General data and fundamental constants

Quantity	Symbol	Value	Power of ten	Units
Speed of light	c	2.997 925 58*	10^{8}	$\mathrm{m\ s^{-1}}$
Elementary charge	e	1.602 176	10^{-19}	C
Faraday constant	$F = N_A e$	9.648 53	10^{4}	$\mathrm{C\ mol^{-1}}$
Boltzmann constant	k	1.380 65	10^{-23}	$\mathrm{J\ K^{-1}}$
Gas constant	$R = N_A k$	8.314 47		$\mathrm{J\ K^{-1}\ mol^{-1}}$
		8.314 47	10^{-2}	$\mathrm{L\ bar\ K^{-1}\ mol^{-1}}$
		8.205 74	10^{-2}	$\mathrm{L\ atm\ K^{-1}\ mol^{-1}}$
		6.236 37	10	$\mathrm{L\ Torr\ K^{-1}\ mol^{-1}}$
Planck's constant	h	6.626 08	10^{-34}	J s
	$\hbar = h/2\pi$	1.054 57	10^{-34}	J s
Avogadro's constant	N_A	6.022 14	10^{23}	$\mathrm{mol^{-1}}$
Atomic mass unit	u	1.660 54	10^{-27}	kg
Mass				
electron	m_e	9.109 38	10^{-31}	kg
proton	m_p	1.672 62	10^{-27}	kg
neutron	m_n	1.674 93	10^{-27}	kg
Vacuum permittivity	$\varepsilon_0 = 1/c^2\mu_0$	8.854 19	10^{-12}	$\mathrm{J^{-1}\ C^2\ m^{-1}}$
	$4\pi\varepsilon_0$	1.112 65	10^{-10}	$\mathrm{J^{-1}\ C^2\ m^{-1}}$
Vacuum permeability	μ_0	4π	10^{-7}	$\mathrm{J\ s^2\ C^{-2}\ m^{-1}\ (= T^2\ J^{-1}\ m^3)}$
Magneton				
Bohr	$\mu_B = e\hbar/2m_e$	9.274 01	10^{-24}	$\mathrm{J\ T^{-1}}$
nuclear	$\mu_N = e\hbar/2m_p$	5.050 78	10^{-27}	$\mathrm{J\ T^{-1}}$
g value	g_e	2.002 32		
Bohr radius	$a_0 = 4\pi\varepsilon_0\hbar^2/m_e e^2$	5.291 77	10^{-11}	m
Fine-structure constant	$\alpha = \mu_0 e^2 c/2h$	7.297 35	10^{-3}	
	α^{-1}	1.370 36	10^{2}	
Second radiation constant	$c_2 = hc/k$	1.438 78	10^{-2}	m K
Stefan-Boltzmann constant	$\sigma = 2\pi^5k^4/15h^3c^2$	5.670 51	10^{-8}	$\mathrm{W\ m^{-2}\ K^{-4}}$
Rydberg constant	$R = m_e e^4/8h^3c\varepsilon_0^2$	1.097 37	10^{5}	$\mathrm{cm^{-1}}$
Standard acceleration of free fall	g	9.806 65*		$\mathrm{m\ s^{-2}}$
Gravitational constant	G	6.673	10^{-11}	$\mathrm{N\ m^2\ kg^{-2}}$

**Exact value*

The Greek alphabet

A, α	alpha	H, η	eta	N, ν	nu	Υ, υ	upsilon
B, β	beta	Θ, θ	theta	Ξ, ξ	xi	Φ, ϕ	phi
Γ, γ	gamma	I, ι	iota	Π, π	pi	X, χ	chi
Δ, δ	delta	K, κ	kappa	P, ρ	rho	Ψ, ψ	psi
E, ε	epsilon	Λ, λ	lambda	Σ, σ	sigma	Ω, ω	omega
Z, ζ	zeta	M, μ	mu	T, τ	tau		

ATKINS'

PHYSICAL CHEMISTRY

ATKINS' PHYSICAL CHEMISTRY

Eighth Edition

Peter Atkins
Professor of Chemistry,
University of Oxford,
and Fellow of Lincoln College, Oxford

Julio de Paula
Professor and Dean of the College of Arts and Sciences,
Lewis and Clark College,
Portland, Oregon

OXFORD
UNIVERSITY PRESS

OXFORD
UNIVERSITY PRESS

Great Clarendon Street, Oxford OX2 6DP

Oxford University Press is a department of the University of Oxford.
It furthers the University's objective of excellence in research, scholarship,
and education by publishing worldwide in

Oxford New York

Auckland Cape Town Dar es Salaam Hong Kong Karachi
Kuala Lumpur Madrid Melbourne Mexico City Nairobi
New Delhi Shanghai Taipei Toronto

With offices in

Argentina Austria Brazil Chile Czech Republic France Greece
Guatemala Hungary Italy Japan Poland Portugal Singapore
South Korea Switzerland Thailand Turkey Ukraine Vietnam

Published in the United States and Canada
by W. H. Freeman and Company

First published 2006

British Library Cataloguing in Publication Data
Data available

Library of Congress Cataloging in Publication Data
Data available

Typeset by Graphicraft Limited, Hong Kong
Printed in China
on acid-free paper by Asia Pacific Offset

ISBN 9780198700722

5 7 9 10 8 6 4

Preface

We have taken the opportunity to refresh both the content and presentation of this text while—as for all its editions—keeping it flexible to use, accessible to students, broad in scope, and authoritative. The bulk of textbooks is a perennial concern: we have sought to tighten the presentation in this edition. However, it should always be borne in mind that much of the bulk arises from the numerous pedagogical features that we include (such as *Worked examples* and the *Data section*), not necessarily from density of information.

The most striking change in presentation is the use of colour. We have made every effort to use colour systematically and pedagogically, not gratuitously, seeing it as a medium for making the text more attractive but using it to convey concepts and data more clearly. The text is still divided into three parts, but material has been moved between chapters and the chapters have been reorganized. We have responded to the shift in emphasis away from classical thermodynamics by combining several chapters in Part 1 (Equilibrium), bearing in mind that some of the material will already have been covered in earlier courses. We no longer make a distinction between 'concepts' and 'machinery', and as a result have provided a more compact presentation of thermodynamics with fewer artificial divisions between the approaches. Similarly, equilibrium electrochemistry now finds a home within the chapter on chemical equilibrium, where space has been made by reducing the discussion of acids and bases.

In Part 2 (Structure) the principal changes are within the chapters, where we have sought to bring into the discussion contemporary techniques of spectroscopy and approaches to computational chemistry. In recognition of the major role that physical chemistry plays in materials science, we have a short sequence of chapters on materials, which deal respectively with hard and soft matter. Moreover, we have introduced concepts of nanoscience throughout much of Part 2.

Part 3 has lost its chapter on dynamic electrochemistry, but not the material. We regard this material as highly important in a contemporary context, but as a final chapter it rarely received the attention it deserves. To make it more readily accessible within the context of courses and to acknowledge that the material it covers is at home intellectually with other material in the book, the description of electron transfer reactions is now a part of the sequence on chemical kinetics and the description of processes at electrodes is now a part of the general discussion of solid surfaces.

We have discarded the Boxes of earlier editions. They have been replaced by more fully integrated and extensive *Impact* sections, which show how physical chemistry is applied to biology, materials, and the environment. By liberating these topics from their boxes, we believe they are more likely to be used and read; there are end-of-chapter problems on most of the material in these sections.

In the preface to the seventh edition we wrote that there was vigorous discussion in the physical chemistry community about the choice of a 'quantum first' or a 'thermodynamics first' approach. That discussion continues. In response we have paid particular attention to making the organization flexible. The strategic aim of this revision is to make it possible to work through the text in a variety of orders, and at the end of this Preface we once again include two suggested road maps.

The concern expressed in the seventh edition about the level of mathematical ability has not evaporated, of course, and we have developed further our strategies for showing the absolute centrality of mathematics to physical chemistry and to make it accessible. Thus, we give more help with the development of equations, motivate them, justify them, and comment on the steps. We have kept in mind the struggling student, and have tried to provide help at every turn.

We are, of course, alert to the developments in electronic resources and have made a special effort in this edition to encourage the use of the resources in our online resource centre (at www.oxfordtextbooks.co.uk/orc/pchem8e/) where you can also access the eBook. In particular, we think it important to encourage students to use the *Living graphs* and their considerable extension as *Explorations in Physical Chemistry*. To do so, wherever we call out a *Living graph* (by an icon attached to a graph in the text), we include an *Exploration* in the figure legend, suggesting how to explore the consequences of changing parameters.

Overall, we have taken this opportunity to refresh the text thoroughly, to integrate applications, to encourage the use of electronic resources, and to make the text even more flexible and up to date.

Oxford P.W.A.
Portland J.de P.

Traditional approach

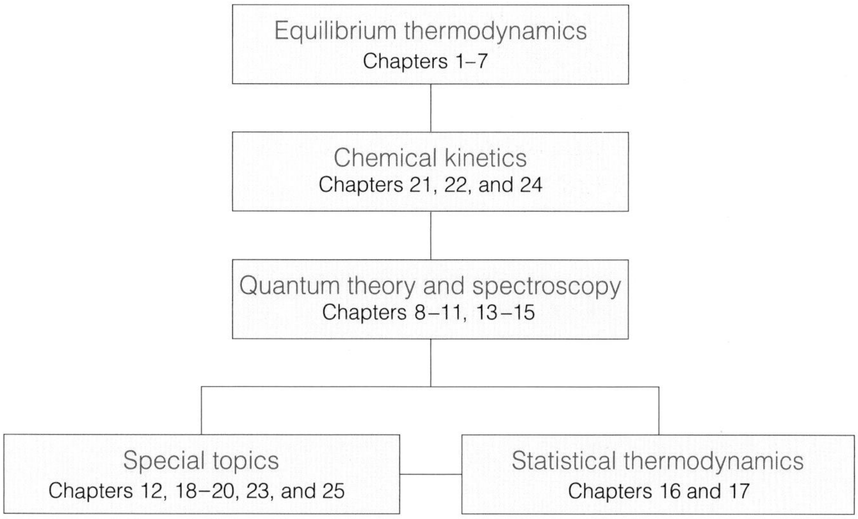

Molecular approach

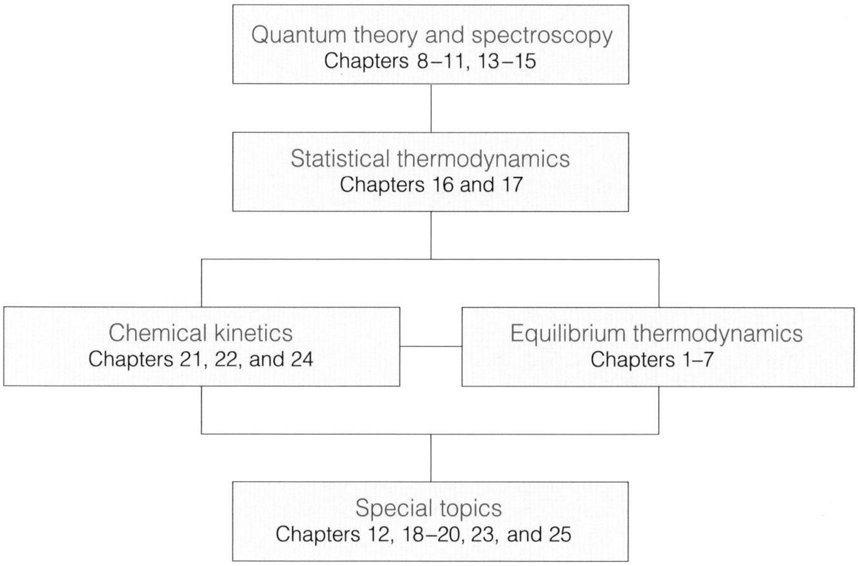

About the book

There are numerous features in this edition that are designed to make learning physical chemistry more effective and more enjoyable. One of the problems that make the subject daunting is the sheer amount of information: we have introduced several devices for organizing the material: see ***Organizing the information***. We appreciate that mathematics is often troublesome, and therefore have taken care to give help with this enormously important aspect of physical chemistry: see ***Mathematics and Physics support***. Problem solving—especially, 'where do I start?'—is often a challenge, and we have done our best to help overcome this first hurdle: see ***Problem solving***. Finally, the web is an extraordinary resource, but it is necessary to know where to start, or where to go for a particular piece of information; we have tried to indicate the right direction: see ***About the Online Resource Centre***. The following paragraphs explain the features in more detail.

Organizing the information

Checklist of key ideas

☐ 1. A gas is a form of matter that fills any container it occupies.

☐ 2. An equation of state interrelates pressure, volume, temperature, and amount of substance: $p = f(T,V,n)$.

☐ 3. The pressure is the force divided by the area to which the force is applied. The standard pressure is $p^{\ominus} = 1$ bar (10^5 Pa).

☐ 4. Mechanical equilibrium is the condition of equality of pressure on either side of a movable wall.

☐ 5. Temperature is the property that indicates the direction of the flow of energy through a thermally conducting, rigid wall.

☐ 6. A diathermic boundary is a boundary that permits the passage of energy as heat. An adiabatic boundary is a boundary that prevents the passage of energy as heat.

☐ 7. Thermal equilibrium is a condition in which no change of state occurs when two objects A and B are in contact through a diathermic boundary.

☐ 8. The Zeroth Law of thermodynamics states that, if A is in thermal equilibrium with B, and B is in thermal equilibrium with C, then C is also in thermal equilibrium with A.

☐ 9. The Celsius and thermodynamic temperature scales are related by $T/\text{K} = \theta/°\text{C} + 273.15$.

☐ 10. A perfect gas obeys the perfect gas equation, $pV = nRT$, exactly

☐ 12. The partial pressure of any gas i
$x_J = n_J/n$ is its mole fraction in a
pressure.

☐ 13. In real gases, molecular interact
state; the true equation of state
coefficients $B, C, \ldots : pV_m = RT$

☐ 14. The vapour pressure is the pres
with its condensed phase.

☐ 15. The critical point is the point at
end of the horizontal part of th
a single point. The critical cons
pressure, molar volume, and te
critical point.

☐ 16. A supercritical fluid is a dense f
temperature and pressure.

☐ 17. The van der Waals equation of
the true equation of state in wh
by a parameter a and repulsion
parameter b: $p = nRT/(V - nb)$

☐ 18. A reduced variable is the actual
corresponding critical constant

Checklist of key ideas

Here we collect together the major concepts introduced in the chapter. We suggest checking off the box that precedes each entry when you feel confident about the topic.

IMPACT ON NANOSCIENCE
I20.2 Nanowires

We have already remarked (*Impacts* I9.1, I9.2, and I19.3) that research on nanometre-sized materials is motivated by the possibility that they will form the basis for cheaper and smaller electronic devices. The synthesis of *nanowires*, nanometre-sized atomic assemblies that conduct electricity, is a major step in the fabrication of nanodevices. An important type of nanowire is based on carbon nanotubes, which, like graphite, can conduct electrons through delocalized π molecular orbitals that form from unhybridized $2p$ orbitals on carbon. Recent studies have shown a correlation between structure and conductivity in single-walled nanotubes (SWNTs) that does not occur in graphite. The SWNT in Fig. 20.45 is a semiconductor. If the hexagons are rotated by 60° about their sixfold axis, the resulting SWNT is a metallic conductor.

Carbon nanotubes are promising building blocks not only because they have useful electrical properties but also because they have unusual mechanical properties. For example, an SWNT has a Young's modulus that is approximately five times larger and a tensile strength that is approximately 375 times larger than that of steel.

Silicon nanowires can be made by focusing a pulsed laser beam on to a solid target composed of silicon and iron. The laser ejects Fe and Si atoms from the surface of the

Impact sections

Where appropriate, we have separated the principles from their applications: the principles are constant and straightforward; the applications come and go as the subject progresses. The *Impact* sections show how the principles developed in the chapter are currently being applied in a variety of modern contexts.

Notes on good practice

Science is a precise activity and its language should be used accurately. We have used this feature to help encourage the use of the language and procedures of science in conformity to international practice and to help avoid common mistakes.

A note on good practice We write $T = 0$, not $T = 0$ K for the zero temperature on the thermodynamic temperature scale. This scale is absolute, and the lowest temperature is 0 regardless of the size of the divisions on the scale (just as we write $p = 0$ for zero pressure, regardless of the size of the units we adopt, such as bar or pascal). However, we write 0°C because the Celsius scale is not absolute.

Justifications

On first reading it might be sufficient to appreciate the 'bottom line' rather than work through detailed development of a mathematical expression. However, mathematical development is an intrinsic part of physical chemistry, and it is important to see how a particular expression is obtained. The *Justifications* let you adjust the level of detail that you require to your current needs, and make it easier to review material.

5.8 **The activities of regular solutions**

The material on regular solutions presented in Section 5.4 gives further insight into the origin of deviations from Raoult's law and its relation to activity coefficients. The starting point is the expression for the Gibbs energy of mixing for a regular solution (eqn 5.31). We show in the following *Justification* that eqn 5.31 implies that the activity coefficients are given by expressions of the form

$$\ln \gamma_A = \beta x_B^2 \qquad \ln \gamma_B = \beta x_A^2 \tag{5.57}$$

These relations are called the **Margules equations**.

Justification 5.4 *The Margules equations*

The Gibbs energy of mixing to form a nonideal solution is

$$\Delta_{mix}G = nRT\{x_A \ln a_A + x_B \ln a_B\}$$

This relation follows from the derivation of eqn 5.31 with activities in place of mole fractions. If each activity is replaced by γx, this expression becomes

$$\Delta_{mix}G = nRT\{x_A \ln x_A + x_B \ln x_B + x_A \ln \gamma_A + x_B \ln \gamma_B\}$$

Now we introduce the two expressions in eqn 5.57, and use $x_A + x_B = 1$, which gives

$$\begin{aligned}\Delta_{mix}G &= nRT\{x_A \ln x_A + x_B \ln x_B + \beta x_A x_B^2 + \beta x_B x_A^2\}\\ &= nRT\{x_A \ln x_A + x_B \ln x_B + \beta x_A x_B(x_A + x_B)\}\\ &= nRT\{x_A \ln x_A + x_B \ln x_B + \beta x_A x_B\}\end{aligned}$$

as required by eqn 5.31. Note, moreover, that the activity coefficients behave correctly for dilute solutions: $\gamma_A \to 1$ as $x_B \to 0$ and $\gamma_B \to 1$ as $x_A \to 0$.

Molecular interpretation sections

Historically, much of the material in the first part of the text was developed before the emergence of detailed models of atoms, molecules, and molecular assemblies. The *Molecular interpretation* sections enhance and enrich coverage of that material by explaining how it can be understood in terms of the behaviour of atoms and molecules.

Molecular interpretation 5.2 *The lowering of vapour pressure of a solvent in a mixture*

The molecular origin of the lowering of the chemical potential is not the energy of interaction of the solute and solvent particles, because the lowering occurs even in an ideal solution (for which the enthalpy of mixing is zero). If it is not an enthalpy effect, it must be an entropy effect.

The pure liquid solvent has an entropy that reflects the number of microstates available to its molecules. Its vapour pressure reflects the tendency of the solution towards greater entropy, which can be achieved if the liquid vaporizes to form a gas. When a solute is present, there is an additional contribution to the entropy of the liquid, even in an ideal solution. Because the entropy of the liquid is already higher than that of the pure liquid, there is a weaker tendency to form the gas (Fig. 5.22). The effect of the solute appears as a lowered vapour pressure, and hence a higher boiling point.

Similarly, the enhanced molecular randomness of the solution opposes the tendency to freeze. Consequently, a lower temperature must be reached before equilibrium between solid and solution is achieved. Hence, the freezing point is lowered.

Further information

Further information 5.1 *The Debye–Hückel theory of ionic solutions*

Imagine a solution in which all the ions have their actual positions, but in which their Coulombic interactions have been turned off. The difference in molar Gibbs energy between the ideal and real solutions is equal to w_e, the electrical work of charging the system in this arrangement. For a salt M_pX_q, we write

$$w_e = \overbrace{(p\mu_+ + q\mu_-)}^{G_m} - \overbrace{(p\mu_+^{\text{ideal}} + q\mu_-^{\text{ideal}})}^{G_m^{\text{ideal}}}$$
$$= p(\mu_+ - \mu_+^{\text{ideal}}) + q(\mu_- - \mu_-^{\text{ideal}})$$

From eqn 5.64 we write

$$\mu_+ - \mu_+^{\text{ideal}} = \mu_- - \mu_-^{\text{ideal}} = RT \ln \gamma_\pm$$

So it follows that

$$\ln \gamma_\pm = \frac{w_e}{sRT} \qquad s = p + q \qquad (5.73)$$

This equation tells us that we must first find the final distribution of the ions and then the work of charging them in that distribution.

The Coulomb potential at a distance r from an isolated ion of charge $z_i e$ in a medium of permittivity ε is

$$\phi_i = \frac{Z_i}{r} \qquad Z_i = \frac{z_i e}{4\pi\varepsilon} \qquad (5.74)$$

The ionic atmosphere causes the potential to decay with distance more sharply than this expression implies. Such shielding is a familiar problem in electrostatics, and its effect is taken into account by replacing the Coulomb potential by the **shielded Coulomb potential**, an expression of the form

$$\phi_i = \frac{Z_i}{r} e^{-r/r_D} \qquad (5.75)$$

where r_D is called the **Debye length**. W
potential is virtually the same as the un
small, the shielded potential is much sm
potential, even for short distances (Fig.

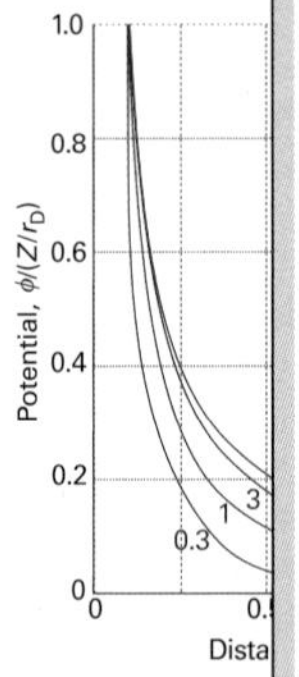

Fig. 5.36 The variation of the shielded C
distance for different values of the Deb
Debye length, the more sharply the pot
case, a is an arbitrary unit of length.

Exploration Write an expression
unshielded and shielded Coulon
Then plot this expression against r_D an
interpretation for the shape of the plot.

Further information

In some cases, we have judged that a derivation is too long, too detailed, or too different in level for it to be included in the text. In these cases, the derivations will be found less obtrusively at the end of the chapter.

966 Appendix 2 MATHEMATICAL TECHNIQUES

A2.6 **Partial derivatives**

A **partial derivative** of a function of more than one variabl
of the function with respect to one of the variables, all th
constant (see Fig. 2.*). Although a partial derivative sho
when one variable changes, it may be used to determine
when more than one variable changes by an infinitesimal
tion of x and y, then when x and y change by dx and dy, re

$$df = \left(\frac{\partial f}{\partial x}\right)_y dx + \left(\frac{\partial f}{\partial y}\right)_x dy$$

where the symbol ∂ is used (instead of d) to denote a part
df is also called the **differential** of f. For example, if $f = ax^3$

$$\left(\frac{\partial f}{\partial x}\right)_y = 3ax^2y \qquad \left(\frac{\partial f}{\partial y}\right)_x = ax^3 + 2by$$

Appendices

Physical chemistry draws on a lot of background material, especially in mathematics and physics. We have included a set of *Appendices* to provide a quick survey of some of the information relating to units, physics, and mathematics that we draw on in the text.

1000 DATA SECTION

Table 2.8 Expansion coefficients, α, and isothermal compressibilities, κ_T

	$\alpha/(10^{-4}\,K^{-1})$	$\kappa_T/(10^{-6}\,atm^{-1})$
Liquids		
Benzene	12.4	92.1
Carbon tetrachloride	12.4	90.5
Ethanol	11.2	76.8
Mercury	1.82	38.7
Water	2.1	49.6
Solids		
Copper	0.501	0.735
Diamond	0.030	0.187
Iron	0.354	0.589
Lead	0.861	2.21

The values refer to 20°C.
Data: AIP(α), KL(κ_T).

Table 2.9 Inversion temperatures, n
points, and Joule–Thomson coefficien

	T_I/K	T_f/K
Air	603	
Argon	723	83.8
Carbon dioxide	1500	194.7s
Helium	40	
Hydrogen	202	14.0
Krypton	1090	116.6
Methane	968	90.6
Neon	231	24.5
Nitrogen	621	63.3
Oxygen	764	54.8

s: sublimes.
Data: AIP, JL, and M.W. Zemansky, *Heat and*
New York (1957).

Synoptic tables and the Data section

Long tables of data are helpful for assembling and solving exercises and problems, but can break up the flow of the text. We provide a lot of data in the *Data section* at the end of the text and short extracts in the *Synoptic tables* in the text itself to give an idea of the typical values of the physical quantities we are introducing.

Mathematics and Physics support

Comment 1.2
A hyperbola is a curve obtained by plotting y against x with xy = constant.

Comment 2.5
The partial-differential operation $(\partial z/\partial x)_y$ consists of taking the first derivative of $z(x,y)$ with respect to x, treating y as a constant. For example, if $z(x,y) = x^2y$, then

$$\left(\frac{\partial z}{\partial x}\right)_y = \left(\frac{\partial [x^2y]}{\partial x}\right)_y = y\frac{dx^2}{dx} = 2yx$$

Partial derivatives are reviewed in *Appendix 2.*

Comments

A topic often needs to draw on a mathematical procedure or a concept of physics; a *Comment* is a quick reminder of the procedure or concept.

978 Appendix 3 ESSENTIAL CONCEPTS OF PHYSICS

Classical mechanics

Classical mechanics describes the behaviour of objects in
expresses the fact that the total energy is constant in the a
other expresses the response of particles to the forces act

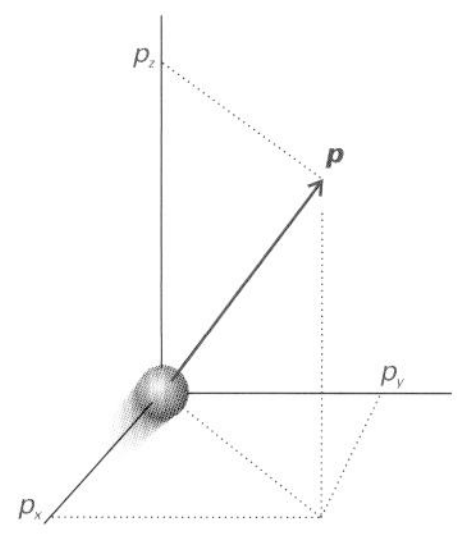

A3.1 The linear momentum of a particle is a vector property and points in the direction of motion.

A3.3 **The trajectory in terms of the energy**

The **velocity**, v, of a particle is the rate of change of its po

$$v = \frac{dr}{dt}$$

The velocity is a vector, with both direction and magni
velocity is the **speed**, v. The **linear momentum**, p, of a pa
its velocity, v, by

$$p = mv$$

Like the velocity vector, the linear momentum vector po
of the particle (Fig. A3.1). In terms of the linear moment
ticle is

Appendices

There is further information on mathematics and physics in Appendices 2 and 3, respectively. These appendices do not go into great detail, but should be enough to act as reminders of topics learned in other courses.

Problem solving

Illustration 5.2 *Using Henry's law*

To estimate the molar solubility of oxygen in water at 25°C and a partial pressure of 21 kPa, its partial pressure in the atmosphere at sea level, we write

$$b_{O_2} = \frac{p_{O_2}}{K_{O_2}} = \frac{21\ \text{kPa}}{7.9\times10^4\ \text{kPa kg mol}^{-1}} = 2.9\times10^{-4}\ \text{mol kg}^{-1}$$

The molality of the saturated solution is therefore 0.29 mmol kg^{-1}. To convert this quantity to a molar concentration, we assume that the mass density of this dilute solution is essentially that of pure water at 25°C, or $\rho_{H_2O} = 0.99709$ kg dm^{-3}. It follows that the molar concentration of oxygen is

$$[O_2] = b_{O_2}\times\rho_{H_2O} = 0.29\ \text{mmol kg}^{-1}\times0.99709\ \text{kg dm}^{-3} = 0.29\ \text{mmol dm}^{-3}$$

A note on good practice The number of significant figures in the result of a calculation should not exceed the number in the data (only two in this case).

Self-test 5.5 Calculate the molar solubility of nitrogen in water exposed to air at 25°C; partial pressures were calculated in *Example 1.3.* [0.51 mmol dm^{-3}]

Illustrations

An *Illustration* (don't confuse this with a diagram!) is a short example of how to use an equation that has just been introduced in the text. In particular, we show how to use data and how to manipulate units correctly.

Example 8.1 *Calculating the number of photons*

Calculate the number of photons emitted by a 100 W yellow lamp in 1.0 s. Take the wavelength of yellow light as 560 nm and assume 100 per cent efficiency.

Method Each photon has an energy $h\nu$, so the total number of photons needed to produce an energy E is $E/h\nu$. To use this equation, we need to know the frequency of the radiation (from $\nu = c/\lambda$) and the total energy emitted by the lamp. The latter is given by the product of the power (P, in watts) and the time interval for which the lamp is turned on ($E = P\Delta t$).

Answer The number of photons is

$$N = \frac{E}{h\nu} = \frac{P\Delta t}{h(c/\lambda)} = \frac{\lambda P \Delta t}{hc}$$

Substitution of the data gives

$$N = \frac{(5.60 \times 10^{-7}\ \text{m}) \times (100\ \text{J s}^{-1}) \times (1.0\ \text{s})}{(6.626 \times 10^{-34}\ \text{J s}) \times (2.998 \times 10^{8}\ \text{m s}^{-1})} = 2.8 \times 10^{20}$$

Note that it would take nearly 40 min to produce 1 mol of these photons.

A note on good practice To avoid rounding and other numerical errors, it is best to carry out algebraic mainpulations first, and to substitute numerical values into a single, final formula. Moreover, an analytical result may be used for other data without having to repeat the entire calculation.

Self-test 8.1 How many photons does a monochromatic (single frequency) infrared rangefinder of power 1 mW and wavelength 1000 nm emit in 0.1 s?
[5×10^{14}]

Worked examples

A *Worked example* is a much more structured form of *Illustration*, often involving a more elaborate procedure. Every *Worked example* has a Method section to suggest how to set up the problem (another way might seem more natural: setting up problems is a highly personal business). Then there is the worked-out Answer.

Self-test 3.12 Calculate the change in G_m for ice at −10°C, with density 917 kg m^{-3}, when the pressure is increased from 1.0 bar to 2.0 bar. [+2.0 J mol^{-1}]

Self-tests

Each *Worked example*, and many of the *Illustrations*, has a *Self-test*, with the answer provided as a check that the procedure has been mastered. There are also free-standing *Self-tests* where we thought it a good idea to provide a question to check understanding. Think of *Self-tests* as in-chapter *Exercises* designed to help monitor your progress.

Discussion questions

1.1 Explain how the perfect gas equation of state arises by combination of Boyle's law, Charles's law, and Avogadro's principle.

1.2 Explain the term 'partial pressure' and explain why Dalton's law is a limiting law.

1.3 Explain how the compression factor varies with pressure and temperature and describe how it reveals information about intermolecular interactions in real gases.

1.4 What is the significance of the critical c

1.5 Describe the formulation of the van de rationale for one other equation of state in

1.6 Explain how the van der Waals equatio behaviour.

Discussion questions

The end-of-chapter material starts with a short set of questions that are intended to encourage reflection on the material and to view it in a broader context than is obtained by solving numerical problems.

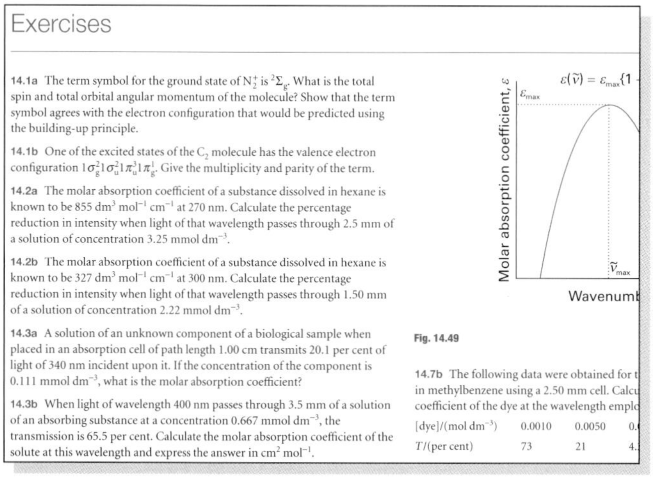

Exercises

14.1a The term symbol for the ground state of N_2^+ is $^2\Sigma_g$. What is the total spin and total orbital angular momentum of the molecule? Show that the term symbol agrees with the electron configuration that would be predicted using the building-up principle.

14.1b One of the excited states of the C_2 molecule has the valence electron configuration $1\sigma_g^2 1\sigma_u^2 1\pi_u^3 1\pi_g^1$. Give the multiplicity and parity of the term.

14.2a The molar absorption coefficient of a substance dissolved in hexane is known to be 855 $dm^3\ mol^{-1}\ cm^{-1}$ at 270 nm. Calculate the percentage reduction in intensity when light of that wavelength passes through 2.5 mm of a solution of concentration 3.25 mmol dm^{-3}.

14.2b The molar absorption coefficient of a substance dissolved in hexane is known to be 327 $dm^3\ mol^{-1}\ cm^{-1}$ at 300 nm. Calculate the percentage reduction in intensity when light of that wavelength passes through 1.50 mm of a solution of concentration 2.22 mmol dm^{-3}.

14.3a A solution of an unknown component of a biological sample when placed in an absorption cell of path length 1.00 cm transmits 20.1 per cent of light of 340 nm incident upon it. If the concentration of the component is 0.111 mmol dm^{-3}, what is the molar absorption coefficient?

14.3b When light of wavelength 400 nm passes through 3.5 mm of a solution of an absorbing substance at a concentration 0.667 mmol dm^{-3}, the transmission is 65.5 per cent. Calculate the molar absorption coefficient of the solute at this wavelength and express the answer in $cm^2\ mol^{-1}$.

Fig. 14.49

14.7b The following data were obtained for t in methylbenzene using a 2.50 mm cell. Calcu coefficient of the dye at the wavelength emplo

[dye]/(mol dm^{-3})	0.0010	0.0050	0.
T/(per cent)	73	21	4.

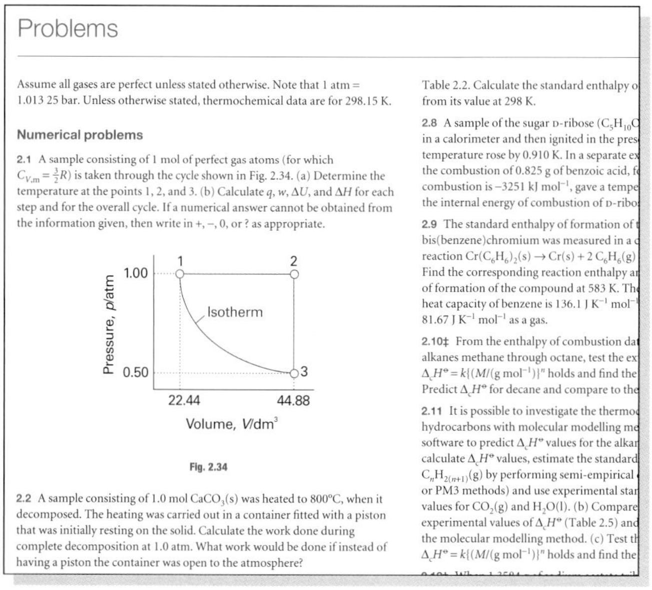

Problems

Assume all gases are perfect unless stated otherwise. Note that 1 atm = 1.013 25 bar. Unless otherwise stated, thermochemical data are for 298.15 K.

Numerical problems

2.1 A sample consisting of 1 mol of perfect gas atoms (for which $C_{V,m} = \frac{3}{2}R$) is taken through the cycle shown in Fig. 2.34. (a) Determine the temperature at the points 1, 2, and 3. (b) Calculate q, w, ΔU, and ΔH for each step and for the overall cycle. If a numerical answer cannot be obtained from the information given, then write in +, −, 0, or ? as appropriate.

Fig. 2.34

2.2 A sample consisting of 1.0 mol $CaCO_3(s)$ was heated to 800°C, when it decomposed. The heating was carried out in a container fitted with a piston that was initially resting on the solid. Calculate the work done during complete decomposition at 1.0 atm. What work would be done if instead of having a piston the container was open to the atmosphere?

Table 2.2. Calculate the standard enthalpy o from its value at 298 K.

2.8 A sample of the sugar D-ribose (C_5H_{10}C in a calorimeter and then ignited in the pres temperature rose by 0.910 K. In a separate ex the combustion of 0.825 g of benzoic acid, f combustion is −3251 kJ mol^{-1}, gave a tempe the internal energy of combustion of D-ribo

2.9 The standard enthalpy of formation of t bis(benzene)chromium was measured in a c reaction $Cr(C_6H_6)_2(s) \rightarrow Cr(s) + 2\,C_6H_6(g)$ Find the corresponding reaction enthalpy an of formation of the compound at 583 K. Th heat capacity of benzene is 136.1 J $K^{-1}\ mol^{-1}$ 81.67 J $K^{-1}\ mol^{-1}$ as a gas.

2.10‡ From the enthalpy of combustion dat alkanes methane through octane, test the ex $\Delta_c H^\ominus = k\{(M/(\text{g mol}^{-1}))\}^n$ holds and find the Predict $\Delta_c H^\ominus$ for decane and compare to the

2.11 It is possible to investigate the thermod hydrocarbons with molecular modelling me software to predict $\Delta_c H^\ominus$ values for the alkan calculate $\Delta_c H^\ominus$ values, estimate the standard $C_nH_{2(n+1)}(g)$ by performing semi-empirical or PM3 methods) and use experimental stan values for $CO_2(g)$ and $H_2O(l)$. (b) Compare experimental values of $\Delta_c H^\ominus$ (Table 2.5) and the molecular modelling method. (c) Test th $\Delta_c H^\ominus = k\{(M/(\text{g mol}^{-1}))\}^n$ holds and find the

Exercises and Problems

The real core of testing understanding is the collection of end-of-chapter *Exercises* and *Problems*. The *Exercises* are straightforward numerical tests that give practice with manipulating numerical data. The *Problems* are more searching. They are divided into 'numerical', where the emphasis is on the manipulation of data, and 'theoretical', where the emphasis is on the manipulation of equations before (in some cases) using numerical data. At the end of the *Problems* are collections of problems that focus on practical applications of various kinds, including the material covered in the *Impact* sections.

About the Online Resource Centre

The Online Resource Centre provides teaching and learning resources. It is free of charge, complements the textbook, and offers additional materials which can be downloaded. The resources it provides are fully customizable and can be incorporated into a virtual learning environment. Go to:

www.oxfordtextbooks.co.uk/orc/pchem8e/

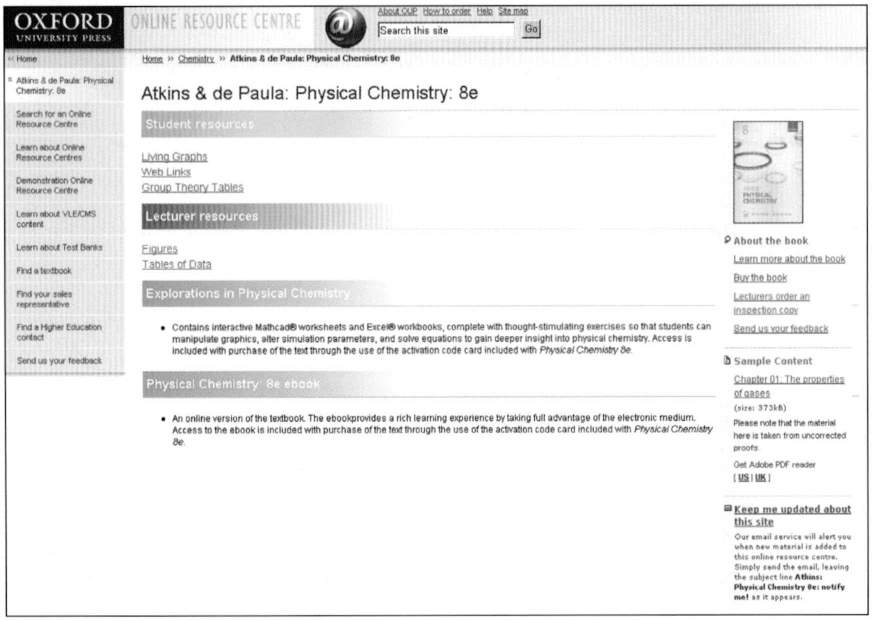

online
resource
centre

The material includes:

Living graphs

A *Living graph* is indicated in the text by the icon attached to a graph. This feature can be used to explore how a property changes as a variety of parameters are changed. To encourage the use of this resource (and the more extensive *Explorations in Physical Chemistry*) we have added a question to each figure where a *Living graph* is called out.

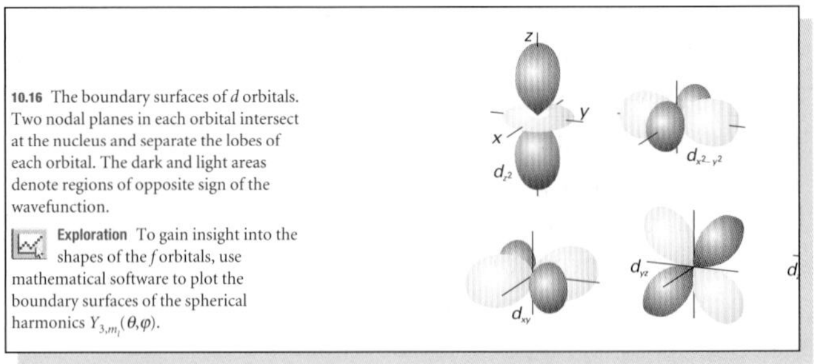
10.16 The boundary surfaces of d orbitals. Two nodal planes in each orbital intersect at the nucleus and separate the lobes of each orbital. The dark and light areas denote regions of opposite sign of the wavefunction.
Exploration To gain insight into the shapes of the f orbitals, use mathematical software to plot the boundary surfaces of the spherical harmonics $Y_{3,m_l}(\theta,\varphi)$.
z
y
x
d_{z^2}
$d_{x^2-y^2}$
d_{yz}
d_{xy}

Artwork

An instructor may wish to use the illustrations from this text in a lecture. Almost all the illustrations are available and can be used for lectures without charge (but not for commercial purposes without specific permission). This edition is in full colour: we have aimed to use colour systematically and helpfully, not just to make the page prettier.

Tables of data

All the tables of data that appear in the chapter text are available and may be used under the same conditions as the figures.

Web links

There is a huge network of information available about physical chemistry, and it can be bewildering to find your way to it. Also, a piece of information may be needed that we have not included in the text. The web site might suggest where to find the specific data or indicate where additional data can be found.

Group theory tables

Comprehensive group theory tables are available for downloading.

Explorations in Physical Chemistry

Physical chemistry describes the dynamic processes that shape the world around us; it is far removed from the perception of abstract theories and relationships held by so many students. But how can students make the jump from abstract equation to the reality of physical chemistry in action? *Explorations in Physical Chemistry* offers a unique way to bring physical chemistry to life.

Explorations in Physical Chemistry consists of interactive Mathcad® worksheets and interactive Excel® workbooks, complete with thought-stimulating exercises. They motivate students to simulate physical, chemical, and biochemical phenomena with their personal computers. Harnessing the computational power of Mathcad® by Mathsoft, Inc. and Excel® by Microsoft Corporation, students can manipulate over 75 graphics, alter simulation parameters, and solve equations to gain deeper insight into physical chemistry.

Explorations in Physical Chemistry is available on-line at www.oxfordtextbooks.co.uk/orc/explorations/, using the activation code card included with *Atkins' Physical Chemistry* 8e. It can also be purchased as a CD-ROM (978-019-928894-6).

Physical Chemistry, Eighth Edition eBook

A complete online version of the textbook. The eBook provides a rich learning experience by taking full advantage of the electronic medium integrating all student media resources and adds features unique to the eBook. The eBook also offers lecturers unparalleled flexibility and customization options not previously possible with any printed textbook. Access to the eBook is included with purchase of the text through the use of the activation code card included with *Atkins' Physical Chemistry* 8e.

Key features of the eBook include:

- Easy access from any Internet-connected computer via a standard Web browser.
- Quick, intuitive navigation to any section or subsection, as well as any printed book page number.
- Integration of all Living Graph animations.
- Text **highlighting**, down to the level of individual phrases.
- A **book marking** feature that allows for quick reference to any page.
- A powerful **Notes** feature that allows students or instructors to add notes to any page.
- A full **index**.
- **Full-text search**, including an option to also search the glossary and index.
- Automatic saving of all notes, highlighting, and bookmarks.

Additional features for lecturers:

- Custom chapter selection: Lecturers can choose the chapters that correspond with their syllabus, and students will get a custom version of the eBook with the selected chapters only.
- Instructor notes: Lecturers can choose to create an annotated version of the eBook with their notes on any page. When students in their course log in, they will see the lecturer's version.
- Custom content: Lecturer notes can include text, web links, and even images, allowing lecturers to place any content they choose exactly where they want it.

Solutions manuals

As with previous editions, Charles Trapp, Carmen Giunta, and Marshall Cady have produced the solutions manuals to accompany this book. A *Student's Solutions Manual* (978-019-928858-8) provides full solutions to the 'a' exercises and the odd-numbered problems. An *Instructor's Solutions Manual* (978-019-928857-1) provides full solutions to the 'b' exercises and the even-numbered problems.

About the authors

Julio de Paula is Professor of Chemistry and Dean of the College of Arts & Sciences at Lewis & Clark College. A native of Brazil, Professor de Paula received a B.A. degree in chemistry from Rutgers, The State University of New Jersey, and a Ph.D. in biophysical chemistry from Yale University. His research activities encompass the areas of molecular spectroscopy, biophysical chemistry, and nanoscience. He has taught courses in general chemistry, physical chemistry, biophysical chemistry, instrumental analysis, and writing.

Peter Atkins is Professor of Chemistry at Oxford University, a fellow of Lincoln College, and the author of more than fifty books for students and a general audience. His texts are market leaders around the globe. A frequent lecturer in the United States and throughout the world, he has held visiting prefessorships in France, Israel, Japan, China, and New Zealand. He was the founding chairman of the Committee on Chemistry Education of the International Union of Pure and Applied Chemistry and a member of IUPAC's Physical and Biophysical Chemistry Division.

Acknowledgements

A book as extensive as this could not have been written without significant input from many individuals. We would like to reiterate our thanks to the hundreds of people who contributed to the first seven editions. Our warm thanks go Charles Trapp, Carmen Giunta, and Marshall Cady who have produced the *Solutions manuals* that accompany this book.

Many people gave their advice based on the seventh edition, and others reviewed the draft chapters for the eighth edition as they emerged. We therefore wish to thank the following colleagues most warmly:

Joe Addison, Governors State University
Joseph Alia, University of Minnesota Morris
David Andrews, University of East Anglia
Mike Ashfold, University of Bristol
Daniel E. Autrey, Fayetteville State University
Jeffrey Bartz, Kalamazoo College
Martin Bates, University of Southampton
Roger Bickley, University of Bradford
E.M. Blokhuis, Leiden University
Jim Bowers, University of Exeter
Mark S. Braiman, Syracuse University
Alex Brown, University of Alberta
David E. Budil, Northeastern University
Dave Cook, University of Sheffield
Ian Cooper, University of Newcastle-upon-Tyne
T. Michael Duncan, Cornell University
Christer Elvingson, Uppsala University
Cherice M. Evans, Queens College—CUNY
Stephen Fletcher, Loughborough University
Alyx S. Frantzen, Stephen F. Austin State University
David Gardner, Lander University
Roberto A. Garza-López, Pomona College
Robert J. Gordon, University of Illinois at Chicago
Pete Griffiths, Cardiff University
Robert Haines, University of Prince Edward Island
Ron Haines, University of New South Wales
Arthur M. Halpern, Indiana State University
Tom Halstead, University of York
Todd M. Hamilton, Adrian College
Gerard S. Harbison, University Nebraska at Lincoln
Ulf Henriksson, Royal Institute of Technology, Sweden
Mike Hey, University of Nottingham
Paul Hodgkinson, University of Durham
Robert E. Howard, University of Tulsa
Mike Jezercak, University of Central Oklahoma
Clarence Josefson, Millikin University
Pramesh N. Kapoor, University of Delhi
Peter Karadakov, University of York
Miklos Kertesz, Georgetown University
Neil R. Kestner, Louisiana State University
Sanjay Kumar, Indian Institute of Technology
Jeffry D. Madura, Duquesne University
Andrew Masters, University of Manchester
Paul May, University of Bristol
Mitchell D. Menzmer, Southwestern Adventist University
David A. Micha, University of Florida
Sergey Mikhalovsky, University of Brighton
Jonathan Mitschele, Saint Joseph's College
Vicki D. Moravec, Tri-State University
Gareth Morris, University of Manchester
Tony Morton-Blake, Trinity College, Dublin
Andy Mount, University of Edinburgh
Maureen Kendrick Murphy, Huntingdon College
John Parker, Heriot Watt University
Jozef Peeters, University of Leuven
Michael J. Perona, CSU Stanislaus
Nils-Ola Persson, Linköping University
Richard Pethrick, University of Strathclyde
John A. Pojman, The University of Southern Mississippi
Durga M. Prasad, University of Hyderabad
Steve Price, University College London
S. Rajagopal, Madurai Kamaraj University
R. Ramaraj, Madurai Kamaraj University
David Ritter, Southeast Missouri State University
Bent Ronsholdt, Aalborg University
Stephen Roser, University of Bath
Kathryn Rowberg, Purdue University Calumet
S.A. Safron, Florida State University
Kari Salmi, Espoo-Vantaa Institute of Technology
Stephan Sauer, University of Copenhagen
Nicholas Schlotter, Hamline University
Roseanne J. Sension, University of Michigan
A.J. Shaka, University of California
Joe Shapter, Flinders University of South Australia
Paul D. Siders, University of Minnesota, Duluth
Harjinder Singh, Panjab University
Steen Skaarup, Technical University of Denmark
David Smith, University of Exeter
Patricia A. Snyder, Florida Atlantic University
Olle Söderman, Lund University
Peter Stilbs, Royal Institute of Technology, Sweden
Svein Stølen, University of Oslo
Fu-Ming Tao, California State University, Fullerton
Eimer Tuite, University of Newcastle
Eric Waclawik, Queensland University of Technology
Yan Waguespack, University of Maryland Eastern Shore
Terence E. Warner, University of Southern Denmark

Richard Wells, University of Aberdeen
Ben Whitaker, University of Leeds
Christopher Whitehead, University of Manchester
Mark Wilson, University College London
Kazushige Yokoyama, State University of New York at Geneseo
Nigel Young, University of Hull
Sidney H. Young, University of South Alabama

We also thank Fabienne Meyers (of the IUPAC Secretariat) for helping us to bring colour to most of the illustrations and doing so on a very short timescale. We would also like to thank our two publishers, Oxford University Press and W.H. Freeman & Co., for their constant encouragement, advice, and assistance, and in particular our editors Jonathan Crowe, Jessica Fiorillo, and Ruth Hughes. Authors could not wish for a more congenial publishing environment.

Summary of contents

Contents

List of impact sections

PART 1 Equilibrium

Part 1 of the text develops the concepts that are needed for the discussion of equilibria in chemistry. Equilibria include physical change, such as fusion and vaporization, and chemical change, including electrochemistry. The discussion is in terms of thermodynamics, and particularly in terms of enthalpy and entropy. We see that we can obtain a unified view of equilibrium and the direction of spontaneous change in terms of the chemical potentials of substances. The chapters in Part 1 deal with the bulk properties of matter; those of Part 2 will show how these properties stem from the behaviour of individual atoms.

The properties of gases

1

This chapter establishes the properties of gases that will be used throughout the text. It begins with an account of an idealized version of a gas, a perfect gas, and shows how its equation of state may be assembled experimentally. We then see how the properties of real gases differ from those of a perfect gas, and construct an equation of state that describes their properties.

The simplest state of matter is a **gas**, a form of matter that fills any container it occupies. Initially we consider only pure gases, but later in the chapter we see that the same ideas and equations apply to mixtures of gases too.

The perfect gas

We shall find it helpful to picture a gas as a collection of molecules (or atoms) in continuous random motion, with average speeds that increase as the temperature is raised. A gas differs from a liquid in that, except during collisions, the molecules of a gas are widely separated from one another and move in paths that are largely unaffected by intermolecular forces.

1.1 The states of gases

The **physical state** of a sample of a substance, its physical condition, is defined by its physical properties. Two samples of a substance that have the same physical properties are in the same state. The state of a pure gas, for example, is specified by giving its volume, V, amount of substance (number of moles), n, pressure, p, and temperature, T. However, it has been established experimentally that it is sufficient to specify only three of these variables, for then the fourth variable is fixed. That is, it is an experimental fact that each substance is described by an **equation of state**, an equation that interrelates these four variables.

The general form of an equation of state is

$$p = f(T,V,n) \tag{1.1}$$

This equation tells us that, if we know the values of T, V, and n for a particular substance, then the pressure has a fixed value. Each substance is described by its own equation of state, but we know the explicit form of the equation in only a few special cases. One very important example is the equation of state of a 'perfect gas', which has the form $p = nRT/V$, where R is a constant. Much of the rest of this chapter will examine the origin of this equation of state and its applications.

Table 1.1 Pressure units

Name	Symbol	Value
pascal	1 Pa	$1\ N\ m^{-2}$, $1\ kg\ m^{-1}\ s^{-2}$
bar	1 bar	10^5 Pa
atmosphere	1 atm	101.325 kPa
torr	1 Torr	(101 325/760) Pa = 133.32 . . . Pa
millimetres of mercury	1 mmHg	133.322 . . . Pa
pound per square inch	1 psi	6.894 757 . . . kPa

(a) Pressure

Comment 1.1

The International System of units (SI, from the French *Système International d'Unités*) is discussed in *Appendix* 1.

Pressure is defined as force divided by the area to which the force is applied. The greater the force acting on a given area, the greater the pressure. The origin of the force exerted by a gas is the incessant battering of the molecules on the walls of its container. The collisions are so numerous that they exert an effectively steady force, which is experienced as a steady pressure.

The SI unit of pressure, the *pascal* (Pa), is defined as 1 newton per metre-squared:

$$1\ Pa = 1\ N\ m^{-2} \qquad [1.2a]$$

In terms of base units,

$$1\ Pa = 1\ kg\ m^{-1}\ s^{-2} \qquad [1.2b]$$

Several other units are still widely used (Table 1.1); of these units, the most commonly used are atmosphere (1 atm = $1.013\ 25 \times 10^5$ Pa exactly) and bar (1 bar = 10^5 Pa). A pressure of 1 bar is the **standard pressure** for reporting data; we denote it $p^{\ominus}$.

Self-test 1.1 Calculate the pressure (in pascals and atmospheres) exerted by a mass of 1.0 kg pressing through the point of a pin of area 1.0×10^{-2} mm^2 at the surface of the Earth. *Hint.* The force exerted by a mass *m* due to gravity at the surface of the Earth is *mg*, where *g* is the acceleration of free fall (see endpaper 2 for its standard value). [0.98 GPa, 9.7×10^3 atm]

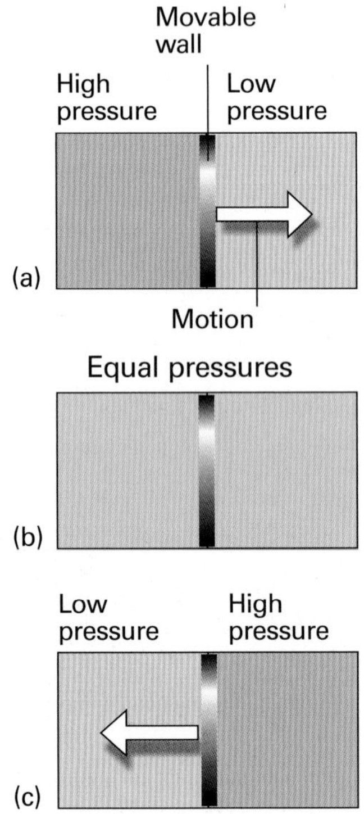

Fig. 1.1 When a region of high pressure is separated from a region of low pressure by a movable wall, the wall will be pushed into one region or the other, as in (a) and (c). However, if the two pressures are identical, the wall will not move (b). The latter condition is one of mechanical equilibrium between the two regions.

If two gases are in separate containers that share a common movable wall (Fig. 1.1), the gas that has the higher pressure will tend to compress (reduce the volume of) the gas that has lower pressure. The pressure of the high-pressure gas will fall as it expands and that of the low-pressure gas will rise as it is compressed. There will come a stage when the two pressures are equal and the wall has no further tendency to move. This condition of equality of pressure on either side of a movable wall (a 'piston') is a state of **mechanical equilibrium** between the two gases. The pressure of a gas is therefore an indication of whether a container that contains the gas will be in mechanical equilibrium with another gas with which it shares a movable wall.

(b) The measurement of pressure

The pressure exerted by the atmosphere is measured with a **barometer**. The original version of a barometer (which was invented by Torricelli, a student of Galileo) was an inverted tube of mercury sealed at the upper end. When the column of mercury is in mechanical equilibrium with the atmosphere, the pressure at its base is equal to that exerted by the atmosphere. It follows that the height of the mercury column is proportional to the external pressure.

Example 1.1 *Calculating the pressure exerted by a column of liquid*

Derive an equation for the pressure at the base of a column of liquid of mass density ρ (rho) and height h at the surface of the Earth.

Method Pressure is defined as $p = F/A$ where F is the force applied to the area A, and $F = mg$. To calculate F we need to know the mass m of the column of liquid, which is its mass density, ρ, multiplied by its volume, V: $m = \rho V$. The first step, therefore, is to calculate the volume of a cylindrical column of liquid.

Answer Let the column have cross-sectional area A; then its volume is Ah and its mass is $m = \rho Ah$. The force the column of this mass exerts at its base is

$$F = mg = \rho Ahg$$

The pressure at the base of the column is therefore

$$p = \frac{F}{A} = \frac{\rho Ahg}{A} = \rho gh \qquad (1.3)$$

Note that the pressure is independent of the shape and cross-sectional area of the column. The mass of the column of a given height increases as the area, but so does the area on which the force acts, so the two cancel.

Self-test 1.2 Derive an expression for the pressure at the base of a column of liquid of length l held at an angle θ (theta) to the vertical (**1**). [$p = \rho gl \cos \theta$]

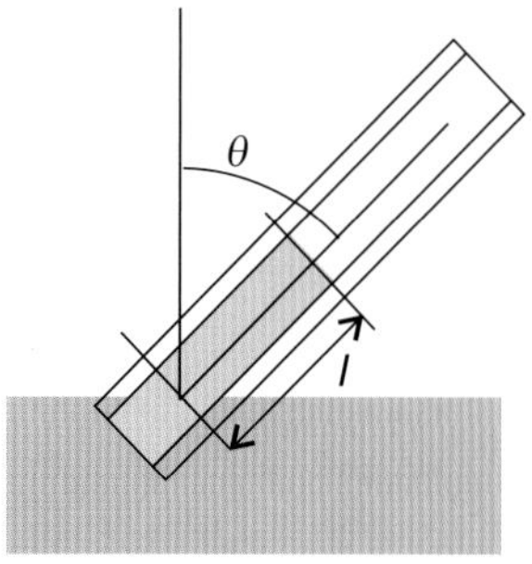

1

The pressure of a sample of gas inside a container is measured by using a pressure gauge, which is a device with electrical properties that depend on the pressure. For instance, a *Bayard–Alpert pressure gauge* is based on the ionization of the molecules present in the gas and the resulting current of ions is interpreted in terms of the pressure. In a *capacitance manometer*, the deflection of a diaphragm relative to a fixed electrode is monitored through its effect on the capacitance of the arrangement. Certain semiconductors also respond to pressure and are used as transducers in solid-state pressure gauges.

(c) Temperature

The concept of temperature springs from the observation that a change in physical state (for example, a change of volume) can occur when two objects are in contact with one another, as when a red-hot metal is plunged into water. Later (Section 2.1) we shall see that the change in state can be interpreted as arising from a flow of energy as heat from one object to another. The **temperature**, T, is the property that indicates the direction of the flow of energy through a thermally conducting, rigid wall. If energy flows from A to B when they are in contact, then we say that A has a higher temperature than B (Fig. 1.2).

It will prove useful to distinguish between two types of boundary that can separate the objects. A boundary is **diathermic** (thermally conducting) if a change of state is observed when two objects at different temperatures are brought into contact.[1] A

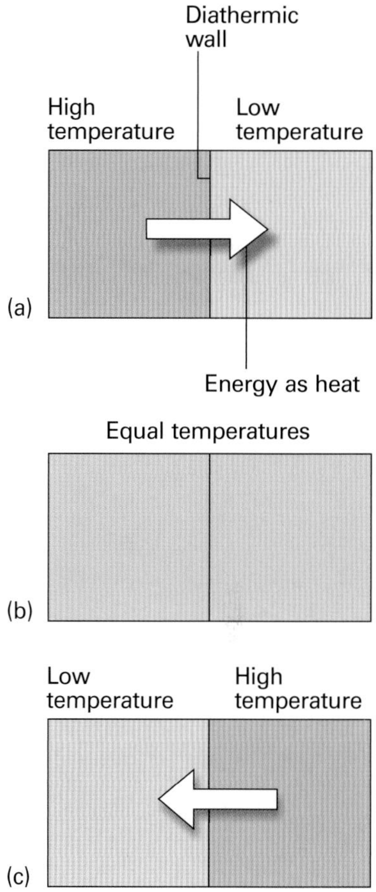

Fig. 1.2 Energy flows as heat from a region at a higher temperature to one at a lower temperature if the two are in contact through a diathermic wall, as in (a) and (c). However, if the two regions have identical temperatures, there is no net transfer of energy as heat even though the two regions are separated by a diathermic wall (b). The latter condition corresponds to the two regions being at thermal equilibrium.

[1] The word dia is from the Greek for 'through'.

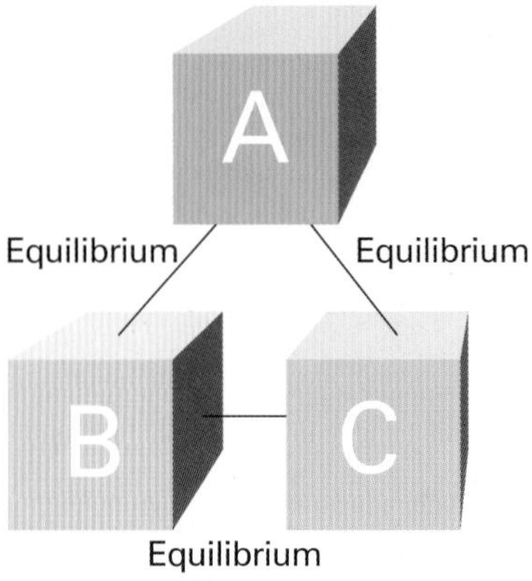

Fig. 1.3 The experience summarized by the Zeroth Law of thermodynamics is that, if an object A is in thermal equilibrium with B and B is in thermal equilibrium with C, then C is in thermal equilibrium with A.

metal container has diathermic walls. A boundary is **adiabatic** (thermally insulating) if no change occurs even though the two objects have different temperatures. A vacuum flask is an approximation to an adiabatic container.

The temperature is a property that indicates whether two objects would be in 'thermal equilibrium' if they were in contact through a diathermic boundary. **Thermal equilibrium** is established if no change of state occurs when two objects A to B are in contact through a diathermic boundary. Suppose an object A (which we can think of as a block of iron) is in thermal equilibrium with an object B (a block of copper), and that B is also in thermal equilibrium with another object C (a flask of water). Then it has been found experimentally that A and C will also be in thermal equilibrium when they are put in contact (Fig. 1.3). This observation is summarized by the **Zeroth Law of thermodynamics**:

> If A is in thermal equilibrium with B, and B is in thermal equilibrium with C, then C is also in thermal equilibrium with A.

The Zeroth Law justifies the concept of temperature and the use of a **thermometer**, a device for measuring the temperature. Thus, suppose that B is a glass capillary containing a liquid, such as mercury, that expands significantly as the temperature increases. Then, when A is in contact with B, the mercury column in the latter has a certain length. According to the Zeroth Law, if the mercury column in B has the same length when it is placed in thermal contact with another object C, then we can predict that no change of state of A and C will occur when they are in thermal contact. Moreover, we can use the length of the mercury column as a measure of the temperatures of A and C.

In the early days of thermometry (and still in laboratory practice today), temperatures were related to the length of a column of liquid, and the difference in lengths shown when the thermometer was first in contact with melting ice and then with boiling water was divided into 100 steps called 'degrees', the lower point being labelled 0. This procedure led to the **Celsius scale** of temperature. In this text, temperatures on the Celsius scale are denoted θ and expressed in *degrees Celsius* (°C). However, because different liquids expand to different extents, and do not always expand uniformly over a given range, thermometers constructed from different materials showed different numerical values of the temperature between their fixed points. The pressure of a gas, however, can be used to construct a **perfect-gas temperature scale** that is independent of the identity of the gas. The perfect-gas scale turns out to be identical to the **thermodynamic temperature scale** to be introduced in Section 3.2c, so we shall use the latter term from now on to avoid a proliferation of names. On the thermodynamic temperature scale, temperatures are denoted T and are normally reported in kelvins, K (not °K). Thermodynamic and Celsius temperatures are related by the exact expression

$$T/\text{K} = \theta/^\circ\text{C} + 273.15 \tag{1.4}$$

This relation, in the form $\theta/^\circ\text{C} = T/\text{K} - 273.15$, is the current definition of the Celsius scale in terms of the more fundamental **Kelvin scale**. It implies that a difference in temperature of 1°C is equivalent to a difference of 1 K.

> ***A note on good practice*** We write $T = 0$, not $T = 0$ K for the zero temperature on the thermodynamic temperature scale. This scale is absolute, and the lowest temperature is 0 regardless of the size of the divisions on the scale (just as we write $p = 0$ for zero pressure, regardless of the size of the units we adopt, such as bar or pascal). However, we write 0°C because the Celsius scale is not absolute.

Illustration 1.1 *Converting temperatures*

To express 25.00°C as a temperature in kelvins, we use eqn 1.4 to write

$$T/\text{K} = (25.00°\text{C})/°\text{C} + 273.15 = 25.00 + 273.15 = 298.15$$

Note how the units (in this case, °C) are cancelled like numbers. This is the procedure called 'quantity calculus' in which a physical quantity (such as the temperature) is the product of a numerical value (25.00) and a unit (1°C). Multiplication of both sides by the unit K then gives $T = 298.15$ K.

A note on good practice When the units need to be specified in an equation, the approved procedure, which avoids any ambiguity, is to write (physical quantity)/units, which is a dimensionless number, just as (25.00°C)/°C = 25.00 in this *Illustration*. Units may be multiplied and cancelled just like numbers.

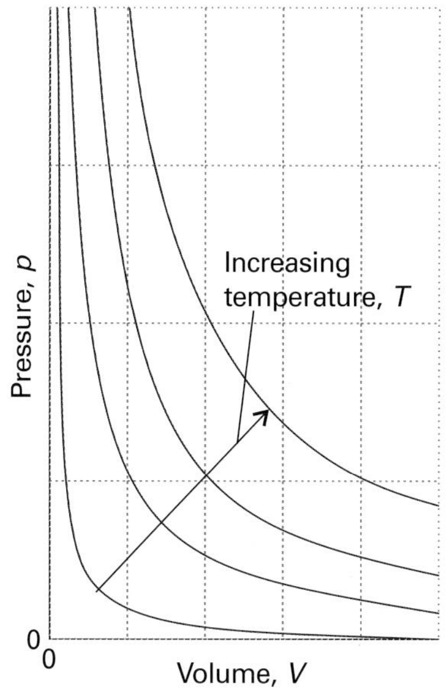

Fig. 1.4 The pressure–volume dependence of a fixed amount of perfect gas at different temperatures. Each curve is a hyperbola (pV = constant) and is called an *isotherm*.

Exploration[3] Explore how the pressure of 1.5 mol CO_2(g) varies with volume as it is compressed at (a) 273 K, (b) 373 K from 30 dm^3 to 15 dm^3.

1.2 The gas laws

The equation of state of a gas at low pressure was established by combining a series of empirical laws.

(a) The perfect gas law

We assume that the following individual gas laws are familiar:

Boyle's law: $pV = \text{constant}$, at constant n, T (1.5)°

Charles's law: $V = \text{constant} \times T$, at constant n, p (1.6a)°

$p = \text{constant} \times T$, at constant n, V (1.6b)°

Avogadro's principle:[2] $V = \text{constant} \times n$ at constant p, T (1.7)°

Boyle's and Charles's laws are examples of a **limiting law**, a law that is strictly true only in a certain limit, in this case $p \to 0$. Equations valid in this limiting sense will be signalled by a ° on the equation number, as in these expressions. Avogadro's principle is commonly expressed in the form 'equal volumes of gases at the same temperature and pressure contain the same numbers of molecules'. In this form, it is increasingly true as $p \to 0$. Although these relations are strictly true only at $p = 0$, they are reasonably reliable at normal pressures ($p \approx 1$ bar) and are used widely throughout chemistry.

Figure 1.4 depicts the variation of the pressure of a sample of gas as the volume is changed. Each of the curves in the graph corresponds to a single temperature and hence is called an **isotherm**. According to Boyle's law, the isotherms of gases are hyperbolas. An alternative depiction, a plot of pressure against 1/volume, is shown in Fig. 1.5. The linear variation of volume with temperature summarized by Charles's law is illustrated in Fig. 1.6. The lines in this illustration are examples of **isobars**, or lines showing the variation of properties at constant pressure. Figure 1.7 illustrates the linear variation of pressure with temperature. The lines in this diagram are **isochores**, or lines showing the variation of properties at constant volume.

Comment 1.2

A hyperbola is a curve obtained by plotting y against x with xy = constant.

[2] Avogadro's principle is a principle rather than a law (a summary of experience) because it depends on the validity of a model, in this case the existence of molecules. Despite there now being no doubt about the existence of molecules, it is still a model-based principle rather than a law.

[3] To solve this and other *Explorations*, use either mathematical software or the *Living graphs* from the text's web site.

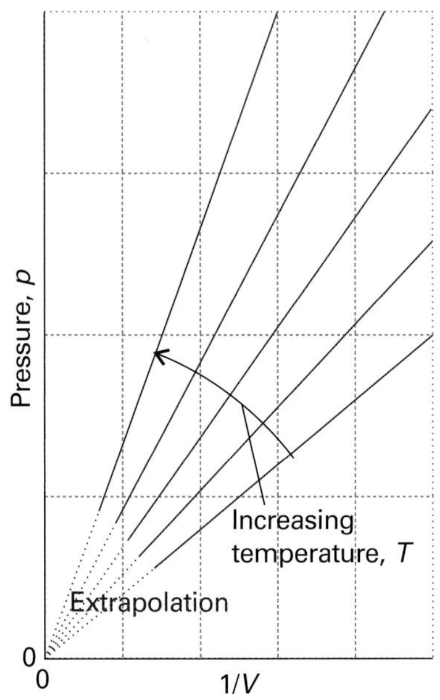

Fig. 1.5 Straight lines are obtained when the pressure is plotted against $1/V$ at constant temperature.

Exploration Repeat *Exploration 1.4*, but plot the data as p against $1/V$.

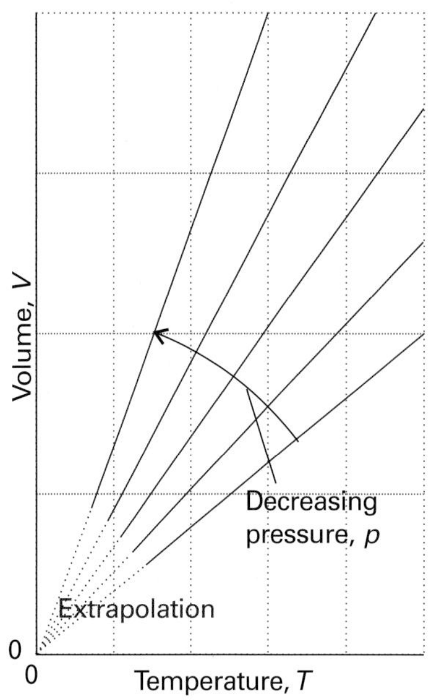

Fig. 1.6 The variation of the volume of a fixed amount of gas with the temperature at constant pressure. Note that in each case the isobars extrapolate to zero volume at $T = 0$, or $\theta = -273°C$.

Exploration Explore how the volume of 1.5 mol $CO_2(g)$ in a container maintained at (a) 1.00 bar, (b) 0.50 bar varies with temperature as it is cooled from 373 K to 273 K.

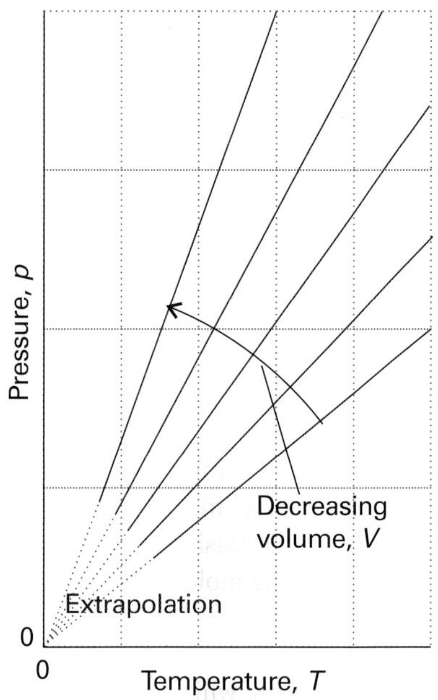

Fig. 1.7 The pressure also varies linearly with the temperature at constant volume, and extrapolates to zero at $T = 0$ (−273°C).

Exploration Explore how the pressure of 1.5 mol $CO_2(g)$ in a container of volume (a) 30 dm^3, (b) 15 dm^3 varies with temperature as it is cooled from 373 K to 273 K.

A note on good practice To test the validity of a relation between two quantities, it is best to plot them in such a way that they should give a straight line, for deviations from a straight line are much easier to detect than deviations from a curve.

The empirical observations summarized by eqns 1.5–7 can be combined into a single expression:

$$pV = \text{constant} \times nT$$

This expression is consistent with Boyle's law ($pV = \text{constant}$) when n and T are constant, with both forms of Charles's law ($p \propto T$, $V \propto T$) when n and either V or p are held constant, and with Avogadro's principle ($V \propto n$) when p and T are constant. The constant of proportionality, which is found experimentally to be the same for all gases, is denoted R and called the **gas constant**. The resulting expression

$$pV = nRT \qquad (1.8)°$$

is the **perfect gas equation**. It is the approximate equation of state of any gas, and becomes increasingly exact as the pressure of the gas approaches zero. A gas that obeys eqn 1.8 exactly under all conditions is called a **perfect gas** (or *ideal gas*). A **real gas**, an actual gas, behaves more like a perfect gas the lower the pressure, and is described exactly by eqn 1.8 in the limit of $p \to 0$. The gas constant R can be determined by evaluating $R = pV/nT$ for a gas in the limit of zero pressure (to guarantee that it is behaving perfectly). However, a more accurate value can be obtained by measuring the speed of sound in a low-pressure gas (argon is used in practice) and extrapolating its value to zero pressure. Table 1.2 lists the values of R in a variety of units.

Table 1.2 The gas constant

R	
8.314 47	$\mathrm{J\,K^{-1}\,mol^{-1}}$
$8.205\,74 \times 10^{-2}$	$\mathrm{dm^3\,atm\,K^{-1}\,mol^{-1}}$
$8.314\,47 \times 10^{-2}$	$\mathrm{dm^3\,bar\,K^{-1}\,mol^{-1}}$
8.314 47	$\mathrm{Pa\,m^3\,K^{-1}\,mol^{-1}}$
1 62.364	$\mathrm{dm^3\,Torr\,K^{-1}\,mol^{-1}}$
1.987 21	$\mathrm{cal\,K^{-1}\,mol^{-1}}$

Molecular interpretation 1.1 *The kinetic model of gases*

The molecular explanation of Boyle's law is that, if a sample of gas is compressed to half its volume, then twice as many molecules strike the walls in a given period of time than before it was compressed. As a result, the average force exerted on the walls is doubled. Hence, when the volume is halved the pressure of the gas is doubled, and $p \times V$ is a constant. Boyle's law applies to all gases regardless of their chemical identity (provided the pressure is low) because at low pressures the average separation of molecules is so great that they exert no influence on one another and hence travel independently. The molecular explanation of Charles's law lies in the fact that raising the temperature of a gas increases the average speed of its molecules. The molecules collide with the walls more frequently and with greater impact. Therefore they exert a greater pressure on the walls of the container.

These qualitative concepts are expressed quantitatively in terms of the kinetic model of gases, which is described more fully in Chapter 21. Briefly, the kinetic model is based on three assumptions:

1. The gas consists of molecules of mass m in ceaseless random motion.
2. The size of the molecules is negligible, in the sense that their diameters are much smaller than the average distance travelled between collisions.
3. The molecules interact only through brief, infrequent, and elastic collisions.

An *elastic collision* is a collision in which the total translational kinetic energy of the molecules is conserved. From the very economical assumptions of the kinetic model, it can be deduced (as we shall show in detail in Chapter 21) that the pressure and volume of the gas are related by

$$pV = \tfrac{1}{3}nMc^2 \qquad (1.9)^\circ$$

where $M = mN_A$, the molar mass of the molecules, and c is the *root mean square speed* of the molecules, the square root of the mean of the squares of the speeds, v, of the molecules:

$$c = \langle v^2 \rangle^{1/2} \qquad (1.10)$$

We see that, if the root mean square speed of the molecules depends only on the temperature, then at constant temperature

$$pV = \text{constant}$$

which is the content of Boyle's law. Moreover, for eqn 1.9 to be the equation of state of a perfect gas, its right-hand side must be equal to nRT. It follows that the root mean square speed of the molecules in a gas at a temperature T must be

$$c = \left(\frac{3RT}{M}\right)^{1/2} \qquad (1.11)^\circ$$

We can conclude that *the root mean square speed of the molecules of a gas is proportional to the square root of the temperature and inversely proportional to the square root of the molar mass.* That is, the higher the temperature, the higher the root mean square speed of the molecules, and, at a given temperature, heavy molecules travel more slowly than light molecules. The root mean square speed of N_2 molecules, for instance, is found from eqn 1.11 to be 515 m s^{-1} at 298 K.

Comment 1.3
For an object of mass m moving at a speed v, the kinetic energy is $E_K = \tfrac{1}{2}mv^2$. The potential energy, E_P or V, of an object is the energy arising from its position (not speed). No universal expression for the potential energy can be given because it depends on the type of interaction the object experiences.

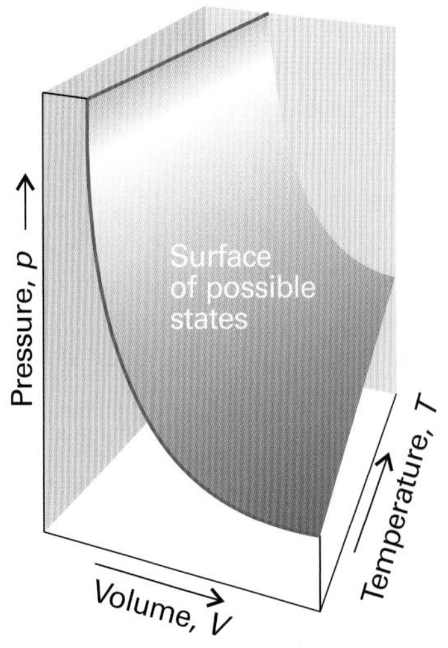

Fig. 1.8 A region of the p,V,T surface of a fixed amount of perfect gas. The points forming the surface represent the only states of the gas that can exist.

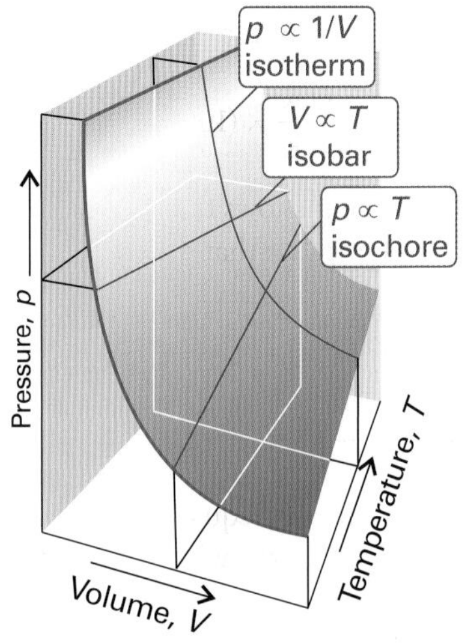

Fig. 1.9 Sections through the surface shown in Fig. 1.8 at constant temperature give the isotherms shown in Fig. 1.4 and the isobars shown in Fig. 1.6.

The surface in Fig. 1.8 is a plot of the pressure of a fixed amount of perfect gas against its volume and thermodynamic temperature as given by eqn 1.8. The surface depicts the only possible states of a perfect gas: the gas cannot exist in states that do not correspond to points on the surface. The graphs in Figs. 1.4 and 1.6 correspond to the sections through the surface (Fig. 1.9).

Example 1.2 *Using the perfect gas equation*

In an industrial process, nitrogen is heated to 500 K in a vessel of constant volume. If it enters the vessel at 100 atm and 300 K, what pressure would it exert at the working temperature if it behaved as a perfect gas?

Method We expect the pressure to be greater on account of the increase in temperature. The perfect gas law in the form $PV/nT = R$ implies that, if the conditions are changed from one set of values to another, then because PV/nT is equal to a constant, the two sets of values are related by the 'combined gas law':

$$\frac{p_1V_1}{n_1T_1} = \frac{p_2V_2}{n_2T_2} \qquad (1.12)°$$

The known and unknown data are summarized in (**2**).

	n	p	V	T
Initial	Same	100	Same	300
Final	Same	?	Same	500

2

Answer Cancellation of the volumes (because $V_1 = V_2$) and amounts (because $n_1 = n_2$) on each side of the combined gas law results in

$$\frac{p_1}{T_1} = \frac{p_2}{T_2}$$

which can be rearranged into

$$p_2 = \frac{T_2}{T_1} \times p_1$$

Substitution of the data then gives

$$p_2 = \frac{500\ \text{K}}{300\ \text{K}} \times (100\ \text{atm}) = 167\ \text{atm}$$

Experiment shows that the pressure is actually 183 atm under these conditions, so the assumption that the gas is perfect leads to a 10 per cent error.

Self-test 1.3 What temperature would result in the same sample exerting a pressure of 300 atm? [900 K]

The perfect gas equation is of the greatest importance in physical chemistry because it is used to derive a wide range of relations that are used throughout thermodynamics. However, it is also of considerable practical utility for calculating the properties of a gas under a variety of conditions. For instance, the molar volume, $V_m = V/n$, of a perfect gas under the conditions called **standard ambient temperature and pressure** (SATP), which means 298.15 K and 1 bar (that is, exactly 10^5 Pa), is easily calculated from $V_m = RT/p$ to be 24.789 $dm^3\ mol^{-1}$. An earlier definition, **standard temperature and pressure** (STP), was 0°C and 1 atm; at STP, the molar volume of a perfect gas is 22.414 $dm^3\ mol^{-1}$. Among other applications, eqn 1.8 can be used to discuss processes in the atmosphere that give rise to the weather.

IMPACT ON ENVIRONMENTAL SCIENCE

I1.1 The gas laws and the weather

The biggest sample of gas readily accessible to us is the atmosphere, a mixture of gases with the composition summarized in Table 1.3. The composition is maintained moderately constant by diffusion and convection (winds, particularly the local turbulence called *eddies*) but the pressure and temperature vary with altitude and with the local conditions, particularly in the troposphere (the 'sphere of change'), the layer extending up to about 11 km.

Table 1.3 The composition of dry air at sea level

Component	Percentage	
	By volume	By mass
Nitrogen, N_2	78.08	75.53
Oxygen, O_2	20.95	23.14
Argon, Ar	0.93	1.28
Carbon dioxide, CO_2	0.031	0.047
Hydrogen, H_2	5.0×10^{-3}	2.0×10^{-4}
Neon, Ne	1.8×10^{-3}	1.3×10^{-3}
Helium, He	5.2×10^{-4}	7.2×10^{-5}
Methane, CH_4	2.0×10^{-4}	1.1×10^{-4}
Krypton, Kr	1.1×10^{-4}	3.2×10^{-4}
Nitric oxide, NO	5.0×10^{-5}	1.7×10^{-6}
Xenon, Xe	8.7×10^{-6}	1.2×10^{-5}
Ozone, O_3: summer	7.0×10^{-6}	1.2×10^{-5}
winter	2.0×10^{-6}	3.3×10^{-6}

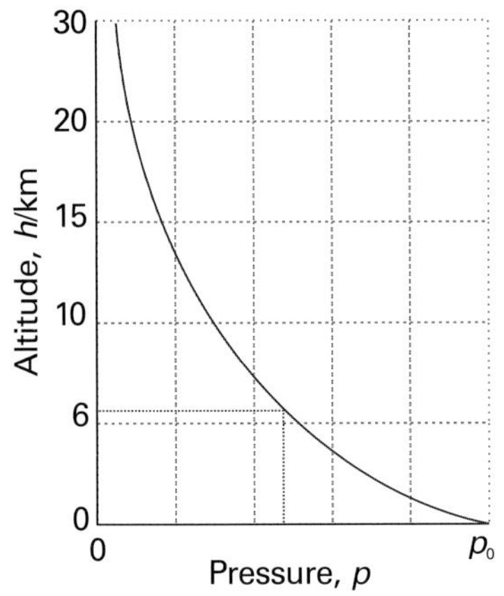

Fig. 1.10 The variation of atmospheric pressure with altitude, as predicted by the barometric formula and as suggested by the 'US Standard Atmosphere', which takes into account the variation of temperature with altitude.

Exploration How would the graph shown in the illustration change if the temperature variation with altitude were taken into account? Construct a graph allowing for a linear decrease in temperature with altitude.

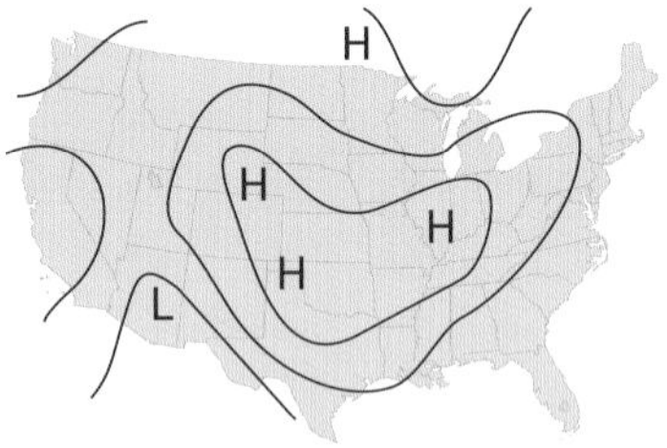

Fig. 1.11 A typical weather map; in this case, for the United States on 1 January 2000.

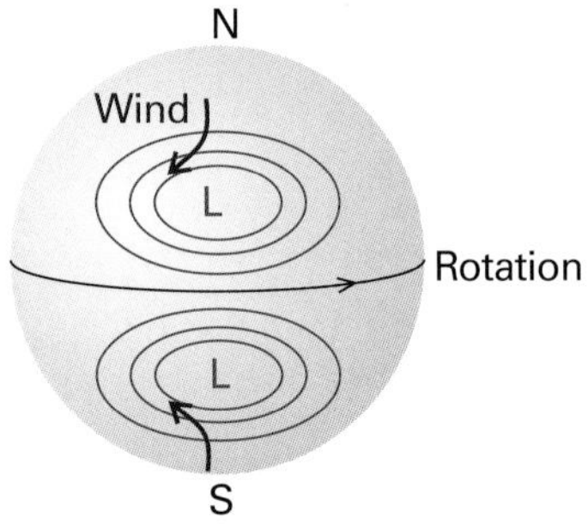

Fig. 1.12 The flow of air ('wind') around regions of high and low pressure in the Northern and Southern hemispheres.

In the troposphere the average temperature is 15°C at sea level, falling to −57°C at the bottom of the tropopause at 11 km. This variation is much less pronounced when expressed on the Kelvin scale, ranging from 288 K to 216 K, an average of 268 K. If we suppose that the temperature has its average value all the way up to the tropopause, then the pressure varies with altitude, *h*, according to the *barometric formula:*

$$p = p_0 e^{-h/H}$$

where p_0 is the pressure at sea level and H is a constant approximately equal to 8 km. More specifically, $H = RT/Mg$, where M is the average molar mass of air and T is the temperature. The barometric formula fits the observed pressure distribution quite well even for regions well above the troposphere (see Fig. 1.10). It implies that the pressure of the air and its density fall to half their sea-level value at $h = H \ln 2$, or 6 km.

Local variations of pressure, temperature, and composition in the troposphere are manifest as 'weather'. A small region of air is termed a *parcel*. First, we note that a parcel of warm air is less dense than the same parcel of cool air. As a parcel rises, it expands adiabatically (that is, without transfer of heat from its surroundings), so it cools. Cool air can absorb lower concentrations of water vapour than warm air, so the moisture forms clouds. Cloudy skies can therefore be associated with rising air and clear skies are often associated with descending air.

The motion of air in the upper altitudes may lead to an accumulation in some regions and a loss of molecules from other regions. The former result in the formation of regions of high pressure ('highs' or anticyclones) and the latter result in regions of low pressure ('lows', depressions, or cyclones). These regions are shown as H and L on the accompanying weather map (Fig. 1.11). The lines of constant pressure—differing by 4 mbar (400 Pa, about 3 Torr)—marked on it are called *isobars*. The elongated regions of high and low pressure are known, respectively, as *ridges* and *troughs*.

In meteorology, large-scale vertical movement is called *convection*. Horizontal pressure differentials result in the flow of air that we call *wind* (see Fig.1.12). Winds coming from the north in the Northern hemisphere and from the south in the Southern hemisphere are deflected towards the west as they migrate from a region where the Earth is rotating slowly (at the poles) to where it is rotating most rapidly (at the equator). Winds travel nearly parallel to the isobars, with low pressure to their left in the Northern hemisphere and to the right in the Southern hemisphere. At the surface, where wind speeds are lower, the winds tend to travel perpendicular to the isobars from high to low pressure. This differential motion results in a spiral outward flow of air clockwise in the Northern hemisphere around a high and an inward counter-clockwise flow around a low.

The air lost from regions of high pressure is restored as an influx of air converges into the region and descends. As we have seen, descending air is associated with clear skies. It also becomes warmer by compression as it descends, so regions of high pressure are associated with high surface temperatures. In winter, the cold surface air may prevent the complete fall of air, and result in a temperature *inversion*, with a layer of warm air over a layer of cold air. Geographical conditions may also trap cool air, as in Los Angeles, and the photochemical pollutants we know as *smog* may be trapped under the warm layer.

(b) Mixtures of gases

When dealing with gaseous mixtures, we often need to know the contribution that each component makes to the total pressure of the sample. The **partial pressure**, p_J, of a gas J in a mixture (any gas, not just a perfect gas), is defined as

$$p_J = x_J p \qquad [1.13]$$

where x_J is the **mole fraction** of the component J, the amount of J expressed as a fraction of the total amount of molecules, n, in the sample:

$$x_J = \frac{n_J}{n} \qquad n = n_A + n_B + \cdots \qquad [1.14]$$

When no J molecules are present, $x_J = 0$; when only J molecules are present, $x_J = 1$. It follows from the definition of x_J that, whatever the composition of the mixture, $x_A + x_B + \cdots = 1$ and therefore that the sum of the partial pressures is equal to the total pressure:

$$p_A + p_B + \cdots = (x_A + x_B + \cdots)p = p \qquad (1.15)$$

This relation is true for both real and perfect gases.

When all the gases are perfect, the partial pressure as defined in eqn 1.13 is also the pressure that each gas would occupy if it occupied the same container alone at the same temperature. The latter is the original meaning of 'partial pressure'. That identification was the basis of the original formulation of **Dalton's law**:

> The pressure exerted by a mixture of gases is the sum of the pressures that each one would exist if it occupied the container alone.

Now, however, the relation between partial pressure (as defined in eqn 1.13) and total pressure (as given by eqn 1.15) is true for all gases and the identification of partial pressure with the pressure that the gas would exert on its own is valid only for a perfect gas.

Example 1.3 *Calculating partial pressures*

The mass percentage composition of dry air at sea level is approximately N_2: 75.5; O_2: 23.2; Ar: 1.3. What is the partial pressure of each component when the total pressure is 1.00 atm?

Method We expect species with a high mole fraction to have a proportionally high partial pressure. Partial pressures are defined by eqn 1.13. To use the equation, we need the mole fractions of the components. To calculate mole fractions, which are defined by eqn 1.14, we use the fact that the amount of molecules J of molar mass M_J in a sample of mass m_J is $n_J = m_J/M_J$. The mole fractions are independent of the total mass of the sample, so we can choose the latter to be 100 g (which makes the conversion from mass percentages very easy). Thus, the mass of N_2 present is 75.5 per cent of 100 g, which is 75.5 g.

Answer The amounts of each type of molecule present in 100 g of air, in which the masses of N_2, O_2, and Ar are 75.5 g, 23.2 g, and 1.3 g, respectively, are

$$n(N_2) = \frac{75.5\ \text{g}}{28.02\ \text{g mol}^{-1}} = \frac{75.5}{28.02}\ \text{mol}$$

$$n(O_2) = \frac{23.2\ \text{g}}{32.00\ \text{g mol}^{-1}} = \frac{23.2}{32.00}\ \text{mol}$$

$$n(\text{Ar}) = \frac{1.3\ \text{g}}{39.95\ \text{g mol}^{-1}} = \frac{1.3}{39.95}\ \text{mol}$$

These three amounts work out as 2.69 mol, 0.725 mol, and 0.033 mol, respectively, for a total of 3.45 mol. The mole fractions are obtained by dividing each of the above amounts by 3.45 mol and the partial pressures are then obtained by multiplying the mole fraction by the total pressure (1.00 atm):

	N_2	O_2	Ar
Mole fraction:	0.780	0.210	0.0096
Partial pressure/atm:	0.780	0.210	0.0096

We have not had to assume that the gases are perfect: partial pressures are defined as $p_J = x_J p$ for any kind of gas.

Self-test 1.4 When carbon dioxide is taken into account, the mass percentages are 75.52 (N_2), 23.15 (O_2), 1.28 (Ar), and 0.046 (CO_2). What are the partial pressures when the total pressure is 0.900 atm? [0.703, 0.189, 0.0084, 0.00027 atm]

Real gases

Real gases do not obey the perfect gas law exactly. Deviations from the law are particularly important at high pressures and low temperatures, especially when a gas is on the point of condensing to liquid.

1.3 Molecular interactions

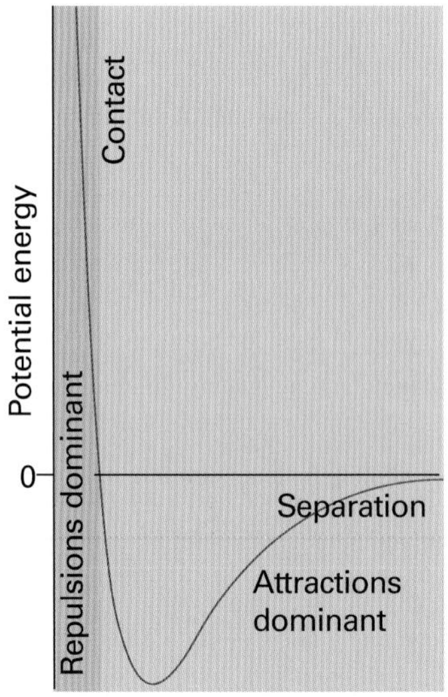

Fig. 1.13 The variation of the potential energy of two molecules on their separation. High positive potential energy (at very small separations) indicates that the interactions between them are strongly repulsive at these distances. At intermediate separations, where the potential energy is negative, the attractive interactions dominate. At large separations (on the right) the potential energy is zero and there is no interaction between the molecules.

Real gases show deviations from the perfect gas law because molecules interact with one another. Repulsive forces between molecules assist expansion and attractive forces assist compression.

Repulsive forces are significant only when molecules are almost in contact: they are short-range interactions, even on a scale measured in molecular diameters (Fig. 1.13). Because they are short-range interactions, repulsions can be expected to be important only when the average separation of the molecules is small. This is the case at high pressure, when many molecules occupy a small volume. On the other hand, attractive intermolecular forces have a relatively long range and are effective over several molecular diameters. They are important when the molecules are fairly close together but not necessarily touching (at the intermediate separations in Fig. 1.13). Attractive forces are ineffective when the molecules are far apart (well to the right in Fig. 1.13). Intermolecular forces are also important when the temperature is so low that the molecules travel with such low mean speeds that they can be captured by one another.

At low pressures, when the sample occupies a large volume, the molecules are so far apart for most of the time that the intermolecular forces play no significant role, and the gas behaves virtually perfectly. At moderate pressures, when the average separation of the molecules is only a few molecular diameters, the attractive forces dominate the repulsive forces. In this case, the gas can be expected to be more compressible than a perfect gas because the forces help to draw the molecules together. At high pressures, when the average separation of the molecules is small, the repulsive forces dominate and the gas can be expected to be less compressible because now the forces help to drive the molecules apart.

(a) The compression factor

The **compression factor**, Z, of a gas is the ratio of its measured molar volume, $V_m = V/n$, to the molar volume of a perfect gas, V_m^o, at the same pressure and temperature:

$$Z = \frac{V_m}{V_m^\circ} \quad [1.16]$$

Because the molar volume of a perfect gas is equal to RT/p, an equivalent expression is $Z = RT/pV_m^\circ$, which we can write as

$$pV_m = RTZ \quad (1.17)$$

Because for a perfect gas $Z = 1$ under all conditions, deviation of Z from 1 is a measure of departure from perfect behaviour.

Some experimental values of Z are plotted in Fig. 1.14. At very low pressures, all the gases shown have $Z \approx 1$ and behave nearly perfectly. At high pressures, all the gases have $Z > 1$, signifying that they have a larger molar volume than a perfect gas. Repulsive forces are now dominant. At intermediate pressures, most gases have $Z < 1$, indicating that the attractive forces are reducing the molar volume relative to that of a perfect gas.

(b) Virial coefficients

Figure 1.15 shows the experimental isotherms for carbon dioxide. At large molar volumes and high temperatures the real-gas isotherms do not differ greatly from perfect-gas isotherms. The small differences suggest that the perfect gas law is in fact the first term in an expression of the form

$$pV_m = RT(1 + B'p + C'p^2 + \cdots) \quad (1.18)$$

This expression is an example of a common procedure in physical chemistry, in which a simple law that is known to be a good first approximation (in this case $pV = nRT$) is

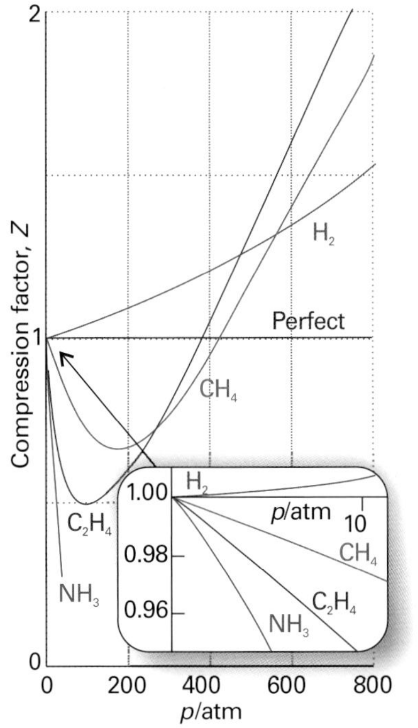

Fig. 1.14 The variation of the compression factor, Z, with pressure for several gases at 0°C. A perfect gas has $Z = 1$ at all pressures. Notice that, although the curves approach 1 as $p \to 0$, they do so with different slopes.

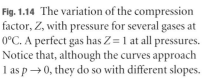

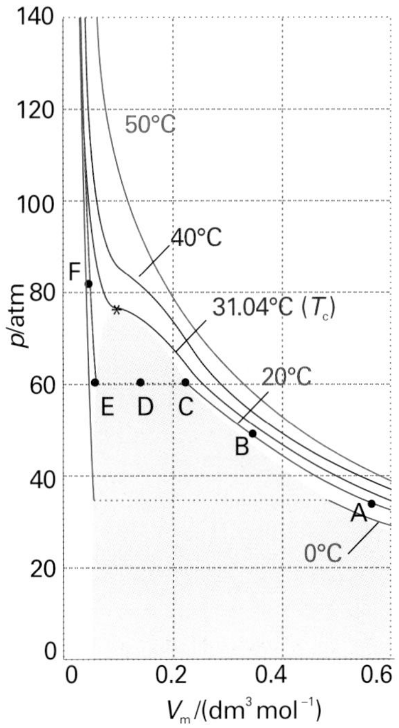

Fig. 1.15 Experimental isotherms of carbon dioxide at several temperatures. The 'critical isotherm', the isotherm at the critical temperature, is at 31.04°C. The critical point is marked with a star.

Comment 1.4
Series expansions are discussed in *Appendix 2.*

treated as the first term in a series in powers of a variable (in this case p). A more convenient expansion for many applications is

$$pV_m = RT\left(1 + \frac{B}{V_m} + \frac{C}{V_m^2} + \cdots\right) \quad (1.19)$$

These two expressions are two versions of the **virial equation of state.**[4] By comparing the expression with eqn 1.17 we see that the term in parentheses can be identified with the compression factor, Z.

The coefficients $B, C, \ldots$, which depend on the temperature, are the second, third, ... **virial coefficients** (Table 1.4); the first virial coefficient is 1. The third virial coefficient, C, is usually less important than the second coefficient, B, in the sense that at typical molar volumes $C/V_m^2 \ll B/V_m$.

Synoptic Table 1.4* Second virial coefficients, $B/(\text{cm}^3\ \text{mol}^{-1})$

	Temperature	
	273 K	600 K
Ar	−21.7	11.9
CO_2	−149.7	−12.4
N_2	−10.5	21.7
Xe	−153.7	−19.6

* More values are given in the *Data section.*

We can use the virial equation to demonstrate the important point that, although the equation of state of a real gas may coincide with the perfect gas law as $p \to 0$, not all its properties necessarily coincide with those of a perfect gas in that limit. Consider, for example, the value of dZ/dp, the slope of the graph of compression factor against pressure. For a perfect gas $dZ/dp = 0$ (because $Z = 1$ at all pressures), but for a real gas from eqn 1.18 we obtain

$$\frac{dZ}{dp} = B' + 2pC' + \cdots \to B' \quad \text{as} \quad p \to 0 \quad (1.20a)$$

However, B' is not necessarily zero, so the slope of Z with respect to p does not necessarily approach 0 (the perfect gas value), as we can see in Fig. 1.14. Because several physical properties of gases depend on derivatives, the properties of real gases do not always coincide with the perfect gas values at low pressures. By a similar argument,

$$\frac{dZ}{d(1/V_m)} \to B \text{ as } V_m \to \infty, \quad \text{corresponding to} \quad p \to 0 \quad (1.20b)$$

Because the virial coefficients depend on the temperature, there may be a temperature at which $Z \to 1$ with zero slope at low pressure or high molar volume (Fig. 1.16). At this temperature, which is called the **Boyle temperature**, T_B, the properties of the real gas do coincide with those of a perfect gas as $p \to 0$. According to eqn 1.20b, Z has zero slope as $p \to 0$ if $B = 0$, so we can conclude that $B = 0$ at the Boyle temperature. It then follows from eqn 1.19 that $pV_m \approx RT_B$ over a more extended range of pressures than at other temperatures because the first term after 1 (that is, B/V_m) in the virial equation is zero and C/V_m^2 and higher terms are negligibly small. For helium $T_B = 22.64$ K; for air $T_B = 346.8$ K; more values are given in Table 1.5.

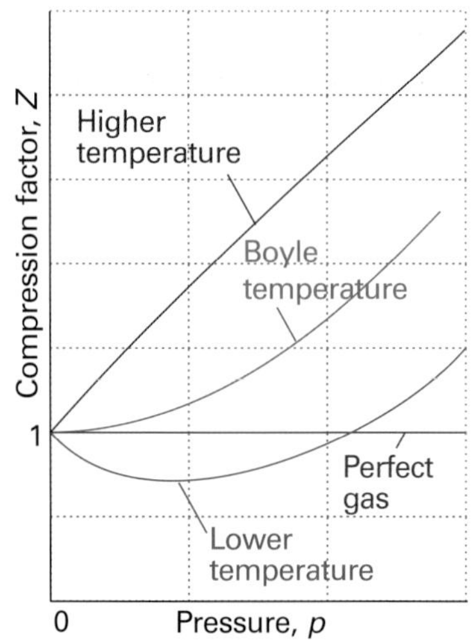

Fig. 1.16 The compression factor, Z, approaches 1 at low pressures, but does so with different slopes. For a perfect gas, the slope is zero, but real gases may have either positive or negative slopes, and the slope may vary with temperature. At the Boyle temperature, the slope is zero and the gas behaves perfectly over a wider range of conditions than at other temperatures.

Synoptic Table 1.5* Critical constants of gases

	p_c/atm	$V_c/(\text{cm}^3\ \text{mol}^{-1})$	T_c/K	Z_c	T_B/K
Ar	48.0	75.3	150.7	0.292	411.5
CO_2	72.9	94.0	304.2	0.274	714.8
He	2.26	57.8	5.2	0.305	22.64
O_2	50.14	78.0	154.8	0.308	405.9

* More values are given in the *Data section.*

[4] The name comes from the Latin word for force. The coefficients are sometimes denoted $B_2, B_3, \ldots$.

(c) Condensation

Now consider what happens when we compress a sample of gas initially in the state marked A in Fig. 1.15 at constant temperature (by pushing in a piston). Near A, the pressure of the gas rises in approximate agreement with Boyle's law. Serious deviations from that law begin to appear when the volume has been reduced to B.

At C (which corresponds to about 60 atm for carbon dioxide), all similarity to perfect behaviour is lost, for suddenly the piston slides in without any further rise in pressure: this stage is represented by the horizontal line CDE. Examination of the contents of the vessel shows that just to the left of C a liquid appears, and there are two phases separated by a sharply defined surface. As the volume is decreased from C through D to E, the amount of liquid increases. There is no additional resistance to the piston because the gas can respond by condensing. The pressure corresponding to the line CDE, when both liquid and vapour are present in equilibrium, is called the **vapour pressure** of the liquid at the temperature of the experiment.

At E, the sample is entirely liquid and the piston rests on its surface. Any further reduction of volume requires the exertion of considerable pressure, as is indicated by the sharply rising line to the left of E. Even a small reduction of volume from E to F requires a great increase in pressure.

(d) Critical constants

The isotherm at the temperature T_c (304.19 K, or 31.04°C for CO_2) plays a special role in the theory of the states of matter. An isotherm slightly below T_c behaves as we have already described: at a certain pressure, a liquid condenses from the gas and is distinguishable from it by the presence of a visible surface. If, however, the compression takes place at T_c itself, then a surface separating two phases does not appear and the volumes at each end of the horizontal part of the isotherm have merged to a single point, the **critical point** of the gas. The temperature, pressure, and molar volume at the critical point are called the **critical temperature**, T_c, **critical pressure**, p_c, and **critical molar volume**, V_c, of the substance. Collectively, p_c, V_c, and T_c are the **critical constants** of a substance (Table 1.5).

Comment 1.5
The web site contains links to online databases of properties of gases.

At and above T_c, the sample has a single phase that occupies the entire volume of the container. Such a phase is, by definition, a gas. Hence, the liquid phase of a substance does not form above the critical temperature. The critical temperature of oxygen, for instance, signifies that it is impossible to produce liquid oxygen by compression alone if its temperature is greater than 155 K: to liquefy oxygen—to obtain a fluid phase that does not occupy the entire volume—the temperature must first be lowered to below 155 K, and then the gas compressed isothermally. The single phase that fills the entire volume when $T > T_c$ may be much denser than we normally consider typical of gases, and the name **supercritical fluid** is preferred.

1.4 The van der Waals equation

We can draw conclusions from the virial equations of state only by inserting specific values of the coefficients. It is often useful to have a broader, if less precise, view of all gases. Therefore, we introduce the approximate equation of state suggested by J.D. van der Waals in 1873. This equation is an excellent example of an expression that can be obtained by thinking scientifically about a mathematically complicated but physically simple problem, that is, it is a good example of 'model building'.

The **van der Waals equation** is

$$p = \frac{nRT}{V - nb} - a\left(\frac{n}{V}\right)^2 \qquad (1.21a)$$

Synoptic Table 1.6* van der Waals coefficients

	$a/(\text{atm dm}^6\text{ mol}^{-2})$	$b/(10^{-2}\text{ dm}^3\text{ mol}^{-1})$
Ar	1.337	3.20
CO_2	3.610	4.29
He	0.0341	2.38
Xe	4.137	5.16

* More values are given in the *Data section*.

and a derivation is given in *Justification 1.1*. The equation is often written in terms of the molar volume $V_m = V/n$ as

$$p = \frac{RT}{V_m - b} - \frac{a}{V_m^2} \tag{1.21b}$$

The constants a and b are called the **van der Waals coefficients**. They are characteristic of each gas but independent of the temperature (Table 1.6).

Justification 1.1 *The van der Waals equation of state*

The repulsive interactions between molecules are taken into account by supposing that they cause the molecules to behave as small but impenetrable spheres. The non-zero volume of the molecules implies that instead of moving in a volume V they are restricted to a smaller volume $V - nb$, where nb is approximately the total volume taken up by the molecules themselves. This argument suggests that the perfect gas law $p = nRT/V$ should be replaced by

$$p = \frac{nRT}{V - nb}$$

when repulsions are significant. The closest distance of two hard-sphere molecules of radius r, and volume $V_{\text{molecule}} = \frac{4}{3}\pi r^3$, is $2r$, so the volume excluded is $\frac{4}{3}\pi(2r)^3$, or $8V_{\text{molecule}}$. The volume excluded per molecule is one-half this volume, or $4V_{\text{molecule}}$, so $b \approx 4V_{\text{molecule}}N_A$.

The pressure depends on both the frequency of collisions with the walls and the force of each collision. Both the frequency of the collisions and their force are reduced by the attractive forces, which act with a strength proportional to the molar concentration, n/V, of molecules in the sample. Therefore, because both the frequency and the force of the collisions are reduced by the attractive forces, the pressure is reduced in proportion to the square of this concentration. If the reduction of pressure is written as $-a(n/V)^2$, where a is a positive constant characteristic of each gas, the combined effect of the repulsive and attractive forces is the van der Waals equation of state as expressed in eqn 1.21.

In this *Justification* we have built the van der Waals equation using vague arguments about the volumes of molecules and the effects of forces. The equation can be derived in other ways, but the present method has the advantage that it shows how to derive the form of an equation out of general ideas. The derivation also has the advantage of keeping imprecise the significance of the coefficients a and b: they are much better regarded as empirical parameters than as precisely defined molecular properties.

Example 1.4 *Using the van der Waals equation to estimate a molar volume*

Estimate the molar volume of CO_2 at 500 K and 100 atm by treating it as a van der Waals gas.

Method To express eqn 1.21b as an equation for the molar volume, we multiply both sides by $(V_m - b)V_m^2$, to obtain

$$(V_m - b)V_m^2 p = RTV_m^2 - (V_m - b)a$$

and, after division by p, collect powers of V_m to obtain

$$V_m^3 - \left(b + \frac{RT}{p}\right)V_m^2 + \left(\frac{a}{p}\right)V_m - \frac{ab}{p} = 0$$

Although closed expressions for the roots of a cubic equation can be given, they are very complicated. Unless analytical solutions are essential, it is usually more expedient to solve such equations with commercial software.

Answer According to Table 1.6, $a = 3.592$ dm^6 atm mol^{-2} and $b = 4.267 \times 10^{-2}$ dm^3 mol^{-1}. Under the stated conditions, $RT/p = 0.410$ dm^3 mol^{-1}. The coefficients in the equation for V_m are therefore

$$b + RT/p = 0.453 \text{ dm}^3 \text{ mol}^{-1}$$
$$a/p = 3.61 \times 10^{-2} \text{ (dm}^3 \text{ mol}^{-1})^2$$
$$ab/p = 1.55 \times 10^{-3} \text{ (dm}^3 \text{ mol}^{-1})^3$$

Therefore, on writing $x = V_m/(\text{dm}^3 \text{ mol}^{-1})$, the equation to solve is

$$x^3 - 0.453x^2 + (3.61 \times 10^{-2})x - (1.55 \times 10^{-3}) = 0$$

The acceptable root is $x = 0.366$, which implies that $V_m = 0.366$ dm^3 mol^{-1}. For a perfect gas under these conditions, the molar volume is 0.410 dm^3 mol^{-1}.

Self-test 1.5 Calculate the molar volume of argon at 100°C and 100 atm on the assumption that it is a van der Waals gas. [0.298 dm^3 mol^{-1}]

(a) The reliability of the equation

We now examine to what extent the van der Waals equation predicts the behaviour of real gases. It is too optimistic to expect a single, simple expression to be the true equation of state of all substances, and accurate work on gases must resort to the virial equation, use tabulated values of the coefficients at various temperatures, and analyse the systems numerically. The advantage of the van der Waals equation, however, is that it is analytical (that is, expressed symbolically) and allows us to draw some general conclusions about real gases. When the equation fails we must use one of the other equations of state that have been proposed (some are listed in Table 1.7), invent a new one, or go back to the virial equation.

That having been said, we can begin to judge the reliability of the equation by comparing the isotherms it predicts with the experimental isotherms in Fig. 1.15. Some

Table 1.7 Selected equations of state

	Equation	Reduced form*	Critical constants		
			p_c	V_c	T_c
Perfect gas	$p = \dfrac{RT}{V_m}$				
van der Waals	$p = \dfrac{RT}{V_m - b} - \dfrac{a}{V_m^2}$	$p = \dfrac{8T_r}{3V_r - 1} - \dfrac{3}{V_r^2}$	$\dfrac{a}{27b^2}$	$3b$	$\dfrac{8a}{27bR}$
Berthelot	$p = \dfrac{RT}{V_m - b} - \dfrac{a}{TV_m^2}$	$p = \dfrac{8T_r}{3V_r - 1} - \dfrac{3}{T_rV_r^2}$	$\dfrac{1}{12}\left(\dfrac{2aR}{3b^3}\right)^{1/2}$	$3b$	$\dfrac{2}{3}\left(\dfrac{2a}{3bR}\right)^{1/2}$
Dieterici	$p = \dfrac{RTe^{-a/RTV_m}}{V_m - b}$	$p = \dfrac{e^2T_re^{-2/T_rV_r}}{2V_r - 1}$	$\dfrac{a}{4e^2b^2}$	$2b$	$\dfrac{a}{4bR}$
Virial	$p = \dfrac{RT}{V_m}\left\{1 + \dfrac{B(T)}{V_m} + \dfrac{C(T)}{V_m^2} + \cdots\right\}$				

* Reduced variables are defined in Section 1.5.

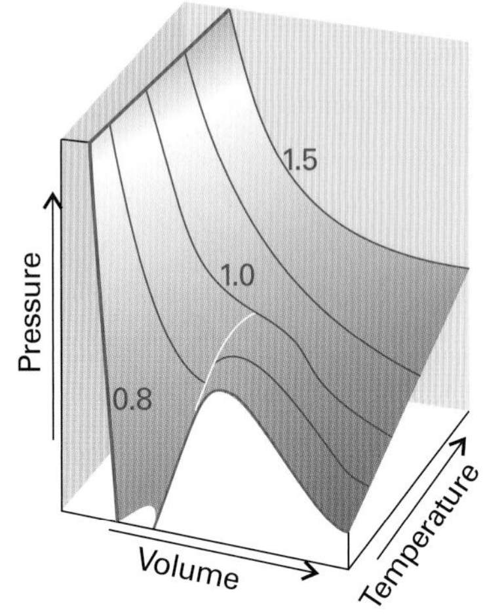

Fig. 1.17 The surface of possible states allowed by the van der Waals equation. Compare this surface with that shown in Fig. 1.8.

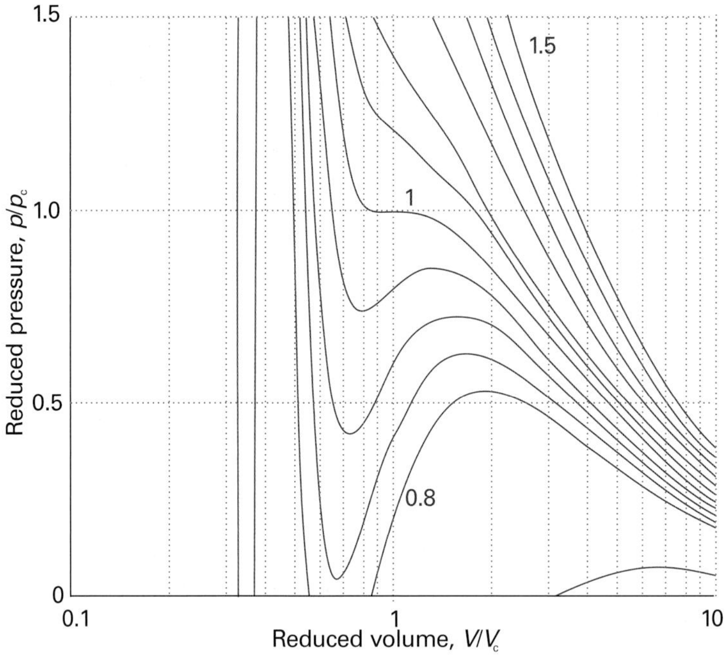

Fig. 1.18 Van der Waals isotherms at several values of T/T_c. Compare these curves with those in Fig. 1.15. The van der Waals loops are normally replaced by horizontal straight lines. The critical isotherm is the isotherm for $T/T_c = 1$.

Exploration Calculate the molar volume of chlorine gas on the basis of the van der Waals equation of state at 250 K and 150 kPa and calculate the percentage difference from the value predicted by the perfect gas equation.

calculated isotherms are shown in Figs. 1.17 and 1.18. Apart from the oscillations below the critical temperature, they do resemble experimental isotherms quite well. The oscillations, the **van der Waals loops**, are unrealistic because they suggest that under some conditions an increase of pressure results in an increase of volume. Therefore they are replaced by horizontal lines drawn so the loops define equal areas above and below the lines: this procedure is called the **Maxwell construction** (**3**). The van der Waals coefficients, such as those in Table 1.7, are found by fitting the calculated curves to the experimental curves.

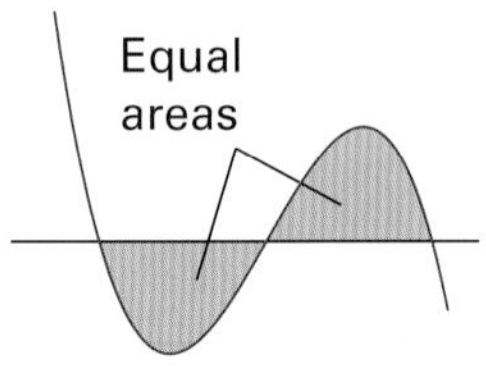

3

(b) The features of the equation

The principal features of the van der Waals equation can be summarized as follows.

(1) Perfect gas isotherms are obtained at high temperatures and large molar volumes.

When the temperature is high, RT may be so large that the first term in eqn 1.21b greatly exceeds the second. Furthermore, if the molar volume is large in the sense $V_m \gg b$, then the denominator $V_m - b \approx V_m$. Under these conditions, the equation reduces to $p = RT/V_m$, the perfect gas equation.

(2) Liquids and gases coexist when cohesive and dispersing effects are in balance.

The van der Waals loops occur when both terms in eqn 1.21b have similar magnitudes. The first term arises from the kinetic energy of the molecules and their repulsive interactions; the second represents the effect of the attractive interactions.

(3) The critical constants are related to the van der Waals coefficients.

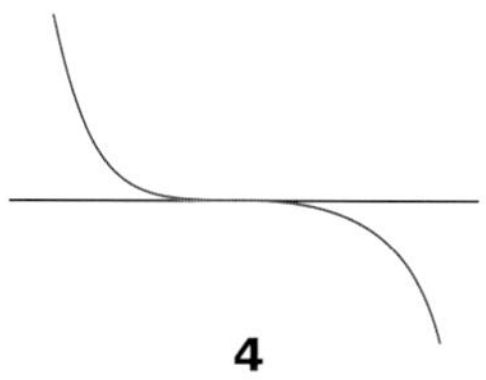

4

For $T < T_c$, the calculated isotherms oscillate, and each one passes through a minimum followed by a maximum. These extrema converge as $T \to T_c$ and coincide at $T = T_c$; at the critical point the curve has a flat inflexion (**4**). From the properties of curves, we know that an inflexion of this type occurs when both the first and second derivatives are zero. Hence, we can find the critical constants by calculating these derivatives and setting them equal to zero:

$$\frac{dp}{dV_m} = -\frac{RT}{(V_m - b)^2} + \frac{2a}{V_m^3} = 0$$

$$\frac{d^2p}{dV_m^2} = \frac{2RT}{(V_m - b)^3} - \frac{6a}{V_m^4} = 0$$

at the critical point. The solutions of these two equations (and using eqn 1.21b to calculate p_c from V_c and T_c) are

$$V_c = 3b \qquad p_c = \frac{a}{27b^2} \qquad T_c = \frac{8a}{27Rb} \tag{1.22}$$

These relations provide an alternative route to the determination of a and b from the values of the critical constants. They can be tested by noting that the **critical compression factor**, Z_c, is predicted to be equal to

$$Z_c = \frac{p_c V_c}{RT_c} = \frac{3}{8} \tag{1.23}$$

for all gases. We see from Table 1.5 that, although $Z_c < \frac{3}{8} = 0.375$, it is approximately constant (at 0.3) and the discrepancy is reasonably small.

1.5 The principle of corresponding states

An important general technique in science for comparing the properties of objects is to choose a related fundamental property of the same kind and to set up a relative scale on that basis. We have seen that the critical constants are characteristic properties of gases, so it may be that a scale can be set up by using them as yardsticks. We therefore introduce the dimensionless **reduced variables** of a gas by dividing the actual variable by the corresponding critical constant:

$$p_r = \frac{p}{p_c} \qquad V_r = \frac{V_m}{V_c} \qquad T_r = \frac{T}{T_c} \qquad [1.24]$$

If the reduced pressure of a gas is given, we can easily calculate its actual pressure by using $p = p_r p_c$, and likewise for the volume and temperature. Van der Waals, who first tried this procedure, hoped that gases confined to the same reduced volume, V_r, at the same reduced temperature, T_r, would exert the same reduced pressure, p_r. The hope was largely fulfilled (Fig. 1.19). The illustration shows the dependence of the compression factor on the reduced pressure for a variety of gases at various reduced temperatures. The success of the procedure is strikingly clear: compare this graph with Fig. 1.14, where similar data are plotted without using reduced variables. The observation that real gases at the same reduced volume and reduced temperature exert the same reduced pressure is called the **principle of corresponding states**. The principle is only an approximation. It works best for gases composed of spherical molecules; it fails, sometimes badly, when the molecules are non-spherical or polar.

The van der Waals equation sheds some light on the principle. First, we express eqn 1.21b in terms of the reduced variables, which gives

$$p_r p_c = \frac{RT_r T_c}{V_r V_c - b} - \frac{a}{V_r^2 V_c^2}$$

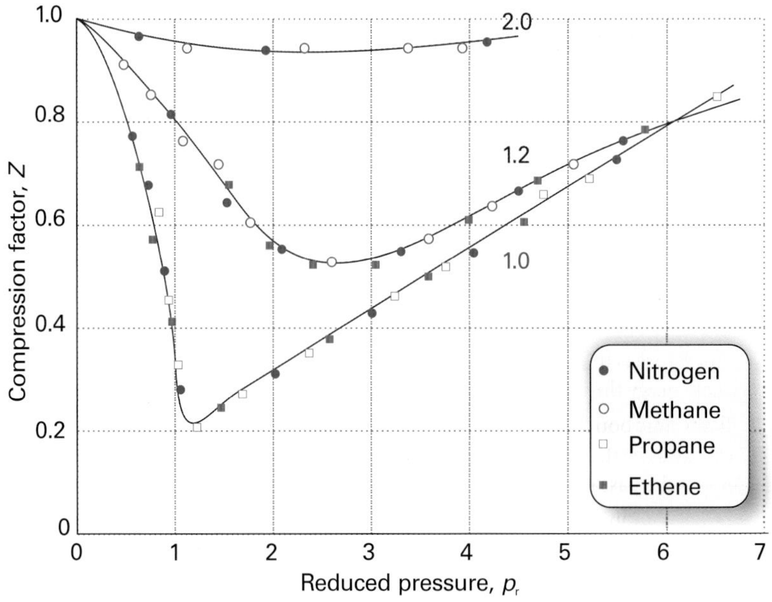

Fig. 1.19 The compression factors of four of the gases shown in Fig. 1.14 plotted using reduced variables. The curves are labelled with the reduced temperature $T_r = T/T_c$. The use of reduced variables organizes the data on to single curves.

Exploration Is there a set of conditions at which the compression factor of a van der Waals gas passes through a minimum? If so, how does the location and value of the minimum value of Z depend on the coefficients a and b?

Then we express the critical constants in terms of a and b by using eqn 1.22:

$$\frac{ap_r}{27b^2} = \frac{8aT_r}{27b(3bV_r - b)} - \frac{a}{9b^2V_r^2}$$

which can be reorganized into

$$p_r = \frac{8T_r}{3V_r - 1} - \frac{3}{V_r^2} \qquad (1.25)$$

This equation has the same form as the original, but the coefficients a and b, which differ from gas to gas, have disappeared. It follows that if the isotherms are plotted in terms of the reduced variables (as we did in fact in Fig. 1.18 without drawing attention to the fact), then the same curves are obtained whatever the gas. This is precisely the content of the principle of corresponding states, so the van der Waals equation is compatible with it.

Looking for too much significance in this apparent triumph is mistaken, because other equations of state also accommodate the principle (Table 1.7). In fact, all we need are two parameters playing the roles of a and b, for then the equation can always be manipulated into reduced form. The observation that real gases obey the principle approximately amounts to saying that the effects of the attractive and repulsive interactions can each be approximated in terms of a single parameter. The importance of the principle is then not so much its theoretical interpretation but the way that it enables the properties of a range of gases to be coordinated on to a single diagram (for example, Fig. 1.19 instead of Fig. 1.14).

Checklist of key ideas

- ☐ 1. A gas is a form of matter that fills any container it occupies.
- ☐ 2. An equation of state interrelates pressure, volume, temperature, and amount of substance: $p = f(T,V,n)$.
- ☐ 3. The pressure is the force divided by the area to which the force is applied. The standard pressure is $p^\ominus = 1$ bar (10^5 Pa).
- ☐ 4. Mechanical equilibrium is the condition of equality of pressure on either side of a movable wall.
- ☐ 5. Temperature is the property that indicates the direction of the flow of energy through a thermally conducting, rigid wall.
- ☐ 6. A diathermic boundary is a boundary that permits the passage of energy as heat. An adiabatic boundary is a boundary that prevents the passage of energy as heat.
- ☐ 7. Thermal equilibrium is a condition in which no change of state occurs when two objects A and B are in contact through a diathermic boundary.
- ☐ 8. The Zeroth Law of thermodynamics states that, if A is in thermal equilibrium with B, and B is in thermal equilibrium with C, then C is also in thermal equilibrium with A.
- ☐ 9. The Celsius and thermodynamic temperature scales are related by $T/\text{K} = \theta/°\text{C} + 273.15$.
- ☐ 10. A perfect gas obeys the perfect gas equation, $pV = nRT$, exactly under all conditions.
- ☐ 11. Dalton's law states that the pressure exerted by a mixture of gases is the sum of the partial pressures of the gases.
- ☐ 12. The partial pressure of any gas is defined as $p_J = x_J p$, where $x_J = n_J/n$ is its mole fraction in a mixture and p is the total pressure.
- ☐ 13. In real gases, molecular interactions affect the equation of state; the true equation of state is expressed in terms of virial coefficients $B, C, \ldots : pV_m = RT(1 + B/V_m + C/V_m^2 + \cdots)$.
- ☐ 14. The vapour pressure is the pressure of a vapour in equilibrium with its condensed phase.
- ☐ 15. The critical point is the point at which the volumes at each end of the horizontal part of the isotherm have merged to a single point. The critical constants p_c, V_c, and T_c are the pressure, molar volume, and temperature, respectively, at the critical point.
- ☐ 16. A supercritical fluid is a dense fluid phase above its critical temperature and pressure.
- ☐ 17. The van der Waals equation of state is an approximation to the true equation of state in which attractions are represented by a parameter a and repulsions are represented by a parameter b: $p = nRT/(V - nb) - a(n/V)^2$.
- ☐ 18. A reduced variable is the actual variable divided by the corresponding critical constant.
- ☐ 19. According to the principle of corresponding states, real gases at the same reduced volume and reduced temperature exert the same reduced pressure.

Further reading

Articles and texts

J.L. Pauley and E.H. Davis, *P-V-T* isotherms of real gases: Experimental versus calculated values. *J. Chem. Educ.* **63**, 466 (1986).

M. Ross, Equations of state. In *Encyclopedia of applied physics* (ed. G.L. Trigg), **6**, 291. VCH, New York (1993).

A. J. Walton, *Three phases of matter*. Oxford University Press (1983).

R.P. Wayne, *Chemistry of atmospheres, an introduction to the chemistry of atmospheres of earth, the planets, and their satellites.* Oxford University Press (2000).

Sources of data and information

J.H. Dymond and E.B. Smith, *The virial coefficients of pure gases and mixtures.* Oxford University Press (1980).

A.D. McNaught and A. Wilkinson, *Compendium of chemical terminology.* Blackwell Scientific, Oxford (1997).

Discussion questions

1.1 Explain how the perfect gas equation of state arises by combination of Boyle's law, Charles's law, and Avogadro's principle.

1.2 Explain the term 'partial pressure' and explain why Dalton's law is a limiting law.

1.3 Explain how the compression factor varies with pressure and temperature and describe how it reveals information about intermolecular interactions in real gases.

1.4 What is the significance of the critical constants?

1.5 Describe the formulation of the van der Waals equation and suggest a rationale for one other equation of state in Table 1.7.

1.6 Explain how the van der Waals equation accounts for critical behaviour.

Exercises

1.1(a) (a) Could 131 g of xenon gas in a vessel of volume 1.0 dm^3 exert a pressure of 20 atm at 25°C if it behaved as a perfect gas? If not, what pressure would it exert? (b) What pressure would it exert if it behaved as a van der Waals gas?

1.1(b) (a) Could 25 g of argon gas in a vessel of volume 1.5 dm^3 exert a pressure of 2.0 bar at 30°C if it behaved as a perfect gas? If not, what pressure would it exert? (b) What pressure would it exert if it behaved as a van der Waals gas?

1.2(a) A perfect gas undergoes isothermal compression, which reduces its volume by 2.20 dm^3. The final pressure and volume of the gas are 5.04 bar and 4.65 dm^3, respectively. Calculate the original pressure of the gas in (a) bar, (b) atm.

1.2(b) A perfect gas undergoes isothermal compression, which reduces its volume by 1.80 dm^3. The final pressure and volume of the gas are 1.97 bar and 2.14 dm^3, respectively. Calculate the original pressure of the gas in (a) bar, (b) Torr.

1.3(a) A car tyre (i.e. an automobile tire) was inflated to a pressure of 24 lb in^{-2} (1.00 atm = 14.7 lb in^{-2}) on a winter's day when the temperature was −5°C. What pressure will be found, assuming no leaks have occurred and that the volume is constant, on a subsequent summer's day when the temperature is 35°C? What complications should be taken into account in practice?

1.3(b) A sample of hydrogen gas was found to have a pressure of 125 kPa when the temperature was 23°C. What can its pressure be expected to be when the temperature is 11°C?

1.4(a) A sample of 255 mg of neon occupies 3.00 dm^3 at 122 K. Use the perfect gas law to calculate the pressure of the gas.

1.4(b) A homeowner uses 4.00×10^3 m^3 of natural gas in a year to heat a home. Assume that natural gas is all methane, CH_4, and that methane is a perfect gas for the conditions of this problem, which are 1.00 atm and 20°C. What is the mass of gas used?

1.5(a) A diving bell has an air space of 3.0 m^3 when on the deck of a boat. What is the volume of the air space when the bell has been lowered to a depth of 50 m? Take the mean density of sea water to be 1.025 g cm^{-3} and assume that the temperature is the same as on the surface.

1.5(b) What pressure difference must be generated across the length of a 15 cm vertical drinking straw in order to drink a water-like liquid of density 1.0 g cm^{-3}?

1.6(a) A manometer consists of a U-shaped tube containing a liquid. One side is connected to the apparatus and the other is open to the atmosphere. The pressure inside the apparatus is then determined from the difference in heights of the liquid. Suppose the liquid is water, the external pressure is 770 Torr, and the open side is 10.0 cm lower than the side connected to the apparatus. What is the pressure in the apparatus? (The density of water at 25°C is 0.997 07 g cm^{-3}.)

1.6(b) A manometer like that described in Exercise 1.6a contained mercury in place of water. Suppose the external pressure is 760 Torr, and the open side is 10.0 cm higher than the side connected to the apparatus. What is the pressure in the apparatus? (The density of mercury at 25°C is 13.55 g cm^{-3}.)

1.7(a) In an attempt to determine an accurate value of the gas constant, R, a student heated a container of volume 20.000 dm^3 filled with 0.251 32 g of helium gas to 500°C and measured the pressure as 206.402 cm of water in a manometer at 25°C. Calculate the value of R from these data. (The density of water at 25°C is 0.997 07 g cm^{-3}; the construction of a manometer is described in Exercise 1.6a.)

1.7(b) The following data have been obtained for oxygen gas at 273.15 K. Calculate the best value of the gas constant R from them and the best value of the molar mass of O_2.

p/atm	0.750 000	0.500 000	0.250 000
$V_m/(dm^3\ mol^{-1})$	29.9649	44.8090	89.6384
$\rho/(g\ dm^{-3})$	1.07144	0.714110	0.356975

1.8(a) At 500°C and 93.2 kPa, the mass density of sulfur vapour is 3.710 kg m^{-3}. What is the molecular formula of sulfur under these conditions?

1.8(b) At 100°C and 1.60 kPa, the mass density of phosphorus vapour is 0.6388 kg m^{-3}. What is the molecular formula of phosphorus under these conditions?

1.9(a) Calculate the mass of water vapour present in a room of volume 400 m^3 that contains air at 27°C on a day when the relative humidity is 60 per cent.

1.9(b) Calculate the mass of water vapour present in a room of volume 250 m^3 that contains air at 23°C on a day when the relative humidity is 53 per cent.

1.10(a) Given that the density of air at 0.987 bar and 27°C is 1.146 kg m^{-3}, calculate the mole fraction and partial pressure of nitrogen and oxygen assuming that (a) air consists only of these two gases, (b) air also contains 1.0 mole per cent Ar.

1.10(b) A gas mixture consists of 320 mg of methane, 175 mg of argon, and 225 mg of neon. The partial pressure of neon at 300 K is 8.87 kPa. Calculate (a) the volume and (b) the total pressure of the mixture.

1.11(a) The density of a gaseous compound was found to be 1.23 kg m^{-3} at 330 K and 20 kPa. What is the molar mass of the compound?

1.11(b) In an experiment to measure the molar mass of a gas, 250 cm^3 of the gas was confined in a glass vessel. The pressure was 152 Torr at 298 K and, after correcting for buoyancy effects, the mass of the gas was 33.5 mg. What is the molar mass of the gas?

1.12(a) The densities of air at −85°C, 0°C, and 100°C are 1.877 g dm^{-3}, 1.294 g dm^{-3}, and 0.946 g dm^{-3}, respectively. From these data, and assuming that air obeys Charles's law, determine a value for the absolute zero of temperature in degrees Celsius.

1.12(b) A certain sample of a gas has a volume of 20.00 dm^3 at 0°C and 1.000 atm. A plot of the experimental data of its volume against the Celsius temperature, θ, at constant p, gives a straight line of slope 0.0741 dm^3 $(°C)^{-1}$. From these data alone (without making use of the perfect gas law), determine the absolute zero of temperature in degrees Celsius.

1.13(a) Calculate the pressure exerted by 1.0 mol C_2H_6 behaving as (a) a perfect gas, (b) a van der Waals gas when it is confined under the following conditions: (i) at 273.15 K in 22.414 dm^3, (ii) at 1000 K in 100 cm^3. Use the data in Table 1.6.

1.13(b) Calculate the pressure exerted by 1.0 mol H_2S behaving as (a) a perfect gas, (b) a van der Waals gas when it is confined under the following conditions: (i) at 273.15 K in 22.414 dm^3, (ii) at 500 K in 150 cm^3. Use the data in Table 1.6.

1.14(a) Express the van der Waals parameters $a = 0.751$ atm dm^6 mol^{-2} and $b = 0.0226$ dm^3 mol^{-1} in SI base units.

1.14(b) Express the van der Waals parameters $a = 1.32$ atm dm^6 mol^{-2} and $b = 0.0436$ dm^3 mol^{-1} in SI base units.

1.15(a) A gas at 250 K and 15 atm has a molar volume 12 per cent smaller than that calculated from the perfect gas law. Calculate (a) the compression factor under these conditions and (b) the molar volume of the gas. Which are dominating in the sample, the attractive or the repulsive forces?

1.15(b) A gas at 350 K and 12 atm has a molar volume 12 per cent larger than that calculated from the perfect gas law. Calculate (a) the compression factor under these conditions and (b) the molar volume of the gas. Which are dominating in the sample, the attractive or the repulsive forces?

1.16(a) In an industrial process, nitrogen is heated to 500 K at a constant volume of 1.000 m^3. The gas enters the container at 300 K and 100 atm. The mass of the gas is 92.4 kg. Use the van der Waals equation to determine the approximate pressure of the gas at its working temperature of 500 K. For nitrogen, $a = 1.352\ dm^6\ atm\ mol^{-2}$, $b = 0.0387\ dm^3\ mol^{-1}$.

1.16(b) Cylinders of compressed gas are typically filled to a pressure of 200 bar. For oxygen, what would be the molar volume at this pressure and 25°C based on (a) the perfect gas equation, (b) the van der Waals equation. For oxygen, $a = 1.364\ dm^6\ atm\ mol^{-2}$, $b = 3.19 \times 10^{-2}\ dm^3\ mol^{-1}$.

1.17(a) Suppose that 10.0 mol C_2H_6(g) is confined to 4.860 dm^3 at 27°C. Predict the pressure exerted by the ethane from (a) the perfect gas and (b) the van der Waals equations of state. Calculate the compression factor based on these calculations. For ethane, $a = 5.507\ dm^6\ atm\ mol^{-2}$, $b = 0.0651\ dm^3\ mol^{-1}$.

1.17(b) At 300 K and 20 atm, the compression factor of a gas is 0.86. Calculate (a) the volume occupied by 8.2 mmol of the gas under these conditions and (b) an approximate value of the second virial coefficient B at 300 K.

1.18(a) A vessel of volume 22.4 dm^3 contains 2.0 mol H_2 and 1.0 mol N_2 at 273.15 K. Calculate (a) the mole fractions of each component, (b) their partial pressures, and (c) their total pressure.

1.18(b) A vessel of volume 22.4 dm^3 contains 1.5 mol H_2 and 2.5 mol N_2 at 273.15 K. Calculate (a) the mole fractions of each component, (b) their partial pressures, and (c) their total pressure.

1.19(a) The critical constants of methane are $p_c = 45.6$ atm, $V_c = 98.7\ cm^3\ mol^{-1}$, and $T_c = 190.6$ K. Calculate the van der Waals parameters of the gas and estimate the radius of the molecules.

1.19(b) The critical constants of ethane are $p_c = 48.20$ atm, $V_c = 148\ cm^3\ mol^{-1}$, and $T_c = 305.4$ K. Calculate the van der Waals parameters of the gas and estimate the radius of the molecules.

1.20(a) Use the van der Waals parameters for chlorine to calculate approximate values of (a) the Boyle temperature of chlorine and (b) the radius of a Cl_2 molecule regarded as a sphere.

1.20(b) Use the van der Waals parameters for hydrogen sulfide to calculate approximate values of (a) the Boyle temperature of the gas and (b) the radius of a H_2S molecule regarded as a sphere ($a = 4.484\ dm^6\ atm\ mol^{-2}$, $b = 0.0434\ dm^3\ mol^{-1}$).

1.21(a) Suggest the pressure and temperature at which 1.0 mol of (a) NH_3, (b) Xe, (c) He will be in states that correspond to 1.0 mol H_2 at 1.0 atm and 25°C.

1.21(b) Suggest the pressure and temperature at which 1.0 mol of (a) H_2S, (b) CO_2, (c) Ar will be in states that correspond to 1.0 mol N_2 at 1.0 atm and 25°C.

1.22(a) A certain gas obeys the van der Waals equation with $a = 0.50\ m^6\ Pa\ mol^{-2}$. Its volume is found to be $5.00 \times 10^{-4}\ m^3\ mol^{-1}$ at 273 K and 3.0 MPa. From this information calculate the van der Waals constant b. What is the compression factor for this gas at the prevailing temperature and pressure?

1.22(b) A certain gas obeys the van der Waals equation with $a = 0.76\ m^6\ Pa\ mol^{-2}$. Its volume is found to be $4.00 \times 10^{-4}\ m^3\ mol^{-1}$ at 288 K and 4.0 MPa. From this information calculate the van der Waals constant b. What is the compression factor for this gas at the prevailing temperature and pressure?

Problems*

Numerical problems

1.1 Recent communication with the inhabitants of Neptune have revealed that they have a Celsius-type temperature scale, but based on the melting point (0°N) and boiling point (100°N) of their most common substance, hydrogen. Further communications have revealed that the Neptunians know about perfect gas behaviour and they find that, in the limit of zero pressure, the value of pV is 28 dm^3 atm at 0°N and 40 dm^3 atm at 100°N. What is the value of the absolute zero of temperature on their temperature scale?

1.2 Deduce the relation between the pressure and mass density, ρ, of a perfect gas of molar mass M. Confirm graphically, using the following data on dimethyl ether at 25°C, that perfect behaviour is reached at low pressures and find the molar mass of the gas.

p/kPa	12.223	25.20	36.97	60.37	85.23	101.3
$\rho/(\text{kg m}^{-3})$	0.225	0.456	0.664	1.062	1.468	1.734

1.3 Charles's law is sometimes expressed in the form $V = V_0(1 + \alpha\theta)$, where θ is the Celsius temperature, α is a constant, and V_0 is the volume of the sample at 0°C. The following values for α have been reported for nitrogen at 0°C:

p/Torr	749.7	599.6	333.1	98.6
$10^3\alpha/(°\text{C})^{-1}$	3.6717	3.6697	3.6665	3.6643

For these data calculate the best value for the absolute zero of temperature on the Celsius scale.

1.4 The molar mass of a newly synthesized fluorocarbon was measured in a gas microbalance. This device consists of a glass bulb forming one end of a beam, the whole surrounded by a closed container. The beam is pivoted, and the balance point is attained by raising the pressure of gas in the container, so increasing the buoyancy of the enclosed bulb. In one experiment, the balance point was reached when the fluorocarbon pressure was 327.10 Torr; for the same setting of the pivot, a balance was reached when CHF_3 ($M = 70.014\ g\ mol^{-1}$) was introduced at 423.22 Torr. A repeat of the experiment with a different setting of the pivot required a pressure of 293.22 Torr of the fluorocarbon and 427.22 Torr of the CHF_3. What is the molar mass of the fluorocarbon? Suggest a molecular formula.

1.5 A constant-volume perfect gas thermometer indicates a pressure of 6.69 kPa at the triple point temperature of water (273.16 K). (a) What change of pressure indicates a change of 1.00 K at this temperature? (b) What pressure indicates a temperature of 100.00°C? (c) What change of pressure indicates a change of 1.00 K at the latter temperature?

1.6 A vessel of volume 22.4 dm^3 contains 2.0 mol H_2 and 1.0 mol N_2 at 273.15 K initially. All the H_2 reacted with sufficient N_2 to form NH_3. Calculate the partial pressures and the total pressure of the final mixture.

1.7 Calculate the molar volume of chlorine gas at 350 K and 2.30 atm using (a) the perfect gas law and (b) the van der Waals equation. Use the answer to (a) to calculate a first approximation to the correction term for attraction and then use successive approximations to obtain a numerical answer for part (b).

* Problems denoted with the symbol ‡ were supplied by Charles Trapp, Carmen Giunta, and Marshall Cady.

1.8 At 273 K measurements on argon gave $B = -21.7\ \text{cm}^3\ \text{mol}^{-1}$ and $C = 1200\ \text{cm}^6\ \text{mol}^{-2}$, where B and C are the second and third virial coefficients in the expansion of Z in powers of $1/V_m$. Assuming that the perfect gas law holds sufficiently well for the estimation of the second and third terms of the expansion, calculate the compression factor of argon at 100 atm and 273 K. From your result, estimate the molar volume of argon under these conditions.

1.9 Calculate the volume occupied by 1.00 mol N_2 using the van der Waals equation in the form of a virial expansion at (a) its critical temperature, (b) its Boyle temperature, and (c) its inversion temperature. Assume that the pressure is 10 atm throughout. At what temperature is the gas most perfect? Use the following data: $T_c = 126.3$ K, $a = 1.352\ \text{dm}^6\ \text{atm}\ \text{mol}^{-2}$, $b = 0.0387\ \text{dm}^3\ \text{mol}^{-1}$.

1.10‡ The second virial coefficient of methane can be approximated by the empirical equation $B'(T) = a + be^{-c/T^2}$, where $a = -0.1993\ \text{bar}^{-1}$, $b = 0.2002\ \text{bar}^{-1}$, and $c = 1131\ \text{K}^2$ with $300\ \text{K} < T < 600\ \text{K}$. What is the Boyle temperature of methane?

1.11 The mass density of water vapour at 327.6 atm and 776.4 K is $133.2\ \text{kg m}^{-3}$. Given that for water $T_c = 647.4$ K, $p_c = 218.3$ atm, $a = 5.464\ \text{dm}^6\ \text{atm}\ \text{mol}^{-2}$, $b = 0.03049\ \text{dm}^3\ \text{mol}^{-1}$, and $M = 18.02\ \text{g mol}^{-1}$, calculate (a) the molar volume. Then calculate the compression factor (b) from the data, (c) from the virial expansion of the van der Waals equation.

1.12 The critical volume and critical pressure of a certain gas are $160\ \text{cm}^3\ \text{mol}^{-1}$ and 40 atm, respectively. Estimate the critical temperature by assuming that the gas obeys the Berthelot equation of state. Estimate the radii of the gas molecules on the assumption that they are spheres.

1.13 Estimate the coefficients a and b in the Dieterici equation of state from the critical constants of xenon. Calculate the pressure exerted by 1.0 mol Xe when it is confined to $1.0\ \text{dm}^3$ at 25°C.

Theoretical problems

1.14 Show that the van der Waals equation leads to values of $Z < 1$ and $Z > 1$, and identify the conditions for which these values are obtained.

1.15 Express the van der Waals equation of state as a virial expansion in powers of $1/V_m$ and obtain expressions for B and C in terms of the parameters a and b. The expansion you will need is $(1 - x)^{-1} = 1 + x + x^2 + \cdots$. Measurements on argon gave $B = -21.7\ \text{cm}^3\ \text{mol}^{-1}$ and $C = 1200\ \text{cm}^6\ \text{mol}^{-2}$ for the virial coefficients at 273 K. What are the values of a and b in the corresponding van der Waals equation of state?

1.16‡ Derive the relation between the critical constants and the Dieterici equation parameters. Show that $Z_c = 2e^{-2}$ and derive the reduced form of the Dieterici equation of state. Compare the van der Waals and Dieterici predictions of the critical compression factor. Which is closer to typical experimental values?

1.17 A scientist proposed the following equation of state:

$$p = \frac{RT}{V_m} - \frac{B}{V_m^2} + \frac{C}{V_m^3}$$

Show that the equation leads to critical behaviour. Find the critical constants of the gas in terms of B and C and an expression for the critical compression factor.

1.18 Equations 1.18 and 1.19 are expansions in p and $1/V_m$, respectively. Find the relation between B, C and B', C'.

1.19 The second virial coefficient B' can be obtained from measurements of the density ρ of a gas at a series of pressures. Show that the graph of p/ρ against p should be a straight line with slope proportional to B'. Use the data on dimethyl ether in Problem 1.2 to find the values of B' and B at 25°C.

1.20 The equation of state of a certain gas is given by $p = RT/V_m + (a + bT)/V_m^2$, where a and b are constants. Find $(\partial V/\partial T)_p$.

1.21 The following equations of state are occasionally used for approximate calculations on gases: (gas A) $pV_m = RT(1 + b/V_m)$, (gas B) $p(V_m - b) = RT$. Assuming that there were gases that actually obeyed these equations of state, would it be possible to liquefy either gas A or B? Would they have a critical temperature? Explain your answer.

1.22 Derive an expression for the compression factor of a gas that obeys the equation of state $p(V - nb) = nRT$, where b and R are constants. If the pressure and temperature are such that $V_m = 10b$, what is the numerical value of the compression factor?

1.23‡ The discovery of the element argon by Lord Rayleigh and Sir William Ramsay had its origins in Rayleigh's measurements of the density of nitrogen with an eye toward accurate determination of its molar mass. Rayleigh prepared some samples of nitrogen by chemical reaction of nitrogen-containing compounds; under his standard conditions, a glass globe filled with this 'chemical nitrogen' had a mass of 2.2990 g. He prepared other samples by removing oxygen, carbon dioxide, and water vapour from atmospheric air; under the same conditions, this 'atmospheric nitrogen' had a mass of 2.3102 g (Lord Rayleigh, *Royal Institution Proceedings* **14,** 524 (1895)). With the hindsight of knowing accurate values for the molar masses of nitrogen and argon, compute the mole fraction of argon in the latter sample on the assumption that the former was pure nitrogen and the latter a mixture of nitrogen and argon.

1.24‡ A substance as elementary and well known as argon still receives research attention. Stewart and Jacobsen have published a review of thermodynamic properties of argon (R.B. Stewart and R.T. Jacobsen, *J. Phys. Chem. Ref. Data* **18**, 639 (1989)) that included the following 300 K isotherm.

p/MPa	0.4000	0.5000	0.6000	0.8000	1.000
$V_m/(\text{dm}^3\ \text{mol}^{-1})$	6.2208	4.9736	4.1423	3.1031	2.4795
p/MPa	1.500	2.000	2.500	3.000	4.000
$V_m/(\text{dm}^3\ \text{mol}^{-1})$	1.6483	1.2328	0.98357	0.81746	0.60998

(a) Compute the second virial coefficient, B, at this temperature. (b) Use non-linear curve-fitting software to compute the third virial coefficient, C, at this temperature.

Applications: to environmental science

1.25 Atmospheric pollution is a problem that has received much attention. Not all pollution, however, is from industrial sources. Volcanic eruptions can be a significant source of air pollution. The Kilauea volcano in Hawaii emits 200–300 t of SO_2 per day. If this gas is emitted at 800°C and 1.0 atm, what volume of gas is emitted?

1.26 Ozone is a trace atmospheric gas that plays an important role in screening the Earth from harmful ultraviolet radiation, and the abundance of ozone is commonly reported in *Dobson units*. One Dobson unit is the thickness, in thousandths of a centimetre, of a column of gas if it were collected as a pure gas at 1.00 atm and 0°C. What amount of O_3 (in moles) is found in a column of atmosphere with a cross-sectional area of $1.00\ \text{dm}^2$ if the abundance is 250 Dobson units (a typical mid-latitude value)? In the seasonal Antarctic ozone hole, the column abundance drops below 100 Dobson units; how many moles of ozone are found in such a column of air above a $1.00\ \text{dm}^2$ area? Most atmospheric ozone is found between 10 and 50 km above the surface of the earth. If that ozone is spread uniformly through this portion of the atmosphere, what is the average molar concentration corresponding to (a) 250 Dobson units, (b) 100 Dobson units?

1.27 The barometric formula relates the pressure of a gas of molar mass M at an altitude h to its pressure p_0 at sea level. Derive this relation by showing that the change in pressure dp for an infinitesimal change in altitude dh where the density is ρ is $dp = -\rho g dh$. Remember that ρ depends on the pressure. Evaluate (a) the pressure difference between the top and bottom of a laboratory vessel of height 15 cm, and (b) the external atmospheric pressure at a typical cruising altitude of an aircraft (11 km) when the pressure at ground level is 1.0 atm.

1.28 Balloons are still used to deploy sensors that monitor meteorological phenomena and the chemistry of the atmosphere. It is possible to investigate some of the technicalities of ballooning by using the perfect gas law. Suppose your balloon has a radius of 3.0 m and that it is spherical. (a) What amount of H_2 (in moles) is needed to inflate it to 1.0 atm in an ambient temperature of 25°C at sea level? (b) What mass can the balloon lift at sea level, where the density of air is 1.22 kg m^{-3}? (c) What would be the payload if He were used instead of H_2?

1.29‡ The preceding problem is most readily solved (see the *Solutions manual*) with the use of the Archimedes principle, which states that the lifting force is equal to the difference between the weight of the displaced air and the weight of the balloon. Prove the Archimedes principle for the atmosphere from the barometric formula. *Hint*. Assume a simple shape for the balloon, perhaps a right circular cylinder of cross–sectional area A and height h.

1.30 ‡ Chlorofluorocarbons such as CCl_3F and CCl_2F_2 have been linked to ozone depletion in Antarctica. As of 1994, these gases were found in quantities of 261 and 509 parts per trillion (10^{12}) by volume (World Resources Institute, *World resources* 1996–97). Compute the molar concentration of these gases under conditions typical of (a) the mid-latitude troposphere (10°C and 1.0 atm) and (b) the Antarctic stratosphere (200 K and 0.050 atm).

2 The First Law

This chapter introduces some of the basic concepts of thermodynamics. It concentrates on the conservation of energy—the experimental observation that energy can be neither created nor destroyed—and shows how the principle of the conservation of energy can be used to assess the energy changes that accompany physical and chemical processes. Much of this chapter examines the means by which a system can exchange energy with its surroundings in terms of the work it may do or the heat that it may produce. The target concept of the chapter is enthalpy, which is a very useful book-keeping property for keeping track of the heat output (or requirements) of physical processes and chemical reactions at constant pressure. We also begin to unfold some of the power of thermodynamics by showing how to establish relations between different properties of a system. We shall see that one very useful aspect of thermodynamics is that a property can be measured indirectly by measuring others and then combining their values. The relations we derive also enable us to discuss the liquefaction of gases and to establish the relation between the heat capacities of a substance under different conditions.

The release of energy can be used to provide heat when a fuel burns in a furnace, to produce mechanical work when a fuel burns in an engine, and to generate electrical work when a chemical reaction pumps electrons through a circuit. In chemistry, we encounter reactions that can be harnessed to provide heat and work, reactions that liberate energy which is squandered (often to the detriment of the environment) but which give products we require, and reactions that constitute the processes of life. **Thermodynamics**, the study of the transformations of energy, enables us to discuss all these matters quantitatively and to make useful predictions.

The basic concepts

For the purposes of physical chemistry, the universe is divided into two parts, the system and its surroundings. The **system** is the part of the world in which we have a special interest. It may be a reaction vessel, an engine, an electrochemical cell, a biological cell, and so on. The **surroundings** comprise the region outside the system and are where we make our measurements. The type of system depends on the characteristics of the boundary that divides it from the surroundings (Fig. 2.1). If matter can be transferred through the boundary between the system and its surroundings the system is classified as **open**. If matter cannot pass through the boundary the system is classified as **closed**. Both open and closed systems can exchange energy with their surroundings. For example, a closed system can expand and thereby raise a weight in the surroundings; it may also transfer energy to them if they are at a lower temperature.

An **isolated system** is a closed system that has neither mechanical nor thermal contact with its surroundings.

2.1 Work, heat, and energy

The fundamental physical property in thermodynamics is work: **work** is motion against an opposing force. Doing work is equivalent to raising a weight somewhere in the surroundings. An example of doing work is the expansion of a gas that pushes out a piston and raises a weight. A chemical reaction that drives an electric current through a resistance also does work, because the same current could be driven through a motor and used to raise a weight.

The **energy** of a system is its capacity to do work. When work is done on an otherwise isolated system (for instance, by compressing a gas or winding a spring), the capacity of the system to do work is increased; in other words, the energy of the system is increased. When the system does work (when the piston moves out or the spring unwinds), the energy of the system is reduced and it can do less work than before.

Experiments have shown that the energy of a system may be changed by means other than work itself. When the energy of a system changes as a result of a temperature difference between the system and its surroundings we say that energy has been transferred as **heat**. When a heater is immersed in a beaker of water (the system), the capacity of the system to do work increases because hot water can be used to do more work than the same amount of cold water. Not all boundaries permit the transfer of energy even though there is a temperature difference between the system and its surroundings.

An **exothermic process** is a process that releases energy as heat into its surroundings. All combustion reactions are exothermic. An **endothermic process** is a process in which energy is acquired from its surroundings as heat. An example of an endothermic process is the vaporization of water. To avoid a lot of awkward circumlocution, we say that in an exothermic process energy is transferred 'as heat' to the surroundings and in an endothermic process energy is transferred 'as heat' from the surroundings into the system. However, it must never be forgotten that heat is a process (the transfer of energy as a result of a temperature difference), not an entity. An endothermic process in a diathermic container results in energy flowing into the system as heat. An exothermic process in a similar diathermic container results in a release of energy as heat into the surroundings. When an endothermic process takes place in an adiabatic container, it results in a lowering of temperature of the system; an exothermic process results in a rise of temperature. These features are summarized in Fig. 2.2.

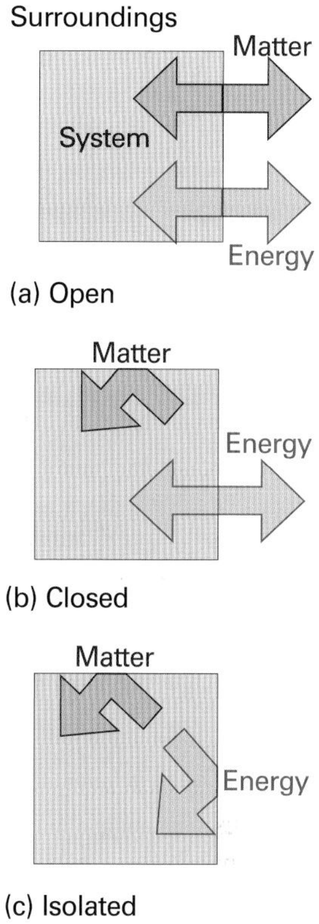

Fig. 2.1 (a) An open system can exchange matter and energy with its surroundings. (b) A closed system can exchange energy with its surroundings, but it cannot exchange matter. (c) An isolated system can exchange neither energy nor matter with its surroundings.

Molecular interpretation 2.1 *Heat and work*

In molecular terms, heating is the transfer of energy that makes use of *disorderly molecular motion*. The disorderly motion of molecules is called **thermal motion**. The thermal motion of the molecules in the hot surroundings stimulates the molecules in the cooler system to move more vigorously and, as a result, the energy of the system is increased. When a system heats its surroundings, molecules of the system stimulate the thermal motion of the molecules in the surroundings (Fig. 2.3).

In contrast, *work is the transfer of energy that makes use of organized motion* (Fig. 2.4). When a weight is raised or lowered, its atoms move in an organized way (up or down). The atoms in a spring move in an orderly way when it is wound; the

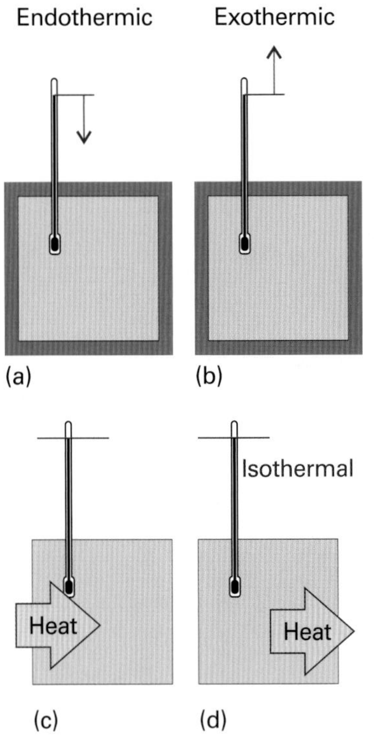

Fig. 2.2 (a) When an endothermic process occurs in an adiabatic system, the temperature falls; (b) if the process is exothermic, then the temperature rises. (c) When an endothermic process occurs in a diathermic container, energy enters as heat from the surroundings, and the system remains at the same temperature. (d) If the process is exothermic, then energy leaves as heat, and the process is isothermal.

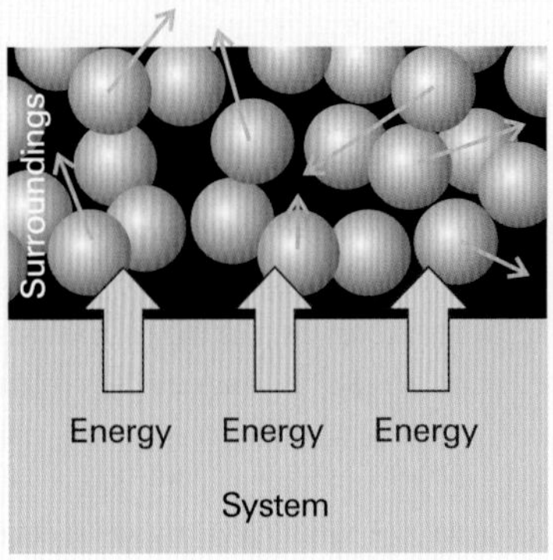

Fig. 2.3 When energy is transferred to the surroundings as heat, the transfer stimulates random motion of the atoms in the surroundings. Transfer of energy from the surroundings to the system makes use of random motion (thermal motion) in the surroundings.

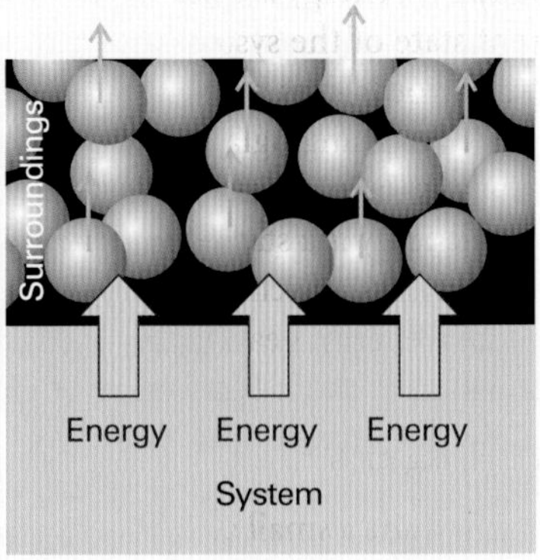

Fig. 2.4 When a system does work, it stimulates orderly motion in the surroundings. For instance, the atoms shown here may be part of a weight that is being raised. The ordered motion of the atoms in a falling weight does work on the system.

electrons in an electric current move in an orderly direction when it flows. When a system does work it causes atoms or electrons in its surroundings to move in an organized way. Likewise, when work is done on a system, molecules in the surroundings are used to transfer energy to it in an organized way, as the atoms in a weight are lowered or a current of electrons is passed.

The distinction between work and heat is made in the surroundings. The fact that a falling weight may stimulate thermal motion in the system is irrelevant to the distinction between heat and work: work is identified as energy transfer making use of the organized motion of atoms in the surroundings, and heat is identified as energy transfer making use of thermal motion in the surroundings. In the compression of a gas, for instance, work is done as the atoms of the compressing weight descend in an orderly way, but the effect of the incoming piston is to accelerate the gas molecules to higher average speeds. Because collisions between molecules quickly randomize their directions, the orderly motion of the atoms of the weight is in effect stimulating thermal motion in the gas. We observe the falling weight, the orderly descent of its atoms, and report that work is being done even though it is stimulating thermal motion.

2.2 The internal energy

In thermodynamics, the total energy of a system is called its **internal energy**, U. The internal energy is the total kinetic and potential energy of the molecules in the system (see *Comment 1.3* for the definitions of kinetic and potential energy).[1] We denote by ΔU the change in internal energy when a system changes from an initial state i with internal energy U_i to a final state f of internal energy U_f:

$$\Delta U = U_f - U_i \qquad [2.1]$$

[1] The internal energy does not include the kinetic energy arising from the motion of the system as a whole, such as its kinetic energy as it accompanies the Earth on its orbit round the Sun.

The internal energy is a **state function** in the sense that its value depends only on the current state of the system and is independent of how that state has been prepared. In other words, it is a function of the properties that determine the current state of the system. Changing any one of the state variables, such as the pressure, results in a change in internal energy. The internal energy is an extensive property. That the internal energy is a state function has consequences of the greatest importance, as we start to unfold in Section 2.10.

Internal energy, heat, and work are all measured in the same units, the joule (J). The joule, which is named after the nineteenth-century scientist J.P. Joule, is defined as

$$1\ \text{J} = 1\ \text{kg m}^2\ \text{s}^{-2}$$

A joule is quite a small unit of energy: for instance, each beat of the human heart consumes about 1 J. Changes in molar internal energy, ΔU_m, are typically expressed in kilojoules per mole (kJ mol^{-1}). Certain other energy units are also used, but are more common in fields other than thermodynamics. Thus, 1 electronvolt (1 eV) is defined as the kinetic energy acquired when an electron is accelerated from rest through a potential difference of 1 V; the relation between electronvolts and joules is 1 eV ≈ 0.16 aJ (where 1 aJ = 10^{-18} J). Many processes in chemistry have an energy of several electronvolts. Thus, the energy to remove an electron from a sodium atom is close to 5 eV. Calories (cal) and kilocalories (kcal) are still encountered. The current definition of the calorie in terms of joules is

$$1\ \text{cal} = 4.184\ \text{J exactly}$$

An energy of 1 cal is enough to raise the temperature of 1 g of water by 1°C.

Comment 2.1

An extensive property is a property that depends on the amount of substance in the sample. An intensive property is a property that is independent of the amount of substance in the sample. Two examples of extensive properties are mass and volume. Examples of intensive properties are temperature, mass density (mass divided by volume), and pressure.

Molecular interpretation 2.2 *The internal energy of a gas*

A molecule has a certain number of degrees of freedom, such as the ability to translate (the motion of its centre of mass through space), rotate around its centre of mass, or vibrate (as its bond lengths and angles change). Many physical and chemical properties depend on the energy associated with each of these modes of motion. For example, a chemical bond might break if a lot of energy becomes concentrated in it.

The *equipartition theorem* of classical mechanics is a useful guide to the average energy associated with each degree of freedom when the sample is at a temperature *T*. First, we need to know that a 'quadratic contribution' to the energy means a contribution that can be expressed as the square of a variable, such as the position or the velocity. For example, the kinetic energy an atom of mass *m* as it moves through space is

$$E_K = \tfrac{1}{2}mv_x^2 + \tfrac{1}{2}mv_y^2 + \tfrac{1}{2}mv_z^2$$

and there are three quadratic contributions to its energy. The equipartition theorem then states that, for a collection of particles at thermal equilibrium at a temperature *T*, *the average value of each quadratic contribution to the energy is the same and equal to* $\tfrac{1}{2}kT$, where *k* is Boltzmann's constant ($k = 1.381 \times 10^{-23}$ J K^{-1}).

The equipartition theorem is a conclusion from classical mechanics and is applicable only when the effects of quantization can be ignored (see Chapters 16 and 17). In practice, it can be used for molecular translation and rotation but not vibration. At 25°C, $\tfrac{1}{2}kT = 2$ zJ (where 1 zJ = 10^{-21} J), or about 13 meV.

According to the equipartition theorem, the average energy of each term in the expression above is $\tfrac{1}{2}kT$. Therefore, the mean energy of the atoms is $\tfrac{3}{2}kT$ and the

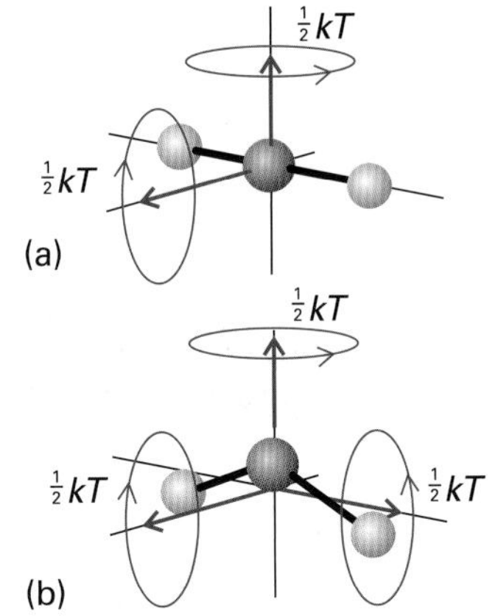

Fig. 2.5 The rotational modes of molecules and the corresponding average energies at a temperature T. (a) A linear molecule can rotate about two axes perpendicular to the line of the atoms. (b) A nonlinear molecule can rotate about three perpendicular axes.

total energy of the gas (there being no potential energy contribution) is $\frac{3}{2}NkT$, or $\frac{3}{2}nRT$ (because $N = nN_A$ and $R = N_Ak$). We can therefore write

$$U_m = U_m(0) + \tfrac{3}{2}RT$$

where $U_m(0)$ is the molar internal energy at $T = 0$, when all translational motion has ceased and the sole contribution to the internal energy arises from the internal structure of the atoms. This equation shows that the internal energy of a perfect gas increases linearly with temperature. At 25°C, $\frac{3}{2}RT = 3.7$ kJ mol^{-1}, so translational motion contributes about 4 kJ mol^{-1} to the molar internal energy of a gaseous sample of atoms or molecules (the remaining contribution arises from the internal structure of the atoms and molecules).

When the gas consists of polyatomic molecules, we need to take into account the effect of rotation and vibration. A linear molecule, such as N_2 and CO_2, can rotate around two axes perpendicular to the line of the atoms (Fig. 2.5), so it has two rotational modes of motion, each contributing a term $\frac{1}{2}kT$ to the internal energy. Therefore, the mean rotational energy is kT and the rotational contribution to the molar internal energy is RT. By adding the translational and rotational contributions, we obtain

$$U_m = U_m(0) + \tfrac{5}{2}RT \qquad \text{(linear molecule, translation and rotation only)}$$

A nonlinear molecule, such as CH_4 or water, can rotate around three axes and, again, each mode of motion contributes a term $\frac{1}{2}kT$ to the internal energy. Therefore, the mean rotational energy is $\frac{3}{2}kT$ and there is a rotational contribution of $\frac{3}{2}RT$ to the molar internal energy of the molecule. That is,

$$U_m = U_m(0) + 3RT \qquad \text{(nonlinear molecule, translation and rotation only)}$$

The internal energy now increases twice as rapidly with temperature compared with the monatomic gas.

The internal energy of interacting molecules in condensed phases also has a contribution from the potential energy of their interaction. However, no simple expressions can be written down in general. Nevertheless, the crucial molecular point is that, as the temperature of a system is raised, the internal energy increases as the various modes of motion become more highly excited.

It has been found experimentally that the internal energy of a system may be changed either by doing work on the system or by heating it. Whereas we may know how the energy transfer has occurred (because we can see if a weight has been raised or lowered in the surroundings, indicating transfer of energy by doing work, or if ice has melted in the surroundings, indicating transfer of energy as heat), the system is blind to the mode employed. *Heat and work are equivalent ways of changing a system's internal energy*. A system is like a bank: it accepts deposits in either currency, but stores its reserves as internal energy. It is also found experimentally that, if a system is isolated from its surroundings, then no change in internal energy takes place. This summary of observations is now known as the **First Law of thermodynamics** and expressed as follows:

The internal energy of an isolated system is constant.

We cannot use a system to do work, leave it isolated for a month, and then come back expecting to find it restored to its original state and ready to do the same work again. The evidence for this property is that no 'perpetual motion machine' (a machine that does work without consuming fuel or some other source of energy) has ever been built.

These remarks may be summarized as follows. If we write w for the work done on a system, q for the energy transferred as heat to a system, and ΔU for the resulting change in internal energy, then it follows that

$$\Delta U = q + w \tag{2.2}$$

Equation 2.2 is the mathematical statement of the First Law, for it summarizes the equivalence of heat and work and the fact that the internal energy is constant in an isolated system (for which $q = 0$ and $w = 0$). The equation states that the change in internal energy of a closed system is equal to the energy that passes through its boundary as heat or work. It employs the 'acquisitive convention', in which $w > 0$ or $q > 0$ if energy is transferred to the system as work or heat and $w < 0$ or $q < 0$ if energy is lost from the system as work or heat. In other words, we view the flow of energy as work or heat from the system's perspective.

Illustration 2.1 *The sign convention in thermodynamics*

If an electric motor produced 15 kJ of energy each second as mechanical work and lost 2 kJ as heat to the surroundings, then the change in the internal energy of the motor each second is

$$\Delta U = -2\ \text{kJ} - 15\ \text{kJ} = -17\ \text{kJ}$$

Suppose that, when a spring was wound, 100 J of work was done on it but 15 J escaped to the surroundings as heat. The change in internal energy of the spring is

$$\Delta U = +100\ \text{kJ} - 15\ \text{kJ} = +85\ \text{kJ}$$

2.3 Expansion work

The way can now be opened to powerful methods of calculation by switching attention to infinitesimal changes of state (such as infinitesimal change in temperature) and infinitesimal changes in the internal energy $\mathrm{d}U$. Then, if the work done on a system is $\mathrm{d}w$ and the energy supplied to it as heat is $\mathrm{d}q$, in place of eqn 2.2 we have

$$\mathrm{d}U = \mathrm{d}q + \mathrm{d}w \tag{2.3}$$

To use this expression we must be able to relate $\mathrm{d}q$ and $\mathrm{d}w$ to events taking place in the surroundings.

We begin by discussing **expansion work**, the work arising from a change in volume. This type of work includes the work done by a gas as it expands and drives back the atmosphere. Many chemical reactions result in the generation or consumption of gases (for instance, the thermal decomposition of calcium carbonate or the combustion of octane), and the thermodynamic characteristics of a reaction depend on the work it can do. The term 'expansion work' also includes work associated with negative changes of volume, that is, compression.

(a) The general expression for work

The calculation of expansion work starts from the definition used in physics, which states that the work required to move an object a distance $\mathrm{d}z$ against an opposing force of magnitude F is

$$\mathrm{d}w = -F\mathrm{d}z \qquad [2.4]$$

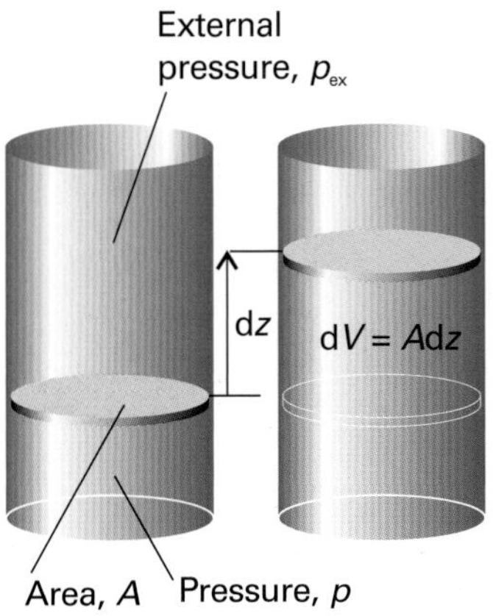

Fig. 2.6 When a piston of area A moves out through a distance dz, it sweeps out a volume $dV = Adz$. The external pressure p_{ex} is equivalent to a weight pressing on the piston, and the force opposing expansion is $F = p_{ex}A$.

The negative sign tells us that, when the system moves an object against an opposing force, the internal energy of the system doing the work will decrease. Now consider the arrangement shown in Fig. 2.6, in which one wall of a system is a massless, frictionless, rigid, perfectly fitting piston of area A. If the external pressure is p_{ex}, the magnitude of the force acting on the outer face of the piston is $F = p_{ex}A$. When the system expands through a distance dz against an external pressure p_{ex}, it follows that the work done is $dw = -p_{ex}Adz$. But Adz is the change in volume, dV, in the course of the expansion. Therefore, the work done when the system expands by dV against a pressure p_{ex} is

$$dw = -p_{ex}dV \tag{2.5}$$

To obtain the total work done when the volume changes from V_i to V_f we integrate this expression between the initial and final volumes:

$$w = -\int_{V_i}^{V_f} p_{ex}dV \tag{2.6}$$

The force acting on the piston, $p_{ex}A$, is equivalent to a weight that is raised as the system expands.

If the system is compressed instead, then the same weight is lowered in the surroundings and eqn 2.6 can still be used, but now $V_f < V_i$. It is important to note that it is still the external pressure that determines the magnitude of the work. This somewhat perplexing conclusion seems to be inconsistent with the fact that the gas *inside* the container is opposing the compression. However, when a gas is compressed, the ability of the *surroundings* to do work is diminished by an amount determined by the weight that is lowered, and it is this energy that is transferred into the system.

Other types of work (for example, electrical work), which we shall call either **non-expansion work** or **additional work**, have analogous expressions, with each one the product of an intensive factor (the pressure, for instance) and an extensive factor (the change in volume). Some are collected in Table 2.1. For the present we continue with the work associated with changing the volume, the expansion work, and see what we can extract from eqns 2.5 and 2.6.

(b) Free expansion

By **free expansion** we mean expansion against zero opposing force. It occurs when $p_{ex} = 0$. According to eqn 2.5, $dw = 0$ for each stage of the expansion. Hence, overall:

$$\textit{Free expansion:} \quad w = 0 \tag{2.7}$$

Table 2.1 Varieties of work*

Type of work	dw	Comments	Units†
Expansion	$-p_{ex}dV$	p_{ex} is the external pressure dV is the change in volume	Pa m^3
Surface expansion	$\gamma d\sigma$	γ is the surface tension $d\sigma$ is the change in area	$N\,m^{-1}$ m^2
Extension	fdl	f is the tension dl is the change in length	N m
Electrical	ϕdQ	ϕ is the electric potential dQ is the change in charge	V C

* In general, the work done on a system can be expressed in the form $dw = -Fdz$, where F is a 'generalized force' and dz is a 'generalized displacement'.
† For work in joules (J). Note that $1\,N\,m = 1\,J$ and $1\,V\,C = 1\,J$.

That is, no work is done when a system expands freely. Expansion of this kind occurs when a system expands into a vacuum.

(c) Expansion against constant pressure

Now suppose that the external pressure is constant throughout the expansion. For example, the piston may be pressed on by the atmosphere, which exerts the same pressure throughout the expansion. A chemical example of this condition is the expansion of a gas formed in a chemical reaction. We can evaluate eqn 2.6 by taking the constant p_{ex} outside the integral:

$$w = -p_{ex}\int_{V_i}^{V_f} dV = -p_{ex}(V_f - V_i)$$

Therefore, if we write the change in volume as $\Delta V = V_f - V_i$,

$$w = -p_{ex}\Delta V \qquad (2.8)$$

This result is illustrated graphically in Fig. 2.7, which makes use of the fact that an integral can be interpreted as an area. The magnitude of w, denoted $|w|$, is equal to the area beneath the horizontal line at $p = p_{ex}$ lying between the initial and final volumes. A p,V-graph used to compute expansion work is called an **indicator diagram**; James Watt first used one to indicate aspects of the operation of his steam engine.

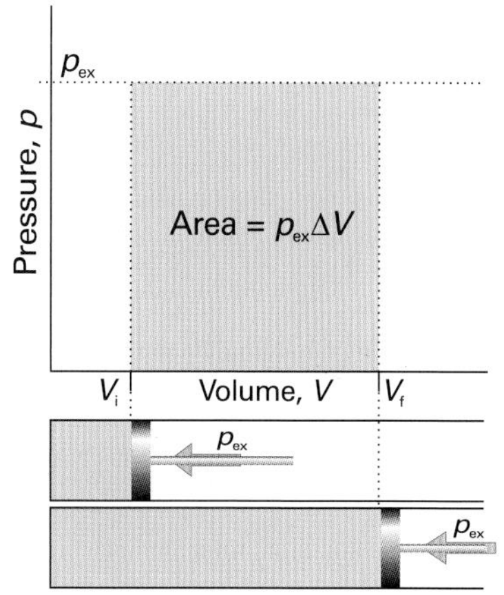

Fig. 2.7 The work done by a gas when it expands against a constant external pressure, p_{ex}, is equal to the shaded area in this example of an indicator diagram.

Comment 2.2

The value of the integral $\int_a^b f(x)dx$ is equal to the area under the graph of $f(x)$ between $x = a$ and $x = b$. For instance, the area under the curve $f(x) = x^2$ shown in the illustration that lies between $x = 1$ and 3 is

$$\int_1^3 x^2 dx = (\tfrac{1}{3}x^3 + \text{constant})\Big|_1^3 = \tfrac{1}{3}(3^3 - 1^3) = \tfrac{26}{3} \approx 8.67$$

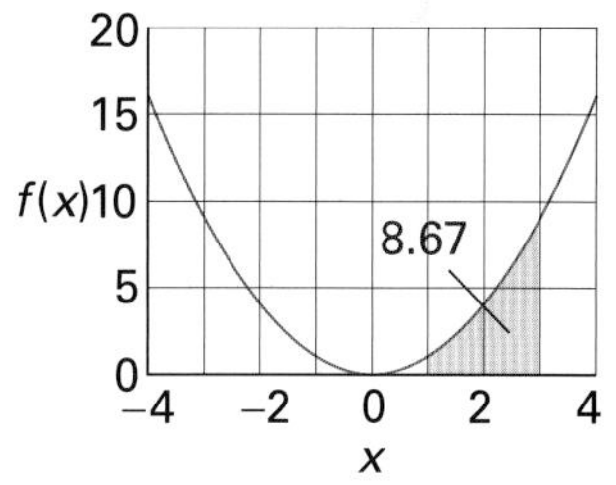

(d) Reversible expansion

A **reversible change** in thermodynamics is a change that can be reversed by an infinitesimal modification of a variable. The key word 'infinitesimal' sharpens the everyday meaning of the word 'reversible' as something that can change direction. We say that a system is in **equilibrium** with its surroundings if an infinitesimal change in the conditions in opposite directions results in opposite changes in its state. One example of reversibility that we have encountered already is the thermal equilibrium of two systems with the same temperature. The transfer of energy as heat between the two is reversible because, if the temperature of either system is lowered infinitesimally, then energy flows into the system with the lower temperature. If the temperature of either system at thermal equilibrium is raised infinitesimally, then energy flows out of the hotter system.

Suppose a gas is confined by a piston and that the external pressure, p_{ex}, is set equal to the pressure, p, of the confined gas. Such a system is in mechanical equilibrium with its surroundings (as illustrated in Section 1.1) because an infinitesimal change in the external pressure in either direction causes changes in volume in opposite directions. If the external pressure is reduced infinitesimally, then the gas expands slightly. If the external pressure is increased infinitesimally, then the gas contracts slightly. In either case the change is reversible in the thermodynamic sense. If, on the other hand, the external pressure differs measurably from the internal pressure, then changing p_{ex} infinitesimally will not decrease it below the pressure of the gas, so will not change the direction of the process. Such a system is not in mechanical equilibrium with its surroundings and the expansion is thermodynamically irreversible.

To achieve reversible expansion we set p_{ex} equal to p at each stage of the expansion. In practice, this equalization could be achieved by gradually removing weights from the piston so that the downward force due to the weights always matched the changing upward force due to the pressure of the gas. When we set $p_{ex} = p$, eqn 2.5 becomes

$$dw = -p_{ex}dV = -pdV \qquad (2.9)_{rev}$$

(Equations valid only for reversible processes are labelled with a subscript rev.) Although the pressure inside the system appears in this expression for the work, it does so only because p_{ex} has been set equal to p to ensure reversibility. The total work of reversible expansion is therefore

$$w = -\int_{V_i}^{V_f} p\,dV \qquad (2.10)_{rev}$$

We can evaluate the integral once we know how the pressure of the confined gas depends on its volume. Equation 2.10 is the link with the material covered in Chapter 1 for, if we know the equation of state of the gas, then we can express p in terms of V and evaluate the integral.

(e) Isothermal reversible expansion

Consider the isothermal, reversible expansion of a perfect gas. The expansion is made isothermal by keeping the system in thermal contact with its surroundings (which may be a constant-temperature bath). Because the equation of state is $pV = nRT$, we know that at each stage $p = nRT/V$, with V the volume at that stage of the expansion. The temperature T is constant in an isothermal expansion, so (together with n and R) it may be taken outside the integral. It follows that the work of reversible isothermal expansion of a perfect gas from V_i to V_f at a temperature T is

Comment 2.3

An integral that occurs throughout thermodynamics is

$$\int_a^b \frac{1}{x}\,dx = (\ln x + \text{constant})\Big|_a^b = \ln\frac{b}{a}$$

$$w = -nRT\int_{V_i}^{V_f} \frac{dV}{V} = -nRT \ln \frac{V_f}{V_i} \qquad (2.11)^{\circ}_{rev}$$

When the final volume is greater than the initial volume, as in an expansion, the logarithm in eqn 2.11 is positive and hence $w < 0$. In this case, the system has done work on the surroundings and the internal energy of the system has decreased as a result.[2] The equations also show that more work is done for a given change of volume when the temperature is increased. The greater pressure of the confined gas then needs a higher opposing pressure to ensure reversibility.

We can express the result of the calculation as an indicator diagram, for the magnitude of the work done is equal to the area under the isotherm $p = nRT/V$ (Fig. 2.8). Superimposed on the diagram is the rectangular area obtained for irreversible expansion against constant external pressure fixed at the same final value as that reached in the reversible expansion. More work is obtained when the expansion is reversible (the area is greater) because matching the external pressure to the internal pressure at each stage of the process ensures that none of the system's pushing power is wasted. We cannot obtain more work than for the reversible process because increasing the external pressure even infinitesimally at any stage results in compression. We may infer from this discussion that, because some pushing power is wasted when $p > p_{ex}$, the maximum work available from a system operating between specified initial and final states and passing along a specified path is obtained when the change takes place reversibly.

We have introduced the connection between reversibility and maximum work for the special case of a perfect gas undergoing expansion. Later (in Section 3.5) we shall see that it applies to all substances and to all kinds of work.

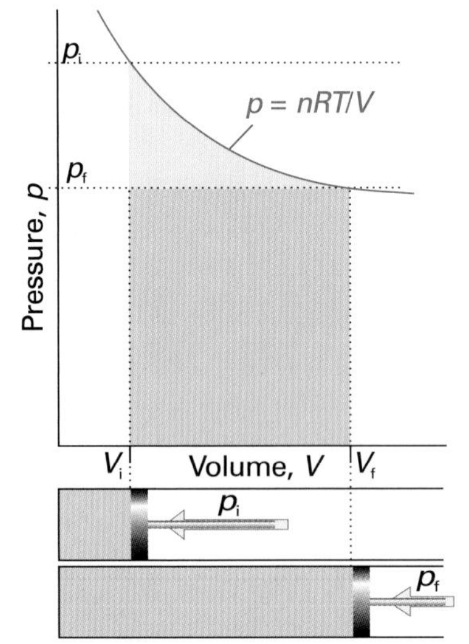

Fig. 2.8 The work done by a perfect gas when it expands reversibly and isothermally is equal to the area under the isotherm $p = nRT/V$. The work done during the irreversible expansion against the same final pressure is equal to the rectangular area shown slightly darker. Note that the reversible work is greater than the irreversible work.

Exploration Calculate the work of isothermal reversible expansion of 1.0 mol CO_2(g) at 298 K from 1.0 m^3 to 3.0 m^3 on the basis that it obeys the van der Waals equation of state.

Example 2.1 *Calculating the work of gas production*

Calculate the work done when 50 g of iron reacts with hydrochloric acid in (a) a closed vessel of fixed volume, (b) an open beaker at 25°C.

[2] We shall see later that there is a compensating influx of energy as heat, so overall the internal energy is constant for the isothermal expansion of a perfect gas.

Method We need to judge the magnitude of the volume change and then to decide how the process occurs. If there is no change in volume, there is no expansion work however the process takes place. If the system expands against a constant external pressure, the work can be calculated from eqn 2.8. A general feature of processes in which a condensed phase changes into a gas is that the volume of the former may usually be neglected relative to that of the gas it forms.

Answer In (a) the volume cannot change, so no expansion work is done and $w = 0$. In (b) the gas drives back the atmosphere and therefore $w = -p_{ex}\Delta V$. We can neglect the initial volume because the final volume (after the production of gas) is so much larger and $\Delta V = V_f - V_i \approx V_f = nRT/p_{ex}$, where n is the amount of H_2 produced. Therefore,

$$w = -p_{ex}\Delta V \approx -p_{ex} \times \frac{nRT}{p_{ex}} = -nRT$$

Because the reaction is $Fe(s) + 2\,HCl(aq) \rightarrow FeCl_2(aq) + H_2(g)$, we know that 1 mol H_2 is generated when 1 mol Fe is consumed, and n can be taken as the amount of Fe atoms that react. Because the molar mass of Fe is 55.85 g mol^{-1}, it follows that

$$\begin{aligned} w &\approx -\frac{50\text{ g}}{55.85\text{ g mol}^{-1}} \times (8.3145\text{ J K}^{-1}\text{ mol}^{-1}) \times (298\text{ K}) \\ &\approx -2.2\text{ kJ} \end{aligned}$$

The system (the reaction mixture) does 2.2 kJ of work driving back the atmosphere. Note that (for this perfect gas system) the magnitude of the external pressure does not affect the final result: the lower the pressure, the larger the volume occupied by the gas, so the effects cancel.

Self-test 2.1 Calculate the expansion work done when 50 g of water is electrolysed under constant pressure at 25°C. [−10 kJ]

2.4 Heat transactions

In general, the change in internal energy of a system is

$$dU = dq + dw_{exp} + dw_e \tag{2.12}$$

where dw_e is work in addition (e for 'extra') to the expansion work, dw_{exp}. For instance, dw_e might be the electrical work of driving a current through a circuit. A system kept at constant volume can do no expansion work, so $dw_{exp} = 0$. If the system is also incapable of doing any other kind of work (if it is not, for instance, an electrochemical cell connected to an electric motor), then $dw_e = 0$ too. Under these circumstances:

$$dU = dq \qquad \text{(at constant volume, no additional work)} \tag{2.13a}$$

We express this relation by writing $dU = dq_V$, where the subscript implies a change at constant volume. For a measurable change,

$$\Delta U = q_V \tag{2.13b}$$

It follows that, by measuring the energy supplied to a constant-volume system as heat ($q > 0$) or obtained from it as heat ($q < 0$) when it undergoes a change of state, we are in fact measuring the change in its internal energy.

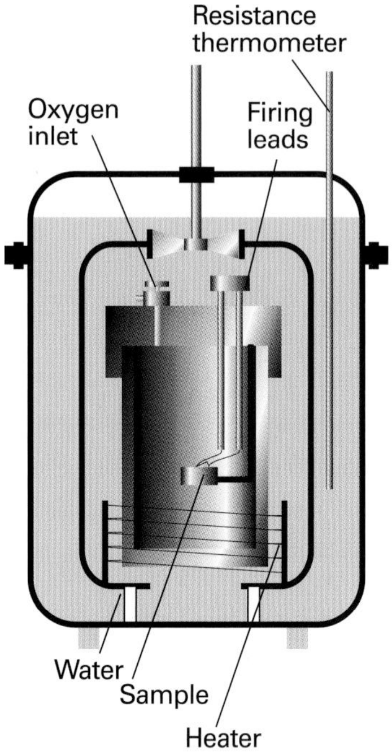

Fig. 2.9 A constant-volume bomb calorimeter. The 'bomb' is the central vessel, which is strong enough to withstand high pressures. The calorimeter (for which the heat capacity must be known) is the entire assembly shown here. To ensure adiabaticity, the calorimeter is immersed in a water bath with a temperature continuously readjusted to that of the calorimeter at each stage of the combustion.

(a) Calorimetry

Calorimetry is the study of heat transfer during physical and chemical processes. A **calorimeter** is a device for measuring energy transferred as heat. The most common device for measuring ΔU is an **adiabatic bomb calorimeter** (Fig. 2.9). The process we wish to study—which may be a chemical reaction—is initiated inside a constant-volume container, the 'bomb'. The bomb is immersed in a stirred water bath, and the whole device is the calorimeter. The calorimeter is also immersed in an outer water bath. The water in the calorimeter and of the outer bath are both monitored and adjusted to the same temperature. This arrangement ensures that there is no net loss of heat from the calorimeter to the surroundings (the bath) and hence that the calorimeter is adiabatic.

The change in temperature, ΔT, of the calorimeter is proportional to the heat that the reaction releases or absorbs. Therefore, by measuring ΔT we can determine q_V and hence find ΔU. The conversion of ΔT to q_V is best achieved by calibrating the calorimeter using a process of known energy output and determining the **calorimeter constant**, the constant C in the relation

$$q = C\Delta T \qquad (2.14a)$$

The calorimeter constant may be measured electrically by passing a constant current, I, from a source of known potential difference, $\mathcal{V}$, through a heater for a known period of time, t, for then

$$q = I\mathcal{V}t \qquad (2.14b)$$

Alternatively, C may be determined by burning a known mass of substance (benzoic acid is often used) that has a known heat output. With C known, it is simple to interpret an observed temperature rise as a release of heat.

Illustration 2.2 *The calibration of a calorimeter*

If we pass a current of 10.0 A from a 12 V supply for 300 s, then from eqn 2.14b the energy supplied as heat is

$$q = (10.0\ \text{A}) \times (12\ \text{V}) \times (300\ \text{s}) = 3.6 \times 10^4\ \text{A V s} = 36\ \text{kJ}$$

because 1 A V s = 1 J. If the observed rise in temperature is 5.5 K, then the calorimeter constant is $C = (36\ \text{kJ})/(5.5\ \text{K}) = 6.5\ \text{kJ K}^{-1}$.

(b) Heat capacity

The internal energy of a substance increases when its temperature is raised. The increase depends on the conditions under which the heating takes place and for the present we suppose that the sample is confined to a constant volume. For example, the sample may be a gas in a container of fixed volume. If the internal energy is plotted against temperature, then a curve like that in Fig. 2.10 may be obtained. The slope of the tangent to the curve at any temperature is called the **heat capacity** of the system at that temperature. The **heat capacity at constant volume** is denoted C_V and is defined formally as[3]

Comment 2.4

Electrical charge is measured in *coulombs*, C. The motion of charge gives rise to an electric current, I, measured in coulombs per second, or *amperes*, A, where $1\ \text{A} = 1\ \text{C s}^{-1}$. If a constant current I flows through a potential difference $\mathcal{V}$ (measured in volts, V), the total energy supplied in an interval t is

$$\textit{Energy supplied} = I\mathcal{V}t$$

Because $1\ \text{A V s} = 1\ (\text{C s}^{-1})\ \text{V s} = 1\ \text{C V} = 1\ \text{J}$, the energy is obtained in joules with the current in amperes, the potential difference in volts, and the time in seconds. We write the electrical power, P, as

$$P = (\textit{energy supplied})/(\textit{time interval}) = I\mathcal{V}t/t = I\mathcal{V}$$

[3] If the system can change its composition, it is necessary to distinguish between equilibrium and fixed-composition values of C_V. All applications in this chapter refer to a single substance, so this complication can be ignored.

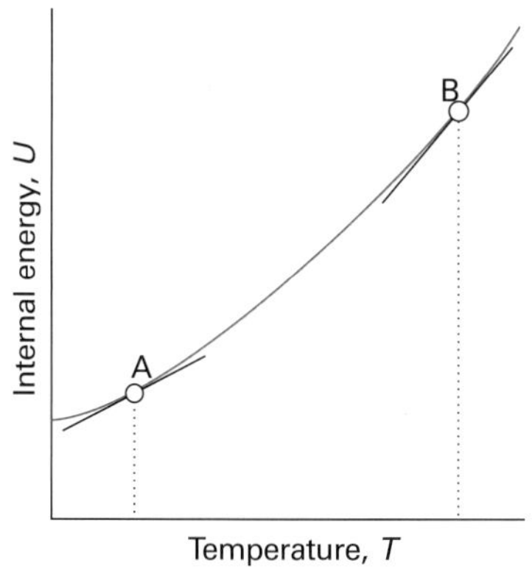

Fig. 2.10 The internal energy of a system increases as the temperature is raised; this graph shows its variation as the system is heated at constant volume. The slope of the tangent to the curve at any temperature is the heat capacity at constant volume at that temperature. Note that, for the system illustrated, the heat capacity is greater at B than at A.

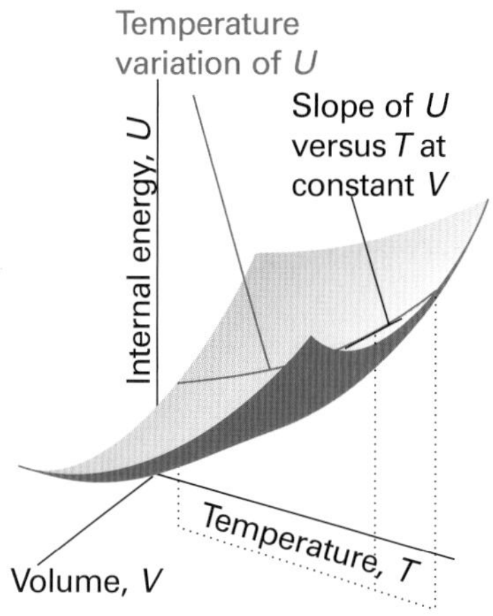

Fig. 2.11 The internal energy of a system varies with volume and temperature, perhaps as shown here by the surface. The variation of the internal energy with temperature at one particular constant volume is illustrated by the curve drawn parallel to T. The slope of this curve at any point is the partial derivative $(\partial U/\partial T)_V$.

$$C_V = \left(\frac{\partial U}{\partial T}\right)_V \qquad [2.15]$$

In this case, the internal energy varies with the temperature and the volume of the sample, but we are interested only in its variation with the temperature, the volume being held constant (Fig. 2.11).

Illustration 2.3 *Estimating a constant-volume heat capacity*

The heat capacity of a monatomic perfect gas can be calculated by inserting the expression for the internal energy derived in *Molecular interpretation 2.2*. There we saw that $U_m = U_m(0) + \frac{3}{2}RT$, so from eqn 2.15

$$C_{V,m} = \frac{\partial}{\partial T}(U_m(0) + \tfrac{3}{2}RT) = \tfrac{3}{2}R$$

The numerical value is 12.47 J K^{-1} mol^{-1}.

Heat capacities are extensive properties: 100 g of water, for instance, has 100 times the heat capacity of 1 g of water (and therefore requires 100 times the energy as heat to bring about the same rise in temperature). The **molar heat capacity at constant volume**, $C_{V,m} = C_V/n$, is the heat capacity per mole of material, and is an intensive property (all molar quantities are intensive). Typical values of $C_{V,m}$ for polyatomic gases are close to 25 J K^{-1} mol^{-1}. For certain applications it is useful to know the **specific heat capacity** (more informally, the 'specific heat') of a substance, which is the heat capacity of the sample divided by the mass, usually in grams: $C_{V,s} = C_V/m$. The specific heat capacity of water at room temperature is close to 4 J K^{-1} g^{-1}. In general,

Comment 2.5

The partial-differential operation $(\partial z/\partial x)_y$ consists of taking the first derivative of $z(x,y)$ with respect to x, treating y as a constant. For example, if $z(x,y) = x^2y$, then

$$\left(\frac{\partial z}{\partial x}\right)_y = \left(\frac{\partial[x^2y]}{\partial x}\right)_y = y\frac{dx^2}{dx} = 2yx$$

Partial derivatives are reviewed in *Appendix 2*.

heat capacities depend on the temperature and decrease at low temperatures. However, over small ranges of temperature at and above room temperature, the variation is quite small and for approximate calculations heat capacities can be treated as almost independent of temperature.

The heat capacity is used to relate a change in internal energy to a change in temperature of a constant-volume system. It follows from eqn 2.15 that

$$dU = C_V dT \qquad \text{(at constant volume)} \tag{2.16a}$$

That is, at constant volume, an infinitesimal change in temperature brings about an infinitesimal change in internal energy, and the constant of proportionality is C_V. If the heat capacity is independent of temperature over the range of temperatures of interest, a measurable change of temperature, ΔT, brings about a measurable increase in internal energy, ΔU, where

$$\Delta U = C_V \Delta T \qquad \text{(at constant volume)} \tag{2.16b}$$

Because a change in internal energy can be identified with the heat supplied at constant volume (eqn 2.13b), the last equation can be written

$$q_V = C_V \Delta T \tag{2.17}$$

This relation provides a simple way of measuring the heat capacity of a sample: a measured quantity of energy is transferred as heat to the sample (electrically, for example), and the resulting increase in temperature is monitored. The ratio of the energy transferred as heat to the temperature rise it causes ($q_V/\Delta T$) is the constant-volume heat capacity of the sample.

A large heat capacity implies that, for a given quantity of energy transferred as heat, there will be only a small increase in temperature (the sample has a large capacity for heat). An infinite heat capacity implies that there will be no increase in temperature however much energy is supplied as heat. At a phase transition, such as at the boiling point of water, the temperature of a substance does not rise as energy is supplied as heat: the energy is used to drive the endothermic transition, in this case to vaporize the water, rather than to increase its temperature. Therefore, at the temperature of a phase transition, the heat capacity of a sample is infinite. The properties of heat capacities close to phase transitions are treated more fully in Section 4.7.

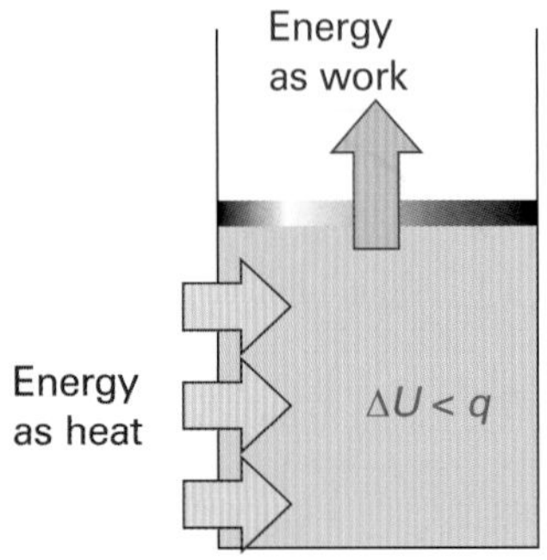

Fig. 2.12 When a system is subjected to constant pressure and is free to change its volume, some of the energy supplied as heat may escape back into the surroundings as work. In such a case, the change in internal energy is smaller than the energy supplied as heat.

2.5 Enthalpy

The change in internal energy is not equal to the energy transferred as heat when the system is free to change its volume. Under these circumstances some of the energy supplied as heat to the system is returned to the surroundings as expansion work (Fig. 2.12), so dU is less than dq. However, we shall now show that in this case the energy supplied as heat at constant pressure is equal to the change in another thermodynamic property of the system, the enthalpy.

(a) The definition of enthalpy

The **enthalpy**, H, is defined as

$$H = U + pV \tag*{[2.18]}$$

where p is the pressure of the system and V is its volume. Because U, p, and V are all state functions, the enthalpy is a state function too. As is true of any state function, the change in enthalpy, ΔH, between any pair of initial and final states is independent of the path between them.

Although the definition of enthalpy may appear arbitrary, it has important implications for thermochemisty. For instance, we show in the following *Justification* that eqn 2.18 implies that *the change in enthalpy is equal to the energy supplied as heat at constant pressure* (provided the system does no additional work):

$$dH = dq \qquad \text{(at constant pressure, no additional work)} \qquad (2.19a)$$

For a measurable change,

$$\Delta H = q_p \qquad (2.19b)$$

Justification 2.1 *The relation* $\Delta H = q_p$

For a general infinitesimal change in the state of the system, U changes to $U + dU$, p changes to $p + dp$, and V changes to $V + dV$, so from the definition in eqn 2.18, H changes from $U + pV$ to

$$\begin{aligned} H + dH &= (U + dU) + (p + dp)(V + dV) \\ &= U + dU + pV + pdV + Vdp + dpdV \end{aligned}$$

The last term is the product of two infinitesimally small quantities and can therefore be neglected. As a result, after recognizing $U + pV = H$ on the right, we find that H changes to

$$H + dH = H + dU + pdV + Vdp$$

and hence that

$$dH = dU + pdV + Vdp$$

If we now substitute $dU = dq + dw$ into this expression, we get

$$dH = dq + dw + pdV + Vdp$$

If the system is in mechanical equilibrium with its surroundings at a pressure p and does only expansion work, we can write $dw = -pdV$ and obtain

$$dH = dq + Vdp$$

Now we impose the condition that the heating occurs at constant pressure by writing $dp = 0$. Then

$$dH = dq \qquad \text{(at constant pressure, no additional work)}$$

as in eqn 2.19a.

The result expressed in eqn 2.19 states that, when a system is subjected to a constant pressure, and only expansion work can occur, the change in enthalpy is equal to the energy supplied as heat. For example, if we supply 36 kJ of energy through an electric heater immersed in an open beaker of water, then the enthalpy of the water increases by 36 kJ and we write $\Delta H = +36$ kJ.

(b) The measurement of an enthalpy change

An enthalpy change can be measured calorimetrically by monitoring the temperature change that accompanies a physical or chemical change occurring at constant pressure. A calorimeter for studying processes at constant pressure is called an **isobaric calorimeter**. A simple example is a thermally insulated vessel open to the atmosphere: the heat released in the reaction is monitored by measuring the change in temperature

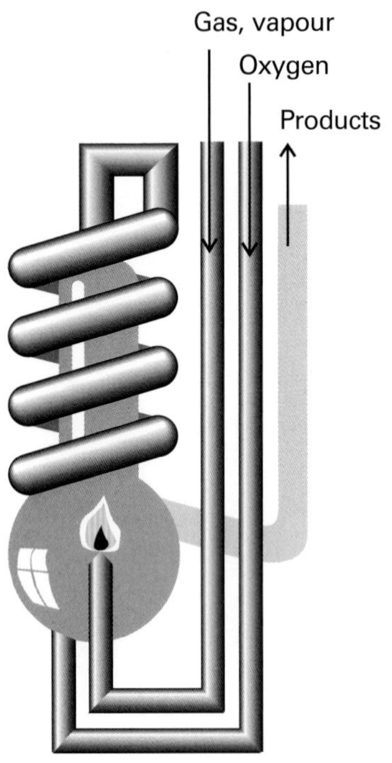

Fig. 2.13 A constant-pressure flame calorimeter consists of this component immersed in a stirred water bath. Combustion occurs as a known amount of reactant is passed through to fuel the flame, and the rise of temperature is monitored.

of the contents. For a combustion reaction an **adiabatic flame calorimeter** may be used to measure ΔT when a given amount of substance burns in a supply of oxygen (Fig. 2.13). Another route to ΔH is to measure the internal energy change by using a bomb calorimeter, and then to convert ΔU to ΔH. Because solids and liquids have small molar volumes, for them pV_m is so small that the molar enthalpy and molar internal energy are almost identical ($H_m = U_m + pV_m \approx U_m$). Consequently, if a process involves only solids or liquids, the values of ΔH and ΔU are almost identical. Physically, such processes are accompanied by a very small change in volume, the system does negligible work on the surroundings when the process occurs, so the energy supplied as heat stays entirely within the system. The most sophisticated way to measure enthalpy changes, however, is to use a **differential scanning calorimeter** (DSC). Changes in enthalpy and internal energy may also be measured by noncalorimetric methods (see Chapter 7).

Example 2.2 *Relating ΔH and ΔU*

The internal energy change when 1.0 mol $CaCO_3$ in the form of calcite converts to aragonite is +0.21 kJ. Calculate the difference between the enthalpy change and the change in internal energy when the pressure is 1.0 bar given that the densities of the solids are 2.71 g cm^{-3} and 2.93 g cm^{-3}, respectively.

Method The starting point for the calculation is the relation between the enthalpy of a substance and its internal energy (eqn 2.18). The difference between the two quantities can be expressed in terms of the pressure and the difference of their molar volumes, and the latter can be calculated from their molar masses, M, and their mass densities, ρ, by using $\rho = M/V_m$.

Answer The change in enthalpy when the transition occurs is

$$\begin{aligned}\Delta H &= H(\text{aragonite}) - H(\text{calcite}) \\ &= \{U(\text{a}) + pV(\text{a})\} - \{U(\text{c}) + pV(\text{c})\} \\ &= \Delta U + p\{V(\text{a}) - V(\text{c})\} = \Delta U + p\Delta V\end{aligned}$$

The volume of 1.0 mol $CaCO_3$ (100 g) as aragonite is 34 cm^3, and that of 1.0 mol $CaCO_3$ as calcite is 37 cm^3. Therefore,

$$p\Delta V = (1.0 \times 10^5\ \text{Pa}) \times (34 - 37) \times 10^{-6}\ \text{m}^3 = -0.3\ \text{J}$$

(because 1 Pa m^3 = 1 J). Hence,

$$\Delta H - \Delta U = -0.3\ \text{J}$$

which is only 0.1 per cent of the value of ΔU. We see that it is usually justifiable to ignore the difference between the enthalpy and internal energy of condensed phases, except at very high pressures, when pV is no longer negligible.

Self-test 2.2 Calculate the difference between ΔH and ΔU when 1.0 mol Sn(s, grey) of density 5.75 g cm^{-3} changes to Sn(s, white) of density 7.31 g cm^{-3} at 10.0 bar. At 298 K, ΔH = +2.1 kJ. [$\Delta H - \Delta U$ = −4.4 J]

The enthalpy of a perfect gas is related to its internal energy by using $pV = nRT$ in the definition of H:

$$H = U + pV = U + nRT \tag{2.20°}$$

This relation implies that the change of enthalpy in a reaction that produces or consumes gas is

$$\Delta H = \Delta U + \Delta n_g RT \qquad (2.21)°$$

where Δn_g is the change in the amount of gas molecules in the reaction.

Illustration 2.4 *The relation between ΔH and ΔU for gas-phase reactions*

In the reaction 2 $H_2(g)$ + $O_2(g) \rightarrow$ 2 $H_2O(l)$, 3 mol of gas-phase molecules is replaced by 2 mol of liquid-phase molecules, so $\Delta n_g = -3$ mol. Therefore, at 298 K, when $RT = 2.5$ kJ mol^{-1}, the enthalpy and internal energy changes taking place in the system are related by

$$\Delta H - \Delta U = (-3\text{ mol}) \times RT \approx -7.4\text{ kJ}$$

Note that the difference is expressed in kilojoules, not joules as in Example 2.2. The enthalpy change is smaller (in this case, less negative) than the change in internal energy because, although heat escapes from the system when the reaction occurs, the system contracts when the liquid is formed, so energy is restored to it from the surroundings.

Example 2.3 *Calculating a change in enthalpy*

Water is heated to boiling under a pressure of 1.0 atm. When an electric current of 0.50 A from a 12 V supply is passed for 300 s through a resistance in thermal contact with it, it is found that 0.798 g of water is vaporized. Calculate the molar internal energy and enthalpy changes at the boiling point (373.15 K).

Method Because the vaporization occurs at constant pressure, the enthalpy change is equal to the heat supplied by the heater. Therefore, the strategy is to calculate the energy supplied as heat (from $q = I\mathcal{V}t$), express that as an enthalpy change, and then convert the result to a molar enthalpy change by division by the amount of H_2O molecules vaporized. To convert from enthalpy change to internal energy change, we assume that the vapour is a perfect gas and use eqn 2.21.

Answer The enthalpy change is

$$\Delta H = q_p = (0.50\text{ A}) \times (12\text{ V}) \times (300\text{ s}) = +(0.50 \times 12 \times 300)\text{ J}$$

Here we have used 1 A V s = 1 J (see *Comment 2.4*). Because 0.798 g of water is (0.798 g)/(18.02 g mol^{-1}) = (0.798/18.02) mol H_2O, the enthalpy of vaporization per mole of H_2O is

$$\Delta H_m = +\frac{0.50 \times 12 \times 300\text{ J}}{(0.798/18.02)\text{ mol}} = +41\text{ kJ mol}^{-1}$$

In the process $H_2O(l) \rightarrow H_2O(g)$ the change in the amount of gas molecules is $\Delta n_g = +1$ mol, so

$$\Delta U_m = \Delta H_m - RT = +38\text{ kJ mol}^{-1}$$

The plus sign is added to positive quantities to emphasize that they represent an increase in internal energy or enthalpy. Notice that the internal energy change is smaller than the enthalpy change because energy has been used to drive back the surrounding atmosphere to make room for the vapour.

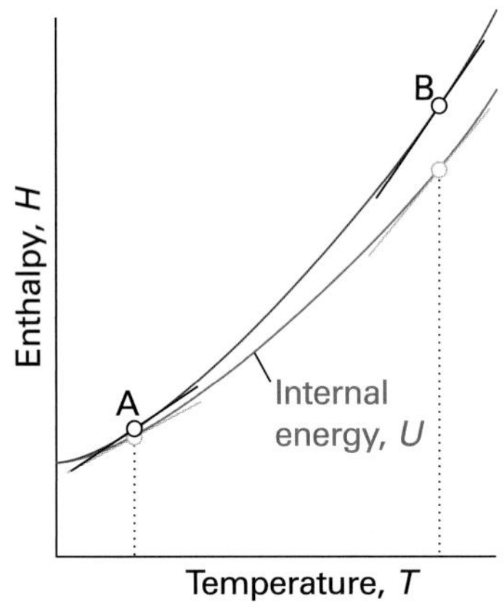

Fig. 2.14 The slope of the tangent to a curve of the enthalpy of a system subjected to a constant pressure plotted against temperature is the constant-pressure heat capacity. The slope may change with temperature, in which case the heat capacity varies with temperature. Thus, the heat capacities at A and B are different. For gases, at a given temperature the slope of enthalpy versus temperature is steeper than that of internal energy versus temperature, and $C_{p,\mathrm{m}}$ is larger than $C_{V,\mathrm{m}}$.

Self-test 2.3 The molar enthalpy of vaporization of benzene at its boiling point (353.25 K) is 30.8 kJ mol^{-1}. What is the molar internal energy change? For how long would the same 12 V source need to supply a 0.50 A current in order to vaporize a 10 g sample? [+27.9 kJ mol^{-1}, 660 s]

(c) The variation of enthalpy with temperature

The enthalpy of a substance increases as its temperature is raised. The relation between the increase in enthalpy and the increase in temperature depends on the conditions (for example, constant pressure or constant volume). The most important condition is constant pressure, and the slope of the tangent to a plot of enthalpy against temperature at constant pressure is called the **heat capacity at constant pressure**, C_p, at a given temperature (Fig. 2.14). More formally:

$$C_p = \left(\frac{\partial H}{\partial T}\right)_p \qquad [2.22]$$

The heat capacity at constant pressure is the analogue of the heat capacity at constant volume, and is an extensive property.[4] The **molar heat capacity at constant pressure**, $C_{p,\mathrm{m}}$, is the heat capacity per mole of material; it is an intensive property.

The heat capacity at constant pressure is used to relate the change in enthalpy to a change in temperature. For infinitesimal changes of temperature,

$$\mathrm{d}H = C_p\mathrm{d}T \qquad \text{(at constant pressure)} \qquad (2.23\text{a})$$

If the heat capacity is constant over the range of temperatures of interest, then for a measurable increase in temperature

$$\Delta H = C_p\Delta T \qquad \text{(at constant pressure)} \qquad (2.23\text{b})$$

Because an increase in enthalpy can be equated with the energy supplied as heat at constant pressure, the practical form of the latter equation is

$$q_p = C_p\Delta T \qquad (2.24)$$

This expression shows us how to measure the heat capacity of a sample: a measured quantity of energy is supplied as heat under conditions of constant pressure (as in a sample exposed to the atmosphere and free to expand), and the temperature rise is monitored.

The variation of heat capacity with temperature can sometimes be ignored if the temperature range is small; this approximation is highly accurate for a monatomic perfect gas (for instance, one of the noble gases at low pressure). However, when it is necessary to take the variation into account, a convenient approximate empirical expression is

$$C_{p,\mathrm{m}} = a + bT + \frac{c}{T^2} \qquad (2.25)$$

The empirical parameters a, b, and c are independent of temperature (Table 2.2).

[4] As in the case of C_V, if the system can change its composition it is necessary to distinguish between equilibrium and fixed-composition values. All applications in this chapter refer to pure substances, so this complication can be ignored.

Synoptic Table 2.2* Temperature variation of molar heat capacities, $C_{p,m}/(\mathrm{J\,K^{-1}\,mol^{-1}}) = a + bT + c/T^2$

	a	$b/(10^{-3}\,\mathrm{K})$	$c/(10^5\,\mathrm{K}^2)$
C(s, graphite)	16.86	4.77	−8.54
CO_2(g)	44.22	8.79	−8.62
H_2O(l)	75.29	0	0
N_2(g)	28.58	3.77	−0.50

* More values are given in the *Data section*.

Example 2.4 *Evaluating an increase in enthalpy with temperature*

What is the change in molar enthalpy of N_2 when it is heated from 25°C to 100°C? Use the heat capacity information in Table 2.2.

Method The heat capacity of N_2 changes with temperature, so we cannot use eqn 2.23b (which assumes that the heat capacity of the substance is constant). Therefore, we must use eqn 2.23a, substitute eqn 2.25 for the temperature dependence of the heat capacity, and integrate the resulting expression from 25°C to 100°C.

Answer For convenience, we denote the two temperatures T_1 (298 K) and T_2 (373 K). The integrals we require are

$$\int_{H(T_1)}^{H(T_2)} \mathrm{d}H = \int_{T_1}^{T_2} \left(a + bT + \frac{c}{T^2}\right)\mathrm{d}T$$

Notice how the limits of integration correspond on each side of the equation: the integration over H on the left ranges from $H(T_1)$, the value of H at T_1, up to $H(T_2)$, the value of H at T_2, while on the right the integration over the temperature ranges from T_1 to T_2. Now we use the integrals

$$\int \mathrm{d}x = x + \text{constant} \qquad \int x\,\mathrm{d}x = \tfrac{1}{2}x^2 + \text{constant} \qquad \int \frac{\mathrm{d}x}{x^2} = -\frac{1}{x} + \text{constant}$$

to obtain

$$H(T_2) - H(T_1) = a(T_2 - T_1) + \tfrac{1}{2}b(T_2^2 - T_1^2) - c\left(\frac{1}{T_2} - \frac{1}{T_1}\right)$$

Substitution of the numerical data results in

$$H(373\,\mathrm{K}) = H(298\,\mathrm{K}) + 2.20\,\mathrm{kJ\,mol^{-1}}$$

If we had assumed a constant heat capacity of 29.14 $\mathrm{J\,K^{-1}\,mol^{-1}}$ (the value given by eqn 2.25 at 25°C), we would have found that the two enthalpies differed by 2.19 $\mathrm{kJ\,mol^{-1}}$.

Self-test 2.4 At very low temperatures the heat capacity of a solid is proportional to T^3, and we can write $C_p = aT^3$. What is the change in enthalpy of such a substance when it is heated from 0 to a temperature T (with T close to 0)? [$\Delta H = \tfrac{1}{4}aT^4$]

Comment 2.6

Integrals commonly encountered in physical chemistry are listed inside the front cover.

Most systems expand when heated at constant pressure. Such systems do work on the surroundings and therefore some of the energy supplied to them as heat escapes back to the surroundings. As a result, the temperature of the system rises less than when the heating occurs at constant volume. A smaller increase in temperature implies a larger heat capacity, so we conclude that in most cases the heat capacity at constant pressure of a system is larger than its heat capacity at constant volume. We show later (Section 2.11) that there is a simple relation between the two heat capacities of a perfect gas:

$$C_p - C_V = nR \qquad (2.26)°$$

It follows that the molar heat capacity of a perfect gas is about 8 J K^{-1} mol^{-1} larger at constant pressure than at constant volume. Because the heat capacity at constant volume of a monatomic gas is about 12 J K^{-1} mol^{-1}, the difference is highly significant and must be taken into account.

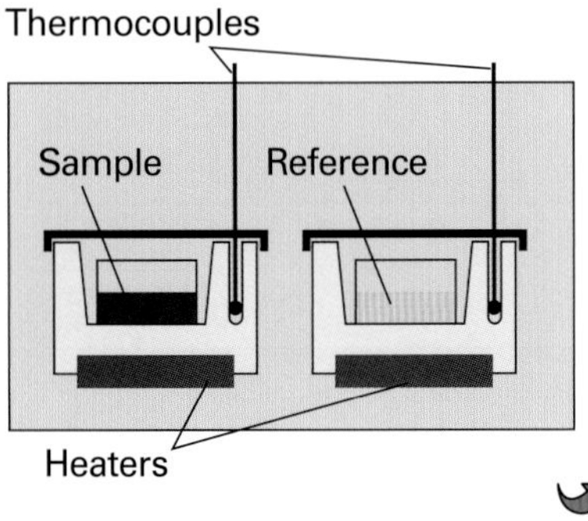

Fig. 2.15 A differential scanning calorimeter. The sample and a reference material are heated in separate but identical metal heat sinks. The output is the difference in power needed to maintain the heat sinks at equal temperatures as the temperature rises.

IMPACT ON BIOCHEMISTRY AND MATERIALS SCIENCE

I2.1 Differential scanning calorimetry

A *differential scanning calorimeter* (DSC) measures the energy transferred as heat to or from a sample at constant pressure during a physical or chemical change. The term 'differential' refers to the fact that the behaviour of the sample is compared to that of a reference material which does not undergo a physical or chemical change during the analysis. The term 'scanning' refers to the fact that the temperatures of the sample and reference material are increased, or scanned, during the analysis.

A DSC consists of two small compartments that are heated electrically at a constant rate. The temperature, T, at time t during a linear scan is $T = T_0 + \alpha t$, where T_0 is the initial temperature and α is the temperature scan rate (in kelvin per second, K s^{-1}). A computer controls the electrical power output in order to maintain the same temperature in the sample and reference compartments throughout the analysis (see Fig. 2.15).

The temperature of the sample changes significantly relative to that of the reference material if a chemical or physical process involving the transfer of energy as heat occurs in the sample during the scan. To maintain the same temperature in both compartments, excess energy is transferred as heat to or from the sample during the process. For example, an endothermic process lowers the temperature of the sample relative to that of the reference and, as a result, the sample must be heated more strongly than the reference in order to maintain equal temperatures.

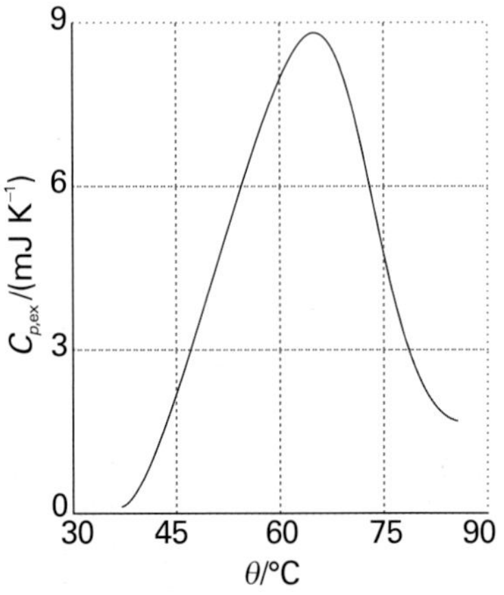

Fig. 2.16 A thermogram for the protein ubiquitin at pH = 2.45. The protein retains its native structure up to about 45°C and then undergoes an endothermic conformational change. (Adapted from B. Chowdhry and S. LeHarne, *J. Chem. Educ.* **74**, 236 (1997).)

If no physical or chemical change occurs in the sample at temperature T, we write the heat transferred to the sample as $q_p = C_p\Delta T$, where $\Delta T = T - T_0$ and we have assumed that C_p is independent of temperature. The chemical or physical process requires the transfer of $q_p + q_{p,ex}$, where $q_{p,ex}$ is excess energy transferred as heat, to attain the same change in temperature of the sample. We interpret $q_{p,ex}$ in terms of an apparent change in the heat capacity at constant pressure of the sample, C_p, during the temperature scan. Then we write the heat capacity of the sample as $C_p + C_{p,ex}$, and

$$q_p + q_{p,ex} = (C_p + C_{p,ex})\Delta T$$

It follows that

$$C_{p,ex} = \frac{q_{p,ex}}{\Delta T} = \frac{q_{p,ex}}{\alpha t} = \frac{P_{ex}}{\alpha}$$

where $P_{ex} = q_{p,ex}/t$ is the excess electrical power necessary to equalize the temperature of the sample and reference compartments.

A DSC trace, also called a *thermogram*, consists of a plot of P_{ex} or $C_{p,ex}$ against T (see Fig. 2.16). Broad peaks in the thermogram indicate processes requiring transfer of energy as heat. From eqn 2.23a, the enthalpy change associated with the process is

$$\Delta H = \int_{T_1}^{T_2} C_{p,\text{ex}} \mathrm{d}T$$

where T_1 and T_2 are, respectively, the temperatures at which the process begins and ends. This relation shows that the enthalpy change is then the area under the curve of $C_{p,\text{ex}}$ against T. With a DSC, enthalpy changes may be determined in samples of masses as low as 0.5 mg, which is a significant advantage over bomb or flame calorimeters, which require several grams of material.

Differential scanning calorimetry is used in the chemical industry to characterize polymers and in the biochemistry laboratory to assess the stability of proteins, nucleic acids, and membranes. Large molecules, such as synthetic or biological polymers, attain complex three-dimensional structures due to intra- and intermolecular interactions, such as hydrogen bonding and hydrophobic interactions (Chapter 18). Disruption of these interactions is an endothermic process that can be studied with a DSC. For example, the thermogram shown in the illustration indicated that the protein ubiquitin retains its native structure up to about 45°C. At higher temperatures, the protein undergoes an endothermic conformational change that results in the loss of its three-dimensional structure. The same principles also apply to the study of structural integrity and stability of synthetic polymers, such as plastics.

2.6 Adiabatic changes

We are now equipped to deal with the changes that occur when a perfect gas expands adiabatically. A decrease in temperature should be expected: because work is done but no heat enters the system, the internal energy falls, and therefore the temperature of the working gas also falls. In molecular terms, the kinetic energy of the molecules falls as work is done, so their average speed decreases, and hence the temperature falls.

The change in internal energy of a perfect gas when the temperature is changed from T_i to T_f and the volume is changed from V_i to V_f can be expressed as the sum of two steps (Fig. 2.17). In the first step, only the volume changes and the temperature is held constant at its initial value. However, because the internal energy of a perfect gas is independent of the volume the molecules occupy, the overall change in internal energy arises solely from the second step, the change in temperature at constant volume. Provided the heat capacity is independent of temperature, this change is

$$\Delta U = C_V(T_\text{f} - T_\text{i}) = C_V\Delta T$$

Because the expansion is adiabatic, we know that $q = 0$; because $\Delta U = q + w$, it then follows that $\Delta U = w_\text{ad}$. The subscript 'ad' denotes an adiabatic process. Therefore, by equating the two values we have obtained for ΔU, we obtain

$$w_\text{ad} = C_V\Delta T \qquad (2.27)$$

That is, the work done during an adiabatic expansion of a perfect gas is proportional to the temperature difference between the initial and final states. That is exactly what we expect on molecular grounds, because the mean kinetic energy is proportional to T, so a change in internal energy arising from temperature alone is also expected to be proportional to ΔT. In *Further information 2.1* we show that the initial and final temperatures of a perfect gas that undergoes reversible adiabatic expansion (reversible expansion in a thermally insulated container) can be calculated from

$$T_\text{f} = T_\text{i}\left(\frac{V_\text{i}}{V_\text{f}}\right)^{1/c} \qquad (2.28\text{a})^\circ_\text{rev}$$

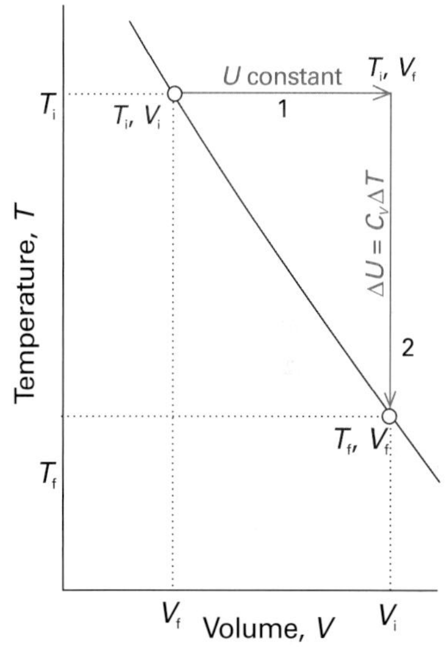

Fig. 2.17 To achieve a change of state from one temperature and volume to another temperature and volume, we may consider the overall change as composed of two steps. In the first step, the system expands at constant temperature; there is no change in internal energy if the system consists of a perfect gas. In the second step, the temperature of the system is reduced at constant volume. The overall change in internal energy is the sum of the changes for the two steps.

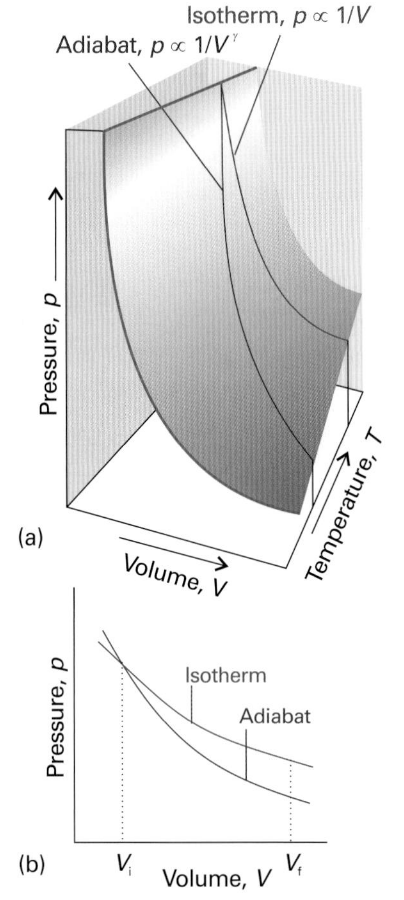

Fig. 2.18 An adiabat depicts the variation of pressure with volume when a gas expands adiabatically. (a) An adiabat for a perfect gas undergoing reversible expansion. (b) Note that the pressure declines more steeply for an adiabat than it does for an isotherm because the temperature decreases in the former.

Exploration Explore how the parameter γ affects the dependence of the pressure on the volume. Does the pressure–volume dependence become stronger or weaker with increasing volume?

where $c = C_{V,\mathrm{m}}/R$, or equivalently

$$V_\mathrm{i}T_\mathrm{i}^c = V_\mathrm{f}T_\mathrm{f}^c \qquad (2.28\mathrm{b})^\circ_\mathrm{rev}$$

This result is often summarized in the form $VT^c = \text{constant}$.

Illustration 2.5 *Work of adiabatic expansion*

Consider the adiabatic, reversible expansion of 0.020 mol Ar, initially at 25°C, from 0.50 dm^3 to 1.00 dm^3. The molar heat capacity of argon at constant volume is 12.48 J K^{-1} mol^{-1}, so $c = 1.501$. Therefore, from eqn 2.28a,

$$T_\mathrm{f} = (298\ \mathrm{K}) \times \left(\frac{0.50\ \mathrm{dm}^3}{1.00\ \mathrm{dm}^3}\right)^{1/1.501} = 188\ \mathrm{K}$$

It follows that $\Delta T = -110$ K, and therefore, from eqn 2.27, that

$$w = \{(0.020\ \mathrm{mol}) \times (12.48\ \mathrm{J\ K^{-1}\ mol^{-1}})\} \times (-110\ \mathrm{K}) = -27\ \mathrm{J}$$

Note that temperature change is independent of the amount of gas but the work is not.

Self-test 2.5 Calculate the final temperature, the work done, and the change of internal energy when ammonia is used in a reversible adiabatic expansion from 0.50 dm^3 to 2.00 dm^3, the other initial conditions being the same.

[195 K, −56 J, −56 J]

We also show in *Further information 2.1* that the pressure of a perfect gas that undergoes reversible adiabatic expansion from a volume V_i to a volume V_f is related to its initial pressure by

$$p_\mathrm{f}V_\mathrm{f}^\gamma = p_\mathrm{i}V_\mathrm{i}^\gamma \qquad (2.29)^\circ_\mathrm{rev}$$

where $\gamma = C_{p,\mathrm{m}}/C_{V,\mathrm{m}}$. This result is summarized in the form $pV^\gamma = \text{constant}$. For a monatomic perfect gas, $C_{V,\mathrm{m}} = \frac{3}{2}R$ (see *Illustration 2.3*), and from eqn 2.26 $C_{p,\mathrm{m}} = \frac{5}{2}R$; so $\gamma = \frac{5}{3}$. For a gas of nonlinear polyatomic molecules (which can rotate as well as translate), $C_{V,\mathrm{m}} = 3R$, so $\gamma = \frac{4}{3}$. The curves of pressure versus volume for adiabatic change are known as **adiabats**, and one for a reversible path is illustrated in Fig. 2.18. Because $\gamma > 1$, an adiabat falls more steeply ($p \propto 1/V^\gamma$) than the corresponding isotherm ($p \propto 1/V$). The physical reason for the difference is that, in an isothermal expansion, energy flows into the system as heat and maintains the temperature; as a result, the pressure does not fall as much as in an adiabatic expansion.

Illustration 2.6 *The pressure change accompanying adiabatic expansion*

When a sample of argon (for which $\gamma = \frac{5}{3}$) at 100 kPa expands reversibly and adiabatically to twice its initial volume the final pressure will be

$$p_\mathrm{f} = \left(\frac{V_\mathrm{i}}{V_\mathrm{f}}\right)^\gamma p_\mathrm{i} = \left(\frac{1}{2}\right)^{5/3} \times (100\ \mathrm{kPa}) = 32\ \mathrm{kPa}$$

For an isothermal doubling of volume, the final pressure would be 50 kPa.

Thermochemistry

The study of the energy transferred as heat during the course of chemical reactions is called **thermochemistry**. Thermochemistry is a branch of thermodynamics because a reaction vessel and its contents form a system, and chemical reactions result in the exchange of energy between the system and the surroundings. Thus we can use calorimetry to measure the energy supplied or discarded as heat by a reaction, and can identify q with a change in internal energy (if the reaction occurs at constant volume) or a change in enthalpy (if the reaction occurs at constant pressure). Conversely, if we know ΔU or ΔH for a reaction, we can predict the energy (transferred as heat) the reaction can produce.

We have already remarked that a process that releases energy by heating the surroundings is classified as exothermic and one that absorbs energy by cooling the surroundings is classified as endothermic. Because the release of energy by heating the surroundings signifies a decrease in the enthalpy of a system (at constant pressure), we can now see that an exothermic process at constant pressure is one for which $\Delta H < 0$. Conversely, because the absorption of energy by cooling the surroundings results in an increase in enthalpy, an endothermic process at constant pressure has $\Delta H > 0$.

2.7 Standard enthalpy changes

Changes in enthalpy are normally reported for processes taking place under a set of standard conditions. In most of our discussions we shall consider the **standard enthalpy change**, $\Delta H^{\ominus}$, the change in enthalpy for a process in which the initial and final substances are in their standard states:

> The **standard state** of a substance at a specified temperature is its pure form at 1 bar.[5]

For example, the standard state of liquid ethanol at 298 K is pure liquid ethanol at 298 K and 1 bar; the standard state of solid iron at 500 K is pure iron at 500 K and 1 bar. The standard enthalpy change for a reaction or a physical process is the difference between the products in their standard states and the reactants in their standard states, all at the same specified temperature.

As an example of a standard enthalpy change, the *standard enthalpy of vaporization*, $\Delta_{vap}H^{\ominus}$, is the enthalpy change per mole when a pure liquid at 1 bar vaporizes to a gas at 1 bar, as in

$$H_2O(l) \rightarrow H_2O(g) \qquad \Delta_{vap}H^{\ominus}(373\text{ K}) = +40.66\text{ kJ mol}^{-1}$$

As implied by the examples, standard enthalpies may be reported for any temperature. However, the conventional temperature for reporting thermodynamic data is 298.15 K (corresponding to 25.00°C). Unless otherwise mentioned, all thermodynamic data in this text will refer to this conventional temperature.

A note on good practice The attachment of the name of the transition to the symbol Δ, as in $\Delta_{vap}H$, is the modern convention. However, the older convention, ΔH_{vap}, is still widely used. The new convention is more logical because the subscript identifies the type of change, not the physical observable related to the change.

[5] The definition of standard state is more sophisticated for a real gas (*Further information 3.2*) and for solutions (Sections 5.6 and 5.7).

Synoptic Table 2.3* Standard enthalpies of fusion and vaporization at the transition temperature, $\Delta_{trs}H^{\ominus}/(\text{kJ mol}^{-1})$

	T_f/K	Fusion	T_b/K	Vaporization
Ar	83.81	1.188	87.29	6.506
C_6H_6	278.61	10.59	353.2	30.8
H_2O	273.15	6.008	373.15	40.656 (44.016 at 298 K)
He	3.5	0.021	4.22	0.084

* More values are given in the *Data section*.

(a) Enthalpies of physical change

The standard enthalpy change that accompanies a change of physical state is called the **standard enthalpy of transition** and is denoted $\Delta_{trs}H^{\ominus}$ (Table 2.3). The **standard enthalpy of vaporization**, $\Delta_{vap}H^{\ominus}$, is one example. Another is the **standard enthalpy of fusion**, $\Delta_{fus}H^{\ominus}$, the standard enthalpy change accompanying the conversion of a solid to a liquid, as in

$$H_2O(s) \rightarrow H_2O(l) \qquad \Delta_{fus}H^{\ominus}(273\text{ K}) = +6.01\text{ kJ mol}^{-1}$$

As in this case, it is sometimes convenient to know the standard enthalpy change at the transition temperature as well as at the conventional temperature.

Because enthalpy is a state function, a change in enthalpy is independent of the path between the two states. This feature is of great importance in thermochemistry, for it implies that the same value of $\Delta H^{\ominus}$ will be obtained however the change is brought about between the same initial and final states. For example, we can picture the conversion of a solid to a vapour either as occurring by sublimation (the direct conversion from solid to vapour),

$$H_2O(s) \rightarrow H_2O(g) \qquad \Delta_{sub}H^{\ominus}$$

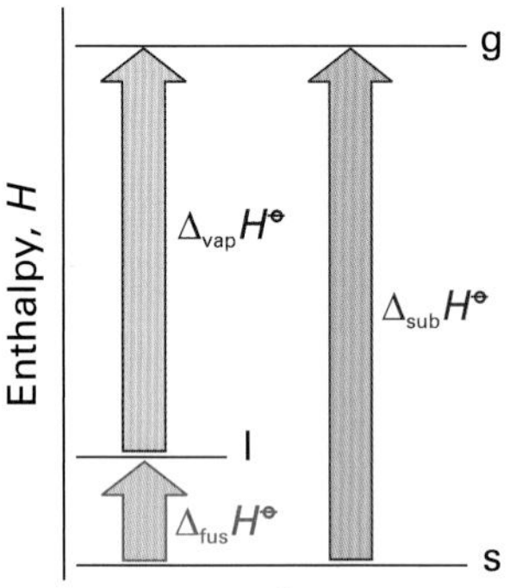

1

or as occurring in two steps, first fusion (melting) and then vaporization of the resulting liquid:

$$\begin{array}{ll} H_2O(s) \rightarrow H_2O(l) & \Delta_{fus}H^{\ominus} \\ H_2O(l) \rightarrow H_2O(g) & \Delta_{vap}H^{\ominus} \\ \text{Overall: } H_2O(s) \rightarrow H_2O(g) & \Delta_{fus}H^{\ominus} + \Delta_{vap}H^{\ominus} \end{array}$$

Because the overall result of the indirect path is the same as that of the direct path, the overall enthalpy change is the same in each case (**1**), and we can conclude that (for processes occurring at the same temperature)

$$\Delta_{sub}H^{\ominus} = \Delta_{fus}H^{\ominus} + \Delta_{vap}H^{\ominus} \tag{2.30}$$

An immediate conclusion is that, because all enthalpies of fusion are positive, the enthalpy of sublimation of a substance is greater than its enthalpy of vaporization (at a given temperature).

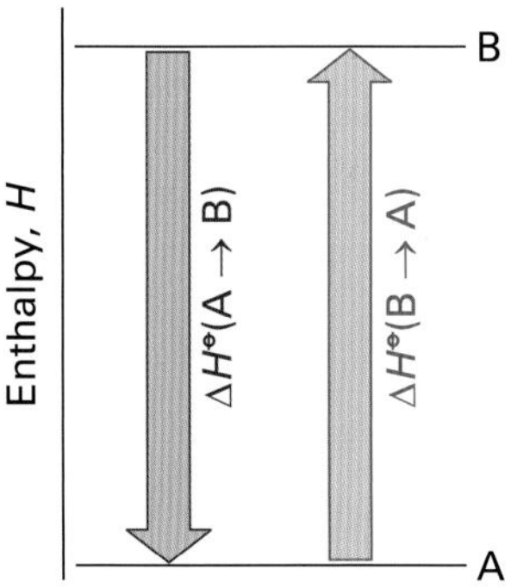

2

Another consequence of H being a state function is that the standard enthalpy changes of a forward process and its reverse differ in sign (**2**):

$$\Delta H^{\ominus}(A \rightarrow B) = -\Delta H^{\ominus}(B \rightarrow A) \tag{2.31}$$

For instance, because the enthalpy of vaporization of water is +44 kJ mol^{-1} at 298 K, its enthalpy of condensation at that temperature is −44 kJ mol^{-1}.

Table 2.4 Enthalpies of transition

Transition	Process	Symbol*
Transition	Phase $\alpha \rightarrow$ phase β	$\Delta_{trs}H$
Fusion	$s \rightarrow l$	$\Delta_{fus}H$
Vaporization	$l \rightarrow g$	$\Delta_{vap}H$
Sublimation	$s \rightarrow g$	$\Delta_{sub}H$
Mixing	Pure $\rightarrow$ mixture	$\Delta_{mix}H$
Solution	Solute $\rightarrow$ solution	$\Delta_{sol}H$
Hydration	$X^{\pm}(g) \rightarrow X^{\pm}(aq)$	$\Delta_{hyd}H$
Atomization	Species(s, l, g) $\rightarrow$ atoms(g)	$\Delta_{at}H$
Ionization	$X(g) \rightarrow X^{+}(g) + e^{-}(g)$	$\Delta_{ion}H$
Electron gain	$X(g) + e^{-}(g) \rightarrow X^{-}(g)$	$\Delta_{eg}H$
Reaction	Reactants $\rightarrow$ products	$\Delta_{r}H$
Combustion	Compounds(s, l, g) + $O_2(g) \rightarrow CO_2(g), H_2O(l, g)$	$\Delta_{c}H$
Formation	Elements $\rightarrow$ compound	$\Delta_{f}H$
Activation	Reactants $\rightarrow$ activated complex	$\Delta^{\ddagger}H$

* IUPAC recommendations. In common usage, the transition subscript is often attached to ΔH, as in ΔH_{trs}.

The different types of enthalpies encountered in thermochemistry are summarized in Table 2.4. We shall meet them again in various locations throughout the text.

(b) Enthalpies of chemical change

Now we consider enthalpy changes that accompany chemical reactions. There are two ways of reporting the change in enthalpy that accompanies a chemical reaction. One is to write the **thermochemical equation**, a combination of a chemical equation and the corresponding change in standard enthalpy:

$$CH_4(g) + 2\,O_2(g) \rightarrow CO_2(g) + 2\,H_2O(l) \qquad \Delta H^{\ominus} = -890\ \text{kJ}$$

$\Delta H^{\ominus}$ is the change in enthalpy when reactants in their standard states change to products in their standard states:

Pure, separate reactants in their standard states
→ pure, separate products in their standard states

Except in the case of ionic reactions in solution, the enthalpy changes accompanying mixing and separation are insignificant in comparison with the contribution from the reaction itself. For the combustion of methane, the standard value refers to the reaction in which 1 mol CH_4 in the form of pure methane gas at 1 bar reacts completely with 2 mol O_2 in the form of pure oxygen gas to produce 1 mol CO_2 as pure carbon dioxide at 1 bar and 2 mol H_2O as pure liquid water at 1 bar; the numerical value is for the reaction at 298 K.

Alternatively, we write the chemical equation and then report the **standard reaction enthalpy**, $\Delta_r H^{\ominus}$. Thus, for the combustion of reaction, we write

$$CH_4(g) + 2\,O_2(g) \rightarrow CO_2(g) + 2\,H_2O(l) \qquad \Delta_r H^{\ominus} = -890\ \text{kJ mol}^{-1}$$

For the reaction

$$2\,A + B \rightarrow 3\,C + D$$

Synoptic Table 2.5* Standard enthalpies of formation and combustion of organic compounds at 298 K

	$\Delta_f H^{\ominus}/(\text{kJ mol}^{-1})$	$\Delta_c H^{\ominus}/(\text{kJ mol}^{-1})$
Benzene, $C_6H_6(l)$	+49.0	−3268
Ethane, $C_2H_6(g)$	−84.7	−1560
Glucose, $C_6H_{12}O_6(s)$	−1274	−2808
Methane, $CH_4(g)$	−74.8	−890
Methanol, $CH_3OH(l)$	−238.7	−721

* More values are given in the *Data section*.

the standard reaction enthalpy is

$$\Delta_r H^{\ominus} = \{3H_m^{\ominus}(C) + H_m^{\ominus}(D)\} - \{2H_m^{\ominus}(A) + H_m^{\ominus}(B)\}$$

where $H_m^{\ominus}(J)$ is the standard molar enthalpy of species J at the temperature of interest. Note how the 'per mole' of $\Delta_r H^{\ominus}$ comes directly from the fact that molar enthalpies appear in this expression. We interpret the 'per mole' by noting the stoichiometic coefficients in the chemical equation. In this case 'per mole' in $\Delta_r H^{\ominus}$ means 'per 2 mol A', 'per mole B', 'per 3 mol C', or 'per mol D'. In general,

$$\Delta_r H^{\ominus} = \sum_{\text{Products}} \nu H_m^{\ominus} - \sum_{\text{Reactants}} \nu H_m^{\ominus} \tag{2.32}$$

where in each case the molar enthalpies of the species are multiplied by their stoichiometric coefficients, ν.[6]

Some standard reaction enthalpies have special names and a particular significance. For instance, the **standard enthalpy of combustion**, $\Delta_c H^{\ominus}$, is the standard reaction enthalpy for the complete oxidation of an organic compound to CO_2 gas and liquid H_2O if the compound contains C, H, and O, and to N_2 gas if N is also present. An example is the combustion of glucose:

$$C_6H_{12}O_6(s) + 6\,O_2(g) \rightarrow 6\,CO_2(g) + 6\,H_2O(l) \qquad \Delta_c H^{\ominus} = -2808 \text{ kJ mol}^{-1}$$

The value quoted shows that 2808 kJ of heat is released when 1 mol $C_6H_{12}O_6$ burns under standard conditions (at 298 K). Some further values are listed in Table 2.5.

IMPACT ON BIOLOGY

I2.2 Food and energy reserves

The thermochemical properties of fuels Table 2.6 and foods are commonly discussed in terms of their *specific enthalpy*, the enthalpy of combustion per gram of material. Thus, if the standard enthalpy of combustion is $\Delta_c H^{\ominus}$ and the molar mass of the compound is M, then the specific enthalpy is $\Delta_c H^{\ominus}/M$. Table 2.6 lists the specific enthalpies of several fuels.

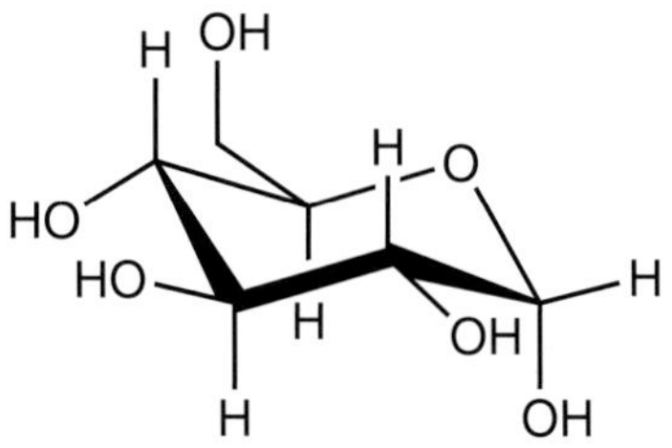

3

A typical 18–20 year old man requires a daily input of about 12 MJ; a woman of the same age needs about 9 MJ. If the entire consumption were in the form of glucose (**3**; which has a specific enthalpy of 16 kJ g^{-1}), that would require the consumption of 750 g of glucose for a man and 560 g for a woman. In fact, digestible carbohydrates have a slightly higher specific enthalpy (17 kJ g^{-1}) than glucose itself, so a carbohydrate

6 In this and similar expressions, all stoichiometric coefficients are positive. For a more sophisticated way of writing eqn 2.32, see Section 7.2.

Table 2.6 Thermochemical properties of some fuels

Fuel	Combustion equation	$\Delta_c H^\ominus$/(kJ mol^{-1})	Specific enthalpy/(kJ g^{-1})	Enthalpy density/(kJ dm^{-3})
Hydrogen	$H_2(g) + \frac{1}{2}O_2(g) \rightarrow H_2O(l)$	−286	142	13
Methane	$CH_4(g) + 2\,O_2(g) \rightarrow CO_2(g) + 2\,H_2O(l)$	−890	55	40
Octane	$C_8H_{18}(l) + \frac{25}{2}O_2(g) \rightarrow 8\,CO_2(g) + 9\,H_2O(l)$	−5471	48	3.8×10^4
Methanol	$CH_3OH(l) + \frac{3}{2}O_2(g) \rightarrow CO_2(g) + 2\,H_2O(l)$	−726	23	1.8×10^4

diet is slightly less daunting than a pure glucose diet, as well as being more appropriate in the form of fibre, the indigestible cellulose that helps move digestion products through the intestine.

The specific enthalpy of fats, which are long-chain esters like tristearin (beef fat), is much greater than that of carbohydrates, at around 38 kJ g^{-1}, slightly less than the value for the hydrocarbon oils used as fuel (48 kJ g^{-1}). Fats are commonly used as an energy store, to be used only when the more readily accessible carbohydrates have fallen into short supply. In Arctic species, the stored fat also acts as a layer of insulation; in desert species (such as the camel), the fat is also a source of water, one of its oxidation products.

Proteins are also used as a source of energy, but their components, the amino acids, are often too valuable to squander in this way, and are used to construct other proteins instead. When proteins are oxidized (to urea, $CO(NH_2)_2$), the equivalent enthalpy density is comparable to that of carbohydrates.

The heat released by the oxidation of foods needs to be discarded in order to maintain body temperature within its typical range of 35.6–37.8°C. A variety of mechanisms contribute to this aspect of homeostasis, the ability of an organism to counteract environmental changes with physiological responses. The general uniformity of temperature throughout the body is maintained largely by the flow of blood. When heat needs to be dissipated rapidly, warm blood is allowed to flow through the capillaries of the skin, so producing flushing. Radiation is one means of discarding heat; another is evaporation and the energy demands of the enthalpy of vaporization of water. Evaporation removes about 2.4 kJ per gram of water perspired. When vigorous exercise promotes sweating (through the influence of heat selectors on the hypothalamus), 1–2 dm^3 of perspired water can be produced per hour, corresponding to a heat loss of 2.4–5.0 MJ h^{-1}.

(c) Hess's law

Standard enthalpies of individual reactions can be combined to obtain the enthalpy of another reaction. This application of the First Law is called **Hess's law**:

> The standard enthalpy of an overall reaction is the sum of the standard enthalpies of the individual reactions into which a reaction may be divided.

The individual steps need not be realizable in practice: they may be hypothetical reactions, the only requirement being that their chemical equations should balance. The thermodynamic basis of the law is the path-independence of the value of $\Delta_r H^\ominus$ and the implication that we may take the specified reactants, pass through any (possibly hypothetical) set of reactions to the specified products, and overall obtain the same change of enthalpy. The importance of Hess's law is that information about a reaction of interest, which may be difficult to determine directly, can be assembled from information on other reactions.

Example 2.5 *Using Hess's law*

The standard reaction enthalpy for the hydrogenation of propene,

$$CH_2{=}CHCH_3(g) + H_2(g) \rightarrow CH_3CH_2CH_3(g)$$

is -124 kJ mol^{-1}. The standard reaction enthalpy for the combustion of propane,

$$CH_3CH_2CH_3(g) + 5\,O_2(g) \rightarrow 3\,CO_2(g) + 4\,H_2O(l)$$

is -2220 kJ mol^{-1}. Calculate the standard enthalpy of combustion of propene.

Method The skill to develop is the ability to assemble a given thermochemical equation from others. Add or subtract the reactions given, together with any others needed, so as to reproduce the reaction required. Then add or subtract the reaction enthalpies in the same way. Additional data are in Table 2.5.

Answer The combustion reaction we require is

$$C_3H_6(g) + \tfrac{9}{2}\,O_2(g) \rightarrow 3\,CO_2(g) + 3\,H_2O(l)$$

This reaction can be recreated from the following sum:

	$\Delta_r H^{\ominus}/(\text{kJ mol}^{-1})$
$C_3H_6(g) + H_2(g) \rightarrow C_3H_8(g)$	-124
$C_3H_8(g) + 5\,O_2(g) \rightarrow 3\,CO_2(g) + 4\,H_2O(l)$	-2220
$H_2O(l) \rightarrow H_2(g) + \tfrac{1}{2}\,O_2(g)$	$+286$
$C_3H_6(g) + \tfrac{9}{2}\,O_2(g) \rightarrow 3\,CO_2(g) + 3\,H_2O(l)$	-2058

Self-test 2.6 Calculate the enthalpy of hydrogenation of benzene from its enthalpy of combustion and the enthalpy of combustion of cyclohexane. [-205 kJ mol^{-1}]

Synoptic Table 2.7* Standard enthalpies of formation of inorganic compounds at 298 K

	$\Delta_f H^{\ominus}/(\text{kJ mol}^{-1})$
$H_2O(l)$	-285.83
$H_2O(g)$	-187.78
$NH_3(g)$	-46.11
$N_2H_4(l)$	$+50.63$
$NO_2(g)$	33.18
$N_2O_4(g)$	$+9.16$
NaCl(s)	-411.15
KCl(s)	-436.75

* More values are given in the *Data section*.

2.8 Standard enthalpies of formation

The **standard enthalpy of formation**, $\Delta_f H^{\ominus}$, of a substance is the standard reaction enthalpy for the formation of the compound from its elements in their reference states. The **reference state** of an element is its most stable state at the specified temperature and 1 bar. For example, at 298 K the reference state of nitrogen is a gas of N_2 molecules, that of mercury is liquid mercury, that of carbon is graphite, and that of tin is the white (metallic) form. There is one exception to this general prescription of reference states: the reference state of phosphorus is taken to be white phosphorus despite this allotrope not being the most stable form but simply the more reproducible form of the element. Standard enthalpies of formation are expressed as enthalpies per mole of molecules or (for ionic substances) formula units of the compound. The standard enthalpy of formation of liquid benzene at 298 K, for example, refers to the reaction

$$6\,C(s, \text{graphite}) + 3\,H_2(g) \rightarrow C_6H_6(l)$$

and is $+49.0$ kJ mol^{-1}. The standard enthalpies of formation of elements in their reference states are zero at all temperatures because they are the enthalpies of such 'null' reactions as $N_2(g) \rightarrow N_2(g)$. Some enthalpies of formation are listed in Tables 2.5 and 2.7.

Comment 2.7

The NIST WebBook listed in the web site for this book links to online databases of thermochemical data.

The standard enthalpy of formation of ions in solution poses a special problem because it is impossible to prepare a solution of cations alone or of anions alone. This problem is solved by defining one ion, conventionally the hydrogen ion, to have zero standard enthalpy of formation at all temperatures:

$$\Delta_f H^{\ominus}(H^+, aq) = 0 \qquad [2.33]$$

Thus, if the enthalpy of formation of HBr(aq) is found to be −122 kJ mol^{-1}, then the whole of that value is ascribed to the formation of Br^-(aq), and we write $\Delta_f H^{\ominus}(Br^-, aq) = -122$ kJ mol^{-1}. That value may then be combined with, for instance, the enthalpy formation of AgBr(aq) to determine the value of $\Delta_f H^{\ominus}(Ag^+, aq)$, and so on. In essence, this definition adjusts the actual values of the enthalpies of formation of ions by a fixed amount, which is chosen so that the standard value for one of them, H^+(aq), has the value zero.

(a) The reaction enthalpy in terms of enthalpies of formation

Conceptually, we can regard a reaction as proceeding by decomposing the reactants into their elements and then forming those elements into the products. The value of $\Delta_r H^{\ominus}$ for the overall reaction is the sum of these 'unforming' and forming enthalpies. Because 'unforming' is the reverse of forming, the enthalpy of an unforming step is the negative of the enthalpy of formation (**4**). Hence, in the enthalpies of formation of substances, we have enough information to calculate the enthalpy of any reaction by using

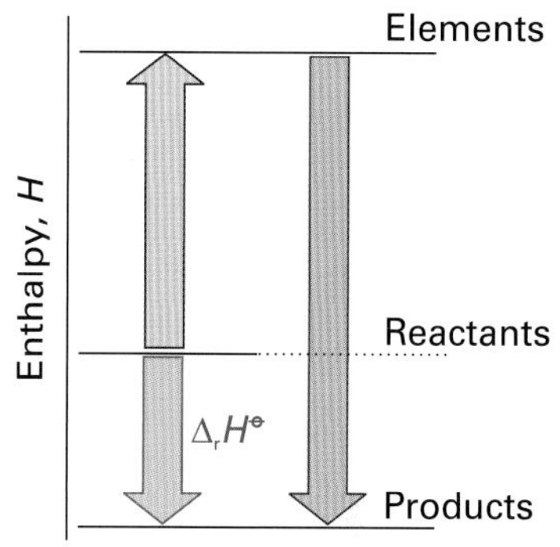

4

$$\Delta_r H^{\ominus} = \sum_{\text{Products}} \nu \Delta_f H^{\ominus} - \sum_{\text{Reactants}} \nu \Delta_f H^{\ominus} \qquad (2.34)$$

where in each case the enthalpies of formation of the species that occur are multiplied by their stoichiometric coefficients.

Illustration 2.7 *Using standard enthalpies of formation*

The standard reaction enthalpy of $2\,HN_3(l) + 2\,NO(g) \rightarrow H_2O_2(l) + 4\,N_2(g)$ is calculated as follows:

$$\begin{aligned}\Delta_r H^{\ominus} &= \{\Delta_f H^{\ominus}(H_2O_2,l) + 4\Delta_f H^{\ominus}(N_2,g)\} - \{2\Delta_f H^{\ominus}(HN_3,l) + 2\Delta_f H^{\ominus}(NO,g)\}\\ &= \{-187.78 + 4(0)\}\ \text{kJ mol}^{-1} - \{2(264.0) + 2(90.25)\}\ \text{kJ mol}^{-1}\\ &= -896.3\ \text{kJ mol}^{-1}\end{aligned}$$

(b) Enthalpies of formation and molecular modelling

We have seen how to construct standard reaction enthalpies by combining standard enthalpies of formation. The question that now arises is whether we can construct standard enthalpies of formation from a knowledge of the chemical constitution of the species. The short answer is that there is no thermodynamically exact way of expressing enthalpies of formation in terms of contributions from individual atoms and bonds. In the past, approximate procedures based on **mean bond enthalpies**, ΔH(A—B), the average enthalpy change associated with the breaking of a specific A—B bond,

$$\text{A—B(g)} \rightarrow \text{A(g)} + \text{B(g)} \qquad \Delta H(\text{A—B})$$

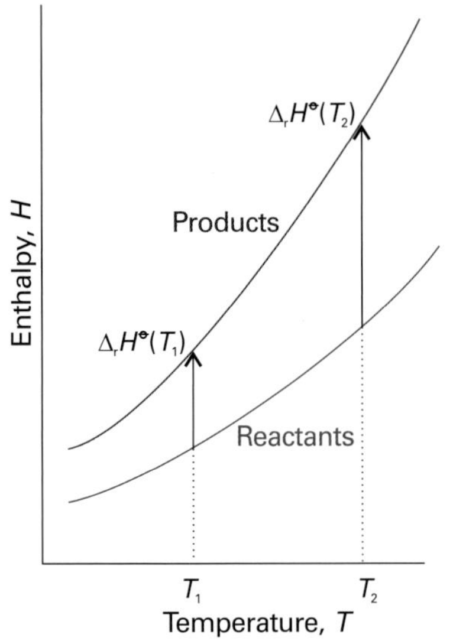

Fig. 2.19 An illustration of the content of Kirchhoff's law. When the temperature is increased, the enthalpy of the products and the reactants both increase, but may do so to different extents. In each case, the change in enthalpy depends on the heat capacities of the substances. The change in reaction enthalpy reflects the difference in the changes of the enthalpies.

have been used. However, this procedure is notoriously unreliable, in part because the ΔH(A—B) are average values for a series of related compounds. Nor does the approach distinguish between geometrical isomers, where the same atoms and bonds may be present but experimentally the enthalpies of formation might be significantly different.

Computer-aided molecular modelling has largely displaced this more primitive approach. Commercial software packages use the principles developed in Chapter 11 to calculate the standard enthalpy of formation of a molecule drawn on the computer screen. These techniques can be applied to different conformations of the same molecule. In the case of methylcyclohexane, for instance, the calculated conformational energy difference ranges from 5.9 to 7.9 kJ mol^{-1}, with the equatorial conformer having the lower standard enthalpy of formation. These estimates compare favourably with the experimental value of 7.5 kJ mol^{-1}. However, good agreement between calculated and experimental values is relatively rare. Computational methods almost always predict correctly which conformer is more stable but do not always predict the correct magnitude of the conformational energy difference.

2.9 The temperature-dependence of reaction enthalpies

The standard enthalpies of many important reactions have been measured at different temperatures. However, in the absence of this information, standard reaction enthalpies at different temperatures may be calculated from heat capacities and the reaction enthalpy at some other temperature (Fig. 2.19). In many cases heat capacity data are more accurate than reaction enthalpies so, providing the information is available, the procedure we are about to describe is more accurate than a direct measurement of a reaction enthalpy at an elevated temperature.

It follows from eqn 2.23a that, when a substance is heated from T_1 to T_2, its enthalpy changes from $H(T_1)$ to

$$H(T_2) = H(T_1) + \int_{T_1}^{T_2} C_p \mathrm{d}T \tag{2.35}$$

(We have assumed that no phase transition takes place in the temperature range of interest.) Because this equation applies to each substance in the reaction, the standard reaction enthalpy changes from $\Delta_r H^\ominus(T_1)$ to

$$\Delta_r H^\ominus(T_2) = \Delta_r H^\ominus(T_1) + \int_{T_1}^{T_2} \Delta_r C_p^\ominus \mathrm{d}T \tag{2.36}$$

where $\Delta_r C_p^\ominus$ is the difference of the molar heat capacities of products and reactants under standard conditions weighted by the stoichiometric coefficients that appear in the chemical equation:

$$\Delta_r C_p^\ominus = \sum_{\text{Products}} \nu C_{p,\mathrm{m}}^\ominus - \sum_{\text{Reactants}} \nu C_{p,\mathrm{m}}^\ominus \qquad [2.37]$$

Equation 2.36 is known as **Kirchhoff's law**. It is normally a good approximation to assume that $\Delta_r C_p$ is independent of the temperature, at least over reasonably limited ranges, as illustrated in the following example. Although the individual heat capacities may vary, their difference varies less significantly. In some cases the temperature dependence of heat capacities is taken into account by using eqn 2.25.

Example 2.6 *Using Kirchhoff's law*

The standard enthalpy of formation of gaseous H_2O at 298 K is -241.82 kJ mol^{-1}. Estimate its value at 100°C given the following values of the molar heat capacities at constant pressure: $H_2O(g)$: 33.58 J K^{-1} mol^{-1}; $H_2(g)$: 28.84 J K^{-1} mol^{-1}; $O_2(g)$: 29.37 J K^{-1} mol^{-1}. Assume that the heat capacities are independent of temperature.

Method When $\Delta C_p^{\ominus}$ is independent of temperature in the range T_1 to T_2, the integral in eqn 2.36 evaluates to $(T_2 - T_1)\Delta_r C_p^{\ominus}$. Therefore,

$$\Delta_r H^{\ominus}(T_2) = \Delta_r H^{\ominus}(T_1) + (T_2 - T_1)\Delta_r C_p^{\ominus}$$

To proceed, write the chemical equation, identify the stoichiometric coefficients, and calculate $\Delta_r C_p^{\ominus}$ from the data.

Answer The reaction is $H_2(g) + \frac{1}{2}O_2(g) \rightarrow H_2O(g)$, so

$$\Delta_r C_p^{\ominus} = C_{p,m}^{\ominus}(H_2O, g) - \{C_{p,m}^{\ominus}(H_2, g) + \tfrac{1}{2}C_{p,m}^{\ominus}(O_2, g)\} = -9.94 \text{ J K}^{-1} \text{ mol}^{-1}$$

It then follows that

$$\Delta_f H^{\ominus}(373 \text{ K}) = -241.82 \text{ kJ mol}^{-1} + (75 \text{ K}) \times (-9.94 \text{ J K}^{-1} \text{ mol}^{-1}) = -242.6 \text{ kJ mol}^{-1}$$

Self-test 2.7 Estimate the standard enthalpy of formation of cyclohexene at 400 K from the data in Table 2.5. [-163 kJ mol^{-1}]

State functions and exact differentials

We saw in Section 2.2 that a 'state function' is a property that is independent of how a sample is prepared. In general, such properties are functions of variables that define the current state of the system, such as pressure and temperature. The internal energy and enthalpy are examples of state functions, for they depend on the current state of the system and are independent of its previous history. Processes that describe the preparation of the state are called **path functions**. Examples of path functions are the work and heating that are done when preparing a state. We do not speak of a system in a particular state as possessing work or heat. In each case, the energy transferred as work or heat relates to the path being taken between states, not the current state itself.

We can use the mathematical properties of state functions to draw far-reaching conclusions about the relations between physical properties and establish connections that may be completely unexpected. The practical importance of these results is that we can combine measurements of different properties to obtain the value of a property we require.

2.10 Exact and inexact differentials

Consider a system undergoing the changes depicted in Fig. 2.20. The initial state of the system is i and in this state the internal energy is U_i. Work is done by the system as it expands adiabatically to a state f. In this state the system has an internal energy U_f and the work done on the system as it changes along Path 1 from i to f is w. Notice our use of language: U is a property of the state; w is a property of the path. Now consider another process, Path 2, in which the initial and final states are the same as those in Path 1 but in which the expansion is not adiabatic. The internal energy of both the

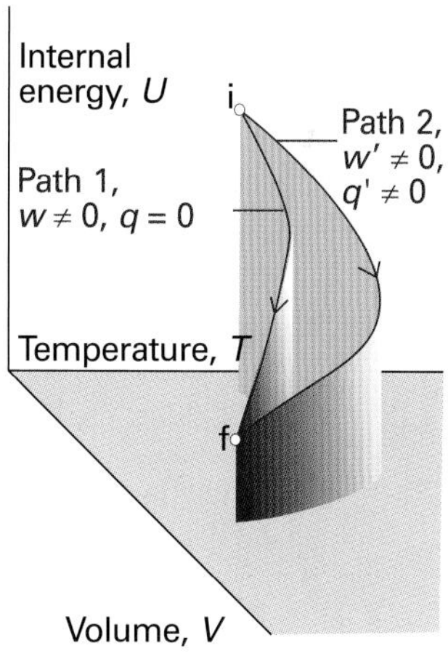

Fig. 2.20 As the volume and temperature of a system are changed, the internal energy changes. An adiabatic and a non-adiabatic path are shown as Path 1 and Path 2, respectively: they correspond to different values of q and w but to the same value of ΔU.

initial and the final states are the same as before (because U is a state function). However, in the second path an energy q' enters the system as heat and the work w' is not the same as w. The work and the heat are path functions. In terms of the mountaineering analogy in Section 2.2, the change in altitude (a state function) is independent of the path, but the distance travelled (a path function) does depend on the path taken between the fixed endpoints.

If a system is taken along a path (for example, by heating it), U changes from U_i to U_f, and the overall change is the sum (integral) of all the infinitesimal changes along the path:

$$\Delta U = \int_i^f dU \tag{2.38}$$

The value of ΔU depends on the initial and final states of the system but is independent of the path between them. This path-independence of the integral is expressed by saying that dU is an 'exact differential'. In general, an **exact differential** is an infinitesimal quantity that, when integrated, gives a result that is independent of the path between the initial and final states.

When a system is heated, the total energy transferred as heat is the sum of all individual contributions at each point of the path:

$$q = \int_{i,\,\text{path}}^f dq \tag{2.39}$$

Notice the difference between this equation and eqn 2.38. First, we do not write Δq, because q is not a state function and the energy supplied as heat cannot be expressed as $q_f - q_i$. Secondly, we must specify the path of integration because q depends on the path selected (for example, an adiabatic path has $q = 0$, whereas a nonadiabatic path between the same two states would have $q \neq 0$). This path-dependence is expressed by saying that dq is an 'inexact differential'. In general, an **inexact differential** is an infinitesimal quantity that, when integrated, gives a result that depends on the path between the initial and final states. Often dq is written $đq$ to emphasize that it is inexact and requires the specification of a path.

The work done on a system to change it from one state to another depends on the path taken between the two specified states; for example, in general the work is different if the change takes place adiabatically and non-adiabatically. It follows that dw is an inexact differential. It is often written $đw$.

Example 2.7 *Calculating work, heat, and internal energy*

Consider a perfect gas inside a cylinder fitted with a piston. Let the initial state be T, V_i and the final state be T, V_f. The change of state can be brought about in many ways, of which the two simplest are the following: Path 1, in which there is free expansion against zero external pressure; Path 2, in which there is reversible, isothermal expansion. Calculate w, q, and ΔU for each process.

Method To find a starting point for a calculation in thermodynamics, it is often a good idea to go back to first principles, and to look for a way of expressing the quantity we are asked to calculate in terms of other quantities that are easier to calculate. We saw in *Molecular interpretation 2.2* that the internal energy of a perfect gas depends only on the temperature and is independent of the volume those molecules occupy, so for any isothermal change, $\Delta U = 0$. We also know that in general $\Delta U = q + w$. The question depends on being able to combine the two expressions. In this chapter, we derived a number of expressions for the work done in a variety of processes, and here we need to select the appropriate ones.

Answer Because $\Delta U = 0$ for both paths and $\Delta U = q + w$, in each case $q = -w$. The work of free expansion is zero (Section 2.3b); so in Path 1, $w = 0$ and $q = 0$. For Path 2, the work is given by eqn 2.11, so $w = -nRT \ln(V_f/V_i)$ and consequently $q = nRT \ln(V_f/V_i)$. These results are consequences of the path independence of U, a state function, and the path dependence of q and w, which are path functions.

Self-test 2.8 Calculate the values of q, w, and ΔU for an irreversible isothermal expansion of a perfect gas against a constant nonzero external pressure.

[$q = p_{ex}\Delta V$, $w = -p_{ex}\Delta V$, $\Delta U = 0$]

2.11 Changes in internal energy

We begin to unfold the consequences of dU being an exact differential by exploring a closed system of constant composition (the only type of system considered in the rest of this chapter). The internal energy U can be regarded as a function of V, T, and p, but, because there is an equation of state, stating the values of two of the variables fixes the value of the third. Therefore, it is possible to write U in terms of just two independent variables: V and T, p and T, or p and V. Expressing U as a function of volume and temperature fits the purpose of our discussion.

(a) General considerations

When V changes to $V + \mathrm{d}V$ at constant temperature, U changes to

$$U' = U + \left(\frac{\partial U}{\partial V}\right)_T \mathrm{d}V$$

The coefficient $(\partial U/\partial V)_T$, the slope of a plot of U against V at constant temperature, is the partial derivative of U with respect to V (Fig. 2.21). If, instead, T changes to $T + \mathrm{d}T$ at constant volume (Fig. 2.22), then the internal energy changes to

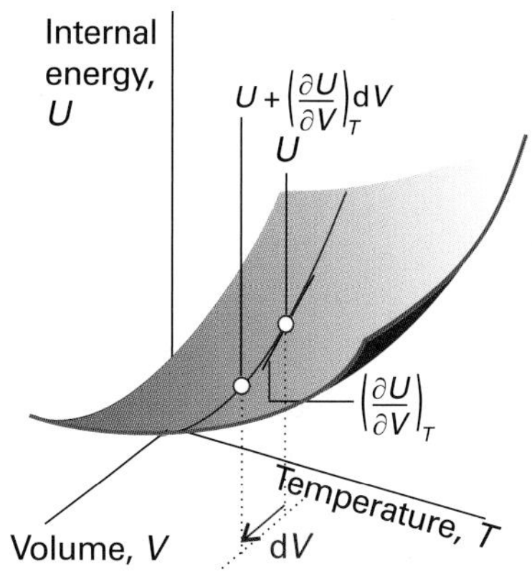

Fig. 2.21 The partial derivative $(\partial U/\partial V)_T$ is the slope of U with respect to V with the temperature T held constant.

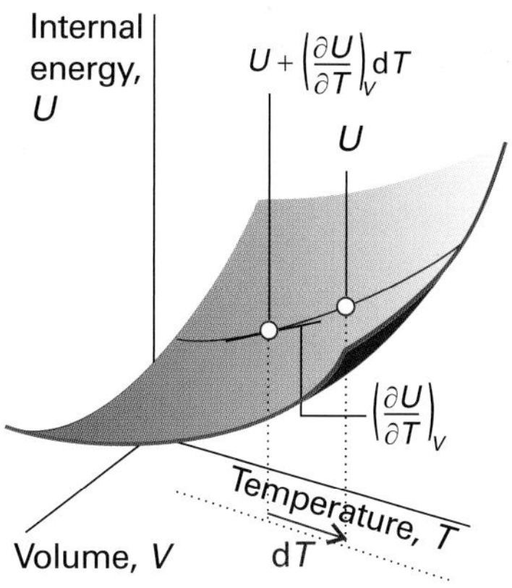

Fig. 2.22 The partial derivative $(\partial U/\partial T)_V$ is the slope of U with respect to T with the volume V held constant.

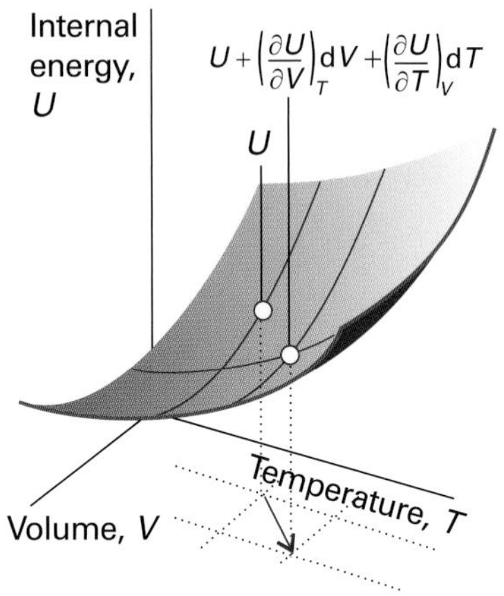

Fig. 2.23 An overall change in U, which is denoted dU, arises when both V and T are allowed to change. If second-order infinitesimals are ignored, the overall change is the sum of changes for each variable separately.

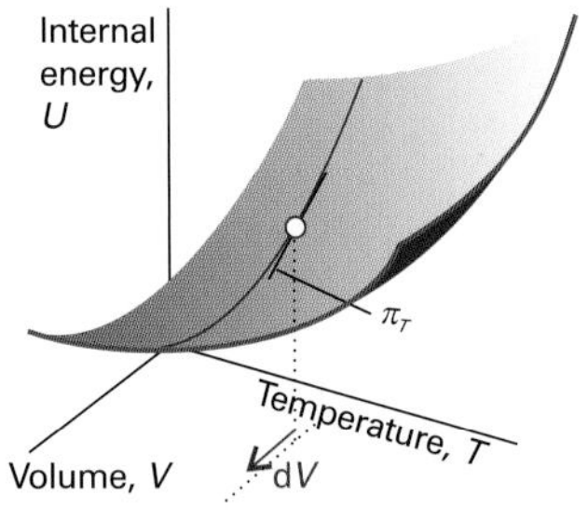

Fig. 2.24 The internal pressure, π_T, is the slope of U with respect to V with the temperature T held constant.

$$U' = U + \left(\frac{\partial U}{\partial T}\right)_V dT$$

Now suppose that V and T both change infinitesimally (Fig. 2.23). The new internal energy, neglecting second-order infinitesimals (those proportional to $dVdT$), is the sum of the changes arising from each increment:

$$U' = U + \left(\frac{\partial U}{\partial V}\right)_T dV + \left(\frac{\partial U}{\partial T}\right)_V dT$$

As a result of the infinitesimal changes in conditions, the internal energy U' differs from U by the infinitesimal amount dU, so we can write $U' = U + dU$. Therefore, from the last equation we obtain the very important result that

$$dU = \left(\frac{\partial U}{\partial V}\right)_T dV + \left(\frac{\partial U}{\partial T}\right)_V dT \qquad (2.40)$$

The interpretation of this equation is that, in a closed system of constant composition, any infinitesimal change in the internal energy is proportional to the infinitesimal changes of volume and temperature, the coefficients of proportionality being the two partial derivatives.

In many cases partial derivatives have a straightforward physical interpretation, and thermodynamics gets shapeless and difficult only when that interpretation is not kept in sight. In the present case, we have already met $(\partial U/\partial T)_V$ in eqn 2.15, where we saw that it is the constant-volume heat capacity, C_V. The other coefficient, $(\partial U/\partial V)_T$, plays a major role in thermodynamics because it is a measure of the variation of the internal energy of a substance as its volume is changed at constant temperature (Fig. 2.24). We shall denote it π_T and, because it has the same dimensions as pressure, call it the **internal pressure**:

$$\pi_T = \left(\frac{\partial U}{\partial V}\right)_T \qquad [2.41]$$

In terms of the notation C_V and π_T, eqn 2.40 can now be written

$$dU = \pi_T dV + C_V dT \qquad (2.42)$$

(b) The Joule experiment

When there are no interactions between the molecules, the internal energy is independent of their separation and hence independent of the volume of the sample (see *Molecular interpretation 2.2*). Therefore, for a perfect gas we can write $\pi_T = 0$. The statement $\pi_T = 0$ (that is, the internal energy is independent of the volume occupied by the sample) can be taken to be the definition of a perfect gas, for later we shall see that it implies the equation of state $pV = nRT$. If the internal energy increases ($dU > 0$) as the volume of the sample expands isothermally ($dV > 0$), which is the case when there are attractive forces between the particles, then a plot of internal energy against volume slopes upwards and $\pi_T > 0$ (Fig. 2.25).

James Joule thought that he could measure π_T by observing the change in temperature of a gas when it is allowed to expand into a vacuum. He used two metal vessels immersed in a water bath (Fig. 2.26). One was filled with air at about 22 atm and the other was evacuated. He then tried to measure the change in temperature of the water of the bath when a stopcock was opened and the air expanded into a vacuum. He observed no change in temperature.

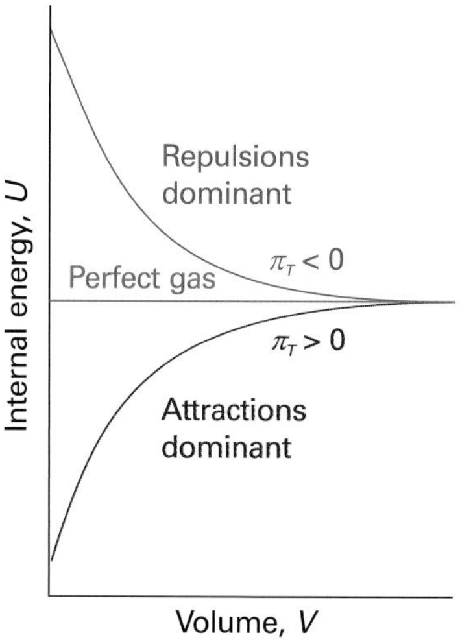

Fig. 2.25 For a perfect gas, the internal energy is independent of the volume (at constant temperature). If attractions are dominant in a real gas, the internal energy increases with volume because the molecules become farther apart on average. If repulsions are dominant, the internal energy decreases as the gas expands.

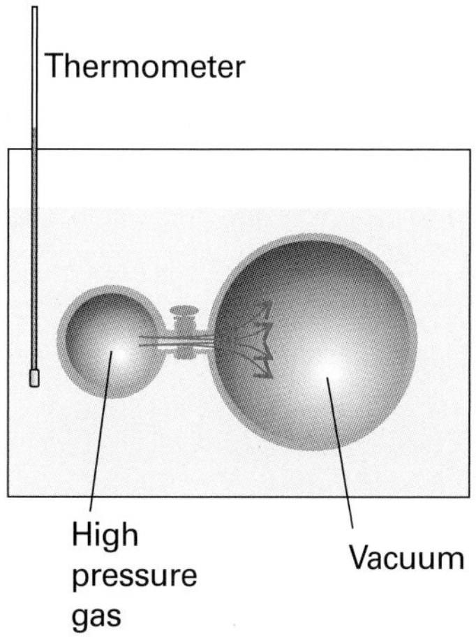

Fig. 2.26 A schematic diagram of the apparatus used by Joule in an attempt to measure the change in internal energy when a gas expands isothermally. The heat absorbed by the gas is proportional to the change in temperature of the bath.

The thermodynamic implications of the experiment are as follows. No work was done in the expansion into a vacuum, so $w = 0$. No energy entered or left the system (the gas) as heat because the temperature of the bath did not change, so $q = 0$. Consequently, within the accuracy of the experiment, $\Delta U = 0$. It follows that U does not change much when a gas expands isothermally and therefore that $\pi_T = 0$.

Joule's experiment was crude. In particular, the heat capacity of the apparatus was so large that the temperature change that gases do in fact cause was too small to measure. From his experiment Joule extracted an essential limiting property of a gas, a property of a perfect gas, without detecting the small deviations characteristic of real gases.

(c) Changes in internal energy at constant pressure

Partial derivatives have many useful properties and some that we shall draw on frequently are reviewed in *Appendix 2*. Skilful use of them can often turn some unfamiliar quantity into a quantity that can be recognized, interpreted, or measured.

As an example, suppose we want to find out how the internal energy varies with temperature when the pressure of the system is kept constant. If we divide both sides of eqn 2.42 by $\mathrm{d}T$ and impose the condition of constant pressure on the resulting differentials, so that $\mathrm{d}U/\mathrm{d}T$ on the left becomes $(\partial U/\partial T)_p$, we obtain

$$\left(\frac{\partial U}{\partial T}\right)_p = \pi_T\left(\frac{\partial V}{\partial T}\right)_p + C_V$$

It is usually sensible in thermodynamics to inspect the output of a manipulation like this to see if it contains any recognizable physical quantity. The partial derivative on the right in this expression is the slope of the plot of volume against temperature (at

Synoptic Table 2.8* Expansion coefficients (α) and isothermal compressibilities (κ_T) at 298 K

	$\alpha/(10^{-4}\,\mathrm{K}^{-1})$	$\kappa_T/(10^{-6}\,\mathrm{bar}^{-1})$
Benzene	12.4	90.9
Diamond	0.030	0.185
Lead	0.861	2.18
Water	2.1	49.0

* More values are given in the *Data section*.

constant pressure). This property is normally tabulated as the **expansion coefficient**, α, of a substance,[7] which is defined as

$$\alpha = \frac{1}{V}\left(\frac{\partial V}{\partial T}\right)_p \qquad [2.43]$$

and physically is the fractional change in volume that accompanies a rise in temperature. A large value of α means that the volume of the sample responds strongly to changes in temperature. Table 2.8 lists some experimental values of α and of the **isothermal compressibility**, κ_T (kappa), which is defined as

$$\kappa_T = -\frac{1}{V}\left(\frac{\partial V}{\partial p}\right)_T \qquad [2.44]$$

The isothermal compressibility is a measure of the fractional change in volume when the pressure is increased by a small amount; the negative sign in the definition ensures that the compressibility is a positive quantity, because an increase of pressure, implying a positive dp, brings about a reduction of volume, a negative dV.

Example 2.8 *Calculating the expansion coefficient of a gas*

Derive an expression for the expansion coefficient of a perfect gas.

Method The expansion coefficient is defined in eqn 2.43. To use this expression, substitute the expression for V in terms of T obtained from the equation of state for the gas. As implied by the subscript in eqn 2.43, the pressure, p, is treated as a constant.

Answer Because $pV = nRT$, we can write

$$\alpha = \frac{1}{V}\left(\frac{\partial(nRT/p)}{\partial T}\right)_p = \frac{1}{V} \times \frac{nR}{p}\frac{\mathrm{d}T}{\mathrm{d}T} = \frac{nR}{pV} = \frac{1}{T}$$

The higher the temperature, the less responsive is the volume of a perfect gas to a change in temperature.

Self-test 2.9 Derive an expression for the isothermal compressibility of a perfect gas. [$\kappa_T = 1/p$]

When we introduce the definition of α into the equation for $(\partial U/\partial T)_p$, we obtain

$$\left(\frac{\partial U}{\partial T}\right)_p = \alpha\pi_T V + C_V \qquad (2.45)$$

This equation is entirely general (provided the system is closed and its composition is constant). It expresses the dependence of the internal energy on the temperature at constant pressure in terms of C_V, which can be measured in one experiment, in terms of α, which can be measured in another, and in terms of the quantity π_T. For a perfect gas, $\pi_T = 0$, so then

$$\left(\frac{\partial U}{\partial T}\right)_p = C_V \qquad (2.46)^\circ$$

[7] As for heat capacities, the expansion coefficients of a mixture depends on whether or not the composition is allowed to change. Throughout this chapter, we deal only with pure substances, so this complication can be disregarded.

That is, although the constant-volume heat capacity of a perfect gas is defined as the slope of a plot of internal energy against temperature at constant volume, for a perfect gas C_V is also the slope at constant pressure.

Equation 2.46 provides an easy way to derive the relation between C_p and C_V for a perfect gas expressed in eqn 2.26. Thus, we can use it to express both heat capacities in terms of derivatives at constant pressure:

$$C_p - C_V = \left(\frac{\partial H}{\partial T}\right)_p - \left(\frac{\partial U}{\partial T}\right)_p \qquad (2.47)^\circ$$

Then we introduce $H = U + pV = U + nRT$ into the first term, which results in

$$C_p - C_V = \left(\frac{\partial U}{\partial T}\right)_p + nR - \left(\frac{\partial U}{\partial T}\right)_p = nR \qquad (2.48)^\circ$$

which is eqn 2.26. We show in *Further information 2.2* that in general

$$C_p - C_V = \frac{\alpha^2 TV}{\kappa_T} \qquad (2.49)$$

Equation 2.49 applies to any substance (that is, it is 'universally true'). It reduces to eqn 2.48 for a perfect gas when we set $\alpha = 1/T$ and $\kappa_T = 1/p$. Because expansion coefficients α of liquids and solids are small, it is tempting to deduce from eqn 2.49 that for them $C_p \approx C_V$. But this is not always so, because the compressibility κ_T might also be small, so α^2/κ_T might be large. That is, although only a little work need be done to push back the atmosphere, a great deal of work may have to be done to pull atoms apart from one another as the solid expands. As an illustration, for water at 25°C, eqn 2.49 gives $C_{p,\mathrm{m}} = 75.3\ \mathrm{J\,K^{-1}\,mol^{-1}}$ compared with $C_{V,\mathrm{m}} = 74.8\ \mathrm{J\,K^{-1}\,mol^{-1}}$. In some cases, the two heat capacities differ by as much as 30 per cent.

2.12 The Joule–Thomson effect

We can carry out a similar set of operations on the enthalpy, $H = U + pV$. The quantities U, p, and V are all state functions; therefore H is also a state function and $\mathrm{d}H$ is an exact differential. It turns out that H is a useful thermodynamic function when the pressure is under our control: we saw a sign of that in the relation $\Delta H = q_p$ (eqn 2.19). We shall therefore regard H as a function of p and T, and adapt the argument in Section 2.10 to find an expression for the variation of H with temperature at constant volume. As set out in *Justification 2.2*, we find that for a closed system of constant composition,

$$\mathrm{d}H = -\mu C_p \mathrm{d}p + C_p \mathrm{d}T \qquad (2.50)$$

where the **Joule–Thomson coefficient**, μ (mu), is defined as

$$\mu = \left(\frac{\partial T}{\partial p}\right)_H \qquad [2.51]$$

This relation will prove useful for relating the heat capacities at constant pressure and volume and for a discussion of the liquefaction of gases.

Justification 2.2 *The variation of enthalpy with pressure and temperature*

By the same argument that led to eqn 2.40 but with H regarded as a function of p and T we can write

$$\mathrm{d}H = \left(\frac{\partial H}{\partial p}\right)_T \mathrm{d}p + \left(\frac{\partial H}{\partial T}\right)_p \mathrm{d}T \qquad (2.52)$$

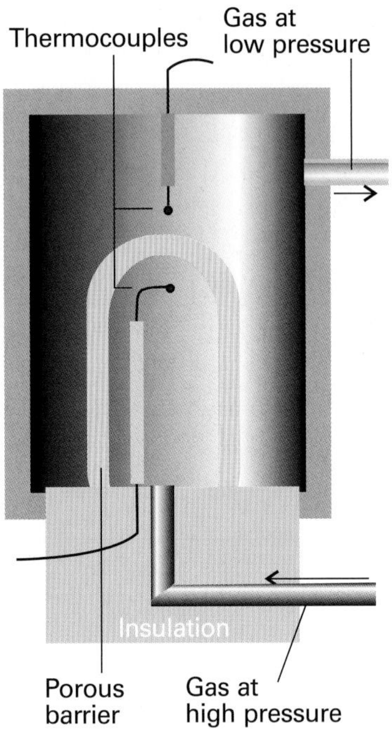

Fig. 2.27 The apparatus used for measuring the Joule–Thomson effect. The gas expands through the porous barrier, which acts as a throttle, and the whole apparatus is thermally insulated. As explained in the text, this arrangement corresponds to an isenthalpic expansion (expansion at constant enthalpy). Whether the expansion results in a heating or a cooling of the gas depends on the conditions.

The second partial derivative is C_p; our task here is to express $(\partial H/\partial p)_T$ in terms of recognizable quantities. The chain relation (see *Further information 2.2*) lets us write

$$\left(\frac{\partial H}{\partial p}\right)_T = -\frac{1}{(\partial p/\partial T)_H(\partial T/\partial H)_p}$$

and both partial derivatives can be brought into the numerator by using the reciprocal identity (see *Further information 2.2*) twice:

$$\left(\frac{\partial H}{\partial p}\right)_T = -\frac{(\partial T/\partial p)_H}{(\partial T/\partial H)_p} = \left(\frac{\partial T}{\partial p}\right)_H\left(\frac{\partial H}{\partial T}\right)_p = -\mu C_p \qquad (2.53)$$

We have used the definitions of the constant-pressure heat capacity, C_p, and the Joule–Thomson coefficient, μ (eqn 2.51). Equation 2.50 now follows directly.

The analysis of the Joule–Thomson coefficient is central to the technological problems associated with the liquefaction of gases. We need to be able to interpret it physically and to measure it. As shown in the *Justification* below, the cunning required to impose the constraint of constant enthalpy, so that the process is **isenthalpic**, was supplied by Joule and William Thomson (later Lord Kelvin). They let a gas expand through a porous barrier from one constant pressure to another, and monitored the difference of temperature that arose from the expansion (Fig. 2.27). The whole apparatus was insulated so that the process was adiabatic. They observed a lower temperature on the low pressure side, the difference in temperature being proportional to the pressure difference they maintained. This cooling by isenthalpic expansion is now called the **Joule–Thomson effect**.

Justification 2.3 *The Joule–Thomson effect*

Here we show that the experimental arrangement results in expansion at constant enthalpy. Because all changes to the gas occur adiabatically,

$$q = 0, \text{ which implies } \Delta U = w$$

Consider the work done as the gas passes through the barrier. We focus on the passage of a fixed amount of gas from the high pressure side, where the pressure is p_i, the temperature T_i, and the gas occupies a volume V_i (Fig. 2.28). The gas emerges on the low pressure side, where the same amount of gas has a pressure p_f, a temperature T_f, and occupies a volume V_f. The gas on the left is compressed isothermally by the upstream gas acting as a piston. The relevant pressure is p_i and the volume changes from V_i to 0; therefore, the work done on the gas is

$$w_1 = -p_i(0 - V_i) = p_iV_i$$

The gas expands isothermally on the right of the barrier (but possibly at a different constant temperature) against the pressure p_f provided by the downstream gas acting as a piston to be driven out. The volume changes from 0 to V_f, so the work done on the gas in this stage is

$$w_2 = -p_f(V_f - 0) = -p_fV_f$$

The total work done on the gas is the sum of these two quantities, or

$$w = w_1 + w_2 = p_iV_i - p_fV_f$$

It follows that the change of internal energy of the gas as it moves adiabatically from one side of the barrier to the other is

$$U_f - U_i = w = p_iV_i - p_fV_f$$

Reorganization of this expression gives

$$U_f + p_fV_f = U_i + p_iV_i, \text{ or } H_f = H_i$$

Therefore, the expansion occurs without change of enthalpy.

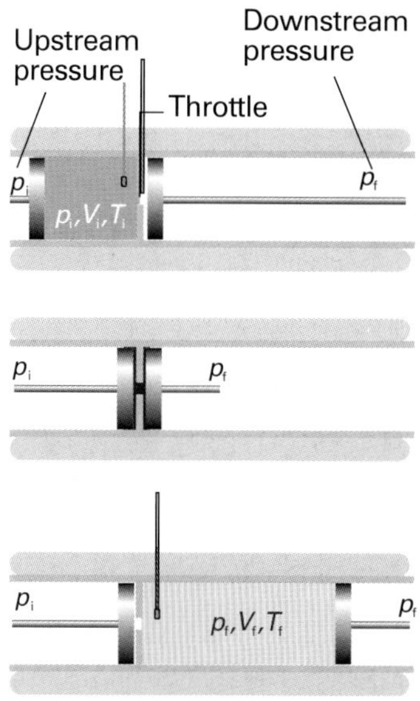

Fig. 2.28 The thermodynamic basis of Joule–Thomson expansion. The pistons represent the upstream and downstream gases, which maintain constant pressures either side of the throttle. The transition from the top diagram to the bottom diagram, which represents the passage of a given amount of gas through the throttle, occurs without change of enthalpy.

The property measured in the experiment is the ratio of the temperature change to the change of pressure, $\Delta T/\Delta p$. Adding the constraint of constant enthalpy and taking the limit of small Δp implies that the thermodynamic quantity measured is $(\partial T/\partial p)_H$, which is the Joule–Thomson coefficient, μ. In other words, the physical interpretation of μ is that it is the ratio of the change in temperature to the change in pressure when a gas expands under conditions that ensure there is no change in enthalpy.

The modern method of measuring μ is indirect, and involves measuring the **isothermal Joule–Thomson coefficient**, the quantity

$$\mu_T = \left(\frac{\partial H}{\partial p}\right)_T \qquad [2.54]$$

which is the slope of a plot of enthalpy against pressure at constant temperature (Fig. 2.29). Comparing eqns 2.53 and 2.54, we see that the two coefficients are related by:

$$\mu_T = -C_p\mu \qquad (2.55)$$

To measure μ_T, the gas is pumped continuously at a steady pressure through a heat exchanger (which brings it to the required temperature), and then through a porous plug inside a thermally insulated container. The steep pressure drop is measured, and the cooling effect is exactly offset by an electric heater placed immediately after the plug (Fig. 2.30). The energy provided by the heater is monitored. Because the energy transferred as heat can be identified with the value of ΔH for the gas (because

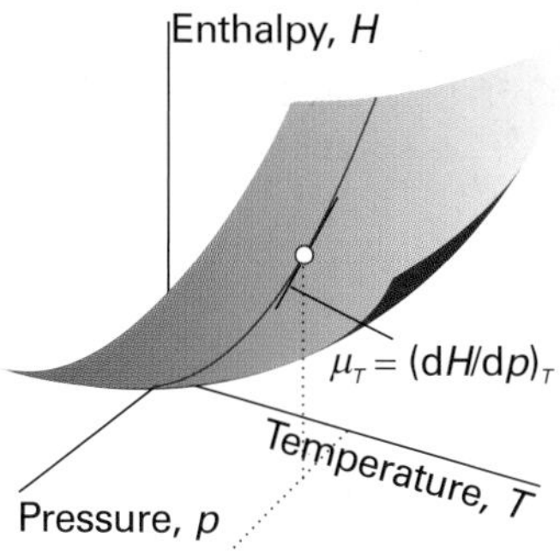

Fig. 2.29 The isothermal Joule–Thomson coefficient is the slope of the enthalpy with respect to changing pressure, the temperature being held constant.

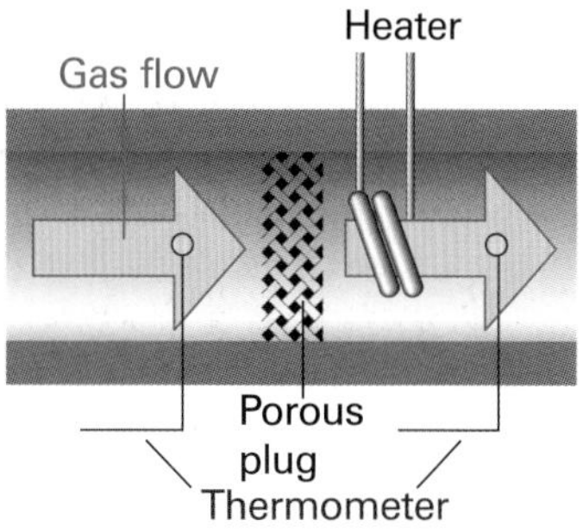

Fig. 2.30 A schematic diagram of the apparatus used for measuring the isothermal Joule–Thomson coefficient. The electrical heating required to offset the cooling arising from expansion is interpreted as ΔH and used to calculate $(\partial H/\partial p)_T$, which is then converted to μ as explained in the text.

Synoptic Table 2.9* Inversion temperatures (T_I), normal freezing (T_f) and boiling (T_b) points, and Joule–Thomson coefficient (μ) at 1 atm and 298 K

	T_I/K	T_f/K	T_b/K	μ/(K bar^{-1})
Ar	723	83.8	87.3	
CO_2	1500		194.7	+1.10
He	40		4.2	−0.060
N_2	621	63.3	77.4	+0.25

* More values are given in the *Data section.*

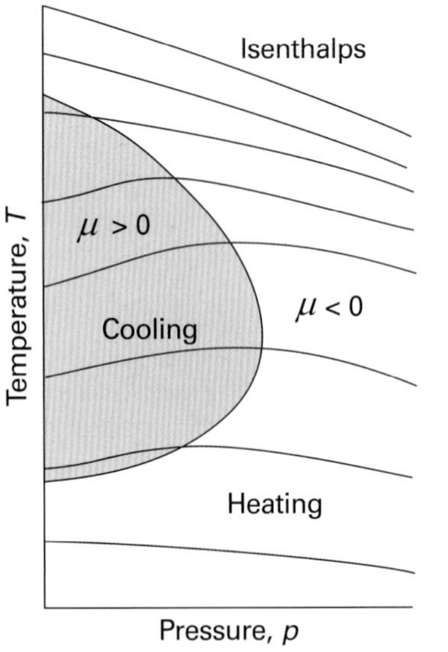

Fig. 2.31 The sign of the Joule–Thomson coefficient, μ, depends on the conditions. Inside the boundary, the shaded area, it is positive and outside it is negative. The temperature corresponding to the boundary at a given pressure is the 'inversion temperature' of the gas at that pressure. For a given pressure, the temperature must be below a certain value if cooling is required but, if it becomes too low, the boundary is crossed again and heating occurs. Reduction of pressure under adiabatic conditions moves the system along one of the isenthalps, or curves of constant enthalpy. The inversion temperature curve runs through the points of the isenthalps where their slope changes from negative to positive.

$\Delta H = q_p$), and the pressure change Δp is known, we can find μ_T from the limiting value of $\Delta H/\Delta p$ as $\Delta p \to 0$, and then convert it to μ. Table 2.9 lists some values obtained in this way.

Real gases have nonzero Joule–Thomson coefficients. Depending on the identity of the gas, the pressure, the relative magnitudes of the attractive and repulsive intermolecular forces (see *Molecular interpretation 2.1*), and the temperature, the sign of the coefficient may be either positive or negative (Fig. 2.31). A positive sign implies that dT is negative when dp is negative, in which case the gas cools on expansion. Gases that show a heating effect ($\mu < 0$) at one temperature show a cooling effect ($\mu > 0$) when the temperature is below their upper **inversion temperature**, T_I (Table 2.9, Fig. 2.32). As indicated in Fig. 2.32, a gas typically has two inversion temperatures, one at high temperature and the other at low.

The 'Linde refrigerator' makes use of Joule–Thompson expansion to liquefy gases (Fig. 2.33). The gas at high pressure is allowed to expand through a throttle; it cools and is circulated past the incoming gas. That gas is cooled, and its subsequent expansion cools it still further. There comes a stage when the circulating gas becomes so cold that it condenses to a liquid.

For a perfect gas, $\mu = 0$; hence, the temperature of a perfect gas is unchanged by Joule–Thomson expansion.[8] This characteristic points clearly to the involvement of intermolecular forces in determining the size of the effect. However, the Joule–Thomson coefficient of a real gas does not necessarily approach zero as the pressure is reduced even though the equation of state of the gas approaches that of a perfect gas. The coefficient behaves like the properties discussed in Section 1.3b in the sense that it depends on derivatives and not on p, V, and T themselves.

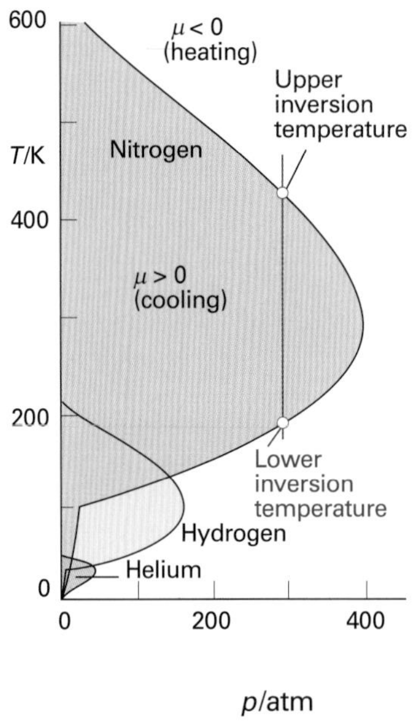

Fig. 2.32 The inversion temperatures for three real gases, nitrogen, hydrogen, and helium.

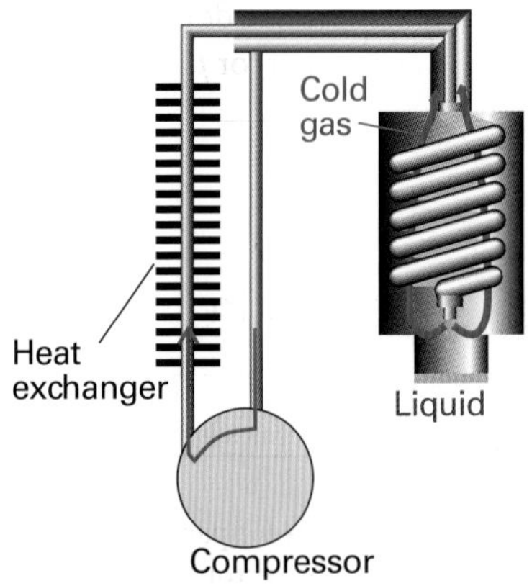

Fig. 2.33 The principle of the Linde refrigerator is shown in this diagram. The gas is recirculated, and so long as it is beneath its inversion temperature it cools on expansion through the throttle. The cooled gas cools the high-pressure gas, which cools still further as it expands. Eventually liquefied gas drips from the throttle.

[8] Simple adiabatic expansion does cool a perfect gas, because the gas does work; recall Section 2.6.

Molecular interpretation 2.3 *Molecular interactions and the Joule–Thomson effect*

The kinetic model of gases (*Molecular interpretation 1.1*) and the equipartition theorem (*Molecular interpretation 2.2*) imply that the mean kinetic energy of molecules in a gas is proportional to the temperature. It follows that reducing the average speed of the molecules is equivalent to cooling the gas. If the speed of the molecules can be reduced to the point that neighbours can capture each other by their intermolecular attractions, then the cooled gas will condense to a liquid.

To slow the gas molecules, we make use of an effect similar to that seen when a ball is thrown into the air: as it rises it slows in response to the gravitational attraction of the Earth and its kinetic energy is converted into potential energy. We saw in Section 1.3 that molecules in a real gas attract each other (the attraction is not gravitational, but the effect is the same). It follows that, if we can cause the molecules to move apart from each other, like a ball rising from a planet, then they should slow. It is very easy to move molecules apart from each other: we simply allow the gas to expand, which increases the average separation of the molecules. To cool a gas, therefore, we allow it to expand without allowing any energy to enter from outside as heat. As the gas expands, the molecules move apart to fill the available volume, struggling as they do so against the attraction of their neighbours. Because some kinetic energy must be converted into potential energy to reach greater separations, the molecules travel more slowly as their separation increases. This sequence of molecular events explains the Joule–Thomson effect: the cooling of a real gas by adiabatic expansion. The cooling effect, which corresponds to $\mu > 0$, is observed under conditions when attractive interactions are dominant ($Z < 1$, eqn 1.17), because the molecules have to climb apart against the attractive force in order for them to travel more slowly. For molecules under conditions when repulsions are dominant ($Z > 1$), the Joule–Thomson effect results in the gas becoming warmer, or $\mu < 0$.

Checklist of key ideas

- ☐ 1. Thermodynamics is the study of the transformations of energy.
- ☐ 2. The system is the part of the world in which we have a special interest. The surroundings is the region outside the system where we make our measurements.
- ☐ 3. An open system has a boundary through which matter can be transferred. A closed system has a boundary through which matter cannot be transferred. An isolated system has a boundary through which neither matter nor energy can be transferred.
- ☐ 4. Energy is the capacity to do work. The internal energy is the total energy of a system.
- ☐ 5. Work is the transfer of energy by motion against an opposing force, $dw = -F dz$. Heat is the transfer of energy as a result of a temperature difference between the system and the surroundings.
- ☐ 6. An exothermic process releases energy as heat to the surroundings. An endothermic process absorbs energy as heat from the surroundings.
- ☐ 7. A state function is a property that depends only on the current state of the system and is independent of how that state has been prepared.
- ☐ 8. The First Law of thermodynamics states that the internal energy of an isolated system is constant, $\Delta U = q + w$.
- ☐ 9. Expansion work is the work of expansion (or compression) of a system, $dw = -p_{ex} dV$. The work of free expansion is $w = 0$. The work of expansion against a constant external pressure is $w = -p_{ex}\Delta V$. The work of isothermal reversible expansion of a perfect gas is $w = -nRT \ln(V_f/V_i)$.
- ☐ 10. A reversible change is a change that can be reversed by an infinitesimal modification of a variable.
- ☐ 11. Maximum work is achieved in a reversible change.

☐ 12. Calorimetry is the study of heat transfers during physical and chemical processes.

☐ 13. The heat capacity at constant volume is defined as $C_V = (\partial U/\partial T)_V$. The heat capacity at constant pressure is $C_p = (\partial H/\partial T)_p$. For a perfect gas, the heat capacities are related by $C_p - C_V = nR$.

☐ 14. The enthalpy is defined as $H = U + pV$. The enthalpy change is the energy transferred as heat at constant pressure, $\Delta H = q_p$.

☐ 15. During a reversible adiabatic change, the temperature of a perfect gas varies according to $T_f = T_i(V_i/V_f)^{1/c}$, $c = C_{V,m}/R$. The pressure and volume are related by $pV^\gamma = \text{constant}$, with $\gamma = C_{p,m}/C_{V,m}$.

☐ 16. The standard enthalpy change is the change in enthalpy for a process in which the initial and final substances are in their standard states. The standard state is the pure substance at 1 bar.

☐ 17. Enthalpy changes are additive, as in $\Delta_{sub}H^\ominus = \Delta_{fus}H^\ominus + \Delta_{vap}H^\ominus$.

☐ 18. The enthalpy change for a process and its reverse are related by $\Delta_{forward}H^\ominus = -\Delta_{reverse}H^\ominus$.

☐ 19. The standard enthalpy of combustion is the standard reaction enthalpy for the complete oxidation of an organic compound to CO_2 gas and liquid H_2O if the compound contains C, H, and O, and to N_2 gas if N is also present.

☐ 20. Hess's law states that the standard enthalpy of an overall reaction is the sum of the standard enthalpies of the individual reactions into which a reaction may be divided.

☐ 21. The standard enthalpy of formation ($\Delta_f H^\ominus$) is the standard reaction enthalpy for the formation of the compound from its elements in their reference states. The reference state is the most stable state of an element at the specified temperature and 1 bar.

☐ 22. The standard reaction enthalpy may be estimated by combining enthalpies of formation, $\Delta_r H^\ominus = \sum_{Products} \nu\Delta_f H^\ominus - \sum_{Reactants} \nu\Delta_f H^\ominus$.

☐ 23. The temperature dependence of the reaction enthalpy is given by Kirchhoff's law, $\Delta_r H^\ominus(T_2) = \Delta_r H^\ominus(T_1) + \int_{T_1}^{T_2} \Delta_r C_p^\ominus \mathrm{d}T$.

☐ 24. An exact differential is an infinitesimal quantity that, when integrated, gives a result that is independent of the path between the initial and final states. An inexact differential is an infinitesimal quantity that, when integrated, gives a result that depends on the path between the initial and final states.

☐ 25. The internal pressure is defined as $\pi_T = (\partial U/\partial V)_T$. For a perfect gas, $\pi_T = 0$.

☐ 26. The Joule–Thomson effect is the cooling of a gas by isenthalpic expansion.

☐ 27. The Joule–Thomson coefficient is defined as $\mu = (\partial T/\partial p)_H$. The isothermal Joule–Thomson coefficient is defined as $\mu_T = (\partial H/\partial p)_T = -C_p\mu$.

☐ 28. The inversion temperature is the temperature at which the Joule–Thomson coefficient changes sign.

Further reading

Articles and texts

P.W. Atkins and J.C. de Paula, *Physical chemistry for the life sciences.* W.H. Freeman, New York (2005).

G.A. Estèvez, K. Yang, and B.B. Dasgupta, Thermodynamic partial derivatives and experimentally measurable quantities. *J. Chem. Educ.* **66**, 890 (1989).

I.M. Klotz and R.M. Rosenberg, *Chemical thermodynamics: basic theory and methods.* Wiley–Interscience, New York (2000).

G.N. Lewis and M. Randall, *Thermodynamics.* Revised by K.S. Pitzer and L. Brewer. McGraw–Hill, New York (1961).

J. Wisniak, The Joule–Thomson coefficient for pure gases and their mixtures. *J. Chem. Educ.* **4**, 51 (1999).

Sources of data and information

M.W. Chase, Jr. (ed.), *NIST–JANAF thermochemical tables.* Published as *J. Phys. Chem. Ref. Data, Monograph no. 9.* American Institute of Physics, New York (1998).

J.D. Cox, D.D. Wagman, and V.A. Medvedev, *CODATA key values for thermodynamics.* Hemisphere Publishing Corp., New York (1989).

D.B. Wagman, W.H. Evans, V.B. Parker, R.H. Schumm, I. Halow, S.M. Bailey, K.L. Churney, and R.L. Nuttall, *The NBS tables of chemical thermodynamic properties.* Published as *J. Phys. Chem. Ref. Data* **11**, Supplement 2 (1982).

R.C. Weast (ed.), *Handbook of chemistry and physics*, Vol. 81. CRC Press, Boca Raton (2000).

M. Zabransky, V. Ruzicka Jr., V. Majer, and E. S. Domalski. *Heat capacity of liquids.* Published as *J. Phys. Chem. Ref. Data, Monograph no. 6.* American Institute of Physics, New York (1996).

Further information

Further information 2.1 *Adiabatic processes*

Consider a stage in a reversible adiabatic expansion when the pressure inside and out is p. The work done when the gas expands by dV is $dw = -pdV$; however, for a perfect gas, $dU = C_V dT$.

Therefore, because for an adiabatic change ($dq = 0$) $dU = dw + dq = dw$, we can equate these two expressions for dU and write

$$C_V dT = -pdV$$

We are dealing with a perfect gas, so we can replace p by nRT/V and obtain

$$\frac{C_V dT}{T} = -\frac{nR dV}{V}$$

To integrate this expression we note that T is equal to T_i when V is equal to V_i, and is equal to T_f when V is equal to V_f at the end of the expansion. Therefore,

$$C_V \int_{T_i}^{T_f} \frac{dT}{T} = -nR \int_{V_i}^{V_f} \frac{dV}{V}$$

(We are taking C_V to be independent of temperature.) Then, because $\int dx/x = \ln x +$ constant, we obtain

$$C_V \ln \frac{T_f}{T_i} = -nR \ln \frac{V_f}{V_i}$$

Because $\ln(x/y) = -\ln(y/x)$, this expression rearranges to

$$\frac{C_V}{nR} \ln \frac{T_f}{T_i} = \ln \frac{V_i}{V_f}$$

With $c = C_V/nR$ we obtain (because $\ln x^a = a \ln x$)

$$\ln\left(\frac{T_f}{T_i}\right)^c = \ln\left(\frac{V_i}{V_f}\right)$$

which implies that $(T_f/T_i)^c = (V_i/V_f)$ and, upon rearrangement, eqn 2.28.

The initial and final states of a perfect gas satisfy the perfect gas law regardless of how the change of state takes place, so we can use $pV = nRT$ to write

$$\frac{p_i V_i}{p_f V_f} = \frac{T_i}{T_f}$$

However, we have just shown that

$$\frac{T_i}{T_f} = \left(\frac{V_f}{V_i}\right)^{1/c} = \left(\frac{V_f}{V_i}\right)^{\gamma-1}$$

where we use the definition of the heat capacity ratio where $\gamma = C_{p,m}/C_{V,m}$ and the fact that, for a perfect gas, $C_{p,m} - C_{V,m} = R$ (the molar version of eqn 2.26). Then we combine the two expressions, to obtain

$$\frac{p_i}{p_f} = \frac{V_f}{V_i} \times \left(\frac{V_f}{V_i}\right)^{\gamma-1} = \left(\frac{V_f}{V_i}\right)^{\gamma}$$

which rearranges to $p_i V_i^{\gamma} = p_f V_f^{\gamma}$, which is eqn 2.29.

Further information 2.2 *The relation between heat capacities*

A useful rule when doing a problem in thermodynamics is to go back to first principles. In the present problem we do this twice, first by expressing C_p and C_V in terms of their definitions and then by inserting the definition $H = U + pV$:

$$C_p - C_V = \left(\frac{\partial H}{\partial T}\right)_p - \left(\frac{\partial U}{\partial T}\right)_V$$

$$= \left(\frac{\partial U}{\partial T}\right)_p + \left(\frac{\partial (pV)}{\partial T}\right)_p - \left(\frac{\partial U}{\partial T}\right)_V$$

We have already calculated the difference of the first and third terms on the right, and eqn 2.45 lets us write this difference as $\alpha\pi_T V$. The factor αV gives the change in volume when the temperature is raised, and $\pi_T = (\partial U/\partial V)_T$ converts this change in volume into a change in internal energy. We can simplify the remaining term by noting that, because p is constant,

$$\left(\frac{\partial (pV)}{\partial T}\right)_p = p\left(\frac{\partial V}{\partial T}\right)_p = \alpha p V$$

The middle term of this expression identifies it as the contribution to the work of pushing back the atmosphere: $(\partial V/\partial T)_p$ is the change of volume caused by a change of temperature, and multiplication by p converts this expansion into work.

Collecting the two contributions gives

$$C_p - C_V = \alpha(p + \pi_T)V \qquad (2.56)$$

As just remarked, the first term on the right, αpV, is a measure of the work needed to push back the atmosphere; the second term on the right, $\alpha\pi_T V$, is the work required to separate the molecules composing the system.

At this point we can go further by using the result we prove in Section 3.8 that

$$\pi_T = T\left(\frac{\partial p}{\partial T}\right)_V - p$$

When this expression is inserted in the last equation we obtain

$$C_p - C_V = \alpha T V\left(\frac{\partial p}{\partial T}\right)_V \qquad (2.57)$$

We now transform the remaining partial derivative. It follows from Euler's chain relation that

$$\left(\frac{\partial p}{\partial T}\right)_V \left(\frac{\partial T}{\partial V}\right)_p \left(\frac{\partial V}{\partial p}\right)_T = -1$$

Comment 2.8

The Euler chain relation states that, for a differentiable function $z = z(x,y)$,

$$\left(\frac{\partial y}{\partial x}\right)_z \left(\frac{\partial x}{\partial z}\right)_y \left(\frac{\partial z}{\partial y}\right)_x = -1$$

For instance, if $z(x,y) = x^2 y$,

$$\left(\frac{\partial y}{\partial x}\right)_z = \left(\frac{\partial(z/x^2)}{\partial x}\right)_z = z\frac{\mathrm{d}(1/x^2)}{\mathrm{d}x} = -\frac{2z}{x^3}$$

$$\left(\frac{\partial x}{\partial z}\right)_y = \left(\frac{\partial(z/y)^{1/2}}{\partial z}\right)_y = \frac{1}{y^{1/2}}\frac{\mathrm{d}z^{1/2}}{\mathrm{d}z} = \frac{1}{2(yz)^{1/2}}$$

$$\left(\frac{\partial z}{\partial y}\right)_x = \left(\frac{\partial(x^2 y)}{\partial y}\right)_x = x^2\frac{\mathrm{d}y}{\mathrm{d}y} = x^2$$

Multiplication of the three terms together gives the result −1.

and therefore that

$$\left(\frac{\partial p}{\partial T}\right)_V = -\frac{1}{(\partial T/\partial V)_p(\partial V/\partial p)_T}$$

Unfortunately, $(\partial T/\partial V)_p$ occurs instead of $(\partial V/\partial T)_p$. However, the 'reciprocal identity' allows us to invert partial derivatives and to write

$$\left(\frac{\partial p}{\partial T}\right)_V = -\frac{(\partial V/\partial T)_p}{(\partial V/\partial p)_T} = \frac{\alpha}{\kappa_T}$$

Comment 2.9

The reciprocal identity states that

$$\left(\frac{\partial y}{\partial x}\right)_z = \frac{1}{(\partial x/\partial y)_z}$$

For example, for the function $z(x,y) = x^2 y$,

$$\left(\frac{\partial y}{\partial x}\right)_z = \left(\frac{\partial(z/x^2)}{\partial x}\right)_z = z\frac{\mathrm{d}(1/x^2)}{\mathrm{d}x} = -\frac{2z}{x^3}$$

We can also write $x = (z/y)^{1/2}$, in which case

$$\left(\frac{\partial x}{\partial y}\right)_z = \left(\frac{\partial(z/y)^{1/2}}{\partial y}\right)_z = z^{1/2}\frac{\mathrm{d}(1/y^{1/2})}{\mathrm{d}y}$$

$$= -\frac{z^{1/2}}{2y^{3/2}} = -\frac{z^{1/2}}{2(z/x^2)^{3/2}} = -\frac{x^3}{2z}$$

which is the reciprocal of the coefficient derived above.

Insertion of this relation into eqn 2.57 produces eqn 2.49.

Discussion questions

2.1 Provide mechanical and molecular definitions of work and heat.

2.2 Consider the reversible expansion of a perfect gas. Provide a physical interpretation for the fact that pV^γ = constant for an adiabatic change, whereas pV = constant for an isothermal change.

2.3 Explain the difference between the change in internal energy and the change in enthalpy accompanying a chemical or physical process.

2.4 Explain the significance of a physical observable being a state function and compile a list of as many state functions as you can identify.

2.5 Explain the significance of the Joule and Joule–Thomson experiments. What would Joule observe in a more sensitive apparatus?

2.6 Suggest (with explanation) how the internal energy of a van der Waals gas should vary with volume at constant temperature.

2.7 In many experimental thermograms, such as that shown in Fig. 2.16, the baseline below T_1 is at a different level from that above T_2. Explain this observation.

Exercises

Assume all gases are perfect unless stated otherwise. Unless otherwise stated, thermochemical data are for 298.15 K.

2.1(a) Calculate the work needed for a 65 kg person to climb through 4.0 m on the surface of (a) the Earth and (b) the Moon (g = 1.60 m s^{-2}).

2.1(b) Calculate the work needed for a bird of mass 120 g to fly to a height of 50 m from the surface of the Earth.

2.2(a) A chemical reaction takes place in a container of cross-sectional area 100 cm^2. As a result of the reaction, a piston is pushed out through 10 cm against an external pressure of 1.0 atm. Calculate the work done by the system.

2.2(b) A chemical reaction takes place in a container of cross-sectional area 50.0 cm^2. As a result of the reaction, a piston is pushed out through 15 cm against an external pressure of 121 kPa. Calculate the work done by the system.

2.3(a) A sample consisting of 1.00 mol Ar is expanded isothermally at 0°C from 22.4 dm^3 to 44.8 dm^3 (a) reversibly, (b) against a constant external pressure equal to the final pressure of the gas, and (c) freely (against zero external pressure). For the three processes calculate q, w, ΔU, and ΔH.

2.3(b) A sample consisting of 2.00 mol He is expanded isothermally at 22°C from 22.8 dm^3 to 31.7 dm^3 (a) reversibly, (b) against a constant external pressure equal to the final pressure of the gas, and (c) freely (against zero external pressure). For the three processes calculate q, w, ΔU, and ΔH.

2.4(a) A sample consisting of 1.00 mol of perfect gas atoms, for which $C_{V,\mathrm{m}} = \frac{3}{2}R$, initially at p_1 = 1.00 atm and T_1 = 300 K, is heated reversibly to 400 K at constant volume. Calculate the final pressure, ΔU, q, and w.

2.4(b) A sample consisting of 2.00 mol of perfect gas molecules, for which $C_{V,\mathrm{m}} = \frac{5}{2}R$, initially at p_1 = 111 kPa and T_1 = 277 K, is heated reversibly to 356 K at constant volume. Calculate the final pressure, ΔU, q, and w.

2.5(a) A sample of 4.50 g of methane occupies 12.7 dm^3 at 310 K. (a) Calculate the work done when the gas expands isothermally against a constant external pressure of 200 Torr until its volume has increased by 3.3 dm^3. (b) Calculate the work that would be done if the same expansion occurred reversibly.

2.5(b) A sample of argon of mass 6.56 g occupies 18.5 dm^3 at 305 K. (a) Calculate the work done when the gas expands isothermally against a constant external pressure of 7.7 kPa until its volume has increased by 2.5 dm^3. (b) Calculate the work that would be done if the same expansion occurred reversibly.

2.6(a) A sample of 1.00 mol $H_2O(g)$ is condensed isothermally and reversibly to liquid water at 100°C. The standard enthalpy of vaporization of water at 100°C is 40.656 kJ mol^{-1}. Find w, q, ΔU, and ΔH for this process.

2.6(b) A sample of 2.00 mol $CH_3OH(g)$ is condensed isothermally and reversibly to liquid at 64°C. The standard enthalpy of vaporization of methanol at 64°C is 35.3 kJ mol^{-1}. Find w, q, ΔU, and ΔH for this process.

2.7(a) A strip of magnesium of mass 15 g is dropped into a beaker of dilute hydrochloric acid. Calculate the work done by the system as a result of the reaction. The atmospheric pressure is 1.0 atm and the temperature 25°C.

2.7(b) A piece of zinc of mass 5.0 g is dropped into a beaker of dilute hydrochloric acid. Calculate the work done by the system as a result of the reaction. The atmospheric pressure is 1.1 atm and the temperature 23°C.

2.8(a) The constant-pressure heat capacity of a sample of a perfect gas was found to vary with temperature according to the expression $C_p/(\mathrm{J\,K^{-1}}) = 20.17 + 0.3665(T/\mathrm{K})$. Calculate q, w, ΔU, and ΔH when the temperature is raised from 25°C to 200°C (a) at constant pressure, (b) at constant volume.

2.8(b) The constant-pressure heat capacity of a sample of a perfect gas was found to vary with temperature according to the expression $C_p/(\mathrm{J\,K^{-1}}) = 20.17 + 0.4001(T/\mathrm{K})$. Calculate q, w, ΔU, and ΔH when the temperature is raised from 0°C to 100°C (a) at constant pressure, (b) at constant volume.

2.9(a) Calculate the final temperature of a sample of argon of mass 12.0 g that is expanded reversibly and adiabatically from 1.0 dm^3 at 273.15 K to 3.0 dm^3.

2.9(b) Calculate the final temperature of a sample of carbon dioxide of mass 16.0 g that is expanded reversibly and adiabatically from 500 cm^3 at 298.15 K to 2.00 dm^3.

2.10(a) A sample of carbon dioxide of mass 2.45 g at 27.0°C is allowed to expand reversibly and adiabatically from 500 cm^3 to 3.00 dm^3. What is the work done by the gas?

2.10(b) A sample of nitrogen of mass 3.12 g at 23.0°C is allowed to expand reversibly and adiabatically from 400 cm^3 to 2.00 dm^3. What is the work done by the gas?

2.11(a) Calculate the final pressure of a sample of carbon dioxide that expands reversibly and adiabatically from 57.4 kPa and 1.0 dm^3 to a final volume of 2.0 dm^3. Take $\gamma = 1.4$.

2.11(b) Calculate the final pressure of a sample of water vapour that expands reversibly and adiabatically from 87.3 Torr and 500 cm^3 to a final volume of 3.0 dm^3. Take $\gamma = 1.3$.

2.12(a) When 229 J of energy is supplied as heat to 3.0 mol Ar(g), the temperature of the sample increases by 2.55 K. Calculate the molar heat capacities at constant volume and constant pressure of the gas.

2.12(b) When 178 J of energy is supplied as heat to 1.9 mol of gas molecules, the temperature of the sample increases by 1.78 K. Calculate the molar heat capacities at constant volume and constant pressure of the gas.

2.13(a) When 3.0 mol O_2 is heated at a constant pressure of 3.25 atm, its temperature increases from 260 K to 285 K. Given that the molar heat capacity of O_2 at constant pressure is 29.4 J K^{-1} mol^{-1}, calculate q, ΔH, and ΔU.

2.13(b) When 2.0 mol CO_2 is heated at a constant pressure of 1.25 atm, its temperature increases from 250 K to 277 K. Given that the molar heat capacity of CO_2 at constant pressure is 37.11 J K^{-1} mol^{-1}, calculate q, ΔH, and ΔU.

2.14(a) A sample of 4.0 mol O_2 is originally confined in 20 dm^3 at 270 K and then undergoes adiabatic expansion against a constant pressure of 600 Torr until the volume has increased by a factor of 3.0. Calculate q, w, ΔT, ΔU, and ΔH. (The final pressure of the gas is not necessarily 600 Torr.)

2.14(b) A sample of 5.0 mol CO_2 is originally confined in 15 dm^3 at 280 K and then undergoes adiabatic expansion against a constant pressure of 78.5 kPa until the volume has increased by a factor of 4.0. Calculate q, w, ΔT, ΔU, and ΔH. (The final pressure of the gas is not necessarily 78.5 kPa.)

2.15(a) A sample consisting of 1.0 mol of perfect gas molecules with C_V = 20.8 J K^{-1} is initially at 3.25 atm and 310 K. It undergoes reversible adiabatic expansion until its pressure reaches 2.50 atm. Calculate the final volume and temperature and the work done.

2.15(b) A sample consisting of 1.5 mol of perfect gas molecules with $C_{p,\mathrm{m}}$ = 20.8 J K^{-1} mol^{-1} is initially at 230 kPa and 315 K. It undergoes reversible adiabatic expansion until its pressure reaches 170 kPa. Calculate the final volume and temperature and the work done.

2.16(a) A certain liquid has $\Delta_{\mathrm{vap}}H^{\ominus}$ = 26.0 kJ mol^{-1}. Calculate q, w, ΔH, and ΔU when 0.50 mol is vaporized at 250 K and 750 Torr.

2.16(b) A certain liquid has $\Delta_{\mathrm{vap}}H^{\ominus}$ = 32.0 kJ mol^{-1}. Calculate q, w, ΔH, and ΔU when 0.75 mol is vaporized at 260 K and 765 Torr.

2.17(a) The standard enthalpy of formation of ethylbenzene is −12.5 kJ mol^{-1}. Calculate its standard enthalpy of combustion.

2.17(b) The standard enthalpy of formation of phenol is −165.0 kJ mol^{-1}. Calculate its standard enthalpy of combustion.

2.18(a) The standard enthalpy of combustion of cyclopropane is −2091 kJ mol^{-1} at 25°C. From this information and enthalpy of formation data for $CO_2(g)$ and $H_2O(g)$, calculate the enthalpy of formation of cyclopropane. The enthalpy of formation of propene is +20.42 kJ mol^{-1}. Calculate the enthalpy of isomerization of cyclopropane to propene.

2.18(b) From the following data, determine $\Delta_f H^{\ominus}$ for diborane, $B_2H_6(g)$, at 298 K:

(1)	$B_2H_6(g) + 3\,O_2(g) \rightarrow B_2O_3(s) + 3\,H_2O(g)$	$\Delta_r H^{\ominus} = -1941$ kJ mol^{-1}
(2)	$2\,B(s) + \frac{3}{2}\,O_2(g) \rightarrow B_2O_3(s)$	$\Delta_r H^{\ominus} = -2368$ kJ mol^{-1}
(3)	$H_2(g) + \frac{1}{2}\,O_2(g) \rightarrow H_2O(g)$	$\Delta_r H^{\ominus} = -241.8$ kJ mol^{-1}

2.19(a) When 120 mg of naphthalene, $C_{10}H_8(s)$, was burned in a bomb calorimeter the temperature rose by 3.05 K. Calculate the calorimeter constant. By how much will the temperature rise when 10 mg of phenol, $C_6H_5OH(s)$, is burned in the calorimeter under the same conditions?

2.19(b) When 2.25 mg of anthracene, $C_{14}H_{10}(s)$, was burned in a bomb calorimeter the temperature rose by 1.35 K. Calculate the calorimeter constant. By how much will the temperature rise when 135 mg of phenol, $C_6H_5OH(s)$, is burned in the calorimeter under the same conditions? ($\Delta_c H^{\ominus}(C_{14}H_{10}, s) = -7061$ kJ mol^{-1}.)

2.20(a) Calculate the standard enthalpy of solution of AgCl(s) in water from the enthalpies of formation of the solid and the aqueous ions.

2.20(b) Calculate the standard enthalpy of solution of AgBr(s) in water from the enthalpies of formation of the solid and the aqueous ions.

2.21(a) The standard enthalpy of decomposition of the yellow complex H_3NSO_2 into NH_3 and SO_2 is +40 kJ mol^{-1}. Calculate the standard enthalpy of formation of H_3NSO_2.

2.21(b) Given that the standard enthalpy of combustion of graphite is −393.51 kJ mol^{-1} and that of diamond is −395.41 kJ mol^{-1}, calculate the enthalpy of the graphite-to-diamond transition.

2.22(a) Given the reactions (1) and (2) below, determine (a) $\Delta_r H^{\ominus}$ and $\Delta_r U^{\ominus}$ for reaction (3), (b) $\Delta_f H^{\ominus}$ for both HCl(g) and $H_2O(g)$ all at 298 K.

(1)	$H_2(g) + Cl_2(g) \rightarrow 2\,HCl(g)$	$\Delta_r H^{\ominus} = -184.62$ kJ mol^{-1}
(2)	$2\,H_2(g) + O_2(g) \rightarrow 2\,H_2O(g)$	$\Delta_r H^{\ominus} = -483.64$ kJ mol^{-1}
(3)	$4\,HCl(g) + O_2(g) \rightarrow Cl_2(g) + 2\,H_2O(g)$	

2.22(b) Given the reactions (1) and (2) below, determine (a) $\Delta_r H^\ominus$ and $\Delta_r U^\ominus$ for reaction (3), (b) $\Delta_f H^\ominus$ for both HI(g) and H_2O(g) all at 298 K.

(1) $H_2(g) + I_2(s) \rightarrow 2\,HI(g)$ $\Delta_r H^\ominus = +52.96$ kJ mol^{-1}

(2) $2\,H_2(g) + O_2(g) \rightarrow 2\,H_2O(g)$ $\Delta_r H^\ominus = -483.64$ kJ mol^{-1}

(3) $4\,HI(g) + O_2(g) \rightarrow 2\,I_2(s) + 2\,H_2O(g)$

2.23(a) For the reaction $C_2H_5OH(l) + 3\,O_2(g) \rightarrow 2\,CO_2(g) + 3\,H_2O(g)$, $\Delta_r U^\ominus = -1373$ kJ mol^{-1} at 298 K. Calculate $\Delta_r H^\ominus$.

2.23(b) For the reaction $2\,C_6H_5COOH(s) + 13\,O_2(g) \rightarrow 12\,CO_2(g) + 6\,H_2O(g)$, $\Delta_r U^\ominus = -772.7$ kJ mol^{-1} at 298 K. Calculate $\Delta_r H^\ominus$.

2.24(a) Calculate the standard enthalpies of formation of (a) $KClO_3$(s) from the enthalpy of formation of KCl, (b) $NaHCO_3$(s) from the enthalpies of formation of CO_2 and NaOH together with the following information:

$2\,KClO_3(s) \rightarrow 2\,KCl(s) + 3\,O_2(g)$ $\Delta_r H^\ominus = -89.4$ kJ mol^{-1}

$NaOH(s) + CO_2(g) \rightarrow NaHCO_3(s)$ $\Delta_r H^\ominus = -127.5$ kJ mol^{-1}

2.24(b) Calculate the standard enthalpy of formation of NOCl(g) from the enthalpy of formation of NO given in Table 2.5, together with the following information:

$2\,NOCl(g) \rightarrow 2\,NO(g) + Cl_2(g)$ $\Delta_r H^\ominus = +75.5$ kJ mol^{-1}

2.25(a) Use the information in Table 2.5 to predict the standard reaction enthalpy of $2\,NO_2(g) \rightarrow N_2O_4(g)$ at 100°C from its value at 25°C.

2.25(b) Use the information in Table 2.5 to predict the standard reaction enthalpy of $2\,H_2(g) + O_2(g) \rightarrow 2\,H_2O(l)$ at 100°C from its value at 25°C.

2.26(a) From the data in Table 2.5, calculate $\Delta_r H^\ominus$ and $\Delta_r U^\ominus$ at (a) 298 K, (b) 378 K for the reaction $C(\text{graphite}) + H_2O(g) \rightarrow CO(g) + H_2(g)$. Assume all heat capacities to be constant over the temperature range of interest.

2.26(b) Calculate $\Delta_r H^\ominus$ and $\Delta_r U^\ominus$ at 298 K and $\Delta_r H^\ominus$ at 348 K for the hydrogenation of ethyne (acetylene) to ethene (ethylene) from the enthalpy of combustion and heat capacity data in Tables 2.5 and 2.7. Assume the heat capacities to be constant over the temperature range involved.

2.27(a) Calculate $\Delta_r H^\ominus$ for the reaction $Zn(s) + CuSO_4(aq) \rightarrow ZnSO_4(aq) + Cu(s)$ from the information in Table 2.7 in the *Data section*.

2.27(b) Calculate $\Delta_r H^\ominus$ for the reaction $NaCl(aq) + AgNO_3(aq) \rightarrow AgCl(s) + NaNO_3(aq)$ from the information in Table 2.7 in the *Data section*.

2.28(a) Set up a thermodynamic cycle for determining the enthalpy of hydration of Mg^{2+} ions using the following data: enthalpy of sublimation of Mg(s), +167.2 kJ mol^{-1}; first and second ionization enthalpies of Mg(g), 7.646 eV and 15.035 eV; dissociation enthalpy of Cl_2(g), +241.6 kJ mol^{-1}; electron gain enthalpy of Cl(g), −3.78 eV; enthalpy of solution of $MgCl_2$(s), −150.5 kJ mol^{-1}; enthalpy of hydration of Cl^-(g), −383.7 kJ mol^{-1}.

2.28(b) Set up a thermodynamic cycle for determining the enthalpy of hydration of Ca^{2+} ions using the following data: enthalpy of sublimation of Ca(s), +178.2 kJ mol^{-1}; first and second ionization enthalpies of Ca(g), 589.7 kJ mol^{-1} and 1145 kJ mol^{-1}; enthalpy of vaporization of bromine, +30.91 kJ mol^{-1}; dissociation enthalpy of Br_2(g), +192.9 kJ mol^{-1}; electron gain enthalpy of Br(g), −331.0 kJ mol^{-1}; enthalpy of solution of $CaBr_2$(s), −103.1 kJ mol^{-1}; enthalpy of hydration of Br^-(g), −337 kJ mol^{-1}.

2.29(a) When a certain freon used in refrigeration was expanded adiabatically from an initial pressure of 32 atm and 0°C to a final pressure of 1.00 atm, the temperature fell by 22 K. Calculate the Joule–Thomson coefficient, μ, at 0°C, assuming it remains constant over this temperature range.

2.29(b) A vapour at 22 atm and 5°C was allowed to expand adiabatically to a final pressure of 1.00 atm; the temperature fell by 10 K. Calculate the Joule–Thomson coefficient, μ, at 5°C, assuming it remains constant over this temperature range.

2.30(a) For a van der Waals gas, $\pi_T = a/V_m^2$. Calculate ΔU_m for the isothermal expansion of nitrogen gas from an initial volume of 1.00 dm^3 to 24.8 dm^3 at 298 K. What are the values of q and w?

2.30(b) Repeat Exercise 2.30(a) for argon, from an initial volume of 1.00 dm^3 to 22.1 dm^3 at 298 K.

2.31(a) The volume of a certain liquid varies with temperature as

$$V = V'\{0.75 + 3.9 \times 10^{-4}(T/\text{K}) + 1.48 \times 10^{-6}(T/\text{K})^2\}$$

where V' is its volume at 300 K. Calculate its expansion coefficient, α, at 320 K.

2.31(b) The volume of a certain liquid varies with temperature as

$$V = V'\{0.77 + 3.7 \times 10^{-4}(T/\text{K}) + 1.52 \times 10^{-6}(T/\text{K})^2\}$$

where V' is its volume at 298 K. Calculate its expansion coefficient, α, at 310 K.

2.32(a) The isothermal compressibility of copper at 293 K is 7.35×10^{-7} atm^{-1}. Calculate the pressure that must be applied in order to increase its density by 0.08 per cent.

2.32(b) The isothermal compressibility of lead at 293 K is 2.21×10^{-6} atm^{-1}. Calculate the pressure that must be applied in order to increase its density by 0.08 per cent.

2.33(a) Given that $\mu = 0.25$ K atm^{-1} for nitrogen, calculate the value of its isothermal Joule–Thomson coefficient. Calculate the energy that must be supplied as heat to maintain constant temperature when 15.0 mol N_2 flows through a throttle in an isothermal Joule–Thomson experiment and the pressure drop is 75 atm.

2.33(b) Given that $\mu = 1.11$ K atm^{-1} for carbon dioxide, calculate the value of its isothermal Joule–Thomson coefficient. Calculate the energy that must be supplied as heat to maintain constant temperature when 12.0 mol CO_2 flows through a throttle in an isothermal Joule–Thomson experiment and the pressure drop is 55 atm.

Problems*

Assume all gases are perfect unless stated otherwise. Note that 1 atm = 1.013 25 bar. Unless otherwise stated, thermochemical data are for 298.15 K.

Numerical problems

2.1 A sample consisting of 1 mol of perfect gas atoms (for which $C_{V,\text{m}} = \frac{3}{2}R$) is taken through the cycle shown in Fig. 2.34. (a) Determine the temperature at the points 1, 2, and 3. (b) Calculate q, w, ΔU, and ΔH for each step and for the overall cycle. If a numerical answer cannot be obtained from the information given, then write in +, −, 0, or ? as appropriate.

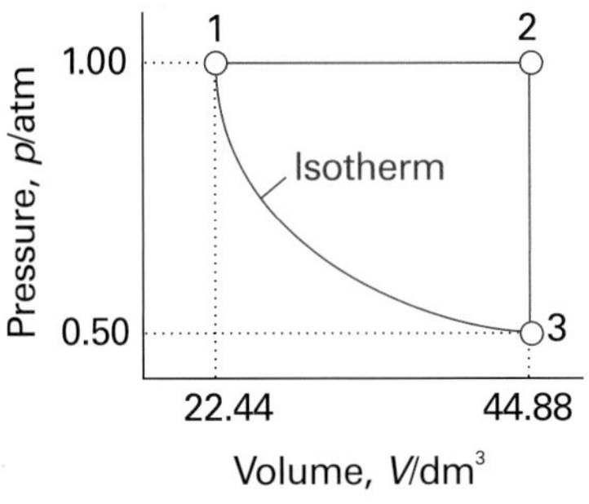

Fig. 2.34

2.2 A sample consisting of 1.0 mol $CaCO_3$(s) was heated to 800°C, when it decomposed. The heating was carried out in a container fitted with a piston that was initially resting on the solid. Calculate the work done during complete decomposition at 1.0 atm. What work would be done if instead of having a piston the container was open to the atmosphere?

2.3 A sample consisting of 2.0 mol CO_2 occupies a fixed volume of 15.0 dm^3 at 300 K. When it is supplied with 2.35 kJ of energy as heat its temperature increases to 341 K. Assume that CO_2 is described by the van der Waals equation of state, and calculate w, ΔU, and ΔH.

2.4 A sample of 70 mmol Kr(g) expands reversibly and isothermally at 373 K from 5.25 cm^3 to 6.29 cm^3, and the internal energy of the sample is known to increase by 83.5 J. Use the virial equation of state up to the second coefficient $B = -28.7\ \text{cm}^3\ \text{mol}^{-1}$ to calculate w, q, and ΔH for this change of state.

2.5 A sample of 1.00 mol perfect gas molecules with $C_{p,\text{m}} = \frac{7}{2}R$ is put through the following cycle: (a) constant-volume heating to twice its initial volume, (b) reversible, adiabatic expansion back to its initial temperature, (c) reversible isothermal compression back to 1.00 atm. Calculate q, w, ΔU, and ΔH for each step and overall.

2.6 Calculate the work done during the isothermal reversible expansion of a van der Waals gas. Account physically for the way in which the coefficients a and b appear in the final expression. Plot on the same graph the indicator diagrams for the isothermal reversible expansion of (a) a perfect gas, (b) a van der Waals gas in which $a = 0$ and $b = 5.11 \times 10^{-2}\ \text{dm}^3\ \text{mol}^{-1}$, and (c) $a = 4.2\ \text{dm}^6\ \text{atm mol}^{-2}$ and $b = 0$. The values selected exaggerate the imperfections but give rise to significant effects on the indicator diagrams. Take $V_\text{i} = 1.0\ \text{dm}^3$, $n = 1.0$ mol, and $T = 298$ K.

2.7 The molar heat capacity of ethane is represented in the temperature range 298 K to 400 K by the empirical expression $C_{p,\text{m}}/(\text{J K}^{-1}\ \text{mol}^{-1}) = 14.73 + 0.1272(T/\text{K})$. The corresponding expressions for C(s) and H_2(g) are given in Table 2.2. Calculate the standard enthalpy of formation of ethane at 350 K from its value at 298 K.

2.8 A sample of the sugar D-ribose ($C_5H_{10}O_5$) of mass 0.727 g was placed in a calorimeter and then ignited in the presence of excess oxygen. The temperature rose by 0.910 K. In a separate experiment in the same calorimeter, the combustion of 0.825 g of benzoic acid, for which the internal energy of combustion is −3251 kJ mol^{-1}, gave a temperature rise of 1.940 K. Calculate the internal energy of combustion of D-ribose and its enthalpy of formation.

2.9 The standard enthalpy of formation of the metallocene bis(benzene)chromium was measured in a calorimeter. It was found for the reaction $Cr(C_6H_6)_2(s) \rightarrow Cr(s) + 2\,C_6H_6(g)$ that $\Delta_\text{r}U^\ominus(583\ \text{K}) = +8.0\ \text{kJ mol}^{-1}$. Find the corresponding reaction enthalpy and estimate the standard enthalpy of formation of the compound at 583 K. The constant-pressure molar heat capacity of benzene is 136.1 J K^{-1} mol^{-1} in its liquid range and 81.67 J K^{-1} mol^{-1} as a gas.

2.10‡ From the enthalpy of combustion data in Table 2.5 for the alkanes methane through octane, test the extent to which the relation $\Delta_\text{c}H^\ominus = k\{(M/(\text{g mol}^{-1})\}^n$ holds and find the numerical values for k and n. Predict $\Delta_\text{c}H^\ominus$ for decane and compare to the known value.

2.11 It is possible to investigate the thermochemical properties of hydrocarbons with molecular modelling methods. (a) Use electronic structure software to predict $\Delta_\text{c}H^\ominus$ values for the alkanes methane through pentane. To calculate $\Delta_\text{c}H^\ominus$ values, estimate the standard enthalpy of formation of $C_nH_{2(n+1)}$(g) by performing semi-empirical calculations (for example, AM1 or PM3 methods) and use experimental standard enthalpy of formation values for CO_2(g) and H_2O(l). (b) Compare your estimated values with the experimental values of $\Delta_\text{c}H^\ominus$ (Table 2.5) and comment on the reliability of the molecular modelling method. (c) Test the extent to which the relation $\Delta_\text{c}H^\ominus = k\{(M/(\text{g mol}^{-1})\}^n$ holds and find the numerical values for k and n.

2.12‡ When 1.3584 g of sodium acetate trihydrate was mixed into 100.0 cm^3 of 0.2000 M HCl(aq) at 25°C in a solution calorimeter, its temperature fell by 0.397°C on account of the reaction:

$$H_3O^+(aq) + NaCH_3CO_2 \cdot 3\,H_2O(s) \rightarrow Na^+(aq) + CH_3COOH(aq) + 4\,H_2O(l).$$

The heat capacity of the calorimeter is 91.0 J K^{-1} and the heat capacity density of the acid solution is 4.144 J K^{-1} cm^{-3}. Determine the standard enthalpy of formation of the aqueous sodium cation. The standard enthalpy of formation of sodium acetate trihydrate is −1064 kJ mol^{-1}.

2.13‡ Since their discovery in 1985, fullerenes have received the attention of many chemical researchers. Kolesov *et al.* reported the standard enthalpy of combustion and of formation of crystalline C_{60} based on calorimetric measurements (V.P. Kolesov, S.M. Pimenova, V.K. Pavlovich, N.B. Tamm, and A.A. Kurskaya, *J. Chem. Thermodynamics* **28**, 1121 (1996)). In one of their runs, they found the standard specific internal energy of combustion to be −36.0334 kJ g^{-1} at 298.15 K Compute $\Delta_\text{c}H^\ominus$ and $\Delta_\text{f}H^\ominus$ of C_{60}.

2.14‡ A thermodynamic study of $DyCl_3$ (E.H.P. Cordfunke, A.S. Booji, and M. Yu. Furkaliouk, *J. Chem. Thermodynamics* **28**, 1387 (1996)) determined its standard enthalpy of formation from the following information

(1) $DyCl_3(s) \rightarrow DyCl_3$(aq, in 4.0 M HCl) $\quad \Delta_\text{r}H^\ominus = -180.06\ \text{kJ mol}^{-1}$

(2) $Dy(s) + 3\,HCl$(aq, 4.0 M) $\rightarrow$ $DyCl_3$(aq, in 4.0 M HCl(aq)) $+ \frac{3}{2}H_2(g)$ $\quad \Delta_\text{r}H^\ominus = -699.43\ \text{kJ mol}^{-1}$

(3) $\frac{1}{2}H_2(g) + \frac{1}{2}Cl_2(g) \rightarrow$ HCl(aq, 4.0 M) $\quad \Delta_\text{r}H^\ominus = -158.31\ \text{kJ mol}^{-1}$

Determine $\Delta_\text{f}H^\ominus(DyCl_3, s)$ from these data.

* Problems denoted with the symbol ‡ were supplied by Charles Trapp, Carmen Giunta, and Marshall Cady.

2.15‡ Silylene (SiH_2) is a key intermediate in the thermal decomposition of silicon hydrides such as silane (SiH_4) and disilane (Si_2H_6). Moffat *et al.* (H.K. Moffat, K.F. Jensen, and R.W. Carr, *J. Phys. Chem.* **95**, 145 (1991)) report $\Delta_f H^\ominus(SiH_2) = +274\ kJ\ mol^{-1}$. If $\Delta_f H^\ominus(SiH_4) = +34.3\ kJ\ mol^{-1}$ and $\Delta_f H^\ominus(Si_2H_6) = +80.3\ kJ\ mol^{-1}$ (*CRC Handbook* (2004)), compute the standard enthalpies of the following reactions:

(a) $SiH_4(g) \rightarrow SiH_2(g) + H_2(g)$

(b) $Si_2H_6(g) \rightarrow SiH_2(g) + SiH_4(g)$

2.16‡ Silanone (SiH_2O) and silanol (SiH_3OH) are species believed to be important in the oxidation of silane (SiH_4). These species are much more elusive than their carbon counterparts. C.L. Darling and H.B. Schlegel (*J. Phys. Chem.* **97**, 8207 (1993)) report the following values (converted from calories) from a computational study: $\Delta_f H^\ominus(SiH_2O) = -98.3\ kJ\ mol^{-1}$ and $\Delta_f H^\ominus(SiH_3OH) = -282\ kJ\ mol^{-1}$. Compute the standard enthalpies of the following reactions:

(a) $SiH_4(g) + \frac{1}{2}O_2(g) \rightarrow SiH_3OH(g)$

(b) $SiH_4(g) + O_2(g) \rightarrow SiH_2O(g) + H_2O(l)$

(c) $SiH_3OH(g) \rightarrow SiH_2O(g) + H_2(g)$

Note that $\Delta_f H^\ominus(SiH_4, g) = +34.3\ kJ\ mol^{-1}$ (*CRC Handbook* (2004)).

2.17 The constant-volume heat capacity of a gas can be measured by observing the decrease in temperature when it expands adiabatically and reversibly. If the decrease in pressure is also measured, we can use it to infer the value of $\gamma = C_p/C_V$ and hence, by combining the two values, deduce the constant-pressure heat capacity. A fluorocarbon gas was allowed to expand reversibly and adiabatically to twice its volume; as a result, the temperature fell from 298.15 K to 248.44 K and its pressure fell from 202.94 kPa to 81.840 kPa. Evaluate C_p.

2.18 A sample consisting of 1.00 mol of a van der Waals gas is compressed from 20.0 dm^3 to 10.0 dm^3 at 300 K. In the process, 20.2 kJ of work is done on the gas. Given that $\mu = \{(2a/RT) - b\}/C_{p,m}$, with $C_{p,m} = 38.4\ J\ K^{-1}\ mol^{-1}$, $a = 3.60\ dm^6\ atm\ mol^{-2}$, and $b = 0.44\ dm^3\ mol^{-1}$, calculate ΔH for the process.

2.19 Take nitrogen to be a van der Waals gas with $a = 1.352\ dm^6\ atm\ mol^{-2}$ and $b = 0.0387\ dm^3\ mol^{-1}$, and calculate ΔH_m when the pressure on the gas is decreased from 500 atm to 1.00 atm at 300 K. For a van der Waals gas, $\mu = \{(2a/RT) - b\}/C_{p,m}$. Assume $C_{p,m} = \frac{7}{2}R$.

Theoretical problems

2.20 Show that the following functions have exact differentials: (a) $x^2y + 3y^2$, (b) $x \cos xy$, (c) x^3y^2, (d) $t(t + e^s) + s$.

2.21 (a) What is the total differential of $z = x^2 + 2y^2 - 2xy + 2x - 4y - 8$? (b) Show that $\partial^2 z/\partial y\partial x = \partial^2 z/\partial x\partial y$ for this function. (c) Let $z = xy - y + \ln x + 2$. Find dz and show that it is exact.

2.22 (a) Express $(\partial C_V/\partial V)_T$ as a second derivative of U and find its relation to $(\partial U/\partial V)_T$ and $(\partial C_p/\partial p)_T$ as a second derivative of H and find its relation to $(\partial H/\partial p)_T$. (b) From these relations show that $(\partial C_V/\partial V)_T = 0$ and $(\partial C_p/\partial p)_T = 0$ for a perfect gas.

2.23 (a) Derive the relation $C_V = -(\partial U/\partial V)_T(\partial V/\partial T)_U$ from the expression for the total differential of $U(T,V)$ and (b) starting from the expression for the total differential of $H(T,p)$, express $(\partial H/\partial p)_T$ in terms of C_p and the Joule–Thomson coefficient, μ.

2.24 Starting from the expression $C_p - C_V = T(\partial p/\partial T)_V(\partial V/\partial T)_p$, use the appropriate relations between partial derivatives to show that

$$C_p - C_V = \frac{T(\partial V/\partial T)_p^2}{(\partial V/\partial T)_T}$$

Evaluate $C_p - C_V$ for a perfect gas.

2.25 (a) By direct differentiation of $H = U + pV$, obtain a relation between $(\partial H/\partial U)_p$ and $(\partial U/\partial V)_p$. (b) Confirm that $(\partial H/\partial U)_p = 1 + p(\partial V/\partial U)_p$ by expressing $(\partial H/\partial U)_p$ as the ratio of two derivatives with respect to volume and then using the definition of enthalpy.

2.26 (a) Write expressions for dV and dp given that V is a function of p and T and p is a function of V and T. (b) Deduce expressions for $d\ln V$ and $d\ln p$ in terms of the expansion coefficient and the isothermal compressibility.

2.27 Calculate the work done during the isothermal reversible expansion of a gas that satisfies the virial equation of state, eqn 1.19. Evaluate (a) the work for 1.0 mol Ar at 273 K (for data, see Table 1.3) and (b) the same amount of a perfect gas. Let the expansion be from 500 cm^3 to 1000 cm^3 in each case.

2.28 Express the work of isothermal reversible expansion of a van der Waals gas in reduced variables and find a definition of reduced work that makes the overall expression independent of the identity of the gas. Calculate the work of isothermal reversible expansion along the critical isotherm from V_c to xV_c.

2.29‡ A gas obeying the equation of state $p(V - nb) = nRT$ is subjected to a Joule–Thomson expansion. Will the temperature increase, decrease, or remain the same?

2.30 Use the fact that $(\partial U/\partial V)_T = a/V_m^2$ for a van der Waals gas to show that $\mu C_{p,m} \approx (2a/RT) - b$ by using the definition of μ and appropriate relations between partial derivatives. (*Hint.* Use the approximation $pV_m \approx RT$ when it is justifiable to do so.)

2.31 Rearrange the van der Waals equation of state to give an expression for T as a function of p and V (with n constant). Calculate $(\partial T/\partial p)_V$ and confirm that $(\partial T/\partial p)_V = 1/(\partial p/\partial T)_V$. Go on to confirm Euler's chain relation.

2.32 Calculate the isothermal compressibility and the expansion coefficient of a van der Waals gas. Show, using Euler's chain relation, that $\kappa_T R = \alpha(V_m - b)$.

2.33 Given that $\mu C_p = T(\partial V/\partial T)_p - V$, derive an expression for μ in terms of the van der Waals parameters a and b, and express it in terms of reduced variables. Evaluate μ at 25°C and 1.0 atm, when the molar volume of the gas is 24.6 $dm^3\ mol^{-1}$. Use the expression obtained to derive a formula for the inversion temperature of a van der Waals gas in terms of reduced variables, and evaluate it for the xenon sample.

2.34 The thermodynamic equation of state $(\partial U/\partial V)_T = T(\partial p/\partial T)_V - p$ was quoted in the chapter. Derive its partner

$$\left(\frac{\partial H}{\partial p}\right)_T = -T\left(\frac{\partial V}{\partial T}\right)_p + V$$

from it and the general relations between partial differentials.

2.35 Show that for a van der Waals gas,

$$C_{p,m} - C_{V,m} = \lambda R \qquad \frac{1}{\lambda} = 1 - \frac{(3V_r - 1)^2}{4V_r^3 T_r}$$

and evaluate the difference for xenon at 25°C and 10.0 atm.

2.36 The speed of sound, c_s, in a gas of molar mass M is related to the ratio of heat capacities γ by $c_s = (\gamma RT/M)^{1/2}$. Show that $c_s = (\gamma p/\rho)^{1/2}$, where ρ is the mass density of the gas. Calculate the speed of sound in argon at 25°C.

2.37‡ A gas obeys the equation of state $V_m = RT/p + aT^2$ and its constant-pressure heat capacity is given by $C_{p,m} = A + BT + Cp$, where a, A, B, and C are constants independent of T and p. Obtain expressions for (a) the Joule–Thomson coefficient and (b) its constant-volume heat capacity.

Applications: to biology, materials science, and the environment

2.38 It is possible to see with the aid of a powerful microscope that a long piece of double-stranded DNA is flexible, with the distance between the ends of the chain adopting a wide range of values. This flexibility is important because it allows DNA to adopt very compact conformations as it is packaged in a chromosome (see Chapter 18). It is convenient to visualize a long piece of DNA as a *freely jointed chain*, a chain of N small, rigid units of length l that are free to make any angle with respect to each other. The length l, the *persistence length*, is approximately 45 nm, corresponding to approximately 130 base pairs. You will now explore the work associated with extending a DNA molecule. (a) Suppose that a DNA molecule resists being extended from an equilibrium, more compact conformation with a *restoring force* $F = -k_F x$, where x is the difference in the end-to-end distance of the chain from an equilibrium value and k_F is the *force constant*. Systems showing this behaviour are said to obey *Hooke's law*. (i) What are the limitations of this model of the DNA molecule? (ii) Using this model, write an expression for the work that must be done to extend a DNA molecule by x. Draw a graph of your conclusion. (b) A better model of a DNA molecule is the *one-dimensional freely jointed chain*, in which a rigid unit of length l can only make an angle of 0° or 180° with an adjacent unit. In this case, the restoring force of a chain extended by $x = nl$ is given by

$$F = \frac{kT}{2l} \ln\left(\frac{1+\nu}{1-\nu}\right) \qquad \nu = n/N$$

where $k = 1.381 \times 10^{-23}$ J K^{-1} is *Boltzmann's constant* (not a force constant). (i) What are the limitations of this model? (ii) What is the magnitude of the force that must be applied to extend a DNA molecule with $N = 200$ by 90 nm? (iii) Plot the restoring force against ν, noting that ν can be either positive or negative. How is the variation of the restoring force with end-to-end distance different from that predicted by Hooke's law? (iv) Keeping in mind that the difference in end-to-end distance from an equilibrium value is $x = nl$ and, consequently, $dx = l dn = Nl d\nu$, write an expression for the work of extending a DNA molecule. (v) Calculate the work of extending a DNA molecule from $\nu = 0$ to $\nu = 1.0$. *Hint.* You must integrate the expression for w. The task can be accomplished easily with mathematical software. (c) Show that for small extensions of the chain, when $\nu \ll 1$, the restoring force is given by

$$F \approx \frac{\nu kT}{l} = \frac{nkT}{Nl}$$

Hint. See *Appendix* 2 for a review of series expansions of functions. (d) Is the variation of the restoring force with extension of the chain given in part (c) different from that predicted by Hooke's law? Explain your answer.

2.39 There are no dietary recommendations for consumption of carbohydrates. Some nutritionists recommend diets that are largely devoid of carbohydrates, with most of the energy needs being met by fats. However, the most common recommendation is that at least 65 per cent of our food calories should come from carbohydrates. A $\frac{3}{4}$-cup serving of pasta contains 40 g of carbohydrates. What percentage of the daily calorie requirement for a person on a 2200 Calorie diet (1 Cal = 1 kcal) does this serving represent?

2.40 An average human produces about 10 MJ of heat each day through metabolic activity. If a human body were an isolated system of mass 65 kg with the heat capacity of water, what temperature rise would the body experience? Human bodies are actually open systems, and the main mechanism of heat loss is through the evaporation of water. What mass of water should be evaporated each day to maintain constant temperature?

2.41 Glucose and fructose are simple sugars with the molecular formula $C_6H_{12}O_6$. Sucrose, or table sugar, is a complex sugar with molecular formula $C_{12}H_{22}O_{11}$ that consists of a glucose unit covalently bound to a fructose unit (a water molecule is given off as a result of the reaction between glucose and fructose to form sucrose). (a) Calculate the energy released as heat when a typical table sugar cube of mass 1.5 g is burned in air. (b) To what height could you climb on the energy a table sugar cube provides assuming 25 per cent of the energy is available for work? (c) The mass of a typical glucose tablet is 2.5 g. Calculate the energy released as heat when a glucose tablet is burned in air. (d) To what height could you climb on the energy a cube provides assuming 25 per cent of the energy is available for work?

2.42 In biological cells that have a plentiful supply of O_2, glucose is oxidized completely to CO_2 and H_2O by a process called *aerobic oxidation*. Muscle cells may be deprived of O_2 during vigorous exercise and, in that case, one molecule of glucose is converted to two molecules of lactic acid ($CH_3CH(OH)COOH$) by a process called *anaerobic glycolysis* (see *Impact I7.2*). (a) When 0.3212 g of glucose was burned in a bomb calorimeter of calorimeter constant 641 J K^{-1} the temperature rose by 7.793 K. Calculate (i) the standard molar enthalpy of combustion, (ii) the standard internal energy of combustion, and (iii) the standard enthalpy of formation of glucose. (b) What is the biological advantage (in kilojoules per mole of energy released as heat) of complete aerobic oxidation compared with anaerobic glycolysis to lactic acid?

2.43 You have at your disposal a sample of pure polymer P and a sample of P that has just been synthesized in a large chemical reactor and that may contain impurities. Describe how you would use differential scanning calorimetry to determine the mole percentage composition of P in the allegedly impure sample.

2.44‡ Alkyl radicals are important intermediates in the combustion and atmospheric chemistry of hydrocarbons. Seakins *et al.* (P.W. Seakins, M.J. Pilling, J.T. Niiranen, D. Gutman, and L.N. Krasnoperov, *J. Phys. Chem.* **96**, 9847 (1992)) report $\Delta_f H^\ominus$ for a variety of alkyl radicals in the gas phase, information that is applicable to studies of pyrolysis and oxidation reactions of hydrocarbons. This information can be combined with thermodynamic data on alkenes to determine the reaction enthalpy for possible fragmentation of a large alkyl radical into smaller radicals and alkenes. Use the following set of data to compute the standard reaction enthalpies for three possible fates of the *tert*-butyl radical, namely, (a) *tert*-$C_4H_9 \rightarrow$ *sec*-C_4H_9, (b) *tert*-$C_4H_9 \rightarrow C_3H_6 + CH_3$, (c) *tert*-$C_4H_9 \rightarrow C_2H_4 + C_2H_5$.

Species:	C_2H_5	*sec*-C_4H_9	*tert*-C_4H_9
$\Delta_f H^\ominus$/(kJ mol^{-1})	+121.0	+67.5	+51.3

2.45‡ In 1995, the Intergovernmental Panel on Climate Change (IPCC) considered a global average temperature rise of 1.0–3.5°C likely by the year 2100, with 2.0°C its best estimate. Predict the average rise in sea level due to thermal expansion of sea water based on temperature rises of 1.0°C, 2.0°C, and 3.5°C given that the volume of the Earth's oceans is 1.37×10^9 km^3 and their surface area is 361×10^6 km^2, and state the approximations that go into the estimates.

2.46‡ Concerns over the harmful effects of chlorofluorocarbons on stratospheric ozone have motivated a search for new refrigerants. One such alternative is 2,2-dichloro-1,1,1-trifluoroethane (refrigerant 123). Younglove and McLinden published a compendium of thermophysical properties of this substance (B.A. Younglove and M. McLinden, *J. Phys. Chem. Ref. Data* **23**, 7 (1994)), from which properties such as the Joule–Thomson coefficient μ can be computed. (a) Compute μ at 1.00 bar and 50°C given that $(\partial H/\partial p)_T = -3.29 \times 10^3$ J MPa^{-1} mol^{-1} and $C_{p,m} = 110.0$ J K^{-1} mol^{-1}. (b) Compute the temperature change that would accompany adiabatic expansion of 2.0 mol of this refrigerant from 1.5 bar to 0.5 bar at 50°C.

2.47‡ Another alternative refrigerant (see preceding problem) is 1,1,1,2-tetrafluoroethane (refrigerant HFC-134a). Tillner-Roth and Baehr published a compendium of thermophysical properties of this substance (R. Tillner-Roth and H.D. Baehr, *J. Phys. Chem. Ref. Data* **23**, 657 (1994)), from which properties such as the Joule–Thomson coefficient μ can be computed. (a) Compute μ at 0.100 MPa and 300 K from the following data (all referring to 300 K):

p/MPa	0.080	0.100	0.12
Specific enthalpy/(kJ kg^{-1})	426.48	426.12	425.76

(The specific constant-pressure heat capacity is 0.7649 kJ K^{-1} kg^{-1}.)
(b) Compute μ at 1.00 MPa and 350 K from the following data (all referring to 350 K):

p/MPa	0.80	1.00	1.2
Specific enthalpy/(kJ kg^{-1})	461.93	459.12	456.15

(The specific constant-pressure heat capacity is 1.0392 kJ K^{-1} kg^{-1}.)

3 The Second Law

The purpose of this chapter is to explain the origin of the spontaneity of physical and chemical change. We examine two simple processes and show how to define, measure, and use a property, the entropy, to discuss spontaneous changes quantitatively. The chapter also introduces a major subsidiary thermodynamic property, the Gibbs energy, which lets us express the spontaneity of a process in terms of the properties of a system. The Gibbs energy also enables us to predict the maximum non-expansion work that a process can do. As we began to see in Chapter 2, one application of thermodynamics is to find relations between properties that might not be thought to be related. Several relations of this kind can be established by making use of the fact that the Gibbs energy is a state function. We also see how to derive expressions for the variation of the Gibbs energy with temperature and pressure and how to formulate expressions that are valid for real gases. These expressions will prove useful later when we discuss the effect of temperature and pressure on equilibrium constants.

Some things happen naturally; some things don't. A gas expands to fill the available volume, a hot body cools to the temperature of its surroundings, and a chemical reaction runs in one direction rather than another. Some aspect of the world determines the **spontaneous** direction of change, the direction of change that does not require work to be done to bring it about. A gas can be confined to a smaller volume, an object can be cooled by using a refrigerator, and some reactions can be driven in reverse (as in the electrolysis of water). However, none of these processes is spontaneous; each one must be brought about by doing work. An important point, though, is that throughout this text 'spontaneous' must be interpreted as a natural *tendency* that may or may not be realized in practice. Thermodynamics is silent on the rate at which a spontaneous change in fact occurs, and some spontaneous processes (such as the conversion of diamond to graphite) may be so slow that the tendency is never realized in practice whereas others (such as the expansion of a gas into a vacuum) are almost instantaneous.

The recognition of two classes of process, spontaneous and non-spontaneous, is summarized by the **Second Law of thermodynamics**. This law may be expressed in a variety of equivalent ways. One statement was formulated by Kelvin:

> No process is possible in which the sole result is the absorption of heat from a reservoir and its complete conversion into work.

For example, it has proved impossible to construct an engine like that shown in Fig. 3.1, in which heat is drawn from a hot reservoir and completely converted into work. All real heat engines have both a hot source and a cold sink; some energy is always discarded into the cold sink as heat and not converted into work. The Kelvin statement is a generalization of another everyday observation, that a ball at rest on a surface has never been observed to leap spontaneously upwards. An upward leap of the ball would be equivalent to the conversion of heat from the surface into work.

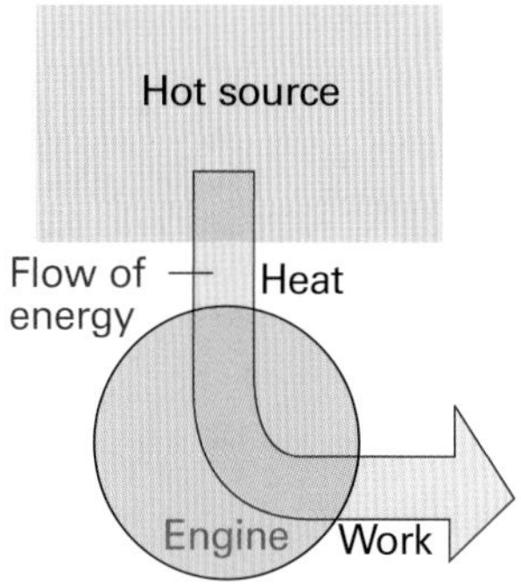

Fig. 3.1 The Kelvin statement of the Second Law denies the possibility of the process illustrated here, in which heat is changed completely into work, there being no other change. The process is not in conflict with the First Law because energy is conserved.

The direction of spontaneous change

What determines the direction of spontaneous change? It is not the total energy of the isolated system. The First Law of thermodynamics states that energy is conserved in any process, and we cannot disregard that law now and say that everything tends towards a state of lower energy: the total energy of an isolated system is constant.

Is it perhaps the energy of the *system* that tends towards a minimum? Two arguments show that this cannot be so. First, a perfect gas expands spontaneously into a vacuum, yet its internal energy remains constant as it does so. Secondly, if the energy of a system does happen to decrease during a spontaneous change, the energy of its surroundings must increase by the same amount (by the First Law). The increase in energy of the surroundings is just as spontaneous a process as the decrease in energy of the system.

When a change occurs, the total energy of an isolated system remains constant but it is parcelled out in different ways. Can it be, therefore, that the direction of change is related to the *distribution* of energy? We shall see that this idea is the key, and that spontaneous changes are always accompanied by a dispersal of energy.

3.1 The dispersal of energy

We can begin to understand the role of the distribution of energy by thinking about a ball (the system) bouncing on a floor (the surroundings). The ball does not rise as high after each bounce because there are inelastic losses in the materials of the ball and floor. The kinetic energy of the ball's overall motion is spread out into the energy of thermal motion of its particles and those of the floor that it hits. The direction of spontaneous change is towards a state in which the ball is at rest with all its energy dispersed into random thermal motion of molecules in the air and of the atoms of the virtually infinite floor (Fig. 3.2).

A ball resting on a warm floor has never been observed to start bouncing. For bouncing to begin, something rather special would need to happen. In the first place, some of the thermal motion of the atoms in the floor would have to accumulate in a single, small object, the ball. This accumulation requires a spontaneous localization of energy from the myriad vibrations of the atoms of the floor into the much smaller number of atoms that constitute the ball (Fig. 3.3). Furthermore, whereas the thermal motion is random, for the ball to move upwards its atoms must all move in the same direction. The localization of random, disorderly motion as concerted, ordered motion is so unlikely that we can dismiss it as virtually impossible.[1]

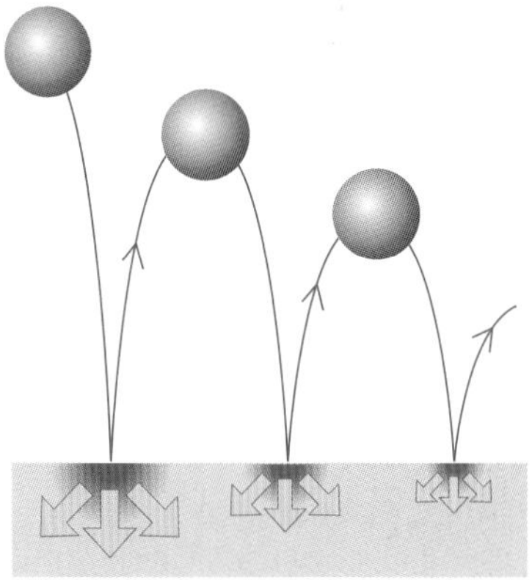

Fig. 3.2 The direction of spontaneous change for a ball bouncing on a floor. On each bounce some of its energy is degraded into the thermal motion of the atoms of the floor, and that energy disperses. The reverse has never been observed to take place on a macroscopic scale.

We appear to have found the signpost of spontaneous change: *we look for the direction of change that leads to dispersal of the total energy of the isolated system.* This principle accounts for the direction of change of the bouncing ball, because its energy is spread out as thermal motion of the atoms of the floor. The reverse process is not spontaneous because it is highly improbable that energy will become localized, leading to uniform motion of the ball's atoms. A gas does not contract spontaneously because

[1] Concerted motion, but on a much smaller scale, is observed as *Brownian motion*, the jittering motion of small particles suspended in water.

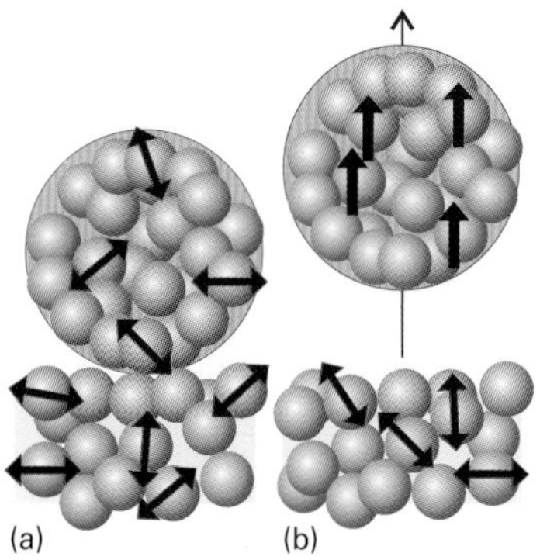

Fig. 3.3 The molecular interpretation of the irreversibility expressed by the Second Law. (a) A ball resting on a warm surface; the atoms are undergoing thermal motion (vibration, in this instance), as indicated by the arrows. (b) For the ball to fly upwards, some of the random vibrational motion would have to change into coordinated, directed motion. Such a conversion is highly improbable.

to do so the random motion of its molecules, which spreads out the distribution of kinetic energy throughout the container, would have to take them all into the same region of the container, thereby localizing the energy. The opposite change, spontaneous expansion, is a natural consequence of energy becoming more dispersed as the gas molecules occupy a larger volume. An object does not spontaneously become warmer than its surroundings because it is highly improbable that the jostling of randomly vibrating atoms in the surroundings will lead to the localization of thermal motion in the object. The opposite change, the spreading of the object's energy into the surroundings as thermal motion, is natural.

It may seem very puzzling that the spreading out of energy and matter, the collapse into disorder, can lead to the formation of such ordered structures as crystals or proteins. Nevertheless, in due course, we shall see that dispersal of energy and matter accounts for change in all its forms.

3.2 Entropy

The First Law of thermodynamics led to the introduction of the internal energy, U. The internal energy is a state function that lets us assess whether a change is permissible: only those changes may occur for which the internal energy of an isolated system remains constant. The law that is used to identify the signpost of spontaneous change, the Second Law of thermodynamics, may also be expressed in terms of another state function, the **entropy**, S. We shall see that the entropy (which we shall define shortly, but is a measure of the energy dispersed in a process) lets us assess whether one state is accessible from another by a spontaneous change. The First Law uses the internal energy to identify *permissible* changes; the Second Law uses the entropy to identify the *spontaneous changes* among those permissible changes.

The Second Law of thermodynamics can be expressed in terms of the entropy:

> The entropy of an isolated system increases in the course of a spontaneous change: $\Delta S_{tot} > 0$

where S_{tot} is the total entropy of the system and its surroundings. Thermodynamically irreversible processes (like cooling to the temperature of the surroundings and the free expansion of gases) are spontaneous processes, and hence must be accompanied by an increase in total entropy.

(a) The thermodynamic definition of entropy

The thermodynamic definition of entropy concentrates on the change in entropy, dS, that occurs as a result of a physical or chemical change (in general, as a result of a 'process'). The definition is motivated by the idea that a change in the extent to which energy is dispersed depends on how much energy is transferred as heat. As we have remarked, heat stimulates random motion in the surroundings. On the other hand, work stimulates uniform motion of atoms in the surroundings and so does not change their entropy.

The thermodynamic definition of entropy is based on the expression

$$\mathrm{d}S = \frac{\mathrm{d}q_{\mathrm{rev}}}{T} \qquad [3.1]$$

For a measurable change between two states i and f this expression integrates to

$$\Delta S = \int_{\mathrm{i}}^{\mathrm{f}} \frac{\mathrm{d}q_{\mathrm{rev}}}{T} \qquad (3.2)$$

That is, to calculate the difference in entropy between any two states of a system, we find a *reversible* path between them, and integrate the energy supplied as heat at each stage of the path divided by the temperature at which heating occurs.

Example 3.1 *Calculating the entropy change for the isothermal expansion of a perfect gas*

Calculate the entropy change of a sample of perfect gas when it expands isothermally from a volume V_i to a volume V_f.

Method The definition of entropy instructs us to find the energy supplied as heat for a reversible path between the stated initial and final states regardless of the actual manner in which the process takes place. A simplification is that the expansion is isothermal, so the temperature is a constant and may be taken outside the integral in eqn 3.2. The energy absorbed as heat during a reversible isothermal expansion of a perfect gas can be calculated from $\Delta U = q + w$ and $\Delta U = 0$, which implies that $q = -w$ in general and therefore that $q_{rev} = -w_{rev}$ for a reversible change. The work of reversible isothermal expansion was calculated in Section 2.3.

Answer Because the temperature is constant, eqn 3.2 becomes

$$\Delta S = \frac{1}{T}\int_i^f dq_{rev} = \frac{q_{rev}}{T}$$

From eqn 2.11, we know that

$$q_{rev} = -w_{rev} = nRT \ln \frac{V_f}{V_i}$$

It follows that

$$\Delta S = nR \ln \frac{V_f}{V_i}$$

As an illustration of this formula, when the volume occupied by 1.00 mol of any perfect gas molecules is doubled at any constant temperature, $V_f/V_i = 2$ and

$$\Delta S = (1.00 \text{ mol}) \times (8.3145 \text{ J K}^{-1} \text{ mol}^{-1}) \times \ln 2 = +5.76 \text{ J K}^{-1}$$

A note on good practice According to eqn 3.2, when the energy transferred as heat is expressed in joules and the temperature is in kelvins, the units of entropy are joules per kelvin (J K^{-1}). Entropy is an extensive property. Molar entropy, the entropy divided by the amount of substance, is expressed in joules per kelvin per mole ($\text{J K}^{-1} \text{ mol}^{-1}$).[2] The molar entropy is an intensive property.

Self-test 3.1 Calculate the change in entropy when the pressure of a perfect gas is changed isothermally from p_i to p_f. [$\Delta S = nR \ln(p_i/p_f)$]

We can use the definition in eqn 3.1 to formulate an expression for the change in entropy of the surroundings, ΔS_{sur}. Consider an infinitesimal transfer of heat dq_{sur} to the surroundings. The surroundings consist of a reservoir of constant volume, so the energy supplied to them by heating can be identified with the change in their

[2] The units of entropy are the same as those of the gas constant, R, and molar heat capacities.

internal energy, dU_{sur}.[3] The internal energy is a state function, and dU_{sur} is an exact differential. As we have seen, these properties imply that dU_{sur} is independent of how the change is brought about and in particular is independent of whether the process is reversible or irreversible. The same remarks therefore apply to dq_{sur}, to which dU_{sur} is equal. Therefore, we can adapt the definition in eqn 3.1 to write

$$dS_{sur} = \frac{dq_{sur,rev}}{T_{sur}} = \frac{dq_{sur}}{T_{sur}} \tag{3.3a}$$

Furthermore, because the temperature of the surroundings is constant whatever the change, for a measurable change

$$\Delta S_{sur} = \frac{q_{sur}}{T_{sur}} \tag{3.3b}$$

That is, regardless of how the change is brought about in the system, reversibly or irreversibly, we can calculate the change of entropy of the surroundings by dividing the heat transferred by the temperature at which the transfer takes place.

Equation 3.3 makes it very simple to calculate the changes in entropy of the surroundings that accompany any process. For instance, for any adiabatic change, $q_{sur} = 0$, so

For an adiabatic change: $\Delta S_{sur} = 0$ (3.4)

This expression is true however the change takes place, reversibly or irreversibly, provided no local hot spots are formed in the surroundings. That is, it is true so long as the surroundings remain in internal equilibrium. If hot spots do form, then the localized energy may subsequently disperse spontaneously and hence generate more entropy.

Illustration 3.1 *Calculating the entropy change in the surroundings*

To calculate the entropy change in the surroundings when 1.00 mol $H_2O(l)$ is formed from its elements under standard conditions at 298 K, we use $\Delta H^{\ominus} = -286$ kJ from Table 2.7. The energy released as heat is supplied to the surroundings, now regarded as being at constant pressure, so $q_{sur} = +286$ kJ. Therefore,

$$\Delta S_{sur} = \frac{2.86 \times 10^5 \text{ J}}{298 \text{ K}} = +960 \text{ J K}^{-1}$$

This strongly exothermic reaction results in an increase in the entropy of the surroundings as energy is released as heat into them.

Self-test 3.2 Calculate the entropy change in the surroundings when 1.00 mol $N_2O_4(g)$ is formed from 2.00 mol $NO_2(g)$ under standard conditions at 298 K.
[-192 J K^{-1}]

Molecular interpretation 3.1 *The statistical view of entropy*

The entry point into the molecular interpretation of the Second Law of thermodynamics is the realization that an atom or molecule can possess only certain energies, called its 'energy levels'. The continuous thermal agitation that molecules

[3] Alternatively, the surroundings can be regarded as being at constant pressure, in which case we could equate dq_{sur} to dH_{sur}.

experience in a sample at $T > 0$ ensures that they are distributed over the available energy levels. One particular molecule may be in one low energy state at one instant, and then be excited into a high energy state a moment later. Although we cannot keep track of the energy state of a single molecule, we can speak of the **population** of the state, the average number of molecules in each state; these populations are constant in time provided the temperature remains the same.

Only the lowest energy state is occupied at $T = 0$. Raising the temperature excites some molecules into higher energy states, and more and more states become accessible as the temperature is raised further (Fig. 3.4). Nevertheless, whatever the temperature, there is always a higher population in a state of low energy than one of high energy. The only exception occurs when the temperature is infinite: then all states of the system are equally populated. These remarks were summarized quantitatively by the Austrian physicist Ludwig Boltzmann in the *Boltzmann distribution*:

$$N_i = \frac{N\mathrm{e}^{-E_i/kT}}{\sum_i \mathrm{e}^{-E_i/kT}}$$

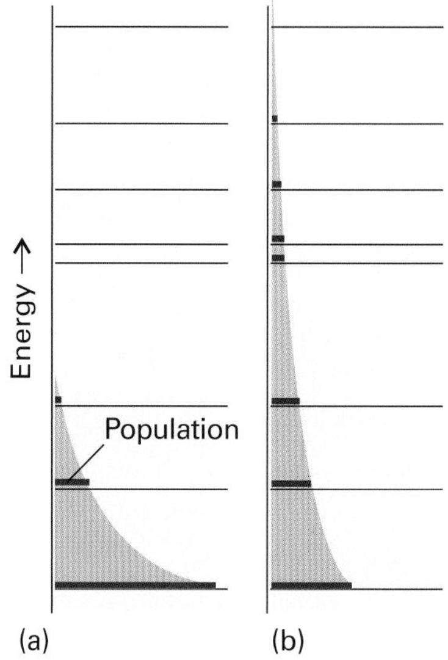

Fig. 3.4 The Boltzmann distribution predicts that the population of a state decreases exponentially with the energy of the state. (a) At low temperatures, only the lowest states are significantly populated; (b) at high temperatures, there is significant population in high-energy states as well as in low-energy states. At infinite temperature (not shown), all states are equally populated.

where $k = 1.381 \times 10^{-23}$ J K^{-1} and N_i is the number of molecules in a sample of N molecules that will be found in a state with an energy E_i when it is part of a system in thermal equilibrium at a temperature T. Care must be taken with the exact interpretation, though, because more than one state may correspond to the same energy: that is, an energy level may consist of several states.

Boltzmann also made the link between the distribution of molecules over energy levels and the entropy. He proposed that the entropy of a system is given by

$$S = k \ln W \qquad (3.5)$$

where W is the number of *microstates*, the ways in which the molecules of a system can be arranged while keeping the total energy constant. Each microstate lasts only for an instant and has a distinct distribution of molecules over the available energy levels. When we measure the properties of a system, we are measuring an average taken over the many microstates the system can occupy under the conditions of the experiment. The concept of the number of microstates makes quantitative the ill-defined qualitative concepts of 'disorder' and 'the dispersal of matter and energy' that are used widely to introduce the concept of entropy: a more 'disorderly' distribution of energy and matter corresponds to a greater number of microstates associated with the same total energy.

Equation 3.5 is known as the **Boltzmann formula** and the entropy calculated from it is sometimes called the **statistical entropy**. We see that if $W = 1$, which corresponds to one microstate (only one way of achieving a given energy, all molecules in exactly the same state), then $S = 0$ because $\ln 1 = 0$. However, if the system can exist in more than one microstate, then $W > 1$ and $S > 0$. But, if more molecules can participate in the distribution of energy, then there are more microstates for a given total energy and the entropy is greater than when the energy is confined so a smaller number of molecules. Therefore, the statistical view of entropy summarized by the Boltzmann formula is consistent with our previous statement that the entropy is related to the dispersal of energy.

The molecular interpretation of entropy advanced by Boltzmann also suggests the thermodynamic definition given by eqn 3.1. To appreciate this point, consider that molecules in a system at high temperature can occupy a large number of the available energy levels, so a small additional transfer of energy as heat will lead to a relatively small change in the number of accessible energy levels. Consequently, the

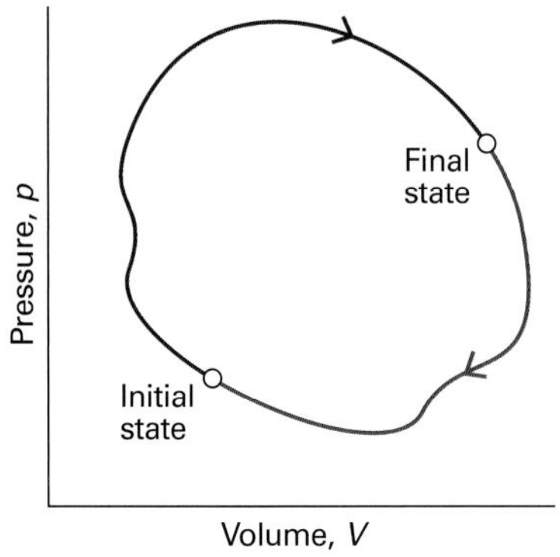

Fig. 3.5 In a thermodynamic cycle, the overall change in a state function (from the initial state to the final state and then back to the initial state again) is zero.

number of microstates does not increase appreciably and neither does the entropy of the system. In contrast, the molecules in a system at low temperature have access to far fewer energy levels (at $T=0$, only the lowest level is accessible), and the transfer of the same quantity of energy by heating will increase the number of accessible energy levels and the number of microstates rather significantly. Hence, the change in entropy upon heating will be greater when the energy is transferred to a cold body than when it is transferred to a hot body. This argument suggests that the change in entropy should be inversely proportional to the temperature at which the transfer takes place, as in eqn 3.1.

(b) The entropy as a state function

Entropy is a state function. To prove this assertion, we need to show that the integral of dS is independent of path. To do so, it is sufficient to prove that the integral of eqn 3.1 around an arbitrary cycle is zero, for that guarantees that the entropy is the same at the initial and final states of the system regardless of the path taken between them (Fig. 3.5). That is, we need to show that

$$\oint \frac{\mathrm{d}q_{\mathrm{rev}}}{T} = 0 \tag{3.6}$$

where the symbol $\oint$ denotes integration around a closed path. There are three steps in the argument:

1. First, to show that eqn 3.6 is true for a special cycle (a 'Carnot cycle') involving a perfect gas.
2. Then to show that the result is true whatever the working substance.
3. Finally, to show that the result is true for any cycle.

A **Carnot cycle**, which is named after the French engineer Sadi Carnot, consists of four reversible stages (Fig. 3.6):

1. Reversible isothermal expansion from A to B at T_h; the entropy change is q_h/T_h, where q_h is the energy supplied to the system as heat from the hot source.
2. Reversible adiabatic expansion from B to C. No energy leaves the system as heat, so the change in entropy is zero. In the course of this expansion, the temperature falls from T_h to T_c, the temperature of the cold sink.
3. Reversible isothermal compression from C to D at T_c. Energy is released as heat to the cold sink; the change in entropy of the system is q_c/T_c; in this expression q_c is negative.
4. Reversible adiabatic compression from D to A. No energy enters the system as heat, so the change in entropy is zero. The temperature rises from T_c to T_h.

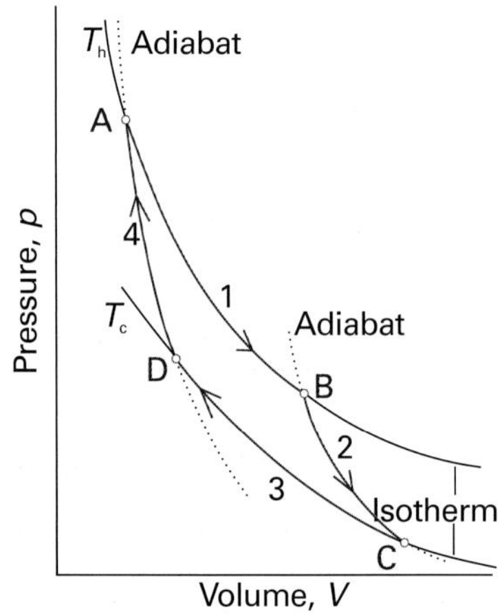

Fig. 3.6 The basic structure of a Carnot cycle. In Step 1, there is isothermal reversible expansion at the temperature T_h. Step 2 is a reversible adiabatic expansion in which the temperature falls from T_h to T_c. In Step 3 there is an isothermal reversible compression at T_c, and that isothermal step is followed by an adiabatic reversible compression, which restores the system to its initial state.

The total change in entropy around the cycle is

$$\oint \mathrm{d}S = \frac{q_h}{T_h} + \frac{q_c}{T_c}$$

However, we show in *Justification 3.1* that, for a perfect gas:

$$\frac{q_h}{q_c} = -\frac{T_h}{T_c} \tag{3.7}_{\mathrm{rev}}$$

Substitution of this relation into the preceding equation gives zero on the right, which is what we wanted to prove.

Justification 3.1 *Heating accompanying reversible adiabatic expansion*

This *Justification* is based on the fact that the two temperatures in eqn 3.7 lie on the same adiabat in Fig. 3.6. As explained in Example 3.1, for a perfect gas:

$$q_h = nRT_h \ln\frac{V_B}{V_A} \qquad q_c = nRT_c \ln\frac{V_D}{V_C}$$

From the relations between temperature and volume for reversible adiabatic processes (eqn 2.28):

$$V_A T_h^c = V_D T_c^c \qquad V_C T_c^c = V_B T_h^c$$

Multiplication of the first of these expressions by the second gives

$$V_A V_C T_h^c T_c^c = V_D V_B T_h^c T_c^c$$

which simplifies to

$$\frac{V_A}{V_B} = \frac{V_D}{V_C}$$

Consequently,

$$q_c = nRT_c \ln\frac{V_D}{V_C} = nRT_c \ln\frac{V_A}{V_B} = -nRT_c \ln\frac{V_B}{V_A}$$

and therefore

$$\frac{q_h}{q_c} = \frac{nRT_h \ln(V_B/V_A)}{-nRT_c \ln(V_B/V_A)} = -\frac{T_h}{T_c}$$

as in eqn 3.7.

In the second step we need to show that eqn 3.7 applies to any material, not just a perfect gas (which is why, in anticipation, we have not labelled it with a °). We begin this step of the argument by introducing the **efficiency**, ε (epsilon), of a heat engine:

$$\varepsilon = \frac{\text{work performed}}{\text{heat absorbed}} = \frac{|w|}{q_h} \qquad [3.8]$$

The definition implies that, the greater the work output for a given supply of heat from the hot reservoir, the greater is the efficiency of the engine. We can express the definition in terms of the heat transactions alone, because (as shown in Fig. 3.7) the energy supplied as work by the engine is the difference between the energy supplied as heat by the hot reservoir and returned to the cold reservoir:

$$\varepsilon = \frac{q_h + q_c}{q_h} = 1 + \frac{q_c}{q_h} \qquad (3.9)$$

(Remember that $q_c < 0$.) It then follows from eqn 3.7 that

$$\varepsilon_{rev} = 1 - \frac{T_c}{T_h} \qquad (3.10)_{rev}$$

Now we are ready to generalize this conclusion. The Second Law of thermodynamics implies that *all reversible engines have the same efficiency regardless of their construction.* To see the truth of this statement, suppose two reversible engines are coupled together and run between the same two reservoirs (Fig. 3.8). The working substances and details of construction of the two engines are entirely arbitrary. Initially, suppose that

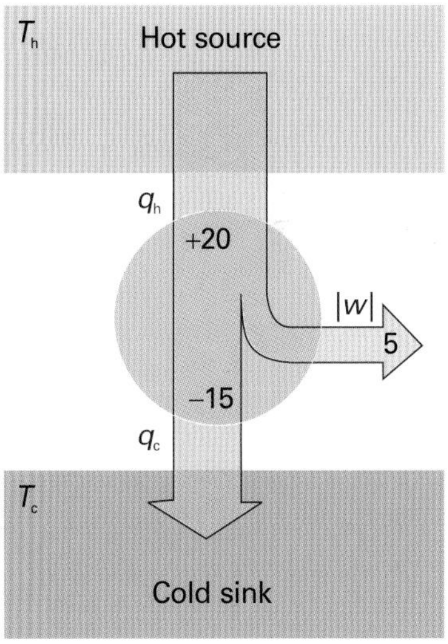

Fig. 3.7 Suppose an energy q_h (for example, 20 kJ) is supplied to the engine and q_c is lost from the engine (for example, $q_c = -15$ kJ) and discarded into the cold reservoir. The work done by the engine is equal to $q_h + q_c$ (for example, 20 kJ + (−15 kJ) = 5 kJ). The efficiency is the work done divided by the energy supplied as heat from the hot source.

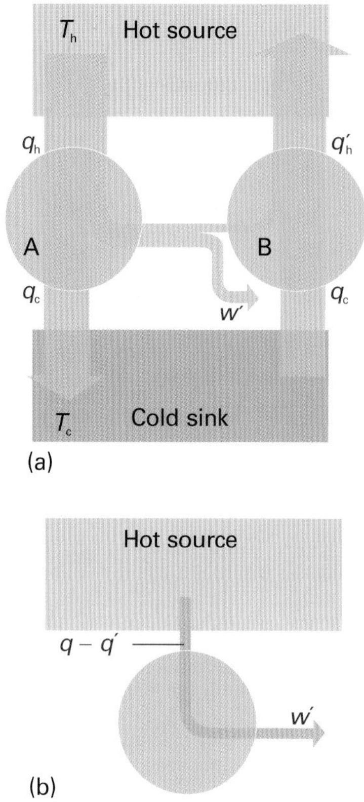

Fig. 3.8 (a) The demonstration of the equivalence of the efficiencies of all reversible engines working between the same thermal reservoirs is based on the flow of energy represented in this diagram. (b) The net effect of the processes is the conversion of heat into work without there being a need for a cold sink: this is contrary to the Kelvin statement of the Second Law.

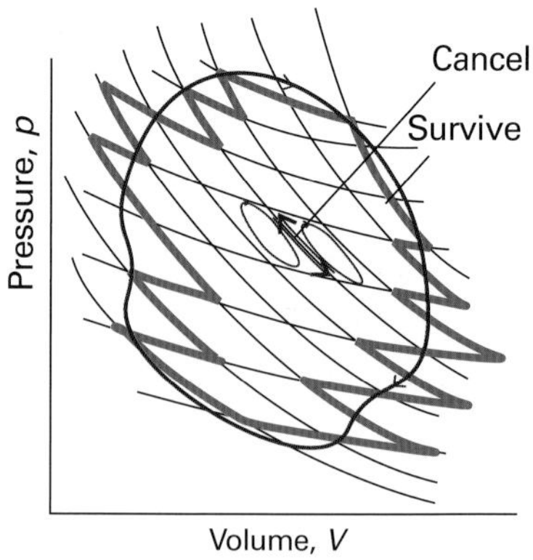

Fig. 3.9 A general cycle can be divided into small Carnot cycles. The match is exact in the limit of infinitesimally small cycles. Paths cancel in the interior of the collection, and only the perimeter, an increasingly good approximation to the true cycle as the number of cycles increases, survives. Because the entropy change around every individual cycle is zero, the integral of the entropy around the perimeter is zero too.

engine A is more efficient than engine B and that we choose a setting of the controls that causes engine B to acquire energy as heat q_c from the cold reservoir and to release a certain quantity of energy as heat into the hot reservoir. However, because engine A is more efficient than engine B, not all the work that A produces is needed for this process, and the difference can be used to do work. The net result is that the cold reservoir is unchanged, work has been done, and the hot reservoir has lost a certain amount of energy. This outcome is contrary to the Kelvin statement of the Second Law, because some heat has been converted directly into work. In molecular terms, the random thermal motion of the hot reservoir has been converted into ordered motion characteristic of work. Because the conclusion is contrary to experience, the initial assumption that engines A and B can have different efficiencies must be false. It follows that the relation between the heat transfers and the temperatures must also be independent of the working material, and therefore that eqn 3.7 is always true for any substance involved in a Carnot cycle.

For the final step in the argument, we note that any reversible cycle can be approximated as a collection of Carnot cycles and the cyclic integral around an arbitrary path is the sum of the integrals around each of the Carnot cycles (Fig. 3.9). This approximation becomes exact as the individual cycles are allowed to become infinitesimal. The entropy change around each individual cycle is zero (as demonstrated above), so the sum of entropy changes for all the cycles is zero. However, in the sum, the entropy change along any individual path is cancelled by the entropy change along the path it shares with the neighbouring cycle. Therefore, all the entropy changes cancel except for those along the perimeter of the overall cycle. That is,

$$\sum_{\text{all}} \frac{q_{\text{rev}}}{T} = \sum_{\text{perimeter}} \frac{q_{\text{rev}}}{T} = 0$$

In the limit of infinitesimal cycles, the non-cancelling edges of the Carnot cycles match the overall cycle exactly, and the sum becomes an integral. Equation 3.6 then follows immediately. This result implies that dS is an exact differential and therefore that S is a state function.

IMPACT ON ENGINEERING
I3.1 Refrigeration

The discussion of the text is the basis of the thermodynamic assessment of the power needed to cool objects in refrigerators. First, we consider the work required to cool an object, and refer to Fig. 3.10.

When an energy $|q_c|$ is removed from a cool source at a temperature T_c and then deposited in a warmer sink at a temperature T_h, as in a typical refrigerator, the change in entropy is

$$\Delta S = -\frac{|q_c|}{T_c} + \frac{|q_c|}{T_h} < 0$$

The process is not spontaneous because not enough entropy is generated in the warm sink to overcome the entropy loss from the cold source. To generate more entropy, energy must be added to the stream that enters the warm sink. Our task is to find the minimum energy that needs to be supplied. The outcome is expressed as the *coefficient of performance*, c:

$$c = \frac{\text{energy transferred as heat}}{\text{energy transferred as work}} = \frac{|q_c|}{|w|}$$

The less the work that is required to achieve a given transfer, the greater the coefficient of performance and the more efficient the refrigerator.

Because $|q_c|$ is removed from the cold source, and the work $|w|$ is added to the energy stream, the energy deposited as heat in the hot sink is $|q_h| = |q_c| + |w|$. Therefore,

$$\frac{1}{c} = \frac{|q_h| - |q_c|}{|q_c|} = \frac{|q_h|}{|q_c|} - 1$$

We can now use eqn 3.7 to express this result in terms of the temperatures alone, which is possible if the transfer is performed reversibly. This substitution leads to

$$c = \frac{T_c}{T_h - T_c}$$

for the thermodynamically optimum coefficient of performance. For a refrigerator withdrawing heat from ice-cold water (T_c = 273 K) in a typical environment (T_h = 293 K), c = 14, so, to remove 10 kJ (enough to freeze 30 g of water), requires transfer of at least 0.71 kJ as work. Practical refrigerators, of course, have a lower coefficient of performance.

The work to *maintain* a low temperature is also relevant to the design of refrigerators. No thermal insulation is perfect, so there is always a flow of energy as heat into the sample at a rate proportional to the temperature difference. If the rate at which energy leaks in is written $A(T_h - T_c)$, where A is a constant that depends on the size of the sample and the details of the insulation, then the minimum power, P, required to maintain the original temperature difference by pumping out that energy by heating the surroundings is

$$P = \frac{1}{c} \times A(T_h - T_c) = A \times \frac{(T_h - T_c)^2}{T_c}$$

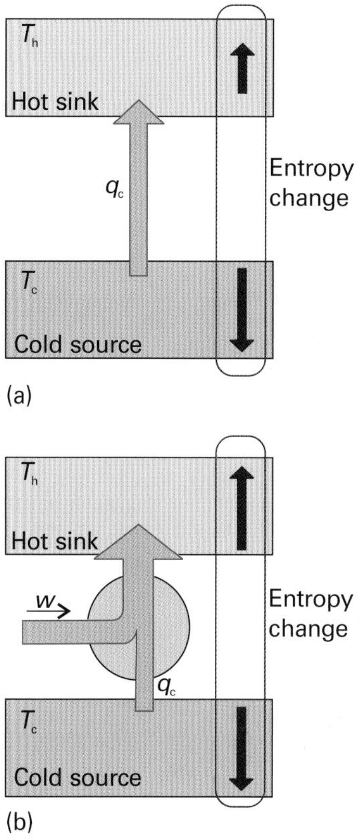

Fig. 3.10 (a) The flow of energy as heat from a cold source to a hot sink is not spontaneous. As shown here, the entropy increase of the hot sink is smaller than the entropy increase of the cold source, so there is a net decrease in entropy. (b) The process becomes feasible if work is provided to add to the energy stream. Then the increase in entropy of the hot sink can be made to cancel the entropy decrease of the hot source.

We see that the power increases as the square of the temperature difference we are trying to maintain. For this reason, air-conditioners are much more expensive to run on hot days than on mild days.

(c) The thermodynamic temperature

Suppose we have an engine that is working reversibly between a hot source at a temperature T_h and a cold sink at a temperature T; then we know from eqn 3.10 that

$$T = (1 - \varepsilon)T_h \qquad (3.11)$$

Comment 3.1

The triple point of a substance represents the set of conditions at which the three phases coexist in equilibrium. For water, the triple point occurs at 273.16 K and 611 Pa. See Section 4.2 for details.

This expression enabled Kelvin to define the **thermodynamic temperature scale** in terms of the efficiency of a heat engine. The zero of the scale occurs for a Carnot efficiency of 1. The size of the unit is entirely arbitrary, but on the Kelvin scale is defined by setting the temperature of the triple point of water as 273.16 K exactly. Then, if the heat engine has a hot source at the triple point of water, the temperature of the cold sink (the object we want to measure) is found by measuring the efficiency of the engine. This result is independent of the working substance.

(d) The Clausius inequality

We now show that the definition of entropy is consistent with the Second Law. To begin, we recall that more energy flows as work under reversible conditions than under irreversible conditions. That is, $-dw_{rev} \geq -dw$, or $dw - dw_{rev} \geq 0$. Because the internal energy is a state function, its change is the same for irreversible and reversible paths between the same two states, so we can also write:

$$dU = dq + dw = dq_{rev} + dw_{rev}$$

It follows that $dq_{rev} - dq = dw - dw_{rev} \geq 0$, or $dq_{rev} \geq dq$, and therefore that $dq_{rev}/T \geq dq/T$. Now we use the thermodynamic definition of the entropy (eqn 3.1; $dS = dq_{rev}/T$) to write

$$dS \geq \frac{dq}{T} \qquad (3.12)$$

This expression is the **Clausius inequality**. It will prove to be of great importance for the discussion of the spontaneity of chemical reactions, as we shall see in Section 3.5.

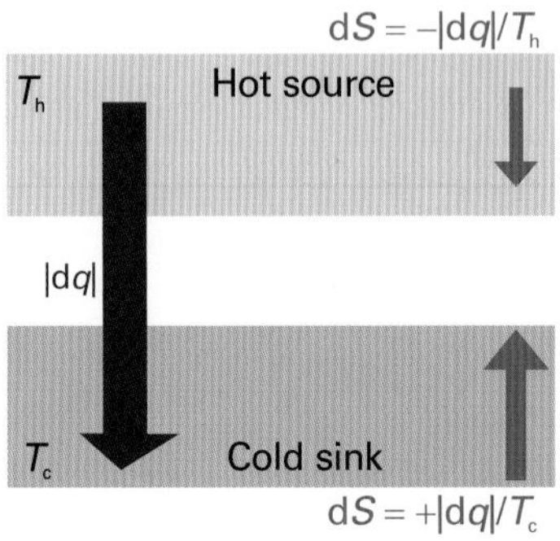

Fig. 3.11 When energy leaves a hot reservoir as heat, the entropy of the reservoir decreases. When the same quantity of energy enters a cooler reservoir, the entropy increases by a larger amount. Hence, overall there is an increase in entropy and the process is spontaneous. Relative changes in entropy are indicated by the sizes of the arrows.

Illustration 3.2 *Spontaneous cooling*

Consider the transfer of energy as heat from one system—the hot source—at a temperature T_h to another system—the cold sink—at a temperature T_c (Fig. 3.11). When $|dq|$ leaves the hot source (so $dq_h < 0$), the Clausius inequality implies that $dS \geq dq_h/T_h$. When $|dq|$ enters the cold sink the Clausius inequality implies that $dS \geq dq_c/T_c$ (with $dq_c > 0$). Overall, therefore,

$$dS \geq \frac{dq_h}{T_h} + \frac{dq_c}{T_c}$$

However, $dq_h = -dq_c$, so

$$dS \geq -\frac{dq_c}{T_h} + \frac{dq_c}{T_c} = dq_c\left(\frac{1}{T_c} - \frac{1}{T_h}\right)$$

which is positive (because $dq_c > 0$ and $T_h \geq T_c$). Hence, cooling (the transfer of heat from hot to cold) is spontaneous, as we know from experience.

We now suppose that the system is isolated from its surroundings, so that $dq = 0$. The Clausius inequality implies that

$$dS \geq 0$$

and we conclude that *in an isolated system the entropy cannot decrease when a spontaneous change occurs.* This statement captures the content of the Second Law.

3.3 Entropy changes accompanying specific processes

We now see how to calculate the entropy changes that accompany a variety of basic processes.

(a) Expansion

We established in Example 3.1 that the change in entropy of a perfect gas that expands isothermally from V_i to V_f is

$$\Delta S = nR \ln \frac{V_f}{V_i} \qquad (3.13)^{\circ}$$

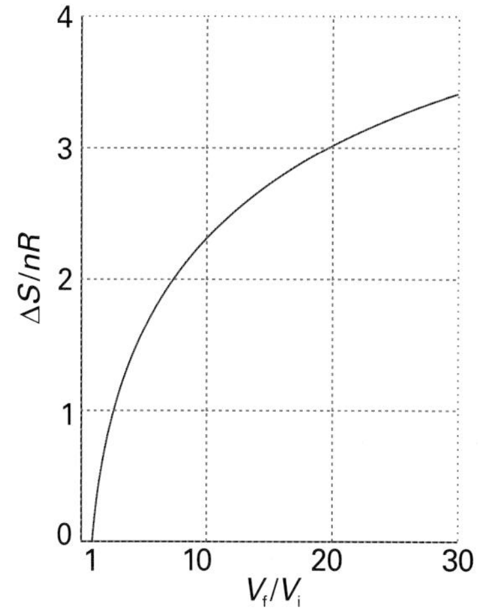

Fig. 3.12 The logarithmic increase in entropy of a perfect gas as it expands isothermally.

Exploration Evaluate the change in entropy that accompanies the expansion of 1.00 mol CO_2(g) from 0.001 m^3 to 0.010 m^3 at 298 K, treated as a van der Waals gas.

Because S is a state function, the value of ΔS *of the system* is independent of the path between the initial and final states, so this expression applies whether the change of state occurs reversibly or irreversibly. The logarithmic dependence of entropy on volume is illustrated in Fig. 3.12.

The *total* change in entropy, however, does depend on how the expansion takes place. For any process $dq_{sur} = -dq$, and for a reversible change we use the expression in Example 3.1; consequently, from eqn 3.3b

$$\Delta S_{sur} = \frac{q_{sur}}{T} = -\frac{q_{rev}}{T} = -nR \ln \frac{V_f}{V_i} \qquad (3.14)^{\circ}_{rev}$$

This change is the negative of the change in the system, so we can conclude that $\Delta S_{tot} = 0$, which is what we should expect for a reversible process. If the isothermal expansion occurs freely ($w = 0$) and irreversibly, then $q = 0$ (because $\Delta U = 0$). Consequently, $\Delta S_{sur} = 0$, and the total entropy change is given by eqn 3.13 itself:

$$\Delta S_{tot} = nR \ln \frac{V_f}{V_i} \qquad (3.15)^{\circ}$$

In this case, $\Delta S_{tot} > 0$, as we expect for an irreversible process.

(b) Phase transition

The degree of dispersal of matter and energy changes when a substance freezes or boils as a result of changes in the order with which the molecules pack together and the extent to which the energy is localized or dispersed. Therefore, we should expect the transition to be accompanied by a change in entropy. For example, when a substance vaporizes, a compact condensed phase changes into a widely dispersed gas and we can expect the entropy of the substance to increase considerably. The entropy of a solid also increases when it melts to a liquid and when that liquid turns into a gas.

Consider a system and its surroundings at the **normal transition temperature**, T_{trs}, the temperature at which two phases are in equilibrium at 1 atm. This temperature is 0°C (273 K) for ice in equilibrium with liquid water at 1 atm, and 100°C (373 K) for water in equilibrium with its vapour at 1 atm. At the transition temperature, any transfer of energy as heat between the system and its surroundings is reversible

Synoptic Table 3.1* Standard entropies (and temperatures) of phase transitions, $\Delta_{trs}S^{\ominus}/(J\ K^{-1}\ mol^{-1})$

	Fusion (at T_f)	Vaporization (at T_b)
Argon, Ar	14.17 (at 83.8 K)	74.53 (at 87.3 K)
Benzene, C_6H_6	38.00 (at 279 K)	87.19 (at 353 K)
Water, H_2O	22.00 (at 273.15 K)	109.0 (at 373.15 K)
Helium, He	4.8 (at 8 K and 30 bar)	19.9 (at 4.22K)

* More values are given in the *Data section*.

Synoptic Table 3.2* The standard entropies of vaporization of liquids

	$\Delta_{vap}H^{\ominus}/(kJ\ mol^{-1})$	$\theta_b/°C$	$\Delta_{vap}S^{\ominus}/(J\ K^{-1}\ mol^{-1})$
Benzene	30.8	80.1	87.2
Carbon tetrachloride	30	76.7	85.8
Cyclohexane	30.1	80.7	85.1
Hydrogen sulfide	18.7	−60.4	87.9
Methane	8.18	−161.5	73.2
Water	40.7	100.0	109.1

* More values are given in the *Data section*.

because the two phases in the system are in equilibrium. Because at constant pressure $q=\Delta_{trs}H$, the change in molar entropy of the system is[4]

$$\Delta_{trs}S = \frac{\Delta_{trs}H}{T_{trs}} \tag{3.16}$$

If the phase transition is exothermic ($\Delta_{trs}H<0$, as in freezing or condensing), then the entropy change is negative. This decrease in entropy is consistent with localization of matter and energy that accompanies the formation of a solid from a liquid or a liquid from a gas. If the transition is endothermic ($\Delta_{trs}H>0$, as in melting and vaporization), then the entropy change is positive, which is consistent with dispersal of energy and matter in the system.

Table 3.1 lists some experimental entropies of transition. Table 3.2 lists in more detail the standard entropies of vaporization of several liquids at their boiling points. An interesting feature of the data is that a wide range of liquids give approximately the same standard entropy of vaporization (about 85 $J\ K^{-1}\ mol^{-1}$): this empirical observation is called **Trouton's rule**.

Molecular interpretation 3.2 *Trouton's rule*

The explanation of Trouton's rule is that a comparable change in volume occurs (with an accompanying change in the number of accessible microstates) when any liquid evaporates and becomes a gas. Hence, all liquids can be expected to have similar standard entropies of vaporization.

[4] Recall from Section 2.7 that $\Delta_{trs}H$ is an enthalpy change per mole of substance; so $\Delta_{trs}S$ is also a molar quantity.

Liquids that show significant deviations from Trouton's rule do so on account of strong molecular interactions that restrict molecular motion. As a result, there is a greater dispersal of energy and matter when the liquid turns into a vapour than would occur for a liquid in which molcular motion is less restricted. An example is water, where the large entropy of vaporization reflects the presence of structure arising from hydrogen-bonding in the liquid. Hydrogen bonds tend to organize the molecules in the liquid so that they are less random than, for example, the molecules in liquid hydrogen sulfide (in which there is no hydrogen bonding).

Methane has an unusually low entropy of vaporization. A part of the reason is that the entropy of the gas itself is slightly low (186 J K^{-1} mol^{-1} at 298 K); the entropy of N_2 under the same conditions is 192 J K^{-1} mol^{-1}. As we shall see in Chapter 13, small molecules are difficult to excite into rotation; as a result, only a few rotational states are accessible at room temperature and, consequently, the number of rotational energy levels among which energy can be dispersed is low.

Illustration 3.3 *Using Trouton's rule*

There is no hydrogen bonding in liquid bromine and Br_2 is a heavy molecule that is unlikely to display unusual behaviour in the gas phase, so it is probably safe to use Trouton's rule. To predict the standard molar enthalpy of vaporization of bromine given that it boils at 59.2°C, we use the rule in the form

$$\Delta_{\text{vap}}H^{\ominus} = T_{\text{b}} \times (85 \text{ J K}^{-1} \text{ mol}^{-1})$$

Substitution of the data then gives

$$\Delta_{\text{vap}}H^{\ominus} = (332.4 \text{ K}) \times (85 \text{ J K}^{-1} \text{ mol}^{-1}) = +2.8 \times 10^3 \text{ J mol}^{-1} = +28 \text{ kJ mol}^{-1}$$

The experimental value is +29.45 kJ mol^{-1}.

Self-test 3.3 Predict the enthalpy of vaporization of ethane from its boiling point, −88.6°C. [16 kJ mol^{-1}]

(c) Heating

We can use eqn 3.2 to calculate the entropy of a system at a temperature T_f from a knowledge of its entropy at a temperature T_i and the heat supplied to change its temperature from one value to the other:

$$S(T_{\text{f}}) = S(T_{\text{i}}) + \int_{T_{\text{i}}}^{T_{\text{f}}} \frac{\text{d}q_{\text{rev}}}{T} \qquad (3.17)$$

We shall be particularly interested in the entropy change when the system is subjected to constant pressure (such as from the atmosphere) during the heating. Then, from the definition of constant-pressure heat capacity (eqn 2.22), $\text{d}q_{\text{rev}} = C_p\text{d}T$ provided the system is doing no non-expansion work. Consequently, at constant pressure:

$$S(T_{\text{f}}) = S(T_{\text{i}}) + \int_{T_{\text{i}}}^{T_{\text{f}}} \frac{C_p\text{d}T}{T} \qquad (3.18)$$

The same expression applies at constant volume, but with C_p replaced by C_V. When C_p is independent of temperature in the temperature range of interest, it can be taken outside the integral and we obtain

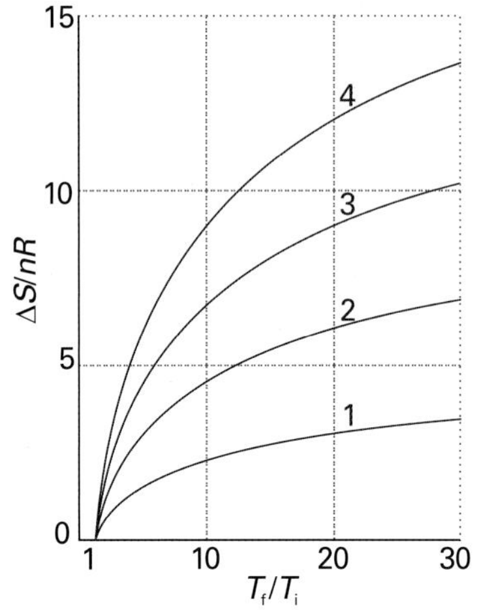

Fig. 3.13 The logarithmic increase in entropy of a substance as it is heated at constant volume. Different curves correspond to different values of the constant-volume heat capacity (which is assumed constant over the temperature range) expressed as $C_{V,m}/R$.

Exploration Plot the change in entropy of a perfect gas of (a) atoms, (b) linear rotors, (c) nonlinear rotors as the sample is heated over the same range under conditions of (i) constant volume, (ii) constant pressure.

$$S(T_f) = S(T_i) + C_p \int_{T_i}^{T_f} \frac{dT}{T} = S(T_i) + C_p \ln \frac{T_f}{T_i} \qquad (3.19)$$

with a similar expression for heating at constant volume. The logarithmic dependence of entropy on temperature is illustrated in Fig. 3.13.

Example 3.2 *Calculating the entropy change*

Calculate the entropy change when argon at 25°C and 1.00 bar in a container of volume 0.500 dm^3 is allowed to expand to 1.000 dm^3 and is simultaneously heated to 100°C.

Method Because S is a state function, we are free to choose the most convenient path from the initial state. One such path is reversible isothermal expansion to the final volume, followed by reversible heating at constant volume to the final temperature. The entropy change in the first step is given by eqn 3.13 and that of the second step, provided C_V is independent of temperature, by eqn 3.19 (with C_V in place of C_p). In each case we need to know n, the amount of gas molecules, and can calculate it from the perfect gas equation and the data for the initial state from $n = p_iV_i/RT_i$. The heat capacity at constant volume is given by the equipartition theorem as $\frac{3}{2}R$. (The equipartition theorem is reliable for monatomic gases: for others and in general use experimental data like that in Table 2.7, converting to the value at constant volume by using the relation $C_{p,m} - C_{V,m} = R$.)

Answer Because $n = p_iV_i/RT_i$, from eqn 3.13

$$\Delta S(\text{Step 1}) = \left(\frac{p_iV_i}{RT_i}\right) \times R \ln \frac{V_f}{V_i} = \frac{p_iV_i}{T_i} \ln \frac{V_f}{V_i}$$

The entropy change in the second step, from 298 K to 373 K at constant volume, is

$$\Delta S(\text{Step 2}) = \left(\frac{p_iV_i}{RT_i}\right) \times \tfrac{3}{2}R \ln \frac{T_f}{T_i} = \frac{p_iV_i}{T_i} \ln \left(\frac{T_f}{T_i}\right)^{3/2}$$

The overall entropy change, the sum of these two changes, is

$$\Delta S = \frac{p_iV_i}{T_i} \ln \frac{V_f}{V_i} + \frac{p_iV_i}{T_i} \ln \left(\frac{T_f}{T_i}\right)^{3/2} = \frac{p_iV_i}{T_i} \ln \left\{ \frac{V_f}{V_i}\left(\frac{T_f}{T_i}\right)^{3/2} \right\}$$

At this point we substitute the data and obtain (by using 1 Pa m^3 = 1 J)

$$\Delta S = \frac{(1.00 \times 10^5 \text{ Pa}) \times (0.500 \times 10^{-3} \text{ m}^3)}{298 \text{ K}} \ln \left\{ \frac{1.000}{0.500}\left(\frac{373}{298}\right)^{3/2} \right\}$$

$$= +0.173 \text{ J K}^{-1}$$

A note on good practice It is sensible to proceed as generally as possible before inserting numerical data so that, if required, the formula can be used for other data and to avoid rounding errors.

Self-test 3.4 Calculate the entropy change when the same initial sample is compressed to 0.0500 dm^3 and cooled to −25°C. [−0.44 J K^{-1}]

(d) The measurement of entropy

The entropy of a system at a temperature T is related to its entropy at $T = 0$ by measuring its heat capacity C_p at different temperatures and evaluating the integral in eqn 3.18, taking care to add the entropy of transition ($\Delta_{trs}H/T_{trs}$) for each phase transition between $T = 0$ and the temperature of interest. For example, if a substance melts at T_f and boils at T_b, then its entropy above its boiling temperature is given by

$$S(T) = S(0) + \int_0^{T_f} \frac{C_p(s)dT}{T} + \frac{\Delta_{fus}H}{T_f} + \int_{T_f}^{T_b} \frac{C_p(l)dT}{T} + \frac{\Delta_{vap}H}{T_b} + \int_{T_b}^{T} \frac{C_p(g)dT}{T} \qquad (3.20)$$

All the properties required, except $S(0)$, can be measured calorimetrically, and the integrals can be evaluated either graphically or, as is now more usual, by fitting a polynomial to the data and integrating the polynomial analytically. The former procedure is illustrated in Fig. 3.14: the area under the curve of C_p/T against T is the integral required. Because $dT/T = d\ln T$, an alternative procedure is to evaluate the area under a plot of C_p against $\ln T$.

One problem with the determination of entropy is the difficulty of measuring heat capacities near $T = 0$. There are good theoretical grounds for assuming that the heat capacity is proportional to T^3 when T is low (see Section 8.1), and this dependence is the basis of the **Debye extrapolation**. In this method, C_p is measured down to as low a temperature as possible, and a curve of the form aT^3 is fitted to the data. That fit determines the value of a, and the expression $C_p = aT^3$ is assumed valid down to $T = 0$.

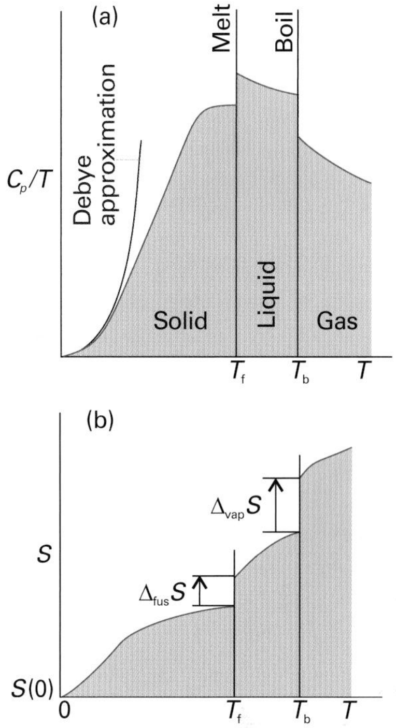

Fig. 3.14 The calculation of entropy from heat capacity data. (a) The variation of C_p/T with the temperature for a sample. (b) The entropy, which is equal to the area beneath the upper curve up to the corresponding temperature, plus the entropy of each phase transition passed.

Exploration Allow for the temperature dependence of the heat capacity by writing $C = a + bT + c/T^2$, and plot the change in entropy for different values of the three coefficients (including negative values of c).

Illustration 3.4 *Calculating a standard molar entropy*

The standard molar entropy of nitrogen gas at 25°C has been calculated from the following data:

	$S_m^\ominus/(J\,K^{-1}\,mol^{-1})$
Debye extrapolation	1.92
Integration, from 10 K to 35.61 K	25.25
Phase transition at 35.61 K	6.43
Integration, from 35.61 K to 63.14 K	23.38
Fusion at 63.14 K	11.42
Integration, from 63.14 K to 77.32 K	11.41
Vaporization at 77.32 K	72.13
Integration, from 77.32 K to 298.15 K	39.20
Correction for gas imperfection	0.92
Total	192.06

Therefore,

$$S_m(298.15\,K) = S_m(0) + 192.1\,J\,K^{-1}\,mol^{-1}$$

Example 3.3 *Calculating the entropy at low temperatures*

The molar constant-pressure heat capacity of a certain solid at 4.2 K is 0.43 J K^{-1} mol^{-1}. What is its molar entropy at that temperature?

Method Because the temperature is so low, we can assume that the heat capacity varies with temperature as aT^3, in which case we can use eqn 3.18 to calculate the entropy at a temperature T in terms of the entropy at $T = 0$ and the constant a. When the integration is carried out, it turns out that the result can be expressed in terms of the heat capacity at the temperature T, so the data can be used directly to calculate the entropy.

Answer The integration required is

$$S(T) = S(0) + \int_0^T \frac{aT^3\mathrm{d}T}{T} = S(0) + a\int_0^T T^2\mathrm{d}T = S(0) + \tfrac{1}{3}aT^3$$

However, because aT^3 is the heat capacity at the temperature T,

$$S(T) = S(0) + \tfrac{1}{3}C_p(T)$$

from which it follows that

$$S_\mathrm{m}(4.2\ \mathrm{K}) = S_\mathrm{m}(0) + 0.14\ \mathrm{J\ K^{-1}\ mol^{-1}}$$

Self-test 3.5 For metals, there is also a contribution to the heat capacity from the electrons which is linearly proportional to T when the temperature is low. Find its contribution to the entropy at low temperatures. [$S(T) = S(0) + C_p(T)$]

3.4 The Third Law of thermodynamics

At $T = 0$, all energy of thermal motion has been quenched, and in a perfect crystal all the atoms or ions are in a regular, uniform array. The localization of matter and the absence of thermal motion suggest that such materials also have zero entropy. This conclusion is consistent with the molecular interpretation of entropy, because $S = 0$ if there is only one way of arranging the molecules and only one microstate is accessible (the ground state).

(a) The Nernst heat theorem

The experimental observation that turns out to be consistent with the view that the entropy of a regular array of molecules is zero at $T = 0$ is summarized by the **Nernst heat theorem**:

> The entropy change accompanying any physical or chemical transformation approaches zero as the temperature approaches zero: $\Delta S \to 0$ as $T \to 0$ provided all the substances involved are perfectly crystalline.

Illustration 3.5 *Using the Nernst heat theorem*

Consider the entropy of the transition between orthorhombic sulfur, S(α), and monoclinic sulfur, S(β), which can be calculated from the transition enthalpy ($-402\ \mathrm{J\ mol^{-1}}$) at the transition temperature (369 K):

$$\Delta_\mathrm{trs}S = S_\mathrm{m}(\alpha) - S_\mathrm{m}(\beta) = \frac{(-402\ \mathrm{J\ mol^{-1}})}{369\ \mathrm{K}} = -1.09\ \mathrm{J\ K^{-1}\ mol^{-1}}$$

The two individual entropies can also be determined by measuring the heat capacities from $T = 0$ up to $T = 369$ K. It is found that $S_\mathrm{m}(\alpha) = S_\mathrm{m}(\alpha,0) + 37\ \mathrm{J\ K^{-1}\ mol^{-1}}$

and $S_m(\beta) = S_m(\beta,0) + 38\ \mathrm{J\ K^{-1}\ mol^{-1}}$. These two values imply that at the transition temperature

$$\Delta_{trs}S = S_m(\alpha,0) - S_m(\beta,0) = -1\ \mathrm{J\ K^{-1}\ mol^{-1}}$$

On comparing this value with the one above, we conclude that $S_m(\alpha,0) - S_m(\beta,0) \approx 0$, in accord with the theorem.

It follows from the Nernst theorem that, if we arbitrarily ascribe the value zero to the entropies of elements in their perfect crystalline form at $T = 0$, then all perfect crystalline compounds also have zero entropy at $T = 0$ (because the change in entropy that accompanies the formation of the compounds, like the entropy of all transformations at that temperature, is zero). This conclusion is summarized by the **Third Law of thermodynamics**:

The entropy of all perfect crystalline substances is zero at $T = 0$.

As far as thermodynamics is concerned, choosing this common value as zero is then a matter of convenience. The molecular interpretation of entropy, however, justifies the value $S = 0$ at $T = 0$.

Molecular interpretation 3.3 *The statistical view of the Third Law of thermodynamics*

We saw in *Molecular interpretation 3.1* that, according to the Boltzmann formula, the entropy is zero if there is only one accessible microstate ($W = 1$). In most cases, $W = 1$ at $T = 0$ because there is only one way of achieving the lowest total energy: put all the molecules into the same, lowest state. Therefore, $S = 0$ at $T = 0$, in accord with the Third Law of thermodynamics. In certain cases, though, W may differ from 1 at $T = 0$. This is the case if there is no energy advantage in adopting a particular orientation even at absolute zero. For instance, for a diatomic molecule AB there may be almost no energy difference between the arrangements . . . AB AB AB . . . and . . . BA AB BA . . . , so $W > 1$ even at $T = 0$. If $S > 0$ at $T = 0$ we say that the substance has a **residual entropy**. Ice has a residual entropy of $3.4\ \mathrm{J\ K^{-1}\ mol^{-1}}$. It stems from the arrangement of the hydrogen bonds between neighbouring water molecules: a given O atom has two short O—H bonds and two long O···H bonds to its neighbours, but there is a degree of randomness in which two bonds are short and which two are long.

(b) Third-Law entropies

Entropies reported on the basis that $S(0) = 0$ are called **Third-Law entropies** (and often just 'entropies'). When the substance is in its standard state at the temperature T, the **standard (Third-Law) entropy** is denoted $S^\ominus(T)$. A list of values at 298 K is given in Table 3.3.

The **standard reaction entropy**, $\Delta_r S^\ominus$, is defined, like the standard reaction enthalpy, as the difference between the molar entropies of the pure, separated products and the pure, separated reactants, all substances being in their standard states at the specified temperature:

$$\Delta_r S^\ominus = \sum_{\text{Products}} \nu S_m^\ominus - \sum_{\text{Reactants}} \nu S_m^\ominus \qquad (3.21)$$

In this expression, each term is weighted by the appropriate stoichiometric coefficient. Standard reaction entropies are likely to be positive if there is a net formation of gas in a reaction, and are likely to be negative if there is a net consumption of gas.

Synoptic Table 3.3* Standard Third-Law entropies at 298 K

	$S_m^\ominus/(\mathrm{J\ K^{-1}\ mol^{-1}})$
Solids	
Graphite, C(s)	5.7
Diamond, C(s)	2.4
Sucrose, $C_{12}H_{22}O_{11}$(s)	360.2
Iodine, I_2(s)	116.1
Liquids	
Benzene, C_6H_6(l)	173.3
Water, H_2O(l)	69.9
Mercury, Hg(l)	76.0
Gases	
Methane, CH_4(g)	186.3
Carbon dioxide, CO_2(g)	213.7
Hydrogen, H_2(g)	130.7
Helium, He	126.2
Ammonia, NH_3(g)	192.4

* More values are given in the *Data section*.

Illustration 3.6 *Calculating a standard reaction entropy*

To calculate the standard reaction entropy of $H_2(g) + \frac{1}{2}O_2(g) \rightarrow H_2O(l)$ at 25°C, we use the data in Table 2.7 of the *Data Section* to write

$$\begin{aligned}\Delta_r S^{\ominus} &= S_m^{\ominus}(H_2O, l) - \{S_m^{\ominus}(H_2, g) + \tfrac{1}{2}S_m^{\ominus}(O_2, g)\}\\ &= 69.9\ \mathrm{J\ K^{-1}\ mol^{-1}} - \{130.7 + \tfrac{1}{2}(205.0)\}\ \mathrm{J\ K^{-1}\ mol^{-1}}\\ &= -163.4\ \mathrm{J\ K^{-1}\ mol^{-1}}\end{aligned}$$

The negative value is consistent with the conversion of two gases to a compact liquid.

A note on good practice Do not make the mistake of setting the standard molar entropies of elements equal to zero: they have non-zero values (provided $T > 0$), as we have already discussed.

Self-test 3.6 Calculate the standard reaction entropy for the combustion of methane to carbon dioxide and liquid water at 25°C. [$-243\ \mathrm{J\ K^{-1}\ mol^{-1}}$]

Just as in the discussion of enthalpies in Section 2.8, where we acknowledged that solutions of cations cannot be prepared in the absence of anions, the standard molar entropies of ions in solution are reported on a scale in which the standard entropy of the H^+ ions in water is taken as zero at all temperatures:

$$S^{\ominus}(H^+, aq) = 0 \qquad [3.22]$$

The values based on this choice are listed in Table 2.7 in the *Data section*.[5] Because the entropies of ions in water are values relative to the hydrogen ion in water, they may be either positive or negative. A positive entropy means that an ion has a higher molar entropy than H^+ in water and a negative entropy means that the ion has a lower molar entropy than H^+ in water. For instance, the standard molar entropy of $Cl^-(aq)$ is +57 $\mathrm{J\ K^{-1}\ mol^{-1}}$ and that of $Mg^{2+}(aq)$ is $-128\ \mathrm{J\ K^{-1}\ mol^{-1}}$. Ion entropies vary as expected on the basis that they are related to the degree to which the ions order the water molecules around them in the solution. Small, highly charged ions induce local structure in the surrounding water, and the disorder of the solution is decreased more than in the case of large, singly charged ions. The absolute, Third-Law standard molar entropy of the proton in water can be estimated by proposing a model of the structure it induces, and there is some agreement on the value $-21\ \mathrm{J\ K^{-1}\ mol^{-1}}$. The negative value indicates that the proton induces order in the solvent.

Concentrating on the system

Entropy is the basic concept for discussing the direction of natural change, but to use it we have to analyse changes in both the system and its surroundings. We have seen that it is always very simple to calculate the entropy change in the surroundings, and we shall now see that it is possible to devise a simple method for taking that contribution into account automatically. This approach focuses our attention on the system

[5] In terms of the language to be introduced in Section 5.1, the entropies of ions in solution are actually *partial molar entropies*, for their values include the consequences of their presence on the organization of the solvent molecules around them.

and simplifies discussions. Moreover, it is the foundation of all the applications of chemical thermodynamics that follow.

3.5 The Helmholtz and Gibbs energies

Consider a system in thermal equilibrium with its surroundings at a temperature T. When a change in the system occurs and there is a transfer of energy as heat between the system and the surroundings, the Clausius inequality, eqn 3.12, reads

$$dS - \frac{dq}{T} \geq 0 \qquad (3.23)$$

We can develop this inequality in two ways according to the conditions (of constant volume or constant pressure) under which the process occurs.

(a) Criteria for spontaneity

First, consider heating at constant volume. Then, in the absence of non-expansion work, we can write $dq_V = dU$; consequently

$$dS - \frac{dU}{T} \geq 0 \qquad (3.24)$$

The importance of the inequality in this form is that it expresses the criterion for spontaneous change solely in terms of the state functions of the system. The inequality is easily rearranged to

$$TdS \geq dU \qquad \text{(constant } V\text{, no additional work)}^{6} \qquad (3.25)$$

At either constant internal energy ($dU = 0$) or constant entropy ($dS = 0$), this expression becomes, respectively,

$$dS_{U,V} \geq 0 \qquad dU_{S,V} \leq 0 \qquad (3.26)$$

where the subscripts indicate the constant conditions.

Equation 3.26 expresses the criteria for spontaneous change in terms of properties relating to the system. The first inequality states that, in a system at constant volume and constant internal energy (such as an isolated system), the entropy increases in a spontaneous change. That statement is essentially the content of the Second Law. The second inequality is less obvious, for it says that, if the entropy and volume of the system are constant, then the internal energy must decrease in a spontaneous change. Do not interpret this criterion as a tendency of the system to sink to lower energy. It is a disguised statement about entropy, and should be interpreted as implying that, if the entropy of the system is unchanged, then there must be an increase in entropy of the surroundings, which can be achieved only if the energy of the system decreases as energy flows out as heat.

When energy is transferred as heat at constant pressure, and there is no work other than expansion work, we can write $dq_p = dH$ and obtain

$$TdS \geq dH \qquad \text{(constant } p\text{, no additional work)} \qquad (3.27)$$

At either constant enthalpy or constant entropy this inequality becomes, respectively,

$$dS_{H,p} \geq 0 \qquad dH_{S,p} \leq 0 \qquad (3.28)$$

The interpretations of these inequalities are similar to those of eqn 3.26. The entropy of the system at constant pressure must increase if its enthalpy remains constant (for

[6] Recall that 'additional work' is work other than expansion work.

there can then be no change in entropy of the surroundings). Alternatively, the enthalpy must decrease if the entropy of the system is constant, for then it is essential to have an increase in entropy of the surroundings.

Because eqns 3.25 and 3.27 have the forms $\mathrm{d}U - T\mathrm{d}S \le 0$ and $\mathrm{d}H - T\mathrm{d}S \le 0$, respectively, they can be expressed more simply by introducing two more thermodynamic quantities. One is the **Helmholtz energy**, A, which is defined as

$$A = U - TS \qquad [3.29]$$

The other is the **Gibbs energy**, G:

$$G = H - TS \qquad [3.30]$$

All the symbols in these two definitions refer to the system.

When the state of the system changes at constant temperature, the two properties change as follows:

$$\text{(a)}\ \mathrm{d}A = \mathrm{d}U - T\mathrm{d}S \qquad \text{(b)}\ \mathrm{d}G = \mathrm{d}H - T\mathrm{d}S \qquad (3.31)$$

When we introduce eqns 3.25 and 3.27, respectively, we obtain the criteria of spontaneous change as

$$\text{(a)}\ \mathrm{d}A_{T,V} \le 0 \qquad \text{(b)}\ \mathrm{d}G_{T,p} \le 0 \qquad (3.32)$$

These inequalities are the most important conclusions from thermodynamics for chemistry. They are developed in subsequent sections and chapters.

(b) Some remarks on the Helmholtz energy

A change in a system at constant temperature and volume is spontaneous if $\mathrm{d}A_{T,V} \le 0$. That is, a change under these conditions is spontaneous if it corresponds to a decrease in the Helmholtz energy. Such systems move spontaneously towards states of lower A if a path is available. The criterion of equilibrium, when neither the forward nor reverse process has a tendency to occur, is

$$\mathrm{d}A_{T,V} = 0 \qquad (3.33)$$

The expressions $\mathrm{d}A = \mathrm{d}U - T\mathrm{d}S$ and $\mathrm{d}A < 0$ are sometimes interpreted as follows. A negative value of $\mathrm{d}A$ is favoured by a negative value of $\mathrm{d}U$ and a positive value of $T\mathrm{d}S$. This observation suggests that the tendency of a system to move to lower A is due to its tendency to move towards states of lower internal energy and higher entropy. However, this interpretation is false (even though it is a good rule of thumb for remembering the expression for $\mathrm{d}A$) because the tendency to lower A is solely a tendency towards states of greater overall entropy. Systems change spontaneously if in doing so the total entropy of the system and its surroundings increases, not because they tend to lower internal energy. The form of $\mathrm{d}A$ may give the impression that systems favour lower energy, but that is misleading: $\mathrm{d}S$ is the entropy change of the system, $-\mathrm{d}U/T$ is the entropy change of the surroundings (when the volume of the system is constant), and their total tends to a maximum.

(c) Maximum work

It turns out that A carries a greater significance than being simply a signpost of spontaneous change: *the change in the Helmholtz function is equal to the maximum work accompanying a process*:

$$\mathrm{d}w_{\max} = \mathrm{d}A \qquad (3.34)$$

As a result, A is sometimes called the 'maximum work function', or the 'work function'.[7]

[7] *Arbeit* is the German word for work; hence the symbol A.

Justification 3.2 *Maximum work*

To demonstrate that maximum work can be expressed in terms of the changes in Helmholtz energy, we combine the Clausius inequality $dS \geq dq/T$ in the form $TdS \geq dq$ with the First Law, $dU = dq + dw$, and obtain

$$dU \leq TdS + dw$$

(dU is smaller than the term on the right because we are replacing dq by TdS, which in general is larger.) This expression rearranges to

$$dw \geq dU - TdS$$

It follows that the most negative value of dw, and therefore the maximum energy that can be obtained from the system as work, is given by

$$dw_{max} = dU - TdS$$

and that this work is done only when the path is traversed reversibly (because then the equality applies). Because at constant temperature $dA = dU - TdS$, we conclude that $dw_{max} = dA$.

When a macroscopic isothermal change takes place in the system, eqn 3.34 becomes

$$w_{max} = \Delta A \qquad (3.35)$$

with

$$\Delta A = \Delta U - T\Delta S \qquad (3.36)$$

This expression shows that in some cases, depending on the sign of $T\Delta S$, not all the change in internal energy may be available for doing work. If the change occurs with a decrease in entropy (of the system), in which case $T\Delta S < 0$, then the right-hand side of this equation is not as negative as ΔU itself, and consequently the maximum work is less than ΔU. For the change to be spontaneous, some of the energy must escape as heat in order to generate enough entropy in the surroundings to overcome the reduction in entropy in the system (Fig. 3.15). In this case, Nature is demanding a tax on the internal energy as it is converted into work. This is the origin of the alternative name 'Helmholtz free energy' for A, because ΔA is that part of the change in internal energy that we are free to use to do work.

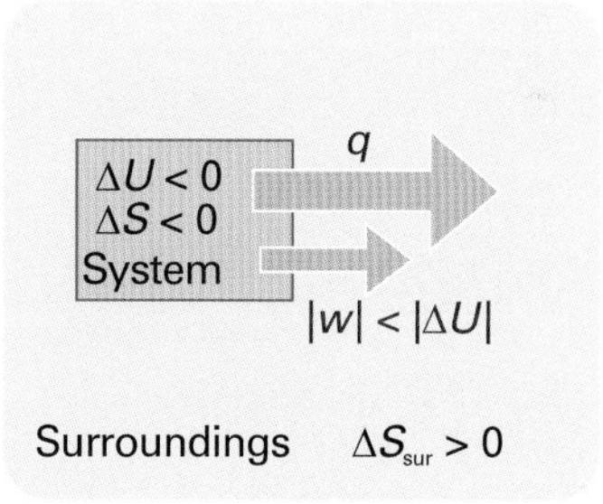

Fig. 3.15 In a system not isolated from its surroundings, the work done may be different from the change in internal energy. Moreover, the process is spontaneous if overall the entropy of the global, isolated system increases. In the process depicted here, the entropy of the system decreases, so that of the surroundings must increase in order for the process to be spontaneous, which means that energy must pass from the system to the surroundings as heat. Therefore, less work than ΔU can be obtained.

Molecular interpretation 3.4 *Maximum work and the Helmholtz energy*

Further insight into the relation between the work that a system can do and the Helmholtz energy is obtained by recalling that work is energy transferred to the surroundings as the uniform motion of atoms. We can interpret the expression $A = U - TS$ as showing that A is the total internal energy of the system, U, less a contribution that is stored as energy of thermal motion (the quantity TS). Because energy stored in random thermal motion cannot be used to achieve uniform motion in the surroundings, only the part of U that is not stored in that way, the quantity $U - TS$, is available for conversion into work.

If the change occurs with an increase of entropy of the system (in which case $T\Delta S > 0$), the right-hand side of the equation is more negative than ΔU. In this case, the maximum work that can be obtained from the system is greater than ΔU. The explanation of this apparent paradox is that the system is not isolated and energy may

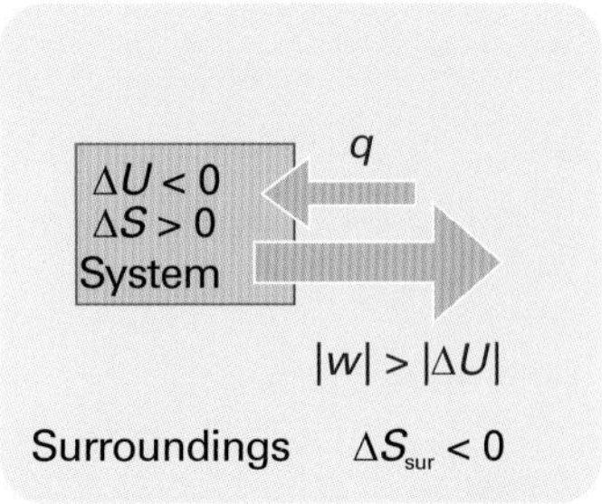

Fig. 3.16 In this process, the entropy of the system increases; hence we can afford to lose some entropy of the surroundings. That is, some of their energy may be lost as heat to the system. This energy can be returned to them as work. Hence the work done can exceed ΔU.

flow in as heat as work is done. Because the entropy of the system increases, we can afford a reduction of the entropy of the surroundings yet still have, overall, a spontaneous process. Therefore, some energy (no more than the value of $T\Delta S$) may leave the surroundings as heat and contribute to the work the change is generating (Fig. 3.16). Nature is now providing a tax refund.

Example 3.4 *Calculating the maximum available work*

When 1.000 mol $C_6H_{12}O_6$ (glucose) is oxidized to carbon dioxide and water at 25°C according to the equation $C_6H_{12}O_6(s) + 6\ O_2(g) \rightarrow 6\ CO_2(g) + 6\ H_2O(l)$, calorimetric measurements give $\Delta_r U^{\ominus} = -2808$ kJ mol^{-1} and $\Delta_r S = +182.4$ J K^{-1} mol^{-1} at 25°C. How much of this energy change can be extracted as (a) heat at constant pressure, (b) work?

Method We know that the heat released at constant pressure is equal to the value of ΔH, so we need to relate $\Delta_r H^{\ominus}$ to $\Delta_r U^{\ominus}$, which is given. To do so, we suppose that all the gases involved are perfect, and use eqn 2.21 in the form $\Delta_r H = \Delta_r U + \Delta\nu_g RT$. For the maximum work available from the process we use eqn 3.34.

Answer (a) Because $\Delta\nu_g = 0$, we know that $\Delta_r H^{\ominus} = \Delta_r U^{\ominus} = -2808$ kJ mol^{-1}. Therefore, at constant pressure, the energy available as heat is 2808 kJ mol^{-1}. (b) Because $T = 298$ K, the value of $\Delta_r A^{\ominus}$ is

$$\Delta_r A^{\ominus} = \Delta_r U^{\ominus} - T\Delta_r S^{\ominus} = -2862 \text{ kJ mol}^{-1}$$

Therefore, the combustion of 1.000 mol $C_6H_{12}O_6$ can be used to produce up to 2862 kJ of work. The maximum work available is greater than the change in internal energy on account of the positive entropy of reaction (which is partly due to the generation of a large number of small molecules from one big one). The system can therefore draw in energy from the surroundings (so reducing their entropy) and make it available for doing work.

Self-test 3.7 Repeat the calculation for the combustion of 1.000 mol $CH_4(g)$ under the same conditions, using data from Table 2.5. [$|q_p| = 890$ kJ, $|w_{max}| = 813$ kJ]

(d) Some remarks on the Gibbs energy

The Gibbs energy (the 'free energy') is more common in chemistry than the Helmholtz energy because, at least in laboratory chemistry, we are usually more interested in changes occurring at constant pressure than at constant volume. The criterion $dG_{T,p} \leq 0$ carries over into chemistry as the observation that, *at constant temperature and pressure, chemical reactions are spontaneous in the direction of decreasing Gibbs energy*. Therefore, if we want to know whether a reaction is spontaneous, the pressure and temperature being constant, we assess the change in the Gibbs energy. If G decreases as the reaction proceeds, then the reaction has a spontaneous tendency to convert the reactants into products. If G increases, then the reverse reaction is spontaneous.

The existence of spontaneous endothermic reactions provides an illustration of the role of G. In such reactions, H increases, the system rises spontaneously to states of higher enthalpy, and $dH > 0$. Because the reaction is spontaneous we know that $dG < 0$ despite $dH > 0$; it follows that the entropy of the system increases so much that TdS outweighs dH in $dG = dH - TdS$. Endothermic reactions are therefore driven by the increase of entropy of the system, and this entropy change overcomes the reduction of entropy brought about in the surroundings by the inflow of heat into the system ($dS_{sur} = -dH/T$ at constant pressure).

(e) Maximum non-expansion work

The analogue of the maximum work interpretation of ΔA, and the origin of the name 'free energy', can be found for ΔG. In the *Justification* below, we show that, at constant temperature and pressure, the maximum additional (non-expansion) work, $w_{\text{add,max}}$, is given by the change in Gibbs energy:

$$\mathrm{d}w_{\text{add,max}} = \mathrm{d}G \tag{3.37}$$

The corresponding expression for a measurable change is

$$w_{\text{add,max}} = \Delta G \tag{3.38}$$

This expression is particularly useful for assessing the electrical work that may be produced by fuel cells and electrochemical cells, and we shall see many applications of it.

Justification 3.3 *Maximum non-expansion work*

Because $H = U + pV$, for a general change in conditions, the change in enthalpy is

$$\mathrm{d}H = \mathrm{d}q + \mathrm{d}w + \mathrm{d}(pV)$$

The corresponding change in Gibbs energy ($G = H - TS$) is

$$\mathrm{d}G = \mathrm{d}H - T\mathrm{d}S - S\mathrm{d}T = \mathrm{d}q + \mathrm{d}w + \mathrm{d}(pV) - T\mathrm{d}S - S\mathrm{d}T$$

When the change is isothermal we can set $\mathrm{d}T = 0$; then

$$\mathrm{d}G = \mathrm{d}q + \mathrm{d}w + \mathrm{d}(pV) - T\mathrm{d}S$$

When the change is reversible, $\mathrm{d}w = \mathrm{d}w_{\text{rev}}$ and $\mathrm{d}q = \mathrm{d}q_{\text{rev}} = T\mathrm{d}S$, so for a reversible, isothermal process

$$\mathrm{d}G = T\mathrm{d}S + \mathrm{d}w_{\text{rev}} + \mathrm{d}(pV) - T\mathrm{d}S = \mathrm{d}w_{\text{rev}} + \mathrm{d}(pV)$$

The work consists of expansion work, which for a reversible change is given by $-p\mathrm{d}V$, and possibly some other kind of work (for instance, the electrical work of pushing electrons through a circuit or of raising a column of liquid); this additional work we denote $\mathrm{d}w_{\text{add}}$. Therefore, with $\mathrm{d}(pV) = p\mathrm{d}V + V\mathrm{d}p$,

$$\mathrm{d}G = (-p\mathrm{d}V + \mathrm{d}w_{\text{add,rev}}) + p\mathrm{d}V + V\mathrm{d}p = \mathrm{d}w_{\text{add,rev}} + V\mathrm{d}p$$

If the change occurs at constant pressure (as well as constant temperature), we can set $\mathrm{d}p = 0$ and obtain $\mathrm{d}G = \mathrm{d}w_{\text{add,rev}}$. Therefore, at constant temperature and pressure, $\mathrm{d}w_{\text{add,rev}} = \mathrm{d}G$. However, because the process is reversible, the work done must now have its maximum value, so eqn 3.37 follows.

Example 3.5 *Calculating the maximum non-expansion work of a reaction*

How much energy is available for sustaining muscular and nervous activity from the combustion of 1.00 mol of glucose molecules under standard conditions at 37°C (blood temperature)? The standard entropy of reaction is $+182.4\ \mathrm{J\,K^{-1}\,mol^{-1}}$.

Method The non-expansion work available from the reaction is equal to the change in standard Gibbs energy for the reaction ($\Delta_r G^{\ominus}$, a quantity defined more fully below). To calculate this quantity, it is legitimate to ignore the temperature-dependence of the reaction enthalpy, to obtain $\Delta_r H^{\ominus}$ from Table 2.5, and to substitute the data into $\Delta_r G^{\ominus} = \Delta_r H^{\ominus} - T\Delta_r S^{\ominus}$.

Answer Because the standard reaction enthalpy is $-2808\ \mathrm{kJ\,mol^{-1}}$, it follows that the standard reaction Gibbs energy is

$$\Delta_r G^{\ominus} = -2808\ \mathrm{kJ\,mol^{-1}} - (310\ \mathrm{K}) \times (182.4\ \mathrm{J\,K^{-1}\,mol^{-1}}) = -2865\ \mathrm{kJ\,mol^{-1}}$$

Therefore, $w_{\text{add,max}} = -2865$ kJ for the combustion of 1 mol glucose molecules, and the reaction can be used to do up to 2865 kJ of non-expansion work. To place this result in perspective, consider that a person of mass 70 kg needs to do 2.1 kJ of work to climb vertically through 3.0 m; therefore, at least 0.13 g of glucose is needed to complete the task (and in practice significantly more).

Self-test 3.8 How much non-expansion work can be obtained from the combustion of 1.00 mol $CH_4(g)$ under standard conditions at 298 K? Use $\Delta_r S^{\ominus} = -243\ \text{J K}^{-1}\ \text{mol}^{-1}$. [818 kJ]

3.6 Standard reaction Gibbs energies

Standard entropies and enthalpies of reaction can be combined to obtain the **standard Gibbs energy of reaction** (or 'standard reaction Gibbs energy'), $\Delta_r G^{\ominus}$:

$$\Delta_r G^{\ominus} = \Delta_r H^{\ominus} - T\Delta_r S^{\ominus} \qquad [3.39]$$

The standard Gibbs energy of reaction is the difference in standard molar Gibbs energies of the products and reactants in their standard states at the temperature specified for the reaction as written. As in the case of standard reaction enthalpies, it is convenient to define the **standard Gibbs energies of formation**, $\Delta_f G^{\ominus}$, the standard reaction Gibbs energy for the formation of a compound from its elements in their reference states.[8] Standard Gibbs energies of formation of the elements in their reference states are zero, because their formation is a 'null' reaction. A selection of values for compounds is given in Table 3.4. From the values there, it is a simple matter to obtain the standard Gibbs energy of reaction by taking the appropriate combination:

$$\Delta_r G^{\ominus} = \sum_{\text{Products}} \nu\Delta_f G^{\ominus} - \sum_{\text{Reactants}} \nu\Delta_f G^{\ominus} \qquad (3.40)$$

with each term weighted by the appropriate stoichiometric coefficient.

Illustration 3.7 *Calculating a standard Gibbs energy of reaction*

To calculate the standard Gibbs energy of the reaction $CO(g) + \frac{1}{2}O_2(g) \rightarrow CO_2(g)$ at 25°C, we write

$$\begin{aligned}\Delta_r G^{\ominus} &= \Delta_f G^{\ominus}(CO_2, g) - \{\Delta_f G^{\ominus}(CO, g) + \tfrac{1}{2}\Delta_f G^{\ominus}(O_2, g)\}\\ &= -394.4\ \text{kJ mol}^{-1} - \{(-137.2) + \tfrac{1}{2}(0)\}\ \text{kJ mol}^{-1}\\ &= -257.2\ \text{kJ mol}^{-1}\end{aligned}$$

Self-test 3.9 Calculate the standard reaction Gibbs energy for the combustion of $CH_4(g)$ at 298 K. [−818 kJ mol^{-1}]

Just as we did in Section 2.8, where we acknowledged that solutions of cations cannot be prepared without their accompanying anions, we define one ion, conventionally the hydrogen ion, to have zero standard Gibbs energy of formation at all temperatures:

$$\Delta_f G^{\ominus}(H^+, aq) = 0 \qquad [3.41]$$

[8] The reference state of an element was defined in Section 2.7.

Synoptic Table 3.4* Standard Gibbs energies of formation (at 298 K)

	$\Delta_f G^{\ominus}/(\text{kJ mol}^{-1})$
Diamond, C(s)	+2.9
Benzene, $C_6H_6(l)$	+124.3
Methane, $CH_4(g)$	−50.7
Carbon dioxide, $CO_2(g)$	−394.4
Water, $H_2O(l)$	−237.1
Ammonia, $NH_3(g)$	−16.5
Sodium chloride, NaCl(s)	−384.1

* More values are given in the *Data section*.

In essence, this definition adjusts the actual values of the Gibbs energies of formation of ions by a fixed amount, which is chosen so that the standard value for one of them, $H^+(aq)$, has the value zero. Then for the reaction

$$\tfrac{1}{2}\,H_2(g) + \tfrac{1}{2}\,Cl_2(g) \rightarrow H^+(aq) + Cl^-(aq) \qquad \Delta_r G^{\ominus} = -131.23\ \text{kJ mol}^{-1}$$

we can write

$$\Delta_r G^{\ominus} = \Delta_f G^{\ominus}(H^+, aq) + \Delta_f G^{\ominus}(Cl^-, aq) = \Delta_f G^{\ominus}(Cl^-, aq)$$

and hence identify $\Delta_f G^{\ominus}(Cl^-, aq)$ as -131.23 kJ mol^{-1}. All the Gibbs energies of formation of ions tabulated in the *Data section* were calculated in the same way.

Illustration 3.8 *Calculating the standard Gibbs energy of formation of an ion*

With the value of $\Delta_f G^{\ominus}(Cl^-, aq)$ established, we can find the value of $\Delta_f G^{\ominus}(Ag^+, aq)$ from

$$Ag(s) + \tfrac{1}{2}\,Cl_2(g) \rightarrow Ag^+(aq) + Cl^-(aq) \qquad \Delta_r G^{\ominus} = -54.12\ \text{kJ mol}^{-1}$$

which leads to $\Delta_f G^{\ominus}(Ag^+, aq) = +77.11$ kJ mol^{-1}.

The factors responsible for the magnitude of the Gibbs energy of formation of an ion in solution can be identified by analysing it in terms of a thermodynamic cycle. As an illustration, we consider the standard Gibbs energies of formation of Cl^- in water, which is -131 kJ mol^{-1}. We do so by treating the formation reaction

$$\tfrac{1}{2}\,H_2(g) + \tfrac{1}{2}\,X_2(g) \rightarrow H^+(aq) + X^-(aq)$$

as the outcome of the sequence of steps shown in Fig. 3.17 (with values taken from the *Data section*). The sum of the Gibbs energies for all the steps around a closed cycle is zero, so

$$\Delta_f G^{\ominus}(Cl^-, aq) = 1272\ \text{kJ mol}^{-1} + \Delta_{solv}G^{\ominus}(H^+) + \Delta_{solv}G^{\ominus}(Cl^-)$$

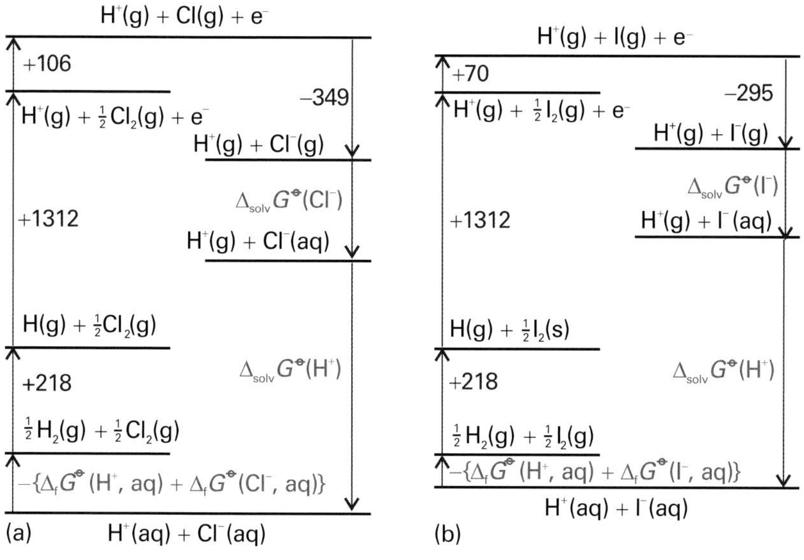

Fig. 3.17 The thermodynamic cycles for the discussion of the Gibbs energies of solvation (hydration) and formation of (a) chloride ions, (b) iodide ions in aqueous solution. The sum of the changes in Gibbs energies around the cycle sum to zero because G is a state function.

Comment 3.2

The standard Gibbs energies of formation of the gas-phase ions are unknown. We have therefore used ionization energies (the energies associated with the removal of electrons from atoms or cations in the gas phase) or electron affinities (the energies associated with the uptake of electrons by atoms or anions in the gas phase) and have assumed that any differences from the Gibbs energies arising from conversion to enthalpy and the inclusion of entropies to obtain Gibbs energies in the formation of H^+ are cancelled by the corresponding terms in the electron gain of X. The conclusions from the cycles are therefore only approximate.

An important point to note is that the value of $\Delta_f G^\ominus$ of an ion X is not determined by the properties of X alone but includes contributions from the dissociation, ionization, and hydration of hydrogen.

Gibbs energies of solvation of individual ions may be estimated from an equation derived by Max Born, who identified $\Delta_{solv}G^\ominus$ with the electrical work of transferring an ion from a vacuum into the solvent treated as a continuous dielectric of relative permittivity ε_r. The resulting **Born equation**, which is derived in *Further information 3.1*, is

$$\Delta_{solv}G^\ominus = -\frac{z_i^2 e^2 N_A}{8\pi\varepsilon_0 r_i}\left(1 - \frac{1}{\varepsilon_r}\right) \qquad (3.42a)$$

where z_i is the charge number of the ion and r_i its radius (N_A is Avogadro's constant). Note that $\Delta_{solv}G^\ominus < 0$, and that $\Delta_{solv}G^\ominus$ is strongly negative for small, highly charged ions in media of high relative permittivity. For water at 25°C,

$$\Delta_{solv}G^\ominus = -\frac{z_i^2}{(r_i/\text{pm})} \times (6.86 \times 10^4 \text{ kJ mol}^{-1}) \qquad (3.42b)$$

Illustration 3.9 *Using the Born equation*

To see how closely the Born equation reproduces the experimental data, we calculate the difference in the values of $\Delta_f G^\ominus$ for Cl^- and I^- in water, for which $\varepsilon_r = 78.54$ at 25°C, given their radii as 181 pm and 220 pm (Table 20.3), respectively, is

$$\Delta_{solv}G^\ominus(Cl^-) - \Delta_{solv}G^\ominus(I^-) = -\left(\frac{1}{181} - \frac{1}{220}\right) \times (6.86 \times 10^4 \text{ kJ mol}^{-1})$$

$$= -67 \text{ kJ mol}^{-1}$$

This estimated difference is in good agreement with the experimental difference, which is -61 kJ mol^{-1}.

Self-test 3.10 Estimate the value of $\Delta_{solv}G^\ominus(Cl^-, aq) - \Delta_{solv}G^\ominus(Br^-, aq)$ from experimental data and from the Born equation.

[-26 kJ mol^{-1} experimental; -29 kJ mol^{-1} calculated]

Comment 3.3
The *NIST WebBook* is a good source of links to online databases of thermochemical data.

Calorimetry (for ΔH directly, and for S via heat capacities) is only one of the ways of determining Gibbs energies. They may also be obtained from equilibrium constants and electrochemical measurements (Chapter 7), and for gases they may be calculated using data from spectroscopic observations (Chapter 17).

Combining the First and Second Laws

The First and Second Laws of thermodynamics are both relevant to the behaviour of matter, and we can bring the whole force of thermodynamics to bear on a problem by setting up a formulation that combines them.

3.7 The fundamental equation

We have seen that the First Law of thermodynamics may be written $dU = dq + dw$. For a reversible change in a closed system of constant composition, and in the absence of any additional (non-expansion) work, we may set $\mathrm{d}w_{\mathrm{rev}} = -p\mathrm{d}V$ and (from the definition of entropy) $\mathrm{d}q_{\mathrm{rev}} = T\mathrm{d}S$, where p is the pressure of the system and T its temperature. Therefore, for a reversible change in a closed system,

$$\mathrm{d}U = T\mathrm{d}S - p\mathrm{d}V \tag{3.43}$$

However, because $\mathrm{d}U$ is an exact differential, its value is independent of path. Therefore, the same value of $\mathrm{d}U$ is obtained whether the change is brought about irreversibly or reversibly. Consequently, *eqn 3.43 applies to any change—reversible or irreversible—of a closed system that does no additional (non-expansion) work*. We shall call this combination of the First and Second Laws the **fundamental equation**.

The fact that the fundamental equation applies to both reversible and irreversible changes may be puzzling at first sight. The reason is that only in the case of a reversible change may $T\mathrm{d}S$ be identified with $\mathrm{d}q$ and $-p\mathrm{d}V$ with $\mathrm{d}w$. When the change is irreversible, $T\mathrm{d}S > \mathrm{d}q$ (the Clausius inequality) and $-p\mathrm{d}V > \mathrm{d}w$. The sum of $\mathrm{d}w$ and $\mathrm{d}q$ remains equal to the sum of $T\mathrm{d}S$ and $-p\mathrm{d}V$, provided the composition is constant.

3.8 Properties of the internal energy

Equation 3.43 shows that the internal energy of a closed system changes in a simple way when either S or V is changed ($\mathrm{d}U \propto \mathrm{d}S$ and $\mathrm{d}U \propto \mathrm{d}V$). These simple proportionalities suggest that U should be regarded as a function of S and V. We could regard U as a function of other variables, such as S and p or T and V, because they are all interrelated; but the simplicity of the fundamental equation suggests that $U(S,V)$ is the best choice.

The *mathematical* consequence of U being a function of S and V is that we can express an infinitesimal change $\mathrm{d}U$ in terms of changes $\mathrm{d}S$ and $\mathrm{d}V$ by

$$\mathrm{d}U = \left(\frac{\partial U}{\partial S}\right)_V \mathrm{d}S + \left(\frac{\partial U}{\partial V}\right)_S \mathrm{d}V \tag{3.44}$$

Comment 3.4

Partial derivatives were introduced in *Comment 2.5* and are reviewed in *Appendix 2*. The type of result in eqn 3.44 was first obtained in Section 2.11, where we treated U as a function of T and V.

The two partial derivatives are the slopes of the plots of U against S and V, respectively. When this expression is compared to the *thermodynamic* relation, eqn 3.43, we see that, for systems of constant composition,

$$\left(\frac{\partial U}{\partial S}\right)_V = T \qquad \left(\frac{\partial U}{\partial V}\right)_S = -p \tag{3.45}$$

The first of these two equations is a purely thermodynamic definition of temperature (a Zeroth-Law concept) as the ratio of the changes in the internal energy (a First-Law concept) and entropy (a Second-Law concept) of a constant-volume, closed, constant-composition system. We are beginning to generate relations between the properties of a system and to discover the power of thermodynamics for establishing unexpected relations.

(a) The Maxwell relations

An infinitesimal change in a function $f(x,y)$ can be written $\mathrm{d}f = g\mathrm{d}x + h\mathrm{d}y$ where g and h are functions of x and y. The mathematical criterion for $\mathrm{d}f$ being an exact differential (in the sense that its integral is independent of path) is that

$$\left(\frac{\partial g}{\partial y}\right)_x = \left(\frac{\partial h}{\partial x}\right)_y \tag{3.46}$$

Comment 3.5

To illustrate the criterion set by eqn 3.46, let's test whether $\mathrm{d}f = 2xy\mathrm{d}x + x^2\mathrm{d}y$ is an exact differential. We identify $g = 2xy$ and $h = x^2$ and form

$$\left(\frac{\partial g}{\partial y}\right)_x = \left(\frac{\partial (2xy)}{\partial y}\right)_x = 2x$$

$$\left(\frac{\partial h}{\partial x}\right)_y = \left(\frac{\partial x^2}{\partial x}\right)_y = 2x$$

Because these two coefficients are equal, $\mathrm{d}f$ is exact.

Because the fundamental equation, eqn 3.43, is an expression for an exact differential, the functions multiplying $\mathrm{d}S$ and $\mathrm{d}V$ (namely T and $-p$) must pass this test. Therefore, it must be the case that

Table 3.5 The Maxwell relations

From U:	$\left(\frac{\partial T}{\partial V}\right)_S = -\left(\frac{\partial p}{\partial S}\right)_V$
From H:	$\left(\frac{\partial T}{\partial p}\right)_S = \left(\frac{\partial V}{\partial S}\right)_p$
From A:	$\left(\frac{\partial p}{\partial T}\right)_V = \left(\frac{\partial S}{\partial V}\right)_T$
From G:	$\left(\frac{\partial V}{\partial T}\right)_p = -\left(\frac{\partial S}{\partial p}\right)_T$

$$\left(\frac{\partial T}{\partial V}\right)_S = -\left(\frac{\partial p}{\partial S}\right)_V \tag{3.47}$$

We have generated a relation between quantities that, at first sight, would not seem to be related.

Equation 3.47 is an example of a **Maxwell relation**. However, apart from being unexpected, it does not look particularly interesting. Nevertheless, it does suggest that there may be other similar relations that are more useful. Indeed, we can use the fact that H, G, and A are all state functions to derive three more Maxwell relations. The argument to obtain them runs in the same way in each case: because H, G, and A are state functions, the expressions for dH, dG, and dA satisfy relations like eqn 3.47. All four relations are listed in Table 3.5 and we put them to work later in the chapter.

(b) The variation of internal energy with volume

The quantity $\pi_T = (\partial U/\partial V)_T$, which represents how the internal energy changes as the volume of a system is changed isothermally, played a central role in the manipulation of the First Law, and in *Further information 2.2* we used the relation

$$\pi_T = T\left(\frac{\partial p}{\partial T}\right)_V - p \tag{3.48}$$

This relation is called a **thermodynamic equation of state** because it is an expression for pressure in terms of a variety of thermodynamic properties of the system. We are now ready to derive it by using a Maxwell relation.

Justification 3.4 *The thermodynamic equation of state*

We obtain an expression for the coefficient π_T by dividing both sides of eqn 3.43 by dV, imposing the constraint of constant temperature, which gives

$$\left(\frac{\partial U}{\partial V}\right)_T = \left(\frac{\partial U}{\partial S}\right)_V\left(\frac{\partial S}{\partial V}\right)_T + \left(\frac{\partial U}{\partial V}\right)_S$$

Next, we introduce the two relations in eqn 3.45 and the definition of π_T to obtain

$$\pi_T = T\left(\frac{\partial S}{\partial V}\right)_T - p$$

The third Maxwell relation in Table 3.5 turns $(\partial S/\partial V)_T$ into $(\partial p/\partial T)_V$, which completes the proof of eqn 3.48.

Example 3.6 *Deriving a thermodynamic relation*

Show thermodynamically that $\pi_T = 0$ for a perfect gas, and compute its value for a van der Waals gas.

Method Proving a result 'thermodynamically' means basing it entirely on general thermodynamic relations and equations of state, without drawing on molecular arguments (such as the existence of intermolecular forces). We know that for a perfect gas, $p = nRT/V$, so this relation should be used in eqn 3.48. Similarly, the van der Waals equation is given in Table 1.7, and for the second part of the question it should be used in eqn 3.48.

Answer For a perfect gas we write

$$\left(\frac{\partial p}{\partial T}\right)_V = \left(\frac{\partial(nRT/V)}{\partial T}\right)_V = \frac{nR}{V}$$

Then, eqn 3.48 becomes

$$\pi_T = \frac{nRT}{V} - p = 0$$

The equation of state of a van der Waals gas is

$$p = \frac{nRT}{V - nb} - a\frac{n^2}{V^2}$$

Because a and b are independent of temperature,

$$\left(\frac{\partial p}{\partial T}\right)_V = \frac{nR}{V - nb}$$

Therefore, from eqn 3.48,

$$\pi_T = \frac{nRT}{V - nb} - \frac{nRT}{V - nb} + a\frac{n^2}{V^2} = a\frac{n^2}{V^2}$$

This result for π_T implies that the internal energy of a van der Waals gas increases when it expands isothermally (that is, $(\partial U/\partial V)_T > 0$), and that the increase is related to the parameter a, which models the attractive interactions between the particles. A larger molar volume, corresponding to a greater average separation between molecules, implies weaker mean intermolecular attractions, so the total energy is greater.

Self-test 3.11 Calculate π_T for a gas that obeys the virial equation of state (Table 1.7). $[\pi_T = RT^2(\partial B/\partial T)_V/V_m^2 + \cdots]$

3.9 Properties of the Gibbs energy

The same arguments that we have used for U can be used for the Gibbs energy $G = H - TS$. They lead to expressions showing how G varies with pressure and temperature that are important for discussing phase transitions and chemical reactions.

(a) General considerations

When the system undergoes a change of state, G may change because H, T, and S all change. As in *Justification 2.1*, we write for infinitesimal changes in each property

$$dG = dH - d(TS) = dH - TdS - SdT$$

Because $H = U + pV$, we know that

$$dH = dU + d(pV) = dU + pdV + Vdp$$

and therefore

$$dG = dU + pdV + Vdp - TdS - SdT$$

For a closed system doing no non-expansion work, we can replace dU by the fundamental equation $dU = TdS - pdV$ and obtain

$$dG = TdS - pdV + pdV + Vdp - TdS - SdT$$

Four terms now cancel on the right, and we conclude that, for a closed system in the absence of non-expansion work and at constant composition,

$$dG = Vdp - SdT \tag{3.49}$$

This expression, which shows that a change in G is proportional to a change in p or T, suggests that G may be best regarded as a function of p and T. It confirms that G is an important quantity in chemistry because the pressure and temperature are usually the variables under our control. In other words, G carries around the combined consequences of the First and Second Laws in a way that makes it particularly suitable for chemical applications.

The same argument that led to eqn 3.45, when applied to the exact differential $dG = Vdp - SdT$, now gives

$$\left(\frac{\partial G}{\partial T}\right)_p = -S \qquad \left(\frac{\partial G}{\partial p}\right)_T = V \tag{3.50}$$

These relations show how the Gibbs energy varies with temperature and pressure (Fig. 3.18). The first implies that:

- Because $S > 0$ for all substances, G always *decreases* when the temperature is raised (at constant pressure and composition).
- Because $(\partial G/\partial T)_p$ becomes more negative as S increases, G decreases most sharply when the entropy of the system is large.

Therefore, the Gibbs energy of the gaseous phase of a substance, which has a high molar entropy, is more sensitive to temperature than its liquid and solid phases (Fig. 3.19). Similarly, the second relation implies that:

- Because $V > 0$ for all substances, G always *increases* when the pressure of the system is increased (at constant temperature and composition).

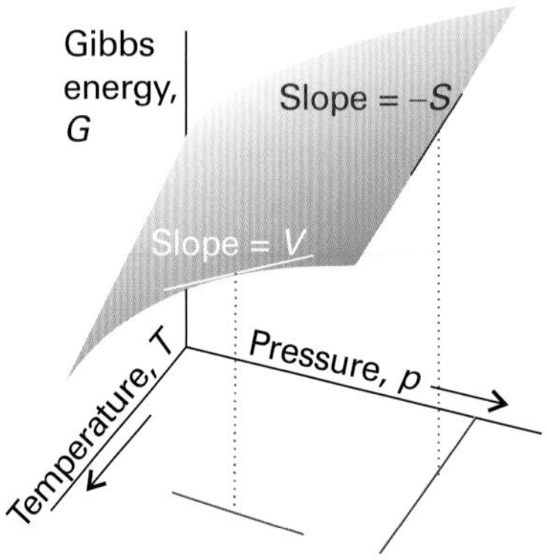

Fig. 3.18 The variation of the Gibbs energy of a system with (a) temperature at constant pressure and (b) pressure at constant temperature. The slope of the former is equal to the negative of the entropy of the system and that of the latter is equal to the volume.

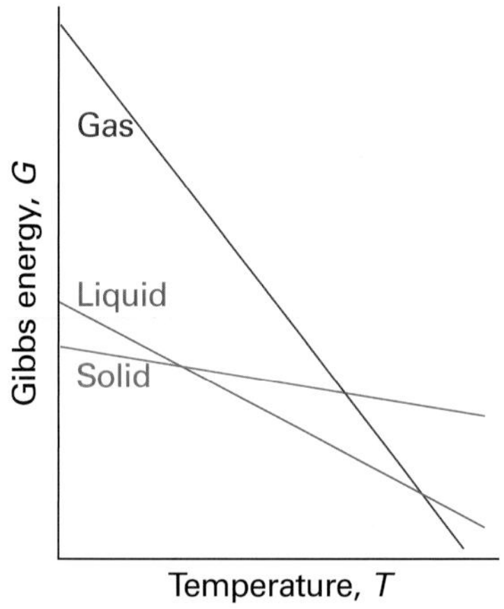

Fig. 3.19 The variation of the Gibbs energy with the temperature is determined by the entropy. Because the entropy of the gaseous phase of a substance is greater than that of the liquid phase, and the entropy of the solid phase is smallest, the Gibbs energy changes most steeply for the gas phase, followed by the liquid phase, and then the solid phase of the substance.

• Because $(\partial G/\partial p)_T$ increases with V, G is more sensitive to pressure when the volume of the system is large.

Because the molar volume of the gaseous phase of a substance is greater than that of its condensed phases, the molar Gibbs energy of a gas is more sensitive to pressure than its liquid and solid phases (Fig. 3.20).

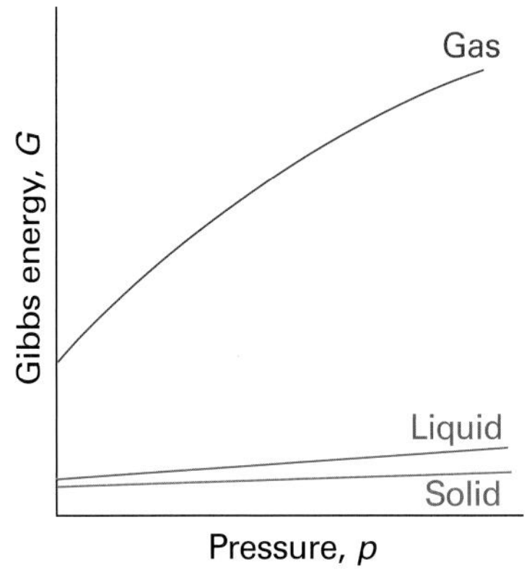

Fig. 3.20 The variation of the Gibbs energy with the pressure is determined by the volume of the sample. Because the volume of the gaseous phase of a substance is greater than that of the same amount of liquid phase, and the entropy of the solid phase is smallest (for most substances), the Gibbs energy changes most steeply for the gas phase, followed by the liquid phase, and then the solid phase of the substance. Because the volumes of the solid and liquid phases of a substance are similar, their molar Gibbs energies vary by similar amounts as the pressure is changed.

(b) The variation of the Gibbs energy with temperature

As we remarked in the introduction, because the equilibrium composition of a system depends on the Gibbs energy, to discuss the response of the composition to temperature we need to know how G varies with temperature.

The first relation in eqn 3.50, $(\partial G/\partial T)_p = -S$, is our starting point for this discussion. Although it expresses the variation of G in terms of the entropy, we can express it in terms of the enthalpy by using the definition of G to write $S = (H - G)/T$. Then

$$\left(\frac{\partial G}{\partial T}\right)_p = \frac{G-H}{T} \qquad (3.51)$$

We shall see later that the equilibrium constant of a reaction is related to G/T rather than to G itself,[9] and it is easy to deduce from the last equation (see the *Justification* below) that

$$\left(\frac{\partial}{\partial T}\frac{G}{T}\right)_p = -\frac{H}{T^2} \qquad (3.52)$$

This expression is called the **Gibbs–Helmholtz equation**. It shows that if we know the enthalpy of the system, then we know how G/T varies with temperature.

Justification 3.5 *The Gibbs–Helmholtz equation*

First, we note that

$$\left(\frac{\partial}{\partial T}\frac{G}{T}\right)_p = \frac{1}{T}\left(\frac{\partial G}{\partial T}\right)_p + G\frac{\mathrm{d}}{\mathrm{d}T}\frac{1}{T} = \frac{1}{T}\left(\frac{\partial G}{\partial T}\right)_p - \frac{G}{T^2} = \frac{1}{T}\left\{\left(\frac{\partial G}{\partial T}\right)_p - \frac{G}{T}\right\}$$

Then we use eqn 3.51 in the form

$$\left(\frac{\partial G}{\partial T}\right)_p - \frac{G}{T} = -\frac{H}{T}$$

It follows that

$$\left(\frac{\partial}{\partial T}\frac{G}{T}\right)_p = \frac{1}{T}\left\{-\frac{H}{T}\right\} = -\frac{H}{T^2}$$

which is eqn 3.52.

Comment 3.6

For this step, we use the rule for differentiating a product of functions (which is valid for partial derivatives as well as ordinary derivatives):

$$\frac{\mathrm{d}uv}{\mathrm{d}x} = u\frac{\mathrm{d}v}{\mathrm{d}x} + v\frac{\mathrm{d}u}{\mathrm{d}x}$$

For instance, to differentiate x^2e^{ax}, we write

$$\frac{\mathrm{d}(\overbrace{x^2}^{u}\overbrace{e^{ax}}^{v})}{\mathrm{d}x} = x^2\frac{\mathrm{d}e^{ax}}{\mathrm{d}x} + e^{ax}\frac{\mathrm{d}x^2}{\mathrm{d}x} = ax^2e^{ax} + 2xe^{ax}$$

The Gibbs–Helmholtz equation is most useful when it is applied to changes, including changes of physical state and chemical reactions at constant pressure. Then, because $\Delta G = G_f - G_i$ for the change of Gibbs energy between the final and initial states and because the equation applies to both G_f and G_i, we can write

[9] In Section 7.2b we derive the result that the equilibrium constant for a reaction is related to its standard reaction Gibbs energy by $\Delta_r G^\ominus/T = -R \ln K$.

$$\left(\frac{\partial}{\partial T}\frac{\Delta G}{T}\right)_p = -\frac{\Delta H}{T^2} \quad (3.53)$$

This equation shows that, if we know the change in enthalpy of a system that is undergoing some kind of transformation (such as vaporization or reaction), then we know how the corresponding change in Gibbs energy varies with temperature. As we shall see, this is a crucial piece of information in chemistry.

(c) The variation of the Gibbs energy with pressure

To find the Gibbs energy at one pressure in terms of its value at another pressure, the temperature being constant, we set $dT = 0$ in eqn 3.49, which gives $dG = Vdp$, and integrate:

$$G(p_f) = G(p_i) + \int_{p_i}^{p_f} V\,dp \quad (3.54a)$$

For molar quantities,

$$G_m(p_f) = G_m(p_i) + \int_{p_i}^{p_f} V_m\,dp \quad (3.54b)$$

This expression is applicable to any phase of matter, but to evaluate it we need to know how the molar volume, V_m, depends on the pressure.

The molar volume of a condensed phase changes only slightly as the pressure changes (Fig. 3.21), so we can treat V_m as a constant and take it outside the integral:

$$G_m(p_f) = G_m(p_i) + V_m\int_{p_i}^{p_f} dp = G_m(p_i) + (p_f - p_i)V_m \quad (3.55)$$

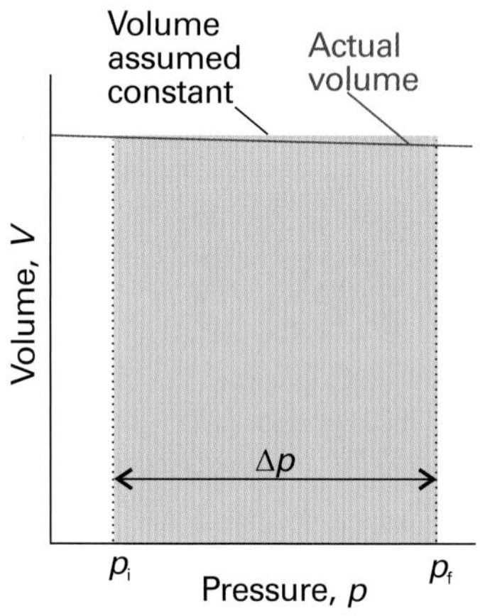

Fig. 3.21 The difference in Gibbs energy of a solid or liquid at two pressures is equal to the rectangular area shown. We have assumed that the variation of volume with pressure is negligible.

Self-test 3.12 Calculate the change in G_m for ice at −10°C, with density 917 kg m^{-3}, when the pressure is increased from 1.0 bar to 2.0 bar. [+2.0 J mol^{-1}]

Under normal laboratory conditions $(p_f - p_i)V_m$ is very small and may be neglected. Hence, we may usually suppose that the Gibbs energies of solids and liquids are independent of pressure. However, if we are interested in geophysical problems, then because pressures in the Earth's interior are huge, their effect on the Gibbs energy cannot be ignored. If the pressures are so great that there are substantial volume changes over the range of integration, then we must use the complete expression, eqn 3.54.

Illustration 3.10 *Gibbs energies at high pressures*

Suppose that for a certain phase transition of a solid $\Delta_{trs}V = +1.0$ cm^3 mol^{-1} independent of pressure. Then, for an increase in pressure to 3.0 Mbar (3.0×10^{11} Pa) from 1.0 bar (1.0×10^5 Pa), the Gibbs energy of the transition changes from $\Delta_{trs}G$(1 bar) to

$$\begin{aligned}\Delta_{trs}G(3\,\text{Mbar}) &= \Delta_{trs}G(1\,\text{bar}) + (1.0\times10^{-6}\,\text{m}^3\,\text{mol}^{-1})\times(3.0\times10^{11}\,\text{Pa} - 1.0\times10^5\,\text{Pa})\\ &= \Delta_{trs}G(1\,\text{bar}) + 3.0\times10^2\,\text{kJ mol}^{-1}\end{aligned}$$

where we have used 1 Pa m^3 = 1 J.

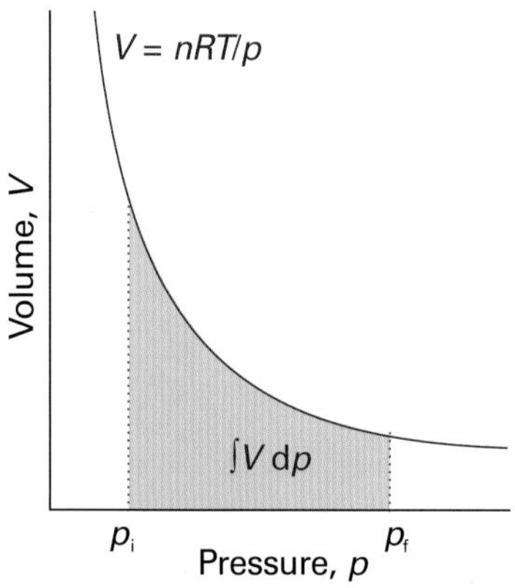

Fig. 3.22 The difference in Gibbs energy for a perfect gas at two pressures is equal to the area shown below the perfect-gas isotherm.

The molar volumes of gases are large, so the Gibbs energy of a gas depends strongly on the pressure. Furthermore, because the volume also varies markedly with the pressure, we cannot treat it as a constant in the integral in eqn 3.54b (Fig. 3.22).

For a perfect gas we substitute $V_m = RT/p$ into the integral, treat RT as a constant, and find

$$G_m(p_f) = G_m(p_i) + RT\int_{p_i}^{p_f}\frac{dp}{p} = G_m(p_i) + RT\ln\frac{p_f}{p_i} \qquad (3.56)°$$

This expression shows that when the pressure is increased tenfold at room temperature, the molar Gibbs energy increases by $RT \ln 10 \approx 6$ kJ mol^{-1}. It also follows from this equation that, if we set $p_i = p^{\ominus}$ (the standard pressure of 1 bar), then the molar Gibbs energy of a perfect gas at a pressure p (set $p_f = p$) is related to its standard value by

$$G_m(p) = G_m^{\ominus} + RT\ln\frac{p}{p^{\ominus}} \qquad (3.57)°$$

Self-test 3.13 Calculate the change in the molar Gibbs energy of water vapour (treated as a perfect gas) when the pressure is increased isothermally from 1.0 bar to 2.0 bar at 298 K. Note that, whereas the change in molar Gibbs energy for a condensed phase (Self-test 3.12) is a few joules per mole, the answer you should get for a gas is of the order of kilojoules per mole. [+1.7 kJ mol^{-1}]

The logarithmic dependence of the molar Gibbs energy on the pressure predicted by eqn 3.57 is illustrated in Fig. 3.23. This very important expression, the consequences of which we unfold in the following chapters, applies to perfect gases (which is usually a good enough approximation). *Further information 3.2* describes how to take into account gas imperfections.

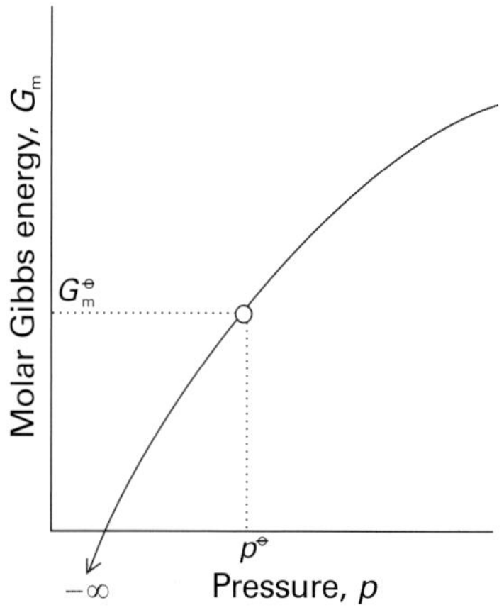

Fig. 3.23 The molar Gibbs energy of a perfect gas is proportional to ln p, and the standard state is reached at $p^{\ominus}$. Note that, as $p \to 0$, the molar Gibbs energy becomes negatively infinite.

Exploration Show how the first derivative of G, $(\partial G/\partial p)_T$, varies with pressure, and plot the resulting expression over a pressure range. What is the physical significance of $(\partial G/\partial p)_T$?

Checklist of key ideas

- ☐ 1. Kelvin statement of the Second Law of thermodynamics: No process is possible in which the sole result is the absorption of heat from a reservoir and its complete conversion into work.
- ☐ 2. The Second Law in terms of entropy: The entropy of an isolated system increases in the course of a spontaneous change: $\Delta S_{tot} > 0$.
- ☐ 3. The thermodynamic definition of entropy is $dS = dq_{rev}/T$. The statistical definition of entropy is given by the Boltzmann formula, $S = k \ln W$.
- ☐ 4. A Carnot cycle is a cycle composed of a sequence of isothermal and adiabatic reversible expansions and compressions.
- ☐ 5. The efficiency of a heat engine is $\varepsilon = |w|/q_h$. The Carnot efficiency is $\varepsilon_{rev} = 1 - T_c/T_h$.
- ☐ 6. The Kelvin scale is a thermodynamic temperature scale in which the triple point of water defines the point 273.16 K.
- ☐ 7. The Clausius inequality is $dS \geq dq/T$.
- ☐ 8. The normal transition temperature, T_{trs}, is the temperature at which two phases are in equilibrium at 1 atm. The entropy of transition at the transition temperature, $\Delta_{trs}S = \Delta_{trs}H/T_{trs}$.
- ☐ 9. Trouton's rule states that many normal liquids have approximately the same standard entropy of vaporization (about 85 J K^{-1} mol^{-1}).
- ☐ 10. The variation of entropy with temperature is given by $$S(T_f) = S(T_i) + \int_{T_i}^{T_f}(C_p/T)dT.$$
- ☐ 11. The entropy of a substance is measured from the area under a graph of C_p/T against T, using the Debye extrapolation at low temperatures, $C_p = aT^3$ as $T \to 0$.
- ☐ 12. The Nernst heat theorem states that the entropy change accompanying any physical or chemical transformation approaches zero as the temperature approaches zero: $\Delta S \to 0$ as $T \to 0$ provided all the substances involved are perfectly ordered.
- ☐ 13. Third Law of thermodynamics: The entropy of all perfect crystalline substances is zero at $T = 0$.
- ☐ 14. The standard reaction entropy is calculated from $\Delta_r S^{\ominus} = \sum_{Products}\nu S_m^{\ominus} - \sum_{Reactants}\nu S_m^{\ominus}$.
- ☐ 15. The standard molar entropies of ions in solution are reported on a scale in which $S^{\ominus}(H^+, aq) = 0$ at all temperatures.

☐ 16. The Helmholtz energy is $A = U - TS$. The Gibbs energy is $G = H - TS$.

☐ 17. The criteria of spontaneity may be written as: (a) $dS_{U,V} \geq 0$ and $dU_{S,V} \leq 0$, or (b) $dA_{T,V} \leq 0$ and $dG_{T,p} \leq 0$.

☐ 18. The criterion of equilibrium at constant temperature and volume, $dA_{T,V} = 0$. The criterion of equilibrium at constant temperature and pressure, $dG_{T,p} = 0$.

☐ 19. The maximum work and the Helmholtz energy are related by $w_{max} = \Delta A$. The maximum additional (non-expansion) work and the Gibbs energy are related by $w_{add,max} = \Delta G$.

☐ 20. The standard Gibbs energy of reaction is given by $\Delta_r G^\ominus = \Delta_r H^\ominus - T\Delta_r S^\ominus = \sum_{Products} \nu G^\ominus_m - \sum_{Reactants} \nu G^\ominus_m$.

☐ 21. The standard Gibbs energy of formation ($\Delta_f G^\ominus$) is the standard reaction Gibbs energy for the formation of a compound from its elements in their reference states.

☐ 22. The standard Gibbs energy of reaction may be expressed in terms of $\Delta_f G^\ominus$, $\Delta_r G^\ominus = \sum_{Products} \nu \Delta_f G^\ominus - \sum_{Reactants} \nu \Delta_f G^\ominus$.

☐ 23. The standard Gibbs energies of formation of ions are reported on a scale in which $\Delta_f G^\ominus(H^+, aq) = 0$ at all temperatures.

☐ 24. The fundamental equation is $dU = TdS - pdV$.

☐ 25. The Maxwell relations are listed in Table 3.5.

☐ 26. A thermodynamic equation of state is an expression for pressure in terms of thermodynamic quantities, $\pi_T = T(\partial p/\partial T)_V - p$.

☐ 27. The Gibbs energy is best described as a function of pressure and temperature, $dG = Vdp - SdT$. The variation of Gibbs energy with pressure and temperature are, respectively, $(\partial G/\partial p)_T = V$ and $(\partial G/\partial T)_p = -S$.

☐ 28. The temperature dependence of the Gibbs energy is given by the Gibbs–Helmholtz equation, $(\partial(G/T)/\partial T)_p = -H/T^2$.

☐ 29. For a condensed phase, the Gibbs energy varies with pressure as $G(p_f) = G(p_i) + V_m\Delta p$. For a perfect gas, $G(p_f) = G(p_i) + nRT\ln(p_f/p_i)$.

Further reading[10]

Articles and texts

N.C. Craig, Entropy analyses of four familiar processes. *J. Chem. Educ.* **65**, 760 (1988).

J.B. Fenn, *Engines, energy, and entropy*. W.H. Freeman and Co., New York (1982).

F.J. Hale, Heat engines and refrigerators. In *Encyclopedia of applied physics* (ed. G.L. Trigg), **7**, 303. VCH, New York (1993).

D. Kondepudi and I. Prigogine, *Modern thermodynamics: from heat engines to dissipative structures.* Wiley, New York (1998).

P.G. Nelson, Derivation of the Second Law of thermodynamics from Boltzmann's distribution law. *J. Chem. Educ.* **65**, 390 (1988).

Sources of data and information

M.W. Chase, Jr. (ed.), *NIST–JANAF thermochemical tables*. Published as *J. Phys. Chem. Ref. Data, Monograph no. 9*. American Institute of Physics, New York (1998).

R.C. Weast (ed.), *Handbook of chemistry and physics*, Vol. 81. CRC Press, Boca Raton (2004).

Further information

Further information 3.1 *The Born equation*

The electrical concepts required in this derivation are reviewed in *Appendix 3*. The strategy of the calculation is to identify the Gibbs energy of solvation with the work of transferring an ion from a vacuum into the solvent. That work is calculated by taking the difference of the work of charging an ion when it is in the solution and the work of charging the same ion when it is in a vacuum.

The Coulomb interaction between two charges q_1 and q_2 separated by a distance r is described by the *Coulombic potential energy*:

$$V = \frac{q_1 q_2}{4\pi\varepsilon r}$$

where ε is the medium's permittivity. The permittivity of vacuum is $\varepsilon_0 = 8.854 \times 10^{-12}\ J^{-1}\ C^2\ m^{-1}$. The relative permittivity (formerly called the 'dielectric constant') of a substance is defined as $\varepsilon_r = \varepsilon/\varepsilon_0$. Ions do not interact as strongly in a solvent of high relative permittivity (such as water, with $\varepsilon_r = 80$ at 293 K) as they do in a solvent of lower relative permittivity (such as ethanol, with $\varepsilon_r = 25$ at 293 K). See Chapter 18 for more details. The potential energy of a charge q_1 in the presence of a charge q_2 can be expressed in terms of the *Coulomb potential*, ϕ:

$$V = q_1\phi \qquad \phi = \frac{q_2}{4\pi\varepsilon r}$$

We model an ion as a sphere of radius r_i immersed in a medium of permittivity ε. It turns out that, when the charge of the sphere is q, the electric potential, ϕ, at its surface is the same as the potential due to a point charge at its centre, so we can use the last expression and write

[10] See *Further reading* in Chapter 2 for additional articles, texts, and sources of thermochemical data.

$$\phi = \frac{q}{4\pi\varepsilon r_i}$$

The work of bringing up a charge dq to the sphere is ϕdq. Therefore, the total work of charging the sphere from 0 to $z_i e$ is

$$w = \int_0^{z_i e} \phi \, dq = \frac{1}{4\pi\varepsilon r_i}\int_0^{z_i e} q \, dq = \frac{z_i^2 e^2}{8\pi\varepsilon r_i}$$

This electrical work of charging, when multiplied by Avogadro's constant, is the molar Gibbs energy for charging the ions.

The work of charging an ion in a vacuum is obtained by setting $\varepsilon = \varepsilon_0$, the vacuum permittivity. The corresponding value for charging the ion in a medium is obtained by setting $\varepsilon = \varepsilon_r \varepsilon_0$, where ε_r is the relative permittivity of the medium. It follows that the change in molar Gibbs energy that accompanies the transfer of ions from a vacuum to a solvent is the difference of these two quantities:

$$\Delta_{solv}G^{\ominus} = \frac{z_i^2 e^2 N_A}{8\pi\varepsilon r_i} - \frac{z_i^2 e^2 N_A}{8\pi\varepsilon_0 r_i} = \frac{z_i^2 e^2 N_A}{8\pi\varepsilon_r\varepsilon_0 r_i} - \frac{z_i^2 e^2 N_A}{8\pi\varepsilon_0 r_i} = -\frac{z_i^2 e^2 N_A}{8\pi\varepsilon_0 r_i}\left(1 - \frac{1}{\varepsilon_r}\right)$$

which is eqn 3.42.

Further information 3.2 *Real gases: the fugacity*

At various stages in the development of physical chemistry it is necessary to switch from a consideration of idealized systems to real systems. In many cases it is desirable to preserve the form of the expressions that have been derived for an idealized system. Then deviations from the idealized behaviour can be expressed most simply. For instance, the pressure-dependence of the molar Gibbs energy of a real gas might resemble that shown in Fig. 3.24. To adapt eqn 3.57 to this case, we replace the true pressure, p, by an effective pressure, called the **fugacity**,[11] f, and write

$$G_m = G_m^{\ominus} + RT \ln \frac{f}{p^{\ominus}} \qquad [3.58]$$

The fugacity, a function of the pressure and temperature, is defined so that this relation is exactly true. Although thermodynamic expressions in terms of fugacities derived from this expression are exact, they are useful only if we know how to interpret fugacities in terms of actual pressures. To develop this relation we write the fugacity as

$$f = \phi p \qquad [3.59]$$

where ϕ is the dimensionless **fugacity coefficient**, which in general depends on the temperature, the pressure, and the identity of the gas.

Equation 3.54b is true for all gases whether real or perfect. Expressing it in terms of the fugacity by using eqn 3.58 turns it into

$$\int_{p'}^{p} V_m dp = G_m(p) - G_m(p') = \left\{G_m^{\ominus} + RT\ln\frac{f}{p^{\ominus}}\right\} - \left\{G_m^{\ominus} + RT\ln\frac{f'}{p^{\ominus}}\right\}$$
$$= RT\ln\frac{f}{f'}$$

In this expression, f is the fugacity when the pressure is p and f' is the fugacity when the pressure is p'. If the gas were perfect, we would write

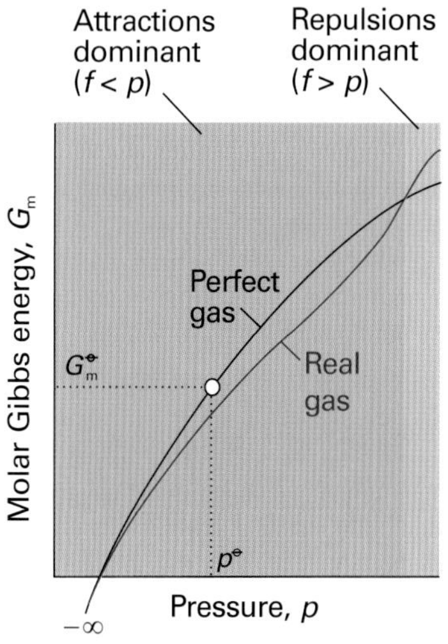

Fig. 3.24 The molar Gibbs energy of a real gas. As $p \to 0$, the molar Gibbs energy coincides with the value for a perfect gas (shown by the black line). When attractive forces are dominant (at intermediate pressures), the molar Gibbs energy is less than that of a perfect gas and the molecules have a lower 'escaping tendency'. At high pressures, when repulsive forces are dominant, the molar Gibbs energy of a real gas is greater than that of a perfect gas. Then the 'escaping tendency' is increased.

$$\int_{p'}^{p} V_{perfect,\,m} dp = RT\ln\frac{p}{p'}$$

The difference between the two equations is

$$\int_{p'}^{p}(V_m - V_{perfect,\,m})dp = RT\left(\ln\frac{f}{f'} - \ln\frac{p}{p'}\right) = RT\ln\left(\frac{f/f'}{p/p'}\right)$$
$$= RT\ln\left(\frac{f}{f'}\times\frac{p'}{p}\right)$$

which can be rearranged into

$$\ln\left(\frac{f}{p}\times\frac{p'}{f'}\right) = \frac{1}{RT}\int_{p'}^{p}(V_m - V_{perfect,\,m})dp$$

When $p' \to 0$, the gas behaves perfectly and f' becomes equal to the pressure, p'. Therefore, $f'/p' \to 1$ as $p' \to 0$. If we take this limit, which means setting $f'/p' = 1$ on the left and $p' = 0$ on the right, the last equation becomes

$$\ln\frac{f}{p} = \frac{1}{RT}\int_0^p (V_m - V_{perfect,\,m})dp$$

Then, with $\phi = f/p$,

$$\ln\phi = \frac{1}{RT}\int_0^p (V_m - V_{perfect,m})dp$$

For a perfect gas, $V_{perfect,m} = RT/p$. For a real gas, $V_m = RTZ/p$, where Z is the compression factor of the gas (Section 1.3). With these two substitutions, we obtain

[11] The name 'fugacity' comes from the Latin for 'fleetness' in the sense of 'escaping tendency'; fugacity has the same dimensions as pressure.

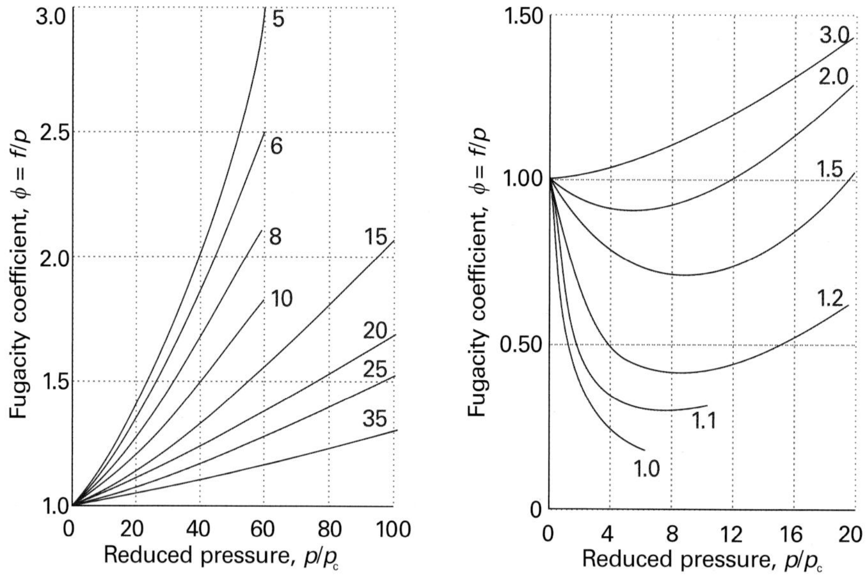

Fig. 3.25 The fugacity coefficient of a van der Waals gas plotted using the reduced variables of the gas. The curves are labelled with the reduced temperature $T_r = T/T_c$.

Exploration Evaluate the fugacity coefficient as a function of the reduced volume of a van der Waals gas and plot the outcome for a selection of reduced temperatures over the range $0.8 \le V_r \le 3$.

$$\ln \phi = \int_0^p \frac{Z-1}{p} \mathrm{d}p \qquad (3.60)$$

Provided we know how Z varies with pressure up to the pressure of interest, this expression enable us to determine the fugacity coefficient and hence, through eqn 3.59, to relate the fugacity to the pressure of the gas.

We see from Fig. 1.14 that for most gases $Z < 1$ up to moderate pressures, but that $Z > 1$ at higher pressures. If $Z < 1$ throughout the range of integration, then the integrand in eqn 3.60 is negative and $\phi < 1$. This value implies that $f < p$ (the molecules tend to stick together) and that the molar Gibbs energy of the gas is less than that of a perfect gas. At higher pressures, the range over which $Z > 1$ may dominate the range over which $Z < 1$. The integral is then positive, $\phi > 1$, and $f > p$ (the repulsive interactions are dominant and tend to drive the particles apart). Now the molar Gibbs energy of the gas is greater than that of the perfect gas at the same pressure.

Figure 3.25, which has been calculated using the full van der Waals equation of state, shows how the fugacity coefficient depends on the pressure in terms of the reduced variables (Section 1.5). Because critical constants are available in Table 1.6, the graphs can be used for quick estimates of the fugacities of a wide range of gases. Table 3.6 gives some explicit values for nitrogen.

Synoptic table 3.6* The fugacity of nitrogen at 273 K

p/atm	f/atm
1	0.999 55
10	9.9560
100	97.03
1000	1839

* More values are given in the *Data section*.

Discussion questions

3.1 The evolution of life requires the organization of a very large number of molecules into biological cells. Does the formation of living organisms violate the Second Law of thermodynamics? State your conclusion clearly and present detailed arguments to support it.

3.2 You received an unsolicited proposal from a self-declared inventor who is seeking investors for the development of his latest idea: a device that uses heat extracted from the ground by a heat pump to boil water into steam that is used to heat a home and to power a steam engine that drives the heat pump. This procedure is potentially very lucrative because, after an initial extraction of energy from the ground, no fossil fuels would be required to keep the device running indefinitely. Would you invest in this idea? State your conclusion clearly and present detailed arguments to support it.

3.3 The following expressions have been used to establish criteria for spontaneous change: $\Delta S_{tot} > 0$, $\mathrm{d}S_{U,V} \ge 0$ and $\mathrm{d}U_{S,V} \le 0$, $\mathrm{d}A_{T,V} \le 0$, and $\mathrm{d}G_{T,p} \le 0$. Discuss the origin, significance, and applicability of each criterion.

3.4 The following expressions have been used to establish criteria for reversibility: $\mathrm{d}A_{T,V} = 0$ and $\mathrm{d}G_{T,p} = 0$. Discuss the origin, significance, and applicability of each criterion.

3.5 Discuss the physical interpretation of any one Maxwell relation.

3.6 Account for the dependence of π_T of a van der Waals gas in terms of the significance of the parameters a and b.

3.7 Suggest a physical interpretation of the dependence of the Gibbs energy on the pressure.

3.8 Suggest a physical interpretation of the dependence of the Gibbs energy on the temperature.

Exercises

Assume that all gases are perfect and that data refer to 298.15 K unless otherwise stated.

3.1(a) Calculate the change in entropy when 25 kJ of energy is transferred reversibly and isothermally as heat to a large block of iron at (a) 0°C, (b) 100°C.

3.1(b) Calculate the change in entropy when 50 kJ of energy is transferred reversibly and isothermally as heat to a large block of copper at (a) 0°C, (b) 70°C.

3.2(a) Calculate the molar entropy of a constant-volume sample of neon at 500 K given that it is 146.22 J K^{-1} mol^{-1} at 298 K.

3.2(b) Calculate the molar entropy of a constant-volume sample of argon at 250 K given that it is 154.84 J K^{-1} mol^{-1} at 298 K.

3.3(a) Calculate ΔS (for the system) when the state of 3.00 mol of perfect gas atoms, for which $C_{p,\mathrm{m}} = \frac{5}{2}R$, is changed from 25°C and 1.00 atm to 125°C and 5.00 atm. How do you rationalize the sign of ΔS?

3.3(b) Calculate ΔS (for the system) when the state of 2.00 mol diatomic perfect gas molecules, for which $C_{p,\mathrm{m}} = \frac{7}{2}R$, is changed from 25°C and 1.50 atm to 135°C and 7.00 atm. How do you rationalize the sign of ΔS?

3.4(a) A sample consisting of 3.00 mol of diatomic perfect gas molecules at 200 K is compressed reversibly and adiabatically until its temperature reaches 250 K. Given that $C_{V,\mathrm{m}} = 27.5$ J K^{-1} mol^{-1}, calculate q, w, ΔU, ΔH, and ΔS.

3.4(b) A sample consisting of 2.00 mol of diatomic perfect gas molecules at 250 K is compressed reversibly and adiabatically until its temperature reaches 300 K. Given that $C_{V,\mathrm{m}} = 27.5$ J K^{-1} mol^{-1}, calculate q, w, ΔU, ΔH, and ΔS.

3.5(a) Calculate ΔH and ΔS_{tot} when two copper blocks, each of mass 10.0 kg, one at 100°C and the other at 0°C, are placed in contact in an isolated container. The specific heat capacity of copper is 0.385 J K^{-1} g^{-1} and may be assumed constant over the temperature range involved.

3.5(b) Calculate ΔH and ΔS_{tot} when two iron blocks, each of mass 1.00 kg, one at 200°C and the other at 25°C, are placed in contact in an isolated container. The specific heat capacity of iron is 0.449 J K^{-1} g^{-1} and may be assumed constant over the temperature range involved.

3.6(a) Consider a system consisting of 2.0 mol $CO_2(g)$, initially at 25°C and 10 atm and confined to a cylinder of cross-section 10.0 cm^2. It is allowed to expand adiabatically against an external pressure of 1.0 atm until the piston has moved outwards through 20 cm. Assume that carbon dioxide may be considered a perfect gas with $C_{V,\mathrm{m}} = 28.8$ J K^{-1} mol^{-1} and calculate (a) q, (b) w, (c) ΔU, (d) ΔT, (e) ΔS.

3.6(b) Consider a system consisting of 1.5 mol $CO_2(g)$, initially at 15°C and 9.0 atm and confined to a cylinder of cross-section 100.0 cm^2. The sample is allowed to expand adiabatically against an external pressure of 1.5 atm until the piston has moved outwards through 15 cm. Assume that carbon dioxide may be considered a perfect gas with $C_{V,\mathrm{m}} = 28.8$ J K^{-1} mol^{-1}, and calculate (a) q, (b) w, (c) ΔU, (d) ΔT, (e) ΔS.

3.7(a) The enthalpy of vaporization of chloroform ($CHCl_3$) is 29.4 kJ mol^{-1} at its normal boiling point of 334.88 K. Calculate (a) the entropy of vaporization of chloroform at this temperature and (b) the entropy change of the surroundings.

3.7(b) The enthalpy of vaporization of methanol is 35.27 kJ mol^{-1} at its normal boiling point of 64.1°C. Calculate (a) the entropy of vaporization of methanol at this temperature and (b) the entropy change of the surroundings.

3.8(a) Calculate the standard reaction entropy at 298 K of

(a) $2\,CH_3CHO(g) + O_2(g) \rightarrow 2\,CH_3COOH(l)$

(b) $2\,AgCl(s) + Br_2(l) \rightarrow 2\,AgBr(s) + Cl_2(g)$

(c) $Hg(l) + Cl_2(g) \rightarrow HgCl_2(s)$

3.8(b) Calculate the standard reaction entropy at 298 K of

(a) $Zn(s) + Cu^{2+}(aq) \rightarrow Zn^{2+}(aq) + Cu(s)$

(b) $C_{12}H_{22}O_{11}(s) + 12\,O_2(g) \rightarrow 12\,CO_2(g) + 11\,H_2O(l)$

3.9(a) Combine the reaction entropies calculated in Exercise 3.8a with the reaction enthalpies, and calculate the standard reaction Gibbs energies at 298 K.

3.9(b) Combine the reaction entropies calculated in Exercise 3.8b with the reaction enthalpies, and calculate the standard reaction Gibbs energies at 298 K.

3.10(a) Use standard Gibbs energies of formation to calculate the standard reaction Gibbs energies at 298 K of the reactions in Exercise 3.8a.

3.10(b) Use standard Gibbs energies of formation to calculate the standard reaction Gibbs energies at 298 K of the reactions in Exercise 3.8b.

3.11(a) Calculate the standard Gibbs energy of the reaction $4\,HCl(g) + O_2(g) \rightarrow 2\,Cl_2(g) + 2\,H_2O(l)$ at 298 K, from the standard entropies and enthalpies of formation given in the *Data section*.

3.11(b) Calculate the standard Gibbs energy of the reaction $CO(g) + CH_3OH(l) \rightarrow CH_3COOH(l)$ at 298 K, from the standard entropies and enthalpies of formation given in the *Data section*.

3.12(a) The standard enthalpy of combustion of solid phenol (C_6H_5OH) is −3054 kJ mol^{-1} at 298 K and its standard molar entropy is 144.0 J K^{-1} mol^{-1}. Calculate the standard Gibbs energy of formation of phenol at 298 K.

3.12(b) The standard enthalpy of combustion of solid urea ($CO(NH_2)_2$) is −632 kJ mol^{-1} at 298 K and its standard molar entropy is 104.60 J K^{-1} mol^{-1}. Calculate the standard Gibbs energy of formation of urea at 298 K.

3.13(a) Calculate the change in the entropies of the system and the surroundings, and the total change in entropy, when a sample of nitrogen gas of mass 14 g at 298 K and 1.00 bar doubles its volume in (a) an isothermal reversible expansion, (b) an isothermal irreversible expansion against $p_{\mathrm{ex}} = 0$, and (c) an adiabatic reversible expansion.

3.13(b) Calculate the change in the entropies of the system and the surroundings, and the total change in entropy, when the volume of a sample of argon gas of mass 21 g at 298 K and 1.50 bar increases from 1.20 dm^3 to 4.60 dm^3 in (a) an isothermal reversible expansion, (b) an isothermal irreversible expansion against $p_{\mathrm{ex}} = 0$, and (c) an adiabatic reversible expansion.

3.14(a) Calculate the maximum non-expansion work per mole that may be obtained from a fuel cell in which the chemical reaction is the combustion of methane at 298 K.

3.14(b) Calculate the maximum non-expansion work per mole that may be obtained from a fuel cell in which the chemical reaction is the combustion of propane at 298 K.

3.15(a) (a) Calculate the Carnot efficiency of a primitive steam engine operating on steam at 100°C and discharging at 60°C. (b) Repeat the calculation for a modern steam turbine that operates with steam at 300°C and discharges at 80°C.

3.15(b) A certain heat engine operates between 1000 K and 500 K. (a) What is the maximum efficiency of the engine? (b) Calculate the maximum work that can be done by for each 1.0 kJ of heat supplied by the hot source. (c) How much heat is discharged into the cold sink in a reversible process for each 1.0 kJ supplied by the hot source?

3.16(a) Suppose that 3.0 mmol $N_2(g)$ occupies 36 cm^3 at 300 K and expands to 60 cm^3. Calculate ΔG for the process.

3.16(b) Suppose that 2.5 mmol Ar(g) occupies 72 dm^3 at 298 K and expands to 100 dm^3. Calculate ΔG for the process.

3.17(a) The change in the Gibbs energy of a certain constant-pressure process was found to fit the expression $\Delta G/J = -85.40 + 36.5(T/K)$. Calculate the value of ΔS for the process.

3.17(b) The change in the Gibbs energy of a certain constant-pressure process was found to fit the expression $\Delta G/J = -73.1 + 42.8(T/K)$. Calculate the value of ΔS for the process.

3.18(a) Calculate the change in Gibbs energy of 35 g of ethanol (mass density 0.789 g cm^{-3}) when the pressure is increased isothermally from 1 atm to 3000 atm.

3.18(b) Calculate the change in Gibbs energy of 25 g of methanol (mass density 0.791 g cm^{-3}) when the pressure is increased isothermally from 100 kPa to 100 MPa.

3.19(a) Calculate the change in chemical potential of a perfect gas when its pressure is increased isothermally from 1.8 atm to 29.5 atm at 40°C.

3.19(b) Calculate the change in chemical potential of a perfect gas when its pressure is increased isothermally from 92.0 kPa to 252.0 kPa at 50°C.

3.20(a) The fugacity coefficient of a certain gas at 200 K and 50 bar is 0.72. Calculate the difference of its molar Gibbs energy from that of a perfect gas in the same state.

3.20(b) The fugacity coefficient of a certain gas at 290 K and 2.1 MPa is 0.68. Calculate the difference of its molar Gibbs energy from that of a perfect gas in the same state.

3.21(a) Estimate the change in the Gibbs energy of 1.0 dm^3 of benzene when the pressure acting on it is increased from 1.0 atm to 100 atm.

3.21(b) Estimate the change in the Gibbs energy of 1.0 dm^3 of water when the pressure acting on it is increased from 100 kPa to 300 kPa.

3.22(a) Calculate the change in the molar Gibbs energy of hydrogen gas when its pressure is increased isothermally from 1.0 atm to 100.0 atm at 298 K.

3.22(b) Calculate the change in the molar Gibbs energy of oxygen when its pressure is increased isothermally from 50.0 kPa to 100.0 kPa at 500 K.

Problems*

Assume that all gases are perfect and that data refer to 298 K unless otherwise stated.

Numerical problems

3.1 Calculate the difference in molar entropy (a) between liquid water and ice at −5°C, (b) between liquid water and its vapour at 95°C and 1.00 atm. The differences in heat capacities on melting and on vaporization are 37.3 J K^{-1} mol^{-1} and −41.9 J K^{-1} mol^{-1}, respectively. Distinguish between the entropy changes of the sample, the surroundings, and the total system, and discuss the spontaneity of the transitions at the two temperatures.

3.2 The heat capacity of chloroform (trichloromethane, $CHCl_3$) in the range 240 K to 330 K is given by $C_{p,m}/(J\,K^{-1}\,mol^{-1}) = 91.47 + 7.5 \times 10^{-2}\,(T/K)$. In a particular experiment, 1.00 mol $CHCl_3$ is heated from 273 K to 300 K. Calculate the change in molar entropy of the sample.

3.3 A block of copper of mass 2.00 kg ($C_{p,m} = 24.44$ J K^{-1} mol^{-1}) and temperature 0°C is introduced into an insulated container in which there is 1.00 mol $H_2O(g)$ at 100°C and 1.00 atm. (a) Assuming all the steam is condensed to water, what will be the final temperature of the system, the heat transferred from water to copper, and the entropy change of the water, copper, and the total system? (b) In fact, some water vapour is present at equilibrium. From the vapour pressure of water at the temperature calculated in (a), and assuming that the heat capacities of both gaseous and liquid water are constant and given by their values at that temperature, obtain an improved value of the final temperature, the heat transferred, and the various entropies. (*Hint.* You will need to make plausible approximations.)

3.4 Consider a perfect gas contained in a cylinder and separated by a frictionless adiabatic piston into two sections A and B. All changes in B is isothermal; that is, a thermostat surrounds B to keep its temperature constant. There is 2.00 mol of the gas in each section. Initially, $T_A = T_B = 300$ K, $V_A = V_B = 2.00$ dm^3. Energy is supplied as heat to Section A and the piston moves to the right reversibly until the final volume of Section B is 1.00 dm^3. Calculate (a) ΔS_A and ΔS_B, (b) ΔA_A and ΔA_B, (c) ΔG_A and ΔG_B, (d) ΔS of the total system and its surroundings. If numerical values cannot be obtained, indicate whether the values should be positive, negative, or zero or are indeterminate from the information given. (Assume $C_{V,m} = 20$ J K^{-1} mol^{-1}.)

3.5 A Carnot cycle uses 1.00 mol of a monatomic perfect gas as the working substance from an initial state of 10.0 atm and 600 K. It expands isothermally to a pressure of 1.00 atm (Step 1), and then adiabatically to a temperature of 300 K (Step 2). This expansion is followed by an isothermal compression (Step 3), and then an adiabatic compression (Step 4) back to the initial state. Determine the values of q, w, ΔU, ΔH, ΔS, ΔS_{tot}, and ΔG for each stage of the cycle and for the cycle as a whole. Express your answer as a table of values.

3.6 1.00 mol of perfect gas molecules at 27°C is expanded isothermally from an initial pressure of 3.00 atm to a final pressure of 1.00 atm in two ways: (a) reversibly, and (b) against a constant external pressure of 1.00 atm. Determine the values of q, w, ΔU, ΔH, ΔS, ΔS_{sur}, ΔS_{tot} for each path.

3.7 The standard molar entropy of $NH_3(g)$ is 192.45 J K^{-1} mol^{-1} at 298 K, and its heat capacity is given by eqn 2.25 with the coefficients given in Table 2.2. Calculate the standard molar entropy at (a) 100°C and (b) 500°C.

3.8 A block of copper of mass 500 g and initially at 293 K is in thermal contact with an electric heater of resistance 1.00 kΩ and negligible mass. A current of 1.00 A is passed for 15.0 s. Calculate the change in entropy of the copper, taking $C_{p,m} = 24.4$ J K^{-1} mol^{-1}. The experiment is then repeated with the copper immersed in a stream of water that maintains its temperature at 293 K. Calculate the change in entropy of the copper and the water in this case.

3.9 Find an expression for the change in entropy when two blocks of the same substance and of equal mass, one at the temperature T_h and the other at T_c, are brought into thermal contact and allowed to reach equilibrium. Evaluate the

* Problems denoted with the symbol ‡ were supplied by Charles Trapp, Carmen Giunta, and Marshall Cady.

change for two blocks of copper, each of mass 500 g, with $C_{p,m} = 24.4$ J K^{-1} mol^{-1}, taking $T_h = 500$ K and $T_c = 250$ K.

3.10 A gaseous sample consisting of 1.00 mol molecules is described by the equation of state $pV_m = RT(1 + Bp)$. Initially at 373 K, it undergoes Joule–Thomson expansion from 100 atm to 1.00 atm. Given that $C_{p,m} = \frac{5}{2}R$, $\mu = 0.21$ K atm^{-1}, $B = -0.525(\text{K}/T)$ atm^{-1}, and that these are constant over the temperature range involved, calculate ΔT and ΔS for the gas.

3.11 The molar heat capacity of lead varies with temperature as follows:

T/K	10	15	20	25	30	50
$C_{p,m}$/(J K^{-1} mol^{-1})	2.8	7.0	10.8	14.1	16.5	21.4
T/K	70	100	150	200	250	298
$C_{p,m}$/(J K^{-1} mol^{-1})	23.3	24.5	25.3	25.8	26.2	26.6

Calculate the standard Third-Law entropy of lead at (a) 0°C and (b) 25°C.

3.12 From standard enthalpies of formation, standard entropies, and standard heat capacities available from tables in the *Data section*, calculate the standard enthalpies and entropies at 298 K and 398 K for the reaction $CO_2(g) + H_2(g) \rightarrow CO(g) + H_2O(g)$. Assume that the heat capacities are constant over the temperature range involved.

3.13 The heat capacity of anhydrous potassium hexacyanoferrate(II) varies with temperature as follows:

T/K	$C_{p,m}$/(J K^{-1} mol^{-1})	T/K	$C_{p,m}$/(J K^{-1} mol^{-1})
10	2.09	100	179.6
20	14.43	110	192.8
30	36.44	150	237.6
40	62.55	160	247.3
50	87.03	170	256.5
60	111.0	180	265.1
70	131.4	190	273.0
80	149.4	200	280.3
90	165.3		

Calculate the molar enthalpy relative to its value at $T = 0$ and the Third-Law entropy at each of these temperatures.

3.14 The compound 1,3,5-trichloro-2,4,6-trifluorobenzene is an intermediate in the conversion of hexachlorobenzene to hexafluorobenzene, and its thermodynamic properties have been examined by measuring its heat capacity over a wide temperature range (R.L. Andon and J.F. Martin, *J. Chem. Soc. Faraday Trans. I.* 871 (1973)). Some of the data are as follows:

T/K	14.14	16.33	20.03	31.15	44.08	64.81
$C_{p,m}$/(J K^{-1} mol^{-1})	9.492	12.70	18.18	32.54	46.86	66.36
T/K	100.90	140.86	183.59	225.10	262.99	298.06
$C_{p,m}$/(J K^{-1} mol^{-1})	95.05	121.3	144.4	163.7	180.2	196.4

Calculate the molar enthalpy relative to its value at $T = 0$ and the Third-Law molar entropy of the compound at these temperatures.

3.15‡ Given that $S^{\ominus}_m = 29.79$ J K^{-1} mol^{-1} for bismuth at 100 K and the following tabulated heat capacities data (D.G. Archer, *J. Chem. Eng. Data* **40**, 1015 (1995)), compute the standard molar entropy of bismuth at 200 K.

T/K	100	120	140	150	160	180	200
$C_{p,m}$/(J K^{-1} mol^{-1})	23.00	23.74	24.25	24.44	24.61	24.89	25.11

Compare the value to the value that would be obtained by taking the heat capacity to be constant at 24.44 J K^{-1} mol^{-1} over this range.

3.16 Calculate $\Delta_r G^{\ominus}(375\text{ K})$ for the reaction $2\,CO(g) + O_2(g) \rightarrow 2\,CO_2(g)$ from the value of $\Delta_r G^{\ominus}(298\text{ K})$, $\Delta_r H^{\ominus}(298\text{ K})$, and the Gibbs–Helmholtz equation.

3.17 Estimate the standard reaction Gibbs energy of $N_2(g) + 3\,H_2(g) \rightarrow 2\,NH_3(g)$ at (a) 500 K, (b) 1000 K from their values at 298 K.

3.18 At 200 K, the compression factor of oxygen varies with pressure as shown below. Evaluate the fugacity of oxygen at this temperature and 100 atm.

p/atm	1.0000	4.00000	7.00000	10.0000	40.00	70.00	100.0
Z	0.9971	0.98796	0.97880	0.96956	0.8734	0.7764	0.6871

Theoretical problems

3.19 Represent the Carnot cycle on a temperature–entropy diagram and show that the area enclosed by the cycle is equal to the work done.

3.20 Prove that two reversible adiabatic paths can never cross. Assume that the energy of the system under consideration is a function of temperature only. (*Hint.* Suppose that two such paths can intersect, and complete a cycle with the two paths plus one isothermal path. Consider the changes accompanying each stage of the cycle and show that they conflict with the Kelvin statement of the Second Law.)

3.21 Prove that the perfect gas temperature scale and the thermodynamic temperature scale based on the Second Law of thermodynamics differ from each other by at most a constant numerical factor.

3.22 The molar Gibbs energy of a certain gas is given by $G_m = RT \ln p + A + Bp + \frac{1}{2}Cp^2 + \frac{1}{3}Dp^3$, where A, B, C, and D are constants. Obtain the equation of state of the gas.

3.23 Evaluate $(\partial S/\partial V)_T$ for (a) a van der Waals gas, (b) a Dieterici gas (Table 1.7). For an isothermal expansion, for which kind of gas (and a perfect gas) will ΔS be greatest? Explain your conclusion.

3.24 Show that, for a perfect gas, $(\partial U/\partial S)_V = T$ and $(\partial U/\partial V)_S = -p$.

3.25 Two of the four Maxwell relations were derived in the text, but two were not. Complete their derivation by showing that $(\partial S/\partial V)_T = (\partial p/\partial T)_V$ and $(\partial T/\partial p)_S = (\partial V/\partial S)_p$.

3.26 Use the Maxwell relations to express the derivatives (a) $(\partial S/\partial V)_T$ and $(\partial V/\partial S)_p$ and (b) $(\partial p/\partial S)_V$ and $(\partial V/\partial S)_p$ in terms of the heat capacities, the expansion coefficient α, and the isothermal compressibility, κ_T.

3.27 Use the Maxwell relations to show that the entropy of a perfect gas depends on the volume as $S \propto R \ln V$.

3.28 Derive the thermodynamic equation of state

$$\left(\frac{\partial H}{\partial p}\right)_T = V - T\left(\frac{\partial V}{\partial T}\right)_p$$

Derive an expression for $(\partial H/\partial p)_T$ for (a) a perfect gas and (b) a van der Waals gas. In the latter case, estimate its value for 1.0 mol Ar(g) at 298 K and 10 atm. By how much does the enthalpy of the argon change when the pressure is increased isothermally to 11 atm?

3.29 Show that if $B(T)$ is the second virial coefficient of a gas, and $\Delta B = B(T'') - B(T')$, $\Delta T = T'' - T'$, and T is the mean of T'' and T', then $\pi_T \approx RT^2\Delta B/V_m^2\Delta T$. Estimate π_T for argon given that $B(250\text{ K}) = -28.0$ cm^3 mol^{-1} and $B(300\text{ K}) = -15.6$ cm^3 mol^{-1} at 275 K at (a) 1.0 atm, (b) 10.0 atm.

3.30 The Joule coefficient, μ_J, is defined as $\mu_J = (\partial T/\partial V)_U$. Show that $\mu_J C_V = p - \alpha T/\kappa_T$.

3.31 Evaluate π_T for a Dieterici gas (Table 1.7). Justify physically the form of the expression obtained.

3.32 The adiabatic compressibility, κ_S, is defined like κ_T (eqn 2.44) but at constant entropy. Show that for a perfect gas $p\gamma\kappa_S = 1$ (where γ is the ratio of heat capacities).

3.33 Suppose that S is regarded as a function of p and T. Show that $TdS = C_p dT - \alpha TV dp$. Hence, show that the energy transferred as heat when the pressure on an incompressible liquid or solid is increased by Δp is equal to $-\alpha TV\Delta p$. Evaluate q when the pressure acting on 100 cm^3 of mercury at 0°C is increased by 1.0 kbar. ($\alpha = 1.82 \times 10^{-4}$ K^{-1}.)

3.34 Suppose that (a) the attractive interactions between gas particles can be neglected, (b) the attractive interaction is dominant in a van der Waals gas, and the pressure is low enough to make the approximation $4ap/(RT)^2 \ll 1$. Find expressions for the fugacity of a van der Waals gas in terms of the pressure and estimate its value for ammonia at 10.00 atm and 298.15 K in each case.

3.35 Find an expression for the fugacity coefficient of a gas that obeys the equation of state $pV_m = RT(1 + B/V_m + C/V_m^2)$. Use the resulting expression to estimate the fugacity of argon at 1.00 atm and 100 K using $B = -21.13$ cm^3 mol^{-1} and $C = 1054$ cm^6 mol^{-2}.

Applications: to biology, environmental science, polymer science, and engineering

3.36 The protein lysozyme unfolds at a transition temperature of 75.5°C and the standard enthalpy of transition is 509 kJ mol^{-1}. Calculate the entropy of unfolding of lysozyme at 25.0°C, given that the difference in the constant-pressure heat capacities upon unfolding is 6.28 kJ K^{-1} mol^{-1} and can be assumed to be independent of temperature. *Hint.* Imagine that the transition at 25.0°C occurs in three steps: (i) heating of the folded protein from 25.0°C to the transition temperature, (ii) unfolding at the transition temperature, and (iii) cooling of the unfolded protein to 25.0°C. Because the entropy is a state function, the entropy change at 25.0°C is equal to the sum of the entropy changes of the steps.

3.37 At 298 K the standard enthalpy of combustion of sucrose is −5797 kJ mol^{-1} and the standard Gibbs energy of the reaction is −6333 kJ mol^{-1}. Estimate the additional non-expansion work that may be obtained by raising the temperature to blood temperature, 37°C.

3.38 In biological cells, the energy released by the oxidation of foods (*Impact on Biology I2.2*) is stored in adenosine triphosphate (ATP or ATP^{4-}). The essence of ATP's action is its ability to lose its terminal phosphate group by hydrolysis and to form adenosine diphosphate (ADP or ADP^{3-}):

$$ATP^{4-}(aq) + H_2O(l) \rightarrow ADP^{3-}(aq) + HPO_4^{2-}(aq) + H_3O^+(aq)$$

At pH = 7.0 and 37°C (310 K, blood temperature) the enthalpy and Gibbs energy of hydrolysis are $\Delta_r H = -20$ kJ mol^{-1} and $\Delta_r G = -31$ kJ mol^{-1}, respectively. Under these conditions, the hydrolysis of 1 mol ATP^{4-}(aq) results in the extraction of up to 31 kJ of energy that can be used to do non-expansion work, such as the synthesis of proteins from amino acids, muscular contraction, and the activation of neuronal circuits in our brains. (a) Calculate and account for the sign of the entropy of hydrolysis of ATP at pH = 7.0 and 310 K. (b) Suppose that the radius of a typical biological cell is 10 µm and that inside it 10^6 ATP molecules are hydrolysed each second. What is the power density of the cell in watts per cubic metre (1 W = 1 J s^{-1})? A computer battery delivers about 15 W and has a volume of 100 cm^3. Which has the greater power density, the cell or the battery? (c) The formation of glutamine from glutamate and ammonium ions requires 14.2 kJ mol^{-1} of energy input. It is driven by the hydrolysis of ATP to ADP mediated by the enzyme glutamine synthetase. How many moles of ATP must be hydrolysed to form 1 mol glutamine?

3.39‡ In 1995, the Intergovernmental Panel on Climate Change (IPCC) considered a global average temperature rise of 1.0–3.5°C likely by the year 2100, with 2.0°C its best estimate. Because water vapour is itself a greenhouse gas, the increase in water vapour content of the atmosphere is of some concern to climate change experts. Predict the relative increase in water vapour in the atmosphere based on a temperature rises of 2.0 K, assuming that the relative humidity remains constant. (The present global mean temperature is 290 K, and the equilibrium vapour pressure of water at that temperature is 0.0189 bar.)

3.40‡ Nitric acid hydrates have received much attention as possible catalysts for heterogeneous reactions that bring about the Antarctic ozone hole. Worsnop *et al.* investigated the thermodynamic stability of these hydrates under conditions typical of the polar winter stratosphere (D. R. Worsnop, L.E. Fox, M.S. Zahniser, and S.C. Wofsy, *Science* **259**, 71 (1993)). They report thermodynamic data for the sublimation of mono-, di-, and trihydrates to nitric acid and water vapours, $HNO_3 \cdot nH_2O(s) \rightarrow HNO_3(g) + nH_2O(g)$, for $n = 1, 2$, and 3. Given $\Delta_r G^\ominus$ and $\Delta_r H^\ominus$ for these reactions at 220 K, use the Gibbs–Helmholtz equation to compute $\Delta_r G^\ominus$ at 190 K.

n	1	2	3
$\Delta_r G^\ominus$/(kJ mol^{-1})	46.2	69.4	93.2
$\Delta_r H^\ominus$/(kJ mol^{-1})	127	188	237

3.41‡ J. Gao and J. H. Weiner in their study of the origin of stress on the atomic level in dense polymer systems (*Science* **266**, 748 (1994)), observe that the tensile force required to maintain the length, l, of a long linear chain of N freely jointed links each of length a, can be interpreted as arising from an entropic spring. For such a chain, $S(l) = -3kl^2/2Na^2 + C$, where k is the Boltzmann constant and C is a constant. Using thermodynamic relations of this and previous chapters, show that the tensile force obeys Hooke's law, $f = -k_f l$, if we assume that the energy U is independent of l.

3.42 Suppose that an internal combustion engine runs on octane, for which the enthalpy of combustion is −5512 kJ mol^{-1} and take the mass of 1 gallon of fuel as 3 kg. What is the maximum height, neglecting all forms of friction, to which a car of mass 1000 kg can be driven on 1.00 gallon of fuel given that the engine cylinder temperature is 2000°C and the exit temperature is 800°C?

3.43 The cycle involved in the operation of an internal combustion engine is called the *Otto cycle*. Air can be considered to be the working substance and can be assumed to be a perfect gas. The cycle consists of the following steps: (1) reversible adiabatic compression from A to B, (2) reversible constant-volume pressure increase from B to C due to the combustion of a small amount of fuel, (3) reversible adiabatic expansion from C to D, and (4) reversible and constant-volume pressure decrease back to state A. Determine the change in entropy (of the system and of the surroundings) for each step of the cycle and determine an expression for the efficiency of the cycle, assuming that the heat is supplied in Step 2. Evaluate the efficiency for a compression ratio of 10:1. Assume that, in state A, $V = 4.00$ dm^3, $p = 1.00$ atm, and $T = 300$ K, that $V_A = 10V_B$, $p_C/p_B = 5$, and that $C_{p,m} = \frac{7}{2}R$.

3.44 To calculate the work required to lower the temperature of an object, we need to consider how the coefficient of performance changes with the temperature of the object. (a) Find an expression for the work of cooling an object from T_i to T_f when the refrigerator is in a room at a temperature T_h. *Hint.* Write $dw = dq/c(T)$, relate dq to dT through the heat capacity C_p, and integrate the resulting expression. Assume that the heat capacity is independent of temperature in the range of interest. (b) Use the result in part (a) to calculate the work needed to freeze 250 g of water in a refrigerator at 293 K. How long will it take when the refrigerator operates at 100 W?

3.45 The expressions that apply to the treatment of refrigerators also describe the behaviour of heat pumps, where warmth is obtained from the back of a refrigerator while its front is being used to cool the outside world. Heat pumps are popular home heating devices because they are very efficient. Compare heating of a room at 295 K by each of two methods: (a) direct conversion of 1.00 kJ of electrical energy in an electrical heater, and (b) use of 1.00 kJ of electrical energy to run a reversible heat pump with the outside at 260 K. Discuss the origin of the difference in the energy delivered to the interior of the house by the two methods.

Physical transformations of pure substances

4

The discussion of the phase transitions of pure substances is among the simplest applications of thermodynamics to chemistry. We shall see that a phase diagram is a map of the pressures and temperatures at which each phase of a substance is the most stable. First, we describe the interpretation of empirically determined phase diagrams for a selection of materials. Then we turn to a consideration of the factors that determine the positions and shapes of the boundaries between the regions on a phase diagram. The practical importance of the expressions we derive is that they show how the vapour pressure of a substance varies with temperature and how the melting point varies with pressure. We shall see that the transitions between phases can be classified by noting how various thermodynamic functions change when the transition occurs. This chapter also introduces the chemical potential, a property that is at the centre of discussions of phase transitions and chemical reactions.

Vaporization, melting, and the conversion of graphite to diamond are all examples of changes of phase without change of chemical composition. In this chapter we describe such processes thermodynamically, using as the guiding principle the tendency of systems at constant temperature and pressure to minimize their Gibbs energy.

Phase diagrams

One of the most succinct ways of presenting the physical changes of state that a substance can undergo is in terms of its phase diagram. We present the concept in this section.

4.1 The stabilities of phases

A **phase** of a substance is a form of matter that is uniform throughout in chemical composition and physical state. Thus, we speak of solid, liquid, and gas phases of a substance, and of its various solid phases, such as the white and black allotropes of phosphorus. A **phase transition**, the spontaneous conversion of one phase into another phase, occurs at a characteristic temperature for a given pressure. Thus, at 1 atm, ice is the stable phase of water below 0°C, but above 0°C liquid water is more stable. This difference indicates that below 0°C the Gibbs energy decreases as liquid water changes into ice and that above 0°C the Gibbs energy decreases as ice changes into liquid water. The **transition temperature**, T_{trs}, is the temperature at which the two phases are in equilibrium and the Gibbs energy is minimized at the prevailing pressure.

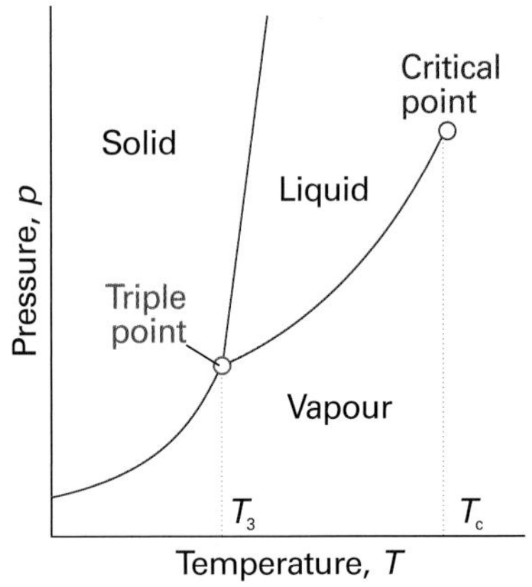

Fig. 4.1 The general regions of pressure and temperature where solid, liquid, or gas is stable (that is, has minimum molar Gibbs energy) are shown on this phase diagram. For example, the solid phase is the most stable phase at low temperatures and high pressures. In the following paragraphs we locate the precise boundaries between the regions.

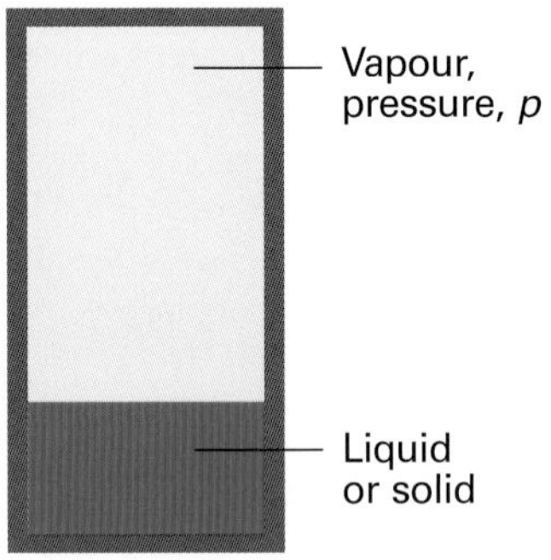

Fig. 4.2 The vapour pressure of a liquid or solid is the pressure exerted by the vapour in equilibrium with the condensed phase.

Comment 4.1

The *NIST Chemistry WebBook* is a good source of links to online databases of data on phase transitions.

As we stressed at the beginning of Chapter 3, we must distinguish between the thermodynamic description of a phase transition and the rate at which the transition occurs. A transition that is predicted from thermodynamics to be spontaneous may occur too slowly to be significant in practice. For instance, at normal temperatures and pressures the molar Gibbs energy of graphite is lower than that of diamond, so there is a thermodynamic tendency for diamond to change into graphite. However, for this transition to take place, the C atoms must change their locations, which is an immeasurably slow process in a solid except at high temperatures. The discussion of the rate of attainment of equilibrium is a kinetic problem and is outside the range of thermodynamics. In gases and liquids the mobilities of the molecules allow phase transitions to occur rapidly, but in solids thermodynamic instability may be frozen in. Thermodynamically unstable phases that persist because the transition is kinetically hindered are called **metastable phases**. Diamond is a metastable phase of carbon under normal conditions.

4.2 Phase boundaries

The **phase diagram** of a substance shows the regions of pressure and temperature at which its various phases are thermodynamically stable (Fig. 4.1). The lines separating the regions, which are called **phase boundaries**, show the values of p and T at which two phases coexist in equilibrium.

Consider a liquid sample of a pure substance in a closed vessel. The pressure of a vapour in equilibrium with the liquid is called the **vapour pressure** of the substance (Fig. 4.2). Therefore, the liquid–vapour phase boundary in a phase diagram shows how the vapour pressure of the liquid varies with temperature. Similarly, the solid–vapour phase boundary shows the temperature variation of the **sublimation vapour pressure**, the vapour pressure of the solid phase. The vapour pressure of a substance increases with temperature because at higher temperatures more molecules have sufficient energy to escape from their neighbours.

(a) Critical points and boiling points

When a liquid is heated in an open vessel, the liquid vaporizes from its surface. At the temperature at which its vapour pressure would be equal to the external pressure, vaporization can occur throughout the bulk of the liquid and the vapour can expand freely into the surroundings. The condition of free vaporization throughout the liquid is called **boiling**. The temperature at which the vapour pressure of a liquid is equal to the external pressure is called the **boiling temperature** at that pressure. For the special case of an external pressure of 1 atm, the boiling temperature is called the **normal boiling point**, T_b. With the replacement of 1 atm by 1 bar as standard pressure, there is some advantage in using the **standard boiling point** instead: this is the temperature at which the vapour pressure reaches 1 bar. Because 1 bar is slightly less than 1 atm (1.00 bar = 0.987 atm), the standard boiling point of a liquid is slightly lower than its normal boiling point. The normal boiling point of water is 100.0°C; its standard boiling point is 99.6°C.

Boiling does not occur when a liquid is heated in a rigid, closed vessel. Instead, the vapour pressure, and hence the density of the vapour, rise as the temperature is raised (Fig. 4.3). At the same time, the density of the liquid decreases slightly as a result of its expansion. There comes a stage when the density of the vapour is equal to that of the remaining liquid and the surface between the two phases disappears. The temperature at which the surface disappears is the **critical temperature**, T_c, of the substance. We first encountered this property in Section 1.3d. The vapour pressure at the critical temperature is called the **critical pressure**, p_c. At and above the critical temperature, a single uniform phase called a **supercritical fluid** fills the container and an interface no longer exists. That is, above the critical temperature, the liquid phase of the substance does not exist.

(b) Melting points and triple points

The temperature at which, under a specified pressure, the liquid and solid phases of a substance coexist in equilibrium is called the **melting temperature**. Because a substance melts at exactly the same temperature as it freezes, the melting temperature of a substance is the same as its **freezing temperature**. The freezing temperature when the pressure is 1 atm is called the **normal freezing point**, T_f, and its freezing point when the pressure is 1 bar is called the **standard freezing point**. The normal and standard freezing points are negligibly different for most purposes. The normal freezing point is also called the **normal melting point**.

There is a set of conditions under which three different phases of a substance (typically solid, liquid, and vapour) all simultaneously coexist in equilibrium. These conditions are represented by the **triple point**, a point at which the three phase boundaries meet. The temperature at the triple point is denoted T_3. The triple point of a pure substance is outside our control: it occurs at a single definite pressure and temperature characteristic of the substance. The triple point of water lies at 273.16 K and 611 Pa (6.11 mbar, 4.58 Torr), and the three phases of water (ice, liquid water, and water vapour) coexist in equilibrium at no other combination of pressure and temperature. This invariance of the triple point is the basis of its use in the definition of the thermodynamic temperature scale (Section 3.2c).

As we can see from Fig. 4.1, the triple point marks the lowest pressure at which a liquid phase of a substance can exist. If (as is common) the slope of the solid–liquid phase boundary is as shown in the diagram, then the triple point also marks the lowest temperature at which the liquid can exist; the critical temperature is the upper limit.

Fig. 4.3 (a) A liquid in equilibrium with its vapour. (b) When a liquid is heated in a sealed container, the density of the vapour phase increases and that of the liquid decreases slightly. There comes a stage, (c), at which the two densities are equal and the interface between the fluids disappears. This disappearance occurs at the critical temperature. The container needs to be strong: the critical temperature of water is 374°C and the vapour pressure is then 218 atm.

IMPACT ON CHEMICAL ENGINEERING AND TECHNOLOGY
I4.1 Supercritical fluids

Supercritical carbon dioxide, $scCO_2$, is the centre of attention for an increasing number of solvent-based processes. The critical temperature of CO_2, 304.2 K (31.0°C) and its critical pressure, 72.9 atm, are readily accessible, it is cheap, and it can readily be recycled. The density of $scCO_2$ at its critical point is 0.45 g cm^{-3}. However, the transport properties of any supercritical fluid depend strongly on its density, which in turn is sensitive to the pressure and temperature. For instance, densities may be adjusted from a gas-like 0.1 g cm^{-3} to a liquid-like 1.2 g cm^{-3}. A useful rule of thumb is that the solubility of a solute is an exponential function of the density of the supercritical fluid, so small increases in pressure, particularly close to the critical point, can have very large effects on solubility.

A great advantage of $scCO_2$ is that there are no noxious residues once the solvent has been allowed to evaporate, so, coupled with its low critical temperature, $scCO_2$ is ideally suited to food processing and the production of pharmaceuticals. It is used, for instance, to remove caffeine from coffee. The supercritical fluid is also increasingly being used for dry cleaning, which avoids the use of carcinogenic and environmentally deleterious chlorinated hydrocarbons.

Supercritical CO_2 has been used since the 1960s as a mobile phase in *supercritical fluid chromatography* (SFC), but it fell out of favour when the more convenient technique of high-performance liquid chromatography (HPLC) was introduced. However, interest in SFC has returned, and there are separations possible in SFC that cannot easily be achieved by HPLC, such as the separation of lipids and of phospholipids. Samples as small as 1 pg can be analysed. The essential advantage of SFC is that diffusion coefficients in supercritical fluids are an order of magnitude greater than in

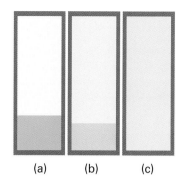

liquids, so there is less resistance to the transfer of solutes through the column, with the result that separations may be effected rapidly or with high resolution.

The principal problem with $scCO_2$, though, is that the fluid is not a very good solvent and surfactants are needed to induce many potentially interesting solutes to dissolve. Indeed, $scCO_2$-based dry cleaning depends on the availability of cheap surfactants; so too does the use of $scCO_2$ as a solvent for homogeneous catalysts, such as metal complexes. There appear to be two principal approaches to solving the solubilization problem. One solution is to use fluorinated and siloxane-based polymeric stabilizers, which allow polymerization reactions to proceed in $scCO_2$. The disadvantage of these stabilizers for commercial use is their great expense. An alternative and much cheaper approach is poly(ether-carbonate) copolymers. The copolymers can be made more soluble in $scCO_2$ by adjusting the ratio of ether and carbonate groups.

The critical temperature of water is 374°C and its pressure is 218 atm. The conditions for using scH_2O are therefore much more demanding than for $scCO_2$ and the properties of the fluid are highly sensitive to pressure. Thus, as the density of scH_2O decreases, the characteristics of a solution change from those of an aqueous solution through those of a non-aqueous solution and eventually to those of a gaseous solution. One consequence is that reaction mechanisms may change from those involving ions to those involving radicals.

4.3 Three typical phase diagrams

We shall now see how these general features appear in the phase diagrams of pure substances.

(a) Carbon dioxide

The phase diagram for carbon dioxide is shown in Fig. 4.4. The features to notice include the positive slope of the solid–liquid boundary (the direction of this line is characteristic of most substances), which indicates that the melting temperature of solid carbon dioxide rises as the pressure is increased. Notice also that, as the triple point lies above 1 atm, the liquid cannot exist at normal atmospheric pressures whatever the temperature, and the solid sublimes when left in the open (hence the name 'dry ice'). To obtain the liquid, it is necessary to exert a pressure of at least 5.11 atm. Cylinders of carbon dioxide generally contain the liquid or compressed gas; at 25°C that implies a vapour pressure of 67 atm if both gas and liquid are present in equilibrium. When the gas squirts through the throttle it cools by the Joule–Thomson effect, so when it emerges into a region where the pressure is only 1 atm, it condenses into a finely divided snow-like solid.

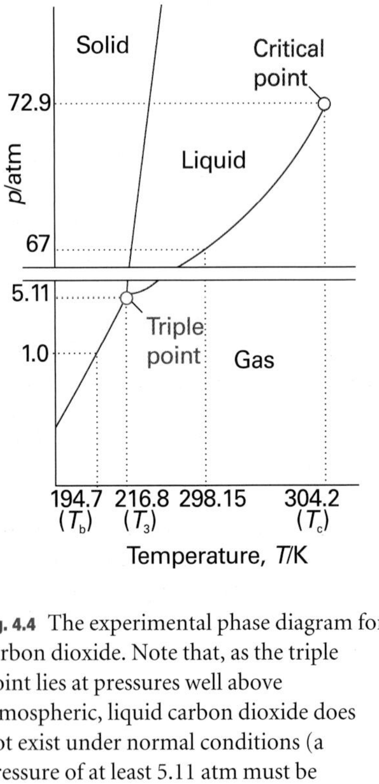

Fig. 4.4 The experimental phase diagram for carbon dioxide. Note that, as the triple point lies at pressures well above atmospheric, liquid carbon dioxide does not exist under normal conditions (a pressure of at least 5.11 atm must be applied).

(b) Water

Figure 4.5 is the phase diagram for water. The liquid–vapour boundary in the phase diagram summarizes how the vapour pressure of liquid water varies with temperature. It also summarizes how the boiling temperature varies with pressure: we simply read off the temperature at which the vapour pressure is equal to the prevailing atmospheric pressure. The solid–liquid boundary shows how the melting temperature varies with the pressure. Its very steep slope indicates that enormous pressures are needed to bring about significant changes. Notice that the line has a negative slope up to 2 kbar, which means that the melting temperature falls as the pressure is raised. The reason for this almost unique behaviour can be traced to the decrease in volume that occurs on melting, and hence it being more favourable for the solid to transform into the liquid as the pressure is raised. The decrease in volume is a result of the very

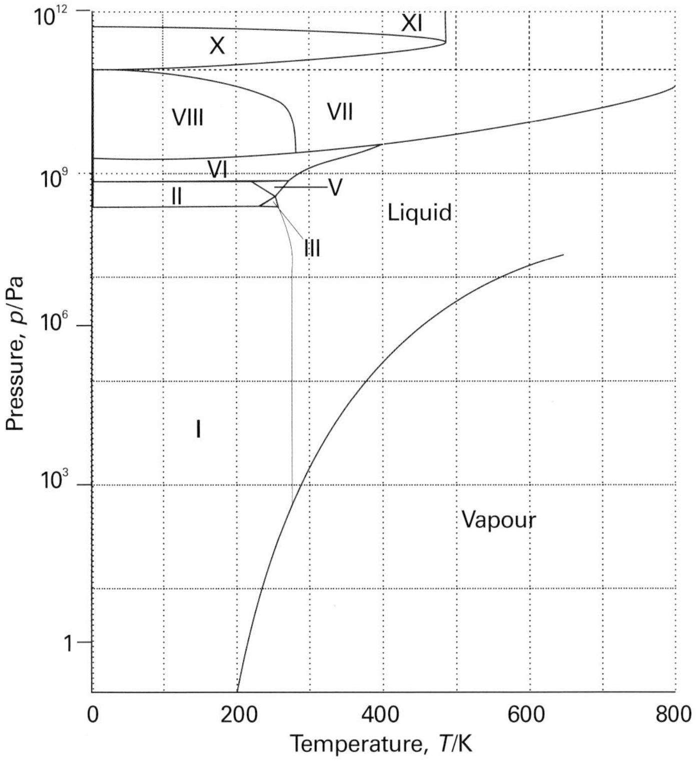

Fig. 4.5 The experimental phase diagram for water showing the different solid phases.

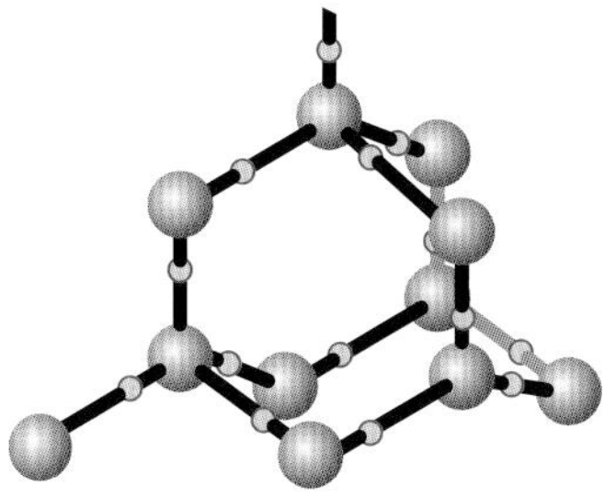

Fig. 4.6 A fragment of the structure of ice (ice-I). Each O atom is linked by two covalent bonds to H atoms and by two hydrogen bonds to a neighbouring O atom, in a tetrahedral array.

open molecular structure of ice: as shown in Fig 4.6, the water molecules are held apart, as well as together, by the hydrogen bonds between them but the structure partially collapses on melting and the liquid is denser than the solid.

Figure 4.5 shows that water has one liquid phase but many different solid phases other than ordinary ice ('ice I', shown in Fig. 4.5). Some of these phases melt at high temperatures. Ice VII, for instance, melts at 100°C but exists only above 25 kbar. Note that five more triple points occur in the diagram other than the one where vapour, liquid, and ice I coexist. Each one occurs at a definite pressure and temperature that cannot be changed. The solid phases of ice differ in the arrangement of the water molecules: under the influence of very high pressures, hydrogen bonds buckle and the H_2O molecules adopt different arrangements. These **polymorphs**, or different solid phases, of ice may be responsible for the advance of glaciers, for ice at the bottom of glaciers experiences very high pressures where it rests on jagged rocks.

(c) Helium

Figure 4.7 shows the phase diagram of helium. Helium behaves unusually at low temperatures. For instance, the solid and gas phases of helium are never in equilibrium however low the temperature: the atoms are so light that they vibrate with a large-amplitude motion even at very low temperatures and the solid simply shakes itself apart. Solid helium can be obtained, but only by holding the atoms together by applying pressure.

When considering helium at low temperatures it is necessary to distinguish between the isotopes ^{3}He and ^{4}He. Pure helium-4 has two liquid phases. The phase marked He-I in the diagram behaves like a normal liquid; the other phase, He-II, is a **superfluid**;

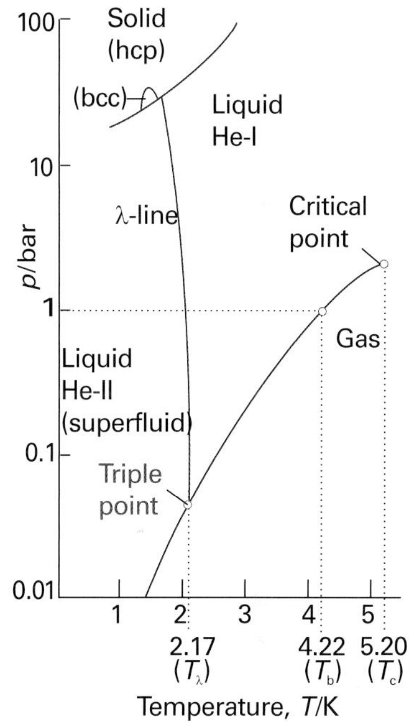

Fig. 4.7 The phase diagram for helium (^{4}He). The λ-line marks the conditions under which the two liquid phases are in equilibrium. Helium-II is the superfluid phase. Note that a pressure of over 20 bar must be exerted before solid helium can be obtained. The labels hcp and bcc denote different solid phases in which the atoms pack together differently: hcp denotes hexagonal closed packing and bcc denotes body-centred cubic (see Section 20.1 for a description of these structures).

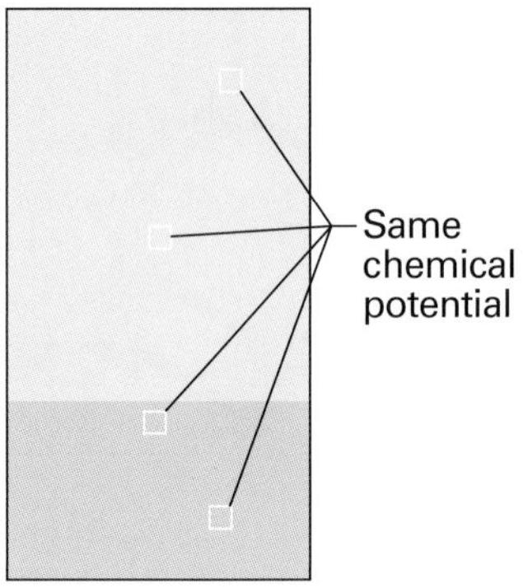

Fig. 4.8 When two or more phases are in equilibrium, the chemical potential of a substance (and, in a mixture, a component) is the same in each phase and is the same at all points in each phase.

it is so called because it flows without viscosity.[1] Provided we discount the liquid crystalline substances discussed in Section 6.6, helium is the only known substance with a liquid–liquid boundary, shown as the **λ-line** (lambda line) in Fig. 4.7. The phase diagram of helium-3 differs from the phase diagram of helium-4, but it also possesses a superfluid phase. Helium-3 is unusual in that the entropy of the liquid is lower than that of the solid, and melting is exothermic.

Phase stability and phase transitions

We shall now see how thermodynamic considerations can account for the features of the phase diagrams we have just described. All our considerations will be based on the Gibbs energy of a substance, and in particular on its molar Gibbs energy, G_m. In fact, this quantity will play such an important role in this chapter and the rest of the text that we give it a special name and symbol, the **chemical potential**, μ (mu). For a one-component system, 'molar Gibbs energy' and 'chemical potential' are synonyms, so $\mu = G_m$, but in Chapter 5 we shall see that chemical potential has a broader significance and a more general definition. The name 'chemical potential' is also instructive: as we develop the concept, we shall see that μ is a measure of the potential that a substance has for undergoing change in a system. In this chapter, it reflects the potential of a substance to undergo physical change. In Chapter 7 we shall see that μ is the potential of a substance to undergo chemical change.

4.4 The thermodynamic criterion of equilibrium

We base our discussion on the following consequence of the Second Law: *at equilibrium, the chemical potential of a substance is the same throughout a sample, regardless of how many phases are present*. When the liquid and solid phases of a substance are in equilibrium, the chemical potential of the substance is the same throughout the system (Fig. 4.8).

To see the validity of this remark, consider a system in which the chemical potential of a substance is μ_1 at one location and μ_2 at another location. The locations may be in the same or in different phases. When an amount dn of the substance is transferred from one location to the other, the Gibbs energy of the system changes by $-\mu_1 \mathrm{d}n$ when material is removed from location 1, and it changes by $+\mu_2 \mathrm{d}n$ when that material is added to location 2. The overall change is therefore $\mathrm{d}G = (\mu_2 - \mu_1)\mathrm{d}n$. If the chemical potential at location 1 is higher than that at location 2, the transfer is accompanied by a decrease in G, and so has a spontaneous tendency to occur. Only if $\mu_1 = \mu_2$ is there no change in G, and only then is the system at equilibrium. We conclude that the transition temperature, T_{trs}, is the temperature at which the chemical potentials of two phases are equal.

4.5 The dependence of stability on the conditions

At low temperatures and provided the pressure is not too low, the solid phase of a substance has the lowest chemical potential and is therefore the most stable phase. However, the chemical potentials of different phases change with temperature in different ways, and above a certain temperature the chemical potential of another phase (perhaps another solid phase, a liquid, or a gas) may turn out to be the lowest. When that happens, a transition to the second phase is spontaneous and occurs if it is kinetically feasible to do so.

[1] Recent work has suggested that water may also have a superfluid liquid phase.

(a) The temperature dependence of phase stability

The temperature dependence of the Gibbs energy is expressed in terms of the entropy of the system by eqn 3.50 ($(\partial G/\partial T)_p = -S$). Because the chemical potential of a pure substance is just another name for its molar Gibbs energy, it follows that

$$\left(\frac{\partial \mu}{\partial T}\right)_p = -S_m \qquad (4.1)$$

This relation shows that, as the temperature is raised, the chemical potential of a pure substance decreases: $S_m > 0$ for all substances, so the slope of a plot of μ against T is negative.

Equation 4.1 implies that the slope of a plot of μ against temperature is steeper for gases than for liquids, because $S_m(g) > S_m(l)$. The slope is also steeper for a liquid than the corresponding solid, because $S_m(l) > S_m(s)$ almost always. These features are illustrated in Fig. 4.9. The steep negative slope of $\mu(l)$ results in its falling below $\mu(s)$ when the temperature is high enough, and then the liquid becomes the stable phase: the solid melts. The chemical potential of the gas phase plunges steeply downwards as the temperature is raised (because the molar entropy of the vapour is so high), and there comes a temperature at which it lies lowest. Then the gas is the stable phase and vaporization is spontaneous.

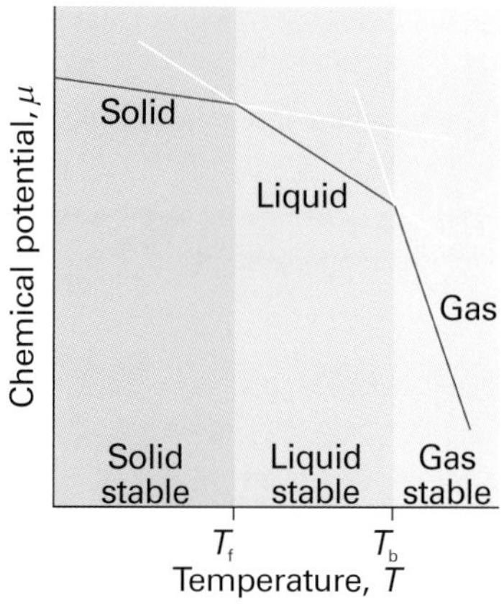

Fig. 4.9 The schematic temperature dependence of the chemical potential of the solid, liquid, and gas phases of a substance (in practice, the lines are curved). The phase with the lowest chemical potential at a specified temperature is the most stable one at that temperature. The transition temperatures, the melting and boiling temperatures (T_f and T_b, respectively), are the temperatures at which the chemical potentials of the two phases are equal.

(b) The response of melting to applied pressure

Most substances melt at a higher temperature when subjected to pressure. It is as though the pressure is preventing the formation of the less dense liquid phase. Exceptions to this behaviour include water, for which the liquid is denser than the solid. Application of pressure to water encourages the formation of the liquid phase. That is, water freezes at a lower temperature when it is under pressure.

We can rationalize the response of melting temperatures to pressure as follows. The variation of the chemical potential with pressure is expressed (from the second of eqn 3.50) by

$$\left(\frac{\partial \mu}{\partial p}\right)_T = V_m \qquad (4.2)$$

This equation shows that the slope of a plot of chemical potential against pressure is equal to the molar volume of the substance. An increase in pressure raises the chemical potential of any pure substance (because $V_m > 0$). In most cases, $V_m(l) > V_m(s)$ and the equation predicts that an increase in pressure increases the chemical potential of the liquid more than that of the solid. As shown in Fig. 4.10a, the effect of pressure in such

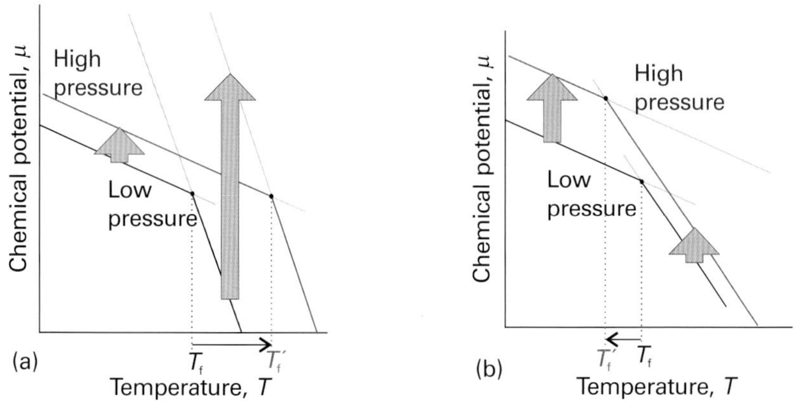

Fig. 4.10 The pressure dependence of the chemical potential of a substance depends on the molar volume of the phase. The lines show schematically the effect of increasing pressure on the chemical potential of the solid and liquid phases (in practice, the lines are curved), and the corresponding effects on the freezing temperatures. (a) In this case the molar volume of the solid is smaller than that of the liquid and $\mu(s)$ increases less than $\mu(l)$. As a result, the freezing temperature rises. (b) Here the molar volume is greater for the solid than the liquid (as for water), $\mu(s)$ increases more strongly than $\mu(l)$, and the freezing temperature is lowered.

a case is to raise the melting temperature slightly. For water, however, $V_m(l) < V_m(s)$, and an increase in pressure increases the chemical potential of the solid more than that of the liquid. In this case, the melting temperature is lowered slightly (Fig. 4.10b).

Example 4.1 *Assessing the effect of pressure on the chemical potential*

Calculate the effect on the chemical potentials of ice and water of increasing the pressure from 1.00 bar to 2.00 bar at 0°C. The density of ice is 0.917 g cm^{-3} and that of liquid water is 0.999 g cm^{-3} under these conditions.

Method From eqn 4.2, we know that the change in chemical potential of an incompressible substance when the pressure is changed by Δp is $\Delta\mu = V_m\Delta p$. Therefore, to answer the question, we need to know the molar volumes of the two phases of water. These values are obtained from the mass density, ρ, and the molar mass, M, by using $V_m = M/\rho$. We therefore use the expression $\Delta\mu = M\Delta p/\rho$.

Answer The molar mass of water is 18.02 g mol^{-1} (1.802×10^{-2} kg mol^{-1}); therefore,

$$\Delta\mu(\text{ice}) = \frac{(1.802\times10^{-2}\ \text{kg mol}^{-1})\times(1.00\times10^{5}\ \text{Pa})}{917\ \text{kg m}^{-3}} = +1.97\ \text{J mol}^{-1}$$

$$\Delta\mu(\text{water}) = \frac{(1.802\times10^{-2}\ \text{kg mol}^{-1})\times(1.00\times10^{5}\ \text{Pa})}{999\ \text{kg m}^{-3}} = +1.80\ \text{J mol}^{-1}$$

We interpret the numerical results as follows: the chemical potential of ice rises more sharply than that of water, so if they are initially in equilibrium at 1 bar, then there will be a tendency for the ice to melt at 2 bar.

Self-test 4.1 Calculate the effect of an increase in pressure of 1.00 bar on the liquid and solid phases of carbon dioxide (of molar mass 44.0 g mol^{-1}) in equilibrium with densities 2.35 g cm^{-3} and 2.50 g cm^{-3}, respectively.

[$\Delta\mu(l) = +1.87$ J mol^{-1}, $\Delta\mu(s) = +1.76$ J mol^{-1}; solid forms]

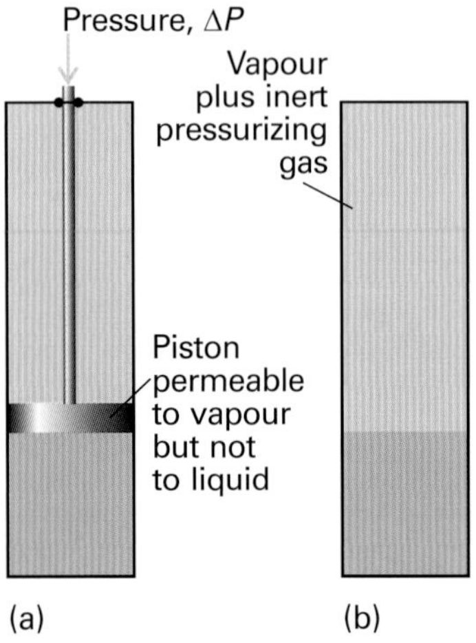

Fig. 4.11 Pressure may be applied to a condensed phases either (a) by compressing the condensed phase or (b) by subjecting it to an inert pressurizing gas. When pressure is applied, the vapour pressure of the condensed phase increases.

(c) The effect of applied pressure on vapour pressure

When pressure is applied to a condensed phase, its vapour pressure rises: in effect, molecules are squeezed out of the phase and escape as a gas. Pressure can be exerted on the condensed phases mechanically or by subjecting it to the applied pressure of an inert gas (Fig. 4.11); in the latter case, the vapour pressure is the partial pressure of the vapour in equilibrium with the condensed phase, and we speak of the **partial vapour pressure** of the substance. One complication (which we ignore here) is that, if the condensed phase is a liquid, then the pressurizing gas might dissolve and change the properties of the liquid. Another complication is that the gas phase molecules might attract molecules out of the liquid by the process of **gas solvation**, the attachment of molecules to gas phase species.

As shown in the following *Justification*, the quantitative relation between the vapour pressure, p, when a pressure ΔP is applied and the vapour pressure, p^*, of the liquid in the absence of an additional pressure is

$$p = p^* e^{V_m(l)\Delta P/RT} \tag{4.3}$$

This equation shows how the vapour pressure increases when the pressure acting on the condensed phase is increased.

Justification 4.1 *The vapour pressure of a pressurized liquid*

We calculate the vapour pressure of a pressurized liquid by using the fact that at equilibrium the chemical potentials of the liquid and its vapour are equal: $\mu(\mathrm{l}) = \mu(\mathrm{g})$. It follows that, for any change that preserves equilibrium, the resulting change in $\mu(\mathrm{l})$ must be equal to the change in $\mu(\mathrm{g})$; therefore, we can write $\mathrm{d}\mu(\mathrm{g}) = \mathrm{d}\mu(\mathrm{l})$. When the pressure P on the liquid is increased by $\mathrm{d}P$, the chemical potential of the liquid changes by $\mathrm{d}\mu(\mathrm{l}) = V_\mathrm{m}(\mathrm{l})\mathrm{d}P$. The chemical potential of the vapour changes by $\mathrm{d}\mu(\mathrm{g}) = V_\mathrm{m}(\mathrm{g})\mathrm{d}p$ where $\mathrm{d}p$ is the change in the vapour pressure we are trying to find. If we treat the vapour as a perfect gas, the molar volume can be replaced by $V_\mathrm{m}(\mathrm{g}) = RT/p$, and we obtain

$$\mathrm{d}\mu(\mathrm{g}) = \frac{RT\mathrm{d}p}{p}$$

Next, we equate the changes in chemical potentials of the vapour and the liquid:

$$\frac{RT\mathrm{d}p}{p} = V_\mathrm{m}(\mathrm{l})\mathrm{d}P$$

We can integrate this expression once we know the limits of integration.

When there is no additional pressure acting on the liquid, P (the pressure experienced by the liquid) is equal to the normal vapour pressure p^*, so when $P = p^*$, $p = p^*$ too. When there is an additional pressure ΔP on the liquid, with the result that $P = p + \Delta P$, the vapour pressure is p (the value we want to find). Provided the effect of pressure on the vapour pressure is small (as will turn out to be the case) a good approximation is to replace the p in $p + \Delta P$ by p^* itself, and to set the upper limit of the integral to $p^* + \Delta P$. The integrations required are therefore as follows:

$$RT\int_{p^*}^{p} \frac{\mathrm{d}p}{p} = \int_{p^*}^{p^*+\Delta P} V_\mathrm{m}(\mathrm{l})\mathrm{d}P$$

We now divide both sides by RT and assume that the molar volume of the liquid is the same throughout the small range of pressures involved:

$$\int_{p^*}^{p} \frac{\mathrm{d}p}{p} = \frac{V_\mathrm{m}(\mathrm{l})}{RT}\int_{p^*}^{p^*+\Delta P} \mathrm{d}P$$

Then both integrations are straightforward, and lead to

$$\ln\frac{p}{p^*} = \frac{V_\mathrm{m}(\mathrm{l})}{RT}\Delta P$$

which rearranges to eqn 4.3 because $\mathrm{e}^{\ln x} = x$.

Illustration 4.1 *The effect of applied pressure on the vapour pressure of liquid water*

For water, which has density 0.997 g cm^{-3} at 25°C and therefore molar volume 18.1 cm^3 mol^{-1}, when the pressure is increased by 10 bar (that is, $\Delta P = 1.0 \times 10^5$ Pa)

$$\frac{V_\mathrm{m}(\mathrm{l})\Delta P}{RT} = \frac{(1.81 \times 10^{-5}\ \mathrm{m^3\ mol^{-1}}) \times (1.0 \times 10^6\ \mathrm{Pa})}{(8.3145\ \mathrm{J\ K^{-1}\ mol^{-1}}) \times (298\ \mathrm{K})} = \frac{1.81 \times 1.0 \times 10}{8.3145 \times 298}$$

where we have used 1 J = 1 Pa m^3. It follows that $p = 1.0073p^*$, an increase of 0.73 per cent.

Self-test 4.2 Calculate the effect of an increase in pressure of 100 bar on the vapour pressure of benzene at 25°C, which has density 0.879 g cm^{-3}. [43 per cent]

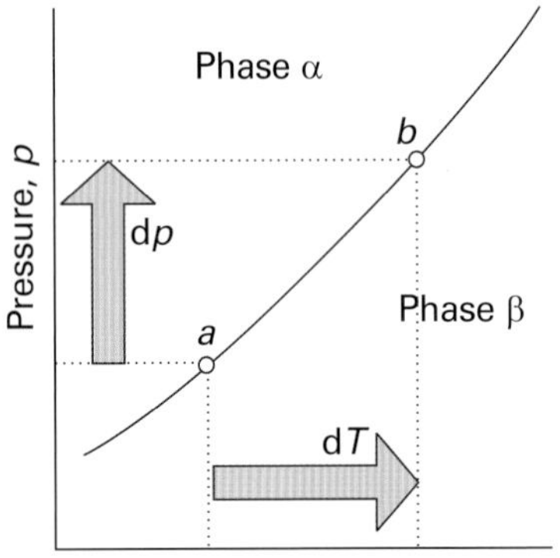

Fig. 4.12 When pressure is applied to a system in which two phases are in equilibrium (at a), the equilibrium is disturbed. It can be restored by changing the temperature, so moving the state of the system to b. It follows that there is a relation between dp and dT that ensures that the system remains in equilibrium as either variable is changed.

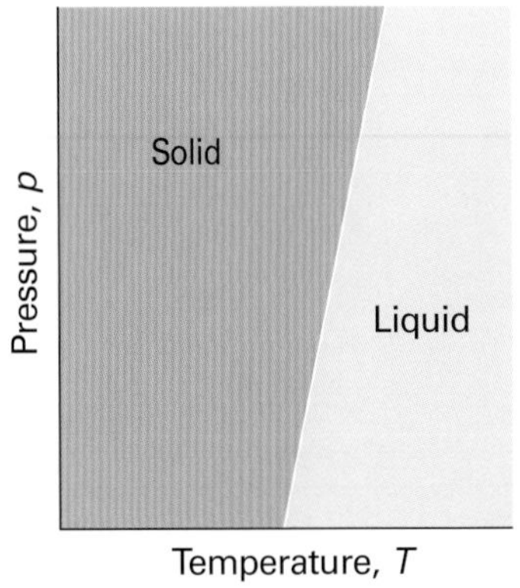

Fig. 4.13 A typical solid–liquid phase boundary slopes steeply upwards. This slope implies that, as the pressure is raised, the melting temperature rises. Most substances behave in this way.

4.6 The location of phase boundaries

We can find the precise locations of the phase boundaries—the pressures and temperatures at which two phases can coexist—by making use of the fact that, when two phases are in equilibrium, their chemical potentials must be equal. Therefore, where the phases α and β are in equilibrium,

$$\mu_\alpha(p,T) = \mu_\beta(p,T) \qquad (4.4)$$

By solving this equation for p in terms of T, we get an equation for the phase boundary.

(a) The slopes of the phase boundaries

It turns out to be simplest to discuss the phase boundaries in terms of their slopes, $\mathrm{d}p/\mathrm{d}T$. Let p and T be changed infinitesimally, but in such a way that the two phases α and β remain in equilibrium. The chemical potentials of the phases are initially equal (the two phases are in equilibrium). They remain equal when the conditions are changed to another point on the phase boundary, where the two phases continue to be in equilibrium (Fig. 4.12). Therefore, the changes in the chemical potentials of the two phases must be equal and we can write $\mathrm{d}\mu_\alpha = \mathrm{d}\mu_\beta$. Because, from eqn 3.49 ($\mathrm{d}G = V\mathrm{d}p - S\mathrm{d}T$), we know that $\mathrm{d}\mu = -S_\mathrm{m}\mathrm{d}T + V_\mathrm{m}\mathrm{d}p$ for each phase, it follows that

$$-S_{\alpha,\mathrm{m}}\mathrm{d}T + V_{\alpha,\mathrm{m}}\mathrm{d}p = -S_{\beta,\mathrm{m}}\mathrm{d}T + V_{\beta,\mathrm{m}}\mathrm{d}p$$

where $S_{\alpha,\mathrm{m}}$ and $S_{\beta,\mathrm{m}}$ are the molar entropies of the phases and $V_{\alpha,\mathrm{m}}$ and $V_{\beta,\mathrm{m}}$ are their molar volumes. Hence

$$(V_{\beta,\mathrm{m}} - V_{\alpha,\mathrm{m}})\mathrm{d}p = (S_{\beta,\mathrm{m}} - S_{\alpha,\mathrm{m}})\mathrm{d}T \qquad (4.5)$$

which rearranges into the **Clapeyron equation**:

$$\frac{\mathrm{d}p}{\mathrm{d}T} = \frac{\Delta_\mathrm{trs}S}{\Delta_\mathrm{trs}V} \qquad (4.6)$$

In this expression $\Delta_\mathrm{trs}S = S_{\beta,\mathrm{m}} - S_{\alpha,\mathrm{m}}$ and $\Delta_\mathrm{trs}V = V_{\beta,\mathrm{m}} - V_{\alpha,\mathrm{m}}$ are the entropy and volume of transition, respectively. The Clapeyron equation is an exact expression for the slope of the phase boundary and applies to any phase equilibrium of any pure substance. It implies that we can use thermodynamic data to predict the appearance of phase diagrams and to understand their form. A more practical application is to the prediction of the response of freezing and boiling points to the application of pressure.

(b) The solid–liquid boundary

Melting (fusion) is accompanied by a molar enthalpy change $\Delta_\mathrm{fus}H$ and occurs at a temperature T. The molar entropy of melting at T is therefore $\Delta_\mathrm{fus}H/T$ (Section 3.3), and the Clapeyron equation becomes

$$\frac{\mathrm{d}p}{\mathrm{d}T} = \frac{\Delta_\mathrm{fus}H}{T\Delta_\mathrm{fus}V} \qquad (4.7)$$

where $\Delta_\mathrm{fus}V$ is the change in molar volume that occurs on melting. The enthalpy of melting is positive (the only exception is helium-3) and the volume change is usually positive and always small. Consequently, the slope $\mathrm{d}p/\mathrm{d}T$ is steep and usually positive (Fig. 4.13).

We can obtain the formula for the phase boundary by integrating $\mathrm{d}p/\mathrm{d}T$, assuming that $\Delta_\mathrm{fus}H$ and $\Delta_\mathrm{fus}V$ change so little with temperature and pressure that they can be treated as constant. If the melting temperature is T^* when the pressure is p^*, and T when the pressure is p, the integration required is

$$\int_{p^*}^{p} \mathrm{d}p = \frac{\Delta_{\mathrm{fus}}H}{\Delta_{\mathrm{fus}}V}\int_{T^*}^{T}\frac{\mathrm{d}T}{T}$$

Therefore, the approximate equation of the solid–liquid boundary is

$$p \approx p^* + \frac{\Delta_{\mathrm{fus}}H}{\Delta_{\mathrm{fus}}V}\ln\frac{T}{T^*} \qquad (4.8)$$

This equation was originally obtained by yet another Thomson—James, the brother of William, Lord Kelvin. When T is close to T^*, the logarithm can be approximated by using

$$\ln\frac{T}{T^*} = \ln\left(1 + \frac{T-T^*}{T^*}\right) \approx \frac{T-T^*}{T^*}$$

therefore,

$$p \approx p^* + \frac{\Delta_{\mathrm{fus}}H}{T^*\Delta_{\mathrm{fus}}V}(T-T^*) \qquad (4.9)$$

This expression is the equation of a steep straight line when p is plotted against T (as in Fig. 4.13).

Comment 4.2

Calculations involving natural logarithms often become simpler if we note that, provided $-1 < x < 1$, $\ln(1+x) = x - \frac{1}{2}x^2 + \frac{1}{3}x^3 \cdots$. If $x \ll 1$, a good approximation is $\ln(1+x) \approx x$.

(c) The liquid–vapour boundary

The entropy of vaporization at a temperature T is equal to $\Delta_{\mathrm{vap}}H/T$; the Clapeyron equation for the liquid–vapour boundary is therefore

$$\frac{\mathrm{d}p}{\mathrm{d}T} = \frac{\Delta_{\mathrm{vap}}H}{T\Delta_{\mathrm{vap}}V} \qquad (4.10)$$

The enthalpy of vaporization is positive; $\Delta_{\mathrm{vap}}V$ is large and positive. Therefore, $\mathrm{d}p/\mathrm{d}T$ is positive, but it is much smaller than for the solid–liquid boundary. It follows that $\mathrm{d}T/\mathrm{d}p$ is large, and hence that the boiling temperature is more responsive to pressure than the freezing temperature.

Example 4.2 *Estimating the effect of pressure on the boiling temperature*

Estimate the typical size of the effect of increasing pressure on the boiling point of a liquid.

Method To use eqn 4.10 we need to estimate the right-hand side. At the boiling point, the term $\Delta_{\mathrm{vap}}H/T$ is Trouton's constant (Section 3.3b). Because the molar volume of a gas is so much greater than the molar volume of a liquid, we can write

$$\Delta_{\mathrm{vap}}V = V_{\mathrm{m}}(\mathrm{g}) - V_{\mathrm{m}}(\mathrm{l}) \approx V_{\mathrm{m}}(\mathrm{g})$$

and take for $V_{\mathrm{m}}(\mathrm{g})$ the molar volume of a perfect gas (at low pressures, at least).

Answer Trouton's constant has the value 85 J K^{-1} mol^{-1}. The molar volume of a perfect gas is about 25 dm^3 mol^{-1} at 1 atm and near but above room temperature. Therefore,

$$\frac{\mathrm{d}p}{\mathrm{d}T} \approx \frac{85\ \mathrm{J\ K^{-1}\ mol^{-1}}}{2.5\times10^{-2}\ \mathrm{m^3\ mol^{-1}}} = 3.4\times10^{3}\ \mathrm{Pa\ K^{-1}}$$

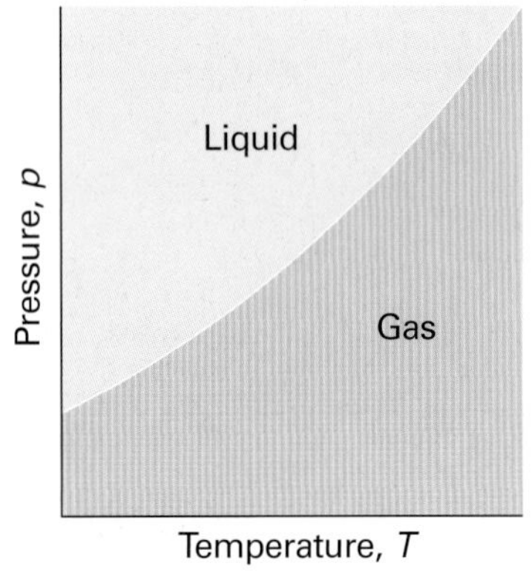

Fig. 4.14 A typical liquid–vapour phase boundary. The boundary can be regarded as a plot of the vapour pressure against the temperature. Note that, in some depictions of phase diagrams in which a logarithmic pressure scale is used, the phase boundary has the opposite curvature (see Fig. 4.7). This phase boundary terminates at the critical point (not shown).

We have used 1 J = 1 Pa m^3. This value corresponds to 0.034 atm K^{-1}, and hence to dT/dp = 29 K atm^{-1}. Therefore, a change of pressure of +0.1 atm can be expected to change a boiling temperature by about +3 K.

Self-test 4.3 Estimate dT/dp for water at its normal boiling point using the information in Table 3.2 and $V_m(g) = RT/p$. [28 K atm^{-1}]

Because the molar volume of a gas is so much greater than the molar volume of a liquid, we can write $\Delta_{vap}V \approx V_m(g)$ (as in *Example 4.2*). Moreover, if the gas behaves perfectly, $V_m(g) = RT/p$. These two approximations turn the exact Clapeyron equation into

$$\frac{dp}{dT} = \frac{\Delta_{vap}H}{T(RT/p)}$$

which rearranges into the **Clausius–Clapeyron equation** for the variation of vapour pressure with temperature:

$$\frac{d\ln p}{dT} = \frac{\Delta_{vap}H}{RT^2} \qquad (4.11)°$$

(We have used $dx/x = d\ln x$.) Like the Clapeyron equation, the Clausius–Clapeyron equation is important for understanding the appearance of phase diagrams, particularly the location and shape of the liquid–vapour and solid–vapour phase boundaries. It lets us predict how the vapour pressure varies with temperature and how the boiling temperature varies with pressure. For instance, if we also assume that the enthalpy of vaporization is independent of temperature, this equation can be integrated as follows:

$$\int_{\ln p^*}^{\ln p} d\ln p = \frac{\Delta_{vap}H}{R}\int_{T^*}^{T}\frac{dT}{T^2} = -\frac{\Delta_{vap}H}{R}\left(\frac{1}{T}-\frac{1}{T^*}\right)$$

where p^* is the vapour pressure when the temperature is T^* and p the vapour pressure when the temperature is T. Therefore, because the integral on the left evaluates to $\ln(p/p^*)$, the two vapour pressures are related by

$$p = p^* e^{-\chi} \qquad \chi = \frac{\Delta_{vap}H}{R}\left(\frac{1}{T}-\frac{1}{T^*}\right) \qquad (4.12)°$$

Equation 4.12 is plotted as the liquid–vapour boundary in Fig. 4.14. The line does not extend beyond the critical temperature T_c, because above this temperature the liquid does not exist.

Illustration 4.2 *The effect of temperature on the vapour pressure of a liquid*

Equation 4.12 can be used to estimate the vapour pressure of a liquid at any temperature from its normal boiling point, the temperature at which the vapour pressure is 1.00 atm (101 kPa). Thus, because the normal boiling point of benzene is 80°C (353 K) and (from Table 2.3), $\Delta_{vap}H^{\ominus} = 30.8$ kJ mol^{-1}, to calculate the vapour pressure at 20°C (293 K), we write

$$\chi = \frac{3.08\times10^4\ \text{J mol}^{-1}}{8.3145\ \text{J K}^{-1}\ \text{mol}^{-1}}\left(\frac{1}{293\ \text{K}}-\frac{1}{353\ \text{K}}\right) = \frac{3.08\times10^4}{8.3145}\left(\frac{1}{293}-\frac{1}{353}\right)$$

and substitute this value into eqn 4.12 with $p^* = 101$ kPa. The result is 12 kPa. The experimental value is 10 kPa.

A note on good practice Because exponential functions are so sensitive, it is good practice to carry out numerical calculations like this without evaluating the intermediate steps and using rounded values.

(d) The solid–vapour boundary

The only difference between this case and the last is the replacement of the enthalpy of vaporization by the enthalpy of sublimation, $\Delta_{sub}H$. Because the enthalpy of sublimation is greater than the enthalpy of vaporization ($\Delta_{sub}H = \Delta_{fus}H + \Delta_{vap}H$), the equation predicts a steeper slope for the sublimation curve than for the vaporization curve at similar temperatures, which is near where they meet at the triple point (Fig. 4.15).

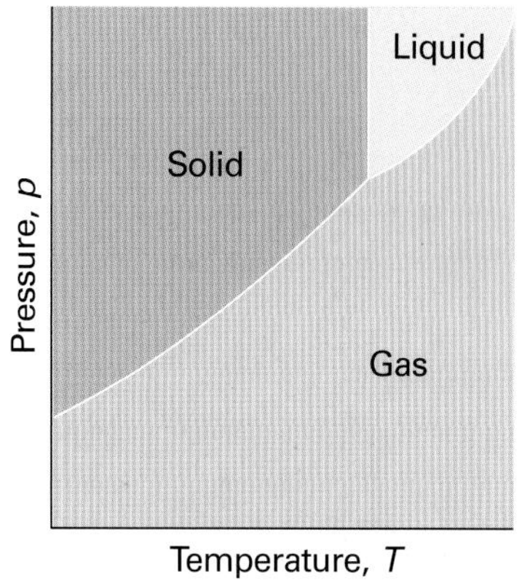

Fig. 4.15 Near the point where they coincide (at the triple point), the solid–gas boundary has a steeper slope than the liquid–gas boundary because the enthalpy of sublimation is greater than the enthalpy of vaporization and the temperatures that occur in the Clausius–Clapeyron equation for the slope have similar values.

4.7 The Ehrenfest classification of phase transitions

There are many different types of phase transition, including the familiar examples of fusion and vaporization and the less familiar examples of solid–solid, conducting–superconducting, and fluid–superfluid transitions. We shall now see that it is possible to use thermodynamic properties of substances, and in particular the behaviour of the chemical potential, to classify phase transitions into different types. The classification scheme was originally proposed by Paul Ehrenfest, and is known as the **Ehrenfest classification**.

Many familiar phase transitions, like fusion and vaporization, are accompanied by changes of enthalpy and volume. These changes have implications for the slopes of the chemical potentials of the phases at either side of the phase transition. Thus, at the transition from a phase α to another phase β,

$$\left(\frac{\partial \mu_\beta}{\partial p}\right)_T - \left(\frac{\partial \mu_\alpha}{\partial p}\right)_T = V_{\beta,m} - V_{\alpha,m} = \Delta_{trs}V$$
$$\left(\frac{\partial \mu_\beta}{\partial T}\right)_p - \left(\frac{\partial \mu_\alpha}{\partial T}\right)_p = -S_{\beta,m} + S_{\alpha,m} = \Delta_{trs}S = \frac{\Delta_{trs}H}{T_{trs}} \qquad (4.13)$$

Because $\Delta_{trs}V$ and $\Delta_{trs}H$ are non-zero for melting and vaporization, it follows that for such transitions the slopes of the chemical potential plotted against either pressure or temperature are different on either side of the transition (Fig. 4.16a). In other words, the first derivatives of the chemical potentials with respect to pressure and temperature are discontinuous at the transition.

A transition for which the first derivative of the chemical potential with respect to temperature is discontinuous is classified as a **first-order phase transition**. The constant-pressure heat capacity, C_p, of a substance is the slope of a plot of the enthalpy with respect to temperature. At a first-order phase transition, H changes by a finite amount for an infinitesimal change of temperature. Therefore, at the transition the heat capacity is infinite. The physical reason is that heating drives the transition rather than raising the temperature. For example, boiling water stays at the same temperature even though heat is being supplied.

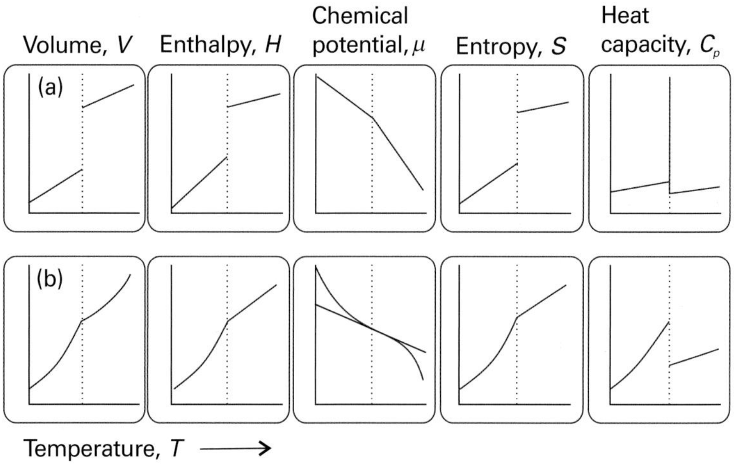

Fig. 4.16 The changes in thermodynamic properties accompanying (a) first-order and (b) second-order phase transitions.

A **second-order phase transition** in the Ehrenfest sense is one in which the first derivative of μ with respect to temperature is continuous but its second derivative is discontinuous. A continuous slope of μ (a graph with the same slope on either side of the transition) implies that the volume and entropy (and hence the enthalpy) do not change at the transition (Fig. 4.16b). The heat capacity is discontinuous at the transition but does not become infinite there. An example of a second-order transition is the conducting–superconducting transition in metals at low temperatures.[2]

The term **λ-transition** is applied to a phase transition that is not first-order yet the heat capacity becomes infinite at the transition temperature. Typically, the heat capacity of a system that shows such a transition begins to increase well before the transition (Fig. 4.17), and the shape of the heat capacity curve resembles the Greek letter lambda. This type of transition includes order–disorder transitions in alloys, the onset of ferromagnetism, and the fluid–superfluid transition of liquid helium.

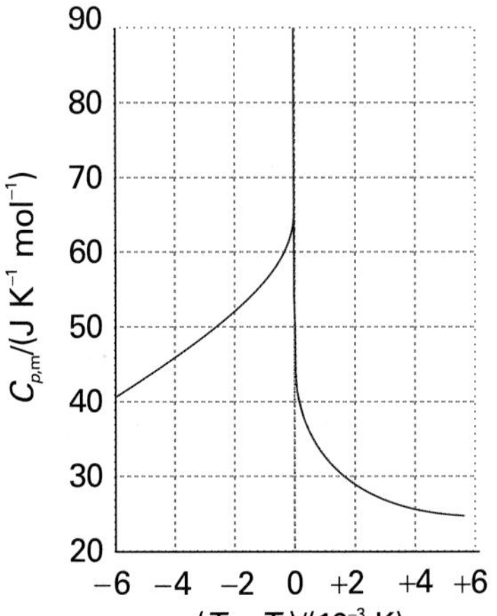

Fig. 4.17 The λ-curve for helium, where the heat capacity rises to infinity. The shape of this curve is the origin of the name λ-transition.

Molecular interpretation 4.1 *Second-order phase transitions and λ-transitions*

One type of second-order transition is associated with a change in symmetry of the crystal structure of a solid. Thus, suppose the arrangement of atoms in a solid is like that represented in Fig. 4.18a, with one dimension (technically, of the unit cell) longer than the other two, which are equal. Such a crystal structure is classified as tetragonal (see Section 20.1). Moreover, suppose the two shorter dimensions increase more than the long dimension when the temperature is raised. There may come a stage when the three dimensions become equal. At that point the crystal has cubic symmetry (Fig. 4.18b), and at higher temperatures it will expand equally in all three directions (because there is no longer any distinction between them). The tetragonal → cubic phase transition has occurred, but as it has not involved a discontinuity in the interaction energy between the atoms or the volume they occupy, the transition is not first-order.

2 A metallic conductor is a substance with an electrical conductivity that decreases as the temperature increases. A superconductor is a solid that conducts electricity without resistance. See Chapter 20 for more details.

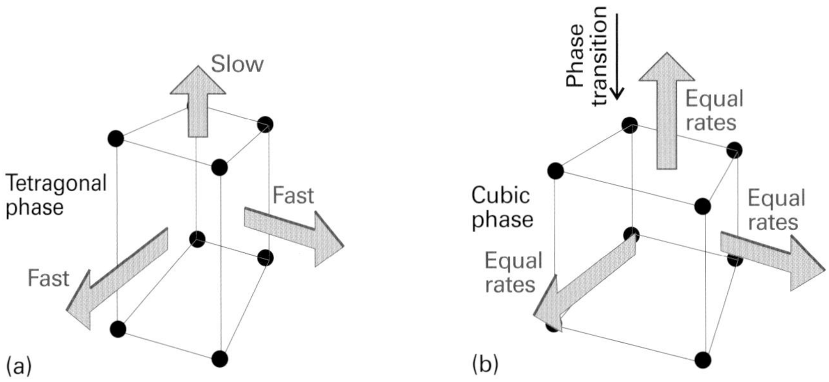

Fig. 4.18 One version of a second-order phase transition in which (a) a tetragonal phase expands more rapidly in two directions than a third, and hence becomes a cubic phase, which (b) expands uniformly in three directions as the temperature is raised. There is no rearrangement of atoms at the transition temperature, and hence no enthalpy of transition.

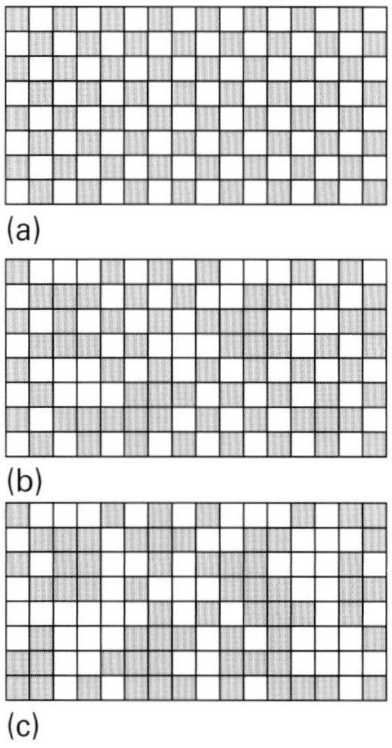

Fig. 4.19 An order–disorder transition. (a) At $T = 0$, there is perfect order, with different kinds of atoms occupying alternate sites. (b) As the temperature is increased, atoms exchange locations and islands of each kind of atom form in regions of the solid. Some of the original order survives. (c) At and above the transition temperature, the islands occur at random throughout the sample.

The order–disorder transition in β-brass (CuZn) is an example of a λ-transition. The low-temperature phase is an orderly array of alternating Cu and Zn atoms. The high-temperature phase is a random array of the atoms (Fig. 4.19). At $T = 0$ the order is perfect, but islands of disorder appear as the temperature is raised. The islands form because the transition is cooperative in the sense that, once two atoms have exchanged locations, it is easier for their neighbours to exchange their locations. The islands grow in extent, and merge throughout the crystal at the transition temperature (742 K). The heat capacity increases as the transition temperature is approached because the cooperative nature of the transition means that it is increasingly easy for the heat supplied to drive the phase transition rather than to be stored as thermal motion.

Checklist of key ideas

- ☐ 1. A phase is a form of matter that is uniform throughout in chemical composition and physical state.
- ☐ 2. A transition temperature is the temperature at which the two phases are in equilibrium.
- ☐ 3. A metastable phase is a thermodynamically unstable phase that persists because the transition is kinetically hindered.
- ☐ 4. A phase diagram is a diagram showing the regions of pressure and temperature at which its various phases are thermodynamically stable.
- ☐ 5. A phase boundary is a line separating the regions in a phase diagram showing the values of p and T at which two phases coexist in equilibrium.
- ☐ 6. The vapour pressure is the pressure of a vapour in equilibrium with the condensed phase.
- ☐ 7. Boiling is the condition of free vaporization throughout the liquid.
- ☐ 8. The boiling temperature is the temperature at which the vapour pressure of a liquid is equal to the external pressure.
- ☐ 9. The critical temperature is the temperature at which a liquid surface disappears and above which a liquid does not exist whatever the pressure. The critical pressure is the vapour pressure at the critical temperature.
- ☐ 10. A supercritical fluid is a dense fluid phase above the critical temperature.
- ☐ 11. The melting temperature (or freezing temperature) is the temperature at which, under a specified pressure, the liquid and solid phases of a substance coexist in equilibrium.
- ☐ 12. The triple point is a point on a phase diagram at which the three phase boundaries meet and all three phases are in mutual equilibrium.
- ☐ 13. The chemical potential μ of a pure substance is the molar Gibbs energy of the substance.

☐ 14. The chemical potential is uniform throughout a system at equilibrium.

☐ 15. The chemical potential varies with temperature as $(\partial\mu/\partial T)_p = -S_m$ and with pressure as $(\partial\mu/\partial p)_T = V_m$.

☐ 16. The vapour pressure in the presence of applied pressure is given by $p = p^{*}e^{V_m\Delta P/RT}$.

☐ 17. The temperature dependence of the vapour pressure is given by the Clapeyron equation, $dp/dT = \Delta_{trs}S/\Delta_{trs}V$.

☐ 18. The temperature dependence of the vapour pressure of a condensed phase is given by the Clausius–Clapeyron equation, $d\ln p/dT = \Delta_{vap}H/RT^2$.

☐ 19. The Ehrenfest classification is a classification of phase transitions based on the behaviour of the chemical potential.

Further reading

Articles and texts

E.K.H. Salje, Phase transitions, structural. In *Encyclopedia of applied physics* (ed. G.L. Trigg), **13**, 373. VCH, New York (1995).

J.M. Sanchez, Order–disorder transitions. In *Encyclopedia of applied physics* (ed. G.L. Trigg), **13**, 1. VCH, New York (1995).

K.M. Scholsky, Supercritical phase transitions at very high pressure. *J. Chem. Educ.* **66**, 989 (1989).

W.D. Callister, Jr., *Materials science and engineering, an introduction.* Wiley, New York (2000).

Sources of data and information[3]

T. Boublik, V. Fried, and E. Hála, *The vapor pressures of pure substances.* Elsevier, Amsterdam (1984).

R.C. Weast (ed.), *Handbook of chemistry and physics*, Vol. 81. CRC Press, Boca Raton (2004).

Discussion questions

4.1 Discuss the implications for phase stability of the variation of chemical potential with temperature and pressure.

4.2 Suggest a physical interpretation of the phenomena of superheating and supercooling.

4.3 Discuss what would be observed as a sample of water is taken along a path that encircles and is close to its critical point.

4.4 Use the phase diagram in Fig. 4.4 to state what would be observed when a sample of carbon dioxide, initially at 1.0 atm and 298 K, is subjected to the following cycle: (a) isobaric (constant–pressure) heating to 320 K, (b) isothermal compression to 100 atm, (c) isobaric cooling to 210 K, (d) isothermal decompression to 1.0 atm, (e) isobaric heating to 298 K.

4.5 The use of supercritical fluids for the extraction of a component from a complicated mixture is not confined to the decaffeination of coffee. Consult library and internet resources and prepare a discussion of the principles, advantages, disadvantages, and current uses of supercritical fluid extraction technology.

4.6 Explain the significance of the Clapeyron equation and of the Clausius–Clapeyron equation.

4.7 Distinguish between a first-order phase transition, a second-order phase transition, and a λ-transition at both molecular and macroscopic levels.

Exercises

4.1(a) The vapour pressure of dichloromethane at 24.1°C is 53.3 kPa and its enthalpy of vaporization is 28.7 kJ mol^{-1}. Estimate the temperature at which its vapour pressure is 70.0 kPa.

4.1(b) The vapour pressure of a substance at 20.0°C is 58.0 kPa and its enthalpy of vaporization is 32.7 kJ mol^{-1}. Estimate the temperature at which its vapour pressure is 66.0 kPa.

4.2(a) The molar volume of a certain solid is 161.0 cm^3 mol^{-1} at 1.00 atm and 350.75 K, its melting temperature. The molar volume of the liquid at this temperature and pressure is 163.3 cm^3 mol^{-1}. At 100 atm the melting temperature changes to 351.26 K. Calculate the enthalpy and entropy of fusion of the solid.

4.2(b) The molar volume of a certain solid is 142.0 cm^3 mol^{-1} at 1.00 atm and 427.15 K, its melting temperature. The molar volume of the liquid at this temperature and pressure is 152.6 cm^3 mol^{-1}. At 1.2 MPa the melting temperature changes to 429.26 K. Calculate the enthalpy and entropy of fusion of the solid.

4.3(a) The vapour pressure of a liquid in the temperature range 200 K to 260 K was found to fit the expression $\ln(p/\text{Torr}) = 16.255 - 2501.8/(T/\text{K})$. Calculate the enthalpy of vaporization of the liquid.

4.3(b) The vapour pressure of a liquid in the temperature range 200 K to 260 K was found to fit the expression $\ln(p/\text{Torr}) = 18.361 - 3036.8/(T/\text{K})$. Calculate the enthalpy of vaporization of the liquid.

[3] See *Further reading* in Chapter 2 for additional sources of thermochemical data.

4.4(a) The vapour pressure of benzene between 10°C and 30°C fits the expression $\log(p/\text{Torr}) = 7.960 - 1780/(T/\text{K})$. Calculate (a) the enthalpy of vaporization and (b) the normal boiling point of benzene.

4.4(b) The vapour pressure of a liquid between 15°C and 35°C fits the expression $\log(p/\text{Torr}) = 8.750 - 1625/(T/\text{K})$. Calculate (a) the enthalpy of vaporization and (b) the normal boiling point of the liquid.

4.5(a) When benzene freezes at 5.5°C its density changes from 0.879 g cm^{-3} to 0.891 g cm^{-3}. Its enthalpy of fusion is 10.59 kJ mol^{-1}. Estimate the freezing point of benzene at 1000 atm.

4.5(b) When a certain liquid freezes at −3.65°C its density changes from 0.789 g cm^{-3} to 0.801 g cm^{-3}. Its enthalpy of fusion is 8.68 kJ mol^{-1}. Estimate the freezing point of the liquid at 100 MPa.

4.6(a) In July in Los Angeles, the incident sunlight at ground level has a power density of 1.2 kW m^{-2} at noon. A swimming pool of area 50 m^2 is directly exposed to the sun. What is the maximum rate of loss of water? Assume that all the radiant energy is absorbed.

4.6(b) Suppose the incident sunlight at ground level has a power density of 0.87 kW m^{-2} at noon. What is the maximum rate of loss of water from a lake of area 1.0 ha? (1 ha = 10^4 m^2.) Assume that all the radiant energy is absorbed.

4.7(a) An open vessel containing (a) water, (b) benzene, (c) mercury stands in a laboratory measuring 5.0 m × 5.0 m × 3.0 m at 25°C. What mass of each substance will be found in the air if there is no ventilation? (The vapour pressures are (a) 3.2 kPa, (b) 13.1 kPa, (c) 0.23 Pa.)

4.7(b) On a cold, dry morning after a frost, the temperature was −5°C and the partial pressure of water in the atmosphere fell to 0.30 kPa. Will the frost sublime? What partial pressure of water would ensure that the frost remained?

4.8(a) Naphthalene, $C_{10}H_8$, melts at 80.2°C. If the vapour pressure of the liquid is 1.3 kPa at 85.8°C and 5.3 kPa at 119.3°C, use the Clausius–Clapeyron equation to calculate (a) the enthalpy of vaporization, (b) the normal boiling point, and (c) the enthalpy of vaporization at the boiling point.

4.8(b) The normal boiling point of hexane is 69.0°C. Estimate (a) its enthalpy of vaporization and (b) its vapour pressure at 25°C and 60°C.

4.9(a) Calculate the melting point of ice under a pressure of 50 bar. Assume that the density of ice under these conditions is approximately 0.92 g cm^{-3} and that of liquid water is 1.00 g cm^{-3}.

4.9(b) Calculate the melting point of ice under a pressure of 10 MPa. Assume that the density of ice under these conditions is approximately 0.915 g cm^{-3} and that of liquid water is 0.998 g cm^{-3}.

4.10(a) What fraction of the enthalpy of vaporization of water is spent on expanding the water vapour?

4.10(b) What fraction of the enthalpy of vaporization of ethanol is spent on expanding its vapour?

Problems*

Numerical problems

4.1 The temperature dependence of the vapour pressure of solid sulfur dioxide can be approximately represented by the relation $\log(p/\text{Torr}) = 10.5916 - 1871.2/(T/\text{K})$ and that of liquid sulfur dioxide by $\log(p/\text{Torr}) = 8.3186 - 1425.7/(T/\text{K})$. Estimate the temperature and pressure of the triple point of sulfur dioxide.

4.2 Prior to the discovery that freon-12 (CF_2Cl_2) was harmful to the Earth's ozone layer, it was frequently used as the dispersing agent in spray cans for hair spray, etc. Its enthalpy of vaporization at its normal boiling point of −29.2°C is 20.25 kJ mol^{-1}. Estimate the pressure that a can of hair spray using freon-12 had to withstand at 40°C, the temperature of a can that has been standing in sunlight. Assume that $\Delta_{\text{vap}}H$ is a constant over the temperature range involved and equal to its value at −29.2°C.

4.3 The enthalpy of vaporization of a certain liquid is found to be 14.4 kJ mol^{-1} at 180 K, its normal boiling point. The molar volumes of the liquid and the vapour at the boiling point are 115 cm^3 mol^{-1} and 14.5 dm^3 mol^{-1}, respectively. (a) Estimate $\text{d}p/\text{d}T$ from the Clapeyron equation and (b) the percentage error in its value if the Clausius–Clapeyron equation is used instead.

4.4 Calculate the difference in slope of the chemical potential against temperature on either side of (a) the normal freezing point of water and (b) the normal boiling point of water. (c) By how much does the chemical potential of water supercooled to −5.0°C exceed that of ice at that temperature?

4.5 Calculate the difference in slope of the chemical potential against pressure on either side of (a) the normal freezing point of water and (b) the normal boiling point of water. The densities of ice and water at 0°C are 0.917 g cm^{-3} and 1.000 g cm^{-3}, and those of water and water vapour at 100°C are 0.958 g cm^{-3} and 0.598 g dm^{-3}, respectively. By how much does the chemical potential of water vapour exceed that of liquid water at 1.2 atm and 100°C?

4.6 The enthalpy of fusion of mercury is 2.292 kJ mol^{-1}, and its normal freezing point is 234.3 K with a change in molar volume of +0.517 cm^{-3} mol^{-1} on melting. At what temperature will the bottom of a column of mercury (density 13.6 g cm^{-3}) of height 10.0 m be expected to freeze?

4.7 50.0 dm^3 of dry air was slowly bubbled through a thermally insulated beaker containing 250 g of water initially at 25°C. Calculate the final temperature. (The vapour pressure of water is approximately constant at 3.17 kPa throughout, and its heat capacity is 75.5 J K^{-1} mol^{-1}. Assume that the air is not heated or cooled and that water vapour is a perfect gas.)

4.8 The vapour pressure, p, of nitric acid varies with temperature as follows:

θ/°C	0	20	40	50	70	80	90	100
p/kPa	1.92	6.38	17.7	27.7	62.3	89.3	124.9	170.9

What are (a) the normal boiling point and (b) the enthalpy of vaporization of nitric acid?

4.9 The vapour pressure of the ketone carvone (M = 150.2 g mol^{-1}), a component of oil of spearmint, is as follows:

θ/°C	57.4	100.4	133.0	157.3	203.5	227.5
p/Torr	1.00	10.0	40.0	100	400	760

What are (a) the normal boiling point and (b) the enthalpy of vaporization of carvone?

* Problems denoted by the symbol ‡ were supplied by Charles Trapp, Carmen Giunta, and Marshall Cady.

4.10 Construct the phase diagram for benzene near its triple point at 36 Torr and 5.50°C using the following data: $\Delta_{fus}H = 10.6$ kJ mol^{-1}, $\Delta_{vap}H = 30.8$ kJ mol^{-1}, $\rho(s) = 0.891$ g cm^{-3}, $\rho(l) = 0.879$ g cm^{-3}.

4.11‡ In an investigation of thermophysical properties of toluene (R.D. Goodwin *J. Phys. Chem. Ref. Data* **18**, 1565 (1989)) presented expressions for two coexistence curves (phase boundaries). The solid–liquid coexistence curve is given by

$$p/\text{bar} = p_3/\text{bar} + 1000 \times (5.60 + 11.727x)x$$

where $x = T/T_3 - 1$ and the triple point pressure and temperature are $p_3 = 0.4362$ μbar and $T_3 = 178.15$ K. The liquid–vapour curve is given by:

$$\ln(p/\text{bar}) = -10.418/y + 21.157 - 15.996y + 14.015y^2 - 5.0120y^3 + 4.7224(1-y)^{1.70}$$

where $y = T/T_c = T/(593.95\text{ K})$. (a) Plot the solid–liquid and liquid–vapour phase boundaries. (b) Estimate the standard melting point of toluene. (c) Estimate the standard boiling point of toluene. (d) Compute the standard enthalpy of vaporization of toluene, given that the molar volumes of the liquid and vapour at the normal boiling point are 0.12 dm^3 mol^{-1} and 30.3 dm^3 mol^{-1}, respectively.

4.12‡ In a study of the vapour pressure of chloromethane, A. Bah and N. Dupont-Pavlovsky (*J. Chem. Eng. Data* **40**, 869 (1995)) presented data for the vapour pressure over solid chloromethane at low temperatures. Some of that data is shown below:

T/K	145.94	147.96	149.93	151.94	153.97	154.94
p/Pa	13.07	18.49	25.99	36.76	50.86	59.56

Estimate the standard enthalpy of sublimation of chloromethane at 150 K. (Take the molar volume of the vapour to be that of a perfect gas, and that of the solid to be negligible.)

Theoretical problems

4.13 Show that, for a transition between two incompressible solid phases, ΔG is independent of the pressure.

4.14 The change in enthalpy is given by $dH = C_p dT + V dp$. The Clapeyron equation relates dp and dT at equilibrium, and so in combination the two equations can be used to find how the enthalpy changes along a phase boundary as the temperature changes and the two phases remain in equilibrium. Show that $d(\Delta H/T) = \Delta C_p \, d\ln T$.

4.15 In the 'gas saturation method' for the measurement of vapour pressure, a volume V of gas (as measured at a temperature T and a pressure p) is bubbled slowly through the liquid that is maintained at the temperature T, and a mass loss m is measured. Show that the vapour pressure, p, of the liquid is related to its molar mass, M, by $p = AmP/(1 + Am)$, where $A = RT/MPV$. The vapour pressure of geraniol ($M = 154.2$ g mol^{-1}), which is a component of oil of roses, was measured at 110°C. It was found that, when 5.00 dm^3 of nitrogen at 760 Torr was passed slowly through the heated liquid, the loss of mass was 0.32 g. Calculate the vapour pressure of geraniol.

4.16 Combine the barometric formula (stated in *Impact I1.1*) for the dependence of the pressure on altitude with the Clausius–Clapeyron equation, and predict how the boiling temperature of a liquid depends on the altitude and the ambient temperature. Take the mean ambient temperature as 20°C and predict the boiling temperature of water at 3000 m.

4.17 Figure 4.9 gives a schematic representation of how the chemical potentials of the solid, liquid, and gaseous phases of a substance vary with temperature. All have a negative slope, but it is unlikely that they are truly straight lines as indicated in the illustration. Derive an expression for the curvatures (specifically, the second derivatives with respect to temperature) of these lines. Is there a restriction on the curvature of these lines? Which state of matter shows the greatest curvature?

4.18 The Clapeyron equation does not apply to second-order phase transitions, but there are two analogous equations, the *Ehrenfest equations*, that do. They are:

$$\frac{dp}{dT} = \frac{\alpha_2 - \alpha_1}{\kappa_{T,2} - \kappa_{T,1}} \qquad \frac{dp}{dT} = \frac{C_{p,m2} - C_{p,m1}}{TV_m(\alpha_2 - \alpha_1)}$$

where α is the expansion coefficient, κ_T the isothermal compressibility, and the subscripts 1 and 2 refer to two different phases. Derive these two equations. Why does the Clapeyron equation not apply to second-order transitions?

4.19 For a first-order phase transition, to which the Clapeyron equation does apply, prove the relation

$$C_S = C_p - \frac{\alpha V \Delta_{trs}H}{\Delta_{trs}V}$$

where $C_S = (\partial q/\partial T)_S$ is the heat capacity along the coexistence curve of two phases.

Applications: to biology and engineering

4.20 Proteins are *polypeptides*, polymers of amino acids that can exist in ordered structures stabilized by a variety of molecular interactions. However, when certain conditions are changed, the compact structure of a polypeptide chain may collapse into a random coil. This structural change may be regarded as a phase transition occurring at a characteristic transition temperature, the *melting temperature*, T_m, which increases with the strength and number of intermolecular interactions in the chain. A thermodynamic treatment allows predictions to be made of the temperature T_m for the unfolding of a helical polypeptide held together by hydrogen bonds into a random coil. If a polypeptide has n amino acids, $n - 4$ hydrogen bonds are formed to form an α-helix, the most common type of helix in naturally occurring proteins (see Chapter 19). Because the first and last residues in the chain are free to move, $n - 2$ residues form the compact helix and have restricted motion. Based on these ideas, the molar Gibbs energy of unfolding of a polypeptide with $n \geq 5$ may be written as

$$\Delta G_m = (n-4)\Delta_{hb}H_m - (n-2)T\Delta_{hb}S_m$$

where $\Delta_{hb}H_m$ and $\Delta_{hb}S_m$ are, respectively, the molar enthalpy and entropy of dissociation of hydrogen bonds in the polypeptide. (a) Justify the form of the equation for the Gibbs energy of unfolding. That is, why are the enthalpy and entropy terms written as $(n-4)\Delta_{hb}H_m$ and $(n-2)\Delta_{hb}S_m$, respectively? (b) Show that T_m may be written as

$$T_m = \frac{(n-4)\Delta_{hb}H_m}{(n-2)\Delta_{hb}S_m}$$

(c) Plot $T_m/(\Delta_{hb}H_m/\Delta_{hb}S_m)$ for $5 \leq n \leq 20$. At what value of n does T_m change by less than 1% when n increases by one?

4.21‡ The use of supercritical fluids as mobile phases in SFC depends on their properties as nonpolar solvents. The solubility parameter, δ, is defined as $(\Delta U_{cohesive}/V_m)^{1/2}$, where $\Delta U_{cohesive}$ is the cohesive energy of the solvent, the energy per mole needed to increase the volume isothermally to an infinite value. Diethyl ether, carbon tetrachloride, and dioxane have solubility parameter ranges of 7–8, 8–9, and 10–11, respectively. (a) Derive a practical equation for the computation of the isotherms for the reduced internal energy change, $\Delta U_r(T_r,V_r)$ defined as

$$\Delta U_r(T_r,V_r) = \frac{U_r(T_r,V_r) - U_r(T_r,\infty)}{p_c V_c}$$

(b) Draw a graph of ΔU_r against p_r for the isotherms $T_r = 1, 1.2$, and 1.5 in the reduced pressure range for which $0.7 \leq V_r \leq 2$. (c) Draw a graph of δ against p_r for the carbon dioxide isotherms $T_r = 1$ and 1.5 in the reduced pressure range for which $1 \leq V_r \leq 3$. In what pressure range at $T_f = 1$ will carbon dioxide have solvent properties similar to those of liquid carbon tetrachloride? *Hint*. Use mathematical software or a spreadsheet.

4.22‡ A substance as well–known as methane still receives research attention because it is an important component of natural gas, a commonly used fossil fuel . Friend *et al.* have published a review of thermophysical properties of methane (D.G. Friend, J.F. Ely, and H. Ingham, *J. Phys. Chem. Ref. Data* **18**, 583 (1989)), which included the following data describing the liquid–vapour phase boundary.

T/K	100	108	110	112	114	120	130	140	150	160	170	190
p/MPa	0.034	0.074	0.088	0.104	0.122	0.192	0.368	0.642	1.041	1.593	2.329	4.521

(a) Plot the liquid–vapour phase boundary. (b) Estimate the standard boiling point of methane. (c) Compute the standard enthalpy of vaporization of methane, given that the molar volumes of the liquid and vapour at the standard boiling point are 3.80×10^{-2} and 8.89 dm^3 mol^{-1}, respectively.

4.23‡ Diamond, an allotrope of carbon, is the hardest substance and the best conductor of heat yet characterized. For these reasons, diamond is used widely in industrial applications that require a strong abrasive. Unfortunately, it is difficult to synthesize diamond from the more readily available allotropes of carbon, such as graphite. To illustrate this point, calculate the pressure required to convert graphite into diamond at 25°C. The following data apply to 25°C and 100 kPa. Assume the specific volume, V_s, and κ_T are constant with respect to pressure changes.

	Graphite	Diamond
$\Delta_f G^{\ominus}/(\text{kJ mol}^{-1})$	0	+2.8678
$V_s/(\text{cm}^3\ \text{g}^{-1})$	0.444	0.284
κ_T/kPa	3.04×10^{-8}	0.187×10^{-8}

5 Simple mixtures

This chapter begins by developing the concept of chemical potential to show that it is a particular case of a class of properties called partial molar quantities. Then it explores how to use the chemical potential of a substance to describe the physical properties of mixtures. The underlying principle to keep in mind is that at equilibrium the chemical potential of a species is the same in every phase. We see, by making use of the experimental observations known as Raoult's and Henry's laws, how to express the chemical potential of a substance in terms of its mole fraction in a mixture. With this result established, we can calculate the effect of a solute on certain thermodynamic properties of a solution. These properties include the lowering of vapour pressure of the solvent, the elevation of its boiling point, the depression of its freezing point, and the origin of osmotic pressure. Finally, we see how to express the chemical potential of a substance in a real mixture in terms of a property known as the activity. We see how the activity may be measured, and conclude with a discussion of how the standard states of solutes and solvents are defined and how ion–ion interactions are taken into account in electrolyte solutions.

Chemistry deals with mixtures, including mixtures of substances that can react together. Therefore, we need to generalize the concepts introduced so far to deal with substances that are mingled together. As a first step towards dealing with chemical reactions (which are treated in Chapter 7), here we consider mixtures of substances that do not react together. At this stage we deal mainly with *binary mixtures*, which are mixtures of two components, A and B. We shall therefore often be able to simplify equations by making use of the relation $x_A + x_B = 1$.

The thermodynamic description of mixtures

We have already seen that the partial pressure, which is the contribution of one component to the total pressure, is used to discuss the properties of mixtures of gases. For a more general description of the thermodynamics of mixtures we need to introduce other analogous 'partial' properties.

5.1 Partial molar quantities

The easiest partial molar property to visualize is the 'partial molar volume', the contribution that a component of a mixture makes to the total volume of a sample.

(a) Partial molar volume

Imagine a huge volume of pure water at 25°C. When a further 1 mol H_2O is added, the volume increases by 18 cm^3 and we can report that 18 $cm^3\,mol^{-1}$ is the molar volume of pure water. However, when we add 1 mol H_2O to a huge volume of pure ethanol, the volume increases by only 14 cm^3. The reason for the different increase in volume is that the volume occupied by a given number of water molecules depends on the identity of the molecules that surround them. In the latter case there is so much ethanol present that each H_2O molecule is surrounded by ethanol molecules, and the packing of the molecules results in the H_2O molecules increasing the volume by only 14 cm^3. The quantity 14 $cm^3\ mol^{-1}$ is the partial molar volume of water in pure ethanol. In general, the **partial molar volume** of a substance A in a mixture is the change in volume per mole of A added to a large volume of the mixture.

The partial molar volumes of the components of a mixture vary with composition because the environment of each type of molecule changes as the composition changes from pure A to pure B. It is this changing molecular environment, and the consequential modification of the forces acting between molecules, that results in the variation of the thermodynamic properties of a mixture as its composition is changed. The partial molar volumes of water and ethanol across the full composition range at 25°C are shown in Fig. 5.1.

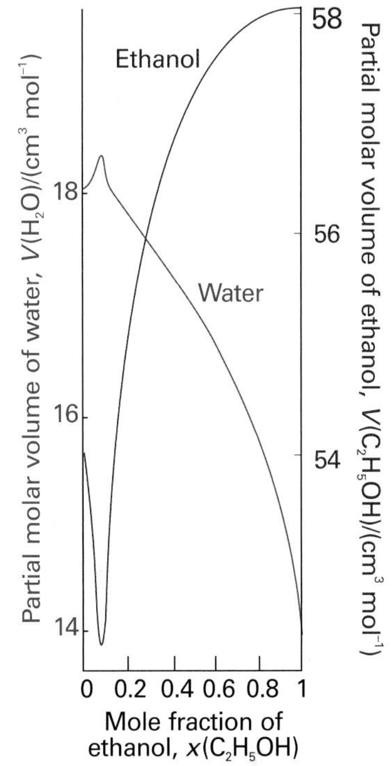

Fig. 5.1 The partial molar volumes of water and ethanol at 25°C. Note the different scales (water on the left, ethanol on the right).

The partial molar volume, V_J, of a substance J at some general composition is defined formally as follows:

$$V_J = \left(\frac{\partial V}{\partial n_J}\right)_{p,T,n'} \qquad [5.1]$$

where the subscript n' signifies that the amounts of all other substances present are constant.[1] The partial molar volume is the slope of the plot of the total volume as the amount of J is changed, the pressure, temperature, and amount of the other components being constant (Fig. 5.2). Its value depends on the composition, as we saw for water and ethanol. The definition in eqn 5.1 implies that, when the composition of the mixture is changed by the addition of dn_A of A and dn_B of B, then the total volume of the mixture changes by

$$dV = \left(\frac{\partial V}{\partial n_A}\right)_{p,T,n_B} dn_A + \left(\frac{\partial V}{\partial n_B}\right)_{p,T,n_A} dn_B = V_A dn_A + V_B dn_B \qquad (5.2)$$

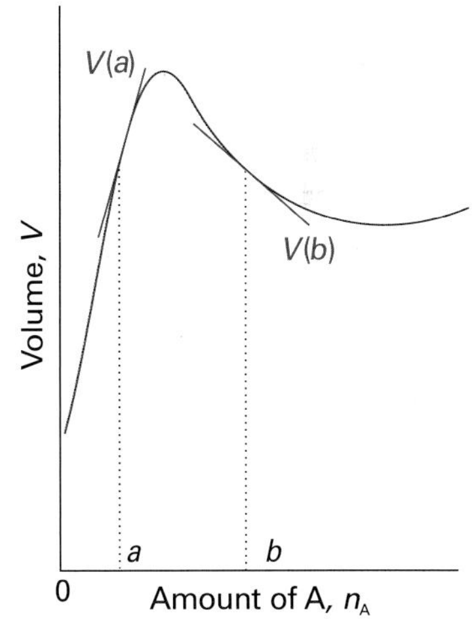

Fig. 5.2 The partial molar volume of a substance is the slope of the variation of the total volume of the sample plotted against the composition. In general, partial molar quantities vary with the composition, as shown by the different slopes at the compositions *a* and *b*. Note that the partial molar volume at *b* is negative: the overall volume of the sample decreases as A is added.

Provided the composition is held constant as the amounts of A and B are increased, the final volume of a mixture can be calculated by integration. Because the partial molar volumes are constant (provided the composition is held constant throughout the integration) we can write

$$V = \int_0^{n_A} V_A dn_A + \int_0^{n_B} V_B dn_B = V_A \int_0^{n_A} dn_A + V_B \int_0^{n_B} dn_B$$
$$= V_A n_A + V_B n_B \qquad (5.3)$$

Although we have envisaged the two integrations as being linked (in order to preserve constant composition), because V is a state function the final result in eqn 5.3 is valid however the solution is in fact prepared.

Partial molar volumes can be measured in several ways. One method is to measure the dependence of the volume on the composition and to fit the observed volume to a function of the amount of the substance. Once the function has been found, its slope can be determined at any composition of interest by differentiation.

[1] The IUPAC recommendation is to denote a partial molar quantity by $\bar{X}$, but only when there is the possibility of confusion with the quantity X. For instance, the partial molar volume of NaCl in water could be written $\bar{V}$(NaCl, aq) to distinguish it from the volume of the solution, V(NaCl, aq).

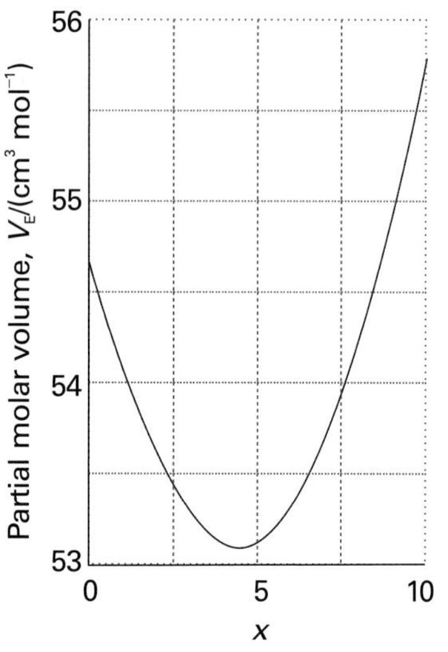

Fig. 5.3 The partial molar volume of ethanol as expressed by the polynomial in *Illustration 5.1*.

Exploration Using the data from *Illustration 5.1*, determine the value of b at which V_E has a minimum value.

Illustration 5.1 *The determination of partial molar volume*

A polynomial fit to measurements of the total volume of a water/ethanol mixture at 25°C that contains 1.000 kg of water is

$$v = 1002.93 + 54.6664x - 0.363\,94x^2 + 0.028\,256x^3$$

where $v = V/\text{cm}^3$, $x = n_E/\text{mol}$, and n_E is the amount of CH_3CH_2OH present. The partial molar volume of ethanol, V_E, is therefore

$$V_E = \left(\frac{\partial V}{\partial n_E}\right)_{p,T,n_W} = \left(\frac{\partial(V/\text{cm}^3)}{\partial(n_E/\text{mol})}\right)_{p,T,n_W} \text{cm}^3\,\text{mol}^{-1} = \left(\frac{\partial v}{\partial x}\right)_{p,T,n_W} \text{cm}^3\,\text{mol}^{-1}$$

Then, because

$$\frac{\text{d}v}{\text{d}x} = 54.6664 - 2(0.36394)x + 3(0.028256)x^2$$

we can conclude that

$$V_E/(\text{cm}^3\,\text{mol}^{-1}) = 54.6664 - 0.72788x + 0.084768x^2$$

Figure 5.3 is a graph of this function.

Self-test 5.1 At 25°C, the density of a 50 per cent by mass ethanol/water solution is 0.914 g cm^{-3}. Given that the partial molar volume of water in the solution is 17.4 cm^3 mol^{-1}, what is the partial molar volume of the ethanol?

[56.4 cm^3 mol^{-1}]

Molar volumes are always positive, but partial molar quantities need not be. For example, the limiting partial molar volume of $MgSO_4$ in water (its partial molar volume in the limit of zero concentration) is -1.4 cm^3 mol^{-1}, which means that the addition of 1 mol $MgSO_4$ to a large volume of water results in a decrease in volume of 1.4 cm^3. The mixture contracts because the salt breaks up the open structure of water as the ions become hydrated, and it collapses slightly.

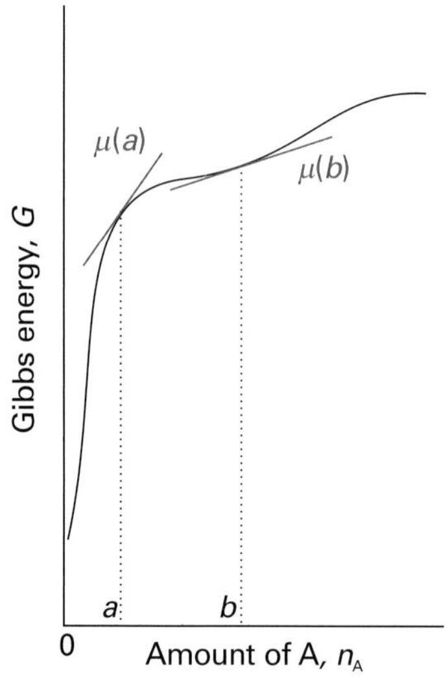

Fig. 5.4 The chemical potential of a substance is the slope of the total Gibbs energy of a mixture with respect to the amount of substance of interest. In general, the chemical potential varies with composition, as shown for the two values at a and b. In this case, both chemical potentials are positive.

(b) Partial molar Gibbs energies

The concept of a partial molar quantity can be extended to any extensive state function. For a substance in a mixture, the chemical potential is *defined* as the partial molar Gibbs energy:

$$\mu_J = \left(\frac{\partial G}{\partial n_J}\right)_{p,T,n'} \qquad [5.4]$$

That is, the chemical potential is the slope of a plot of Gibbs energy against the amount of the component J, with the pressure and temperature (and the amounts of the other substances) held constant (Fig. 5.4). For a pure substance we can write $G = n_J G_{J,m}$, and from eqn 5.4 obtain $\mu_J = G_{J,m}$: in this case, the chemical potential is simply the molar Gibbs energy of the substance, as we used in Chapter 4.

By the same argument that led to eqn 5.3, it follows that the total Gibbs energy of a binary mixture is

$$G = n_A\mu_A + n_B\mu_B \qquad (5.5)$$

where μ_A and μ_B are the chemical potentials at the composition of the mixture. That is, the chemical potential of a substance in a mixture is the contribution of that substance to the total Gibbs energy of the mixture. Because the chemical potentials depend on composition (and the pressure and temperature), the Gibbs energy of a mixture may change when these variables change, and for a system of components A, B, etc., the equation $dG = Vdp - SdT$ becomes

$$dG = Vdp - SdT + \mu_A dn_A + \mu_B dn_B + \cdots \tag{5.6}$$

This expression is the **fundamental equation of chemical thermodynamics**. Its implications and consequences are explored and developed in this and the next two chapters.

At constant pressure and temperature, eqn 5.6 simplifies to

$$dG = \mu_A dn_A + \mu_B dn_B + \cdots \tag{5.7}$$

We saw in Section 3.5e that under the same conditions $dG = dw_{add,max}$. Therefore, at constant temperature and pressure,

$$dw_{add,max} = \mu_A dn_A + \mu_B dn_B + \cdots \tag{5.8}$$

That is, additional (non-expansion) work can arise from the changing composition of a system. For instance, in an electrochemical cell, the chemical reaction is arranged to take place in two distinct sites (at the two electrodes). The electrical work the cell performs can be traced to its changing composition as products are formed from reactants.

(c) The wider significance of the chemical potential

The chemical potential does more than show how G varies with composition. Because $G = U + pV - TS$, and therefore $U = -pV + TS + G$, we can write a general infinitesimal change in U for a system of variable composition as

$$\begin{aligned} dU &= -pdV - Vdp + SdT + TdS + dG \\ &= -pdV - Vdp + SdT + TdS + (Vdp - SdT + \mu_A dn_A + \mu_B dn_B + \cdots) \\ &= -pdV + TdS + \mu_A dn_A + \mu_B dn_B + \cdots \end{aligned}$$

This expression is the generalization of eqn 3.43 (that $dU = TdS - pdV$) to systems in which the composition may change. It follows that at constant volume and entropy,

$$dU = \mu_A dn_A + \mu_B dn_B + \cdots \tag{5.9}$$

and hence that

$$\mu_J = \left(\frac{\partial U}{\partial n_J}\right)_{S,V,n'} \tag{5.10}$$

Therefore, not only does the chemical potential show how G changes when the composition changes, it also shows how the internal energy changes too (but under a different set of conditions). In the same way it is easy to deduce that

$$\text{(a)}\ \mu_J = \left(\frac{\partial H}{\partial n_J}\right)_{S,p,n'} \qquad \text{(b)}\ \mu_J = \left(\frac{\partial A}{\partial n_J}\right)_{V,T,n'} \tag{5.11}$$

Thus we see that the μ_J shows how all the extensive thermodynamic properties U, H, A, and G depend on the composition. This is why the chemical potential is so central to chemistry.

(d) The Gibbs–Duhem equation

Because the total Gibbs energy of a binary mixture is given by eqn 5.5 and the chemical potentials depend on the composition, when the compositions are changed infinitesimally we might expect G of a binary system to change by

$$dG = \mu_A dn_A + \mu_B dn_B + n_A d\mu_A + n_B d\mu_B$$

However, we have seen that at constant pressure and temperature a change in Gibbs energy is given by eqn 5.7. Because G is a state function, these two equations must be equal, which implies that at constant temperature and pressure

$$n_A d\mu_A + n_B d\mu_B = 0 \tag{5.12a}$$

This equation is a special case of the **Gibbs–Duhem equation**:

$$\sum_J n_J d\mu_J = 0 \tag{5.12b}$$

The significance of the Gibbs–Duhem equation is that the chemical potential of one component of a mixture cannot change independently of the chemical potentials of the other components. In a binary mixture, if one partial molar quantity increases, then the other must decrease, with the two changes related by

$$d\mu_B = -\frac{n_A}{n_B} d\mu_A \tag{5.13}$$

The same line of reasoning applies to all partial molar quantities. We can see in Fig. 5.1, for example, that, where the partial molar volume of water increases, that of ethanol decreases. Moreover, as eqn 5.13 shows, and as we can see from Fig. 5.1, a small change in the partial molar volume of A corresponds to a large change in the partial molar volume of B if n_A/n_B is large, but the opposite is true when this ratio is small. In practice, the Gibbs–Duhem equation is used to determine the partial molar volume of one component of a binary mixture from measurements of the partial molar volume of the second component.

Comment 5.1

The *molar concentration* (colloquially, the 'molarity', [J] or c_J) is the amount of solute divided by the volume of the solution and is usually expressed in moles per cubic decimetre (mol dm^{-3}). We write $c^\ominus = 1$ mol dm^{-3}. The term *molality*, *b*, is the amount of solute divided by the mass of solvent and is usually expressed in moles per kilogram of solvent (mol kg^{-1}). We write $b^\ominus =$ 1 mol kg^{-1}.

Example 5.1 *Using the Gibbs–Duhem equation*

The experimental values of the partial molar volume of K_2SO_4(aq) at 298 K are found to fit the expression

$$v_B = 32.280 + 18.216x^{1/2}$$

where $v_B = V_{K_2SO_4}/(\text{cm}^3\ \text{mol}^{-1})$ and x is the numerical value of the molality of K_2SO_4 ($x = b/b^\ominus$; see *Comment 5.1*). Use the Gibbs–Duhem equation to derive an equation for the molar volume of water in the solution. The molar volume of pure water at 298 K is 18.079 cm^3 mol^{-1}.

Method Let A denote H_2O, the solvent, and B denote K_2SO_4, the solute. The Gibbs–Duhem equation for the partial molar volumes of two components is $n_A dV_A + n_B dV_B = 0$. This relation implies that $dv_A = -(n_B/n_A)dv_B$, and therefore that v_A can be found by integration:

$$v_A = v_A^* - \int \frac{n_B}{n_A} dv_B$$

where $v_A^* = V_A/(\text{cm}^3\ \text{mol}^{-1})$ is the numerical value of the molar volume of pure A. The first step is to change the variable v_B to $x = b/b^\ominus$ and then to integrate the right-hand side between $x = 0$ (pure B) and the molality of interest.

Answer It follows from the information in the question that, with B = K_2SO_4, $dv_B/dx = 9.108x^{-1/2}$. Therefore, the integration required is

$$v_B = v_B^* - 9.108\int_0^{b/b^\ominus} \frac{n_B}{n_A} x^{-1/2} dx$$

However, the ratio of amounts of A (H_2O) and B (K_2SO_4) is related to the molality of B, $b = n_B/(1\text{ kg water})$ and $n_A = (1\text{ kg water})/M_A$ where M_A is the molar mass of water, by

$$\frac{n_B}{n_A} = \frac{n_B}{(1\text{ kg})/M_A} = \frac{n_B M_A}{1\text{ kg}} = bM_A = xb^{\ominus}M_A$$

and hence

$$v_A = v_A^{\star} - 9.108M_A b^{\ominus}\int_0^{b/b^{\ominus}} x^{1/2}\mathrm{d}x = v_A^{\star} - \tfrac{2}{3}\{9.108M_A b^{\ominus}(b/b^{\ominus})^{3/2}\}$$

It then follows, by substituting the data (including $M_A = 1.802 \times 10^{-2}$ kg mol^{-1}, the molar mass of water), that

$$V_A/(\text{cm}^3\text{ mol}^{-1}) = 18.079 - 0.1094(b/b^{\ominus})^{3/2}$$

The partial molar volumes are plotted in Fig. 5.5.

Self-test 5.2 Repeat the calculation for a salt B for which $V_B/(\text{cm}^3\text{ mol}^{-1}) = 6.218 + 5.146b - 7.147b^2$. [$V_A/(\text{cm}^3\text{ mol}^{-1}) = 18.079 - 0.0464b^2 + 0.0859b^3$]

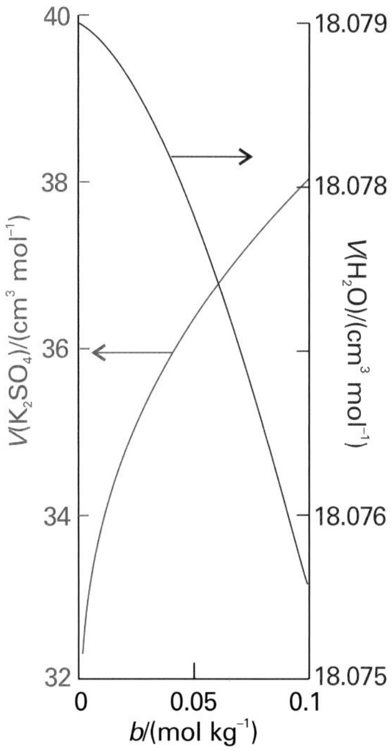

Fig. 5.5 The partial molar volumes of the components of an aqueous solution of potassium sulfate.

5.2 The thermodynamics of mixing

The dependence of the Gibbs energy of a mixture on its composition is given by eqn 5.5, and we know that at constant temperature and pressure systems tend towards lower Gibbs energy. This is the link we need in order to apply thermodynamics to the discussion of spontaneous changes of composition, as in the mixing of two substances. One simple example of a spontaneous mixing process is that of two gases introduced into the same container. The mixing is spontaneous, so it must correspond to a decrease in G. We shall now see how to express this idea quantitatively.

(a) The Gibbs energy of mixing of perfect gases

Let the amounts of two perfect gases in the two containers be n_A and n_B; both are at a temperature T and a pressure p (Fig. 5.6). At this stage, the chemical potentials of the two gases have their 'pure' values, which are obtained by applying the definition $\mu = G_m$ to eqn 3.57:

$$\mu = \mu^{\ominus} + RT\ln\frac{p}{p^{\ominus}} \qquad (5.14a)^{\circ}$$

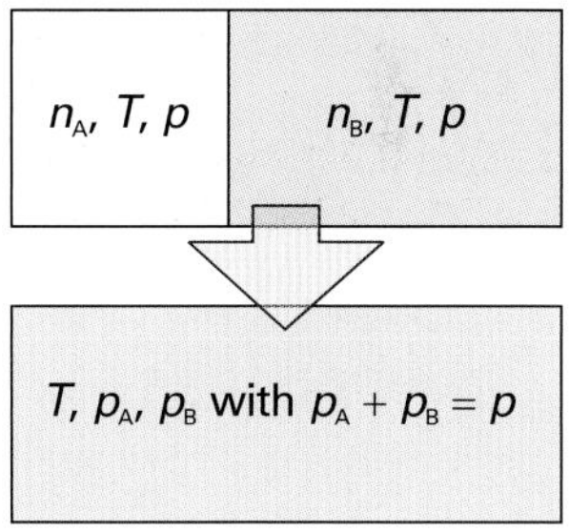

Fig. 5.6 The arrangement for calculating the thermodynamic functions of mixing of two perfect gases.

where $\mu^{\ominus}$ is the **standard chemical potential**, the chemical potential of the pure gas at 1 bar. It will be much simpler notationally if we agree to let p denote the pressure relative to $p^{\ominus}$; that is, to replace $p/p^{\ominus}$ by p, for then we can write

$$\mu = \mu^{\ominus} + RT\ln p \qquad \{5.14b\}^{\circ}$$

Equations for which this convention is used will be labelled {1}, {2}, . . . ; to use the equations, we have to remember to replace p by $p/p^{\ominus}$ again. In practice, that simply means using the numerical value of p in bars. The Gibbs energy of the total system is then given by eqn 5.5 as

$$G_i = n_A\mu_A + n_B\mu_B = n_A(\mu_A^{\ominus} + RT\ln p) + n_B(\mu_B^{\ominus} + RT\ln p) \qquad \{5.15\}^{\circ}$$

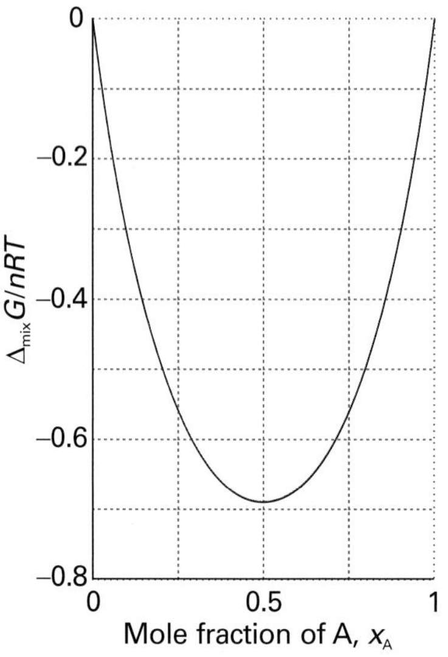

Fig. 5.7 The Gibbs energy of mixing of two perfect gases and (as discussed later) of two liquids that form an ideal solution. The Gibbs energy of mixing is negative for all compositions and temperatures, so perfect gases mix spontaneously in all proportions.

Exploration Draw graphs of $\Delta_{mix}G$ against x_A at different temperatures in the range 298 K to 500 K. For what value of x_A does $\Delta_{mix}G$ depend on temperature most strongly?

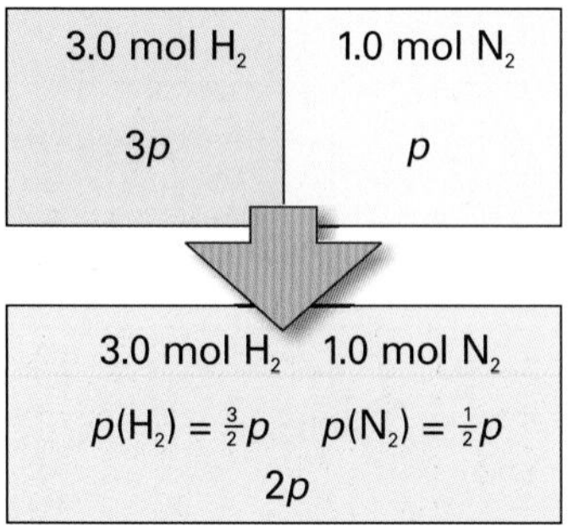

Fig. 5.8 The initial and final states considered in the calculation of the Gibbs energy of mixing of gases at different initial pressures.

After mixing, the partial pressures of the gases are p_A and p_B, with $p_A + p_B = p$. The total Gibbs energy changes to

$$G_f = n_A(\mu_A^{\ominus} + RT\ln p_A) + n_B(\mu_B^{\ominus} + RT\ln p_B) \qquad \{5.16\}^{\circ}$$

The difference $G_f - G_i$, the **Gibbs energy of mixing**, $\Delta_{mix}G$, is therefore

$$\Delta_{mix}G = n_ART\ln\frac{p_A}{p} + n_BRT\ln\frac{p_B}{p} \qquad (5.17)^{\circ}$$

At this point we may replace n_J by x_Jn, where n is the total amount of A and B, and use the relation between partial pressure and mole fraction (Section 1.2b) to write $p_J/p = x_J$ for each component, which gives

$$\Delta_{mix}G = nRT(x_A\ln x_A + x_B\ln x_B) \qquad (5.18)^{\circ}$$

Because mole fractions are never greater than 1, the logarithms in this equation are negative, and $\Delta_{mix}G < 0$ (Fig. 5.7). The conclusion that $\Delta_{mix}G$ is negative for all compositions confirms that perfect gases mix spontaneously in all proportions. However, the equation extends common sense by allowing us to discuss the process quantitatively.

Example 5.2 *Calculating a Gibbs energy of mixing*

A container is divided into two equal compartments (Fig. 5.8). One contains 3.0 mol H_2(g) at 25°C; the other contains 1.0 mol N_2(g) at 25°C. Calculate the Gibbs energy of mixing when the partition is removed. Assume perfect behaviour.

Method Equation 5.18 cannot be used directly because the two gases are initially at different pressures. We proceed by calculating the initial Gibbs energy from the chemical potentials. To do so, we need the pressure of each gas. Write the pressure of nitrogen as p; then the pressure of hydrogen as a multiple of p can be found from the gas laws. Next, calculate the Gibbs energy for the system when the partition is removed. The volume occupied by each gas doubles, so its initial partial pressure is halved.

Answer Given that the pressure of nitrogen is p, the pressure of hydrogen is $3p$; therefore, the initial Gibbs energy is

$$G_i = (3.0\text{ mol})\{\mu^{\ominus}(H_2) + RT\ln 3p\} + (1.0\text{ mol})\{\mu^{\ominus}(N_2) + RT\ln p\}$$

When the partition is removed and each gas occupies twice the original volume, the partial pressure of nitrogen falls to $\frac{1}{2}p$ and that of hydrogen falls to $\frac{3}{2}p$. Therefore, the Gibbs energy changes to

$$G_f = (3.0\text{ mol})\{\mu^{\ominus}(H_2) + RT\ln \tfrac{3}{2}p\} + (1.0\text{ mol})\{\mu^{\ominus}(N_2) + RT\ln \tfrac{1}{2}p\}$$

The Gibbs energy of mixing is the difference of these two quantities:

$$\begin{aligned}\Delta_{mix}G &= (3.0\text{ mol})RT\ln\left(\frac{\frac{3}{2}p}{3p}\right) + (1.0\text{ mol})RT\ln\left(\frac{\frac{1}{2}p}{p}\right)\\ &= -(3.0\text{ mol})RT\ln 2 - (1.0\text{ mol})RT\ln 2\\ &= -(4.0\text{ mol})RT\ln 2 = -6.9\text{ kJ}\end{aligned}$$

In this example, the value of $\Delta_{mix}G$ is the sum of two contributions: the mixing itself, and the changes in pressure of the two gases to their final total pressure, $2p$. When 3.0 mol H_2 mixes with 1.0 mol N_2 at the same pressure, with the volumes of the vessels adjusted accordingly, the change of Gibbs energy is −5.6 kJ.

Self-test 5.3 Suppose that 2.0 mol H_2 at 2.0 atm and 25°C and 4.0 mol N_2 at 3.0 atm and 25°C are mixed at constant volume. Calculate $\Delta_{mix}G$. What would be the value of $\Delta_{mix}G$ had the pressures been identical initially? [−9.7 kJ, −9.5 kJ]

(b) Other thermodynamic mixing functions

Because $(\partial G/\partial T)_{p,n} = -S$, it follows immediately from eqn 5.18 that, for a mixture of perfect gases initially at the same pressure, the entropy of mixing, $\Delta_{mix}S$, is

$$\Delta_{mix}S = \left(\frac{\partial \Delta_{mix}G}{\partial T}\right)_{p,n_A,n_B} = -nR(x_A \ln x_A + x_B \ln x_B) \qquad (5.19)^\circ$$

Because $\ln x < 0$, it follows that $\Delta_{mix}S > 0$ for all compositions (Fig. 5.9). For equal amounts of gas, for instance, we set $x_A = x_B = \frac{1}{2}$, and obtain $\Delta_{mix}S = nR \ln 2$, with n the total amount of gas molecules. This increase in entropy is what we expect when one gas disperses into the other and the disorder increases.

We can calculate the isothermal, isobaric (constant pressure) **enthalpy of mixing**, $\Delta_{mix}H$, the enthalpy change accompanying mixing, of two perfect gases from $\Delta G = \Delta H - T\Delta S$. It follows from eqns 5.18 and 5.19 that

$$\Delta_{mix}H = 0 \qquad (5.20)^\circ$$

The enthalpy of mixing is zero, as we should expect for a system in which there are no interactions between the molecules forming the gaseous mixture. It follows that the whole of the driving force for mixing comes from the increase in entropy of the system, because the entropy of the surroundings is unchanged.

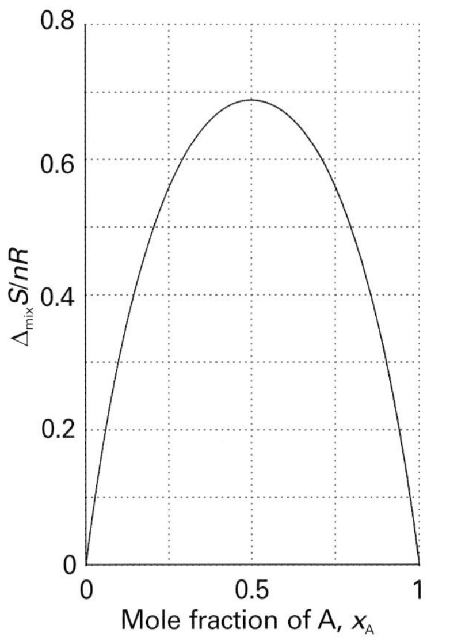

Fig. 5.9 The entropy of mixing of two perfect gases and (as discussed later) of two liquids that form an ideal solution. The entropy increases for all compositions and temperatures, so perfect gases mix spontaneously in all proportions. Because there is no transfer of heat to the surroundings when perfect gases mix, the entropy of the surroundings is unchanged. Hence, the graph also shows the total entropy of the system plus the surroundings when perfect gases mix.

5.3 The chemical potentials of liquids

To discuss the equilibrium properties of liquid mixtures we need to know how the Gibbs energy of a liquid varies with composition. To calculate its value, we use the fact that, at equilibrium, the chemical potential of a substance present as a vapour must be equal to its chemical potential in the liquid.

(a) Ideal solutions

We shall denote quantities relating to pure substances by a superscript *, so the chemical potential of pure A is written μ_A^*, and as $\mu_A^*(l)$ when we need to emphasize that A is a liquid. Because the vapour pressure of the pure liquid is p_A^*, it follows from eqn 5.14 that the chemical potential of A in the vapour (treated as a perfect gas) is $\mu_A^\ominus + RT \ln p_A^*$ (with p_A to be interpreted as the relative pressure $p_A/p^\ominus$). These two chemical potentials are equal at equilibrium (Fig. 5.10), so we can write

$$\mu_A^* = \mu_A^\ominus + RT \ln p_A^* \qquad \{5.21\}$$

If another substance, a solute, is also present in the liquid, the chemical potential of A in the liquid is changed to μ_A and its vapour pressure is changed to p_A. The vapour and solvent are still in equilibrium, so we can write

$$\mu_A = \mu_A^\ominus + RT \ln p_A \qquad \{5.22\}$$

Next, we combine these two equations to eliminate the standard chemical potential of the gas. To do so, we write eqn 5.21 as $\mu_A^\ominus = \mu_A^* - RT \ln p_A^*$ and substitute this expression into eqn 5.22 to obtain

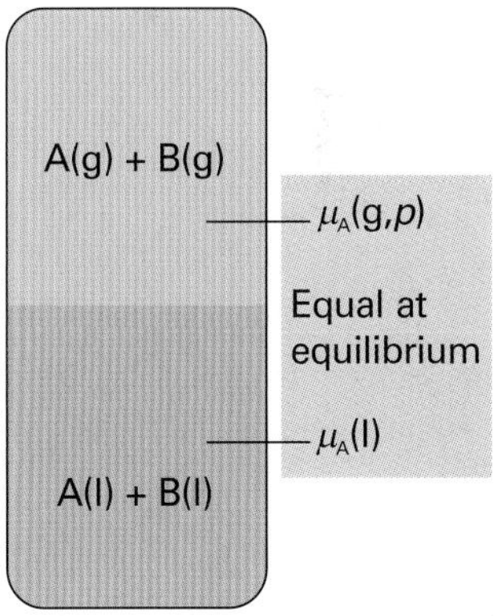

Fig. 5.10 At equilibrium, the chemical potential of the gaseous form of a substance A is equal to the chemical potential of its condensed phase. The equality is preserved if a solute is also present. Because the chemical potential of A in the vapour depends on its partial vapour pressure, it follows that the chemical potential of liquid A can be related to its partial vapour pressure.

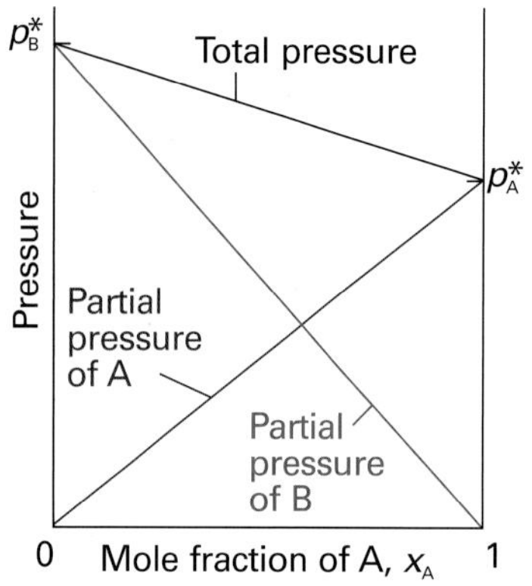

Fig. 5.11 The total vapour pressure and the two partial vapour pressures of an ideal binary mixture are proportional to the mole fractions of the components.

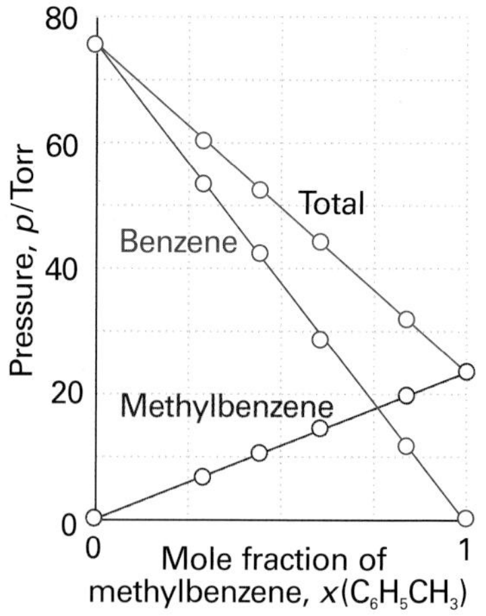

Fig. 5.12 Two similar liquids, in this case benzene and methylbenzene (toluene), behave almost ideally, and the variation of their vapour pressures with composition resembles that for an ideal solution.

$$\mu_A = \mu_A^* - RT\ln p_A^* + RT\ln p_A = \mu_A^* + RT\ln\frac{p_A}{p_A^*} \qquad (5.23)$$

In the final step we draw on additional experimental information about the relation between the ratio of vapour pressures and the composition of the liquid. In a series of experiments on mixtures of closely related liquids (such as benzene and methylbenzene), the French chemist François Raoult found that the ratio of the partial vapour pressure of each component to its vapour pressure as a pure liquid, p_A/p_A^*, is approximately equal to the mole fraction of A in the liquid mixture. That is, he established what we now call **Raoult's law**:

$$p_A = x_A p_A^* \qquad (5.24)°$$

This law is illustrated in Fig. 5.11. Some mixtures obey Raoult's law very well, especially when the components are structurally similar (Fig. 5.12). Mixtures that obey the law throughout the composition range from pure A to pure B are called **ideal solutions**. When we write equations that are valid only for ideal solutions, we shall label them with a superscript °, as in eqn 5.24.

For an ideal solution, it follows from eqns 5.23 and 5.24 that

$$\mu_A = \mu_A^* + RT\ln x_A \qquad (5.25)°$$

This important equation can be used as the *definition* of an ideal solution (so that it implies Raoult's law rather than stemming from it). It is in fact a better definition than eqn 5.24 because it does not assume that the vapour is a perfect gas.

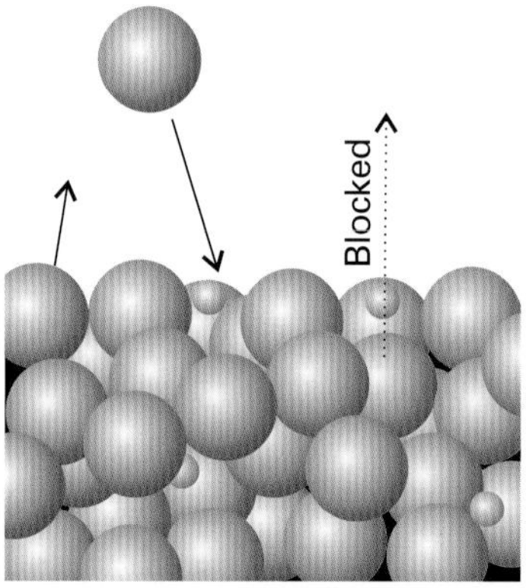

Fig. 5.13 A pictorial representation of the molecular basis of Raoult's law. The large spheres represent solvent molecules at the surface of a solution (the uppermost line of spheres), and the small spheres are solute molecules. The latter hinder the escape of solvent molecules into the vapour, but do not hinder their return.

Molecular interpretation 5.1 *The molecular origin of Raoult's law*

The origin of Raoult's law can be understood in molecular terms by considering the rates at which molecules leave and return to the liquid. The law reflects the fact that the presence of a second component reduces the rate at which A molecules leave the surface of the liquid but does not inhibit the rate at which they return (Fig. 5.13).

The rate at which A molecules leave the surface is proportional to the number of them at the surface, which in turn is proportional to the mole fraction of A:

$$\text{rate of vaporization} = kx_A$$

where k is a constant of proportionality. The rate at which molecules condense is proportional to their concentration in the gas phase, which in turn is proportional to their partial pressure:

$$\text{rate of condensation} = k'p_A$$

At equilibrium, the rates of vaporization and condensation are equal, so $k'p_A = kx_A$. It follows that

$$p_A = \frac{k}{k'}x_A$$

For the pure liquid, $x_A = 1$; so in this special case $p_A^* = k/k'$. Equation 5.24 then follows by substitution of this relation into the line above.

Some solutions depart significantly from Raoult's law (Fig. 5.14). Nevertheless, even in these cases the law is obeyed increasingly closely for the component in excess (the solvent) as it approaches purity. The law is therefore a good approximation for the properties of the solvent if the solution is dilute.

(b) Ideal-dilute solutions

In ideal solutions the solute, as well as the solvent, obeys Raoult's law. However, the English chemist William Henry found experimentally that, for real solutions at low concentrations, although the vapour pressure of the solute is proportional to its mole fraction, the constant of proportionality is not the vapour pressure of the pure substance (Fig. 5.15). **Henry's law** is:

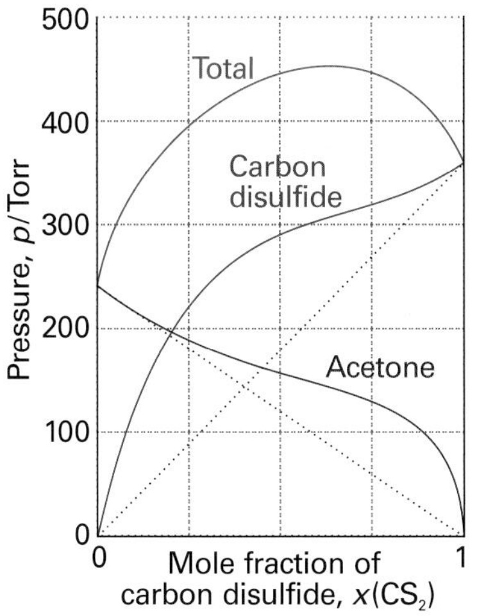

Fig. 5.14 Strong deviations from ideality are shown by dissimilar liquids (in this case carbon disulfide and acetone, propanone).

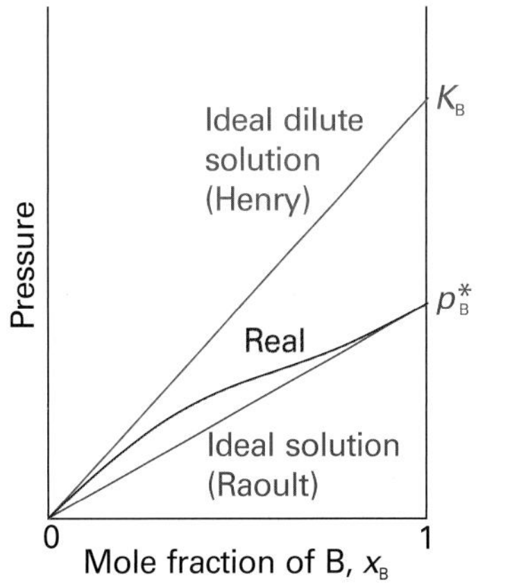

Fig. 5.15 When a component (the solvent) is nearly pure, it has a vapour pressure that is proportional to mole fraction with a slope p_B^* (Raoult's law). When it is the minor component (the solute) its vapour pressure is still proportional to the mole fraction, but the constant of proportionality is now K_B (Henry's law).

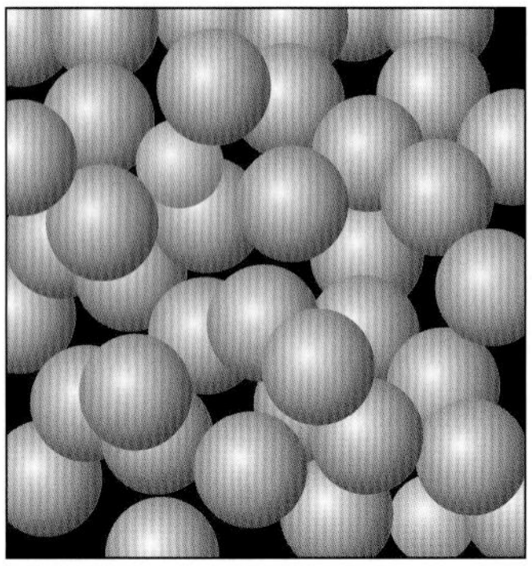

Fig. 5.16 In a dilute solution, the solvent molecules (the purple spheres) are in an environment that differs only slightly from that of the pure solvent. The solute particles, however, are in an environment totally unlike that of the pure solute.

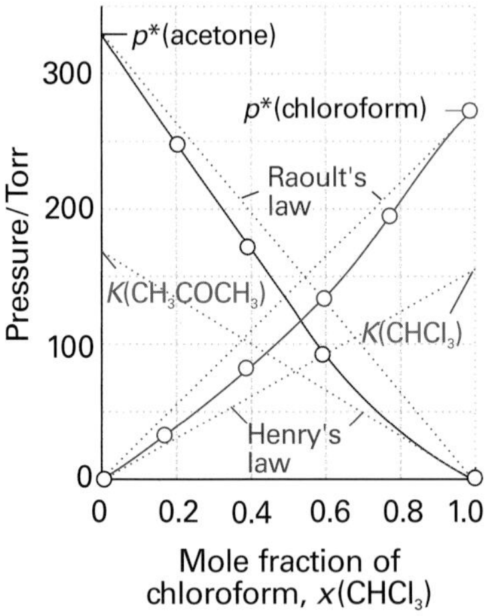

Fig. 5.17 The experimental partial vapour pressures of a mixture of chloroform (trichloromethane) and acetone (propanone) based on the data in *Example 5.3*. The values of K are obtained by extrapolating the dilute solution vapour pressures as explained in the *Example*.

$$p_B = x_B K_B \qquad (5.26)°$$

In this expression x_B is the mole fraction of the solute and K_B is an empirical constant (with the dimensions of pressure) chosen so that the plot of the vapour pressure of B against its mole fraction is tangent to the experimental curve at $x_B = 0$.

Mixtures for which the solute obeys Henry's law and the solvent obeys Raoult's law are called **ideal-dilute solutions**. We shall also label equations with a superscript ° when they have been derived from Henry's law. The difference in behaviour of the solute and solvent at low concentrations (as expressed by Henry's and Raoult's laws, respectively) arises from the fact that in a dilute solution the solvent molecules are in an environment very much like the one they have in the pure liquid (Fig. 5.16). In contrast, the solute molecules are surrounded by solvent molecules, which is entirely different from their environment when pure. Thus, the solvent behaves like a slightly modified pure liquid, but the solute behaves entirely differently from its pure state unless the solvent and solute molecules happen to be very similar. In the latter case, the solute also obeys Raoult's law.

Example 5.3 *Investigating the validity of Raoult's and Henry's laws*

The vapour pressures of each component in a mixture of propanone (acetone, A) and trichloromethane (chloroform, C) were measured at 35°C with the following results:

x_C	0	0.20	0.40	0.60	0.80	1
p_C/kPa	0	4.7	11	18.9	26.7	36.4
p_A/kPa	46.3	33.3	23.3	12.3	4.9	0

Confirm that the mixture conforms to Raoult's law for the component in large excess and to Henry's law for the minor component. Find the Henry's law constants.

Method Both Raoult's and Henry's laws are statements about the form of the graph of partial vapour pressure against mole fraction. Therefore, plot the partial vapour pressures against mole fraction. Raoult's law is tested by comparing the data with the straight line $p_J = x_J p_J^*$ for each component in the region in which it is in excess (and acting as the solvent). Henry's law is tested by finding a straight line $p_J = x_J K_J$ that is tangent to each partial vapour pressure at low x, where the component can be treated as the solute.

Answer The data are plotted in Fig. 5.17 together with the Raoult's law lines. Henry's law requires K = 23.3 kPa for propanone and K = 22.0 kPa for trichloromethane. Notice how the system deviates from both Raoult's and Henry's laws even for quite small departures from $x = 1$ and $x = 0$, respectively. We deal with these deviations in Section 5.5.

Self-test 5.4 The vapour pressure of chloromethane at various mole fractions in a mixture at 25°C was found to be as follows:

x	0.005	0.009	0.019	0.024
p/kPa	27.3	48.4	101	126

Estimate Henry's law constant. [5 MPa]

For practical applications, Henry's law is expressed in terms of the molality, b, of the solute,

$$p_B = b_B K_B$$

Some Henry's law data for this convention are listed in Table 5.1. As well as providing a link between the mole fraction of solute and its partial pressure, the data in the table may also be used to calculate gas solubilities. A knowledge of Henry's law constants for gases in blood and fats is important for the discussion of respiration, especially when the partial pressure of oxygen is abnormal, as in diving and mountaineering, and for the discussion of the action of gaseous anaesthetics.

Synoptic Table 5.1* Henry's law constants for gases in water at 298 K

	$K/(\text{kPa kg mol}^{-1})$
CO_2	3.01×10^3
H_2	1.28×10^5
N_2	1.56×10^5
O_2	7.92×10^4

* More values are given in the *Data section*.

Illustration 5.2 *Using Henry's law*

To estimate the molar solubility of oxygen in water at 25°C and a partial pressure of 21 kPa, its partial pressure in the atmosphere at sea level, we write

$$b_{O_2} = \frac{p_{O_2}}{K_{O_2}} = \frac{21\ \text{kPa}}{7.9 \times 10^4\ \text{kPa kg mol}^{-1}} = 2.9 \times 10^{-4}\ \text{mol kg}^{-1}$$

The molality of the saturated solution is therefore 0.29 mmol kg^{-1}. To convert this quantity to a molar concentration, we assume that the mass density of this dilute solution is essentially that of pure water at 25°C, or $\rho_{H_2O} = 0.99709$ kg dm^{-3}. It follows that the molar concentration of oxygen is

$$[O_2] = b_{O_2} \times \rho_{H_2O} = 0.29\ \text{mmol kg}^{-1} \times 0.99709\ \text{kg dm}^{-3} = 0.29\ \text{mmol dm}^{-3}$$

A note on good practice The number of significant figures in the result of a calculation should not exceed the number in the data (only two in this case).

Self-test 5.5 Calculate the molar solubility of nitrogen in water exposed to air at 25°C; partial pressures were calculated in *Example 1.3*. [0.51 mmol dm^{-3}]

IMPACT ON BIOLOGY

I5.1 Gas solubility and breathing

We inhale about 500 cm^3 of air with each breath we take. The influx of air is a result of changes in volume of the lungs as the diaphragm is depressed and the chest expands, which results in a decrease in pressure of about 100 Pa relative to atmospheric pressure. Expiration occurs as the diaphragm rises and the chest contracts, and gives rise to a differential pressure of about 100 Pa above atmospheric pressure. The total volume of air in the lungs is about 6 dm^3, and the additional volume of air that can be exhaled forcefully after normal expiration is about 1.5 dm^3. Some air remains in the lungs at all times to prevent the collapse of the alveoli.

Comment 5.2
The web site contains links to online databases of Henry's law constants.

A knowledge of Henry's law constants for gases in fats and lipids is important for the discussion of respiration. The effect of gas exchange between blood and air inside the alveoli of the lungs means that the composition of the air in the lungs changes throughout the breathing cycle. Alveolar gas is in fact a mixture of newly inhaled air and air about to be exhaled. The concentration of oxygen present in arterial blood is equivalent to a partial pressure of about 40 Torr (5.3 kPa), whereas the partial pressure of freshly inhaled air is about 104 Torr (13.9 kPa). Arterial blood remains in the capillary passing through the wall of an alveolus for about 0.75 s, but such is the steepness of the pressure gradient that it becomes fully saturated with oxygen in about 0.25 s. If the lungs collect fluids (as in pneumonia), then the respiratory membrane thickens, diffusion is greatly slowed, and body tissues begin to suffer from oxygen starvation. Carbon dioxide moves in the opposite direction across the respiratory tissue, but the partial pressure gradient is much less, corresponding to about 5 Torr (0.7 kPa) in blood and 40 Torr (5.3 kPa) in air at equilibrium. However, because carbon dioxide is much more soluble in the alveolar fluid than oxygen is, equal amounts of oxygen and carbon dioxide are exchanged in each breath.

A hyperbaric oxygen chamber, in which oxygen is at an elevated partial pressure, is used to treat certain types of disease. Carbon monoxide poisoning can be treated in this way as can the consequences of shock. Diseases that are caused by anaerobic bacteria, such as gas gangrene and tetanus, can also be treated because the bacteria cannot thrive in high oxygen concentrations.

In scuba diving (where *scuba* is an acronym formed from 'self-contained underwater breathing apparatus'), air is supplied at a higher pressure, so that the pressure within the diver's chest matches the pressure exerted by the surrounding water. The latter increases by about 1 atm for each 10 m of descent. One unfortunate consequence of breathing air at high pressures is that nitrogen is much more soluble in fatty tissues than in water, so it tends to dissolve in the central nervous system, bone marrow, and fat reserves. The result is *nitrogen narcosis*, with symptoms like intoxication. If the diver rises too rapidly to the surface, the nitrogen comes out of its lipid solution as bubbles, which causes the painful and sometimes fatal condition known as *the bends*. Many cases of scuba drowning appear to be consequences of arterial embolisms (obstructions in arteries caused by gas bubbles) and loss of consciousness as the air bubbles rise into the head.

The properties of solutions

In this section we consider the thermodynamics of mixing of liquids. First, we consider the simple case of mixtures of liquids that mix to form an ideal solution. In this way, we identify the thermodynamic consequences of molecules of one species mingling randomly with molecules of the second species. The calculation provides a background for discussing the deviations from ideal behaviour exhibited by real solutions.

5.4 Liquid mixtures

Thermodynamics can provide insight into the properties of liquid mixtures, and a few simple ideas can bring the whole field of study together.

(a) Ideal solutions

The Gibbs energy of mixing of two liquids to form an ideal solution is calculated in exactly the same way as for two gases (Section 5.2). The total Gibbs energy before liquids are mixed is

$$G_i = n_A\mu_A^* + n_B\mu_B^*$$

When they are mixed, the individual chemical potentials are given by eqn 5.25 and the total Gibbs energy is

$$G_f = n_A\{\mu_A^* + RT\ln x_A\} + n_B\{\mu_B^* + RT\ln x_B\}$$

Consequently, the Gibbs energy of mixing is

$$\Delta_{mix}G = nRT\{x_A\ln x_A + x_B\ln x_B\} \quad (5.27)^\circ$$

where $n = n_A + n_B$. As for gases, it follows that the ideal entropy of mixing of two liquids is

$$\Delta_{mix}S = -nR\{x_A\ln x_A + x_B\ln x_B\} \quad (5.28)^\circ$$

and, because $\Delta_{mix}H = \Delta_{mix}G + T\Delta_{mix}S = 0$, the ideal enthalpy of mixing is zero. The ideal volume of mixing, the change in volume on mixing, is also zero because it follows from eqn 3.50 ($(\partial G/\partial p)_T = V$) that $\Delta_{mix}V = (\partial\Delta_{mix}G/\partial p)_T$, but $\Delta_{mix}G$ in eqn 5.27 is independent of pressure, so the derivative with respect to pressure is zero.

Equation 5.27 is the same as that for two perfect gases and all the conclusions drawn there are valid here: the driving force for mixing is the increasing entropy of the system as the molecules mingle and the enthalpy of mixing is zero. It should be noted, however, that solution ideality means something different from gas perfection. In a perfect gas there are no forces acting between molecules. In ideal solutions there are interactions, but the average energy of A—B interactions in the mixture is the same as the average energy of A—A and B—B interactions in the pure liquids.[2] The variation of the Gibbs energy of mixing with composition is the same as that already depicted for gases in Fig. 5.7; the same is true of the entropy of mixing, Fig. 5.9.

Real solutions are composed of particles for which A—A, A—B, and B—B interactions are all different. Not only may there be enthalpy and volume changes when liquids mix, but there may also be an additional contribution to the entropy arising from the way in which the molecules of one type might cluster together instead of mingling freely with the others. If the enthalpy change is large and positive or if the entropy change is adverse (because of a reorganization of the molecules that results in an orderly mixture), then the Gibbs energy might be positive for mixing. In that case, separation is spontaneous and the liquids may be immiscible. Alternatively, the liquids might be **partially miscible**, which means that they are miscible only over a certain range of compositions.

(b) Excess functions and regular solutions

The thermodynamic properties of real solutions are expressed in terms of the **excess functions**, X^E, the difference between the observed thermodynamic function of mixing and the function for an ideal solution. The **excess entropy**, S^E, for example, is defined as

$$S^E = \Delta_{mix}S - \Delta_{mix}S^{ideal} \qquad [5.29]$$

where $\Delta_{mix}S^{ideal}$ is given by eqn 5.28. The excess enthalpy and volume are both equal to the observed enthalpy and volume of mixing, because the ideal values are zero in each case.

Deviations of the excess energies from zero indicate the extent to which the solutions are nonideal. In this connection a useful model system is the **regular solution**, a solution for which $H^E \neq 0$ but $S^E = 0$. We can think of a regular solution as one in which the two kinds of molecules are distributed randomly (as in an ideal solution) but have different energies of interactions with each other. Figure 5.18 shows two examples of the composition dependence of molar excess functions.

We can make this discussion more quantitative by supposing that the excess enthalpy depends on composition as

$$H^E = n\beta RTx_Ax_B \qquad (5.30)$$

where β is a dimensionless parameter that is a measure of the energy of AB interactions relative to that of the AA and BB interactions. The function given by eqn 5.30 is plotted in Fig. 5.19, and we see it resembles the experimental curve in Fig. 5.18. If $\beta < 0$, mixing is exothermic and the solute–solvent interactions are more favourable than the solvent–solvent and solute–solute interactions. If $\beta > 0$, then the mixing is

[2] It is on the basis of this distinction that the term 'perfect gas' is preferable to the more common 'ideal gas'.

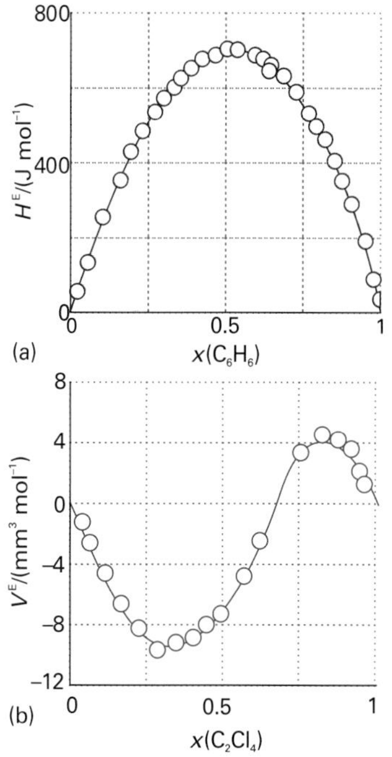

Fig. 5.18 Experimental excess functions at 25°C. (a) H^E for benzene/cyclohexane; this graph shows that the mixing is endothermic (because $\Delta_{mix}H = 0$ for an ideal solution). (b) The excess volume, V^E, for tetrachloroethene/cyclopentane; this graph shows that there is a contraction at low tetrachloroethane mole fractions, but an expansion at high mole fractions (because $\Delta_{mix}V = 0$ for an ideal mixture).

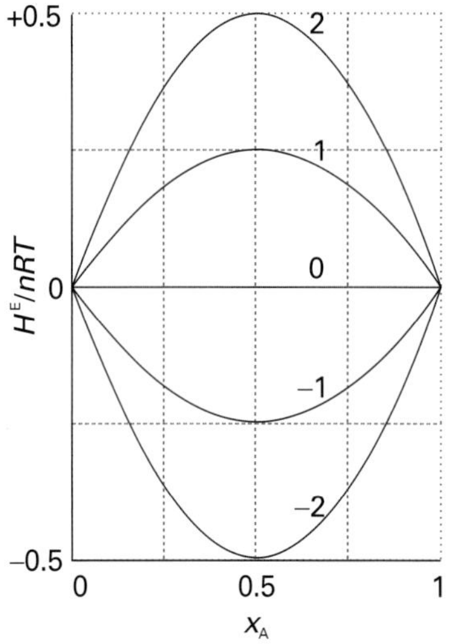

Fig. 5.19 The excess enthalpy according to a model in which it is proportional to $\beta x_A x_B$, for different values of the parameter β.

Exploration Using the graph above, fix β and vary the temperature. For what value of x_A does the excess enthalpy depend on temperature most strongly?

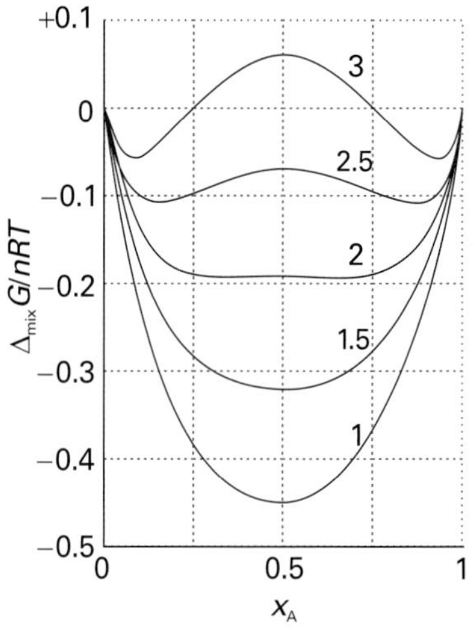

Fig. 5.20 The Gibbs energy of mixing for different values of the parameter β.

Exploration Using the graph above, fix β at 1.5 and vary the temperature. Is there a range of temperatures over which you observe phase separation?

endothermic. Because the entropy of mixing has its ideal value for a regular solution, the excess Gibbs energy is equal to the excess enthalpy, and the Gibbs energy of mixing is

$$\Delta_{mix}G = nRT\{x_A \ln x_A + x_B \ln x_B + \beta x_A x_B\} \tag{5.31}$$

Figure 5.20 shows how $\Delta_{mix}G$ varies with composition for different values of β. The important feature is that for $\beta > 2$ the graph shows two minima separated by a maximum. The implication of this observation is that, provided $\beta > 2$, then the system will separate spontaneously into two phases with compositions corresponding to the two minima, for that separation corresponds to a reduction in Gibbs energy. We develop this point in Sections 5.8 and 6.5.

5.5 Colligative properties

The properties we now consider are the lowering of vapour pressure, the elevation of boiling point, the depression of freezing point, and the osmotic pressure arising from the presence of a solute. In dilute solutions these properties depend only on the number of solute particles present, not their identity. For this reason, they are called **colligative properties** (denoting 'depending on the collection').

We assume throughout the following that the solute is not volatile, so it does not contribute to the vapour. We also assume that the solute does not dissolve in the solid solvent: that is, the pure solid solvent separates when the solution is frozen. The latter assumption is quite drastic, although it is true of many mixtures; it can be avoided at the expense of more algebra, but that introduces no new principles.

(a) The common features of colligative properties

All the colligative properties stem from the reduction of the chemical potential of the liquid solvent as a result of the presence of solute. For an ideal-dilute solution, the reduction is from μ_A^* for the pure solvent to $\mu_A^* + RT \ln x_A$ when a solute is present ($\ln x_A$ is negative because $x_A < 1$). There is no direct influence of the solute on the chemical potential of the solvent vapour and the solid solvent because the solute appears in neither the vapour nor the solid. As can be seen from Fig. 5.21, the reduction in chemical potential of the solvent implies that the liquid–vapour equilibrium occurs at a higher temperature (the boiling point is raised) and the solid–liquid equilibrium occurs at a lower temperature (the freezing point is lowered).

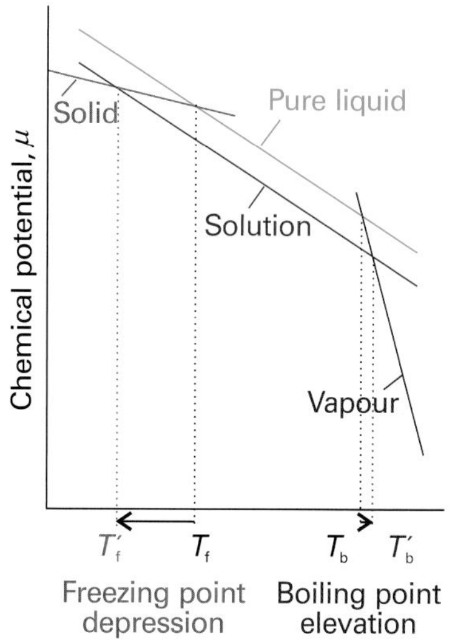

Fig. 5.21 The chemical potential of a solvent in the presence of a solute. The lowering of the liquid's chemical potential has a greater effect on the freezing point than on the boiling point because of the angles at which the lines intersect.

Molecular interpretation 5.2 *The lowering of vapour pressure of a solvent in a mixture*

The molecular origin of the lowering of the chemical potential is not the energy of interaction of the solute and solvent particles, because the lowering occurs even in an ideal solution (for which the enthalpy of mixing is zero). If it is not an enthalpy effect, it must be an entropy effect.

The pure liquid solvent has an entropy that reflects the number of microstates available to its molecules. Its vapour pressure reflects the tendency of the solution towards greater entropy, which can be achieved if the liquid vaporizes to form a gas. When a solute is present, there is an additional contribution to the entropy of the liquid, even in an ideal solution. Because the entropy of the liquid is already higher than that of the pure liquid, there is a weaker tendency to form the gas (Fig. 5.22). The effect of the solute appears as a lowered vapour pressure, and hence a higher boiling point.

Similarly, the enhanced molecular randomness of the solution opposes the tendency to freeze. Consequently, a lower temperature must be reached before equilibrium between solid and solution is achieved. Hence, the freezing point is lowered.

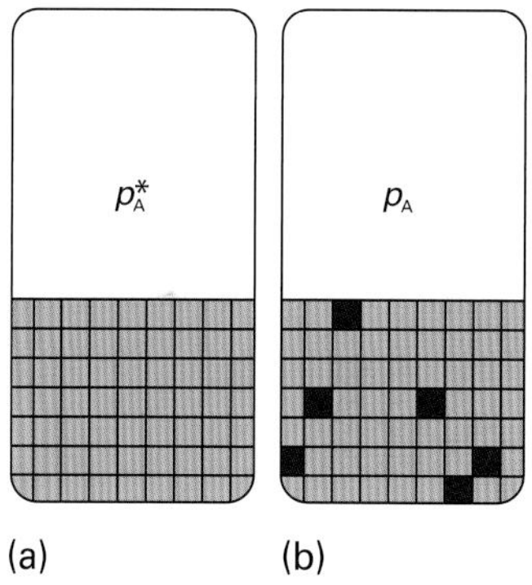

Fig. 5.22 The vapour pressure of a pure liquid represents a balance between the increase in disorder arising from vaporization and the decrease in disorder of the surroundings. (a) Here the structure of the liquid is represented highly schematically by the grid of squares. (b) When solute (the dark squares) is present, the disorder of the condensed phase is higher than that of the pure liquid, and there is a decreased tendency to acquire the disorder characteristic of the vapour.

The strategy for the quantitative discussion of the elevation of boiling point and the depression of freezing point is to look for the temperature at which, at 1 atm, one phase (the pure solvent vapour or the pure solid solvent) has the same chemical potential as the solvent in the solution. This is the new equilibrium temperature for the phase transition at 1 atm, and hence corresponds to the new boiling point or the new freezing point of the solvent.

(b) The elevation of boiling point

The heterogeneous equilibrium of interest when considering boiling is between the solvent vapour and the solvent in solution at 1 atm (Fig. 5.23). We denote the solvent by A and the solute by B. The equilibrium is established at a temperature for which

$$\mu_A^*(g) = \mu_A^*(l) + RT \ln x_A \qquad (5.32)^\circ$$

(The pressure of 1 atm is the same throughout, and will not be written explicitly.) We show in the *Justification* below that this equation implies that the presence of a solute at a mole fraction x_B causes an increase in normal boiling point from T^* to $T^* + \Delta T$, where

$$\Delta T = Kx_B \qquad K = \frac{RT^{*2}}{\Delta_{vap}H} \qquad (5.33)^\circ$$

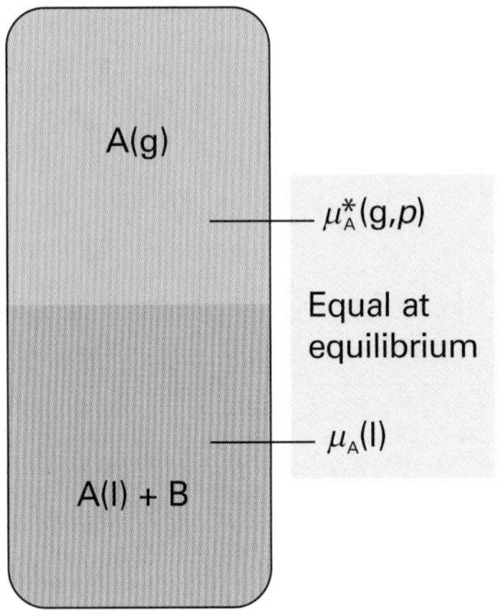

Fig. 5.23 The heterogeneous equilibrium involved in the calculation of the elevation of boiling point is between A in the pure vapour and A in the mixture, A being the solvent and B an involatile solute.

Justification 5.1 *The elevation of the boiling point of a solvent*

Equation 5.32 can be rearranged into

$$\ln x_A = \frac{\mu_A^*(g) - \mu_A^*(l)}{RT} = \frac{\Delta_{vap}G}{RT}$$

where $\Delta_{vap}G$ is the Gibbs energy of vaporization of the pure solvent (A). First, to find the relation between a change in composition and the resulting change in boiling temperature, we differentiate both sides with respect to temperature and use the Gibbs–Helmholtz equation (eqn 3.52, $(\partial(G/T)/\partial T)_p = -H/T^2$) to express the term on the right:

$$\frac{d\ln x_A}{dT} = \frac{1}{R}\frac{d(\Delta_{vap}G/T)}{dT} = -\frac{\Delta_{vap}H}{RT^2}$$

Now multiply both sides by dT and integrate from $x_A = 1$, corresponding to $\ln x_A = 0$ (and when $T = T^*$, the boiling point of pure A) to x_A (when the boiling point is T):

$$\int_0^{\ln x_A} d\ln x_A = -\frac{1}{R}\int_{T^*}^{T}\frac{\Delta_{vap}H}{T^2}\,dT$$

The left-hand side integrates to $\ln x_A$, which is equal to $\ln(1 - x_B)$. The right-hand side can be integrated if we assume that the enthalpy of vaporization is a constant over the small range of temperatures involved and can be taken outside the integral. Thus, we obtain

$$\ln(1 - x_B) = \frac{\Delta_{vap}H}{R}\left(\frac{1}{T} - \frac{1}{T^*}\right)$$

We now suppose that the amount of solute present is so small that $x_B \ll 1$. We can then write $\ln(1 - x_B) \approx -x_B$ and hence obtain

$$x_B = \frac{\Delta_{vap}H}{R}\left(\frac{1}{T^*} - \frac{1}{T}\right)$$

Finally, because $T \approx T^*$, it also follows that

$$\frac{1}{T^*} - \frac{1}{T} = \frac{T - T^*}{TT^*} \approx \frac{\Delta T}{T^{*2}}$$

with $\Delta T = T - T^*$. The previous equation then rearranges into eqn 5.33.

Comment 5.3

The series expansion of a natural logarithm (see *Appendix 2*) is

$$\ln(1 - x) = -x - \tfrac{1}{2}x^2 - \tfrac{1}{3}x^3 \cdots$$

provided that $-1 < x < 1$. If $x \ll 1$, then the terms involving x raised to a power greater than 1 are much smaller than x, so $\ln(1 - x) \approx -x$.

Because eqn 5.33 makes no reference to the identity of the solute, only to its mole fraction, we conclude that the elevation of boiling point is a colligative property. The value of ΔT does depend on the properties of the solvent, and the biggest changes occur for solvents with high boiling points.[3] For practical applications of eqn 5.33, we note that the mole fraction of B is proportional to its molality, b, in the solution, and write

$$\Delta T = K_b b \qquad (5.34)$$

where K_b is the empirical **boiling-point constant** of the solvent (Table 5.2).

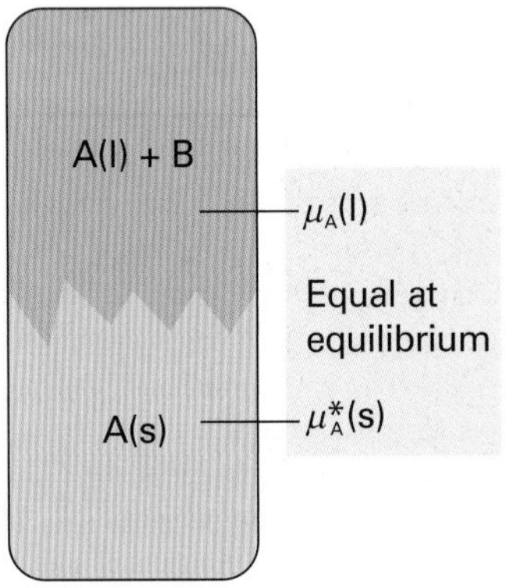

Fig. 5.24 The heterogeneous equilibrium involved in the calculation of the lowering of freezing point is between A in the pure solid and A in the mixture, A being the solvent and B a solute that is insoluble in solid A.

(c) The depression of freezing point

The heterogeneous equilibrium now of interest is between pure solid solvent A and the solution with solute present at a mole fraction x_B (Fig. 5.24). At the freezing point, the chemical potentials of A in the two phases are equal:

[3] By Trouton's rule (Section 3.3b), $\Delta_{vap}H/T^*$ is a constant; therefore eqn 5.33 has the form $\Delta T \propto T^*$ and is independent of $\Delta_{vap}H$ itself.

Synoptic Table 5.2* Freezing-point and boiling-point constants

	$K_f/(\mathrm{K\ kg\ mol^{-1}})$	$K_b/(\mathrm{K\ kg\ mol^{-1}})$
Benzene	5.12	2.53
Camphor	40	
Phenol	7.27	3.04
Water	1.86	0.51

* More values are given in the *Data section*.

$$\mu_A^*(\mathrm{s}) = \mu_A^*(\mathrm{l}) + RT\ln x_A \qquad (5.35)°$$

The only difference between this calculation and the last is the appearance of the solid's chemical potential in place of the vapour's. Therefore we can write the result directly from eqn 5.33:

$$\Delta T = K'x_B \qquad K' = \frac{RT^{*2}}{\Delta_{\mathrm{fus}}H} \qquad (5.36)°$$

where ΔT is the freezing point depression, $T^* - T$, and $\Delta_{\mathrm{fus}}H$ is the enthalpy of fusion of the solvent. Larger depressions are observed in solvents with low enthalpies of fusion and high melting points. When the solution is dilute, the mole fraction is proportional to the molality of the solute, b, and it is common to write the last equation as

$$\Delta T = K_f b \qquad (5.37)$$

where K_f is the empirical **freezing-point constant** (Table 5.2). Once the freezing-point constant of a solvent is known, the depression of freezing point may be used to measure the molar mass of a solute in the method known as **cryoscopy**; however, the technique is of little more than historical interest.

(d) Solubility

Although solubility is not strictly a colligative property (because solubility varies with the identity of the solute), it may be estimated by the same techniques as we have been using. When a solid solute is left in contact with a solvent, it dissolves until the solution is saturated. Saturation is a state of equilibrium, with the undissolved solute in equilibrium with the dissolved solute. Therefore, in a saturated solution the chemical potential of the pure solid solute, $\mu_B^*(\mathrm{s})$, and the chemical potential of B in solution, μ_B, are equal (Fig. 5.25). Because the latter is

$$\mu_B = \mu_B^*(\mathrm{l}) + RT\ln x_B$$

we can write

$$\mu_B^*(\mathrm{s}) = \mu_B^*(\mathrm{l}) + RT\ln x_B \qquad (5.38)°$$

This expression is the same as the starting equation of the last section, except that the quantities refer to the solute B, not the solvent A. We now show in the following *Justification* that

$$\ln x_B = \frac{\Delta_{\mathrm{fus}}H}{R}\left(\frac{1}{T_f} - \frac{1}{T}\right) \qquad (5.39)°$$

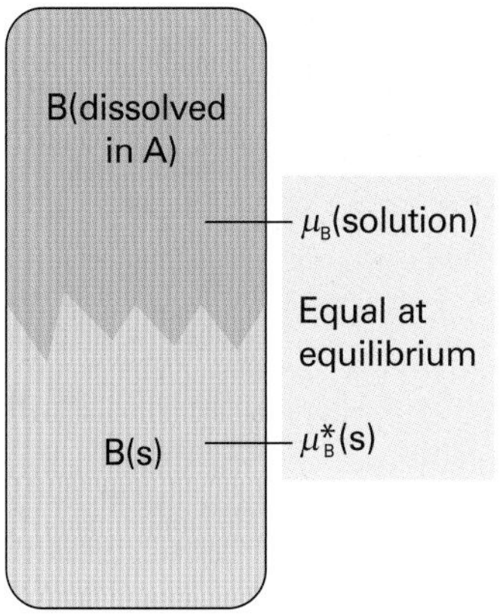

Fig. 5.25 The heterogeneous equilibrium involved in the calculation of the solubility is between pure solid B and B in the mixture.

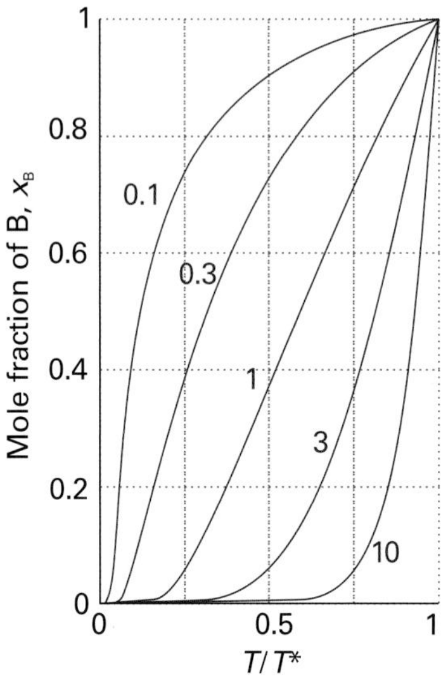

Fig. 5.26 The variation of solubility (the mole fraction of solute in a saturated solution) with temperature (T^* is the freezing temperature of the solute). Individual curves are labelled with the value of $\Delta_{fus}H/RT^*$.

Exploration Derive an expression for the temperature coefficient of the solubility, dx_B/dT, and plot it as a function of temperature for several values of the enthalpy of fusion.

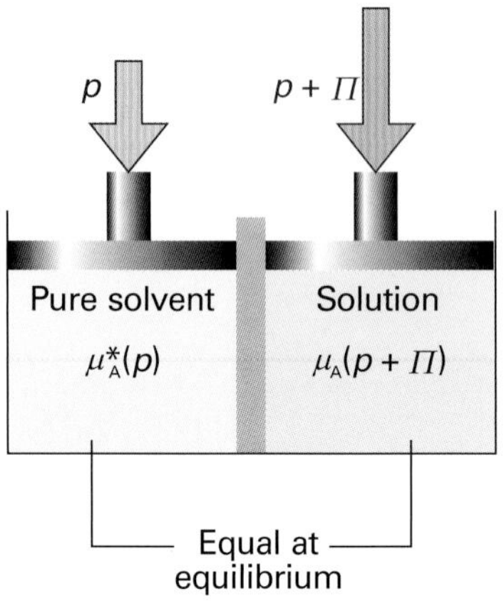

Fig. 5.27 The equilibrium involved in the calculation of osmotic pressure, Π, is between pure solvent A at a pressure p on one side of the semipermeable membrane and A as a component of the mixture on the other side of the membrane, where the pressure is $p + \Pi$.

Justification 5.2 *The solubility of an ideal solute.*

The starting point is the same as in *Justification 5.1* but the aim is different. In the present case, we want to find the mole fraction of B in solution at equilibrium when the temperature is T. Therefore, we start by rearranging eqn 5.38 into

$$\ln x_B = \frac{\mu_B^*(s) - \mu_B^*(l)}{RT} = -\frac{\Delta_{fus}G}{RT}$$

As in *Justification 5.1*, we relate the change in composition $d \ln x_B$ to the change in temperature by differentiation and use of the Gibbs–Helmholtz equation. Then we integrate from the melting temperature of B (when $x_B = 1$ and $\ln x_B = 0$) to the *lower* temperature of interest (when x_B has a value between 0 and 1):

$$\int_0^{\ln x_B} d \ln x_B = \frac{1}{R}\int_{T_f}^{T} \frac{\Delta_{fus}H}{T^2}\,dT$$

If we suppose that the enthalpy of fusion of B is constant over the range of temperatures of interest, it can be taken outside the integral, and we obtain eqn 5.39.

Equation 5.39 is plotted in Fig. 5.26. It shows that the solubility of B decreases exponentially as the temperature is lowered from its melting point. The illustration also shows that solutes with high melting points and large enthalpies of melting have low solubilities at normal temperatures. However, the detailed content of eqn 5.39 should not be treated too seriously because it is based on highly questionable approximations, such as the ideality of the solution. One aspect of its approximate character is that it fails to predict that solutes will have different solubilities in different solvents, for no solvent properties appear in the expression.

(e) Osmosis

The phenomenon of **osmosis** (from the Greek word for 'push') is the spontaneous passage of a pure solvent into a solution separated from it by a **semipermeable membrane**, a membrane permeable to the solvent but not to the solute (Fig. 5.27). The **osmotic pressure**, Π, is the pressure that must be applied to the solution to stop the influx of solvent. Important examples of osmosis include transport of fluids through cell membranes, dialysis and **osmometry**, the determination of molar mass by the measurement of osmotic pressure. Osmometry is widely used to determine the molar masses of macromolecules.

In the simple arrangement shown in Fig. 5.28, the opposing pressure arises from the head of solution that the osmosis itself produces. Equilibrium is reached when the hydrostatic pressure of the column of solution matches the osmotic pressure. The complicating feature of this arrangement is that the entry of solvent into the solution results in its dilution, and so it is more difficult to treat than the arrangement in Fig. 5.27, in which there is no flow and the concentrations remain unchanged.

The thermodynamic treatment of osmosis depends on noting that, at equilibrium, the chemical potential of the solvent must be the same on each side of the membrane. The chemical potential of the solvent is lowered by the solute, but is restored to its 'pure' value by the application of pressure. As shown in the *Justification* below, this equality implies that for dilute solutions the osmotic pressure is given by the **van 't Hoff equation**:

$$\Pi = [B]RT \qquad (5.40)°$$

where $[B] = n_B/V$ is the molar concentration of the solute.

Justification 5.3 *The van 't Hoff equation*

On the pure solvent side the chemical potential of the solvent, which is at a pressure p, is $\mu_A^\star(p)$. On the solution side, the chemical potential is lowered by the presence of the solute, which reduces the mole fraction of the solvent from 1 to x_A. However, the chemical potential of A is raised on account of the greater pressure, $p + \Pi$, that the solution experiences. At equilibrium the chemical potential of A is the same in both compartments, and we can write

$$\mu_A^\star(p) = \mu_A(x_A, p+\Pi)$$

The presence of solute is taken into account in the normal way:

$$\mu_A(x_A, p+\Pi) = \mu_A^\star(p+\Pi) + RT\ln x_A$$

We saw in Section 3.9c (eqn 3.54) how to take the effect of pressure into account:

$$\mu_A^\star(p+\Pi) = \mu_A^\star(p) + \int_p^{p+\Pi} V_m \mathrm{d}p$$

where V_m is the molar volume of the pure solvent A. When these three equations are combined we get

$$-RT\ln x_A = \int_p^{p+\Pi} V_m \mathrm{d}p$$

This expression enables us to calculate the additional pressure Π that must be applied to the solution to restore the chemical potential of the solvent to its 'pure' value and thus to restore equilibrium across the semipermeable membrane. For dilute solutions, $\ln x_A$ may be replaced by $\ln(1-x_B) \approx -x_B$. We may also assume that the pressure range in the integration is so small that the molar volume of the solvent is a constant. That being so, V_m may be taken outside the integral, giving

$$RTx_B = \Pi V_m$$

When the solution is dilute, $x_B \approx n_B/n_A$. Moreover, because $n_A V_m = V$, the total volume of the solvent, the equation simplifies to eqn 5.40.

Because the effect of osmotic pressure is so readily measurable and large, one of the most common applications of osmometry is to the measurement of molar masses of macromolecules, such as proteins and synthetic polymers. As these huge molecules dissolve to produce solutions that are far from ideal, it is assumed that the van 't Hoff equation is only the first term of a virial-like expansion:[4]

$$\Pi = [\mathrm{J}]RT\{1 + B[\mathrm{J}] + \cdots\} \qquad (5.41)$$

The additional terms take the nonideality into account; the empirical constant B is called the **osmotic virial coefficient**.

Example 5.4 *Using osmometry to determine the molar mass of a macromolecule*

The osmotic pressures of solutions of poly(vinyl chloride), PVC, in cyclohexanone at 298 K are given below. The pressures are expressed in terms of the heights of solution (of mass density $\rho = 0.980$ g cm^{-3}) in balance with the osmotic pressure. Determine the molar mass of the polymer.

$c/(\mathrm{g\,dm^{-3}})$	1.00	2.00	4.00	7.00	9.00
h/cm	0.28	0.71	2.01	5.10	8.00

4 We have denoted the solute J to avoid too many different Bs in this expression.

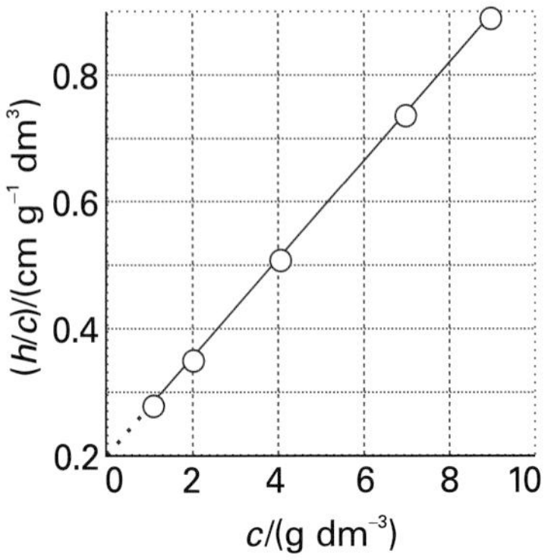

Fig. 5.28 The plot involved in the determination of molar mass by osmometry. The molar mass is calculated from the intercept at $c = 0$; in Chapter 19 we shall see that additional information comes from the slope.

Exploration Calculate the osmotic virial coefficient B from these data.

Method The osmotic pressure is measured at a series of mass concentrations, c, and a plot of Π/c against c is used to determine the molar mass of the polymer. We use eqn 5.41 with $[\mathrm{J}] = c/M$ where c is the mass concentration of the polymer and M is its molar mass. The osmotic pressure is related to the hydrostatic pressure by $\Pi = \rho gh$ (Example 1.2) with $g = 9.81\ \mathrm{m\ s^{-2}}$. With these substitutions, eqn 5.41 becomes

$$\frac{h}{c} = \frac{RT}{\rho gM}\left(1 + \frac{Bc}{M} + \cdots\right) = \frac{RT}{\rho gM} + \left(\frac{RTB}{\rho gM^2}\right)c + \cdots$$

Therefore, to find M, plot h/c against c, and expect a straight line with intercept $RT/\rho gM$ at $c = 0$.

Answer The data give the following values for the quantities to plot:

$c/(\mathrm{g\ dm^{-3}})$	1.00	2.00	4.00	7.00	9.00
$(h/c)/(\mathrm{cm\ g^{-1}\ dm^3})$	0.28	0.36	0.503	0.729	0.889

The points are plotted in Fig. 5.28. The intercept is at 0.21. Therefore,

$$\begin{aligned} M &= \frac{RT}{\rho g} \times \frac{1}{0.21\ \mathrm{cm\ g^{-1}\ dm^3}} \\ &= \frac{(8.3145\ \mathrm{J\ K^{-1}\ mol^{-1}}) \times (298\ \mathrm{K})}{(980\ \mathrm{kg\ m^{-3}}) \times (9.81\ \mathrm{m\ s^{-2}})} \times \frac{1}{2.1 \times 10^{-3}\ \mathrm{m^4\ kg^{-1}}} \\ &= 1.2 \times 10^2\ \mathrm{kg\ mol^{-1}} \end{aligned}$$

where we have used $1\ \mathrm{kg\ m^2\ s^{-2}} = 1\ \mathrm{J}$. Molar masses of macromolecules are often reported in daltons (Da), with $1\ \mathrm{Da} = 1\ \mathrm{g\ mol^{-1}}$. The macromolecule in this example has a molar mass of about 120 kDa. Modern osmometers give readings of osmotic pressure in pascals, so the analysis of the data is more straightforward and eqn 5.41 can be used directly. As we shall see in Chapter 19, the value obtained from osmometry is the 'number average molar mass'.

Self-test 5.6 Estimate the depression of freezing point of the most concentrated of these solutions, taking K_f as about $10\ \mathrm{K/(mol\ kg^{-1})}$. [0.8 mK]

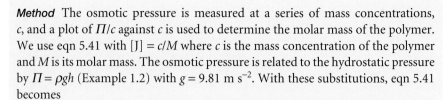

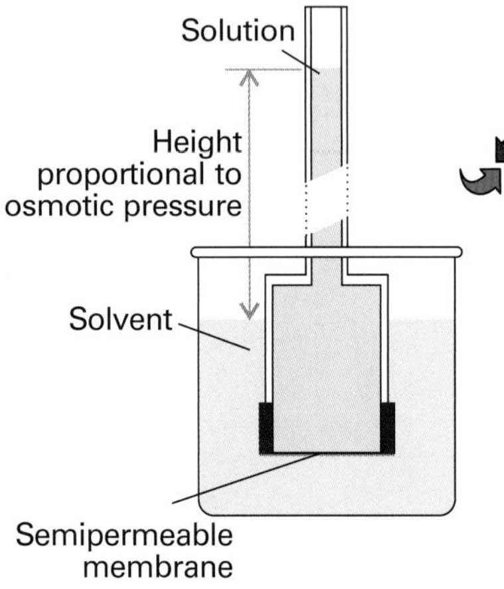

Fig. 5.29 In a simple version of the osmotic pressure experiment, A is at equilibrium on each side of the membrane when enough has passed into the solution to cause a hydrostatic pressure difference.

IMPACT ON BIOLOGY

I5.2 Osmosis in physiology and biochemistry

Osmosis helps biological cells maintain their structure. Cell membranes are semipermeable and allow water, small molecules, and hydrated ions to pass, while blocking the passage of biopolymers synthesized inside the cell. The difference in concentrations of solutes inside and outside the cell gives rise to an osmotic pressure, and water passes into the more concentrated solution in the interior of the cell, carrying small nutrient molecules. The influx of water also keeps the cell swollen, whereas dehydration causes the cell to shrink. These effects are important in everyday medical practice. To maintain the integrity of blood cells, solutions that are injected into the bloodstream for blood transfusions and intravenous feeding must be *isotonic* with the blood, meaning that they must have the same osmotic pressure as blood. If the injected solution is too dilute, or *hypotonic*, the flow of solvent into the cells, required to equalize the osmotic pressure, causes the cells to burst and die by a process called *haemolysis*. If the solution is too concentrated, or *hypertonic*, equalization of the osmotic pressure requires flow of solvent out of the cells, which shrink and die.

Osmosis also forms the basis of *dialysis*, a common technique for the removal of impurities from solutions of biological macromolecules and for the study of binding of small molecules to macromolecules, such as an inhibitor to an enzyme, an antibiotic to DNA, and any other instance of cooperation or inhibition by small molecules attaching to large ones. In a purification experiment, a solution of macromolecules containing impurities, such as ions or small molecules (including small proteins or nucleic acids), is placed in a bag made of a material that acts as a semipermeable membrane and the filled bag is immersed in a solvent. The membrane permits the passage of the small ions and molecules but not the larger macromolecules, so the former migrate through the membrane, leaving the macromolecules behind. In practice, purification of the sample requires several changes of solvent to coax most of the impurities out of the dialysis bag.

In a binding experiment, a solution of macromolecules and smaller ions or molecules is placed in a dialysis bag, which is then immersed in a solvent. Suppose the molar concentration of the macromolecule M is [M] and the total concentration of the small molecule A in the bag is $[A]_{in}$. This total concentration is the sum of the concentrations of free A and bound A, which we write $[A]_{free}$ and $[A]_{bound}$, respectively. At equilibrium, the chemical potential of free A in the macromolecule solution is equal to the chemical potential of A in the solution on the other side of the membrane, where its concentration is $[A]_{out}$. We shall see in Section 5.7 that the equality $\mu_{A,free} = \mu_{A,out}$ implies that $[A]_{free} = [A]_{out}$, provided the activity coefficient of A is the same in both solutions. Therefore, by measuring the concentration of A in the 'outside' solution, we can find the concentration of unbound A in the macromolecule solution and, from the difference $[A]_{in} - [A]_{free}$, which is equal to $[A]_{in} - [A]_{out}$, the concentration of bound A. The average number of A molecules bound to M molecules, ν, is then the ratio

$$\nu = \frac{[A]_{bound}}{[M]} = \frac{[A]_{in} - [A]_{out}}{[M]}$$

The bound and unbound A molecules are in equilibrium, $M + A \rightleftharpoons MA$, so their concentrations are related by an equilibrium constant K, where

$$K = \frac{[MA]}{[M]_{free}[A]_{free}} = \frac{[A]_{bound}}{([M] - [A]_{bound})[A]_{free}}$$

We have used $[MA] = [A]_{bound}$ and $[M]_{free} = [M] - [MA] = [M] - [A]_{bound}$. On division by [M], and replacement of $[A]_{free}$ by $[A]_{out}$, the last expression becomes

$$K = \frac{\nu}{(1-\nu)[A]_{out}}$$

If there are N *identical* and *independent* binding sites on each macromolecule, each macromolecule behaves like N separate smaller macromolecules, with the same value of K for each site. The average number of A molecules per site is ν/N, so the last equation becomes

$$K = \frac{\nu/N}{\left(1 - \dfrac{\nu}{N}\right)[A]_{out}}$$

It then follows that

$$\frac{\nu}{[A]_{out}} = KN - K\nu$$

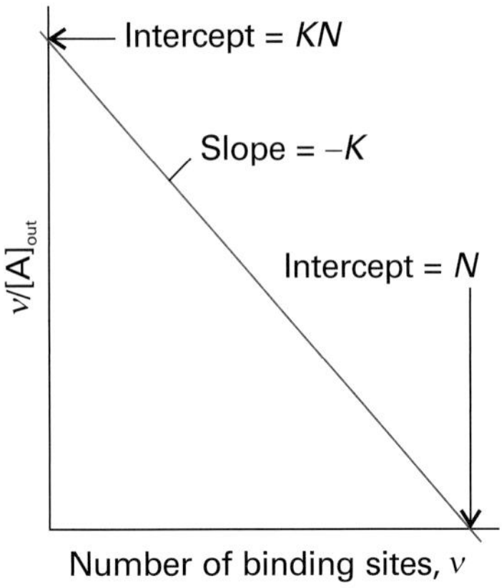

Fig. 5.30 A Scatchard plot of $v/[\mathrm{A}]_{\mathrm{out}}$ against v. The slope is $-K$ and the intercept at $v = 0$ is KN.

Exploration The following tasks will give you an idea of how graphical analysis can distinguish between systems with the same values of K or N. First, display on the same graph several Scatchard plots with varying K values but fixed N. Then repeat the process, this time varying N but fixing K.

This expression is the *Scatchard equation*. It implies that a plot of $v/[\mathrm{A}]_{\mathrm{out}}$ against v should be a straight line of slope $-K$ and intercept KN at $v = 0$ (see Fig. 5.30). From these two quantities, we can find the equilibrium constant for binding and the number of binding sites on each macromolecule. If a straight line is not obtained we can conclude that the binding sites are not equivalent or independent.

Activities

Now we see how to adjust the expressions developed earlier in the chapter to take into account deviations from ideal behaviour. In Chapter 3 (specifically, *Further information 3.2*) we remarked that a quantity called 'fugacity' takes into account the effects of gas imperfections in a manner that resulted in the least upset of the form of equations. Here we see how the expressions encountered in the treatment of ideal solutions can also be preserved almost intact by introducing the concept of 'activity'. It is important to be aware of the different definitions of standard states and activities, and they are summarized in Table 5.3. We shall put them to work in the next few chapters, when we shall see that using them is much easier than defining them.

5.6 The solvent activity

The general form of the chemical potential of a real or ideal solvent is given by a straightforward modification of eqn 5.23 (that $\mu_{\mathrm{A}} = \mu_{\mathrm{A}}^{\star} + RT\ln(p_{\mathrm{A}}/p_{\mathrm{A}}^{\star})$, where $p_{\mathrm{A}}^{\star}$ is the vapour pressure of pure A and p_{A} is the vapour pressure of A when it is a component of a solution. For an ideal solution, as we have seen, the solvent obeys Raoult's law at all concentrations and we can express this relation as eqn 5.25 (that is, as $\mu_{\mathrm{A}} = \mu_{\mathrm{A}}^{\star} + RT\ln x_{\mathrm{A}}$). The form of the this relation can be preserved when the solution does not obey Raoult's law by writing

$$\mu_{\mathrm{A}} = \mu_{\mathrm{A}}^{\star} + RT\ln a_{\mathrm{A}} \qquad [5.42]$$

The quantity a_{A} is the **activity** of A, a kind of 'effective' mole fraction, just as the fugacity is an effective pressure.

Because eqn 5.23 is true for both real and ideal solutions (the only approximation being the use of pressures rather than fugacities), we can conclude by comparing it with eqn 5.42 that

$$a_{\mathrm{A}} = \frac{p_{\mathrm{A}}}{p_{\mathrm{A}}^{\star}} \qquad (5.43)$$

Table 5.3 Standard states

Component	Basis	Standard state	Activity	Limits
Solid or liquid		Pure	$a = 1$	
Solvent	Raoult	Pure solvent	$a = p/p^{\star}$, $a = \gamma x$	$\gamma \to 1$ as $x \to 1$ (pure solvent)
Solute	Henry	(1) A hypothetical state of the pure solute	$a = p/K$, $a = \gamma x$	$\gamma \to 1$ as $x \to 0$
		(2) A hypothetical state of the solute at molality $b^{\ominus}$	$a = \gamma b/b^{\ominus}$	$\gamma \to 1$ as $b \to 0$

In each case, $\mu = \mu^{\ominus} + RT\ln a$.

We see that there is nothing mysterious about the activity of a solvent: it can be determined experimentally simply by measuring the vapour pressure and then using eqn 5.43.

Illustration 5.3 *Calculating the solvent activity*

The vapour pressure of 0.500 M $KNO_3(aq)$ at 100°C is 99.95 kPa, so the activity of water in the solution at this temperature is

$$a_A = \frac{99.95 \text{ kPa}}{101.325 \text{ kPa}} = 0.9864$$

Because all solvents obey Raoult's law (that $p_A/p_A^* = x_A$) increasingly closely as the concentration of solute approaches zero, the activity of the solvent approaches the mole fraction as $x_A \to 1$:

$$a_A \to x_A \quad \text{as} \quad x_A \to 1 \qquad (5.44)$$

A convenient way of expressing this convergence is to introduce the **activity coefficient**, γ, by the definition

$$a_A = \gamma_A x_A \qquad \gamma_A \to 1 \quad \text{as} \quad x_A \to 1 \qquad [5.45]$$

at all temperatures and pressures. The chemical potential of the solvent is then

$$\mu_A = \mu_A^* + RT \ln x_A + RT \ln \gamma_A \qquad (5.46)$$

The standard state of the solvent, the pure liquid solvent at 1 bar, is established when $x_A = 1$.

5.7 The solute activity

The problem with defining activity coefficients and standard states for solutes is that they approach ideal-dilute (Henry's law) behaviour as $x_B \to 0$, not as $x_B \to 1$ (corresponding to pure solute). We shall show how to set up the definitions for a solute that obeys Henry's law exactly, and then show how to allow for deviations.

(a) Ideal-dilute solutions

A solute B that satisfies Henry's law has a vapour pressure given by $p_B = K_B x_B$, where K_B is an empirical constant. In this case, the chemical potential of B is

$$\mu_B = \mu_B^* + RT \ln \frac{p_B}{p_B^*} = \mu_B^* + RT \ln \frac{K_B}{p_B^*} + RT \ln x_B$$

Both K_B and p_B^* are characteristics of the solute, so the second term may be combined with the first to give a new standard chemical potential:

$$\mu_B^{\ominus} = \mu_B^* + RT \ln \frac{K_B}{p_B^*} \qquad [5.47]$$

It then follows that the chemical potential of a solute in an ideal-dilute solution is related to its mole fraction by

$$\mu_B = \mu_B^{\ominus} + RT \ln x_B \qquad (5.48)^\circ$$

If the solution is ideal, $K_B = p_B^*$ and eqn 5.47 reduces to $\mu_B^{\ominus} = \mu_B^*$, as we should expect.

(b) Real solutes

We now permit deviations from ideal-dilute, Henry's law behaviour. For the solute, we introduce a_B in place of x_B in eqn 5.48, and obtain

$$\mu_B = \mu_B^{\ominus} + RT \ln a_B \qquad [5.49]$$

The standard state remains unchanged in this last stage, and all the deviations from ideality are captured in the activity a_B. The value of the activity at any concentration can be obtained in the same way as for the solvent, but in place of eqn 5.43 we use

$$a_B = \frac{p_B}{K_B} \qquad (5.50)$$

As for the solvent, it is sensible to introduce an activity coefficient through

$$a_B = \gamma_B x_B \qquad [5.51]$$

Now all the deviations from ideality are captured in the activity coefficient γ_B. Because the solute obeys Henry's law as its concentration goes to zero, it follows that

$$a_B \rightarrow x_B \quad \text{and} \quad \gamma_B \rightarrow 1 \quad \text{as} \quad x_B \rightarrow 0 \qquad (5.52)$$

at all temperatures and pressures. Deviations of the solute from ideality disappear as zero concentration is approached.

Example 5.5 *Measuring activity*

Use the information in Example 5.3 to calculate the activity and activity coefficient of chloroform in acetone at 25°C, treating it first as a solvent and then as a solute. For convenience, the data are repeated here:

x_C	0	0.20	0.40	0.60	0.80	1
p_C/kPa	0	4.7	11	18.9	26.7	36.4
p_A/kPa	46.3	33.3	23.3	12.3	4.9	0

Method For the activity of chloroform as a solvent (the Raoult's law activity), form $a_C = p_C/p_C^*$ and $\gamma_C = a_C/x_C$. For its activity as a solute (the Henry's law activity), form $a_C = p_C/K_C$ and $\gamma_C = a_C/x_C$.

Answer Because $p_C^* = 36.4$ kPa and $K_C = 22.0$ kPa, we can construct the following tables. For instance, at $x_C = 0.20$, in the Raoult's law case we find $a_C = (4.7\text{ kPa})/(36.4\text{ kPa}) = 0.13$ and $\gamma_C = 0.13/0.20 = 0.65$; likewise, in the Henry's law case, $a_C = (4.7\text{ kPa})/(22.0\text{ kPa}) = 0.21$ and $\gamma_C = 0.21/0.20 = 1.05$.

From Raoult's law (chloroform regarded as the solvent):

a_C	0	0.13	0.30	0.52	0.73	1.00
γ_C		0.65	0.75	0.87	0.91	1.00

From Henry's law (chloroform regarded as the solute):

a_C	0	0.21	0.50	0.86	1.21	1.65
γ_C	1	1.05	1.25	1.43	1.51	1.65

These values are plotted in Fig. 5.31. Notice that $\gamma_C \rightarrow 1$ as $x_C \rightarrow 1$ in the Raoult's law case, but that $\gamma_C \rightarrow 1$ as $x_C \rightarrow 0$ in the Henry's law case.

Self-test 5.7 Calculate the activities and activity coefficients for acetone according to the two conventions.

[At $x_A = 0.60$, for instance $a_R = 0.50$; $\gamma_R = 0.83$; $a_H = 1.00$, $\gamma_H = 1.67$]

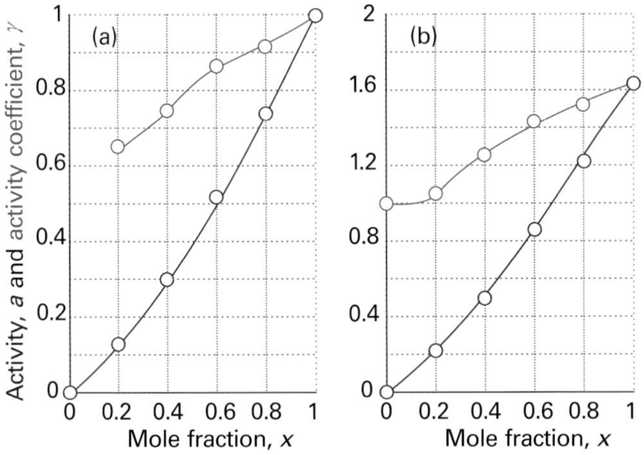

Fig. 5.31 The variation of activity and activity coefficient of chloroform (trichloromethane) and acetone (propanone) with composition according to (a) Raoult's law, (b) Henry's law.

(c) Activities in terms of molalities

The selection of a standard state is entirely arbitrary, so we are free to choose one that best suits our purpose and the description of the composition of the system. In chemistry, compositions are often expressed as molalities, b, in place of mole fractions. It therefore proves convenient to write

$$\mu_B = \mu_B^{\ominus} + RT \ln b_B \qquad \{5.53\}^{\circ}$$

where $\mu^{\ominus}$ has a different value from the standard values introduced earlier. According to this definition, the chemical potential of the solute has its standard value $\mu^{\ominus}$ when the molality of B is equal to $b^{\ominus}$ (that is, at 1 mol kg^{-1}). Note that as $b_B \to 0$, $\mu_B \to -\infty$; that is, as the solution becomes diluted, so the solute becomes increasingly stabilized. The practical consequence of this result is that it is very difficult to remove the last traces of a solute from a solution.

Now, as before, we incorporate deviations from ideality by introducing a dimensionless activity a_B, a dimensionless activity coefficient γ_B, and writing

$$a_B = \gamma_B \frac{b_B}{b^{\ominus}} \quad \text{where} \quad \gamma_B \to 1 \quad \text{as} \quad b_B \to 0 \qquad [5.54]$$

at all temperatures and pressures. The standard state remains unchanged in this last stage and, as before, all the deviations from ideality are captured in the activity coefficient γ_B. We then arrive at the following succinct expression for the chemical potential of a real solute at any molality:

$$\mu = \mu^{\ominus} + RT \ln a \qquad (5.55)$$

(d) The biological standard state

One important illustration of the ability to choose a standard state to suit the circumstances arises in biological applications. The conventional standard state of hydrogen ions (unit activity, corresponding to pH = 0)[5] is not appropriate to normal biological conditions. Therefore, in biochemistry it is common to adopt the **biological standard state**, in which pH = 7 (an activity of 10^{-7}, neutral solution) and to label the corresponding standard thermodynamic functions as $G^{\oplus}$, $H^{\oplus}$, $\mu^{\oplus}$, and $S^{\oplus}$ (some texts use $X^{\ominus\prime}$).

[5] Recall from introductory chemistry courses that pH = –log $a_{H_3O^+}$.

To find the relation between the thermodynamic and biological standard values of the chemical potential of hydrogen ions we need to note from eqn 5.55 that

$$\mu_{H^+} = \mu^{\ominus}_{H^+} + RT \ln a_{H^+} = \mu^{\ominus}_{H^+} - (RT \ln 10) \times \text{pH}$$

It follows that

$$\mu^{\oplus}_{H^+} = \mu^{\ominus}_{H^+} - 7RT \ln 10 \qquad (5.56)$$

At 298 K, $7RT \ln 10 = 39.96$ kJ mol^{-1}, so the two standard values differ by about 40 kJ mol^{-1}.

5.8 The activities of regular solutions

The material on regular solutions presented in Section 5.4 gives further insight into the origin of deviations from Raoult's law and its relation to activity coefficients. The starting point is the expression for the Gibbs energy of mixing for a regular solution (eqn 5.31). We show in the following *Justification* that eqn 5.31 implies that the activity coefficients are given by expressions of the form

$$\ln \gamma_A = \beta x_B^2 \qquad \ln \gamma_B = \beta x_A^2 \qquad (5.57)$$

These relations are called the **Margules equations**.

Justification 5.4 *The Margules equations*

The Gibbs energy of mixing to form a nonideal solution is

$$\Delta_{mix}G = nRT\{x_A \ln a_A + x_B \ln a_B\}$$

This relation follows from the derivation of eqn 5.31 with activities in place of mole fractions. If each activity is replaced by γx, this expression becomes

$$\Delta_{mix}G = nRT\{x_A \ln x_A + x_B \ln x_B + x_A \ln \gamma_A + x_B \ln \gamma_B\}$$

Now we introduce the two expressions in eqn 5.57, and use $x_A + x_B = 1$, which gives

$$\begin{aligned}\Delta_{mix}G &= nRT\{x_A \ln x_A + x_B \ln x_B + \beta x_A x_B^2 + \beta x_B x_A^2\} \\ &= nRT\{x_A \ln x_A + x_B \ln x_B + \beta x_A x_B(x_A + x_B)\} \\ &= nRT\{x_A \ln x_A + x_B \ln x_B + \beta x_A x_B\}\end{aligned}$$

as required by eqn 5.31. Note, moreover, that the activity coefficients behave correctly for dilute solutions: $\gamma_A \to 1$ as $x_B \to 0$ and $\gamma_B \to 1$ as $x_A \to 0$.

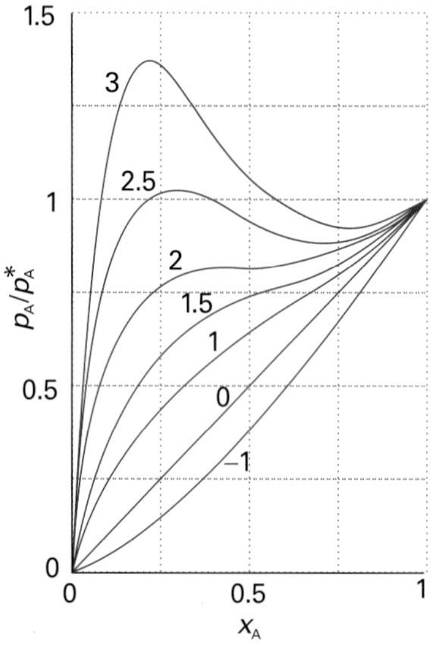

Fig. 5.32 The vapour pressure of a mixture based on a model in which the excess enthalpy is proportional to $\beta x_A x_B$. An ideal solution corresponds to $\beta = 0$ and gives a straight line, in accord with Raoult's law. Positive values of β give vapour pressures higher than ideal. Negative values of β give a lower vapour pressure.

Exploration Plot p_A/p_A^* against x_A with $\beta = 2.5$ by using eqn 5.24 and then eqn 5.59. Above what value of x_A do the values of p_A/p_A^* given by these equations differ by more than 10 per cent?

At this point we can use the Margules equations to write the activity of A as

$$a_A = \gamma_A x_A = x_A e^{\beta x_B^2} = x_A e^{\beta(1-x_A)^2} \qquad (5.58)$$

with a similar expression for a_B. The activity of A, though, is just the ratio of the vapour pressure of A in the solution to the vapour pressure of pure A (eqn 5.43), so we can write

$$p_A = \{x_A e^{\beta(1-x_A)^2}\} p_A^* \qquad (5.59)$$

This function is plotted in Fig. 5.32. We see that $\beta = 0$, corresponding to an ideal solution, gives a straight line, in accord with Raoult's law (indeed, when $\beta = 0$, eqn 5.59 becomes $p_A = x_A p_A^*$, which is Raoult's law). Positive values of β (endothermic mixing, unfavourable solute–solvent interactions) give vapour pressures higher than ideal. Negative values of β (exothermic mixing, favourable solute–solvent interactions) give a lower vapour pressure. All the curves approach linearity and coincide with the Raoult's law line as $x_A \to 1$ and the exponential function in eqn 5.59 approaches 1. When $x_A \ll 1$, eqn 5.59 approaches

$$p_A = x_A e^{\beta} p_A^\star \qquad (5.60)$$

This expression has the form of Henry's law once we identify K with $e^{\beta}p_A^\star$, which is different for each solute–solvent system.

5.9 The activities of ions in solution

Interactions between ions are so strong that the approximation of replacing activities by molalities is valid only in very dilute solutions (less than 10^{-3} mol kg^{-1} in total ion concentration) and in precise work activities themselves must be used. We need, therefore, to pay special attention to the activities of ions in solution, especially in preparation for the discussion of electrochemical phenomena.

(a) Mean activity coefficients

If the chemical potential of a univalent cation M^+ is denoted μ_+ and that of a univalent anion X^- is denoted μ_-, the total molar Gibbs energy of the ions in the electrically neutral solution is the sum of these partial molar quantities. The molar Gibbs energy of an *ideal* solution is

$$G_m^{ideal} = \mu_+^{ideal} + \mu_-^{ideal} \qquad (5.61)^\circ$$

However, for a *real* solution of M^+ and X^- of the same molality,

$$G_m = \mu_+ + \mu_- = \mu_+^{ideal} + \mu_-^{ideal} + RT\ln\gamma_+ + RT\ln\gamma_- = G_m^{ideal} + RT\ln\gamma_+\gamma_- \qquad (5.62)$$

All the deviations from ideality are contained in the last term.

There is no experimental way of separating the product $\gamma_+\gamma_-$ into contributions from the cations and the anions. The best we can do experimentally is to assign responsibility for the nonideality equally to both kinds of ion. Therefore, for a 1,1-electrolyte, we introduce the **mean activity coefficient** as the geometric mean of the individual coefficients:

$$\gamma_\pm = (\gamma_+\gamma_-)^{1/2} \qquad [5.63]$$

Comment 5.4

The geometric mean of x^p and y^q is $(x^p y^q)^{1/(p+q)}$. For example, the geometric mean of x^2 and y^{-3} is $(x^2y^{-3})^{-1}$.

and express the individual chemical potentials of the ions as

$$\mu_+ = \mu_+^{ideal} + RT\ln\gamma_\pm \qquad \mu_- = \mu_-^{ideal} + RT\ln\gamma_\pm \qquad (5.64)$$

The sum of these two chemical potentials is the same as before, eqn 5.62, but now the nonideality is shared equally.

We can generalize this approach to the case of a compound M_pX_q that dissolves to give a solution of p cations and q anions from each formula unit. The molar Gibbs energy of the ions is the sum of their partial molar Gibbs energies:

$$G_m = p\mu_+ + q\mu_- = G_m^{ideal} + pRT\ln\gamma_+ + qRT\ln\gamma_- \qquad (5.65)$$

If we introduce the mean activity coefficient

$$\gamma_\pm = (\gamma_+^p\gamma_-^q)^{1/s} \qquad s = p + q \qquad [5.66]$$

and write the chemical potential of each ion as

$$\mu_i = \mu_i^{ideal} + RT\ln\gamma_\pm \qquad (5.67)$$

we get the same expression as in eqn 5.62 for G_m when we write

$$G = p\mu_+ + q\mu_- \qquad (5.68)$$

However, both types of ion now share equal responsibility for the nonideality.

(b) The Debye–Hückel limiting law

The long range and strength of the Coulombic interaction between ions means that it is likely to be primarily responsible for the departures from ideality in ionic solutions

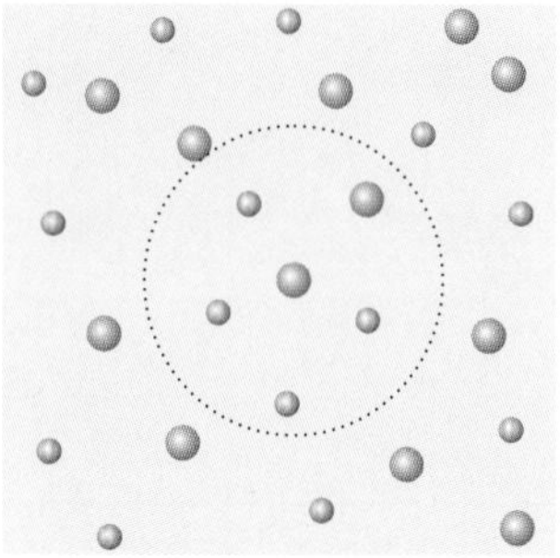

Fig. 5.33 The picture underlying the Debye–Hückel theory is of a tendency for anions to be found around cations, and of cations to be found around anions (one such local clustering region is shown by the circle). The ions are in ceaseless motion, and the diagram represents a snapshot of their motion. The solutions to which the theory applies are far less concentrated than shown here.

and to dominate all the other contributions to nonideality. This domination is the basis of the **Debye–Hückel theory** of ionic solutions, which was devised by Peter Debye and Erich Hückel in 1923. We give here a qualitative account of the theory and its principal conclusions. The calculation itself, which is a profound example of how a seemingly intractable problem can be formulated and then resolved by drawing on physical insight, is described in *Further information 5.1*.

Oppositely charged ions attract one another. As a result, anions are more likely to be found near cations in solution, and vice versa (Fig. 5.33). Overall, the solution is electrically neutral, but near any given ion there is an excess of counter ions (ions of opposite charge). Averaged over time, counter ions are more likely to be found near any given ion. This time-averaged, spherical haze around the central ion, in which counter ions outnumber ions of the same charge as the central ion, has a net charge equal in magnitude but opposite in sign to that on the central ion, and is called its **ionic atmosphere**. The energy, and therefore the chemical potential, of any given central ion is lowered as a result of its electrostatic interaction with its ionic atmosphere. This lowering of energy appears as the difference between the molar Gibbs energy G_m and the ideal value G_m^{ideal} of the solute, and hence can be identified with $RT \ln \gamma_\pm$. The stabilization of ions by their interaction with their ionic atmospheres is part of the explanation why chemists commonly use dilute solutions, in which the stabilization is less important, to achieve precipitation of ions from electrolyte solutions.

The model leads to the result that at very low concentrations the activity coefficient can be calculated from the **Debye–Hückel limiting law**

$$\log \gamma_\pm = -|z_+z_-|AI^{1/2} \qquad (5.69)$$

where $A = 0.509$ for an aqueous solution at 25°C and I is the dimensionless **ionic strength** of the solution:

$$I = \tfrac{1}{2}\sum_i z_i^2(b_i/b^\ominus) \qquad [5.70]$$

In this expression z_i is the charge number of an ion i (positive for cations and negative for anions) and b_i is its molality. The ionic strength occurs widely wherever ionic solutions are discussed, as we shall see. The sum extends over all the ions present in the solution. For solutions consisting of two types of ion at molalities b_+ and b_-,

$$I = \tfrac{1}{2}(b_+z_+^2 + b_-z_-^2)/b^\ominus \qquad (5.71)$$

The ionic strength emphasizes the charges of the ions because the charge numbers occur as their squares. Table 5.4 summarizes the relation of ionic strength and molality in an easily usable form.

Table 5.4 Ionic strength and molality, $I = kb/b^\ominus$

k	X^-	X^{2-}	X^{3-}	X^{4-}
M^+	1	3	6	10
M^{2+}	3	4	15	12
M^{3+}	6	15	9	42
M^{4+}	10	12	42	16

For example, the ionic strength of an M_2X_3 solution of molality b, which is understood to give M^{3+} and X^{2-} ions in solution is $15b/b^\ominus$.

Illustration 5.4 *Using the Debye–Hückel limiting law*

The mean activity coefficient of 5.0×10^{-3} mol kg^{-1} KCl(aq) at 25°C is calculated by writing

$$I = \tfrac{1}{2}(b_+ + b_-)/b^\ominus = b/b^\ominus$$

where b is the molality of the solution (and $b_+ = b_- = b$). Then, from eqn 5.69,

$$\log \gamma_\pm = -0.509 \times (5.0 \times 10^{-3})^{1/2} = -0.036$$

Hence, $\gamma_\pm = 0.92$. The experimental value is 0.927.

Self-test 5.8 Calculate the ionic strength and the mean activity coefficient of 1.00 mmol kg^{-1} $CaCl_2$(aq) at 25°C. [3.00 mmol kg^{-1}, 0.880]

The name 'limiting law' is applied to eqn 5.69 because ionic solutions of moderate molalities may have activity coefficients that differ from the values given by this expression, yet all solutions are expected to conform as $b \to 0$. Table 5.5 lists some experimental values of activity coefficients for salts of various valence types. Figure 5.34 shows some of these values plotted against $I^{1/2}$, and compares them with the theoretical straight lines calculated from eqn 5.69. The agreement at very low molalities (less than about 1 mmol kg^{-1}, depending on charge type) is impressive, and convincing evidence in support of the model. Nevertheless, the departures from the theoretical curves above these molalities are large, and show that the approximations are valid only at very low concentrations.

Synoptic Table 5.5* Mean activity coefficients in water at 298 K

$b/b^{\ominus}$	KCl	$CaCl_2$
0.001	0.966	0.888
0.01	0.902	0.732
0.1	0.770	0.524
1.0	0.607	0.725

* More values are given in the *Data section*.

(c) The extended Debye–Hückel law

When the ionic strength of the solution is too high for the limiting law to be valid, the activity coefficient may be estimated from the **extended Debye–Hückel law**:

$$\log \gamma_{\pm} = -\frac{A\,|z_+z_-|\,I^{1/2}}{1+BI^{1/2}} + CI \qquad (5.72)$$

where B and C are dimensionless constants. Although B can be interpreted as a measure of the closest approach of the ions, it (like C) is best regarded as an adjustable empirical parameter. A curve drawn in this way is shown in Fig. 5.35. It is clear that

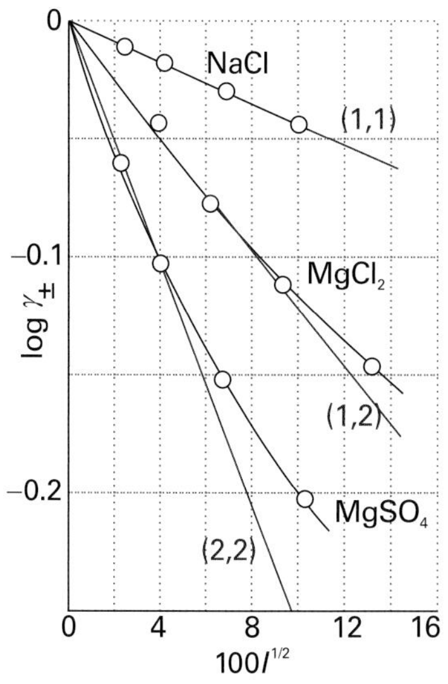

Fig. 5.34 An experimental test of the Debye–Hückel limiting law. Although there are marked deviations for moderate ionic strengths, the limiting slopes as $I \to 0$ are in good agreement with the theory, so the law can be used for extrapolating data to very low molalities.

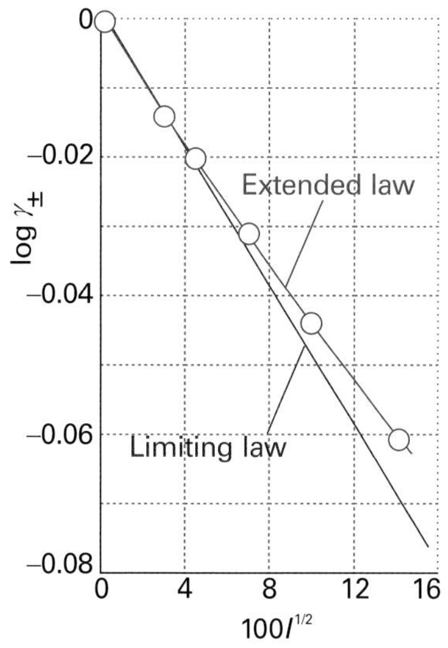

Fig. 5.35 The extended Debye–Hückel law gives agreement with experiment over a wider range of molalities (as shown here for a 1,1-electrolyte), but it fails at higher molalities.

Exploration Consider the plot of $\log \gamma_{\pm}$ against $I^{1/2}$ with $B = 1.50$ and $C = 0$ as a representation of experimental data for a certain 1,1-electrolyte. Over what range of ionic strengths does the application of the limiting law lead to an error in the value of the activity coefficient of less than 10 per cent of the value predicted by the extended law?

eqn 5.72 accounts for some activity coefficients over a moderate range of dilute solutions (up to about 0.1 mol kg^{-1}); nevertheless it remains very poor near 1 mol kg^{-1}.

Current theories of activity coefficients for ionic solutes take an indirect route. They set up a theory for the dependence of the activity coefficient of the solvent on the concentration of the solute, and then use the Gibbs–Duhem equation (eqn 5.12) to estimate the activity coefficient of the solute. The results are reasonably reliable for solutions with molalities greater than about 0.1 mol kg^{-1} and are valuable for the discussion of mixed salt solutions, such as sea water.

Checklist of key ideas

☐ 1. The partial molar volume is the change in volume per mole of A added to a large volume of the mixture: $V_J = (\partial V/\partial n_J)_{p,T,n'}$. The total volume of a mixture is $V = n_A V_A + n_B V_B$.

☐ 2. The chemical potential can be defined in terms of the partial molar Gibbs energy, $\mu_J = (\partial G/\partial n_J)_{p,T,n'}$. The total Gibbs energy of a mixture is $G = n_A\mu_A + n_B\mu_B$.

☐ 3. The fundamental equation of chemical thermodynamics relates the change in Gibbs energy to changes in pressure, temperature, and composition: $dG = Vdp - SdT + \mu_A dn_A + \mu_B dn_B + \cdots$.

☐ 4. The Gibbs–Duhem equation is $\sum_J n_J d\mu_J = 0$.

☐ 5. The chemical potential of a perfect gas is $\mu = \mu^\ominus + RT\ln(p/p^\ominus)$, where $\mu^\ominus$ is the standard chemical potential, the chemical potential of the pure gas at 1 bar.

☐ 6. The Gibbs energy of mixing of two perfect gases is given by $\Delta_{mix}G = nRT(x_A \ln x_A + x_B \ln x_B)$.

☐ 7. The entropy of mixing of two perfect gases is given by $\Delta_{mix}S = -nR(x_A \ln x_A + x_B \ln x_B)$.

☐ 8. The enthalpy of mixing is $\Delta_{mix}H = 0$ for perfect gases.

☐ 9. An ideal solution is a solution in which all components obeys Raoult's law ($p_A = x_A p_A^*$) throughout the composition range.

☐ 10. The chemical potential of a component of an ideal solution is given by $\mu_A = \mu_A^* + RT\ln x_A$.

☐ 11. An ideal-dilute solution is a solution for which the solute obeys Henry's law ($p_B = x_B K_B^*$) and the solvent obeys Raoult's law.

☐ 12. The Gibbs energy of mixing of two liquids that form an ideal solution is given by $\Delta_{mix}G = nRT(x_A \ln x_A + x_B \ln x_B)$.

☐ 13. The entropy of mixing of two liquids that form an ideal solution is given by $\Delta_{mix}S = -nR(x_A \ln x_A + x_B \ln x_B)$.

☐ 14. An excess function (X^E) is the difference between the observed thermodynamic function of mixing and the function for an ideal solution.

☐ 15. A regular solution is a solution for which $H^E \neq 0$ but $S^E = 0$.

☐ 16. A colligative property is a property that depends only on the number of solute particles present, not their identity.

☐ 17. The elevation of boiling point is given by $\Delta T = K_b b$, where K_b is the ebullioscopic constant. The depression of freezing point is given by $\Delta T = K_f b$, where K_f is the cryoscopic constant.

☐ 18. Osmosis is the spontaneous passage of a pure solvent into a solution separated from it by a semipermeable membrane, a membrane permeable to the solvent but not to the solute.

☐ 19. The osmotic pressure is the pressure that must be applied to the solution to stop the influx of solvent.

☐ 20. The van 't Hoff equation for the osmotic pressure is $\Pi = [B]RT$.

☐ 21. The activity is defined as $a_A = p_A/p_A^*$.

☐ 22. The solvent activity is related to its chemical potential by $\mu_A = \mu_A^* + RT\ln a_A$. The activity may be written in terms of the activity coefficient $\gamma_A = a_A/x_A$.

☐ 23. The chemical potential of a solute in an ideal-dilute solution is given by $\mu_B = \mu_B^\ominus + RT\ln a_B$. The activity may be written in terms of the activity coefficient $\gamma_B = a_B/x_B$.

☐ 24. The biological standard state (pH = 7) is related to the thermodynamic standard state by $\mu_{H^+}^\oplus = \mu_{H^+}^\ominus - 7RT\ln 10$.

☐ 25. The mean activity coefficient is the geometric mean of the individual coefficients: $\gamma_\pm = (\gamma_+^p\gamma_-^q)^{1/(p+q)}$.

☐ 26. The Debye–Hückel theory of activity coefficients of electrolyte solutions is based on the assumption that Coulombic interactions between ions are dominant; a key idea of the theory is that of an ionic atmosphere.

☐ 27. The Debye–Hückel limiting law is $\log\gamma_\pm = -|z_+z_-|AI^{1/2}$ where I is the ionic strength, $I = \frac{1}{2}\sum_i z_i^2(b_i/b^\ominus)$.

☐ 28. The extended Debye–Hückel law is $\ln\gamma_\pm = -|z_+z_-|AI^{1/2}/(1 + BI^{1/2}) + CI$.

Further reading

Articles and texts[6]

B. Freeman, Osmosis. In *Encyclopedia of applied physics* (ed. G.L. Trigg), **13**, 59. VCH, New York (1995).

J.N. Murrell and A.D. Jenkins, *Properties of liquids and solutions.* Wiley–Interscience, New York (1994).

J.S. Rowlinson and F.L. Swinton, *Liquids and liquid mixtures.* Butterworths, London (1982).

S. Sattar, Thermodynamics of mixing real gases. *J. Chem. Educ.* **77**, 1361 (2000).

Sources of data and information

M.R.J. Dack, Solutions and solubilities. In *Techniques of chemistry* (ed. A. Weissberger and B.W. Rossiter), **8**. Wiley, New York (1975).

R.C. Weast (ed.), *Handbook of chemistry and physics*, Vol. 81. CRC Press, Boca Raton (2004).

Further information

Further information 5.1 *The Debye–Hückel theory of ionic solutions*

Imagine a solution in which all the ions have their actual positions, but in which their Coulombic interactions have been turned off. The difference in molar Gibbs energy between the ideal and real solutions is equal to w_e, the electrical work of charging the system in this arrangement. For a salt M_pX_q, we write

$$w_e = \overbrace{(p\mu_+ + q\mu_-)}^{G_m} - \overbrace{(p\mu_+^{ideal} + q\mu_-^{ideal})}^{G_m^{ideal}}$$
$$= p(\mu_+ - \mu_+^{ideal}) + q(\mu_- - \mu_-^{ideal})$$

From eqn 5.64 we write

$$\mu_+ - \mu_+^{ideal} = \mu_- - \mu_-^{ideal} = RT \ln \gamma_\pm$$

So it follows that

$$\ln \gamma_\pm = \frac{w_e}{sRT} \qquad s = p + q \tag{5.73}$$

This equation tells us that we must first find the final distribution of the ions and then the work of charging them in that distribution.

The Coulomb potential at a distance r from an isolated ion of charge z_ie in a medium of permittivity ε is

$$\phi_i = \frac{Z_i}{r} \qquad Z_i = \frac{z_t e}{4\pi\varepsilon} \tag{5.74}$$

The ionic atmosphere causes the potential to decay with distance more sharply than this expression implies. Such shielding is a familiar problem in electrostatics, and its effect is taken into account by replacing the Coulomb potential by the **shielded Coulomb potential**, an expression of the form

$$\phi_i = \frac{Z_i}{r} e^{-r/r_D} \tag{5.75}$$

where r_D is called the **Debye length**. When r_D is large, the shielded potential is virtually the same as the unshielded potential. When it is small, the shielded potential is much smaller than the unshielded potential, even for short distances (Fig. 5.36).

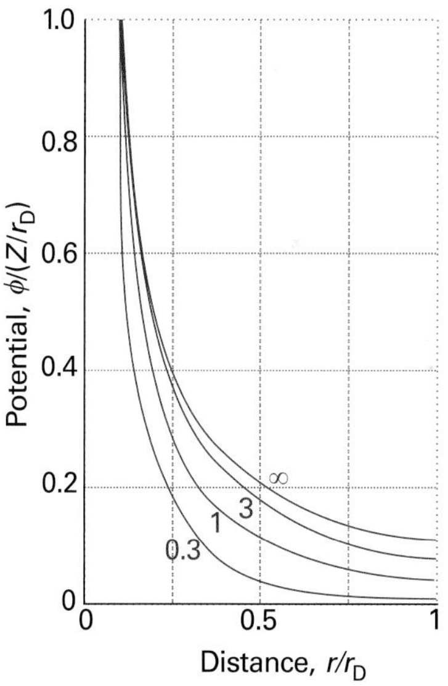

Fig. 5.36 The variation of the shielded Coulomb potential with distance for different values of the Debye length, r_D/a. The smaller the Debye length, the more sharply the potential decays to zero. In each case, a is an arbitrary unit of length.

Exploration Write an expression for the difference between the unshielded and shielded Coulomb potentials evaluated at r_D. Then plot this expression against r_D and provide a physical interpretation for the shape of the plot.

[6] See *Further reading* in Chapter 2 for additional texts on chemical thermodynamics.

To calculate r_D, we need to know how the **charge density**, ρ_i, of the ionic atmosphere, the charge in a small region divided by the volume of the region, varies with distance from the ion. This step draws on another standard result of electrostatics, in which charge density and potential are related by **Poisson's equation** (see *Appendix 3*):

$$\nabla^2\phi = -\frac{\rho}{\varepsilon} \tag{5.76}$$

where $\nabla^2 = (\partial^2/\partial x^2 + \partial^2/\partial y^2 + \partial^2/\partial z^2)$ is called the *laplacian*. Because we are considering only a spherical ionic atmosphere, we can use a simplified form of this equation in which the charge density varies only with distance from the central ion:

$$\frac{1}{r^2}\frac{\mathrm{d}}{\mathrm{d}r}\left(r^2\frac{\mathrm{d}\phi_i}{\mathrm{d}r}\right) = -\frac{\rho_i}{\varepsilon}$$

Comment 5.5

For systems with spherical symmetry, it is best to work in spherical polar coordinates r, θ, and ϕ (see the illustration): $x = r\sin\theta\cos\phi$, $y = r\sin\theta\sin\phi$, and $z = r\cos\theta$. The laplacian in spherical polar coordinates is

$$\nabla^2 = \frac{1}{r^2}\frac{\partial}{\partial r}\left(r^2\frac{\partial}{\partial r}\right) + \frac{1}{r^2\sin\theta}\frac{\partial}{\partial\theta}\left(\sin\theta\frac{\partial}{\partial\theta}\right) + \frac{1}{r^2\sin^2\theta}\frac{\partial^2}{\partial\theta^2}$$

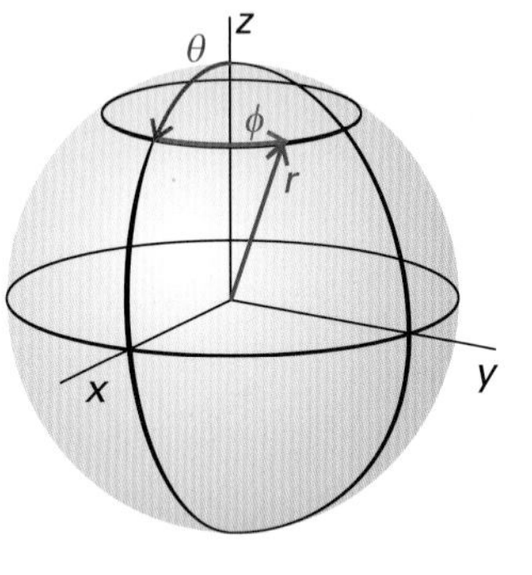

When a function depends only on r, the derivatives in the second and third terms evaluate to zero and the laplacian takes the form

$$\nabla^2 = \frac{1}{r^2}\frac{\mathrm{d}}{\mathrm{d}r}\left(r^2\frac{\mathrm{d}}{\mathrm{d}r}\right)$$

Substitution of the expression for the shielded potential, eqn 5.75, results in

$$r_D^2 = -\frac{\varepsilon\phi_i}{\rho_i} \tag{5.77}$$

To solve this equation we need to relate ρ_i and ϕ_i.

For the next step we draw on the fact that the energy of an ion depends on its closeness to the central ion, and then use the Boltzmann distribution (see Section 16.1) to work out the probability that an ion will be found at each distance. The energy of an ion of charge z_je at a distance where it experiences the potential ϕ_i of the central ion i relative to its energy when it is far away in the bulk solution is its charge times the potential:

$$E = z_je\phi_i$$

Therefore, according to the Boltzmann distribution, the ratio of the molar concentration, c_j, of ions at a distance r and the molar concentration in the bulk, c_j°, where the energy is zero, is:

$$\frac{c_j}{c_j^\circ} = \mathrm{e}^{-E/kT}$$

The charge density, ρ_i, at a distance r from the ion i is the molar concentration of each type of ion multiplied by the charge per mole of ions, z_ieN_A. The quantity eN_A, the magnitude of the charge per mole of electrons, is Faraday's constant, $F = 96.48\ \mathrm{kC\ mol^{-1}}$. It follows that

$$\rho_i = c_+z_+F + c_-z_-F = c_+^\circ z_+F\mathrm{e}^{-z_+e\phi_i/kT} + c_-^\circ z_-F\mathrm{e}^{-z_-e\phi_i/kT} \tag{5.78}$$

At this stage we need to simplify the expression to avoid the awkward exponential terms. Because the average electrostatic interaction energy is small compared with kT we may write eqn 5.78 as

$$\rho_i = c_+^\circ z_+F\left(1 - \frac{z_+e\phi_i}{kT} + \cdots\right) + c_-^\circ z_-F\left(1 - \frac{z_-e\phi_i}{kT} + \cdots\right)$$
$$= (c_+^\circ z_+ + c_-^\circ z_-)F - (c_+^\circ z_+^2 + c_-^\circ z_-^2)F\frac{e\phi_i}{kT} + \cdots$$

Comment 5.6

The expansion of an exponential function used here is $\mathrm{e}^{-x} = 1 - x + \frac{1}{2}x^2 - \cdots$. If $x \ll 1$, then $\mathrm{e}^{-x} \approx 1 - x$.

Replacing e by F/N_A and N_Ak by R results in the following expression:

$$\rho_i = (c_+^\circ z_+ + c_-^\circ z_-)F - (c_+^\circ z_+^2 + c_-^\circ z_-^2)\frac{F^2\phi_i}{RT} + \cdots \tag{5.79}$$

The first term in the expansion is zero because it is the charge density in the bulk, uniform solution, and the solution is electrically neutral. The unwritten terms are assumed to be too small to be significant. The one remaining term can be expressed in terms of the ionic strength, eqn 5.70, by noting that in the dilute aqueous solutions we are considering there is little difference between molality and molar concentration, and $c \approx b\rho$, where ρ is the mass density of the solvent

$$c_+^\circ z_+^2 + c_-^\circ z_-^2 \approx (b_+^\circ z_+^2 + b_-^\circ z_-^2)\rho = 2Ib^\ominus\rho$$

With these approximations, eqn 5.78 becomes

$$\rho_i = -\frac{2\rho F^2Ib^\ominus\phi_i}{RT}$$

We can now solve eqn 5.77 for r_D:

$$r_D = \left(\frac{\varepsilon RT}{2\rho F^2Ib^\ominus}\right)^{1/2} \tag{5.80}$$

To calculate the activity coefficient we need to find the electrical work of charging the central ion when it is surrounded by its atmosphere. To do so, we need to know the potential at the ion due to its atmosphere, ϕ_{atmos}. This potential is the difference between the total potential, given by eqn 5.75, and the potential due to the central ion itself:

$$\phi_{\text{atmos}} = \phi - \phi_{\text{central ion}} = Z_i\left(\frac{\mathrm{e}^{-r/r_D}}{r} - \frac{1}{r}\right)$$

The potential at the central ion (at $r = 0$) is obtained by taking the limit of this expression as $r \to 0$ and is

$$\phi_{\text{atmos}}(0) = \frac{Z_i}{r_D}$$

This expression shows us that the potential of the ionic atmosphere is equivalent to the potential arising from a single charge of equal magnitude but opposite sign to that of the central ion and located at a distance r_D from the ion. If the charge of the central ion were q and not $z_i e$, then the potential due to its atmosphere would be

$$\phi_{atmos}(0) = -\frac{q}{4\pi\varepsilon r_D}$$

The work of adding a charge dq to a region where the electrical potential is $\phi_{atmos}(0)$ is

$$dw_e = \phi_{atmos}(0)dq$$

Therefore, the total molar work of fully charging the ions is

$$w_e = N_A\int_0^{z_i e}\phi_{atmos}(0)\,dq = -\frac{N_A}{4\pi\varepsilon r_D}\int_0^{z_i e} q\,dq$$

$$= -\frac{N_A z_i^2 e^2}{8\pi\varepsilon r_D} = -\frac{z_i^2 F^2}{8\pi\varepsilon N_A r_D}$$

where in the last step we have used $F = N_A e$. It follows from eqn 5.73 that the mean activity coefficient of the ions is

$$\ln\gamma_\pm = \frac{pw_{e,+} + qw_{e,-}}{sRT} = -\frac{(pz_+^2 + qz_-^2)F^2}{8\pi\varepsilon s N_A RT r_D}$$

However, for neutrality $pz_+ + qz_- = 0$; therefore

Comment 5.7

For this step, multiply $pz_+ + qz_- = 0$ by p and also, separately, by q; add the two expressions and rearrange the result by using $p + q = s$ and $z_+z_- = -|z_+z_-|$.

$$\ln\gamma_\pm = -\frac{|z_+z_-|F^2}{8\pi\varepsilon N_A RT r_D}$$

Replacing r_D with the expression in eqn 5.79 gives

$$\ln\gamma_\pm = -\frac{|z_+z_-|F^2}{8\pi\varepsilon N_A RT}\left(\frac{2\rho F^2 I b^\ominus}{\varepsilon RT}\right)^{1/2}$$

$$= -|z_+z_-|\left\{\frac{F^3}{4\pi N_A}\left(\frac{\rho b^\ominus}{2\varepsilon^3 R^3 T^3}\right)^{1/2}\right\} I^{1/2}$$

where we have grouped terms in such a way as to show that this expression is beginning to take the form of eqn 5.69. Indeed, conversion to common logarithms (by using $\ln x = \ln 10 \times \log x$) gives

$$\log\gamma_\pm = -|z_+z_-|\left\{\frac{F^3}{4\pi N_A \ln 10}\left(\frac{\rho b^\ominus}{2\varepsilon^3 R^3 T^3}\right)^{1/2}\right\} I^{1/2}$$

which is eqn 5.69 ($\log\gamma_\pm = -|z_+z_-|AI^{1/2}$) with

$$A = \frac{F^3}{4\pi N_A \ln 10}\left(\frac{\rho b^\ominus}{2\varepsilon^3 R^3 T^3}\right)^{1/2} \qquad (5.81)$$

Discussion questions

5.1 State and justify the thermodynamic criterion for solution–vapour equilibrium.

5.2 How is Raoult's law modified so as to describe the vapour pressure of real solutions?

5.3 Explain how colligative properties are used to determine molar mass.

5.4 Explain the origin of colligative properties.

5.5 Explain what is meant by a regular solution.

5.6 Describe the general features of the Debye–Hückel theory of electrolyte solutions.

Exercises

5.1(a) The partial molar volumes of acetone (propanone) and chloroform (trichloromethane) in a mixture in which the mole fraction of $CHCl_3$ is 0.4693 are 74.166 $cm^3\,mol^{-1}$ and 80.235 $cm^3\,mol^{-1}$, respectively. What is the volume of a solution of mass 1.000 kg?

5.1(b) The partial molar volumes of two liquids A and B in a mixture in which the mole fraction of A is 0.3713 are 188.2 $cm^3\,mol^{-1}$ and 176.14 $cm^3\,mol^{-1}$, respectively. The molar masses of A and B are 241.1 g mol^{-1} and 198.2 g mol^{-1}. What is the volume of a solution of mass 1.000 kg?

5.2(a) At 25°C, the density of a 50 per cent by mass ethanol–water solution is 0.914 g cm^{-3}. Given that the partial molar volume of water in the solution is 17.4 $cm^3\,mol^{-1}$, calculate the partial molar volume of the ethanol.

5.2(b) At 20°C, the density of a 20 per cent by mass ethanol–water solution is 968.7 kg m^{-3}. Given that the partial molar volume of ethanol in the solution is 52.2 $cm^3\,mol^{-1}$, calculate the partial molar volume of the water.

5.3(a) At 300 K, the partial vapour pressures of HCl (that is, the partial pressure of the HCl vapour) in liquid $GeCl_4$ are as follows:

x_{HCl}	0.005	0.012	0.019
p_{HCl}/kPa	32.0	76.9	121.8

Show that the solution obeys Henry's law in this range of mole fractions, and calculate Henry's law constant at 300 K.

5.3(b) At 310 K, the partial vapour pressures of a substance B dissolved in a liquid A are as follows:

x_B	0.010	0.015	0.020
p_B/kPa	82.0	122.0	166.1

Show that the solution obeys Henry's law in this range of mole fractions, and calculate Henry's law constant at 310 K.

5.4(a) Predict the partial vapour pressure of HCl above its solution in liquid germanium tetrachloride of molality 0.10 mol kg^{-1}. For data, see Exercise 5.3a.

5.4(b) Predict the partial vapour pressure of the component B above its solution in A in Exercise 5.3b when the molality of B is 0.25 mol kg^{-1}. The molar mass of A is 74.1 g mol^{-1}.

5.5(a) The vapour pressure of benzene is 53.3 kPa at 60.6°C, but it fell to 51.5 kPa when 19.0 g of an involatile organic compound was dissolved in 500 g of benzene. Calculate the molar mass of the compound.

5.5(b) The vapour pressure of 2-propanol is 50.00 kPa at 338.8°C, but it fell to 49.62 kPa when 8.69 g of an involatile organic compound was dissolved in 250 g of 2-propanol. Calculate the molar mass of the compound.

5.6(a) The addition of 100 g of a compound to 750 g of CCl_4 lowered the freezing point of the solvent by 10.5 K. Calculate the molar mass of the compound.

5.6(b) The addition of 5.00 g of a compound to 250 g of naphthalene lowered the freezing point of the solvent by 0.780 K. Calculate the molar mass of the compound.

5.7(a) The osmotic pressure of an aqueous solution at 300 K is 120 kPa. Calculate the freezing point of the solution.

5.7(b) The osmotic pressure of an aqueous solution at 288 K is 99.0 kPa. Calculate the freezing point of the solution.

5.8(a) Consider a container of volume 5.0 dm^3 that is divided into two compartments of equal size. In the left compartment there is nitrogen at 1.0 atm and 25°C; in the right compartment there is hydrogen at the same temperature and pressure. Calculate the entropy and Gibbs energy of mixing when the partition is removed. Assume that the gases are perfect.

5.8(b) Consider a container of volume 250 cm^3 that is divided into two compartments of equal size. In the left compartment there is argon at 100 kPa and 0°C; in the right compartment there is neon at the same temperature and pressure. Calculate the entropy and Gibbs energy of mixing when the partition is removed. Assume that the gases are perfect.

5.9(a) Air is a mixture with a composition given in *Self-test 1.4*. Calculate the entropy of mixing when it is prepared from the pure (and perfect) gases.

5.9(b) Calculate the Gibbs energy, entropy, and enthalpy of mixing when 1.00 mol C_6H_{14} (hexane) is mixed with 1.00 mol C_7H_{16} (heptane) at 298 K; treat the solution as ideal.

5.10(a) What proportions of hexane and heptane should be mixed (a) by mole fraction, (b) by mass in order to achieve the greatest entropy of mixing?

5.10(b) What proportions of benzene and ethylbenzene should be mixed (a) by mole fraction, (b) by mass in order to achieve the greatest entropy of mixing?

5.11(a) Use Henry's law and the data in Table 5.1 to calculate the solubility (as a molality) of CO_2 in water at 25°C when its partial pressure is (a) 0.10 atm, (b) 1.00 atm.

5.11(b) The mole fractions of N_2 and O_2 in air at sea level are approximately 0.78 and 0.21. Calculate the molalities of the solution formed in an open flask of water at 25°C.

5.12(a) A water carbonating plant is available for use in the home and operates by providing carbon dioxide at 5.0 atm. Estimate the molar concentration of the soda water it produces.

5.12(b) After some weeks of use, the pressure in the water carbonating plant mentioned in the previous exercise has fallen to 2.0 atm. Estimate the molar concentration of the soda water it produces at this stage.

5.13(a) The enthalpy of fusion of anthracene is 28.8 kJ mol^{-1} and its melting point is 217°C. Calculate its ideal solubility in benzene at 25°C.

5.13(b) Predict the ideal solubility of lead in bismuth at 280°C given that its melting point is 327°C and its enthalpy of fusion is 5.2 kJ mol^{-1}.

5.14(a) The osmotic pressure of solutions of polystyrene in toluene were measured at 25°C and the pressure was expressed in terms of the height of the solvent of density 1.004 g cm^{-3}:

$c/(\text{g dm}^{-3})$	2.042	6.613	9.521	12.602
h/cm	0.592	1.910	2.750	3.600

Calculate the molar mass of the polymer.

5.14(b) The molar mass of an enzyme was determined by dissolving it in water, measuring the osmotic pressure at 20°C, and extrapolating the data to zero concentration. The following data were obtained:

$c/(\text{mg cm}^{-3})$	3.221	4.618	5.112	6.722
h/cm	5.746	8.238	9.119	11.990

Calculate the molar mass of the enzyme.

5.15(a) Substances A and B are both volatile liquids with $p_A^* = 300$ Torr, $p_B^* = 250$ Torr, and $K_B = 200$ Torr (concentration expressed in mole fraction). When $x_A = 0.9$, $b_B = 2.22$ mol kg^{-1}, $p_A = 250$ Torr, and $p_B = 25$ Torr. Calculate the activities and activity coefficients of A and B. Use the mole fraction, Raoult's law basis system for A and the Henry's law basis system (both mole fractions and molalities) for B.

5.15(b) Given that $p^*(H_2O) = 0.02308$ atm and $p(H_2O) = 0.02239$ atm in a solution in which 0.122 kg of a non-volatile solute ($M = 241$ g mol^{-1}) is dissolved in 0.920 kg water at 293 K, calculate the activity and activity coefficient of water in the solution.

5.16(a) A dilute solution of bromine in carbon tetrachloride behaves as an ideal-dilute solution. The vapour pressure of pure CCl_4 is 33.85 Torr at 298 K. The Henry's law constant when the concentration of Br_2 is expressed as a mole fraction is 122.36 Torr. Calculate the vapour pressure of each component, the total pressure, and the composition of the vapour phase when the mole fraction of Br_2 is 0.050, on the assumption that the conditions of the ideal-dilute solution are satisfied at this concentration.

5.16(b) Benzene and toluene form nearly ideal solutions. The boiling point of pure benzene is 80.1°C. Calculate the chemical potential of benzene relative to that of pure benzene when $x_{\text{benzene}} = 0.30$ at its boiling point. If the activity coefficient of benzene in this solution were actually 0.93 rather than 1.00, what would be its vapour pressure?

5.17(a) By measuring the equilibrium between liquid and vapour phases of an acetone (A)–methanol (M) solution at 57.2°C at 1.00 atm, it was found that $x_A = 0.400$ when $y_A = 0.516$. Calculate the activities and activity coefficients of both components in this solution on the Raoult's law basis. The vapour pressures of the pure components at this temperature are: $p_A^* = 105$ kPa and $p_M^* = 73.5$ kPa. (x_A is the mole fraction in the liquid and y_A the mole fraction in the vapour.)

5.17(b) By measuring the equilibrium between liquid and vapour phases of a solution at 30°C at 1.00 atm, it was found that $x_A = 0.220$ when $y_A = 0.314$. Calculate the activities and activity coefficients of both components in this solution on the Raoult's law basis. The vapour pressures of the pure components at this temperature are: $p_A^* = 73.0$ kPa and $p_B^* = 92.1$ kPa. (x_A is the mole fraction in the liquid and y_A the mole fraction in the vapour.)

5.18(a) Calculate the ionic strength of a solution that is 0.10 mol kg^{-1} in KCl(aq) and 0.20 mol kg^{-1} in $CuSO_4$(aq).

5.18(b) Calculate the ionic strength of a solution that is 0.040 mol kg^{-1} in $K_3[Fe(CN)_6]$(aq), 0.030 mol kg^{-1} in KCl(aq), and 0.050 mol kg^{-1} in NaBr(aq).

5.19(a) Calculate the masses of (a) $Ca(NO_3)_2$ and, separately, (b) NaCl to add to a 0.150 mol kg^{-1} solution of KNO_3(aq) containing 500 g of solvent to raise its ionic strength to 0.250.

5.19(b) Calculate the masses of (a) KNO_3 and, separately, (b) $Ba(NO_3)_2$ to add to a 0.110 mol kg^{-1} solution of KNO_3(aq) containing 500 g of solvent to raise its ionic strength to 1.00.

5.20(a) Estimate the mean ionic activity coefficient and activity of a solution that is 0.010 mol kg^{-1} $CaCl_2$(aq) and 0.030 mol kg^{-1} NaF(aq).

5.20(b) Estimate the mean ionic activity coefficient and activity of a solution that is 0.020 mol kg^{-1} NaCl(aq) and 0.035 mol kg^{-1} $Ca(NO_3)_2$(aq).

5.21(a) The mean activity coefficients of HBr in three dilute aqueous solutions at 25°C are 0.930 (at 5.0 mmol kg^{-1}), 0.907 (at 10.0 mmol kg^{-1}), and 0.879 (at 20.0 mmol kg^{-1}). Estimate the value of B in the extended Debye–Hückel law.

5.21(b) The mean activity coefficients of KCl in three dilute aqueous solutions at 25°C are 0.927 (at 5.0 mmol kg^{-1}), 0.902 (at 10.0 mmol kg^{-1}), and 0.816 (at 50.0 mmol kg^{-1}). Estimate the value of B in the extended Debye–Hückel law.

Problems*

Numerical problems

5.1 The following table gives the mole fraction of methylbenzene (A) in liquid and gaseous mixtures with butanone at equilibrium at 303.15 K and the total pressure p. Take the vapour to be perfect and calculate the partial pressures of the two components. Plot them against their respective mole fractions in the liquid mixture and find the Henry's law constants for the two components.

x_A	0	0.0898	0.2476	0.3577	0.5194	0.6036
y_A	0	0.0410	0.1154	0.1762	0.2772	0.3393
p/kPa	36.066	34.121	30.900	28.626	25.239	23.402

x_A	0.7188	0.8019	0.9105	1
y_A	0.4450	0.5435	0.7284	1
p/kPa	20.6984	18.592	15.496	12.295

5.2 The volume of an aqueous solution of NaCl at 25°C was measured at a series of molalities b, and it was found that the volume fitted the expression $v = 1003 + 16.62x + 1.77x^{3/2} + 0.12x^2$ where $v = V/\text{cm}^3$, V is the volume of a solution formed from 1.000 kg of water, and $x = b/b^{\ominus}$. Calculate the partial molar volume of the components in a solution of molality 0.100 mol kg^{-1}.

5.3 At 18°C the total volume V of a solution formed from $MgSO_4$ and 1.000 kg of water fits the expression $v = 1001.21 + 34.69(x - 0.070)^2$, where $v = V/\text{cm}^3$ and $x = b/b^{\ominus}$. Calculate the partial molar volumes of the salt and the solvent when in a solution of molality 0.050 mol kg^{-1}.

5.4 The densities of aqueous solutions of copper(II) sulfate at 20°C were measured as set out below. Determine and plot the partial molar volume of $CuSO_4$ in the range of the measurements.

$m(CuSO_4)$/g	5	10	15	20
ρ/(g cm^{-3})	1.051	1.107	1.167	1.230

where $m(CuSO_4)$ is the mass of $CuSO_4$ dissolved in 100 g of solution.

5.5 What proportions of ethanol and water should be mixed in order to produce 100 cm^3 of a mixture containing 50 per cent by mass of ethanol? What change in volume is brought about by adding 1.00 cm^3 of ethanol to the mixture? (Use data from Fig. 5.1.)

5.6 Potassium fluoride is very soluble in glacial acetic acid and the solutions have a number of unusual properties. In an attempt to understand them, freezing point depression data were obtained by taking a solution of known molality and then diluting it several times (J. Emsley, *J. Chem. Soc. A*, 2702 (1971)). The following data were obtained:

b/(mol kg^{-1})	0.015	0.037	0.077	0.295	0.602
ΔT/K	0.115	0.295	0.470	1.381	2.67

Calculate the apparent molar mass of the solute and suggest an interpretation. Use $\Delta_{fus}H = 11.4$ kJ mol^{-1} and $T_f^* = 290$ K.

5.7 In a study of the properties of an aqueous solution of $Th(NO_3)_4$ (by A. Apelblat, D. Azoulay, and A. Sahar, *J. Chem. Soc. Faraday Trans., I*, 1618, (1973)), a freezing point depression of 0.0703 K was observed for an aqueous solution of molality 9.6 mmol kg^{-1}. What is the apparent number of ions per formula unit?

5.8 The table below lists the vapour pressures of mixtures of iodoethane (I) and ethyl acetate (A) at 50°C. Find the activity coefficients of both components on (a) the Raoult's law basis, (b) the Henry's law basis with iodoethane as solute.

x_I	0	0.0579	0.1095	0.1918	0.2353	0.3718
p_I/kPa	0	3.73	7.03	11.7	14.05	20.72
p_A/kPa	37.38	35.48	33.64	30.85	29.44	25.05

x_I	0.5478	0.6349	0.8253	0.9093	1.0000
p_I/kPa	28.44	31.88	39.58	43.00	47.12
p_A/kPa	19.23	16.39	8.88	5.09	0

5.9 Plot the vapour pressure data for a mixture of benzene (B) and acetic acid (A) given below and plot the vapour pressure/composition curve for the mixture at 50°C. Then confirm that Raoult's and Henry's laws are obeyed in the appropriate regions. Deduce the activities and activity coefficients of the components on the Raoult's law basis and then, taking B as the solute, its activity and activity coefficient on a Henry's law basis. Finally, evaluate the excess Gibbs energy of the mixture over the composition range spanned by the data.

x_A	0.0160	0.0439	0.0835	0.1138	0.1714
p_A/kPa	0.484	0.967	1.535	1.89	2.45
p_B/kPa	35.05	34.29	33.28	32.64	30.90

x_A	0.2973	0.3696	0.5834	0.6604	0.8437	0.9931
p_A/kPa	3.31	3.83	4.84	5.36	6.76	7.29
p_B/kPa	28.16	26.08	20.42	18.01	10.0	0.47

5.10‡ Aminabhavi *et al.* examined mixtures of cyclohexane with various long-chain alkanes (T.M. Aminabhavi, V.B. Patil, M.I, Aralaguppi, J.D. Ortego, and K.C. Hansen, *J. Chem. Eng. Data* **41**, 526 (1996)). Among their data are the following measurements of the density of a mixture of cyclohexane and pentadecane as a function of mole fraction of cyclohexane (x_c) at 298.15 K:

x_c	0.6965	0.7988	0.9004
ρ/(g cm^{-3})	0.7661	0.7674	0.7697

* Problems denoted with the symbol ‡ were supplied by Charles Trapp, Carmen Giunta, and Marshall Cady.

Compute the partial molar volume for each component in a mixture that has a mole fraction cyclohexane of 0.7988.

5.11‡ Comelli and Francesconi examined mixtures of propionic acid with various other organic liquids at 313.15 K (F. Comelli and R. Francesconi, *J. Chem. Eng. Data* **41**, 101 (1996)). They report the excess volume of mixing propionic acid with oxane as $V^{E}=x_1x_2\{a_0+a_1(x_1-x_2)\}$, where x_1 is the mole fraction of propionic acid, x_2 that of oxane, $a_0=-2.4697$ cm^3 mol^{-1} and $a_1=0.0608$ cm^3 mol^{-1}. The density of propionic acid at this temperature is 0.97174 g cm^{-3}; that of oxane is 0.86398 g cm^{-3}. (a) Derive an expression for the partial molar volume of each component at this temperature. (b) Compute the partial molar volume for each component in an equimolar mixture.

5.12‡ Francesconi *et al.* studied the liquid–vapour equilibria of trichloromethane and 1,2-epoxybutane at several temperatures (R. Francesconi, B. Lunelli, and F. Comelli, *J. Chem. Eng. Data* **41**, 310 (1996)). Among their data are the following measurements of the mole fractions of trichloromethane in the liquid phase (x_T) and the vapour phase (y_T) at 298.15 K as a function of pressure.

p/kPa	23.40	21.75	20.25	18.75	18.15	20.25	22.50	26.30
x	0	0.129	0.228	0.353	0.511	0.700	0.810	1
y	0	0.065	0.145	0.285	0.535	0.805	0.915	1

Compute the activity coefficients of both components on the basis of Raoult's law.

5.13‡ Chen and Lee studied the liquid–vapour equilibria of cyclohexanol with several gases at elevated pressures (J.-T. Chen and M.-J. Lee, *J. Chem. Eng. Data* **41**, 339 (1996)). Among their data are the following measurements of the mole fractions of cyclohexanol in the vapour phase (y) and the liquid phase (x) at 393.15 K as a function of pressure.

p/bar	10.0	20.0	30.0	40.0	60.0	80.0
y_{cyc}	0.0267	0.0149	0.0112	0.00947	0.00835	0.00921
x_{cyc}	0.9741	0.9464	0.9204	0.892	0.836	0.773

Determine the Henry's law constant of CO_2 in cyclohexanol, and compute the activity coefficient of CO_2.

5.14‡ Equation 5.39 indicates that solubility is an exponential function of temperature. The data in the table below gives the solubility, S, of calcium acetate in water as a function of temperature.

θ/°C	0	20	40	60	80
S/(mol dm^{-3})	36.4	34.9	33.7	32.7	31.7

Determine the extent to which the data fit the exponential $S=S_0e^{\tau/T}$ and obtain values for S_0 and τ. Express these constants in terms of properties of the solute.

5.15 The excess Gibbs energy of solutions of methylcyclohexane (MCH) and tetrahydrofuran (THF) at 303.15 K was found to fit the expression

$$G^{E}=RTx(1-x)\{0.4857-0.1077(2x-1)+0.0191(2x-1)^2\}$$

where x is the mole fraction of the methylcyclohexane. Calculate the Gibbs energy of mixing when a mixture of 1.00 mol of MCH and 3.00 mol of THF is prepared.

5.16 The mean activity coefficients for aqueous solutions of NaCl at 25°C are given below. Confirm that they support the Debye–Hückel limiting law and that an improved fit is obtained with the extended law.

b/(mmol kg^{-1})	1.0	2.0	5.0	10.0	20.0
$\gamma_\pm$	0.9649	0.9519	0.9275	0.9024	0.8712

Theoretical problems

5.17 The excess Gibbs energy of a certain binary mixture is equal to $gRTx(1-x)$ where g is a constant and x is the mole fraction of a solute A. Find an expression for the chemical potential of A in the mixture and sketch its dependence on the composition.

5.18 Use the Gibbs–Duhem equation to derive the Gibbs–Duhem–Margules equation

$$\left(\frac{\partial \ln f_A}{\partial \ln x_A}\right)_{p,T}=\left(\frac{\partial \ln f_B}{\partial \ln x_B}\right)_{p,T}$$

where f is the fugacity. Use the relation to show that, when the fugacities are replaced by pressures, if Raoult's law applies to one component in a mixture it must also apply to the other.

5.19 Use the Gibbs–Duhem equation to show that the partial molar volume (or any partial molar property) of a component B can be obtained if the partial molar volume (or other property) of A is known for all compositions up to the one of interest. Do this by proving that

$$V_B=V_B^*-\int_{V_A^*}^{V_A}\frac{x_A}{1-x_A}\,dV_A$$

Use the following data (which are for 298 K) to evaluate the integral graphically to find the partial molar volume of acetone at $x=0.500$.

$x(CHCl_3)$	0	0.194	0.385	0.559	0.788	0.889	1.000
V_m/(cm^3 mol^{-1})	73.99	75.29	76.50	77.55	79.08	79.82	80.67

5.20 Use the Gibbs–Helmholtz equation to find an expression for d ln x_A in terms of dT. Integrate d ln x_A from $x_A=0$ to the value of interest, and integrate the right–hand side from the transition temperature for the pure liquid A to the value in the solution. Show that, if the enthalpy of transition is constant, then eqns 5.33 and 5.36 are obtained.

5.21 The 'osmotic coefficient', ϕ, is defined as $\phi=-(x_A/x_B)\ln a_A$. By writing $r=x_B/x_A$, and using the Gibbs–Duhem equation, show that we can calculate the activity of B from the activities of A over a composition range by using the formula

$$\ln\left(\frac{a_B}{r}\right)=\phi-\phi(0)+\int_0^r\left(\frac{\phi-1}{r}\right)dr$$

5.22 Show that the osmotic pressure of a real solution is given by $\Pi V=-RT\ln a_A$. Go on to show that, provided the concentration of the solution is low, this expression takes the form $\Pi V=\phi RT[B]$ and hence that the osmotic coefficient, ϕ, (which is defined in Problem 5.21) may be determined from osmometry.

5.23 Show that the freezing-point depression of a real solution in which the solvent of molar mass M has activity a_A obeys

$$\frac{d\ln a_A}{d(\Delta T)}=-\frac{M}{K_f}$$

and use the Gibbs–Duhem equation to show that

$$\frac{d\ln a_B}{d(\Delta T)}=-\frac{1}{b_BK_f}$$

where a_B is the solute activity and b_B is its molality. Use the Debye–Hückel limiting law to show that the osmotic coefficient (ϕ, Problem 5.21) is given by $\phi=1-\frac{1}{3}A'I$ with $A'=2.303A$ and $I=b/b^{\ominus}$.

Applications: to biology and polymer science

5.24 Haemoglobin, the red blood protein responsible for oxygen transport, binds about 1.34 cm^3 of oxygen per gram. Normal blood has a haemoglobin concentration of 150 g dm^{-3}. Haemoglobin in the lungs is about 97 per cent saturated with oxygen, but in the capillary is only about 75 per cent saturated.

What volume of oxygen is given up by 100 cm^3 of blood flowing from the lungs in the capillary?

5.25 For the calculation of the solubility c of a gas in a solvent, it is often convenient to use the expression $c = Kp$, where K is the Henry's law constant. Breathing air at high pressures, such as in scuba diving, results in an increased concentration of dissolved nitrogen. The Henry's law constant for the solubility of nitrogen is 0.18 μg/(g H_2O atm). What mass of nitrogen is dissolved in 100 g of water saturated with air at 4.0 atm and 20°C? Compare your answer to that for 100 g of water saturated with air at 1.0 atm. (Air is 78.08 mole per cent N_2.) If nitrogen is four times as soluble in fatty tissues as in water, what is the increase in nitrogen concentration in fatty tissue in going from 1 atm to 4 atm?

5.26 Ethidium bromide binds to DNA by a process called *intercalation*, in which the aromatic ethidium cation fits between two adjacent DNA base pairs. An equilibrium dialysis experiment was used to study the binding of ethidium bromide (EB) to a short piece of DNA. A 1.00 μmol dm^{-3} aqueous solution of the DNA sample was dialysed against an excess of EB. The following data were obtained for the total concentration of EB:

[EB]/(μmol dm^{-3})					
Side without DNA	0.042	0.092	0.204	0.526	1.150
Side with DNA	0.292	0.590	1.204	2.531	4.150

From these data, make a Scatchard plot and evaluate the intrinsic equilibrium constant, K, and total number of sites per DNA molecule. Is the identical and independent sites model for binding applicable?

5.27 The form of the Scatchard equation given in *Impact I5.2* applies only when the macromolecule has identical and independent binding sites. For non-identical independent binding sites, the Scatchard equation is

$$\frac{\nu}{[\mathrm{A}]_{\mathrm{out}}} = \sum_i \frac{N_i K_i}{1 + K_i[\mathrm{A}]_{\mathrm{out}}}$$

Plot $\nu/[\mathrm{A}]$ for the following cases. (a) There are four independent sites on an enzyme molecule and the intrinsic binding constant is $K = 1.0 \times 10^7$. (b) There are a total of six sites per polymer. Four of the sites are identical and have an intrinsic binding constant of 1×10^5. The binding constants for the other two sites are 2×10^6.

5.28 The addition of a small amount of a salt, such as $(NH_4)_2SO_4$, to a solution containing a charged protein increases the solubility of the protein in water. This observation is called the *salting-in effect*. However, the addition of large amounts of salt can decrease the solubility of the protein to such an extent that the protein precipitates from solution. This observation is called the *salting-out effect* and is used widely by biochemists to isolate and purify proteins. Consider the equilibrium $PX_\nu(s) \rightleftharpoons P^{\nu+}(aq) + \nu X^-(aq)$, where $P^{\nu+}$ is a polycationic protein of charge $+\nu$ and X^- is its counter ion. Use Le Chatelier's principle and the physical principles behind the Debye–Hückel theory to provide a molecular interpretation for the salting-in and salting-out effects.

5.29‡ Polymer scientists often report their data in rather strange units. For example, in the determination of molar masses of polymers in solution by osmometry, osmotic pressures are often reported in grams per square centimetre (g cm^{-2}) and concentrations in grams per cubic centimetre (g cm^{-3}). (a) With these choices of units, what would be the units of R in the van't Hoff equation? (b) The data in the table below on the concentration dependence of the osmotic pressure of polyisobutene in chlorobenzene at 25°C have been adapted from J. Leonard and H. Daoust (*J. Polymer Sci.* **57**, 53 (1962)). From these data, determine the molar mass of polyisobutene by plotting Π/c against c. (c) Theta solvents are solvents for which the second osmotic coefficient is zero; for 'poor' solvents the plot is linear and for good solvents the plot is nonlinear. From your plot, how would you classify chlorobenzene as a solvent for polyisobutene? Rationalize the result in terms of the molecular structure of the polymer and solvent. (d) Determine the second and third osmotic virial coefficients by fitting the curve to the virial form of the osmotic pressure equation. (e) Experimentally, it is often found that the virial expansion can be represented as

$$\Pi/c = RT/M\,(1 + B'c + gB'^2c'^2 + \cdots)$$

and in good solvents, the parameter g is often about 0.25. With terms beyond the second power ignored, obtain an equation for $(\Pi/c)^{1/2}$ and plot this quantity against c. Determine the second and third virial coefficients from the plot and compare to the values from the first plot. Does this plot confirm the assumed value of g?

$10^{-2}(\Pi/c)/(\mathrm{g\ cm^{-2}/g\ cm^{-3}})$	2.6	2.9	3.6	4.3	6.0	12.0
$c/(\mathrm{g\ cm^{-3}})$	0.0050	0.010	0.020	0.033	0.057	0.10
$10^{-2}(\Pi/c)/(\mathrm{g\ cm^{-2}/g\ cm^{-3}})$	19.0	31.0	38.0	52	63	
$c/(\mathrm{g\ cm^{-3}})$	0.145	0.195	0.245	0.27	0.29	

5.30‡ K. Sato, F.R. Eirich, and J.E. Mark (*J. Polymer Sci., Polym. Phys.* **14**, 619 (1976)) have reported the data in the table below for the osmotic pressures of polychloroprene ($\rho = 1.25$ g cm^{-3}) in toluene ($\rho = 0.858$ g cm^{-3}) at 30°C. Determine the molar mass of polychloroprene and its second osmotic virial coefficient.

$c/(\mathrm{mg\ cm^{-3}})$	1.33	2.10	4.52	7.18	9.87
$\Pi/(\mathrm{N\ m^{-2}})$	30	51	132	246	390

6 Phase diagrams

Phase diagrams for pure substances were introduced in Chapter 4. Now we develop their use systematically and show how they are rich summaries of empirical information about a wide range of systems. To set the stage, we introduce the famous phase rule of Gibbs, which shows the extent to which various parameters can be varied yet the equilibrium between phases preserved. With the rule established, we see how it can be used to discuss the phase diagrams that we met in the two preceding chapters. The chapter then introduces systems of gradually increasing complexity. In each case we shall see how the phase diagram for the system summarizes empirical observations on the conditions under which the various phases of the system are stable.

In this chapter we describe a systematic way of discussing the physical changes mixtures undergo when they are heated or cooled and when their compositions are changed. In particular, we see how to use phase diagrams to judge whether two substances are mutually miscible, whether an equilibrium can exist over a range of conditions, or whether the system must be brought to a definite pressure, temperature, and composition before equilibrium is established. Phase diagrams are of considerable commercial and industrial significance, particularly for semiconductors, ceramics, steels, and alloys. They are also the basis of separation procedures in the petroleum industry and of the formulation of foods and cosmetic preparations.

Phases, components, and degrees of freedom

All phase diagrams can be discussed in terms of a relationship, the phase rule, derived by J.W. Gibbs. We shall derive this rule first, and then apply it to a wide variety of systems. The phase rule requires a careful use of terms, so we begin by presenting a number of definitions.

6.1 Definitions

The term **phase** was introduced at the start of Chapter 4, where we saw that it signifies a state of matter that is uniform throughout, not only in chemical composition but also in physical state.[1] Thus we speak of the solid, liquid, and gas phases of a substance, and of its various solid phases (as for black phosphorus and white phosphorus). The number of phases in a system is denoted P. A gas, or a gaseous mixture, is a single phase, a crystal is a single phase, and two totally miscible liquids form a single phase.

[1] The words are Gibbs's.

A solution of sodium chloride in water is a single phase. Ice is a single phase ($P = 1$) even though it might be chipped into small fragments. A slurry of ice and water is a two-phase system ($P = 2$) even though it is difficult to map the boundaries between the phases. A system in which calcium carbonate undergoes thermal decomposition consists of two solid phases (one consisting of calcium carbonate and the other of calcium oxide) and one gaseous phase (consisting of carbon dioxide).

An alloy of two metals is a two-phase system ($P = 2$) if the metals are immiscible, but a single-phase system ($P = 1$) if they are miscible. This example shows that it is not always easy to decide whether a system consists of one phase or of two. A solution of solid B in solid A—a homogeneous mixture of the two substances—is uniform on a molecular scale. In a solution, atoms of A are surrounded by atoms of A and B, and any sample cut from the sample, however small, is representative of the composition of the whole.

A dispersion is uniform on a macroscopic scale but not on a microscopic scale, for it consists of grains or droplets of one substance in a matrix of the other. A small sample could come entirely from one of the minute grains of pure A and would not be representative of the whole (Fig. 6.1). Dispersions are important because, in many advanced materials (including steels), heat treatment cycles are used to achieve the precipitation of a fine dispersion of particles of one phase (such as a carbide phase) within a matrix formed by a saturated solid solution phase. The ability to control this microstructure resulting from phase equilibria makes it possible to tailor the mechanical properties of the materials to a particular application.

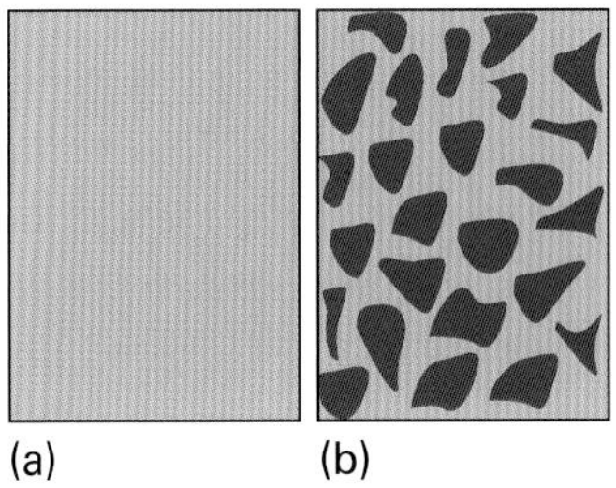

Fig. 6.1 The difference between (a) a single-phase solution, in which the composition is uniform on a microscopic scale, and (b) a dispersion, in which regions of one component are embedded in a matrix of a second component.

By a **constituent** of a system we mean a chemical species (an ion or a molecule) that is present. Thus, a mixture of ethanol and water has two constituents. A solution of sodium chloride has three constituents: water, Na^+ ions, and Cl^- ions. The term constituent should be carefully distinguished from 'component', which has a more technical meaning. A **component** is a *chemically independent* constituent of a system. The number of components, C, in a system is the minimum number of independent species necessary to define the composition of all the phases present in the system.

When no reaction takes place and there are no other constraints (such as charge balance), the number of components is equal to the number of constituents. Thus, pure water is a one-component system ($C = 1$), because we need only the species H_2O to specify its composition. Similarly, a mixture of ethanol and water is a two-component system ($C = 2$): we need the species H_2O and C_2H_5OH to specify its composition. An aqueous solution of sodium chloride has two components because, by charge balance, the number of Na^+ ions must be the same as the number of Cl^- ions.

A system that consists of hydrogen, oxygen, and water at room temperature has three components ($C = 3$), despite it being possible to form H_2O from H_2 and O_2: under the conditions prevailing in the system, hydrogen and oxygen do not react to form water, so they are independent constituents. When a reaction can occur under the conditions prevailing in the system, we need to decide the minimum number of species that, after allowing for reactions in which one species is synthesized from others, can be used to specify the composition of all the phases. Consider, for example, the equilibrium

$$\underset{\text{Phase 1}}{CaCO_3(s)} \rightleftharpoons \underset{\text{Phase 2}}{CaO(s)} + \underset{\text{Phase 3}}{CO_2(g)}$$

in which there are three constituents and three phases. To specify the composition of the gas phase (Phase 3) we need the species CO_2, and to specify the composition of Phase 2 we need the species CaO. However, we do not need an additional species to specify the composition of Phase 1 because its identity ($CaCO_3$) can be expressed in terms of the other two constituents by making use of the stoichiometry of the reaction. Hence, the system has only two components ($C = 2$).

Example 6.1 *Counting components*

How many components are present in a system in which ammonium chloride undergoes thermal decomposition?

Method Begin by writing down the chemical equation for the reaction and identifying the constituents of the system (all the species present) and the phases. Then decide whether, under the conditions prevailing in the system, any of the constituents can be prepared from any of the other constituents. The removal of these constituents leaves the number of independent constituents. Finally, identify the minimum number of these independent constituents that are needed to specify the composition of all the phases.

Answer The chemical reaction is

$$NH_4Cl(s) \rightleftharpoons NH_3(g) + HCl(g)$$

There are three constituents and two phases (one solid, one gas). However, NH_3 and HCl are formed in fixed stoichiometric proportions by the reaction. Therefore, the compositions of both phases can be expressed in terms of the single species NH_4Cl. It follows that there is only one component in the system ($C = 1$). If additional HCl (or NH_3) were supplied to the system, the decomposition of NH_4Cl would not give the correct composition of the gas phase and HCl (or NH_3) would have to be invoked as a second component.

Self-test 6.1 Give the number of components in the following systems: (a) water, allowing for its autoprotolysis, (b) aqueous acetic acid, (c) magnesium carbonate in equilibrium with its decomposition products. [(a) 1, (b) 2, (c) 2]

The **variance**, F, of a system is the number of intensive variables that can be changed independently without disturbing the number of phases in equilibrium. In a single-component, single-phase system ($C = 1$, $P = 1$), the pressure and temperature may be changed independently without changing the number of phases, so $F = 2$. We say that such a system is **bivariant**, or that it has two **degrees of freedom**. On the other hand, if two phases are in equilibrium (a liquid and its vapour, for instance) in a single-component system ($C = 1$, $P = 2$), the temperature (or the pressure) can be changed at will, but the change in temperature (or pressure) demands an accompanying change in pressure (or temperature) to preserve the number of phases in equilibrium. That is, the variance of the system has fallen to 1.

6.2 The phase rule

In one of the most elegant calculations of the whole of chemical thermodynamics, J.W. Gibbs deduced the **phase rule**, which is a general relation between the variance, F, the number of components, C, and the number of phases at equilibrium, P, for a system of any composition:

$$F = C - P + 2 \tag{6.1}$$

Comment 6.1

Josiah Willard Gibbs spent most of his working life at Yale, and may justly be regarded as the originator of chemical thermodynamics. He reflected for years before publishing his conclusions, and then did so in precisely expressed papers in an obscure journal (*The Transactions of the Connecticut Academy of Arts and Sciences*). He needed interpreters before the power of his work was recognized and before it could be applied to industrial processes. He is regarded by many as the first great American theoretical scientist.

Justification 6.1 *The phase rule*

Consider first the special case of a one-component system. For two phases in equilibrium, we can write $\mu_J(\alpha) = \mu_J(\beta)$. Each chemical potential is a function of the pressure and temperature, so

$$\mu_J(\alpha; p,T) = \mu_J(\beta; p,T)$$

This is an equation relating p and T, so only one of these variables is independent (just as the equation $x + y = 2$ is a relation for y in terms of x: $y = 2 - x$). That conclusion is consistent with $F = 1$. For three phases in mutual equilibrium,

$$\mu_J(\alpha; p,T) = \mu_J(\beta; p,T) = \mu_J(\gamma; p,T)$$

This relation is actually two equations for two unknowns ($\mu_J(\alpha; p,T) = \mu_J(\beta; p,T)$ and $\mu_J(\beta; p,T) = \mu_J(\gamma; p,T)$), and therefore has a solution only for a single value of p and T (just as the pair of equations $x + y = 2$ and $3x - y = 4$ has the single solution $x = \frac{3}{2}$ and $y = \frac{1}{2}$). That conclusion is consistent with $F = 0$. Four phases cannot be in mutual equilibrium in a one-component system because the three equalities

$$\mu_J(\alpha; p,T) = \mu_J(\beta; p,T) \qquad \mu_J(\beta; p,T) = \mu_J(\gamma; p,T) \qquad \mu_J(\gamma; p,T) = \mu_J(\delta; p,T)$$

are three equations for two unknowns (p and T) and are not consistent (just as $x + y = 2$, $3x - y = 4$, and $x + 4y = 6$ have no solution).

Now consider the general case. We begin by counting the total number of intensive variables. The pressure, p, and temperature, T, count as 2. We can specify the composition of a phase by giving the mole fractions of $C - 1$ components. We need specify only $C - 1$ and not all C mole fractions because $x_1 + x_2 + \cdots + x_C = 1$, and all mole fractions are known if all except one are specified. Because there are P phases, the total number of composition variables is $P(C - 1)$. At this stage, the total number of intensive variables is $P(C - 1) + 2$.

At equilibrium, the chemical potential of a component J must be the same in every phase (Section 4.4):

$$\mu_J(\alpha) = \mu_J(\beta) = \ldots \qquad \text{for } P \text{ phases}$$

That is, there are $P - 1$ equations of this kind to be satisfied for each component J. As there are C components, the total number of equations is $C(P - 1)$. Each equation reduces our freedom to vary one of the $P(C - 1) + 2$ intensive variables. It follows that the total number of degrees of freedom is

$$F = P(C - 1) + 2 - C(P - 1) = C - P + 2$$

which is eqn 6.1.

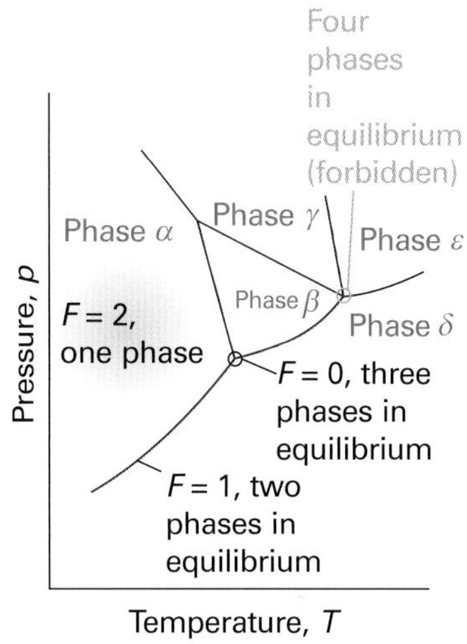

Fig. 6.2 The typical regions of a one-component phase diagram. The lines represent conditions under which the two adjoining phases are in equilibrium. A point represents the unique set of conditions under which three phases coexist in equilibrium. Four phases cannot mutually coexist in equilibrium.

(a) One-component systems

For a one-component system, such as pure water, $F = 3 - P$. When only one phase is present, $F = 2$ and both p and T can be varied independently without changing the number of phases. In other words, a single phase is represented by an *area* on a phase diagram. When two phases are in equilibrium $F = 1$, which implies that pressure is not freely variable if the temperature is set; indeed, at a given temperature, a liquid has a characteristic vapour pressure. It follows that the equilibrium of two phases is represented by a *line* in the phase diagram. Instead of selecting the temperature, we could select the pressure, but having done so the two phases would be in equilibrium at a single definite temperature. Therefore, freezing (or any other phase transition) occurs at a definite temperature at a given pressure.

When three phases are in equilibrium, $F = 0$ and the system is invariant. This special condition can be established only at a definite temperature and pressure that is characteristic of the substance and outside our control. The equilibrium of three phases is therefore represented by a *point*, the triple point, on a phase diagram. Four phases cannot be in equilibrium in a one-component system because F cannot be negative. These features are summarized in Fig. 6.2.

We can identify the features in Fig. 6.2 in the experimentally determined phase diagram for water (Fig. 6.3). This diagram summarizes the changes that take place as a sample, such as that at a, is cooled at constant pressure. The sample remains entirely

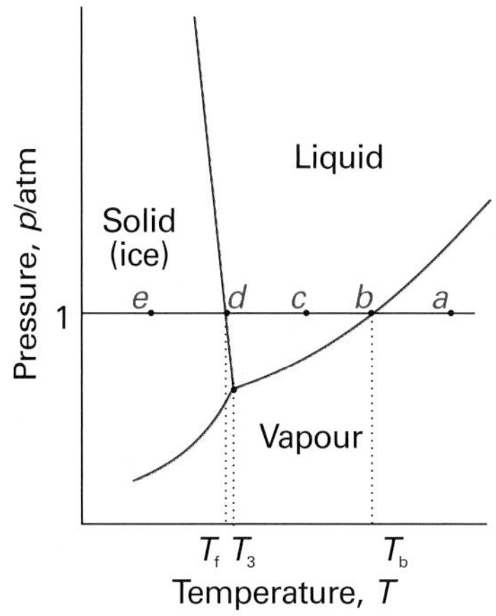

Fig. 6.3 The phase diagram for water, a simplified version of Fig. 4.5. The label T_3 marks the temperature of the triple point, T_b the normal boiling point, and T_f the normal freezing point.

gaseous until the temperature reaches *b*, when liquid appears. Two phases are now in equilibrium and $F = 1$. Because we have decided to specify the pressure, which uses up the single degree of freedom, the temperature at which this equilibrium occurs is not under our control. Lowering the temperature takes the system to *c* in the one-phase, liquid region. The temperature can now be varied around the point *c* at will, and only when ice appears at *d* does the variance become 1 again.

(b) Experimental procedures

Detecting a phase change is not always as simple as seeing water boil in a kettle, so special techniques have been developed. Two techniques are **thermal analysis**, which takes advantage of the effect of the enthalpy change during a first-order transition (Section 4.7), and differential scanning calorimetry (see *Impact I2.1*). They are useful for solid–solid transitions, where simple visual inspection of the sample may be inadequate. In thermal analysis, a sample is allowed to cool and its temperature is monitored. At a first-order transition, heat is evolved and the cooling stops until the transition is complete. The cooling curve along the isobar *cde* in Fig. 6.3 therefore has the shape shown in Fig. 6.4. The transition temperature is obvious, and is used to mark point *d* on the phase diagram.

Modern work on phase transitions often deals with systems at very high pressures, and more sophisticated detection procedures must be adopted. Some of the highest pressures currently attainable are produced in a **diamond-anvil cell** like that illustrated in Fig. 6.5. The sample is placed in a minute cavity between two gem-quality diamonds, and then pressure is exerted simply by turning the screw. The advance in design this represents is quite remarkable for, with a turn of the screw, pressures of up to about 1 Mbar can be reached that a few years ago could not be reached with equipment weighing tons.

The pressure is monitored spectroscopically by observing the shift of spectral lines in small pieces of ruby added to the sample, and the properties of the sample itself are observed optically through the diamond anvils. One application of the technique is to study the transition of covalent solids to metallic solids. Iodine, I_2, for instance, becomes metallic at around 200 kbar and makes a transition to a monatomic metallic

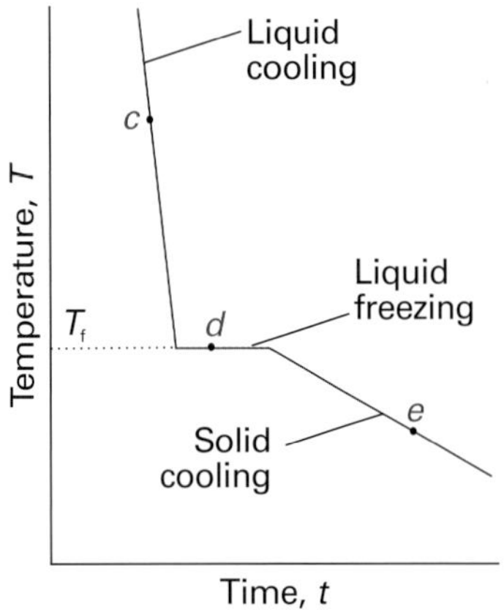

Fig. 6.4 The cooling curve for the isobar *cde* in Fig. 6.3. The halt marked *d* corresponds to the pause in the fall of temperature while the first-order exothermic transition (freezing) occurs. This pause enables T_f to be located even if the transition cannot be observed visually.

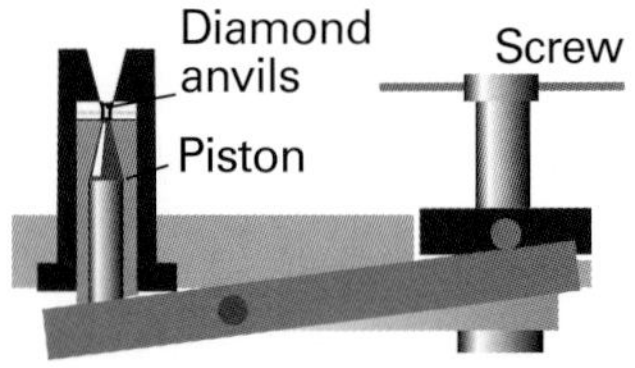

Fig. 6.5 Ultrahigh pressures (up to about 2 Mbar) can be achieved using a diamond anvil. The sample, together with a ruby for pressure measurement and a drop of liquid for pressure transmission, are placed between two gem-quality diamonds. The principle of its action is like that of a nutcracker: the pressure is exerted by turning the screw by hand.

solid at around 210 kbar. Studies such as these are relevant to the structure of material deep inside the Earth (at the centre of the Earth the pressure is around 5 Mbar) and in the interiors of the giant planets, where even hydrogen may be metallic.

Two-component systems

When two components are present in a system, $C=2$ and $F=4-P$. If the temperature is constant, the remaining variance is $F'=3-P$, which has a maximum value of 2. (The prime on F indicates that one of the degrees of freedom has been discarded, in this case the temperature.) One of these two remaining degrees of freedom is the pressure and the other is the composition (as expressed by the mole fraction of one component). Hence, one form of the phase diagram is a map of pressures and compositions at which each phase is stable. Alternatively, the pressure could be held constant and the phase diagram depicted in terms of temperature and composition.

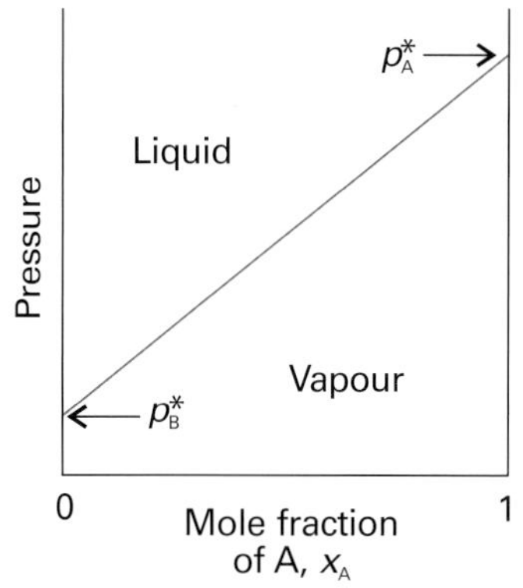

Fig. 6.6 The variation of the total vapour pressure of a binary mixture with the mole fraction of A in the liquid when Raoult's law is obeyed.

6.3 Vapour pressure diagrams

The partial vapour pressures of the components of an ideal solution of two volatile liquids are related to the composition of the liquid mixture by Raoult's law (Section 5.3a)

$$p_A = x_A p_A^* \qquad p_B = x_B p_B^* \tag{6.2}^\circ$$

where p_A^* is the vapour pressure of pure A and p_B^* that of pure B. The total vapour pressure p of the mixture is therefore

$$p = p_A + p_B = x_A p_A^* + x_B p_B^* = p_B^* + (p_A^* - p_B^*)x_A \tag{6.3}^\circ$$

This expression shows that the total vapour pressure (at some fixed temperature) changes linearly with the composition from p_B^* to p_A^* as x_A changes from 0 to 1 (Fig. 6.6).

(a) The composition of the vapour

The compositions of the liquid and vapour that are in mutual equilibrium are not necessarily the same. Common sense suggests that the vapour should be richer in the more volatile component. This expectation can be confirmed as follows. The partial pressures of the components are given by eqn 6.2. It follows from Dalton's law that the mole fractions in the gas, y_A and y_B, are

$$y_A = \frac{p_A}{p} \qquad y_B = \frac{p_B}{p} \tag{6.4}$$

Provided the mixture is ideal, the partial pressures and the total pressure may be expressed in terms of the mole fractions in the liquid by using eqn 6.2 for p_J and eqn 6.3 for the total vapour pressure p, which gives

$$y_A = \frac{x_A p_A^*}{p_B^* + (p_A^* - p_B^*)x_A} \qquad y_B = 1 - y_A \tag{6.5}^\circ$$

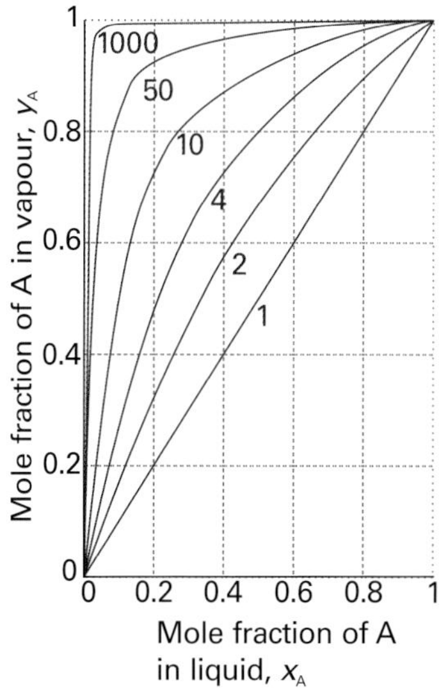

Fig. 6.7 The mole fraction of A in the vapour of a binary ideal solution expressed in terms of its mole fraction in the liquid, calculated using eqn 6.5 for various values of p_A^*/p_B^* (the label on each curve) with A more volatile than B. In all cases the vapour is richer than the liquid in A.

Exploration To reproduce the results of Fig. 6.7, first rearrange eqn 6.5 so that y_A is expressed as a function of x_A and the ratio p_A^*/p_B^*. Then plot y_A against x_A for several values of $p_A^*/p_B^* > 1$.

Figure 6.7 shows the composition of the vapour plotted against the composition of the liquid for various values of $p_A^*/p_B^* > 1$. We see that in all cases $y_A > x_A$, that is, the vapour is richer than the liquid in the more volatile component. Note that if B is non-volatile, so that $p_B^* = 0$ at the temperature of interest, then it makes no contribution to the vapour ($y_B = 0$).

Equation 6.3 shows how the total vapour pressure of the mixture varies with the composition of the liquid. Because we can relate the composition of the liquid to the composition of the vapour through eqn 6.5, we can now also relate the total vapour pressure to the composition of the vapour:

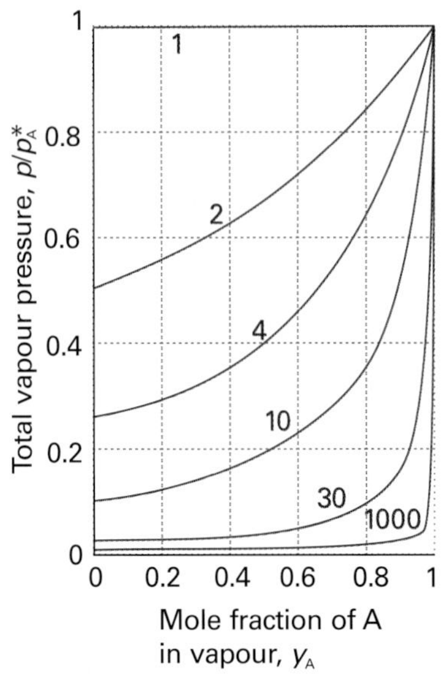

Fig. 6.8 The dependence of the vapour pressure of the same system as in Fig. 6.7, but expressed in terms of the mole fraction of A in the vapour by using eqn 6.6. Individual curves are labelled with the value of p_A^*/p_B^*.

Exploration To reproduce the results of Fig. 6.8, first rearrange eqn 6.6 so that the ratio p_A/p_A^* is expressed as a function of y_A and the ratio p_A^*/p_B^*. Then plot p_A/p_A^* against y_A for several values of $p_A^*/p_B^* > 1$.

$$p = \frac{p_A^* p_B^*}{p_A^* + (p_B^* - p_A^*)y_A} \qquad (6.6)°$$

This expression is plotted in Fig. 6.8.

(b) The interpretation of the diagrams

If we are interested in distillation, both the vapour and the liquid compositions are of equal interest. It is therefore sensible to combine Figs. 6.7 and 6.8 into one (Fig. 6.9). The point *a* indicates the vapour pressure of a mixture of composition x_A, and the point *b* indicates the composition of the vapour that is in equilibrium with the liquid at that pressure. Note that, when two phases are in equilibrium, $P = 2$ so $F' = 1$ (as usual, the prime indicating that one degree of freedom, the temperature, has already been discarded). That is, if the composition is specified (so using up the only remaining degree of freedom), the pressure at which the two phases are in equilibrium is fixed.

A richer interpretation of the phase diagram is obtained if we interpret the horizontal axis as showing the *overall* composition, z_A, of the system. If the horizontal axis of the vapour pressure diagram is labelled with z_A, then all the points down to the solid diagonal line in the graph correspond to a system that is under such high pressure that it contains only a liquid phase (the applied pressure is higher than the vapour pressure), so $z_A = x_A$, the composition of the liquid. On the other hand, all points below the lower curve correspond to a system that is under such low pressure that it contains only a vapour phase (the applied pressure is lower than the vapour pressure), so $z_A = y_A$.

Points that lie between the two lines correspond to a system in which there are two phases present, one a liquid and the other a vapour. To see this interpretation, consider the effect of lowering the pressure on a liquid mixture of overall composition *a* in Fig. 6.10. The lowering of pressure can be achieved by drawing out a piston (Fig. 6.11). This degree of freedom is permitted by the phase rule because $F' = 2$ when $P = 1$, and even if the composition is selected one degree of freedom remains. The changes to the system do not affect the overall composition, so the state of the system moves down the vertical line that passes through *a*. This vertical line is called an

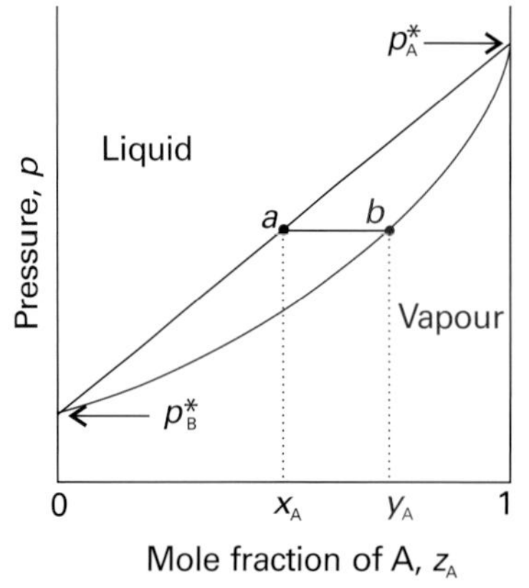

Fig. 6.9 The dependence of the total vapour pressure of an ideal solution on the mole fraction of A in the entire system. A point between the two lines corresponds to both liquid and vapour being present; outside that region there is only one phase present. The mole fraction of A is denoted z_A, as explained below.

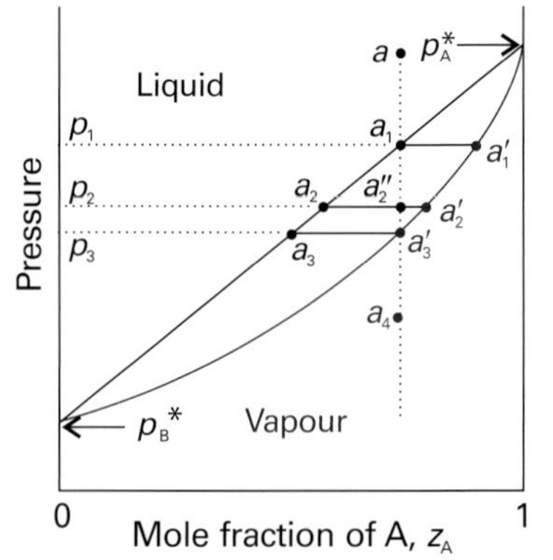

Fig. 6.10 The points of the pressure–composition diagram discussed in the text. The vertical line through *a* is an *isopleth*, a line of constant composition of the entire system.

isopleth, from the Greek words for 'equal abundance'. Until the point a_1 is reached (when the pressure has been reduced to p_1), the sample consists of a single liquid phase. At a_1 the liquid can exist in equilibrium with its vapour. As we have seen, the composition of the vapour phase is given by point a_1'. A line joining two points representing phases in equilibrium is called a **tie line**. The composition of the liquid is the same as initially (a_1 lies on the isopleth through a), so we have to conclude that at this pressure there is virtually no vapour present; however, the tiny amount of vapour that is present has the composition a_1'.

Now consider the effect of lowering the pressure to p_2, so taking the system to a pressure and overall composition represented by the point a_2''. This new pressure is below the vapour pressure of the original liquid, so it vaporizes until the vapour pressure of the remaining liquid falls to p_2. Now we know that the composition of such a liquid must be a_2. Moreover, the composition of the vapour in equilibrium with that liquid must be given by the point a_2' at the other end of the tie line. Note that two phases are now in equilibrium, so $F' = 1$ for all points between the two lines; hence, for a given pressure (such as at p_2) the variance is zero, and the vapour and liquid phases have fixed compositions (Fig. 6.12). If the pressure is reduced to p_3, a similar readjustment in composition takes place, and now the compositions of the liquid and vapour are represented by the points a_3 and a_3', respectively. The latter point corresponds to a system in which the composition of the vapour is the same as the overall composition, so we have to conclude that the amount of liquid present is now virtually zero, but the tiny amount of liquid present has the composition a_3. A further decrease in pressure takes the system to the point a_4; at this stage, only vapour is present and its composition is the same as the initial overall composition of the system (the composition of the original liquid).

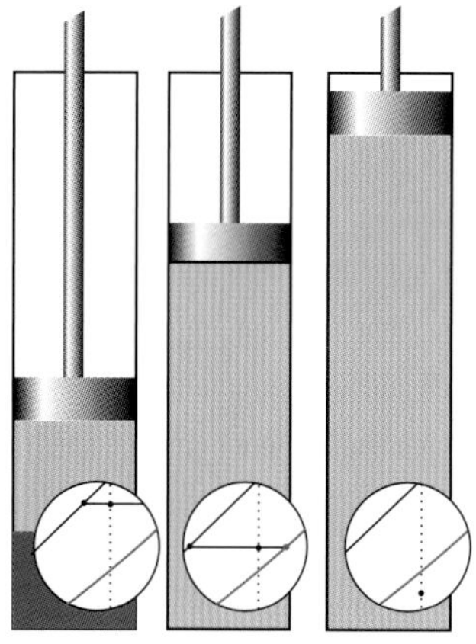

Fig. 6.11 (a) A liquid in a container exists in equilibrium with its vapour. The superimposed fragment of the phase diagram shows the compositions of the two phases and their abundances (by the lever rule). (b) When the pressure is changed by drawing out a piston, the compositions of the phases adjust as shown by the tie line in the phase diagram. (c) When the piston is pulled so far out that all the liquid has vaporized and only the vapour is present, the pressure falls as the piston is withdrawn and the point on the phase diagram moves into the one-phase region.

(c) The lever rule

A point in the two-phase region of a phase diagram indicates not only qualitatively that both liquid and vapour are present, but represents quantitatively the relative amounts of each. To find the relative amounts of two phases α and β that are in equilibrium, we measure the distances l_α and l_β along the horizontal tie line, and then use the **lever rule** (Fig. 6.13):

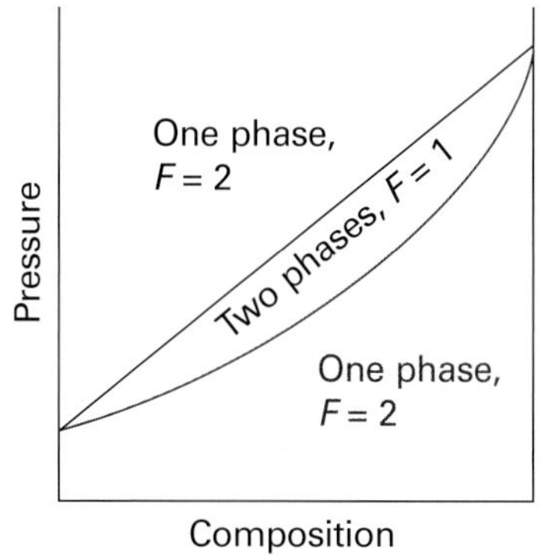

Fig. 6.12 The general scheme of interpretation of a pressure–composition diagram (a vapour pressure diagram).

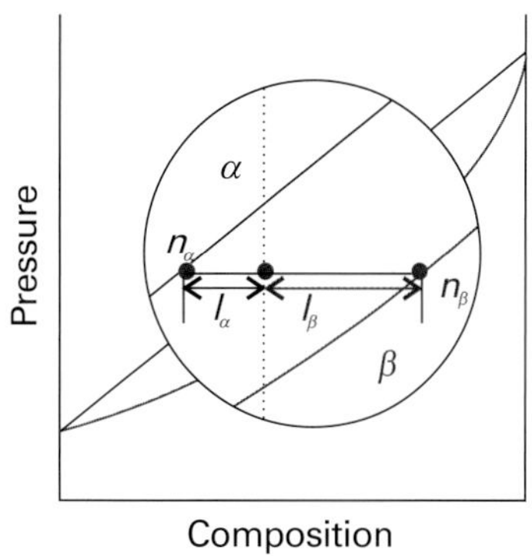

Fig. 6.13 The lever rule. The distances l_α and l_β are used to find the proportions of the amounts of phases α (such as vapour) and β (for example, liquid) present at equilibrium. The lever rule is so called because a similar rule relates the masses at two ends of a lever to their distances from a pivot ($m_\alpha l_\alpha = m_\beta l_\beta$ for balance).

$$n_\alpha l_\alpha = n_\beta l_\beta \tag{6.7}$$

Here n_α is the amount of phase α and n_β the amount of phase β. In the case illustrated in Fig. 6.13, because $l_\beta \approx 2l_\alpha$, the amount of phase α is about twice the amount of phase β.

Justification 6.2 *The lever rule*

To prove the lever rule we write $n = n_\alpha + n_\beta$ and the overall amount of A as nz_A. The overall amount of A is also the sum of its amounts in the two phases:

$$nz_A = n_\alpha x_A + n_\beta y_A$$

Since also

$$nz_A = n_\alpha z_A + n_\beta z_A$$

by equating these two expressions it follows that

$$n_\alpha(x_A - z_A) = n_\beta(z_A - y_A)$$

which corresponds to eqn 6.7.

Illustration 6.1 *Using the lever rule*

At p_1 in Fig. 6.10, the ratio l_{vap}/l_{liq} is almost infinite for this tie line, so n_{liq}/n_{vap} is also almost infinite, and there is only a trace of vapour present. When the pressure is reduced to p_2, the value of l_{vap}/l_{liq} is about 0.3, so $n_{liq}/n_{vap} \approx 0.3$ and the amount of liquid is about 0.3 times the amount of vapour. When the pressure has been reduced to p_3, the sample is almost completely gaseous and because $l_{vap}/l_{liq} \approx 0$ we conclude that there is only a trace of liquid present.

6.4 Temperature–composition diagrams

To discuss distillation we need a **temperature–composition diagram**, a phase diagram in which the boundaries show the composition of the phases that are in equilibrium at various temperatures (and a given pressure, typically 1 atm). An example is shown in Fig. 6.14. Note that the liquid phase now lies in the lower part of the diagram.

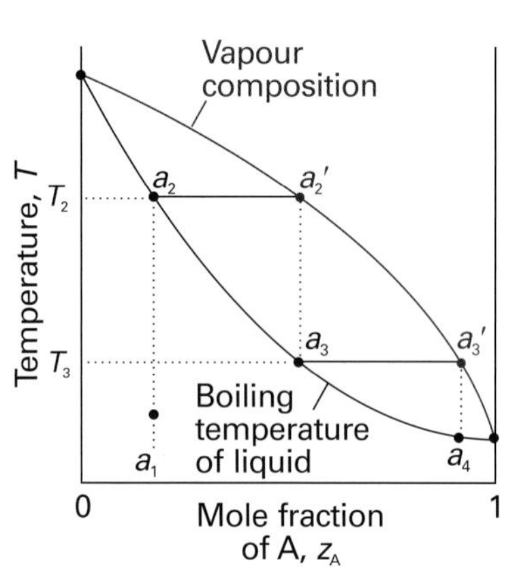

Fig. 6.14 The temperature–composition diagram corresponding to an ideal mixture with the component A more volatile than component B. Successive boilings and condensations of a liquid originally of composition a_1 lead to a condensate that is pure A. The separation technique is called fractional distillation.

(a) The distillation of mixtures

The region between the lines in Fig. 6.14 is a two-phase region where $F' = 1$. As usual, the prime indicates that one degree of freedom has been discarded; in this case, the pressure is being kept fixed, and hence at a given temperature the compositions of the phases in equilibrium are fixed. The regions outside the phase lines correspond to a single phase, so $F' = 2$, and the temperature and composition are both independently variable.

Consider what happens when a liquid of composition a_1 is heated. It boils when the temperature reaches T_2. Then the liquid has composition a_2 (the same as a_1) and the vapour (which is present only as a trace) has composition a_2'. The vapour is richer in the more volatile component A (the component with the lower boiling point). From the location of a_2, we can state the vapour's composition at the boiling point, and from the location of the tie line joining a_2 and a_2' we can read off the boiling temperature (T_2) of the original liquid mixture.

Comment 6.2
The textbook's web site contains links to online databases of phase diagrams.

In a **simple distillation**, the vapour is withdrawn and condensed. This technique is used to separate a volatile liquid from a non-volatile solute or solid. In **fractional distillation**, the boiling and condensation cycle is repeated successively. This technique is used to separate volatile liquids. We can follow the changes that occur by seeing what happens when the first condensate of composition a_3 is reheated. The phase diagram shows that this mixture boils at T_3 and yields a vapour of composition a_3', which is even richer in the more volatile component. That vapour is drawn off, and the first drop condenses to a liquid of composition a_4. The cycle can then be repeated until in due course almost pure A is obtained.

The efficiency of a fractionating column is expressed in terms of the number of **theoretical plates**, the number of effective vaporization and condensation steps that are required to achieve a condensate of given composition from a given distillate. Thus, to achieve the degree of separation shown in Fig. 6.15a, the fractionating column must correspond to three theoretical plates. To achieve the same separation for the system shown in Fig. 6.15b, in which the components have more similar partial pressures, the fractionating column must be designed to correspond to five theoretical plates.

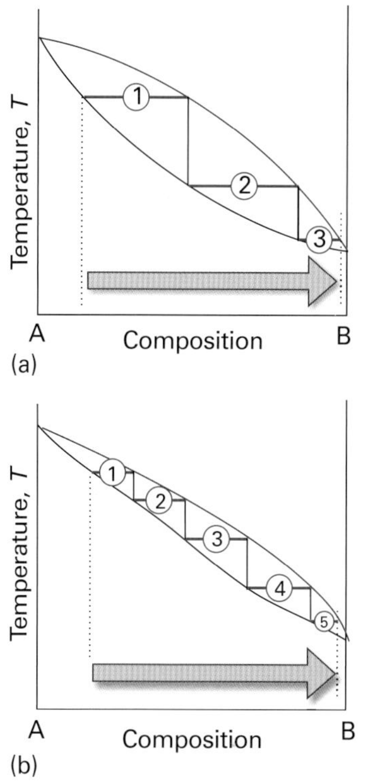

Fig. 6.15 The number of theoretical plates is the number of steps needed to bring about a specified degree of separation of two components in a mixture. The two systems shown correspond to (a) 3, (b) 5 theoretical plates.

(b) Azeotropes

Although many liquids have temperature–composition phase diagrams resembling the ideal version in Fig. 6.14, in a number of important cases there are marked deviations. A maximum in the phase diagram (Fig. 6.16) may occur when the favourable interactions between A and B molecules reduce the vapour pressure of the mixture below the ideal value: in effect, the A–B interactions stabilize the liquid. In such cases the excess Gibbs energy, G^E (Section 5.4), is negative (more favourable to mixing than ideal). Examples of this behaviour include trichloromethane/propanone and nitric acid/water mixtures. Phase diagrams showing a minimum (Fig. 6.17) indicate that the mixture is destabilized relative to the ideal solution, the A–B interactions then being unfavourable. For such mixtures G^E is positive (less favourable to mixing than ideal), and there may be contributions from both enthalpy and entropy effects. Examples include dioxane/water and ethanol/water mixtures.

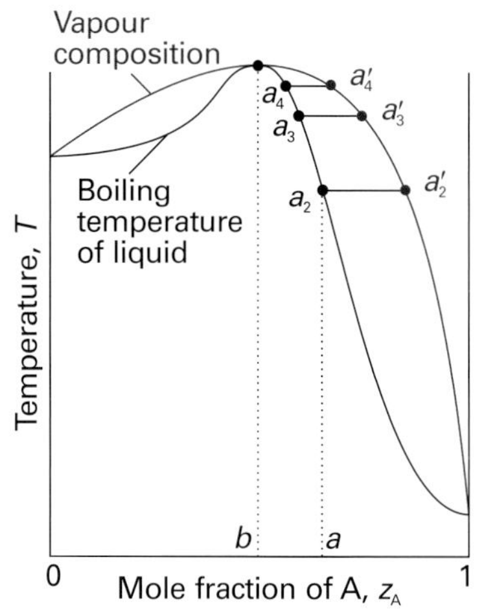

Fig. 6.16 A high-boiling azeotrope. When the liquid of composition a is distilled, the composition of the remaining liquid changes towards b but no further.

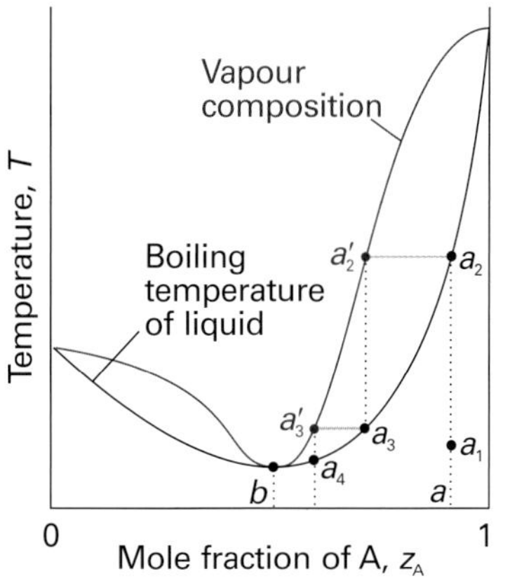

Fig. 6.17 A low-boiling azeotrope. When the mixture at a is fractionally distilled, the vapour in equilibrium in the fractionating column moves towards b and then remains unchanged.

Deviations from ideality are not always so strong as to lead to a maximum or minimum in the phase diagram, but when they do there are important consequences for distillation. Consider a liquid of composition a on the right of the maximum in Fig. 6.16. The vapour (at a_2') of the boiling mixture (at a_2) is richer in A. If that vapour is removed (and condensed elsewhere), then the remaining liquid will move to a composition that is richer in B, such as that represented by a_3, and the vapour in equilibrium with this mixture will have composition a_3'. As that vapour is removed, the composition of the boiling liquid shifts to a point such as a_4, and the composition of the vapour shifts to a_4'. Hence, as evaporation proceeds, the composition of the remaining liquid shifts towards B as A is drawn off. The boiling point of the liquid rises, and the vapour becomes richer in B. When so much A has been evaporated that the liquid has reached the composition b, the vapour has the same composition as the liquid. Evaporation then occurs without change of composition. The mixture is said to form an **azeotrope.**[2] When the azeotropic composition has been reached, distillation cannot separate the two liquids because the condensate has the same composition as the azeotropic liquid. One example of azeotrope formation is hydrochloric acid/water, which is azeotropic at 80 per cent by mass of water and boils unchanged at 108.6°C.

The system shown in Fig. 6.17 is also azeotropic, but shows its azeotropy in a different way. Suppose we start with a mixture of composition a_1, and follow the changes in the composition of the vapour that rises through a fractionating column (essentially a vertical glass tube packed with glass rings to give a large surface area). The mixture boils at a_2 to give a vapour of composition a_2'. This vapour condenses in the column to a liquid of the same composition (now marked a_3). That liquid reaches equilibrium with its vapour at a_3', which condenses higher up the tube to give a liquid of the same composition, which we now call a_4. The fractionation therefore shifts the vapour towards the azeotropic composition at b, but not beyond, and the azeotropic vapour emerges from the top of the column. An example is ethanol/water, which boils unchanged when the water content is 4 per cent by mass and the temperature is 78°C.

(c) Immiscible liquids

Finally we consider the distillation of two immiscible liquids, such as octane and water. At equilibrium, there is a tiny amount of A dissolved in B, and similarly a tiny amount of B dissolved in A: both liquids are saturated with the other component (Fig. 6.18a). As a result, the total vapour pressure of the mixture is close to $p = p_A^* + p_B^*$. If the temperature is raised to the value at which this total vapour pressure is equal to the atmospheric pressure, boiling commences and the dissolved substances are purged from their solution. However, this boiling results in a vigorous agitation of the mixture, so each component is kept saturated in the other component, and the purging continues as the very dilute solutions are replenished. This intimate contact is essential: two immiscible liquids heated in a container like that shown in Fig. 6.18b would not boil at the same temperature. The presence of the saturated solutions means that the 'mixture' boils at a lower temperature than either component would alone because boiling begins when the total vapour pressure reaches 1 atm, not when either vapour pressure reaches 1 atm. This distinction is the basis of **steam distillation**, which enables some heat-sensitive, water-insoluble organic compounds to be distilled at a lower temperature than their normal boiling point. The only snag is that the composition of the condensate is in proportion to the vapour pressures of the components, so oils of low volatility distil in low abundance.

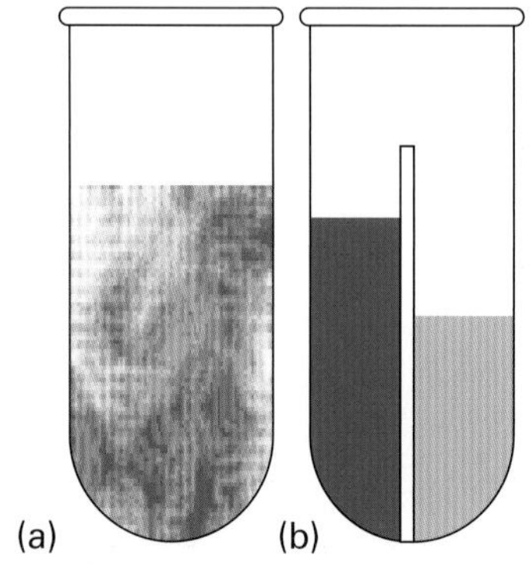

Fig. 6.18 The distillation of (a) two immiscible liquids can be regarded as (b) the joint distillation of the separated components, and boiling occurs when the sum of the partial pressures equals the external pressure.

[2] The name comes from the Greek words for 'boiling without changing'.

6.5 Liquid–liquid phase diagrams

Now we consider temperature–composition diagrams for systems that consist of pairs of **partially miscible** liquids, which are liquids that do not mix in all proportions at all temperatures. An example is hexane and nitrobenzene. The same principles of interpretation apply as to liquid–vapour diagrams. When $P=2$, $F'=1$ (the prime denoting the adoption of constant pressure), and the selection of a temperature implies that the compositions of the immiscible liquid phases are fixed. When $P=1$ (corresponding to a system in which the two liquids are fully mixed), both the temperature and the composition may be adjusted.

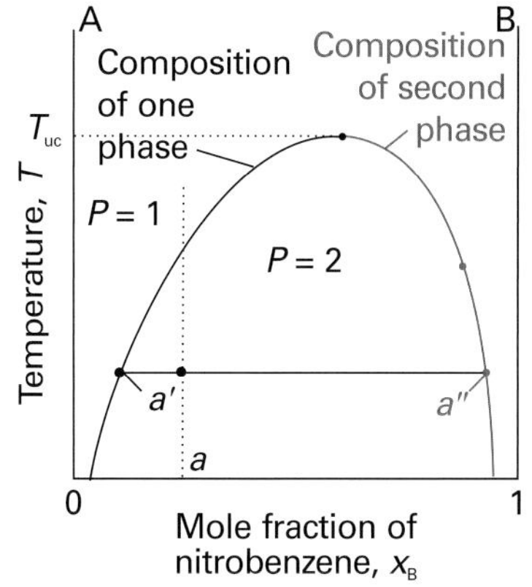

Fig. 6.19 The temperature–composition diagram for hexane and nitrobenzene at 1 atm. The region below the curve corresponds to the compositions and temperatures at which the liquids are partially miscible. The upper critical temperature, T_{uc}, is the temperature above which the two liquids are miscible in all proportions.

(a) Phase separation

Suppose a small amount of a liquid B is added to a sample of another liquid A at a temperature T'. It dissolves completely, and the binary system remains a single phase. As more B is added, a stage comes at which no more dissolves. The sample now consists of two phases in equilibrium with each other ($P = 2$), the most abundant one consisting of A saturated with B, the minor one a trace of B saturated with A. In the temperature–composition diagram drawn in Fig. 6.19, the composition of the former is represented by the point a' and that of the latter by the point a''. The relative abundances of the two phases are given by the lever rule.

When more B is added, A dissolves in it slightly. The compositions of the two phases in equilibrium remain a' and a'' because $P = 2$ implies that $F' = 0$, and hence that the compositions of the phases are invariant at a fixed temperature and pressure. However, the amount of one phase increases at the expense of the other. A stage is reached when so much B is present that it can dissolve all the A, and the system reverts to a single phase. The addition of more B now simply dilutes the solution, and from then on it remains a single phase.

The composition of the two phases at equilibrium varies with the temperature. For hexane and nitrobenzene, raising the temperature increases their miscibility. The two-phase system therefore becomes less extensive, because each phase in equilibrium is richer in its minor component: the A-rich phase is richer in B and the B-rich phase is richer in A. We can construct the entire phase diagram by repeating the observations at different temperatures and drawing the envelope of the two-phase region.

Example 6.2 *Interpreting a liquid–liquid phase diagram*

A mixture of 50 g of hexane (0.59 mol C_6H_{14}) and 50 g of nitrobenzene (0.41 mol $C_6H_5NO_2$) was prepared at 290 K. What are the compositions of the phases, and in what proportions do they occur? To what temperature must the sample be heated in order to obtain a single phase?

Method The compositions of phases in equilibrium are given by the points where the tie-line representing the temperature intersects the phase boundary. Their proportions are given by the lever rule (eqn 6.7). The temperature at which the components are completely miscible is found by following the isopleth upwards and noting the temperature at which it enters the one-phase region of the phase diagram.

Answer We denote hexane by H and nitrobenzene by N; refer to Fig. 6.20, which is a simplified version of Fig. 6.19. The point $x_N = 0.41$, $T = 290$ K occurs in the two-phase region of the phase diagram. The horizontal tie line cuts the phase

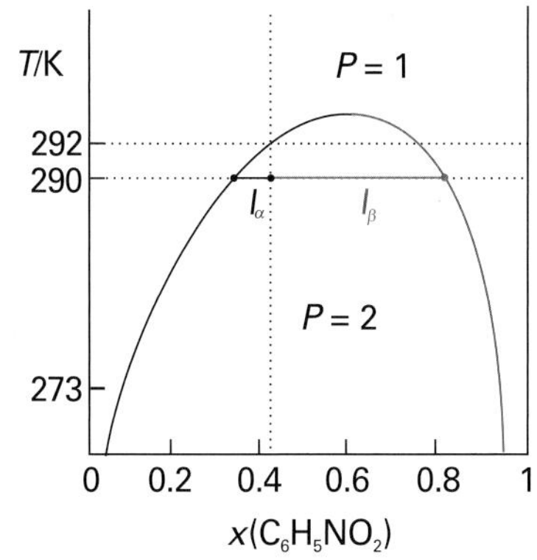

Fig. 6.20 The temperature–composition diagram for hexane and nitrobenzene at 1 atm again, with the points and lengths discussed in the text.

boundary at $x_N = 0.35$ and $x_N = 0.83$, so those are the compositions of the two phases. According to the lever rule, the ratio of amounts of each phase is equal to the ratio of the distances l_α and l_β:

$$\frac{n_\alpha}{n_\beta} = \frac{l_\beta}{l_\alpha} = \frac{0.83 - 0.41}{0.41 - 0.35} = \frac{0.42}{0.06} = 7$$

That is, there is about 7 times more hexane-rich phase than nitrobenzene-rich phase. Heating the sample to 292 K takes it into the single-phase region.

Because the phase diagram has been constructed experimentally, these conclusions are not based on any assumptions about ideality. They would be modified if the system were subjected to a different pressure.

Self-test 6.2 Repeat the problem for 50 g of hexane and 100 g of nitrobenzene at 273 K. [$x_N = 0.09$ and 0.95 in ratio 1:1.3; 294 K]

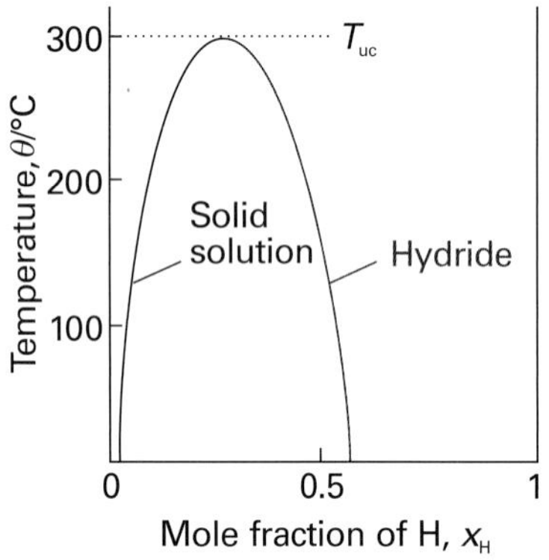

Fig. 6.21 The phase diagram for palladium and palladium hydride, which has an upper critical temperature at 300°C.

(b) Critical solution temperatures

The **upper critical solution temperature**, T_{uc}, is the highest temperature at which phase separation occurs. Above the upper critical temperature the two components are fully miscible. This temperature exists because the greater thermal motion overcomes any potential energy advantage in molecules of one type being close together. One example is the nitrobenzene/hexane system shown in Fig. 6.19. An example of a solid solution is the palladium/hydrogen system, which shows two phases, one a solid solution of hydrogen in palladium and the other a palladium hydride, up to 300°C but forms a single phase at higher temperatures (Fig. 6.21).

The thermodynamic interpretation of the upper critical solution temperature focuses on the Gibbs energy of mixing and its variation with temperature. We saw in Section 5.4 that a simple model of a real solution results in a Gibbs energy of mixing that behaves as shown in Fig. 5.20. Provided the parameter β that was introduced in eqn 5.30 is greater than 2, the Gibbs energy of mixing has a double minimum (Fig. 6.22). As a result, for $\beta > 2$ we can expect phase separation to occur. The same model shows that the compositions corresponding to the minima are obtained by looking for the conditions at which $\partial\Delta_{mix}G/\partial x = 0$, and a simple manipulation of eqn 5.31 shows that we have to solve

$$\ln\frac{x}{1-x} + \beta(1-2x) = 0$$

The solutions are plotted in Fig. 6.23. We see that, as β decreases, which can be interpreted as an increase in temperature provided the intermolecular forces remain constant, then the two minima move together and merge when $\beta = 2$.

Some systems show a **lower critical solution temperature,** T_{lc}, below which they mix in all proportions and above which they form two phases. An example is water and triethylamine (Fig. 6.24). In this case, at low temperatures the two components are more miscible because they form a weak complex; at higher temperatures the complexes break up and the two components are less miscible.

Some systems have both upper and lower critical solution temperatures. They occur because, after the weak complexes have been disrupted, leading to partial miscibility, the thermal motion at higher temperatures homogenizes the mixture again, just as in the case of ordinary partially miscible liquids. The most famous example is nicotine and water, which are partially miscible between 61°C and 210°C (Fig. 6.25).

Comment 6.3

This expression is an example of a *transcendental equation*, an equation that does not have a solution that can be expressed in a closed form. The solutions can be found numerically by using mathematical software or by plotting the first term against the second and identifying the points of intersection as β is changed.

Comment 6.4

The upper critical solution temperature and the lower critical solution temperature are also called the 'upper consolute temperature' and 'lower consolute temperature', respectively.

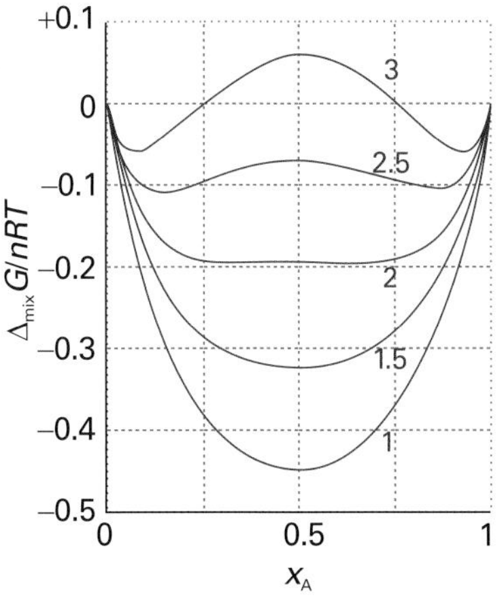

Fig. 6.22 The temperature variation of the Gibbs energy of mixing of a system that is partially miscible at low temperatures. A system of composition in the region $P = 2$ forms two phases with compositions corresponding to the two local minima of the curve. This illustration is a duplicate of Fig. 5.20.

Exploration Working from eqn 5.31, write an expression for T_{min}, the temperature at which $\Delta_{mix}G$ has a minimum, as a function of β and x_A. Then, plot T_{min} against x_A for several values of β. Provide a physical interpretation for any maxima or minima that you observe in these plots.

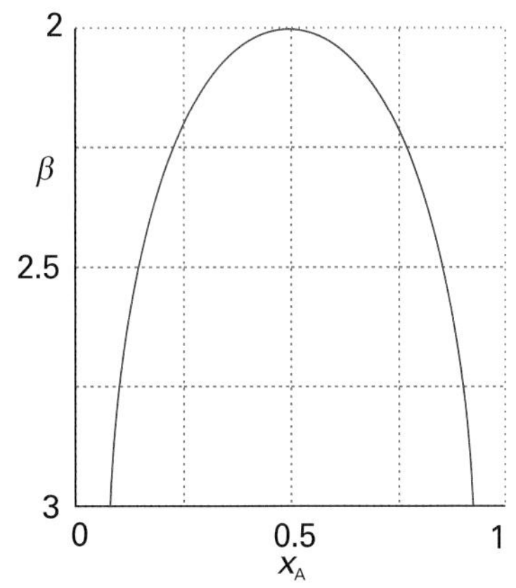

Fig. 6.23 The location of the phase boundary as computed on the basis of the β-parameter model introduced in Section 5.4.

Exploration Using mathematical software or an electronic spreadsheet, generate the plot of β against x_A by one of two methods: (a) solve the transcendental equation $\ln\{(x/(1-x)\} + \beta(1-2x) = 0$ numerically, or (b) plot the first term of the transcendental equation against the second and identify the points of intersection as β is changed.

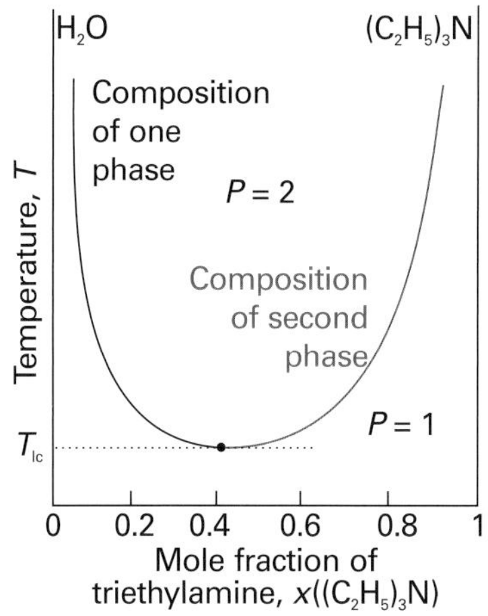

Fig. 6.24 The temperature–composition diagram for water and triethylamine. This system shows a lower critical temperature at 292 K. The labels indicate the interpretation of the boundaries.

(c) The distillation of partially miscible liquids

Consider a pair of liquids that are partially miscible and form a low-boiling azeotrope. This combination is quite common because both properties reflect the tendency of the two kinds of molecule to avoid each other. There are two possibilities: one in which the liquids become fully miscible before they boil; the other in which boiling occurs before mixing is complete.

Figure 6.26 shows the phase diagram for two components that become fully miscible before they boil. Distillation of a mixture of composition a_1 leads to a vapour of composition b_1, which condenses to the completely miscible single-phase solution at b_2. Phase separation occurs only when this distillate is cooled to a point in the two-phase liquid region, such as b_3. This description applies only to the first drop of distillate. If distillation continues, the composition of the remaining liquid changes. In the end, when the whole sample has evaporated and condensed, the composition is back to a_1.

Figure 6.27 shows the second possibility, in which there is no upper critical solution temperature. The distillate obtained from a liquid initially of composition a_1 has composition b_3 and is a two-phase mixture. One phase has composition b_3' and the other has composition b_3''.

The behaviour of a system of composition represented by the isopleth e in Fig. 6.27 is interesting. A system at e_1 forms two phases, which persist (but with changing proportions) up to the boiling point at e_2. The vapour of this mixture has the same

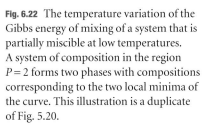

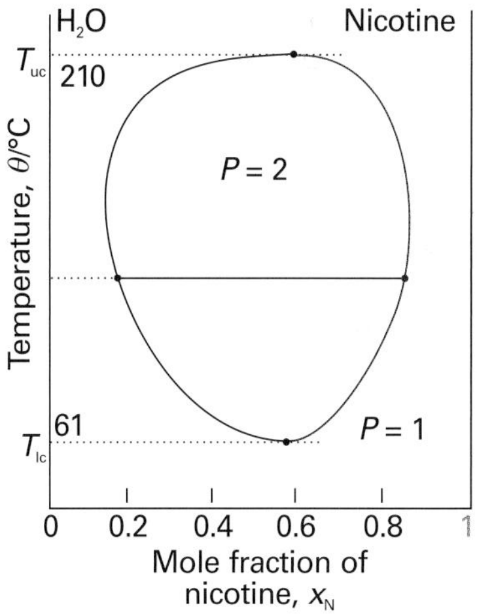

Fig. 6.25 The temperature–composition diagram for water and nicotine, which has both upper and lower critical temperatures. Note the high temperatures for the liquid (especially the water): the diagram corresponds to a sample under pressure.

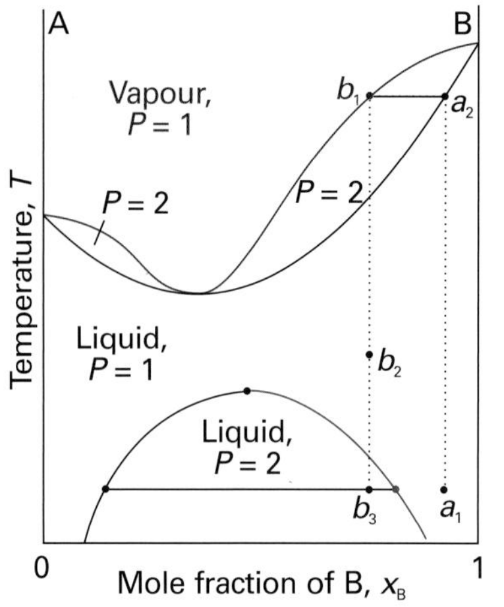

Fig. 6.26 The temperature–composition diagram for a binary system in which the upper critical temperature is less than the boiling point at all compositions. The mixture forms a low-boiling azeotrope.

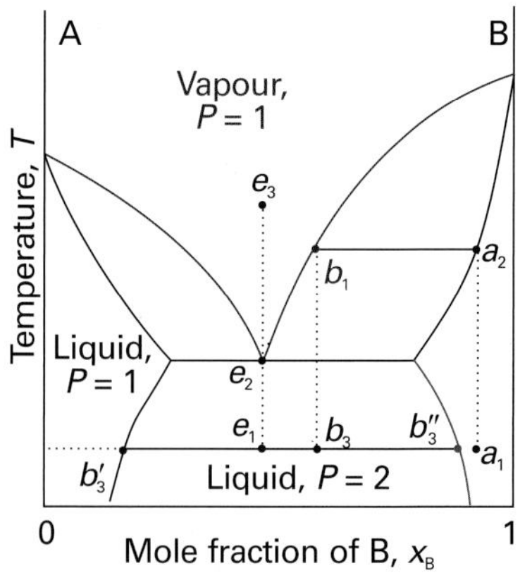

Fig. 6.27 The temperature–composition diagram for a binary system in which boiling occurs before the two liquids are fully miscible.

composition as the liquid (the liquid is an azeotrope). Similarly, condensing a vapour of composition e_3 gives a two-phase liquid of the same overall composition. At a fixed temperature, the mixture vaporizes and condenses like a single substance.

Example 6.3 *Interpreting a phase diagram*

State the changes that occur when a mixture of composition $x_B = 0.95$ (a_1) in Fig. 6.28 is boiled and the vapour condensed.

Method The area in which the point lies gives the number of phases; the compositions of the phases are given by the points at the intersections of the horizontal tie line with the phase boundaries; the relative abundances are given by the lever rule (eqn 6.7).

Answer The initial point is in the one-phase region. When heated it boils at 350 K (a_2) giving a vapour of composition $x_B = 0.66$ (b_1). The liquid gets richer in B, and the last drop (of pure B) evaporates at 390 K. The boiling range of the liquid is therefore 350 to 390 K. If the initial vapour is drawn off, it has a composition $x_B = 0.66$. This composition would be maintained if the sample were very large, but for a finite sample it shifts to higher values and ultimately to $x_B = 0.95$. Cooling the distillate corresponds to moving down the $x_B = 0.66$ isopleth. At 330 K, for instance, the liquid phase has composition $x_B = 0.87$, the vapour $x_B = 0.49$; their relative proportions are 1:3. At 320 K the sample consists of three phases: the vapour and two liquids. One liquid phase has composition $x_B = 0.30$; the other has composition $x_B = 0.80$ in the ratio 0.62:1. Further cooling moves the system into the two-phase region, and at 298 K the compositions are 0.20 and 0.90 in the ratio 0.82:1. As further distillate boils over, the overall composition of the distillate becomes richer in B. When the last drop has been condensed the phase composition is the same as at the beginning.

Self-test 6.3 Repeat the discussion, beginning at the point $x_B = 0.4$, $T = 298$ K.

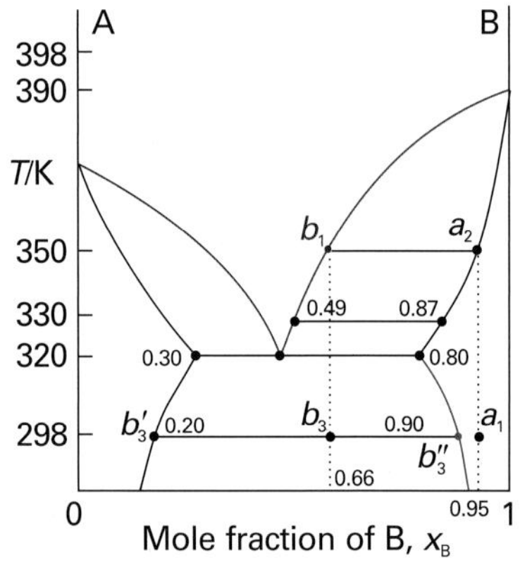

Fig. 6.28 The points of the phase diagram in Fig. 6.27 that are discussed in Example 6.3.

6.6 Liquid–solid phase diagrams

Knowledge of the temperature–composition diagrams for solid mixtures guides the design of important industrial processes, such as the manufacture of liquid crystal displays and semiconductors. In this section, we shall consider systems where solid and liquid phases may both be present at temperatures below the boiling point.

Consider the two-component liquid of composition a_1 in Fig. 6.29. The changes that occur may be expressed as follows.

1. $a_1 \rightarrow a_2$. The system enters the two-phase region labelled 'Liquid + B'. Pure solid B begins to come out of solution and the remaining liquid becomes richer in A.

2. $a_2 \rightarrow a_3$. More of the solid forms, and the relative amounts of the solid and liquid (which are in equilibrium) are given by the lever rule. At this stage there are roughly equal amounts of each. The liquid phase is richer in A than before (its composition is given by b_3) because some B has been deposited.

3. $a_3 \rightarrow a_4$. At the end of this step, there is less liquid than at a_3, and its composition is given by e. This liquid now freezes to give a two-phase system of pure B and pure A.

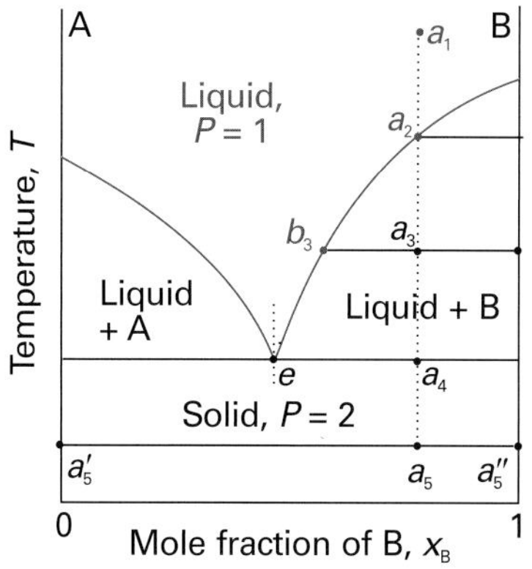

Fig. 6.29 The temperature–composition phase diagram for two almost immiscible solids and their completely miscible liquids. Note the similarity to Fig. 6.27. The isopleth through e corresponds to the eutectic composition, the mixture with lowest melting point.

(a) Eutectics

The isopleth at e in Fig. 6.29 corresponds to the **eutectic** composition, the mixture with the lowest melting point.[3] A liquid with the eutectic composition freezes at a single temperature, without previously depositing solid A or B. A solid with the eutectic composition melts, without change of composition, at the lowest temperature of any mixture. Solutions of composition to the right of e deposit B as they cool, and solutions to the left deposit A: only the eutectic mixture (apart from pure A or pure B) solidifies at a single definite temperature ($F' = 0$ when $C = 2$ and $P = 3$) without gradually unloading one or other of the components from the liquid.

One technologically important eutectic is solder, which has mass composition of about 67 per cent tin and 33 per cent lead and melts at 183°C. The eutectic formed by 23 per cent NaCl and 77 per cent H_2O by mass melts at −21.1°C. When salt is added to ice under isothermal conditions (for example, when spread on an icy road) the mixture melts if the temperature is above −21.1°C (and the eutectic composition has been achieved). When salt is added to ice under adiabatic conditions (for example, when added to ice in a vacuum flask) the ice melts, but in doing so it absorbs heat from the rest of the mixture. The temperature of the system falls and, if enough salt is added, cooling continues down to the eutectic temperature. Eutectic formation occurs in the great majority of binary alloy systems, and is of great importance for the microstructure of solid materials. Although a eutectic solid is a two-phase system, it crystallizes out in a nearly homogeneous mixture of microcrystals. The two microcrystalline phases can be distinguished by microscopy and structural techniques such as X-ray diffraction (Chapter 20).

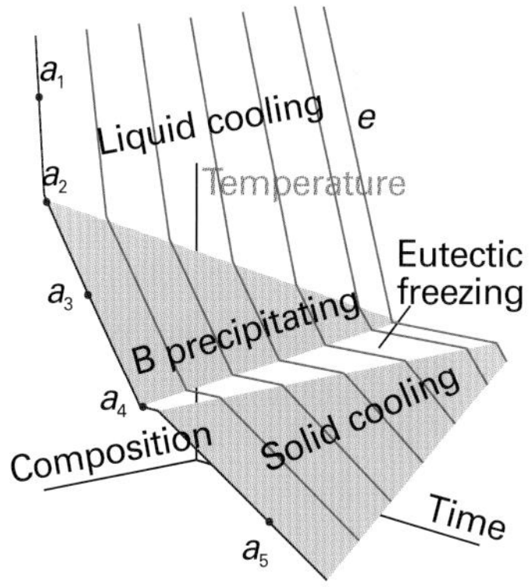

Fig. 6.30 The cooling curves for the system shown in Fig. 6.29. For isopleth a, the rate of cooling slows at a_2 because solid B deposits from solution. There is a complete halt at a_4 while the eutectic solidifies. This halt is longest for the eutectic isopleth, e. The eutectic halt shortens again for compositions beyond e (richer in A). Cooling curves are used to construct the phase diagram.

Thermal analysis is a very useful practical way of detecting eutectics. We can see how it is used by considering the rate of cooling down the isopleth through a_1 in Fig. 6.29. The liquid cools steadily until it reaches a_2, when B begins to be deposited (Fig. 6.30). Cooling is now slower because the solidification of B is exothermic and retards the cooling. When the remaining liquid reaches the eutectic composition, the temperature remains constant ($F' = 0$) until the whole sample has solidified: this region of constant temperature is the eutectic halt. If the liquid has the eutectic composition e initially, the liquid cools steadily down to the freezing temperature of the eutectic,

[3] The name comes from the Greek words for 'easily melted'.

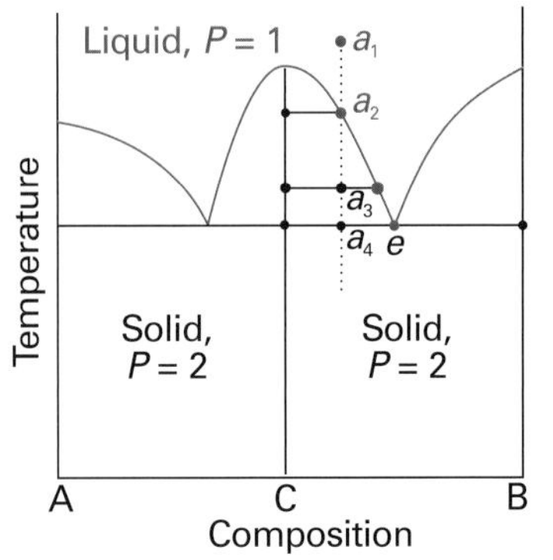

Fig. 6.31 The phase diagram for a system in which A and B react to form a compound C = AB. This resembles two versions of Fig. 6.29 in each half of the diagram. The constituent C is a true compound, not just an equimolar mixture.

when there is a long **eutectic halt** as the entire sample solidifies (like the freezing of a pure liquid).

Monitoring the cooling curves at different overall compositions gives a clear indication of the structure of the phase diagram. The solid–liquid boundary is given by the points at which the rate of cooling changes. The longest eutectic halt gives the location of the eutectic composition and its melting temperature.

(b) Reacting systems

Many binary mixtures react to produce compounds, and technologically important examples of this behaviour include the III/V semiconductors, such as the gallium arsenide system, which forms the compound GaAs. Although three constituents are present, there are only two components because GaAs is formed from the reaction Ga + As $\rightleftharpoons$ GaAs. We shall illustrate some of the principles involved with a system that forms a compound C that also forms eutectic mixtures with the species A and B (Fig. 6.31).

A system prepared by mixing an excess of B with A consists of C and unreacted B. This is a binary B, C system, which we suppose forms a eutectic. The principal change from the eutectic phase diagram in Fig. 6.29 is that the whole of the phase diagram is squeezed into the range of compositions lying between equal amounts of A and B ($x_B = 0.5$, marked C in Fig. 6.31) and pure B. The interpretation of the information in the diagram is obtained in the same way as for Fig. 6.29. The solid deposited on cooling along the isopleth a is the compound C. At temperatures below a_4 there are two solid phases, one consisting of C and the other of B. The pure compound C melts **congruently**, that is, the composition of the liquid it forms is the same as that of the solid compound.

(c) Incongruent melting

In some cases the compound C is not stable as a liquid. An example is the alloy Na_2K, which survives only as a solid (Fig. 6.32). Consider what happens as a liquid at a_1 is cooled:

1. $a_1 \rightarrow a_2$. Some solid Na is deposited, and the remaining liquid is richer in K.
2. $a_2 \rightarrow$ just below a_3. The sample is now entirely solid, and consists of solid Na and solid Na_2K.

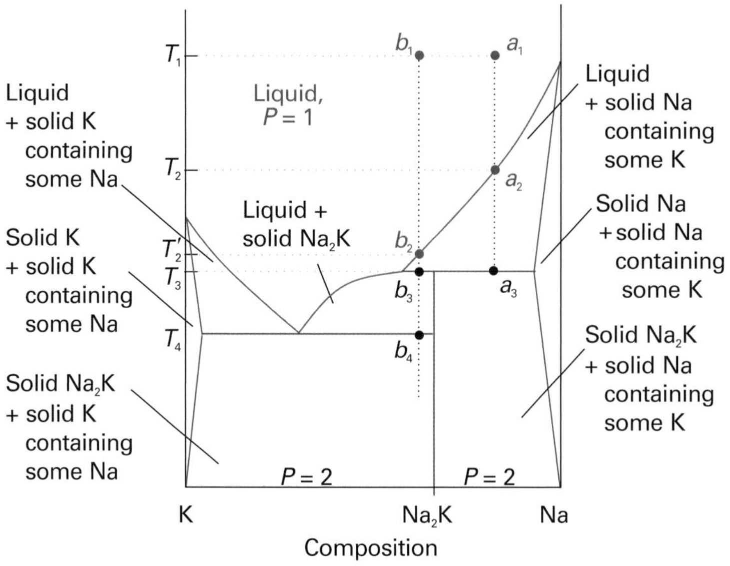

Fig. 6.32 The phase diagram for an actual system (sodium and potassium) like that shown in Fig. 6.31, but with two differences. One is that the compound is Na_2K, corresponding to A_2B and not AB as in that illustration. The second is that the compound exists only as the solid, not as the liquid. The transformation of the compound at its melting point is an example of incongruent melting.

Now consider the isopleth through b_1:

1. $b_1 \rightarrow b_2$. No obvious change occurs until the phase boundary is reached at b_2 when solid Na begins to deposit.

2. $b_2 \rightarrow b_3$. Solid Na deposits, but at b_3 a reaction occurs to form Na_2K: this compound is formed by the K atoms diffusing into the solid Na.

3. b_3. At b_3, three phases are in mutual equilibrium: the liquid, the compound Na_2K, and solid Na. The horizontal line representing this three-phase equilibrium is called a **peritectic line**.

At this stage the liquid Na/K mixture is in equilibrium with a little solid Na_2K, but there is still no liquid compound.

4. $b_3 \rightarrow b_4$. As cooling continues, the amount of solid compound increases until at b_4 the liquid reaches its eutectic composition. It then solidifies to give a two-phase solid consisting of solid K and solid Na_2K.

If the solid is reheated, the sequence of events is reversed. No liquid Na_2K forms at any stage because it is too unstable to exist as a liquid. This behaviour is an example of **incongruent melting**, in which a compound melts into its components and does not itself form a liquid phase.

IMPACT ON MATERIALS SCIENCE

I6.1 Liquid crystals

A *mesophase* is a phase intermediate between solid and liquid. Mesophases are of great importance in biology, for they occur as lipid bilayers and in vesicular systems. A mesophase may arise when molecules have highly non-spherical shapes, such as being long and thin (**1**), or disk-like (**2**). When the solid melts, some aspects of the long-range order characteristic of the solid may be retained, and the new phase may be a *liquid crystal*, a substance having liquid-like imperfect long-range order in at least one direction in space but positional or orientational order in at least one other direction. *Calamitic liquid crystals* (from the Greek word for reed) are made from long and thin molecules, whereas *discotic liquid crystals* are made from disk-like molecules. A

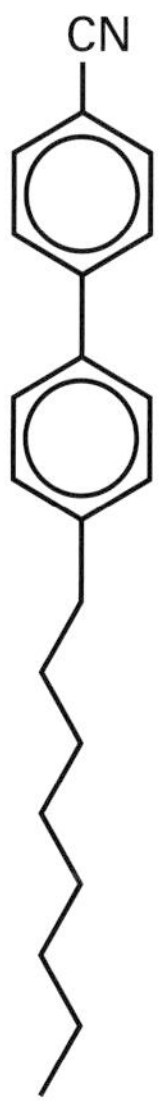

1

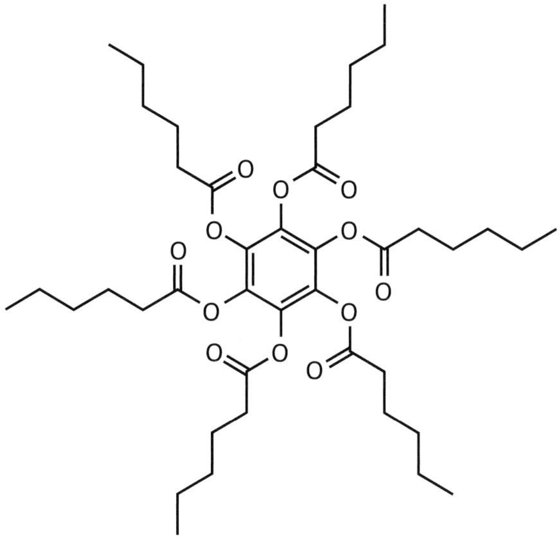

2

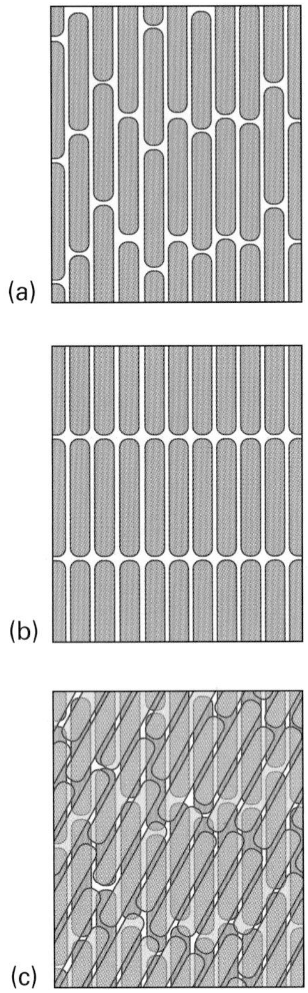

Fig. 6.33 The arrangement of molecules in (a) the nematic phase, (b) the smectic phase, and (c) the cholesteric phase of liquid crystals. In the cholesteric phase, the stacking of layers continues to give a helical arrangement of molecules.

thermotropic liquid crystal displays a transition to the liquid crystalline phase as the temperature is changed. A *lyotropic* liquid crystal is a solution that undergoes a transition to the liquid crystalline phase as the composition is changed.

One type of retained long-range order gives rise to a *smectic phase* (from the Greek word for soapy), in which the molecules align themselves in layers (see Fig. 6.33). Other materials, and some smectic liquid crystals at higher temperatures, lack the layered structure but retain a parallel alignment; this mesophase is called a *nematic phase* (from the Greek for thread, which refers to the observed defect structure of the phase). In the *cholesteric phase* (from the Greek for bile solid) the molecules lie in sheets at angles that change slightly between each sheet. That is, they form helical structures with a pitch that depends on the temperature. As a result, cholesteric liquid crystals diffract light and have colours that depend on the temperature. Disk-like molecules such as (**2**) can form nematic and *columnar* mesophases. In the latter, the aromatic rings stack one on top of the other and are separated by very small distances (less than 0.5 nm).

The optical properties of nematic liquid crystals are anisotropic, meaning that they depend on the relative orientation of the molecular assemblies with respect to the polarization of the incident beam of light. Nematic liquid crystals also respond in special ways to electric fields. Together, these unique optical and electrical properties form the basis of operation of liquid crystal displays (LCDs). In a 'twisted nematic' LCD, the liquid crystal is held between two flat plates about 10 μm apart. The inner surface of each plate is coated with a transparent conducting material, such as indium–tin oxide. The plates also have a surface that causes the liquid crystal to adopt a particular orientation at its interface and are typically set at 90° to each other but 270° in a'supertwist' arrangement. The entire assembly is set between two polarizers, optical filters that allow light of one one specific plane of polarization to pass. The incident light passes through the outer polarizer, then its plane of polarization is rotated as it passes through the twisted nematic and, depending on the setting of the second polarizer, will pass through (if that is how the second polarizer is arranged). When a potential difference is applied across the cell, the helical arrangement is lost and the plane of the light is no longer rotated and will be blocked by the second polarizer.

Although there are many liquid crystalline materials, some difficulty is often experienced in achieving a technologically useful temperature range for the existence of the mesophase. To overcome this difficulty, mixtures can be used. An example of the type of phase diagram that is then obtained is shown in Fig. 6.34. As can be seen, the mesophase exists over a wider range of temperatures than either liquid crystalline material alone.

IMPACT ON MATERIALS SCIENCE

I6.2 Ultrapurity and controlled impurity

Advances in technology have called for materials of extreme purity. For example, semiconductor devices consist of almost perfectly pure silicon or germanium doped to a precisely controlled extent. For these materials to operate successfully, the impurity level must be kept down to less than 1 ppb (1 part in 10^9, which corresponds to 1 mg of impurity in 1 t of material, about a small grain of salt in 5 t of sugar).[4]

In the technique of *zone refining* the sample is in the form of a narrow cylinder. This cylinder is heated in a thin disk-like zone which is swept from one end of the sample to the other. The advancing liquid zone accumulates the impurities as it passes. In practice, a train of hot and cold zones are swept repeatedly from one end to the other

[4] 1 t = 10^3 kg.

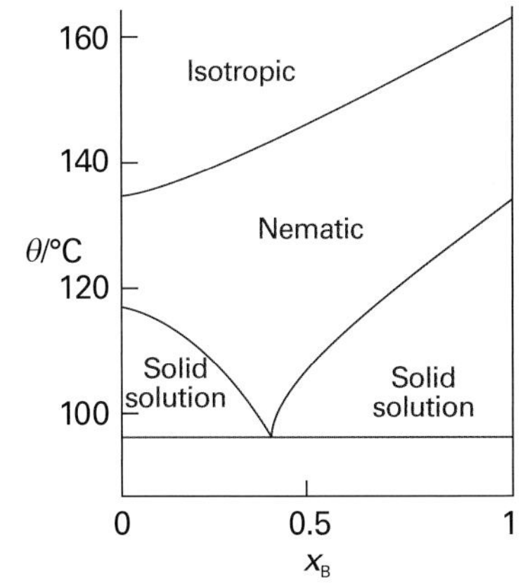

Fig. 6.34 The phase diagram at 1 atm for a binary system of two liquid crystalline materials, 4,4′-dimethoxyazoxybenzene (A) and 4,4′-diethoxyazoxybenzene (B).

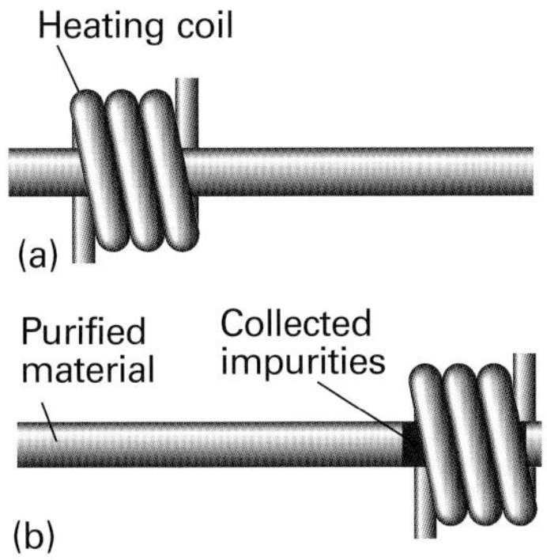

Fig. 6.35 The procedure for zone refining. (a) Initially, impurities are distributed uniformly along the sample. (b) After a molten zone is passed along the rod, the impurities are more concentrated at the right. In practice, a series of molten zones are passed along the rod from left to right.

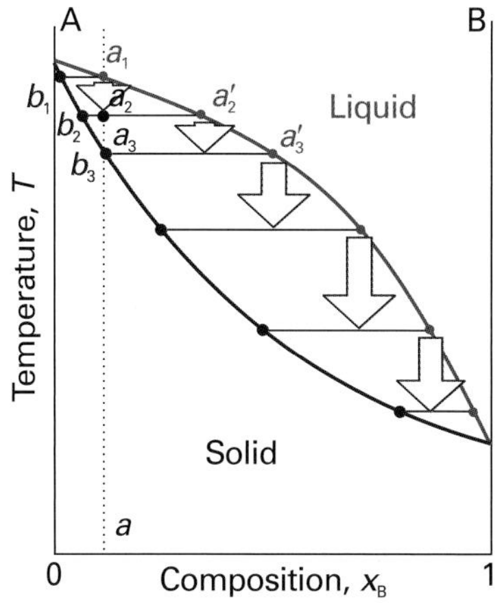

Fig. 6.36 A binary temperature–composition diagram can be used to discuss zone refining, as explained in the text.

as shown in Fig. 6.35. The zone at the end of the sample is the impurity dump: when the heater has gone by, it cools to a dirty solid which can be discarded.

The technique makes use of the non-equilibrium properties of the system. It relies on the impurities being more soluble in the molten sample than in the solid, and sweeps them up by passing a molten zone repeatedly from one end to the other along a sample. The phase diagram in Fig. 6.36 gives some insight into the process. Consider a liquid (this represents the molten zone) on the isopleth through a_1, and let it cool without the entire sample coming to overall equilibrium. If the temperature falls to a_2 a solid of composition b_2 is deposited and the remaining liquid (the zone where the heater has moved on) is at a_2'. Cooling that liquid down an isopleth passing through a_2' deposits solid of composition b_3 and leaves liquid at a_3'. The process continues until the last drop of liquid to solidify is heavily contaminated with B. There is plenty of everyday evidence that impure liquids freeze in this way. For example, an ice cube is clear near the surface but misty in the core: the water used to make ice normally contains dissolved air; freezing proceeds from the outside, and air is accumulated in the retreating liquid phase. It cannot escape from the interior of the cube, and so when that freezes it occludes the air in a mist of tiny bubbles.

A modification of zone refining is *zone levelling*. It is used to introduce controlled amounts of impurity (for example, of indium into germanium). A sample rich in the required dopant is put at the head of the main sample, and made molten. The zone is then dragged repeatedly in alternate directions through the sample, where it deposits a uniform distribution of the impurity.

Checklist of key ideas

☐ 1. A phase is a state of matter that is uniform throughout, not only in chemical composition but also in physical state.

☐ 2. A constituent is a chemical species (an ion or a molecule). A component is a chemically independent constituent of a system.

☐ 3. The variance F, or degree of freedom, is the number of intensive variables that can be changed independently without disturbing the number of phases in equilibrium.

☐ 4. The phase rule states that $F = C - P + 2$.

- ☐ 5. Thermal analysis is a technique for detecting phase transitions that takes advantage of the effect of the enthalpy change during a first-order transition.
- ☐ 6. The vapour pressure of an ideal solution is given by $p = p_B^* + (p_A^* - p_B^*)x_A$. The composition of the vapour, $y_A = x_A p_A^* / \{p_B^* + (p_A^* - p_B^*)x_A\}$, $y_B = 1 - y_A$.
- ☐ 7. The total vapour pressure of a mixture is given by $p = p_A^* p_B^* / \{p_A^* + (p_B^* - p_A^*)y_A\}$.
- ☐ 8. An isopleth is a line of constant composition in a phase diagram. A tie line is a line joining two points representing phases in equilibrium.
- ☐ 9. The lever rule allows for the calculation of the relative amounts of two phases in equilibrium: $n_\alpha l_\alpha = n_\beta l_\beta$.
- ☐ 10. A temperature–composition diagram is a phase diagram in which the boundaries show the composition of the phases that are in equilibrium at various temperatures.
- ☐ 11. An azeotrope is a mixture that boils without change of composition.
- ☐ 12. Partially miscible liquids are liquids that do not mix in all proportions at all temperatures.
- ☐ 13. The upper critical solution temperature is the highest temperature at which phase separation occurs in a binary liquid mixture. The lower critical solution temperature is the temperature below which the components of a binary mixture mix in all proportions and above which they form two phases.
- ☐ 14. A eutectic is the mixture with the lowest melting point; a liquid with the eutectic composition freezes at a single temperature. A eutectic halt is a delay in cooling while the eutectic freezes.
- ☐ 15. Incongruent melting occurs when a compound melts into its components and does not itself form a liquid phase.

Further reading

Articles and texts

J.S. Alper, The Gibbs phase rule revisited: interrelationships between components and phases. *J. Chem. Educ.* **76**, 1567 (1999).

W.D. Callister, Jr., *Materials science and engineering, an introduction.* Wiley, New York (2000).

P.J. Collings and M. Hird, *Introduction to liquid crystals: chemistry and physics.* Taylor & Francis, London (1997).

M. Hillert, *Phase equilibria, phase diagrams and phase transformations: a thermodynamic basis.* Cambridge University Press (1998).

H.-G. Lee, *Chemical thermodynamics for metals and materials.* Imperial College Press, London (1999).

R.J. Stead and K. Stead, Phase diagrams for ternary liquid systems. *J. Chem. Educ.* **67**, 385 (1990).

S.I. Sandler, *Chemical and engineering thermodynamics.* Wiley, New York (1998).

Sources of data and information

A. Alper, *Phase diagrams*, Vols. 1, 2, and 3. Academic Press, New York (1970).

J. Wisniak, *Phase diagrams: a literature source book.* Elsevier, Amsterdam (1981–86).

Discussion questions

6.1 Define the following terms: phase, constituent, component, and degree of freedom.

6.2 What factors determine the number of theoretical plates required to achieve a desired degree of separation in fractional distillation?

6.3 Draw phase diagrams for the following types of systems. Label the regions and intersections of the diagrams, stating what materials (possibly compounds or azeotropes) are present and whether they are solid liquid or gas. (a) One-component, pressure–temperature diagram, liquid density greater than that of solid. (b) Two-component, temperature–composition, solid–liquid diagram, one compound AB formed that melts congruently, negligible solid–solid solubility.

6.4 Draw phase diagrams for the following types of systems. Label the regions and intersections of the diagrams, stating what materials (possibly compounds or azeotropes) are present and whether they are solid liquid or gas. (a) Two-component, temperature–composition, solid–liquid diagram, one compound of formula AB_2 that melts incongruently, negligible solid–solid solubility; (b) two-component, constant temperature–composition, liquid–vapour diagram, formation of an azeotrope at $x_B = 0.333$, complete miscibility.

6.5 Label the regions of the phase diagram in Fig. 6.37. State what substances (if compounds give their formulas) exist in each region. Label each substance in each region as solid, liquid, or gas.

6.6 Label the regions of the phase diagram in Fig. 6.38. State what substances (if compounds give their formulas) exist in each region. Label each substance in each region as solid, liquid, or gas.

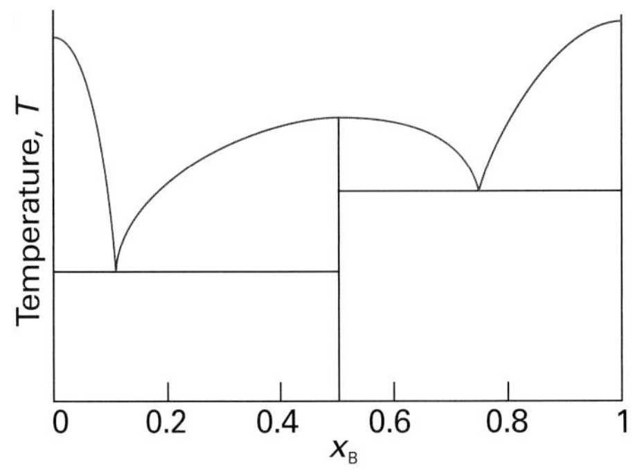

Fig. 6.37

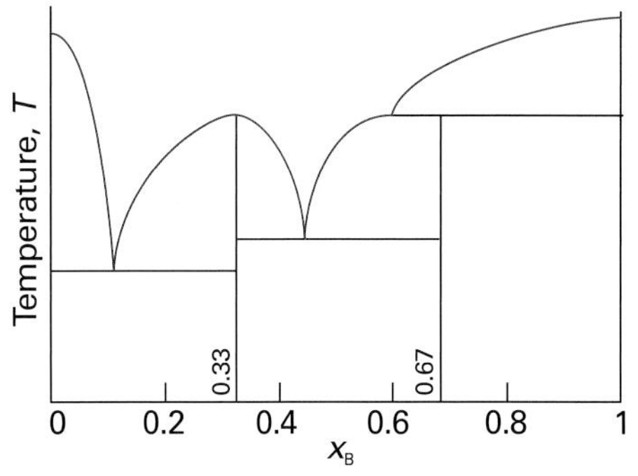

Fig. 6.38

Exercises

6.1(a) At 90°C, the vapour pressure of methylbenzene is 53.3 kPa and that of 1,2-dimethylbenzene is 20.0 kPa. What is the composition of a liquid mixture that boils at 90°C when the pressure is 0.50 atm? What is the composition of the vapour produced?

6.1(b) At 90°C, the vapour pressure of 1,2-dimethylbenzene is 20 kPa and that of 1,3-dimethylbenzene is 18 kPa. What is the composition of a liquid mixture that boils at 90°C when the pressure is 19 kPa? What is the composition of the vapour produced?

6.2(a) The vapour pressure of pure liquid A at 300 K is 76.7 kPa and that of pure liquid B is 52.0 kPa. These two compounds form ideal liquid and gaseous mixtures. Consider the equilibrium composition of a mixture in which the mole fraction of A in the vapour is 0.350. Calculate the total pressure of the vapour and the composition of the liquid mixture.

6.2(b) The vapour pressure of pure liquid A at 293 K is 68.8 kPa and that of pure liquid B is 82.1 kPa. These two compounds form ideal liquid and gaseous mixtures. Consider the equilibrium composition of a mixture in which the mole fraction of A in the vapour is 0.612. Calculate the total pressure of the vapour and the composition of the liquid mixture.

6.3(a) It is found that the boiling point of a binary solution of A and B with $x_A = 0.6589$ is 88°C. At this temperature the vapour pressures of pure A and B are 127.6 kPa and 50.60 kPa, respectively. (a) Is this solution ideal? (b) What is the initial composition of the vapour above the solution?

6.3(b) It is found that the boiling point of a binary solution of A and B with $x_A = 0.4217$ is 96°C. At this temperature the vapour pressures of pure A and B are 110.1 kPa and 76.5 kPa, respectively. (a) Is this solution ideal? (b) What is the initial composition of the vapour above the solution?

6.4(a) Dibromoethene (DE, $p^*_{DE} = 22.9$ kPa at 358 K) and dibromopropene (DP, $p^*_{DP} = 17.1$ kPa at 358 K) form a nearly ideal solution. If $z_{DE} = 0.60$, what is (a) p_{total} when the system is all liquid, (b) the composition of the vapour when the system is still almost all liquid?

6.4(b) Benzene and toluene form nearly ideal solutions. Consider an equimolar solution of benzene and toluene. At 20°C the vapour pressures of pure benzene and toluene are 9.9 kPa and 2.9 kPa, respectively. The solution is boiled by reducing the external pressure below the vapour pressure. Calculate (a) the pressure when boiling begins, (b) the composition of each component in the vapour, and (c) the vapour pressure when only a few drops of liquid remain. Assume that the rate of vaporization is low enough for the temperature to remain constant at 20°C.

6.5(a) The following temperature/composition data were obtained for a mixture of octane (O) and methylbenzene (M) at 1.00 atm, where x is the mole fraction in the liquid and y the mole fraction in the vapour at equilibrium.

θ/°C	110.9	112.0	114.0	115.8	117.3	119.0	121.1	123.0
x_M	0.908	0.795	0.615	0.527	0.408	0.300	0.203	0.097
y_M	0.923	0.836	0.698	0.624	0.527	0.410	0.297	0.164

The boiling points are 110.6°C and 125.6°C for M and O, respectively. Plot the temperature/composition diagram for the mixture. What is the composition of the vapour in equilibrium with the liquid of composition (a) $x_M = 0.250$ and (b) $x_O = 0.250$?

6.5(b) The following temperature/composition data were obtained for a mixture of two liquids A and B at 1.00 atm, where x is the mole fraction in the liquid and y the mole fraction in the vapour at equilibrium.

θ/°C	125	130	135	140	145	150
x_A	0.91	0.65	0.45	0.30	0.18	0.098
y_A	0.99	0.91	0.77	0.61	0.45	0.25

The boiling points are 124°C for A and 155°C for B. Plot the temperature–composition diagram for the mixture. What is the composition of the vapour in equilibrium with the liquid of composition (a) $x_A = 0.50$ and (b) $x_B = 0.33$?

6.6(a) State the number of components in the following systems. (a) NaH_2PO_4 in water at equilibrium with water vapour but disregarding the fact that the salt is ionized. (b) The same, but taking into account the ionization of the salt.

6.6(b) State the number of components for a system in which $AlCl_3$ is dissolved in water, noting that hydrolysis and precipitation of $Al(OH)_3$ occur.

6.7(a) Blue $CuSO_4{\cdot}5H_2O$ crystals release their water of hydration when heated. How many phases and components are present in an otherwise empty heated container?

6.7(b) Ammonium chloride, NH_4Cl, decomposes when it is heated. (a) How many components and phases are present when the salt is heated in an otherwise empty container? (b) Now suppose that additional ammonia is also present. How many components and phases are present?

6.8(a) A saturated solution of Na_2SO_4, with excess of the solid, is present at equilibrium with its vapour in a closed vessel. (a) How many phases and components are present. (b) What is the variance (the number of degrees of freedom) of the system? Identify the independent variables.

6.8(b) Suppose that the solution referred to in Exercise 6.8a is not saturated. (a) How many phases and components are present. (b) What is the variance (the number of degrees of freedom) of the system? Identify the independent variables.

6.9(a) Methylethyl ether (A) and diborane, B_2H_6 (B), form a compound that melts congruently at 133 K. The system exhibits two eutectics, one at 25 mol per cent B and 123 K and a second at 90 mol per cent B and 104 K. The melting points of pure A and B are 131 K and 110 K, respectively. Sketch the phase diagram for this system. Assume negligible solid–solid solubility.

6.9(b) Sketch the phase diagram of the system NH_3/N_2H_4 given that the two substances do not form a compound with each other, that NH_3 freezes at −78°C and N_2H_4 freezes at +2°C, and that a eutectic is formed when the mole fraction of N_2H_4 is 0.07 and that the eutectic melts at −80°C.

6.10(a) Figure 6.39 shows the phase diagram for two partially miscible liquids, which can be taken to be that for water (A) and 2-methyl-1-propanol (B). Describe what will be observed when a mixture of composition $x_B = 0.8$ is heated, at each stage giving the number, composition, and relative amounts of the phases present.

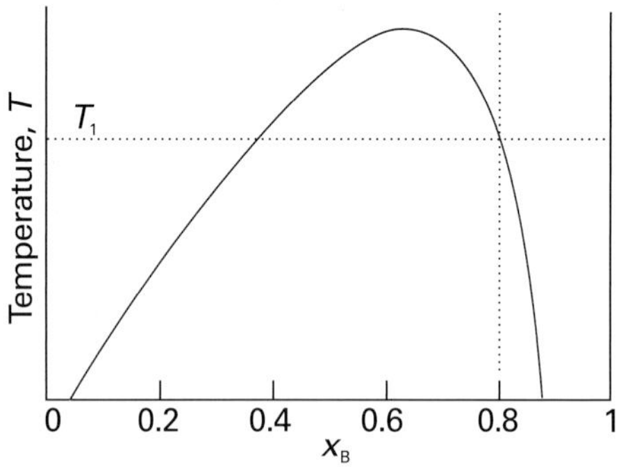

Fig. 6.39

6.10(b) Figure 6.40 is the phase diagram for silver and tin. Label the regions, and describe what will be observed when liquids of compositions a and b are cooled to 200 K.

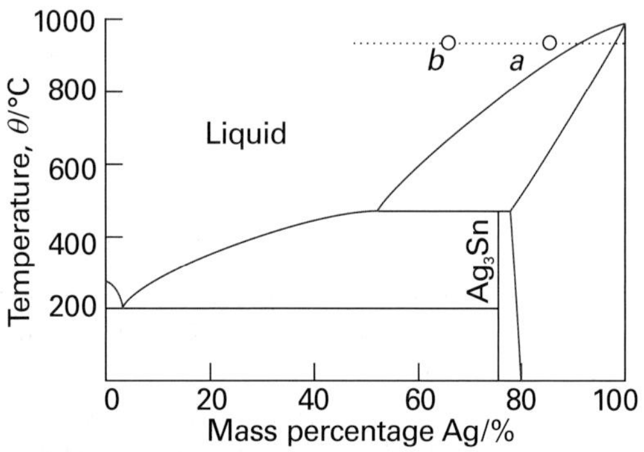

Fig. 6.40

6.11(a) Indicate on the phase diagram in Fig. 6.41 the feature that denotes incongruent melting. What is the composition of the eutectic mixture and at what temperature does it melt?

6.11(b) Indicate on the phase diagram in Fig. 6.42 the feature that denotes incongruent melting. What is the composition of the eutectic mixture and at what temperature does it melt?

6.12(a) Sketch the cooling curves for the isopleths a and b in Fig. 6.41.

6.12(b) Sketch the cooling curves for the isopleths a and b in Fig. 6.42.

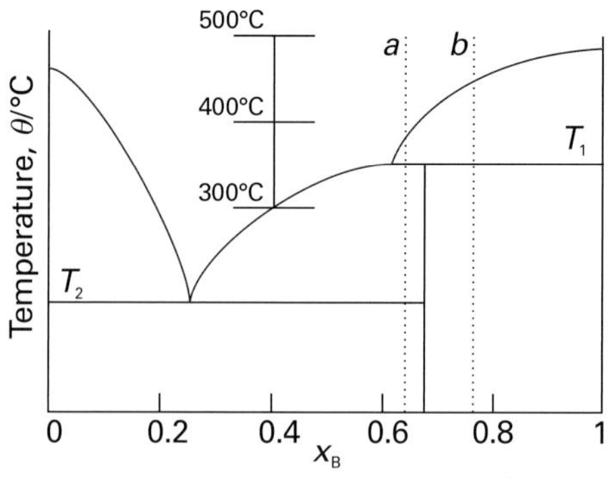

Fig. 6.41

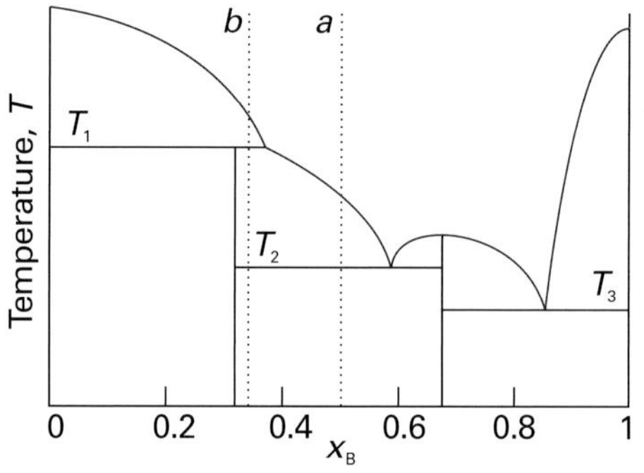

Fig. 6.42

6.13(a) Use the phase diagram in Fig. 6.40 to state (a) the solubility of Ag in Sn at 800°C and (b) the solubility of Ag_3Sn in Ag at 460°C, (c) the solubility of Ag_3Sn in Ag at 300°C.

6.13(b) Use the phase diagram in Fig. 6.41 to state (a) the solubility of B in A at 500°C and (b) the solubility of AB_2 in A at 390°C, (c) the solubility of AB_2 in B at 300°C.

6.14(a) Figure 6.43 shows the experimentally determined phase diagrams for the nearly ideal solution of hexane and heptane. (a) Label the regions of the diagrams as to which phases are present. (b) For a solution containing 1 mol each of hexane and heptane, estimate the vapour pressure at 70°C when vaporization on reduction of the external pressure just begins. (c) What is the vapour pressure of the solution at 70°C when just one drop of liquid remains. (d) Estimate from the figures the mole fraction of hexane in the liquid and vapour phases for the conditions of part b. (e) What are the mole fractions for the conditions of part c? (f) At 85°C and 760 Torr, what are the amounts of substance in the liquid and vapour phases when $z_{heptane} = 0.40$?

6.14(b) Uranium tetrafluoride and zirconium tetrafluoride melt at 1035°C and 912°C, respectively. They form a continuous series of solid solutions with a minimum melting temperature of 765°C and composition $x(ZrF_4) = 0.77$. At 900°C, the liquid solution of composition $x(ZrF_4) = 0.28$ is in equilibrium with a solid solution of composition $x(ZrF_4) = 0.14$. At 850°C the two compositions are 0.87 and 0.90, respectively. Sketch the phase diagram for this system and state what is observed when a liquid of composition $x(ZrF_4) = 0.40$ is cooled slowly from 900°C to 500°C.

6.15(a) Methane (melting point 91 K) and tetrafluoromethane (melting point 89 K) do not form solid solutions with each other, and as liquids they are only partially miscible. The upper critical temperature of the liquid mixture is 94 K at $x(CF_4) = 0.43$ and the eutectic temperature is 84 K at $x(CF_4) = 0.88$. At 86 K, the phase in equilibrium with the tetrafluoromethane-rich

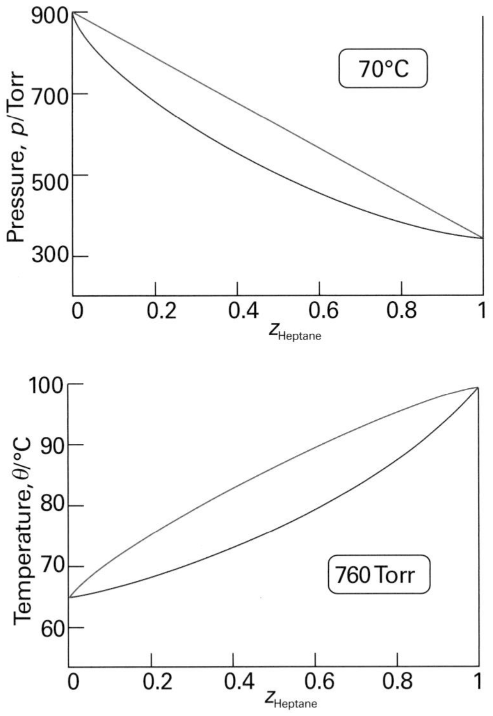

Fig. 6.43

solution changes from solid methane to a methane-rich liquid. At that temperature, the two liquid solutions that are in mutual equilibrium have the compositions $x(CF_4) = 0.10$ and $x(CF_4) = 0.80$. Sketch the phase diagram.

6.15(b) Describe the phase changes that take place when a liquid mixture of 4.0 mol B_2H_6 (melting point 131 K) and 1.0 mol CH_3OCH_3 (melting point 135 K) is cooled from 140 K to 90 K. These substances form a compound $(CH_3)_2OB_2H_6$ that melts congruently at 133 K. The system exhibits one eutectic at $x(B_2H_6) = 0.25$ and 123 K and another at $x(B_2H_6) = 0.90$ and 104 K.

6.16(a) Refer to the information in Exercise 6.15(b) and sketch the cooling curves for liquid mixtures in which $x(B_2H_6)$ is (a) 0.10, (b) 0.30, (c) 0.50, (d) 0.80, and (e) 0.95.

6.16(b) Refer to the information in Exercise 6.15(a) and sketch the cooling curves for liquid mixtures in which $x(CF_4)$ is (a) 0.10, (b) 0.30, (c) 0.50, (d) 0.80, and (e) 0.95.

6.17(a) Hexane and perfluorohexane show partial miscibility below 22.70°C. The critical concentration at the upper critical temperature is $x = 0.355$, where x is the mole fraction of C_6F_{14}. At 22.0°C the two solutions in equilibrium have $x = 0.24$ and $x = 0.48$, respectively, and at 21.5°C the mole fractions are 0.22 and 0.51. Sketch the phase diagram. Describe the phase changes that occur when perfluorohexane is added to a fixed amount of hexane at (a) 23°C, (b) 22°C.

6.17(b) Two liquids, A and B, show partial miscibility below 52.4°C. The critical concentration at the upper critical temperature is $x = 0.459$, where x is the mole fraction of A. At 40.0°C the two solutions in equilibrium have $x = 0.22$ and $x = 0.60$, respectively, and at 42.5°C the mole fractions are 0.24 and 0.48. Sketch the phase diagram. Describe the phase changes that occur when B is added to a fixed amount of A at (a) 48°C, (b) 52.4°C.

Problems*

Numerical problems

6.1‡ 1-Butanol and chlorobenzene form a minimum-boiling azeotropic system. The mole fraction of 1-butanol in the liquid (x) and vapour (y) phases at 1.000 atm is given below for a variety of boiling temperatures (H. Artigas, C. Lafuente, P. Cea, F.M. Royo, and J.S. Urieta, *J. Chem. Eng. Data* **42**, 132 (1997)).

T/K	396.57	393.94	391.60	390.15	389.03	388.66	388.57
x	0.1065	0.1700	0.2646	0.3687	0.5017	0.6091	0.7171
y	0.2859	0.3691	0.4505	0.5138	0.5840	0.6409	0.7070

Pure chlorobenzene boils at 404.86 K. (a) Construct the chlorobenzene-rich portion of the phase diagram from the data. (b) Estimate the temperature at which a solution whose mole fraction of 1-butanol is 0.300 begins to boil. (c) State the compositions and relative proportions of the two phases present after a solution initially 0.300 1-butanol is heated to 393.94 K.

6.2‡ An *et al.* investigated the liquid–liquid coexistence curve of *N,N*-dimethylacetamide and heptane (X. An, H. Zhao, F. Fuguo, and W. Shen, *J. Chem. Thermodynamics* **28**, 1221 (1996)). Mole fractions of *N,N*-dimethylacetamide in the upper (x_1) and lower (x_2) phases of a two-phase region are given below as a function of temperature:

T/K	309.820	309.422	309.031	308.006	306.686
x_1	0.473	0.400	0.371	0.326	0.293
x_2	0.529	0.601	0.625	0.657	0.690
T/K	304.553	301.803	299.097	296.000	294.534
x_1	0.255	0.218	0.193	0.168	0.157
x_2	0.724	0.758	0.783	0.804	0.814

(a) Plot the phase diagram. (b) State the proportions and compositions of the two phases that form from mixing 0.750 mol of *N,N*-dimethylacetamide with 0.250 mol of heptane at 296.0 K. To what temperature must the mixture be heated to form a single-phase mixture?

6.3‡ The following data have been obtained for the liquid–vapour equilibrium compositions of mixtures of nitrogen and oxygen at 100 kPa.

T/K	77.3	78	80	82	84	86	88	90.2
$x(O_2)$	0	10	34	54	70	82	92	100
$y(O_2)$	0	2	11	22	35	52	73	100
$p^*(O_2)$/Torr	154	171	225	294	377	479	601	760

Plot the data on a temperature–composition diagram and determine the extent to which it fits the predictions for an ideal solution by calculating the activity coefficients of O_2 at each composition.

6.4 Phosphorus and sulfur form a series of binary compounds. The best characterized are P_4S_3, P_4S_7, and P_4S_{10}, all of which melt congruently. Assuming that only these three binary compounds of the two elements exist, (a) draw schematically only the P/S phase diagram. Label each region of the diagram with the substance that exists in that region and indicate its phase. Label the horizontal axis as x_S and give the numerical values of x_S that correspond to the compounds. The melting point of pure phosphorus is 44°C and that of pure sulfur is 119°C. (b) Draw, schematically, the cooling curve for a mixture of composition $x_S = 0.28$. Assume that a eutectic occurs at $x_S = 0.2$ and negligible solid–solid solubility.

6.5 The table below gives the break and halt temperatures found in the cooling curves of two metals A and B. Construct a phase diagram consistent

* Problems denoted with the symbol ‡ were supplied by Charles Trapp, Carmen Giunta, and Marshall Cady.

with the data of these curves. Label the regions of the diagram, stating what phases and substances are present. Give the probable formulas of any compounds that form.

$100x_B$	$\theta_{break}/°C$	$\theta_{halt,1}/°C$	$\theta_{halt,2}/°C$
0		1100	
10.0	1060	700	
20.0	1000	700	
30.0	940	700	400
40.0	850	700	400
50.0	750	700	400
60.0	670	400	
70.0	550	400	
80.0		400	
90.0	450	400	
100.0		500	

6.6 Consider the phase diagram in Fig. 6.44, which represents a solid–liquid equilibrium. Label all regions of the diagram according to the chemical species that exist in that region and their phases. Indicate the number of species and phases present at the points labelled *b*, *d*, *e*, *f*, *g*, and *k*. Sketch cooling curves for compositions $x_B = 0.16, 0.23, 0.57, 0.67,$ and 0.84.

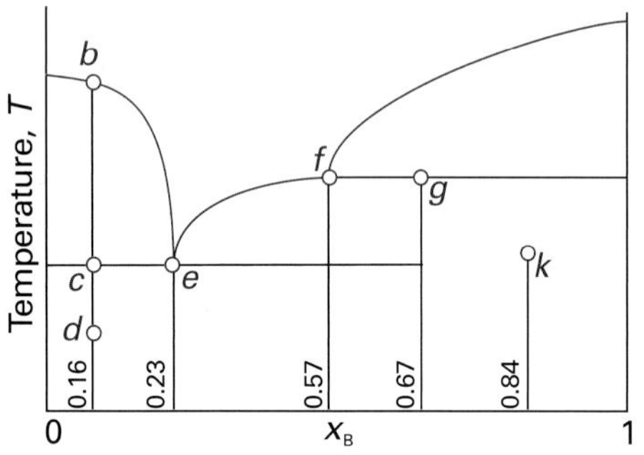

Fig. 6.44

6.7 Sketch the phase diagram for the Mg/Cu system using the following information: $\theta_f(Mg) = 648°C$, $\theta_f(Cu) = 1085°C$; two intermetallic compounds are formed with $\theta_f(MgCu_2) = 800°C$ and $\theta_f(Mg_2Cu) = 580°C$; eutectics of mass percentage Mg composition and melting points 10 per cent (690°C), 33 per cent (560°C), and 65 per cent (380°C). A sample of Mg/Cu alloy containing 25 per cent Mg by mass was prepared in a crucible heated to 800°C in an inert atmosphere. Describe what will be observed if the melt is cooled slowly to room temperature. Specify the composition and relative abundances of the phases and sketch the cooling curve.

6.8‡ Figure 6.45 shows $\Delta_{mix}G(x_{Pb}, T)$ for a mixture of copper and lead. (a) What does the graph reveal about the miscibility of copper and lead and the spontaneity of solution formation? What is the variance (F) at (i) 1500 K, (ii) 1100 K? (b) Suppose that at 1500 K a mixture of composition (i) $x_{Pb} = 0.1$, (ii) $x_{Pb} = 0.7$, is slowly cooled to 1100 K. What is the equilibrium composition of the final mixture? Include an estimate of the relative amounts of each phase. (c) What is the solubility of (i) lead in copper, (ii) copper in lead at 1100 K?

6.9‡ The temperature–composition diagram for the Ca/Si binary system is shown in Fig. 6.46. (a) Identify eutectics, congruent melting compounds, and incongruent melting compounds. (b) If a 20 per cent by atom composition melt of silicon at 1500°C is cooled to 1000°C, what phases (and phase composition) would be at equilibrium? Estimate the relative amounts of each phase. (c) Describe the equilibrium phases observed when an 80 per cent by atom composition Si melt is cooled to 1030°C. What phases, and relative amounts, would be at equilibrium at a temperature (i) slightly higher than 1030°C, (ii) slightly lower than 1030°C? Draw a graph of the mole percentages of both Si(s) and $CaSi_2$(s) as a function of mole percentage of melt that is freezing at 1030°C.

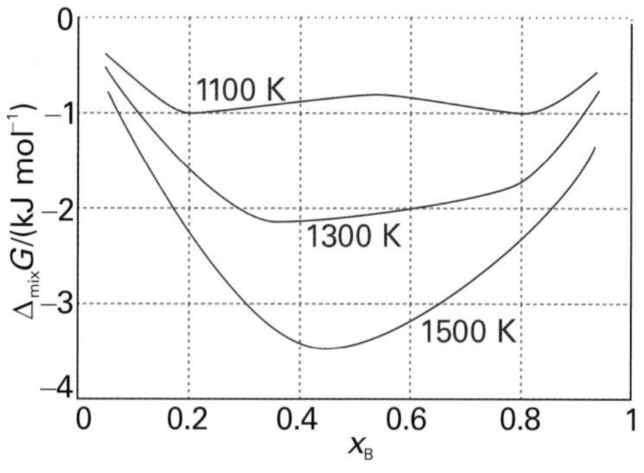

Fig. 6.45

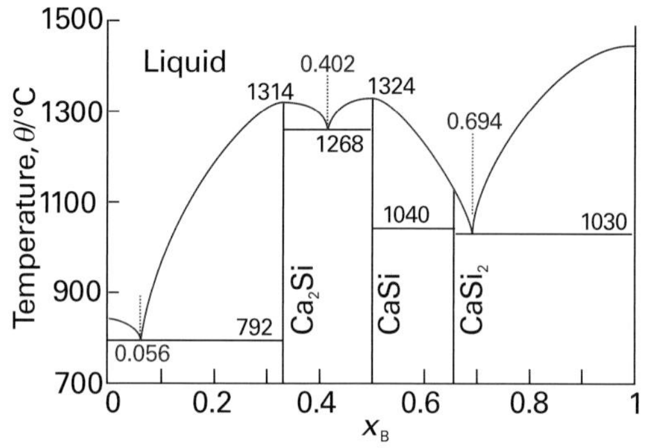

Fig. 6.46

6.10 Iron(II) chloride (melting point 677°C) and potassium chloride (melting point 776°C) form the compounds $KFeCl_3$ and K_2FeCl_4 at elevated temperatures. $KFeCl_3$ melts congruently at 380°C and K_2FeCl_4 melts incongruently at 399°C. Eutectics are formed with compositions $x = 0.38$ (melting point 351°C) and $x = 0.54$ (melting point 393°C), where x is the mole fraction of $FeCl_2$. The KCl solubility curve intersects the K_2FeCl_4 curve at $x = 0.34$. Sketch the phase diagram. State the phases that are in equilibrium when a mixture of composition $x = 0.36$ is cooled from 400°Cto 300°C.

Theoretical problems

6.11 Show that two phases are in thermal equilibrium only if their temperatures are the same.

6.12 Show that two phases are in mechanical equilibrium only if their pressures are equal.

Applications: to biology, materials science, and chemical engineering

6.13 The unfolding, or *denaturation*, of a biological macromolecule may be brought about by treatment with substances, called *denaturants*, that disrupt the intermolecular interactions responsible for the native three-dimensional conformation of the polymer. For example, urea, $CO(NH_2)_2$, competes for NH and CO groups and interferes with hydrogen bonding in a polypeptide.

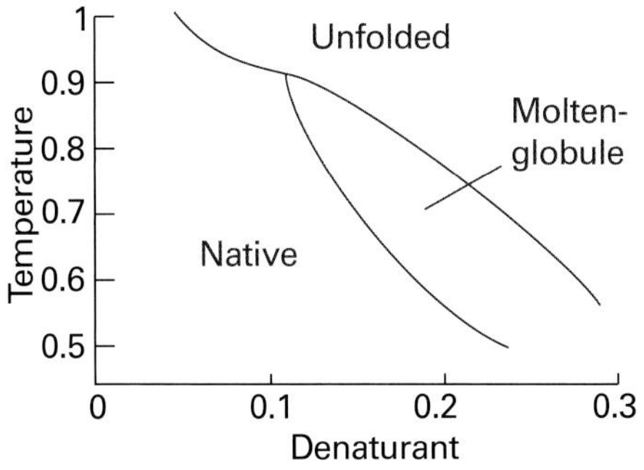

Fig. 6.47

In a theoretical study of a protein, the temperature–composition diagram shown in Fig. 6.47 was obtained. It shows three structural regions: the native form, the unfolded form, and a 'molten globule' form, a partially unfolded but still compact form of the protein. (i) Is the molten globule form ever stable when the denaturant concentration is below 0.1? (ii) Describe what happens to the polymer as the native form is heated in the presence of denaturant at concentration 0.15.

6.14 The basic structural element of a membrane is a phospholipid, such as phosphatidyl choline, which contains long hydrocarbon chains (typically in the range C_{14}–C_{24}) and a variety of polar groups, such as $-CH_2CH_2N(CH_3)_3^+$. The hydrophobic chains stack together to form an extensive bilayer about 5 nm across, leaving the polar groups exposed to the aqueous environment on either side of the membrane (see Chapter 19 for details). All lipid bilayers undergo a transition from a state of low chain mobility (the *gel* form) to high chain mobility (the *liquid crystal* form) at a temperature that depends on the structure of the lipid. Biological cell membranes exist as liquid crystals at physiological temperatures. In an experimental study of membrane-like assemblies, a phase diagram like that shown in Fig. 6.48 was obtained. The two components are dielaidoylphosphatidylcholine (DEL) and dipalmitoylphosphatidylcholine (DPL). Explain what happens as a liquid mixture of composition $x_{DEL} = 0.5$ is cooled from 45°C.

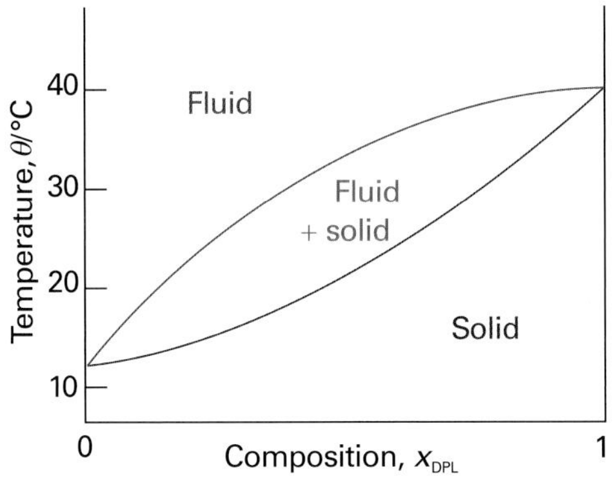

Fig. 6.48

6.15 The compound *p*-azoxyanisole forms a liquid crystal. 5.0 g of the solid was placed in a tube, which was then evacuated and sealed. Use the phase rule to prove that the solid will melt at a definite temperature and that the liquid crystal phase will make a transition to a normal liquid phase at a definite temperature.

6.16 Some polymers can form liquid crystal mesophases with unusual physical properties. For example, liquid crystalline Kevlar (**3**) is strong enough to be the material of choice for bulletproof vests and is stable at temperatures up to 600 K. What molecular interactions contribute to the formation, thermal stability, and mechanical strength of liquid crystal mesophases in Kevlar?

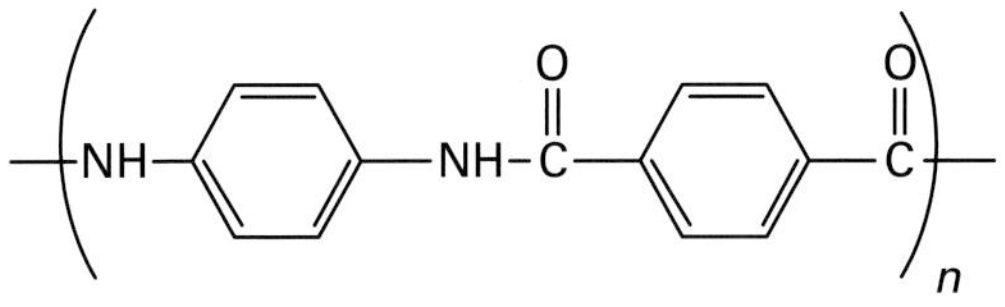

3

6.17 Use a phase diagram like that shown in Fig. 6.36 to indicate how zone levelling may be described.

6.18 The technique of *float zoning*, which is similar to zone refining (*Impact I6.2*), has produced very pure samples of silicon for use in the semiconductor industry. Consult a textbook of materials science or metallurgy and prepare a discussion of the principles, advantages, and disadvantages of float zoning.

6.19 Magnesium oxide and nickel oxide withstand high temperatures. However, they do melt when the temperature is high enough and the behaviour of mixtures of the two is of considerable interest to the ceramics industry. Draw the temperature–composition diagram for the system using the data below, where *x* is the mole fraction of MgO in the solid and *y* its mole fraction in the liquid.

θ/°C	1960	2200	2400	2600	2800
x	0	0.35	0.60	0.83	1.00
y	0	0.18	0.38	0.65	1.00

State (a) the melting point of a mixture with $x = 0.30$, (b) the composition and proportion of the phases present when a solid of composition $x = 0.30$ is heated to 2200°C, (c) the temperature at which a liquid of composition $y = 0.70$ will begin to solidify.

6.20 The bismuth–cadmium phase diagram is of interest in metallurgy, and its general form can be estimated from expressions for the depression of freezing point. Construct the diagram using the following data: $T_f(\text{Bi}) = 544.5$ K, $T_f(\text{Cd}) = 594$ K, $\Delta_{\text{fus}}H(\text{Bi}) = 10.88$ kJ mol^{-1}, $\Delta_{\text{fus}}H(\text{Cd}) = 6.07$ kJ mol^{-1}. The metals are mutually insoluble as solids. Use the phase diagram to state what would be observed when a liquid of composition $x(\text{Bi}) = 0.70$ is cooled slowly from 550 K. What are the relative abundances of the liquid and solid at (a) 460 K and (b) 350 K? Sketch the cooling curve for the mixture.

6.21‡ Carbon dioxide at high pressure is used to separate various compounds in citrus oil. The mole fraction of CO_2 in the liquid (x) and vapour (y) at 323.2 K is given below for a variety of pressures (Y. Iwai, T. Morotomi, K. Sakamoto, Y. Koga, and Y. Arai, *J. Chem. Eng. Data* **41**, 951 (1996)).

p/MPa	3.94	6.02	7.97	8.94	9.27
x	0.2873	0.4541	0.6650	0.7744	0.8338
y	0.9982	0.9980	0.9973	0.9958	0.9922

(a) Plot the portion of the phase diagram represented by these data. (b) State the compositions and relative proportions of the two phases present after an equimolar gas mixture is compressed to 6.02 MPa at 323.2 K.

7 Chemical equilibrium

This chapter develops the concept of chemical potential and shows how it is used to account for the equilibrium composition of chemical reactions. The equilibrium composition corresponds to a minimum in the Gibbs energy plotted against the extent of reaction, and by locating this minimum we establish the relation between the equilibrium constant and the standard Gibbs energy of reaction. The thermodynamic formulation of equilibrium enables us to establish the quantitative effects of changes in the conditions. The principles of thermodynamics established in the preceding chapters can be applied to the description of the thermodynamic properties of reactions that take place in electrochemical cells, in which, as the reaction proceeds, it drives electrons through an external circuit. Thermodynamic arguments can be used to derive an expression for the electric potential of such cells and the potential can be related to their composition. There are two major topics developed in this connection. One is the definition and tabulation of standard potentials; the second is the use of these standard potentials to predict the equilibrium constants and other thermodynamic properties of chemical reactions.

Chemical reactions tend to move towards a dynamic equilibrium in which both reactants and products are present but have no further tendency to undergo net change. In some cases, the concentration of products in the equilibrium mixture is so much greater than that of the unchanged reactants that for all practical purposes the reaction is 'complete'. However, in many important cases the equilibrium mixture has significant concentrations of both reactants and products. In this chapter we see how to use thermodynamics to predict the equilibrium composition under any reaction conditions. Because many reactions of ions involve the transfer of electrons, they can be studied (and utilized) by allowing them to take place in an electrochemical cell. Measurements like those described in this chapter provide data that are very useful for discussing the characteristics of electrolyte solutions and of ionic equilibria in solution.

Spontaneous chemical reactions

We have seen that the direction of spontaneous change at constant temperature and pressure is towards lower values of the Gibbs energy, *G*. The idea is entirely general, and in this chapter we apply it to the discussion of chemical reactions.

7.1 The Gibbs energy minimum

We locate the equilibrium composition of a reaction mixture by calculating the Gibbs energy of the reaction mixture and identifying the composition that corresponds to minimum *G*.

(a) The reaction Gibbs energy

Consider the equilibrium A ⇌ B. Even though this reaction looks trivial, there are many examples of it, such as the isomerization of pentane to 2-methylbutane and the conversion of L-alanine to D-alanine. Suppose an infinitesimal amount $d\xi$ of A turns into B, then the change in the amount of A present is $dn_A = -d\xi$ and the change in the amount of B present is $dn_B = +d\xi$. The quantity ξ (xi) is called the **extent of reaction**; it has the dimensions of amount of substance and is reported in moles. When the extent of reaction changes by a finite amount $\Delta\xi$, the amount of A present changes from $n_{A,0}$ to $n_{A,0} - \Delta\xi$ and the amount of B changes from $n_{B,0}$ to $n_{B,0} + \Delta\xi$. So, if initially 2.0 mol A is present and we wait until $\Delta\xi = +1.5$ mol, then the amount of A remaining will be 0.5 mol.

The **reaction Gibbs energy**, $\Delta_r G$, is defined as the slope of the graph of the Gibbs energy plotted against the extent of reaction:

$$\Delta_r G = \left(\frac{\partial G}{\partial \xi}\right)_{p,T} \qquad [7.1]$$

Although Δ normally signifies a *difference* in values, here Δ_r signifies a *derivative*, the slope of G with respect to ξ. However, to see that there is a close relationship with the normal usage, suppose the reaction advances by $d\xi$. The corresponding change in Gibbs energy is

$$dG = \mu_A dn_A + \mu_B dn_B = -\mu_A d\xi + \mu_B d\xi = (\mu_B - \mu_A)d\xi$$

This equation can be reorganized into

$$\left(\frac{\partial G}{\partial \xi}\right)_{p,T} = \mu_B - \mu_A$$

That is,

$$\Delta_r G = \mu_B - \mu_A \qquad (7.2)$$

We see that $\Delta_r G$ can also be interpreted as the difference between the chemical potentials (the partial molar Gibbs energies) of the reactants and products *at the composition of the reaction mixture*.

Because chemical potential varies with composition, the slope of the plot of Gibbs energy against extent of reaction changes as the reaction proceeds. Moreover, because the reaction runs in the direction of decreasing G (that is, down the slope of G plotted against ξ), we see from eqn 7.2 that the reaction A → B is spontaneous when $\mu_A > \mu_B$, whereas the reverse reaction is spontaneous when $\mu_B > \mu_A$. The slope is zero, and the reaction is spontaneous in neither direction, when

$$\Delta_r G = 0 \qquad (7.3)$$

This condition occurs when $\mu_B = \mu_A$ (Fig. 7.1). It follows that, if we can find the composition of the reaction mixture that ensures $\mu_B = \mu_A$, then we can identify the composition of the reaction mixture at equilibrium.

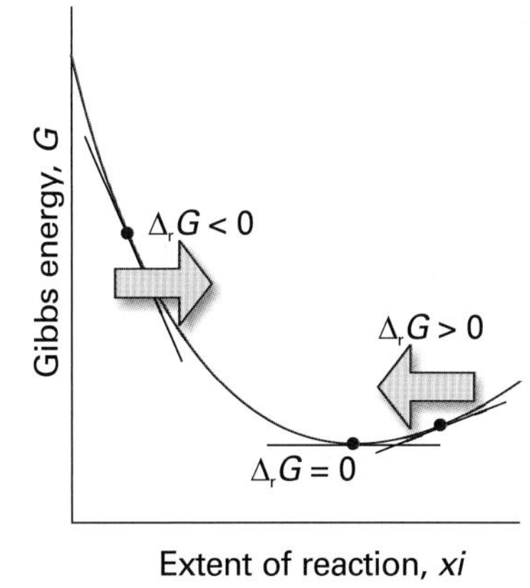

Fig. 7.1 As the reaction advances (represented by motion from left to right along the horizontal axis) the slope of the Gibbs energy changes. Equilibrium corresponds to zero slope, at the foot of the valley.

(b) Exergonic and endergonic reactions

We can express the spontaneity of a reaction at constant temperature and pressure in terms of the reaction Gibbs energy:

If $\Delta_r G < 0$, the forward reaction is spontaneous.

If $\Delta_r G > 0$, the reverse reaction is spontaneous.

If $\Delta_r G = 0$, the reaction is at equilibrium.

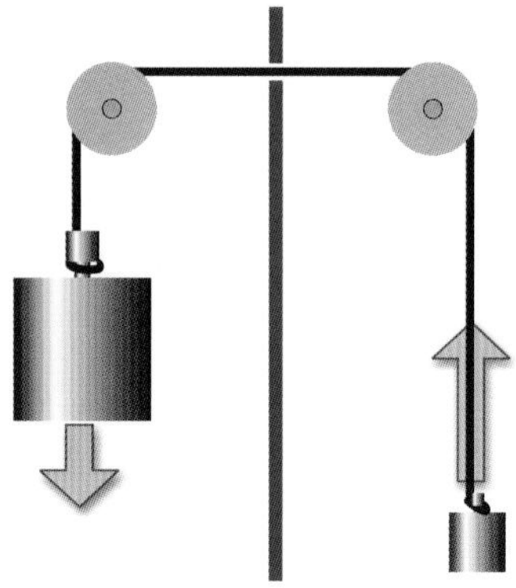

Fig. 7.2 If two weights are coupled as shown here, then the heavier weight will move the lighter weight in its non-spontaneous direction: overall, the process is still spontaneous. The weights are the analogues of two chemical reactions: a reaction with a large negative ΔG can force another reaction with a less ΔG to run in its non-spontaneous direction.

A reaction for which $\Delta_r G < 0$ is called **exergonic** (from the Greek words for work producing). The name signifies that, because the process is spontaneous, it can be used to drive another process, such as another reaction, or used to do non-expansion work. A simple mechanical analogy is a pair of weights joined by a string (Fig. 7.2): the lighter of the pair of weights will be pulled up as the heavier weight falls down. Although the lighter weight has a natural tendency to move downward, its coupling to the heavier weight results in it being raised. In biological cells, the oxidation of carbohydrates act as the heavy weight that drives other reactions forward and results in the formation of proteins from amino acids, muscle contraction, and brain activity. A reaction for which $\Delta_r G > 0$ is called **endergonic** (signifying work consuming). The reaction can be made to occur only by doing work on it, such as electrolysing water to reverse its spontaneous formation reaction. Reactions at equilibrium are spontaneous in neither direction: they are neither exergonic nor endergonic.

7.2 The description of equilibrium

With the background established, we are now ready to see how to apply thermodynamics to the description of chemical equilibrium.

(a) Perfect gas equilibria

When A and B are perfect gases we can use eqn 5.14 ($\mu = \mu^\ominus + RT \ln p$, with p interpreted as $p/p^\ominus$) to write

$$\Delta_r G = \mu_B - \mu_A = (\mu_B^\ominus + RT \ln p_B) - (\mu_A^\ominus + RT \ln p_A)$$

$$= \Delta_r G^\ominus + RT \ln \frac{p_B}{p_A} \qquad (7.4)^\circ$$

If we denote the ratio of partial pressures by Q, we obtain

$$\Delta_r G = \Delta_r G^\ominus + RT \ln Q \qquad Q = \frac{p_B}{p_A} \qquad (7.5)^\circ$$

Comment 7.1

Note that in the definition of $\Delta_r G^\ominus$, the Δ_r has its normal meaning as a difference whereas in the definition of $\Delta_r G$ the Δ_r signifies a derivative.

The ratio Q is an example of a **reaction quotient**. It ranges from 0 when $p_B = 0$ (corresponding to pure A) to infinity when $p_A = 0$ (corresponding to pure B). The **standard reaction Gibbs energy**, $\Delta_r G^\ominus$, is defined (like the standard reaction enthalpy) as the difference in the standard molar Gibbs energies of the reactants and products. For our reaction

$$\Delta_r G^\ominus = G^\ominus_{B,m} - G^\ominus_{A,m} = \mu_B^\ominus - \mu_A^\ominus \qquad (7.6)$$

In Section 3.6 we saw that the difference in standard molar Gibbs energies of the products and reactants is equal to the difference in their standard Gibbs energies of formation, so in practice we calculate $\Delta_r G^\ominus$ from

$$\Delta_r G^\ominus = \Delta_f G^\ominus(B) - \Delta_f G^\ominus(A) \qquad (7.7)$$

At equilibrium $\Delta_r G = 0$. The ratio of partial pressures at equilibrium is denoted K, and eqn 7.5 becomes

$$0 = \Delta_r G^\ominus + RT \ln K$$

This expression rearranges to

$$RT \ln K = -\Delta_r G^\ominus \qquad K = \left(\frac{p_B}{p_A}\right)_{\text{equilibrium}} \qquad (7.8)^\circ$$

This relation is a special case of one of the most important equations in chemical thermodynamics: it is the link between tables of thermodynamic data, such as those in the *Data section* at the end of this volume, and the chemically important **equilibrium constant**, K.

Molecular interpretation 7.1 *The approach to equilibrium*

In molecular terms, the minimum in the Gibbs energy, which corresponds to $\Delta_r G = 0$, stems from the Gibbs energy of mixing of the two gases. Hence, an important contribution to the position of chemical equilibrium is the mixing of the products with the reactants as the products are formed.

Consider a hypothetical reaction in which A molecules change into B molecules without mingling together. The Gibbs energy of the system changes from $G^\ominus(\mathrm{A})$ to $G^\ominus(\mathrm{B})$ in proportion to the amount of B that had been formed, and the slope of the plot of G against the extent of reaction is a constant and equal to $\Delta_r G^\ominus$ at all stages of the reaction (Fig. 7.3). There is no intermediate minimum in the graph. However, in fact, the newly produced B molecules do mix with the surviving A molecules. We have seen that the contribution of a mixing process to the change in Gibbs energy is given by eqn 5.27 ($\Delta_{\mathrm{mix}} G = nRT(x_A \ln x_A + x_B \ln x_B)$). This expression makes a U-shaped contribution to the total change in Gibbs energy. As can be seen from Fig. 7.3, there is now an intermediate minimum in the Gibbs energy, and its position corresponds to the equilibrium composition of the reaction mixture.

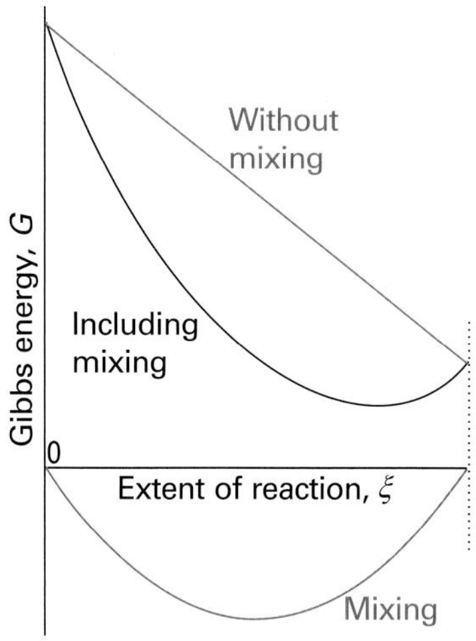

Fig. 7.3 If the mixing of reactants and products is ignored, then the Gibbs energy changes linearly from its initial value (pure reactants) to its final value (pure products) and the slope of the line is $\Delta_r G^\ominus$. However, as products are produced, there is a further contribution to the Gibbs energy arising from their mixing (lowest curve). The sum of the two contributions has a minimum. That minimum corresponds to the equilibrium composition of the system.

We see from eqn 7.8 that, when $\Delta_r G^\ominus > 0$, the equilibrium constant $K < 1$. Therefore, at equilibrium the partial pressure of A exceeds that of B, which means that the reactant A is favoured in the equilibrium. When $\Delta_r G^\ominus < 0$, the equilibrium constant $K > 1$, so at equilibrium the partial pressure of B exceeds that of A. Now the product B is favoured in the equilibrium.

(b) The general case of a reaction

We can easily extend the argument that led to eqn 7.8 to a general reaction. First, we need to generalize the concept of extent of reaction.

Consider the reaction $2\,\mathrm{A} + \mathrm{B} \rightarrow 3\,\mathrm{C} + \mathrm{D}$. A more sophisticated way of expressing the chemical equation is to write it in the symbolic form

$$0 = 3\,\mathrm{C} + \mathrm{D} - 2\,\mathrm{A} - \mathrm{B}$$

by subtracting the reactants from both sides (and replacing the arrow by an equals sign). This equation has the form

$$0 = \sum_J \nu_J \mathrm{J} \qquad (7.9)$$

where J denotes the substances and the ν_J are the corresponding **stoichiometric numbers** in the chemical equation. In our example, these numbers have the values $\nu_A = -2$, $\nu_B = -1$, $\nu_C = +3$, and $\nu_D = +1$. A stoichiometric number is positive for products and negative for reactants. Then we define ξ so that, if it changes by $\Delta\xi$, then the change in the amount of any species J is $\nu_J \Delta\xi$.

Illustration 7.1 *Identifying stoichiometric numbers*

To express the equation

$$\mathrm{N_2(g)} + 3\,\mathrm{H_2(g)} \rightarrow 2\,\mathrm{NH_3(g)} \qquad (7.10)$$

in the notation of eqn 7.9, we rearrange it to

$$0 = 2\,NH_3(g) - N_2(g) + 3\,H_2(g)$$

and then identify the stoichiometric numbers as $\nu_{N_2} = -1$, $\nu_{H_2} = -3$, and $\nu_{NH_3} = +2$. Therefore, if initially there is 10 mol N_2 present, then when the extent of reaction changes from $\xi = 0$ to $\xi = 1$ mol, implying that $\Delta\xi = +1$ mol, the amount of N_2 changes from 10 mol to 9 mol. All the N_2 has been consumed when $\xi = 10$ mol. When $\Delta\xi = +1$ mol, the amount of H_2 changes by $-3 \times (1\text{ mol}) = -3$ mol and the amount of NH_3 changes by $+2 \times (1\text{ mol}) = +2$ mol.

A note on good practice Stoichiometric *numbers* may be positive or negative; stoichiometric *coefficients* are always positive. Few, however, make the distinction between the two types of quantity.

The reaction Gibbs energy, $\Delta_r G$, is defined in the same way as before, eqn 7.1. In the *Justification* below, we show that the Gibbs energy of reaction can always be written

$$\Delta_r G = \Delta_r G^{\ominus} + RT \ln Q \qquad (7.11)$$

with the standard reaction Gibbs energy calculated from

$$\Delta_r G^{\ominus} = \sum_{\text{Products}} \nu \Delta_f G^{\ominus} - \sum_{\text{Reactants}} \nu \Delta_f G^{\ominus} \qquad (7.12a)$$

or, more formally,

$$\Delta_r G^{\ominus} = \sum_J \nu_J \Delta_f G^{\ominus}(J) \qquad (7.12b)$$

The reaction quotient, Q, has the form

$$Q = \frac{\text{activities of products}}{\text{activities of reactants}} \qquad (7.13a)$$

with each species raised to the power given by its stoichiometric coefficient. More formally, to write the general expression for Q we introduce the symbol Π to denote the product of what follows it (just as Σ denotes the sum), and define Q as

$$Q = \prod_J a_J^{\nu_J} \qquad [7.13b]$$

Because reactants have negative stoichiometric numbers, they automatically appear as the denominator when the product is written out explicitly. Recall from Table 5.3 that, for pure solids and liquids, the activity is 1, so such substances make no contribution to Q even though they may appear in the chemical equation.

Illustration 7.2 *Writing a reaction quotient*

Consider the reaction $2\,A + 3\,B \rightarrow C + 2\,D$, in which case $\nu_A = -2$, $\nu_B = -3$, $\nu_C = +1$, and $\nu_D = +2$. The reaction quotient is then

$$Q = a_A^{-2} a_B^{-3} a_C a_D^2 = \frac{a_C a_D^2}{a_A^2 a_B^3}$$

Justification 7.1 *The dependence of the reaction Gibbs energy on the reaction quotient*

Consider the reaction with stoichiometric numbers ν_J. When the reaction advances by $d\xi$, the amounts of reactants and products change by $dn_J = \nu_J d\xi$. The resulting infinitesimal change in the Gibbs energy at constant temperature and pressure is

$$dG = \sum_J \mu_J dn_J = \sum_J \mu_J \nu_J d\xi = \left(\sum_J \nu_J \mu_J\right) d\xi \qquad (7.14)$$

It follows that

$$\Delta_r G = \left(\frac{\partial G}{\partial \xi}\right)_{p,T} = \sum_J \nu_J \mu_J \qquad (7.15)$$

To make further progress, we note that the chemical potential of a species J is related to its activity by eqn 5.25 ($\mu_J = \mu_J^{\ominus} + RT \ln a_J$). When this expression is substituted into eqn 7.15 we obtain

$$\begin{aligned}\Delta_r G &= \overbrace{\sum_J \nu_J \mu_J^{\ominus}}^{\Delta_r G^{\ominus}} + RT\sum_J \nu_J \ln a_J \\ &= \Delta_r G^{\ominus} + RT\sum_J \ln a_J^{\nu_J} = \Delta_r G^{\ominus} + RT \ln \overbrace{\prod_J a_J^{\nu_J}}^{Q} \\ &= \Delta_r G^{\ominus} + RT \ln Q\end{aligned}$$

with Q given by eqn 7.13b.

Comment 7.2
Recall that $a \ln x = \ln x^a$ and $\ln x + \ln y + \cdots = \ln xy\cdots$, so $\sum_i \ln x_i = \ln\left(\prod_i x_i\right)$.

Now we conclude the argument based on eqn 7.11. At equilibrium, the slope of G is zero: $\Delta_r G = 0$. The activities then have their equilibrium values and we can write

$$K = \left(\prod_J a_J^{\nu_J}\right)_{\text{equilibrium}} \qquad [7.16]$$

This expression has the same form as Q, eqn 7.13, but is evaluated using equilibrium activities. From now on, we shall not write the 'equilibrium' subscript explicitly, and will rely on the context to make it clear that for K we use equilibrium values and for Q we use the values at the specified stage of the reaction.

An equilibrium constant K expressed in terms of activities (or fugacities) is called a **thermodynamic equilibrium constant**. Note that, because activities are dimensionless numbers, the thermodynamic equilibrium constant is also dimensionless. In elementary applications, the activities that occur in eqn 7.16 are often replaced by the numerical values of molalities (that is, by replacing a_J by $b_J/b^{\ominus}$, where $b^{\ominus} = 1$ mol kg^{-1}), molar concentrations (that is, as $[J]/c^{\ominus}$, where $c^{\ominus} = 1$ mol dm^{-3}), or the numerical values of partial pressures (that is, by $p_J/p^{\ominus}$, where $p^{\ominus} = 1$ bar). In such cases, the resulting expressions are only approximations. The approximation is particularly severe for electrolyte solutions, for in them activity coefficients differ from 1 even in very dilute solutions (Section 5.9).

Illustration 7.3 *Writing an equilibrium constant.*

The equilibrium constant for the heterogeneous equilibrium $CaCO_3(s) \rightleftharpoons CaO(s) + CO_2(g)$ is

$$K = a_{CaCO_3(s)}^{-1} a_{CaO(s)} a_{CO_2(g)} = \frac{\overbrace{a_{CaO(s)}}^{1} a_{CO_2(g)}}{\underbrace{a_{CaCO_3(s)}}_{1}} = a_{CO_2}$$

(Table 5.3). Provided the carbon dioxide can be treated as a perfect gas, we can go on to write

$$K \approx p_{CO_2}/p^{\ominus}$$

and conclude that in this case the equilibrium constant is the numerical value of the decomposition vapour pressure of calcium carbonate.

Comment 7.3

In Chapter 17 we shall see that the right-hand side of eqn 7.17 may be expressed in terms of spectroscopic data for gas-phase species; so this expression also provides a link between spectroscopy and equilibrium composition.

At this point we set $\Delta_r G = 0$ in eqn 7.11 and replace Q by K. We immediately obtain

$$RT \ln K = -\Delta_r G^{\ominus} \tag{7.17}$$

This is an exact and highly important thermodynamic relation, for it enables us to predict the equilibrium constant of any reaction from tables of thermodynamic data, and hence to predict the equilibrium composition of the reaction mixture.

Example 7.1 *Calculating an equilibrium constant*

Calculate the equilibrium constant for the ammonia synthesis reaction, eqn 7.10, at 298 K and show how K is related to the partial pressures of the species at equilibrium when the overall pressure is low enough for the gases to be treated as perfect.

Method Calculate the standard reaction Gibbs energy from eqn 7.12 and convert it to the value of the equilibrium constant by using eqn 7.17. The expression for the equilibrium constant is obtained from eqn 7.16, and because the gases are taken to be perfect, we replace each activity by the ratio $p/p^{\ominus}$, where p is a partial pressure.

Answer The standard Gibbs energy of the reaction is

$$\begin{aligned}\Delta_r G^{\ominus} &= 2\Delta_f G^{\ominus}(NH_3, g) - \{\Delta_f G^{\ominus}(N_2, g) + 3\Delta_f G^{\ominus}(H_2, g)\}\\ &= 2\Delta_f G^{\ominus}(NH_3, g) = 2 \times (-16.5 \text{ kJ mol}^{-1})\end{aligned}$$

Then,

$$\ln K = -\frac{2 \times (-16.5 \times 10^3 \text{ J mol}^{-1})}{(8.3145 \text{ J K}^{-1} \text{ mol}^{-1}) \times (298 \text{ K})} = \frac{2 \times 16.5 \times 10^3}{8.3145 \times 298}$$

Hence, $K = 6.1 \times 10^5$. This result is thermodynamically exact. The thermodynamic equilibrium constant for the reaction is

$$K = \frac{a_{NH_3}^2}{a_{N_2} a_{H_2}^3}$$

and this ratio has exactly the value we have just calculated. At low overall pressures, the activities can be replaced by the ratios $p/p^{\ominus}$, where p is a partial pressure, and an approximate form of the equilibrium constant is

$$K=\frac{(p_{NH_3}/p^{\ominus})^2}{(p_{N_2}/p^{\ominus})(p_{H_2}/p^{\ominus})^3}=\frac{p_{NH_3}^2 p^{\ominus 2}}{p_{N_2}p_{H_2}^3}$$

Self-test 7.1 Evaluate the equilibrium constant for $N_2O_4(g) \rightleftharpoons 2\,NO_2(g)$ at 298 K. [$K=0.15$]

Example 7.2 *Estimating the degree of dissociation at equilibrium*

The degree of dissociation, α, is defined as the fraction of reactant that has decomposed; if the initial amount of reactant is n and the amount at equilibrium is n_{eq}, then $\alpha=(n-n_{eq})/n$. The standard Gibbs energy of reaction for the decomposition $H_2O(g) \rightarrow H_2(g)+\frac{1}{2}O_2(g)$ is $+118.08$ kJ mol^{-1} at 2300 K. What is the degree of dissociation of H_2O at 2300 K and 1.00 bar?

Method The equilibrium constant is obtained from the standard Gibbs energy of reaction by using eqn 7.17, so the task is to relate the degree of dissociation, α, to K and then to find its numerical value. Proceed by expressing the equilibrium compositions in terms of α, and solve for α in terms of K. Because the standard Gibbs energy of reaction is large and positive, we can anticipate that K will be small, and hence that $\alpha \ll 1$, which opens the way to making approximations to obtain its numerical value.

Answer The equilibrium constant is obtained from eqn 7.17 in the form

$$\ln K=-\frac{\Delta_r G^{\ominus}}{RT}=-\frac{(+118.08\times10^3\ \mathrm{J\ mol^{-1}})}{(8.3145\ \mathrm{J\ K^{-1}\ mol^{-1}})\times(2300\ \mathrm{K})}$$
$$=-\frac{118.08\times10^3}{8.3145\times2300}$$

It follows that $K=2.08\times10^{-3}$. The equilibrium composition can be expressed in terms of α by drawing up the following table:

	H_2O	H_2	O_2	
Initial amount	n	0	0	
Change to reach equilibrium	$-\alpha n$	$+\alpha n$	$+\frac{1}{2}\alpha n$	
Amount at equilibrium	$(1-\alpha)n$	αn	$\frac{1}{2}\alpha n$	Total: $(1+\frac{1}{2}\alpha)n$
Mole fraction, x_J	$\frac{1-\alpha}{1+\frac{1}{2}\alpha}$	$\frac{\alpha}{1+\frac{1}{2}\alpha}$	$\frac{\frac{1}{2}\alpha}{1+\frac{1}{2}\alpha}$	
Partial pressure, p_J	$\frac{(1-\alpha)p}{1+\frac{1}{2}\alpha}$	$\frac{\alpha p}{1+\frac{1}{2}\alpha}$	$\frac{\frac{1}{2}\alpha p}{1+\frac{1}{2}\alpha}$	

where, for the entries in the last row, we have used $p_J=x_J p$ (eqn 1.13). The equilibrium constant is therefore

$$K=\frac{p_{H_2}p_{O_2}^{1/2}}{p_{H_2O}}=\frac{\alpha^{3/2}p^{1/2}}{(1-\alpha)(2+\alpha)^{1/2}}$$

In this expression, we have written p in place of $p/p^{\ominus}$, to keep the notation simple. Now make the approximation that $\alpha \ll 1$, and hence obtain

$$K \approx \frac{\alpha^{3/2} p^{1/2}}{2^{1/2}}$$

Under the stated condition, $p = 1.00$ bar (that is, $p/p^{\ominus} = 1.00$), so $\alpha \approx (2^{1/2}K)^{2/3} = 0.0205$. That is, about 2 per cent of the water has decomposed.

A note on good practice Always check that the approximation is consistent with the final answer. In this case $\alpha \ll 1$ in accord with the original assumption.

Comment 7.4

The textbook's web site contains links to online tools for the estimation of equilibrium constants of gas-phase reactions.

Self-test 7.2 Given that the standard Gibbs energy of reaction at 2000 K is +135.2 kJ mol^{-1} for the same reaction, suppose that steam at 200 kPa is passed through a furnace tube at that temperature. Calculate the mole fraction of O_2 present in the output gas stream. [0.00221]

(c) The relation between equilibrium constants

The only remaining problem is to express the thermodynamic equilibrium constant in terms of the mole fractions, x_J, or molalities, b_J, of the species. To do so, we need to know the activity coefficients, and then to use $a_J = \gamma_J x_J$ or $a_J = \gamma_J b_J/b^{\ominus}$ (recalling that the activity coefficients depend on the choice). For example, in the latter case, for an equilibrium of the form A + B $\rightleftharpoons$ C + D, where all four species are solutes, we write

$$K = \frac{a_C a_D}{a_A a_B} = \frac{\gamma_C \gamma_D}{\gamma_A \gamma_B} \times \frac{b_C b_D}{b_A b_B} = K_\gamma K_b \qquad (7.18)$$

The activity coefficients must be evaluated at the equilibrium composition of the mixture (for instance, by using one of the Debye–Hückel expressions, Section 5.9), which may involve a complicated calculation, because the activity coefficients are known only if the equilibrium composition is already known. In elementary applications, and to begin the iterative calculation of the concentrations in a real example, the assumption is often made that the activity coefficients are all so close to unity that $K_\gamma = 1$. Then we obtain the result widely used in elementary chemistry that $K \approx K_b$, and equilibria are discussed in terms of molalities (or molar concentrations) themselves.

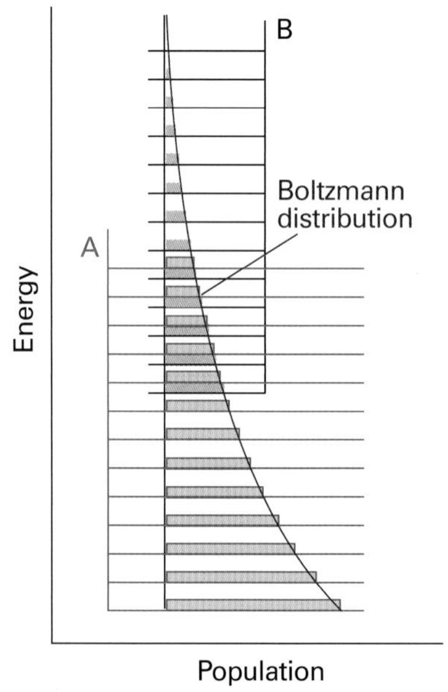

Fig. 7.4 The Boltzmann distribution of populations over the energy levels of two species A and B with similar densities of energy levels; the reaction A → B is endothermic in this example. The bulk of the population is associated with the species A, so that species is dominant at equilibrium.

Molecular interpretation 7.2 *The molecular origin of the equilibrium constant*

We can obtain a deeper insight into the origin and significance of the equilibrium constant by considering the Boltzmann distribution of molecules over the available states of a system composed of reactants and products (recall *Molecular interpretation 3.1*). When atoms can exchange partners, as in a reaction, the available states of the system include arrangements in which the atoms are present in the form of reactants and in the form of products: these arrangements have their characteristic sets of energy levels, but the Boltzmann distribution does not distinguish between their identities, only their energies. The atoms distribute themselves over both sets of energy levels in accord with the Boltzmann distribution (Fig. 7.4). At a given temperature, there will be a specific distribution of populations, and hence a specific composition of the reaction mixture.

It can be appreciated from the illustration that, if the reactants and products both have similar arrays of molecular energy levels, then the dominant species in a reaction mixture at equilibrium will be the species with the lower set of energy levels. However, the fact that the Gibbs energy occurs in the expression is a signal that entropy plays a role as well as energy. Its role can be appreciated by referring to Fig. 7.5. We see that, although the B energy levels lie higher than the A energy levels, in this instance they are much more closely spaced. As a result, their total population may be considerable and B could even dominate in the reaction mixture at equilibrium. Closely spaced energy levels correlate with a high entropy (see *Molecular interpretation 3.1*), so in this case we see that entropy effects dominate adverse energy effects. This competition is mirrored in eqn 7.17, as can be seen most clearly by using $\Delta_r G^\ominus = \Delta_r H^\ominus - T\Delta_r S^\ominus$ and writing it in the form

$$K = e^{-\Delta_r H^\ominus/RT} e^{\Delta_r S^\ominus/R} \qquad (7.19)$$

Note that a positive reaction enthalpy results in a lowering of the equilibrium constant (that is, an endothermic reaction can be expected to have an equilibrium composition that favours the reactants). However, if there is positive reaction entropy, then the equilibrium composition may favour products, despite the endothermic character of the reaction.

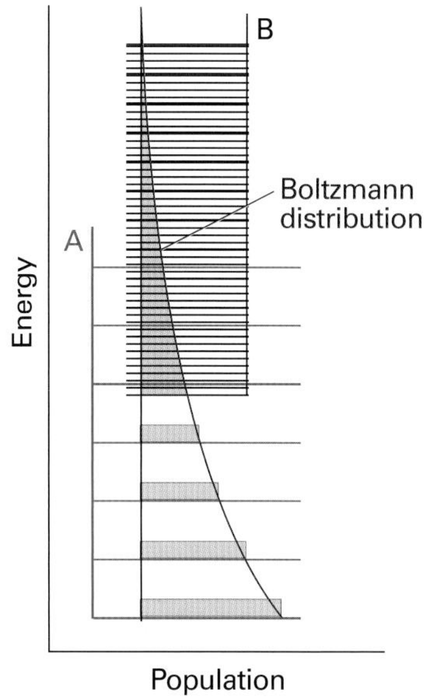

Fig. 7.5 Even though the reaction A → B is endothermic, the density of energy levels in B is so much greater than that in A that the population associated with B is greater than that associated with A, so B is dominant at equilibrium.

(d) Equilibria in biological systems

We saw in Section 5.7 that for biological systems it is appropriate to adopt the biological standard state, in which $a_{H^+} = 10^{-7}$ and $pH = -\log a_{H^+} = 7$. It follows from eqn 5.56 that the relation between the thermodynamic and biological standard Gibbs energies of reaction for a reaction of the form

$$A + \nu H^+(aq) \rightarrow P \qquad (7.20a)$$

is

$$\Delta_r G^\oplus = \Delta_r G^\ominus + 7\nu RT \ln 10 \qquad (7.20b)$$

Note that there is no difference between the two standard values if hydrogen ions are not involved in the reaction ($\nu = 0$).

Illustration 7.4 *Using the biological standard state*

Consider the reaction

$$NADH(aq) + H^+(aq) \rightarrow NAD^+(aq) + H_2(g)$$

at 37°C, for which $\Delta_r G^\ominus = -21.8$ kJ mol^{-1}. NADH is the reduced form of nicotinamide adenine dinucleotide and NAD^+ is its oxidized form; the molecules play an important role in the later stages of the respiratory process. It follows that, because $\nu = 1$ and $7 \ln 10 = 16.1$,

$$\Delta_r G^\oplus = -21.8 \text{ kJ mol}^{-1} + 16.1 \times (8.3145 \times 10^{-3} \text{ kJ K}^{-1} \text{ mol}^{-1}) \times (310 \text{ K})$$
$$= +19.7 \text{ kJ mol}^{-1}$$

Note that the biological standard value is opposite in sign (in this example) to the thermodynamic standard value: the much lower concentration of hydronium ions (by seven orders of magnitude) at pH = 7 in place of pH = 0, has resulted in the reverse reaction becoming spontaneous.

Self-test 7.3 For a particular reaction of the form $A \rightarrow B + 2\,H^+$ in aqueous solution, it was found that $\Delta_r G^\ominus = +20$ kJ mol^{-1} at 28°C. Estimate the value of $\Delta_r G^\oplus$. [−61 kJ mol^{-1}]

The response of equilibria to the conditions

Equilibria respond to changes in pressure, temperature, and concentrations of reactants and products. The equilibrium constant for a reaction is not affected by the presence of a catalyst or an enzyme (a biological catalyst). As we shall see in detail in Sections 22.5 and 25.6, catalysts increase the rate at which equilibrium is attained but do not affect its position. However, it is important to note that in industry reactions rarely reach equilibrium, partly on account of the rates at which reactants mix.

7.3 How equilibria respond to pressure

The equilibrium constant depends on the value of $\Delta_r G^{\ominus}$, which is defined at a single, standard pressure. The value of $\Delta_r G^{\ominus}$, and hence of K, is therefore independent of the pressure at which the equilibrium is actually established. Formally we may express this independence as

$$\left(\frac{\partial K}{\partial p}\right)_T = 0 \tag{7.21}$$

The conclusion that K is independent of pressure does not necessarily mean that the equilibrium composition is independent of the pressure, and its effect depends on how the pressure is applied. The pressure within a reaction vessel can be increased by injecting an inert gas into it. However, so long as the gases are perfect, this addition of gas leaves all the partial pressures of the reacting gases unchanged: the partial pressures of a perfect gas is the pressure it would exert if it were alone in the container, so the presence of another gas has no effect. It follows that pressurization by the addition of an inert gas has no effect on the equilibrium composition of the system (provided the gases are perfect). Alternatively, the pressure of the system may be increased by confining the gases to a smaller volume (that is, by compression). Now the individual partial pressures are changed but their ratio (as it appears in the equilibrium constant) remains the same. Consider, for instance, the perfect gas equilibrium A $\rightleftharpoons$ 2 B, for which the equilibrium constant is

$$K = \frac{p_B^2}{p_A p^{\ominus}}$$

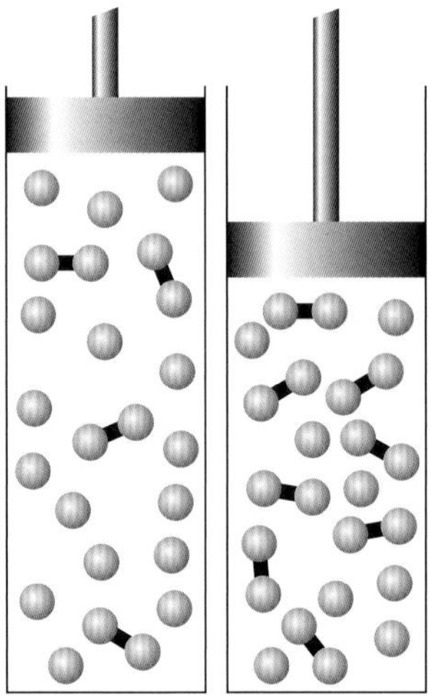

Fig. 7.6 When a reaction at equilibrium is compressed (from *a* to *b*), the reaction responds by reducing the number of molecules in the gas phase (in this case by producing the dimers represented by the linked spheres).

The right-hand side of this expression remains constant only if an increase in p_A cancels an increase in the *square* of p_B. This relatively steep increase of p_A compared to p_B will occur if the equilibrium composition shifts in favour of A at the expense of B. Then the number of A molecules will increase as the volume of the container is decreased and its partial pressure will rise more rapidly than can be ascribed to a simple change in volume alone (Fig. 7.6).

The increase in the number of A molecules and the corresponding decrease in the number of B molecules in the equilibrium A $\rightleftharpoons$ 2 B is a special case of a principle proposed by the French chemist Henri Le Chatelier.[1] **Le Chatelier's principle** states that:

> A system at equilibrium, when subjected to a disturbance, responds in a way that tends to minimize the effect of the disturbance

The principle implies that, if a system at equilibrium is compressed, then the reaction will adjust so as to minimize the increase in pressure. This it can do by reducing the number of particles in the gas phase, which implies a shift A $\leftarrow$ 2 B.

[1] Le Chatelier also invented oxyacetylene welding.

To treat the effect of compression quantitatively, we suppose that there is an amount n of A present initially (and no B). At equilibrium the amount of A is $(1-\alpha)n$ and the amount of B is $2\alpha n$, where α is the extent of dissociation of A into 2B. It follows that the mole fractions present at equilibrium are

$$x_A = \frac{(1-\alpha)n}{(1-\alpha)n+2\alpha n} = \frac{1-\alpha}{1+\alpha} \qquad x_B = \frac{2\alpha}{1+\alpha}$$

The equilibrium constant for the reaction is

$$K = \frac{p_B^2}{p_A p^\ominus} = \frac{x_B^2 p^2}{x_A p p^\ominus} = \frac{4\alpha^2(p/p^\ominus)}{1-\alpha^2}$$

which rearranges to

$$\alpha = \left(\frac{1}{1+4p/Kp^\ominus}\right)^{1/2} \qquad (7.22)$$

This formula shows that, even though K is independent of pressure, the amounts of A and B do depend on pressure (Fig. 7.7). It also shows that as p is increased, α decreases, in accord with Le Chatelier's principle.

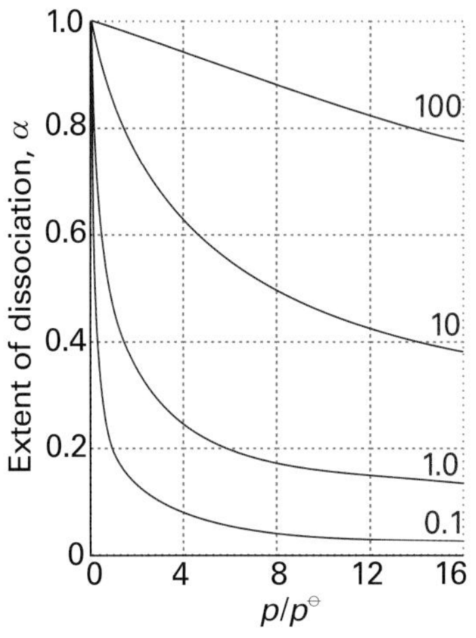

Fig. 7.7 The pressure dependence of the degree of dissociation, α, at equilibrium for an $A(g) \rightleftharpoons 2\,B(g)$ reaction for different values of the equilibrium constant K. The value $\alpha = 0$ corresponds to pure A; $\alpha = 1$ corresponds to pure B.

Exploration Plot x_A and x_B against the pressure p for several values of the equilibrium constant K.

Illustration 7.5 *Predicting the effect of compression*

To predict the effect of an increase in pressure on the composition of the ammonia synthesis at equilibrium, eqn 7.10, we note that the number of gas molecules decreases (from 4 to 2). So, Le Chatelier's principle predicts that an increase in pressure will favour the product. The equilibrium constant is

$$K = \frac{p_{NH_3}^2 p^{\ominus 2}}{p_{N_2} p_{H_2}^3} = \frac{x_{NH_3}^2 p^2 p^{\ominus 2}}{x_{N_2} x_{H_2}^3 p^4} = \frac{K_x p^{\ominus 2}}{p^2}$$

where K_x is the part of the equilibrium constant expression that contains the equilibrium mole fractions of reactants and products (note that, unlike K itself, K_x is not an equilibrium constant). Therefore, doubling the pressure must increase K_x by a factor of 4 to preserve the value of K.

Self-test 7.4 Predict the effect of a tenfold pressure increase on the equilibrium composition of the reaction $3\,N_2(g) + H_2(g) \rightarrow 2\,N_3H(g)$.

[100-fold increase in K_x]

7.4 The response of equilibria to temperature

Le Chatelier's principle predicts that a system at equilibrium will tend to shift in the endothermic direction if the temperature is raised, for then energy is absorbed as heat and the rise in temperature is opposed. Conversely, an equilibrium can be expected to shift in the exothermic direction if the temperature is lowered, for then energy is released and the reduction in temperature is opposed. These conclusions can be summarized as follows:

Exothermic reactions: increased temperature favours the reactants.

Endothermic reactions: increased temperature favours the products.

We shall now justify these remarks and see how to express the changes quantitatively.

(a) The van 't Hoff equation

The **van 't Hoff equation**, which is derived in the *Justification* below, is an expression for the slope of a plot of the equilibrium constant (specifically, ln K) as a function of temperature. It may be expressed in either of two ways:

$$\text{(a)}\quad \frac{\mathrm{d}\ln K}{\mathrm{d}T}=\frac{\Delta_r H^\ominus}{RT^2} \qquad \text{(b)}\quad \frac{\mathrm{d}\ln K}{\mathrm{d}(1/T)}=-\frac{\Delta_r H^\ominus}{R} \qquad (7.23)$$

Justification 7.2 *The van 't Hoff equation*

From eqn 7.17, we know that

$$\ln K=-\frac{\Delta_r G^\ominus}{RT}$$

Differentiation of ln K with respect to temperature then gives

$$\frac{\mathrm{d}\ln K}{\mathrm{d}T}=-\frac{1}{R}\frac{\mathrm{d}(\Delta_r G^\ominus/T)}{\mathrm{d}T}$$

The differentials are complete because K and $\Delta_r G^\ominus$ depend only on temperature, not on pressure. To develop this equation we use the Gibbs–Helmholtz equation (eqn 3.53) in the form

$$\frac{\mathrm{d}(\Delta_r G^\ominus/T)}{\mathrm{d}T}=-\frac{\Delta_r H^\ominus}{T^2}$$

where $\Delta_r H^\ominus$ is the standard reaction enthalpy at the temperature T. Combining the two equations gives the van 't Hoff equation, eqn 7.23a. The second form of the equation is obtained by noting that

$$\frac{\mathrm{d}(1/T)}{\mathrm{d}T}=-\frac{1}{T^2}, \qquad \text{so} \quad \mathrm{d}T=-T^2\mathrm{d}(1/T)$$

It follows that eqn 7.23a can be rewritten as

$$-\frac{\mathrm{d}\ln K}{T^2\mathrm{d}(1/T)}=\frac{\Delta_r H^\ominus}{RT^2}$$

which simplifies into eqn 7.23b.

Equation 7.23a shows that $\mathrm{d}\ln K/\mathrm{d}T<0$ (and therefore that $\mathrm{d}K/\mathrm{d}T<0$) for a reaction that is exothermic under standard conditions ($\Delta_r H^\ominus<0$). A negative slope means that ln K, and therefore K itself, decreases as the temperature rises. Therefore, as asserted above, in the case of an exothermic reaction the equilibrium shifts away from products. The opposite occurs in the case of endothermic reactions.

Some insight into the thermodynamic basis of this behaviour comes from the expression $\Delta_r G^\ominus=\Delta_r H^\ominus-T\Delta_r S^\ominus$ written in the form $-\Delta_r G^\ominus/T=-\Delta_r H^\ominus/T+\Delta_r S^\ominus$. When the reaction is exothermic, $-\Delta_r H^\ominus/T$ corresponds to a positive change of entropy of the surroundings and favours the formation of products. When the temperature is raised, $-\Delta_r H^\ominus/T$ decreases, and the increasing entropy of the surroundings has a less important role. As a result, the equilibrium lies less to the right. When the reaction is endothermic, the principal factor is the increasing entropy of the reaction system. The importance of the unfavourable change of entropy of the surroundings is reduced if the temperature is raised (because then $\Delta_r H^\ominus/T$ is smaller), and the reaction is able to shift towards products.

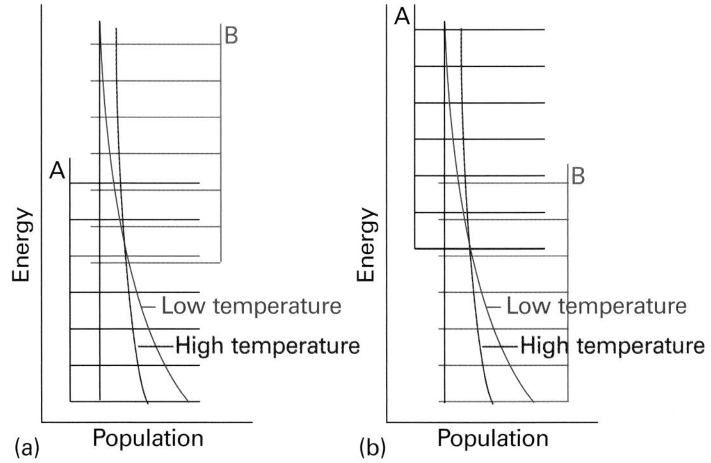

Fig. 7.8 The effect of temperature on a chemical equilibrium can be interpreted in terms of the change in the Boltzmann distribution with temperature and the effect of that change in the population of the species. (a) In an endothermic reaction, the population of B increases at the expense of A as the temperature is raised. (b) In an exothermic reaction, the opposite happens.

Molecular interpretation 7.3 *The temperature dependence of the equilibrium constant*

The typical arrangement of energy levels for an endothermic reaction is shown in Fig. 7.8a. When the temperature is increased, the Boltzmann distribution adjusts and the populations change as shown. The change corresponds to an increased population of the higher energy states at the expense of the population of the lower energy states. We see that the states that arise from the B molecules become more populated at the expense of the A molecules. Therefore, the total population of B states increases, and B becomes more abundant in the equilibrium mixture. Conversely, if the reaction is exothermic (Fig. 7.8b), then an increase in temperature increases the population of the A states (which start at higher energy) at the expense of the B states, so the reactants become more abundant.

Example 7.3 *Measuring a reaction enthalpy*

The data below show the temperature variation of the equilibrium constant of the reaction $Ag_2CO_3(s) \rightleftharpoons Ag_2O(s) + CO_2(g)$. Calculate the standard reaction enthalpy of the decomposition.

T/K	350	400	450	500
K	3.98×10^{-4}	1.41×10^{-2}	1.86×10^{-1}	1.48

Method It follows from eqn 7.23b that, provided the reaction enthalpy can be assumed to be independent of temperature, a plot of $-\ln K$ against $1/T$ should be a straight line of slope $\Delta_r H^\ominus/R$.

Answer We draw up the following table:

T/K	350	400	450	500
$(10^3\ \text{K})/T$	2.86	2.50	2.22	2.00
$-\ln K$	7.83	4.26	1.68	-0.39

These points are plotted in Fig. 7.9. The slope of the graph is $+9.6 \times 10^3$, so

$$\Delta_r H^\ominus = (+9.6 \times 10^3\ \text{K}) \times R = +80\ \text{kJ mol}^{-1}$$

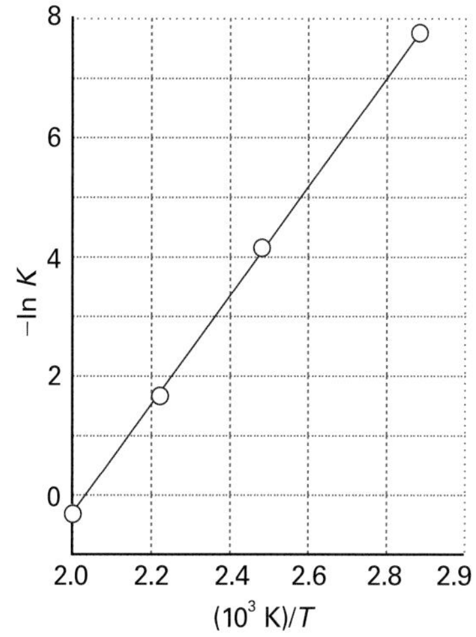

Fig. 7.9 When $-\ln K$ is plotted against $1/T$, a straight line is expected with slope equal to $\Delta_r H^\ominus/R$ if the standard reaction enthalpy does not vary appreciably with temperature. This is a non-calorimetric method for the measurement of reaction enthalpies.

Exploration The equilibrium constant of a reaction is found to fit the expression $\ln K = a + b/(T/K) + c/(T/K)^3$ over a range of temperatures. (a) Write expressions for $\Delta_r H^\ominus$ and $\Delta_r S^\ominus$. (b) Plot $\ln K$ against $1/T$ between 400 K and 600 K for $a = -2.0$, $b = -1.0 \times 10^3$, and $c = 2.0 \times 10^7$.

Self-test 7.5 The equilibrium constant of the reaction $2\ SO_2(g) + O_2(g) \rightleftharpoons 2\ SO_3(g)$ is 4.0×10^{24} at 300 K, 2.5×10^{10} at 500 K, and 3.0×10^{4} at 700 K. Estimate the reaction enthalpy at 500 K. [-200 kJ mol^{-1}]

The temperature dependence of the equilibrium constant provides a non-calorimetric method of determining $\Delta_r H^{\ominus}$. A drawback is that the reaction enthalpy is actually temperature-dependent, so the plot is not expected to be perfectly linear. However, the temperature dependence is weak in many cases, so the plot is reasonably straight. In practice, the method is not very accurate, but it is often the only method available.

(b) The value of K at different temperatures

To find the value of the equilibrium constant at a temperature T_2 in terms of its value K_1 at another temperature T_1, we integrate eqn 7.23b between these two temperatures:

$$\ln K_2 - \ln K_1 = -\frac{1}{R}\int_{1/T_1}^{1/T_2} \Delta_r H^{\ominus} \mathrm{d}(1/T) \tag{7.24}$$

If we suppose that $\Delta_r H^{\ominus}$ varies only slightly with temperature over the temperature range of interest, then we may take it outside the integral. It follows that

$$\ln K_2 - \ln K_1 = -\frac{\Delta_r H^{\ominus}}{R}\left(\frac{1}{T_2} - \frac{1}{T_1}\right) \tag{7.25}$$

Illustration 7.6 *Estimating an equilibrium constant at a different temperature*

To estimate the equilibrium constant for the synthesis of ammonia at 500 K from its value at 298 K (6.1×10^5 for the reaction as written in eqn 7.10) we use the standard reaction enthalpy, which can be obtained from Table 2.7 in the *Data section* by using $\Delta_r H^{\ominus} = 2\Delta_f H^{\ominus}(NH_3, g)$, and assume that its value is constant over the range of temperatures. Then, with $\Delta_r H^{\ominus} = -92.2$ kJ mol^{-1}, from eqn 7.25 we find

$$\ln K_2 = \ln(6.1 \times 10^5) - \frac{(-92.2 \times 10^3\ \mathrm{J\ mol^{-1}})}{8.3145\ \mathrm{J\ K^{-1}\ mol^{-1}}}\left(\frac{1}{500\ \mathrm{K}} - \frac{1}{298\ \mathrm{K}}\right)$$
$$= -1.71$$

It follows that $K_2 = 0.18$, a lower value than at 298 K, as expected for this exothermic reaction.

Self-test 7.6 The equilibrium constant for $N_2O_4(g) \rightleftharpoons 2\ NO_2(g)$ was calculated in Self-test 7.1. Estimate its value at 100°C. [15]

Knowledge of the temperature dependence of the equilibrium constant for a reaction can be useful in the design of laboratory and industrial processes. For example, synthetic chemists can improve the yield of a reaction by changing the temperature of the reaction mixture. Also, reduction of a metal oxide with carbon or carbon monoxide results in the extraction of the metal when the process is carried out at a temperature for which the equilibrium constant for the reaction is much greater than one.

IMPACT ON ENGINEERING

I7.1 The extraction of metals from their oxides

Metals can be obtained from their oxides by reduction with carbon or carbon monoxide if any of the equilibria

$$MO(s) + C(s) \rightleftharpoons M(s) + CO(g)$$
$$MO(s) + \tfrac{1}{2}C(s) \rightleftharpoons M(s) + \tfrac{1}{2}CO_2(g)$$
$$MO(s) + CO(g) \rightleftharpoons M(s) + CO_2(g)$$

lie to the right (that is, have $K > 1$). As we shall see, these equilibria can be discussed in terms of the thermodynamic functions for the reactions

$$\text{(i) } M(s) + \tfrac{1}{2}O_2(g) \rightarrow MO(s)$$
$$\text{(ii) } \tfrac{1}{2}C(s) + \tfrac{1}{2}O_2(g) \rightarrow \tfrac{1}{2}CO_2(g)$$
$$\text{(iii) } C(s) + \tfrac{1}{2}O_2(g) \rightarrow CO(g)$$
$$\text{(iv) } CO(g) + \tfrac{1}{2}O_2(g) \rightarrow CO_2(g)$$

The temperature dependences of the standard Gibbs energies of reactions (i)–(iv) depend on the reaction entropy through $d\Delta_rG^\ominus/dT = -\Delta_rS^\ominus$. Because in reaction (iii) there is a net increase in the amount of gas, the standard reaction entropy is large and positive; therefore, its $\Delta_rG^\ominus$ decreases sharply with increasing temperature. In reaction (iv), there is a similar net decrease in the amount of gas, so $\Delta_rG^\ominus$ increases sharply with increasing temperature. In reaction (ii), the amount of gas is constant, so the entropy change is small and $\Delta_rG^\ominus$ changes only slightly with temperature. These remarks are summarized in Fig. 7.10, which is called an *Ellingham diagram*. Note that $\Delta_rG^\ominus$ decreases upwards!

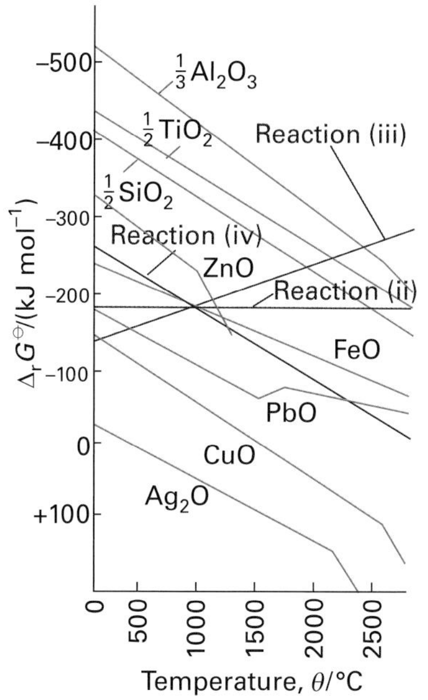

Fig. 7.10 An Ellingham diagram for the discussion of the reduction of metal ores.

At room temperature, $\Delta_rG^\ominus$ is dominated by the contribution of the reaction enthalpy ($T\Delta_rS^\ominus$ being relatively small), so the order of increasing $\Delta_rG^\ominus$ is the same as the order of increasing $\Delta_rH^\ominus$ (Al_2O_3 is most exothermic; Ag_2O is least). The standard reaction entropy is similar for all metals because in each case gaseous oxygen is eliminated and a compact, solid oxide is formed. As a result, the temperature dependence of the standard Gibbs energy of oxidation should be similar for all metals, as is shown by the similar slopes of the lines in the diagram. The kinks at high temperatures correspond to the evaporation of the metals; less pronounced kinks occur at the melting temperatures of the metals and the oxides.

Successful reduction of the oxide depends on the outcome of the competition of the carbon for the oxygen bound to the metal. The standard Gibbs energies for the reductions can be expressed in terms of the standard Gibbs energies for the reactions above:

$$MO(s) + C(s) \rightarrow M(s) + CO(g) \qquad \Delta_rG^\ominus = \Delta_rG^\ominus(\text{iii}) - \Delta_rG^\ominus(\text{i})$$
$$MO(s) + \tfrac{1}{2}C(s) \rightarrow M(s) + \tfrac{1}{2}CO_2(g) \qquad \Delta_rG^\ominus = \Delta_rG^\ominus(\text{ii}) - \Delta_rG^\ominus(\text{i})$$
$$MO(s) + CO(g) \rightarrow M(s) + CO_2(g) \qquad \Delta_rG^\ominus = \Delta_rG^\ominus(\text{iv}) - \Delta_rG^\ominus(\text{i})$$

The equilibrium lies to the right if $\Delta_rG^\ominus < 0$. This is the case when the line for reaction (i) lies below (is more positive than) the line for one of the reactions (ii) to (iv).

The spontaneity of a reduction at any temperature can be predicted simply by looking at the diagram: a metal oxide is reduced by any carbon reaction lying above it, because the overall reaction then has $\Delta_rG^\ominus < 0$. For example, CuO can be reduced to Cu at any temperature above room temperature. Even in the absence of carbon, Ag_2O decomposes when heated above 200°C because then the standard Gibbs energy for reaction (i) becomes positive (and the reverse reaction is then spontaneous). On the other hand, Al_2O_3 is not reduced by carbon until the temperature has been raised to above 2000°C.

Equilibrium electrochemistry

We shall now see how the foregoing ideas, with certain changes of technical detail, can be used to describe the equilibrium properties of reactions taking place in electrochemical cells. The ability to make very precise measurements of currents and potential differences ('voltages') means that electrochemical methods can be used to determine thermodynamic properties of reactions that may be inaccessible by other methods.

An **electrochemical cell** consists of two **electrodes**, or metallic conductors, in contact with an **electrolyte**, an ionic conductor (which may be a solution, a liquid, or a solid). An electrode and its electrolyte comprise an **electrode compartment**. The two electrodes may share the same compartment. The various kinds of electrode are summarized in Table 7.1. Any 'inert metal' shown as part of the specification is present to act as a source or sink of electrons, but takes no other part in the reaction other than acting as a catalyst for it. If the electrolytes are different, the two compartments may be joined by a **salt bridge**, which is a tube containing a concentrated electrolyte solution (almost always potassium chloride in agar jelly) that completes the electrical circuit and enables the cell to function. A **galvanic cell** is an electrochemical cell that produces electricity as a result of the spontaneous reaction occurring inside it. An **electrolytic cell** is an electrochemical cell in which a non-spontaneous reaction is driven by an external source of current.

7.5 Half-reactions and electrodes

It will be familiar from introductory chemistry courses that **oxidation** is the removal of electrons from a species, a **reduction** is the addition of electrons to a species, and a **redox reaction** is a reaction in which there is a transfer of electrons from one species to another. The electron transfer may be accompanied by other events, such as atom or ion transfer, but the net effect is electron transfer and hence a change in oxidation number of an element. The **reducing agent** (or 'reductant') is the electron donor; the **oxidizing agent** (or 'oxidant') is the electron acceptor. It should also be familiar that any redox reaction may be expressed as the difference of two reduction **half-reactions**, which are conceptual reactions showing the gain of electrons. Even reactions that are not redox reactions may often be expressed as the difference of two reduction half-reactions. The reduced and oxidized species in a half-reaction form a **redox couple**. In general we write a couple as Ox/Red and the corresponding reduction half-reaction as

$$\mathrm{Ox} + \nu\,\mathrm{e}^- \rightarrow \mathrm{Red} \tag{7.26}$$

Table 7.1 Varieties of electrode

Electrode type	Designation	Redox couple	Half-reaction
Metal/metal ion	$M(s)\vert M^+(aq)$	M^+/M	$M^+(aq) + e^- \rightarrow M(s)$
Gas	$Pt(s)\vert X_2(g)\vert X^+(aq)$	X^+/X_2	$X^+(aq) + e^- \rightarrow \frac{1}{2}X_2(g)$
	$Pt(s)\vert X_2(g)\vert X^-(aq)$	X_2/X^-	$\frac{1}{2}X_2(g) + e^- \rightarrow X^-(aq)$
Metal/insoluble salt	$M(s)\vert MX(s)\vert X^-(aq)$	$MX/M,X^-$	$MX(s) + e^- \rightarrow M(s) + X^-(aq)$
Redox	$Pt(s)\vert M^+(aq),M^{2+}(aq)$	M^{2+}/M^+	$M^{2+}(aq) + e^- \rightarrow M^+(aq)$

Illustration 7.7 *Expressing a reaction in terms of half-reactions*

The dissolution of silver chloride in water $AgCl(s) \rightarrow Ag^+(aq) + Cl^-(aq)$, which is not a redox reaction, can be expressed as the difference of the following two reduction half-reactions:

$$AgCl(s) + e^- \rightarrow Ag(s) + Cl^-(aq)$$
$$Ag^+(aq) + e^- \rightarrow Ag(s)$$

The redox couples are $AgCl/Ag,Cl^-$ and Ag^+/Ag, respectively.

Self-test 7.7 Express the formation of H_2O from H_2 and O_2 in acidic solution (a redox reaction) as the difference of two reduction half-reactions.
[$4\,H^+(aq) + 4\,e^- \rightarrow 2\,H_2(g)$, $O_2(g) + 4\,H^+(aq) + 4\,e^- \rightarrow 2\,H_2O(l)$]

We shall often find it useful to express the composition of an electrode compartment in terms of the reaction quotient, Q, for the half-reaction. This quotient is defined like the reaction quotient for the overall reaction, but the electrons are ignored.

Illustration 7.8 *Writing the reaction quotient of a half-reaction*

The reaction quotient for the reduction of O_2 to H_2O in acid solution, $O_2(g) + 4\,H^+(aq) + 4\,e^- \rightarrow 2\,H_2O(l)$, is

$$Q = \frac{a_{H_2O}^2}{a_{H^+}^4 a_{O_2}} \approx \frac{p^\ominus}{a_{H^+}^4 p_{O_2}}$$

The approximations used in the second step are that the activity of water is 1 (because the solution is dilute) and the oxygen behaves as a perfect gas, so $a_{O_2} \approx p_{O_2}/p^\ominus$.

Self-test 7.8 Write the half-reaction and the reaction quotient for a chlorine gas electrode. [$Cl_2(g) + 2\,e^- \rightarrow 2\,Cl^-(aq)$, $Q = a_{Cl^-}^2 p^\ominus / p_{Cl_2}$]

The reduction and oxidation processes responsible for the overall reaction in a cell are separated in space: oxidation takes place at one electrode and reduction takes place at the other. As the reaction proceeds, the electrons released in the oxidation $Red_1 \rightarrow Ox_1 + \nu\,e^-$ at one electrode travel through the external circuit and re-enter the cell through the other electrode. There they bring about reduction $Ox_2 + \nu\,e^- \rightarrow Red_2$. The electrode at which oxidation occurs is called the **anode**; the electrode at which reduction occurs is called the **cathode**. In a galvanic cell, the cathode has a higher potential than the anode: the species undergoing reduction, Ox_2, withdraws electrons from its electrode (the cathode, Fig. 7.11), so leaving a relative positive charge on it (corresponding to a high potential). At the anode, oxidation results in the transfer of electrons to the electrode, so giving it a relative negative charge (corresponding to a low potential).

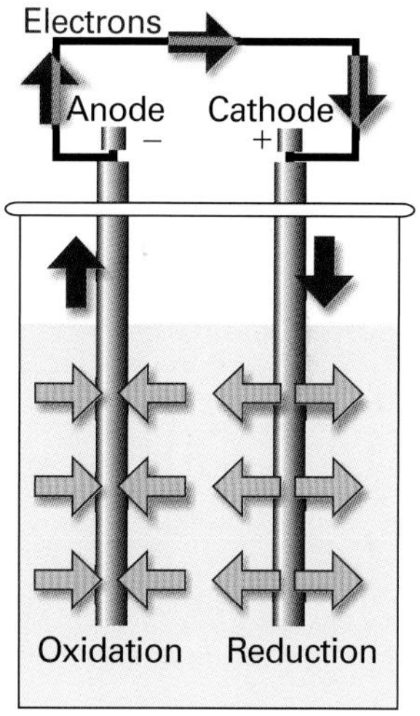

Fig. 7.11 When a spontaneous reaction takes place in a galvanic cell, electrons are deposited in one electrode (the site of oxidation, the anode) and collected from another (the site of reduction, the cathode), and so there is a net flow of current which can be used to do work. Note that the + sign of the cathode can be interpreted as indicating the electrode at which electrons enter the cell, and the − sign of the anode is where the electrons leave the cell.

7.6 **Varieties of cells**

The simplest type of cell has a single electrolyte common to both electrodes (as in Fig. 7.11). In some cases it is necessary to immerse the electrodes in different electrolytes,

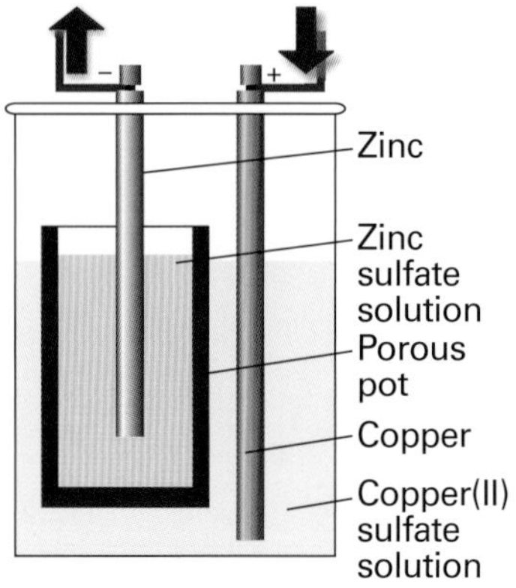

Fig. 7.12 One version of the Daniell cell. The copper electrode is the cathode and the zinc electrode is the anode. Electrons leave the cell from the zinc electrode and enter it again through the copper electrode.

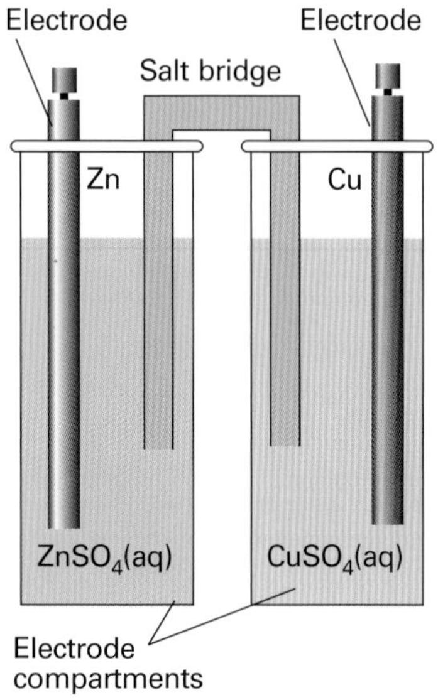

Fig. 7.13 The salt bridge, essentially an inverted U-tube full of concentrated salt solution in a jelly, has two opposing liquid junction potentials that almost cancel.

as in the 'Daniell cell' in which the redox couple at one electrode is Cu^{2+}/Cu and at the other is Zn^{2+}/Zn (Fig. 7.12). In an **electrolyte concentration cell**, the electrode compartments are identical except for the concentrations of the electrolytes. In an **electrode concentration cell** the electrodes themselves have different concentrations, either because they are gas electrodes operating at different pressures or because they are amalgams (solutions in mercury) with different concentrations.

(a) Liquid junction potentials

In a cell with two different electrolyte solutions in contact, as in the Daniell cell, there is an additional source of potential difference across the interface of the two electrolytes. This potential is called the **liquid junction potential**, E_{lj}. Another example of a junction potential is that between different concentrations of hydrochloric acid. At the junction, the mobile H^+ ions diffuse into the more dilute solution. The bulkier Cl^- ions follow, but initially do so more slowly, which results in a potential difference at the junction. The potential then settles down to a value such that, after that brief initial period, the ions diffuse at the same rates. Electrolyte concentration cells always have a liquid junction; electrode concentration cells do not.

The contribution of the liquid junction to the potential can be reduced (to about 1 to 2 mV) by joining the electrolyte compartments through a salt bridge (Fig. 7.13). The reason for the success of the salt bridge is that the liquid junction potentials at either end are largely independent of the concentrations of the two dilute solutions, and so nearly cancel.

(b) Notation

In the notation for cells, phase boundaries are denoted by a vertical bar. For example,

$$\mathrm{Pt(s)|H_2(g)|HCl(aq)|AgCl(s)|Ag(s)}$$

A liquid junction is denoted by ⋮, so the cell in Fig. 7.12, is denoted

$$\mathrm{Zn(s)|ZnSO_4(aq)\vdots CuSO_4(aq)|Cu(s)}$$

A double vertical line, ||, denotes an interface for which it is assumed that the junction potential has been eliminated. Thus the cell in Fig. 7.13 is denoted

$$\mathrm{Zn(s)|ZnSO_4(aq)||CuSO_4(aq)|Cu(s)}$$

An example of an electrolyte concentration cell in which the liquid junction potential is assumed to be eliminated is

$$\mathrm{Pt(s)|H_2(g)|HCl(aq},\, b_1\mathrm{)||HCl(aq},\, b_2\mathrm{)|H_2(g)|Pt(s)}.$$

7.7 The electromotive force

The current produced by a galvanic cell arises from the spontaneous chemical reaction taking place inside it. The **cell reaction** is the reaction in the cell written on the assumption that the right-hand electrode is the cathode, and hence that the spontaneous reaction is one in which reduction is taking place in the right-hand compartment. Later we see how to predict if the right-hand electrode is in fact the cathode; if it is, then the cell reaction is spontaneous as written. If the left-hand electrode turns out to be the cathode, then the reverse of the corresponding cell reaction is spontaneous.

To write the cell reaction corresponding to a cell diagram, we first write the right-hand half-reaction as a reduction (because we have assumed that to be spontaneous). Then we subtract from it the left-hand reduction half-reaction (for, by implication, that electrode is the site of oxidation). Thus, in the cell Zn(s) | $ZnSO_4$(aq) || $CuSO_4$(aq) | Cu(s) the two electrodes and their reduction half-reactions are

Right-hand electrode: $Cu^{2+}(aq) + 2\,e^- \rightarrow Cu(s)$

Left-hand electrode: $Zn^{2+}(aq) + 2\,e^- \rightarrow Zn(s)$

Hence, the overall cell reaction is the difference:

$$Cu^{2+}(aq) + Zn(s) \rightarrow Cu(s) + Zn^{2+}(aq)$$

(a) The Nernst equation

A cell in which the overall cell reaction has not reached chemical equilibrium can do electrical work as the reaction drives electrons through an external circuit. The work that a given transfer of electrons can accomplish depends on the potential difference between the two electrodes. This potential difference is called the **cell potential** and is measured in volts, V ($1\ \text{V} = 1\ \text{J C}^{-1}\ \text{s}$). When the cell potential is large, a given number of electrons travelling between the electrodes can do a large amount of electrical work. When the cell potential is small, the same number of electrons can do only a small amount of work. A cell in which the overall reaction is at equilibrium can do no work, and then the cell potential is zero.

According to the discussion in Section 3.5e, we know that the maximum non-expansion work, which in the current context is electrical work, that a system (the cell) can do is given by eqn 3.38 ($w_{e,max} = \Delta G$), with ΔG identified (as we shall show) with the Gibbs energy of the cell reaction, $\Delta_r G$. It follows that, to draw thermodynamic conclusions from measurements of the work a cell can do, we must ensure that the cell is operating reversibly, for only then is it producing maximum work. Moreover, we saw in Section 7.1a that the reaction Gibbs energy is actually a property relating to a specified composition of the reaction mixture. Therefore, to make use of $\Delta_r G$ we must ensure that the cell is operating reversibly at a specific, constant composition. Both these conditions are achieved by measuring the cell potential when it is balanced by an exactly opposing source of potential so that the cell reaction occurs reversibly, the composition is constant, and no current flows: in effect, the cell reaction is poised for change, but not actually changing. The resulting potential difference is called the **electromotive force** (emf), E, of the cell.

As we show in the *Justification* below, the relation between the reaction Gibbs energy and the emf of the cell is

$$-\nu FE = \Delta_r G \qquad (7.27)$$

where F is Faraday's constant, $F = eN_A$, and ν is the stoichiometric coefficient of the electrons in the half-reactions into which the cell reaction can be divided. This equation is the key connection between electrical measurements on the one hand and thermodynamic properties on the other. It will be the basis of all that follows.

Justification 7.3 *The relation between the electromotive force and the reaction Gibbs energy*

We consider the change in G when the cell reaction advances by an infinitesimal amount $d\xi$ at some composition. From eqn 7.15 we can write (at constant temperature and pressure)

$$dG = \Delta_r G d\xi$$

The maximum non-expansion (electrical) work that the reaction can do as it advances by $d\xi$ at constant temperature and pressure is therefore

$$dw_e = \Delta_r G d\xi$$

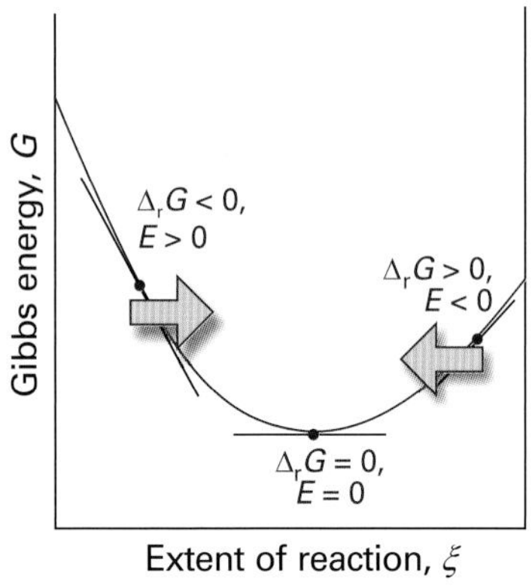

Fig. 7.14 A spontaneous reaction occurs in the direction of decreasing Gibbs energy. When expressed in terms of a cell potential, the spontaneous direction of change can be expressed in terms of the cell emf, E. The reaction is spontaneous as written (from left to right on the illustration) when $E > 0$. The reverse reaction is spontaneous when $E < 0$. When the cell reaction is at equilibrium, the cell potential is zero.

This work is infinitesimal, and the composition of the system is virtually constant when it occurs.

Suppose that the reaction advances by $d\xi$, then $\nu d\xi$ electrons must travel from the anode to the cathode. The total charge transported between the electrodes when this change occurs is $-\nu eN_A d\xi$ (because $\nu d\xi$ is the amount of electrons and the charge per mole of electrons is $-eN_A$). Hence, the total charge transported is $-\nu F d\xi$ because $eN_A = F$. The work done when an infinitesimal charge $-\nu F d\xi$ travels from the anode to the cathode is equal to the product of the charge and the potential difference E (see Table 2.1 and *Appendix 3*):

$$dw_e = -\nu FE d\xi$$

When we equate this relation to the one above ($dw_e = \Delta_r G d\xi$), the advancement $d\xi$ cancels, and we obtain eqn 7.27.

It follows from eqn 7.27 that, by knowing the reaction Gibbs energy at a specified composition, we can state the cell emf at that composition. Note that a negative reaction Gibbs energy, corresponding to a spontaneous cell reaction, corresponds to a positive cell emf. Another way of looking at the content of eqn 7.27 is that it shows that the driving power of a cell (that is, its emf) is proportional to the slope of the Gibbs energy with respect to the extent of reaction. It is plausible that a reaction that is far from equilibrium (when the slope is steep) has a strong tendency to drive electrons through an external circuit (Fig. 7.14). When the slope is close to zero (when the cell reaction is close to equilibrium), the emf is small.

Illustration 7.9 *Converting between the cell emf and the reaction Gibbs energy*

Equation 7.27 provides an electrical method for measuring a reaction Gibbs energy at any composition of the reaction mixture: we simply measure the cell's emf and convert it to $\Delta_r G$. Conversely, if we know the value of $\Delta_r G$ at a particular composition, then we can predict the emf. For example, if $\Delta_r G = -1 \times 10^2$ kJ mol^{-1} and $\nu = 1$, then

$$E = -\frac{\Delta_r G}{\nu F} = -\frac{(-1 \times 10^5 \text{ J mol}^{-1})}{1 \times (9.6485 \times 10^4 \text{ C mol}^{-1})} = 1 \text{ V}$$

where we have used 1 J = 1 C V.

We can go on to relate the emf to the activities of the participants in the cell reaction. We know that the reaction Gibbs energy is related to the composition of the reaction mixture by eqn 7.11 ($\Delta_r G = \Delta_r G^\ominus + RT \ln Q$); it follows, on division of both sides by $-\nu F$, that

$$E = -\frac{\Delta_r G^\ominus}{\nu F} - \frac{RT}{\nu F} \ln Q$$

The first term on the right is written

$$E^\ominus = -\frac{\Delta_r G^\ominus}{\nu F} \qquad [7.28]$$

and called the **standard emf** of the cell. That is, the standard emf is the standard reaction Gibbs energy expressed as a potential (in volts). It follows that

$$E = E^{\ominus} - \frac{RT}{\nu F}\ln Q \qquad (7.29)$$

This equation for the emf in terms of the composition is called the **Nernst equation**; the dependence of cell potential on composition that it predicts is summarized in Fig. 7.15. One important application of the Nernst equation is to the determination of the pH of a solution and, with a suitable choice of electrodes, of the concentration of other ions (Section 7.9c).

We see from eqn 7.29 that the standard emf (which will shortly move to centre stage of the exposition) can be interpreted as the emf when all the reactants and products in the cell reaction are in their standard states, for then all activities are 1, so $Q = 1$ and $\ln Q = 0$. However, the fact that the standard emf is merely a disguised form of the standard reaction Gibbs energy (eqn 7.28) should always be kept in mind and underlies all its applications.

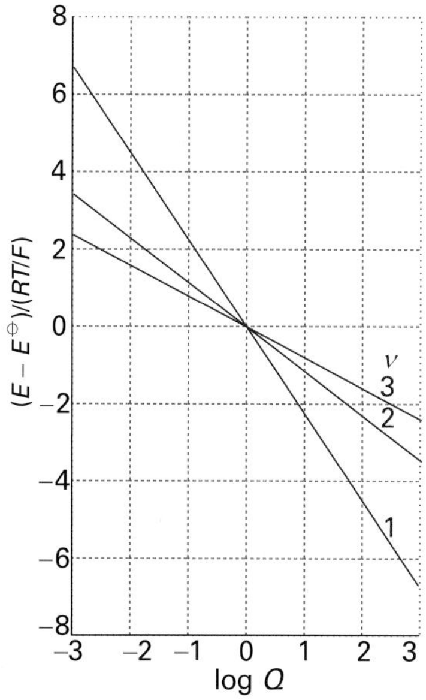

Fig. 7.15 The variation of cell emf with the value of the reaction quotient for the cell reaction for different values of ν (the number of electrons transferred). At 298 K, $RT/F = 25.69$ mV, so the vertical scale refers to multiples of this value.

Exploration Plot the variation of cell emf with the value of the reaction quotient for the cell reaction for different values of the temperature. Does the cell emf become more or less sensitive to composition as the temperature increases?

Illustration 7.10 *Using the Nernst equation*

Because $RT/F = 25.7$ mV at 25°C, a practical form of the Nernst equation is

$$E = E^{\ominus} - \frac{25.7\text{ mV}}{\nu}\ln Q$$

It then follows that, for a reaction in which $\nu = 1$, if Q is increased by a factor of 10, then the emf decreases by 59.2 mV.

(b) Cells at equilibrium

A special case of the Nernst equation has great importance in electrochemistry and provides a link to the earlier part of the chapter. Suppose the reaction has reached equilibrium; then $Q = K$, where K is the equilibrium constant of the cell reaction. However, a chemical reaction at equilibrium cannot do work, and hence it generates zero potential difference between the electrodes of a galvanic cell. Therefore, setting $E = 0$ and $Q = K$ in the Nernst equation gives

$$\ln K = \frac{\nu F E^{\ominus}}{RT} \qquad (7.30)$$

This very important equation (which could also have been obtained more directly by substituting eqn 7.29 into eqn 7.17) lets us predict equilibrium constants from measured standard cell potentials. However, before we use it extensively, we need to establish a further result.

Illustration 7.11 *Calculating an equilibrium constant from a standard cell potential*

Because the standard emf of the Daniell cell is +1.10 V, the equilibrium constant for the cell reaction $Cu^{2+}(aq) + Zn(s) \rightarrow Cu(s) + Zn^{2+}(aq)$, for which $\nu = 2$, is $K = 1.5 \times 10^{37}$ at 298 K. We conclude that the displacement of copper by zinc goes virtually to completion. Note that an emf of about 1 V is easily measurable but corresponds to an equilibrium constant that would be impossible to measure by direct chemical analysis.

7.8 Standard potentials

A galvanic cell is a combination of two electrodes, and each one can be considered as making a characteristic contribution to the overall cell potential. Although it is not possible to measure the contribution of a single electrode, we can define the potential of one of the electrodes as zero and then assign values to others on that basis. The specially selected electrode is the **standard hydrogen electrode** (SHE):

$$\mathrm{Pt(s)|H_2(g)|H^+(aq)} \qquad E^\ominus = 0 \tag{7.31}$$

at all temperatures. To achieve the standard conditions, the activity of the hydrogen ions must be 1 (that is, pH = 0) and the pressure (more precisely, the fugacity) of the hydrogen gas must be 1 bar. The **standard potential**, $E^\ominus$, of another couple is then assigned by constructing a cell in which it is the right-hand electrode and the standard hydrogen electrode is the left-hand electrode.

The procedure for measuring a standard potential can be illustrated by considering a specific case, the silver chloride electrode. The measurement is made on the 'Harned cell':

$$\mathrm{Pt(s)|H_2(g)|HCl(aq)|AgCl(s)|Ag(s)} \qquad \tfrac{1}{2}\mathrm{H_2(g) + AgCl(s) \rightarrow HCl(aq) + Ag(s)}$$

for which the Nernst equation is

$$E = E^\ominus(\mathrm{AgCl/Ag, Cl^-}) - \frac{RT}{F}\ln\frac{a_{\mathrm{H^+}}a_{\mathrm{Cl^-}}}{a_{\mathrm{H_2}}^{1/2}}$$

We shall set $a_{\mathrm{H_2}} = 1$ from now on, and for simplicity write the standard potential as $E^\ominus$; then

$$E = E^\ominus - \frac{RT}{F}\ln a_{\mathrm{H^+}}a_{\mathrm{Cl^-}}$$

The activities can be expressed in terms of the molality b of HCl(aq) through $a_{\mathrm{H^+}} = \gamma_\pm b/b^\ominus$ and $a_{\mathrm{Cl^-}} = \gamma_\pm b/b^\ominus$ as we saw in Section 5.9, so

$$E = E^\ominus - \frac{RT}{F}\ln b^2 - \frac{RT}{F}\ln\gamma_\pm^2$$

where for simplicity we have replaced $b/b^\ominus$ by b. This expression rearranges to

$$E + \frac{2RT}{F}\ln b = E^\ominus - \frac{2RT}{F}\ln\gamma_\pm \tag{7.32}$$

From the Debye–Hückel limiting law for a 1,1-electrolyte (Section 5.9; a 1,1-electrolyte is a solution of singly charged M^+ and X^- ions), we know that $\ln\gamma_\pm \propto -b^{1/2}$. The natural logarithm used here is proportional to the common logarithm that appears in eqn 5.69 (because $\ln x = \ln 10 \log x = 2.303 \log x$). Therefore, with the constant of proportionality in this relation written as $(F/2RT)C$, eqn 7.32 becomes

$$E + \frac{2RT}{F}\ln b = E^\ominus + Cb^{1/2} \tag{7.33}$$

The expression on the left is evaluated at a range of molalities, plotted against $b^{1/2}$, and extrapolated to $b = 0$. The intercept at $b^{1/2} = 0$ is the value of $E^\ominus$ for the silver/silver-chloride electrode. In precise work, the $b^{1/2}$ term is brought to the left, and a higher-order correction term from the extended Debye–Hückel law is used on the right.

Illustration 7.12 *Determining the standard emf of a cell*

The emf of the cell Pt(s) | H_2(g, $p^⊖$) | HCl(aq, b) | AgCl(s) | Ag(s) at 25°C has the following values:

$b/(10^{-3}\,b^⊖)$	3.215	5.619	9.138	25.63
E/V	0.52053	0.49257	0.46860	0.41824

To determine the standard emf of the cell we draw up the following table, using $2RT/F = 0.051\,39$ V:

$b/(10^{-3}\,b^⊖)$	3.215	5.619	9.138	25.63
$\{b/(10^{-3}\,b^⊖)\}^{1/2}$	1.793	2.370	3.023	5.063
E/V	0.52053	0.49257	0.46860	0.41824
$E/\text{V} + 0.051\,39 \ln b$	0.2256	0.2263	0.2273	0.2299

The data are plotted in Fig. 7.16; as can be seen, they extrapolate to $E^⊖ = 0.2232$ V.

Self-test 7.9 The data below are for the cell Pt(s) | H_2(g, $p^⊖$) | HBr(aq, b) | AgBr(s) | Ag(s) at 25°C. Determine the standard emf of the cell.

$b/(10^{-4}\,b^⊖)$	4.042	8.444	37.19
E/V	0.47381	0.43636	0.36173

[0.071 V]

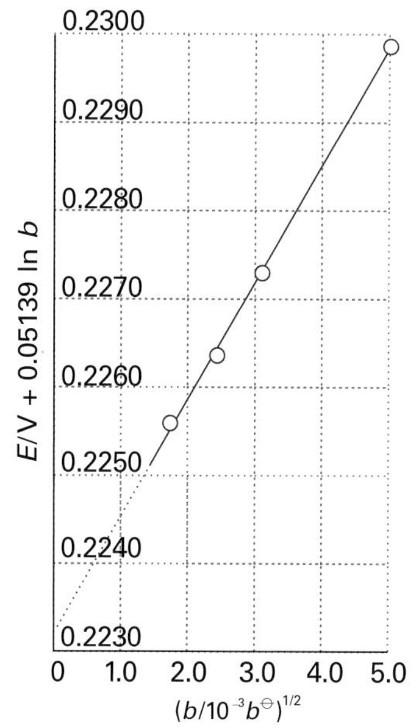

Fig. 7.16 The plot and the extrapolation used for the experimental measurement of a standard cell emf. The intercept at $b^{1/2} = 0$ is $E^⊖$.

Exploration Suppose that the procedure in *Illustration 7.12* results in a plot that deviates from linearity. What might be the cause of this behaviour? How might you modify the procedure to obtain a reliable value of the standard potential?

Table 7.2 lists standard potentials at 298 K. An important feature of standard emf of cells and standard potentials of electrodes is that they are unchanged if the chemical equation for the cell reaction or a half-reaction is multiplied by a numerical factor. A numerical factor increases the value of the standard Gibbs energy for the reaction. However, it also increases the number of electrons transferred by the same factor, and by eqn 7.27 the value of $E^⊖$ remains unchanged. A practical consequence is that a cell emf is independent of the physical size of the cell. In other words, cell emf is an intensive property.

The standard potentials in Table 7.2 may be combined to give values for couples that are not listed there. However, to do so, we must take into account the fact that different couples may correspond to the transfer of different numbers of electrons. The procedure is illustrated in the following *Example*.

Synoptic Table 7.2* Standard potentials at 298 K

Couple	$E^⊖$/V
$Ce^{4+}(aq) + e^- \rightarrow Ce^{3+}(aq)$	+1.61
$Cu^{2+}(aq) + 2\,e^- \rightarrow Cu(s)$	+0.34
$H^+(aq) + e^- \rightarrow \frac{1}{2}H_2(g)$	0
$AgCl(s) + e^- \rightarrow Ag(s) + Cl^-(aq)$	+0.22
$Zn^{2+}(aq) + 2\,e^- \rightarrow Zn(s)$	−0.76
$Na^+(aq) + e^- \rightarrow Na(s)$	−2.71

* More values are given in the *Data section*.

Example 7.4 *Evaluating a standard potential from two others*

Given that the standard potentials of the Cu^{2+}/Cu and Cu^+/Cu couples are +0.340 V and +0.522 V, respectively, evaluate $E^⊖(Cu^{2+},Cu^+)$.

Method First, we note that reaction Gibbs energies may be added (as in a Hess's law analysis of reaction enthalpies). Therefore, we should convert the $E^⊖$ values to $\Delta G^⊖$ values by using eqn 7.27, add them appropriately, and then convert the overall $\Delta G^⊖$ to the required $E^⊖$ by using eqn 7.27 again. This roundabout procedure is necessary because, as we shall see, although the factor F cancels, the factor ν in general does not.

Answer The electrode reactions are as follows:

(a) $Cu^{2+}(aq) + 2\,e^- \rightarrow Cu(s)$ $\quad E^⊖ = +0.340$ V, $\quad$ so $\Delta_r G^⊖ = -2(0.340\text{ V})F$

(b) $Cu^+(aq) + e^- \rightarrow Cu(s)$ $\quad E^⊖ = +0.522$ V, $\quad$ so $\Delta_r G^⊖ = -(0.522\text{ V})F$

The required reaction is

$$\text{(c) } Cu^{2+}(aq) + e^- \rightarrow Cu^+(aq) \qquad E^\ominus = -\Delta_r G^\ominus / F$$

Because (c) = (a) − (b), the standard Gibbs energy of reaction (c) is

$$\Delta_r G^\ominus = \Delta_r G^\ominus(a) - \Delta_r G^\ominus(b) = -(-0.158\ \text{V}) \times F$$

Therefore, $E^\ominus = +0.158$ V. Note that the generalization of the calculation we just performed is

$$\nu_c E^\ominus(c) = \nu_a E^\ominus(a) + \nu_b E^\ominus(b) \qquad (7.34)$$

A note on good practice Whenever combining standard potentials to obtain the standard potential of a third couple, always work via the Gibbs energies because they are additive, whereas, in general, standard potentials are not.

Self-test 7.10 Calculate the standard potential of the Fe^{3+}/Fe couple from the values for the Fe^{3+}/Fe^{2+} and Fe^{2+}/Fe couples. [−0.037 V]

7.9 Applications of standard potentials

Cell emfs are a convenient source of data on equilibrium constants and the Gibbs energies, enthalpies, and entropies of reactions. In practice the standard values of these quantities are the ones normally determined.

(a) The electrochemical series

We have seen that for two redox couples, Ox_1/Red_1 and Ox_2/Red_2, and the cell

$$Red_1,Ox_1 \| Red_2,Ox_2 \qquad E^\ominus = E_2^\ominus - E_1^\ominus \qquad (7.35a)$$

that the cell reaction

$$Red_1 + Ox_2 \rightarrow Ox_1 + Red_2 \qquad (7.35b)$$

is spontaneous as written if $E^\ominus > 0$, and therefore if $E_2^\ominus > E_1^\ominus$. Because in the cell reaction Red_1 reduces Ox_2, we can conclude that

Red_1 has a thermodynamic tendency to reduce Ox_2 if $E_1^\ominus < E_2^\ominus$

More briefly: low reduces high.

Illustration 7.13 *Using the electrochemical series*

Because $E^\ominus(Zn^{2+},Zn) = -0.76\ \text{V} < E^\ominus(Cu^{2+},Cu) = +0.34$ V, zinc has a thermodynamic tendency to reduce Cu^{2+} ions in aqueous solution.

Table 7.3 The electrochemical series of the metals*

Least strongly reducing
Gold
Platinum
Silver
Mercury
Copper
(Hydrogen)
Lead
Tin
Nickel
Iron
Zinc
Chromium
Aluminium
Magnesium
Sodium
Calcium
Potassium
Most strongly reducing

* The complete series can be inferred from Table 7.2.

Table 7.3 shows a part of the **electrochemical series**, the metallic elements (and hydrogen) arranged in the order of their reducing power as measured by their standard potentials in aqueous solution. A metal low in the series (with a lower standard potential) can reduce the ions of metals with higher standard potentials. This conclusion is qualitative. The quantitative value of K is obtained by doing the calculations we have described previously. For example, to determine whether zinc can displace magnesium from aqueous solutions at 298 K, we note that zinc lies above magnesium in the electrochemical series, so zinc cannot reduce magnesium ions in aqueous solution. Zinc can reduce hydrogen ions, because hydrogen lies higher in the series. However, even for reactions that are thermodynamically favourable, there may be kinetic factors that result in very slow rates of reaction.

The reactions of the electron transport chains of respiration are good applications of this principle.

IMPACT ON BIOCHEMISTRY

I7.2 Energy conversion in biological cells

The whole of life's activities depends on the coupling of exergonic and endergonic reactions, for the oxidation of food drives other reactions forward. In biological cells, the energy released by the oxidation of foods is stored in adenosine triphosphate (ATP, **1**). The essence of the action of ATP is its ability to lose its terminal phosphate group by hydrolysis and to form adenosine diphosphate (ADP):

$$\mathrm{ATP(aq) + H_2O(l) \rightarrow ADP(aq) + P_i^-(aq) + H_3O^+(aq)}$$

where P_i^- denotes an inorganic phosphate group, such as $H_2PO_4^-$. The biological standard values for ATP hydrolysis at 37°C (310 K, blood temperature) are $\Delta_r G^{\oplus} = -31$ kJ mol^{-1}, $\Delta_r H^{\oplus} = -20$ kJ mol^{-1}, and $\Delta_r S^{\oplus} = +34$ J K^{-1} mol^{-1}. The hydrolysis is therefore exergonic ($\Delta_r G^{\oplus} < 0$) under these conditions and 31 kJ mol^{-1} is available for driving other reactions. Moreover, because the reaction entropy is large, the reaction Gibbs energy is sensitive to temperature. In view of its exergonicity the ADP—phosphate bond has been called a 'high-energy phosphate bond'. The name is intended to signify a high tendency to undergo reaction, and should not be confused with 'strong' bond. In fact, even in the biological sense it is not of very 'high energy'. The action of ATP depends on it being intermediate in activity. Thus ATP acts as a phosphate donor to a number of acceptors (for example, glucose), but is recharged by more powerful phosphate donors in a number of biochemical processes.

We now use the oxidation of glucose to CO_2 and H_2O by O_2 as an example of how the breakdown of foods is coupled to the formation of ATP in the cell. The process begins with *glycolysis*, a partial oxidation of glucose by nicotinamide adenine dinucleotide (NAD^+, **2**) to pyruvate ion, $CH_3COCO_2^-$, continues with the *citric acid cycle*, which oxidizes pyruvate to CO_2, and ends with *oxidative phosphorylation*, which reduces O_2 to H_2O. Glycolysis is the main source of energy during *anaerobic metabolism*, a form of metabolism in which inhaled O_2 does not play a role. The citric acid cycle and oxidative phosphorylation are the main mechanisms for the extraction of energy from carbohydrates during *aerobic metabolism*, a form of metabolism in which inhaled O_2 does play a role.

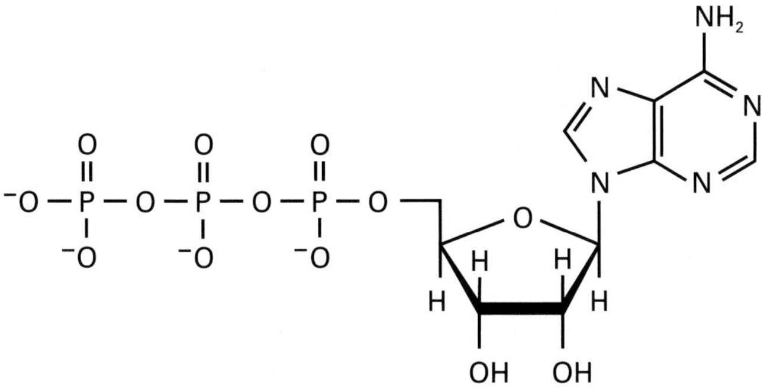

1 ATP

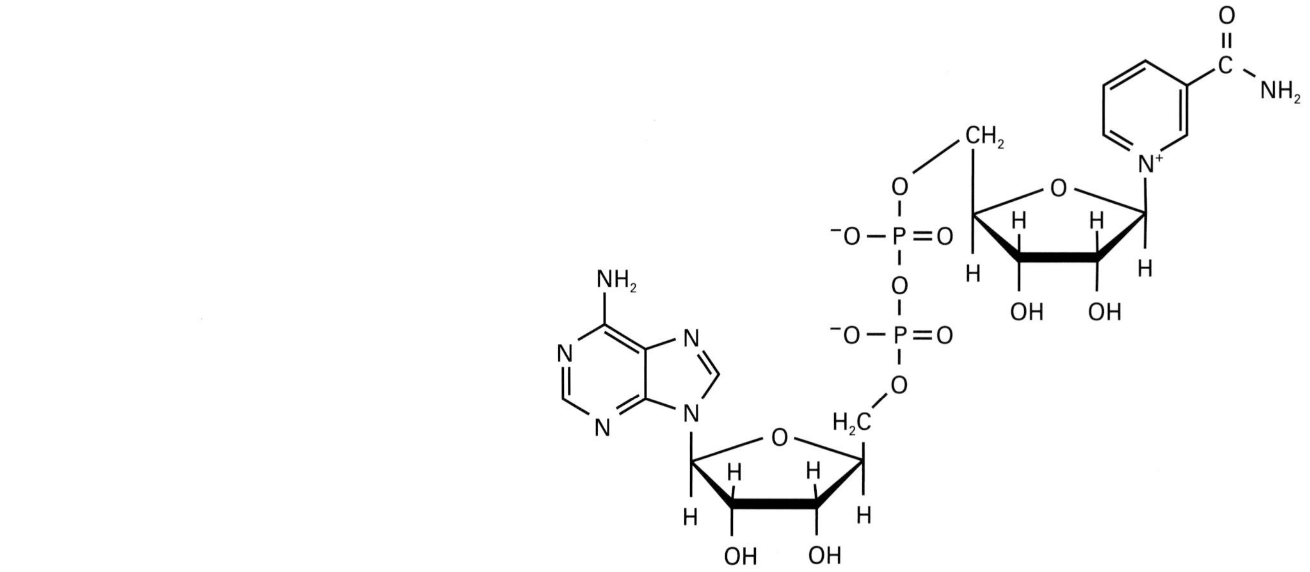

2 NAD+

Glycolysis

Glycolysis occurs in the *cytosol*, the aqueous material encapsulated by the cell membrane, and consists of ten enzyme-catalysed reactions. At blood temperature, $\Delta_r G^{\ominus} = -147\ \text{kJ mol}^{-1}$ for the oxidation of glucose by NAD^+ to pyruvate ions. The oxidation of one glucose molecule is coupled to the conversion of two ADP molecules to two ATP molecules, so the net reaction of glycolysis is:

$$C_6H_{12}O_6(aq) + 2\,NAD^+(aq) + 2\,ADP(aq) + 2\,P_i^-(aq) + 2\,H_2O(l)$$
$$\rightarrow 2\,CH_3COCO_2^-(aq) + 2\,NADH(aq) + 2\,ATP(aq) + 2\,H_3O^+(aq)$$

The standard reaction Gibbs energy is $(-147) - 2(-31)\ \text{kJ mol}^{-1} = -85\ \text{kJ mol}^{-1}$. The reaction is exergonic, and therefore spontaneous: the oxidation of glucose is used to 'recharge' the ATP. In cells that are deprived of O_2, pyruvate ion is reduced to lactate ion, $CH_3C(OH)CO_2^-$, by NADH.[2] Very strenuous exercise, such as bicycle racing, can decrease sharply the concentration of O_2 in muscle cells and the condition known as muscle fatigue results from increased concentrations of lactate ion.

The citric acid cycle

The standard Gibbs energy of combustion of glucose is $-2880\ \text{kJ mol}^{-1}$, so terminating its oxidation at pyruvate is a poor use of resources. In the presence of O_2, pyruvate is oxidized further during the citric acid cycle and oxidative phosphorylation, which occur in a special compartment of the cell called the *mitochondrion*. The citric acid cycle requires eight enzymes that couple the synthesis of ATP to the oxidation of pyruvate by NAD^+ and flavin adenine dinucleotide (FAD, **3**):

$$2\,CH_3COCO_2^-(aq) + 8\,NAD^+(aq) + 2\,FAD(aq) + 2\,ADP(aq) + 2\,P_i(aq) + 8\,H_2O(l)$$
$$\rightarrow 6\,CO_2(g) + 8\,NADH(aq) + 4\,H_3O^+(aq) + 2\,FADH_2(aq) + 2\,ATP(aq)$$

The NADH and $FADH_2$ go on to reduce O_2 during oxidative phosphorylation, which also produces ATP. The citric acid cycle and oxidative phosphorylation generate as many as 38 ATP molecules for each glucose molecule consumed. Each mole of ATP molecules extracts 31 kJ from the 2880 kJ supplied by 1 mol $C_6H_{12}O_6$ (180 g of

[2] In yeast, the terminal products are ethanol and CO_2.

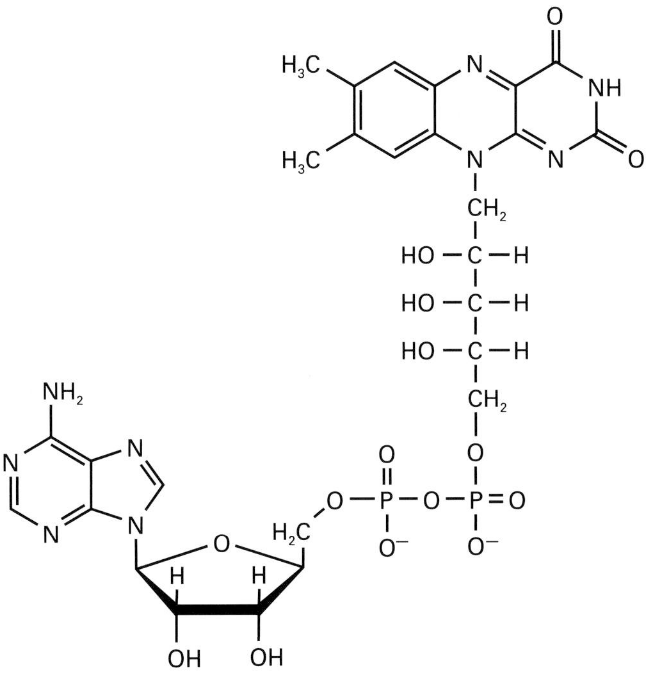

3 FAD

glucose), so 1178 kJ is stored for later use. Therefore, aerobic oxidation of glucose is much more efficient than glycolysis.

In the cell, each ATP molecule can be used to drive an endergonic reaction for which $\Delta_r G^{\ominus}$ does not exceed +31 kJ mol^{-1}. For example, the biosynthesis of sucrose from glucose and fructose can be driven by plant enzymes because the reaction is endergonic to the extent $\Delta_r G^{\ominus} = +23$ kJ mol^{-1}. The biosynthesis of proteins is strongly endergonic, not only on account of the enthalpy change but also on account of the large decrease in entropy that occurs when many amino acids are assembled into a precisely determined sequence. For instance, the formation of a peptide link is endergonic, with $\Delta_r G^{\ominus} = +17$ kJ mol^{-1}, but the biosynthesis occurs indirectly and is equivalent to the consumption of three ATP molecules for each link. In a moderately small protein like myoglobin, with about 150 peptide links, the construction alone requires 450 ATP molecules, and therefore about 12 mol of glucose molecules for 1 mol of protein molecules.

The respiratory chain

In the exergonic oxidation of glucose 24 electrons are transferred from each $C_6H_{12}O_6$ molecule to six O_2 molecules. The half-reactions for the oxidation of glucose and the reduction of O_2 are

$$C_6H_{12}O_6(s) + 6\,H_2O(l) \rightarrow 6\,CO_2(g) + 24\,H^+(aq) + 24\,e^-$$
$$6\,O_2(g) + 24\,H^+(aq) + 24\,e^- \rightarrow 12\,H_2O(l)$$

The electrons do not flow directly from glucose to O_2. We have already seen that, in biological cells, glucose is oxidized to CO_2 by NAD^+ and FAD during glycolysis and the citric acid cycle:

$$C_6H_{12}O_6(s) + 10\,NAD^+ + 2\,FAD + 4\,ADP + 4\,P_i^- + 2\,H_2O$$
$$\rightarrow 6\,CO_2 + 10\,NADH + 2\,FADH_2 + 4\,ATP + 6\,H^+$$

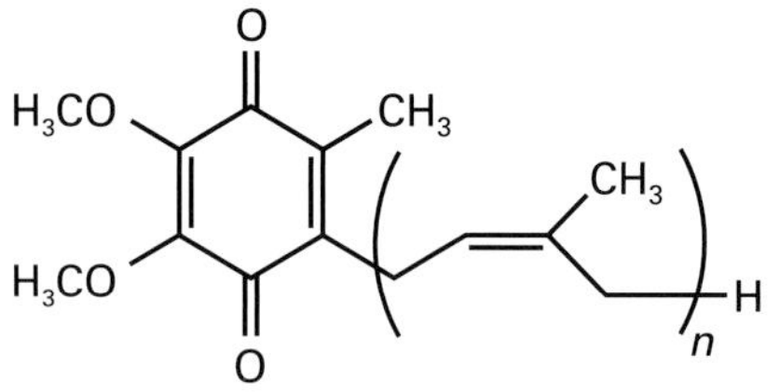

4 Coenzyme Q, Q

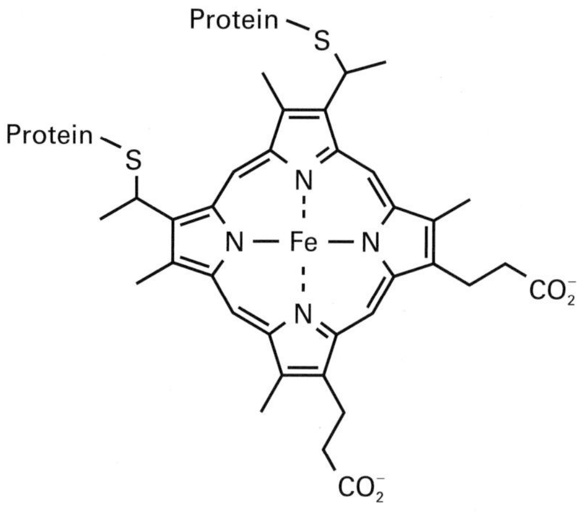

5 Heme *c*

In the *respiratory chain*, electrons from the powerful reducing agents NADH and $FADH_2$ pass through four membrane-bound protein complexes and two mobile electron carriers before reducing O_2 to H_2O. We shall see that the electron transfer reactions drive the synthesis of ATP at three of the membrane protein complexes.

The respiratory chain begins in complex I (NADH-Q oxidoreductase), where NADH is oxidized by coenzyme Q (Q, **4**) in a two-electron reaction:

$$H^+ + NADH + Q \xrightarrow{\text{complex I}} NAD^+ + QH_2 \qquad E^{\oplus} = +0.42\ \text{V}, \quad \Delta_r G^{\oplus} = -81\ \text{kJ mol}^{-1}$$

Additional Q molecules are reduced by $FADH_2$ in complex II (succinate-Q reductase):

$$FADH_2 + Q \xrightarrow{\text{complex II}} FAD + QH_2 \qquad E^{\oplus} = +0.015\ \text{V}, \qquad \Delta_r G^{\oplus} = -2.9\ \text{kJ mol}^{-1}$$

Reduced Q migrates to complex III (Q-cytochrome *c* oxidoreductase), which catalyses the reduction of the protein cytochrome *c* (Cyt *c*). Cytochrome *c* contains the haem *c* group (**5**), the central iron ion of which can exist in oxidation states +3 and +2. The net reaction catalysed by complex III is

$$QH_2 + 2\,Fe^{3+}(\text{Cyt } c) \xrightarrow{\text{complex III}} Q + 2\,Fe^{2+}(\text{Cyt } c) + 2\,H^+$$
$$E^{\oplus} = +0.15\ \text{V}, \qquad \Delta_r G^{\oplus} = -30\ \text{kJ mol}^{-1}$$

Reduced cytochrome *c* carries electrons from complex III to complex IV (cytochrome *c* oxidase), where O_2 is reduced to H_2O:

$$2\,Fe^{2+}(\text{Cyt } c) + 2\,H^+ + \tfrac{1}{2}\,O_2 \xrightarrow{\text{complex IV}} 2\,Fe^{3+}(\text{Cyt } c) + H_2O$$
$$E^{\oplus} = +0.815\ \text{V}, \qquad \Delta_r G^{\oplus} = -109\ \text{kJ mol}^{-1}$$

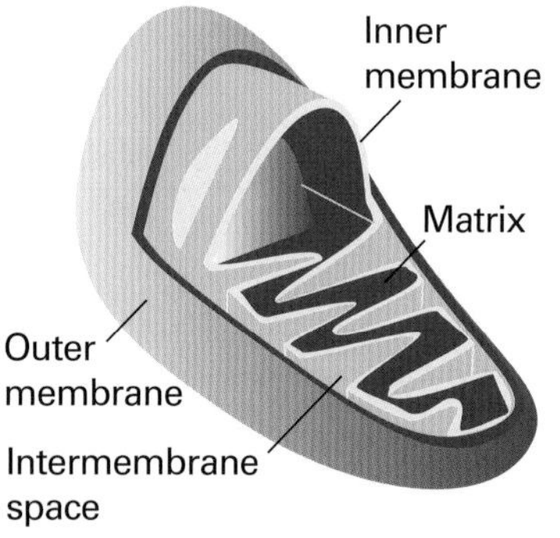

Fig. 7.17 The general features of a typical mitochondrion.

Oxidative phosphorylation

The reactions that occur in complexes I, III, and IV are sufficiently exergonic to drive the synthesis of ATP in the process called *oxidative phosphorylation*:

$$ADP + P_i^- + H^+ \rightarrow ATP \qquad \Delta_r G^{\oplus} = +31\ \text{kJ mol}^{-1}$$

We saw above that the phosphorylation of ADP to ATP can be coupled to the exergonic dephosphorylation of other molecules. Indeed, this is the mechanism by which ATP is synthesized during glycolysis and the citric acid cycle. However, oxidative phosphorylation operates by a different mechanism.

The structure of a mitochondrion is shown in Fig. 7.17. The protein complexes associated with the electron transport chain span the inner membrane and phosphorylation takes place in the matrix. The Gibbs energy of the reactions in complexes I, III, and IV is first used to do the work of moving protons across the mitochondrial membrane. The complexes are oriented asymmetrically in the inner membrane so that the protons abstracted from one side of the membrane can be deposited on the other side. For example, the oxidation of NADH by Q in complex I is coupled to the transfer of four protons across the membrane. The coupling of electron transfer and proton pumping in complexes III and IV contribute further to a gradient of proton concentration across the membrane. Then the enzyme H^+-ATPase uses the energy stored in the proton gradient to phosphorylate ADP to ATP. Experiments show that 11 molecules of ATP are made for every three molecules of NADH and one molecule of $FADH_2$ that are oxidized by the respiratory chain. The ATP is then hydrolysed on demand to perform useful biochemical work throughout the cell.

The *chemiosmotic theory* proposed by Peter Mitchell explains how H^+-ATPases synthesize ATP from ADP. The energy stored in a transmembrane proton gradient come from two contributions. First, the difference in activity of H^+ ion results in a difference in molar Gibbs energy across the mitochrondrial membrane

$$\Delta G_{m,1} = G_{m,in} - G_{m,out} = RT \ln \frac{a_{H^+,in}}{a_{H^+,out}}$$

Second, there is a membrane potential difference $\Delta\phi = \phi_{in} - \phi_{out}$ that arises from differences in Coulombic interactions on each side of the membrane. The charge difference across a membrane per mole of H^+ ions is $N_A e$, or F, where $F = eN_A$. It follows from *Justification 7.3*, that the molar Gibbs energy difference is then $\Delta G_{m,2} = F\Delta\phi$. Adding this contribution to $\Delta G_{m,1}$ gives the total Gibbs energy stored by the combination of an an activity gradient and a membrane potential gradient:

$$\Delta G_m = RT \ln \frac{[H^+]_{in}}{[H^+]_{out}} + F\Delta\phi$$

where we have replaced activities by molar concentrations. This equation also provides an estimate of the Gibbs energy available for phosphorylation of ADP. After using $\ln [H^+] \approx \ln 10 \times \log [H^+]$ and substituting $\Delta pH = pH_{in} - pH_{out} = -\log [H^+]_{in} + \log [H^+]_{out}$, it follows that

$$\Delta G_m = F\Delta\phi - (RT \ln 10)\Delta pH$$

In the mitochondrion, $\Delta pH \approx -1.4$ and $\Delta\phi \approx 0.14$ V, so $\Delta G_m \approx +21.5$ kJ mol^{-1}. Because 31 kJ mol^{-1} is needed for phosphorylation, we conclude that at least 2 mol H^+ (and probably more) must flow through the membrane for the phosphorylation of 1 mol ADP.

(b) The determination of activity coefficients

Once the standard potential of an electrode in a cell is known, we can use it to determine mean activity coefficients by measuring the cell emf with the ions at the concentration of interest. For example, the mean activity coefficient of the ions in hydrochloric acid of molality b is obtained from eqn 7.32 in the form

$$\ln \gamma_\pm = \frac{E^\ominus - E}{2RT/F} - \ln b \qquad \{7.36\}$$

once E has been measured.

(c) The determination of equilibrium constants

The principal use for standard potentials is to calculate the standard emf of a cell formed from any two electrodes. To do so, we subtract the standard potential of the left-hand electrode from the standard potential of the right-hand electrode:

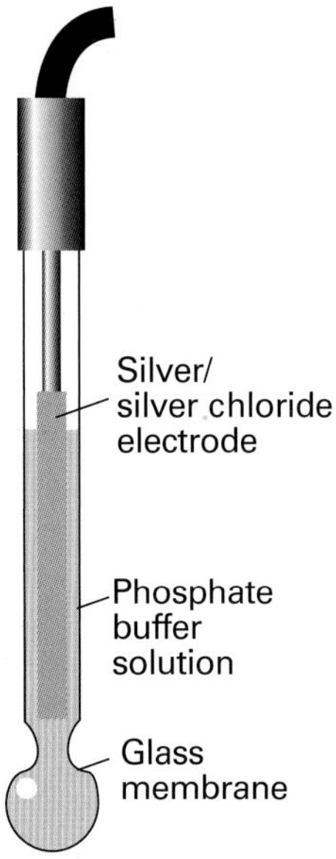

Fig. 7.18 The glass electrode. It is commonly used in conjunction with a calomel electrode that makes contact with the test solution through a salt bridge.

$$E^\ominus = E^\ominus(\text{right}) - E^\ominus(\text{left}) \tag{7.37}$$

Because $\Delta G^\ominus = -\nu FE^\ominus$, it then follows that, if the result gives $E^\ominus > 0$, then the corresponding cell reaction has $K > 1$.

Illustration 7.14 *Calculating an equilibrium constant from standard potentials*

A disproportionation is a reaction in which a species is both oxidized and reduced. To study the disproportionation $2\,Cu^+(aq) \rightarrow Cu(s) + Cu^{2+}(aq)$ we combine the following electrodes:

Right-hand electrode:
$Cu(s)|Cu^+(aq)$ $\quad Cu^+(aq) + e^- \rightarrow Cu(aq)$ $\quad E^\ominus = +0.52\text{ V}$
Left-hand electrode:
$Pt(s)|Cu^{2+}(aq),Cu^+(aq)$ $\quad Cu^{2+}(aq) + e^- \rightarrow Cu^+(s)$ $\quad E^\ominus = +0.16\text{ V}$

where the standard potentials are measured at 298 K. The standard emf of the cell is therefore

$$E^\ominus = +0.52\text{ V} - 0.16\text{ V} = +0.36\text{ V}$$

We can now calculate the equilibrium constant of the cell reaction. Because $\nu = 1$, from eqn 7.30

$$\ln K = \frac{0.36\text{ V}}{0.025693\text{ V}} = \frac{0.36}{0.025693}$$

Hence, $K = 1.2 \times 10^6$.

Self-test 7.11 Calculate the solubility constant (the equilibrium constant for the reaction $Hg_2Cl_2(s) \rightleftharpoons Hg_2^{2+}(aq) + 2\,Cl^-(aq)$) and the solubility of mercury(I) chloride at 298.15 K. *Hint.* The mercury(I) ion is the diatomic species Hg_2^{2+}.

[2.6×10^{-18}, $8.7 \times 10^{-7}\text{ mol kg}^{-1}$]

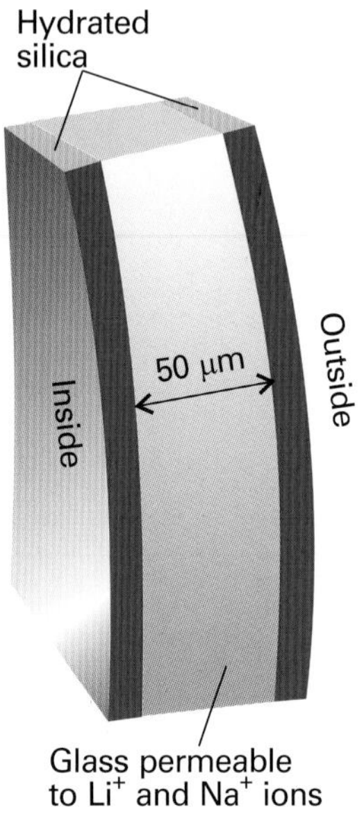

Fig. 7.19 A section through the wall of a glass electrode.

(d) Species-selective electrodes

An **ion-selective electrode** is an electrode that generates a potential in response to the presence of a solution of specific ions. An example is the **glass electrode** (Fig. 7.18), which is sensitive to hydrogen ion activity, and has a potential proportional to pH. It is filled with a phosphate buffer containing Cl^- ions, and conveniently has $E = 0$ when the external medium is at pH = 7. It is necessary to calibrate the glass electrode before use with solutions of known pH.

The responsiveness of a glass electrode to the hydrogen ion activity is a result of complex processes at the interface between the glass membrane and the solutions on either side of it. The membrane itself is permeable to Na^+ and Li^+ ions but not to H^+ ions. Therefore, the potential difference across the glass membrane must arise by a mechanism different from that responsible for biological transmembrane potentials (*Impact on biochemistry 7.2*). A clue to the mechanism comes from a detailed inspection of the glass membrane, for each face is coated with a thin layer of hydrated silica (Fig. 7.19). The hydrogen ions in the test solution modify this layer to an extent that depends on their activity in the solution, and the charge modification of the outside layer is transmitted to the inner layer by the Na^+ and Li^+ ions in the glass. The hydrogen ion activity gives rise to a membrane potential by this indirect mechanism.

Electrodes sensitive to hydrogen ions, and hence to pH, are typically glasses based on lithium silicate doped with heavy-metal oxides. The glass can also be made responsive to Na^+, K^+, and NH_4^+ ions by being doped with Al_2O_3 and B_2O_3.

A suitably adapted glass electrode can be used to detect the presence of certain gases. A simple form of a **gas-sensing electrode** consists of a glass electrode contained in an outer sleeve filled with an aqueous solution and separated from the test solution by a membrane that is permeable to gas. When a gas such as sulfur dioxide or ammonia diffuses into the aqueous solution, it modifies its pH, which in turn affects the potential of the glass electrode. The presence of an enzyme that converts a compound, such as urea or an amino acid, into ammonia, which then affects the pH, can be used to detect these organic compounds.

Somewhat more sophisticated devices are used as ion-selective electrodes that give potentials according to the presence of specific ions present in a test solution. In one arrangement, a porous lipophilic (hydrocarbon-attracting) membrane is attached to a small reservoir of a hydrophobic (water-repelling) liquid, such as dioctylphenylphosphonate, that saturates it (Fig. 7.20). The liquid contains an agent, such as $(RO)_2PO_2^-$ with R a C_8 to C_{18} chain, that acts as a kind of solubilizing agent for the ions with which it can form a complex. The complex's ions are able to migrate through the lipophilic membrane, and hence give rise to a transmembrane potential, which is detected by a silver/silver chloride electrode in the interior of the assembly. Electrodes of this construction can be designed to be sensitive to a variety of ionic species, including calcium, zinc, iron, lead, and copper ions.

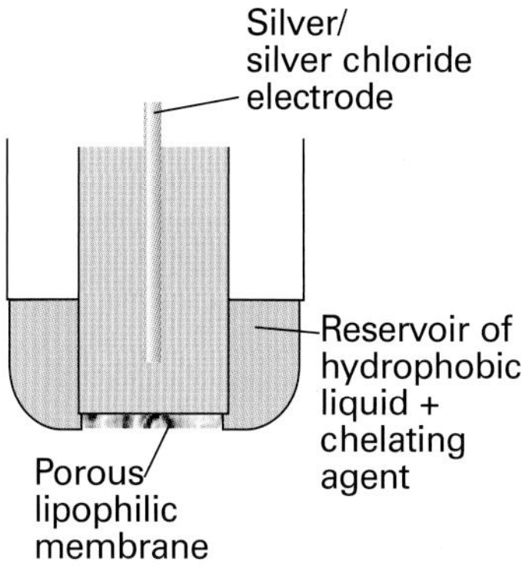

Fig. 7.20 The structure of an ion-selective electrode. Chelated ions are able to migrate through the lipophilic membrane.

In theory, the transmembrane potential should be determined entirely by differences in the activity of the species that the electrode was designed to detect. In practice, a small potential difference, called the **asymmetry potential**, is observed even when the activity of the test species is the same on both sides of the membrane. The asymmetry potential is due to the fact that it is not possible to manufacture a membrane material that has the same structure and the same chemical properties throughout. Furthermore, all species-selective electrodes are sensitive to more than one species. For example, a Na^+ selective electrode also responds, albeit less effectively, to the activity of K^+ ions in the test solution. As a result of these effects, the potential of an electrode sensitive to species X^+ that is also susceptible to interference by species Y^+ is given by a modified form of the Nernst equation:

$$E = E_{ap} + \beta \frac{RT}{F} \ln(a_{X^+} + k_{X,Y} a_{Y^+}) \qquad (7.38)$$

where E_{ap} is the asymmetry potential, β is an experimental parameter that captures deviations from the Nernst equation, and $k_{X,Y}$ is the **selectivity coefficient** of the electrode and is related to the response of the electrode to the interfering species Y^+. A value of $\beta = 1$ indicates that the electrode responds to the activity of ions in solution in a way that is consistent with the Nernst equation and, in practice, most species-selective electrodes of high quality have $\beta \approx 1$. The selectivity coefficient, and hence interference effects, can be minimized when designing and manufacturing a species-selective electrode. For precise work, it is necessary to calibrate the response of the electrode by measuring E_{ap}, β, and $k_{X,Y}$ before performing experiments on solutions of unknown concentration of X^+.

(e) The determination of thermodynamic functions

The standard emf of a cell is related to the standard reaction Gibbs energy through eqn 7.28 ($\Delta_r G^{\ominus} = -\nu F E^{\ominus}$). Therefore, by measuring $E^{\ominus}$ we can obtain this important thermodynamic quantity. Its value can then be used to calculate the Gibbs energy of formation of ions by using the convention explained in Section 3.6.

Illustration 7.15 *Determining the Gibbs energy of formation of an ion electrochemically*

The cell reaction taking place in

$$\mathrm{Pt(s)|H_2|H^+(aq)||Ag^+(aq)|Ag(s)} \qquad E^{\ominus} = +0.7996\ \mathrm{V}$$

is

$$\mathrm{Ag^+(aq)} + \tfrac{1}{2}\mathrm{H_2(g)} \rightarrow \mathrm{H^+(aq)} + \mathrm{Ag(s)} \qquad \Delta_r G^{\ominus} = -\Delta_f G^{\ominus}(\mathrm{Ag^+, aq})$$

Therefore, with $\nu = 1$, we find

$$\Delta_f G^{\ominus}(\mathrm{Ag^+, aq}) = -(-FE^{\ominus}) = +77.15\ \mathrm{kJ\ mol^{-1}}$$

which is in close agreement with the value in Table 2.6 of the *Data section*.

The temperature coefficient of the standard cell emf, $\mathrm{d}E^{\ominus}/\mathrm{d}T$, gives the standard entropy of the cell reaction. This conclusion follows from the thermodynamic relation $(\partial G/\partial T)_p = -S$ and eqn 7.27, which combine to give

$$\frac{\mathrm{d}E^{\ominus}}{\mathrm{d}T} = \frac{\Delta_r S^{\ominus}}{\nu F} \tag{7.39}$$

The derivative is complete because $E^{\ominus}$, like $\Delta_r G^{\ominus}$, is independent of the pressure. Hence we have an electrochemical technique for obtaining standard reaction entropies and through them the entropies of ions in solution.

Finally, we can combine the results obtained so far and use them to obtain the standard reaction enthalpy:

$$\Delta_r H^{\ominus} = \Delta_r G^{\ominus} + T\Delta_r S^{\ominus} = -\nu F\left(E^{\ominus} - T\frac{\mathrm{d}E^{\ominus}}{\mathrm{d}T}\right) \tag{7.40}$$

This expression provides a non-calorimetric method for measuring $\Delta_r H^{\ominus}$ and, through the convention $\Delta_f H^{\ominus}(\mathrm{H^+, aq}) = 0$, the standard enthalpies of formation of ions in solution (Section 2.8). Thus, electrical measurements can be used to calculate all the thermodynamic properties with which this chapter began.

Example 7.5 *Using the temperature coefficient of the cell potential*

The standard emf of the cell $\mathrm{Pt(s)|H_2(g)|HBr(aq)|AgBr(s)|Ag(s)}$ was measured over a range of temperatures, and the data were fitted to the following polynomial:

$$E^{\ominus}/\mathrm{V} = 0.07131 - 4.99\times10^{-4}(T/\mathrm{K} - 298) - 3.45\times10^{-6}(T/\mathrm{K} - 298)^2$$

Evaluate the standard reaction Gibbs energy, enthalpy, and entropy at 298 K.

Method The standard Gibbs energy of reaction is obtained by using eqn 7.28 after evaluating $E^{\ominus}$ at 298 K and by using 1 V C = 1 J. The standard entropy of reaction is obtained by using eqn 7.39, which involves differentiating the polynomial with respect to T and then setting $T = 298$ K. The reaction enthalpy is obtained by combining the values of the standard Gibbs energy and entropy.

Answer At $T = 298$ K, $E^{\ominus} = +0.07131$ V, so

$$\begin{aligned}\Delta_r G^{\ominus} &= -\nu F E^{\ominus} = -(1)\times(9.6485\times10^4\ \mathrm{C\ mol^{-1}})\times(+0.07131\ \mathrm{V})\\ &= -6.880\times10^3\ \mathrm{V\ C\ mol^{-1}} = -6.880\ \mathrm{kJ\ mol^{-1}}\end{aligned}$$

The temperature coefficient of the cell potential is

$$\frac{dE^{\ominus}}{dT} = -4.99\times10^{-4}\ \mathrm{V\ K^{-1}} - 2(3.45\times10^{-6})(T/\mathrm{K}-298)\ \mathrm{V\ K^{-1}}$$

At $T = 298$ K this expression evaluates to

$$\frac{dE}{dT} = -4.99\times10^{-4}\ \mathrm{V\ K^{-1}}$$

So, from eqn 7.39, the reaction entropy is

$$\Delta_r S^{\ominus} = 1\times(9.6485\times10^{4}\ \mathrm{C\ mol^{-1}})\times(-4.99\times10^{-4}\ \mathrm{V\ K^{-1}}) = -48.2\ \mathrm{J\ K^{-1}\ mol^{-1}}$$

It then follows that

$$\Delta_r H^{\ominus} = \Delta_r G^{\ominus} + T\Delta_r S^{\ominus} = -6.880\ \mathrm{kJ\ mol^{-1}} + (298\ \mathrm{K})\times(-0.0482\ \mathrm{kJ\ K^{-1}\ mol^{-1}}) = -21.2\ \mathrm{kJ\ mol^{-1}}$$

One difficulty with this procedure lies in the accurate measurement of small temperature coefficients of cell potential. Nevertheless, it is another example of the striking ability of thermodynamics to relate the apparently unrelated, in this case to relate electrical measurements to thermal properties.

Self-test 7.12 Predict the standard potential of the Harned cell at 303 K from tables of thermodynamic data. [+0.2222 V]

Checklist of key ideas

- ☐ 1. The extent of reaction (ξ) is defined such that, when the extent of reaction changes by a finite amount $\Delta\xi$, the amount of A present changes from $n_{A,0}$ to $n_{A,0} - \Delta\xi$.
- ☐ 2. The reaction Gibbs energy is the slope of the graph of the Gibbs energy plotted against the extent of reaction: $\Delta_r G = (\partial G/\partial\xi)_{p,T}$; at equilibrium, $\Delta_r G = 0$.
- ☐ 3. An exergonic reaction is a reaction for which $\Delta_r G < 0$; such a reaction can be used to drive another process. An endergonic reaction is a reaction for which $\Delta_r G > 0$.
- ☐ 4. The general expression for $\Delta_r G$ at an arbitrary stage of the reaction is $\Delta_r G = \Delta_r G^{\ominus} + RT\ln Q$.
- ☐ 5. The equilibrium constant (K) may be written in terms of $\Delta_r G^{\ominus}$ as $\Delta_r G^{\ominus} = -RT\ln K$.
- ☐ 6. The standard reaction Gibbs energy may be calculated from standard Gibbs energies of formation, $\Delta_r G^{\ominus} = \sum_{\mathrm{Products}} \nu\Delta_f G^{\ominus} - \sum_{\mathrm{Reactants}} \nu\Delta_f G^{\ominus} = \sum_J \nu_J \Delta_f G^{\ominus}(J)$.
- ☐ 7. Thermodynamic equilibrium constant is an equilibrium constant K expressed in terms of activities (or fugacities): $K = \left(\prod_J a_J^{\nu_J}\right)_{\mathrm{equilibrium}}$.
- ☐ 8. A catalyst does not affect the equilibrium constant.
- ☐ 9. Changes in pressure do not affect the equilibrium constant: $(\partial K/\partial p)_T = 0$. However, partial pressures and concentrations can change in response to a change in pressure.
- ☐ 10. Le Chatelier's principle states that a system at equilibrium, when subjected to a disturbance, responds in a way that tends to minimize the effect of the disturbance.
- ☐ 11. Increased temperature favours the reactants in exothermic reactions and the products in endothermic reactions.
- ☐ 12. The temperature dependence of the equilibrium constant is given by the van 't Hoff equation: $d\ln K/dT = \Delta_r H^{\ominus}/RT^2$. To calculate K at one temperature in terms of its value at another temperature, and provided $\Delta_r H^{\ominus}$ is independent of temperature, we use $\ln K_2 - \ln K_1 = -(\Delta_r H^{\ominus}/R)(1/T_2 - 1/T_1)$.
- ☐ 13. A galvanic cell is an electrochemical cell that produces electricity as a result of the spontaneous reaction occurring inside it. An electrolytic cell is an electrochemical cell in which a non-spontaneous reaction is driven by an external source of current.
- ☐ 14. Oxidation is the removal of electrons from a species; reduction is the addition of electrons to a species; a redox

reaction is a reaction in which there is a transfer of electrons from one species to another.

☐ 15. The anode is the electrode at which oxidation occurs. The cathode is the electrode at which reduction occurs.

☐ 16. The electromotive force (emf) is the cell potential when it is balanced by an exactly opposing source of potential so that the cell reaction occurs reversibly, the composition is constant, and no current flows.

☐ 17. The cell potential and the reaction Gibbs energy are related by $-\nu FE = \Delta_r G$.

☐ 18. The standard emf is the standard reaction Gibbs energy expressed as a potential: $E^{\ominus} = \Delta_r G^{\ominus}/\nu F$.

☐ 19. The Nernst equation is the equation for the emf of a cell in terms of the composition: $E = E^{\ominus} - (RT/\nu F) \ln Q$.

☐ 20. The equilibrium constant for a cell reaction is related to the standard emf by $\ln K = \nu FE^{\ominus}/RT$.

☐ 21. The standard potential of a couple ($E^{\ominus}$) is the standard emf of a cell in which a couple forms the right-hand electrode and the standard hydrogen electrode is the left-hand electrode.

☐ 22. To calculate the standard emf, form the difference of electrode potentials: $E^{\ominus} = E^{\ominus}(\text{right}) - E^{\ominus}(\text{left})$.

☐ 23. The temperature coefficient of cell potential is given by $\mathrm{d}E^{\ominus}/\mathrm{d}T = \Delta_r S^{\ominus}/\nu F$.

☐ 24. The standard reaction entropy and enthalpy are calculated from the temperature dependence of the standard emf by: $\Delta_r S^{\ominus} = \nu F\mathrm{d}E^{\ominus}/\mathrm{d}T$, $\Delta_r H^{\ominus} = -\nu(FE^{\ominus} - T\mathrm{d}E^{\ominus}/\mathrm{d}T)$.

Further reading

Articles and texts

P.W. Atkins and J.C. de Paula, *Physical chemistry for the life sciences.* W.H. Freeman and Company, New York (2005).

A.J. Bard and L.R. Faulkner, *Electrochemical methods.* Wiley, New York (2000).

M.J. Blandamer, *Chemical equilibria in solution.* Ellis Horwood/Prentice Hall, Hemel Hempstead (1992).

W.A. Cramer and D.A. Knaff, *Energy transduction in biological membranes, a textbook of bioenergetics.* Springer–Verlag, New York (1990).

D.R. Crow, *Principles and applications of electrochemistry.* Blackie, London (1994).

K. Denbigh, *The principles of chemical equilibrium, with applications in chemistry and chemical engineering.* Cambridge University Press (1981).

C.H. Hamann, A. Hamnett, and W. Vielstich, *Electrochemistry.* Wiley-VCH, Weinheim (1998).

Sources of data and information

M.S. Antelman, *The encyclopedia of chemical electrode potentials*, Plenum, New York (1982).

A.J. Bard, R. Parsons, and J. Jordan (ed.), *Standard potentials in aqueous solution.* Marcel Dekker, New York (1985).

R.N. Goldberg and Y.B. Tewari, Thermodynamics of enzyme-catalyzed reactions. *J. Phys. Chem. Ref. Data.* Part 1: **22**, 515 (1993). Part 2: **23**, 547 (1994). Part 3: **23**, 1035 (1994). Part 4: **24**, 1669 (1995). Part 5: **24**, 1765 (1995).

Discussion questions

7.1 Explain how the mixing of reactants and products affects the position of chemical equilibrium.

7.2 Suggest how the thermodynamic equilibrium constant may respond differently to changes in pressure and temperature from the equilibrium constant expressed in terms of partial pressures.

7.3 Account for Le Chatelier's principle in terms of thermodynamic quantities.

7.4 Explain the molecular basis of the van 't Hoff equation for the temperature dependence of K.

7.5 (a) How may an Ellingham diagram be used to decide whether one metal may be used to reduce the oxide of another metal? (b) Use the Ellingham diagram in Fig. 7.10 to identify the lowest temperature at which zinc oxide can be reduced to zinc metal by carbon.

7.6 Distinguish between cell potential and electromotive force and explain why the latter is related to thermodynamic quantities.

7.7 Describe the contributions to the emf of cells formed by combining the electrodes specified in Table 7.1.

7.8 Describe a method for the determination of a standard potential of a redox couple.

7.9 Devise a method for the determination of the pH of an aqueous solution.

Exercises

7.1(a) At 2257 K and 1.00 atm total pressure, water is 1.77 per cent dissociated at equilibrium by way of the reaction $2\ H_2O(g) \rightleftharpoons 2\ H_2(g) + O_2(g)$. Calculate (a) K, (b) $\Delta_r G^\ominus$, and (c) $\Delta_r G$ at this temperature.

7.1(b) For the equilibrium, $N_2O_4(g) \rightleftharpoons 2\ NO_2(g)$, the degree of dissociation, α_e, at 298 K is 0.201 at 1.00 bar total pressure. Calculate (a) $\Delta_r G$, (2) K, and (3) $\Delta_r G^\ominus$ at 298 K.

7.2(a) Dinitrogen tetroxide is 18.46 per cent dissociated at 25°C and 1.00 bar in the equilibrium $N_2O_4(g) \rightleftharpoons 2\ NO_2(g)$. Calculate (a) K at 25°C, (b) $\Delta_r G^\ominus$, (c) K at 100°C given that $\Delta_r H^\ominus = +57.2\ \text{kJ mol}^{-1}$ over the temperature range.

7.2(b) Molecular bromine is 24 per cent dissociated at 1600 K and 1.00 bar in the equilibrium $Br_2(g) \rightleftharpoons 2\ Br(g)$. Calculate (a) K at 25°C, (b) $\Delta_r G^\ominus$, (c) K at 2000°C given that $\Delta_r H^\ominus = +112\ \text{kJ mol}^{-1}$ over the temperature range.

7.3(a) From information in the *Data section*, calculate the standard Gibbs energy and the equilibrium constant at (a) 298 K and (b) 400 K for the reaction $PbO(s) + CO(g) \rightleftharpoons Pb(s) + CO_2(g)$. Assume that the reaction enthalpy is independent of temperature.

7.3(b) From information in the *Data section*, calculate the standard Gibbs energy and the equilibrium constant at (a) 25°C and (b) 50°C for the reaction $CH_4(g) + 3\ Cl_2(g) \rightleftharpoons CHCl_3(l) + 3\ HCl(g)$. Assume that the reaction enthalpy is independent of temperature.

7.4(a) In the gas-phase reaction $2\ A + B \rightleftharpoons 3\ C + 2\ D$, it was found that, when 1.00 mol A, 2.00 mol B, and 1.00 mol D were mixed and allowed to come to equilibrium at 25°C, the resulting mixture contained 0.90 mol C at a total pressure of 1.00 bar. Calculate (a) the mole fractions of each species at equilibrium, (b) K_x, (c) K, and (d) $\Delta_r G^\ominus$.

7.4(b) In the gas-phase reaction $A + B \rightleftharpoons C + 2\ D$, it was found that, when 2.00 mol A, 1.00 mol B, and 3.00 mol D were mixed and allowed to come to equilibrium at 25°C, the resulting mixture contained 0.79 mol C at a total pressure of 1.00 bar. Calculate (a) the mole fractions of each species at equilibrium, (b) K_x, (c) K, and (d) $\Delta_r G^\ominus$.

7.5(a) The standard reaction enthalpy of $Zn(s) + H_2O(g) \rightarrow ZnO(s) + H_2(g)$ is approximately constant at $+224\ \text{kJ mol}^{-1}$ from 920 K up to 1280 K. The standard reaction Gibbs energy is $+33\ \text{kJ mol}^{-1}$ at 1280 K. Estimate the temperature at which the equilibrium constant becomes greater than 1.

7.5(b) The standard enthalpy of a certain reaction is approximately constant at $+125\ \text{kJ mol}^{-1}$ from 800 K up to 1500 K. The standard reaction Gibbs energy is $+22\ \text{kJ mol}^{-1}$ at 1120 K. Estimate the temperature at which the equilibrium constant becomes greater than 1.

7.6(a) The equilibrium constant of the reaction $2\ C_3H_6(g) \rightleftharpoons C_2H_4(g) + C_4H_8(g)$ is found to fit the expression $\ln K = A + B/T + C/T^2$ between 300 K and 600 K, with $A = -1.04$, $B = -1088$ K, and $C = 1.51 \times 10^5\ \text{K}^2$. Calculate the standard reaction enthalpy and standard reaction entropy at 400 K.

7.6(b) The equilibrium constant of a reaction is found to fit the expression $\ln K = A + B/T + C/T^3$ between 400 K and 500 K with $A = -2.04$, $B = -1176$ K, and $C = 2.1 \times 10^7\ \text{K}^3$. Calculate the standard reaction enthalpy and standard reaction entropy at 450 K.

7.7(a) The standard reaction Gibbs energy of the isomerization of borneol ($C_{10}H_{17}OH$) to isoborneol in the gas phase at 503 K is $+9.4\ \text{kJ mol}^{-1}$. Calculate the reaction Gibbs energy in a mixture consisting of 0.15 mol of borneol and 0.30 mol of isoborneol when the total pressure is 600 Torr.

7.7(b) The equilibrium pressure of H_2 over solid uranium and uranium hydride, UH_3, at 500 K is 139 Pa. Calculate the standard Gibbs energy of formation of $UH_3(s)$ at 500 K.

7.8(a) Calculate the percentage change in K_x for the reaction $H_2CO(g) \rightleftharpoons CO(g) + H_2(g)$ when the total pressure is increased from 1.0 bar to 2.0 bar at constant temperature.

7.8(b) Calculate the percentage change in K_x for the reaction $CH_3OH(g) + NOCl(g) \rightleftharpoons HCl(g) + CH_3NO_2(g)$ when the total pressure is increased from 1.0 bar to 2.0 bar at constant temperature.

7.9(a) The equilibrium constant for the gas-phase isomerization of borneol ($C_{10}H_{17}OH$) to isoborneol at 503 K is 0.106. A mixture consisting of 7.50 g of borneol and 14.0 g of isoborneol in a container of volume 5.0 dm^3 is heated to 503 K and allowed to come to equilibrium. Calculate the mole fractions of the two substances at equilibrium.

7.9(b) The equilibrium constant for the reaction $N_2(g) + O_2(g) \rightleftharpoons 2\ NO(g)$ is 1.69×10^{-3} at 2300 K. A mixture consisting of 5.0 g of nitrogen and 2.0 g of oxygen in a container of volume 1.0 dm^3 is heated to 2300 K and allowed to come to equilibrium. Calculate the mole fraction of NO at equilibrium.

7.10(a) What is the standard enthalpy of a reaction for which the equilibrium constant is (a) doubled, (b) halved when the temperature is increased by 10 K at 298 K?

7.10(b) What is the standard enthalpy of a reaction for which the equilibrium constant is (a) doubled, (b) halved when the temperature is increased by 15 K at 310 K?

7.11(a) The standard Gibbs energy of formation of $NH_3(g)$ is $-16.5\ \text{kJ mol}^{-1}$ at 298 K. What is the reaction Gibbs energy when the partial pressures of the N_2, H_2, and NH_3 (treated as perfect gases) are 3.0 bar, 1.0 bar, and 4.0 bar, respectively? What is the spontaneous direction of the reaction in this case?

7.11(b) The dissociation vapour pressure of NH_4Cl at 427°C is 608 kPa but at 459°C it has risen to 1115 kPa. Calculate (a) the equilibrium constant, (b) the standard reaction Gibbs energy, (c) the standard enthalpy, (d) the standard entropy of dissociation, all at 427°C. Assume that the vapour behaves as a perfect gas and that $\Delta H^\ominus$ and $\Delta S^\ominus$ are independent of temperature in the range given.

7.12(a) Estimate the temperature at which $CaCO_3$(calcite) decomposes.

7.12(b) Estimate the temperature at which $CuSO_4 \cdot 5H_2O$ undergoes dehydration.

7.13(a) For $CaF_2(s) \rightleftharpoons Ca^{2+}(aq) + 2\ F^-(aq)$, $K = 3.9 \times 10^{-11}$ at 25°C and the standard Gibbs energy of formation of $CaF_2(s)$ is $-1167\ \text{kJ mol}^{-1}$. Calculate the standard Gibbs energy of formation of $CaF_2(aq)$.

7.13(b) For $PbI_2(s) \rightleftharpoons Pb^{2+}(aq) + 2\ I^-(aq)$, $K = 1.4 \times 10^{-8}$ at 25°C and the standard Gibbs energy of formation of $PbI_2(s)$ is $-173.64\ \text{kJ mol}^{-1}$. Calculate the standard Gibbs energy of formation of $PbI_2(aq)$.

7.14(a) Write the cell reaction and electrode half-reactions and calculate the standard emf of each of the following cells:

(a) $Zn|ZnSO_4(aq)||AgNO_3(aq)|Ag$

(b) $Cd|CdCl_2(aq)||HNO_3(aq)|H_2(g)|Pt$

(c) $Pt|K_3[Fe(CN)_6](aq),K_4[Fe(CN)_6](aq)||CrCl_3(aq)|Cr$

7.14(b) Write the cell reaction and electrode half-reactions and calculate the standard emf of each the following cells:

(a) $Pt|Cl_2(g)|HCl(aq)||K_2CrO_4(aq)|Ag_2CrO_4(s)|Ag$

(b) $Pt|Fe^{3+}(aq),Fe^{2+}(aq)||Sn^{4+}(aq),Sn^{2+}(aq)|Pt$

(c) $Cu|Cu^{2+}(aq)||Mn^{2+}(aq),H^+(aq)|MnO_2(s)|Pt$

7.15(a) Devise cells in which the following are the reactions and calculate the standard emf in each case:

(a) $Zn(s) + CuSO_4(aq) \rightarrow ZnSO_4(aq) + Cu(s)$

(b) $2\ AgCl(s) + H_2(g) \rightarrow 2\ HCl(aq) + 2\ Ag(s)$

(c) $2\ H_2(g) + O_2(g) \rightarrow 2\ H_2O(l)$

7.15(b) Devise cells in which the following are the reactions and calculate the standard emf in each case:

(a) $2\ Na(s) + 2\ H_2O(l) \rightarrow 2\ NaOH(aq) + H_2(g)$

(b) $H_2(g) + I_2(g) \rightarrow 2\ HI(aq)$

(c) $H_3O^+(aq) + OH^-(aq) \rightarrow 2\ H_2O(l)$

7.16(a) Use the Debye–Hückel limiting law and the Nernst equation to estimate the potential of the cell $Ag|AgBr(s)|KBr(aq, 0.050\ mol\ kg^{-1})\ ||\ Cd(NO_3)_2(aq, 0.010\ mol\ kg^{-1})|Cd$ at 25°C.

7.16(b) Consider the cell $Pt|H_2(g,p^{\ominus})|HCl(aq)|AgCl(s)|Ag$, for which the cell reaction is $2\ AgCl(s) + H_2(g) \rightarrow 2\ Ag(s) + 2\ HCl(aq)$. At 25°C and a molality of HCl of 0.010 mol kg^{-1}, $E = +0.4658$ V. (a) Write the Nernst equation for the cell reaction. (b) Calculate $\Delta_r G$ for the cell reaction. (c) Assuming that the Debye–Hückel limiting law holds at this concentration, calculate $E^{\ominus}(AgCl, Ag)$.

7.17(a) Calculate the equilibrium constants of the following reactions at 25°C from standard potential data:

(a) $Sn(s) + Sn^{4+}(aq) \rightleftharpoons 2\ Sn^{2+}(aq)$

(b) $Sn(s) + 2\ AgCl(s) \rightleftharpoons SnCl_2(aq) + 2\ Ag(s)$

7.17(b) Calculate the equilibrium constants of the following reactions at 25°C from standard potential data:

(a) $Sn(s) + CuSO_4(aq) \rightleftharpoons Cu(s) + SnSO_4(aq)$

(b) $Cu^{2+}(aq) + Cu(s) \rightleftharpoons 2\ Cu^+(aq)$

7.18(a) The emf of the cell $Ag|AgI(s)|AgI(aq)|Ag$ is +0.9509 V at 25°C. Calculate (a) the solubility product of AgI and (b) its solubility.

7.18(b) The emf of the cell $Bi|Bi_2S_3(s)|Bi_2S_3(aq)|Bi$ is −0.96 V at 25°C. Calculate (a) the solubility product of Bi_2S_3 and (b) its solubility.

Problems*

Numerical problems

7.1 The equilibrium constant for the reaction, $I_2(s) + Br_2(g) \rightleftharpoons 2\ IBr(g)$ is 0.164 at 25°C. (a) Calculate $\Delta_r G^{\ominus}$ for this reaction. (b) Bromine gas is introduced into a container with excess solid iodine. The pressure and temperature are held at 0.164 atm and 25°C, respectively. Find the partial pressure of IBr(g) at equilibrium. Assume that all the bromine is in the liquid form and that the vapour pressure of iodine is negligible. (c) In fact, solid iodine has a measurable vapour pressure at 25°C. In this case, how would the calculation have to be modified?

7.2 Consider the dissociation of methane, $CH_4(g)$, into the elements $H_2(g)$ and C(s, graphite). (a) Given that $\Delta_f H^{\ominus}(CH_4, g) = -74.85$ kJ mol^{-1} and that $\Delta_f S^{\ominus}(CH_4, g) = -80.67$ J K^{-1} mol^{-1} at 298 K, calculate the value of the equilibrium constant at 298 K. (b) Assuming that $\Delta_f H^{\ominus}$ is independent of temperature, calculate K at 50°C. (c) Calculate the degree of dissociation, α_e, of methane at 25°C and a total pressure of 0.010 bar. (d) Without doing any numerical calculations, explain how the degree of dissociation for this reaction will change as the pressure and temperature are varied.

7.3 The equilibrium pressure of H_2 over U(s) and $UH_3(s)$ between 450 K and 715 K fits the expression $\ln(p/Pa) = A + B/T + C\ln(T/K)$, with $A = 69.32$, $B = -1.464 \times 10^4$ K, and $C = -5.65$. Find an expression for the standard enthalpy of formation of $UH_3(s)$ and from it calculate $\Delta_r C_p^{\ominus}$.

7.4 The degree of dissociation, α_e, of $CO_2(g)$ into CO(g) and $O_2(g)$ at high temperatures was found to vary with temperature as follows:

T/K	1395	1443	1498
$\alpha_e/10^{-4}$	1.44	2.50	4.71

Assuming $\Delta_r H^{\ominus}$ to be constant over this temperature range, calculate K, $\Delta_r G^{\ominus}$, $\Delta_r H^{\ominus}$, and $\Delta_r S^{\ominus}$. Make any justifiable approximations.

7.5 The standard reaction enthalpy for the decomposition of $CaCl_2{\cdot}NH_3(s)$ into $CaCl_2(s)$ and $NH_3(g)$ is nearly constant at +78 kJ mol^{-1} between 350 K and 470 K. The equilibrium pressure of NH_3 in the presence of $CaCl_2{\cdot}NH_3$ is 1.71 kPa at 400 K. Find an expression for the temperature dependence of $\Delta_r G^{\ominus}$ in the same range.

7.6 Calculate the equilibrium constant of the reaction $CO(g) + H_2(g) \rightleftharpoons H_2CO(g)$ given that, for the production of liquid formaldehyde, $\Delta_r G^{\ominus} = +28.95$ kJ mol^{-1} at 298 K and that the vapour pressure of formaldehyde is 1500 Torr at that temperature.

7.7 Acetic acid was evaporated in container of volume 21.45 cm^3 at 437 K and at an external pressure of 101.9 kPa, and the container was then sealed. The mass of acid present in the sealed container was 0.0519 g. The experiment was repeated with the same container but at 471 K, and it was found that 0.0380 g of acetic acid was present. Calculate the equilibrium constant for the dimerization of the acid in the vapour and the enthalpy of vaporization.

7.8 A sealed container was filled with 0.300 mol $H_2(g)$, 0.400 mol $I_2(g)$, and 0.200 mol HI(g) at 870 K and total pressure 1.00 bar. Calculate the amounts of the components in the mixture at equilibrium given that $K = 870$ for the reaction $H_2(g) + I_2(g) \rightleftharpoons 2\ HI(g)$.

7.9 The dissociation of I_2 can be monitored by measuring the total pressure, and three sets of results are as follows:

T/K	973	1073	1173
$100p$/atm	6.244	7.500	9.181
$10^4 n_I$	2.4709	2.4555	2.4366

where n_I is the amount of I atoms per mole of I_2 molecules in the mixture, which occupied 342.68 cm^3. Calculate the equilibrium constants of the dissociation and the standard enthalpy of dissociation at the mean temperature.

7.10‡ Thorn *et al.* carried out a study of $Cl_2O(g)$ by photoelectron ionization (R.P. Thorn, L.J. Stief, S.-C. Kuo, and R.B. Klemm, *J. Phys. Chem.* **100**, 14178 (1996)). From their measurements, they report $\Delta_f H^{\ominus}(Cl_2O) = +77.2$ kJ mol^{-1}. They combined this measurement with literature data on the reaction $Cl_2O\ (g) + H_2O(g) \rightarrow 2\ HOCl(g)$, for which $K = 8.2 \times 10^{-2}$ and $\Delta_r S^{\ominus} =$

* Problems denoted with the symbol ‡ were supplied by Charles Trapp, Carmen Giunta, and Marshall Cady.

+16.38 J K^{-1} mol^{-1}, and with readily available thermodynamic data on water vapour to report a value for $\Delta_f H^{\ominus}$(HOCl). Calculate that value. All quantities refer to 298 K.

7.11‡ The 1980s saw reports of $\Delta_f H^{\ominus}(SiH_2)$ ranging from 243 to 289 kJ mol^{-1}. For example, the lower value was cited in the review article by R. Walsh (*Acc. Chem. Res.* **14**, 246 (1981)); Walsh later leant towards the upper end of the range (H.M. Frey, R. Walsh, and I.M. Watts, *J. Chem. Soc., Chem. Commun.* 1189 (1986)). The higher value was reported in S.-K. Shin and J.L. Beauchamp, *J. Phys. Chem.* **90**, 1507 (1986). If the standard enthalpy of formation is uncertain by this amount, by what factor is the equilibrium constant for the formation of SiH_2 from its elements uncertain at (a) 298 K, (b) 700 K?

7.12 Fuel cells provide electrical power for spacecraft (as in the NASA space shuttles) and also show promise as power sources for automobiles. Hydrogen and carbon monoxide have been investigated for use in fuel cells, so their solubilities in molten salts are of interest. Their solubilities in a molten $NaNO_3/KNO_3$ mixture were examined (E. Desimoni and P.G. Zambonin, *J. Chem. Soc. Faraday Trans. 1*, 2014 (1973)) with the following results:

$$\log s_{H_2} = -5.39 - \frac{980}{T/K} \qquad \log s_{CO} = -5.98 - \frac{980}{T/K}$$

where s is the solubility in mol cm^{-3} bar^{-1}. Calculate the standard molar enthalpies of solution of the two gases at 570 K.

7.13 Given that $\Delta_r G^{\ominus} = -212.7$ kJ mol^{-1} for the reaction in the Daniell cell at 25°C, and $b(CuSO_4) = 1.0 \times 10^{-3}$ mol kg^{-1} and $b(ZnSO_4) = 3.0 \times 10^{-3}$ mol kg^{-1}, calculate (a) the ionic strengths of the solutions, (b) the mean ionic activity coefficients in the compartments, (c) the reaction quotient, (d) the standard cell potential, and (e) the cell potential. (Take $\gamma_+ = \gamma_- = \gamma_\pm$ in the respective compartments.)

7.14 A fuel cell develops an electric potential from the chemical reaction between reagents supplied from an outside source. What is the emf of a cell fuelled by (a) hydrogen and oxygen, (b) the combustion of butane at 1.0 bar and 298 K?

7.15 Although the hydrogen electrode may be conceptually the simplest electrode and is the basis for our reference state of electrical potential in electrochemical systems, it is cumbersome to use. Therefore, several substitutes for it have been devised. One of these alternatives is the quinhydrone electrode (quinhydrone, $Q \cdot QH_2$, is a complex of quinone, $C_6H_4O_2 = Q$, and hydroquinone, $C_6H_4O_2H_2 = QH_2$). The electrode half-reaction is $Q(aq) + 2H^+(aq) + 2e^- \rightarrow QH_2(aq)$, $E^{\ominus} = +0.6994$ V. If the cell Hg|Hg_2Cl_2(s)|HCl(aq)|$Q \cdot QH_2$|Au is prepared, and the measured cell potential is +0.190 V, what is the pH of the HCl solution? Assume that the Debye–Hückel limiting law is applicable.

7.16 Consider the cell, Zn(s)|$ZnCl_2$ (0.0050 mol kg^{-1})|Hg_2Cl_2(s)|Hg(l), for which the cell reaction is $Hg_2Cl_2(s) + Zn(s) \rightarrow 2\,Hg(l) + 2\,Cl^-(aq) + Zn^{2+}(aq)$. Given that $E^{\ominus}(Zn^{2+},Zn) = -0.7628$ V, $E^{\ominus}(Hg_2Cl_2, Hg) = +0.2676$ V, and that the emf is +1.2272 V, (a) write the Nernst equation for the cell. Determine (b) the standard emf, (c) $\Delta_r G$, $\Delta_r G^{\ominus}$, and K for the cell reaction, (d) the mean ionic activity and activity coefficient of $ZnCl_2$ from the measured cell potential, and (e) the mean ionic activity coefficient of $ZnCl_2$ from the Debye–Hückel limiting law. (f) Given that $(\partial E/\partial T)_p = -4.52 \times 10^{-4}$ V K^{-1}. Calculate ΔS and ΔH.

7.17 The emf of the cell Pt|H_2(g, $p^{\ominus}$)|HCl(aq,b)|Hg_2Cl_2(s)|Hg(l) has been measured with high precision (G.J. Hills and D.J.G. Ives, *J. Chem. Soc.*, 311 (1951)) with the following results at 25°C:

b/(mmol kg^{-1})	1.6077	3.0769	5.0403	7.6938	10.9474
E/V	0.60080	0.56825	0.54366	0.52267	0.50532

Determine the standard emf of the cell and the mean activity coefficient of HCl at these molalities. (Make a least-squares fit of the data to the best straight line.)

7.18 Careful measurements of the emf of the cell Pt|H_2(g, $p^{\ominus}$)|NaOH(aq, 0.0100 mol kg^{-1}), NaCl(aq, 0.01125 mol kg^{-1})|AgCl(s)|Ag have been reported (C.P. Bezboruah, M.F.G.F.C. Camoes, A.K. Covington, and J.V. Dobson, *J. Chem. Soc. Faraday Trans. I* **69**, 949 (1973)). Among the data is the following information:

θ/°C	20.0	25.0	30.0
E/V	1.04774	1.04864	1.04942

Calculate pK_w at these temperatures and the standard enthalpy and entropy of the autoprotolysis of water at 25.0°C.

7.19 Measurements of the emf of cells of the type Ag|AgX(s)|MX(b_1)|M_xHg|MX(b_2)|AgX(s)|Ag, where M_xHg denotes an amalgam and the electrolyte is an alkali metal halide dissolved in ethylene glycol, have been reported (U. Sen, *J. Chem. Soc. Faraday Trans. I* **69**, 2006 (1973)) and some values for LiCl are given below. Estimate the activity coefficient at the concentration marked * and then use this value to calculate activity coefficients from the measured cell potential at the other concentrations. Base your answer on the following version of the extended Debye–Hückel law:

$$\log \gamma = -\frac{AI^{1/2}}{1 - BI^{1/2}} + kI$$

with $A = 1.461$, $B = 1.70$, $k = 0.20$, and $I = b/b^{\ominus}$. For $b_2 = 0.09141$ mol kg^{-1}:

b_1/(mol kg^{-1})	0.0555	0.09141*	0.1652	0.2171	1.040	1.350
E/V	−0.0220	0.0000	0.0263	0.0379	0.1156	0.1336

7.20 The standard potential of the AgCl/Ag,Cl^- couple has been measured very carefully over a range of temperature (R.G. Bates and V.E. Bowers, *J. Res. Nat. Bur. Stand.* **53**, 283 (1954)) and the results were found to fit the expression

$$E^{\ominus}/V = 0.23659 - 4.8564 \times 10^{-4}(\theta/°C) - 3.4205 \times 10^{-6}(\theta/°C)^2 + 5.869 \times 10^{-9}(\theta/°C)^3$$

Calculate the standard Gibbs energy and enthalpy of formation of Cl^-(aq) and its entropy at 298 K.

7.21‡ (a) Derive a general relation for $(\partial E/\partial p)_{T,n}$ for electrochemical cells employing reactants in any state of matter. (b) E. Cohen and K. Piepenbroek (*Z. Physik. Chem.* **167A**, 365 (1933)) calculated the change in volume for the reaction $TlCl(s) + CNS^-(aq) \rightarrow TlCNS(s) + Cl^-(aq)$ at 30°C from density data and obtained $\Delta_r V = -2.666 \pm 0.080$ cm^3 mol^{-1}. They also measured the emf of the cell Tl(Hg)|TlCNS(s)|KCNS:KCl|TlCl|Tl(Hg) at pressures up to 1500 atm. Their results are given in the following table:

p/atm	1.00	250	500	750	1000	1250	1500
E/mV	8.56	9.27	9.98	10.69	11.39	12.11	12.82

From this information, obtain $(\partial E/\partial p)_{T,n}$ at 30°C and compare to the value obtained from $\Delta_r V$. (c) Fit the data to a polynomial for E against p. How constant is $(\partial E/\partial p)_{T,n}$? (d) From the polynomial, estimate an effective isothermal compressibility for the cell as a whole.

7.22‡ The table below summarizes the emf observed for the cell Pd|H_2(g, 1 bar)|BH(aq, b), B(aq, b)|AgCl(s)|Ag. Each measurement is made at equimolar concentrations of 2-aminopyridinium chloride (BH) and 2-aminopyridine (B). The data are for 25°C and it is found that $E^{\ominus} = 0.22251$ V. Use the data to determine pK_a for the acid at 25°C and the mean activity coefficient ($\gamma_\pm$) of BH as a function of molality (b) and ionic strength (I). Use the extended Debye–Hückel equation for the mean activity coefficient in the form

$$-\log \gamma_\pm = \frac{AI^{1/2}}{1 + BI^{1/2}} - kb$$

where $A = 0.5091$ and B and k are parameters that depend upon the ions. Draw a graph of the mean activity coefficient with $b = 0.04$ mol kg^{-1} and $0 \le I \le 0.1$.

b/(mol kg^{-1})	0.01	0.02	0.03	0.04	0.05
E(25°C)/V	0.74452	0.72853	0.71928	0.71314	0.70809
b/(mol kg^{-1})	0.06	0.07	0.08	0.09	0.10
E(25°C)/V	0.70380	0.70059	0.69790	0.69571	0.69338

Hint. Use mathematical software or a spreadsheet.

7.23 Superheavy elements are now of considerable interest, particularly because signs of stability are starting to emerge with element 114, which has recently been made. Shortly before it was (falsely) believed that the first superheavy element had been discovered, an attempt was made to predict the chemical properties of ununpentium (Uup, element 115, O.L. Keller, C.W. Nestor, and B. Fricke, *J. Phys. Chem.* **78**, 1945 (1974)). In one part of the paper the standard enthalpy and entropy of the reaction $\text{Uup}^+(\text{aq}) + \frac{1}{2}\text{H}_2(\text{g}) \rightarrow \text{Uup(s)} + \text{H}^+(\text{aq})$ were estimated from the following data: $\Delta_{\text{sub}}H^{\ominus}(\text{Uup}) = +1.5$ eV, $I(\text{Uup}) = 5.52$ eV, $\Delta_{\text{hyd}}H^{\ominus}(\text{Uup}^+, \text{aq}) = -3.22$ eV, $S^{\ominus}(\text{Uup}^+, \text{aq}) = +1.34$ meV K^{-1}, $S^{\ominus}(\text{Uup, s}) = 0.69$ meV K^{-1}. Estimate the expected standard potential of the Uup$^+$/Uup couple.

7.24 Sodium fluoride is routinely added to public water supplies because it is known that fluoride ion can prevent tooth decay. In a fluoride-selective electrode used in the analysis of water samples a crystal of LaF_3 doped with Eu^{2+}, denoted as Eu^{2+}:LaF_3, provides a semipermeable barrier between the test solution and the solution inside the electrode (the filling solution), which contains 0.1 mol kg^{-1} NaF(aq) and 0.1 mol kg^{-1} NaCl(aq). A silver–silver chloride electrode immersed in the filling solution is connected to a potentiometer and the emf of the cell can be measured against an appropriate reference electrode. It follows that the half-cell for a fluoride-selective electrode is represented by

$$\text{Ag(s)}\,|\,\text{AgCl(s)}\,|\,\text{NaCl}(\text{aq}, b_1), \text{NaF}(\text{aq}, b_1)\,|\,\text{Eu}^{2+}\text{:LaF}_3\text{(s)}\,|\,\text{F}^-(\text{aq}, b_2)$$

where b_1 and b_2 are the molalities of fluoride ion in the filling and test solutions, respectively. (a) Derive an expression for the emf of this half-cell. (b) The fluoride-selective electrode just described is not sensitive to HF(aq). Hydroxide ion is the only interfering species, with $k_{\text{F}^-,\text{OH}^-} = 0.1$. Use this information and the fact that K_a of HF is 3.5×10^{-4} at 298 K to specify a range of pH values in which the electrode responds accurately to the activity of F$^-$ in the test solution at 298 K.

Theoretical problems

7.25 Express the equilibrium constant of a gas-phase reaction $\text{A} + 3\,\text{B} \rightleftharpoons 2\,\text{C}$ in terms of the equilibrium value of the extent of reaction, ξ, given that initially A and B were present in stoichiometric proportions. Find an expression for ξ as a function of the total pressure, p, of the reaction mixture and sketch a graph of the expression obtained.

7.26 Find an expression for the standard reaction Gibbs energy at a temperature T' in terms of its value at another temperature T and the coefficients a, b, and c in the expression for the molar heat capacity listed in Table 2.2. Evaluate the standard Gibbs energy of formation of H_2O(l) at 372 K from its value at 298 K.

7.27 Show that, if the ionic strength of a solution of the sparingly soluble salt MX and the freely soluble salt NX is dominated by the concentration C of the latter, and if it is valid to use the Debye–Hückel limiting law, the solubility S' in the mixed solution is given by

$$S' = \frac{K_s e^{4.606AC^{1/2}}}{C}$$

when K_s is small (in a sense to be specified).

Applications: to biology, environmental science, and chemical engineering

7.28 Here we investigate the molecular basis for the observation that the hydrolysis of ATP is exergonic at pH = 7.0 and 310 K. (a) It is thought that the exergonicity of ATP hydrolysis is due in part to the fact that the standard entropies of hydrolysis of polyphosphates are positive. Why would an increase in entropy accompany the hydrolysis of a triphosphate group into a diphosphate and a phosphate group? (b) Under identical conditions, the Gibbs energies of hydrolysis of H_4ATP and MgATP^{2-}, a complex between the Mg^{2+} ion and ATP^{4-}, are less negative than the Gibbs energy of hydrolysis of ATP^{4-}. This observation has been used to support the hypothesis that electrostatic repulsion between adjacent phosphate groups is a factor that controls the exergonicity of ATP hydrolysis. Provide a rationale for the hypothesis and discuss how the experimental evidence supports it. Do these electrostatic effects contribute to the $\Delta_r H$ or $\Delta_r S$ terms that determine the exergonicity of the reaction? *Hint.* In the MgATP^{2-}complex, the Mg^{2+} ion and ATP^{4-} anion form two bonds: one that involves a negatively charged oxygen belonging to the terminal phosphate group of ATP^{4-} and another that involves a negatively charged oxygen belonging to the phosphate group adjacent to the terminal phosphate group of ATP^{4-}.

7.29 To get a sense of the effect of cellular conditions on the ability of ATP to drive biochemical processes, compare the standard Gibbs energy of hydrolysis of ATP to ADP with the reaction Gibbs energy in an environment at 37°C in which pH = 7.0 and the ATP, ADP, and P_i^- concentrations are all 1.0 μmol dm^{-3}.

7.30 Under biochemical standard conditions, aerobic respiration produces approximately 38 molecules of ATP per molecule of glucose that is completely oxidized. (a) What is the percentage efficiency of aerobic respiration under biochemical standard conditions? (b) The following conditions are more likely to be observed in a living cell: $p_{CO_2} = 5.3 \times 10^{-2}$ atm, $p_{O_2} = 0.132$ atm, [glucose] = 5.6×10^{-2} mol dm^{-3}, [ATP] = [ADP] = [P_i] = 1.0×10^{-4} mol dm^{-3}, pH = 7.4, T = 310 K. Assuming that activities can be replaced by the numerical values of molar concentrations, calculate the efficiency of aerobic respiration under these physiological conditions. (c) A typical diesel engine operates between T_c = 873 K and T_h = 1923 K with an efficiency that is approximately 75 per cent of the theoretical limit of $(1 - T_c/T_h)$ (see Section 3.2). Compare the efficiency of a typical diesel engine with that of aerobic respiration under typical physiological conditions (see part b). Why is biological energy conversion more or less efficient than energy conversion in a diesel engine?

7.31 In anaerobic bacteria, the source of carbon may be a molecule other than glucose and the final electron acceptor is some molecule other than O_2. Could a bacterium evolve to use the ethanol/nitrate pair instead of the glucose/O_2 pair as a source of metabolic energy?

7.32 If the mitochondrial electric potential between matrix and the intermembrane space were 70 mV, as is common for other membranes, how much ATP could be synthesized from the transport of 4 mol H$^+$, assuming the pH difference remains the same?

7.33 The standard potentials of proteins are not commonly measured by the methods described in this chapter because proteins often lose their native structure and function when they react on the surfaces of electrodes. In an alternative method, the oxidized protein is allowed to react with an appropriate electron donor in solution. The standard potential of the protein is then determined from the Nernst equation, the equilibrium concentrations of all species in solution, and the known standard potential of the electron donor. We shall illustrate this method with the protein cytochrome *c*. The one-electron reaction between cytochrome *c*, cyt, and 2,6-dichloroindophenol, D, can be followed spectrophotometrically because each of the four species in solution has a distinct colour, or absorption spectrum. We write the reaction as $\text{cyt}_{\text{ox}} + \text{D}_{\text{red}} \rightleftharpoons \text{cyt}_{\text{red}} + \text{D}_{\text{ox}}$, where the subscripts 'ox' and 'red' refer to oxidized and reduced states, respectively. (a) Consider $E^{\ominus}_{cyt}$ and $E^{\ominus}_{D}$ to be the standard potentials of cytochrome *c* and D, respectively. Show that, at equilibrium ('eq'), a plot of $\ln([D_{ox}]_{eq}/[D_{red}]_{eq})$ versus $\ln([cyt_{ox}]_{eq}/[cyt_{red}]_{eq})$ is linear with slope of 1 and *y*-intercept $F(E^{\ominus}_{cyt} - E^{\ominus}_{D})/RT$, where equilibrium activities are replaced by the numerical values of equilibrium molar concentrations. (b) The following data were obtained for the reaction between oxidized cytochrome *c* and reduced D in a pH 6.5 buffer at 298 K. The ratios $[D_{ox}]_{eq}/[D_{red}]_{eq}$ and $[cyt_{ox}]_{eq}/[cyt_{red}]_{eq}$ were adjusted by titrating a solution containing oxidized cytochrome *c* and reduced D with a solution of sodium ascorbate, which is a strong reductant. From the data and the standard potential of D of 0.237 V, determine the standard potential cytochrome *c* at pH 6.5 and 298K.

$[D_{ox}]_{eq}/[D_{red}]_{eq}$	0.00279	0.00843	0.0257	0.0497	0.0748	0.238	0.534
$[cyt_{ox}]_{eq}/[cyt_{red}]_{eq}$	0.0106	0.0230	0.0894	0.197	0.335	0.809	1.39

7.34‡ The dimerization of ClO in the Antarctic winter stratosphere is believed to play an important part in that region's severe seasonal depletion of ozone. The following equilibrium constants are based on measurements by R.A. Cox and C.A. Hayman (*Nature* **332**, 796 (1988)) on the reaction $2ClO(g) \rightarrow (ClO)_2(g)$.

T/K	233	248	258	268	273	280
K	4.13×10^8	5.00×10^7	1.45×10^7	5.37×10^6	3.20×10^6	9.62×10^5
T/K	288	295	303			
K	4.28×10^5	1.67×10^5	7.02×10^4			

(a) Derive the values of $\Delta_r H^{\ominus}$ and $\Delta_r S^{\ominus}$ for this reaction. (b) Compute the standard enthalpy of formation and the standard molar entropy of $(ClO)_2$ given $\Delta_f H^{\ominus}(ClO) = +101.8$ kJ mol^{-1} and $S^{\ominus}_m(ClO) = 226.6$ J K^{-1} mol^{-1} (*CRC Handbook 2004*).

7.35‡ Nitric acid hydrates have received much attention as possible catalysts for heterogeneous reactions that bring about the Antarctic ozone hole. Worsnop *et al.* investigated the thermodynamic stability of these hydrates under conditions typical of the polar winter stratosphere (D.R. Worsnop, L.E. Fox, M.S. Zahniser, and S.C. Wofsy, *Science* **259**, 71 (1993)). Standard reaction Gibbs energies can be computed for the following reactions at 190 K from their data:

(i) $H_2O(g) \rightarrow H_2O(s)$ $\quad \Delta_r G^{\ominus} = -23.6$ kJ mol^{-1}

(ii) $H_2O(g) + HNO_3(g) \rightarrow HNO_3{\cdot}H_2O(s)$ $\quad \Delta_r G^{\ominus} = -57.2$ kJ mol^{-1}

(iii) $2H_2O(g) + HNO_3(g) \rightarrow HNO_3{\cdot}2H_2O(s)$ $\quad \Delta_r G^{\ominus} = -85.6$ kJ mol^{-1}

(iv) $3H_2O(g) + HNO_3(g) \rightarrow HNO_3{\cdot}3H_2O(s)$ $\quad \Delta_r G^{\ominus} = -112.8$ kJ mol^{-1}

Which solid is thermodynamically most stable at 190 K if $p_{H_2O} = 1.3\times10^{-7}$ bar and $p_{HNO_3} = 4.1\times10^{-10}$ bar? *Hint.* Try computing $\Delta_r G$ for each reaction under the prevailing conditions; if more than one solid forms spontaneously, examine $\Delta_r G$ for the conversion of one solid to another.

7.36‡ Suppose that an iron catalyst at a particular manufacturing plant produces ammonia in the most cost-effective manner at 450°C when the pressure is such that $\Delta_r G$ for the reaction $\frac{1}{2}N_2(g) + \frac{3}{2}H_2(g) \rightarrow NH_3(g)$ is equal to -500 J mol^{-1}. (a) What pressure is needed? (b) Now suppose that a new catalyst is developed that is most cost-effective at 400°C when the pressure gives the same value of $\Delta_r G$. What pressure is needed when the new catalyst is used? What are the advantages of the new catalyst? Assume that (i) all gases are perfect gases or that (ii) all gases are van der Waals gases. Isotherms of $\Delta_r G(T, p)$ in the pressure range 100 atm $\le p \le$ 400 atm are needed to derive the answer. (c) Do the isotherms you plotted confirm Le Chatelier's principle concerning the response of equilibrium changes in temperature and pressure?

PART 2 Structure

In Part 1 we examined the properties of bulk matter from the viewpoint of thermodynamics. In Part 2 we examine the structures and properties of individual atoms and molecules from the viewpoint of quantum mechanics. The two viewpoints merge in Chapters 16 and 17.

Quantum theory: introduction and principles

8

This chapter introduces some of the basic principles of quantum mechanics. First, it reviews the experimental results that overthrew the concepts of classical physics. These experiments led to the conclusion that particles may not have an arbitrary energy and that the classical concepts of 'particle' and 'wave' blend together. The overthrow of classical mechanics inspired the formulation of a new set of concepts and led to the formulation of quantum mechanics. In quantum mechanics, all the properties of a system are expressed in terms of a wavefunction that is obtained by solving the Schrödinger equation. We see how to interpret wavefunctions. Finally, we introduce some of the techniques of quantum mechanics in terms of operators, and see that they lead to the uncertainty principle, one of the most profound departures from classical mechanics.

It was once thought that the motion of atoms and subatomic particles could be expressed using **classical mechanics**, the laws of motion introduced in the seventeenth century by Isaac Newton, for these laws were very successful at explaining the motion of everyday objects and planets. However, towards the end of the nineteenth century, experimental evidence accumulated showing that classical mechanics failed when it was applied to particles as small as electrons, and it took until the 1920s to discover the appropriate concepts and equations for describing them. We describe the concepts of this new mechanics, which is called **quantum mechanics**, in this chapter, and apply them throughout the remainder of the text.

The origins of quantum mechanics

The basic principles of classical mechanics are reviewed in *Appendix* 2. In brief, they show that classical physics (1) predicts a precise trajectory for particles, with precisely specified locations and momenta at each instant, and (2) allows the translational, rotational, and vibrational modes of motion to be excited to any energy simply by controlling the forces that are applied. These conclusions agree with everyday experience. Everyday experience, however, does not extend to individual atoms, and careful experiments of the type described below have shown that classical mechanics fails when applied to the transfers of very small energies and to objects of very small mass.

We shall also investigate the properties of light. In classical physics, light is described as electromagnetic radiation, which is understood in terms of the **electromagnetic field**, an oscillating electric and magnetic disturbance that spreads as a harmonic wave through empty space, the vacuum. Such waves are generated by the acceleration of electric charge, as in the oscillating motion of electrons in the antenna of a radio transmitter. The wave travels at a constant speed called the *speed of light*, c, which

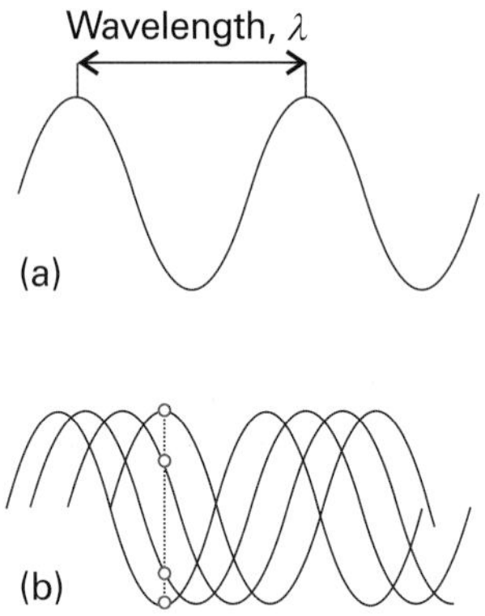

Fig. 8.1 The wavelength, λ, of a wave is the peak-to-peak distance. (b) The wave is shown travelling to the right at a speed c. At a given location, the instantaneous amplitude of the wave changes through a complete cycle (the four dots show half a cycle). The frequency, ν, is the number of cycles per second that occur at a given point.

Comment 8.1

Harmonic waves are waves with displacements that can be expressed as sine or cosine functions. The physics of waves is reviewed in *Appendix 3*.

is about 3×10^8 m s^{-1}. As its name suggests, an electromagnetic field has two components, an **electric field** that acts on charged particles (whether stationary or moving) and a **magnetic field** that acts only on moving charged particles. The electromagnetic field is characterized by a **wavelength**, λ (lambda), the distance between the neighbouring peaks of the wave, and its **frequency**, ν (nu), the number of times per second at which its displacement at a fixed point returns to its original value (Fig. 8.1). The frequency is measured in *hertz*, where 1 Hz = 1 s^{-1}. The wavelength and frequency of an electromagnetic wave are related by

$$\lambda\nu = c \tag{8.1}$$

Therefore, the shorter the wavelength, the higher the frequency. The characteristics of the wave are also reported by giving the **wavenumber**, $\tilde{\nu}$ (nu tilde), of the radiation, where

$$\tilde{\nu} = \frac{\nu}{c} = \frac{1}{\lambda} \qquad [8.2]$$

Wavenumbers are normally reported in reciprocal centimetres (cm^{-1}).

Figure 8.2 summarizes the **electromagnetic spectrum**, the description and classification of the electromagnetic field according to its frequency and wavelength. White light is a mixture of electromagnetic radiation with wavelengths ranging from about 380 nm to about 700 nm (1 nm = 10^{-9} m). Our eyes perceive different wavelengths of radiation in this range as different colours, so it can be said that white light is a mixture of light of all different colours.

The wave model falls short of describing all the properties of radiation. So, just as our view of particles (and in particular small particles) needs to be adjusted, a new view of light also has to be developed.

8.1 The failures of classical physics

In this section we review some of the experimental evidence that showed that several concepts of classical mechanics are untenable. In particular, we shall see that observations of the radiation emitted by hot bodies, heat capacities, and the spectra of atoms and molecules indicate that systems can take up energy only in discrete amounts.

(a) Black-body radiation

A hot object emits electromagnetic radiation. At high temperatures, an appreciable proportion of the radiation is in the visible region of the spectrum, and a higher

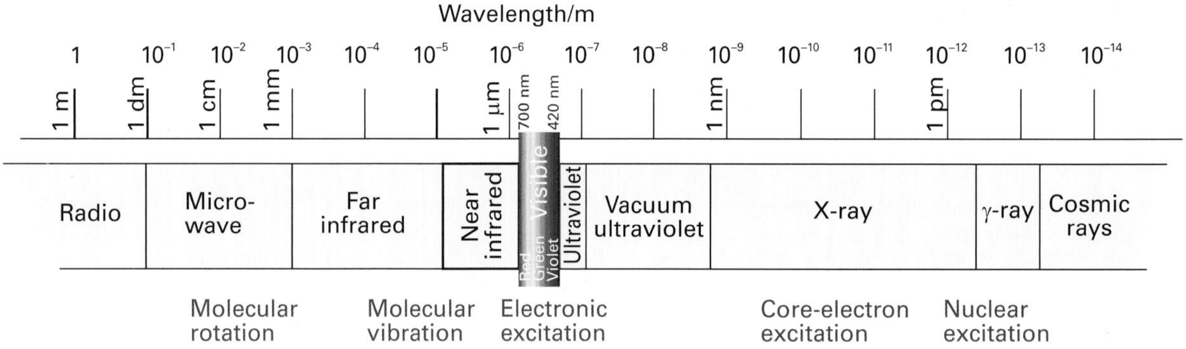

Fig. 8.2 The electromagnetic spectrum and the classification of the spectral regions.

proportion of short-wavelength blue light is generated as the temperature is raised. This behaviour is seen when a heated iron bar glowing red hot becomes white hot when heated further. The dependence is illustrated in Fig. 8.3, which shows how the energy output varies with wavelength at several temperatures. The curves are those of an ideal emitter called a **black body**, which is an object capable of emitting and absorbing all frequencies of radiation uniformly. A good approximation to a black body is a pinhole in an empty container maintained at a constant temperature, because any radiation leaking out of the hole has been absorbed and re-emitted inside so many times that it has come to thermal equilibrium with the walls (Fig. 8.4).

The explanation of black-body radiation was a major challenge for nineteenth-century scientists, and in due course it was found to be beyond the capabilities of classical physics. The physicist Lord Rayleigh studied it theoretically from a classical viewpoint, and thought of the electromagnetic field as a collection of oscillators of all possible frequencies. He regarded the presence of radiation of frequency ν (and therefore of wavelength $\lambda = c/\nu$) as signifying that the electromagnetic oscillator of that frequency had been excited (Fig. 8.5). Rayleigh used the equipartition principle (Section 2.2) to calculate the average energy of each oscillator as kT. Then, with minor help from James Jeans, he arrived at the **Rayleigh–Jeans law** (see *Further reading* for its justification):

$$d\mathcal{E} = \rho d\lambda \qquad \rho = \frac{8\pi kT}{\lambda^4} \tag{8.3}$$

where ρ (rho), the **density of states**, is the proportionality constant between $d\lambda$ and the energy density, $d\mathcal{E}$, in the range of wavelengths between λ and $\lambda + d\lambda$, k is Boltzmann's constant ($k = 1.381 \times 10^{-23}$ J K^{-1}). The units of ρ are typically joules per metre^4 (J m^{-4}), to give an energy density $d\mathcal{E}$ in joules per cubic metre (J m^{-3}) when multiplied by a wavelength range $d\lambda$ in metres. A high density of states at the wavelength λ simply means that there is a lot of energy associated with wavelengths lying between λ and $\lambda + d\lambda$. The total energy density (in joules per cubic metre) in a region is obtained by integrating eqn 8.3 over all wavelengths between zero and infinity, and the total energy (in joules) within the region is obtained by multiplying that total energy density by the volume of the region.

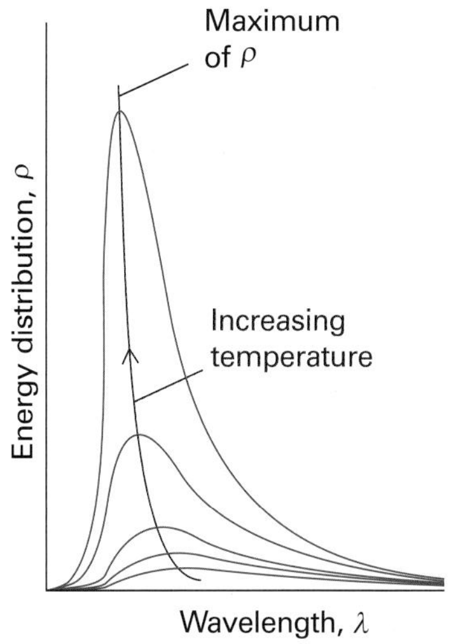

Fig. 8.3 The energy distribution in a black-body cavity at several temperatures. Note how the energy density increases in the region of shorter wavelengths as the temperature is raised, and how the peak shifts to shorter wavelengths. The total energy density (the area under the curve) increases as the temperature is increased (as T^4).

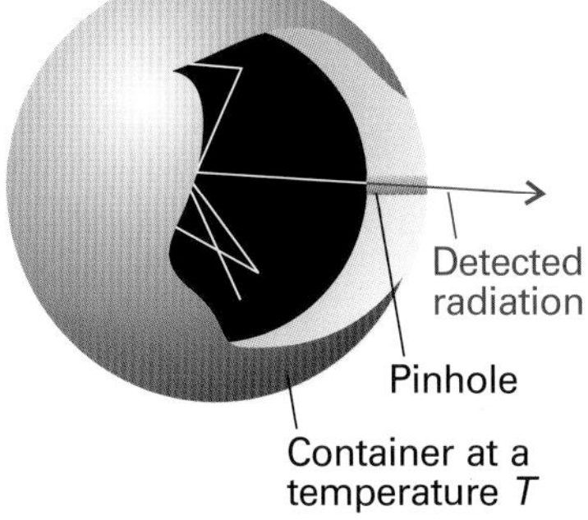

Fig. 8.4 An experimental representation of a black-body is a pinhole in an otherwise closed container. The radiation is reflected many times within the container and comes to thermal equilibrium with the walls at a temperature T. Radiation leaking out through the pinhole is characteristic of the radiation within the container.

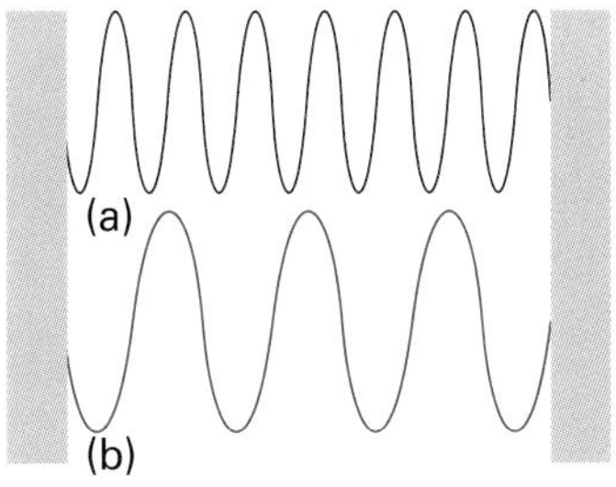

Fig. 8.5 The electromagnetic vacuum can be regarded as able to support oscillations of the electromagnetic field. When a high frequency, short wavelength oscillator (a) is excited, that frequency of radiation is present. The presence of low frequency, long wavelength radiation (b) signifies that an oscillator of the corresponding frequency has been excited.

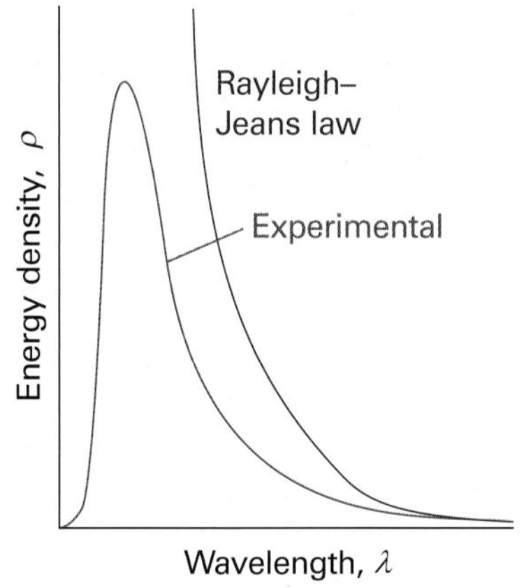

Fig. 8.6 The Rayleigh–Jeans law (eqn 8.3) predicts an infinite energy density at short wavelengths. This approach to infinity is called the *ultraviolet catastrophe*.

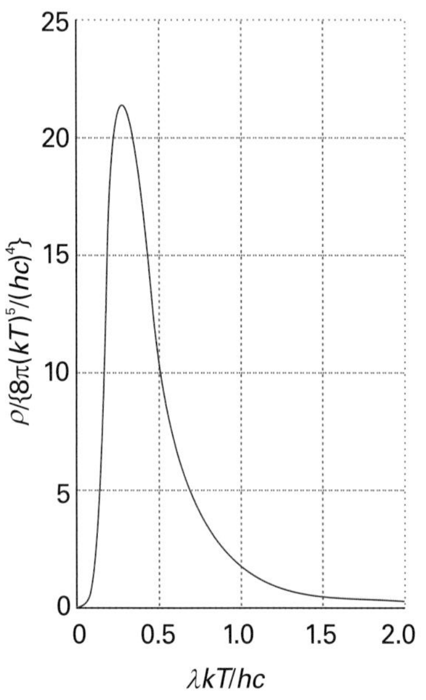

Fig. 8.7 The Planck distribution (eqn 8.5) accounts very well for the experimentally determined distribution of black-body radiation. Planck's quantization hypothesis essentially quenches the contributions of high frequency, short wavelength oscillators. The distribution coincides with the Rayleigh–Jeans distribution at long wavelengths.

Exploration Plot the Planck distribution at several temperatures and confirm that eqn 8.5 predicts the behaviour summarized by Fig. 8.2.

Unfortunately (for Rayleigh, Jeans, and classical physics), although the Rayleigh–Jeans law is quite successful at long wavelengths (low frequencies), it fails badly at short wavelengths (high frequencies). Thus, as λ decreases, ρ increases without going through a maximum (Fig. 8.6). The equation therefore predicts that oscillators of very short wavelength (corresponding to ultraviolet radiation, X-rays, and even γ-rays) are strongly excited even at room temperature. This absurd result, which implies that a large amount of energy is radiated in the high-frequency region of the electromagnetic spectrum, is called the **ultraviolet catastrophe**. According to classical physics, even cool objects should radiate in the visible and ultraviolet regions, so objects should glow in the dark; there should in fact be no darkness.

(b) The Planck distribution

The German physicist Max Planck studied black-body radiation from the viewpoint of thermodynamics. In 1900 he found that he could account for the experimental observations by proposing that the energy of each electromagnetic oscillator is limited to discrete values and cannot be varied arbitrarily. This proposal is quite contrary to the viewpoint of classical physics (on which the equipartition principle used by Rayleigh is based), in which all possible energies are allowed. The limitation of energies to discrete values is called the **quantization of energy**. In particular, Planck found that he could account for the observed distribution of energy if he supposed that the permitted energies of an electromagnetic oscillator of frequency ν are integer multiples of $h\nu$:

$$E = nh\nu \qquad n = 0, 1, 2, \ldots \tag{8.4}$$

where h is a fundamental constant now known as **Planck's constant**. On the basis of this assumption, Planck was able to derive the **Planck distribution**:

$$\mathrm{d}\mathcal{E} = \rho\mathrm{d}\lambda \qquad \rho = \frac{8\pi hc}{\lambda^5(\mathrm{e}^{hc/\lambda kT} - 1)} \tag{8.5}$$

(For references to the derivation of this expression, see *Further reading*.) This expression fits the experimental curve very well at all wavelengths (Fig. 8.7), and the value of h, which is an undetermined parameter in the theory, may be obtained by varying its value until a best fit is obtained. The currently accepted value for h is 6.626×10^{-34} J s.

The Planck distribution resembles the Rayleigh–Jeans law (eqn 8.3) apart from the all-important exponential factor in the denominator. For short wavelengths, $hc/\lambda kT \gg 1$ and $\mathrm{e}^{hc/\lambda kT} \to \infty$ faster than $\lambda^5 \to 0$; therefore $\rho \to 0$ as $\lambda \to 0$ or $\nu \to \infty$. Hence, the energy density approaches zero at high frequencies, in agreement with observation. For long wavelengths, $hc/\lambda kT \ll 1$, and the denominator in the Planck distribution can be replaced by

$$\mathrm{e}^{hc/\lambda kT} - 1 = \left(1 + \frac{hc}{\lambda kT} + \cdots\right) - 1 \approx \frac{hc}{\lambda kT}$$

When this approximation is substituted into eqn 8.5, we find that the Planck distribution reduces to the Rayleigh–Jeans law.

It is quite easy to see why Planck's approach was successful while Rayleigh's was not. The thermal motion of the atoms in the walls of the black body excites the oscillators of the electromagnetic field. According to classical mechanics, all the oscillators of the field share equally in the energy supplied by the walls, so even the highest frequencies are excited. The excitation of very high frequency oscillators results in the ultraviolet catastrophe. According to Planck's hypothesis, however, oscillators are excited only if they can acquire an energy of at least $h\nu$. This energy is too large for the walls to supply in the case of the very high frequency oscillators, so the latter remain unexcited. The effect of quantization is to reduce the contribution from the high frequency oscillators, for they cannot be significantly excited with the energy available.

Comment 8.2

The series expansion of an exponential function is $e^x = 1 + x + \frac{1}{2}x^2 + \cdots$. If $x \ll 1$, a good approximation is $e^x \approx 1 + x$. For example, $e^{0.01} = 1.010\,050\ldots \approx 1 + 0.01$.

(c) Heat capacities

In the early nineteenth century, the French scientists Pierre-Louis Dulong and Alexis-Thérèse Petit determined the heat capacities of a number of monatomic solids. On the basis of some somewhat slender experimental evidence, they proposed that the molar heat capacities of all monatomic solids are the same and (in modern units) close to $25\ \mathrm{J\,K^{-1}\,mol^{-1}}$.

Comment 8.3

The internal energy, U, a concept from thermodynamics (Chapter 2), can be regarded as the total energy of the particles making up a sample of matter. The constant-volume heat capacity is defined as $C_V = (\partial U/\partial T)_V$. A small heat capacity indicates that a large rise in temperature results from a given transfer of energy.

Dulong and Petit's law is easy to justify in terms of classical physics. If classical physics were valid, the equipartition principle could be used to calculate the heat capacity of a solid. According to this principle, the mean energy of an atom as it oscillates about its mean position in a solid is kT for each direction of displacement. As each atom can oscillate in three dimensions, the average energy of each atom is $3kT$; for N atoms the total energy is $3NkT$. The contribution of this motion to the molar internal energy is therefore

$$U_m = 3N_AkT = 3RT$$

because $N_Ak = R$, the gas constant. The molar constant volume heat capacity (see *Comment* 8.3) is then predicted to be

$$C_{V,m} = \left(\frac{\partial U_m}{\partial T}\right)_V = 3R \tag{8.6}$$

This result, with $3R = 24.9\ \mathrm{J\,K^{-1}\,mol^{-1}}$, is in striking accord with Dulong and Petit's value.

Unfortunately (this time, for Dulong and Petit), significant deviations from their law were observed when advances in refrigeration techniques made it possible to measure heat capacities at low temperatures. It was found that the molar heat capacities of all monatomic solids are lower than $3R$ at low temperatures, and that the values approach zero as $T \to 0$. To account for these observations, Einstein (in 1905) assumed that each atom oscillated about its equilibrium position with a single frequency ν. He then invoked Planck's hypothesis to assert that the energy of oscillation is confined to discrete values, and specifically to $nh\nu$, where n is an integer. Einstein first calculated the contribution of the oscillations of the atoms to the total molar energy of the metal (by a method described in Section 16.4) and obtained

$$U_m = \frac{3N_Ah\nu}{e^{h\nu/kT} - 1}$$

in place of the classical expression $3RT$. Then he found the molar heat capacity by differentiating U_m with respect to T. The resulting expression is now known as the **Einstein formula**:

$$C_{V,m} = 3Rf \qquad f = \left(\frac{\theta_E}{T}\right)^2 \left(\frac{e^{\theta_E/2T}}{e^{\theta_E/T} - 1}\right)^2 \tag{8.7}$$

The **Einstein temperature**, $\theta_E = h\nu/k$, is a way of expressing the frequency of oscillation of the atoms as a temperature: a high frequency corresponds to a high Einstein temperature.

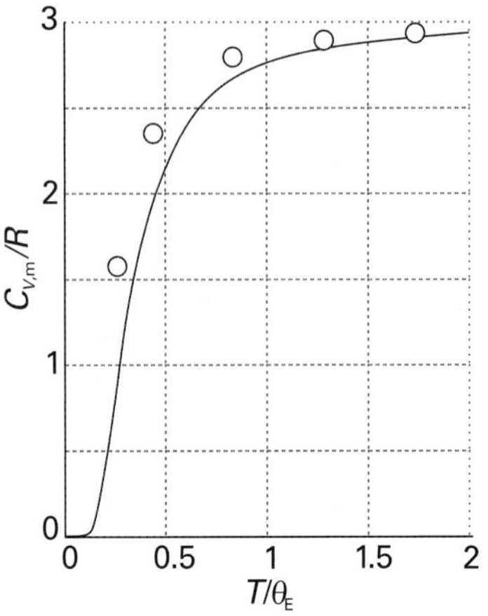

Fig. 8.8 Experimental low-temperature molar heat capacities and the temperature dependence predicted on the basis of Einstein's theory. His equation (eqn 8.7) accounts for the dependence fairly well, but is everywhere too low.

Exploration Using eqn 8.7, plot $C_{V,m}$ against T for several values of the Einstein temperature θ_E. At low temperature, does an increase in θ_E result in an increase or decrease of $C_{V,m}$? Estimate the temperature at which the value of $C_{V,m}$ reaches the classical value given by eqn 8.6.

At high temperatures (when $T \gg \theta_E$) the exponentials in f can be expanded as $1 + \theta_E/T + \cdots$ and higher terms ignored (see *Comment 8.2*). The result is

$$f = \left(\frac{\theta_E}{T}\right)^2 \left\{\frac{1 + \theta_E/2T + \cdots}{(1 + \theta_E/T + \cdots) - 1}\right\}^2 \approx 1 \tag{8.8a}$$

Consequently, the classical result ($C_{V,m} = 3R$) is obtained at high temperatures. At low temperatures, when $T \ll \theta_E$,

$$f \approx \left(\frac{\theta_E}{T}\right)^2 \left(\frac{e^{\theta_E/2T}}{e^{\theta_E/T}}\right)^2 = \left(\frac{\theta_E}{T}\right)^2 e^{-\theta_E/T} \tag{8.8b}$$

The strongly decaying exponential function goes to zero more rapidly than $1/T$ goes to infinity; so $f \to 0$ as $T \to 0$, and the heat capacity therefore approaches zero too. We see that Einstein's formula accounts for the decrease of heat capacity at low temperatures. The physical reason for this success is that at low temperatures only a few oscillators possess enough energy to oscillate significantly. At higher temperatures, there is enough energy available for all the oscillators to become active: all $3N$ oscillators contribute, and the heat capacity approaches its classical value.

Figure 8.8 shows the temperature dependence of the heat capacity predicted by the Einstein formula. The general shape of the curve is satisfactory, but the numerical agreement is in fact quite poor. The poor fit arises from Einstein's assumption that all the atoms oscillate with the same frequency, whereas in fact they oscillate over a range of frequencies from zero up to a maximum value, ν_D. This complication is taken into account by averaging over all the frequencies present, the final result being the **Debye formula**:

$$C_{V,m} = 3Rf \qquad f = 3\left(\frac{T}{\theta_D}\right)^3 \int_0^{\theta_D/T} \frac{x^4 e^x}{(e^x - 1)^2} dx \tag{8.9}$$

where $\theta_D = h\nu_D/k$ is the **Debye temperature** (for a derivation, see *Further reading*). The integral in eqn 8.9 has to be evaluated numerically, but that is simple with mathematical software. The details of this modification, which, as Fig. 8.9 shows, gives improved agreement with experiment, need not distract us at this stage from the main conclusion, which is that quantization must be introduced in order to explain the thermal properties of solids.

Illustration 8.1 *Assessing the heat capacity*

The Debye temperature for lead is 105 K, corresponding to a vibrational frequency of 2.2×10^{12} Hz, whereas that for diamond and its much lighter, more rigidly bonded atoms, is 2230 K, corresponding to 4.6×10^{13} Hz. As we see from Fig. 8.9, $f \approx 1$ for $T > \theta_D$ and the heat capacity is almost classical. For lead at 25°C, corresponding to $T/\theta_D = 2.8$, $f = 0.99$ and the heat capacity has almost its classical value. For diamond at the same temperature, $T/\theta_D = 0.13$, corresponding to $f = 0.15$, and the heat capacity is only 15 per cent of its classical value.

(d) Atomic and molecular spectra

The most compelling evidence for the quantization of energy comes from **spectroscopy**, the detection and analysis of the electromagnetic radiation absorbed, emitted, or scattered by a substance. The record of the intensity of light transmitted

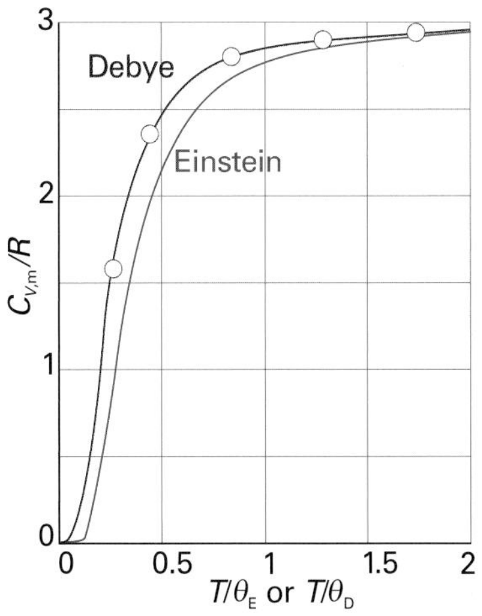

Fig. 8.9 Debye's modification of Einstein's calculation (eqn 8.9) gives very good agreement with experiment. For copper, $T/\theta_D = 2$ corresponds to about 170 K, so the detection of deviations from Dulong and Petit's law had to await advances in low-temperature physics.

Exploration Starting with the Debye formula (eqn 8.9), plot $dC_{V,m}/dT$, the temperature coefficient of $C_{V,m}$, against T for $\theta_D = 400$ K. At what temperature is $C_{V,m}$ most sensitive to temperature?

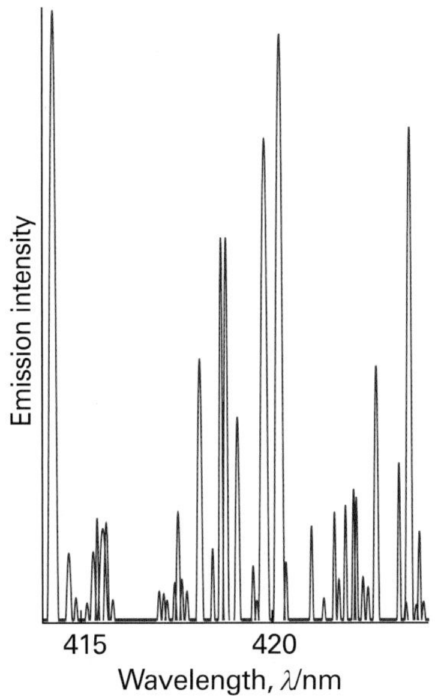

Fig. 8.10 A region of the spectrum of radiation emitted by excited iron atoms consists of radiation at a series of discrete wavelengths (or frequencies).

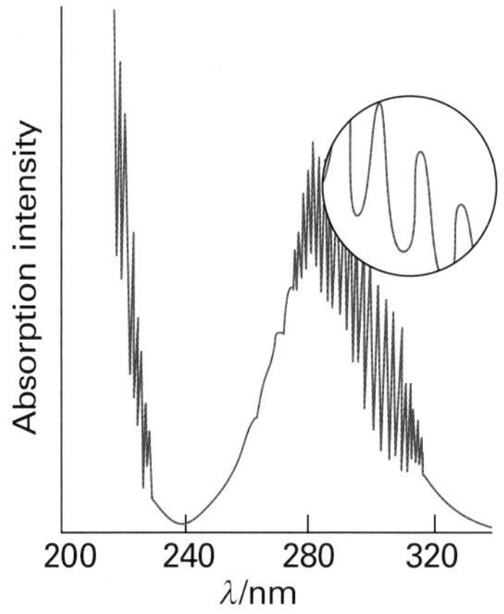

Fig. 8.11 When a molecule changes its state, it does so by absorbing radiation at definite frequencies. This spectrum is part of that due to the electronic, vibrational, and rotational excitation of sulfur dioxide (SO_2) molecules. This observation suggests that molecules can possess only discrete energies, not an arbitrary energy.

or scattered by a molecule as a function of frequency (ν), wavelength (λ), or wavenumber ($\tilde{\nu} = \nu/c$) is called its **spectrum** (from the Latin word for appearance).

A typical atomic spectrum is shown in Fig. 8.10, and a typical molecular spectrum is shown in Fig. 8.11. The obvious feature of both is that radiation is emitted or absorbed at a series of discrete frequencies. This observation can be understood if the energy of the atoms or molecules is also confined to discrete values, for then energy can be discarded or absorbed only in discrete amounts (Fig. 8.12). Then, if the energy of an atom decreases by ΔE, the energy is carried away as radiation of frequency ν, and an emission 'line', a sharply defined peak, appears in the spectrum. We say that a molecule undergoes a **spectroscopic transition**, a change of state, when the **Bohr frequency condition**

$$\Delta E = h\nu \qquad (8.10)$$

is fulfilled. We develop the principles and applications of atomic spectroscopy in Chapter 10 and of molecular spectroscopy in Chapters 13–15.

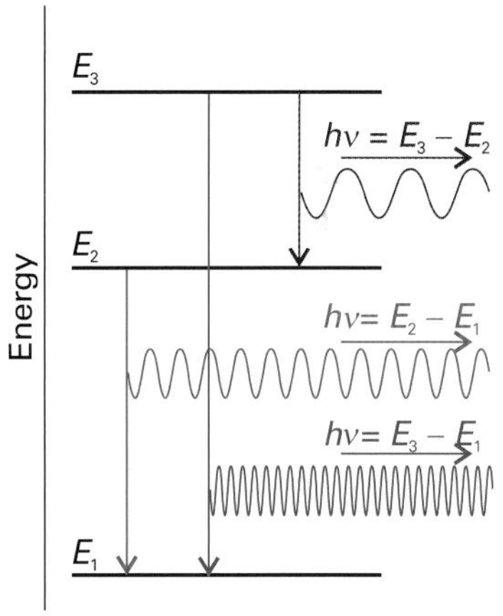

Fig. 8.12 Spectroscopic transitions, such as those shown above, can be accounted for if we assume that a molecule emits a photon as it changes between discrete energy levels. Note that high-frequency radiation is emitted when the energy change is large.

8.2 Wave–particle duality

At this stage we have established that the energies of the electromagnetic field and of oscillating atoms are quantized. In this section we shall see the experimental evidence that led to the revision of two other basic concepts concerning natural phenomena. One experiment shows that electromagnetic radiation—which classical physics treats as wave-like—actually also displays the characteristics of particles. Another experiment shows that electrons—which classical physics treats as particles—also display the characteristics of waves.

(a) The particle character of electromagnetic radiation

The observation that electromagnetic radiation of frequency ν can possess only the energies 0, $h\nu$, $2h\nu$, . . . suggests that it can be thought of as consisting of 0, 1, 2, . . . particles, each particle having an energy $h\nu$. Then, if one of these particles is present, the energy is $h\nu$, if two are present the energy is $2h\nu$, and so on. These particles of electromagnetic radiation are now called **photons**. The observation of discrete spectra from atoms and molecules can be pictured as the atom or molecule generating a photon of energy $h\nu$ when it discards an energy of magnitude ΔE, with $\Delta E = h\nu$.

Example 8.1 *Calculating the number of photons*

Calculate the number of photons emitted by a 100 W yellow lamp in 1.0 s. Take the wavelength of yellow light as 560 nm and assume 100 per cent efficiency.

Method Each photon has an energy $h\nu$, so the total number of photons needed to produce an energy E is $E/h\nu$. To use this equation, we need to know the frequency of the radiation (from $\nu = c/\lambda$) and the total energy emitted by the lamp. The latter is given by the product of the power (P, in watts) and the time interval for which the lamp is turned on ($E = P\Delta t$).

Answer The number of photons is

$$N = \frac{E}{h\nu} = \frac{P\Delta t}{h(c/\lambda)} = \frac{\lambda P\Delta t}{hc}$$

Substitution of the data gives

$$N = \frac{(5.60 \times 10^{-7}\ \text{m}) \times (100\ \text{J s}^{-1}) \times (1.0\ \text{s})}{(6.626 \times 10^{-34}\ \text{J s}) \times (2.998 \times 10^{8}\ \text{m s}^{-1})} = 2.8 \times 10^{20}$$

Note that it would take nearly 40 min to produce 1 mol of these photons.

A note on good practice To avoid rounding and other numerical errors, it is best to carry out algebraic manipulations first, and to substitute numerical values into a single, final formula. Moreover, an analytical result may be used for other data without having to repeat the entire calculation.

Self-test 8.1 How many photons does a monochromatic (single frequency) infrared rangefinder of power 1 mW and wavelength 1000 nm emit in 0.1 s?

[5×10^{14}]

Further evidence for the particle-like character of radiation comes from the measurement of the energies of electrons produced in the **photoelectric effect**. This effect is the ejection of electrons from metals when they are exposed to ultraviolet radiation. The experimental characteristics of the photoelectric effect are as follows:

1 No electrons are ejected, regardless of the intensity of the radiation, unless its frequency exceeds a threshold value characteristic of the metal.

2 The kinetic energy of the ejected electrons increases linearly with the frequency of the incident radiation but is independent of the intensity of the radiation.

3 Even at low light intensities, electrons are ejected immediately if the frequency is above the threshold.

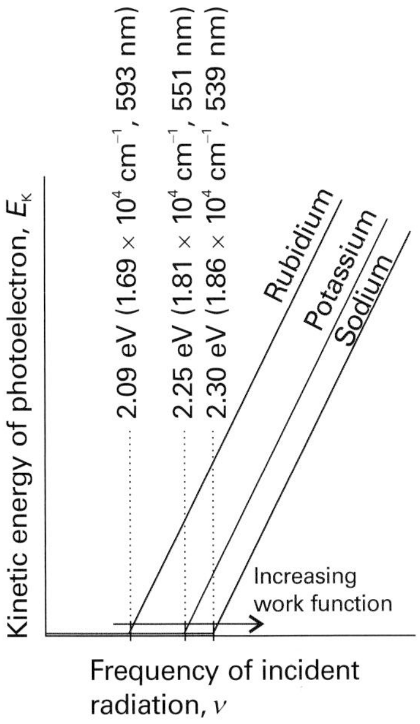

Fig. 8.13 In the photoelectric effect, it is found that no electrons are ejected when the incident radiation has a frequency below a value characteristic of the metal and, above that value, the kinetic energy of the photoelectrons varies linearly with the frequency of the incident radiation.

Exploration Calculate the value of Planck's constant given that the following kinetic energies were observed for photoejected electrons irradiated by radiation of the wavelengths noted.

λ_i/nm	320	330	345	360	385
E_K/eV	1.17	1.05	0.885	0.735	0.511

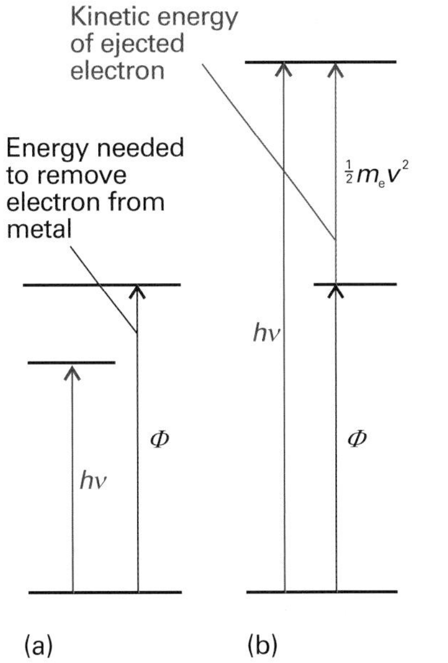

Fig. 8.14 The photoelectric effect can be explained if it is supposed that the incident radiation is composed of photons that have energy proportional to the frequency of the radiation. (a) The energy of the photon is insufficient to drive an electron out of the metal. (b) The energy of the photon is more than enough to eject an electron, and the excess energy is carried away as the kinetic energy of the photoelectron (the ejected electron).

Figure 8.13 illustrates the first and second characteristics.

These observations strongly suggest that the photoelectric effect depends on the ejection of an electron when it is involved in a collision with a particle-like projectile that carries enough energy to eject the electron from the metal. If we suppose that the projectile is a photon of energy $h\nu$, where ν is the frequency of the radiation, then the conservation of energy requires that the kinetic energy of the ejected electron should obey

$$\tfrac{1}{2}m_ev^2 = h\nu - \Phi \qquad (8.11)$$

In this expression Φ is a characteristic of the metal called its **work function**, the energy required to remove an electron from the metal to infinity (Fig. 8.14), the analogue of the ionization energy of an individual atom or molecule. Photoejection cannot occur if $h\nu < \Phi$ because the photon brings insufficient energy: this conclusion accounts for observation (1). Equation 8.11 predicts that the kinetic energy of an ejected electron should increase linearly with frequency, in agreement with observation (2). When a photon collides with an electron, it gives up all its energy, so we should expect electrons to appear as soon as the collisions begin, provided the photons have sufficient energy;

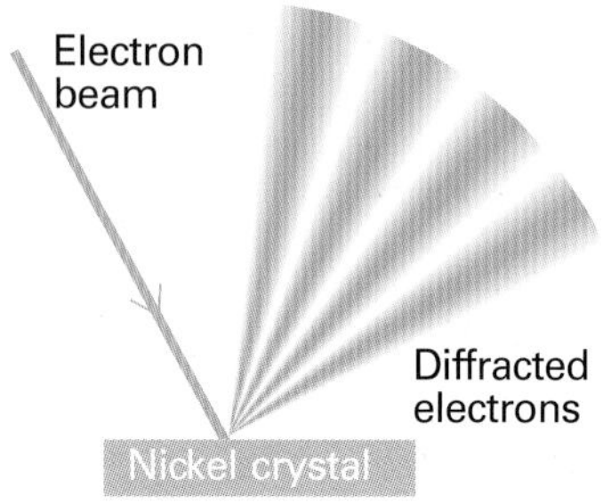

Fig. 8.15 The Davisson–Germer experiment. The scattering of an electron beam from a nickel crystal shows a variation of intensity characteristic of a diffraction experiment in which waves interfere constructively and destructively in different directions.

Comment 8.4

A characteristic property of waves is that they interfere with one another, giving a greater displacement where peaks or troughs coincide, leading to constructive interference, and a smaller displacement where peaks coincide with troughs, leading to destructive interference (see the illustration: (a) constructive, (b) destructive).

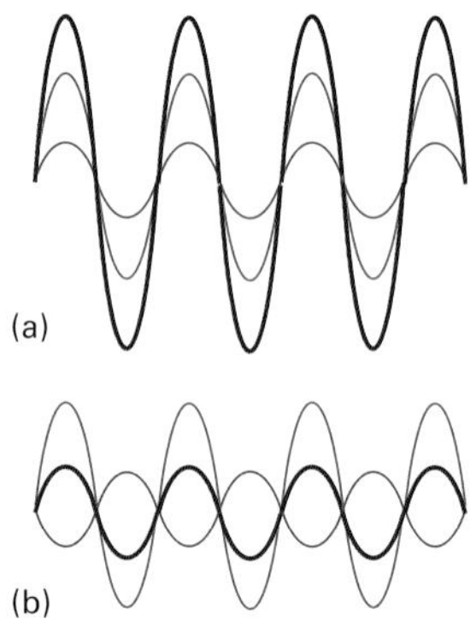

this conclusion agrees with observation (3). A practical application of eqn 8.11 is that it provides a technique for the determination of Planck's constant, for the slopes of the lines in Fig. 8.13 are all equal to h.

(b) The wave character of particles

Although contrary to the long-established wave theory of light, the view that light consists of particles had been held before, but discarded. No significant scientist, however, had taken the view that matter is wave-like. Nevertheless, experiments carried out in 1925 forced people to consider that possibility. The crucial experiment was performed by the American physicists Clinton Davisson and Lester Germer, who observed the diffraction of electrons by a crystal (Fig. 8.15). Diffraction is the interference caused by an object in the path of waves. Depending on whether the interference is constructive or destructive, the result is a region of enhanced or diminished intensity of the wave. Davisson and Germer's success was a lucky accident, because a chance rise of temperature caused their polycrystalline sample to anneal, and the ordered planes of atoms then acted as a diffraction grating. At almost the same time, G.P. Thomson, working in Scotland, showed that a beam of electrons was diffracted when passed through a thin gold foil. Electron diffraction is the basis for special techniques in microscopy used by biologists and materials scientists (*Impact I8.1* and Section 20.4).

The Davisson–Germer experiment, which has since been repeated with other particles (including α particles and molecular hydrogen), shows clearly that particles have wave-like properties, and the diffraction of neutrons is a well-established technique for investigating the structures and dynamics of condensed phases (see Chapter 20). We have also seen that waves of electromagnetic radiation have particle-like properties. Thus we are brought to the heart of modern physics. When examined on an atomic scale, the classical concepts of particle and wave melt together, particles taking on the characteristics of waves, and waves the characteristics of particles.

Some progress towards coordinating these properties had already been made by the French physicist Louis de Broglie when, in 1924, he suggested that any particle, not only photons, travelling with a linear momentum p should have (in some sense) a wavelength given by the **de Broglie relation**:

$$\lambda = \frac{h}{p} \tag{8.12}$$

That is, a particle with a high linear momentum has a short wavelength (Fig. 8.16). Macroscopic bodies have such high momenta (because their mass is so great), even when they are moving slowly, that their wavelengths are undetectably small, and the wave-like properties cannot be observed.

Example 8.2 *Estimating the de Broglie wavelength*

Estimate the wavelength of electrons that have been accelerated from rest through a potential difference of 40 kV.

Method To use the de Broglie relation, we need to know the linear momentum, p, of the electrons. To calculate the linear momentum, we note that the energy acquired by an electron accelerated through a potential difference V is eV, where e is the magnitude of its charge. At the end of the period of acceleration, all the acquired energy is in the form of kinetic energy, $E_K = p^2/2m_e$, so we can determine p by setting $p^2/2m_e$ equal to eV. As before, carry through the calculation algebraically before substituting the data.

Answer The expression $p^2/2m_e = eV$ solves to $p = (2m_eeV)^{1/2}$; then, from the de Broglie relation $\lambda = h/p$,

$$\lambda = \frac{h}{(2m_e eV)^{1/2}}$$

Substitution of the data and the fundamental constants (from inside the front cover) gives

$$\lambda = \frac{6.626 \times 10^{-34}\ \text{J s}}{\{2 \times (9.109 \times 10^{-31}\ \text{kg}) \times (1.609 \times 10^{-19}\ \text{C}) \times (4.0 \times 10^4\ \text{V})\}^{1/2}}$$
$$= 6.1 \times 10^{-12}\ \text{m}$$

where we have used 1 V C = 1 J and 1 J = 1 kg m^2 s^{-2}. The wavelength of 6.1 pm is shorter than typical bond lengths in molecules (about 100 pm). Electrons accelerated in this way are used in the technique of electron diffraction for the determination of molecular structure (see Section 20.4).

Self-test 8.2 Calculate (a) the wavelength of a neutron with a translational kinetic energy equal to kT at 300 K, (b) a tennis ball of mass 57 g travelling at 80 km/h.
[(a) 178 pm, (b) 5.2×10^{-34} m]

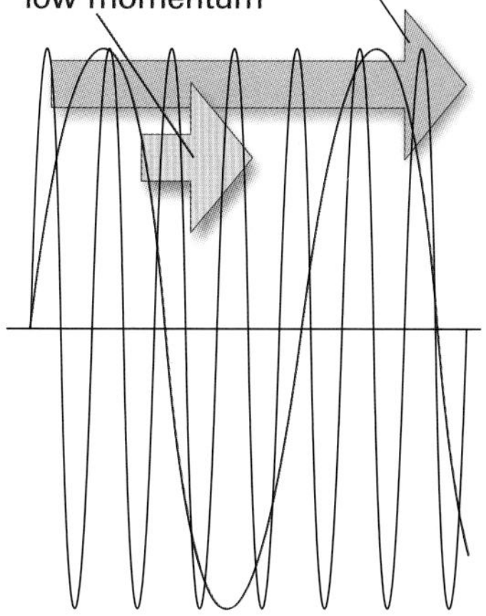

Fig. 8.16 An illustration of the de Broglie relation between momentum and wavelength. The wave is associated with a particle (shortly this wave will be seen to be the wavefunction of the particle). A particle with high momentum has a wavefunction with a short wavelength, and vice versa.

We now have to conclude that, not only has electromagnetic radiation the character classically ascribed to particles, but electrons (and all other particles) have the characteristics classically ascribed to waves. This joint particle and wave character of matter and radiation is called **wave–particle duality**. Duality strikes at the heart of classical physics, where particles and waves are treated as entirely distinct entities. We have also seen that the energies of electromagnetic radiation and of matter cannot be varied continuously, and that for small objects the discreteness of energy is highly significant. In classical mechanics, in contrast, energies could be varied continuously. Such total failure of classical physics for small objects implied that its basic concepts were false. A new mechanics had to be devised to take its place.

IMPACT ON BIOLOGY
I8.1 Electron microscopy

The basic approach of illuminating a small area of a sample and collecting light with a microscope has been used for many years to image small specimens. However, the *resolution* of a microscope, the minimum distance between two objects that leads to two distinct images, is on the order of the wavelength of light used as a probe (see *Impact I13.1*). Therefore, conventional microscopes employing visible light have resolutions in the micrometre range and are blind to features on a scale of nanometres.

There is great interest in the development of new experimental probes of very small specimens that cannot be studied by traditional light microscopy. For example, our understanding of biochemical processes, such as enzymatic catalysis, protein folding, and the insertion of DNA into the cell's nucleus, will be enhanced if it becomes possible to image individual biopolymers—with dimensions much smaller than visible wavelengths—at work. One technique that is often used to image nanometre-sized objects is *electron microscopy*, in which a beam of electrons with a well defined de Broglie wavelength replaces the lamp found in traditional light microscopes. Instead of glass or quartz lenses, magnetic fields are used to focus the beam. In *transmission electron microscopy* (TEM), the electron beam passes through the specimen and the

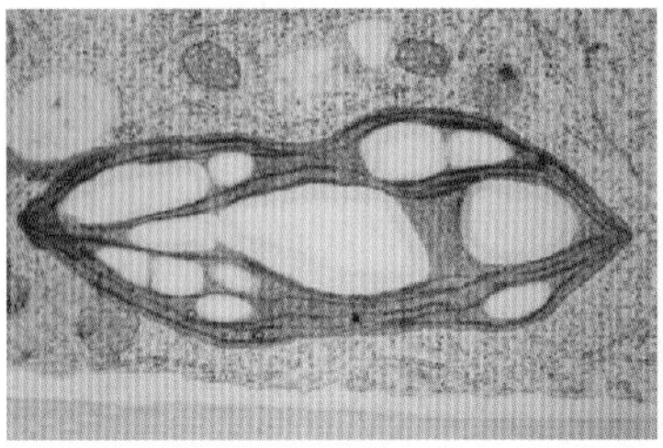

Fig. 8.17 A TEM image of a cross-section of a plant cell showing chloroplasts, organelles responsible for the reactions of photosynthesis (Chapter 23). Chloroplasts are typically 5 μm long. (Image supplied by Brian Bowes.)

image is collected on a screen. In *scanning electron microscopy* (SEM), electrons scattered back from a small irradiated area of the sample are detected and the electrical signal is sent to a video screen. An image of the surface is then obtained by scanning the electron beam across the sample.

As in traditional light microscopy, the wavelength of and the ability to focus the incident beam—in this case a beam of electrons—govern the resolution. Electron wavelengths in typical electron microscopes can be as short as 10 pm, but it is not possible to focus electrons well with magnetic lenses so, in the end, typical resolutions of TEM and SEM instruments are about 2 nm and 50 nm, respectively. It follows that electron microscopes cannot resolve individual atoms (which have diameters of about 0.2 nm). Furthermore, only certain samples can be observed under certain conditions. The measurements must be conducted under high vacuum. For TEM observations, the samples must be very thin cross-sections of a specimen and SEM observations must be made on dry samples. A consequence of these requirements is that neither technique can be used to study living cells. In spite of these limitations, electron microscopy is very useful in studies of the internal structure of cells (Fig. 8.17).

The dynamics of microscopic systems

Quantum mechanics acknowledges the wave–particle duality of matter by supposing that, rather than travelling along a definite path, a particle is distributed through space like a wave. This remark may seem mysterious: it will be interpreted more fully shortly. The mathematical representation of the wave that in quantum mechanics replaces the classical concept of trajectory is called a **wavefunction,** ψ (psi).

8.3 The Schrödinger equation

In 1926, the Austrian physicist Erwin Schrödinger proposed an equation for finding the wavefunction of any system. The **time-independent Schrödinger equation** for a particle of mass m moving in one dimension with energy E is

$$-\frac{\hbar^2}{2m}\frac{d^2\psi}{dx^2} + V(x)\psi = E\psi \qquad (8.13)$$

The factor $V(x)$ is the potential energy of the particle at the point x; because the total energy E is the sum of potential and kinetic energies, the first term must be related (in a manner we explore later) to the kinetic energy of the particle; $\hbar$ (which is read h-cross or h-bar) is a convenient modification of Planck's constant:

$$\hbar = \frac{h}{2\pi} = 1.054\,57 \times 10^{-34}\ \text{J s} \qquad (8.14)$$

For a partial justification of the form of the Schrödinger equation, see the *Justification* below. The discussions later in the chapter will help to overcome the apparent arbitrariness of this complicated expression. For the present, treat the equation as a quantum-mechanical postulate. Various ways of expressing the Schrödinger equation, of incorporating the time-dependence of the wavefunction, and of extending it to more dimensions, are collected in Table 8.1. In Chapter 9 we shall solve the equation for a number of important cases; in this chapter we are mainly concerned with its significance, the interpretation of its solutions, and seeing how it implies that energy is quantized.

Table 8.1 The Schrödinger equation

For one-dimensional systems:

$$-\frac{\hbar^2}{2m}\frac{d^2\psi}{dx^2}+V(x)\psi=E\psi$$

Where $V(x)$ is the potential energy of the particle and E is its total energy. For three-dimensional systems

$$-\frac{\hbar^2}{2m}\nabla^2\psi+V\psi=E\psi$$

where V may depend on position and ∇^2 ('del squared') is

$$\nabla^2=\frac{\partial^2}{\partial x^2}+\frac{\partial^2}{\partial y^2}+\frac{\partial^2}{\partial z^2}$$

In systems with spherical symmetry three equivalent forms are

$$\nabla^2=\frac{1}{r}\frac{\partial^2}{\partial r^2}r+\frac{1}{r^2}\Lambda^2$$

$$=\frac{1}{r^2}\frac{\partial}{\partial r}r^2\frac{\partial}{\partial r}+\frac{1}{r^2}\Lambda^2$$

$$=\frac{\partial^2}{\partial r^2}+\frac{2}{r}\frac{\partial}{\partial r}+\frac{1}{r^2}\Lambda^2$$

where

$$\Lambda^2=\frac{1}{\sin^2\theta}\frac{\partial^2}{\partial\phi^2}+\frac{1}{\sin\theta}\frac{\partial}{\partial\theta}\sin\theta\frac{\partial}{\partial\theta}$$

In the general case the Schrodinger equation is written

$$\hat{H}\psi=E\psi$$

where $\hat{H}$ is the hamiltonian operator for the system:

$$\hat{H}=-\frac{\hbar^2}{2m}\nabla^2+V$$

For the evolution of a system with time, it is necessary to solve the time-dependent Schrödinger equation:

$$\hat{H}\Psi=i\hbar\frac{\partial\Psi}{\partial t}$$

Justification 8.1 *Using the Schrödinger equation to develop the de Broglie relation*

Although the Schrödinger equation should be regarded as a postulate, like Newton's equations of motion, it can be seen to be plausible by noting that it implies the de Broglie relation for a freely moving particle in a region with constant potential energy V. After making the substitution $V(x) = V$, we can rearrange eqn 8.13 into

$$\frac{d^2\psi}{dx^2}=-\frac{2m}{\hbar^2}(E-V)\psi$$

General strategies for solving differential equations of this and other types that occur frequently in physical chemistry are treated in *Appendix 2*. In the case at hand, we note that a solution is

$$\psi=e^{ikx}\qquad k=\left\{\frac{2m(E-V)}{\hbar^2}\right\}^{1/2}$$

Comment 8.5

Complex numbers and functions are discussed in *Appendix 2*. Complex numbers have the form $z = x + iy$, where $i = (-1)^{1/2}$ and the real numbers x and y are the real and imaginary parts of z, denoted $\mathrm{Re}(z)$ and $\mathrm{Im}(z)$, respectively. Similarly, a complex function of the form $f = g + ih$, where g and h are functions of real arguments, has a real part $\mathrm{Re}(f) = g$ and an imaginary part $\mathrm{Im}(f) = h$.

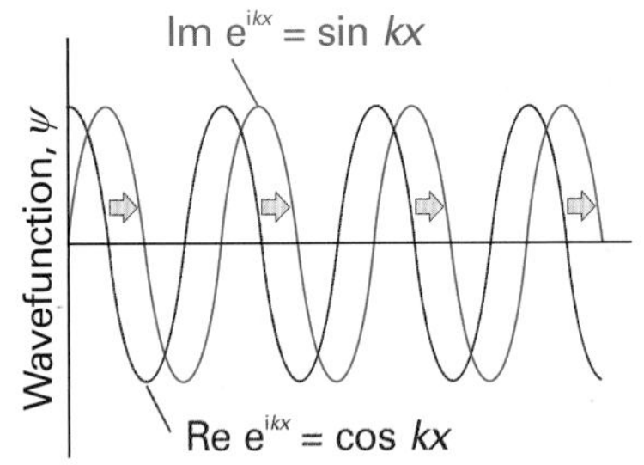

Fig. 8.18 The real (purple) and imaginary (blue) parts of a free particle wavefunction corresponding to motion towards positive x (as shown by the arrow).

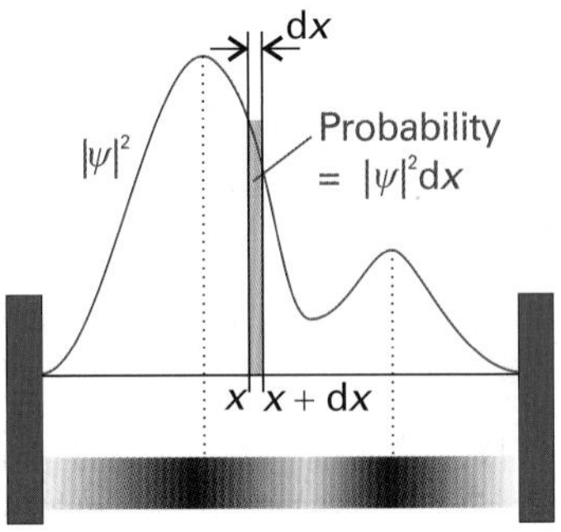

Fig. 8.19 The wavefunction ψ is a probability amplitude in the sense that its square modulus ($\psi^*\psi$ or $|\psi|^2$) is a probability density. The probability of finding a particle in the region dx located at x is proportional to $|\psi|^2$dx. We represent the probability density by the density of shading in the superimposed band.

Comment 8.6

To form the complex conjugate, ψ^*, of a complex function, replace i wherever it occurs by –i. For instance, the complex conjugate of e^{ikx} is e^{-ikx}. If the wavefunction is real, $|\psi|^2 = \psi^2$.

In quantum mechanics, a wavefunction that describes the spatial distribution of a particle (a 'spatial wavefunction') is complex if the particle it describes has a net motion. In the present case, we can use the relation $e^{i\theta} = \cos\theta + i\sin\theta$ to write

$$\psi = \cos kx + i \sin kx$$

The real and imaginary parts of ψ are drawn in Fig. 8.18, and we see that the imaginary component $\mathrm{Im}(\psi) = \sin kx$ is shifted in the direction of the particle's motion. That is, both the real and imaginary parts of the wavefunction are 'real', in the sense of being present, and we express ψ as a complex function simply to help with the visualization of the motion of the particle the wavefunction desribes.

Now we recognize that cos kx (or sin kx) is a wave of wavelength $\lambda = 2\pi/k$, as can be seen by comparing cos kx with the standard form of a harmonic wave, $\cos(2\pi x/\lambda)$. The quantity $E - V$ is equal to the kinetic energy of the particle, E_K, so $k = (2mE_K/\hbar^2)^{1/2}$, which implies that $E_K = k^2\hbar^2/2m$. Because $E_K = p^2/2m$, it follows that

$$p = k\hbar$$

Therefore, the linear momentum is related to the wavelength of the wavefunction by

$$p = \frac{2\pi}{\lambda} \times \frac{h}{2\pi} = \frac{h}{\lambda}$$

which is the de Broglie relation.

8.4 The Born interpretation of the wavefunction

A principal tenet of quantum mechanics is that *the wavefunction contains all the dynamical information about the system it describes.* Here we concentrate on the information it carries about the location of the particle.

The interpretation of the wavefunction in terms of the location of the particle is based on a suggestion made by Max Born. He made use of an analogy with the wave theory of light, in which the square of the amplitude of an electromagnetic wave in a region is interpreted as its intensity and therefore (in quantum terms) as a measure of the probability of finding a photon present in the region. The **Born interpretation** of the wavefunction focuses on the square of the wavefunction (or the square modulus, $|\psi|^2 = \psi^*\psi$, if ψ is complex). It states that the value of $|\psi|^2$ at a point is proportional to the probability of finding the particle in a region around that point. Specifically, for a one-dimensional system (Fig. 8.19):

> If the wavefunction of a particle has the value ψ at some point x, then the probability of finding the particle between x and $x + dx$ is proportional to $|\psi|^2 dx$.

Thus, $|\psi|^2$ is the **probability density**, and to obtain the probability it must be multiplied by the length of the infinitesimal region dx. The wavefunction ψ itself is called the **probability amplitude.** For a particle free to move in three dimensions (for example, an electron near a nucleus in an atom), the wavefunction depends on the point d$\boldsymbol{r}$ with coordinates x, y, and z, and the interpretation of $\psi(\boldsymbol{r})$ is as follows (Fig. 8.20):

> If the wavefunction of a particle has the value ψ at some point $\boldsymbol{r}$, then the probability of finding the particle in an infinitesimal volume $d\tau = dxdydz$ at that point is proportional to $|\psi|^2 d\tau$.

The Born interpretation does away with any worry about the significance of a negative (and, in general, complex) value of ψ because $|\psi|^2$ is real and never negative. There is no *direct* significance in the negative (or complex) value of a wavefunction: only the square modulus, a positive quantity, is directly physically significant, and both negative and positive regions of a wavefunction may correspond to a high probability of finding a particle in a region (Fig. 8.21). However, later we shall see that the presence of positive and negative regions of a wavefunction is of great *indirect* significance, because it gives rise to the possibility of constructive and destructive interference between different wavefunctions.

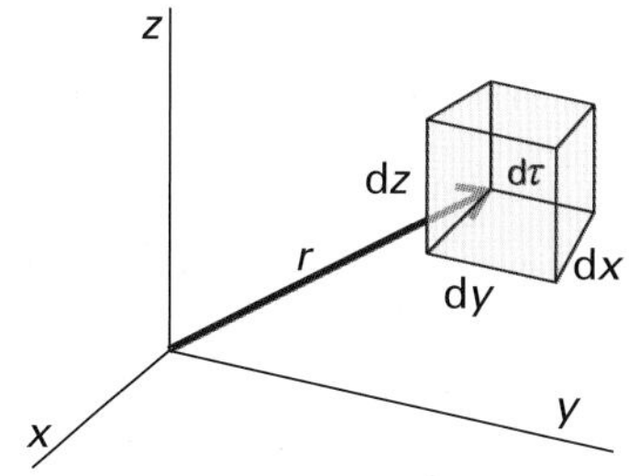

Fig. 8.20 The Born interpretation of the wavefunction in three-dimensional space implies that the probability of finding the particle in the volume element $d\tau = dxdydz$ at some location $\boldsymbol{r}$ is proportional to the product of $d\tau$ and the value of $|\psi|^2$ at that location.

Example 8.3 *Interpreting a wavefunction*

We shall see in Chapter 10 that the wavefunction of an electron in the lowest energy state of a hydrogen atom is proportional to e^{-r/a_0}, with a_0 a constant and r the distance from the nucleus. (Notice that this wavefunction depends only on this distance, not the angular position relative to the nucleus.) Calculate the relative probabilities of finding the electron inside a region of volume 1.0 pm^3, which is small even on the scale of the atom, located at (a) the nucleus, (b) a distance a_0 from the nucleus.

Method The region of interest is so small on the scale of the atom that we can ignore the variation of ψ within it and write the probability, P, as proportional to the probability density (ψ^2; note that ψ is real) evaluated at the point of interest multiplied by the volume of interest, δV. That is, $P \propto \psi^2\delta V$, with $\psi^2 \propto e^{-2r/a_0}$.

Answer In each case $\delta V = 1.0$ pm^3. (a) At the nucleus, $r = 0$, so

$$P \propto e^0 \times (1.0\ pm^3) = (1.0) \times (1.0\ pm^3)$$

(b) At a distance $r = a_0$ in an arbitrary direction,

$$P \propto e^{-2} \times (1.0\ pm^3) = (0.14\ ) \times (1.0\ pm^3)$$

Therefore, the ratio of probabilities is 1.0/0.14 = 7.1. Note that it is more probable (by a factor of 7) that the electron will be found at the nucleus than in a volume element of the same size located at a distance a_0 from the nucleus. The negatively charged electron is attracted to the positively charged nucleus, and is likely to be found close to it.

A note on good practice The square of a wavefunction is not a probability: it is a probability density, and (in three dimensions) has the dimensions of 1/length^3. It becomes a (unitless) probability when multiplied by a volume. In general, we have to take into account the variation of the amplitude of the wavefunction over the volume of interest, but here we are supposing that the volume is so small that the variation of ψ in the region can be ignored.

Self-test 8.3 The wavefunction for the electron in its lowest energy state in the ion He^+ is proportional to e^{-2r/a_0}. Repeat the calculation for this ion. Any comment?

[55; more compact wavefunction]

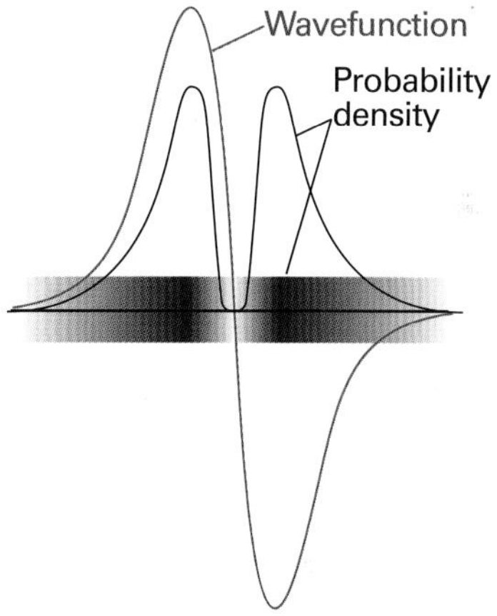

Fig. 8.21 The sign of a wavefunction has no direct physical significance: the positive and negative regions of this wavefunction both correspond to the same probability distribution (as given by the square modulus of ψ and depicted by the density of shading).

(a) Normalization

A mathematical feature of the Schrödinger equation is that, if ψ is a solution, then so is $N\psi$, where N is any constant. This feature is confirmed by noting that ψ occurs in every term in eqn 8.13, so any constant factor can be cancelled. This freedom to vary the wavefunction by a constant factor means that it is always possible to find a **normalization constant**, N, such that the proportionality of the Born interpretation becomes an equality.

We find the normalization constant by noting that, for a normalized wavefunction $N\psi$, the probability that a particle is in the region $\mathrm{d}x$ is equal to $(N\psi^*)(N\psi)\mathrm{d}x$ (we are taking N to be real). Furthermore, the sum over all space of these individual probabilities must be 1 (the probability of the particle being somewhere is 1). Expressed mathematically, the latter requirement is

$$N^2\int_{-\infty}^{\infty}\psi^*\psi\mathrm{d}x=1 \tag{8.15}$$

Almost all wavefunctions go to zero at sufficiently great distances so there is rarely any difficulty with the evaluation of this integral, and wavefunctions for which the integral in eqn 8.15 exists (in the sense of having a finite value) are said to be 'square-integrable'. It follows that

$$N=\frac{1}{\left(\int_{-\infty}^{\infty}\psi^*\psi\mathrm{d}x\right)^{1/2}} \tag{8.16}$$

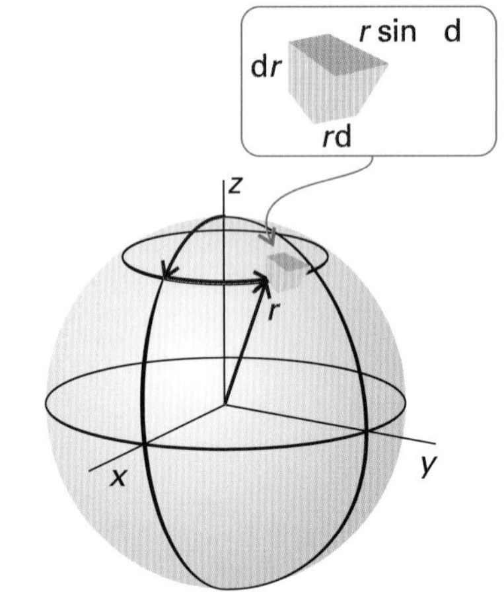

Fig. 8.22 The spherical polar coordinates used for discussing systems with spherical symmetry.

Therefore, by evaluating the integral, we can find the value of N and hence 'normalize' the wavefunction. From now on, unless we state otherwise, we always use wavefunctions that have been normalized to 1; that is, from now on we assume that ψ already includes a factor that ensures that (in one dimension)

$$\int_{-\infty}^{\infty}\psi^*\psi\,\mathrm{d}x=1 \tag{8.17a}$$

In three dimensions, the wavefunction is normalized if

$$\int_{-\infty}^{\infty}\int_{-\infty}^{\infty}\int_{-\infty}^{\infty}\psi^*\psi\,\mathrm{d}x\mathrm{d}y\mathrm{d}z=1 \tag{8.17b}$$

or, more succinctly, if

$$\int\psi^*\psi\,\mathrm{d}\tau=1 \tag{8.17c}$$

where $\mathrm{d}\tau=\mathrm{d}x\mathrm{d}y\mathrm{d}z$ and the limits of this definite integral are not written explicitly: in all such integrals, the integration is over all the space accessible to the particle. For systems with spherical symmetry it is best to work in **spherical polar coordinates** r, θ, and ϕ (Fig. 8.22): $x=r\sin\theta\cos\phi$, $y=r\sin\theta\sin\phi$, $z=r\cos\theta$. The volume element in spherical polar coordinates is $\mathrm{d}\tau=r^2\sin\theta\,\mathrm{d}r\mathrm{d}\theta\mathrm{d}\phi$. To cover all space, the radius r ranges from 0 to ∞, the colatitude, θ, ranges from 0 to π, and the azimuth, ϕ, ranges from 0 to 2π (Fig. 8.23), so the explicit form of eqn 8.17c is

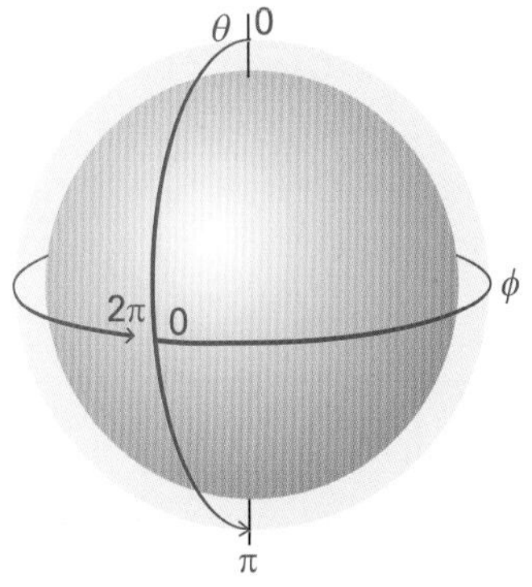

Fig. 8.23 The surface of a sphere is covered by allowing θ to range from 0 to π, and then sweeping that arc around a complete circle by allowing ϕ to range from 0 to 2π.

$$\int_0^{\infty}\int_0^{\pi}\int_0^{2\pi}\psi^*\psi r^2\sin\theta\mathrm{d}r\mathrm{d}\theta\mathrm{d}\phi=1 \tag{8.17d}$$

Example 8.4 *Normalizing a wavefunction*

Normalize the wavefunction used for the hydrogen atom in Example 8.3.

Method We need to find the factor N that guarantees that the integral in eqn 8.17c is equal to 1. Because the system is spherical, it is most convenient to use spherical coordinates and to carry out the integrations specified in eqn 8.17d. Note that the limits on the first integral sign refer to r, those on the second to θ, and those on the third to ϕ. A useful integral for calculations on atomic wavefunctions is

$$\int_0^{\infty} x^n \mathrm{e}^{-ax}\mathrm{d}x = \frac{n!}{a^{n+1}}$$

where $n!$ denotes a factorial: $n! = n(n-1)(n-2)\ldots 1$.

Answer The integration required is the product of three factors:

$$\int \psi^{\star}\psi \mathrm{d}\tau = N^2 \overbrace{\int_0^{\infty} r^2 \mathrm{e}^{-2r/a_0}\mathrm{d}r}^{\frac{1}{4}a_0^3} \overbrace{\int_0^{\pi} \sin\theta\,\mathrm{d}\theta}^{2} \overbrace{\int_0^{2\pi} \mathrm{d}\phi}^{2\pi} = \pi a_0^3 N^2$$

Therefore, for this integral to equal 1, we must set

$$N = \left(\frac{1}{\pi a_0^3}\right)^{1/2}$$

and the normalized wavefunction is

$$\psi = \left(\frac{1}{\pi a_0^3}\right)^{1/2} \mathrm{e}^{-r/a_0}$$

Note that, because a_0 is a length, the dimensions of ψ are 1/length$^{3/2}$ and therefore those of ψ^2 are 1/length^3 (for instance, 1/m^3) as is appropriate for a probability density.

If Example 8.3 is now repeated, we can obtain the actual probabilities of finding the electron in the volume element at each location, not just their relative values. Given (from Section 10.1) that $a_0 = 52.9$ pm, the results are (a) 2.2×10^{-6}, corresponding to 1 chance in about 500 000 inspections of finding the electron in the test volume, and (b) 2.9×10^{-7}, corresponding to 1 chance in 3.4 million.

Self-test 8.4 Normalize the wavefunction given in Self-test 8.3.

[$N = (8/\pi a_0^3)^{1/2}$]

(b) Quantization

The Born interpretation puts severe restrictions on the acceptability of wavefunctions. The principal constraint is that ψ must not be infinite anywhere. If it were, the integral in eqn 8.17 would be infinite (in other words, ψ would not be square-integrable) and the normalization constant would be zero. The normalized function would then be zero everywhere, except where it is infinite, which would be unacceptable. The requirement that ψ is finite everywhere rules out many possible solutions of the Schrödinger equation, because many mathematically acceptable solutions rise to infinity and are therefore physically unacceptable. We shall meet several examples shortly.

The requirement that ψ is finite everywhere is not the only restriction implied by the Born interpretation. We could imagine (and in Section 9.6a will meet) a solution of the Schrödinger equation that gives rise to more than one value of $|\psi|^2$ at a single point. The Born interpretation implies that such solutions are unacceptable, because it would be absurd to have more than one probability that a particle is at the same point. This restriction is expressed by saying that the wavefunction must be *single-valued*; that is, have only one value at each point of space.

Comment 8.7

Infinitely sharp spikes are acceptable provided they have zero width, so it is more appropriate to state that the wavefunction must not be infinite over any finite region. In elementary quantum mechanics the simpler restriction, to finite ψ, is sufficient.

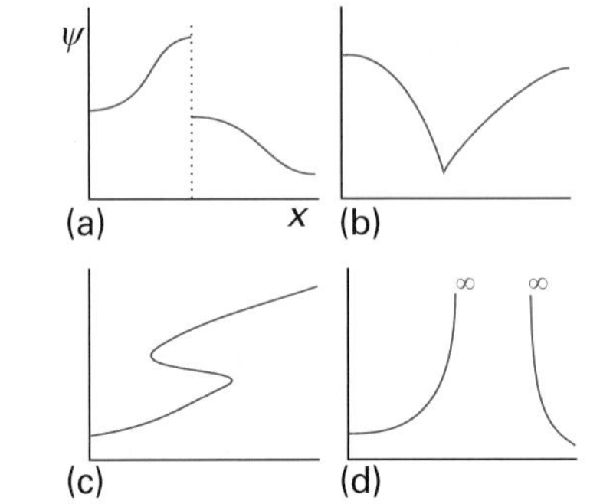

Fig. 8.24 The wavefunction must satisfy stringent conditions for it to be acceptable. (a) Unacceptable because it is not continuous; (b) unacceptable because its slope is discontinuous; (c) unacceptable because it is not single-valued; (d) unacceptable because it is infinite over a finite region.

The Schrödinger equation itself also implies some mathematical restrictions on the type of functions that will occur. Because it is a second-order differential equation, the second derivative of ψ must be well-defined if the equation is to be applicable everywhere. We can take the second derivative of a function only if it is continuous (so there are no sharp steps in it, Fig. 8.24) and if its first derivative, its slope, is continuous (so there are no kinks).

At this stage we see that ψ must be continuous, have a continuous slope, be single-valued, and be square-integrable. An acceptable wavefunction cannot be zero everywhere, because the particle it describes must be somewhere. These are such severe restrictions that acceptable solutions of the Schrödinger equation do not in general exist for arbitrary values of the energy E. In other words, a particle may possess only certain energies, for otherwise its wavefunction would be physically unacceptable. That is, *the energy of a particle is quantized*. We can find the acceptable energies by solving the Schrödinger equation for motion of various kinds, and selecting the solutions that conform to the restrictions listed above. That is the task of the next chapter.

Comment 8.8

There are cases, and we shall meet them, where acceptable wavefunctions have kinks. These cases arise when the potential energy has peculiar properties, such as rising abruptly to infinity. When the potential energy is smoothly well-behaved and finite, the slope of the wavefunction must be continuous; if the potential energy becomes infinite, then the slope of the wavefunction need not be continuous. There are only two cases of this behaviour in elementary quantum mechanics, and the peculiarity will be mentioned when we meet them.

Quantum mechanical principles

We have claimed that a wavefunction contains all the information it is possible to obtain about the dynamical properties of the particle (for example, its location and momentum). We have seen that the Born interpretation tells us as much as we can know about location, but how do we find any additional information?

8.5 The information in a wavefunction

The Schrödinger equation for a particle of mass m free to move parallel to the x-axis with zero potential energy is obtained from eqn 8.13 by setting $V = 0$, and is

$$-\frac{\hbar^2}{2m}\frac{d^2\psi}{dx^2} = E\psi \qquad (8.18)$$

The solutions of this equation have the form

$$\psi = Ae^{ikx} + Be^{-ikx} \qquad E = \frac{k^2\hbar^2}{2m} \qquad (8.19)$$

where A and B are constants. To verify that ψ is a solution of eqn 8.18, we simply substitute it into the left-hand side of the equation and confirm that we obtain $E\psi$:

$$\begin{aligned}
-\frac{\hbar^2}{2m}\frac{d^2\psi}{dx^2} &= -\frac{\hbar^2}{2m}\frac{d^2}{dx^2}(Ae^{ikx} + Be^{-kx}) \\
&= -\frac{\hbar^2}{2m}\{A(ik)^2e^{ikx} + B(-ik)^2e^{-ikx}\} \\
&= \frac{\hbar^2k^2}{2m}(Ae^{ikx} + Be^{-ikx}) = E\psi
\end{aligned}$$

(a) The probability density

We shall see later what determines the values of A and B; for the time being we can treat them as arbitrary constants. Suppose that $B = 0$ in eqn 8.19, then the wavefunction is simply

$$\psi = Ae^{ikx} \qquad (8.20)$$

Where is the particle? To find out, we calculate the probability density:

$$|\psi|^2 = (Ae^{ikx})^*(Ae^{ikx}) = (A^*e^{-ikx})(Ae^{ikx}) = |A|^2 \qquad (8.21)$$

This probability density is independent of x; so, wherever we look along the x-axis, there is an equal probability of finding the particle (Fig. 8.25a). In other words, if the wavefunction of the particle is given by eqn 8.20, then we cannot predict where we will find the particle. The same would be true if the wavefunction in eqn 8.19 had $A = 0$; then the probability density would be $|B|^2$, a constant.

Now suppose that in the wavefunction $A = B$. Then eqn 8.19 becomes

$$\psi = A(e^{ikx} + e^{-ikx}) = 2A \cos kx \qquad (8.22)$$

The probability density now has the form

$$|\psi|^2 = (2A \cos kx)^*(2A \cos kx) = 4|A|^2 \cos^2 kx \qquad (8.23)$$

This function is illustrated in Fig. 8.25b. As we see, the probability density periodically varies between 0 and $4|A|^2$. The locations where the probability density is zero correspond to *nodes* in the wavefunction: particles will never be found at the nodes. Specifically, a **node** is a point where a wavefunction passes *through* zero. The location where a wavefunction approaches zero without actually passing through zero is not a node. Nodes are defined in terms of the probability amplitude, the wavefunction itself. The probability density, of course, never passes through zero because it cannot be negative.

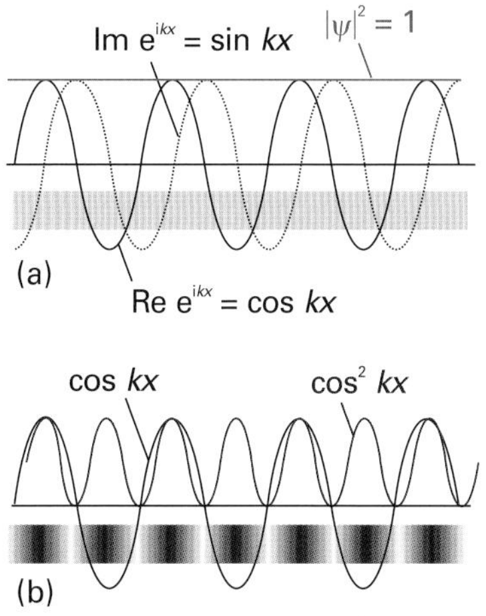

Fig. 8.25 (a) The square modulus of a wavefunction corresponding to a definite state of linear momentum is a constant; so it corresponds to a uniform probability of finding the particle anywhere. (b) The probability distribution corresponding to the superposition of states of equal magnitude of linear momentum but opposite direction of travel.

(b) Operators, eigenvalues, and eigenfunctions

To formulate a systematic way of extracting information from the wavefunction, we first note that any Schrödinger equation (such as those in eqns 8.13 and 8.18) may be written in the succinct form

$$\hat{H}\psi = E\psi \qquad (8.24a)$$

with (in one dimension)

$$\hat{H} = -\frac{\hbar^2}{2m}\frac{d^2}{dx^2} + V(x) \qquad (8.24b)$$

Comment 8.9

If the probability density of a particle is a constant, then it follows that, with x ranging from $-\infty$ to $+\infty$, the normalization constants, A or B, are 0. To avoid this embarrassing problem, x is allowed to range from $-L$ to $+L$, and L is allowed to go to infinity at the end of all calculations. We ignore this complication here.

The quantity $\hat{H}$ is an **operator**, something that carries out a mathematical operation on the function ψ. In this case, the operation is to take the second derivative of ψ and (after multiplication by $-\hbar^2/2m$) to add the result to the outcome of multiplying ψ by V. The operator $\hat{H}$ plays a special role in quantum mechanics, and is called the **hamiltonian operator** after the nineteenth century mathematician William Hamilton, who developed a form of classical mechanics that, it subsequently turned out, is well suited to the formulation of quantum mechanics. The hamiltonian operator is the operator corresponding to the total energy of the system, the sum of the kinetic and potential energies. Consequently, we can infer—as we anticipated in Section 8.3—that the first term in eqn 8.24b (the term proportional to the second derivative) must be the operator for the kinetic energy. When the Schrödinger equation is written as in eqn 8.24a, it is seen to be an **eigenvalue equation**, an equation of the form

$$\text{(Operator)(function)} = \text{(constant factor)} \times \text{(same function)} \qquad (8.25a)$$

If we denote a general operator by $\hat{\Omega}$ (where Ω is uppercase omega) and a constant factor by ω (lowercase omega), then an eigenvalue equation has the form

$$\hat{\Omega}\psi = \omega\psi \qquad (8.25b)$$

The factor ω is called the **eigenvalue** of the operator $\hat{\Omega}$. The eigenvalue in eqn 8.24a is the energy. The function ψ in an equation of this kind is called an **eigenfunction** of the operator $\hat{\Omega}$ and is different for each eigenvalue. The eigenfunction in eqn 8.24a is the wavefunction corresponding to the energy E. It follows that another way of saying 'solve the Schrödinger equation' is to say 'find the eigenvalues and eigenfunctions of the hamiltonian operator for the system'. The wavefunctions are the eigenfunctions of the hamiltonian operator, and the corresponding eigenvalues are the allowed energies.

Example 8.5 *Identifying an eigenfunction*

Show that e^{ax} is an eigenfunction of the operator d/dx, and find the corresponding eigenvalue. Show that e^{ax^2} is not an eigenfunction of d/dx.

Method We need to operate on the function with the operator and check whether the result is a constant factor times the original function.

Answer For $\hat{\Omega} = \mathrm{d}/\mathrm{d}x$ and $\psi = e^{ax}$:

$$\hat{\Omega}\psi = \frac{\mathrm{d}}{\mathrm{d}x}e^{ax} = ae^{ax} = a\psi$$

Therefore e^{ax} is indeed an eigenfunction of d/dx, and its eigenvalue is a. For $\psi = e^{ax^2}$,

$$\hat{\Omega}\psi = \frac{\mathrm{d}}{\mathrm{d}x}e^{ax^2} = 2axe^{ax^2} = 2ax \times \psi$$

which is not an eigenvalue equation even though the same function ψ occurs on the right, because ψ is now multiplied by a variable factor ($2ax$), not a constant factor. Alternatively, if the right-hand side is written $2a(xe^{ax^2})$, we see that it is a constant ($2a$) times a *different* function.

Self-test 8.5 Is the function cos ax an eigenfunction of (a) d/dx, (b) $\mathrm{d}^2/\mathrm{d}x^2$?

[(a) No, (b) yes]

The importance of eigenvalue equations is that the pattern

(Energy operator)ψ = (energy) × ψ

exemplified by the Schrödinger equation is repeated for other **observables**, or measurable properties of a system, such as the momentum or the electric dipole moment. Thus, it is often the case that we can write

(Operator corresponding to an observable)ψ = (value of observable) × ψ

The symbol $\hat{\Omega}$ in eqn 8.25b is then interpreted as an operator (for example, the hamiltonian, $\hat{H}$) corresponding to an observable (for example, the energy), and the eigenvalue ω is the value of that observable (for example, the value of the energy, E). Therefore, if we know both the wavefunction ψ and the operator $\hat{\Omega}$ corresponding to the observable Ω of interest, and the wavefunction is an eigenfunction of the operator $\hat{\Omega}$, then we can predict the outcome of an observation of the property Ω (for example, an atom's energy) by picking out the factor ω in the eigenvalue equation, eqn 8.25b.

A basic postulate of quantum mechanics tells us how to set up the operator corresponding to a given observable:

Observables, Ω, are represented by operators, $\hat{\Omega}$, built from the following position and momentum operators:

$$\hat{x} = x\times \qquad \hat{p}_x = \frac{\hbar}{\mathrm{i}}\frac{\mathrm{d}}{\mathrm{d}x} \qquad [8.26]$$

That is, the operator for location along the x-axis is multiplication (of the wavefunction) by x and the operator for linear momentum parallel to the x-axis is proportional to taking the derivative (of the wavefunction) with respect to x.

Comment 8.10

The rules summarized by eqn 8.26 apply to observables that depend on spatial variables; intrinsic properties, such as spin (see Section 9.8) are treated differently.

Example 8.6 *Determining the value of an observable*

What is the linear momentum of a particle described by the wavefunction in eqn 8.19 with (a) $B = 0$, (b) $A = 0$?

Method We operate on ψ with the operator corresponding to linear momentum (eqn 8.26), and inspect the result. If the outcome is the original wavefunction multiplied by a constant (that is, we generate an eigenvalue equation), then the constant is identified with the value of the observable.

Answer (a) With the wavefunction given in eqn 8.19 with $B = 0$,

$$\hat{p}_x\psi = \frac{\hbar}{\mathrm{i}}\frac{\mathrm{d}\psi}{\mathrm{d}x} = \frac{\hbar}{\mathrm{i}}A\frac{\mathrm{d}\mathrm{e}^{ikx}}{\mathrm{d}x} = \frac{\hbar}{\mathrm{i}}B\times \mathrm{i}k\mathrm{e}^{ikx} = k\hbar B\mathrm{e}^{-ikx} = k\hbar\psi$$

This is an eigenvalue equation, and by comparing it with eqn 8.25b we find that $p_x = +k\hbar$. (b) For the wavefunction with $A = 0$,

$$\hat{p}_x\psi = \frac{\hbar}{\mathrm{i}}\frac{\mathrm{d}\psi}{\mathrm{d}x} = \frac{\hbar}{\mathrm{i}}B\frac{\mathrm{d}\mathrm{e}^{-ikx}}{\mathrm{d}x} = \frac{\hbar}{\mathrm{i}}B\times(-\mathrm{i}k)\mathrm{e}^{-ikx} = -k\hbar B\mathrm{e}^{-ikx} = -k\hbar\psi$$

The magnitude of the linear momentum is the same in each case ($k\hbar$), but the signs are different: In (a) the particle is travelling to the right (positive x) but in (b) it is travelling to the left (negative x).

Self-test 8.6 The operator for the angular momentum of a particle travelling in a circle in the xy-plane is $\hat{l}_z = (\hbar/\mathrm{i})\mathrm{d}/\mathrm{d}\phi$, where ϕ is its angular position. What is the angular momentum of a particle described by the wavefunction $\mathrm{e}^{-2\mathrm{i}\phi}$?

[$l_z = -2\hbar$]

We use the definitions in eqn 8.26 to construct operators for other spatial observables. For example, suppose we wanted the operator for a potential energy of the form $V = \frac{1}{2}kx^2$, with k a constant (later, we shall see that this potential energy describes the vibrations of atoms in molecules). Then it follows from eqn 8.26 that the operator corresponding to V is multiplication by x^2:

$$\hat{V} = \tfrac{1}{2}kx^2\times \qquad (8.27)$$

In normal practice, the multiplication sign is omitted. To construct the operator for kinetic energy, we make use of the classical relation between kinetic energy and linear

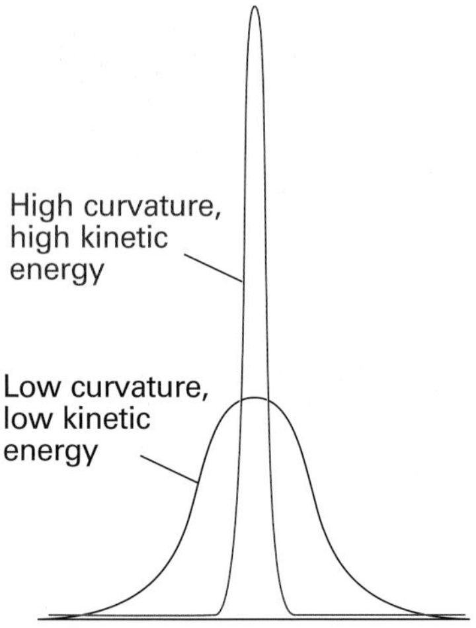

Fig. 8.26 Even if a wavefunction does not have the form of a periodic wave, it is still possible to infer from it the average kinetic energy of a particle by noting its average curvature. This illustration shows two wavefunctions: the sharply curved function corresponds to a higher kinetic energy than the less sharply curved function.

Comment 8.11

We are using the term 'curvature' informally: the precise technical definition of the curvature of a function f is $(d^2f/dx^2)/\{1+(df/dx)^2\}^{3/2}$.

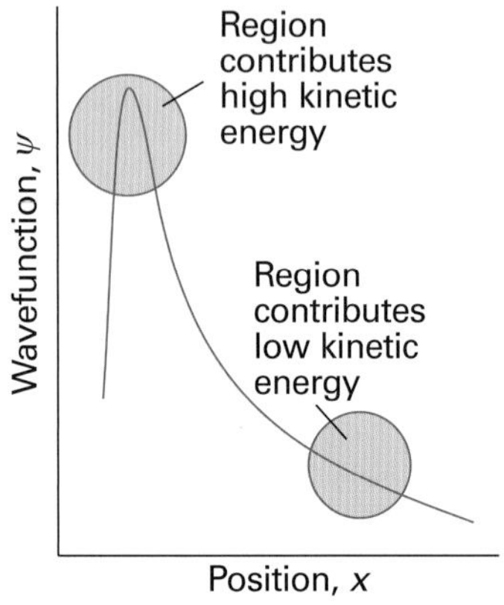

Fig. 8.27 The observed kinetic energy of a particle is an average of contributions from the entire space covered by the wavefunction. Sharply curved regions contribute a high kinetic energy to the average; slightly curved regions contribute only a small kinetic energy.

momentum, which in one dimension is $E_K = p_x^2/2m$. Then, by using the operator for p_x in eqn 8.26 we find:

$$\hat{E}_K = \frac{1}{2m}\left(\frac{\hbar}{i}\frac{d}{dx}\right)\left(\frac{\hbar}{i}\frac{d}{dx}\right) = -\frac{\hbar^2}{2m}\frac{d^2}{dx^2} \tag{8.28}$$

It follows that the operator for the total energy, the hamiltonian operator, is

$$\hat{H} = \hat{E}_K + \hat{V} = -\frac{\hbar^2}{2m}\frac{d^2}{dx^2} + \hat{V} \tag{8.29}$$

with $\hat{V}$ the multiplicative operator in eqn 8.27 (or some other relevant potential energy).

The expression for the kinetic energy operator, eqn 8.28, enables us to develop the point made earlier concerning the interpretation of the Schrödinger equation. In mathematics, the second derivative of a function is a measure of its curvature: a large second derivative indicates a sharply curved function (Fig. 8.26). It follows that a sharply curved wavefunction is associated with a high kinetic energy, and one with a low curvature is associated with a low kinetic energy. This interpretation is consistent with the de Broglie relation, which predicts a short wavelength (a sharply curved wavefunction) when the linear momentum (and hence the kinetic energy) is high. However, it extends the interpretation to wavefunctions that do not spread through space and resemble those shown in Fig. 8.26. The curvature of a wavefunction in general varies from place to place. Wherever a wavefunction is sharply curved, its contribution to the total kinetic energy is large (Fig. 8.27). Wherever the wavefunction is not sharply curved, its contribution to the overall kinetic energy is low. As we shall shortly see, the observed kinetic energy of the particle is an integral of all the contributions of the kinetic energy from each region. Hence, we can expect a particle to have a high kinetic energy if the average curvature of its wavefunction is high. Locally there can be both positive and negative contributions to the kinetic energy (because the curvature can be either positive, ⌣, or negative, ⌢), but the average is always positive (see Problem 8.22).

The association of high curvature with high kinetic energy will turn out to be a valuable guide to the interpretation of wavefunctions and the prediction of their shapes. For example, suppose we need to know the wavefunction of a particle with a given total energy and a potential energy that decreases with increasing x (Fig. 8.28). Because the difference $E - V = E_K$ increases from left to right, the wavefunction must become more sharply curved as x increases: its wavelength decreases as the local contributions to its kinetic energy increase. We can therefore guess that the wavefunction will look like the function sketched in the illustration, and more detailed calculation confirms this to be so.

(c) Hermitian operators

All the quantum mechanical operators that correspond to observables have a very special mathematical property: they are 'hermitian'. An **hermitian operator** is one for which the following relation is true:

$$\text{Hermiticity:} \quad \int \psi_i^* \hat{\Omega} \psi_j \, dx = \left\{ \int \psi_j^* \hat{\Omega} \psi_i \, dx \right\}^* \tag{8.30}$$

It is easy to confirm that the position operator ($x\times$) is hermitian because we are free to change the order of the factors in the integrand:

$$\int_{-\infty}^{\infty} \psi_i^* x \psi_j \, dx = \int_{-\infty}^{\infty} \psi_j x \psi_i^* dx = \left\{ \int_{-\infty}^{\infty} \psi_j^* x \psi_i \, dx \right\}^*$$

The demonstration that the linear momentum operator is hermitian is more involved because we cannot just alter the order of functions we differentiate, but it is hermitian, as we show in the following *Justification*.

Justification 8.2 *The hermiticity of the linear momentum operator*

Our task is to show that

$$\int_{-\infty}^{\infty}\psi_i^*\hat{p}_x\psi_j\mathrm{d}x=\left\{\int_{-\infty}^{\infty}\psi_j^*\hat{p}_x\psi_i\mathrm{d}x\right\}^*$$

with $\hat{p}_x$ given in eqn 8.26. To do so, we use 'integration by parts', the relation

$$\int f\frac{\mathrm{d}g}{\mathrm{d}x}\mathrm{d}x=fg-\int g\frac{\mathrm{d}f}{\mathrm{d}x}\mathrm{d}x$$

In the present case we write

$$\int_{-\infty}^{\infty}\psi_i^*\hat{p}_x\psi_j\mathrm{d}x=\frac{\hbar}{\mathrm{i}}\int_{-\infty}^{\infty}\psi_i^*\frac{\mathrm{d}\psi_j}{\mathrm{d}x}\mathrm{d}x$$

$$=\frac{\hbar}{\mathrm{i}}\psi_i^*\psi_j\Big|_{-\infty}^{\infty}-\frac{\hbar}{\mathrm{i}}\int_{-\infty}^{\infty}\psi_j\frac{\mathrm{d}\psi_i^*}{\mathrm{d}x}\mathrm{d}x$$

The first term on the right is zero, because all wavefunctions are zero at infinity in either direction, so we are left with

$$\int_{-\infty}^{\infty}\psi_i^*\hat{p}_x\psi_j\mathrm{d}x=-\frac{\hbar}{\mathrm{i}}\int_{-\infty}^{\infty}\psi_j\frac{\mathrm{d}\psi_i^*}{\mathrm{d}x}\mathrm{d}x=\left\{\frac{\hbar}{\mathrm{i}}\int_{-\infty}^{\infty}\psi_j^*\frac{\mathrm{d}\psi_i}{\mathrm{d}x}\mathrm{d}x\right\}^*$$

$$=\left\{\int_{-\infty}^{\infty}\psi_j^*\hat{p}_x\psi_i\mathrm{d}x\right\}^*$$

as we set out to prove.

Self-test 8.7 Confirm that the operator $\mathrm{d}^2/\mathrm{d}x^2$ is hermitian.

Hermitian operators are enormously important by virtue of two properties: their eigenvalues are real (as we prove in the *Justification* below), and their eigenfunctions are 'orthogonal'. All observables have real values (in the mathematical sense, such as $x=2$ m and $E=10$ J), so all observables are represented by hermitian operators. To say that two different functions ψ_i and ψ_j are **orthogonal** means that the integral (over all space) of their product is zero:

$$\text{Orthogonality:}\quad \int\psi_i^*\psi_j\,\mathrm{d}\tau=0 \qquad [8.31]$$

For example, the hamiltonian operator is hermitian (it corresponds to an observable, the energy). Therefore, if ψ_1 corresponds to one energy, and ψ_2 corresponds to a different energy, then we know at once that the two functions are orthogonal and that the integral of their product is zero.

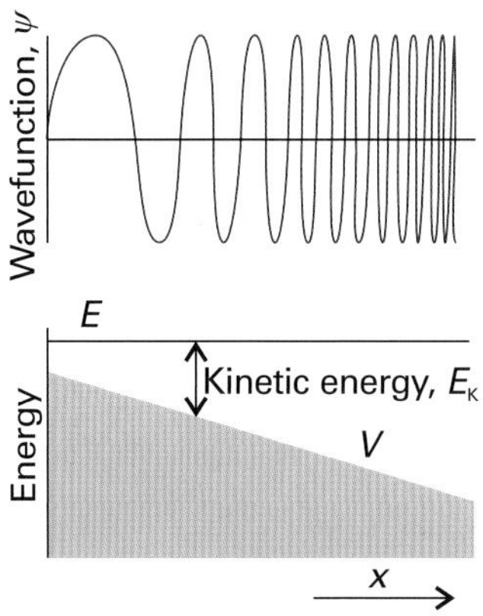

Fig. 8.28 The wavefunction of a particle in a potential decreasing towards the right and hence subjected to a constant force to the right. Only the real part of the wavefunction is shown, the imaginary part is similar, but displaced to the right.

Justification 8.3 *The reality of eigenvalues*

For a wavefunction ψ that is normalized and is an eigenfunction of an hermitian operator $\hat{\Omega}$ with eigenvalue ω, we can write

$$\int \psi^* \hat{\Omega} \psi \mathrm{d}\tau = \int \psi^* \omega \psi \mathrm{d}\tau = \omega \int \psi^* \psi \mathrm{d}\tau = \omega$$

However, by taking the complex conjugate we can write

$$\omega^* = \left\{ \int \psi^* \hat{\Omega} \psi \mathrm{d}\tau \right\}^* = \int \psi^* \hat{\Omega} \psi \mathrm{d}\tau = \omega$$

where in the second equality we have used the hermiticity of $\hat{\Omega}$. The conclusion that $\omega^* = \omega$ confirms that ω is real.

Illustration 8.2 *Confirming orthogonality*

The wavefunctions sin x and sin $2x$ are eigenfunctions of the hermitian operator $\mathrm{d}^2/\mathrm{d}x^2$, with eigenvalues -1 and -4, respectively. To verify that the two wavefunctions are mutually orthogonal, we integrate the product (sin x)(sin $2x$) over all space, which we may take to span from $x = 0$ to $x = 2\pi$, because both functions repeat themselves outside that range. Hence proving that the integral of their product is zero within that range implies that the integral over the whole of space is also zero (Fig. 8.29). A useful integral for this calculation is

$$\int \sin ax \sin bx \, \mathrm{d}x = \frac{\sin(a-b)x}{2(a-b)} - \frac{\sin(a+b)x}{2(a+b)} + \text{constant}, \qquad \text{if } a^2 \neq b^2$$

It follows that, for $a = 1$ and $b = 2$, and given the fact that $\sin 0 = 0$, $\sin 2\pi = 0$, and $\sin 6\pi = 0$,

$$\int_0^{2\pi} (\sin x)(\sin 2x)\mathrm{d}x = 0$$

and the two functions are mutually orthogonal.

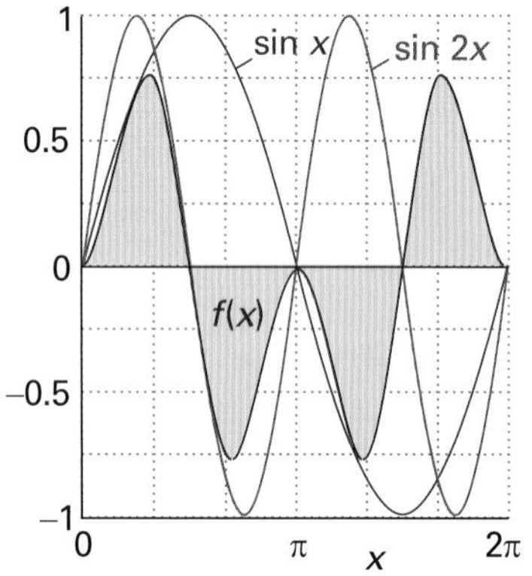

Fig. 8.29 The integral of the function $f(x) = \sin x \sin 2x$ is equal to the area (tinted) below the brown curve, and is zero, as can be inferred by symmetry. The function—and the value of the integral—repeats itself for all replications of the section between 0 and 2π, so the integral from $-\infty$ to ∞ is zero.

Self-test 8.8 Confirm that the functions sin x and sin $3x$ are mutually orthogonal.

$$\left[\int_{-\infty}^{\infty} \sin x \sin 3x \mathrm{d}x = 0 \right]$$

(d) Superpositions and expectation values

Suppose now that the wavefunction is the one given in eqn 8.19 (with $A = B$). What is the linear momentum of the particle it describes? We quickly run into trouble if we use the operator technique. When we operate with p_x, we find

$$\frac{\hbar}{\mathrm{i}}\frac{\mathrm{d}\psi}{\mathrm{d}x} = \frac{2\hbar}{\mathrm{i}} A \frac{\mathrm{d}\cos kx}{\mathrm{d}x} = -\frac{2k\hbar}{\mathrm{i}} A \sin kx \qquad (8.32)$$

This expression is not an eigenvalue equation, because the function on the right (sin kx) is different from that on the left (cos kx).

When the wavefunction of a particle is not an eigenfunction of an operator, the property to which the operator corresponds does not have a definite value. However, in the current example the momentum is not completely indefinite because the cosine wavefunction is a **linear combination**, or sum, of e^{ikx} and e^{-ikx}, and these two functions, as we have seen, individually correspond to definite momentum states. We say that the total wavefunction is a **superposition** of more than one wavefunction. Symbolically we can write the superposition as

$$\psi = \underset{\substack{\text{Particle with}\\ \text{linear}\\ \text{momentum}\\ +k\hbar}}{\psi_{\rightarrow}} + \underset{\substack{\text{Particle with}\\ \text{linear}\\ \text{momentum}\\ -k\hbar}}{\psi_{\leftarrow}}$$

The interpretation of this composite wavefunction is that, if the momentum of the particle is repeatedly measured in a long series of observations, then its magnitude will found to be $k\hbar$ in all the measurements (because that is the value for each component of the wavefunction). However, because the two component wavefunctions occur equally in the superposition, half the measurements will show that the particle is moving to the right ($p_x = +k\hbar$), and half the measurements will show that it is moving to the left ($p_x = -k\hbar$). According to quantum mechanics, we cannot predict in which direction the particle will in fact be found to be travelling; all we can say is that, in a long series of observations, if the particle is described by this wavefunction, then there are equal probabilities of finding the particle travelling to the right and to the left.

The same interpretation applies to any wavefunction written as a linear combination of eigenfunctions of an operator. Thus, suppose the wavefunction is known to be a superposition of many different linear momentum eigenfunctions and written as the linear combination

$$\psi = c_1\psi_1 + c_2\psi_2 + \cdots = \sum_k c_k\psi_k \qquad (8.33)$$

where the c_k are numerical (possibly complex) coefficients and the ψ_k correspond to different momentum states. The functions ψ_k are said to form a **complete set** in the sense that any arbitrary function can be expressed as a linear combination of them. Then according to quantum mechanics:

1 When the momentum is measured, in a single observation one of the eigenvalues corresponding to the ψ_k that contribute to the superposition will be found.

2 The probability of measuring a particular eigenvalue in a series of observations is proportional to the square modulus ($|c_k|^2$) of the corresponding coefficient in the linear combination.

3 The average value of a large number of observations is given by the expectation value, $\langle\Omega\rangle$, of the operator $\hat{\Omega}$ corresponding to the observable of interest.

The **expectation value** of an operator $\hat{\Omega}$ is defined as

$$\langle\Omega\rangle = \int \psi^{\star}\hat{\Omega}\psi\,\mathrm{d}\tau \qquad [8.34]$$

This formula is valid only for normalized wavefunctions. As we see in the *Justification* below, an expectation value is the weighted average of a large number of observations of a property.

Comment 8.12

In general, a linear combination of two functions f and g is $c_1f + c_2g$, where c_1 and c_2 are numerical coefficients, so a linear combination is a more general term than 'sum'. In a sum, $c_1 = c_2 = 1$. A linear combination might have the form $0.567f + 1.234g$, for instance, so it is more general than the simple sum $f + g$.

Justification 8.4 *The expectation value of an operator*

If ψ is an eigenfunction of $\hat{\Omega}$ with eigenvalue ω, the expectation value of $\hat{\Omega}$ is

$$\langle\Omega\rangle = \int \psi^* \overbrace{\hat{\Omega}\psi}^{\omega\psi} \mathrm{d}\tau = \int \psi^* \omega\psi \mathrm{d}\tau = \omega \int \psi^* \psi \mathrm{d}\tau = \omega$$

because ω is a constant and may be taken outside the integral, and the resulting integral is equal to 1 for a normalized wavefunction. The interpretation of this expression is that, because every observation of the property Ω results in the value ω (because the wavefunction is an eigenfunction of $\hat{\Omega}$), the mean value of all the observations is also ω.

A wavefunction that is not an eigenfunction of the operator of interest can be written as a linear combination of eigenfunctions. For simplicity, suppose the wavefunction is the sum of two eigenfunctions (the general case, eqn 8.33, can easily be developed). Then

$$\begin{aligned}\langle\Omega\rangle &= \int (c_1\psi_1 + c_2\psi_2)^* \hat{\Omega}(c_1\psi_1 + c_2\psi_2)\mathrm{d}\tau \\ &= \int (c_1\psi_1 + c_2\psi_2)^* (c_1\hat{\Omega}\psi_1 + c_2\hat{\Omega}\psi_2)\mathrm{d}\tau \\ &= \int (c_1\psi_1 + c_2\psi_2)^* (c_1\omega_1\psi_1 + c_2\omega_2\psi_2)\mathrm{d}\tau \\ &= c_1^* c_1\omega_1 \overbrace{\int \psi_1^*\psi_1\mathrm{d}\tau}^{1} + c_2^* c_2\omega_2 \overbrace{\int \psi_2^*\psi_2\mathrm{d}\tau}^{1} \\ &\quad + c_2^* c_1\omega_1 \overbrace{\int \psi_2^*\psi_1\mathrm{d}\tau}^{0} + c_1^* c_2\omega_2 \overbrace{\int \psi_1^*\psi_2\mathrm{d}\tau}^{0}\end{aligned}$$

The first two integrals on the right are both equal to 1 because the wavefunctions are individually normalized. Because ψ_1 and ψ_2 correspond to different eigenvalues of an hermitian operator, they are orthogonal, so the third and fourth integrals on the right are zero. We can conclude that

$$\langle\Omega\rangle = |c_1|^2\omega_1 + |c_2|^2\omega_2$$

This expression shows that the expectation value is the sum of the two eigenvalues weighted by the probabilities that each one will be found in a series of measurements. Hence, the expectation value is the weighted mean of a series of observations.

Example 8.7 *Calculating an expectation value*

Calculate the average value of the distance of an electron from the nucleus in the hydrogen atom in its state of lowest energy.

Method The average radius is the expectation value of the operator corresponding to the distance from the nucleus, which is multiplication by r. To evaluate $\langle r\rangle$, we need to know the normalized wavefunction (from Example 8.4) and then evaluate the integral in eqn 8.34.

Answer The average value is given by the expectation value

$$\langle r\rangle = \int \psi^* r\psi \,\mathrm{d}\tau$$

which we evaluate by using spherical polar coordinates. Using the normalized function in Example 8.4 gives

$$\langle r\rangle = \frac{1}{\pi a_0^3}\overbrace{\int_0^\infty r^3 \mathrm{e}^{-2r/a_0}\mathrm{d}r}^{3!a_0^3/2^4}\overbrace{\int_0^\pi \sin\theta\,\mathrm{d}\theta}^{2}\overbrace{\int_0^{2\pi}\mathrm{d}\phi}^{2\pi} = \tfrac{3}{2}a_0$$

Because $a_0 = 52.9$ pm (see Section 10.1), $\langle r\rangle = 79.4$ pm. This result means that, if a very large number of measurements of the distance of the electron from the nucleus are made, then their mean value will be 79.4 pm. However, each different observation will give a different and unpredictable individual result because the wavefunction is not an eigenfunction of the operator corresponding to r.

Self-test 8.9 Evaluate the root mean square distance, $\langle r^2\rangle^{1/2}$, of the electron from the nucleus in the hydrogen atom. [$3^{1/2}a_0 = 91.6$ pm]

The mean kinetic energy of a particle in one dimension is the expectation value of the operator given in eqn 8.28. Therefore, we can write

$$\langle E_\mathrm{K}\rangle = \int \psi^* \hat{E}_\mathrm{K}\psi\,\mathrm{d}\tau = -\frac{\hbar^2}{2m}\int \psi^* \frac{\mathrm{d}^2\psi}{\mathrm{d}x^2}\mathrm{d}\tau \qquad (8.35)$$

This conclusion confirms the previous assertion that the kinetic energy is a kind of average over the curvature of the wavefunction: we get a large contribution to the observed value from regions where the wavefunction is sharply curved (so $\mathrm{d}^2\psi/\mathrm{d}x^2$ is large) and the wavefunction itself is large (so that ψ^* is large too).

8.6 The uncertainty principle

We have seen that, if the wavefunction is $A\mathrm{e}^{ikx}$, then the particle it describes has a definite state of linear momentum, namely travelling to the right with momentum $p_x = +k\hbar$. However, we have also seen that the position of the particle described by this wavefunction is completely unpredictable. In other words, if the momentum is specified precisely, it is impossible to predict the location of the particle. This statement is one-half of a special case of the **Heisenberg uncertainty principle**, one of the most celebrated results of quantum mechanics:

> It is impossible to specify simultaneously, with arbitrary precision, both the momentum and the position of a particle.

Before discussing the principle further, we must establish its other half: that, if we know the position of a particle exactly, then we can say nothing about its momentum. The argument draws on the idea of regarding a wavefunction as a superposition of eigenfunctions, and runs as follows.

If we know that the particle is at a definite location, its wavefunction must be large there and zero everywhere else (Fig. 8.30). Such a wavefunction can be created by superimposing a large number of harmonic (sine and cosine) functions, or, equivalently, a number of e^{ikx} functions. In other words, we can create a sharply localized

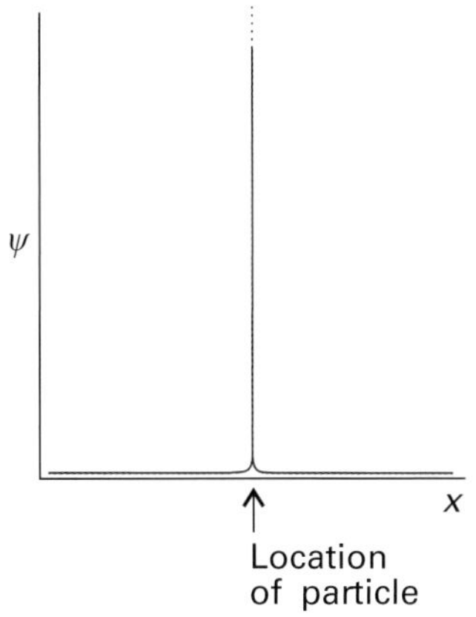

Fig. 8.30 The wavefunction for a particle at a well-defined location is a sharply spiked function that has zero amplitude everywhere except at the particle's position.

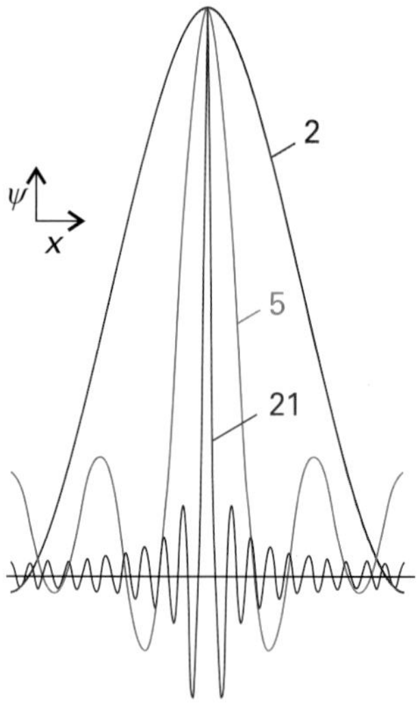

Fig. 8.31 The wavefunction for a particle with an ill-defined location can be regarded as the superposition of several wavefunctions of definite wavelength that interfere constructively in one place but destructively elsewhere. As more waves are used in the superposition (as given by the numbers attached to the curves), the location becomes more precise at the expense of uncertainty in the particle's momentum. An infinite number of waves is needed to construct the wavefunction of a perfectly localized particle.

Exploration Use mathematical software or an electronic spreadsheet to construct superpositions of cosine functions as $\psi(x)=\sum_{k=1}^{N}(1/N)\cos(k\pi x)$, where the constant $1/N$ is introduced to keep the superpositions with the same overall magnitude. Explore how the probability density $\psi^2(x)$ changes with the value of N.

wavefunction, called a **wave packet**, by forming a linear combination of wavefunctions that correspond to many different linear momenta. The superposition of a few harmonic functions gives a wavefunction that spreads over a range of locations (Fig. 8.31). However, as the number of wavefunctions in the superposition increases, the wave packet becomes sharper on account of the more complete interference between the positive and negative regions of the individual waves. When an infinite number of components is used, the wave packet is a sharp, infinitely narrow spike, which corresponds to perfect localization of the particle. Now the particle is perfectly localized. However, we have lost all information about its momentum because, as we saw above, a measurement of the momentum will give a result corresponding to any one of the infinite number of waves in the superposition, and which one it will give is unpredictable. Hence, if we know the location of the particle precisely (implying that its wavefunction is a superposition of an infinite number of momentum eigenfunctions), then its momentum is completely unpredictable.

A quantitative version of this result is

$$\Delta p\Delta q \geq \tfrac{1}{2}\hbar \tag{8.36a}$$

In this expression Δp is the 'uncertainty' in the linear momentum parallel to the axis q, and Δq is the uncertainty in position along that axis. These 'uncertainties' are precisely defined, for they are the root mean square deviations of the properties from their mean values:

$$\Delta p=\{\langle p^2\rangle-\langle p\rangle^2\}^{1/2} \qquad \Delta q=\{\langle q^2\rangle-\langle q\rangle^2\}^{1/2} \tag{8.36b}$$

If there is complete certainty about the position of the particle ($\Delta q=0$), then the only way that eqn 8.36a can be satisfied is for $\Delta p=\infty$, which implies complete uncertainty about the momentum. Conversely, if the momentum parallel to an axis is known exactly ($\Delta p = 0$), then the position along that axis must be completely uncertain ($\Delta q=\infty$).

The p and q that appear in eqn 8.36 refer to the same direction in space. Therefore, whereas simultaneous specification of the position on the x-axis and momentum parallel to the x-axis are restricted by the uncertainty relation, simultaneous location of position on x and motion parallel to y or z are not restricted. The restrictions that the uncertainty principle implies are summarized in Table 8.2.

Example 8.8 *Using the uncertainty principle*

Suppose the speed of a projectile of mass 1.0 g is known to within 1 μm s^{-1}. Calculate the minimum uncertainty in its position.

Method Estimate Δp from $m\Delta v$, where Δv is the uncertainty in the speed; then use eqn 8.36a to estimate the minimum uncertainty in position, Δq.

Answer The minimum uncertainty in position is

$$\Delta q=\frac{\hbar}{2m\Delta v}$$

$$=\frac{1.055\times10^{-34}\,\mathrm{J\,s}}{2\times(1.0\times10^{-3}\,\mathrm{kg})\times(1\times10^{-6}\,\mathrm{m\,s^{-1}})}=5\times10^{-26}\,\mathrm{m}$$

where we have used 1 J = 1 kg m^2 s^{-2}. The uncertainty is completely negligible for all practical purposes concerning macroscopic objects. However, if the mass is that of an electron, then the same uncertainty in speed implies an uncertainty in position far larger than the diameter of an atom (the analogous calculation gives $\Delta q = 60$ m); so the concept of a trajectory, the simultaneous possession of a precise position and momentum, is untenable.

Self-test 8.10 Estimate the minimum uncertainty in the speed of an electron in a one-dimensional region of length $2a_0$. [500 km s^{-1}]

The Heisenberg uncertainty principle is more general than eqn 8.36 suggests. It applies to any pair of observables called **complementary observables**, which are defined in terms of the properties of their operators. Specifically, two observables Ω_1 and Ω_2 are complementary if

$$\hat{\Omega}_1(\hat{\Omega}_2\psi) \neq \hat{\Omega}_2(\hat{\Omega}_1\psi) \tag{8.37}$$

When the effect of two operators depends on their order (as this equation implies), we say that they do not **commute**. The different outcomes of the effect of applying $\hat{\Omega}_1$ and $\hat{\Omega}_2$ in a different order are expressed by introducing the **commutator** of the two operators, which is defined as

$$[\hat{\Omega}_1,\hat{\Omega}_2] = \hat{\Omega}_1\hat{\Omega}_2 - \hat{\Omega}_2\hat{\Omega}_1 \qquad [8.38]$$

We can conclude from *Illustration 8.3* that the commutator of the operators for position and linear momentum is

$$[\hat{x}, \hat{p}_x] = \mathrm{i}\hbar \tag{8.39}$$

Illustration 8.3 *Evaluating a commutator*

To show that the operators for position and momentum do not commute (and hence are complementary observables) we consider the effect of $\hat{x}\hat{p}_x$ (that is, the effect of $\hat{p}_x$ followed by the effect on the outcome of multiplication by x) on a wavefunction ψ:

$$\hat{x}\hat{p}_x\psi = x \times \frac{\hbar}{\mathrm{i}}\frac{\mathrm{d}\psi}{\mathrm{d}x}$$

Next, we consider the effect of $\hat{p}_x\hat{x}$ on the same function (that is, the effect of multiplication by x followed by the effect of $\hat{p}_x$ on the outcome):

$$\hat{p}_x\hat{x}\psi = \frac{\hbar}{\mathrm{i}}\frac{\mathrm{d}}{\mathrm{d}x}x\psi = \frac{\hbar}{\mathrm{i}}\left(\psi + x\frac{\mathrm{d}\psi}{\mathrm{d}x}\right)$$

For this step we have used the standard rule about differentiating a product of functions. The second expression is clearly different from the first, so the two operators do not commute. Subtraction of the second expression from the first gives eqn 8.39.

The commutator in eqn 8.39 is of such vital significance in quantum mechanics that it is taken as a fundamental distinction between classical mechanics and quantum mechanics. In fact, this commutator may be taken as a postulate of quantum mechanics, and is used to justify the choice of the operators for position and linear momentum given in eqn 8.26.

Table 8.2* Constraints of the uncertainty principle

	Variable 1					
Variable 2	x	y	z	p_x	p_y	p_z
x						
y						
z						
p_x						
p_y						
p_z						

* Pairs of observables that cannot be determined simultaneously with arbitrary precision are marked with a white rectangle; all others are unrestricted.

Comment 8.13

For two functions f and g, $\mathrm{d}(fg) = f\mathrm{d}g + g\mathrm{d}f$.

Comment 8.14

The 'modulus' notation |. . .| means take the magnitude of the term the bars enclose: for a real quantity x, $|x|$ is the magnitude of x (its value without its sign); for an imaginary quantity iy, $|iy|$ is the magnitude of y; and—most generally—for a complex quantity $z=x+iy$, $|z|$ is the value of $(z^*z)^{1/2}$. For example, $|-2|=2$, $|3i|=3$, and $|-2+3i|=\{(-2-3i)(-2+3i)\}^{1/2}=13^{1/2}$. Physically, the modulus on the right of eqn 8.40 ensures that the product of uncertainties has a real, non-negative value.

With the concept of commutator established, the Heisenberg uncertainty principle can be given its most general form. For *any* two pairs of observables, Ω_1 and Ω_2, the uncertainties (to be precise, the root mean square deviations of their values from the mean) in simultaneous determinations are related by

$$\Delta\Omega_1\Delta\Omega_2 \geq \tfrac{1}{2}|\langle[\hat{\Omega}_1,\hat{\Omega}_2]\rangle| \qquad (8.40)$$

We obtain the special case of eqn 8.36a when we identify the observables with x and p_x and use eqn 8.39 for their commutator.

Complementary observables are observables with non-commuting operators. With the discovery that some pairs of observables are complementary (we meet more examples in the next chapter), we are at the heart of the difference between classical and quantum mechanics. Classical mechanics supposed, falsely as we now know, that the position and momentum of a particle could be specified simultaneously with arbitrary precision. However, quantum mechanics shows that position and momentum are complementary, and that we have to make a choice: we can specify position at the expense of momentum, or momentum at the expense of position.

The realization that some observables are complementary allows us to make considerable progress with the calculation of atomic and molecular properties, but it does away with some of the most cherished concepts of classical physics.

8.7 The postulates of quantum mechanics

For convenience, we collect here the postulates on which quantum mechanics is based and which have been introduced in the course of this chapter.

The wavefunction. All dynamical information is contained in the wavefunction ψ for the system, which is a mathematical function found by solving the Schrödinger equation for the system. In one dimension:

$$-\frac{\hbar^2}{2m}\frac{d^2\psi}{dx^2}+V(x)\psi=E\psi$$

The Born interpretation. If the wavefunction of a particle has the value ψ at some point $\boldsymbol{r}$, then the probability of finding the particle in an infinitesimal volume $d\tau = dxdydz$ at that point is proportional to $|\psi|^2 d\tau$.

Acceptable wavefunctions. An acceptable wavefunction must be continuous, have a continuous first derivative, be single-valued, and be square-integrable.

Observables. Observables, Ω, are represented by operators, $\hat{\Omega}$, built from position and momentum operators of the form

$$\hat{x}=x\times \qquad \hat{p}_x=\frac{\hbar}{i}\frac{d}{dx}$$

or, more generally, from operators that satisfy the commutation relation $[\hat{x},\hat{p}_x]=i\hbar$.

The Heisenberg uncertainty principle. It is impossible to specify simultaneously, with arbitrary precision, both the momentum and the position of a particle and, more generally, any pair of observable with operators that do not commute.

Checklist of key ideas

☐ 1. In classical physics, radiation is described in terms of an oscillating electromagnetic disturbance that travels through vacuum at a constant speed $c = \lambda\nu$.

☐ 2. A black body is an object that emits and absorbs all frequencies of radiation uniformly.

☐ 3. The variation of the energy output of a black body with wavelength is explained by invoking quantization of energy, the limitation of energies to discrete values, which in turn leads to the Planck distribution, eqn 8.5.

☐ 4. The variation of the molar heat capacity of a solid with temperature is explained by invoking quantization of energy, which leads to the Einstein and Debye formulas, eqns 8.7 and 8.9.

☐ 5. Spectroscopic transitions are changes in populations of quantized energy levels of a system involving the absorption, emission, or scattering of electromagnetic radiation, $\Delta E = h\nu$.

☐ 6. The photoelectric effect is the ejection of electrons from metals when they are exposed to ultraviolet radiation: $\frac{1}{2}m_e v^2 = h\nu - \Phi$, where Φ is the work function, the energy required to remove an electron from the metal to infinity.

☐ 7. The photoelectric effect and electron diffraction are phenomena that confirm wave–particle duality, the joint particle and wave character of matter and radiation.

☐ 8. The de Broglie relation, $\lambda = h/p$, relates the momentum of a particle with its wavelength.

☐ 9. A wavefunction is a mathematical function obtained by solving the Schrödinger equation and which contains all the dynamical information about a system.

☐ 10. The time-independent Schrödinger equation in one dimension is $-(\hbar^2/2m)(d^2\psi/dx^2) + V(x)\psi = E\psi$.

☐ 11. The Born interpretation of the wavefunction states that the value of $|\psi|^2$, the probability density, at a point is proportional to the probability of finding the particle at that point.

☐ 12. Quantization is the confinement of a dynamical observable to discrete values.

☐ 13. An acceptable wavefunction must be continuous, have a continuous first derivative, be single-valued, and be square-integrable.

☐ 14. An operator is something that carries out a mathematical operation on a function. The position and momentum operators are $\hat{x} = x\times$ and $\hat{p}_x = (\hbar/i)d/dx$, respectively.

☐ 15. The hamiltonian operator is the operator for the total energy of a system, $\hat{H}\psi = E\psi$ and is the sum of the operators for kinetic energy and potential energy.

☐ 16. An eigenvalue equation is an equation of the form $\hat{\Omega}\psi = \omega\psi$. The eigenvalue is the constant ω in the eigenvalue equation; the eigenfunction is the function ψ in the eigenvalue equation.

☐ 17. The expectation value of an operator is $\langle\Omega\rangle = \int\psi^*\hat{\Omega}\psi\,d\tau$.

☐ 18. An hermitian operator is one for which $\int\psi_i^*\hat{\Omega}\psi_j dx = (\int\psi_j^*\hat{\Omega}\psi_i dx)^*$. The eigenvalues of hermitian operators are real and correspond to observables, measurable properties of a system. The eigenfunctions of hermitian operations are orthogonal, meaning that $\int\psi_i^*\psi_j\,d\tau = 0$.

☐ 19. The Heisenberg uncertainty principle states that it is impossible to specify simultaneously, with arbitrary precision, both the momentum and the position of a particle; $\Delta p\Delta q \geq \frac{1}{2}\hbar$.

☐ 20. Two operators commute when $[\hat{\Omega}_1,\hat{\Omega}_2] = \hat{\Omega}_1\hat{\Omega}_2 - \hat{\Omega}_2\hat{\Omega}_1 = 0$.

☐ 21. Complementary observables are observables corresponding to non-commuting operators.

☐ 22. The general form of the Heisenberg uncertainty principle is $\Delta\Omega_1\Delta\Omega_2 \geq \frac{1}{2}|\langle[\hat{\Omega}_1,\hat{\Omega}_2]\rangle|$.

Further reading

Articles and texts

P.W. Atkins, *Quanta: A handbook of concepts*. Oxford University Press (1991).

P.W. Atkins and R.S. Friedman, *Molecular quantum mechanics*. Oxford University Press (2005).

D. Bohm, *Quantum theory*. Dover, New York (1989).

R.P. Feynman, R.B. Leighton, and M. Sands, *The Feynman lectures on physics*. Volume III. Addison–Wesley, Reading (1965).

C.S. Johnson, Jr. and L.G. Pedersen, *Problems and solutions in quantum chemistry and physics*. Dover, New York, 1986.

L. Pauling and E.B. Wilson, *Introduction to quantum mechanics with applications to chemistry*. Dover, New York (1985).

Discussion questions

8.1 Summarize the evidence that led to the introduction of quantum mechanics.

8.2 Explain why Planck's introduction of quantization accounted for the properties of black-body radiation.

8.3 Explain why Einstein's introduction of quantization accounted for the properties of heat capacities at low temperatures.

8.4 Describe how a wavefunction determines the dynamical properties of a system and how those properties may be predicted.

8.5 Account for the uncertainty relation between position and linear momentum in terms of the shape of the wavefunction.

8.6 Suggest how the general shape of a wavefunction can be predicted without solving the Schrödinger equation explicitly.

Exercises

8.1a To what speed must an electron be accelerated for it to have a wavelength of 3.0 cm?

8.1b To what speed must a proton be accelerated for it to have a wavelength of 3.0 cm?

8.2a The fine-structure constant, α, plays a special role in the structure of matter; its approximate value is 1/137. What is the wavelength of an electron travelling at a speed αc, where c is the speed of light? (Note that the circumference of the first Bohr orbit in the hydrogen atom is 331 pm.)

8.2b Calculate the linear momentum of photons of wavelength 350 nm. What speed does a hydrogen molecule need to travel to have the same linear momentum?

8.3a The speed of a certain proton is 0.45 Mm s^{-1}. If the uncertainty in its momentum is to be reduced to 0.0100 per cent, what uncertainty in its location must be tolerated?

8.3b The speed of a certain electron is 995 km s^{-1}. If the uncertainty in its momentum is to be reduced to 0.0010 per cent, what uncertainty in its location must be tolerated?

8.4a Calculate the energy per photon and the energy per mole of photons for radiation of wavelength (a) 600 nm (red), (b) 550 nm (yellow), (c) 400 nm (blue).

8.4b Calculate the energy per photon and the energy per mole of photons for radiation of wavelength (a) 200 nm (ultraviolet), (b) 150 pm (X-ray), (c) 1.00 cm (microwave).

8.5a Calculate the speed to which a stationary H atom would be accelerated if it absorbed each of the photons used in Exercise 8.4a.

8.5b Calculate the speed to which a stationary ^{4}He atom (mass 4.0026 u) would be accelerated if it absorbed each of the photons used in Exercise 8.4b.

8.6a A glow-worm of mass 5.0 g emits red light (650 nm) with a power of 0.10 W entirely in the backward direction. To what speed will it have accelerated after 10 y if released into free space and assumed to live?

8.6b A photon-powered spacecraft of mass 10.0 kg emits radiation of wavelength 225 nm with a power of 1.50 kW entirely in the backward direction. To what speed will it have accelerated after 10.0 y if released into free space?

8.7a A sodium lamp emits yellow light (550 nm). How many photons does it emit each second if its power is (a) 1.0 W, (b) 100 W?

8.7b A laser used to read CDs emits red light of wavelength 700 nm. How many photons does it emit each second if its power is (a) 0.10 W, (b) 1.0 W?

8.8a The work function for metallic caesium is 2.14 eV. Calculate the kinetic energy and the speed of the electrons ejected by light of wavelength (a) 700 nm, (b) 300 nm.

8.8b The work function for metallic rubidium is 2.09 eV. Calculate the kinetic energy and the speed of the electrons ejected by light of wavelength (a) 650 nm, (b) 195 nm.

8.9a Calculate the size of the quantum involved in the excitation of (a) an electronic oscillation of period 1.0 fs, (b) a molecular vibration of period 10 fs, (c) a pendulum of period 1.0 s. Express the results in joules and kilojoules per mole.

8.9b Calculate the size of the quantum involved in the excitation of (a) an electronic oscillation of period 2.50 fs, (b) a molecular vibration of period 2.21 fs, (c) a balance wheel of period 1.0 ms. Express the results in joules and kilojoules per mole.

8.10a Calculate the de Broglie wavelength of (a) a mass of 1.0 g travelling at 1.0 cm s^{-1}, (b) the same, travelling at 100 km s^{-1}, (c) an He atom travelling at 1000 m s^{-1} (a typical speed at room temperature).

8.10b Calculate the de Broglie wavelength of an electron accelerated from rest through a potential difference of (a) 100 V, (b) 1.0 kV, (c) 100 kV.

8.11a Confirm that the operator $\hat{l}_z = (\hbar/\mathrm{i})\mathrm{d}/\mathrm{d}\phi$, where ϕ is an angle, is hermitian.

8.11b Show that the linear combinations $\hat{A} + \mathrm{i}\hat{B}$ and $\hat{A} - \mathrm{i}\hat{B}$ are not hermitian if $\hat{A}$ and $\hat{B}$ are hermitian operators.

8.12a Calculate the minimum uncertainty in the speed of a ball of mass 500 g that is known to be within 1.0 μm of a certain point on a bat. What is the minimum uncertainty in the position of a bullet of mass 5.0 g that is known to have a speed somewhere between 350.000 01 m s^{-1} and 350.000 00 m s^{-1}?

8.12b An electron is confined to a linear region with a length of the same order as the diameter of an atom (about 100 pm). Calculate the minimum uncertainties in its position and speed.

8.13a In an X-ray photoelectron experiment, a photon of wavelength 150 pm ejects an electron from the inner shell of an atom and it emerges with a speed of 21.4 Mm s^{-1}. Calculate the binding energy of the electron.

8.13b In an X-ray photoelectron experiment, a photon of wavelength 121 pm ejects an electron from the inner shell of an atom and it emerges with a speed of 56.9 Mm s^{-1}. Calculate the binding energy of the electron.

8.14a Determine the commutators of the operators (a) $\mathrm{d}/\mathrm{d}x$ and $1/x$, (b) $\mathrm{d}/\mathrm{d}x$ and x^2.

8.14b Determine the commutators of the operators a and $a^\dagger$, where $a = (\hat{x} + \mathrm{i}\hat{p})/2^{1/2}$ and $a^\dagger = (\hat{x} - \mathrm{i}\hat{p})/2^{1/2}$.

Problems*

Numerical problems

8.1 The Planck distribution gives the energy in the wavelength range $d\lambda$ at the wavelength λ. Calculate the energy density in the range 650 nm to 655 nm inside a cavity of volume 100 cm^3 when its temperature is (a) 25°C, (b) 3000°C.

8.2 For a black body, the temperature and the wavelength of emission maximum, λ_{max}, are related by Wien's law, $\lambda_{max}T = \frac{1}{5}c_2$, where $c_2 = hc/k$ (see Problem 8.10). Values of λ_{max} from a small pinhole in an electrically heated container were determined at a series of temperatures, and the results are given below. Deduce a value for Planck's constant.

θ/°C	1000	1500	2000	2500	3000	3500
λ_{max}/nm	2181	1600	1240	1035	878	763

8.3 The Einstein frequency is often expressed in terms of an equivalent temperature θ_E, where $\theta_E = h\nu/k$. Confirm that θ_E has the dimensions of temperature, and express the criterion for the validity of the high-temperature form of the Einstein equation in terms of it. Evaluate θ_E for (a) diamond, for which $\nu = 46.5$ THz and (b) for copper, for which $\nu = 7.15$ THz. What fraction of the Dulong and Petit value of the heat capacity does each substance reach at 25°C?

8.4 The ground-state wavefunction for a particle confined to a one-dimensional box of length L is

$$\psi = \left(\frac{2}{L}\right)^{1/2} \sin\left(\frac{\pi x}{L}\right)$$

Suppose the box is 10.0 nm long. Calculate the probability that the particle is (a) between $x = 4.95$ nm and 5.05 nm, (b) between $x = 1.95$ nm and 2.05 nm, (c) between $x = 9.90$ nm and 10.00 nm, (d) in the right half of the box, (e) in the central third of the box.

8.5 The ground-state wavefunction of a hydrogen atom is

$$\psi = \left(\frac{1}{\pi a_0^3}\right)^{1/2} e^{-r/a_0}$$

where $a_0 = 53$ pm (the Bohr radius). (a) Calculate the probability that the electron will be found somewhere within a small sphere of radius 1.0 pm centred on the nucleus. (b) Now suppose that the same sphere is located at $r = a_0$. What is the probability that the electron is inside it?

8.6 The normalized wavefunctions for a particle confined to move on a circle are $\psi(\phi) = (1/2\pi)^{1/2}e^{-im\phi}$, where $m = 0, \pm1, \pm2, \pm3, \ldots$ and $0 \le \phi \le 2\pi$. Determine $\langle\phi\rangle$.

8.7 A particle is in a state described by the wavefunction $\psi(x) = (2a/\pi)^{1/4}e^{-ax^2}$, where a is a constant and $-\infty \le x \le \infty$. Verify that the value of the product $\Delta p\Delta x$ is consistent with the predictions from the uncertainty principle.

8.8 A particle is in a state described by the wavefunction $\psi(x) = a^{1/2}e^{-ax}$, where a is a constant and $0 \le x \le \infty$. Determine the expectation value of the commutator of the position and momentum operators.

Theoretical problems

8.9 Demonstrate that the Planck distribution reduces to the Rayleigh–Jeans law at long wavelengths.

8.10 Derive *Wien's law*, that $\lambda_{max}T$ is a constant, where λ_{max} is the wavelength corresponding to maximum in the Planck distribution at the temperature T, and deduce an expression for the constant as a multiple of the *second radiation constant*, $c_2 = hc/k$.

8.11 Use the Planck distribution to deduce the *Stefan–Boltzmann law* that the total energy density of black-body radiation is proportional to T^4, and find the constant of proportionality.

8.12‡ Prior to Planck's derivation of the distribution law for black-body radiation, Wien found empirically a closely related distribution function which is very nearly but not exactly in agreement with the experimental results, namely, $\rho = (a/\lambda^5)e^{-b/\lambda kT}$. This formula shows small deviations from Planck's at long wavelengths. (a) By fitting Wien's empirical formula to Planck's at short wavelengths determine the constants a and b. (b) Demonstrate that Wien's formula is consistent with Wien's law (Problem 8.10) and with the Stefan–Boltzmann law (Problem 8.11).

8.13 Normalize the following wavefunctions: (a) $\sin(n\pi x/L)$ in the range $0 \le x \le L$, where $n = 1, 2, 3, \ldots$, (b) a constant in the range $-L \le x \le L$, (c) $e^{-r/a}$ in three-dimensional space, (d) $xe^{-r/2a}$ in three-dimensional space. *Hint*: The volume element in three dimensions is $d\tau = r^2 dr \sin\theta\, d\theta\, d\phi$, with $0 \le r < \infty$, $0 \le \theta \le \pi$, $0 \le \phi \le 2\pi$. Use the integral in Example 8.4.

8.14 (a) Two (unnormalized) excited state wavefunctions of the H atom are

$$\text{(i)}\ \psi = \left(2 - \frac{r}{a_0}\right)e^{-r/a_0} \qquad \text{(ii)}\ \psi = r\sin\theta\cos\phi\, e^{-r/2a_0}$$

Normalize both functions to 1. (b) Confirm that these two functions are mutually orthogonal.

8.15 Identify which of the following functions are eigenfunctions of the operator d/dx: (a) e^{ikx}, (b) $\cos kx$, (c) k, (d) kx, (e) $e^{-\alpha x^2}$. Give the corresponding eigenvalue where appropriate.

8.16 Determine which of the following functions are eigenfunctions of the inversion operator $\hat{\imath}$ (which has the effect of making the replacement $x \to -x$): (a) $x^3 - kx$, (b) $\cos kx$, (c) $x^2 + 3x - 1$. State the eigenvalue of $\hat{\imath}$ when relevant.

8.17 Which of the functions in Problem 8.15 are (a) also eigenfunctions of d^2/dx^2 and (b) only eigenfunctions of d^2/dx^2? Give the eigenvalues where appropriate.

8.18 A particle is in a state described by the wavefunction $\psi = (\cos\chi)e^{ikx} + (\sin\chi)e^{-ikx}$, where χ (chi) is a parameter. What is the probability that the particle will be found with a linear momentum (a) $+k\hbar$, (b) $-k\hbar$? What form would the wavefunction have if it were 90 per cent certain that the particle had linear momentum $+k\hbar$?

8.19 Evaluate the kinetic energy of the particle with wavefunction given in Problem 8.18.

8.20 Calculate the average linear momentum of a particle described by the following wavefunctions: (a) e^{ikx}, (b) $\cos kx$, (c) $e^{-\alpha x^2}$, where in each one x ranges from $-\infty$ to $+\infty$.

8.21 Evaluate the expectation values of r and r^2 for a hydrogen atom with wavefunctions given in Problem 8.14.

8.22 Calculate (a) the mean potential energy and (b) the mean kinetic energy of an electron in the ground state of a hydrogenic atom.

* Problems denoted with the symbol ‡ were supplied by Charles Trapp, Carmen Giunta, and Marshall Cady.

8.23 Use mathematical software to construct superpositions of cosine functions and determine the probability that a given momentum will be observed. If you plot the superposition (which you should), set $x = 0$ at the centre of the screen and build the superposition there. Evaluate the root mean square location of the packet, $\langle x^2 \rangle^{1/2}$.

8.24 Show that the expectation value of an operator that can be written as the square of an hermitian operator is positive.

8.25 (a) Given that any operators used to represent observables must satisfy the commutation relation in eqn 8.38, what would be the operator for position if the choice had been made to represent linear momentum parallel to the x-axis by multiplication by the linear momentum? These different choices are all valid 'representations' of quantum mechanics. (b) With the identification of $\hat{x}$ in this representation, what would be the operator for $1/x$? *Hint.* Think of $1/x$ as x^{-1}.

Applications: to biology, environmental science, and astrophysics

8.26‡ The temperature of the Sun's surface is approximately 5800 K. On the assumption that the human eye evolved to be most sensitive at the wavelength of light corresponding to the maximum in the Sun's radiant energy distribution, determine the colour of light to which the eye is most sensitive. *Hint*: See Problem 8.10.

8.27 We saw in *Impact I8.1* that electron microscopes can obtain images with several hundredfold higher resolution than optical microscopes because of the short wavelength obtainable from a beam of electrons. For electrons moving at speeds close to c, the speed of light, the expression for the de Broglie wavelength (eqn 8.12) needs to be corrected for relativistic effects:

$$\lambda = \frac{h}{\left\{2m_e eV\left(1+\frac{eV}{2m_e c^2}\right)\right\}^{1/2}}$$

where c is the speed of light in vacuum and V is the potential difference through which the electrons are accelerated. (a) Use the expression above to calculate the de Broglie wavelength of electrons accelerated through 50 kV. (b) Is the relativistic correction important?

8.28‡ Solar energy strikes the top of the Earth's atmosphere at a rate of 343 W m^{-2}. About 30 per cent of this energy is reflected directly back into space by the Earth or the atmosphere. The Earth–atmosphere system absorbs the remaining energy and re-radiates it into space as black-body radiation. What is the average black-body temperature of the Earth? What is the wavelength of the most plentiful of the Earth's black-body radiation? *Hint.* Use Wien's law, Problem 8.10.

8.29‡ A star too small and cold to shine has been found by S. Kulkarni, K. Matthews, B.R. Oppenheimer, and T. Nakajima (*Science* **270**, 1478 (1995)). The spectrum of the object shows the presence of methane, which, according to the authors, would not exist at temperatures much above 1000 K. The mass of the star, as determined from its gravitational effect on a companion star, is roughly 20 times the mass of Jupiter. The star is considered to be a brown dwarf, the coolest ever found. (a) From available thermodynamic data, test the stability of methane at temperatures above 1000 K. (b) What is λ_{max} for this star? (c) What is the energy density of the star relative to that of the Sun (6000 K)? (d) To determine whether the star will shine, estimate the fraction of the energy density of the star in the visible region of the spectrum.

Quantum theory: techniques and applications

9

To find the properties of systems according to quantum mechanics we need to solve the appropriate Schrödinger equation. This chapter presents the essentials of the solutions for three basic types of motion: translation, vibration, and rotation. We shall see that only certain wavefunctions and their corresponding energies are acceptable. Hence, quantization emerges as a natural consequence of the equation and the conditions imposed on it. The solutions bring to light a number of highly nonclassical, and therefore surprising, features of particles, especially their ability to tunnel into and through regions where classical physics would forbid them to be found. We also encounter a property of the electron, its spin, that has no classical counterpart. The chapter concludes with an introduction to the experimental techniques used to probe the quantization of energy in molecules.

The three basic modes of motion—translation (motion through space), vibration, and rotation—all play an important role in chemistry because they are ways in which molecules store energy. Gas-phase molecules, for instance, undergo translational motion and their kinetic energy is a contribution to the total internal energy of a sample. Molecules can also store energy as rotational kinetic energy and transitions between their rotational energy states can be observed spectroscopically. Energy is also stored as molecular vibration and transitions between vibrational states are responsible for the appearance of infrared and Raman spectra.

Translational motion

Section 8.5 introduced the quantum mechanical description of free motion in one dimension. We saw there that the Schrödinger equation is

$$-\frac{\hbar^2}{2m}\frac{d^2\psi}{dx^2} = E\psi \tag{9.1a}$$

or more succinctly

$$\hat{H}\psi = E\psi \qquad \hat{H} = -\frac{\hbar^2}{2m}\frac{d^2}{dx^2} \tag{9.1b}$$

The general solutions of eqn 9.1 are

$$\psi_k = Ae^{ikx} + Be^{-ikx} \qquad E_k = \frac{k^2\hbar^2}{2m} \tag{9.2}$$

Note that we are now labelling both the wavefunctions and the energies (that is, the eigenfunctions and eigenvalues of $\hat{H}$) with the index k. We can verify that these

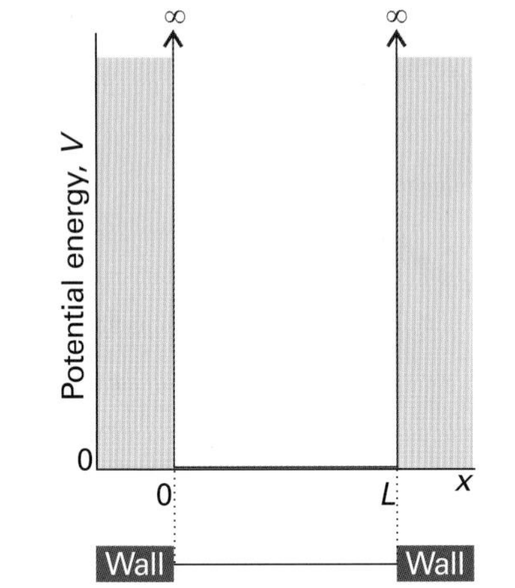

Fig. 9.1 A particle in a one-dimensional region with impenetrable walls. Its potential energy is zero between $x = 0$ and $x = L$, and rises abruptly to infinity as soon as it touches the walls.

functions are solutions by substituting ψ_k into the left-hand side of eqn 9.1a and showing that the result is equal to $E_k\psi_k$. In this case, all values of k, and therefore all values of the energy, are permitted. It follows that the translational energy of a free particle is not quantized.

We saw in Section 8.5b that a wavefunction of the form e^{ikx} describes a particle with linear momentum $p_x = +k\hbar$, corresponding to motion towards positive x (to the right), and that a wavefunction of the form e^{-ikx} describes a particle with the same magnitude of linear momentum but travelling towards negative x (to the left). That is, e^{ikx} is an eigenfunction of the operator $\hat{p}_x$ with eigenvalue $+k\hbar$, and e^{-ikx} is an eigenfunction with eigenvalue $-k\hbar$. In either state, $|\psi|^2$ is independent of x, which implies that the position of the particle is completely unpredictable. This conclusion is consistent with the uncertainty principle because, if the momentum is certain, then the position cannot be specified (the operators $\hat{x}$ and $\hat{p}_x$ do not commute, Section 8.6).

9.1 A particle in a box

In this section, we consider a **particle in a box**, in which a particle of mass m is confined between two walls at $x = 0$ and $x = L$: the potential energy is zero inside the box but rises abruptly to infinity at the walls (Fig. 9.1). This model is an idealization of the potential energy of a gas-phase molecule that is free to move in a one-dimensional container. However, it is also the basis of the treatment of the electronic structure of metals (Chapter 20) and of a primitive treatment of conjugated molecules. The particle in a box is also used in statistical thermodynamics in assessing the contribution of the translational motion of molecules to their thermodynamic properties (Chapter 16).

(a) The acceptable solutions

The Schrödinger equation for the region between the walls (where $V = 0$) is the same as for a free particle (eqn 9.1), so the general solutions given in eqn 9.2 are also the same. However, we can us $e^{\pm ix} = \cos x \pm i \sin x$ to write

$$\psi_k = Ae^{ikx} + Be^{-ikx} = A(\cos kx + i \sin kx) + B(\cos kx - i \sin kx)$$
$$= (A + B)\cos kx + (A - B)i \sin kx$$

If we absorb all numerical factors into two new coefficients C and D, then the general solutions take the form

$$\psi_k(x) = C \sin kx + D \cos kx \qquad E_k = \frac{k^2\hbar^2}{2m} \tag{9.3}$$

For a free particle, any value of E_k corresponds to an acceptable solution. However, when the particle is confined within a region, the acceptable wavefunctions must satisfy certain **boundary conditions**, or constraints on the function at certain locations. As we shall see when we discuss penetration into barriers, a wavefunction decays exponentially with distance inside a barrier, such as a wall, and the decay is infinitely fast when the potential energy is infinite. This behaviour is consistent with the fact that it is physically impossible for the particle to be found with an infinite potential energy. We conclude that the wavefunction must be zero where V is infinite, at $x < 0$ and $x > L$. The continuity of the wavefunction then requires it to vanish just inside the well at $x = 0$ and $x = L$. That is, the boundary conditions are $\psi_k(0) = 0$ and $\psi_k(L) = 0$. These boundary conditions imply quantization, as we show in the following *Justification*.

Justification 9.1 *The energy levels and wavefunctions of a particle in a one-dimensional box*

For an informal demonstration of quantization, we consider each wavefunction to be a de Broglie wave that must fit within the container. The permitted wavelengths satisfy

$$L = n \times \tfrac{1}{2}\lambda \qquad n = 1, 2, \ldots$$

and therefore

$$\lambda = \frac{2L}{n} \qquad \text{with } n = 1, 2, \ldots$$

According to the de Broglie relation, these wavelengths correspond to the momenta

$$p = \frac{h}{\lambda} = \frac{nh}{2L}$$

The particle has only kinetic energy inside the box (where $V = 0$), so the permitted energies are

$$E = \frac{p^2}{2m} = \frac{n^2h^2}{8mL^2} \qquad \text{with } n = 1, 2, \ldots$$

A more formal and widely applicable approach is as follows. Consider the wall at $x = 0$. According to eqn 9.3, $\psi(0) = D$ (because $\sin 0 = 0$ and $\cos 0 = 1$). But because $\psi(0) = 0$ we must have $D = 0$. It follows that the wavefunction must be of the form $\psi_k(x) = C \sin kx$. The value of ψ at the other wall (at $x = L$) is $\psi_k(L) = C \sin kL$, which must also be zero. Taking $C = 0$ would give $\psi_k(x) = 0$ for all x, which would conflict with the Born interpretation (the particle must be somewhere). Therefore, kL must be chosen so that $\sin kL = 0$, which is satisfied by

$$kL = n\pi \qquad n = 1, 2, \ldots$$

The value $n = 0$ is ruled out, because it implies $k = 0$ and $\psi_k(x) = 0$ everywhere (because $\sin 0 = 0$), which is unacceptable. Negative values of n merely change the sign of $\sin kL$ (because $\sin(-x) = -\sin x$). The wavefunctions are therefore

$$\psi_n(x) = C \sin(n\pi x/L) \qquad n = 1, 2, \ldots$$

(At this point we have started to label the solutions with the index n instead of k.) Because k and E_k are related by eqn 9.3, and k and n are related by $kL = n\pi$, it follows that the energy of the particle is limited to $E_n = n^2h^2/8mL^2$, the values obtained by the informal procedure.

We conclude that the energy of the particle in a one-dimensional box is quantized and that this quantization arises from the boundary conditions that ψ must satisfy if it is to be an acceptable wavefunction. This is a general conclusion: *the need to satisfy boundary conditions implies that only certain wavefunctions are acceptable, and hence restricts observables to discrete values.* So far, only energy has been quantized; shortly we shall see that other physical observables may also be quantized.

(b) Normalization

Before discussing the solution in more detail, we shall complete the derivation of the wavefunctions by finding the normalization constant (here written C and regarded as real, that is, does not contain i). To do so, we look for the value of C that ensures that the integral of ψ^2 over all the space available to the particle (that is, from $x = 0$ to $x = L$) is equal to 1:

$$\int_0^L \psi^2\,dx = C^2 \int_0^L \sin^2 \frac{n\pi x}{L} = C^2 \times \frac{L}{2} = 1, \qquad \text{so } C = \left(\frac{2}{L}\right)^{1/2}$$

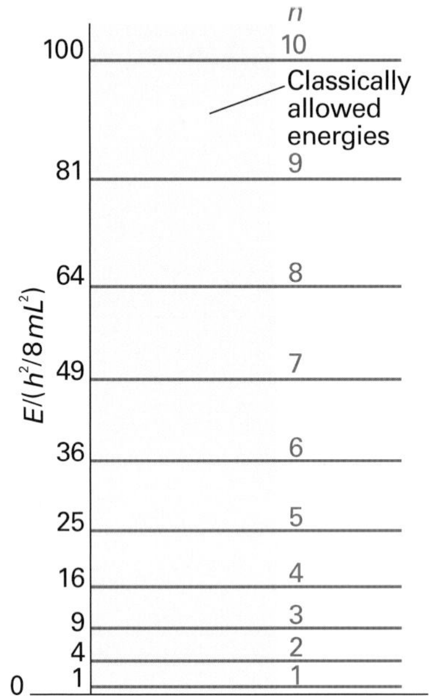

Fig. 9.2 The allowed energy levels for a particle in a box. Note that the energy levels increase as n^2, and that their separation increases as the quantum number increases.

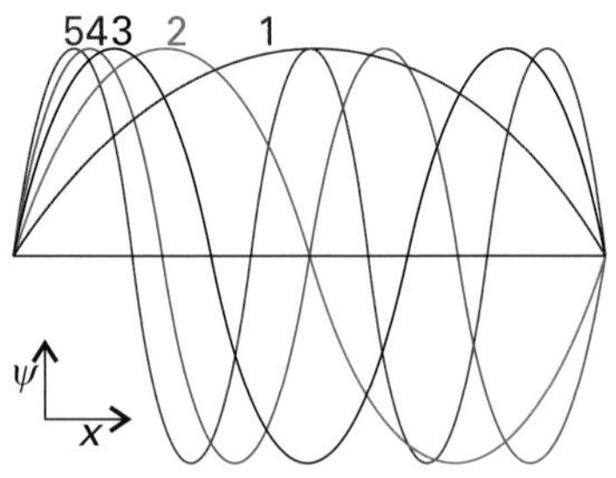

Fig. 9.3 The first five normalized wavefunctions of a particle in a box. Each wavefunction is a standing wave, and successive functions possess one more half wave and a correspondingly shorter wavelength.

Exploration Plot the probability density for a particle in a box with $n = 1, 2, \ldots 5$ and $n = 50$. How do your plots illustrate the correspondence principle?

Comment 9.1

It is sometimes useful to write

$\cos x = (e^{ix} + e^{-ix})/2 \quad \sin x = (e^{ix} - e^{-ix})/2i$

for all n. Therefore, the complete solution to the problem is

$$E_n = \frac{n^2h^2}{8mL^2} \qquad n = 1, 2, \ldots \tag{9.4a}$$

$$\psi_n(x) = \left(\frac{2}{L}\right)^{1/2} \sin\left(\frac{n\pi x}{L}\right) \qquad \text{for } 0 \le x \le L \tag{9.4b}$$

Self-test 9.1 Provide the intermediate steps for the determination of the normalization constant *C*. *Hint.* Use the standard integral $\int \sin^2 ax\, dx = \frac{1}{2}x - (\frac{1}{4}a) \sin 2ax +$ constant and the fact that $\sin 2m\pi = 0$, with $m = 0, 1, 2, \ldots$.

The energies and wavefunctions are labelled with the 'quantum number' n. A **quantum number** is an integer (in some cases, as we shall see, a half-integer) that labels the state of the system. For a particle in a box there is an infinite number of acceptable solutions, and the quantum number n specifies the one of interest (Fig. 9.2). As well as acting as a label, a quantum number can often be used to calculate the energy corresponding to the state and to write down the wavefunction explicitly (in the present example, by using eqn 9.4).

(c) The properties of the solutions

Figure 9.3 shows some of the wavefunctions of a particle in a box: they are all sine functions with the same amplitude but different wavelengths. Shortening the wavelength results in a sharper average curvature of the wavefunction and therefore an increase in the kinetic energy of the particle. Note that the number of nodes (points where the wavefunction passes *through* zero) also increases as n increases, and that the wavefunction ψ_n has $n - 1$ nodes. Increasing the number of nodes between walls of a given separation increases the average curvature of the wavefunction and hence the kinetic energy of the particle.

The linear momentum of a particle in a box is not well defined because the wavefunction $\sin kx$ is a standing wave and, like the example of $\cos kx$ treated in Section 8.5d, not an eigenfunction of the linear momentum operator. However, each wavefunction is a superposition of momentum eigenfunctions:

$$\psi_n = \left(\frac{2}{L}\right)^{1/2} \sin\frac{n\pi x}{L} = \frac{1}{2i}\left(\frac{2}{L}\right)^{1/2} (e^{ikx} - e^{-ikx}) \qquad k = \frac{n\pi}{L} \tag{9.5}$$

It follows that measurement of the linear momentum will give the value $+k\hbar$ for half the measurements of momentum and $-k\hbar$ for the other half. This detection of opposite directions of travel with equal probability is the quantum mechanical version of the classical picture that a particle in a box rattles from wall to wall, and in any given period spends half its time travelling to the left and half travelling to the right.

Self-test 9.2 What is (a) the average value of the linear momentum of a particle in a box with quantum number n, (b) the average value of p^2?

[(a) $\langle p \rangle = 0$, (b) $\langle p^2 \rangle = n^2h^2/4L^2$]

Because n cannot be zero, the lowest energy that the particle may possess is not zero (as would be allowed by classical mechanics, corresponding to a stationary particle) but

$$E_1 = \frac{h^2}{8mL^2} \tag{9.6}$$

This lowest, irremovable energy is called the **zero-point energy**. The physical origin of the zero-point energy can be explained in two ways. First, the uncertainty principle requires a particle to possess kinetic energy if it is confined to a finite region: the location of the particle is not completely indefinite, so its momentum cannot be precisely zero. Hence it has nonzero kinetic energy. Second, if the wavefunction is to be zero at the walls, but smooth, continuous, and not zero everywhere, then it must be curved, and curvature in a wavefunction implies the possession of kinetic energy.

The separation between adjacent energy levels with quantum numbers n and $n+1$ is

$$E_{n+1} - E_n = \frac{(n+1)^2h^2}{8mL^2} - \frac{n^2h^2}{8mL^2} = (2n+1)\frac{h^2}{8mL^2} \tag{9.7}$$

This separation decreases as the length of the container increases, and is very small when the container has macroscopic dimensions. The separation of adjacent levels becomes zero when the walls are infinitely far apart. Atoms and molecules free to move in normal laboratory-sized vessels may therefore be treated as though their translational energy is not quantized. The translational energy of completely free particles (those not confined by walls) is not quantized.

Illustration 9.1 *Accounting for the electronic absorption spectra of polyenes*

β-Carotene (**1**) is a linear polyene in which 10 single and 11 double bonds alternate along a chain of 22 carbon atoms. If we take each CC bond length to be about 140 pm, then the length L of the molecular box in β-carotene is $L = 0.294$ nm. For reasons that will be familiar from introductory chemistry, each C atom contributes one p-electron to the π orbitals and, in the lowest energy state of the molecule, each level up to $n = 11$ is occupied by two electrons. From eqn 9.7 it follows that the separation in energy between the ground state and the state in which one electron is promoted from $n = 11$ to $n = 12$ is

$$\begin{aligned}\Delta E = E_{12} - E_{11} &= (2\times 11+1)\frac{(6.626\times 10^{-34}\ \text{J s})^2}{8\times(9.110\times 10^{-31}\ \text{kg})\times(2.94\times 10^{-10}\ \text{m})^2}\\ &= 1.60\times 10^{-19}\ \text{J}\end{aligned}$$

It follows from the Bohr frequency condition (eqn 8.10, $\Delta E = h\nu$) that the frequency of radiation required to cause this transition is

$$\nu = \frac{\Delta E}{h} = \frac{1.60\times 10^{-19}\ \text{J}}{6.626\times 10^{-34}\ \text{J s}} = 2.41\times 10^{14}\ \text{s}^{-1}$$

The experimental value is $\nu = 6.03\times 10^{14}\ \text{s}^{-1}$ ($\lambda = 497$ nm), corresponding to radiation in the visible range of the electromagnetic spectrum.

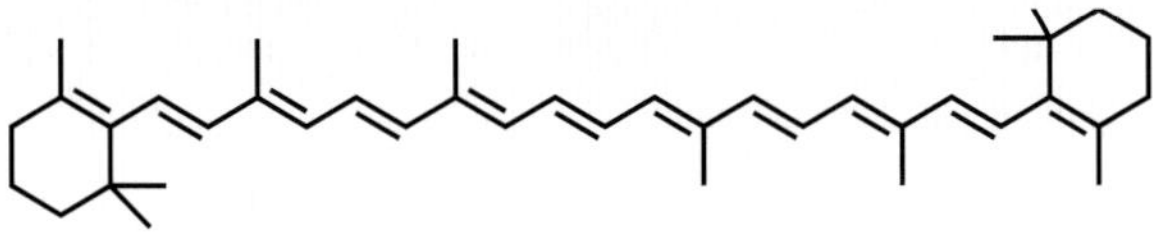

1 β-Carotene

Self-test 9.3 Estimate a typical nuclear excitation energy by calculating the first excitation energy of a proton confined to a square well with a length equal to the diameter of a nucleus (approximately 1 fm). [0.6 GeV]

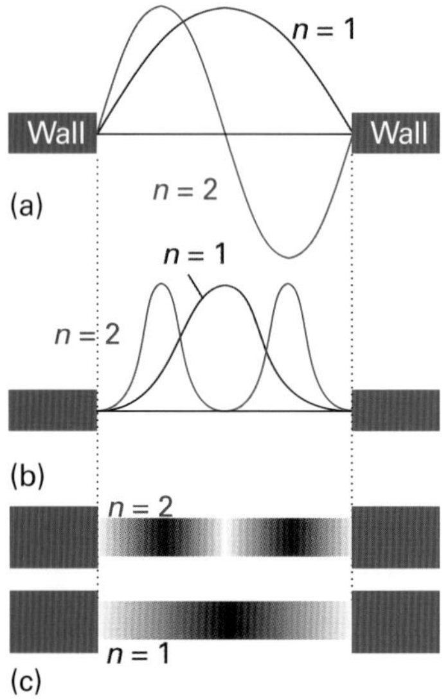

Fig. 9.4 (a) The first two wavefunctions, (b) the corresponding probability distributions, and (c) a representation of the probability distribution in terms of the darkness of shading.

The probability density for a particle in a box is

$$\psi^2(x) = \frac{2}{L}\sin^2\frac{n\pi x}{L} \tag{9.8}$$

and varies with position. The nonuniformity is pronounced when n is small (Fig. 9.4), but—provided we ignore the increasingly rapid oscillations—$\psi^2(x)$ becomes more uniform as n increases. The distribution at high quantum numbers reflects the classical result that a particle bouncing between the walls spends, on the average, equal times at all points. That the quantum result corresponds to the classical prediction at high quantum numbers is an illustration of the **correspondence principle**, which states that classical mechanics emerges from quantum mechanics as high quantum numbers are reached.

Example 9.1 *Using the particle in a box solutions*

The wavefunctions of an electron in a conjugated polyene can be approximated by particle-in-a-box wavefunctions. What is the probability, P, of locating the electron between $x = 0$ (the left-hand end of a molecule) and $x = 0.2$ nm in its lowest energy state in a conjugated molecule of length 1.0 nm?

Method The value of $\psi^2\mathrm{d}x$ is the probability of finding the particle in the small region $\mathrm{d}x$ located at x; therefore, the total probability of finding the electron in the specified region is the integral of $\psi^2\mathrm{d}x$ over that region. The wavefunction of the electron is given in eqn 9.4b with $n = 1$.

Answer The probability of finding the particle in a region between $x = 0$ and $x = l$ is

$$P = \int_0^l \psi_n^2\,\mathrm{d}x = \frac{2}{L}\int_0^l \sin^2\frac{n\pi x}{L}\,\mathrm{d}x = \frac{l}{L} - \frac{1}{2n\pi}\sin\frac{2\pi nl}{L}$$

We then set $n = 1$ and $l = 0.2$ nm, which gives $P = 0.05$. The result corresponds to a chance of 1 in 20 of finding the electron in the region. As n becomes infinite, the sine term, which is multiplied by $1/n$, makes no contribution to P and the classical result, $P = l/L$, is obtained.

Self-test 9.4 Calculate the probability that an electron in the state with $n = 1$ will be found between $x = 0.25L$ and $x = 0.75L$ in a conjugated molecule of length L (with $x = 0$ at the left-hand end of the molecule). [0.82]

(d) Orthogonality

We can now illustrate a property of wavefunctions first mentioned in Section 8.5. Two wavefunctions are **orthogonal** if the integral of their product vanishes. Specifically, the functions ψ_n and $\psi_{n'}$ are orthogonal if

$$\int \psi_n^*\psi_{n'}\,\mathrm{d}\tau = 0 \tag{9.9}$$

where the integration is over all space. A general feature of quantum mechanics, which we prove in the *Justification* below, is that *wavefunctions corresponding to different energies are orthogonal*; therefore, we can be confident that all the wavefunctions of a particle in a box are mutually orthogonal. A more compact notation for integrals of this kind is described in *Further information 9.1*.

Justification 9.2 *The orthogonality of wavefunctions*

Suppose we have two wavefunctions ψ_n and ψ_m corresponding to two different energies E_n and E_m, respectively. Then we can write

$$\hat{H}\psi_n = E_n\psi_n \qquad \hat{H}\psi_m = E_m\psi_m$$

Now multiply the first of these two Schrödinger equations by $\psi_m^\star$ and the second by $\psi_n^\star$ and integrate over all space:

$$\int \psi_m^\star \hat{H}\psi_n \mathrm{d}\tau = E_n \int \psi_m^\star \psi_n \mathrm{d}\tau \qquad \int \psi_n^\star \hat{H}\psi_m \mathrm{d}\tau = E_m \int \psi_n^\star \psi_m \mathrm{d}\tau$$

Next, noting that the energies themselves are real, form the complex conjugate of the second expression (for the state m) and subtract it from the first expression (for the state n):

$$\int \psi_m^\star \hat{H}\psi_n \mathrm{d}\tau - \left(\int \psi_n^\star \hat{H}\psi_m \mathrm{d}\tau\right)^\star = E_n \int \psi_m^\star \psi_n \mathrm{d}\tau - E_m \int \psi_n \psi_m^\star \mathrm{d}\tau$$

By the hermiticity of the hamiltonian (Section 8.5c), the two terms on the left are equal, so they cancel and we are left with

$$0 = (E_n - E_m)\int \psi_m^\star \psi_n \mathrm{d}\tau$$

However, the two energies are different; therefore the integral on the right must be zero, which confirms that two wavefunctions belonging to different energies are orthogonal.

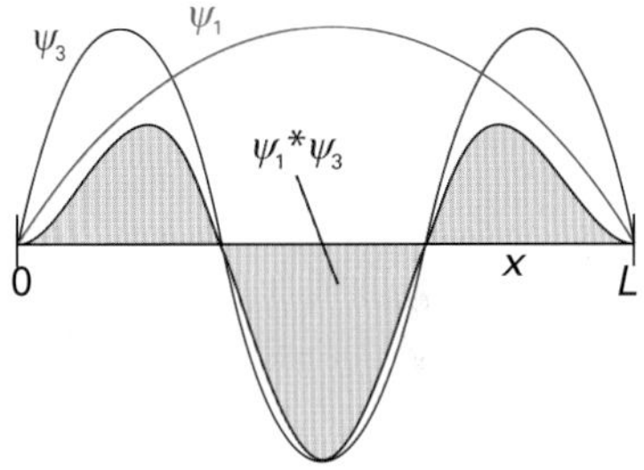

Fig. 9.5 Two functions are orthogonal if the integral of their product is zero. Here the calculation of the integral is illustrated graphically for two wavefunctions of a particle in a square well. The integral is equal to the total area beneath the graph of the product, and is zero.

Illustration 9.2 *Verifying the orthogonality of the wavefunctions for a particle in a box*

We can verify the orthogonality of wavefunctions of a particle in a box with $n = 1$ and $n = 3$ (Fig. 9.5):

$$\int_0^L \psi_1^\star \psi_3 \mathrm{d}x = \frac{2}{L}\int_0^L \sin\frac{\pi x}{L}\sin\frac{3\pi x}{L}\mathrm{d}x = 0$$

We have used the standard integral given in *Illustration 8.2*.

The property of orthogonality is of great importance in quantum mechanics because it enables us to eliminate a large number of integrals from calculations. Orthogonality plays a central role in the theory of chemical bonding (Chapter 11) and spectroscopy (Chapter 14). Sets of functions that are normalized and mutually orthogonal are called **orthonormal**. The wavefunctions in eqn 9.4b are orthonormal.

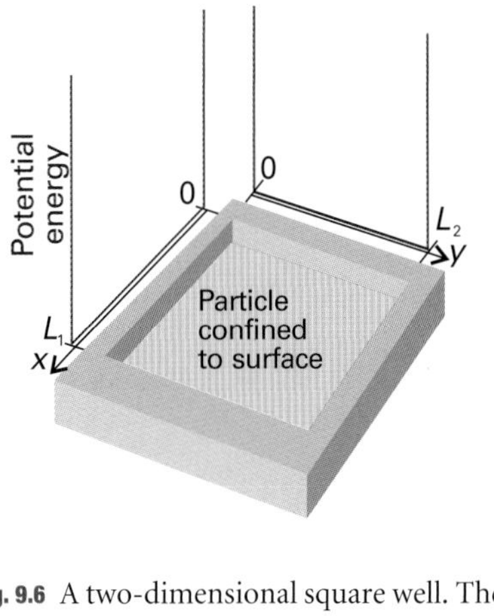

Fig. 9.6 A two-dimensional square well. The particle is confined to the plane bounded by impenetrable walls. As soon as it touches the walls, its potential energy rises to infinity.

9.2 Motion in two and more dimensions

Next, we consider a two-dimensional version of the particle in a box. Now the particle is confined to a rectangular surface of length L_1 in the x-direction and L_2 in the y-direction; the potential energy is zero everywhere except at the walls, where it is infinite (Fig. 9.6). The wavefunction is now a function of both x and y and the Schrödinger equation is

$$-\frac{\hbar^2}{2m}\left(\frac{\partial^2\psi}{\partial x^2}+\frac{\partial^2\psi}{\partial y^2}\right) = E\psi \qquad (9.10)$$

We need to see how to solve this *partial* differential equation, an equation in more than one variable.

(a) Separation of variables

Some partial differential equations can be simplified by the **separation of variables technique**, which divides the equation into two or more ordinary differential equations, one for each variable. An important application of this procedure, as we shall see, is the separation of the Schrödinger equation for the hydrogen atom into equations that describe the radial and angular variation of the wavefunction. The technique is particularly simple for a two-dimensional square well, as can be seen by testing whether a solution of eqn 9.10 can be found by writing the wavefunction as a product of functions, one depending only on x and the other only on y:

$$\psi(x,y) = X(x)Y(y)$$

With this substitution, we show in the *Justification* below that eqn 9.10 separates into two ordinary differential equations, one for each coordinate:

$$-\frac{\hbar^2}{2m}\frac{d^2X}{dx^2} = E_X X \qquad -\frac{\hbar^2}{2m}\frac{d^2Y}{dy^2} = E_Y Y \qquad E = E_X + E_Y \tag{9.11}$$

The quantity E_X is the energy associated with the motion of the particle parallel to the x-axis, and likewise for E_Y and motion parallel to the y-axis.

Justification 9.3 *The separation of variables technique applied to the particle in a two-dimensional box*

The first step in the justification of the separability of the wavefunction into the product of two functions X and Y is to note that, because X is independent of y and Y is independent of x, we can write

$$\frac{\partial^2\psi}{\partial x^2} = \frac{\partial^2 XY}{\partial x^2} = Y\frac{d^2X}{dx^2} \qquad \frac{\partial^2\psi}{\partial y^2} = \frac{\partial^2 XY}{\partial y^2} = X\frac{d^2Y}{dy^2}$$

Then eqn 9.10 becomes

$$-\frac{\hbar^2}{2m}\left(Y\frac{d^2X}{dx^2} + X\frac{d^2Y}{dy^2}\right) = EXY$$

When both sides are divided by XY, we can rearrange the resulting equation into

$$\frac{1}{X}\frac{d^2X}{dx^2} + \frac{1}{Y}\frac{d^2Y}{dy^2} = -\frac{2mE}{\hbar^2}$$

The first term on the left is independent of y, so if y is varied only the second term can change. But the sum of these two terms is a constant given by the right-hand side of the equation; therefore, even the second term cannot change when y is changed. In other words, the second term is a constant, which we write $-2mE_Y/\hbar^2$. By a similar argument, the first term is a constant when x changes, and we write it $-2mE_X/\hbar^2$, and $E = E_X + E_Y$. Therefore, we can write

$$\frac{1}{X}\frac{d^2X}{dx^2} = -\frac{2mE_X}{\hbar^2} \qquad \frac{1}{Y}\frac{d^2Y}{dy^2} = -\frac{2mE_Y}{\hbar^2}$$

which rearrange into the two ordinary (that is, single variable) differential equations in eqn 9.11.

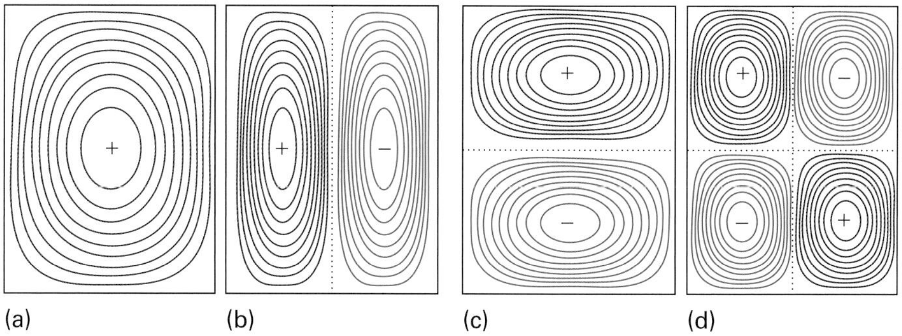

Fig. 9.7 The wavefunctions for a particle confined to a rectangular surface depicted as contours of equal amplitude. (a) $n_1 = 1$, $n_2 = 1$, the state of lowest energy, (b) $n_1 = 1$, $n_2 = 2$, (c) $n_1 = 2$, $n_2 = 1$, and (d) $n_1 = 2$, $n_2 = 2$.

Exploration Use mathematical software to generate three-dimensional plots of the functions in this illustration. Deduce a rule for the number of nodal lines in a wavefunction as a function of the values of n_x and n_y.

Each of the two ordinary differential equations in eqn 9.11 is the same as the one-dimensional square-well Schrödinger equation. We can therefore adapt the results in eqn 9.4 without further calculation:

$$X_{n_1}(x) = \left(\frac{2}{L_1}\right)^{1/2} \sin\frac{n_1\pi x}{L_1} \qquad Y_{n_2}(y) = \left(\frac{2}{L_2}\right)^{1/2} \sin\frac{n_2\pi y}{L_2}$$

Then, because $\psi = XY$ and $E = E_X + E_Y$, we obtain

$$\psi_{n_1,n_2}(x,y) = \frac{2}{(L_1L_2)^{1/2}} \sin\frac{n_1\pi x}{L_1} \sin\frac{n_2\pi y}{L_2} \qquad 0 \le x \le L_1, 0 \le y \le L_2$$

$$E_{n_1n_2} = \left(\frac{n_1^2}{L_1^2} + \frac{n_2^2}{L_2^2}\right)\frac{h^2}{8m} \tag{9.12a}$$

with the quantum numbers taking the values $n_1 = 1, 2, \ldots$ and $n_2 = 1, 2, \ldots$ independently. Some of these functions are plotted in Fig. 9.7. They are the two-dimensional versions of the wavefunctions shown in Fig. 9.3. Note that two quantum numbers are needed in this two-dimensional problem.

We treat a particle in a three-dimensional box in the same way. The wavefunctions have another factor (for the z-dependence), and the energy has an additional term in n_3^2/L_3^2. Solution of the Schrödinger equation by the separation of variables technique then gives

$$\psi_{n_1,n_2,n_3}(x,y,z) = \left(\frac{8}{L_1L_2L_3}\right)^{1/2} \sin\frac{n_1\pi x}{L_1} \sin\frac{n_2\pi y}{L_2} \sin\frac{n_3\pi z}{L_3}$$

$$0 \le x \le L_1, 0 \le y \le L_2, 0 \le z \le L_3 \tag{9.12b}$$

$$E_{n_1n_2n_3} = \left(\frac{n_1^2}{L_1^2} + \frac{n_2^2}{L_2^2} + \frac{n_3^2}{L_3^2}\right)\frac{h^2}{8m}$$

(b) Degeneracy

An interesting feature of the solutions for a particle in a two-dimensional box is obtained when the plane surface is square, with $L_1 = L_2 = L$. Then eqn 9.12a becomes

$$\psi_{n_1,n_2}(x,y) = \frac{2}{L} \sin\frac{n_1\pi x}{L} \sin\frac{n_2\pi y}{L} \qquad E_{n_1n_2} = (n_1^2 + n_2^2)\frac{h^2}{8mL^2} \tag{9.13}$$

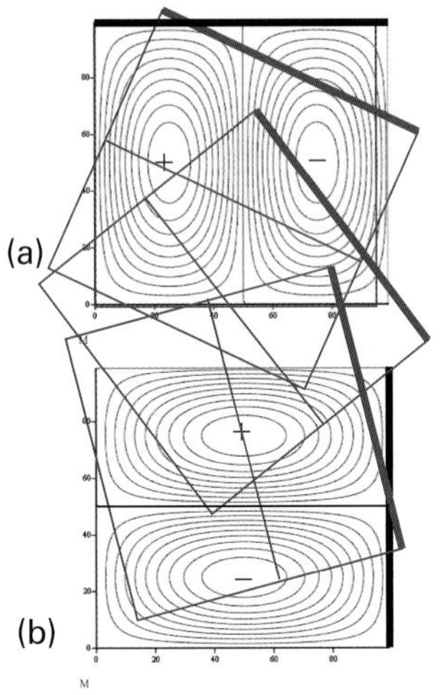

Fig. 9.8 The wavefunctions for a particle confined to a square surface. Note that one wavefunction can be converted into the other by a rotation of the box by 90°. The two functions correspond to the same energy. Degeneracy and symmetry are closely related.

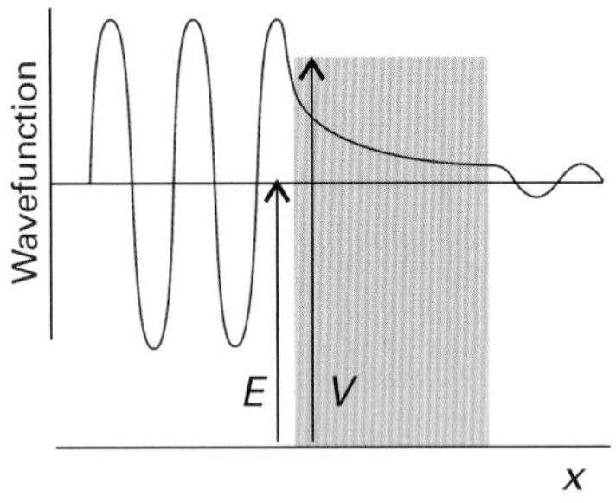

Fig. 9.9 A particle incident on a barrier from the left has an oscillating wave function, but inside the barrier there are no oscillations (for $E < V$). If the barrier is not too thick, the wavefunction is nonzero at its opposite face, and so oscillates begin again there. (Only the real component of the wavefunction is shown.)

Consider the cases $n_1 = 1, n_2 = 2$ and $n_1 = 2, n_2 = 1$:

$$\psi_{1,2} = \frac{2}{L}\sin\frac{\pi x}{L}\sin\frac{2\pi y}{L} \qquad E_{1,2} = \frac{5h^2}{8mL^2}$$

$$\psi_{2,1} = \frac{2}{L}\sin\frac{2\pi x}{L}\sin\frac{\pi y}{L} \qquad E_{2,1} = \frac{5h^2}{8mL^2}$$

We see that, although the wavefunctions are different, they are **degenerate**, meaning that they correspond to the same energy. In this case, in which there are two degenerate wavefunctions, we say that the energy level $5(h^2/8mL^2)$ is 'doubly degenerate'.

The occurrence of degeneracy is related to the symmetry of the system. Figure 9.8 shows contour diagrams of the two degenerate functions $\psi_{1,2}$ and $\psi_{2,1}$. Because the box is square, we can convert one wavefunction into the other simply by rotating the plane by 90°. Interconversion by rotation through 90° is not possible when the plane is not square, and $\psi_{1,2}$ and $\psi_{2,1}$ are then not degenerate. Similar arguments account for the degeneracy of states in a cubic box. We shall see many other examples of degeneracy in the pages that follow (for instance, in the hydrogen atom), and all of them can be traced to the symmetry properties of the system (see Section 12.4b).

9.3 Tunnelling

If the potential energy of a particle does not rise to infinity when it is in the walls of the container, and $E < V$, the wavefunction does not decay abruptly to zero. If the walls are thin (so that the potential energy falls to zero again after a finite distance), then the wavefunction oscillates inside the box, varies smoothly inside the region representing the wall, and oscillates again on the other side of the wall outside the box (Fig. 9.9). Hence the particle might be found on the outside of a container even though according to classical mechanics it has insufficient energy to escape. Such leakage by penetration through a classically forbidden region is called **tunnelling**.

The Schrödinger equation can be used to calculate the probability of tunnelling of a particle of mass m incident on a finite barrier from the left. On the left of the barrier (for $x < 0$) the wavefunctions are those of a particle with $V = 0$, so from eqn 9.2 we can write

$$\psi = Ae^{ikx} + Be^{-ikx} \qquad k\hbar = (2mE)^{1/2} \tag{9.14}$$

The Schrödinger equation for the region representing the barrier (for $0 \le x \le L$), where the potential energy is the constant V, is

$$-\frac{\hbar^2}{2m}\frac{d^2\psi}{dx^2} + V\psi = E\psi \tag{9.15}$$

We shall consider particles that have $E < V$ (so, according to classical physics, the particle has insufficient energy to pass over the barrier), and therefore $V - E$ is positive. The general solutions of this equation are

$$\psi = Ce^{\kappa x} + De^{-\kappa x} \qquad \kappa\hbar = \{2m(V - E)\}^{1/2} \tag{9.16}$$

as we can readily verify by differentiating ψ twice with respect to x. The important feature to note is that the two exponentials are now real functions, as distinct from the complex, oscillating functions for the region where $V = 0$ (oscillating functions would be obtained if $E > V$). To the right of the barrier ($x > L$), where $V = 0$ again, the wavefunctions are

$$\psi = A'e^{ikx} + B'e^{-ikx} \qquad k\hbar = (2mE)^{1/2} \tag{9.17}$$

The complete wavefunction for a particle incident from the left consists of an incident wave, a wave reflected from the barrier, the exponentially changing amplitudes inside the barrier, and an oscillating wave representing the propagation of the particle to the right after tunnelling through the barrier successfully (Fig. 9.10). The acceptable wavefunctions must obey the conditions set out in Section 8.4b. In particular, they must be continuous at the edges of the barrier (at $x = 0$ and $x = L$, remembering that $e^0 = 1$):

$$A + B = C + D \qquad Ce^{\kappa L} + De^{-\kappa L} = A'e^{ikL} + B'e^{-ikL} \tag{9.18}$$

Their slopes (their first derivatives) must also be continuous there (Fig. 9.11):

$$ikA - ikB = \kappa C - \kappa D \qquad \kappa Ce^{\kappa L} - \kappa De^{-\kappa L} = ikA'e^{ikL} - ikB'e^{-ikL} \tag{9.19}$$

At this stage, we have four equations for the six unknown coefficients. If the particles are shot towards the barrier from the left, there can be no particles travelling to the left on the right of the barrier. Therefore, we can set $B' = 0$, which removes one more unknown. We cannot set $B = 0$ because some particles may be reflected back from the barrier toward negative x.

The probability that a particle is travelling towards positive x (to the right) on the left of the barrier is proportional to $|A|^2$, and the probability that it is travelling to the right on the right of the barrier is $|A'|^2$. The ratio of these two probabilities is called the **transmission probability**, T. After some algebra (see Problem 9.9) we find

$$T = \left\{1 + \frac{(e^{\kappa L} - e^{-\kappa L})^2}{16\varepsilon(1-\varepsilon)}\right\}^{-1} \tag{9.20a}$$

where $\varepsilon = E/V$. This function is plotted in Fig. 9.12; the transmission coefficient for $E > V$ is shown there too. For high, wide barriers (in the sense that $\kappa L \gg 1$), eqn 9.20a simplifies to

$$T \approx 16\varepsilon(1-\varepsilon)e^{-2\kappa L} \tag{9.20b}$$

The transmission probability decreases exponentially with the thickness of the barrier and with $m^{1/2}$. It follows that particles of low mass are more able to tunnel through barriers than heavy ones (Fig. 9.13). Tunnelling is very important for electrons and muons, and moderately important for protons; for heavier particles it is less important.

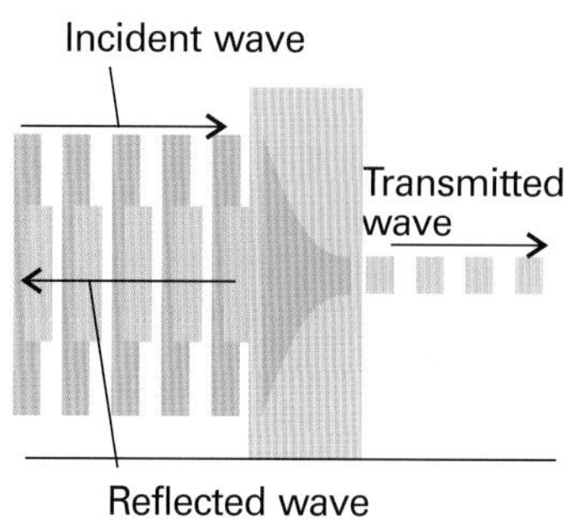

Fig. 9.10 When a particle is incident on a barrier from the left, the wavefunction consists of a wave representing linear momentum to the right, a reflected component representing momentum to the left, a varying but not oscillating component inside the barrier, and a (weak) wave representing motion to the right on the far side of the barrier.

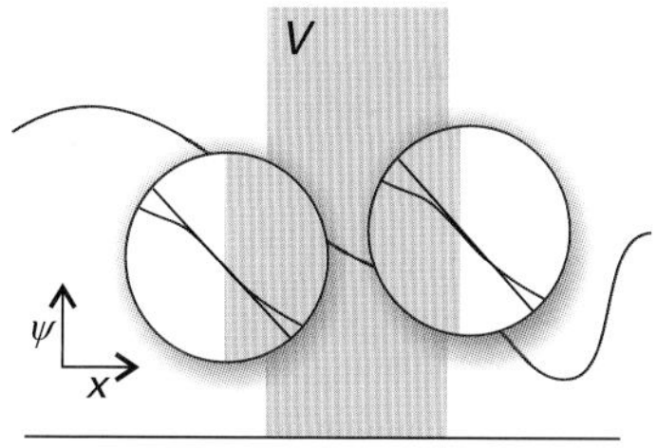

Fig. 9.11 The wavefunction and its slope must be continuous at the edges of the barrier. The conditions for continuity enable us to connect the wavefunctions in the three zones and hence to obtain relations between the coefficients that appear in the solutions of the Schrödinger equation.

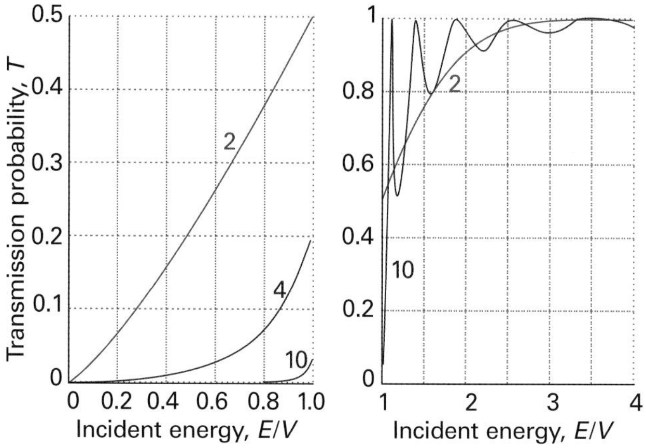

Fig. 9.12 The transmission probabilities for passage through a barrier. The horizontal axis is the energy of the incident particle expressed as a multiple of the barrier height. The curves are labelled with the value of $L(2mV)^{1/2}/\hbar$. The graph on the left is for $E < V$ and that on the right for $E > V$. Note that $T > 0$ for $E < V$ whereas classically T would be zero. However, $T < 1$ for $E > V$, whereas classically T would be 1.

Exploration Plot T against ε for a hydrogen molecule, a proton and an electron.

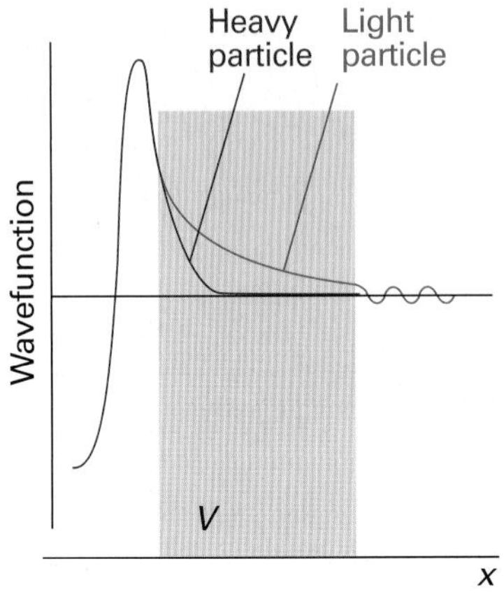

Fig. 9.13 The wavefunction of a heavy particle decays more rapidly inside a barrier than that of a light particle. Consequently, a light particle has a greater probability of tunnelling through the barrier.

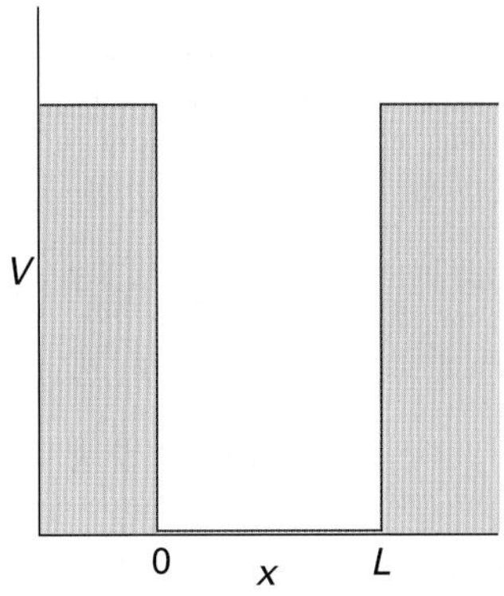

Fig. 9.14 A potential well with a finite depth.

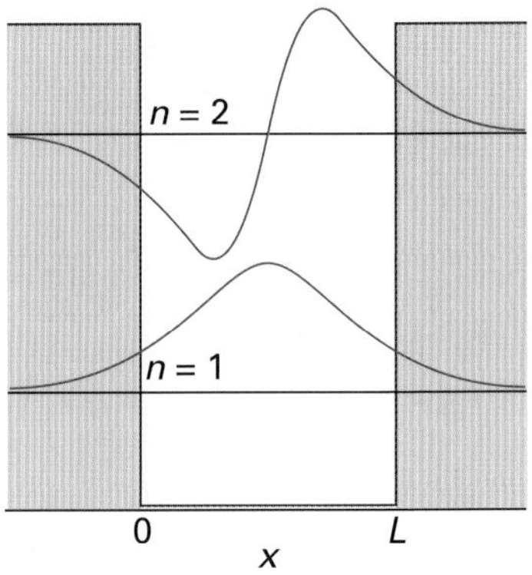

Fig. 9.15 The lowest two bound-state wavefunctions for a particle in the well shown in Fig. 9.14 and one of the wavefunctions corresponding to an unbound state ($E > V$).

A number of effects in chemistry (for example, the isotope-dependence of some reaction rates) depend on the ability of the proton to tunnel more readily than the deuteron. The very rapid equilibration of proton transfer reactions is also a manifestation of the ability of protons to tunnel through barriers and transfer quickly from an acid to a base. Tunnelling of protons between acidic and basic groups is also an important feature of the mechanism of some enzyme-catalysed reactions. As we shall see in Chapters 24 and 25, electron tunnelling is one of the factors that determine the rates of electron transfer reactions at electrodes and in biological systems.

A problem related to the one just considered is that of a particle in a square-well potential of finite depth (Fig. 9.14). In this kind of potential, the wavefunction penetrates into the walls, where it decays exponentially towards zero, and oscillates within the well. The wavefunctions are found by ensuring, as in the discussion of tunnelling, that they and their slopes are continuous at the edges of the potential. Some of the lowest energy solutions are shown in Fig. 9.15. A further difference from the solutions for an infinitely deep well is that there is only a finite number of bound states. Regardless of the depth and length of the well, there is always at least one bound state. Detailed consideration of the Schrödinger equation for the problem shows that in general the number of levels is equal to N, with

$$N - 1 < \frac{(8mVL)^{1/2}}{h} < N \tag{9.21}$$

where V is the depth of the well and L is its length (for a derivation of this expression, see *Further reading*). We see that the deeper and wider the well, the greater the number of bound states. As the depth becomes infinite, so the number of bound states also becomes infinite, as we have already seen.

IMPACT ON NANOSCIENCE

I9.1 Scanning probe microscopy

Nanoscience is the study of atomic and molecular assemblies with dimensions ranging from 1 nm to about 100 nm and *nanotechnology* is concerned with the incorporation of such assemblies into devices. The future economic impact of nanotechnology could be very significant. For example, increased demand for very small digital electronic devices has driven the design of ever smaller and more powerful microprocessors. However, there is an upper limit on the density of electronic circuits that can be incorporated into silicon-based chips with current fabrication technologies. As the ability to process data increases with the number of circuits in a chip, it follows that soon chips and the devices that use them will have to become bigger if processing power is to increase indefinitely. One way to circumvent this problem is to fabricate devices from nanometre-sized components.

We will explore several concepts of nanoscience throughout the text. We begin with the description of *scanning probe microscopy* (SPM), a collection of techniques that can be used to visualize and manipulate objects as small as atoms on surfaces. Consequently, SPM has far better resolution than electron microscopy (*Impact I8.1*).

One modality of SPM is *scanning tunnelling microscopy* (STM), in which a platinum–rhodium or tungsten needle is scanned across the surface of a conducting solid. When the tip of the needle is brought very close to the surface, electrons tunnel across the intervening space (Fig. 9.16). In the constant-current mode of operation, the stylus moves up and down corresponding to the form of the surface, and the topography of the surface, including any adsorbates, can be mapped on an atomic scale. The vertical motion of the stylus is achieved by fixing it to a piezoelectric cylinder, which contracts or expands according to the potential difference it experiences. In the constant-z mode, the vertical position of the stylus is held constant and the current is monitored. Because the tunnelling probability is very sensitive to the size of the gap, the microscope can detect tiny, atom-scale variations in the height of the surface.

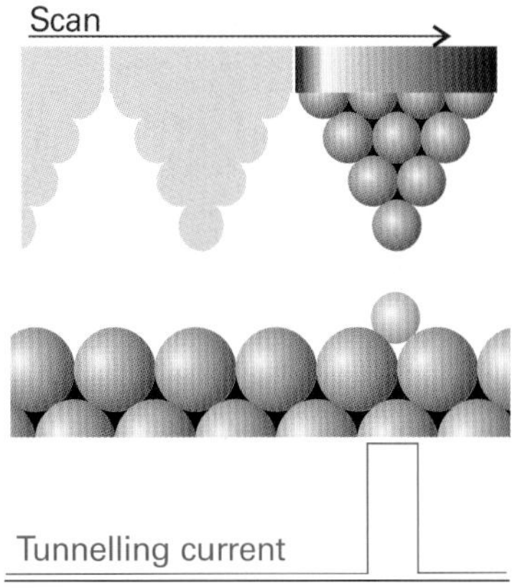

Fig. 9.16 A scanning tunnelling microscope makes use of the current of electrons that tunnel between the surface and the tip. That current is very sensitive to the distance of the tip above the surface.

Figure 9.17 shows an example of the kind of image obtained with a surface, in this case of gallium arsenide, that has been modified by addition of atoms, in this case caesium atoms. Each 'bump' on the surface corresponds to an atom. In a further variation of the STM technique, the tip may be used to nudge single atoms around on the surface, making possible the fabrication of complex and yet very tiny nanometre-sized structures.

In *atomic force microscopy* (AFM) a sharpened stylus attached to a cantilever is scanned across the surface. The force exerted by the surface and any bound species pushes or pulls on the stylus and deflects the cantilever (Fig. 9.18). The deflection is monitored either by interferometry or by using a laser beam. Because no current is needed between the sample and the probe, the technique can be applied to non-conducting surfaces too. A spectacular demonstration of the power of AFM is given in Fig. 9.19, which shows individual DNA molecules on a solid surface.

Fig. 9.17 An STM image of caesium atoms on a gallium arsenide surface.

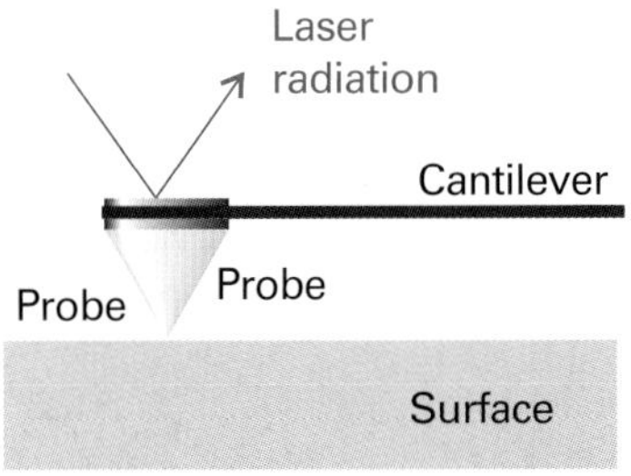

Fig. 9.18 In atomic force microscopy, a laser beam is used to monitor the tiny changes in the position of a probe as it is attracted to or repelled from atoms on a surface.

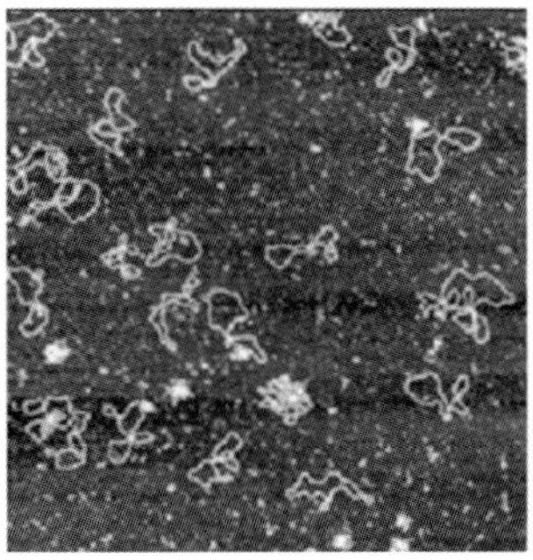

Fig. 9.19 An AFM image of bacterial DNA plasmids on a mica surface. (Courtesy of Veeco Instruments.)

Example 9.2 *Exploring the origin of the current in scanning tunnelling microscopy*

To get an idea of the distance dependence of the tunnelling current in STM, suppose that the wavefunction of the electron in the gap between sample and needle is given by $\psi = Be^{-\kappa x}$, where $\kappa = \{2m_e(V - E)/\hbar^2\}^{1/2}$; take $V - E = 2.0$ eV. By what factor would the current drop if the needle is moved from $L_1 = 0.50$ nm to $L_2 = 0.60$ nm from the surface?

Method We regard the tunnelling current to be proportional to the transmission probability T, so the ratio of the currents is equal to the ratio of the transmission probabilities. To choose between eqn 9.20a or 9.20b for the calculation of T, first calculate κL for the shortest distance L_1: if $\kappa L_1 > 1$, then use eqn 9.20b.

Answer When $L = L_1 = 0.50$ nm and $V - E = 2.0$ eV $= 3.20 \times 10^{-19}$ J the value of κL is

$$\kappa L_1 = \left\{\frac{2m_e(V-E)}{\hbar^2}\right\}^{1/2} L_1$$

$$= \left\{\frac{2 \times (9.109 \times 10^{-31}\,\text{kg}) \times (3.20 \times 10^{-19}\,\text{J})}{(1.054 \times 10^{-34}\,\text{J s})^2}\right\}^{1/2} \times (5.0 \times 10^{-10}\,\text{m})$$

$$= (7.25 \times 10^9\,\text{m}^{-1}) \times (5.0 \times 10^{-10}\,\text{m}) = 3.6$$

Because $\kappa L_1 > 1$, we use eqn 9.20b to calculate the transmission probabilities at the two distances. It follows that

$$\frac{\text{current at } L_2}{\text{current at } L_1} = \frac{T(L_2)}{T(L_1)} = \frac{16\varepsilon(1-\varepsilon)e^{-2\kappa L_2}}{16\varepsilon(1-\varepsilon)e^{-2\kappa L_1}} = e^{-2\kappa(L_2 - L_1)}$$

$$= e^{-2\times(7.25\times10^{-9}\,\text{m}^{-1})\times(1.0\times10^{-10}\,\text{m})} = 0.23$$

We conclude that, at a distance of 0.60 nm between the surface and the needle, the current is 23 per cent of the value measured when the distance is 0.50 nm.

Self-test 9.5 The ability of a proton to tunnel through a barrier contributes to the rapidity of proton transfer reactions in solution and therefore to the properties of acids and bases. Estimate the relative probabilities that a proton and a deuteron can tunnel through the same barrier of height 1.0 eV (1.6×10^{-19} J) and length 100 pm when their energy is 0.9 eV. Any comment?

[$T_H/T_D = 3.7 \times 10^2$; we expect proton transfer reactions to be much faster than deuteron transfer reactions.]

Vibrational motion

A particle undergoes **harmonic motion** if it experiences a restoring force proportional to its displacement:

$$F = -kx \tag{9.22}$$

where k is the **force constant**: the stiffer the 'spring', the greater the value of k. Because force is related to potential energy by $F = -\mathrm{d}V/\mathrm{d}x$ (see *Appendix 3*), the force in eqn 9.22 corresponds to a potential energy

$$V = \tfrac{1}{2}kx^2 \tag{9.23}$$

This expression, which is the equation of a parabola (Fig. 9.20), is the origin of the term 'parabolic potential energy' for the potential energy characteristic of a harmonic oscillator. The Schrödinger equation for the particle is therefore

$$-\frac{\hbar^2}{2m}\frac{d^2\psi}{dx^2}+\tfrac{1}{2}kx^2\psi=E\psi \qquad (9.24)$$

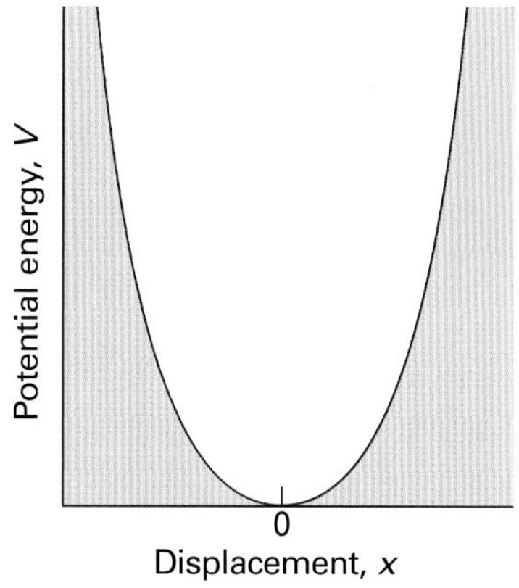

Fig. 9.20 The parabolic potential energy $V=\frac{1}{2}kx^2$ of a harmonic oscillator, where x is the displacement from equilibrium. The narrowness of the curve depends on the force constant k: the larger the value of k, the narrower the well.

9.4 The energy levels

Equation 9.24 is a standard equation in the theory of differential equations and its solutions are well known to mathematicians (for details, see *Further reading*). Quantization of energy levels arises from the boundary conditions: the oscillator will not be found with infinitely large compressions or extensions, so the only allowed solutions are those for which $\psi=0$ at $x=\pm\infty$. The permitted energy levels are

$$E_v=(v+\tfrac{1}{2})\hbar\omega \qquad \omega=\left(\frac{k}{m}\right)^{1/2} \qquad v=0,1,2,\ldots \qquad (9.25)$$

Note that ω (omega) increases with increasing force constant and decreasing mass. It follows that the separation between adjacent levels is

$$E_{v+1}-E_v=\hbar\omega \qquad (9.26)$$

which is the same for all v. Therefore, the energy levels form a uniform ladder of spacing $\hbar\omega$ (Fig. 9.21). The energy separation $\hbar\omega$ is negligibly small for macroscopic objects (with large mass), but is of great importance for objects with mass similar to that of atoms.

Because the smallest permitted value of v is 0, it follows from eqn 9.26 that a harmonic oscillator has a zero-point energy

$$E_0=\tfrac{1}{2}\hbar\omega \qquad (9.27)$$

The mathematical reason for the zero-point energy is that v cannot take negative values, for if it did the wavefunction would be ill-behaved. The physical reason is the same as for the particle in a square well: the particle is confined, its position is not completely uncertain, and therefore its momentum, and hence its kinetic energy, cannot be exactly zero. We can picture this zero-point state as one in which the particle fluctuates incessantly around its equilibrium position; classical mechanics would allow the particle to be perfectly still.

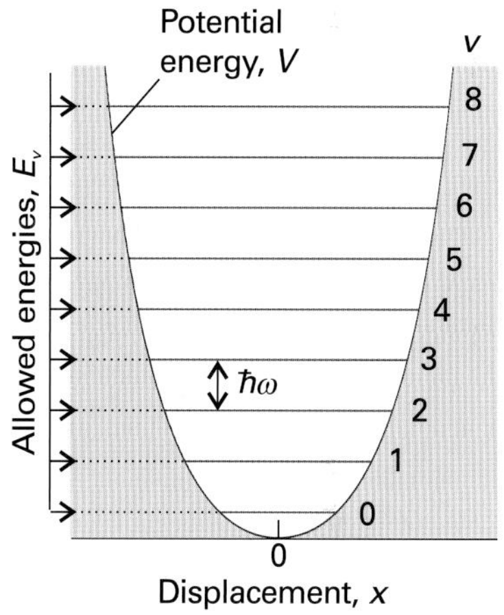

Fig. 9.21 The energy levels of a harmonic oscillator are evenly spaced with separation $\hbar\omega$, with $\omega=(k/m)^{1/2}$. Even in its lowest state, an oscillator has an energy greater than zero.

Illustration 9.3 *Calculating a molecular vibrational absorption frequency*

Atoms vibrate relative to one another in molecules with the bond acting like a spring. Consider an X—H chemical bond, where a heavy X atom forms a stationary anchor for the very light H atom. That is, only the H atom moves, vibrating as a simple harmonic oscillator. Therefore, eqn 9.25 describes the allowed vibrational energy levels of a X—H bond. The force constant of a typical X—H chemical bond is around 500 N m^{-1}. For example $k=516.3$ N m^{-1} for the $^1H^{35}Cl$ bond. Because the mass of a proton is about 1.7×10^{-27} kg, using $k=500$ N m^{-1} in eqn 9.25 gives $\omega\approx5.4\times10^{14}$ s^{-1} (5.4×10^2 THz). It follows from eqn 9.26 that the separation of adjacent levels is $\hbar\omega\approx5.7\times10^{-20}$ J (57 zJ, about 0.36 eV). This energy separation corresponds to 34 kJ mol^{-1}, which is chemically significant. From eqn 9.27, the zero-point energy of this molecular oscillator is about 3 zJ, which corresponds to 0.2 eV, or 15 kJ mol^{-1}.

The excitation of the vibration of the bond from one level to the level immediately above requires 57 zJ. Therefore, if it is caused by a photon, the excitation requires radiation of frequency $\nu = \Delta E/h = 86$ THz and wavelength $\lambda = c/\nu = 3.5$ μm. It follows that transitions between adjacent vibrational energy levels of molecules are stimulated by or emit infrared radiation. We shall see in Chapter 13 that the concepts just described represent the starting point for the interpretation of vibrational spectroscopy, an important technique for the characterization of small and large molecules in the gas phase or in condensed phases.

9.5 The wavefunctions

It is helpful at the outset to identify the similarities between the harmonic oscillator and the particle in a box, for then we shall be able to anticipate the form of the oscillator wavefunctions without detailed calculation. Like the particle in a box, a particle undergoing harmonic motion is trapped in a symmetrical well in which the potential energy rises to large values (and ultimately to infinity) for sufficiently large displacements (compare Figs. 9.1 and 9.20). However, there are two important differences. First, because the potential energy climbs towards infinity only as x^2 and not abruptly, the wavefunction approaches zero more slowly at large displacements than for the particle in a box. Second, as the kinetic energy of the oscillator depends on the displacement in a more complex way (on account of the variation of the potential energy), the curvature of the wavefunction also varies in a more complex way.

(a) The form of the wavefunctions

The detailed solution of eqn 9.24 shows that the wavefunction for a harmonic oscillator has the form

$$\psi(x) = N \times (\text{polynomial in } x) \times (\text{bell-shaped Gaussian function})$$

where N is a normalization constant. A Gaussian function is a function of the form e^{-x^2} (Fig. 9.22). The precise form of the wavefunctions are

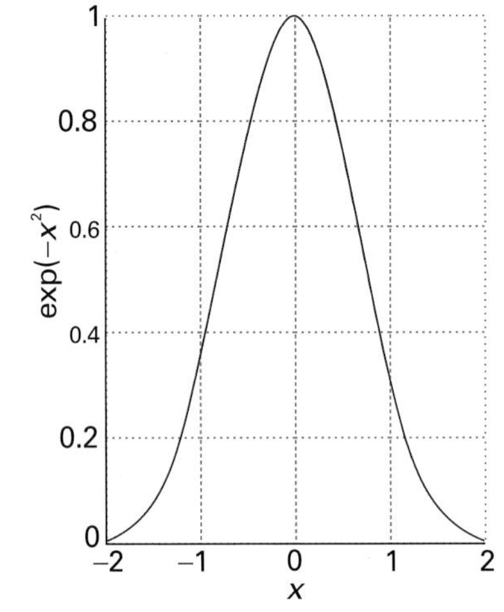

Fig. 9.22 The graph of the Gaussian function, $f(x) = e^{-x^2}$.

$$\psi_v(x) = N_v H_v(y) e^{-y^2/2} \qquad y = \frac{x}{\alpha} \qquad \alpha = \left(\frac{\hbar^2}{mk}\right)^{1/4} \tag{9.28}$$

The factor $H_v(y)$ is a **Hermite polynomial** (Table 9.1). For instance, because $H_0(y) = 1$, the wavefunction for the ground state (the lowest energy state, with $v = 0$) of the harmonic oscillator is

$$\psi_0(x) = N_0 e^{-y^2/2} = N_0 e^{-x^2/2\alpha^2} \tag{9.29a}$$

It follows that the probability density is the bell-shaped Gaussian function

$$\psi_0^2(x) = N_0^2 e^{-x^2/\alpha^2} \tag{9.29b}$$

The wavefunction and the probability distribution are shown in Fig. 9.23. Both curves have their largest values at zero displacement (at $x = 0$), so they capture the classical picture of the zero-point energy as arising from the ceaseless fluctuation of the particle about its equilibrium position.

The wavefunction for the first excited state of the oscillator, the state with $v = 1$, is obtained by noting that $H_1(y) = 2y$ (note that some of the Hermite polynomials are very simple functions!):

$$\psi_1(x) = N_1 \times 2y e^{-y^2/2} \tag{9.30}$$

Comment 9.2

The Hermite polynomials are solutions of the differential equation

$$H_v'' - 2yH_v' + 2vH_v = 0$$

where primes denote differentiation. They satisfy the recursion relation

$$H_{v+1} - 2yH_v + 2vH_{v-1} = 0$$

An important integral is

$$\int_{-\infty}^{\infty} H_{v'} H_v e^{-y^2} dy = \begin{cases} 0 & \text{if } v' \neq v \\ \pi^{1/2} 2^v v! & \text{if } v' = v \end{cases}$$

Hermite polynomials are members of a class of functions called *orthogonal polynomials.* These polynomials have a wide range of important properties that allow a number of quantum mechanical calculations to be done with relative ease. See *Further reading* for a reference to their properties.

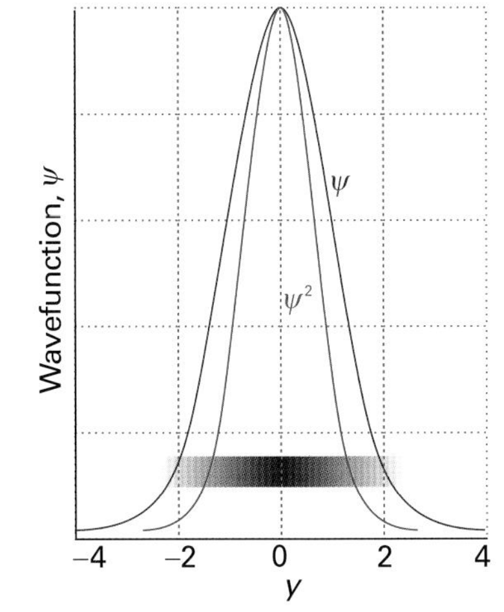

Fig. 9.23 The normalized wavefunction and probability distribution (shown also by shading) for the lowest energy state of a harmonic oscillator.

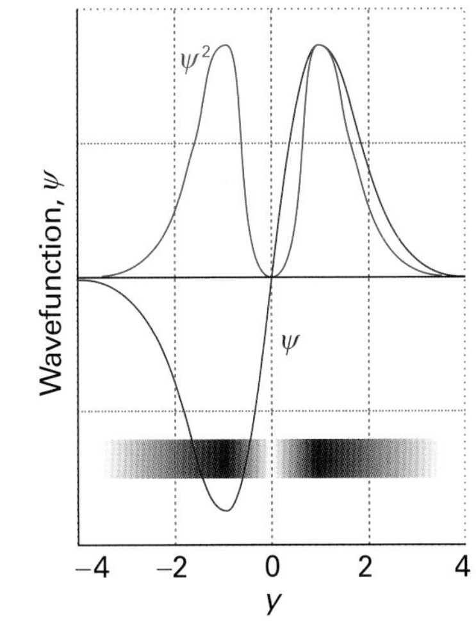

Fig. 9.24 The normalized wavefunction and probability distribution (shown also by shading) for the first excited state of a harmonic oscillator.

Table 9.1 The Hermite polynomials $H_v(y)$

v	$H_v(y)$
0	1
1	$2y$
2	$4y^2 - 2$
3	$8y^3 - 12y$
4	$16y^4 - 48y^2 + 12$
5	$32y^5 - 160y^3 + 120y$
6	$64y^6 - 480y^4 + 720y^2 - 120$

This function has a node at zero displacement ($x = 0$), and the probability density has maxima at $x = \pm\alpha$, corresponding to $y = \pm 1$ (Fig. 9.24).

Once again, we should interpret the mathematical expressions we have derived. In the case of the harmonic oscillator wavefunctions in eqn 9.28, we should note the following:

1. The Gaussian function goes very strongly to zero as the displacement increases (in either direction), so all the wavefunctions approach zero at large displacements.
2. The exponent y^2 is proportional to $x^2 \times (mk)^{1/2}$, so the wavefunctions decay more rapidly for large masses and stiff springs.
3. As v increases, the Hermite polynomials become larger at large displacements (as x^v), so the wavefunctions grow large before the Gaussian function damps them down to zero: as a result, the wavefunctions spread over a wider range as v increases.

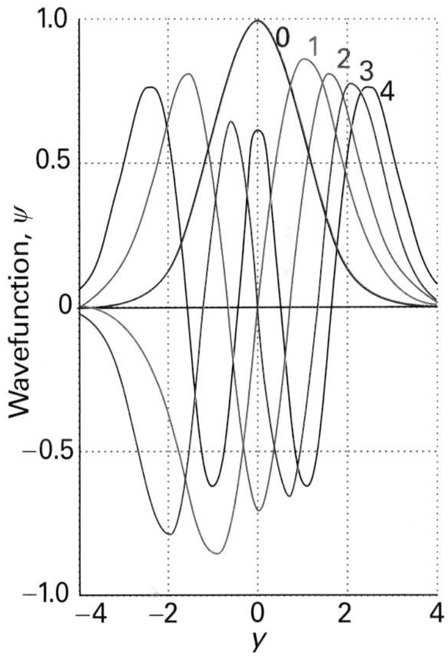

Fig. 9.25 The normalized wavefunctions for the first five states of a harmonic oscillator. Note that the number of nodes is equal to v and that alternate wavefunctions are symmetrical or antisymmetrical about $y = 0$ (zero displacement).

The shapes of several wavefunctions are shown in Fig. 9.25. The shading in Fig. 9.26 that represents the probability density is based on the squares of these functions. At high quantum numbers, harmonic oscillator wavefunctions have their largest amplitudes near the turning points of the classical motion (the locations at which $V = E$, so

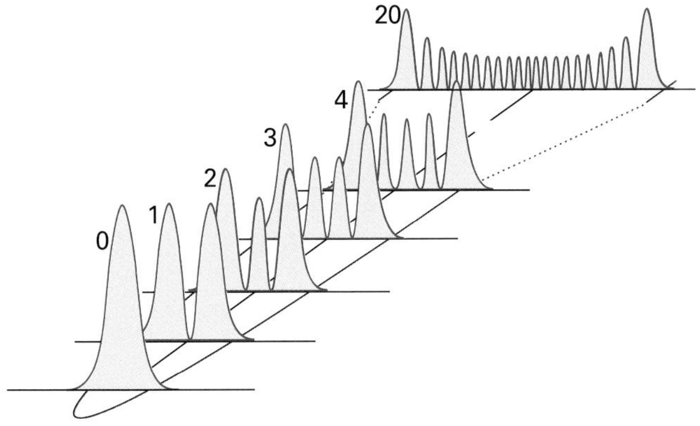

Fig. 9.26 The probability distributions for the first five states of a harmonic oscillator and the state with $v = 20$. Note how the regions of highest probability move towards the turning points of the classical motion as v increases.

Exploration To gain some insight into the origins of the nodes in the harmonic oscillator wavefunctions, plot the Hermite polynomials $H_v(y)$ for $v = 0$ through 5.

the kinetic energy is zero). We see classical properties emerging in the correspondence limit of high quantum numbers, for a classical particle is most likely to be found at the turning points (where it travels most slowly) and is least likely to be found at zero displacement (where it travels most rapidly).

Example 9.3 *Normalizing a harmonic oscillator wavefunction*

Find the normalization constant for the harmonic oscillator wavefunctions.

Method Normalization is always carried out by evaluating the integral of $|\psi|^2$ over all space and then finding the normalization factor from eqn 8.16. The normalized wavefunction is then equal to $N\psi$. In this one-dimensional problem, the volume element is dx and the integration is from $-\infty$ to $+\infty$. The wavefunctions are expressed in terms of the dimensionless variable $y = x/\alpha$, so begin by expressing the integral in terms of y by using $dx = \alpha dy$. The integrals required are given in *Comment 9.2*.

Answer The unnormalized wavefunction is

$$\psi_v(x) = H_v(y)e^{-y^2/2}$$

It follows from the integrals given in *Comment 9.2* that

$$\int_{-\infty}^{\infty} \psi_v^* \psi_v dx = \alpha \int_{-\infty}^{\infty} \psi_v^* \psi_v dy = \alpha \int_{-\infty}^{\infty} H_v^2(y)e^{-y^2}dy = \alpha\pi^{1/2}2^v v!$$

where $v! = v(v-1)(v-2)\ldots 1$. Therefore,

$$N_v = \left(\frac{1}{\alpha\pi^{1/2}2^v v!}\right)^{1/2}$$

Note that for a harmonic oscillator N_v is different for each value of v.

Self-test 9.6 Confirm, by explicit evaluation of the integral, that ψ_0 and ψ_1 are orthogonal.

[Evaluate the integral $\int_{-\infty}^{\infty}\psi_0^*\psi_1\,dx$ by using the information in *Comment 9.2*.]

(b) The properties of oscillators

With the wavefunctions that are available, we can start calculating the properties of a harmonic oscillator. For instance, we can calculate the expectation values of an observable Ω by evaluating integrals of the type

$$\langle\Omega\rangle = \int_{-\infty}^{\infty} \psi_v^* \hat{\Omega} \psi_v dx \qquad (9.31)$$

(Here and henceforth, the wavefunctions are all taken to be normalized to 1.) When the explicit wavefunctions are substituted, the integrals look fearsome, but the Hermite polynomials have many simplifying features. For instance, we show in the following example that the mean displacement, $\langle x\rangle$, and the mean square displacement, $\langle x^2\rangle$, of the oscillator when it is in the state with quantum number v are

$$\langle x\rangle = 0 \qquad \langle x^2\rangle = (v + \tfrac{1}{2})\frac{\hbar}{(mk)^{1/2}} \qquad (9.32)$$

The result for $\langle x\rangle$ shows that the oscillator is equally likely to be found on either side of $x=0$ (like a classical oscillator). The result for $\langle x^2\rangle$ shows that the mean square displacement increases with v. This increase is apparent from the probability densities in Fig. 9.26, and corresponds to the classical amplitude of swing increasing as the oscillator becomes more highly excited.

Example 9.4 *Calculating properties of a harmonic oscillator*

We can imagine the bending motion of a CO_2 molecule as a harmonic oscillation relative to the linear conformation of the molecule. We may be interested in the extent to which the molecule bends. Calculate the mean displacement of the oscillator when it is in a quantum state v.

Method Normalized wavefunctions must be used to calculate the expectation value. The operator for position along x is multiplication by the value of x (Section 8.5b). The resulting integral can be evaluated either by inspection (the integrand is the product of an odd and an even function), or by explicit evaluation using the formulas in *Comment 9.2*. To give practice in this type of calculation, we illustrate the latter procedure. We shall need the relation $x=\alpha y$, which implies that $dx=\alpha dy$.

Answer The integral we require is

$$\langle x\rangle=\int_{-\infty}^{\infty}\psi_v^* x\psi_v \mathrm{d}x = N_v^2\int_{-\infty}^{\infty}(H_v\mathrm{e}^{-y^2/2})x(H_v\mathrm{e}^{-y^2/2})\mathrm{d}x$$

$$=\alpha^2N_v^2\int_{-\infty}^{\infty}(H_v\mathrm{e}^{-y^2/2})y(H_v\mathrm{e}^{-y^2/2})\mathrm{d}y$$

$$=\alpha^2N_v^2\int_{-\infty}^{\infty}H_v yH_v\mathrm{e}^{-y^2}\mathrm{d}y$$

Now use the recursion relation (see *Comment 9.2*) to form

$$yH_v=vH_{v-1}+\tfrac{1}{2}H_{v+1}$$

which turns the integral into

$$\int_{-\infty}^{\infty}H_v yH_v\mathrm{e}^{-y^2}\mathrm{d}y=v\int_{-\infty}^{\infty}H_{v-1}H_v\mathrm{e}^{-y^2}\mathrm{d}y+\tfrac{1}{2}\int_{-\infty}^{\infty}H_{v+1}H_v\mathrm{e}^{-y^2}\mathrm{d}y$$

Both integrals are zero, so $\langle x\rangle=0$. As remarked in the text, the mean displacement is zero because the displacement occurs equally on either side of the equilibrium position. The following Self-test extends this calculation by examining the mean square displacement, which we can expect to be non-zero and to increase with increasing v.

Comment 9.3

An even function is one for which $f(-x)=f(x)$; an odd function is one for which $f(-x)=-f(x)$. The product of an odd and even function is itself odd, and the integral of an odd function over a symmetrical range about $x=0$ is zero.

Self-test 9.7 Calculate the mean square displacement $\langle x^2\rangle$ of the particle from its equilibrium position. (Use the recursion relation twice.) [eqn 9.32]

The mean potential energy of an oscillator, the expectation value of $V=\tfrac{1}{2}kx^2$, can now be calculated very easily:

$$\langle V\rangle=\langle\tfrac{1}{2}kx^2\rangle=\tfrac{1}{2}(v+\tfrac{1}{2})\hbar\left(\frac{k}{m}\right)^{1/2}=\tfrac{1}{2}(v+\tfrac{1}{2})\hbar\omega \qquad (9.33)$$

Because the total energy in the state with quantum number v is $(v+\frac{1}{2})\hbar\omega$, it follows that

$$\langle V\rangle = \tfrac{1}{2}E_v \qquad (9.34a)$$

The total energy is the sum of the potential and kinetic energies, so it follows at once that the mean kinetic energy of the oscillator is

$$\langle E_K\rangle = \tfrac{1}{2}E_v \qquad (9.34b)$$

The result that the mean potential and kinetic energies of a harmonic oscillator are equal (and therefore that both are equal to half the total energy) is a special case of the **virial theorem**:

> If the potential energy of a particle has the form $V = ax^b$, then its mean potential and kinetic energies are related by
>
> $$2\langle E_K\rangle = b\langle V\rangle \qquad (9.35)$$

For a harmonic oscillator $b = 2$, so $\langle E_K\rangle = \langle V\rangle$, as we have found. The virial theorem is a short cut to the establishment of a number of useful results, and we shall use it again.

An oscillator may be found at extensions with $V > E$ that are forbidden by classical physics, for they correspond to negative kinetic energy. For example, it follows from the shape of the wavefunction (see the *Justification* below) that in its lowest energy state there is about an 8 per cent chance of finding an oscillator stretched beyond its classical limit and an 8 per cent chance of finding it with a classically forbidden compression. These tunnelling probabilities are independent of the force constant and mass of the oscillator. The probability of being found in classically forbidden regions decreases quickly with increasing v, and vanishes entirely as v approaches infinity, as we would expect from the correspondence principle. Macroscopic oscillators (such as pendulums) are in states with very high quantum numbers, so the probability that they will be found in a classically forbidden region is wholly negligible. Molecules, however, are normally in their vibrational ground states, and for them the probability is very significant.

Justification 9.4 *Tunnelling in the quantum mechanical harmonic oscillator*

According to classical mechanics, the turning point, x_{tp}, of an oscillator occurs when its kinetic energy is zero, which is when its potential energy $\frac{1}{2}kx^2$ is equal to its total energy E. This equality occurs when

$$x_{tp}^2 = \frac{2E}{k} \qquad \text{or, } x_{tp} = \pm\left(\frac{2E}{k}\right)^{1/2}$$

with E given by eqn 9.25. The probability of finding the oscillator stretched beyond a displacement x_{tp} is the sum of the probabilities $\psi^2\mathrm{d}x$ of finding it in any of the intervals $\mathrm{d}x$ lying between x_{tp} and infinity:

$$P = \int_{x_{tp}}^{\infty} \psi_v^2\,\mathrm{d}x$$

The variable of integration is best expressed in terms of $y = x/\alpha$ with $\alpha = (\hbar^2/mk)^{1/4}$, and then the turning point on the right lies at

$$y_{tp} = \frac{x_{tp}}{\alpha} = \left\{\frac{2(v+\frac{1}{2})\hbar\omega}{\alpha^2 k}\right\}^{1/2} = (2v+1)^{1/2}$$

For the state of lowest energy ($v = 0$), $y_{tp} = 1$ and the probability is

$$P=\int_{x_{tp}}^{\infty}\psi_0^2\mathrm{d}x=\alpha N_0^2\int_1^{\infty}\mathrm{e}^{-y^2}\mathrm{d}y$$

The integral is a special case of the *error function*, erf z, which is defined as follows:

$$\mathrm{erf}\,z=1-\frac{2}{\pi^{1/2}}\int_z^{\infty}\mathrm{e}^{-y^2}\mathrm{d}y$$

The values of this function are tabulated and available in mathematical software packages, and a small selection of values is given in Table 9.2. In the present case

$$P=\tfrac{1}{2}(1-\mathrm{erf}\,1)=\tfrac{1}{2}(1-0.843)=0.079$$

It follows that, in 7.9 per cent of a large number of observations, any oscillator in the state $v=0$ will be found stretched to a classically forbidden extent. There is the same probability of finding the oscillator with a classically forbidden compression. The total probability of finding the oscillator tunnelled into a classically forbidden region (stretched or compressed) is about 16 per cent. A similar calculation for the state with $v=6$ shows that the probability of finding the oscillator outside the classical turning points has fallen to about 7 per cent.

Table 9.2 The error function

z	erf z
0	0
0.01	0.0113
0.05	0.0564
0.10	0.1125
0.50	0.5205
1.00	0.8427
1.50	0.9661
2.00	0.9953

Rotational motion

The treatment of rotational motion can be broken down into two parts. The first deals with motion in two dimensions and the second with rotation in three dimensions. It may be helpful to review the classical description of rotational motion given in *Appendix 3*, particularly the concepts of moment of inertia and angular momentum.

9.6 Rotation in two dimensions: a particle on a ring

We consider a particle of mass m constrained to move in a circular path of radius r in the xy-plane (Fig. 9.27). The total energy is equal to the kinetic energy, because $V=0$ everywhere. We can therefore write $E=p^2/2m$. According to classical mechanics, the **angular momentum**, J_z, around the z-axis (which lies perpendicular to the xy-plane) is $J_z=\pm pr$, so the energy can be expressed as $J_z^2/2mr^2$. Because mr^2 is the **moment of inertia**, I, of the mass on its path, it follows that

$$E=\frac{J_z^2}{2I} \tag{9.36}$$

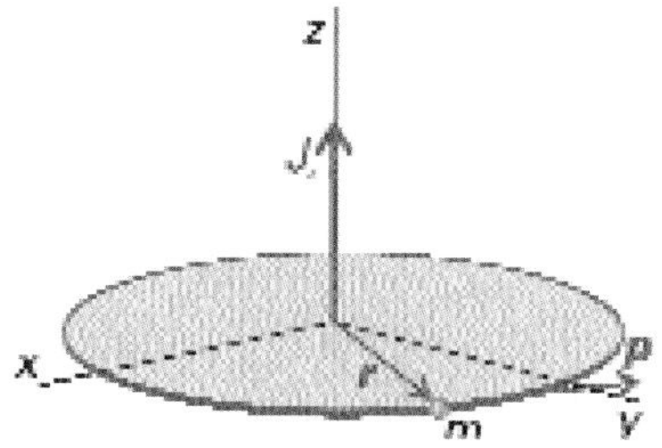

Fig. 9.27 The angular momentum of a particle of mass m on a circular path of radius r in the xy-plane is represented by a vector $\boldsymbol{J}$ with the single non-zero component J_z of magnitude pr perpendicular to the plane.

We shall now see that not all the values of the angular momentum are permitted in quantum mechanics, and therefore that both angular momentum and rotational energy are quantized.

(a) The qualitative origin of quantized rotation

Because $J_z=\pm pr$, and, from the de Broglie relation, $p=h/\lambda$, the angular momentum about the z-axis is

$$J_z=\pm\frac{hr}{\lambda}$$

Opposite signs correspond to opposite directions of travel. This equation shows that the shorter the wavelength of the particle on a circular path of given radius, the greater the angular momentum of the particle. It follows that, if we can see why the

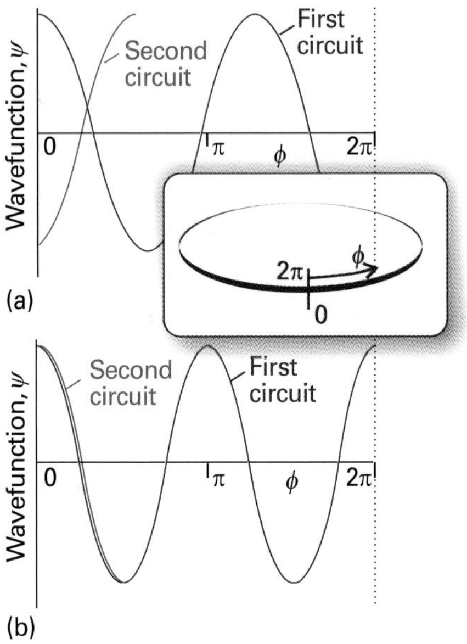

Fig. 9.28 Two solutions of the Schrödinger equation for a particle on a ring. The circumference has been opened out into a straight line; the points at $\phi = 0$ and 2π are identical. The solution in (a) is unacceptable because it is not single-valued. Moreover, on successive circuits it interferes destructively with itself, and does not survive. The solution in (b) is acceptable: it is single-valued, and on successive circuits it reproduces itself.

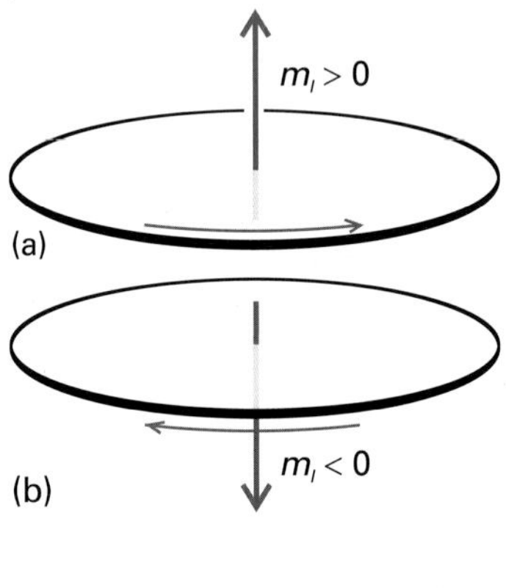

Fig. 9.29 The angular momentum of a particle confined to a plane can be represented by a vector of length $|m_l|$ units along the z-axis and with an orientation that indicates the direction of motion of the particle. The direction is given by the right-hand screw rule.

wavelength is restricted to discrete values, then we shall understand why the angular momentum is quantized.

Suppose for the moment that λ can take an arbitrary value. In that case, the wavefunction depends on the azimuthal angle ϕ as shown in Fig. 9.28a. When ϕ increases beyond 2π, the wavefunction continues to change, but for an arbitrary wavelength it gives rise to a different value at each point, which is unacceptable (Section 8.4b). An acceptable solution is obtained only if the wavefunction reproduces itself on successive circuits, as in Fig. 9.28b. Because only some wavefunctions have this property, it follows that only some angular momenta are acceptable, and therefore that only certain rotational energies exist. Hence, the energy of the particle is quantized. Specifically, the only allowed wavelengths are

$$\lambda = \frac{2\pi r}{m_l}$$

with m_l, the conventional notation for this quantum number, taking integral values including 0. The value $m_l = 0$ corresponds to $\lambda = \infty$; a 'wave' of infinite wavelength has a constant height at all values of ϕ. The angular momentum is therefore limited to the values

$$J_z = \pm\frac{hr}{\lambda} = \frac{m_l hr}{2\pi r} = \frac{m_l h}{2\pi}$$

where we have allowed m_l to have positive or negative values. That is,

$$J_z = m_l\hbar \qquad m_l = 0, \pm1, \pm2, \ldots \tag{9.37}$$

Positive values of m_l correspond to rotation in a clockwise sense around the z-axis (as viewed in the direction of z, Fig. 9.29) and negative values of m_l correspond to counter-clockwise rotation around z. It then follows from eqn 9.36 that the energy is limited to the values

$$E = \frac{J_z^2}{2I} = \frac{m_l^2\hbar^2}{2I} \tag{9.38a}$$

We shall see shortly that the corresponding normalized wavefunctions are

$$\psi_{m_l}(\phi) = \frac{e^{im_l\phi}}{(2\pi)^{1/2}} \tag{9.38b}$$

The wavefunction with $m_l = 0$ is $\psi_0(\phi) = 1/(2\pi)^{1/2}$, and has the same value at all points on the circle.

We have arrived at a number of conclusions about rotational motion by combining some classical notions with the de Broglie relation. Such a procedure can be very useful for establishing the general form (and, as in this case, the exact energies) for a quantum mechanical system. However, to be sure that the correct solutions have been obtained, and to obtain practice for more complex problems where this less formal approach is inadequate, we need to solve the Schrödinger equation explicitly. The formal solution is described in the *Justification* that follows.

Justification 9.5 *The energies and wavefunctions of a particle on a ring*

The hamiltonian for a particle of mass m in a plane (with $V = 0$) is the same as that given in eqn 9.10:

$$\hat{H} = -\frac{\hbar^2}{2m}\left(\frac{\partial^2}{\partial x^2} + \frac{\partial^2}{\partial y^2}\right)$$

and the Schrödinger equation is $H\psi = E\psi$, with the wavefunction a function of the angle ϕ. It is always a good idea to use coordinates that reflect the full symmetry of the system, so we introduce the coordinates r and ϕ (Fig. 9.30), where $x = r\cos\phi$ and $y = r\sin\phi$. By standard manipulations (see *Further reading*) we can write

$$\frac{\partial^2}{\partial x^2}+\frac{\partial^2}{\partial y^2}=\frac{\partial^2}{\partial r^2}+\frac{1}{r}\frac{\partial}{\partial r}+\frac{1}{r^2}\frac{\partial^2}{\partial\phi^2} \qquad (9.39)$$

However, because the radius of the path is fixed, the derivatives with respect to r can be discarded. The hamiltonian then becomes

$$\hat{H}=-\frac{\hbar^2}{2mr^2}\frac{d^2}{d\phi^2}$$

The moment of inertia $I = mr^2$ has appeared automatically, so H may be written

$$\hat{H}=-\frac{\hbar^2}{2I}\frac{d^2}{d\phi^2} \qquad (9.40)$$

and the Schrödinger equation is

$$\frac{d^2\psi}{d\phi^2}=-\frac{2IE}{\hbar^2}\psi \qquad (9.41)$$

The normalized general solutions of the equation are

$$\psi_{m_l}(\phi)=\frac{e^{im_l\phi}}{(2\pi)^{1/2}} \qquad m_l=\pm\frac{(2IE)^{1/2}}{\hbar} \qquad (9.42)$$

The quantity m_l is just a dimensionless number at this stage.

We now select the acceptable solutions from among these general solutions by imposing the condition that the wavefunction should be single-valued. That is, the wavefunction ψ must satisfy a **cyclic boundary condition**, and match at points separated by a complete revolution: $\psi(\phi + 2\pi) = \psi(\phi)$. On substituting the general wavefunction into this condition, we find

$$\psi_{m_l}(\phi+2\pi)=\frac{e^{im_l(\phi+2\pi)}}{(2\pi)^{1/2}}=\frac{e^{im_l\phi}e^{2\pi i m_l}}{(2\pi)^{1/2}}=\psi_{m_l}(\phi)e^{2\pi i m_l}$$

As $e^{i\pi} = -1$, this relation is equivalent to

$$\psi_{m_l}(\phi+2\pi)=(-1)^{2m_l}\psi(\phi) \qquad (9.43)$$

Because we require $(-1)^{2m_l} = 1$, $2m_l$ must be a positive or a negative even integer (including 0), and therefore m_l must be an integer: $m_l = 0, \pm1, \pm2, \ldots$.

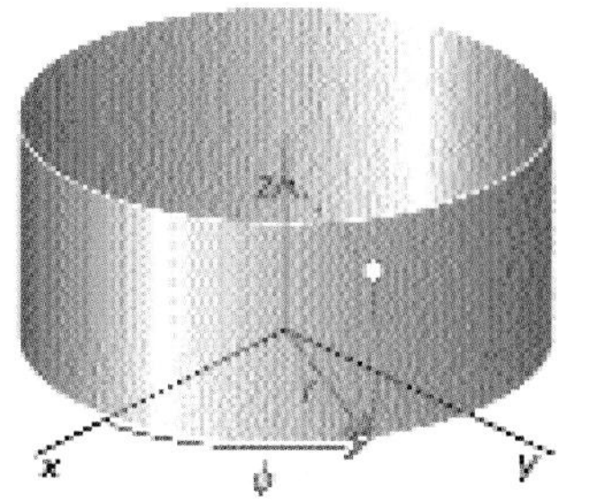

Fig. 9.30 The cylindrical coordinates z, r, and ϕ for discussing systems with axial (cylindrical) symmetry. For a particle confined to the xy-plane, only r and ϕ can change.

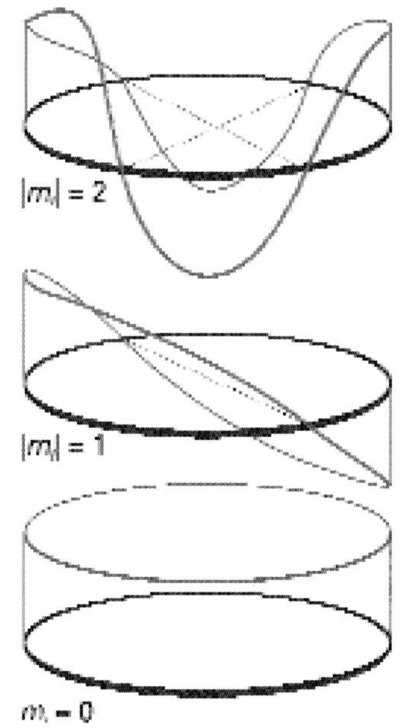

Fig. 9.31 The real parts of the wavefunctions of a particle on a ring. As shorter wavelengths are achieved, the magnitude of the angular momentum around the z-axis grows in steps of $\hbar$.

(b) Quantization of rotation

We can summarize the conclusions so far as follows. The energy is quantized and restricted to the values given in eqn 9.38a ($E = m_l^2\hbar^2/2I$). The occurrence of m_l as its square means that the energy of rotation is independent of the sense of rotation (the sign of m_l), as we expect physically. In other words, states with a given value of $|m_l|$ are doubly degenerate, except for $m_l = 0$, which is non-degenerate. Although the result has been derived for the rotation of a single mass point, it also applies to any body of moment of inertia I constrained to rotate about one axis.

We have also seen that the angular momentum is quantized and confined to the values given in eqn 9.37 ($J_z = m_l\hbar$). The increasing angular momentum is associated with the increasing number of nodes in the real and imaginary parts of the wavefunction: the wavelength decreases stepwise as $|m_l|$ increases, so the momentum with which the particle travels round the ring increases (Fig. 9.31). As shown in the following

Comment 9.4

The complex function $e^{im_l\phi}$ does not have nodes; however, it may be written as $\cos m_l\phi + i\sin m_l\phi$, and the real ($\cos m_l\phi$) and imaginary ($\sin m_l\phi$) components do have nodes.

Comment 9.5

The angular momentum in three dimensions is defined as

$$\boldsymbol{l}=\boldsymbol{r}\times\boldsymbol{p}=\begin{vmatrix} \boldsymbol{i} & \boldsymbol{j} & \boldsymbol{k} \\ x & y & z \\ p_x & p_y & p_z \end{vmatrix}$$
$$=(yp_z-zp_y)\boldsymbol{i}-(xp_z-zp_x)\boldsymbol{j}+(xp_y-yp_x)\boldsymbol{k}$$

where $\boldsymbol{i}$, $\boldsymbol{j}$, and $\boldsymbol{k}$ are unit vectors pointing along the positive directions on the x-, y-, and z-axes. It follows that the z-component of the angular momentum has a magnitude given by eqn 9.44. For more information on vectors, see *Appendix 2.*

Justification, we can come to the same conclusion more formally by using the argument about the relation between eigenvalues and the values of observables established in Section 8.5.

Justification 9.6 *The quantization of angular momentum*

In the discussion of translational motion in one dimension, we saw that the opposite signs in the wavefunctions e^{ikx} and e^{-ikx} correspond to opposite directions of travel, and that the linear momentum is given by the eigenvalue of the linear momentum operator. The same conclusions can be drawn here, but now we need the eigenvalues of the angular momentum operator. In classical mechanics the orbital angular momentum l_z about the z-axis is defined as

$$l_z=xp_y-yp_x \qquad [9.44]$$

where p_x is the component of linear motion parallel to the x-axis and p_y is the component parallel to the y-axis. The operators for the two linear momentum components are given in eqn 8.26, so the operator for angular momentum about the z-axis, which we denote $\hat{l}_z$, is

$$\hat{l}_z=\frac{\hbar}{\mathrm{i}}\left(x\frac{\partial}{\partial y}-y\frac{\partial}{\partial x}\right) \qquad (9.45)$$

When expressed in terms of the coordinates r and ϕ, by standard manipulations this equation becomes

$$\hat{l}_z=\frac{\hbar}{\mathrm{i}}\frac{\partial}{\partial\phi} \qquad (9.46)$$

With the angular momentum operator available, we can test the wavefunction in eqn 9.38b. Disregarding the normalization constant, we find

$$\hat{l}_z\psi_{m_l}=\frac{\hbar}{\mathrm{i}}\frac{\mathrm{d}\psi_{m_l}}{\mathrm{d}\phi}=\mathrm{i}m_l\frac{\hbar}{\mathrm{i}}\mathrm{e}^{\mathrm{i}m_l\phi}=m_l\hbar\psi_{m_l} \qquad (9.47)$$

That is, ψ_{m_l} is an eigenfunction of $\hat{l}_z$, and corresponds to an angular momentum $m_l\hbar$. When m_l is positive, the angular momentum is positive (clockwise when seen from below); when m_l is negative, the angular momentum is negative (counterclockwise when seen from below). These features are the origin of the vector representation of angular momentum, in which the magnitude is represented by the length of a vector and the direction of motion by its orientation (Fig. 9.32).

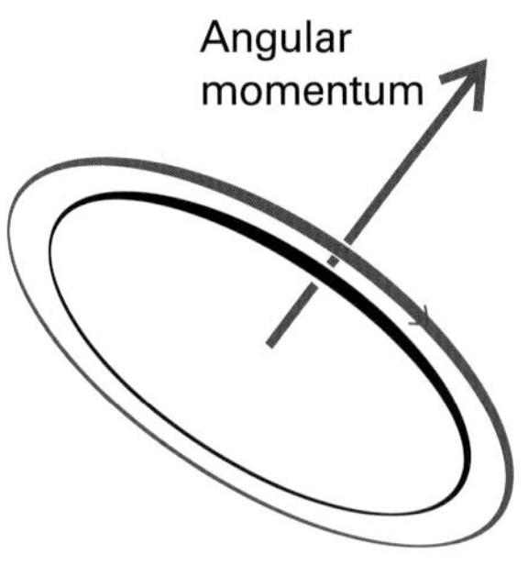

Fig. 9.32 The basic ideas of the vector representation of angular momentum: the magnitude of the angular momentum is represented by the length of the vector, and the orientation of the motion in space by the orientation of the vector (using the right-hand screw rule).

To locate the particle given its wavefunction in eqn 9.38b, we form the probability density:

$$\psi^{*}_{m_l}\psi_{m_l}=\left(\frac{\mathrm{e}^{\mathrm{i}m_l\phi}}{(2\pi)^{1/2}}\right)^{*}\left(\frac{\mathrm{e}^{\mathrm{i}m_l\phi}}{(2\pi)^{1/2}}\right)=\left(\frac{\mathrm{e}^{-\mathrm{i}m_l\phi}}{(2\pi)^{1/2}}\right)\left(\frac{\mathrm{e}^{\mathrm{i}m_l\phi}}{(2\pi)^{1/2}}\right)=\frac{1}{2\pi}$$

Because this probability density is independent of ϕ, the probability of locating the particle somewhere on the ring is also independent of ϕ (Fig. 9.33). Hence the location of the particle is completely indefinite, and knowing the angular momentum precisely eliminates the possibility of specifying the particle's location. Angular momentum and angle are a pair of complementary observables (in the sense defined in Section 8.6), and the inability to specify them simultaneously with arbitrary precision is another example of the uncertainty principle.

9.7 Rotation in three dimensions: the particle on a sphere

We now consider a particle of mass m that is free to move anywhere on the surface of a sphere of radius r. We shall need the results of this calculation when we come to describe rotating molecules and the states of electrons in atoms and in small clusters of atoms. The requirement that the wavefunction should match as a path is traced over the poles as well as round the equator of the sphere surrounding the central point introduces a second cyclic boundary condition and therefore a second quantum number (Fig. 9.34).

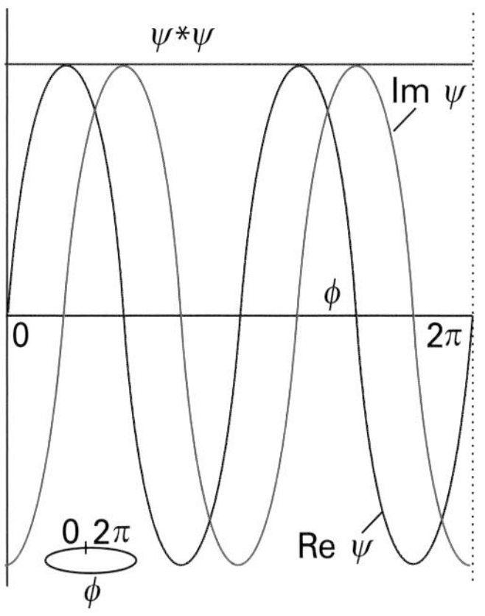

Fig. 9.33 The probability density for a particle in a definite state of angular momentum is uniform, so there is an equal probability of finding the particle anywhere on the ring.

(a) The Schrödinger equation

The hamiltonian for motion in three dimensions (Table 8.1) is

$$\hat{H} = -\frac{\hbar^2}{2m}\nabla^2 + V \qquad \nabla^2 = \frac{\partial^2}{\partial x^2} + \frac{\partial^2}{\partial y^2} + \frac{\partial^2}{\partial z^2} \tag{9.48}$$

The symbol ∇^2 is a convenient abbreviation for the sum of the three second derivatives; it is called the **laplacian**, and read either 'del squared' or 'nabla squared'. For the particle confined to a spherical surface, $V = 0$ wherever it is free to travel, and the radius r is a constant. The wavefunction is therefore a function of the **colatitude**, θ, and the **azimuth**, ϕ (Fig. 9.35), and we write it $\psi(\theta,\phi)$. The Schrödinger equation is

$$-\frac{\hbar^2}{2m}\nabla^2\psi = E\psi \tag{9.49}$$

As shown in the following *Justification*, this partial differential equation can be simplified by the separation of variables procedure by expressing the wavefunction (for constant r) as the product

$$\psi(\theta,\phi) = \Theta(\theta)\Phi(\phi) \tag{9.50}$$

where Θ is a function only of θ and Φ is a function only of ϕ.

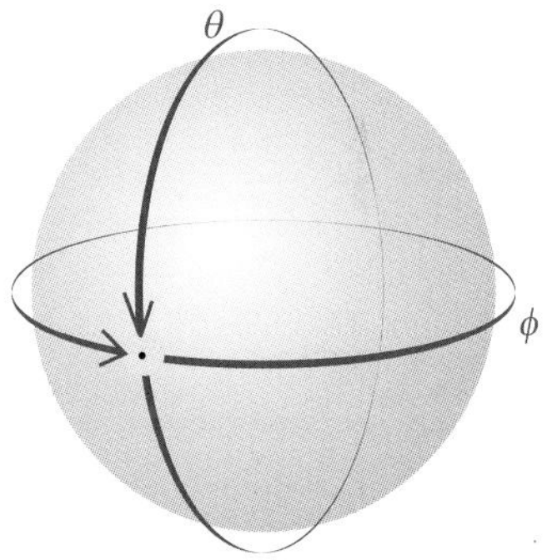

Fig. 9.34 The wavefunction of a particle on the surface of a sphere must satisfy two cyclic boundary conditions; this requirement leads to two quantum numbers for its state of angular momentum.

Justification 9.7 *The separation of variables technique applied to the particle on a sphere*

The laplacian in spherical polar coordinates is (see *Further reading*)

$$\nabla^2 = \frac{\partial^2}{\partial r^2} + \frac{2}{r}\frac{\partial}{\partial r} + \frac{1}{r^2}\Lambda^2 \tag{9.51a}$$

where the **legendrian**, Λ^2, is

$$\Lambda^2 = \frac{1}{\sin^2\theta}\frac{\partial^2}{\partial\phi^2} + \frac{1}{\sin\theta}\frac{\partial}{\partial\theta}\sin\theta\frac{\partial}{\partial\theta} \tag{9.51b}$$

Because r is constant, we can discard the part of the laplacian that involves differentiation with respect to r, and so write the Schrödinger equation as

$$\frac{1}{r^2}\Lambda^2\psi = -\frac{2mE}{\hbar^2}\psi$$

or, because $I = mr^2$, as

$$\Lambda^2\psi = -\varepsilon\psi \qquad \varepsilon = \frac{2IE}{\hbar^2}$$

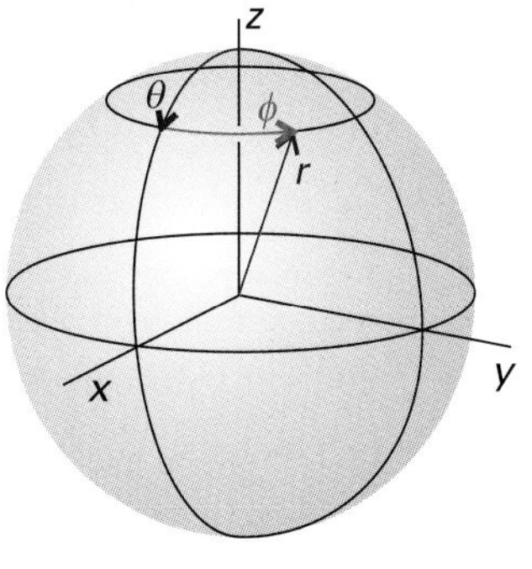

Fig. 9.35 Spherical polar coordinates. For a particle confined to the surface of a sphere, only the colatitude, θ, and the azimuth, ϕ, can change.

Table 9.3 The spherical harmonics

l	m_l	$Y_{l,m_l}(\theta,\varphi)$
0	0	$\left(\frac{1}{4\pi}\right)^{1/2}$
1	0	$\left(\frac{3}{4\pi}\right)^{1/2}\cos\theta$
	±1	$\mp\left(\frac{3}{8\pi}\right)^{1/2}\sin\theta\, e^{\pm i\phi}$
2	0	$\left(\frac{5}{16\pi}\right)^{1/2}(3\cos^2\theta-1)$
	±1	$\mp\left(\frac{15}{8\pi}\right)^{1/2}\cos\theta\sin\theta\, e^{\pm i\phi}$
	±2	$\left(\frac{15}{32\pi}\right)^{1/2}\sin^2\theta\, e^{\pm 2i\phi}$
3	0	$\left(\frac{7}{16\pi}\right)^{1/2}(5\cos^3\theta-3\cos\theta)$
	±1	$\mp\left(\frac{21}{64\pi}\right)^{1/2}(5\cos^2\theta-1)\sin\theta\, e^{\pm i\phi}$
	±2	$\left(\frac{105}{32\pi}\right)^{1/2}\sin^2\theta\cos\theta\, e^{\pm 2i\phi}$
	±3	$\mp\left(\frac{35}{64\pi}\right)^{1/2}\sin^3\theta\, e^{\pm 3i\phi}$

To verify that this expression is separable, we substitute $\psi=\Theta\Phi$:

$$\frac{1}{\sin^2\theta}\frac{\partial^2(\Theta\Phi)}{\partial\phi^2}+\frac{1}{\sin\theta}\frac{\partial}{\partial\theta}\sin\theta\frac{\partial(\Phi\Theta)}{\partial\theta}=-\varepsilon\Theta\Phi$$

We now use the fact that Θ and Φ are each functions of one variable, so the partial derivatives become complete derivatives:

$$\frac{\Theta}{\sin^2\theta}\frac{d^2\Phi}{d\phi^2}+\frac{\Phi}{\sin\theta}\frac{d}{d\theta}\sin\theta\frac{d\Theta}{d\theta}=-\varepsilon\Theta\Phi$$

Division through by $\Theta\Phi$, multiplication by $\sin^2\theta$, and minor rearrangement gives

$$\frac{1}{\Phi}\frac{d^2\Phi}{d\phi^2}+\frac{\sin\theta}{\Theta}\frac{d}{d\theta}\sin\theta\frac{d\Theta}{d\theta}+\varepsilon\sin^2\theta=0$$

The first term on the left depends only on ϕ and the remaining two terms depend only on θ. We met a similar situation when discussing a particle on a rectangular surface (*Justification 9.3*), and by the same argument, the complete equation can be separated. Thus, if we set the first term equal to the numerical constant $-m_l^2$ (using a notation chosen with an eye to the future), the separated equations are

$$\frac{1}{\Phi}\frac{d^2\Phi}{d\phi^2}=-m_l^2 \qquad \frac{\sin\theta}{\Theta}\frac{d}{d\theta}\sin\theta\frac{d\Theta}{d\theta}+\varepsilon\sin^2\theta=m_l^2$$

The first of these two equations is the same as that in *Justification 9.5*, so it has the same solutions (eqn 9.38b). The second is much more complicated to solve, but the solutions are tabulated as the *associated Legendre functions*. The cyclic boundary conditions on Θ result in the introduction of a second quantum number, l, which identifies the acceptable solutions. The presence of the quantum number m_l in the second equation implies, as we see below, that the range of acceptable values of m_l is restricted by the value of l.

As indicated in *Justification 9.7*, solution of the Schrödinger equation shows that the acceptable wavefunctions are specified by two quantum numbers l and m_l that are restricted to the values

$$l=0, 1, 2, \ldots \qquad m_l=l, l-1, \ldots, -l \tag{9.52}$$

Note that the **orbital angular momentum quantum number** l is non-negative and that, for a given value of l, there are $2l+1$ permitted values of the **magnetic quantum number**, m_l. The normalized wavefunctions are usually denoted $Y_{l,m_l}(\theta,\phi)$ and are called the **spherical harmonics** (Table 9.3).

Figure 9.36 is a representation of the spherical harmonics for $l=0$ to 4 and $m_l=0$ which emphasizes how the number of angular nodes (the positions at which the wavefunction passes through zero) increases as the value of l increases. There are no angular nodes around the z-axis for functions with $m_l=0$, which corresponds to there being no component of orbital angular momentum about that axis. Figure 9.37 shows the distribution of the particle of a given angular momentum in more detail. In this representation, the value of $|Y_{l,m_l}|^2$ at each value of θ and ϕ is proportional to the distance of the surface from the origin. Note how, for a given value of l, the most probable location of the particle migrates towards the xy-plane as the value of $|m_l|$ increases.

It also follows from the solution of the Schrödinger equation that the energy E of the particle is restricted to the values

$$E=l(l+1)\frac{\hbar^2}{2I} \qquad l=0, 1, 2, \ldots \tag{9.53}$$

Comment 9.6

The spherical harmonics are orthogonal and normalized in the following sense:

$$\int_0^{\pi}\int_0^{2\pi} Y_{l',m_l'}(\theta,\phi)^* Y_{l,m_l}(\theta,\phi)\sin\theta\, d\theta\, d\phi = \delta_{l'l}\delta_{m_l'm_l}$$

An important 'triple integral' is

$$\int_0^{\pi}\int_0^{2\pi} Y_{l'',m_l''}(\theta,\phi)^* Y_{l',m_l'}(\theta,\phi) Y_{l,m_l}(\theta,\phi) \sin\theta\, d\theta\, d\phi=0$$

unless $m_l''=m_l'+m_l$ and we can form a triangle with sides of lengths l'', l', and l (such as 1, 2, and 3 or 1, 1, and 1, but not 1, 2, and 4).

Comment 9.7

The real and imaginary components of the Φ component of the wavefunctions, $e^{im_l\phi}=\cos m_l\phi+i\sin m_l\phi$, each have $|m_l|$ angular nodes, but these nodes are not seen when we plot the probability density, because $|e^{im_l\phi}|^2=1$.

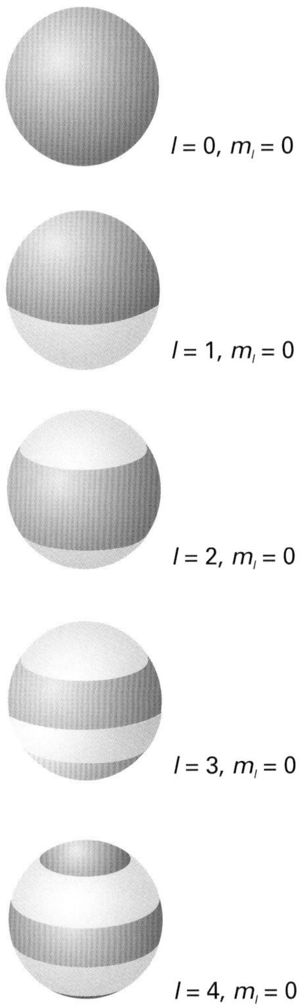

Fig. 9.36 A representation of the wavefunctions of a particle on the surface of a sphere which emphasizes the location of angular nodes: dark and light shading correspond to different signs of the wavefunction. Note that the number of nodes increases as the value of l increases. All these wavefunctions correspond to $m_l = 0$; a path round the vertical z-axis of the sphere does not cut through any nodes.

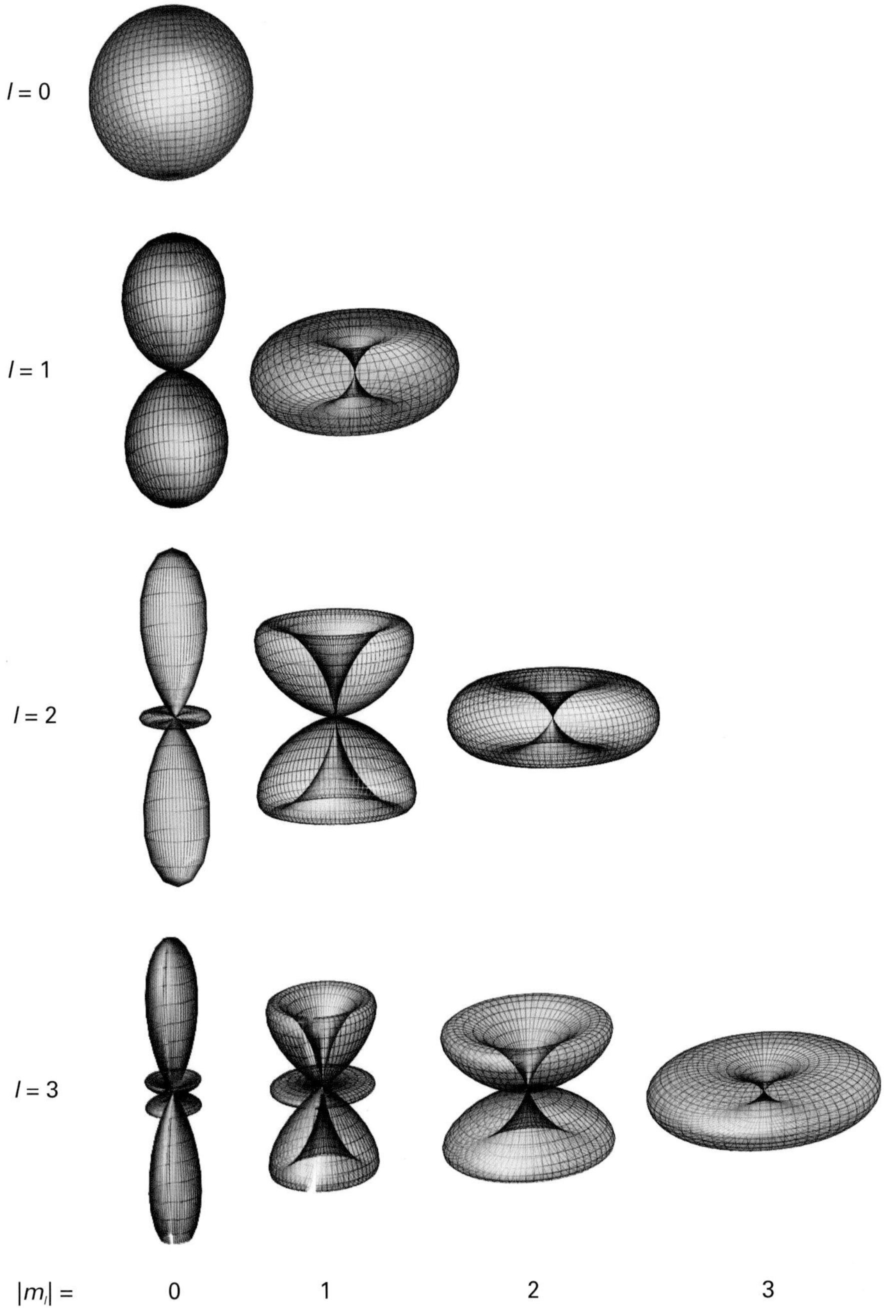

Fig. 9.37 A more complete representation of the wavefunctions for $l = 0$, 1, 2, and 3. The distance of a point on the surface from the origin is proportional to the square modulus of the amplitude of the wavefunction at that point.

Exploration Plot the variation with the radius r of the first ten energy levels of a particle on a sphere. Which of the following statements are true: (a) for a given value of r, the energy separation between adjacent levels decreases with increasing l, (b) increasing r leads to an decrease in the value of the energy for each level, (c) the energy difference between adjacent levels increases as r increases.

We see that the energy is quantized, and that it is independent of m_l. Because there are $2l+1$ different wavefunctions (one for each value of m_l) that correspond to the same energy, it follows that a level with quantum number l is $(2l+1)$-fold degenerate.

(b) Angular momentum

The energy of a rotating particle is related classically to its angular momentum J by $E=J^2/2I$ (see *Appendix 3*). Therefore, by comparing this equation with eqn 9.53, we can deduce that, because the energy is quantized, then so too is the magnitude of the angular momentum, and confined to the values

$$\text{Magnitude of angular momentum} = \{l(l+1)\}^{1/2}\hbar \qquad l=0,1,2\ldots \tag{9.54a}$$

We have already seen (in the context of rotation in a plane) that the angular momentum about the z-axis is quantized, and that it has the values

$$z\text{-Component of angular momentum} = m_l\hbar \qquad m_l=l,l-1,\ldots,-l \tag{9.54b}$$

The fact that the number of nodes in $\psi_{l,m_l}(\theta,\phi)$ increases with l reflects the fact that higher angular momentum implies higher kinetic energy, and therefore a more sharply buckled wavefunction. We can also see that the states corresponding to high angular momentum around the z-axis are those in which most nodal lines cut the equator: a high kinetic energy now arises from motion parallel to the equator because the curvature is greatest in that direction.

Illustration 9.4 *Calculating the frequency of a molecular rotational transition*

Under certain circumstances, the particle on a sphere is a reasonable model for the description of the rotation of diatomic molecules. Consider, for example, the rotation of a $^1H^{127}I$ molecule: because of the large difference in atomic masses, it is appropriate to picture the ^1H atom as orbiting a stationary ^{127}I atom at a distance $r=160$ pm, the equilibrium bond distance. The moment of inertia of $^1H^{127}I$ is then $I=m_Hr^2=4.288\times10^{-47}$ kg m^2. It follows that

$$\frac{\hbar^2}{2I}=\frac{(1.054\,57\times10^{-34}\text{ J s})^2}{2\times(4.288\times10^{-47}\text{ kg m}^2)}=1.297\times10^{-22}\text{ J}$$

or 0.1297 zJ. This energy corresponds to 78.09 J mol^{-1}. From eqn 9.53, the first few rotational energy levels are therefore 0 ($l=0$), 0.2594 zJ ($l=1$), 0.7782 zJ ($l=2$), and 1.556 zJ ($l=3$). The degeneracies of these levels are 1, 3, 5, and 7, respectively (from $2l+1$), and the magnitudes of the angular momentum of the molecule are 0, $2^{1/2}\hbar$, $6^{1/2}\hbar$, and $(12)^{1/2}\hbar$ (from eqn 9.54a). It follows from our calculations that the $l=0$ and $l=1$ levels are separated by $\Delta E=0.2594$ zJ. A transition between these two rotational levels of the molecule can be brought about by the emission or absorption of a photon with a frequency given by the Bohr frequency condition (eqn 8.10):

$$\nu=\frac{\Delta E}{h}=\frac{2.594\times10^{-22}\text{ J}}{6.626\times10^{-34}\text{ J s}}=3.915\times10^{11}\text{ Hz}=391.5\text{ GHz}$$

Radiation with this frequency belongs to the microwave region of the electromagnetic spectrum, so microwave spectroscopy is a convenient method for the study of molecular rotations. Because the transition energies depend on the moment of inertia, microwave spectroscopy is a very accurate technique for the determination of bond lengths. We discuss rotational spectra further in Chapter 13.

Self-test 9.8 Repeat the calculation for a $^2H^{127}I$ molecule (same bond length as $^1H^{127}I$).

[Energies are smaller by a factor of two; same angular momenta and numbers of components]

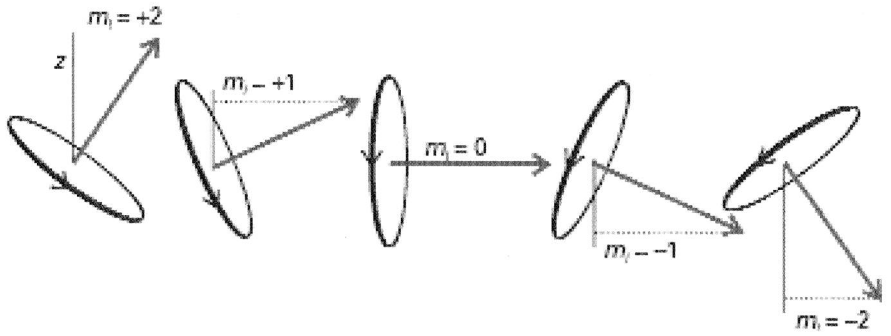

Fig. 9.38 The permitted orientations of angular momentum when $l = 2$. We shall see soon that this representation is too specific because the azimuthal orientation of the vector (its angle around z) is indeterminate.

(c) Space quantization

The result that m_l is confined to the values $l, l-1, \ldots, -l$ for a given value of l means that the component of angular momentum about the z-axis may take only $2l + 1$ values. If the angular momentum is represented by a vector of length proportional to its magnitude (that is, of length $\{l(l+1)\}^{1/2}$ units), then to represent correctly the value of the component of angular momentum, the vector must be oriented so that its projection on the z-axis is of length m_l units. In classical terms, this restriction means that the plane of rotation of the particle can take only a discrete range of orientations (Fig. 9.38). The remarkable implication is that *the orientation of a rotating body is quantized.*

The quantum mechanical result that a rotating body may not take up an arbitrary orientation with respect to some specified axis (for example, an axis defined by the direction of an externally applied electric or magnetic field) is called **space quantization.** It was confirmed by an experiment first performed by Otto Stern and Walther Gerlach in 1921, who shot a beam of silver atoms through an inhomogeneous magnetic field (Fig. 9.39). The idea behind the experiment was that a rotating, charged body behaves like a magnet and interacts with the applied field. According to classical mechanics, because the orientation of the angular momentum can take any value, the associated magnet can take any orientation. Because the direction in which the magnet is driven by the inhomogeneous field depends on the magnet's orientation, it follows that a broad band of atoms is expected to emerge from the region where the magnetic field acts. According to quantum mechanics, however, because the angular momentum is quantized, the associated magnet lies in a number of discrete orientations, so several sharp bands of atoms are expected.

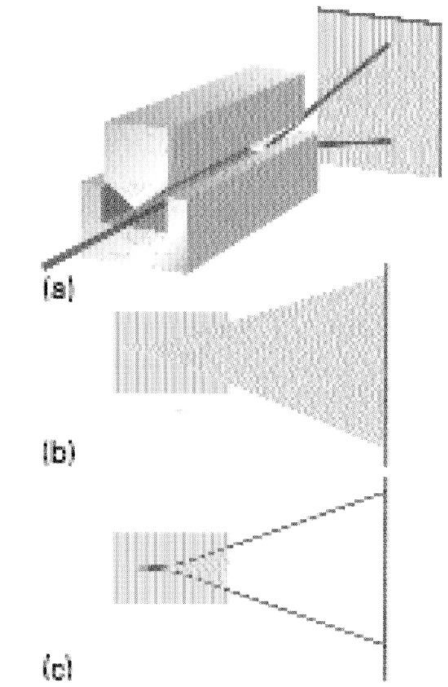

Fig. 9.39 (a) The experimental arrangement for the Stern–Gerlach experiment: the magnet provides an inhomogeneous field. (b) The classically expected result. (c) The observed outcome using silver atoms.

In their first experiment, Stern and Gerlach appeared to confirm the classical prediction. However, the experiment is difficult because collisions between the atoms in the beam blur the bands. When the experiment was repeated with a beam of very low intensity (so that collisions were less frequent) they observed discrete bands, and so confirmed the quantum prediction.

(d) The vector model

Throughout the preceding discussion, we have referred to the z-component of angular momentum (the component about an arbitrary axis, which is conventionally denoted z), and have made no reference to the x- and y-components (the components

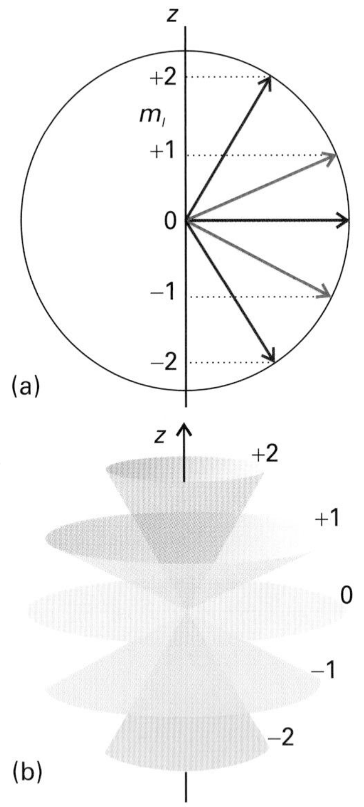

Fig. 9.40 (a) A summary of Fig. 9.38. However, because the azimuthal angle of the vector around the z-axis is indeterminate, a better representation is as in (b), where each vector lies at an unspecified azimuthal angle on its cone.

about the two axes perpendicular to z). The reason for this omission is found by examining the operators for the three components, each one being given by a term like that in eqn 9.45:

$$\hat{l}_x = \frac{\hbar}{\mathrm{i}}\left(y\frac{\partial}{\partial z} - z\frac{\partial}{\partial y}\right) \qquad \hat{l}_y = \frac{\hbar}{\mathrm{i}}\left(z\frac{\partial}{\partial x} - x\frac{\partial}{\partial z}\right) \qquad \hat{l}_z = \frac{\hbar}{\mathrm{i}}\left(x\frac{\partial}{\partial y} - y\frac{\partial}{\partial x}\right) \tag{9.55}$$

As you are invited to show in Problem 9.27, these three operators do not commute with one another:

$$[\hat{l}_x,\hat{l}_y] = \mathrm{i}\hbar\hat{l}_z \qquad [\hat{l}_y,\hat{l}_z] = \mathrm{i}\hbar\hat{l}_x \qquad [\hat{l}_z,\hat{l}_x] = \mathrm{i}\hbar\hat{l}_y \tag{9.56a}$$

Therefore, we cannot specify more than one component (unless $l = 0$). In other words, l_x, l_y, and l_z are complementary observables. On the other hand, the operator for the square of the magnitude of the angular momentum is

$$\hat{l}^2 = \hat{l}_x^2 + \hat{l}_y^2 + \hat{l}_z^2 = \hbar^2\Lambda^2 \tag{9.56b}$$

where Λ^2 is the legendrian in eqn 9.51b. This operator does commute with all three components:

$$[\hat{l}^2,\hat{l}_q] = 0 \qquad q = x, y, \text{and } z \tag{9.56c}$$

(See Problem 9.29.) Therefore, although we may specify the magnitude of the angular momentum and any of its components, if l_z is known, then it is impossible to ascribe values to the other two components. It follows that the illustration in Fig. 9.38, which is summarized in Fig. 9.40a, gives a false impression of the state of the system, because it suggests definite values for the x- and y-components. A better picture must reflect the impossibility of specifying l_x and l_y if l_z is known.

The **vector model** of angular momentum uses pictures like that in Fig. 9.40b. The cones are drawn with side $\{l(l+1)\}^{1/2}$ units, and represent the magnitude of the angular momentum. Each cone has a definite projection (of m_l units) on the z-axis, representing the system's precise value of l_z. The l_x and l_y projections, however, are indefinite. The vector representing the state of angular momentum can be thought of as lying with its tip on any point on the mouth of the cone. At this stage it should not be thought of as sweeping round the cone; that aspect of the model will be added later when we allow the picture to convey more information.

IMPACT ON NANOSCIENCE
I9.2 Quantum dots

In *Impact I9.1* we outlined some advantages of working in the nanometre regime. Another is the possibility of using quantum mechanical effects that render the properties of an assembly dependent on its size. Here we focus on the origins and consequences of these quantum mechanical effects.

Consider a sample of a metal, such as copper or gold. It carries an electrical current because the electrons are delocalized over all the atomic nuclei. That is, we may treat the movement of electrons in metals with a particle in a box model, though it is necessary to imagine that the electrons move independently of each other. Immediately, we predict from eqn 9.6 that the energy levels of the electrons in a large box, such as a copper wire commonly used to make electrical connections, form a continuum so we are justified in neglecting quantum mechanical effects on the properties of the material. However, consider a *nanocrystal*, a small cluster of atoms with dimensions in the nanometre scale. Again using eqn 9.6, we predict that the electronic energies are quantized and that the separation between energy levels decreases with increasing size of the cluster. This quantum mechanical effect can be observed in 'boxes' of any shape. For example, you are invited to show in Problem 9.39 that the energy levels of an electron *in* a sphere of radius R are given by

$$E_n = \frac{n^2h^2}{8m_eR^2} \tag{9.57}$$

The quantization of energy in nanocrystals has important technological implications when the material is a semiconductor, in which the electrical conductivity increases with increasing temperature or upon excitation by light (see Chapter 20 for a more detailed discussion). Transfer of energy to a semiconductor increases the mobility of electrons in the material. However, for every electron that moves to a different site in the sample, a unit of positive charge, called a *hole*, is left behind. The holes are also mobile, so to describe electrical conductivity in semiconductors we need to consider the movement of electron–hole pairs, also called **excitons**, in the material.

The electrons and holes may be regarded as particles trapped in a box, so eqn 9.6 can give us qualitative insight into the origins of conductivity in semiconductors. We conclude as before that only in nanocrystals are the energies of the charge carriers quantized. Now we explore the impact of energy quantization on the optical and electronic properties of semiconducting nanocrystals.

Three-dimensional nanocrystals of semiconducting materials containing 10^3 to 10^5 atoms are called **quantum dots**. They can be made in solution or by depositing atoms on a surface, with the size of the nanocrystal being determined by the details of the synthesis (see, for example, *Impact I20.2*). A quantitative but approximate treatment that leads to the energy of the exciton begins with the following hamiltonian for a spherical quantum dot of radius R:

$$\hat{H} = -\frac{\hbar^2}{2m_e}\nabla_e^2 - \frac{\hbar^2}{2m_h}\nabla_h^2 + V(\boldsymbol{r}_e,\boldsymbol{r}_h) \tag{9.58}$$

where the first two terms are the kinetic energy operators for the electron and hole (with masses m_e and m_h, respectively), and the third term is the potential energy of interaction between electron and hole, which are located at positions $\boldsymbol{r}_e$ and $\boldsymbol{r}_h$ from the centre of the sphere. Taking into account only the Coulomb attraction between the hole, with charge $+e$, and the electron, with charge $-e$, we write (see Chapter 9 and *Appendix 3* for details):

$$V(\boldsymbol{r}_e,\boldsymbol{r}_h) = -\frac{e^2}{4\pi\varepsilon|\boldsymbol{r}_e - \boldsymbol{r}_h|} \tag{9.59}$$

where $|\boldsymbol{r}_e - \boldsymbol{r}_h|$ is the distance between the electron and hole and ε is the permittivity of the medium (we are ignoring the effect of polarization of the medium due to the presence of charges). Solving the Schrödinger equation in this case is not a trivial task, but the final expression for the energy of the exciton, E_{ex}, is relatively simple (see *Further reading* for details):

$$E_{ex} = \frac{h^2}{8R^2}\left(\frac{1}{m_e} + \frac{1}{m_h}\right) - \frac{1.8e^2}{4\pi\varepsilon R} \tag{9.60}$$

As expected, we see that the energy of the exciton decreases with increasing radius of the quantum dot. Moreover, for small R, the second term on the right of the preceding equation is smaller than the first term and the energy of the exciton is largely kinetic, with the resulting expression resembling the case for a particle in a sphere.

The expression for E_{ex} has important consequences for the optical properties of quantum dots. First, we see that the energy required to create mobile charge carriers and to induce electrical conductivity depends on the size of the quantum dot. The electrical properties of large, macroscopic samples of semiconductors cannot be tuned in this way. Second, in many quantum dots, such as the nearly spherical nanocrystals of cadmium selenide (CdSe), the exciton can be generated by absorption of visible light. Therefore, we predict that, as the radius of the quantum dot decreases, the excitation wavelength decreases. That is, as the size of the quantum dot varies, so does the colour of the material. This phenomenon is indeed observed in suspensions of CdSe quantum dots of different sizes.

Because quantum dots are semiconductors with tunable electrical properties, it is easy to imagine uses for these materials in the manufacture of transistors. But the special optical properties of quantum dots can also be exploited. Just as the generation of an electron–hole pair requires absorption of light of a specific wavelength, so does recombination of the pair result in the emission of light of a specific wavelength. This property forms the basis for the use of quantum dots in the visualization of biological cells at work. For example, a CdSe quantum dot can be modified by covalent attachment of an organic spacer to its surface. When the other end of the spacer reacts specifically with a cellular component, such as a protein, nucleic acid, or membrane, the cell becomes labelled with a light-emitting quantum dot. The spatial distribution of emission intensity and, consequently, of the labelled molecule can then be measured with a microscope. Though this technique has been used extensively with organic molecules as labels, quantum dots are more stable and are stronger light emitters.

9.8 Spin

Stern and Gerlach observed *two* bands of Ag atoms in their experiment. This observation seems to conflict with one of the predictions of quantum mechanics, because an angular momentum l gives rise to $2l+1$ orientations, which is equal to 2 only if $l=\frac{1}{2}$, contrary to the conclusion that l must be an integer. The conflict was resolved by the suggestion that the angular momentum they were observing was not due to orbital angular momentum (the motion of an electron around the atomic nucleus) but arose instead from the motion of the electron about its own axis. This intrinsic angular momentum of the electron is called its **spin**. The explanation of the existence of spin emerged when Dirac combined quantum mechanics with special relativity and established the theory of relativistic quantum mechanics.

The spin of an electron about its own axis does not have to satisfy the same boundary conditions as those for a particle circulating around a central point, so the quantum number for spin angular momentum is subject to different restrictions. To distinguish this spin angular momentum from orbital angular momentum we use the **spin quantum number** s (in place of l; like l, s is a non-negative number) and m_s, the **spin magnetic quantum number**, for the projection on the z-axis. The magnitude of the spin angular momentum is $\{s(s+1)\}^{1/2}\hbar$ and the component $m_s\hbar$ is restricted to the $2s+1$ values

$$m_s = s, s-1, \ldots, -s \tag{9.61}$$

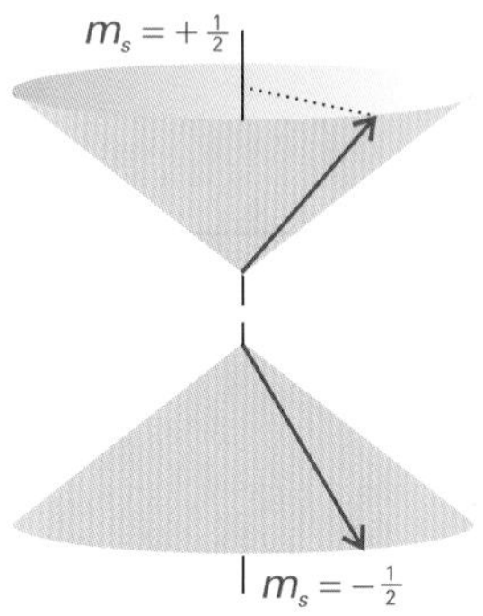

Fig. 9.41 An electron spin ($s=\frac{1}{2}$) can take only two orientations with respect to a specified axis. An α electron (top) is an electron with $m_s=+\frac{1}{2}$; a β electron (bottom) is an electron with $m_s=-\frac{1}{2}$. The vector representing the spin angular momentum lies at an angle of 55° to the z-axis (more precisely, the half-angle of the cones is $\arccos(1/3^{1/2})$).

The detailed analysis of the spin of a particle is sophisticated and shows that the property should not be taken to be an actual spinning motion. It is better to regard 'spin' as an intrinsic property like mass and charge. However, the picture of an actual spinning motion can be very useful when used with care. For an electron it turns out that only one value of s is allowed, namely $s=\frac{1}{2}$, corresponding to an angular momentum of magnitude $(\frac{3}{4})^{1/2}\hbar = 0.866\hbar$. This spin angular momentum is an intrinsic property of the electron, like its rest mass and its charge, and every electron has exactly the same value: the magnitude of the spin angular momentum of an electron cannot be changed. The spin may lie in $2s+1=2$ different orientations (Fig. 9.41).

One orientation corresponds to $m_s = +\frac{1}{2}$ (this state is often denoted α or $\uparrow$); the other orientation corresponds to $m_s = -\frac{1}{2}$ (this state is denoted β or $\downarrow$).

The outcome of the Stern–Gerlach experiment can now be explained if we suppose that each Ag atom possesses an angular momentum due to the spin of a single electron, because the two bands of atoms then correspond to the two spin orientations. Why the atoms behave like this is explained in Chapter 10 (but it is already probably familiar from introductory chemistry that the ground-state configuration of a silver atom is [Kr]$4d^{10}5s^1$, a single unpaired electron outside a closed shell).

Like the electron, other elementary particles have characteristic spin. For example, protons and neutrons are spin-$\frac{1}{2}$ particles (that is, $s = \frac{1}{2}$) and invariably spin with angular momentum $(\frac{3}{4})^{1/2}\hbar = 0.866\hbar$. Because the masses of a proton and a neutron are so much greater than the mass of an electron, yet they all have the same spin angular momentum, the classical picture would be of these two particles spinning much more slowly than an electron. Some elementary particles have $s = 1$, and so have an intrinsic angular momentum of magnitude $2^{1/2}\hbar$. Some mesons are spin-1 particles (as are some atomic nuclei), but for our purposes the most important spin-1 particle is the photon. From the discussion in this chapter, we see that the photon has zero rest mass, zero charge, an energy $h\nu$, a linear momentum h/λ or $h\nu/c$, an intrinsic angular momentum of $2^{1/2}\hbar$, and travels at the speed c. We shall see the importance of photon spin in the next chapter.

Particles with half-integral spin are called **fermions** and those with integral spin (including 0) are called **bosons**. Thus, electrons and protons are fermions and photons are bosons. It is a very deep feature of nature that all the elementary particles that constitute matter are fermions whereas the elementary particles that are responsible for the forces that bind fermions together are all bosons. Photons, for example, transmit the electromagnetic force that binds together electrically charged particles. Matter, therefore, is an assembly of fermions held together by forces conveyed by bosons.

The properties of angular momentum that we have developed are set out in Table 9.4. As mentioned there, when we use the quantum numbers l and m_l we shall mean orbital angular momentum (circulation in space). When we use s and m_s we shall mean spin angular momentum (intrinsic angular momentum). When we use j and m_j we shall mean either (or, in some contexts to be described in Chapter 10, a combination of orbital and spin momenta).

Table 9.4 Properties of angular momentum

Quantum number	Symbol	Values*	Specifies
Orbital angular momentum	l	$0, 1, 2, \ldots$	Magnitude, $\{l(l+1)\}^{1/2}\hbar$
Magnetic	m_l	$l, l-1, \ldots, -l$	Component on z-axis, $m_l\hbar$
Spin	s	$\frac{1}{2}$	Magnitude, $\{s(s+1)\}^{1/2}\hbar$
Spin magnetic	m_s	$\pm\frac{1}{2}$	Component on z-axis, $m_s\hbar$
Total	j	$l+s, l+s-1, \ldots, \lvert l-s \rvert$	Magnitude, $\{j(j+1)\}^{1/2}\hbar$
Total magnetic	m_j	$j, j-1, \ldots, -j$	Component on z-axis, $m_j\hbar$

To combine two angular momenta, use the Clebsch–Gordan series:

$$j = j_1 + j_2, j_1 + j_2 - 1, \ldots, \lvert j_1 - j_2 \rvert$$

For many-electron systems, the quantum numbers are designated by uppercase letters (L, M_L, S, M_S, etc.).

*Note that the quantum numbers for magnitude (l, s, j, etc.) are never negative.

Techniques of approximation

All the applications treated so far have had exact solutions. However, many problems —and almost all the problems of interest in chemistry—do not have exact solutions. To make progress with these problems we need to develop techniques of approximation. There are two major approaches, *variation theory* and *perturbation theory.* Variation theory is most commonly encountered in the context of molecular orbital theory, and we consider it there (Chapter 11). Here, we concentrate on perturbation theory.

9.9 Time-independent perturbation theory

In **perturbation theory**, we suppose that the hamiltonian for the problem we are trying to solve, $\hat{H}$, can be expressed as the sum of a simple hamiltonian, $\hat{H}^{(0)}$, which has known eigenvalues and eigenfunctions, and a contribution, $\hat{H}^{(1)}$, which represents the extent to which the true hamiltonian differs from the 'model' hamiltonian:

$$\hat{H}=\hat{H}^{(0)}+\hat{H}^{(1)} \tag{9.62}$$

In **time-independent perturbation theory**, the perturbation is always present and unvarying. For example, it might represent a dip in the potential energy of a particle in a box in some region along the length of the box.

In time-independent perturbation theory, we suppose that the true energy of the system differs from the energy of the simple system, and that we can write

$$E=E^{(0)}+E^{(1)}+E^{(2)}+\dots \tag{9.63}$$

where $E^{(1)}$ is the 'first-order correction' to the energy, a contribution proportional to $\hat{H}^{(1)}$, and $E^{(2)}$ is the 'second-order correction' to the energy, a contribution proportional to $\hat{H}^{(1)2}$, and so on. The true wavefunction also differs from the 'simple' wavefunction, and we write

$$\psi=\psi^{(0)}+\psi^{(1)}+\psi^{(2)}+\dots \tag{9.64}$$

In practice, we need to consider only the 'first-order correction' to the wavefunction, $\psi^{(1)}$. As we show in *Further information 9.2*, the first- and second-order corrections to the energy of the ground state (with the wavefunction ψ_0 and energy E_0), are

$$E_0^{(1)}=\int\psi_0^{(0)\star}\hat{H}^{(1)}\psi_0^{(0)}\mathrm{d}\tau=H_{00}^{(1)} \tag{9.65a}$$

and

$$E_0^{(2)}=\sum_{n\neq 0}\frac{\left|\int\psi_n^{(0)\star}\hat{H}^{(1)}\psi_0^{(0)}\mathrm{d}\tau\right|^2}{E_0^{(0)}-E_n^{(0)}}=\sum_{n\neq 0}\frac{|H_{n0}^{(1)}|^2}{E_0^{(0)}-E_n^{(0)}} \tag{9.65b}$$

where we have introduced the **matrix element**

$$\Omega_{nm}=\int\psi_n^{\star}\hat{\Omega}\psi_m\mathrm{d}\tau \qquad [9.65c]$$

in a convenient compact notation for integrals that we shall use frequently.

As usual, it is important to be able to interpret these equations physically. We can interpret $E^{(1)}$ as the average value of the perturbation, calculated by using the unperturbed wavefunction. An analogy is the shift in energy of vibration of a violin string when small weights are hung along its length. The weights hanging close to the nodes have little effect on its energy of vibration. Those hanging at the antinodes, however, have a pronounced effect (Fig. 9.42a). The second-order energy represents a similar average of the perturbation, but now the average is taken over the *perturbed* wavefunctions. In terms of the violin analogy, the average is now taken over the distorted waveform of the vibrating string, in which the nodes and antinodes are slightly shifted (Fig. 9.42b).

We should note the following three features of eqn 9.65b:

1. Because $E_n(0) > E_0(0)$, all the terms in the denominator are negative and, because the numerators are all positive, the second-order correction is negative, which represents a *lowering* of the energy of the ground state.

2. The perturbation appears (as its square) in the numerator; so the stronger the perturbation, the greater the lowering of the ground-state energy.

3. If the energy levels of the system are widely spaced, all the denominators are large, so the sum is likely to be small; in which case the perturbation has little effect on the energy of the system: the system is 'stiff', and unresponsive to perturbations. The opposite is true when the energy levels lie close together.

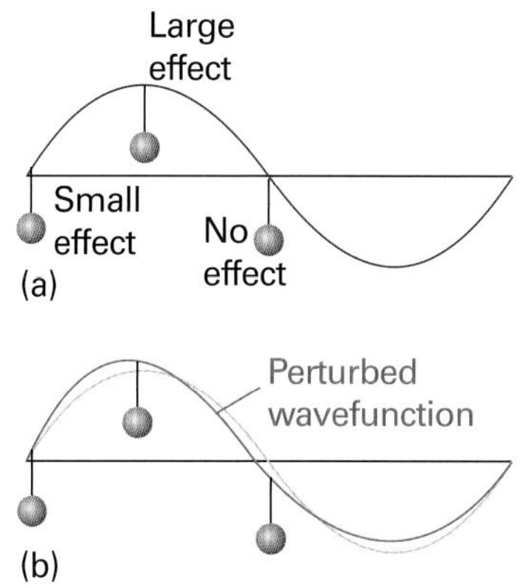

Fig. 9.42 (a) The first-order energy is an average of the perturbation (represented by the hanging weights) over the unperturbed wavefunction. (b) The second-order energy is a similar average, but over the distortion induced by the perturbation.

Example 9.5 *Using perturbation theory*

Find the first-order correction to the ground-state energy for a particle in a well with a variation in the potential of the form $V = -\varepsilon \sin(\pi x/L)$, as in Fig. 9.43.

Method Identify the first-order perturbation hamiltonian and evaluate $E_0^{(1)}$ from eqn 9.65a. We can expect a small lowering of the energy because the average potential energy of the particle is lower in the distorted box.

Answer The perturbation hamiltonian is

$$\hat{H}^{(1)} = -\varepsilon \sin(\pi x/L)$$

Therefore, the first-order correction to the energy is

$$E_0^{(1)} = \int_0^L \psi_1 \hat{H}^{(1)} \psi_1 \mathrm{d}x = -\frac{2\varepsilon}{L} \overbrace{\int_0^L \sin^3 \frac{\pi x}{L} \mathrm{d}x}^{4L/3\pi} = -\frac{8\varepsilon}{3\pi}$$

Note that the energy is lowered by the perturbation, as would be expected for the shape shown in Fig. 9.43.

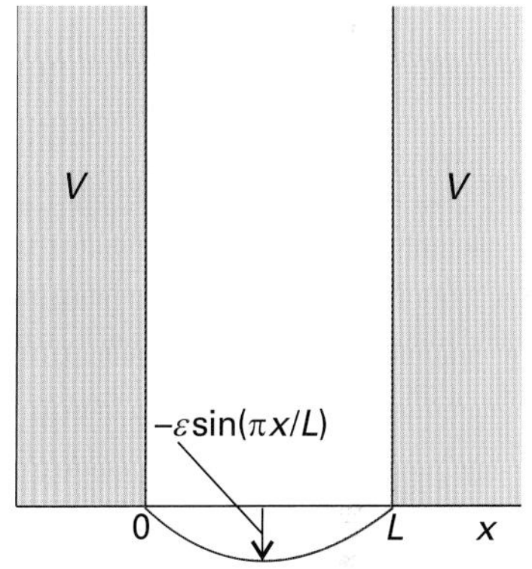

Fig. 9.43 The potential energy for a particle in a box with a potential that varies as $-\varepsilon \sin(\pi x/L)$ across the floor of the box. We can expect the particle to accumulate more in the centre of the box (in the ground state at least) than in the unperturbed box.

Self-test 9.9 Suppose that only ψ_3 contributes to the distortion of the wavefunction: calculate the coefficient c_3 and the second-order correction to the energy by using eqn 9.65b and eqn 9.76 in *Further information 9.2*.

$[c_3 = -8\varepsilon mL^2/15\pi h^2, E_0^{(2)} = -64\varepsilon^2 mL^2/225\pi^2 h^2]$

9.10 Time-dependent perturbation theory

In **time-dependent perturbation theory**, the perturbation is either switched on and allowed to rise to its final value or is varying in time. Many of the perturbations encountered in chemistry are time-dependent. The most important is the effect of an oscillating electromagnetic field, which is responsible for spectroscopic transitions between quantized energy levels in atoms and molecules.

Classically, for a molecule to be able to interact with the electromagnetic field and absorb or emit a photon of frequency ν, it must possess, at least transiently, a dipole oscillating at that frequency. In this section, we develop the quantum mechanical view and begin by writing the hamiltonian for the system as

$$\hat{H} = \hat{H}^{(0)} + \hat{H}^{(1)}(t) \tag{9.66}$$

where $\hat{H}^{(1)}(t)$ is the time-dependent perturbation. Because the perturbation arises from the effect of an oscillating electric field with the electric dipole moment, we write

$$\hat{H}^{(1)}(t) = -\mu_z \mathcal{E} \cos \omega t \tag{9.67}$$

where ω is the frequency of the field and $\mathcal{E}$ is its amplitude. We suppose that the perturbation is absent until $t = 0$, and then it is turned on.

Comment 9.8

An electric dipole consists of two electric charges $+q$ and $-q$ separated by a distance R. The electric dipole moment vector $\boldsymbol{\mu}$ has a magnitude $\mu = qR$.

We show in *Further information 9.2* that the rate of change of population of the state ψ_f due to transitions from state ψ_i, $w_{\mathrm{f}\leftarrow\mathrm{i}}$, is proportional to the square modulus of the matrix element of the perturbation between the two states:

$$w_{\mathrm{f}\leftarrow\mathrm{i}} \propto |H^{(1)}_{\mathrm{fi}}|^2 \tag{9.68}$$

Because in our case the perturbation is that of the interaction of the electromagnetic field with a molecule (eqn 9.67), we conclude that

$$w_{\mathrm{f}\leftarrow\mathrm{i}} \propto |\mu_{z,\mathrm{fi}}|^2 \mathcal{E}^2 \tag{9.69}$$

Therefore, the rate of transition, and hence the intensity of absorption of the incident radiation, is proportional to the square of the **transition dipole moment**:

$$\mu_{z,\mathrm{fi}} = \int \psi_\mathrm{f}^* \mu_z \psi_\mathrm{i} \mathrm{d}\tau \qquad [9.70]$$

The size of the transition dipole can be regarded as a measure of the charge redistribution that accompanies a transition.

The rate of transition is also proportional to $\mathcal{E}^2$, and therefore the intensity of the incident radiation (because the intensity is proportional to $\mathcal{E}^2$; see *Appendix 3*). This result will be the basis of most of our subsequent discussion of spectroscopy in Chapters 10 and 13–15 and of the kinetics of electron transfer in Chapter 24.

Checklist of key ideas

- ☐ 1. The wavefunction of a free particle is $\psi_k = A\mathrm{e}^{\mathrm{i}kx} + B\mathrm{e}^{-\mathrm{i}kx}$, $E_k = k^2\hbar^2/2m$.
- ☐ 2. The wavefunctions and energies of a particle in a one-dimensional box of length L are, respectively, $\psi_n(x) = (2/L)^{1/2} \sin(n\pi x/L)$ and $E_n = n^2h^2/8mL^2$, $n = 1,2,\ldots$. The zero-point energy, the lowest possible energy is $E_1 = h^2/8mL^2$.
- ☐ 3. The correspondence principle states that classical mechanics emerges from quantum mechanics as high quantum numbers are reached.
- ☐ 4. The functions ψ_n and $\psi_{n'}$ are orthogonal if $\int \psi_n^* \psi_{n'} \mathrm{d}\tau = 0$; all wavefunctions corresponding to different energies of a system are orthogonal. Orthonormal functions are sets of functions that are normalized and mutually orthogonal.
- ☐ 5. The wavefunctions and energies of a particle in a two-dimensional box are given by eqn 9.12a.
- ☐ 6. Degenerate wavefunctions are different wavefunctions corresponding to the same energy.
- ☐ 7. Tunnelling is the penetration into or through classically forbidden regions. The transmission probability is given by eqn 9.20a.
- ☐ 8. Harmonic motion is the motion in the presence of a restoring force proportional to the displacement, $F = -kx$, where k is the force constant. As a consequence, $V = \frac{1}{2}kx^2$.
- ☐ 9. The wavefunctions and energy levels of a quantum mechanical harmonic oscillator are given by eqns 9.28 and 9.25, respectively.
- ☐ 10. The virial theorem states that, if the potential energy of a particle has the form $V = ax^b$, then its mean potential and kinetic energies are related by $2\langle E_\mathrm{K}\rangle = b\langle V\rangle$.
- ☐ 11. Angular momentum is the moment of linear momentum around a point.

☐ 12. The wavefunctions and energies of a particle on a ring are, respectively, $\psi_{m_l}(\phi) = (1/2\pi)^{1/2}e^{im_l\phi}$ and $E = m_l^2\hbar^2/2I$, with $I = mr^2$ and $m_l = 0, \pm1, \pm2, \ldots$.

☐ 13. The wavefunctions of a particle on a sphere are the spherical harmonics, the functions $Y_{l,m_l}(\theta,\phi)$ (Table 9.3). The energies are $E = l(l+1)\hbar^2/2I$, $l = 0, 1, 2, \ldots$.

☐ 14. For a particle on a sphere, the magnitude of the angular momentum is $\{l(l+1)\}^{1/2}\hbar$ and the z-component of the angular momentum is $m_l\hbar$, $m_l = l, l-1, \ldots, -l$.

☐ 15. Space quantization is the restriction of the component of angular momentum around an axis to discrete values.

☐ 16. Spin is an intrinsic angular momentum of a fundamental particle. A fermion is a particle with a half-integral spin quantum number; a boson is a particle with an integral spin quantum number.

☐ 17. For an electron, the spin quantum number is $s = \frac{1}{2}$.

☐ 18. The spin magnetic quantum number is $m_s = s, s-1, \ldots, -s$; for an electron, $m_s = +\frac{1}{2}, -\frac{1}{2}$.

☐ 19. Perturbation theory is a technique that supplies approximate solutions to the Schrödinger equation and in which the hamiltonian for the problem is expressed as a sum of simpler hamiltonians.

☐ 20. In time-independent perturbation theory, the perturbation is always present and unvarying. The first- and second-order corrections to the energy are given by eqns 9.65a and 9.65b, respectively. In time-dependent perturbation theory, the perturbation is either switched on and allowed to rise to its final value or is varying in time.

☐ 21. The rate of change of population of the state ψ_f due to transitions from state ψ_i is $w_{f\leftarrow i} \propto |\mu_{z,fi}|^2\mathcal{E}^2$, where $\mu_{z,fi} = \int\psi_f^*\mu_z\psi_i d\tau$ is the transition dipole moment.

Further reading

Articles and texts

P.W. Atkins and R.S. Friedman, *Molecular quantum mechanics.* Oxford University Press (2005).

C.S. Johnson, Jr. and L.G. Pedersen, *Problems and solutions in quantum chemistry and physics.* Dover, New York (1986).

I.N. Levine, *Quantum chemistry.* Prentice–Hall, Upper Saddle River (2000).

D.A. McQuarrie, *Mathematical methods for scientists and engineers.* University Science Books, Mill Valley (2003).

J.J.C. Mulder, Closed-form spherical harmonics: explicit polynomial expression for the associated legendre functions. *J. Chem. Educ.* **77**, 244 (2000).

L. Pauling and E.B. Wilson, *Introduction to quantum mechanics with applications to chemistry.* Dover, New York (1985).

Further information

Further information 9.1 *Dirac notation*

The integral in eqn 9.9 is often written

$$\langle n|n'\rangle = 0 \qquad (n' \neq n)$$

This **Dirac bracket notation** is much more succinct than writing out the integral in full. It also introduces the words 'bra' and 'ket' into the language of quantum mechanics. Thus, the **bra** $\langle n|$ corresponds to ψ_n^* and the **ket** $|n'\rangle$ corresponds to the wavefunction $\psi_{n'}$. When the bra and ket are put together as in this expression, the integration over all space is understood. Similarly, the normalization condition in eqn 8.17c becomes simply

$$\langle n|n\rangle = 1$$

in bracket notation. These two expressions can be combined into one:

$$\langle n|n'\rangle = \delta_{nn'} \qquad (9.71)$$

Here $\delta_{nn'}$, which is called the **Kronecker delta**, is 1 when $n' = n$ and 0 when $n' \neq n$.

Integrals of the form $\int\psi_n^*\hat{\Omega}\psi_m d\tau$, which we first encounter in connection with perturbation theory (Section 9.9) and which are commonly called 'matrix elements', are incorporated into the bracket notation by writing

$$\langle n|\hat{\Omega}|m\rangle = \int\psi_n^*\hat{\Omega}\psi_m d\tau \qquad [9.72]$$

Note how the operator stands between the bra and the ket (which may denote different states), in the place of the c in ⟨bra|c|ket⟩. An integration is implied whenever a complete bracket is written. In this notation, an expectation value is

$$\langle\Omega\rangle = \langle n|\hat{\Omega}|n\rangle \qquad (9.73)$$

with the bra and the ket corresponding to the same state (with quantum number n and wavefunction ψ_n). In this notation, an operator is hermitian (eqn 8.30) if

$$\langle n|\hat{\Omega}|m\rangle = \langle m|\hat{\Omega}|n\rangle^* \qquad (9.74)$$

Further information 9.2 *Perturbation theory*

Here we treat perturbation theory in detail. Our first task is to develop the results of time-independent perturbation theory, in which a system is subjected to a perturbation that does not vary with time. Then, we go on to discuss time-dependent perturbation theory, in which a perturbation is turned on at a specific time and the system is allowed to evolve.

1 Time-independent perturbation theory

To develop expressions for the corrections to the wavefunction and energy of a system subjected to a time-independent perturbation, we write

$$\psi=\psi^{(0)}+\lambda\psi^{(1)}+\lambda^2\psi^{(2)}+\ldots$$

where the power of λ indicates the order of the correction. Likewise, we write

$$\hat{H}=\hat{H}^{(0)}+\lambda\hat{H}^{(1)}$$

and

$$E=E^{(0)}+\lambda E^{(1)}+\lambda^2E^{(2)}+\ldots$$

When these expressions are inserted into the Schrödinger equation, $\hat{H}\psi=E\psi$, we obtain

$$\begin{aligned}&(\hat{H}^{(0)}+\lambda\hat{H}^{(1)})(\psi^{(0)}+\lambda\psi^{(1)}+\lambda^2\psi^{(2)}+\ldots)\\&\quad=(E^{(0)}+\lambda E^{(1)}+\lambda^2E^{(2)}+\ldots)(\psi^{(0)}+\lambda\psi^{(1)}+\lambda^2\psi^{(2)}+\ldots)\end{aligned}$$

which we can rewrite as

$$\begin{aligned}&\hat{H}^{(0)}\psi^{(0)}+\lambda(\hat{H}^{(1)}\psi^{(0)}+\hat{H}^{(0)}\psi^{(1)})+\lambda^2(\hat{H}^{(0)}\psi^{(2)}+\hat{H}^{(1)}\psi^{(1)})+\ldots\\&\quad=E^{(0)}\psi^{(0)}+\lambda(E^{(0)}\psi^{(1)}+E^{(1)}\psi^{(0)})+\lambda^2(E^{(2)}\psi^{(0)}+E^{(1)}\psi^{(1)}\\&\quad\quad+E^{(0)}\psi^{(2)})+\ldots\end{aligned}$$

By comparing powers of λ, we find

Terms in λ^0: $\hat{H}^{(0)}\psi^{(0)}=E^{(0)}\psi^{(0)}$

Terms in λ: $\hat{H}^{(1)}\psi^{(0)}+\hat{H}^{(0)}\psi^{(1)}=E^{(0)}\psi^{(1)}+E^{(1)}\psi^{(0)}$

Terms in λ^2: $\hat{H}^{(0)}\psi^{(2)}+\hat{H}^{(1)}\psi^{(1)}=E^{(2)}\psi^{(0)}+E^{(1)}\psi^{(1)}+E^{(0)}\psi^{(2)}$

and so on.

The equations we have derived are applicable to any state of the system. From now on we shall consider only the ground state ψ_0 with energy E_0. The first equation, which we now write

$$\hat{H}^{(0)}\psi_0^{(0)}=E_0^{(0)}\psi_0^{(0)}$$

is the Schrödinger equation for the ground state of the unperturbed system, which we assume we can solve (for instance, it might be the equation for the ground state of the particle in a box, with the solutions given in eqn 9.7). To solve the next equation, which is now written

$$\hat{H}^{(1)}\psi_0^{(0)}+\hat{H}^{(0)}\psi_0^{(1)}=E_0^{(0)}\psi^{(1)}+E_0^{(1)}\psi_0^{(0)}$$

we suppose that the first-order correction to the wavefunction can be expressed as a linear combination of the wavefunctions of the unperturbed system, and write

$$\psi_0^{(1)}=\sum_n c_n\psi_n^{(0)} \tag{9.75}$$

Substitution of this expression gives

$$\hat{H}^{(1)}\psi_0^{(0)}+\sum_n c_n\hat{H}^{(0)}\psi_n^{(0)}=\sum_n c_nE_0^{(0)}\psi_n^{(0)}+E_0^{(1)}\psi_0^{(0)}$$

We can isolate the term in $E_0^{(1)}$ by making use of the fact that the $\psi_n^{(0)}$ form a complete orthogonal and normalized set in the sense that

$$\int\psi_0^{(0)*}\psi_n^{(0)}\mathrm{d}\tau=0 \quad \text{if } n\neq0, \quad \text{but 1 if } n=0$$

Therefore, when we multiply through by $\psi_0^{(0)*}$ and integrate over all space, we get

$$\begin{aligned}&\int\psi_0^{(0)*}\hat{H}^{(1)}\psi_0^{(0)}\mathrm{d}\tau+\sum_n c_n\overbrace{\int\psi_0^{(0)*}\hat{H}^{(0)}\psi_n^{(0)}\mathrm{d}\tau}^{E_0^{(0)}\text{ if }n=0,\text{ 0 otherwise}}\\&\quad=\sum_n c_nE_0^{(0)}\overbrace{\int\psi_0^{(0)*}\psi_n^{(0)}\mathrm{d}\tau}^{1\text{ if }n=0,\text{ 0 otherwise}}+E_0^{(1)}\overbrace{\int\psi_0^{(0)*}\psi_0^{(0)}\mathrm{d}\tau}^{1}\end{aligned}$$

That is,

$$\int\psi_0^{(0)*}\hat{H}^{(1)}\psi_0^{(0)}\mathrm{d}\tau=E_0^{(1)}$$

which is eqn 9.65a.

To find the coefficients c_n, we multiply the same expression through by $\psi_k^{(0)*}$, where now $k\neq0$, which gives

$$\begin{aligned}&\int\psi_k^{(0)*}\hat{H}^{(1)}\psi_0^{(0)}\mathrm{d}\tau+\sum_n c_n\overbrace{\int\psi_k^{(0)*}\hat{H}^{(0)}\psi_n^{(0)}\mathrm{d}\tau}^{E_k^{(0)}\delta_{kn}}\\&\quad=\sum_n c_nE_0^{(0)}\overbrace{\int\psi_k^{(0)*}\psi_n^{(0)}\mathrm{d}\tau}^{1\text{ if }n=k,\text{ 0 otherwise}}+E_0^{(1)}\overbrace{\int\psi_k^{(0)*}\psi_0^{(0)}\mathrm{d}\tau}^{0}\end{aligned}$$

That is,

$$\int\psi_k^{(0)*}\hat{H}^{(1)}\psi_0^{(0)}\mathrm{d}\tau+c_kE_k^{(0)}=c_kE_0^{(0)}$$

which we can rearrange into

$$c_k=-\frac{\int\psi_k^{(0)*}\hat{H}^{(1)}\psi_0^{(0)}\mathrm{d}\tau}{E_k^{(0)}-E_0^{(0)}} \tag{9.76}$$

The second-order energy is obtained starting from the second-order expression, which for the ground state is

$$\hat{H}^{(0)}\psi_0^{(2)}+\hat{H}^{(1)}\psi_0^{(1)}=E_0^{(2)}\psi_0^{(0)}+E_0^{(1)}\psi_0^{(1)}+E_0^{(0)}\psi_0^{(2)}$$

To isolate the term $E_0^{(2)}$ we multiply both sides by $\psi_0^{(0)*}$, integrate over all space, and obtain

$$\begin{aligned}&\overbrace{\int\psi_0^{(0)*}\hat{H}^{(0)}\psi_0^{(2)}\mathrm{d}\tau}^{E_0^{(0)}\int\psi_0^{(0)*}\psi_0^{(2)}\mathrm{d}\tau}+\int\psi_0^{(0)*}\hat{H}^{(1)}\psi_0^{(1)}\mathrm{d}\tau\\&\quad=E_0^{(2)}\underbrace{\int\psi_0^{(0)*}\psi_0^{(0)}\mathrm{d}\tau}_{1}+E_0^{(1)}\int\psi_0^{(0)*}\psi_0^{(1)}\mathrm{d}\tau+E_0^{(0)}\int\psi_0^{(0)*}\psi_0^{(2)}\mathrm{d}\tau\end{aligned}$$

The first and last terms cancel, and we are left with

$$E_0^{(2)}=\int\psi_0^{(0)*}\hat{H}^{(1)}\psi_0^{(1)}\mathrm{d}\tau-E_0^{(1)}\int\psi_0^{(0)*}\psi_0^{(1)}\mathrm{d}\tau$$

We have already found the first-order corrections to the energy and the wavefunction, so this expression could be regarded as an explicit expression for the second-order energy. However, we can go one step further by substituting eqn 9.75:

$$E_0^{(2)} = \sum_n c_n \overbrace{\int \psi_0^{(0)\star} \hat{H}^{(1)} \psi_n^{(0)} \mathrm{d}\tau}^{H_{0n}^{(1)}} - \sum_n c_n E_0^{(1)} \overbrace{\int \psi_0^{(0)\star} \psi_n^{(0)} \mathrm{d}\tau}^{\delta_{0n}}$$

$$= \sum_n c_n H_{0n}^{(1)} - c_0 E_0^{(1)}$$

The final term cancels the term $c_0 H_{00}^{(1)}$ in the sum, and we are left with

$$E_0^{(2)} = \sum_{n \neq 0} c_n H_{0n}^{(1)}$$

Substitution of the expression for c_n in eqn 9.76 now produces the final result, eqn 9.65b.

2 Time-dependent perturbation theory

To cope with a perturbed wavefunction that evolves with time, we need to solve the time-dependent Schrödinger equation,

$$\hat{H}\Psi = \mathrm{i}\hbar \frac{\partial \Psi}{\partial t} \quad (9.77)$$

We confirm below, that if we write the first-order correction to the wavefunction as

$$\Psi_0^{(1)}(t) = \sum_n c_n(t) \Psi_n(t) = \sum_n c_n(t) \psi_n^{(0)} \mathrm{e}^{-\mathrm{i}E_n^{(0)}t/\hbar} \quad (9.78\mathrm{a})$$

then the coefficients in this expansion are given by

$$c_n(t) = \frac{1}{\mathrm{i}\hbar} \int_0^t H_{n0}^{(1)}(t) \mathrm{e}^{\mathrm{i}\omega_{n0}t} \mathrm{d}t \quad (9.78\mathrm{b})$$

The formal demonstration of eqn 9.78 is quite lengthy (see *Further reading*). Here we shall show that, given eqn 9.78b, a perturbation that is switched on very slowly to a constant value gives the same expression for the coefficients as we obtained for time-independent perturbation theory. For such a perturbation, we write

$$\hat{H}^{(1)}(t) = \hat{H}^{(1)}(1 - \mathrm{e}^{-t/\tau})$$

and take the time constant τ to be very long (Fig. 9.44). Substitution of this expression into eqn 9.78b gives

$$c_n(t) = \frac{1}{\mathrm{i}\hbar} H_{n0}^{(1)} \int_0^t (1 - \mathrm{e}^{-t/\tau}) \mathrm{e}^{\mathrm{i}\omega_{n0}t} \mathrm{d}t$$

$$= \frac{1}{\mathrm{i}\hbar} H_{n0}^{(1)} \left\{ \frac{\mathrm{e}^{\mathrm{i}\omega_{n0}t} - 1}{\mathrm{i}\omega_{n0}} - \frac{\mathrm{e}^{(\mathrm{i}\omega_{n0} - 1/\tau)t} - 1}{\mathrm{i}\omega_{n0} - 1/\tau} \right\}$$

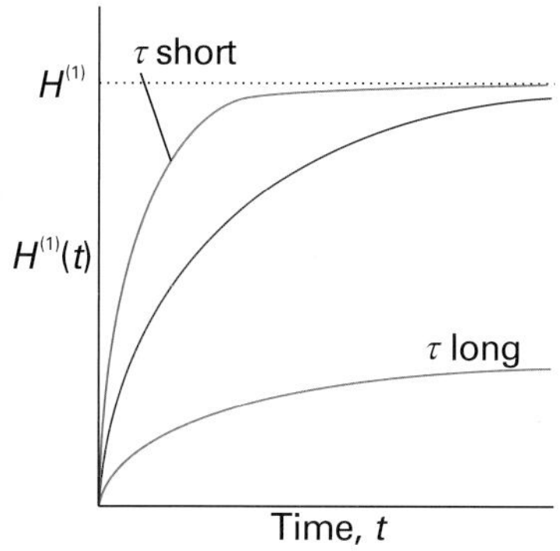

Fig. 9.44 The time-dependence of a slowly switched perturbation. A large value of τ corresponds to very slow switching.

At this point we suppose that the perturbation is switched slowly, in the sense that $\tau \gg 1/\omega_{n0}$ (so that the $1/\tau$ in the second denominator can be ignored). We also suppose that we are interested in the coefficients long after the perturbation has settled down into its final value, when $t \gg \tau$ (so that the exponential in the second numerator is close to zero and can be ignored). Under these conditions,

$$c_n(t) = -\frac{H_{n0}^{(0)}}{\hbar\omega_{n0}} \mathrm{e}^{\mathrm{i}\omega_{n0}t}$$

Now we recognize that $\hbar\omega_{n0} = E_n^{(0)} - E_0^{(0)}$, which gives

$$c_n(t) = -\frac{H_{n0}^{(0)}}{E_n^{(0)} - E_0^{(0)}} \mathrm{e}^{\mathrm{i}E_n^{(0)}t} \mathrm{e}^{-\mathrm{i}E_0^{(0)}t}$$

When this expression is substituted into eqn 9.78a, we obtain the time-independent expression, eqn 9.76 (apart from an irrelevant overall phase factor).

In accord with the general rules for the interpretation of wavefunctions, the probability that the system will be found in the state n is proportional to the square modulus of the coefficient of the state, $|c_n(t)|^2$. Therefore, the rate of change of population of a final state ψ_f due to transitions from an initial state ψ_i is

$$w_{\mathrm{f}\leftarrow\mathrm{i}} = \frac{\mathrm{d}|c_\mathrm{f}|^2}{\mathrm{d}t} = \frac{\mathrm{d}c_\mathrm{f}^\star c_\mathrm{f}}{\mathrm{d}t} = c_\mathrm{f}\frac{\mathrm{d}c_\mathrm{f}^\star}{\mathrm{d}t} + c_\mathrm{f}^\star\frac{\mathrm{d}c_\mathrm{f}}{\mathrm{d}t}$$

Because the coefficients are proportional to the matrix elements of the perturbation, $w_{\mathrm{f}\leftarrow\mathrm{i}}$ is proportional to the square modulus of the matrix element of the perturbation between the two states:

$$w_{\mathrm{f}\leftarrow\mathrm{i}} \propto |H_{\mathrm{fi}}^{(1)}|^2$$

which is eqn 9.68.

Discussion questions

9.1 Discuss the physical origin of quantization energy for a particle confined to moving inside a one-dimensional box or on a ring.

9.2 Discuss the correspondence principle and provide two examples.

9.3 Define, justify, and provide examples of zero-point energy.

9.4 Discuss the physical origins of quantum mechanical tunnelling. Why is tunnelling more likely to contribute to the mechanisms of electron transfer and proton transfer processes than to mechanisms of group transfer reactions, such as AB + C → A + BC (where A, B, and C are large molecular groups)?

9.5 Distinguish between a fermion and a boson. Provide examples of each type of particle.

9.6 Describe the features that stem from nanometre-scale dimensions that are not found in macroscopic objects.

Exercises

9.1a Calculate the energy separations in joules, kilojoules per mole, electronvolts, and reciprocal centimetres between the levels (a) $n = 2$ and $n = 1$, (b) $n = 6$ and $n = 5$ of an electron in a box of length 1.0 nm.

9.1b Calculate the energy separations in joules, kilojoules per mole, electronvolts, and reciprocal centimetres between the levels (a) $n = 3$ and $n = 1$, (b) $n = 7$ and $n = 6$ of an electron in a box of length 1.50 nm.

9.2a Calculate the probability that a particle will be found between $0.49L$ and $0.51L$ in a box of length L when it has (a) $n = 1$, (b) $n = 2$. Take the wavefunction to be a constant in this range.

9.2b Calculate the probability that a particle will be found between $0.65L$ and $0.67L$ in a box of length L when it has (a) $n = 1$, (b) $n = 2$. Take the wavefunction to be a constant in this range.

9.3a Calculate the expectation values of p and p^2 for a particle in the state $n = 1$ in a square-well potential.

9.3b Calculate the expectation values of p and p^2 for a particle in the state $n = 2$ in a square-well potential.

9.4a An electron is confined to a a square well of length L. What would be the length of the box such that the zero-point energy of the electron is equal to its rest mass energy, m_ec^2? Express your answer in terms of the parameter $\lambda_C = h/m_ec$, the 'Compton wavelength' of the electron.

9.4b Repeat Exercise 9.4a for a general particle of mass m in a cubic box.

9.5a What are the most likely locations of a particle in a box of length L in the state $n = 3$?

9.5b What are the most likely locations of a particle in a box of length L in the state $n = 5$?

9.6a Consider a particle in a cubic box. What is the degeneracy of the level that has an energy three times that of the lowest level?

9.6b Consider a particle in a cubic box. What is the degeneracy of the level that has an energy $\frac{14}{3}$ times that of the lowest level?

9.7a Calculate the percentage change in a given energy level of a particle in a cubic box when the length of the edge of the cube is decreased by 10 per cent in each direction.

9.7b A nitrogen molecule is confined in a cubic box of volume 1.00 m^3. Assuming that the molecule has an energy equal to $\frac{3}{2}kT$ at $T = 300$ K, what is the value of $n = (n_x^2 + n_y^2 + n_z^2)^{1/2}$ for this molecule? What is the energy separation between the levels n and $n + 1$? What is its de Broglie wavelength? Would it be appropriate to describe this particle as behaving classically?

9.8a Calculate the zero-point energy of a harmonic oscillator consisting of a particle of mass 2.33×10^{-26} kg and force constant 155 N m^{-1}.

9.8b Calculate the zero-point energy of a harmonic oscillator consisting of a particle of mass 5.16×10^{-26} kg and force constant 285 N m^{-1}.

9.9a For a harmonic oscillator of effective mass 1.33×10^{-25} kg, the difference in adjacent energy levels is 4.82 zJ. Calculate the force constant of the oscillator.

9.9b For a harmonic oscillator of effective mass 2.88×10^{-25} kg, the difference in adjacent energy levels is 3.17 zJ. Calculate the force constant of the oscillator.

9.10a Calculate the wavelength of a photon needed to excite a transition between neighbouring energy levels of a harmonic oscillator of effective mass equal to that of a proton (1.0078 u) and force constant 855 N m^{-1}.

9.10b Calculate the wavelength of a photon needed to excite a transition between neighbouring energy levels of a harmonic oscillator of effective mass equal to that of an oxygen atom (15.9949 u) and force constant 544 N m^{-1}.

9.11a Refer to Exercise 9.10a and calculate the wavelength that would result from doubling the effective mass of the oscillator.

9.11b Refer to Exercise 9.10b and calculate the wavelength that would result from doubling the effective mass of the oscillator.

9.12a Calculate the minimum excitation energies of (a) a pendulum of length 1.0 m on the surface of the Earth, (b) the balance-wheel of a clockwork watch ($\nu = 5$ Hz).

9.12b Calculate the minimum excitation energies of (a) the 33 kHz quartz crystal of a watch, (b) the bond between two O atoms in O_2, for which $k = 1177$ N m^{-1}.

9.13a Confirm that the wavefunction for the ground state of a one-dimensional linear harmonic oscillator given in Table 9.1 is a solution of the Schrödinger equation for the oscillator and that its energy is $\frac{1}{2}\hbar\omega$.

9.13b Confirm that the wavefunction for the first excited state of a one-dimensional linear harmonic oscillator given in Table 9.1 is a solution of the Schrödinger equation for the oscillator and that its energy is $\frac{3}{2}\hbar\omega$.

9.14a Locate the nodes of the harmonic oscillator wavefunction with $v = 4$.

9.14b Locate the nodes of the harmonic oscillator wavefunction with $v = 5$.

9.15a Assuming that the vibrations of a $^{35}Cl_2$ molecule are equivalent to those of a harmonic oscillator with a force constant $k = 329$ N m^{-1}, what is the zero-point energy of vibration of this molecule? The mass of a ^{35}Cl atom is 34.9688 u.

9.15b Assuming that the vibrations of a $^{14}N_2$ molecule are equivalent to those of a harmonic oscillator with a force constant $k = 2293.8$ N m^{-1}, what is the zero-point energy of vibration of this molecule? The mass of a ^{14}N atom is 14.0031 u.

9.16a The wavefunction, $\psi(\phi)$, for the motion of a particle in a ring is of the form $\psi = Ne^{im_l\phi}$. Determine the normalization constant, N.

9.16b Confirm that wavefunctions for a particle in a ring with different values of the quantum number m_l are mutually orthogonal.

9.17a A point mass rotates in a circle with $l = 1$, Calculate the magnitude of its angular momentum and the possible projections of the angular momentum on an arbitrary axis.

9.17b A point mass rotates in a circle with $l = 2$, Calculate the magnitude of its angular momentum and the possible projections of the angular momentum on an arbitrary axis.

9.18a Draw scale vector diagrams to represent the states (a) $s = \frac{1}{2}$, $m_s = +\frac{1}{2}$, (b) $l = 1$, $m_l = +1$, (c) $l = 2$, $m_l = 0$.

9.18b Draw the vector diagram for all the permitted states of a particle with $l = 6$.

Problems*

Numerical problems

9.1 Calculate the separation between the two lowest levels for an O_2 molecule in a one-dimensional container of length 5.0 cm. At what value of n does the energy of the molecule reach $\frac{1}{2}kT$ at 300 K, and what is the separation of this level from the one immediately below?

9.2 The mass to use in the expression for the vibrational frequency of a diatomic molecule is the effective mass $\mu = m_A m_B/(m_A + m_B)$, where m_A and m_B are the masses of the individual atoms. The following data on the infrared absorption wavenumbers (in cm^{-1}) of molecules are taken from *Spectra of diatomic molecules*, G. Herzberg, van Nostrand (1950):

$H^{35}Cl$	$H^{81}Br$	HI	CO	NO
2990	2650	2310	2170	1904

Calculate the force constants of the bonds and arrange them in order of increasing stiffness.

9.3 The rotation of an $^1H^{127}I$ molecule can be pictured as the orbital motion of an H atom at a distance 160 pm from a stationary I atom. (This picture is quite good; to be precise, both atoms rotate around their common centre of mass, which is very close to the I nucleus.) Suppose that the molecule rotates only in a plane. Calculate the energy needed to excite the molecule into rotation. What, apart from 0, is the minimum angular momentum of the molecule?

9.4 Calculate the energies of the first four rotational levels of $^1H^{127}I$ free to rotate in three dimensions, using for its moment of inertia $I = \mu R^2$, with $\mu = m_H m_I/(m_H + m_I)$ and $R = 160$ pm.

9.5 A small step in the potential energy is introduced into the one-dimensional square-well problem as in Fig. 9.45. (a) Write a general expression for the first-order correction to the ground-state energy, $E_0^{(1)}$. (b) Evaluate the energy correction for $a = L/10$ (so the blip in the potential occupies the central 10 per cent of the well), with $n = 1$.

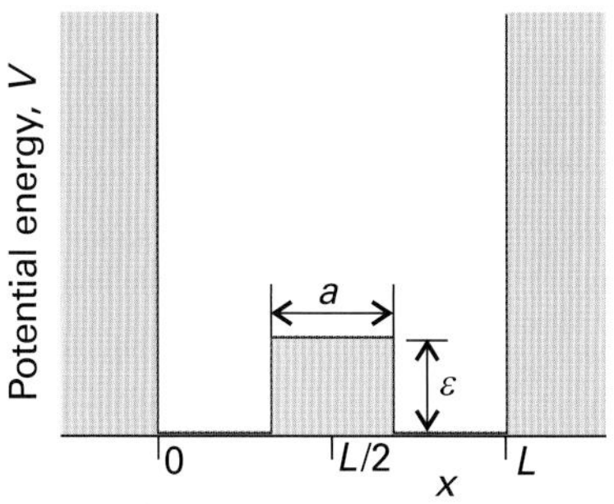

Fig. 9.45

9.6 We normally think of the one-dimensional well as being horizontal. Suppose it is vertical; then the potential energy of the particle depends on x because of the presence of the gravitational field. Calculate the first-order correction to the zero-point energy, and evaluate it for an electron in a box on the surface of the Earth. Account for the result. *Hint.* The energy of the particle depends on its height as mgh, where $g = 9.81$ m s^{-2}. Because g is so small, the energy correction is small; but it would be significant if the box were near a very massive star.

9.7 Calculate the second-order correction to the energy for the system described in Problem 9.6 and calculate the ground-state wavefunction. Account for the shape of the distortion caused by the perturbation. *Hint.* The following integrals are useful

$$\int x \sin ax \sin bx\,dx = -\frac{d}{da}\int \cos ax \sin bx\,dx$$

$$\int \cos ax \sin bx\,dx = \frac{\cos(a-b)x}{2(a-b)} - \frac{\cos(a+b)x}{2(a+b)} + \text{constant}$$

Theoretical problems

9.8 Suppose that 1.0 mol perfect gas molecules all occupy the lowest energy level of a cubic box. How much work must be done to change the volume of the box by ΔV? Would the work be different if the molecules all occupied a state $n \neq 1$? What is the relevance of this discussion to the expression for the expansion work discussed in Chapter 2? Can you identify a distinction between adiabatic and isothermal expansion?

9.9 Derive eqn 9.20a, the expression for the transmission probability.

9.10‡ Consider the one-dimensional space in which a particle can experience one of three potentials depending upon its position. They are: $V = 0$ for $-\infty < x \leq 0$, 0, $V = V_2$ for $0 \leq x \leq L$, and $V = V_3$ for $L \leq x < \infty$. The particle wavefunction is to have both a component e^{ik_1x} that is incident upon the barrier V_2 and a reflected component e^{-ik_1x} in region 1 ($-\infty < x \leq 0$). In region 3 the wavefunction has only a forward component, e^{ik_3x}, which represents a particle that has traversed the barrier. The energy of the particle, E, is somewhere in the range of the $V_2 > E > V_3$. The transmission probability, T, is the ratio of the square modulus of the region 3 amplitude to the square modulus of the incident amplitude. (a) Base your calculation on the continuity of the amplitudes and the slope of the wavefunction at the locations of the zone boundaries and derive a general equation for T. (b) Show that the general equation for T reduces to eqn 9.20b in the high, wide barrier limit when

* Problems denoted with the symbol ‡ were supplied by Charles Trapp, Carmen Giunta, and Marshall Cady.

$V_1 = V_3 = 0$. (c) Draw a graph of the probability of proton tunnelling when $V_3 = 0$, $L = 50$ pm, and $E = 10$ kJ mol^{-1} in the barrier range $E < V_2 < 2E$.

9.11 The wavefunction inside a long barrier of height V is $\psi = Ne^{-\kappa x}$. Calculate (a) the probability that the particle is inside the barrier and (b) the average penetration depth of the particle into the barrier.

9.12 Confirm that a function of the form e^{-gx^2} is a solution of the Schrödinger equation for the ground state of a harmonic oscillator and find an expression for g in terms of the mass and force constant of the oscillator.

9.13 Calculate the mean kinetic energy of a harmonic oscillator by using the relations in Table 9.1.

9.14 Calculate the values of $\langle x^3 \rangle$ and $\langle x^4 \rangle$ for a harmonic oscillator by using the relations in Table 9.1.

9.15 Determine the values of $\delta x = (\langle x^2 \rangle - \langle x \rangle^2)^{1/2}$ and $\delta p = (\langle p^2 \rangle - \langle p \rangle^2)^{1/2}$ for (a) a particle in a box of length L and (b) a harmonic oscillator. Discuss these quantities with reference to the uncertainty principle.

9.16 We shall see in Chapter 13 that the intensities of spectroscopic transitions between the vibrational states of a molecule are proportional to the square of the integral $\int \psi_{v'} x \psi_v \mathrm{d}x$ over all space. Use the relations between Hermite polynomials given in Table 9.1 to show that the only permitted transitions are those for which $v' = v \pm 1$ and evaluate the integral in these cases.

9.17 The potential energy of the rotation of one CH_3 group relative to its neighbour in ethane can be expressed as $V(\varphi) = V_0 \cos 3\varphi$. Show that for small displacements the motion of the group is harmonic and calculate the energy of excitation from $v = 0$ to $v = 1$. What do you expect to happen to the energy levels and wavefunctions as the excitation increases?

9.18 Show that, whatever superposition of harmonic oscillator states is used to construct a wavepacket, it is localized at the same place at the times 0, T, $2T$, . . . , where T is the classical period of the oscillator.

9.19 Use the virial theorem to obtain an expression for the relation between the mean kinetic and potential energies of an electron in a hydrogen atom.

9.20 Evaluate the z-component of the angular momentum and the kinetic energy of a particle on a ring that is described by the (unnormalized) wavefunctions (a) $e^{i\phi}$, (b) $e^{-2i\phi}$, (c) $\cos \phi$, and (d) $(\cos \chi)e^{i\phi} + (\sin \chi)e^{-i\phi}$.

9.21 Is the Schrödinger equation for a particle on an elliptical ring of semimajor axes a and b separable? *Hint.* Although r varies with angle φ, the two are related by $r^2 = a^2 \sin^2\phi + b^2 \cos^2\phi$.

9.22 Use mathematical software to construct a wavepacket of the form

$$\Psi(\phi,t) = \sum_{m_l=0}^{m_{l,\max}} c_{m_l} e^{i(m_l\phi - E_{m_l}t/\hbar)} \qquad E_{m_l} = m_l^2\hbar^2/2I$$

with coefficients c of your choice (for example, all equal). Explore how the wavepacket migrates on the ring but spreads with time.

9.23 Confirm that the spherical harmonics (a) $Y_{0,0}$, (b) $Y_{2,-1}$, and (c) $Y_{3,+3}$ satisfy the Schrödinger equation for a particle free to rotate in three dimensions, and find its energy and angular momentum in each case.

9.24 Confirm that $Y_{3,+3}$ is normalized to 1. (The integration required is over the surface of a sphere.)

9.25 Derive an expression in terms of l and m_l for the half-angle of the apex of the cone used to represent an angular momentum according to the vector model. Evaluate the expression for an α spin. Show that the minimum possible angle approaches 0 as $l \to \infty$.

9.26 Show that the function $f = \cos ax \cos by \cos cz$ is an eigenfunction of ∇^2, and determine its eigenvalue.

9.27 Derive (in Cartesian coordinates) the quantum mechanical operators for the three components of angular momentum starting from the classical definition of angular momentum, $\boldsymbol{l} = \boldsymbol{r} \times \boldsymbol{p}$. Show that any two of the components do not mutually commute, and find their commutator.

9.28 Starting from the operator $l_z = xp_y - yp_x$, prove that in spherical polar coordinates $l_z = -i\hbar\partial/\partial\phi$.

9.29 Show that the commutator $[l^2, l_z] = 0$, and then, without further calculation, justify the remark that $[l^2, l_q] = 0$ for all $q = x, y$, and z.

9.30‡ A particle is confined to move in a one-dimensional box of length L. (a) If the particle is classical, show that the average value of x is $\frac{1}{2}L$ and that the root-mean square value is $L/3^{1/2}$. (b) Show that for large values of n, a quantum particle approaches the classical values. This result is an example of the correspondence principle, which states that, for very large values of the quantum numbers, the predictions of quantum mechanics approach those of classical mechanics.

Applications: to biology and nanotechnology

9.31 When β-carotene is oxidized *in vivo*, it breaks in half and forms two molecules of retinal (vitamin A), which is a precursor to the pigment in the retina responsible for vision (*Impact I14.1*). The conjugated system of retinal consists of 11 C atoms and one O atom. In the ground state of retinal, each level up to $n = 6$ is occupied by two electrons. Assuming an average internuclear distance of 140 pm, calculate (a) the separation in energy between the ground state and the first excited state in which one electron occupies the state with $n = 7$, and (b) the frequency of the radiation required to produce a transition between these two states. (c) Using your results and *Illustration 9.1*, choose among the words in parentheses to generate a rule for the prediction of frequency shifts in the absorption spectra of linear polyenes:

> The absorption spectrum of a linear polyene shifts to (higher/lower) frequency as the number of conjugated atoms (increases/decreases).

9.32 Many biological electron transfer reactions, such as those associated with biological energy conversion, may be visualized as arising from electron tunnelling between protein-bound co-factors, such as cytochromes, quinones, flavins, and chlorophylls. This tunnelling occurs over distances that are often greater than 1.0 nm, with sections of protein separating electron donor from acceptor. For a specific combination of donor and acceptor, the rate of electron tunnelling is proportional to the transmission probability, with $\kappa \approx 7$ nm^{-1} (eqn 9.20). By what factor does the rate of electron tunnelling between two co-factors increase as the distance between them changes from 2.0 nm to 1.0 nm?

9.33 Carbon monoxide binds strongly to the Fe^{2+} ion of the haem group of the protein myoglobin. Estimate the vibrational frequency of CO bound to myoglobin by using the data in Problem 9.2 and by making the following assumptions: the atom that binds to the haem group is immobilized, the protein is infinitely more massive than either the C or O atom, the C atom binds to the Fe^{2+} ion, and binding of CO to the protein does not alter the force constant of the C≡O bond.

9.34 Of the four assumptions made in Problem 9.33, the last two are questionable. Suppose that the first two assumptions are still reasonable and that you have at your disposal a supply of myoglobin, a suitable buffer in which to suspend the protein, $^{12}C^{16}O$, $^{13}C^{16}O$, $^{12}C^{18}O$, $^{13}C^{18}O$, and an infrared spectrometer. Assuming that isotopic substitution does not affect the force constant of the C≡O bond, describe a set of experiments that: (a) proves which atom, C or O, binds to the haem group of myoglobin, and (b) allows for the determination of the force constant of the C≡O bond for myoglobin-bound carbon monoxide.

9.35 The particle on a ring is a useful model for the motion of electrons around the porphine ring (**2**), the conjugated macrocycle that forms the

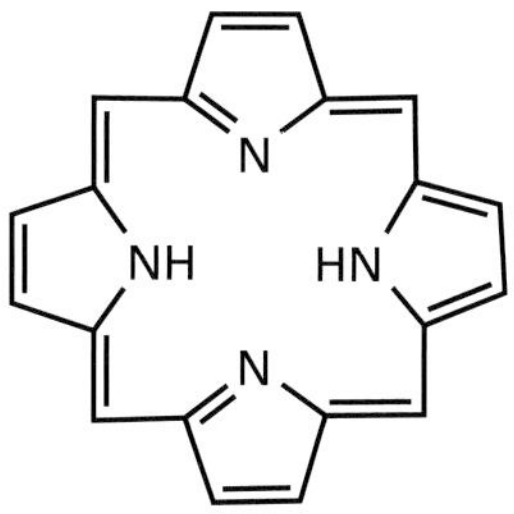

2 Porphine (free base form)

structural basis of the haem group and the chlorophylls. We may treat the group as a circular ring of radius 440 pm, with 22 electrons in the conjugated system moving along the perimeter of the ring. As in *Illustration 9.1*, we assume that in the ground state of the molecule each state is occupied by two electrons. (a) Calculate the energy and angular momentum of an electron in the highest occupied level. (b) Calculate the frequency of radiation that can induce a transition between the highest occupied and lowest unoccupied levels.

9.36 When in Chapter 19 we come to study macromolecules, such as synthetic polymers, proteins, and nucleic acids, we shall see that one conformation is that of a random coil. For a one-dimensional random coil of N units, the restoring force at small displacements and at a temperature T is

$$F = -\frac{kT}{2l}\ln\left(\frac{N+n}{N-n}\right)$$

where l is the length of each monomer unit and nl is the distance between the ends of the chain (see Section 19.8). Show that for small extensions ($n \ll N$) the restoring force is proportional to n and therefore the coil undergoes harmonic oscillation with force constant kT/Nl^2. Suppose that the mass to use for the vibrating chain is its total mass Nm, where m is the mass of one monomer unit, and deduce the root mean square separation of the ends of the chain due to quantum fluctuations in its vibrational ground state.

9.37 The forces measured by AFM arise primarily from interactions between electrons of the stylus and on the surface. To get an idea of the magnitudes of these forces, calculate the force acting between two electrons separated by 2.0 nm. *Hints*. The Coulombic potential energy of a charge q_1 at a distance r from another charge q_2 is

$$V = \frac{q_1 q_2}{4\pi\varepsilon_0 r}$$

where $\varepsilon_0 = 8.854 \times 10^{-12}\ \mathrm{C^2\,J^{-1}\,m^{-1}}$ is the vacuum permittivity. To calculate the force between the electrons, note that $F = -\mathrm{d}V/\mathrm{d}r$.

9.38 Here we explore further the idea introduced in *Impact I9.2* that quantum mechanical effects need to be invoked in the description of the electronic properties of metallic nanocrystals, here modelled as three-dimensional boxes. (a) Set up the Schrödinger equation for a particle of mass m in a three-dimensional rectangular box with sides L_1, L_2, and L_3. Show that the Schrödinger equation is separable. (b) Show that the wavefunction and the energy are defined by three quantum numbers. (c) Specialize the result from part (b) to an electron moving in a cubic box of side $L = 5$ nm and draw an energy diagram resembling Fig. 9.2 and showing the first 15 energy levels. Note that each energy level may consist of degenerate energy states. (d) Compare the energy level diagram from part (c) with the energy level diagram for an electron in a one-dimensional box of length $L = 5$ nm. Are the energy levels become more or less sparsely distributed in the cubic box than in the one-dimensional box?

9.39 We remarked in *Impact I9.2* that the particle *in* a sphere is a reasonable starting point for the discussion of the electronic properties of spherical metal nanoparticles. Here, we justify eqn 9.54, which shows that the energy of an electron in a sphere is quantized. (a) The Hamiltonian for a particle free to move inside a sphere of radius R is

$$\hat{H} = -\frac{\hbar}{2m}\nabla^2$$

Show that the Schrödinger equation is separable into radial and angular components. That is, begin by writing $\psi(r,\theta,\phi) = X(r)Y(\theta,\phi)$, where $X(r)$ depends only on the distance of the particle away from the centre of the sphere, and $Y(\theta,\phi)$ is a spherical harmonic. Then show that the Schrödinger equation can be separated into two equations, one for X, the radial equation, and the other for Y, the angular equation:

$$-\frac{\hbar^2}{2m}\left(\frac{\mathrm{d}^2X(r)}{\mathrm{d}r^2} + \frac{2}{r}\frac{\mathrm{d}X(r)}{\mathrm{d}r}\right) + \frac{l(l+1)\hbar^2}{2mr^2}X(r) = EX(r)$$

$$\Lambda^2 Y = -l(l+1)Y$$

You may wish to consult *Further information 10.1* for additional help.
(b) Consider the case $l = 0$. Show by differentiation that the solution of the radial equation has the form

$$X(r) = (2\pi R)^{-1/2}\frac{\sin(n\pi r/R)}{r}$$

(c) Now go on to show that the allowed energies are given by:

$$E_n = \frac{n^2h^2}{8mR^2}$$

This result for the energy (which is eqn 9.54 after substituting m_e for m) also applies when $l \neq 0$.

10 Atomic structure and atomic spectra

We now use the principles of quantum mechanics introduced in the preceding two chapters to describe the internal structures of atoms. We see what experimental information is available from a study of the spectrum of atomic hydrogen. Then we set up the Schrödinger equation for an electron in an atom and separate it into angular and radial parts. The wavefunctions obtained are the 'atomic orbitals' of hydrogenic atoms. Next, we use these hydrogenic atomic orbitals to describe the structures of many-electron atoms. In conjunction with the Pauli exclusion principle, we account for the periodicity of atomic properties. The spectra of many-electron atoms are more complicated than those of hydrogen, but the same principles apply. We see in the closing sections of the chapter how such spectra are described by using term symbols, and the origin of the finer details of their appearance.

In this chapter we see how to use quantum mechanics to describe the **electronic structure** of an atom, the arrangement of electrons around a nucleus. The concepts we meet are of central importance for understanding the structures and reactions of atoms and molecules, and hence have extensive chemical applications. We need to distinguish between two types of atoms. A **hydrogenic atom** is a one-electron atom or ion of general atomic number Z; examples of hydrogenic atoms are H, He^+, Li^{2+}, O^{7+}, and even U^{91+}. A **many-electron atom** (or *polyelectronic atom*) is an atom or ion with more than one electron; examples include all neutral atoms other than H. So even He, with only two electrons, is a many-electron atom. Hydrogenic atoms are important because their Schrödinger equations can be solved exactly. They also provide a set of concepts that are used to describe the structures of many-electron atoms and, as we shall see in the next chapter, the structures of molecules too.

The structure and spectra of hydrogenic atoms

When an electric discharge is passed through gaseous hydrogen, the H_2 molecules are dissociated and the energetically excited H atoms that are produced emit light of discrete frequencies, producting a spectrum of a series of 'lines' (Fig. 10.1). The Swedish spectroscopist Johannes Rydberg noted (in 1890) that all of them are described by the expression

$$\tilde{\nu} = R_H\left(\frac{1}{n_1^2} - \frac{1}{n_2^2}\right) \qquad R_H = 109\,677\ \text{cm}^{-1} \tag{10.1}$$

with $n_1 = 1$ (the *Lyman series*), 2 (the *Balmer series*), and 3 (the *Paschen series*), and that in each case $n_2 = n_1 + 1, n_1 + 2, \ldots$. The constant R_H is now called the **Rydberg constant** for the hydrogen atom.

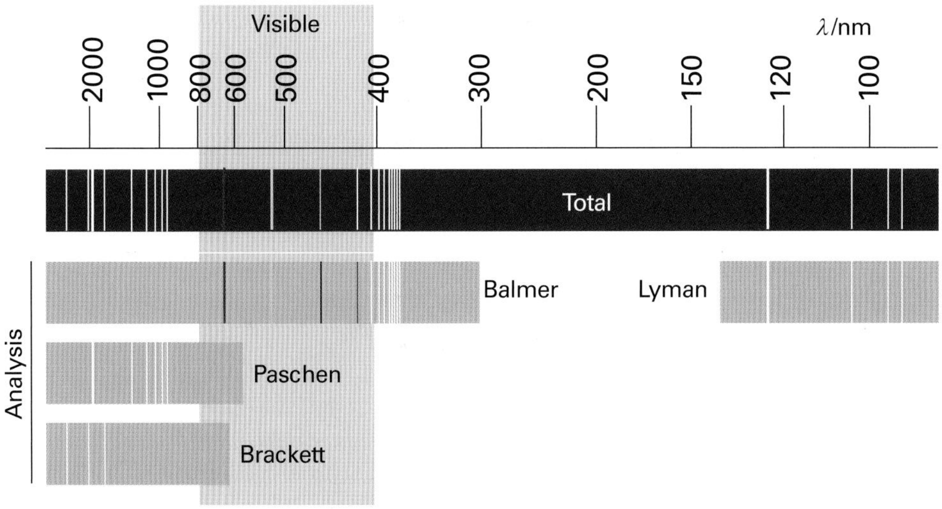

Fig. 10.1 The spectrum of atomic hydrogen. Both the observed spectrum and its resolution into overlapping series are shown. Note that the Balmer series lies in the visible region.

Self-test 10.1 Calculate the shortest wavelength line in the Paschen series.
[821 nm]

The form of eqn 10.1 strongly suggests that the wavenumber of each spectral line can be written as the difference of two **terms**, each of the form

$$T_n = \frac{R_H}{n^2} \tag{10.2}$$

The **Ritz combination principle** states that *the wavenumber of any spectral line is the difference between two terms*. We say that two terms T_1 and T_2 'combine' to produce a spectral line of wavenumber

$$\tilde{\nu} = T_1 - T_2 \tag{10.3}$$

Thus, if each spectroscopic term represents an energy hcT, the difference in energy when the atom undergoes a transition between two terms is $\Delta E = hcT_1 - hcT_2$ and, according to the Bohr frequency conditions (Section 8.1d), the frequency of the radiation emitted is given by $\nu = cT_1 - cT_2$. This expression rearranges into the Ritz formula when expressed in terms of wavenumbers (on division by c). The Ritz combination principle applies to all types of atoms and molecules, but only for hydrogenic atoms do the terms have the simple form (constant)/n^2.

Because spectroscopic observations show that electromagnetic radiation is absorbed and emitted by atoms only at certain wavenumbers, it follows that only certain energy states of atoms are permitted. Our tasks in the first part of this chapter are to determine the origin of this energy quantization, to find the permitted energy levels, and to account for the value of R_H.

10.1 The structure of hydrogenic atoms

The Coulomb potential energy of an electron in a hydrogenic atom of atomic number Z (and nuclear charge Ze) is

$$V = -\frac{Ze^2}{4\pi\varepsilon_0 r} \tag{10.4}$$

where r is the distance of the electron from the nucleus and ε_0 is the vacuum permittivity. The hamiltonian for the electron and a nucleus of mass m_N is therefore

$$\begin{aligned}\hat{H} &= \hat{E}_{K,\text{electron}} + \hat{E}_{K,\text{nucleus}} + \hat{V} \\ &= -\frac{\hbar^2}{2m_e}\nabla_e^2 - \frac{\hbar^2}{2m_N}\nabla_N^2 - \frac{Ze^2}{4\pi\varepsilon_0 r}\end{aligned} \tag{10.5}$$

The subscripts on ∇^2 indicate differentiation with respect to the electron or nuclear coordinates.

(a) The separation of variables

Physical intuition suggests that the full Schrödinger equation ought to separate into two equations, one for the motion of the atom as a whole through space and the other for the motion of the electron relative to the nucleus. We show in *Further information 10.1* how this separation is achieved, and that the Schrödinger equation for the internal motion of the electron relative to the nucleus is

$$-\frac{\hbar^2}{2\mu}\nabla^2\psi - \frac{Ze^2}{4\pi\varepsilon_0 r}\psi = E\psi \qquad \frac{1}{\mu} = \frac{1}{m_e} + \frac{1}{m_N} \tag{10.6}$$

where differentiation is now with respect to the coordinates of the electron relative to the nucleus. The quantity μ is called the **reduced mass**. The reduced mass is very similar to the electron mass because m_N, the mass of the nucleus, is much larger than the mass of an electron, so $1/\mu \approx 1/m_e$. In all except the most precise work, the reduced mass can be replaced by m_e.

Because the potential energy is centrosymmetric (independent of angle), we can suspect that the equation is separable into radial and angular components. Therefore, we write

$$\psi(r,\theta,\phi) = R(r)Y(\theta,\phi) \tag{10.7}$$

and examine whether the Schrödinger equation can be separated into two equations, one for R and the other for Y. As shown in *Further information 10.1*, the equation does separate, and the equations we have to solve are

$$\Lambda^2 Y = -l(l+1)Y \tag{10.8}$$

$$-\frac{\hbar^2}{2\mu}\frac{d^2u}{dr^2} + V_{\text{eff}}u = Eu \qquad u = rR \tag{10.9}$$

where

$$V_{\text{eff}} = -\frac{Ze^2}{4\pi\varepsilon_0 r} + \frac{l(l+1)\hbar^2}{2\mu r^2} \tag{10.10}$$

Equation 10.8 is the same as the Schrödinger equation for a particle free to move round a central point, and we considered it in Section 9.7. The solutions are the spherical harmonics (Table 9.3), and are specified by the quantum numbers l and m_l. We consider them in more detail shortly. Equation 10.9 is called the **radial wave equation**. The radial wave equation is the description of the motion of a particle of mass μ in a one-dimensional region $0 \le r < \infty$ where the potential energy is V_{eff}.

(b) The radial solutions

We can anticipate some features of the shapes of the radial wavefunctions by analysing the form of V_{eff}. The first term in eqn 10.10 is the Coulomb potential energy of the electron in the field of the nucleus. The second term stems from what in classical physics would be called the centrifugal force that arises from the angular momentum of the electron around the nucleus. When $l = 0$, the electron has no angular momentum, and the effective potential energy is purely Coulombic and attractive at all radii (Fig. 10.2). When $l \neq 0$, the centrifugal term gives a positive (repulsive) contribution to the effective potential energy. When the electron is close to the nucleus ($r \approx 0$), this repulsive term, which is proportional to $1/r^2$, dominates the attractive Coulombic component, which is proportional to $1/r$, and the net effect is an effective repulsion of the electron from the nucleus. The two effective potential energies, the one for $l = 0$ and the one for $l \neq 0$, are qualitatively very different close to the nucleus. However, they are similar at large distances because the centrifugal contribution tends to zero more rapidly (as $1/r^2$) than the Coulombic contribution (as $1/r$). Therefore, we can expect the solutions with $l = 0$ and $l \neq 0$ to be quite different near the nucleus but similar far away from it. We show in the *Justification* below that close to the nucleus the radial wavefunction is proportional to r^l, and the higher the orbital angular momentum, the less likely the electron is to be found (Fig. 10.3). We also show that far from the nucleus all wavefunctions approach zero exponentially.

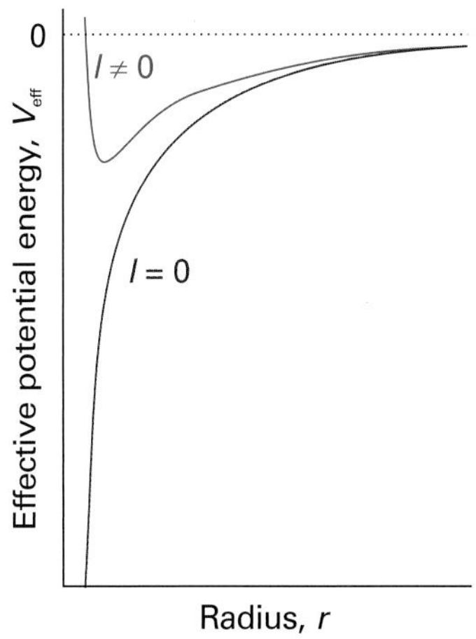

Fig. 10.2 The effective potential energy of an electron in the hydrogen atom. When the electron has zero orbital angular momentum, the effective potential energy is the Coulombic potential energy. When the electron has nonzero orbital angular momentum, the centrifugal effect gives rise to a positive contribution that is very large close to the nucleus. We can expect the $l = 0$ and $l \neq 0$ wavefunctions to be very different near the nucleus.

Exploration Plot the effective potential energy against r for several nonzero values of the orbital angular momentum l. How does the location of the minimum in the effective potential energy vary with l?

Justification 10.1 *The shape of the radial wavefunction*

When r is very small (close to the nucleus), $u \approx 0$, so the right-hand side of eqn 10.9 is zero; we can also ignore all but the largest terms (those depending on $1/r^2$) in eqn 10.9 and write

$$-\frac{d^2u}{dr^2} + \frac{l(l+1)}{r^2}u \approx 0$$

The solution of this equation (for $r \approx 0$) is

$$u \approx Ar^{l+1} + \frac{B}{r^l}$$

Because $R = u/r$, and R cannot be infinite at $r = 0$, we must set $B = 0$, and hence obtain $R \approx Ar^l$.

Far from the nucleus, when r is very large, we can ignore all terms in $1/r$ and eqn 10.9 becomes

$$-\frac{\hbar^2}{2\mu}\frac{d^2u}{dr^2} \simeq Eu$$

where $\simeq$ means 'asymptotically equal to'. Because

$$\frac{d^2u}{dr^2} = \frac{d^2(rR)}{dr^2} = r\frac{d^2R}{dr^2} + 2\frac{dR}{dr} \simeq r\frac{d^2R}{dr^2}$$

this equation has the form

$$-\frac{\hbar^2}{2\mu}\frac{d^2R}{dr^2} \simeq ER$$

The acceptable (finite) solution of this equation (for r large) is

$$R \simeq e^{-(2\mu|E|/\hbar^2)r}$$

and the wavefunction decays exponentially towards zero as r increases.

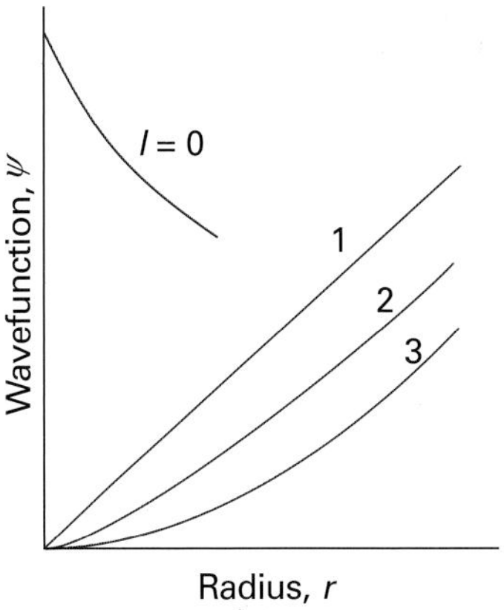

Fig. 10.3 Close to the nucleus, orbitals with $l = 1$ are proportional to r, orbitals with $l = 2$ are proportional to r^2, and orbitals with $l = 3$ are proportional to r^3. Electrons are progressively excluded from the neighbourhood of the nucleus as l increases. An orbital with $l = 0$ has a finite, nonzero value at the nucleus.

We shall not go through the technical steps of solving the radial equation for the full range of radii, and see how the form r^l close to the nucleus blends into the exponentially decaying form at great distances (see *Further reading*). It is sufficient to know that the two limits can be matched only for integral values of a quantum number n, and that the allowed energies corresponding to the allowed solutions are

$$E_n = -\frac{Z^2\mu e^4}{32\pi^2\varepsilon_0^2\hbar^2 n^2} \tag{10.11}$$

with $n = 1, 2, \ldots$ Likewise, the radial wavefunctions depend on the values of both n and l (but not on m_l), and all of them have the form

$$R(r) = (\text{polynomial in } r) \times (\text{decaying exponential in } r) \tag{10.12}$$

These functions are most simply written in terms of the dimensionless quantity ρ (rho), where

$$\rho = \frac{2Zr}{na_0} \qquad a_0 = \frac{4\pi\varepsilon_0\hbar^2}{m_e e^2} \tag{10.13}$$

The **Bohr radius**, a_0, has the value 52.9 pm; it is so called because the same quantity appeared in Bohr's early model of the hydrogen atom as the radius of the electron orbit of lowest energy. Specifically, the radial wavefunctions for an electron with quantum numbers n and l are the (real) functions

$$R_{n,l}(r) = N_{n,l}\rho^l L_{n+1}^{2l+1}(\rho)e^{-\rho/2} \tag{10.14}$$

where L is a polynomial in ρ called an *associated Laguerre polynomial*: it links the $r \approx 0$ solutions on its left (corresponding to $R \propto \rho^l$) to the exponentially decaying function on its right. The notation might look fearsome, but the polynomials have quite simple forms, such as 1, ρ, and $2 - \rho$ (they can be picked out in Table 10.1). Specifically, we can interpret the components of this expression as follows:

1 The exponential factor ensures that the wavefunction approaches zero far from the nucleus.

Table 10.1 Hydrogenic radial wavefunctions

Orbital	n	l	$R_{n,l}$
1s	1	0	$2\left(\frac{Z}{a}\right)^{3/2} e^{-\rho/2}$
2s	2	0	$\frac{1}{8^{1/2}}\left(\frac{Z}{a}\right)^{3/2}(2-\rho)e^{-\rho/2}$
2p	2	1	$\frac{1}{24^{1/2}}\left(\frac{Z}{a}\right)^{3/2}\rho e^{-\rho/2}$
3s	3	0	$\frac{1}{243^{1/2}}\left(\frac{Z}{a}\right)^{3/2}(6-6\rho+\rho^2)e^{-\rho/2}$
3p	3	1	$\frac{1}{486^{1/2}}\left(\frac{Z}{a}\right)^{3/2}(4-\rho)\rho e^{-\rho/2}$
3d	3	2	$\frac{1}{2430^{1/2}}\left(\frac{Z}{a}\right)^{3/2}\rho^2 e^{-\rho/2}$

$\rho = (2Z/na)r$ with $a = 4\pi\varepsilon_0\hbar^2/\mu e^2$. For an infinitely heavy nucleus (or one that may be assumed to be so), $\mu = m_e$ and $a = a_0$, the Bohr radius. The full wavefunction is obtained by multiplying R by the appropriate Y given in Table 9.3.

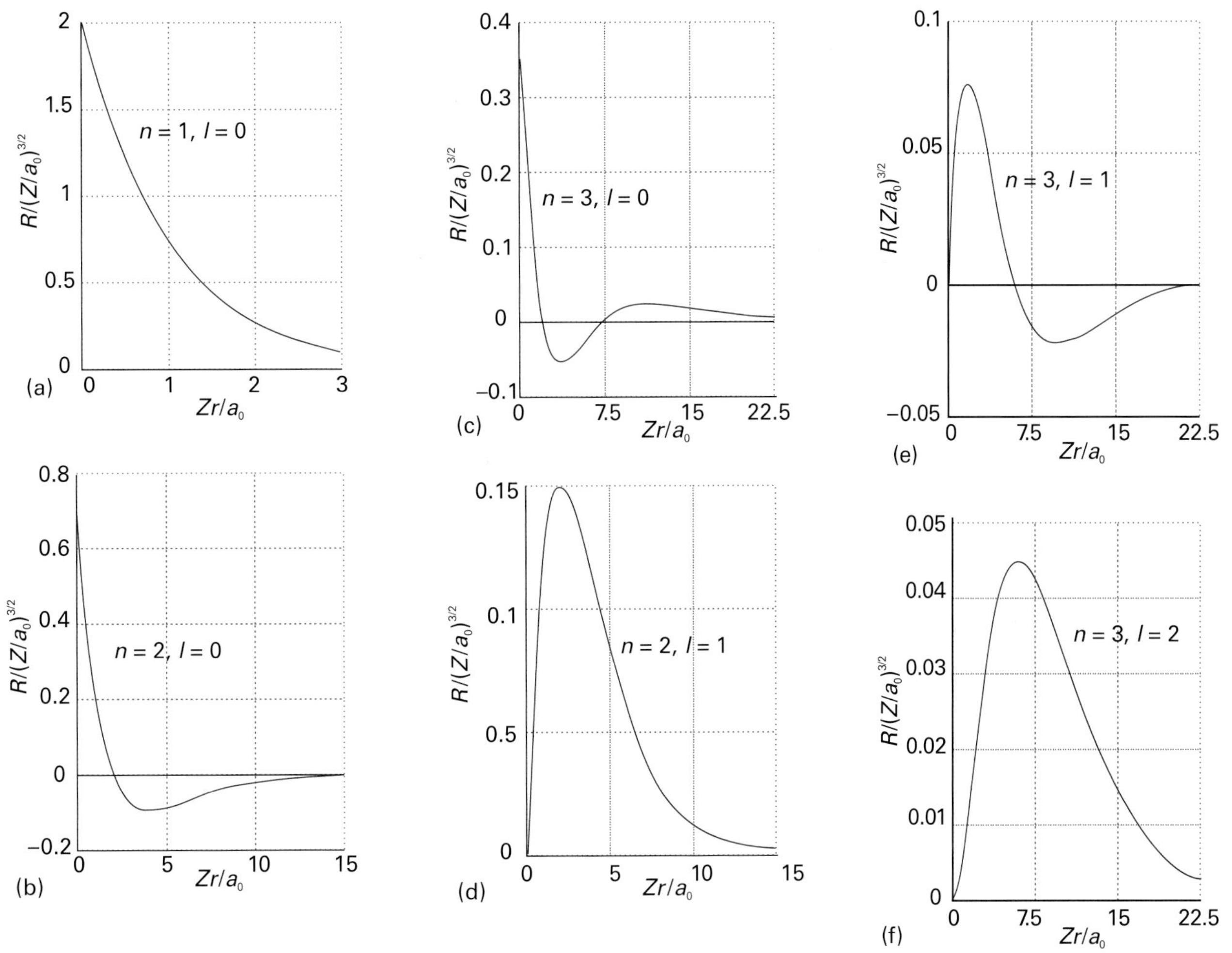

Fig. 10.4 The radial wavefunctions of the first few states of hydrogenic atoms of atomic number Z. Note that the orbitals with $l = 0$ have a nonzero and finite value at the nucleus. The horizontal scales are different in each case: orbitals with high principal quantum numbers are relatively distant from the nucleus.

Exploration Use mathematical software to find the locations of the radial nodes in hydrogenic wavefunctions with n up to 3.

2 The factor ρ^l ensures that (provided $l > 0$) the wavefunction vanishes at the nucleus.

3 The associated Laguerre polynomial is a function that oscillates from positive to negative values and accounts for the presence of radial nodes.

Expressions for some radial wavefunctions are given in Table 10.1 and illustrated in Fig. 10.4.

Comment 10.1

The zero at $r = 0$ is not a radial node because the radial wavefunction does not pass through zero at that point (because r cannot be negative). Nodes at the nucleus are all angular nodes.

Illustration 10.1 *Calculating a probability density*

To calculate the probability density at the nucleus for an electron with $n = 1, l = 0$, and $m_l = 0$, we evaluate ψ at $r = 0$:

$$\psi_{1,0,0}(0,\theta,\phi) = R_{1,0}(0)Y_{0,0}(\theta,\phi) = 2\left(\frac{Z}{a_0}\right)^{3/2}\left(\frac{1}{4\pi}\right)^{1/2}$$

The probability density is therefore

$$\psi_{1,0,0}(0,\theta,\phi)^2 = \frac{Z^3}{\pi a_0^3}$$

which evaluates to 2.15×10^{-6} pm^{-3} when $Z = 1$.

Self-test 10.2 Evaluate the probability density at the nucleus of the electron for an electron with $n = 2$, $l = 0$, $m_l = 0$. $[(Z/a_0)^3/8\pi]$

10.2 Atomic orbitals and their energies

An **atomic orbital** is a one-electron wavefunction for an electron in an atom. Each hydrogenic atomic orbital is defined by three quantum numbers, designated n, l, and m_l. When an electron is described by one of these wavefunctions, we say that it 'occupies' that orbital. We could go on to say that the electron is in the state $|n,l,m_l\rangle$. For instance, an electron described by the wavefunction $\psi_{1,0,0}$ and in the state $|1,0,0\rangle$ is said to occupy the orbital with $n = 1$, $l = 0$, and $m_l = 0$.

The quantum number n is called the **principal quantum number**; it can take the values $n = 1, 2, 3, \ldots$ and determines the energy of the electron:

An electron in an orbital with quantum number n has an energy given by eqn 10.11.

The two other quantum numbers, l and m_l, come from the angular solutions, and specify the angular momentum of the electron around the nucleus:

An electron in an orbital with quantum number l has an angular momentum of magnitude $\{l(l+1)\}^{1/2}\hbar$, with $l = 0, 1, 2, \ldots, n-1$.

An electron in an orbital with quantum number m_l has a z-component of angular momentum $m_l\hbar$, with $m_l = 0, \pm 1, \pm 2, \ldots, \pm l$.

Note how the value of the principal quantum number, n, controls the maximum value of l and l controls the range of values of m_l.

To define the state of an electron in a hydrogenic atom fully we need to specify not only the orbital it occupies but also its spin state. We saw in Section 9.8 that an electron possesses an intrinsic angular momentum that is described by the two quantum numbers s and m_s (the analogues of l and m_l). The value of s is fixed at $\frac{1}{2}$ for an electron, so we do not need to consider it further at this stage. However, m_s may be either $+\frac{1}{2}$ or $-\frac{1}{2}$, and to specify the electron's state in a hydrogenic atom we need to specify which of these values describes it. It follows that, to specify the state of an electron in a hydrogenic atom, we need to give the values of four quantum numbers, namely n, l, m_l, and m_s.

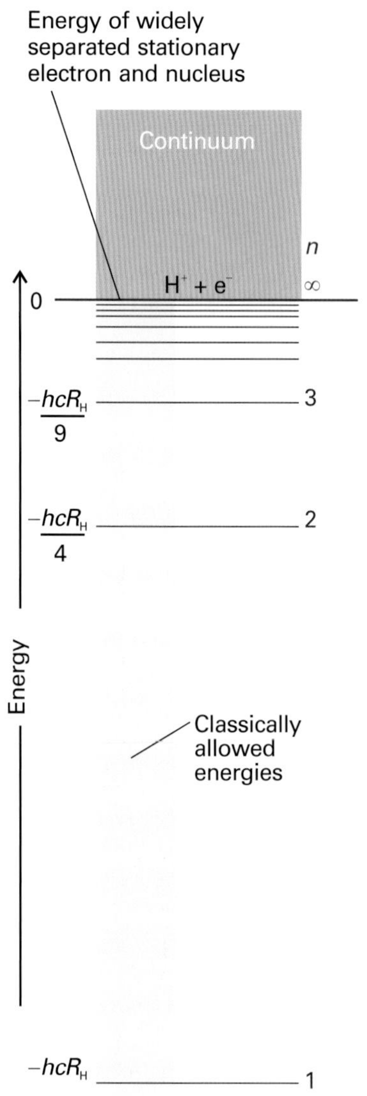

Fig. 10.5 The energy levels of a hydrogen atom. The values are relative to an infinitely separated, stationary electron and a proton.

(a) The energy levels

The energy levels predicted by eqn 10.11 are depicted in Fig. 10.5. The energies, and also the separation of neighbouring levels, are proportional to Z^2, so the levels are four times as wide apart (and the ground state four times deeper in energy) in He^+ ($Z = 2$) than in H ($Z = 1$). All the energies given by eqn 10.11 are negative. They refer to the **bound states** of the atom, in which the energy of the atom is lower than that of the infinitely separated, stationary electron and nucleus (which corresponds to the zero of energy). There are also solutions of the Schrödinger equation with positive energies. These solutions correspond to **unbound states** of the electron, the states to which an electron is raised when it is ejected from the atom by a high-energy collision or photon. The energies of the unbound electron are not quantized and form the continuum states of the atom.

Equation 10.11 is consistent with the spectroscopic result summarized by eqn 10.1, and we can identify the Rydberg constant for hydrogen ($Z = 1$) as

$$hcR_{\mathrm{H}} = \frac{\mu_{\mathrm{H}}e^4}{32\pi^2\varepsilon_0^2\hbar^2} \qquad [10.15]$$

where μ_{H} is the reduced mass for hydrogen. The **Rydberg constant** itself, R, is defined by the same expression except for the replacement of μ_{H} by the mass of an electron, m_{e}, corresponding to a nucleus of infinite mass:

$$R_{\mathrm{H}} = \frac{\mu_{\mathrm{H}}}{m_{\mathrm{e}}}R \qquad R = \frac{m_{\mathrm{e}}e^4}{8\varepsilon_0^2h^3c} \qquad [10.16]$$

Insertion of the values of the fundamental constants into the expression for R_{H} gives almost exact agreement with the experimental value. The only discrepancies arise from the neglect of relativistic corrections (in simple terms, the increase of mass with speed), which the non-relativistic Schrödinger equation ignores.

Comment 10.2

The particle in a finite well, discussed in Section 9.3, is a primitive but useful model that gives insight into the bound and unbound states of the electron in a hydrogenic atom. Figure 9.15 shows that the energies of a particle (for example, an electron in a hydrogenic atom) are quantized when its total energy, E, is lower than its potential energy, V (the Coulomb interaction energy between the electron and the nucleus). When $E > V$, the particle can escape from the well (the atom is ionized) and its energies are no longer quantized, forming a continuum.

(b) Ionization energies

The **ionization energy**, I, of an element is the minimum energy required to remove an electron from the ground state, the state of lowest energy, of one of its atoms. Because the ground state of hydrogen is the state with $n = 1$, with energy $E_1 = -hcR_{\mathrm{H}}$ and the atom is ionized when the electron has been excited to the level corresponding to $n = \infty$ (see Fig. 10.5), the energy that must be supplied is

$$I = hcR_{\mathrm{H}} \qquad (10.17)$$

The value of I is 2.179 aJ (a, for atto, is the prefix that denotes 10^{-18}), which corresponds to 13.60 eV.

Example 10.1 *Measuring an ionization energy spectroscopically*

The emission spectrum of atomic hydrogen shows lines at 82 259, 97 492, 102 824, 105 292, 106 632, and 107 440 cm^{-1}, which correspond to transitions to the same lower state. Determine (a) the ionization energy of the lower state, (b) the value of the Rydberg constant.

Method The spectroscopic determination of ionization energies depends on the determination of the series limit, the wavenumber at which the series terminates and becomes a continuum. If the upper state lies at an energy $-hcR_{\mathrm{H}}/n^2$, then, when the atom makes a transition to E_{lower}, a photon of wavenumber

$$\tilde{\nu} = -\frac{R_{\mathrm{H}}}{n^2} - \frac{E_{\text{lower}}}{hc}$$

is emitted. However, because $I = -E_{\text{lower}}$, it follows that

$$\tilde{\nu} = \frac{I}{hc} - \frac{R_{\mathrm{H}}}{n^2}$$

A plot of the wavenumbers against $1/n^2$ should give a straight line of slope $-R_{\mathrm{H}}$ and intercept I/hc. Use a computer to make a least-squares fit of the data to get a result that reflects the precision of the data.

Answer The wavenumbers are plotted against $1/n^2$ in Fig. 10.6. The (least-squares) intercept lies at 109 679 cm^{-1}, so the ionization energy is 2.1788 aJ (1312.1 kJ mol^{-1}).

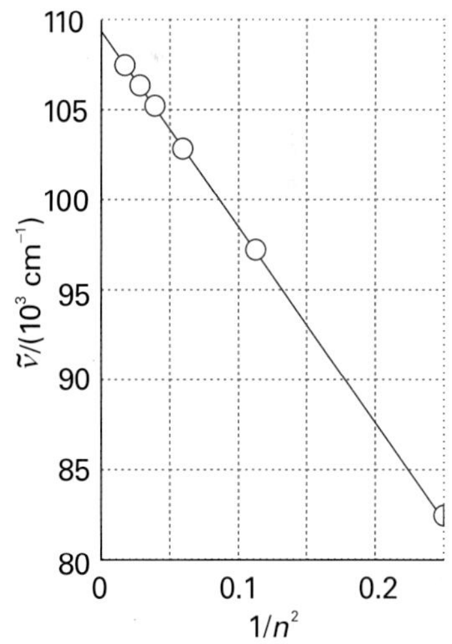

Fig. 10.6 The plot of the data in Example 10.1 used to determine the ionization energy of an atom (in this case, of H).

Exploration The initial value of n was not specified in Example 10.1. Show that the correct value can be determined by making several choices and selecting the one that leads to a straight line.

The slope is, in this instance, numerically the same, so $R_H = 109\,679\ cm^{-1}$. A similar extrapolation procedure can be used for many-electron atoms (see Section 10.5).

Self-test 10.3 The emission spectrum of atomic deuterium shows lines at 15 238, 20 571, 23 039, and 24 380 cm^{-1}, which correspond to transitions to the same lower state. Determine (a) the ionization energy of the lower state, (b) the ionization energy of the ground state, (c) the mass of the deuteron (by expressing the Rydberg constant in terms of the reduced mass of the electron and the deuteron, and solving for the mass of the deuteron).

[(a) 328.1 kJ mol^{-1}, (b) 1312.4 kJ mol^{-1}, (c) 2.8×10^{-27} kg, a result very sensitive to R_D]

(c) Shells and subshells

All the orbitals of a given value of n are said to form a single **shell** of the atom. In a hydrogenic atom, all orbitals of given n, and therefore belonging to the same shell, have the same energy. It is common to refer to successive shells by letters:

$n =$	1	2	3	4 . . .
	K	L	M	N . . .

Thus, all the orbitals of the shell with $n = 2$ form the L shell of the atom, and so on.

The orbitals with the same value of n but different values of l are said to form a **subshell** of a given shell. These subshells are generally referred to by letters:

$l =$	0	1	2	3	4	5	6 . . .
	s	p	d	f	g	h	i . . .

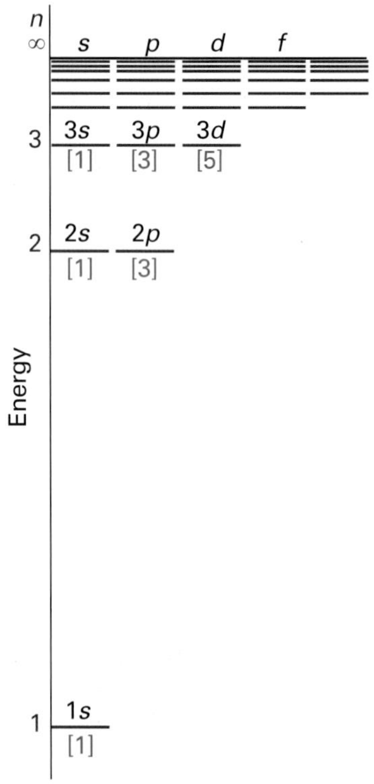

Fig. 10.7 The energy levels of the hydrogen atom showing the subshells and (in square brackets) the numbers of orbitals in each subshell. In hydrogenic atoms, all orbitals of a given shell have the same energy.

The letters then run alphabetically (j is not used). Figure 10.7 is a version of Fig. 10.5 which shows the subshells explicitly. Because l can range from 0 to $n - 1$, giving n values in all, it follows that there are n subshells of a shell with principal quantum number n. Thus, when $n = 1$, there is only one subshell, the one with $l = 0$. When $n = 2$, there are two subshells, the 2s subshell (with $l = 0$) and the 2p subshell (with $l = 1$).

When $n = 1$ there is only one subshell, that with $l = 0$, and that subshell contains only one orbital, with $m_l = 0$ (the only value of m_l permitted). When $n = 2$, there are four orbitals, one in the s subshell with $l = 0$ and $m_l = 0$, and three in the $l = 1$ subshell with $m_l = +1, 0, -1$. When $n = 3$ there are nine orbitals (one with $l = 0$, three with $l = 1$, and five with $l = 2$). The organization of orbitals in the shells is summarized in Fig. 10.8. In general, the number of orbitals in a shell of principal quantum number n is n^2, so in a hydrogenic atom each energy level is n^2-fold degenerate.

(d) Atomic orbitals

The orbital occupied in the ground state is the one with $n = 1$ (and therefore with $l = 0$ and $m_l = 0$, the only possible values of these quantum numbers when $n = 1$). From Table 10.1 we can write (for $Z = 1$):

$$\psi = \frac{1}{(\pi a_0^3)^{1/2}} e^{-r/a_0} \tag{10.18}$$

This wavefunction is independent of angle and has the same value at all points of constant radius; that is, the 1s orbital is *spherically symmetrical.* The wavefunction decays exponentially from a maximum value of $1/(\pi a_0^3)^{1/2}$ at the nucleus (at $r = 0$). It follows that the most probable point at which the electron will be found is at the nucleus itself.

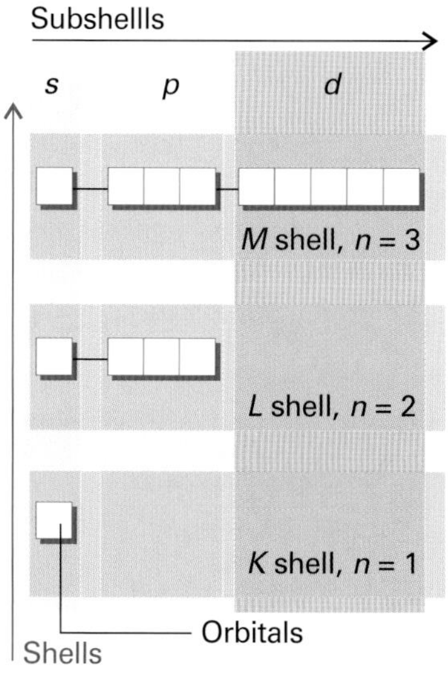

Fig. 10.8 The organization of orbitals (white squares) into subshells (characterized by l) and shells (characterized by n).

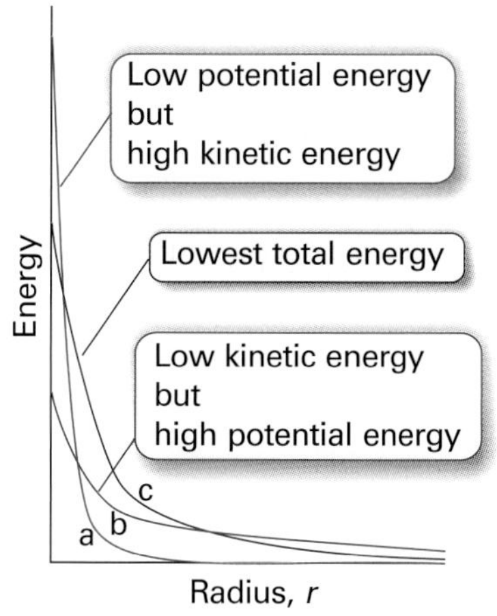

Fig. 10.9 The balance of kinetic and potential energies that accounts for the structure of the ground state of hydrogen (and similar atoms). (a) The sharply curved but localized orbital has high mean kinetic energy, but low mean potential energy; (b) the mean kinetic energy is low, but the potential energy is not very favourable; (c) the compromise of moderate kinetic energy and moderately favourable potential energy.

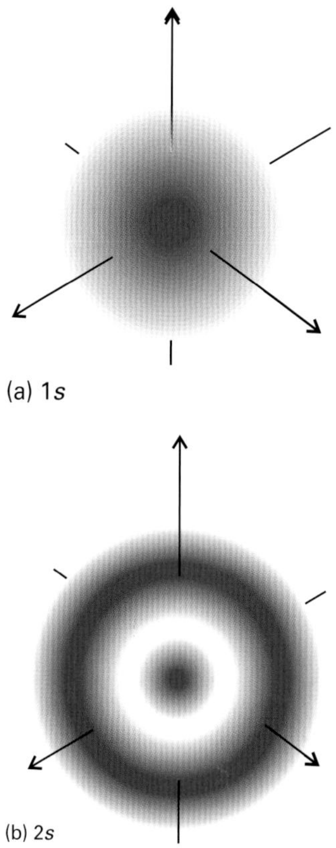

Fig. 10.10 Representations of the 1s and 2s hydrogenic atomic orbitals in terms of their electron densities (as represented by the density of shading)

We can understand the general form of the ground-state wavefunction by considering the contributions of the potential and kinetic energies to the total energy of the atom. The closer the electron is to the nucleus on average, the lower its average potential energy. This dependence suggests that the lowest potential energy should be obtained with a sharply peaked wavefunction that has a large amplitude at the nucleus and is zero everywhere else (Fig. 10.9). However, this shape implies a high kinetic energy, because such a wavefunction has a very high average curvature. The electron would have very low kinetic energy if its wavefunction had only a very low average curvature. However, such a wavefunction spreads to great distances from the nucleus and the average potential energy of the electron will be correspondingly high. The actual ground-state wavefunction is a compromise between these two extremes: the wavefunction spreads away from the nucleus (so the expectation value of the potential energy is not as low as in the first example, but nor is it very high) and has a reasonably low average curvature (so the expectation of the kinetic energy is not very low, but nor is it as high as in the first example).

The energies of ns orbitals increase (become less negative; the electron becomes less tightly bound) as n increases because the average distance of the electron from the nucleus increases. By the virial theorem with $b=-1$ (eqn 9.35), $\langle E_K\rangle = -\frac{1}{2}\langle V\rangle$ so, even though the average kinetic energy decreases as n increases, the total energy is equal to $\frac{1}{2}\langle V\rangle$, which becomes less negative as n increases.

One way of depicting the probability density of the electron is to represent $|\psi|^2$ by the density of shading (Fig. 10.10). A simpler procedure is to show only the **boundary surface**, the surface that captures a high proportion (typically about 90 per cent) of the electron probability. For the 1s orbital, the boundary surface is a sphere centred on the nucleus (Fig. 10.11).

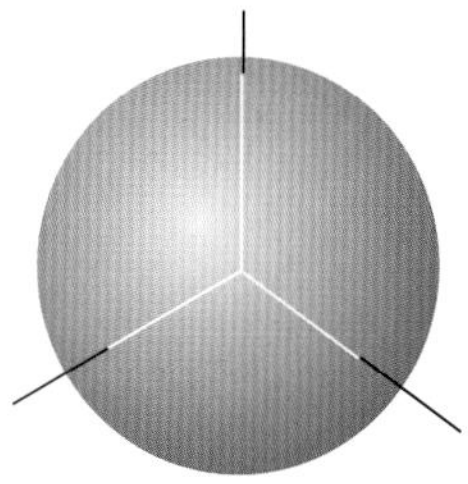

Fig. 10.11 The boundary surface of an s orbital, within which there is a 90 per cent probability of finding the electron.

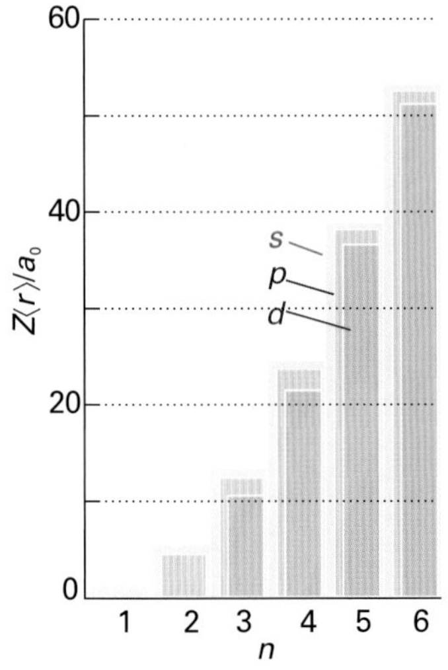

Fig. 10.12 The variation of the mean radius of a hydrogenic atom with the principal and orbital angular momentum quantum numbers. Note that the mean radius lies in the order $d < p < s$ for a given value of n.

The general expression for the mean radius of an orbital with quantum numbers l and n is

$$r_{n,l} = n^2\left\{1 + \tfrac{1}{2}\left(1 - \frac{l(l+1)}{n^2}\right)\right\}\frac{a_0}{Z} \tag{10.19}$$

The variation with n and l is shown in Fig. 10.12. Note that, for a given principal quantum number, the mean radius decreases as l increases, so the average distance of an electron from the nucleus is less when it is in a $2p$ orbital, for instance, than when it is in a $2s$ orbital.

Example 10.2 *Calculating the mean radius of an orbital*

Use hydrogenic orbitals to calculate the mean radius of a $1s$ orbital.

Method The mean radius is the expectation value

$$\langle r\rangle = \int \psi^* r\psi\, d\tau = \int r|\psi|^2 d\tau$$

We therefore need to evaluate the integral using the wavefunctions given in Table 10.1 and $d\tau = r^2 dr \sin\theta\, d\theta\, d\phi$. The angular parts of the wavefunction are normalized in the sense that

$$\int_0^{\pi}\int_0^{2\pi} |Y_{l,m_l}|^2 \sin\theta d\theta\, d\phi = 1$$

The integral over r required is given in Example 8.7.

Answer With the wavefunction written in the form $\psi = RY$, the integration is

$$\langle r\rangle = \int_0^{\infty}\int_0^{\pi}\int_0^{2\pi} rR_{n,l}^2 |Y_{l,m_l}|^2 r^2\, dr \sin\theta\, d\theta\, d\phi = \int_0^{\infty} r^3 R_{n,l}^2 dr$$

For a $1s$ orbital,

$$R_{1,0} = 2\left(\frac{Z}{a_0^3}\right)^{1/2} e^{-Zr/a_0}$$

Hence

$$\langle r\rangle = \frac{4Z}{a_0^3}\int_0^{\infty} r^3 e^{-2Zr/a_0} dr = \frac{3a_0}{2Z}$$

Self-test 10.4 Evaluate the mean radius (a) of a $3s$ orbital by integration, and (b) of a $3p$ orbital by using the general formula, eqn 10.19. [(a) $27a_0/2Z$; (b) $25a_0/2Z$]

All s-orbitals are spherically symmetric, but differ in the number of radial nodes. For example, the $1s$, $2s$, and $3s$ orbitals have 0, 1, and 2 radial nodes, respectively. In general, an ns orbital has $n - 1$ radial nodes.

Self-test 10.5 (a) Use the fact that a $2s$ orbital has radial nodes where the polynomial factor (Table 10.1) is equal to zero, and locate the radial node at $2a_0/Z$ (see Fig. 10.4). (b) Similarly, locate the two nodes of a $3s$ orbital.
[(a) $2a_0/Z$; (b)$1.90a_0/Z$ and $7.10a_0/Z$]

(e) Radial distribution functions

The wavefunction tells us, through the value of $|\psi|^2$, the probability of finding an electron in any region. We can imagine a probe with a volume $d\tau$ and sensitive to electrons, and which we can move around near the nucleus of a hydrogen atom. Because the probability density in the ground state of the atom is $|\psi|^2 \propto e^{-2Zr/a_0}$, the reading from the detector decreases exponentially as the probe is moved out along any radius but is constant if the probe is moved on a circle of constant radius (Fig. 10.13).

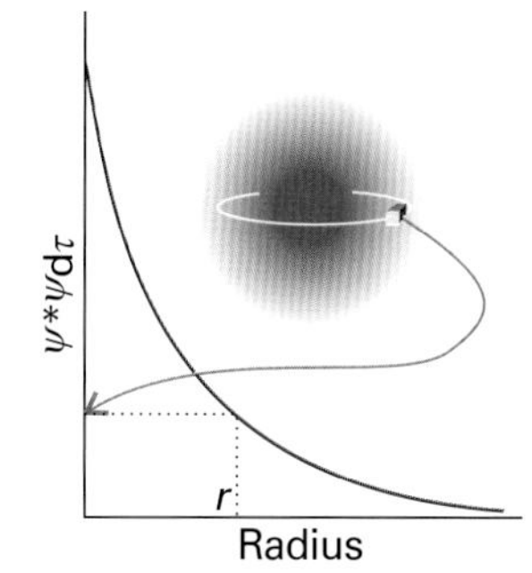

Fig. 10.13 A constant-volume electron-sensitive detector (the small cube) gives its greatest reading at the nucleus, and a smaller reading elsewhere. The same reading is obtained anywhere on a circle of given radius: the *s* orbital is spherically symmetrical.

Now consider the probability of finding the electron *anywhere* between the two walls of a spherical shell of thickness dr at a radius r. The sensitive volume of the probe is now the volume of the shell (Fig. 10.14), which is $4\pi r^2 dr$ (the product of its surface area, $4\pi r^2$, and its thickness, dr). The probability that the electron will be found between the inner and outer surfaces of this shell is the probability density at the radius r multiplied by the volume of the probe, or $|\psi|^2 \times 4\pi r^2 dr$. This expression has the form $P(r)dr$, where

$$P(r) = 4\pi r^2 \psi^2 \qquad (10.20)$$

The more general expression, which also applies to orbitals that are not spherically symmetrical, is

$$P(r) = r^2 R(r)^2 \qquad (10.21)$$

where $R(r)$ is the radial wavefunction for the orbital in question.

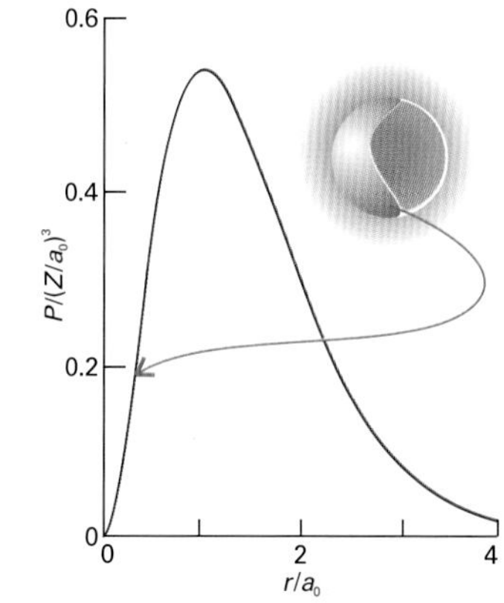

Fig. 10.14 The radial distribution function P gives the probability that the electron will be found anywhere in a shell of radius r. For a 1*s* electron in hydrogen, P is a maximum when r is equal to the Bohr radius a_0. The value of P is equivalent to the reading that a detector shaped like a spherical shell would give as its radius is varied.

Justification 10.2 *The general form of the radial distribution function*

The probability of finding an electron in a volume element $d\tau$ when its wavefunction is $\psi = RY$ is $|RY|^2 d\tau$ with $d\tau = r^2 dr \sin\theta d\theta d\phi$. The total probability of finding the electron at any angle at a constant radius is the integral of this probability over the surface of a sphere of radius r, and is written $P(r)dr$; so

$$P(r)dr = \int_0^{\pi}\int_0^{2\pi} \overbrace{R(r)^2|Y(\theta,\phi)|^2}^{\psi^2} \overbrace{r^2 dr \sin\theta d\theta d\phi}^{d\tau}$$

$$= r^2 R(r)^2 dr \overbrace{\int_0^{\pi}\int_0^{2\pi} |Y(\theta,\phi)|^2 \sin\theta d\theta d\phi}^{1} = r^2 R(r)^2 dr$$

The last equality follows from the fact that the spherical harmonics are normalized to 1 (see Example 10.2). It follows that $P(r) = r^2 R(r)^2$, as stated in the text.

The **radial distribution function**, $P(r)$, is a probability density in the sense that, when it is multiplied by dr, it gives the probability of finding the electron anywhere between the two walls of a spherical shell of thickness dr at the radius r. For a 1*s* orbital,

$$P(r) = \frac{4Z^3}{a_0^3} r^2 e^{-2Zr/a_0} \qquad (10.22)$$

Let's interpret this expression:

1 Because $r^2 = 0$ at the nucleus, at the nucleus $P(0) = 0$.

2 As $r \to \infty$, $P(r) \to 0$ on account of the exponential term.

3 The increase in r^2 and the decrease in the exponential factor means that P passes through a maximum at an intermediate radius (see Fig. 10.14).

The maximum of $P(r)$, which can be found by differentiation, marks the most probable radius at which the electron will be found, and for a 1*s* orbital in hydrogen occurs at $r = a_0$, the Bohr radius. When we carry through the same calculation for the radial distribution function of the 2*s* orbital in hydrogen, we find that the most probable radius is $5.2a_0 = 275$ pm. This larger value reflects the expansion of the atom as its energy increases.

Example 10.3 *Calculating the most probable radius*

Calculate the most probable radius, r^*, at which an electron will be found when it occupies a 1*s* orbital of a hydrogenic atom of atomic number Z, and tabulate the values for the one-electron species from H to Ne^{9+}.

Method We find the radius at which the radial distribution function of the hydrogenic 1*s* orbital has a maximum value by solving $dP/dr = 0$. If there are several maxima, then we choose the one corresponding to the greatest amplitude (the outermost one).

Answer The radial distribution function is given in eqn 10.22. It follows that

$$\frac{dP}{dr} = \frac{4Z^3}{a_0^3}\left(2r - \frac{2Zr^2}{a_0}\right)e^{-2Zr/a_0}$$

This function is zero where the term in parentheses is zero, which is at

$$r^* = \frac{a_0}{Z}$$

Then, with $a_0 = 52.9$ pm, the radial node lies at

	H	He^+	Li^{2+}	Be^{3+}	B^{4+}	C^{5+}	N^{6+}	O^{7+}	F^{8+}	Ne^{9+}
r^*/pm	52.9	26.5	17.6	13.2	10.6	8.82	7.56	6.61	5.88	5.29

Notice how the 1*s* orbital is drawn towards the nucleus as the nuclear charge increases. At uranium the most probable radius is only 0.58 pm, almost 100 times closer than for hydrogen. (On a scale where $r^* = 10$ cm for H, $r^* = 1$ mm for U.) The electron then experiences strong accelerations and relativistic effects are important.

Self-test 10.6 Find the most probable distance of a 2*s* electron from the nucleus in a hydrogenic atom. $[(3 + 5^{1/2})a_0/Z]$

(f) *p* Orbitals

The three 2*p* orbitals are distinguished by the three different values that m_l can take when $l = 1$. Because the quantum number m_l tells us the orbital angular momentum around an axis, these different values of m_l denote orbitals in which the electron has different orbital angular momenta around an arbitrary *z*-axis but the same magnitude of that momentum (because l is the same for all three). The orbital with $m_l = 0$, for instance, has zero angular momentum around the *z*-axis. Its angular variation is proportional to $\cos\theta$, so the probability density, which is proportional to $\cos^2\theta$, has its maximum value on either side of the nucleus along the *z*-axis (at $\theta = 0$ and 180°). The wavefunction of a 2*p*-orbital with $m_l = 0$ is

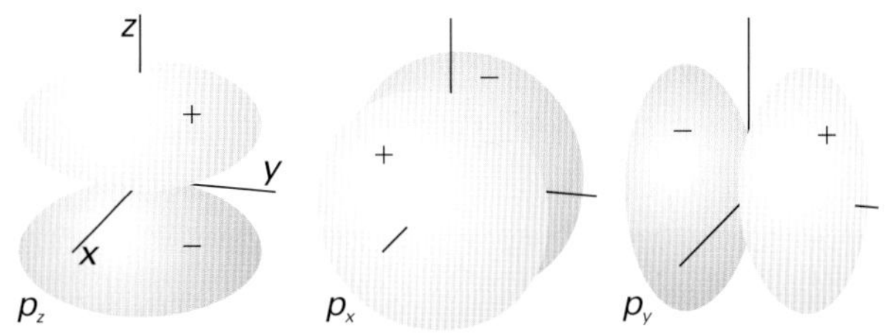

Fig. 10.15 The boundary surfaces of p orbitals. A nodal plane passes through the nucleus and separates the two lobes of each orbital. The dark and light areas denote regions of opposite sign of the wavefunction.

Exploration Use mathematical software to plot the boundary surfaces of the real parts of the spherical harmonics $Y_{1,m_l}(\theta,\phi)$. The resulting plots are not strictly the p orbital boundary surfaces, but sufficiently close to be reasonable representations of the shapes of hydrogenic orbitals.

$$\psi_{p_0} = R_{2,1}(r)Y_{1,0}(\theta,\phi) = \frac{1}{4(2\pi)^{1/2}}\left(\frac{Z}{a_0}\right)^{5/2} r\cos\theta e^{-Zr/2a_0}$$
$$= r\cos\theta f(r)$$

where $f(r)$ is a function only of r. Because in spherical polar coordinates $z = r\cos\theta$, this wavefunction may also be written

$$\psi_{p_z} = zf(r) \tag{10.23}$$

All p orbitals with $m_l = 0$ have wavefunctions of this form regardless of the value of n. This way of writing the orbital is the origin of the name 'p_z orbital': its boundary surface is shown in Fig. 10.15. The wavefunction is zero everywhere in the xy-plane, where $z = 0$, so the xy-plane is a **nodal plane** of the orbital: the wavefunction changes sign on going from one side of the plane to the other.

The wavefunctions of $2p$ orbitals with $m_l = \pm 1$ have the following form:

$$\psi_{p_{\pm 1}} = R_{2,1}(r)Y_{1,\pm 1}(\theta,\phi) = \mp\frac{1}{8\pi^{1/2}}\left(\frac{Z}{a_0}\right)^{5/2} re^{-Zr/2a_0}\sin\theta e^{\pm i\phi}$$
$$= \mp\frac{1}{2^{1/2}} r\sin\theta e^{\pm i\phi} f(r)$$

We saw in Chapter 8 that a moving particle can be described by a complex wavefunction. In the present case, the functions correspond to non-zero angular momentum about the z-axis: $e^{+i\phi}$ corresponds to clockwise rotation when viewed from below, and $e^{-i\phi}$ corresponds to counter-clockwise rotation (from the same viewpoint). They have zero amplitude where $\theta = 0$ and $180°$ (along the z-axis) and maximum amplitude at $90°$, which is in the xy-plane. To draw the functions it is usual to represent them as standing waves. To do so, we take the real linear combinations

$$\begin{aligned}\psi_{p_x} &= -\frac{1}{2^{1/2}}(p_{+1} - p_{-1}) = r\sin\theta\cos\phi f(r) = xf(r)\\ \psi_{p_y} &= \frac{i}{2^{1/2}}(p_{+1} + p_{-1}) = r\sin\theta\sin\phi f(r) = yf(r)\end{aligned} \tag{10.24}$$

These linear combinations are indeed standing waves with no net orbital angular momentum around the z-axis, as they are superpositions of states with equal and opposite values of m_l. The p_x orbital has the same shape as a p_z orbital, but it is directed along the x-axis (see Fig. 10.15); the p_y orbital is similarly directed along the y-axis. The wavefunction of any p orbital of a given shell can be written as a product of x, y, or z and the same radial function (which depends on the value of n).

Justification 10.3 *The linear combination of degenerate wavefunctions*

We justify here the step of taking linear combinations of degenerate orbitals when we want to indicate a particular point. The freedom to do so rests on the fact that, whenever two or more wavefunctions correspond to the same energy, any linear combination of them is an equally valid solution of the Schrödinger equation.

Suppose ψ_1 and ψ_2 are both solutions of the Schrödinger equation with energy E; then we know that

$$H\psi_1 = E\psi_1 \qquad H\psi_2 = E\psi_2$$

Now consider the linear combination

$$\psi = c_1\psi_1 + c_2\psi_2$$

where c_1 and c_2 are arbitrary coefficients. Then it follows that

$$H\psi = H(c_1\psi_1 + c_2\psi_2) = c_1H\psi_1 + c_2H\psi_2 = c_1E\psi_1 + c_2E\psi_2 = E\psi$$

Hence, the linear combination is also a solution corresponding to the same energy E.

(g) *d* Orbitals

When $n = 3$, l can be 0, 1, or 2. As a result, this shell consists of one 3s orbital, three 3p orbitals, and five 3d orbitals. The five d orbitals have $m_l = +2, +1, 0, -1, -2$ and correspond to five different angular momenta around the z-axis (but the same *magnitude* of angular momentum, because $l = 2$ in each case). As for the p orbitals, d orbitals with opposite values of m_l (and hence opposite senses of motion around the z-axis) may be combined in pairs to give real standing waves, and the boundary surfaces of the resulting shapes are shown in Fig. 10.16. The real combinations have the following forms:

$$d_{xy} = xyf(r) \qquad d_{yz} = yzf(r) \qquad d_{zx} = zxf(r)$$
$$d_{x^2-y^2} = \tfrac{1}{2}(x^2 - y^2)f(r) \qquad d_{z^2} = (\tfrac{1}{2}\sqrt{3})(3z^2 - r^2)f(r) \qquad (10.25)$$

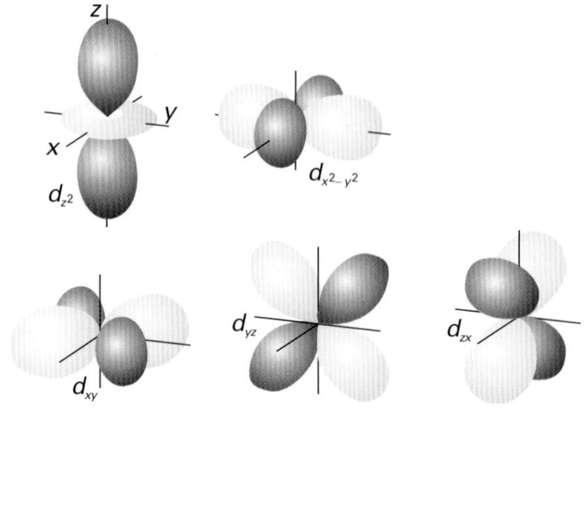

Fig. 10.16 The boundary surfaces of d orbitals. Two nodal planes in each orbital intersect at the nucleus and separate the lobes of each orbital. The dark and light areas denote regions of opposite sign of the wavefunction.

Exploration To gain insight into the shapes of the f orbitals, use mathematical software to plot the boundary surfaces of the spherical harmonics $Y_{3,m_l}(\theta,\phi)$.

10.3 Spectroscopic transitions and selection rules

The energies of the hydrogenic atoms are given by eqn 10.11. When the electron undergoes a **transition**, a change of state, from an orbital with quantum numbers n_1, l_1, m_{l1} to another (lower energy) orbital with quantum numbers n_2, l_2, m_{l2}, it undergoes a change of energy ΔE and discards the excess energy as a photon of electromagnetic radiation with a frequency ν given by the Bohr frequency condition (eqn 8.10).

It is tempting to think that all possible transitions are permissible, and that a spectrum arises from the transition of an electron from any initial orbital to any other orbital. However, this is not so, because a photon has an intrinsic spin angular momentum corresponding to $s = 1$ (Section 9.8). The change in angular momentum of the electron must compensate for the angular momentum carried away by the photon. Thus, an electron in a d orbital ($l = 2$) cannot make a transition into an s orbital ($l = 0$) because the photon cannot carry away enough angular momentum. Similarly, an s electron cannot make a transition to another s orbital, because there would then be no change in the electron's angular momentum to make up for the angular momentum carried away by the photon. It follows that some spectroscopic transitions are **allowed**, meaning that they can occur, whereas others are **forbidden**, meaning that they cannot occur.

A **selection rule** is a statement about which transitions are allowed. They are derived (for atoms) by identifying the transitions that conserve angular momentum when a photon is emitted or absorbed. The selection rules for hydrogenic atoms are

$$\Delta l = \pm 1 \qquad \Delta m_l = 0, \pm 1 \tag{10.26}$$

The principal quantum number n can change by any amount consistent with the Δl for the transition, because it does not relate directly to the angular momentum.

Justification 10.4 *The identification of selection rules*

We saw in Section 9.10 that the rate of transition between two states is proportional to the square of the transition dipole moment, $\boldsymbol{\mu}_{\rm fi}$, between the initial and final states, where (using the notation introduced in *Further information 9.1*)

$$\boldsymbol{\mu}_{\rm fi} = \langle \mathrm{f}|\boldsymbol{\mu}|\mathrm{i}\rangle \qquad [10.27]$$

and $\boldsymbol{\mu}$ is the electric dipole moment operator. For a one-electron atom $\boldsymbol{\mu}$ is multiplication by $-e\boldsymbol{r}$ with components $\mu_x = -ex$, $\mu_y = -ey$, and $\mu_z = -ez$. If the transition dipole moment is zero, the transition is forbidden; the transition is allowed if the transition moment is non-zero. Physically, the transition dipole moment is a measure of the dipolar 'kick' that the electron gives to or receives from the electromagnetic field.

To evaluate a transition dipole moment, we consider each component in turn. For example, for the z-component,

$$\mu_{z,\rm fi} = -e\langle \mathrm{f}|z|\mathrm{i}\rangle = -e\int \psi_{\rm f}^{*} z \psi_{\rm i}\,\mathrm{d}\tau \tag{10.28}$$

To evaluate the integral, we note from Table 9.3 that $z = (4\pi/3)^{1/2} r Y_{1,0}$, so

$$\int \psi_{\rm f}^{*} z \psi_{\rm i}\,\mathrm{d}\tau = \int_0^{\infty}\int_0^{\pi}\int_0^{2\pi} \overbrace{R_{n_{\rm f},l_{\rm f}} Y_{l_{\rm f},m_{l,\rm f}}^{*}}^{\psi_{\rm f}} \overbrace{\left(\frac{4\pi}{3}\right)^{1/2} r Y_{1,0}}^{z} \overbrace{R_{n_{\rm i},l_{\rm i}} Y_{l_{\rm i},m_{l,\rm i}}}^{\psi_{\rm i}} \overbrace{r^2\mathrm{d}r \sin\theta \mathrm{d}\theta \mathrm{d}\phi}^{\mathrm{d}\tau}$$

This multiple integral is the product of three factors, an integral over r and two integrals over the angles, so the factors on the right can be grouped as follows:

$$\int \psi_{\rm f}^{*} z \psi_{\rm i}\,\mathrm{d}\tau = \left(\frac{4\pi}{3}\right)^{1/2} \int_0^{\infty} R_{n_{\rm f},l_{\rm f}} r R_{n_{\rm i},l_{\rm i}} r^2\mathrm{d}r \int_0^{\pi}\int_0^{2\pi} Y_{l_{\rm f},m_{l,\rm f}}^{*} Y_{1,0} Y_{l_{\rm i},m_{l,\rm i}} \sin\theta \mathrm{d}\theta \mathrm{d}\phi$$

It follows from the properties of the spherical harmonics (Comment 9.6) that the integral

$$\int_0^{\pi}\int_0^{2\pi} Y^*_{l_f,m_{l_f}} Y_{1,m} Y_{l_i,m_{l_i}} \sin\theta \mathrm{d}\theta\mathrm{d}\phi$$

is zero unless $l_f = l_i \pm 1$ and $m_{l,f} = m_{l,i} + m$. Because $m = 0$ in the present case, the angular integral, and hence the z-component of the transition dipole moment, is zero unless $\Delta l = \pm 1$ and $\Delta m_l = 0$, which is a part of the set of selection rules. The same procedure, but considering the x- and y-components, results in the complete set of rules.

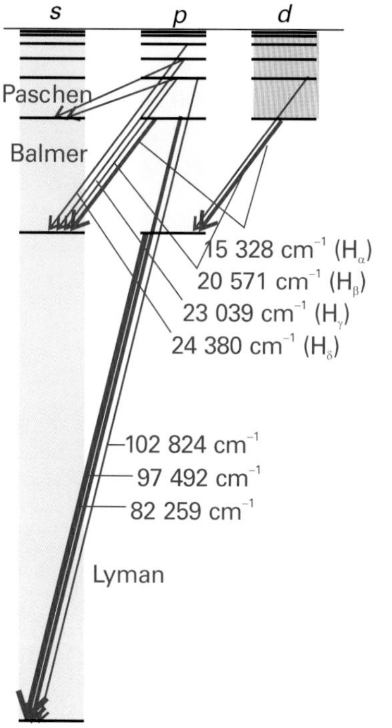

Fig. 10.17 A Grotrian diagram that summarizes the appearance and analysis of the spectrum of atomic hydrogen. The thicker the line, the more intense the transition.

Illustration 10.2 *Applying selection rules*

To identify the orbitals to which a 4d electron may make radiative transitions, we first identify the value of l and then apply the selection rule for this quantum number. Because $l = 2$, the final orbital must have $l = 1$ or 3. Thus, an electron may make a transition from a 4d orbital to any np orbital (subject to $\Delta m_l = 0, \pm 1$) and to any nf orbital (subject to the same rule). However, it cannot undergo a transition to any other orbital, so a transition to any ns orbital or to another nd orbital is forbidden.

Self-test 10.7 To what orbitals may a 4s electron make electric-dipole allowed radiative transitions? [to np orbitals only]

The selection rules and the atomic energy levels jointly account for the structure of a **Grotrian diagram** (Fig. 10.17), which summarizes the energies of the states and the transitions between them. The thicknesses of the transition lines in the diagram denote their relative intensities in the spectrum; we see how to determine transition intensities in Section 13.2.

The structures of many-electron atoms

The Schrödinger equation for a many-electron atom is highly complicated because all the electrons interact with one another. Even for a helium atom, with its two electrons, no analytical expression for the orbitals and energies can be given, and we are forced to make approximations. We shall adopt a simple approach based on what we already know about the structure of hydrogenic atoms. Later we shall see the kind of numerical computations that are currently used to obtain accurate wavefunctions and energies.

10.4 The orbital approximation

The wavefunction of a many-electron atom is a very complicated function of the coordinates of all the electrons, and we should write it $\psi(\boldsymbol{r}_1,\boldsymbol{r}_2,\ldots)$, where $\boldsymbol{r}_i$ is the vector from the nucleus to electron i. However, in the **orbital approximation** we suppose that a reasonable first approximation to this exact wavefunction is obtained by thinking of each electron as occupying its 'own' orbital, and write

$$\psi(\boldsymbol{r}_1,\boldsymbol{r}_2,\ldots) = \psi(\boldsymbol{r}_1)\psi(\boldsymbol{r}_2)\ldots \qquad (10.29)$$

We can think of the individual orbitals as resembling the hydrogenic orbitals, but corresponding to nuclear charges modified by the presence of all the other electrons in the atom. This description is only approximate, but it is a useful model for discussing the chemical properties of atoms, and is the starting point for more sophisticated descriptions of atomic structure.

Justification 10.5 *The orbital approximation*

The orbital approximation would be exact if there were no interactions between electrons. To demonstrate the validity of this remark, we need to consider a system in which the hamiltonian for the energy is the sum of two contributions, one for electron 1 and the other for electron 2:

$$\hat{H} = \hat{H}_1 + \hat{H}_2$$

In an actual atom (such as helium atom), there is an additional term corresponding to the interaction of the two electrons, but we are ignoring that term. We shall now show that if $\psi(\boldsymbol{r}_1)$ is an eigenfunction of $\hat{H}_1$ with energy E_1, and $\psi(\boldsymbol{r}_2)$ is an eigenfunction of $\hat{H}_2$ with energy E_2, then the product $\psi(\boldsymbol{r}_1,\boldsymbol{r}_2) = \psi(\boldsymbol{r}_1)\psi(\boldsymbol{r}_2)$ is an eigenfunction of the combined hamiltonian $\hat{H}$. To do so we write

$$\begin{aligned}\hat{H}\psi(\boldsymbol{r}_1,\boldsymbol{r}_2) &= (\hat{H}_1 + \hat{H}_2)\psi(\boldsymbol{r}_1)\psi(\boldsymbol{r}_2) = \hat{H}_1\psi(\boldsymbol{r}_1)\psi(\boldsymbol{r}_2) + \psi(\boldsymbol{r}_1)\hat{H}_2\psi(\boldsymbol{r}_2)\\ &= E_1\psi(\boldsymbol{r}_1)\psi(\boldsymbol{r}_2) + \psi(\boldsymbol{r}_1)E_2\psi(\boldsymbol{r}_2) = (E_1 + E_2)\psi(\boldsymbol{r}_1)\psi(\boldsymbol{r}_2)\\ &= E\psi(\boldsymbol{r}_1,\boldsymbol{r}_2)\end{aligned}$$

where $E = E_1 + E_2$. This is the result we need to prove. However, if the electrons interact (as they do in fact), then the proof fails.

(a) The helium atom

The orbital approximation allows us to express the electronic structure of an atom by reporting its **configuration**, the list of occupied orbitals (usually, but not necessarily, in its ground state). Thus, as the ground state of a hydrogenic atom consists of the single electron in a 1*s* orbital, we report its configuration as $1s^1$.

The He atom has two electrons. We can imagine forming the atom by adding the electrons in succession to the orbitals of the bare nucleus (of charge 2*e*). The first electron occupies a 1*s* hydrogenic orbital, but because $Z = 2$ that orbital is more compact than in H itself. The second electron joins the first in the 1*s* orbital, so the electron configuration of the ground state of He is $1s^2$.

(b) The Pauli principle

Lithium, with $Z = 3$, has three electrons. The first two occupy a 1*s* orbital drawn even more closely than in He around the more highly charged nucleus. The third electron, however, does not join the first two in the 1*s* orbital because that configuration is forbidden by the **Pauli exclusion principle**:

> No more than two electrons may occupy any given orbital, and if two do occupy one orbital, then their spins must be paired.

Electrons with paired spins, denoted ↑↓, have zero net spin angular momentum because the spin of one electron is cancelled by the spin of the other. Specifically, one electron has $m_s = +\frac{1}{2}$, the other has $m_s = -\frac{1}{2}$ and they are orientated on their respective cones so that the resultant spin is zero (Fig. 10.18). The exclusion principle is the key to the structure of complex atoms, to chemical periodicity, and to molecular structure. It was proposed by Wolfgang Pauli in 1924 when he was trying to account for the

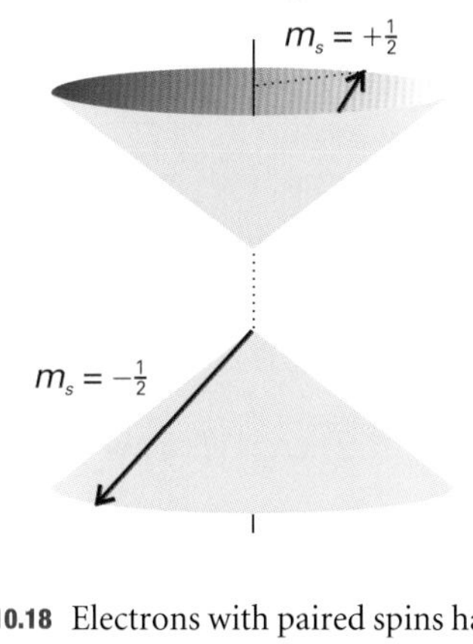

Fig. 10.18 Electrons with paired spins have zero resultant spin angular momentum. They can be represented by two vectors that lie at an indeterminate position on the cones shown here, but wherever one lies on its cone, the other points in the opposite direction; their resultant is zero.

absence of some lines in the spectrum of helium. Later he was able to derive a very general form of the principle from theoretical considerations.

The Pauli exclusion principle in fact applies to any pair of identical fermions (particles with half integral spin). Thus it applies to protons, neutrons, and ^{13}C nuclei (all of which have spin $\frac{1}{2}$) and to ^{35}Cl nuclei (which have spin $\frac{3}{2}$). It does not apply to identical bosons (particles with integral spin), which include photons (spin 1), ^{12}C nuclei (spin 0). Any number of identical bosons may occupy the same state (that is, be described by the same wavefunction).

The Pauli exclusion principle is a special case of a general statement called the **Pauli principle**:

> When the labels of any two identical fermions are exchanged, the total wavefunction changes sign; when the labels of any two identical bosons are exchanged, the total wavefunction retains the same sign.

By 'total wavefunction' is meant the entire wavefunction, including the spin of the particles. To see that the Pauli principle implies the Pauli exclusion principle, we consider the wavefunction for two electrons $\psi(1,2)$. The Pauli principle implies that it is a fact of nature (which has its roots in the theory of relativity) that the wavefunction must change sign if we interchange the labels 1 and 2 wherever they occur in the function:

$$\psi(2,1) = -\psi(1,2) \tag{10.30}$$

Suppose the two electrons in an atom occupy an orbital ψ, then in the orbital approximation the overall wavefunction is $\psi(1)\psi(2)$. To apply the Pauli principle, we must deal with the total wavefunction, the wavefunction including spin. There are several possibilities for two spins: both α, denoted $\alpha(1)\alpha(2)$, both β, denoted $\beta(1)\beta(2)$, and one α the other β, denoted either $\alpha(1)\beta(2)$ or $\alpha(2)\beta(1)$. Because we cannot tell which electron is α and which is β, in the last case it is appropriate to express the spin states as the (normalized) linear combinations

$$\sigma_+(1,2) = (1/2^{1/2})\{\alpha(1)\beta(2) + \beta(1)\alpha(2)\}$$
$$\sigma_-(1,2) = (1/2^{1/2})\{\alpha(1)\beta(2) - \beta(1)\alpha(2)\} \tag{10.31}$$

because these combinations allow one spin to be α and the other β with equal probability. The total wavefunction of the system is therefore the product of the orbital part and one of the four spin states:

$$\psi(1)\psi(2)\alpha(1)\alpha(2) \quad \psi(1)\psi(2)\beta(1)\beta(2) \quad \psi(1)\psi(2)\sigma_+(1,2) \quad \psi(1)\psi(2)\sigma_-(1,2)$$

The Pauli principle says that for a wavefunction to be acceptable (for electrons), it must change sign when the electrons are exchanged. In each case, exchanging the labels 1 and 2 converts the factor $\psi(1)\psi(2)$ into $\psi(2)\psi(1)$, which is the same, because the order of multiplying the functions does not change the value of the product. The same is true of $\alpha(1)\alpha(2)$ and $\beta(1)\beta(2)$. Therefore, the first two overall products are not allowed, because they do not change sign. The combination $\sigma_+(1,2)$ changes to

$$\sigma_+(2,1) = (1/2^{1/2})\{\alpha(2)\beta(1) + \beta(2)\alpha(1)\} = \sigma_+(1,2)$$

because it is simply the original function written in a different order. The third overall product is therefore also disallowed. Finally, consider $\sigma_-(1,2)$:

$$\sigma_-(2,1) = (1/2^{1/2})\{\alpha(2)\beta(1) - \beta(2)\alpha(1)\}$$
$$= -(1/2^{1/2})\{\alpha(1)\beta(2) - \beta(1)\alpha(2)\} = -\sigma_-(1,2)$$

This combination does change sign (it is 'antisymmetric'). The product $\psi(1)\psi(2)$ $\sigma_-(1,2)$ also changes sign under particle exchange, and therefore it is acceptable.

Comment 10.3

A stronger justification for taking linear combinations in eqn 10.31 is that they correspond to eigenfunctions of the total spin operators S^2 and S_z, with $M_S = 0$ and, respectively, $S = 1$ and 0. See Section 10.7.

Now we see that only one of the four possible states is allowed by the Pauli principle, and the one that survives has paired α and β spins. This is the content of the Pauli exclusion principle. The exclusion principle is irrelevant when the orbitals occupied by the electrons are different, and both electrons may then have (but need not have) the same spin state. Nevertheless, even then the overall wavefunction must still be antisymmetric overall, and must still satisfy the Pauli principle itself.

A final point in this connection is that the acceptable product wavefunction $\psi(1)\psi(2)\sigma_-(1,2)$ can be expressed as a determinant:

$$\frac{1}{2^{1/2}}\begin{vmatrix}\psi(1)\alpha(1) & \psi(2)\alpha(2)\\ \psi(1)\beta(1) & \psi(2)\beta(2)\end{vmatrix}=\frac{1}{2^{1/2}}\{\psi(1)\alpha(1)\psi(2)\beta(2)-\psi(2)\alpha(2)\psi(1)\beta(1)\}$$
$$=\psi(1)\psi(2)\sigma_-(1,2)$$

Any acceptable wavefunction for a closed-shell species can be expressed as a **Slater determinant**, as such determinants are known. In general, for N electrons in orbitals $\psi_a, \psi_b, \ldots$

$$\psi(1,2,\ldots,N)=\frac{1}{(N!)^{1/2}}\begin{vmatrix}\psi_a(1)\alpha(1) & \psi_a(2)\alpha(2) & \psi_a(3)\alpha(3) & \ldots & \psi_a(N)\alpha(N)\\ \psi_a(1)\beta(1) & \psi_a(2)\beta(2) & \psi_a(3)\beta(3) & \ldots & \psi_a(N)\beta(N)\\ \psi_b(1)\alpha(1) & \psi_b(2)\alpha(2) & \psi_b(3)\alpha(3) & \ldots & \psi_b(N)\alpha(N)\\ \vdots & \vdots & \vdots & \vdots & \vdots\\ \psi_z(1)\beta(1) & \psi_z(2)\beta(2) & \psi_z(3)\beta(3) & \ldots & \psi_z(N)\beta(N)\end{vmatrix} \qquad [10.32]$$

Writing a many-electron wavefunction in this way ensures that it is antisymmetric under the interchange of any pair of electrons, as is explored in Problem 10.23.

Now we can return to lithium. In Li ($Z=3$), the third electron cannot enter the 1*s* orbital because that orbital is already full: we say the *K* shell is *complete* and that the two electrons form a **closed shell**. Because a similar closed shell is characteristic of the He atom, we denote it [He]. The third electron is excluded from the *K* shell and must occupy the next available orbital, which is one with $n=2$ and hence belonging to the *L* shell. However, we now have to decide whether the next available orbital is the 2*s* orbital or a 2*p* orbital, and therefore whether the lowest energy configuration of the atom is [He]$2s^1$ or [He]$2p^1$.

(c) Penetration and shielding

Unlike in hydrogenic atoms, the 2*s* and 2*p* orbitals (and, in general, all subshells of a given shell) are not degenerate in many-electron atoms. As will be familiar from introductory chemistry, an electron in a many-electron atom experiences a Coulombic repulsion from all the other electrons present. If it is at a distance *r* from the nucleus, it experiences an average repulsion that can be represented by a point negative charge located at the nucleus and equal in magnitude to the total charge of the electrons within a sphere of radius *r* (Fig. 10.19). The effect of this point negative charge, when averaged over all the locations of the electron, is to reduce the full charge of the nucleus from Ze to $Z_{\text{eff}}e$, the **effective nuclear charge**. In everyday parlance, Z_{eff} itself is commonly referred to as the 'effective nuclear charge'. We say that the electron experiences a **shielded** nuclear charge, and the difference between Z and Z_{eff} is called the **shielding constant**, σ:

$$Z_{\text{eff}}=Z-\sigma \qquad [10.33]$$

The electrons do not actually 'block' the full Coulombic attraction of the nucleus: the shielding constant is simply a way of expressing the net outcome of the nuclear

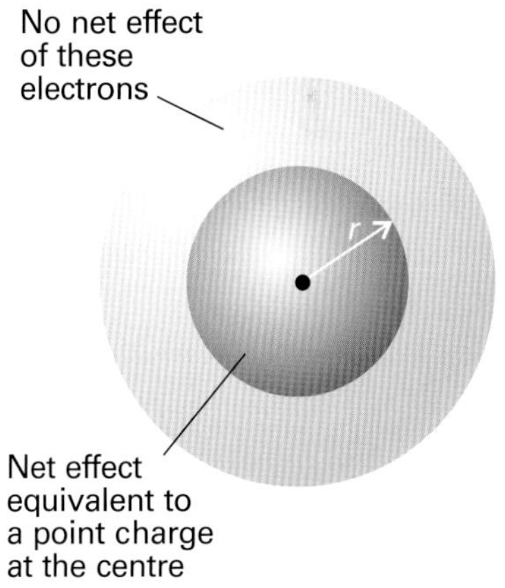

Fig. 10.19 An electron at a distance *r* from the nucleus experiences a Coulombic repulsion from all the electrons within a sphere of radius *r* and which is equivalent to a point negative charge located on the nucleus. The negative charge reduces the effective nuclear charge of the nucleus from Ze to $Z_{\text{eff}}e$.

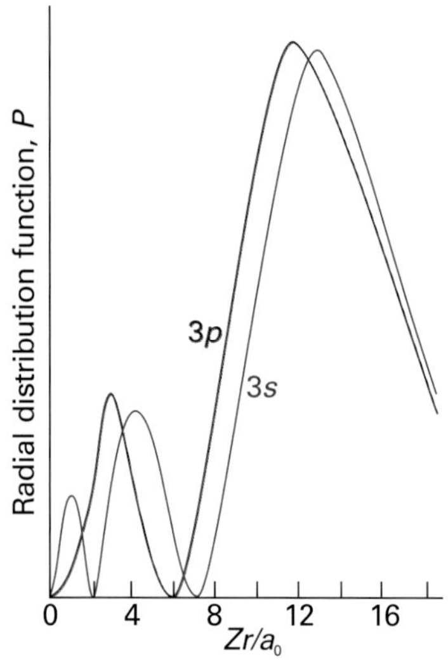

Fig. 10.20 An electron in an *s* orbital (here a 3*s* orbital) is more likely to be found close to the nucleus than an electron in a *p* orbital of the same shell (note the closeness of the innermost peak of the 3*s* orbital to the nucleus at $r = 0$). Hence an *s* electron experiences less shielding and is more tightly bound than a *p* electron.

Exploration Calculate and plot the graphs given above for $n = 4$.

Synoptic table 10.2* Effective nuclear charge, $Z_{eff} = Z - \sigma$

Element	Z	Orbital	Z_{eff}
He	2	1*s*	1.6875
C	6	1*s*	5.6727
		2*s*	3.2166
		2*p*	3.1358

* More values are given in the *Data section*.

attraction and the electronic repulsions in terms of a single equivalent charge at the centre of the atom.

The shielding constant is different for *s* and *p* electrons because they have different radial distributions (Fig. 10.20). An *s* electron has a greater **penetration** through inner shells than a *p* electron, in the sense that it is more likely to be found close to the nucleus than a *p* electron of the same shell (the wavefunction of a *p* orbital, remember, is zero at the nucleus). Because only electrons inside the sphere defined by the location of the electron (in effect, the core electrons) contribute to shielding, an *s* electron experiences less shielding than a *p* electron. Consequently, by the combined effects of penetration and shielding, an *s* electron is more tightly bound than a *p* electron of the same shell. Similarly, a *d* electron penetrates less than a *p* electron of the same shell (recall that the wavefunction of a *d* orbital varies as r^2 close to the nucleus, whereas a *p* orbital varies as *r*), and therefore experiences more shielding.

Shielding constants for different types of electrons in atoms have been calculated from their wavefunctions obtained by numerical solution of the Schrödinger equation for the atom (Table 10.2). We see that, in general, valence-shell *s* electrons do experience higher effective nuclear charges than *p* electrons, although there are some discrepancies. We return to this point shortly.

The consequence of penetration and shielding is that the energies of subshells of a shell in a many-electron atom in general lie in the order

$$s < p < d < f$$

The individual orbitals of a given subshell remain degenerate because they all have the same radial characteristics and so experience the same effective nuclear charge.

We can now complete the Li story. Because the shell with $n = 2$ consists of two nondegenerate subshells, with the 2*s* orbital lower in energy than the three 2*p* orbitals, the third electron occupies the 2*s* orbital. This occupation results in the ground-state configuration $1s^22s^1$, with the central nucleus surrounded by a complete helium-like shell of two 1*s* electrons, and around that a more diffuse 2*s* electron. The electrons in the outermost shell of an atom in its ground state are called the **valence electrons** because they are largely responsible for the chemical bonds that the atom forms. Thus, the valence electron in Li is a 2*s* electron and its other two electrons belong to its core.

(d) The building-up principle

The extension of this argument is called the **building-up principle**, or the *Aufbau principle*, from the German word for building up, which will be familiar from introductory courses. In brief, we imagine the bare nucleus of atomic number *Z*, and then feed into the orbitals *Z* electrons in succession. The order of occupation is

$$1s \quad 2s \quad 2p \quad 3s \quad 3p \quad 4s \quad 3d \quad 4p \quad 5s \quad 4d \quad 5p \quad 6s$$

and each orbital may accommodate up to two electrons. As an example, consider the carbon atom, for which $Z = 6$ and there are six electrons to accommodate. Two electrons enter and fill the 1*s* orbital, two enter and fill the 2*s* orbital, leaving two electrons to occupy the orbitals of the 2*p* subshell. Hence the ground-state configuration of C is $1s^22s^22p^2$, or more succinctly $[He]2s^22p^2$, with [He] the helium-like $1s^2$ core. However, we can be more precise: we can expect the last two electrons to occupy different 2*p* orbitals because they will then be further apart on average and repel each other less than if they were in the same orbital. Thus, one electron can be thought of as occupying the $2p_x$ orbital and the other the $2p_y$ orbital (the *x*, *y*, *z* designation is arbitrary, and it would be equally valid to use the complex forms of these orbitals), and the lowest energy configuration of the atom is $[He]2s^22p_x^12p_y^1$. The same rule applies whenever degenerate orbitals of a subshell are available for occupation. Thus, another rule of the building-up principle is:

> Electrons occupy different orbitals of a given subshell before doubly occupying any one of them.

For instance, nitrogen ($Z=7$) has the configuration [He]$2s^2 2p_x^1 2p_y^1 2p_z^1$, and only when we get to oxygen ($Z = 8$) is a $2p$ orbital doubly occupied, giving [He]$2s^2 2p_x^2 2p_y^1 2p_z^1$. When electrons occupy orbitals singly we invoke **Hund's maximum multiplicity rule**:

> An atom in its ground state adopts a configuration with the greatest number of unpaired electrons.

The explanation of Hund's rule is subtle, but it reflects the quantum mechanical property of **spin correlation**, that electrons with parallel spins behave as if they have a tendency to stay well apart, and hence repel each other less. In essence, the effect of spin correlation is to allow the atom to shrink slightly, so the electron–nucleus interaction is improved when the spins are parallel. We can now conclude that, in the ground state of the carbon atom, the two $2p$ electrons have the same spin, that all three $2p$ electrons in the N atoms have the same spin, and that the two $2p$ electrons in different orbitals in the O atom have the same spin (the two in the $2p_x$ orbital are necessarily paired).

Justification 10.6 *Spin correlation*

Suppose electron 1 is described by a wavefunction $\psi_a(\boldsymbol{r}_1)$ and electron 2 is described by a wavefunction $\psi_b(\boldsymbol{r}_2)$; then, in the orbital approximation, the joint wavefunction of the electrons is the product $\psi = \psi_a(\boldsymbol{r}_1)\psi_b(\boldsymbol{r}_2)$. However, this wavefunction is not acceptable, because it suggests that we know which electron is in which orbital, whereas we cannot keep track of electrons. According to quantum mechanics, the correct description is either of the two following wavefunctions:

$$\psi_{\pm} = (1/2^{1/2})\{\psi_a(\boldsymbol{r}_1)\psi_b(\boldsymbol{r}_2) \pm \psi_b(\boldsymbol{r}_1)\psi_a(\boldsymbol{r}_2)\}$$

According to the Pauli principle, because ψ_+ is symmetrical under particle interchange, it must be multiplied by an antisymmetric spin function (the one denoted σ_-). That combination corresponds to a spin-paired state. Conversely, ψ_- is antisymmetric, so it must be multiplied by one of the three symmetric spin states. These three symmetric states correspond to electrons with parallel spins (see Section 10.7 for an explanation).

Now consider the values of the two combinations when one electron approaches another, and $\boldsymbol{r}_1 = \boldsymbol{r}_2$. We see that ψ_- vanishes, which means that there is zero probability of finding the two electrons at the same point in space when they have parallel spins. The other combination does not vanish when the two electrons are at the same point in space. Because the two electrons have different relative spatial distributions depending on whether their spins are parallel or not, it follows that their Coulombic interaction is different, and hence that the two states have different energies.

Neon, with $Z = 10$, has the configuration [He]$2s^2 2p^6$, which completes the L shell. This closed-shell configuration is denoted [Ne], and acts as a core for subsequent elements. The next electron must enter the $3s$ orbital and begin a new shell, so an Na atom, with $Z = 11$, has the configuration [Ne]$3s^1$. Like lithium with the configuration [He]$2s^1$, sodium has a single s electron outside a complete core. This analysis has brought us to the origin of chemical periodicity. The L shell is completed by eight electrons, so the element with $Z = 3$ (Li) should have similar properties to the element with $Z = 11$ (Na). Likewise, Be ($Z = 4$) should be similar to $Z = 12$ (Mg), and so on, up to the noble gases He ($Z = 2$), Ne ($Z = 10$), and Ar ($Z = 18$).

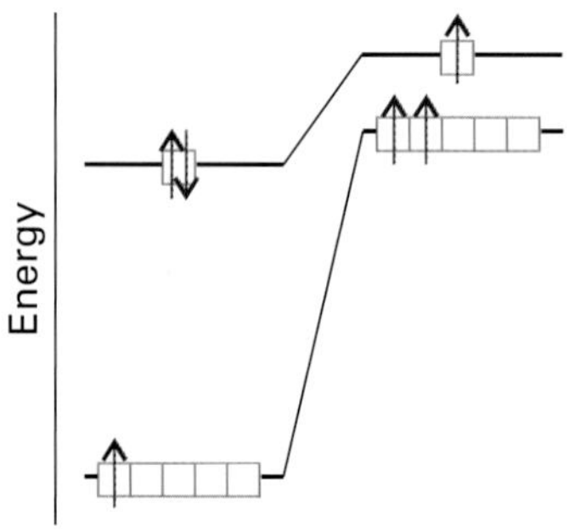

Fig. 10.21 Strong electron–electron repulsions in the $3d$ orbitals are minimized in the ground state of Sc if the atom has the configuration $[Ar]3d^14s^2$ (shown on the left) instead of $[Ar]3d^24s^1$ (shown on the right). The total energy of the atom is lower when it has the $[Ar]3d^14s^2$ configuration despite the cost of populating the high energy $4s$ orbital.

Ten electrons can be accommodated in the five $3d$ orbitals, which accounts for the electron configurations of scandium to zinc. Calculations of the type discussed in Section 10.5 show that for these atoms the energies of the $3d$ orbitals are always lower than the energy of the $4s$ orbital. However, spectroscopic results show that Sc has the configuration $[Ar]3d^14s^2$, instead of $[Ar]3d^3$ or $[Ar]3d^24s^1$. To understand this observation, we have to consider the nature of electron–electron repulsions in $3d$ and $4s$ orbitals. The most probable distance of a $3d$ electron from the nucleus is less than that for a $4s$ electron, so two $3d$ electrons repel each other more strongly than two $4s$ electrons. As a result, Sc has the configuration $[Ar]3d^14s^2$ rather than the two alternatives, for then the strong electron–electron repulsions in the $3d$ orbitals are minimized. The total energy of the atom is least despite the cost of allowing electrons to populate the high energy $4s$ orbital (Fig. 10.21). The effect just described is generally true for scandium through zinc, so their electron configurations are of the form $[Ar]3d^n4s^2$, where $n = 1$ for scandium and $n = 10$ for zinc. Two notable exceptions, which are observed experimentally, are Cr, with electron configuration $[Ar]3d^54s^1$, and Cu, with electron configuration $[Ar]3d^{10}4s^1$ (see *Further reading* for a discussion of the theoretical basis for these exceptions).

At gallium, the building-up principle is used in the same way as in preceding periods. Now the $4s$ and $4p$ subshells constitute the valence shell, and the period terminates with krypton. Because 18 electrons have intervened since argon, this period is the first 'long period' of the periodic table. The existence of the d-block elements (the 'transition metals') reflects the stepwise occupation of the $3d$ orbitals, and the subtle shades of energy differences and effects of electron–electron repulsion along this series gives rise to the rich complexity of inorganic d-metal chemistry. A similar intrusion of the f orbitals in Periods 6 and 7 accounts for the existence of the f block of the periodic table (the lanthanoids and actinoids).

We derive the configurations of cations of elements in the s, p, and d blocks of the periodic table by removing electrons from the ground-state configuration of the neutral atom in a specific order. First, we remove valence p electrons, then valence s electrons, and then as many d electrons as are necessary to achieve the specified charge. For instance, because the configuration of V is $[Ar]3d^34s^2$, the V^{2+} cation has the configuration $[Ar]3d^3$. It is reasonable that we remove the more energetic $4s$ electrons in order to form the cation, but it is not obvious why the $[Ar]3d^3$ configuration is preferred in V^{2+} over the $[Ar]3d^14s^2$ configuration, which is found in the isoelectronic Sc atom. Calculations show that the energy difference between $[Ar]3d^3$ and $[Ar]3d^14s^2$ depends on Z_{eff}. As Z_{eff} increases, transfer of a $4s$ electron to a $3d$ orbital becomes more favourable because the electron–electron repulsions are compensated by attractive interactions between the nucleus and the electrons in the spatially compact $3d$ orbital. Indeed, calculations reveal that, for a sufficiently large Z_{eff}, $[Ar]3d^3$ is lower in energy than $[Ar]3d^14s^2$. This conclusion explains why V^{2+} has a $[Ar]3d^3$ configuration and also accounts for the observed $[Ar]4s^03d^n$ configurations of the M^{2+} cations of Sc through Zn.

The configurations of anions of the p-block elements are derived by continuing the building-up procedure and adding electrons to the neutral atom until the configuration of the next noble gas has been reached. Thus, the configuration of the O^{2-} ion is achieved by adding two electrons to $[He]2s^22p^4$, giving $[He]2s^22p^6$, the same as the configuration of neon.

Comment 10.4

The web site for this text contains links to databases of atomic properties.

(e) Ionization energies and electron affinities

The minimum energy necessary to remove an electron from a many-electron atom in the gas phase is the **first ionization energy**, I_1, of the element. The **second ionization energy**, I_2, is the minimum energy needed to remove a second electron (from the

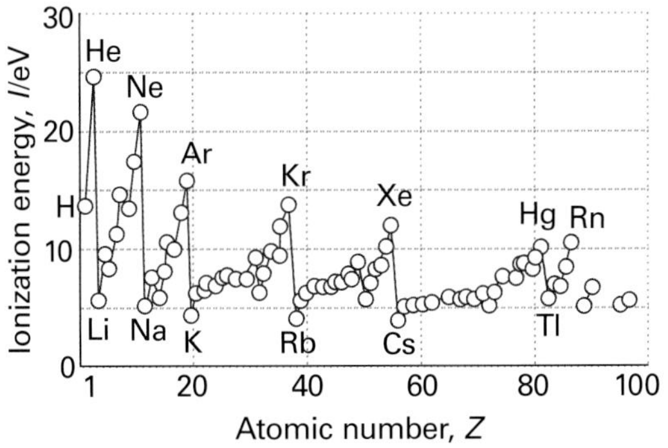

Fig. 10.22 The first ionization energies of the elements plotted against atomic number.

singly charged cation). The variation of the first ionization energy through the periodic table is shown in Fig. 10.22 and some numerical values are given in Table 10.3. In thermodynamic calculations we often need the **standard enthalpy of ionization**, $\Delta_{\text{ion}}H^{\ominus}$. As shown in the *Justification* below, the two are related by

$$\Delta_{\text{ion}}H^{\ominus}(T) = I + \tfrac{5}{2}RT \qquad (10.34)$$

At 298 K, the difference between the ionization enthalpy and the corresponding ionization energy is 6.20 kJ mol^{-1}.

Justification 10.7 *The ionization enthalpy and the ionization energy*

It follows from Kirchhoff's law (Section 2.9 and eqn 2.36) that the reaction enthalpy for

$$\text{M(g)} \rightarrow \text{M}^+\text{(g)} + \text{e}^-\text{(g)}$$

at a temperature T is related to the value at $T = 0$ by

$$\Delta_r H^{\ominus}(T) = \Delta_r H^{\ominus}(0) + \int_0^T \Delta_r C_p^{\ominus} \mathrm{d}T$$

The molar constant-pressure heat capacity of each species in the reaction is $\tfrac{5}{2}R$, so $\Delta_r C_p^{\ominus} = +\tfrac{5}{2}R$. The integral in this expression therefore evaluates to $+\tfrac{5}{2}RT$. The reaction enthalpy at $T = 0$ is the same as the (molar) ionization energy, I. Equation 10.33 then follows. The same expression applies to each successive ionization step, so the overall ionization enthalpy for the formation of M^{2+} is

$$\Delta_r H^{\ominus}(T) = I_1 + I_2 + 5RT$$

Synoptic table 10.3* First and second ionization energies

Element	I_1/(kJ mol^{-1})	I_2/(kJ mol^{-1})
H	1312	
He	2372	5251
Mg	738	1451
Na	496	4562

* More values are given in the *Data section*.

Synoptic table 10.4* Electron affinities, E_a/(kJ mol^{-1})

Cl	349		
F	322		
H	73		
O	141	O^-	−844

* More values are given in the *Data section*.

The **electron affinity**, E_{ea}, is the energy released when an electron attaches to a gas-phase atom (Table 10.4). In a common, logical, but not universal convention (which we adopt), the electron affinity is positive if energy is released when the electron attaches to the atom (that is, $E_{\text{ea}} > 0$ implies that electron attachment is exothermic). It follows from a similar argument to that given in the *Justification* above that the **standard enthalpy of electron gain**, $\Delta_{\text{eg}}H^{\ominus}$, at a temperature T is related to the electron affinity by

$$\Delta_{\text{eg}}H^{\ominus}(T) = -E_{\text{ea}} - \tfrac{5}{2}RT \qquad (10.35)$$

Note the change of sign. In typical thermodynamic cycles the $\tfrac{5}{2}RT$ that appears in eqn 10.35 cancels that in eqn 10.34, so ionization energies and electron affinities can be used directly. A final preliminary point is that the electron-gain enthalpy of a species X is the negative of the ionization enthalpy of its negative ion:

$$\Delta_{\text{eg}}H^{\ominus}(\text{X}) = -\Delta_{\text{ion}}H^{\ominus}(\text{X}^-) \qquad (10.36)$$

As ionization energy is often easier to measure than electron affinity, this relation can be used to determine numerical values of the latter.

As will be familiar from introductory chemistry, ionization energies and electron affinities show periodicities. The former is more regular and we concentrate on it. Lithium has a low first ionization energy because its outermost electron is well shielded from the nucleus by the core ($Z_{eff} = 1.3$, compared with $Z = 3$). The ionization energy of beryllium ($Z = 4$) is greater but that of boron is lower because in the latter the outermost electron occupies a 2*p* orbital and is less strongly bound than if it had been a 2*s* electron. The ionization energy increases from boron to nitrogen on account of the increasing nuclear charge. However, the ionization energy of oxygen is less than would be expected by simple extrapolation. The explanation is that at oxygen a 2*p* orbital must become doubly occupied, and the electron–electron repulsions are increased above what would be expected by simple extrapolation along the row. In addition, the loss of a 2*p* electron results in a configuration with a half-filled subshell (like that of N), which is an arrangement of low energy, so the energy of $O^+ + e^-$ is lower than might be expected, and the ionization energy is correspondingly low too. (The kink is less pronounced in the next row, between phosphorus and sulfur because their orbitals are more diffuse.) The values for oxygen, fluorine, and neon fall roughly on the same line, the increase of their ionization energies reflecting the increasing attraction of the more highly charged nuclei for the outermost electrons.

The outermost electron in sodium is 3*s*. It is far from the nucleus, and the latter's charge is shielded by the compact, complete neon-like core. As a result, the ionization energy of sodium is substantially lower than that of neon. The periodic cycle starts again along this row, and the variation of the ionization energy can be traced to similar reasons.

Electron affinities are greatest close to fluorine, for the incoming electron enters a vacancy in a compact valence shell and can interact strongly with the nucleus. The attachment of an electron to an anion (as in the formation of O^{2-} from O^-) is invariably endothermic, so E_{ea} is negative. The incoming electron is repelled by the charge already present. Electron affinities are also small, and may be negative, when an electron enters an orbital that is far from the nucleus (as in the heavier alkali metal atoms) or is forced by the Pauli principle to occupy a new shell (as in the noble gas atoms).

10.5 Self-consistent field orbitals

The central difficulty of the Schrödinger equation is the presence of the electron–electron interaction terms. The potential energy of the electrons is

$$V = -\sum_i \frac{Ze^2}{4\pi\varepsilon_0 r_i} + \frac{1}{2}\sum_{i,j}{}' \frac{e^2}{4\pi\varepsilon_0 r_{ij}} \qquad (10.37)$$

The prime on the second sum indicates that $i \neq j$, and the factor of one-half prevents double-counting of electron pair repulsions (1 with 2 is the same as 2 with 1). The first term is the total attractive interaction between the electrons and the nucleus. The second term is the total repulsive interaction between the electrons; r_{ij} is the distance between electrons i and j. It is hopeless to expect to find analytical solutions of a Schrödinger equation with such a complicated potential energy term, but computational techniques are available that give very detailed and reliable numerical solutions for the wavefunctions and energies. The techniques were originally introduced by D.R. Hartree (before computers were available) and then modified by V. Fock to take into account the Pauli principle correctly. In broad outline, the **Hartree–Fock self-consistent field** (HF-SCF) procedure is as follows.

Imagine that we have a rough idea of the structure of the atom. In the Ne atom, for instance, the orbital approximation suggests the configuration $1s^22s^22p^6$ with the orbitals approximated by hydrogenic atomic orbitals. Now consider one of the 2*p* electrons. A Schrödinger equation can be written for this electron by ascribing to it a potential energy due to the nuclear attraction and the repulsion from the other electrons. This equation has the form

$$H(1)\psi_{2p}(1) + V(\text{other electrons})\psi_{2p}(1) - V(\text{exchange correction})\psi_{2p}(1) = E_{2p}\psi_{2p}(1) \quad (10.38)$$

A similar equation can be written for the 1*s* and 2*s* orbitals in the atom. The various terms are as follows:

1 The first term on the left is the contribution of the kinetic energy and the attraction of the electron to the nucleus, just as in a hydrogenic atom.

2 The second takes into account the potential energy of the electron of interest due to the electrons in the other occupied orbitals.

3 The third term takes into account the spin correlation effects discussed earlier.

Although the equation is for the 2*p* orbital in neon, it depends on the wavefunctions of all the other occupied orbitals in the atom.

There is no hope of solving eqn 10.38 analytically. However, it can be solved numerically if we guess an approximate form of the wavefunctions of all the orbitals except 2*p*. The procedure is then repeated for the other orbitals in the atom, the 1*s* and 2*s* orbitals. This sequence of calculations gives the form of the 2*p*, 2*s*, and 1*s* orbitals, and in general they will differ from the set used initially to start the calculation. These improved orbitals can be used in another cycle of calculation, and a second improved set of orbitals is obtained. The recycling continues until the orbitals and energies obtained are insignificantly different from those used at the start of the current cycle. The solutions are then self-consistent and accepted as solutions of the problem.

Figure 10.23 shows plots of some of the HF-SCF radial distribution functions for sodium. They show the grouping of electron density into shells, as was anticipated by the early chemists, and the differences of penetration as discussed above. These SCF calculations therefore support the qualitative discussions that are used to explain chemical periodicity. They also considerably extend that discussion by providing detailed wavefunctions and precise energies.

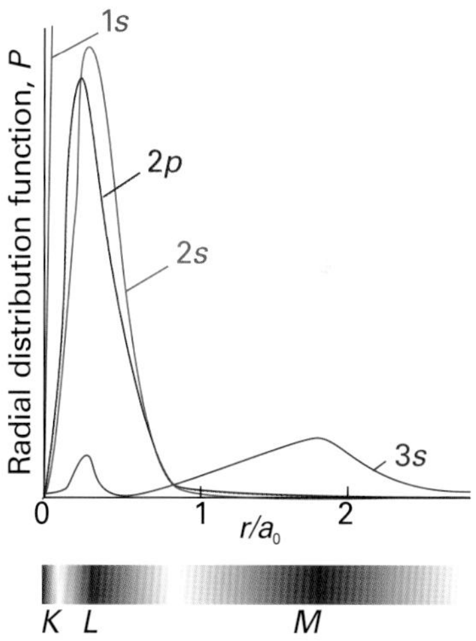

Fig. 10.23 The radial distribution functions for the orbitals of Na based on SCF calculations. Note the shell-like structure, with the 3*s* orbital outside the inner *K* and *L* shells.

The spectra of complex atoms

The spectra of atoms rapidly become very complicated as the number of electrons increases, but there are some important and moderately simple features that make atomic spectroscopy useful in the study of the composition of samples as large and as complex as stars (*Impact I10.1*). The general idea is straightforward: lines in the spectrum (in either emission or absorption) occur when the atom undergoes a transition with a change of energy $|\Delta E|$, and emits or absorbs a photon of frequency $\nu = |\Delta E|/h$ and $\tilde{\nu} = |\Delta E|/hc$. Hence, we can expect the spectrum to give information about the energies of electrons in atoms. However, the actual energy levels are not given solely by the energies of the orbitals, because the electrons interact with one another in various ways, and there are contributions to the energy in addition to those we have already considered.

Comment 10.5

The web site for this text contains links to databases of atomic spectra.

IMPACT ON ASTROPHYSICS

I10.1 Spectroscopy of stars

The bulk of stellar material consists of neutral and ionized forms of hydrogen and helium atoms, with helium being the product of 'hydrogen burning' by nuclear fusion. However, nuclear fusion also makes heavier elements. It is generally accepted that the outer layers of stars are composed of lighter elements, such as H, He, C, N, O, and Ne in both neutral and ionized forms. Heavier elements, including neutral and ionized forms of Si, Mg, Ca, S, and Ar, are found closer to the stellar core. The core itself contains the heaviest elements and ^{56}Fe is particularly abundant because it is a very stable nuclide. All these elements are in the gas phase on account of the very high temperatures in stellar interiors. For example, the temperature is estimated to be 3.6 MK half way to the centre of the Sun.

Astronomers use spectroscopic techniques to determine the chemical composition of stars because each element, and indeed each isotope of an element, has a characteristic spectral signature that is transmitted through space by the star's light. To understand the spectra of stars, we must first know why they shine. Nuclear reactions in the dense stellar interior generate radiation that travels to less dense outer layers. Absorption and re-emission of photons by the atoms and ions in the interior give rise to a quasi-continuum of radiation energy that is emitted into space by a thin layer of gas called the *photosphere*. To a good approximation, the distribution of energy emitted from a star's photosphere resembles the Planck distribution for a very hot black body (Section 8.1). For example, the energy distribution of our Sun's photosphere may be modelled by a Planck distribution with an effective temperature of 5.8 kK. Superimposed on the black-body radiation continuum are sharp absorption and emission lines from neutral atoms and ions present in the photosphere. Analysis of stellar radiation with a spectrometer mounted on to a telescope yields the chemical composition of the star's photosphere by comparison with known spectra of the elements. The data can also reveal the presence of small molecules, such as CN, C_2, TiO, and ZrO, in certain 'cold' stars, which are stars with relatively low effective temperatures.

The two outermost layers of a star are the *chromosphere*, a region just above the photosphere, and the *corona*, a region above the chromosphere that can be seen (with proper care) during eclipses. The photosphere, chromosphere, and corona comprise a star's 'atmosphere'. Our Sun's chromosphere is much less dense than its photosphere and its temperature is much higher, rising to about 10 kK. The reasons for this increase in temperature are not fully understood. The temperature of our Sun's corona is very high, rising up to 1.5 MK, so black-body emission is strong from the X-ray to the radiofrequency region of the spectrum. The spectrum of the Sun's corona is dominated by emission lines from electronically excited species, such as neutral atoms and a number of highly ionized species . The most intense emission lines in the visible range are from the Fe^{13+} ion at 530.3 nm, the Fe^{9+} ion at 637.4 nm, and the Ca^{4+} ion at 569.4 nm.

Because only light from the photosphere reaches our telescopes, the overall chemical composition of a star must be inferred from theoretical work on its interior and from spectral analysis of its atmosphere. Data on the Sun indicate that it is 92 per cent hydrogen and 7.8 per cent helium. The remaining 0.2 per cent is due to heavier elements, among which C, N, O, Ne, and Fe are the most abundant. More advanced analysis of spectra also permit the determination of other properties of stars, such as their relative speeds (Problem 10.27) and their effective temperatures (Problem 13.29).

10.6 Quantum defects and ionization limits

One application of atomic spectroscopy is to the determination of ionization energies. However, we cannot use the procedure illustrated in Example 10.1 indiscriminately because the energy levels of a many-electron atom do not in general vary as $1/n^2$. If we confine attention to the outermost electrons, then we know that, as a result of penetration and shielding, they experience a nuclear charge of slightly more than $1e$ because in a neutral atom the other $Z - 1$ electrons cancel all but about one unit of nuclear charge. Typical values of Z_{eff} are a little more than 1, so we expect binding energies to be given by a term of the form $-hcR/n^2$, but lying slightly lower in energy than this formula predicts. We therefore introduce a **quantum defect**, δ, and write the energy as $-hcR/(n - \delta)^2$. The quantum defect is best regarded as a purely empirical quantity.

There are some excited states that are so diffuse that the $1/n^2$ variation is valid: these states are called **Rydberg states**. In such cases we can write

$$\tilde{\nu} = \frac{I}{hc} - \frac{R}{n^2} \qquad (10.39)$$

and a plot of wavenumber against $1/n^2$ can be used to obtain I by extrapolation; in practice, one would use a linear regression fit using a computer. If the lower state is not the ground state (a possibility if we wish to generalize the concept of ionization energy), the ionization energy of the ground state can be determined by adding the appropriate energy difference to the ionization energy obtained as described here.

10.7 Singlet and triplet states

Suppose we were interested in the energy levels of a He atom, with its two electrons. We know that the ground-state configuration is $1s^2$, and can anticipate that an excited configuration will be one in which one of the electrons has been promoted into a $2s$ orbital, giving the configuration $1s^12s^1$. The two electrons need not be paired because they occupy different orbitals. According to Hund's maximum multiplicity rule, the state of the atom with the spins parallel lies lower in energy than the state in which they are paired. Both states are permissible, and can contribute to the spectrum of the atom.

Parallel and antiparallel (paired) spins differ in their overall spin angular momentum. In the paired case, the two spin momenta cancel each other, and there is zero net spin (as was depicted in Fig. 10.18). The paired-spin arrangement is called a **singlet**. Its spin state is the one we denoted σ_- in the discussion of the Pauli principle:

$$\sigma_-(1,2) = (1/2^{1/2})\{\alpha(1)\beta(2) - \beta(1)\alpha(2)\} \qquad (10.40a)$$

The angular momenta of two parallel spins add together to give a nonzero total spin, and the resulting state is called a **triplet**. As illustrated in Fig. 10.24, there are three ways of achieving a nonzero total spin, but only one way to achieve zero spin. The three spin states are the symmetric combinations introduced earlier:

$$\alpha(1)\alpha(2) \qquad \sigma_+(1,2) = (1/2^{1/2})\{\alpha(1)\beta(2) + \beta(1)\alpha(2)\} \qquad \beta(1)\beta(2) \qquad (10.40b)$$

The fact that the parallel arrangement of spins in the $1s^12s^1$ configuration of the He atom lies lower in energy than the antiparallel arrangement can now be expressed by saying that the triplet state of the $1s^12s^1$ configuration of He lies lower in energy than the singlet state. This is a general conclusion that applies to other atoms (and molecules) and, *for states arising from the same configuration, the triplet state generally lies lower than the singlet state.* The origin of the energy difference lies in the effect of spin correlation on the Coulombic interactions between electrons, as we saw in the case of Hund's rule for ground-state configurations. Because the Coulombic interaction between electrons in an atom is strong, the difference in energies between singlet and triplet states of the same configuration can be large. The two states of $1s^12s^1$ He, for instance, differ by 6421 cm^{-1} (corresponding to 0.80 eV).

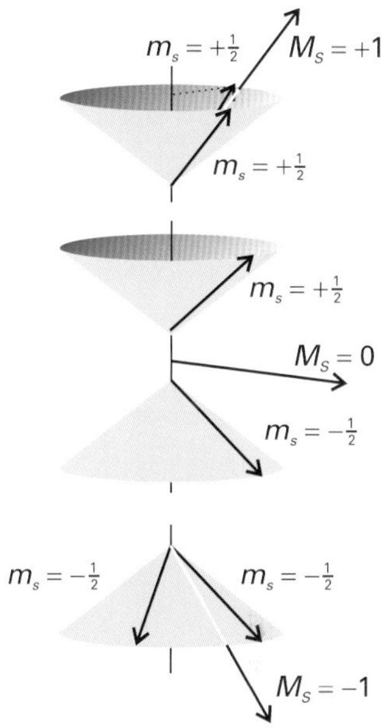

Fig. 10.24 When two electrons have parallel spins, they have a nonzero total spin angular momentum. There are three ways of achieving this resultant, which are shown by these vector representations. Note that, although we cannot know the orientation of the spin vectors on the cones, the angle between the vectors is the same in all three cases, for all three arrangements have the same total spin angular momentum (that is, the resultant of the two vectors has the same length in each case, but points in different directions). Compare this diagram with Fig. 10.18, which shows the antiparallel case. Note that, whereas two paired spins are precisely antiparallel, two 'parallel' spins are not strictly parallel.

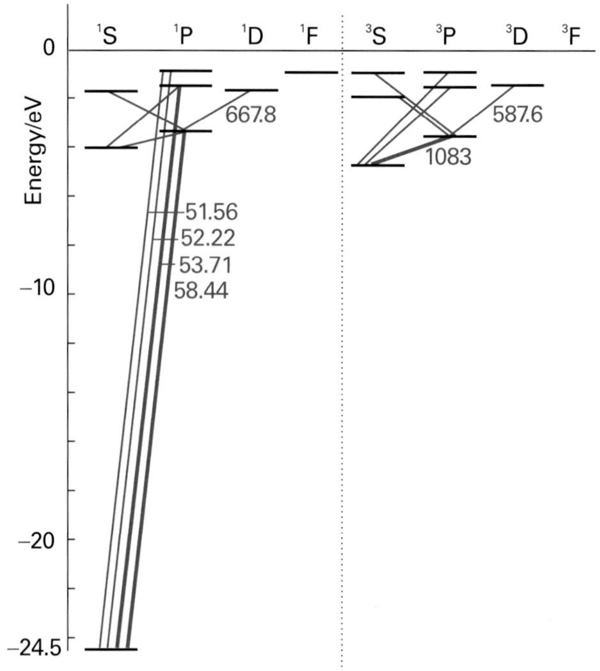

Fig. 10.25 Part of the Grotrian diagram for a helium atom. Note that there are no transitions between the singlet and triplet levels.

The spectrum of atomic helium is more complicated than that of atomic hydrogen, but there are two simplifying features. One is that the only excited configurations it is necessary to consider are of the form $1s^1nl^1$: that is, only one electron is excited. Excitation of two electrons requires an energy greater than the ionization energy of the atom, so the He^+ ion is formed instead of the doubly excited atom. Second, no radiative transitions take place between singlet and triplet states because the relative orientation of the two electron spins cannot change during a transition. Thus, there is a spectrum arising from transitions between singlet states (including the ground state) and between triplet states, but not between the two. Spectroscopically, helium behaves like two distinct species, and the early spectroscopists actually thought of helium as consisting of 'parahelium' and 'orthohelium'. The Grotrian diagram for helium in Fig. 10.25 shows the two sets of transitions.

Comment 10.6

We have already remarked that the electron's spin is a purely quantum mechanical phenomenon that has no classical counterpart. However a classical model can give us partial insight into the origin of an electron's magnetic moment. Namely, the magnetic field generated by a spinning electron, regarded classically as a moving charge, induces a magnetic moment. This model is merely a visualization aid and cannot be used to explain the magnitude of the magnetic moment of the electron or the origin of spin magnetic moments in electrically neutral particles, such as the neutron.

10.8 Spin–orbit coupling

An electron has a magnetic moment that arises from its spin (Fig. 10.26). Similarly, an electron with orbital angular momentum (that is, an electron in an orbital with $l > 0$) is in effect a circulating current, and possesses a magnetic moment that arises from its orbital momentum. The interaction of the spin magnetic moment with the magnetic field arising from the orbital angular momentum is called **spin–orbit coupling**. The strength of the coupling, and its effect on the energy levels of the atom, depend on the relative orientations of the spin and orbital magnetic moments, and therefore on the relative orientations of the two angular momenta (Fig. 10.27).

(a) The total angular momentum

One way of expressing the dependence of the spin–orbit interaction on the relative orientation of the spin and orbital momenta is to say that it depends on the total angular momentum of the electron, the vector sum of its spin and orbital momenta. Thus, when the spin and orbital angular momenta are nearly parallel, the total angular

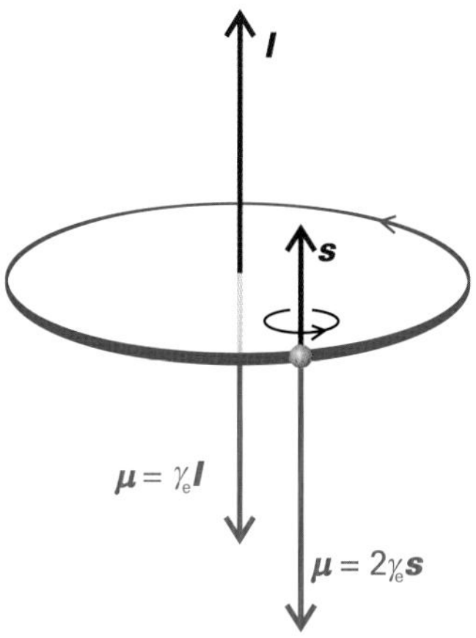

Fig. 10.26 Angular momentum gives rise to a magnetic moment ($\boldsymbol{\mu}$). For an electron, the magnetic moment is antiparallel to the orbital angular momentum, but proportional to it. For spin angular momentum, there is a factor 2, which increases the magnetic moment to twice its expected value (see Section 10.10).

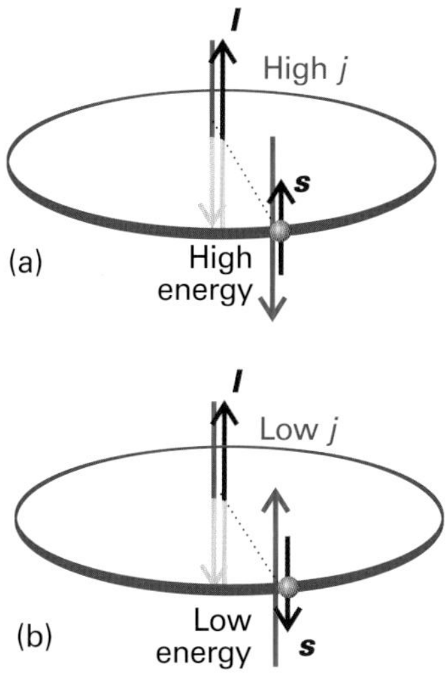

Fig. 10.27 Spin–orbit coupling is a magnetic interaction between spin and orbital magnetic moments. When the angular momenta are parallel, as in (a), the magnetic moments are aligned unfavourably; when they are opposed, as in (b), the interaction is favourable. This magnetic coupling is the cause of the splitting of a configuration into levels.

momentum is high; when the two angular momenta are opposed, the total angular momentum is low.

The total angular momentum of an electron is described by the quantum numbers j and m_j, with $j = l + \frac{1}{2}$ (when the two angular momenta are in the same direction) or $j = l - \frac{1}{2}$ (when they are opposed, Fig. 10.28). The different values of j that can arise for a given value of l label **levels** of a term. For $l = 0$, the only permitted value is $j = \frac{1}{2}$ (the total angular momentum is the same as the spin angular momentum because there is no other source of angular momentum in the atom). When $l = 1$, j may be either $\frac{3}{2}$ (the spin and orbital angular momenta are in the same sense) or $\frac{1}{2}$ (the spin and angular momenta are in opposite senses).

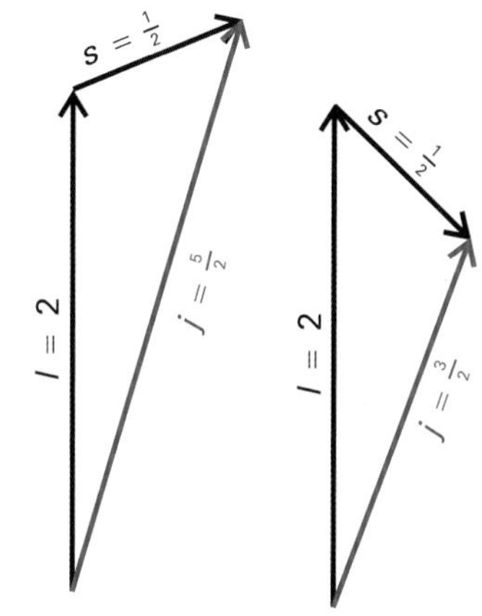

Fig. 10.28 The coupling of the spin and orbital angular momenta of a d electron ($l = 2$) gives two possible values of j depending on the relative orientations of the spin and orbital angular momenta of the electron.

Example 10.4 *Identifying the levels of a configuration*

Identify the levels that may arise from the configurations (a) d^1, (b) s^1.

Method In each case, identify the value of l and then the possible values of j. For these one-electron systems, the total angular momentum is the sum and difference of the orbital and spin momenta.

Answer (a) For a d electron, $l = 2$ and there are two levels in the configuration, one with $j = 2 + \frac{1}{2} = \frac{5}{2}$ and the other with $j = 2 - \frac{1}{2} = \frac{3}{2}$. (b) For an s electron $l = 0$, so only one level is possible, and $j = \frac{1}{2}$.

Self-test 10.8 Identify the levels of the configurations (a) p^1 and (b) f^1.

[(a) $\frac{3}{2}, \frac{1}{2}$; (b) $\frac{7}{2}, \frac{5}{2}$]

The dependence of the spin–orbit interaction on the value of j is expressed in terms of the **spin–orbit coupling constant**, A (which is typically expressed as a wavenumber). A quantum mechanical calculation leads to the result that the energies of the levels with quantum numbers s, l, and j are given by

$$E_{l,s,j} = \tfrac{1}{2}hcA\{j(j+1) - l(l+1) - s(s+1)\} \qquad (10.41)$$

Justification 10.8 *The energy of spin–orbit interaction*

The energy of a magnetic moment $\boldsymbol{\mu}$ in a magnetic field $\boldsymbol{B}$ is equal to their scalar product $-\boldsymbol{\mu}\cdot\boldsymbol{B}$. If the magnetic field arises from the orbital angular momentum of the electron, it is proportional to $\boldsymbol{l}$; if the magnetic moment $\boldsymbol{\mu}$ is that of the electron spin, then it is proportional to $\boldsymbol{s}$. It then follows that the energy of interaction is proportional to the scalar product $\boldsymbol{s}\cdot\boldsymbol{l}$:

$$\text{Energy of interaction} = -\boldsymbol{\mu}\cdot\boldsymbol{B} \propto \boldsymbol{s}\cdot\boldsymbol{l}$$

We take this expression to be the first-order perturbation contribution to the hamiltonian. Next, we note that the total angular momentum is the vector sum of the spin and orbital momenta: $\boldsymbol{j} = \boldsymbol{l} + \boldsymbol{s}$. The magnitude of the vector $\boldsymbol{j}$ is calculated by evaluating

$$\boldsymbol{j}\cdot\boldsymbol{j} = (\boldsymbol{l}+\boldsymbol{s})\cdot(\boldsymbol{l}+\boldsymbol{s}) = \boldsymbol{l}\cdot\boldsymbol{l} + \boldsymbol{s}\cdot\boldsymbol{s} + 2\boldsymbol{s}\cdot\boldsymbol{l}$$

That is,

$$\boldsymbol{s}\cdot\boldsymbol{l} = \tfrac{1}{2}\{j^2 - l^2 - s^2\}$$

where we have used the fact that the scalar product of two vectors $\boldsymbol{u}$ and $\boldsymbol{v}$ is $\boldsymbol{u}\cdot\boldsymbol{v} = uv\cos\theta$, from which it follows that $\boldsymbol{u}\cdot\boldsymbol{u} = u^2$.

The preceding equation is a classical result. To make the transition to quantum mechanics, we treat all the quantities as operators, and write

$$\hat{\boldsymbol{s}}\cdot\hat{\boldsymbol{l}} = \tfrac{1}{2}\{\hat{\boldsymbol{j}}^2 - \hat{\boldsymbol{l}}^2 - \hat{\boldsymbol{s}}^2\} \qquad (10.42)$$

At this point, we calculate the first-order correction to the energy by evaluating the expectation value:

$$\langle j,l,s|\hat{\boldsymbol{s}}\cdot\hat{\boldsymbol{l}}|j,l,s\rangle = \tfrac{1}{2}\langle j,l,s|\hat{\boldsymbol{j}}^2 - \hat{\boldsymbol{l}}^2 - \hat{\boldsymbol{s}}^2|j,l,s\rangle = \tfrac{1}{2}\{j(j+1) - l(l+1) - s(s+1)\}\hbar^2 \qquad (10.43)$$

Then, by inserting this expression into the formula for the energy of interaction ($E \propto \boldsymbol{s}\cdot\boldsymbol{l}$), and writing the constant of proportionality as $hcA/\hbar^2$, we obtain eqn 10.42. The calculation of A is much more complicated: see *Further reading*.

Comment 10.7

The scalar product (or dot product) $\boldsymbol{u}\cdot\boldsymbol{v}$ of two vectors $\boldsymbol{u}$ and $\boldsymbol{v}$ with magnitudes u and v is $\boldsymbol{u}\cdot\boldsymbol{v} = uv\cos\theta$, where θ is the angle between the two vectors.

Illustration 10.3 *Calculating the energies of levels*

The unpaired electron in the ground state of an alkali metal atom has $l = 0$, so $j = \tfrac{1}{2}$. Because the orbital angular momentum is zero in this state, the spin–orbit coupling energy is zero (as is confirmed by setting $j = s$ and $l = 0$ in eqn 10.41). When the electron is excited to an orbital with $l = 1$, it has orbital angular momentum and can give rise to a magnetic field that interacts with its spin. In this configuration the electron can have $j = \tfrac{3}{2}$ or $j = \tfrac{1}{2}$, and the energies of these levels are

$$E_{3/2} = \tfrac{1}{2}hcA\{\tfrac{3}{2}\times\tfrac{5}{2} - 1\times 2 - \tfrac{1}{2}\times\tfrac{3}{2}\} = \tfrac{1}{2}hcA$$
$$E_{1/2} = \tfrac{1}{2}hcA\{\tfrac{1}{2}\times\tfrac{3}{2} - 1\times 2 - \tfrac{1}{2}\times\tfrac{3}{2}\} = -hcA$$

The corresponding energies are shown in Fig. 10.29. Note that the baricentre (the 'centre of gravity') of the levels is unchanged, because there are four states of energy $\tfrac{1}{2}hcA$ and two of energy $-hcA$.

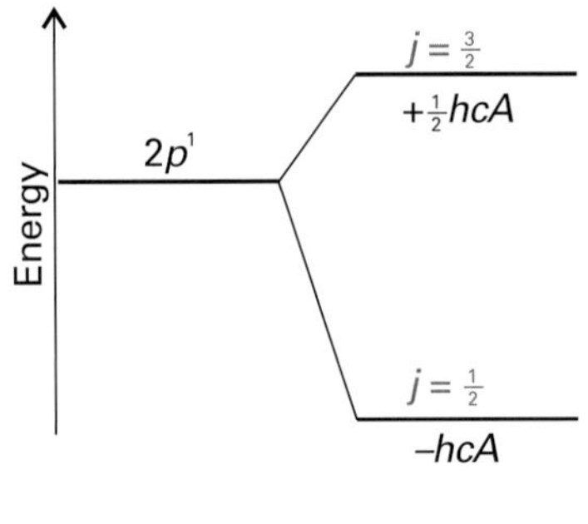

Fig. 10.29 The levels of a ^{2}P term arising from spin–orbit coupling. Note that the low-j level lies below the high-j level in energy.

The strength of the spin–orbit coupling depends on the nuclear charge. To understand why this is so, imagine riding on the orbiting electron and seeing a charged nucleus apparently orbiting around us (like the Sun rising and setting). As a result, we find ourselves at the centre of a ring of current. The greater the nuclear charge, the greater this current, and therefore the stronger the magnetic field we detect. Because the spin magnetic moment of the electron interacts with this orbital magnetic field, it follows that the greater the nuclear charge, the stronger the spin–orbit interaction. The coupling increases sharply with atomic number (as Z^4). Whereas it is only small in H (giving rise to shifts of energy levels of no more than about 0.4 cm^{-1}), in heavy atoms like Pb it is very large (giving shifts of the order of thousands of reciprocal centimetres).

(b) Fine structure

Two spectral lines are observed when the p electron of an electronically excited alkali metal atom undergoes a transition and falls into a lower s orbital. One line is due to a transition starting in a $j=\frac{3}{2}$ level and the other line is due to a transition starting in the $j=\frac{1}{2}$ level of the same configuration. The two lines are an example of the **fine structure** of a spectrum, the structure in a spectrum due to spin–orbit coupling. Fine structure can be clearly seen in the emission spectrum from sodium vapour excited by an electric discharge (for example, in one kind of street lighting). The yellow line at 589 nm (close to 17 000 cm^{-1}) is actually a doublet composed of one line at 589.76 nm (16 956.2 cm^{-1}) and another at 589.16 nm (16 973.4 cm^{-1}); the components of this doublet are the 'D lines' of the spectrum (Fig. 10.30). Therefore, in Na, the spin–orbit coupling affects the energies by about 17 cm^{-1}.

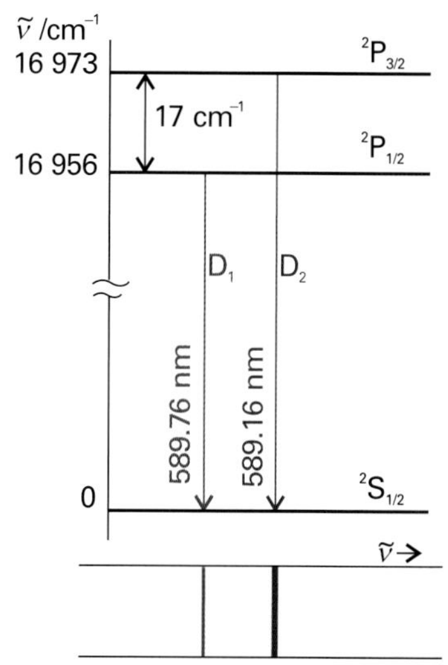

Fig. 10.30 The energy-level diagram for the formation of the sodium D lines. The splitting of the spectral lines (by 17 cm^{-1}) reflects the splitting of the levels of the ^2P term.

Example 10.5 *Analysing a spectrum for the spin–orbit coupling constant*

The origin of the D lines in the spectrum of atomic sodium is shown in Fig. 10.30. Calculate the spin–orbit coupling constant for the upper configuration of the Na atom.

Method We see from Fig. 10.30 that the splitting of the lines is equal to the energy separation of the $j=\frac{3}{2}$ and $\frac{1}{2}$ levels of the excited configuration. This separation can be expressed in terms of A by using eqn 10.40. Therefore, set the observed splitting equal to the energy separation calculated from eqn 10.40 and solve the equation for A.

Answer The two levels are split by

$$\Delta\tilde{\nu} = A\tfrac{1}{2}\{\tfrac{3}{2}(\tfrac{3}{2}+1) - \tfrac{1}{2}(\tfrac{1}{2}+1)\} = \tfrac{3}{2}A$$

The experimental value is 17.2 cm^{-1}; therefore

$$A = \tfrac{2}{3}\times(17.2\ \mathrm{cm^{-1}}) = 11.5\ \mathrm{cm^{-1}}$$

The same calculation repeated for the other alkali metal atoms gives Li: 0.23 cm^{-1}, K: 38.5 cm^{-1}, Rb: 158 cm^{-1}, Cs: 370 cm^{-1}. Note the increase of A with atomic number (but more slowly than Z^4 for these many-electron atoms).

Self-test 10.9 The configuration . . . $4p^65d^1$ of rubidium has two levels at 25 700.56 cm^{-1} and 25 703.52 cm^{-1} above the ground state. What is the spin–orbit coupling constant in this excited state? [1.18 cm^{-1}]

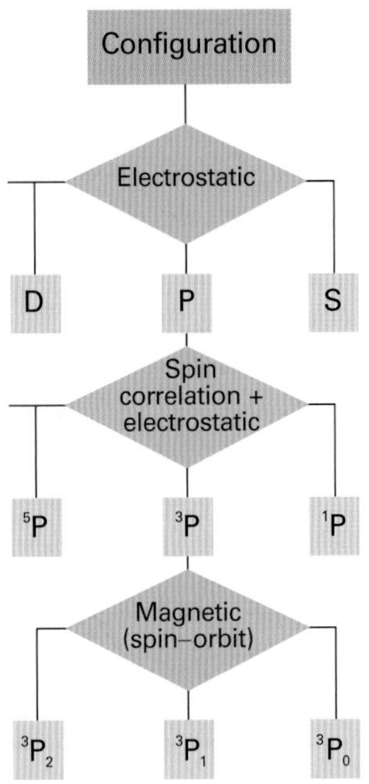

Fig. 10.31 A summary of the types of interaction that are responsible for the various kinds of splitting of energy levels in atoms. For light atoms, magnetic interactions are small, but in heavy atoms they may dominate the electrostatic (charge–charge) interactions.

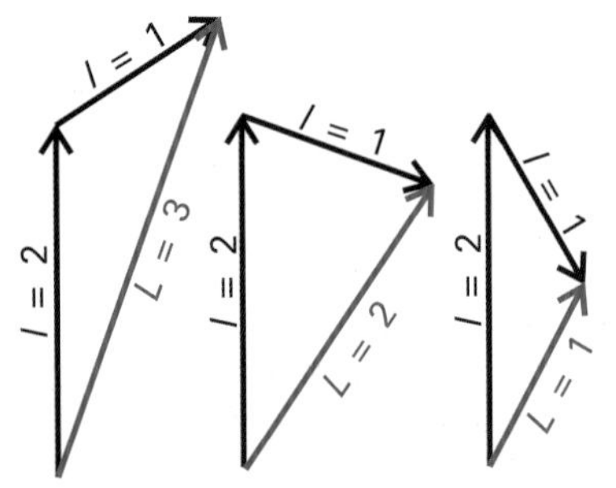

Fig. 10.32 The total angular orbital momenta of a p electron and a d electron correspond to $L = 3$, 2, and 1 and reflect the different relative orientations of the two momenta.

10.9 Term symbols and selection rules

We have used expressions such as 'the $j = \frac{3}{2}$ level of a configuration'. A **term symbol**, which is a symbol looking like $^2P_{3/2}$ or $^3D_2$, conveys this information much more succinctly. The convention of using lowercase letters to label orbitals and uppercase letters to label overall states applies throughout spectroscopy, not just to atoms.

A term symbol gives three pieces of information:

1 The letter (P or D in the examples) indicates the total orbital angular momentum quantum number, L.

2 The left superscript in the term symbol (the 2 in $^2P_{3/2}$) gives the multiplicity of the term.

3 The right subscript on the term symbol (the $\frac{3}{2}$ in $^2P_{3/2}$) is the value of the total angular momentum quantum number, J.

We shall now say what each of these statements means; the contributions to the energies which we are about to discuss are summarized in Fig. 10.31.

(a) The total orbital angular momentum

When several electrons are present, it is necessary to judge how their individual orbital angular momenta add together or oppose each other. The **total orbital angular momentum quantum number**, L, tells us the magnitude of the angular momentum through $\{L(L+1)\}^{1/2}\hbar$. It has $2L+1$ orientations distinguished by the quantum number M_L, which can take the values $L, L-1, \ldots, -L$. Similar remarks apply to the **total spin quantum number**, S, and the quantum number M_S, and the **total angular momentum quantum number**, J, and the quantum number M_J.

The value of L (a non-negative integer) is obtained by coupling the individual orbital angular momenta by using the **Clebsch–Gordan series:**

$$L = l_1 + l_2, l_1 + l_2 - 1, \ldots, |l_1 - l_2| \tag{10.44}$$

The modulus signs are attached to $l_1 - l_2$ because L is non-negative. The maximum value, $L = l_1 + l_2$, is obtained when the two orbital angular momenta are in the same direction; the lowest value, $|l_1 - l_2|$, is obtained when they are in opposite directions. The intermediate values represent possible intermediate relative orientations of the two momenta (Fig. 10.32). For two p electrons (for which $l_1 = l_2 = 1$), $L = 2, 1, 0$. The code for converting the value of L into a letter is the same as for the $s, p, d, f, \ldots$ designation of orbitals, but uses uppercase Roman letters:

L:	0	1	2	3	4	5	6...
	S	P	D	F	G	H	I...

Thus, a p^2 configuration can give rise to D, P, and S terms. The terms differ in energy on account of the different spatial distribution of the electrons and the consequent differences in repulsion between them.

A closed shell has zero orbital angular momentum because all the individual orbital angular momenta sum to zero. Therefore, when working out term symbols, we need consider only the electrons of the unfilled shell. In the case of a single electron outside a closed shell, the value of L is the same as the value of l; so the configuration [Ne]$3s^1$ has only an S term.

Example 10.6 *Deriving the total orbital angular momentum of a configuration*

Find the terms that can arise from the configurations (a) d^2, (b) p^3.

Method Use the Clebsch–Gordan series and begin by finding the minimum value of L (so that we know where the series terminates). When there are more than two electrons to couple together, use two series in succession: first couple two electrons, and then couple the third to each combined state, and so on.

Answer (a) The minimum value is $|l_1 - l_2| = |2 - 2| = 0$. Therefore,

$$L = 2 + 2, 2 + 2 - 1, \ldots, 0 = 4, 3, 2, 1, 0$$

corresponding to G, F, D, P, S terms, respectively. (b) Coupling two electrons gives a minimum value of $|1 - 1| = 0$. Therefore,

$$L' = 1 + 1, 1 + 1 - 1, \ldots, 0 = 2, 1, 0$$

Now couple l_3 with $L' = 2$, to give $L = 3, 2, 1$; with $L' = 1$, to give $L = 2, 1, 0$; and with $L' = 0$, to give $L = 1$. The overall result is

$$L = 3, 2, 2, 1, 1, 1, 0$$

giving one F, two D, three P, and one S term.

Self-test 10.10 Repeat the question for the configurations (a) f^1d^1 and (b) d^3.
[(a) H, G, F, D, P; (b) I, 2H, 3G, 4F, 5D, 3P, S]

Comment 10.8

Throughout our discussion of atomic spectroscopy, distinguish italic S, the total spin quantum number, from Roman S, the term label.

(b) The multiplicity

When there are several electrons to be taken into account, we must assess their total spin angular momentum quantum number, S (a non-negative integer or half integer). Once again, we use the Clebsch–Gordan series in the form

$$S = s_1 + s_2, s_1 + s_2 - 1, \ldots, |s_1 - s_2| \qquad (10.45)$$

to decide on the value of S, noting that each electron has $s = \frac{1}{2}$, which gives $S = 1, 0$ for two electrons (Fig. 10.33). If there are three electrons, the total spin angular momentum is obtained by coupling the third spin to each of the values of S for the first two spins, which results in $S = \frac{3}{2}$, and $S = \frac{1}{2}$.

The **multiplicity** of a term is the value of $2S + 1$. When $S = 0$ (as for a closed shell, like $1s^2$) the electrons are all paired and there is no net spin: this arrangement gives a singlet term, ^{1}S. A single electron has $S = s = \frac{1}{2}$, so a configuration such as [Ne]$3s^1$ can give rise to a doublet term, ^{2}S. Likewise, the configuration [Ne]$3p^1$ is a doublet, ^{2}P. When there are two unpaired electrons $S = 1$, so $2S + 1 = 3$, giving a triplet term, such as ^{3}D. We discussed the relative energies of singlets and triplets in Section 10.7 and saw that their energies differ on account of the different effects of spin correlation.

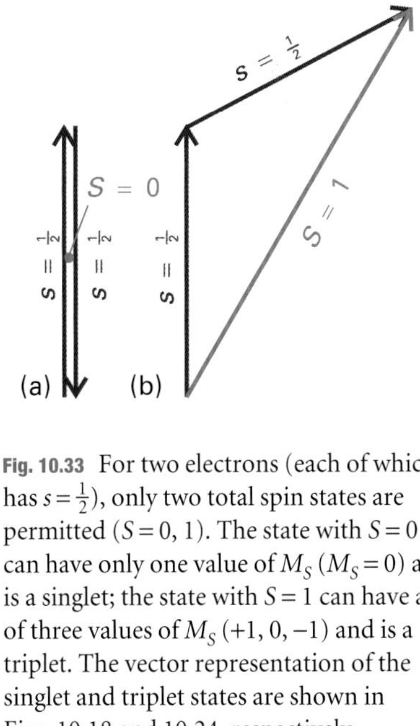

Fig. 10.33 For two electrons (each of which has $s = \frac{1}{2}$), only two total spin states are permitted ($S = 0, 1$). The state with $S = 0$ can have only one value of M_S ($M_S = 0$) and is a singlet; the state with $S = 1$ can have any of three values of M_S (+1, 0, −1) and is a triplet. The vector representation of the singlet and triplet states are shown in Figs. 10.18 and 10.24, respectively.

(c) The total angular momentum

As we have seen, the quantum number j tells us the relative orientation of the spin and orbital angular momenta of a single electron. The **total angular momentum quantum number**, J (a non-negative integer or half integer), does the same for several electrons. If there is a single electron outside a closed shell, $J = j$, with j either $l + \frac{1}{2}$ or $|l - \frac{1}{2}|$. The [Ne]$3s^1$ configuration has $j = \frac{1}{2}$ (because $l = 0$ and $s = \frac{1}{2}$), so the 2s term has a single level, which we denote ^{2}S$_{1/2}$. The [Ne]$3p^1$ configuration has $l = 1$; therefore

$j = \frac{3}{2}$ and $\frac{1}{2}$; the ^{2}P term therefore has two levels, $^2P_{3/2}$ and $^2P_{1/2}$. These levels lie at different energies on account of the magnetic spin–orbit interaction.

If there are several electrons outside a closed shell we have to consider the coupling of all the spins and all the orbital angular momenta. This complicated problem can be simplified when the spin–orbit coupling is weak (for atoms of low atomic number), for then we can use the **Russell–Saunders coupling** scheme. This scheme is based on the view that, if spin–orbit coupling is weak, then it is effective only when all the orbital momenta are operating cooperatively. We therefore imagine that all the orbital angular momenta of the electrons couple to give a total L, and that all the spins are similarly coupled to give a total S. Only at this stage do we imagine the two kinds of momenta coupling through the spin–orbit interaction to give a total J. The permitted values of J are given by the Clebsch–Gordan series

$$J = L + S, L + S - 1, \ldots, |L - S| \tag{10.46}$$

For example, in the case of the ^{3}D term of the configuration [Ne]$2p^13p^1$, the permitted values of J are 3, 2, 1 (because ^{3}D has $L = 2$ and $S = 1$), so the term has three levels, $^3D_3$, $^3D_2$, and $^3D_1$.

When $L \geq S$, the multiplicity is equal to the number of levels. For example, a ^{2}P term has the two levels $^2P_{3/2}$ and $^2P_{1/2}$, and ^{3}D has the three levels $^3D_3$, $^3D_2$, and $^3D_1$. However, this is not the case when $L < S$: the term ^{2}S, for example, has only the one level $^2S_{1/2}$.

Example 10.7 *Deriving term symbols*

Write the term symbols arising from the ground-state configurations of (a) Na and (b) F, and (c) the excited configuration $1s^22s^22p^13p^1$ of C.

Method Begin by writing the configurations, but ignore inner closed shells. Then couple the orbital momenta to find L and the spins to find S. Next, couple L and S to find J. Finally, express the term as $^{2S+1}\{L\}_J$, where $\{L\}$ is the appropriate letter. For F, for which the valence configuration is $2p^5$, treat the single gap in the closed-shell $2p^6$ configuration as a single particle.

Answer (a) For Na, the configuration is [Ne]$3s^1$, and we consider the single $3s$ electron. Because $L = l = 0$ and $S = s = \frac{1}{2}$, it is possible for $J = j = s = \frac{1}{2}$ only. Hence the term symbol is $^2S_{1/2}$. (b) For F, the configuration is [He]$2s^22p^5$, which we can treat as [Ne]$2p^{-1}$ (where the notation $2p^{-1}$ signifies the absence of a $2p$ electron). Hence $L = 1$, and $S = s = \frac{1}{2}$. Two values of $J = j$ are allowed: $J = \frac{3}{2}, \frac{1}{2}$. Hence, the term symbols for the two levels are $^2P_{3/2}$, $^2P_{1/2}$. (c) We are treating an excited configuration of carbon because, in the ground configuration, $2p^2$, the Pauli principle forbids some terms, and deciding which survive (^{1}D, ^{3}P, ^{1}S, in fact) is quite complicated. That is, there is a distinction between 'equivalent electrons', which are electrons that occupy the same orbitals, and 'inequivalent electrons', which are electrons that occupy different orbitals. For information about how to deal with equivalent electrons, see *Further reading*. The excited configuration of C under consideration is effectively $2p^13p^1$. This is a two-electron problem, and $l_1 = l_2 = 1$, $s_1 = s_2 = \frac{1}{2}$. It follows that $L = 2, 1, 0$ and $S = 1, 0$. The terms are therefore ^{3}D and ^{1}D, ^{3}P and ^{1}P, and ^{3}S and ^{1}S. For ^{3}D, $L = 2$ and $S = 1$; hence $J = 3, 2, 1$ and the levels are $^3D_3$, $^3D_2$, and $^3D_1$. For ^{1}D, $L = 2$ and $S = 0$, so the single level is $^1D_2$. The triplet of levels of ^{3}P is $^3P_2$, $^3P_1$, and $^3P_0$, and the singlet is $^1P_1$. For the ^{3}S term there is only one level, $^3S_1$ (because $J = 1$ only), and the singlet term is $^1S_0$.

Self-test 10.11 Write down the terms arising from the configurations (a) $2s^1 2p^1$, (b) $2p^1 3d^1$.

[(a) $^3P_2$, $^3P_1$, $^3P_0$, $^1P_1$;
(b) $^3F_4$, $^3F_3$, $^3F_2$, $^1F_3$, $^3D_3$, $^3D_2$, $^3D_1$, $^1D_2$, $^3P_1$, $^3P_0$, $^1P_1$]

Russell–Saunders coupling fails when the spin–orbit coupling is large (in heavy atoms). In that case, the individual spin and orbital momenta of the electrons are coupled into individual j values; then these momenta are combined into a grand total, J. This scheme is called ***jj*-coupling**. For example, in a p^2 configuration, the individual values of j are $\frac{3}{2}$ and $\frac{1}{2}$ for each electron. If the spin and the orbital angular momentum of each electron are coupled together strongly, it is best to consider each electron as a particle with angular momentum $j = \frac{3}{2}$ or $\frac{1}{2}$. These individual total momenta then couple as follows:

$j_1 = \frac{3}{2}$ and $j_2 = \frac{3}{2}$ $\quad J = 3, 2, 1, 0$
$j_1 = \frac{3}{2}$ and $j_2 = \frac{1}{2}$ $\quad J = 2, 1$
$j_1 = \frac{1}{2}$ and $j_2 = \frac{3}{2}$ $\quad J = 2, 1$
$j_1 = \frac{1}{2}$ and $j_2 = \frac{1}{2}$ $\quad J = 1, 0$

For heavy atoms, in which *jj*-coupling is appropriate, it is best to discuss their energies using these quantum numbers.

Although *jj*-coupling should be used for assessing the energies of heavy atoms, the term symbols derived from Russell–Saunders coupling can still be used as labels. To see why this procedure is valid, we need to examine how the energies of the atomic states change as the spin–orbit coupling increases in strength. Such a **correlation diagram** is shown in Fig. 10.34. It shows that there is a correspondence between the low spin–orbit coupling (Russell–Saunders coupling) and high spin–orbit coupling (*jj*-coupling) schemes, so the labels derived by using the Russell–Saunders scheme can be used to label the states of the *jj*-coupling scheme.

(d) Selection rules

Any state of the atom, and any spectral transition, can be specified by using term symbols. For example, the transitions giving rise to the yellow sodium doublet (which were shown in Fig. 10.30) are

$$3p^1\ ^2P_{3/2} \rightarrow 3s^1\ ^2S_{1/2} \qquad 3p^1\ ^2P_{1/2} \rightarrow 3s^1\ ^2S_{1/2}$$

By convention, the upper term precedes the lower. The corresponding absorptions are therefore denoted

$$^2P_{3/2} \leftarrow {}^2S_{1/2} \qquad ^2P_{1/2} \leftarrow {}^2S_{1/2}$$

(The configurations have been omitted.)

We have seen that selection rules arise from the conservation of angular momentum during a transition and from the fact that a photon has a spin of 1. They can therefore be expressed in terms of the term symbols, because the latter carry information about angular momentum. A detailed analysis leads to the following rules:

$$\Delta S = 0 \qquad \Delta L = 0, \pm 1 \qquad \Delta l = \pm 1 \qquad \Delta J = 0, \pm 1, \text{ but } J = 0 \leftarrow\!|\!\rightarrow J = 0 \qquad (10.47)$$

where the symbol $\leftarrow\!|\!\rightarrow$ denotes a forbidden transition. The rule about ΔS (no change of overall spin) stems from the fact that the light does not affect the spin directly. The rules about ΔL and Δl express the fact that the orbital angular momentum of an

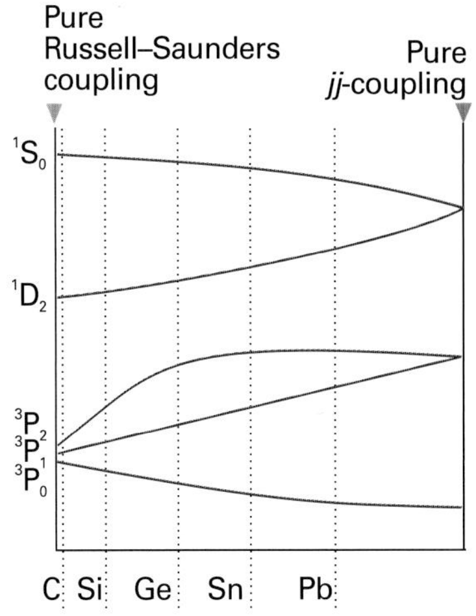

Fig. 10.34 The correlation diagram for some of the states of a two-electron system. All atoms lie between the two extremes, but the heavier the atom, the closer it lies to the pure *jj*-coupling case.

individual electron must change (so $\Delta l = \pm 1$), but whether or not this results in an overall change of orbital momentum depends on the coupling.

The selection rules given above apply when Russell–Saunders coupling is valid (in light atoms). If we insist on labelling the terms of heavy atoms with symbols like ^3D, then we shall find that the selection rules progressively fail as the atomic number increases because the quantum numbers S and L become ill defined as jj-coupling becomes more appropriate. As explained above, Russell–Saunders term symbols are only a convenient way of labelling the terms of heavy atoms: they do not bear any direct relation to the actual angular momenta of the electrons in a heavy atom. For this reason, transitions between singlet and triplet states (for which $\Delta S = \pm 1$), while forbidden in light atoms, are allowed in heavy atoms.

Checklist of key ideas

☐ 1. A hydrogenic atom is a one-electron atom or ion of general atomic number Z. A many-electron (polyelectronic) atom is an atom or ion with more than one electron.

☐ 2. The Lyman, Balmer, and Paschen series in the spectrum of atomic hydrogen arise, respectively, from the transitions $n \to 1$, $n \to 2$, and $n \to 3$.

☐ 3. The wavenumbers of all the spectral lines of a hydrogen atom can be expressed as $\tilde{\nu} = R_H(1/n_1^2 - 1/n_2^2)$, where R_H is the Rydberg constant for hydrogen.

☐ 4. The Ritz combination principle states that the wavenumber of any spectral line is the difference between two spectroscopic energy levels, or terms: $\tilde{\nu} = T_1 - T_2$.

☐ 5. The wavefunction of the hydrogen atom is the product of a radial wavefunction and an angular wavefunction: $\psi(r,\theta,\phi) = R(r)Y(\theta,\phi)$.

☐ 6. An atomic orbital is a one-electron wavefunction for an electron in an atom.

☐ 7. The energies of an electron in a hydrogen atom are given by $E_n = -Z^2\mu e^4/32\pi^2\varepsilon_0^2\hbar^2 n^2$, where n is the principal quantum number, $n = 1, 2, \ldots$.

☐ 8. All the orbitals of a given value of n belong to a given shell; orbitals with the same value of n but different values of l belong to different subshells.

☐ 9. The radial distribution function is a probability density that, when it is multiplied by dr, gives the probability of finding the electron anywhere in a shell of thickness dr at the radius r; $P(r) = r^2R(r)^2$.

☐ 10. A selection rule is a statement about which spectroscopic transitions are allowed; a Grotrian diagram is a diagram summarizing the energies of the states and atom and the transitions between them.

☐ 11. In the orbital approximation it is supposed that each electron occupies its 'own' orbital, $\psi(\boldsymbol{r}_1,\boldsymbol{r}_2,\ldots) = \psi(\boldsymbol{r}_1)\psi(\boldsymbol{r}_2)\ldots$.

☐ 12. The Pauli exclusion principle states that no more than two electrons may occupy any given orbital and, if two do occupy one orbital, then their spins must be paired.

☐ 13. The Pauli principle states that, when the labels of any two identical fermions are exchanged, the total wavefunction changes sign; when the labels of any two identical bosons are exchanged, the total wavefunction retains the same sign.

☐ 14. The effective nuclear charge Z_{eff} is the net charge experienced by an electron allowing for electron–electron repulsions.

☐ 15. Shielding is the effective reduction in charge of a nucleus by surrounding electrons; the shielding constant σ is given by $Z_{\text{eff}} = Z - \sigma$.

☐ 16. Penetration is the ability of an electron to be found inside inner shells and close to the nucleus.

☐ 17. The building-up (*Aufbau*) principle is the procedure for filling atomic orbitals that leads to the ground-state configuration of an atom.

☐ 18. Hund's maximum multiplicity rule states that an atom in its ground state adopts a configuration with the greatest number of unpaired electrons.

☐ 19. The first ionization energy I_1 is the minimum energy necessary to remove an electron from a many-electron atom in the gas phase; the second ionization energy I_2 is the minimum energy necessary to remove an electron from an ionized many-electron atom in the gas phase.

☐ 20. The electron affinity E_{ea} is the energy released when an electron attaches to a gas-phase atom.

☐ 21. A singlet term has $S = 0$; a triplet term has $S = 1$.

☐ 22. Spin–orbit coupling is the interaction of the spin magnetic moment with the magnetic field arising from the orbital angular momentum.

☐ 23. Fine structure is the structure in a spectrum due to spin–orbit coupling.

☐ 24. A term symbol is a symbolic specification of the state of an atom, $^{2S+1}\{L\}_J$.

☐ 25. The allowed values of a combined angular momenta are obtained by using the Clebsch–Gordan series: $J = j_1 + j_2, j_1 + j_2 - 1, \ldots |j_1 - j_2|$.

☐ 26. The multiplicity of a term is the value of $2S+1$; provided $L \geq S$, the multiplicity is the number of levels of the term.

☐ 27. A level is a group of states with a common value of J.

☐ 28. Russell–Saunders coupling is a coupling scheme based on the view that, if spin–orbit coupling is weak, then it is effective only when all the orbital momenta are operating cooperatively.

☐ 29. *jj*-Coupling is a coupling scheme based on the view that the individual spin and orbital momenta of the electrons are coupled into individual j values and these momenta are combined into a grand total, J.

☐ 30. The selection rules for spectroscopic transitions in polyelectronic atoms are: $\Delta S=0$, $\Delta L=0, \pm 1$, $\Delta l=\pm 1$, $\Delta J=0, \pm 1$, but $J=0 \leftarrow\!|\!\rightarrow J=0$.

Further reading

Articles and texts

P.W. Atkins, *Quanta: a handbook of concepts*. Oxford University Press (1991).

P.F. Bernath, *Spectra of atoms and molecules*. Oxford University Press (1995).

K. Bonin and W. Happer, Atomic spectroscopy. In *Encyclopedia of applied physics* (ed. G.L. Trigg), **2**, 245. VCH, New York (1991).

E.U. Condon and H. Odabaçi, *Atomic structure*. Cambridge University Press (1980).

C.W. Haigh, The theory of atomic spectroscopy: *jj* coupling, intermediate coupling, and configuration interaction. *J. Chem. Educ.* **72**, 206 (1995).

C.S. Johnson, Jr. and L.G. Pedersen, *Problems and solutions in quantum chemistry and physics*. Dover, New York (1986).

J.C. Morrison, A.W. Weiss, K. Kirby, and D. Cooper, Electronic structure of atoms and molecules. In *Encyclopedia of applied physics* (ed. G.L. Trigg), **6**, 45. VCH, New York (1993).

N. Shenkuan, The physical basis of Hund's rule: orbital contraction effects. *J. Chem. Educ.* **69**, 800 (1992).

L.G. Vanquickenborne, K. Pierloot, and D. Devoghel, Transition metals and the *Aufbau* principle. *J. Chem. Educ.* **71**, 469 (1994).

Sources of data and information

S. Bashkin and J.O. Stonor, Jr., *Atomic energy levels and Grotrian diagrams*. North-Holland, Amsterdam (1975–1982).

D.R. Lide (ed.), *CRC handbook of chemistry and physics*, Section 10, CRC Press, Boca Raton (2000).

Further information

Further information 10.1 *The separation of motion*

The separation of internal and external motion

Consider a one-dimensional system in which the potential energy depends only on the separation of the two particles. The total energy is

$$E=\frac{p_1^2}{2m_1}+\frac{p_2^2}{2m_2}+V$$

where $p_1=m_1\dot{x}_1$ and $p_2=m_2\dot{x}_2$, the dot signifying differentiation with respect to time. The centre of mass (Fig. 10.35) is located at

$$X=\frac{m_1}{m}x_1+\frac{m_2}{m}x_2 \qquad m=m_1+m_2$$

and the separation of the particles is $x=x_1-x_2$. It follows that

$$x_1=X+\frac{m_2}{m}x \qquad x_2=X-\frac{m_1}{m}x$$

The linear momenta of the particles can be expressed in terms of the rates of change of x and X:

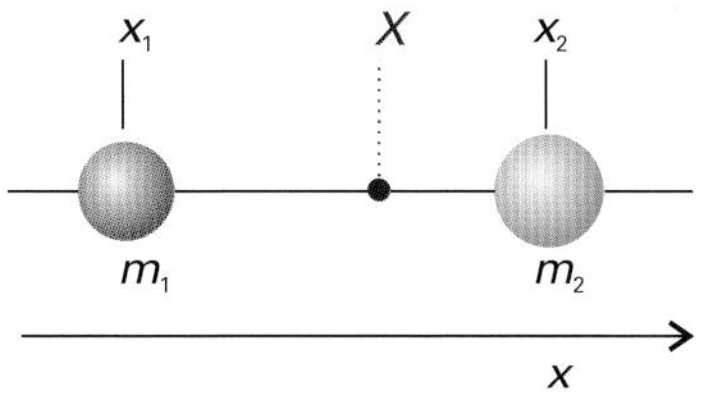

Fig. 10.35 The coordinates used for discussing the separation of the relative motion of two particles from the motion of the centre of mass.

$$p_1=m_1\dot{x}_1=m_1\dot{X}+\frac{m_1m_2}{m}\dot{x} \qquad p_2=m_2\dot{x}_2=m_2\dot{X}-\frac{m_1m_2}{m}\dot{x}$$

Then it follows that

$$\frac{p_1^2}{2m_1}+\frac{p_2^2}{2m_2}=\tfrac{1}{2}m\dot{X}^2+\tfrac{1}{2}\mu\dot{x}^2$$

where μ is given in eqn 10.6. By writing $P=m\dot{X}$ for the linear momentum of the system as a whole and defining p as $\mu\dot{x}$, we find

$$E=\frac{P^2}{2m}+\frac{p^2}{2\mu}+V$$

The corresponding hamiltonian (generalized to three dimensions) is therefore

$$\hat{H}=-\frac{\hbar^2}{2m}\nabla^2_{\text{c.m.}}-\frac{\hbar^2}{2\mu}\nabla^2+V$$

where the first term differentiates with respect to the centre of mass coordinates and the second with respect to the relative coordinates.

Now we write the overall wavefunction as the product $\psi_{\text{total}}=\psi_{\text{c.m.}}\psi$, where the first factor is a function of only the centre of mass coordinates and the second is a function of only the relative coordinates. The overall Schrödinger equation, $H\psi_{\text{total}}=E_{\text{total}}\psi_{\text{total}}$, then separates by the argument that we have used in Sections 9.2a and 9.7, with $E_{\text{total}}=E_{\text{c.m.}}+E$.

The separation of angular and radial motion

The laplacian in three dimensions is given in eqn 9.51a. It follows that the Schrödinger equation in eqn 10.6 is

$$-\frac{\hbar^2}{2\mu}\left(\frac{\partial^2}{\partial r^2}+\frac{2}{r}\frac{\partial}{\partial r}+\frac{1}{r^2}\Lambda^2\right)RY+VRY=ERY$$

Because R depends only on r and Y depends only on the angular coordinates, this equation becomes

$$-\frac{\hbar^2}{2\mu}\left(Y\frac{\mathrm{d}^2R}{\mathrm{d}r^2}+\frac{2Y}{r}\frac{\mathrm{d}R}{\mathrm{d}r}+\frac{R}{r^2}\Lambda^2Y\right)+VRY=ERY$$

If we multiply through by r^2/RY, we obtain

$$-\frac{\hbar^2}{2\mu R}\left(r^2\frac{\mathrm{d}^2R}{\mathrm{d}r^2}+2r\frac{\mathrm{d}R}{\mathrm{d}r}\right)+Vr^2-\frac{\hbar^2}{2\mu Y}\Lambda^2Y=Er^2$$

At this point we employ the usual argument. The term in Y is the only one that depends on the angular variables, so it must be a constant. When we write this constant as $\hbar^2l(l+1)/2\mu$, eqn 10.10 follows immediately.

Discussion questions

10.1 Describe the separation of variables procedure as it is applied to simplify the description of a hydrogenic atom free to move through space.

10.2 List and describe the significance of the quantum numbers needed to specify the internal state of a hydrogenic atom.

10.3 Specify and account for the selection rules for transitions in hydrogenic atoms.

10.4 Explain the significance of (a) a boundary surface and (b) the radial distribution function for hydrogenic orbitals.

10.5 Outline the electron configurations of many-electron atoms in terms of their location in the periodic table.

10.6 Describe and account for the variation of first ionization energies along Period 2 of the periodic table.

10.7 Describe the orbital approximation for the wavefunction of a many-electron atom. What are the limitations of the approximation?

10.8 Explain the origin of spin–orbit coupling and how it affects the appearance of a spectrum.

Exercises

10.1a When ultraviolet radiation of wavelength 58.4 nm from a helium lamp is directed on to a sample of krypton, electrons are ejected with a speed of 1.59 Mm s^{-1}. Calculate the ionization energy of krypton.

10.1b When ultraviolet radiation of wavelength 58.4 nm from a helium lamp is directed on to a sample of xenon, electrons are ejected with a speed of 1.79 Mm s^{-1}. Calculate the ionization energy of xenon.

10.2a By differentiation of the 2*s* radial wavefunction, show that it has two extrema in its amplitude, and locate them.

10.2b By differentiation of the 3*s* radial wavefunction, show that it has three extrema in its amplitude, and locate them.

10.3a Locate the radial nodes in the 3*s* orbital of an H atom.

10.3b Locate the radial nodes in the 4*p* orbital of an H atom where, in the notation of Table 10.1, the radial wavefunction is proportional to $20-10\rho+\rho^2$.

10.4a The wavefunction for the ground state of a hydrogen atom is Ne^{-r/a_0}. Determine the normalization constant N.

10.4b The wavefunction for the 2*s* orbital of a hydrogen atom is $N(2-r/a_0)\mathrm{e}^{-r/2a_0}$. Determine the normalization constant N.

10.5a Calculate the average kinetic and potential energies of an electron in the ground state of a hydrogen atom.

10.5b Calculate the average kinetic and potential energies of a 2*s* electron in a hydrogenic atom of atomic number Z.

10.6a Write down the expression for the radial distribution function of a 2*s* electron in a hydrogenic atom and determine the radius at which the electron is most likely to be found.

10.6b Write down the expression for the radial distribution function of a 3*s* electron in a hydrogenic atom and determine the radius at which the electron is most likely to be found.

10.7a Write down the expression for the radial distribution function of a 2*p* electron in a hydrogenic atom and determine the radius at which the electron is most likely to be found.

10.7b Write down the expression for the radial distribution function of a $3p$ electron in a hydrogenic atom and determine the radius at which the electron is most likely to be found.

10.8a What is the orbital angular momentum of an electron in the orbitals (a) $1s$, (b) $3s$, (c) $3d$? Give the numbers of angular and radial nodes in each case.

10.8b What is the orbital angular momentum of an electron in the orbitals (a) $4d$, (b) $2p$, (c) $3p$? Give the numbers of angular and radial nodes in each case.

10.9a Calculate the permitted values of j for (a) a d electron, (b) an f electron.

10.9b Calculate the permitted values of j for (a) a p electron, (b) an h electron.

10.10a An electron in two different states of an atom is known to have $j = \frac{3}{2}$ and $\frac{1}{2}$. What is its orbital angular momentum quantum number in each case?

10.10b What are the allowed total angular momentum quantum numbers of a composite system in which $j_1 = 5$ and $j_2 = 3$?

10.11a State the orbital degeneracy of the levels in a hydrogen atom that have energy (a) $-hcR_H$; (b) $-\frac{1}{9}hcR_H$; (c) $-\frac{1}{25}hcR_H$.

10.11b State the orbital degeneracy of the levels in a hydrogenic atom (Z in parentheses) that have energy (a) $-4hcR_{atom}$ (2); (b) $-\frac{1}{4}hcR_{atom}$ (4), and (c) $-hcR_{atom}$ (5).

10.12a What information does the term symbol $^1D_2$ provide about the angular momentum of an atom?

10.12b What information does the term symbol $^3F_4$ provide about the angular momentum of an atom?

10.13a At what radius does the probability of finding an electron at a point in the H atom fall to 50 per cent of its maximum value?

10.13b At what radius in the H atom does the radial distribution function of the ground state have (a) 50 per cent, (b) 75 per cent of its maximum value?

10.14a Which of the following transitions are allowed in the normal electronic emission spectrum of an atom: (a) $2s \rightarrow 1s$, (b) $2p \rightarrow 1s$, (c) $3d \rightarrow 2p$?

10.14b Which of the following transitions are allowed in the normal electronic emission spectrum of an atom: (a) $5d \rightarrow 2s$, (b) $5p \rightarrow 3s$, (c) $6p \rightarrow 4f$?

10.15a (a) Write the electronic configuration of the Ni^{2+} ion. (b) What are the possible values of the total spin quantum numbers S and M_S for this ion?

10.15b (a) Write the electronic configuration of the V^{2+} ion. (b) What are the possible values of the total spin quantum numbers S and M_S for this ion?

10.16a Suppose that an atom has (a) 2, (b) 3 electrons in different orbitals. What are the possible values of the total spin quantum number S? What is the multiplicity in each case?

10.16b Suppose that an atom has (a) 4, (b) 5 electrons in different orbitals. What are the possible values of the total spin quantum number S? What is the multiplicity in each case?

10.17a What atomic terms are possible for the electron configuration ns^1nd^1? Which term is likely to lie lowest in energy?

10.17b What atomic terms are possible for the electron configuration np^1nd^1? Which term is likely to lie lowest in energy?

10.18a What values of J may occur in the terms (a) ^1S, (b) ^2P, (c) ^3P? How many states (distinguished by the quantum number M_J) belong to each level?

10.18b What values of J may occur in the terms (a) ^3D, (b) ^4D, (c) ^2G? How many states (distinguished by the quantum number M_J) belong to each level?

10.19a Give the possible term symbols for (a) Li [He]$2s^1$, (b) Na [Ne]$3p^1$.

10.19b Give the possible term symbols for (a) Sc [Ar]$3d^14s^2$, (b) Br [Ar]$3d^{10}4s^24p^5$.

Problems*

Numerical problems

10.1 The *Humphreys series* is a group of lines in the spectrum of atomic hydrogen. It begins at 12 368 nm and has been traced to 3281.4 nm. What are the transitions involved? What are the wavelengths of the intermediate transitions?

10.2 A series of lines in the spectrum of atomic hydrogen lies at 656.46 nm, 486.27 nm, 434.17 nm, and 410.29 nm. What is the wavelength of the next line in the series? What is the ionization energy of the atom when it is in the lower state of the transitions?

10.3 The Li^{2+} ion is hydrogenic and has a Lyman series at 740 747 cm^{-1}, 877 924 cm^{-1}, 925 933 cm^{-1}, and beyond. Show that the energy levels are of the form $-hcR/n^2$ and find the value of R for this ion. Go on to predict the wavenumbers of the two longest-wavelength transitions of the Balmer series of the ion and find the ionization energy of the ion.

10.4 A series of lines in the spectrum of neutral Li atoms rise from combinations of $1s^22p^1\ ^2P$ with $1s^2nd^1\ ^2D$ and occur at 610.36 nm, 460.29 nm, and 413.23 nm. The d orbitals are hydrogenic. It is known that the ^2P term lies at 670.78 nm above the ground state, which is $1s^22s^1\ ^2S$. Calculate the ionization energy of the ground-state atom.

10.5‡ W.P. Wijesundera, S.H. Vosko, and F.A. Parpia (*Phys. Rev.* A **51**, 278 (1995)) attempted to determine the electron configuration of the ground state of lawrencium, element 103. The two contending configurations are [Rn]$5f^{14}7s^27p^1$ and [Rn]$5f^{14}6d7s^2$. Write down the term symbols for each of these configurations, and identify the lowest level within each configuration. Which level would be lowest according to a simple estimate of spin–orbit coupling?

10.6 The characteristic emission from K atoms when heated is purple and lies at 770 nm. On close inspection, the line is found to have two closely spaced components, one at 766.70 nm and the other at 770.11 nm. Account for this observation, and deduce what information you can.

10.7 Calculate the mass of the deuteron given that the first line in the Lyman series of H lies at 82 259.098 cm^{-1} whereas that of D lies at 82 281.476 cm^{-1}. Calculate the ratio of the ionization energies of H and D.

10.8 Positronium consists of an electron and a positron (same mass, opposite charge) orbiting round their common centre of mass. The broad features of the spectrum are therefore expected to be hydrogen-like, the differences arising largely from the mass differences. Predict the wavenumbers of the first three lines of the Balmer series of positronium. What is the binding energy of the ground state of positronium?

* Problems denoted with the symbol ‡ were supplied by Charles Trapp, Carmen Giunta, and Marshall Cady.

10.9 The *Zeeman effect* is the modification of an atomic spectrum by the application of a strong magnetic field. It arises from the interaction between applied magnetic fields and the magnetic moments due to orbital and spin angular momenta (recall the evidence provided for electron spin by the Stern–Gerlach experiment, Section 9.8). To gain some appreciation for the so-called *normal Zeeman effect*, which is observed in transitions involving singlet states, consider a *p* electron, with $l = 1$ and $m_l = 0, \pm 1$. In the absence of a magnetic field, these three states are degenerate. When a field of magnitude B is present, the degeneracy is removed and it is observed that the state with $m_l = +1$ moves up in energy by $\mu_B B$, the state with $m_l = 0$ is unchanged, and the state with $m_l = -1$ moves down in energy by $\mu_B B$, where $\mu_B = e\hbar/2m_e = 9.274 \times 10^{-24}$ J T^{-1} is the Bohr magneton (see Section 15.1). Therefore, a transition between a $^1S_0$ term and a $^1P_1$ term consists of three spectral lines in the presence of a magnetic field where, in the absence of the magnetic field, there is only one. (a) Calculate the splitting in reciprocal centimetres between the three spectral lines of a transition between a $^1S_0$ term and a $^1P_1$ term in the presence of a magnetic field of 2 T (where 1 T = 1 kg s^{-2} A^{-1}). (b) Compare the value you calculated in (a) with typical optical transition wavenumbers, such as those for the Balmer series of the H atom. Is the line splitting caused by the normal Zeeman effect relatively small or relatively large?

10.10 In 1976 it was mistakenly believed that the first of the 'superheavy' elements had been discovered in a sample of mica. Its atomic number was believed to be 126. What is the most probable distance of the innermost electrons from the nucleus of an atom of this element? (In such elements, relativistic effects are very important, but ignore them here.)

Theoretical problems

10.11 What is the most probable point (not radius) at which a 2*p* electron will be found in the hydrogen atom?

10.12 Show by explicit integration that (a) hydrogenic 1*s* and 2*s* orbitals, (b) $2p_x$ and $2p_y$ orbitals are mutually orthogonal.

10.13‡ Explicit expressions for hydrogenic orbitals are given in Tables 10.1 and 9.3. (a) Verify both that the $3p_x$ orbital is normalized (to 1) and that $3p_x$ and $3d_{xy}$ are mutually orthogonal. (b) Determine the positions of both the radial nodes and nodal planes of the 3*s*, $3p_x$, and $3d_{xy}$ orbitals. (c) Determine the mean radius of the 3*s* orbital. (d) Draw a graph of the radial distribution function for the three orbitals (of part (b)) and discuss the significance of the graphs for interpreting the properties of many-electron atoms. (e) Create both *xy*-plane polar plots and boundary surface plots for these orbitals. Construct the boundary plots so that the distance from the origin to the surface is the absolute value of the angular part of the wavefunction. Compare the *s*, *p*, and *d* boundary surface plots with that of an *f*-orbital; e.g. $\psi_f \propto x(5z^2 - r^2) \propto \sin\theta\,(5\cos^2\theta - 1)\cos\phi$.

10.14 Determine whether the p_x and p_y orbitals are eigenfunctions of l_z. If not, does a linear combination exist that is an eigenfunction of l_z?

10.15 Show that l_z and l^2 both commute with the hamiltonian for a hydrogen atom. What is the significance of this result?

10.16 The 'size' of an atom is sometimes considered to be measured by the radius of a sphere that contains 90 per cent of the charge density of the electrons in the outermost occupied orbital. Calculate the 'size' of a hydrogen atom in its ground state according to this definition. Go on to explore how the 'size' varies as the definition is changed to other percentages, and plot your conclusion.

10.17 Some atomic properties depend on the average value of $1/r$ rather than the average value of r itself. Evaluate the expectation value of $1/r$ for (a) a hydrogen 1*s* orbital, (b) a hydrogenic 2*s* orbital, (c) a hydrogenic 2*p* orbital.

10.18 One of the most famous of the obsolete theories of the hydrogen atom was proposed by Bohr. It has been replaced by quantum mechanics but, by a remarkable coincidence (not the only one where the Coulomb potential is concerned), the energies it predicts agree exactly with those obtained from the Schrödinger equation. In the Bohr atom, an electron travels in a circle around the nucleus. The Coulombic force of attraction ($Ze^2/4\pi\varepsilon_0 r^2$) is balanced by the centrifugal effect of the orbital motion. Bohr proposed that the angular momentum is limited to integral values of $\hbar$. When the two forces are balanced, the atom remains in a stationary state until it makes a spectral transition. Calculate the energies of a hydrogenic atom using the Bohr model.

10.19 The Bohr model of the atom is specified in Problem 10.18. What features of it are untenable according to quantum mechanics? How does the Bohr ground state differ from the actual ground state? Is there an experimental distinction between the Bohr and quantum mechanical models of the ground state?

10.20 Atomic units of length and energy may be based on the properties of a particular atom. The usual choice is that of a hydrogen atom, with the unit of length being the Bohr radius, a_0, and the unit of energy being the (negative of the) energy of the 1*s* orbital. If the positronium atom (e^+,e^-) were used instead, with analogous definitions of units of length and energy, what would be the relation between these two sets of atomic units?

10.21 Some of the selection rules for hydrogenic atoms were derived in *Justification* 10.4. Complete the derivation by considering the *x*- and *y*-components of the electric dipole moment operator.

10.22‡ Stern–Gerlach splittings of atomic beams are small and require either large magnetic field gradients or long magnets for their observation. For a beam of atoms with zero orbital angular momentum, such as H or Ag, the deflection is given by $x = \pm(\mu_B L^2/4E_K)\mathrm{d}B/\mathrm{d}z$, where μ_B is the Bohr magneton (Problem 10.9), L is the length of the magnet, E_K is the average kinetic energy of the atoms in the beam, and $\mathrm{d}B/\mathrm{d}z$ is the magnetic field gradient across the beam. (a) Use the Maxwell–Boltzmann velocity distribution to show that the average translational kinetic energy of the atoms emerging as a beam from a pinhole in an oven at temperature T is $2kT$. (b) Calculate the magnetic field gradient required to produce a splitting of 1.00 mm in a beam of Ag atoms from an oven at 1000 K with a magnet of length 50 cm.

10.23 The wavefunction of a many-electron closed-shell atom can expressed as a Slater determinant (Section 10.4b). A useful property of determinants is that interchanging any two rows or columns changes their sign and therefore, if any two rows or columns are identical, then the determinant vanishes. Use this property to show that (a) the wavefunction is antisymmetric under particle exchange, (b) no two electrons can occupy the same orbital with the same spin.

Applications: to astrophysics and biochemistry

10.24 Hydrogen is the most abundant element in all stars. However, neither absorption nor emission lines due to neutral hydrogen are found in the spectra of stars with effective temperatures higher than 25 000 K. Account for this observation.

10.25 The distribution of isotopes of an element may yield clues about the nuclear reactions that occur in the interior of a star. Show that it is possible to use spectroscopy to confirm the presence of both $^4He^+$ and $^3He^+$ in a star by calculating the wavenumbers of the $n = 3 \rightarrow n = 2$ and of the $n = 2 \rightarrow n = 1$ transitions for each isotope.

10.26‡ Highly excited atoms have electrons with large principal quantum numbers. Such *Rydberg atoms* have unique properties and are of interest to astrophysicists . For hydrogen atoms with large *n*, derive a relation for the separation of energy levels. Calculate this separation for $n = 100$; also calculate the average radius, the geometric cross-section, and the ionization energy. Could a thermal collision with another hydrogen atom ionize this Rydberg atom? What minimum velocity of the second atom is required? Could a normal-sized neutral H atom simply pass through the Rydberg atom leaving it undisturbed? What might the radial wavefunction for a 100*s* orbital be like?

10.27 The spectrum of a star is used to measure its *radial velocity* with respect to the Sun, the component of the star's velocity vector that is parallel to a vector connecting the star's centre to the centre of the Sun. The measurement relies on the *Doppler effect*, in which radiation is shifted in frequency when the source is moving towards or away from the observer. When a star emitting electromagnetic radiation of frequency ν moves with a speed s relative to an observer, the observer detects radiation of frequency $\nu_{\text{receding}} = \nu f$ or $\nu_{\text{approaching}} = \nu/f$, where $f = \{(1 - s/c)/(1 + s/c)\}^{1/2}$ and c is the speed of light. It is easy to see that $\nu_{\text{receding}} < \nu$ and a receding star is characterized by a *red shift* of its spectrum with respect to the spectrum of an identical, but stationary source. Furthermore, $\nu_{\text{approaching}} > \nu$ and an approaching star is characterized by a *blue shift* of its spectrum with respect to the spectrum of an identical, but stationary source. In a typical experiment, ν is the frequency of a spectral line of an element measured in a stationary Earth-bound laboratory from a calibration source, such as an arc lamp. Measurement of the same spectral line in a star gives ν_{star} and the speed of recession or approach may be calculated from the value of ν and the equations above. (a) Three Fe I lines of the star HDE 271 182, which belongs to the Large Magellanic Cloud, occur at 438.882 nm, 441.000 nm, and 442.020 nm. The same lines occur at 438.392 nm, 440.510 nm, and 441.510 nm in the spectrum of an Earth-bound iron arc. Determine whether HDE 271 182 is receding from or approaching the Earth and estimate the star's radial speed with respect to the Earth. (b) What additional information would you need to calculate the radial velocity of HDE 271 182 with respect to the Sun?

10.28 The *d*-metals iron, copper, and manganese form cations with different oxidation states. For this reason, they are found in many oxidoreductases and in several proteins of oxidative phosphorylation and photosynthesis (*Impact* I7.2 and I23.2). Explain why many *d*-metals form cations with different oxidation states.

10.29 Thallium, a neurotoxin, is the heaviest member of Group 13 of the periodic table and is found most usually in the +1 oxidation state. Aluminium, which causes anaemia and dementia, is also a member of the group but its chemical properties are dominated by the +3 oxidation state. Examine this issue by plotting the first, second, and third ionization energies for the Group 13 elements against atomic number. Explain the trends you observe. *Hints*. The third ionization energy, I_3, is the minimum energy needed to remove an electron from the doubly charged cation: $E^{2+}(g) \rightarrow E^{3+}(g) + e^-(g)$, $I_3 = E(E^{3+}) - E(E^{2+})$. For data, see the links to databases of atomic properties provided in the text's web site.

11 Molecular structure

The concepts developed in Chapter 10, particularly those of orbitals, can be extended to a description of the electronic structures of molecules. There are two principal quantum mechanical theories of molecular electronic structure. In valence-bond theory, the starting point is the concept of the shared electron pair. We see how to write the wavefunction for such a pair, and how it may be extended to account for the structures of a wide variety of molecules. The theory introduces the concepts of σ and π bonds, promotion, and hybridization that are used widely in chemistry. In molecular orbital theory (with which the bulk of the chapter is concerned), the concept of atomic orbital is extended to that of molecular orbital, which is a wavefunction that spreads over all the atoms in a molecule.

In this chapter we consider the origin of the strengths, numbers, and three-dimensional arrangement of chemical bonds between atoms. The quantum mechanical description of chemical bonding has become highly developed through the use of computers, and it is now possible to consider the structures of molecules of almost any complexity. We shall concentrate on the quantum mechanical description of the **covalent bond**, which was identified by G.N. Lewis (in 1916, before quantum mechanics was fully established) as an electron pair shared between two neighbouring atoms. We shall see, however, that the other principal type of bond, an **ionic bond**, in which the cohesion arises from the Coulombic attraction between ions of opposite charge, is also captured as a limiting case of a covalent bond between dissimilar atoms. In fact, although the Schrödinger equation might shroud the fact in mystery, all chemical bonding can be traced to the interplay between the attraction of opposite charges, the repulsion of like charges, and the effect of changing kinetic energy as the electrons are confined to various regions when bonds form.

There are two major approaches to the calculation of molecular structure, **valence-bond theory** (VB theory) and **molecular orbital theory** (MO theory). Almost all modern computational work makes use of MO theory, and we concentrate on that theory in this chapter. Valence-bond theory, though, has left its imprint on the language of chemistry, and it is important to know the significance of terms that chemists use every day. Therefore, our discussion is organized as follows. First, we set out the concepts common to all levels of description. Then we present VB theory, which gives us a simple qualitative understanding of bond formation. Next, we present the basic ideas of MO theory. Finally, we see how computational techniques pervade all current discussions of molecular structure, including the prediction of chemical reactivity.

The Born–Oppenheimer approximation

All theories of molecular structure make the same simplification at the outset. Whereas the Schrödinger equation for a hydrogen atom can be solved exactly, an exact solution is not possible for any molecule because the simplest molecule consists of three particles (two nuclei and one electron). We therefore adopt the **Born–Oppenheimer approximation** in which it is supposed that the nuclei, being so much heavier than an electron, move relatively slowly and may be treated as stationary while the electrons move in their field. We can therefore think of the nuclei as being fixed at arbitrary locations, and then solve the Schrödinger equation for the wavefunction of the electrons alone.

The approximation is quite good for ground-state molecules, for calculations suggest that the nuclei in H_2 move through only about 1 pm while the electron speeds through 1000 pm, so the error of assuming that the nuclei are stationary is small. Exceptions to the approximation's validity include certain excited states of polyatomic molecules and the ground states of cations; both types of species are important when considering photoelectron spectroscopy (Section 11.4) and mass spectrometry.

The Born–Oppenheimer approximation allows us to select an internuclear separation in a diatomic molecule and then to solve the Schrödinger equation for the electrons at that nuclear separation. Then we choose a different separation and repeat the calculation, and so on. In this way we can explore how the energy of the molecule varies with bond length (in polyatomic molecules, with angles too) and obtain a **molecular potential energy curve** (Fig. 11.1). When more than one molecular parameter is changed in a polyatomic molecule, we obtain a potential energy surface. It is called a *potential* energy curve because the kinetic energy of the stationary nuclei is zero. Once the curve has been calculated or determined experimentally (by using the spectroscopic techniques described in Chapters 13 and 14), we can identify the **equilibrium bond length**, R_e, the internuclear separation at the minimum of the curve, and the **bond dissociation energy**, D_0, which is closely related to the depth, D_e, of the minimum below the energy of the infinitely widely separated and stationary atoms.

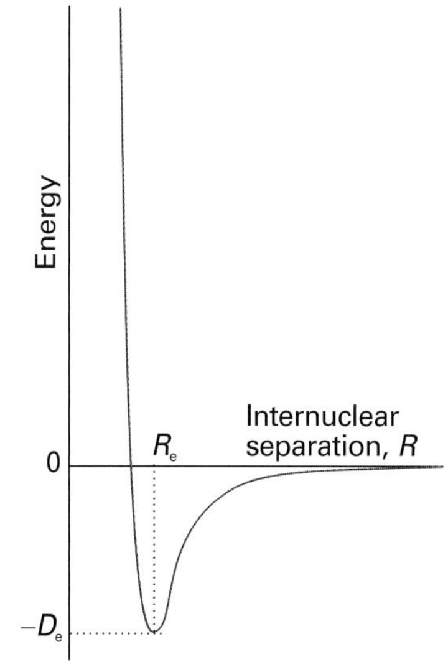

Fig. 11.1 A molecular potential energy curve. The equilibrium bond length corresponds to the energy minimum.

Comment 11.1

The dissociation energy differs from the depth of the well by an energy equal to the zero-point vibrational energy of the bonded atoms: $D_0 = D_e - \frac{1}{2}\hbar\omega$, where ω is the vibrational frequency of the bond (Section 13.9).

Valence-bond theory

Valence-bond theory was the first quantum mechanical theory of bonding to be developed. The language it introduced, which includes concepts such as spin pairing, orbital overlap, σ and π bonds, and hybridization, is widely used throughout chemistry, especially in the description of the properties and reactions of organic compounds. Here we summarize essential topics of VB theory that are familiar from introductory chemistry and set the stage for the development of MO theory.

11.1 Homonuclear diatomic molecules

In VB theory, a bond is regarded as forming when an electron in an atomic orbital on one atom pairs its spin with that of an electron in an atomic orbital on another atom. To understand why this pairing leads to bonding, we have to examine the wavefunction for the two electrons that form the bond. We begin by considering the simplest possible chemical bond, the one in molecular hydrogen, H_2.

The spatial wavefunction for an electron on each of two widely separated H atoms is

$$\psi = \chi_{\mathrm{H1}s_{\mathrm{A}}}(\boldsymbol{r}_1)\chi_{\mathrm{H1}s_{\mathrm{B}}}(\boldsymbol{r}_2)$$

if electron 1 is on atom A and electron 2 is on atom B; in this chapter we use χ (chi) to denote atomic orbitals. For simplicity, we shall write this wavefunction as $\psi = A(1)B(2)$. When the atoms are close, it is not possible to know whether it is electron 1 that is on A or electron 2. An equally valid description is therefore $\psi = A(2)B(1)$, in which electron 2 is on A and electron 1 is on B. When two outcomes are equally probable, quantum mechanics instructs us to describe the true state of the system as a superposition of the wavefunctions for each possibility (Section 8.5d), so a better description of the molecule than either wavefunction alone is the (unnormalized) linear combination

$$\psi = A(1)B(2) \pm A(2)B(1) \tag{11.1}$$

It turns out that the combination with lower energy is the one with a + sign, so the valence-bond wavefunction of the H_2 molecule is

$$\psi = A(1)B(2) + A(2)B(1) \tag{11.2}$$

The formation of the bond in H_2 can be pictured as due to the high probability that the two electrons will be found between the two nuclei and hence will bind them together. More formally, the wave pattern represented by the term $A(1)B(2)$ interferes constructively with the wave pattern represented by the contribution $A(2)B(1)$, and there is an enhancement in the value of the wavefunction in the internuclear region (Fig. 11.2).

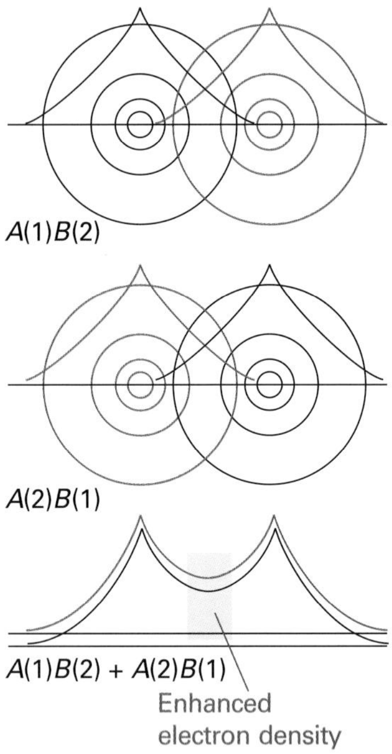

Fig. 11.2 It is very difficult to represent valence-bond wavefunctions because they refer to two electrons simultaneously. However, this illustration is an attempt. The atomic orbital for electron 1 is represented by the black contours, and that of electron 2 is represented by the blue contours. The top illustration represents $A(1)B(2)$, and the middle illustration represents the contribution $A(2)B(1)$. When the two contributions are superimposed, there is interference between the black contributions and between the blue contributions, resulting in an enhanced (two-electron) density in the internuclear region.

The electron distribution described by the wavefunction in eqn 11.2 is called a **σ bond**. A σ bond has cylindrical symmetry around the internuclear axis, and is so called because, when viewed along the internuclear axis, it resembles a pair of electrons in an *s* orbital (and σ is the Greek equivalent of s).

A chemist's picture of a covalent bond is one in which the spins of two electrons pair as the atomic orbitals overlap. The origin of the role of spin is that the wavefunction given in eqn 11.2 can be formed only by a pair of electrons with opposed spins. Spin pairing is not an end in itself: it is a means of achieving a wavefunction (and the probability distribution it implies) that corresponds to a low energy.

Justification 11.1 *Electron pairing in VB theory*

The Pauli principle requires the wavefunction of two electrons to change sign when the labels of the electrons are interchanged (see Section 10.4b). The total VB wavefunction for two electrons is

$$\psi(1,2) = \{A(1)B(2) + A(2)B(1)\}\sigma(1,2)$$

where σ represents the spin component of the wavefunction. When the labels 1 and 2 are interchanged, this wavefunction becomes

$$\psi(2,1) = \{A(2)B(1) + A(1)B(2)\}\sigma(2,1) = \{A(1)B(2) + A(2)B(1)\}\sigma(2,1)$$

The Pauli principle requires that $\psi(2,1) = -\psi(1,2)$, which is satisfied only if $\sigma(2,1) = -\sigma(1,2)$. The combination of two spins that has this property is

$$\sigma_-(1,2) = (1/2^{1/2})\{\alpha(1)\beta(2) - \alpha(2)\beta(1)\}$$

which corresponds to paired electron spins (Section 10.7). Therefore, we conclude that the state of lower energy (and hence the formation of a chemical bond) is achieved if the electron spins are paired.

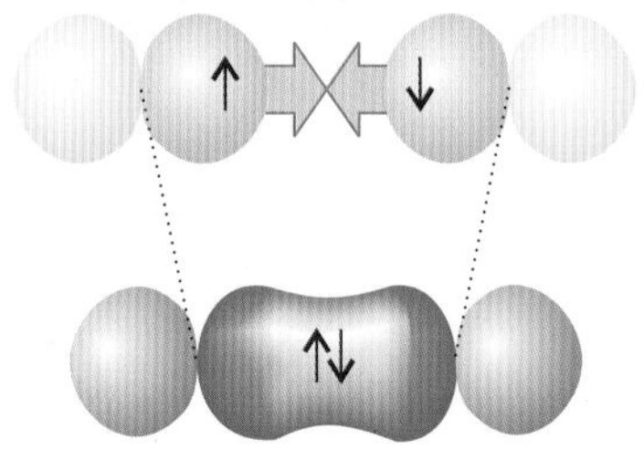

Fig. 11.3 The orbital overlap and spin pairing between electrons in two collinear *p* orbitals that results in the formation of a σ bond.

The VB description of H_2 can be applied to other homonuclear diatomic molecules, such as nitrogen, N_2. To construct the valence bond description of N_2, we consider the valence electron configuration of each atom, which is $2s^2 2p_x^1 2p_y^1 2p_z^1$. It is conventional to take the *z*-axis to be the internuclear axis, so we can imagine each atom as having a $2p_z$ orbital pointing towards a $2p_z$ orbital on the other atom (Fig. 11.3), with the $2p_x$ and $2p_y$ orbitals perpendicular to the axis. A σ bond is then formed by spin pairing between the two electrons in the two $2p_z$ orbitals. Its spatial wavefunction is given by eqn 11.2, but now *A* and *B* stand for the two $2p_z$ orbitals.

The remaining $2p$ orbitals cannot merge to give σ bonds as they do not have cylindrical symmetry around the internuclear axis. Instead, they merge to form two π bonds. A **π bond** arises from the spin pairing of electrons in two p orbitals that approach side-by-side (Fig. 11.4). It is so called because, viewed along the internuclear axis, a π bond resembles a pair of electrons in a p orbital (and π is the Greek equivalent of p).

There are two π bonds in N_2, one formed by spin pairing in two neighbouring $2p_x$ orbitals and the other by spin pairing in two neighbouring $2p_y$ orbitals. The overall bonding pattern in N_2 is therefore a σ bond plus two π bonds (Fig. 11.5), which is consistent with the Lewis structure :N≡N: for nitrogen.

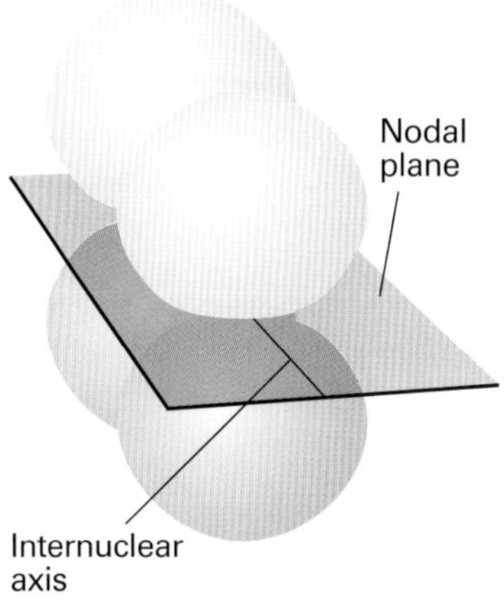

Fig. 11.4 A π bond results from orbital overlap and spin pairing between electrons in p orbitals with their axes perpendicular to the internuclear axis. The bond has two lobes of electron density separated by a nodal plane.

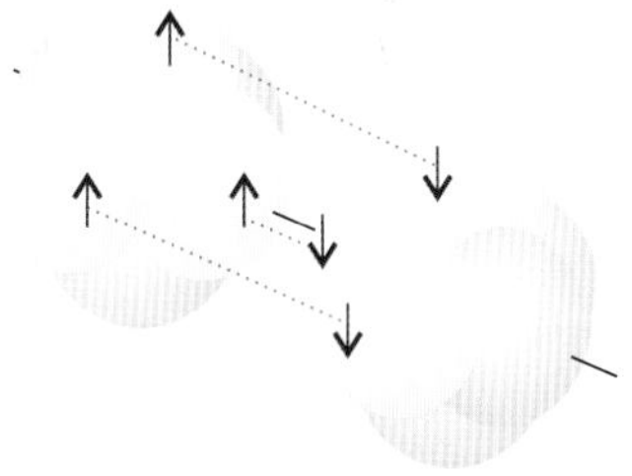

Fig. 11.5 The structure of bonds in a nitrogen molecule: there is one σ bond and two π bonds. As explained later, the overall electron density has cylindrical symmetry around the internuclear axis.

11.2 Polyatomic molecules

Each σ bond in a polyatomic molecule is formed by the spin pairing of electrons in atomic orbitals with cylindrical symmetry about the relevant internuclear axis. Likewise, π bonds are formed by pairing electrons that occupy atomic orbitals of the appropriate symmetry.

The VB description of H_2O will make this clear. The valence electron configuration of an O atom is $2s^22p_x^22p_y^12p_z^1$. The two unpaired electrons in the O2p orbitals can each pair with an electron in an H1s orbital, and each combination results in the formation of a σ bond (each bond has cylindrical symmetry about the respective O—H internuclear axis). Because the $2p_y$ and $2p_z$ orbitals lie at 90° to each other, the two σ bonds also lie at 90° to each other (Fig. 11.6). We can predict, therefore, that H_2O should be an angular molecule, which it is. However, the theory predicts a bond angle of 90°, whereas the actual bond angle is 104.5°.

Self-test 11.1 Use valence-bond theory to suggest a shape for the ammonia molecule, NH_3.

[A trigonal pyramidal molecule with each N—H bond 90°; experimental: 107°]

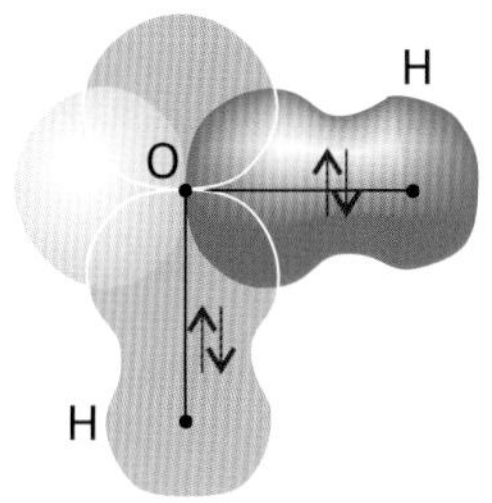

Fig. 11.6 A first approximation to the valence-bond description of bonding in an H_2O molecule. Each σ bond arises from the overlap of an H1s orbital with one of the O2p orbitals. This model suggests that the bond angle should be 90°, which is significantly different from the experimental value.

Another deficiency of VB theory is its inability to account for carbon's tetravalence (its ability to form four bonds). The ground-state configuration of C is $2s^22p_x^12p_y^1$, which suggests that a carbon atom should be capable of forming only two bonds, not four. This deficiency is overcome by allowing for **promotion**, the excitation of an electron to an orbital of higher energy. In carbon, for example, the promotion of a $2s$ electron to a $2p$ orbital can be thought of as leading to the configuration $2s^12p_x^12p_y^12p_z^1$, with four unpaired electrons in separate orbitals. These electrons may pair with four electrons in orbitals provided by four other atoms (such as four H1s orbitals if the molecule is CH_4), and hence form four σ bonds. Although energy was required to promote the electron, it is more than recovered by the promoted atom's ability to form four bonds in place of the two bonds of the unpromoted atom. Promotion, and the formation of four bonds, is a characteristic feature of carbon because the promotion energy is quite small: the promoted electron leaves a doubly occupied $2s$ orbital and enters a vacant $2p$ orbital, hence significantly relieving the electron–electron repulsion it experiences in the former. However, we need to remember that promotion is not a 'real' process in which an atom somehow becomes excited and then forms bonds: it is a notional contribution to the overall energy change that occurs when bonds form.

The description of the bonding in CH_4 (and other alkanes) is still incomplete because it implies the presence of three σ bonds of one type (formed from H1s and C2p

Comment 11.2

A characteristic property of waves is that they interfere with one another, resulting in a greater displacement where peaks or troughs coincide, giving rise to *constructive interference*, and a smaller displacement where peaks coincide with troughs, giving rise to *destructive interference*. The physics of waves is reviewed in *Appendix 3*.

orbitals) and a fourth σ bond of a distinctly different character (formed from H1*s* and C2*s*). This problem is overcome by realizing that the electron density distribution in the promoted atom is equivalent to the electron density in which each electron occupies a **hybrid orbital** formed by interference between the C2*s* and C2*p* orbitals. The origin of the hybridization can be appreciated by thinking of the four atomic orbitals centred on a nucleus as waves that interfere destructively and constructively in different regions, and give rise to four new shapes.

The specific linear combinations that give rise to four equivalent hybrid orbitals are

$$h_1 = s + p_x + p_y + p_z \qquad h_2 = s - p_x - p_y + p_z$$
$$h_3 = s - p_x + p_y - p_z \qquad h_4 = s + p_x - p_y - p_z \qquad (11.3)$$

As a result of the interference between the component orbitals, each hybrid orbital consists of a large lobe pointing in the direction of one corner of a regular tetrahedron (Fig. 11.7). The angle between the axes of the hybrid orbitals is the tetrahedral angle, 109.47°. Because each hybrid is built from one *s* orbital and three *p* orbitals, it is called an ***sp*^{3} hybrid orbital**.

It is now easy to see how the valence-bond description of the CH_4 molecule leads to a tetrahedral molecule containing four equivalent C—H bonds. Each hybrid orbital of the promoted C atom contains a single unpaired electron; an H1*s* electron can pair with each one, giving rise to a σ bond pointing in a tetrahedral direction. For example, the (un-normalized) wavefunction for the bond formed by the hybrid orbital h_1 and the $1s_A$ orbital (with wavefunction that we shall denote A) is

$$\psi = h_1(1)A(2) + h_1(2)A(1)$$

Because each sp^3 hybrid orbital has the same composition, all four σ bonds are identical apart from their orientation in space (Fig. 11.8).

A hybrid orbital has enhanced amplitude in the internuclear region, which arises from the constructive interference between the *s* orbital and the positive lobes of the *p* orbitals (Fig. 11.9). As a result, the bond strength is greater than for a bond formed

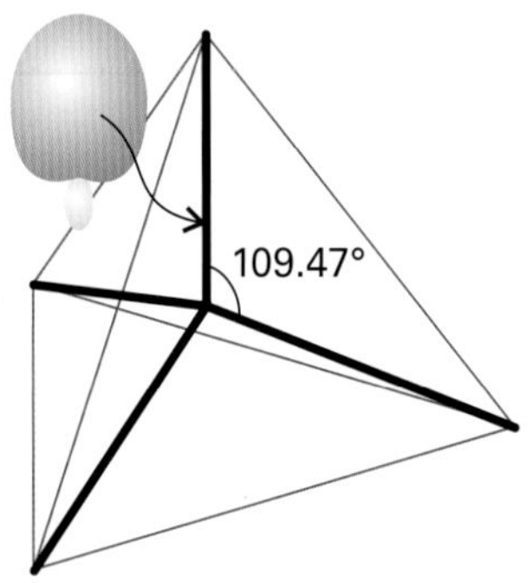

Fig. 11.7 An sp^3 hybrid orbital formed from the superposition of *s* and *p* orbitals on the same atom. There are four such hybrids: each one points towards the corner of a regular tetrahedron. The overall electron density remains spherically symmetrical.

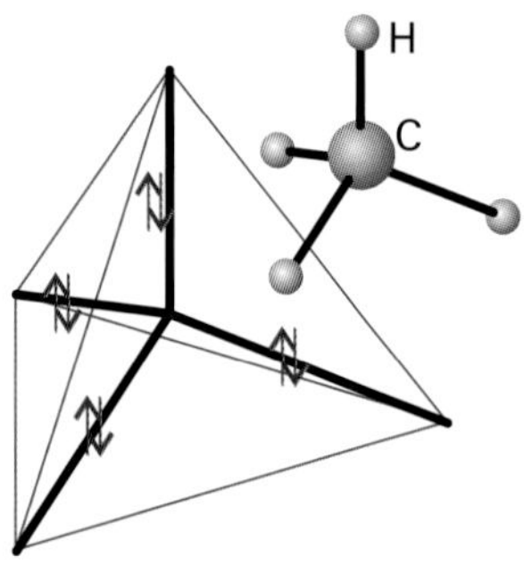

Fig. 11.8 Each sp^3 hybrid orbital forms a σ bond by overlap with an H1*s* orbital located at the corner of the tetrahedron. This model accounts for the equivalence of the four bonds in CH_4.

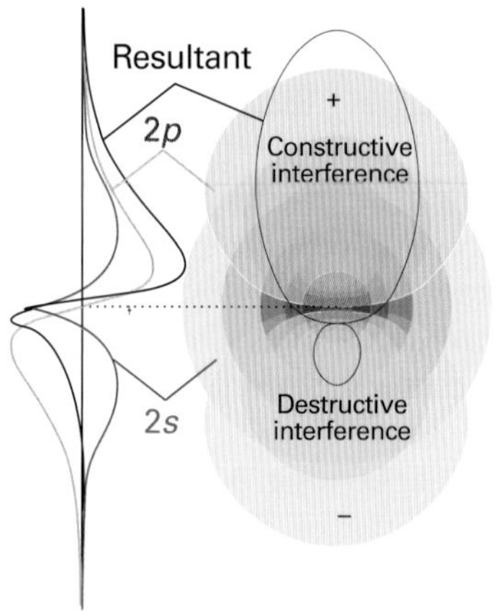

Fig. 11.9 A more detailed representation of the formation of an sp^3 hybrid by interference between wavefunctions centred on the same atomic nucleus. (To simplify the representation, we have ignored the radial node of the 2*s* orbital.)

from an s or p orbital alone. This increased bond strength is another factor that helps to repay the promotion energy.

Hybridization can also be used to describe the structure of an ethene molecule, $H_2C{=}CH_2$, and the torsional rigidity of double bonds. An ethene molecule is planar, with HCH and HCC bond angles close to 120°. To reproduce the σ bonding structure, we promote each C atom to a $2s^12p^3$ configuration. However, instead of using all four orbitals to form hybrids, we form ***sp*² hybrid orbitals**:

$$h_1 = s + 2^{1/2}p_y \qquad h_2 = s + (\tfrac{3}{2})^{1/2}p_x - (\tfrac{1}{2})^{1/2}p_y \qquad h_3 = s - (\tfrac{3}{2})^{1/2}p_x - (\tfrac{1}{2})^{1/2}p_y \quad (11.4)$$

that lie in a plane and point towards the corners of an equilateral triangle (Fig. 11.10). The third $2p$ orbital ($2p_z$) is not included in the hybridization; its axis is perpendicular to the plane in which the hybrids lie. As always in superpositions, the proportion of each orbital in the mixture is given by the *square* of the corresponding coefficient. Thus, in the first of these hybrids the ratio of s to p contributions is 1:2. Similarly, the total p contribution in each of h_2 and h_3 is $\frac{3}{2}+\frac{1}{2}=2$, so the ratio for these orbitals is also 1:2. The different signs of the coefficients ensure that constructive interference takes place in different regions of space, so giving the patterns in the illustration.

We can now describe the structure of $CH_2{=}CH_2$ as follows. The sp^2-hybridized C atoms each form three σ bonds by spin pairing with either the h_1 hybrid of the other C atom or with H1s orbitals. The σ framework therefore consists of C—H and C—C σ bonds at 120° to each other. When the two CH_2 groups lie in the same plane, the two electrons in the unhybridized p orbitals can pair and form a π bond (Fig. 11.11). The formation of this π bond locks the framework into the planar arrangement, for any rotation of one CH_2 group relative to the other leads to a weakening of the π bond (and consequently an increase in energy of the molecule).

A similar description applies to ethyne, HC≡CH, a linear molecule. Now the C atoms are ***sp* hybridized**, and the σ bonds are formed using hybrid atomic orbitals of the form

$$h_1 = s + p_z \qquad h_2 = s - p_z \qquad (11.5)$$

These two orbitals lie along the internuclear axis. The electrons in them pair either with an electron in the corresponding hybrid orbital on the other C atom or with an electron in one of the H1s orbitals. Electrons in the two remaining p orbitals on each atom, which are perpendicular to the molecular axis, pair to form two perpendicular π bonds (Fig. 11.12).

Self-test 11.2 Hybrid orbitals do not always form bonds. They may also contain lone pairs of electrons. Use valence-bond theory to suggest possible shapes for the hydrogen peroxide molecule, H_2O_2.

[Each H—O—O bond angle is predicted to be approximately 109° (experimental: 94.8°); rotation around the O—O bond is possible, so the molecule interconverts between planar and non-planar geometries at high temperatures.]

Other hybridization schemes, particularly those involving d orbitals, are often invoked in elementary work to be consistent with other molecular geometries (Table 11.1). The hybridization of N atomic orbitals always results in the formation of N hybrid orbitals, which may either form bonds or may contain lone pairs of electrons. For example, sp^3d^2 hybridization results in six equivalent hybrid orbitals pointing towards the corners of a regular octahedron and is sometimes invoked to account for the structure of octahedral molecules, such as SF_6.

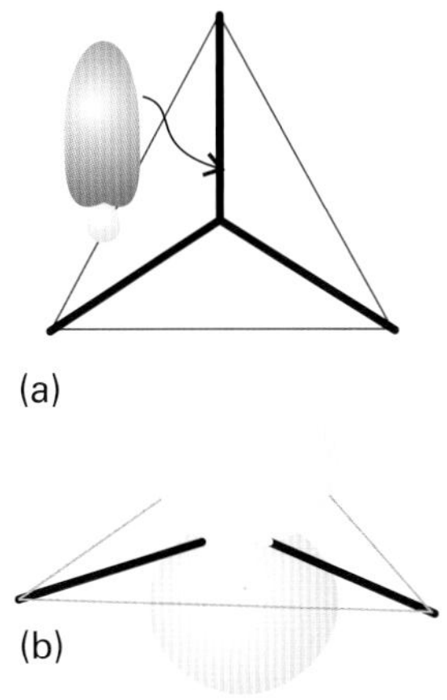

Fig. 11.10 (a) An s orbital and two p orbitals can be hybridized to form three equivalent orbitals that point towards the corners of an equilateral triangle. (b) The remaining, unhybridized p orbital is perpendicular to the plane.

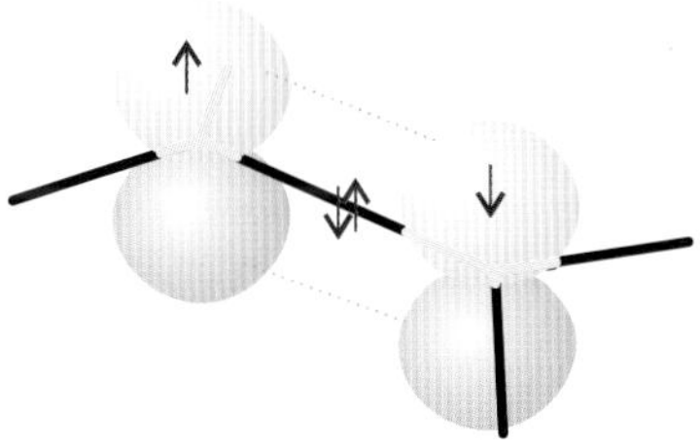

Fig. 11.11 A representation of the structure of a double bond in ethene; only the π bond is shown explicitly.

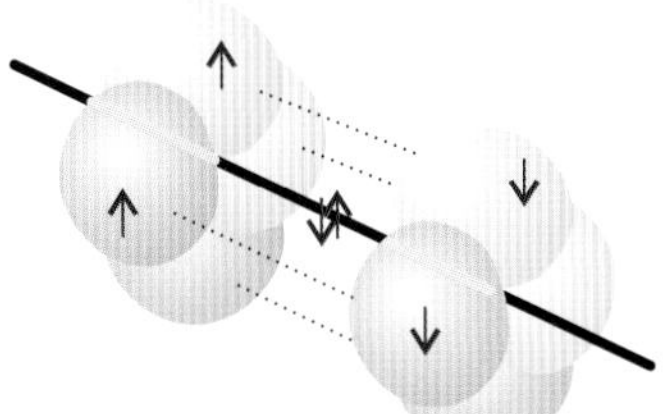

Fig. 11.12 A representation of the structure of a triple bond in ethyne; only the π bonds are shown explicitly. The overall electron density has cylindrical symmetry around the axis of the molecule.

Table 11.1* Some hybridization schemes

Coordination number	Arrangement	Composition
2	Linear	*sp*, *pd*, *sd*
	Angular	*sd*
3	Trigonal planar	sp^2, p^2d
	Unsymmetrical planar	*spd*
	Trigonal pyramidal	pd^2
4	Tetrahedral	sp^3, sd^3
	Irregular tetrahedral	spd^2, p^3d, dp^3
	Square planar	p^2d^2, sp^2d
5	Trigonal bipyramidal	sp^3d, spd^2
	Tetragonal pyramidal	sp^2d^2, sd^4, pd^4, p^3d^2
	Pentagonal planar	p^2d^3
6	Octahedral	sp^3d^2
	Trigonal prismatic	spd^4, pd^5
	Trigonal antiprismatic	p^3d^2

* Source: H. Eyring, J. Walter, and G.E. Kimball, *Quantum chemistry*, Wiley (1944).

Molecular orbital theory

In MO theory, it is accepted that electrons should not be regarded as belonging to particular bonds but should be treated as spreading throughout the entire molecule. This theory has been more fully developed than VB theory and provides the language that is widely used in modern discussions of bonding. To introduce it, we follow the same strategy as in Chapter 10, where the one-electron H atom was taken as the fundamental species for discussing atomic structure and then developed into a description of many-electron atoms. In this chapter we use the simplest molecular species of all, the hydrogen molecule-ion, H_2^+, to introduce the essential features of bonding, and then use it as a guide to the structures of more complex systems. To that end, we will progress to **homonuclear diatomic molecules**, which, like the H_2^+ molecule-ion, are formed from two atoms of the same element, then describe **heteronuclear diatomic molecules**, which are diatomic molecules formed from atoms of two different elements (such as CO and HCl), and end with a treatment of polyatomic molecules that forms the basis for modern computational models of molecular structure and chemical reactivity.

11.3 The hydrogen molecule-ion

The hamiltonian for the single electron in H_2^+ is

$$H=-\frac{\hbar^2}{2m_e}\nabla_1^2+V \qquad V=-\frac{e^2}{4\pi\varepsilon_0}\left(\frac{1}{r_{A1}}+\frac{1}{r_{B1}}-\frac{1}{R}\right) \tag{11.6}$$

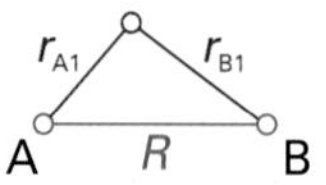

1

where r_{A1} and r_{B1} are the distances of the electron from the two nuclei (**1**) and R is the distance between the two nuclei. In the expression for V, the first two terms in parentheses are the attractive contribution from the interaction between the electron and the nuclei; the remaining term is the repulsive interaction between the nuclei.

The one-electron wavefunctions obtained by solving the Schrödinger equation $H\psi = E\psi$ are called **molecular orbitals** (MO). A molecular orbital ψ gives, through the value of $|\psi|^2$, the distribution of the electron in the molecule. A molecular orbital is like an atomic orbital, but spreads throughout the molecule.

The Schrödinger equation can be solved analytically for H_2^+ (within the Born–Oppenheimer approximation), but the wavefunctions are very complicated functions; moreover, the solution cannot be extended to polyatomic systems. Therefore, we adopt a simpler procedure that, while more approximate, can be extended readily to other molecules.

(a) Linear combinations of atomic orbitals

If an electron can be found in an atomic orbital belonging to atom A and also in an atomic orbital belonging to atom B, then the overall wavefunction is a superposition of the two atomic orbitals:

$$\psi_{\pm} = N(A \pm B) \tag{11.7}$$

where, for H_2^+, A denotes $\chi_{\mathrm{H1s_A}}$, B denotes $\chi_{\mathrm{H1s_B}}$, and N is a normalization factor. The technical term for the superposition in eqn 11.7 is a **linear combination of atomic orbitals** (LCAO). An approximate molecular orbital formed from a linear combination of atomic orbitals is called an **LCAO-MO**. A molecular orbital that has cylindrical symmetry around the internuclear axis, such as the one we are discussing, is called a **σ orbital** because it resembles an *s* orbital when viewed along the axis and, more precisely, because it has zero orbital angular momentum around the internuclear axis.

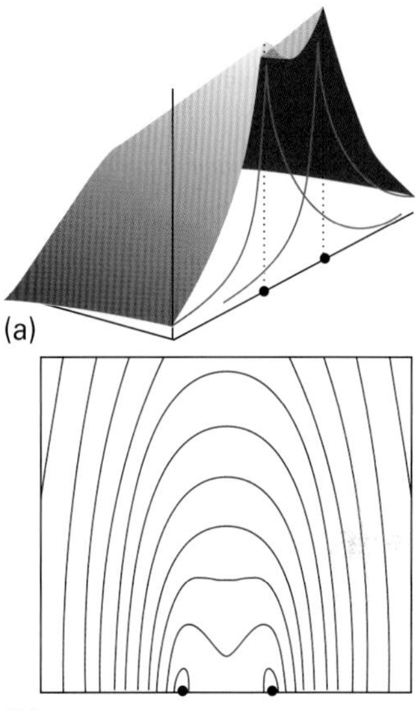

Fig. 11.13 (a) The amplitude of the bonding molecular orbital in a hydrogen molecule-ion in a plane containing the two nuclei and (b) a contour representation of the amplitude.

Exploration Plot the 1σ orbital for different values of the internuclear distance. Point to the features of the 1σ orbital that lead to bonding.

Example 11.1 *Normalizing a molecular orbital*

Normalize the molecular orbital ψ_+ in eqn 11.7.

Method We need to find the factor N such that

$$\int \psi^* \psi \,\mathrm{d}\tau = 1$$

To proceed, substitute the LCAO into this integral, and make use of the fact that the atomic orbitals are individually normalized.

Answer When we substitute the wavefunction, we find

$$\int \psi^* \psi \,\mathrm{d}\tau = N^2 \left\{ \int A^2 \mathrm{d}\tau + \int B^2 \mathrm{d}\tau + 2\int AB \,\mathrm{d}\tau \right\} = N^2(1 + 1 + 2S)$$

where $S = \int AB \,\mathrm{d}\tau$. For the integral to be equal to 1, we require

$$N = \frac{1}{\{2(1+S)\}^{1/2}}$$

In H_2^+, $S \approx 0.59$, so $N = 0.56$.

Self-test 11.3 Normalize the orbital ψ_- in eqn 11.7.

[$N = 1/\{2(1-S)\}^{1/2}$, so $N = 1.10$]

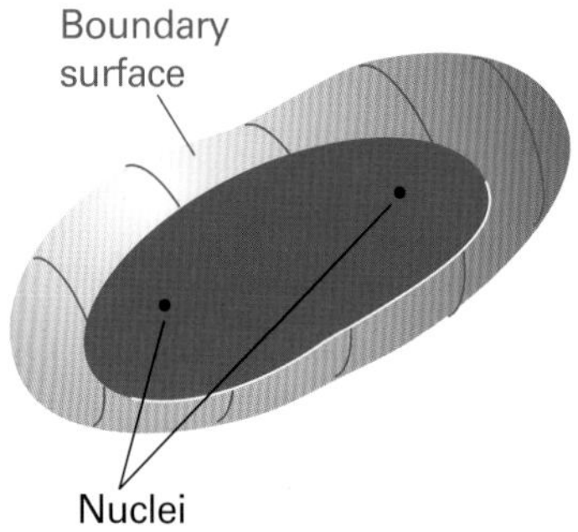

Fig. 11.14 A general indication of the shape of the boundary surface of a σ orbital.

Figure 11.13 shows the contours of constant amplitude for the two molecular orbitals in eqn 11.7, and Fig. 11.14 shows their boundary surfaces. Plots like these are readily obtained using commercially available software. The calculation is quite straightforward, because all we need do is feed in the mathematical forms of the two atomic orbitals and then let the program do the rest. In this case, we use

Comment 11.3

The *law of cosines* states that for a triangle such as that shown in (2) with sides r_A, r_B, and R, and angle θ facing side r_B we may write: $r_B^2 = r_A^2 + R^2 - 2r_AR\cos\theta$.

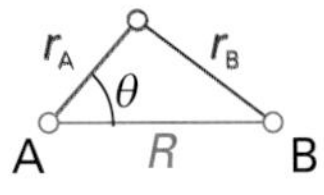

2

$$A = \frac{e^{-r_A/a_0}}{(\pi a_0^3)^{1/2}} \qquad B = \frac{e^{-r_B/a_0}}{(\pi a_0^3)^{1/2}} \tag{11.8}$$

and note that r_A and r_B are not independent (**2**), but related by the *law of cosines* (see *Comment 11.3*):

$$r_B = \{r_A^2 + R^2 - 2r_AR\cos\theta\}^{1/2} \tag{11.9}$$

To make this plot, we have taken $N^2 = 0.31$ (Example 11.1).

(b) Bonding orbitals

According to the Born interpretation, the probability density of the electron in H_2^+ is proportional to the square modulus of its wavefunction. The probability density corresponding to the (real) wavefunction ψ_+ in eqn 11.7 is

$$\psi_+^2 = N^2(A^2 + B^2 + 2AB) \tag{11.10}$$

This probability density is plotted in Fig. 11.15.

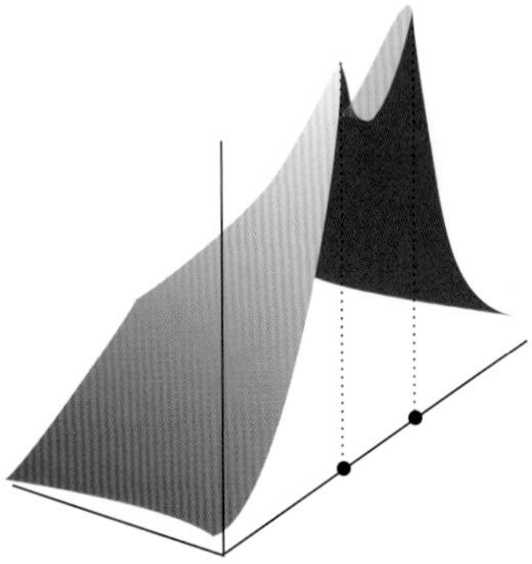

Fig. 11.15 The electron density calculated by forming the square of the wavefunction used to construct Fig. 11.13. Note the accumulation of electron density in the internuclear region.

An important feature of the probability density becomes apparent when we examine the internuclear region, where both atomic orbitals have similar amplitudes. According to eqn 11.10, the total probability density is proportional to the sum of

1 A^2, the probability density if the electron were confined to the atomic orbital A.

2 B^2, the probability density if the electron were confined to the atomic orbital B.

3 $2AB$, an extra contribution to the density.

This last contribution, the **overlap density**, is crucial, because it represents an enhancement of the probability of finding the electron in the internuclear region. The enhancement can be traced to the constructive interference of the two atomic orbitals: each has a positive amplitude in the internuclear region, so the total amplitude is greater there than if the electron were confined to a single atomic orbital.

We shall frequently make use of the result that *electrons accumulate in regions where atomic orbitals overlap and interfere constructively*. The conventional explanation is based on the notion that accumulation of electron density between the nuclei puts the electron in a position where it interacts strongly with both nuclei. Hence, the energy of the molecule is lower than that of the separate atoms, where each electron can interact strongly with only one nucleus. This conventional explanation, however, has been called into question, because shifting an electron away from a nucleus into the internuclear region *raises* its potential energy. The modern (and still controversial) explanation does not emerge from the simple LCAO treatment given here. It seems that, at the same time as the electron shifts into the internuclear region, the atomic orbitals shrink. This orbital shrinkage improves the electron–nucleus attraction more than it is decreased by the migration to the internuclear region, so there is a net lowering of potential energy. The kinetic energy of the electron is also modified because the curvature of the wavefunction is changed, but the change in kinetic energy is dominated by the change in potential energy. Throughout the following discussion we ascribe the strength of chemical bonds to the accumulation of electron density in the internuclear region. We leave open the question whether in molecules more complicated than H_2^+ the true source of energy lowering is that accumulation itself or some indirect but related effect.

The σ orbital we have described is an example of a **bonding orbital**, an orbital which, if occupied, helps to bind two atoms together. Specifically, we label it 1σ as it is the σ orbital of lowest energy. An electron that occupies a σ orbital is called a **σ electron**, and if that is the only electron present in the molecule (as in the ground state of H_2^+), then we report the configuration of the molecule as $1\sigma^1$.

The energy $E_{1\sigma}$ of the 1σ orbital is (see Problem 11.23):

$$E_{1\sigma} = E_{\text{H1}s} + \frac{e^2}{4\pi\varepsilon_0 R} - \frac{j+k}{1+S} \tag{11.11}$$

where

$$S = \int AB\,\mathrm{d}\tau = \left\{1 + \frac{R}{a_0} + \frac{1}{3}\left(\frac{R}{a_0}\right)^2\right\}\mathrm{e}^{-R/a_0} \tag{11.12a}$$

$$j = \frac{e^2}{4\pi\varepsilon_0}\int \frac{A^2}{r_\text{B}}\mathrm{d}\tau = \frac{e^2}{4\pi\varepsilon_0 R}\left\{1 - \left(1 + \frac{R}{a_0}\right)\mathrm{e}^{-2R/a_0}\right\} \tag{11.12b}$$

$$k = \frac{e^2}{4\pi\varepsilon_0}\int \frac{AB}{r_\text{B}}\mathrm{d}\tau = \frac{e^2}{4\pi\varepsilon_0 a_0}\left(1 + \frac{R}{a_0}\right)\mathrm{e}^{-R/a_0} \tag{11.12c}$$

We can interpret the preceding integrals as follows:

1 All three integrals are positive and decline towards zero at large internuclear separations (S and k on account of the exponential term, j on account of the factor $1/R$).

2 The integral j is a measure of the interaction between a nucleus and electron density centred on the other nucleus.

3 The integral k is a measure of the interaction between a nucleus and the excess probability in the internuclear region arising from overlap.

Figure 11.16 is a plot of $E_{1\sigma}$ against R relative to the energy of the separated atoms. The energy of the 1σ orbital decreases as the internuclear separation decreases from large values because electron density accumulates in the internuclear region as the constructive interference between the atomic orbitals increases (Fig. 11.17). However, at small separations there is too little space between the nuclei for significant accumulation of electron density there. In addition, the nucleus–nucleus repulsion (which is proportional to $1/R$) becomes large. As a result, the energy of the molecule rises at short distances, and there is a minimum in the potential energy curve. Calculations on H_2^+ give $R_e = 130$ pm and $D_e = 1.77$ eV (171 kJ mol^{-1}); the experimental values are 106 pm and 2.6 eV, so this simple LCAO-MO description of the molecule, while inaccurate, is not absurdly wrong.

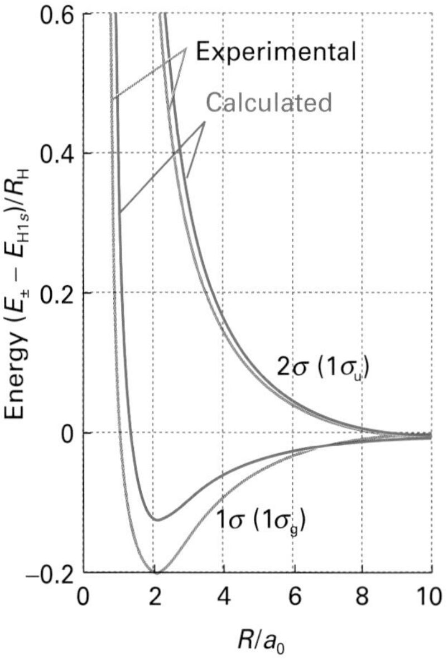

Fig. 11.16 The calculated and experimental molecular potential energy curves for a hydrogen molecule-ion showing the variation of the energy of the molecule as the bond length is changed. The alternative g,u notation is introduced in Section 11.3c.

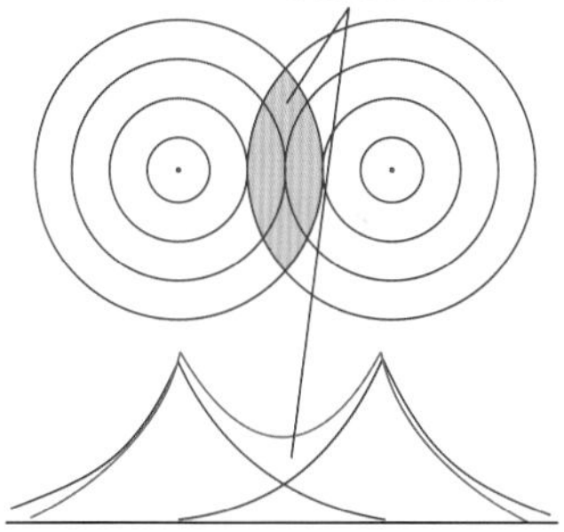

Fig. 11.17 A representation of the constructive interference that occurs when two H1s orbitals overlap and form a bonding σ orbital.

(c) Antibonding orbitals

The linear combination ψ_- in eqn 11.7 corresponds to a higher energy than that of ψ_+. Because it is also a σ orbital we label it 2σ. This orbital has an internuclear nodal plane where A and B cancel exactly (Figs. 11.18 and 11.19). The probability density is

$$\psi_-^2 = N^2(A^2 + B^2 - 2AB) \tag{11.13}$$

There is a reduction in probability density between the nuclei due to the $-2AB$ term (Fig. 11.20); in physical terms, there is destructive interference where the two atomic orbitals overlap. The 2σ orbital is an example of an **antibonding orbital**, an orbital that, if occupied, contributes to a reduction in the cohesion between two atoms and helps to raise the energy of the molecule relative to the separated atoms.

The energy $E_{2\sigma}$ of the 2σ antibonding orbital is given by (see Problem 11.23)

$$E_{2\sigma} = E_{\text{H1}s} + \frac{e^2}{4\pi\varepsilon_0 R} - \frac{j-k}{1-S} \tag{11.14}$$

where the integrals S, j, and k are given by eqn 11.12. The variation of $E_{2\sigma}$ with R is shown in Fig. 11.16, where we see the destabilizing effect of an antibonding electron.

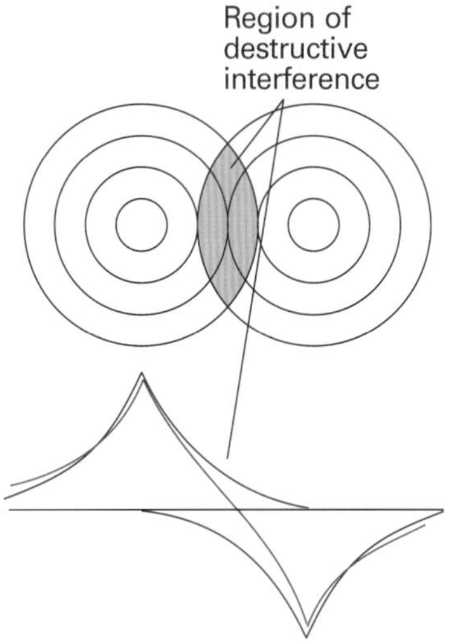

Fig. 11.18 A representation of the destructive interference that occurs when two H1*s* orbitals overlap and form an antibonding 2σ orbital.

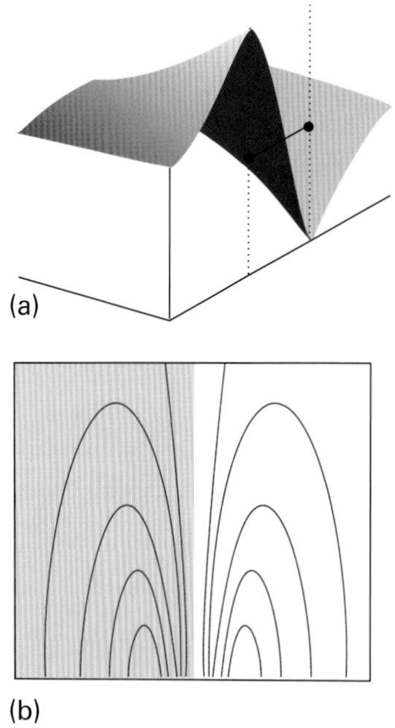

Fig. 11.19 (a) The amplitude of the antibonding molecular orbital in a hydrogen molecule-ion in a plane containing the two nuclei and (b) a contour representation of the amplitude. Note the internuclear node.

Exploration Plot the 2σ orbital for different values of the internuclear distance. Point to the features of the 2σ orbital that lead to antibonding.

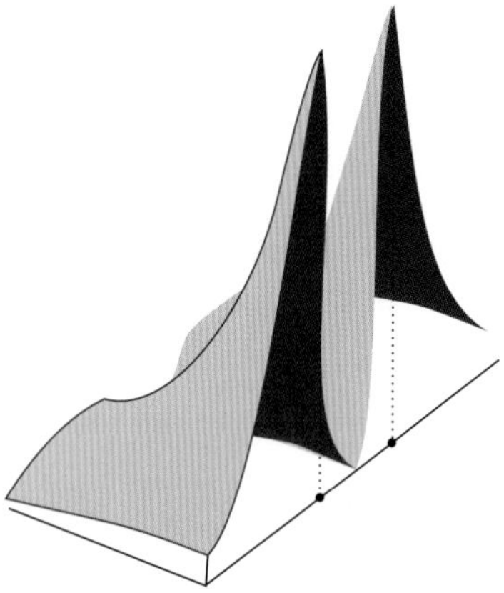

Fig. 11.20 The electron density calculated by forming the square of the wavefunction used to construct Fig. 11.19. Note the elimination of electron density from the internuclear region.

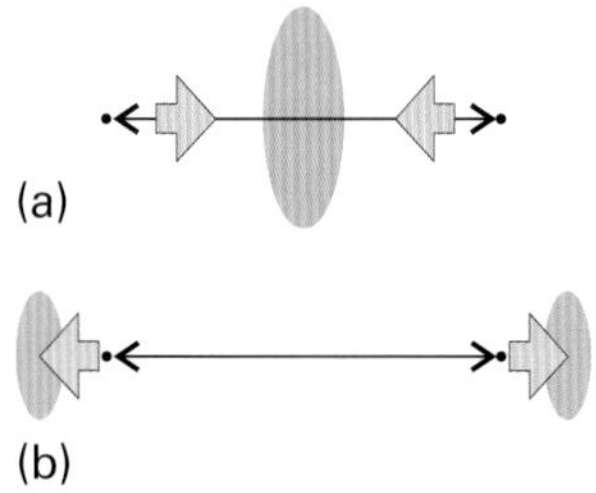

Fig. 11.21 A partial explanation of the origin of bonding and antibonding effects. (a) In a bonding orbital, the nuclei are attracted to the accumulation of electron density in the internuclear region. (b) In an antibonding orbital, the nuclei are attracted to an accumulation of electron density outside the internuclear region.

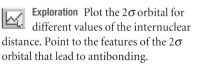

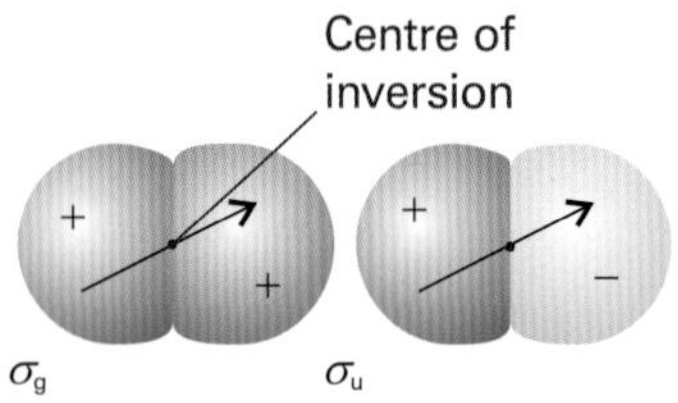

Fig. 11.22 The parity of an orbital is even (g) if its wavefunction is unchanged under inversion through the centre of symmetry of the molecule, but odd (u) if the wavefunction changes sign. Heteronuclear diatomic molecules do not have a centre of inversion, so for them the g, u classification is irrelevant.

The effect is partly due to the fact that an antibonding electron is excluded from the internuclear region, and hence is distributed largely outside the bonding region. In effect, whereas a bonding electron pulls two nuclei together, an antibonding electron pulls the nuclei apart (Fig. 11.21). Figure 11.16 also shows another feature that we draw on later: $|E_- - E_{\mathrm{H1}s}| > |E_+ - E_{\mathrm{H1}s}|$, which indicates that *the antibonding orbital is more antibonding than the bonding orbital is bonding*. This important conclusion stems in part from the presence of the nucleus–nucleus repulsion ($e^2/4\pi\varepsilon_0 R$): this contribution raises the energy of both molecular orbitals. Antibonding orbitals are often labelled with an asterisk (*), so the 2σ orbital could also be denoted $2\sigma^*$ (and read '2 sigma star').

For homonuclear diatomic molecules, it is helpful to describe a molecular orbital by identifying its **inversion symmetry**, the behaviour of the wavefunction when it is inverted through the centre (more formally, the centre of inversion) of the molecule. Thus, if we consider any point on the bonding σ orbital, and then project it through the centre of the molecule and out an equal distance on the other side, then we arrive at an identical value of the wavefunction (Fig. 11.22). This so-called **gerade symmetry** (from the German word for 'even') is denoted by a subscript g, as in σ_g. On the other hand, the same procedure applied to the antibonding 2σ orbital results in the same size but opposite sign of the wavefunction. This **ungerade symmetry** ('odd symmetry') is denoted by a subscript u, as in σ_u. This inversion symmetry classification is not applicable to diatomic molecules formed by atoms from two different elements (such as CO) because these molecules do not have a centre of inversion. When using the g,u notation, each set of orbitals of the same inversion symmetry are labelled separately so, whereas 1σ becomes $1\sigma_g$, its antibonding partner, which so far we have called 2σ, is the first orbital of a different symmetry, and is denoted $1\sigma_u$. The general rule is that *each set of orbitals of the same symmetry designation is labelled separately*.

Comment 11.4

When treating homonuclear diatomic molecules, we shall favour the more modern notation that focuses attention on the symmetry properties of the orbital. For all other molecules, we shall use asterisks from time to time to denote antibonding orbitals.

11.4 **Homonuclear diatomic molecules**

In Chapter 10 we used the hydrogenic atomic orbitals and the building-up principle to deduce the ground electronic configurations of many-electron atoms. We now do the same for many-electron diatomic molecules by using the H_2^+ molecular orbitals. The general procedure is to construct molecular orbitals by combining the available atomic orbitals. The electrons supplied by the atoms are then accommodated in the orbitals so as to achieve the lowest overall energy subject to the constraint of the Pauli exclusion principle, that no more than two electrons may occupy a single orbital (and then must be paired). As in the case of atoms, if several degenerate molecular orbitals are available, we add the electrons singly to each individual orbital before doubly occupying any one orbital (because that minimizes electron–electron repulsions). We also take note of Hund's maximum multiplicity rule (Section 10.4) that, if electrons do occupy different degenerate orbitals, then a lower energy is obtained if they do so with parallel spins.

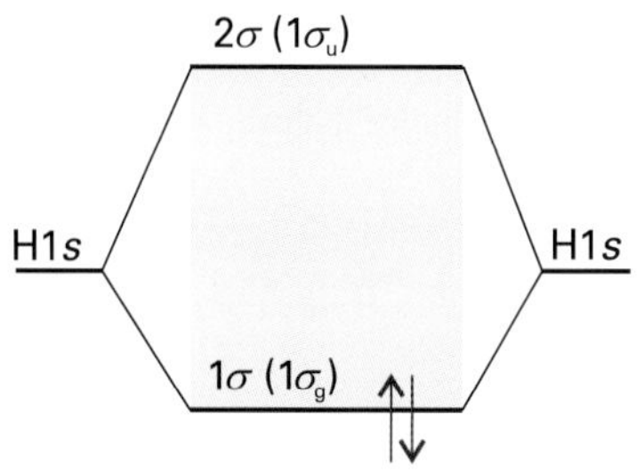

Fig. 11.23 A molecular orbital energy level diagram for orbitals constructed from the overlap of H1*s* orbitals; the separation of the levels corresponds to that found at the equilibrium bond length. The ground electronic configuration of H_2 is obtained by accommodating the two electrons in the lowest available orbital (the bonding orbital).

(a) σ orbitals

Consider H_2, the simplest many-electron diatomic molecule. Each H atom contributes a 1*s* orbital (as in H_2^+), so we can form the $1\sigma_g$ and $1\sigma_u$ orbitals from them, as we have seen already. At the experimental internuclear separation these orbitals will have the energies shown in Fig. 11.23, which is called a **molecular orbital energy level diagram**. Note that from two atomic orbitals we can build two molecular orbitals. In general, from *N* atomic orbitals we can build *N* molecular orbitals.

There are two electrons to accommodate, and both can enter $1\sigma_g$ by pairing their spins, as required by the Pauli principle (see the following *Justification*). The ground-state configuration is therefore $1\sigma_g^2$ and the atoms are joined by a bond consisting of an electron pair in a bonding σ orbital. This approach shows that an electron pair, which was the focus of Lewis's account of chemical bonding, represents the maximum number of electrons that can enter a bonding molecular orbital.

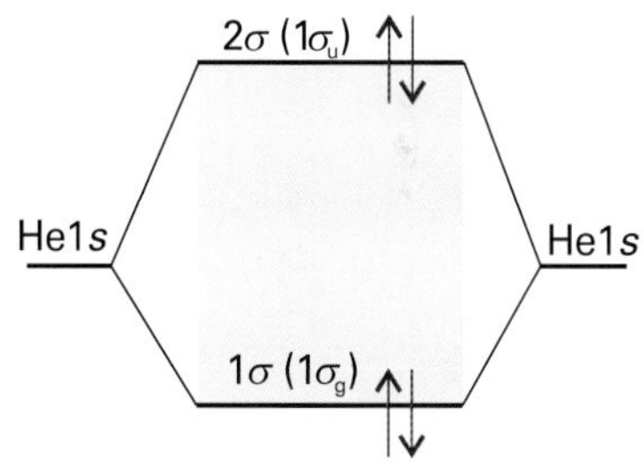

Fig. 11.24 The ground electronic configuration of the hypothetical four-electron molecule He_2 has two bonding electrons and two antibonding electrons. It has a higher energy than the separated atoms, and so is unstable.

Justification 11.2 *Electron pairing in MO theory*

The spatial wavefunction for two electrons in a bonding molecular orbital ψ such as the bonding orbital in eqn 11.7, is $\psi(1)\psi(2)$. This two-electron wavefunction is obviously symmetric under interchange of the electron labels. To satisfy the Pauli principle, it must be multiplied by the antisymmetric spin state $\alpha(1)\beta(2) - \beta(1)\alpha(2)$ to give the overall antisymmetric state

$$\psi(1,2) = \psi(1)\psi(2)\{\alpha(1)\beta(2) - \beta(1)\alpha(2)\}$$

Because $\alpha(1)\beta(2) - \beta(1)\alpha(2)$ corresponds to paired electron spins, we see that two electrons can occupy the same molecular orbital (in this case, the bonding orbital) only if their spins are paired.

The same argument shows why He does not form diatomic molecules. Each He atom contributes a 1*s* orbital, so $1\sigma_g$ and $1\sigma_u$ molecular orbitals can be constructed. Although these orbitals differ in detail from those in H_2, the general shape is the same, and we can use the same qualitative energy level diagram in the discussion. There are four electrons to accommodate. Two can enter the $1\sigma_g$ orbital, but then it is full, and the next two must enter the $1\sigma_u$ orbital (Fig. 11.24). The ground electronic configuration

Comment 11.5

Diatomic helium 'molecules' have been prepared: they consist of pairs of atoms held together by weak van der Waals forces of the type described in Chapter 18.

of He_2 is therefore $1\sigma_g^2 1\sigma_u^2$. We see that there is one bond and one antibond. Because an antibond is slightly more antibonding than a bond is bonding, an He_2 molecule has a higher energy than the separated atoms, so it is unstable relative to the individual atoms.

We shall now see how the concepts we have introduced apply to homonuclear diatomic molecules in general. In elementary treatments, only the orbitals of the valence shell are used to form molecular orbitals so, for molecules formed with atoms from Period 2 elements, only the 2*s* and 2*p* atomic orbitals are considered.

A general principle of molecular orbital theory is that *all orbitals of the appropriate symmetry contribute to a molecular orbital.* Thus, to build σ orbitals, we form linear combinations of all atomic orbitals that have cylindrical symmetry about the internuclear axis. These orbitals include the 2*s* orbitals on each atom and the $2p_z$ orbitals on the two atoms (Fig. 11.25). The general form of the σ orbitals that may be formed is therefore

$$\psi = c_{A2s}\chi_{A2s} + c_{B2s}\chi_{B2s} + c_{A2p_z}\chi_{A2p_z} + c_{B2p_z}\chi_{B2p_z} \qquad (11.15)$$

From these four atomic orbitals we can form four molecular orbitals of σ symmetry by an appropriate choice of the coefficients *c*.

The procedure for calculating the coefficients will be described in Section 11.6. At this stage we adopt a simpler route, and suppose that, because the 2*s* and $2p_z$ orbitals have distinctly different energies, they may be treated separately. That is, the four σ orbitals fall approximately into two sets, one consisting of two molecular orbitals of the form

$$\psi = c_{A2s}\chi_{A2s} + c_{B2s}\chi_{B2s} \qquad (11.16a)$$

and another consisting of two orbitals of the form

$$\psi = c_{A2p_z}\chi_{A2p_z} + c_{B2p_z}\chi_{B2p_z} \qquad (11.16b)$$

Because atoms A and B are identical, the energies of their 2*s* orbitals are the same, so the coefficients are equal (apart from a possible difference in sign); the same is true of the $2p_z$ orbitals. Therefore, the two sets of orbitals have the form $\chi_{A2s} \pm \chi_{B2s}$ and $\chi_{A2p_z} \pm \chi_{B2p_z}$.

The 2*s* orbitals on the two atoms overlap to give a bonding and an antibonding σ orbital ($1\sigma_g$ and $1\sigma_u$, respectively) in exactly the same way as we have already seen for 1*s* orbitals. The two $2p_z$ orbitals directed along the internuclear axis overlap strongly. They may interfere either constructively or destructively, and give a bonding or antibonding σ orbital, respectively (Fig. 11.26). These two σ orbitals are labelled $2\sigma_g$ and $2\sigma_u$, respectively. In general, note how the numbering follows the order of increasing energy.

(b) π orbitals

Now consider the $2p_x$ and $2p_y$ orbitals of each atom. These orbitals are perpendicular to the internuclear axis and may overlap broadside-on. This overlap may be constructive or destructive, and results in a bonding or an antibonding **π orbital** (Fig. 11.27). The notation π is the analogue of *p* in atoms, for when viewed along the axis of the molecule, a π orbital looks like a *p* orbital, and has one unit of orbital angular momentum around the internuclear axis. The two $2p_x$ orbitals overlap to give a bonding and antibonding π_x orbital, and the two $2p_y$ orbitals overlap to give two π_y orbitals. The π_x and π_y bonding orbitals are degenerate; so too are their antibonding partners. We also see from Fig. 11.27 that a bonding π orbital has odd parity and is denoted π_u and an antibonding π orbital has even parity, denoted π_g.

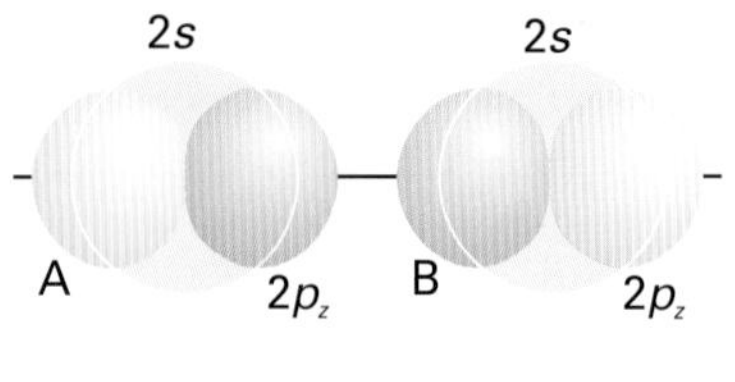

Fig. 11.25 According to molecular orbital theory, σ orbitals are built from all orbitals that have the appropriate symmetry. In homonuclear diatomic molecules of Period 2, that means that two 2*s* and two $2p_z$ orbitals should be used. From these four orbitals, four molecular orbitals can be built.

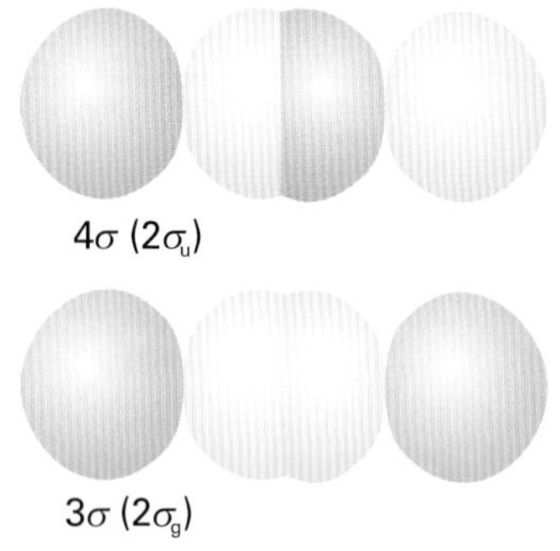

Fig. 11.26 A representation of the composition of bonding and antibonding σ orbitals built from the overlap of *p* orbitals. These illustrations are schematic.

Comment 11.6

Note that we number only the molecular orbitals formed from atomic orbitals in the valence shell. In an alternative system of notation, $1\sigma_g$ and $1\sigma_u$ are used to designate the molecular orbitals formed from the core 1*s* orbitals of the atoms; the orbitals we are considering would then be labelled starting from 2.

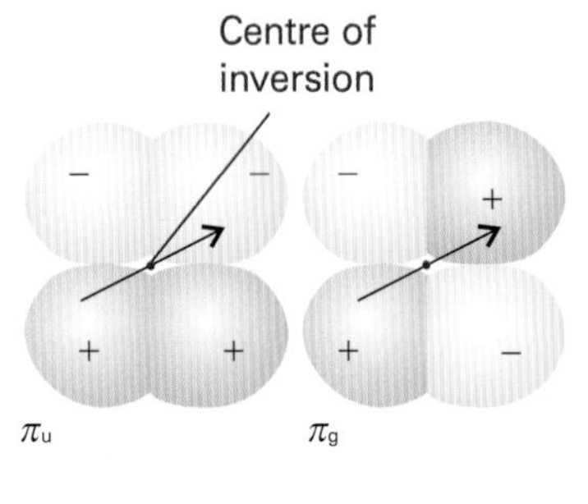

Fig. 11.27 A schematic representation of the structure of π bonding and antibonding molecular orbitals. The figure also shows that the bonding π orbital has odd parity, whereas the antiboding π orbital has even parity.

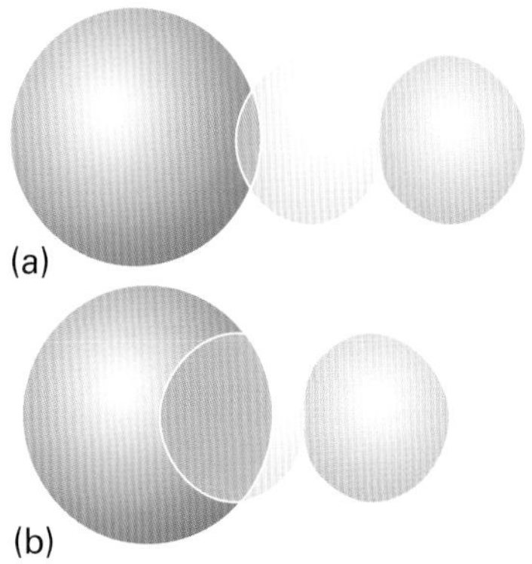

Fig. 11.28 (a) When two orbitals are on atoms that are far apart, the wavefunctions are small where they overlap, so S is small. (b) When the atoms are closer, both orbitals have significant amplitudes where they overlap, and S may approach 1. Note that S will decrease again as the two atoms approach more closely than shown here, because the region of negative amplitude of the p orbital starts to overlap the positive overlap of the s orbital. When the centres of the atoms coincide, $S = 0$.

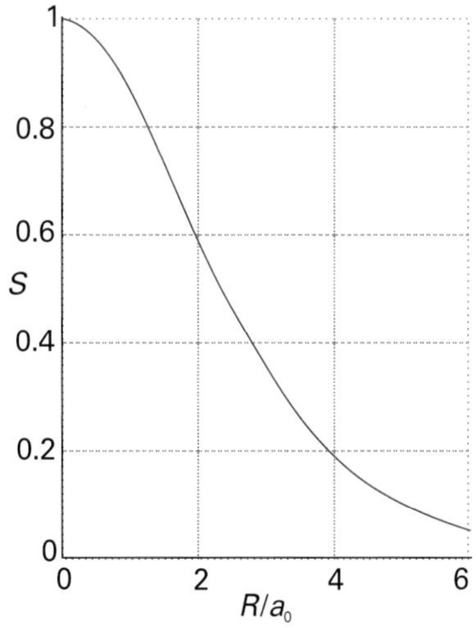

Fig. 11.29 The overlap integral, S, between two H1s orbitals as a function of their separation R.

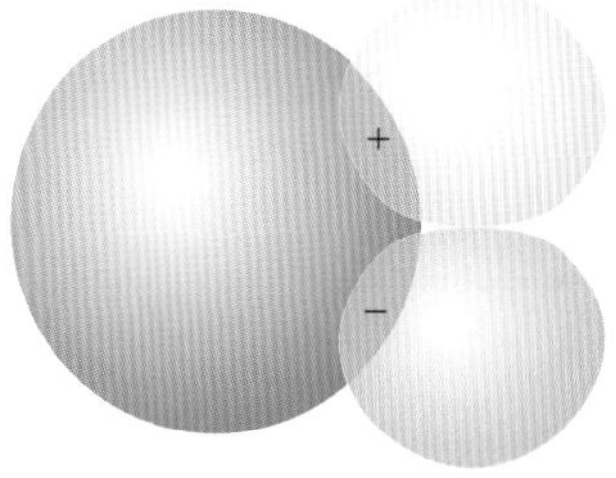

Fig. 11.30 A p orbital in the orientation shown here has zero net overlap ($S = 0$) with the s orbital at all internuclear separations.

(c) The overlap integral

The extent to which two atomic orbitals on different atoms overlap is measured by the **overlap integral**, S:

$$S = \int \chi_A^* \chi_B \, d\tau \qquad [11.17]$$

If the atomic orbital χ_A on A is small wherever the orbital χ_B on B is large, or vice versa, then the product of their amplitudes is everywhere small and the integral—the sum of these products—is small (Fig. 11.28). If χ_A and χ_B are simultaneously large in some region of space, then S may be large. If the two normalized atomic orbitals are identical (for instance, 1s orbitals on the same nucleus), then $S = 1$. In some cases, simple formulas can be given for overlap integrals and the variation of S with bond length plotted (Fig. 11.29). It follows that $S = 0.59$ for two H1s orbitals at the equilibrium bond length in H_2^+, which is an unusually large value. Typical values for orbitals with $n = 2$ are in the range 0.2 to 0.3.

Now consider the arrangement in which an s orbital is superimposed on a p_x orbital of a different atom (Fig. 11.30). The integral over the region where the product of orbitals is positive exactly cancels the integral over the region where the product of orbitals is negative, so overall $S = 0$ exactly. Therefore, there is no net overlap between the s and p orbitals in this arrangement.

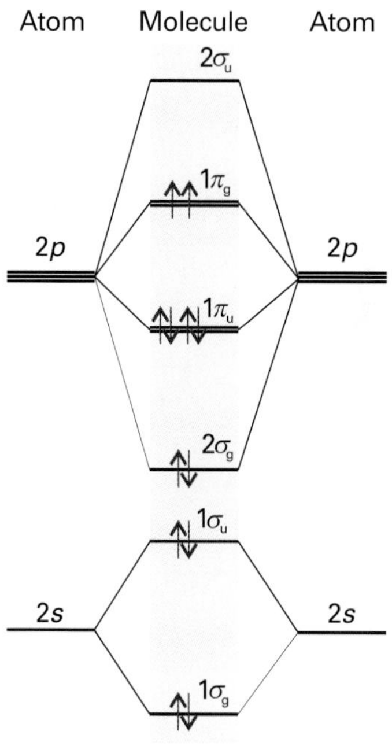

Fig. 11.31 The molecular orbital energy level diagram for homonuclear diatomic molecules. The lines in the middle are an indication of the energies of the molecular orbitals that can be formed by overlap of atomic orbitals. As remarked in the text, this diagram should be used for O_2 (the configuration shown) and F_2.

(d) The electronic structures of homonuclear diatomic molecules

To construct the molecular orbital energy level diagram for Period 2 homonuclear diatomic molecules, we form eight molecular orbitals from the eight valence shell orbitals (four from each atom). In some cases, π orbitals are less strongly bonding than σ orbitals because their maximum overlap occurs off-axis. This relative weakness suggests that the molecular orbital energy level diagram ought to be as shown in Fig. 11.31. However, we must remember that we have assumed that 2s and 2p_z orbitals

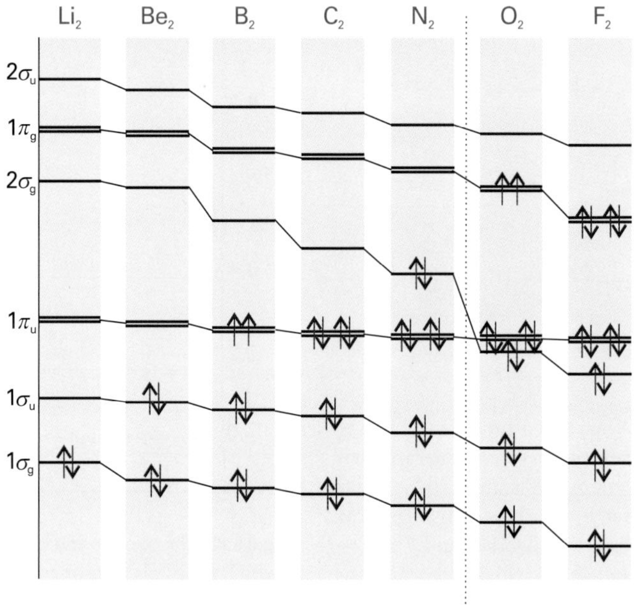

Fig. 11.32 The variation of the orbital energies of Period 2 homonuclear diatomics.

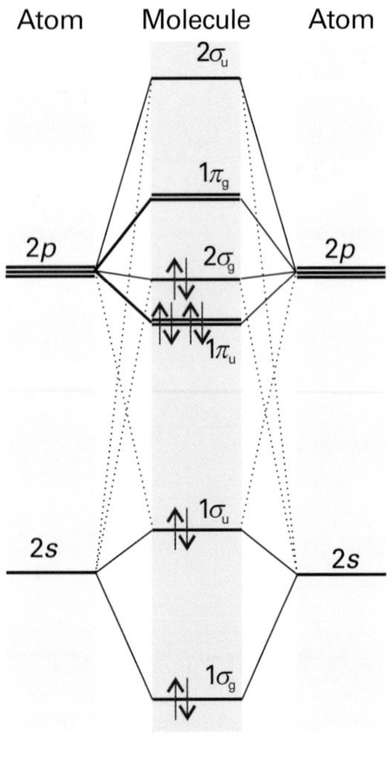

Fig. 11.33 An alternative molecular orbital energy level diagram for homonuclear diatomic molecules. As remarked in the text, this diagram should be used for diatomics up to and including N_2 (the configuration shown).

contribute to different sets of molecular orbitals whereas in fact all four atomic orbitals contribute jointly to the four σ orbitals. Hence, there is no guarantee that this order of energies should prevail, and it is found experimentally (by spectroscopy) and by detailed calculation that the order varies along Period 2 (Fig. 11.32). The order shown in Fig. 11.33 is appropriate as far as N_2, and Fig. 11.31 applies for O_2 and F_2. The relative order is controlled by the separation of the 2*s* and 2*p* orbitals in the atoms, which increases across the group. The consequent switch in order occurs at about N_2.

With the orbitals established, we can deduce the ground configurations of the molecules by adding the appropriate number of electrons to the orbitals and following the building-up rules. Anionic species (such as the peroxide ion, O_2^{2-}) need more electrons than the parent neutral molecules; cationic species (such as O_2^+) need fewer.

Consider N_2, which has 10 valence electrons. Two electrons pair, occupy, and fill the $1\sigma_g$ orbital; the next two occupy and fill the $1\sigma_u$ orbital. Six electrons remain. There are two $1\pi_u$ orbitals, so four electrons can be accommodated in them. The last two enter the $2\sigma_g$ orbital. Therefore, the ground-state configuration of N_2 is $1\sigma_g^2 1\sigma_u^2 1\pi_u^4 2\sigma_g^2$.

A measure of the net bonding in a diatomic molecule is its **bond order**, *b*:

$$b = \tfrac{1}{2}(n - n^*) \qquad [11.18]$$

where n is the number of electrons in bonding orbitals and n^* is the number of electrons in antibonding orbitals. Thus each electron pair in a bonding orbital increases the bond order by 1 and each pair in an antibonding orbital decreases b by 1. For H_2, $b = 1$, corresponding to a single bond, H—H, between the two atoms. In He_2, $b = 0$, and there is no bond. In N_2, $b = \tfrac{1}{2}(8 - 2) = 3$. This bond order accords with the Lewis structure of the molecule (:N≡N:).

The ground-state electron configuration of O_2, with 12 valence electrons, is based on Fig. 11.31, and is $1\sigma_g^2 1\sigma_u^2 2\sigma_g^2 1\pi_u^4 1\pi_g^2$. Its bond order is 2. According to the building-up principle, however, the two $1\pi_g$ electrons occupy different orbitals: one will enter $1\pi_{g,x}$ and the other will enter $1\pi_{g,y}$. Because the electrons are in different orbitals, they will have parallel spins. Therefore, we can predict that an O_2 molecule will have a net spin angular momentum $S = 1$ and, in the language introduced in Section 10.7, be in a triplet state. Because electron spin is the source of a magnetic moment, we can go on to predict that oxygen should be paramagnetic. This prediction, which VB theory does not make, is confirmed by experiment.

Synoptic table 11.2* Bond lengths

Bond	Order	R_e/pm
HH	1	74.14
NN	3	109.76
HCl	1	127.45
CH	1	114
CC	1	*154*
CC	2	*134*
CC	3	*120*

* More values will be found in the *Data section*. Numbers in italics are mean values for polyatomic molecules.

Synoptic table 11.3* Bond dissociation energies

Bond	Order	D_0/(kJ mol^{-1})
HH	1	432.1
NN	3	941.7
HCl	1	427.7
CH	1	*435*
CC	1	*368*
CC	2	*720*
CC	3	*962*

* More values will be found in the *Data section*. Numbers in italics are mean values for polyatomic molecules.

Comment 11.7

A paramagnetic substance tends to move into a magnetic field; a diamagnetic substance tends to move out of one. Paramagnetism, the rarer property, arises when the molecules have unpaired electron spins. Both properties are discussed in more detail in Chapter 20.

An F_2 molecule has two more electrons than an O_2 molecule. Its configuration is therefore $1\sigma_g^2 1\sigma_u^2 2\sigma_g^2 1\pi_u^4 1\pi_g^4$ and $b = 1$. We conclude that F_2 is a singly-bonded molecule, in agreement with its Lewis structure. The hypothetical molecule dineon, Ne_2, has two further electrons: its configuration is $1\sigma_g^2 1\sigma_u^2 2\sigma_g^2 1\pi_u^4 1\pi_g^4 2\sigma_u^2$ and $b = 0$. The zero bond order is consistent with the monatomic nature of Ne.

The bond order is a useful parameter for discussing the characteristics of bonds, because it correlates with bond length and bond strength. For bonds between atoms of a given pair of elements:

1 The greater the bond order, the shorter the bond.

2 The greater the bond order, the greater the bond strength.

Table 11.2 lists some typical bond lengths in diatomic and polyatomic molecules. The strength of a bond is measured by its bond dissociation energy, D_e, the energy required to separate the atoms to infinity. Table 11.3 lists some experimental values of dissociation energies.

Comment 11.8

Bond dissociation energies are commonly used in thermodynamic cycles, where bond enthalpies, $\Delta_{bond}H^{\ominus}$, should be used instead. It follows from the same kind of argument used in *Justification 10.7* concerning ionization enthalpies, that

$$X_2(g) \rightarrow 2\,X(g) \quad \Delta_{bond}H^{\ominus}(T) = D_e + \tfrac{3}{2}RT$$

To derive this relation, we have supposed that the molar constant-pressure heat capacity of X_2 is $\tfrac{7}{2}R$ (*Molecular interpretation 2.2*) for there is a contribution from two rotational modes as well as three translational modes.

Example 11.2 *Judging the relative bond strengths of molecules and ions*

Judge whether N_2^+ is likely to have a larger or smaller dissociation energy than N_2.

Method Because the molecule with the larger bond order is likely to have the larger dissociation energy, compare their electronic configurations and assess their bond orders.

Answer From Fig. 11.33, the electron configurations and bond orders are

$$\begin{array}{lll} N_2 & 1\sigma_g^2 1\sigma_u^2 1\pi_u^4 2\sigma_g^2 & b=3 \\ N_2^+ & 1\sigma_g^2 1\sigma_u^2 1\pi_u^4 2\sigma_g^1 & b=2\tfrac{1}{2} \end{array}$$

Because the cation has the smaller bond order, we expect it to have the smaller dissociation energy. The experimental dissociation energies are 945 kJ mol^{-1} for N_2 and 842 kJ mol^{-1} for N_2^+.

Self-test 11.4 Which can be expected to have the higher dissociation energy, F_2 or F_2^+? [F_2^+]

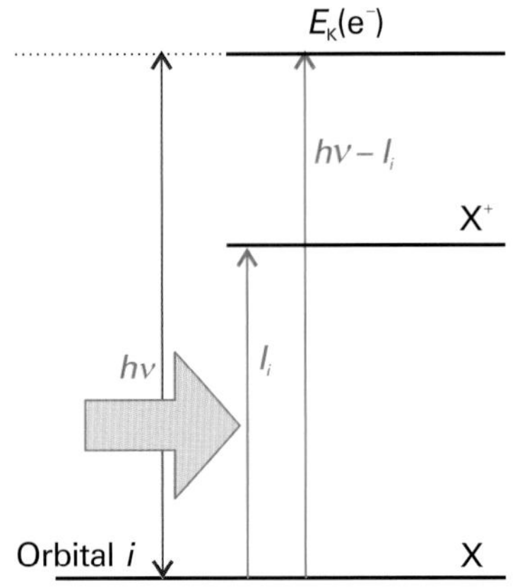

Fig. 11.34 An incoming photon carries an energy $h\nu$; an energy I_i is needed to remove an electron from an orbital i, and the difference appears as the kinetic energy of the electron.

(e) Photoelectron spectroscopy

So far we have treated molecular orbitals as purely theoretical constructs, but is there experimental evidence for their existence? **Photoelectron spectroscopy** (PES) measures the ionization energies of molecules when electrons are ejected from different orbitals by absorption of a photon of the proper energy, and uses the information to infer the energies of molecular orbitals. The technique is also used to study solids, and in Chapter 25 we shall see the important information that it gives about species at or on surfaces.

Because energy is conserved when a photon ionizes a sample, the energy of the incident photon $h\nu$ must be equal to the sum of the ionization energy, I, of the sample and the kinetic energy of the **photoelectron**, the ejected electron (Fig. 11.34):

$$h\nu = \tfrac{1}{2}m_e v^2 + I \tag{11.19}$$

This equation (which is like the one used for the photoelectric effect, Section 8.2a) can be refined in two ways. First, photoelectrons may originate from one of a number of different orbitals, and each one has a different ionization energy. Hence, a series of different kinetic energies of the photoelectrons will be obtained, each one satisfying

$$h\nu = \tfrac{1}{2}m_e v^2 + I_i \tag{11.20}$$

where I_i is the ionization energy for ejection of an electron from an orbital i. Therefore, by measuring the kinetic energies of the photoelectrons, and knowing ν, these ionization energies can be determined. Photoelectron spectra are interpreted in terms of an approximation called **Koopmans' theorem**, which states that the ionization energy I_i is equal to the orbital energy of the ejected electron (formally: $I_i = -\varepsilon_i$). That is, we can identify the ionization energy with the energy of the orbital from which it is ejected. Similarly, the energy of unfilled ('virtual orbitals') is related to the electron affinity. The theorem is only an approximation because it ignores the fact that the remaining electrons adjust their distributions when ionization occurs.

The ionization energies of molecules are several electronvolts even for valence electrons, so it is essential to work in at least the ultraviolet region of the spectrum and with wavelengths of less than about 200 nm. Much work has been done with radiation generated by a discharge through helium: the He(I) line ($1s^1 2p^1 \rightarrow 1s^2$) lies at 58.43 nm, corresponding to a photon energy of 21.22 eV. Its use gives rise to the technique of **ultraviolet photoelectron spectroscopy** (UPS). When core electrons are being studied, photons of even higher energy are needed to expel them: X–rays are used, and the technique is denoted XPS.

The kinetic energies of the photoelectrons are measured using an electrostatic deflector that produces different deflections in the paths of the photoelectrons as they pass between charged plates (Fig. 11.35). As the field strength is increased, electrons of different speeds, and therefore kinetic energies, reach the detector. The electron flux can be recorded and plotted against kinetic energy to obtain the photoelectron spectrum.

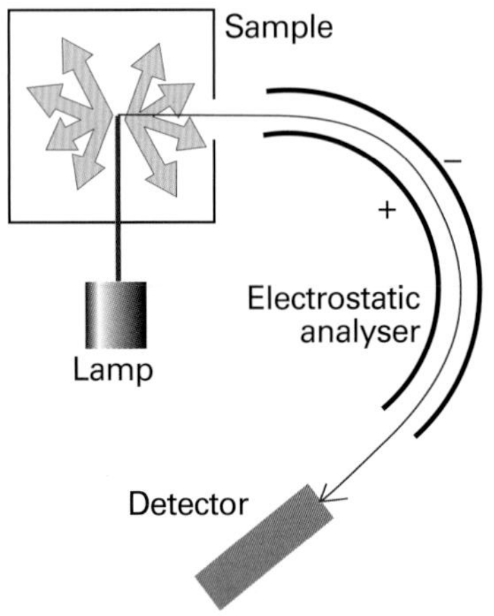

Fig. 11.35 A photoelectron spectrometer consists of a source of ionizing radiation (such as a helium discharge lamp for UPS and an X-ray source for XPS), an electrostatic analyser, and an electron detector. The deflection of the electron path caused by the analyser depends on their speed.

Illustration 11.1 *Interpreting a photoelectron spectrum*

Photoelectrons ejected from N_2 with He(I) radiation had kinetic energies of 5.63 eV (1 eV = 8065.5 cm^{-1}). Helium(I) radiation of wavelength 58.43 nm has wavenumber 1.711×10^5 cm^{-1} and therefore corresponds to an energy of 21.22 eV. Then, from eqn 11.20, 21.22 eV = 5.63 eV + I_i, so I_i = 15.59 eV. This ionization energy is the energy needed to remove an electron from the occupied molecular orbital with the highest energy of the N_2 molecule, the $2\sigma_g$ bonding orbital (see Fig. 11.33).

Self-test 11.5 Under the same circumstances, photoelectrons are also detected at 4.53 eV. To what ionization energy does that correspond? Suggest an origin.

[16.7 eV, $1\pi_u$]

11.5 Heteronuclear diatomic molecules

The electron distribution in the covalent bond between the atoms in a heteronuclear diatomic molecule is not shared evenly because it is energetically favourable for the electron pair to be found closer to one atom than the other. This imbalance results in a **polar bond**, a covalent bond in which the electron pair is shared unequally by the two atoms. The bond in HF, for instance, is polar, with the electron pair closer to the F atom. The accumulation of the electron pair near the F atom results in that atom having a net negative charge, which is called a **partial negative charge** and denoted $\delta-$. There is a matching **partial positive charge**, $\delta+$, on the H atom.

(a) Polar bonds

A polar bond consists of two electrons in an orbital of the form

$$\psi = c_A A + c_B B \qquad (11.21)$$

with unequal coefficients. The proportion of the atomic orbital A in the bond is $|c_A|^2$ and that of B is $|c_B|^2$. A nonpolar bond has $|c_A|^2 = |c_B|^2$ and a pure ionic bond has one coefficient zero (so the species A^+B^- would have $c_A = 0$ and $c_B = 1$). The atomic orbital with the lower energy makes the larger contribution to the bonding molecular orbital. The opposite is true of the antibonding orbital, for which the dominant component comes from the atomic orbital with higher energy.

These points can be illustrated by considering HF, and judging the energies of the atomic orbitals from the ionization energies of the atoms. The general form of the molecular orbitals is

$$\psi = c_H \chi_H + c_F \chi_F \qquad (11.22)$$

where χ_H is an H1s orbital and χ_F is an F2p orbital. The H1s orbital lies 13.6 eV below the zero of energy (the separated proton and electron) and the F2p orbital lies at 18.6 eV (Fig. 11.36). Hence, the bonding σ orbital in HF is mainly F2p and the antibonding σ orbital is mainly H1s orbital in character. The two electrons in the bonding orbital are most likely to be found in the F2p orbital, so there is a partial negative charge on the F atom and a partial positive charge on the H atom.

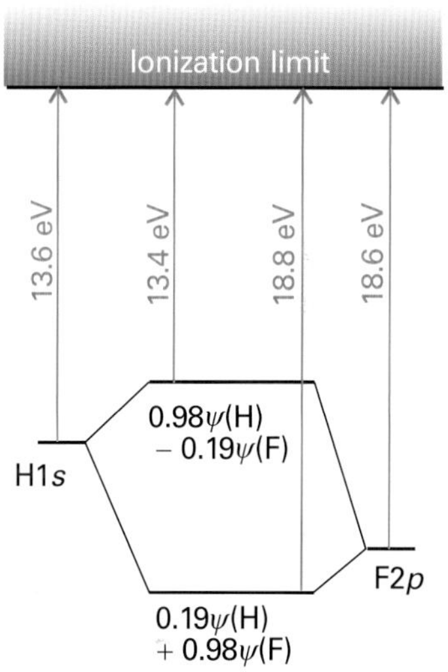

Fig. 11.36 The atomic orbital energy levels of H and F atoms and the molecular orbitals they form.

(b) Electronegativity

The charge distribution in bonds is commonly discussed in terms of the **electronegativity**, χ, of the elements involved (there should be little danger of confusing this use of χ with its use to denote an atomic orbital, which is another common convention). The electronegativity is a parameter introduced by Linus Pauling as a measure of the power of an atom to attract electrons to itself when it is part of a compound. Pauling used valence-bond arguments to suggest that an appropriate numerical scale of electronegativities could be defined in terms of bond dissociation energies, D, in electronvolts and proposed that the difference in electronegativities could be expressed as

$$|\chi_A - \chi_B| = 0.102\{D(A{-}B) - \tfrac{1}{2}[D(A{-}A) + D(B{-}B)]\}^{1/2} \qquad [11.23]$$

Electronegativities based on this definition are called **Pauling electronegativities.** A list of Pauling electronegativities is given in Table 11.4. The most electronegative

Synoptic table 11.4* Pauling electronegativities

Element	χ_P
H	2.2
C	2.6
N	3.0
O	3.4
F	4.0
Cl	3.2
Cs	0.79

* More values will be found in the *Data section*.

elements are those close to fluorine; the least are those close to caesium. It is found that the greater the difference in electronegativities, the greater the polar character of the bond. The difference for HF, for instance, is 1.78; a C—H bond, which is commonly regarded as almost nonpolar, has an electronegativity difference of 0.51.

The spectroscopist Robert Mulliken proposed an alternative definition of electronegativity. He argued that an element is likely to be highly electronegative if it has a high ionization energy (so it will not release electrons readily) and a high electron affinity (so it is energetically favorable to acquire electrons). The **Mulliken electronegativity scale** is therefore based on the definition

$$\chi_M = \tfrac{1}{2}(I + E_{ea}) \qquad [11.24]$$

where I is the ionization energy of the element and E_{ea} is its electron affinity (both in electronvolts, Section 10.4e). The Mulliken and Pauling scales are approximately in line with each other. A reasonably reliable conversion between the two is $\chi_P = 1.35\chi_M^{1/2} - 1.37$.

(c) The variation principle

A more systematic way of discussing bond polarity and finding the coefficients in the linear combinations used to build molecular orbitals is provided by the **variation principle**:

> If an arbitrary wavefunction is used to calculate the energy, the value calculated is never less than the true energy.

This principle is the basis of all modern molecular structure calculations (Section 11.7). The arbitrary wavefunction is called the **trial wavefunction**. The principle implies that, if we vary the coefficients in the trial wavefunction until the lowest energy is achieved (by evaluating the expectation value of the hamiltonian for each wavefunction), then those coefficients will be the best. We might get a lower energy if we use a more complicated wavefunction (for example, by taking a linear combination of several atomic orbitals on each atom), but we shall have the optimum (minimum energy) molecular orbital that can be built from the chosen **basis set**, the given set of atomic orbitals.

Comment 11.9
The name 'secular' is derived from the Latin word for age or generation. The term comes via astronomy, where the same equations appear in connection with slowly accumulating modifications of planetary orbits.

The method can be illustrated by the trial wavefunction in eqn 11.21. We show in the *Justification* below that the coefficients are given by the solutions of the two **secular equations**

$$(\alpha_A - E)c_A + (\beta - ES)c_B = 0 \qquad (11.25a)$$

$$(\beta - ES)c_A + (\alpha_B - E)c_B = 0 \qquad (11.25b)$$

The parameter α is called a **Coulomb integral**. It is negative and can be interpreted as the energy of the electron when it occupies A (for α_A) or B (for α_B). In a homonuclear diatomic molecule, $\alpha_A = \alpha_B$. The parameter β is called a **resonance integral** (for classical reasons). It vanishes when the orbitals do not overlap, and at equilibrium bond lengths it is normally negative.

Justification 11.3 *The variation principle applied to a heteronuclear diatomic molecule*

The trial wavefunction in eqn 11.21 is real but not normalized because at this stage the coefficients can take arbitrary values. Therefore, we can write $\psi^* = \psi$ but do not assume that $\int\psi^2 d\tau = 1$. The energy of the trial wavefunction is the expectation value of the energy operator (the hamiltonian, $\hat{H}$, Section 8.5):

$$E = \frac{\int \psi^* \hat{H} \psi \, \mathrm{d}\tau}{\int \psi^* \psi \, \mathrm{d}\tau} \tag{11.26}$$

We must search for values of the coefficients in the trial function that minimize the value of E. This is a standard problem in calculus, and is solved by finding the coefficients for which

$$\frac{\partial E}{\partial c_{\mathrm{A}}} = 0 \qquad \frac{\partial E}{\partial c_{\mathrm{B}}} = 0$$

The first step is to express the two integrals in terms of the coefficients. The denominator is

$$\begin{aligned}\int \psi^2 \, \mathrm{d}\tau &= \int (c_{\mathrm{A}}A + c_{\mathrm{B}}B)^2 \, \mathrm{d}\tau \\ &= c_{\mathrm{A}}^2 \int A^2 \, \mathrm{d}\tau + c_{\mathrm{B}}^2 \int B^2 \, \mathrm{d}\tau + 2c_{\mathrm{A}}c_{\mathrm{B}} \int AB \, \mathrm{d}\tau \\ &= c_{\mathrm{A}}^2 + c_{\mathrm{B}}^2 + 2c_{\mathrm{A}}c_{\mathrm{B}}S\end{aligned}$$

because the individual atomic orbitals are normalized and the third integral is the overlap integral S (eqn 11.17). The numerator is

$$\begin{aligned}\int \psi \hat{H} \psi \, \mathrm{d}\tau &= \int (c_{\mathrm{A}}A + c_{\mathrm{B}}B)\hat{H}(c_{\mathrm{A}}A + c_{\mathrm{B}}B) \, \mathrm{d}\tau \\ &= c_{\mathrm{A}}^2 \int A\hat{H}A \, \mathrm{d}\tau + c_{\mathrm{B}}^2 \int B\hat{H}B \, \mathrm{d}\tau + c_{\mathrm{A}}c_{\mathrm{B}} \int A\hat{H}B \, \mathrm{d}\tau + c_{\mathrm{A}}c_{\mathrm{B}} \int B\hat{H}A \, \mathrm{d}\tau\end{aligned}$$

There are some complicated integrals in this expression, but we can combine them all into the parameters

$$\alpha_{\mathrm{A}} = \int A\hat{H}A \, \mathrm{d}\tau \qquad \alpha_{\mathrm{B}} = \int B\hat{H}B \, \mathrm{d}\tau \tag*{[11.27]}$$

$$\beta = \int A\hat{H}B \, \mathrm{d}\tau = \int B\hat{H}A \, \mathrm{d}\tau \text{ (by the hermiticity of } \hat{H}\text{)}$$

Then

$$\int \psi \hat{H} \psi \, \mathrm{d}\tau = c_{\mathrm{A}}^2 \alpha_{\mathrm{A}} + c_{\mathrm{B}}^2 \alpha_{\mathrm{B}} + 2c_{\mathrm{A}}c_{\mathrm{B}}\beta$$

The complete expression for E is

$$E = \frac{c_{\mathrm{A}}^2 \alpha_{\mathrm{A}} + c_{\mathrm{B}}^2 \alpha_{\mathrm{B}} + 2c_{\mathrm{A}}c_{\mathrm{B}}\beta}{c_{\mathrm{A}}^2 + c_{\mathrm{B}}^2 + 2c_{\mathrm{A}}c_{\mathrm{B}}S} \tag{11.28}$$

Its minimum is found by differentiation with respect to the two coefficients and setting the results equal to 0. After a bit of work, we obtain

$$\frac{\partial E}{\partial c_{\mathrm{A}}} = \frac{2 \times (c_{\mathrm{A}}\alpha_{\mathrm{A}} - c_{\mathrm{A}}E + c_{\mathrm{B}}\beta - c_{\mathrm{B}}SE)}{c_{\mathrm{A}}^2 + c_{\mathrm{B}}^2 + 2c_{\mathrm{A}}c_{\mathrm{B}}S} = 0$$

$$\frac{\partial E}{\partial c_{\mathrm{B}}} = \frac{2 \times (c_{\mathrm{B}}\alpha_{\mathrm{B}} - c_{\mathrm{B}}E + c_{\mathrm{A}}\beta - c_{\mathrm{A}}SE)}{c_{\mathrm{A}}^2 + c_{\mathrm{B}}^2 + 2c_{\mathrm{A}}c_{\mathrm{B}}S} = 0$$

For the derivatives to vanish, the numerators of the expressions above must vanish. That is, we must find values of c_A and c_B that satisfy the conditions

$$c_A\alpha_A - c_AE + c_B\beta - c_BSE = (\alpha_A - E)c_A + (\beta - ES)c_B = 0$$
$$c_A\beta - c_ASE + c_B\alpha_B - c_BE = (\beta - ES)c_A + (\alpha_B - E)c_B = 0$$

which are the secular equations (eqn 11.25).

To solve the secular equations for the coefficients we need to know the energy E of the orbital. As for any set of simultaneous equations, the secular equations have a solution if the **secular determinant**, the determinant of the coefficients, is zero; that is, if

$$\begin{vmatrix} \alpha_A - E & \beta - ES \\ \beta - ES & \alpha_B - E \end{vmatrix} = 0 \qquad (11.29)$$

Comment 11.10
We need to know that a 2×2 determinant expands as follows:

$$\begin{vmatrix} a & b \\ c & d \end{vmatrix} = ad - bc$$

This determinant expands to a quadratic equation in E (see *Illustration 11.2*). Its two roots give the energies of the bonding and antibonding molecular orbitals formed from the atomic orbitals and, according to the variation principle, the lower root is the best energy achievable with the given basis set.

Illustration 11.2 *Using the variation principle (1)*

To find the energies E of the bonding and antibonding orbitals of a homonuclear diatomic molecule set with $\alpha_A = \alpha_B = \alpha$ in eqn 11.29 and get

$$\begin{vmatrix} \alpha - E & \beta - ES \\ \beta - ES & \alpha - E \end{vmatrix} = (\alpha - E)^2 - (\beta - ES)^2 = 0$$

The solutions of this equation are

$$E_\pm = \frac{\alpha \pm \beta}{1 \pm S}$$

The values of the coefficients in the linear combination are obtained by solving the secular equations using the two energies obtained from the secular determinant. The lower energy (E_+ in the *Illustration*) gives the coefficients for the bonding molecular orbital, the upper energy (E_-) the coefficients for the antibonding molecular orbital. The secular equations give expressions for the ratio of the coefficients in each case, so we need a further equation in order to find their individual values. This equation is obtained by demanding that the best wavefunction should also be normalized. This condition means that, at this final stage, we must also ensure that

$$\int \psi^2\, d\tau = c_A^2 + c_B^2 + 2c_Ac_BS = 1 \qquad (11.30)$$

Illustration 11.3 *Using the variation principle (2)*

To find the values of the coefficients c_A and c_B in the linear combination that corresponds to the energy E_+ from *Illustration 11.2*, we use eqn 11.28 (with $\alpha_A = \alpha_B = \alpha$) to write

$$E_+ = \frac{\alpha + \beta}{1 + S} = \frac{c_A^2\alpha + c_B^2\alpha + 2c_Ac_B\beta}{c_A^2 + c_B^2 + 2c_Ac_BS}$$

Now we use the normalization condition, eqn 11.30, to set $c_A^2 + c_B^2 + 2c_Ac_BS = 1$, and so write

$$\frac{\alpha+\beta}{1+S} = (c_A^2 + c_B^2)\alpha + 2c_Ac_B\beta$$

This expression implies that

$$c_A^2 + c_B^2 = 2c_Ac_B = \frac{1}{1+S} \qquad \text{and} \qquad |c_A| = \frac{1}{\{2(1+S)\}^{1/2}} \qquad c_B = c_A$$

Proceeding in a similar way to find the coefficients in the linear combination that corresponds to the energy E_-, we write

$$E_- = \frac{\alpha-\beta}{1-S} = (c_A^2 + c_B^2)\alpha + 2c_Ac_B\beta$$

which implies that

$$c_A^2 + c_B^2 = -2c_Ac_B = \frac{1}{1-S} \qquad \text{and} \qquad |c_A| = \frac{1}{\{2(1-S)\}^{1/2}} \qquad c_B = -c_A$$

(d) Two simple cases

The complete solutions of the secular equations are very cumbersome, even for 2×2 determinants, but there are two cases where the roots can be written down very simply.

We saw in *Illustrations* 11.2 and 11.3 that, when the two atoms are the same, and we can write $\alpha_A = \alpha_B = \alpha$, the solutions are

$$E_+ = \frac{\alpha+\beta}{1+S} \qquad c_A = \frac{1}{\{2(1+S)\}^{1/2}} \qquad c_B = c_A \tag{11.31a}$$

$$E_- = \frac{\alpha-\beta}{1-S} \qquad c_A = \frac{1}{\{2(1-S)\}^{1/2}} \qquad c_B = -c_A \tag{11.31b}$$

In this case, the bonding orbital has the form

$$\psi_+ = \frac{A+B}{\{2(1+S)\}^{1/2}} \tag{11.32a}$$

and the corresponding antibonding orbital is

$$\psi_- = \frac{A-B}{\{2(1-S)\}^{1/2}} \tag{11.32b}$$

in agreement with the discussion of homonuclear diatomics we have already given, but now with the normalization constant in place.

The second simple case is for a heteronuclear diatomic molecule but with $S = 0$ (a common approximation in elementary work). The secular determinant is then

$$\begin{vmatrix} \alpha_A - E & \beta \\ \beta & \alpha_B - E \end{vmatrix} = (\alpha_A - E)(\alpha_B - E) - \beta^2 = 0$$

The solutions can be expressed in terms of the parameter ζ (zeta), with

$$\zeta = \tfrac{1}{2}\arctan\frac{2|\beta|}{\alpha_B - \alpha_A} \tag{11.33}$$

and are

$$E_- = \alpha_B - \beta \tan \zeta \qquad \psi_- = -A \sin \zeta + B \cos \zeta \tag{11.34a}$$

$$E_+ = \alpha_A + \beta \tan \zeta \qquad \psi_+ = A \cos \zeta + B \sin \zeta \tag{11.34b}$$

An important feature revealed by these solutions is that as the energy difference $|\alpha_B - \alpha_A|$ between the interacting atomic orbitals increases, the value of ζ decreases. We show in the following *Justification* that, when the energy difference is very large, in the sense that $|\alpha_B - \alpha_A| \gg 2|\beta|$, the energies of the resulting molecular orbitals differ only slightly from those of the atomic orbitals, which implies in turn that the bonding and antibonding effects are small. That is, the strongest bonding and antibonding effects are obtained when the two contributing orbitals have closely similar energies. The difference in energy between core and valence orbitals is the justification for neglecting the contribution of core orbitals to bonding. The core orbitals of one atom have a similar energy to the core orbitals of the other atom; but core–core interaction is largely negligible because the overlap between them (and hence the value of β) is so small.

Justification 11.4 *Bonding and antibonding effects in heteronuclear diatomic molecules*

When $|\alpha_B - \alpha_A| \gg 2|\beta|$ and $2|\beta|/|\alpha_B - \alpha_A| \ll 1$, we can write $\arctan 2|\beta|/|\alpha_B - \alpha_A| \approx 2|\beta|/|\alpha_B - \alpha_A|$ and, from eqn 11.33, $\zeta \approx |\beta|/(\alpha_B - \alpha_A)$. It follows that $\tan \zeta \approx |\beta|/(\alpha_B - \alpha_A)$. Noting that β is normally a negative number, so that $\beta/|\beta| = -1$, we can use eqn 11.34 to write

Comment 11.11

For $x \ll 1$, we can write: $\sin x \approx x$, $\cos x \approx 1$, $\tan x \approx x$, and $\arctan x = \tan^{-1} x \approx x$.

$$E_- = \alpha_B + \frac{\beta^2}{\alpha_B - \alpha_A} \qquad E_+ = \alpha_A - \frac{\beta^2}{\alpha_B - \alpha_A}$$

(In Problem 11.25 you are invited to derive these expressions via a different route.) It follows that, when the energy difference between the atomic orbitals is so large that $|\alpha_B - \alpha_A| \gg 2|\beta|$, the energies of the two molecular orbitals are $E_- \approx \alpha_B$ and $E_+ \approx \alpha_A$.

Now we consider the behaviour of the wavefunctions in the limit of large $|\alpha_B - \alpha_A|$, when $\zeta \ll 1$. In this case, $\sin \zeta \approx \zeta$ and $\cos \zeta \approx 1$ and, from eqn 11.34, we write $\psi_- \approx B$ and $\psi_+ \approx A$. That is, the molecular orbitals are respectively almost pure B and almost pure A.

Example 11.3 *Calculating the molecular orbitals of HF*

Calculate the wavefunctions and energies of the σ orbitals in the HF molecule, taking $\beta = -1.0$ eV and the following ionization energies: H1*s*: 13.6 eV, F2*s*: 40.2 eV, F2*p*: 17.4 eV.

Method Because the F2*p* and H1*s* orbitals are much closer in energy than the F2*s* and H1*s* orbitals, to a first approximation neglect the contribution of the F2*s* orbital. To use eqn 11.34, we need to know the values of the Coulomb integrals α_H and α_F. Because these integrals represent the energies of the H1*s* and F2*p* electrons, respectively, they are approximately equal to (the negative of) the ionization energies of the atoms. Calculate ζ from eqn 11.33 (with A identified as F and B as H), and then write the wavefunctions by using eqn 11.34.

Answer Setting $\alpha_H = -13.6$ eV and $\alpha_F = -17.4$ eV gives $\tan 2\zeta = 0.58$; so $\zeta = 13.9°$. Then

$E_- = -13.4$ eV $\quad \psi_- = 0.97\chi_H - 0.24\chi_F$

$E_+ = -17.6$ eV $\quad \psi_+ = 0.24\chi_H + 0.97\chi_F$

Notice how the lower energy orbital (the one with energy −17.6 eV) has a composition that is more F2*p* orbital than H1*s*, and that the opposite is true of the higher energy, antibonding orbital.

Self-test 11.6 The ionization energy of Cl is 13.1 eV; find the form and energies of the σ orbitals in the HCl molecule using $\beta = -1.0$ eV.
[$E_- = -12.8$ eV, $\psi_- = -0.62\chi_H + 0.79\chi_{Cl}$; $E_+ = -13.9$ eV, $\psi_+ = 0.79\chi_H + 0.62\chi_{Cl}$]

IMPACT ON BIOCHEMISTRY
I11.1 The biochemical reactivity of O_2, N_2, and NO

We can now see how some of these concepts are applied to diatomic molecules that play a vital biochemical role. At sea level, air contains approximately 23.1 per cent O_2 and 75.5 per cent N_2 by mass. Molecular orbital theory predicts—correctly—that O_2 has unpaired electron spins and, consequently, is a reactive component of the Earth's atmosphere; its most important biological role is as an oxidizing agent. By contrast N_2, the major component of the air we breathe, is so stable (on account of the triple bond connecting the atoms) and unreactive that *nitrogen fixation*, the reduction of atmospheric N_2 to NH_3, is among the most thermodynamically demanding of biochemical reactions, in the sense that it requires a great deal of energy derived from metabolism. So taxing is the process that only certain bacteria and archaea are capable of carrying it out, making nitrogen available first to plants and other microorganisms in the form of ammonia. Only after incorporation into amino acids by plants does nitrogen adopt a chemical form that, when consumed, can be used by animals in the synthesis of proteins and other nitrogen-containing molecules.

The reactivity of O_2, while important for biological energy conversion, also poses serious physiological problems. During the course of metabolism, some electrons escape from complexes I, II, and III of the respiratory chain and reduce O_2 to superoxide ion, O_2^-. The ground-state electronic configuration of O_2^- is $1\sigma_g^2 1\sigma_u^2 2\sigma_g^2 1\pi_u^4 1\pi_g^3$, so the ion is a radical with a bond order $b = \frac{3}{2}$. We predict that the superoxide ion is a reactive species that must be scavenged to prevent damage to cellular components. The enzyme superoxide dismutase protects cells by catalysing the disproportionation (or dismutation) of O_2^- into O_2 and H_2O_2:

$$2\,O_2^- + 2\,H^+ \rightarrow H_2O_2 + O_2$$

However, H_2O_2 (hydrogen peroxide), formed by the reaction above and by leakage of electrons out of the respiratory chain, is a powerful oxidizing agent and also harmful to cells. It is metabolized further by catalases and peroxidases. A catalase catalyses the reaction

$$2\,H_2O_2 \rightarrow 2\,H_2O + O_2$$

and a peroxidase reduces hydrogen peroxide to water by oxidizing an organic molecule. For example, the enzyme glutathione peroxidase catalyses the reaction

$$2\,\text{glutathione}_{\text{red}} + H_2O_2 \rightarrow \text{glutathione}_{\text{ox}} + 2\,H_2O$$

There is growing evidence for the involvement of the damage caused by reactive oxygen species (ROS), such as O_2^-, H_2O_2, and ·OH (the hydroxyl radical), in the mechanism of ageing and in the development of cardiovascular disease, cancer,

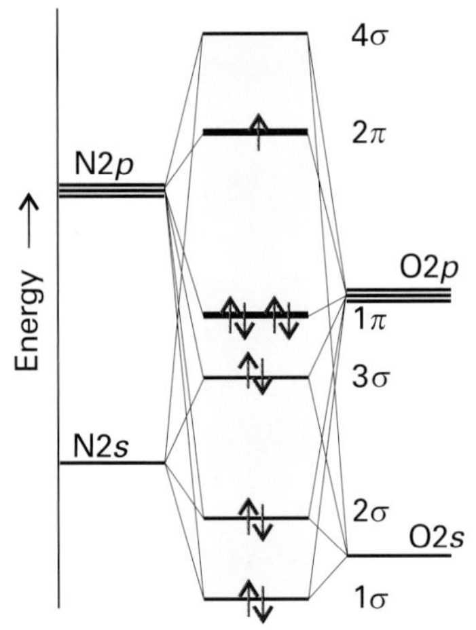

Fig. 11.37 The molecular orbital energy level diagram for NO.

stroke, inflammatory disease, and other conditions. For this reason, much effort has been expended on studies of the biochemistry of *antioxidants*, substances that can either deactivate ROS directly (as glutathione does) or halt the progress of cellular damage through reactions with radicals formed by processes initiated by ROS. Important examples of antioxidants are vitamin C (ascorbic acid), vitamin E (α-tocopherol), and uric acid.

Nitric oxide (nitrogen monoxide, NO) is a small molecule that diffuses quickly between cells, carrying chemical messages that help initiate a variety of processes, such as regulation of blood pressure, inhibition of platelet aggregation, and defence against inflammation and attacks to the immune system. The molecule is synthesized from the amino acid arginine in a series of reactions catalysed by nitric oxide synthase and requiring O_2 and NADPH.

Figure 11.37 shows the bonding scheme in NO and illustrates a number of points we have made about heteronuclear diatomic molecules. The ground configuration is $1\sigma^2 2\sigma^2 3\sigma^2 1\pi^4 2\pi^1$. The 3σ and 1π orbitals are predominantly of O character as that is the more electronegative element. The highest-energy occupied orbital is 2π, contains one electron, and has more N character than O character. It follows that NO is a radical with an unpaired electron that can be regarded as localized more on the N atom than on the O atom. The lowest-energy unoccupied orbital is 4σ, which is also localized predominantly on N.

Because NO is a radical, we expect it to be reactive. Its half-life is estimated at approximately 1–5 s, so it needs to be synthesized often in the cell. As we saw above, there is a biochemical price to be paid for the reactivity of biological radicals. Like O_2, NO participates in some reactions that are not beneficial to the cell. Indeed, the radicals O_2^- and NO combine to form the peroxynitrite ion:

$$NO\cdot + O_2^-\cdot \rightarrow ONOO^-$$

where we have shown the unpaired electrons explicitly. The peroxynitrite ion is a reactive oxygen species that damages proteins, DNA, and lipids, possibly leading to heart disease, amyotrophic lateral sclerosis (Lou Gehrig's disease), Alzheimer's disease, and multiple sclerosis. Note that the structure of the ion is consistent with the bonding scheme in Fig. 11.37: because the unpaired electron in NO is slightly more localized on the N atom, we expect that atom to form a bond with an O atom from the O_2^- ion.

Molecular orbitals for polyatomic systems

The molecular orbitals of polyatomic molecules are built in the same way as in diatomic molecules, the only difference being that we use more atomic orbitals to construct them. As for diatomic molecules, polyatomic molecular orbitals spread over the entire molecule. A molecular orbital has the general form

$$\psi = \sum_i c_i \chi_i \tag{11.35}$$

where χ_i is an atomic orbital and the sum extends over all the valence orbitals of all the atoms in the molecule. To find the coefficients, we set up the secular equations and the secular determinant, just as for diatomic molecules, solve the latter for the energies, and then use these energies in the secular equations to find the coefficients of the atomic orbitals for each molecular orbital.

The principal difference between diatomic and polyatomic molecules lies in the greater range of shapes that are possible: a diatomic molecule is necessarily linear, but a triatomic molecule, for instance, may be either linear or angular with a characteristic bond angle. The shape of a polyatomic molecule—the specification of its bond lengths and its bond angles—can be predicted by calculating the total energy of the molecule for a variety of nuclear positions, and then identifying the conformation that corresponds to the lowest energy.

11.6 The Hückel approximation

Molecular orbital theory takes large molecules and extended aggregates of atoms, such as solid materials, in its stride. First we shall consider conjugated molecules, in which there is an alternation of single and double bonds along a chain of carbon atoms. Although the classification of an orbital as σ or π is strictly valid only in linear molecules, as will be familiar from introductory chemistry courses, it is also used to denote the local symmetry with respect to a given A—B bond axis.

The π molecular orbital energy level diagrams of conjugated molecules can be constructed using a set of approximations suggested by Erich Hückel in 1931. In his approach, the π orbitals are treated separately from the σ orbitals, and the latter form a rigid framework that determines the general shape of the molecule. All the C atoms are treated identically, so all the Coulomb integrals α for the atomic orbitals that contribute to the π orbitals are set equal. For example, in ethene, we take the σ bonds as fixed, and concentrate on finding the energies of the single π bond and its companion antibond.

(a) Ethene and frontier orbitals

We express the π orbitals as LCAOs of the C2p orbitals that lie perpendicular to the molecular plane. In ethene, for instance, we would write

$$\psi = c_A A + c_B B \qquad (11.36)$$

where the A is a C2p orbital on atom A, and so on. Next, the optimum coefficients and energies are found by the variation principle as explained in Section 11.5. That is, we have to solve the secular determinant, which in the case of ethene is eqn 11.29 with $\alpha_A = \alpha_B = \alpha$:

$$\begin{vmatrix} \alpha - E & \beta - ES \\ \beta - ES & \alpha - E \end{vmatrix} = 0 \qquad (11.37)$$

The roots of this determinant can be found very easily (they are the same as those in *Illustration 11.2*). In a modern computation all the resonance integrals and overlap integrals would be included, but an indication of the molecular orbital energy level diagram can be obtained very readily if we make the following additional **Hückel approximations**:

1 All overlap integrals are set equal to zero.

2 All resonance integrals between non-neighbours are set equal to zero.

3 All remaining resonance integrals are set equal (to β).

These approximations are obviously very severe, but they let us calculate at least a general picture of the molecular orbital energy levels with very little work. The assumptions result in the following structure of the secular determinant:

1 All diagonal elements: $\alpha - E$.

2 Off-diagonal elements between neighbouring atoms: β.

3 All other elements: 0.

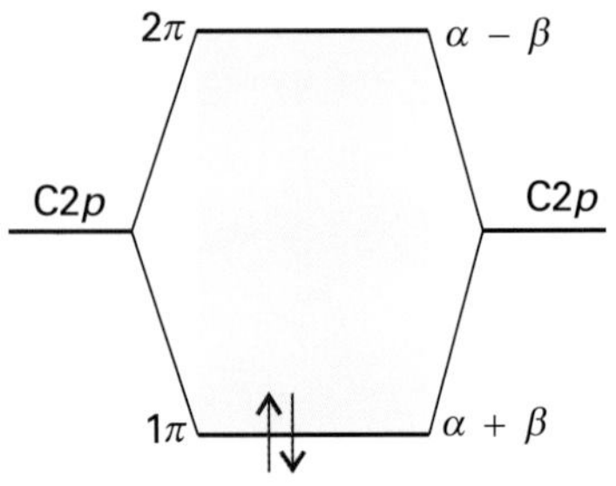

Fig. 11.38 The Hückel molecular orbital energy levels of ethene. Two electrons occupy the lower π orbital.

These approximations lead to

$$\begin{vmatrix} \alpha - E & \beta \\ \beta & \alpha - E \end{vmatrix} = (\alpha - E)^2 - \beta^2 = 0 \qquad (11.38)$$

The roots of the equation are

$$E_{\pm} = \alpha \pm \beta \qquad (11.39)$$

The + sign corresponds to the bonding combination (β is negative) and the − sign corresponds to the antibonding combination (Fig. 11.38). We see the effect of neglecting overlap by comparing this result with eqn 11.31.

The building-up principle leads to the configuration $1\pi^2$, because each carbon atom supplies one electron to the π system. The **highest occupied molecular orbital** in ethene, its HOMO, is the 1π orbital; the **lowest unfilled molecular orbital**, its LUMO, is the 2π orbital (or, as it is sometimes denoted, the $2\pi^*$ orbital). These two orbitals jointly form the **frontier orbitals** of the molecule. The frontier orbitals are important because they are largely responsible for many of the chemical and spectroscopic properties of the molecule. For example, we can estimate that $2|\beta|$ is the $\pi^* \leftarrow \pi$ excitation energy of ethene, the energy required to excite an electron from the 1π to the 2π orbital. The constant β is often left as an adjustable parameter; an approximate value for π bonds formed from overlap of two C2p atomic orbitals is about −2.4 eV (−230 kJ mol^{-1}).

(b) The matrix formulation of the Hückel method

In preparation for making Hückel theory more sophisticated and readily applicable to bigger molecules, we need to reformulate it in terms of matrices and vectors (see *Appendix 2*). We have seen that the secular equations that we have to solve for a two-atom system have the form

$$(H_{AA} - E_i S_{AA})c_{i,A} + (H_{AB} - E_i S_{AB})c_{i,B} = 0 \qquad (11.40a)$$

$$(H_{BA} - E_i S_{BA})c_{i,A} + (H_{BB} - E_i S_{BB})c_{i,B} = 0 \qquad (11.40b)$$

where the eigenvalue E_i corresponds to a wavefunction of the form $\psi_i = c_{i,A}A + c_{i,B}B$. (These expressions generalize eqn 11.25). There are two atomic orbitals, two eigenvalues, and two wavefunctions, so there are two pairs of secular equations, with the first corresponding to E_1 and ψ_1:

$$(H_{AA} - E_1 S_{AA})c_{1,A} + (H_{AB} - E_1 S_{AB})c_{1,B} = 0 \qquad (11.41a)$$

$$(H_{BA} - E_1 S_{BA})c_{1,A} + (H_{BB} - E_1 S_{BB})c_{1,B} = 0 \qquad (11.41b)$$

and another corresponding to E_2 and ψ_2:

$$(H_{AA} - E_2 S_{AA})c_{2,A} + (H_{AB} - E_2 S_{AB})c_{2,B} = 0 \qquad (11.41c)$$

$$(H_{BA} - E_2 S_{BA})c_{2,A} + (H_{BB} - E_2 S_{BB})c_{2,B} = 0 \qquad (11.41d)$$

If we introduce the following matrices and column vectors

$$\boldsymbol{H} = \begin{pmatrix} H_{AA} & H_{AB} \\ H_{BA} & H_{BB} \end{pmatrix} \quad \boldsymbol{S} = \begin{pmatrix} S_{AA} & S_{AB} \\ S_{BA} & S_{BB} \end{pmatrix} \quad \boldsymbol{c}_i = \begin{pmatrix} c_{i,A} \\ c_{i,B} \end{pmatrix} \qquad (11.42)$$

then each pair of equations may be written more succinctly as

$$(\boldsymbol{H} - E_i \boldsymbol{S})\boldsymbol{c}_i = 0 \quad \text{or} \quad \boldsymbol{H}\boldsymbol{c}_i = \boldsymbol{S}\boldsymbol{c}_i E_i \qquad (11.43)$$

where $\boldsymbol{H}$ is the hamiltonian matrix and $\boldsymbol{S}$ is the overlap matrix. To proceed with the calculation of the eigenvalues and coefficients, we introduce the matrices

$$\boldsymbol{C} = (\boldsymbol{c}_1 \quad \boldsymbol{c}_2) = \begin{pmatrix} c_{1,A} & c_{2,A} \\ c_{1,B} & c_{2,B} \end{pmatrix} \quad \boldsymbol{E} = \begin{pmatrix} E_1 & 0 \\ 0 & E_2 \end{pmatrix} \qquad [11.44]$$

for then the entire set of equations we have to solve can be expressed as

$$\boldsymbol{HC} = \boldsymbol{SCE} \tag{11.45}$$

Self-test 11.7 Show by carrying out the necessary matrix operations that eqn 11.45 is a representation of the system of equations consisting of eqns 11.41(a)–(d).

In the Hückel approximation, $H_{AA} = H_{BB} = \alpha$, $H_{AB} = H_{BA} = \beta$, and we neglect overlap, setting $\boldsymbol{S} = \mathbf{1}$, the unit matrix (with 1 on the diagonal and 0 elsewhere). Then

$$\boldsymbol{HC} = \boldsymbol{CE}$$

At this point, we multiply from the left by the inverse matrix $\boldsymbol{C}^{-1}$, and find

$$\boldsymbol{C}^{-1}\boldsymbol{HC} = \boldsymbol{E} \tag{11.46}$$

where we have used $\boldsymbol{C}^{-1}\boldsymbol{C} = \mathbf{1}$. In other words, to find the eigenvalues E_i, we have to find a transformation of $\boldsymbol{H}$ that makes it diagonal. This procedure is called **matrix diagonalization**. The diagonal elements then correspond to the eigenvalues E_i and the columns of the matrix $\boldsymbol{C}$ that brings about this diagonalization are the coefficients of the members of the **basis set**, the set of atomic orbitals used in the calculation, and hence give us the composition of the molecular orbitals. If there are N orbitals in the basis set (there are only two in our example), then there are N eigenvalues E_i and N corresponding column vectors $\boldsymbol{c}_i$. As a result, we have to solve N equations of the form $\boldsymbol{Hc}_i = \boldsymbol{Sc}_iE_i$ by diagonalization of the $N \times N$ matrix $\boldsymbol{H}$, as directed by eqn 11.46.

Example 11.4 *Finding the molecular orbitals by matrix diagonalization*

Set up and solve the matrix equations within the Hückel approximation for the π-orbitals of butadiene (**3**).

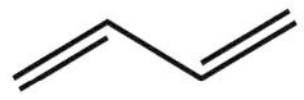

3

Method The matrices will be four-dimensional for this four-atom system. Ignore overlap, and construct the matrix $\boldsymbol{H}$ by using the Hückel values α and β. Find the matrix $\boldsymbol{C}$ that diagonalizes $\boldsymbol{H}$: for this step, use mathematical software. Full details are given in *Appendix 2*.

Solution

$$\boldsymbol{H} = \begin{pmatrix} H_{11} & H_{12} & H_{13} & H_{14} \\ H_{21} & H_{22} & H_{23} & H_{24} \\ H_{31} & H_{32} & H_{33} & H_{34} \\ H_{41} & H_{42} & H_{43} & H_{44} \end{pmatrix} = \begin{pmatrix} \alpha & \beta & 0 & 0 \\ \beta & \alpha & \beta & 0 \\ 0 & \beta & \alpha & \beta \\ 0 & 0 & \beta & \alpha \end{pmatrix}$$

Mathematical software then diagonalizes this matrix to

$$\boldsymbol{E} = \begin{pmatrix} \alpha + 1.62\beta & 0 & 0 & 0 \\ 0 & \alpha + 0.62\beta & 0 & 0 \\ 0 & 0 & \alpha - 0.62\beta & 0 \\ 0 & 0 & 0 & \alpha - 1.62\beta \end{pmatrix}$$

and the matrix that achieves the diagonalization is

$$\boldsymbol{C} = \begin{pmatrix} 0.372 & 0.602 & 0.602 & -0.372 \\ 0.602 & 0.372 & -0.372 & 0.602 \\ 0.602 & -0.372 & -0.372 & -0.602 \\ 0.372 & -0.602 & 0.602 & 0.372 \end{pmatrix}$$

We can conclude that the energies and molecular orbitals are

$$\begin{aligned}
E_1 &= \alpha + 1.62\beta & \psi_1 &= 0.372\chi_A + 0.602\chi_B + 0.602\chi_C + 0.372\chi_D \\
E_2 &= \alpha + 0.62\beta & \psi_2 &= 0.602\chi_A + 0.372\chi_B - 0.372\chi_C - 0.602\chi_D \\
E_3 &= \alpha - 0.62\beta & \psi_3 &= 0.602\chi_A - 0.372\chi_B - 0.372\chi_C + 0.602\chi_D \\
E_4 &= \alpha - 1.62\beta & \psi_4 &= -0.372\chi_A + 0.602\chi_B - 0.602\chi_C - 0.372\chi_D
\end{aligned}$$

where the C2*p* atomic orbitals are denoted by $\chi_A, \ldots, \chi_D$. Note that the orbitals are mutually orthogonal and, with overlap neglected, normalized.

Self-test 11.8 Repeat the exercise for the allyl radical, $\cdot CH_2{-}CH{=}CH_2$.

[$E = \alpha + 2^{1/2}\beta, \alpha, \alpha - 2^{1/2}\beta$; $\psi_1 = \frac{1}{2}\chi_A + (\frac{1}{2})^{1/2}\chi_B + \frac{1}{2}\chi_C$, $\psi_2 = (\frac{1}{2})^{1/2}\chi_A - (\frac{1}{2})^{1/2}\chi_C$, $\psi_3 = \frac{1}{2}\chi_A - (\frac{1}{2})^{1/2}\chi_B + \frac{1}{2}\chi_C$]

(c) Butadiene and π-electron binding energy

As we saw in the preceding example, the energies of the four LCAO-MOs for butadiene are

$$E = \alpha \pm 1.62\beta, \qquad \alpha \pm 0.62\beta \tag{11.47}$$

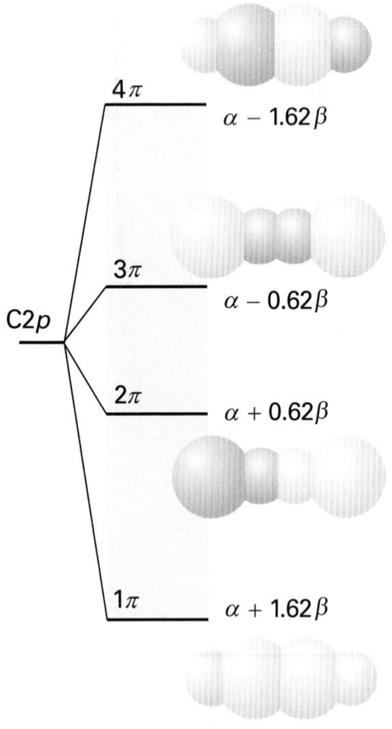

Fig. 11.39 The Hückel molecular orbital energy levels of butadiene and the top view of the corresponding π orbitals. The four *p* electrons (one supplied by each C) occupy the two lower π orbitals. Note that the orbitals are delocalized.

These orbitals and their energies are drawn in Fig. 11.39. Note that the greater the number of internuclear nodes, the higher the energy of the orbital. There are four electrons to accommodate, so the ground-state configuration is $1\pi^2 2\pi^2$. The frontier orbitals of butadiene are the 2π orbital (the HOMO, which is largely bonding) and the 3π orbital (the LUMO, which is largely antibonding). 'Largely' bonding means that an orbital has both bonding and antibonding interactions between various neighbours, but the bonding effects dominate. 'Largely antibonding' indicates that the antibonding effects dominate.

An important point emerges when we calculate the total **π-electron binding energy**, E_π, the sum of the energies of each π electron, and compare it with what we find in ethene. In ethene the total energy is

$$E_\pi = 2(\alpha + \beta) = 2\alpha + 2\beta$$

In butadiene it is

$$E_\pi = 2(\alpha + 1.62\beta) + 2(\alpha + 0.62\beta) = 4\alpha + 4.48\beta$$

Therefore, the energy of the butadiene molecule lies lower by 0.48β (about 110 kJ mol^{-1}) than the sum of two individual π bonds. This extra stabilization of a conjugated system is called the **delocalization energy**. A closely related quantity is the **π-bond formation energy**, the energy released when a π bond is formed. Because the contribution of α is the same in the molecule as in the atoms, we can find the π-bond formation energy from the π-electron binding energy by writing

$$E_{bf} = E_\pi - N\alpha \tag{11.48}$$

where *N* is the number of carbon atoms in the molecule. The π-bond formation energy in butadiene, for instance, is 4.48β.

Example 11.5 *Estimating the delocalization energy*

Use the Hückel approximation to find the energies of the π orbitals of cyclobutadiene, and estimate the delocalization energy.

Method Set up the secular determinant using the same basis as for butadiene, but note that atoms A and D are also now neighbours. Then solve for the roots of the secular equation and assess the total π-bond energy. For the delocalization energy, subtract from the total π-bond energy the energy of two π-bonds.

Answer The hamiltonian matrix is

$$\boldsymbol{H}=\begin{pmatrix}\alpha & \beta & 0 & \beta\\ \beta & \alpha & \beta & 0\\ 0 & \beta & \alpha & \beta\\ \beta & 0 & \beta & \alpha\end{pmatrix}$$

Diagonalization gives the energies of the orbitals as

$$E=\alpha+2\beta, \qquad \alpha, \qquad \alpha, \qquad \alpha-2\beta$$

Four electrons must be accommodated. Two occupy the lowest orbital (of energy $\alpha+2\beta$), and two occupy the doubly degenerate orbitals (of energy α). The total energy is therefore $4\alpha+4\beta$. Two isolated π bonds would have an energy $4\alpha+4\beta$; therefore, in this case, the delocalization energy is zero.

Self-test 11.9 Repeat the calculation for benzene. [See next subsection]

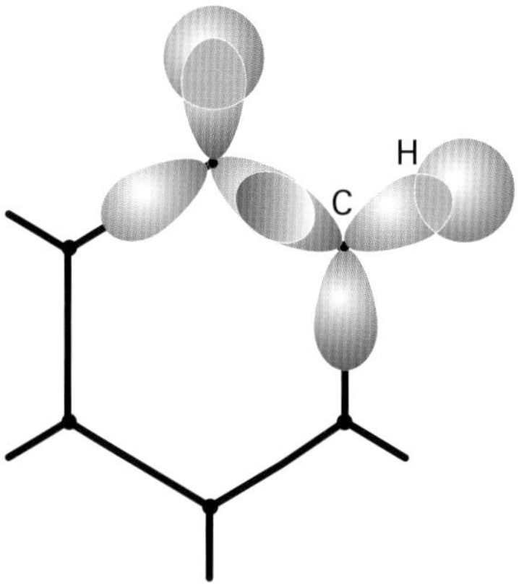

Fig. 11.40 The σ framework of benzene is formed by the overlap of Csp^2 hybrids, which fit without strain into a hexagonal arrangement.

(d) Benzene and aromatic stability

The most notable example of delocalization conferring extra stability is benzene and the aromatic molecules based on its structure. Benzene is often expressed in a mixture of valence-bond and molecular orbital terms, with typically valence-bond language used for its σ framework and molecular orbital language used to describe its π electrons.

First, the valence-bond component. The six C atoms are regarded as sp^2 hybridized, with a single unhydridized perpendicular $2p$ orbital. One H atom is bonded by (Csp^2,H1s) overlap to each C carbon, and the remaining hybrids overlap to give a regular hexagon of atoms (Fig. 11.40). The internal angle of a regular hexagon is 120°, so sp^2 hybridization is ideally suited for forming σ bonds. We see that benzene's hexagonal shape permits strain-free σ bonding.

Now consider the molecular orbital component of the description. The six C2p orbitals overlap to give six π orbitals that spread all round the ring. Their energies are calculated within the Hückel approximation by diagonalizing the hamiltonian matrix

$$\boldsymbol{H}=\begin{pmatrix}\alpha & \beta & 0 & 0 & 0 & \beta\\ \beta & \alpha & \beta & 0 & 0 & 0\\ 0 & \beta & \alpha & \beta & 0 & 0\\ 0 & 0 & \beta & \alpha & \beta & 0\\ 0 & 0 & 0 & \beta & \alpha & \beta\\ \beta & 0 & 0 & 0 & \beta & \alpha\end{pmatrix}$$

The MO energies, the eigenvalues of this matrix, are simply

$$E=\alpha\pm 2\beta, \alpha\pm\beta, \alpha\pm\beta \qquad (11.49)$$

as shown in Fig. 11.41. The orbitals there have been given symmetry labels that we explain in Chapter 12. Note that the lowest energy orbital is bonding between all neighbouring atoms, the highest energy orbital is antibonding between each pair of neighbours, and the intermediate orbitals are a mixture of bonding, nonbonding, and antibonding character between adjacent atoms.

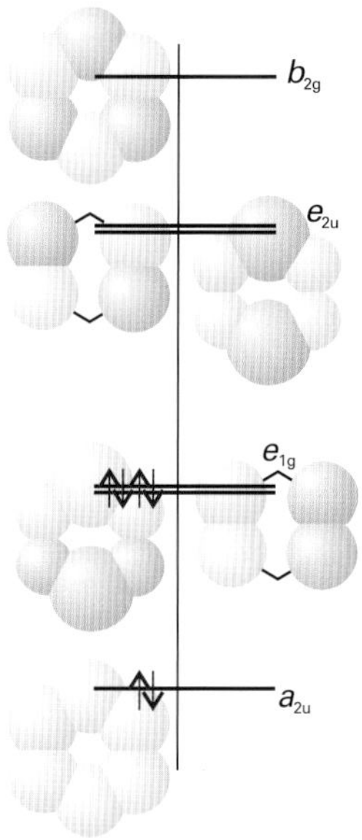

Fig. 11.41 The Hückel orbitals of benzene and the corresponding energy levels. The symmetry labels are explained in Chapter 12. The bonding and antibonding character of the delocalized orbitals reflects the numbers of nodes between the atoms. In the ground state, only the bonding orbitals are occupied.

We now apply the building-up principle to the π system. There are six electrons to accommodate (one from each C atom), so the three lowest orbitals (a_{2u} and the doubly-degenerate pair e_{1g}) are fully occupied, giving the ground-state configuration $a_{2u}^2 e_{1g}^4$. A significant point is that the only molecular orbitals occupied are those with net bonding character.

The π-electron energy of benzene is

$$E_\pi = 2(\alpha + 2\beta) + 4(\alpha + \beta) = 6\alpha + 8\beta$$

If we ignored delocalization and thought of the molecule as having three isolated π bonds, it would be ascribed a π-electron energy of only $3(2\alpha + 2\beta) = 6\alpha + 6\beta$. The delocalization energy is therefore $2\beta \approx -460$ kJ mol^{-1}, which is considerably more than for butadiene. The π-bond formation energy in benzene is 8β.

This discussion suggests that aromatic stability can be traced to two main contributions. First, the shape of the regular hexagon is ideal for the formation of strong σ bonds: the σ framework is relaxed and without strain. Second, the π orbitals are such as to be able to accommodate all the electrons in bonding orbitals, and the delocalization energy is large.

11.7 Computational chemistry

The difficulties arising from the severe assumptions of Hückel method have been overcome by more sophisticated theories that not only calculate the shapes and energies of molecular orbitals but also predict with reasonable accuracy the structure and reactivity of molecules. The full treatment of molecular electronic structure is quite easy to formulate but difficult to implement. However, it has received an enormous amount of attention by chemists, and has become a keystone of modern chemical research. John Pople and Walter Kohn were awarded the Nobel Prize in Chemistry for 1998 for their contributions to the development of computational techniques for the elucidation of molecular structure and reactivity.

Comment 11.12

The web site contains links to sites where you may perform semi-empirical and *ab initio* calculations on simple molecules directly from your web browser.

(a) The Hartree–Fock equations

The starting point is to write down the many-electron wavefunction as a product of one-electron wavefunctions:

$$\Psi = \psi_{a,\alpha}(1)\psi_{a,\beta}(2)\ldots\psi_{z,\beta}(N)$$

This is the wavefunction for an N-electron closed-shell molecule in which electron 1 occupies molecular orbital ψ_a with spin α, electron 2 occupies molecular orbital ψ_a with spin β, and so on. However, the wavefunction must satisfy the Pauli principle and change sign under the permutation of any pair of electrons. To achieve this behaviour, we write the wavefunction as a sum of all possible permutations with the appropriate sign:

$$\Psi = \psi_{a,\alpha}(1)\psi_{a,\beta}(2)\ldots\psi_{z,\beta}(N) - \psi_{a,\alpha}(2)\psi_{a,\beta}(1)\ldots\psi_{z,\beta}(N) + \cdots$$

There are $N!$ terms in this sum, and the entire sum can be written as a determinant:

$$\Psi = \frac{1}{\sqrt{N!}}\begin{vmatrix} \psi_{a,\alpha}(1) & \psi_{a,\beta}(1) & \cdots & \cdots & \psi_{z,\beta}(1) \\ \psi_{a,\alpha}(2) & \psi_{a,\beta}(2) & \cdots & \cdots & \psi_{z,\beta}(2) \\ \vdots & \vdots & & & \vdots \\ \vdots & \vdots & & & \vdots \\ \psi_{a,\alpha}(N) & \psi_{a,\beta}(N) & \cdots & \cdots & \psi_{z,\beta}(N) \end{vmatrix} \qquad (11.50a)$$

The initial factor ensures that the wavefunction is normalized if the component molecular orbitals are normalized. To save the tedium of writing out large determinants, the wavefunction is normally written simply as

$$\Psi = (1/N!)^{1/2}\det|\psi_{a,\alpha}(1)\psi_{a,\beta}(2)\ldots\psi_{z,\beta}(N)| \qquad (11.50b)$$

When the determinantal wavefunction is combined with the variation principle (Section 11.5c), the optimum wavefunctions, in the sense of corresponding to the lowest total energy, must satisfy the **Hartree–Fock equations:**

$$f_1\psi_{a,\sigma}(1) = \varepsilon\psi_{a,\sigma}(1) \qquad (11.51)$$

where σ is either α or β. The **Fock operator** f_1 is

$$f_1 = h_1 + \Sigma_j\{2J_j(1) - K_j(1)\} \qquad (11.52)$$

The three terms in this expression are the **core hamiltonian**

$$h_1 = -\frac{\hbar^2}{2m_e}\nabla_1^2 - \sum_n \frac{Z_n e^2}{4\pi\varepsilon_0 r_{ni}} \qquad [11.53a]$$

the **Coulomb operator** J, where

$$J_j(1)\psi_a(1) = \int \psi_j^*(2)\psi_j(2)\left(\frac{e^2}{4\pi\varepsilon_0 r_{12}}\right)\psi_a(1)\mathrm{d}\tau_2 \qquad [11.53b]$$

and the **exchange operator**, K, where

$$K_j(1)\psi_a(1) = \int \psi_j^*(2)\psi_a(2)\left(\frac{e^2}{4\pi\varepsilon_0 r_{12}}\right)\psi_j(1)\mathrm{d}\tau_2 \qquad [11.53c]$$

Although the Hartree–Fock equations look deceptively simple, with the Fock operator looking like a hamiltonian, we see from these definitions that f actually depends on the wavefunctions of all the electrons. To proceed, we have to guess the initial form of the wavefunctions, use them in the definition of the Coulomb and exchange operators, and solve the Hartree–Fock equations. That process is then continued using the newly found wavefunctions until each cycle of calculation leaves the energies and wavefunctions unchanged to within a chosen criterion. This is the origin of the term **self-consistent field** (SCF) for this type of procedure.

The difficulty in this procedure is in the solution of the Hartree–Fock equations. To make progress, we have to express the wavefunctions as linear combinations of M atomic orbitals χ_i, and write

$$\psi_a = \sum_{i=1}^{M} c_{ia}\chi_i$$

As we show in the *Justification* below, the use of a linear combination like this leads to a set of equations that can be expressed in a matrix form known as the **Roothaan equations:**

$$\boldsymbol{FC} = \boldsymbol{SC\varepsilon} \qquad (11.54)$$

where $\boldsymbol{F}$ is the matrix formed from the Fock operator:

$$F_{ij} = \int \chi_i^*(1)f_1\chi_j(1)\mathrm{d}\tau \qquad (11.55a)$$

and $\boldsymbol{S}$ is the matrix of overlap integrals

$$S_{ij} = \int \chi_i^*(1)\chi_j(1)\mathrm{d}\tau \qquad (11.55b)$$

Justification 11.5 *The Roothaan equations*

To construct the Roothaan equations we substitute the linear combination of atomic orbitals into eqn 11.51, which gives

$$f_1 \sum_{i=1}^{M} c_{i\alpha}\chi_i(1) = \varepsilon_\alpha \sum_{i=1}^{M} c_{i\alpha}\chi_i(1)$$

Now multiply from the left by $\chi_j^*(1)$ and integrate over the coordinates of electron 1:

$$\sum_{i=1}^{M} c_{i\alpha} \overbrace{\int \chi_j(1)^* f(1)\chi_i(1)\mathrm{d}\boldsymbol{r}_1}^{F_{ji}} = \varepsilon_\alpha \sum_{i=1}^{M} c_{i\alpha} \overbrace{\int \chi_j(1)^* \chi_i(1)\mathrm{d}\boldsymbol{r}_1}^{S_{ji}}$$

That is,

$$\sum_{i=1}^{M} F_{ji}c_{i\alpha} = \varepsilon_\alpha \sum_{i=1}^{M} S_{ji}c_{i\alpha}$$

This expression has the form of the matrix equation in eqn 11.54.

(b) Semi-empirical and *ab initio* methods

There are two main strategies for continuing the calculation from this point. In the **semi-empirical methods**, many of the integrals are estimated by appealing to spectroscopic data or physical properties such as ionization energies, and using a series of rules to set certain integrals equal to zero. In the ***ab initio*** **methods**, an attempt is made to calculate all the integrals that appear in the Fock and overlap matrices. Both procedures employ a great deal of computational effort and, along with cryptanalysts and meteorologists, theoretical chemists are among the heaviest users of the fastest computers.

The Fock matrix has elements that consist of integrals of the form

$$(AB|CD) = \int A(1)B(1)\left(\frac{e^2}{4\pi\varepsilon_0 r_{12}}\right)C(2)D(2)\mathrm{d}\tau_1\mathrm{d}\tau_2 \qquad (11.56)$$

where A, B, C, and D are atomic orbitals that in general may be centred on different nuclei. It can be appreciated that, if there are several dozen atomic orbitals used to build the molecular orbitals, then there will be tens of thousands of integrals of this form to evaluate (the number of integrals increases as the fourth power of the number of atomic orbitals in the basis). One severe approximation is called **complete neglect of differential overlap** (CNDO), in which all integrals are set to zero unless A and B are the same orbitals centred on the same nucleus, and likewise for C and D. The surviving integrals are then adjusted until the energy levels are in good agreement with experiment. The more recent semi-empirical methods make less draconian decisions about which integrals are to be ignored, but they are all descendants of the early CNDO technique. These procedures are now readily available in commercial software packages and can be used with very little detailed knowledge of their mode of calculation. The packages also have sophisticated graphical output procedures, which enable one to analyse the shapes of orbitals and the distribution of electric charge in molecules. The latter is important when assessing, for instance, the likelihood that a given molecule will bind to an active site in an enzyme.

Commercial packages are also available for *ab initio* calculations. Here the problem is to evaluate as efficiently as possible thousands of integrals. This task is greatly facilitated by expressing the atomic orbitals used in the LCAOs as linear combinations of

Gaussian orbitals. A **Gaussian type orbital** (GTO) is a function of the form $e^{-\zeta r^2}$. The advantage of GTOs over the correct orbitals (which for hydrogenic systems are proportional to $e^{-\zeta r}$) is that the product of two Gaussian functions is itself a Gaussian function that lies between the centres of the two contributing functions (Fig. 11.42). In this way, the four-centre integrals like that in eqn 11.56 become two-centre integrals of the form

$$(AB|CD) = \int X(1)\left(\frac{e^2}{4\pi\varepsilon_0 r_{12}}\right)Y(2)\mathrm{d}\tau_1\mathrm{d}\tau_2 \qquad (11.57)$$

where X is the Gaussian corresponding to the product AB and Y is the corresponding Gaussian from CD. Integrals of this form are much easier and faster to evaluate numerically than the original four-centre integrals. Although more GTOs have to be used to simulate the atomic orbitals, there is an overall increase in speed of computation.

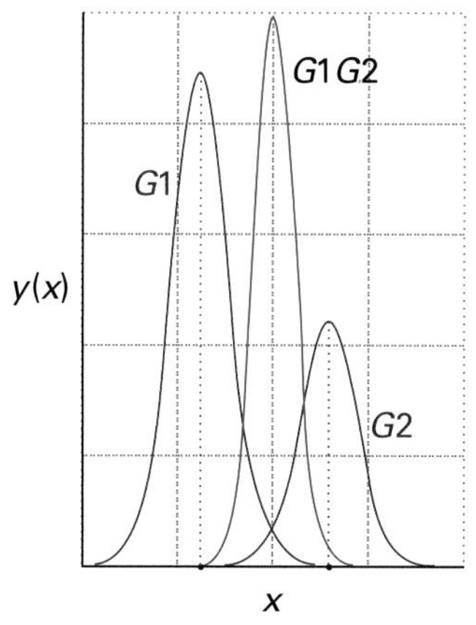

Fig. 11.42 The product of two Gaussian functions (the purple curves) is itself a Gaussian function located between the two contributing Gaussians.

(c) Density functional theory

A technique that has gained considerable ground in recent years to become one of the most widely used techniques for the calculation of molecular structure is **density functional theory** (DFT). Its advantages include less demanding computational effort, less computer time, and—in some cases (particularly d-metal complexes)—better agreement with experimental values than is obtained from Hartree–Fock procedures.

The central focus of DFT is the electron density, ρ, rather than the wavefunction ψ. The 'functional' part of the name comes from the fact that the energy of the molecule is a function of the electron density, written $E[\rho]$, and the electron density is itself a function of position, $\rho(\boldsymbol{r})$, and in mathematics a function of a function is called a *functional*. The exact ground-state energy of an n-electron molecule is

$$E[\rho] = E_{\mathrm{K}} + E_{\mathrm{P;e,N}} + E_{\mathrm{P;e,e}} + E_{\mathrm{XC}}[\rho] \qquad (11.58)$$

where E_{K} is the total electron kinetic energy, $E_{\mathrm{P;e,N}}$ the electron–nucleus potential energy, $E_{\mathrm{P;e,e}}$ the electron–electron potential energy, and $E_{\mathrm{XC}}[\rho]$ the **exchange–correlation energy**, which takes into account all the effects due to spin. The orbitals used to construct the electron density from

$$\rho(\boldsymbol{r}) = \sum_{i=1}^{N} |\psi_i(\boldsymbol{r})|^2 \qquad (11.59)$$

are calculated from the **Kohn–Sham equations**, which are found by applying the variation principle to the electron energy, and are like the Hartree–Fock equations except for a term V_{XC}, which is called the **exchange–correlation potential**:

$$\Bigg\{\underbrace{-\frac{\hbar^2}{2m_\mathrm{e}}\nabla_1^2}_{\text{Kinetic energy}} - \underbrace{\sum_{j=1}^{N}\frac{Z_j e^2}{4\pi\varepsilon_0 r_{j1}}}_{\text{Electron–nucleus attraction}} + \underbrace{\int\frac{\rho(\boldsymbol{r}_2)e^2}{4\pi\varepsilon_0 r_{12}}\mathrm{d}\boldsymbol{r}_2}_{\text{Electron–electron repulsion}} + \underbrace{V_{\mathrm{XC}}(\boldsymbol{r}_1)}_{\text{Exchange–correlation}}\Bigg\}\psi_i(\boldsymbol{r}_1) = \varepsilon_i\psi_i(\boldsymbol{r}_1) \qquad (11.60)$$

The exchange–correlation potential is the 'functional derivative' of the exchange–correlation energy:

$$V_{\mathrm{XC}}[\rho] = \frac{\delta E_{\mathrm{XC}}[\rho]}{\delta\rho} \qquad (11.61)$$

The Kohn–Sham equations are solved iteratively and self-consistently. First, we guess the electron density. For this step it is common to use a superposition of atomic

Comment 11.13

Consider the functional $G[f]$ where f is a function of x. When x changes to $x + \delta x$, the function changes to $f + \delta f$ and the functional changes to $G[f + \delta f]$. By analogy with the derivative of a function, the functional derivative is then defined as

$$\frac{\delta G}{\delta f} = \lim_{\delta f\to 0}\frac{G[f+\delta f] - G[f]}{\delta f}$$

where the manner in which δf goes to zero must be specified explicitly. See *Appendix 2* for more details and examples.

electron densities. Then the exchange–correlation potential is calculated by assuming an approximate form of the dependence of the exchange–correlation energy on the electron density and evaluating the functional derivative in eqn 11.61. For this step, the simplest approximation is the **local-density approximation** and to write

$$E_{XC}[\rho] = \int \rho(\boldsymbol{r})\varepsilon_{XC}[\rho(\boldsymbol{r})]\mathrm{d}\boldsymbol{r} \qquad (11.62)$$

where ε_{XC} is the exchange–correlation energy per electron in a homogeneous gas of constant density. Next, the Kohn–Sham equations are solved to obtain an initial set of orbitals. This set of orbitals is used to obtain a better approximation to the electron density (from eqn 11.59) and the process is repeated until the density and the exchange–correlation energy are constant to within some tolerance.

11.8 The prediction of molecular properties

The results of molecular orbital calculations are only approximate, with deviations from experimental values increasing with the size of the molecule. Therefore, one goal of computational chemistry is to gain insight into trends in properties of molecules, without necessarily striving for ultimate accuracy. In the next sections we give a brief summary of strategies used by computational chemists for the prediction of molecular properties.

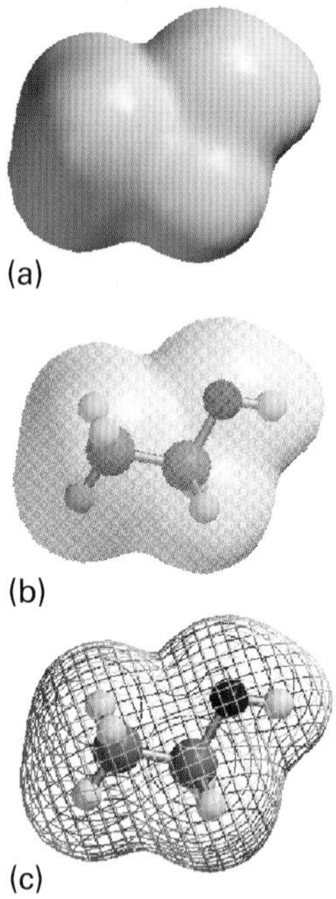

Fig. 11.43 Various representations of an isodensity surface of ethanol (a) solid surface, (b) transparent surface, and (c) mesh surface.

(a) Electron density and the electrostatic potential surfaces

One of the most significant developments in computational chemistry has been the introduction of graphical representations of molecular orbitals and electron densities. The raw output of a molecular structure calculation is a list of the coefficients of the atomic orbitals in each molecular orbital and the energies of these orbitals. The graphical representation of a molecular orbital uses stylized shapes to represent the basis set, and then scales their size to indicate the coefficient in the linear combination. Different signs of the wavefunctions are represented by different colours.

Once the coefficients are known, we can build up a representation of the electron density in the molecule by noting which orbitals are occupied and then forming the squares of those orbitals. The total electron density at any point is then the sum of the squares of the wavefunctions evaluated at that point. The outcome is commonly represented by a **isodensity surface**, a surface of constant total electron density (Fig. 11.43). As shown in the illustration, there are several styles of representing an isodensity surface, as a solid form, as a transparent form with a ball-and-stick representation of the molecule within, or as a mesh. A related representation is a **solvent-accessible surface** in which the shape represents the shape of the molecule by imagining a sphere representing a solvent molecule rolling across the surface and plotting the locations of the centre of that sphere.

One of the most important aspects of a molecule other than its geometrical shape is the distribution of charge over its surface. The net charge at each point on an isodensity surface can be calculated by subtracting the charge due to the electron density at that point form the charge due to the nuclei: the result is an **electrostatic potential surface** (an 'elpot surface') in which net positive charge is shown in one colour and net negative charge is shown in another, with intermediate gradations of colour (Fig. 11.44).

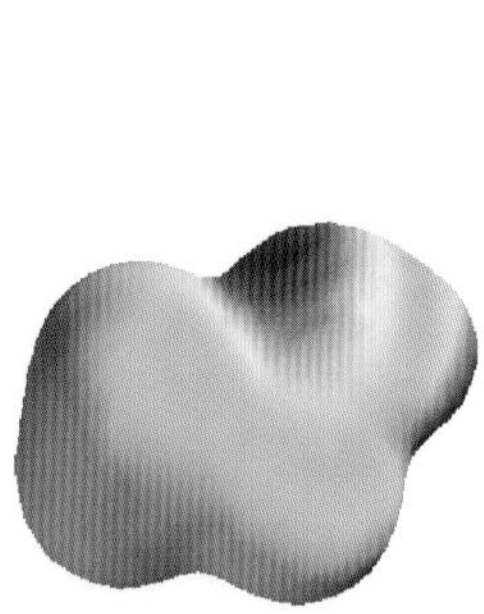

Fig. 11.44 An elpot diagram of ethanol.

Representations such as those we have illustrated are of critical importance in a number of fields. For instance, they may be used to identify an electron-poor region of a molecule that is susceptible to association with or chemical attack by an electron-rich region of another molecule. Such considerations are important for assessing the pharmacological activity of potential drugs.

(b) Thermodynamic and spectroscopic properties

We already saw in Section 2.8 that computational chemistry is becoming the technique of choice for estimating standard enthalpies of formation of molecules with complex three-dimensional structures. The computational approach also makes it possible to gain insight into the effect of solvation on the enthalpy of formation without conducting experiments. A calculation performed in the absence of solvent molecules estimates the properties of the molecule of interest in the gas phase. Computational methods are available that allow for the inclusion of several solvent molecules around a solute molecule, thereby taking into account the effect of molecular interactions with the solvent on the enthalpy of formation of the solute. Again, the numerical results are only estimates and the primary purpose of the calculation is to predict whether interactions with the solvent increase or decrease the enthalpy of formation. As an example, consider the amino acid glycine, which can exist in a neutral (**4**) or zwitterionic (**5**) form, in which the amino group is protonated and the carboxyl group is deprotonated. It is possible to show computationally that in the gas phase the neutral form has a lower enthalpy of formation than the zwitterionic form. However, in water the opposite is true because of strong interactions between the polar solvent and the charges in the zwitterion.

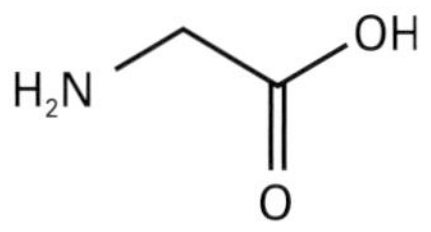

4

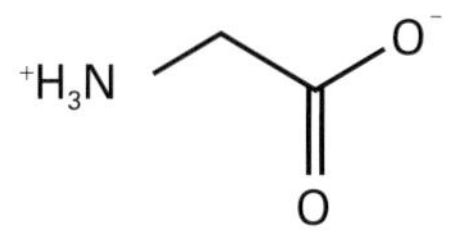

5

Molecular orbital calculations can also be used to predict trends in electrochemical properties, such as standard potentials (Chapter 7). Several experimental and computational studies of aromatic hydrocarbons indicate that decreasing the energy of the LUMO enhances the ability of a molecule to accept an electron into the LUMO, with an attendant increase in the value of the molecule's standard potential. The effect is also observed in quinones and flavins, co-factors involved in biological electron transfer reactions. For example, stepwise substitution of the hydrogen atoms in *p*-benzoquinone by methyl groups ($-CH_3$) results in a systematic increase in the energy of the LUMO and a decrease in the standard potential for formation of the semiquinone radical (**6**):

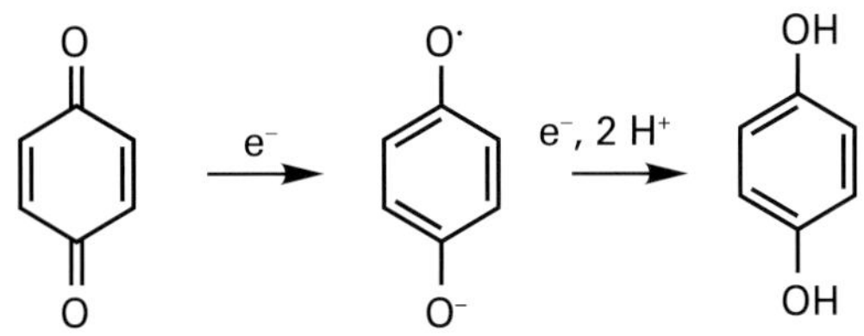

6

The standard potentials of naturally occurring quinones are also modified by the presence of different substituents, a strategy that imparts specific functions to specific quinones. For example, the substituents in coenzyme Q are largely responsible for poising its standard potential so that the molecule can function as an electron shuttle between specific electroactive proteins in the respiratory chain (*Impact* I7.2).

We remarked in Chapter 8 that a molecule can absorb or emit a photon of energy hc/λ, resulting in a transition between two quantized molecular energy levels. The transition of lowest energy (and longest wavelength) occurs between the HOMO and LUMO. We can use calculations based on semi-empirical, *ab initio*, and DFT methods to correlate the calculated HOMO–LUMO energy gap with the wavelength of absorption. For example, consider the linear polyenes shown in Table 11.5: ethene (C_2H_4), butadiene (C_4H_6), hexatriene (C_6H_8), and octatetraene (C_8H_{10}), all of which absorb in the ultraviolet region of the spectrum. The table also shows that, as expected, the wavelength of the lowest-energy electronic transition decreases as the energy separation between the HOMO and LUMO increases. We also see that the smallest HOMO–LUMO gap and longest transition wavelength correspond to octatetraene, the longest

Table 11.5 *Ab initio* calculations and spectroscopic data

Polyene	$\{E(\text{HOMO}) - E(\text{LUMO})\}/\text{eV}$	λ/nm
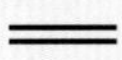 (C_2H_4)	18.1	163
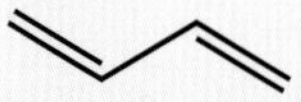	14.5	217
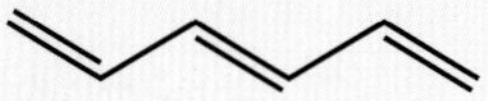	12.7	252
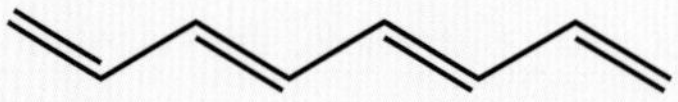	11.8	304

polyene in the group. It follows that the wavelength of the transition increases with increasing number of conjugated double bonds in linear polyenes. Extrapolation of the trend suggests that a sufficiently long linear polyene should absorb light in the visible region of the electromagnetic spectrum. This is indeed the case for β-carotene (7), which absorbs light with $\lambda \approx 450$ nm. The ability of β-carotene to absorb visible light is part of the strategy employed by plants to harvest solar energy for use in photosynthesis (Chapter 23).

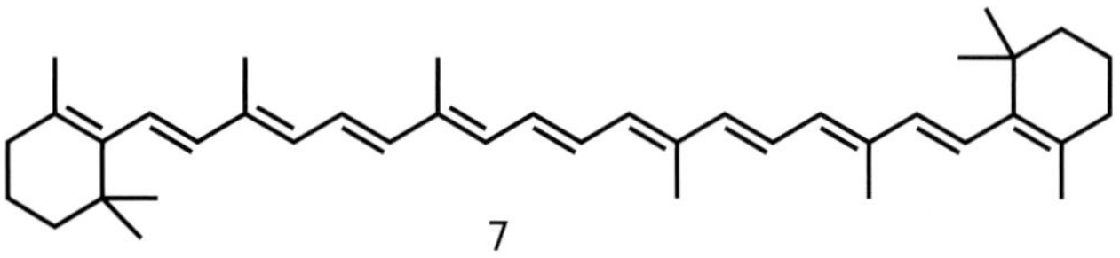

7

Checklist of key ideas

☐ 1. In the Born–Oppenheimer approximation, nuclei are treated as stationary while electrons move around them.

☐ 2. In valence-bond theory (VB theory), a bond is regarded as forming when an electron in an atomic orbital on one atoms pairs its spin with that of an electron in an atomic orbital on another atom.

☐ 3. A valence bond wavefunction with cylindrical symmetry around the internuclear axis is a σ bond. A π bond arises from the merging of two *p* orbitals that approach side-by-side and the pairing of electrons that they contain.

☐ 4. Hybrid orbitals are mixtures of atomic orbitals on the same atom and are invoked in VB theory to explain molecular geometries.

☐ 5. In molecular orbital theory (MO theory), electrons are treated as spreading throughout the entire molecule.

☐ 6. A bonding orbital is a molecular orbital that, if occupied, contributes to the strength of a bond between two atoms. An antibonding orbital is a molecular orbital that, if occupied, decreases the strength of a bond between two atoms.

☐ 7. A σ molecular orbital has zero orbital angular momentum about the internuclear axis. A π molecular orbital has one unit of angular momentum around the internuclear axis; in a nonlinear molecule, it has a nodal plane that includes the internucelar axis.

☐ 8. The electron configurations of homonuclear diatomic molecules are shown in Figs. 11.31 and 11.33.

☐ 9. When constructing molecular orbitals, we need to consider only combinations of atomic orbitals of similar energies and of the same symmetry around the internuclear axis.

☐ 10. The bond order of a diatomic molecule is $b = \frac{1}{2}(n - n^*)$, where n and n^* are the numbers of electrons in bonding and antibonding orbitals, respectively.

☐ 11. The electronegativity, χ, of an element is the power of its atoms to draw electrons to itself when it is part of a compound.

☐ 12. In a bond between dissimilar atoms, the atomic orbital belonging to the more electronegative atom makes the larger contribution to the molecular orbital with the lowest energy. For the molecular orbital with the highest energy, the principal contribution comes from the atomic orbital belonging to the less electronegative atom.

☐ 13. The hamiltonian matrix, $\boldsymbol{H}$, is formed of all integrals $H_{ij} = \int \psi_i^* \hat{H} \psi_j \mathrm{d}\tau$. The overlap matrix, $\boldsymbol{S}$, is formed of all $S_{ij} = \int \psi_i^* \psi_j \mathrm{d}\tau$.

☐ 14. The variation principle states that if an arbitrary wavefunction is used to calculate the energy, the value calculated is never less than the true energy.

☐ 15. In the Hückel method, all Coulomb integrals H_{ii} are set equal (to α), all overlap integrals are set equal to zero, all resonance integrals H_{ij} between non-neighbours are set equal to zero, and all remaining resonance integrals are set equal (to β).

☐ 16. The π-electron binding energy is the sum of the energies of each π electron. The π-bond formation energy is the energy released when a π bond is formed. The delocalization energy is the extra stabilization of a conjugated system.

☐ 17. In the self-consistent field procedure, an initial guess about the composition of the molecular orbitals is successively refined until the solution remains unchanged in a cycle of calculations.

☐ 18. In semi-empirical methods for the determination of electronic structure, the Schrödinger equation is written in terms of parameters chosen to agree with selected experimental quantities. In *ab initio* and density functional methods, the Schrödinger equation is solved numerically, without the need of parameters that appeal to experimental data.

Further reading

Articles and texts

T.A. Albright and J.K. Burdett, *Problems in molecular orbital theory*. Oxford University Press (1992).

P.W. Atkins and R.S. Friedman, *Molecular quantum mechanics*. Oxford University Press (2005).

I.N. Levine, *Quantum chemistry*. Prentice–Hall, Upper Saddle River (2000).

D.A. McQuarrie, *Mathematical methods for scientists and engineers*. University Science Books, Mill Valley (2003).

R.C. Mebane, S.A. Schanley, T.R. Rybolt, and C.D. Bruce, The correlation of physical properties of organic molecules with computed molecular surface areas. *J. Chem. Educ.* **76**, 688 (1999).

L. Pauling, *The nature of the chemical bond*. Cornell University Press, Ithaca (1960).

C.M. Quinn, *Computational quantum chemistry: an interactive guide to basis set theory*. Academic Press, San Diego (2002).

Sources of data and information

D.R. Lide (ed.), *CRC handbook of chemistry and physics*, Section 9, CRC Press, Boca Raton (2000).

P.R. Scott and W.G. Richards, *Energy levels in atoms and molecules*. Oxford Chemistry Primers, Oxford University Press (1994).

Discussion questions

11.1 Compare the approximations built into valence-bond theory and molecular-orbital theory.

11.2 Discuss the steps involved in the construction of sp^3, sp^2, and sp hybrid orbitals.

11.3 Distinguish between the Pauling and Mulliken electronegativity scales.

11.4 Discuss the steps involved in the calculation of the energy of a system by using the variation principle.

11.5 Discuss the approximations built into the Hückel method.

11.6 Distinguish between delocalization energy, π-electron binding energy, and π-bond formation energy.

11.7 Use concepts of molecular orbital theory to describe the biochemical reactivity of O_2, N_2, and NO.

11.8 Distinguish between semi-empirical, *ab initio*, and density functional theory methods of electronic structure determination.

Exercises

11.1a Give the ground-state electron configurations and bond orders of (a) Li_2, (b) Be_2, and (c) C_2.

11.1b Give the ground-state electron configurations of (a) H_2^-, (b) N_2, and (c) O_2.

11.2a Give the ground-state electron configurations of (a) CO, (b) NO, and (c) CN^-.

11.2b Give the ground-state electron configurations of (a) ClF, (b) CS, and (c) $O_2^{-\cdot}$.

11.3a From the ground-state electron configurations of B_2 and C_2, predict which molecule should have the greater bond dissociation energy.

11.3b Which of the molecules N_2, NO, O_2, C_2, F_2, and CN would you expect to be stabilized by (a) the addition of an electron to form AB^-, (b) the removal of an electron to form AB^+?

11.4a Sketch the molecular orbital energy level diagram for XeF and deduce its ground-state electron configurations. Is XeF likely to have a shorter bond length than XeF^+?

11.4b Sketch the molecular orbital energy level diagrams for BrCl and deduce its ground-state electron configurations. Is BrCl likely to have a shorter bond length than $BrCl^-$?

11.5a Use the electron configurations of NO and N_2 to predict which is likely to have the shorter bond length.

11.5b Arrange the species O_2^+, O_2, O_2^-, O_2^{2-} in order of increasing bond length.

11.6a Show that the sp^2 hybrid orbital $(s + 2^{1/2}p)/3^{1/2}$ is normalized to 1 if the s and p orbitals are normalized to 1.

11.6b Normalize the molecular orbital $\psi_A + \lambda\psi_B$ in terms of the parameter λ and the overlap integral S.

11.7a Confirm that the bonding and antibonding combinations $\psi_A \pm \psi_B$ are mutually orthogonal in the sense that their mutual overlap is zero.

11.7b Suppose that a molecular orbital has the form $N(0.145A + 0.844B)$. Find a linear combination of the orbitals A and B that is orthogonal to this combination.

11.8a Can the function $\psi = x(L - x)$ be used as a trial wavefunction for the $n = 1$ state of a particle with mass m in a one-dimensional box of length L? If the answer is yes, then express the energy of this trial wavefunction in terms of h, m, and L and compare it with the exact result (eqn 9.4). If the answer is no, explain why this is not a suitable trial wavefunction.

11.8b Can the function $\psi = x^2(L - 2x)$ be used as a trial wavefunction for the $n = 1$ state of a particle with mass m in a one-dimensional box of length L? If the answer is yes, then express the energy of this trial wavefunction in terms of h, m, and L and compare it with the exact result (eqn 9.4). If the answer is no, explain why this is not a suitable trial wavefunction.

11.9a Suppose that the function $\psi = Ae^{-ar^2}$, with A being the normalization constant and a being an adjustable parameter, is used as a trial wavefunction for the 1s orbital of the hydrogen atom. Express the energy of this trial wavefunction as a function of the h, a, e, the electron charge, and μ, the effective mass of the H atom.

11.9b Suppose that the function $\psi = Ae^{-ar^2}$, with A being the normalization constant and a being an adjustable parameter, is used as a trial wavefunction for the 1s orbital of the hydrogen atom. The energy of this trial wavefunction is

$$E = \frac{3a\hbar^2}{2\mu} - \frac{e^2}{\varepsilon_0}\left(\frac{a}{2\pi^3}\right)^{1/2}$$

where e is the electron charge, and μ is the effective mass of the H atom. What is the minimum energy associated with this trial wavefunction?

11.10a What is the energy of an electron that has been ejected from an orbital of ionization energy 11.0 eV by a photon of radiation of wavelength 100 nm?

11.10b What is the energy of an electron that has been ejected from an orbital of ionization energy 4.69 eV by a photon of radiation of wavelength 584 pm?

11.11a Construct the molecular orbital energy level diagrams of ethene on the basis that the molecule is formed from the appropriately hybridized CH_2 or CH fragments.

11.11b Construct the molecular orbital energy level diagrams of ethyne (acetylene) on the basis that the molecule is formed from the appropriately hybridized CH_2 or CH fragments.

11.12a Write down the secular determinants for (a) linear H_3, (b) cyclic H_3 within the Hückel approximation.

11.12b Predict the electronic configurations of (a) the benzene anion, (b) the benzene cation. Estimate the π-electron binding energy in each case.

11.13a Write down the secular determinants (a) anthracene (**8**), (b) phenanthrene (**9**) within the Hückel approximation and using the C2p orbitals as the basis set.

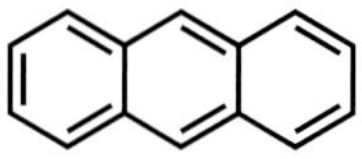

8

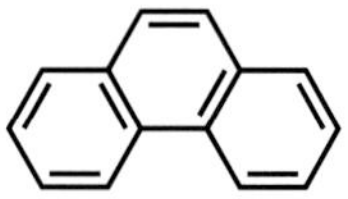

9

11.13b Use mathematical software to estimate the π-electron binding energy of (a) anthracene (**8**), (b) phenanthrene (**9**) within the Hückel approximation.

Problems*

Numerical problems

11.1 Show that, if a wave cos kx centred on A (so that x is measured from A) interferes with a similar wave cos $k'x$ centred on B (with x measured from B) a distance R away, then constructive interference occurs in the intermediate region when $k = k' = \pi/2R$ and destructive interference if $kR = \frac{1}{2}\pi$ and $k'R = \frac{3}{2}\pi$.

11.2 The overlap integral between two H1s orbitals on nuclei separated by a distance R is $S = \{1 + (R/a_0) + \frac{1}{3}(R/a_0)^2\}e^{-R/a_0}$. Plot this function for $0 \le R < \infty$.

11.3 Before doing the calculation below, sketch how the overlap between a 1s orbital and a 2p orbital can be expected to depend on their separation. The overlap integral between an H1s orbital and an H2p orbital on nuclei separated by a distance R and forming a σ orbital is $S = (R/a_0)\{1 + (R/a_0) + \frac{1}{3}(R/a_0)^2\}e^{-R/a_0}$. Plot this function, and find the separation for which the overlap is a maximum.

11.4 Calculate the total amplitude of the normalized bonding and antibonding LCAO-MOs that may be formed from two H1s orbitals at a separation of 106 pm. Plot the two amplitudes for positions along the molecular axis both inside and outside the internuclear region.

11.5 Repeat the calculation in Problem 11.4 but plot the probability densities of the two orbitals. Then form the difference density, the difference between ψ^2 and $\frac{1}{2}\{\psi_A^2 + \psi_B^2\}$.

* Problems denoted with the symbol ‡ were supplied by Charles Trapp, Carmen Giunta, and Marshall Cady.

11.6‡ Use the $2p_x$ and $2p_z$ hydrogenic atomic orbitals to construct simple LCAO descriptions of $2p\sigma$ and $2p\pi$ molecular orbitals. (a) Make a probability density plot, and both surface and contour plots of the *xz*-plane amplitudes of the $2p_z\sigma$ and $2p_z\sigma^*$ molecular orbitals. (b) Make surface and contour plots of the *xz*-plane amplitudes of the $2p_x\pi$ and $2p_x\pi^*$ molecular orbitals. Include plots for both internuclear distances, *R*, of $10a_0$ and $3a_0$, where $a_0 = 52.9$ pm. Interpret the graphs, and describe why this graphical information is useful.

11.7 Imagine a small electron-sensitive probe of volume 1.00 pm^3 inserted into an H_2^+ molecule-ion in its ground state. Calculate the probability that it will register the presence of an electron at the following positions: (a) at nucleus A, (b) at nucleus B, (c) half-way between A and B, (c) at a point 20 pm along the bond from A and 10 pm perpendicularly. Do the same for the molecule-ion the instant after the electron has been excited into the antibonding LCAO-MO.

11.8 The energy of H_2^+ with internuclear separation *R* is given by the expression

$$E = E_H + \frac{e^2}{4\pi\varepsilon_0 R} - \frac{V_1 + V_2}{1+S}$$

where E_H is the energy of an isolated H atom, V_1 is the attractive potential energy between the electron centred on one nucleus and the charge of the other nucleus, V_2 is the attraction between the overlap density and one of the nuclei, *S* is the overlap integral. The values are given below. Plot the molecular potential energy curve and find the bond dissociation energy (in electronvolts) and the equilibrium bond length.

R/a_0	0	1	2	3	4
V_1/E_h	11.000	10.729	10.473	10.330	10.250
V_2/E_h	11.000	10.736	10.406	10.199	10.092
S	1.000	0.858	0.587	0.349	0.189

where $E_h = 27.3$ eV and $a_0 = 52.9$ pm and $E_H = -\tfrac{1}{2}E_h$.

11.9 The same data as in Problem 11.8 may be used to calculate the molecular potential energy curve for the antibonding orbital, which is given by

$$E = E_H + \frac{e^2}{4\pi\varepsilon_0 R} - \frac{V_1 - V_2}{1-S}$$

Plot the curve.

11.10‡ J.G. Dojahn, E.C.M. Chen, and W.E. Wentworth (*J. Phys. Chem.* **100**, 9649 (1996)) characterized the potential energy curves of homonuclear diatomic halogen molecules and molecular anions. Among the properties they report are the equilibrium internuclear distance R_e, the vibrational wavenumber, $\tilde{\nu}$, and the dissociation energy, D_e:

Species	R_e	$\tilde{\nu}/\text{cm}^{-1}$	D_e/eV
F_2	1.411	916.6	1.60
F_2^-	1.900	450.0	1.31

Rationalize these data in terms of molecular orbital configurations.

11.11‡ *Rydberg molecules* are molecules with an electron in an atomic orbital with principal quantum number *n* one higher than the valence shells of the constituent atoms. Speculate about the existence of 'hyper Rydberg' H_2 formed from two H atoms with 100*s* electrons. Make reasonable guesses about the binding energy, the equilibrium internuclear separation, the vibrational force constant, and the rotational constant. Is such a molecule likely to exist under any circumstances?

11.12 In a particular photoelectron spectrum using 21.21 eV photons, electrons were ejected with kinetic energies of 11.01 eV, 8.23 eV, and 5.22 eV. Sketch the molecular orbital energy level diagram for the species, showing the ionization energies of the three identifiable orbitals.

11.13‡ Set up and solve the Hückel secular equations for the π electrons of NO_3^-. Express the energies in terms of the Coulomb integrals α_O and α_N and the resonance integral β. Determine the delocalization energy of the ion.

11.14 In the 'free electron molecular orbital' (FEMO) theory, the electrons in a conjugated molecule are treated as independent particles in a box of length *L*. Sketch the form of the two occupied orbitals in butadiene predicted by this model and predict the minimum excitation energy of the molecule. The tetraene $CH_2{=}CHCH{=}CHCH{=}CHCH{=}CH_2$ can be treated as a box of length 8*R*, where $R \approx 140$ pm (as in this case, an extra half bond-length is often added at each end of the box). Calculate the minimum excitation energy of the molecule and sketch the HOMO and LUMO. Estimate the colour a sample of the compound is likely to appear in white light.

11.15 The FEMO theory (Problem 11.14) of conjugated molecules is rather crude and better results are obtained with simple Hückel theory. (a) For a linear conjugated polyene with each of *N* carbon atoms contributing an electron in a 2*p* orbital, the energies E_k of the resulting π molecular orbitals are given by (see also Section 20.9):

$$E_k = \alpha + 2\beta\cos\frac{k\pi}{N+1} \qquad k = 1, 2, 3, \ldots, N$$

Use this expression to determine a reasonable empirical estimate of the resonance integral β for the homologous series consisting of ethene, butadiene, hexatriene, and octatetraene given that $\pi^*\leftarrow\pi$ ultraviolet absorptions from the HOMO to the LUMO occur at 61 500, 46 080, 39 750, and 32 900 cm^{-1}, respectively. (b) Calculate the π-electron delocalization energy, $E_{\text{deloc}} = E_\pi - n(\alpha+\beta)$, of octatetraene, where E_π is the total π-electron binding energy and *n* is the total number of π-electrons. (c) In the context of this Hückel model, the π molecular orbitals are written as linear combinations of the carbon 2*p* orbitals. The coefficient of the *j*th atomic orbital in the *k*th molecular orbital is given by:

$$c_{kj} = \left(\frac{2}{N+1}\right)^{1/2}\sin\frac{jk\pi}{N+1} \qquad j = 1, 2, 3, \ldots, N$$

Determine the values of the coefficients of each of the six 2*p* orbitals in each of the six π molecular orbitals of hexatriene. Match each set of coefficients (that is, each molecular orbital) with a value of the energy calculated with the expression given in part (a) of the molecular orbital. Comment on trends that relate the energy of a molecular orbital with its 'shape', which can be inferred from the magnitudes and signs of the coefficients in the linear combination that describes the molecular orbital.

11.16 For monocyclic conjugated polyenes (such as cyclobutadiene and benzene) with each of *N* carbon atoms contributing an electron in a 2*p* orbital, simple Hückel theory gives the following expression for the energies E_k of the resulting π molecular orbitals:

$$E_k = \alpha + 2\beta\cos\frac{2k\pi}{N} \qquad k = 0, \pm1, \pm2, \ldots, \pm N/2 \text{ (even } N) \qquad k = 0, \pm1, \pm2, \ldots, \pm(N-1)/2 \text{ (odd } N)$$

(a) Calculate the energies of the π molecular orbitals of benzene and cyclooctatetraene. Comment on the presence or absence of degenerate energy levels. (b) Calculate and compare the delocalization energies of benzene (using the expression above) and hexatriene (see Problem 11.15a). What do you conclude from your results? (c) Calculate and compare the delocalization energies of cyclooctaene and octatetraene. Are your conclusions for this pair of molecules the same as for the pair of molecules investigated in part (b)?

11.17 If you have access to mathematical software that can perform matrix diagonalization, use it to solve Problems 11.15 and 11.16, disregarding the expressions for the energies and coefficients given there.

11.18 Molecular orbital calculations based on semi-empirical, *ab initio*, and DFT methods describe the spectroscopic properties of conjugated molecules a bit better than simple Hückel theory. (a) Using molecular modelling software[2] and the computational method of your choice (semi-empirical, *ab initio*, or density functioncal methods), calculate the energy separation between the HOMO and LUMO of ethene, butadiene, hexatriene, and octatetraene. (b) Plot the HOMO–LUMO energy separations against the experimental frequencies for $\pi^*\leftarrow\pi$ ultraviolet absorptions for these molecules (Problem 11.15). Use mathematical software to find the polynomial equation that best fits the data. (c) Use your polynomial fit from part (b) to estimate the frequency of the $\pi^*\leftarrow\pi$ ultraviolet absorption of decapentaene from the calculated HOMO–LUMO energy separation. (d) Discuss why the calibration procedure of part (b) is necessary.

11.19 Electronic excitation of a molecule may weaken or strengthen some bonds because bonding and antibonding characteristics differ between the HOMO and the LUMO. For example, a carbon–carbon bond in a linear polyene may have bonding character in the HOMO and antibonding character in the LUMO. Therefore, promotion of an electron from the HOMO to the LUMO weakens this carbon–carbon bond in the excited electronic state, relative to the ground electronic state. Display the HOMO and LUMO of each molecule in Problem 11.15 and discuss in detail any changes in bond order that accompany the $\pi^*\leftarrow\pi$ ultraviolet absorptions in these molecules.

11.20 As mentioned in Section 2.8, molecular electronic structure methods may be used to estimate the standard enthalpy of formation of molecules in the gas phase. (a) Using molecular modelling software and a semi-empirical method of your choice, calculate the standard enthalpy of formation of ethene, butadiene, hexatriene, and octatetraene in the gas phase. (b) Consult a database of thermochemical data, such as the online sources listed in this textbook's web site, and, for each molecule in part (a), calculate the relative error between the calculated and experimental values of the standard enthalpy of formation. (c) A good thermochemical database will also report the uncertainty in the experimental value of the standard enthalpy of formation. Compare experimental uncertainties with the relative errors calculated in part (b) and discuss the reliability of your chosen semi-empirical method for the estimation of thermochemical properties of linear polyenes.

Theoretical problems

11.21 An sp^2 hybrid orbital that lies in the xy-plane and makes an angle of 120° to the x-axis has the form

$$\psi=\frac{1}{3^{1/2}}\left(s-\frac{1}{2^{1/2}}p_x+\frac{3^{1/2}}{2^{1/2}}p_y\right)$$

Use hydrogenic atomic orbitals to write the explicit form of the hybrid orbital. Show that it has its maximum amplitude in the direction specified.

11.22 Use the expressions in Problems 11.8 and 11.9 to show that the antibonding orbital is more antibonding than the bonding orbital is bonding at most internuclear separations.

11.23 Derive eqns 11.11 and 11.14 by working with the normalized LCAO-MOs for the H_2^+ molecule-ion (Section 11.3a). Proceed by evaluating the expectation value of the hamiltonian for the ion. Make use of the fact that A and B each individually satisfy the Schrödinger equation for an isolated H atom.

11.24 Take as a trial function for the ground state of the hydrogen atom (a) e^{-kr}, (b) e^{-kr^2} and use the variation principle to find the optimum value of k in each case. Identify the better wavefunction. The only part of the laplacian that need be considered is the part that involves radial derivatives (eqn 9.5).

11.25 We saw in Section 11.5 that, to find the energies of the bonding and antibonding orbitals of a heteronuclear diatomic molecule, we need to solve the secular determinant

$$\begin{vmatrix}\alpha_A-E & B\\ \beta & \alpha_B-E\end{vmatrix}=0$$

where $\alpha_A\neq\alpha_B$ and we have taken $S=0$. Equations 11.34a and 11.34b give the general solution to this problem. Here, we shall develop the result for the case $(\alpha_B-\alpha_A)^2 >> \beta^2$. (a) Begin by showing that

$$E_\pm=\frac{\alpha_A+\alpha_B}{2}\pm\frac{\alpha_A-\alpha_B}{2}\left[1+\frac{4\beta^2}{(\alpha_A-\alpha_B)^2}\right]^{1/2}$$

where E_+ and E_- are the energies of the bonding and antibonding molecular orbitals, respectively. (b) Now use the expansion

$$(1+x)^{1/2}=1+\frac{x}{2}-\frac{x^3}{8}+\cdots$$

to show that

$$E_-=\alpha_B+\frac{\beta^2}{\alpha_B-\alpha_A}\qquad E_+=\alpha_A-\frac{\beta^2}{\alpha_B-\alpha_A}$$

which is the limiting result used in *Justification 11.4*.

Applications: to astrophysics and biology

11.26‡ In Exercise 11.12a you were invited to set up the Hückel secular determinant for linear and cyclic H_3. The same secular determinant applies to the molecular ions H_3^+ and D_3^+. The molecular ion H_3^+ was discovered as long ago as 1912 by J.J. Thomson, but only more recently has the equivalent equilateral triangular structure been confirmed by M.J. Gaillard *et al.* (*Phys. Rev.* **A17**, 1797 (1978)). The molecular ion H_3^+ is the simplest polyatomic species with a confirmed existence and plays an important role in chemical reactions occurring in interstellar clouds that may lead to the formation of water, carbon monoxide, and ethyl alcohol. The H_3^+ ion has also been found in the atmospheres of Jupiter, Saturn, and Uranus. (a) Solve the Hückel secular equations for the energies of the H_3 system in terms of the parameters α and β, draw an energy level diagram for the orbitals, and determine the binding energies of H_3^+, H_3, and H_3^-. (b) Accurate quantum mechanical calculations by G.D. Carney and R.N. Porter (*J. Chem. Phys.* **65**, 3547 (1976)) give the dissociation energy for the process $H_3^+\rightarrow H+H+H^+$ as 849 kJ mol^{-1}. From this information and data in Table 11.3, calculate the enthalpy of the reaction $H^+(g)+H_2(g)\rightarrow H_3^+(g)$. (c) From your equations and the information given, calculate a value for the resonance integral β in H_3^+. Then go on to calculate the bind energies of the other H_3 species in (a).

11.27‡ There is some indication that other hydrogen ring compounds and ions in addition to H_3 and D_3 species may play a role in interstellar chemistry. According to J.S. Wright and G.A. DiLabio (*J. Phys. Chem.* **96**, 10793 (1992)), H_5^-, H_6, and H_7^+ are particularly stable whereas H_4 and H_5^+ are not. Confirm these statements by Hückel calculations.

11.28 Here we develop a molecular orbital theory treatment of the peptide group (**10**), which links amino acids in proteins. Specifically, we shall describe the factors that stabilize the planar conformation of the peptide group.

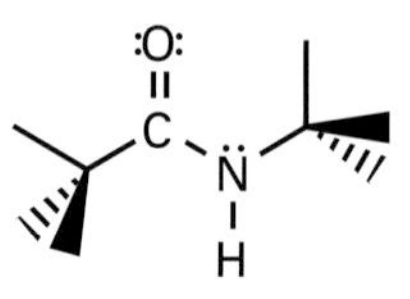

10

[2] The web site contains links to molecular modelling freeware and to other sites where you may perform molecular orbital calculations directly from your web browser.

(a) It will be familiar from introductory chemistry that valence bond theory explains the planar conformation of the peptide group by invoking delocalization of the π bond between the oxygen, carbon, and nitrogen atoms (**11**, **12**):

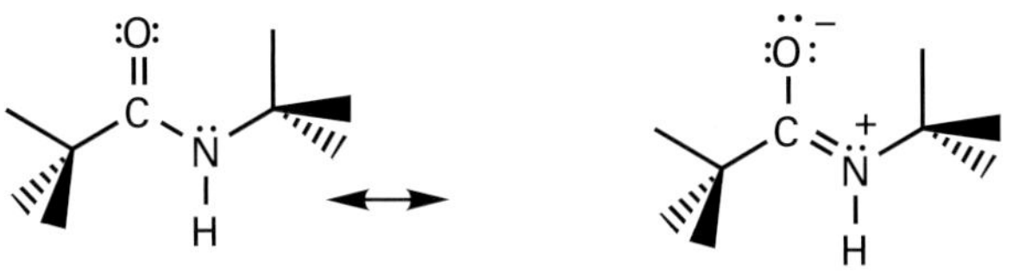

11 **12**

It follows that we can model the peptide group with molecular orbital theory by making LCAO-MOs from $2p$ orbitals perpendicular to the plane defined by the O, C, and N atoms. The three combinations have the form:

$$\psi_1 = a\psi_O + b\psi_C + c\psi_N \qquad \psi_2 = d\psi_O - e\psi_N \qquad \psi_3 = f\psi_O - g\psi_C + h\psi_N$$

where the coefficients a through h are all positive. Sketch the orbitals ψ_1, ψ_2, and ψ_3 and characterize them as bonding, non-bonding, or antibonding molecular orbitals. In a non-bonding molecular orbital, a pair of electrons resides in an orbital confined largely to one atom and not appreciably involved in bond formation. (b) Show that this treatment is consistent only with a planar conformation of the peptide link. (c) Draw a diagram showing the relative energies of these molecular orbitals and determine the occupancy of the orbitals. *Hint.* Convince yourself that there are four electrons to be distributed among the molecular orbitals. (d) Now consider a non-planar conformation of the peptide link, in which the O2p and C2p orbitals are perpendicular to the plane defined by the O, C, and N atoms, but the N2p orbital lies on that plane. The LCAO-MOs are given by

$$\psi_4 = a\psi_O + b\psi_C \qquad \psi_5 = e\psi_N \qquad \psi_6 = f\psi_O - g\psi_C$$

Just as before, sketch these molecular orbitals and characterize them as bonding, non-bonding, or antibonding. Also, draw an energy level diagram and determine the occupancy of the orbitals. (e) Why is this arrangement of atomic orbitals consistent with a non-planar conformation for the peptide link? (f) Does the bonding MO associated with the planar conformation have the same energy as the bonding MO associated with the non-planar conformation? If not, which bonding MO is lower in energy? Repeat the analysis for the non-bonding and anti-bonding molecular orbitals. (g) Use your results from parts (a)–(f) to construct arguments that support the planar model for the peptide link.

11.29 Molecular orbital calculations may be used to predict trends in the standard potentials of conjugated molecules, such as the quinones and flavins, that are involved in biological electron transfer reactions (*Impact I7.2*). It is commonly assumed that decreasing the energy of the LUMO enhances the ability of a molecule to accept an electron into the LUMO, with an attendant increase in the value of the molecule's standard potential. Furthermore, a number of studies indicate that there is a linear correlation between the LUMO energy and the reduction potential of aromatic hydrocarbons (see, for example, J.P. Lowe, *Quantum chemistry*, Chapter 8, Academic Press (1993)). (a) The standard potentials at pH = 7 for the one-electron reduction of methyl-substituted 1,4-benzoquinones (**13**) to their respective semiquinone radical anions are:

R_2	R_3	R_5	R_6	$E^\ominus$/V
H	H	H	H	0.078
CH_3	H	H	H	0.023
CH_3	H	CH_3	H	−0.067
CH_3	CH_3	CH_3	H	−0.165
CH_3	CH_3	CH_3	CH_3	−0.260

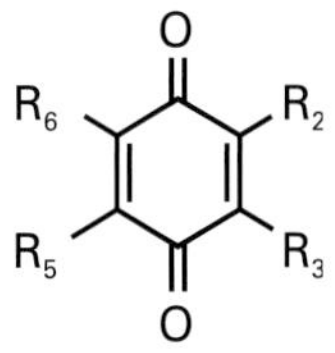

13

Using molecular modelling software and the computational method of your choice (semi-empirical, *ab initio*, or density functional theory methods), calculate E_{LUMO}, the energy of the LUMO of each substituted 1,4-benzoquinone, and plot E_{LUMO} against $E^\ominus$. Do your calculations support a linear relation between E_{LUMO} and $E^\ominus$? (b) The 1,4-benzoquinone for which $R_2 = R_3 = CH_3$ and $R_5 = R_6 = OCH_3$ is a suitable model of ubiquinone, a component of the respiratory electron transport chain (*Impact I7.2*). Determine E_{LUMO} of this quinone and then use your results from part (a) to estimate its standard potential. (c) The 1,4-benzoquinone for which $R_2 = R_3 = R_5 = CH_3$ and $R_6 = H$ is a suitable model of plastoquinone, a component of the photosynthetic electron transport chain (*Impact I7.2*). Determine E_{LUMO} of this quinone and then use your results from part (a) to estimate its standard potential. Is plastoquinone expected to be a better or worse oxidizing agent than ubiquinone? (d) Based on your predictions and on basic concepts of biological electron transport (*Impact I7.2 and I23.2*), suggest a reason why ubiquinone is used in respiration and plastoquinone is used in photosynthesis.

12 Molecular symmetry

In this chapter we sharpen the concept of 'shape' into a precise definition of 'symmetry', and show that symmetry may be discussed systematically. We see how to classify any molecule according to its symmetry and how to use this classification to discuss molecular properties. After describing the symmetry properties of molecules themselves, we turn to a consideration of the effect of symmetry transformations on orbitals and see that their transformation properties can be used to set up a labelling scheme. These symmetry labels are used to identify integrals that necessarily vanish. One important integral is the overlap integral between two orbitals. By knowing which atomic orbitals may have nonzero overlap, we can decide which ones can contribute to molecular orbitals. We also see how to select linear combinations of atomic orbitals that match the symmetry of the nuclear framework. Finally, by considering the symmetry properties of integrals, we see that it is possible to derive the selection rules that govern spectroscopic transitions.

The systematic discussion of symmetry is called **group theory**. Much of group theory is a summary of common sense about the symmetries of objects. However, because group theory is systematic, its rules can be applied in a straightforward, mechanical way. In most cases the theory gives a simple, direct method for arriving at useful conclusions with the minimum of calculation, and this is the aspect we stress here. In some cases, though, it leads to unexpected results.

The symmetry elements of objects

Some objects are 'more symmetrical' than others. A sphere is more symmetrical than a cube because it looks the same after it has been rotated through any angle about any diameter. A cube looks the same only if it is rotated through certain angles about specific axes, such as 90°, 180°, or 270° about an axis passing through the centres of any of its opposite faces (Fig. 12.1), or by 120° or 240° about an axis passing through any of its opposite corners. Similarly, an NH_3 molecule is 'more symmetrical' than an H_2O molecule because NH_3 looks the same after rotations of 120° or 240° about the axis shown in Fig. 12.2, whereas H_2O looks the same only after a rotation of 180°.

An action that leaves an object looking the same after it has been carried out is called a **symmetry operation**. Typical symmetry operations include rotations, reflections, and inversions. There is a corresponding **symmetry element** for each symmetry operation, which is the point, line, or plane with respect to which the symmetry operation is performed. For instance, a rotation (a symmetry operation) is carried out around an axis (the corresponding symmetry element). We shall see that we can classify molecules by identifying all their symmetry elements, and grouping together molecules that

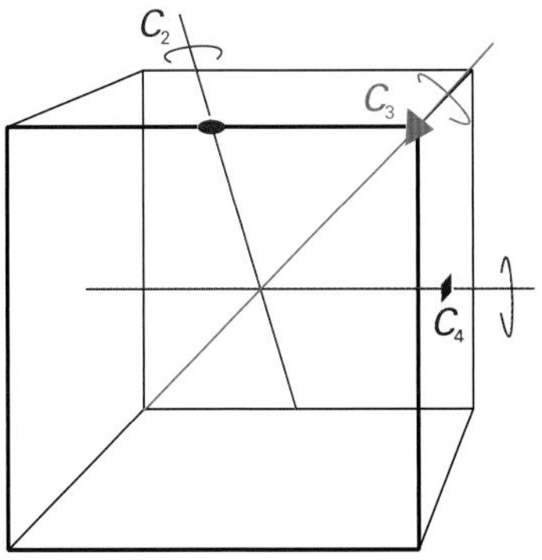

Fig. 12.1 Some of the symmetry elements of a cube. The twofold, threefold, and fourfold axes are labelled with the conventional symbols.

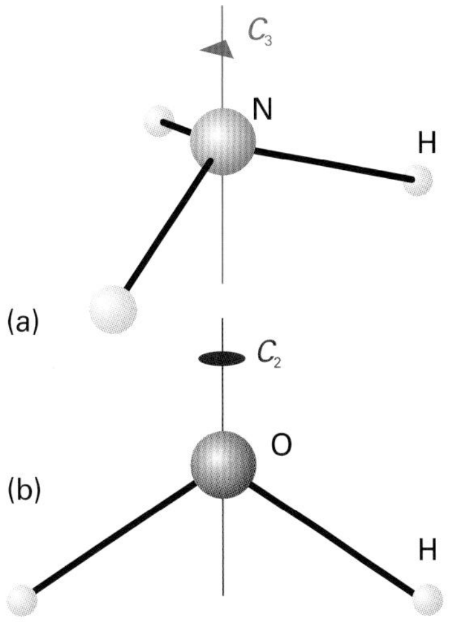

Fig. 12.2 (a) An NH_3 molecule has a threefold (C_3) axis and (b) an H_2O molecule has a twofold (C_2) axis. Both have other symmetry elements too.

possess the same set of symmetry elements. This procedure, for example, puts the trigonal pyramidal species NH_3 and SO_3^{2-} into one group and the angular species H_2O and SO_2 into another group.

12.1 Operations and symmetry elements

The classification of objects according to symmetry elements corresponding to operations that leave at least one common point unchanged gives rise to the **point groups**. There are five kinds of symmetry operation (and five kinds of symmetry element) of this kind. When we consider crystals (Chapter 20), we shall meet symmetries arising from translation through space. These more extensive groups are called **space groups**.

The **identity**, E, consists of doing nothing; the corresponding symmetry element is the entire object. Because every molecule is indistinguishable from itself if nothing is done to it, every object possesses at least the identity element. One reason for including the identity is that some molecules have only this symmetry element (**1**); another reason is technical and connected with the detailed formulation of group theory.

An ***n*-fold rotation** (the operation) about an ***n*-fold axis of symmetry**, C_n (the corresponding element) is a rotation through 360°/n. The operation C_1 is a rotation through 360°, and is equivalent to the identity operation E. An H_2O molecule has one twofold axis, C_2. An NH_3 molecule has one threefold axis, C_3, with which is associated two symmetry operations, one being 120° rotation in a clockwise sense and the other 120° rotation in a counter-clockwise sense. A pentagon has a C_5 axis, with two (clockwise and counterclockwise) rotations through 72° associated with it. It also has an axis denoted C_5^2, corresponding to two successive C_5 rotations; there are two such operations, one through 144° in a clockwise sense and the other through 144° in a counter-clockwise sense. A cube has three C_4 axes, four C_3 axes, and six C_2 axes. However, even this high symmetry is exceeded by a sphere, which possesses an infinite number of symmetry axes (along any diameter) of all possible integral values of n. If a molecule possesses several rotation axes, then the one (or more) with the greatest value of n is called the **principal axis**. The principal axis of a benzene molecule is the sixfold axis perpendicular to the hexagonal ring (**2**).

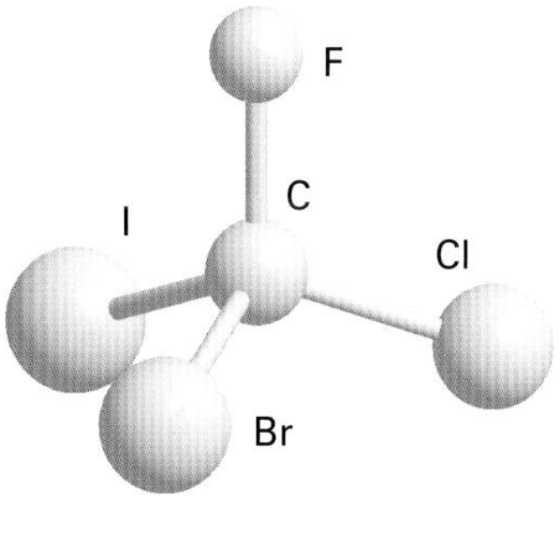

1 CBrClFI

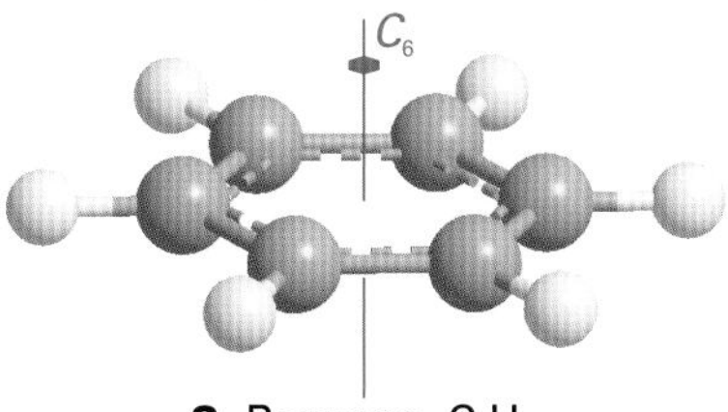

2 Benzene, C_6H_6

Comment 12.1

There is only one twofold rotation associated with a C_2 axis because clockwise and counter-clockwise 180° rotations are identical.

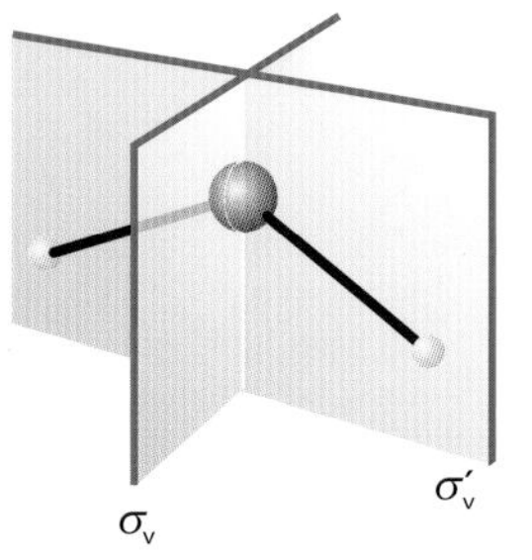

Fig. 12.3 An H_2O molecule has two mirror planes. They are both vertical (i.e. contain the principal axis), so are denoted σ_v and σ_v'.

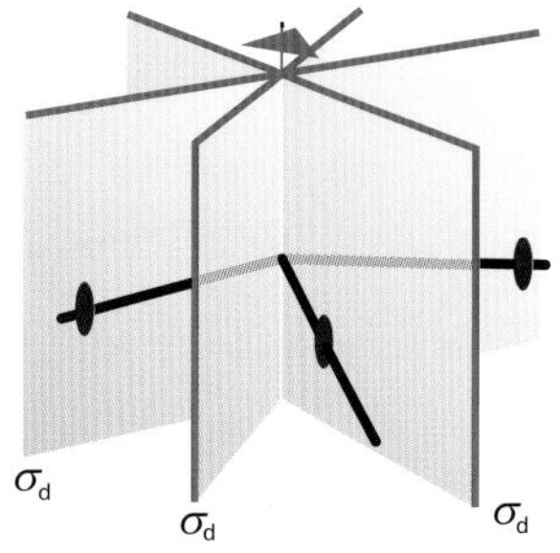

Fig. 12.4 Dihedral mirror planes (σ_d) bisect the C_2 axes perpendicular to the principal axis.

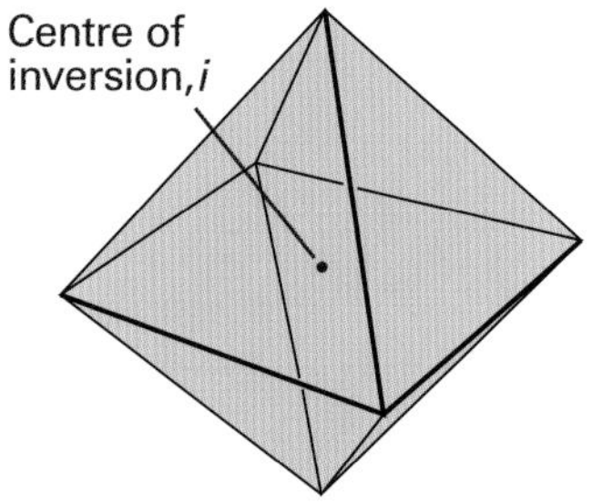

Fig. 12.5 A regular octahedron has a centre of inversion (i).

A **reflection** (the operation) in a **mirror plane**, σ (the element), may contain the principal axis of a molecule or be perpendicular to it. If the plane is parallel to the principal axis, it is called 'vertical' and denoted σ_v. An H_2O molecule has two vertical planes of symmetry (Fig. 12.3) and an NH_3 molecule has three. A vertical mirror plane that bisects the angle between two C_2 axes is called a 'dihedral plane' and is denoted σ_d (Fig. 12.4). When the plane of symmetry is perpendicular to the principal axis it is called 'horizontal' and denoted σ_h. A C_6H_6 molecule has a C_6 principal axis and a horizontal mirror plane (as well as several other symmetry elements).

In an **inversion** (the operation) through a **centre of symmetry**, i (the element), we imagine taking each point in a molecule, moving it to the centre of the molecule, and then moving it out the same distance on the other side; that is, the point (x, y, z) is taken into the point $(-x, -y, -z)$. Neither an H_2O molecule nor an NH_3 molecule has a centre of inversion, but a sphere and a cube do have one. A C_6H_6 molecule does have a centre of inversion, as does a regular octahedron (Fig. 12.5); a regular tetrahedron and a CH_4 molecule do not.

An ***n*-fold improper rotation** (the operation) about an ***n*-fold axis of improper rotation** or an ***n*-fold improper rotation axis**, S_n, (the symmetry element) is composed of two successive transformations. The first component is a rotation through $360°/n$, and the second is a reflection through a plane perpendicular to the axis of that rotation; neither operation alone needs to be a symmetry operation. A CH_4 molecule has three S_4 axes (Fig. 12.6).

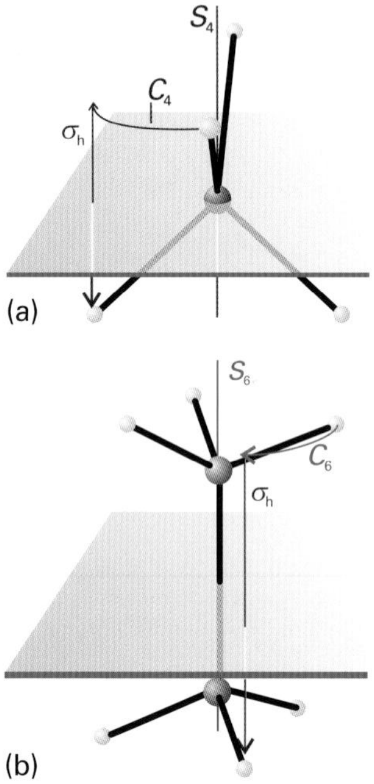

Fig. 12.6 (a) A CH_4 molecule has a fourfold improper rotation axis (S_4): the molecule is indistinguishable after a 90° rotation followed by a reflection across the horizontal plane, but neither operation alone is a symmetry operation. (b) The staggered form of ethane has an S_6 axis composed of a 60° rotation followed by a reflection.

12.2 The symmetry classification of molecules

To classify molecules according to their symmetries, we list their symmetry elements and collect together molecules with the same list of elements. This procedure puts CH_4 and CCl_4, which both possess the same symmetry elements as a regular tetrahedron, into the same group, and H_2O into another group.

The name of the group to which a molecule belongs is determined by the symmetry elements it possesses. There are two systems of notation (Table 12.1). The **Schoenflies system** (in which a name looks like C_{4v}) is more common for the discussion of individual molecules, and the **Hermann–Mauguin system**, or **International system** (in which a name looks like $4mm$), is used almost exclusively in the discussion of crystal symmetry. The identification of a molecule's point group according to the Schoenflies system is simplified by referring to the flow diagram in Fig. 12.7 and the shapes shown in Fig. 12.8.

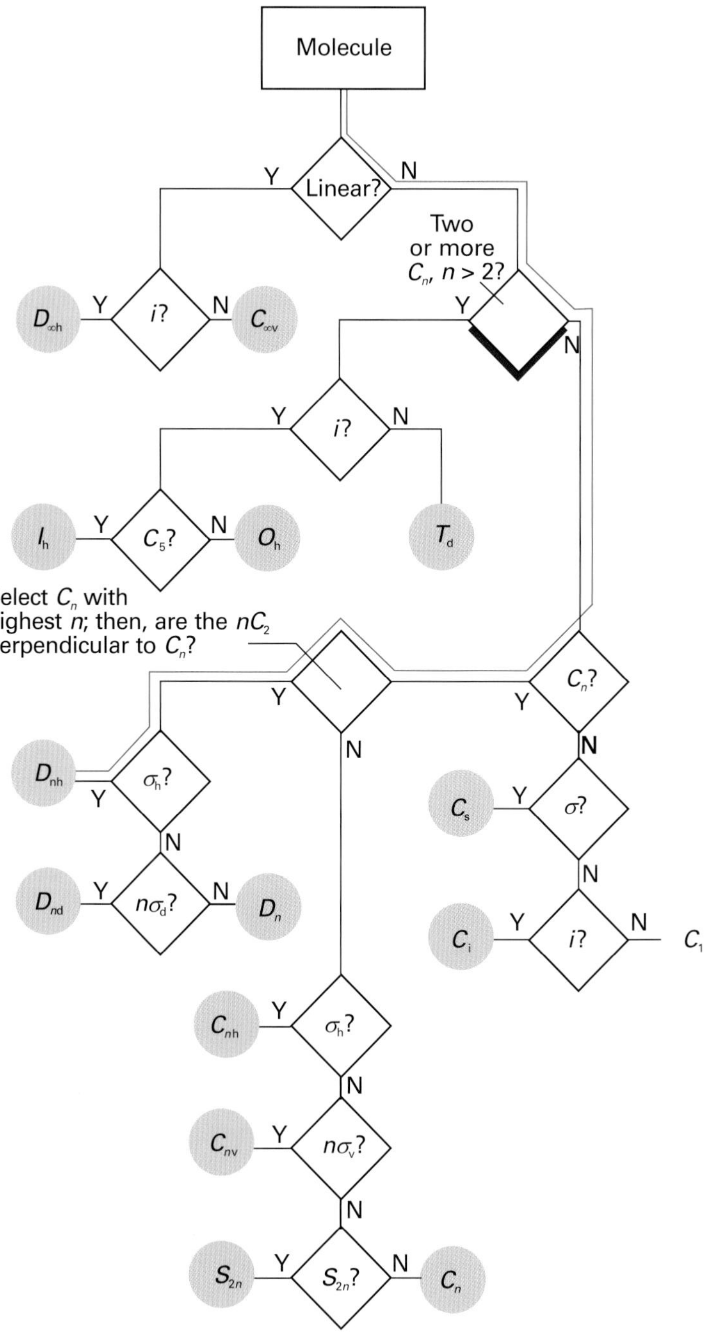

Fig. 12.7 A flow diagram for determining the point group of a molecule. Start at the top and answer the question posed in each diamond (Y = yes, N = no).

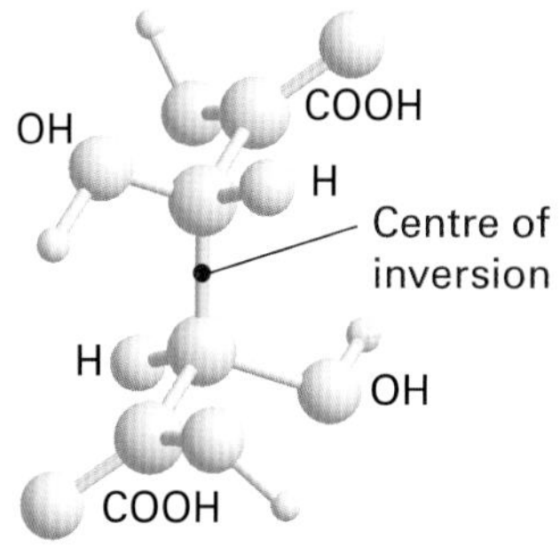

3 Meso-tartaric acid, HOOCCH(OH)CH(OH)COOH

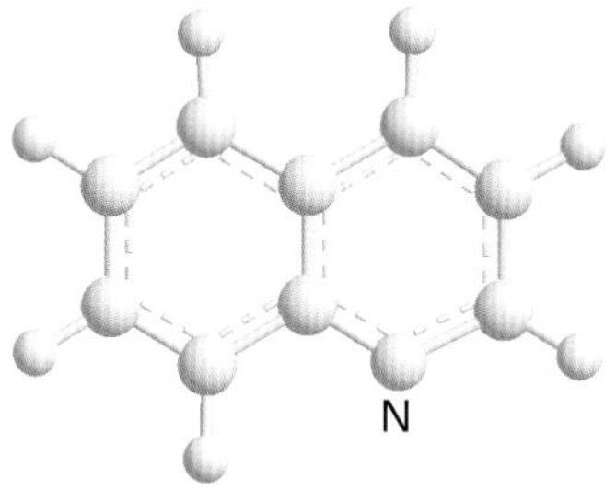

4 Quinoline, C_9H_7N

(a) The groups C_1, C_i, and C_s

A molecule belongs to the group C_1 if it has no element other than the identity, as in (**1**). It belongs to C_i if it has the identity and the inversion alone (**3**), and to C_s if it has the identity and a mirror plane alone (**4**).

Table 12.1 The notation for point groups*

C_i	$\bar{1}$								
C_s	m								
C_1	1	C_2	2	C_3	3	C_4	4	C_6	6
		C_{2v}	$2mm$	C_{3v}	$3m$	C_{4v}	$4mm$	C_{6v}	$6mm$
		C_{2h}	$2/m$	C_{3h}	$\bar{6}$	C_{4h}	$4/m$	C_{6h}	$6/m$
		D_2	222	D_3	32	D_4	422	D_6	622
		D_{2h}	mmm	D_{3h}	$\bar{6}2m$	D_{4h}	$4/mmm$	D_{6h}	$6/mmm$
		D_{2d}	$\bar{4}2m$	D_{3d}	$\bar{3}m$	S_4	$\bar{4}/m$	S_6	$\bar{3}$
T	23	T_d	$\bar{4}3m$	T_h	$m3$				
O	432	O_h	$m3m$						

* In the International system (or Hermann–Mauguin system) for point groups, a number n denotes the presence of an n-fold axis and m denotes a mirror plane. A slash (/) indicates that the mirror plane is perpendicular to the symmetry axis. It is important to distinguish symmetry elements of the same type but of different classes, as in $4/mmm$, in which there are three classes of mirror plane. A bar over a number indicates that the element is combined with an inversion. The only groups listed here are the so-called 'crystallographic point groups' (Section 20.1).

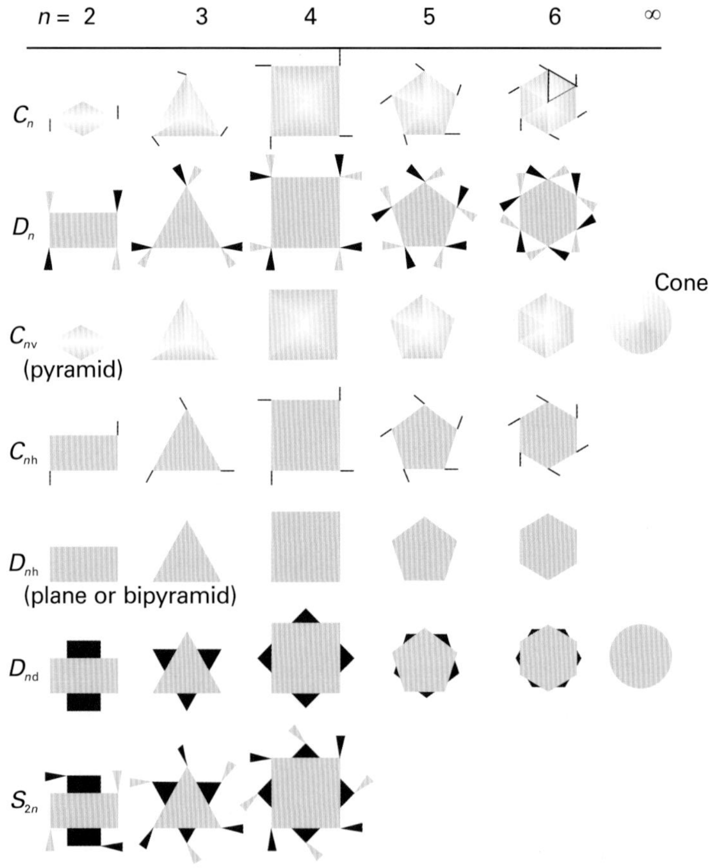

Fig. 12.8 A summary of the shapes corresponding to different point groups. The group to which a molecule belongs can often be identified from this diagram without going through the formal procedure in Fig. 12.7.

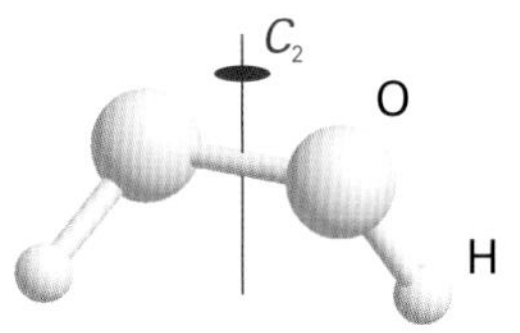

5 Hydrogen peroxide, H_2O_2

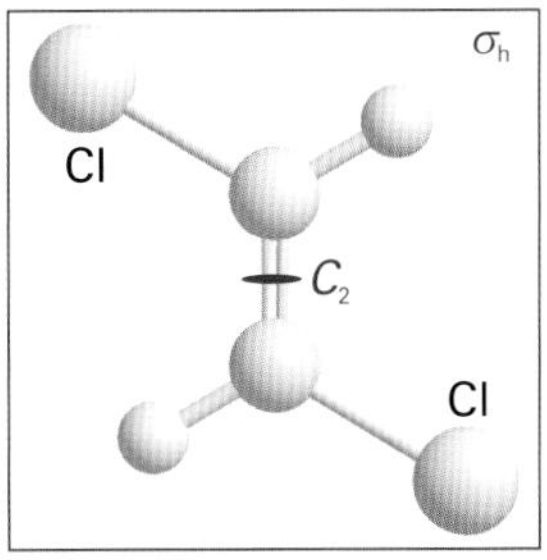

6 *trans*-CHCl=CHCl

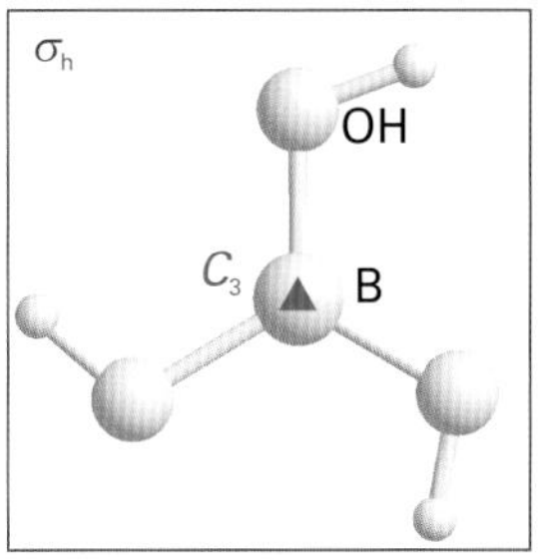

7 $B(OH)_3$

(b) The groups C_n, C_{nv}, and C_{nh}

A molecule belongs to the group C_n if it possesses an n-fold axis. Note that symbol C_n is now playing a triple role: as the label of a symmetry element, a symmetry operation, and a group. For example, an H_2O_2 molecule has the elements E and C_2 (**5**), so it belongs to the group C_2.

If in addition to the identity and a C_n axis a molecule has n vertical mirror planes σ_v, then it belongs to the group C_{nv}. An H_2O molecule, for example, has the symmetry elements E, C_2, and $2\sigma_v$, so it belongs to the group C_{2v}. An NH_3 molecule has the elements E, C_3, and $3\sigma_v$, so it belongs to the group C_{3v}. A heteronuclear diatomic molecule such as HCl belongs to the group $C_{\infty v}$ because all rotations around the axis and reflections across the axis are symmetry operations. Other members of the group $C_{\infty v}$ include the linear OCS molecule and a cone.

Objects that in addition to the identity and an n-fold principal axis also have a horizontal mirror plane σ_h belong to the groups C_{nh}. An example is *trans*-CHCl=CHCl (**6**), which has the elements E, C_2, and σ_h, so belongs to the group C_{2h}; the molecule $B(OH)_3$ in the conformation shown in (**7**) belongs to the group C_{3h}. The presence of certain symmetry elements may be implied by the presence of others: thus, in C_{2h} the operations C_2 and σ_h jointly imply the presence of a centre of inversion (Fig. 12.9).

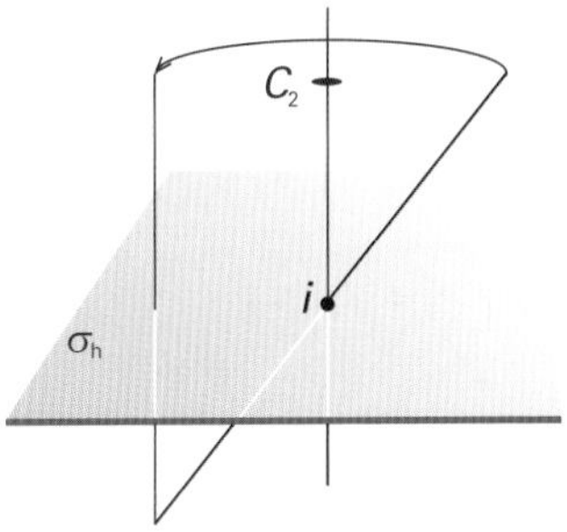

Fig. 12.9 The presence of a twofold axis and a horizontal mirror plane jointly imply the presence of a centre of inversion in the molecule.

(c) The groups D_n, D_{nh}, and D_{nd}

We see from Fig. 12.7 that a molecule that has an n-fold principal axis and n twofold axes perpendicular to C_n belongs to the group D_n. A molecule belongs to D_{nh} if it also possesses a horizontal mirror plane. The planar trigonal BF_3 molecule has the elements E, C_3, $3C_2$, and σ_h (with one C_2 axis along each B—F bond), so belongs to D_{3h} (**8**). The C_6H_6 molecule has the elements E, C_6, $3C_2$, $3C_2'$, and σ_h together with some others that these elements imply, so it belongs to D_{6h}. All homonuclear diatomic molecules, such as N_2, belong to the group $D_{\infty h}$ because all rotations around the axis are symmetry operations, as are end-to-end rotation and end-to-end reflection; $D_{\infty h}$ is also the group of the linear OCO and HCCH molecules and of a uniform cylinder. Other examples of D_{nh} molecules are shown in (**9**), (**10**), and (**11**).

A molecule belongs to the group D_{nd} if in addition to the elements of D_n it possesses n dihedral mirror planes σ_d. The twisted, 90° allene (**12**) belongs to D_{2d}, and the staggered conformation of ethane (**13**) belongs to D_{3d}.

Comment 12.2

The prime on $3C_2'$ indicates that the three C_2 axes are different from the other three C_2 axes. In benzene, three of the C_2 axes bisect C—C bonds and the other three pass through vertices of the hexagon formed by the carbon framework of the molecule.

(d) The groups S_n

Molecules that have not been classified into one of the groups mentioned so far, but that possess one S_n axis, belong to the group S_n. An example is tetraphenylmethane, which belongs to the point group S_4 (**14**). Molecules belonging to S_n with $n > 4$ are rare. Note that the group S_2 is the same as C_i, so such a molecule will already have been classified as C_i.

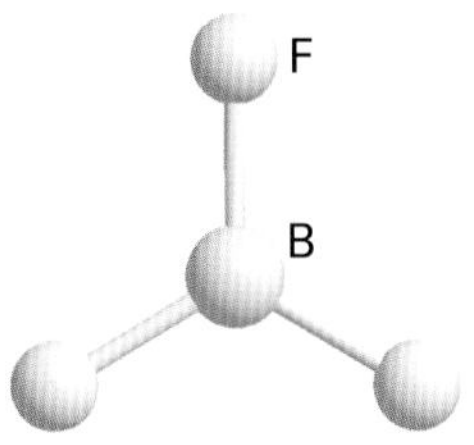

8 Boron trifluoride, BF_3

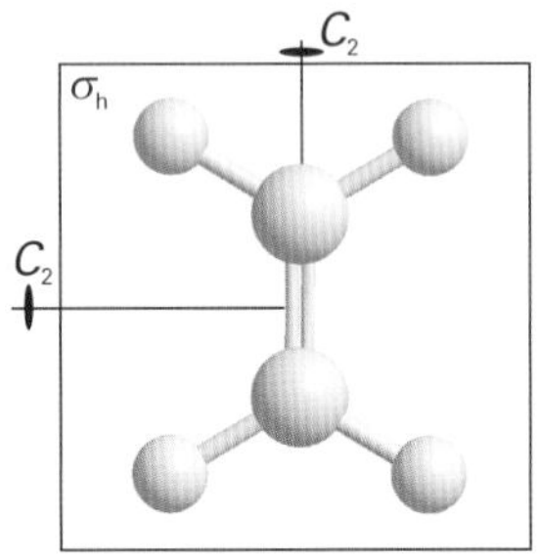

9 Ethene, $CH_2{=}CH_2$ (D_{2h})

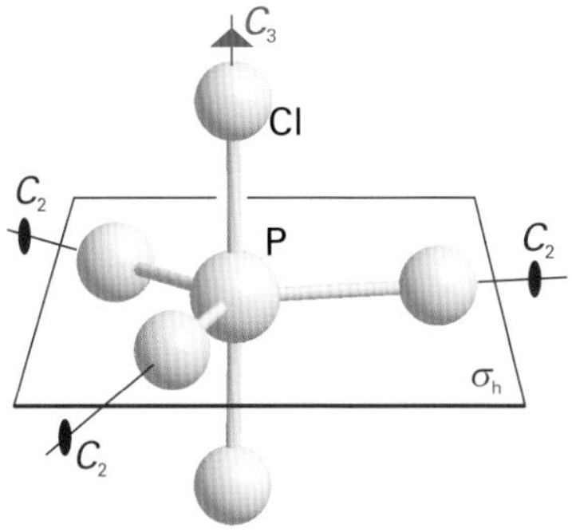

10 Phosphorus pentachloride, PCl_5 (D_{3h})

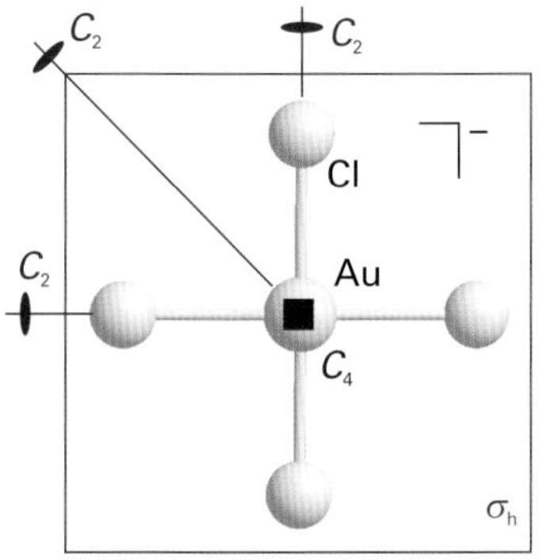

11 Tetrachloroaurate(III) ion, $[AuCl_4]^-$ (D_{4h})

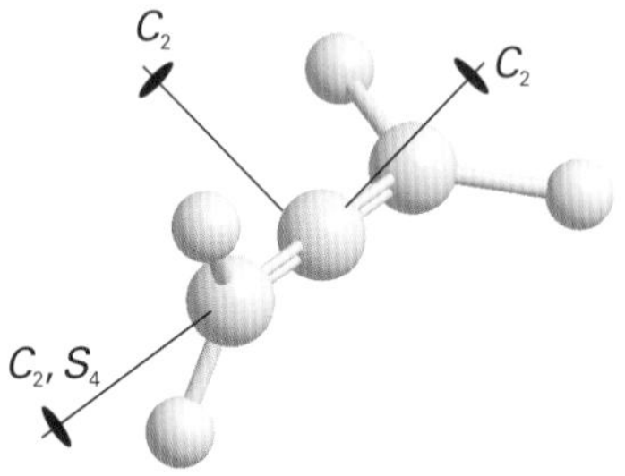

12 Allene, C_3H_4 (D_{2d})

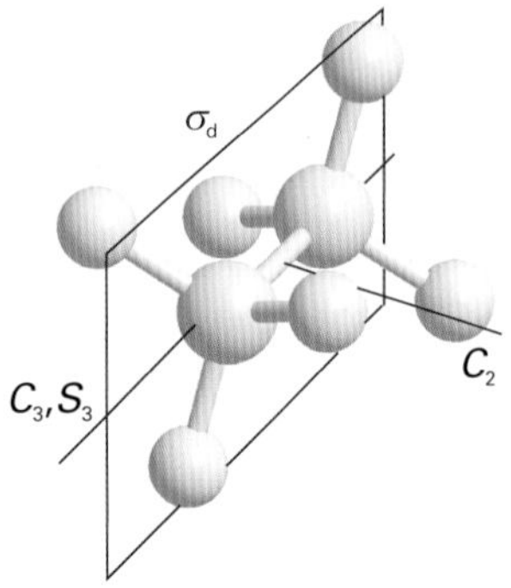

13 Ethane, C_2H_6 (D_{3d})

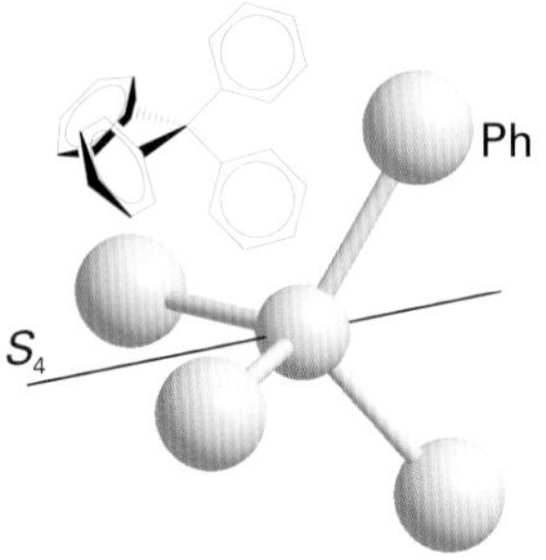

14 Tetraphenylmethane, $C(C_6H_5)_4$ (S_4)

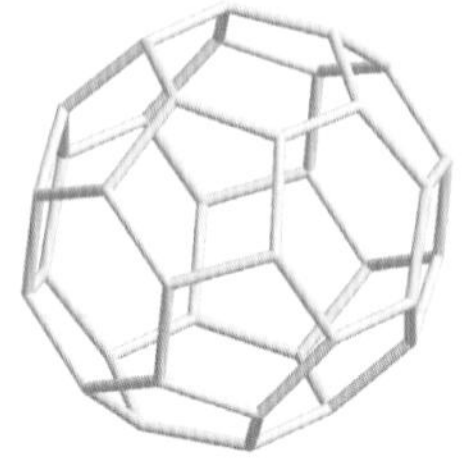

15 Buckminsterfullerene, C_{60} (I)

(e) The cubic groups

A number of very important molecules (e.g. CH_4 and SF_6) possess more than one principal axis. Most belong to the **cubic groups**, and in particular to the **tetrahedral groups** T, T_d, and T_h (Fig. 12.10a) or to the **octahedral groups** O and O_h (Fig. 12.10b). A few icosahedral (20-faced) molecules belonging to the **icosahedral group**, I (Fig. 12.10c), are also known: they include some of the boranes and buckminsterfullerene, C_{60} (**15**). The groups T_d and O_h are the groups of the regular tetrahedron (for instance, CH_4) and the regular octahedron (for instance, SF_6), respectively. If the object possesses the rotational symmetry of the tetrahedron or the octahedron, but none of their planes of reflection, then it belongs to the simpler groups T or O (Fig. 12.11). The group T_h is based on T but also contains a centre of inversion (Fig. 12.12).

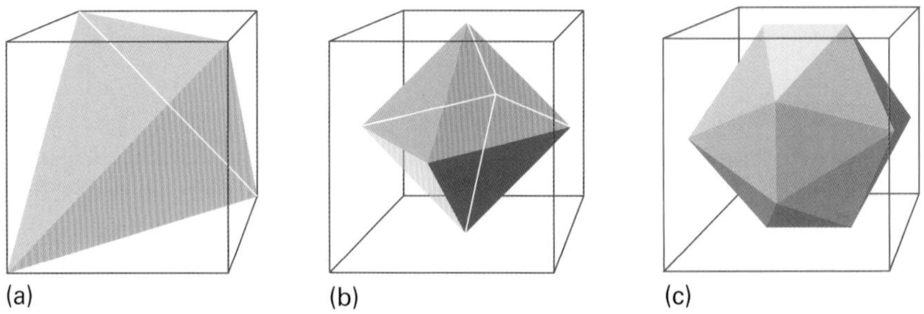

Fig. 12.10 (a) Tetrahedral, (b) octahedral, and (c) icosahedral molecules are drawn in a way that shows their relation to a cube: they belong to the cubicgroups T_d, O_h, and I_h, respectively.

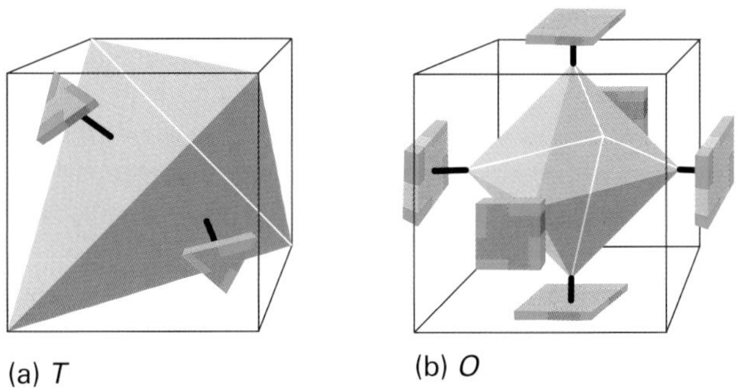

Fig. 12.11 Shapes corresponding to the point groups (a) T and (b) O. The presence of the decorated slabs reduces the symmetry of the object from T_d and O_h, respectively.

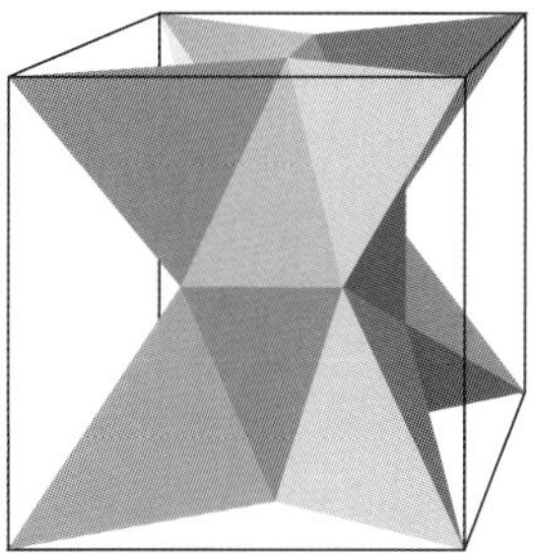

Fig. 12.12 The shape of an object belonging to the group T_h.

(f) The full rotation group

The **full rotation group**, R_3 (the 3 refers to rotation in three dimensions), consists of an infinite number of rotation axes with all possible values of n. A sphere and an atom belong to R_3, but no molecule does. Exploring the consequences of R_3 is a very important way of applying symmetry arguments to atoms, and is an alternative approach to the theory of orbital angular momentum.

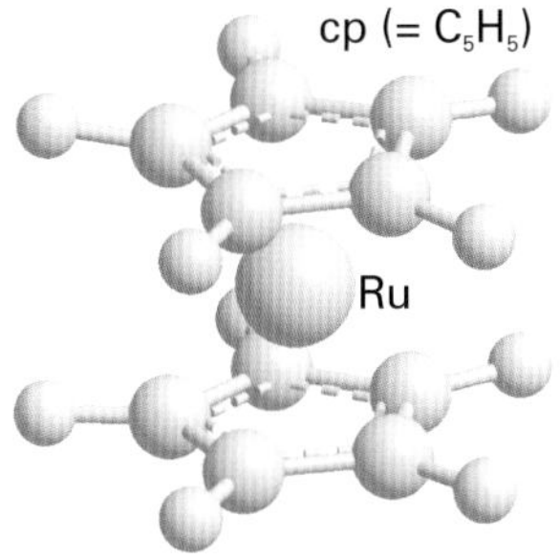

16 Ruthenocene, $Ru(cp)_2$

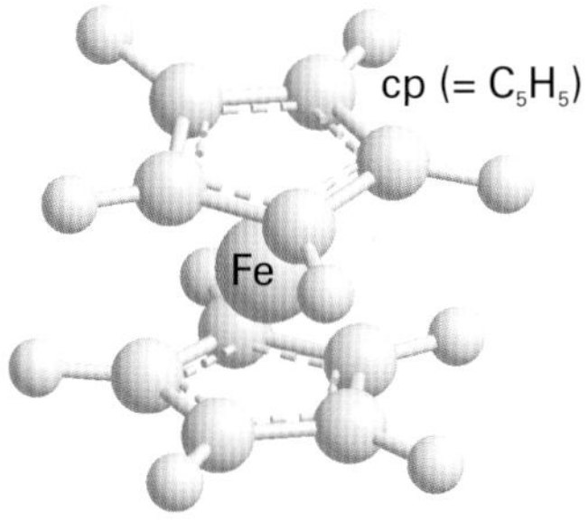

17 Ferrocene, $Fe(cp)_2$

Example 12.1 *Identifying a point group of a molecule*

Identify the point group to which a ruthenocene molecule (**16**) belongs.

Method Use the flow diagram in Fig. 12.7.

Answer The path to trace through the flow diagram in Fig. 12.7 is shown by a green line; it ends at D_{nh}. Because the molecule has a fivefold axis, it belongs to the group D_{5h}. If the rings were staggered, as they are in an excited state of ferrocene that lies 4 kJ mol^{-1} above the ground state (**17**), the horizontal reflection plane would be absent, but dihedral planes would be present.

Self-test 12.1 Classify the pentagonal antiprismatic excited state of ferrocene (**17**). [D_{5d}]

12.3 Some immediate consequences of symmetry

Some statements about the properties of a molecule can be made as soon as its point group has been identified.

Comment 12.3
The web site contains links to interactive tutorials, where you use your web browser to assign point groups of molecules.

(a) Polarity

A **polar molecule** is one with a permanent electric dipole moment (HCl, O_3, and NH_3 are examples). If the molecule belongs to the group C_n with $n > 1$, it cannot possess a charge distribution with a dipole moment perpendicular to the symmetry axis because the symmetry of the molecule implies that any dipole that exists in one direction perpendicular to the axis is cancelled by an opposing dipole (Fig. 12.13a). For example, the perpendicular component of the dipole associated with one O—H bond in H_2O is cancelled by an equal but opposite component of the dipole of the second O—H bond, so any dipole that the molecule has must be parallel to the twofold symmetry axis.

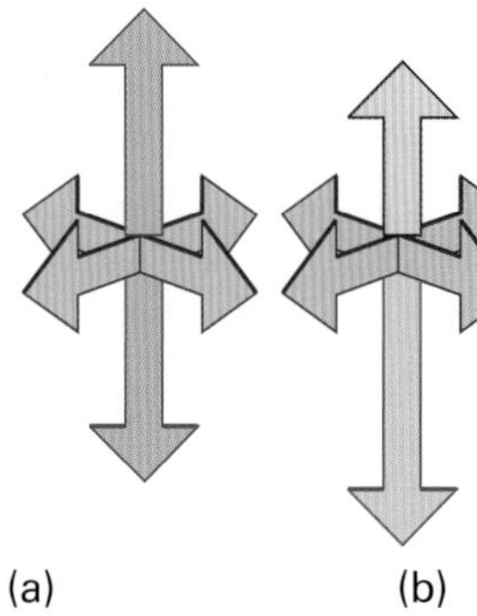

Fig. 12.13 (a) A molecule with a C_n axis cannot have a dipole perpendicular to the axis, but (b) it may have one parallel to the axis. The arrows represent local contributions to the overall electric dipole, such as may arise from bonds between pairs of neighbouring atoms with different electronegativities.

However, as the group makes no reference to operations relating the two ends of the molecule, a charge distribution may exist that results in a dipole along the axis (Fig. 12.13b), and H_2O has a dipole moment parallel to its twofold symmetry axis. The same remarks apply generally to the group C_{nv}, so molecules belonging to any of the C_{nv} groups may be polar. In all the other groups, such as C_{3h}, D, etc., there are symmetry operations that take one end of the molecule into the other. Therefore, as well as having no dipole perpendicular to the axis, such molecules can have none along the axis, for otherwise these additional operations would not be symmetry operations. We can conclude that *only molecules belonging to the groups* C_n, C_{nv}, *and* C_s *may have a permanent electric dipole moment.*

For C_n and C_{nv}, that dipole moment must lie along the symmetry axis. Thus ozone, O_3, which is angular and belongs to the group C_{2v}, may be polar (and is), but carbon dioxide, CO_2, which is linear and belongs to the group $D_{\infty h}$, is not.

(b) Chirality

A **chiral molecule** (from the Greek word for 'hand') is a molecule that cannot be superimposed on its mirror image. An **achiral molecule** is a molecule that can be superimposed on its mirror image. Chiral molecules are **optically active** in the sense that they rotate the plane of polarized light (a property discussed in more detail in Appendix 3). A chiral molecule and its mirror-image partner constitute an **enantiomeric pair** of isomers and rotate the plane of polarization in equal but opposite directions.

A molecule may be chiral, and therefore optically active, only if it does not possess an axis of improper rotation, S_n. However, we need to be aware that such an axis may be present under a different name, and be implied by other symmetry elements that are present. For example, molecules belonging to the groups C_{nh} possess an S_n axis implicitly because they possess both C_n and σ_h, which are the two components of an improper rotation axis. Any molecule containing a centre of inversion, i, also possesses an S_2 axis, because i is equivalent to C_2 in conjunction with σ_h, and that combination of elements is S_2 (Fig. 12.14). It follows that all molecules with centres of inversion are achiral and hence optically inactive. Similarly, because $S_1 = \sigma$, it follows that any molecule with a mirror plane is achiral.

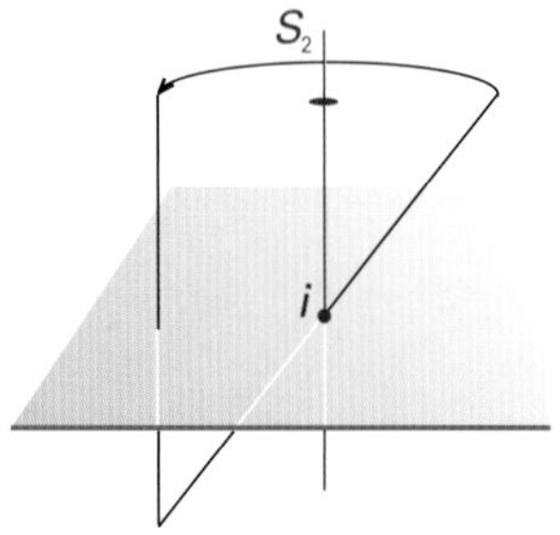

Fig. 12.14 Some symmetry elements are implied by the other symmetry elements in a group. Any molecule containing an inversion also possesses at least an S_2 element because i and S_2 are equivalent.

A molecule may be chiral if it does not have a centre of inversion or a mirror plane, which is the case with the amino acid alanine (**18**), but not with glycine (**19**). However, a molecule may be achiral even though it does not have a centre of inversion. For example, the S_4 species (**20**) is achiral and optically inactive: though it lacks i (that is, S_2) it does have an S_4 axis.

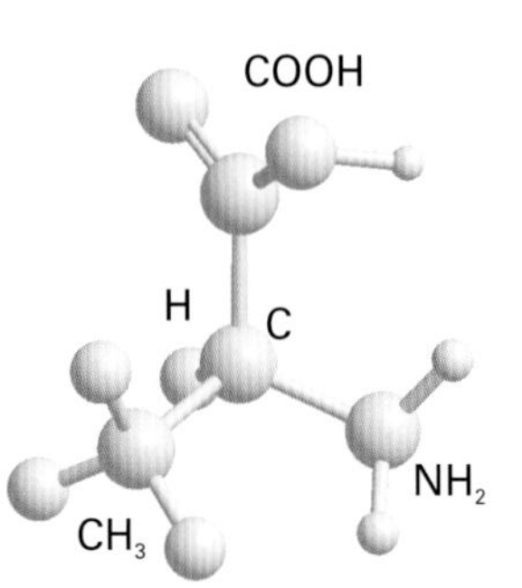

18 L-Alanine, $NH_2CH(CH_3)COOH$

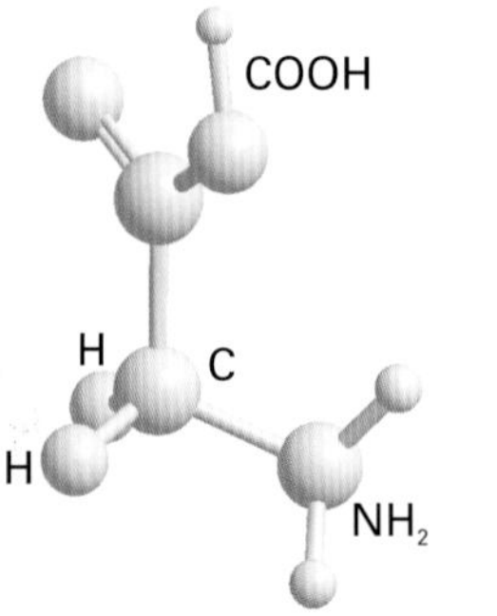

19 Glycine, NH_2CH_2COOH

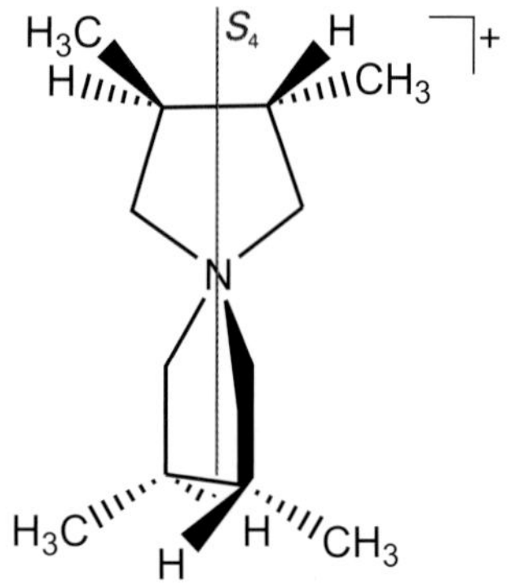

20 $N(CH_2CH(CH_3)CH(CH_3)CH_2)_2^+$

Applications to molecular orbital theory and spectroscopy

We shall now turn our attention away from the symmetries of molecules themselves and direct it towards the symmetry characteristics of orbitals that belong to the various atoms in a molecule. This material will enable us to discuss the formulation and labelling of molecular orbitals and selection rules in spectroscopy.

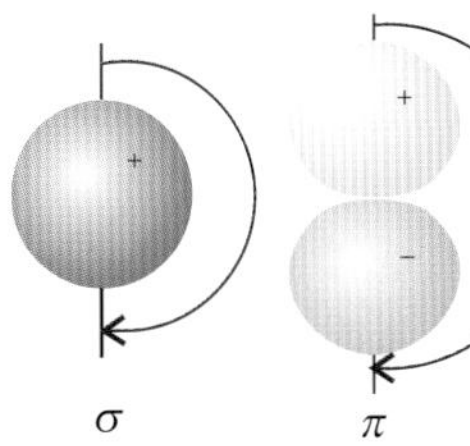

Fig. 12.15 A rotation through 180° about the internuclear axis (perpendicular to the page) leaves the sign of a σ orbital unchanged but the sign of a π orbital is changed. In the language introduced in this chapter, the characters of the C_2 rotation are +1 and −1 for the σ and π orbitals, respectively.

12.4 Character tables and symmetry labels

We saw in Chapter 11 that molecular orbitals of diatomic and linear polyatomic molecules are labelled σ, π, etc. These labels refer to the symmetries of the orbitals with respect to rotations around the principal symmetry axis of the molecule. Thus, a σ orbital does not change sign under a rotation through any angle, a π orbital changes sign when rotated by 180°, and so on (Fig. 12.15). The symmetry classifications σ and π can also be assigned to individual atomic orbitals in a linear molecule. For example, we can speak of an individual p_z orbital as having σ symmetry if the z-axis lies along the bond, because p_z is cylindrically symmetrical about the bond. This labelling of orbitals according to their behaviour under rotations can be generalized and extended to nonlinear polyatomic molecules, where there may be reflections and inversions to take into account as well as rotations.

(a) Representations and characters

Labels analogous to σ and π are used to denote the symmetries of orbitals in polyatomic molecules. These labels look like a, a_1, e, e_g, and we first encountered them in Fig. 11.4 in connection with the molecular orbitals of benzene. As we shall see, these labels indicate the behaviour of the orbitals under the symmetry operations of the relevant point group of the molecule.

A label is assigned to an orbital by referring to the **character table** of the group, a table that characterizes the different symmetry types possible in the point group. Thus, to assign the labels σ and π, we use the table shown in the margin. This table is a fragment of the full character table for a linear molecule. The entry +1 shows that the orbital remains the same and the entry −1 shows that the orbital changes sign under the operation C_2 at the head of the column (as illustrated in Fig. 12.15). So, to assign the label σ or π to a particular orbital, we compare the orbital's behaviour with the information in the character table.

	C_2	(i.e. rotation by 180°)
σ	+1	(i.e. no change of sign)
π	−1	(i.e. change of sign)

The entries in a complete character table are derived by using the formal techniques of group theory and are called **characters**, χ (chi). These numbers characterize the essential features of each symmetry type in a way that we can illustrate by considering the C_{2v} molecule SO_2 and the valence p_x orbitals on each atom, which we shall denote p_S, p_A, and p_B (Fig. 12.16).

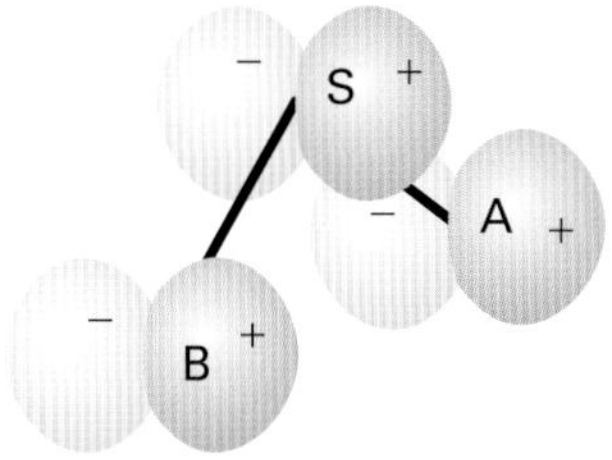

Fig. 12.16 The three p_x orbitals that are used to illustrate the construction of a matrix representation in a C_{2v} molecule (SO_2).

Under σ_v, the change $(p_S, p_B, p_A) \leftarrow (p_S, p_A, p_B)$ takes place. We can express this transformation by using matrix multiplication:

$$(p_S, p_B, p_A) = (p_S, p_A, p_B)\begin{pmatrix} 1 & 0 & 0 \\ 0 & 0 & 1 \\ 0 & 1 & 0 \end{pmatrix} = (p_S, p_A, p_B)\boldsymbol{D}(\sigma_v) \qquad (12.1)$$

The matrix $\boldsymbol{D}(\sigma_v)$ is called a **representative** of the operation σ_v. Representatives take different forms according to the **basis**, the set of orbitals, that has been adopted.

Comment 12.4

See *Appendix 2* for a summary of the rules of matrix algebra.

We can use the same technique to find matrices that reproduce the other symmetry operations. For instance, C_2 has the effect $(-p_S, -p_B, -p_A) \leftarrow (p_S, p_A, p_B)$, and its representative is

$$\boldsymbol{D}(C_2) = \begin{pmatrix} -1 & 0 & 0 \\ 0 & 0 & -1 \\ 0 & -1 & 0 \end{pmatrix} \tag{12.2}$$

The effect of σ'_v is $(-p_S, -p_A, -p_B) \leftarrow (p_S, p_A, p_B)$, and its representative is

$$\boldsymbol{D}(\sigma'_v) = \begin{pmatrix} -1 & 0 & 0 \\ 0 & -1 & 0 \\ 0 & 0 & -1 \end{pmatrix} \tag{12.3}$$

The identity operation has no effect on the basis, so its representative is the 3×3 unit matrix:

$$\boldsymbol{D}(E) = \begin{pmatrix} 1 & 0 & 0 \\ 0 & 1 & 0 \\ 0 & 0 & 1 \end{pmatrix} \tag{12.4}$$

The set of matrices that represents *all* the operations of the group is called a **matrix representation**, Γ (uppercase gamma), of the group for the particular basis we have chosen. We denote this three-dimensional representation by $\Gamma^{(3)}$. The discovery of a matrix representation of the group means that we have found a link between symbolic manipulations of operations and algebraic manipulations of numbers.

The character of an operation in a particular matrix representation is the sum of the diagonal elements of the representative of that operation. Thus, in the basis we are illustrating, the characters of the representatives are

$\boldsymbol{D}(E)$	$\boldsymbol{D}(C_2)$	$\boldsymbol{D}(\sigma_v)$	$\boldsymbol{D}(\sigma'_v)$
3	−1	1	−3

The character of an operation depends on the basis.

Inspection of the representatives shows that they are all of **block-diagonal form**:

$$\boldsymbol{D} = \begin{pmatrix} [\blacksquare] & 0 & 0 \\ 0 & [\blacksquare] & [\blacksquare] \\ 0 & [\blacksquare] & [\blacksquare] \end{pmatrix}$$

The block-diagonal form of the representatives show us that the symmetry operations of C_{2v} never mix p_S with the other two functions. Consequently, the basis can be cut into two parts, one consisting of p_S alone and the other of (p_A, p_B). It is readily verified that the p_S orbital itself is a basis for the one-dimensional representation

$$\boldsymbol{D}(E) = 1 \qquad \boldsymbol{D}(C_2) = -1 \qquad \boldsymbol{D}(\sigma_v) = 1 \qquad \boldsymbol{D}(\sigma'_v) = -1$$

which we shall call $\Gamma^{(1)}$. The remaining two basis functions are a basis for the two-dimensional representation $\Gamma^{(2)}$:

$$\boldsymbol{D}(E) = \begin{pmatrix} 1 & 0 \\ 0 & 1 \end{pmatrix} \quad \boldsymbol{D}(C_2) = \begin{pmatrix} 0 & -1 \\ -1 & 0 \end{pmatrix} \quad \boldsymbol{D}(\sigma_v) = \begin{pmatrix} 0 & 1 \\ 1 & 0 \end{pmatrix} \quad \boldsymbol{D}(\sigma'_v) = \begin{pmatrix} -1 & 0 \\ 0 & -1 \end{pmatrix}$$

These matrices are the same as those of the original three-dimensional representation, except for the loss of the first row and column. We say that the original three-dimensional representation has been **reduced** to the 'direct sum' of a one-dimensional representation 'spanned' by p_S, and a two-dimensional representation spanned by (p_A, p_B). This reduction is consistent with the common sense view that the central orbital plays a role different from the other two. We denote the reduction symbolically by writing

$$\Gamma^{(3)} = \Gamma^{(1)} + \Gamma^{(2)} \tag{12.5}$$

The one-dimensional representation $\Gamma^{(1)}$ cannot be reduced any further, and is called an **irreducible representation** of the group (an 'irrep'). We can demonstrate that the two-dimensional representation $\Gamma^{(2)}$ is reducible (for this basis in this group) by switching attention to the linear combinations $p_1 = p_A + p_B$ and $p_2 = p_A - p_B$. These combinations are sketched in Fig. 12.17. The representatives in the new basis can be constructed from the old by noting, for example, that under σ_v, $(p_B, p_A) \leftarrow (p_A, p_B)$. In this way we find the following representation in the new basis:

$$\boldsymbol{D}(E) = \begin{pmatrix} 1 & 0 \\ 0 & 1 \end{pmatrix} \quad \boldsymbol{D}(C_2) = \begin{pmatrix} -1 & 0 \\ 0 & 1 \end{pmatrix} \quad \boldsymbol{D}(\sigma_v) = \begin{pmatrix} 1 & 0 \\ 0 & -1 \end{pmatrix} \quad \boldsymbol{D}(\sigma_v') = \begin{pmatrix} -1 & 0 \\ 0 & -1 \end{pmatrix}$$

The new representatives are all in block-diagonal form, and the two combinations are not mixed with each other by any operation of the group. We have therefore achieved the reduction of $\Gamma^{(2)}$ to the sum of two one-dimensional representations. Thus, p_1 spans

$$D(E) = 1 \quad D(C_2) = -1 \quad D(\sigma_v) = 1 \quad D(\sigma_v') = -1$$

which is the same one-dimensional representation as that spanned by p_S, and p_2 spans

$$D(E) = 1 \quad D(C_2) = 1 \quad D(\sigma_v) = -1 \quad D(\sigma_v') = -1$$

which is a different one-dimensional representation; we shall denote it $\Gamma^{(1)\prime}$.

At this point we have found two irreducible representations of the group C_{2v} (Table 12.2). The two irreducible representations are normally labelled B_1 and A_2, respectively. An A or a B is used to denote a one-dimensional representation; A is used if the character under the principal rotation is +1, and B is used if the character is −1. Subscripts are used to distinguish the irreducible representations if there is more than one of the same type: A_1 is reserved for the representation with character 1 for all operations. When higher dimensional irreducible representations are permitted, E denotes a two-dimensional irreducible representation and T a three-dimensional irreducible representation; all the irreducible representations of C_{2v} are one-dimensional.

There are in fact only two more species of irreducible representations of this group, for a surprising theorem of group theory states that

$$\text{Number of symmetry species} = \text{number of classes} \tag{12.6}$$

Symmetry operations fall into the same **class** if they are of the same type (for example, rotations) and can be transformed into one another by a symmetry operation of the

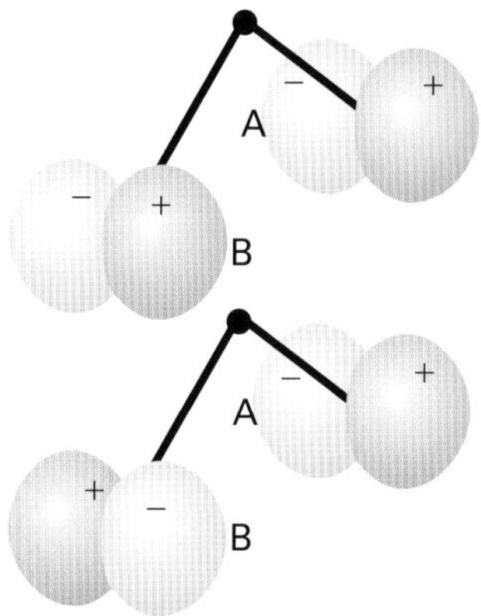

Fig. 12.17 Two symmetry-adapted linear combinations of the basis orbitals shown in Fig. 12.16. The two combinations each span a one-dimensional irreducible representation, and their symmetry species are different.

Table 12.2* The C_{2v} character table

C_{2v}, 2*mm*	E	C_2	σ_v	σ_v'	$h=4$	
A_1	1	1	1	1	z	z^2, y^2, x^2
A_2	1	1	−1	−1		xy
B_1	1	−1	1	−1	x	zx
B_2	1	−1	−1	1	y	yz

* More character tables are given at the end of the *Data section*.

Table 12.3* The C_{3v} character table

C_{3v}, $3m$	E	$2C_3$	$3\sigma_v$	$h=6$	
A_1	1	1	1	z	z^2, x^2+y^2
A_2	1	1	−1		
E	2	−1	0	(x, y)	$(xy, x^2-y^2), (yz, zx)$

* More character tables are given at the end of the *Data section.*

group. In C_{2v}, for instance, there are four classes (four columns in the character table), so there are only four species of irreducible representation. The character table in Table 12.2 therefore shows the characters of all the irreducible representations of this group.

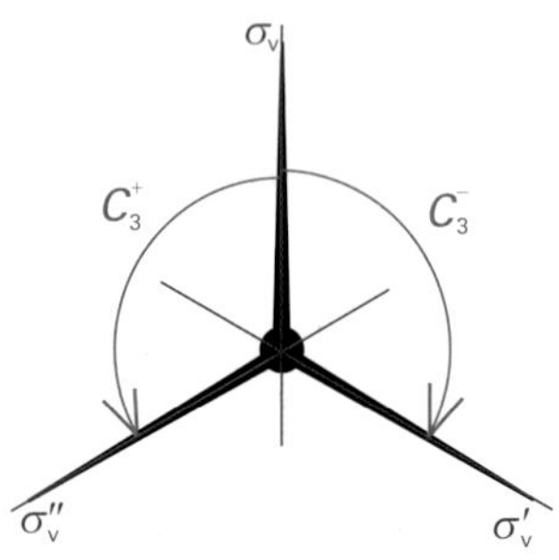

Fig. 12.18 Symmetry operations in the same class are related to one another by the symmetry operations of the group. Thus, the three mirror frames shown here are related by threefold rotations, and the two rotations shown here are related by reflection in σ_v.

(b) The structure of character tables

In general, the columns in a character table are labelled with the symmetry operations of the group. For instance, for the group C_{3v} the columns are headed E, C_3, and σ_v (Table 12.3). The numbers multiplying each operation are the numbers of members of each class. In the C_{3v} character table we see that the two threefold rotations (clockwise and counter-clockwise rotations by 120°) belong to the same class: they are related by a reflection (Fig. 12.18). The three reflections (one through each of the three vertical mirror planes) also lie in the same class: they are related by the threefold rotations. The two reflections of the group C_{2v} fall into different classes: although they are both reflections, one cannot be transformed into the other by any symmetry operation of the group.

The total number of operations in a group is called the **order**, h, of the group. The order of the group C_{3v}, for instance, is 6.

The rows under the labels for the operations summarize the symmetry properties of the orbitals. They are labelled with the **symmetry species** (the analogues of the labels σ and π). More formally, the symmetry species label the irreducible representations of the group, which are the basic types of behaviour that orbitals may show when subjected to the symmetry operations of the group, as we have illustrated for the group C_{2v}. By convention, irreducible representations are labelled with upper case Roman letters (such as A_1 and E) but the orbitals to which they apply are labelled with the lower case italic equivalents (so an orbital of symmetry species A_1 is called an a_1 orbital). Examples of each type of orbital are shown in Fig. 12.19.

Comment 12.5

Note that care must be taken to distinguish the identity element E (italic, a column heading) from the symmetry label E (roman, a row label).

(c) Character tables and orbital degeneracy

The character of the identity operation E tells us the degeneracy of the orbitals. Thus, in a C_{3v} molecule, any orbital with a symmetry label a_1 or a_2 is nondegenerate. Any doubly degenerate pair of orbitals in C_{3v} must be labelled e because, in this group, only E symmetry species have characters greater than 1.

Because there are no characters greater than 2 in the column headed E in C_{3v}, we know that there can be no triply degenerate orbitals in a C_{3v} molecule. This last point is a powerful result of group theory, for it means that, with a glance at the character table of a molecule, we can state the maximum possible degeneracy of its orbitals.

Example 12.2 *Using a character table to judge degeneracy*

Can a trigonal planar molecule such as BF_3 have triply degenerate orbitals? What is the minimum number of atoms from which a molecule can be built that does display triple degeneracy?

Method First, identify the point group, and then refer to the corresponding character table in the *Data section*. The maximum number in the column headed by the identity E is the maximum orbital degeneracy possible in a molecule of that point group. For the second part, consider the shapes that can be built from two, three, etc. atoms, and decide which number can be used to form a molecule that can have orbitals of symmetry species T.

Answer Trigonal planar molecules belong to the point group D_{3h}. Reference to the character table for this group shows that the maximum degeneracy is 2, as no character exceeds 2 in the column headed E. Therefore, the orbitals cannot be triply degenerate. A tetrahedral molecule (symmetry group T) has an irreducible representation with a T symmetry species. The minimum number of atoms needed to build such a molecule is four (as in P_4, for instance).

Self-test 12.2 A buckminsterfullerene molecule, C_{60} (**15**), belongs to the icosahedral point group. What is the maximum possible degree of degeneracy of its orbitals? [5]

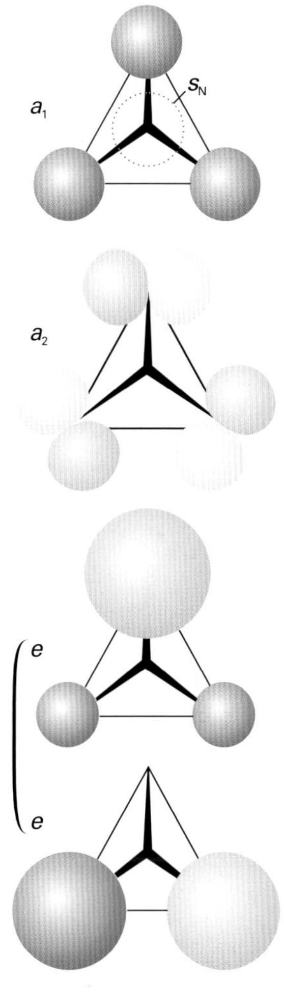

Fig. 12.19 Typical symmetry-adapted linear combinations of orbitals in a C_{3v} molecule.

(d) Characters and operations

The characters in the rows labelled A and B and in the columns headed by symmetry operations other than the identity E indicate the behaviour of an orbital under the corresponding operations: a +1 indicates that an orbital is unchanged, and a −1 indicates that it changes sign. It follows that we can identify the symmetry label of the orbital by comparing the changes that occur to an orbital under each operation, and then comparing the resulting +1 or −1 with the entries in a row of the character table for the point group concerned.

For the rows labelled E or T (which refer to the behaviour of sets of doubly and triply degenerate orbitals, respectively), the characters in a row of the table are the sums of the characters summarizing the behaviour of the individual orbitals in the basis. Thus, if one member of a doubly degenerate pair remains unchanged under a symmetry operation but the other changes sign (Fig. 12.20), then the entry is reported as $\chi = 1 - 1 = 0$. Care must be exercised with these characters because the transformations of orbitals can be quite complicated; nevertheless, the sums of the individual characters are integers.

As an example, consider the O2p_x orbital in H_2O. Because H_2O belongs to the point group C_{2v}, we know by referring to the C_{2v} character table (Table 12.2) that the labels available for the orbitals are a_1, a_2, b_1, and b_2. We can decide the appropriate label for O2p_x by noting that under a 180° rotation (C_2) the orbital changes sign (Fig. 12.21), so it must be either B_1 or B_2, as only these two symmetry types have character −1 under C_2. The O2p_x orbital also changes sign under the reflection σ_v', which identifies it as B_1. As we shall see, any molecular orbital built from this atomic orbital will also be a b_1 orbital. Similarly, O2p_y changes sign under C_2 but not under σ_v'; therefore, it can contribute to b_2 orbitals.

The behaviour of s, p, and d orbitals on a central atom under the symmetry operations of the molecule is so important that the symmetry species of these orbitals

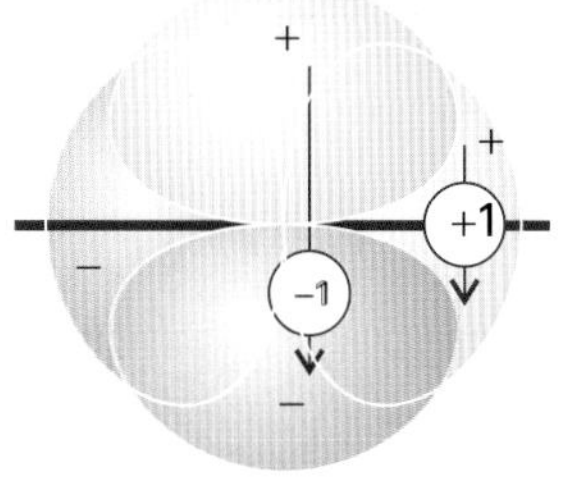

Fig. 12.20 The two orbitals shown here have different properties under reflection through the mirror plane: one changes sign (character −1), the other does not (character +1).

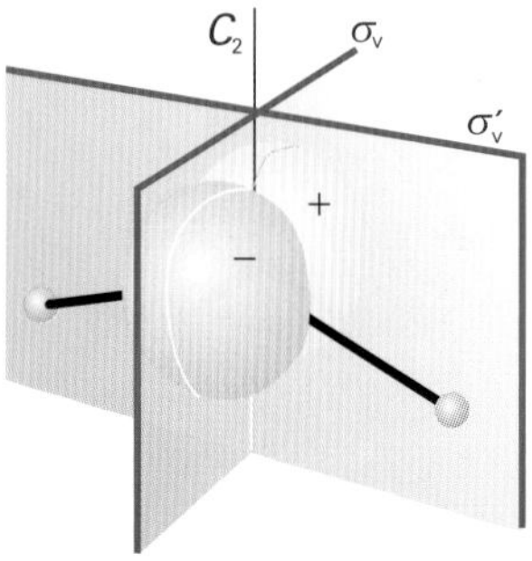

Fig. 12.21 A p_x orbital on the central atom of a C_{2v} molecule and the symmetry elements of the group.

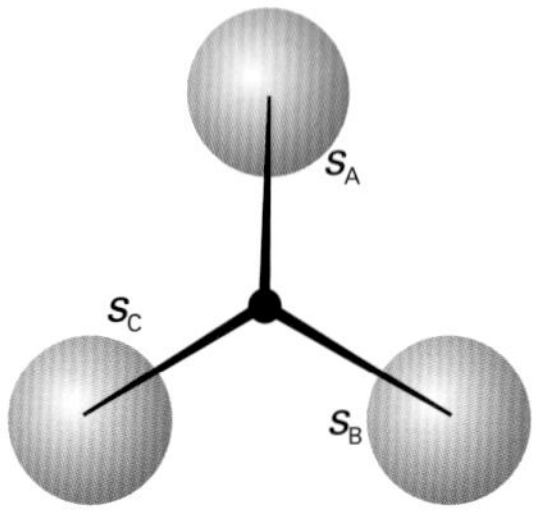

Fig. 12.22 The three H1*s* orbitals used to construct symmetry-adapted linear combinations in a C_{3v} molecule such as NH_3.

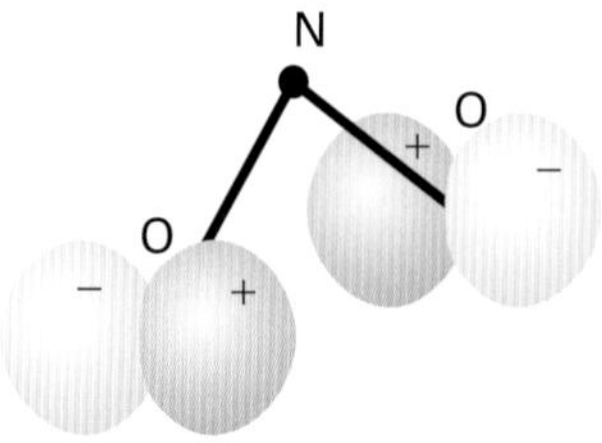

Fig. 12.23 One symmetry-adapted linear combination of O2p_x orbitals in the C_{2v} NO_2^- ion.

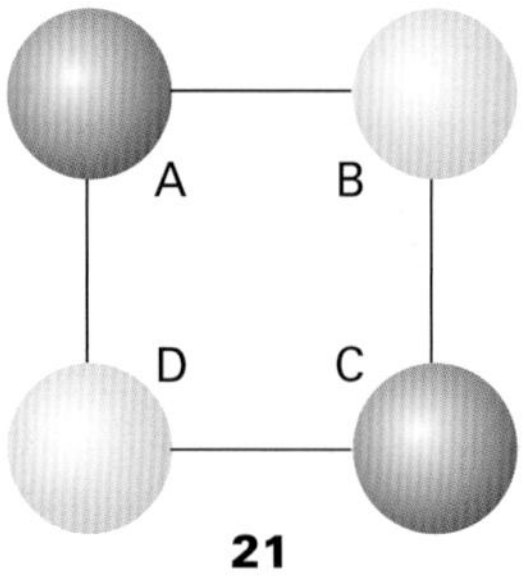

21

are generally indicated in a character table. To make these allocations, we look at the symmetry species of *x*, *y*, and *z*, which appear on the right-hand side of the character table. Thus, the position of *z* in Table 12.3 shows that p_z (which is proportional to $zf(r)$), has symmetry species A_1 in C_{3v}, whereas p_x and p_y (which are proportional to $xf(r)$ and $yf(r)$, respectively) are jointly of E symmetry. In technical terms, we say that p_x and p_y jointly **span** an irreducible representation of symmetry species E. An *s* orbital on the central atom always spans the fully symmetrical irreducible representation (typically labelled A_1 but sometimes A_1') of a group as it is unchanged under all symmetry operations.

The five *d* orbitals of a shell are represented by *xy* for d_{xy}, etc, and are also listed on the right of the character table. We can see at a glance that in C_{3v}, d_{xy} and $d_{x^2-y^2}$ on a central atom jointly belong to E and hence form a doubly degenerate pair.

(e) The classification of linear combinations of orbitals

So far, we have dealt with the symmetry classification of individual orbitals. The same technique may be applied to linear combinations of orbitals on atoms that are related by symmetry transformations of the molecule, such as the combination $\psi_1 = \psi_A + \psi_B + \psi_C$ of the three H1*s* orbitals in the C_{3v} molecule NH_3 (Fig. 12.22). This combination remains unchanged under a C_3 rotation and under any of the three vertical reflections of the group, so its characters are

$$\chi(E) = 1 \qquad \chi(C_3) = 1 \qquad \chi(\sigma_v) = 1$$

Comparison with the C_{3v} character table shows that ψ_1 is of symmetry species A_1, and therefore that it contributes to a_1 molecular orbitals in NH_3.

Example 12.3 *Identifying the symmetry species of orbitals*

Identify the symmetry species of the orbital $\psi = \psi_A - \psi_B$ in a C_{2v} NO_2 molecule, where ψ_A is an O2p_x orbital on one O atom and ψ_B that on the other O atom.

Method The negative sign in ψ indicates that the sign of ψ_B is opposite to that of ψ_A. We need to consider how the combination changes under each operation of the group, and then write the character as +1, −1, or 0 as specified above. Then we compare the resulting characters with each row in the character table for the point group, and hence identify the symmetry species.

Answer The combination is shown in Fig. 12.23. Under C_2, ψ changes into itself, implying a character of +1. Under the reflection σ_v, both orbitals change sign, so $\psi \to -\psi$, implying a character of −1. Under σ_v', $\psi \to -\psi$, so the character for this operation is also −1. The characters are therefore

$$\chi(E) = 1 \qquad \chi(C_2) = 1 \qquad \chi(\sigma_v) = -1 \qquad \chi(\sigma_v') = -1$$

These values match the characters of the A_2 symmetry species, so ψ can contribute to an a_2 orbital.

Self-test 12.3 Consider $PtCl_4^-$, in which the Cl ligands form a square planar array of point group D_{4h} (**21**). Identify the symmetry type of the combination $\psi_A - \psi_B + \psi_C - \psi_D$. [$B_{2g}$]

12.5 Vanishing integrals and orbital overlap

Suppose we had to evaluate the integral

$$I = \int f_1 f_2 \, d\tau \qquad (12.7)$$

where f_1 and f_2 are functions. For example, f_1 might be an atomic orbital A on one atom and f_2 an atomic orbital B on another atom, in which case I would be their overlap integral. If we knew that the integral is zero, we could say at once that a molecular orbital does not result from (A,B) overlap in that molecule. We shall now see that character tables provide a quick way of judging whether an integral is necessarily zero.

(a) The criteria for vanishing integrals

The key point in dealing with the integral I is that the value of any integral, and of an overlap integral in particular, is independent of the orientation of the molecule (Fig. 12.24). In group theory we express this point by saying that *I is invariant under any symmetry operation of the molecule*, and that each operation brings about the trivial transformation $I \rightarrow I$. Because the volume element $d\tau$ is invariant under any symmetry operation, it follows that the integral is nonzero only if the integrand itself, the product $f_1 f_2$, is unchanged by any symmetry operation of the molecular point group. If the integrand changed sign under a symmetry operation, the integral would be the sum of equal and opposite contributions, and hence would be zero. It follows that the only contribution to a nonzero integral comes from functions for which under any symmetry operation of the molecular point group $f_1 f_2 \rightarrow f_1 f_2$, and hence for which the characters of the operations are all equal to +1. Therefore, for I not to be zero, *the integrand $f_1 f_2$ must have symmetry species A_1* (or its equivalent in the specific molecular point group).

We use the following procedure to deduce the symmetry species spanned by the product $f_1 f_2$ and hence to see whether it does indeed span A_1.

1 Decide on the symmetry species of the individual functions f_1 and f_2 by reference to the character table, and write their characters in two rows in the same order as in the table.

2 Multiply the numbers in each column, writing the results in the same order.

3 Inspect the row so produced, and see if it can be expressed as a sum of characters from each column of the group. The integral must be zero if this sum does not contain A_1.

For example, if f_1 is the s_N orbital in NH_3 and f_2 is the linear combination $s_3 = s_B - s_C$ (Fig. 12.25), then, because s_N spans A_1 and s_3 is a member of the basis spanning E, we write

f_1:	1	1	1
f_2:	2	−1	0
$f_1 f_2$:	2	−1	0

The characters 2, −1, 0 are those of E alone, so the integrand does not span A_1. It follows that the integral must be zero. Inspection of the form of the functions (see Fig. 12.25) shows why this is so: s_3 has a node running through s_N. Had we taken $f_1 = s_N$ and $f_2 = s_1$ instead, where $s_1 = s_A + s_B + s_C$, then because each spans A_1 with characters 1,1,1:

f_1:	1	1	1
f_2:	1	1	1
$f_1 f_2$:	1	1	1

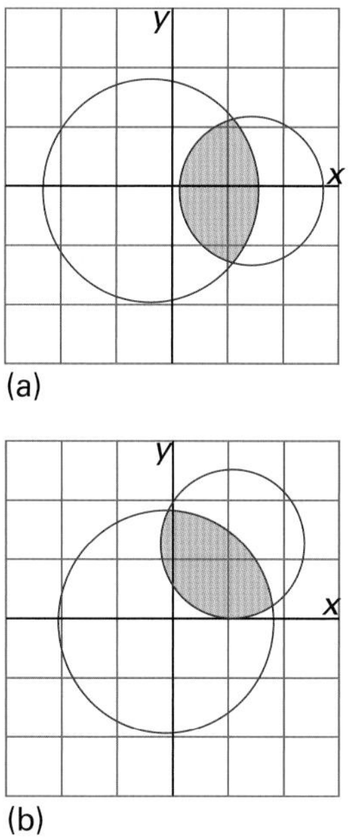

Fig. 12.24 The value of an integral I (for example, an area) is independent of the coordinate system used to evaluate it. That is, I is a basis of a representation of symmetry species A_1 (or its equivalent).

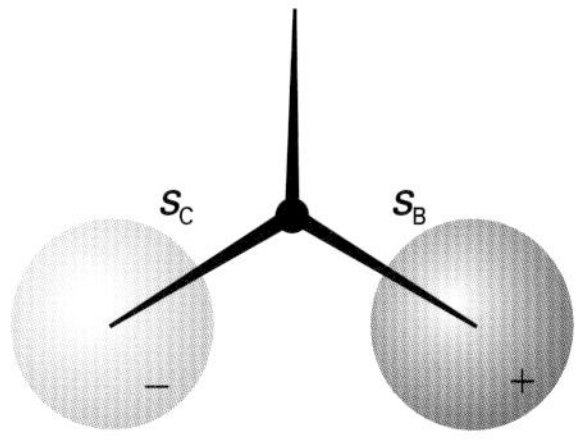

Fig. 12.25 A symmetry-adapted linear combination that belongs to the symmetry species E in a C_{3v} molecule such as NH_3. This combination can form a molecular orbital by overlapping with the p_x orbital on the central atom (the orbital with its axis parallel to the width of the page; see Fig. 12.28c).

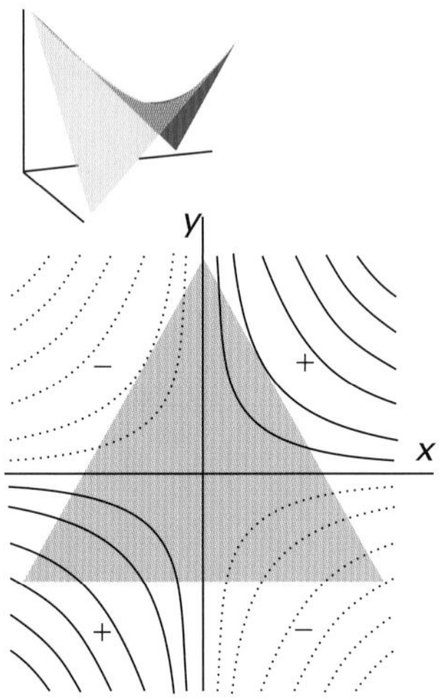

Fig. 12.26 The integral of the function $f = xy$ over the tinted region is zero. In this case, the result is obvious by inspection, but group theory can be used to establish similar results in less obvious cases. The insert shows the shape of the function in three dimensions.

The characters of the product are those of A_1 itself. Therefore, s_1 and s_N may have nonzero overlap. A shortcut that works when f_1 and f_2 are bases for irreducible representations of a group is to note their symmetry species: if they are different, then the integral of their product must vanish; if they are the same, then the integral may be nonzero.

It is important to note that group theory is specific about when an integral must be zero, but integrals that it allows to be nonzero may be zero for reasons unrelated to symmetry. For example, the N—H distance in ammonia may be so great that the (s_1, s_N) overlap integral is zero simply because the orbitals are so far apart.

Example 12.4 *Deciding if an integral must be zero (1)*

May the integral of the function $f = xy$ be nonzero when evaluated over a region the shape of an equilateral triangle centred on the origin (Fig. 12.26)?

Method First, note that an integral over a single function f is included in the previous discussion if we take $f_1 = f$ and $f_2 = 1$ in eqn 12.7. Therefore, we need to judge whether f alone belongs to the symmetry species A_1 (or its equivalent) in the point group of the system. To decide that, we identify the point group and then examine the character table to see whether f belongs to A_1 (or its equivalent).

Answer An equilateral triangle has the point-group symmetry D_{3h}. If we refer to the character table of the group, we see that xy is a member of a basis that spans the irreducible representation E′. Therefore, its integral must be zero, because the integrand has no component that spans A_1'.

Self-test 12.4 Can the function $x^2 + y^2$ have a nonzero integral when integrated over a regular pentagon centred on the origin? [Yes, Fig. 12.27]

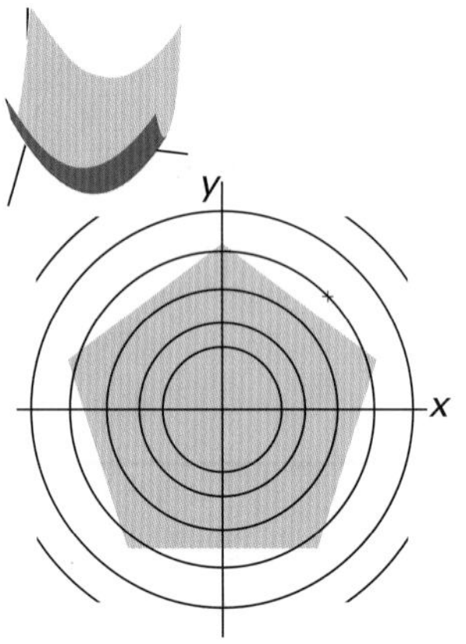

Fig. 12.27 The integration of a function over a pentagonal region. The insert shows the shape of the function in three dimensions.

In many cases, the product of functions f_1 and f_2 spans a sum of irreducible representations. For instance, in C_{2v} we may find the characters 2, 0, 0, −2 when we multiply the characters of f_1 and f_2 together. In this case, we note that these characters are the sum of the characters for A_2 and B_1:

	E	C_2	σ_v	σ_v'
A_2	1	1	−1	−1
B_1	1	−1	1	−1
$A_2 + B_1$	2	0	0	−2

To summarize this result we write the symbolic expression $A_2 \times B_1 = A_2 + B_1$, which is called the **decomposition of a direct product**. This expression is symbolic. The × and + signs in this expression are not ordinary multiplication and addition signs: formally, they denote technical procedures with matrices called a 'direct product' and a 'direct sum'. Because the sum on the right does not include a component that is a basis for an irreducible representation of symmetry species A_1, we can conclude that the integral of $f_1 f_2$ over all space is zero in a C_{2v} molecule.

Whereas the decomposition of the characters 2, 0, 0, −2 can be done by inspection in this simple case, in other cases and more complex groups the decomposition is often far from obvious. For example, if we found the characters 8, −2, −6, 4, it would not be obvious that the sum contains A_1. Group theory, however, provides a systematic way of using the characters of the representation spanned by a product to find the symmetry species of the irreducible representations. The recipe is as follows:

1 Write down a table with columns headed by the symmetry operations of the group.

2 In the first row write down the characters of the symmetry species we want to analyse.

3 In the second row, write down the characters of the irreducible representation Γ we are interested in.

4 Multiply the two rows together, add the products together, and divide by the order of the group.

The resulting number is the number of times Γ occurs in the decomposition.

Illustration 12.1 *To find whether A_1 occurs in a direct product*

To find whether A_1 does indeed occur in the product with characters 8, −2, −6, 4 in C_{2v}, we draw up the following table:

	E	C_2	σ_v	σ_v'	
					$h=4$ (the order of the group)
f_1f_2	8	−2	−6	4	(the characters of the product)
A_1	1	1	1	1	(the symmetry species we are interested in)
	8	−2	−6	4	(the product of the two sets of characters)

The sum of the numbers in the last line is 4; when that number is divided by the order of the group, we get 1, so A_1 occurs once in the decomposition. When the procedure is repeated for all four symmetry species, we find that f_1f_2 spans $A_1 + 2A_2 + 5B_2$.

Self-test 12.5 Does A_2 occur among the symmetry species of the irreducible representations spanned by a product with characters 7, −3, −1, 5 in the group C_{2v}?

[No]

(b) Orbitals with nonzero overlap

The rules just given let us decide which atomic orbitals may have nonzero overlap in a molecule. We have seen that s_N may have nonzero overlap with s_1 (the combination $s_A + s_B + s_C$), so bonding and antibonding molecular orbitals can form from (s_N, s_1) overlap (Fig. 12.28). The general rule is that *only orbitals of the same symmetry species may have nonzero overlap*, so only orbitals of the same symmetry species form bonding and antibonding combinations. It should be recalled from Chapter 11 that the selection of atomic orbitals that had mutual nonzero overlap is the central and initial step in the construction of molecular orbitals by the LCAO procedure. We are therefore at the point of contact between group theory and the material introduced in that chapter. The molecular orbitals formed from a particular set of atomic orbitals with nonzero overlap are labelled with the lowercase letter corresponding to the symmetry species. Thus, the (s_N, s_1)-overlap orbitals are called a_1 orbitals (or a_1^*, if we wish to emphasize that they are antibonding).

The linear combinations $s_2 = 2s_a - s_b - s_c$ and $s_3 = s_b - s_c$ have symmetry species E. Does the N atom have orbitals that have nonzero overlap with them (and give rise to e molecular orbitals)? Intuition (as supported by Figs. 12.28b and c) suggests that N2p_x and N2p_y should be suitable. We can confirm this conclusion by noting that the character table shows that, in C_{3v}, the functions x and y jointly belong to the symmetry species E. Therefore, N2p_x and N2p_y also belong to E, so may have nonzero overlap with s_2 and s_3. This conclusion can be verified by multiplying the characters

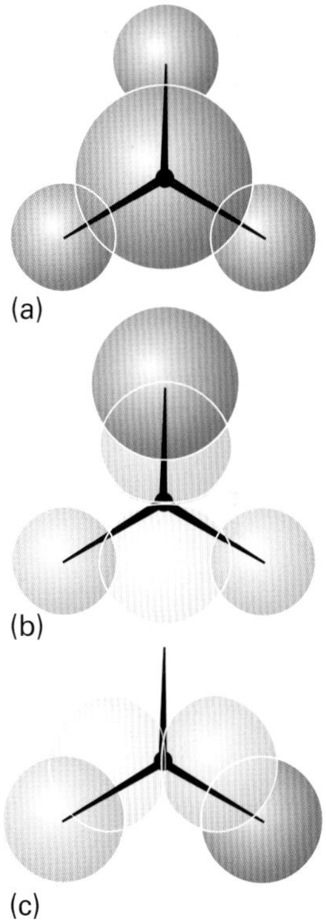

Fig. 12.28 Orbitals of the same symmetry species may have non-vanishing overlap. This diagram illustrates the three bonding orbitals that may be constructed from (N2s, H1s) and (N2p, H1s) overlap in a C_{3v} molecule. (a) a_1; (b) and (c) the two components of the doubly degenerate e orbitals. (There are also three antibonding orbitals of the same species.)

and finding that the product of characters can be expressed as the decomposition $E \times E = A_1 + A_2 + E$. The two *e* orbitals that result are shown in Fig. 12.28 (there are also two antibonding *e* orbitals).

We can see the power of the method by exploring whether any *d* orbitals on the central atom can take part in bonding. As explained earlier, reference to the C_{3v} character table shows that d_{z^2} has A_1 symmetry and that the pairs $(d_{x^2-y^2}, d_{xy})$ and (d_{yz}, d_{zx}) each transform as E. It follows that molecular orbitals may be formed by (s_1, d_{z^2}) overlap and by overlap of the s_2, s_3 combinations with the E *d* orbitals. Whether or not the *d* orbitals are in fact important is a question group theory cannot answer because the extent of their involvement depends on energy considerations, not symmetry.

Example 12.5 *Determining which orbitals can contribute to bonding*

The four H1*s* orbitals of methane span $A_1 + T_2$. With which of the C atom orbitals can they overlap? What bonding pattern would be possible if the C atom had *d* orbitals available?

Method Refer to the T_d character table (in the *Data section*) and look for *s*, *p*, and *d* orbitals spanning A_1 or T_2.

Answer An *s* orbital spans A_1, so it may have nonzero overlap with the A_1 combination of H1*s* orbitals. The C2*p* orbitals span T_2, so they may have nonzero overlap with the T_2 combination. The d_{xy}, d_{yz}, and d_{zx} orbitals span T_2, so they may overlap the same combination. Neither of the other two *d* orbitals span A_1 (they span E), so they remain nonbonding orbitals. It follows that in methane there are (C2*s*,H1*s*)-overlap a_1 orbitals and (C2*p*,H1*s*)-overlap t_2 orbitals. The C3*d* orbitals might contribute to the latter. The lowest energy configuration is probably $a_1^2 t_2^6$, with all bonding orbitals occupied.

Self-test 12.6 Consider the octahedral SF_6 molecule, with the bonding arising from overlap of S orbitals and a 2*p* orbital on each F directed towards the central S atom. The latter span $A_{1g} + E_g + T_{1u}$. What *s* orbitals have nonzero overlap? Suggest what the ground-state configuration is likely to be.

[3*s*(A_{1g}), 3*p*(T_{1u}), 3*d*(E_g); $a_{1g}^2 t_{1u}^6 e_g^4$]

(c) Symmetry-adapted linear combinations

So far, we have only asserted the forms of the linear combinations (such as s_1, etc.) that have a particular symmetry. Group theory also provides machinery that takes an arbitrary **basis**, or set of atomic orbitals (s_A, etc.), as input and generates combinations of the specified symmetry. Because these combinations are adapted to the symmetry of the molecule, they are called **symmetry-adapted linear combinations** (SALC). Symmetry-adapted linear combinations are the building blocks of LCAO molecular orbitals, for they include combinations such as those used to construct molecular orbitals in benzene. The construction of SALCs is the first step in any molecular orbital treatment of molecules.

The technique for building SALCs is derived by using the full power of group theory. We shall not show the derivation (see *Further reading*), which is very lengthy, but present the main conclusions as a set of rules:

1 Construct a table showing the effect of each operation on each orbital of the original basis.

2 To generate the combination of a specified symmetry species, take each column in turn and:

(i) Multiply each member of the column by the character of the corresponding operation.

(ii) Add together all the orbitals in each column with the factors as determined in (i).

(iii) Divide the sum by the order of the group.

	Original basis			
	s_N	s_A	s_B	s_C
Under E	s_N	s_A	s_B	s_C
C_3^+	s_N	s_B	s_C	s_A
C_3^-	s_N	s_C	s_A	s_B
σ_v	s_N	s_A	s_C	s_B
σ_v'	s_N	s_B	s_A	s_C
σ_v''	s_N	s_C	s_B	s_A

For example, from the (s_N,s_A,s_B,s_C) basis in NH_3 we form the table shown in the margin. To generate the A_1 combination, we take the characters for A_1 (1,1,1,1,1,1); then rules (i) and (ii) lead to

$$\psi \propto s_N + s_N + \cdots = 6s_N$$

The order of the group (the number of elements) is 6, so the combination of A_1 symmetry that can be generated from s_N is s_N itself. Applying the same technique to the column under s_A gives

$$\psi = \tfrac{1}{6}(s_A + s_B + s_C + s_A + s_B + s_C) = \tfrac{1}{3}(s_A + s_B + s_C)$$

The same combination is built from the other two columns, so they give no further information. The combination we have just formed is the s_1 combination we used before (apart from the numerical factor).

We now form the overall molecular orbital by forming a linear combination of all the SALCs of the specified symmetry species. In this case, therefore, the a_1 molecular orbital is

$$\psi = c_N s_N + c_1 s_1$$

This is as far as group theory can take us. The coefficients are found by solving the Schrödinger equation; they do not come directly from the symmetry of the system.

We run into a problem when we try to generate an SALC of symmetry species E, because, for representations of dimension 2 or more, the rules generate sums of SALCs. This problem can be illustrated as follows. In C_{3v}, the E characters are 2, −1, −1, 0, 0, 0, so the column under s_N gives

$$\psi = \tfrac{1}{6}(2s_N - s_N - s_N + 0 + 0 + 0) = 0$$

The other columns give

$$\tfrac{1}{6}(2s_A - s_B - s_C) \qquad \tfrac{1}{6}(2s_B - s_A - s_C) \qquad \tfrac{1}{6}(2s_C - s_B - s_A)$$

However, any one of these three expressions can be expressed as a sum of the other two (they are not 'linearly independent'). The difference of the second and third gives $\tfrac{1}{2}(s_B - s_C)$, and this combination and the first, $\tfrac{1}{6}(2s_A - s_B - s_C)$ are the two (now linearly independent) SALCs we have used in the discussion of e orbitals.

12.6 **Vanishing integrals and selection rules**

Integrals of the form

$$I = \int f_1 f_2 f_3 \, d\tau \tag{12.8}$$

are also common in quantum mechanics for they include matrix elements of operators (Section 8.5d), and it is important to know when they are necessarily zero. For the integral to be nonzero, *the product $f_1 f_2 f_3$ must span A_1 (or its equivalent) or contain a component that spans A_1*. To test whether this is so, the characters of all three functions are multiplied together in the same way as in the rules set out above.

Example 12.6 *Deciding if an integral must be zero (2)*

Does the integral $\int(3d_{z^2})x(3d_{xy})\mathrm{d}\tau$ vanish in a C_{2v} molecule?

Method We must refer to the C_{2v} character table (Table 12.2) and the characters of the irreducible representations spanned by $3z^2 - r^2$ (the form of the d_{z^2} orbital), x, and xy; then we can use the procedure set out above (with one more row of multiplication).

Answer We draw up the following table:

	E	C_2	σ_v	σ'_v	
$f_3 = d_{xy}$	1	1	−1	−1	A_2
$f_2 = x$	1	−1	1	−1	B_1
$f_1 = d_{z^2}$	1	1	1	1	A_1
$f_1f_2f_3$	1	−1	−1	1	

The characters are those of B_2. Therefore, the integral is necessarily zero.

Self-test 12.7 Does the integral $\int(2p_x)(2p_y)(2p_z)\mathrm{d}\tau$ necessarily vanish in an octahedral environment? [No]

We saw in Chapters 9 and 10, and will see in more detail in Chapters 13 and 14, that the intensity of a spectral line arising from a molecular transition between some initial state with wavefunction ψ_i and a final state with wavefunction ψ_f depends on the (electric) transition dipole moment, $\boldsymbol{\mu}_{fi}$. The z-component of this vector is defined through

$$\mu_{z,fi} = -e\int \psi_f^* z \psi_i \,\mathrm{d}\tau \qquad [12.9]$$

where $-e$ is the charge of the electron. The transition moment has the form of the integral in eqn 12.8, so, once we know the symmetry species of the states, we can use group theory to formulate the selection rules for the transitions.

As an example, we investigate whether an electron in an a_1 orbital in H_2O (which belongs to the group C_{2v}) can make an electric dipole transition to a b_1 orbital (Fig. 12.29). We must examine all three components of the transition dipole moment, and take f_2 in eqn 12.8 as x, y, and z in turn. Reference to the C_{2v} character table shows that these components transform as B_1, B_2, and A_1, respectively. The three calculations run as follows:

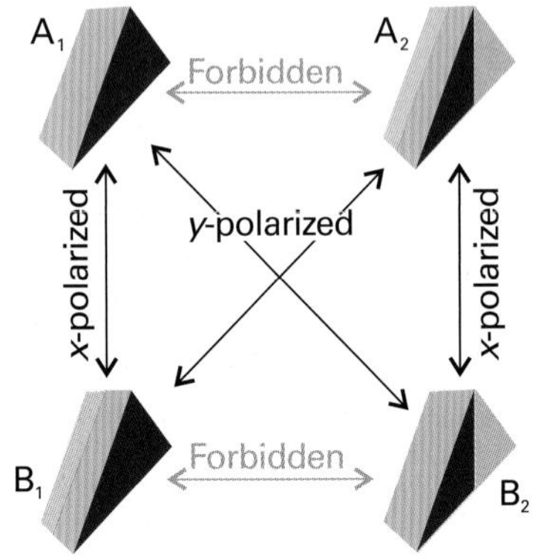

Fig. 12.29 The polarizations of the allowed transitions in a C_{2v} molecule. The shading indicates the structure of the orbitals of the specified symmetry species. The perspective view of the molecule makes it look rather like a door stop; however, from the side, each 'door stop' is in fact an isosceles triangle.

	x-component				*y*-component				*z*-component			
	E	C_2	σ_v	σ'_v	E	C_2	σ_v	σ'_v	E	C_2	σ_v	σ'_v
f_3	1	−1	1	−1	1	−1	1	−1	1	−1	1	−1B_1
f_2	1	−1	1	−1	1	−1	−1	1	1	1	1	1
f_1	1	1	1	1	1	1	1	1	1	1	1	1A_1
$f_1f_2f_3$	1	1	1	1	1	1	−1	−1	1	−1	1	−1

Only the first product (with $f_2 = x$) spans A_1, so only the x-component of the transition dipole moment may be nonzero. Therefore, we conclude that the electric dipole transitions between a_1 and b_1 are allowed. We can go on to state that the radiation emitted (or absorbed) is x-polarized and has its electric field vector in the x-direction, because that form of radiation couples with the x–component of a transition dipole.

Example 12.7 *Deducing a selection rule*

Is $p_x \rightarrow p_y$ an allowed transition in a tetrahedral environment?

Method We must decide whether the product $p_y q p_x$, with $q = x, y$, or z, spans A_1 by using the T_d character table.

Answer The procedure works out as follows:

	E	$8C_3$	$3C_2$	$6\sigma_d$	$6S_4$	
$f_3(p_y)$	3	0	−1	−1	1	T_2
$f_2(q)$	3	0	−1	−1	1	T_2
$f_1(p_x)$	3	0	−1	−1	1	T_2
$f_1 f_2 f_3$	27	0	−1	−1	1	

We can use the decomposition procedure described in Section 12.5a to deduce that A_1 occurs (once) in this set of characters, so $p_x \rightarrow p_y$ is allowed.

A more detailed analysis (using the matrix representatives rather than the characters) shows that only $q = z$ gives an integral that may be nonzero, so the transition is z-polarized. That is, the electromagnetic radiation involved in the transition has its electric vector aligned in the z-direction.

Self-test 12.8 What are the allowed transitions, and their polarizations, of a b_1 electron in a C_{4v} molecule? $[b_1 \rightarrow b_1(z); b_1 \rightarrow e(x,y)]$

The following chapters will show many more examples of the systematic use of symmetry. We shall see that the techniques of group theory greatly simplify the analysis of molecular structure and spectra.

Checklist of key ideas

- ☐ 1. A symmetry operation is an action that leaves an object looking the same after it has been carried out.
- ☐ 2. A symmetry element is a point, line, or plane with respect to which a symmetry operation is performed.
- ☐ 3. A point group is a group of symmetry operations that leaves at least one common point unchanged. A space group, a group of symmetry operations that includes translation through space.
- ☐ 4. The notation for point groups commonly used for molecules and solids is summarized in Table 12.1.
- ☐ 5. To be polar, a molecule must belong to C_n, C_{nv}, or C_s (and have no higher symmetry).
- ☐ 6. A molecule may be chiral only if it does not possess an axis of improper rotation, S_n.
- ☐ 7. A representative $\boldsymbol{D}(X)$ is a matrix that brings about the transformation of the basis under the operation X. The basis is the set of functions on which the representative acts.
- ☐ 8. A character, χ, is the sum of the diagonal elements of a matrix representative.
- ☐ 9. A character table characterizes the different symmetry types possible in the point group.
- ☐ 10. In a reduced representation all the matrices have block-diagonal form. An irreducible representation cannot be reduced further.
- ☐ 11. Symmetry species are the labels for the irreducible representations of a group.
- ☐ 12. Decomposition of the direct product is the reduction of a product of symmetry species to a sum of symmetry species, $\Gamma \times \Gamma' = \Gamma^{(1)} + \Gamma^{(2)} + \cdots$

☐ 13. For an integral $\int f_1 f_2\,d\tau$ to be nonzero, the integrand $f_1 f_2$ must have the symmetry species A_1 (or its equivalent in the specific molecular point group).

☐ 14. A symmetry-adapted linear combination (SALC) is a combination of atomic orbitals adapted to the symmetry of the molecule and used as the building blocks for LCAO molecular orbitals.

☐ 15. Allowed and forbidden spectroscopic transitions can be identified by considering the symmetry criteria for the non-vanishing of the transition moment between the initial and final states.

Further reading

Articles and texts

P.W. Atkins and R.S. Friedman, *Molecular quantum mechanics.* Oxford University Press (2005).

F.A. Cotton, *Chemical applications of group theory.* Wiley, New York (1990).

R. Drago, *Physical methods for chemists.* Saunders, Philadelphia (1992).

D.C. Harris and M.D. Bertollucci, *Symmetry and spectroscopy: an introduction to vibrational and electronic spectroscopy.* Dover, New York (1989).

S.F.A. Kettle, *Symmetry and structure: readable group theory for chemists.* Wiley, New York (1995).

Sources of data and information

G.L. Breneman, Crystallographic symmetry point group notation flow chart. *J. Chem. Educ.* **64**, 216 (1987).

P.W. Atkins, M.S. Child, and C.S.G. Phillips, *Tables for group theory.* Oxford University Press (1970).

Discussion questions

12.1 Explain how a molecule is assigned to a point group.

12.2 List the symmetry operations and the corresponding symmetry elements of the point groups.

12.3 Explain the symmetry criteria that allow a molecule to be polar.

12.4 Explain the symmetry criteria that allow a molecule to be optically active.

12.5 Explain what is meant by (a) a representative and (b) a representation in the context of group theory.

12.6 Explain the construction and content of a character table.

12.7 Explain how spectroscopic selection rules arise and how they are formulated by using group theory.

12.8 Outline how a direct product is expressed as a direct sum and how to decide whether the totally symmetric irreducible representation is present in the direct product.

Exercises

12.1a The CH_3Cl molecule belongs to the point group C_{3v}. List the symmetry elements of the group and locate them in the molecule.

12.1b The CCl_4 molecule belongs to the point group T_d. List the symmetry elements of the group and locate them in the molecule.

12.2a Which of the following molecules may be polar? (a) pyridine (C_{2v}), (b) nitroethane (C_s), (c) gas-phase $HgBr_2$ ($D_{\infty h}$), (d) $B_3N_3N_6$ (D_{3h}).

12.2b Which of the following molecules may be polar? (a) CH_3Cl (C_{3v}), (b) $HW_2(CO)_{10}$ (D_{4h}), (c) $SnCl_4$ (T_d).

12.3a Use symmetry properties to determine whether or not the integral $\int p_x z p_z\,d\tau$ is necessarily zero in a molecule with symmetry C_{4v}.

12.3b Use symmetry properties to determine whether or not the integral $\int p_x z p_z\,d\tau$ is necessarily zero in a molecule with symmetry D_{6h}.

12.4a Show that the transition $A_1 \rightarrow A_2$ is forbidden for electric dipole transitions in a C_{3v} molecule.

12.4b Is the transition $A_{1g} \rightarrow E_{2u}$ forbidden for electric dipole transitions in a D_{6h} molecule?

12.5a Show that the function xy has symmetry species B_2 in the group C_{4v}.

12.5b Show that the function xyz has symmetry species A_1 in the group D_2.

12.6a Molecules belonging to the point groups D_{2h} or C_{3h} cannot be chiral. Which elements of these groups rule out chirality?

12.6b Molecules belonging to the point groups T_h or T_d cannot be chiral. Which elements of these groups rule out chirality?

12.7a The group D_2 consists of the elements E, C_2, C_2', and C_2'', where the three twofold rotations are around mutually perpendicular axes. Construct the group multiplication table.

12.7b The group C_{4v} consists of the elements E, $2C_4$, C_2, and $2\sigma_v$, $2\sigma_d$. Construct the group multiplication table.

12.8a Identify the point groups to which the following objects belong: (a) a sphere, (b) an isosceles triangle, (c) an equilateral triangle, (d) an unsharpened cylindrical pencil.

12.8b Identify the point groups to which the following objects belong: (a) a sharpened cylindrical pencil, (b) a three-bladed propellor, (c) a four-legged table, (d) yourself (approximately).

12.9a List the symmetry elements of the following molecules and name the point groups to which they belong: (a) NO_2, (b) N_2O, (c) $CHCl_3$, (d) $CH_2{=}CH_2$, (e) *cis*-CHBr=CHBr, (f) *trans*-CHCl=CHCl.

12.9b List the symmetry elements of the following molecules and name the point groups to which they belong: (a) naphthalene, (b) anthracene, (c) the three dichlorobenzenes.

12.10a Assign (a) *cis*-dichloroethene and (b) *trans*-dichloroethene to point groups.

12.10b Assign the following molecules to point groups: (a) HF, (b) IF_7 (pentagonal bipyramid), (c) XeO_2F_2 (see-saw), (d) $Fe_2(CO)_9$ (**22**), (e) cubane, C_8H_8, (f) tetrafluorocubane, $C_8H_4F_4$ (**23**).

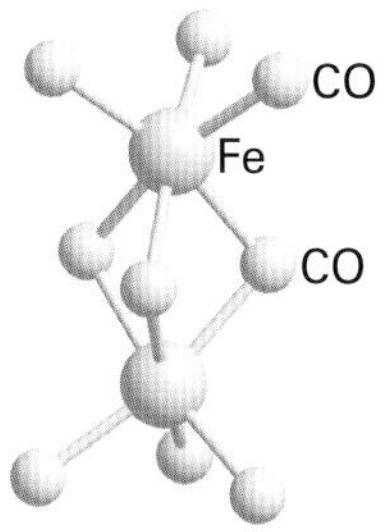

22

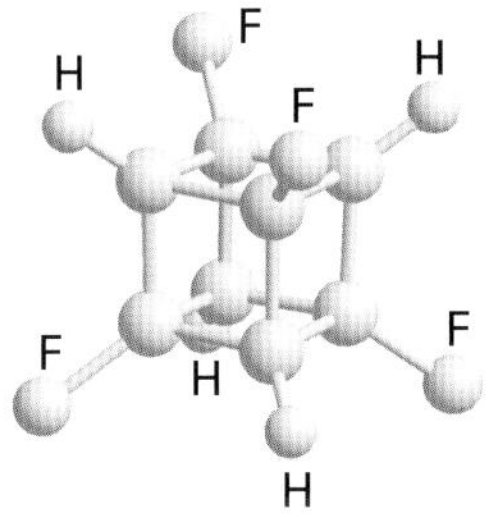

23

12.11a Which of the molecules in Exercises 12.9a and 12.10a can be (a) polar, (b) chiral?

12.11b Which of the molecules in Exercises 12.9b and 12.10b can be (a) polar, (b) chiral?

12.12a Consider the C_{2v} molecule NO_2. The combination $p_x(A) - p_x(B)$ of the two O atoms (with x perpendicular to the plane) spans A_2. Is there any orbital of the central N atom that can have a nonzero overlap with that combination of O orbitals? What would be the case in SO_2, where $3d$ orbitals might be available?

12.12b Consider the C_{3v} ion NO_3^-. Is there any orbital of the central N atom that can have a nonzero overlap with the combination $2p_z(A) - p_z(B) - p_z(C)$ of the three O atoms (with z perpendicular to the plane). What would be the case in SO_3, where $3d$ orbitals might be available?

12.13a The ground state of NO_2 is A_1 in the group C_{2v}. To what excited states may it be excited by electric dipole transitions, and what polarization of light is it necessary to use?

12.13b The ClO_2 molecule (which belongs to the group C_{2v}) was trapped in a solid. Its ground state is known to be B_1. Light polarized parallel to the y-axis (parallel to the OO separation) excited the molecule to an upper state. What is the symmetry of that state?

12.14a What states of (a) benzene, (b) naphthalene may be reached by electric dipole transitions from their (totally symmetrical) ground states?

12.14b What states of (a) anthracene, (b) coronene (**24**) may be reached by electric dipole transitions from their (totally symmetrical) ground states?

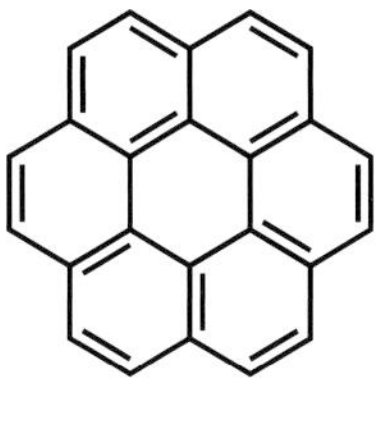

24 Coronene

12.15a Write $f_1 = \sin\theta$ and $f_2 = \cos\theta$, and show by symmetry arguments using the group C_s that the integral of their product over a symmetrical range around $\theta = 0$ is zero.

12.15b Determine whether the integral over f_1 and f_2 in Exercise 12.15a is zero over a symmetrical range about $\theta = 0$ in the group C_{3v}.

Problems*

12.1 List the symmetry elements of the following molecules and name the point groups to which they belong: (a) staggered CH_3CH_3, (b) chair and boat cyclohexane, (c) B_2H_6, (d) $[Co(en)_3]^{3+}$, where en is ethylenediamine (ignore its detailed structure), (e) crown-shaped S_8. Which of these molecules can be (i) polar, (ii) chiral?

12.2 The group C_{2h} consists of the elements E, C_2, σ_h, i. Construct the group multiplication table and find an example of a molecule that belongs to the group.

12.3 The group D_{2h} has a C_2 axis perpendicular to the principal axis and a horizontal mirror plane. Show that the group must therefore have a centre of inversion.

12.4 Consider the H_2O molecule, which belongs to the group C_{2v}. Take as a basis the two H1s orbitals and the four valence orbital of the O atom and set up the 6×6 matrices that represent the group in this basis. Confirm by explicit matrix multiplication that the group multiplications (a) $C_2\sigma_v = \sigma_v'$ and (b) $\sigma_v\sigma_v' = C_2$. Confirm, by calculating the traces of the matrices: (a) that symmetry elements in the same class have the same character, (b) that the representation is reducible, and (c) that the basis spans $3A_1 + B_1 + 2B_2$.

12.5 Confirm that the z-component of orbital angular momentum is a basis for an irreducible representation of A_2 symmetry in C_{3v}.

12.6 The (one-dimensional) matrices $\boldsymbol{D}(C_3) = \mathbf{1}$ and $\boldsymbol{D}(C_2) = \mathbf{1}$, and $\boldsymbol{D}(C_3) = \mathbf{1}$ and $\boldsymbol{D}(C_2) = \mathbf{-1}$ both represent the group multiplication $C_3C_2 = C_6$ in the group C_{6v} with $\boldsymbol{D}(C_6) = +1$ and -1, respectively. Use the character table to confirm these remarks. What are the representatives of σ_v and σ_d in each case?

12.7 Construct the multiplication table of the Pauli spin matrices, $\boldsymbol{\sigma}$, and the 2×2 unit matrix:

* Problems denoted with the symbol ‡ were supplied by Charles Trapp and Carmen Giunta.

$$\sigma_x = \begin{pmatrix} 0 & 1 \\ 1 & 0 \end{pmatrix} \qquad \sigma_y = \begin{pmatrix} 0 & -\mathrm{i} \\ \mathrm{i} & 0 \end{pmatrix} \qquad \sigma_z = \begin{pmatrix} 1 & 0 \\ 0 & -1 \end{pmatrix} \qquad \sigma_0 = \begin{pmatrix} 1 & 0 \\ 0 & 1 \end{pmatrix}$$

Do the four matrices form a group under multiplication?

12.8 What irreducible representations do the four H1*s* orbitals of CH_4 span? Are there *s* and *p* orbitals of the central C atom that may form molecular orbitals with them? Could *d* orbitals, even if they were present on the C atom, play a role in orbital formation in CH_4?

12.9 Suppose that a methane molecule became distorted to (a) C_{3v} symmetry by the lengthening of one bond, (b) C_{2v} symmetry, by a kind of scissors action in which one bond angle opened and another closed slightly. Would more *d* orbitals become available for bonding?

12.10‡ B.A. Bovenzi and G.A. Pearse, Jr. (*J. Chem. Soc. Dalton Trans.*, 2763 (1997)) synthesized coordination compounds of the tridentate ligand pyridine-2,6-diamidoxime ($C_7H_9N_5O_2$, **25**). Reaction with $NiSO_4$ produced a complex in which two of the essentially planar ligands are bonded at right angles to a single Ni atom. Name the point group and the symmetry operations of the resulting $[Ni(C_7H_9N_5O_2)_2]^{2+}$ complex cation.

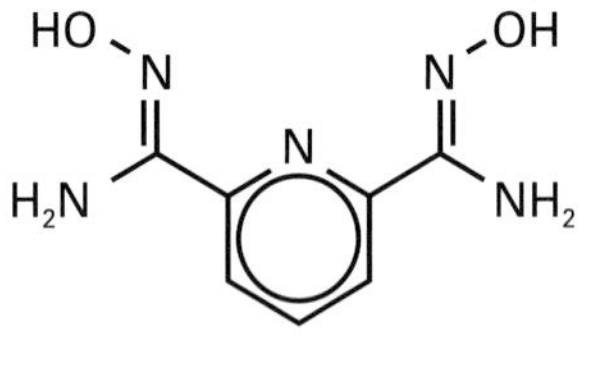

25

12.11‡ R. Eujen, B. Hoge, and D.J. Brauer (*Inorg. Chem.* **36**, 1464 (1997)) prepared and characterized several square-planar Ag(III) complex anions. In the complex anion $[trans\text{-}Ag(CF_3)_2(CN)_2]^-$, the Ag—CN groups are collinear. (a) Assuming free rotation of the CF_3 groups (that is, disregarding the AgCF and AgCH angles), name the point group of this complex anion. (b) Now suppose the CF_3 groups cannot rotate freely (because the ion was in a solid, for example). Structure (**26**) shows a plane that bisects the NC—Ag—CN axis and is perpendicular to it. Name the point group of the complex if each CF_3 group has a CF bond in that plane (so the CF_3 groups do not point to either CN group preferentially) and the CF_3 groups are (i) staggered, (ii) eclipsed.

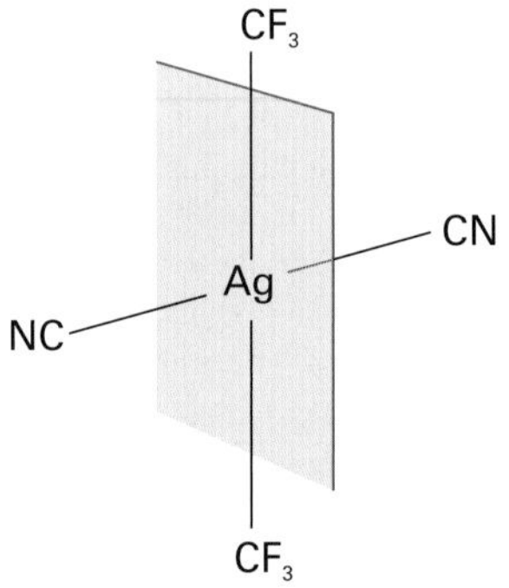

26

12.12‡ A computational study by C.J. Marsden (*Chem. Phys. Lett.* **245**, 475 (1995)) of AM_x compounds, where A is in Group 14 of the periodic table and M is an alkali metal, shows several deviations from the most symmetric structures for each formula. For example, most of the AM_4 structures were not tetrahedral but had two distinct values for MAM bond angles. They could be

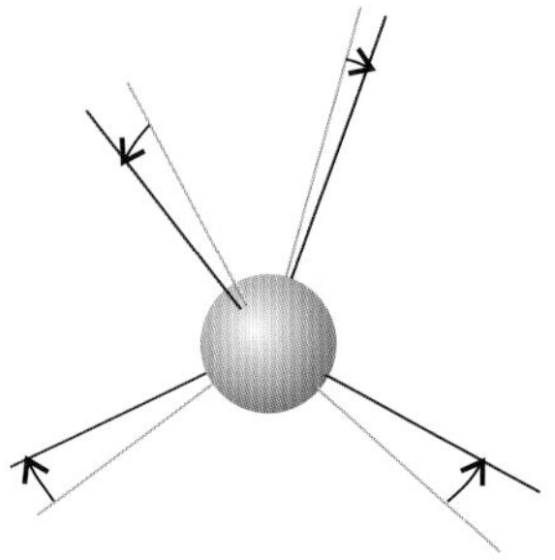

27

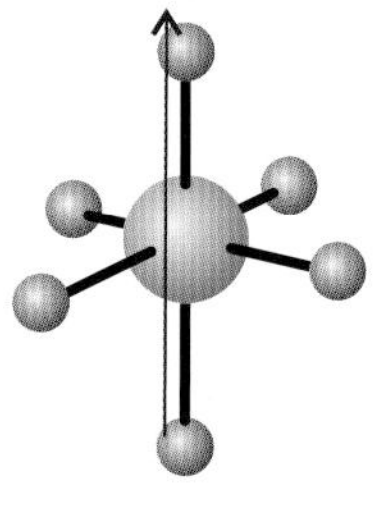

28

derived from a tetrahedron by a distortion shown in (**27**). (a) What is the point group of the distorted tetrahedron? (b) What is the symmetry species of the distortion considered as a vibration in the new, less symmetric group? Some AM_6 structures are not octahedral, but could be derived from an octahedron by translating a C—M—C axis as in (**28**). (c) What is the point group of the distorted octahedron? (d) What is the symmetry species of the distortion considered as a vibration in the new, less symmetric group?

12.13 The algebraic forms of the *f* orbitals are a radial function multiplied by one of the factors (a) $z(5z^2 - 3r^2)$, (b) $y(5y^2 - 3r^2)$, (c) $x(5x^2 - 3r^2)$, (d) $z(x^2 - y^2)$, (e) $y(x^2 - z^2)$, (f) $x(z^2 - y^2)$, (g) xyz. Identify the irreducible representations spanned by these orbitals in (a) C_{2v}, (b) C_{3v}, (c) T_d, (d) O_h. Consider a lanthanide ion at the centre of (a) a tetrahedral complex, (b) an octahedral complex. What sets of orbitals do the seven *f* orbitals split into?

12.14 Does the product *xyz* necessarily vanish when integrated over (a) a cube, (b) a tetrahedron, (c) a hexagonal prism, each centred on the origin?

12.15 The NO_2 molecule belongs to the group C_{2v}, with the C_2 axis bisecting the ONO angle. Taking as a basis the N2*s*, N2*p*, and O2*p* orbitals, identify the irreducible representations they span, and construct the symmetry-adapted linear combinations.

12.16 Construct the symmetry-adapted linear combinations of C2p_z orbitals for benzene, and use them to calculate the Hückel secular determinant. This procedure leads to equations that are much easier to solve than using the original orbitals, and show that the Hückel orbitals are those specified in Section 11.6d.

12.17 The phenanthrene molecule (**29**) belongs to the group C_{2v} with the C_2 axis in the plane of the molecule. (a) Classify the irreducible representations spanned by the carbon 2p_z orbitals and find their symmetry-adapted linear combinations. (b) Use your results from part (a) to calculate the Hückel secular determinant. (c) What states of phenanthrene may be reached by electric dipole transitions from its (totally symmetrical) ground state?

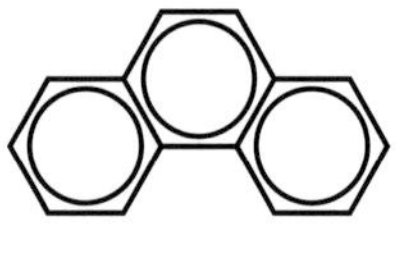

29 Phenanthrene

12.18‡ In a spectroscopic study of C_{60}, F. Negri, G. Orlandi, and F. Zerbetto (*J. Phys. Chem.* **100**, 10849 (1996)) assigned peaks in the fluorescence spectrum. The molecule has icosahedral symmetry (I_h). The ground electronic state is A_{1g}, and the lowest-lying excited states are T_{1g} and G_g. (a) Are photon-induced transitions allowed from the ground state to either of these excited states? Explain your answer. (b) What if the transition is accompanied by a vibration that breaks the parity?

Applications: to astrophysics and biology

12.19‡ The H_3^+ molecular ion, which plays an important role in chemical reactions occurring in interstellar clouds, is known to be equilateral triangular. (a) Identify the symmetry elements and determine the point group of this molecule. (b) Take as a basis for a representation of this molecule the three H1*s* orbitals and set up the matrices that group in this basis. (c) Obtain the group multiplication table by explicit multiplication of the matrices. (d) Determine if the representation is reducible and, if so, give the irreducible representations obtained.

12.20‡ The H_3^+ molecular ion has recently been found in the interstellar medium and in the atmospheres of Jupiter, Saturn, and Uranus. The H_4 analogues have not yet been found, and the square planar structure is thought to be unstable with respect to vibration. Take as a basis for a representation of the point group of this molecule the four H1*s* orbitals and determine if this representation is reducible.

12.21 Some linear polyenes, of which β-carotene is an example, are important biological co-factors that participate in processes as diverse as the absorption of solar energy in photosynthesis (*Impact I23.2*) and protection against harmful biological oxidations. Use as a model of β-carotene a linear polyene containing 22 conjugated C atoms. (a) To what point group does this model of β-carotene belong? (b) Classify the irreducible representations spanned by the carbon $2p_z$ orbitals and find their symmetry-adapted linear combinations. (c) Use your results from part (b) to calculate the Hückel secular determinant. (d) What states of this model of β-carotene may be reached by electric dipole transitions from its (totally symmetrical) ground state?

12.22 The chlorophylls that participate in photosynthesis (*Impact I24.2*) and the haem groups of cytochromes (*Impact I7.2*) are derived from the porphine dianion group (**30**), which belongs to the D_{4h} point group. The ground electronic state is A_{1g} and the lowest-lying excited state is E_u. Is a photon-induced transition allowed from the ground state to the excited state? Explain your answer.

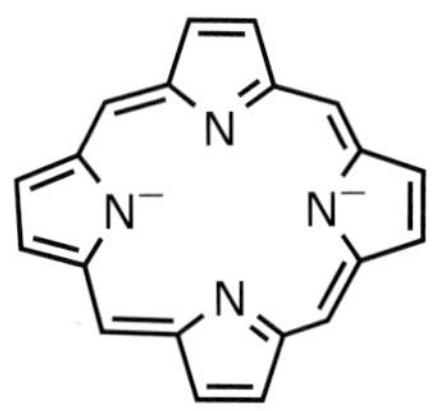

30

13 Molecular spectroscopy 1: rotational and vibrational spectra

The general strategy we adopt in the chapter is to set up expressions for the energy levels of molecules and then apply selection rules and considerations of populations to infer the form of spectra. Rotational energy levels are considered first, and we see how to derive expressions for their values and how to interpret rotational spectra in terms of molecular dimensions. Not all molecules can occupy all rotational states: we see the experimental evidence for this restriction and its explanation in terms of nuclear spin and the Pauli principle. Next, we consider the vibrational energy levels of diatomic molecules, and see that we can use the properties of harmonic oscillators developed in Chapter 9. Then we consider polyatomic molecules and find that their vibrations may be discussed as though they consisted of a set of independent harmonic oscillators, so the same approach as employed for diatomic molecules may be used. We also see that the symmetry properties of the vibrations of polyatomic molecules are helpful for deciding which modes of vibration can be studied spectroscopically.

The origin of spectral lines in molecular spectroscopy is the absorption, emission, or scattering of a photon when the energy of a molecule changes. The difference from atomic spectroscopy is that the energy of a molecule can change not only as a result of electronic transitions but also because it can undergo changes of rotational and vibrational state. Molecular spectra are therefore more complex than atomic spectra. However, they also contain information relating to more properties, and their analysis leads to values of bond strengths, lengths, and angles. They also provide a way of determining a variety of molecular properties, particularly molecular dimensions, shapes, and dipole moments. Molecular spectroscopy is also useful to astrophysicists and environmental scientists, for the chemical composition of interstellar space and of planetary atmospheres can be inferred from their rotational, vibrational, and electronic spectra.

Pure rotational spectra, in which only the rotational state of a molecule changes, can be observed in the gas phase. Vibrational spectra of gaseous samples show features that arise from rotational transitions that accompany the excitation of vibration. Electronic spectra, which are described in Chapter 14, show features arising from simultaneous vibrational and rotational transitions. The simplest way of dealing with these complexities is to tackle each type of transition in turn, and then to see how simultaneous changes affect the appearance of spectra.

General features of spectroscopy

All types of spectra have some features in common, and we examine these first. In **emission spectroscopy**, a molecule undergoes a transition from a state of high energy E_1 to a state of lower energy E_2 and emits the excess energy as a photon. In **absorption spectroscopy**, the net absorption of nearly monochromatic (single frequency) incident radiation is monitored as the radiation is swept over a range of frequencies. We say *net* absorption, because it will become clear that, when a sample is irradiated, both absorption and emission at a given frequency are stimulated, and the detector measures the difference, the net absorption.

The energy, $h\nu$, of the photon emitted or absorbed, and therefore the frequency ν of the radiation emitted or absorbed, is given by the Bohr frequency condition, $h\nu = |E_1 - E_2|$ (eqn 8.10). Emission and absorption spectroscopy give the same information about energy level separations, but practical considerations generally determine which technique is employed. We shall discuss emission spectroscopy in Chapter 14; here we focus on absorption spectroscopy, which is widely employed in studies of electronic transitions, molecular rotations, and molecular vibrations.

In Chapter 9 we saw that transitions between electronic energy levels are stimulated by or emit ultraviolet, visible, or near-infrared radiation. Vibrational and rotational transitions, the focus of the discussion in this chapter, can be induced in two ways. First, the direct absorption or emission of infrared radiation can cause changes in vibrational energy levels, whereas absorption or emission of microwave radiation gives information about rotational energy levels. Second, vibrational and rotational energy levels can be explored by examining the frequencies present in the radiation scattered by molecules in **Raman spectroscopy**. About 1 in 10^7 of the incident photons collide with the molecules, give up some of their energy, and emerge with a lower energy. These scattered photons constitute the lower-frequency **Stokes radiation** from the sample (Fig. 13.1). Other incident photons may collect energy from the molecules (if they are already excited), and emerge as higher-frequency **anti-Stokes radiation**. The component of radiation scattered without change of frequency is called **Rayleigh radiation**.

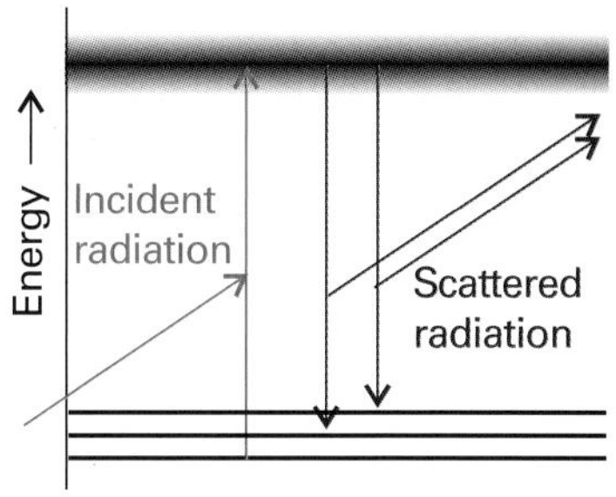

Fig. 13.1 In Raman spectroscopy, an incident photon is scattered from a molecule with either an increase in frequency (if the radiation collects energy from the molecule) or—as shown here for the case of scattered Stokes radiation—with a lower frequency if it loses energy to the molecule. The process can be regarded as taking place by an excitation of the molecule to a wide range of states (represented by the shaded band), and the subsequent return of the molecule to a lower state; the net energy change is then carried away by the photon.

13.1 Experimental techniques

A **spectrometer** is an instrument that detects the characteristics of light scattered, emitted, or absorbed by atoms and molecules. Figure 13.2 shows the general layouts of absorption and emission spectrometers operating in the ultraviolet and visible ranges. Radiation from an appropriate source is directed toward a sample. In most

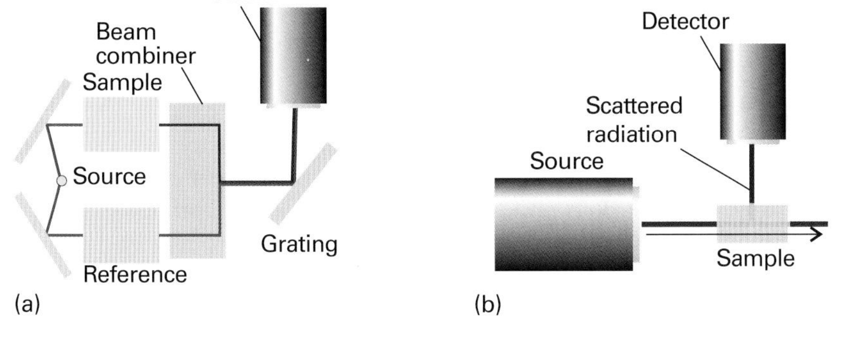

Fig. 13.2 Two examples of spectrometers: (a) the layout of an absorption spectrometer, used primarily for studies in the ultraviolet and visible ranges, in which the exciting beams of radiation pass alternately through a sample and a reference cell, and the detector is synchronized with them so that the relative absorption can be determined, and (b) a simple emission spectrometer, where light emitted or scattered by the sample is detected at right angles to the direction of propagation of an incident beam of radiation.

Comment 13.1

The principles of operation of radiation sources, dispersing elements, Fourier transform spectrometers, and detectors are described in *Further information 13.1*.

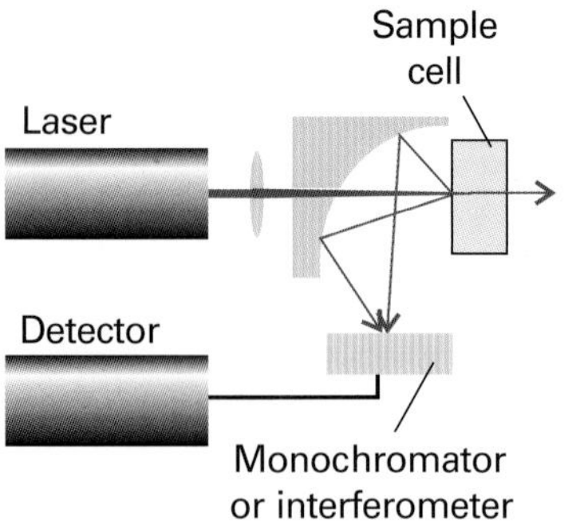

Fig. 13.3 A common arrangement adopted in Raman spectroscopy. A laser beam first passes through a lens and then through a small hole in a mirror with a curved reflecting surface. The focused beam strikes the sample and scattered light is both deflected and focused by the mirror. The spectrum is analysed by a monochromator or an interferometer.

spectrometers, light transmitted, emitted, or scattered by the sample is collected by mirrors or lenses and strikes a **dispersing element** that separates radiation into different frequencies. The intensity of light at each frequency is then analysed by a suitable detector. In a typical Raman spectroscopy experiment, a monochromatic incident laser beam is passed through the sample and the radiation scattered from the front face of the sample is monitored (Fig. 13.3). This detection geometry allows for the study of gases, pure liquids, solutions, suspensions, and solids.

Modern spectrometers, particularly those operating in the infrared and near-infrared, now almost always use **Fourier transform techniques** of spectral detection and analysis. The heart of a Fourier transform spectrometer is a *Michelson interferometer*, a device for analysing the frequencies present in a composite signal. The total signal from a sample is like a chord played on a piano, and the Fourier transform of the signal is equivalent to the separation of the chord into its individual notes, its spectrum.

13.2 The intensities of spectral lines

The ratio of the transmitted intensity, I, to the incident intensity, I_0, at a given frequency is called the **transmittance**, T, of the sample at that frequency:

$$T = \frac{I}{I_0} \qquad [13.1]$$

It is found empirically that the transmitted intensity varies with the length, l, of the sample and the molar concentration, [J], of the absorbing species J in accord with the **Beer–Lambert law**:

$$I = I_0 10^{-\varepsilon[\mathrm{J}]l} \qquad (13.2)$$

The quantity ε is called the **molar absorption coefficient** (formerly, and still widely, the 'extinction coefficient'). The molar absorption coefficient depends on the frequency of the incident radiation and is greatest where the absorption is most intense. Its dimensions are 1/(concentration × length), and it is normally convenient to express it in cubic decimetres per mole per centimetre ($\mathrm{dm^3\,mol^{-1}\,cm^{-1}}$). Alternative units are square centimetres per mole ($\mathrm{cm^2\,mol^{-1}}$). This change of units demonstrates that ε may be regarded as a molar cross-section for absorption and, the greater the cross-sectional area of the molecule for absorption, the greater its ability to block the passage of the incident radiation.

To simplify eqn 13.2, we introduce the **absorbance**, A, of the sample at a given wavenumber as

$$A = \log\frac{I_0}{I} \qquad \text{or} \qquad A = -\log T \qquad [13.3]$$

Then the Beer–Lambert law becomes

$$A = \varepsilon[\mathrm{J}]l \qquad (13.4)$$

The product $\varepsilon[\mathrm{J}]l$ was known formerly as the *optical density* of the sample. Equation 13.4 suggests that, to achieve sufficient absorption, path lengths through gaseous samples must be very long, of the order of metres, because concentrations are low. Long path lengths are achieved by multiple passage of the beam between parallel mirrors at each end of the sample cavity. Conversely, path lengths through liquid samples can be significantly shorter, of the order of millimetres or centimetres.

Justification 13.1 *The Beer–Lambert law*

The Beer–Lambert law is an empirical result. However, it is simple to account for its form. The reduction in intensity, dI, that occurs when light passes through a layer of thickness dl containing an absorbing species J at a molar concentration [J] is proportional to the thickness of the layer, the concentration of J, and the intensity, I, incident on the layer (because the rate of absorption is proportional to the intensity, see below). We can therefore write

$$dI = -\kappa[J]I dl$$

where κ (kappa) is the proportionality coefficient, or equivalently

$$\frac{dI}{I} = -\kappa[J]dl$$

This expression applies to each successive layer into which the sample can be regarded as being divided. Therefore, to obtain the intensity that emerges from a sample of thickness l when the intensity incident on one face of the sample is I_0, we sum all the successive changes:

$$\int_{I_0}^{I} \frac{dI}{I} = -\kappa \int_0^l [J]dl$$

If the concentration is uniform, [J] is independent of location, and the expression integrates to

$$\ln\frac{I}{I_0} = -\kappa[J]l$$

This expression gives the Beer–Lambert law when the logarithm is converted to base 10 by using $\ln x = (\ln 10)\log x$ and replacing κ by $\varepsilon \ln 10$.

Illustration 13.1 *Using the Beer–Lambert law*

The Beer–Lambert law implies that the intensity of electromagnetic radiation transmitted through a sample at a given wavenumber decreases exponentially with the sample thickness and the molar concentration. If the transmittance is 0.1 for a path length of 1 cm (corresponding to a 90 per cent reduction in intensity), then it would be $(0.1)^2 = 0.01$ for a path of double the length (corresponding to a 99 per cent reduction in intensity overall).

The maximum value of the molar absorption coefficient, ε_{max}, is an indication of the intensity of a transition. However, as absorption bands generally spread over a range of wavenumbers, quoting the absorption coefficient at a single wavenumber might not give a true indication of the intensity of a transition. The **integrated absorption coefficient**, $\mathcal{A}$, is the sum of the absorption coefficients over the entire band (Fig. 13.4), and corresponds to the area under the plot of the molar absorption coefficient against wavenumber:

$$\mathcal{A} = \int_{\text{band}} \varepsilon(\tilde{\nu})\, d\tilde{\nu} \qquad [13.5]$$

For lines of similar widths, the integrated absorption coefficients are proportional to the heights of the lines.

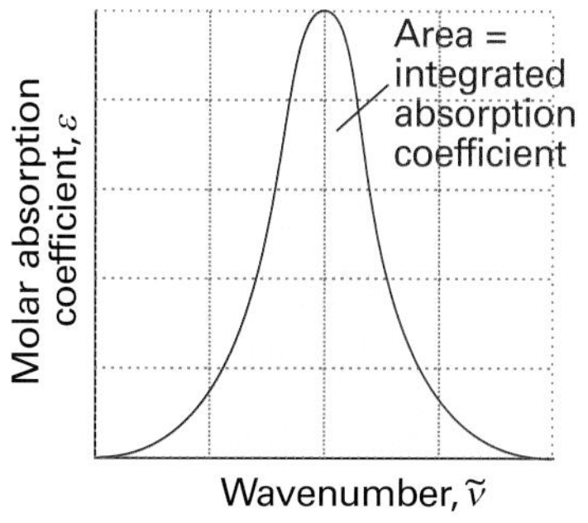

Fig. 13.4 The integrated absorption coefficient of a transition is the area under a plot of the molar absorption coefficient against the wavenumber of the incident radiation.

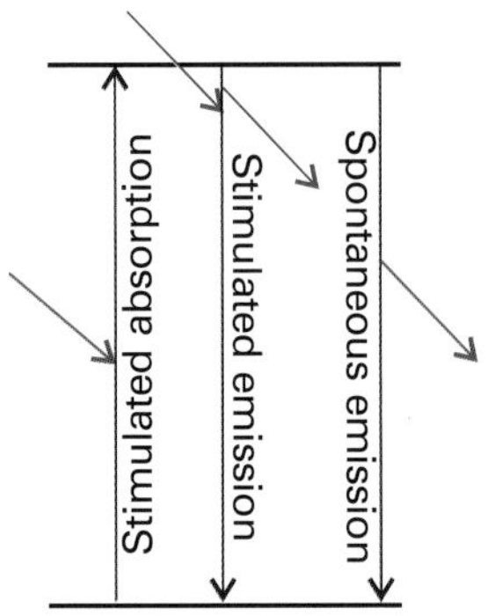

Fig. 13.5 The processes that account for absorption and emission of radiation and the attainment of thermal equilibrium. The excited state can return to the lower state spontaneously as well as by a process stimulated by radiation already present at the transition frequency.

Comment 13.2

The slight difference between the forms of the Planck distribution shown in eqns 8.5 and 13.7 stems from the fact that it is written here as $\rho \mathrm{d}\nu$, and $\mathrm{d}\lambda = (c/\nu^2)\mathrm{d}\nu$.

(a) Absorption intensities

Einstein identified three contributions to the transitions between states. **Stimulated absorption** is the transition from a low energy state to one of higher energy that is driven by the electromagnetic field oscillating at the transition frequency. We saw in Section 9.10 that the transition rate, w, is the rate of change of probability of the molecule being found in the upper state. We also saw that the more intense the electromagnetic field (the more intense the incident radiation), the greater the rate at which transitions are induced and hence the stronger the absorption by the sample (Fig. 13.5). Einstein wrote the transition rate as

$$w = B\rho \tag{13.6}$$

The constant B is the **Einstein coefficient of stimulated absorption** and $\rho \mathrm{d}\nu$ is the energy density of radiation in the frequency range ν to $\nu + \mathrm{d}\nu$, where ν is the frequency of the transition. When the molecule is exposed to black-body radiation from a source of temperature T, ρ is given by the Planck distribution (eqn 8.5):

$$\rho = \frac{8\pi h\nu^3/c^3}{e^{h\nu/kT} - 1} \tag{13.7}$$

For the time being, we can treat B as an empirical parameter that characterizes the transition: if B is large, then a given intensity of incident radiation will induce transitions strongly and the sample will be strongly absorbing. The **total rate of absorption**, W, the number of molecules excited during an interval divided by the duration of the interval, is the transition rate of a single molecule multiplied by the number of molecules N in the lower state: $W = Nw$.

Einstein considered that the radiation was also able to induce the molecule in the upper state to undergo a transition to the lower state, and hence to generate a photon of frequency ν. Thus, he wrote the rate of this stimulated emission as

$$w' = B'\rho \tag{13.8}$$

where B' is the **Einstein coefficient of stimulated emission**. Note that only radiation of the same frequency as the transition can stimulate an excited state to fall to a lower state. However, he realized that stimulated emission was not the only means by which the excited state could generate radiation and return to the lower state, and suggested that an excited state could undergo **spontaneous emission** at a rate that was independent of the intensity of the radiation (of any frequency) that is already present. Einstein therefore wrote the total rate of transition from the upper to the lower state as

$$w' = A + B'\rho \tag{13.9}$$

The constant A is the **Einstein coefficient of spontaneous emission**. The overall rate of emission is

$$W' = N'(A + B'\rho) \tag{13.10}$$

where N' is the population of the upper state.

As demonstrated in the *Justification* below, Einstein was able to show that the two coefficients of stimulated absorption and emission are equal, and that the coefficient of spontaneous emission is related to them by

$$A = \left(\frac{8\pi h\nu^3}{c^3}\right)B \tag{13.11}$$

Justification 13.2 *The relation between the Einstein coefficients*

At thermal equilibrium, the total rates of emission and absorption are equal, so

$$NB\rho = N'(A + B'\rho)$$

This expression rearranges into

$$\rho = \frac{N'A}{NB - N'B'} = \frac{A/B}{N/N' - B'/B} = \frac{A/B}{e^{h\nu/kT} - B'/B}$$

We have used the Boltzmann expression (*Molecular interpretation* 3.1) for the ratio of populations of states of energies E and E' in the last step:

$$\frac{N'}{N} = e^{-h\nu/kT} \qquad h\nu = E' - E$$

This result has the same form as the Planck distribution (eqn 13.7), which describes the radiation density at thermal equilibrium. Indeed, when we compare the two expressions for ρ, we can conclude that $B' = B$ and that A is related to B by eqn 13.11.

The growth of the importance of spontaneous emission with increasing frequency is a very important conclusion, as we shall see when we consider the operation of lasers (Section 14.5). The equality of the coefficients of stimulated emission and absorption implies that, if two states happen to have equal populations, then the rate of stimulated emission is equal to the rate of stimulated absorption, and there is then no net absorption.

Spontaneous emission can be largely ignored at the relatively low frequencies of rotational and vibrational transitions, and the intensities of these transitions can be discussed in terms of stimulated emission and absorption. Then the net rate of absorption is given by

$$W_{\text{net}} = NB\rho - N'B'\rho = (N - N')B\rho \qquad (13.12)$$

and is proportional to the population difference of the two states involved in the transition.

(b) Selection rules and transition moments

We met the concept of a 'selection rule' in Sections 10.3 and 12.6 as a statement about whether a transition is forbidden or allowed. Selection rules also apply to molecular spectra, and the form they take depends on the type of transition. The underlying classical idea is that, for the molecule to be able to interact with the electromagnetic field and absorb or create a photon of frequency ν, it must possess, at least transiently, a dipole oscillating at that frequency. We saw in Section 9.10 that this transient dipole is expressed quantum mechanically in terms of the transition dipole moment, $\boldsymbol{\mu}_{\text{fi}}$, between states ψ_{i} and ψ_{f}:

$$\boldsymbol{\mu}_{\text{fi}} = \int \psi_{\text{f}}^{*}\hat{\boldsymbol{\mu}}\psi_{\text{i}}\,d\tau \qquad [13.13]$$

where $\hat{\boldsymbol{\mu}}$ is the electric dipole moment operator. The size of the transition dipole can be regarded as a measure of the charge redistribution that accompanies a transition: a transition will be active (and generate or absorb photons) only if the accompanying charge redistribution is dipolar (Fig. 13.6).

We know from time-dependent perturbation theory (Section 9.10) that the transition rate is proportional to $|\mu_{\text{fi}}|^2$. It follows that the coefficient of stimulated absorption

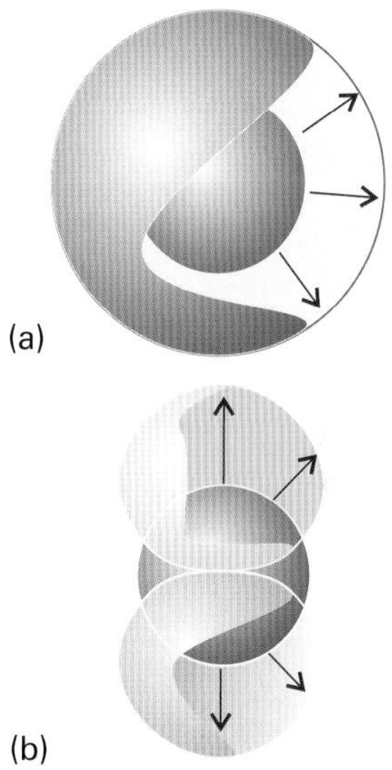

Fig. 13.6 (a) When a 1*s* electron becomes a 2*s* electron, there is a spherical migration of charge; there is no dipole moment associated with this migration of charge; this transition is electric-dipole forbidden. (b) In contrast, when a 1*s* electron becomes a 2*p* electron, there is a dipole associated with the charge migration; this transition is allowed. (There are subtle effects arising from the sign of the wavefunction that give the charge migration a dipolar character, which this diagram does not attempt to convey.)

(and emission), and therefore the intensity of the transition, is also proportional to $|\mu_{fi}|^2$. A detailed analysis gives

$$B = \frac{|\mu_{fi}|^2}{6\varepsilon_0\hbar^2} \tag{13.14}$$

Only if the transition moment is nonzero does the transition contribute to the spectrum. It follows that, to identify the selection rules, we must establish the conditions for which $\boldsymbol{\mu}_{fi} \neq 0$.

A **gross selection rule** specifies the general features a molecule must have if it is to have a spectrum of a given kind. For instance, we shall see that a molecule gives a rotational spectrum only if it has a permanent electric dipole moment. This rule, and others like it for other types of transition, will be explained in the relevant sections of the chapter. A detailed study of the transition moment leads to the **specific selection rules** that express the allowed transitions in terms of the changes in quantum numbers. We have already encountered examples of specific selection rules when discussing atomic spectra (Section 10.3), such as the rule $\Delta l = \pm 1$ for the angular momentum quantum number.

13.3 Linewidths

A number of effects contribute to the widths of spectroscopic lines. Some contributions to linewidths can be modified by changing the conditions, and to achieve high resolutions we need to know how to minimize these contributions. Other contributions cannot be changed, and represent an inherent limitation on resolution.

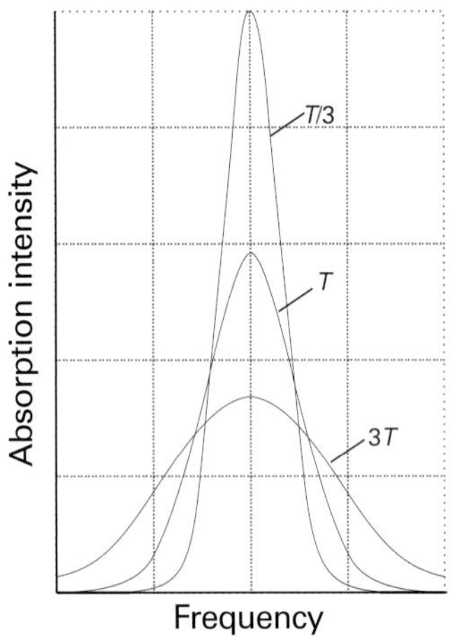

Fig. 13.7 The Gaussian shape of a Doppler-broadened spectral line reflects the Maxwell distribution of speeds in the sample at the temperature of the experiment. Notice that the line broadens as the temperature is increased.

Exploration In a spectrometer that makes use of *phase-sensitive detection* the output signal is proportional to the first derivative of the signal intensity, $dI/d\nu$. Plot the resulting line shape for various temperatures. How is the separation of the peaks related to the temperature?

(a) Doppler broadening

The study of gaseous samples is very important, as it can inform our understanding of atmospheric chemistry. In some cases, meaningful spectroscopic data can be obtained only from gaseous samples. For example, they are essential for rotational spectroscopy, for only in gases can molecules rotate freely.

One important broadening process in gaseous samples is the **Doppler effect**, in which radiation is shifted in frequency when the source is moving towards or away from the observer. When a source emitting electromagnetic radiation of frequency ν moves with a speed s relative to an observer, the observer detects radiation of frequency

$$\nu_{\text{receding}} = \nu\left(\frac{1-s/c}{1+s/c}\right)^{1/2} \qquad \nu_{\text{approaching}} = \nu\left(\frac{1+s/c}{1-s/c}\right)^{1/2} \tag{13.15}$$

where c is the speed of light (see *Further reading* for derivations). For nonrelativistic speeds ($s \ll c$), these expressions simplify to

$$\nu_{\text{receding}} \approx \frac{\nu}{1+s/c} \qquad \nu_{\text{approaching}} \approx \frac{\nu}{1-s/c} \tag{13.16}$$

Molecules reach high speeds in all directions in a gas, and a stationary observer detects the corresponding Doppler-shifted range of frequencies. Some molecules approach the observer, some move away; some move quickly, others slowly. The detected spectral 'line' is the absorption or emission profile arising from all the resulting Doppler shifts. As shown in the following *Justification*, the profile reflects the distribution of molecular velocities parallel to the line of sight, which is a bell-shaped Gaussian curve. The Doppler line shape is therefore also a Gaussian (Fig. 13.7), and we show in the

Justification that, when the temperature is T and the mass of the molecule is m, then the observed width of the line at half-height (in terms of frequency or wavelength) is

$$\delta\nu_{obs} = \frac{2\nu}{c}\left(\frac{2kT\ln 2}{m}\right)^{1/2} \qquad \delta\lambda_{obs} = \frac{2\lambda}{c}\left(\frac{2kT\ln 2}{m}\right)^{1/2} \qquad (13.17)$$

For a molecule like N_2 at room temperature ($T \approx 300$ K), $\delta\nu/\nu \approx 2.3 \times 10^{-6}$. For a typical rotational transition wavenumber of 1 cm^{-1} (corresponding to a frequency of 30 GHz), the linewidth is about 70 kHz. Doppler broadening increases with temperature because the molecules acquire a wider range of speeds. Therefore, to obtain spectra of maximum sharpness, it is best to work with cold samples.

Comment 13.3

A Gaussian function of the general form $y(x) = ae^{-(x-b)^2/2\sigma^2}$, where a, b, and σ are constants, has a maximum $y(b) = a$ and a width at half-height $\delta x = 2\sigma(2\ln 2)^{1/2}$.

A note on good practice You will often hear people speak of 'a frequency as so many wavenumbers'. This usage is doubly wrong. First, *frequency* and *wavenumber* are two distinct physical observables with different units, and should be distinguished. Second, 'wavenumber' is not a unit, it is an observable with the dimensions of 1/length and commonly reported in reciprocal centimetres (cm^{-1}).

Justification 13.3 *Doppler broadening*

We know from the Boltzmann distribution (*Molecular interpretation 3.1*) that the probability that a gas molecule of mass m and speed s in a sample with temperature T has kinetic energy $E_K = \frac{1}{2}ms^2$ is proportional to $e^{-ms^2/2kT}$. The observed frequencies, ν_{obs}, emitted or absorbed by the molecule are related to its speed by eqn 13.16:

$$\nu_{obs} = \nu\left(\frac{1}{1 \pm s/c}\right)$$

where ν is the unshifted frequency. When $s \ll c$, the Doppler shift in the frequency is

$$\nu_{obs} - \nu \approx \pm\nu s/c$$

which implies a symmetrical distribution of observed frequencies with respect to molecular speeds. More specifically, the intensity I of a transition at ν_{obs} is proportional to the probability of finding the molecule that emits or absorbs at ν_{obs}, so it follows from the Boltzmannm distribution and the expression for the Doppler shift that

$$I(\nu_{obs}) \propto e^{-mc^2(\nu_{obs}-\nu)^2/2\nu^2kT}$$

which has the form of a Gaussian function. The width at half-height can be calculated directly from the exponent (see *Comment 13.3*) to give eqn 13.17.

(b) Lifetime broadening

It is found that spectroscopic lines from gas-phase samples are not infinitely sharp even when Doppler broadening has been largely eliminated by working at low temperatures. The same is true of the spectra of samples in condensed phases and solution. This residual broadening is due to quantum mechanical effects. Specifically, when the Schrödinger equation is solved for a system that is changing with time, it is found that it is impossible to specify the energy levels exactly. If on average a system survives in a state for a time τ (tau), the **lifetime** of the state, then its energy levels are blurred to an extent of order δE, where

$$\delta E \approx \frac{\hbar}{\tau} \tag{13.18}$$

This expression is reminiscent of the Heisenberg uncertainty principle (eqn 8.40), and consequently this **lifetime broadening** is often called 'uncertainty broadening'. When the energy spread is expressed as a wavenumber through $\delta E = hc\delta\tilde{\nu}$, and the values of the fundamental constants introduced, this relation becomes

$$\delta\tilde{\nu} \approx \frac{5.3\ \mathrm{cm}^{-1}}{\tau/\mathrm{ps}} \tag{13.19}$$

No excited state has an infinite lifetime; therefore, all states are subject to some lifetime broadening and, the shorter the lifetimes of the states involved in a transition, the broader the corresponding spectral lines.

Two processes are responsible for the finite lifetimes of excited states. The dominant one for low frequency transitions is **collisional deactivation**, which arises from collisions between molecules or with the walls of the container. If the **collisional lifetime**, the mean time between collisions, is τ_{col}, the resulting collisional linewidth is $\delta E_{\mathrm{col}} \approx \hbar/\tau_{\mathrm{col}}$. Because $\tau_{\mathrm{col}} = 1/z$, where z is the collision frequency, and from the kinetic model of gases (Section 1.3) we know that z is proportional to the pressure, we see that the collisional linewidth is proportional to the pressure. The collisional linewidth can therefore be minimized by working at low pressures.

The rate of spontaneous emission cannot be changed. Hence it is a natural limit to the lifetime of an excited state, and the resulting lifetime broadening is the **natural linewidth** of the transition. The natural linewidth is an intrinsic property of the transition, and cannot be changed by modifying the conditions. Natural linewidths depend strongly on the transition frequency (they increase with the coefficient of spontaneous emission A and therefore as ν^3), so low frequency transitions (such as the microwave transitions of rotational spectroscopy) have very small natural linewidths, and collisional and Doppler line-broadening processes are dominant. The natural lifetimes of electronic transitions are very much shorter than for vibrational and rotational transitions, so the natural linewidths of electronic transitions are much greater than those of vibrational and rotational transitions. For example, a typical electronic excited state natural lifetime is about 10^{-8} s (10 ns), corresponding to a natural width of about 5×10^{-4} cm^{-1} (15 MHz). A typical rotational state natural lifetime is about 10^3 s, corresponding to a natural linewidth of only 5×10^{-15} cm^{-1} (of the order of 10^{-4} Hz).

IMPACT ON ASTROPHYSICS
I13.1 Rotational and vibrational spectroscopy of interstellar space

Observations by the Cosmic Background Explorer (COBE) satellite support the long-held hypothesis that the distribution of energy in the current Universe can be modelled by a Planck distribution (eqn. 8.5) with $T = 2.726 \pm 0.001$ K, the bulk of the radiation spanning the microwave region of the spectrum. This *cosmic microwave background radiation* is the residue of energy released during the Big Bang, the event that brought the Universe into existence. Very small fluctuations in the background temperature are believed to account for the large-scale structure of the Universe.

The interstellar space in our galaxy is a little warmer than the cosmic background and consists largely of dust grains and gas clouds. The dust grains are carbon-based compounds and silicates of aluminium, magnesium, and iron, in which are embedded trace amounts of methane, water, and ammonia. Interstellar clouds are significant because it is from them that new stars, and consequently new planets, are formed. The hottest clouds are plasmas with temperatures of up to 10^6 K and densities of only about 3×10^3 particles m^{-3}. Colder clouds range from 0.1 to 1000 solar masses (1 solar

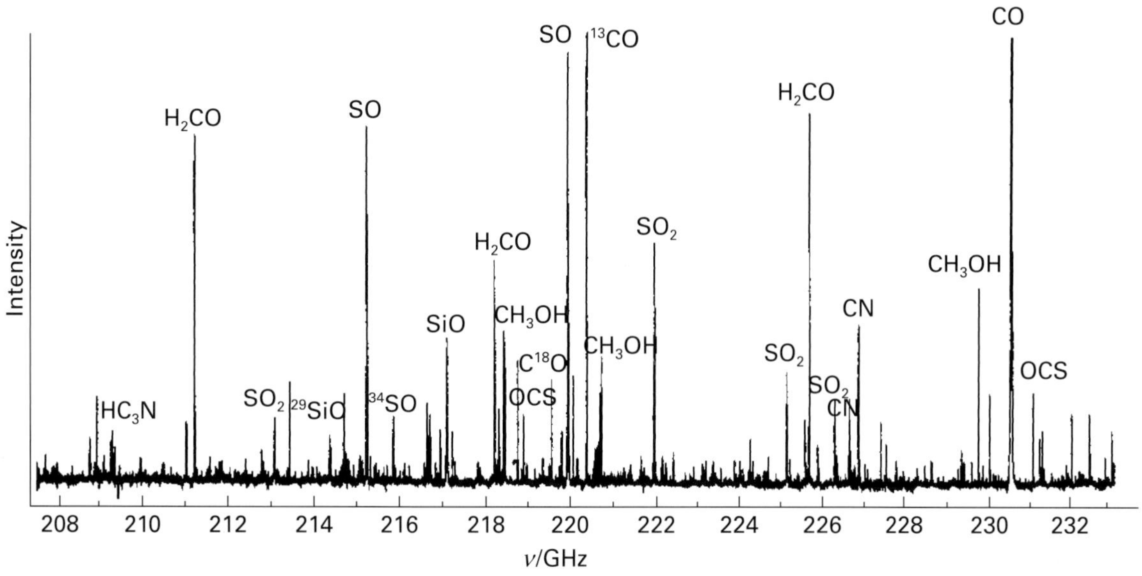

Fig. 13.8 Rotational spectrum of the Orion nebula, showing spectral fingerprints of diatomic and polyatomic molecules present in the interstellar cloud. Adapted from G.A. Blake *et al.*, *Astrophys. J.* **315**, 621 (1987).

mass = 2×10^{30} kg), have a density of about 5×10^5 particles m^{-3}, consist largely of hydrogen atoms, and have a temperature of about 80 K. There are also colder and denser clouds, some with masses greater than 500 000 solar masses, densities greater than 10^9 particles m^{-3}, and temperatures that can be lower than 10 K. They are also called *molecular clouds*, because they are composed primarily of H_2 and CO gas in a proportion of about 10^5 to 1. There are also trace amounts of larger molecules. To place the densities in context, the density of liquid water at 298 K and 1 bar is about 3×10^{28} particles m^{-3}.

It follows from the the Boltzmann distribution and the low temperature of a molecular cloud that the vast majority of a cloud's molecules are in their vibrational and electronic ground states. However, rotational excited states are populated at 10–100 K and decay by spontaneous emission. As a result, the spectrum of the cloud in the radiofrequency and microwave regions consists of sharp lines corresponding to rotational transitions (Fig. 13.8). The emitted light is collected by Earth-bound or space-borne radiotelescopes, telescopes with antennas and detectors optimized for the collection and analysis of radiation in the microwave–radiowave range of the spectrum. Earth-bound radiotelescopes are often located at the tops of high mountains, as atmospheric water vapour can reabsorb microwave radiation from space and hence interfere with the measurement.

Over 100 interstellar molecules have been identified by their rotational spectra, often by comparing radiotelescope data with spectra obtained in the laboratory or calculated by computational methods. The experiments have revealed the presence of trace amounts (with abundances of less than 10^{-8} relative to hydrogen) of neutral molecules, ions, and radicals. Examples of neutral molecules include hydrides, oxides (including water), sulfides, halogenated compounds, nitriles, hydrocarbons, aldehydes, alcohols, ethers, ketones, and amides. The largest molecule detected by rotational spectroscopy is the nitrile $HC_{11}N$.

Interstellar space can also be investigated with vibrational spectroscopy by using a combination of telescopes and infrared detectors. The experiments are conducted

Table 13.1 Moments of inertia*

1. *Diatomic molecules*

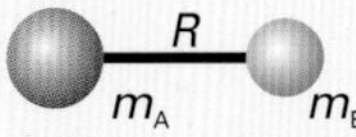

$$I = \mu R^2 \qquad \mu = \frac{m_A m_B}{m}$$

2. *Triatomic linear rotors*

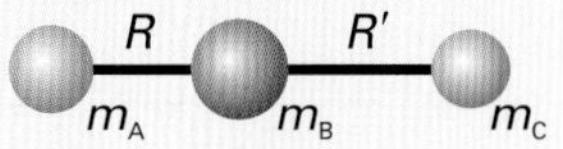

$$I = m_A R^2 + m_C R'^2 - \frac{(m_A R - m_C R')^2}{m}$$

$$I = 2m_A R^2$$

3. *Symmetric rotors*

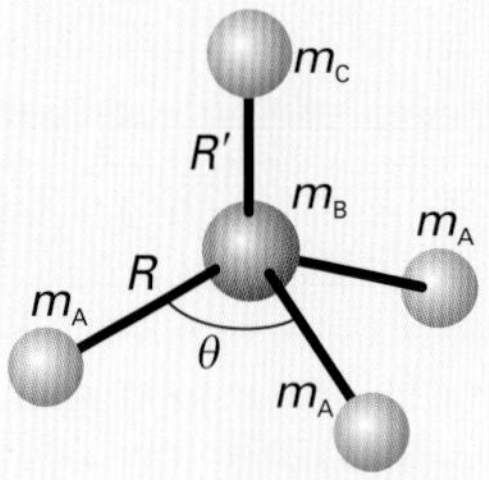

$$I_\parallel = 2m_A(1-\cos\theta)R^2$$

$$I_\perp = m_A(1-\cos\theta)R^2 + \frac{m_A}{m}(m_B + m_C)(1+2\cos\theta)R^2 + \frac{m_C}{m}\{(3m_A + m_B)R' + 6m_A R[\tfrac{1}{3}(1+2\cos\theta)]^{1/2}\}R'$$

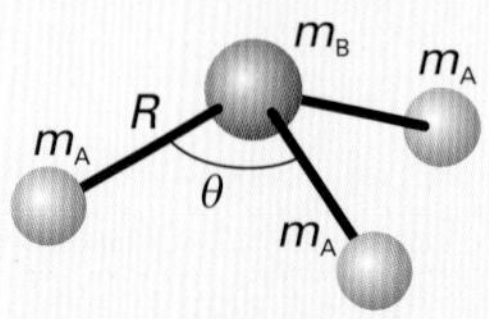

$$I_\parallel = 2m_A(1-\cos\theta)R^2$$

$$I_\perp = m_A(1-\cos\theta)R^2 + \frac{m_A m_B}{m}(1+2\cos\theta)R^2$$

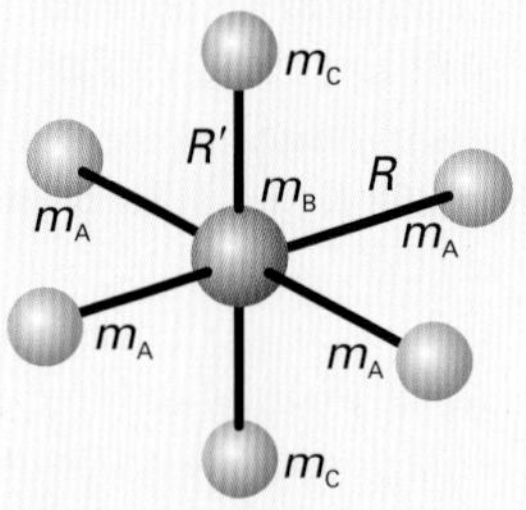

$$I_\parallel = 4m_A R^2$$

$$I_\perp = 2m_A R^2 + 2m_C R'^2$$

4. *Spherical rotors*

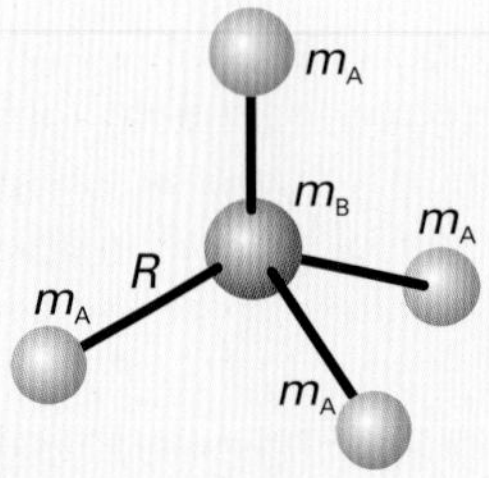

$$I = \tfrac{8}{3} m_A R^2$$

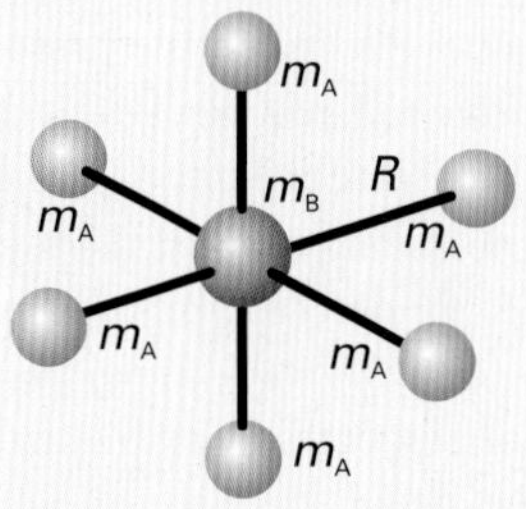

$$I = 4m_A R^2$$

* In each case, m is the total mass of the molecule.

primarily in space-borne telescopes because the Earth's atmosphere absorbs a great deal of infrared radiation (see *Impact I13.2*). In most cases, absorption by an interstellar species is detected against the background of infrared radiation emitted by a nearby star. The data can detect the presence of gas and solid water, CO, and CO_2 in molecular clouds. In certain cases, infrared emission can be detected, but these events are rare because interstellar space is too cold and does not provide enough energy to promote a significant number of molecules to vibrational excited states. However, infrared emissions can be observed if molecules are occasionally excited by high-energy photons emitted by hot stars in the vicinity of the cloud. For example, the polycyclic aromatic hydrocarbons hexabenzocoronene ($C_{48}H_{24}$) and circumcoronene ($C_{54}H_{18}$) have been identified from characteristic infrared emissions.

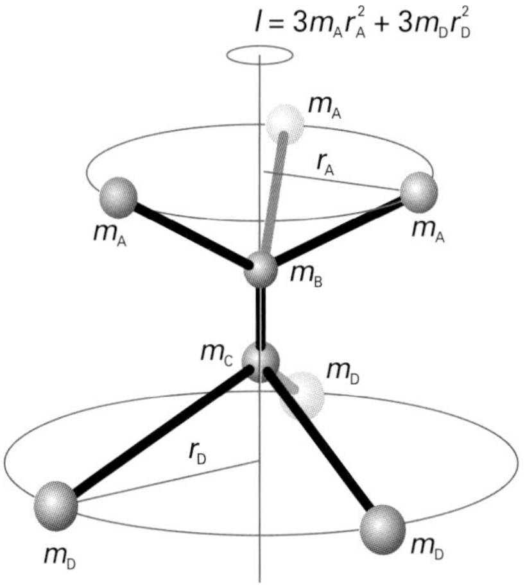

Fig. 13.9 The definition of moment of inertia. In this molecule there are three identical atoms attached to the B atom and three different but mutually identical atoms attached to the C atom. In this example, the centre of mass lies on an axis passing through the B and C atom, and the perpendicular distances are measured from this axis.

Pure rotation spectra

The general strategy we adopt for discussing molecular spectra and the information they contain is to find expressions for the energy levels of molecules and then to calculate the transition frequencies by applying the selection rules. We then predict the appearance of the spectrum by taking into account the transition moments and the populations of the states. In this section we illustrate the strategy by considering the rotational states of molecules.

13.4 Moments of inertia

The key molecular parameter we shall need is the **moment of inertia**, I, of the molecule (Section 9.6). The moment of inertia of a molecule is defined as the mass of each atom multiplied by the square of its distance from the rotational axis through the centre of mass of the molecule (Fig. 13.9):

$$I=\sum_i m_i r_i^2 \qquad [13.20]$$

where r_i is the perpendicular distance of the atom i from the axis of rotation. The moment of inertia depends on the masses of the atoms present and the molecular geometry, so we can suspect (and later shall see explicitly) that rotational spectroscopy will give information about bond lengths and bond angles.

In general, the rotational properties of any molecule can be expressed in terms of the moments of inertia about three perpendicular axes set in the molecule (Fig. 13.10). The convention is to label the moments of inertia I_a, I_b, and I_c, with the axes chosen so that $I_c \geq I_b \geq I_a$. For linear molecules, the moment of inertia around the internuclear axis is zero. The explicit expressions for the moments of inertia of some symmetrical molecules are given in Table 13.1.

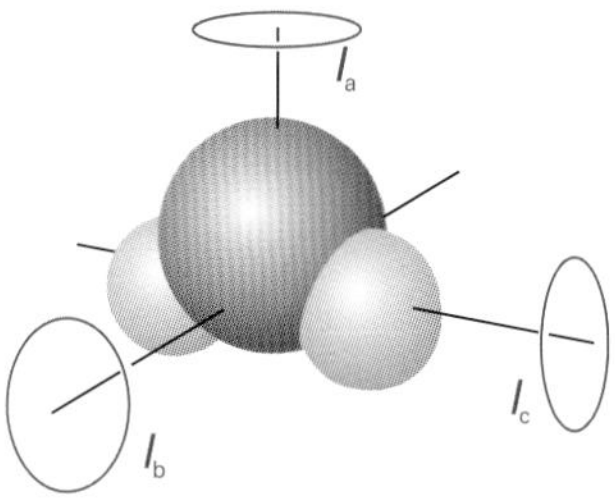

Fig. 13.10 An asymmetric rotor has three different moments of inertia; all three rotation axes coincide at the centre of mass of the molecule.

Example 13.1 *Calculating the moment of inertia of a molecule*

Calculate the moment of inertia of an H_2O molecule around the axis defined by the bisector of the HOH angle (**1**). The HOH bond angle is 104.5° and the bond length is 95.7 pm.

Method According to eqn 13.20, the moment of inertia is the sum of the masses multiplied by the squares of their distances from the axis of rotation. The latter can be expressed by using trigonometry and the bond angle and bond length.

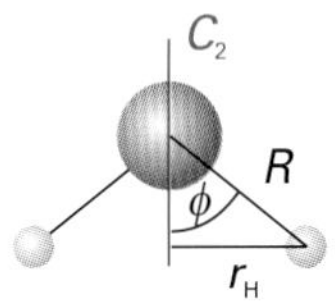

1

Answer From eqn 13.20,

$$I=\sum_i m_i r_i^2 = m_{\rm H}r_{\rm H}^2 + 0 + m_{\rm H}r_{\rm H}^2 = 2m_{\rm H}r_{\rm H}^2$$

If the bond angle of the molecule is denoted 2ϕ and the bond length is R, trigonometry gives $r_{\rm H} = R\sin\phi$. It follows that

$$I = 2m_{\rm H}R^2\sin^2\phi$$

Substitution of the data gives

$$I = 2\times(1.67\times10^{-27}\,{\rm kg})\times(9.57\times10^{-11}\,{\rm m})^2\times\sin^2 52.3° = 1.91\times10^{-47}\,{\rm kg\,m^2}$$

Note that the mass of the O atom makes no contribution to the moment of inertia for this mode of rotation as the atom is immobile while the H atoms circulate around it.

A note on good practice The mass to use in the calculation of the moment of inertia is the actual atomic mass, not the element's molar mass; don't forget to convert from atomic mass units (u, formerly amu) to kilograms.

Self-test 13.1 Calculate the moment of inertia of a $CH^{35}Cl_3$ molecule around a rotational axis that contains the C—H bond. The C—Cl bond length is 177 pm and the HCCl angle is 107°; $m(^{35}Cl) = 34.97$ u. [4.99×10^{-45} kg m^2]

We shall suppose initially that molecules are **rigid rotors**, bodies that do not distort under the stress of rotation. Rigid rotors can be classified into four types (Fig. 13.11):

Spherical rotors have three equal moments of inertia (examples: CH_4, SiH_4, and SF_6).

Symmetric rotors have two equal moments of inertia (examples: NH_3, CH_3Cl, and CH_3CN).

Linear rotors have one moment of inertia (the one about the molecular axis) equal to zero (examples: CO_2, HCl, OCS, and HC≡CH).

Asymmetric rotors have three different moments of inertia (examples: H_2O, H_2CO, and CH_3OH).

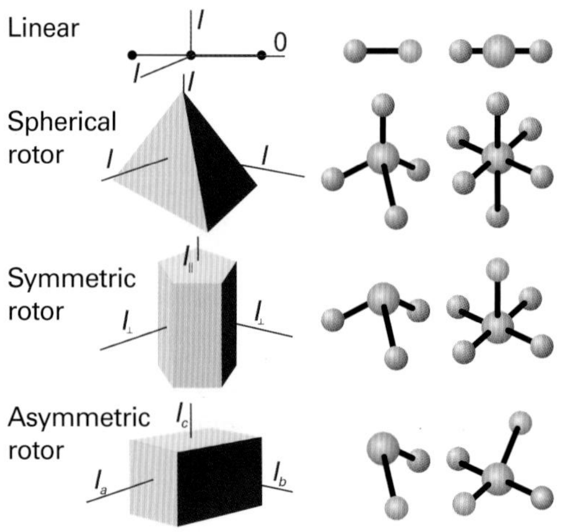

Fig. 13.11 A schematic illustration of the classification of rigid rotors.

13.5 The rotational energy levels

The rotational energy levels of a rigid rotor may be obtained by solving the appropriate Schrödinger equation. Fortunately, however, there is a much less onerous short cut to the exact expressions that depends on noting the classical expression for the energy of a rotating body, expressing it in terms of the angular momentum, and then importing the quantum mechanical properties of angular momentum into the equations.

The classical expression for the energy of a body rotating about an axis a is

$$E_a = \tfrac{1}{2}I_a\omega_a^2 \qquad (13.21)$$

where ω_a is the angular velocity (in radians per second, rad s^{-1}) about that axis and I_a is the corresponding moment of inertia. A body free to rotate about three axes has an energy

$$E = \tfrac{1}{2}I_a\omega_a^2 + \tfrac{1}{2}I_b\omega_b^2 + \tfrac{1}{2}I_c\omega_c^2$$

Because the classical angular momentum about the axis a is $J_a = I_a\omega_a$, with similar expressions for the other axes, it follows that

$$E = \frac{J_a^2}{2I_a} + \frac{J_b^2}{2I_b} + \frac{J_c^2}{2I_c} \qquad (13.22)$$

This is the key equation. We described the quantum mechanical properties of angular momentum in Section 9.7b, and can now make use of them in conjunction with this equation to obtain the rotational energy levels.

(a) Spherical rotors

When all three moments of inertia are equal to some value I, as in CH_4 and SF_6, the classical expression for the energy is

$$E = \frac{J_a^2 + J_b^2 + J_c^2}{2I} = \frac{\mathcal{J}^2}{2I}$$

where $\mathcal{J}^2 = J_a^2 + J_b^2 + J_c^2$ is the square of the magnitude of the angular momentum. We can immediately find the quantum expression by making the replacement

$$\mathcal{J}^2 \rightarrow J(J+1)\hbar^2 \qquad J = 0, 1, 2, \ldots$$

Therefore, the energy of a spherical rotor is confined to the values

$$E_J = J(J+1)\frac{\hbar^2}{2I} \qquad J = 0, 1, 2, \ldots \qquad (13.23)$$

The resulting ladder of energy levels is illustrated in Fig. 13.12. The energy is normally expressed in terms of the **rotational constant**, B, of the molecule, where

$$hcB = \frac{\hbar^2}{2I} \quad \text{so} \quad B = \frac{\hbar}{4\pi cI} \qquad [13.24]$$

The expression for the energy is then

$$E_J = hcBJ(J+1) \qquad J = 0, 1, 2, \ldots \qquad (13.25)$$

The rotational constant as defined by eqn 13.25 is a wavenumber. The energy of a rotational state is normally reported as the **rotational term**, $F(J)$, a wavenumber, by division by hc:

$$F(J) = BJ(J+1) \qquad (13.26)$$

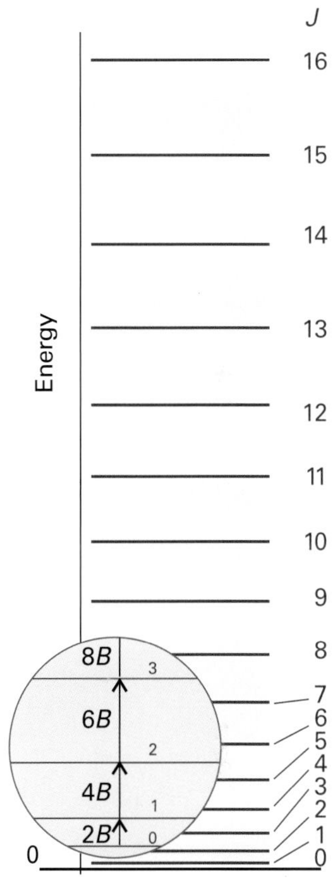

Fig. 13.12 The rotational energy levels of a linear or spherical rotor. Note that the energy separation between neighbouring levels increases as J increases.

Comment 13.4

The definition of B as a wavenumber is convenient when we come to vibration–rotation spectra. However, for pure rotational spectroscopy it is more common to define B as a frequency. Then $B = \hbar/4\pi I$ and the energy is $E = hBJ(J+1)$.

The separation of adjacent levels is

$$F(J) - F(J-1) = 2BJ \tag{13.27}$$

Because the rotational constant decreases as I increases, we see that large molecules have closely spaced rotational energy levels. We can estimate the magnitude of the separation by considering CCl_4: from the bond lengths and masses of the atoms we find $I = 4.85 \times 10^{-45}$ kg m^2, and hence $B = 0.0577$ cm^{-1}.

(b) Symmetric rotors

In symmetric rotors, two moments of inertia are equal but different from the third (as in CH_3Cl, NH_3, and C_6H_6); the unique axis of the molecule is its **principal axis** (or *figure axis*). We shall write the unique moment of inertia (that about the principal axis) as $I_{\parallel}$ and the other two as $I_{\perp}$. If $I_{\parallel} > I_{\perp}$, the rotor is classified as **oblate** (like a pancake, and C_6H_6); if $I_{\parallel} < I_{\perp}$ it is classified as **prolate** (like a cigar, and CH_3Cl). The classical expression for the energy, eqn 13.22, becomes

$$E = \frac{J_b^2 + J_c^2}{2I_{|}} + \frac{J_a^2}{2I_{\parallel}}$$

Again, this expression can be written in terms of $\mathcal{J}^2 = J_a^2 + J_b^2 + J_c^2$:

$$E = \frac{\mathcal{J}^2 - J_a^2}{2I_{\perp}} + \frac{J_a^2}{2I_{\parallel}} = \frac{\mathcal{J}^2}{2I} + \left(\frac{1}{2I_{\parallel}} - \frac{1}{2I_{\perp}}\right)J_a^2 \tag{13.28}$$

Now we generate the quantum expression by replacing $\mathcal{J}^2$ by $J(J+1)\hbar^2$, where J is the angular momentum quantum number. We also know from the quantum theory of angular momentum (Section 9.7b) that the component of angular momentum about any axis is restricted to the values $K\hbar$, with $K = 0, \pm 1, \ldots, \pm J$. ($K$ is the quantum number used to signify a component on the principal axis; M_J is reserved for a component on an externally defined axis.) Therefore, we also replace J_a^2 by $K^2\hbar^2$. It follows that the rotational terms are

$$F(J,K) = BJ(J+1) + (A-B)K^2 \qquad J = 0, 1, 2, \ldots \qquad K = 0, \pm 1, \ldots, \pm J \tag{13.29}$$

with

$$A = \frac{\hbar}{4\pi c I_{\parallel}} \qquad B = \frac{\hbar}{4\pi c I_{\perp}} \qquad [13.30]$$

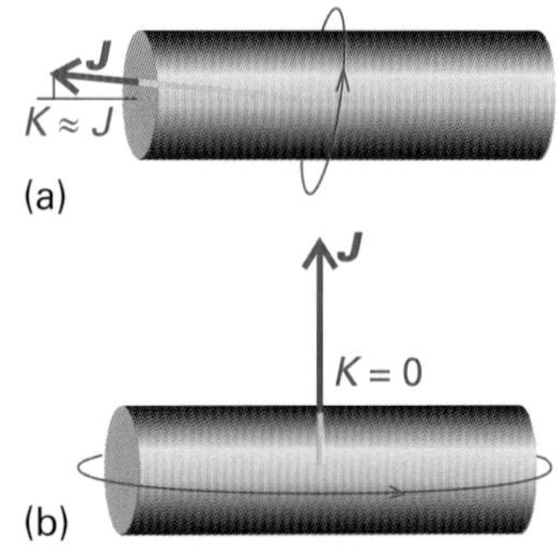

Fig. 13.13 The significance of the quantum number K. (a) When $|K|$ is close to its maximum value, J, most of the molecular rotation is around the figure axis. (b) When $K = 0$ the molecule has no angular momentum about its principal axis: it is undergoing end-over-end rotation.

Equation 13.29 matches what we should expect for the dependence of the energy levels on the two distinct moments of inertia of the molecule. When $K = 0$, there is no component of angular momentum about the principal axis, and the energy levels depend only on $I_{\perp}$ (Fig. 13.13). When $K = \pm J$, almost all the angular momentum arises from rotation around the principal axis, and the energy levels are determined largely by $I_{\parallel}$. The sign of K does not affect the energy because opposite values of K correspond to opposite senses of rotation, and the energy does not depend on the sense of rotation.

Example 13.2 *Calculating the rotational energy levels of a molecule*

A $^{14}NH_3$ molecule is a symmetric rotor with bond length 101.2 pm and HNH bond angle 106.7°. Calculate its rotational terms.

A note on good practice To calculate moments of inertia precisely, we need to specify the nuclide.

Method Begin by calculating the rotational constants A and B by using the expressions for moments of inertia given in Table 13.1. Then use eqn 13.29 to find the rotational terms.

Answer Substitution of $m_A = 1.0078$ u, $m_B = 14.0031$ u, $R = 101.2$ pm, and $\theta = 106.7°$ into the second of the symmetric rotor expressions in Table 13.1 gives $I_{\parallel} = 4.4128 \times 10^{-47}$ kg m^2 and $I_{\perp} = 2.8059 \times 10^{-47}$ kg m^2. Hence, $A = 6.344$ cm^{-1} and $B = 9.977$ cm^{-1}. It follows from eqn 13.29 that

$$F(J,K)/\text{cm}^{-1} = 9.977J(J+1) - 3.633K^2$$

Upon multiplication by c, $F(J,K)$ acquires units of frequency:

$$F(J,K)/\text{GHz} = 299.1J(J+1) - 108.9K^2$$

For $J = 1$, the energy needed for the molecule to rotate mainly about its figure axis ($K = \pm J$) is equivalent to 16.32 cm^{-1} (489.3 GHz), but end-over-end rotation ($K = 0$) corresponds to 19.95 cm^{-1} (598.1 GHz).

Self-test 13.2 A $CH_3{}^{35}Cl$ molecule has a C—Cl bond length of 178 pm, a C—H bond length of 111 pm, and an HCH angle of 110.5°. Calculate its rotational energy terms.

[$F(J,K)/\text{cm}^{-1} = 0.444J(J+1) + 4.58K^2$; also $F(J,K)/\text{GHz} = 13.3J(J+1) + 137K^2$]

(c) Linear rotors

For a linear rotor (such as CO_2, HCl, and C_2H_2), in which the nuclei are regarded as mass points, the rotation occurs only about an axis perpendicular to the line of atoms and there is zero angular momentum around the line. Therefore, the component of angular momentum around the figure axis of a linear rotor is identically zero, and $K \equiv 0$ in eqn 13.29. The rotational terms of a linear molecule are therefore

$$F(J) = BJ(J+1) \qquad J = 0, 1, 2, \ldots \tag{13.31}$$

This expression is the same as eqn 13.26 but we have arrived at it in a significantly different way: here $K \equiv 0$ but for a spherical rotor $A = B$.

(d) Degeneracies and the Stark effect

The energy of a symmetric rotor depends on J and K, and each level except those with $K = 0$ is doubly degenerate: the states with K and $-K$ have the same energy. However, we must not forget that the angular momentum of the molecule has a component on an external, laboratory-fixed axis. This component is quantized, and its permitted values are $M_J\hbar$, with $M_J = 0, \pm 1, \ldots, \pm J$, giving $2J + 1$ values in all (Fig. 13.14). The quantum number M_J does not appear in the expression for the energy, but it is necessary for a complete specification of the state of the rotor. Consequently, all $2J + 1$ orientations of the rotating molecule have the same energy. It follows that a symmetric rotor level is $2(2J + 1)$-fold degenerate for $K \neq 0$ and $(2J + 1)$-fold degenerate for $K = 0$. A linear rotor has K fixed at 0, but the angular momentum may still have $2J + 1$ components on the laboratory axis, so its degeneracy is $2J + 1$.

A spherical rotor can be regarded as a version of a symmetric rotor in which A is equal to B: The quantum number K may still take any one of $2J + 1$ values, but the energy is independent of which value it takes. Therefore, as well as having a $(2J + 1)$-fold degeneracy arising from its orientation in space, the rotor also has a $(2J + 1)$-fold degeneracy arising from its orientation with respect to an arbitrary axis in the

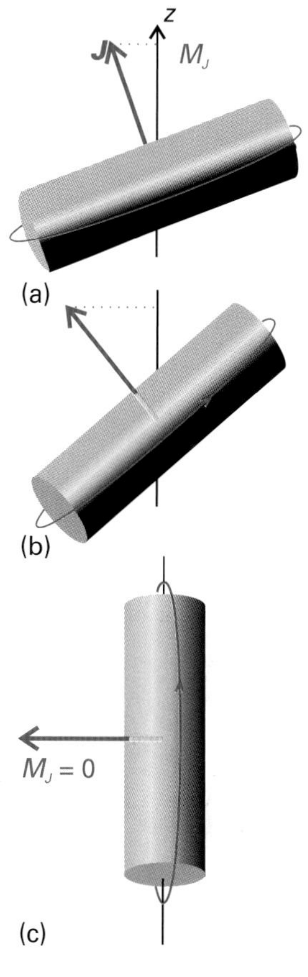

Fig. 13.14 The significance of the quantum number M_J. (a) When M_J is close to its maximum value, J, most of the molecular rotation is around the laboratory z-axis. (b) An intermediate value of M_J. (c) When $M_J = 0$ the molecule has no angular momentum about the z-axis All three diagrams correspond to a state with $K = 0$; there are corresponding diagrams for different values of K, in which the angular momentum makes a different angle to the molecule's principal axis.

molecule. The overall degeneracy of a symmetric rotor with quantum number J is therefore $(2J+1)^2$. This degeneracy increases very rapidly: when $J = 10$, for instance, there are 441 states of the same energy.

The degeneracy associated with the quantum number M_J (the orientation of the rotation in space) is partly removed when an electric field is applied to a polar molecule (e.g. HCl or NH_3), as illustrated in Fig. 13.15. The splitting of states by an electric field is called the **Stark effect**. For a linear rotor in an electric field $\mathcal{E}$, the energy of the state with quantum numbers J and M_J is given by

$$E(J,M_J) = hcBJ(J+1) + a(J,M_J)\mu^2\mathcal{E}^2 \quad (13.32a)$$

where (see *Further reading* for a derivation)

$$a(J,M_J) = \frac{J(J+1) - 3M_J^2}{2hcBJ(J+1)(2J-1)(2J+3)} \quad (13.32b)$$

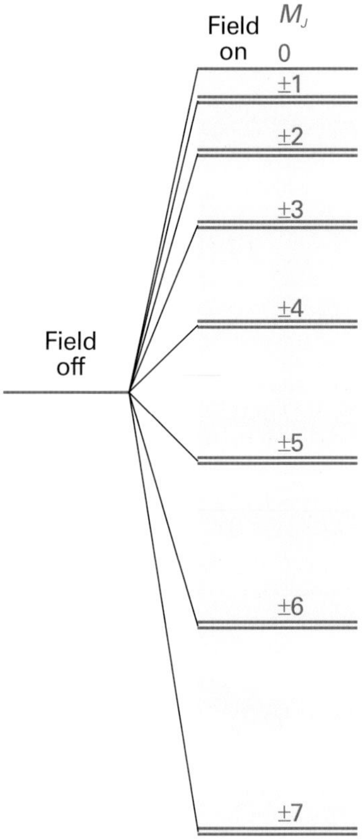

Fig. 13.15 The effect of an electric field on the energy levels of a polar linear rotor. All levels are doubly degenerate except that with $M_J = 0$.

Note that the energy of a state with quantum number M_J depends on the square of the permanent electric dipole moment, μ. The observation of the Stark effect can therefore be used to measure this property, but the technique is limited to molecules that are sufficiently volatile to be studied by rotational spectroscopy. However, as spectra can be recorded for samples at pressures of only about 1 Pa and special techniques (such as using an intense laser beam or an electrical discharge) can be used to vaporize even some quite nonvolatile substances, a wide variety of samples may be studied. Sodium chloride, for example, can be studied as diatomic NaCl molecules at high temperatures.

(e) Centrifugal distortion

We have treated molecules as rigid rotors. However, the atoms of rotating molecules are subject to centrifugal forces that tend to distort the molecular geometry and change the moments of inertia (Fig. 13.16). The effect of centrifugal distortion on a diatomic molecule is to stretch the bond and hence to increase the moment of inertia. As a result, centrifugal distortion reduces the rotational constant and consequently the energy levels are slightly closer than the rigid-rotor expressions predict. The effect is usually taken into account largely empirically by subtracting a term from the energy and writing

$$F(J) = BJ(J+1) - D_J J^2(J+1)^2 \quad (13.33)$$

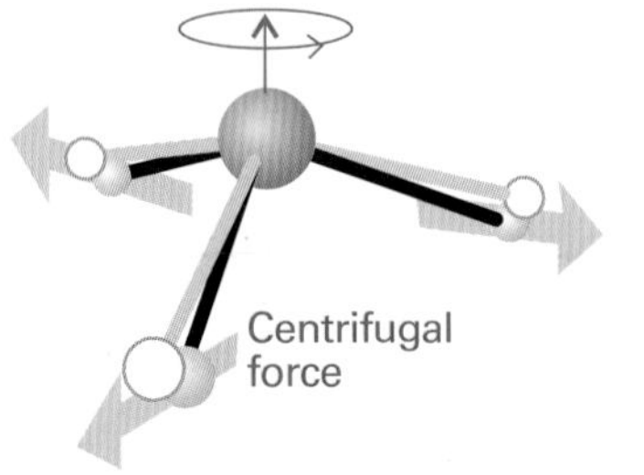

Fig. 13.16 The effect of rotation on a molecule. The centrifugal force arising from rotation distorts the molecule, opening out bond angles and stretching bonds slightly. The effect is to increase the moment of inertia of the molecule and hence to decrease its rotational constant.

The parameter D_J is the **centrifugal distortion constant**. It is large when the bond is easily stretched. The centrifugal distortion constant of a diatomic molecule is related to the vibrational wavenumber of the bond, $\tilde{\nu}$ (which, as we shall see later, is a measure of its stiffness), through the approximate relation (see Problem 13.22)

$$D_J = \frac{4B^3}{\tilde{\nu}^2} \quad (13.34)$$

Hence the observation of the convergence of the rotational levels as J increases can be interpreted in terms of the rigidity of the bond.

13.6 Rotational transitions

Typical values of B for small molecules are in the region of 0.1 to10 cm^{-1} (for example, 0.356 cm^{-1} for NF_3 and 10.59 cm^{-1} for HCl), so rotational transitions lie in the microwave region of the spectrum. The transitions are detected by monitoring the net absorption of microwave radiation. Modulation of the transmitted intensity can be achieved by varying the energy levels with an oscillating electric field. In this **Stark**

modulation, an electric field of about 10^5 V m^{-1} and a frequency of between 10 and 100 kHz is applied to the sample.

(a) Rotational selection rules

We have already remarked (Section 13.2) that the gross selection rule for the observation of a pure rotational spectrum is that a molecule must have a permanent electric dipole moment. That is, *for a molecule to give a pure rotational spectrum, it must be polar*. The classical basis of this rule is that a polar molecule appears to possess a fluctuating dipole when rotating, but a nonpolar molecule does not (Fig.13.17). The permanent dipole can be regarded as a handle with which the molecule stirs the electromagnetic field into oscillation (and vice versa for absorption). Homonuclear diatomic molecules and symmetrical linear molecules such as CO_2 are rotationally inactive. Spherical rotors cannot have electric dipole moments unless they become distorted by rotation, so they are also inactive except in special cases. An example of a spherical rotor that does become sufficiently distorted for it to acquire a dipole moment is SiH_4, which has a dipole moment of about 8.3 μD by virtue of its rotation when $J \approx 10$ (for comparison, HCl has a permanent dipole moment of 1.1 D; molecular dipole moments and their units are discussed in Section 18.1). The pure rotational spectrum of SiH_4 has been detected by using long path lengths (10 m) through high-pressure (4 atm) samples.

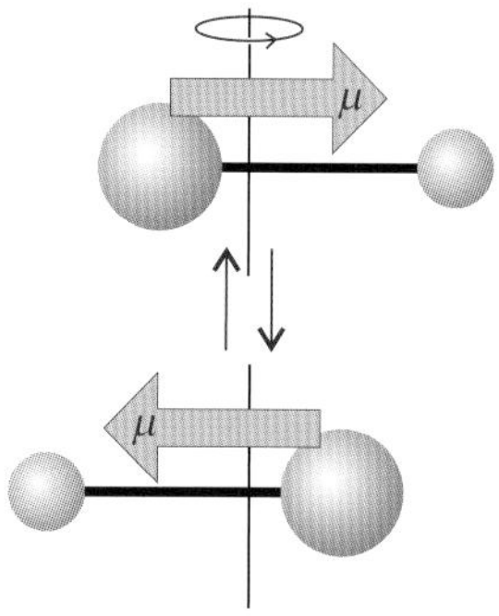

Fig. 13.17 To a stationary observer, a rotating polar molecule looks like an oscillating dipole that can stir the electromagnetic field into oscillation (and vice versa for absorption). This picture is the classical origin of the gross selection rule for rotational transitions.

Illustration 13.2 *Identifying rotationally active molecules*

Of the molecules N_2, CO_2, OCS, H_2O, $CH_2{=}CH_2$, C_6H_6, only OCS and H_2O are polar, so only these two molecules have microwave spectra.

Self-test 13.3 Which of the molecules H_2, NO, N_2O, CH_4 can have a pure rotational spectrum? [NO, N_2O]

The specific rotational selection rules are found by evaluating the transition dipole moment between rotational states. We show in *Further information 13.2* that, for a linear molecule, the transition moment vanishes unless the following conditions are fulfilled:

$$\Delta J = \pm 1 \qquad \Delta M_J = 0, \pm 1 \tag{13.35}$$

The transition $\Delta J = +1$ corresponds to absorption and the transition $\Delta J = -1$ corresponds to emission. The allowed change in J in each case arises from the conservation of angular momentum when a photon, a spin-1 particle, is emitted or absorbed (Fig. 13.18).

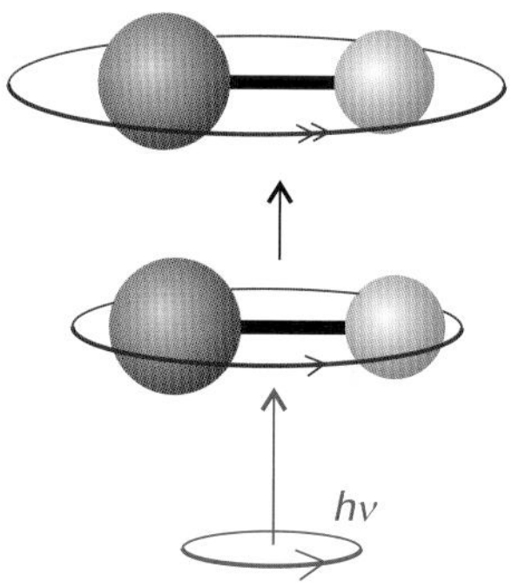

Fig. 13.18 When a photon is absorbed by a molecule, the angular momentum of the combined system is conserved. If the molecule is rotating in the same sense as the spin of the incoming photon, then J increases by 1.

When the transition moment is evaluated for all possible relative orientations of the molecule to the line of flight of the photon, it is found that the total $J + 1 \leftrightarrow J$ transition intensity is proportional to

$$|\mu_{J+1,J}|^2 = \left(\frac{J+1}{2J+1}\right)\mu_0^2 \rightarrow \tfrac{1}{2}\mu_0^2 \qquad \text{for } J \gg 1 \tag{13.36}$$

where μ_0 is the permanent electric dipole moment of the molecule. The intensity is proportional to the square of the permanent electric dipole moment, so strongly polar molecules give rise to much more intense rotational lines than less polar molecules.

For symmetric rotors, an additional selection rule states that $\Delta K = 0$. To understand this rule, consider the symmetric rotor NH_3, where the electric dipole moment lies parallel to the figure axis. Such a molecule cannot be accelerated into different states of rotation around the figure axis by the absorption of radiation, so $\Delta K = 0$.

(b) The appearance of rotational spectra

When these selection rules are applied to the expressions for the energy levels of a rigid symmetric or linear rotor, it follows that the wavenumbers of the allowed $J+1 \leftarrow J$ absorptions are

$$\tilde{\nu}(J+1 \leftarrow J) = 2B(J+1) \qquad J = 0, 1, 2, \ldots \tag{13.37}$$

When centrifugal distortion is taken into account, the corresponding expression is

$$\tilde{\nu}(J+1 \leftarrow J) = 2B(J+1) - 4D_J(J+1)^3 \tag{13.38}$$

However, because the second term is typically very small compared with the first, the appearance of the spectrum closely resembles that predicted from eqn 13.37.

Example 13.3 *Predicting the appearance of a rotational spectrum*

Predict the form of the rotational spectrum of $^{14}NH_3$.

Method We calculated the energy levels in Example 13.2. The $^{14}NH_3$ molecule is a polar symmetric rotor, so the selection rules $\Delta J = \pm 1$ and $\Delta K = 0$ apply. For absorption, $\Delta J = +1$ and we can use eqn 13.37. Because $B = 9.977$ cm^{-1}, we can draw up the following table for the $J+1 \leftarrow J$ transitions.

J	0	1	2	3	...
$\tilde{\nu}$/cm^{-1}	19.95	39.91	59.86	79.82	...
ν/GHz	598.1	1197	1795	2393	...

The line spacing is 19.95 cm^{-1} (598.1 GHz).

Self-test 13.4 Repeat the problem for $C^{35}ClH_3$ (see Self-test 13.2 for details).
[Lines of separation 0.888 cm^{-1} (26.6 GHz)]

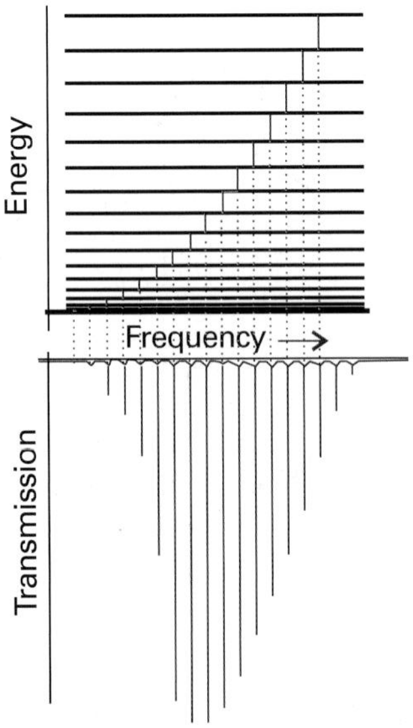

Fig. 13.19 The rotational energy levels of a linear rotor, the transitions allowed by the selection rule $\Delta J = \pm 1$, and a typical pure rotational absorption spectrum (displayed here in terms of the radiation transmitted through the sample). The intensities reflect the populations of the initial level in each case and the strengths of the transition dipole moments.

The form of the spectrum predicted by eqn 13.37 is shown in Fig. 13.19. The most significant feature is that it consists of a series of lines with wavenumbers $2B$, $4B$, $6B$, ... and of separation $2B$. The measurement of the line spacing gives B, and hence the moment of inertia perpendicular to the principal axis of the molecule. Because the masses of the atoms are known, it is a simple matter to deduce the bond length of a diatomic molecule. However, in the case of a polyatomic molecule such as OCS or NH_3, the analysis gives only a single quantity, $I_\perp$, and it is not possible to infer both bond lengths (in OCS) or the bond length and bond angle (in NH_3). This difficulty can be overcome by using isotopically substituted molecules, such as ABC and A′BC; then, by assuming that R(A—B) = R(A′—B), both A—B and B—C bond lengths can be extracted from the two moments of inertia. A famous example of this procedure is the study of OCS; the actual calculation is worked through in Problem 13.10. The assumption that bond lengths are unchanged by isotopic substitution is only an approximation, but it is a good approximation in most cases.

The intensities of spectral lines increase with increasing J and pass through a maximum before tailing off as J becomes large. The most important reason for the maximum in intensity is the existence of a maximum in the population of rotational levels. The Boltzmann distribution (*Molecular interpretation 3.1*) implies that the population of each state decays exponentially with increasing J, but the degeneracy of the levels increases. Specifically, the population of a rotational energy level J is given by the Boltzmann expression

$$N_J \propto N g_J e^{-E_J/kT}$$

where N is the total number of molecules and g_J is the degeneracy of the level J. The value of J corresponding to a maximum of this expression is found by treating J as a continuous variable, differentiating with respect to J, and then setting the result equal to zero. The result is (see Problem 13.24)

$$J_{\max} \approx \left(\frac{kT}{2hcB}\right)^{1/2} - \tfrac{1}{2} \qquad (13.39)$$

For a typical molecule (for example, OCS, with $B = 0.2\ \text{cm}^{-1}$) at room temperature, $kT \approx 1000hcB$, so $J_{\max} \approx 30$. However, it must be recalled that the intensity of each transition also depends on the value of J (eqn 13.36) and on the population difference between the two states involved in the transition (see Section 13.2). Hence the value of J corresponding to the most intense line is not quite the same as the value of J for the most highly populated level.

13.7 Rotational Raman spectra

The gross selection rule for rotational Raman transitions is *that the molecule must be anisotropically polarizable*. We begin by explaining what this means. A formal derivation of this rule is given in *Further information 13.2*.

The distortion of a molecule in an electric field is determined by its polarizability, α (Section 18.2). More precisely, if the strength of the field is $\mathcal{E}$, then the molecule acquires an induced dipole moment of magnitude

$$\mu = \alpha\mathcal{E} \qquad (13.40)$$

in addition to any permanent dipole moment it may have. An atom is isotropically polarizable. That is, the same distortion is induced whatever the direction of the applied field. The polarizability of a spherical rotor is also isotropic. However, non-spherical rotors have polarizabilities that do depend on the direction of the field relative to the molecule, so these molecules are anisotropically polarizable (Fig. 13.20). The electron distribution in H_2, for example, is more distorted when the field is applied parallel to the bond than when it is applied perpendicular to it, and we write $\alpha_{\parallel} > \alpha_{\perp}$.

All linear molecules and diatomics (whether homonuclear or heteronuclear) have anisotropic polarizabilities, and so are rotationally Raman active. This activity is one reason for the importance of rotational Raman spectroscopy, for the technique can be used to study many of the molecules that are inaccessible to microwave spectroscopy. Spherical rotors such as CH_4 and SF_6, however, are rotationally Raman inactive as well as microwave inactive. This inactivity does not mean that such molecules are never found in rotationally excited states. Molecular collisions do not have to obey such restrictive selection rules, and hence collisions between molecules can result in the population of any rotational state.

We show in *Further information 13.2* that the specific rotational Raman selection rules are

$$\text{Linear rotors:} \quad \Delta J = 0, \pm 2 \qquad (13.41)$$

$$\text{Symmetric rotors:} \quad \Delta J = 0, \pm 1, \pm 2; \quad \Delta K = 0$$

The $\Delta J = 0$ transitions do not lead to a shift of the scattered photon's frequency in pure rotational Raman spectroscopy, and contribute to the unshifted Rayleigh radiation.

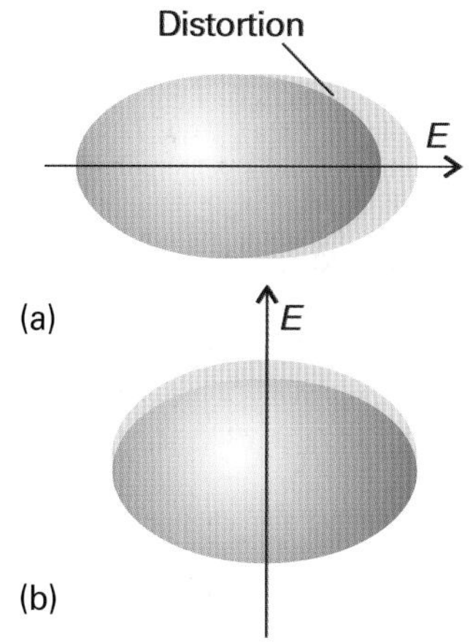

Fig. 13.20 An electric field applied to a molecule results in its distortion, and the distorted molecule acquires a contribution to its dipole moment (even if it is nonpolar initially). The polarizability may be different when the field is applied (a) parallel or (b) perpendicular to the molecular axis (or, in general, in different directions relative to the molecule); if that is so, then the molecule has an anisotropic polarizability.

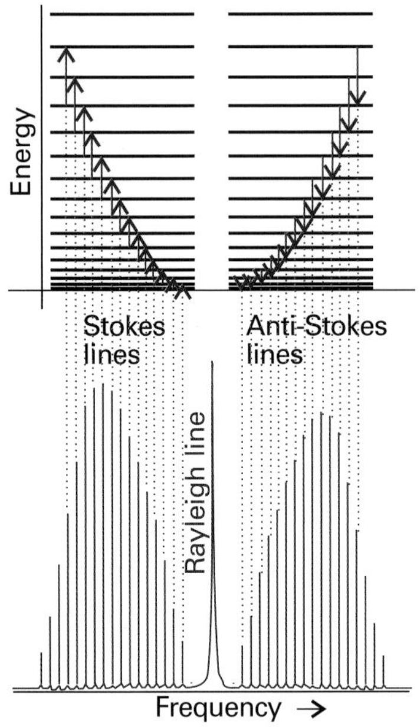

Fig. 13.21 The rotational energy levels of a linear rotor and the transitions allowed by the $\Delta J = \pm 2$ Raman selection rules. The form of a typical rotational Raman spectrum is also shown. The Rayleigh line is much stronger than depicted in the figure; it is shown as a weaker line to improve visualization of the Raman lines.

We can predict the form of the Raman spectrum of a linear rotor by applying the selection rule $\Delta J = \pm 2$ to the rotational energy levels (Fig. 13.21). When the molecule makes a transition with $\Delta J = +2$, the scattered radiation leaves the molecule in a higher rotational state, so the wavenumber of the incident radiation, initially $\tilde{\nu}_i$, is decreased. These transitions account for the Stokes lines in the spectrum:

$$\tilde{\nu}(J+2 \leftarrow J) = \tilde{\nu}_i - \{F(J+2) - F(J)\} = \tilde{\nu}_i - 2B(2J+3) \qquad (13.42a)$$

The Stokes lines appear to low frequency of the incident radiation and at displacements $6B$, $10B$, $14B$, . . . from $\tilde{\nu}_i$ for $J = 0, 1, 2, \ldots$. When the molecule makes a transition with $\Delta J = -2$, the scattered photon emerges with increased energy. These transitions account for the anti-Stokes lines of the spectrum:

$$\tilde{\nu}(J-2 \leftarrow J) = \tilde{\nu}_i + \{F(J) - F(J-2)\} = \tilde{\nu}_i + 2B(2J-1) \qquad (13.42b)$$

The anti-Stokes lines occur at displacements of $6B$, $10B$, $14B$, . . . (for $J = 2, 3, 4, \ldots$; $J = 2$ is the lowest state that can contribute under the selection rule $\Delta J = -2$) to high frequency of the incident radiation. The separation of adjacent lines in both the Stokes and the anti-Stokes regions is $4B$, so from its measurement $I_\perp$ can be determined and then used to find the bond lengths exactly as in the case of microwave spectroscopy.

Example 13.4 *Predicting the form of a Raman spectrum*

Predict the form of the rotational Raman spectrum of $^{14}N_2$, for which $B = 1.99\ cm^{-1}$, when it is exposed to monochromatic 336.732 nm laser radiation.

Method The molecule is rotationally Raman active because end-over-end rotation modulates its polarizability as viewed by a stationary observer. The Stokes and anti-Stokes lines are given by eqn 13.42.

Answer Because $\lambda_i = 336.732$ nm corresponds to $\tilde{\nu}_i = 29\,697.2\ cm^{-1}$, eqns 13.42a and 13.42b give the following line positions:

J	0	1	2	3
Stokes lines				
$\tilde{\nu}/cm^{-1}$	29 685.3	29 677.3	29 669.3	29 661.4
λ/nm	336.868	336.958	337.048	337.139
Anti-Stokes lines				
$\tilde{\nu}/cm^{-1}$			29 709.1	29 717.1
λ/nm			336.597	336.507

There will be a strong central line at 336.732 nm accompanied on either side by lines of increasing and then decreasing intensity (as a result of transition moment and population effects). The spread of the entire spectrum is very small, so the incident light must be highly monochromatic.

Self-test 13.5 Repeat the calculation for the rotational Raman spectrum of NH_3 ($B = 9.977\ cm^{-1}$).

[Stokes lines at 29 637.3, 29 597.4, 29 557.5, 29 517.6 cm^{-1}, anti-Stokes lines at 29 757.1, 29 797.0 cm^{-1}.]

13.8 Nuclear statistics and rotational states

If eqn 13.42 is used in conjunction with the rotational Raman spectrum of CO_2, the rotational constant is inconsistent with other measurements of C—O bond lengths.

The results are consistent only if it is supposed that the molecule can exist in states with even values of J, so the Stokes lines are $2 \leftarrow 0, 4 \leftarrow 2, \ldots$ and not $5 \leftarrow 3, 3 \leftarrow 1, \ldots$.

The explanation of the missing lines is the Pauli principle and the fact that O nuclei are spin-0 bosons: just as the Pauli principle excludes certain electronic states, so too does it exclude certain molecular rotational states. The form of the Pauli principle given in *Justification 10.4* states that, when two identical bosons are exchanged, the overall wavefunction must remain unchanged in every respect, including sign. In particular, when a CO_2 molecule rotates through 180°, two identical O nuclei are interchanged, so the overall wavefunction of the molecule must remain unchanged. However, inspection of the form of the rotational wavefunctions (which have the same form as the *s*, *p*, etc. orbitals of atoms) shows that they change sign by $(-1)^J$ under such a rotation (Fig. 13.22). Therefore, only even values of J are permissible for CO_2, and hence the Raman spectrum shows only alternate lines.

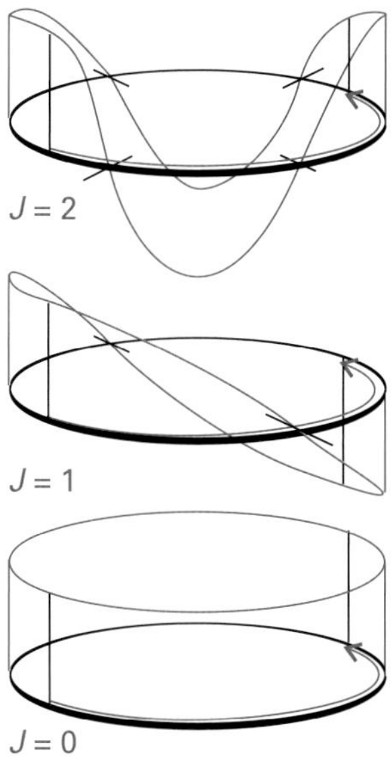

Fig. 13.22 The symmetries of rotational wavefunctions (shown here, for simplicity as a two-dimensional rotor) under a rotation through 180°. Wavefunctions with J even do not change sign; those with J odd do change sign.

The selective occupation of rotational states that stems from the Pauli principle is termed **nuclear statistics**. Nuclear statistics must be taken into account whenever a rotation interchanges equivalent nuclei. However, the consequences are not always as simple as for CO_2 because there are complicating features when the nuclei have nonzero spin: there may be several different relative nuclear spin orientations consistent with even values of J and a different number of spin orientations consistent with odd values of J. For molecular hydrogen and fluorine, for instance, with their two identical spin-$\frac{1}{2}$ nuclei, we show in the *Justification* below that there are three times as many ways of achieving a state with odd J than with even J, and there is a corresponding 3:1 alternation in intensity in their rotational Raman spectra (Fig. 13.23). In general, for a homonuclear diatomic molecule with nuclei of spin I, the numbers of ways of achieving states of odd and even J are in the ratio

$$\frac{\text{Number of ways of achieving odd } J}{\text{Number of ways of achieving even } J} = \begin{cases} (I+1)/I \text{ for half-integral spin nuclei} \\ I/(I+1) \text{ for integral spin nuclei} \end{cases} \tag{13.43}$$

For hydrogen, $I = \frac{1}{2}$, and the ratio is 3:1. For N_2, with $I = 1$, the ratio is 1:2.

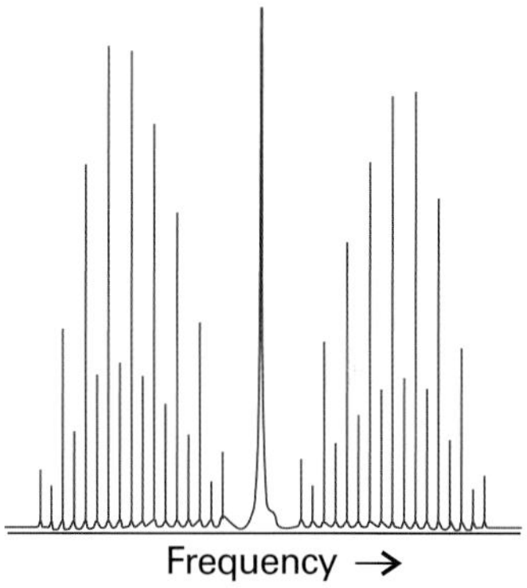

Fig. 13.23 The rotational Raman spectrum of a diatomic molecule with two identical spin-$\frac{1}{2}$ nuclei shows an alternation in intensity as a result of nuclear statistics. The Rayleigh line is much stronger than depicted in the figure; it is shown as a weaker line to improve visualization of the Raman lines.

Justification 13.4 *The effect of nuclear statistics on rotational spectra*

Hydrogen nuclei are fermions, so the Pauli principle requires the overall wavefunction to change sign under particle interchange. However, the rotation of an H_2 molecule through 180° has a more complicated effect than merely relabelling the nuclei, because it interchanges their spin states too if the nuclear spins are paired (↑↓) but not if they are parallel (↑↑).

For the overall wavefunction of the molecule to change sign when the spins are parallel, the rotational wavefunction must change sign. Hence, only odd values of J are allowed. In contrast, if the nuclear spins are paired, their wavefunction is $\alpha(A)\beta(B) - \alpha(B)\beta(A)$, which changes sign when α and β are exchanged in order to bring about a simple $A \leftrightarrow B$ interchange overall (Fig. 13.24). Therefore, for the overall wavefunction to change sign in this case requires the rotational wavefunction *not* to change sign. Hence, only even values of J are allowed if the nuclear spins are paired.

As there are three nuclear spin states with parallel spins (just like the triplet state of two parallel electrons, as in Fig. 10.24), but only one state with paired spins (the analogue of the singlet state of two electrons, see Fig. 10.18), it follows that the populations of the odd J and even J states should be in the ratio of 3:1, and hence the intensities of transitions originating in these levels will be in the same ratio.

Different relative nuclear spin orientations change into one another only very slowly, so an H_2 molecule with parallel nuclear spins remains distinct from one with

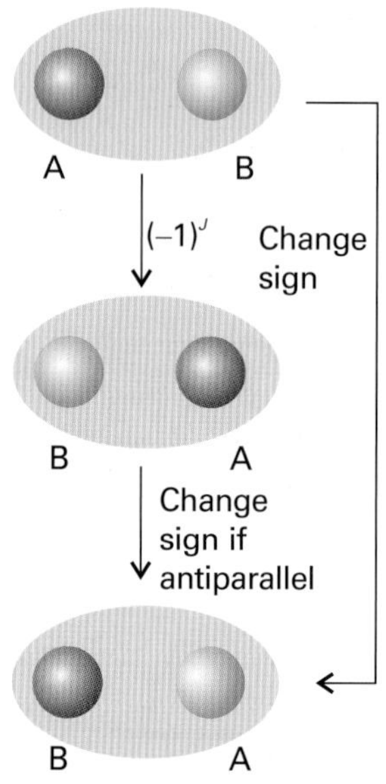

Fig. 13.24 The interchange of two identical fermion nuclei results in the change in sign of the overall wavefunction. The relabelling can be thought of as occurring in two steps: the first is a rotation of the molecule; the second is the interchange of unlike spins (represented by the different colours of the nuclei). The wavefunction changes sign in the second step if the nuclei have antiparallel spins.

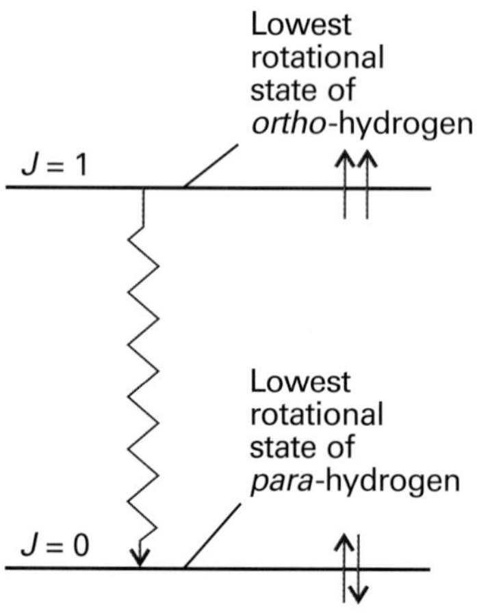

Fig. 13.25 When hydrogen is cooled, the molecules with parallel nuclear spins accumulate in their lowest available rotational state, the one with $J = 1$. They can enter the lowest rotational state ($J = 0$) only if the spins change their relative orientation and become antiparallel. This is a slow process under normal circumstances, so energy is slowly released.

paired nuclear spins for long periods. The two forms of hydrogen can be separated by physical techniques, and stored. The form with parallel nuclear spins is called ***ortho*-hydrogen** and the form with paired nuclear spins is called ***para*-hydrogen**. Because *ortho*-hydrogen cannot exist in a state with $J = 0$, it continues to rotate at very low temperatures and has an effective rotational zero-point energy (Fig. 13.25). This energy is of some concern to manufacturers of liquid hydrogen, for the slow conversion of *ortho*-hydrogen into *para*-hydrogen (which can exist with $J = 0$) as nuclear spins slowly realign releases rotational energy, which vaporizes the liquid. Techniques are used to accelerate the conversion of *ortho*-hydrogen to *para*-hydrogen to avoid this problem. One such technique is to pass hydrogen over a metal surface: the molecules adsorb on the surface as atoms, which then recombine in the lower energy *para*-hydrogen form.

The vibrations of diatomic molecules

In this section, we adopt the same strategy of finding expressions for the energy levels, establishing the selection rules, and then discussing the form of the spectrum. We shall also see how the simultaneous excitation of rotation modifies the appearance of a vibrational spectrum.

13.9 Molecular vibrations

We base our discussion on Fig. 13.26, which shows a typical potential energy curve (as in Fig. 11.1) of a diatomic molecule. In regions close to R_e (at the minimum of the curve) the potential energy can be approximated by a parabola, so we can write

$$V = \tfrac{1}{2}kx^2 \qquad x = R - R_e \tag{13.44}$$

where k is the **force constant** of the bond. The steeper the walls of the potential (the stiffer the bond), the greater the force constant.

To see the connection between the shape of the molecular potential energy curve and the value of k, note that we can expand the potential energy around its minimum by using a Taylor series:

$$V(x) = V(0) + \left(\frac{dV}{dx}\right)_0 x + \tfrac{1}{2}\left(\frac{d^2V}{dx^2}\right)_0 x^2 + \cdots \tag{13.45}$$

The term $V(0)$ can be set arbitrarily to zero. The first derivative of V is 0 at the minimum. Therefore, the first surviving term is proportional to the square of the displacement. For small displacements we can ignore all the higher terms, and so write

$$V(x) \approx \tfrac{1}{2}\left(\frac{d^2V}{dx^2}\right)_0 x^2 \tag{13.46}$$

Therefore, the first approximation to a molecular potential energy curve is a parabolic potential, and we can identify the force constant as

$$k = \left(\frac{d^2V}{dx^2}\right)_0 \qquad [13.47]$$

We see that if the potential energy curve is sharply curved close to its minimum, then k will be large. Conversely, if the potential energy curve is wide and shallow, then k will be small (Fig. 13.27).

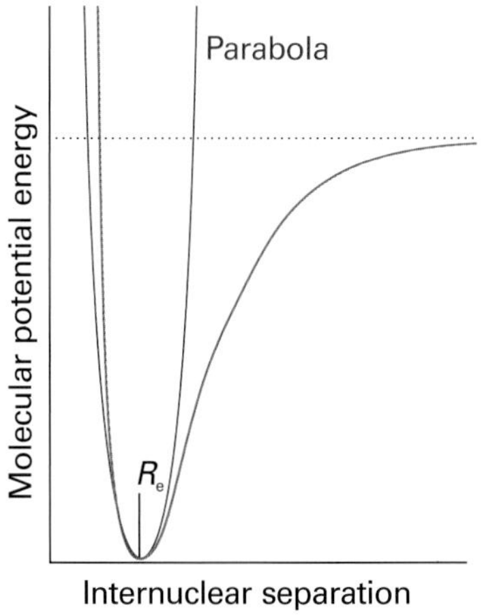

Fig. 13.26 A molecular potential energy curve can be approximated by a parabola near the bottom of the well. The parabolic potential leads to harmonic oscillations. At high excitation energies the parabolic approximation is poor (the true potential is less confining), and it is totally wrong near the dissociation limit.

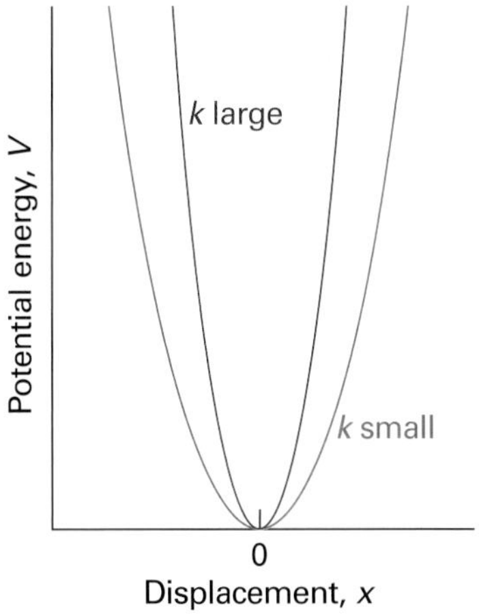

Fig. 13.27 The force constant is a measure of the curvature of the potential energy close to the equilibrium extension of the bond. A strongly confining well (one with steep sides, a stiff bond) corresponds to high values of *k*.

Comment 13.5

It is often useful to express a function $f(x)$ in the vicinity of $x = a$ as an infinite Taylor series of the form:

$$f(x) = f(a) + \left(\frac{df}{dx}\right)_a (x-a) + \frac{1}{2!}\left(\frac{d^2f}{dx^2}\right)_a (x-a)^2 + \cdots + \frac{1}{n!}\left(\frac{d^nf}{dx^n}\right)_a (x-a)^n + \cdots$$

where $n = 0, 1, 2, \ldots$.

The Schrödinger equation for the relative motion of two atoms of masses m_1 and m_2 with a parabolic potential energy is

$$-\frac{\hbar^2}{2m_{\text{eff}}}\frac{d^2\psi}{dx^2} + \tfrac{1}{2}kx^2\psi = E\psi \tag{13.48}$$

where m_{eff} is the **effective mass**:

$$m_{\text{eff}} = \frac{m_1 m_2}{m_1 + m_2} \qquad [13.49]$$

These equations are derived in the same way as in *Further information 10.1*, but here the separation of variables procedure is used to separate the relative motion of the atoms from the motion of the molecule as a whole. In that context, the effective mass is called the 'reduced mass', and the name is widely used in this context too.

The Schrödinger equation in eqn 13.48 is the same as eqn 9.24 for a particle of mass m undergoing harmonic motion. Therefore, we can use the results of Section 9.4 to write down the permitted vibrational energy levels:

$$E_v = (v + \tfrac{1}{2})\hbar\omega \qquad \omega = \left(\frac{k}{m_{\text{eff}}}\right)^{1/2} \qquad v = 0, 1, 2, \ldots \tag{13.50}$$

The **vibrational terms** of a molecule, the energies of its vibrational states expressed in wavenumbers, are denoted $G(v)$, with $E_v = hcG(v)$, so

$$G(v) = (v + \tfrac{1}{2})\tilde{\nu} \qquad \tilde{\nu} = \frac{1}{2\pi c}\left(\frac{k}{m_{\text{eff}}}\right)^{1/2} \tag{13.51}$$

The vibrational wavefunctions are the same as those discussed in Section 9.5.

It is important to note that the vibrational terms depend on the effective mass of the molecule, not directly on its total mass. This dependence is physically reasonable, for if atom 1 were as heavy as a brick wall, then we would find $m_{\text{eff}} \approx m_2$, the mass of the lighter atom. The vibration would then be that of a light atom relative to that of a stationary wall (this is approximately the case in HI, for example, where the I atom barely moves and $m_{\text{eff}} \approx m_{\text{H}}$). For a homonuclear diatomic molecule $m_1 = m_2$, and the effective mass is half the total mass: $m_{\text{eff}} = \frac{1}{2}m$.

Illustration 13.3 *Calculating a vibrational wavenumber*

An HCl molecule has a force constant of 516 N m^{-1}, a reasonably typical value for a single bond. The effective mass of $^1H^{35}Cl$ is 1.63×10^{-27} kg (note that this mass is very close to the mass of the hydrogen atom, 1.67×10^{-27} kg, so the Cl atom is acting like a brick wall). These values imply $\omega = 5.63 \times 10^{14}$ s^{-1}, $\nu = 89.5$ THz (1 THz $= 10^{12}$ Hz), $\tilde{\nu} = 2990$ cm^{-1}, $\lambda = 3.35$ μm. These characteristics correspond to electromagnetic radiation in the infrared region.

13.10 Selection rules

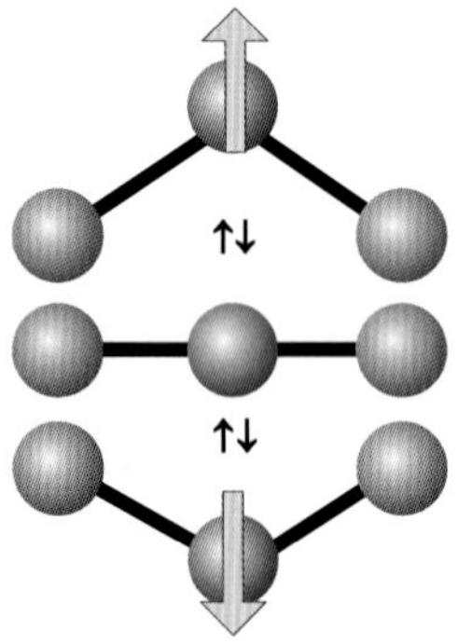

Fig. 13.28 The oscillation of a molecule, even if it is nonpolar, may result in an oscillating dipole that can interact with the electromagnetic field.

The gross selection rule for a change in vibrational state brought about by absorption or emission of radiation is that *the electric dipole moment of the molecule must change when the atoms are displaced relative to one another*. Such vibrations are said to be **infrared active**. The classical basis of this rule is that the molecule can shake the electromagnetic field into oscillation if its dipole changes as it vibrates, and vice versa (Fig. 13.28); its formal basis is given in *Further information 13.2*. Note that the molecule need not have a permanent dipole: the rule requires only a change in dipole moment, possibly from zero. Some vibrations do not affect the molecule's dipole moment (e.g. the stretching motion of a homonuclear diatomic molecule), so they neither absorb nor generate radiation: such vibrations are said to be **infrared inactive**. Homonuclear diatomic molecules are infrared inactive because their dipole moments remain zero however long the bond; heteronuclear diatomic molecules are infrared active.

Illustration 13.4 *Identifying infrared active molecules*

Of the molecules N_2, CO_2, OCS, H_2O, $CH_2{=}CH_2$, and C_6H_6, all except N_2 possess at least one vibrational mode that results in a change of dipole moment, so all except N_2 can show a vibrational absorption spectrum. Not all the modes of complex molecules are vibrationally active. For example, the symmetric stretch of CO_2, in which the O—C—O bonds stretch and contract symmetrically is inactive because it leaves the dipole moment unchanged (at zero).

Self-test 13.6 Which of the molecules H_2, NO, N_2O, and CH_4 have infrared active vibrations? [NO, N_2O, CH_4]

The specific selection rule, which is obtained from an analysis of the expression for the transition moment and the properties of integrals over harmonic oscillator wavefunctions (as shown in *Further information 13.2*), is

$$\Delta v = \pm 1 \tag{13.52}$$

Transitions for which $\Delta v = +1$ correspond to absorption and those with $\Delta v = -1$ correspond to emission.

It follows from the specific selection rules that the wavenumbers of allowed vibrational transitions, which are denoted $\Delta G_{v+\frac{1}{2}}$ for the transition $v+1 \leftarrow v$, are

$$\Delta G_{v+\frac{1}{2}} = G(v+1) - G(v) = \tilde{\nu} \tag{13.53}$$

As we have seen, $\tilde{\nu}$ lies in the infrared region of the electromagnetic spectrum, so vibrational transitions absorb and generate infrared radiation.

At room temperature $kT/hc \approx 200\ \text{cm}^{-1}$, and most vibrational wavenumbers are significantly greater than $200\ \text{cm}^{-1}$. It follows from the Boltzmann distribution that almost all the molecules will be in their vibrational ground states initially. Hence, the dominant spectral transition will be the **fundamental transition**, $1 \leftarrow 0$. As a result, the spectrum is expected to consist of a single absorption line. If the molecules are formed in a vibrationally excited state, such as when vibrationally excited HF molecules are formed in the reaction $H_2 + F_2 \rightarrow 2\,HF^*$, the transitions $5 \rightarrow 4, 4 \rightarrow 3, \ldots$ may also appear (in emission). In the harmonic approximation, all these lines lie at the same frequency, and the spectrum is also a single line. However, as we shall now show, the breakdown of the harmonic approximation causes the transitions to lie at slightly different frequencies, so several lines are observed.

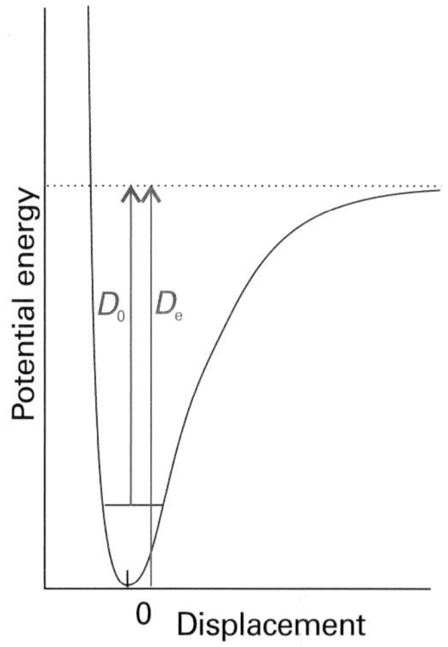

Fig. 13.29 The dissociation energy of a molecule, D_0, differs from the depth of the potential well, D_e, on account of the zero-point energy of the vibrations of the bond.

13.11 Anharmonicity

The vibrational terms in eqn 13.53 are only approximate because they are based on a parabolic approximation to the actual potential energy curve. A parabola cannot be correct at all extensions because it does not allow a bond to dissociate. At high vibrational excitations the swing of the atoms (more precisely, the spread of the vibrational wavefunction) allows the molecule to explore regions of the potential energy curve where the parabolic approximation is poor and additional terms in the Taylor expansion of V (eqn 13.45) must be retained. The motion then becomes **anharmonic**, in the sense that the restoring force is no longer proportional to the displacement. Because the actual curve is less confining than a parabola, we can anticipate that the energy levels become less widely spaced at high excitations.

(a) The convergence of energy levels

One approach to the calculation of the energy levels in the presence of anharmonicity is to use a function that resembles the true potential energy more closely. The **Morse potential energy** is

$$V = hcD_e\{1 - e^{-a(R-R_e)}\}^2 \qquad a = \left(\frac{m_{\text{eff}}\omega^2}{2hcD_e}\right)^{1/2} \tag{13.54}$$

where D_e is the depth of the potential minimum (Fig. 13.29). Near the well minimum the variation of V with displacement resembles a parabola (as can be checked by expanding the exponential as far as the first term) but, unlike a parabola, eqn 13.54 allows for dissociation at large displacements. The Schrödinger equation can be solved for the Morse potential and the permitted energy levels are

$$G(v) = (v+\tfrac{1}{2})\tilde{\nu} - (v+\tfrac{1}{2})^2 x_e\tilde{\nu} \qquad x_e = \frac{a^2\hbar}{2m_{\text{eff}}\omega} = \frac{\tilde{\nu}}{4D_e} \tag{13.55}$$

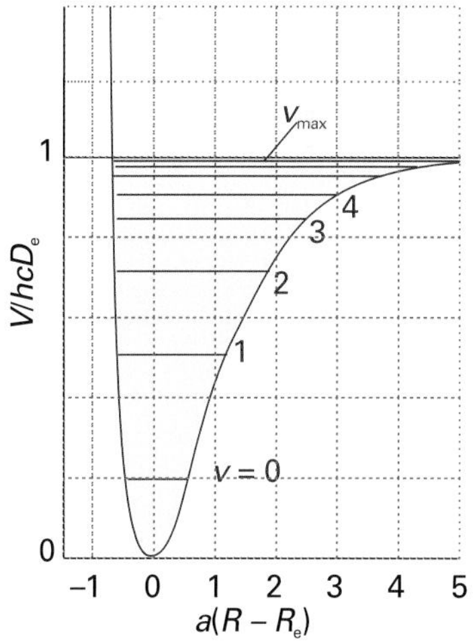

Fig. 13.30 The Morse potential energy curve reproduces the general shape of a molecular potential energy curve. The corresponding Schrödinger equation can be solved, and the values of the energies obtained. The number of bound levels is finite.

The parameter x_e is called the **anharmonicity constant**. The number of vibrational levels of a Morse oscillator is finite, and $v = 0, 1, 2, \ldots, v_{\text{max}}$, as shown in Fig. 13.30 (see also Problem 13.26). The second term in the expression for G subtracts from the first with increasing effect as v increases, and hence gives rise to the convergence of the levels at high quantum numbers.

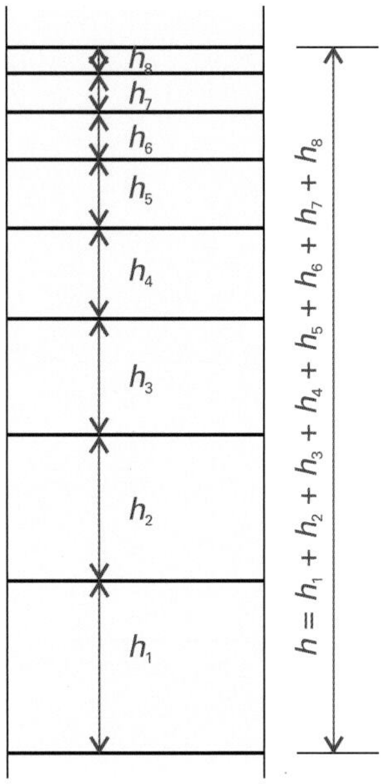

Fig. 13.31 The dissociation energy is the sum of the separations of the vibrational energy levels up to the dissociation limit just as the length of a ladder is the sum of the separations of its rungs.

Although the Morse oscillator is quite useful theoretically, in practice the more general expression

$$G(v) = (v + \tfrac{1}{2})\tilde{\nu} - (v + \tfrac{1}{2})^2 x_e\tilde{\nu} + (v + \tfrac{1}{2})^3 y_e\tilde{\nu} + \cdots \quad (13.56)$$

where x_e, y_e, . . . are empirical dimensionless constants characteristic of the molecule, is used to fit the experimental data and to find the dissociation energy of the molecule. When anharmonicities are present, the wavenumbers of transitions with $\Delta v = +1$ are

$$\Delta G_{v+\frac{1}{2}} = \tilde{\nu} - 2(v + 1)x_e\tilde{\nu} + \cdots \quad (13.57)$$

Equation 13.57 shows that, when $x_e > 0$, the transitions move to lower wavenumbers as v increases.

Anharmonicity also accounts for the appearance of additional weak absorption lines corresponding to the transitions $2 \leftarrow 0$, $3 \leftarrow 0$, . . . , even though these first, second, . . . **overtones** are forbidden by the selection rule $\Delta v = \pm 1$. The first overtone, for example, gives rise to an absorption at

$$G(v + 2) - G(v) = 2\tilde{\nu} - 2(2v + 3)x_e\tilde{\nu} + \cdots \quad (13.58)$$

The reason for the appearance of overtones is that the selection rule is derived from the properties of harmonic oscillator wavefunctions, which are only approximately valid when anharmonicity is present. Therefore, the selection rule is also only an approximation. For an anharmonic oscillator, all values of Δv are allowed, but transitions with $\Delta v > 1$ are allowed only weakly if the anharmonicity is slight.

(b) The Birge–Sponer plot

When several vibrational transitions are detectable, a graphical technique called a **Birge–Sponer plot** may be used to determine the dissociation energy, D_0, of the bond. The basis of the Birge–Sponer plot is that the sum of successive intervals $\Delta G_{v+\frac{1}{2}}$ from the zero-point level to the dissociation limit is the dissociation energy:

$$D_0 = \Delta G_{1/2} + \Delta G_{3/2} + \cdots = \sum_v \Delta G_{v+\frac{1}{2}} \quad (13.59)$$

just as the height of the ladder is the sum of the separations of its rungs (Fig. 13.31). The construction in Fig. 13.32 shows that the area under the plot of $\Delta G_{v+\frac{1}{2}}$ against $v + \frac{1}{2}$ is equal to the sum, and therefore to D_0. The successive terms decrease linearly when only the x_e anharmonicity constant is taken into account and the inaccessible part of the spectrum can be estimated by linear extrapolation. Most actual plots differ from the linear plot as shown in Fig. 13.32, so the value of D_0 obtained in this way is usually an overestimate of the true value.

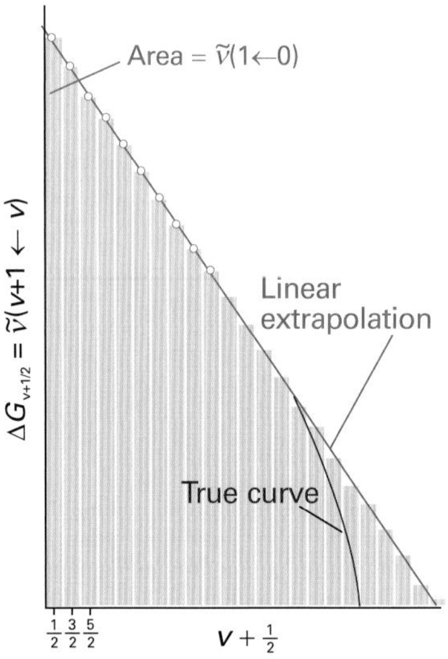

Fig. 13.32 The area under a plot of transition wavenumber against vibrational quantum number is equal to the dissociation energy of the molecule. The assumption that the differences approach zero linearly is the basis of the Birge–Sponer extrapolation.

Example 13.5 *Using a Birge–Sponer plot*

The observed vibrational intervals of H_2^+ lie at the following values for $1 \leftarrow 0$, $2 \leftarrow 1$, . . . respectively (in cm^{-1}): 2191, 2064, 1941, 1821, 1705, 1591, 1479, 1368, 1257, 1145, 1033, 918, 800, 677, 548, 411. Determine the dissociation energy of the molecule.

Method Plot the separations against $v + \frac{1}{2}$, extrapolate linearly to the point cutting the horizontal axis, and then measure the area under the curve.

Answer The points are plotted in Fig. 13.33, and a linear extrapolation is shown as a dotted line. The area under the curve (use the formula for the area of a triangle or count the squares) is 214. Each square corresponds to 100 cm^{-1} (refer to the scale of the vertical axis); hence the dissociation energy is 21 400 cm^{-1} (corresponding to 256 kJ mol^{-1}).

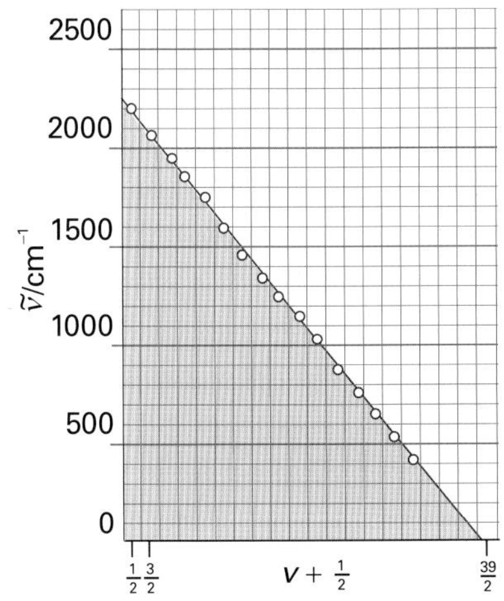

Fig. 13.33 The Birge–Sponer plot used in Example 13.5. The area is obtained simply by counting the squares beneath the line or using the formula for the area of a right triangle.

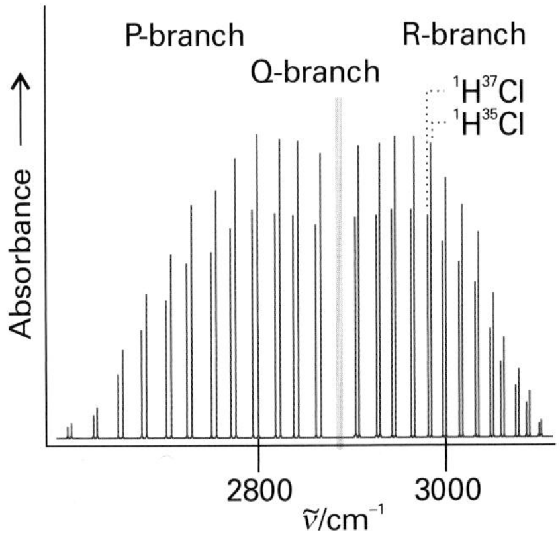

Fig. 13.34 A high-resolution vibration–rotation spectrum of HCl. The lines appear in pairs because $H^{35}Cl$ and $H^{37}Cl$ both contribute (their abundance ratio is 3:1). There is no Q branch, because $\Delta J = 0$ is forbidden for this molecule.

Self-test 13.7 The vibrational levels of HgH converge rapidly, and successive intervals are 1203.7 (which corresponds to the $1 \leftarrow 0$ transition), 965.6, 632.4, and 172 cm^{-1}. Estimate the dissociation energy. [35.6 kJ mol^{-1}]

13.12 Vibration–rotation spectra

Each line of the high resolution vibrational spectrum of a gas-phase heteronuclear diatomic molecule is found to consist of a large number of closely spaced components (Fig. 13.34). Hence, molecular spectra are often called **band spectra**. The separation between the components is less than 10 cm^{-1}, which suggests that the structure is due to rotational transitions accompanying the vibrational transition. A rotational change should be expected because classically we can think of the transition as leading to a sudden increase or decrease in the instantaneous bond length. Just as ice-skaters rotate more rapidly when they bring their arms in, and more slowly when they throw them out, so the molecular rotation is either accelerated or retarded by a vibrational transition.

(a) Spectral branches

A detailed analysis of the quantum mechanics of simultaneous vibrational and rotational changes shows that the rotational quantum number J changes by ± 1 during the vibrational transition of a diatomic molecule. If the molecule also possesses angular momentum about its axis, as in the case of the electronic orbital angular momentum of the paramagnetic molecule NO, then the selection rules also allow $\Delta J = 0$.

The appearance of the vibration–rotation spectrum of a diatomic molecule can be discussed in terms of the combined vibration–rotation terms, S:

$$S(v,J) = G(v) + F(J) \tag{13.60}$$

If we ignore anharmonicity and centrifugal distortion,

$$S(v,J) = (v + \tfrac{1}{2})\tilde{\nu} + BJ(J+1) \tag{13.61}$$

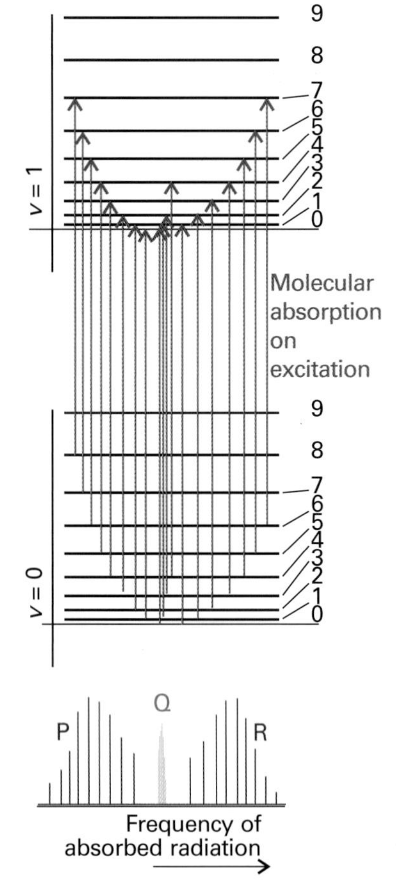

Fig. 13.35 The formation of P, Q, and R branches in a vibration–rotation spectrum. The intensities reflect the populations of the initial rotational levels.

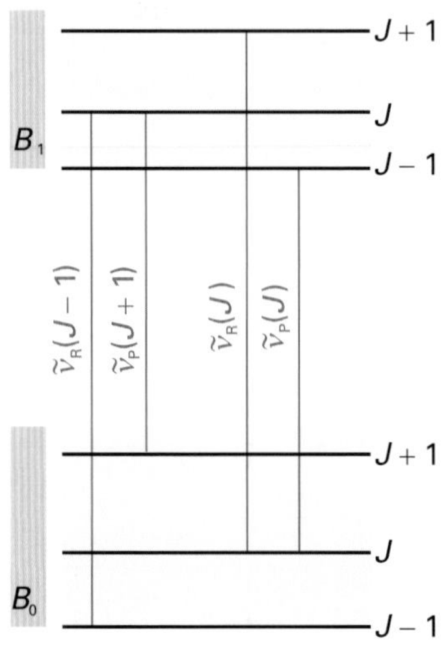

Fig. 13.36 The method of combination differences makes use of the fact that some transitions share a common level.

In a more detailed treatment, B is allowed to depend on the vibrational state because, as v increases, the molecule swells slightly and the moment of inertia changes. We shall continue with the simple expression initially.

When the vibrational transition $v+1 \leftarrow v$ occurs, J changes by ± 1 and in some cases by 0 (when $\Delta J = 0$ is allowed). The absorptions then fall into three groups called **branches** of the spectrum. The **P branch** consists of all transitions with $\Delta J = -1$:

$$\tilde{\nu}_{\mathrm{P}}(J) = S(v+1, J-1) - S(v,J) = \tilde{\nu} - 2BJ \qquad (13.62a)$$

This branch consists of lines at $\tilde{\nu} - 2B$, $\tilde{\nu} - 4B$, . . . with an intensity distribution reflecting both the populations of the rotational levels and the magnitude of the $J-1 \leftarrow J$ transition moment (Fig. 13.35). The **Q branch** consists of all lines with $\Delta J = 0$, and its wavenumbers are all

$$\tilde{\nu}_{\mathrm{Q}}(J) = S(v+1, J) - S(v,J) = \tilde{\nu} \qquad (13.62b)$$

for all values of J. This branch, when it is allowed (as in NO), appears at the vibrational transition wavenumber. In Fig. 13.35 there is a gap at the expected location of the Q branch because it is forbidden in HCl. The **R branch** consists of lines with $\Delta J = +1$:

$$\tilde{\nu}_{\mathrm{R}}(J) = S(v+1, J+1) - S(v,J) = \tilde{\nu} + 2B(J+1) \qquad (13.62c)$$

This branch consists of lines displaced from $\tilde{\nu}$ to high wavenumber by $2B$, $4B$, ...

The separation between the lines in the P and R branches of a vibrational transition gives the value of B. Therefore, the bond length can be deduced without needing to take a pure rotational microwave spectrum. However, the latter is more precise.

(b) Combination differences

The rotational constant of the vibrationally excited state, B_1 (in general, B_v), is in fact slightly smaller than that of the ground vibrational state, B_0, because the anharmonicity of the vibration results in a slightly extended bond in the upper state. As a result, the Q branch (if it exists) consists of a series of closely spaced lines. The lines of the R branch converge slightly as J increases; and those of the P branch diverge:

$$\begin{aligned}
\tilde{\nu}_{\mathrm{P}}(J) &= \tilde{\nu} - (B_1 + B_0)J + (B_1 - B_0)J^2 \\
\tilde{\nu}_{\mathrm{Q}}(J) &= \tilde{\nu} + (B_1 - B_0)J(J+1) \\
\tilde{\nu}_{\mathrm{R}}(J) &= \tilde{\nu} + (B_1 + B_0)(J+1) + (B_1 - B_0)(J+1)^2
\end{aligned} \qquad (13.63)$$

To determine the two rotational constants individually, we use the method of **combination differences.** This procedure is used widely in spectroscopy to extract information about a particular state. It involves setting up expressions for the difference in the wavenumbers of transitions to a common state; the resulting expression then depends solely on properties of the other state.

As can be seen from Fig. 13.36, the transitions $\tilde{\nu}_{\mathrm{R}}(J-1)$ and $\tilde{\nu}_{\mathrm{P}}(J+1)$ have a common upper state, and hence can be anticipated to depend on B_0. Indeed, it is easy to show from eqn 13.63 that

$$\tilde{\nu}_{\mathrm{R}}(J-1) - \tilde{\nu}_{\mathrm{P}}(J+1) = 4B_0(J+\tfrac{1}{2}) \qquad (13.64a)$$

Therefore, a plot of the combination difference against $J+\frac{1}{2}$ should be a straight line of slope $4B_0$, so the rotational constant of the molecule in the state $v = 0$ can be determined. (Any deviation from a straight line is a consequence of centrifugal distortion, so that effect can be investigated too.) Similarly, $\tilde{\nu}_{\mathrm{R}}(J)$ and $\tilde{\nu}_{\mathrm{P}}(J)$ have a common lower state, and hence their combination difference gives information about the upper state:

$$\tilde{\nu}_R(J) - \tilde{\nu}_P(J) = 4B_1(J + \tfrac{1}{2}) \qquad (13.64b)$$

The two rotational constants of $^1H^{35}Cl$ found in this way are $B_0 = 10.440\ \text{cm}^{-1}$ and $B_1 = 10.136\ \text{cm}^{-1}$.

13.13 Vibrational Raman spectra of diatomic molecules

The gross selection rule for vibrational Raman transitions is that *the polarizability should change as the molecule vibrates.* As homonuclear and heteronuclear diatomic molecules swell and contract during a vibration, the control of the nuclei over the electrons varies, and hence the molecular polarizability changes. Both types of diatomic molecule are therefore vibrationally Raman active. The specific selection rule for vibrational Raman transitions in the harmonic approximation is $\Delta v = \pm 1$. The formal basis for the gross and specific selection rules is given in *Further information 13.2*.

The lines to high frequency of the incident radiation, the anti-Stokes lines, are those for which $\Delta v = -1$. The lines to low frequency, the Stokes lines, correspond to $\Delta v = +1$. The intensities of the anti-Stokes and Stokes lines are governed largely by the Boltzmann populations of the vibrational states involved in the transition. It follows that anti-Stokes lines are usually weak because very few molecules are in an excited vibrational state initially.

In gas-phase spectra, the Stokes and anti-Stokes lines have a branch structure arising from the simultaneous rotational transitions that accompany the vibrational excitation (Fig. 13.37). The selection rules are $\Delta J = 0, \pm 2$ (as in pure rotational Raman spectroscopy), and give rise to the **O branch** ($\Delta J = -2$), the **Q branch** ($\Delta J = 0$), and the **S branch** ($\Delta J = +2$):

$$\begin{aligned}\tilde{\nu}_O(J) &= \tilde{\nu}_i - \tilde{\nu} - 2B + 4BJ \\ \tilde{\nu}_Q(J) &= \tilde{\nu}_i - \tilde{\nu} \\ \tilde{\nu}_S(J) &= \tilde{\nu}_i - \tilde{\nu} - 6B - 4BJ\end{aligned} \qquad (13.65)$$

Note that, unlike in infrared spectroscopy, a Q branch is obtained for all linear molecules. The spectrum of CO, for instance, is shown in Fig. 13.38: the structure of the Q branch arises from the differences in rotational constants of the upper and lower vibrational states.

The information available from vibrational Raman spectra adds to that from infrared spectroscopy because homonuclear diatomics can also be studied. The spectra can be interpreted in terms of the force constants, dissociation energies, and bond lengths, and some of the information obtained is included in Table 13.2.

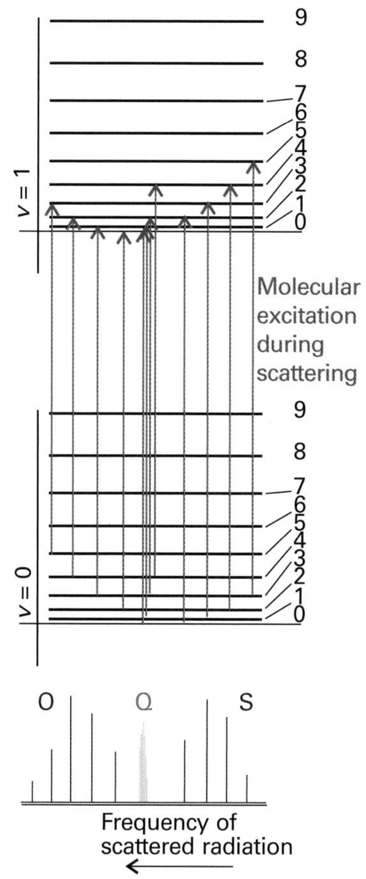

Fig. 13.37 The formation of O, Q, and S branches in a vibration–rotation Raman spectrum of a linear rotor. Note that the frequency scale runs in the opposite direction to that in Fig. 13.35, because the higher energy transitions (on the right) extract more energy from the incident beam and leave it at lower frequency.

Synoptic table 13.2* Properties of diatomic molecules

	$\tilde{\nu}/\text{cm}^{-1}$	R_e/pm	B/cm^{-1}	$k/(\text{N m}^{-1})$	$D_e/(\text{kJ mol}^{-1})$
$^1H_2$	4401	74	60.86	575	432
$^1H^{35}Cl$	2991	127	10.59	516	428
$^1H^{127}I$	2308	161	6.61	314	295
$^{35}Cl_2$	560	199	0.244	323	239

* More values are given in the *Data section*.

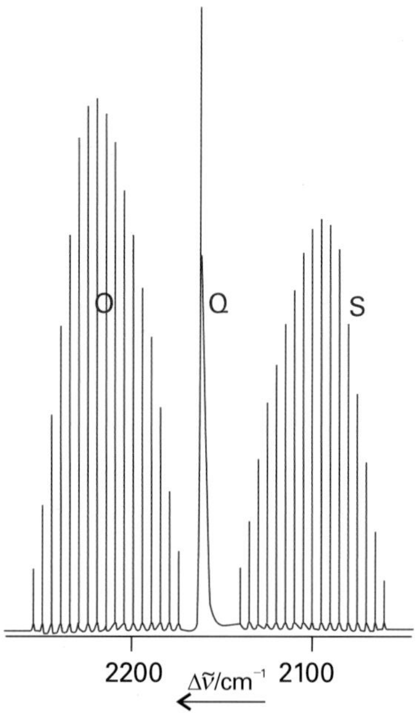

Fig. 13.38 The structure of a vibrational line in the vibrational Raman spectrum of carbon monoxide, showing the O, Q, and S branches.

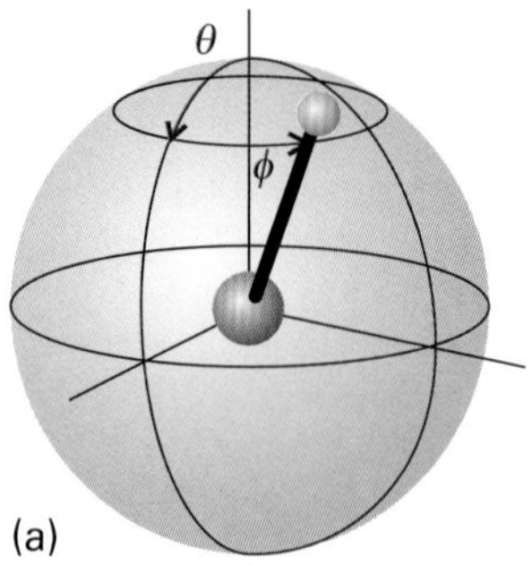

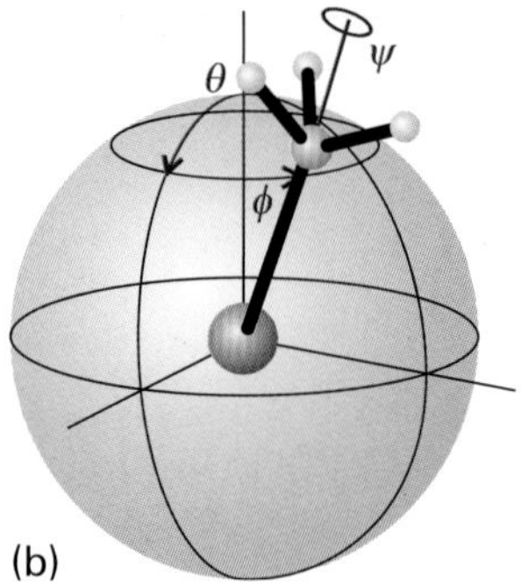

Fig. 13.39 (a) The orientation of a linear molecule requires the specification of two angles. (b) The orientation of a nonlinear molecule requires the specification of three angles.

The vibrations of polyatomic molecules

There is only one mode of vibration for a diatomic molecule, the bond stretch. In polyatomic molecules there are several modes of vibration because all the bond lengths and angles may change and the vibrational spectra are very complex. Nonetheless, we shall see that infrared and Raman spectroscopy can be used to obtain information about the structure of systems as large as animal and plant tissues (see *Impact I13.3*).

13.14 Normal modes

We begin by calculating the total number of vibrational modes of a polyatomic molecule. We then see that we can choose combinations of these atomic displacements that give the simplest description of the vibrations.

As shown in the *Justification* below, for a nonlinear molecule that consists of N atoms, there are $3N - 6$ independent modes of vibration. If the molecule is linear, there are $3N - 5$ independent vibrational modes.

Justification 13.5 *The number of vibrational modes*

The total number of coordinates needed to specify the locations of N atoms is $3N$. Each atom may change its location by varying one of its three coordinates (x, y, and z), so the total number of displacements available is $3N$. These displacements can be grouped together in a physically sensible way. For example, three coordinates are needed to specify the location of the centre of mass of the molecule, so three of the $3N$ displacements correspond to the translational motion of the molecule as a whole. The remaining $3N - 3$ are non-translational 'internal' modes of the molecule.

Two angles are needed to specify the orientation of a linear molecule in space: in effect, we need to give only the latitude and longitude of the direction in which the molecular axis is pointing (Fig. 13.39a). However, three angles are needed for a nonlinear molecule because we also need to specify the orientation of the molecule around the direction defined by the latitude and longitude (Fig. 13.39b). Therefore, two (linear) or three (nonlinear) of the $3N - 3$ internal displacements are rotational. This leaves $3N - 5$ (linear) or $3N - 6$ (nonlinear) displacements of the atoms relative to one another: these are the vibrational modes. It follows that the number of modes of vibration N_{vib} is $3N - 5$ for linear molecules and $3N - 6$ for nonlinear molecules.

Illustration 13.5 *Determining the number of vibrational modes*

Water, H_2O, is a nonlinear triatomic molecule, and has three modes of vibration (and three modes of rotation); CO_2 is a linear triatomic molecule, and has four modes of vibration (and only two modes of rotation). Even a middle-sized molecule such as naphthalene ($C_{10}H_8$) has 48 distinct modes of vibration.

The next step is to find the best description of the modes. One choice for the four modes of CO_2, for example, might be the ones in Fig. 13.40a. This illustration shows the stretching of one bond (the mode ν_L), the stretching of the other (ν_R), and the two perpendicular bending modes (ν_2). The description, while permissible, has a disadvantage: when one CO bond vibration is excited, the motion of the C atom sets the other CO bond in motion, so energy flows backwards and forwards between ν_L and ν_R. Moreover, the position of the centre of mass of the molecule varies in the course of either vibration.

The description of the vibrational motion is much simpler if linear combinations of ν_L and ν_R are taken. For example, one combination is ν_1 in Fig. 13.40b: this mode is the **symmetric stretch**. In this mode, the C atom is buffeted simultaneously from each side and the motion continues indefinitely. Another mode is ν_3, the **antisymmetric stretch**, in which the two O atoms always move in the same direction and opposite to that of the C atom. Both modes are independent in the sense that, if one is excited, then it does not excite the other. They are two of the 'normal modes' of the molecule, its independent, collective vibrational displacements. The two other normal modes are the bending modes ν_2. In general, a **normal mode** is an independent, synchronous motion of atoms or groups of atoms that may be excited without leading to the excitation of any other normal mode and without involving translation or rotation of the molecule as a whole.

The four normal modes of CO_2, and the N_{vib} normal modes of polyatomics in general, are the key to the description of molecular vibrations. Each normal mode, q, behaves like an independent harmonic oscillator (if anharmonicities are neglected), so each has a series of terms

$$G_q(v) = (v + \tfrac{1}{2})\tilde{\nu}_q \qquad \tilde{\nu}_q = \frac{1}{2\pi c}\left(\frac{k_q}{m_q}\right)^{1/2} \tag{13.66}$$

where $\tilde{\nu}_q$ is the wavenumber of mode q and depends on the force constant k_q for the mode and on the effective mass m_q of the mode. The effective mass of the mode is a measure of the mass that is swung about by the vibration and in general is a complicated function of the masses of the atoms. For example, in the symmetric stretch of CO_2, the C atom is stationary, and the effective mass depends on the masses of only the O atoms. In the antisymmetric stretch and in the bends, all three atoms move, so all contribute to the effective mass. The three normal modes of H_2O are shown in Fig. 13.41: note that the predominantly bending mode (ν_2) has a lower frequency than the others, which are predominantly stretching modes. It is generally the case that the frequencies of bending motions are lower than those of stretching modes. One point that must be appreciated is that only in special cases (such as the CO_2 molecule) are the normal modes purely stretches or purely bends. In general, a normal mode is a composite motion of simultaneous stretching and bending of bonds. Another point in this connection is that heavy atoms generally move less than light atoms in normal modes.

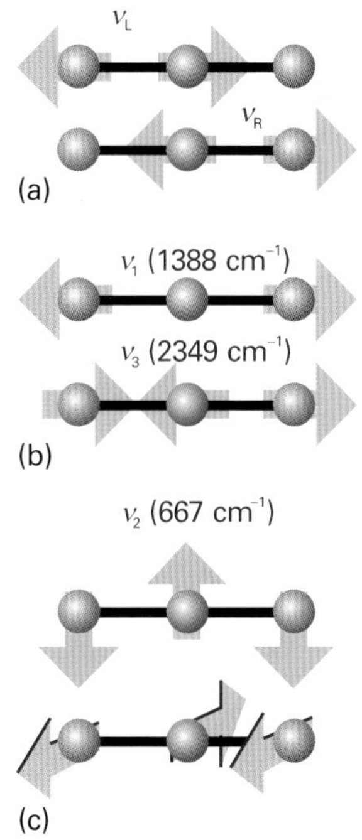

Fig. 13.40 Alternative descriptions of the vibrations of CO_2. (a) The stretching modes are not independent, and if one C—O group is excited the other begins to vibrate. They are not normal modes of vibration of the molecule. (b) The symmetric and antisymmetric stretches are independent, and one can be excited without affecting the other: they are normal modes. (c) The two perpendicular bending motions are also normal modes.

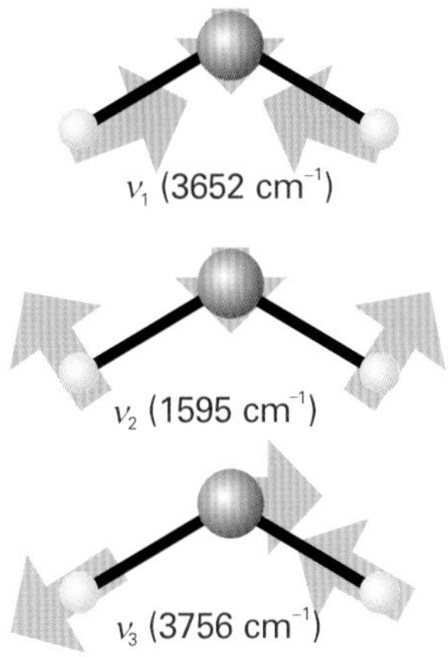

Fig. 13.41 The three normal modes of H_2O. The mode ν_2 is predominantly bending, and occurs at lower wavenumber than the other two.

13.15 Infrared absorption spectra of polyatomic molecules

The gross selection rule for infrared activity is that *the motion corresponding to a normal mode should be accompanied by a change of dipole moment.* Deciding whether this is so can sometimes be done by inspection. For example, the symmetric stretch of CO_2 leaves the dipole moment unchanged (at zero, see Fig. 13.40), so this mode is infrared inactive. The antisymmetric stretch, however, changes the dipole moment because the molecule becomes unsymmetrical as it vibrates, so this mode is infrared active. Because the dipole moment change is parallel to the principal axis, the transitions arising from this mode are classified as **parallel bands** in the spectrum. Both bending modes are infrared active: they are accompanied by a changing dipole perpendicular to the principal axis, so transitions involving them lead to a **perpendicular band** in the spectrum. The latter bands eliminate the linearity of the molecule, and as a result a Q branch is observed; a parallel band does not have a Q branch.

The active modes are subject to the specific selection rule $\Delta v_q = \pm 1$ in the harmonic approximation, so the wavenumber of the fundamental transition (the 'first harmonic') of each active mode is $\tilde{\nu}_q$. From the analysis of the spectrum, a picture may be constructed of the stiffness of various parts of the molecule, that is, we can establish its **force field**, the set of force constants corresponding to all the displacements of the atoms. The force field may also be estimated by using the semi-empirical, *ab initio*, and DFT computational techniques described in Section 11.7. Superimposed on the simple force field scheme are the complications arising from anharmonicities and the effects of molecular rotation. Very often the sample is a liquid or a solid, and the molecules are unable to rotate freely. In a liquid, for example, a molecule may be able to rotate through only a few degrees before it is struck by another, so it changes its rotational state frequently. This random changing of orientation is called **tumbling**.

Comment 13.6

The web site for this text contains links to sites where you can perform quantum mechanical calculations of frequencies and atomic displacements of normal modes of simple molecules.

Comment 13.7

The web site for this text contains links to databases of infrared spectra.

The lifetimes of rotational states in liquids are very short, so in most cases the rotational energies are ill-defined. Collisions occur at a rate of about 10^{13} s^{-1} and, even allowing for only a 10 per cent success rate in knocking the molecule into another rotational state, a lifetime broadening (eqn 13.19) of more than 1 cm^{-1} can easily result. The rotational structure of the vibrational spectrum is blurred by this effect, so the infrared spectra of molecules in condensed phases usually consist of broad lines spanning the entire range of the resolved gas-phase spectrum, and showing no branch structure.

Synoptic table 13.3* Typical vibrational wavenumbers

Vibration type	$\tilde{\nu}$/cm^{-1}
C—H stretch	2850–2960
C—H bend	1340–1465
C—C stretch, bend	700–1250
C=C stretch	1620–1680

* More values are given in the *Data section*.

One very important application of infrared spectroscopy to condensed phase samples, and for which the blurring of the rotational structure by random collisions is a welcome simplification, is to chemical analysis. The vibrational spectra of different groups in a molecule give rise to absorptions at characteristic frequencies because a normal mode of even a very large molecule is often dominated by the motion of a small group of atoms. The intensities of the vibrational bands that can be identified with the motions of small groups are also transferable between molecules. Consequently, the molecules in a sample can often be identified by examining its infrared spectrum and referring to a table of characteristic frequencies and intensities (Table 13.3).

IMPACT ON ENVIRONMENTAL SCIENCE

I13.2 Global warming[1]

Solar energy strikes the top of the Earth's atmosphere at a rate of 343 W m^{-2}. About 30 per cent of this energy is reflected back into space by the Earth or the atmosphere. The Earth–atmosphere system absorbs the remaining energy and re-emits it into space as black-body radiation, with most of the intensity being carried by infrared radiation in the range 200–2500 cm^{-1} (4–50 μm). The Earth's average temperature is maintained by an energy balance between solar radiation absorbed by the Earth and black-body radiation emitted by the Earth.

The trapping of infrared radiation by certain gases in the atmosphere is known as the *greenhouse effect*, so called because it warms the Earth as if the planet were enclosed in a huge greenhouse. The result is that the natural greenhouse effect raises the average surface temperature well above the freezing point of water and creates an environment in which life is possible. The major constituents to the Earth's atmosphere, O_2 and N_2, do not contribute to the greenhouse effect because homonuclear diatomic molecules cannot absorb infrared radiation. However, the minor atmospheric gases, water vapour and CO_2, do absorb infrared radiation and hence are responsible for the

[1] This section is based on a similar contribution initially prepared by Loretta Jones and appearing in *Chemical principles*, Peter Atkins and Loretta Jones, W.H. Freeman and Co., New York (2005).

greenhouse effect (Fig. 13.42). Water vapour absorbs strongly in the ranges 1300–1900 cm^{-1} (5.3–7.7 μm) and 3550–3900 cm^{-1} (2.6–2.8 μm), whereas CO_2 shows strong absorption in the ranges 500–725 cm^{-1} (14–20 μm) and 2250–2400 cm^{-1} (4.2–4.4 μm).

Increases in the levels of greenhouse gases, which also include methane, dinitrogen oxide, ozone, and certain chlorofluorocarbons, as a result of human activity have the potential to enhance the natural greenhouse effect, leading to significant warming of the planet. This problem is referred to as *global warming*, which we now explore in some detail.

The concentration of water vapour in the atmosphere has remained steady over time, but concentrations of some other greenhouse gases are rising. From about the year 1000 until about 1750, the CO_2 concentration remained fairly stable, but, since then, it has increased by 28 per cent. The concentration of methane, CH_4, has more than doubled during this time and is now at its highest level for 160 000 years (160 ka; a is the SI unit denoting 1 year). Studies of air pockets in ice cores taken from Antarctica show that increases in the concentration of both atmospheric CO_2 and CH_4 over the past 160 ka correlate well with increases in the global surface temperature.

Human activities are primarily responsible for the rising concentrations of atmospheric CO_2 and CH_4. Most of the atmospheric CO_2 comes from the burning of hydrocarbon fuels, which began on a large scale with the Industrial Revolution in the middle of the nineteenth century. The additional methane comes mainly from the petroleum industry and from agriculture.

The temperature of the surface of the Earth has increased by about 0.5 K since the late nineteenth century (Fig. 13.43). If we continue to rely on hydrocarbon fuels and current trends in population growth and energy are not reversed, then by the middle of the twenty-first century, the concentration of CO_2 in the atmosphere will be about twice its value prior to the Industrial Revolution. The Intergovernmental Panel on Climate Change (IPCC) estimated in 1995 that, by the year 2100, the Earth will undergo an increase in temperature of 3 K. Furthermore, the rate of temperature change is likely to be greater than at any time in the last 10 ka. To place a temperature rise of 3 K in perspective, it is useful to consider that the average temperature of the Earth during the last ice age was only 6 K colder than at present. Just as cooling the planet (for example, during an ice age) can lead to detrimental effects on ecosystems, so too can a dramatic warming of the globe. One example of a significant change in the

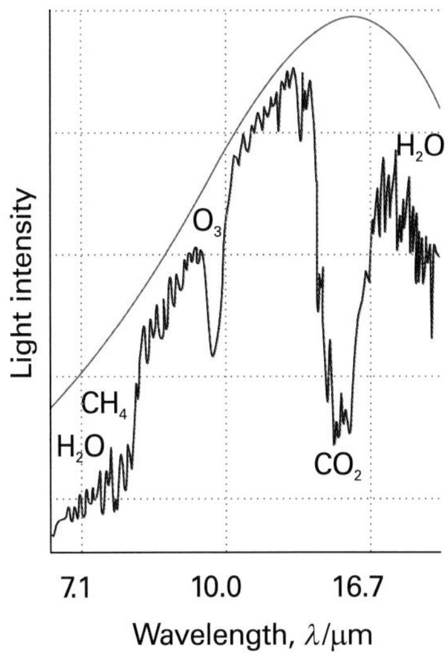

Fig. 13.42 The intensity of infrared radiation that would be lost from Earth in the absence of greenhouse gases is shown by the blue line. The purple line is the intensity of the radiation actually emitted. The maximum wavelength of radiation absorbed by each greenhouse gas is indicated.

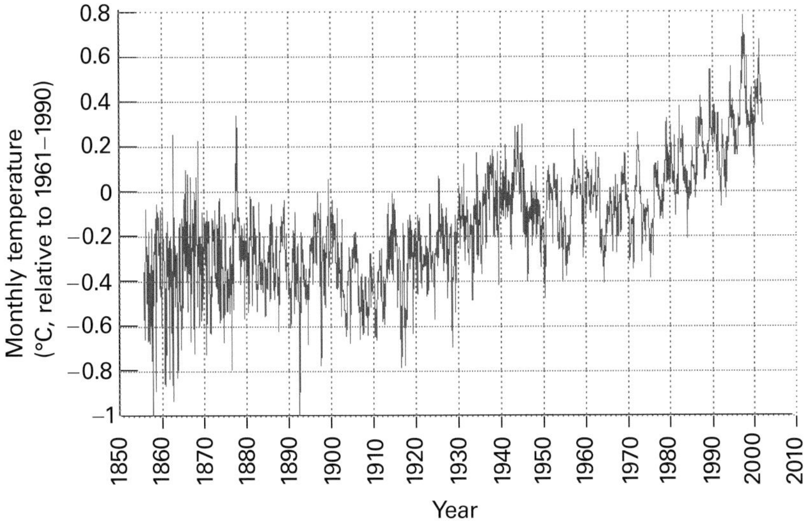

Fig. 13.43 The average change in surface temperature of the Earth from 1855 to 2002.

environment caused by a temperature increase of 3 K is a rise in sea level by about 0.5 m, which is sufficient to alter weather patterns and submerge currently coastal ecosystems.

Computer projections for the next 200 years predict further increases in atmospheric CO_2 levels and suggest that, to maintain CO_2 at its current concentration, we would have to reduce hydrocarbon fuel consumption immediately by about 50 per cent. Clearly, in order to reverse global warming trends, we need to develop alternatives to fossil fuels, such as hydrogen (which can be used in fuel cells, *Impact I25.3*) and solar energy technologies.

13.16 **Vibrational Raman spectra of polyatomic molecules**

The normal modes of vibration of molecules are Raman active if they are accompanied by a changing polarizability. It is sometimes quite difficult to judge by inspection when this is so. The symmetric stretch of CO_2, for example, alternately swells and contracts the molecule: this motion changes the polarizability of the molecule, so the mode is Raman active. The other modes of CO_2 leave the polarizability unchanged, so they are Raman inactive.

A more exact treatment of infrared and Raman activity of normal modes leads to the **exclusion rule**:

> If the molecule has a centre of symmetry, then no modes can be both infrared and Raman active.

(A mode may be inactive in both.) Because it is often possible to judge intuitively if a mode changes the molecular dipole moment, we can use this rule to identify modes that are not Raman active. The rule applies to CO_2 but to neither H_2O nor CH_4 because they have no centre of symmetry. In general, it is necessary to use group theory to predict whether a mode is infrared or Raman active (Section 13.17).

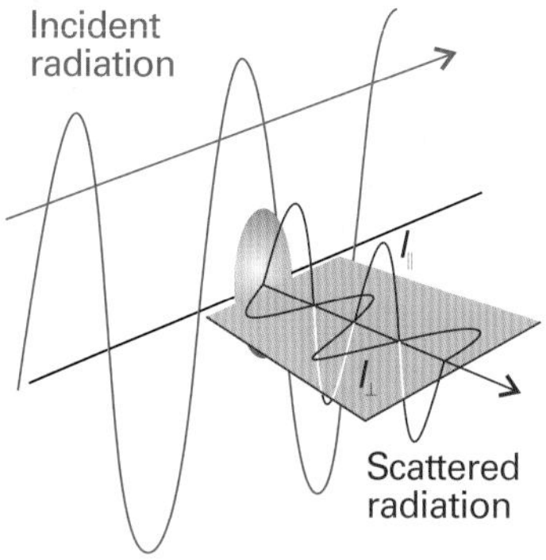

Fig. 13.44 The definition of the planes used for the specification of the depolarization ratio, ρ, in Raman scattering.

(a) Depolarization

The assignment of Raman lines to particular vibrational modes is aided by noting the state of polarization of the scattered light. The **depolarization ratio**, ρ, of a line is the ratio of the intensities, I, of the scattered light with polarizations perpendicular and parallel to the plane of polarization of the incident radiation:

$$\rho = \frac{I_{\perp}}{I_{\parallel}} \qquad [13.67]$$

To measure ρ, the intensity of a Raman line is measured with a polarizing filter (a 'half-wave plate') first parallel and then perpendicular to the polarization of the incident beam. If the emergent light is not polarized, then both intensities are the same and ρ is close to 1; if the light retains its initial polarization, then $I_{\perp} = 0$, so $\rho = 0$ (Fig. 13.44). A line is classified as **depolarized** if it has ρ close to or greater than 0.75 and as **polarized** if $\rho < 0.75$. Only totally symmetrical vibrations give rise to polarized lines in which the incident polarization is largely preserved. Vibrations that are not totally symmetrical give rise to depolarized lines because the incident radiation can give rise to radiation in the perpendicular direction too.

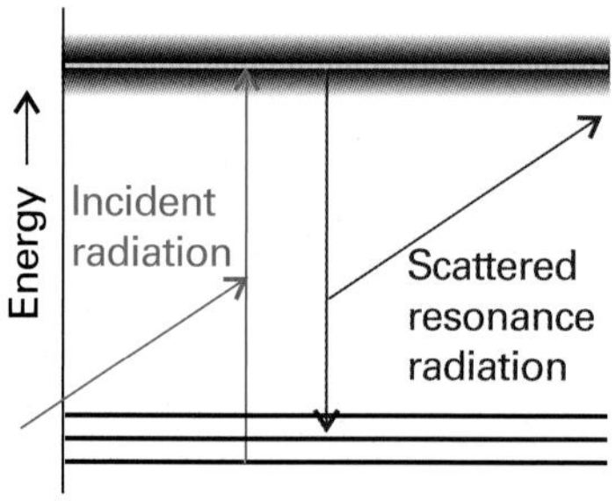

Fig. 13.45 In the *resonance Raman effect*, the incident radiation has a frequency corresponding to an actual electronic excitation of the molecule. A photon is emitted when the excited state returns to a state close to the ground state.

(b) Resonance Raman spectra

A modification of the basic Raman effect involves using incident radiation that nearly coincides with the frequency of an electronic transition of the sample (Fig. 13.45). The technique is then called **resonance Raman spectroscopy**. It is characterized by a much greater intensity in the scattered radiation. Furthermore, because it is often the case that only a few vibrational modes contribute to the more intense scattering, the spectrum is greatly simplified.

Resonance Raman spectroscopy is used to study biological molecules that absorb strongly in the ultraviolet and visible regions of the spectrum. Examples include the pigments β-carotene and chlorophyll, which capture solar energy during plant photosynthesis (see *Impact I23.2*). The resonance Raman spectra of Fig. 13.46 show vibrational transitions from only the few pigment molecules that are bound to very large proteins dissolved in an aqueous buffer solution. This selectivity arises from the fact that water (the solvent), amino acid residues, and the peptide group do not have electronic transitions at the laser wavelengths used in the experiment, so their *conventional* Raman spectra are weak compared to the enhanced spectra of the pigments. Comparison of the spectra in Figs. 13.46a and 13.46b also shows that, with proper choice of excitation wavelength, it is possible to examine individual classes of pigments bound to the same protein: excitation at 488 nm, where β-carotene absorbs strongly, shows vibrational bands from β-carotene only, whereas excitation at 442 nm, where chlorophyll *a* and β-carotene absorb, reveals features from both types of pigments.

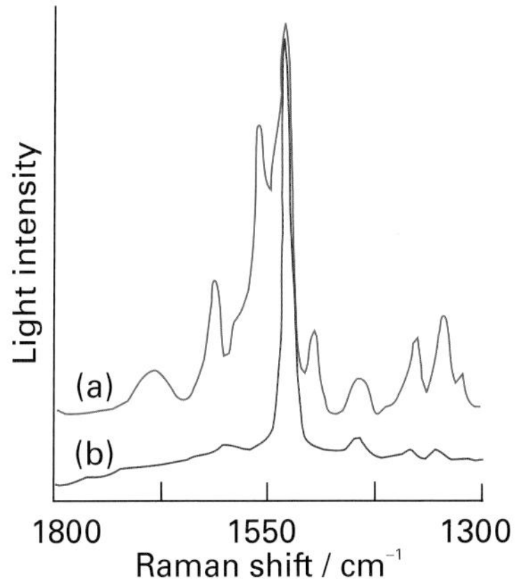

Fig. 13.46 The resonance Raman spectra of a protein complex that is responsible for some of the initial electron transfer events in plant photosynthesis. (a) Laser excitation of the sample at 407 nm shows Raman bands due to both chlorophyll *a* and β-carotene bound to the protein because both pigments absorb light at this wavelength. (b) Laser excitation at 488 nm shows Raman bands from β-carotene only because chlorophyll *a* does not absorb light very strongly at this wavelength. (Adapted from D.F. Ghanotakis *et al.*, *Biochim. Biophys. Acta* **974**, 44 (1989).)

(c) Coherent anti-Stokes Raman spectroscopy

The intensity of Raman transitions may be enhanced by **coherent anti-Stokes Raman spectroscopy** (CARS, Fig. 13.47). The technique relies on the fact that, if two laser beams of frequencies ν_1 and ν_2 pass through a sample, then they may mix together and give rise to coherent radiation of several different frequencies, one of which is

$$\nu' = 2\nu_1 - \nu_2 \tag{13.68}$$

Suppose that ν_2 is varied until it matches any Stokes line from the sample, such as the one with frequency $\nu_1 - \Delta\nu$; then the coherent emission will have frequency

$$\nu' = 2\nu_1 - (\nu_1 - \Delta\nu) = \nu_1 + \Delta\nu \tag{13.69}$$

which is the frequency of the corresponding anti-Stokes line. This coherent radiation forms a narrow beam of high intensity.

An advantage of CARS is that it can be used to study Raman transitions in the presence of competing incoherent background radiation, and so can be used to observe the Raman spectra of species in flames. One example is the vibration–rotation CARS spectrum of N_2 gas in a methane–air flame shown in Fig 13.48.

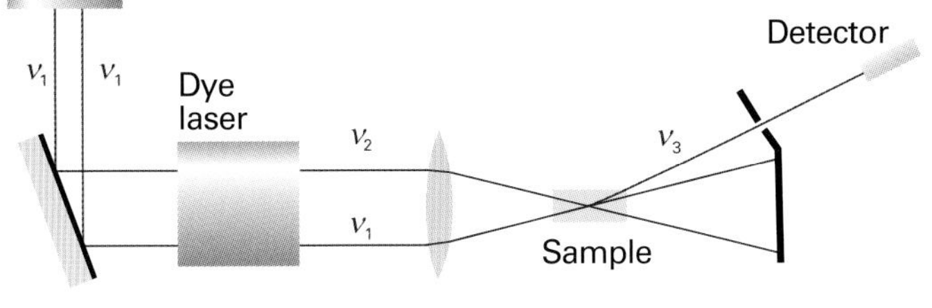

Fig. 13.47 The experimental arrangement for the CARS experiment.

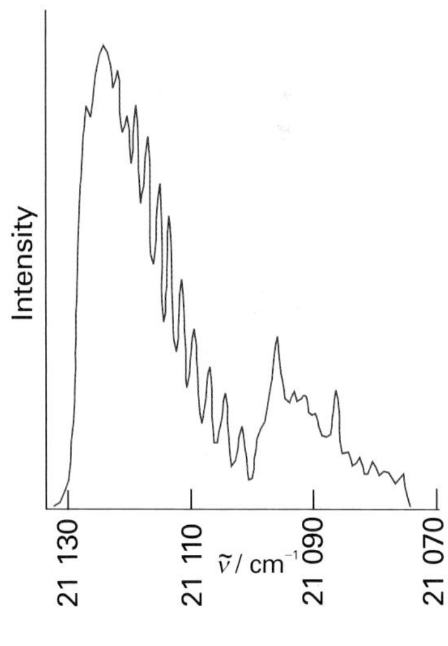

Fig. 13.48 CARS spectrum of a methane–air flame at 2104 K. The peaks correspond to the Q branch of the vibration–rotation spectrum of N_2 gas. (Adapted from J.F. Verdieck *et al.*, *J. Chem. Educ.* **59**, 495 (1982).)

IMPACT ON BIOCHEMISTRY

I13.3 Vibrational microscopy

Optical microscopes can now be combined with infrared and Raman spectrometers and the vibrational spectra of specimens as small as single biological cells obtained. The techniques of *vibrational microscopy* provide details of cellular events that cannot be observed with traditional light or electron microscopy.

The principles behind the operation of infrared and Raman microscopes are simple: radiation illuminates a small area of the sample, and the transmitted, reflected, or scattered light is first collected by a microscope and then analysed by a spectrometer. The sample is then moved by very small increments along a plane perpendicular to the direction of illumination and the process is repeated until vibrational spectra for all sections the sample are obtained. The size of a sample that can be studied by vibrational microscopy depends on a number of factors, such as the area of illumination and the excitance and wavelength of the incident radiation. Up to a point, the smaller the area that is illuminated, the smaller the area from which a spectrum can be obtained. High excitance is required to increase the rate of arrival of photons at the detector from small illuminated areas. For this reason, lasers and synchrotron radiation (see *Further information 13.1*) are the preferred radiation sources.

In a conventional light microscope, an image is constructed from a pattern of diffracted light waves that emanate from the illuminated object. As a result, some information about the specimen is lost by destructive interference of scattered light waves. Ultimately, this *diffraction limit* prevents the study of samples that are much smaller than the wavelength of light used as a probe. In practice, two objects will appear as distinct images under a microscope if the distance between their centres is greater than the *Airy radius*, $r_{\text{Airy}} = 0.61\lambda/a$, where λ is the wavelength of the incident beam of radiation and a is the numerical aperture of the *objective lens*, the lens that collects light scattered by the object. The numerical aperture of the objective lens is defined as $a = n_{\text{r}} \sin \alpha$, where n_{r} is the refractive index of the lens material (the greater the refractive index, the greater the bending of a ray of light by the lens) and the angle α is the half-angle of the widest cone of scattered light that can collected by the lens (so the lens collects light beams sweeping a cone with angle 2α). Use of the best equipment makes it possible to probe areas as small as 9 μm^2 by vibrational microscopy.

Figure 13.49 shows the infrared spectra of a single mouse cell, living and dying. Both spectra have features at 1545 cm^{-1} and 1650 cm^{-1} that are due to the peptide carbonyl groups of proteins and a feature at 1240 cm^{-1} that is due to the phosphodiester (PO_2^-) groups of lipids. The dying cell shows an additional absorption at 1730 cm^{-1}, which is due to the ester carbonyl group from an unidentified compound. From a plot of the intensities of individual absorption features as a function of position in the cell, it has been possible to map the distribution of proteins and lipids during cell division and cell death.

Vibrational microscopy has also been used in biomedical and pharmaceutical laboratories. Examples include the determination of the size and distribution of a drug in a tablet, the observation of conformational changes in proteins of cancerous cells upon administration of anti-tumour drugs, and the measurement of differences between diseased and normal tissue, such as diseased arteries and the white matter from brains of multiple sclerosis patients.

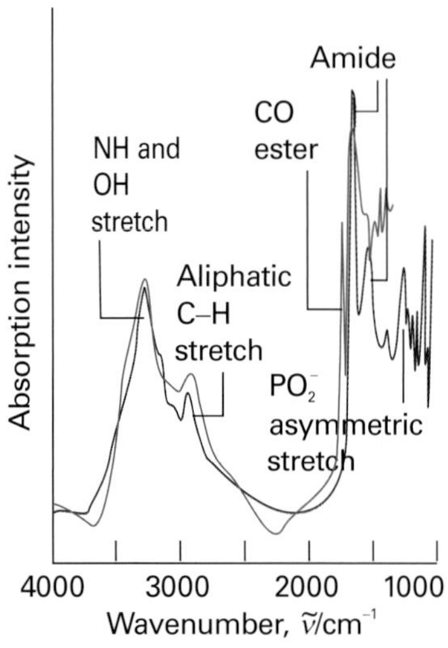

Fig. 13.49 Infrared absorption spectra of a single mouse cell: (purple) living cell, (blue) dying cell. Adapted from N. Jamin *et al., Proc. Natl. Acad. Sci.USA* **95**, 4837 (1998).

13.17 **Symmetry aspects of molecular vibrations**

One of the most powerful ways of dealing with normal modes, especially of complex molecules, is to classify them according to their symmetries. Each normal mode must belong to one of the symmetry species of the molecular point group, as discussed in Chapter 12.

Example 13.6 *Identifying the symmetry species of a normal mode*

Establish the symmetry species of the normal mode vibrations of CH_4, which belongs to the group T_d.

Method The first step in the procedure is to identify the symmetry species of the irreducible representations spanned by all the $3N$ displacements of the atoms, using the characters of the molecular point group. Find these characters by counting 1 if the displacement is unchanged under a symmetry operation, −1 if it changes sign, and 0 if it is changed into some other displacement. Next, subtract the symmetry species of the translations. Translational displacements span the same symmetry species as x, y, and z, so they can be obtained from the right-most column of the character table. Finally, subtract the symmetry species of the rotations, which are also given in the character table (and denoted there by R_x, R_y, or R_z).

Answer There are $3 \times 5 = 15$ degrees of freedom, of which $(3 \times 5) - 6 = 9$ are vibrations. Refer to Fig. 13.50. Under E, no displacement coordinates are changed, so the character is 15. Under C_3, no displacements are left unchanged, so the character is 0. Under the C_2 indicated, the z-displacement of the central atom is left unchanged, whereas its x- and y-components both change sign. Therefore $\chi(C_2) = 1 - 1 - 1 + 0 + 0 + \cdots = -1$. Under the S_4 indicated, the z-displacement of the central atom is reversed, so $\chi(S_4) = -1$. Under σ_d, the x- and z-displacements of C, H_3, and H_4 are left unchanged and the y-displacements are reversed; hence $\chi(\sigma_d) = 3 + 3 - 3 = 3$. The characters are therefore 15, 0, −1, −1, 3. By decomposing the direct product (Section 12.5a), we find that this representation spans $A_1 + E + T_1 + 3T_2$. The translations span T_2; the rotations span T_1. Hence, the nine vibrations span $A_1 + E + 2T_2$.

The modes themselves are shown in Fig. 13.51. We shall see that symmetry analysis gives a quick way of deciding which modes are active.

Self-test 13.8 Establish the symmetry species of the normal modes of H_2O.

$[2A_1 + B_2]$

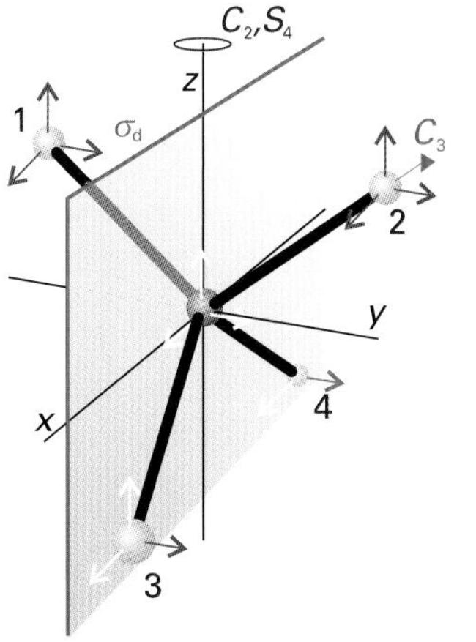

Fig. 13.50 The atomic displacements of CH_4 and the symmetry elements used to calculate the characters.

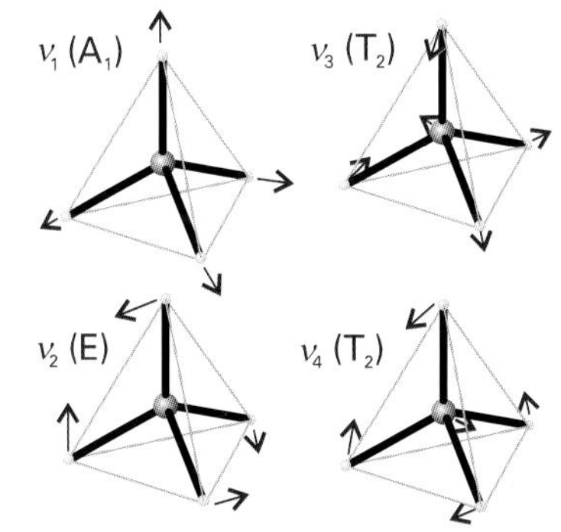

Fig. 13.51 Typical normal modes of vibration of a tetrahedral molecule. There are in fact two modes of symmetry species E and three modes of each T_2 symmetry species.

(a) Infrared activity of normal modes

It is best to use group theory to judge the activities of more complex modes of vibration. This is easily done by checking the character table of the molecular point group for the symmetry species of the irreducible representations spanned by x, y, and z, for their species are also the symmetry species of the components of the electric dipole moment. Then apply the following rule:

> If the symmetry species of a normal mode is the same as any of the symmetry species of x, y, or z, then the mode is infrared active.

Justification 13.6 *Using group theory to identify infrared active normal modes*

The rule hinges on the form of the transition dipole moment between the ground-state vibrational wavefunction, ψ_0, and that of the first excited state, ψ_1. The x-component is

$$\mu_{x,10} = -e\int \psi_1^{\star} x \psi_0 \, d\tau \tag{13.70}$$

with similar expressions for the two other components of the transition moment. The ground-state vibrational wavefunction is a Gaussian function of the form e^{-x^2}, so it is symmetrical in x. The wavefunction for the first excited state gives a non-vanishing integral only if it is proportional to x, for then the integrand is proportional to x^2 rather than to xy or xz. Consequently, the excited state wavefunction must have the same symmetry as the displacement x.

Example 13.7 *Identifying infrared active modes*

Which modes of CH_4 are infrared active?

Method Refer to the T_d character table to establish the symmetry species of x, y, and z for this molecule, and then use the rule given above.

Answer The functions x, y, and z span T_2. We found in Example13.6 that the symmetry species of the normal modes are $A_1 + E + 2T_2$. Therefore, only the T_2 modes are infrared active. The distortions accompanying these modes lead to a changing dipole moment. The A_1 mode, which is inactive, is the symmetrical 'breathing' mode of the molecule.

Self-test 13.9 Which of the normal modes of H_2O are infrared active? [All three]

(b) Raman activity of normal modes

Group theory provides an explicit recipe for judging the Raman activity of a normal mode. In this case, the symmetry species of the quadratic forms (x^2, xy, etc.) listed in the character table are noted (they transform in the same way as the polarizability), and then we use the following rule:

> If the symmetry species of a normal mode is the same as the symmetry species of a quadratic form, then the mode is Raman active.

Illustration 13.6 *Identifying Raman active modes*

To decide which of the vibrations of CH_4 are Raman active, refer to the T_d character table. It was established in Example 13.6 that the symmetry species of the normal modes are $A_1 + E + 2T_2$. Because the quadratic forms span $A_1 + E + T_2$, all the normal modes are Raman active. By combining this information with that in Example 13.6, we see how the infrared and Raman spectra of CH_4 are assigned. The assignment of spectral features to the T_2 modes is straightforward because these are the only modes that are both infrared and Raman active. This leaves the A_1 and E modes to be assigned in the Raman spectrum. Measurement of the depolarization ratio distinguishes between these modes because the A_1 mode, being totally symmetric, is polarized and the E mode is depolarized.

Self-test 13.10 Which of the vibrational modes of H_2O are Raman active? [All three]

Checklist of key ideas

☐ 1. Emission spectroscopy is based on the detection of a transition from a state of high energy to a state of lower energy; absorption spectroscopy is based on the detection of the net absorption of nearly monochromatic incident radiation as the radiation is swept over a range of frequencies.

☐ 2. In Raman spectroscopy molecular energy levels are explored by examining the frequencies present in scattered radiation. Stokes and anti-Stokes radiation are scattered radiation at a lower and higher frequency, respectively, than the incident radiation. Rayleigh radiation is the component of radiation scattered into the forward direction without change of frequency.

☐ 3. The Beer–Lambert law is $I(\tilde{\nu}) = I_0(\tilde{\nu})10^{-\varepsilon(\tilde{\nu})[J]l}$, where $I(\tilde{\nu})$ is the transmitted intensity, $I_0(\tilde{\nu})$ is the incident intensity, and $\varepsilon(\tilde{\nu})$ is the molar absorption coefficient.

☐ 4. The transmittance, $T = I/I_0$, and the absorbance, A, of a sample at a given wavenumber are related by $A = -\log T$.

☐ 5. The integrated absorption coefficient, $\mathcal{A}$, is the sum of the absorption coefficients over the entire band, $\mathcal{A} = \int_{\text{band}} \varepsilon(\tilde{\nu})\text{d}\tilde{\nu}$.

☐ 6. Stimulated absorption is the radiation-driven transition from a low energy state to one of higher energy. Stimulated emission is the radiation-driven transition from a high energy state to one of lower energy. Spontaneous emission is radiative emission independent of the intensity of the radiation (of any frequency) that is already present.

☐ 7. The natural linewidth of a spectral line is due to spontaneous emission. Spectral lines are affected by Doppler broadening, lifetime broadening, and collisional deactivation of excited states.

☐ 8. A rigid rotor is a body that does not distort under the stress of rotation. A spherical rotor is a rigid rotor with three equal moments of inertia. A symmetric rotor is a rigid rotor with two equal moments of inertia. A linear rotor is a rigid rotor with one moment of inertia equal to zero. An asymmetric rotor is a rigid rotor with three different moments of inertia.

☐ 9. The rotational terms of a spherical rotor are $F(J) = BJ(J+1)$ with $B = \hbar/4\pi cI$, $J = 0, 1, 2, \ldots$, and are $(2J+1)^2$-fold degenerate.

☐ 10. The principal axis (figure axis) is the unique axis of a symmetric top. In an oblate top, $I_{\parallel} > I_{\perp}$. In a prolate top, $I_{\parallel} < I_{\perp}$.

☐ 11. The rotational terms of a symmetric rotor are $F(J,K) = BJ(J+1) + (A-B)K^2$, $J = 0, 1, 2, \ldots$, $K = 0, \pm1, \ldots, \pm J$.

☐ 12. The rotational terms of a linear rotor are $F(J) = BJ(J+1)$, $J = 0, 1, 2, \ldots$ and are $(2J+1)$-fold degenerate.

☐ 13. The centrifugal distortion constant, D_J, is the empirical constant in the expression $F(J) = BJ(J+1) - D_J J^2(J+1)^2$ that takes into account centrifugal distortion, $D_J \approx 4B^3/\tilde{\nu}^2$.

☐ 14. The gross rotational selection rule for microwave spectra is: for a molecule to give a pure rotational spectrum, it must be polar. The specific rotational selection rule is: $\Delta J = \pm1$, $\Delta M_J = 0, \pm1$, $\Delta K = 0$. The rotational wavenumbers in the absence and presence of centrifugal distortion are given by eqns 13.37 and 13.38, respectively.

☐ 15. The gross selection rule for rotational Raman spectra is: the molecule must be anisotropically polarizable. The specific selection rules are: (i) linear rotors, $\Delta J = 0, \pm2$; (ii) symmetric rotors, $\Delta J = 0, \pm1, \pm2$; $\Delta K = 0$.

☐ 16. The appearance of rotational spectra is affected by nuclear statistics, the selective occupation of rotational states that stems from the Pauli principle.

☐ 17. The vibrational energy levels of a diatomic molecule modelled as a harmonic oscillator are $E_v = (v + \frac{1}{2})\hbar\omega$, $\omega = (k/m_{\text{eff}})^{1/2}$; the vibrational terms are $G(v) = (v + \frac{1}{2})\tilde{\nu}$, $\tilde{\nu} = (1/2\pi c)(k/m_{\text{eff}})^{1/2}$.

☐ 18. The gross selection rule for infrared spectra is: the electric dipole moment of the molecule must change when the atoms are displaced relative to one another. The specific selection rule is: $\Delta v = \pm1$.

☐ 19. Morse potential energy, eqn 13.54, describes anharmonic motion, oscillatory motion in which the restoring force is not proportional to the displacement; the vibrational terms of a Morse oscillator are given by eqn 13.55.

☐ 20. A Birge–Sponer plot is a graphical procedure for determining the dissociation energy of a bond.

☐ 21. The P branch consists of vibration–rotation infrared transitions with $\Delta J = -1$; the Q branch has transitions with $\Delta J = 0$; the R branch has transitions with $\Delta J = +1$.

☐ 22. The gross selection rule for vibrational Raman spectra is: the polarizability must change as the molecule vibrates. The specific selection rule is: $\Delta v = \pm1$.

☐ 23. A normal mode is an independent, synchronous motion of atoms or groups of atoms that may be excited without leading to the excitation of any other normal mode. The number of normal modes is $3N-6$ (for nonlinear molecules) or $3N-5$ (linear molecules).

☐ 24. A symmetric stretch is a symmetrically stretching vibrational mode. An antisymmetric stretch is a stretching mode, one half of which is the mirror image of the other half.

☐ 25. The exclusion rule states that, if the molecule has a centre of symmetry, then no modes can be both infrared and Raman active.

☐ 26. The depolarization ratio, ρ, the ratio of the intensities, I, of the scattered light with polarizations perpendicular and parallel to the plane of polarization of the incident radiation, $\rho = I_{\perp}/I_{\parallel}$. A depolarized line is a line with ρ close to or greater than 0.75. A polarized line is a line with $\rho < 0.75$.

☐ 27. Resonance Raman spectroscopy is a Raman technique in which the frequency of the incident radiation nearly coincides with the frequency of an electronic transition of the sample. Coherent anti-Stokes Raman spectroscopy (CARS) is a

Raman technique that relies on the use of two incident beams of radiation.

☐ 28. A normal mode is infrared active if its symmetry species is the same as any of the symmetry species of *x*, *y*, or *z*, then the mode is infrared active. A normal mode is Raman active if its symmetry species is the same as the symmetry species of a quadratic form.

Further reading

Articles and texts

L. Glasser, Fourier transforms for chemists. Part I. Introduction to the Fourier transform. *J. Chem. Educ.* **64**, A228 (1987). Part II. Fourier transforms in chemistry and spectroscopy. *J. Chem. Educ.* **64**, A260 (1987).

H.-U. Gremlich and B. Yan, *Infrared and Raman spectroscopy of biological materials.* Marcel Dekker, New York (2001).

G. Herzberg, *Molecular spectra and molecular structure I. Spectra of diatomic molecules.* Krieger, Malabar (1989).

G. Herzberg, *Molecular spectra and molecular structure II. Infrared and Raman spectra of polyatomic molecules.* Van Nostrand–Reinhold, New York (1945).

J.C. Lindon, G.E. Tranter, and J.L. Holmes (ed.), *Encyclopedia of spectroscopy and spectrometry*. Academic Press, San Diego (2000).

D.P. Strommen, Specific values of the depolarization ratio in Raman spectroscopy: Their origins and significance. *J. Chem. Educ.* **69**, 803 (1992).

E.B. Wilson, J.C. Decius, and P.C. Cross, *Molecular vibrations.* Dover, New York (1980).

J.M. Brown and A. Carrington, *Rotational spectroscopy of diatomic molecules.* Cambridge University Press (2003).

Sources of data and information

M.E. Jacox, *Vibrational and electronic energy levels of polyatomic transient molecules.* Journal of Physical and Chemical Reference Data, Monograph No. 3 (1994).

K.P. Huber and G. Herzberg, *Molecular spectra and molecular structure IV. Constants of diatomic molecules.* Van Nostrand-Reinhold, New York (1979).

B. Schrader, *Raman/IR atlas of organic compounds.* VCH, New York (1989).

G. Socrates, *Infrared and Raman characteristic group frequencies: tables and charts.* Wiley, New York (2000).

Further information

Further information 13.1 *Spectrometers*

Here we provide additional detail on the principles of operation of spectrometers, describing radiation sources, dispersing elements, detectors, and Fourier transform techniques.

Sources of radiation

Sources of radiation are either *monochromatic*, those spanning a very narrow range of frequencies around a central value, or *polychromatic*, those spanning a wide range of frequencies. Monochromatic sources that can be tuned over a range of frequencies include the *klystron* and the *Gunn diode*, which operate in the microwave range, and lasers, which are discussed in Chapter 14.

Polychromatic sources that take advantage of black-body radiation from hot materials. For far infrared radiation with 35 cm^{-1} $< \tilde{\nu} <$ 200 cm^{-1}, a typical source is a mercury arc inside a quartz envelope, most of the radiation being generated by the hot quartz. Either a *Nernst filament* or a *globar* is used as a source of mid-infrared radiation with 200 cm^{-1} $< \tilde{\nu} <$ 4000 cm^{-1}. The Nernst filament consists of a ceramic filament of lanthanoid oxides that is heated to temperatures ranging from 1200 to 2000 K. The globar consists of a rod of silicon carbide, which is heated electrically to about 1500 K.

A *quartz–tungsten–halogen lamp* consists of a tungsten filament that, when heated to about 3000 K, emits light in the range 320 nm $< \lambda <$ 2500 nm. Near the surface of the lamp's quartz envelope, iodine atoms and tungsten atoms ejected from the filament combine to make a variety of tungsten–iodine compounds that decompose at the hot filament, replenishing it with tungsten atoms.

A *gas discharge lamp* is a common source of ultraviolet and visible radiation. In a *xenon discharge lamp*, an electrical discharge excites xenon atoms to excited states, which then emit ultraviolet radiation. At pressures exceeding 1 kPa, the output consists of sharp lines on a broad, intense background due to emission from a mixture of ions formed by the electrical discharge. These high-pressure xenon lamps have emission profiles similar to that of a black body heated to 6000 K. In a *deuterium lamp*, excited D_2 molecules dissociate into electronically excited D atoms, which emit intense radiation between 200–400 nm.

For certain applications, synchrotron radiation is generated in a *synchrotron storage ring*, which consists of an electron beam travelling in a circular path with circumferences of up to several hunderd metres. As electrons travelling in a circle are constantly accelerated by the forces that constrain them to their path, they generate radiation (Fig. 13.52). Synchrotron radiation spans a wide range of frequencies,

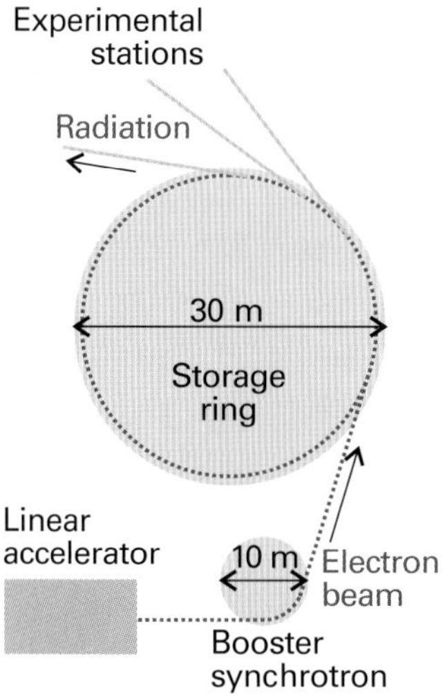

Fig. 13.52 A synchrotron storage ring. The electrons injected into the ring from the linear accelerator and booster synchrotron are accelerated to high speed in the main ring. An electron in a curved path is subject to constant acceleration, and an accelerated charge radiates electromagnetic energy.

including the infrared and X-rays. Except in the microwave region, synchrotron radiation is much more intense than can be obtained by most conventional sources.

The dispersing element

The dispersing element in most absorption spectrometers operating in the ultraviolet to near-infrared region of the spectrum is a **diffraction grating**, which consists of a glass or ceramic plate into which fine grooves have been cut and covered with a reflective aluminium coating. The grating causes interference between waves reflected from its surface, and constructive interference occurs when

$$n\lambda = d(\sin\theta - \sin\phi) \tag{13.71}$$

where $n = 1, 2, \ldots$ is the *diffraction order*, λ is the wavelength of the diffracted radiation, d is the distance between grooves, θ is the angle of incidence of the beam, and ϕ is the angle of emergence of the beam (Fig. 13.53). For given values of n and θ, larger differences in ϕ are

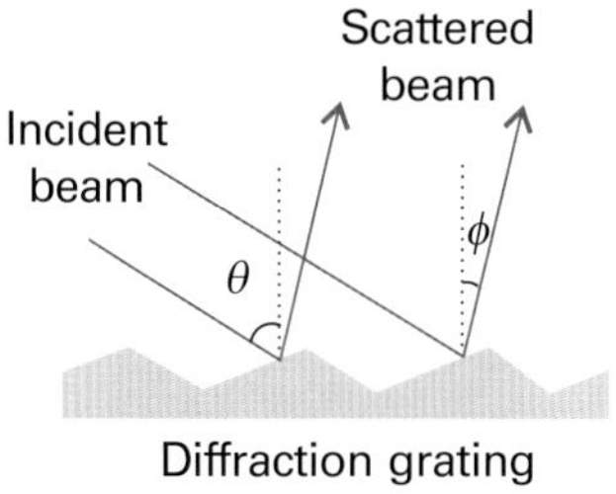

Fig. 13.53 One common dispersing element is a diffraction grating, which separates wavelengths spatially as a result of the scattering of light by fine grooves cut into a coated piece of glass. When a polychromatic light beam strikes the surface at an angle θ, several light beams of different wavelengths emerge at different angles ϕ (eqn 13.71).

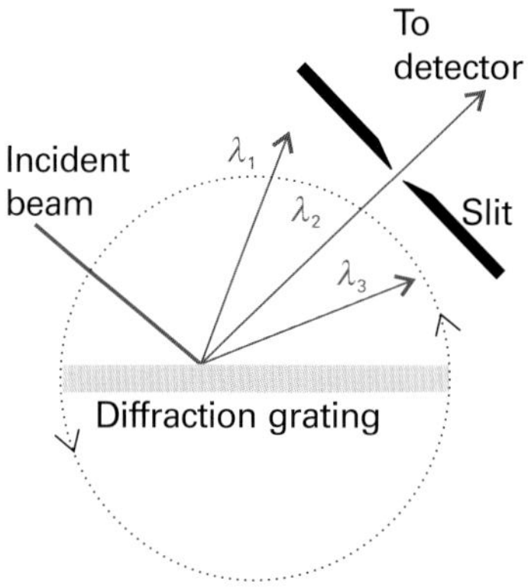

Fig. 13.54 A polychromatic beam is dispersed by a diffraction grating into three component wavelengths λ_1, λ_2, and λ_3. In the configuration shown, only radiation with λ_2 passes through a narrow slit and reaches the detector. Rotating the diffraction grating as shown by the double arrows allows λ_1 or λ_3 to reach the detector.

observed for different wavelengths when d is similar to the wavelength of radiation being analysed. Wide angular separation results in wide spatial separation between wavelengths some distance away from the grating, where a detector is placed.

In a **monochromator**, a narrow exit slit allows only a narrow range of wavelengths to reach the detector (Fig. 13.54). Turning the grating around an axis perpendicular to the incident and diffracted beams allows different wavelengths to be analysed; in this way, the absorption spectrum is built up one narrow wavelength range at a time. Typically, the grating is swept through an angle that investigates only the first order of diffraction ($n = 1$). In a **polychromator**, there is no slit and a broad range of wavelengths can be analysed simultaneously by *array detectors*, such as those discussed below.

Fourier transform techniques

In a Fourier transform instrument, the diffraction grating is replaced by a Michelson interferometer, which works by splitting the beam from the sample into two and introducing a varying path difference, p, into one of them (Fig. 13.55). When the two components

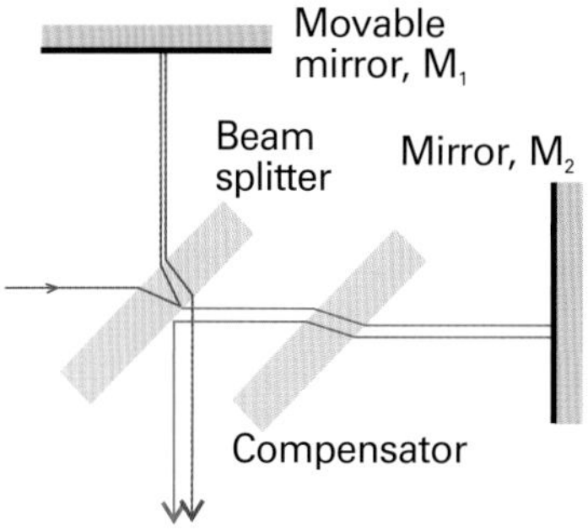

Fig. 13.55 A Michelson interferometer. The beam-splitting element divides the incident beam into two beams with a path difference that depends on the location of the mirror M_1. The compensator ensures that both beams pass through the same thickness of material.

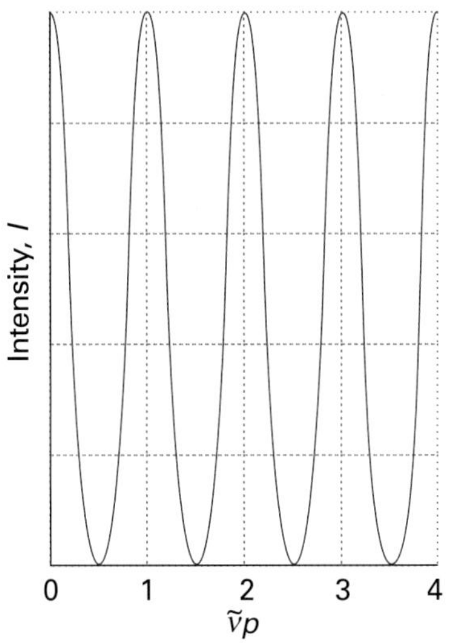

Fig. 13.56 An interferogram produced as the path length p is changed in the interferometer shown in Fig. 13.55. Only a single frequency component is present in the signal, so the graph is a plot of the function $I(p) = I_0(1 + \cos 2\pi\tilde{\nu}p)$, where I_0 is the intensity of the radiation.

Exploration Referring to Fig. 13.55, the mirror M_1 moves in finite distance increments, so the path difference p is also incremented in finite steps. Explore the effect of increasing the step size on the shape of the interferogram for a monochromatic beam of wavenumber $\tilde{\nu}$ and intensity I_0. That is, draw plots of $I(p)/I_0$ against $\tilde{\nu}p$, each with a different number of data points spanning the same total distance path taken by the movable mirror M_1.

recombine, there is a phase difference between them, and they interfere either constructively or destructively depending on the difference in path lengths. The detected signal oscillates as the two components alternately come into and out of phase as the path difference is changed (Fig. 13.56). If the radiation has wavenumber $\tilde{\nu}$, the intensity of the detected signal due to radiation in the range of wavenumbers $\tilde{\nu}$ to $\tilde{\nu} + d\tilde{\nu}$, which we denote $I(p,\tilde{\nu})d\tilde{\nu}$, varies with p as

$$I(p,\tilde{\nu})d\tilde{\nu} = I(\tilde{\nu})(1 + \cos 2\pi\tilde{\nu}p)d\tilde{\nu} \qquad (13.72)$$

Hence, the interferometer converts the presence of a particular wavenumber component in the signal into a variation in intensity of the radiation reaching the detector. An actual signal consists of radiation spanning a large number of wavenumbers, and the total intensity at the detector, which we write $I(p)$, is the sum of contributions from all the wavenumbers present in the signal (Fig. 13.57):

$$I(p) = \int_0^\infty I(p,\tilde{\nu})d\tilde{\nu} = \int_0^\infty I(\tilde{\nu})(1 + \cos 2\pi\tilde{\nu}p)d\tilde{\nu} \qquad (13.73)$$

The problem is to find $I(\tilde{\nu})$, the variation of intensity with wavenumber, which is the spectrum we require, from the record of values of $I(p)$. This step is a standard technique of mathematics, and is the 'Fourier transformation' step from which this form of spectroscopy takes its name. Specifically:

$$I(\tilde{\nu}) = 4\int_0^\infty \{I(p) - \tfrac{1}{2}I(0)\} \cos 2\pi\tilde{\nu}p\, dp \qquad (13.74)$$

where $I(0)$ is given by eqn 13.73 with $p = 0$. This integration is carried out numerically in a computer connected to the spectrometer, and the output, $I(\tilde{\nu})$, is the transmission spectrum of the sample (Fig. 13.58).

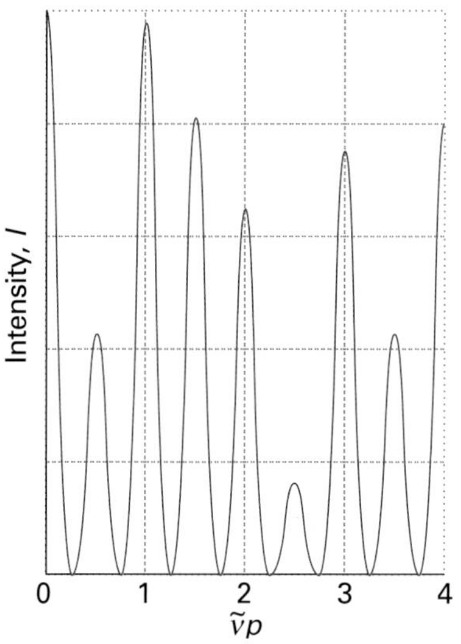

Fig. 13.57 An interferogram obtained when several (in this case, three) frequencies are present in the radiation.

Exploration For a signal consisting of only a few monochromatic beams, the integral in eqn 13.73 can be replaced by a sum over the finite number of wavenumbers. Use this information to draw your own version of Fig. 13.57. Then, go on to explore the effect of varying the wavenumbers and intensities of the three components of the radiation on the shape of the interferogram.

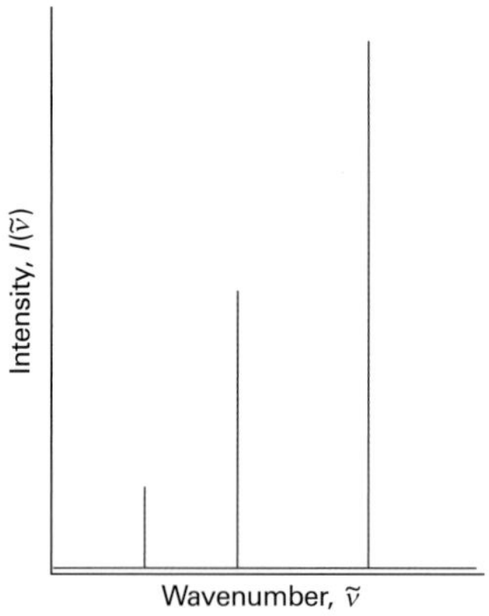

Fig. 13.58 The three frequency components and their intensities that account for the appearance of the interferogram in Fig. 13.57. This spectrum is the Fourier transform of the interferogram, and is a depiction of the contributing frequencies.

Exploration Calculate the Fourier transforms of the functions you generated in the previous *Exploration*.

A major advantage of the Fourier transform procedure is that all the radiation emitted by the source is monitored continuously. This is in contrast to a spectrometer in which a monochromator discards most of the generated radiation. As a result, Fourier transform spectrometers have a higher sensitivity than conventional spectrometers.

Detectors

A **detector** is a device that converts radiation into an electric current or voltage for appropriate signal processing and display. Detectors may consist of a single radiation sensing element or of several small elements arranged in one or two-dimensional arrays.

A microwave detector is typically a *crystal diode* consisting of a tungsten tip in contact with a semiconductor. The most common detectors found in commercial infrared spectrometers are sensitive in the mid-infrared region. In a *photovoltaic device* the potential difference changes upon exposure to infrared radiation. In a *pyroelectric device* the capacitance is sensitive to temperature and hence the presence of infrared radiation.

A common detector for work in the ultraviolet and visible ranges is the *photomultiplier tube* (PMT), in which the photoelectric effect (Section 8.2a) is used to generate an electrical signal proportional to the intensity of light that strikes the detector. In a PMT, photons first strike the *photocathode*, usually a metallic surface to which a large negative potential is applied. Each electron ejected from the photocathode is accelerated by a potential difference to another metallic surface, called the *dynode*, from which more electrons are ejected. After travelling through a chain of dynodes, with each dynode at a more positive potential than the preceding dynode in the chain, all the ejected electrons are collected at a final metallic surface called the *anode*. Depending on how the detector is constructed, a PMT can produce up to 10^8 electrons per photon that strikes the photocathode. This sensitivity is important for the detection of light from weak sources, but can pose problems as well. At room temperature, a small number of electrons on the surfaces of the photocathode and dynodes have sufficient energy to be ejected even in the dark. When amplified through the dynode chain, these electrons give rise to a *dark current*, which interferes with measurements on the sample of interest. To minimize the dark current, it is common to lower the temperature of the PMT detector.

A common, but less sensitive, alternative to the PMT is the *photodiode*, a solid-state device that conducts electricity when struck by photons because light-induced electron transfer reactions in the detector material create mobile charge carriers (negatively charged electrons and positively charged 'holes'). In an *avalanche photodiode*, the photo-generated electrons are accelerated through a very large electrical potential difference. The high-energy electrons then collide with other atoms in the solid and ionize them, thus creating an avalanche of secondary charge carriers and increasing the sensitivity of the device toward photons.

The *charge-coupled device* (CCD) is a two-dimensional array of several million small photodiode detectors. With a CCD, a wide range of wavelengths that emerge from a polychromator are detected simultaneously, thus eliminating the need to measure light intensity one narrow wavelength range at a time. CCD detectors are the imaging devices in digital cameras, but are also used widely in spectroscopy to measure absorption, emission, and Raman scattering. In Raman microscopy, a CCD detector can be used in a variation of the technique known as *Raman imaging*: a special optical filter allows only one Stokes line to reach the two-dimensional detector, which then contains a map of the distribution of the intensity of that line in the illuminated area.

Resolution

A number of factors determine a spectrometer's resolution, the smallest observable separation between two closely spaced spectral bands. We have already seen that the ability of a diffraction grating to disperse light depends on the distance between the grating's grooves and on the wavelength of the incident radiation. Furthermore, the distance between the grating and the slit placed in front of a detector must be long enough and the slit's width must be narrow enough so full advantage can be taken of the grating's dispersing ability (Fig. 13.54). It follows that a bad choice of grating, slit width, and detector placement may result in the failure to distinguish between closely spaced lines or to measure the actual linewidth of any one band in the spectrum.

The resolution of Fourier transform spectrometers is determined by the maximum path length difference, $p_{\max}$, of the interferometer:

$$\Delta\tilde{\nu} = \frac{1}{2p_{\max}} \qquad (13.75)$$

To achieve a resolution of 0.1 cm^{-1} requires a maximum path length difference of 5 cm.

Assuming that all instrumental factors have been optimized, the highest resolution is obtained when the sample is gaseous and at such low pressure that collisions between the molecules are infrequent (see Section 13.3b). In liquids and solids, the actual linewidths can be so broad that the sample itself can limit the resolution.

Further information 13.2 *Selection rules for rotational and vibrational spectroscopy*

Here we derive the gross and specific selection rules for microwave, infrared, and rotational and vibrational Raman spectroscopy. The starting point for our discussion is the total wavefunction for a molecule, which can be written as

$$\psi_{\text{total}} = \psi_{\text{c.m.}}\psi$$

where $\psi_{\text{c.m.}}$ describes the motion of the centre of mass and ψ describes the internal motion of the molecule. If we neglect the effect of electron spin, the Born–Oppenheimer approximation allows us to write ψ as the product of an electronic part, $|\varepsilon\rangle$, a vibrational part, $|\upsilon\rangle$, and a rotational part, which for a diatomic molecule can be represented by the spherical harmonics $Y_{J,M_J}(\theta,\phi)$ (Section 9.7). To simplify the form of the integrals that will soon follow, we are using the Dirac bracket notation introduced in *Further information 9.1*. The transition dipole moment for a spectroscopic transition can now be written as:

$$\boldsymbol{\mu}_{\text{fi}} = \langle \varepsilon_{\text{f}}\upsilon_{\text{f}}Y_{J,\text{f},M_J,\text{f}}|\hat{\boldsymbol{\mu}}|\varepsilon_{\text{i}}\upsilon_{\text{i}}Y_{J,\text{i},M_J,\text{i}}\rangle \qquad (13.76)$$

and our task is to explore conditions for which this integral vanishes or has a non-zero value.

Microwave spectra

During a pure rotational transition the molecule does not change electronic or vibrational states, so that $\langle \varepsilon_f v_f| = \langle \varepsilon_i v_i| = \langle \varepsilon v|$ and we identify $\boldsymbol{\mu}_{\varepsilon v} = \langle \varepsilon v|\hat{\boldsymbol{\mu}}|\varepsilon v\rangle$ with the *permanent* electric dipole moment of the molecule in the state εv. Equation 13.76 becomes

$$\boldsymbol{\mu}_{fi} = \langle Y_{J,f,M_{J,f}}|\hat{\boldsymbol{\mu}}_{\varepsilon v}|Y_{J,i,M_{J,i}}\rangle$$

The electric dipole moment has components $\mu_{\varepsilon v,x}$, $\mu_{\varepsilon v,y}$, and $\mu_{\varepsilon v,z}$, which, in spherical polar coordinates, are written in terms of μ_0, the magnitude of the vector, and the angles θ and ϕ as

$$\mu_{\varepsilon v,x} = \mu_0 \sin\theta\cos\phi \qquad \mu_{\varepsilon v,y} = \mu_0 \sin\theta\sin\phi \qquad \mu_{\varepsilon v,z} = \mu_0\cos\theta$$

Here, we have taken the z-axis to be coincident with the figure axis. The transition dipole moment has three components, given by:

$$\mu_{fi,x} = \mu_0\langle Y_{J,f,M_{J,f}}|\sin\theta\cos\phi|Y_{J,i,M_{J,i}}\rangle$$
$$\mu_{fi,y} = \mu_0\langle Y_{J,f,M_{J,f}}|\sin\theta\sin\phi|Y_{J,i,M_{J,i}}\rangle$$
$$\mu_{fi,z} = \mu_0\langle Y_{J,f,M_{J,f}}|\cos\theta|Y_{J,i,M_{J,i}}\rangle$$

We see immediately that the molecule must have a permanent dipole moment in order to have a microwave spectrum. This is the gross selection rule for microwave spectroscopy.

For the specific selection rules we need to examine the conditions for which the integrals do not vanish, and we must consider each component. For the z-component, we simplify the integral by using $\cos\theta \propto Y_{1,0}$ (Table 9.3). It follows that

$$\mu_{fi,z} \propto \langle Y_{J,f,M_{J,f}}|Y_{1,0}|Y_{J,i,M_{J,i}}\rangle$$

According to the properties of the spherical harmonics (*Comment* 13.8), this integral vanishes unless $J_f - J_i = \pm 1$ and $M_{J,f} - M_{J,i} = 0$. These are two of the selection rules stated in eqn 13.35.

Comment 13.8

An important 'triple integral' involving the spherical harmonics is

$$\int_0^{\pi}\int_0^{2\pi} Y_{l'',m_l''}(\theta,\phi)^* Y_{l',m_l'}(\theta,\phi) Y_{l,m_l}(\theta,\phi)\sin\theta\,d\theta\,d\phi = 0$$

unless $m_l'' = m_l' + m_l$ and lines of length l'', l', and l can form a triangle.

For the x- and y-components, we use $\cos\phi = \frac{1}{2}(e^{i\phi} + e^{-i\phi})$ and $\sin\phi = -\frac{1}{2}i(e^{i\phi} - e^{-i\phi})$ to write $\sin\theta\cos\phi \propto Y_{1,1} + Y_{1,-1}$ and $\sin\theta\sin\phi \propto Y_{1,1} - Y_{1,-1}$. It follows that

$$\mu_{fi,x} \propto \langle Y_{J,f,M_{J,f}}|(Y_{1,1} + Y_{1,-1})|Y_{J,i,M_{J,i}}\rangle$$
$$\mu_{fi,y} \propto \langle Y_{J,f,M_{J,f}}|(Y_{1,1} - Y_{1,-1})|Y_{J,i,M_{J,i}}\rangle$$

According to the properties of the spherical harmonics, these integrals vanish unless $J_f - J_i = \pm 1$ and $M_{J,f} - M_{J,i} = \pm 1$. This completes the selection rules of eqn 13.35.

Rotational Raman spectra

We understand the origin of the gross and specific selection rules for rotational Raman spectroscopy by using a diatomic molecule as an example. The incident electric field of a wave of electromagnetic radiation of frequency ω_i induces a molecular dipole moment that is given by

$$\mu_{ind} = \alpha\mathcal{E}(t) = \alpha\mathcal{E}\cos\omega_i t$$

If the molecule is rotating at a circular frequency ω_R, to an external observer its polarizability is also time dependent (if it is anisotropic), and we can write

$$\alpha = \alpha_0 + \Delta\alpha\cos 2\omega_R t$$

where $\Delta\alpha = \alpha_\parallel - \alpha_\perp$ and α ranges from $\alpha_0 + \Delta\alpha$ to $\alpha_0 - \Delta\alpha$ as the molecule rotates. The 2 appears because the polarizability returns to its initial value twice each revolution (Fig. 13.59). Substituting this expression into the expression for the induced dipole moment gives

$$\begin{aligned}\mu_{ind} &= (\alpha_0 + \Delta\alpha\cos 2\omega_R t)\times(\mathcal{E}\cos\omega_i t)\\ &= \alpha_0\mathcal{E}\cos\omega_i t + \mathcal{E}\Delta\alpha\cos 2\omega_R t\cos\omega_i t\\ &= \alpha_0\mathcal{E}\cos\omega_i t + \tfrac{1}{2}\mathcal{E}\Delta\alpha\{\cos(\omega_i + 2\omega_R)t + \cos(\omega_i - 2\omega_R)t\}\end{aligned}$$

This calculation shows that the induced dipole has a component oscillating at the incident frequency (which generates Rayleigh radiation), and that it also has two components at $\omega_i \pm 2\omega_R$, which give rise to the shifted Raman lines. These lines appear only if $\Delta\alpha \neq 0$; hence the polarizability must be anisotropic for there to be Raman lines. This is the gross selection rule for rotational Raman spectroscopy. We also see that the distortion induced in the molecule by the incident electric field returns to its initial value after a rotation of 180° (that is, twice a revolution). This is the origin of the specific selection rule $\Delta J = \pm 2$.

We now use a quantum mechanical formalism to understand the selection rules. First, we write the x-, y-, and z-components of the induced dipole moment as

$$\mu_{ind,x} = \mu_x\sin\theta\cos\phi \qquad \mu_{ind,y} = \mu_y\sin\theta\sin\phi \qquad \mu_{ind,z} = \mu_z\cos\theta$$

where μ_x, μ_y, and μ_z are the components of the electric dipole moment of the molecule and the z-axis is coincident with the molecular figure axis. The incident electric field also has components along the x-, y-, and z-axes:

$$\mathcal{E}_x = \mathcal{E}\sin\theta\cos\phi \qquad \mathcal{E}_y = \mathcal{E}\sin\theta\sin\phi \qquad \mathcal{E}_z = \mathcal{E}\cos\theta$$

Using eqn 13.40 and the preceding equations, it follows that

$$\begin{aligned}\mu_{ind} &= \alpha_\perp\mathcal{E}_x\sin\theta\cos\phi + \alpha_\perp\mathcal{E}_y\sin\theta\sin\phi + \alpha_\parallel\mathcal{E}\cos\theta\\ &= \alpha_\perp\mathcal{E}\sin^2\theta + \alpha_\parallel\mathcal{E}\cos^2\theta\end{aligned}$$

By using the spherical harmonic $Y_{2,0}(\theta,\phi) = (5/16\pi)^{1/2}(3\cos^2\theta - 1)$ and the relation $\sin^2\theta = 1 - \cos^2\theta$, it follows that:

$$\mu_{ind} = \left\{\tfrac{1}{3}\alpha_\parallel + \tfrac{2}{3}\alpha_\perp + \tfrac{4}{3}\left(\frac{\pi}{5}\right)^{1/2}\Delta\alpha Y_{2,0}(\theta,\phi)\right\}\mathcal{E}$$

For a transition between two rotational states, we calculate the integral $\langle Y_{J,f,M_{J,f}}|\mu_{ind}|Y_{J,i,M_{J,i}}\rangle$, which has two components:

$$(\tfrac{1}{3}\alpha_\parallel + \tfrac{2}{3}\alpha_\perp)\langle Y_{J_f,M_{J,f}}|(Y_{J_i,M_{J,i}}\rangle \quad \text{and} \quad \mathcal{E}\Delta\alpha\langle Y_{J_f,M_{J,f}}|Y_{2,0}Y_{J_i,M_{J,i}}\rangle$$

According to the properties of the spherical harmonics (Table 9.3), the first integral vanishes unless $J_f - J_i = 0$ and the second integral vanishes unless $J_f - J_i = \pm 2$ and $\Delta\alpha \neq 0$. These are the gross and specific selection rules for linear rotors.

Infrared spectra

The gross selection rule for infrared spectroscopy is based on an analysis of the transition dipole moment $\langle v_f|\hat{\boldsymbol{\mu}}|v_i\rangle$, which arises from eqn 13.76 when the molecule does not change electronic or rotational

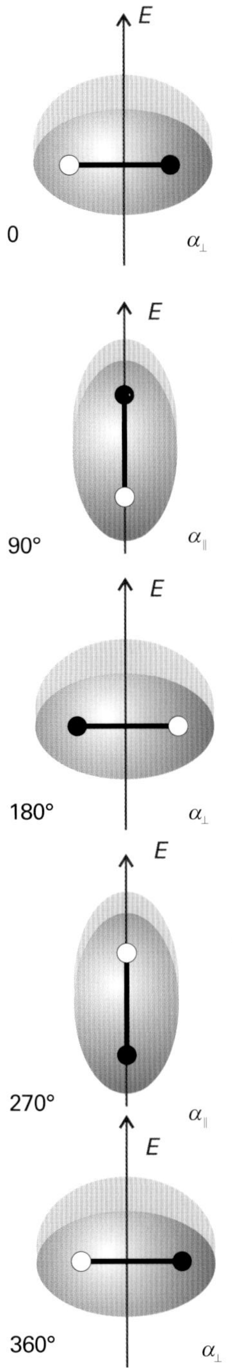

Fig. 13.59 The distortion induced in a molecule by an applied electric field returns to its initial value after a rotation of only 180° (that is, twice a revolution). This is the origin of the $\Delta J = \pm 2$ selection rule in rotational Raman spectroscopy.

states. For simplicity, we shall consider a one-dimensional oscillator (like a diatomic molecule). The electric dipole moment operator depends on the location of all the electrons and all the nuclei in the molecule, so it varies as the internuclear separation changes

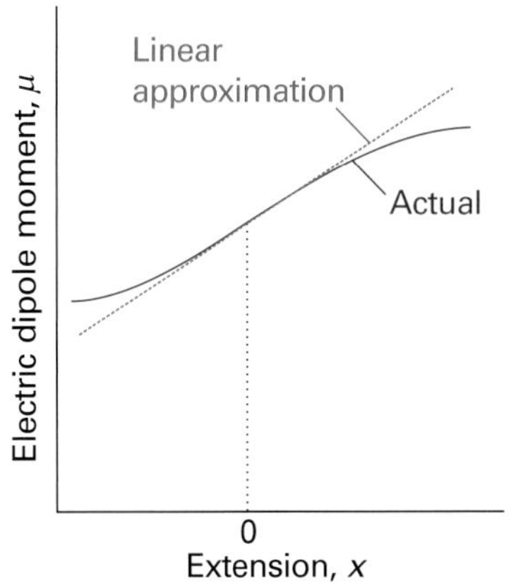

Fig. 13.60 The electric dipole moment of a heteronuclear diatomic molecule varies as shown by the purple curve. For small displacements the change in dipole moment is proportional to the displacement.

(Fig. 13.60). If we think of the dipole moment as arising from two partial charges $\pm\delta q$ separated by a distance $R = R_e + x$, we can write its variation with displacement from the equilibrium separation, x, as

$$\mu = R\delta q = R_e\delta q + x\delta q = \mu_0 + x\delta q$$

where μ_0 is the electric dipole moment operator when the nuclei have their equilibrium separation. It then follows that, with $\mathrm{f} \neq \mathrm{i}$,

$$\langle v_f|\hat{\mu}|v_i\rangle = \mu_0\langle v_f|v_i\rangle + \delta q\langle v_f|x|v_i\rangle$$

The term proportional to μ_0 is zero because the states with different values of v are orthogonal. It follows that the transition dipole moment is

$$\langle v_f|\hat{\mu}|v_i\rangle = \langle v_f|x|v_i\rangle\delta q$$

Because

$$\delta q = \frac{d\mu}{dx}$$

we can write the transition dipole moment more generally as

$$\langle v_f|\hat{\mu}|v_i\rangle = \langle v_f|x|v_i\rangle\left(\frac{d\mu}{dx}\right)$$

and we see that the right-hand side is zero unless the dipole moment varies with displacement. This is the gross selection rule for infrared spectroscopy.

The specific selection rule is determined by considering the value of $\langle v_f|x|v_i\rangle$. We need to write out the wavefunctions in terms of the Hermite polynomials given in Section 9.5 and then to use their properties (Example 9.4 should be reviewed, for it gives further details of the calculation). We note that $x = \alpha y$ with $\alpha = (\hbar^2/m_{\text{eff}}k)^{1/4}$ (eqn 9.28; note that in this context α is not the polarizability). Then we write

$$\langle v_f|x|v_i\rangle = N_{v_f}N_{v_i}\int_{-\infty}^{\infty} H_{v_f}xH_{v_i}e^{-y^2}dx = \alpha^2 N_{v_f}N_{v_i}\int_{-\infty}^{\infty} H_{v_f}yH_{v_i}e^{-y^2}dy$$

To evaluate the integral we use the recursion relation

$$yH_v = vH_{v-1} + \tfrac{1}{2}H_{v+1}$$

which turns the matrix element into

$$\langle v_f|x|v_i\rangle = \alpha^2 N_{v_f} N_{v_i}\left\{v_i\int_{-\infty}^{\infty} H_{v_f}H_{v_i-1}\mathrm{e}^{-y^2}\mathrm{d}y + \tfrac{1}{2}\int_{-\infty}^{\infty} H_{v_f}H_{v_i+1}\mathrm{e}^{-y^2}\mathrm{d}y\right\}$$

The first integral is zero unless $v_f = v_i - 1$ and that the second is zero unless $v_f = v_i + 1$. It follows that the transition dipole moment is zero unless $\Delta v = \pm 1$.

Comment 13.9

An important integral involving Hermite polynomials is

$$\int_{-\infty}^{\infty} H_{v'}H_v\mathrm{e}^{-y^2}\mathrm{d}y = \begin{cases} 0 & \text{if } v' \neq v \\ \pi^{1/2}2^v v! & \text{if } v' = v \end{cases}$$

Vibrational Raman spectra

The gross selection rule for vibrational Raman spectroscopy is based on an analysis of the transition dipole moment $\langle \varepsilon v_f|\hat{\mu}|\varepsilon v_i\rangle$, which is written from eqn 13.76 by using the Born–Oppenheimer approximation and neglecting the effect of rotation and electron spin. For simplicity, we consider a one-dimensional harmonic oscillator (like a diatomic molecule).

First, we use eqn 13.40 to write the transition dipole moment as

$$\mu_{fi} = \langle \varepsilon v_f|\hat{\mu}|\varepsilon v_i\rangle = \langle \varepsilon v_f|\alpha|\varepsilon v_i\rangle\mathcal{E} = \langle v_f|\alpha(x)|v_i\rangle\mathcal{E}$$

where $\alpha(x) = \langle \varepsilon|\alpha|\varepsilon\rangle$ is the polarizability of the molecule, which we expect to be a function of small displacements x from the equilibrium bond length of the molecule (Section 13.13). Next, we expand $\alpha(x)$ as a Taylor series, so the transition dipole moment becomes

$$\mu_{fi} = \left\langle v_f \;\; \alpha(0) + \left(\frac{\mathrm{d}\alpha}{\mathrm{d}x}\right)_0 x + \cdots \;\; v_i\right\rangle\mathcal{E}$$

$$= \langle v_f|v_i\rangle\alpha(0)\mathcal{E} + \left(\frac{\mathrm{d}\alpha}{\mathrm{d}x}\right)_0 \langle v_f|x|v_i\rangle\mathcal{E} + \cdots$$

The term containing $\langle v_f|v_i\rangle$ vanishes for $f \neq i$ because the harmonic oscillator wavefunctions are orthogonal. Therefore, the vibration is Raman active if $(\mathrm{d}\alpha/\mathrm{d}x)_0 \neq 0$ and $\langle v_f|x|v_i\rangle \neq 0$. Therefore, the polarizability of the molecule must change during the vibration; this is the gross selection rule of Raman spectroscopy. Also, we already know that $\langle v_f|x|v_i\rangle \neq 0$ if $v_f - v_i = \pm 1$; this is the specific selection rule of Raman spectroscopy.

Discussion questions

13.1 Describe the physical origins of linewidths in the absorption and emission spectra of gases, liquids, and solids.

13.2 Discuss the physical origins of the gross selection rules for microwave and infrared spectroscopy.

13.3 Discuss the physical origins of the gross selection rules for rotational and vibrational Raman spectroscopy.

13.4 Consider a diatomic molecule that is highly susceptible to centrifugal distortion in its ground vibrational state. Do you expect excitation to high rotational energy levels to change the equilibrium bond length of this molecule? Justify your answer.

13.5 Suppose that you wish to characterize the normal modes of benzene in the gas phase. Why is it important to obtain both infrared absorption and Raman spectra of your sample?

Exercises

13.1a Calculate the ratio of the Einstein coefficients of spontaneous and stimulated emission, A and B, for transitions with the following characteristics: (a) 70.8 pm X-rays, (b) 500 nm visible light, (c) 3000 cm^{-1} infrared radiation.

13.1b Calculate the ratio of the Einstein coefficients of spontaneous and stimulated emission, A and B, for transitions with the following characteristics: (a) 500 MHz radiofrequency radiation, (e) 3.0 cm microwave radiation.

13.2a What is the Doppler-shifted wavelength of a red (660 nm) traffic light approached at 80 km h^{-1}?

13.2b At what speed of approach would a red (660 nm) traffic light appear green (520 nm)?

13.3a Estimate the lifetime of a state that gives rise to a line of width (a) 0.10 cm^{-1}, (b) 1.0 cm^{-1}.

13.3b Estimate the lifetime of a state that gives rise to a line of width (a) 100 MHZ, (b) 2.14 cm^{-1}.

13.4a A molecule in a liquid undergoes about 1.0×10^{13} collisions in each second. Suppose that (a) every collision is effective in deactivating the molecule vibrationally and (b) that one collision in 100 is effective. Calculate the width (in cm^{-1}) of vibrational transitions in the molecule.

13.4b A molecule in a gas undergoes about 1.0×10^9 collisions in each second. Suppose that (a) every collision is effective in deactivating the molecule rotationally and (b) that one collision in 10 is effective. Calculate the width (in hertz) of rotational transitions in the molecule.

13.5a Calculate the frequency of the $J = 4 \leftarrow 3$ transition in the pure rotational spectrum of $^{14}N^{16}O$. The equilibrium bond length is 115 pm.

13.5b Calculate the frequency of the $J = 3 \leftarrow 2$ transition in the pure rotational spectrum of $^{12}C^{16}O$. The equilibrium bond length is 112.81 pm.

13.6a If the wavenumber of the $J = 3 \leftarrow 2$ rotational transition of $^1H^{35}Cl$ considered as a rigid rotator is 63.56 cm^{-1}, what is (a) the moment of inertia of the molecule, (b) the bond length?

13.6b If the wavenumber of the $J = 1 \leftarrow 0$ rotational transition of $^1H^{81}Br$ considered as a rigid rotator is 16.93 cm^{-1}, what is (a) the moment of inertia of the molecule, (b) the bond length?

13.7a Given that the spacing of lines in the microwave spectrum of $^{27}Al^1H$ is constant at 12.604 cm^{-1}, calculate the moment of inertia and bond length of the molecule ($m(^{27}Al) = 26.9815$ u).

13.7b Given that the spacing of lines in the microwave spectrum of $^{35}Cl^{19}F$ is constant at 1.033 cm^{-1}, calculate the moment of inertia and bond length of the molecule ($m(^{35}Cl) = 34.9688$ u, $m(^{19}F) = 18.9984$ u).

13.8a The rotational constant of $^{127}I^{35}Cl$ is 0.1142 cm^{-1}. Calculate the ICl bond length ($m(^{35}Cl) = 34.9688$ u, $m(^{127}I) = 126.9045$ u).

13.8b The rotational constant of $^{12}C^{16}O_2$ is 0.39021 cm^{-1}. Calculate the bond length of the molecule ($m(^{12}C) = 12$ u exactly, $m(^{16}O) = 15.9949$ u).

13.9a Determine the HC and CN bond lengths in HCN from the rotational constants $B(^1H^{12}C^{14}N) = 44.316$ GHz and $B(^2H^{12}C^{14}N) = 36.208$ GHz.

13.9b Determine the CO and CS bond lengths in OCS from the rotational constants $B(^{16}O^{12}C^{32}S) = 6081.5$ MHz, $B(^{16}O^{12}C^{34}S) = 5932.8$ MHz.

13.10a The wavenumber of the incident radiation in a Raman spectrometer is 20 487 cm^{-1}. What is the wavenumber of the scattered Stokes radiation for the $J = 2 \leftarrow 0$ transition of $^{14}N_2$?

13.10b The wavenumber of the incident radiation in a Raman spectrometer is 20 623 cm^{-1}. What is the wavenumber of the scattered Stokes radiation for the $J = 4 \leftarrow 2$ transition of $^{16}O_2$?

13.11a The rotational Raman spectrum of $^{35}Cl_2$ ($m(^{35}Cl) = 34.9688$ u) shows a series of Stokes lines separated by 0.9752 cm^{-1} and a similar series of anti-Stokes lines. Calculate the bond length of the molecule.

13.11b The rotational Raman spectrum of $^{19}F_2$ ($m(^{19}F) = 18.9984$ u) shows a series of Stokes lines separated by 3.5312 cm^{-1} and a similar series of anti-Stokes lines. Calculate the bond length of the molecule.

13.12a Which of the following molecules may show a pure rotational microwave absorption spectrum: (a) H_2, (b) HCl, (c) CH_4, (d) CH_3Cl, (e) CH_2Cl_2?

13.12b Which of the following molecules may show a pure rotational microwave absorption spectrum: (a) H_2O, (b) H_2O_2, (c) NH_3, (d) N_2O?

13.13a Which of the following molecules may show a pure rotational Raman spectrum: (a) H_2, (b) HCl, (c) CH_4, (d) CH_3Cl?

13.13b Which of the following molecules may show a pure rotational Raman spectrum: (a) CH_2Cl_2, (b) CH_3CH_3, (c) SF_6, (d) N_2O?

13.14a An object of mass 1.0 kg suspended from the end of a rubber band has a vibrational frequency of 2.0 Hz. Calculate the force constant of the rubber band.

13.14b An object of mass 2.0 g suspended from the end of a spring has a vibrational frequency of 3.0 Hz. Calculate the force constant of the spring.

13.15a Calculate the percentage difference in the fundamental vibration wavenumber of $^{23}Na^{35}Cl$ and $^{23}Na^{37}Cl$ on the assumption that their force constants are the same.

13.15b Calculate the percentage difference in the fundamental vibration wavenumber of $^1H^{35}Cl$ and $^2H^{37}Cl$ on the assumption that their force constants are the same.

13.16a The wavenumber of the fundamental vibrational transition of $^{35}Cl_2$ is 564.9 cm^{-1}. Calculate the force constant of the bond ($m(^{35}Cl) = 34.9688$ u).

13.16b The wavenumber of the fundamental vibrational transition of $^{79}Br^{81}Br$ is 323.2 cm^{-1}. Calculate the force constant of the bond ($m(^{79}Br) = 78.9183$ u, $m(^{81}Br) = 80.9163$ u).

13.17a Calculate the relative numbers of Cl_2 molecules ($\tilde{\nu} = 559.7$ cm^{-1}) in the ground and first excited vibrational states at (a) 298 K, (b) 500 K.

13.17b Calculate the relative numbers of Br_2 molecules ($\tilde{\nu} = 321$ cm^{-1}) in the second and first excited vibrational states at (a) 298 K, (b) 800 K.

13.18a The hydrogen halides have the following fundamental vibrational wavenumbers: 4141.3 cm^{-1} (HF); 2988.9 cm^{-1} ($H^{35}Cl$); 2649.7 cm^{-1} ($H^{81}Br$); 2309.5 cm^{-1} ($H^{127}I$). Calculate the force constants of the hydrogen–halogen bonds.

13.18b From the data in Exercise 13.18a, predict the fundamental vibrational wavenumbers of the deuterium halides.

13.19a For $^{16}O_2$, ΔG values for the transitions $v = 1 \leftarrow 0$, $2 \leftarrow 0$, and $3 \leftarrow 0$ are, respectively, 1556.22, 3088.28, and 4596.21 cm^{-1}. Calculate $\tilde{\nu}$ and x_e. Assume y_e to be zero.

13.19b For $^{14}N_2$, ΔG values for the transitions $v = 1 \leftarrow 0$, $2 \leftarrow 0$, and $3 \leftarrow 0$ are, respectively, 2345.15, 4661.40, and 6983.73 cm^{-1}. Calculate $\tilde{\nu}$ and x_e. Assume y_e to be zero.

13.20a The first five vibrational energy levels of HCl are at 1481.86, 4367.50, 7149.04, 9826.48, and 12 399.8 cm^{-1}. Calculate the dissociation energy of the molecule in reciprocal centimetres and electronvolts.

13.20b The first five vibrational energy levels of HI are at 1144.83, 3374.90, 5525.51, 7596.66, and 9588.35 cm^{-1}. Calculate the dissociation energy of the molecule in reciprocal centimetres and electronvolts.

13.21a Infrared absorption by $^1H^{81}Br$ gives rise to an R branch from $v = 0$. What is the wavenumber of the line originating from the rotational state with $J = 2$? Use the information in Table 13.2.

13.21b Infrared absorption by $^1H^{127}I$ gives rise to an R branch from $v = 0$. What is the wavenumber of the line originating from the rotational state with $J = 2$? Use the information in Table 13.2.

13.22a Which of the following molecules may show infrared absorption spectra: (a) H_2, (b) HCl, (c) CO_2, (d) H_2O?

13.22b Which of the following molecules may show infrared absorption spectra: (a) CH_3CH_3, (b) CH_4, (c) CH_3Cl, (d) N_2?

13.23a How many normal modes of vibration are there for the following molecules: (a) H_2O, (b) H_2O_2, (c) C_2H_4?

13.23b How many normal modes of vibration are there for the following molecules: (a) C_6H_6, (b) $C_6H_6CH_3$, (c) HC≡C—C≡CH.

13.24a Which of the three vibrations of an AB_2 molecule are infrared or Raman active when it is (a) angular, (b) linear?

13.24b Which of the vibrations of an AB_3 molecule are infrared or Raman active when it is (a) trigonal planar, (b) trigonal pyramidal?

13.25a Consider the vibrational mode that corresponds to the uniform expansion of the benzene ring. Is it (a) Raman, (b) infrared active?

13.25b Consider the vibrational mode that corresponds to the boat-like bending of a benzene ring. Is it (a) Raman, (b) infrared active?

13.26a The molecule CH_2Cl_2 belongs to the point group C_{2v}. The displacements of the atoms span $5A_1 + 2A_2 + 4B_1 + 4B_2$. What are the symmetries of the normal modes of vibration?

13.26b A carbon disulfide molecule belongs to the point group $D_{\infty h}$. The nine displacements of the three atoms span $A_{1g} + A_{1u} + A_{2g} + 2E_{1u} + E_{1g}$. What are the symmetries of the normal modes of vibration?

Problems*

Numerical problems

13.1 Use mathematical software to evaluate the Planck distribution at any temperature and wavelength or frequency, and evaluate integrals for the energy density of the radiation between any two wavelengths. Calculate the total energy density in the visible region (700 nm to 400 nm) for a black body at (a) 1500 K, a typical operating temperature for globars, (b) 2500 K, a typical operating temperature for tungsten filament lamps, (c) 5800 K, the surface temperature of the Sun. What are the classical values at these temperatures?

13.2 Calculate the Doppler width (as a fraction of the transition wavelength) for any kind of transition in (a) HCl, (b) ICl at 25°C. What would be the widths of the rotational and vibrational transitions in these molecules (in MHz and cm^{-1}, respectively), given $B(ICl) = 0.1142\ cm^{-1}$ and $\tilde{\nu}(ICl) = 384\ cm^{-1}$ and additional information in Table 13.2.

13.3 The collision frequency z of a molecule of mass m in a gas at a pressure p is $z = 4\sigma(kT/\pi m)^{1/2}p/kT$, where σ is the collision cross-section. Find an expression for the collision-limited lifetime of an excited state assuming that every collision is effective. Estimate the width of rotational transition in HCl ($\sigma = 0.30\ nm^2$) at 25°C and 1.0 atm. To what value must the pressure of the gas be reduced in order to ensure that collision broadening is less important than Doppler broadening?

13.4 The rotational constant of NH_3 is equivalent to 298 GHz. Compute the separation of the pure rotational spectrum lines in GHz, cm^{-1}, and mm, and show that the value of B is consistent with an N—H bond length of 101.4 pm and a bond angle of 106.78°.

13.5 The rotational constant for CO is $1.9314\ cm^{-1}$ and $1.6116\ cm^{-1}$ in the ground and first excited vibrational states, respectively. By how much does the internuclear distance change as a result of this transition?

13.6 Pure rotational Raman spectra of gaseous C_6H_6 and C_6D_6 yield the following rotational constants: $B(C_6H_6) = 0.189\ 60\ cm^{-1}$, $B(C_6D_6) = 0.156\ 81\ cm^{-1}$. The moments of inertia of the molecules about any axis perpendicular to the C_6 axis were calculated from these data as $I(C_6H_6) = 1.4759 \times 10^{-45}\ kg\ m^2$, $I(C_6D_6) = 1.7845 \times 10^{-45}\ kg\ m^2$. Calculate the CC, CH, and CD bond lengths.

13.7 Rotational absorption lines from $^1H^{35}Cl$ gas were found at the following wavenumbers (R.L. Hausler and R.A. Oetjen, *J. Chem. Phys.* **21**, 1340 (1953)): 83.32, 104.13, 124.73, 145.37, 165.89, 186.23, 206.60, 226.86 cm^{-1}. Calculate the moment of inertia and the bond length of the molecule. Predict the positions of the corresponding lines in $^2H^{35}Cl$.

13.8 Is the bond length in HCl the same as that in DCl? The wavenumbers of the $J = 1 \leftarrow 0$ rotational transitions for $H^{35}Cl$ and $^2H^{35}Cl$ are 20.8784 and 10.7840 cm^{-1}, respectively. Accurate atomic masses are 1.007825 u and 2.0140 u for ^1H and ^2H, respectively. The mass of ^{35}Cl is 34.96885 u. Based on this information alone, can you conclude that the bond lengths are the same or different in the two molecules?

13.9 Thermodynamic considerations suggest that the copper monohalides CuX should exist mainly as polymers in the gas phase, and indeed it proved difficult to obtain the monomers in sufficient abundance to detect spectroscopically. This problem was overcome by flowing the halogen gas over copper heated to 1100 K (E.L. Manson, F.C. de Lucia, and W. Gordy, *J. Chem. Phys.* **63**, 2724 (1975)). For CuBr the $J = 13$–14, 14–15, and 15–16 transitions occurred at 84 421.34, 90 449.25, and 96 476.72 MHz, respectively. Calculate the rotational constant and bond length of CuBr.

13.10 The microwave spectrum of $^{16}O^{12}CS$ (C.H. Townes, A.N. Holden, and F.R. Merritt, *Phys. Rev.* **74**, 1113 (1948)) gave absorption lines (in GHz) as follows:

J	1	2	3	4
^{32}S	24.325 92	36.488 82	48.651 64	60.814 08
^{34}S	23.732 33		47.462 40	

Use the expressions for moments of inertia in Table 13.1 and assume that the bond lengths are unchanged by substitution; calculate the CO and CS bond lengths in OCS.

13.11‡ In a study of the rotational spectrum of the linear FeCO radical, K. Tanaka, M. Shirasaka, and T. Tanaka (*J. Chem. Phys.* **106**, 6820 (1997)) report the following $J + 1 \leftarrow J$ transitions:

J	24	25	26	27	28	29
$\tilde{\nu}/m^{-1}$	214 777.7	223 379.0	231 981.2	240 584.4	249 188.5	257 793.5

Evaluate the rotational constant of the molecule. Also, estimate the value of J for the most highly populated rotational energy level at 298 K and at 100 K.

13.12 The vibrational energy levels of NaI lie at the wavenumbers 142.81, 427.31, 710.31, and 991.81 cm^{-1}. Show that they fit the expression $(v + \frac{1}{2})\tilde{\nu} - (v + \frac{1}{2})^2 x\tilde{\nu}$, and deduce the force constant, zero-point energy, and dissociation energy of the molecule.

13.13 Predict the shape of the nitronium ion, NO_2^+, from its Lewis structure and the VSEPR model. It has one Raman active vibrational mode at 1400 cm^{-1}, two strong IR active modes at 2360 and 540 cm^{-1}, and one weak IR mode at 3735 cm^{-1}. Are these data consistent with the predicted shape of the molecule? Assign the vibrational wavenumbers to the modes from which they arise.

13.14 At low resolution, the strongest absorption band in the infrared absorption spectrum of $^{12}C^{16}O$ is centred at 2150 cm^{-1}. Upon closer examination at higher resolution, this band is observed to be split into two sets of closely spaced peaks, one on each side of the centre of the spectrum at 2143.26 cm^{-1}. The separation between the peaks immediately to the right and left of the centre is 7.655 cm^{-1}. Make the harmonic oscillator and rigid rotor approximations and calculate from these data: (a) the vibrational wavenumber of a CO molecule, (b) its molar zero-point vibrational energy, (c) the force constant of the CO bond, (d) the rotational constant B, and (e) the bond length of CO.

13.15 The HCl molecule is quite well described by the Morse potential with $D_e = 5.33$ eV, $\tilde{\nu} = 2989.7\ cm^{-1}$, and $x\tilde{\nu} = 52.05\ cm^{-1}$. Assuming that the potential is unchanged on deuteration, predict the dissociation energies (D_0) of (a) HCl, (b) DCl.

13.16 The Morse potential (eqn 13.54) is very useful as a simple representation of the actual molecular potential energy. When RbH was studied, it was found that $\tilde{\nu} = 936.8\ cm^{-1}$ and $x\tilde{\nu} = 14.15\ cm^{-1}$. Plot the potential energy curve from 50 pm to 800 pm around $R_e = 236.7$ pm. Then go on to explore how the rotation of a molecule may weaken its bond by allowing for the kinetic energy of rotation of a molecule and plotting $V^* = V + hcBJ(J+1)$ with $B = \hbar/4\pi c\mu R^2$. Plot these curves on the same diagram for $J = 40$, 80, and 100, and observe how the dissociation energy is affected by the

* Problems denoted with the symbol ‡ were supplied by Charles Trapp, Carmen Giunta, and Marshall Cady.

rotation. (Taking $B = 3.020\ \text{cm}^{-1}$ at the equilibrium bond length will greatly simplify the calculation.)

13.17‡ F. Luo, G.C. McBane, G. Kim, C.F. Giese, and W.R. Gentry (*J. Chem. Phys.* **98**, 3564 (1993)) reported experimental observation of the He_2 complex, a species that had escaped detection for a long time. The fact that the observation required temperatures in the neighbourhood of 1 mK is consistent with computational studies that suggest that hcD_e for He_2 is about 1.51×10^{-23} J, hcD_0 about 2×10^{-26} J, and R_e about 297 pm. (a) Estimate the fundamental vibrational wavenumber, force constant, moment of inertia, and rotational constant based on the harmonic oscillator and rigid-rotor approximations. (b) Such a weakly bound complex is hardly likely to be rigid. Estimate the vibrational wavenumber and anharmonicity constant based on the Morse potential.

13.18 As mentioned in Section 13.15, the semi-empirical, *ab initio*, and DFT methods discussed in Chapter 11 can be used to estimate the force field of a molecule. The molecule's vibrational spectrum can be simulated, and it is then possible to determine the correspondence between a vibrational frequency and the atomic displacements that give rise to a normal mode. (a) Using molecular modelling software[3] and the computational method of your choice (semi-empirical, *ab initio*, or DFT methods), calculate the fundamental vibrational wavenumbers and visualize the vibrational normal modes of SO_2 in the gas phase. (b) The experimental values of the fundamental vibrational wavenumbers of SO_2 in the gas phase are 525 cm^{-1}, 1151 cm^{-1}, and 1336 cm^{-1}. Compare the calculated and experimental values. Even if agreement is poor, is it possible to establish a correlation between an experimental value of the vibrational wavenumber with a specific vibrational normal mode?

13.19 Consider the molecule CH_3Cl. (a) To what point group does the molecule belong? (b) How many normal modes of vibration does the molecule have? (c) What are the symmetries of the normal modes of vibration for this molecule? (d) Which of the vibrational modes of this molecule are infrared active? (e) Which of the vibrational modes of this molecule are Raman active?

13.20 Suppose that three conformations are proposed for the nonlinear molecule H_2O_2 (**2**, **3**, and **4**). The infrared absorption spectrum of gaseous H_2O_2 has bands at 870, 1370, 2869, and 3417 cm^{-1}. The Raman spectrum of the same sample has bands at 877, 1408, 1435, and 3407 cm^{-1}. All bands correspond to fundamental vibrational wavenumbers and you may assume that: (i) the 870 and 877 cm^{-1} bands arise from the same normal mode, and (ii) the 3417 and 3407 cm^{-1} bands arise from the same normal mode. (a) If H_2O_2 were linear, how many normal modes of vibration would it have? (b) Give the symmetry point group of each of the three proposed conformations of nonlinear H_2O_2. (c) Determine which of the proposed conformations is inconsistent with the spectroscopic data. Explain your reasoning.

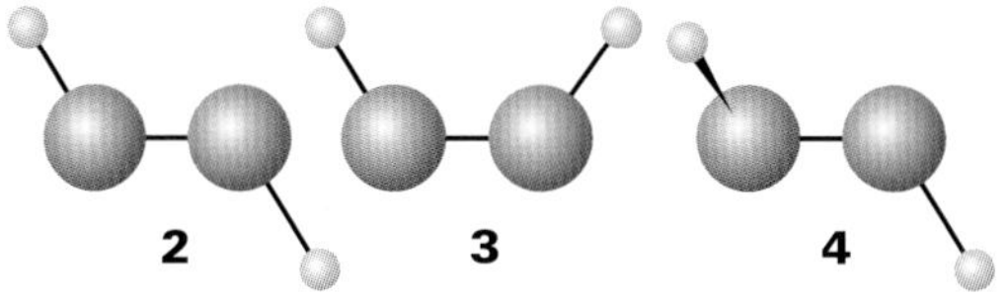

2 **3** **4**

Theoretical problems

13.21 Show that the moment of inertia of a diatomic molecule composed of atoms of masses m_A and m_B and bond length R is equal to $m_{eff}R^2$, where $m_{eff} = m_A m_B/(m_A + m_B)$.

13.22 Derive eqn 13.34 for the centrifugal distortion constant D_J of a diatomic molecule of effective mass m_{eff}. Treat the bond as an elastic spring with force constant k and equilibrium length r_e that is subjected to a centrifugal distortion to a new length r_c. Begin the derivation by letting the particles experience a restoring force of magnitude $k(r_c - r_e)$ that is countered perfectly by a centrifugal force $m_{eff}\omega^2 r_c$, where ω is the angular velocity of the rotating molecule. Then introduce quantum mechanical effects by writing the angular momentum as $\{J(J+1)\}^{1/2}\hbar$. Finally, write an expression for the energy of the rotating molecule, compare it with eqn 13.33, and write an expression for D_J. For help with the classical aspects of this derivation, see *Appendix 3*.

13.23 In the group theoretical language developed in Chapter 12, a spherical rotor is a molecule that belongs to a cubic or icosahedral point group, a symmetric rotor is a molecule with at least a threefold axis of symmetry, and an asymmetric rotor is a molecule without a threefold (or higher) axis. Linear molecules are linear rotors. Classify each of the following molecules as a spherical, symmetric, linear, or asymmetric rotor and justify your answers with group theoretical arguments: (a) CH_4, (b) CH_3CN, (c) CO_2, (d) CH_3OH, (e) benzene, (f) pyridine.

13.24 Derive an expression for the value of J corresponding to the most highly populated rotational energy level of a diatomic rotor at a temperature T remembering that the degeneracy of each level is $2J + 1$. Evaluate the expression for ICl (for which $B = 0.1142\ \text{cm}^{-1}$) at 25°C. Repeat the problem for the most highly populated level of a spherical rotor, taking note of the fact that each level is $(2J + 1)^2$-fold degenerate. Evaluate the expression for CH_4 (for which $B = 5.24\ \text{cm}^{-1}$) at 25°C.

13.25 The moments of inertia of the linear mercury(II) halides are very large, so the O and S branches of their vibrational Raman spectra show little rotational structure. Nevertheless, the peaks of both branches can be identified and have been used to measure the rotational constants of the molecules (R.J.H. Clark and D.M. Rippon, *J. Chem. Soc. Faraday Soc. II*, **69**, 1496 (1973)). Show, from a knowledge of the value of J corresponding to the intensity maximum, that the separation of the peaks of the O and S branches is given by the Placzek–Teller relation $\delta\tilde{\nu} = (32BkT/hc)^{1/2}$. The following widths were obtained at the temperatures stated:

	$HgCl_2$	$HgBr_2$	HgI_2
θ/°C	282	292	292
$\delta\tilde{\nu}$/cm^{-1}	23.8	15.2	11.4

Calculate the bond lengths in the three molecules.

13.26 Confirm that a Morse oscillator has a finite number of bound states, the states with $V < hcD_e$. Determine the value of v_{max} for the highest bound state.

Applications: to biology, environmental science, and astrophysics

13.27 The protein haemerythrin is responsible for binding and carrying O_2 in some invertebrates. Each protein molecule has two Fe^{2+} ions that are in very close proximity and work together to bind one molecule of O_2. The Fe_2O_2 group of oxygenated haemerythrin is coloured and has an electronic absorption band at 500 nm. The resonance Raman spectrum of oxygenated haemerythrin obtained with laser excitation at 500 nm has a band at 844 cm^{-1} that has been attributed to the O—O stretching mode of bound $^{16}O_2$. (a) Why is resonance Raman spectroscopy and not infrared spectroscopy the method of choice for the study of the binding of O_2 to haemerythrin? (b) Proof that the 844 cm^{-1} band arises from a bound O_2 species may be obtained by conducting experiments on samples of haemerythrin that have been mixed with $^{18}O_2$, instead of $^{16}O_2$. Predict the fundamental vibrational wavenumber of the ^{18}O—^{18}O stretching mode in a sample of haemerythrin that has been treated with $^{18}O_2$. (c) The fundamental vibrational wavenumbers for the

[3] The web site contains links to molecular modelling freeware and to other sites where you may perform molecular orbital calculations directly from your web browser.

O—O stretching modes of O_2, O_2^- (superoxide anion), and O_2^{2-} (peroxide anion) are 1555, 1107, and 878 cm^{-1}, respectively. Explain this trend in terms of the electronic structures of O_2, O_2^-, and O_2^{2-}. *Hint:* Review Section 11.4. What are the bond orders of O_2, O_2^-, and O_2^{2-}? (d) Based on the data given above, which of the following species best describes the Fe_2O_2 group of haemerythrin: $Fe_2^{2+}O_2$, $Fe^{2+}Fe^{3+}O_2^-$, or $Fe_2^{3+}O_2^{2-}$? Explain your reasoning. (e) The resonance Raman spectrum of haemerythrin mixed with $^{16}O^{18}O$ has two bands that can be attributed to the O—O stretching mode of bound oxygen. Discuss how this observation may be used to exclude one or more of the four proposed schemes (**5–8**) for binding of O_2 to the Fe_2 site of haemerythrin.

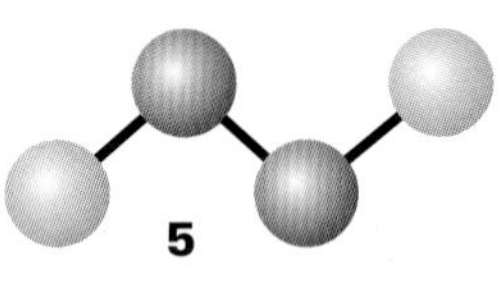
5

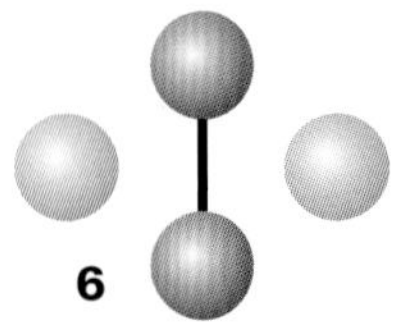
6

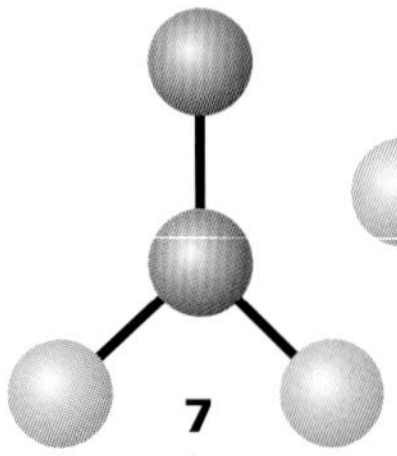
7

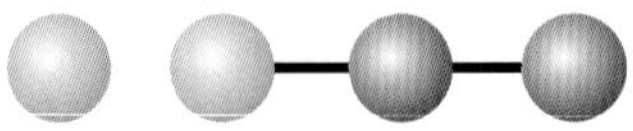
8

13.28‡ A mixture of carbon dioxide (2.1 per cent) and helium, at 1.00 bar and 298 K in a gas cell of length 10 cm has an infrared absorption band centred at 2349 cm^{-1} with absorbances, $A(\tilde{\nu})$, described by:

$$A(\tilde{\nu}) = \frac{a_1}{1 + a_2(\tilde{\nu} - a_3)^2} + \frac{a_4}{1 + a_5(\tilde{\nu} - a_6)^2}$$

where the coefficients are $a_1 = 0.932$, $a_2 = 0.005050$ cm^2, $a_3 = 2333$ cm^{-1}, $a_4 = 1.504$, $a_5 = 0.01521$ cm^2, $a_6 = 2362$ cm^{-1}. (a) Draw graphs of $A(\tilde{\nu})$ and $\varepsilon(\tilde{\nu})$. What is the origin of both the band and the band width? What are the allowed and forbidden transitions of this band? (b) Calculate the transition wavenumbers and absorbances of the band with a simple harmonic oscillator-rigid rotor model and compare the result with the experimental spectra. The CO bond length is 116.2 pm. (c) Within what height, h, is basically all the infrared emission from the Earth in this band absorbed by atmospheric carbon dioxide? The mole fraction of CO_2 in the atmosphere is 3.3×10^{-4} and $T/K = 288 - 0.0065(h/m)$ below 10 km. Draw a surface plot of the atmospheric transmittance of the band as a function of both height and wavenumber.

13.29 In Problem 10.27, we saw that Doppler shifts of atomic spectral lines are used to estimate the speed of recession or approach of a star. From the discussion in Section 13.3a, it is easy to see that Doppler broadening of an atomic spectral line depends on the temperature of the star that emits the radiation. A spectral line of $^{48}Ti^{8+}$ (of mass 47.95 u) in a distant star was found to be shifted from 654.2 nm to 706.5 nm and to be broadened to 61.8 pm. What is the speed of recession and the surface temperature of the star?

13.30 A. Dalgarno, in Chemistry in the interstellar medium, *Frontiers of Astrophysics*, E.H. Avrett (ed.), Harvard University Press, Cambridge (1976), notes that although both CH and CN spectra show up strongly in the interstellar medium in the constellation Ophiuchus, the CN spectrum has become the standard for the determination of the temperature of the cosmic microwave background radiation. Demonstrate through a calculation why CH would not be as useful for this purpose as CN. The rotational constant B_0 for CH is 14.190 cm^{-1}.

13.31‡ There is a gaseous interstellar cloud in the constellation Ophiuchus that is illuminated from behind by the star ζ-Ophiuci. Analysis of the electronic–vibrational–rotational absorption lines obtained by H.S. Uhler and R.A. Patterson (*Astrophys. J.* **42**, 434 (1915)) shows the presence of CN molecules in the interstellar medium. A strong absorption line in the ultraviolet region at $\lambda = 387.5$ nm was observed corresponding to the transition $J = 0$–1. Unexpectedly, a second strong absorption line with 25 per cent of the intensity of the first was found at a slightly longer wavelength ($\Delta\lambda = 0.061$ nm) corresponding to the transition $J = 1$–1 (here allowed). Calculate the temperature of the CN molecules. Gerhard Herzberg, who was later to receive the Nobel Prize for his contributions to spectroscopy, calculated the temperature as 2.3 K. Although puzzled by this result, he did not realize its full significance. If he had, his prize might have been for the discovery of the cosmic microwave background radiation.

13.32‡ The H_3^+ ion has recently been found in the interstellar medium and in the atmospheres of Jupiter, Saturn, and Uranus. The rotational energy levels of H_3^+, an oblate symmetric rotor, are given by eqn 13.29, with C replacing A, when centrifugal distortion and other complications are ignored. Experimental values for vibrational–rotational constants are $\tilde{\nu}(E') = 2521.6$ cm^{-1}, $B = 43.55$ cm^{-1}, and $C = 20.71$ cm^{-1}. (a) Show that, for a nonlinear planar molecule (such as H_3^+), $I_C = 2I_B$. The rather large discrepancy with the experimental values is due to factors ignored in eqn 13.29. (b) Calculate an approximate value of the H—H bond length in H_3^+. (c) The value of R_e obtained from the best quantum mechanical calculations by J.B. Anderson (*J. Chem. Phys.* **96**, 3702 (1991)) is 87.32 pm. Use this result to calculate the values of the rotational constants B and C. (d) Assuming that the geometry and force constants are the same in D_3^+ and H_3^+, calculate the spectroscopic constants of D_3^+. The molecular ion D_3^+ was first produced by J.T. Shy, J.W. Farley, W.E. Lamb Jr, and W.H. Wing (*Phys. Rev. Lett* **45**, 535 (1980)) who observed the $\nu_2(E')$ band in the infrared.

13.33 The space immediately surrounding stars, also called the *circumstellar space*, is significantly warmer because stars are very intense black-body emitters with temperatures of several thousand kelvin. Discuss how such factors as cloud temperature, particle density, and particle velocity may affect the rotational spectrum of CO in an interstellar cloud. What new features in the spectrum of CO can be observed in gas ejected from and still near a star with temperatures of about 1000 K, relative to gas in a cloud with temperature of about 10 K? Explain how these features may be used to distinguish between circumstellar and interstellar material on the basis of the rotational spectrum of CO.

Molecular spectroscopy 2: electronic transitions

14

Simple analytical expressions for the electronic energy levels of molecules cannot be given, so this chapter concentrates on the qualitative features of electronic transitions. A common theme throughout the chapter is that electronic transitions occur within a stationary nuclear framework. We pay particular attention to spontaneous radiative decay processes, which include fluorescence and phosphorescence. A specially important example of stimulated radiative decay is that responsible for the action of lasers, and we see how this stimulated emission may be achieved and employed.

The energies needed to change the electron distributions of molecules are of the order of several electronvolts (1 eV is equivalent to about 8000 cm^{-1} or 100 kJ mol^{-1}). Consequently, the photons emitted or absorbed when such changes occur lie in the visible and ultraviolet regions of the spectrum (Table 14.1).

One of the revolutions that has occurred in physical chemistry in recent years is the application of lasers to spectroscopy and kinetics. Lasers have brought unprecedented precision to spectroscopy, made Raman spectroscopy a widely useful technique, and have made it possible to study chemical reactions on a femtosecond time scale. We shall see the principles of their action in this chapter and their applications throughout the rest of the book.

The characteristics of electronic transitions

In the ground state of a molecule the nuclei are at equilibrium in the sense that they experience no net force from the electrons and other nuclei in the molecule. Immediately after an electronic transition they are subjected to different forces and

Synoptic table 14.1* Colour, frequency, and energy of light

Colour	λ/nm	$\nu/(10^{14}$ Hz)	E/(kJ mol^{-1})
Infrared	>1000	<3.0	<120
Red	700	4.3	170
Yellow	580	5.2	210
Blue	470	6.4	250
Ultraviolet	<300	>10	>400

* More values are given in the *Data section*.

the molecule may respond by starting to vibrate. The resulting vibrational structure of electronic transitions can be resolved for gaseous samples, but in a liquid or solid the lines usually merge together and result in a broad, almost featureless band (Fig. 14.1). Superimposed on the vibrational transitions that accompany the electronic transition of a molecule in the gas phase is an additional branch structure that arises from rotational transitions. The electronic spectra of gaseous samples are therefore very complicated but rich in information.

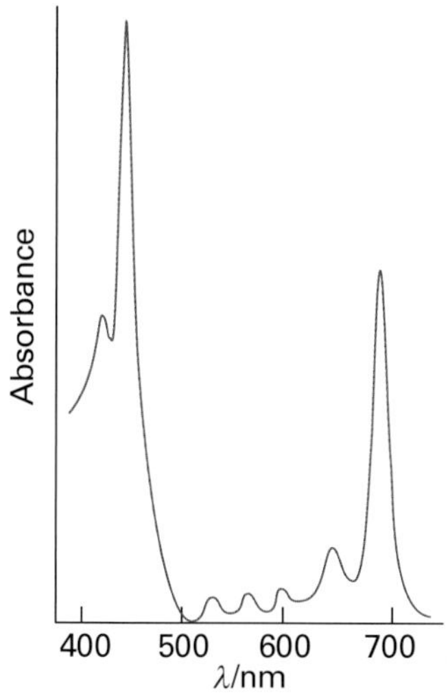

Fig. 14.1 The absorption spectrum of chlorophyll in the visible region. Note that it absorbs in the red and blue regions, and that green light is not absorbed.

14.1 The electronic spectra of diatomic molecules

We examine some general features of electronic transitions by using diatomic molecules as examples. We begin by assigning term symbols to ground and excited electronic states. Then we use the symmetry designations to formulate selection rules. Finally, we examine the origin of vibrational structure in electronic spectra.

(a) Term symbols

The term symbols of linear molecules (the analogues of the symbols ^{2}P, etc. for atoms) are constructed in a similar way to those for atoms, but now we must pay attention to the component of total orbital angular momentum about the internuclear axis, $\Lambda\hbar$. The value of $|\Lambda|$ is denoted by the symbols Σ, Π, Δ, . . . for $|\Lambda| = 0, 1, 2 \ldots$, respectively. These labels are the analogues of S, P, D, . . . for atoms. The value of Λ is the sum of the values of λ, the quantum number for the component $\lambda\hbar$ of orbital angular momentum of an individual electron around the internuclear axis. A single electron in a σ orbital has $\lambda = 0$: the orbital is cylindrically symmetrical and has no angular nodes when viewed along the internuclear axis. Therefore, if that is the only electron present, $\Lambda = 0$. The term symbol for H_2^+ is therefore Σ.

As in atoms, we use a superscript with the value of $2S + 1$ to denote the multiplicity of the term. The component of total spin angular momentum about the internuclear axis is denoted Σ, where $\Sigma = S, S-1, S-2, \ldots, -S$. For H_2^+, because there is only one electron, $S = s = \frac{1}{2}$ $(\Sigma = \pm\frac{1}{2})$ and the term symbol is $^2\Sigma$, a doublet term. The overall parity of the term is added as a right subscript. For H_2^+, the parity of the only occupied orbital is g (Section 11.3c), so the term itself is also g, and in full dress is $^2\Sigma_g$. If there are several electrons, the overall parity is calculated by using

Comment 14.1

It is important to distinguish between the (upright) term symbol Σ and the (sloping) quantum number Σ.

$$\mathrm{g \times g = g \qquad u \times u = g \qquad u \times g = u} \tag{14.1}$$

These rules are generated by interpreting g as +1 and u as −1. The term symbol for the ground state of any closed-shell homonuclear diatomic molecule is $^1\Sigma_g$ because the spin is zero (a singlet term in which all electrons paired), there is no orbital angular momentum from a closed shell, and the overall parity is g.

A π electron in a diatomic molecule has one unit of orbital angular momentum about the internuclear axis ($\lambda = \pm 1$), and if it is the only electron outside a closed shell, gives rise to a Π term. If there are two π electrons (as in the ground state of O_2, with configuration $1\pi_u^4 1\pi_g^2$) then the term symbol may be either Σ (if the electrons are travelling in opposite directions, which is the case if they occupy different π orbitals, one with $\lambda = +1$ and the other with $\lambda = -1$) or Δ (if they are travelling in the same direction, which is the case if they occupy the same π orbital, both $\lambda = +1$, for instance). For O_2, the two π electrons occupy different orbitals with parallel spins (a triplet term), so the ground term is $^3\Sigma$. The overall parity of the molecule is

$$\text{(closed shell)} \times \mathrm{g} \times \mathrm{g} = \mathrm{g}$$

The term symbol is therefore $^3\Sigma_g$.

Table 14.2 Properties of O_2 in its lower electronic states*

Configuration†	Term	Relative energy/cm^{-1}	$\tilde{\nu}$/cm^{-1}	R_e/pm
$\pi_u^2\pi_u^2\pi_g^1\pi_g^1$	$^3\Sigma_g^-$	0	1580	120.74
$\pi_u^2\pi_u^2\pi_g^2\pi_g^0$	$^1\Delta_g$	7 882.39	1509	121.55
$\pi_u^2\pi_u^2\pi_g^1\pi_g^1$	$^1\Sigma_g^+$	13 120.9	1433	122.68
$\pi_u^2\pi_u^1\pi_g^2\pi_g^1$	$^3\Sigma_u^+$	35 713	819	142
$\pi_u^2\pi_u^1\pi_g^2\pi_g^1$	$^3\Sigma_u^-$	49 363	700	160

* Adapted from G. Herzberg, *Spectra of diatomic molecules*, Van Nostrand, New York (1950) and D.C. Harris and M.D. Bertolucci, *Symmetry and spectroscopy: an introduction to vibrational and electronic spectroscopy*, Dover, New York (1989).

† The configuration $\pi_u^2\pi_u^1\pi_g^2\pi_g^1$ should also give rise to a $^3\Delta_u$ term, but electronic transitions to or from this state have not been observed.

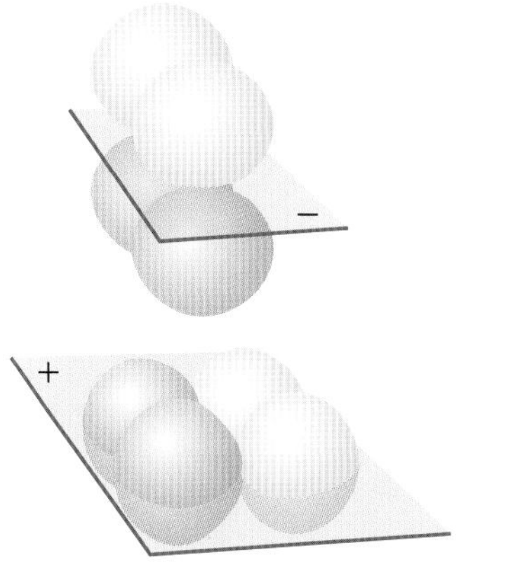

Fig. 14.2 The + or – on a term symbol refers to the overall symmetry of a configuration under reflection in a plane containing the two nuclei.

For Σ terms, a ± superscript denotes the behaviour of the molecular wavefunction under reflection in a plane containing the nuclei (Fig. 14.2). If, for convenience, we think of O_2 as having one electron in $1\pi_{g,x}$, which changes sign under reflection in the *yz*-plane, and the other electron in $1\pi_{g,y}$, which does not change sign under reflection in the same plane, then the overall reflection symmetry is

$$(\text{closed shell}) \times (+) \times (-) = (-)$$

and the full term symbol of the ground electronic state of O_2 is $^3\Sigma_g^-$.

The term symbols of excited electronic states are constructed in a similar way. For example, the term symbol for the excited state of O_2 formed by placing two electrons in a $1\pi_{g,x}$ (or in a $1\pi_{g,y}$) orbital is $^1\Delta_g$ because $|\Lambda| = 2$ (two electrons in the same π orbital), the spin is zero (all electrons are paired), and the overall parity is (closed shell) × g × g = g. Table 14.2 and Fig. 14.3 summarize the configurations, term symbols, and energies of the ground and some excited states of O_2.

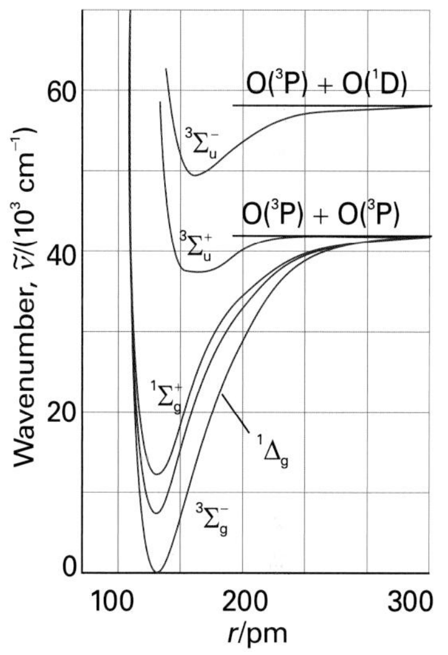

Fig. 14.3 The electronic states of dioxygen.

(b) Selection rules

A number of selection rules govern which transitions will be observed in the electronic spectrum of a molecule. The selection rules concerned with changes in angular momentum are

$$\Delta\Lambda = 0, \pm 1 \qquad \Delta S = 0 \qquad \Delta\Sigma = 0 \qquad \Delta\Omega = 0, \pm 1$$

where $\Omega = \Lambda + \Sigma$ is the quantum number for the component of total angular momentum (orbital and spin) around the internuclear axis (Fig. 14.4). As in atoms (Section 10.9), the origins of these rules are conservation of angular momentum during a transition and the fact that a photon has a spin of 1.

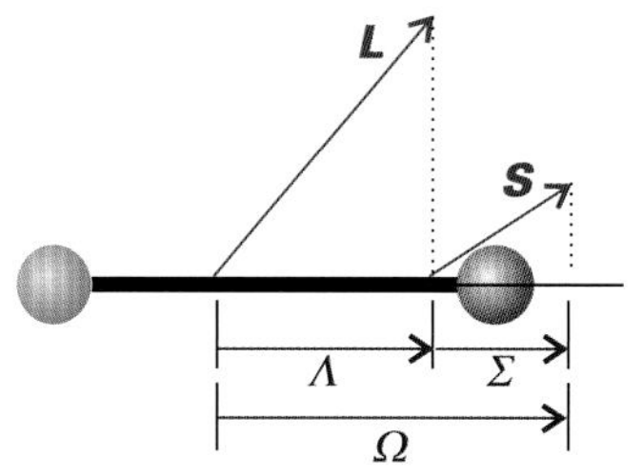

Fig. 14.4 The coupling of spin and orbital angular momenta in a linear molecule: only the components along the internuclear axis are conserved.

There are two selection rules concerned with changes in symmetry. First, for Σ terms, only $\Sigma^+ \leftrightarrow \Sigma^+$ and $\Sigma^- \leftrightarrow \Sigma^-$ transitions are allowed. Second, the **Laporte selection rule** for centrosymmetric molecules (those with a centre of inversion) and atoms states that:

> The only allowed transitions are transitions that are accompanied by a change of parity.

That is, u → g and g → u transitions are allowed, but g → g and u → u transitions are forbidden.

Justification 14.1 *The Laporte selection rule*

The last two selection rules result from the fact that the electric-dipole transition moment

$$\boldsymbol{\mu}_{\mathrm{fi}} = \int \psi_{\mathrm{f}}^{*}\hat{\mu}\psi_{\mathrm{i}}\,\mathrm{d}\tau$$

vanishes unless the integrand is invariant under all symmetry operations of the molecule. The three components of the dipole moment operator transform like x, y, and z, and are all u. Therefore, for a g → g transition, the overall parity of the transition dipole moment is g × u × g = u, so it must be zero. Likewise, for a u → u transition, the overall parity is u × u × u = u, so the transition dipole moment must also vanish. Hence, transitions without a change of parity are forbidden. The z-component of the dipole moment operator, the only component of $\boldsymbol{\mu}$ responsible for $\Sigma \leftrightarrow \Sigma$ transitions, has (+) symmetry. Therefore, for a (+) ↔ (−) transition, the overall symmetry of the transition dipole moment is (+) × (+) × (−) = (−), so it must be zero. Therefore, for Σ terms, $\Sigma^+ \leftrightarrow \Sigma^-$ transitions are not allowed.

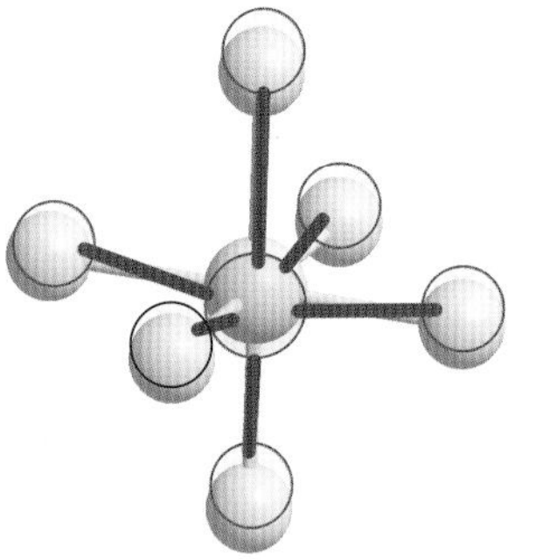

Fig. 14.5 A *d–d* transition is parity-forbidden because it corresponds to a g–g transition. However, a vibration of the molecule can destroy the inversion symmetry of the molecule and the g,u classification no longer applies. The removal of the centre of symmetry gives rise to a vibronically allowed transition.

A forbidden g → g transition can become allowed if the centre of symmetry is eliminated by an asymmetrical vibration, such as the one shown in Fig. 14.5. When the centre of symmetry is lost, g → g and u → u transitions are no longer parity-forbidden and become weakly allowed. A transition that derives its intensity from an asymmetrical vibration of a molecule is called a **vibronic transition**.

Self-test 14.1 Which of the following electronic transitions are allowed in O_2: ${}^3\Sigma_g^- \leftrightarrow {}^1\Delta_g$, ${}^3\Sigma_g^- \leftrightarrow {}^1\Sigma_g^+$, ${}^3\Sigma_g^- \leftrightarrow {}^3\Delta_u$, ${}^3\Sigma_g^- \leftrightarrow {}^3\Sigma_u^+$, ${}^3\Sigma_g^- \leftrightarrow {}^3\Sigma_u^-$? [${}^3\Sigma_g^- \leftrightarrow {}^3\Sigma_u^-$]

(c) Vibrational structure

To account for the vibrational structure in electronic spectra of molecules (Fig. 14.6), we apply the **Franck–Condon principle**:

> Because the nuclei are so much more massive than the electrons, an electronic transition takes place very much faster than the nuclei can respond.

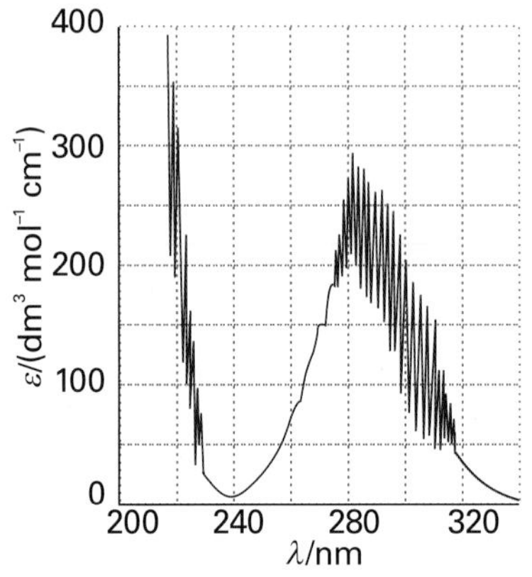

Fig. 14.6 The electronic spectra of some molecules show significant vibrational structure. Shown here is the ultraviolet spectrum of gaseous SO_2 at 298 K. As explained in the text, the sharp lines in this spectrum are due to transitions from a lower electronic state to different vibrational levels of a higher electronic state.

As a result of the transition, electron density is rapidly built up in new regions of the molecule and removed from others. The initially stationary nuclei suddenly experience a new force field, to which they respond by beginning to vibrate and (in classical terms) swing backwards and forwards from their original separation (which was maintained during the rapid electronic excitation). The stationary equilibrium separation of the nuclei in the initial electronic state therefore becomes a stationary turning point in the final electronic state (Fig. 14.7).

The quantum mechanical version of the Franck–Condon principle refines this picture. Before the absorption, the molecule is in the lowest vibrational state of its lowest electronic state (Fig. 14.8); the most probable location of the nuclei is at their equilibrium separation, R_e. The electronic transition is most likely to take place when the nuclei have this separation. When the transition occurs, the molecule is excited to the state represented by the upper curve. According to the Franck–Condon principle, the nuclear framework remains constant during this excitation, so we may imagine the transition as being up the vertical line in Fig. 14.7. The vertical line is the origin of the expression **vertical transition**, which is used to denote an electronic transition that occurs without change of nuclear geometry.

The vertical transition cuts through several vibrational levels of the upper electronic state. The level marked * is the one in which the nuclei are most probably at the same initial separation R_e (because the vibrational wavefunction has maximum amplitude there), so this vibrational state is the most probable state for the termination of the transition. However, it is not the only accessible vibrational state because several nearby states have an appreciable probability of the nuclei being at the separation R_e. Therefore, transitions occur to all the vibrational states in this region, but most intensely to the state with a vibrational wavefunction that peaks most strongly near R_e.

The vibrational structure of the spectrum depends on the relative horizontal position of the two potential energy curves, and a long **vibrational progression**, a lot of vibrational structure, is stimulated if the upper potential energy curve is appreciably displaced horizontally from the lower. The upper curve is usually displaced to greater equilibrium bond lengths because electronically excited states usually have more antibonding character than electronic ground states.

The separation of the vibrational lines of an electronic absorption spectrum depends on the vibrational energies of the *upper* electronic state. Hence, electronic absorption spectra may be used to assess the force fields and dissociation energies of electronically excited molecules (for example, by using a Birge–Sponer plot, as in Problem 14.2).

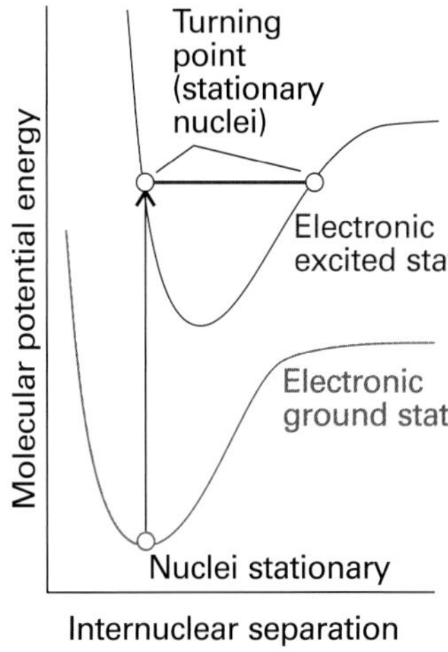

Fig. 14.7 According to the Franck–Condon principle, the most intense vibronic transition is from the ground vibrational state to the vibrational state lying vertically above it. Transitions to other vibrational levels also occur, but with lower intensity.

(d) Franck–Condon factors

The quantitative form of the Franck–Condon principle is derived from the expression for the transition dipole moment, $\boldsymbol{\mu}_{fi} = \langle f|\boldsymbol{\mu}|i\rangle$. The dipole moment operator is a sum over all nuclei and electrons in the molecule:

$$\hat{\mu} = -e\sum_i \boldsymbol{r}_i + e\sum_I Z_I \boldsymbol{R}_I \tag{14.2}$$

where the vectors are the distances from the centre of charge of the molecule. The intensity of the transition is proportional to the square modulus, $|\boldsymbol{\mu}_{fi}|^2$, of the magnitude of the transition dipole moment (eqn 9.70), and we show in the *Justification* below that this intensity is proportional to the square modulus of the overlap integral, $S(v_f,v_i)$, between the vibrational states of the initial and final electronic states. This overlap integral is a measure of the match between the vibrational wavefunctions in the upper and lower electronic states: $S = 1$ for a perfect match and $S = 0$ when there is no similarity.

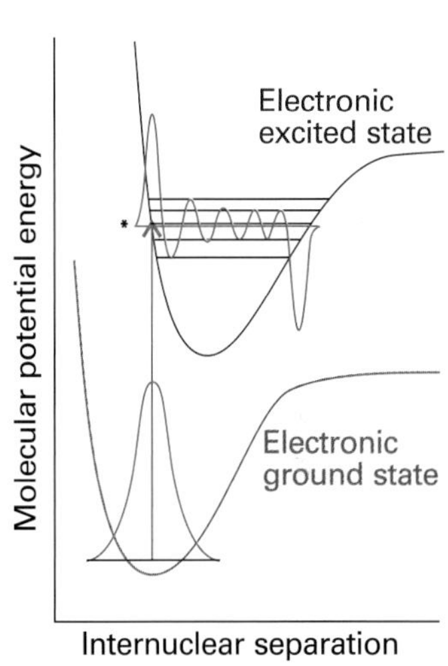

Fig. 14.8 In the quantum mechanical version of the Franck–Condon principle, the molecule undergoes a transition to the upper vibrational state that most closely resembles the vibrational wavefunction of the vibrational ground state of the lower electronic state. The two wavefunctions shown here have the greatest overlap integral of all the vibrational states of the upper electronic state and hence are most closely similar.

Justification 14.2 *The Franck–Condon approximation*

The overall state of the molecule consists of an electronic part, $|\varepsilon\rangle$, and a vibrational part, $|v\rangle$. Therefore, within the Born–Oppenheimer approximation, the transition dipole moment factorizes as follows:

$$\boldsymbol{\mu}_{fi} = \langle \varepsilon_f v_f|\{-e\sum_i \boldsymbol{r}_i + e\sum_I Z_I \boldsymbol{R}_I\}|\varepsilon_i v_i\rangle$$

$$= -e\sum_i \langle \varepsilon_f|\boldsymbol{r}_i|\varepsilon_i\rangle\langle v_f|v_i\rangle + e\sum_I Z_r\langle \varepsilon_f|\varepsilon_i\rangle\langle v_f|\boldsymbol{R}_I|v_i\rangle$$

The second term on the right of the second row is zero, because $\langle \varepsilon_f|\varepsilon_i\rangle = 0$ for two different electronic states (they are orthogonal). Therefore,

$$\boldsymbol{\mu}_{fi} = -e\sum_i \langle \varepsilon_f|\boldsymbol{r}_i|\varepsilon_i\rangle\langle v_f|v_i\rangle = \boldsymbol{\mu}_{\varepsilon_f,\varepsilon_i} S(v_f,v_i) \tag{14.3}$$

where

$$\boldsymbol{\mu}_{\varepsilon_f,\varepsilon_i} = -e\sum_i \langle \varepsilon_f|\boldsymbol{r}_i|\varepsilon_i\rangle \quad S(v_f,v_i) = \langle v_f|v_i\rangle \tag{14.4}$$

The matrix element $\boldsymbol{\mu}_{\varepsilon_f,\varepsilon_i}$ is the electric-dipole transition moment arising from the redistribution of electrons (and a measure of the 'kick' this redistribution gives to the electromagnetic field, and vice versa for absorption). The factor $S(v_f,v_i)$, is the overlap integral between the vibrational state $|v_i\rangle$ in the initial electronic state of the molecule, and the vibrational state $|v_f\rangle$ in the final electronic state of the molecule.

Because the transition intensity is proportional to the square of the magnitude of the transition dipole moment, the intensity of an absorption is proportional to $|S(v_f,v_i)|^2$, which is known as the **Franck–Condon factor** for the transition. It follows that, the greater the overlap of the vibrational state wavefunction in the upper electronic state with the vibrational wavefunction in the lower electronic state, the greater the absorption intensity of that particular simultaneous electronic and vibrational transition. This conclusion is the basis of the illustration in Fig. 14.8, where we see that the vibrational wavefunction of the ground state has the greatest overlap with the vibrational states that have peaks at similar bond lengths in the upper electronic state.

Example 14.1 *Calculating a Franck–Condon factor*

Consider the transition from one electronic state to another, their bond lengths being R_e and R'_e and their force constants equal. Calculate the Franck–Condon factor for the 0–0 transition and show that the transition is most intense when the bond lengths are equal.

Method We need to calculate $S(0,0)$, the overlap integral of the two ground-state vibrational wavefunctions, and then take its square. The difference between harmonic and anharmonic vibrational wavefunctions is negligible for $v = 0$, so harmonic oscillator wavefunctions can be used (Table 9.1).

Answer We use the (real) wavefunctions

$$\psi_0 = \left(\frac{1}{\alpha\pi^{1/2}}\right)^{1/2} e^{-x^2/2\alpha^2} \qquad \psi'_0 = \left(\frac{1}{\alpha\pi^{1/2}}\right)^{1/2} e^{-x'^2/2\alpha^2}$$

where $x = R - R_e$ and $x' = R - R'_e$, with $\alpha = (\hbar^2/mk)^{1/4}$ (Section 9.5a). The overlap integral is

$$S(0,0) = \langle 0|0\rangle = \int_{-\infty}^{\infty} \psi'_0\psi_0 \mathrm{d}R = \frac{1}{\alpha\pi^{1/2}}\int_{-\infty}^{\infty} e^{-(x^2+x'^2)/2\alpha^2}\mathrm{d}x$$

We now write $\alpha z = R - \frac{1}{2}(R_e + R'_e)$, and manipulate this expression into

$$S(0,0) = \frac{1}{\pi^{1/2}} e^{-(R_e-R'_e)^2/4\alpha^2}\int_{-\infty}^{\infty} e^{-z^2}\mathrm{d}z$$

The value of the integral is $\pi^{1/2}$. Therefore, the overlap integral is

$$S(0,0) = e^{-(R_e-R'_e)^2/4\alpha^2}$$

and the Franck–Condon factor is

$$S(0,0)^2 = e^{-(R_e-R'_e)^2/2\alpha^2}$$

This factor is equal to 1 when $R'_e = R_e$ and decreases as the equilibrium bond lengths diverge from each other (Fig. 14.9).

For Br_2, $R_e = 228$ pm and there is an upper state with $R'_e = 266$ pm. Taking the vibrational wavenumber as 250 cm^{-1} gives $S(0,0)^2 = 5.1 \times 10^{-10}$, so the intensity of the 0–0 transition is only 5.1×10^{-10} what it would have been if the potential curves had been directly above each other.

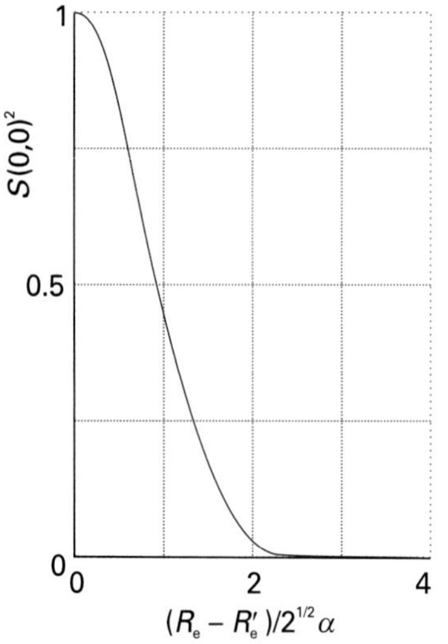

Fig. 14.9 The Franck–Condon factor for the arrangement discussed in Example 14.1.

Self-test 14.2 Suppose the vibrational wavefunctions can be approximated by rectangular functions of width W and W', centred on the equilibrium bond lengths (Fig. 14.10). Find the corresponding Franck–Condon factors when the centres are coincident and $W' < W$. [$S^2 = W'/W$]

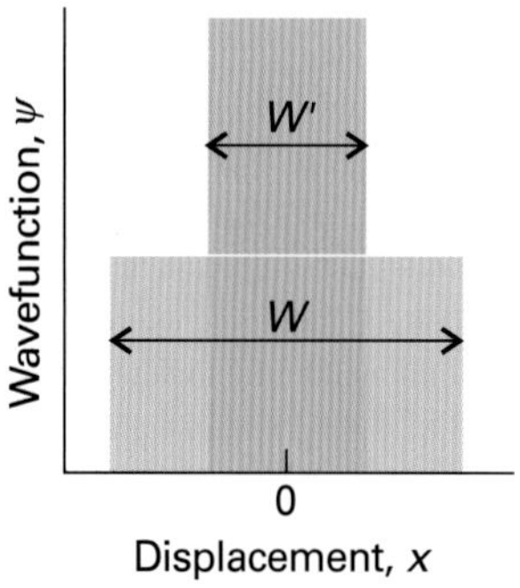

Fig. 14.10 The model wavefunctions used in Self-test 14.2.

(e) Rotational structure

Just as in vibrational spectroscopy, where a vibrational transition is accompanied by rotational excitation, so rotational transitions accompany the excitation of the vibrational excitation that accompanies electronic excitation. We therefore see P, Q, and R branches for each vibrational transition, and the electronic transition has a very rich structure. However, the principal difference is that electronic excitation can result in much larger changes in bond length than vibrational excitation causes alone, and the rotational branches have a more complex structure than in vibration–rotation spectra.

We suppose that the rotational constants of the electronic ground and excited states are B and B', respectively. The rotational energy levels of the initial and final states are

$$E(J) = hcBJ(J+1) \qquad E(J') = hcB'J'(J'+1)$$

and the rotational transitions occur at the following positions relative to the vibrational transition of wavenumber $\tilde{\nu}$ that they accompany:

$$\text{P branch } (\Delta J = -1): \quad \tilde{\nu}_{\mathrm{P}}(J) = \tilde{\nu} - (B'+B)J + (B'-B)J^2 \qquad (14.5a)$$

$$\text{Q branch } (\Delta J = 0): \quad \tilde{\nu}_{\mathrm{Q}}(J) = \tilde{\nu} + (B'-B)J(J+1) \qquad (14.5b)$$

$$\text{R branch } (\Delta J = +1): \quad \tilde{\nu}_{\mathrm{R}}(J) = \tilde{\nu} + (B'+B)(J+1) + (B'-B)(J+1)^2 \qquad (14.5c)$$

(These are the analogues of eqn 13.63.) First, suppose that the bond length in the electronically excited state is greater than that in the ground state; then $B' < B$ and $B' - B$ is negative. In this case the lines of the R branch converge with increasing J and when J is such that $|B' - B|(J+1) > B' + B$ the lines start to appear at successively decreasing wavenumbers. That is, the R branch has a **band head** (Fig. 14.11a). When the bond is shorter in the excited state than in the ground state, $B' > B$ and $B' - B$ is positive. In this case, the lines of the P branch begin to converge and go through a head when J is such that $(B' - B)J > B' + B$ (Fig. 14.11b).

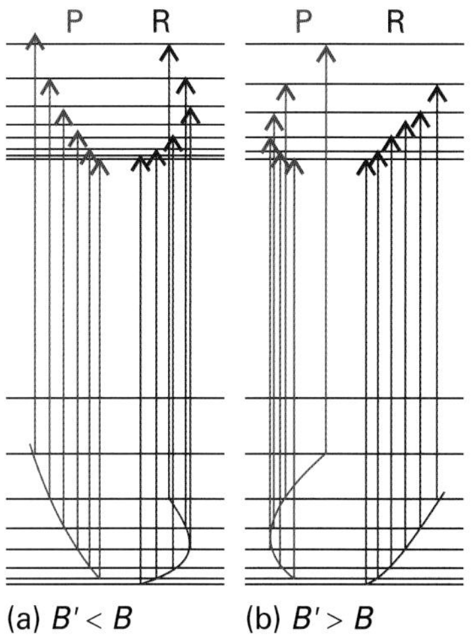

Fig. 14.11 When the rotational constants of a diatomic molecule differ significantly in the initial and final states of an electronic transition, the P and R branches show a head. (a) The formation of a head in the R branch when $B' < B$; (b) the formation of a head in the P branch when $B' > B$.

14.2 The electronic spectra of polyatomic molecules

The absorption of a photon can often be traced to the excitation of specific types of electrons or to electrons that belong to a small group of atoms in a polyatomic molecule. For example, when a carbonyl group (>C=O) is present, an absorption at about 290 nm is normally observed, although its precise location depends on the nature of the rest of the molecule. Groups with characteristic optical absorptions are called **chromophores** (from the Greek for 'colour bringer'), and their presence often accounts for the colours of substances (Table 14.3).

Comment 14.2

The web site for this text contains links to databases of electronic spectra.

(a) *d*–*d* transitions

In a free atom, all five *d* orbitals of a given shell are degenerate. In a *d*-metal complex, where the immediate environment of the atom is no longer spherical, the *d* orbitals

Synoptic table 14.3* Absorption characteristics ofsome groups and molecules

Group	$\tilde{\nu}/\mathrm{cm^{-1}}$	$\lambda_{max}/\mathrm{nm}$	$\varepsilon/(\mathrm{dm^3\,mol^{-1}\,cm^{-1}})$
C=C ($\pi^* \leftarrow \pi$)	61 000	163	5 000
	57 300	174	15 500
C=O ($\pi^* \leftarrow n$)	35 000–37 000	270–290	10–20
H_2O ($\pi^* \leftarrow n$)	60 000	167	7 000

* More values are given in the *Data section*.

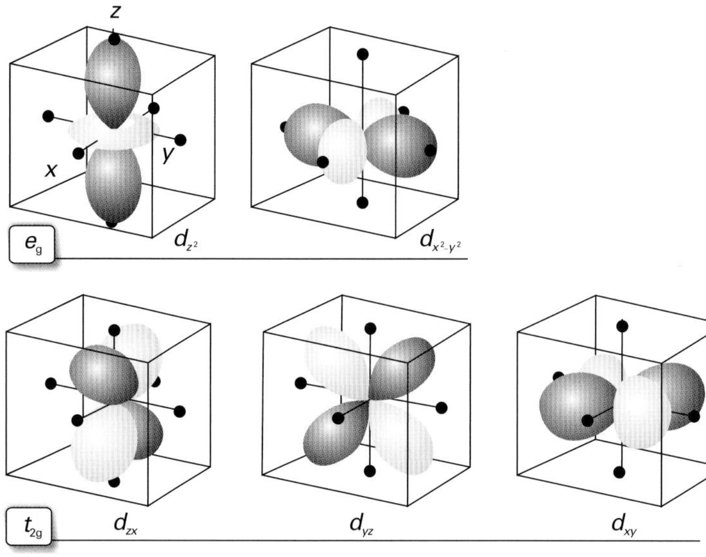

Fig. 14.12 The classification of *d*-orbitals in an octahedral environment.

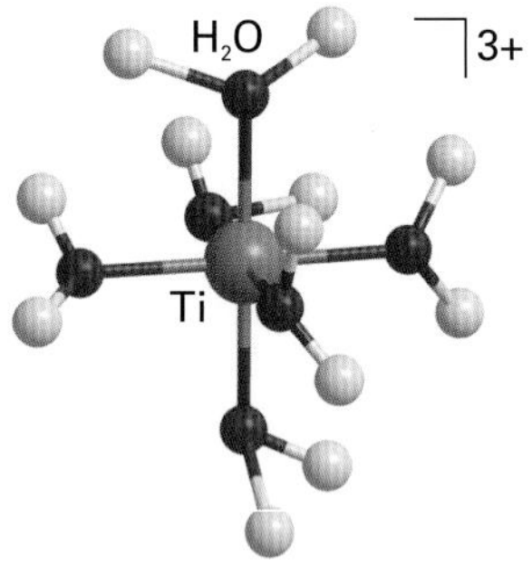

1 $[Ti(OH_2)_6]^{3+}$

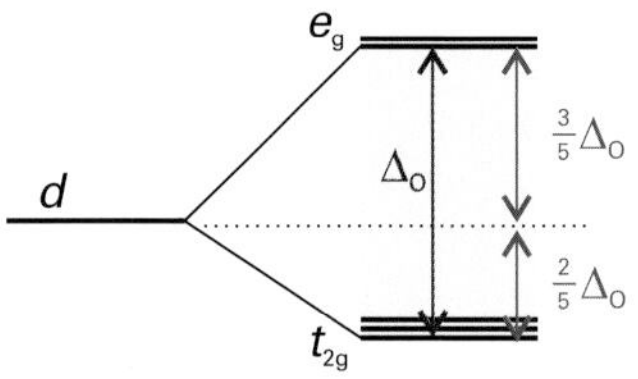

2

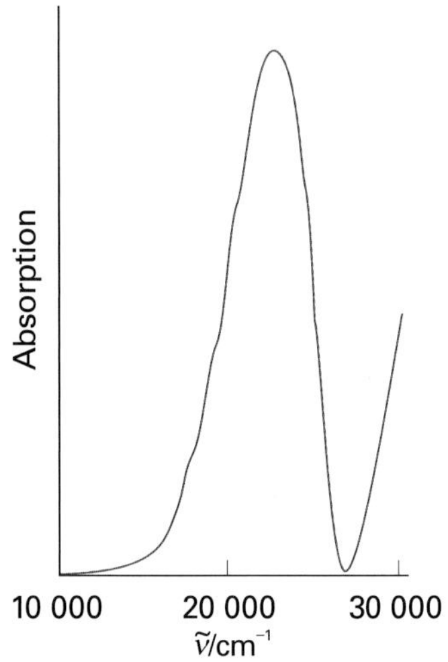

Fig. 14.13 The electronic absorption spectrum of $[Ti(OH_2)_6]^{3+}$ in aqueous solution.

are not all degenerate, and electrons can absorb energy by making transitions between them. We show in the following *Justification* that in an octahedral complex, such as $[Ti(OH_2)_6]^{3+}$ (**1**), the five *d* orbitals of the central atom are split into two sets (**2**), a triply degenerate set labelled t_{2g} and a doubly degenerate set labelled e_g. The three t_{2g} orbitals lie below the two e_g orbitals; the difference in energy is denoted Δ_O and called the **ligand-field splitting parameter** (the O denoting octahedral symmetry).

Justification 14.3 *The splitting of* d*-orbitals in an octahedral* d*-metal complex*

In an octahedral *d*-metal complex, six identical ions or molecules, the *ligands*, are at the vertices of a regular octahedron, with the metal ion at its centre. The ligands can be regarded as point negative charges that are repelled by the *d*-electrons of the central ion. Figure 14.12 shows the consequence of this arrangement: the five *d*-orbitals fall into two groups, with $d_{x^2-y^2}$ and d_{z^2} pointing directly towards the ligand positions, and d_{xy}, d_{yz}, and d_{zx} pointing between them. An electron occupying an orbital of the former group has a less favourable potential energy than when it occupies any of the three orbitals of the other group, and so the *d*-orbitals split into the two sets shown in (**2**) with an energy difference Δ_O: a triply degenerate set comprising the d_{xy}, d_{yz}, and d_{zx} orbitals and labelled t_{2g}, and a doubly degenerate set comprising the $d_{x^2-y^2}$ and d_{z^2} orbitals and labelled e_g.

The *d*-orbitals also divide into two sets in a tetrahedral complex, but in this case the *e* orbitals lie below the t_2 orbitals and their separation is written Δ_T. Neither Δ_O nor Δ_T is large, so transitions between the two sets of orbitals typically occur in the visible region of the spectrum. The transitions are responsible for many of the colours that are so characteristic of *d*-metal complexes. As an example, the spectrum of $[Ti(OH_2)_6]^{3+}$ near 20 000 cm^{-1} (500 nm) is shown in Fig. 14.13, and can be ascribed to the promotion of its single *d* electron from a t_{2g} orbital to an e_g orbital. The wavenumber of the absorption maximum suggests that $\Delta_O \approx 20\ 000$ cm^{-1} for this complex, which corresponds to about 2.5 eV.

According to the Laporte rule (Section 14.1b), *d–d* transitions are parity-forbidden in octahedral complexes because they are g → g transitions (more specifically $e_g \leftarrow t_{2g}$ transitions). However, *d–d* transitions become weakly allowed as vibronic transitions as a result of coupling to asymmetrical vibrations such as that shown in Fig. 14.5.

(b) Charge-transfer transitions

A complex may absorb radiation as a result of the transfer of an electron from the ligands into the *d*-orbitals of the central atom, or vice versa. In such **charge-transfer transitions** the electron moves through a considerable distance, which means that the transition dipole moment may be large and the absorption is correspondingly intense. This mode of chromophore activity is shown by the permanganate ion, MnO_4^-, and accounts for its intense violet colour (which arises from strong absorption within the range 420–700 nm). In this oxoanion, the electron migrates from an orbital that is largely confined to the O atom ligands to an orbital that is largely confined to the Mn atom. It is therefore an example of a **ligand-to-metal charge-transfer transition** (LMCT). The reverse migration, a **metal-to-ligand charge-transfer transition** (MLCT), can also occur. An example is the transfer of a *d* electron into the antibonding π orbitals of an aromatic ligand. The resulting excited state may have a very long lifetime if the electron is extensively delocalized over several aromatic rings, and such species can participate in photochemically induced redox reactions (see Section 23.7).

The intensities of charge-transfer transitions are proportional to the square of the transition dipole moment, in the usual way. We can think of the transition moment as a measure of the distance moved by the electron as it migrates from metal to ligand or vice versa, with a large distance of migration corresponding to a large transition dipole moment and therefore a high intensity of absorption. However, because the integrand in the transition dipole is proportional to the product of the initial and final wavefunctions, it is zero unless the two wavefunctions have nonzero values in the same region of space. Therefore, although large distances of migration favour high intensities, the diminished overlap of the initial and final wavefunctions for large separations of metal and ligands favours low intensities (see Problem 14.17). We encounter similar considerations when we examine electron transfer reactions (Chapter 24), which can be regarded as a special type of charge-transfer transition.

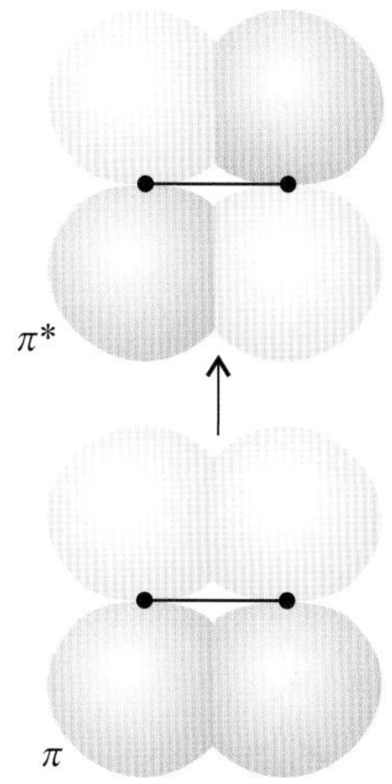

Fig. 14.14 A C=C double bond acts as a chromophore. One of its important transitions is the $\pi^* \leftarrow \pi$ transition illustrated here, in which an electron is promoted from a π orbital to the corresponding antibonding orbital.

(c) $\pi^* \leftarrow \pi$ and $\pi^* \leftarrow n$ transitions

Absorption by a C=C double bond results in the excitation of a π electron into an antibonding π^* orbital (Fig. 14.14). The chromophore activity is therefore due to a $\pi^* \leftarrow \pi$ transition (which is normally read 'π to π-star transition'). Its energy is about 7 eV for an unconjugated double bond, which corresponds to an absorption at 180 nm (in the ultraviolet). When the double bond is part of a conjugated chain, the energies of the molecular orbitals lie closer together and the $\pi^* \leftarrow \pi$ transition moves to longer wavelength; it may even lie in the visible region if the conjugated system is long enough. An important example of an $\pi^* \leftarrow \pi$ transition is provided by the photochemical mechanism of vision (*Impact I14.1*).

The transition responsible for absorption in carbonyl compounds can be traced to the lone pairs of electrons on the O atom. The Lewis concept of a 'lone pair' of electrons is represented in molecular orbital theory by a pair of electrons in an orbital confined largely to one atom and not appreciably involved in bond formation. One of these electrons may be excited into an empty π^* orbital of the carbonyl group (Fig. 14.15), which gives rise to a $\pi^* \leftarrow n$ transition (an 'n to π-star transition'). Typical absorption energies are about 4 eV (290 nm). Because $\pi^* \leftarrow n$ transitions in carbonyls are symmetry forbidden, the absorptions are weak.

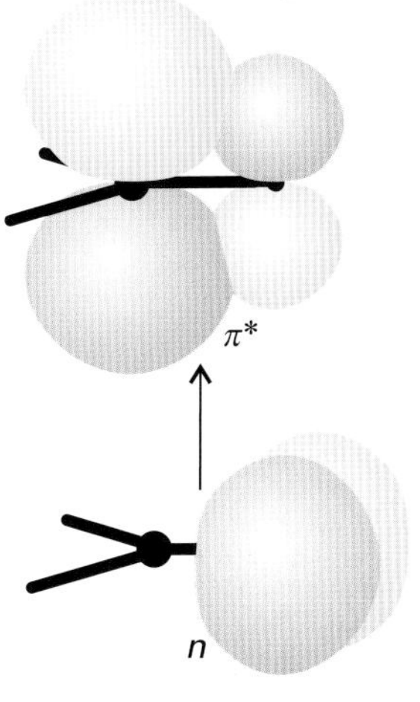

Fig. 14.15 A carbonyl group (C=O) acts as a chromophore primarily on account of the excitation of a nonbonding O lone-pair electron to an antibonding CO π orbital.

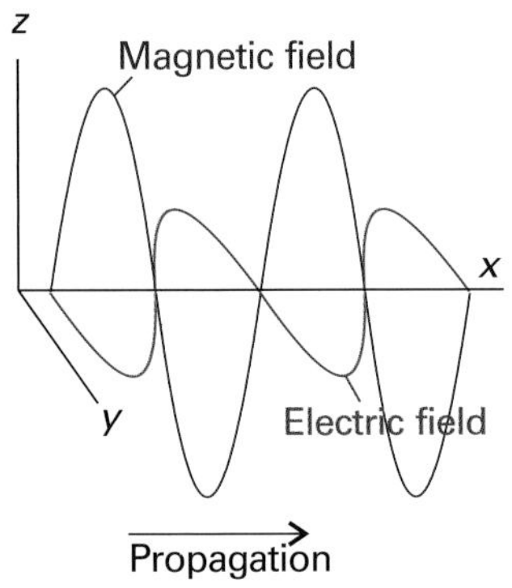

Fig. 14.16 Electromagnetic radiation consists of a wave of electric and magnetic fields perpendicular to the direction of propagation (in this case the *x*-direction), and mutually perpendicular to each other. This illustration shows a plane-polarized wave, with the electric and magnetic fields oscillating in the *xy* and *xz* planes, respectively.

(d) Circular dichroism spectroscopy

Electronic spectra can reveal additional details of molecular structure when experiments are conducted with **polarized light**, electromagnetic radiation with electric and magnetic fields that oscillate only in certain directions. Light is **plane polarized** when the electric and magnetic fields each oscillate in a single plane (Fig. 14.16). The plane of polarization may be oriented in any direction around the direction of propagation (the *x*-direction in Fig. 14.16), with the electric and magnetic fields perpendicular to that direction (and perpendicular to each other). An alternative mode of polarization is **circular polarization**, in which the electric and magnetic fields rotate around the direction of propagation in either a clockwise or a counter-clockwise sense but remain perpendicular to it and each other.

When plane-polarized radiation passes through samples of certain kinds of matter, the plane of polarization is rotated around the direction of propagation. This rotation is the familiar phenomenon of optical activity, observed when the molecules in the sample are chiral (Section 12.3b). Chiral molecules have a second characteristic: they absorb left and right circularly polarized light to different extents. In a circularly polarized ray of light, the electric field describes a helical path as the wave travels through space (Fig. 14.17), and the rotation may be either clockwise or counterclockwise. The differential absorption of left- and right-circularly polarized light is called **circular dichroism**. In terms of the absorbances for the two components, A_L and A_R, the circular dichroism of a sample of molar concentration [J] is reported as

$$\Delta\varepsilon = \varepsilon_L - \varepsilon_R = \frac{A_L - A_R}{[J]l} \tag{14.6}$$

where l is the path length of the sample.

Circular dichroism is a useful adjunct to visible and UV spectroscopy. For example, the CD spectra of the enantiomeric pairs of chiral *d*-metal complexes are distinctly different, whereas there is little difference between their absorption spectra (Fig. 14.18). Moreover, CD spectra can be used to assign the absolute configuration of complexes by comparing the observed spectrum with the CD spectrum of a similar complex of known handedness. We shall see in Chapter 19 that the CD spectra of biological polymers, such as proteins and nucleic acids, give similar structural information. In these cases the spectrum of the polymer chain arises from the chirality of individual monomer units and, in addition, a contribution from the three-dimensional structure of the polymer itself.

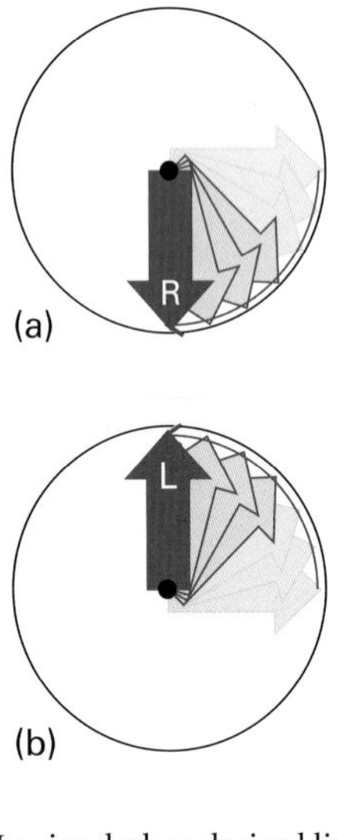

Fig. 14.17 In circularly polarized light, the electric field at different points along the direction of propagation rotates. The arrays of arrows in these illustrations show the view of the electric field when looking toward the oncoming ray: (a) right-circularly polarized, (b) left-circularly polarized light.

IMPACT ON BIOCHEMISTRY

I14.1 Vision

The eye is an exquisite photochemical organ that acts as a transducer, converting radiant energy into electrical signals that travel along neurons. Here we concentrate on the events taking place in the human eye, but similar processes occur in all animals. Indeed, a single type of protein, rhodopsin, is the primary receptor for light throughout the animal kingdom, which indicates that vision emerged very early in evolutionary history, no doubt because of its enormous value for survival.

Photons enter the eye through the cornea, pass through the ocular fluid that fills the eye, and fall on the retina. The ocular fluid is principally water, and passage of light through this medium is largely responsible for the *chromatic aberration* of the eye, the blurring of the image as a result of different frequencies being brought to slightly different focuses. The chromatic aberration is reduced to some extent by the tinted region called the *macular pigment* that covers part of the retina. The pigments in this region are the carotene-like xanthophylls (**3**), which absorb some of the blue light and

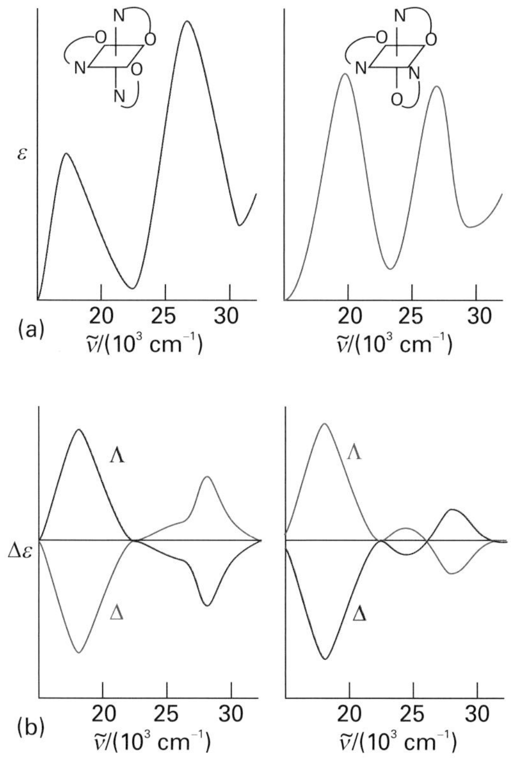

Fig. 14.18 (a) The absorption spectra of two isomers, denoted mer and fac, of $[Co(ala)_3]$, where ala is the conjugate base of alanine, and (b) the corresponding CD spectra. The left- and right-handed forms of these isomers give identical absorption spectra. However, the CD spectra are distinctly different, and the absolute configurations (denoted Λ and Δ) have been assigned by comparison with the CD spectra of a complex of known absolute configuration.

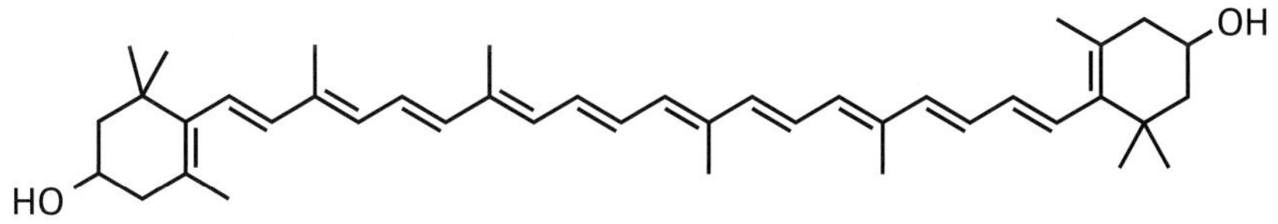

3 A xanthophyll

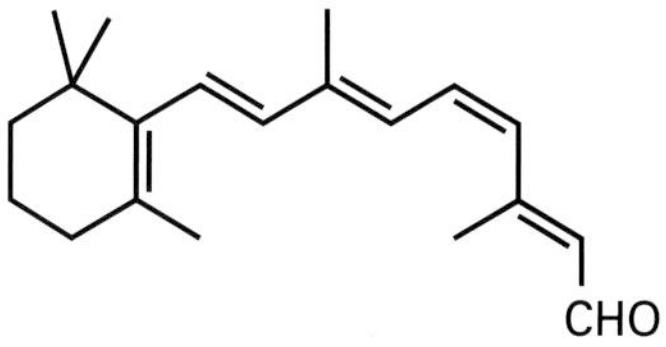

4 11-*cis*-retinal

hence help to sharpen the image. They also protect the photoreceptor molecules from too great a flux of potentially dangerous high energy photons. The xanthophylls have delocalized electrons that spread along the chain of conjugated double bonds, and the $\pi^* \leftarrow \pi$ transition lies in the visible.

About 57 per cent of the photons that enter the eye reach the retina; the rest are scattered or absorbed by the ocular fluid. Here the primary act of vision takes place, in which the chromophore of a rhodopsin molecule absorbs a photon in another $\pi^* \leftarrow \pi$ transition. A rhodopsin molecule consists of an opsin protein molecule to which is attached a 11-*cis*-retinal molecule (**4**). The latter resembles half a carotene molecule, showing Nature's economy in its use of available materials. The attachment is by the formation of a protonated Schiff's base, utilizing the —CHO group of the chromophore and the terminal NH_2 group of the sidechain, a lysine residue from opsin. The free 11-*cis*-retinal molecule absorbs in the ultraviolet, but attachment to the opsin protein molecule shifts the absorption into the visible region. The rhodopsin molecules are situated in the membranes of special cells (the 'rods' and the 'cones') that cover the retina. The opsin molecule is anchored into the cell membrane by two hydrophobic groups and largely surrounds the chromophore (Fig. 14.19).

Immediately after the absorption of a photon, the 11-*cis*-retinal molecule undergoes photoisomerization into all-*trans*-retinal (**5**). Photoisomerization takes about

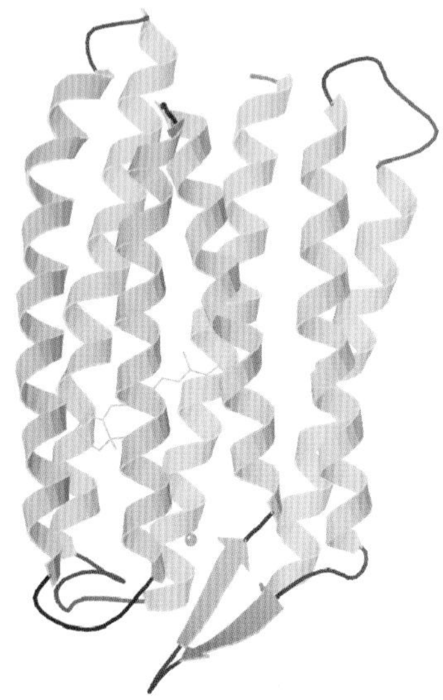

Fig. 14.19 The structure of the rhodopsin molecule, consisting of an opsin protein to which is attached an 11-*cis*-retinal molecule embedded in the space surrounded by the helical regions.

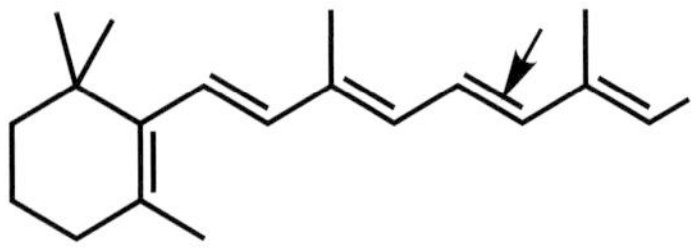

5 All-*trans*-retinal

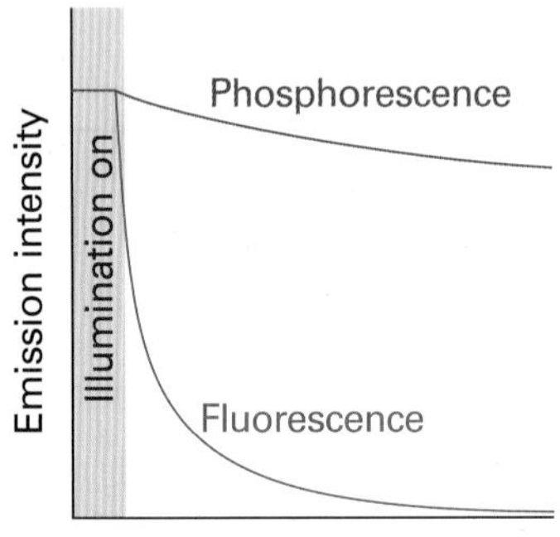

Fig. 14.20 The empirical (observation-based) distinction between fluorescence and phosphorescence is that the former is extinguished very quickly after the exciting source is removed, whereas the latter continues with relatively slowly diminishing intensity.

200 fs and about 67 pigment molecules isomerize for every 100 photons that are absorbed. The process occurs because the $\pi^* \leftarrow \pi$ excitation of an electron loosens one of the π-bonds (the one indicated by the arrow in **5**), its torsional rigidity is lost, and one part of the molecule swings round into its new position. At that point, the molecule returns to its ground state, but is now trapped in its new conformation. The straightened tail of all-*trans*-retinal results in the molecule taking up more space than 11-*cis*-retinal did, so the molecule presses against the coils of the opsin molecule that surrounds it. In about 0.25–0.50 ms from the initial absorption event, the rhodopsin molecule is activated both by the isomerization of retinal and deprotonation of its Schiff's base tether to opsin, forming an intermediate known as *metarhodopsin II*.

In a sequence of biochemical events known as the *biochemical cascade*, metarhodopsin II activates the protein transducin, which in turn activates a phosphodiesterase enzyme that hydrolyses cyclic guanine monophosphate (cGMP) to GMP. The reduction in the concentration of cGMP causes ion channels, proteins that mediate the movement of ions across biological membranes, to close and the result is a sizable change in the transmembrane potential (see *Impact I7.2* for a discussion of transmembrane potentials). The pulse of electric potential travels through the optical nerve and into the optical cortex, where it is interpreted as a signal and incorporated into the web of events we call 'vision'.

The resting state of the rhodopsin molecule is restored by a series of nonradiative chemical events powered by ATP. The process involves the escape of all-*trans*-retinal as all-*trans*-retinol (in which —CHO has been reduced to $-CH_2OH$) from the opsin molecule by a process catalysed by the enzyme rhodopsin kinase and the attachment of another protein molecule, arrestin. The free all-*trans*-retinol molecule now undergoes enzyme-catalysed isomerization into 11-*cis*-retinol followed by dehydrogenation to form 11-*cis*-retinal, which is then delivered back into an opsin molecule. At this point, the cycle of excitation, photoisomerization, and regeneration is ready to begin again.

The fates of electronically excited states

A **radiative decay process** is a process in which a molecule discards its excitation energy as a photon. A more common fate is **nonradiative decay**, in which the excess energy is transferred into the vibration, rotation, and translation of the surrounding molecules. This thermal degradation converts the excitation energy completely into thermal motion of the environment (that is, to 'heat'). An excited molecule may also take part in a chemical reaction, as we discuss in Chapter 23.

14.3 Fluorescence and phosphorescence

In **fluorescence**, spontaneous emission of radiation occurs within a few nanoseconds after the exciting radiation is extinguished (Fig. 14.20). In **phosphorescence**, the spontaneous emission may persist for long periods (even hours, but characteristically seconds or fractions of seconds). The difference suggests that fluorescence is a fast conversion of absorbed radiation into re-emitted energy, and that phosphorescence involves the storage of energy in a reservoir from which it slowly leaks.

(a) Fluorescence

Figure 14.21 shows the sequence of steps involved in fluorescence. The initial absorption takes the molecule to an excited electronic state, and if the absorption spectrum were monitored it would look like the one shown in Fig. 14.22a. The excited molecule

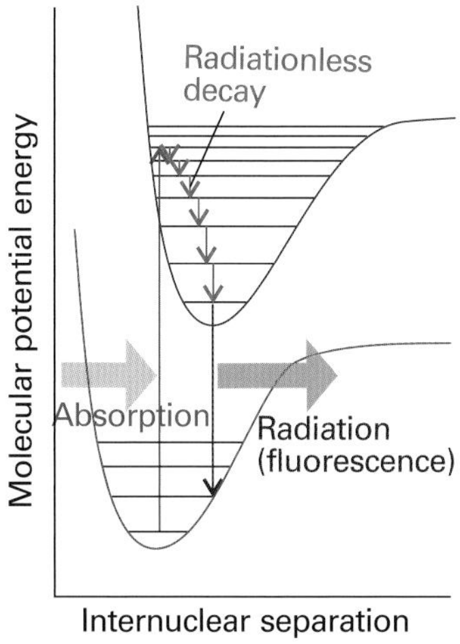

Fig. 14.21 The sequence of steps leading to fluorescence. After the initial absorption, the upper vibrational states undergo radiationless decay by giving up energy to the surroundings. A radiative transition then occurs from the vibrational ground state of the upper electronic state.

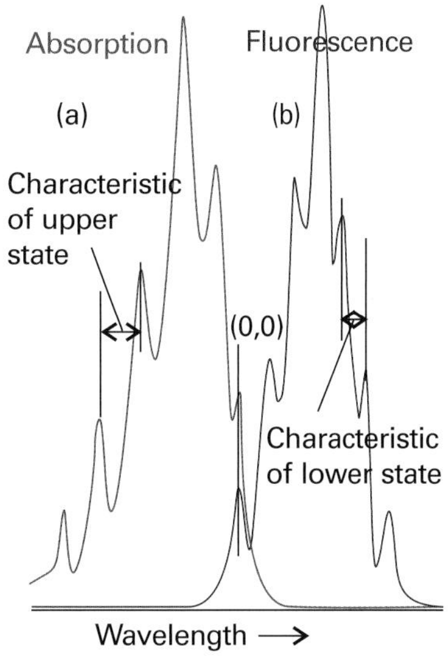

Fig. 14.22 An absorption spectrum (a) shows a vibrational structure characteristic of the upper state. A fluorescence spectrum (b) shows a structure characteristic of the lower state; it is also displaced to lower frequencies (but the 0–0 transitions are coincident) and resembles a mirror image of the absorption.

is subjected to collisions with the surrounding molecules, and as it gives up energy nonradiatively it steps down the ladder of vibrational levels to the lowest vibrational level of the electronically excited molecular state. The surrounding molecules, however, might now be unable to accept the larger energy difference needed to lower the molecule to the ground electronic state. It might therefore survive long enough to undergo spontaneous emission, and emit the remaining excess energy as radiation. The downward electronic transition is vertical (in accord with the Franck–Condon principle) and the fluorescence spectrum has a vibrational structure characteristic of the *lower* electronic state (Fig. 14.22b).

Provided they can be seen, the 0–0 absorption and fluorescence transitions can be expected to be coincident. The absorption spectrum arises from 1–0, 2–0, . . . transitions that occur at progressively higher wavenumber and with intensities governed by the Franck–Condon principle. The fluorescence spectrum arises from 0–0, 0–1, . . . *downward* transitions that hence occur with decreasing wavenumbers. The 0–0 absorption and fluorescence peaks are not always exactly coincident, however, because the solvent may interact differently with the solute in the ground and excited states (for instance, the hydrogen bonding pattern might differ). Because the solvent molecules do not have time to rearrange during the transition, the absorption occurs in an environment characteristic of the solvated ground state; however, the fluorescence occurs in an environment characteristic of the solvated excited state (Fig. 14.23).

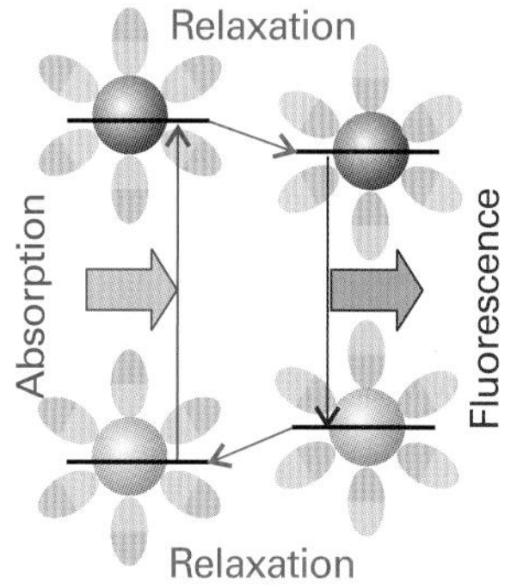

Fig. 14.23 The solvent can shift the fluorescence spectrum relative to the absorption spectrum. On the left we see that the absorption occurs with the solvent (the ellipses) in the arrangement characteristic of the ground electronic state of the molecule (the sphere). However, before fluorescence occurs, the solvent molecules relax into a new arrangement, and that arrangement is preserved during the subsequent radiative transition.

Fluorescence occurs at lower frequencies (longer wavelengths) than the incident radiation because the emissive transition occurs after some vibrational energy has been discarded into the surroundings. The vivid oranges and greens of fluorescent dyes are an everyday manifestation of this effect: they absorb in the ultraviolet and blue, and fluoresce in the visible. The mechanism also suggests that the intensity of the fluorescence ought to depend on the ability of the solvent molecules to accept the

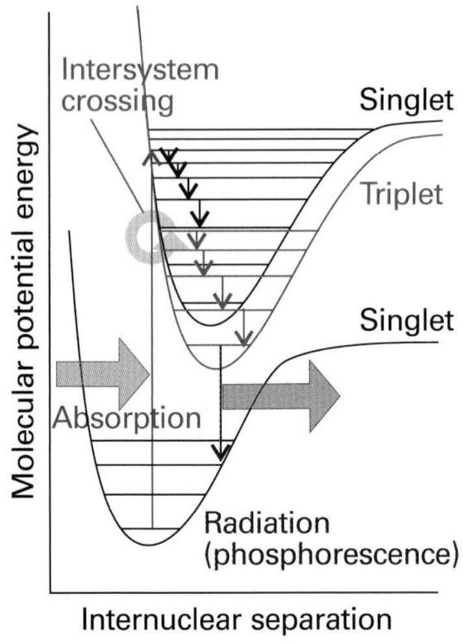

Fig. 14.24 The sequence of steps leading to phosphorescence. The important step is the intersystem crossing, the switch from a singlet state to a triplet state brought about by spin–orbit coupling. The triplet state acts as a slowly radiating reservoir because the return to the ground state is spin-forbidden.

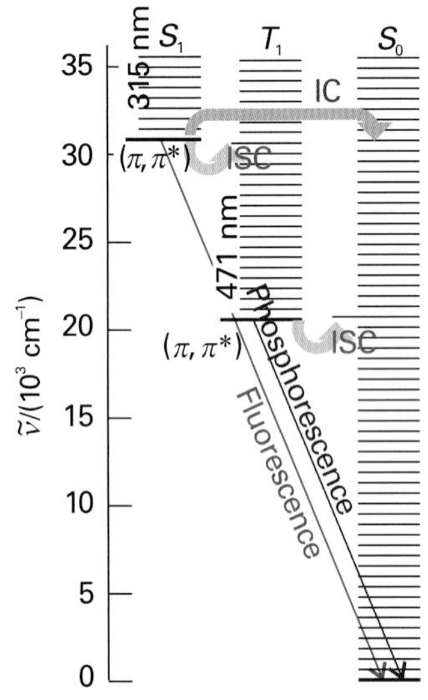

Fig. 14.25 A Jablonski diagram (here, for naphthalene) is a simplified portrayal of the relative positions of the electronic energy levels of a molecule. Vibrational levels of states of a given electronic state lie above each other, but the relative horizontal locations of the columns bear no relation to the nuclear separations in the states. The ground vibrational states of each electronic state are correctly located vertically but the other vibrational states are shown only schematically. (IC: internal conversion; ISC: intersystem crossing.)

electronic and vibrational quanta. It is indeed found that a solvent composed of molecules with widely spaced vibrational levels (such as water) can in some cases accept the large quantum of electronic energy and so extinguish, or 'quench', the fluorescence. We examine the mechanisms of fluorescence quenching in Chapter 23.

(b) Phosphorescence

Figure 14.24 shows the sequence of events leading to phosphorescence for a molecule with a singlet ground state. The first steps are the same as in fluorescence, but the presence of a triplet excited state plays a decisive role. The singlet and triplet excited states share a common geometry at the point where their potential energy curves intersect. Hence, if there is a mechanism for unpairing two electron spins (and achieving the conversion of ↑↓ to ↑↑), the molecule may undergo **intersystem crossing**, a nonradiative transition between states of different multiplicity, and become a triplet state. We saw in the discussion of atomic spectra (Section 10.9d) that singlet–triplet transitions may occur in the presence of spin–orbit coupling, and the same is true in molecules. We can expect intersystem crossing to be important when a molecule contains a moderately heavy atom (such as S), because then the spin–orbit coupling is large.

If an excited molecule crosses into a triplet state, it continues to deposit energy into the surroundings. However, it is now stepping down the triplet's vibrational ladder, and at the lowest energy level it is trapped because the triplet state is at a lower energy than the corresponding singlet (recall Hund's rule, Section 13.7). The solvent cannot absorb the final, large quantum of electronic excitation energy, and the molecule cannot radiate its energy because return to the ground state is spin-forbidden (Section 14.1). The radiative transition, however, is not totally forbidden because the spin–orbit coupling that was responsible for the intersystem crossing also breaks the selection rule. The molecules are therefore able to emit weakly, and the emission may continue long after the original excited state was formed.

The mechanism accounts for the observation that the excitation energy seems to get trapped in a slowly leaking reservoir. It also suggests (as is confirmed experimentally) that phosphorescence should be most intense from solid samples: energy transfer is then less efficient and intersystem crossing has time to occur as the singlet excited state steps slowly past the intersection point. The mechanism also suggests that the phosphorescence efficiency should depend on the presence of a moderately heavy atom (with strong spin–orbit coupling), which is in fact the case. The confirmation of the mechanism is the experimental observation (using the sensitive magnetic resonance techniques described in Chapter 15) that the sample is paramagnetic while the reservoir state, with its unpaired electron spins, is populated.

The various types of nonradiative and radiative transitions that can occur in molecules are often represented on a schematic **Jablonski diagram** of the type shown in Fig. 14.25.

IMPACT ON BIOCHEMISTRY

I14.2 Fluorescence microscopy

Apart from a small number of co-factors, such as the chlorophylls and flavins, the majority of the building blocks of proteins and nucleic acids do not fluoresce strongly. Four notable exceptions are the amino acids tryptophan ($\lambda_{abs} \approx 280$ nm and $\lambda_{fluor} \approx 348$ nm in water), tyrosine ($\lambda_{abs} \approx 274$ nm and $\lambda_{fluor} \approx 303$ nm in water), and phenylalanine ($\lambda_{abs} \approx 257$ nm and $\lambda_{fluor} \approx 282$ nm in water), and the oxidized form of the sequence serine–tyrosine–glycine (**6**) found in the green fluorescent protein (GFP) of certain jellyfish. The wild type of GFP from *Aequora victoria* absorbs strongly at 395 nm and emits maximally at 509 nm.

In **fluorescence microscopy**, images of biological cells at work are obtained by attaching a large number of fluorescent molecules to proteins, nucleic acids, and membranes and then measuring the distribution of fluorescence intensity within the illuminated area. A common fluorescent label is GFP. With proper filtering to remove light due to Rayleigh scattering of the incident beam, it is possible to collect light from the sample that contains only fluorescence from the label. However, great care is required to eliminate fluorescent impurities from the sample.

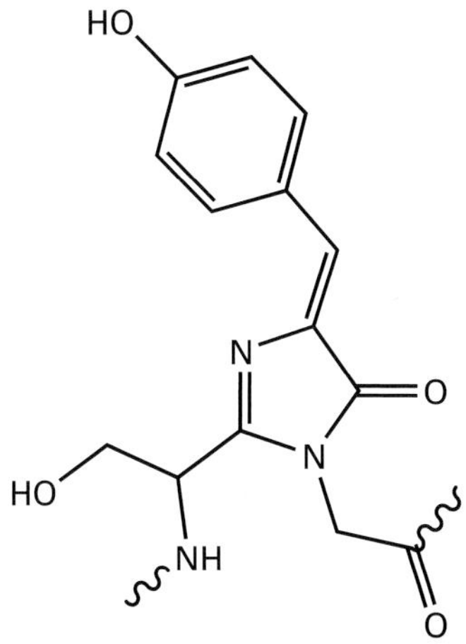

6 The chromophore of GFP

14.4 Dissociation and predissociation

Another fate for an electronically excited molecule is **dissociation**, the breaking of bonds (Fig. 14.26). The onset of dissociation can be detected in an absorption spectrum by seeing that the vibrational structure of a band terminates at a certain energy. Absorption occurs in a continuous band above this **dissociation limit** because the final state is an unquantized translational motion of the fragments. Locating the dissociation limit is a valuable way of determining the bond dissociation energy.

In some cases, the vibrational structure disappears but resumes at higher photon energies. This **predissociation** can be interpreted in terms of the molecular potential energy curves shown in Fig. 14.27. When a molecule is excited to a vibrational level, its electrons may undergo a redistribution that results in it undergoing an **internal conversion**, a radiationless conversion to another state of the same multiplicity. An internal conversion occurs most readily at the point of intersection of the two molecular potential energy curves, because there the nuclear geometries of the two states are the same. The state into which the molecule converts may be dissociative, so the states near the intersection have a finite lifetime, and hence their energies are imprecisely defined. As a result, the absorption spectrum is blurred in the vicinity of the intersection. When the incoming photon brings enough energy to excite the molecule

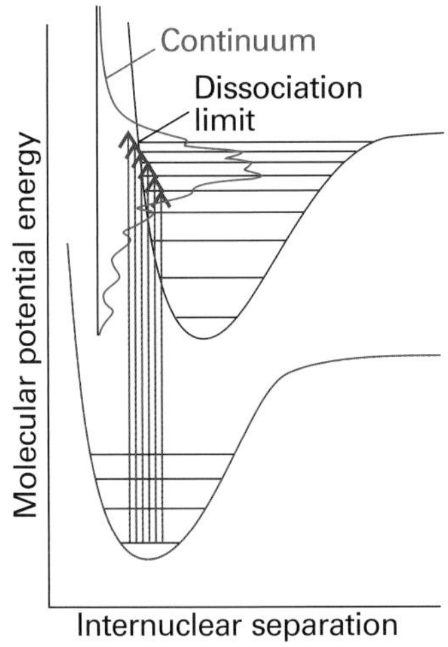

Fig. 14.26 When absorption occurs to unbound states of the upper electronic state, the molecule dissociates and the absorption is a continuum. Below the dissociation limit the electronic spectrum shows a normal vibrational structure.

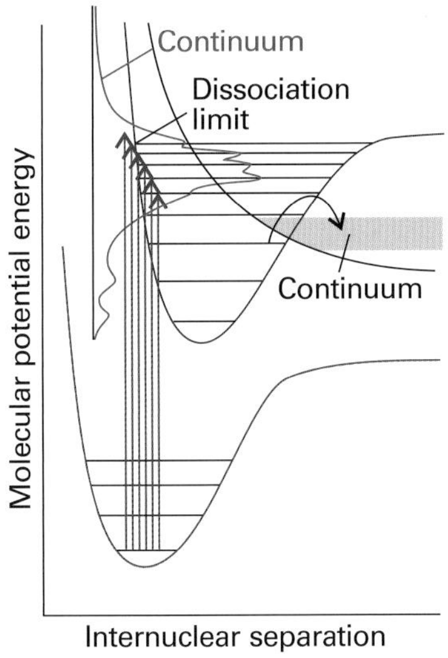

Fig. 14.27 When a dissociative state crosses a bound state, as in the upper part of the illustration, molecules excited to levels near the crossing may dissociate. This process is called predissociation, and is detected in the spectrum as a loss of vibrational structure that resumes at higher frequencies.

to a vibrational level high above the intersection, the internal conversion does not occur (the nuclei are unlikely to have the same geometry). Consequently, the levels resume their well-defined, vibrational character with correspondingly well-defined energies, and the line structure resumes on the high-frequency side of the blurred region.

Lasers

Lasers have transformed chemistry as much as they have transformed the everyday world. They lie very much on the frontier of physics and chemistry, for their operation depends on details of optics and, in some cases, of solid-state processes. In this section, we discuss the mechanisms of laser action, and then explore their applications in chemistry. In *Further information 14.1*, we discuss the modes of operation of a number of commonly available laser systems.

14.5 General principles of laser action

The word laser is an acronym formed from **l**ight **a**mplification by **s**timulated **e**mission of **r**adiation. In stimulated emission, an excited state is stimulated to emit a photon by radiation of the same frequency; the more photons that are present, the greater the probability of the emission. The essential feature of laser action is positive-feedback: the more photons present of the appropriate frequency, the more photons of that frequency that will be stimulated to form.

(a) Population inversion

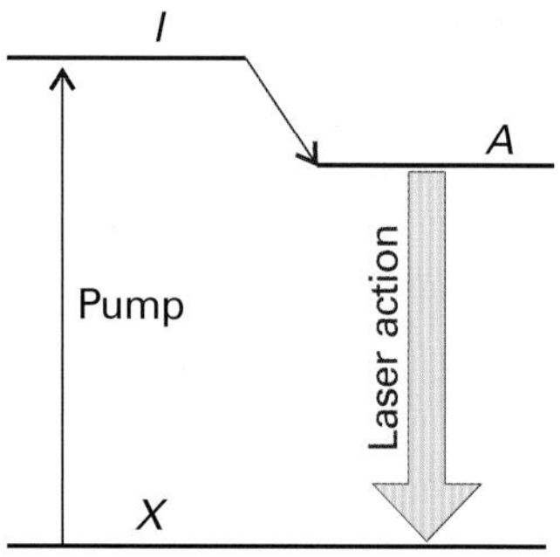

Fig. 14.28 The transitions involved in one kind of three-level laser. The pumping pulse populates the intermediate state *I*, which in turn populates the laser state *A*. The laser transition is the stimulated emission $A \rightarrow X$.

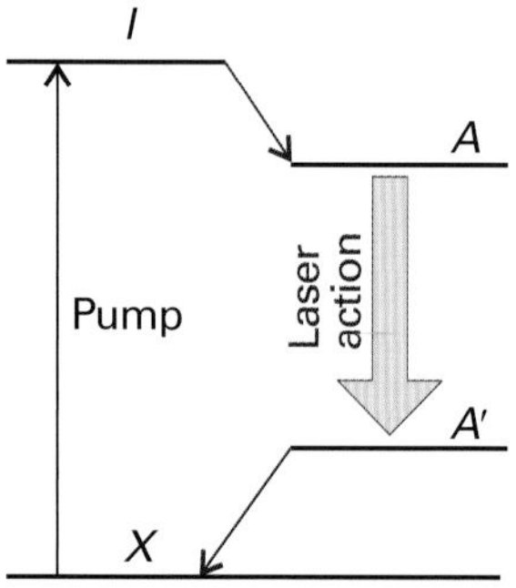

Fig. 14.29 The transitions involved in a four-level laser. Because the laser transition terminates in an excited state (A'), the population inversion between *A* and A' is much easier to achieve.

One requirement of laser action is the existence of a **metastable excited state**, an excited state with a long enough lifetime for it to participate in stimulated emission. Another requirement is the existence of a greater population in the metastable state than in the lower state where the transition terminates, for then there will be a net emission of radiation. Because at thermal equilibrium the opposite is true, it is necessary to achieve a **population inversion** in which there are more molecules in the upper state than in the lower.

One way of achieving population inversion is illustrated in Fig. 14.28. The molecule is excited to an intermediate state *I*, which then gives up some of its energy nonradiatively and changes into a lower state *A*; the laser transition is the return of *A* to the ground state *X*. Because three energy levels are involved overall, this arrangement leads to a **three-level laser**. In practice, *I* consists of many states, all of which can convert to the upper of the two laser states *A*. The $I \leftarrow X$ transition is stimulated with an intense flash of light in the process called **pumping**. The pumping is often achieved with an electric discharge through xenon or with the light of another laser. The conversion of *I* to *A* should be rapid, and the laser transitions from *A* to *X* should be relatively slow.

The disadvantage of this three-level arrangement is that it is difficult to achieve population inversion, because so many ground-state molecules must be converted to the excited state by the pumping action. The arrangement adopted in a **four-level laser** simplifies this task by having the laser transition terminate in a state A' other than the ground state (Fig. 14.29). Because A' is unpopulated initially, any population in *A* corresponds to a population inversion, and we can expect laser action if *A* is sufficiently metastable. Moreover, this population inversion can be maintained if the $A' \leftarrow X$ transitions are rapid, for these transitions will deplete any population in A' that stems from the laser transition, and keep the state A' relatively empty.

(b) Cavity and mode characteristics

The laser medium is confined to a cavity that ensures that only certain photons of a particular frequency, direction of travel, and state of polarization are generated abundantly. The cavity is essentially a region between two mirrors, which reflect the light back and forth. This arrangement can be regarded as a version of the particle in a box, with the particle now being a photon. As in the treatment of a particle in a box (Section 9.1), the only wavelengths that can be sustained satisfy

$$n\times\tfrac{1}{2}\lambda=L \qquad (14.7)$$

where n is an integer and L is the length of the cavity. That is, only an integral number of half-wavelengths fit into the cavity; all other waves undergo destructive interference with themselves. In addition, not all wavelengths that can be sustained by the cavity are amplified by the laser medium (many fall outside the range of frequencies of the laser transitions), so only a few contribute to the laser radiation. These wavelengths are the **resonant modes** of the laser.

Photons with the correct wavelength for the resonant modes of the cavity and the correct frequency to stimulate the laser transition are highly amplified. One photon might be generated spontaneously, and travel through the medium. It stimulates the emission of another photon, which in turn stimulates more (Fig. 14.30). The cascade of energy builds up rapidly, and soon the cavity is an intense reservoir of radiation at all the resonant modes it can sustain. Some of this radiation can be withdrawn if one of the mirrors is partially transmitting.

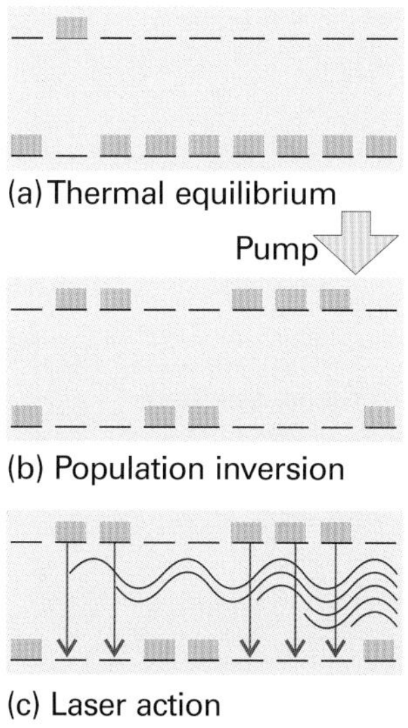

Fig. 14.30 A schematic illustration of the steps leading to laser action. (a) The Boltzmann population of states (see *Molecular interpretation* 3.1), with more atoms in the ground state. (b) When the initial state absorbs, the populations are inverted (the atoms are pumped to the excited state). (c) A cascade of radiation then occurs, as one emitted photon stimulates another atom to emit, and so on. The radiation is coherent (phases in step).

The resonant modes of the cavity have various natural characteristics, and to some extent may be selected. Only photons that are travelling strictly parallel to the axis of the cavity undergo more than a couple of reflections, so only they are amplified, all others simply vanishing into the surroundings. Hence, laser light generally forms a beam with very low divergence. It may also be polarized, with its electric vector in a particular plane (or in some other state of polarization), by including a polarizing filter into the cavity or by making use of polarized transitions in a solid medium.

Laser radiation is **coherent** in the sense that the electromagnetic waves are all in step. In **spatial coherence** the waves are in step across the cross-section of the beam emerging from the cavity. In **temporal coherence** the waves remain in step along the beam. The latter is normally expressed in terms of a **coherence length**, l_C, the distance over which the waves remain coherent, and is related to the range of wavelengths, $\Delta\lambda$ present in the beam:

$$l_C=\frac{\lambda^2}{2\Delta\lambda} \qquad (14.8)$$

If the beam were perfectly monochromatic, with strictly one wavelength present, $\Delta\lambda$ would be zero and the waves would remain in step for an infinite distance. When many wavelengths are present, the waves get out of step in a short distance and the coherence length is small. A typical light bulb gives out light with a coherence length of only about 400 nm; a He–Ne laser with $\Delta\lambda\approx 2$ pm has a coherence length of about 10 cm.

(c) *Q*-switching

A laser can generate radiation for as long as the population inversion is maintained. A laser can operate continuously when heat is easily dissipated, for then the population of the upper level can be replenished by pumping. When overheating is a problem, the laser can be operated only in pulses, perhaps of microsecond or millisecond duration, so that the medium has a chance to cool or the lower state discard its population.

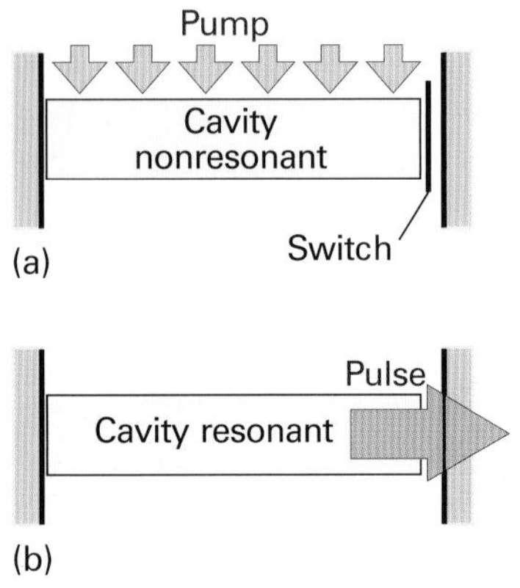

Fig. 14.31 The principle of Q-switching. The excited state is populated while the cavity is nonresonant. Then the resonance characteristics are suddenly restored, and the stimulated emission emerges in a giant pulse.

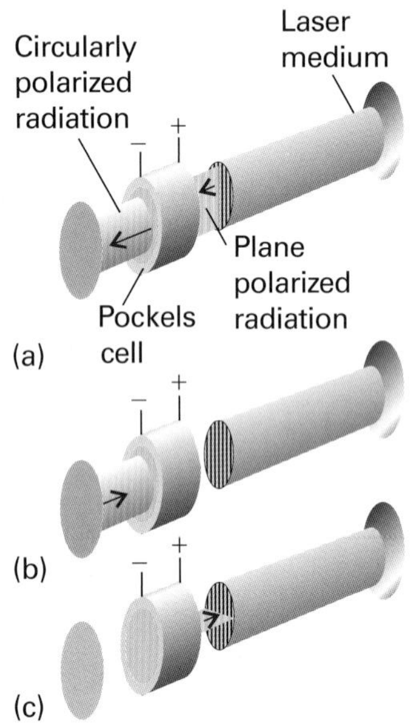

Fig. 14.32 The principle of a Pockels cell. When light passes through a cell that is 'on', its plane of polarization is rotated and so the laser cavity is non-resonant (its Q-factor is reduced). In this sequence, (a) the plane polarized ray becomes circularly polarized, (b) is reflected, and (c) emerges from the Pockels cell with perpendicular plane polarization. When the cell is turned off, no change of polarization occurs, and the cavity becomes resonant.

However, it is sometimes desirable to have pulses of radiation rather than a continuous output, with a lot of power concentrated into a brief pulse. One way of achieving pulses is by ***Q*-switching**, the modification of the resonance characteristics of the laser cavity. The name comes from the 'Q-factor' used as a measure of the quality of a resonance cavity in microwave engineering.

Example 14.2 *Relating the power and energy of a laser*

A laser rated at 0.10 J can generate radiation in 3.0 ns pulses at a pulse repetition rate of 10 Hz. Assuming that the pulses are rectangular, calculate the peak power output and the average power output of this laser.

Method The power output is the energy released in an interval divided by the duration of the interval, and is expressed in watts ($1\ \text{W} = 1\ \text{J s}^{-1}$). To calculate the peak power output, P_{peak}, we divide the energy released during the pulse divided by the duration of the pulse. The average power output, P_{average}, is the total energy released by a large number of pulses divided by the duration of the time interval over which the total energy was measured. So, the average power is simply the energy released by one pulse multiplied by the pulse repetition rate.

Answer From the data,

$$P_{\text{peak}} = \frac{0.10\ \text{J}}{3.0 \times 10^{-9}\ \text{s}} = 3.3 \times 10^{7}\ \text{J s}^{-1}$$

That is, the peak power output is 33 MW. The pulse repetition rate is 10 Hz, so ten pulses are emitted by the laser for every second of operation. It follows that the average power output is

$$P_{\text{average}} = 0.10\ \text{J} \times 10\ \text{s}^{-1} = 1.0\ \text{J s}^{-1} = 1.0\ \text{W}$$

The peak power is much higher than the average power because this laser emits light for only 30 ns during each second of operation.

Self-test 14.3 Calculate the peak power and average power output of a laser with a pulse energy of 2.0 mJ, a pulse duration of 30 ps, and a pulse repetition rate of 38 MHz. [$P_{\text{peak}} = 67$ MW, $P_{\text{average}} = 76$ kW]

The aim of Q-switching is to achieve a healthy population inversion in the absence of the resonant cavity, then to plunge the population-inverted medium into a cavity, and hence to obtain a sudden pulse of radiation. The switching may be achieved by impairing the resonance characteristics of the cavity in some way while the pumping pulse is active, and then suddenly to improve them (Fig. 14.31). One technique is to use a **Pockels cell**, which is an electro-optical device based on the ability of some crystals, such as those of potassium dihydrogenphosphate (KH_2PO_4), to convert plane-polarized light to circularly polarized light when an electrical potential difference is applied. If a Pockels cell is made part of a laser cavity, then its action and the change in polarization that occurs when light is reflected from a mirror convert light polarized in one plane into reflected light polarized in the perpendicular plane (Fig. 14.32). As a result, the reflected light does not stimulate more emission. However, if the cell is suddenly turned off, the polarization effect is extinguished and all the energy stored in the cavity can emerge as an intense pulse of stimulated emission. An alternative technique is to use a **saturable absorber**, typically a solution of a dye that loses its ability to absorb when many of its molecules have been excited by intense radiation. The dye then suddenly becomes transparent and the cavity becomes resonant. In practice, Q-switching can give pulses of about 5 ns duration.

(d) Mode locking

The technique of **mode locking** can produce pulses of picosecond duration and less. A laser radiates at a number of different frequencies, depending on the precise details of the resonance characteristics of the cavity and in particular on the number of half-wavelengths of radiation that can be trapped between the mirrors (the cavity modes). The resonant modes differ in frequency by multiples of $c/2L$ (as can be inferred from eqn 14.8 with $\nu = c/\lambda$). Normally, these modes have random phases relative to each other. However, it is possible to lock their phases together. Then interference occurs to give a series of sharp peaks, and the energy of the laser is obtained in short bursts (Fig. 14.33). The sharpness of the peaks depends on the range of modes superimposed, and the wider the range, the narrower the pulses. In a laser with a cavity of length 30 cm, the peaks are separated by 2 ns. If 1000 modes contribute, the width of the pulses is 4 ps.

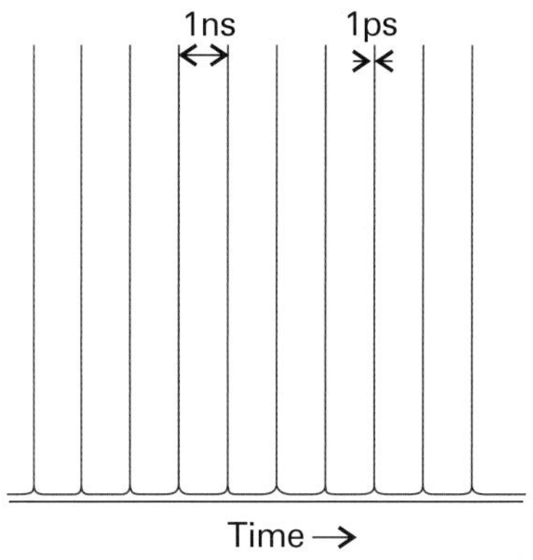

Fig. 14.33 The output of a mode-locked laser consists of a stream of very narrow pulses separated by an interval equal to the time it takes for light to make a round trip inside the cavity.

Justification 14.4 *The origin of mode locking*

The general expression for a (complex) wave of amplitude $\mathcal{E}_0$ and frequency ω is $\mathcal{E}_0 e^{i\omega t}$. Therefore, each wave that can be supported by a cavity of length L has the form

$$\mathcal{E}_n(t) = \mathcal{E}_0 e^{2\pi(\nu + nc/2L)t}$$

where ν is the lowest frequency. A wave formed by superimposing N modes with $n = 0, 1, \ldots, N-1$ has the form

$$\mathcal{E}(t) = \sum_{n=0}^{N-1} \mathcal{E}_n(t) = \mathcal{E}_0 e^{2\pi i\nu t} \sum_{n=0}^{N-1} e^{i\pi nct/L}$$

The sum is a geometrical progression:

$$\sum_{n=0}^{N-1} e^{i\pi nct/L} = 1 + e^{i\pi ct/L} + e^{2i\pi ct/L} + \cdots$$
$$= \frac{\sin(N\pi ct/2L)}{\sin(\pi ct/2L)} \times e^{(N-1)i\pi ct/2L}$$

The intensity, I, of the radiation is proportional to the square modulus of the total amplitude, so

$$I \propto \mathcal{E}^*\mathcal{E} = \mathcal{E}_0^2 \frac{\sin^2(N\pi ct/2L)}{\sin^2(\pi ct/2L)}$$

This function is shown in Fig. 14.34. We see that it is a series of peaks with maxima separated by $t = 2L/c$, the round-trip transit time of the light in the cavity, and that the peaks become sharper as N is increased.

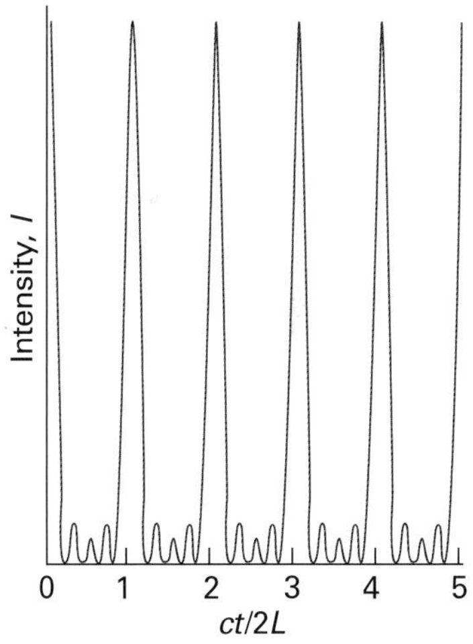

Fig. 14.34 The function derived in *Justification 14.4* showing in more detail the structure of the pulses generated by a mode-locked laser.

Mode locking is achieved by varying the Q-factor of the cavity periodically at the frequency $c/2L$. The modulation can be pictured as the opening of a shutter in synchrony with the round-trip travel time of the photons in the cavity, so only photons making the journey in that time are amplified. The modulation can be achieved by linking a prism in the cavity to a transducer driven by a radiofrequency source at a frequency $c/2L$. The transducer sets up standing-wave vibrations in the prism and modulates the

Table 14.4 Characteristics of laser radiation and their chemical applications

Characteristic	Advantage	Application
High power	Multiphoton process	Nonlinear spectroscopy
		Saturation spectroscopy
	Low detector noise	Improved sensitivity
	High scattering intensity	Raman spectroscopy
Monochromatic	High resolution	Spectroscopy
	State selection	Isotope separation
		Photochemically precise
		State-to-state reaction dynamics
Collimated beam	Long path lengths	Sensitivitv
	Forward-scattering observable	Nonlinear Raman spectroscopy
Coherent	Interference between separate beams	CARS
Pulsed	Precise timing of excitation	Fast reactions
		Relaxation
		Energy transfer

loss it introduces into the cavity. We also see in Section 20.10c that the unique optical properties of some materials can be exploited to bring about mode-locking.

14.6 Applications of lasers in chemistry

Laser radiation has five striking characteristics (Table 14.4). Each of them (sometimes in combination with the others) opens up interesting opportunities in spectroscopy, giving rise to 'laser spectroscopy' and, in photochemistry, giving rise to 'laser photochemistry'. What follows is only an initial listing of applications of lasers to chemistry. We see throughout the text how lasers are also used in the study of macromolecules (Chapter 19) and reaction dynamics (Chapter 24).

(a) Multiphoton spectroscopy

The large number of photons in an incident beam generated by a laser gives rise to a qualitatively different branch of spectroscopy, for the photon density is so high that more than one photon may be absorbed by a single molecule and give rise to **multiphoton processes**. One application of multiphoton processes is that states inaccessible by conventional one-photon spectroscopy become observable because the overall transition occurs with no change of parity. For example, in one-photon spectroscopy, only $g \leftrightarrow u$ transitions are observable; in two-photon spectroscopy, however, the overall outcome of absorbing two photons is a $g \rightarrow g$ or a $u \rightarrow u$ transition.

(b) Raman spectroscopy

Raman spectroscopy was revitalized by the introduction of lasers. An intense excitation beam increases the intensity of scattered radiation, so the use of laser sources increases the sensitivity of Raman spectroscopy. A well-defined beam also implies that the detector can be designed to collect only the radiation that has passed through a sample, and can be screened much more effectively against stray scattered light, which can obscure the Raman signal. The monochromaticity of laser radiation is also a great advantage, for it makes possible the observation of scattered light that differs by only fractions of reciprocal centimetres from the incident radiation. Such high resolution is particularly useful for observing the rotational structure of Raman lines because rotational transitions are of the order of a few reciprocal centimetres. Monochromaticity also allows observations to be made very close to absorption frequencies, giving rise to the techniques of Fourier-transform Raman spectroscopy (Section 13.1) and resonance Raman spectroscopy (Section 13.16b).

The availability of nondivergent beams makes possible a qualitatively different kind of spectroscopy. The beam is so well-defined that it is possible to observe Raman transitions very close to the direction of propagation of the incident beam. This configuration is employed in the technique called **stimulated Raman spectroscopy**. In this form of spectroscopy, the Stokes and anti-Stokes radiation in the forward direction are powerful enough to undergo more scattering and hence give up or acquire more quanta of energy from the molecules in the sample. This multiple scattering results in lines of frequency $\nu_i \pm 2\nu_M$, $\nu_i \pm 3\nu_M$, and so on, where ν_i is the frequency of the incident radiation and ν_M the frequency of a molecular excitation.

(c) Precision-specified transitions

The monochromatic character of laser radiation is a very powerful characteristic because it allows us to excite specific states with very high precision. One consequence of state-specificity for photochemistry is that the illumination of a sample may be photochemically precise and hence efficient in stimulating a reaction, because its frequency can be tuned exactly to an absorption. The specific excitation of a particular excited state of a molecule may greatly enhance the rate of a reaction even at low temperatures. The rate of a reaction is generally increased by raising the temperature because the energies of the various modes of motion of the molecule are enhanced. However, this enhancement increases the energy of all the modes, even those that do not contribute appreciably to the reaction rate. With a laser we can excite the kinetically significant mode, so rate enhancement is achieved most efficiently. An example is the reaction

$$BCl_3 + C_6H_6 \rightarrow C_6H_5{-}BCl_2 + HCl$$

which normally proceeds only above 600°C in the presence of a catalyst; exposure to 10.6 μm CO_2 laser radiation results in the formation of products at room temperature without a catalyst. The commercial potential of this procedure is considerable (provided laser photons can be produced sufficiently cheaply), because heat-sensitive compounds, such as pharmaceuticals, may perhaps be made at lower temperatures than in conventional reactions.

A related application is the study of **state-to-state reaction dynamics**, in which a specific state of a reactant molecule is excited and we monitor not only the rate at which it forms products but also the states in which they are produced. Studies such as these give highly detailed information about the deployment of energy in chemical reactions (Chapter 24).

(d) Isotope separation

The precision state-selectivity of lasers is also of considerable potential for laser isotope separation. Isotope separation is possible because two **isotopomers**, or species that differ only in their isotopic composition, have slightly different energy levels and hence slightly different absorption frequencies.

One approach is to use **photoionization**, the ejection of an electron by the absorption of electromagnetic radiation. Direct photoionization by the absorption of a single photon does not distinguish between isotopomers because the upper level belongs to a continuum; to distinguish isotopomers it is necessary to deal with discrete states. At least two absorption processes are required. In the first step, a photon excites an atom to a higher state; in the second step, a photon achieves photoionization from that state (Fig. 14.35). The energy separation between the two states involved in the first step

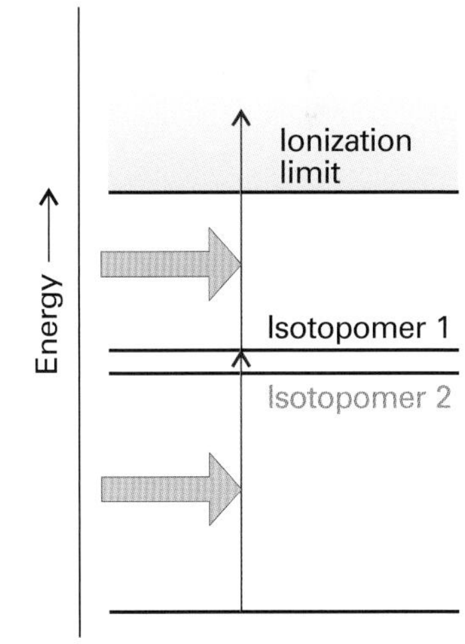

Fig. 14.35 In one method of isotope separation, one photon excites an isotopomer to an excited state, and then a second photon achieves photoionization. The success of the first step depends on the nuclear mass.

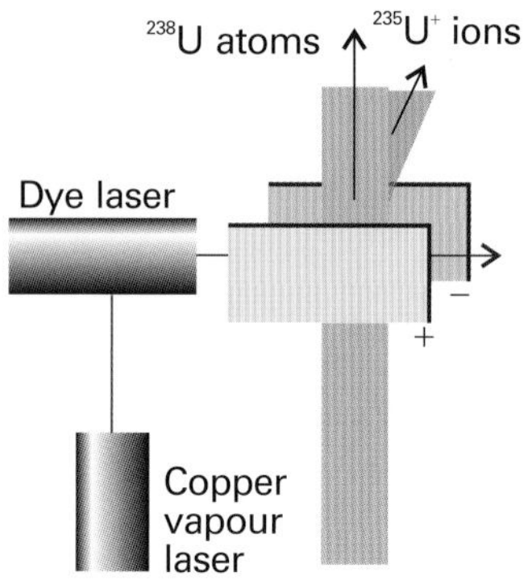

Fig. 14.36 An experimental arrangement for isotope separation. The dye laser, which is pumped by a copper-vapour laser, photoionizes the U atoms selectively according to their mass, and the ions are deflected by the electric field applied between the plates.

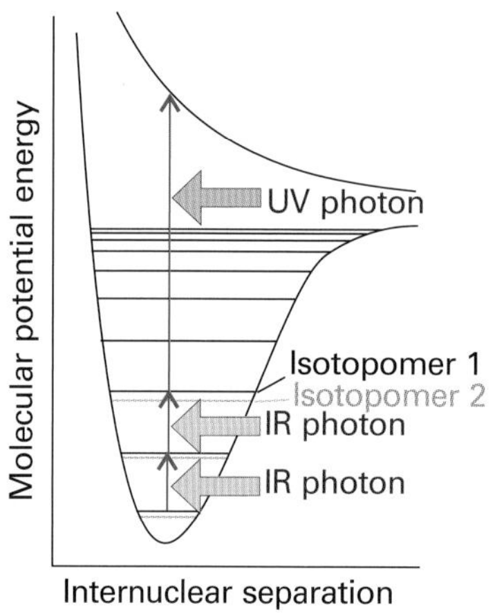

Fig. 14.37 Isotopomers may be separated by making use of their selective absorption of infrared photons followed by photodissociation with an ultraviolet photon.

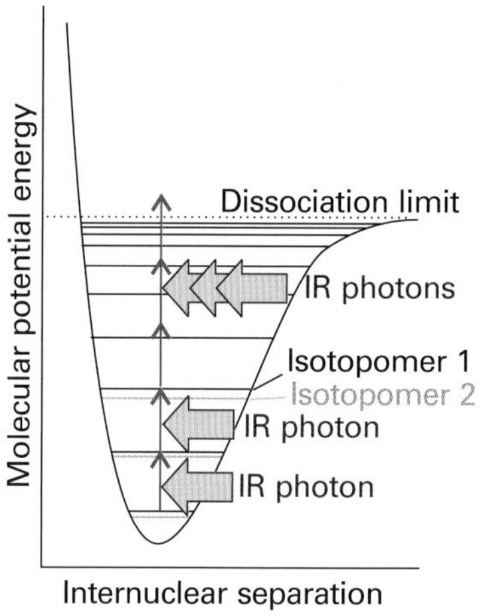

Fig. 14.38 In an alternative scheme for separating isotopomers, multiphoton absorption of infrared photons is used to reach the dissociation limit of a ground electronic state.

depends on the nuclear mass. Therefore, if the laser radiation is tuned to the appropriate frequency, only one of the isotopomers will undergo excitation and hence be available for photoionization in the second step. An example of this procedure is the photoionization of uranium vapour, in which the incident laser is tuned to excite ^{235}U but not ^{238}U. The ^{235}U atoms in the atomic beam are ionized in the two-step process; they are then attracted to a negatively charged electrode, and may be collected (Fig. 14.36). This procedure is being used in the latest generation of uranium separation plants.

Molecular isotopomers are used in techniques based on **photodissociation**, the fragmentation of a molecule following absorption of electromagnetic radiation. The key problem is to achieve both mass selectivity (which requires excitation to take place between discrete states) and dissociation (which requires excitation to continuum states). In one approach, two lasers are used: an infrared photon excites one isotopomer selectively to a higher vibrational level, and then an ultraviolet photon completes the process of photodissociation (Fig. 14.37). An alternative procedure is to make use of multiphoton absorption within the ground electronic state (Fig. 14.38); the efficiency of absorption of the first few photons depends on the match of their frequency to the energy level separations, so it is sensitive to nuclear mass. The absorbed photons open the door to a subsequent influx of enough photons to complete the dissociation process. The isotopomers $^{32}SF_6$ and $^{34}SF_6$ have been separated in this way.

In a third approach, a selectively vibrationally excited species may react with another species and give rise to products that can be separated chemically. This procedure has been employed successfully to separate isotopes of B, N, O, and, most efficiently, H. A variation on this procedure is to achieve selective **photoisomerization**, the conversion of a species to one of its isomers (particularly a geometrical isomer) on absorption of electromagnetic radiation. Once again, the initial absorption, which is isotope selective, opens the way to subsequent further absorption and the formation of a geometrical isomer that can be separated chemically. The approach has been used with the photoisomerization of CH_3NC to CH_3CN.

A different, more physical approach, that of **photodeflection**, is based on the recoil that occurs when a photon is absorbed by an atom and the linear momentum of the photon (which is equal to h/λ) is transferred to the atom. The atom is deflected from its original path only if the absorption actually occurs, and the incident radiation can be tuned to a particular isotope. The deflection is very small, so an atom must absorb dozens of photons before its path is changed sufficiently to allow collection. For instance, if a Ba atom absorbs about 50 photons of 550 nm light, it will be deflected by only about 1 mm after a flight of 1 m.

(e) Time-resolved spectroscopy

The ability of lasers to produce pulses of very short duration is particularly useful in chemistry when we want to monitor processes in time. *Q*-switched lasers produce nanosecond pulses, which are generally fast enough to study reactions with rates controlled by the speed with which reactants can move through a fluid medium. However, when we want to study the rates at which energy is converted from one mode to another within a molecule, we need femtosecond and picosecond pulses. These timescales are available from mode-locked lasers.

In **time-resolved spectroscopy**, laser pulses are used to obtain the absorption, emission, or Raman spectrum of reactants, intermediates, products, and even transition states of reactions. It is also possible to study energy transfer, molecular rotations, vibrations, and conversion from one mode of motion to another. We shall see some of the information obtained from time-resolved spectroscopy in Chapters 22 to 24. Here, we describe some of the experimental techniques that employ pulsed lasers.

The arrangement shown in Fig. 14.39 is often used to study ultrafast chemical reactions that can be initiated by light, such as the initial events of vision (*Impact I14.1*). A strong and short laser pulse, the *pump*, promotes a molecule A to an excited electronic state A* that can either emit a photon (as fluorescence or phosphorescence) or react with another species B to yield a product C:

$$\begin{array}{ll} \mathrm{A} + h\nu \rightarrow \mathrm{A}^* & \text{(absorption)} \\ \mathrm{A}^* \rightarrow \mathrm{A} & \text{(emission)} \\ \mathrm{A}^* + \mathrm{B} \rightarrow [\mathrm{AB}] \rightarrow \mathrm{C} & \text{(reaction)} \end{array}$$

Here [AB] denotes either an intermediate or an activated complex. The rates of appearance and disappearance of the various species are determined by observing time-dependent changes in the absorption spectrum of the sample during the course of the reaction. This monitoring is done by passing a weak pulse of white light, the *probe*, through the sample at different times after the laser pulse. Pulsed 'white' light can be generated directly from the laser pulse by the phenomenon of **continuum generation**, in which focusing an ultrafast laser pulse on a vessel containing a liquid such as water, carbon tetrachloride, CaF, or sapphire results in an outgoing beam with a wide distribution of frequencies. A time delay between the strong laser pulse and the 'white' light pulse can be introduced by allowing one of the beams to travel a longer distance before reaching the sample. For example, a difference in travel distance of $\Delta d = 3$ mm corresponds to a time delay $\Delta t = \Delta d/c \approx 10$ ps between two beams, where c is the speed of light. The relative distances travelled by the two beams in Fig. 14.39 are controlled by directing the 'white' light beam to a motorized stage carrying a pair of mirrors.

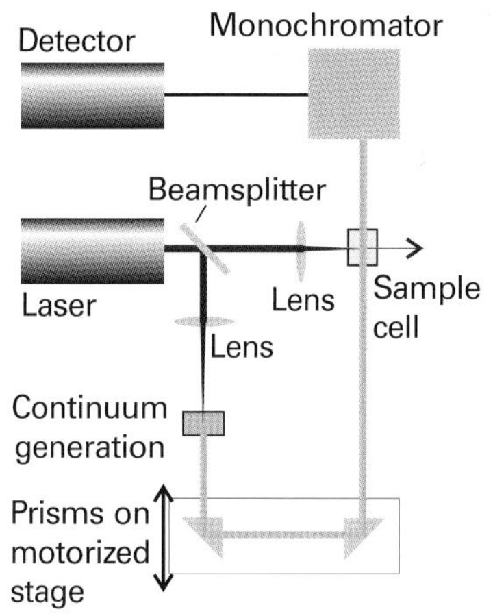

Fig. 14.39 A configuration used for time-resolved absorption spectroscopy, in which the same pulsed laser is used to generate a monochromatic pump pulse and, after continuum generation in a suitable liquid, a 'white' light probe pulse. The time delay between the pump and probe pulses may be varied by moving the motorized stage in the direction shown by the double arrow.

Variations of the arrangement in Fig. 14.39 allow for the observation of fluorescence decay kinetics of A* and time-resolved Raman spectra during the course of the reaction. The fluorescence lifetime of A* can be determined by exciting A as before and measuring the decay of the fluorescence intensity after the pulse with a fast photodetector system. In this case, continuum generation is not necessary. Time-resolved resonance Raman spectra of A, A*, B, [AB], or C can be obtained by initiating the reaction with a strong laser pulse of a certain wavelength and then, some time later, irradiating the sample with another laser pulse that can excite the resonance Raman spectrum of the desired species. Also in this case continuum generation is not necessary. Instead, the Raman excitation beam may be generated in a dye laser (see *Further information 14.1*) or by stimulated Raman scattering of the laser pulse in a medium such as $H_2(g)$ or $CH_4(g)$.

(f) Spectroscopy of single molecules

There is great interest in the development of new experimental probes of very small specimens. On the one hand, our understanding of biochemical processes, such as enzymatic catalysis, protein folding, and the insertion of DNA into the cell's nucleus, will be enhanced if it is possible to visualize individual biopolymers at work. On the other hand, techniques that can probe the structure, dynamics, and reactivity of single molecules will be needed to advance research on nanometre-sized materials (*Impact I20.2*).

We saw in *Impact I13.3* that it is possible to obtain the vibrational spectrum of samples with areas of more than 10 μm^2. Fluorescence microscopy (*Impact I14.2*) has also been used for many years to image biological cells, but the diffraction limit prevents the visualization of samples that are smaller than the wavelength of light used as a probe (*Impact I13.3*). Most molecules—including biological polymers—have dimensions that are much smaller than visible wavelengths, so special techniques had to be developed to make single-molecule spectroscopy possible.

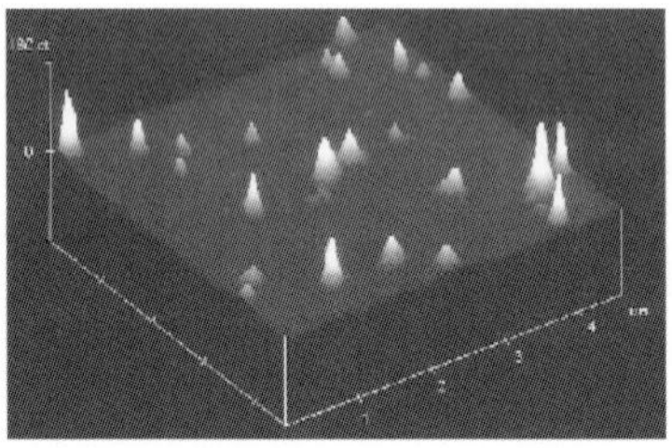

Fig. 14.40 Image of a 4.5 μm × 4.5 μm sample of oxazine-720 dye molecules embedded in a polymer film and obtained with NSOM. Each peak corresponds to a single dye molecule. Reproduced with permission from X.S. Xie. *Acc. Chem. Res.* 1996, **29**, 598.

The bulk of the work done in the field of single-molecule spectroscopy is based on fluorescence microscopy with laser excitation. The laser is the radiation source of choice because it provides the high excitance required to increase the rate of arrival of photons on to the detector from small illuminated areas. Two techniques are commonly used to circumvent the diffraction limit. First, the concentration of the sample is kept so low that, on average, only one fluorescent molecule is in the illuminated area. Second, special strategies are used to illuminate very small volumes. In **near-field optical microscopy** (NSOM), a very thin metal-coated optical fibre is used to deliver light to a small area. It is possible to construct fibres with tip diameters in the range of 50 to 100 nm, which are indeed smaller than visible wavelengths. The fibre tip is placed very close to the sample, in a region known as the *near field*, where, according to classical physics, photons do not diffract. Figure 14.40 shows the image of a 4.5 μm × 4.5 μm sample of oxazine 720 dye molecules embedded in a polymer film and obtained with NSOM by measuring the fluorescence intensity as the tip is scanned over the film surface. Each peak corresponds to a single dye molecule.

In **far-field confocal microscopy**, laser light focused by an objective lens is used to illuminate about 1 μm^3 of a very dilute sample placed beyond the near field. This illumination scheme is limited by diffraction and, as a result, data from far-field microscopy have less structural detail than data from NSOM. However, far-field microscopes are very easy to construct and the technique can be used to probe single molecules as long as there is one molecule, on average, in the illuminated area.

In the **wide-field epifluorescence method**, a two-dimensional array detector (*Further information 13.1*) detects fluorescence excited by a laser and scattered scattered back from the sample (Fig. 14.41a). If the fluorescing molecules are well separated in the specimen, then it is possible to obtain a map of the distribution of fluorescent molecules in the illuminated area. For example, Fig. 14.41b shows how epifluorescence microscopy can be used to observe single molecules of the major histocompatibility (MHC) protein on the surface of a cell.

Though still a relatively new technique, single-molecule spectroscopy has already been used to address important problems in chemistry and biology. Nearly all the techniques discussed in this text measure the average value of a property in a large ensemble of molecules. Single-molecule methods allow a chemist to study the nature

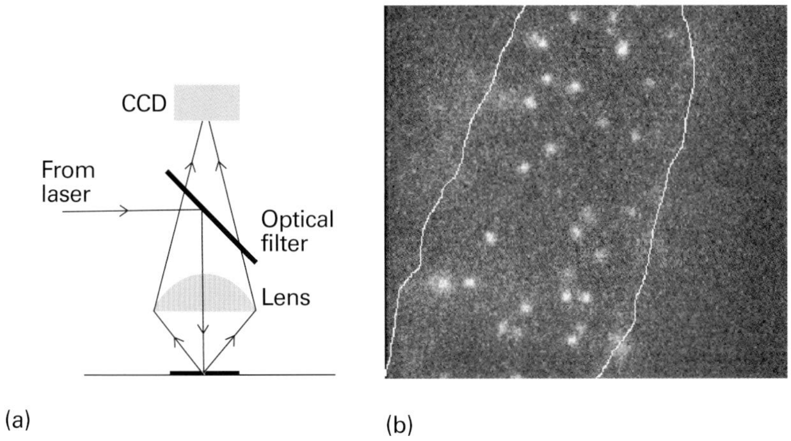

Fig. 14.41 (a) Layout of an epifluorescence microscope. Laser radiation is diverted to a sample by a special optical filter that reflects radiation with a specified wavelength (in this case the laser excitation wavelength) but transmits radiation with other wavelengths (in this case, wavelengths at which the fluorescent label emits). A CCD detector (see *Further information 13.1*) analyses the spatial distribution of the fluorescence signal from the illuminated area. (b) Observation of fluorescence from single MHC proteins that have been labelled with a fluorescent marker and are bound to the surface of a cell (the area shown has dimensions of 12 μm × 12 μm). Image provided by Professor W.E. Moerner, Stanford University, USA.

of distributions of physical and chemical properties in an ensemble of molecules. For example, it is possible to measure the fluorescence lifetime of a molecule by moving the laser focus to a location on the sample that contains a molecule and then measuring the decay of fluorescence intensity after excitation with a pulsed laser. Such studies have shown that not every molecule in a sample has the same fluorescence lifetime, probably because each molecule interacts with its immediate environment in a slightly different way. These details are not apparent from conventional measurements of fluorescence lifetimes, in which many molecules are excited electronically and only an average lifetime for the ensemble can be measured.

Checklist of key ideas

☐ 1. The selection rules for electronic transitions that are concerned with changes in angular momentum are: $\Delta\Lambda = 0, \pm 1, \Delta S = 0, \Delta\Sigma = 0, \Delta\Omega = 0, \pm 1$.

☐ 2. The Laporte selection rule (for centrosymmetric molecules) states that the only allowed transitions are transitions that are accompanied by a change of parity.

☐ 3. The Franck–Condon principle states that, because the nuclei are so much more massive than the electrons, an electronic transition takes place very much faster than the nuclei can respond.

☐ 4. The intensity of an electronic transition is proportional to the Franck–Condon factor, the quantity $|S(v_f,v_i)|^2$, with $S(v_f,v_i) = \langle v_f|v_i\rangle$.

☐ 5. Examples of electronic transitions include *d–d* transitions in *d*-metal complexes, charge-transfer transitions (a transition in which an electron moves from metal to ligand or from ligand to metal in a complex), $\pi^* \leftarrow \pi$, and $\pi^* \leftarrow n$ transitions.

☐ 6. A Jablonski diagram is a schematic diagram of the various types of nonradiative and radiative transitions that can occur in molecules.

☐ 7. Fluorescence is the spontaneous emission of radiation arising from a transition between states of the same multiplicity.

☐ 8. Phosphorescence is the spontaneous emission of radiation arising from a transition between states of different multiplicity.

☐ 9. Intersystem crossing is a nonradiative transition between states of different multiplicity.

☐ 10. Internal conversion is a nonradiative transition between states of the same multiplicity.

☐ 11. Laser action depends on the achievement of population inversion, an arrangement in which there are more molecules in an upper state than in a lower state, and the stimulated emission of radiation.

☐ 12. Resonant modes are the wavelengths that can be sustained by an optical cavity and contribute to the laser action. *Q*-switching is the modification of the resonance characteristics of the laser cavity and, consequently, of the laser output.

☐ 13. Mode locking is a technique for producing pulses of picosecond duration and less by matching the phases of many resonant cavity modes.

☐ 14. Applications of lasers in chemistry include multiphoton spectroscopy, Raman spectroscopy, precision-specified transitions, isotope separation, time-resolved spectroscopy, and single-molecule spectroscopy.

Further reading

Articles and texts

G. Herzberg, *Molecular spectra and molecular structure I. Spectra of diatomic molecules*. Krieger, Malabar (1989).

G. Herzberg, *Molecular spectra and molecular structure III. Electronic spectra and electronic structure of polyatomic molecules*. Van Nostrand–Reinhold, New York (1966).

J.R. Lakowicz, *Principles of fluorescence spectroscopy*. Kluwer/Plenum, New York (1999).

J.C. Lindon, G.E. Tranter, and J.L. Holmes (ed.), *Encyclopedia of spectroscopy and spectrometry*. Academic Press, San Diego (2000).

G.R. van Hecke and K.K. Karukstis, *A guide to lasers in chemistry*. Jones and Bartlett, Boston (1998).

G. Steinmeyer, D.H. Sutter, L. Gallmann, N. Matuschek, and U. Keller, Frontiers in ultrashort pulse generation: pushing the limits in linear and nonlinear optics. *Science* 1999, **286**, 1507.

Sources of data and information

M.E. Jacox, *Vibrational and electronic energy levels of polyatomic transient molecules*. Journal of Physical and Chemical Reference Data , Monograph No. 3 (1994).

D.R. Lide (ed.), *CRC Handbook of Chemistry and Physics*, Sections 9 and 10, CRC Press, Boca Raton (2000).

Further information

Further information 14.1 *Examples of practical lasers*

Figure 14.42 summarizes the requirements for an efficient laser. In practice, the requirements can be satisfied by using a variety of different systems, and this section reviews some that are commonly available. We also include some lasers that operate by using other than electronic transitions. Noticeably absent from this discussion are solid state lasers (including the ubiquitous diode lasers), which we discuss in Chapter 20.

Comment 14.3

The web site for this text contains links to databases on the optical properties of laser materials.

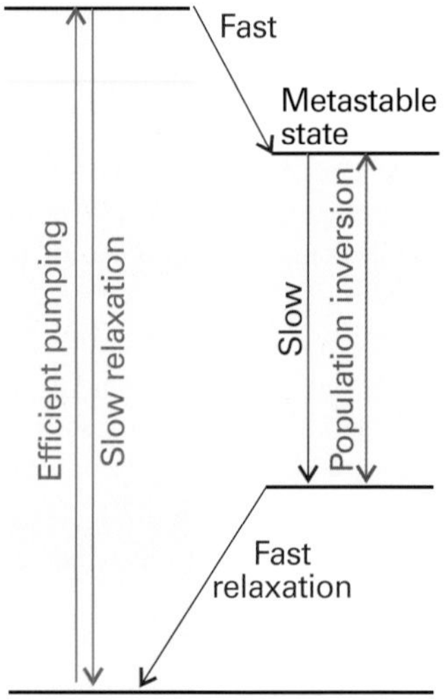

Fig. 14.42 A summary of the features needed for efficient laser action.

Gas lasers

Because gas lasers can be cooled by a rapid flow of the gas through the cavity, they can be used to generate high powers. The pumping is normally achieved using a gas that is different from the gas responsible for the laser emission itself.

In the **helium–neon laser** the active medium is a mixture of helium and neon in a mole ratio of about 5:1 (Fig. 14.43). The initial step is the excitation of an He atom to the metastable $1s^12s^1$

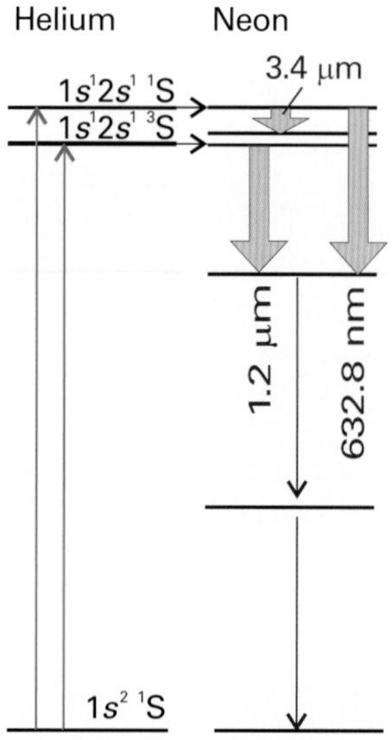

Fig. 14.43 The transitions involved in a helium–neon laser. The pumping (of the neon) depends on a coincidental matching of the helium and neon energy separations, so excited He atoms can transfer their excess energy to Ne atoms during a collision.

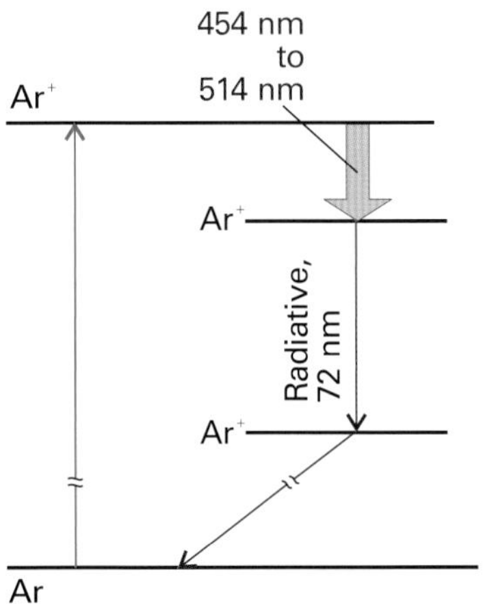

Fig. 14.44 The transitions involved in an argon-ion laser.

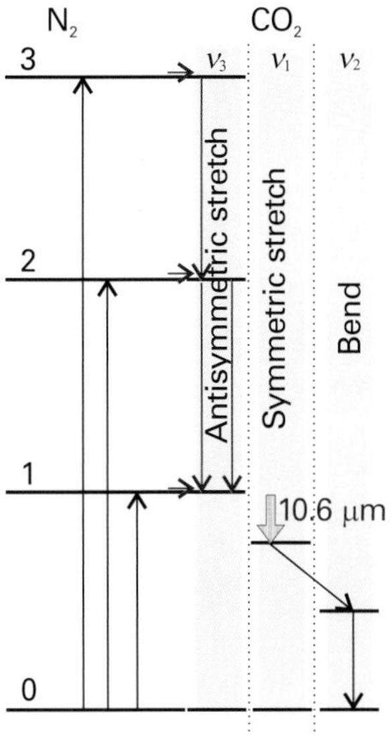

Fig. 14.45 The transitions involved in a carbon dioxide laser. The pumping also depends on the coincidental matching of energy separations; in this case the vibrationally excited N_2 molecules have excess energies that correspond to a vibrational excitation of the antisymmetric stretch of CO_2. The laser transition is from $\nu_3 = 1$ to $\nu_1 = 1$.

configuration by using an electric discharge (the collisions of electrons and ions cause transitions that are not restricted by electric-dipole selection rules). The excitation energy of this transition happens to match an excitation energy of neon, and during an He–Ne collision efficient transfer of energy may occur, leading to the production of highly excited, metastable Ne atoms with unpopulated intermediate states. Laser action generating 633 nm radiation (among about 100 other lines) then occurs.

The **argon-ion laser** (Fig.14.44), one of a number of 'ion lasers', consists of argon at about 1 Torr, through which is passed an electric discharge. The discharge results in the formation of Ar^+ and Ar^{2+} ions in excited states, which undergo a laser transition to a lower state. These ions then revert to their ground states by emitting hard ultraviolet radiation (at 72 nm), and are then neutralized by a series of electrodes in the laser cavity. One of the design problems is to find materials that can withstand this damaging residual radiation. There are many lines in the laser transition because the excited ions may make transitions to many lower states, but two strong emissions from Ar^+ are at 488 nm (blue) and 514 nm (green); other transitions occur elsewhere in the visible region, in the infrared, and in the ultraviolet. The **krypton-ion laser** works similarly. It is less efficient, but gives a wider range of wavelengths, the most intense being at 647 nm (red), but it can also generate yellow, green, and violet lines. Both lasers are widely used in laser light shows (for this application argon and krypton are often used simultaneously in the same cavity) as well as laboratory sources of high-power radiation.

The **carbon dioxide laser** works on a slightly different principle (Fig. 14.45), for its radiation (between 9.2 μm and 10.8 μm, with the strongest emission at 10.6 μm, in the infrared) arises from vibrational transitions. Most of the working gas is nitrogen, which becomes vibrationally excited by electronic and ionic collisions in an electric discharge. The vibrational levels happen to coincide with the ladder of antisymmetric stretch (ν_3, see Fig. 13.40) energy levels of CO_2, which pick up the energy during a collision. Laser action then occurs from the lowest excited level of ν_3 to the lowest excited level of the symmetric stretch (ν_1), which has remained unpopulated during the collisions. This transition is allowed by anharmonicities in the molecular potential energy. Some helium is included in the gas to help remove energy from this state and maintain the population inversion.

In a **nitrogen laser**, the efficiency of the stimulated transition (at 337 nm, in the ultraviolet, the transition $C^3\Pi_u \rightarrow B^3\Pi_g$) is so great that a single passage of a pulse of radiation is enough to generate laser radiation and mirrors are unnecessary: such lasers are said to be **superradiant**.

Chemical and exciplex lasers

Chemical reactions may also be used to generate molecules with nonequilibrium, inverted populations. For example, the photolysis of Cl_2 leads to the formation of Cl atoms which attack H_2 molecules in the mixture and produce HCl and H. The latter then attacks Cl_2 to produce vibrationally excited ('hot') HCl molecules. Because the newly formed HCl molecules have nonequilibrium vibrational populations, laser action can result as they return to lower states. Such processes are remarkable examples of the direct conversion of chemical energy into coherent electromagnetic radiation.

The population inversion needed for laser action is achieved in a more underhand way in **exciplex lasers**, for in these (as we shall see) the lower state does not effectively exist. This odd situation is achieved by forming an **exciplex**, a combination of two atoms that survives only in an excited state and which dissociates as soon as the excitation energy has been discarded. An exciplex can be formed in a mixture of xenon, chlorine, and neon (which acts as a buffer gas). An electric discharge through the mixture produces excited Cl atoms, which attach to the Xe atoms to give the exciplex XeCl*. The exciplex survives for about 10 ns, which is time for it to participate in laser action at 308 nm (in the ultraviolet). As soon as XeCl* has discarded a

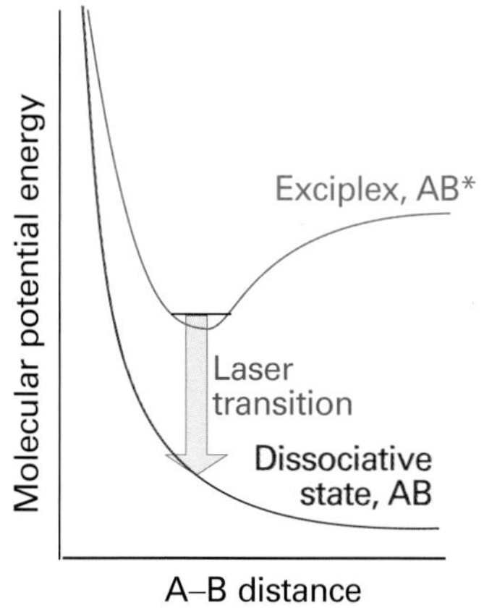

Fig. 14.46 The molecular potential energy curves for an exciplex. The species can survive only as an excited state, because on discarding its energy it enters the lower, dissociative state. Because only the upper state can exist, there is never any population in the lower state.

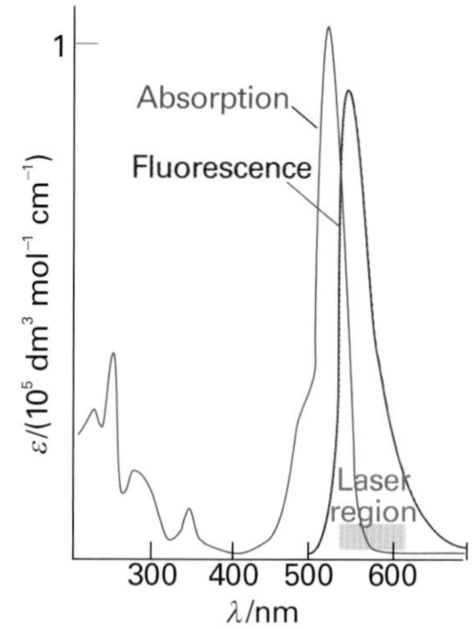

Fig. 14.47 The optical absorption spectrum of the dye Rhodamine 6G and the region used for laser action.

photon, the atoms separate because the molecular potential energy curve of the ground state is dissociative, and the ground state of the exciplex cannot become populated (Fig. 14.46). The KrF* exciplex laser is another example: it produces radiation at 249 nm.

Comment 14.4

The term 'excimer laser' is also widely encountered and used loosely when 'exciplex laser' is more appropriate. An exciplex has the form AB* whereas an excimer, an excited dimer, is AA*.

Dye lasers

Gas lasers and most solid state lasers operate at discrete frequencies and, although the frequency required may be selected by suitable optics, the laser cannot be tuned continuously. The tuning problem is overcome by using a titanium sapphire laser (see above) or a **dye laser**, which has broad spectral characteristics because the solvent broadens the vibrational structure of the transitions into bands. Hence, it is possible to scan the wavelength continuously (by rotating the diffraction grating in the cavity) and achieve laser action at any chosen wavelength. A commonly used dye is rhodamine 6G in methanol (Fig. 14.47). As the gain is very high, only a short length of the optical path need be through the dye. The excited states of the active medium, the dye, are sustained by another laser or a flash lamp, and the dye solution is flowed through the laser cavity to avoid thermal degradation (Fig. 14.48).

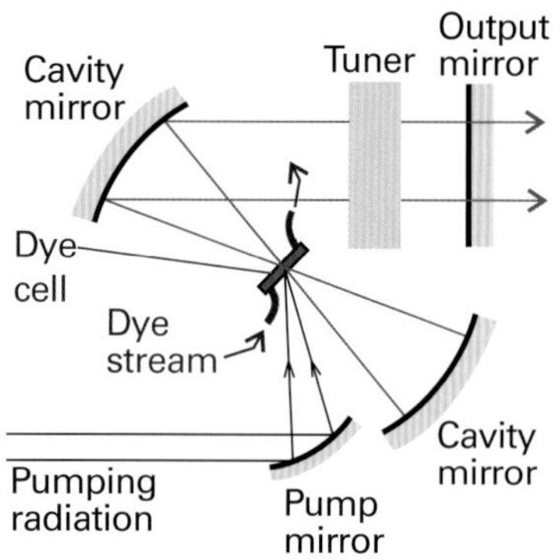

Fig. 14.48 The configuration used for a dye laser. The dye is flowed through the cell inside the laser cavity. The flow helps to keep it cool and prevents degradation.

Discussion questions

14.1 Explain the origin of the term symbol $^3\Sigma_g^-$ for the ground state of dioxygen.

14.2 Explain the basis of the Franck–Condon principle and how it leads to the formation of a vibrational progression.

14.3 How do the band heads in P and R branches arise? Could the Q branch show a head?

14.4 Explain how colour can arise from molecules.

14.5 Describe the mechanism of fluorescence. To what extent is a fluorescence spectrum not the exact mirror image of the corresponding absorption spectrum?

14.6 What is the evidence for the correctness of the mechanism of fluorescence?

14.7 Describe the principles of laser action, with actual examples.

14.8 What features of laser radiation are applied in chemistry? Discuss two applications of lasers in chemistry.

Exercises

14.1a The term symbol for the ground state of N_2^+ is $^2\Sigma_g^+$. What is the total spin and total orbital angular momentum of the molecule? Show that the term symbol agrees with the electron configuration that would be predicted using the building-up principle.

14.1b One of the excited states of the C_2 molecule has the valence electron configuration $1\sigma_g^2 1\sigma_u^2 1\pi_u^3 1\pi_g^1$. Give the multiplicity and parity of the term.

14.2a The molar absorption coefficient of a substance dissolved in hexane is known to be 855 $dm^3\ mol^{-1}\ cm^{-1}$ at 270 nm. Calculate the percentage reduction in intensity when light of that wavelength passes through 2.5 mm of a solution of concentration 3.25 mmol dm^{-3}.

14.2b The molar absorption coefficient of a substance dissolved in hexane is known to be 327 $dm^3\ mol^{-1}\ cm^{-1}$ at 300 nm. Calculate the percentage reduction in intensity when light of that wavelength passes through 1.50 mm of a solution of concentration 2.22 mmol dm^{-3}.

14.3a A solution of an unknown component of a biological sample when placed in an absorption cell of path length 1.00 cm transmits 20.1 per cent of light of 340 nm incident upon it. If the concentration of the component is 0.111 mmol dm^{-3}, what is the molar absorption coefficient?

14.3b When light of wavelength 400 nm passes through 3.5 mm of a solution of an absorbing substance at a concentration 0.667 mmol dm^{-3}, the transmission is 65.5 per cent. Calculate the molar absorption coefficient of the solute at this wavelength and express the answer in $cm^2\ mol^{-1}$.

14.4a The molar absorption coefficient of a solute at 540 nm is 286 $dm^3\ mol^{-1}\ cm^{-1}$. When light of that wavelength passes through a 6.5 mm cell containing a solution of the solute, 46.5 per cent of the light is absorbed. What is the concentration of the solution?

14.4b The molar absorption coefficient of a solute at 440 nm is 323 $dm^3\ mol^{-1}\ cm^{-1}$. When light of that wavelength passes through a 7.50 mm cell containing a solution of the solute, 52.3 per cent of the light is absorbed. What is the concentration of the solution?

14.5a The absorption associated with a particular transition begins at 230 nm, peaks sharply at 260 nm, and ends at 290 nm. The maximum value of the molar absorption coefficient is 1.21×10^4 $dm^3\ mol^{-1}\ cm^{-1}$. Estimate the integrated absorption coefficient of the transition assuming a triangular lineshape (see eqn 13.5).

14.5b The absorption associated with a certain transition begins at 199 nm, peaks sharply at 220 nm, and ends at 275 nm. The maximum value of the molar absorption coefficient is 2.25×10^4 $dm^3\ mol^{-1}\ cm^{-1}$. Estimate the integrated absorption coefficient of the transition assuming an inverted parabolic lineshape (Fig. 14.49; use eqn 13.5).

14.6a The two compounds, 2,3-dimethyl-2-butene and 2,5-dimethyl-2,4-hexadiene, are to be distinguished by their ultraviolet absorption spectra. The maximum absorption in one compound occurs at 192 nm and in the other at 243 nm. Match the maxima to the compounds and justify the assignment.

14.6b 1,3,5-hexatriene (a kind of 'linear' benzene) was converted into benzene itself. On the basis of a free-electron molecular orbital model (in which hexatriene is treated as a linear box and benzene as a ring), would you expect the lowest energy absorption to rise or fall in energy?

14.7a The following data were obtained for the absorption by Br_2 in carbon tetrachloride using a 2.0 mm cell. Calculate the molar absorption coefficient of bromine at the wavelength employed:

$[Br_2]/(mol\ dm^{-3})$	0.0010	0.0050	0.0100	0.0500
$T/(\text{per cent})$	81.4	35.6	12.7	3.0×10^{-3}

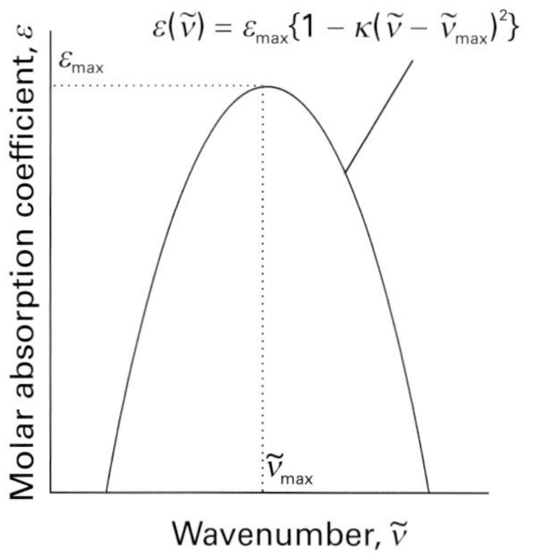

Fig. 14.49

14.7b The following data were obtained for the absorption by a dye dissolved in methylbenzene using a 2.50 mm cell. Calculate the molar absorption coefficient of the dye at the wavelength employed:

$[\text{dye}]/(mol\ dm^{-3})$	0.0010	0.0050	0.0100	0.0500
$T/(\text{per cent})$	73	21	4.2	1.33×10^{-5}

14.8a A 2.0-mm cell was filled with a solution of benzene in a non-absorbing solvent. The concentration of the benzene was 0.010 mol dm^{-3} and the wavelength of the radiation was 256 nm (where there is a maximum in the absorption). Calculate the molar absorption coefficient of benzene at this wavelength given that the transmission was 48 per cent. What will the transmittance be in a 4.0-mm cell at the same wavelength?

14.8b A 2.50-mm cell was filled with a solution of a dye. The concentration of the dye was 15.5 mmol dm^{-3}. Calculate the molar absorption coefficient of benzene at this wavelength given that the transmission was 32 per cent. What will the transmittance be in a 4.50-mm cell at the same wavelength?

14.9a A swimmer enters a gloomier world (in one sense) on diving to greater depths. Given that the mean molar absorption coefficient of sea water in the visible region is 6.2×10^{-3} $dm^3\ mol^{-1}\ cm^{-1}$, calculate the depth at which a diver will experience (a) half the surface intensity of light, (b) one tenth the surface intensity.

14.9b Given that the maximum molar absorption coefficient of a molecule containing a carbonyl group is 30 $dm^3\ mol^{-1}\ cm^{-1}$ near 280 nm, calculate the thickness of a sample that will result in (a) half the initial intensity of radiation, (b) one-tenth the initial intensity.

14.10a The electronic absorption bands of many molecules in solution have half-widths at half-height of about 5000 cm^{-1}. Estimate the integrated absorption coefficients of bands for which (a) $\varepsilon_{max} \approx 1\times10^4$ $dm^3\ mol^{-1}\ cm^{-1}$, (b) $\varepsilon_{max} \approx 5\times10^2$ $dm^3\ mol^{-1}\ cm^{-1}$.

14.10b The electronic absorption band of a compound in solution had a Gaussian lineshape and a half-width at half-height of 4233 cm^{-1} and $\varepsilon_{max} = 1.54\times10^4$ $dm^3\ mol^{-1}\ cm^{-1}$. Estimate the integrated absorption coefficient.

14.11a The photoionization of H_2 by 21 eV photons produces H_2^+. Explain why the intensity of the $v = 2 \leftarrow 0$ transition is stronger than that of the $0 \leftarrow 0$ transition.

14.11b The photoionization of F_2 by 21 eV photons produces F_2^+. Would you expect the $2 \leftarrow 0$ transition to be weaker or stronger than the $0 \leftarrow 0$ transition? Justify your answer.

Problems*

Numerical problems

14.1 The vibrational wavenumber of the oxygen molecule in its electronic ground state is 1580 cm^{-1}, whereas that in the first excited state (B $^3\Sigma_u^-$), to which there is an allowed electronic transition, is 700 cm^{-1}. Given that the separation in energy between the minima in their respective potential energy curves of these two electronic states is 6.175 eV, what is the wavenumber of the lowest energy transition in the band of transitions originating from the $v = 0$ vibrational state of the electronic ground state to this excited state? Ignore any rotational structure or anharmonicity.

14.2 A Birge–Sponer extrapolation yields 7760 cm^{-1} as the area under the curve for the B state of the oxygen molecule described in Problem 14.1. Given that the B state dissociates to ground-state atoms (at zero energy, ^{3}P) and 15 870 cm^{-1} (^{1}D) and the lowest vibrational state of the B state is 49 363 cm^{-1} above the lowest vibrational state of the ground electronic state, calculate the dissociation energy of the molecular ground state to the ground-state atoms.

14.3 The electronic spectrum of the IBr molecule shows two low-lying, well defined convergence limits at 14 660 and 18 345 cm^{-1}. Energy levels for the iodine and bromine atoms occur at 0, 7598; and 0, 3685 cm^{-1}, respectively. Other atomic levels are at much higher energies. What possibilities exist for the numerical value of the dissociation energy of IBr? Decide which is the correct possibility by calculating this quantity from $\Delta_f H^\ominus$(IBr, g) = +40.79 kJ mol^{-1} and the dissociation energies of I_2(g) and Br_2(g) which are 146 and 190 kJ mol^{-1}, respectively.

14.4 In many cases it is possible to assume that an absorption band has a Gaussian lineshape (one proportional to e^{-x^2}) centred on the band maximum. Assume such a line shape, and show that $A \approx 1.0645\varepsilon_{max}\Delta\tilde{\nu}_{1/2}$, where $\Delta\tilde{\nu}_{1/2}$ is the width at half-height. The absorption spectrum of azoethane ($CH_3CH_2N_2$) between 24 000 cm^{-1} and 34 000 cm^{-1} is shown in Fig. 14.50. First, estimate A for the band by assuming that it is Gaussian. Then integrate the absorption band graphically. The latter can be done either by ruling and counting squares, or by tracing the lineshape on to paper and weighing. A more sophisticated procedure would be to use mathematical software to fit a polynomial to the absorption band (or a Gaussian), and then to integrate the result analytically.

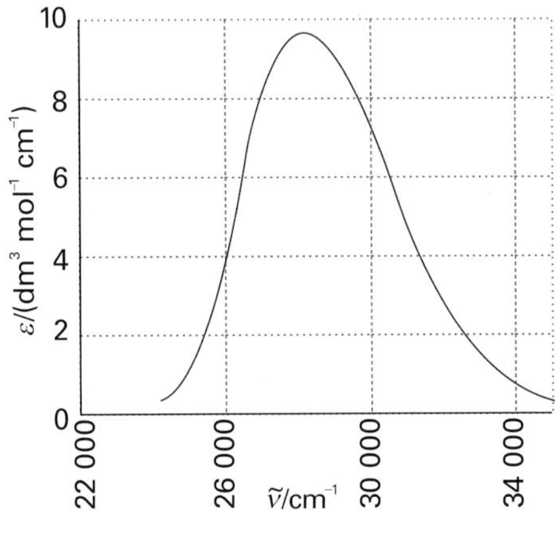

Fig. 14.50

14.5 A lot of information about the energy levels and wavefunctions of small inorganic molecules can be obtained from their ultraviolet spectra. An example of a spectrum with considerable vibrational structure, that of gaseous SO_2 at 25°C, is shown in Fig. 14.6. Estimate the integrated absorption coefficient for the transition. What electronic states are accessible from the A_1 ground state of this C_{2v} molecule by electric dipole transitions?

14.6‡ J.G. Dojahn, E.C.M. Chen, and W.E. Wentworth (*J. Phys. Chem.* **100**, 9649 (1996)) characterized the potential energy curves of the ground and electronic states of homonuclear diatomic halogen anions. These anions have a $^2\Sigma_u^+$ ground state and $^2\Pi_g$, $^2\Pi_u$, and $^2\Sigma_g^+$ excited states. To which of the excited states are transitions by absorption of photons allowed? Explain.

14.7 A transition of particular importance in O_2 gives rise to the 'Schumann–Runge band' in the ultraviolet region. The wavenumbers (in cm^{-1}) of transitions from the ground state to the vibrational levels of the first excited state ($^3\Sigma_u^-$) are 50 062.6, 50 725.4, 51 369.0, 51 988.6, 52 579.0, 53 143.4, 53 679.6, 54 177.0, 54 641.8, 55 078.2, 55 460.0, 55 803.1, 56 107.3, 56 360.3, 56 570.6. What is the dissociation energy of the upper electronic state? (Use a Birge–Sponer plot.) The same excited state is known to dissociate into one ground-state O atom and one excited-state atom with an energy 190 kJ mol^{-1} above the ground state. (This excited atom is responsible for a great deal of photochemical mischief in the atmosphere.) Ground-state O_2 dissociates into two ground-state atoms. Use this information to calculate the dissociation energy of ground-state O_2 from the Schumann–Runge data.

14.8 The compound $CH_3CH{=}CHCHO$ has a strong absorption in the ultraviolet at 46 950 cm^{-1} and a weak absorption at 30 000 cm^{-1}. Justify these features in terms of the structure of the molecule.

14.9 Aromatic hydrocarbons and I_2 form complexes from which charge-transfer electronic transitions are observed. The hydrocarbon acts as an electron donor and I_2 as an electron acceptor. The energies $h\nu_{max}$ of the charge-transfer transitions for a number of hydrocarbon–I_2 complexes are given below:

Hydrocarbon	benzene	biphenyl	naphthalene	phenanthrene	pyrene	anthracene
$h\nu_{max}$/eV	4.184	3.654	3.452	3.288	2.989	2.890

Investigate the hypothesis that there is a correlation between the energy of the HOMO of the hydrocarbon (from which the electron comes in the charge–transfer transition) and $h\nu_{max}$. Use one of the molecular electronic structure methods discussed in Chapter 11 to determine the energy of the HOMO of each hydrocarbon in the data set.[1]

14.10 A certain molecule fluoresces at a wavelength of 400 nm with a half-life of 1.0 ns. It phosphoresces at 500 nm. If the ratio of the transition probabilities for stimulated emission for the S* → S to the T → S transitions is 1.0×10^5, what is the half-life of the phosphorescent state?

14.11 Consider some of the precautions that must be taken when conducting single-molecule spectroscopy experiments. (a) What is the molar concentration of a solution in which there is, on average, one solute molecule in 1.0 μm^3 (1.0 fL) of solution? (b) It is important to use pure solvents in single-molecule spectroscopy because optical signals from fluorescent

* Problems denoted with the symbol ‡ were supplied by Charles Trapp and Carmen Giunta.

1 The web site contains links to molecular modelling freeware and to other sites where you may perform molecular orbital calculations directly from your web browser.

impurities in the solvent may mask optical signals from the solute. Suppose that water containing a fluorescent impurity of molar mass 100 g mol^{-1} is used as solvent and that analysis indicates the presence of 0.10 mg of impurity per 1.0 kg of solvent. On average, how many impurity molecules will be present in 1.0 μm^3 of solution? You may take the density of water as 1.0 g cm^{-3}. Comment on the suitability of this solvent for single-molecule spectroscopy experiments.

14.12 Light-induced degradation of molecules, also called *photobleaching*, is a serious problem in single-molecule spectroscopy. A molecule of a fluorescent dye commonly used to label biopolymers can withstand about 10^6 excitations by photons before light-induced reactions destroy its π system and the molecule no longer fluoresces. For how long will a single dye molecule fluoresce while being excited by 1.0 mW of 488 nm radiation from a continuous-wave argon ion laser? You may assume that the dye has an absorption spectrum that peaks at 488 nm and that every photon delivered by the laser is absorbed by the molecule.

Theoretical problems

14.13 Assume that the electronic states of the π electrons of a conjugated molecule can be approximated by the wavefunctions of a particle in a one-dimensional box, and that the dipole moment can be related to the displacement along this length by $\mu = -ex$. Show that the transition probability for the transition $n = 1 \to n = 2$ is nonzero, whereas that for $n = 1 \to n = 3$ is zero. *Hint*. The following relations will be useful:

$$\sin x \sin y = \tfrac{1}{2}\cos(x-y) - \tfrac{1}{2}\cos(x+y)$$

$$\int x \cos ax\,dx = \frac{1}{a^2}\cos ax + \frac{x}{a}\sin ax$$

14.14 Use a group theoretical argument to decide which of the following transitions are electric-dipole allowed: (a) the $\pi^* \leftarrow \pi$ transition in ethene, (b) the $\pi^* \leftarrow n$ transition in a carbonyl group in a C_{2v} environment.

14.15 Suppose that you are a colour chemist and had been asked to intensify the colour of a dye without changing the type of compound, and that the dye in question was a polyene. Would you choose to lengthen or to shorten the chain? Would the modification to the length shift the apparent colour of the dye towards the red or the blue?

14.16 One measure of the intensity of a transition of frequency ν is the *oscillator strength*, f, which is defined as

$$f = \frac{8\pi^2 m_e \nu |\mu_{fi}|^2}{3he^2}$$

Consider an electron in an atom to be oscillating harmonically in one dimension (the three-dimensional version of this model was used in early attempts to describe atomic structure). The wavefunctions for such an electron are those in Table 9.1. Show that the oscillator strength for the transition of this electron from its ground state is exactly $\tfrac{1}{3}$.

14.17 Estimate the oscillator strength (see Problem 14.16) of a charge-transfer transition modelled as the migration of an electron from an H1*s* orbital on one atom to another H1*s* orbital on an atom a distance *R* away. Approximate the transition moment by $-eRS$ where *S* is the overlap integral of the two orbitals. Sketch the oscillator strength as a function of *R* using the curve for *S* given in Fig. 11.29. Why does the intensity fall to zero as *R* approaches 0 and infinity?

14.18 The line marked A in Fig. 14.51 is the fluorescence spectrum of benzophenone in solid solution in ethanol at low temperatures observed when the sample is illuminated with 360 nm light. What can be said about the vibrational energy levels of the carbonyl group in (a) its ground electronic state and (b) its excited electronic state? When naphthalene is illuminated with 360 nm light it does not absorb, but the line marked B in the illustration is the phosphorescence spectrum of a solid solution of a mixture of naphthalene and benzophenone in ethanol. Now a component of fluorescence from naphthalene can be detected. Account for this observation.

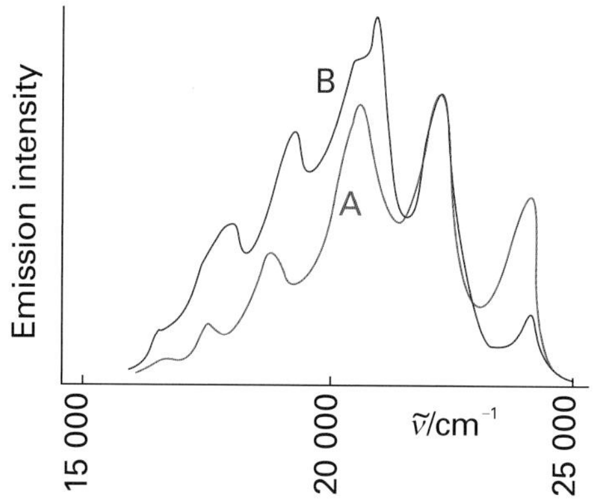

Fig. 14.51

14.19 The fluorescence spectrum of anthracene vapour shows a series of peaks of increasing intensity with individual maxima at 440 nm, 410 nm, 390 nm, and 370 nm followed by a sharp cut-off at shorter wavelengths. The absorption spectrum rises sharply from zero to a maximum at 360 nm with a trail of peaks of lessening intensity at 345 nm, 330 nm, and 305 nm. Account for these observations.

14.20 The Beer–Lambert law states that the absorbance of a sample at a wavenumber $\tilde{\nu}$ is proportional to the molar concentration [J] of the absorbing species J and to the length *l* of the sample (eqn 13.4). In this problem you will show that the intensity of fluorescence emission from a sample of J is also proportional to [J] and *l*. Consider a sample of J that is illuminated with a beam of intensity $I_0(\tilde{\nu})$ at the wavenumber $\tilde{\nu}$. Before fluorescence can occur, a fraction of $I_0(\tilde{\nu})$ must be absorbed and an intensity $I(\tilde{\nu})$ will be transmitted. However, not all of the absorbed intensity is emitted and the intensity of fluorescence depends on the fluorescence quantum yield, ϕ_f, the efficiency of photon emission. The fluorescence quantum yield ranges from 0 to 1 and is proportional to the ratio of the integral of the fluorescence spectrum over the integrated absorption coefficient. Because of a Stokes shift of magnitude $\Delta\tilde{\nu}_{Stokes}$, fluorescence occurs at a wavenumber $\tilde{\nu}_f$, with $\tilde{\nu}_f + \Delta\tilde{\nu}_{Stokes} = \tilde{\nu}$. It follows that the fluorescence intensity at $\tilde{\nu}_f$, $I_f(\tilde{\nu}_f)$, is proportional to ϕ_f and to the intensity of exciting radiation that is absorbed by J, $I_{abs}(\tilde{\nu}) = I_0(\tilde{\nu}) - I(\tilde{\nu})$. (a) Use the Beer–Lambert law to express $I_{abs}(\tilde{\nu})$ in terms of $I_0(\tilde{\nu})$, [J], *l*, and $\varepsilon(\tilde{\nu})$, the molar absorption coefficient of J at $\tilde{\nu}$. (b) Use your result from part (a) to show that $I_f(\tilde{\nu}_f) \propto I_0(\tilde{\nu})\varepsilon(\tilde{\nu})\phi_f[J]l$.

14.21 Spin angular momentum is conserved when a molecule dissociates into atoms. What atom multiplicities are permitted when (a) an O_2 molecule, (b) an N_2 molecule dissociates into atoms?

Applications: to biochemistry, environmental science, and astrophysics

14.22 The protein haemerythrin (Her) is responsible for binding and carrying O_2 in some invertebrates. Each protein molecule has two Fe^{2+} ions that are in very close proximity and work together to bind one molecule of O_2. The Fe_2O_2 group of oxygenated haemerythrin is coloured and has an electronic absorption band at 500 nm. Figure 14.52 shows the UV-visible absorption spectrum of a derivative of haemerythrin in the presence of different concentrations of CNS^- ions. What may be inferred from the spectrum?

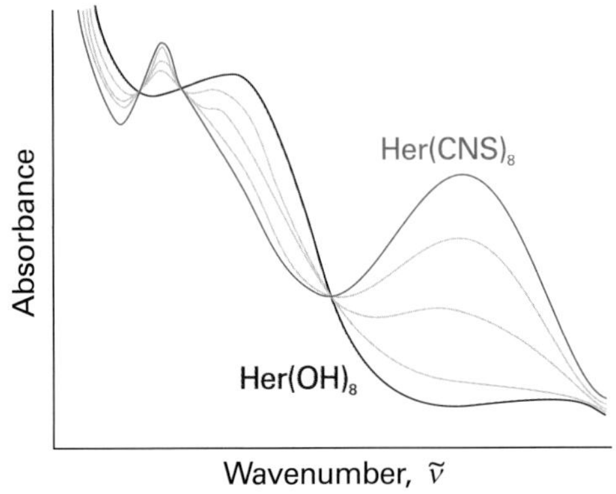

Fig. 14.52

14.23 The flux of visible photons reaching Earth from the North Star is about $4 \times 10^3\ mm^{-2}\ s^{-1}$. Of these photons, 30 per cent are absorbed or scattered by the atmosphere and 25 per cent of the surviving photons are scattered by the surface of the cornea of the eye. A further 9 per cent are absorbed inside the cornea. The area of the pupil at night is about 40 mm^2 and the response time of the eye is about 0.1 s. Of the photons passing through the pupil, about 43 per cent are absorbed in the ocular medium. How many photons from the North Star are focused on to the retina in 0.1 s? For a continuation of this story, see R.W. Rodieck, *The first steps in seeing*, Sinauer, Sunderland (1998).

14.24 Use molecule (**7**) as a model of the *trans* conformation of the chromophore found in rhodopsin. In this model, the methyl group bound to the nitrogen atom of the protonated Schiff's base replaces the protein. (a) Using molecular modelling software and the computational method of your instructor's choice, calculate the energy separation between the HOMO and LUMO of (**7**). (b) Repeat the calculation for the 11-*cis* form of (**7**). (c) Based on your results from parts (a) and (b), do you expect the experimental frequency for the $\pi^* \leftarrow \pi$ visible absorption of the *trans* form of (**7**) to be higher or lower than that for the 11-*cis* form of (**7**)?

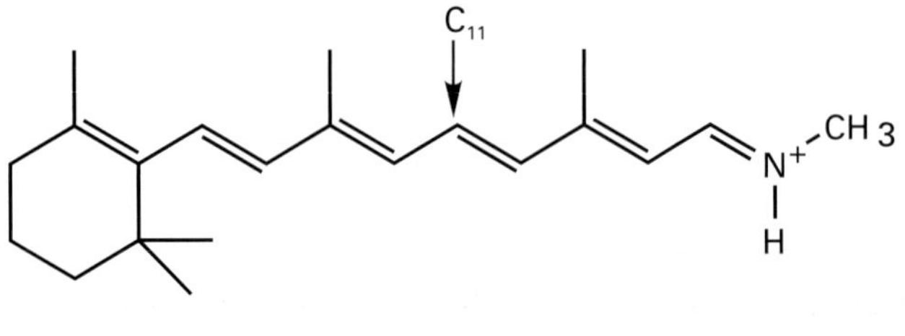

7

14.25‡ Ozone absorbs ultraviolet radiation in a part of the electromagnetic spectrum energetic enough to disrupt DNA in biological organisms and that is absorbed by no other abundant atmospheric constituent. This spectral range, denoted UV-B, spans the wavelengths of about 290 nm to 320 nm. The molar extinction coefficient of ozone over this range is given in the table below (W.B. DeMore, S.P. Sander, D.M. Golden, R.F. Hampson, M.J. Kurylo, C.J. Howard, A.R. Ravishankara, C.E. Kolb, and M.J. Molina, *Chemical kinetics and photochemical data for use in stratospheric modeling: Evaluation Number 11*, JPL Publication 94–26 (1994).

λ/nm	292.0	296.3	300.8	305.4	310.1	315.0	320.0
$\varepsilon/(dm^3\ mol^{-1}\ cm^{-1})$	1512	865	477	257	135.9	69.5	34.5

Compute the integrated absorption coefficient of ozone over the wavelength range 290–320 nm. (*Hint.* $\varepsilon(\tilde{\nu})$ can be fitted to an exponential function quite well.)

14.26‡ The abundance of ozone is typically inferred from measurements of UV absorption and is often expressed in terms of *Dobson units* (DU): 1 DU is equivalent to a layer of pure ozone 10^{-3} cm thick at 1 atm and 0°C. Compute the absorbance of UV radiation at 300 nm expected for an ozone abundance of 300 DU (a typical value) and 100 DU (a value reached during seasonal Antarctic ozone depletions) given a molar absorption coefficient of 476 $dm^3\ mol^{-1}\ cm^{-1}$.

14.27‡ G.C.G. Wachewsky, R. Horansky, and V. Vaida (*J. Phys. Chem.* **100**, 11559 (1996)) examined the UV absorption spectrum of CH_3I, a species of interest in connection with stratospheric ozone chemistry. They found the integrated absorption coefficient to be dependent on temperature and pressure to an extent inconsistent with internal structural changes in isolated CH_3I molecules; they explained the changes as due to dimerization of a substantial fraction of the CH_3I, a process that would naturally be pressure and temperature dependent. (a) Compute the integrated absorption coefficient over a triangular lineshape in the range 31 250 to 34 483 cm^{-1} and a maximal molar absorption coefficient of 150 $dm^3\ mol^{-1}\ cm^{-1}$ at 31 250 cm^{-1}. (b) Suppose 1 per cent of the CH_3I units in a sample at 2.4 Torr and 373 K exists as dimers. Compute the absorbance expected at 31 250 cm^{-1} in a sample cell of length 12.0 cm. (c) Suppose 18 per cent of the CH_3I units in a sample at 100 Torr and 373 K exists as dimers. Compute the absorbance expected at 31 250 cm^{-1} in a sample cell of length 12.0 cm; compute the molar absorption coefficient that would be inferred from this absorbance if dimerization was not considered.

14.28‡ The molecule Cl_2O_2 is believed to participate in the seasonal depletion of ozone over Antarctica. M. Schwell, H.-W. Jochims, B. Wassermann, U. Rockland, R. Flesch, and E. Rühl (*J. Phys. Chem.* **100**, 10070 (1996)) measured the ionization energies of Cl_2O_2 by photoelectron spectroscopy in which the ionized fragments were detected using a mass spectrometer. From their data, we can infer that the ionization enthalpy of Cl_2O_2 is 11.05 eV and the enthalpy of the dissociative ionization $Cl_2O_2 \rightarrow Cl + OClO^+ + e^-$ is 10.95 eV. They used this information to make some inferences about the structure of Cl_2O_2. Computational studies had suggested that the lowest energy isomer is ClOOCl, but that $ClClO_2$ (C_{2v}) and ClOClO are not very much higher in energy. The Cl_2O_2 in the photoionization step is the lowest energy isomer, whatever its structure may be, and its enthalpy of formation had previously been reported as +133 kJ mol^{-1}. The Cl_2O_2 in the dissociative ionization step is unlikely to be ClOOCl, for the product can be derived from it only with substantial rearrangement. Given $\Delta_f H^{\ominus}(OClO^+) = +1096$ kJ mol^{-1} and $\Delta_f H^{\ominus}(e^-) = 0$, determine whether the Cl_2O_2 in the dissociative ionization is the same as that in the photoionization. If different, how much greater is its $\Delta_f H^{\ominus}$? Are these results consistent with or contradictory to the computational studies?

14.29‡ One of the principal methods for obtaining the electronic spectra of unstable radicals is to study the spectra of comets, which are almost entirely due to radicals. Many radical spectra have been found in comets, including that due to CN. These radicals are produced in comets by the absorption of far ultraviolet solar radiation by their parent compounds. Subsequently, their fluorescence is excited by sunlight of longer wavelength. The spectra of comet Hale–Bopp (C/1995 O1) have been the subject of many recent studies. One such study is that of the fluorescence spectrum of CN in the comet at large heliocentric distances by R.M. Wagner and D.G. Schleicher (*Science* **275**, 1918 (1997)), in which the authors determine the spatial distribution and rate of production of CN in the coma. The (0–0) vibrational band is centred on 387.6 nm and the weaker (1–1) band with relative intensity 0.1 is centred on 386.4 nm. The band heads for (0–0) and (0–1) are known to be 388.3 and 421.6 nm, respectively. From these data, calculate the energy of the excited S_1 state relative to the ground S_0 state, the vibrational wavenumbers and the difference in the vibrational wavenumbers of the two states, and the relative populations of the $v = 0$ and $v = 1$ vibrational levels of the S_1 state. Also estimate the effective temperature of the molecule in the excited S_1 state. Only eight rotational levels of the S_1 state are thought to be populated. Is that observation consistent with the effective temperature of the S_1 state?

Molecular spectroscopy 3: magnetic resonance

15

One of the most widely used spectroscopic procedures in chemistry makes use of the classical concept of resonance. The chapter begins with an account of conventional nuclear magnetic resonance, which shows how the resonance frequency of a magnetic nucleus is affected by its electronic environment and the presence of magnetic nuclei in its vicinity. Then we turn to the modern versions of NMR, which are based on the use of pulses of electromagnetic radiation and the processing of the resulting signal by Fourier transform techniques. The experimental techniques for electron paramagnetic resonance resemble those used in the early days of NMR. The information obtained is very useful for the determination of the properties of species with unpaired electrons.

When two pendulums share a slightly flexible support and one is set in motion, the other is forced into oscillation by the motion of the common axle. As a result, energy flows between the two pendulums. The energy transfer occurs most efficiently when the frequencies of the two pendulums are identical. The condition of strong effective coupling when the frequencies of two oscillators are identical is called **resonance**. Resonance is the basis of a number of everyday phenomena, including the response of radios to the weak oscillations of the electromagnetic field generated by a distant transmitter. In this chapter we explore some spectroscopic applications that, as originally developed (and in some cases still), depend on matching a set of energy levels to a source of monochromatic radiation and observing the strong absorption that occurs at resonance.

The effect of magnetic fields on electrons and nuclei

The Stern–Gerlach experiment (Section 9.8) provided evidence for electron spin. It turns out that many nuclei also possess spin angular momentum. Orbital and spin angular momenta give rise to magnetic moments, and to say that electrons and nuclei have magnetic moments means that, to some extent, they behave like small bar magnets. First, we establish how the energies of electrons and nuclei depend on the strength of an external field. Then we see how to use this dependence to study the structure and dynamics of complex molecules.

15.1 The energies of electrons in magnetic fields

Classically, the energy of a magnetic moment $\boldsymbol{\mu}$ in a magnetic field $\boldsymbol{\mathcal{B}}$ is equal to the scalar product

Comment 15.1

More formally, $\mathcal{B}$ is the magnetic induction and is measured in tesla, T; $1\ \mathrm{T} = 1\ \mathrm{kg\ s^{-2}\ A^{-1}}$. The (non-SI) unit gauss, G, is also occasionally used: $1\ \mathrm{T} = 10^4\ \mathrm{G}$.

Comment 15.2

The scalar product (or 'dot product') of two vectors $\boldsymbol{a}$ and $\boldsymbol{b}$ is given by $\boldsymbol{a}\cdot\boldsymbol{b} = ab\cos\theta$, where a and b are the magnitudes of $\boldsymbol{a}$ and $\boldsymbol{b}$, respectively, and θ is the angle between them.

$$E = -\boldsymbol{\mu}\cdot\boldsymbol{\mathcal{B}} \tag{15.1}$$

Quantum mechanically, we write the hamiltonian as

$$\hat{H} = -\hat{\boldsymbol{\mu}}\cdot\boldsymbol{\mathcal{B}} \tag{15.2}$$

To write an expression for $\hat{\boldsymbol{\mu}}$, we recall from Section 9.8 that the magnetic moment is proportional to the angular momentum. For an electron possessing orbital angular momentum we write

$$\hat{\boldsymbol{\mu}} = \gamma_e\hat{\boldsymbol{l}} \qquad \text{and} \qquad \hat{H} = -\gamma_e\boldsymbol{\mathcal{B}}\cdot\hat{\boldsymbol{l}} \tag{15.3}$$

where $\hat{\boldsymbol{l}}$ is the orbital angular momentum operator and

$$\gamma_e = -\frac{e}{2m_e} \tag{15.4}$$

γ_e is called the **magnetogyric ratio** of the electron: The negative sign (arising from the sign of the electron's charge) shows that the orbital moment is antiparallel to its orbital angular momentum (as was depicted in Fig 10.26).

For a magnetic field $\mathcal{B}_0$ along the z-direction, eqn 15.3 becomes

$$\hat{\mu}_z = \gamma_e\hat{l}_z \qquad \text{and} \qquad \hat{H} = -\gamma_e\mathcal{B}_0\hat{l}_z = -\hat{\mu}_z\mathcal{B}_0 \tag{15.5a}$$

Because the eigenvalues of the operator $\hat{\boldsymbol{l}}_z$ are $m_l\hbar$, the z-component of the orbital magnetic moment and the energy of interaction are, respectively,

$$\mu_z = \gamma_e m_l\hbar \qquad \text{and} \qquad E = -\gamma_e m_l\hbar\mathcal{B}_0 = m_l\mu_B\mathcal{B}_0 \tag{15.5b}$$

where the **Bohr magneton**, μ_B, is

$$\mu_B = -\gamma_e\hbar = \frac{e\hbar}{2m_e} = 9.724\times10^{-24}\ \mathrm{J\ T^{-1}} \tag*{[15.6]}$$

The Bohr magneton is often regarded as the fundamental quantum of magnetic moment.

The spin magnetic moment of an electron, which has a spin quantum number $s = \frac{1}{2}$ (Section 9.8), is also proportional to its spin angular momentum. However, instead of eqn 15.3, the spin magnetic moment and hamiltonian operators are, respectively,

$$\hat{\boldsymbol{\mu}} = g_e\gamma_e\hat{\boldsymbol{s}} \qquad \text{and} \qquad \hat{H} = -g_e\gamma_e\boldsymbol{\mathcal{B}}\cdot\hat{\boldsymbol{s}} \tag{15.7}$$

where $\hat{\boldsymbol{s}}$ is the spin angular momentum operator and the extra factor g_e is called the **g-value** of the electron: $g_e = 2.002\,319\ldots$. The g-value arises from relativistic effects and from interactions of the electron with the electromagnetic fluctuations of the vacuum that surrounds the electron. For a magnetic field $\mathcal{B}_0$ in the z-direction,

$$\hat{\mu}_z = g_e\gamma_e\hat{s}_z \qquad \text{and} \qquad \hat{H} = -g_e\gamma_e\mathcal{B}_0\hat{s}_z \tag{15.8a}$$

Because the eigenvalues of the operator $\hat{s}_z$ are $m_s\hbar$ with $m_s = +\frac{1}{2}(\alpha)$ and $m_s = -\frac{1}{2}(\beta)$, it follows that the energies of an electron spin in a magnetic field are

$$\mu_z = g_e\gamma_e m_s\hbar \qquad \text{and} \qquad E_{m_s} = -g_e\gamma_e m_s\hbar\mathcal{B}_0 = g_e\mu_B m_s\mathcal{B}_0 \tag{15.8b}$$

with $m_s = \pm\frac{1}{2}$.

In the absence of a magnetic field, the states with different values of m_l and m_s are degenerate. When a field is present, the degeneracy is removed: the state with $m_s = +\frac{1}{2}$ moves up in energy by $\frac{1}{2}g_e\mu_B\mathcal{B}_0$ and the state with $m_s = -\frac{1}{2}$ moves down by $\frac{1}{2}g_e\mu_B\mathcal{B}_0$. The different energies arising from an interaction with an external field are sometimes represented on the vector model by picturing the vectors as **precessing**, or sweeping

Table 15.1 Nuclear constitution and the nuclear spin quantum number*

Number of protons	Number of neutrons	I
even	even	0
odd	odd	integer (1, 2, 3, . . .)
even	odd	half-integer ($\frac{1}{2}, \frac{3}{2}, \frac{5}{2}, \ldots$)
odd	even	half-integer ($\frac{1}{2}, \frac{3}{2}, \frac{5}{2}, \ldots$)

* The spin of a nucleus may be different if it is in an excited state; throughout this chapter we deal only with the ground state of nuclei.

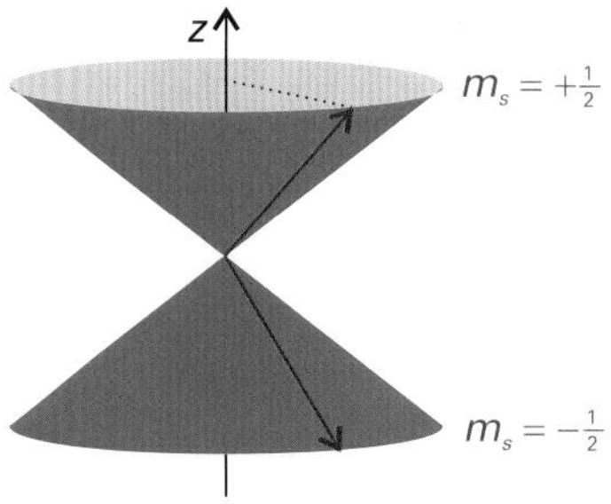

Fig. 15.1 The interactions between the m_s states of an electron and an external magnetic field may be visualized as the precession of the vectors representing the angular momentum.

round their cones (Fig. 15.1), with the rate of precession equal to the **Larmor frequency**, ν_L:

$$\nu_L = \frac{\gamma_e \mathcal{B}_0}{2\pi} \tag{15.9}$$

Equation 15.9 shows that the Larmor frequency increases with the strength of the magnetic field. For a field of 1 T, the Larmor frequency is 30 GHz.

15.2 The energies of nuclei in magnetic fields

The spin quantum number, I, of a nucleus is a fixed characteristic property of a nucleus and is either an integer or a half-integer (Table 15.1). A nucleus with spin quantum number I has the following properties:

1. An angular momentum of magnitude $\{I(I+1)\}^{1/2}\hbar$.
2. A component of angular momentum $m_I\hbar$ on a specified axis ('the z-axis'), where $m_I = I, I-1, \ldots, -I$.
3. If $I > 0$, a magnetic moment with a constant magnitude and an orientation that is determined by the value of m_I.

According to the second property, the spin, and hence the magnetic moment, of the nucleus may lie in $2I+1$ different orientations relative to an axis. A proton has $I = \frac{1}{2}$ and its spin may adopt either of two orientations; a ^{14}N nucleus has $I = 1$ and its spin may adopt any of three orientations; both ^{12}C and ^{16}O have $I = 0$ and hence zero magnetic moment.

The energy of interaction between a nucleus with a magnetic moment $\boldsymbol{\mu}$ and an external magnetic field $\boldsymbol{\mathcal{B}}$ may be calculated by using operators analogous to those of eqn 15.3:

$$\hat{\boldsymbol{\mu}} = \gamma \hat{\boldsymbol{I}} \quad \text{and} \quad H = -\gamma \boldsymbol{\mathcal{B}} \cdot \hat{\boldsymbol{I}} \tag{15.10a}$$

where γ is the magnetogyric ratio of the specified nucleus, an empirically determined characteristic arising from the internal structure of the nucleus (Table 15.2). The corresponding energies are

$$E_{m_I} = -\mu_z \mathcal{B}_0 = -\gamma\hbar\mathcal{B}_0 m_I \tag{15.10b}$$

As for electrons, the nuclear spin may be pictured as precessing around the direction of the applied field at a rate proportional to the applied field. For protons, a field of 1 T corresponds to a Larmor frequency (eqn 15.9, with γ_e replaced by γ) of about 40 MHz.

Synoptic table 15.2* Nuclear spin properties

Nuclide	Natural abundance/%	Spin I	g-factor, g_I	Magnetogyric ratio, $\gamma/(10^7\ \mathrm{T^{-1}\ s^{-1}})$	NMR frequency at 1 T, ν/MHz
^{1}n		$\frac{1}{2}$	−3.826	−18.32	29.165
^{1}H	99.98	$\frac{1}{2}$	5.586	26.75	42.577
^{2}H	0.02	1	0.857	4.10	6.536
^{13}C	1.11	$\frac{1}{2}$	1.405	6.73	10.705
^{14}N	99.64	1	0.404	1.93	3.076

* More values are given in the *Data section*.

The magnetic moment of a nucleus is sometimes expressed in terms of the **nuclear g-factor**, g_I, a characteristic of the nucleus, and the **nuclear magneton**, μ_N, a quantity independent of the nucleus, by using

$$\gamma\hbar = g_I\mu_N \qquad \mu_N = \frac{e\hbar}{2m_p} = 5.051\times10^{-27}\ \mathrm{J\ T^{-1}} \qquad [15.11]$$

where m_p is the mass of the proton. Nuclear g-factors vary between −6 and +6 (see Table 15.2): positive values of g_I and γ denote a magnetic moment that is parallel to the spin; negative values indicate that the magnetic moment and spin are antiparallel. For the remainder of this chapter we shall assume that γ is positive, as is the case for the majority of nuclei. In such cases, the states with negative values of m_I lie above states with positive values of m_I. The nuclear magneton is about 2000 times smaller than the Bohr magneton, so nuclear magnetic moments—and consequently the energies of interaction with magnetic fields—are about 2000 times weaker than the electron spin magnetic moment.

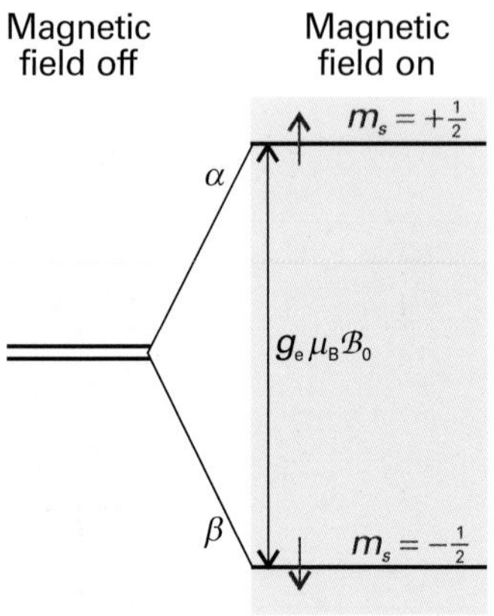

Fig. 15.2 Electron spin levels in a magnetic field. Note that the β state is lower in energy than the α state (because the magnetogyric ratio of an electron is negative). Resonance is achieved when the frequency of the incident radiation matches the frequency corresponding to the energy separation.

15.3 Magnetic resonance spectroscopy

In its original form, the magnetic resonance experiment is the resonant absorption of radiation by nuclei or unpaired electrons in a magnetic field. From eqn 15.8b, the separation between the $m_s = -\frac{1}{2}$ and $m_s = +\frac{1}{2}$ levels of an electron spin in a magnetic field $\mathcal{B}_0$ is

$$\Delta E = E_\alpha - E_\beta = g_e\mu_B\mathcal{B}_0 \qquad (15.12a)$$

If the sample is exposed to radiation of frequency ν, the energy separations come into resonance with the radiation when the frequency satisfies the **resonance condition** (Fig. 15.2):

$$h\nu = g_e\mu_B\mathcal{B}_0 \qquad (15.12b)$$

At resonance there is strong coupling between the electron spins and the radiation, and strong absorption occurs as the spins make the transition $\beta \rightarrow \alpha$. **Electron paramagnetic resonance** (EPR), or **electron spin resonance** (ESR), is the study of molecules and ions containing unpaired electrons by observing the magnetic fields at which they come into resonance with monochromatic radiation. Magnetic fields of about 0.3 T (the value used in most commercial EPR spectrometers) correspond to resonance with an electromagnetic field of frequency 10 GHz (10^{10} Hz) and wavelength 3 cm. Because 3 cm radiation falls in the microwave region of the electromagnetic spectrum, EPR is a microwave technique.

The energy separation between the $m_I = +\frac{1}{2}$ ($\uparrow$ or α) and the $m_I = -\frac{1}{2}$ ($\downarrow$ or β) states of **spin-$\frac{1}{2}$ nuclei**, which are nuclei with $I = \frac{1}{2}$, is

$$\Delta E = E_\beta - E_\alpha = \tfrac{1}{2}\gamma\hbar\mathcal{B}_0 - (-\tfrac{1}{2}\gamma\hbar\mathcal{B}_0) = \gamma\hbar\mathcal{B}_0 \quad (15.13a)$$

and resonant absorption occurs when the resonance condition (Fig. 15.3)

$$h\nu = \gamma\hbar\mathcal{B}_0 \quad (15.13b)$$

is fulfilled. Because $\gamma\hbar\mathcal{B}_0/h$ is the Larmor frequency of the nucleus, this resonance occurs when the frequency of the electromagnetic field matches the Larmor frequency ($\nu = \nu_L$). In its simplest form, **nuclear magnetic resonance** (NMR) is the study of the properties of molecules containing magnetic nuclei by applying a magnetic field and observing the frequency of the resonant electromagnetic field. Larmor frequencies of nuclei at the fields normally employed (about 12 T) typically lie in the radiofrequency region of the electromagnetic spectrum (close to 500 MHz), so NMR is a radiofrequency technique.

For much of this chapter we consider spin-$\frac{1}{2}$ nuclei, but NMR is applicable to nuclei with any non-zero spin. As well as protons, which are the most common nuclei studied by NMR, spin-$\frac{1}{2}$ nuclei include ^{13}C, ^{19}F, and ^{31}P.

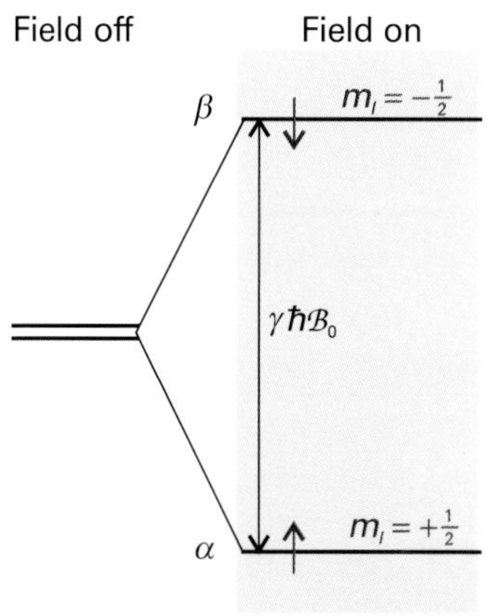

Fig. 15.3 The nuclear spin energy levels of a spin-$\frac{1}{2}$ nucleus with positive magnetogyric ratio (for example, ^1H or ^{13}C) in a magnetic field. Resonance occurs when the energy separation of the levels matches the energy of the photons in the electromagnetic field.

Nuclear magnetic resonance

Although the NMR technique is simple in concept, NMR spectra can be highly complex. However, they have proved invaluable in chemistry, for they reveal so much structural information. A magnetic nucleus is a very sensitive, non-invasive probe of the surrounding electronic structure.

15.4 The NMR spectrometer

An NMR spectrometer consists of the appropriate sources of radiofrequency electromagnetic radiation and a magnet that can produce a uniform, intense field. In simple instruments, the magnetic field is provided by a permanent magnet. For serious work, a **superconducting magnet** capable of producing fields of the order of 10 T and more is used (Fig. 15.4). The sample is placed in the cylindrically wound magnet. In some cases the sample is rotated rapidly to average out magnetic inhomogeneities. However, sample spinning can lead to irreproducible results, and is often avoided. Although a superconducting magnet operates at the temperature of liquid helium (4 K), the sample itself is normally at room temperature.

The intensity of an NMR transition depends on a number of factors. We show in the following *Justification* that

$$\textit{Intensity} \propto (N_\alpha - N_\beta)\mathcal{B}_0 \quad (15.14a)$$

where

$$N_\alpha - N_\beta \approx \frac{N\gamma\hbar\mathcal{B}_0}{2kT} \quad (15.14b)$$

with N the total number of spins ($N = N_\alpha + N_\beta$). It follows that decreasing the temperature increases the intensity by increasing the population difference. By combining these two equations we see that the intensity is proportional to $\mathcal{B}_0^2$, so NMR transitions can be enhanced significantly by increasing the strength of the applied magnetic field. We shall also see (Section 15.6) that the use of high magnetic fields simplifies the

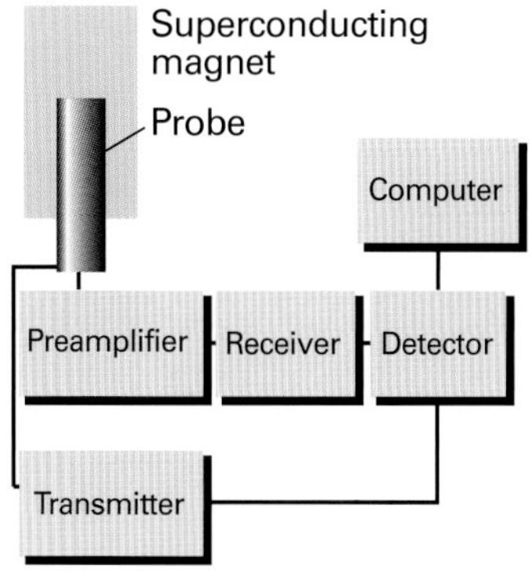

Fig. 15.4 The layout of a typical NMR spectrometer. The link from the transmitter to the detector indicates that the high frequency of the transmitter is subtracted from the high frequency received signal to give a low frequency signal for processing.

appearance of spectra and so allows them to be interpreted more readily. We also conclude that absorptions of nuclei with large magnetogyric ratios (^{1}H, for instance) are more intense than those with small magnetogyric ratios (^{13}C, for instance)

Justification 15.1 *Intensities in NMR spectra*

From the general considerations of transition intensities in Section 13.2, we know that the rate of absorption of electromagnetic radiation is proportional to the population of the lower energy state (N_α in the case of a proton NMR transition) and the rate of stimulated emission is proportional to the population of the upper state (N_β). At the low frequencies typical of magnetic resonance, we can neglect spontaneous emission as it is very slow. Therefore, the net rate of absorption is proportional to the difference in populations, and we can write

$$\text{Rate of absorption} \propto N_\alpha - N_\beta$$

The intensity of absorption, the rate at which energy is absorbed, is proportional to the product of the rate of absorption (the rate at which photons are absorbed) and the energy of each photon, and the latter is proportional to the frequency ν of the incident radiation (through $E = h\nu$). At resonance, this frequency is proportional to the applied magnetic field (through $\nu = \nu_L = \gamma\mathcal{B}_0/2\pi$), so we can write

$$\text{Intensity of absorption} \propto (N_\alpha - N_\beta)\mathcal{B}_0$$

as in eqn 15.14a. To write an expression for the population difference, we use the Boltzmann distribution (*Molecular interpretation* 3.1) to write the ratio of populations as

$$\frac{N_\beta}{N_\alpha} = e^{-\Delta E/kT} \approx 1 - \frac{\Delta E}{kT} = 1 - \frac{\gamma\mathcal{B}_0}{kT}$$

Comment 15.3

The expansion of an exponential function used here is $e^{-x} = 1 - x + \frac{1}{2}x^2 - \dots$. If $x \ll 1$, then $e^{-x} \approx 1 - x$.

where $\Delta E = E_\beta - E_\alpha$. The expansion of the exponential term is appropriate for $\Delta E \ll kT$, a condition usually met for nuclear spins. It follows after rearrangement that

$$\frac{N_\alpha - N_\beta}{N_\alpha + N_\beta} = \frac{N_\alpha(1 - N_\beta/N_\alpha)}{N_\alpha(1 + N_\beta/N_\alpha)} = \frac{1 - N_\beta/N_\alpha}{1 + N_\beta/N_\alpha}$$

$$\approx \frac{1 - (1 - \gamma\hbar\mathcal{B}_0/kT)}{1 + (1 - \gamma\hbar\mathcal{B}_0/kT)} \approx \frac{\gamma\hbar\mathcal{B}_0/kT}{2}$$

Then, with $N_\alpha + N_\beta = N$, the total number of spins, we obtain eqn 15.14b.

15.5 The chemical shift

Nuclear magnetic moments interact with the *local* magnetic field. The local field may differ from the applied field because the latter induces electronic orbital angular momentum (that is, the circulation of electronic currents) which gives rise to a small additional magnetic field $\delta\mathcal{B}$ at the nuclei. This additional field is proportional to the applied field, and it is conventional to write

$$\delta\mathcal{B} = -\sigma\mathcal{B}_0 \qquad [15.15]$$

where the dimensionless quantity σ is called the **shielding constant** of the nucleus (σ is usually positive but may be negative). The ability of the applied field to induce an electronic current in the molecule, and hence affect the strength of the resulting local magnetic field experienced by the nucleus, depends on the details of the electronic structure near the magnetic nucleus of interest, so nuclei in different chemical groups have different shielding constants. The calculation of reliable values of the shielding constant is very difficult, but trends in it are quite well understood and we concentrate on them.

(a) The δ scale of chemical shifts

Because the total local field is

$$\mathcal{B}_{loc} = \mathcal{B}_0 + \delta\mathcal{B} = (1 - \sigma)\mathcal{B}_0 \tag{15.16}$$

the nuclear Larmor frequency is

$$\nu_L = \frac{\gamma\mathcal{B}_{loc}}{2\pi} = (1 - \sigma)\frac{\gamma\mathcal{B}_0}{2\pi} \tag{15.17}$$

This frequency is different for nuclei in different environments. Hence, different nuclei, even of the same element, come into resonance at different frequencies.

It is conventional to express the resonance frequencies in terms of an empirical quantity called the **chemical shift**, which is related to the difference between the resonance frequency, ν, of the nucleus in question and that of a reference standard, ν°:

$$\delta = \frac{\nu - \nu^\circ}{\nu^\circ} \times 10^6 \qquad [15.18]$$

The standard for protons is the proton resonance in tetramethylsilane ($Si(CH_3)_4$, commonly referred to as TMS), which bristles with protons and dissolves without reaction in many liquids. Other references are used for other nuclei. For ^{13}C, the reference frequency is the ^{13}C resonance in TMS; for ^{31}P it is the ^{31}P resonance in 85 per cent H_3PO_4(aq). The advantage of the δ-scale is that shifts reported on it are independent of the applied field (because both numerator and denominator are proportional to the applied field).

Illustration 15.1 *Using the chemical shift*

From eqn 15.18,

$$\nu - \nu^\circ = \nu^\circ\delta \times 10^{-6}$$

A nucleus with $\delta = 1.00$ in a spectrometer operating at 500 MHz will have a shift relative to the reference equal to

$$\nu - \nu^\circ = (500\text{ MHz}) \times 1.00 \times 10^{-6} = 500\text{ Hz}$$

In a spectrometer operating at 100 MHz, the shift relative to the reference would be only 100 Hz.

A note on good practice In much of the literature, chemical shifts are reported in 'parts per million', ppm, in recognition of the factor of 10^6 in the definition. This practice is unnecessary.

The relation between δ and σ is obtained by substituting eqn 15.17 into eqn 15.18:

$$\delta = \frac{(1-\sigma)\mathcal{B}_0 - (1-\sigma^\circ)\mathcal{B}_0}{(1-\sigma^\circ)\mathcal{B}_0} \times 10^6 = \frac{\sigma^\circ - \sigma}{1 - \sigma^\circ} \times 10^6 \approx (\sigma^\circ - \sigma) \times 10^6 \tag{15.19}$$

As the shielding, σ, gets smaller, δ *increases*. Therefore, we speak of nuclei with large chemical shift as being strongly **deshielded**. Some typical chemical shifts are given in Fig. 15.5. As can be seen from the illustration, the nuclei of different elements have

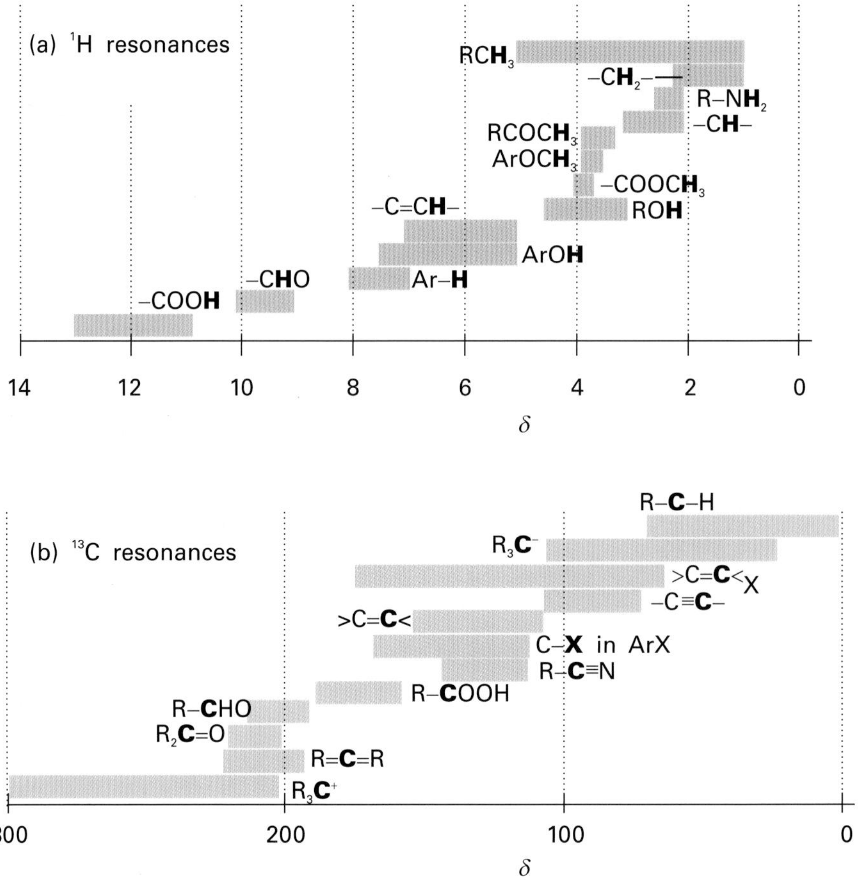

Fig. 15.5 The range of typical chemical shifts for (a) ^1H resonances and (b) ^{13}C resonances.

very different ranges of chemical shifts. The ranges exhibit the variety of electronic environments of the nuclei in molecules: the heavier the element, the greater the number of electrons around the nucleus and hence the greater the range of shieldings.

By convention, NMR spectra are plotted with δ increasing from right to left. Consequently, in a given applied magnetic field the Larmor frequency also increases from right to left. In the original continuous wave (CW) spectrometers, in which the radiofrequency was held constant and the magnetic field varied (a 'field sweep experiment'), the spectrum was displayed with the applied magnetic field increasing from left to right: a nucleus with a small chemical shift experiences a relatively low local magnetic field, so it needs a higher applied magnetic field to bring it into resonance with the radiofrequency field. Consequently, the right-hand (low chemical shift) end of the spectrum became known as the 'high field end' of the spectrum.

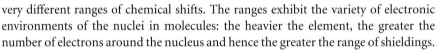

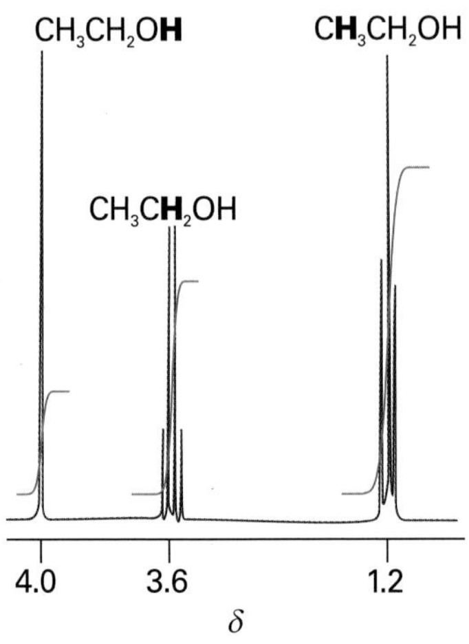

Fig. 15.6 The ^1H-NMR spectrum of ethanol. The bold letters denote the protons giving rise to the resonance peak, and the step-like curve is the integrated signal.

(b) Resonance of different groups of nuclei

The existence of a chemical shift explains the general features of the spectrum of ethanol shown in Fig.15.6. The CH_3 protons form one group of nuclei with $\delta \approx 1$. The two CH_2 protons are in a different part of the molecule, experience a different local magnetic field, and resonate at $\delta \approx 3$. Finally, the OH proton is in another environment, and has a chemical shift of $\delta \approx 4$. The increasing value of δ (that is, the decrease in shielding) is consistent with the electron-withdrawing power of the O atom: it reduces the electron density of the OH proton most, and that proton is strongly deshielded. It reduces the electron density of the distant methyl protons least, and those nuclei are least deshielded.

The relative intensities of the signals (the areas under the absorption lines) can be used to help distinguish which group of lines corresponds to which chemical group. The determination of the area under an absorption line is referred to as the **integration** of the signal (just as any area under a curve may be determined by mathematical integration). Spectrometers can integrate the absorption automatically (as indicated in Fig. 15.6). In ethanol the group intensities are in the ratio 3:2:1 because there are three CH_3 protons, two CH_2 protons, and one OH proton in each molecule. Counting the number of magnetic nuclei as well as noting their chemical shifts helps to identify a compound present in a sample.

(c) The origin of shielding constants

The calculation of shielding constants is difficult, even for small molecules, for it requires detailed information about the distribution of electron density in the ground and excited states and the excitation energies of the molecule. Nevertheless, considerable success has been achieved with the calculation for diatomic molecules and small molecules such as H_2O and CH_4 and even large molecules, such as proteins, are within the scope of some types of calculation. Nevertheless, it is easier to understand the different contributions to chemical shifts by studying the large body of empirical information now available for large molecules.

The empirical approach supposes that the observed shielding constant is the sum of three contributions:

$$\sigma = \sigma(\text{local}) + \sigma(\text{neighbour}) + \sigma(\text{solvent}) \qquad (15.20)$$

The **local contribution**, σ(local), is essentially the contribution of the electrons of the atom that contains the nucleus in question. The **neighbouring group contribution**, σ(neighbour), is the contribution from the groups of atoms that form the rest of the molecule. The **solvent contribution**, σ(solvent), is the contribution from the solvent molecules.

(d) The local contribution

It is convenient to regard the local contribution to the shielding constant as the sum of a **diamagnetic contribution**, σ_d, and a **paramagnetic contribution**, σ_p:

$$\sigma(\text{local}) = \sigma_d + \sigma_p \qquad (15.21)$$

A diamagnetic contribution to σ(local) opposes the applied magnetic field and shields the nucleus in question. A paramagnetic contribution to σ(local) reinforces the applied magnetic field and deshields the nucleus in question. Therefore, $\sigma_d > 0$ and $\sigma_p < 0$. The total local contribution is positive if the diamagnetic contribution dominates, and is negative if the paramagnetic contribution dominates.

The diamagnetic contribution arises from the ability of the applied field to generate a circulation of charge in the ground-state electron distribution of the atom. The circulation generates a magnetic field that opposes the applied field and hence shields the nucleus. The magnitude of σ_d depends on the electron density close to the nucleus and can be calculated from the **Lamb formula** (see *Further reading* for a derivation):

$$\sigma_d = \frac{e^2\mu_0}{12\pi m_e}\left\langle \frac{1}{r} \right\rangle \qquad (15.22)$$

where μ_0 is the vacuum permeability (a fundamental constant, see inside the front cover) and r is the electron–nucleus distance.

Illustration 15.2 *Calculating the diamagnetic contribution to the chemical shift of a proton*

To calculate σ_d for the proton in a free H atom, we need to calculate the expectation value of $1/r$ for a hydrogen 1*s* orbital. Wavefunctions are given in Table 10.1, and a useful integral is given in Example 8.7. Because $d\tau = r^2 dr \sin\theta\, d\theta d\phi$, we can write

$$\left\langle \frac{1}{r} \right\rangle = \int \frac{\psi^*\psi}{r} d\tau = \frac{1}{\pi a_0^3} \int_0^{2\pi} d\phi \int_0^{\pi} \sin\theta d\theta \int_0^{\infty} re^{-2r/a_0} dr = \frac{4}{a_0^3} \int_0^{\infty} re^{-2r/a_0} dr = \frac{1}{a_0}$$

Therefore,

$$\sigma_d = \frac{e^2\mu_0}{12\pi m_e a_0}$$

With the values of the fundamental constants inside the front cover, this expression evaluates to 1.78×10^{-5}.

The diamagnetic contribution is the only contribution in closed-shell free atoms. It is also the only contribution to the local shielding for electron distributions that have spherical or cylindrical symmetry. Thus, it is the only contribution to the local shielding from inner cores of atoms, for cores remain spherical even though the atom may be a component of a molecule and its valence electron distribution highly distorted. The diamagnetic contribution is broadly proportional to the electron density of the atom containing the nucleus of interest. It follows that the shielding is decreased if the electron density on the atom is reduced by the influence of an electronegative atom nearby. That reduction in shielding translates into an increase in deshielding, and hence to an increase in the chemical shift δ as the electronegativity of a neighbouring atom increases (Fig. 15.7). That is, as the electronegativity increases, δ decreases.

The local paramagnetic contribution, σ_p, arises from the ability of the applied field to force electrons to circulate through the molecule by making use of orbitals that are unoccupied in the ground state. It is zero in free atoms and around the axes of linear molecules (such as ethyne, HC≡CH) where the electrons can circulate freely and a field applied along the internuclear axis is unable to force them into other orbitals. We can expect large paramagnetic contributions from small atoms in molecules with low-lying excited states. In fact, the paramagnetic contribution is the dominant local contribution for atoms other than hydrogen.

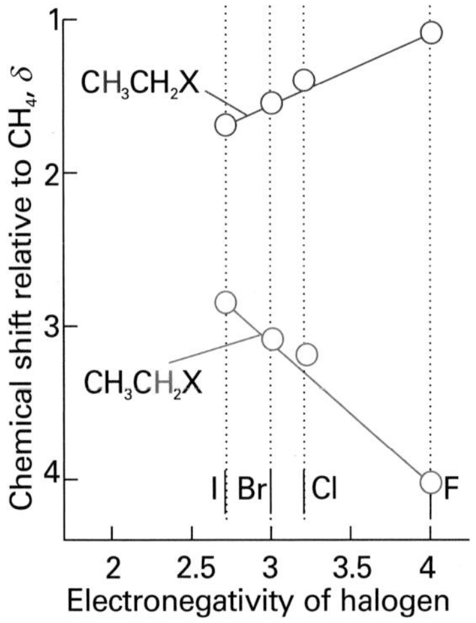

Fig. 15.7 The variation of chemical shielding with electronegativity. The shifts for the methylene protons agree with the trend expected with increasing electronegativity. However, to emphasize that chemical shifts are subtle phenomena, notice that the trend for the methyl protons is opposite to that expected. For these protons another contribution (the magnetic anisotropy of C—H and C—X bonds) is dominant.

(e) Neighbouring group contributions

The neighbouring group contribution arises from the currents induced in nearby groups of atoms. Consider the influence of the neighbouring group X on the proton H in a molecule such as H—X. The applied field generates currents in the electron distribution of X and gives rise to an induced magnetic moment proportional to the applied field; the constant of proportionality is the magnetic susceptibility, χ (chi), of the group X. The proton H is affected by this induced magnetic moment in two ways. First, the strength of the additional magnetic field the proton experiences is inversely proportional to the cube of the distance *r* between H and X. Second, the field at H depends on the anisotropy of the magnetic susceptibility of X, the variation of χ with the angle that X makes to the applied field. We assume that the magnetic susceptibility of X has two components, $\chi_\parallel$ and $\chi_\perp$, which are parallel and perpendicular to the axis of symmetry of X, respectively. The axis of symmetry of X makes an angle θ to the

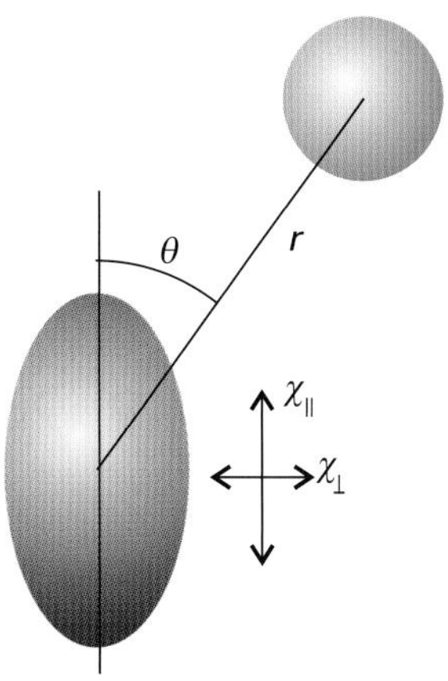

1

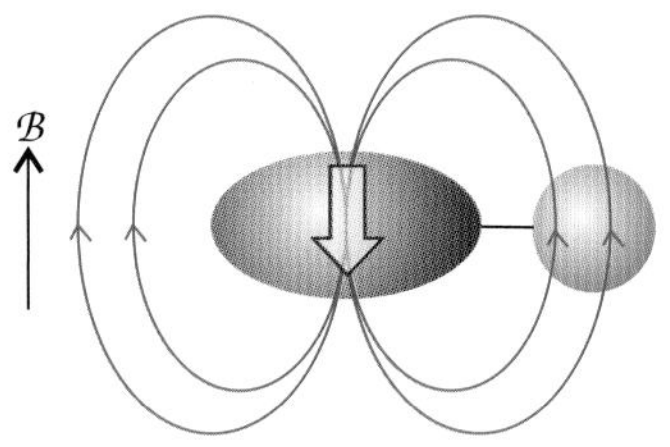

2

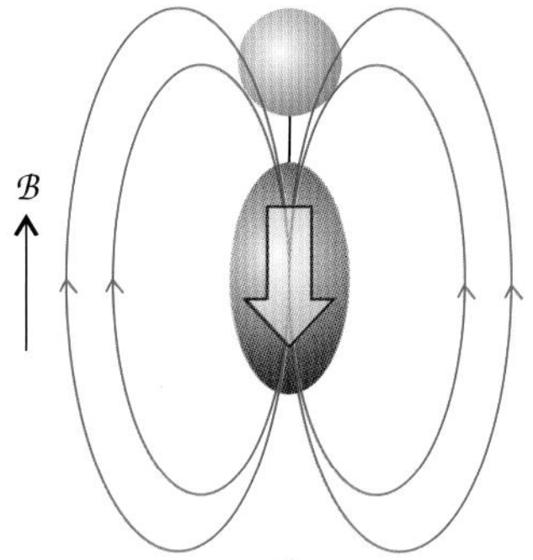

3

vector connecting X to H (**1**, where X is represented by the ellipse and H is represented by the circle).

To examine the effect of anisotropy of the magnetic susceptibility of X on the shielding constant, consider the case $\theta = 0$ for a molecule H—X that is free to tumble (**2** and **3**). Some of the time the H—X axis will be perpendicular to the applied field and then only $\chi_\perp$ will contribute to the induced magnetic moment that shields X from the applied field. The result is deshielding of the proton H, or $\sigma(\text{neighbour}) < 0$ (**2**). When the applied field is parallel to the H—X axis, only $\chi_\parallel$ contributes to the induced magnetic moment at X. The result is shielding of the proton H (**3**). We conclude that, as the molecule tumbles and the H—X axis takes all possible angles with respect to the applied field, the effects of anisotropic magnetic susceptibility do not average to zero because $\chi_\parallel \neq \chi_\perp$.

Self-test 15.1 For a tumbling H—X molecule, show that when $\theta = 90°$: (a) contributions from the $\chi_\perp$ component lead to shielding of H, or $\sigma(\text{neighbour}) > 0$, and (b) contributions from the $\chi_\parallel$ component lead to deshielding of H, or $\sigma(\text{neighbour}) < 0$. Comparison between the $\theta = 0$ and $\theta = 90°$ cases shows that the patterns of shielding and deshielding by neighbouring groups depend not only on differences between $\chi_\parallel$ and $\chi_\perp$, but also the angle θ.

[Draw diagrams similar to **2** and **3** where the $\chi_\perp$ component is parallel to the H—X axis and then analyse the problem as above.]

To a good approximation, the shielding constant $\sigma(\text{neighbour})$ depends on the distance r, the difference $\chi_\parallel - \chi_\perp$, as (see *Further reading* for a derivation)

$$\sigma(\text{neighbour}) \propto (\chi_\parallel - \chi_\perp)\left(\frac{1 - 3\cos^2\theta}{r^3}\right) \qquad (15.23)$$

where $\chi_\parallel$ and $\chi_\perp$ are both negative for a diamagnetic group X. Equation 15.23 shows that the neighbouring group contribution may be positive or negative according to the relative magnitudes of the two magnetic susceptibilities and the relative orientation of the nucleus with respect to X. The latter effect is easy to anticipate: if $54.7° < \theta < 125.3°$, then $1 - 3\cos^2\theta$ is positive, but it is negative otherwise (Fig. 15.8).

A special case of a neighbouring group effect is found in aromatic compounds. The strong anisotropy of the magnetic susceptibility of the benzene ring is ascribed to the

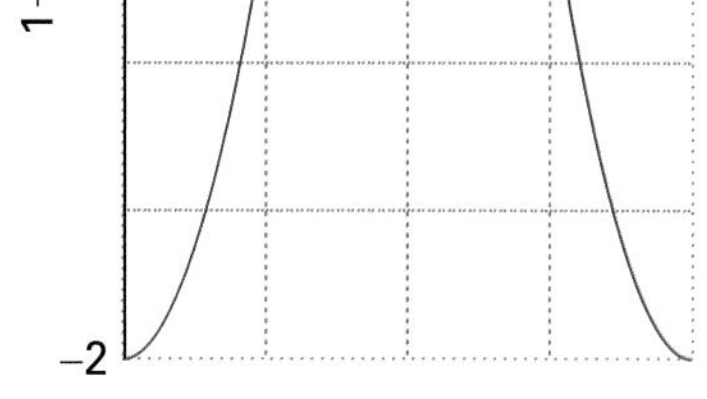

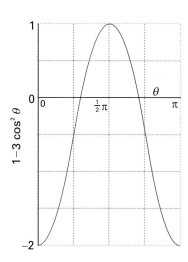

Fig. 15.8 The variation of the function $1 - 3\cos^2\theta$ with the angle θ.

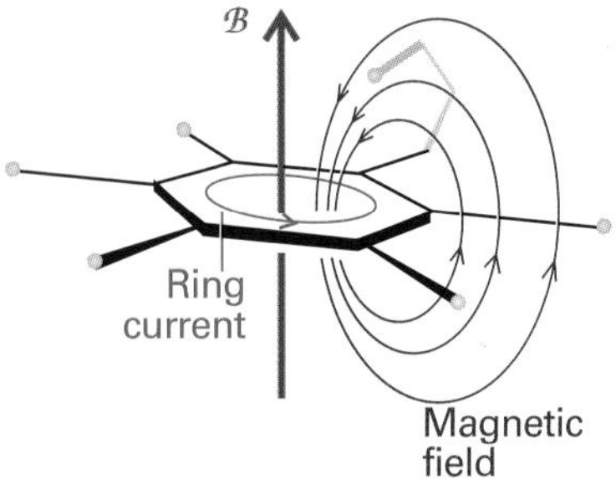

Fig. 15.9 The shielding and deshielding effects of the ring current induced in the benzene ring by the applied field. Protons attached to the ring are deshielded but a proton attached to a substituent that projects above the ring is shielded.

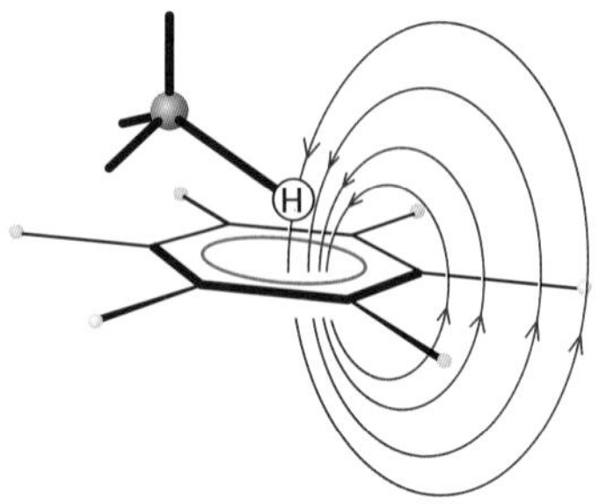

Fig. 15.10 An aromatic solvent (benzene here) can give rise to local currents that shield or deshield a proton in a solute molecule. In this relative orientation of the solvent and solute, the proton on the solute molecule is shielded.

ability of the field to induce a **ring current**, a circulation of electrons around the ring, when it is applied perpendicular to the molecular plane. Protons in the plane are deshielded (Fig. 15.9), but any that happen to lie above or below the plane (as members of substituents of the ring) are shielded.

(f) The solvent contribution

A solvent can influence the local magnetic field experienced by a nucleus in a variety of ways. Some of these effects arise from specific interactions between the solute and the solvent (such as hydrogen-bond formation and other forms of Lewis acid–base complex formation). The anisotropy of the magnetic susceptibility of the solvent molecules, especially if they are aromatic, can also be the source of a local magnetic field. Moreover, if there are steric interactions that result in a loose but specific interaction between a solute molecule and a solvent molecule, then protons in the solute molecule may experience shielding or deshielding effects according to their location relative to the solvent molecule (Fig. 15.10). We shall see that the NMR spectra of species that contain protons with widely different chemical shifts are easier to interpret than those in which the shifts are similar, so the appropriate choice of solvent may help to simplify the appearance and interpretation of a spectrum.

15.6 The fine structure

The splitting of resonances into individual lines in Fig. 15.6 is called the **fine structure** of the spectrum. It arises because each magnetic nucleus may contribute to the local field experienced by the other nuclei and so modify their resonance frequencies. The strength of the interaction is expressed in terms of the **scalar coupling constant**, J, and reported in hertz (Hz). The scalar coupling constant is so called because the energy of interaction it describes is proportional to the scalar product of the two interacting spins: $E \propto \boldsymbol{I}_1 \cdot \boldsymbol{I}_2$. The constant of proportionality in this expression is $hJ/\hbar^2$, because each angular momentum is proportional to $\hbar$.

Spin coupling constants are independent of the strength of the applied field because they do not depend on the latter for their ability to generate local fields. If the resonance line of a particular nucleus is split by a certain amount by a second nucleus, then the resonance line of the second nucleus is split by the first to the same extent.

(a) The energy levels of coupled systems

It will be useful for later discussions to consider an NMR spectrum in terms of the energy levels of the nuclei and the transitions between them. In NMR, letters far apart in the alphabet (typically A and X) are used to indicate nuclei with very different chemical shifts; letters close together (such as A and B) are used for nuclei with similar chemical shifts. We shall consider first an AX system, a molecule that contains two spin-$\frac{1}{2}$ nuclei A and X with very different chemical shifts in the sense that the difference in chemical shift corresponds to a frequency that is large compared to their spin–spin coupling.

The energy level diagram for a single spin-$\frac{1}{2}$ nucleus and its single transition were shown in Fig. 15.3, and nothing more needs to be said. For a spin-$\frac{1}{2}$ AX system there are four spin states:

$$\alpha_A\alpha_X \qquad \alpha_A\beta_X \qquad \beta_A\alpha_X \quad \beta_A\beta_X$$

The energy depends on the orientation of the spins in the external magnetic field, and if spin–spin coupling is neglected

$$E = -\gamma\hbar(1-\sigma_A)\mathcal{B}m_A - \gamma\hbar(1-\sigma_X)\mathcal{B}m_X = -h\nu_A m_A - h\nu_X m_X \tag{15.24}$$

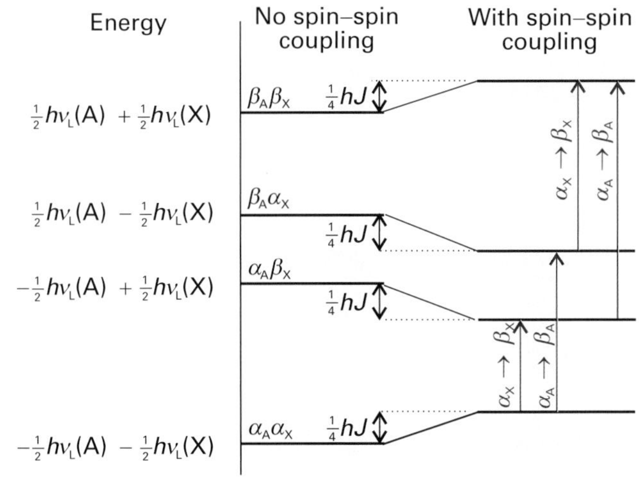

Fig. 15.11 The energy levels of an AX system. The four levels on the left are those of the two spins in the absence of spin–spin coupling. The four levels on the right show how a positive spin–spin coupling constant affects the energies. The transitions shown are for $\beta \leftarrow \alpha$ of A or X, the other nucleus (X or A, respectively) remaining unchanged. We have exaggerated the effect for clarity; in practice, the splitting caused by spin–spin coupling is much smaller than that caused by the applied field.

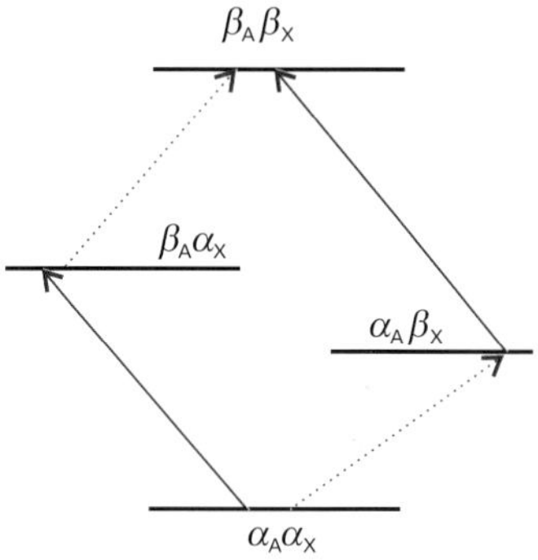

Fig. 15.12 An alternative depiction of the energy levels and transitions shown in Fig. 15.11. Once again, we have exaggerated the effect of spin–spin coupling.

where ν_A and ν_X are the Larmor frequencies of A and X and m_A and m_X are their quantum numbers. This expression gives the four lines on the left of Fig. 15.11. The spin–spin coupling depends on the relative orientation of the two nuclear spins, so it is proportional to the product $m_A m_X$. Therefore, the energy including spin–spin coupling is

$$E = -h\nu_A m_A - h\nu_X m_X + hJm_A m_X \tag{15.25}$$

If $J > 0$, a lower energy is obtained when $m_A m_X < 0$, which is the case if one spin is α and the other is β. A higher energy is obtained if both spins are α or both spins are β. The opposite is true if $J < 0$. The resulting energy level diagram (for $J > 0$) is shown on the right of Fig. 15.11. We see that the $\alpha\alpha$ and $\beta\beta$ states are both raised by $\frac{1}{4}hJ$ and that the $\alpha\beta$ and $\beta\alpha$ states are both lowered by $\frac{1}{4}hJ$.

When a transition of nucleus A occurs, nucleus X remains unchanged. Therefore, the A resonance is a transition for which $\Delta m_A = +1$ and $\Delta m_X = 0$. There are two such transitions, one in which $\beta_A \leftarrow \alpha_A$ occurs when the X nucleus is α_X, and the other in which $\beta_A \leftarrow \alpha_A$ occurs when the X nucleus is β_X. They are shown in Fig. 15.11 and in a slightly different form in Fig. 15.12. The energies of the transitions are

$$\Delta E = h\nu_A \pm \tfrac{1}{2}hJ \tag{15.26a}$$

Therefore, the A resonance consists of a doublet of separation J centred on the chemical shift of A (Fig. 15.13). Similar remarks apply to the X resonance, which consists of two transitions according to whether the A nucleus is α or β (as shown in Fig. 15.12). The transition energies are

$$\Delta E = h\nu_X \pm \tfrac{1}{2}hJ \tag{15.26b}$$

It follows that the X resonance also consists of two lines of separation J, but they are centred on the chemical shift of X (as shown in Fig. 15.13).

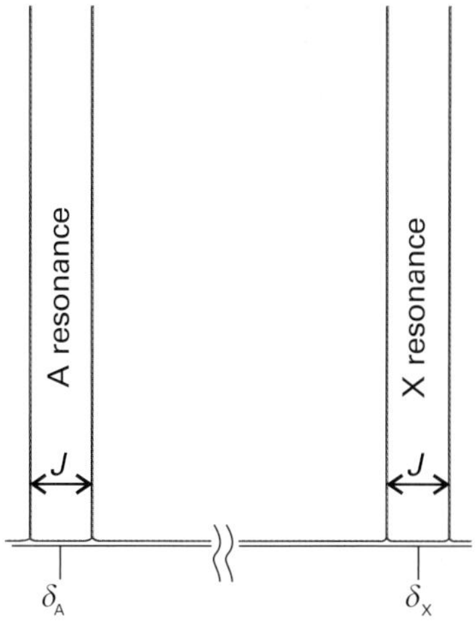

Fig. 15.13 The effect of spin–spin coupling on an AX spectrum. Each resonance is split into two lines separated by J. The pairs of resonances are centred on the chemical shifts of the protons in the absence of spin–spin coupling.

(b) Patterns of coupling

We have seen that in an AX system, spin–spin coupling will result in four lines in the NMR spectrum. Instead of a single line from A, we get a doublet of lines separated by J and centred on the chemical shift characteristic of A. The same splitting occurs in the X resonance: instead of a single line, the resonance is a doublet with splitting J (the same value as for the splitting of A) centred on the chemical shift characteristic of X. These features are summarized in Fig. 15.13.

A subtle point is that the X resonance in an AX_n species (such as an AX_2 or AX_3 species) is also a doublet with splitting J. As we shall explain below, *a group of equivalent nuclei resonates like a single nucleus*. The only difference for the X resonance of an AX_n species is that the intensity is n times as great as that of an AX species (Fig. 15.14). The A resonance in an AX_n species, though, is quite different from the A resonance in an AX species. For example, consider an AX_2 species with two equivalent X nuclei. The resonance of A is split into a doublet of separation J by one X, and each line of that doublet is split again by the same amount by the second X (Fig. 15.15). This splitting results in three lines in the intensity ratio 1:2:1 (because the central frequency can be obtained in two ways). The A resonance of an A_nX_2 species would also be a 1:2:1 triplet of splitting J, the only difference being that the intensity of the A resonance would be n times as great as that of AX_2.

Three equivalent X nuclei (an AX_3 species) split the resonance of A into four lines of intensity ratio 1:3:3:1 and separation J (Fig. 15.16). The X resonance, though, is still a doublet of separation J. In general, n equivalent spin-$\frac{1}{2}$ nuclei split the resonance of a nearby spin or group of equivalent spins into $n + 1$ lines with an intensity distribution given by 'Pascal's triangle' in which each entry is the sum of the two entries immediately above (**4**). The easiest way of constructing the pattern of fine structure is to draw a diagram in which successive rows show the splitting of a subsequent proton. The procedure is illustrated in Fig. 15.17 and was used in Figs. 15.15 and 15.16. It is easily extended to molecules containing nuclei with $I > \frac{1}{2}$ (Fig. 15.18).

1
1 1
1 2 1
1 3 3 1
1 4 6 4 1
1 5 10 10 5 1

4

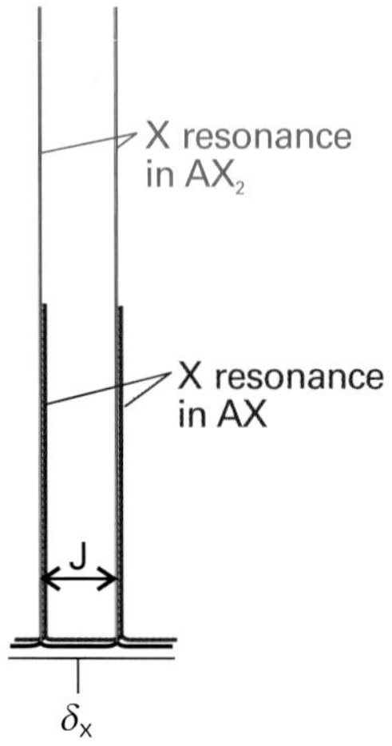

Fig. 15.14 The X resonance of an AX_2 species is also a doublet, because the two equivalent X nuclei behave like a single nucleus; however, the overall absorption is twice as intense as that of an AX species.

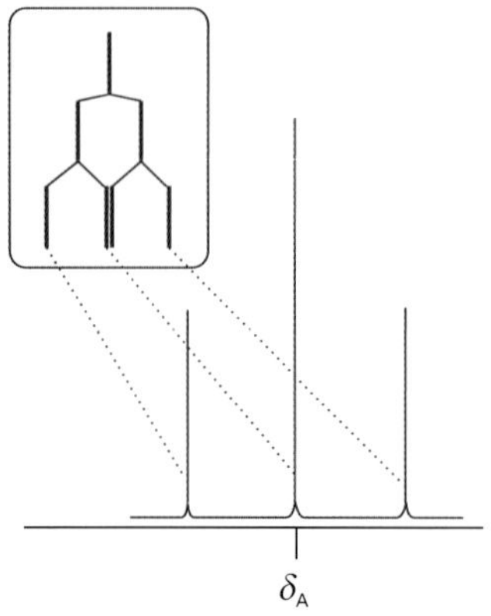

Fig. 15.15 The origin of the 1:2:1 triplet in the A resonance of an AX_2 species. The resonance of A is split into two by coupling with one X nucleus (as shown in the inset), and then each of those two lines is split into two by coupling to the second X nucleus. Because each X nucleus causes the same splitting, the two central transitions are coincident and give rise to an absorption line of double the intensity of the outer lines.

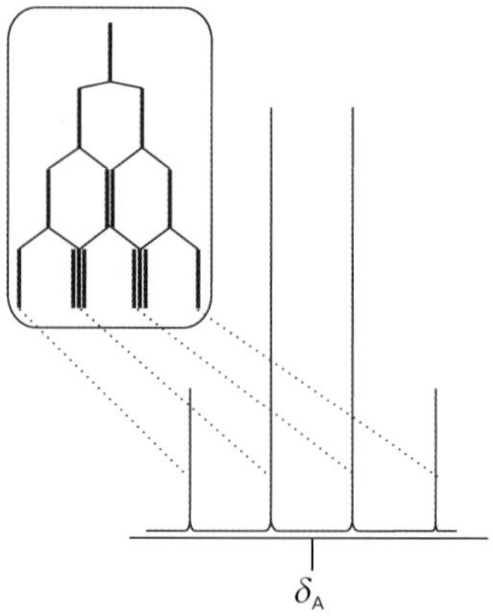

Fig. 15.16 The origin of the 1:3:3:1 quartet in the A resonance of an AX_3 species. The third X nucleus splits each of the lines shown in Fig. 15.15 for an AX_2 species into a doublet, and the intensity distribution reflects the number of transitions that have the same energy.

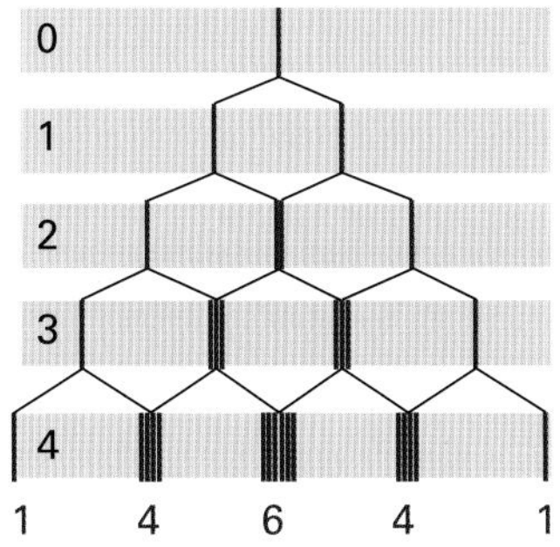

Fig. 15.17 The intensity distribution of the A resonance of an AX_n resonance can be constructed by considering the splitting caused by 1, 2, . . . n protons, as in Figs. 15.15 and 15.16. The resulting intensity distribution has a binomial distribution and is given by the integers in the corresponding row of Pascal's triangle. Note that, although the lines have been drawn side-by-side for clarity, the members of each group are coincident. Four protons, in AX_4, split the A resonance into a 1:4:6:4:1 quintet.

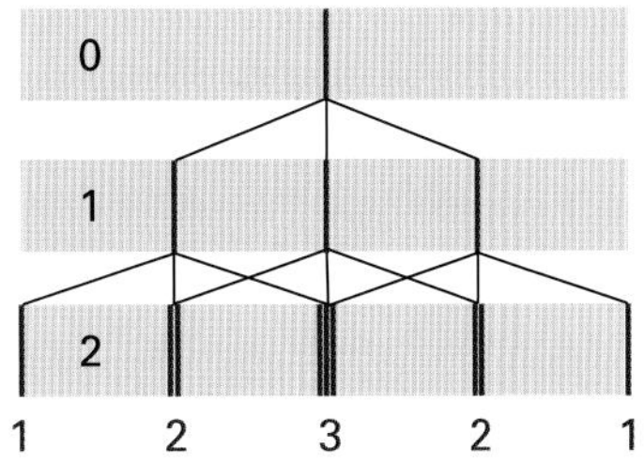

Fig. 15.18 The intensity distribution arising from spin–spin interaction with nuclei with $I = 1$ can be constructed similarly, but each successive nucleus splits the lines into three equal intensity components. Two equivalent spin-1 nuclei give rise to a 1:2:3:2:1 quintet.

Example 15.1 *Accounting for the fine structure in a spectrum*

Account for the fine structure in the NMR spectrum of the C—H protons of ethanol.

Method Consider how each group of equivalent protons (for instance, three methyl protons) split the resonances of the other groups of protons. There is no splitting within groups of equivalent protons. Each splitting pattern can be decided by referring to Pascal's triangle.

Answer The three protons of the CH_3 group split the resonance of the CH_2 protons into a 1:3:3:1 quartet with a splitting J. Likewise, the two protons of the CH_2 group split the resonance of the CH_3 protons into a 1:2:1 triplet with the same splitting J. The CH_2 and CH_3 protons all interact with the OH proton, but these couplings do not cause any splitting because the OH protons migrate rapidly from molecule to molecule and their effect averages to zero.

Self-test 15.2 What fine-structure can be expected for the protons in $^{14}NH_4^+$? The spin quantum number of nitrogen is 1. [1:1:1 triplet from N]

(c) The magnitudes of coupling constants

The scalar coupling constant of two nuclei separated by N bonds is denoted ^NJ, with subscripts for the types of nuclei involved. Thus, $^1J_{CH}$ is the coupling constant for a proton joined directly to a ^{13}C atom, and $^2J_{CH}$ is the coupling constant when the same two nuclei are separated by two bonds (as in ^{13}C—C—H). A typical value of $^1J_{CH}$ is in the range 120 to 250 Hz; $^2J_{CH}$ is between −10 and +20 Hz. Both ^3J and ^4J can give detectable effects in a spectrum, but couplings over larger numbers of bonds can generally

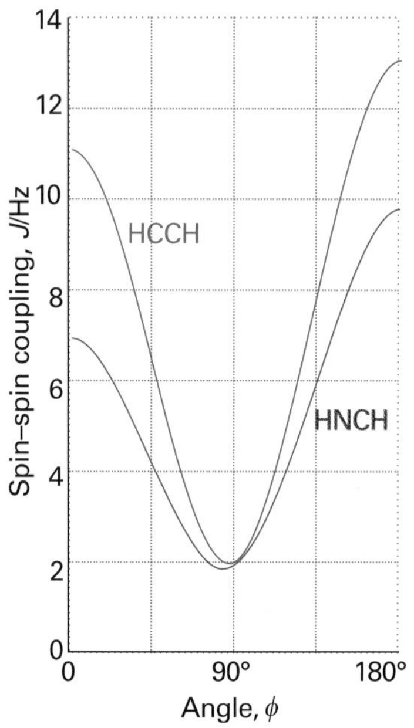

Fig. 15.19 The variation of the spin–spin coupling constant with angle predicted by the Karplus equation for an HCCH group and an HNCH group.

Exploration Draw a family of curves showing the variation of $^3J_{HH}$ with ϕ for which $A = +7.0$ Hz, $B = -1.0$ Hz, and C varies slightly from a typical value of +5.0 Hz. What is the effect of changing the value of the parameter C on the shape of the curve? In a similar fashion, explore the effect of the values of A and B on the shape of the curve.

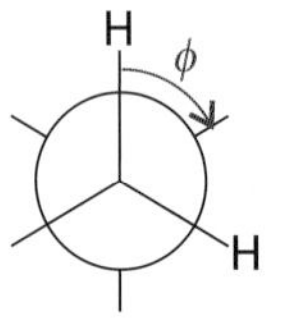

5

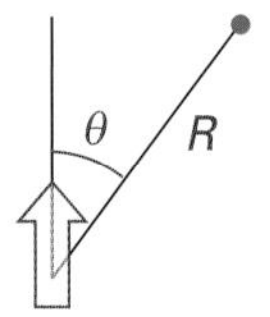

6

Comment 15.4

The average (or mean value) of a function $f(x)$ over the range $x = a$ to $x = b$ is $\int_a^b f(x)\mathrm{d}x/(b - a)$. The volume element in polar coordinates is proportional to $\sin\theta\,\mathrm{d}\theta$, and θ ranges from 0 to π. Therefore the average value of $(1 - 3\cos^2\theta)$ is $\int_0^\pi (1 - 3\cos^2\theta)\sin\theta\,\mathrm{d}\theta/\pi = 0$.

be ignored. One of the longest range couplings that has been detected is $^9J_{HH} = 0.4$ Hz between the CH_3 and CH_2 protons in $CH_3C{\equiv}CC{\equiv}CC{\equiv}CCH_2OH$.

The sign of J_{XY} indicates whether the energy of two spins is lower when they are parallel ($J < 0$) or when they are antiparallel ($J > 0$). It is found that $^1J_{CH}$ is often positive, $^2J_{HH}$ is often negative, $^3J_{HH}$ is often positive, and so on. An additional point is that J varies with the angle between the bonds (Fig. 15.19). Thus, a $^3J_{HH}$ coupling constant is often found to depend on the dihedral angle ϕ (**5**) according to the **Karplus equation**:

$$J = A + B\cos\phi + C\cos 2\phi \qquad (15.27)$$

with A, B, and C empirical constants with values close to +7 Hz, −1 Hz, and +5 Hz, respectively, for an HCCH fragment. It follows that the measurement of $^3J_{HH}$ in a series of related compounds can be used to determine their conformations. The coupling constant $^1J_{CH}$ also depends on the hybridization of the C atom, as the following values indicate:

	sp	sp^2	sp^3
$^1J_{CH}$/Hz:	250	160	125

(d) The origin of spin–spin coupling

Spin–spin coupling is a very subtle phenomenon, and it is better to treat J as an empirical parameter than to use calculated values. However, we can get some insight into its origins, if not its precise magnitude—or always reliably its sign—by considering the magnetic interactions within molecules.

A nucleus with spin projection m_I gives rise to a magnetic field with z-component $\mathcal{B}_{nuc}$ at a distance R, where, to a good approximation,

$$\mathcal{B}_{nuc} = -\frac{\gamma\hbar\mu_0}{4\pi R^3}(1 - 3\cos^2\theta)m_I \qquad (15.28)$$

The angle θ is defined in (**6**). The magnitude of this field is about 0.1 mT when R = 0.3 nm, corresponding to a splitting of resonance signal of about 10^4 Hz, and is of the order of magnitude of the splitting observed in solid samples (see Section 15.3a).

In a liquid, the angle θ sweeps over all values as the molecule tumbles, and $1-3\cos^2\theta$ averages to zero. Hence the direct dipolar interaction between spins cannot account for the fine structure of the spectra of rapidly tumbling molecules. The direct interaction does make an important contribution to the spectra of solid samples and is a very useful indirect source of structure information through its involvement in spin relaxation (Section 15.11).

Spin–spin coupling in molecules in solution can be explained in terms of the **polarization mechanism**, in which the interaction is transmitted through the bonds. The simplest case to consider is that of $^1J_{XY}$ where X and Y are spin-$\frac{1}{2}$ nuclei joined by an electron-pair bond (Fig. 15.20). The coupling mechanism depends on the fact that in some atoms it is favourable for the nucleus and a nearby electron spin to be parallel (both α or both β), but in others it is favourable for them to be antiparallel (one α and the other β). The electron–nucleus coupling is magnetic in origin, and may be either a dipolar interaction between the magnetic moments of the electron and nuclear spins or a **Fermi contact interaction**. A pictorial description of the Fermi contact interaction is as follows. First, we regard the magnetic moment of the nucleus as arising from the circulation of a current in a tiny loop with a radius similar to that of the nucleus (Fig. 15.21). Far from the nucleus the field generated by this loop is indistinguishable from the field generated by a point magnetic dipole. Close to the loop, however, the field differs from that of a point dipole. The magnetic interaction between this non-dipolar field and the electron's magnetic moment is the contact

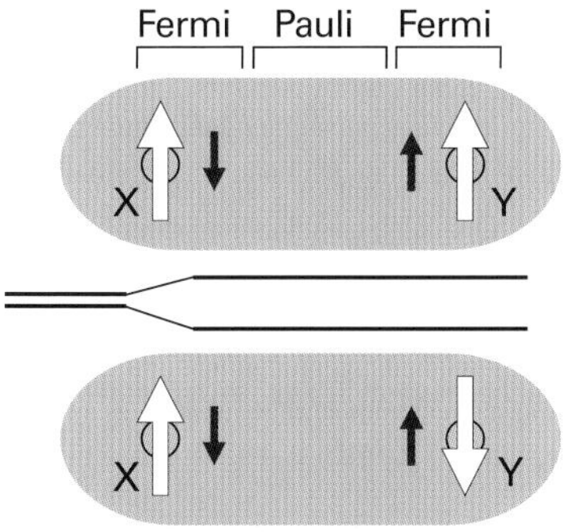

Fig. 15.20 The polarization mechanism for spin–spin coupling ($^1J_{HH}$). The two arrangements have slightly different energies. In this case, J is positive, corresponding to a lower energy when the nuclear spins are antiparallel.

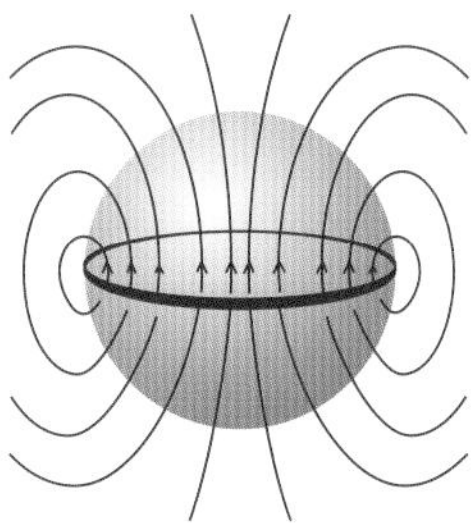

Fig. 15.21 The origin of the Fermi contact interaction. From far away, the magnetic field pattern arising from a ring of current (representing the rotating charge of the nucleus, the pale green sphere) is that of a point dipole. However, if an electron can sample the field close to the region indicated by the sphere, the field distribution differs significantly from that of a point dipole. For example, if the electron can penetrate the sphere, then the spherical average of the field it experiences is not zero.

interaction. The lines of force depicted in Fig. 15.21 correspond to those for a proton with α spin. The lower energy state of an electron spin in such a field is the β state. In conclusion, the contact interaction depends on the very close approach of an electron to the nucleus and hence can occur only if the electron occupies an *s* orbital (which is the reason why $^1J_{CH}$ depends on the hybridization ratio). We shall suppose that it is energetically favourable for an electron spin and a nuclear spin to be antiparallel (as is the case for a proton and an electron in a hydrogen atom).

If the X nucleus is α, a β electron of the bonding pair will tend to be found nearby (because that is energetically favourable for it). The second electron in the bond, which must have α spin if the other is β, will be found mainly at the far end of the bond (because electrons tend to stay apart to reduce their mutual repulsion). Because it is energetically favourable for the spin of Y to be antiparallel to an electron spin, a Y nucleus with β spin has a lower energy, and hence a lower Larmor frequency, than a Y nucleus with α spin. The opposite is true when X is β, for now the α spin of Y has the lower energy. In other words, the antiparallel arrangement of nuclear spins lies lower in energy than the parallel arrangement as a result of their magnetic coupling with the bond electrons. That is, $^1J_{HH}$ is positive.

To account for the value of $^2J_{XY}$, as in H—C—H, we need a mechanism that can transmit the spin alignments through the central C atom (which may be ^{12}C, with no nuclear spin of its own). In this case (Fig. 15.22), an X nucleus with α spin polarizes the electrons in its bond, and the α electron is likely to be found closer to the C nucleus. The more favourable arrangement of two electrons on the same atom is with their spins parallel (Hund's rule, Section 10.4d), so the more favourable arrangement is for the α electron of the neighbouring bond to be close to the C nucleus. Consequently, the β electron of that bond is more likely to be found close to the Y nucleus, and therefore that nucleus will have a lower energy if it is α. Hence, according to this mechanism, the lower Larmor frequency of Y will be obtained if its spin is parallel to that of X. That is, $^2J_{HH}$ is negative.

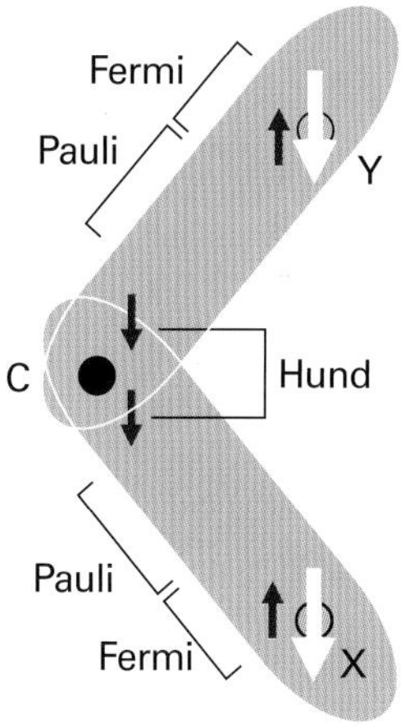

Fig. 15.22 The polarization mechanism for $^2J_{HH}$ spin–spin coupling. The spin information is transmitted from one bond to the next by a version of the mechanism that accounts for the lower energy of electrons with parallel spins in different atomic orbitals (Hund's rule of maximum multiplicity). In this case, $J < 0$, corresponding to a lower energy when the nuclear spins are parallel.

The coupling of nuclear spin to electron spin by the Fermi contact interaction is most important for proton spins, but it is not necessarily the most important mechanism for other nuclei. These nuclei may also interact by a dipolar mechanism with the electron magnetic moments and with their orbital motion, and there is no simple way of specifying whether J will be positive or negative.

(e) Equivalent nuclei

A group of nuclei are **chemically equivalent** if they are related by a symmetry operation of the molecule and have the same chemical shifts. Chemically equivalent nuclei are nuclei that would be regarded as 'equivalent' according to ordinary chemical criteria. Nuclei are **magnetically equivalent** if, as well as being chemically equivalent, they also have identical spin–spin interactions with any other magnetic nuclei in the molecule.

The difference between chemical and magnetic equivalence is illustrated by CH_2F_2 and $H_2C{=}CF_2$. In each of these molecules the protons are chemically equivalent: they are related by symmetry and undergo the same chemical reactions. However, although the protons in CH_2F_2 are magnetically equivalent, those in $CH_2{=}CF_2$ are not. One proton in the latter has a *cis* spin-coupling interaction with a given F nucleus whereas the other proton has a *trans* interaction with it. In constrast, in CH_2F_2 both protons are connected to a given F nucleus by identical bonds, so there is no distinction between them. Strictly speaking, the CH_3 protons in ethanol (and other compounds) are magnetically inequivalent on account of their different interactions with the CH_2 protons in the next group. However, they are in practice made magnetically equivalent by the rapid rotation of the CH_3 group, which averages out any differences. Magnetically inequivalent species can give very complicated spectra (for instance, the proton and ^{19}F spectra of $H_2C{=}CF_2$ each consist of 12 lines), and we shall not consider them further.

An important feature of chemically equivalent magnetic nuclei is that, although they do couple together, the coupling has no effect on the appearance of the spectrum. The reason for the invisibility of the coupling is set out in the following *Justification*, but qualitatively it is that all allowed nuclear spin transitions are *collective* reorientations of groups of equivalent nuclear spins that do not change the relative orientations of the spins within the group (Fig. 15.23). Then, because the relative orientations of nuclear spins are not changed in any transition, the magnitude of the coupling between them is undetectable. Hence, an isolated CH_3 group gives a single, unsplit line because all the allowed transitions of the group of three protons occur without change of their relative orientations.

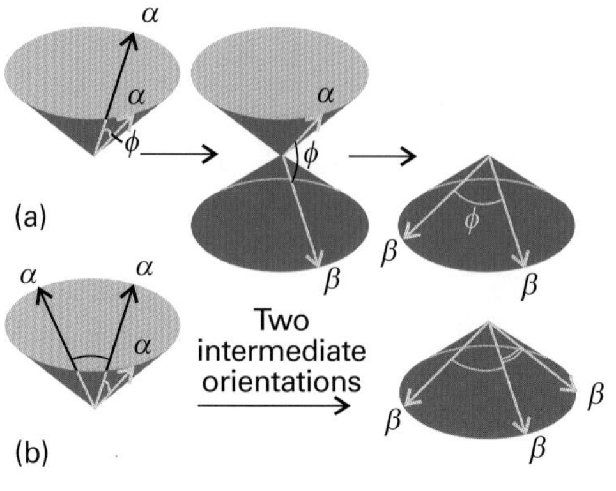

Fig. 15.23 (a) A group of two equivalent nuclei realigns as a group, without change of angle between the spins, when a resonant absorption occurs. Hence it behaves like a single nucleus and the spin–spin coupling between the individual spins of the group is undetectable. (b) Three equivalent nuclei also realign as a group without change of their relative orientations.

Justification 15.2 *The energy levels of an A_2 system*

Consider an A_2 system of two chemically equivalent spin-$\frac{1}{2}$ nuclei. First, consider the energy levels in the absence of spin–spin coupling. There are four spin states that (just as for two electrons) can be classified according to their total spin I (the analogue of S for two electrons) and their total projection M_I on the z-axis. The states are analogous to those we developed for two electrons in singlet and triplet states:

Spins parallel, $I=1$:	$M_I=+1$	$\alpha\alpha$
	$M_I=0$	$(1/2^{1/2})\{\alpha\beta+\beta\alpha\}$
	$M_I=-1$	$\beta\beta$
Spins paired, $I=0$:	$M_I=0$	$(1/2^{1/2})\{\alpha\beta-\beta\alpha\}$

The effect of a magnetic field on these four states is shown in Fig. 15.24: the energies of the two states with $M_I=0$ are unchanged by the field because they are composed of equal proportions of α and β spins.

As remarked in Section 15.6a, the spin–spin coupling energy is proportional to the scalar product of the vectors representing the spins, $E=(hJ/\hbar^2)\boldsymbol{I}_1\cdot\boldsymbol{I}_2$. The scalar product can be expressed in terms of the total nuclear spin by noting that

$$I^2=(\boldsymbol{I}_1+\boldsymbol{I}_2)\cdot(\boldsymbol{I}_1+\boldsymbol{I}_2)=I_1^2+I_2^2+2\boldsymbol{I}_1\cdot\boldsymbol{I}_2$$

rearranging this expression to

$$\boldsymbol{I}_1\cdot\boldsymbol{I}_2=\tfrac{1}{2}\{I^2-I_1^2-I_2^2\}$$

and replacing the magnitudes by their quantum mechanical values:

$$\boldsymbol{I}_1\cdot\boldsymbol{I}_2=\tfrac{1}{2}\{I(I+1)-I_1(I_1+1)-I_2(I_2+1)\}\hbar^2$$

Then, because $I_1=I_2=\frac{1}{2}$, it follows that

$$E=\tfrac{1}{2}hJ\{I(I+1)-\tfrac{3}{2}\}$$

For parallel spins, $I=1$ and $E=+\frac{1}{4}hJ$; for antiparallel spins $I=0$ and $E=-\frac{3}{4}hJ$, as in Fig. 15.24. We see that three of the states move in energy in one direction and the fourth (the one with antiparallel spins) moves three times as much in the opposite direction. The resulting energy levels are shown on the right in Fig. 15.24.

Comment 15.5

As in Section 10.7, the states we have selected in *Justification* 15.2 are those with a definite resultant, and hence a well defined value of I. The + sign in $\alpha\beta+\beta\alpha$ signifies an in-phase alignment of spins and $I=1$; the − sign in $\alpha\beta-\beta\alpha$ signifies an alignment out of phase by π, and hence $I=0$. See Fig. 10.24.

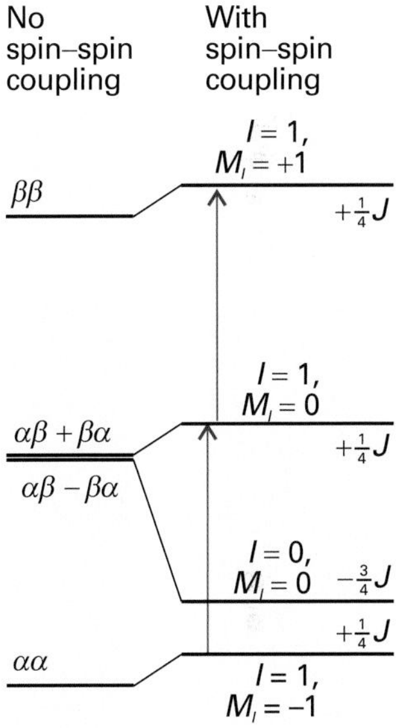

Fig. 15.24 The energy levels of an A_2 system in the absence of spin–spin coupling are shown on the left. When spin–spin coupling is taken into account, the energy levels on the right are obtained. Note that the three states with total nuclear spin $I=1$ correspond to parallel spins and give rise to the same increase in energy (J is positive); the one state with $I=0$ (antiparallel nuclear spins) has a lower energy in the presence of spin–spin coupling. The only allowed transitions are those that preserve the angle between the spins, and so take place between the three states with $I=1$. They occur at the same resonance frequency as they would have in the absence of spin–spin coupling.

The NMR spectrum of the A_2 species arises from transitions between the levels. However, the radiofrequency field affects the two equivalent protons equally, so it cannot change the orientation of one proton relative to the other; therefore, the transitions take place within the set of states that correspond to parallel spin (those labelled $I=1$), and no spin-parallel state can change to a spin-antiparallel state (the state with $I=0$). Put another way, the allowed transitions are subject to the selection rule $\Delta I=0$. This selection rule is in addition to the rule $\Delta M_I=\pm1$ that arises from the conservation of angular momentum and the unit spin of the photon. The allowed transitions are shown in Fig. 15.24: we see that there are only two transitions, and that they occur at the same resonance frequency that the nuclei would have in the absence of spin–spin coupling. Hence, the spin–spin coupling interaction does not affect the appearance of the spectrum.

(f) Strongly coupled nuclei

NMR spectra are usually much more complex than the foregoing simple analysis suggests. We have described the extreme case in which the differences in chemical shifts are much greater than the spin–spin coupling constants. In such cases it is simple to identify groups of magnetically equivalent nuclei and to think of the groups of nuclear spins as reorientating relative to each other. The spectra that result are called **first-order spectra**.

Transitions cannot be allocated to definite groups when the differences in their chemical shifts are comparable to their spin–spin coupling interactions. The complicated spectra that are then obtained are called **strongly coupled spectra** (or 'second-order spectra') and are much more difficult to analyse (Fig. 15.25). Because the difference in resonance frequencies increases with field, but spin–spin coupling constants are independent of it, a second-order spectrum may become simpler (and first-order) at high fields because individual groups of nuclei become identifiable again.

A clue to the type of analysis that is appropriate is given by the notation for the types of spins involved. Thus, an AX spin system (which consists of two nuclei with a large chemical shift difference) has a first-order spectrum. An AB system, on the other hand (with two nuclei of similar chemical shifts), gives a spectrum typical of a strongly coupled system. An AX system may have widely different Larmor frequencies because A and X are nuclei of different elements (such as ^{13}C and ^{1}H), in which case they form a **heteronuclear spin system**. AX may also denote a **homonuclear spin system** in which the nuclei are of the same element but in markedly different environments.

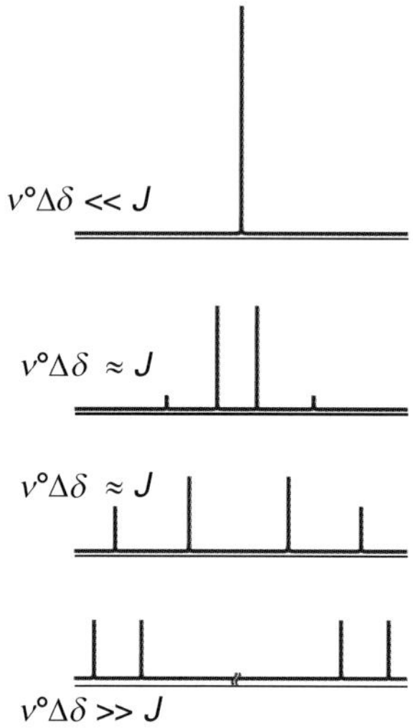

Fig. 15.25 The NMR spectra of an A_2 system (top) and an AX system (bottom) are simple 'first-order' spectra. At intermediate relative values of the chemical shift difference and the spin–spin coupling, complex 'strongly coupled' spectra are obtained. Note how the inner two lines of the bottom spectrum move together, grow in intensity, and form the single central line of the top spectrum. The two outer lines diminish in intensity and are absent in the top spectrum.

15.7 Conformational conversion and exchange processes

The appearance of an NMR spectrum is changed if magnetic nuclei can jump rapidly between different environments. Consider a molecule, such as *N*,*N*-dimethylformamide, that can jump between conformations; in its case, the methyl shifts depend on whether they are *cis* or *trans* to the carbonyl group (Fig. 15.26). When the jumping rate is low, the spectrum shows two sets of lines, one each from molecules in each conformation. When the interconversion is fast, the spectrum shows a single line at the mean of the two chemical shifts. At intermediate inversion rates, the line is very broad. This maximum broadening occurs when the lifetime, τ, of a conformation gives rise to a linewidth that is comparable to the difference of resonance frequencies, $\delta\nu$, and both broadened lines blend together into a very broad line. Coalescence of the two lines occurs when

$$\tau = \frac{\sqrt{2}}{\pi\delta\nu} \tag{15.29}$$

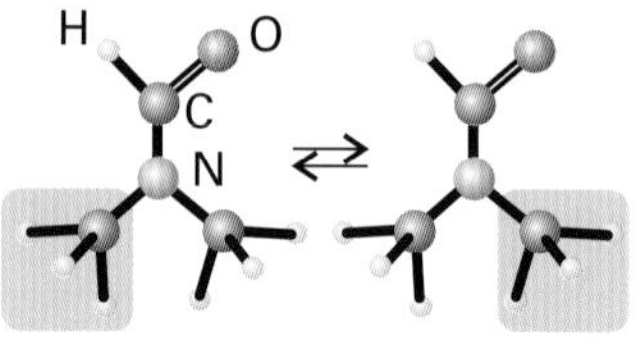

Fig. 15.26 When a molecule changes from one conformation to another, the positions of its protons are interchanged and jump between magnetically distinct environments.

Example 15.2 *Interpreting line broadening*

The NO group in *N*,*N*-dimethylnitrosamine, $(CH_3)_2N{-}NO$, rotates about the N—N bond and, as a result, the magnetic environments of the two CH_3 groups are interchanged. The two CH_3 resonances are separated by 390 Hz in a 600 MHz spectrometer. At what rate of interconversion will the resonance collapse to a single line?

Method Use eqn 15.29 for the average lifetimes of the conformations. The rate of interconversion is the inverse of their lifetime.

Answer With $\delta\nu = 390$ Hz,

$$\tau = \frac{\sqrt{2}}{\pi \times (390\ \mathrm{s}^{-1})} = 1.2\ \mathrm{ms}$$

It follows that the signal will collapse to a single line when the interconversion rate exceeds about 830 s^{-1}. The dependence of the rate of exchange on the temperature is used to determine the energy barrier to interconversion.

Self-test 15.3 What would you deduce from the observation of a single line from the same molecule in a 300 MHz spectrometer?

[Conformation lifetime less than 2.3 ms]

A similar explanation accounts for the loss of fine structure in solvents able to exchange protons with the sample. For example, hydroxyl protons are able to exchange with water protons. When this **chemical exchange** occurs, a molecule ROH with an α-spin proton (we write this ROH_α) rapidly converts to ROH_β and then perhaps to ROH_α again because the protons provided by the solvent molecules in successive exchanges have random spin orientations. Therefore, instead of seeing a spectrum composed of contributions from both ROH_α and ROH_β molecules (that is, a spectrum showing a doublet structure due to the OH proton) we see a spectrum that shows no splitting caused by coupling of the OH proton (as in Fig. 15.6). The effect is observed when the lifetime of a molecule due to this chemical exchange is so short that the lifetime broadening is greater than the doublet splitting. Because this splitting is often very small (a few hertz), a proton must remain attached to the same molecule for longer than about 0.1 s for the splitting to be observable. In water, the exchange rate is much faster than that, so alcohols show no splitting from the OH protons. In dry dimethylsulfoxide (DMSO), the exchange rate may be slow enough for the splitting to be detected.

Pulse techniques in NMR

Modern methods of detecting the energy separation between nuclear spin states are more sophisticated than simply looking for the frequency at which resonance occurs. One of the best analogies that has been suggested to illustrate the difference between the old and new ways of observing an NMR spectrum is that of detecting the spectrum of vibrations of a bell. We could stimulate the bell with a gentle vibration at a gradually increasing frequency, and note the frequencies at which it resonated with the stimulation. A lot of time would be spent getting zero response when the stimulating frequency was between the bell's vibrational modes. However, if we were simply to hit the bell with a hammer, we would immediately obtain a clang composed of all the frequencies that the bell can produce. The equivalent in NMR is to monitor the radiation nuclear spins emit as they return to equilibrium after the appropriate stimulation. The resulting **Fourier-transform NMR** gives greatly increased sensitivity, so opening up the entire periodic table to the technique. Moreover, multiple-pulse FT-NMR gives chemists unparalleled control over the information content and display of spectra. We need to understand how the equivalent of the hammer blow is delivered and how the signal is monitored and interpreted. These features are generally expressed in terms of the vector model of angular momentum introduced in Section 9.7d.

15.8 The magnetization vector

Consider a sample composed of many identical spin-$\frac{1}{2}$ nuclei. As we saw in Section 9.7d, an angular momentum can be represented by a vector of length $\{I(I+1)\}^{1/2}$ units

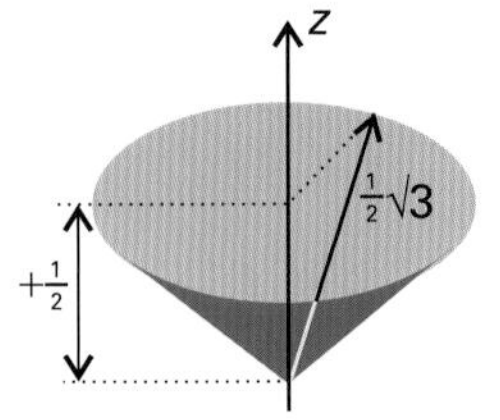

Fig. 15.27 The vector model of angular momentum for a single spin-$\frac{1}{2}$ nucleus. The angle around the z-axis is indeterminate.

with a component of length m_I units along the z-axis. As the uncertainty principle does not allow us to specify the x- and y-components of the angular momentum, all we know is that the vector lies somewhere on a cone around the z-axis. For $I=\frac{1}{2}$, the length of the vector is $\frac{1}{2}\sqrt{3}$ and it makes an angle of 55° to the z-axis (Fig. 15.27).

In the absence of a magnetic field, the sample consists of equal numbers of α and β nuclear spins with their vectors lying at random angles on the cones. These angles are unpredictable, and at this stage we picture the spin vectors as stationary. The **magnetization, *M***, of the sample, its net nuclear magnetic moment, is zero (Fig. 15.28a).

(a) The effect of the static field

Two changes occur in the magnetization when a magnetic field is present. First, the energies of the two orientations change, the α spins moving to low energy and the β spins to high energy (provided $\gamma > 0$). At 10 T, the Larmor frequency for protons is 427 MHz, and in the vector model the individual vectors are pictured as precessing at this rate. This motion is a pictorial representation of the difference in energy of the spin states (it is not an actual representation of reality). As the field is increased, the Larmor frequency increases and the precession becomes faster. Secondly, the populations of the two spin states (the numbers of α and β spins) at thermal equilibrium change, and there will be more α spins than β spins. Because $h\nu_L/kT \approx 7\times10^{-5}$ for protons at 300 K and 10 T, it follows from the Boltzmann distribution that $N_\beta/N_\alpha = e^{-h\nu_L/kT}$ is only slightly less than 1. That is, there is only a tiny imbalance of populations, and it is even smaller for other nuclei with their smaller magnetogyric ratios. However, despite its smallness, the imbalance means that there is a net magnetization that we can represent by a vector $\boldsymbol{M}$ pointing in the z-direction and with a length proportional to the population difference (Fig. 15.28b).

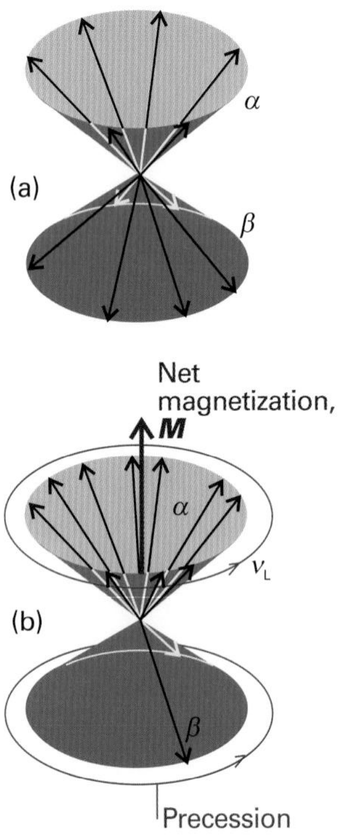

Fig. 15.28 The magnetization of a sample of spin-$\frac{1}{2}$ nuclei is the resultant of all their magnetic moments. (a) In the absence of an externally applied field, there are equal numbers of α and β spins at random angles around the z-axis (the field direction) and the magnetization is zero. (b) In the presence of a field, the spins precess around their cones (that is, there is an energy difference between the α and β states) and there are slightly more α spins than β spins. As a result, there is a net magnetization along the z-axis.

(b) The effect of the radiofrequency field

We now consider the effect of a radiofrequency field circularly polarized in the xy-plane, so that the magnetic component of the electromagnetic field (the only component we need to consider) is rotating around the z-direction, the direction of the applied field $\mathcal{B}_0$, in the same sense as the Larmor precession. The strength of the rotating magnetic field is $\mathcal{B}_1$. Suppose we choose the frequency of this field to be equal to the Larmor frequency of the spins, $\nu_L = (\gamma/2\pi)\mathcal{B}_0$; this choice is equivalent to selecting the resonance condition in the conventional experiment. The nuclei now experience a steady $\mathcal{B}_1$ field because the rotating magnetic field is in step with the precessing spins (Fig. 15.29a). Just as the spins precess about the strong static field $\mathcal{B}_0$ at a frequency $\gamma\mathcal{B}_0/2\pi$, so under the influence of the field $\mathcal{B}_1$ they precess about $\mathcal{B}_1$ at a frequency $\gamma\mathcal{B}_1/2\pi$.

To interpret the effects of radiofrequency pulses on the magnetization, it is often useful to look at the spin system from a different perspective. If we were to imagine stepping on to a platform, a so-called **rotating frame**, that rotates around the direction of the applied field at the radiofrequency, then the nuclear magnetization appears stationary if the radiofrequency is the same as the Larmor frequency (Fig. 15.29b). If the $\mathcal{B}_1$ field is applied in a pulse of duration $\pi/2\gamma\mathcal{B}_1$, the magnetization tips through 90° in the rotating frame and we say that we have applied a **90° pulse**, or a '$\pi/2$ pulse' (Fig. 15.30a). The duration of the pulse depends on the strength of the $\mathcal{B}_1$ field, but is typically of the order of microseconds.

Now imagine stepping out of the rotating frame. To an external observer (the role played by a radiofrequency coil) in this stationary frame, the magnetization vector is now rotating at the Larmor frequency in the xy-plane (Fig. 15.30b). The rotating magnetization induces in the coil a signal that oscillates at the Larmor frequency and that

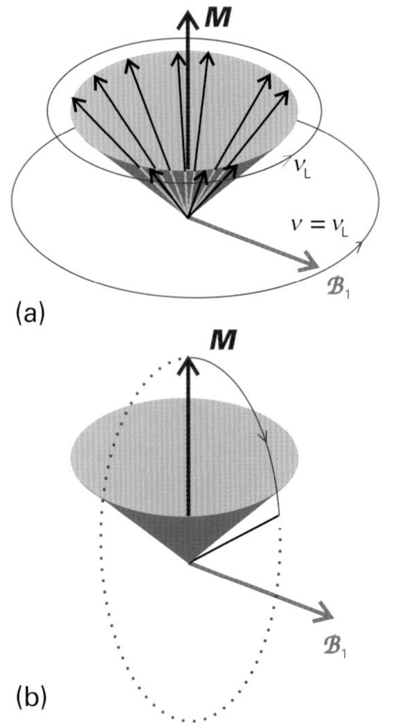

Fig. 15.29 (a) In a resonance experiment, a circularly polarized radiofrequency magnetic field $\mathcal{B}_1$ is applied in the *xy*-plane (the magnetization vector lies along the *z*-axis). (b) If we step into a frame rotating at the radiofrequency, $\mathcal{B}_1$ appears to be stationary, as does the magnetization ***M*** if the Larmor frequency is equal to the radiofrequency. When the two frequencies coincide, the magnetization vector of the sample rotates around the direction of the $\mathcal{B}_1$ field.

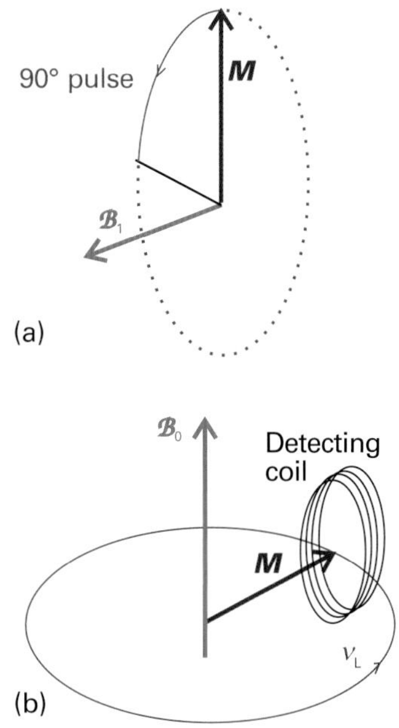

Fig. 15.30 (a) If the radiofrequency field is applied for a certain time, the magnetization vector is rotated into the *xy*-plane. (b) To an external stationary observer (the coil), the magnetization vector is rotating at the Larmor frequency, and can induce a signal in the coil.

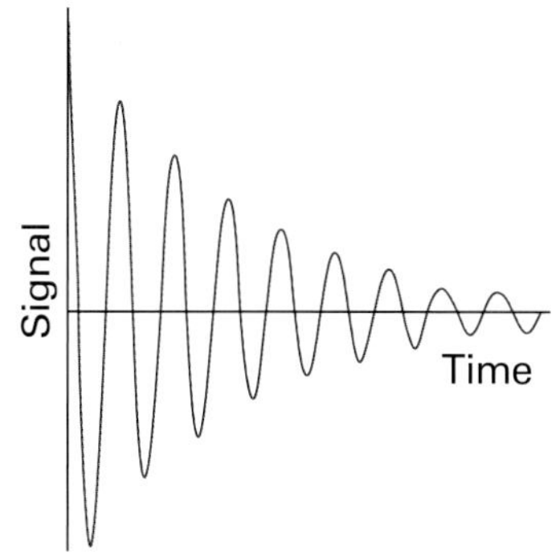

Fig. 15.31 A simple free-induction decay of a sample of spins with a single resonance frequency.

can be amplified and processed. In practice, the processing takes place after subtraction of a constant high frequency component (the radiofrequency used for $\mathcal{B}_1$), so that all the signal manipulation takes place at frequencies of a few kilohertz.

As time passes, the individual spins move out of step (partly because they are precessing at slightly different rates, as we shall explain later), so the magnetization vector shrinks exponentially with a time constant T_2 and induces an ever weaker signal in the detector coil. The form of the signal that we can expect is therefore the oscillating-decaying **free-induction decay** (FID) shown in Fig. 15.31. The *y*-component of the magnetization varies as

$$M_y(t) = M_0 \cos(2\pi\nu_L t)\, e^{-t/T_2} \qquad (15.30)$$

We have considered the effect of a pulse applied at exactly the Larmor frequency. However, virtually the same effect is obtained off resonance, provided that the pulse is applied close to ν_L. If the difference in frequency is small compared to the inverse of the duration of the 90° pulse, the magnetization will end up in the *xy*-plane. Note that we do not need to know the Larmor frequency beforehand: the short pulse is the analogue of the hammer blow on the bell, exciting a range of frequencies. The detected signal shows that a particular resonant frequency is present.

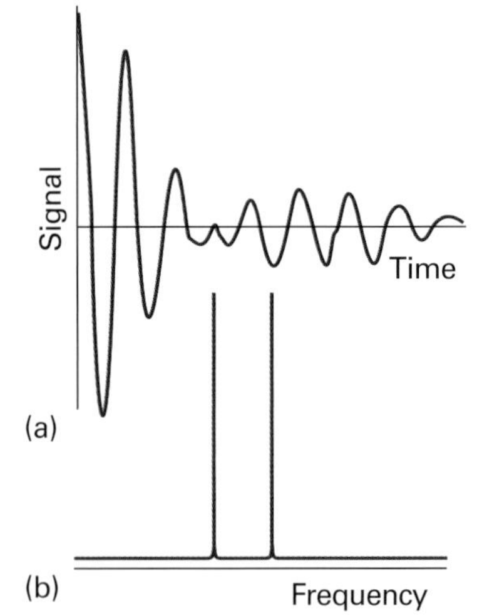

Fig. 15.32 (a) A free induction decay signal of a sample of AX species and (b) its analysis into its frequency components.

Exploration The *Living graphs* section of the text's web site has an applet that allows you to calculate and display the FID curve from an AX system. Explore the effect on the shape of the FID curve of changing the chemical shifts (and therefore the Larmor frequencies) of the A and X nuclei.

Comment 15.6

The web site for this text contains links to databases of NMR spectra and to sites that allow for interactive simulation of NMR spectra.

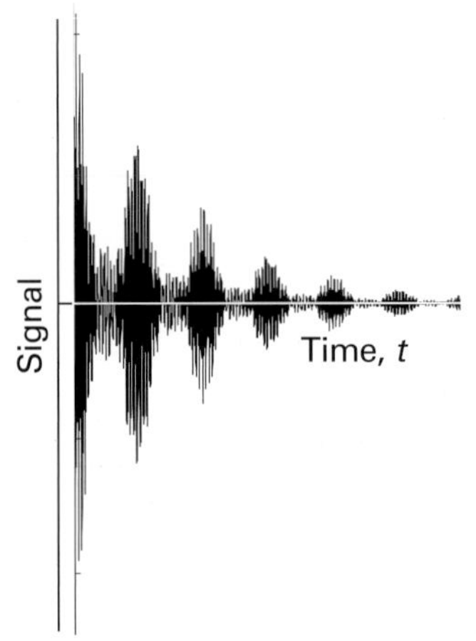

Fig. 15.33 A free induction decay signal of a sample of ethanol. Its Fourier transform is the frequency-domain spectrum shown in Fig. 15.6. The total length of the image corresponds to about 1 s.

(c) Time- and frequency-domain signals

We can think of the magnetization vector of a homonuclear AX spin system with $J = 0$ as consisting of two parts, one formed by the A spins and the other by the X spins. When the 90° pulse is applied, both magnetization vectors are rotated into the *xy*-plane. However, because the A and X nuclei precess at different frequencies, they induce two signals in the detector coils, and the overall FID curve may resemble that in Fig. 15.32a. The composite FID curve is the analogue of the struck bell emitting a rich tone composed of all the frequencies at which it can vibrate.

The problem we must address is how to recover the resonance frequencies present in a free-induction decay. We know that the FID curve is a sum of oscillating functions, so the problem is to analyse it into its component frequencies by carrying out a Fourier transformation (*Further information* 13.2 and 15.1). When the signal in Fig.15.32a is transformed in this way, we get the frequency-domain spectrum shown in Fig. 15.32b. One line represents the Larmor frequency of the A nuclei and the other that of the X nuclei.

The FID curve in Fig. 15.33 is obtained from a sample of ethanol. The frequency-domain spectrum obtained from it by Fourier transformation is the one that we have already discussed (Fig. 15.6). We can now see why the FID curve in Fig. 15.33 is so complex: it arises from the precession of a magnetization vector that is composed of eight components, each with a characteristic frequency.

15.9 Spin relaxation

There are two reasons why the component of the magnetization vector in the *xy*-plane shrinks. Both reflect the fact that the nuclear spins are not in thermal equilibrium with their surroundings (for then $\boldsymbol{M}$ lies parallel to z). The return to equilibrium is the process called **spin relaxation.**

(a) Longitudinal and transverse relaxation

At thermal equilibrium the spins have a Boltzmann distribution, with more α spins than β spins; however, a magnetization vector in the *xy*-plane immediately after a 90° pulse has equal numbers of α and β spins.

Now consider the effect of a **180° pulse**, which may be visualized in the rotating frame as a flip of the net magnetization vector from one direction along the z-axis to the opposite direction. That is, the 180° pulse leads to population inversion of the spin system, which now has more β spins than α spins. After the pulse, the populations revert to their thermal equilibrium values exponentially. As they do so, the z-component of magnetization reverts to its equilibrium value M_0 with a time constant called the **longitudinal relaxation time,** T_1 (Fig. 15.34):

$$M_z(t) - M_0 \propto e^{-t/T_1} \tag{15.31}$$

Because this relaxation process involves giving up energy to the surroundings (the 'lattice') as β spins revert to α spins, the time constant T_1 is also called the **spin–lattice relaxation time.** Spin–lattice relaxation is caused by local magnetic fields that fluctuate at a frequency close to the resonance frequency of the $\alpha \rightarrow \beta$ transition. Such fields can arise from the tumbling motion of molecules in a fluid sample. If molecular tumbling is too slow or too fast compared to the resonance frequency, it will give rise to a fluctuating magnetic field with a frequency that is either too low or too high to stimulate a spin change from β to α, so T_1 will be long. Only if the molecule tumbles at about the resonance frequency will the fluctuating magnetic field be able to induce spin changes effectively, and only then will T_1 be short. The rate of molecular tumbling

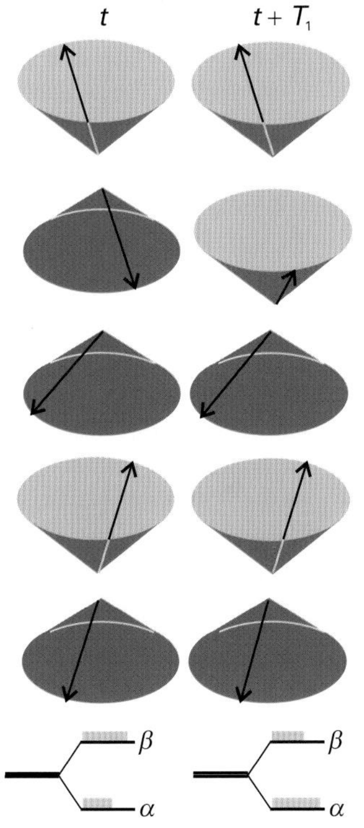

Fig. 15.34 In longitudinal relaxation the spins relax back towards their thermal equilibrium populations. On the left we see the precessional cones representing spin-$\frac{1}{2}$ angular momenta, and they do not have their thermal equilibrium populations (there are more α-spins than β-spins). On the right, which represents the sample a long time after a time T_1 has elapsed, the populations are those characteristic of a Boltzmann distribution (see *Molecular interpretation 3.1*). In actuality, T_1 is the time constant for relaxation to the arrangement on the right and $T_1 \ln 2$ is the half-life of the arrangement on the left.

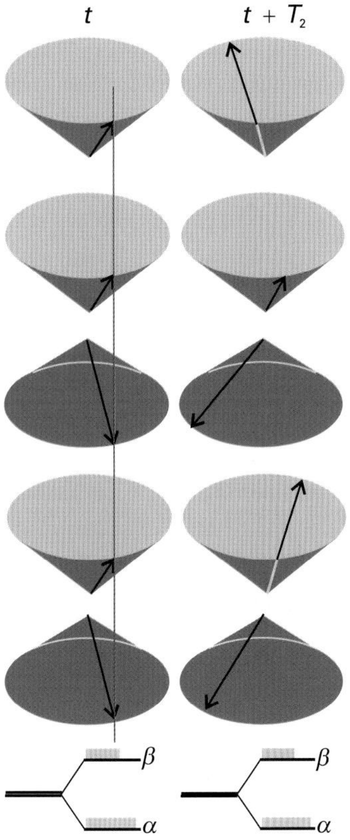

Fig. 15.36 The transverse relaxation time, T_2, is the time constant for the phases of the spins to become randomized (another condition for equilibrium) and to change from the orderly arrangement shown on the left to the disorderly arrangement on the right (long after a time T_2 has elapsed). Note that the populations of the states remain the same; only the relative phase of the spins relaxes. In actuality, T_2 is the time constant for relaxation to the arrangement on the right and $T_2 \ln 2$ is the half-life of the arrangement on the left.

increases with temperature and with reducing viscosity of the solvent, so we can expect a dependence like that shown in Fig. 15.35.

A second aspect of spin relaxation is the fanning-out of the spins in the *xy*-plane if they precess at different rates (Fig. 15.36). The magnetization vector is large when all the spins are bunched together immediately after a 90° pulse. However, this orderly bunching of spins is not at equilibrium and, even if there were no spin–lattice relaxation, we would expect the individual spins to spread out until they were uniformly distributed with all possible angles around the *z*-axis. At that stage, the component of magnetization vector in the plane would be zero. The randomization of the spin directions occurs exponentially with a time constant called the **transverse relaxation time**, T_2:

$$M_y(t) \propto e^{-t/T_2} \qquad (15.32)$$

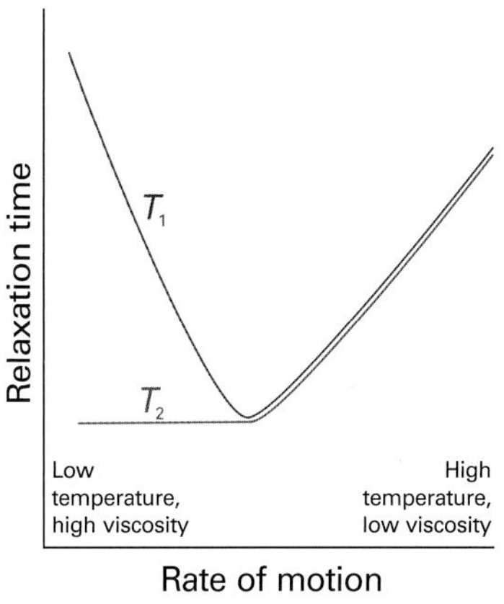

Fig. 15.35 The variation of the two relaxation times with the rate at which the molecules move (either by tumbling or migrating through the solution). The horizontal axis can be interpreted as representing temperature or viscosity. Note that, at rapid rates of motion, the two relaxation times coincide.

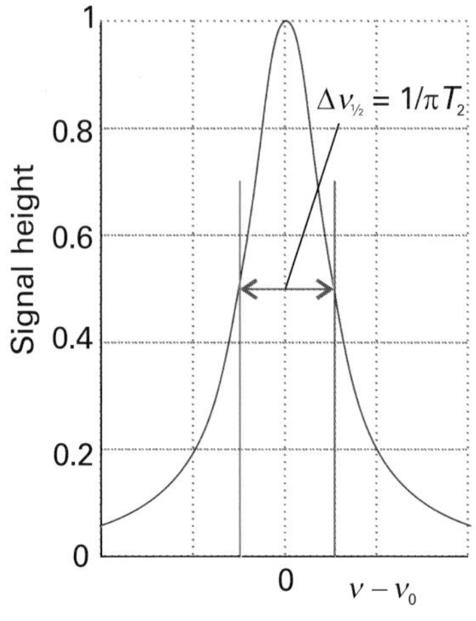

Fig. 15.37 A Lorentzian absorption line. The width at half-height is inversely proportional to the parameter T_2 and the longer the transverse relaxation time, the narrower the line.

Exploration The *Living graphs* section of the text's web site has an applet that allows you to calculate and display Lorenztian absorption lines. Explore the effect of the parameter T_2 on the width and the maximal intensity of a Lorentzian line. Rationalize your observations.

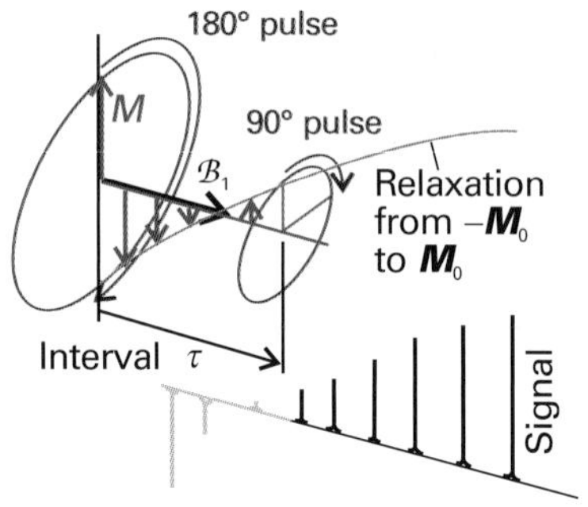

Fig. 15.38 The result of applying a 180° pulse to the magnetization in the rotating frame and the effect of a subsequent 90° pulse. The amplitude of the frequency-domain spectrum varies with the interval between the two pulses because spin–lattice relaxation has time to occur.

Because the relaxation involves the relative orientation of the spins, T_2 is also known as the **spin–spin relaxation time**. Any relaxation process that changes the balance between α and β spins will also contribute to this randomization, so the time constant T_2 is almost always less than or equal to T_1.

Local magnetic fields also affect spin–spin relaxation. When the fluctuations are slow, each molecule lingers in its local magnetic environment and the spin orientations randomize quickly around the applied field direction. If the molecules move rapidly from one magnetic environment to another, the effects of differences in local magnetic field average to zero: individual spins do not precess at very different rates, they can remain bunched for longer, and spin–spin relaxation does not take place as quickly. In other words, slow molecular motion corresponds to short T_2 and fast motion corresponds to long T_2 (as shown in Fig. 15.35). Calculations show that, when the motion is fast, $T_2 \approx T_1$.

If the y-component of magnetization decays with a time constant T_2, the spectral line is broadened (Fig. 15.37), and its width at half-height becomes

$$\Delta \nu_{1/2} = \frac{1}{\pi T_2} \tag{15.33}$$

Typical values of T_2 in proton NMR are of the order of seconds, so linewidths of around 0.1 Hz can be anticipated, in broad agreement with observation.

So far, we have assumed that the equipment, and in particular the magnet, is perfect, and that the differences in Larmor frequencies arise solely from interactions within the sample. In practice, the magnet is not perfect, and the field is different at different locations in the sample. The inhomogeneity broadens the resonance, and in most cases this **inhomogeneous broadening** dominates the broadening we have discussed so far. It is common to express the extent of inhomogeneous broadening in terms of an **effective transverse relaxation time**, $T_2^\star$, by using a relation like eqn 15.33, but writing

$$T_2^\star = \frac{1}{\pi \Delta \nu_{1/2}} \tag{15.34}$$

where $\Delta \nu_{1/2}$ is the observed width at half-height of a line with a Lorenztian shape of the form $I \propto 1/(1 + \nu^2)$. As an example, consider a line in a spectrum with a width of 10 Hz. It follows from eqn 15.34 that the effective transverse relaxation time is

$$T_2^\star = \frac{1}{\pi \times (10 \text{ s}^{-1})} = 32 \text{ ms}$$

(b) The measurement of T_1

The longitudinal relaxation time T_1 can be measured by the **inversion recovery technique**. The first step is to apply a 180° pulse to the sample. A 180° pulse is achieved by applying the $\mathcal{B}_1$ field for twice as long as for a 90° pulse, so the magnetization vector precesses through 180° and points in the $-z$-direction (Fig. 15.38). No signal can be seen at this stage because there is no component of magnetization in the xy-plane (where the coil can detect it). The β spins begin to relax back into α spins, and the magnetization vector first shrinks exponentially, falling through zero to its thermal equilibrium value, M_z. After an interval τ, a 90° pulse is applied that rotates the magnetization into the xy-plane, where it generates an FID signal. The frequency-domain spectrum is then obtained by Fourier transformation.

The intensity of the spectrum obtained in this way depends on the length of the magnetization vector that is rotated into the xy-plane. The length of that vector changes exponentially as the interval between the two pulses is increased, so the intensity of the spectrum also changes exponentially with increasing τ. We can therefore measure T_1 by fitting an exponential curve to the series of spectra obtained with different values of τ.

(c) Spin echoes

The measurement of T_2 (as distinct from T_2^*) depends on being able to eliminate the effects of inhomogeneous broadening. The cunning required is at the root of some of the most important advances that have been made in NMR since its introduction.

A **spin echo** is the magnetic analogue of an audible echo: transverse magnetization is created by a radiofrequency pulse, decays away, is reflected by a second pulse, and grows back to form an echo. The sequence of events is shown in Fig. 15.39. We can consider the overall magnetization as being made up of a number of different magnetizations, each of which arises from a **spin packet** of nuclei with very similar precession frequencies. The spread in these frequencies arises because the applied field $\mathcal{B}_0$ is inhomogeneous, so different parts of the sample experience different fields. The precession frequencies also differ if there is more than one chemical shift present. As will be seen, the importance of a spin echo is that it can suppress the effects of both field inhomogeneities and chemical shifts.

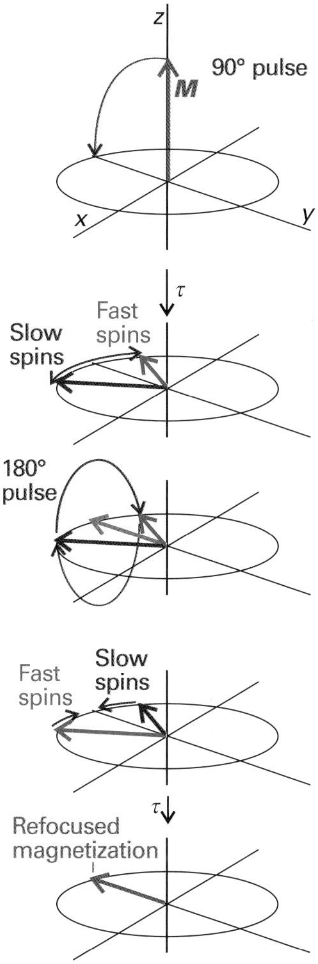

Fig. 15.39 The sequence of pulses leading to the observation of a spin echo.

First, a 90° pulse is applied to the sample. We follow events by using the rotating frame, in which $\mathcal{B}_1$ is stationary along the x-axis and causes the magnetization to be into the xy-plane. The spin packets now begin to fan out because they have different Larmor frequencies, with some above the radiofrequency and some below. The detected signal depends on the resultant of the spin-packet magnetization vectors, and decays with a time-constant T_2^* because of the combined effects of field inhomogeneity and spin–spin relaxation.

After an evolution period τ, a 180° pulse is applied to the sample; this time, about the y-axis of the rotating frame (the axis of the pulse is changed from x to y by a 90° phase shift of the radiofrequency radiation). The pulse rotates the magnetization vectors of the faster spin packets into the positions previously occupied by the slower spin packets, and vice versa. Thus, as the vectors continue to precess, the fast vectors are now behind the slow; the fan begins to close up again, and the resultant signal begins to grow back into an echo. At time 2τ, all the vectors will once more be aligned along the y-axis, and the fanning out caused by the field inhomogeneity is said to have been **refocused**: the spin echo has reached its maximum. Because the effects of field inhomogeneities have been suppressed by the refocusing, the echo signal will have been attenuated by the factor $e^{-2\tau/T_2}$ caused by spin–spin relaxation alone. After the time 2τ, the magnetization will continue to precess, fanning out once again, giving a resultant that decays with time constant T_2^*.

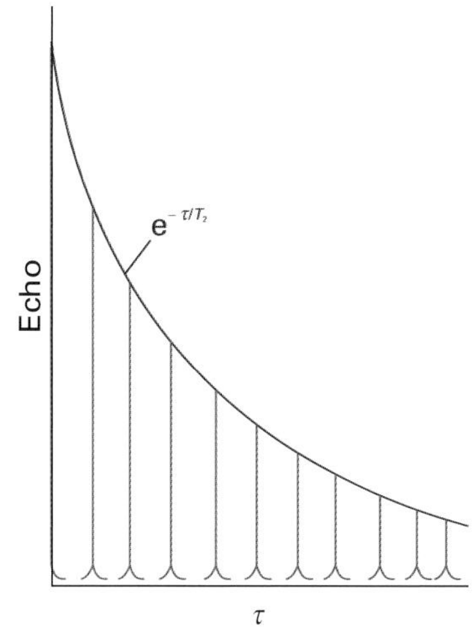

Fig. 15.40 The exponential decay of spin echoes can be used to determine the transverse relaxation time.

The important feature of the technique is that the size of the echo is independent of any local fields that remain constant during the two τ intervals. If a spin packet is 'fast' because it happens to be composed of spins in a region of the sample that experiences higher than average fields, then it remains fast throughout both intervals, and what it gains on the first interval it loses on the second interval. Hence, the size of the echo is independent of inhomogeneities in the magnetic field, for these remain constant. The true transverse relaxation arises from fields that vary on a molecular distance scale, and there is no guarantee that an individual 'fast' spin will remain 'fast' in the refocusing phase: the spins within the packets therefore spread with a time constant T_2. Hence, the effects of the true relaxation are not refocused, and the size of the echo decays with the time constant T_2 (Fig. 15.40).

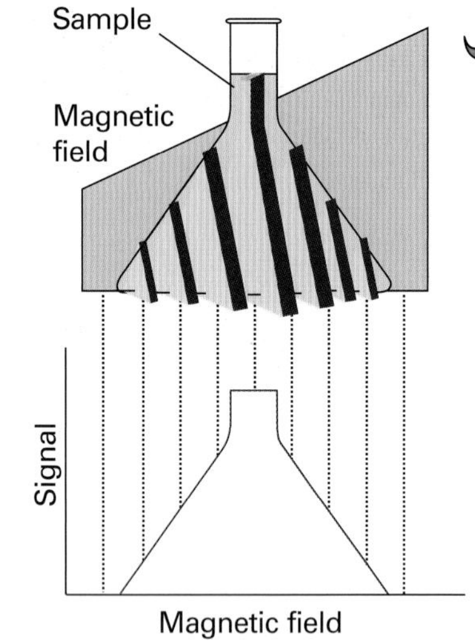

Fig. 15.41 In a magnetic field that varies linearly over a sample, all the protons within a given slice (that is, at a given field value) come into resonance and give a signal of the corresponding intensity. The resulting intensity pattern is a map of the numbers in all the slices, and portrays the shape of the sample. Changing the orientation of the field shows the shape along the corresponding direction, and computer manipulation can be used to build up the three-dimensional shape of the sample.

IMPACT ON MEDICINE

I15.1 Magnetic resonance imaging

One of the most striking applications of nuclear magnetic resonance is in medicine. *Magnetic resonance imaging* (MRI) is a portrayal of the concentrations of protons in a solid object. The technique relies on the application of specific pulse sequences to an object in an inhomogeneous magnetic field.

If an object containing hydrogen nuclei (a tube of water or a human body) is placed in an NMR spectrometer and exposed to a *homogeneous* magnetic field, then a single resonance signal will be detected. Now consider a flask of water in a magnetic field that varies linearly in the z-direction according to $\mathcal{B}_0 + G_z z$, where G_z is the field gradient along the z-direction (Fig. 15.41). Then the water protons will be resonant at the frequencies

$$\nu_L(z) = \frac{\gamma}{2\pi}(\mathcal{B}_0 + G_z z) \qquad (15.35)$$

(Similar equations may be written for gradients along the x- and y-directions.) Application of a 90° radiofrequency pulse with $\nu = \nu_L(z)$ will result in a signal with an intensity that is proportional to the numbers of protons at the position z. This is an example of *slice selection*, the application of a selective 90° pulse that excites nuclei in a specific region, or slice, of the sample. It follows that the intensity of the NMR signal will be a projection of the numbers of protons on a line parallel to the field gradient. The image of a three-dimensional object such as a flask of water can be obtained if the slice selection technique is applied at different orientations (see Fig. 15.41). In *projection reconstruction*, the projections can be analysed on a computer to reconstruct the three-dimensional distribution of protons in the object.

In practice, the NMR signal is not obtained by direct analysis of the FID curve after application of a single 90° pulse. Instead, spin echoes are often detected with several variations of the 90°–τ–180° pulse sequence (Section 15.9c). In *phase encoding*, field gradients are applied during the evolution period and the detection period of a spin-echo pulse sequence. The first step consists of a 90° pulse that results in slice selection along the z-direction. The second step consists of application of a *phase gradient*, a field gradient along the y-direction, during the evolution period. At each position along the gradient, a spin packet will precess at a different Larmor frequency due to chemical shift effects and the field inhomogeneity, so each packet will dephase to a different extent by the end of the evolution period. We can control the extent of dephasing by changing the duration of the evolution period, so Fourier transformation on τ gives information about the location of a proton along the y-direction.[1] For each value of τ, the next steps are application of the 180° pulse and then of a *read gradient*, a field gradient along the x-direction, during detection of the echo. Protons at different positions along x experience different fields and will resonate at different frequencies. Therefore Fourier transformation of the FID gives different signals for protons at different positions along x.

A common problem with the techniques described above is image contrast, which must be optimized in order to show spatial variations in water content in the sample. One strategy for solving this problem takes advantage of the fact that the relaxation times of water protons are shorter for water in biological tissues than for the pure liquid. Furthermore, relaxation times from water protons are also different in healthy and diseased tissues. A *T_1-weighted image* is obtained by repeating the spin echo sequence

[1] For technical reasons, it is more common to vary the magnitude of the phase gradient. See *Further reading* for details.

before spin–lattice relaxation can return the spins in the sample to equilibrium. Under these conditions, differences in signal intensities are directly related to differences in T_1. A *T_2-weighted image* is obtained by using an evolution period τ that is relatively long. Each point on the image is an echo signal that behaves in the manner shown in Fig. 15.40, so signal intensities are strongly dependent on variations in T_2. However, allowing so much of the decay to occur leads to weak signals even for those protons with long spin–spin relaxation times. Another strategy involves the use of *contrast agents*, paramagnetic compounds that shorten the relaxation times of nearby protons. The technique is particularly useful in enhancing image contrast and in diagnosing disease if the contrast agent is distributed differently in healthy and diseased tissues.

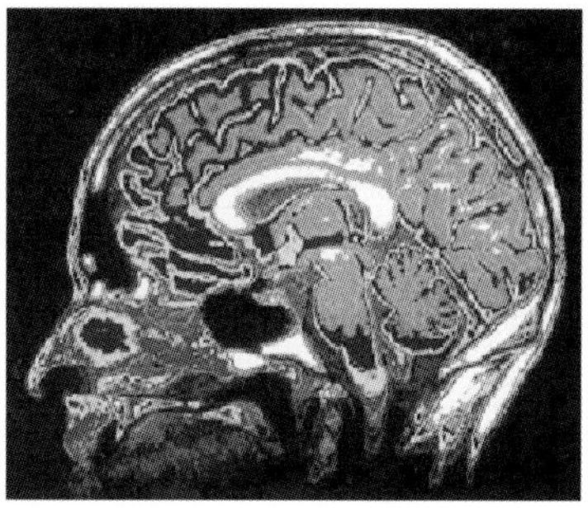

Fig. 15.42 The great advantage of MRI is that it can display soft tissue, such as in this cross-section through a patient's head. (Courtesy of the University of Manitoba.)

The MRI technique is used widely to detect physiological abnormalities and to observe metabolic processes. With *functional MRI*, blood flow in different regions of the brain can be studied and related to the mental activities of the subject. The technique is based on differences in the magnetic properties of deoxygenated and oxygenated haemoglobin, the iron-containing protein that transports O_2 in red blood cells. The more paramagnetic deoxygenated haemoglobin affects the proton resonances of tissue differently from the oxygenated protein. Because there is greater blood flow in active regions of the brain than in inactive regions, changes in the intensities of proton resonances due to changes in levels of oxygenated haemoglobin can be related to brain activity.

The special advantage of MRI is that it can image *soft* tissues (Fig. 15.42), whereas X-rays are largely used for imaging hard, bony structures and abnormally dense regions, such as tumours. In fact, the invisibility of hard structures in MRI is an advantage, as it allows the imaging of structures encased by bone, such as the brain and the spinal cord. X-rays are known to be dangerous on account of the ionization they cause; the high magnetic fields used in MRI may also be dangerous but, apart from anecdotes about the extraction of loose fillings from teeth, there is no convincing evidence of their harmfulness, and the technique is considered safe.

15.10 Spin decoupling

Carbon-13 is a **dilute-spin species** in the sense that it is unlikely that more than one ^{13}C nucleus will be found in any given small molecule (provided the sample has not been enriched with that isotope; the natural abundance of ^{13}C is only 1.1 per cent). Even in large molecules, although more than one ^{13}C nucleus may be present, it is unlikely that they will be close enough to give an observable splitting. Hence, it is not normally necessary to take into account $^{13}C—^{13}C$ spin–spin coupling within a molecule.

Protons are **abundant-spin species** in the sense that a molecule is likely to contain many of them. If we were observing a ^{13}C-NMR spectrum, we would obtain a very complex spectrum on account of the coupling of the one ^{13}C nucleus with many of the protons that are present. To avoid this difficulty, ^{13}C-NMR spectra are normally observed using the technique of **proton decoupling**. Thus, if the CH_3 protons of ethanol are irradiated with a second, strong, resonant radiofrequency pulse, they undergo rapid spin reorientations and the ^{13}C nucleus senses an average orientation. As a result, its resonance is a single line and not a 1:3:3:1 quartet. Proton decoupling has the additional advantage of enhancing sensitivity, because the intensity is concentrated into a single transition frequency instead of being spread over several transition frequencies (see Section 15.11). If care is taken to ensure that the other parameters on which the strength of the signal depends are kept constant, the intensities of proton-decoupled spectra are proportional to the number of ^{13}C nuclei present. The technique is widely used to characterize synthetic polymers.

Comment 15.7

In a dipole–dipole interaction between two nuclei, one nucleus influences the behaviour of another nucleus in much the same way that the orientation of a bar magnet is influenced by the presence of another bar magnet nearby. Dipole–dipole interactions are discussed in Chapter 18.

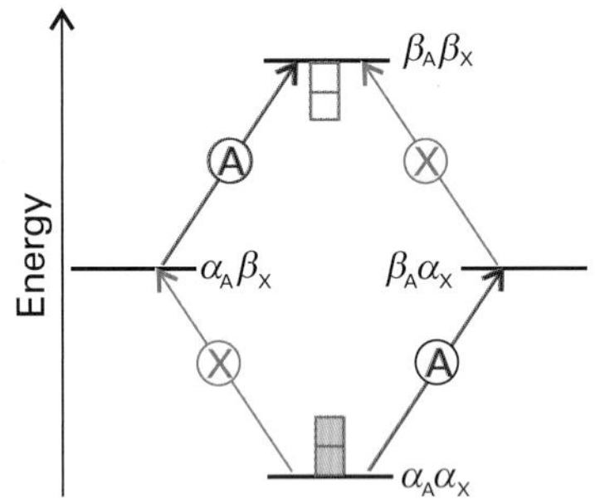

Fig. 15.43 The energy levels of an AX system and an indication of their relative populations. Each grey square above the line represents an excess population and each white square below the line represents a population deficit. The transitions of A and X are marked.

15.11 The nuclear Overhauser effect

We have seen already that one advantage of protons in NMR is their high magnetogyric ratio, which results in relatively large Boltzmann population differences and hence greater resonance intensities than for most other nuclei. In the steady-state **nuclear Overhauser effect** (NOE), spin relaxation processes involving internuclear dipole–dipole interactions are used to transfer this population advantage to another nucleus (such as ^{13}C or another proton), so that the latter's resonances are modified.

To understand the effect, we consider the populations of the four levels of a homonuclear (for instance, proton) AX system; these were shown in Fig. 15.12. At thermal equilibrium, the population of the $\alpha_A\alpha_X$ level is the greatest, and that of the $\beta_A\beta_X$ level is the least; the other two levels have the same energy and an intermediate population. The thermal equilibrium absorption intensities reflect these populations as shown in Fig. 15.43. Now consider the combined effect of spin relaxation and keeping the X spins saturated. When we saturate the X transition, the populations of the X levels are equalized ($N_{\alpha X} = N_{\beta X}$) and all transitions involving $\alpha_X \leftrightarrow \beta_X$ spin flips are no longer observed. At this stage there is no change in the populations of the A levels. If that were all there were to happen, all we would see would be the loss of the X resonance and no effect on the A resonance.

Now consider the effect of spin relaxation. Relaxation can occur in a variety of ways if there is a dipolar interaction between the A and X spins. One possibility is for the magnetic field acting between the two spins to cause them both to flop from β to α, so the $\alpha_A\alpha_X$ and $\beta_A\beta_X$ states regain their thermal equilibrium populations. However, the populations of the $\alpha_A\beta_X$ and $\beta_A\alpha_X$ levels remain unchanged at the values characteristic of saturation. As we see from Fig. 15.44, the population difference between the states joined by transitions of A is now greater than at equilibrium, so the resonance absorption is enhanced. Another possibility is for the dipolar interaction between the two spins to cause α to flip to β and β to flop to α. This transition equilibrates the populations of $\alpha_A\beta_X$ and $\beta_A\alpha_X$ but leaves the $\alpha_A\alpha_X$ and $\beta_A\beta_X$ populations unchanged. Now we see from the illustration that the population differences in the states involved in the A transitions are decreased, so the resonance absorption is diminished.

Which effect wins? Does the NOE enhance the A absorption or does it diminish it? As in the discussion of relaxation times in Section 15.9, the efficiency of the intensity-enhancing $\beta_A\beta_X \leftrightarrow \alpha_A\alpha_X$ relaxation is high if the dipole field oscillates at a frequency close to the transition frequency, which in this case is about 2ν; likewise, the efficiency

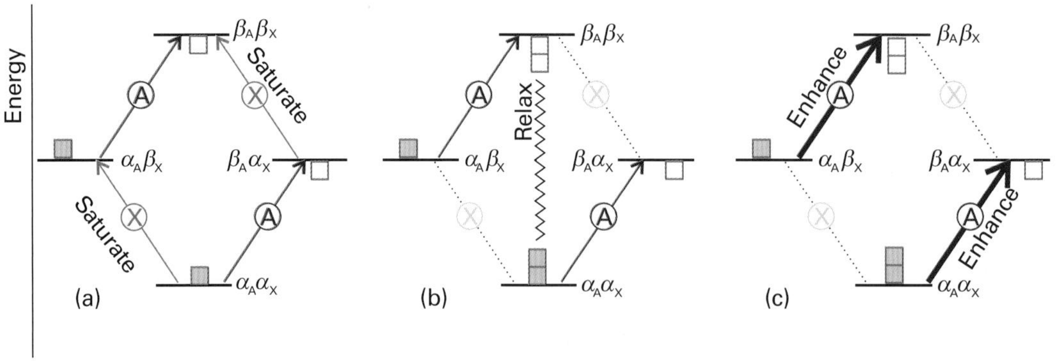

Fig. 15.44 (a) When the X transition is saturated, the populations of its two states are equalized and the population excess and deficit become as shown (using the same symbols as in Fig. 15.43). (b) Dipole–dipole relaxation relaxes the populations of the highest and lowest states, and they regain their original populations. (c) The A transitions reflect the difference in populations resulting from the preceding changes, and are enhanced compared with those shown in Fig. 15.43.

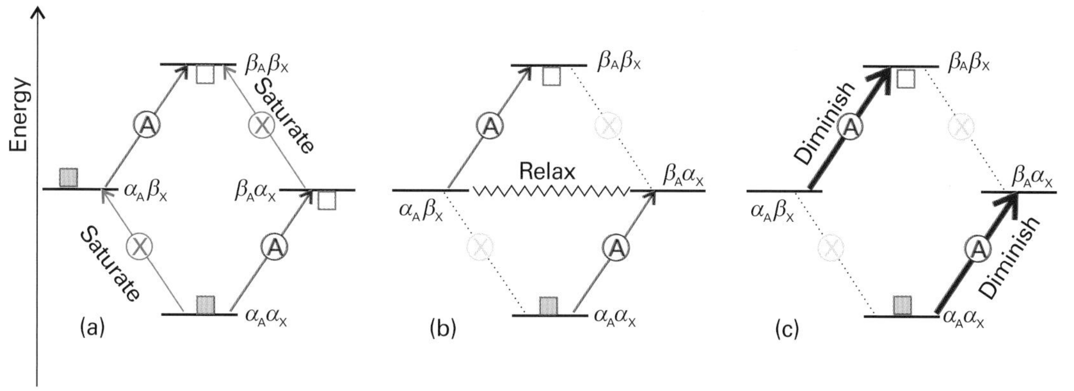

Fig. 15.45 (a) When the X transition is saturated, just as in Fig. 15.44 the populations of its two states are equalized and the population excess and deficit become as shown. (b) Dipole–dipole relaxation relaxes the populations of the two intermediate states, and they regain their original populations. (c) The A transitions reflect the difference in populations resulting from the preceding changes, and are diminished compared with those shown in Fig. 15.41.

of the intensity-diminishing $\alpha_A\beta_X \leftrightarrow \beta_A\alpha_X$ relaxation is high if the dipole field is stationary (as there is no frequency difference between the initial and final states). A large molecule rotates so slowly that there is very little motion at 2ν, so we expect an intensity decrease (Fig. 15.45). A small molecule rotating rapidly can be expected to have substantial motion at 2ν, and a consequent enhancement of the signal. In practice, the enhancement lies somewhere between the two extremes and is reported in terms of the parameter η (eta), where

$$\eta = \frac{I_A - I_A^\circ}{I_A^\circ} \qquad [15.36]$$

Here I_A° and I_A are the intensities of the NMR signals due to nucleus A before and after application of the long ($> T_1$) radiofrequency pulse that saturates transitions due to the X nucleus. When A and X are nuclei of the same species, such as protons, η lies between -1 (diminution) and $+\frac{1}{2}$ (enhancement). However, η also depends on the values of the magnetogyric ratios of A and X. In the case of maximal enhancement it is possible to show that

$$\eta = \frac{\gamma_X}{2\gamma_A} \qquad (15.37)$$

where γ_A and γ_X are the magnetogyric ratios of nuclei A and X, respectively. For ^{13}C close to a saturated proton, the ratio evaluates to 1.99, which shows that an enhancement of about a factor of 2 can be achieved.

The NOE is also used to determine interproton distances. The Overhauser enhancement of a proton A generated by saturating a spin X depends on the fraction of A's spin–lattice relaxation that is caused by its dipolar interaction with X. Because the dipolar field is proportional to r^{-3}, where r is the internuclear distance, and the relaxation effect is proportional to the square of the field, and therefore to r^{-6}, the NOE may be used to determine the geometries of molecules in solution. The determination of the structure of a small protein in solution involves the use of several hundred NOE measurements, effectively casting a net over the protons present. The enormous importance of this procedure is that we can determine the conformation of biological macromolecules in an aqueous environment and do not need to try to make the single crystals that are essential for an X-ray diffraction investigation (Chapter 20).

15.12 Two-dimensional NMR

An NMR spectrum contains a great deal of information and, if many protons are present, is very complex. Even a first-order spectrum is complex, for the fine structure of different groups of lines can overlap. The complexity would be reduced if we could use two axes to display the data, with resonances belonging to different groups lying at different locations on the second axis. This separation is essentially what is achieved in **two-dimensional NMR**.

All two-dimensional NMR experiments use the **PEMD pulse structure**, which consists of:

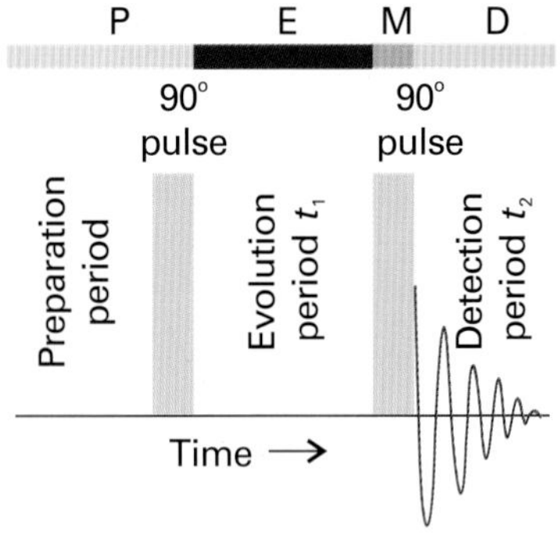

Fig. 15.46 The pulse sequence used in correlation spectroscopy (COSY). The preparation period is much longer than either T_1 or T_2, so the spins have time to relax before the next cycle of pulses begins. Acquisitions of free-induction decays are taken during t_2 for a set of different evolution times t_1. Fourier transformation on both variables t_1 and t_2 results in a two-dimensional spectrum, such as that shown in Fig 15.52.

P: a *preparation period*, in which the spins first return to thermal equilibrium and then are excited by one or more radiofrequency pulses

E: an *evolution period* of duration t_1, during which the spins precess under the influence of their chemical shifts and spin–spin couplings

M: a *mixing period*, in which pulses may be used to transfer information between spins

D: a *detection period* of duration t_2, during which the FID is recorded.

Now we shall see how the PEMD pulse structure can be used to devise experiments that reveal spin–spin couplings and internuclear distances in small and large molecules.

(a) Correlation spectroscopy

Much modern NMR work makes use of techniques such as **correlation spectroscopy** (COSY) in which a clever choice of pulses and Fourier transformation techniques makes it possible to determine all spin–spin couplings in a molecule. The basic COSY experiment uses the simplest of all two-dimensional pulse sequences, consisting of two consecutive 90° pulses (Fig. 15.46).

To see how we can obtain a two-dimensional spectrum from a COSY experiment, we consider a trivial but illustrative example: the spectrum of a compound containing one proton, such as trichloromethane (chloroform, $CHCl_3$). Figure 15.47 shows the effect of the pulse sequence on the magnetization of the sample, which is aligned initially along the z-axis with a magnitude M_0. A 90° pulse applied in the x-direction (in the stationary frame) tilts the magnetization vector toward the y-axis. Then, during the evolution period, the magnetization vector rotates in the xy-plane with a frequency ν. At a time t_1 the vector will have swept through an angle $2\pi\nu t_1$ and the magnitude of the magnetization will have decayed by spin–spin relaxation to $M = M_0 e^{-t_1/T_2}$. By trigonometry, the magnitudes of the components of the magnetization vector are:

$$M_x = M \sin 2\pi\nu t_1 \qquad M_y = M \cos 2\pi\nu t_1 \qquad M_z = 0 \tag{15.38a}$$

Application of the second 90° pulse parallel to the x-axis tilts the magnetization again and the resulting vector has components with magnitudes (once again, in the stationary frame)

$$M_x = M \sin 2\pi\nu t_1 \qquad M_y = 0 \qquad M_z = M \cos 2\pi\nu t_1 \tag{15.38b}$$

The FID is detected over a period t_2 and Fourier transformation yields a signal over a frequency range ν_2 with a peak at ν, the resonance frequency of the proton. The signal intensity is related to M_x, the magnitude of the magnetization that is rotating around the xy-plane at the time of application of the detection pulse, so it follows that the signal strength varies sinusoidally with the duration of the evolution period. That is, if we were to acquire a series of spectra at different evolution times t_1, then we would obtain data as shown in Fig. 15.48a.

Comment 15.8

A vector, v, of length v, in the xy-plane and its two components, v_x and v_y, can be thought of as forming a right-angled triangle, with v the length of the hypotenuse (see the illustration). If θ is the angle that v_y makes with v, then it follows that $v_x = v \sin \theta$ and $v_y = v \cos \theta$.

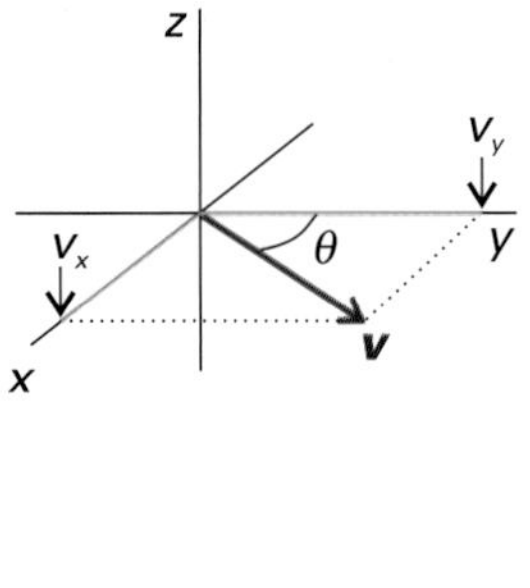

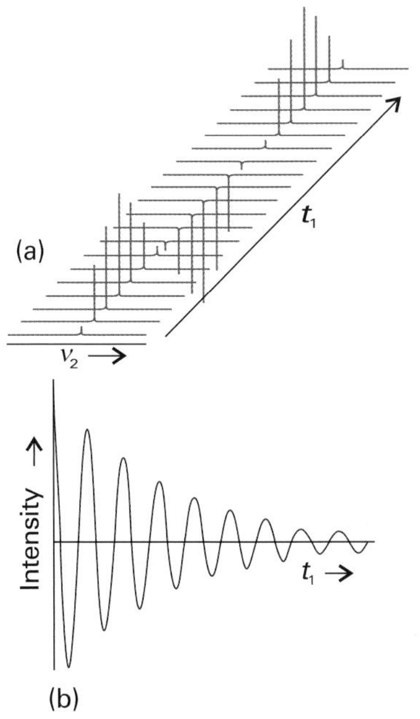

Fig. 15.48 (a) Spectra acquired for different evolution times t_1 between two 90° pulses. (b) A plot of the maximum intensity of each absorption line against t_1. Fourier transformation of this plot leads to a spectrum centred at ν, the resonance frequency of the protons in the sample.

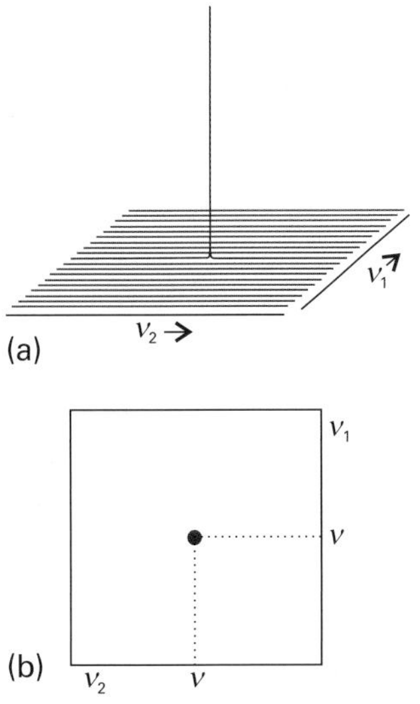

Fig. 15.49 (a) The two-dimensional NMR spectrum of the sample discussed in Figs. 15.47 and 15.48. See the text for an explanation of how the spectrum is obtained from a series of Fourier transformations of the data. (b) The contour plot of the spectrum in (a).

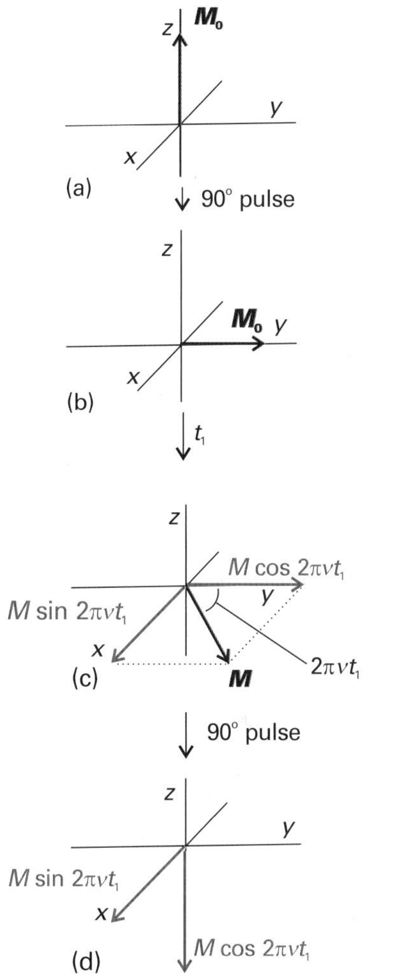

Fig. 15.47 (a) The effect of the pulse sequence shown in Fig. 15.46 on the magnetization $\boldsymbol{M}_0$ of a sample of a compound with only one proton. (b) A 90° pulse applied in the x-direction tilts the magnetization vector toward the y-axis. (c) After a time t_1 has elapsed, the vector will have swept through an angle $2\pi\nu t_1$ and the magnitude of the magnetization will have decayed to $\boldsymbol{M}$. The magnitudes of the components of $\boldsymbol{M}$ are $M_x = M\sin 2\pi\nu t_1$, $M_y = M\cos 2\pi\nu t_1$, and $M_z = 0$. (d) Application of the second 90° pulse parallel to the x-axis tilts the magnetization again and the resulting vector has components with magnitude $M_x = M\sin 2\pi\nu t_1$, $M_y = 0$, and $M_z = M\cos 2\pi\nu t_1$. The FID is detected at this stage of the experiment.

A plot of the maximum intensity of each absorption band in Fig. 15.48a against t_1 has the form shown in Fig. 15.48b. The plot resembles an FID curve with the oscillating component having a frequency ν, so Fourier transformation yields a signal over a frequency range ν_1 with a peak at ν. If we continue the process by first plotting signal intensity against t_1 for several frequencies along the ν_2 axis and then carrying out Fourier transformations, we generate a family of curves that can be pooled together into a three-dimensional plot of $I(\nu_1,\nu_2)$, the signal intensity as a function of the frequencies ν_1 and ν_2 (Fig. 15.49a). This plot is referred to as a *two-dimensional NMR spectrum* because Fourier transformations were performed in two variables. The most common representation of the data is as a contour plot, such as the one shown in Fig. 15.49b.

The experiment described above is not necessary for as simple a system as chloroform because the information contained in the two-dimensional spectrum could have been obtained much more quickly through the conventional, one-dimensional approach. However, when the one-dimensional spectrum is complex, the COSY experiment shows which spins are related by spin–spin coupling. To justify this statement, we now examine a spin-coupled AX system.

From our discussion so far, we know that the key to the COSY technique is the effect of the second 90° pulse. In this more complex example we consider its role for the four energy levels of an AX system (as shown in Fig. 15.12). At thermal equilibrium, the population of the $\alpha_A\alpha_X$ level is the greatest, and that of the $\beta_A\beta_X$ level is the least; the other two levels have the same energy and an intermediate population. After the first 90° pulse, the spins are no longer at thermal equilibrium. If a second 90° pulse

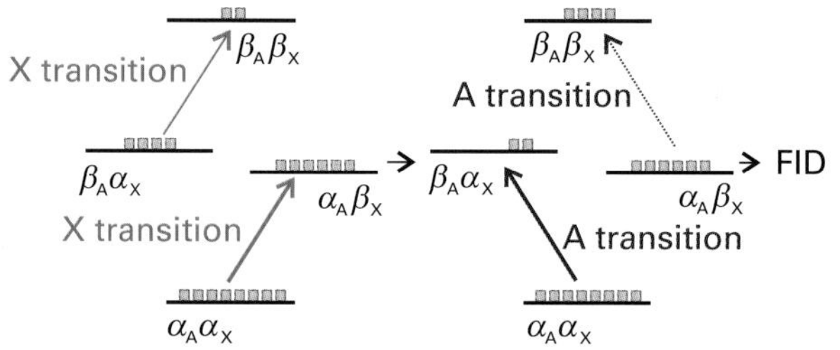

Fig. 15.50 An example of the change in the population of energy levels of an AX spin system that results from the second 90° pulse of a COSY experiment. Each square represents the same large number of spins. In this example, we imagine that the pulse affects the X spins first, and then the A spins. Excitation of the X spins inverts the populations of the $\beta_A\beta_X$ and $\beta_A\alpha_X$ levels and does not affect the populations of the $\alpha_A\alpha_X$ and $\alpha_A\beta_X$ levels. As a result, excitation of the A spins by the pulse generates an FID in which one of the two A transitions has increased in intensity and the other has decreased. That is, magnetization has been transferred from the X spins to the A spins. Similar schemes can be written to show that magnetization can be transferred from the A spins to the X spins.

is applied at a time t_1 that is short compared to the spin–lattice relaxation time T_1, the extra input of energy causes further changes in the populations of the four states. The changes in populations of the four states of the AX system will depend on how far the individual magnetizations have precessed during the evolution period. It is difficult to visualize these changes because the A spins are affecting the X spins and vice-versa.

For simplicity, we imagine that the second pulse induces X and A transitions sequentially. Depending on the evolution time t_1, the 90° pulse may leave the population differences across each of the two X transitions unchanged, inverted, or somewhere in between. Consider the extreme case in which one population difference is inverted and the other unchanged (Fig. 15.50). Excitation of the A transitions will now generate an FID in which one of the two A transitions has increased in intensity (because the population difference is now greater), and the other has decreased (because the population difference is now smaller). The overall effect is that precession of the X spins during the evolution period determines the amplitudes of the signals from the A spins obtained during the detection period. As the evolution time t_1 is increased, the intensities of the signals from A spins oscillate with frequencies determined by the frequencies of the two X transitions. Of course, it is just as easy to turn our scenario around and to conclude that the intensities of signals from X spins oscillate with frequencies determined by the frequencies of the A transitions.

This transfer of information between spins is at the heart of two-dimensional NMR spectroscopy: it leads to the *correlation* between different signals in a spectrum. In this case, information transfer tells us that there is spin–spin coupling between A and X. So, just as before, if we conduct a series of experiments in which t_1 is incremented, Fourier transformation of the FIDs on t_2 yields a set of spectra $I(t_1,F_2)$ in which the signal amplitudes oscillate as a function of t_1. A second Fourier transformation, now on t_1, converts these oscillations into a two-dimensional spectrum $I(F_1,F_2)$. The signals are spread out in F_1 according to their precession frequencies during the detection period. Thus, if we apply the COSY pulse sequence (Fig. 15.46) to the AX spin system, the result is a two-dimensional spectrum that contains four groups of signals in F_1 and F_2 centred on the two chemical shifts (Fig. 15.51). Each group consists of a block of four signals separated by J. The **diagonal peaks** are signals centred on (δ_A,δ_A) and (δ_X,δ_X) and lie along the diagonal $F_1 = F_2$. That is, the spectrum along the diagonal is equivalent to the one-dimensional spectrum obtained with the conventional NMR technique (Fig. 15.13). The **cross-peaks** (or *off-diagonal peaks*) are signals centred on (δ_A,δ_X) and (δ_X,δ_A) and owe their existence to the coupling between A and X.

Although information from two-dimensional NMR spectroscopy is trivial in an AX system, it can be of enormous help in the interpretation of more complex spectra, leading to a map of the couplings between spins and to the determination of the bonding network in complex molecules. Indeed, the spectrum of a synthetic or biological polymer that would be impossible to interpret in one-dimensional NMR but can often be interpreted reasonably rapidly by two-dimensional NMR. Below we illustrate the procedure by assigning the resonances in the COSY spectrum of an amino acid.

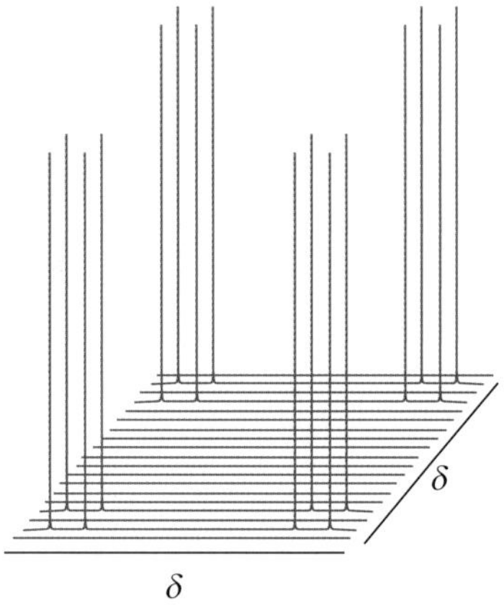

Fig. 15.51 A representation of the two-dimensional NMR spectrum obtained by application of the COSY pulse sequence to an AX spin system.

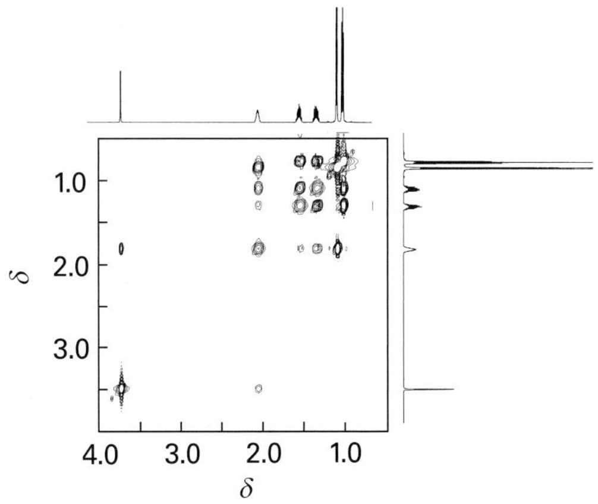

Fig. 15.52 Proton COSY spectrum of isoleucine. (Adapted from K.E. van Holde, W.C. Johnson, and P.S. Ho, *Principles of physical biochemistry*, p. 508, Prentice Hall, Upper Saddle River (1998).)

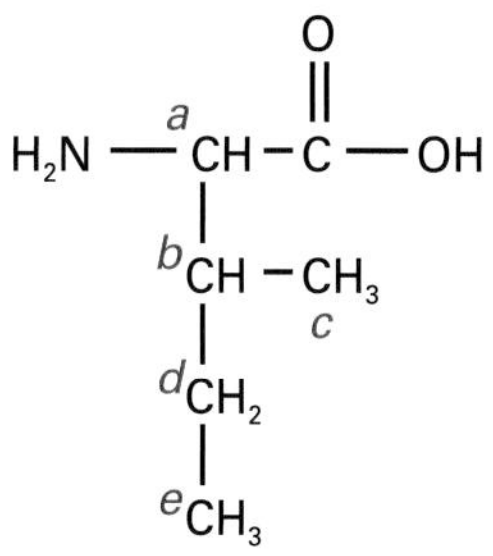

7 Isoleucine

Illustration 15.3 *The COSY spectrum of isoleucine*

Figure 15.52 is a portion of the COSY spectrum of the amino acid isoleucine (**7**), showing the resonances associated with the protons bound to the carbon atoms. We begin the assignment process by considering which protons should be interacting by spin–spin coupling. From the known molecular structure, we conclude that:

1. The C_a—H proton is coupled only to the C_b—H proton.
2. The C_b—H protons are coupled to the C_a—H, C_c—H, and C_d—H protons.
3. The inequivalent C_d—H protons are coupled to the C_b—H and C_e—H protons.

We now note that:

- The resonance with $\delta = 3.6$ shares a cross-peak with only one other resonance at $\delta = 1.9$, which in turn shares cross-peaks with resonances at $\delta = 1.4$, 1.2, and 0.9. This identification is consistent with the resonances at $\delta = 3.6$ and 1.9 corresponding to the C_a—H and C_b—H protons, respectively.
- The proton with resonance at $\delta = 0.8$ is not coupled to the C_b—H protons, so we assign the resonance at $\delta = 0.8$ to the C_e—H protons.
- The resonances at $\delta = 1.4$ and 1.2 do not share cross-peaks with the resonance at $\delta = 0.9$.

In the light of the expected couplings, we assign the resonance at $\delta = 0.9$ to the C_c—H protons and the resonances at $\delta = 1.4$ and 1.2 to the inequivalent C_d—H protons.

Our simplified description of the COSY experiment does not reveal some important details. For example, the second 90° pulse actually mixes the spin state transitions caused by the first 90° pulse (hence the term 'mixing period'). Each of the four transitions (two for A and two for X) generated by the first pulse can be converted into any of the other three, or into formally forbidden **multiple quantum transitions**, which have $|\Delta m| > 1$. The latter transitions cannot generate any signal in the receiver coil of the spectrometer, but their existence can be demonstrated by applying a third pulse to mix them back into the four observable single quantum transitions. Many modern NMR experiments exploit multiple quantum transitions to filter out unwanted signals and to simplify spectra for interpretation.

(b) Nuclear Overhauser effect spectroscopy

Many different two-dimensional NMR experiments are based on the PEMD pulse structure. We have seen that the steady-state nuclear Overhauser effect can provide information about internuclear distances through analysis of enhancement patterns in the NMR spectrum before and after saturation of selected resonances. In its dynamic analogue, **nuclear Overhauser effect spectroscopy** (NOESY), the second 90° pulse of the COSY experiment is replaced by two 90° pulses separated by a time delay during which dipole–dipole interactions cause magnetization to be exchanged between neighbouring spins (Fig. 15.50). The results of double Fourier transformation is a spectrum in which the cross-peaks form a map of all the NOE interactions in a molecule. Because the nuclear Overhauser effect depends on the inverse sixth power of the separation between nuclei (see Section 15.11), NOESY data reveal internuclear distances up to about 0.5 nm.

15.13 Solid-state NMR

The principal difficulty with the application of NMR to solids is the low resolution characteristic of solid samples. Nevertheless, there are good reasons for seeking to overcome these difficulties. They include the possibility that a compound of interest is unstable in solution or that it is insoluble, so conventional solution NMR cannot be employed. Moreover, many species are intrinsically interesting as solids, and it is important to determine their structures and dynamics. Synthetic polymers are particularly interesting in this regard, and information can be obtained about the arrangement of molecules, their conformations, and the motion of different parts of the chain. This kind of information is crucial to an interpretation of the bulk properties of the polymer in terms of its molecular characteristics. Similarly, inorganic substances, such as the zeolites that are used as molecular sieves and shape-selective catalysts, can be studied using solid-state NMR, and structural problems can be resolved that cannot be tackled by X-ray diffraction.

Problems of resolution and linewidth are not the only features that plague NMR studies of solids, but the rewards are so great that considerable efforts have been made to overcome them and have achieved notable success. Because molecular rotation has almost ceased (except in special cases, including 'plastic crystals' in which the molecules continue to tumble), spin–lattice relaxation times are very long but spin–spin relaxation times are very short. Hence, in a pulse experiment, there is a need for lengthy delays—of several seconds—between successive pulses so that the spin system has time to revert to equilibrium. Even gathering the murky information may therefore be a lengthy process. Moreover, because lines are so broad, very high powers of radiofrequency radiation may be required to achieve saturation. Whereas solution pulse NMR uses transmitters of a few tens of watts, solid-state NMR may require transmitters rated at several hundreds of watts.

(a) The origins of linewidths in solids

There are two principal contributions to the linewidths of solids. One is the direct magnetic dipolar interaction between nuclear spins. As we saw in the discussion of spin–spin coupling, a nuclear magnetic moment will give rise to a local magnetic field, which points in different directions at different locations around the nucleus. If we are interested only in the component parallel to the direction of the applied magnetic field (because only this component has a significant effect), then we can use a classical expression to write the magnitude of the local magnetic field as

$$\mathcal{B}_{\text{loc}} = -\frac{\gamma \hbar \mu_0 m_I}{4\pi R^3}(1 - 3\cos^2\theta) \tag{15.39}$$

Unlike in solution, this field is not motionally averaged to zero. Many nuclei may contribute to the total local field experienced by a nucleus of interest, and different nuclei in a sample may experience a wide range of fields. Typical dipole fields are of the order of 10^{-3} T, which corresponds to splittings and linewidths of the order of 10^4 Hz.

A second source of linewidth is the anisotropy of the chemical shift. We have seen that chemical shifts arise from the ability of the applied field to generate electron currents in molecules. In general, this ability depends on the orientation of the molecule relative to the applied field. In solution, when the molecule is tumbling rapidly, only the average value of the chemical shift is relevant. However, the anisotropy is not averaged to zero for stationary molecules in a solid, and molecules in different orientations have resonances at different frequencies. The chemical shift anisotropy also varies with the angle between the applied field and the principal axis of the molecule as $1 - 3\cos^2\theta$.

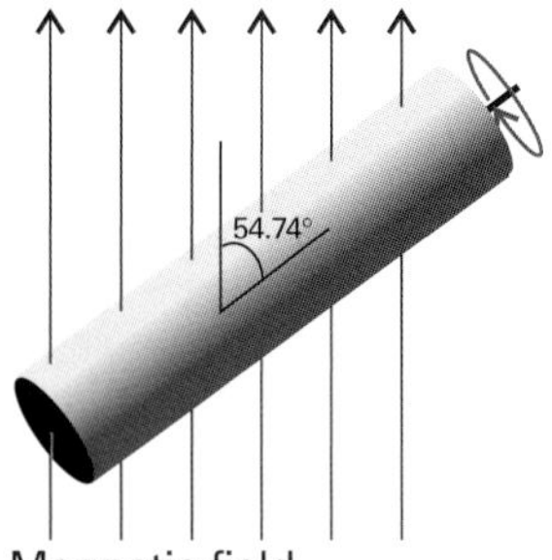

Fig. 15.53 In magic angle spinning, the sample spins at 54.74° (that is, arccos $(\frac{1}{3})^{1/2}$) to the applied magnetic field. Rapid motion at this angle averages dipole–dipole interactions and chemical shift anisotropies to zero.

(b) The reduction of linewidths

Fortunately, there are techniques available for reducing the linewidths of solid samples. One technique, **magic-angle spinning** (MAS), takes note of the $1 - 3\cos^2\theta$ dependence of both the dipole–dipole interaction and the chemical shift anisotropy. The 'magic angle' is the angle at which $1 - 3\cos^2\theta = 0$, and corresponds to 54.74°. In the technique, the sample is spun at high speed at the magic angle to the applied field (Fig. 15.53). All the dipolar interactions and the anisotropies average to the value they would have at the magic angle, but at that angle they are zero. The difficulty with MAS is that the spinning frequency must not be less than the width of the spectrum, which is of the order of kilohertz. However, gas-driven sample spinners that can be rotated at up to 25 kHz are now routinely available, and a considerable body of work has been done.

Pulsed techniques similar to those described in the previous section may also be used to reduce linewidths. The dipolar field of protons, for instance, may be reduced by a decoupling procedure. However, because the range of coupling strengths is so large, radiofrequency power of the order of 1 kW is required. Elaborate pulse sequences have also been devised that reduce linewidths by averaging procedures that make use of twisting the magnetization vector through an elaborate series of angles.

Electron paramagnetic resonance

Electron paramagnetic resonance (EPR) is less widely applicable than NMR because it cannot be detected in normal, spin-paired molecules and the sample must possess unpaired electron spins. It is used to study radicals formed during chemical reactions or by radiation, radicals that act as probes of biological structure, many *d*-metal complexes, and molecules in triplet states (such as those involved in phosphorescence, Section 14.3b). The sample may be a gas, a liquid, or a solid, but the free rotation of molecules in the gas phase gives rise to complications.

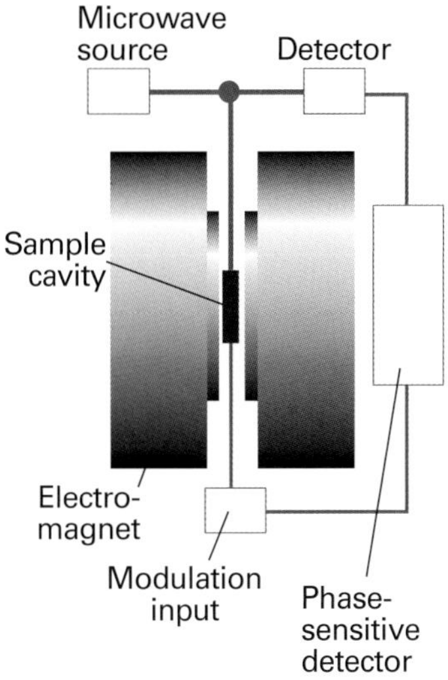

Fig. 15.54 The layout of a continuous-wave EPR spectrometer. A typical magnetic field is 0.3 T, which requires 9 GHz (3 cm) microwaves for resonance.

15.14 The EPR spectrometer

Both Fourier-transform (FT) and continuous wave (CW) EPR spectrometers are available. The FT-EPR instrument is based on the concepts developed in Section 15.8, except that pulses of microwaves are used to excite electron spins in the sample. The layout of the more common CW-EPR spectrometer is shown in Fig. 15.54. It consists

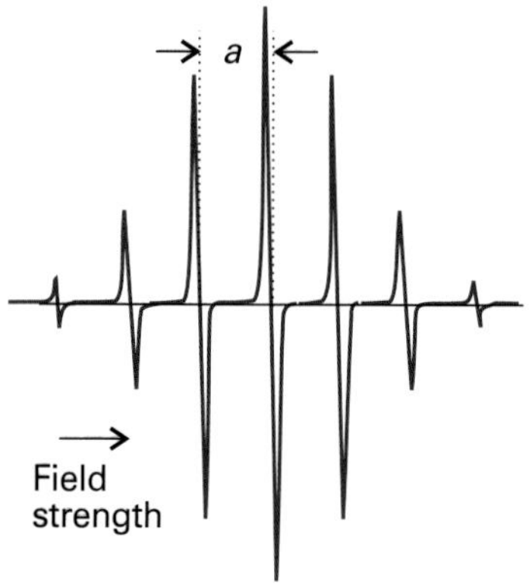

Fig. 15.55 The EPR spectrum of the benzene radical anion, $C_6H_6^-$, in fluid solution. *a* is the hyperfine splitting of the spectrum; the centre of the spectrum is determined by the *g*-value of the radical.

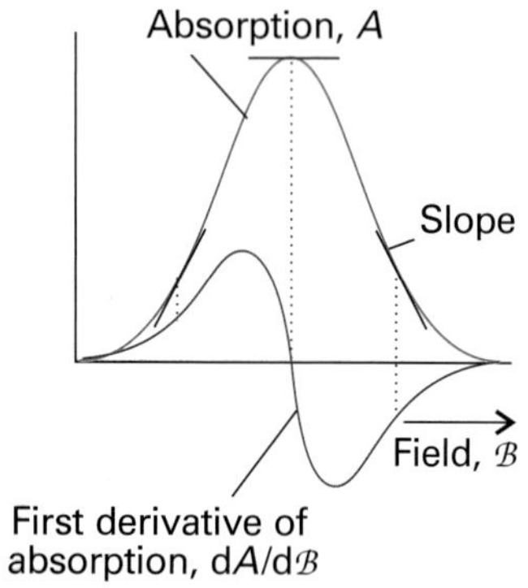

Fig. 15.56 When phase-sensitive detection is used, the signal is the first derivative of the absorption intensity. Note that the peak of the absorption corresponds to the point where the derivative passes through zero.

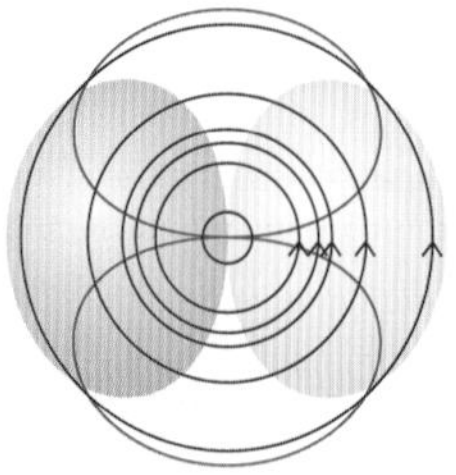

Fig. 15.57 An applied magnetic field can induce circulation of electrons that makes use of excited state orbitals (shown as a green outline).

of a microwave source (a klystron or a Gunn oscillator), a cavity in which the sample is inserted in a glass or quartz container, a microwave detector, and an electromagnet with a field that can be varied in the region of 0.3 T. The EPR spectrum is obtained by monitoring the microwave absorption as the field is changed, and a typical spectrum (of the benzene radical anion, $C_6H_6^-$) is shown in Fig. 15.55. The peculiar appearance of the spectrum, which is in fact the first-derivative of the absorption, arises from the detection technique, which is sensitive to the slope of the absorption curve (Fig. 15.56).

15.15 The *g*-value

Equation 15.13b gives the resonance frequency for a transition between the $m_s = -\frac{1}{2}$ and the $m_s = +\frac{1}{2}$ levels of a 'free' electron in terms of the *g*-value $g_e \approx 2.0023$. The magnetic moment of an unpaired electron in a radical also interacts with an external field, but the *g*-value is different from that for a free electron because of local magnetic fields induced by the molecular framework of the radical. Consequently, the resonance condition is normally written as

$$h\nu = g\mu_B\mathcal{B}_0 \qquad (15.40)$$

where *g* is the ***g*-value** of the radical.

Illustration 15.4 *Calculating the g-value of an organic radical*

The centre of the EPR spectrum of the methyl radical occurred at 329.40 mT in a spectrometer operating at 9.2330 GHz (radiation belonging to the X band of the microwave region). Its *g*-value is therefore

$$g = \frac{h\nu}{\mu_B\mathcal{B}} = \frac{(6.626\,08\times10^{-34}\,\text{J s})\times(9.2330\times10^{9}\,\text{s}^{-1})}{(9.2740\times10^{-24}\,\text{J T}^{-1})\times(0.329\,40\,\text{T})} = 2.0027$$

Self-test 15.4 At what magnetic field would the methyl radical come into resonance in a spectrometer operating at 34.000 GHz (radiation belonging to the Q band of the microwave region)? [1.213 T]

The *g*-value in a molecular environment (a radical or a *d*-metal complex) is related to the ease with which the applied field can stir up currents through the molecular framework and the strength of the magnetic field the currents generate. Therefore, the *g*-value gives some information about electronic structure and plays a similar role in EPR to that played by shielding constants in NMR.

Electrons can migrate through the molecular framework by making use of excited states (Fig. 15.57). This additional path for circulation of electrons gives rise to a local magnetic field that adds to the applied field. Therefore, we expect that the ease of stirring up currents to be inversely proportional to the separation of energy levels, ΔE, in the molecule. As we saw in Section 10.8, the strength of the field generated by electronic currents in atoms (and analogously in molecules) is related to the extent of coupling between spin and orbital angular momenta. That is, the local field strength is proportional to the molecular spin–orbit coupling constant, ξ.

We can conclude from the discussion above that the *g*-value of a radical or *d*-metal complex differs from g_e, the 'free-electron' *g*-value, by an amount that is proportional to $\xi/\Delta E$. This proportionality is widely observed. Many organic radicals have *g*-values close to 2.0027 and inorganic radicals have *g*-values typically in the range 1.9 to 2.1. The *g*-values of paramagnetic *d*-metal complexes often differ considerably from g_e, varying from 0 to 6, because in them ΔE is small (on account of the splitting of *d*-orbitals brought about by interactions with ligands, as we saw in Section 14.2).

Just as in the case of the chemical shift in NMR spectroscopy, the *g*-value is anisotropic: that is, its magnitude depends on the orientation of the radical with respect to the applied field. In solution, when the molecule is tumbling rapidly, only the average value of the *g*-value is observed. Therefore, anisotropy of the *g*-value is observed only for radicals trapped in solids.

15.16 Hyperfine structure

The most important feature of EPR spectra is their **hyperfine structure**, the splitting of individual resonance lines into components. In general in spectroscopy, the term 'hyperfine structure' means the structure of a spectrum that can be traced to interactions of the electrons with nuclei other than as a result of the latter's point electric charge. The source of the hyperfine structure in EPR is the magnetic interaction between the electron spin and the magnetic dipole moments of the nuclei present in the radical.

(a) The effects of nuclear spin

Consider the effect on the EPR spectrum of a single H nucleus located somewhere in a radical. The proton spin is a source of magnetic field, and depending on the orientation of the nuclear spin, the field it generates adds to or subtracts from the applied field. The total local field is therefore

$$\mathcal{B}_{\text{loc}} = \mathcal{B} + am_I \qquad m_I = \pm\tfrac{1}{2} \tag{15.41}$$

where a is the **hyperfine coupling constant**. Half the radicals in a sample have $m_I = +\frac{1}{2}$, so half resonate when the applied field satisfies the condition

$$h\nu = g\mu_{\text{B}}(\mathcal{B} + \tfrac{1}{2}a), \qquad \text{or} \qquad \mathcal{B} = \frac{h\nu}{g\mu_{\text{B}}} - \tfrac{1}{2}a \tag{15.42a}$$

The other half (which have $m_I = -\frac{1}{2}$) resonate when

$$h\nu = g\mu_{\text{B}}(\mathcal{B} - \tfrac{1}{2}a), \qquad \text{or} \qquad \mathcal{B} = \frac{h\nu}{g\mu_{\text{B}}} + \tfrac{1}{2}a \tag{15.42b}$$

Therefore, instead of a single line, the spectrum shows two lines of half the original intensity separated by a and centred on the field determined by g (Fig. 15.58).

If the radical contains an ^{14}N atom ($I = 1$), its EPR spectrum consists of three lines of equal intensity, because the ^{14}N nucleus has three possible spin orientations, and each spin orientation is possessed by one-third of all the radicals in the sample. In general, a spin-I nucleus splits the spectrum into $2I + 1$ hyperfine lines of equal intensity.

When there are several magnetic nuclei present in the radical, each one contributes to the hyperfine structure. In the case of equivalent protons (for example, the two CH_2 protons in the radical CH_3CH_2) some of the hyperfine lines are coincident. It is not hard to show that, if the radical contains N equivalent protons, then there are $N + 1$ hyperfine lines with a binomial intensity distribution (the intensity distribution given by Pascal's triangle). The spectrum of the benzene radical anion in Fig. 15.55, which has seven lines with intensity ratio 1:6:15:20:15:6:1, is consistent with a radical containing six equivalent protons. More generally, if the radical contains N equivalent nuclei with spin quantum number I, then there are $2NI + 1$ hyperfine lines with an intensity distribution based on a modified version of Pascal's triangle as shown in the following *Example*.

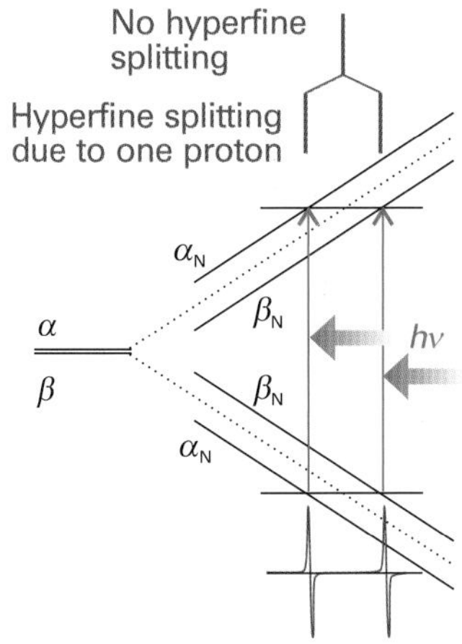

Fig. 15.58 The hyperfine interaction between an electron and a spin-$\frac{1}{2}$ nucleus results in four energy levels in place of the original two. As a result, the spectrum consists of two lines (of equal intensity) instead of one. The intensity distribution can be summarized by a simple stick diagram. The diagonal lines show the energies of the states as the applied field is increased, and resonance occurs when the separation of states matches the fixed energy of the microwave photon.

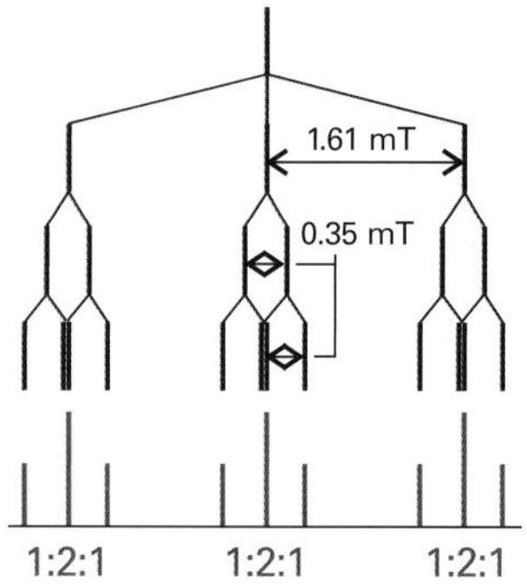

Fig. 15.59 The analysis of the hyperfine structure of radicals containing one ^{14}N nucleus ($I = 1$) and two equivalent protons.

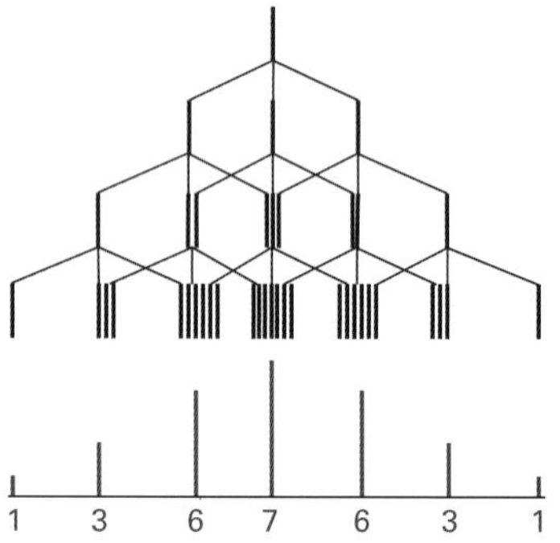

Fig. 15.60 The analysis of the hyperfine structure of radicals containing three equivalent ^{14}N nuclei.

Example 15.3 *Predicting the hyperfine structure of an EPR spectrum*

A radical contains one ^{14}N nucleus ($I = 1$) with hyperfine constant 1.61 mT and two equivalent protons ($I = \frac{1}{2}$) with hyperfine constant 0.35 mT. Predict the form of the EPR spectrum.

Method We should consider the hyperfine structure that arises from each type of nucleus or group of equivalent nuclei in succession. So, split a line with one nucleus, then each of those lines is split by a second nucleus (or group of nuclei), and so on. It is best to start with the nucleus with the largest hyperfine splitting; however, any choice could be made, and the order in which nuclei are considered does not affect the conclusion.

Answer The ^{14}N nucleus gives three hyperfine lines of equal intensity separated by 1.61 mT. Each line is split into doublets of spacing 0.35 mT by the first proton, and each line of these doublets is split into doublets with the same 0.35 mT splitting (Fig. 15.59). The central lines of each split doublet coincide, so the proton splitting gives 1:2:1 triplets of internal splitting 0.35 mT. Therefore, the spectrum consists of three equivalent 1:2:1 triplets.

Self-test 15.5 Predict the form of the EPR spectrum of a radical containing three equivalent ^{14}N nuclei. [Fig. 15.60]

The hyperfine structure of an EPR spectrum is a kind of fingerprint that helps to identify the radicals present in a sample. Moreover, because the magnitude of the splitting depends on the distribution of the unpaired electron near the magnetic nuclei present, the spectrum can be used to map the molecular orbital occupied by the unpaired electron. For example, because the hyperfine splitting in $C_6H_6^-$ is 0.375 mT, and one proton is close to a C atom with one-sixth the unpaired electron spin density (because the electron is spread uniformly around the ring), the hyperfine splitting caused by a proton in the electron spin entirely confined to a single adjacent C atom should be $6 \times 0.375\ \text{mT} = 2.25\ \text{mT}$. If in another aromatic radical we find a hyperfine splitting constant a, then the **spin density**, ρ, the probability that an unpaired electron is on the atom, can be calculated from the **McConnell equation**:

$$a = Q\rho \qquad (15.43)$$

with $Q = 2.25$ mT. In this equation, ρ is the spin density on a C atom and a is the hyperfine splitting observed for the H atom to which it is attached.

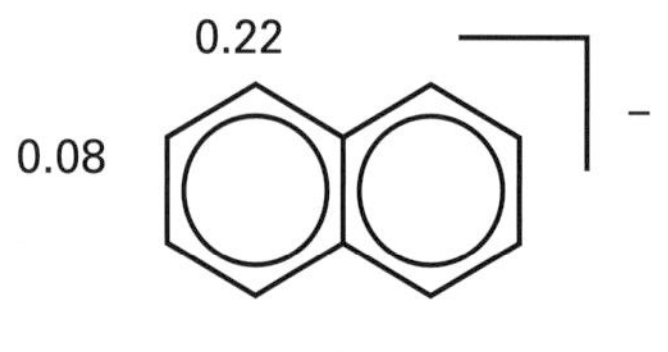

8

Illustration 15.5 *Using the McConnell equation*

The hyperfine structure of the EPR spectrum of the radical anion (naphthalene)$^-$ can be interpreted as arising from two groups of four equivalent protons. Those at the 1, 4, 5, and 8 positions in the ring have $a = 0.490$ mT and for those in the 2, 3, 6, and 7 positions have $a = 0.183$ mT. The densities obtained by using the McConnell equation are 0.22 and 0.08, respectively (**8**).

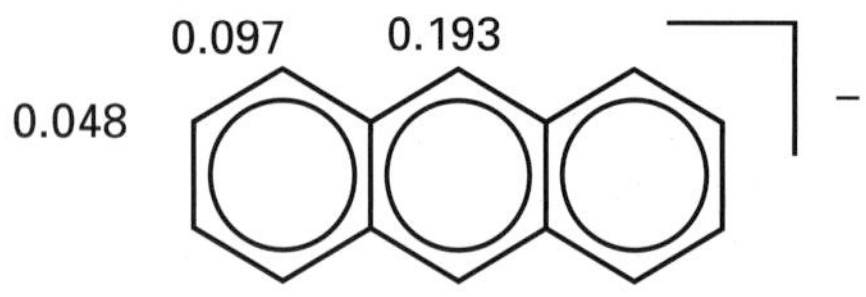

9

Self-test 15.6 The spin density in (anthracene)$^-$ is shown in (**9**). Predict the form of its EPR spectrum.

[A 1:2:1 triplet of splitting 0.43 mT split into a 1:4:6:4:1 quintet of splitting 0.22 mT, split into a 1:4:6:4:1 quintet of plitting 0.11 mT, $3 \times 5 \times 5 = 75$ lines in all]

(b) The origin of the hyperfine interaction

The hyperfine interaction is an interaction between the magnetic moments of the unpaired electron and the nuclei. There are two contributions to the interaction.

An electron in a *p* orbital does not approach the nucleus very closely, so it experiences a field that appears to arise from a point magnetic dipole. The resulting interaction is called the **dipole–dipole interaction**. The contribution of a magnetic nucleus to the local field experienced by the unpaired electron is given by an expression like that in eqn 15.28. A characteristic of this type of interaction is that it is anisotropic. Furthermore, just as in the case of NMR, the dipole–dipole interaction averages to zero when the radical is free to tumble. Therefore, hyperfine structure due to the dipole–dipole interaction is observed only for radicals trapped in solids.

An *s* electron is spherically distributed around a nucleus and so has zero average dipole–dipole interaction with the nucleus even in a solid sample. However, because an *s* electron has a nonzero probability of being at the nucleus, it is incorrect to treat the interaction as one between two point dipoles. An *s* electron has a Fermi contact interaction with the nucleus, which as we saw in Section 15.6d is a magnetic interaction that occurs when the point dipole approximation fails. The contact interaction is isotropic (that is, independent of the radical's orientation), and consequently is shown even by rapidly tumbling molecules in fluids (provided the spin density has some *s* character).

The dipole–dipole interactions of *p* electrons and the Fermi contact interaction of *s* electrons can be quite large. For example, a 2*p* electron in a nitrogen atom experiences an average field of about 3.4 mT from the ^{14}N nucleus. A 1*s* electron in a hydrogen atom experiences a field of about 50 mT as a result of its Fermi contact interaction with the central proton. More values are listed in Table 15.3. The magnitudes of the contact interactions in radicals can be interpreted in terms of the *s* orbital character of the molecular orbital occupied by the unpaired electron, and the dipole–dipole interaction can be interpreted in terms of the *p* character. The analysis of hyperfine structure therefore gives information about the composition of the orbital, and especially the hybridization of the atomic orbitals (see Problem 15.11).

We still have the source of the hyperfine structure of the $C_6H_6^-$ anion and other aromatic radical anions to explain. The sample is fluid, and as the radicals are tumbling the hyperfine structure cannot be due to the dipole–dipole interaction. Moreover, the protons lie in the nodal plane of the π orbital occupied by the unpaired electron, so the structure cannot be due to a Fermi contact interaction. The explanation lies in a **polarization mechanism** similar to the one responsible for spin–spin coupling in NMR. There is a magnetic interaction between a proton and the α electrons ($m_s = +\frac{1}{2}$) which results in one of the electrons tending to be found with a greater probability nearby (Fig. 15.61). The electron with opposite spin is therefore more likely to be close to the C atom at the other end of the bond. The unpaired electron on the C atom has a lower energy if it is parallel to that electron (Hund's rule favours parallel electrons on atoms), so the unpaired electron can detect the spin of the proton indirectly. Calculation using this model leads to a hyperfine interaction in agreement with the observed value of 2.25 mT.

Synoptic table 15.3* Hyperfine coupling constants for atoms, a/mT

Nuclide	Isotropic coupling	Anisotropic coupling
^1H	50.8 (1*s*)	
^2H	7.8 (1*s*)	
^{14}N	55.2 (2*s*)	3.4 (2*p*)
^{19}F	1720 (2*s*)	108.4 (2*p*)

* More values are given in the *Data section*.

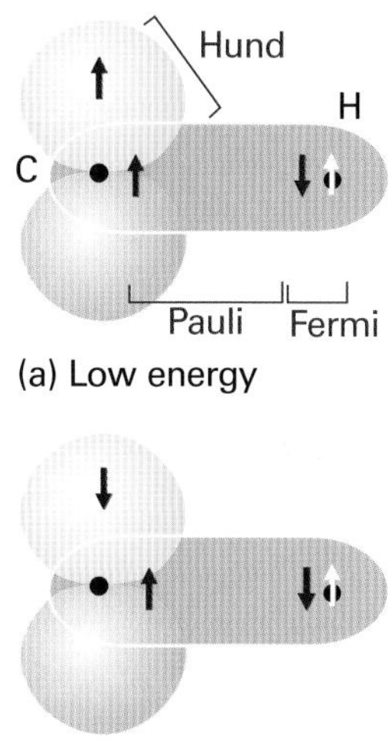

Fig. 15.61 The polarization mechanism for the hyperfine interaction in π-electron radicals. The arrangement in (a) is lower in energy than that in (b), so there is an effective coupling between the unpaired electron and the proton.

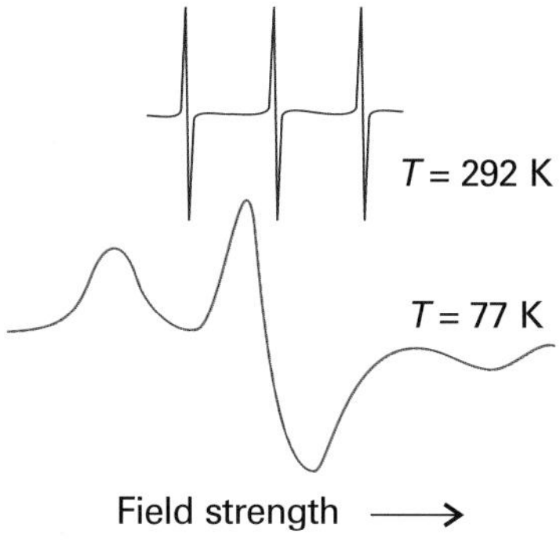

Fig. 15.62 EPR spectra of the di-*tert*-butyl nitroxide radical at 292 K (top) and 77 K (bottom). Adapted from J.R. Bolton, in *Biological applications of electron spin resonance*, H.M. Swartz, J.R. Bolton, and D.C. Borg (ed.), Wiley, New York (1972).

IMPACT ON BIOCHEMISTRY
I15.2 Spin probes

We saw in Sections 15.14 and 15.15 that anisotropy of the *g*-value and of the nuclear hyperfine interactions can be observed when a radical is immobilized in a solid. Figure 15.62 shows the variation of the lineshape of the EPR spectrum of the di-*tert*-butyl nitroxide radical (**10**) with temperature. At 292 K, the radical tumbles freely and isotropic hyperfine coupling to the ^{14}N nucleus gives rise to three sharp peaks. At 77 K,

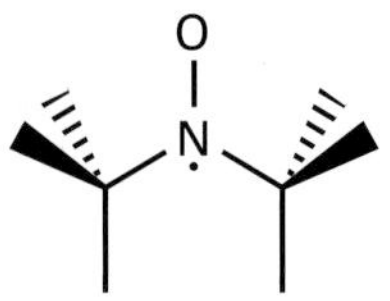

10

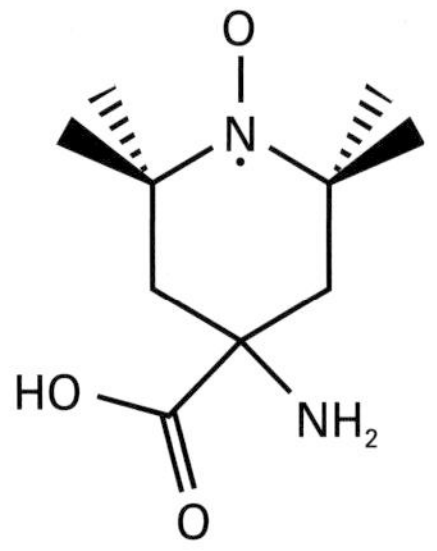

11

motion of the radical is restricted. Both isotropic and anisotropic hyperfine couplings determine the appearance of the spectrum, which now consists of three broad peaks.

A *spin probe* (or *spin label*) is a radical that interacts with a biopolymer and with an EPR spectrum that reports on the dynamical properties of the biopolymer. The ideal spin probe is one with a spectrum that broadens significantly as its motion is restricted to a relatively small extent. Nitroxide spin probes have been used to show that the hydrophobic interiors of biological membranes, once thought to be rigid, are in fact very fluid and individual lipid molecules move laterally through the sheet-like structure of the membrane.

Just as chemical exchange can broaden proton NMR spectra (Section 15.7), electron exchange between two radicals can broaden EPR spectra. Therefore, the distance between two spin probe molecules may be measured from the linewidths of their EPR spectra. The effect can be used in a number of biochemical studies. For example, the kinetics of association of two polypeptides labelled with the synthetic amino acid 2,2,6,6,-tetramethylpiperidine-1-oxyl-4-amino-4-carboxylic acid (**11**) may be studied by measuring the change in linewidth of the label with time. Alternatively, the thermodynamics of association may be studied by examining the temperature dependence of the linewidth.

Checklist of key ideas

☐ 1. The energy of an electron in a magnetic field $\mathcal{B}_0$ is $E_{m_s} = -g_e\gamma_e\hbar\mathcal{B}_0 m_s$, where γ_e is the magnetogyric ratio of the electron. The energy of a nucleus in a magnetic field $\mathcal{B}_0$ is $E_{m_I} = -\gamma\hbar\mathcal{B}_0 m_I$, where γ is the nuclear magnetogyric ratio.

☐ 2. The resonance condition for an electron in a magnetic field is $h\nu = g_e\mu_B\mathcal{B}_0$. The resonance condition for a nucleus in a magnetic field is $h\nu = \gamma\hbar\mathcal{B}_0$.

☐ 3. Nuclear magnetic resonance (NMR) is the observation of the frequency at which magnetic nuclei in molecules come into resonance with an electromagnetic field when the molecule is exposed to a strong magnetic field; NMR is a radiofrequency technique.

☐ 4. Electron paramagnetic resonance (EPR) is the observation of the frequency at which an electron spin comes into resonance with an electromagnetic field when the molecule is exposed to a strong magnetic field; EPR is a microwave technique.

☐ 5. The intensity of an NMR or EPR transition increases with the difference in population of α and β states and the strength of the applied magnetic field (as $\mathcal{B}_0^2$).

☐ 6. The chemical shift of a nucleus is the difference between its resonance frequency and that of a reference standard; chemical shifts are reported on the δ scale, in which $\delta = (\nu - \nu^\circ) \times 10^6/\nu^\circ$.

☐ 7. The observed shielding constant is the sum of a local contribution, a neighbouring group contribution, and a solvent contribution.

☐ 8. The fine structure of an NMR spectrum is the splitting of the groups of resonances into individual lines; the strength of the interaction is expressed in terms of the spin–spin coupling constant, J.

☐ 9. N equivalent spin-$\frac{1}{2}$ nuclei split the resonance of a nearby spin or group of equivalent spins into $N+1$ lines with an intensity distribution given by Pascal's triangle.

☐ 10. Spin–spin coupling in molecules in solution can be explained in terms of the polarization mechanism, in which the interaction is transmitted through the bonds.

☐ 11. The Fermi contact interaction is a magnetic interaction that depends on the very close approach of an electron to the nucleus and can occur only if the electron occupies an *s* orbital.

☐ 12. Coalescence of the two lines occurs in conformational interchange or chemical exchange when the lifetime, τ, of the states is related to their resonance frequency difference, $\delta\nu$, by $\tau = 2^{1/2}/\pi\delta\nu$.

☐ 13. In Fourier-transform NMR, the spectrum is obtained by mathematical analysis of the free-induction decay of magnetization, the response of nuclear spins in a sample to the application of one or more pulses of radiofrequency radiation.

☐ 14. Spin relaxation is the nonradiative return of a spin system to an equilibrium distribution of populations in which the transverse spin orientations are random; the system returns exponentially to the equilibrium population distribution with a time constant called the spin–lattice relaxation time, T_1.

☐ 15. The spin–spin relaxation time, T_2, is the time constant for the exponential return of the system to random transverse spin orientations.

☐ 16. In proton decoupling of ^{13}C-NMR spectra, protons are made to undergo rapid spin reorientations and the ^{13}C nucleus senses an average orientation. As a result, its resonance is a single line and not a group of lines.

☐ 17. The nuclear Overhauser effect (NOE) is the modification of one resonance by the saturation of another.

☐ 18. In two-dimensional NMR, spectra are displayed in two axes, with resonances belonging to different groups lying at different locations on the second axis. An example of a two-dimensional NMR technique is correlation spectroscopy (COSY), in which all spin–spin couplings in a molecule are determined. Another example is nuclear Overhauser effect spectroscopy (NOESY), in which internuclear distances up to about 0.5 nm are determined.

☐ 19. Magic-angle spinning (MAS) is technique in which the NMR linewidths in a solid sample are reduced by spinning at an angle of 54.74° to the applied magnetic field.

☐ 20. The EPR resonance condition is written $h\nu = g\mu_B\mathcal{B}$, where g is the g-value of the radical; the deviation of g from $g_e = 2.0023$ depends on the ability of the applied field to induce local electron currents in the radical.

☐ 21. The hyperfine structure of an EPR spectrum is its splitting of individual resonance lines into components by the magnetic interaction between the electron and nuclei with spin.

☐ 22. If a radical contains N equivalent nuclei with spin quantum number I, then there are $2NI + 1$ hyperfine lines with an intensity distribution given by a modified version of Pascal's triangle.

☐ 23. The hyperfine structure due to a hydrogen attached to an aromatic ring is converted to spin density, ρ, on the neighbouring carbon atom by using the McConnell equation: $a = Q\rho$ with $Q = 2.25$ mT.

Further reading

Articles and texts

N.M. Atherton, *Principles of electron spin resonance.* Ellis Horwood/Prentice Hall, Hemel Hempstead (1993).

E.D. Becker, *High resolution NMR: theory and chemical applications.* Academic Press, San Diego (2000).

R. Freeman, *Spin choreography: basic steps in high resolution NMR.* Oxford University Press (1998).

J.C. Lindon, G.E. Tranter, and J.L. Holmes (ed.), *Encyclopedia of spectroscopy and spectrometry.* Academic Press, San Diego (2000).

R.W. King and K.R. Williams. The Fourier transform in chemistry. Part 1. Nuclear magnetic resonance: Introduction. *J. Chem. Educ.* **66**, A213 (1989); Part 2. Nuclear magnetic resonance: The single pulse experiment. *Ibid.* A243; Part 3. Multiple-pulse experiments. *J. Chem. Educ.* **67**, A93 (1990); Part 4. NMR: Two-dimensional methods. *Ibid.* A125; A glossary of NMR terms. *Ibid.* A100.

M.T. Vlaardingerbroek and J.A. de Boer, *Magnetic resonance imaging: theory and practice.* Springer, Berlin (1999).

M.H. Levitt, *Spin dynamics.* Wiley (2001).

Sources of data and information

D.R. Lide (ed.), *CRC Handbook of Chemistry and Physics*, Section 9, CRC Press, Boca Raton (2000).

C.P. Poole, Jr. and H.A. Farach (ed.), *Handbook of electron spin resonance: data sources, computer technology, relaxation, and ENDOR.* Vols. 1–2. Springer, Berlin (1997, 1999).

Further information

Further information 15.1 *Fourier transformation of the FID curve*

The analysis of the FID curve is achieved by the standard mathematical technique of Fourier transformation, which we explored in the context of FT infrared spectroscopy (*Further information* 13.2). We start by noting that the signal $S(t)$ in the time domain, the total FID curve, is the sum (more precisely, the integral) over all the contributing frequencies

$$S(t) = \int_{-\infty}^{\infty} I(\nu)e^{-2\pi i\nu t}d\nu \qquad (15.44)$$

Because $e^{2\pi i\nu t} = \cos(2\pi\nu t) + i\sin(2\pi\nu t)$, the expression above is a sum over harmonically oscillating functions, with each one weighted by the intensity $I(\nu)$.

We need $I(\nu)$, the spectrum in the frequency domain; it is obtained by evaluating the integral

$$I(\nu) = 2\,\mathrm{Re}\int_{0}^{\infty} S(t)e^{2\pi i\nu t}dt \qquad (15.45)$$

where Re means take the real part of the following expression. This integral is very much like an overlap integral: it gives a nonzero value if $S(t)$ contains a component that matches the oscillating function $e^{2i\pi\nu t}$. The integration is carried out at a series of frequencies ν on a computer that is built into the spectrometer.

Discussion questions

15.1 Discuss in detail the origins of the local, neighbouring group, and solvent contributions to the shielding constant.

15.2 Discuss in detail the effects of a 90° pulse and of a 180° pulse on a system of spin-$\frac{1}{2}$ nuclei in a static magnetic field.

15.3 Suggest a reason why the relaxation times of ^{13}C nuclei are typically much longer than those of ^{1}H nuclei.

15.4 Discuss the origins of diagonal and cross peaks in the COSY spectrum of an AX system.

15.5 Discuss how the Fermi contact interaction and the polarization mechanism contribute to spin–spin couplings in NMR and hyperfine interactions in EPR.

15.6 Suggest how spin probes could be used to estimate the depth of a crevice in a biopolymer, such as the active site of an enzyme.

Exercises

15.1a What is the resonance frequency of a proton in a magnetic field of 14.1 T?

15.1b What is the resonance frequency of a ^{19}F nucleus in a magnetic field of 16.2 T?

15.2a ^{33}S has a nuclear spin of $\frac{3}{2}$ and a nuclear g-factor of 0.4289. Calculate the energies of the nuclear spin states in a magnetic field of 7.500 T.

15.2b ^{14}N has a nuclear spin of 1 and a nuclear g-factor of 0.404. Calculate the energies of the nuclear spin states in a magnetic field of 11.50 T.

15.3a Calculate the frequency separation of the nuclear spin levels of a ^{13}C nucleus in a magnetic field of 14.4 T given that the magnetogyric ratio is $6.73 \times 10^7\ \mathrm{T^{-1}\,s^{-1}}$.

15.3b Calculate the frequency separation of the nuclear spin levels of a ^{14}N nucleus in a magnetic field of 15.4 T given that the magnetogyric ratio is $1.93 \times 10^7\ \mathrm{T^{-1}\,s^{-1}}$.

15.4a In which of the following systems is the energy level separation the largest? (a) A proton in a 600 MHz NMR spectrometer, (b) a deuteron in the same spectrometer.

15.4b In which of the following systems is the energy level separation the largest? (a) A ^{14}N nucleus in (for protons) a 600 MHz NMR spectrometer, (b) an electron in a radical in a field of 0.300 T.

15.5a Calculate the energy difference between the lowest and highest nuclear spin states of a ^{14}N nucleus in a 15.00 T magnetic field.

15.5b Calculate the magnetic field needed to satisfy the resonance condition for unshielded protons in a 150.0 MHz radiofrequency field.

15.6a Use Table 15.2 to predict the magnetic fields at which (a) ^{1}H, (b) ^{2}H, (c) ^{13}C come into resonance at (i) 250 MHz, (ii) 500 MHz.

15.6b Use Table 15.2 to predict the magnetic fields at which (a) ^{14}N, (b) ^{19}F, and (c) ^{31}P come into resonance at (i) 300 MHz, (ii) 750 MHz.

15.7a Calculate the relative population differences ($\delta N/N$) for protons in fields of (a) 0.30 T, (b) 1.5 T, and (c) 10 T at 25°C.

15.7b Calculate the relative population differences ($\delta N/N$) for ^{13}C nuclei in fields of (a) 0.50 T, (b) 2.5 T, and (c) 15.5 T at 25°C.

15.8a The first generally available NMR spectrometers operated at a frequency of 60 MHz; today it is not uncommon to use a spectrometer that operates at 800 MHz. What are the relative population differences of ^{13}C spin states in these two spectrometers at 25°C?

15.8b What are the relative values of the chemical shifts observed for nuclei in the spectrometers mentioned in Exercise 15.8a in terms of (a) δ values, (b) frequencies?

15.9a The chemical shift of the CH_3 protons in acetaldehyde (ethanal) is $\delta = 2.20$ and that of the CHO proton is 9.80. What is the difference in local magnetic field between the two regions of the molecule when the applied field is (a) 1.5 T, (b) 15 T?

15.9b The chemical shift of the CH_3 protons in diethyl ether is $\delta = 1.16$ and that of the CH_2 protons is 3.36. What is the difference in local magnetic field between the two regions of the molecule when the applied field is (a) 1.9 T, (b) 16.5 T?

15.10a Sketch the appearance of the ^{1}H-NMR spectrum of acetaldehyde (ethanal) using $J = 2.90$ Hz and the data in Exercise 15.9a in a spectrometer operating at (a) 250 MHz, (b) 500 MHz.

15.10b Sketch the appearance of the ^{1}H-NMR spectrum of diethyl ether using $J = 6.97$ Hz and the data in Exercise 15.9b in a spectrometer operating at (a) 350 MHz, (b) 650 MHz.

15.11a Two groups of protons are made equivalent by the isomerization of a fluxional molecule. At low temperatures, where the interconversion is slow, one group has $\delta = 4.0$ and the other has $\delta = 5.2$. At what rate of interconversion will the two signals merge in a spectrometer operating at 250 MHz?

15.11b Two groups of protons are made equivalent by the isomerization of a fluxional molecule. At low temperatures, where the interconversion is slow, one group has $\delta = 5.5$ and the other has $\delta = 6.8$. At what rate of interconversion will the two signals merge in a spectrometer operating at 350 MHz?

15.12a Sketch the form of the ^{19}F-NMR spectra of a natural sample of tetrafluoroborate ions, BF_4^-, allowing for the relative abundances of $^{10}BF_4^-$ and $^{11}BF_4^-$.

15.12b From the data in Table 15.2, predict the frequency needed for ^{31}P-NMR in an NMR spectrometer designed to observe proton resonance at 500 MHz. Sketch the proton and ^{31}P resonances in the NMR spectrum of PH_4^+.

15.13a Sketch the form of an $A_3M_2X_4$ spectrum, where A, M, and X are protons with distinctly different chemical shifts and $J_{AM} > J_{AX} > J_{MX}$.

15.13b Sketch the form of an $A_2M_2X_5$ spectrum, where A, M, and X are protons with distinctly different chemical shifts and $J_{AM} > J_{AX} > J_{MX}$.

15.14a Which of the following molecules have sets of nuclei that are chemically but not magnetically equivalent? (a) CH_3CH_3, (b) $CH_2{=}CH_2$.

15.14b Which of the following molecules have sets of nuclei that are chemically but not magnetically equivalent? (a) $CH_2{=}C{=}CF_2$, (b) *cis*- and *trans*-$[Mo(CO)_4(PH_3)_2]$.

15.15a The duration of a 90° or 180° pulse depends on the strength of the $\mathcal{B}_1$ field. If a 90° pulse requires 10 μs, what is the strength of the $\mathcal{B}_1$ field? How long would the corresponding 180° pulse require?

15.15b The duration of a 90° or 180° pulse depends on the strength of the $\mathcal{B}_1$ field. If a 180° pulse requires 12.5 μs, what is the strength of the $\mathcal{B}_1$ field? How long would the corresponding 90° pulse require?

15.16a What magnetic field would be required in order to use an EPR X–band spectrometer (9 GHz) to observe ^{1}H-NMR and a 300 MHz spectrometer to observe EPR?

15.16b Some commercial EPR spectrometers use 8 mm microwave radiation (the Q band). What magnetic field is needed to satisfy the resonance condition?

15.17a The centre of the EPR spectrum of atomic hydrogen lies at 329.12 mT in a spectrometer operating at 9.2231 GHz. What is the g-value of the electron in the atom?

15.17b The centre of the EPR spectrum of atomic deuterium lies at 330.02 mT in a spectrometer operating at 9.2482 GHz. What is the g-value of the electron in the atom?

15.18a A radical containing two equivalent protons shows a three-line spectrum with an intensity distribution 1:2:1. The lines occur at 330.2 mT, 332.5 mT, and 334.8 mT. What is the hyperfine coupling constant for each proton? What is the g-value of the radical given that the spectrometer is operating at 9.319 GHz?

15.18b A radical containing three equivalent protons shows a four–line spectrum with an intensity distribution 1:3:3:1. The lines occur at 331.4 mT, 333.6 mT, 335.8 mT, and 338.0 mT. What is the hyperfine coupling constant for each proton? What is the g-value of the radical given that the spectrometer is operating at 9.332 GHz?

15.19a A radical containing two inequivalent protons with hyperfine constants 2.0 mT and 2.6 mT gives a spectrum centred on 332.5 mT. At what fields do the hyperfine lines occur and what are their relative intensities?

15.19b A radical containing three inequivalent protons with hyperfine constants 2.11 mT, 2.87 mT, and 2.89 mT gives a spectrum centred on 332.8 mT. At what fields do the hyperfine lines occur and what are their relative intensities?

15.20a Predict the intensity distribution in the hyperfine lines of the EPR spectra of (a) $\cdot CH_3$, (b) $\cdot CD_3$.

15.20b Predict the intensity distribution in the hyperfine lines of the EPR spectra of (a) $\cdot CH_2H_3$, (b) $\cdot CD_2CD_3$.

15.21a The benzene radical anion has $g = 2.0025$. At what field should you search for resonance in a spectrometer operating at (a) 9.302 GHz, (b) 33.67 GHz?

15.21b The naphthalene radical anion has $g = 2.0024$. At what field should you search for resonance in a spectrometer operating at (a) 9.312 GHz, (b) 33.88 GHz?

15.22a The EPR spectrum of a radical with a single magnetic nucleus is split into four lines of equal intensity. What is the nuclear spin of the nucleus?

15.22b The EPR spectrum of a radical with two equivalent nuclei of a particular kind is split into five lines of intensity ratio 1:2:3:2:1. What is the spin of the nuclei?

15.23a Sketch the form of the hyperfine structures of radicals XH_2 and XD_2, where the nucleus X has $I = \frac{5}{2}$.

15.23b Sketch the form of the hyperfine structures of radicals XH_3 and XD_3, where the nucleus X has $I = \frac{3}{2}$.

Problems*

Numerical problems

15.1 A scientist investigates the possibility of neutron spin resonance, and has available a commercial NMR spectrometer operating at 300 MHz. What field is required for resonance? What is the relative population difference at room temperature? Which is the lower energy spin state of the neutron?

15.2 Two groups of protons have $\delta = 4.0$ and $\delta = 5.2$ and are interconverted by a conformational change of a fluxional molecule. In a 60 MHz spectrometer the spectrum collapsed into a single line at 280 K but at 300 MHz the collapse did not occur until the temperature had been raised to 300 K. What is the activation energy of the interconversion?

15.3‡ Suppose that the FID in Fig. 15.31 was recorded in a 300 MHz spectrometer, and that the interval between maxima in the oscillations in the FID is 0.10 s. What is the Larmor frequency of the nuclei and the spin–spin relaxation time?

15.4‡ In a classic study of the application of NMR to the measurement of rotational barriers in molecules, P.M. Nair and J.D. Roberts (*J. Am. Chem. Soc.* **79**, 4565 (1957)) obtained the 40 MHz ^{19}F-NMR spectrum of $F_2BrCCBrCl_2$. Their spectra are reproduced in Fig. 15.63. At 193 K the spectrum shows five resonance peaks. Peaks I and III are separated by 160 Hz, as are IV and V. The ratio of the integrated intensities of peak II to peaks I, III, IV, and V is approximately 10 to 1. At 273 K, the five peaks have collapsed into

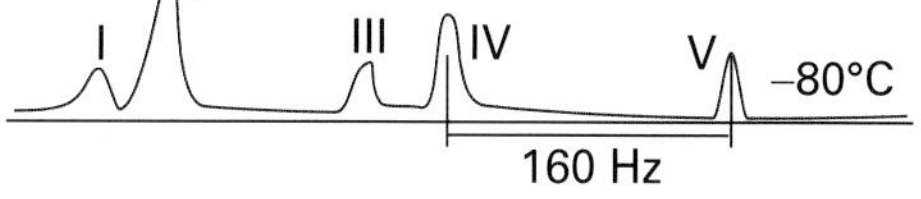

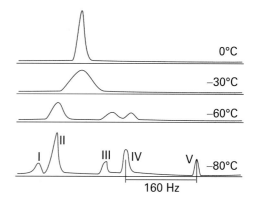

Fig. 15.63

* Problems denoted with the symbol ‡ were supplied by Charles Trapp and Carmen Giunta.

one. Explain the spectrum and its change with temperature. At what rate of interconversion will the spectrum collapse to a single line? Calculate the rotational energy barrier between the rotational isomers on the assumption that it is related to the rate of interconversion between the isomers.

15.5‡ Various versions of the Karplus equation (eqn 15.27) have been used to correlate data on vicinal proton coupling constants in systems of the type $R_1R_2CHCHR_3R_4$. The original version, (M. Karplus, *J. Am. Chem. Soc.* **85**, 2870 (1963)), is $^3J_{HH} = A\cos^2\phi_{HH} + B$. When $R_3 = R_4 = H$, $^3J_{HH} = 7.3$ Hz; when $R_3 = CH_3$ and $R_4 = H$, $^3J_{HH} = 8.0$ Hz; when $R_3 = R_4 = CH_3$, $^3J_{HH} = 11.2$ Hz. Assume that only staggered conformations are important and determine which version of the Karplus equation fits the data better.

15.6‡ It might be unexpected that the Karplus equation, which was first derived for $^3J_{HH}$ coupling constants, should also apply to vicinal coupling between the nuclei of metals such as tin. T.N. Mitchell and B. Kowall (*Magn. Reson. Chem.* **33**, 325 (1995)) have studied the relation between $^3J_{HH}$ and $^3J_{SnSn}$ in compounds of the type $Me_3SnCH_2CHRSnMe_3$ and find that $^3J_{SnSn} = 78.86\,^3J_{HH} + 27.84$ Hz. (a) Does this result support a Karplus type equation for tin? Explain your reasoning. (b) Obtain the Karplus equation for $^3J_{SnSn}$ and plot it as a function of the dihedral angle. (c) Draw the preferred conformation.

15.7 Figure 15.64 shows the proton COSY spectrum of 1-nitropropane. Account for the appearance of off-diagonal peaks in the spectrum.

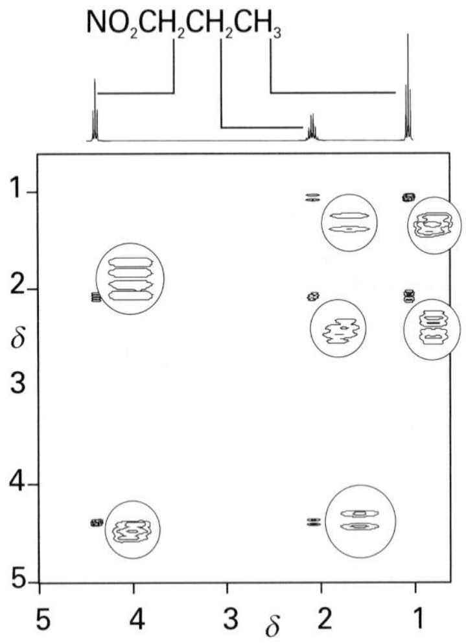

Fig. 15.64 The COSY spectrum of 1-nitropropane ($NO_2CH_2CH_2CH_3$). The circles show enhanced views of the spectral features. (Spectrum provided by Prof. G. Morris.)

15.8 The angular NO_2 molecule has a single unpaired electron and can be trapped in a solid matrix or prepared inside a nitrite crystal by radiation damage of NO_2^- ions. When the applied field is parallel to the OO direction the centre of the spectrum lies at 333.64 mT in a spectrometer operating at 9.302 GHz. When the field lies along the bisector of the ONO angle, the resonance lies at 331.94 mT. What are the *g*-values in the two orientations?

15.9 The hyperfine coupling constant in $\cdot CH_3$ is 2.3 mT. Use the information in Table 15.3 to predict the splitting between the hyperfine lines of the spectrum of $\cdot CD_3$. What are the overall widths of the hyperfine spectra in each case?

15.10 The *p*-dinitrobenzene radical anion can be prepared by reduction of *p*-dinitrobenzene. The radical anion has two equivalent N nuclei ($I = 1$) and four equivalent protons. Predict the form of the EPR spectrum using $a(N) = 0.148$ mT and $a(H) = 0.112$ mT.

15.11 When an electron occupies a 2*s* orbital on an N atom it has a hyperfine interaction of 55.2 mT with the nucleus. The spectrum of NO_2 shows an isotropic hyperfine interaction of 5.7 mT. For what proportion of its time is the unpaired electron of NO_2 occupying a 2*s* orbital? The hyperfine coupling constant for an electron in a 2*p* orbital of an N atom is 3.4 mT. In NO_2 the anisotropic part of the hyperfine coupling is 1.3 mT. What proportion of its time does the unpaired electron spend in the 2*p* orbital of the N atom in NO_2? What is the total probability that the electron will be found on (a) the N atoms, (b) the O atoms? What is the hybridization ratio of the N atom? Does the hybridization support the view that NO_2 is angular?

15.12 The hyperfine coupling constants observed in the radical anions (**12**), (**13**), and (**14**) are shown (in millitesla, mT). Use the value for the benzene radical anion to map the probability of finding the unpaired electron in the π orbital on each C atom.

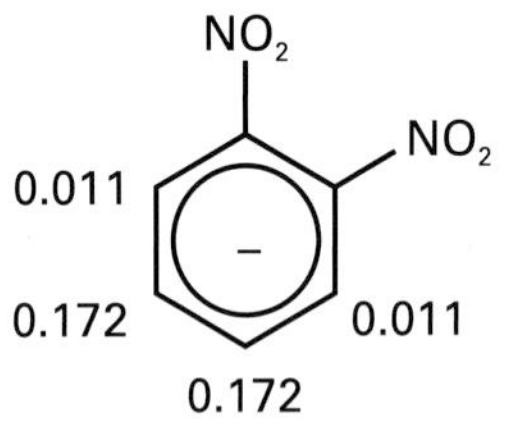

12

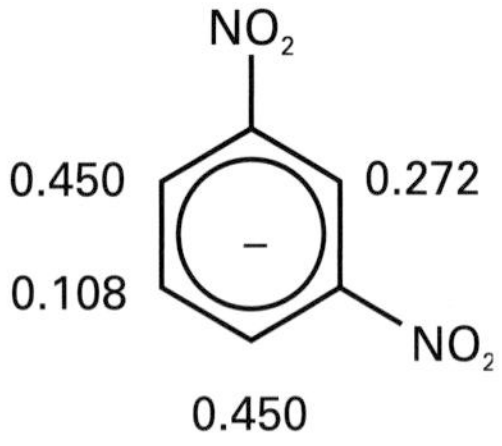

13

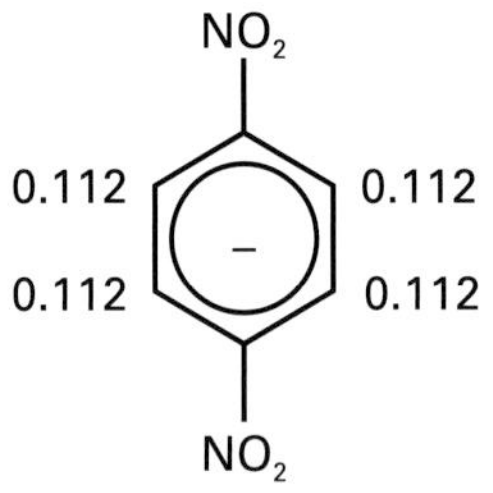

14

Theoretical problems

15.13 Calculate σ_d for a hydrogenic atom with atomic number *Z*.

15.14 In this problem you will use the molecular electronic structure methods described in Chapter 11 to investigate the hypothesis that the magnitude of the ^{13}C chemical shift correlates with the net charge on a ^{13}C atom. (a) Using molecular modelling software[3] and the computational method of your choice, calculate the net charge at the C atom *para* to the substituents in this series of molecules: benzene, phenol, toluene, trifluorotoluene, benzonitrile, and nitrobenzene. (b) The ^{13}C chemical shifts of the *para* C atoms in each of the molecules that you examined in part (a) are given below:

Substituent	OH	CH_3	H	CF_3	CN	NO_2
δ	130.1	128.4	128.5	128.9	129.1	129.4

Is there a linear correlation between net charge and ^{13}C chemical shift of the *para* C atom in this series of molecules? (c) If you did find a correlation in part (b), use the concepts developed in this chapter to explain the physical origins of the correlation.

[3] The web site contains links to molecular modelling freeware and to other sites where you may perform molecular orbital calculations directly from your web browser.

15.15 The z-component of the magnetic field at a distance R from a magnetic moment parallel to the z-axis is given by eqn 15.28. In a solid, a proton at a distance R from another can experience such a field and the measurement of the splitting it causes in the spectrum can be used to calculate R. In gypsum, for instance, the splitting in the H_2O resonance can be interpreted in terms of a magnetic field of 0.715 mT generated by one proton and experienced by the other. What is the separation of the protons in the H_2O molecule?

15.16 In a liquid, the dipolar magnetic field averages to zero: show this result by evaluating the average of the field given in eqn 15.28. *Hint.* The volume element is $\sin\theta\,d\theta d\phi$ in polar coordinates.

15.17 The shape of a spectral line, $I(\omega)$, is related to the free induction decay signal $G(t)$ by

$$I(\omega) = a\,\mathrm{Re}\int_0^{\infty} G(t)e^{i\omega t}dt$$

where a is a constant and 'Re' means take the real part of what follows. Calculate the lineshape corresponding to an oscillating, decaying function $G(t) = \cos\omega_0 t\, e^{-t/\tau}$.

15.18 In the language of Problem 15.17, show that, if $G(t) = (a\cos\omega_1 t + b\cos\omega_2 t)e^{-t/\tau}$, then the spectrum consists of two lines with intensities proportional to a and b and located at $\omega = \omega_1$ and ω_2, respectively.

15.19 EPR spectra are commonly discussed in terms of the parameters that occur in the *spin-hamiltonian*, a hamiltonian operator that incorporates various effects involving spatial operators (like the orbital angular momentum) into operators that depend on the spin alone. Show that, if you use $H = -g_e\gamma_e\mathcal{B}_0 s_z - \gamma_e\mathcal{B}_0 l_z$ as the true hamiltonian, then from second-order perturbation theory (and specifically eqn 9.65), the eigenvalues of the spin are the same as those of the spin-hamiltonian $H_{spin} = -g\gamma_e\mathcal{B}_0 s_z$ (note the g in place of g_e) and find an expression for g.

Applications: to biochemistry and medicine

15.20 Interpret the following features of the NMR spectra of hen lysozyme: (a) saturation of a proton resonance assigned to the side chain of methionine-105 changes the intensities of proton resonances assigned to the side chains of tryptophan-28 and tyrosine-23; (b) saturation of proton resonances assigned to tryptophan-28 did not affect the spectrum of tyrosine-23.

15.21 When interacting with a large biopolymer or even larger organelle, a small molecule might not rotate freely in all directions and the dipolar interaction might not average to zero. Suppose a molecule is bound so that, although the vector separating two protons may rotate freely around the z-axis, the colatitude may vary only between 0 and θ'. Average the dipolar field over this restricted range of orientations and confirm that the average vanishes when $\theta' = \pi$ (corresponding to rotation over an entire sphere). What is the average value of the local dipolar field for the H_2O molecule in Problem 15.15 if it is bound to a biopolymer that enables it to rotate up to $\theta' = 30°$?

15.22 Suggest a reason why the spin–lattice relaxation time of benzene (a small molecule) in a mobile, deuterated hydrocarbon solvent increases with temperature whereas that of an oligonucleotide (a large molecule) decreases.

15.23 NMR spectroscopy may be used to determine the equilibrium constant for dissociation of a complex between a small molecule, such as an enzyme inhibitor I, and a protein, such as an enzyme E:

$$EI \rightleftharpoons E + I \qquad K_I = [E][I]/[EI]$$

In the limit of slow chemical exchange, the NMR spectrum of a proton in I would consist of two resonances: one at ν_I for free I and another at ν_{EI} for bound I. When chemical exchange is fast, the NMR spectrum of the same proton in I consists of a single peak with a resonance frequency ν given by $\nu = f_I\nu_I + f_{EI}\nu_{EI}$, where $f_I = [I]/([I] + [EI])$ and $f_{EI} = [EI]/([I] + [EI])$ are, respectively, the fractions of free I and bound I. For the purposes of analysing the data, it is also useful to define the frequency differences $\delta\nu = \nu - \nu_I$ and $\Delta\nu = \nu_{EI} - \nu_I$. Show that, when the initial concentration of I, $[I]_0$, is much greater than the initial concentration of E, $[E]_0$, a plot of $[I]_0$ against $\delta\nu^{-1}$ is a straight line with slope $[E]_0\Delta\nu$ and y-intercept $-K_I$.

15.24 The molecular electronic structure methods described in Chapter 11 may be used to predict the spin density distribution in a radical. Recent EPR studies have shown that the amino acid tyrosine participates in a number of biological electron transfer reactions, including the processes of water oxidation to O_2 in plant photosystem II and of O_2 reduction to water in cytochrome *c* oxidase (*Impact I17.2*). During the course of these electron transfer reactions, a tyrosine radical forms, with spin density delocalized over the side chain of the amino acid. (a) The phenoxy radical shown in (**15**) is a suitable model of the tyrosine radical. Using molecular modelling software and the computational method of your choice (semi-empirical or *ab initio* methods), calculate the spin densities at the O atom and at all of the C atoms in (**15**). (b) Predict the form of the EPR spectrum of (**15**).

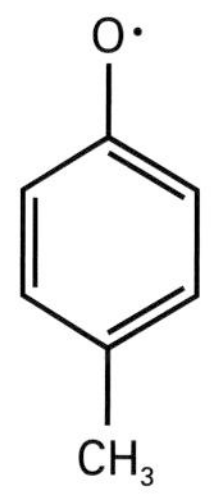

15

15.25 Sketch the EPR spectra of the di-*tert*-butyl nitroxide radical (**10**) at 292 K in the limits of very low concentration (at which electron exchange is negligible), moderate concentration (at which electron exchange effects begin to be observed), and high concentration (at which electron exchange effects predominate). Discuss how the observation of electron exchange between nitroxide spin probes can inform the study of lateral mobility of lipids in a biological membrane.

15.26 You are designing an MRI spectrometer. What field gradient (in microtesla per metre, $\mu T\ m^{-1}$) is required to produce a separation of 100 Hz between two protons separated by the long diameter of a human kidney (taken as 8 cm) given that they are in environments with $\delta = 3.4$? The radiofrequency field of the spectrometer is at 400 MHz and the applied field is 9.4 T.

15.27 Suppose a uniform disk-shaped organ is in a linear field gradient, and that the MRI signal is proportional to the number of protons in a slice of width δx at each horizontal distance x from the centre of the disk. Sketch the shape of the absorption intensity for the MRI image of the disk before any computer manipulation has been carried out.

16 Statistical thermodynamics 1: the concepts

Statistical thermodynamics provides the link between the microscopic properties of matter and its bulk properties. Two key ideas are introduced in this chapter. The first is the Boltzmann distribution, which is used to predict the populations of states in systems at thermal equilibrium. In this chapter we see its derivation in terms of the distribution of particles over available states. The derivation leads naturally to the introduction of the partition function, which is the central mathematical concept of this and the next chapter. We see how to interpret the partition function and how to calculate it in a number of simple cases. We then see how to extract thermodynamic information from the partition function. In the final part of the chapter, we generalize the discussion to include systems that are composed of assemblies of interacting particles. Very similar equations are developed to those in the first part of the chapter, but they are much more widely applicable.

The preceding chapters of this part of the text have shown how the energy levels of molecules can be calculated, determined spectroscopically, and related to their structures. The next major step is to see how a knowledge of these energy levels can be used to account for the properties of matter in bulk. To do so, we now introduce the concepts of **statistical thermodynamics**, the link between individual molecular properties and bulk thermodynamic properties.

The crucial step in going from the quantum mechanics of individual molecules to the thermodynamics of bulk samples is to recognize that the latter deals with the *average* behaviour of large numbers of molecules. For example, the pressure of a gas depends on the average force exerted by its molecules, and there is no need to specify which molecules happen to be striking the wall at any instant. Nor is it necessary to consider the fluctuations in the pressure as different numbers of molecules collide with the wall at different moments. The fluctuations in pressure are very small compared with the steady pressure: it is highly improbable that there will be a sudden lull in the number of collisions, or a sudden surge. Fluctuations in other thermodynamic properties also occur, but for large numbers of particles they are negligible compared to the mean values.

This chapter introduces statistical thermodynamics in two stages. The first, the derivation of the Boltzmann distribution for individual particles, is of restricted applicability, but it has the advantage of taking us directly to a result of central importance in a straightforward and elementary way. We can *use* statistical thermodynamics once we have deduced the Boltzmann distribution. Then (in Section 16.5) we extend the arguments to systems composed of interacting particles.

The distribution of molecular states

We consider a closed system composed of N molecules. Although the total energy is constant at E, it is not possible to be definite about how that energy is shared between the molecules. Collisions result in the ceaseless redistribution of energy not only between the molecules but also among their different modes of motion. The closest we can come to a description of the distribution of energy is to report the **population** of a state, the average number of molecules that occupy it, and to say that on average there are n_i molecules in a state of energy ε_i. The populations of the states remain almost constant, but the precise identities of the molecules in each state may change at every collision.

The problem we address in this section is the calculation of the populations of states for any type of molecule in any mode of motion at any temperature. The only restriction is that the molecules should be independent, in the sense that the total energy of the system is a sum of their individual energies. We are discounting (at this stage) the possibility that in a real system a contribution to the total energy may arise from interactions between molecules. We also adopt the **principle of equal *a priori* probabilities**, the assumption that all possibilities for the distribution of energy are equally probable. *A priori* means in this context loosely 'as far as one knows'. We have no reason to presume otherwise than that, for a collection of molecules at thermal equilibrium, vibrational states of a certain energy, for instance, are as likely to be populated as rotational states of the same energy.

One very important conclusion that will emerge from the following analysis is that the populations of states depend on a single parameter, the 'temperature'. That is, statistical thermodynamics provides a molecular justification for the concept of temperature and some insight into this crucially important quantity.

16.1 Configurations and weights

Any individual molecule may exist in states with energies $\varepsilon_0, \varepsilon_1, \ldots$. We shall always take ε_0, the lowest state, as the zero of energy ($\varepsilon_0 = 0$), and measure all other energies relative to that state. To obtain the actual internal energy, U, we may have to add a constant to the calculated energy of the system. For example, if we are considering the vibrational contribution to the internal energy, then we must add the total zero-point energy of any oscillators in the sample.

(a) Instantaneous configurations

At any instant there will be n_0 molecules in the state with energy ε_0, n_1 with ε_1, and so on. The specification of the set of populations $n_0, n_1, \ldots$ in the form $\{n_0, n_1, \ldots\}$ is a statement of the instantaneous **configuration** of the system. The instantaneous configuration fluctuates with time because the populations change. We can picture a large number of different instantaneous configurations. One, for example, might be $\{N,0,0,\ldots\}$, corresponding to every molecule being in its ground state. Another might be $\{N-2,2,0,0,\ldots\}$, in which two molecules are in the first excited state. The latter configuration is intrinsically more likely to be found than the former because it can be achieved in more ways: $\{N,0,0,\ldots\}$ can be achieved in only one way, but $\{N-2,2,0,\ldots\}$ can be achieved in $\frac{1}{2}N(N-1)$ different ways (Fig. 16.1; see *Justification* 16.1). At this stage in the argument, we are ignoring the requirement that the total energy of the system should be constant (the second configuration has a higher energy than the first). The constraint of total energy is imposed later in this section.

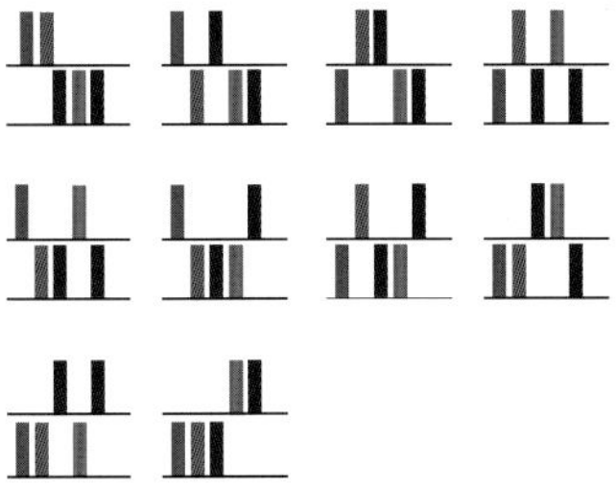

Fig. 16.1 Whereas a configuration $\{5,0,0,\ldots\}$ can be achieved in only one way, a configuration $\{3,2,0,\ldots\}$ can be achieved in the ten different ways shown here, where the tinted blocks represent different molecules.

Fig. 16.2 The 18 molecules shown here can be distributed into four receptacles (distinguished by the three vertical lines) in 18! different ways. However, 3! of the selections that put three molecules in the first receptacle are equivalent, 6! that put six molecules into the second receptacle are equivalent, and so on. Hence the number of distinguishable arrangements is 18!/3!6!5!4!.

If, as a result of collisions, the system were to fluctuate between the configurations $\{N,0,0,\ldots\}$ and $\{N-2,2,0,\ldots\}$, it would almost always be found in the second, more likely state (especially if N were large). In other words, a system free to switch between the two configurations would show properties characteristic almost exclusively of the second configuration. A general configuration $\{n_0,n_1,\ldots\}$ can be achieved in W different ways, where W is called the **weight** of the configuration. The weight of the configuration $\{n_0,n_1,\ldots\}$ is given by the expression

$$W=\frac{N!}{n_0!n_1!n_2!\ldots} \qquad (16.1)$$

Equation 16.1 is a generalization of the formula $W=\frac{1}{2}N(N-1)$, and reduces to it for the configuration $\{N-2,2,0,\ldots\}$.

Comment 16.1

More formally, W is called the *multinomial coefficient* (see *Appendix* 2). In eqn 16.1, $x!$, x factorial, denotes $x(x-1)(x-2)\ldots 1$, and by definition $0!=1$.

Justification 16.1 *The weight of a configuration*

First, consider the weight of the configuration $\{N-2,2,0,0,\ldots\}$. One candidate for promotion to an upper state can be selected in N ways. There are $N-1$ candidates for the second choice, so the total number of choices is $N(N-1)$. However, we should not distinguish the choice (Jack, Jill) from the choice (Jill, Jack) because they lead to the same configurations. Therefore, only half the choices lead to distinguishable configurations, and the total number of distinguishable choices is $\frac{1}{2}N(N-1)$.

Now we generalize this remark. Consider the number of ways of distributing N balls into bins. The first ball can be selected in N different ways, the next ball in $N-1$ different ways for the balls remaining, and so on. Therefore, there are $N(N-1)\ldots 1 = N!$ ways of selecting the balls for distribution over the bins. However, if there are n_0 balls in the bin labelled ε_0, there would be $n_0!$ different ways in which the same balls could have been chosen (Fig. 16.2). Similarly, there are $n_1!$ ways in which the n_1 balls in the bin labelled ε_1 can be chosen, and so on. Therefore, the total number of distinguishable ways of distributing the balls so that there are n_0 in bin ε_0, n_1 in bin ε_1, etc. regardless of the order in which the balls were chosen is $N!/n_0!n_1!\ldots$, which is the content of eqn 16.1.

Illustration 16.1 *Calculating the weight of a distribution*

To calculate the number of ways of distributing 20 identical objects with the arrangement 1, 0, 3, 5, 10, 1, we note that the configuration is $\{1,0,3,5,10,1\}$ with $N=20$; therefore the weight is

$$W=\frac{20!}{1!0!3!5!10!1!}=9.31\times 10^{8}$$

Self-test 16.1 Calculate the weight of the configuration in which 20 objects are distributed in the arrangement 0, 1, 5, 0, 8, 0, 3, 2, 0, 1. [4.19×10^{10}]

It will turn out to be more convenient to deal with the natural logarithm of the weight, ln W, rather than with the weight itself. We shall therefore need the expression

$$\begin{aligned}\ln W &= \ln\frac{N!}{n_0!n_1!n_2!\ldots} = \ln N! - \ln(n_0!n_1!n_2!\cdots)\\ &= \ln N! - (\ln n_0! + \ln n_1! + \ln n_2! + \cdots)\\ &= \ln N! - \sum_i \ln n_i!\end{aligned}$$

where in the first line we have used $\ln(x/y) = \ln x - \ln y$ and in the second $\ln xy = \ln x + \ln y$. One reason for introducing ln W is that it is easier to make approximations. In particular, we can simplify the factorials by using **Stirling's approximation** in the form

$$\ln x! \approx x\ln x - x \qquad (16.2)$$

Comment 16.2

A more accurate form of Stirling's approximation is

$$x! \approx (2\pi)^{1/2}x^{x+\frac{1}{2}}e^{-x}$$

and is in error by less than 1 per cent when x is greater than about 10. We deal with far larger values of x, and the simplified version in eqn 16.2 is adequate.

Then the approximate expression for the weight is

$$\ln W = (N\ln N - N) - \sum_i (n_i \ln n_i - n_i) = N\ln N - \sum_i n_i \ln n_i \qquad (16.3)$$

The final form of eqn 16.3 is derived by noting that the sum of n_i is equal to N, so the second and fourth terms in the second expression cancel.

(b) The Boltzmann distribution

We have seen that the configuration $\{N-2,2,0,\ldots\}$ dominates $\{N,0,0,\ldots\}$, and it should be easy to believe that there may be other configurations that have a much greater weight than both. We shall see, in fact, that there is a configuration with so great a weight that it overwhelms all the rest in importance to such an extent that the system will almost always be found in it. The properties of the system will therefore be characteristic of that particular dominating configuration. This dominating configuration can be found by looking for the values of n_i that lead to a maximum value of W. Because W is a function of all the n_i, we can do this search by varying the n_i and looking for the values that correspond to $dW = 0$ (just as in the search for the maximum of any function), or equivalently a maximum value of ln W. However, there are two difficulties with this procedure.

The first difficulty is that the only permitted configurations are those corresponding to the specified, constant, total energy of the system. This requirement rules out many configurations; $\{N,0,0,\ldots\}$ and $\{N-2,2,0,\ldots\}$, for instance, have different energies, so both cannot occur in the same isolated system. It follows that, in looking for the configuration with the greatest weight, we must ensure that the configuration also satisfies the condition

$$\text{Constant total energy:} \qquad \sum_i n_i\varepsilon_i = E \qquad (16.4)$$

where E is the total energy of the system.

The second constraint is that, because the total number of molecules present is also fixed (at N), we cannot arbitrarily vary all the populations simultaneously. Thus, increasing the population of one state by 1 demands that the population of another state must be reduced by 1. Therefore, the search for the maximum value of W is also subject to the condition

$$\text{Constant total number of molecules:} \qquad \sum_i n_i = N \qquad (16.5)$$

We show in *Further information 16.1* that the populations in the configuration of greatest weight, subject to the two constraints in eqns 16.4 and 16.5, depend on the energy of the state according to the **Boltzmann distribution**:

$$\frac{n_i}{N} = \frac{e^{-\beta\varepsilon_i}}{\sum_i e^{-\beta\varepsilon_i}} \tag{16.6a}$$

where $\varepsilon_0 \le \varepsilon_1 \le \varepsilon_2 \ldots$. Equation 16.6a is the justification of the remark that a single parameter, here denoted β, determines the most probable populations of the states of the system. We shall see in Section 16.3b that

$$\beta = \frac{1}{kT} \tag{16.6b}$$

where T is the thermodynamic temperature and k is Boltzmann's constant. In other words, *the thermodynamic temperature is the unique parameter that governs the most probable populations of states of a system at thermal equilibrium.* In *Further information* 16.3, moreover, we see that β is a more natural measure of temperature than T itself.

16.2 The molecular partition function

From now on we write the Boltzmann distribution as

$$p_i = \frac{e^{-\beta\varepsilon_i}}{q} \tag{16.7}$$

where p_i is the fraction of molecules in the state i, $p_i = n_i/N$, and q is the **molecular partition function**:

$$q = \sum_i e^{-\beta\varepsilon_i} \qquad [16.8]$$

The sum in q is sometimes expressed slightly differently. It may happen that several states have the same energy, and so give the same contribution to the sum. If, for example, g_i states have the same energy ε_i (so the level is g_i-fold degenerate), we could write

$$q = \sum_{\text{levels } i} g_i e^{-\beta\varepsilon_i} \tag{16.9}$$

where the sum is now over energy levels (sets of states with the same energy), not individual states.

Example 16.1 *Writing a partition function*

Write an expression for the partition function of a linear molecule (such as HCl) treated as a rigid rotor.

Method To use eqn 16.9 we need to know (a) the energies of the levels, (b) the degeneracies, the number of states that belong to each level. Whenever calculating a partition function, the energies of the levels are expressed relative to 0 for the state of lowest energy. The energy levels of a rigid linear rotor were derived in Section 13.5c.

Answer From eqn 13.31, the energy levels of a linear rotor are $hcBJ(J+1)$, with $J = 0, 1, 2, \ldots$. The state of lowest energy has zero energy, so no adjustment need be made to the energies given by this expression. Each level consists of $2J + 1$ degenerate states. Therefore,

$$q = \sum_{J=0}^{\infty} \overbrace{(2J+1)}^{g_J} e^{\overbrace{-\beta hcBJ(J+1)}^{\varepsilon_J}}$$

The sum can be evaluated numerically by supplying the value of B (from spectroscopy or calculation) and the temperature. For reasons explained in Section 17.2b, this expression applies only to unsymmetrical linear rotors (for instance, HCl, not CO_2).

Self-test 16.2 Write the partition function for a two-level system, the lower state (at energy 0) being nondegenerate, and the upper state (at an energy ε) doubly degenerate. [$q = 1 + 2e^{-\beta\varepsilon}$]

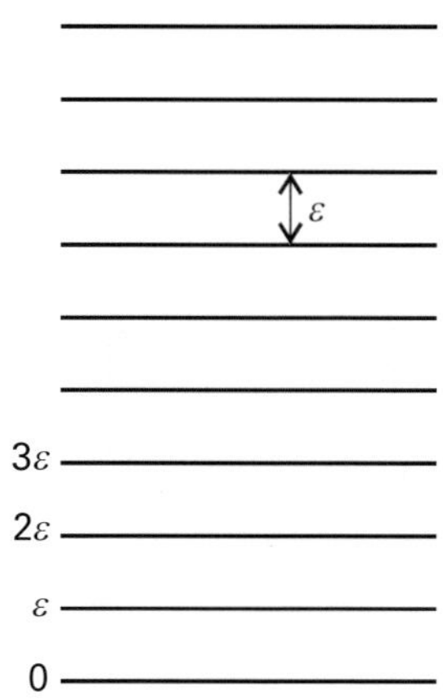

Fig. 16.3 The equally spaced infinite array of energy levels used in the calculation of the partition function. A harmonic oscillator has the same spectrum of levels.

(a) An interpretation of the partition function

Some insight into the significance of a partition function can be obtained by considering how q depends on the temperature. When T is close to zero, the parameter $\beta = 1/kT$ is close to infinity. Then every term except one in the sum defining q is zero because each one has the form e^{-x} with $x \to \infty$. The exception is the term with $\varepsilon_0 \equiv 0$ (or the g_0 terms at zero energy if the ground state is g_0-fold degenerate), because then $\varepsilon_0/kT \equiv 0$ whatever the temperature, including zero. As there is only one surviving term when $T = 0$, and its value is g_0, it follows that

$$\lim_{T\to 0} q = g_0 \qquad (16.10)$$

That is, at $T = 0$, the partition function is equal to the degeneracy of the ground state.

Now consider the case when T is so high that for each term in the sum $\varepsilon_j/kT \approx 0$. Because $e^{-x} = 1$ when $x = 0$, each term in the sum now contributes 1. It follows that the sum is equal to the number of molecular states, which in general is infinite:

$$\lim_{T\to \infty} q = \infty \qquad (16.11)$$

In some idealized cases, the molecule may have only a finite number of states; then the upper limit of q is equal to the number of states. For example, if we were considering only the spin energy levels of a radical in a magnetic field, then there would be only two states ($m_s = \pm\frac{1}{2}$). The partition function for such a system can therefore be expected to rise towards 2 as T is increased towards infinity.

We see that *the molecular partition function gives an indication of the number of states that are thermally accessible to a molecule at the temperature of the system*. At $T = 0$, only the ground level is accessible and $q = g_0$. At very high temperatures, virtually all states are accessible, and q is correspondingly large.

Comment 16.3

The sum of the infinite series $S = 1 + x + x^2 + \cdots$ is obtained by multiplying both sides by x, which gives $xS = x + x^2 + x^3 + \cdots = S - 1$ and hence $S = 1/(1 - x)$.

Example 16.2 *Evaluating the partition function for a uniform ladder of energy levels*

Evaluate the partition function for a molecule with an infinite number of equally spaced nondegenerate energy levels (Fig. 16.3). These levels can be thought of as the vibrational energy levels of a diatomic molecule in the harmonic approximation.

Method We expect the partition function to increase from 1 at $T = 0$ and approach infinity as T to ∞. To evaluate eqn 16.8 explicitly, note that

$$1 + x + x^2 + \cdots = \frac{1}{1 - x}$$

Answer If the separation of neighbouring levels is ε, the partition function is

$$q = 1 + e^{-\beta\varepsilon} + e^{-2\beta\varepsilon} + \cdots = 1 + e^{-\beta\varepsilon} + (e^{-\beta\varepsilon})^2 + \cdots = \frac{1}{1 - e^{-\beta\varepsilon}}$$

This expression is plotted in Fig. 16.4: notice that, as anticipated, q rises from 1 to infinity as the temperature is raised.

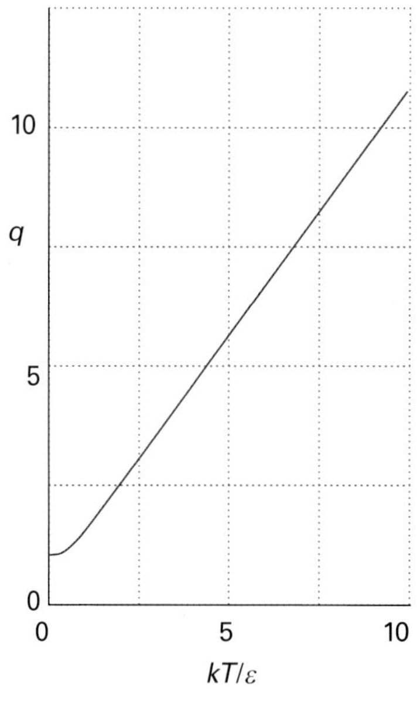

Fig. 16.4 The partition function for the system shown in Fig.16.3 (a harmonic oscillator) as a function of temperature.

Exploration Plot the partition function of a harmonic oscillator against temperature for several values of the energy separation ε. How does q vary with temperature when T is high, in the sense that $kT \gg \varepsilon$ (or $\beta\varepsilon \ll 1$)?

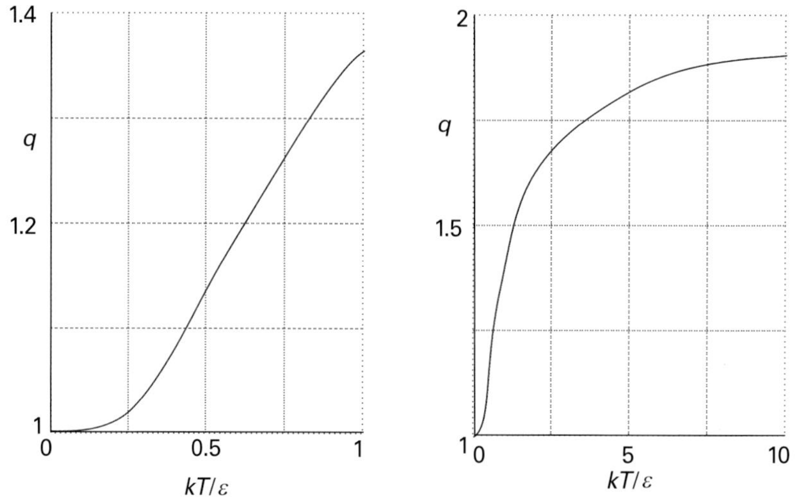

Fig. 16.5 The partition function for a two-level system as a function of temperature. The two graphs differ in the scale of the temperature axis to show the approach to 1 as $T \to 0$ and the slow approach to 2 as $T \to \infty$.

Exploration Consider a three-level system with levels 0, ε, and 2ε. Plot the partition function against kT/ε.

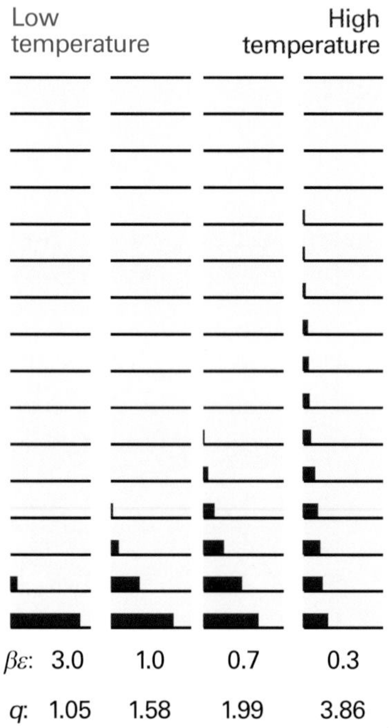

Fig. 16.6 The populations of the energy levels of the system shown in Fig.16.3 at different temperatures, and the corresponding values of the partition function calculated in Example 16.2. Note that $\beta = 1/kT$.

Exploration To visualize the content of Fig. 16.6 in a different way, plot the functions p_0, p_1, p_2, and p_3 against kT/ε.

Self-test 16.3 Find and plot an expression for the partition function of a system with one state at zero energy and another state at the energy ε.

[$q = 1 + e^{-\beta\varepsilon}$, Fig. 16.5]

It follows from eqn 16.8 and the expression for q derived in Example 16.2 for a uniform ladder of states of spacing ε,

$$q = \frac{1}{1 - e^{-\beta\varepsilon}} \tag{16.12}$$

that the fraction of molecules in the state with energy ε_i is

$$p_i = \frac{e^{-\beta\varepsilon_i}}{q} = (1 - e^{-\beta\varepsilon})e^{-\beta\varepsilon_i} \tag{16.13}$$

Figure 16.6 shows how p_i varies with temperature. At very low temperatures, where q is close to 1, only the lowest state is significantly populated. As the temperature is raised, the population breaks out of the lowest state, and the upper states become progressively more highly populated. At the same time, the partition function rises from 1 and its value gives an indication of the range of states populated. The name 'partition function' reflects the sense in which q measures how the total number of molecules is distributed—partitioned—over the available states.

The corresponding expressions for a two-level system derived in Self-test 16.3 are

$$p_0 = \frac{1}{1 + e^{-\beta\varepsilon}} \qquad p_1 = \frac{e^{-\beta\varepsilon}}{1 + e^{-\beta\varepsilon}} \tag{16.14}$$

These functions are plotted in Fig. 16.7. Notice how the populations tend towards equality ($p_0 = \tfrac{1}{2}, p_1 = \tfrac{1}{2}$) as $T \to \infty$. A common error is to suppose that all the molecules in the system will be found in the upper energy state when $T = \infty$; however, we see

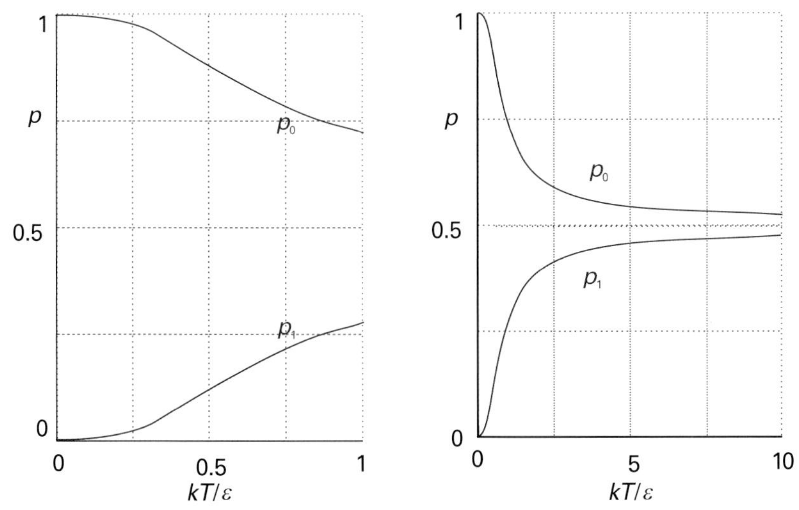

Fig. 16.7 The fraction of populations of the two states of a two-level system as a function of temperature (eqn 16.14). Note that, as the temperature approaches infinity, the populations of the two states become equal (and the fractions both approach 0.5).

Exploration Consider a three-level system with levels 0, ε, and 2ε. Plot the functions p_0, p_1, and p_2 against kT/ε.

from eqn 16.14 that, as $T \to \infty$, the populations of states become equal. The same conclusion is true of multi-level systems too: as $T \to \infty$, all states become equally populated.

Example 16.3 *Using the partition function to calculate a population*

Calculate the proportion of I_2 molecules in their ground, first excited, and second excited vibrational states at 25°C. The vibrational wavenumber is 214.6 cm^{-1}.

Method Vibrational energy levels have a constant separation (in the harmonic approximation, Section 13.9), so the partition function is given by eqn 16.12 and the populations by eqn 16.13. To use the latter equation, we identify the index i with the quantum number v, and calculate p_v for $v = 0$, 1, and 2. At 298.15 K, $kT/hc = 207.226$ cm^{-1}.

Answer First, we note that

$$\beta\varepsilon = \frac{hc\tilde{\nu}}{kT} = \frac{214.6\ \text{cm}^{-1}}{207.226\ \text{cm}^{-1}} = 1.036$$

Then it follows from eqn 16.13 that the populations are

$$p_v = (1 - e^{-\beta\varepsilon})e^{-v\beta\varepsilon} = 0.645e^{-1.036v}$$

Therefore, $p_0 = 0.645$, $p_1 = 0.229$, $p_2 = 0.081$. The I—I bond is not stiff and the atoms are heavy: as a result, the vibrational energy separations are small and at room temperature several vibrational levels are significantly populated. The value of the partition function, $q = 1.55$, reflects this small but significant spread of populations.

Self-test 16.4 At what temperature would the $v = 1$ level of I_2 have (a) half the population of the ground state, (b) the same population as the ground state?

[(a) 445 K, (b) infinite]

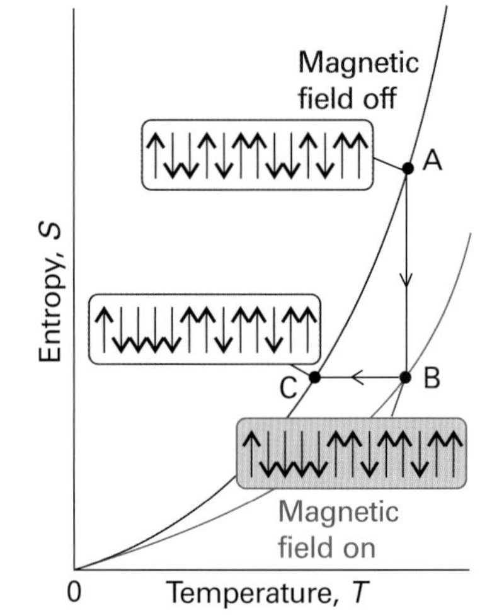

Fig. 16.8 The technique of adiabatic demagnetization is used to attain very low temperatures. The upper curve shows that variation of the entropy of a paramagnetic system in the absence of an applied field. The lower curve shows that variation in entropy when a field is applied and has made the electron magnets more orderly. The isothermal magnetization step is from A to B; the adiabatic demagnetization step (at constant entropy) is from B to C.

It follows from our discussion of the partition function that to reach low temperatures it is necessary to devise strategies that populate the low energy levels of a system at the expense of high energy levels. Common methods used to reach very low temperatures include **optical trapping** and **adiabatic demagnetization**. In optical trapping, atoms in the gas phase are cooled by inelastic collisions with photons from intense laser beams, which act as walls of a very small container. Adiabatic demagnetization is based on the fact that, in the absence of a magnetic field, the unpaired electrons of a paramagnetic material are orientated at random, but in the presence of a magnetic field there are more β spins ($m_s = -\frac{1}{2}$) than α spins ($m_s = +\frac{1}{2}$). In thermodynamic terms, the application of a magnetic field lowers the entropy of a sample and, at a given temperature, the entropy of a sample is lower when the field is on than when it is off. Even lower temperatures can be reached if nuclear spins (which also behave like small magnets) are used instead of electron spins in the technique of **adiabatic nuclear demagnetization**, which has been used to cool a sample of silver to about 280 pK. In certain circumstances it is possible to achieve negative temperatures, and the equations derived later in this chapter can be extended to $T < 0$ with interesting consequences (see *Further information* 16.3).

Illustration 16.2 *Cooling a sample by adiabatic demagnetization*

Consider the situation summarized by Fig. 16.8. A sample of paramagnetic material, such as a *d*- or *f*-metal complex with several unpaired electrons, is cooled to about 1 K by using helium. The sample is then exposed to a strong magnetic field while it is surrounded by helium, which provides thermal contact with the cold reservoir. This magnetization step is isothermal, and energy leaves the system as heat while the electron spins adopt the lower energy state (AB in the illustration). Thermal contact between the sample and the surroundings is now broken by pumping away the helium and the magnetic field is reduced to zero. This step is adiabatic and effectively reversible, so the state of the sample changes from B to C. At the end of this step the sample is the same as it was at A except that it now has a lower entropy. That lower entropy in the absence of a magnetic field corresponds to a lower temperature. That is, adiabatic demagnetization has cooled the sample.

(b) Approximations and factorizations

In general, exact analytical expressions for partition functions cannot be obtained. However, closed approximate expressions can often be found and prove to be very important in a number of chemical and biochemical applications (*Impact* 16.1). For instance, the expression for the partition function for a particle of mass m free to move in a one-dimensional container of length X can be evaluated by making use of the fact that the separation of energy levels is very small and that large numbers of states are accessible at normal temperatures. As shown in the *Justification* below, in this case

$$q_X = \left(\frac{2\pi m}{h^2\beta}\right)^{1/2} X \qquad (16.15)$$

This expression shows that the partition function for translational motion increases with the length of the box and the mass of the particle, for in each case the separation of the energy levels becomes smaller and more levels become thermally accessible. For a given mass and length of the box, the partition function also increases with increasing temperature (decreasing β), because more states become accessible.

Justification 16.2 *The partition function for a particle in a one-dimensional box*

The energy levels of a molecule of mass m in a container of length X are given by eqn 9.4a with $L = X$:

$$E_n = \frac{n^2h^2}{8mX^2} \qquad n = 1, 2, \ldots$$

The lowest level ($n = 1$) has energy $h^2/8mX^2$, so the energies relative to that level are

$$\varepsilon_n = (n^2 - 1)\varepsilon \qquad \varepsilon = h^2/8mX^2$$

The sum to evaluate is therefore

$$q_X = \sum_{n=1}^{\infty} \mathrm{e}^{-(n^2-1)\beta\varepsilon}$$

The translational energy levels are very close together in a container the size of a typical laboratory vessel; therefore, the sum can be approximated by an integral:

$$q_X = \int_1^{\infty} \mathrm{e}^{-(n^2-1)\beta\varepsilon}\mathrm{d}n \approx \int_0^{\infty} \mathrm{e}^{-n^2\beta\varepsilon}\mathrm{d}n$$

The extension of the lower limit to $n = 0$ and the replacement of $n^2 - 1$ by n^2 introduces negligible error but turns the integral into standard form. We make the substitution $x^2 = n^2\beta\varepsilon$, implying $\mathrm{d}n = \mathrm{d}x/(\beta\varepsilon)^{1/2}$, and therefore that

$$q_X = \left(\frac{1}{\beta\varepsilon}\right)^{1/2} \overbrace{\int_0^{\infty} \mathrm{e}^{-x^2}\mathrm{d}x}^{\pi^{1/2}/2} = \left(\frac{1}{\beta\varepsilon}\right)^{1/2} \left(\frac{\pi^{1/2}}{2}\right) = \left(\frac{2\pi m}{h^2\beta}\right)^{1/2} X$$

Another useful feature of partition functions is used to derive expressions when the energy of a molecule arises from several different, independent sources: if the energy is a sum of contributions from independent modes of motion, then the partition function is a product of partition functions for each mode of motion. For instance, suppose the molecule we are considering is free to move in three dimensions. We take the length of the container in the y-direction to be Y and that in the z-direction to be Z. The total energy of a molecule ε is the sum of its translational energies in all three directions:

$$\varepsilon_{n_1n_2n_3} = \varepsilon_{n_1}^{(X)} + \varepsilon_{n_2}^{(Y)} + \varepsilon_{n_3}^{(Z)} \tag{16.16}$$

where n_1, n_2, and n_3 are the quantum numbers for motion in the x-, y-, and z-directions, respectively. Therefore, because $\mathrm{e}^{a+b+c} = \mathrm{e}^a\mathrm{e}^b\mathrm{e}^c$, the partition function factorizes as follows:

$$\begin{aligned} q &= \sum_{\text{all } n} \mathrm{e}^{-\beta\varepsilon_{n_1}^{(X)}-\beta\varepsilon_{n_2}^{(Y)}-\beta\varepsilon_{n_3}^{(Z)}} = \sum_{\text{all } n} \mathrm{e}^{-\beta\varepsilon_{n_1}^{(X)}}\mathrm{e}^{-\beta\varepsilon_{n_2}^{(Y)}}\mathrm{e}^{-\beta\varepsilon_{n_3}^{(Z)}} \\ &= \left(\sum_{n_1} \mathrm{e}^{-\beta\varepsilon_{n_1}^{(X)}}\right)\left(\sum_{n_2} \mathrm{e}^{-\beta\varepsilon_{n_2}^{(Y)}}\right)\left(\sum_{n_3} \mathrm{e}^{-\beta\varepsilon_{n_3}^{(Z)}}\right) \\ &= q_Xq_Yq_Z \end{aligned} \tag{16.17}$$

It is generally true that, if the energy of a molecule can be written as the sum of independent terms, then the partition function is the corresponding product of individual contributions.

Equation 16.15 gives the partition function for translational motion in the x-direction. The only change for the other two directions is to replace the length X by the lengths Y or Z. Hence the partition function for motion in three dimensions is

$$q = \left(\frac{2\pi m}{h^2\beta}\right)^{3/2} XYZ \qquad (16.18)$$

The product of lengths XYZ is the volume, V, of the container, so we can write

$$q = \frac{V}{\Lambda^3} \qquad \Lambda = h\left(\frac{\beta}{2\pi m}\right)^{1/2} = \frac{h}{(2\pi mkT)^{1/2}} \qquad (16.19)$$

The quantity Λ has the dimensions of length and is called the **thermal wavelength** (sometimes the *thermal de Broglie wavelength*) of the molecule. The thermal wavelength decreases with increasing mass and temperature. As in the one-dimensional case, the partition function increases with the mass of the particle (as $m^{3/2}$) and the volume of the container (as V); for a given mass and volume, the partition function increases with temperature (as $T^{3/2}$).

Illustration 16.3 *Calculating the translational partition function*

To calculate the translational partition function of an H_2 molecule confined to a 100 cm^3 vessel at 25°C we use $m = 2.016$ u; then

$$\Lambda = \frac{6.626\times10^{-34}\ \text{J s}}{\{2\pi\times(2.016\times1.6605\times10^{-27}\ \text{kg})\times(1.38\times10^{-23}\ \text{J K}^{-1})\times(298\ \text{K})\}^{1/2}}$$
$$= 7.12\times10^{-11}\ \text{m}$$

where we have used 1 J = 1 kg m^2 s^{-2}. Therefore,

$$q = \frac{1.00\times10^{-4}\ \text{m}^3}{(7.12\times10^{-11}\ \text{m})^3} = 2.77\times10^{26}$$

About 10^{26} quantum states are thermally accessible, even at room temperature and for this light molecule. Many states are occupied if the thermal wavelength (which in this case is 71.2 pm) is small compared with the linear dimensions of the container.

Self-test 16.5 Calculate the translational partition function for a D_2 molecule under the same conditions. [$q = 7.8\times10^{26}$, $2^{3/2}$ times larger]

The validity of the approximations that led to eqn 16.19 can be expressed in terms of the average separation of the particles in the container, d. We do not have to worry about the role of the Pauli principle on the occupation of states if there are many states available for each molecule. Because q is the total number of accessible states, the average number of states per molecule is q/N. For this quantity to be large, we require $V/N\Lambda^3 \gg 1$. However, V/N is the volume occupied by a single particle, and therefore the average separation of the particles is $d = (V/N)^{1/3}$. The condition for there being many states available per molecule is therefore $d^3/\Lambda^3 \gg 1$, and therefore $d \gg \Lambda$. That is, for eqn 16.19 to be valid, *the average separation of the particles must be much greater than their thermal wavelength.* For H_2 molecules at 1 bar and 298 K, the average separation is 3 nm, which is significantly larger than their thermal wavelength (71.2 pm, *Illustration* 16.3).

IMPACT ON BIOCHEMISTRY

I16.1 The helix–coil transition in polypeptides

Proteins are polymers that attain well defined three-dimensional structures both in solution and in biological cells. They are *polypeptides* formed from different amino acids strung together by the *peptide link*, —CONH—. Hydrogen bonds between amino acids of a polypeptide give rise to stable helical or sheet structures, which may collapse into a random coil when certain conditions are changed. The unwinding of a helix into a random coil is a *cooperative transition*, in which the polymer becomes increasingly more susceptible to structural changes once the process has begun. We examine here a model grounded in the principles of statistical thermodynamics that accounts for the cooperativity of the helix–coil transition in polypeptides.

To calculate the fraction of polypeptide molecules present as helix or coil we need to set up the partition function for the various states of the molecule. To illustrate the approach, consider a short polypeptide with four amino acid residues, each labelled *h* if it contributes to a helical region and *c* if it contributes to a random coil region. We suppose that conformations *hhhh* and *cccc* contribute terms q_0 and q_4, respectively, to the partition function q. Then we assume that each of the four conformations with one *c* amino acid (such as *hchh*) contributes q_1. Similarly, each of the six states with two *c* amino acids contributes a term q_2, and each of the four states with three *c* amino acids contributes a term q_3. The partition function is then

$$q = q_0 + 4q_1 + 6q_2 + 4q_3 + q_4 = q_0\left(1 + \frac{4q_1}{q_0} + \frac{6q_2}{q_0} + \frac{4q_3}{q_0} + \frac{q_4}{q_0}\right)$$

We shall now suppose that each partition function differs from q_0 only by the energy of each conformation relative to *hhhh*, and write

$$\frac{q_i}{q_0} = e^{-(\varepsilon_i - \varepsilon_0)/kT}$$

Next, we suppose that the conformational transformations are non-cooperative, in the sense that the energy associated with changing one *h* amino acid into one *c* amino acid has the same value regardless of how many *h* or *c* amino acid residues are in the reactant or product state and regardless of where in the chain the conversion occurs. That is, we suppose that the difference in energy between c^ih^{4-i} and $c^{i+1}h^{3-i}$ has the same value γ for all i. This assumption implies that $\varepsilon_i - \varepsilon_0 = i\gamma$ and therefore that

$$q = q_0(1 + 4s + 6s^2 + 4s^3 + s^4) \qquad s = e^{-\Gamma/RT} \tag{16.20}$$

where $\Gamma = N_A\gamma$ and s is called the *stability parameter*. The term in parentheses has the form of the binomial expansion of $(1 + s)^4$.

$$\frac{q}{q_0} = \sum_{i=0}^{4} C(4,i)s^i \quad \text{with} \quad C(4,i) = \frac{4!}{(4-i)!i!} \tag{16.21}$$

which we interpret as the number of ways in which a state with i *c* amino acids can be formed.

The extension of this treatment to take into account a longer chain of residues is now straightforward: we simply replace the upper limit of 4 in the sum by n:

$$\frac{q}{q_0} = \sum_{i=0}^{n} C(n,i)s^i \tag{16.22}$$

A cooperative transformation is more difficult to accommodate, and depends on building a model of how neighbours facilitate each other's conformational change. In

Comment 16.4

The binomial expansion of $(1 + x)^n$ is

$$(1 + x)^n = \sum_{i=0}^{n} C(n,i)x^i,$$

$$\text{with } C(n,i) = \frac{n!}{(n-i)!i!}$$

the simple *zipper model*, conversion from h to c is allowed only if a residue adjacent to the one undergoing the conversion is already a c residue. Thus, the zipper model allows a transition of the type . . .*hhhch*. . . → . . .*hhhcc*. . . , but not a transition of the type . . .*hhhch*. . . → . . .*hchch*. . . . The only exception to this rule is, of course, the very first conversion from h to c in a fully helical chain. Cooperativity is included in the zipper model by assuming that the first conversion from h to c, called the *nucleation step*, is less favourable than the remaining conversions and replacing s for that step by σs, where $\sigma \ll 1$. Each subsequent step is called a *propagation step* and has a stability parameter s. In Problem 16.24, you are invited to show that the partition function is:

$$q = 1 + \sum_{i=1}^{n} Z(n,i)\sigma s^i \tag{16.23}$$

where $Z(n,i)$ is the number of ways in which a state with a number i of c amino acids can be formed under the strictures of the zipper model. Because $Z(n,i) = n - i + 1$ (see Problem 16.24),

$$q = 1 + \sigma(n+1)\sum_{i=1}^{n} s^i - \sigma\sum_{i=1}^{n} is^i \tag{16.24}$$

After evaluating both geometric series by using the two relations

$$\sum_{i=1}^{n} x^i = \frac{x^{n+1} - x}{x-1} \qquad \sum_{i=1}^{n} ix^i = \frac{x}{(x-1)^2}[nx^{n+1} - (n+1)x^n + 1]$$

we find

$$q = 1 + \frac{\sigma s[s^{n+1} - (n+1)s^n + 1]}{(s-1)^2}$$

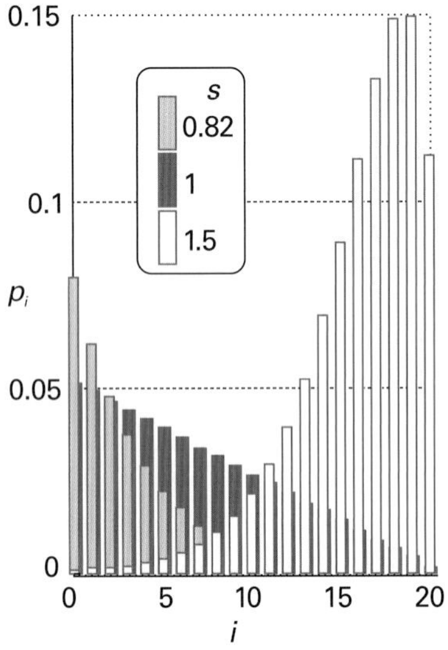

Fig. 16.9 The distribution of p_i, the fraction of molecules that has a number i of c amino acids for $s = 0.8$ ($\langle i \rangle = 1.1$), 1.0 ($\langle i \rangle = 3.8$), and 1.5 ($\langle i \rangle = 15.9$), with $\sigma = 5.0 \times 10^{-3}$.

The fraction $p_i = q_i/q$ of molecules that has a number i of c amino acids is $p_i = [(n-i+1)\sigma s^i]/q$ and the mean value of i is then $\langle i \rangle = \sum_i ip_i$. Figure 16.9 shows the distribution of p_i for various values of s with $\sigma = 5.0 \times 10^{-3}$. We see that most of the polypeptide chains remain largely helical when $s < 1$ and that most of the chains exist largely as random coils when $s > 1$. When $s = 1$, there is a more widespread distribution of length of random coil segments. Because the *degree of conversion*, θ, of a polypeptide with n amino acids to a random coil is defined as $\theta = \langle i \rangle/n$, it is possible to show (see Problem 16.24) that

$$\theta = \frac{1}{n}\frac{\mathrm{d}}{\mathrm{d}(\ln s)}\ln q \tag{16.25}$$

This is a general result that applies to any model of the helix–coil transition in which the partition function q is expressed as a function of the stability parameter s.

A more sophisticated model for the helix–coil transition must allow for helical segments to form in different regions of a long polypeptide chain, with the nascent helices being separated by shrinking coil segments. Calculations based on this more complete *Zimm–Bragg model* give

$$\theta = \tfrac{1}{2}\left(1 + \frac{(s-1) + 2\sigma}{[(s-1)^2 + 4s\sigma]^{1/2}}\right) \tag{16.26}$$

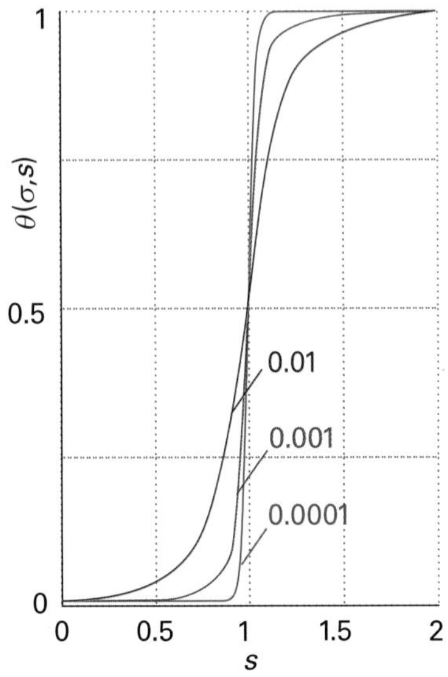

Fig. 16.10 Plots of the degree of conversion θ, against s for several values of σ. The curves show the sigmoidal shape characteristics of cooperative behaviour.

Figure 16.10 shows plots of θ against s for several values of σ. The curves show the sigmoidal shape characteristic of cooperative behaviour. There is a sudden surge of transition to a random coil as s passes through 1 and, the smaller the parameter σ, the greater the sharpness and hence the greater the cooperativity of the transition. That is, the harder it is to get coil formation started, the sharper the transition from helix to coil.

The internal energy and the entropy

The importance of the molecular partition function is that it contains all the information needed to calculate the thermodynamic properties of a system of independent particles. In this respect, q plays a role in statistical thermodynamics very similar to that played by the wavefunction in quantum mechanics: q is a kind of thermal wavefunction.

16.3 The internal energy

We shall begin to unfold the importance of q by showing how to derive an expression for the internal energy of the system.

(a) The relation between U and q

The total energy of the system relative to the energy of the lowest state is

$$E=\sum_i n_i\varepsilon_i \tag{16.27}$$

Because the most probable configuration is so strongly dominating, we can use the Boltzmann distribution for the populations and write

$$E=\frac{N}{q}\sum_i \varepsilon_i e^{-\beta\varepsilon_i} \tag{16.28}$$

To manipulate this expression into a form involving only q we note that

$$\varepsilon_i e^{-\beta\varepsilon_i}=-\frac{d}{d\beta}e^{-\beta\varepsilon_i}$$

It follows that

$$E=-\frac{N}{q}\sum_i\frac{d}{d\beta}e^{-\beta\varepsilon_i}=-\frac{N}{q}\frac{d}{d\beta}\sum_i e^{-\beta\varepsilon_i}=-\frac{N}{q}\frac{dq}{d\beta} \tag{16.29}$$

Illustration 16.4 *The energy of a two-level system*

From the two-level partition function $q=1+e^{-\beta\varepsilon}$, we can deduce that the total energy of N two-level systems is

$$E=-\left(\frac{N}{1+e^{-\beta\varepsilon}}\right)\frac{d}{d\beta}(1+e^{-\beta\varepsilon})=\frac{N\varepsilon e^{-\beta\varepsilon}}{1+e^{-\beta\varepsilon}}=\frac{N\varepsilon}{1+e^{\beta\varepsilon}}$$

This function is plotted in Fig. 16.11. Notice how the energy is zero at $T=0$, when only the lower state (at the zero of energy) is occupied, and rises to $\frac{1}{2}N\varepsilon$ as $T\to\infty$, when the two levels become equally populated.

There are several points in relation to eqn 16.29 that need to be made. Because $\varepsilon_0=0$ (remember that we measure all energies from the lowest available level), E should be interpreted as the value of the internal energy relative to its value at $T=0$, $U(0)$. Therefore, to obtain the conventional internal energy U, we must add the internal energy at $T=0$:

$$U=U(0)+E \tag{16.30}$$

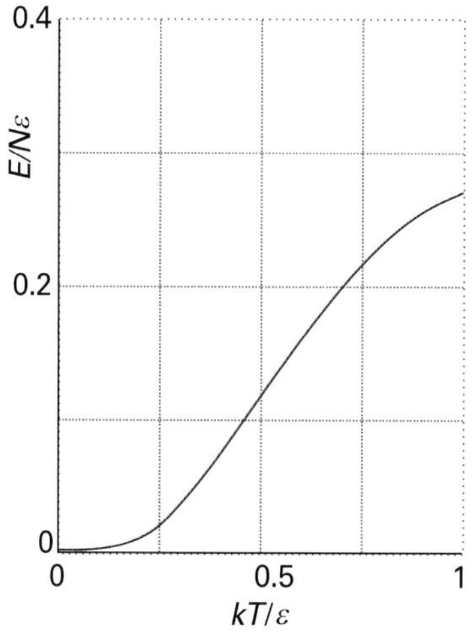

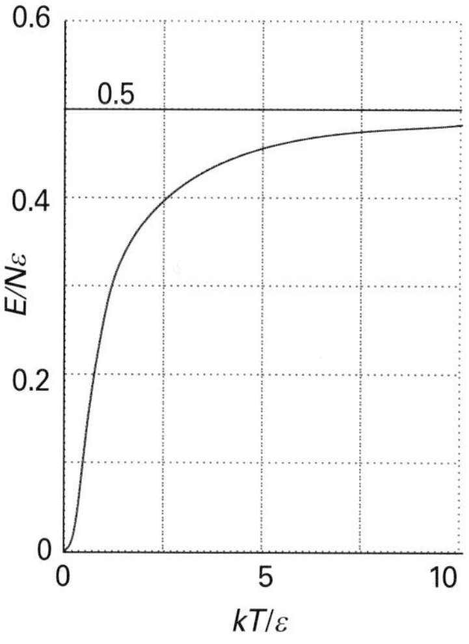

Fig. 16.11 The total energy of a two-level system (expressed as a multiple of $N\varepsilon$) as a function of temperature, on two temperature scales. The graph at the top shows the slow rise away from zero energy at low temperatures; the slope of the graph at $T=0$ is 0 (that is, the heat capacity is zero at $T=0$). The graph below shows the slow rise to 0.5 as $T\to\infty$ as both states become equally populated (see Fig. 16.7).

Exploration Draw graphs similar to those in Fig. 16.11 for a three-level system with levels 0, ε, and 2ε.

Secondly, because the partition function may depend on variables other than the temperature (for example, the volume), the derivative with respect to β in eqn 16.29 is actually a *partial* derivative with these other variables held constant. The complete expression relating the molecular partition function to the thermodynamic internal energy of a system of independent molecules is therefore

$$U = U(0) - \frac{N}{q}\left(\frac{\partial q}{\partial \beta}\right)_V \qquad (16.31a)$$

An equivalent form is obtained by noting that $dx/x = d\ln x$:

$$U = U(0) - N\left(\frac{\partial \ln q}{\partial \beta}\right)_V \qquad (16.31b)$$

These two equations confirm that we need know only the partition function (as a function of temperature) to calculate the internal energy relative to its value at $T = 0$.

(b) The value of β

We now confirm that the parameter β, which we have anticipated is equal to $1/kT$, does indeed have that value. To do so, we compare the equipartition expression for the internal energy of a monatomic perfect gas, which from *Molecular interpretation* 2.2 we know to be

$$U = U(0) + \tfrac{3}{2}nRT \qquad (16.32a)$$

with the value calculated from the translational partition function (see the following *Justification*), which is

$$U = U(0) + \frac{3N}{2\beta} \qquad (16.32b)$$

It follows by comparing these two expressions that

$$\beta = \frac{N}{nRT} = \frac{nN_A}{nN_AkT} = \frac{1}{kT} \qquad (16.33)$$

(We have used $N = nN_A$, where n is the amount of gas molecules, N_A is Avogadro's constant, and $R = N_Ak$.) Although we have proved that $\beta = 1/kT$ by examining a very specific example, the translational motion of a perfect monatomic gas, the result is general (see *Example* 17.1 and *Further reading*).

Justification 16.3 *The internal energy of a perfect gas*

To use eqn 16.31, we introduce the translational partition function from eqn 16.19:

$$\left(\frac{\partial q}{\partial \beta}\right)_V = \left(\frac{\partial}{\partial \beta}\frac{V}{\Lambda^3}\right)_V = V\frac{d}{d\beta}\frac{1}{\Lambda^3} = -3\frac{V}{\Lambda^4}\frac{d\Lambda}{d\beta}$$

Then we note from the formula for Λ in eqn 16.19 that

$$\frac{d\Lambda}{d\beta} = \frac{d}{d\beta}\left\{\frac{h\beta^{1/2}}{(2\pi m)^{1/2}}\right\} = \frac{1}{2\beta^{1/2}} \times \frac{h}{(2\pi m)^{1/2}} = \frac{\Lambda}{2\beta}$$

and so obtain

$$\left(\frac{\partial q}{\partial \beta}\right)_V = -\frac{3V}{2\beta\Lambda^3}$$

Then, by eqn 16.31a,

$$U = U(0) - N\left(\frac{\Lambda^3}{V}\right)\left(-\frac{3V}{2\beta\Lambda^3}\right) = U(0) + \frac{3N}{2\beta}$$

as in eqn 16.32b.

16.4 The statistical entropy

If it is true that the partition function contains all thermodynamic information, then it must be possible to use it to calculate the entropy as well as the internal energy. Because we know (from Section 3.2) that entropy is related to the dispersal of energy and that the partition function is a measure of the number of thermally accessible states, we can be confident that the two are indeed related.

We shall develop the relation between the entropy and the partition function in two stages. In *Further information* 16.2, we justify one of the most celebrated equations in statistical thermodynamics, the **Boltzmann formula** for the entropy:

$$S = k \ln W \qquad [16.34]$$

In this expression, W is the weight of the most probable configuration of the system. In the second stage, we express W in terms of the partition function.

The statistical entropy behaves in exactly the same way as the thermodynamic entropy. Thus, as the temperature is lowered, the value of W, and hence of S, decreases because fewer configurations are compatible with the total energy. In the limit $T \rightarrow 0$, $W = 1$, so $\ln W = 0$, because only one configuration (every molecule in the lowest level) is compatible with $E = 0$. It follows that $S \rightarrow 0$ as $T \rightarrow 0$, which is compatible with the Third Law of thermodynamics, that the entropies of all perfect crystals approach the same value as $T \rightarrow 0$ (Section 3.4).

Now we relate the Boltzmann formula for the entropy to the partition function. To do so, we substitute the expression for $\ln W$ given in eqn 16.3 into eqn 16.34 and, as shown in the *Justification* below, obtain

$$S = \frac{U - U(0)}{T} + Nk \ln q \qquad (16.35)$$

Justification 16.4 *The statistical entropy*

The first stage is to use eqn 16.3 ($\ln W = N \ln N - \sum_i n_i \ln n_i$) and $N = \sum_i n_i$ to write

$$S = k\sum_i (n_i \ln N - n_i \ln n_i) = -k\sum_i n_i \ln \frac{n_i}{N} = -Nk\sum_i p_i \ln p_i$$

where $p_i = n_i/N$, the fraction of molecules in state i. It follows from eqn 16.7 that

$$\ln p_i = -\beta\varepsilon_i - \ln q$$

and therefore that

$$S = -Nk(-\beta\sum_i p_i\varepsilon_i - \sum_i p_i \ln q) = k\beta\{U - U(0)\} + Nk \ln q$$

We have used the fact that the sum over the p_i is equal to 1 and that (from eqns 16.27 and 16.30)

$$N\sum_i p_i\varepsilon_i = \sum_i Np_i\varepsilon_i = \sum_i Np_i\varepsilon_i = \sum_i n_i\varepsilon_i = E = U - U(0)$$

We have already established that $\beta = 1/kT$, so eqn 16.35 immediately follows.

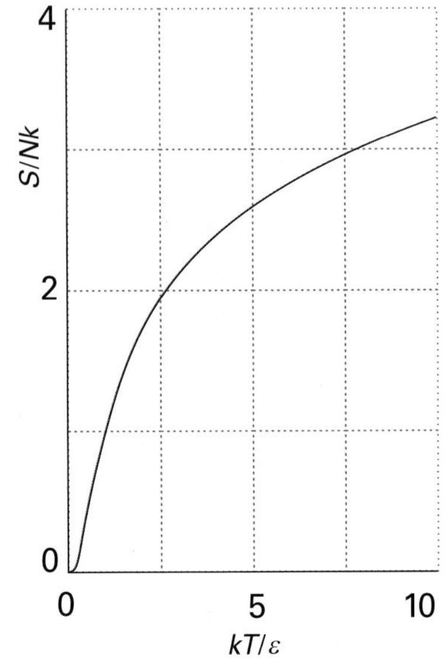

Fig. 16.12 The temperature variation of the entropy of the system shown in Fig. 16.3 (expressed here as a multiple of Nk). The entropy approaches zero as $T \to 0$, and increases without limit as $T \to \infty$.

Exploration Plot the function dS/dT, the temperature coefficient of the entropy, against kT/ε. Is there a temperature at which this coefficient passes through a maximum? If you find a maximum, explain its physical origins.

Example 16.4 *Calculating the entropy of a collection of oscillators*

Calculate the entropy of a collection of N independent harmonic oscillators, and evaluate it using vibrational data for I_2 vapour at 25°C (Example 16.3).

Method To use eqn 16.35, we use the partition function for a molecule with evenly spaced vibrational energy levels, eqn 16.12. With the partition function available, the internal energy can be found by differentiation (as in eqn 16.31a), and the two expressions then combined to give S.

Answer The molecular partition function as given in eqn 16.12 is

$$q = \frac{1}{1-e^{-\beta\varepsilon}}$$

The internal energy is obtained by using eqn 16.31a:

$$U - U(0) = -\frac{N}{q}\left(\frac{\partial q}{\partial \beta}\right)_V = \frac{N\varepsilon e^{-\beta\varepsilon}}{1-e^{-\beta\varepsilon}} = \frac{N\varepsilon}{e^{-\beta\varepsilon}-1}$$

The entropy is therefore

$$S = Nk\left\{\frac{\beta\varepsilon}{e^{\beta\varepsilon}-1} - \ln(1-e^{\beta\varepsilon})\right\}$$

This function is plotted in Fig. 16.12. For I_2 at 25°C, $\beta\varepsilon = 1.036$ (Example 16.3), so $S_m = 8.38\ \mathrm{J\ K^{-1}\ mol^{-1}}$.

Self-test 16.6 Evaluate the molar entropy of N two-level systems and plot the resulting expression. What is the entropy when the two states are equally thermally accessible?

[$S/Nk = \beta\varepsilon/(1+e^{\beta\varepsilon}) + \ln(1+e^{-\beta\varepsilon})$; see Fig. 16.13; $S = Nk \ln 2$]

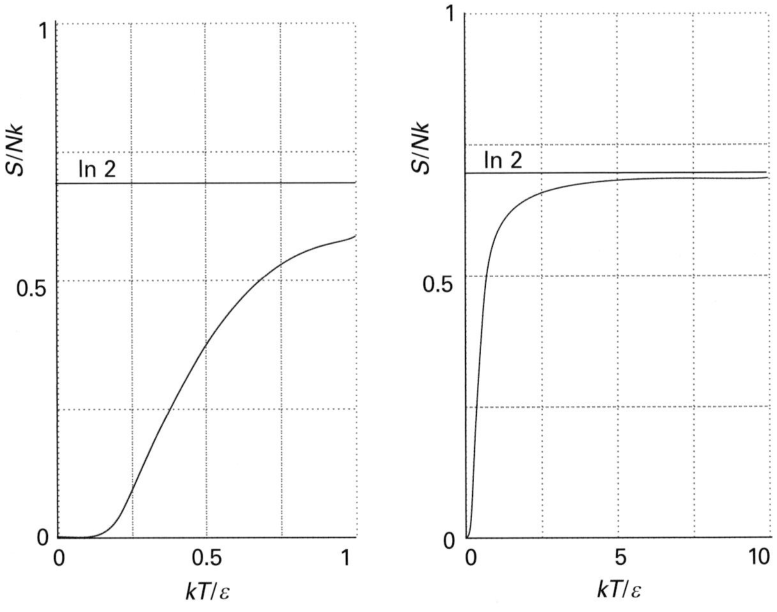

Fig. 16.13 The temperature variation of the entropy of a two-level system (expressed as a multiple of Nk). As $T \to \infty$, the two states become equally populated and S approaches $Nk \ln 2$.

Exploration Draw graphs similar to those in Fig. 16.13 for a three-level system with levels 0, ε, and 2ε.

The canonical partition function

In this section we see how to generalize our conclusions to include systems composed of interacting molecules. We shall also see how to obtain the molecular partition function from the more general form of the partition function developed here.

16.5 The canonical ensemble

The crucial new concept we need when treating systems of interacting particles is the 'ensemble'. Like so many scientific terms, the term has basically its normal meaning of 'collection', but it has been sharpened and refined into a precise significance.

(a) The concept of ensemble

To set up an ensemble, we take a closed system of specified volume, composition, and temperature, and think of it as replicated $\tilde{N}$ times (Fig. 16.14). All the identical closed systems are regarded as being in thermal contact with one another, so they can exchange energy. The total energy of all the systems is $\tilde{E}$ and, because they are in thermal equilibrium with one another, they all have the same temperature, T. This imaginary collection of replications of the actual system with a common temperature is called the **canonical ensemble**. The word 'canon' means 'according to a rule'.

There are two other important ensembles. In the **microcanonical ensemble** the condition of constant temperature is replaced by the requirement that all the systems should have exactly the same energy: each system is individually isolated. In the **grand canonical ensemble** the volume and temperature of each system is the same, but they are open, which means that matter can be imagined as able to pass between the systems; the composition of each one may fluctuate, but now the chemical potential is the same in each system:

Microcanonical ensemble: N, V, E common

Canonical ensemble: N, V, T common

Grand canonical ensemble: μ, V, T common

The important point about an ensemble is that it is a collection of *imaginary* replications of the system, so we are free to let the number of members be as large as we like; when appropriate, we can let $\tilde{N}$ become infinite. The number of members of the ensemble in a state with energy E_i is denoted $\tilde{n}_i$, and we can speak of the configuration of the ensemble (by analogy with the configuration of the system used in Section 16.1) and its weight, $\tilde{W}$. Note that $\tilde{N}$ is unrelated to N, the number of molecules in the actual system; $\tilde{N}$ is the number of imaginary replications of that system.

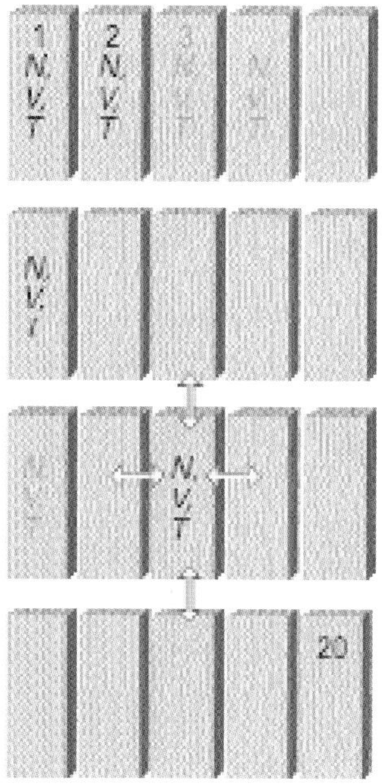

Fig. 16.14 A representation of the canonical ensemble, in this case for $\tilde{N} = 20$. The individual replications of the actual system all have the same composition and volume. They are all in mutual thermal contact, and so all have the same temperature. Energy may be transferred between them as heat, and so they do not all have the same energy. The total energy $\tilde{E}$ of all 20 replications is a constant because the ensemble is isolated overall.

(b) Dominating configurations

Just as in Section 16.1, some of the configurations of the ensemble will be very much more probable than others. For instance, it is very unlikely that the whole of the total energy, $\tilde{E}$, will accumulate in one system. By analogy with the earlier discussion, we can anticipate that there will be a dominating configuration, and that we can evaluate the thermodynamic properties by taking the average over the ensemble using that single, most probable, configuration. In the **thermodynamic limit** of $\tilde{N} \rightarrow \infty$, this dominating configuration is overwhelmingly the most probable, and it dominates the properties of the system virtually completely.

The quantitative discussion follows the argument in Section 16.1 with the modification that N and n_i are replaced by $\tilde{N}$ and $\tilde{n}_i$. The weight of a configuration $\{\tilde{n}_0, \tilde{n}_1, \ldots\}$ is

$$\tilde{W} = \frac{\tilde{N}!}{\tilde{n}_0!\tilde{n}_1!\ldots} \tag{16.36}$$

The configuration of greatest weight, subject to the constraints that the total energy of the ensemble is constant at $\tilde{E}$ and that the total number of members is fixed at $\tilde{N}$, is given by the **canonical distribution**:

$$\frac{\tilde{n}_i}{\tilde{N}} = \frac{e^{-\beta E_i}}{Q} \qquad Q = \sum_i e^{-\beta E_i} \tag{16.37}$$

The quantity Q, which is a function of the temperature, is called the **canonical partition function**.

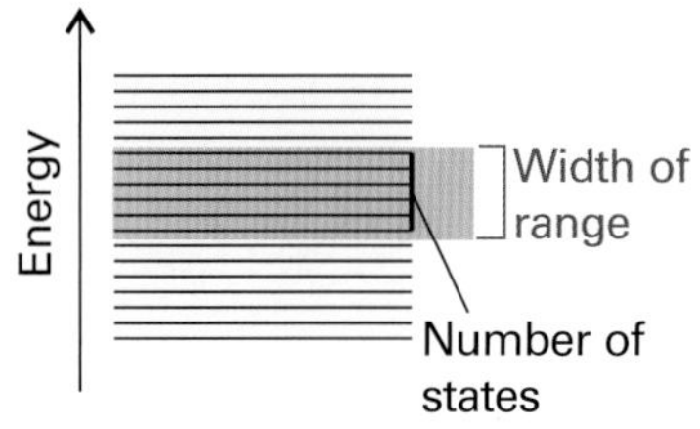

Fig. 16.15 The energy density of states is the number of states in an energy range divided by the width of the range.

(c) Fluctuations from the most probable distribution

The canonical distribution in eqn 16.37 is only apparently an exponentially decreasing function of the energy of the system. We must appreciate that eqn 16.37 gives the probability of occurrence of members in a single state i of the entire system of energy E_i. There may in fact be numerous states with almost identical energies. For example, in a gas the identities of the molecules moving slowly or quickly can change without necessarily affecting the total energy. The density of states, the number of states in an energy range divided by the width of the range (Fig. 16.15), is a very sharply increasing function of energy. It follows that the probability of a member of an ensemble having a specified energy (as distinct from being in a specified state) is given by eqn 16.37, a sharply decreasing function, multiplied by a sharply increasing function (Fig. 16.16). Therefore, the overall distribution is a sharply peaked function. We conclude that most members of the ensemble have an energy very close to the mean value.

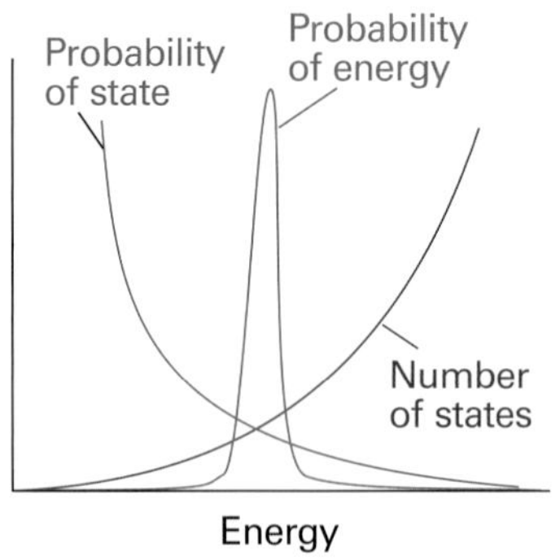

Fig. 16.16 To construct the form of the distribution of members of the canonical ensemble in terms of their energies, we multiply the probability that any one is in a state of given energy, eqn 16.39, by the number of states corresponding to that energy (a steeply rising function). The product is a sharply peaked function at the mean energy, which shows that almost all the members of the ensemble have that energy.

16.6 The thermodynamic information in the partition function

Like the molecular partition function, the canonical partition function carries all the thermodynamic information about a system. However, Q is more general than q because it does not assume that the molecules are independent. We can therefore use Q to discuss the properties of condensed phases and real gases where molecular interactions are important.

(a) The internal energy

If the total energy of the ensemble is $\tilde{E}$, and there are $\tilde{N}$ members, the average energy of a member is $E = \tilde{E}/\tilde{N}$. We use this quantity to calculate the internal energy of the system in the limit of $\tilde{N}$ (and $\tilde{E}$) approaching infinity:

$$U = U(0) + E = U(0) + \tilde{E}/\tilde{N} \qquad \text{as} \qquad \tilde{N} \to \infty \tag{16.38}$$

The fraction, $\tilde{p}_i$, of members of the ensemble in a state i with energy E_i is given by the analogue of eqn 16.7 as

$$\tilde{p}_i = \frac{e^{-\beta E_i}}{Q} \tag{16.39}$$

It follows that the internal energy is given by

$$U = U(0) + \sum_i \tilde{p}_i E_i = U(0) + \frac{1}{Q}\sum_i E_i e^{-\beta E_i} \tag{16.40}$$

By the same argument that led to eqn 16.31,

$$U = U(0) - \frac{1}{Q}\left(\frac{\partial Q}{\partial \beta}\right)_V = U(0) - \left(\frac{\partial \ln Q}{\partial \beta}\right)_V \tag{16.41}$$

(b) The entropy

The total weight, $\tilde{W}$, of a configuration of the ensemble is the product of the average weight W of each member of the ensemble, $\tilde{W} = W^{\tilde{N}}$. Hence, we can calculate S from

$$S = k \ln W = k \ln \tilde{W}^{1/\tilde{N}} = \frac{k}{\tilde{N}} \ln \tilde{W} \tag{16.42}$$

It follows, by the same argument used in Section 16.4, that

$$S = \frac{U - U(0)}{T} + k \ln Q \tag{16.43}$$

16.7 Independent molecules

We shall now see how to recover the molecular partition function from the more general canonical partition function when the molecules are independent. When the molecules are independent and distinguishable (in the sense to be described), the relation between Q and q is

$$Q = q^N \tag{16.44}$$

Justification 16.5 *The relation between Q and q*

The total energy of a collection of N independent molecules is the sum of the energies of the molecules. Therefore, we can write the total energy of a state i of the system as

$$E_i = \varepsilon_i(1) + \varepsilon_i(2) + \cdots + \varepsilon_i(N)$$

In this expression, $\varepsilon_i(1)$ is the energy of molecule 1 when the system is in the state i, $\varepsilon_i(2)$ the energy of molecule 2 when the system is in the same state i, and so on. The canonical partition function is then

$$Q = \sum_i e^{-\beta\varepsilon_i(1) - \beta\varepsilon_i(2) - \cdots - \beta\varepsilon_i(N)}$$

The sum over the states of the system can be reproduced by letting each molecule enter all its own individual states (although we meet an important proviso shortly). Therefore, instead of summing over the states i of the system, we can sum over all the individual states i of molecule 1, all the states i of molecule 2, and so on. This rewriting of the original expression leads to

$$Q = \left(\sum_i e^{-\beta\varepsilon_i}\right)\left(\sum_i e^{-\beta\varepsilon_i}\right)\cdots\left(\sum_i e^{-\beta\varepsilon_i}\right) = \left(\sum_i e^{-\beta\varepsilon_i}\right)^N = q^N$$

(a) Distinguishable and indistinguishable molecules

If all the molecules are identical and free to move through space, we cannot distinguish them and the relation $Q = q^N$ is not valid. Suppose that molecule 1 is in some state a, molecule 2 is in b, and molecule 3 is in c, then one member of the ensemble has an energy $E = \varepsilon_a + \varepsilon_b + \varepsilon_c$. This member, however, is indistinguishable from one formed by putting molecule 1 in state b, molecule 2 in state c, and molecule 3 in state a, or some other permutation. There are six such permutations in all, and $N!$ in general. In the case of indistinguishable molecules, it follows that we have counted too many states in going from the sum over system states to the sum over molecular states, so writing $Q = q^N$ overestimates the value of Q. The detailed argument is quite involved, but at all except very low temperatures it turns out that the correction factor is $1/N!$. Therefore:

- For distinguishable independent molecules: $Q = q^N$ (16.45a)
- For indistinguishable independent molecules: $Q = q^N/N!$ (16.45b)

For molecules to be indistinguishable, they must be of the same kind: an Ar atom is never indistinguishable from a Ne atom. Their identity, however, is not the only criterion. Each identical molecule in a crystal lattice, for instance, can be 'named' with a set of coordinates. Identical molecules in a lattice can therefore be treated as distinguishable because their sites are distinguishable, and we use eqn 16.45a. On the other hand, identical molecules in a gas are free to move to different locations, and there is no way of keeping track of the identity of a given molecule; we therefore use eqn 16.45b.

(b) The entropy of a monatomic gas

An important application of the previous material is the derivation (as shown in the *Justification* below) of the **Sackur–Tetrode equation** for the entropy of a monatomic gas:

$$S = nR\ln\left(\frac{e^{5/2}V}{nN_A\Lambda^3}\right) \qquad \Lambda = \frac{h}{(2\pi mkT)^{1/2}} \tag{16.46a}$$

This equation implies that the molar entropy of a perfect gas of high molar mass is greater than one of low molar mass under the same conditions (because the former has more thermally accessible translational states). Because the gas is perfect, we can use the relation $V = nRT/p$ to express the entropy in terms of the pressure as

$$S = nR\ln\left(\frac{e^{5/2}kT}{p\Lambda^3}\right) \tag{16.46b}$$

Justification 16.6 *The Sackur–Tetrode equation*

For a gas of independent molecules, Q may be replaced by $q^N/N!$, with the result that eqn 16.43 becomes

$$S = \frac{U - U(0)}{T} + Nk\ln q - k\ln N!$$

Because the number of molecules ($N = nN_A$) in a typical sample is large, we can use Stirling's approximation (eqn 16.2) to write

$$S = \frac{U - U(0)}{T} + nR\ln q - nR\ln N + nR$$

The only mode of motion for a gas of atoms is translation, and the partition function is $q = V/\Lambda^3$ (eqn 16.19), where Λ is the thermal wavelength. The internal energy is given by eqn 16.32, so the entropy is

$$S = \tfrac{3}{2}nR + nR\left(\ln\frac{V}{\Lambda^3} - \ln nN_A + 1\right) = nR\left(\ln e^{3/2} + \ln\frac{V}{\Lambda^3} - \ln nN_A + \ln e\right)$$

which rearranges into eqn 16.46.

Example 16.5 *Using the Sackur–Tetrode equation*

Calculate the standard molar entropy of gaseous argon at 25°C.

Method To calculate the molar entropy, S_m, from eqn 16.46b, divide both sides by n. To calculate the standard molar entropy, $S_m^\ominus$, set $p = p^\ominus$ in the expression for S_m:

$$S_m^\ominus = R\ln\left(\frac{e^{5/2}kT}{p^\ominus \Lambda^3}\right)$$

Answer The mass of an Ar atom is $m = 39.95$ u. At 25°C, its thermal wavelength is 16.0 pm (by the same kind of calculation as in *Illustration* 16.3). Therefore,

$$S_m^\ominus = R\ln\left\{\frac{e^{5/2}\times(4.12\times10^{-21}\,\text{J})}{(10^5\,\text{N m}^{-2})\times(1.60\times10^{-11}\,\text{m})^3}\right\} = 18.6R = 155\ \text{J K}^{-1}\,\text{mol}^{-1}$$

We can anticipate, on the basis of the number of accessible states for a lighter molecule, that the standard molar entropy of Ne is likely to be smaller than for Ar; its actual value is 17.60R at 298 K.

Self-test 16.7 Calculate the translational contribution to the standard molar entropy of H_2 at 25°C. [14.2R]

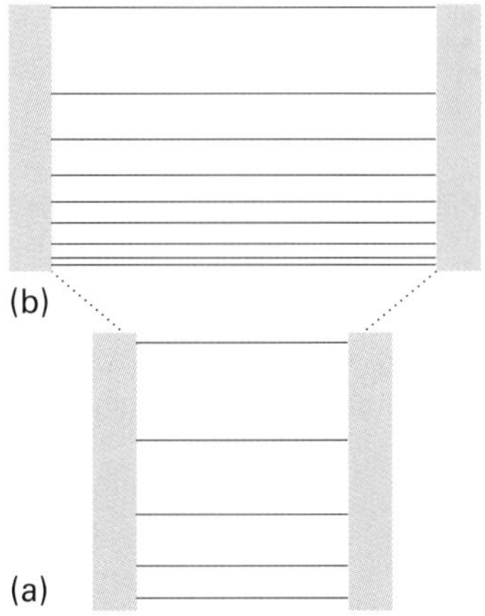

Fig. 16.17 As the width of a container is increased (going from (a) to (b)), the energy levels become closer together (as $1/L^2$), and as a result more are thermally accessible at a given temperature. Consequently, the entropy of the system rises as the container expands.

The Sackur–Tetrode equation implies that, when a monatomic perfect gas expands isothermally from V_i to V_f, its entropy changes by

$$\Delta S = nR\ln(aV_f) - nR\ln(aV_i) = nR\ln\frac{V_f}{V_i} \qquad (16.47)$$

where aV is the collection of quantities inside the logarithm of eqn 16.46a. This is exactly the expression we obtained by using classical thermodynamics (Example 3.1). Now, though, we see that that classical expression is in fact a consequence of the increase in the number of accessible translational states when the volume of the container is increased (Fig. 16.17).

Checklist of key ideas

☐ 1. The instantaneous configuration of a system of N molecules is the specification of the set of populations $n_0, n_1, \ldots$ of the energy levels $\varepsilon_0, \varepsilon_1, \ldots$. The weight W of a configuration is given by $W = N!/n_0!n_1!\ldots$.

☐ 2. The Boltzmann distribution gives the numbers of molecules in each state of a system at any temperature: $N_i = Ne^{-\beta\varepsilon_i}/q$, $\beta = 1/kT$.

☐ 3. The partition function is defined as $q = \sum_j e^{-\beta\varepsilon_j}$ and is an indication of the number of thermally accessible states at the temperature of interest.

☐ 4. The internal energy is $U(T) = U(0) + E$, with $E = -(N/q)(\partial q/\partial\beta)_V = -N(\partial \ln q/\partial\beta)_V$.

☐ 5. The Boltzmann formula for the entropy is $S = k\ln W$, where W is the number of different ways in which the molecules of a system can be arranged while keeping the same total energy.

☐ 6. The entropy in terms of the partition function is $S = \{U - U(0)\}/T + Nk\ln q$ (distinguishable molecules) or $S = \{U - U(0)\}/T + Nk\ln q - Nk(\ln N - 1)$ (indistinguishable molecules).

☐ 7. The canonical ensemble is an imaginary collection of replications of the actual system with a common temperature.

☐ 8. The canonical distribution is given by $\tilde{n}_i/\tilde{N} = e^{-\beta E_i}/\sum_j e^{-\beta E_j}$. The canonical partition function, $Q = \sum_i e^{-\beta E_i}$.

☐ 9. The internal energy and entropy of an ensemble are, respectively, $U = U(0) - (\partial \ln Q/\partial\beta)_V$ and $S = \{U - U(0)\}/T + k\ln Q$.

☐ 10. For distinguishable independent molecules we write $Q = q^N$. For indistinguishable independent molecules we write $Q = q^N/N!$.

☐ 11. The Sackur–Tetrode equation, eqn 16.46, is an expression for the entropy of a monatomic gas.

Further reading

Articles and texts

D. Chandler, *Introduction to modern statistical mechanics*. Oxford University Press (1987).

D.A. McQuarrie and J.D. Simon, *Molecular thermodynamics*. University Science Books, Sausalito (1999).

K.E. van Holde, W.C. Johnson, and P.S. Ho, *Principles of physical biochemistry*. Prentice Hall, Upper Saddle River (1998).

J. Wisniak, Negative absolute temperatures, a novelty. *J. Chem. Educ.* **77**, 518 (2000).

Further information

Further information 16.1 *The Boltzmann distribution*

We remarked in Section 16.1 that ln W is easier to handle than W. Therefore, to find the form of the Boltzmann distribution, we look for the condition for ln W being a maximum rather than dealing directly with W. Because ln W depends on all the n_i, when a configuration changes and the n_i change to $n_i + dn_i$, the function ln W changes to ln W + d ln W, where

$$\mathrm{d}\ln W = \sum_i \left(\frac{\partial \ln W}{\partial n_i}\right)\mathrm{d}n_i$$

All this expression states is that a change in ln W is the sum of contributions arising from changes in each value of n_i. At a maximum, d ln W = 0. However, when the n_i change, they do so subject to the two constraints

$$\sum_i \varepsilon_i \mathrm{d}n_i = 0 \qquad \sum_i \mathrm{d}n_i = 0 \qquad (16.48)$$

The first constraint recognizes that the total energy must not change, and the second recognizes that the total number of molecules must not change. These two constraints prevent us from solving d ln W = 0 simply by setting all $(\partial \ln W/\partial n_i) = 0$ because the $\mathrm{d}n_i$ are not all independent.

The way to take constraints into account was devised by the French mathematician Lagrange, and is called the **method of undetermined multipliers**. The technique is described in *Appendix* 2. All we need here is the rule that a constraint should be multiplied by a constant and then added to the main variation equation. The variables are then treated as though they were all independent, and the constants are evaluated at the end of the calculation.

We employ the technique as follows. The two constraints in eqn 16.48 are multiplied by the constants $-\beta$ and α, respectively (the minus sign in $-\beta$ has been included for future convenience), and then added to the expression for d ln W:

$$\mathrm{d}\ln W = \sum_i \left(\frac{\partial \ln W}{\partial n_i}\right)\mathrm{d}n_i + \alpha\sum_i \mathrm{d}n_i - \beta\sum_i \varepsilon_i \mathrm{d}n_i$$

$$= \sum_i \left\{\left(\frac{\partial \ln W}{\partial n_i}\right) + \alpha - \beta\varepsilon_i\right\}\mathrm{d}n_i$$

All the $\mathrm{d}n_i$ are now treated as independent. Hence the only way of satisfying d ln W = 0 is to require that, for each i,

$$\frac{\partial \ln W}{\partial n_i} + \alpha - \beta\varepsilon_i = 0 \qquad (16.49)$$

when the n_i have their most probable values.

Differentiation of ln W as given in eqn 16.3 with respect to n_i gives

$$\frac{\partial \ln W}{\partial n_i} = \frac{\partial(N\ln N)}{\partial n_i} - \sum_j \frac{\partial(n_j \ln n_j)}{\partial n_i}$$

The derivative of the first term is obtained as follows:

$$\frac{\partial(N\ln N)}{\partial n_i} = \left(\frac{\partial N}{\partial n_i}\right)\ln N + N\left(\frac{\partial \ln N}{\partial n_i}\right)$$

$$= \ln N + \frac{\partial N}{\partial n_i} = \ln N + 1$$

The ln N in the first term on the right in the second line arises because $N = n_1 + n_2 + \cdots$ and so the derivative of N with respect to any of the n_i is 1: that is, $\partial N/\partial n_i = 1$. The second term on the right in the second line arises because $\partial(\ln N)/\partial n_i = (1/N)\partial N/\partial n_i$. The final 1 is then obtained in the same way as in the preceding remark, by using $\partial N/\partial n_i = 1$.

For the derivative of the second term we first note that

$$\frac{\partial \ln n_j}{\partial n_i} = \frac{1}{n_j}\left(\frac{\partial n_j}{\partial n_i}\right)$$

Morever, if $i \neq j$, n_j is independent of n_i, so $\partial n_j/\partial n_i = 0$. However, if $i = j$,

$$\frac{\partial n_j}{\partial n_i} = \frac{\partial n_j}{\partial n_j} = 1$$

Therefore,

$$\frac{\partial n_j}{\partial n_i} = \delta_{ij}$$

with δ_{ij} the Kronecker delta ($\delta_{ij} = 1$ if $i = j$, $\delta_{ij} = 0$ otherwise). Then

$$\sum_j \frac{\partial(n_j \ln n_j)}{\partial n_i} = \sum_j \left\{ \left(\frac{\partial n_j}{\partial n_i}\right) \ln n_j + n_j \left(\frac{\partial \ln n_j}{\partial n_i}\right) \right\}$$

$$= \sum_j \left\{ \left(\frac{\partial n_j}{\partial n_i}\right) \ln n_j + \left(\frac{\partial n_j}{\partial n_i}\right) \right\}$$

$$= \sum_j \left(\frac{\partial n_j}{\partial n_i}\right)(\ln n_j + 1)$$

$$= \sum_j \delta_{ij}(\ln n_j + 1) = \ln n_i + 1$$

and therefore

$$\frac{\partial \ln W}{\partial n_i} = -(\ln n_i + 1) + (\ln N + 1) = -\ln \frac{n_i}{N}$$

It follows from eqn 16.49 that

$$-\ln \frac{n_i}{N} + \alpha - \beta\varepsilon_i = 0$$

and therefore that

$$\frac{n_i}{N} = \mathrm{e}^{\alpha - \beta\varepsilon_i}$$

At this stage we note that

$$N = \sum_i n_i = \sum_i N\mathrm{e}^{\alpha-\beta\varepsilon_i} = N\mathrm{e}^{\alpha} \sum_i \mathrm{e}^{\beta\varepsilon_i}$$

Because the N cancels on each side of this equality, it follows that

$$\mathrm{e}^{\alpha} = \frac{1}{\sum_j \mathrm{e}^{-\beta\varepsilon_j}} \qquad (16.50)$$

and

$$\frac{n_i}{N} = \mathrm{e}^{\alpha-\beta\varepsilon_i} = \mathrm{e}^{\alpha}\mathrm{e}^{-\beta\varepsilon_i} = \frac{1}{\sum_j \mathrm{e}^{-\beta\varepsilon_j}} \mathrm{e}^{-\beta\varepsilon_i}$$

which is eqn 16.6a.

Further information 16.2 *The Boltzmann formula*

A change in the internal energy

$$U = U(0) + \sum_i n_i\varepsilon_i \qquad (16.51)$$

may arise from either a modification of the energy levels of a system (when ε_i changes to $\varepsilon_i + \mathrm{d}\varepsilon_i$) or from a modification of the populations (when n_i changes to $n_i + \mathrm{d}n_i$). The most general change is therefore

$$\mathrm{d}U = \mathrm{d}U(0) + \sum_i n_i \mathrm{d}\varepsilon_i + \sum_i \varepsilon_i \mathrm{d}n_i \qquad (16.52)$$

Because the energy levels do not change when a system is heated at constant volume (Fig. 16.18), in the absence of all changes other than heating

$$\mathrm{d}U = \sum_i \varepsilon_i \mathrm{d}n_i$$

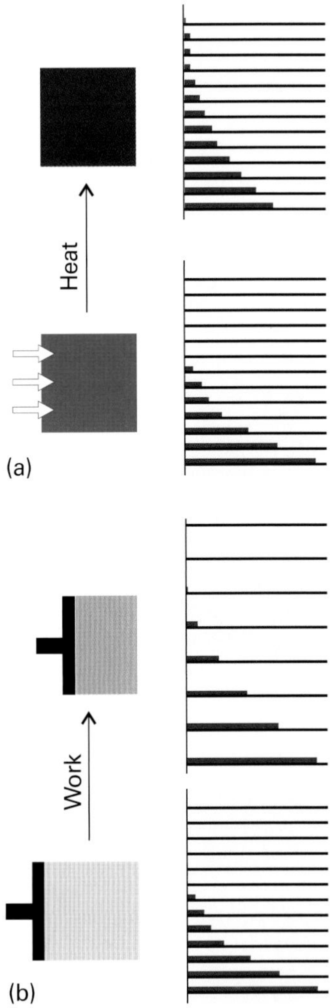

Fig. 16.18 (a) When a system is heated, the energy levels are unchanged but their populations are changed. (b) When work is done on a system, the energy levels themselves are changed. The levels in this case are the one-dimensional particle-in-a-box energy levels of Chapter 9: they depend on the size of the container and move apart as its length is decreased.

We know from thermodynamics (and specifically from eqn 3.43) that under the same conditions

$$\mathrm{d}U = \mathrm{d}q_{\mathrm{rev}} = T\mathrm{d}S$$

Therefore,

$$\mathrm{d}S = \frac{\mathrm{d}U}{T} = k\beta \sum_i \varepsilon_i \mathrm{d}n_i \qquad (16.53)$$

For changes in the most probable configuration (the only one we need consider), we rearrange eqn 16.49 to

$$\beta\varepsilon_i = \frac{\partial \ln W}{\partial n_i} + \alpha$$

and find that

$$dS = k\sum_i \left(\frac{\partial \ln W}{\partial n_i}\right) dn_i + k\alpha \sum_i dn_i$$

But because the number of molecules is constant, the sum over the dn_i is zero. Hence

$$dS = k\sum_i \left(\frac{\partial \ln W}{\partial n_i}\right) dn_i = k(d \ln W)$$

This relation strongly suggests the definition $S = k \ln W$, as in eqn 16.34.

Further information 16.3 *Temperatures below zero*

The Boltzmann distribution tells us that the ratio of populations in a two-level system at a temperature T is

$$\frac{N_+}{N_-} = e^{-\varepsilon/kT} \qquad (16.54)$$

where ε is the separation of the upper state N_+ and the lower state N_-. It follows that, if we can contrive the population of the upper state to exceed that of the lower state, then the temperature must have a negative value. Indeed, for a general population,

$$T = \frac{\varepsilon/k}{\ln(N_-/N_+)} \qquad (16.55)$$

and the temperature is formally negative for all $N_+ > N_-$.

All the statistical thermodynamic expressions we have derived apply to $T < 0$ as well as to $T > 0$, the difference being that states with $T < 0$ are not in thermal equilibrium and therefore have to be achieved by techniques that do not rely on the equalization of temperatures of the system and its surroundings. The Third Law of thermodynamics prohibits the achievement of absolute zero in a finite number of steps. However, it is possible to circumvent this restriction in systems that have a finite number of levels or in systems that are effectively finite because they have such weak coupling to their surroundings. The practical realization of such a system is a group of spin-$\frac{1}{2}$ nuclei that have very long relaxation times, such as the ^{19}F nuclei in cold solid LiF. Pulse techniques in NMR can achieve non-equilibrium populations (Section 15.8) as can pumping procedures in laser technologies (Section 14.5). From now on, we shall suppose that these non-equilibrium distributions have been achieved, and will concentrate on the consequences.

The expressions for q, U, and S that we have derived in this chapter are applicable to $T < 0$ as well as to $T > 0$, and are shown in Fig. 16.19. We see that q and U show sharp discontinuities on passing through zero, and $T = +0$ (corresponding to all population in the lower state) is quite distinct from $T = -0$, where all the population is in the upper state. The entropy S is continuous at $T = 0$. But all these functions are continuous if we use $\beta = 1/kT$ as the dependent variable (Fig. 16.20), which shows that β is a more natural, if less familiar, variable than T. Note that $U \rightarrow 0$ as $\beta \rightarrow \infty$ (that is, as $T \rightarrow 0$, when only the lower state is occupied) and $U \rightarrow N\varepsilon$ as $\beta \rightarrow -\infty$ (that is, as $T \rightarrow -0$);

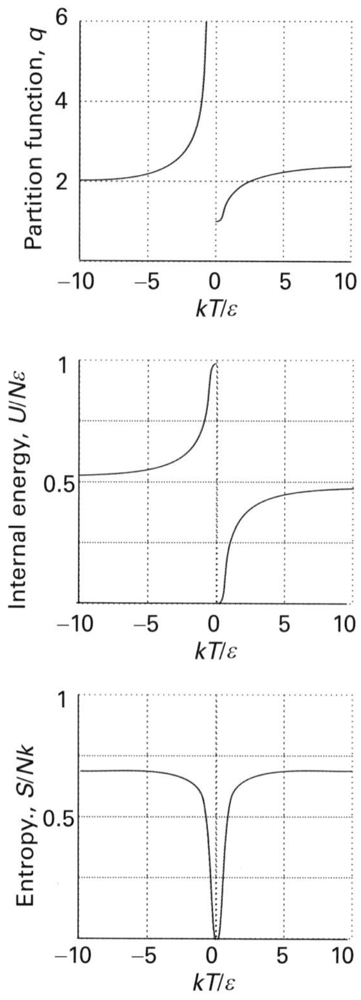

Fig. 16.19 The partition function, internal energy, and entropy of a two-level system extended to negative temperatures.

we see that a state with $T = -0$ is 'hotter' than one with $T = +0$. The entropy of the system is zero on either side of $T = 0$, and rises to $Nk \ln 2$ as $T \rightarrow \pm\infty$. At $T = +0$ only one state is accessible (the lower state), only the upper state is accessible, so the entropy is zero in each case.

We get more insight into the dependence of thermodynamic properties on temperature by noting the thermodynamic result (Section 3.8) that $T = (\partial S/\partial U)_T$. When S is plotted against U for a two-level system (Fig. 16.21), we see that the entropy rises as energy is supplied to the system (as we would expect) provided that $T > 0$ (the thermal equilibrium regime). However, the entropy decreases as energy is supplied when $T < 0$. This conclusion is consistent with the thermodynamic definition of entropy, $dS = dq_{rev}/T$ (where, of course, q denotes heat and not the partition function). Physically, the increase in entropy for $T > 0$ corresponds to the increasing accessibility of the upper state, and the decrease for $T < 0$ corresponds

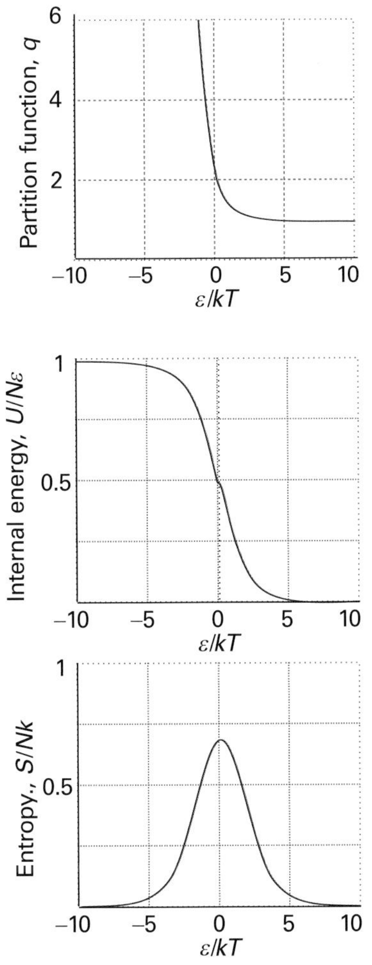

Fig. 16.20 The partition function, internal energy, and entropy of a two-level system extended to negative temperatures but plotted against $\beta = 1/kT$ (modified here to the dimensionless quantity ε/kT).

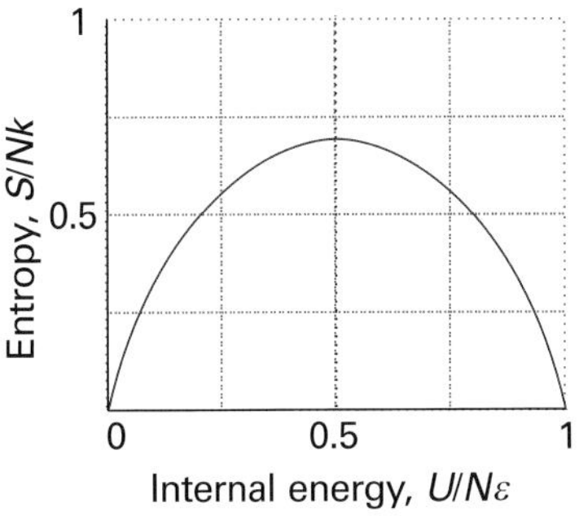

Fig. 16.21 The variation of the entropy with internal energy for a two-level system extended to negative temperatures.

to the shift towards population of the upper state alone as more energy is packed into the system.

The phenomenological laws of thermodynamics survive largely intact at negative temperatures. The First Law (in essence, the conservation of energy) is robust, and independent of how populations are distributed over states. The Second Law survives because the definition of entropy survives (as we have seen above). The efficiency of heat engines (Section 3.2), which is a direct consequence of the Second Law, is still given by $1 - T_{\text{cold}}/T_{\text{hot}}$. However, if the temperature of the cold reservoir is negative, then the efficiency of the engine may be greater than 1. This condition corresponds to the amplification of signals achieved in lasers. Alternatively, an efficiency greater than 1 implies that heat *can* be converted completely into work provided the heat is withdrawn from a reservoir at $T < 0$. If both reservoirs are at negative temperatures, then the efficiency is less than 1, as in the thermal equilibrium case treated in Chapter 3. The Third Law requires a slight amendment on account of the discontinuity of the populations across $T = 0$: it is impossible in a finite number of steps to cool any system down to +0 or to heat any system above −0.

Discussion questions

16.1 Describe the physical significance of the partition function.

16.2 Explain how the internal energy and entropy of a system composed of two levels vary with temperature.

16.3 Enumerate the ways by which the parameter β may be identified with $1/kT$.

16.4 Distinguish between the zipper and Zimm–Bragg models of the helix–coil transition.

16.5 Explain what is meant by an *ensemble* and why it is useful in statistical thermodynamics.

16.6 Under what circumstances may identical particles be regarded as distinguishable?

Exercises

16.1a What are the relative populations of the states of a two-level system when the temperature is infinite?

16.1b What is the temperature of a two-level system of energy separation equivalent to 300 cm^{-1} when the population of the upper state is one-half that of the lower state?

16.2a Calculate the translational partition function at (a) 300 K and (b) 600 K of a molecule of molar mass 120 g mol^{-1} in a container of volume 2.00 cm^3.

16.2b Calculate (a) the thermal wavelength, (b) the translational partition function of an Ar atom in a cubic box of side 1.00 cm at (i) 300 K and (ii) 3000 K.

16.3a Calculate the ratio of the translational partition functions of D_2 and H_2 at the same temperature and volume.

16.3b Calculate the ratio of the translational partition functions of xenon and helium at the same temperature and volume.

16.4a A certain atom has a threefold degenerate ground level, a non-degenerate electronically excited level at 3500 cm^{-1}, and a threefold degenerate level at 4700 cm^{-1}. Calculate the partition function of these electronic states at 1900 K.

16.4b A certain atom has a doubly degenerate ground level, a triply degenerate electronically excited level at 1250 cm^{-1}, and a doubly degenerate level at 1300 cm^{-1}. Calculate the partition function of these electronic states at 2000 K.

16.5a Calculate the electronic contribution to the molar internal energy at 1900 K for a sample composed of the atoms specified in Exercise 16.4a.

16.5b Calculate the electronic contribution to the molar internal energy at 2000 K for a sample composed of the atoms specified in Exercise 16.4b.

16.6a A certain molecule has a non-degenerate excited state lying at 540 cm^{-1} above the non-degenerate ground state. At what temperature will 10 per cent of the molecules be in the upper state?

16.6b A certain molecule has a doubly degenerate excited state lying at 360 cm^{-1} above the non-degenerate ground state. At what temperature will 15 per cent of the molecules be in the upper state?

16.7a An electron spin can adopt either of two orientations in a magnetic field, and its energies are $\pm\mu_B\mathcal{B}$, where μ_B is the Bohr magneton. Deduce an expression for the partition function and mean energy of the electron and sketch the variation of the functions with $\mathcal{B}$. Calculate the relative populations of the spin states at (a) 4.0 K, (b) 298 K when $\mathcal{B}$ = 1.0 T.

16.7b A nitrogen nucleus spin can adopt any of three orientations in a magnetic field, and its energies are 0, $\pm\gamma_N\hbar\mathcal{B}$, where γ_N is the magnetogyric ratio of the nucleus. Deduce an expression for the partition function and mean energy of the nucleus and sketch the variation of the functions with $\mathcal{B}$. Calculate the relative populations of the spin states at (a) 1.0 K, (b) 298 K when $\mathcal{B}$ = 20.0 T.

16.8a Consider a system of distinguishable particles having only two non-degenerate energy levels separated by an energy that is equal to the value of kT at 10 K. Calculate (a) the ratio of populations in the two states at (1) 1.0 K, (2) 10 K, and (3) 100 K, (b) the molecular partition function at 10 K, (c) the molar energy at 10 K, (d) the molar heat capacity at 10 K, (e) the molar entropy at 10 K.

16.8b Consider a system of distinguishable particles having only three non-degenerate energy levels separated by an energy which is equal to the value of kT at 25.0 K. Calculate (a) the ratio of populations in the states at (1) 1.00 K, (2) 25.0 K, and (3) 100 K, (b) the molecular partition function at 25.0 K, (c) the molar energy at 25.0 K, (d) the molar heat capacity at 25.0 K, (e) the molar entropy at 25.0 K.

16.9a At what temperature would the population of the first excited vibrational state of HCl be 1/e times its population of the ground state?

16.9b At what temperature would the population of the first excited rotational level of HCl be 1/e times its population of the ground state?

16.10a Calculate the standard molar entropy of neon gas at (a) 200 K, (b) 298.15 K.

16.10b Calculate the standard molar entropy of xenon gas at (a) 100 K, (b) 298.15 K.

16.11a Calculate the vibrational contribution to the entropy of Cl_2 at 500 K given that the wavenumber of the vibration is 560 cm^{-1}.

16.11b Calculate the vibrational contribution to the entropy of Br_2 at 600 K given that the wavenumber of the vibration is 321 cm^{-1}.

16.12a Identify the systems for which it is essential to include a factor of $1/N!$ on going from Q to q: (a) a sample of helium gas, (b) a sample of carbon monoxide gas, (c) a solid sample of carbon monoxide, (d) water vapour.

16.12b Identify the systems for which it is essential to include a factor of $1/N!$ on going from Q to q: (a) a sample of carbon dioxide gas, (b) a sample of graphite, (c) a sample of diamond, (d) ice.

Problems*

Numerical problems

16.1‡ Consider a system A consisting of subsystems A_1 and A_2, for which $W_1 = 1 \times 10^{20}$ and $W_2 = 2 \times 10^{20}$. What is the number of configurations available to the combined system? Also, compute the entropies S, S_1, and S_2. What is the significance of this result?

16.2‡ Consider 1.00×10^{22} ^{4}He atoms in a box of dimensions 1.0 cm × 1.0 cm × 1.0 cm. Calculate the occupancy of the first excited level at 1.0 mK, 2.0 K, and 4.0 K. Do the same for ^{3}He. What conclusions might you draw from the results of your calculations?

* Problems denoted with the symbol ‡ were supplied by Charles Trapp and Carmen Giunta.

16.3‡ By what factor does the number of available configurations increase when 100 J of energy is added to a system containing 1.00 mol of particles at constant volume at 298 K?

16.4‡ By what factor does the number of available configurations increase when 20 m^3 of air at 1.00 atm and 300 K is allowed to expand by 0.0010 per cent at constant temperature?

16.5 Explore the conditions under which the 'integral' approximation for the translational partition function is not valid by considering the translational partition function of an Ar atom in a cubic box of side 1.00 cm. Estimate the temperature at which, according to the integral approximation, $q = 10$ and evaluate the exact partition function at that temperature.

16.6 A certain atom has a doubly degenerate ground level pair and an upper level of four degenerate states at 450 cm^{-1} above the ground level. In an atomic beam study of the atoms it was observed that 30 per cent of the atoms were in the upper level, and the translational temperature of the beam was 300 K. Are the electronic states of the atoms in thermal equilibrium with the translational states?

16.7 (a) Calculate the electronic partition function of a tellurium atom at (i) 298 K, (ii) 5000 K by direct summation using the following data:

Term	Degeneracy	Wavenumber/cm^{-1}
Ground	5	0
1	1	4 707
2	3	4 751
3	5	10 559

(b) What proportion of the Te atoms are in the ground term and in the term labelled 2 at the two temperatures? (c) Calculate the electronic contribution to the standard molar entropy of gaseous Te atoms.

16.8 The four lowest electronic levels of a Ti atom are: $^3F_2$, $^3F_3$, $^3F_4$, and $^5F_1$, at 0, 170, 387, and 6557 cm^{-1}, respectively. There are many other electronic states at higher energies. The boiling point of titanium is 3287°C. What are the relative populations of these levels at the boiling point? *Hint*. The degeneracies of the levels are $2J + 1$.

16.9 The NO molecule has a doubly degenerate excited electronic level 121.1 cm^{-1} above the doubly degenerate electronic ground term. Calculate and plot the electronic partition function of NO from $T = 0$ to 1000 K. Evaluate (a) the term populations and (b) the electronic contribution to the molar internal energy at 300 K. Calculate the electronic contribution to the molar entropy of the NO molecule at 300 K and 500 K.

16.10‡ J. Sugar and A. Musgrove (*J. Phys. Chem. Ref. Data* **22**, 1213 (1993)) have published tables of energy levels for germanium atoms and cations from Ge^+ to Ge^{+31}. The lowest-lying energy levels in neutral Ge are as follows:

	$^3P_0$	$^3P_1$	$^3P_2$	$^1D_2$	$^1S_0$
E/cm^{-1}	0	557.1	1410.0	7125.3	16 367.3

Calculate the electronic partition function at 298 K and 1000 K by direct summation. *Hint*. The degeneracy of a level is $2J + 1$.

16.11 Calculate, by explicit summation, the vibrational partition function and the vibrational contribution to the molar internal energy of I_2 molecules at (a) 100 K, (b) 298 K given that its vibrational energy levels lie at the following wavenumbers above the zero-point energy level: 0, 213.30, 425.39, 636.27, 845.93 cm^{-1}. What proportion of I_2 molecules are in the ground and first two excited levels at the two temperatures? Calculate the vibrational contribution to the molar entropy of I_2 at the two temperatures.

16.12‡ (a) The standard molar entropy of graphite at 298, 410, and 498 K is 5.69, 9.03, and 11.63 J K^{-1} mol^{-1}, respectively. If 1.00 mol C(graphite) at 298 K is surrounded by thermal insulation and placed next to 1.00 mol C(graphite) at 498 K, also insulated, how many configurations are there altogether for the combined but independent systems? (b) If the same two samples are now placed in thermal contact and brought to thermal equilibrium, the final temperature will be 410 K. (Why might the final temperature not be the average?) How many configurations are there now in the combined system? Neglect any volume changes. (c) Demonstrate that this process is spontaneous.

Theoretical problems

16.13 A sample consisting of five molecules has a total energy 5ε. Each molecule is able to occupy states of energy $j\varepsilon$, with $j = 0, 1, 2, \ldots$. (a) Calculate the weight of the configuration in which the molecules are distributed evenly over the available states. (b) Draw up a table with columns headed by the energy of the states and write beneath them all configurations that are consistent with the total energy. Calculate the weights of each configuration and identify the most probable configurations.

16.14 A sample of nine molecules is numerically tractable but on the verge of being thermodynamically significant. Draw up a table of configurations for $N = 9$, total energy 9ε in a system with energy levels $j\varepsilon$ (as in Problem 16.13). Before evaluating the weights of the configurations, guess (by looking for the most 'exponential' distribution of populations) which of the configurations will turn out to be the most probable. Go on to calculate the weights and identify the most probable configuration.

16.15 The most probable configuration is characterized by a parameter we know as the 'temperature'. The temperatures of the system specified in Problems 16.13 and 16.14 must be such as to give a mean value of ε for the energy of each molecule and a total energy $N\varepsilon$ for the system. (a) Show that the temperature can be obtained by plotting p_j against j, where p_j is the (most probable) fraction of molecules in the state with energy $j\varepsilon$. Apply the procedure to the system in Problem 16.14. What is the temperature of the system when ε corresponds to 50 cm^{-1}? (b) Choose configurations other than the most probable, and show that the same procedure gives a worse straight line, indicating that a temperature is not well-defined for them.

16.16 A certain molecule can exist in either a non-degenerate singlet state or a triplet state (with degeneracy 3). The energy of the triplet exceeds that of the singlet by ε. Assuming that the molecules are distinguishable (localized) and independent, (a) obtain the expression for the molecular partition function. (b) Find expressions in terms of ε for the molar energy, molar heat capacity, and molar entropy of such molecules and calculate their values at $T = \varepsilon/k$.

16.17 Consider a system with energy levels $\varepsilon_j = j\varepsilon$ and N molecules. (a) Show that if the mean energy per molecule is $a\varepsilon$, then the temperature is given by

$$\beta = \frac{1}{\varepsilon}\ln\left(1+\frac{1}{a}\right)$$

Evaluate the temperature for a system in which the mean energy is ε, taking ε equivalent to 50 cm^{-1}. (b) Calculate the molecular partition function q for the system when its mean energy is $a\varepsilon$. (c) Show that the entropy of the system is

$$S/k = (1 + a)\ln(1 + a) - a\ln a$$

and evaluate this expression for a mean energy ε.

16.18 Consider Stirling's approximation for ln $N!$ in the derivation of the Boltzmann distribution. What difference would it make if (a) a cruder approximation, $N! = N^N$, (b) the better approximation in *Comment* 16.2 were used instead?

16.19‡ For gases, the canonical partition function, Q, is related to the molecular partition function q by $Q = q^N/N!$. Use the expression for q and general thermodynamic relations to derive the perfect gas law $pV = nRT$.

Applications: to atmospheric science, astrophysics, and biochemistry

16.20‡ Obtain the barometric formula (Problem 1.27) from the Boltzmann distribution. Recall that the potential energy of a particle at height h above the surface of the Earth is mgh. Convert the barometric formula from pressure to number density, $\mathcal{N}$. Compare the relative number densities, $\mathcal{N}(h)/\mathcal{N}(0)$, for O_2 and H_2O at $h = 8.0$ km, a typical cruising altitude for commercial aircraft.

16.21‡ Planets lose their atmospheres over time unless they are replenished. A complete analysis of the overall process is very complicated and depends upon the radius of the planet, temperature, atmospheric composition, and other factors. Prove that the atmosphere of planets cannot be in an equilibrium state by demonstrating that the Boltzmann distribution leads to a uniform finite number density as $r \rightarrow \infty$. *Hint.* Recall that in a gravitational field the potential energy is $V(r) = -GMm/r$, where G is the gravitational constant, M is the mass of the planet, and m the mass of the particle.

16.22‡ Consider the electronic partition function of a perfect atomic hydrogen gas at a density of 1.99×10^{-4} kg m^{-3} and 5780 K. These are the mean conditions within the Sun's photosphere, the surface layer of the Sun that is about 190 km thick. (a) Show that this partition function, which involves a sum over an infinite number of quantum states that are solutions to the Schrödinger equation for an isolated atomic hydrogen atom, is infinite. (b) Develop a theoretical argument for truncating the sum and estimate the maximum number of quantum states that contribute to the sum. (c) Calculate the equilibrium probability that an atomic hydrogen electron is in each quantum state. Are there any general implications concerning electronic states that will be observed for other atoms and molecules? Is it wise to apply these calculations in the study of the Sun's photosphere?

16.23 Consider a protein P with four distinct sites, with each site capable of binding one ligand L. Show that the possible varieties (configurations) of the species PL_i (with PL_0 denoting P) are given by the binomial coefficients $C(4,i)$.

16.24 Complete some of the derivations in the discussion of the helix–coil transition in polypeptides (*Impact* I16.1). (a) Show that, within the tenets of the zipper model,

$$q = 1 + \sum_{i=1}^{n} Z(n,i)\sigma s^i$$

and that $Z(n,i) = n - i + 1$ is the number of ways in which an allowed state with a number i of c amino acids can be formed. (b) Using the zipper model, show that $\theta = (1/n)\mathrm{d}(\ln q)/\mathrm{d}(\ln s)$. *Hint.* As a first step, show that $\sum_i i(n - i + 1)\sigma s^i = s(\mathrm{d}q/\mathrm{d}s)$.

16.25 Here you will use the zipper model discussed in *Impact* I16.1 to explore the helix–coil transition in polypeptides.(a) Investigate the effect of the parameter s on the distribution of random coil segments in a polypeptide with $n = 20$ by plotting p_i, the fraction of molecules with a number i of amino acids in a coil region, against i for $s = 0.8$, 1.0, and 1.5, with $\sigma = 5.0 \times 10^{-2}$. Discuss the significance of any effects you discover. (b) The average value of i given by $\langle i \rangle = \sum_i i p_i$. Use the results of the zipper model to calculate $\langle i \rangle$ for all the combinations of s and σ used in Fig. 16.10 and part (a).

Statistical thermodynamics 2: applications

17

In this chapter we apply the concepts of statistical thermodynamics to the calculation of chemically significant quantities. First, we establish the relations between thermodynamic functions and partition functions. Next, we show that the molecular partition function can be factorized into contributions from each mode of motion and establish the formulas for the partition functions for translational, rotational, and vibrational modes of motion and the contribution of electronic excitation. These contributions can be calculated from spectroscopic data. Finally, we turn to specific applications, which include the mean energies of modes of motion, the heat capacities of substances, and residual entropies. In the final section, we see how to calculate the equilibrium constant of a reaction and through that calculation understand some of the molecular features that determine the magnitudes of equilibrium constants and their variation with temperature.

A partition function is the bridge between thermodynamics, spectroscopy, and quantum mechanics. Once it is known, a partition function can be used to calculate thermodynamic functions, heat capacities, entropies, and equilibrium constants. It also sheds light on the significance of these properties.

Fundamental relations

In this section we see how to obtain any thermodynamic function once we know the partition function. Then we see how to calculate the molecular partition function, and through that the thermodynamic functions, from spectroscopic data.

17.1 The thermodynamic functions

We have already derived (in Chapter 16) the two expressions for calculating the internal energy and the entropy of a system from its canonical partition function, Q:

$$U - U(0) = -\left(\frac{\partial \ln Q}{\partial \beta}\right)_V \qquad S = \frac{U - U(0)}{T} + k \ln Q \tag{17.1}$$

where $\beta = 1/kT$. If the molecules are independent, we can go on to make the substitutions $Q = q^N$ (for distinguishable molecules, as in a solid) or $Q = q^N/N!$ (for indistinguishable molecules, as in a gas). All the thermodynamic functions introduced in Part 1 are related to U and S, so we have a route to their calculation from Q.

(a) The Helmholtz energy

The Helmholtz energy, A, is defined as $A = U - TS$. This relation implies that $A(0) = U(0)$, so substitution for U and S by using eqn 17.1 leads to the very simple expression

$$A - A(0) = -kT \ln Q \qquad (17.2)$$

(b) The pressure

By an argument like that leading to eqn 3.31, it follows from $A = U - TS$ that $dA = -pdV - SdT$. Therefore, on imposing constant temperature, the pressure and the Helmholtz energy are related by $p = -(\partial A/\partial V)_T$. It then follows from eqn 17.2 that

$$p = kT\left(\frac{\partial \ln Q}{\partial V}\right)_T \qquad (17.3)$$

This relation is entirely general, and may be used for any type of substance, including perfect gases, real gases, and liquids. Because Q is in general a function of the volume, temperature, and amount of substance, eqn 17.3 is an equation of state.

Example 17.1 *Deriving an equation of state*

Derive an expression for the pressure of a gas of independent particles.

Method We should suspect that the pressure is that given by the perfect gas law. To proceed systematically, substitute the explicit formula for Q for a gas of independent, indistinguishable molecules (see eqn 16.45 and Table 17.3 at the end of the chapter) into eqn 17.3.

Answer For a gas of independent molecules, $Q = q^N/N!$ with $q = V/\Lambda^3$:

$$p = kT\left(\frac{\partial \ln Q}{\partial V}\right)_T = \frac{kT}{Q}\left(\frac{\partial Q}{\partial V}\right)_T = \frac{NkT}{q}\left(\frac{\partial q}{\partial V}\right)_T$$

$$= \frac{NkT\Lambda^3}{V} \times \frac{1}{\Lambda^3} = \frac{NkT}{V} = \frac{nRT}{V}$$

To derive this relation, we have used

$$\left(\frac{\partial q}{\partial V}\right)_T = \left(\frac{\partial (V/\Lambda^3)}{\partial V}\right)_T = \frac{1}{\Lambda^3}$$

and $NkT = nN_AkT = nRT$. The calculation shows that the equation of state of a gas of independent particles is indeed the perfect gas law.

Self-test 17.1 Derive the equation of state of a sample for which $Q = q^N f/N!$, with $q = V/\Lambda^3$, where f depends on the volume. $[p = nRT/V + kT(\partial \ln f/\partial V)_T]$

(c) The enthalpy

At this stage we can use the expressions for U and p in the definition $H = U + pV$ to obtain an expression for the enthalpy, H, of any substance:

$$H - H(0) = -\left(\frac{\partial \ln Q}{\partial \beta}\right)_V + kTV\left(\frac{\partial \ln Q}{\partial V}\right)_T \qquad (17.4)$$

We have already seen that $U - U(0) = \frac{3}{2}nRT$ for a gas of independent particles (eqn 16.32a), and have just shown that $pV = nRT$. Therefore, for such a gas,

$$H - H(0) = \tfrac{5}{2}nRT \qquad (17.5)°$$

(d) The Gibbs energy

One of the most important thermodynamic functions for chemistry is the Gibbs energy, $G = H - TS = A + pV$. We can now express this function in terms of the partition function by combining the expressions for A and p:

$$G - G(0) = -kT \ln Q + kTV\left(\frac{\partial \ln Q}{\partial V}\right)_T \qquad (17.6)$$

This expression takes a simple form for a gas of independent molecules because pV in the expression $G = A + pV$ can be replaced by nRT:

$$G - G(0) = -kT \ln Q + nRT \qquad (17.7)°$$

Furthermore, because $Q = q^N/N!$, and therefore $\ln Q = N \ln q - \ln N!$, it follows by using Stirling's approximation ($\ln N! \approx N \ln N - N$) that we can write

$$\begin{aligned} G - G(0) &= -NkT \ln q + kT \ln N! + nRT \\ &= -nRT \ln q + kT(N \ln N - N) + nRT \\ &= -nRT \ln \frac{q}{N} \end{aligned} \qquad (17.8)°$$

with $N = nN_A$. Now we see another interpretation of the Gibbs energy: it is proportional to the logarithm of the average number of thermally accessible states per molecule.

It will turn out to be convenient to define the **molar partition function**, $q_m = q/n$ (with units mol^{-1}), for then

$$G - G(0) = -nRT \ln \frac{q_m}{N_A} \qquad (17.9)°$$

17.2 The molecular partition function

The energy of a molecule is the sum of contributions from its different modes of motion:

$$\varepsilon_i = \varepsilon_i^T + \varepsilon_i^R + \varepsilon_i^V + \varepsilon_i^E \qquad (17.10)$$

where T denotes translation, R rotation, V vibration, and E the electronic contribution. The electronic contribution is not actually a 'mode of motion', but it is convenient to include it here. The separation of terms in eqn 17.10 is only approximate (except for translation) because the modes are not completely independent, but in most cases it is satisfactory. The separation of the electronic and vibrational motions is justified provided only the ground electronic state is occupied (for otherwise the vibrational characteristics depend on the electronic state) and, for the electronic ground state, that the Born–Oppenheimer approximation is valid (Chapter 11). The separation of the vibrational and rotational modes is justified to the extent that the rotational constant is independent of the vibrational state.

Given that the energy is a sum of independent contributions, the partition function factorizes into a product of contributions (recall Section 16.2b):

$$
\begin{aligned}
q &= \sum_i e^{-\beta\varepsilon_i} = \sum_{i\,(\text{all states})} e^{-\beta\varepsilon_i^{T}-\beta\varepsilon_i^{R}-\beta\varepsilon_i^{V}-\beta\varepsilon_i^{E}} \\
&= \sum_{i\,(\text{translational})}\ \sum_{i\,(\text{rotational})}\ \sum_{i\,(\text{vibrational})}\ \sum_{i\,(\text{electronic})} e^{-\beta\varepsilon_i^{T}-\beta\varepsilon_i^{R}-\beta\varepsilon_i^{V}-\beta\varepsilon_i^{E}} \\
&= \left(\sum_{i\,(\text{translational})} e^{-\beta\varepsilon_i^{T}}\right)\left(\sum_{i\,(\text{rotational})} e^{-\beta\varepsilon_i^{R}}\right)\left(\sum_{i\,(\text{vibrational})} e^{-\beta\varepsilon_i^{V}}\right)\left(\sum_{i\,(\text{electronic})} e^{-\beta\varepsilon_i^{E}}\right) \\
&= q^{T}q^{R}q^{V}q^{E}
\end{aligned} \qquad (17.11)
$$

This factorization means that we can investigate each contribution separately.

(a) The translational contribution

The translational partition function of a molecule of mass m in a container of volume V was derived in Section 16.2:

$$q^{T} = \frac{V}{\Lambda^3} \qquad \Lambda = h\left(\frac{\beta}{2\pi m}\right)^{1/2} = \frac{h}{(2\pi mkT)^{1/2}} \qquad (17.12)$$

Notice that $q^{T} \to \infty$ as $T \to \infty$ because an infinite number of states becomes accessible as the temperature is raised. Even at room temperature $q^{T} \approx 2 \times 10^{28}$ for an O_2 molecule in a vessel of volume 100 cm^3.

The thermal wavelength, Λ, lets us judge whether the approximations that led to the expression for q^{T} are valid. The approximations are valid if many states are occupied, which requires V/Λ^3 to be large. That will be so if Λ is small compared with the linear dimensions of the container. For H_2 at 25°C, $\Lambda = 71$ pm, which is far smaller than any conventional container is likely to be (but comparable to pores in zeolites or cavities in clathrates). For O_2, a heavier molecule, $\Lambda = 18$ pm. We saw in Section 16.2 that an equivalent criterion of validity is that Λ should be much less than the average separation of the molecules in the sample.

(b) The rotational contribution

As demonstrated in Example 16.1, the partition function of a nonsymmetrical (AB) linear rotor is

$$q^{R} = \sum_J (2J+1)e^{-\beta hcBJ(J+1)} \qquad (17.13)$$

The direct method of calculating q^{R} is to substitute the experimental values of the rotational energy levels into this expression and to sum the series numerically.

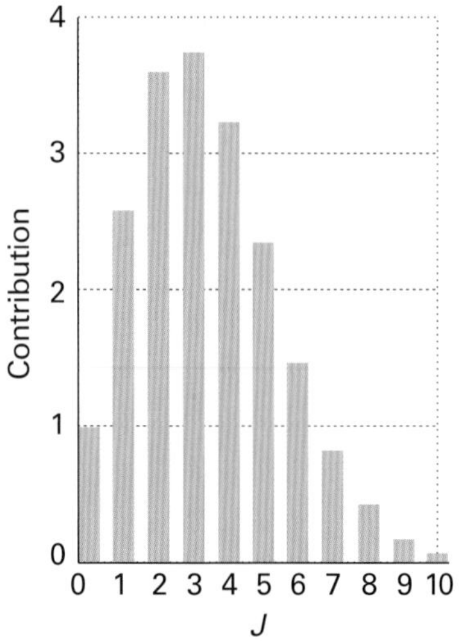

Fig. 17.1 The contributions to the rotational partition function of an HCl molecule at 25°C. The vertical axis is the value of $(2J+1)e^{-\beta hcBJ(J+1)}$. Successive terms (which are proportional to the populations of the levels) pass through a maximum because the population of individual states decreases exponentially, but the degeneracy of the levels increases with J.

Example 17.2 *Evaluating the rotational partition function explicitly*

Evaluate the rotational partition function of $^{1}H^{35}Cl$ at 25°C, given that $B = 10.591$ cm^{-1}.

Method We use eqn 17.13 and evaluate it term by term. A useful relation is $kT/hc = 207.22$ cm^{-1} at 298.15 K. The sum is readily evaluated by using mathematical software.

Answer To show how successive terms contribute, we draw up the following table by using $kT/hcB = 0.051\,11$ (Fig. 17.1):

J	0	1	2	3	4	...	10
$(2J+1)e^{-0.0511J(J+1)}$	1	2.71	3.68	3.79	3.24	...	0.08

The sum required by eqn 17.13 (the sum of the numbers in the second row of the table) is 19.9, hence $q^R = 19.9$ at this temperature. Taking J up to 50 gives $q^R = 19.902$. Notice that about ten J-levels are significantly populated but the number of populated *states* is larger on account of the $(2J + 1)$-fold degeneracy of each level. We shall shortly encounter the approximation that $q^R \approx kT/hcB$, which in the present case gives $q^R = 19.6$, in good agreement with the exact value and with much less work.

Self-test 17.2 Evaluate the rotational partition function for HCl at 0°C. [18.26]

At room temperature $kT/hc \approx 200\ \text{cm}^{-1}$. The rotational constants of many molecules are close to $1\ \text{cm}^{-1}$ (Table 13.2) and often smaller (though the very light H_2 molecule, for which $B = 60.9\ \text{cm}^{-1}$, is one exception). It follows that many rotational levels are populated at normal temperatures. When this is the case, the partition function may be approximated by

Linear rotors: $$q^R = \frac{kT}{hcB} \tag{17.14a}$$

Nonlinear rotors: $$q^R = \left(\frac{kT}{hc}\right)^{3/2}\left(\frac{\pi}{ABC}\right)^{1/2} \tag{17.14b}$$

where A, B, and C are the rotational constants of the molecule. However, before using these expressions, read on (to eqns 17.15 and 17.16).

Justification 17.1 *The rotational contribution to the molecular partition function*

When many rotational states are occupied and kT is much larger than the separation between neighbouring states, the sum in the partition function can be approximated by an integral, much as we did for translational motion in *Justification* 16.2:

$$q^R = \int_0^\infty (2J+1)e^{-\beta hcBJ(J+1)}dJ$$

Although this integral looks complicated, it can be evaluated without much effort by noticing that because

$$\frac{d}{dJ}e^{aJ(J+1)} = \left\{\frac{d}{dJ}aJ(J+1)\right\}e^{aJ(J+1)} = a(2J+1)e^{aJ(J+1)}$$

it can also be written as

$$q^R = \frac{1}{\beta hcB}\int_0^\infty\left(\frac{d}{dJ}e^{-\beta hcBJ(J+1)}\right)dJ$$

Then, because the integral of a derivative of a function is the function itself, we obtain

$$q^R = -\frac{1}{\beta hcB}e^{-\beta hcBJ(J+1)}\Big|_0^\infty = \frac{1}{\beta hcB}$$

which (because $\beta = 1/kT$) is eqn 17.14a.

The calculation for a nonlinear molecule is along the same lines, but slightly trickier. First, we note that the energies of a symmetric rotor are

$$E_{J,K,M_J} = hcBJ(J+1) + hc(A-B)K^2$$

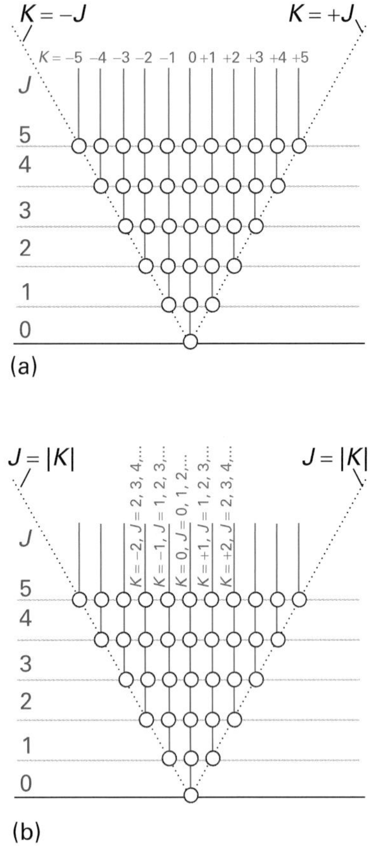

Fig. 17.2 (a) The sum over $J = 0, 1, 2, \ldots$ and $K = J, J-1, \ldots, -J$ (depicted by the circles) can be covered (b) by allowing K to range from $-\infty$ to ∞, with J confined to $|K|$, $|K|+1, \ldots, \infty$ for each value of K.

with $J = 0, 1, 2, \ldots, K = J, J-1, \ldots, -J$, and $M_J = J, J-1, \ldots, -J$. Instead of considering these ranges, we can cover the same values by allowing K to range from $-\infty$ to ∞, with J confined to $|K|, |K|+1, \ldots, \infty$ for each value of K (Fig. 17.2). Because the energy is independent of M_J, and there are $2J+1$ values of M_J for each value of J, each value of J is $2J+1$-fold degenerate. It follows that the partition function

$$q = \sum_{J=0}^{\infty}\sum_{K=-J}^{J}\sum_{M_J=-J}^{J} e^{-E_{JKM_J}/kT}$$

can be written equivalently as

$$\begin{aligned} q &= \sum_{K=-\infty}^{\infty}\sum_{J=|K|}^{\infty}(2J+1)e^{-E_{JKM_J}/kT} \\ &= \sum_{K=-\infty}^{\infty}\sum_{J=|K|}^{\infty}(2J+1)e^{-hc\{BJ(J+1)+(A-B)K^2\}/kT} \\ &= \sum_{K=-\infty}^{\infty} e^{-\{hc(A-B)/kT\}K^2}\sum_{J=|K|}^{\infty}(2J+1)e^{-hcBJ(J+1)/kT} \end{aligned}$$

Now we assume that the temperature is so high that numerous states are occupied and that the sums may be approximated by integrals. Then

$$q = \int_{-\infty}^{\infty} e^{-\{hc(A-B)/kT\}K^2}\int_{|K|}^{\infty}(2J+1)e^{-hcBJ(J+1)/kT}\,dJ\,dK$$

As before, the integral over J can be recognized as the integral of the derivative of a function, which is the function itself, so

$$\begin{aligned} \int_{|K|}^{\infty}(2J+1)e^{-hcBJ(J+1)/kT}dJ &= \int_{|K|}^{\infty}\left(-\frac{kT}{hcB}\right)\frac{d}{dJ}e^{-hcBJ(J+1)/kT}dJ \\ &= \left(-\frac{kT}{hcB}\right)e^{-hcBJ(J+1)/kT}\Big|_{|K|}^{\infty} = \left(\frac{kT}{hcB}\right)e^{-hcB|K|(|K|+1)/kT} \\ &\approx \left(\frac{kT}{hcB}\right)e^{-hcBK^2/kT} \end{aligned}$$

In the last line we have supposed that $|K| \gg 1$ for most contributions. Now we can write

$$\begin{aligned} q &= \frac{kT}{hcB}\int_{-\infty}^{\infty} e^{-\{hc(A-B)/kT\}K^2}e^{-hcBK^2/kT}dK \\ &= \frac{kT}{hcB}\int_{-\infty}^{\infty} e^{-\{hcA/kT\}K^2}dK = \left(\frac{kT}{hcB}\right)\left(\frac{kT}{hcA}\right)^{1/2}\overbrace{\int_{-\infty}^{\infty} e^{-x^2}dx}^{\pi^{1/2}} \\ &= \left(\frac{kT}{hc}\right)^{3/2}\left(\frac{\pi}{AB^2}\right)^{1/2} \end{aligned}$$

For an asymmetric rotor, one of the Bs is replaced by C, to give eqn 17.14b.

Synoptic table 17.1* Rotational and vibrational temperatures

Molecule	Mode	θ_V/K	θ_R/K
H_2		6330	88
HCl		4300	9.4
I_2		309	0.053
CO_2	ν_1	1997	0.561
	ν_2	3380	
	ν_3	960	

* For more values, see Table 13.2 in the *Data section* and use $hc/k = 1.439$ K cm.

A useful way of expressing the temperature above which the rotational approximation is valid is to introduce the **characteristic rotational temperature**, $\theta_R = hcB/k$. Then 'high temperature' means $T \gg \theta_R$ and under these conditions the rotational partition function of a linear molecule is simply T/θ_R. Some typical values of θ_R are shown in Table 17.1. The value for H_2 is abnormally high and we must be careful with the approximation for this molecule.

The general conclusion at this stage is that molecules with large moments of inertia (and hence small rotational constants and low characteristic rotational temperatures) have large rotational partition functions. The large value of q^R reflects the closeness in energy (compared with kT) of the rotational levels in large, heavy molecules, and the large number of them that are accessible at normal temperatures.

We must take care, however, not to include too many rotational states in the sum. For a homonuclear diatomic molecule or a symmetrical linear molecule (such as CO_2 or HC≡CH), a rotation through 180° results in an indistinguishable state of the molecule. Hence, the number of thermally accessible states is only half the number that can be occupied by a heteronuclear diatomic molecule, where rotation through 180° does result in a distinguishable state. Therefore, for a symmetrical linear molecule,

$$q^R = \frac{kT}{2hcB} = \frac{T}{2\theta_R} \tag{17.15a}$$

The equations for symmetrical and nonsymmetrical molecules can be combined into a single expression by introducing the **symmetry number**, σ, which is the number of indistinguishable orientations of the molecule. Then

$$q^R = \frac{kT}{\sigma hcB} = \frac{T}{\sigma\theta_R} \tag{17.15b}$$

For a heteronuclear diatomic molecule $\sigma = 1$; for a homonuclear diatomic molecule or a symmetrical linear molecule, $\sigma = 2$.

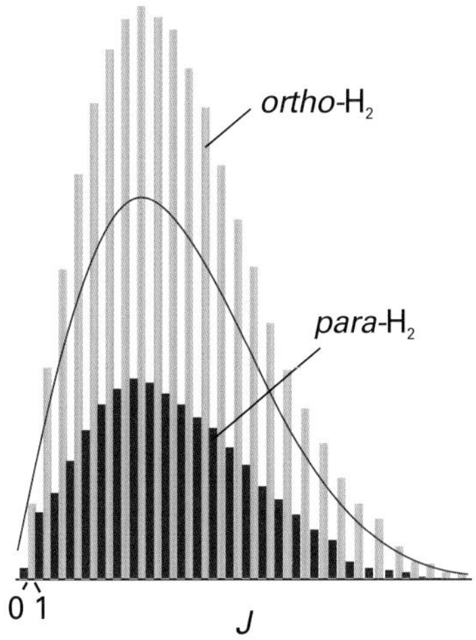

Fig. 17.3 The values of the individual terms $(2J+1)e^{-\beta hcBJ(J+1)}$ contributing to the mean partition function of a 3:1 mixture of *ortho*- and *para*-H_2. The partition function is the sum of all these terms. At high temperatures, the sum is approximately equal to the sum of the terms over all values of J, each with a weight of $\frac{1}{2}$. This is the sum of the contributions indicated by the curve.

Justification 17.2 *The origin of the symmetry number*

The quantum mechanical origin of the symmetry number is the Pauli principle, which forbids the occupation of certain states. We saw in Section 13.8, for example, that H_2 may occupy rotational states with even J only if its nuclear spins are paired (*para*-hydrogen), and odd J states only if its nuclear spins are parallel (*ortho*-hydrogen). There are three states of *ortho*-H_2 to each value of J (because there are three parallel spin states of the two nuclei).

To set up the rotational partition function we note that 'ordinary' molecular hydrogen is a mixture of one part *para*-H_2 (with only its even-J rotational states occupied) and three parts *ortho*-H_2 (with only its odd-J rotational states occupied). Therefore, the average partition function per molecule is

$$q^R = \tfrac{1}{4}\sum_{\text{even } J}(2J+1)e^{-\beta hcBJ(J+1)} + \tfrac{3}{4}\sum_{\text{odd } J}(2J+1)e^{-\beta hcBJ(J+1)}$$

The odd-J states are more heavily weighted than the even-J states (Fig. 17.3). From the illustration we see that we would obtain approximately the same answer for the partition function (the sum of all the populations) if each J term contributed half its normal value to the sum. That is, the last equation can be approximated as

$$q^R = \tfrac{1}{2}\sum_{J}(2J+1)e^{-\beta hcBJ(J+1)}$$

and this approximation is very good when many terms contribute (at high temperatures).

The same type of argument may be used for linear symmetrical molecules in which identical bosons are interchanged by rotation (such as CO_2). As pointed out in Section 13.8, if the nuclear spin of the bosons is 0, then only even-J states are admissible. Because only half the rotational states are occupied, the rotational partition function is only half the value of the sum obtained by allowing all values of J to contribute (Fig. 17.4).

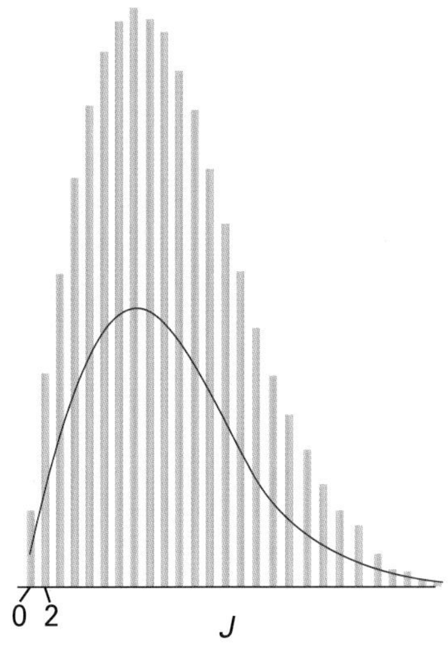

Fig. 17.4 The relative populations of the rotational energy levels of CO_2. Only states with even J values are occupied. The full line shows the smoothed, averaged population of levels.

Synoptic table 17.2* Symmetry numbers

Molecule	σ
H_2O	2
NH_3	3
CH_4	12
C_6H_6	12

* For more values, see Table 13.2 in the *Data section*.

The same care must be exercised for other types of symmetrical molecule, and for a nonlinear molecule we write

$$q^R = \frac{1}{\sigma}\left(\frac{kT}{hc}\right)^{3/2}\left(\frac{\pi}{ABC}\right)^{1/2} \qquad (17.16)$$

Some typical values of the symmetry numbers required are given in Table 17.2. The value $\sigma(H_2O) = 2$ reflects the fact that a 180° rotation about the bisector of the H—O—H angle interchanges two indistinguishable atoms. In NH_3, there are three indistinguishable orientations around the axis shown in (**1**). For CH_4, any of three 120° rotations about any of its four C—H bonds leaves the molecule in an indistinguishable state, so the symmetry number is $3 \times 4 = 12$. For benzene, any of six orientations around the axis perpendicular to the plane of the molecule leaves it apparently unchanged, as does a rotation of 180° around any of six axes in the plane of the molecule (three of which pass along each C—H bond and the remaining three pass through each C—C bond in the plane of the molecule). For the way that group theory is used to identify the value of the symmetry number, see Problem 17.17.

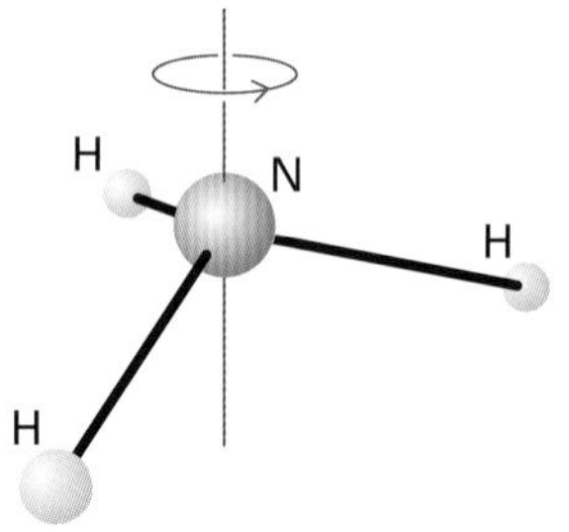

1

(c) The vibrational contribution

The vibrational partition function of a molecule is calculated by substituting the measured vibrational energy levels into the exponentials appearing in the definition of q^V, and summing them numerically. In a polyatomic molecule each normal mode (Section 13.14) has its own partition function (provided the anharmonicities are so small that the modes are independent). The overall vibrational partition function is the product of the individual partition functions, and we can write $q^V = q^V(1)q^V(2)\ldots$, where $q^V(K)$ is the partition function for the Kth normal mode and is calculated by direct summation of the observed spectroscopic levels.

If the vibrational excitation is not too great, the harmonic approximation may be made, and the vibrational energy levels written as

$$E_v = (v + \tfrac{1}{2})hc\tilde{\nu} \qquad v = 0, 1, 2, \ldots \qquad (17.17)$$

If, as usual, we measure energies from the zero-point level, then the permitted values are $\varepsilon_v = vhc\tilde{\nu}$ and the partition function is

$$q^V = \sum_v e^{-\beta vhc\tilde{\nu}} = \sum_v (e^{-\beta hc\tilde{\nu}})^v \qquad (17.18)$$

(because $e^{ax} = (e^x)^a$). We met this sum in Example 16.2 (which is no accident: the ladder-like array of levels in Fig. 16.3 is exactly the same as that of a harmonic oscillator). The series can be summed in the same way, and gives

$$q^V = \frac{1}{1 - e^{-\beta hc\tilde{\nu}}} \qquad (17.19)$$

This function is plotted in Fig. 17.5. In a polyatomic molecule, each normal mode gives rise to a partition function of this form.

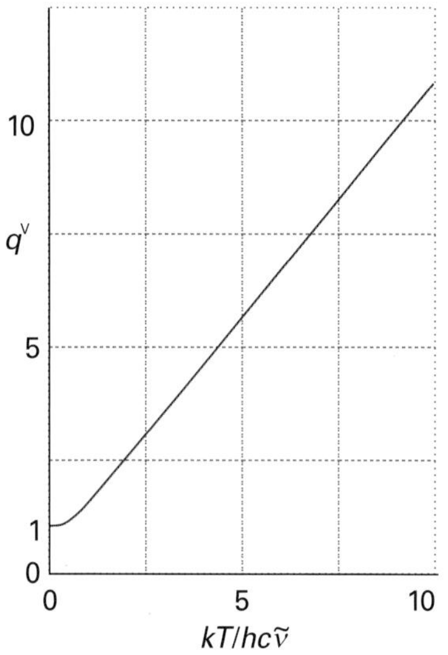

Fig. 17.5 The vibrational partition function of a molecule in the harmonic approximation. Note that the partition function is linearly proportional to the temperature when the temperature is high ($T \gg \theta_V$).

Exploration Plot the temperature dependence of the vibrational contribution to the molecular partition function for several values of the vibrational wavenumber. Estimate from your plots the temperature above which the harmonic oscillator is in the 'high temperature' limit.

Example 17.3 *Calculating a vibrational partition function*

The wavenumbers of the three normal modes of H_2O are 3656.7 cm^{-1}, 1594.8 cm^{-1}, and 3755.8 cm^{-1}. Evaluate the vibrational partition function at 1500 K.

Method Use eqn 17.19 for each mode, and then form the product of the three contributions. At 1500 K, $kT/hc = 1042.6$ cm^{-1}.

Answer We draw up the following table displaying the contributions of each mode:

Mode:	1	2	3
$\tilde{\nu}/\text{cm}^{-1}$	3656.7	1594.8	3755.8
$hc\tilde{\nu}/kT$	3.507	1.530	3.602
q^V	1.031	1.276	1.028

The overall vibrational partition function is therefore

$$q^V = 1.031 \times 1.276 \times 1.028 = 1.353$$

The three normal modes of H_2O are at such high wavenumbers that even at 1500 K most of the molecules are in their vibrational ground state. However, there may be so many normal modes in a large molecule that their excitation may be significant even though each mode is not appreciably excited. For example, a nonlinear molecule containing 10 atoms has $3N - 6 = 24$ normal modes (Section 13.14). If we assume a value of about 1.1 for the vibrational partition function of one normal mode, the overall vibrational partition function is about $q^V \approx (1.1)^{24} = 9.8$, which indicates significant vibrational excitation relative to a smaller molecule, such as H_2O.

Self-test 17.3 Repeat the calculation for CO_2, where the vibrational wavenumbers are 1388 cm^{-1}, 667.4 cm^{-1}, and 2349 cm^{-1}, the second being the doubly degenerate bending mode. [6.79]

In many molecules the vibrational wavenumbers are so great that $\beta hc\tilde{\nu} > 1$. For example, the lowest vibrational wavenumber of CH_4 is 1306 cm^{-1}, so $\beta hc\tilde{\nu} = 6.3$ at room temperature. C—H stretches normally lie in the range 2850 to 2960 cm^{-1}, so for them $\beta hc\tilde{\nu} \approx 14$. In these cases, $e^{-\beta hc\tilde{\nu}}$ in the denominator of q^V is very close to zero (for example, $e^{-6.3} = 0.002$), and the vibrational partition function for a single mode is very close to 1 ($q^V = 1.002$ when $\beta hc\tilde{\nu} = 6.3$), implying that only the zero-point level is significantly occupied.

Now consider the case of bonds so weak that $\beta hc\tilde{\nu} \ll kT$. When this condition is satisfied, the partition function may be approximated by expanding the exponential ($e^x = 1 + x + \cdots$):

$$q^V = \frac{1}{1-(1-\beta hc\tilde{\nu} + \cdots)} \tag{17.20}$$

That is, for weak bonds at high temperatures,

$$q^V = \frac{1}{\beta hc\tilde{\nu}} = \frac{kT}{hc\tilde{\nu}} \tag{17.21}$$

The temperatures for which eqn 17.21 is valid can be expressed in terms of the **characteristic vibrational temperature**, $\theta_V = hc\tilde{\nu}/k$ (Table 17.1). The value for H_2 is abnormally high because the atoms are so light and the vibrational frequency is correspondingly high. In terms of the vibrational temperature, 'high temperature' means $T \gg \theta_V$ and, when this condition is satisfied, $q^V = T/\theta_V$ (the analogue of the rotational expression).

(d) The electronic contribution

Electronic energy separations from the ground state are usually very large, so for most cases $q^E = 1$. An important exception arises in the case of atoms and molecules having

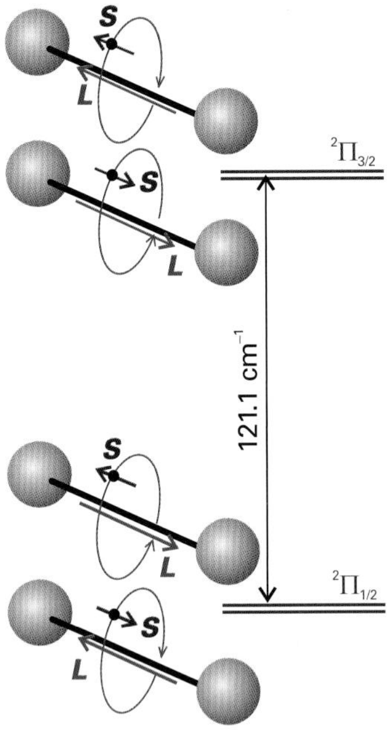

Fig. 17.6 The doubly degenerate ground electronic level of NO (with the spin and orbital angular momentum around the axis in opposite directions) and the doubly degenerate first excited level (with the spin and orbital momenta parallel). The upper level is thermally accessible at room temperature.

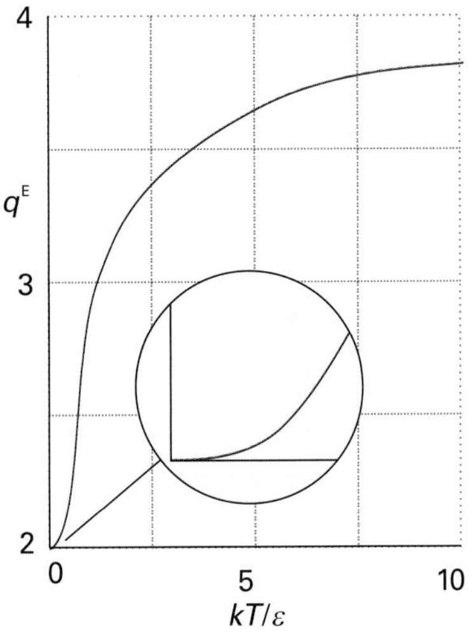

Fig. 17.7 The variation with temperature of the electronic partition function of an NO molecule. Note that the curve resembles that for a two-level system (Fig.16.5), but rises from 2 (the degeneracy of the lower level) and approaches 4 (the total number of states) at high temperatures.

Exploration Plot the temperature dependence of the electronic partition function for several values of the energy separation ε between two doubly degenerate levels. From your plots, estimate the temperature at which the population of the excited level begins to increase sharply.

electronically degenerate ground states, in which case $q^E = g^E$, where g^E is the degeneracy of the electronic ground state. Alkali metal atoms, for example, have doubly degenerate ground states (corresponding to the two orientations of their electron spin), so $q^E = 2$.

Some atoms and molecules have low-lying electronically excited states. (At high enough temperatures, all atoms and molecules have thermally accessible excited states.) An example is NO, which has a configuration of the form . . . π^1 (see *Impact* I11.1). The orbital angular momentum may take two orientations with respect to the molecular axis (corresponding to circulation clockwise or counter-clockwise around the axis), and the spin angular momentum may also take two orientations, giving four states in all (Fig. 17.6). The energy of the two states in which the orbital and spin momenta are parallel (giving the $^2\Pi_{3/2}$ term) is slightly greater than that of the two other states in which they are antiparallel (giving the $^2\Pi_{1/2}$ term). The separation, which arises from spin–orbit coupling (Section 10.8), is only 121 cm^{-1}. Hence, at normal temperatures, all four states are thermally accessible. If we denote the energies of the two levels as $E_{1/2} = 0$ and $E_{3/2} = \varepsilon$, the partition function is

$$q^E = \sum_{\text{energy levels}} g_j e^{-\beta\varepsilon_j} = 2 + 2e^{-\beta\varepsilon} \qquad (17.22)$$

Figure 17.7 shows the variation of this function with temperature. At $T = 0$, $q^E = 2$, because only the doubly degenerate ground state is accessible. At high temperatures, q^E approaches 4 because all four states are accessible. At 25°C, $q^E = 3.1$.

(e) The overall partition function

The partition functions for each mode of motion of a molecule are collected in Table 17.3 at the end of the chapter. The overall partition function is the product of each contribution. For a diatomic molecule with no low-lying electronically excited states and $T \gg \theta_R$,

$$q = g^E\left(\frac{V}{\Lambda^3}\right)\left(\frac{T}{\sigma\theta_R}\right)\left(\frac{1}{1-e^{-T/\theta_V}}\right) \qquad (17.23)$$

Example 17.4 *Calculating a thermodynamic function from spectroscopic data*

Calculate the value of $G_m^{\ominus} - G_m^{\ominus}(0)$ for $H_2O(g)$ at 1500 K given that $A = 27.8778\ cm^{-1}$, $B = 14.5092\ cm^{-1}$, and $C = 9.2869\ cm^{-1}$ and the information in Example 17.3.

Method The starting point is eqn 17.9. For the standard value, we evaluate the translational partition function at $p^{\ominus}$ (that is, at 10^5 Pa exactly). The vibrational partition function was calculated in Example 17.3. Use the expressions in Table 17.3 for the other contributions.

Answer Because $m = 18.015$ u, it follows that $q_m^{T\ominus}/N_A = 1.706 \times 10^8$. For the vibrational contribution we have already found that $q^V = 1.352$. From Table 17.2 we see that $\sigma = 2$, so the rotational contribution is $q^R = 486.7$. Therefore,

$$\begin{aligned} G_m^{\ominus} - G_m^{\ominus}(0) &= -(8.3145\ J\ K^{-1}\ mol^{-1}) \times (1500\ K) \\ &\quad \times \ln\{(1.706 \times 10^8) \times 486.7 \times 1.352\} \\ &= -317.3\ kJ\ mol^{-1} \end{aligned}$$

Self-test 17.4 Repeat the calculation for CO_2. The vibrational data are given in Self-test 17.3; $B = 0.3902\ cm^{-1}$. [$-366.6\ kJ\ mol^{-1}$]

Comment 17.1
The text's web site contains links to on-line databases of atomic and molecular spectra.

Overall partition functions obtained from eqn 17.23 are approximate because they assume that the rotational levels are very close together and that the vibrational levels are harmonic. These approximations are avoided by using the energy levels identified spectroscopically and evaluating the sums explicitly.

Using statistical thermodynamics

We can now calculate any thermodynamic quantity from a knowledge of the energy levels of molecules: we have merged thermodynamics and spectroscopy. In this section, we indicate how to do the calculations for four important properties.

17.3 Mean energies

It is often useful to know the mean energy, $\langle\varepsilon\rangle$, of various modes of motion. When the molecular partition function can be factorized into contributions from each mode, the mean energy of each mode M (from eqn 16.29) is

$$\langle\varepsilon^M\rangle = -\frac{1}{q^M}\left(\frac{\partial q^M}{\partial\beta}\right)_V \qquad M = T, R, V, \text{ or } E \qquad (17.24)$$

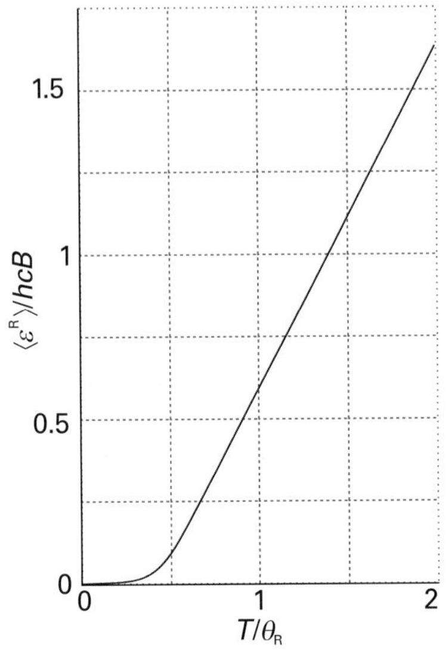

Fig. 17.8 The mean rotational energy of a nonsymmetrical linear rotor as a function of temperature. At high temperatures ($T \gg \theta_R$), the energy is linearly proportional to the temperature, in accord with the equipartition theorem.

Exploration Plot the temperature dependence of the mean rotational energy for several values of the rotational constant (for reasonable values of the rotational constant, see the *Data section*). From your plots, estimate the temperature at which the mean rotational energy begins to increase sharply.

(a) The mean translational energy

To see a pattern emerging, we consider first a one-dimensional system of length X, for which $q^T = X/\Lambda$, with $\Lambda = h(\beta/2\pi m)^{1/2}$. Then, if we note that Λ is a constant times $\beta^{1/2}$,

$$\langle\varepsilon^T\rangle = -\frac{\Lambda}{X}\left(\frac{\partial}{\partial\beta}\frac{X}{\Lambda}\right)_V = -\beta^{1/2}\frac{d}{d\beta}\left(\frac{1}{\beta^{1/2}}\right) = \frac{1}{2\beta} = \tfrac{1}{2}kT \qquad (17.25a)$$

For a molecule free to move in three dimensions, the analogous calculation leads to

$$\langle\varepsilon^T\rangle = \tfrac{3}{2}kT \qquad (17.25b)$$

Both conclusions are in agreement with the classical equipartition theorem (see *Molecular interpretation 2.2*) that the mean energy of each quadratic contribution to the energy is $\tfrac{1}{2}kT$. Furthermore, the fact that the mean energy is independent of the size of the container is consistent with the thermodynamic result that the internal energy of a perfect gas is independent of its volume (*Molecular interpretation 2.2*).

(b) The mean rotational energy

The mean rotational energy of a linear molecule is obtained from the partition function given in eqn 17.13. When the temperature is low ($T < \theta_R$), the series must be summed term by term, which gives

$$q^R = 1 + 3e^{-2\beta hcB} + 5e^{-6\beta hcB} + \cdots$$

Hence

$$\langle\varepsilon^R\rangle = \frac{hcB(6e^{-2\beta hcB} + 30e^{-6\beta hcB} + \cdots)}{1 + 3e^{-2\beta hcB} + 5e^{-6\beta hcB} + \cdots} \qquad (17.26a)$$

This function is plotted in Fig. 17.8. At high temperatures ($T \gg \theta_R$), q^R is given by eqn 17.15, and

$$\langle\varepsilon^R\rangle = -\frac{1}{q^R}\frac{dq^R}{d\beta} = -\sigma hc\beta B\frac{d}{d\beta}\frac{1}{\sigma hc\beta B} = \frac{1}{\beta} = kT \qquad (17.26b)$$

(q^R is independent of V, so the partial derivatives have been replaced by complete derivatives.) The high-temperature result is also in agreement with the equipartition theorem, for the classical expression for the energy of a linear rotor is $E_K = \tfrac{1}{2}I_\perp\omega_a^2 + \tfrac{1}{2}I_\perp\omega_b^2$. (There is no rotation around the line of atoms.) It follows from the equipartition theorem that the mean rotational energy is $2 \times \tfrac{1}{2}kT = kT$.

(c) The mean vibrational energy

The vibrational partition function in the harmonic approximation is given in eqn 17.19. Because q^V is independent of the volume, it follows that

$$\frac{dq^V}{d\beta} = \frac{d}{d\beta}\left(\frac{1}{1 - e^{-\beta hc\tilde{\nu}}}\right) = -\frac{hc\tilde{\nu}e^{-\beta hc\tilde{\nu}}}{(1 - e^{-\beta hc\tilde{\nu}})^2} \qquad (17.27)$$

and hence from

$$\langle\varepsilon^V\rangle = -\frac{1}{q^V}\frac{dq^V}{d\beta} = -(1 - e^{-\beta hc\tilde{\nu}})\left\{-\frac{hc\tilde{\nu}e^{-\beta hc\tilde{\nu}}}{(1 - e^{-\beta hc\tilde{\nu}})^2}\right\} = \frac{hc\tilde{\nu}e^{-\beta hc\tilde{\nu}}}{1 - e^{-\beta hc\tilde{\nu}}}$$

that

$$\langle\varepsilon^V\rangle = \frac{hc\tilde{\nu}}{e^{\beta hc\tilde{\nu}} - 1} \qquad (17.28)$$

The zero-point energy, $\frac{1}{2}hc\tilde{\nu}$, can be added to the right-hand side if the mean energy is to be measured from 0 rather than the lowest attainable level (the zero-point level). The variation of the mean energy with temperature is illustrated in Fig. 17.9. At high temperatures, when $T \gg \theta_V$, or $\beta hc\tilde{\nu} \ll 1$, the exponential functions can be expanded ($e^x = 1 + x + \cdots$) and all but the leading terms discarded. This approximation leads to

$$\langle \varepsilon^V \rangle = \frac{hc\tilde{\nu}}{(1 + \beta hc\tilde{\nu} + \cdots) - 1} \approx \frac{1}{\beta} = kT \qquad (17.29)$$

This result is in agreement with the value predicted by the classical equipartition theorem, because the energy of a one-dimensional oscillator is $E = \frac{1}{2}mv_x^2 + \frac{1}{2}kx^2$ and the mean energy of each quadratic term is $\frac{1}{2}kT$.

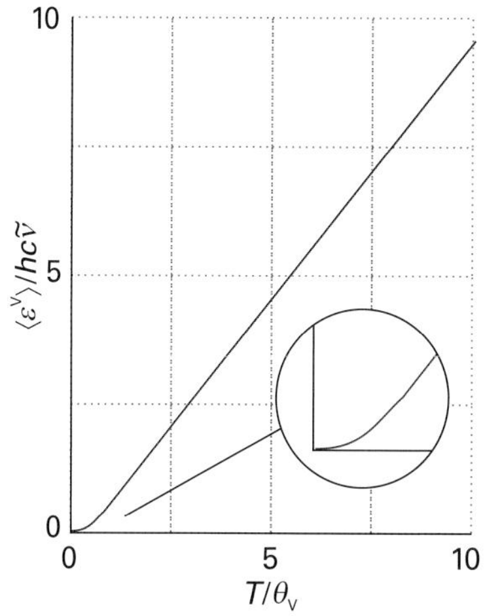

Fig. 17.9 The mean vibrational energy of a molecule in the harmonic approximation as a function of temperature. At high temperatures ($T \gg \theta_V$), the energy is linearly proportional to the temperature, in accord with the equipartition theorem.

Exploration Plot the temperature dependence of the mean vibrational energy for several values of the vibrational wavenumber (for reasonable values of the vibrational wavenumber, see the *Data section*). From your plots, estimate the temperature at which the mean vibrational energy begins to increase sharply.

17.4 Heat capacities

The constant-volume heat capacity is defined as $C_V = (\partial U/\partial T)_V$. The derivative with respect to T is converted into a derivative with respect to β by using

$$\frac{d}{dT} = \frac{d\beta}{dT}\frac{d}{d\beta} = -\frac{1}{kT^2}\frac{d}{d\beta} = -k\beta^2\frac{d}{d\beta} \qquad (17.30)$$

It follows that

$$C_V = -k\beta^2\left(\frac{\partial U}{\partial \beta}\right)_V \qquad (17.31a)$$

Because the internal energy of a perfect gas is a sum of contributions, the heat capacity is also a sum of contributions from each mode. The contribution of mode M is

$$C_V^M = N\left(\frac{\partial \langle \varepsilon^M \rangle}{\partial T}\right)_V = -Nk\beta^2\left(\frac{\partial \langle \varepsilon^M \rangle}{\partial \beta}\right)_V \qquad (17.31b)$$

(a) The individual contributions

The temperature is always high enough (provided the gas is above its condensation temperature) for the mean translational energy to be $\frac{3}{2}kT$, the equipartition value. Therefore, the molar constant-volume heat capacity is

$$C_{V,m}^T = N_A \frac{d(\frac{3}{2}kT)}{dT} = \frac{3}{2}R \qquad (17.32)$$

Translation is the only mode of motion for a monatomic gas, so for such a gas $C_{V,m} = \frac{3}{2}R = 12.47\ J\ K^{-1}\ mol^{-1}$. This result is very reliable: helium, for example, has this value over a range of 2000 K. We saw in Section 2.5 that $C_{p,m} - C_{V,m} = R$, so for a monatomic perfect gas $C_{p,m} = \frac{5}{2}R$, and therefore

$$\gamma = \frac{C_p}{C_V} = \frac{5}{3} \qquad (17.33)°$$

When the temperature is high enough for the rotations of the molecules to be highly excited (when $T \gg \theta_R$), we can use the equipartition value kT for the mean rotational energy (for a linear rotor) to obtain $C_{V,m} = R$. For nonlinear molecules, the mean rotational energy rises to $\frac{3}{2}kT$, so the molar rotational heat capacity rises to $\frac{3}{2}R$ when $T \gg \theta_R$. Only the lowest rotational state is occupied when the temperature is very low, and then rotation does not contribute to the heat capacity. We can

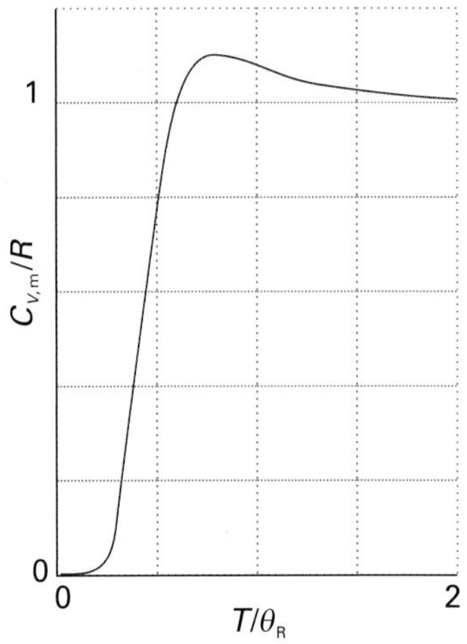

Fig. 17.10 The temperature dependence of the rotational contribution to the heat capacity of a linear molecule.

Exploration The *Living graphs* section of the text's web site has applets for the calculation of the temperature dependence of the rotational contribution to the heat capacity. Explore the effect of the rotational constant on the plot of $C^R_{V,m}$ against T.

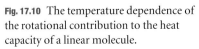

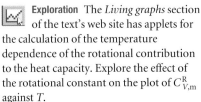

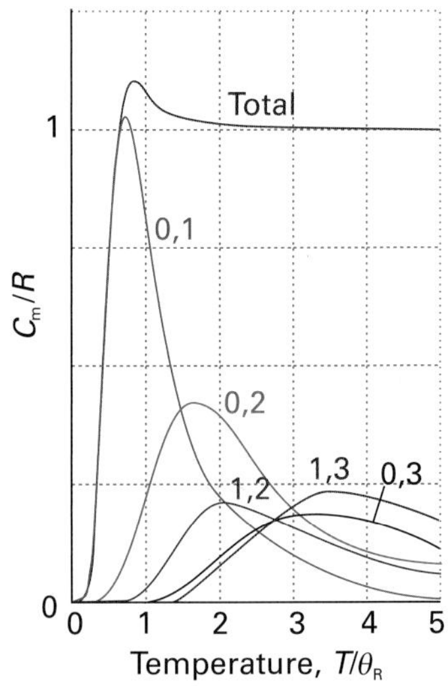

Fig. 17.11 The rotational heat capacity of a linear molecule can be regarded as the sum of contributions from a collection of two-level systems, in which the rise in temperature stimulates transitions between J levels, some of which are shown here. The calculation on which this illustration is based is sketched in Problem 17.19.

calculate the rotational heat capacity at intermediate temperatures by differentiating the equation for the mean rotational energy (eqn 17.26). The resulting (untidy) expression, which is plotted in Fig. 17.10, shows that the contribution rises from zero (when $T = 0$) to the equipartition value (when $T \gg \theta_R$). Because the translational contribution is always present, we can expect the molar heat capacity of a gas of diatomic molecules ($C^T_{V,m} + C^R_{V,m}$) to rise from $\frac{3}{2}R$ to $\frac{5}{2}R$ as the temperature is increased above θ_R. Problem 17.19 explores how the overall shape of the curve can be traced to the sum of thermal excitations between all the available rotational energy levels (Fig. 17.11).

Molecular vibrations contribute to the heat capacity, but only when the temperature is high enough for them to be significantly excited. The equipartition mean energy is kT for each mode, so the maximum contribution to the molar heat capacity is R. However, it is very unusual for the vibrations to be so highly excited that equipartition is valid, and it is more appropriate to use the full expression for the vibrational heat capacity, which is obtained by differentiating eqn 17.28:

Comment 17.2

Equation 17.34 is essentially the same as the Einstein formula for the heat capacity of a solid (eqn 8.7) with θ_V the Einstein temperature, θ_E. The only difference is that vibrations can take place in three dimensions in a solid.

$$C^V_{V,m} = Rf \qquad f = \left(\frac{\theta_V}{T}\right)^2 \left(\frac{e^{-\theta_V/2T}}{1 - e^{-\theta_V/T}}\right)^2 \tag{17.34}$$

where $\theta_V = hc\tilde{\nu}/k$ is the characteristic vibrational temperature. The curve in Fig. 17.12 shows how the vibrational heat capacity depends on temperature. Note that even when the temperature is only slightly above θ_V the heat capacity is close to its equipartition value.

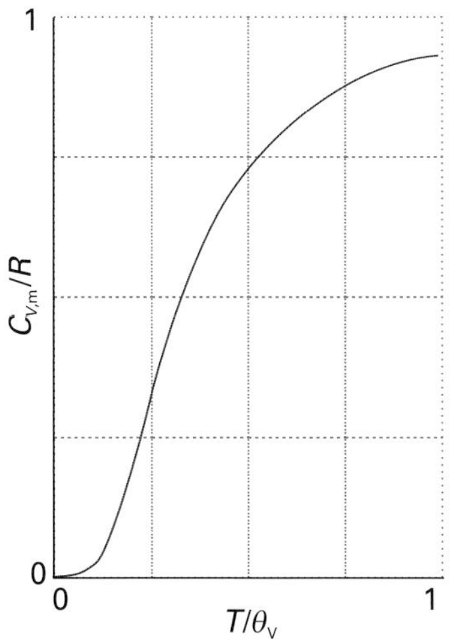

Fig. 17.12 The temperature dependence of the vibrational heat capacity of a molecule in the harmonic approximation calculated by using eqn 17.34. Note that the heat capacity is within 10 per cent of its classical value for temperatures greater than θ_V.

Exploration The *Living graphs* section of the text's web site has applets for the calculation of the temperature dependence of the vibrational contribution to the heat capacity. Explore the effect of the vibrational wavenumber on the plot of $C_{V,m}^{V}$ against T.

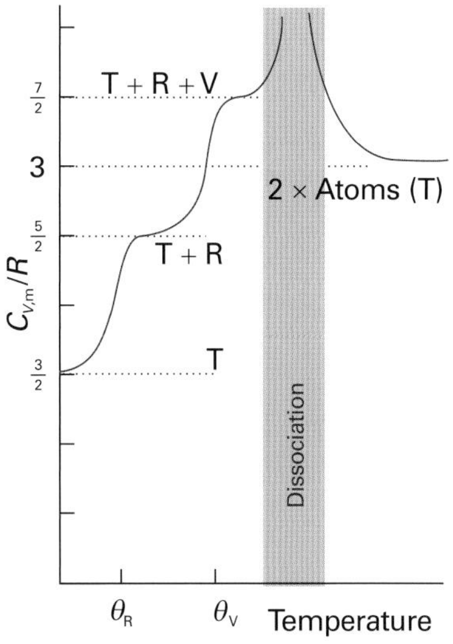

Fig. 17.13 The general features of the temperature dependence of the heat capacity of diatomic molecules are as shown here. Each mode becomes active when its characteristic temperature is exceeded. The heat capacity becomes very large when the molecule dissociates because the energy is used to cause dissociation and not to raise the temperature. Then it falls back to the translation-only value of the atoms.

(b) The overall heat capacity

The total heat capacity of a molecular substance is the sum of each contribution (Fig. 17.13). When equipartition is valid (when the temperature is well above the characteristic temperature of the mode, $T \gg \theta_M$) we can estimate the heat capacity by counting the numbers of modes that are active. In gases, all three translational modes are always active and contribute $\frac{3}{2}R$ to the molar heat capacity. If we denote the number of active rotational modes by ν_R^* (so for most molecules at normal temperatures $\nu_R^* = 2$ for linear molecules, and 3 for nonlinear molecules), then the rotational contribution is $\frac{1}{2}\nu_R^* R$. If the temperature is high enough for ν_V^* vibrational modes to be active, the vibrational contribution to the molar heat capacity is $\nu_V^* R$. In most cases $\nu_V^* \approx 0$. It follows that the total molar heat capacity is

$$C_{V,m} = \tfrac{1}{2}(3 + \nu_R^* + 2\nu_V^*)R \qquad (17.35)$$

Example 17.5 *Estimating the molar heat capacity of a gas*

Estimate the molar constant-volume heat capacity of water vapour at 100°C. Vibrational wavenumbers are given in Example 17.3; the rotational constants of an H_2O molecule are 27.9, 14.5, and 9.3 cm^{-1}.

Method We need to assess whether the rotational and vibrational modes are active by computing their characteristic temperatures from the data (to do so, use $hc/k = 1.439$ cm K).

Answer The characteristic temperatures (in round numbers) of the vibrations are 5300 K, 2300 K, and 5400 K; the vibrations are therefore not excited at 373 K. The three rotational modes have characteristic temperatures 40 K, 21 K, and 13 K, so they are fully excited, like the three translational modes. The translational contribution is $\frac{3}{2}R = 12.5$ J K^{-1} mol^{-1}. Fully excited rotations contribute a further 12.5 J K^{-1} mol^{-1}. Therefore, a value close to 25 J K^{-1} mol^{-1} is predicted. The experimental value is 26.1 J K^{-1} mol^{-1}. The discrepancy is probably due to deviations from perfect gas behaviour.

Self-test 17.5 Estimate the molar constant-volume heat capacity of gaseous I_2 at 25°C ($B = 0.037$ cm^{-1}; see Table 13.2 for more data). [29 J K^{-1} mol^{-1}]

17.5 Equations of state

The relation between p and Q in eqn 17.3 is a very important route to the equations of state of real gases in terms of intermolecular forces, for the latter can be built into Q. We have already seen (Example 17.1) that the partition function for a gas of independent particles leads to the perfect gas equation of state, $pV = nRT$. Real gases differ from perfect gases in their equations of state and we saw in Section 1.3 that their equations of state may be written

$$\frac{pV_{\mathrm{m}}}{RT} = 1 + \frac{B}{V_{\mathrm{m}}} + \frac{C}{V_{\mathrm{m}}^2} + \cdots \qquad (17.36)$$

where B is the second virial coefficient and C is the third virial coefficient.

The total kinetic energy of a gas is the sum of the kinetic energies of the individual molecules. Therefore, even in a real gas the canonical partition function factorizes into a part arising from the kinetic energy, which is the same as for the perfect gas, and a factor called the **configuration integral**, Z, which depends on the intermolecular potentials. We therefore write

$$Q = \frac{Z}{\Lambda^{3N}} \qquad (17.37)$$

By comparing this equation with eqn 16.45 ($Q = q^N/N!$, with $q = V/\Lambda^3$), we see that for a perfect gas of atoms (with no contributions from rotational or vibrational modes)

$$Z = \frac{V^N}{N!} \qquad (17.38)$$

For a real gas of atoms (for which the intermolecular interactions are isotropic), Z is related to the total potential energy E_{P} of interaction of all the particles by

$$Z = \frac{1}{N!}\int \mathrm{e}^{-\beta E_{\mathrm{P}}}\mathrm{d}\tau_1 \mathrm{d}\tau_2 \cdots \mathrm{d}\tau_N \qquad (17.39)$$

where $\mathrm{d}\tau_i$ is the volume element for atom i. The physical origin of this term is that the probability of occurrence of each arrangement of molecules possible in the sample is given by a Boltzmann distribution in which the exponent is given by the potential energy corresponding to that arrangement.

Illustration 17.1 *Calculating a configuration integral*

When the molecules do not interact with one another, $E_P = 0$ and hence $e^{-\beta E_P} = 1$. Then

$$Z = \frac{1}{N!}\int d\tau_1 d\tau_2 \cdots d\tau_N = \frac{V^N}{N!}$$

because $\int d\tau = V$, where V is the volume of the container. This result coincides with eqn 17.39.

When we consider only interactions between pairs of particles the configuration integral simplifies to

$$Z = \tfrac{1}{2}\int e^{-\beta E_P} d\tau_1 d\tau_2 \qquad (17.40)$$

The second virial coefficient then turns out to be

$$B = -\frac{N_A}{2V}\int f\, d\tau_1 d\tau_2 \qquad (17.41)$$

The quantity f is the **Mayer *f*-function**: it goes to zero when the two particles are so far apart that $E_P = 0$. When the intermolecular interaction depends only on the separation r of the particles and not on their relative orientation or their absolute position in space, as in the interaction of closed-shell atoms in a uniform sample, the volume element simplifies to $4\pi r^2 dr$ (because the integrals over the angular variables in $d\tau = r^2 dr \sin\theta\, d\theta d\phi$ give a factor of 4π) and eqn 17.41 becomes

$$B = -2\pi N_A \int_0^\infty f r^2 dr \qquad f = e^{-\beta E_P} - 1 \qquad (17.42)$$

The integral can be evaluated (usually numerically) by substituting an expression for the intermolecular potential energy.

Intermolecular potential energies are discussed in more detail in Chapter 18, where several expressions are developed for them. At this stage, we can illustrate how eqn 17.42 is used by considering the **hard-sphere potential**, which is infinite when the separation of the two molecules, r, is less than or equal to a certain value σ, and is zero for greater separations. Then

$$e^{-\beta E_P} = 0 \qquad f = -1 \quad \text{when} \quad r \le \sigma \quad (\text{and } E_P = \infty) \qquad (17.43a)$$

$$e^{-\beta E_P} = 1 \qquad f = 0 \quad \text{when} \quad r > \sigma \quad (\text{and } E_P = 0) \qquad (17.43b)$$

It follows from eqn 17.42 that the second virial coefficient is

$$B = 2\pi N_A \int_0^\sigma r^2 dr = \tfrac{2}{3}\pi N_A \sigma^3 \qquad (17.44)$$

This calculation of B raises the question as to whether a potential can be found that, when the virial coefficients are evaluated, gives the van der Waals equation of state. Such a potential can be found for weak attractive interactions ($a \ll RT$): it consists of a hard-sphere repulsive core and a long-range, shallow attractive region (see Problem 17.15). A further point is that, once a second virial coefficient has been calculated for a given intermolecular potential, it is possible to calculate other thermodynamic properties that depend on the form of the potential. For example, it is possible to calculate the isothermal Joule–Thomson coefficient, μ_T (Section 3.8), from the thermodynamic relation

$$\lim_{p\to 0} \mu_T = B - T\frac{\mathrm{d}B}{\mathrm{d}T} \tag{17.45}$$

and from the result calculate the Joule–Thomson coefficient itself by using eqn 3.48.

17.6 Molecular interactions in liquids

The starting point for the discussion of solids is the well ordered structure of a perfect crystal, which will be discussed in Chapter 20. The starting point for the discussion of gases is the completely disordered distribution of the molecules of a perfect gas, as we saw in Chapter 1. Liquids lie between these two extremes. We shall see that the structural and thermodynamic properties of liquids depend on the nature of intermolecular interactions and that an equation of state can be built in a similar way to that just demonstrated for real gases.

(a) The radial distribution function

The average relative locations of the particles of a liquid are expressed in terms of the **radial distribution function**, $g(r)$. This function is defined so that $g(r)r^2\mathrm{d}r$ is the probability that a molecule will be found in the range $\mathrm{d}r$ at a distance r from another molecule. In a perfect crystal, $g(r)$ is a periodic array of sharp spikes, representing the certainty (in the absence of defects and thermal motion) that molecules (or ions) lie at definite locations. This regularity continues out to the edges of the crystal, so we say that crystals have **long-range order**. When the crystal melts, the long-range order is lost and, wherever we look at long distances from a given molecule, there is equal probability of finding a second molecule. Close to the first molecule, though, the nearest neighbours might still adopt approximately their original relative positions and, even if they are displaced by newcomers, the new particles might adopt their vacated positions. It is still possible to detect a sphere of nearest neighbours at a distance r_1, and perhaps beyond them a sphere of next-nearest neighbours at r_2. The existence of this **short-range order** means that the radial distribution function can be expected to oscillate at short distances, with a peak at r_1, a smaller peak at r_2, and perhaps some more structure beyond that.

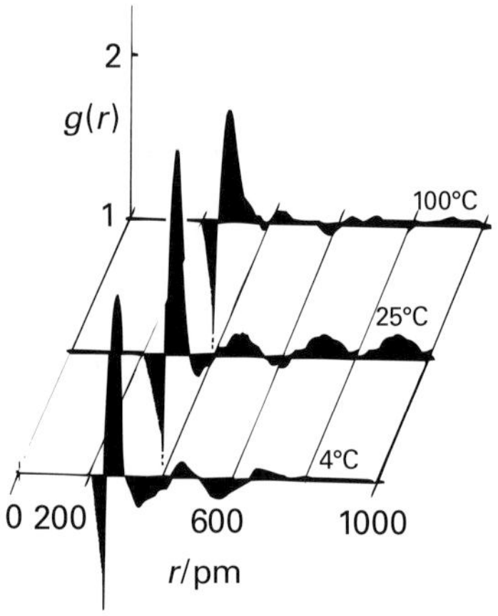

Fig. 17.14 The radial distribution function of the oxygen atoms in liquid water at three temperatures. Note the expansion as the temperature is raised. (A.H. Narten, M.D. Danford, and H.A. Levy, *Discuss. Faraday. Soc.* **43**, 97 (1967).)

The radial distribution function of the oxygen atoms in liquid water is shown in Fig. 17.14. Closer analysis shows that any given H_2O molecule is surrounded by other molecules at the corners of a tetrahedron. The form of $g(r)$ at 100°C shows that the intermolecular interactions (in this case, principally by hydrogen bonds) are strong enough to affect the local structure right up to the boiling point. Raman spectra indicate that in liquid water most molecules participate in either three or four hydrogen bonds. Infrared spectra show that about 90 per cent of hydrogen bonds are intact at the melting point of ice, falling to about 20 per cent at the boiling point.

The formal expression for the radial distribution function for molecules 1 and 2 in a fluid consisting of N particles is the somewhat fearsome equation

$$g(r_{12}) = \frac{\displaystyle\iint\cdots\int \mathrm{e}^{-\beta V_N}\mathrm{d}\tau_3\mathrm{d}\tau_4\ldots\mathrm{d}\tau_N}{\displaystyle N^2\iint\cdots\int \mathrm{e}^{-\beta V_N}\mathrm{d}\tau_1\mathrm{d}\tau_2\ldots\mathrm{d}\tau_N} \tag{17.46}$$

where $\beta = 1/kT$ and V_N is the N-particle potential energy. Although fearsome, this expression is nothing more than the Boltzmann distribution for the relative locations of two molecules in a field provided by all the other molecules in the system.

(b) The calculation of $g(r)$

Because the radial distribution function can be calculated by making assumptions about the intermolecular interactions, it can be used to test theories of liquid structure. However, even a fluid of hard spheres without attractive interactions (a collection of ball-bearings in a container) gives a function that oscillates near the origin (Fig. 17.15), and one of the factors influencing, and sometimes dominating, the structure of a liquid is the geometrical problem of stacking together reasonably hard spheres. Indeed, the radial distribution function of a liquid of hard spheres shows more pronounced oscillations at a given temperature than that of any other type of liquid. The attractive part of the potential modifies this basic structure, but sometimes only quite weakly. One of the reasons behind the difficulty of describing liquids theoretically is the similar importance of both the attractive and repulsive (hard core) components of the potential.

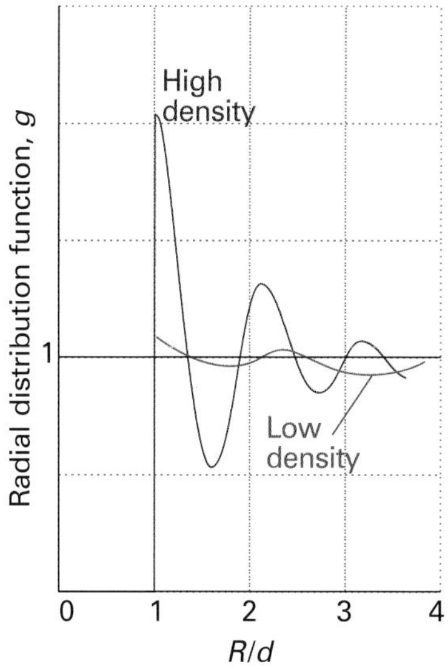

Fig. 17.15 The radial distribution function for a simulation of a liquid using impenetrable hard spheres (ball bearings).

There are several ways of building the intermolecular potential into the calculation of $g(r)$. Numerical methods take a box of about 10^3 particles (the number increases as computers grow more powerful), and the rest of the liquid is simulated by surrounding the box with replications of the original box (Fig. 17.16). Then, whenever a particle leaves the box through one of its faces, its image arrives through the opposite face. When calculating the interactions of a molecule in a box, it interacts with all the molecules in the box and all the periodic replications of those molecules and itself in the other boxes.

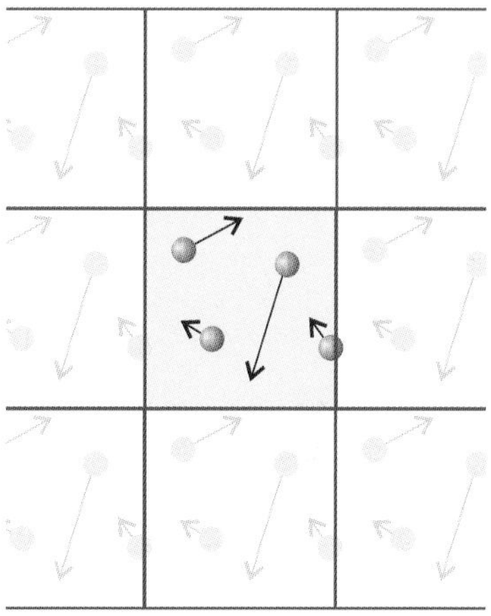

Fig. 17.16 In a two-dimensional simulation of a liquid that uses periodic boundary conditions, when one particle leaves the cell its mirror image enters through the opposite face.

In the **Monte Carlo method**, the particles in the box are moved through small but otherwise random distances, and the change in total potential energy of the N particles in the box, ΔV_N, is calculated using one of the intermolecular potentials discussed in Section 18.4. Whether or not this new configuration is accepted is then judged from the following rules:

1 If the potential energy is not greater than before the change, then the configuration is accepted.

If the potential energy is greater than before the change, then it is necessary to check if the new configuration is reasonable and can exist in equilibrium with configurations of lower potential energy at a given temperature. To make progress, we use the result that, at equilibrium, the ratio of populations of two states with energy separation ΔV_N is $e^{-\Delta V_N/kT}$. Because we are testing the viability of a configuration with a higher potential energy than the previous configuration in the calculation, $\Delta V_N > 0$ and the exponential factor varies between 0 and 1. In the Monte Carlo method, the second rule, therefore, is:

2 The exponential factor is compared with a random number between 0 and 1; if the factor is larger than the random number, then the configuration is accepted; if the factor is not larger, the configuration is rejected.

The configurations generated with Monte Carlo calculations can be used to construct $g(r)$ simply by counting the number of pairs of particles with a separation r and averaging the result over the whole collection of configurations.

In the **molecular dynamics** approach, the history of an initial arrangement is followed by calculating the trajectories of all the particles under the influence of the intermolecular potentials. To appreciate what is involved, we consider the motion of a particle in one dimension. We show in the following *Justification* that, after a time interval Δt, the position of a particle changes from x_{i-1} to a new value x_i given by

$$x_i = x_{i-1} + v_{i-1}\Delta t \tag{17.47}$$

where v_{i-1} is the velocity of the atom when it was at x_{i-1}, its location at the start of the interval. The velocity at x_i is related to v_{i-1}, the velocity at the start of the interval, by

$$v_i = v_{i-1} - m^{-1}\left(\frac{\mathrm{d}V_N(x)}{\mathrm{d}x}\right)_{x_{i-1}}\Delta t \qquad (17.48)$$

where the derivative of the potential energy $V_N(x)$ is evaluated at x_{i-1}. The time interval Δt is approximately 1 fs (10^{-15} s), which is shorter than the average time between collisions. The calculation of x_i and v_i is then repeated for tens of thousands of such steps. The time-consuming part of the calculation is the evaluation of the net force on the molecule arising from all the other molecules present in the system.

Justification 17.3 *Particle trajectories according to molecular dynamics*

Consider a particle of mass m moving along the x direction with an initial velocity v_1 given by

$$v_1 = \frac{\Delta x}{\Delta t}$$

If the initial and new positions of the atom are x_1 and x_2, then $\Delta x = x_2 - x_1$ and

$$x_2 = x_1 + v_1\Delta t$$

The particle moves under the influence of a force arising from interactions with other atoms in the molecule. From Newton's second law of motion, we write the force F_1 at x_1 as

$$F_1 = ma_1$$

where the acceleration a_1 at x_1 is given by $a_1 = \Delta v/\Delta t$. If the initial and new velocities are v_1 and v_2, then $\Delta v = v_2 - v_1$ and

$$v_2 = v_1 + a_1\Delta t = v_1 + \frac{F_1}{m}\Delta t$$

Because $F = -\mathrm{d}V/\mathrm{d}x$, the force acting on the atom is related to the potential energy of interaction with other nearby atoms, the potential energy $V_N(x)$, by

$$F_1 = -\left(\frac{\mathrm{d}V_N(x)}{\mathrm{d}x}\right)_{x_1}$$

where the derivative is evaluated at x_1. It follows that

$$v_2 = v_1 - m^{-1}\left(\frac{\mathrm{d}V_N(x)}{\mathrm{d}x}\right)_{x_1}\Delta t$$

This expression generalizes to eqn 17.48 for the calculation of a velocity v_i from a previous velocity v_{i-1}.

Self-test 17.6 Consider a particle of mass m connected to a stationary wall with a spring of force constant k. Write an expression for the velocity of this particle once it is set into motion in the x direction from an equilibrium position x_0.

[$v_i = v_{i-1} + (k/m)(x_{i-1} - x_0)$]

A molecular dynamics calculation gives a series of snapshots of the liquid, and $g(r)$ can be calculated as before. The temperature of the system is inferred by computing the mean kinetic energy of the particles and using the equipartition result that

$$\langle \tfrac{1}{2}mv_q^2\rangle = \tfrac{1}{2}kT \qquad (17.49)$$

for each coordinate q.

(c) The thermodynamic properties of liquids

Once $g(r)$ is known it can be used to calculate the thermodynamic properties of liquids. For example, the contribution of the pairwise additive intermolecular potential, V_2, to the internal energy is given by the integral

$$U=\frac{2\pi N^2}{V}\int_0^\infty g(r)V_2 r^2\mathrm{d}r \qquad (17.50)$$

That is, U is essentially the average two-particle potential energy weighted by $g(r)r^2\mathrm{d}r$, which is the probability that the pair of particles have a separation between r and $r+\mathrm{d}r$. Likewise, the contribution that pairwise interactions make to the pressure is

$$\frac{pV}{nRT}=1-\frac{2\pi N}{kTV}\int_0^\infty g(r)v_2 r^2\mathrm{d}r \qquad v_2=r\frac{\mathrm{d}V_2}{\mathrm{d}r} \qquad (17.51a)$$

The quantity v_2 is called the **virial** (hence the term 'virial equation of state'). To understand the physical content of this expression, we rewrite it as

$$p=\frac{nRT}{V}-2\pi\left(\frac{N}{V}\right)^2\int_0^\infty g(r)v_2 r^2\mathrm{d}r \qquad (17.51b)$$

The first term on the right is the **kinetic pressure**, the contribution to the pressure from the impact of the molecules in free flight. The second term is essentially the internal pressure, $\pi_T=(\partial U/\partial V)_T$, introduced in Section 2.11, representing the contribution to the pressure from the intermolecular forces. To see the connection, we should recognize $-\mathrm{d}V_2/\mathrm{d}r$ (in v_2) as the force required to move two molecules apart, and therefore $-r(\mathrm{d}V_2/\mathrm{d}r)$ as the work required to separate the molecules through a distance r. The second term is therefore the average of this work over the range of pairwise separations in the liquid as represented by the probability of finding two molecules at separations between r and $r+\mathrm{d}r$, which is $g(r)r^2\mathrm{d}r$. In brief, the integral, when multiplied by the square of the number density, is the change in internal energy of the system as it expands, and therefore is equal to the internal pressure.

17.7 Residual entropies

Entropies may be calculated from spectroscopic data; they may also be measured experimentally (Section 3.3). In many cases there is good agreement, but in some the experimental entropy is less than the calculated value. One possibility is that the experimental determination failed to take a phase transition into account (and a contribution of the form $\Delta_{\mathrm{trs}}H/T_{\mathrm{trs}}$ incorrectly omitted from the sum). Another possibility is that some disorder is present in the solid even at $T=0$. The entropy at $T=0$ is then greater than zero and is called the **residual entropy**.

The origin and magnitude of the residual entropy can be explained by considering a crystal composed of AB molecules, where A and B are similar atoms (such as CO, with its very small electric dipole moment). There may be so little energy difference between . . .AB AB AB AB. . . , . . .AB BA BA AB. . . , and other arrangements that the molecules adopt the orientations AB and BA at random in the solid. We can readily calculate the entropy arising from residual disorder by using the Boltzmann formula $S=k\ln W$. To do so, we suppose that two orientations are equally probable, and that the sample consists of N molecules. Because the same energy can be achieved in 2^N different ways (because each molecule can take either of two orientations), the total number of ways of achieving the same energy is $W=2^N$. It follows that

$$S=k\ln 2^N=Nk\ln 2=nR\ln 2 \qquad (17.52a)$$

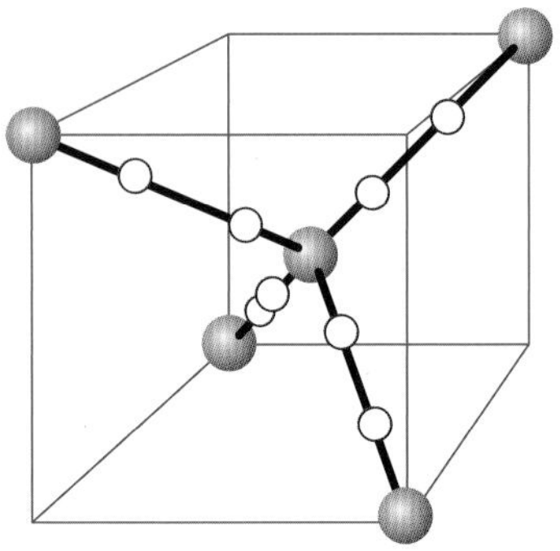

Fig. 17.17 The possible locations of H atoms around a central O atom in an ice crystal are shown by the white spheres. Only one of the locations on each bond may be occupied by an atom, and two H atoms must be close to the O atom and two H atoms must be distant from it.

We can therefore expect a residual molar entropy of $R \ln 2 = 5.8\ \mathrm{J\ K^{-1}\ mol^{-1}}$ for solids composed of molecules that can adopt either of two orientations at $T = 0$. If s orientations are possible, the residual molar entropy will be

$$S_m = R \ln s \qquad (17.52b)$$

An $FClO_3$ molecule, for example, can adopt four orientations with about the same energy (with the F atom at any of the four corners of a tetrahedron), and the calculated residual molar entropy of $R \ln 4 = 11.5\ \mathrm{J\ K^{-1}\ mol^{-1}}$ is in good agreement with the experimental value ($10.1\ \mathrm{J\ K^{-1}\ mol^{-1}}$). For CO, the measured residual entropy is $5\ \mathrm{J\ K^{-1}\ mol^{-1}}$, which is close to $R \ln 2$, the value expected for a random structure of the form . . .CO CO OC CO OC OC. . . .

Illustration 17.2 *Calculating a residual entropy*

Consider a sample of ice with N H_2O molecules. Each O atom is surrounded tetrahedrally by four H atoms, two of which are attached by short σ bonds, the other two being attached by long hydrogen bonds (Fig. 17.17). It follows that each of the $2N$ H atoms can be in one of two positions (either close to or far from an O atom as shown in Fig. 17.18), resulting in 2^{2N} possible arrangements. However, not all these arrangements are acceptable. Indeed, of the $2^4 = 16$ ways of arranging four H atoms around one O atom, only 6 have two short and two long OH distances and hence are acceptable. Therefore, the number of permitted arrangements is

$$W = 2^{2N}\left(\tfrac{6}{16}\right)^N = \left(\tfrac{3}{2}\right)^N$$

It then follows that the residual molar entropy is

$$S_m(0) \approx k \ln\left(\tfrac{3}{2}\right)^{N_A} = N_A k \ln\left(\tfrac{3}{2}\right) = R \ln\left(\tfrac{3}{2}\right) = 3.4\ \mathrm{J\ K^{-1}\ mol^{-1}}$$

which is in good agreement with the experimental value of $3.4\ \mathrm{J\ K^{-1}\ mol^{-1}}$. The model, however, is not exact because it ignores the possibility that next-nearest neighbours and those beyond can influence the local arrangement of bonds.

17.8 Equilibrium constants

The Gibbs energy of a gas of independent molecules is given by eqn 17.9 in terms of the molar partition function, $q_m = q/n$. The equilibrium constant K of a reaction is related to the standard Gibbs energy of reaction by $\Delta_r G^\ominus = -RT \ln K$. To calculate the equilibrium constant, we need to combine these two equations. We shall consider gas phase reactions in which the equilibrium constant is expressed in terms of the partial pressures of the reactants and products.

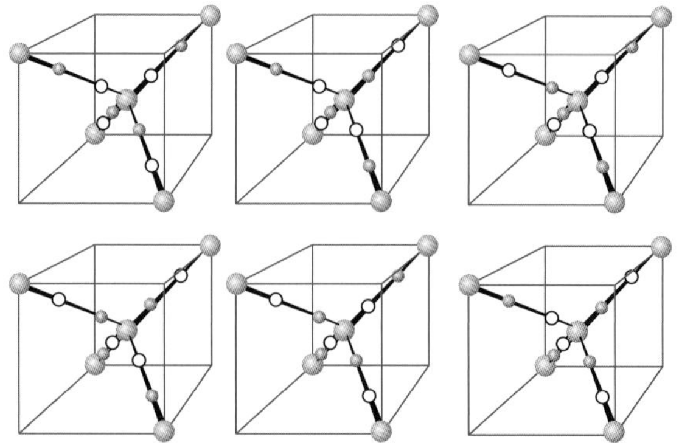

Fig. 17.18 The six possible arrangements of H atoms in the locations identified in Fig. 17.17. Occupied locations are denoted by red spheres and unoccupied locations by white spheres.

(a) The relation between *K* and the partition function

To find an expression for the standard reaction Gibbs energy we need expressions for the standard molar Gibbs energies, $G^{\ominus}/n$, of each species. For these expressions, we need the value of the molar partition function when $p = p^{\ominus}$ (where $p^{\ominus} = 1$ bar): we denote this **standard molar partition function** $q^{\ominus}_{\mathrm{m}}$. Because only the translational component depends on the pressure, we can find $q^{\ominus}_{\mathrm{m}}$ by evaluating the partition function with V replaced by $V^{\ominus}_{\mathrm{m}}$, where $V^{\ominus}_{\mathrm{m}} = RT/p^{\ominus}$. For a species J it follows that

$$G^{\ominus}_{\mathrm{m}}(\mathrm{J}) = G^{\ominus}_{\mathrm{m}}(\mathrm{J},0) - RT\ln\frac{q^{\ominus}_{\mathrm{J,m}}}{N_{\mathrm{A}}} \qquad (17.53)^{\circ}$$

where $q^{\ominus}_{\mathrm{J,m}}$ is the standard molar partition function of J. By combining expressions like this one (as shown in the *Justification* below), the equilibrium constant for the reaction

$$a\,\mathrm{A} + b\,\mathrm{B} \rightarrow c\,\mathrm{C} + d\,\mathrm{D}$$

is given by the expression

$$K = \frac{(q^{\ominus}_{\mathrm{C,m}}/N_{\mathrm{A}})^{c}(q^{\ominus}_{\mathrm{D,m}}/N_{\mathrm{A}})^{d}}{(q^{\ominus}_{\mathrm{A,m}}/N_{\mathrm{A}})^{a}(q^{\ominus}_{\mathrm{B,m}}/N_{\mathrm{A}})^{b}}\mathrm{e}^{-\Delta_{\mathrm{r}}E_0/RT} \qquad (17.54\mathrm{a})$$

where $\Delta_{\mathrm{r}}E_0$ is the difference in molar energies of the ground states of the products and reactants (this term is defined more precisely in the *Justification*), and is calculated from the bond dissociation energies of the species (Fig. 17.19). In terms of the stoichiometric numbers introduced in Section 7.2, we would write

$$K = \left\{\prod_{\mathrm{J}}\left(\frac{q^{\ominus}_{\mathrm{J,m}}}{N_{\mathrm{A}}}\right)^{\nu_{\mathrm{J}}}\right\}\mathrm{e}^{-\Delta_{\mathrm{r}}E_0/RT} \qquad (17.54\mathrm{b})$$

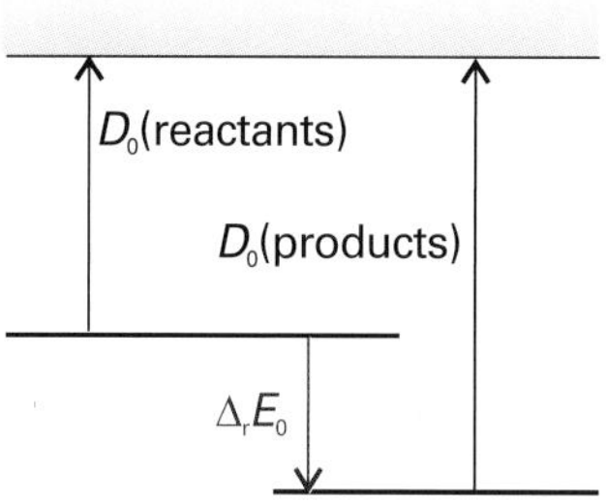

Fig. 17.19 The definition of $\Delta_{\mathrm{r}}E_0$ for the calculation of equilibrium constants.

Justification 17.4 *The equilibrium constant in terms of the partition function 1*

The standard molar reaction Gibbs energy for the reaction is

$$\begin{aligned}\Delta_{\mathrm{r}}G^{\ominus} &= cG^{\ominus}_{\mathrm{m}}(\mathrm{C}) + dG^{\ominus}_{\mathrm{m}}(\mathrm{D}) - aG^{\ominus}_{\mathrm{m}}(\mathrm{A}) - bG^{\ominus}_{\mathrm{m}}(\mathrm{B})\\ &= cG^{\ominus}_{\mathrm{m}}(\mathrm{C},0) + dG^{\ominus}_{\mathrm{m}}(\mathrm{D},0) - aG^{\ominus}_{\mathrm{m}}(\mathrm{A},0) - bG^{\ominus}_{\mathrm{m}}(\mathrm{B},0)\\ &\quad - RT\left\{c\ln\frac{q^{\ominus}_{\mathrm{C,m}}}{N_{\mathrm{A}}} + d\ln\frac{q^{\ominus}_{\mathrm{D,m}}}{N_{\mathrm{A}}} - a\ln\frac{q^{\ominus}_{\mathrm{A,m}}}{N_{\mathrm{A}}} - b\ln\frac{q^{\ominus}_{\mathrm{B,m}}}{N_{\mathrm{A}}}\right\}\end{aligned}$$

Because $G(0) = U(0)$, the first term on the right is

$$\Delta_{\mathrm{r}}E_0 = cU^{\ominus}_{\mathrm{m}}(\mathrm{C},0) + dU^{\ominus}_{\mathrm{m}}(\mathrm{D},0) - aU^{\ominus}_{\mathrm{m}}(\mathrm{A},0) - bU^{\ominus}_{\mathrm{m}}(\mathrm{B},0) \qquad (17.55)$$

the reaction internal energy at $T = 0$ (a molar quantity).

Now we can write

$$\begin{aligned}\Delta_{\mathrm{r}}G^{\ominus} &= \Delta_{\mathrm{r}}E_0 - RT\left\{\ln\left(\frac{q^{\ominus}_{\mathrm{C,m}}}{N_{\mathrm{A}}}\right)^{c} + \ln\left(\frac{q^{\ominus}_{\mathrm{D,m}}}{N_{\mathrm{A}}}\right)^{d} - \ln\left(\frac{q^{\ominus}_{\mathrm{A,m}}}{N_{\mathrm{A}}}\right)^{a} - \ln\left(\frac{q^{\ominus}_{\mathrm{B,m}}}{N_{\mathrm{A}}}\right)^{b}\right\}\\ &= \Delta_{\mathrm{r}}E_0 - RT\ln\frac{(q^{\ominus}_{\mathrm{C,m}}/N_{\mathrm{A}})^{c}(q^{\ominus}_{\mathrm{D,m}}/N_{\mathrm{A}})^{d}}{(q^{\ominus}_{\mathrm{A,m}}/N_{\mathrm{A}})^{a}(q^{\ominus}_{\mathrm{B,m}}/N_{\mathrm{A}})^{b}}\\ &= -RT\left\{\frac{\Delta_{\mathrm{r}}E_0}{RT} + \ln\frac{(q^{\ominus}_{\mathrm{C,m}}/N_{\mathrm{A}})^{c}(q^{\ominus}_{\mathrm{D,m}}/N_{\mathrm{A}})^{d}}{(q^{\ominus}_{\mathrm{A,m}}/N_{\mathrm{A}})^{a}(q^{\ominus}_{\mathrm{B,m}}/N_{\mathrm{A}})^{b}}\right\}\end{aligned}$$

At this stage we can pick out an expression for K by comparing this equation with $\Delta_{\mathrm{r}}G^{\ominus} = -RT\ln K$, which gives

$$\ln K = -\frac{\Delta_r E_0}{RT} + \ln \frac{(q^{\ominus}_{C,m}/N_A)^c (q^{\ominus}_{D,m}/N_A)^d}{(q^{\ominus}_{A,m}/N_A)^a (q^{\ominus}_{B,m}/N_A)^b}$$

This expression is easily rearranged into eqn 17.54a by forming the exponential of both sides.

(b) A dissociation equilibrium

We shall illustrate the application of eqn 17.54 to an equilibrium in which a diatomic molecule X_2 dissociates into its atoms:

$$X_2(g) \rightleftharpoons 2\,X(g) \qquad K = \frac{p_X^2}{p_{X_2} p^{\ominus}}$$

According to eqn 17.54 (with $a = 1$, $b = 0$, $c = 2$, and $d = 0$):

$$K = \frac{(q^{\ominus}_{X,m}/N_A)^2}{q^{\ominus}_{X_2,m}/N_A} e^{-\Delta_r E_0/RT} = \frac{(q^{\ominus}_{X,m})^2}{q^{\ominus}_{X_2,m} N_A} e^{-\Delta_r E_0/RT} \tag{17.56a}$$

with

$$\Delta_r E_0 = 2U^{\ominus}_m(X,0) - U^{\ominus}_m(X_2,0) = D_0(X\text{—}X) \tag{17.56b}$$

where $D_0(X\text{—}X)$ is the dissociation energy of the X—X bond. The standard molar partition functions of the atoms X are

$$q^{\ominus}_{X,m} = g_X \left(\frac{V^{\ominus}_m}{\Lambda_X^3} \right) = \frac{RTg_X}{p^{\ominus} \Lambda_X^3}$$

where g_X is the degeneracy of the electronic ground state of X and we have used $V^{\ominus}_m = RT/p^{\ominus}$. The diatomic molecule X_2 also has rotational and vibrational degrees of freedom, so its standard molar partition function is

$$q^{\ominus}_{X_2,m} = g_{X_2} \left(\frac{V^{\ominus}_m}{\Lambda_{X_2}^3} \right) q^R_{X_2} q^V_{X_2} = \frac{RTg_{X_2} q^R_{X_2} q^V_{X_2}}{p^{\ominus} \Lambda_{X_2}^3}$$

where g_{X_2} is the degeneracy of the electronic ground state of X_2. It follows from eqn 17.54 that the equilibrium constant is

$$K = \frac{kTg_X^2 \Lambda_{X_2}^3}{p^{\ominus} g_{X_2} q^R_{X_2} q^V_{X_2} \Lambda_X^6} e^{-D_0/RT} \tag{17.57}$$

where we have used $R/N_A = k$. All the quantities in this expression can be calculated from spectroscopic data. The Λs are defined in Table 17.3 and depend on the masses of the species and the temperature; the expressions for the rotational and vibrational partition functions are also available in Table 17.3 and depend on the rotational constant and vibrational wavenumber of the molecule.

Example 17.6 *Evaluating an equilibrium constant*

Evaluate the equilibrium constant for the dissociation $Na_2(g) \rightleftharpoons 2\,Na(g)$ at 1000 K from the following data: $B = 0.1547\ cm^{-1}$, $\tilde{\nu} = 159.2\ cm^{-1}$, $D_0 = 70.4\ kJ\ mol^{-1}$. The Na atoms have doublet ground terms.

Method The partition functions required are specified in eqn 17.54. They are evaluated by using the expressions in Table 17.3. For a homonuclear diatomic molecule, $\sigma = 2$. In the evaluation of $kT/p^{\ominus}$ use $p^{\ominus} = 10^5$ Pa and 1 Pa m^3 = 1 J.

Answer The partition functions and other quantities required are as follows:

$\Lambda(Na_2) = 8.14$ pm $\quad \Lambda(Na) = 11.5$ pm
$q^R(Na_2) = 2246 \quad q^V(Na_2) = 4.885$
$g(Na) = 2 \quad g(Na_2) = 1$

Then, from eqn 17.54,

$$K = \frac{(1.38\times10^{-23}\ \mathrm{J\,K^{-1}})\times(1000\ \mathrm{K})\times4\times(8.14\times10^{-12}\ \mathrm{m})^3}{(10^5\ \mathrm{Pa})\times2246\times4.885\times(1.15\times10^{-11}\ \mathrm{m})^6}\times e^{-8.47}$$

$$= 2.42$$

where we have used 1 J = 1 kg m^2 s^{-2} and 1 Pa = 1 kg m^{-1} s^{-1}.

Self-test 17.7 Evaluate K at 1500 K. [52]

(c) Contributions to the equilibrium constant

We are now in a position to appreciate the physical basis of equilibrium constants. To see what is involved, consider a simple R $\rightleftharpoons$ P gas-phase equilibrium (R for reactants, P for products).

Figure 17.20 shows two sets of energy levels; one set of states belongs to R, and the other belongs to P. The populations of the states are given by the Boltzmann distribution, and are independent of whether any given state happens to belong to R or to P. We can therefore imagine a single Boltzmann distribution spreading, without distinction, over the two sets of states. If the spacings of R and P are similar (as in Fig. 17.20), and P lies above R, the diagram indicates that R will dominate in the equilibrium mixture. However, if P has a high density of states (a large number of states in a given energy range, as in Fig. 17.21), then, even though its zero-point energy lies above that of R, the species P might still dominate at equilibrium.

It is quite easy to show (see the *Justification* below) that the ratio of numbers of R and P molecules at equilibrium is given by

$$\frac{N_P}{N_R} = \frac{q_P}{q_R} e^{-\Delta_r E_0/RT} \qquad (17.58a)$$

and therefore that the equilibrium constant for the reaction is

$$K = \frac{q_P}{q_R} e^{-\Delta_r E_0/RT} \qquad (17.58b)$$

just as would be obtained from eqn 17.54.

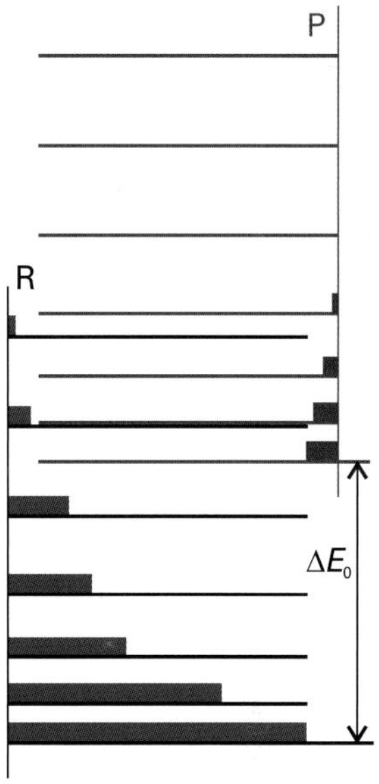

Fig. 17.20 The array of R(eactants) and P(roducts) energy levels. At equilibrium all are accessible (to differing extents, depending on the temperature), and the equilibrium composition of the system reflects the overall Bolzmann distribution of populations. As ΔE_0 increases, R becomes dominant.

Justification 17.5 *The equilibrium constant in terms of the partition function 2*

The population in a state i of the composite (R,P) system is

$$n_i = \frac{Ne^{-\beta\varepsilon_i}}{q}$$

where N is the total number of molecules. The total number of R molecules is the sum of these populations taken over the states belonging to R; these states we label r with energies ε_r. The total number of P molecules is the sum over the states

Comment 17.3

For an R $\rightleftharpoons$ P equilibrium, the V factors in the partition functions cancel, so the appearance of q in place of $q^{\ominus}$ has no effect. In the case of a more general reaction, the conversion from q to $q^{\ominus}$ comes about at the stage of converting the pressures that occur in K to numbers of molecules.

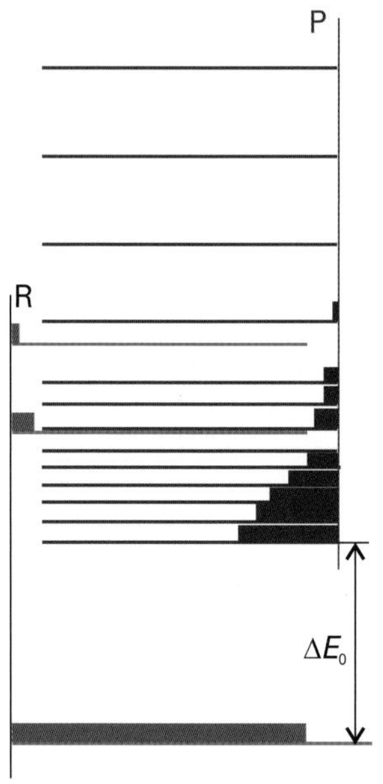

Fig. 17.21 It is important to take into account the densities of states of the molecules. Even though P might lie well above R in energy (that is, ΔE_0 is large and positive), P might have so many states that its total population dominates in the mixture. In classical thermodynamic terms, we have to take entropies into account as well as enthalpies when considering equilibria.

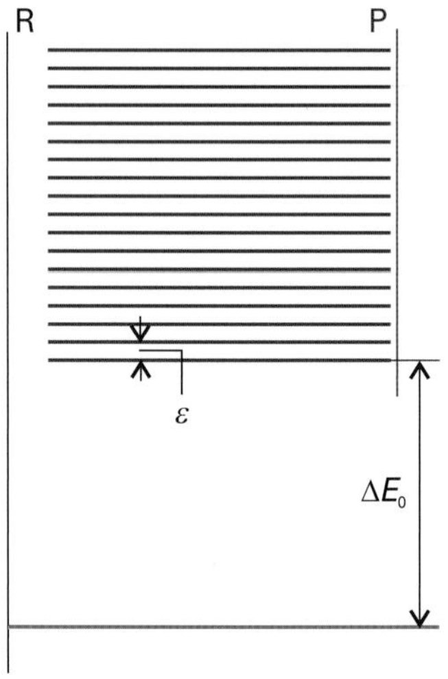

Fig. 17.22 The model used in the text for exploring the effects of energy separations and densities of states on equilibria. The products P can dominate provided ΔE_0 is not too large and P has an appreciable density of states.

belonging to P; these states we label p with energies ε'_p (the prime is explained in a moment):

$$N_\text{R}=\sum_\text{r} n_\text{r}=\frac{N}{q}\sum_\text{r} \text{e}^{-\beta\varepsilon_\text{r}} \qquad N_\text{P}=\sum_\text{p} n_\text{p}=\frac{N}{q}\sum_\text{p} \text{e}^{-\beta\varepsilon'_\text{p}}$$

The sum over the states of R is its partition function, q_R, so

$$N_\text{R}=\frac{Nq_\text{R}}{q}$$

The sum over the states of P is also a partition function, but the energies are measured from the ground state of the combined system, which is the ground state of R. However, because $\varepsilon'_\text{p}=\varepsilon_\text{p}+\Delta\varepsilon_0$, where $\Delta\varepsilon_0$ is the separation of zero-point energies (as in Fig. 17.21),

$$N_\text{P}=\frac{N}{q}\sum_\text{p} \text{e}^{-\beta(\varepsilon_\text{p}+\Delta\varepsilon_0)}=\frac{N}{q}\left(\sum_\text{p} \text{e}^{-\beta\varepsilon_\text{p}}\right)\text{e}^{-\beta\Delta\varepsilon_0}=\frac{Nq_\text{P}}{q}\text{e}^{-\Delta_\text{r}E_0/RT}$$

The switch from $\Delta\varepsilon_0/k$ to $\Delta_\text{r}E_0/R$ in the last step is the conversion of molecular energies to molar energies.

The equilibrium constant of the R ⇌ P reaction is proportional to the ratio of the numbers of the two types of molecule. Therefore,

$$K=\frac{N_\text{P}}{N_\text{R}}=\frac{q_\text{P}}{q_\text{R}}\text{e}^{-\Delta_\text{r}E_0/RT}$$

as in eqn 17.58b.

The content of eqn 17.58 can be seen most clearly by exaggerating the molecular features that contribute to it. We shall suppose that R has only a single accessible level, which implies that $q_\text{R}=1$. We also suppose that P has a large number of evenly, closely spaced levels (Fig. 17.22). The partition function of P is then $q_\text{P}=kT/\varepsilon$. In this model system, the equilibrium constant is

$$K=\frac{kT}{\varepsilon}\text{e}^{-\Delta_\text{r}E_0/RT} \qquad (17.59)$$

When $\Delta_\text{r}E_0$ is very large, the exponential term dominates and $K \ll 1$, which implies that very little P is present at equilibrium. When $\Delta_\text{r}E_0$ is small but still positive, K can exceed 1 because the factor kT/ε may be large enough to overcome the small size of the exponential term. The size of K then reflects the predominance of P at equilibrium on account of its high density of states. At low temperatures $K \ll 1$ and the system consists entirely of R. At high temperatures the exponential function approaches 1 and the pre-exponential factor is large. Hence P becomes dominant. We see that, in this endothermic reaction (endothermic because P lies above R), a rise in temperature favours P, because its states become accessible. This behaviour is what we saw, from the outside, in Chapter 7.

The model also shows why the Gibbs energy, G, and not just the enthalpy, determines the position of equilibrium. It shows that the density of states (and hence the entropy) of each species as well as their relative energies controls the distribution of populations and hence the value of the equilibrium constant.

Checklist of key ideas

☐ 1. The molecular partition function can be written as $q = q^T q^R q^V q^E$, with the contributions summarized in Table 17.3.

☐ 2. Thermodynamic functions can be expressed in terms of the partition function as summarized in Table 17.4.

☐ 3. The mean energy of a mode is $\langle \varepsilon^M \rangle = -(1/q^M)(\partial q^M/\partial\beta)_V$, with the contributions from each mode summarized in Table 17.5.

☐ 4. The contribution of a mode M to the constant-volume heat capacity is $C_V^M = -Nk\beta(\partial\langle\varepsilon^M\rangle/\partial\beta)_V$, with the contributions from each mode summarized in Table 17.5.

☐ 5. The overall heat capacity is written as $C_{V,m} = \frac{1}{2}(3 + \nu_R^* + 2\nu_V^*)R$

☐ 6. The canonical partition function of a gas is $Q = Z/\Lambda^{3N}$, where Z is the configuration integral: $Z = V^N/N!$ for a perfect gas, and $Z = (1/N!)\int e^{-\beta E_P}\,d\tau_1 d\tau_2 \ldots d\tau_N$ for a real gas.

☐ 7. In the virial equation of state, the second virial coefficient can be written as $B = -(N_A/2V)\int f\,d\tau_1 d\tau_2$, where the Mayer f-function is $f = e^{-\beta E_P} - 1$.

☐ 8. The radial distribution function, $g(r)$, where $g(r)r^2dr$, is the probability that a molecule will be found in the range dr at a distance r from another molecule. The internal energy and pressure of a fluid may be expressed in terms of the radial distribution function (eqns 17.50 and 17.51, respectively).

☐ 9. The residual entropy is a non-zero entropy at $T = 0$ arising from molecular disorder.

☐ 10. The equilibrium constant can be written in terms of the partition function (eqn 17.54).

Further reading

Articles and texts

D. Chandler, *Introduction to modern statistical mechanics.* Oxford University Press (1987).

K.A. Dill and S. Bromberg, *Molecular driving forces: statistical thermodynamics in chemistry and biology.* Garland Publishing (2002).

T.L. Hill, *An introduction to statistical thermodynamics.* Dover, New York (1986).

D.A. McQuarrie and J.D. Simon, *Molecular thermodynamics.* University Science Books, Sausalito (1999).

B. Widom, *Statistical mechanics: a concise introduction for chemists.* Cambridge University Press (2002).

Table 17.3 Contributions to the molecular partition function

Mode	Expression	Value
Translation	$q^T = \dfrac{V}{\Lambda^3}$ $\quad \Lambda = \dfrac{h}{(2\pi mkT)^{1/2}}$	$\Lambda/\text{pm} = \dfrac{1749}{(T/\text{K})^{1/2}(M/\text{g mol}^{-1})^{1/2}}$
	$\dfrac{q_m^{T\ominus}}{N_A} = \dfrac{kT}{p^\ominus\Lambda^3}$	$\dfrac{q_m^{T\ominus}}{N_A} = 2.561\times10^{-2}(T/\text{K})^{5/2}(M/\text{g mol}^{-1})^{3/2}$
Rotation		
Linear molecules	$q^R = \dfrac{kT}{\sigma hcB} = \dfrac{T}{\theta_R}$ $\quad \theta_R = \dfrac{hcB}{k}$	$q^R = \dfrac{0.6950}{\sigma}\times\dfrac{T/\text{K}}{(B/\text{cm}^{-1})}$
Nonlinear molecules	$q^R = \dfrac{1}{\sigma}\left(\dfrac{kT}{hc}\right)^{3/2}\left(\dfrac{\pi}{ABC}\right)^{1/2}$	$q^R = \dfrac{1.0270}{\sigma}\times\dfrac{(T/\text{K})^{3/2}}{(ABC/\text{cm}^{-3})^{1/2}}$
Vibration	$q^V = \dfrac{1}{1-e^{-hc\tilde{\nu}/kT}} = \dfrac{1}{1-e^{-\theta_V/T}}$	
	$\theta_V = \dfrac{hc\tilde{\nu}}{k} = \dfrac{h\nu}{k}$	
	For $T \gg \theta_V$, $q^V = \dfrac{kT}{hc\tilde{\nu}} = \dfrac{T}{\theta_V}$	$q^V = 0.695\times\dfrac{T/\text{K}}{\tilde{\nu}/\text{cm}^{-1}}$
Electronic	$q^E = g_0$ [+ higher terms] where g_0 is the degeneracy of the electronic ground state	

Note that $\beta = 1/kT$.

Table 17.4 Thermodynamic functions in terms of the partition function

Function	Expression		
	General case	Independent molecules*	
Internal energy	$U(T)-U(0)=-\left(\dfrac{\partial \ln Q}{\partial \beta}\right)_V$	$U(T)-U(0)=-N\left(\dfrac{\partial \ln q}{\partial \beta}\right)_V$	
Entropy	$S=\dfrac{U(T)-U(0)}{T}+k\ln Q$	$S=\dfrac{U(T)-U(0)}{T}+Nk\ln q$	(a)
		$S=\dfrac{U(T)-U(0)}{T}+Nk\ln\dfrac{eq}{N}$	(b)
Helmholtz energy	$A(T)-A(0)=-kT\ln Q$	$A(T)-A(0)=-NkT\ln q$	(a)
		$A(T)-A(0)=-NkT\ln\dfrac{eq}{N}$	(b)
Pressure	$p=kT\left(\dfrac{\partial \ln Q}{\partial V}\right)_T$	$p=NkT\left(\dfrac{\partial \ln q}{\partial V}\right)_T$	(b)
Enthalpy	$H(T)-H(0)=-\left(\dfrac{\partial \ln Q}{\partial \beta}\right)_V+kTV\left(\dfrac{\partial \ln Q}{\partial V}\right)_T$	$H(T)-H(0)=-N\left(\dfrac{\partial \ln q}{\partial \beta}\right)_V+NkTV\left(\dfrac{\partial \ln q}{\partial V}\right)_T$	
Gibbs energy	$G(T)-G(0)=-kT\ln Q+kTV\left(\dfrac{\partial \ln Q}{\partial V}\right)_T$	$G(T)-G(0)=-NkT\ln q+NkTV\left(\dfrac{\partial \ln q}{\partial V}\right)_T$	(a)
		$G(T)-G(0)=-NkT\ln\dfrac{eq}{N}+NkTV\left(\dfrac{\partial \ln q}{\partial V}\right)_T$	(b)

* (a) is for distinguishable particles (from $Q=q^N$), (b) for indistinquishable particles (from $Q=q^N/N!$).

Table 17.5 Contributions to mean energies and heat capacities

Mode	Expression	
	Mean energy	Heat capacity*
General mode, M	$\langle\varepsilon^{M}\rangle=-\left(\dfrac{\partial \ln q^{M}}{\partial \beta}\right)_V=-\dfrac{1}{q^{M}}\left(\dfrac{\partial q^{M}}{\partial \beta}\right)_V$	$C_V^{M}=-Nk\beta^2\left(\dfrac{\partial\langle\varepsilon^{M}\rangle}{\partial \beta}\right)_V$
Translation	$\langle\varepsilon^{T}\rangle=\frac{3}{2}kT$	$C_V^{T}=\frac{3}{2}nR$
Rotation ($T\gg\theta_R$)	$\langle\varepsilon^{R}\rangle=kT$, *linear molecules*	$C^{R}=nR$, *linear molecules*
	$\langle\varepsilon^{R}\rangle=\frac{3}{2}kT$, *nonlinear molecules*	$C^{R}=\frac{3}{2}nR$, *nonlinear molecules*
Vibration	$\langle\varepsilon^{V}\rangle=\dfrac{hc\nu}{e^{-\theta_V/T}-1}=\dfrac{h\nu}{e^{-\theta_V/T}-1}$	$C^{V}=nRf$,
		$f=\left(\dfrac{\theta_V}{T}\right)^2\dfrac{e^{-\theta_V/T}}{(1-e^{-\theta_V/T})^2}$
Vibration ($T\gg\theta_V$)	$\langle\varepsilon^{V}\rangle=kT$	$C^{V}=nR$

* No distinction need be made between C_V and C_p for internal modes.

Discussion questions

17.1 Discuss the limitations of the expressions $q^R = kT/hcB$, $q^V = kT/hc\tilde{\nu}$, and $q^E = g^E$.

17.2 Explain the origin of the symmetry number.

17.3 Explain the origin of residual entropy.

17.4 Use concepts of statistical thermodynamics to describe the molecular features that determine the magnitudes of the constant-volume molar heat capacity of a molecular substance.

17.5 Use concepts of statistical thermodynamics to describe the molecular features that lead to the equations of state of perfect and real gases.

17.6 Describe how liquids are investigated by using concepts of statistical thermodynamics.

17.7 Use concepts of statistical thermodynamics to describe the molecular features that determine the magnitudes of equilibrium constants and their variation with temperature.

Exercises

17.1a Use the equipartition theorem to estimate the constant-volume molar heat capacity of (a) I_2, (b) CH_4, (c) C_6H_6 in the gas phase at 25°C.

17.1b Use the equipartition theorem to estimate the constant-volume molar heat capacity of (a) O_3, (b) C_2H_6, (c) CO_2 in the gas phase at 25°C.

17.2a Estimate the values of $\gamma = C_p/C_V$ for gaseous ammonia and methane. Do this calculation with and without the vibrational contribution to the energy. Which is closer to the expected experimental value at 25°C?

17.2b Estimate the value of $\gamma = C_p/C_V$ for carbon dioxide. Do this calculation with and without the vibrational contribution to the energy. Which is closer to the expected experimental value at 25°C?

17.3a Estimate the rotational partition function of HCl at (a) 25°C and (b) 250°C.

17.3b Estimate the rotational partition function of O_2 at (a) 25°C and (b) 250°C.

17.4a Give the symmetry number for each of the following molecules: (a) CO, (b) O_2, (c) H_2S, and (d) SiH_4, (e) $CHCl_3$.

17.4b Give the symmetry number for each of the following molecules: (a) CO_2, (b) O_3, (c) SO_3, (d) SF_6, and (e) Al_2Cl_6.

17.5a Calculate the rotational partition function of H_2O at 298 K from its rotational constants 27.878 cm^{-1}, 14.509 cm^{-1}, and 9.287 cm^{-1}. Above what temperature is the high-temperature approximation valid to within 10 per cent of the true value?

17.5b Calculate the rotational partition function of SO_2 at 298 K from its rotational constants 2.027 36 cm^{-1}, 0.344 17 cm^{-1}, and 0.293 535 cm^{-1}. Above what temperature is the high-temperature approximation valid to within 10 per cent of the true value?

17.6a From the results of Exercise 17.5a, calculate the rotational contribution to the molar entropy of gaseous water at 25°C.

17.6b From the results of Exercise 17.5b, calculate the rotational contribution to the molar entropy of sulfur dioxide at 25°C.

17.7a Calculate the rotational partition function of CH_4 (a) by direct summation of the energy levels at 298 K and 500 K, and (b) by the high-temperature approximation. Take $B = 5.2412$ cm^{-1}.

17.7b Calculate the rotational partition function of CH_3CN (a) by direct summation of the energy levels at 298 K and 500 K, and (b) by the high-temperature approximation. Take $A = 5.28$ cm^{-1} and $B = 0.307$ cm^{-1}.

17.8a The bond length of O_2 is 120.75 pm. Use the high-temperature approximation to calculate the rotational partition function of the molecule at 300 K.

17.8b The NOF molecule is an asymmetric rotor with rotational constants 3.1752 cm^{-1}, 0.3951 cm^{-1}, and 0.3505 cm^{-1}. Calculate the rotational partition function of the molecule at (a) 25°C, (b) 100°C.

17.9a Plot the molar heat capacity of a collection of harmonic oscillators as a function of T/θ_V, and predict the vibrational heat capacity of ethyne at (a) 298 K, (b) 500 K. The normal modes (and their degeneracies in parentheses) occur at wavenumbers 612(2), 729(2), 1974, 3287, and 3374 cm^{-1}.

17.9b Plot the molar entropy of a collection of harmonic oscillators as a function of T/θ_V, and predict the standard molar entropy of ethyne at (a) 298 K, (b) 500 K. For data, see the preceding exercise.

17.10a A CO_2 molecule is linear, and its vibrational wavenumbers are 1388.2 cm^{-1}, 667.4 cm^{-1}, and 2349.2 cm^{-1}, the last being doubly degenerate and the others non-degenerate. The rotational constant of the molecule is 0.3902 cm^{-1}. Calculate the rotational and vibrational contributions to the molar Gibbs energy at 298 K.

17.10b An O_3 molecule is angular, and its vibrational wavenumbers are 1110 cm^{-1}, 705 cm^{-1}, and 1042 cm^{-1}. The rotational constants of the molecule are 3.553 cm^{-1}, 0.4452 cm^{-1}, and 0.3948 cm^{-1}. Calculate the rotational and vibrational contributions to the molar Gibbs energy at 298 K.

17.11a The ground level of Cl is $^2P_{3/2}$ and a $^2P_{1/2}$ level lies 881 cm^{-1} above it. Calculate the electronic contribution to the heat capacity of Cl atoms at (a) 500 K and (b) 900 K.

17.11b The first electronically excited state of O_2 is $^1\Delta_g$ and lies 7918.1 cm^{-1} above the ground state, which is $^3\Sigma_g^-$. Calculate the electronic contribution to the molar Gibbs energy of O_2 at 400 K.

17.12a The ground state of the Co^{2+} ion in $CoSO_4{\cdot}7H_2O$ may be regarded as $^4T_{9/2}$. The entropy of the solid at temperatures below 1 K is derived almost entirely from the electron spin. Estimate the molar entropy of the solid at these temperatures.

17.12b Estimate the contribution of the spin to the molar entropy of a solid sample of a *d*-metal complex with $S = \frac{5}{2}$.

17.13a Calculate the residual molar entropy of a solid in which the molecules can adopt (a) three, (b) five, (c) six orientations of equal energy at $T = 0$.

17.13b Suppose that the hexagonal molecule $C_6H_nF_{6-n}$ has a residual entropy on account of the similarity of the H and F atoms. Calculate the residual for each value of n.

17.14a Calculate the equilibrium constant of the reaction $I_2(g) \rightleftharpoons 2\,I(g)$ at 1000 K from the following data for I_2: $\tilde{\nu} = 214.36$ cm^{-1}, $B = 0.0373$ cm^{-1}, $D_e = 1.5422$ eV. The ground state of the I atoms is $^2P_{3/2}$, implying fourfold degeneracy.

17.14b Calculate the value of K at 298 K for the gas-phase isotopic exchange reaction $2\,{}^{79}Br^{81}Br\;{}^{79}Br^{79} \rightleftharpoons {}^{79}Br^{79}Br + {}^{81}Br^{81}Br$. The Br_2 molecule has a non-degenerate ground state, with no other electronic states nearby. Base the calculation on the wavenumber of the vibration of ${}^{79}Br^{81}Br$, which is 323.33 cm^{-1}.

Problems*

Numerical problems

17.1 The NO molecule has a doubly degenerate electronic ground state and a doubly degenerate excited state at 121.1 cm^{-1}. Calculate the electronic contribution to the molar heat capacity of the molecule at (a) 50 K, (b) 298 K, and (c) 500 K.

17.2 Explore whether a magnetic field can influence the heat capacity of a paramagnetic molecule by calculating the electronic contribution to the heat capacity of an NO_2 molecule in a magnetic field. Estimate the total constant-volume heat capacity using equipartition, and calculate the percentage change in heat capacity brought about by a 5.0 T magnetic field at (a) 50 K, (b) 298 K.

17.3 The energy levels of a CH_3 group attached to a larger fragment are given by the expression for a particle on a ring, provided the group is rotating freely. What is the high-temperature contribution to the heat capacity and entropy of such a freely rotating group at 25°C? The moment of inertia of CH_3 about its three-fold rotation axis (the axis that passes through the C atom and the centre of the equilateral triangle formed by the H atoms) is 5.341×10^{-47} kg m^2).

17.4 Calculate the temperature dependence of the heat capacity of p-H_2 (in which only rotational states with even values of J are populated) at low temperatures on the basis that its rotational levels $J = 0$ and $J = 2$ constitute a system that resembles a two-level system except for the degeneracy of the upper level. Use $B = 60.864$ cm^{-1} and sketch the heat capacity curve. The experimental heat capacity of p-H_2 does in fact show a peak at low temperatures.

17.5 The pure rotational microwave spectrum of HCl has absorption lines at the following wavenumbers (in cm^{-1}): 21.19, 42.37, 63.56, 84.75, 105.93, 127.12 148.31 169.49, 190.68, 211.87, 233.06, 254.24, 275.43, 296.62, 317.80, 338.99, 360.18, 381.36, 402.55, 423.74, 444.92, 466.11, 487.30, 508.48. Calculate the rotational partition function at 25°C by direct summation.

17.6 Calculate the standard molar entropy of N_2(g) at 298 K from its rotational constant $B = 1.9987$ cm^{-1} and its vibrational wavenumber $\tilde{\nu} = 2358$ cm^{-1}. The thermochemical value is 192.1 J K^{-1} mol^{-1}. What does this suggest about the solid at $T = 0$?

17.7‡ J.G. Dojahn, E.C.M. Chen, and W.E. Wentworth (*J. Phys. Chem.* **100**, 9649 (1996)) characterized the potential energy curves of the ground and electronic states of homonuclear diatomic halogen anions. The ground state of F_2^- is $^2\Sigma_u^+$ with a fundamental vibrational wavenumber of 450.0 cm^{-1} and equilibrium internuclear distance of 190.0 pm. The first two excited states are at 1.609 and 1.702 eV above the ground state. Compute the standard molar entropy of F_2^- at 298 K.

17.8‡ In a spectroscopic study of buckminsterfullerene C_{60}, F. Negri, G. Orlandi, and F. Zerbetto (*J. Phys. Chem.* **100**, 10849 (1996)) reviewed the wavenumbers of all the vibrational modes of the molecule. The wavenumber for the single A_u mode is 976 cm^{-1}; wavenumbers for the four threefold degenerate T_{1u} modes are 525, 578, 1180, and 1430 cm^{-1}; wavenumbers for the five threefold degenerate T_{2u} modes are 354, 715, 1037, 1190, and 1540 cm^{-1}; wavenumbers for the six fourfold degenerate G_u modes are 345, 757, 776, 963, 1315, and 1410 cm^{-1}; and wavenumbers for the seven fivefold degenerate H_u modes are 403, 525, 667, 738, 1215, 1342, and 1566 cm^{-1}. How many modes have a vibrational temperature θ_V below 1000 K? Estimate the molar constant-volume heat capacity of C_{60} at 1000 K, counting as active all modes with θ_V below this temperature.

17.9‡ Treat carbon monoxide as a perfect gas and apply equilibrium statistical thermodynamics to the study of its properties, as specified below, in the temperature range 100–1000 K at 1 bar. $\tilde{\nu} = 2169.8$ cm^{-1}, $B = 1.931$ cm^{-1}, and $D_0 = 11.09$ eV; neglect anharmonicity and centrifugal distortion. (a) Examine the probability distribution of molecules over available rotational and vibrational states. (b) Explore numerically the differences, if any, between the rotational molecular partition function as calculated with the discrete energy distribution and that calculated with the classical, continuous energy distribution. (c) Calculate the individual contributions to $U_m(T) - U_m(100\text{ K})$, $C_{V,m}(T)$, and $S_m(T) - S_m(100\text{ K})$ made by the translational, rotational, and vibrational degrees of freedom.

17.10 Calculate and plot as a function of temperature, in the range 300 K to 1000 K, the equilibrium constant for the reaction $CD_4(g) + HCl(g) \rightleftharpoons CHD_3(g) + DCl(g)$ using the following data (numbers in parentheses are degeneracies): $\tilde{\nu}(CHD_3)/cm^{-1} = 2993(1), 2142(1), 1003(3), 1291(2), 1036(2)$; $\tilde{\nu}(CD_4)/cm^{-1} = 2109(1), 1092(2), 2259(3), 996(3)$; $\tilde{\nu}(HCl)/cm^{-1} = 2991$; $\tilde{\nu}(DCl)/cm^{-1} = 2145$; $B(HCl)/cm^{-1} = 10.59$; $B(DCl)/cm^{-1} = 5.445$; $B(CHD_3)/cm^{-1} = 3.28$; $A(CHD_3)/cm^{-1} = 2.63$, $B(CD_4)/cm^{-1} = 2.63$.

17.11 The exchange of deuterium between acid and water is an important type of equilibrium, and we can examine it using spectroscopic data on the molecules. Calculate the equilibrium constant at (a) 298 K and (b) 800 K for the gas-phase exchange reaction $H_2O + DCl \rightleftharpoons HDO + HCl$ from the following data: $\tilde{\nu}(H_2O)/cm^{-1} = 3656.7, 1594.8, 3755.8$; $\tilde{\nu}(HDO)/cm^{-1} = 2726.7, 1402.2, 3707.5$; $A(H_2O)/cm^{-1} = 27.88$; $B(H_2O)/cm^{-1} = 14.51$; $C(H_2O)/cm^{-1} = 9.29$; $A(HDO)/cm^{-1} = 23.38$; $B(HDO)/cm^{-1} = 9.102$; $C(HDO)/cm^{-1} = 6.417$; $B(HCl)/cm^{-1} = 10.59$; $B(DCl)/cm^{-1} = 5.449$; $\tilde{\nu}(HCl)/cm^{-1} = 2991$; $\tilde{\nu}(DCl)/cm^{-1} = 2145$.

Theoretical problems

17.12 Derive the Sackur–Tetrode equation for a monatomic gas confined to a two-dimensional surface, and hence derive an expression for the standard molar entropy of condensation to form a mobile surface film.

17.13‡ For H_2 at very low temperatures, only translational motion contributes to the heat capacity. At temperatures above $\theta_R = hcB/k$, the rotational contribution to the heat capacity becomes significant. At still higher temperatures, above $\theta_V = h\nu/k$, the vibrations contribute. But at this latter temperature, dissociation of the molecule into the atoms must be considered. (a) Explain the origin of the expressions for θ_R and θ_V, and calculate their values for hydrogen. (b) Obtain an expression for the molar constant-pressure heat capacity of hydrogen at all temperatures taking into account the dissociation of hydrogen. (c) Make a plot of the molar constant-pressure heat capacity as a function of temperature in the high-temperature region where dissociation of the molecule is significant.

17.14 Derive expressions for the internal energy, heat capacity, entropy, Helmholtz energy, and Gibbs energy of a harmonic oscillator. Express the results in terms of the vibrational temperature, θ_V and plot graphs of each property against T/θ_V.

17.15 Suppose that an intermolecular potential has a hard-sphere core of radius r_1 and a shallow attractive well of uniform depth ε out to a distance r_2. Show, by using eqn 17.42 and the condition $\varepsilon \ll kT$, that such a model is approximately consistent with a van der Waals equation of state when $b \ll V_m$, and relate the van der Waals parameters and the Joule–Thomson coefficient to the parameters in this model.

* Problems denoted with the symbol ‡ were supplied by Charles Trapp, Carmen Giunta, and Marshall Cady.

17.16‡ (a) Show that the number of molecules in any given rotational state of a linear molecule is given by $N_J = C(2J+1)e^{-hcBJ(J+1)/kT}$, where C is independent of J. (b) Use this result to derive eqn 13.39 for the value of J of the most highly populated rotational level. (c) Estimate the temperature at which the spectrum of HCl shown in Fig. 13.44 was obtained. (d) What is the most highly populated level of a spherical rotor at a temperature T?

17.17 A more formal way of arriving at the value of the symmetry number is to note that σ is the order (the number of elements) of the *rotational subgroup* of the molecule, the point group of the molecule with all but the identity and the rotations removed. The rotational subgroup of H_2O is $\{E, C_2\}$, so $\sigma = 2$. The rotational subgroup of NH_3 is $\{E, 2C_3\}$, so $\sigma = 3$. This recipe makes it easy to find the symmetry numbers for more complicated molecules. The rotational subgroup of CH_4 is obtained from the T character table as $\{E, 8C_3, 3C_2\}$, so $\sigma = 12$. For benzene, the rotational subgroup of D_{6h} is $\{E, 2C_6, 2C_3, C_2, 3C_2', 3C_2''\}$, so $\sigma = 12$. (a) Estimate the rotational partition function of ethene at 25°C given that $A = 4.828\ \text{cm}^{-1}$, $B = 1.0012\ \text{cm}^{-1}$, and $C = 0.8282\ \text{cm}^{-1}$. (b) Evaluate the rotational partition function of pyridine, C_5H_5N, at room temperature ($A = 0.2014\ \text{cm}^{-1}$, $B = 0.1936\ \text{cm}^{-1}$, $C = 0.0987\ \text{cm}^{-1}$).

17.18 Although expressions like $\langle\varepsilon\rangle = -\text{d}\ln q/\text{d}\beta$ are useful for formal manipulations in statistical thermodynamics, and for expressing thermodynamic functions in neat formulas, they are sometimes more trouble than they are worth in practical applications. When presented with a table of energy levels, it is often much more convenient to evaluate the following sums directly:

$$q = \sum_j e^{-\beta\varepsilon_j} \qquad \dot{q} = \sum_j \beta\varepsilon_j e^{-\beta\varepsilon_j} \qquad \ddot{q} = \sum_j (\beta\varepsilon_j)^2 e^{-\beta\varepsilon_j}$$

(a) Derive expressions for the internal energy, heat capacity, and entropy in terms of these three functions. (b) Apply the technique to the calculation of the electronic contribution to the constant-volume molar heat capacity of magnesium vapour at 5000 K using the following data:

Term	^{1}S	^{3}P$_0$	^{3}P$_1$	^{3}P$_2$	^{1}P$_1$	^{3}S$_1$
Degeneracy	1	1	3	5	3	3
$\tilde{\nu}$/cm^{-1}	0	21 850	21 870	21 911	35 051	41 197

17.19 Show how the heat capacity of a linear rotor is related to the following sum:

$$\zeta(\beta) = \frac{1}{q^2}\sum_{J,J'}\{\varepsilon(J) - \varepsilon(J')\}^2 g(J)g(J')e^{-\beta\{\varepsilon(J)+\varepsilon(J')\}}$$

by

$$C = \tfrac{1}{2}Nk\beta^2\zeta(\beta)$$

where the $\varepsilon(J)$ are the rotational energy levels and $g(J)$ their degeneracies. Then go on to show graphically that the total contribution to the heat capacity of a linear rotor can be regarded as a sum of contributions due to transitions 0→1, 0→2, 1→2, 1→3, etc. In this way, construct Fig. 17.11 for the rotational heat capacities of a linear molecule.

17.20 Set up a calculation like that in Problem 17.19 to analyse the vibrational contribution to the heat capacity in terms of excitations between levels and illustrate your results graphically in terms of a diagram like that in Fig. 17.11.

17.21 Determine whether a magnetic field can influence the value of an equilibrium constant. Consider the equilibrium $I_2(g) \rightleftharpoons 2\,I(g)$ at 1000 K, and calculate the ratio of equilibrium constants $K(\mathcal{B})/K$, where $K(\mathcal{B})$ is the equilibrium constant when a magnetic field $\mathcal{B}$ is present and removes the degeneracy of the four states of the $^2P_{3/2}$ level. Data on the species are given in Exercise 17.14a. The electronic g value of the atoms is $\frac{4}{3}$. Calculate the field required to change the equilibrium constant by 1 per cent.

17.22 The heat capacity ratio of a gas determines the speed of sound in it through the formula $c_s = (\gamma RT/M)^{1/2}$, where $\gamma = C_p/C_V$ and M is the molar mass of the gas. Deduce an expression for the speed of sound in a perfect gas of (a) diatomic, (b) linear triatomic, (c) nonlinear triatomic molecules at high temperatures (with translation and rotation active). Estimate the speed of sound in air at 25°C.

Applications: to biology, materials science, environmental science, and astrophysics

17.23 An average human DNA molecule has 5×10^8 binucleotides (rungs on the DNA ladder) of four different kinds. If each rung were a random choice of one of these four possibilities, what would be the residual entropy associated with this typical DNA molecule?

17.24 It is possible to write an approximate expression for the partition function of a protein molecule by including contributions from only two states: the native and denatured forms of the polymer. Proceeding with this crude model gives us insight into the contribution of denaturation to the heat capacity of a protein. It follows from *Illustration* 16.4 that the total energy of a system of N protein molecules is

$$E = \frac{N\varepsilon e^{-\varepsilon/kT}}{1 + e^{-\varepsilon/kT}}$$

where ε is the energy separation between the denatured and native forms. (a) Show that the constant-volume molar heat capacity is

$$C_{V,\text{m}} = \frac{R(\varepsilon_\text{m}/RT)^2 e^{-\varepsilon_\text{m}/RT}}{(1 + e^{-\varepsilon_\text{m}/RT})^2}$$

Hint. For two functions f and g, the quotient rule of differentiation states that $\text{d}(f/g)/\text{d}x = (1/g)\text{d}f/\text{d}x - (f/g^2)\text{d}g/\text{d}x$. (b) Plot the variation of $C_{V,\text{m}}$ with temperature. (c) If the function $C_{V,\text{m}}(T)$ has a maximum or minimum, derive an expression for the temperature at which it occurs.

17.25‡ R. Viswanathan, R.W. Schmude, Jr., and K.A. Gingerich (*J. Phys. Chem.* **100**, 10784 (1996)) studied thermodynamic properties of several boron–silicon gas-phase species experimentally and theoretically. These species can occur in the high-temperature chemical vapour deposition (CVD) of silicon-based semiconductors. Among the computations they reported was computation of the Gibbs energy of BSi(g) at several temperatures based on a $^4\Sigma^-$ ground state with equilibrium internuclear distance of 190.5 pm and fundamental vibrational wavenumber of 772 cm^{-1} and a $^2P_0$ first excited level 8000 cm^{-1} above the ground level. Compute the standard molar Gibbs energy $G_\text{m}^\ominus(2000\ \text{K}) - G_\text{m}^\ominus(0)$.

17.26‡ The molecule Cl_2O_2, which is believed to participate in the seasonal depletion of ozone over Antarctica, has been studied by several means. M. Birk, R.R. Friedl, E.A. Cohen, H.M. Pickett, and S.P. Sander (*J. Chem. Phys.* **91**, 6588 (1989)) report its rotational constants (actually cB) as 13 109.4, 2409.8, and 2139.7 MHz. They also report that its rotational spectrum indicates a molecule with a symmetry number of 2. J. Jacobs, M. Kronberg, H.S.P. Möller, and H. Willner (*J. Amer. Chem. Soc.* **116**, 1106 (1994)) report its vibrational wavenumbers as 753, 542, 310, 127, 646, and 419 cm^{-1}. Compute $G_\text{m}^\ominus(200\ \text{K}) - G_\text{m}^\ominus(0)$ of Cl_2O_2.

17.27‡ J. Hutter, H.P. Lüthi, and F. Diederich (*J. Amer. Chem. Soc.* **116**, 750 (1994)) examined the geometric and vibrational structure of several carbon molecules of formula C_n. Given that the ground state of C_3, a molecule found in interstellar space and in flames, is an angular singlet with moments of inertia 39.340, 39.032, and 0.3082 u Å^2 (where 1 Å = 10^{-10} m) and with vibrational wavenumbers of 63.4, 1224.5, and 2040 cm^{-1}, compute $G_\text{m}^\ominus(10.00\ \text{K}) - G_\text{m}^\ominus(0)$ and $G_\text{m}^\ominus(1000\ \text{K}) - G_\text{m}^\ominus(0)$ for C_3.

18 Molecular interactions

In this chapter we examine molecular interactions in gases and liquids and interpret them in terms of electric properties of molecules, such as electric dipole moments and polarizabilities. All these properties reflect the degree to which the nuclei of atoms exert control over the electrons in a molecule, either by causing electrons to accumulate in particular regions, or by permitting them to respond more or less strongly to the effects of external electric fields.

Molecular interactions are responsible for the unique properties of substances as simple as water and as complex as polymers. We begin our examination of molecular interactions by describing the electric properties of molecules, which may be interpreted in terms of concepts in electronic structure introduced in Chapter 11. We shall see that small imbalances of charge distributions in molecules allow them to interact with one another and with externally applied fields. One result of this interaction is the cohesion of molecules to form the bulk phases of matter. These interactions are also important for understanding the shapes adopted by biological and synthetic macromolecules, as we shall see in Chapter 19. The interaction between ions is treated in Chapter 5 (for solutions) and Chapter 20 (for solids).

Electric properties of molecules

Many of the electric properties of molecules can be traced to the competing influences of nuclei with different charges or the competition between the control exercised by a nucleus and the influence of an externally applied field. The former competition may result in an electric dipole moment. The latter may result in properties such as refractive index and optical activity.

18.1 Electric dipole moments

An **electric dipole** consists of two electric charges $+q$ and $-q$ separated by a distance $\boldsymbol{R}$. This arrangement of charges is represented by a vector $\boldsymbol{\mu}$ (**1**). The magnitude of $\boldsymbol{\mu}$ is $\mu = qR$ and, although the SI unit of dipole moment is coulomb metre (C m), it is still commonly reported in the non-SI unit debye, D, named after Peter Debye, a pioneer in the study of dipole moments of molecules, where

$$1\ \text{D} = 3.335\,64 \times 10^{-30}\ \text{C m} \qquad (18.1)$$

The dipole moment of a pair of charges $+e$ and $-e$ separated by 100 pm is 1.6×10^{-29} C m, corresponding to 4.8 D. Dipole moments of small molecules are typically about 1 D. The conversion factor in eqn 18.1 stems from the original definition of the debye in terms of c.g.s. units: 1 D is the dipole moment of two equal and opposite charges of magnitude 1 e.s.u. separated by 1 Å.

(a) Polar molecules

A **polar molecule** is a molecule with a permanent electric dipole moment. The permanent dipole moment stems from the partial charges on the atoms in the molecule that arise from differences in electronegativity or other features of bonding (Section 11.6). Nonpolar molecules acquire an induced dipole moment in an electric field on account of the distortion the field causes in their electronic distributions and nuclear positions; however, this induced moment is only temporary, and disappears as soon as the perturbing field is removed. Polar molecules also have their existing dipole moments temporarily modified by an applied field.

The Stark effect (Section 13.5) is used to measure the electric dipole moments of molecules for which a rotational spectrum can be observed. In many cases microwave spectroscopy cannot be used because the sample is not volatile, decomposes on vaporization, or consists of molecules that are so complex that their rotational spectra cannot be interpreted. In such cases the dipole moment may be obtained by measurements on a liquid or solid bulk sample using a method explained later. Computational software is now widely available, and typically computes electric dipole moments by assessing the electron density at each point in the molecule and its coordinates relative to the centroid of the molecule; however, it is still important to be able to formulate simple models of the origin of these moments and to understand how they arise. The following paragraphs focus on this aspect.

All heteronuclear diatomic molecules are polar, and typical values of μ include 1.08 D for HCl and 0.42 D for HI (Table 18.1). Molecular symmetry is of the greatest importance in deciding whether a polyatomic molecule is polar or not. Indeed, molecular symmetry is more important than the question of whether or not the atoms in the molecule belong to the same element. Homonuclear polyatomic molecules may be polar if they have low symmetry and the atoms are in inequivalent positions. For instance, the angular molecule ozone, O_3 (**2**), is homonuclear; however, it is polar because the central O atom is different from the outer two (it is bonded to two atoms, they are bonded only to one); moreover, the dipole moments associated with each bond make an angle to each other and do not cancel. Heteronuclear polyatomic molecules may be nonpolar if they have high symmetry, because individual bond dipoles may then cancel. The heteronuclear linear triatomic molecule CO_2, for example, is nonpolar because, although there are partial charges on all three atoms, the dipole moment associated with the OC bond points in the opposite direction to the dipole moment associated with the CO bond, and the two cancel (**3**).

To a first approximation, it is possible to resolve the dipole moment of a polyatomic molecule into contributions from various groups of atoms in the molecule and the directions in which these individual contributions lie (Fig. 18.1). Thus,

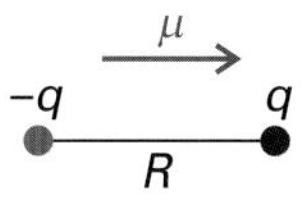

1 Electric dipole

Comment 18.1

In elementary chemistry, an electric dipole moment is represented by the arrow +→ added to the Lewis structure for the molecule, with the + marking the positive end. Note that the direction of the arrow is opposite to that of $\boldsymbol{\mu}$.

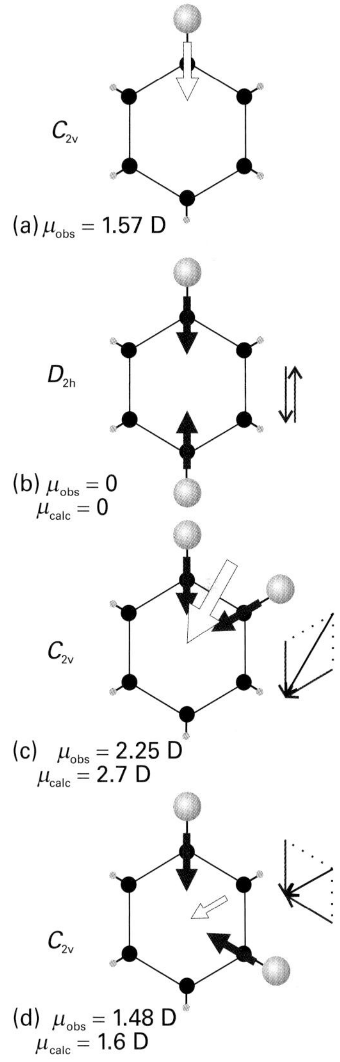

Fig. 18.1 The resultant dipole moments (pale yellow) of the dichlorobenzene isomers (b to d) can be obtained approximately by vectorial addition of two chlorobenzene dipole moments (1.57 D), purple.

Synoptic table 18.1* Dipole moments (μ) and polarizability volumes (α')

	μ/D	$\alpha'/(10^{-30}\ m^3)$
CCl_4	0	10.5
H_2	0	0.819
H_2O	1.85	1.48
HCl	1.08	2.63
HI	0.42	5.45

* More values are given in the *Data section*.

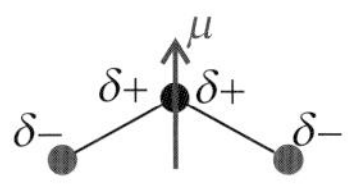

2 Ozone, O_3

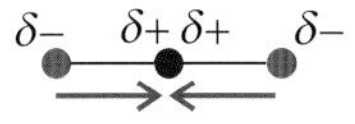

3 Carbon dioxide, CO_2

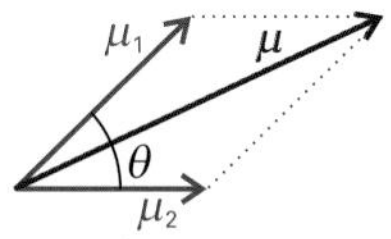

4 Addition of dipole moments

Comment 18.2

Operations involving vectors are described in *Appendix* 2, where eqn 18.2 is also derived.

Comment 18.3

In three dimensions, a vector μ has components μ_x, μ_y, and μ_z along the x-, y-, and z-axes, respectively, as shown in the illustration. The direction of each of the components is denoted with a plus sign or minus sign. For example, if $\mu_x = -1.0$ D, the x-component of the vector μ has a magnitude of 1.0 D and points in the $-x$ direction.

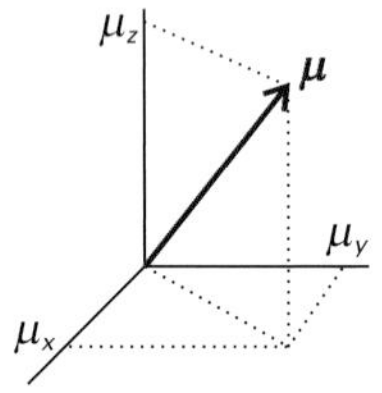

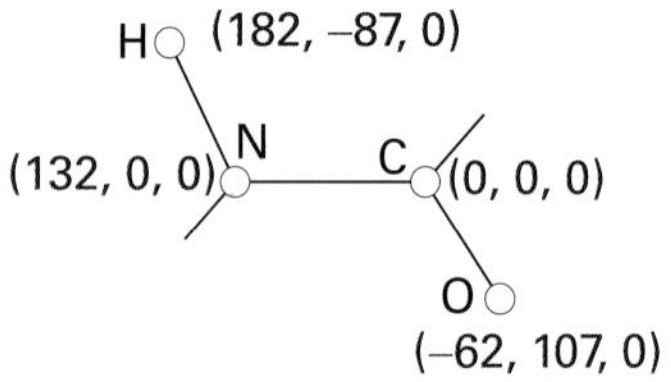

5 Amide group

Table 18.2 Partial charges in polypeptides

Atom	Partial charge/e
C(=O)	+0.45
C(—CO)	+0.06
H(—C)	+0.02
H(—N)	+0.18
H(—O)	+0.42
N	−0.36
O	−0.38

1,4-dichlorobenzene is nonpolar by symmetry on account of the cancellation of two equal but opposing C—Cl moments (exactly as in carbon dioxide). 1,2-Dichlorobenzene, however, has a dipole moment which is approximately the resultant of two chlorobenzene dipole moments arranged at 60° to each other. This technique of 'vector addition' can be applied with fair success to other series of related molecules, and the resultant μ_{res} of two dipole moments μ_1 and μ_2 that make an angle θ to each other (**4**) is approximately

$$\mu_{res} \approx (\mu_1^2 + \mu_2^2 + 2\mu_1\mu_2 \cos\theta)^{1/2} \tag{18.2a}$$

When the two dipole moments have the same magnitude (as in the dichlorobenzenes), this equation simplifies to

$$\mu_{res} \approx 2\mu_1 \cos\tfrac{1}{2}\theta \tag{18.2b}$$

Self-test 18.1 Estimate the ratio of the electric dipole moments of *ortho* (1,2-) and *meta* (1,3-) disubstituted benzenes. [$\mu(ortho)/\mu(meta) = 1.7$]

A better approach to the calculation of dipole moments is to take into account the locations and magnitudes of the partial charges on all the atoms. These partial charges are included in the output of many molecular structure software packages. To calculate the x-component, for instance, we need to know the partial charge on each atom and the atom's x-coordinate relative to a point in the molecule and form the sum

$$\mu_x = \sum_J q_J x_J \tag{18.3a}$$

Here q_J is the partial charge of atom J, x_J is the x-coordinate of atom J, and the sum is over all the atoms in the molecule. Analogous expressions are used for the y- and z-components. For an electrically neutral molecule, the origin of the coordinates is arbitrary, so it is best chosen to simplify the measurements. In common with all vectors, the magnitude of μ is related to the three components μ_x, μ_y, and μ_z by

$$\mu = (\mu_x^2 + \mu_y^2 + \mu_z^2)^{1/2} \tag{18.3b}$$

Example 18.1 *Calculating a molecular dipole moment*

Estimate the electric dipole moment of the amide group shown in (**5**) by using the partial charges (as multiples of e) in Table 18.2 and the locations of the atoms shown.

Method We use eqn 18.3a to calculate each of the components of the dipole moment and then eqn 18.3b to assemble the three components into the magnitude of the dipole moment. Note that the partial charges are multiples of the fundamental charge, $e = 1.609 \times 10^{-19}$ C.

Answer The expression for μ_x is

$$\begin{aligned}\mu_x &= (-0.36e) \times (132\text{ pm}) + (0.45e) \times (0\text{ pm}) + (0.18e) \times (182\text{ pm}) \\ &\quad + (-0.38e) \times (-62.0\text{ pm}) \\ &= 8.8e\text{ pm} \\ &= 8.8 \times (1.609 \times 10^{-19}\text{ C}) \times (10^{-12}\text{ m}) = 1.4 \times 10^{-30}\text{ C m}\end{aligned}$$

corresponding to $\mu_x = 0.42$ D. The expression for μ_y is:

$$\mu_y = (-0.36e) \times (0\text{ pm}) + (0.45e) \times (0\text{ pm}) + (0.18e) \times (-86.6\text{ pm}) + (-0.38e) \times (107\text{ pm})$$
$$= -56e\text{ pm} = -9.1 \times 10^{-30}\text{ C m}$$

It follows that $\mu_y = -2.7$ D. Therefore, because $\mu_z = 0$,

$$\mu = \{(0.42\text{ D})^2 + (-2.7\text{ D})^2\}^{1/2} = 2.7\text{ D}$$

We can find the orientation of the dipole moment by arranging an arrow of length 2.7 units of length to have *x*, *y*, and *z* components of 0.42, −2.7, and 0 units; the orientation is superimposed on (**6**).

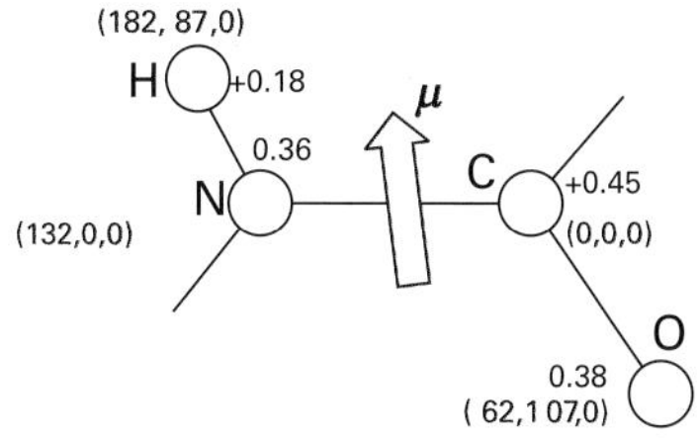

6

Self-test 18.2 Calculate the electric dipole moment of formaldehyde, using the information in (**7**). [−3.2 D]

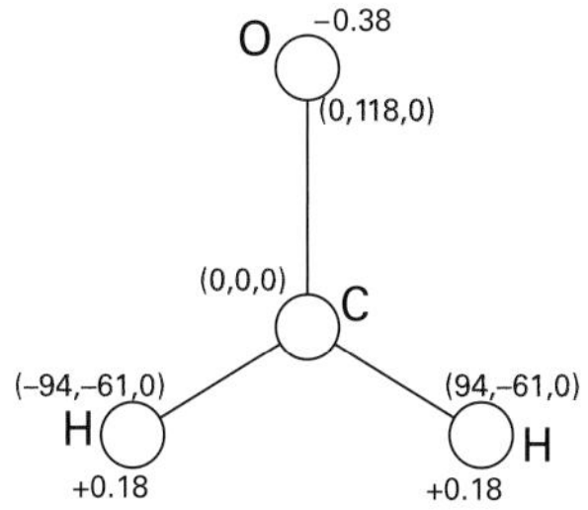

7

(b) Polarization

The **polarization**, *P*, of a sample is the electric dipole moment density, the mean electric dipole moment of the molecules, $\langle\mu\rangle$, multiplied by the number density, $\mathcal{N}$:

$$P = \langle\mu\rangle\mathcal{N} \qquad (18.4)$$

In the following pages we refer to the sample as a **dielectric**, by which is meant a polarizable, nonconducting medium.

The polarization of an isotropic fluid sample is zero in the absence of an applied field because the molecules adopt random orientations, so $\langle\mu\rangle = 0$. In the presence of a field, the dipoles become partially aligned because some orientations have lower energies than others. As a result, the electric dipole moment density is nonzero. We show in the *Justification* below that, at a temperature *T*

$$\langle\mu_z\rangle = \frac{\mu^2\mathcal{E}}{3kT} \qquad (18.5)$$

where *z* is the direction of the applied field $\mathcal{E}$. Moreover, as we shall see, there is an additional contribution from the dipole moment induced by the field.

Justification 18.1 *The thermally averaged dipole moment*

The probability d*p* that a dipole has an orientation in the range θ to $\theta + \mathrm{d}\theta$ is given by the Boltzmann distribution (Section 16.1b), which in this case is

$$\mathrm{d}p = \frac{\mathrm{e}^{-E(\theta)/kT}\sin\theta\,\mathrm{d}\theta}{\int_0^\pi \mathrm{e}^{-E(\theta)/kT}\sin\theta\,\mathrm{d}\theta}$$

where $E(\theta)$ is the energy of the dipole in the field: $E(\theta) = -\mu\mathcal{E}\cos\theta$, with $0 \le \theta \le \pi$. The average value of the component of the dipole moment parallel to the applied electric field is therefore

$$\langle\mu_z\rangle = \int \mu\cos\theta\,\mathrm{d}p = \mu\int\cos\theta\,\mathrm{d}p = \frac{\mu\int_0^\pi \mathrm{e}^{x\cos\theta}\cos\theta\sin\theta\,\mathrm{d}\theta}{\int_0^\pi \mathrm{e}^{x\cos\theta}\sin\theta\,\mathrm{d}\theta}$$

with $x=\mu\mathcal{E}/kT$. The integral takes on a simpler appearance when we write $y=\cos\theta$ and note that $dy=-\sin\theta\,d\theta$.

$$\langle\mu_z\rangle=\frac{\mu\int_{-1}^{1} y e^{xy}dy}{\int_{-1}^{1} e^{xy}dy}$$

At this point we use

$$\int_{-1}^{1} e^{xy}dy=\frac{e^x-e^{-x}}{x} \qquad \int_{-1}^{1} ye^{xy}dy=\frac{e^x+e^{-x}}{x}-\frac{e^x-e^{-x}}{x^2}$$

It is now straightforward algebra to combine these two results and to obtain

$$\langle\mu_z\rangle=\mu L(x) \qquad L(x)=\frac{e^x+e^{-x}}{e^x-e^{-x}}-\frac{1}{x} \qquad x=\frac{\mu\mathcal{E}}{kT} \qquad (18.6)$$

$L(x)$ is called the **Langevin function**.

Under most circumstances, x is very small (for example, if $\mu=1$ D and $T=300$ K, then x exceeds 0.01 only if the field strength exceeds 100 kV cm^{-1}, and most measurements are done at much lower strengths). When $x \ll 1$, the exponentials in the Langevin function can be expanded, and the largest term that survives is

$$L(x)=\tfrac{1}{3}x+\cdots \qquad (18.7)$$

Therefore, the average molecular dipole moment is given by eqn 18.6.

Comment 18.4

When x is small, it is possible to simplify expressions by using the expansion $e^x = 1+x+\frac{1}{2}x^2+\frac{1}{6}x^3+\cdots$; it is important when developing approximations that all terms of the same order are retained because low-order terms might cancel.

18.2 Polarizabilities

An applied electric field can distort a molecule as well as align its permanent electric dipole moment. The **induced dipole moment**, μ^*, is generally proportional to the field strength, $\mathcal{E}$, and we write

$$\mu^*=\alpha\mathcal{E} \qquad (18.8)$$

(See Section 20.10 for exceptions to eqn 18.8.) The constant of proportionality α is the **polarizability** of the molecule. The greater the polarizability, the larger is the induced dipole moment for a given applied field. In a formal treatment, we should use vector quantities and allow for the possibility that the induced dipole moment might not lie parallel to the applied field, but for simplicity we discuss polarizabilities in terms of (scalar) magnitudes.

(a) Polarizability volumes

Polarizability has the units (coulomb metre)^2 per joule ($C^2\,m^2\,J^{-1}$). That collection of units is awkward, so α is often expressed as a **polarizability volume**, α', by using the relation

$$\alpha'=\frac{\alpha}{4\pi\varepsilon_0} \qquad [18.9]$$

where ε_0 is the vacuum permittivity. Because the units of $4\pi\varepsilon_0$ are coulomb-squared per joule per metre ($C^2\,J^{-1}\,m^{-1}$), it follows that α' has the dimensions of volume (hence its name). Polarizability volumes are similar in magnitude to actual molecular volumes (of the order of 10^{-30} m^3, 10^{-3} nm^3, 1 Å^3).

Comment 18.5

When using older compilations of data, it is useful to note that polarizability volumes have the same numerical values as the 'polarizabilities' reported using c.g.s. electrical units, so the tabulated values previously called 'polarizabilities' can be used directly.

Some experimental polarizability volumes of molecules are given in Table 18.1. As shown in the *Justification* below, polarizability volumes correlate with the HOMO–LUMO separations in atoms and molecules. The electron distribution can be distorted readily if the LUMO lies close to the HOMO in energy, so the polarizability is then large. If the LUMO lies high above the HOMO, an applied field cannot perturb the electron distribution significantly, and the polarizability is low. Molecules with small HOMO–LUMO gaps are typically large, with numerous electrons.

Justification 18.2 *Polarizabilities and molecular structures*

When an electric field is increased by $d\mathcal{E}$, the energy of a molecule changes by $-\mu d\mathcal{E}$, and if the molecule is polarizable, we interpret μ as μ^* (eqn 18.8). Therefore, the change in energy when the field is increased from 0 to $\mathcal{E}$ is

$$\Delta E = -\int_0^{\mathcal{E}} \mu^* d\mathcal{E} = -\int_0^{\mathcal{E}} \alpha\mathcal{E}\, d\mathcal{E} = -\tfrac{1}{2}\alpha\mathcal{E}^2$$

The contribution to the hamiltonian when a dipole moment is exposed to an electric field $\mathcal{E}$ in the z-direction is

$$H^{(1)} = -\mu_z\mathcal{E}$$

Comparison of these two expressions suggests that we should use second-order perturbation theory to calculate the energy of the system in the presence of the field, because then we shall obtain an expression proportional to $\mathcal{E}^2$. According to eqn 9.65b, the second-order contribution to the energy is

$$E^{(2)} = \sum_n{}' \frac{\left|\int \psi_n^* H^{(1)} \psi_0 d\tau\right|^2}{E_0^{(0)} - E_n^{(0)}} = \mathcal{E}^2 \sum_n{}' \frac{\left|\int \psi_n^* \mu_z \psi_0 d\tau\right|^2}{E_0^{(0)} - E_n^{(0)}} = \mathcal{E}^2 \sum_n{}' \frac{|\mu_{z,0n}|^2}{E_0^{(0)} - E_n^{(0)}}$$

where $\mu_{z,0n}$ is the *transition* electric dipole moment in the z-direction (eqn 9.70). By comparing the two expressions for the energy, we conclude that the polarizability of the molecule in the z-direction is

$$\alpha = 2\sum_n{}' \frac{|\mu_{z,0n}|^2}{E_n^{(0)} - E_0^{(0)}} \qquad (18.10)$$

The content of eqn 18.10 can be appreciated by approximating the excitation energies by a mean value ΔE (an indication of the HOMO–LUMO separation), and supposing that the most important transition dipole moment is approximately equal to the charge of an electron multiplied by the radius, R, of the molecule. Then

$$\alpha \approx \frac{2e^2R^2}{\Delta E}$$

This expression shows that α increases with the size of the molecule and with the ease with which it can be excited (the smaller the value of ΔE).

If the excitation energy is approximated by the energy needed to remove an electron to infinity from a distance R from a single positive charge, we can write $\Delta E \approx e^2/4\pi\varepsilon_0 R$. When this expression is substituted into the equation above, both sides are divided by $4\pi\varepsilon_0$, and the factor of 2 ignored in this approximation, we obtain $\alpha' \approx R^3$, which is of the same order of magnitude as the molecular volume.

For most molecules, the polarizability is anisotropic, by which is meant that its value depends on the orientation of the molecule relative to the field. The polarizability volume of benzene when the field is applied perpendicular to the ring is 0.0067 nm^3

and it is 0.0123 nm^3 when the field is applied in the plane of the ring. The anisotropy of the polarizability determines whether a molecule is rotationally Raman active (Section 13.7).

(b) Polarization at high frequencies

When the applied field changes direction slowly, the permanent dipole moment has time to reorientate—the whole molecule rotates into a new direction—and follow the field. However, when the frequency of the field is high, a molecule cannot change direction fast enough to follow the change in direction of the applied field and the dipole moment then makes no contribution to the polarization of the sample. Because a molecule takes about 1 ps to turn through about 1 radian in a fluid, the loss of this contribution to the polarization occurs when measurements are made at frequencies greater than about 10^{11} Hz (in the microwave region). We say that the **orientation polarization**, the polarization arising from the permanent dipole moments, is lost at such high frequencies.

The next contribution to the polarization to be lost as the frequency is raised is the **distortion polarization**, the polarization that arises from the distortion of the positions of the nuclei by the applied field. The molecule is bent and stretched by the applied field, and the molecular dipole moment changes accordingly. The time taken for a molecule to bend is approximately the inverse of the molecular vibrational frequency, so the distortion polarization disappears when the frequency of the radiation is increased through the infrared. The disappearance of polarization occurs in stages: as shown in the *Justification* below, each successive stage occurs as the incident frequency rises above the frequency of a particular mode of vibration.

At even higher frequencies, in the visible region, only the electrons are mobile enough to respond to the rapidly changing direction of the applied field. The polarization that remains is now due entirely to the distortion of the electron distribution, and the surviving contribution to the molecular polarizability is called the **electronic polarizability**.

Justification 18.3 *The frequency-dependence of polarizabilities*

The quantum mechanical expression for the polarizability of a molecule in the presence of an electric field that is oscillating at a frequency ω in the z-direction is obtained by using time-dependent perturbation theory (*Further information 9.2*) and is

$$\alpha(\omega) = \frac{2}{\hbar}\sum_n{}' \frac{\omega_{n0}|\mu_{z,0n}|^2}{\omega_{n0}^2 - \omega^2} \qquad (18.11)$$

The quantities in this expression (which is valid provided that ω is not close to ω_{n0}) are the same as those in the previous *Justification*, with $\hbar\omega_{n0} = E_n^{(0)} - E_0^{(0)}$. As $\omega \to 0$, the equation reduces to eqn 18.10 for the static polarizability. As ω becomes very high (and much higher than any excitation frequency of the molecule so that the ω_{n0}^2 in the denominator can be ignored), the polarizability becomes

$$\alpha(\omega) = -\frac{2}{\hbar\omega^2}\sum_n \omega_{n0}|\mu_{0n}|^2 \to 0 \qquad \text{as} \qquad \omega \to \infty$$

That is, when the incident frequency is much higher than any excitation frequency, the polarizability becomes zero. The argument applies to each type of excitation, vibrational as well as electronic, and accounts for the successive decreases in polarizability as the frequency is increased.

18.3 Relative permittivities

When two charges q_1 and q_2 are separated by a distance r in a vacuum, the potential energy of their interaction is (see *Appendix* 3):

$$V=\frac{q_1q_2}{4\pi\varepsilon_0 r} \tag{18.12a}$$

When the same two charges are immersed in a medium (such as air or a liquid), their potential energy is reduced to

$$V=\frac{q_1q_2}{4\pi\varepsilon r} \tag{18.12b}$$

where ε is the **permittivity** of the medium. The permittivity is normally expressed in terms of the dimensionless **relative permittivity**, ε_r, (formerly and still widely called the dielectric constant) of the medium:

$$\varepsilon_r=\frac{\varepsilon}{\varepsilon_0} \qquad [18.13]$$

The relative permittivity can have a very significant effect on the strength of the interactions between ions in solution. For instance, water has a relative permittivity of 78 at 25°C, so the interionic Coulombic interaction energy is reduced by nearly two orders of magnitude from its vacuum value. Some of the consequences of this reduction for electrolyte solutions were explored in Chapter 5.

The relative permittivity of a substance is large if its molecules are polar or highly polarizable. The quantitative relation between the relative permittivity and the electric properties of the molecules is obtained by considering the polarization of a medium, and is expressed by the **Debye equation** (for the derivation of this and the following equations, see *Further reading*):

$$\frac{\varepsilon_r-1}{\varepsilon_r+2}=\frac{\rho P_m}{M} \tag{18.14}$$

where ρ is the mass density of the sample, M is the molar mass of the molecules, and P_m is the **molar polarization**, which is defined as

$$P_m=\frac{N_A}{3\varepsilon_0}\left(\alpha+\frac{\mu^2}{3kT}\right) \qquad [18.15]$$

The term $\mu^2/3kT$ stems from the thermal averaging of the electric dipole moment in the presence of the applied field (eqn 18.5). The corresponding expression without the contribution from the permanent dipole moment is called the **Clausius–Mossotti equation**:

$$\frac{\varepsilon_r-1}{\varepsilon_r+2}=\frac{\rho N_A\alpha}{3M\varepsilon_0} \tag{18.16}$$

The Clausius–Mossotti equation is used when there is no contribution from permanent electric dipole moments to the polarization, either because the molecules are nonpolar or because the frequency of the applied field is so high that the molecules cannot orientate quickly enough to follow the change in direction of the field.

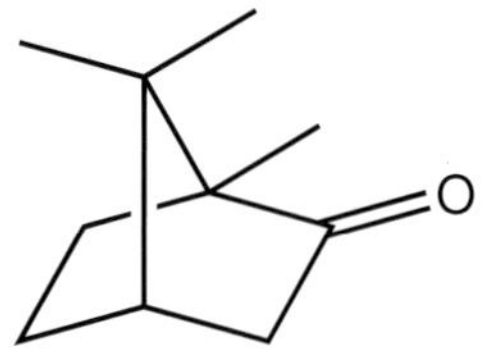

8 Camphor

Example 18.2 *Determining dipole moment and polarizability*

The relative permittivity of a substance is measured by comparing the capacitance of a capacitor with and without the sample present (C and C_0, respectively) and using $\varepsilon_r = C/C_0$. The relative permittivity of camphor (**8**) was measured at a series of temperatures with the results given below. Determine the dipole moment and the polarizability volume of the molecule.

θ/°C	ρ/(g cm^{-3})	ε_r
0	0.99	12.5
20	0.99	11.4
40	0.99	10.8
60	0.99	10.0
80	0.99	9.50
100	0.99	8.90
120	0.97	8.10
140	0.96	7.60
160	0.95	7.11
200	0.91	6.21

Method Equation 18.14 implies that the polarizability and permanent electric dipole moment of the molecules in a sample can be determined by measuring ε_r at a series of temperatures, calculating P_m, and plotting it against $1/T$. The slope of the graph is $N_A\mu^2/9\varepsilon_0 k$ and its intercept at $1/T = 0$ is $N_A\alpha/3\varepsilon_0$. We need to calculate $(\varepsilon_r - 1)/(\varepsilon_r + 2)$ at each temperature, and then multiply by M/ρ to form P_m.

Answer For camphor, $M = 152.23$ g mol^{-1}. We can therefore use the data to draw up the following table:

θ/°C	$(10^3$ K$)/T$	ε_r	$(\varepsilon_r - 1)/(\varepsilon_r + 2)$	P_m/(cm^3 mol^{-1})
0	3.66	12.5	0.793	122
20	3.41	11.4	0.776	119
40	3.19	10.8	0.766	118
60	3.00	10.0	0.750	115
80	2.83	9.50	0.739	114
100	2.68	8.90	0.725	111
120	2.54	8.10	0.703	110
140	2.42	7.60	0.688	109
160	2.31	7.11	0.670	107
200	2.11	6.21	0.634	106

The points are plotted in Fig. 18.2. The intercept lies at 82.7, so $\alpha' = 3.3 \times 10^{-23}$ cm^3. The slope is 10.9, so $\mu = 4.46 \times 10^{-30}$ C m, corresponding to 1.34 D. Because the Debye equation describes molecules that are free to rotate, the data show that camphor, which does not melt until 175°C, is rotating even in the solid. It is an approximately spherical molecule.

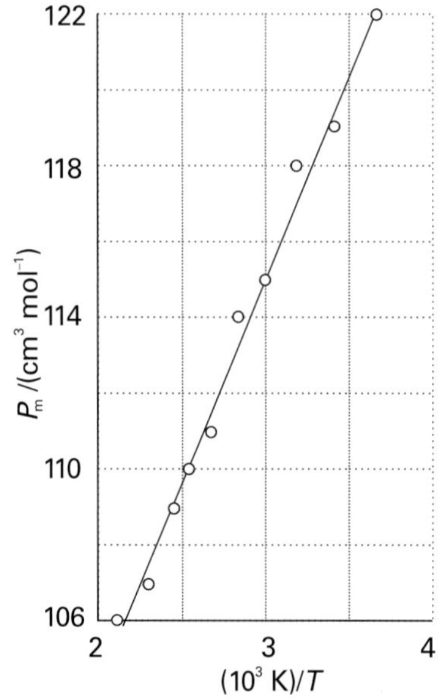

Fig. 18.2 The plot of P_m/(cm^3 mol^{-1}) against $(10^3$ K$)/T$ used in *Example 18.2* for the determination of the polarizability and dipole moment of camphor.

Self-test 18.3 The relative permittivity of chlorobenzene is 5.71 at 20°C and 5.62 at 25°C. Assuming a constant density (1.11 g cm^{-3}), estimate its polarizability volume and dipole moment. [1.4×10^{-23} cm^3, 1.2 D]

The Maxwell equations that describe the properties of electromagnetic radiation (see *Further reading*) relate the refractive index at a (visible or ultraviolet) specified wavelength to the relative permittivity at that frequency:

$$n_r = \varepsilon_r^{1/2} \tag{18.17}$$

Therefore, the molar polarization, P_m, and the molecular polarizability, α, can be measured at frequencies typical of visible light (about 10^{15} to 10^{16} Hz) by measuring the refractive index of the sample and using the Clausius–Mossotti equation.

Comment 18.6

The refractive index, n_r, of the medium is the ratio of the speed of light in a vacuum, c, to its speed c' in the medium: $n_r = c/c'$. A beam of light changes direction ('bends') when it passes from a region of one refractive index to a region with a different refractive index. See *Appendix* 3 for details.

Interactions between molecules

A **van der Waals interaction** is the attractive interaction between closed-shell molecules that depends on the distance between the molecules as $1/r^6$. In addition, there are interactions between ions and the partial charges of polar molecules and repulsive interactions that prevent the complete collapse of matter to nuclear densities. The repulsive interactions arise from Coulombic repulsions and, indirectly, from the Pauli principle and the exclusion of electrons from regions of space where the orbitals of neighbouring species overlap.

18.4 Interactions between dipoles

Most of the discussion in this section is based on the Coulombic potential energy of interaction between two charges (eqn 18.12a). We can easily adapt this expression to find the potential energy of a point charge and a dipole and to extend it to the interaction between two dipoles.

(a) The potential energy of interaction

We show in the *Justification* below that the potential energy of interaction between a point dipole $\mu_1 = q_1 l$ and the point charge q_2 in the arrangement shown in (**9**) is

$$V = -\frac{\mu_1 q_2}{4\pi\varepsilon_0 r^2} \tag{18.18}$$

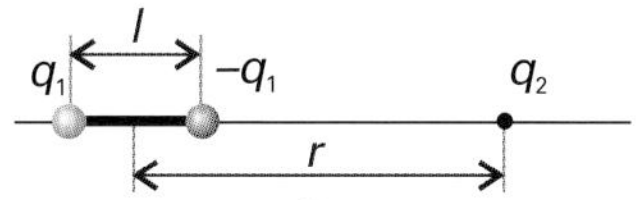

9

With μ in coulomb metres, q_2 in coulombs, and r in metres, V is obtained in joules. A **point dipole** is a dipole in which the separation between the charges is much smaller than the distance at which the dipole is being observed, $l \ll r$. The potential energy rises towards zero (the value at infinite separation of the charge and the dipole) more rapidly (as $1/r^2$) than that between two point charges (which varies as $1/r$) because, from the viewpoint of the point charge, the partial charges of the dipole seem to merge and cancel as the distance r increases (Fig. 18.3).

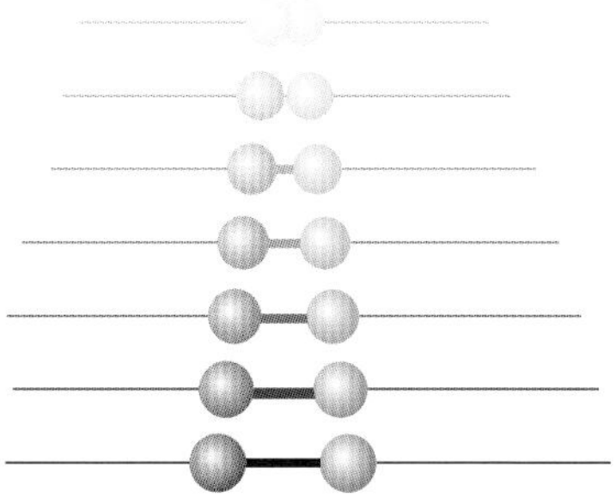

Fig. 18.3 There are two contributions to the diminishing field of an electric dipole with distance (here seen from the side). The potentials of the charges decrease (shown here by a fading intensity) and the two charges appear to merge, so their combined effect approaches zero more rapidly than by the distance effect alone.

Justification 18.4 *The interaction between a point charge and a point dipole*

The sum of the potential energies of repulsion between like charges and attraction between opposite charges in the orientation shown in (**9**) is

$$V = \frac{1}{4\pi\varepsilon_0}\left(-\frac{q_1 q_2}{r-\frac{1}{2}l} + \frac{q_1 q_2}{r+\frac{1}{2}l}\right) = \frac{q_1 q_2}{4\pi\varepsilon_0 r}\left(-\frac{1}{1-x} + \frac{1}{1+x}\right)$$

where $x = l/2r$. Because $l \ll r$ for a point dipole, this expression can be simplified by expanding the terms in x and retaining only the leading term:

$$V = \frac{q_1 q_2}{4\pi\varepsilon_0 r}\{-(1+x+\cdots)+(1-x+\cdots)\} \approx -\frac{2xq_1q_2}{4\pi\varepsilon_0 r} = -\frac{q_1 q_2 l}{4\pi\varepsilon_0 r^2}$$

With $\mu_1 = q_1 l$, this expression becomes eqn 18.18. This expression should be multiplied by $\cos\theta$ when the point charge lies at an angle θ to the axis of the dipole.

Comment 18.7

The following expansions are often useful:

$$\frac{1}{1+x} = 1 - x + x^2 - \cdots$$

$$\frac{1}{1-x} = 1 + x + x^2 + \cdots$$

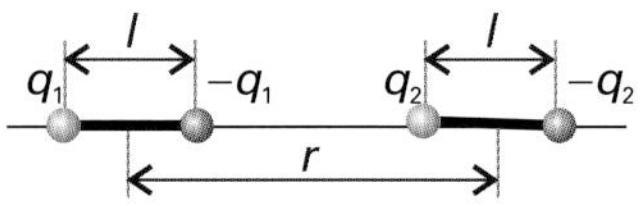

10

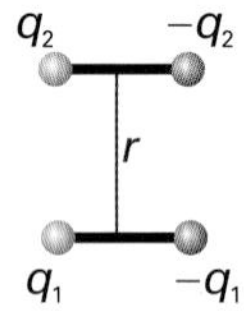

11

Example 18.3 *Calculating the interaction energy of two dipoles*

Calculate the potential energy of interaction of two dipoles in the arrangement shown in (**10**) when their separation is r.

Method We proceed in exactly the same way as in *Justification* 18.4, but now the total interaction energy is the sum of four pairwise terms, two attractions between opposite charges, which contribute negative terms to the potential energy, and two repulsions between like charges, which contribute positive terms.

Answer The sum of the four contributions is

$$V = \frac{1}{4\pi\varepsilon_0}\left(-\frac{q_1q_2}{r+l} + \frac{q_1q_2}{r} + \frac{q_1q_2}{r} - \frac{q_1q_2}{r-l}\right) = -\frac{q_1q_2}{4\pi\varepsilon_0 r}\left(\frac{1}{1+x} - 2 + \frac{1}{1-x}\right)$$

with $x = l/r$. As before, provided $l \ll r$ we can expand the two terms in x and retain only the first surviving term, which is equal to $2x^2$. This step results in the expression

$$V = -\frac{2x^2 q_1 q_2}{4\pi\varepsilon_0 r}$$

Therefore, because $\mu_1 = q_1 l$ and $\mu_2 = q_2 l$, the potential energy of interaction in the alignment shown in the illustration is

$$V = -\frac{\mu_1\mu_2}{2\pi\varepsilon_0 r^3}$$

This interaction energy approaches zero more rapidly (as $1/r^3$) than for the previous case: now both interacting entities appear neutral to each other at large separations. See *Further information* 18.1 for the general expression.

Self-test 18.4 Derive an expression for the potential energy when the dipoles are in the arrangement shown in (**11**). $[V = \mu_1\mu_2/4\pi\varepsilon_0 r^3]$

Table 18.3 summarizes the various expressions for the interaction of charges and dipoles. It is quite easy to extend the formulas given there to obtain expressions for the energy of interaction of higher **multipoles**, or arrays of point charges (Fig. 18.4). Specifically, an ***n*-pole** is an array of point charges with an n-pole moment but no lower moment. Thus, a **monopole** ($n = 1$) is a point charge, and the monopole moment is what we normally call the overall charge. A dipole ($n = 2$), as we have seen, is an array

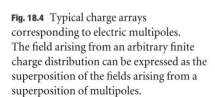

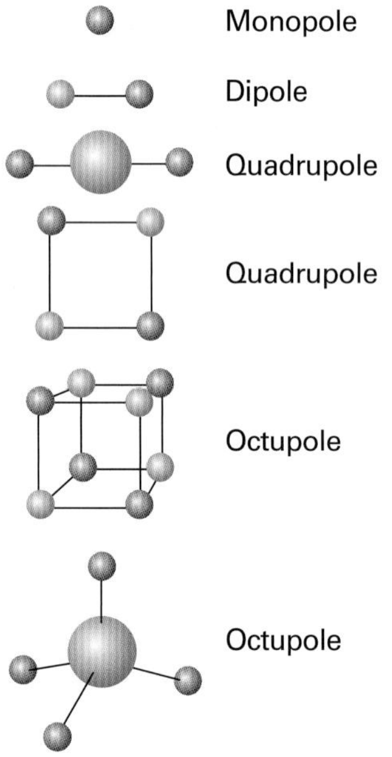

Fig. 18.4 Typical charge arrays corresponding to electric multipoles. The field arising from an arbitrary finite charge distribution can be expressed as the superposition of the fields arising from a superposition of multipoles.

Table 18.3 Multipole interaction potential energies

Interaction type	Distance dependence of potential energy	Typical energy/ (kJ mol^{-1})	Comment
Ion–ion	$1/r$	250	Only between ions*
Ion–dipole	$1/r^2$	15	
Dipole–dipole	$1/r^3$	2	Between stationary polar molecules
	$1/r^6$	0.6	Between rotating polar molecules
London (dispersion)	$1/r^6$	2	Between all types of molecules

The energy of a hydrogen bond A—H···B is typically 20 kJ mol^{-1} and occurs on contact for A, B = O, N, or F.
* Electrolyte solutions are treated in Chapter 5, ionic solids in Chapter 20.

of charges that has no monopole moment (no net charge). A **quadrupole** ($n = 3$) consists of an array of point charges that has neither net charge nor dipole moment (as for CO_2 molecules, **3**). An **octupole** ($n = 4$) consists of an array of point charges that sum to zero and which has neither a dipole moment nor a quadrupole moment (as for CH_4 molecules, **12**). The feature to remember is that the interaction energy falls off more rapidly the higher the order of the multipole. For the interaction of an n-pole with an m-pole, the potential energy varies with distance as

$$V \propto \frac{1}{r^{n+m-1}} \qquad (18.19)$$

The reason for the even steeper decrease with distance is the same as before: the array of charges appears to blend together into neutrality more rapidly with distance the higher the number of individual charges that contribute to the multipole. Note that a given molecule may have a charge distribution that corresponds to a superposition of several different multipoles.

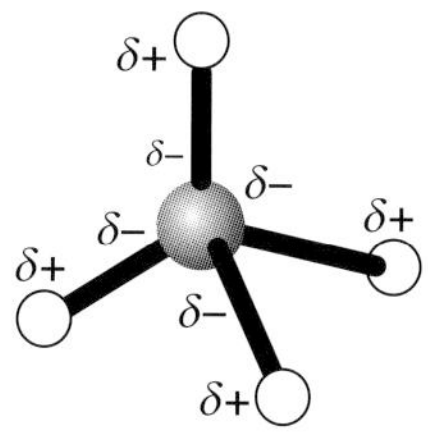

12 CH_4

(b) The electric field

The same kind of argument as that used to derive expressions for the potential energy can be used to establish the distance dependence of the strength of the electric field generated by a dipole. We shall need this expression when we calculate the dipole moment induced in one molecule by another.

The starting point for the calculation is the strength of the electric field generated by a point electric charge:

$$\mathcal{E} = \frac{q}{4\pi\varepsilon_0 r^2} \qquad (18.20)$$

The field generated by a dipole is the sum of the fields generated by each partial charge. For the point-dipole arrangement shown in Fig. 18.5, the same procedure that was used to derive the potential energy gives

$$\mathcal{E} = \frac{\mu}{2\pi\varepsilon_0 r^3} \qquad (18.21)$$

The electric field of a multipole (in this case a dipole) decreases more rapidly with distance (as $1/r^3$ for a dipole) than a monopole (a point charge).

Comment 18.8

The electric field is actually a vector, and we cannot simply add and subtract magnitudes without taking into account the directions of the fields. In the cases we consider, this will not be a complication because the two charges of the dipoles will be collinear and give rise to fields in the same direction. Be careful, though, with more general arrangements of charges.

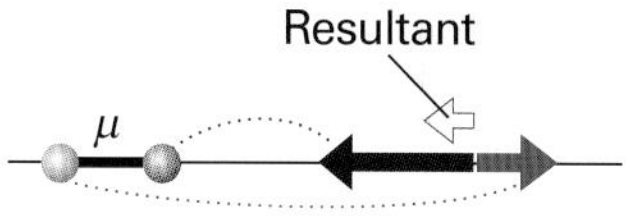

Fig. 18.5 The electric field of a dipole is the sum of the opposing fields from the positive and negative charges, each of which is proportional to $1/r^2$. The difference, the net field, is proportional to $1/r^3$.

(c) Dipole–dipole interactions

The potential energy of interaction between two polar molecules is a complicated function of their relative orientation. When the two dipoles are parallel (as in **13**), the potential energy is simply (see *Further information* 18.1)

$$V = \frac{\mu_1\mu_2 f(\theta)}{4\pi\varepsilon_0 r^3} \qquad f(\theta) = 1 - 3\cos^2\theta \qquad (18.22)$$

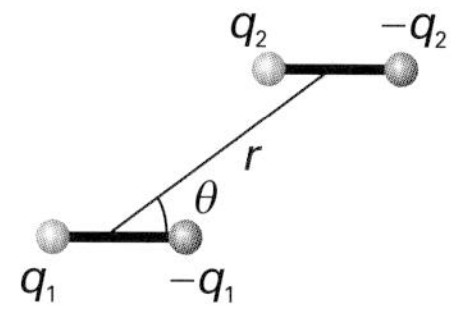

13

This expression applies to polar molecules in a fixed, parallel, orientation in a solid.

In a fluid of freely rotating molecules, the interaction between dipoles averages to zero because $f(\theta)$ changes sign as the orientation changes, and its average value is zero. Physically, the like partial charges of two freely rotating molecules are close together as much as the two opposite charges, and the repulsion of the former is cancelled by the attraction of the latter.

The interaction energy of two *freely* rotating dipoles is zero. However, because their mutual potential energy depends on their relative orientation, the molecules do not in fact rotate completely freely, even in a gas. In fact, the lower energy orientations are

Comment 18.9

The average (or mean value) of a function $f(x)$ over the range from $x = a$ to $x = b$ is

$$\langle f \rangle = \frac{1}{b-a}\int_a^b f(x)\,dx$$

The volume element in polar coordinates is proportional to $\sin\theta\,d\theta$, and θ ranges from 0 to π. Therefore the average value of $(1 - 3\cos^2\theta)$ is

$$(1/\pi)\int_0^\pi (1 - 3\cos^2\theta)\sin\theta\,d\theta = 0.$$

marginally favoured, so there is a nonzero average interaction between polar molecules. We show in the following *Justification* that the average potential energy of two rotating molecules that are separated by a distance r is

$$\langle V\rangle = -\frac{C}{r^6} \qquad C = \frac{2\mu_1^2\mu_2^2}{3(4\pi\varepsilon_0)^2kT} \tag{18.23}$$

This expression describes the **Keesom interaction**, and is the first of the contributions to the van der Waals interaction.

Justification 18.5 *The Keesom interaction*

The detailed calculation of the Keesom interaction energy is quite complicated, but the form of the final answer can be constructed quite simply. First, we note that the average interaction energy of two polar molecules rotating at a fixed separation r is given by

$$\langle V\rangle = \frac{\mu_1\mu_2\langle f\rangle}{4\pi\varepsilon_0 r^3}$$

where $\langle f\rangle$ now includes a weighting factor in the averaging that is equal to the probability that a particular orientation will be adopted. This probability is given by the Boltzmann distribution $p \propto e^{-E/kT}$, with E interpreted as the potential energy of interaction of the two dipoles in that orientation. That is,

$$p \propto e^{-V/kT} \qquad V = \frac{\mu_1\mu_2 f}{4\pi\varepsilon_0 r^3}$$

When the potential energy of interaction of the two dipoles is very small compared with the energy of thermal motion, we can use $V \ll kT$, expand the exponential function in p, and retain only the first two terms:

$$p \propto 1 - V/kT + \cdots$$

The weighted average of f is therefore

$$\langle f\rangle = \langle f\rangle_0 - \frac{\mu_1\mu_2}{4\pi\varepsilon_0 kTr^3}\langle f^2\rangle_0 + \cdots$$

where $\langle\cdots\rangle_0$ denotes an unweighted spherical average. The spherical average of f is zero, so the first term vanishes. However, the average value of f^2 is nonzero because f^2 is positive at all orientations, so we can write

$$\langle V\rangle = -\frac{\mu_1^2\mu_2^2\langle f^2\rangle_0}{(4\pi\varepsilon_0)^2kTr^6}$$

The average value $\langle f^2\rangle_0$ turns out to be $\frac{2}{3}$ when the calculation is carried through in detail. The final result is that quoted in eqn 18.23.

The important features of eqn 18.23 are its negative sign (the average interaction is attractive), the dependence of the average interaction energy on the inverse sixth power of the separation (which identifies it as a van der Waals interaction), and its inverse dependence on the temperature. The last feature reflects the way that the greater thermal motion overcomes the mutual orientating effects of the dipoles at higher temperatures. The inverse sixth power arises from the inverse third power of the interaction potential energy that is weighted by the energy in the Boltzmann term, which is also proportional to the inverse third power of the separation.

At 25°C the average interaction energy for pairs of molecules with $\mu = 1$ D is about -0.07 kJ mol^{-1} when the separation is 0.5 nm. This energy should be compared with the average molar kinetic energy of $\frac{3}{2}RT = 3.7$ kJ mol^{-1} at the same temperature. The interaction energy is also much smaller than the energies involved in the making and breaking of chemical bonds.

(d) Dipole–induced-dipole interactions

A polar molecule with dipole moment μ_1 can induce a dipole μ_2^* in a neighbouring polarizable molecule (Fig. 18.6). The induced dipole interacts with the permanent dipole of the first molecule, and the two are attracted together. The average interaction energy when the separation of the molecules is r is (for a derivation, see *Further reading*)

$$V = -\frac{C}{r^6} \qquad C = \frac{\mu_1^2 \alpha_2'}{4\pi\varepsilon_0} \tag{18.24}$$

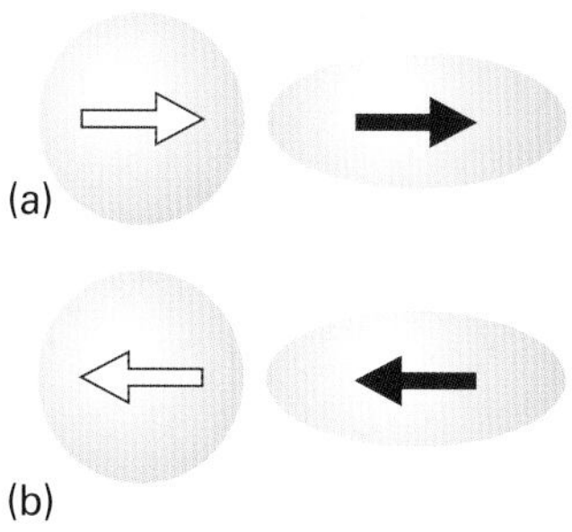

Fig. 18.6 (a) A polar molecule (purple arrow) can induce a dipole (white arrow) in a nonpolar molecule, and (b) the latter's orientation follows the former's, so the interaction does not average to zero.

where α_2' is the polarizability volume of molecule 2 and μ_1 is the permanent dipole moment of molecule 1. Note that the C in this expression is different from the C in eqn 18.23 and other expressions below: we are using the same symbol in C/r^6 to emphasize the similarity of form of each expression.

The dipole–induced-dipole interaction energy is independent of the temperature because thermal motion has no effect on the averaging process. Moreover, like the dipole–dipole interaction, the potential energy depends on $1/r^6$: this distance dependence stems from the $1/r^3$ dependence of the field (and hence the magnitude of the induced dipole) and the $1/r^3$ dependence of the potential energy of interaction between the permanent and induced dipoles. For a molecule with $\mu = 1$ D (such as HCl) near a molecule of polarizability volume $\alpha' = 10 \times 10^{-30}$ m^3 (such as benzene, Table 18.1), the average interaction energy is about -0.8 kJ mol^{-1} when the separation is 0.3 nm.

(e) Induced-dipole–induced-dipole interactions

Nonpolar molecules (including closed-shell atoms, such as Ar) attract one another even though neither has a permanent dipole moment. The abundant evidence for the existence of interactions between them is the formation of condensed phases of nonpolar substances, such as the condensation of hydrogen or argon to a liquid at low temperatures and the fact that benzene is a liquid at normal temperatures.

The interaction between nonpolar molecules arises from the transient dipoles that all molecules possess as a result of fluctuations in the instantaneous positions of electrons. To appreciate the origin of the interaction, suppose that the electrons in one molecule flicker into an arrangement that gives the molecule an instantaneous dipole moment μ_1^*. This dipole generates an electric field that polarizes the other molecule, and induces in that molecule an instantaneous dipole moment μ_2^*. The two dipoles attract each other and the potential energy of the pair is lowered. Although the first molecule will go on to change the size and direction of its instantaneous dipole, the electron distribution of the second molecule will follow; that is, the two dipoles are correlated in direction (Fig. 18.7). Because of this correlation, the attraction between the two instantaneous dipoles does not average to zero, and gives rise to an induced-dipole–induced-dipole interaction. This interaction is called either the **dispersion interaction** or the **London interaction** (for Fritz London, who first described it).

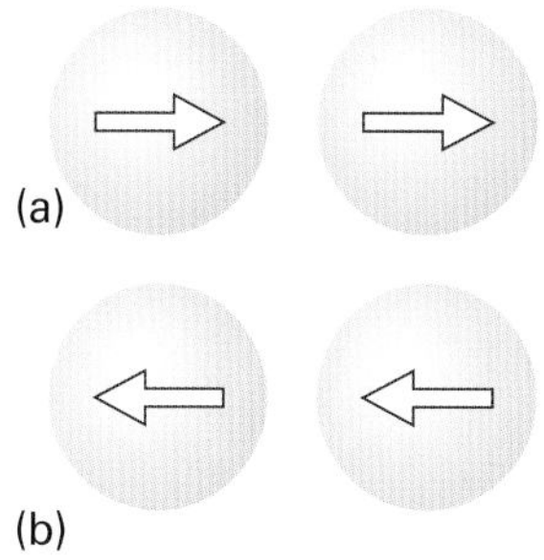

Fig. 18.7 (a) In the dispersion interaction, an instantaneous dipole on one molecule induces a dipole on another molecule, and the two dipoles then interact to lower the energy. (b) The two instantaneous dipoles are correlated and, although they occur in different orientations at different instants, the interaction does not average to zero.

Polar molecules also interact by a dispersion interaction: such molecules also possess instantaneous dipoles, the only difference being that the time average of each fluctuating dipole does not vanish, but corresponds to the permanent dipole. Such molecules therefore interact both through their permanent dipoles and through the correlated, instantaneous fluctuations in these dipoles.

The strength of the dispersion interaction depends on the polarizability of the first molecule because the instantaneous dipole moment μ_1^* depends on the looseness of the control that the nuclear charge exercises over the outer electrons. The strength of the interaction also depends on the polarizability of the second molecule, for that polarizability determines how readily a dipole can be induced by another molecule. The actual calculation of the dispersion interaction is quite involved, but a reasonable approximation to the interaction energy is given by the **London formula:**

$$V=-\frac{C}{r^6} \qquad C=\tfrac{3}{2}\alpha_1'\alpha_2'\frac{I_1I_2}{I_1+I_2} \tag{18.25}$$

where I_1 and I_2 are the ionization energies of the two molecules (Table 10.4). This interaction energy is also proportional to the inverse sixth power of the separation of the molecules, which identifies it as a third contribution to the van der Waals interaction. The dispersion interaction generally dominates all the interactions between molecules other than hydrogen bonds.

Illustration 18.1 *Calculating the strength of the dispersion interaction*

For two CH_4 molecules, we can substitute $\alpha' = 2.6\times10^{-30}$ m^3 and $I\approx 700$ kJ mol^{-1} to obtain $V=-2$ kJ mol^{-1} for $r=0.3$ nm. A very rough check on this figure is the enthalpy of vaporization of methane, which is 8.2 kJ mol^{-1}. However, this comparison is insecure, partly because the enthalpy of vaporization is a many-body quantity and partly because the long-distance assumption breaks down.

(f) Hydrogen bonding

The interactions described so far are universal in the sense that they are possessed by all molecules independent of their specific identity. However, there is a type of interaction possessed by molecules that have a particular constitution. A **hydrogen bond** is an attractive interaction between two species that arises from a link of the form A—H···B, where A and B are highly electronegative elements and B possesses a lone pair of electrons. Hydrogen bonding is conventionally regarded as being limited to N, O, and F but, if B is an anionic species (such as Cl^-), it may also participate in hydrogen bonding. There is no strict cutoff for an ability to participate in hydrogen bonding, but N, O, and F participate most effectively.

The formation of a hydrogen bond can be regarded either as the approach between a partial positive charge of H and a partial negative charge of B or as a particular example of delocalized molecular orbital formation in which A, H, and B each supply one atomic orbital from which three molecular orbitals are constructed (Fig. 18.8). Thus, if the A—H bond is regarded as formed from the overlap of an orbital on A, ψ_A, and a hydrogen 1*s* orbital, ψ_H, and the lone pair on B occupies an orbital on B, ψ_B, then, when the two molecules are close together, we can build three molecular orbitals from the three basis orbitals:

$$\psi=c_1\psi_A+c_2\psi_H+c_3\psi_B$$

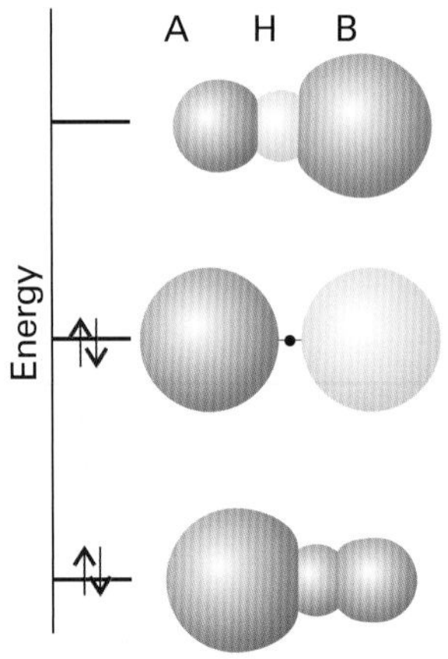

Fig. 18.8 The molecular orbital interpretation of the formation of an A—H···B hydrogen bond. From the three A, H, and B orbitals, three molecular orbitals can be formed (their relative contributions are represented by the sizes of the spheres. Only the two lower energy orbitals are occupied, and there may therefore be a net lowering of energy compared with the separate AH and B species.

One of the molecular orbitals is bonding, one almost nonbonding, and the third antibonding. These three orbitals need to accommodate four electrons (two from the original A—H bond and two from the lone pair of B), so two enter the bonding orbital and two enter the nonbonding orbital. Because the antibonding orbital remains empty, the net effect—depending on the precise location of the almost nonbonding orbital—may be a lowering of energy.

In practice, the strength of the bond is found to be about 20 kJ mol^{-1}. Because the bonding depends on orbital overlap, it is virtually a contact-like interaction that is turned on when AH touches B and is zero as soon as the contact is broken. If hydrogen bonding is present, it dominates the other intermolecular interactions. The properties of liquid and solid water, for example, are dominated by the hydrogen bonding between H_2O molecules. The structure of DNA and hence the transmission of genetic information is crucially dependent on the strength of hydrogen bonds between base pairs. The structural evidence for hydrogen bonding comes from noting that the internuclear distance between formally non-bonded atoms is less than their van der Waals contact distance, which suggests that a dominating attractive interaction is present. For example, the O—O distance in O—H···O is expected to be 280 pm on the basis of van der Waals radii, but is found to be 270 pm in typical compounds. Moreover, the H···O distance is expected to be 260 pm but is found to be only 170 pm.

Hydrogen bonds may be either symmetric or unsymmetric. In a symmetric hydrogen bond, the H atoms lies midway between the two other atoms. This arrangement is rare, but occurs in $F—H\cdots F^-$, where both bond lengths are 120 pm. More common is the unsymmetrical arrangement, where the A—H bond is shorter than the H···B bond. Simple electrostatic arguments, treating A—H···B as an array of point charges (partial negative charges on A and B, partial positive on H) suggest that the lowest energy is achieved when the bond is linear, because then the two partial negative charges are furthest apart. The experimental evidence from structural studies support a linear or near-linear arrangement.

(g) The hydrophobic interaction

Nonpolar molecules do dissolve slightly in polar solvents, but strong interactions between solute and solvent are not possible and as a result it is found that each individual solute molecule is surrounded by a solvent cage (Fig. 18.9). To understand the consequences of this effect, consider the thermodynamics of transfer of a nonpolar hydrocarbon solute from a nonpolar solvent to water, a polar solvent. Experiments indicate that the process is endergonic ($\Delta_{\text{transfer}}G > 0$), as expected on the basis of the increase in polarity of the solvent, but exothermic ($\Delta_{\text{transfer}}H < 0$). Therefore, it is a large decrease in the entropy of the system ($\Delta_{\text{transfer}}S < 0$) that accounts for the positive Gibbs energy of transfer. For example, the process

$$CH_4(\text{in } CCl_4) \rightarrow CH_4(\text{aq})$$

has $\Delta_{\text{transfer}}G = +12\text{ kJ mol}^{-1}$, $\Delta_{\text{transfer}}H = -10\text{ kJ mol}^{-1}$, and $\Delta_{\text{transfer}}S = -75\text{ J K}^{-1}\text{ mol}^{-1}$ at 298 K. Substances characterized by a positive Gibbs energy of transfer from a nonpolar to a polar solvent are called **hydrophobic**.

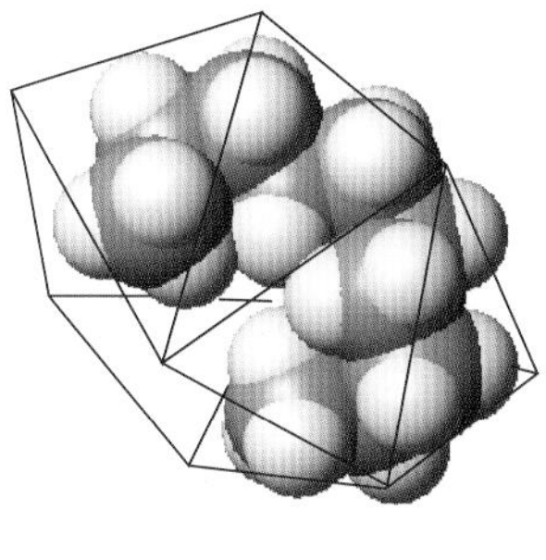

Fig. 18.9 When a hydrocarbon molecule is surrounded by water, the H_2O molecules form a clathrate cage. As a result of this acquisition of structure, the entropy of the water decreases, so the dispersal of the hydrocarbon into the water is entropy-opposed; its coalescence is entropy-favoured.

It is possible to quantify the hydrophobicity of a small molecular group R by defining the **hydrophobicity constant**, π, as

$$\pi = \log \frac{S}{S_0} \qquad [18.26]$$

where S is the ratio of the molar solubility of the compound R—A in octanol, a nonpolar solvent, to that in water, and S_0 is the ratio of the molar solubility of the compound H—A in octanol to that in water. Therefore, positive values of π indicate hydrophobicity and negative values of π indicate hydrophilicity, the thermodynamic preference for water as a solvent. It is observed experimentally that the π values of most groups do not depend on the nature of A. However, measurements do suggest group additivity of π values. For example, π for R = CH_3, CH_2CH_3, $(CH_2)_2CH_3$, $(CH_2)_3CH_3$, and $(CH_2)_4CH_3$ is, respectively, 0.5, 1.0, 1.5, 2.0, and 2.5 and we conclude that acyclic saturated hydrocarbons become more hydrophobic as the carbon chain length increases. This trend can be rationalized by $\Delta_{\text{transfer}}H$ becoming more positive and $\Delta_{\text{transfer}}S$ more negative as the number of carbon atoms in the chain increases.

At the molecular level, formation of a solvent cage around a hydrophobic molecule involves the formation of new hydrogen bonds among solvent molecules. This is an exothermic process and accounts for the negative values of $\Delta_{\text{transfer}}H$. On the other hand, the increase in order associated with formation of a very large number of small solvent cages decreases the entropy of the system and accounts for the negative values of $\Delta_{\text{transfer}}S$. However, when many solute molecules cluster together, fewer (albeit larger) cages are required and more solvent molecules are free to move. The net effect of formation of large clusters of hydrophobic molecules is then a decrease in the organization of the solvent and therefore a net *increase* in entropy of the system. This increase in entropy of the solvent is large enough to render spontaneous the association of hydrophobic molecules in a polar solvent.

The increase in entropy that results from fewer structural demands on the solvent placed by the clustering of nonpolar molecules is the origin of the **hydrophobic interaction**, which tends to stabilize groupings of hydrophobic groups in micelles and biopolymers (Chapter 19). The hydrophobic interaction is an example of an ordering process that is stabilized by a tendency toward greater disorder of the solvent.

(h) The total attractive interaction

We shall consider molecules that are unable to participate in hydrogen bond formation. The total attractive interaction energy between rotating molecules is then the sum of the three van der Waals contributions discussed above. (Only the dispersion interaction contributes if both molecules are nonpolar.) In a fluid phase, all three contributions to the potential energy vary as the inverse sixth power of the separation of the molecules, so we may write

$$V = -\frac{C_6}{r^6} \tag{18.27}$$

where C_6 is a coefficient that depends on the identity of the molecules.

Although attractive interactions between molecules are often expressed as in eqn 18.27, we must remember that this equation has only limited validity. First, we have taken into account only dipolar interactions of various kinds, for they have the longest range and are dominant if the average separation of the molecules is large. However, in a complete treatment we should also consider quadrupolar and higher-order multipole interactions, particularly if the molecules do not have permanent electric dipole moments. Secondly, the expressions have been derived by assuming that the molecules can rotate reasonably freely. That is not the case in most solids, and in rigid media the dipole–dipole interaction is proportional to $1/r^3$ because the Boltzmann averaging procedure is irrelevant when the molecules are trapped into a fixed orientation.

A different kind of limitation is that eqn 18.27 relates to the interactions of pairs of molecules. There is no reason to suppose that the energy of interaction of three (or more) molecules is the sum of the pairwise interaction energies alone. The total dispersion energy of three closed-shell atoms, for instance, is given approximately by the **Axilrod–Teller formula**:

$$V = -\frac{C_6}{r_{AB}^6} - \frac{C_6}{r_{BC}^6} - \frac{C_6}{r_{CA}^6} + \frac{C'}{(r_{AB}r_{BC}r_{CA})^3} \tag{18.28a}$$

where

$$C' = a(3\cos\theta_A \cos\theta_B \cos\theta_C + 1) \tag{18.28b}$$

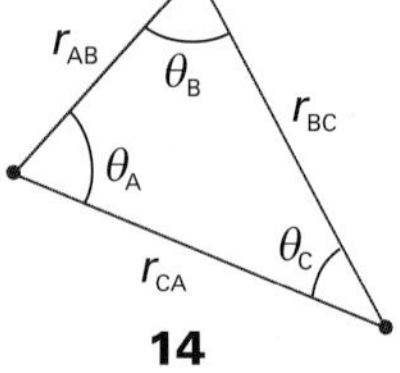

14

The parameter a is approximately equal to $\frac{3}{4}\alpha' C_6$; the angles θ are the internal angles of the triangle formed by the three atoms (**14**). The term in C' (which represents the non-additivity of the pairwise interactions) is negative for a linear arrangement of atoms (so that arrangement is stabilized) and positive for an equilateral triangular cluster. It is found that the three-body term contributes about 10 per cent of the total interaction energy in liquid argon.

18.5 Repulsive and total interactions

When molecules are squeezed together, the nuclear and electronic repulsions and the rising electronic kinetic energy begin to dominate the attractive forces. The repulsions increase steeply with decreasing separation in a way that can be deduced only by very extensive, complicated molecular structure calculations of the kind described in Chapter 11 (Fig. 18.10).

In many cases, however, progress can be made by using a greatly simplified representation of the potential energy, where the details are ignored and the general features expressed by a few adjustable parameters. One such approximation is the **hard-sphere potential**, in which it is assumed that the potential energy rises abruptly to infinity as soon as the particles come within a separation d:

$$V=\infty \quad \text{for} \quad r\le d \qquad V=0 \quad \text{for} \quad r>d \tag{18.29}$$

This very simple potential is surprisingly useful for assessing a number of properties. Another widely used approximation is the **Mie potential**:

$$V=\frac{C_n}{r^n}-\frac{C_m}{r^m} \tag{18.30}$$

with $n > m$. The first term represents repulsions and the second term attractions. The **Lennard-Jones potential** is a special case of the Mie potential with $n = 12$ and $m = 6$ (Fig. 18.11); it is often written in the form

$$V=4\varepsilon\left\{\left(\frac{r_0}{r}\right)^{12}-\left(\frac{r_0}{r}\right)^{6}\right\} \tag{18.31}$$

The two parameters are ε, the depth of the well (not to be confused with the symbol of the permittivity of a medium used in Section 18.3), and r_0, the separation at which $V = 0$ (Table 18.4). The well minimum occurs at $r_e = 2^{1/6}r_0$. Although the Lennard-Jones potential has been used in many calculations, there is plenty of evidence to show that $1/r^{12}$ is a very poor representation of the repulsive potential, and that an exponential form, e^{-r/r_0}, is greatly superior. An exponential function is more faithful to the exponential decay of atomic wavefunctions at large distances, and hence to the overlap that is responsible for repulsion. The potential with an exponential repulsive term and a $1/r^6$ attractive term is known as an **exp-6 potential**. These potentials can be used to calculate the virial coefficients of gases, as explained in Section 17.5, and through them various properties of real gases, such as the Joule–Thompson coefficient. The potentials are also used to model the structures of condensed fluids.

With the advent of **atomic force microscopy** (AFM), in which the force between a molecular sized probe and a surface is monitored (see *Impact I9.1*), it has become possible to measure directly the forces acting between molecules. The force, F, is the negative slope of potential, so for a Lennard-Jones potential between individual molecules we write

$$F=-\frac{\mathrm{d}V}{\mathrm{d}r}=\frac{24\varepsilon}{r_0}\left\{2\left(\frac{r_0}{r}\right)^{13}-\left(\frac{r_0}{r}\right)^{7}\right\} \tag{18.32}$$

The net attractive force is greatest (from $\mathrm{d}F/\mathrm{d}r = 0$) at $r = (\frac{26}{7})^{1/6}r_0$, or $1.244r_0$, and at that distance is equal to $-144(\frac{7}{26})^{7/6}\varepsilon/13r_0$, or $-2.397\varepsilon/r_0$. For typical parameters, the magnitude of this force is about 10 pN.

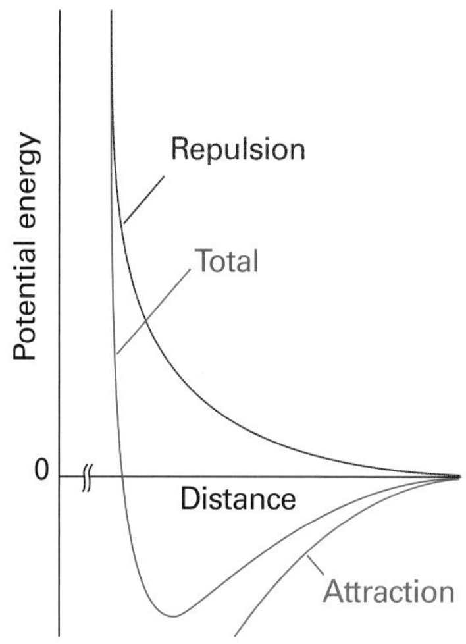

Fig. 18.10 The general form of an intermolecular potential energy curve. At long range the interaction is attractive, but at close range the repulsions dominate.

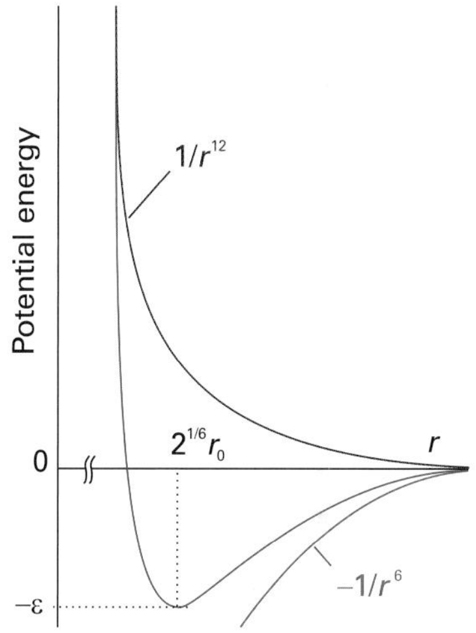

Fig. 18.11 The Lennard-Jones potential, and the relation of the parameters to the features of the curve. The green and purple lines are the two contributions.

Synoptic table 18.4* Lennard-Jones (12,6) parameters

	(ε/k)/K	r_0/pm
Ar	111.84	362.3
CCl_4	376.86	624.1
N_2	91.85	391.9
Xe	213.96	426.0

* More values are given in the *Data section*.

IMPACT ON MEDICINE

I18.1 Molecular recognition and drug design

A drug is a small molecule or protein that binds to a specific receptor site of a target molecule, such as a larger protein or nucleic acid, and inhibits the progress of disease. To devise efficient therapies, we need to know how to characterize and optimize molecular interactions between drug and target.

Molecular interactions are responsible for the assembly of many biological structures. Hydrogen bonding and hydrophobic interactions are primarily responsible for the three-dimensional structures of biopolymers, such as proteins, nucleic acids, and cell membranes. The binding of a ligand, or *guest*, to a biopolymer, or *host*, is also governed by molecular interactions. Examples of biological *host–guest complexes* include enzyme–substrate complexes, antigen–antibody complexes, and drug–receptor complexes. In all these cases, a site on the guest contains functional groups that can interact with complementary functional groups of the host. For example, a hydrogen bond donor group of the guest must be positioned near a hydrogen bond acceptor group of the host for tight binding to occur. It is generally true that many specific intermolecular contacts must be made in a biological host–guest complex and, as a result, a guest binds only hosts that are chemically similar. The strict rules governing molecular recognition of a guest by a host control every biological process, from metabolism to immunological response, and provide important clues for the design of effective drugs for the treatment of disease.

Interactions between nonpolar groups can be important in the binding of a guest to a host. For example, many enzyme active sites have hydrophobic pockets that bind nonpolar groups of a substrate. In addition to dispersion, repulsive, and hydrophobic interactions, π stacking interactions are also possible, in which the planar π systems of aromatic macrocycles lie one on top of the other, in a nearly parallel orientation. Such interactions are responsible for the stacking of hydrogen-bonded base pairs in DNA (Fig. 18.12). Some drugs with planar π systems, shown as a green rectangle in Fig. 18.12, are effective because they intercalate between base pairs through π stacking interactions, causing the helix to unwind slightly and altering the function of DNA.

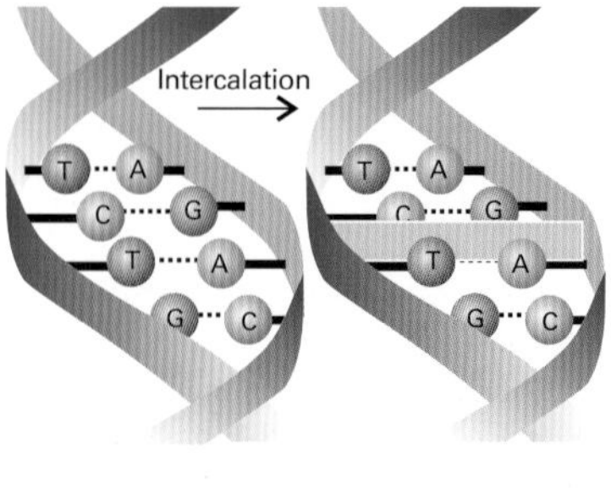

Fig. 18.12 Some drugs with planar π systems, shown as a green rectangle, intercalate between base pairs of DNA.

Coulombic interactions can be important in the interior of a biopolymer host, where the relative permittivity can be much lower than that of the aqueous exterior. For example, at physiological pH, amino acid side chains containing carboxylic acid or amine groups are negatively and positively charged, respectively, and can attract each other. Dipole–dipole interactions are also possible because many of the building blocks of biopolymers are polar, including the peptide link, —CONH— (see *Example* 18.1). However, hydrogen bonding interactions are by far the most prevalent in a biological host–guest complexes. Many effective drugs bind tightly and inhibit the action of enzymes that are associated with the progress of a disease. In many cases, a successful inhibitor will be able to form the same hydrogen bonds with the binding site that the normal substrate of the enzyme can form, except that the drug is chemically inert toward the enzyme.

There are two main strategies for the discovery of a drug. In *structure-based design*, new drugs are developed on the basis of the known structure of the receptor site of a known target. However, in many cases a number of so-called *lead compounds* are known to have some biological activity but little information is available about the target. To design a molecule with improved pharmacological efficacy, **quantitative structure–activity relationships** (QSAR) are often established by correlating data on activity of lead compounds with molecular properties, also called *molecular descriptors*, which can be determined either experimentally or computationally.

In broad terms, the first stage of the QSAR method consists of compiling molecular descriptors for a very large number of lead compounds. Descriptors such as molar mass, molecular dimensions and volume, and relative solubility in water and nonpolar solvents are available from routine experimental procedures. Quantum mechanical descriptors determined by semi-empirical and *ab initio* calculations include bond orders and HOMO and LUMO energies.

In the second stage of the process, biological activity is expressed as a function of the molecular descriptors. An example of a QSAR equation is:

$$Activity = c_0 + c_1 d_1 + c_2 d_1^2 + c_3 d_2 + c_4 d_2^2 + \ldots \quad (18.33)$$

where d_i is the value of the descriptor and c_i is a coefficient calculated by fitting the data by regression analysis. The quadratic terms account for the fact that biological activity can have a maximum or minimum value at a specific descriptor value. For example, a molecule might not cross a biological membrane and become available for binding to targets in the interior of the cell if it is too hydrophilic (water-loving), in which case it will not partition into the hydrophobic layer of the cell membrane (see Section 19.14 for details of membrane structure), or too hydrophobic (water-repelling), for then it may bind too tightly to the membrane. It follows that the activity will peak at some intermediate value of a parameter that measures the relative solubility of the drug in water and organic solvents.

In the final stage of the QSAR process, the activity of a drug candidate can be estimated from its molecular descriptors and the QSAR equation either by interpolation or extrapolation of the data. The predictions are more reliable when a large number of lead compounds and molecular descriptors are used to generate the QSAR equation.

The traditional QSAR technique has been refined into 3D QSAR, in which sophisticated computational methods are used to gain further insight into the three-dimensional features of drug candidates that lead to tight binding to the receptor site of a target. The process begins by using a computer to superimpose three-dimensional structural models of lead compounds and looking for common features, such as similarities in shape, location of functional groups, and electrostatic potential plots, which can be obtained from molecular orbital calculations. The key assumption of the method is that common structural features are indicative of molecular properties that enhance binding of the drug to the receptor. The collection of superimposed molecules is then placed inside a three-dimensional grid of points. An atomic probe, typically an sp^3-hybridized carbon atom, visits each grid point and two energies of interaction are calculated: E_{steric}, the steric energy reflecting interactions between the probe and electrons in uncharged regions of the drug, and E_{elec}, the electrostatic energy arising from interactions between the probe and a region of the molecule carrying a partial charge. The measured equilibrium constant for binding of the drug to the target, K_{bind}, is then assumed to be related to the interaction energies at each point r by the 3D QSAR equation

$$\log K_{\text{bind}} = c_0 + \sum_r \{c_S(\boldsymbol{r})E_{\text{steric}}(\boldsymbol{r}) + c_E(\boldsymbol{r})E_{\text{elec}}(\boldsymbol{r})\} \quad (18.34)$$

where the $c(\boldsymbol{r})$ are coefficients calculated by regression analysis, with the coefficients c_S and c_E reflecting the relative importance of steric and electrostatic interactions, respectively, at the grid point $\boldsymbol{r}$. Visualization of the regression analysis is facilitated by colouring each grid point according to the magnitude of the coefficients. Figure 18.13 shows results of a 3D QSAR analysis of the binding of steroids, molecules with the carbon skeleton shown, to human corticosteroid-binding globulin (CBG). Indeed, we see that the technique lives up to the promise of opening a window into the chemical nature of the binding site even when its structure is not known.

The QSAR and 3D QSAR methods, though powerful, have limited power: the predictions are only as good as the data used in the correlations are both reliable and

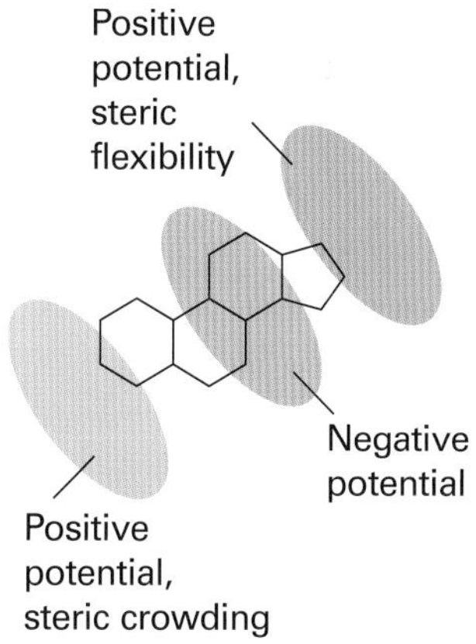

Fig. 18.13 A 3D QSAR analysis of the binding of steroids, molecules with the carbon skeleton shown, to human corticosteroid-binding globulin (CBG). The ellipses indicate areas in the protein's binding site with positive or negative electrostatic potentials and with little or much steric crowding. It follows from the calculations that addition of large substituents near the left-hand side of the molecule (as it is drawn on the page) leads to poor affinity of the drug to the binding site. Also, substituents that lead to the accumulation of negative electrostatic potential at either end of the drug are likely to show enhanced affinity for the binding site. (Adapted from P. Krogsgaard-Larsen, T. Liljefors, U. Madsen (ed.), *Textbook of drug design and discovery*, Taylor & Francis, London (2002).)

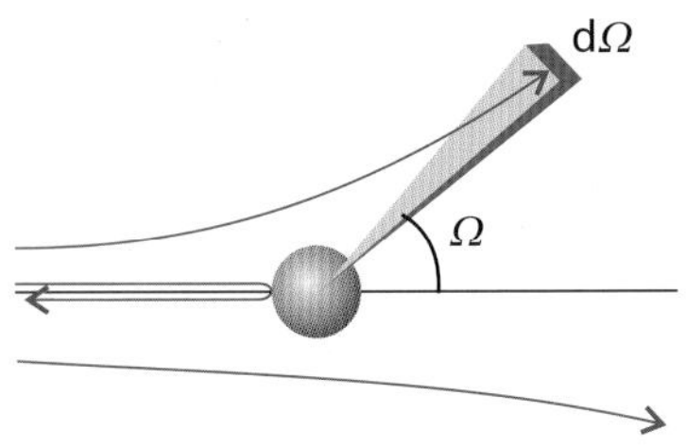

Fig. 18.14 The definition of the solid angle, dΩ, for scattering.

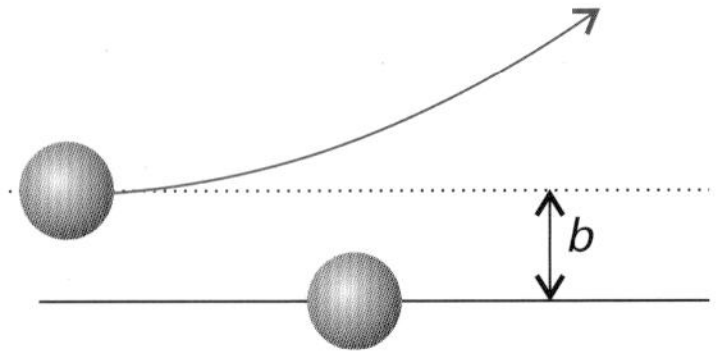

Fig. 18.15 The definition of the impact parameter, b, as the perpendicular separation of the initial paths of the particles.

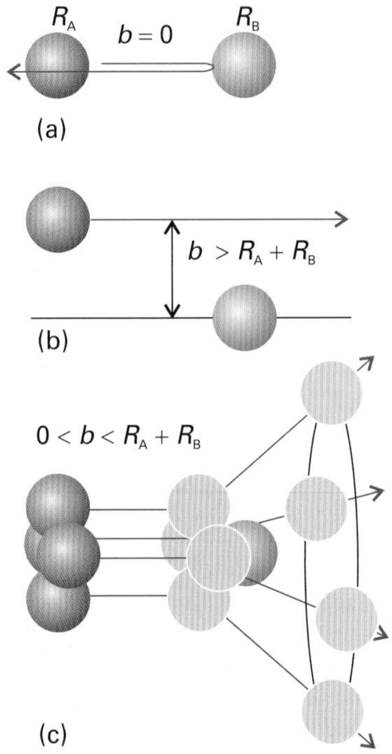

Fig. 18.16 Three typical cases for the collisions of two hard spheres: (a) $b = 0$, giving backward scattering; (b) $b > R_A + R_B$, giving forward scattering; (c) $0 < b < R_A + R_B$, leading to scattering into one direction on a ring of possibilities. (The target molecule is taken to be so heavy that it remains virtually stationary.)

abundant. However, the techniques have been used successfully to identify compounds that deserve further synthetic elaboration, such as addition or removal of functional groups, and testing.

Gases and liquids

The form of matter with the least order is a gas. In a perfect gas there are no intermolecular interactions and the distribution of molecules is completely random. In a real gas there are weak attractions and repulsions that have minimal effect on the relative locations of the molecules but that cause deviations from the perfect gas law for the dependence of pressure on the volume, temperature, and amount (Section 1.3).

The attractions between molecules are responsible for the condensation of gases into liquids at low temperatures. At low enough temperatures the molecules of a gas have insufficient kinetic energy to escape from each other's attraction and they stick together. Second, although molecules attract each other when they are a few diameters apart, as soon as they come into contact they repel each other. This repulsion is responsible for the fact that liquids and solids have a definite bulk and do not collapse to an infinitesimal point. The molecules are held together by molecular interactions, but their kinetic energies are comparable to their potential energies. As a result, we saw in Section 17.6 that, although the molecules of a liquid are not free to escape completely from the bulk, the whole structure is very mobile and we can speak only of the *average* relative locations of molecules. In the following sections we build on those concepts and add thermodynamic arguments to describe the surface of a liquid and the condensation of a gas into a liquid.

18.6 Molecular interactions in gases

Molecular interactions in the gas phase can be studied in **molecular beams**, which consist of a collimated, narrow stream of molecules travelling though an evacuated vessel. The beam is directed towards other molecules, and the scattering that occurs on impact is related to the intermolecular interactions.

The primary experimental information from a molecular beam experiment is the fraction of the molecules in the incident beam that are scattered into a particular direction. The fraction is normally expressed in terms of dI, the rate at which molecules are scattered into a cone that represents the area covered by the 'eye' of the detector (Fig. 18.14). This rate is reported as the **differential scattering cross-section**, σ, the constant of proportionality between the value of dI and the intensity, I, of the incident beam, the number density of target molecules, $\mathcal{N}$, and the infinitesimal path length dx through the sample:

$$\mathrm{d}I = \sigma I \mathcal{N} \mathrm{d}x \qquad (18.35)$$

The value of σ (which has the dimensions of area) depends on the **impact parameter**, b, the initial perpendicular separation of the paths of the colliding molecules (Fig. 18.15), and the details of the intermolecular potential. The role of the impact parameter is most easily seen by considering the impact of two hard spheres (Fig. 18.16). If $b = 0$, the lighter projectile is on a trajectory that leads to a head-on collision, so the only scattering intensity is detected when the detector is at $\theta = \pi$. When the impact parameter is so great that the spheres do not make contact ($b > R_A + R_B$), there is no scattering and the scattering cross-section is zero at all angles except $\theta = 0$. Glancing blows, with $0 < b \leq R_A + R_B$, lead to scattering intensity in cones around the forward direction.

The scattering pattern of real molecules, which are not hard spheres, depends on the details of the intermolecular potential, including the anisotropy that is present when the molecules are non-spherical. The scattering also depends on the relative speed of approach of the two particles: a very fast particle might pass through the interaction region without much deflection, whereas a slower one on the same path might be temporarily captured and undergo considerable deflection (Fig. 18.17). The variation of the scattering cross-section with the relative speed of approach should therefore give information about the strength and range of the intermolecular potential.

A further point is that the outcome of collisions is determined by quantum, not classical, mechanics. The wave nature of the particles can be taken into account, at least to some extent, by drawing all classical trajectories that take the projectile particle from source to detector, and then considering the effects of interference between them.

Two quantum mechanical effects are of great importance. A particle with a certain impact parameter might approach the attractive region of the potential in such a way that the particle is deflected towards the repulsive core (Fig. 18.18), which then repels it out through the attractive region to continue its flight in the forward direction. Some molecules, however, also travel in the forward direction because they have impact parameters so large that they are undeflected. The wavefunctions of the particles that take the two types of path interfere, and the intensity in the forward direction is modified. The effect is called **quantum oscillation**. The same phenomenon accounts for the optical 'glory effect', in which a bright halo can sometimes be seen surrounding an illuminated object. (The coloured rings around the shadow of an aircraft cast on clouds by the Sun, and often seen in flight, is an example of an optical glory.)

The second quantum effect we need consider is the observation of a strongly enhanced scattering in a non-forward direction. This effect is called **rainbow scattering** because the same mechanism accounts for the appearance of an optical rainbow. The origin of the phenomenon is illustrated in Fig. 18.19. As the impact parameter decreases, there comes a stage at which the scattering angle passes through a maximum and the interference between the paths results in a strongly scattered beam. The **rainbow angle**, θ_r, is the angle for which $d\theta/db = 0$ and the scattering is strong.

Another phenomenon that can occur in certain beams is the capturing of one species by another. The vibrational temperature in supersonic beams is so low that **van der Waals molecules** may be formed, which are complexes of the form AB in which A and B are held together by van der Waals forces or hydrogen bonds. Large numbers of such molecules have been studied spectroscopically, including ArHCl, $(HCl)_2$, $ArCO_2$, and $(H_2O)_2$. More recently, van der Waals clusters of water molecules have been pursued as far as $(H_2O)_6$. The study of their spectroscopic properties gives detailed information about the intermolecular potentials involved.

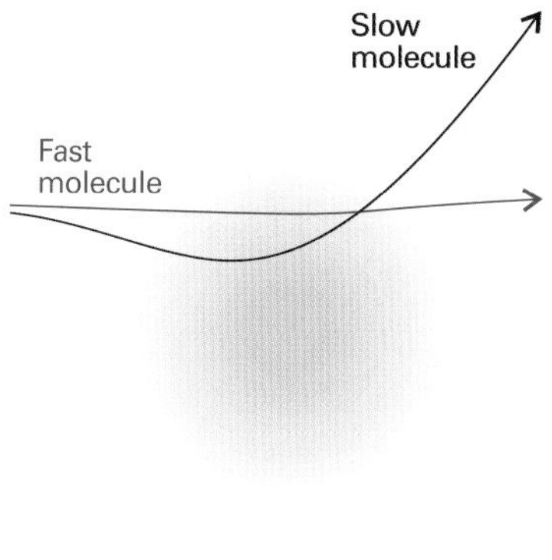

Fig. 18.17 The extent of scattering may depend on the relative speed of approach as well as the impact parameter. The darker central zone represents the repulsive core; the fuzzy outer zone represents the long-range attractive potential.

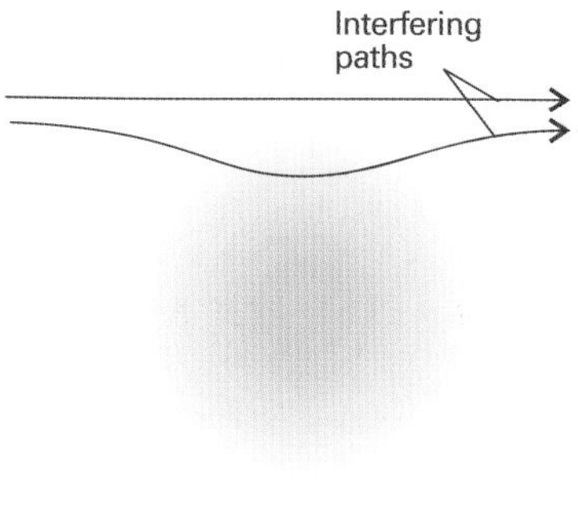

Fig. 18.18 Two paths leading to the same destination will interfere quantum mechanically; in this case they give rise to quantum oscillations in the forward direction.

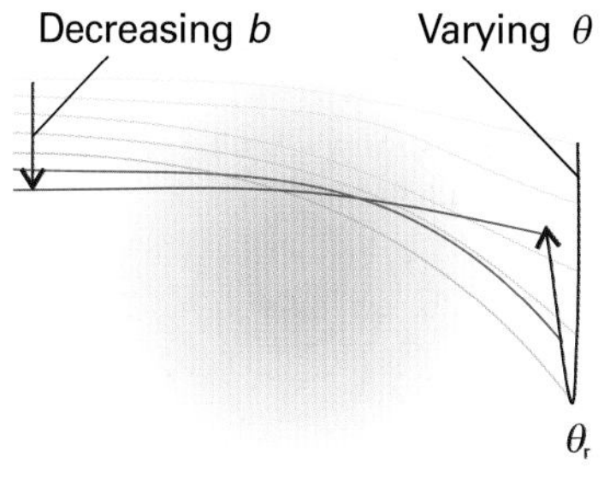

Fig. 18.19 The interference of paths leading to rainbow scattering. The rainbow angle, θ_r, is the maximum scattering angle reached as b is decreased. Interference between the numerous paths at that angle modifies the scattering intensity markedly.

18.7 The liquid–vapour interface

So far, we have concentrated on the properties of gases. In Section 17.6, we described the structure of liquids. Now we turn our attention to the physical boundary between phases, such as the surface where solid is in contact with liquid or liquid is in contact with its vapour, has interesting properties. In this section we concentrate on the liquid–vapour interface, which is interesting because it is so mobile. Chapter 25 deals with solid surfaces and their important role in catalysis.

(a) Surface tension

Liquids tend to adopt shapes that minimize their surface area, for then the maximum number of molecules are in the bulk and hence surrounded by and interacting with neighbours. Droplets of liquids therefore tend to be spherical, because a sphere is the

Synoptic table 18.5* Surface tensions of liquids at 293 K

	$\gamma/(\mathrm{mN\ m^{-1}})^{\dagger}$
Benzene	28.88
Mercury	472
Methanol	22.6
Water	72.75

* More values are given in the *Data section.*
† Note that $1\ \mathrm{N\ m^{-1}} = 1\ \mathrm{J\ m^{-2}}$.

shape with the smallest surface-to-volume ratio. However, there may be other forces present that compete against the tendency to form this ideal shape, and in particular gravity may flatten spheres into puddles or oceans.

Surface effects may be expressed in the language of Helmholtz and Gibbs energies (Chapter 3). The link between these quantities and the surface area is the work needed to change the area by a given amount, and the fact that dA and dG are equal (under different conditions) to the work done in changing the energy of a system. The work needed to change the surface area, σ, of a sample by an infinitesimal amount dσ is proportional to dσ, and we write

$$\mathrm{d}w = \gamma \mathrm{d}\sigma \qquad [18.36]$$

The constant of proportionality, γ, is called the **surface tension**; its dimensions are energy/area and its units are typically joules per metre squared (J m^{-2}). However, as in Table 18.5, values of γ are usually reported in newtons per metre (N m^{-1}, because 1 J = 1 N m). The work of surface formation at constant volume and temperature can be identified with the change in the Helmholtz energy, and we can write

$$\mathrm{d}A = \gamma \mathrm{d}\sigma \qquad (18.37)$$

Because the Helmholtz energy decreases (d$A < 0$) if the surface area decreases (d$\sigma < 0$), surfaces have a natural tendency to contract. This is a more formal way of expressing what we have already described.

Example 18.4 *Using the surface tension*

Calculate the work needed to raise a wire of length l and to stretch the surface of a liquid through a height h in the arrangement shown in Fig. 18.20. Disregard gravitational potential energy.

Method According to eqn 18.36, the work required to create a surface area given that the surface tension does not vary as the surface is formed is $w = \gamma\sigma$. Therefore, all we need do is to calculate the surface area of the two-sided rectangle formed as the frame is withdrawn from the liquid.

Answer When the wire of length l is raised through a height h it increases the area of the liquid by twice the area of the rectangle (because there is a surface on each side). The total increase is therefore $2lh$ and the work done is $2\gamma lh$. The work can be expressed as a force × distance by writing it as $2\gamma l \times h$, and identifying γl as the opposing force on the wire of length l. This is why γ is called a tension and why its units are often chosen to be newtons per metre (N m^{-1}, so γl is a force in newtons).

Self-test 18.5 Calculate the work of creating a spherical cavity of radius r in a liquid of surface tension γ. [$4\pi r^2\gamma$]

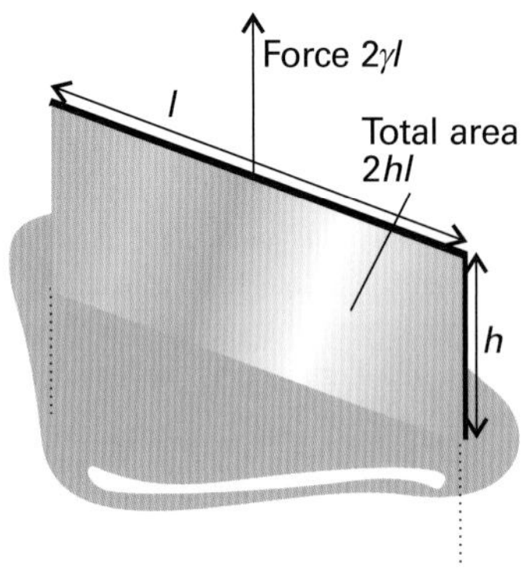

Fig. 18.20 The model used for calculating the work of forming a liquid film when a wire of length l is raised and pulls the surface with it through a height h.

(b) Curved surfaces

The minimization of the surface area of a liquid may result in the formation of a curved surface. A **bubble** is a region in which vapour (and possibly air too) is trapped by a thin film; a **cavity** is a vapour-filled hole in a liquid. What are widely called 'bubbles' in liquids are therefore strictly cavities. True bubbles have two surfaces (one on each side of the film); cavities have only one. The treatments of both are similar, but a factor of 2 is required for bubbles to take into account the doubled surface area. A **droplet** is a small volume of liquid at equilibrium surrounded by its vapour (and possibly also air).

The pressure on the concave side of an interface, p_{in}, is always greater than the pressure on the convex side, p_{out}. This relation is expressed by the **Laplace equation**, which is derived in the following *Justification*:

$$p_{in} = p_{out} + \frac{2\gamma}{r} \tag{18.38}$$

Justification 18.6 *The Laplace equation*

The cavities in a liquid are at equilibrium when the tendency for their surface area to decrease is balanced by the rise of internal pressure which would then result. When the pressure inside a cavity is p_{in} and its radius is r, the outward force is

$$\text{pressure} \times \text{area} = 4\pi r^2 p_{in}$$

The force inwards arises from the external pressure and the surface tension. The former has magnitude $4\pi r^2 p_{out}$. The latter is calculated as follows. The change in surface area when the radius of a sphere changes from r to $r + dr$ is

$$d\sigma = 4\pi(r + dr)^2 - 4\pi r^2 = 8\pi r dr$$

(The second-order infinitesimal, $(dr)^2$, is ignored.) The work done when the surface is stretched by this amount is therefore

$$dw = 8\pi\gamma r dr$$

As force × distance is work, the force opposing stretching through a distance dr when the radius is r is

$$F = 8\pi\gamma r$$

The total inward force is therefore $4\pi r^2 p_{out} + 8\pi\gamma r$. At equilibrium, the outward and inward forces are balanced, so we can write

$$4\pi r^2 p_{in} = 4\pi r^2 p_{out} + 8\pi\gamma r$$

which rearranges into eqn 18.38.

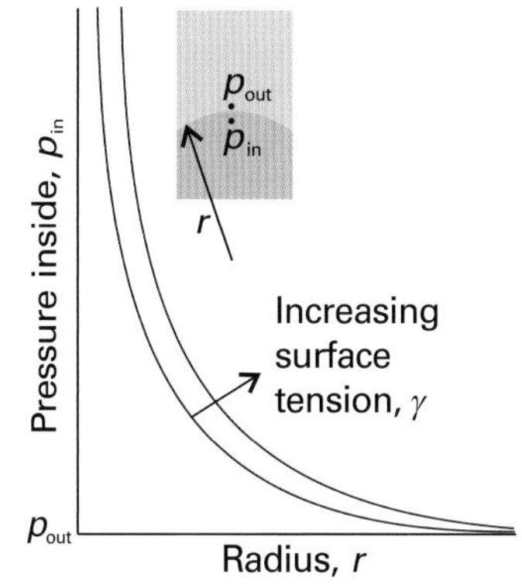

Fig. 18.21 The dependence of the pressure inside a curved surface on the radius of the surface, for two different values of the surface tension.

The Laplace equation shows that the difference in pressure decreases to zero as the radius of curvature becomes infinite (when the surface is flat, Fig. 18.21). Small cavities have small radii of curvature, so the pressure difference across their surface is quite large. For instance, a 'bubble' (actually, a cavity) of radius 0.10 mm in champagne implies a pressure difference of 1.5 kPa, which is enough to sustain a column of water of height 15 cm.

(c) Capillary action

The tendency of liquids to rise up capillary tubes (tubes of narrow bore), which is called **capillary action**, is a consequence of surface tension. Consider what happens when a glass capillary tube is first immersed in water or any liquid that has a tendency to adhere to the walls. The energy is lowest when a thin film covers as much of the glass as possible. As this film creeps up the inside wall it has the effect of curving the surface of the liquid inside the tube. This curvature implies that the pressure just beneath the curving meniscus is less than the atmospheric pressure by approximately $2\gamma/r$, where r is the radius of the tube and we assume a hemispherical surface. The pressure immediately under the flat surface outside the tube is p, the atmospheric pressure; but inside the tube under the curved surface it is only $p - 2\gamma/r$. The excess external pressure presses the liquid up the tube until hydrostatic equilibrium (equal pressures at equal depths) has been reached (Fig. 18.22).

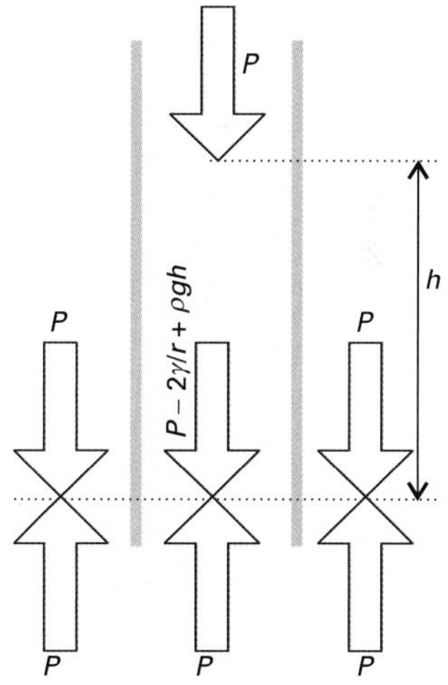

Fig. 18.22 When a capillary tube is first stood in a liquid, the latter climbs up the walls, so curving the surface. The pressure just under the meniscus is less than that arising from the atmosphere by $2\gamma/r$. The pressure is equal at equal heights throughout the liquid provided the hydrostatic pressure (which is equal to ρgh) cancels the pressure difference arising from the curvature.

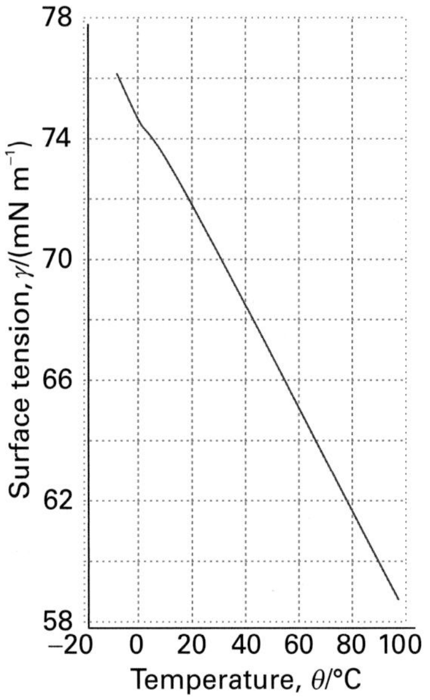

Fig. 18.23 The variation of the surface tension of water with temperature.

To calculate the height to which the liquid rises, we note that the pressure exerted by a column of liquid of mass density ρ and height h is

$$p = \rho g h \tag{18.39}$$

This hydrostatic pressure matches the pressure difference $2\gamma/r$ at equilibrium. Therefore, the height of the column at equilibrium is obtained by equating $2\gamma/r$ and ρgh, which gives

$$h = \frac{2\gamma}{\rho g r} \tag{18.40}$$

This simple expression provides a reasonably accurate way of measuring the surface tension of liquids. Surface tension decreases with increasing temperature (Fig. 18.23).

Illustration 18.2 *Calculating the surface tension of a liquid from its capillary rise*

If water at 25°C rises through 7.36 cm in a capillary of radius 0.20 mm, its surface tension at that temperature is

$$\begin{aligned}\gamma &= \tfrac{1}{2}\rho g h r \\ &= \tfrac{1}{2} \times (997.1\ \mathrm{kg\ m^{-3}}) \times (9.81\ \mathrm{m\ s^{-2}}) \times (7.36 \times 10^{-2}\ \mathrm{m}) \times (2.0 \times 10^{-4}\ \mathrm{m}) \\ &= 72\ \mathrm{mN\ m^{-1}}\end{aligned}$$

where we have used $1\ \mathrm{kg\ m\ s^{-2}} = 1\ \mathrm{N}$.

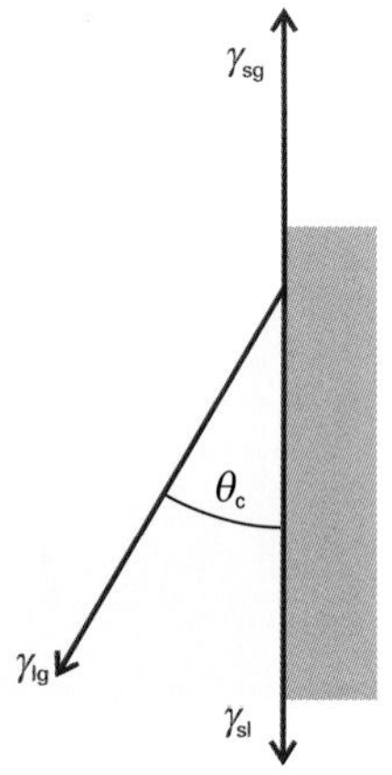

Fig. 18.24 The balance of forces that results in a contact angle, θ_c.

When the adhesive forces between the liquid and the material of the capillary wall are weaker than the cohesive forces within the liquid (as for mercury in glass), the liquid in the tube retracts from the walls. This retraction curves the surface with the concave, high pressure side downwards. To equalize the pressure at the same depth throughout the liquid the surface must fall to compensate for the heightened pressure arising from its curvature. This compensation results in a capillary depression.

In many cases there is a nonzero angle between the edge of the meniscus and the wall. If this contact angle is θ_c, then eqn 18.40 should be modified by multiplying the right-hand side by $\cos\theta_c$. The origin of the contact angle can be traced to the balance of forces at the line of contact between the liquid and the solid (Fig. 18.24). If the solid–gas, solid–liquid, and liquid–gas surface tensions (essentially the energy needed to create unit area of each of the interfaces) are denoted γ_{sg}, γ_{sl}, and γ_{lg}, respectively, then the vertical forces are in balance if

$$\gamma_{sg} = \gamma_{sl} + \gamma_{lg}\cos\theta_c \tag{18.41}$$

This expression solves to

$$\cos\theta_c = \frac{\gamma_{sg} - \gamma_{sl}}{\gamma_{lg}} \tag{18.42}$$

If we note that the superficial work of adhesion of the liquid to the solid (the work of adhesion divided by the area of contact) is

$$w_{ad} = \gamma_{sg} + \gamma_{lg} - \gamma_{sl} \tag{18.43}$$

eqn 18.42 can be written

$$\cos\theta_c = \frac{w_{ad}}{\gamma_{lg}} - 1 \tag{18.44}$$

We now see that the liquid 'wets' (spreads over) the surface, corresponding to $0 < \theta_c < 90°$, when $1 < w_{ad}/\gamma_{lg} < 2$ (Fig. 18.25). The liquid does not wet the surface, corresponding to $90° < \theta_c < 180°$, when $0 < w_{ad}/\gamma_{lg} < 1$. For mercury in contact with glass, $\theta_c = 140°$, which corresponds to $w_{ad}/\gamma_{lg} = 0.23$, indicating a relatively low work of adhesion of the mercury to glass on account of the strong cohesive forces within mercury.

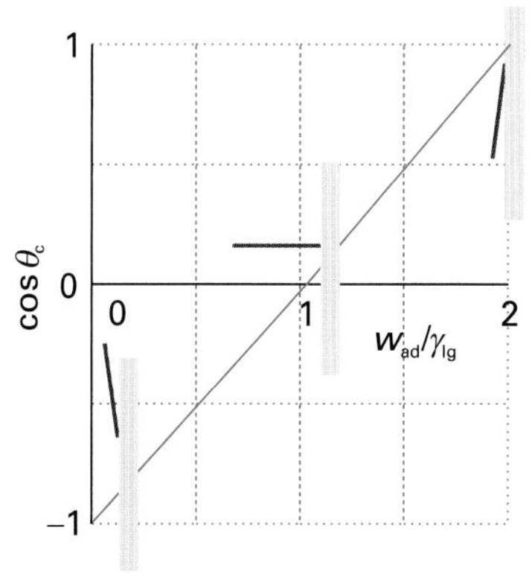

Fig. 18.25 The variation of contact angle (shown by the semaphore-like object) as the ratio w_{ad}/γ_{lg} changes.

18.8 Condensation

We now bring together concepts from this chapter and Chapter 4 to explain the condensation of a gas to a liquid. We saw in Section 4.5 that the vapour pressure of a liquid depends on the pressure applied to the liquid. Because curving a surface gives rise to a pressure differential of $2\gamma/r$, we can expect the vapour pressure above a curved surface to be different from that above a flat surface. By substituting this value of the pressure difference into eqn 4.3 ($p = p^\star e^{V_m\Delta P/RT}$, where $p^\star$ is the vapour pressure when the pressure difference is zero) we obtain the **Kelvin equation** for the vapour pressure of a liquid when it is dispersed as droplets of radius r:

$$p = p^\star e^{2\gamma V_m/rRT} \qquad (18.45)$$

The analogous expression for the vapour pressure inside a cavity can be written at once. The pressure of the liquid outside the cavity is less than the pressure inside, so the only change is in the sign of the exponent in the last expression.

For droplets of water of radius 1 μm and 1 nm the ratios $p/p^\star$ at 25°C are about 1.001 and 3, respectively. The second figure, although quite large, is unreliable because at that radius the droplet is less than about 10 molecules in diameter and the basis of the calculation is suspect. The first figure shows that the effect is usually small; nevertheless it may have important consequences.

Consider, for example, the formation of a cloud. Warm, moist air rises into the cooler regions higher in the atmosphere. At some altitude the temperature is so low that the vapour becomes thermodynamically unstable with respect to the liquid and we expect it to condense into a cloud of liquid droplets. The initial step can be imagined as a swarm of water molecules congregating into a microscopic droplet. Because the initial droplet is so small it has an enhanced vapour pressure. Therefore, instead of growing it evaporates. This effect stabilizes the vapour because an initial tendency to condense is overcome by a heightened tendency to evaporate. The vapour phase is then said to be **supersaturated**. It is thermodynamically unstable with respect to the liquid but not unstable with respect to the small droplets that need to form before the bulk liquid phase can appear, so the formation of the latter by a simple, direct mechanism is hindered.

Clouds do form, so there must be a mechanism. Two processes are responsible. The first is that a sufficiently large number of molecules might congregate into a droplet so big that the enhanced evaporative effect is unimportant. The chance of one of these **spontaneous nucleation centres** forming is low, and in rain formation it is not a dominant mechanism. The more important process depends on the presence of minute dust particles or other kinds of foreign matter. These **nucleate** the condensation (that is, provide centres at which it can occur) by providing surfaces to which the water molecules can attach.

Liquids may be **superheated** above their boiling temperatures and **supercooled** below their freezing temperatures. In each case the thermodynamically stable phase is not achieved on account of the kinetic stabilization that occurs in the absence of nucleation centres. For example, superheating occurs because the vapour pressure inside a cavity is artificially low, so any cavity that does form tends to collapse. This instability is encountered when an unstirred beaker of water is heated, for its temperature may be raised above its boiling point. Violent bumping often ensues as spontaneous nucleation leads to bubbles big enough to survive. To ensure smooth boiling at the true boiling temperature, nucleation centres, such as small pieces of sharp-edged glass or bubbles (cavities) of air, should be introduced.

Checklist of key ideas

☐ 1. A polar molecule is a molecule with a permanent electric dipole moment; the magnitude of a dipole moment is the product of the partial charge and the separation.

☐ 2. The polarization is the electric dipole moment density, $P = \langle \mu \rangle \mathcal{N}$. Orientation polarization is the polarization arising from the permanent dipole moments. Distortion polarization is the polarization arising from the distortion of the positions of the nuclei by the applied field.

☐ 3. The polarizability is a measure of the ability of an electric field to induce a dipole moment in a molecule ($\mu = \alpha \mathcal{E}$). Electronic polarizability is the polarizability due to the distortion of the electron distribution.

☐ 4. The permittivity is the quantity ε in the Coulomb potential energy, $V = q_1 q_2 / 4\pi\varepsilon r$.

☐ 5. The relative permittivity is given by $\varepsilon_r = \varepsilon/\varepsilon_0$ and may be calculated from electric properties by using the Debye equation (eqn 18.14) or the Clausius–Mossotti equation (eqn 18.16).

☐ 6. A van der Waals interaction between closed-shell molecules is inversely proportional to the sixth power of their separation.

☐ 7. The potential energy of the dipole–dipole interaction between two fixed (non-rotating) molecules is proportional to $\mu_1\mu_2/r^3$ and that between molecules that are free to rotate is proportional to $\mu_1^2\mu_2^2/kTr^6$.

☐ 8. The dipole–induced-dipole interaction between two molecules is proportional to $\mu_1^2\alpha_2/r^6$, where α is the polarizability.

☐ 9. The potential energy of the dispersion (or London) interaction is proportional to $\alpha_1\alpha_2/r^6$.

☐ 10. A hydrogen bond is an interaction of the form A—H···B, where A and B are N, O, or F.

☐ 11. A hydrophobic interaction is an interaction that favours formation of clusters of hydrophobic groups in aqueous environments and that stems from changes in entropy of water molecules.

☐ 12. The Lennard-Jones (12,6) potential, $V = 4\varepsilon\{(r_0/r)^{12} - (r_0/r)^6\}$, is a model of the total intermolecular potential energy.

☐ 13. A molecular beam is a collimated, narrow stream of molecules travelling though an evacuated vessel. Molecular beam techniques are used to investigate molecular interactions in gases.

☐ 14. The work of forming a liquid surface is $dw = \gamma d\sigma$, where γ is the surface tension.

☐ 15. The Laplace equation for the vapour pressure at a curved surface is $p_{in} = p_{out} + 2\gamma/r$.

☐ 16. The Kelvin equation for the vapour pressure of droplets is $p = p^* e^{2\gamma V_m/rRT}$.

☐ 17. Capillary action is the tendency of liquids to rise up capillary tubes.

☐ 18. Nucleation provides surfaces to which molecules can attach and thereby induce condensation.

Further reading

Articles and texts

P.W. Atkins and R.S. Friedman, *Molecular quantum mechanics*. Oxford University Press (2005).

M.A.D. Fluendy and K.P. Lawley, *Chemical applications of molecular beam scattering*. Chapman and Hall, London (1973).

J.N. Israelachvili, *Intermolecular and surface forces*. Academic Press, New York (1998).

G.A. Jeffrey, *An introduction to hydrogen bonding*. Oxford University Press (1997).

H.-J. Schneider and A. Yatsimirsky, *Principles and methods in supramolecular chemistry*. Wiley, Chichester (1999).

Sources of data and information

J.J. Jasper, The surface tension of pure liquid compounds. *J. Phys. Chem. Ref. Data* **1**, 841 (1972).

D.R. Lide (ed.), *CRC Handbook of Chemistry and Physics*, Sections 3, 4, 6, 9, 10, 12, and 13. CRC Press, Boca Raton (2000).

Further information

Further information 18.1 *The dipole–dipole interaction*

An important problem in physical chemistry is the calculation of the potential energy of interaction between two point dipoles with moments $\boldsymbol{\mu}_1$ and $\boldsymbol{\mu}_2$, separated by a vector $\boldsymbol{r}$. From classical electromagnetic theory, the potential energy of $\boldsymbol{\mu}_2$ in the electric field $\boldsymbol{\mathcal{E}}_1$ generated by $\boldsymbol{\mu}_1$ is given by the dot (scalar) product

$$V = -\boldsymbol{\mathcal{E}}_1 \cdot \boldsymbol{\mu}_2 \qquad (18.46)$$

To calculate $\mathcal{E}_1$, we consider a distribution of point charges q_i located at x_i, y_i, and z_i from the origin. The Coulomb potential ϕ due to this distribution at a point with coordinates x, y, and z is:

$$\phi = \sum_i \frac{q_i}{4\pi\varepsilon_0} \frac{1}{\{(x-x_i)^2+(y-y_i)^2+(z-z_i)^2\}^{1/2}} \qquad (18.47)$$

Comment 18.10

The potential energy of a charge q_1 in the presence of another charge q_2 may be written as $V = q_1\phi$, where $\phi = q_2/4\pi\varepsilon_0 r$ is the Coulomb potential. If there are several charges $q_2, q_3, \ldots$ present in the system, then the total potential experienced by the charge q_1 is the sum of the potential generated by each charge: $\phi = \phi_2 + \phi_3 + \cdots$. The electric field strength is the negative gradient of the electric potential: $\mathcal{E} = -\nabla\phi$. See *Appendix* 3 for more details.

where r is the location of the point of interest and the $\boldsymbol{r}_i$ are the locations of the charges q_i. If we suppose that all the charges are close to the origin (in the sense that $r_i \ll r$), we can use a Taylor expansion to write

$$\phi(\boldsymbol{r}) = \sum_i \frac{q_i}{4\pi\varepsilon_0}\left\{\frac{1}{r} + \left(\frac{\partial\{(x-x_i)^2+(y-y_i)^2+(z-z_i)^2\}^{1/2}}{\partial x_i}\right)_{x_i=0} x_i + \cdots\right\}$$

$$= \sum_i \frac{q_i}{4\pi\varepsilon_0}\left\{\frac{1}{r} + \frac{xx_i}{r^3} + \cdots\right\} \qquad (18.48)$$

where the ellipses include the terms arising from derivatives with respect to y_i and z_i and higher derivatives. If the charge distribution is electrically neutral, the first term disappears because $\Sigma_i q_i = 0$. Next we note that $\Sigma_i q_i x_i = \mu_x$, and likewise for the y- and z-components. That is,

$$\phi = \frac{1}{4\pi\varepsilon_0 r^3}(\mu_x x + \mu_y y + \mu_z z) = \frac{1}{4\pi\varepsilon_0 r^3}\boldsymbol{\mu}_1\cdot\boldsymbol{r} \qquad (18.49)$$

The electric field strength is (see *Comment* 18.10)

$$\mathcal{E}_1 = \frac{1}{4\pi\varepsilon_0}\nabla\frac{\boldsymbol{\mu}_1\cdot\boldsymbol{r}}{r^3} = -\frac{\boldsymbol{\mu}_1}{4\pi\varepsilon_0 r^3} - \frac{\boldsymbol{\mu}_1\cdot\boldsymbol{r}}{4\pi\varepsilon_0}\nabla\frac{1}{r^3} \qquad (18.50)$$

It follows from eqns 18.46 and 18.50 that

$$V = \frac{\boldsymbol{\mu}_1\cdot\boldsymbol{\mu}_2}{4\pi\varepsilon_0 r^3} - 3\frac{(\boldsymbol{\mu}_1\cdot\boldsymbol{r})(\boldsymbol{\mu}_2\cdot\boldsymbol{r})}{4\pi\varepsilon_0 r^5} \qquad (18.51)$$

For the arrangement shown in (**13**), in which $\boldsymbol{\mu}_1\cdot\boldsymbol{r} = \mu_1 r\cos\theta$ and $\boldsymbol{\mu}_2\cdot\boldsymbol{r} = \mu_2 r\cos\theta$, eqn 18.51 becomes:

$$V = \frac{\mu_1\mu_2 f(\theta)}{4\pi\varepsilon_0 r^3} \qquad f(\theta) = 1 - 3\cos^2\theta \qquad (18.52)$$

which is eqn 18.22.

Further information 18.2 *The basic principles of molecular beams*

The basic arrangement for a molecular beam experiment is shown in Fig. 18.26. If the pressure of vapour in the source is increased so that the mean free path of the molecules in the emerging beam is much shorter than the diameter of the pinhole, many collisions take place even outside the source. The net effect of these collisions, which give rise to **hydrodynamic flow**, is to transfer momentum into the direction of the beam. The molecules in the beam then travel with very similar speeds, so further downstream few collisions take place between them. This condition is called **molecular flow**. Because the spread in speeds is so small, the molecules are effectively in a state of very low translational temperature (Fig. 18.27). The translational temperature may reach as low as 1 K. Such jets are called **supersonic** because the average speed of the molecules in the jet is much greater than the speed of sound for the molecules that are not part of the jet.

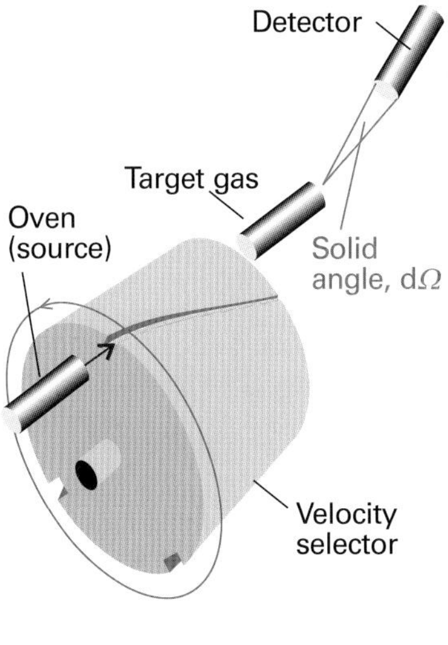

Fig. 18.26 The basic arrangement of a molecular beam apparatus. The atoms or molecules emerge from a heated source, and pass through the velocity selector, a rotating slotted cylinder such as that discussed in Section 1.3a. The scattering occurs from the target gas (which might take the form of another beam), and the flux of particles entering the detector set at some angle is recorded.

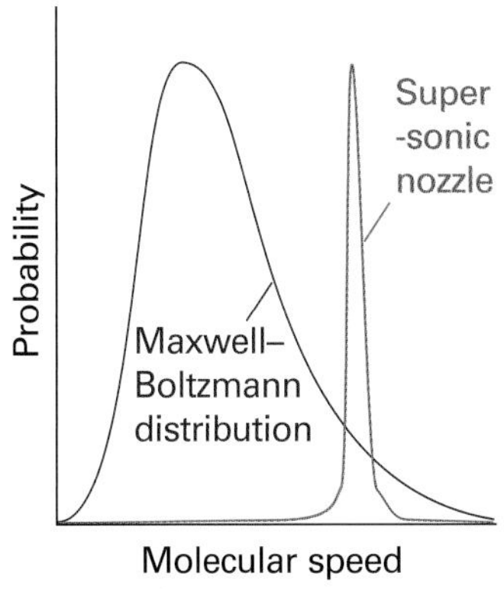

Fig. 18.27 The shift in the mean speed and the width of the distribution brought about by use of a supersonic nozzle.

A supersonic jet can be converted into a more parallel **supersonic beam** if it is 'skimmed' in the region of hydrodynamic flow and the excess gas pumped away. A skimmer consists of a conical nozzle shaped to avoid any supersonic shock waves spreading back into the gas and so increasing the translational temperature (Fig. 18.28). A jet or beam may also be formed by using helium or neon as the principal

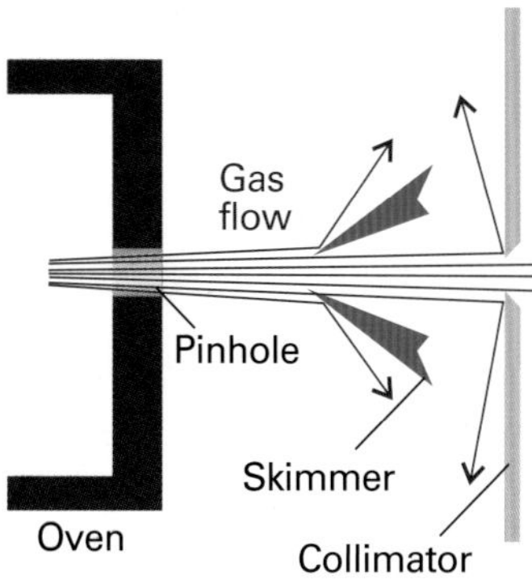

Fig. 18.28 A supersonic nozzle skims off some of the molecules of the beam and leads to a beam with well defined velocity.

gas, and injecting molecules of interest into it in the hydrodynamic region of flow.

The low translational temperature of the molecules is reflected in the low rotational and vibrational temperatures of the molecules. In this context, a rotational or vibrational temperature means the temperature that should be used in the Boltzmann distribution to reproduce the observed populations of the states. However, as rotational modes equilibrate more slowly, and vibrational modes equilibrate even more slowly, the rotational and vibrational populations of the species correspond to somewhat higher temperatures, of the order of 10 K for rotation and 100 K for vibrations.

The target gas may be either a bulk sample or another molecular beam. The latter **crossed beam technique** gives a lot of information because the states of both the target and projectile molecules may be controlled. The intensity of the incident beam is measured by the **incident beam flux**, I, which is the number of particles passing through a given area in a given interval divided by the area and the duration of the interval.

The detectors may consist of a chamber fitted with a sensitive pressure gauge, a bolometer (a detector that responds to the incident energy by making use of the temperature-dependence of resistance), or an ionization detector, in which the incoming molecule is first ionized and then detected electronically. The state of the scattered molecules may also be determined spectroscopically, and is of interest when the collisions change their vibrational or rotational states.

Discussion questions

18.1 Explain how the permanent dipole moment and the polarizability of a molecule arise.

18.2 Explain why the polarizability of a molecule decreases at high frequencies.

18.3 Describe the experimental procedures available for determining the electric dipole moment of a molecule.

18.4 Account for the theoretical conclusion that many attractive interactions between molecules vary with their separation as $1/r^6$.

18.5 Describe the formation of a hydrogen bond in terms of molecular orbitals.

18.6 Account for the hydrophobic interaction and discuss its manifestations.

18.7 Describe how molecular beams are used to investigate intermolecular potentials.

Exercises

18.1a Which of the following molecules may be polar: ClF_3, O_3, H_2O_2?

18.1b Which of the following molecules may be polar: SO_3, XeF_4, SF_4?

18.2a The electric dipole moment of toluene (methylbenzene) is 0.4 D. Estimate the dipole moments of the three xylenes (dimethylbenzene). Which answer can you be sure about?

18.2b Calculate the resultant of two dipole moments of magnitude 1.5 D and 0.80 D that make an angle of 109.5° to each other.

18.3a Calculate the magnitude and direction of the dipole moment of the following arrangement of charges in the xy-plane: $3e$ at (0,0), $-e$ at (0.32 nm, 0), and $-2e$ at an angle of 20° from the x-axis and a distance of 0.23 nm from the origin.

18.3b Calculate the magnitude and direction of the dipole moment of the following arrangement of charges in the xy-plane: $4e$ at (0, 0), $-2e$ at (162 pm, 0), and $-2e$ at an angle of 30° from the x-axis and a distance of 143 pm from the origin.

18.4a The molar polarization of fluorobenzene vapour varies linearly with T^{-1}, and is 70.62 $cm^3\ mol^{-1}$ at 351.0 K and 62.47 $cm^3\ mol^{-1}$ at 423.2 K. Calculate the polarizability and dipole moment of the molecule.

18.4b The molar polarization of the vapour of a compound was found to vary linearly with T^{-1}, and is 75.74 $cm^3\ mol^{-1}$ at 320.0 K and 71.43 $cm^3\ mol^{-1}$ at 421.7 K. Calculate the polarizability and dipole moment of the molecule.

18.5a At 0°C, the molar polarization of liquid chlorine trifluoride is 27.18 $cm^3\ mol^{-1}$ and its density is 1.89 $g\ cm^{-3}$. Calculate the relative permittivity of the liquid.

18.5b At 0°C, the molar polarization of a liquid is 32.16 $cm^3\ mol^{-1}$ and its density is 1.92 $g\ cm^{-3}$. Calculate the relative permittivity of the liquid. Take $M = 55.0\ g\ mol^{-1}$.

18.6a The polarizability volume of H_2O is $1.48 \times 10^{-24}\ cm^3$; calculate the dipole moment of the molecule (in addition to the permanent dipole moment) induced by an applied electric field of strength 1.0 $kV\ cm^{-1}$.

18.6b The polarizability volume of NH_3 is $2.22 \times 10^{-30}\ m^3$; calculate the dipole moment of the molecule (in addition to the permanent dipole moment) induced by an applied electric field of strength 15.0 $kV\ m^{-1}$.

18.7a The refractive index of CH_2I_2 is 1.732 for 656 nm light. Its density at 20°C is 3.32 $g\ cm^{-3}$. Calculate the polarizability of the molecule at this wavelength.

18.7b The refractive index of a compound is 1.622 for 643 nm light. Its density at 20°C is 2.99 g cm^{-3}. Calculate the polarizability of the molecule at this wavelength. Take $M = 65.5$ g mol^{-1}.

18.8a The polarizability volume of H_2O at optical frequencies is 1.5×10^{-24} cm^3: estimate the refractive index of water. The experimental value is 1.33; what may be the origin of the discrepancy?

18.8b The polarizability volume of a liquid of molar mass 72.3 g mol^{-1} and density 865 kg mol^{-1} at optical frequencies is 2.2×10^{-30} m^3: estimate the refractive index of the liquid.

18.9a The dipole moment of chlorobenzene is 1.57 D and its polarizability volume is 1.23×10^{-23} cm^3. Estimate its relative permittivity at 25°C, when its density is 1.173 g cm^{-3}.

18.9b The dipole moment of bromobenzene is 5.17×10^{-30} C m and its polarizability volume is approximately 1.5×10^{-29} m^3. Estimate its relative permittivity at 25°C, when its density is 1491 kg m^{-3}.

18.10a Calculate the vapour pressure of a spherical droplet of water of radius 10 nm at 20°C. The vapour pressure of bulk water at that temperature is 2.3 kPa and its density is 0.9982 g cm^{-3}.

18.10b Calculate the vapour pressure of a spherical droplet of water of radius 20.0 nm at 35.0°C. The vapour pressure of bulk water at that temperature is 5.623 kPa and its density is 994.0 kg m^{-3}.

18.11a The contact angle for water on clean glass is close to zero. Calculate the surface tension of water at 20°C given that at that temperature water climbs to a height of 4.96 cm in a clean glass capillary tube of internal radius 0.300 mm. The density of water at 20°C is 998.2 kg m^{-3}.

18.11b The contact angle for water on clean glass is close to zero. Calculate the surface tension of water at 30°C given that at that temperature water climbs to a height of 9.11 cm in a clean glass capillary tube of internal radius 0.320 mm. The density of water at 30°C is 0.9956 g cm^{-3}.

18.12a Calculate the pressure differential of water across the surface of a spherical droplet of radius 200 nm at 20°C.

18.12b Calculate the pressure differential of ethanol across the surface of a spherical droplet of radius 220 nm at 20°C. The surface tension of ethanol at that temperature is 22.39 mN m^{-1}.

Problems*

Numerical problems

18.1 Suppose an H_2O molecule ($\mu = 1.85$ D) approaches an anion. What is the favourable orientation of the molecule? Calculate the electric field (in volts per metre) experienced by the anion when the water dipole is (a) 1.0 nm, (b) 0.3 nm, (c) 30 nm from the ion.

18.2 An H_2O molecule is aligned by an external electric field of strength 1.0 kV m^{-1} and an Ar atom ($\alpha' = 1.66 \times 10^{-24}$ cm^3) is brought up slowly from one side. At what separation is it energetically favourable for the H_2O molecule to flip over and point towards the approaching Ar atom?

18.3 The relative permittivity of chloroform was measured over a range of temperatures with the following results:

θ/°C	−80	−70	−60	−40	−20	0	20
ε_r	3.1	3.1	7.0	6.5	6.0	5.5	5.0
ρ/(g cm^{-3})	1.65	1.64	1.64	1.61	1.57	1.53	1.50

The freezing point of chloroform is –64°C. Account for these results and calculate the dipole moment and polarizability volume of the molecule.

18.4 The relative permittivities of methanol (m.p. –95°C) corrected for density variation are given below. What molecular information can be deduced from these values? Take $\rho = 0.791$ g cm^{-3} at 20°C.

θ/°C	−185	−170	−150	−140	−110	−80	−50	−20	0	20
ε_r	3.2	3.6	4.0	5.1	67	57	4	43	38	34

18.5 In his classic book *Polar molecules*, Debye reports some early measurements of the polarizability of ammonia. From the selection below, determine the dipole moment and the polarizability volume of the molecule.

T/K	292.2	309.0	333.0	387.0	413.0	446.0
P_m/(cm^3 mol^{-1})	57.57	55.01	51.22	44.99	42.51	39.59

The refractive index of ammonia at 273 K and 100 kPa is 1.000 379 (for yellow sodium light). Calculate the molar polarizability of the gas at this temperature and at 292.2 K. Combine the value calculated with the static molar polarizability at 292.2 K and deduce from this information alone the molecular dipole moment.

18.6 Values of the molar polarization of gaseous water at 100 kPa as determined from capacitance measurements are given below as a function of temperature.

T/K	384.3	420.1	444.7	484.1	522.0
P_m/(cm^3 mol^{-1})	57.4	53.5	50.1	46.8	43.1

Calculate the dipole moment of H_2O and its polarizability volume.

18.7‡ F. Luo, G.C. McBane, O. Kim, C.F. Giese, and W.R. Gentry (*J. Chem. Phys.* **98**, 3564 (1993)) reported experimental observation of the He_2 complex, a species that had escaped detection for a long time. The fact that the observation required temperatures in the neighbourhood of 1 mK is consistent with computational studies which suggest that hcD_e, for He_2 is about 1.51×10^{-23} J, hcD_0 about 2×10^{-26} J, and R about 297 pm. (a) Determine the Lennard-Jones parameters r_0, and ε and plot the Lennard-Jones potential for He–He interactions. (b) Plot the Morse potential given that $a = 5.79 \times 10^{10}$ m^{-1}.

18.8‡ D.D. Nelson, G.T. Fraser, and W. Klemperer (*Science* **238**, 1670 (1987)) examined several weakly bound gas-phase complexes of ammonia in search of examples in which the H atoms in NH_3 formed hydrogen bonds, but found none. For example, they found that the complex of NH_3 and CO_2 has the carbon atom nearest the nitrogen (299 pm away): the CO_2 molecule is at right angles to the C—N 'bond', and the H atoms of NH_3 are pointing away from the CO_2. The permanent dipole moment of this complex is reported as 1.77 D. If the N and C atoms are the centres of the negative and positive charge distributions, respectively, what is the magnitude of those partial charges (as multiples of *e*)?

18.9‡ From data in Table 18.1 calculate the molar polarization, relative permittivity, and refractive index of methanol at 20°C. Its density at that temperature is 0.7914 g cm^{-3}.

* Problems denoted with the symbol‡ were supplied by Charles Trapp and Carmen Giunta.

Theoretical problems

18.10 Calculate the potential energy of the interaction between two linear quadrupoles when they are (a) collinear, (b) parallel and separated by a distance *r*.

18.11 Show that, in a gas (for which the refractive index is close to 1), the refractive index depends on the pressure as $n_r = 1 + \text{const} \times p$, and find the constant of proportionality. Go on to show how to deduce the polarizability volume of a molecule from measurements of the refractive index of a gaseous sample.

18.12 Acetic acid vapour contains a proportion of planar, hydrogen-bonded dimers. The relative permittivity of pure liquid acetic acid is 7.14 at 290 K and increases with increasing temperature. Suggest an interpretation of the latter observation. What effect should isothermal dilution have on the relative permittivity of solutions of acetic acid in benzene?

18.13 Show that the mean interaction energy of *N* atoms of diameter *d* interacting with a potential energy of the form C_6/R^6 is given by $U = -2N^2C_6/3Vd^3$, where *V* is the volume in which the molecules are confined and all effects of clustering are ignored. Hence, find a connection between the van der Waals parameter *a* and C_6, from $n^2a/V^2 = (\partial U/\partial V)_T$.

18.14 Suppose the repulsive term in a Lennard-Jones (12,6)-potential is replaced by an exponential function of the form $e^{-r/d}$. Sketch the form of the potential energy and locate the distance at which it is a minimum.

18.15 The *cohesive energy density*, $\mathcal{V}$, is defined as U/V, where *U* is the mean potential energy of attraction within the sample and *V* its volume. Show that $\mathcal{V} = \frac{1}{2}\mathcal{N}\int V(R)\mathrm{d}\tau$, where $\mathcal{N}$ is the number density of the molecules and $V(R)$ is their attractive potential energy and where the integration ranges from *d* to infinity and over all angles. Go on to show that the cohesive energy density of a uniform distribution of molecules that interact by a van der Waals attraction of the form $-C_6/R^6$ is equal to $(2\pi/3)(N_A^2/d^3M^2)\rho^2C_6$, where ρ is the mass density of the solid sample and *M* is the molar mass of the molecules.

18.16 Consider the collision between a hard-sphere molecule of radius R_1 and mass *m*, and an infinitely massive impenetrable sphere of radius R_2. Plot the scattering angle θ as a function of the impact parameter *b*. Carry out the calculation using simple geometrical considerations.

18.17 The dependence of the scattering characteristics of atoms on the energy of the collision can be modelled as follows. We suppose that the two colliding atoms behave as impenetrable spheres, as in Problem 18.16, but that the effective radius of the heavy atoms depends on the speed *v* of the light atom. Suppose its effective radius depends on *v* as R_2e^{-v/v^*}, where v^* is a constant. Take $R_1 = \frac{1}{2}R_2$ for simplicity and an impact parameter $b = \frac{1}{2}R_2$, and plot the scattering angle as a function of (a) speed, (b) kinetic energy of approach.

Applications: to biochemistry

18.18 Phenylalanine (Phe, **15**) is a naturally occurring amino acid. What is the energy of interaction between its phenyl group and the electric dipole moment of a neighbouring peptide group? Take the distance between the groups as 4.0 nm and treat the phenyl group as a benzene molecule. The dipole moment of the peptide group is $\mu = 2.7$ D and the polarizability volume of benzene is $\alpha' = 1.04 \times 10^{-29}$ m^3.

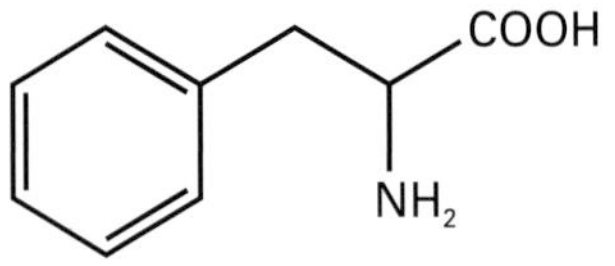

15

18.19 Now consider the London interaction between the phenyl groups of two Phe residues (see Problem 18.18). (a) Estimate the potential energy of interaction between two such rings (treated as benzene molecules) separated by 4.0 nm. For the ionization energy, use $I = 5.0$ eV. (b) Given that force is the negative slope of the potential, calculate the distance dependence of the force acting between two nonbonded groups of atoms, such as the phenyl groups of Phe, in a polypeptide chain that can have a London dispersion interaction with each other. What is the separation at which the force between the phenyl groups (treated as benzene molecules) of two Phe residues is zero? *Hint.* Calculate the slope by considering the potential energy at *r* and $r + \delta r$, with $\delta r \ll r$, and evaluating $\{V(r + \delta r) - V(r)\}/\delta r$. At the end of the calculation, let δr become vanishingly small.

18.20 Molecular orbital calculations may be used to predict structures of intermolecular complexes. Hydrogen bonds between purine and pyrimidine bases are responsible for the double helix structure of DNA (see Chapter 19). Consider methyl-adenine (**16**, with $R = CH_3$) and methyl-thymine (**17**, with $R = CH_3$) as models of two bases that can form hydrogen bonds in DNA. (a) Using molecular modelling software and the computational method of your choice, calculate the atomic charges of all atoms in methyl-adenine and methyl-thymine. (b) Based on your tabulation of atomic charges, identify the atoms in methyl-adenine and methyl-thymine that are likely to participate in hydrogen bonds. (c) Draw all possible adenine–thymine pairs that can be linked by hydrogen bonds, keeping in mind that linear arrangements of the A—H···B fragments are preferred in DNA. For this step, you may want to use your molecular modelling software to align the molecules properly. (d) Consult Chapter 19 and determine which of the pairs that you drew in part (c) occur naturally in DNA molecules. (e) Repeat parts (a)–(d) for cytosine and guanine, which also form base pairs in DNA (see Chapter 19 for the structures of these bases).

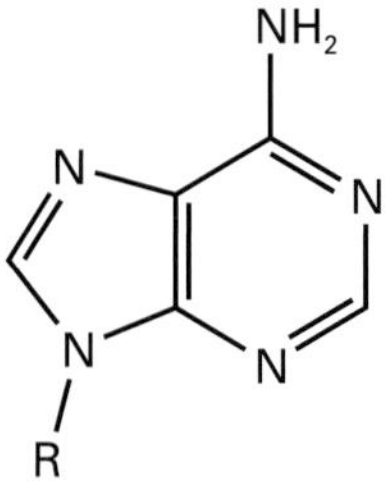

16

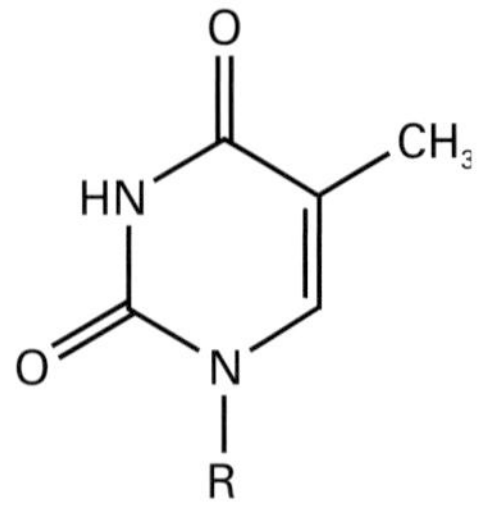

17

18.21 Molecular orbital calculations may be used to predict the dipole moments of molecules. (a) Using molecular modelling software and the computational method of your choice, calculate the dipole moment of the peptide link, modelled as a *trans-N*-methylacetamide (**18**). Plot the energy of interaction between these dipoles against the angle θ for $r = 3.0$ nm (see eqn 18.22). (b) Compare the maximum value of the dipole–dipole interaction energy from part (a) to 20 kJ mol^{-1}, a typical value for the energy of a hydrogen-bonding interaction in biological systems.

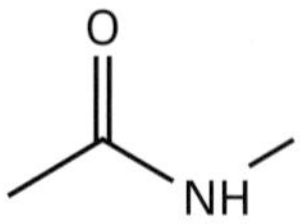

18

18.22 This problem gives a simple example of a quantitative structure–activity relation (QSAR). The binding of nonpolar groups of amino acid to hydrophobic sites in the interior of proteins is governed largely by hydrophobic interactions. (a) Consider a family of hydrocarbons R—H. The

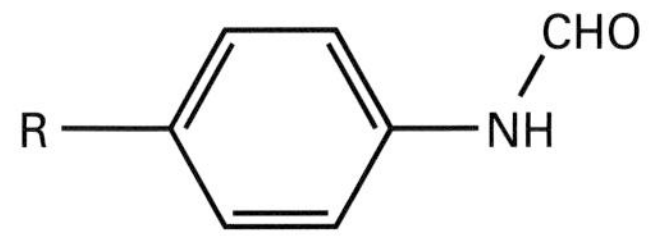

19

hydrophobicity constants, π, for R = CH_3, CH_2CH_3, $(CH_2)_2CH_3$, $(CH_2)_3CH_3$, and $(CH_2)_4CH_3$ are, respectively, 0.5, 1.0, 1.5, 2.0, and 2.5. Use these data to predict the π value for $(CH_2)_6CH_3$. (b) The equilibrium constants K_I for the dissociation of inhibitors (**19**) from the enzyme chymotrypsin were measured for different substituents R:

R	CH_3CO	CN	NO_2	CH_3	Cl
π	−0.20	−0.025	0.33	0.5	0.9
log K_I	−1.73	−1.90	−2.43	−2.55	−3.40

Plot log K_I against π. Does the plot suggest a linear relationship? If so, what are the slope and intercept to the log K_I axis of the line that best fits the data? (c) Predict the value of K_I for the case R = H.

18.23 Derivatives of the compound TIBO (**20**) inhibit the enzyme reverse transcriptase, which catalyses the conversion of retroviral RNA to DNA. A QSAR analysis of the activity A of a number of TIBO derivatives suggests the following equation:

$$\log A = b_0 + b_1S + b_2W$$

where S is a parameter related to the drug's solubility in water and W is a parameter related to the width of the first atom in a substituent X shown in **20**.

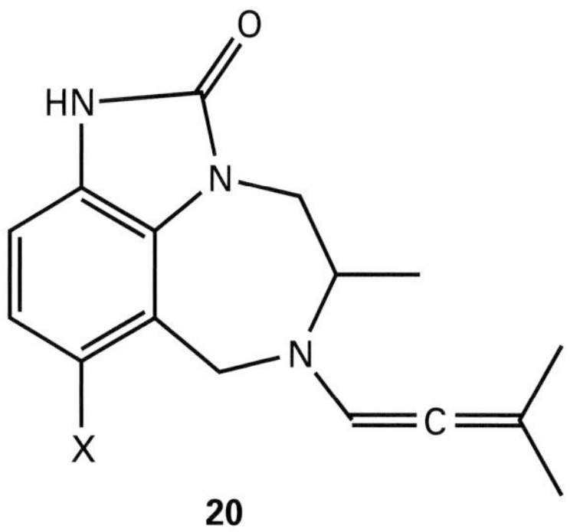

20

(a) Use the following data to determine the values of b_0, b_1, and b_2. *Hint.* The QSAR equation relates one dependent variable, log A, to two independent variables, S and W. To fit the data, you must use the mathematical procedure of *multiple regression*, which can be performed with mathematical software or an electronic spreadsheet.

X	H	Cl	SCH_3	OCH_3	CN	CHO	Br	CH_3	CCH
log A	7.36	8.37	8.3	7.47	7.25	6.73	8.52	7.87	7.53
S	3.53	4.24	4.09	3.45	2.96	2.89	4.39	4.03	3.80
W	1.00	1.80	1.70	1.35	1.60	1.60	1.95	1.60	1.60

(b) What should be the value of W for a drug with S = 4.84 and log A = 7.60?

19 Materials 1: macromolecules and aggregates

Macromolecules exhibit a range of properties and problems that illustrate a wide variety of physical chemical principles. They need to be characterized in terms of their molar mass, their size, and their shape. However, the molecules are so large and the solutions they form depart so strongly from ideality, that techniques for accommodating these departures from ideality need to be developed. Another major problem concerns the influences that determine the shapes of the molecules. We consider a range of influences in this chapter, beginning with a structureless random coil and ending with the structurally precise forces that operate in polypeptides and nucleic acids. Atoms, small molecules, and macromolecules can form large assemblies that are held together by one or more of the molecular interactions described in Chapter 18. These assemblies, which include colloids and biological membranes, exhibit some of the typical properties of molecules but have their own characteristic features.

There are macromolecules everywhere, inside us and outside us. Some are natural: they include polysaccharides such as cellulose, polypeptides such as protein enzymes, and polynucleotides such as deoxyribonucleic acid (DNA). Others are synthetic: they include **polymers** such as nylon and polystyrene that are manufactured by stringing together and (in some cases) cross-linking smaller units known as **monomers**. Life in all its forms, from its intrinsic nature to its technological interaction with its environment, is the chemistry of macromolecules.

Macromolecules give rise to special problems that include the shapes and the lengths of polymer chains, the determination of their sizes, and the large deviations from ideality of their solutions. Natural macromolecules differ in certain respects from synthetic macromolecules, particularly in their composition and the resulting structure, but the two share a number of common properties. We concentrate on these common properties here. Another level of complexity arises when small molecules aggregate into large particles in a process that is called 'self-assembly' and give rise to aggregates. One example is the assembly of haemoglobin from four myoglobin-like polypeptides. A similar type of aggregation gives rise to a variety of disperse phases, which include colloids. The properties of these disperse phases resemble to a certain extent the properties of solutions of macromolecules, and we describe their common attributes in the final part of this chapter.

Determination of size and shape

X-ray diffraction techniques (Chapter 20) can reveal the position of almost every heavy atom (that is, every atom other than hydrogen) even in very large molecules. However, there are several reasons why other techniques must also be used. In the first place, the sample might be a mixture of molecules with different chain lengths and extents of cross-linking, in which case sharp X-ray images are not obtained. Even if all the molecules in the sample are identical, it might prove impossible to obtain a single crystal, which is essential for diffraction studies because only then does the electron density (which is responsible for the scattering) have a large-scale periodic variation. Furthermore, although work on proteins and DNA has shown how immensely interesting and motivating the data can be, the information is incomplete. For instance, what can be said about the shape of the molecule in its natural environment, a biological cell? What can be said about the response of its shape to changes in its environment?

19.1 Mean molar masses

A pure protein is **monodisperse**, meaning that it has a single, definite molar mass (although there may be small variations, such as one amino acid replacing another, depending on the source of the sample). A synthetic polymer, however, is **polydisperse**, in the sense that a sample is a mixture of molecules with various chain lengths and molar masses. The various techniques that are used to measure molar masses result in different types of mean values of polydisperse systems.

The mean obtained from the determination of molar mass by osmometry (Section 5.5) is the **number-average molar mass**, $\bar{M}_\mathrm{n}$, which is the value obtained by weighting each molar mass by the number of molecules of that mass present in the sample:

$$\bar{M}_\mathrm{n} = \frac{1}{N}\sum_i N_i M_i \tag{19.1}$$

where N_i is the number of molecules with molar mass M_i and there are N molecules in all. Viscosity measurements give the **viscosity-average molar mass**, $\bar{M}_\mathrm{v}$, light-scattering experiments give the **weight-average molar mass**, $\bar{M}_\mathrm{w}$, and sedimentation experiments give the ***Z*-average molar mass**, $\bar{M}_Z$. (The name is derived from the *z*-coordinate used to depict data in a procedure for determining the average.) Although such averages are often best left as empirical quantities, some may be interpreted in terms of the composition of the sample. Thus, the weight-average molar mass is the average calculated by weighting the molar masses of the molecules by the mass of each one present in the sample:

$$\bar{M}_\mathrm{w} = \frac{1}{m}\sum_i m_i M_i \tag{19.2}$$

In this expression, m_i is the total mass of molecules of molar mass M_i and m is the total mass of the sample. Because $m_i = N_i M_i / N_\mathrm{A}$, we can also express this average as

$$\bar{M}_\mathrm{w} = \frac{\sum_i N_i M_i^2}{\sum_i N_i M_i} \tag{19.3}$$

This expression shows that the weight-average molar mass is proportional to the mean square molar mass. Similarly, the *Z*-average molar mass can be interpreted in terms of the mean cubic molar mass:

$$\bar{M}_Z = \frac{\sum_i N_i M_i^3}{\sum_i N_i M_i^2} \tag{19.4}$$

Example 19.1 *Calculating number and mass averages*

Determine the number-average and the weight-average molar masses for a sample of poly(vinyl chloride) from the following data:

Molar mass interval/ $(\mathrm{kg\,mol^{-1}})$	Average molar mass within interval/ $(\mathrm{kg\,mol^{-1}})$	Mass of sample within interval/g
5–10	7.5	9.6
10–15	12.5	8.7
15–20	17.5	8.9
20–25	22.5	5.6
25–30	27.5	3.1
30–35	32.5	1.7

Method The relevant equations are eqns 19.1 and 19.2. Calculate the two averages by weighting the molar mass within each interval by the number and mass, respectively, of the molecule in each interval. Obtain the numbers in each interval by dividing the mass of the sample in each interval by the average molar mass for that interval. Because the number of molecules is proportional to the amount of substance (the number of moles), the number-weighted average can be obtained directly from the amounts in each interval.

Answer The amounts in each interval are as follows:

Interval	5–10	10–15	15–20	20–25	25–30	30–35
Molar mass/$(\mathrm{kg\,mol^{-1}})$	7.5	12.5	17.5	22.5	27.5	32.5
Amount/mol	1.3	0.70	0.51	0.25	0.11	0.052
					Total:	2.92

The number-average molar mass is therefore

$$\begin{aligned}\bar{M}_{\mathrm{n}}/(\mathrm{kg\,mol^{-1}}) &= \frac{1}{2.92}(1.3\times 7.5+0.70\times 12.5+0.51\times 17.5+0.25\times 22.5\\ &\quad +0.11\times 27.5+0.052\times 32.5)\\ &=13\end{aligned}$$

The weight-average molar mass is calculated directly from the data after noting that the total mass of the sample is 37.6 g:

$$\begin{aligned}\bar{M}_{\mathrm{w}}/(\mathrm{kg\,mol^{-1}}) &= \frac{1}{37.6}(9.6\times 7.5+8.7\times 12.5+8.9\times 17.5+5.6\times 22.5\\ &\quad +3.1\times 27.5+1.7\times 32.5)\\ &=16\end{aligned}$$

Note the significantly different values of the two averages. In this instance, $\bar{M}_{\mathrm{w}}/\bar{M}_{\mathrm{n}}=1.2$.

Self-test 19.1 Evaluate the Z–average molar mass of the sample. [$19\ \mathrm{kg\,mol^{-1}}$]

The ratio $\bar{M}_{\mathrm{w}}/\bar{M}_{\mathrm{n}}$ is called the **heterogeneity index** (or 'polydispersity index'). In the determination of protein molar masses we expect the various averages to be the same because the sample is monodisperse (unless there has been degradation). A synthetic polymer normally spans a range of molar masses and the different averages yield different values. Typical synthetic materials have $\bar{M}_w/\bar{M}_n \approx 4$. The term 'monodisperse' is conventionally applied to synthetic polymers in which this index is less than 1.1; commercial polyethylene samples might be much more heterogeneous, with a ratio close to 30. One consequence of a narrow molar mass distribution for synthetic polymers is often a higher degree of three-dimensional long-range order in the solid and therefore higher density and melting point. The spread of values is controlled by the choice of catalyst and reaction conditions. In practice, it is found that long-range order is determined more by structural factors (branching, for instance) than by molar mass.

Average molar masses may be determined by osmotic pressure of polymer solutions . The upper limit for the reliability of membrane osmometry is about 1000 kg mol^{-1}. A major problem for macromolecules of relatively low molar mass (less than about 10 kg mol^{-1}) is their ability to percolate through the membrane. One consequence of this partial permeability is that membrane osmometry tends to overestimate the average molar mass of a polydisperse mixture. Several techniques for the determination of molar mass and polydispersity that are not so limited include mass spectrometry, laser light scattering, ultracentrifugation, electrophoresis, and viscosity measurements.

19.2 Mass spectrometry

Mass spectrometry is among the most accurate techniques for the determination of molar masses. The procedure consists of ionizing the sample in the gas phase and then measuring the mass-to-charge number ratio (m/z) of all ions. Macromolecules present a challenge because it is difficult to produce gaseous ions of large species without fragmentation. However, two new techniques have emerged that circumvent this problem: **matrix-assisted laser desorption/ionization** (MALDI) and **electrospray ionization**. We shall discuss **MALDI-TOF mass spectrometry**, so called because the MALDI technique is coupled to a time-of-flight (TOF) ion detector.

Figure 19.1 shows a schematic view of a MALDI-TOF mass spectrometer. The macromolecule is first embedded in a solid matrix that often consists of an organic material such as *trans*-3-indoleacrylic acid and inorganic salts such as sodium chloride or silver trifluoroacetate. This sample is then irradiated with a pulsed laser, such as a nitrogen laser. The laser energy ejects electronically excited matrix ions, cations, and neutral macromolecules, thus creating a dense gas plume above the sample surface. The macromolecule is ionized by collisions and complexation with small cations, such as H^+, Na^+, and Ag^+.

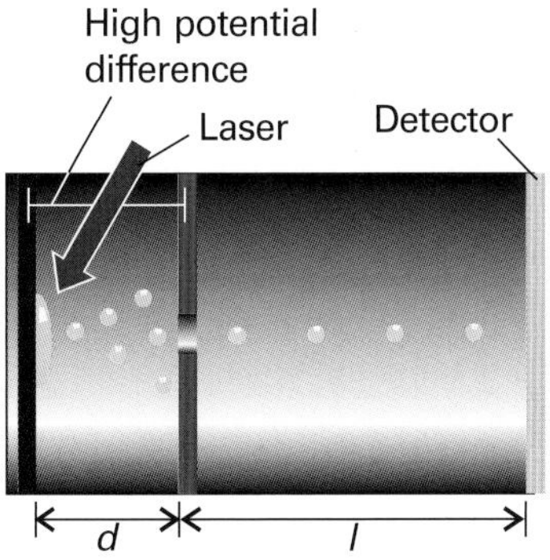

Fig. 19.1 A matrix-assisted laser desorption/ionization time-of-flight (MALDI-TOF) mass spectrometer. A laser beam ejects macromolecules and ions from the solid matrix. The ionized macromolecules are accelerated by an electrical potential difference over a distance d and then travel through a drift region of length l. Ions with the smallest mass to charge ratio (m/z) reach the detector first.

In the TOF spectrometer, the ions are accelerated over a short distance d by an electrical field of strength $\mathcal{E}$ and then travel through a drift region of length l. The time, t, required for an ion of mass m and charge number z to reach the detector at the end of the drift region is (see the *Justification*):

$$t = l\left(\frac{m}{2ze\mathcal{E}d}\right)^{1/2} \tag{19.5}$$

where e is the fundamental charge. Because d, l, and $\mathcal{E}$ are fixed for a given experiment, the time of flight, t, of the ion is a direct measure of its m/z ratio, which is given by:

$$\frac{m}{z} = 2e\mathcal{E}d\left(\frac{t}{l}\right)^2 \tag{19.6}$$

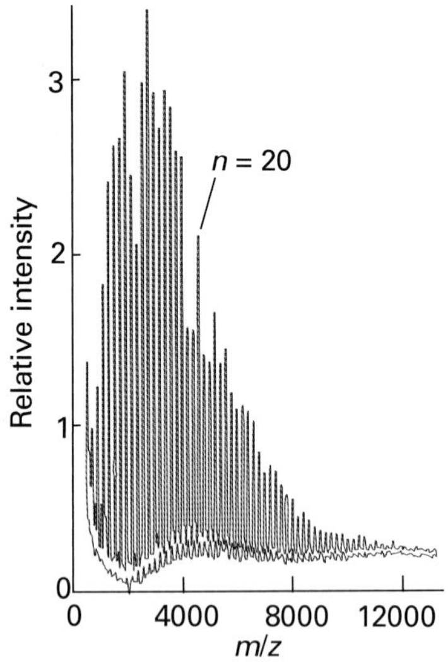

Fig. 19.2 MALDI-TOF spectrum of a sample of poly(butylene adipate) with $\bar{M}_n = 4525$ g mol^{-1} (Adapted from Mudiman *et al.*, *J. Chem. Educ.*, **74**, 1288 (1997).)

Justification 19.1 *The time of flight of an ion in a mass spectrometer*

Consider an ion of charge ze and mass m that is accelerated from rest by an electric field of strength $\mathcal{E}$ applied over a distance d. The kinetic energy, E_K, of the ion is

$$E_K = \tfrac{1}{2}mv^2 = ze\mathcal{E}d$$

where v is the speed of the ion. The drift region, l, and the time of flight, t, in the mass spectrometer are both sufficiently short that we can ignore acceleration and write $v = l/t$. Then substitution into this equation gives

$$\tfrac{1}{2}m\left(\frac{l}{t}\right)^2 = ze\mathcal{E}d$$

Rearrangement of this equation gives eqn 19.6.

Figure 19.2 shows the MALDI-TOF mass spectrum of a polydisperse sample of poly(butylene adipate) (PBA, **1**). The MALDI technique produces mostly singly charged molecular ions that are not fragmented. Therefore, the multiple peaks in the spectrum arise from polymers of different lengths, with the intensity of each peak being proportional to the abundance of each polymer in the sample. Values of $\bar{M}_n$, $\bar{M}_w$, and the heterogeneity index can be calculated from the data. It is also possible to use the mass spectrum to verify the structure of a polymer, as shown in the following example.

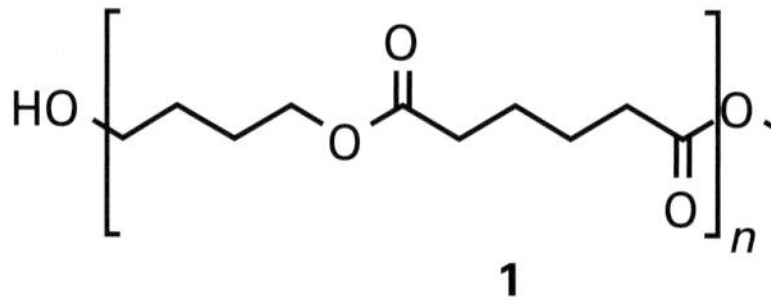

1

Example 19.2 *Interpreting the mass spectrum of a polymer*

The mass spectrum in Fig. 19.2 consists of peaks spaced by 200 g mol^{-1}. The peak at 4113 g mol^{-1} corresponds to the polymer for which $n = 20$. From these data, verify that the sample consists of polymers with the general structure given by (**1**).

Method Because each peak corresponds to a different value of n, the molar mass difference, ΔM, between peaks corresponds to the molar mass, M, of the repeating unit (the group inside the brackets in **1**). Furthermore, the molar mass of the terminal groups (the groups outside the brackets in **1**) may be obtained from the molar mass of any peak by using

$$M(\text{terminal groups}) = M(\text{polymer with } n \text{ repeating units}) - n\Delta M - M(\text{cation})$$

where the last term corresponds to the molar mass of the cation that attaches to the macromolecule during ionization.

Answer The value of ΔM is consistent with the molar mass of the repeating unit shown in (**1**), which is 200 g mol^{-1}. The molar mass of the terminal group is calculated by recalling that Na^+ is the cation in the matrix:

$$M(\text{terminal group}) = 4113\text{ g mol}^{-1} - 20(200\text{ g mol}^{-1}) - 23\text{ g mol}^{-1} = 90\text{ g mol}^{-1}$$

The result is consistent with the molar mass of the —O$(CH_2)_4$OH terminal group (89 g mol^{-1}) plus the molar mass of the —H terminal group (1 g mol^{-1}).

Self-test 19.2 What would be the molar mass of the $n = 20$ polymer if silver trifluoroacetate were used instead of NaCl in the preparation of the matrix? [4198 g mol^{-1}]

19.3 Laser light scattering

Large particles scatter light very efficiently. A familiar example is the light scattered by specks of dust in a sunbeam. Therefore, light scattering is a convenient method for the characterization of polymers, large aggregates (such as colloids), and biological systems from proteins to viruses. Unlike mass spectrometry, laser light scattering measurements may be performed in nearly intact samples; often the only preparation required is filtration of the sample.

(a) General principles of light scattering

When the oscillating electric field of electromagnetic radiation interacts with the electrons in a particle, an oscillating dipole moment develops with a magnitude proportional to the polarizability of the particle and the strength of the field (Section 18.1). Elastic light scattering is observed as the oscillating dipoles in the particle radiate at the same frequency as the frequency of the exciting electromagnetic radiation. The term *elastic* refers to the fact that the incident and scattered photons have the same frequency and hence the same energy. If the medium is perfectly homogeneous, as in a perfect crystal, the scattered waves interfere destructively in all directions except the direction of propagation of the exciting radiation. If the medium is inhomogeneous, as in an imperfect crystal or a solution of macromolecules, radiation is scattered into other directions as well.

Scattering of light by particles with diameters much smaller than the wavelength of the incident radiation is called **Rayleigh scattering** (Fig. 19.3). This type of scattering has several characteristic features.

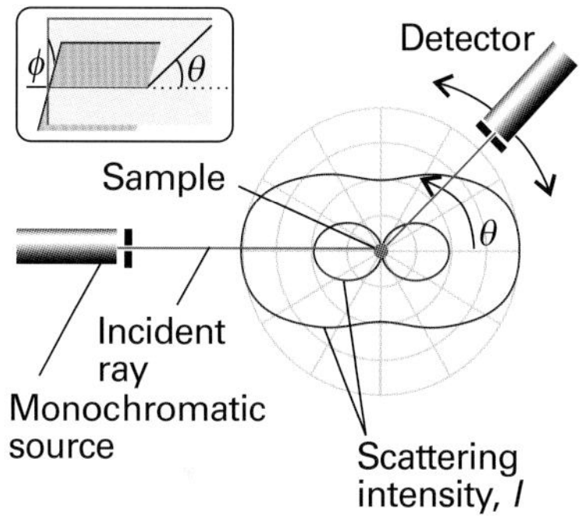

Fig. 19.3 Rayleigh scattering from a sample of point-like particles. The intensity of scattered light depends on the angle θ between the incident and scattered beams. The inset shows the angle ϕ between the plane of polarization of the incident beam and the plane defined by the incident and scattered beams. In a typical experimental arrangement, $\phi = 90°$.

1 The intensity of scattered light is proportional to λ^{-4}, so shorter wavelength radiation is scattered more intensely than longer wavelengths.

2 The intensity of scattered light is proportional to the molar mass of the particle.

3 The intensity of scattered light depends on the scattering angle θ (Fig. 19.3). In practice, data are collected at several angles to the incident laser beam (Example 19.3).

4 For very dilute solutions excited by plane-polarized light, the **Rayleigh ratio**, R_θ, a measure of the intensity of scattered light at a given scattering angle θ, is defined as

$$R_\theta = \frac{I}{I_0} \times \frac{r^2}{\sin^2\phi} \qquad (19.7)$$

where I is the intensity of scattered light, I_0 is the intensity of incident light, r is the distance between the sample and the detector, ϕ is the angle between the plane of polarization of the incident beam and the plane defined by the incident and scattered beams (see the inset in Fig. 19.3).

For a solution of a polymer of mass concentration c_P, the Rayleigh ratio may be written as

$$R_\theta = K P_\theta c_\mathrm{P} \bar{M}_\mathrm{w}, \text{ with } K = \frac{4\pi^2 n_{\mathrm{r},0}^2 V (\mathrm{d}n_\mathrm{r}/\mathrm{d}c_\mathrm{P})^2}{\lambda^4 N_\mathrm{A}} \qquad (19.8)$$

Here $n_{\mathrm{r},0}$ is the refractive index of the pure solvent (see *Comment* 18.6 and *Appendix* 3), $(\mathrm{d}n/\mathrm{d}c_\mathrm{P})$ is the change in refractive index of the solution with concentration of polymer, V is the volume of the sample, and N_A is Avogadro's constant. The parameter P_θ is the **structure factor**, which takes into account the fact that scattering may occur from different sites of the same molecule and interference between scattered rays becomes important when the wavelength of the incident radiation is comparable to the size of the scattering particles. When the molecule is much smaller than the wavelength of incident radiation, $P_\theta \approx 1$. However, when the size of the molecule is

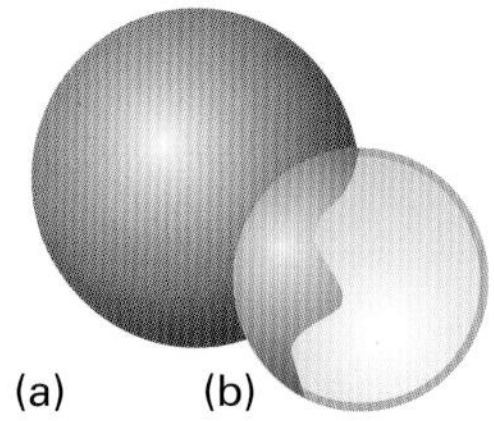

Fig. 19.4 (a) A spherical molecule and (b) the hollow spherical shell that has the same rotational characteristics. The radius of the hollow shell is the radius of gyration of the molecule.

Synoptic table 19.1* Radius of gyration

	$M/(\text{kg mol}^{-1})$	R_g/nm
Serum albumin	66	2.98
Polystyrene	3.2×10^3	50†
DNA	4×10^3	117

* More values are given in the *Data section*.
† In a poor solvent.

about one-tenth the wavelength of the incident radiation, we show in *Further information* 19.1 that

$$P_\theta \approx 1 - \frac{16\pi^2 R_g^2 \sin^2 \frac{1}{2}\theta}{3\lambda^2} \qquad (19.9)$$

where R_g is the **radius of gyration** of the macromolecule, the radius of a thin hollow spherical shell of the same mass and moment of inertia as the molecule (Fig. 19.4 and Section 19.8). Table 19.1 lists some experimental values of R_g.

Illustration 19.1 *Why is the sky blue whereas clouds are white?*

We expect from eqn 19.9 that when the particles are very small and $P_\theta \approx 1$, the medium scatters light of shorter wavelengths much more efficiently than light of longer wavelengths. This effect accounts for the colour of a cloudless sky: the N_2 and O_2 molecules in the atmosphere are much smaller than the wavelengths of visible electromagnetic radiation, so blue light is scattered preferentially. We also see clouds because light scatters from them, but they look white, not blue. In clouds, the water molecules group together into droplets of a size comparable to the wavelength of light, and scatter cooperatively. Although blue light scatters more strongly, more molecules can contribute cooperatively when the wavelength is longer (as for red light), so the net result is uniform scattering for all wavelengths: white light scatters as white light. This paper looks white for the same reason. As a result, the scattering intensity is distorted from the form characteristic of small-particle, Rayleigh scattering, and the distortion is taken into account by values of P_θ that differ from 1.

(b) Scattering by non-ideal solutions of polymers

The preceding discussion shows that structural properties, such as size and the weight-average molar mass of a macromolecule, can be obtained from measurements of light scattering by a sample at several angles θ relative to the direction of propagation on an incident laser beam. However, eqn 19.8 applies only to ideal solutions. In practice, even relatively dilute polymer dispersions can deviate considerably from ideality. Being so large, macromolecules displace a large quantity of solvent instead of replacing individual solvent molecules with negligible disturbance. In thermodynamic terms, the displacement and reorganization of solvent molecules implies that the entropy change is especially important when a macromolecule dissolves. Furthermore, its great bulk means that a macromolecule is unable to move freely through the solution because the molecule is excluded from the regions occupied by other solute molecules. There are also significant contributions to the Gibbs energy from the enthalpy of solution, largely because solvent–solvent interactions are more favourable than the macromolecule–solvent interactions that replace them. To take deviations from ideality into account, it is common to rewrite eqn 19.8 as

$$\frac{Kc_P}{R_\theta} = \frac{1}{P_\theta \bar{M}_w} + Bc_P \qquad (19.10)$$

where B is an empirical constant analogous to the osmotic virial coefficient and indicative of the effect of excluded volume.

For most solute–solvent systems there is a unique temperature (which is not always experimentally attainable) at which the effects leading to non-ideal behaviour cancel and the solution is virtually ideal. This temperature (the analogue of the Boyle temperature for real gases) is called the **θ-temperature** (theta temperature). At this temperature, B is zero. As an example, for polystyrene in cyclohexane the θ-temperature is approximately 306 K, the exact value depending on the average molar mass of the polymer. A solution at its θ-temperature is called a **θ-solution**. Because a θ-solution behaves nearly ideally, its thermodynamic and structural properties are easier to describe even though the molar concentration is not low. In molecular terms, in a theta solution the molecules are in an unperturbed condition, whereas in other solutions expansion of the coiled molecule takes place as a result of interactions with the solvent.

Example 19.3 *Determining the size of a polymer by light scattering*

The following data for a sample of polystyrene in butanone were obtained at 20°C with plane-polarized light at $\lambda = 546$ nm.

$\theta/°$	26.0	36.9	66.4	90.0	113.6
R_θ/m^2	19.7	18.8	17.1	16.0	14.4

In separate experiments, it was determined that $K = 6.42 \times 10^{-5}$ mol m^5 kg^{-2}. From this information, calculate R_g and $\bar{M}_w$ for the sample. Assume that B is negligibly small and that the polymer is small enough that eqn 19.9 holds.

Method Substituting the result of eqn 19.9 into eqn 19.8 we obtain, after some rearrangement:

$$\frac{1}{R_\theta} = \frac{1}{Kc_P\bar{M}_w} + \left(\frac{16\pi^2 R_g^2}{3\lambda^2}\right)\left(\frac{1}{R_\theta}\sin^2\tfrac{1}{2}\theta\right)$$

Hence, a plot of $1/R_\theta$ against $(1/R_\theta)\sin^2\frac{1}{2}\theta$ should be a straight line with slope $16\pi^2R_g^2/3\lambda^2$ and y-intercept $1/Kc_P\bar{M}_w$.

Answer We construct a table of values of $1/R_\theta$ and $(1/R_\theta)\sin^2\frac{1}{2}\theta$ and plot the data (Fig. 19.5).

$10^2 \times R_\theta^{-1}/\text{m}^{-2}$	5.06	5.32	5.83	6.25	6.96
$(10^3/R_\theta)\sin^2(\frac{1}{2}\theta)/\text{m}^{-2}$	2.56	5.33	17.5	31.3	48.7

The best straight line through the data has a slope of 0.391 and a y-intercept of 5.06×10^{-2}. From these values and the value of K, we calculate $R_g = 4.71 \times 10^{-8}$ m = 47.1 nm and $\bar{M}_w = 987$ kg mol^{-1}.

A more accurate method for more concentrated samples consists of conducting a series of experiments where R_θ against θ data are obtained for several c_P values. From analysis of the entire data set, the R_g, $\bar{M}_w$, and B values are obtained.

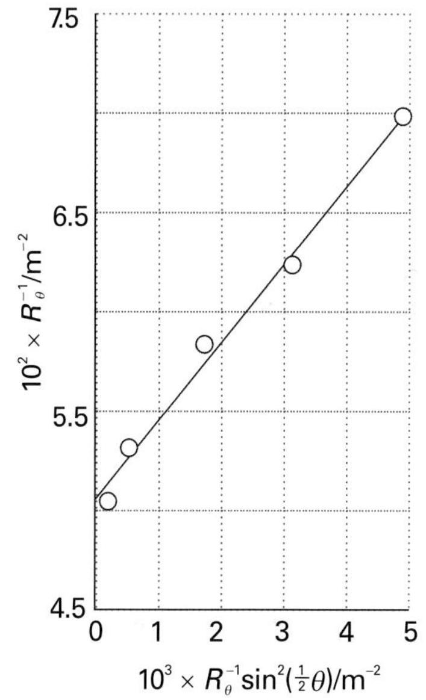

Fig. 19.5 Plot of the data for Example 19.3.

Self-test 19.3 The following data for an aqueous solution of a protein with $c_P = 2.0$ kg m^{-3} were obtained at 20°C with laser light at $\lambda = 532$ nm:

$\theta/°$	15.0	45.0	70.0	85.0	90.0
R_θ/m^2	23.8	22.9	21.6	20.7	20.4

In a separate experiment, it was determined that $K = 2.40 \times 10^{-2}$ mol m^5 kg^{-2}. From this information, calculate the radius of gyration and the molar mass of the protein. Assume that B is negligibly small and that the protein is small enough that eqn 19.9 holds. [$R_g = 39.8$ nm; $M = 498$ kg mol^{-1}]

Synoptic table 19.2* Diffusion coefficients in water at 20°C

	$M/(\text{kg mol}^{-1})$	$D/(\text{m}^2\,\text{s}^{-1})$
Sucrose	0.342	4.59×10^{-10}
Lysozyme	14.1	1.04×10^{-10}
Haemoglobin	68	6.9×10^{-11}
Collagen	345	6.9×10^{-12}

* More values are given in the *Data section*.

Synoptic table 19.3* Frictional coefficients and molecular geometry†

a/b	Prolate	Oblate
2	1.04	1.04
3	1.18	1.17
6	1.31	1.28
8	1.43	1.37
10	1.54	1.46

* More values and analytical expressions are given in the *Data section*.
† Entries are the ratio f/f_0, where $f_0 = 6\pi\eta c$, where $c = (ab^2)^{1/3}$ for prolate ellipsoids and $c = (a^2b)^{1/3}$ for oblate ellipsoids; $2a$ is the major axis and $2b$ is the minor axis.

(c) Dynamic light scattering

A special laser scattering technique, **dynamic light scattering**, can be used to investigate the diffusion of polymers in solution. Consider two polymer molecules being irradiated by a laser beam. Suppose that at a time t the scattered waves from these particles interfere constructively at the detector, leading to a large signal. However, as the molecules move through the solution, the scattered waves may interfere destructively at another time t' and result in no signal. When this behaviour is extended to a very large number of molecules in solution, it results in fluctuations in light intensity that depend on the diffusion coefficient, D, which is a measure of the rate of molecular motion and is given by the **Stokes–Einstein relation** (which is discussed further in Section 21.9e):

$$D = \frac{kT}{f} \tag{19.11}$$

where f is the **frictional coefficient**, a measure of the forces that retard a molecule's motion. Table 19.2 lists some typical values of D. For a spherical particle of radius a in a solvent of viscosity η (see Section 19.6), the frictional coefficient is given by **Stokes's relation**:

$$f = 6\pi a\eta \tag{19.12}$$

If the molecule is not spherical, we use appropriate values of f given in Table 19.3. Hence, dynamic light scattering measurements give the diffusion coefficient and molecular size, in cases where the molecular shape is known. For dilute monodisperse systems of random coils, it has been found empirically that D is related to the molar mass M of the polymer by:

$$D = \beta_D M^{-0.6} \tag{19.13}$$

The coefficient β_D is obtained by determining D at fixed viscosity and temperature for a variety of standard samples with known molar masses. As we should expect, bulky polymers of high molar mass migrate more slowly (have a lower diffusion coefficient) through a solvent than polymers of low molar mass.

19.4 Ultracentrifugation

In a gravitational field, heavy particles settle towards the foot of a column of solution by the process called **sedimentation**. The rate of sedimentation depends on the strength of the field and on the masses and shapes of the particles. Spherical molecules (and compact molecules in general) sediment faster than rod-like and extended molecules. When the sample is at equilibrium, the particles are dispersed over a range of heights in accord with the Boltzmann distribution (because the gravitational field competes with the stirring effect of thermal motion). The spread of heights depends on the masses of the molecules, so the equilibrium distribution is another way to determine molar mass.

Sedimentation is normally very slow, but it can be accelerated by **ultracentrifugation**, a technique that replaces the gravitational field with a centrifugal field. The effect can be achieved in an ultracentrifuge, which is essentially a cylinder that can be rotated at high speed about its axis with a sample in a cell near its periphery (Fig. 19.6). Modern ultracentrifuges can produce accelerations equivalent to about 10^5 that of gravity ('10^5 g'). Initially the sample is uniform, but the 'top' (innermost) boundary of the solute moves outwards as sedimentation proceeds.

(a) The rate of sedimentation

A solute particle of mass m has an effective mass $m_{\text{eff}} = bm$ on account of the buoyancy of the medium, with

$$b = 1 - \rho v_s \tag{19.14}$$

where ρ is the solution density, v_s is the partial specific volume of the solute ($v_s = (\partial V/\partial m_B)_T$, with m_B the total mass of solute), and ρv_s is the mass of solvent displaced per gram of solute. The solute particles at a distance r from the axis of a rotor spinning at an angular velocity ω experience a centrifugal force of magnitude $m_{\text{eff}} r\omega^2$. The acceleration outwards is countered by a frictional force proportional to the speed, s, of the particles through the medium. This force is written fs, where f is the frictional coefficient (Section 19.3). The particles therefore adopt a **drift speed**, a constant speed through the medium, which is found by equating the two forces $m_{\text{eff}} r\omega^2$ and fs. The forces are equal when

$$s = \frac{m_{\text{eff}} r\omega^2}{f} = \frac{bmr\omega^2}{f} \tag{19.15}$$

The drift speed depends on the angular velocity and the radius, and it is convenient to define the **sedimentation constant**, S, as

$$S = \frac{s}{r\omega^2} \tag{19.16}$$

Then, because the average molecular mass is related to the average molar mass $\bar{M}_n$ through $m = \bar{M}_n/N_A$,

$$S = \frac{b\bar{M}_n}{fN_A} \tag{19.17}$$

On substituting the Stokes relation for spherical molecules (eqn 19.12), we obtain

$$S = \frac{b\bar{M}_n}{6\pi a\eta N_A} \tag{19.18}$$

and S may be used to determine either $\bar{M}_n$ or a. Again, if the molecules are not spherical, we use the appropriate value of f given in Table 19.3. As always when dealing with macromolecules, the measurements must be carried out at a series of concentrations and then extrapolated to zero concentration to avoid the complications that arise from the interference between bulky molecules.

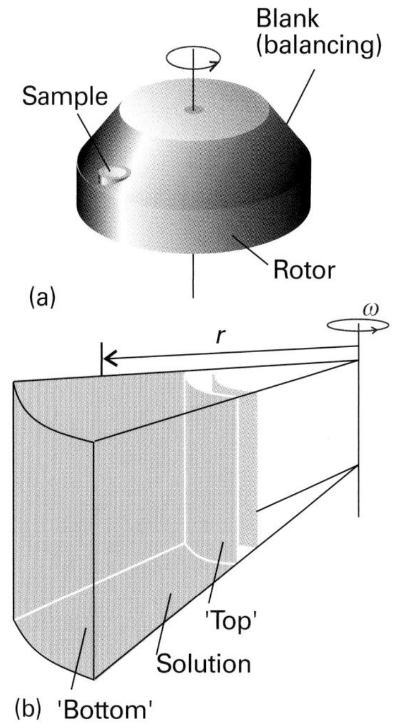

Fig. 19.6 (a) An ultracentrifuge head. The sample on one side is balanced by a blank diametrically opposite. (b) Detail of the sample cavity: the 'top' surface is the inner surface, and the centrifugal force causes sedimentation towards the outer surface; a particle at a radius r experiences a force of magnitude $mr\omega^2$.

Example 19.4 *Determining a sedimentation constant*

The sedimentation of the protein bovine serum albumin (BSA) was monitored at 25°C. The initial location of the solute surface was at 5.50 cm from the axis of rotation, and during centrifugation at 56 850 r.p.m. it receded as follows:

t/s	0	500	1000	2000	3000	4000	5000
r/cm	5.50	5.55	5.60	5.70	5.80	5.91	6.01

Calculate the sedimentation coefficient.

Method Equation 19.16 can be interpreted as a differential equation for $s = dr/dt$ in terms of r; so integrate it to obtain a formula for r in terms of t. The integrated expression, an expression for r as a function of t, will suggest how to plot the data and obtain from it the sedimentation constant.

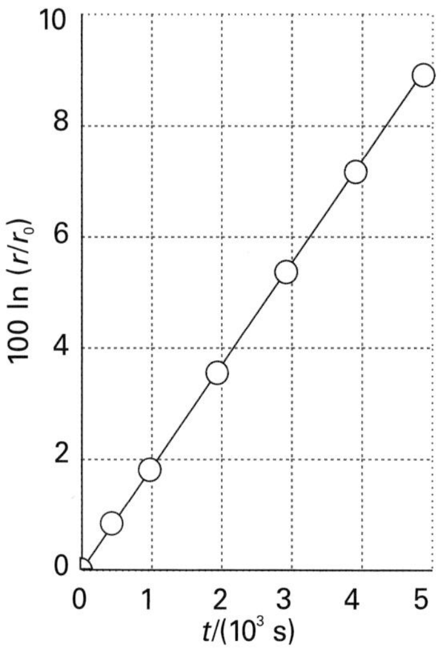

Fig. 19.7 A plot of the data in Example 19.4.

Answer Equation 19.16 may be written

$$\frac{dr}{dt} = r\omega^2 S$$

This equation integrates to

$$\ln\frac{r}{r_0} = \omega^2 St$$

It follows that a plot of $\ln(r/r_0)$ against t should be a straight line of slope $\omega^2 S$. Use $\omega = 2\pi\nu$, where ν is in cycles per second, and draw up the following table:

t/s	0	500	1000	2000	3000	4000	5000
$10^2 \ln(r/r_0)$	0	0.905	1.80	3.57	5.31	7.19	8.87

The straight-line graph (Fig. 19.7) has slope 1.78×10^{-5}; so $\omega^2 S = 1.79 \times 10^{-5}\ \mathrm{s}^{-1}$. Because $\omega = 2\pi \times (56\,850/60)\ \mathrm{s}^{-1} = 5.95 \times 10^3\ \mathrm{s}^{-1}$, it follows that $S = 5.02 \times 10^{-13}$ s. The unit 10^{-13} s is sometimes called a 'svedberg' and denoted Sv; in this case $S = 5.02$ Sv.

Self-test 19.4 Calculate the sedimentation constant given the following data (the other conditions being the same as above):

t/s	0	500	1000	2000	3000	4000	5000
r/cm	5.65	5.68	5.71	5.77	5.84	5.9	5.97

[3.11 Sv]

At this stage, it appears that we need to know the molecular radius a to obtain the molar mass from the value of S. Fortunately, this requirement can be avoided by drawing on the Stokes–Einstein relation (eqn 19.11) between f and the diffusion coefficient, D. The average molar mass is then:

$$\bar{M} = \frac{SRT}{bD} \qquad (19.19)$$

where we are not specifying which mean molar mass because the average obtained depends on technical details of the experiment. The result in eqn 19.19 is independent of the shape of the solute molecules. It follows that we can find the molar mass by combining measurements of S and D by ultracentrifugation and dynamic light scattering, respectively.

(b) Sedimentation equilibria

The difficulty with using sedimentation rates to measure molar masses lies in the inaccuracies inherent in the determination of diffusion coefficients of polydisperse systems. This problem can be avoided by allowing the system to reach equilibrium, for the transport property D is then no longer relevant. As we show in the *Justification* below, the weight-average molar mass can be obtained from the ratio of concentrations of the macromolecules at two different radii in a centrifuge operating at angular frequency ω:

$$\bar{M}_w = \frac{2RT}{(r_2^2 - r_1^2)b\omega^2}\ln\frac{c_2}{c_1} \qquad (19.20)$$

An alternative treatment of the data leads to the Z–average molar mass. The centrifuge is run more slowly in this technique than in the sedimentation rate method to avoid having all the solute pressed in a thin film against the bottom of the cell. At these slower speeds, several days may be needed for equilibrium to be reached.

Justification 19.2 *The weight-average molar mass from sedimentation experiments*

The distribution of particles is the outcome of the balance between the effect of the centrifugal force and the dispersing effect of diffusion down a concentration gradient. The kinetic energy of a particle of effective mass m at a radius r in a rotor spinning at a frequency ω is $\frac{1}{2}m\omega^2r^2$, so the total chemical potential at a radius r is $\bar{\mu}(r) = \mu(r) - \frac{1}{2}m\omega^2r^2$, where $\mu(r)$ is the contribution that depends on the concentration of solute. The condition for equilibrium is that the chemical potential is uniform, so

$$\left(\frac{\partial\bar{\mu}}{\partial r}\right)_T = \left(\frac{\partial\mu}{\partial r}\right)_T - M\omega^2 r = 0$$

To evaluate the partial derivative of μ, we write

$$\left(\frac{\partial\mu}{\partial r}\right)_T = \left(\frac{\partial\mu}{\partial p}\right)_{T,c}\left(\frac{\partial p}{\partial r}\right)_{T,c} + \left(\frac{\partial\mu}{\partial c}\right)_{T,p}\left(\frac{\partial c}{\partial r}\right)_{T,p} = Mv\omega^2 r\rho + RT\left(\frac{\partial \ln c}{\partial r}\right)_{T,p}$$

The first result follows from the fact that $(\partial\mu/\partial p)_T = V_m$, the partial molar volume, and $V_m = Mv$. It also makes use of the fact that the hydrostatic pressure at r is $p(r) = p(r_0) + \frac{1}{2}\rho\omega^2(r^2 - r_0^2)$, where r_0 is the radius of the surface of the liquid in the sample holder (that is, the location of its meniscus), with ρ the mass density of the solution. The concentration term stems from the expression $\mu = \mu^\ominus + RT \ln c$. The condition for equilibrium is therefore

$$Mr\omega^2(1 - v\rho) - RT\left(\frac{\partial \ln c}{\partial r}\right)_{T,p} = 0$$

and therefore, at constant temperature,

$$\mathrm{d}\ln c = \frac{Mr\omega^2(1 - v\rho)\mathrm{d}r}{RT}$$

This expression integrates to eqn 19.20.

19.5 Electrophoresis

Many macromolecules, such as DNA, are charged and move in response to an electric field. This motion is called **electrophoresis**. Electrophoretic mobility is a result of a constant drift speed, s, reached by an ion when the driving force $ze\mathcal{E}$ (where, as usual, ze is the net charge and $\mathcal{E}$ is the field strength) is matched by the frictional force fs. The drift speed (which is treated in detail in Section 21.7) is then:

$$s = \frac{ze\mathcal{E}}{f} \qquad (19.21)$$

Therefore, the mobility of a macromolecule in an electric field depends on its net charge, size (and hence molar mass), and shape. The latter two factors are implied by the dependence of s on f.

The drift speeds attained by polymers in traditional electrophoresis methods are rather low; as a result, several hours are often necessary to effect good separation of complex mixtures. According to eqn 19.21, one way to increase the drift speed is to increase the electric field strength. However, there are limits to this strategy because very large electric fields can heat the large surfaces of an electrophoresis apparatus unevenly, leading to a non-uniform distribution of electrophoretic mobilities and poor separation.

In **capillary electrophoresis**, the sample is dispersed in a medium (such as methylcellulose) and held in a thin glass or plastic tube with diameters ranging from 20 to 100 μm. The small size of the apparatus makes it easy to dissipate heat when large electric fields are applied. Excellent separations may be effected in minutes rather than hours. Each polymer fraction emerging from the capillary can be characterized further by other techniques, such as MALDI-TOF.

IMPACT ON BIOCHEMISTRY

I19.1 Gel electrophoresis in genomics and proteomics

Advances in biotechnology are linked strongly to the development of physical techniques. The ongoing effort to characterize the entire genetic material, or **genome**, of organisms as simple as bacteria and as complex as *Homo sapiens* will lead to important new insights into the molecular mechanisms of disease, primarily through the discovery of previously unknown proteins encoded by the deoxyribonucleic acid (DNA) in genes. However, decoding genomic DNA will not always lead to accurate predictions of the amino acids present in biologically active proteins. Many proteins undergo chemical modification, such as cleavage into smaller proteins, after being synthesized in the cell. Moreover, it is known that one piece of DNA may encode more than one active protein. It follows that it is also important to describe the **proteome**, the full complement of functional proteins of an organism, by characterizing directly the proteins after they have been synthesized and processed in the cell.

The procedures of **genomics** and **proteomics**, the analysis of the genome and proteome, of complex organisms are time-consuming because of the very large number of molecules that must be characterized. For example, the human genome contains about 30 000 genes and the number of active proteins is likely to be much larger. Success in the characterization of the genome and proteome of any organism will depend on the deployment of very rapid techniques for the determination of the order in which molecular building blocks are linked covalently in DNA and proteins.

An important tool in genomics and proteomics is **gel electrophoresis**, in which biopolymers are separated on a slab of a porous gel, a semirigid dispersion of a solid in a liquid. Because the molecules must pass through the pores in the gel, the larger the macromolecule the less mobile it is in the electric field and, conversely, the smaller the macromolecule the more swiftly it moves through the pores. In this way, gel electrophoresis allows for the separation of components of a mixture according to their molar masses. Two common gel materials for the study of proteins and nucleic acids are agarose and cross-linked polyacrylamide. Agarose has large pores and is better suited for the study of large macromolecules, such as DNA and enzyme complexes. Polyacrylamide gels with varying pore sizes can be made by changing the concentration of acrylamide in the polymerization solution. In general, smaller pores form as the concentration of acrylamide is increased, making possible the separation of relatively small macromolecules by **polyacrylamide gel electrophoresis** (PAGE).

The separation of very large pieces of DNA, such as chromosomes, by conventional gel electrophoresis is not effective, making the analysis of genomic material rather difficult. Double-stranded DNA molecules are thin enough to pass through gel pores, but long and flexible DNA coils can become trapped in the pores and the result is impaired mobility along the direction of the applied electric field. This problem can be avoided with **pulsed-field electrophoresis**, in which a brief burst of the electric field is applied first along one direction and then along a perpendicular direction. In response to the switching back and forth between field directions, the DNA coils writhe about and eventually pass through the gel pores. In this way, the mobility of the macromolecule can be related to its molar mass.

We have seen that charge also determines the drift speed. For example, proteins of the same size but different net charge travel along the slab at different speeds. One way to avoid this problem and to achieve separation by molar mass is to denature the proteins in a controlled way. Sodium dodecyl sulfate is an anionic detergent that is very useful in this respect: it denatures proteins, whatever their initial shapes, into rods by forming a complex with them. Moreover, most protein molecules bind a constant number of ions, so the net charge per protein is well regulated. Under these conditions, different proteins in a mixture may be separated according to size only. The molar mass of each constituent protein is estimated by comparing its mobility in its rod-like complex form with that of a standard sample of known molar mass. However, molar masses obtained by this method, often referred to as **SDS-PAGE** when polyacrylamide gels are used, are not as accurate as those obtained by MALDI-TOF or ultracentrifugation.

Another technique that deals with the effect of charge on drift speed takes advantage of the fact that the overall charge of proteins and other biopolymers depends on the pH of the medium. For instance, in acidic environments protons attach to basic groups and the net charge is positive; in basic media the net charge is negative as a result of proton loss. At the **isoelectric point**, the pH is such that there is no net charge on the biopolymer. Consequently, the drift speed of a biopolymer depends on the pH of the medium, with $s = 0$ at the isoelectric point (Fig. 19.8). **Isoelectric focusing** is an electrophoresis method that exploits the dependence of drift speed on pH. In this technique, a mixture of proteins is dispersed in a medium with a pH gradient along the direction of an applied electric field. Each protein in the mixture will stop moving at a position in the gradient where the pH is equal to the isoelectric point. In this manner, the protein mixture can be separated into its components.

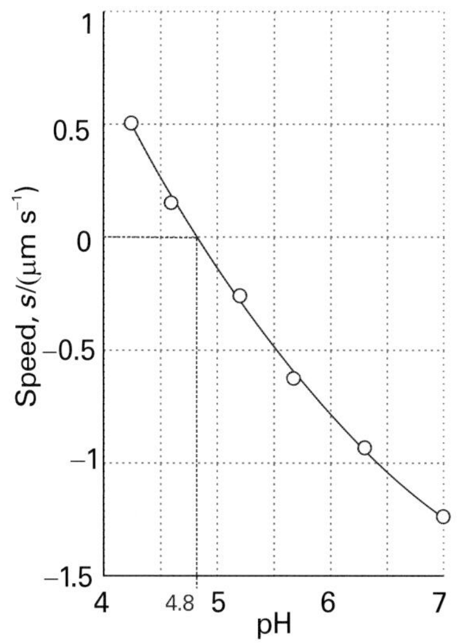

Fig. 19.8 The plot of drift speed of the protein bovine serum albumin in water against pH. The isoelectric point of the macromolecule corresponds to the pH at which the drift speed in the presence of an electric field is zero.

The separation of complicated mixtures of macromolecules may be difficult by SDS-PAGE or isoelectric focusing alone. However, the two techniques can be combined in **two-dimensional (2D) electrophoresis.** In a typical experiment, a protein mixture is separated first by isoelectric focusing, yielding a pattern of bands in a gel slab such as the one shown in Fig. 19.9a. To improve the separation of closely spaced bands, the first slab is attached to a second slab and SDS-PAGE is performed with the electric field being applied in a direction that is perpendicular to the direction in which isoelectric focusing was performed. The macromolecules separate according to their molar masses along this second dimension of the experiment, and the result is that spots are spread widely over the surface of the slab, leading to enhanced separation of the mixture's components (Fig. 19.9b).

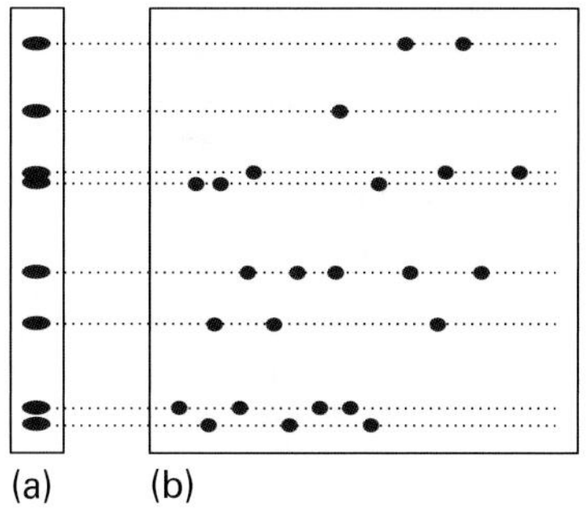

Fig. 19.9 The experimental steps taken during separation of a mixture of biopolymers by two-dimensional electrophoresis. (a) Isoelectric focusing is performed on a thin gel slab, resulting in separation along the vertical direction of the illustration. (b) The first slab is attached to a second, larger slab and SDS-PAGE is performed with the electric field oriented in the horizontal direction of the illustration, resulting in further separation by molar mass. The dashed horizontal lines show how the bands in the two-dimensional gel correspond to the bands in the gel on which isoelectric focusing was performed.

19.6 Viscosity

The formal definition of viscosity is given in Section 21.4; for now, we need to know that highly viscous liquids flow slowly and retard the motion of objects through them. The presence of a macromolecular solute increases the viscosity of a solution. The effect is large even at low concentration, because big molecules affect the fluid flow over an extensive region surrounding them. At low concentrations the viscosity, η, of the solution is related to the viscosity of the pure solvent, η_0, by

$$\eta = \eta_0(1 + [\eta]c + \cdots) \tag{19.22}$$

The **intrinsic viscosity**, $[\eta]$, is the analogue of a virial coefficient (and has dimensions of 1/concentration). It follows from eqn 19.22 that

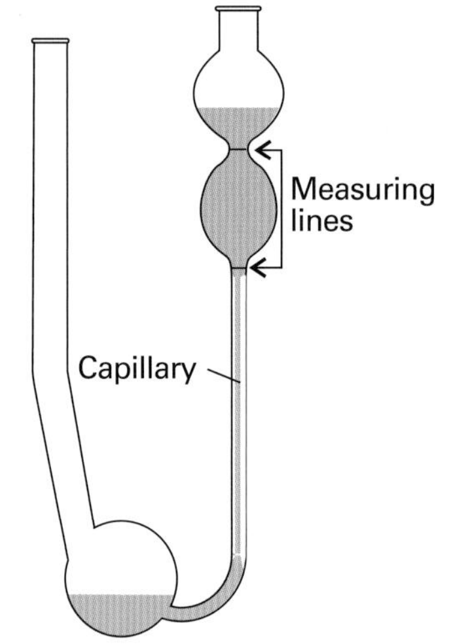

Fig. 19.10 An Ostwald viscometer. The viscosity is measured by noting the time required for the liquid to drain between the two marks.

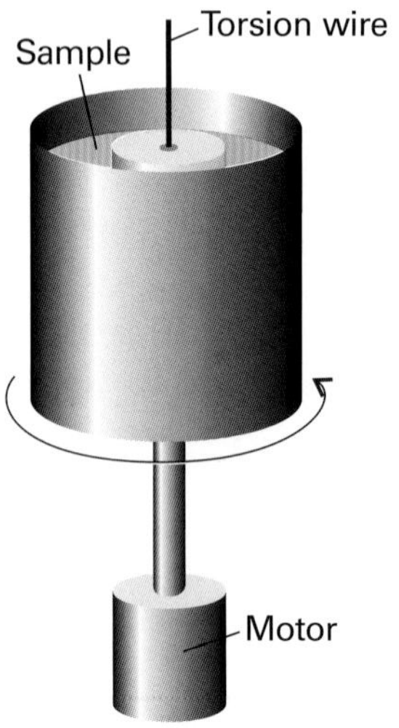

Fig. 19.11 A rotating rheometer. The torque on the inner drum is observed when the outer container is rotated.

Synoptic table 19.4* Intrinsic viscosity

	Solvent	θ/°C	K/(cm^3 g^{-1})	a
Polystyrene	Benzene	25	9.5×10^{-3}	0.74
Polyisobutylene	Benzene	23	8.3×10^{-2}	0.50
Various proteins	Guanidine hydrochloride + $HSCH_2CH_2OH$		7.2×10^{-3}	0.66

* More values are given in the *Data section*.

$$[\eta]=\lim_{c\to0}\left(\frac{\eta-\eta_0}{c\eta_0}\right)=\lim_{c\to0}\left(\frac{\eta/\eta_0-1}{c}\right) \tag{19.23}$$

Viscosities are measured in several ways. In the **Ostwald viscometer** shown in Fig. 19.10, the time taken for a solution to flow through the capillary is noted, and compared with a standard sample. The method is well suited to the determination of $[\eta]$ because the ratio of the viscosities of the solution and the pure solvent is proportional to the drainage time t and t_0 after correcting for different densities ρ and ρ_0:

$$\frac{\eta}{\eta_0}=\frac{t}{t_0}\times\frac{\rho}{\rho_0} \tag{19.24}$$

(In practice, the two densities are only rarely significantly different.) This ratio can be used directly in eqn 19.23. Viscometers in the form of rotating concentric cylinders are also used (Fig. 19.11), and the torque on the inner cylinder is monitored while the outer one is rotated. Such **rotating rheometers** (some instruments for the measurement of viscosity are also called rheometers) have the advantage over the Ostwald viscometer that the shear gradient between the cylinders is simpler than in the capillary and effects of the kind discussed shortly can be studied more easily.

There are many complications in the interpretation of viscosity measurements. Much of the work is based on empirical observations, and the determination of molar mass is usually based on comparisons with standard, nearly monodisperse sample. Some regularities are observed that help in the determination. For example, it is found that θ solutions of macromolecules often fit the **Mark–Kuhn–Houwink–Sakurada equation:**

$$[\eta]=K\bar{M}_v^a \tag{19.25}$$

where K and a are constants that depend on the solvent and type of macromolecule (Table 19.4); the viscosity-average molar mass, $\bar{M}_v$, appears in this expression.

Example 19.5 *Using intrinsic viscosity to measure molar mass*

The viscosities of a series of solutions of polystyrene in toluene were measured at 25°C with the following results:

c/(g dm^{-3})	0	2	4	6	8	10
η/(10^{-4} kg m^{-1} s^{-1})	5.58	6.15	6.74	7.35	7.98	8.64

Calculate the intrinsic viscosity and estimate the molar mass of the polymer by using eqn 19.25 with $K=3.80\times10^{-5}$ dm^3 g^{-1} and $a=0.63$.

Method The intrinsic viscosity is defined in eqn 19.23; therefore, form this ratio at the series of data points and extrapolate to $c = 0$. Interpret $\bar{M}_v$ as $\bar{M}_v/(\text{g mol}^{-1})$ in eqn 19.25.

Answer We draw up the following table:

$c/(\text{g dm}^{-3})$	0	2	4	6	8	10
η/η_0	1	1.102	1.208	1.317	1.43	1.549
$100[(\eta/\eta_0) - 1]/(c/\text{g dm}^{-3})$		5.11	5.2	5.28	5.38	5.49

The points are plotted in Fig. 19.12. The extrapolated intercept at $c = 0$ is 0.0504, so $[\eta] = 0.0504 \text{ dm}^3 \text{ g}^{-1}$. Therefore,

$$\bar{M}_v = \left(\frac{[\eta]}{K}\right)^{1/a} = 9.0 \times 10^4 \text{ g mol}^{-1}$$

Self-test 19.5 Evaluate the viscosity-average molar mass by using the second plotting technique. [90 kg mol^{-1}]

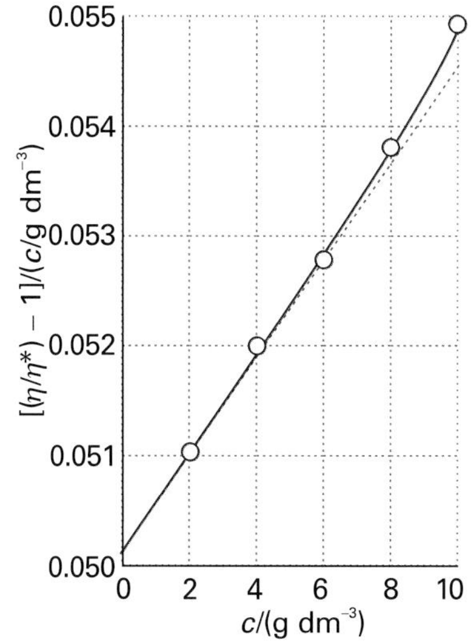

Fig. 19.12 The plot used for the determination of intrinsic viscosity, which is taken from the intercept at $c = 0$; see Example 19.5.

In some cases, the flow is non-Newtonian in the sense that the viscosity of the solution changes as the rate of flow increases. A decrease in viscosity with increasing rate of flow indicates the presence of long rod-like molecules that are orientated by the flow and hence slide past each other more freely. In some somewhat rare cases the stresses set up by the flow are so great that long molecules are broken up, with further consequences for the viscosity.

Structure and dynamics

The concept of the 'structure' of a macromolecule takes on different meanings at the different levels at which we think about the arrangement of the chain or network of monomers. The term **configuration** refers to the structural features that can be changed only by breaking chemical bonds and forming new ones. Thus, the chains —A—B—C— and —A—C—B— have different configurations. The term **conformation** refers to the spatial arrangement of the different parts of a chain, and one conformation can be changed into another by rotating one part of a chain around a bond.

19.7 The different levels of structure

The **primary structure** of a macromolecule is the sequence of small molecular residues making up the polymer. The residues may form either a chain, as in polyethylene, or a more complex network in which cross-links connect different chains, as in cross-linked polyacrylamide. In a synthetic polymer, virtually all the residues are identical and it is sufficient to name the monomer used in the synthesis. Thus, the repeating unit of polyethylene is $-CH_2CH_2-$, and the primary structure of the chain is specified by denoting it as $-(CH_2CH_2)_n-$.

The concept of primary structure ceases to be trivial in the case of synthetic copolymers and biological macromolecules, for in general these substances are chains formed from different molecules. For example, proteins are **polypeptides** formed from different amino acids (about twenty occur naturally) strung together by the **peptide link**, —CONH—. The determination of the primary structure is then a highly

Comment 19.1

More rigorously, the repeating unit in polyethylene is $-CH_2-$ and the substance is polymethylene. However, the advantage of regarding the repeating unit as $-CH_2CH_2-$ and naming it after its monomer is that derivatives, $-CHXCH_2-$, are seen to belong to the same family.

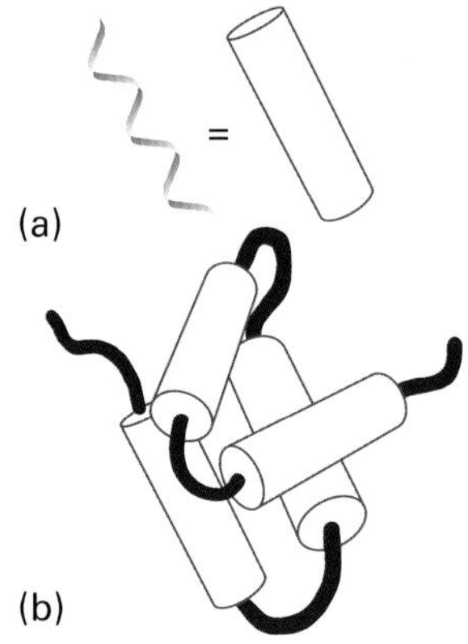

Fig. 19.13 (a) A polymer adopts a highly organized helical conformation, an example of a secondary structure. The helix is represented as a cylinder. (b) Several helical segments connected by short random coils pack together, providing an example of tertiary structure.

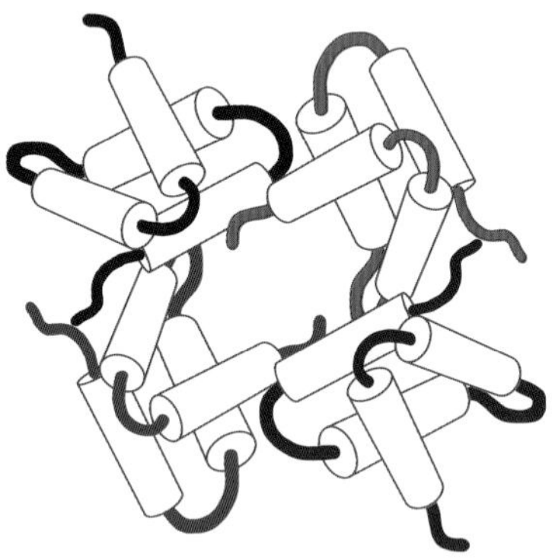

Fig. 19.14 Several subunits with specific tertiary structures pack together, providing an example of quaternary structure.

complex problem of chemical analysis called **sequencing**. The **degradation** of a polymer is a disruption of its primary structure, when the chain breaks into shorter components.

The **secondary structure** of a macromolecule is the (often local) spatial arrangement of a chain. The secondary structure of an isolated molecule of polyethylene is a random coil, whereas that of a protein is a highly organized arrangement determined largely by hydrogen bonds, and taking the form of random coils, helices (Fig. 19.13a), or sheets in various segments of the molecule. The loss of secondary structure is called **denaturation**. When the hydrogen bonds in a protein are destroyed (for instance, by heating, as when cooking an egg) the structure denatures into a random coil.

The **tertiary structure** is the overall three-dimensional structure of a macromolecule. For instance, the hypothetical protein shown in Fig. 19.13b has helical regions connected by short random-coil sections. The helices interact to form a compact tertiary structure.

The **quaternary structure** of a macromolecule is the manner in which large molecules are formed by the aggregation of others. Figure 19.14 shows how four molecular subunits, each with a specific tertiary structure, aggregate together. Quaternary structure can be very important in biology. For example, the oxygen-transport protein haemoglobin consists of four subunits that work together to take up and release O_2.

19.8 Random coils

The most likely conformation of a chain of identical units not capable of forming hydrogen bonds or any other type of specific bond is a **random coil**. Polyethylene is a simple example. The random coil model is a helpful starting point for estimating the orders of magnitude of the hydrodynamic properties of polymers and denatured proteins in solution.

The simplest model of a random coil is a **freely jointed chain**, in which any bond is free to make any angle with respect to the preceding one (Fig. 19.15). We assume that the residues occupy zero volume, so different parts of the chain can occupy the same region of space. The model is obviously an oversimplification because a bond is actually constrained to a cone of angles around a direction defined by its neighbour (Fig. 19.16). In a hypothetical one-dimensional freely jointed chain all the residues lie in a straight line, and the angle between neighbours is either 0° or 180°. The residues in a three-dimensional freely jointed chain are not restricted to lie in a line or a plane.

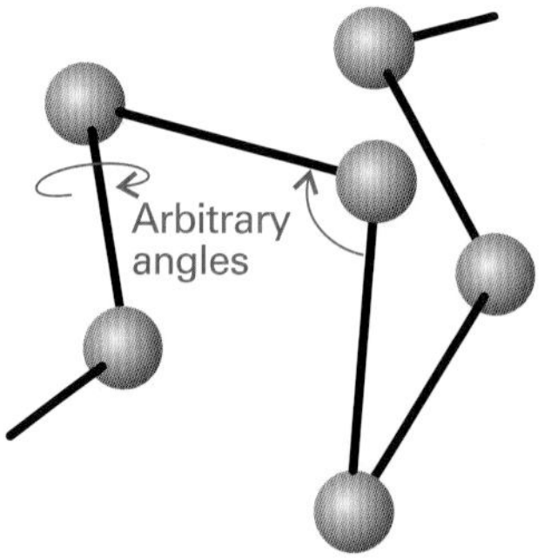

Fig. 19.15 A freely jointed chain is like a three-dimensional random walk, each step being in an arbitrary direction but of the same length.

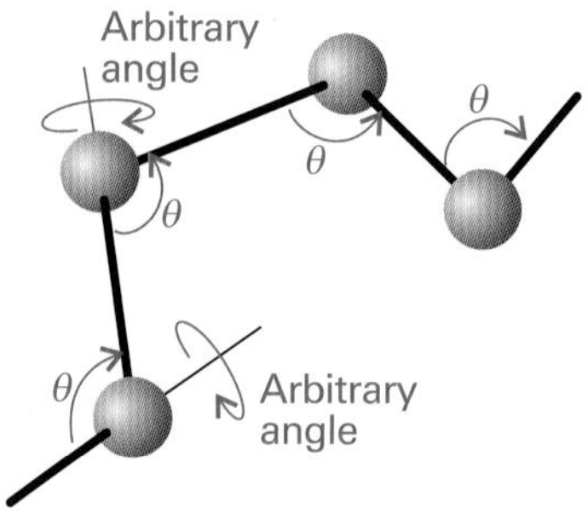

Fig. 19.16 A better description is obtained by fixing the bond angle (for example, at the tetrahedral angle) and allowing free rotation about a bond direction.

(a) Measures of size

As shown in the following *Justification*, we can deduce the probability, P, that the ends of a one-dimensional freely jointed chain composed of N units of length l are a distance nl apart:

$$P = \left(\frac{2}{\pi N}\right)^{1/2} e^{-n^2/2N} \qquad (19.26)$$

This function is plotted in Fig. 19.17 and can be used to calculate the probability that the ends of a three-dimensional freely jointed chain lie in the range r to $r + dr$. We write this probability as $f(r)dr$, where

$$f(r) = 4\pi\left(\frac{a}{\pi^{1/2}}\right)^3 r^2 e^{-a^2r^2} \qquad a = \left(\frac{3}{2Nl^2}\right)^{1/2} \qquad (19.27)$$

In some coils, the ends may be far apart whereas in others their separation is small. Here and elsewhere we are ignoring the fact that the chain cannot be longer than Nl. Although eqn 19.29 gives a nonzero probability for $r > Nl$, the values are so small that the errors in pretending that r can range up to infinity are negligible.

An alternative interpretation of eqn 19.27 is to regard each coil in a sample as ceaselessly writhing from one conformation to another; then $f(r)dr$ is the probability that at any instant the chain will be found with the separation of its ends between r and $r + dr$.

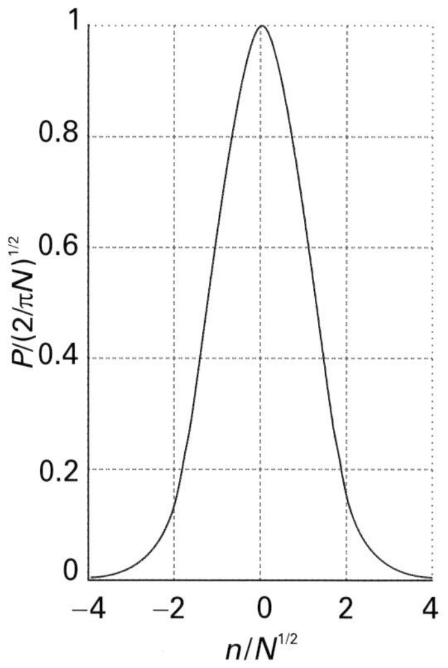

Fig. 19.17 The probability distribution for the separation of the ends of a one-dimensional random coil. The separation of the ends is nl, where l is the bond length.

Justification 19.3 *The one-dimensional freely jointed chain*

Consider a one-dimensional freely jointed polymer. We can specify the conformation of a molecule by stating the number of bonds pointing to the right (N_R) and the number pointing to the left (N_L). The distance between the ends of the chain is $(N_R - N_L)l$, where l is the length of an individual bond. We write $n = N_R - N_L$ and the total number of bonds as $N = N_R + N_L$.

The number of ways W of forming a chain with a given end-to-end distance nl is the number of ways of having N_R right-pointing and N_L left-pointing bonds. There are $N(N-1)(N-2)\ldots 1 = N!$ ways of selecting whether a step should be to the right or the left. If N_L steps are to the left, $N_R = N - N_L$ will be to the right. However, we end up at the same point for all $N_L!$ and $N_R!$ choices of which step is to the left and which to the right. Therefore

$$W = \frac{N!}{N_L!N_R!} = \frac{N!}{\{\tfrac{1}{2}(N+n)\}!\{\tfrac{1}{2}(N-n)\}!} \qquad (19.28)$$

The probability that the separation is nl is

$$P = \frac{\text{number of polymers with } N_R \text{ bonds to the right}}{\text{total number of arrangements of bonds}}$$

$$= \frac{N!/N_R!(N-N_R)!}{2^N} = \frac{N!}{\{\tfrac{1}{2}(N+n)\}!\{\tfrac{1}{2}(N-n)\}!2^N}$$

When the chain is compact in the sense that $n \ll N$, it is more convenient to evaluate $\ln P$: the factorials are then large and we can use Stirling's approximation (Section 16.1a) in the form

$$\ln x! \approx \ln(2\pi)^{1/2} + (x + \tfrac{1}{2})\ln x - x$$

The result, after quite a lot of algebra, is

$$\ln P = \ln\left(\frac{2}{\pi N}\right)^{1/2} - \tfrac{1}{2}(N+n+1)\ln(1+\nu) - \tfrac{1}{2}(N-n+1)\ln(1-\nu) \qquad (19.29)$$

where $\nu = n/N$. For a compact coil ($\nu \ll 1$) we use the approximation $\ln(1 \pm \nu) \approx \pm\nu - \frac{1}{2}\nu^2$ and so obtain

$$\ln P \approx \ln\left(\frac{2}{\pi N}\right)^{1/2} - \tfrac{1}{2}N\nu^2$$

which rearranges into eqn 19.26.

Self-test 19.6 Provide the algebraic steps that lead from eqn 19.28 to eqn 19.29.

There are several measures of the geometrical size of a random coil. The **contour length**, R_c, is the length of the macromolecule measured along its backbone from atom to atom. For a polymer of N monomer units each of length l, the contour length is

$$R_c = Nl \tag{19.30}$$

The **root mean square separation**, R_{rms}, is a measure of the average separation of the ends of a random coil: it is the square root of the mean value of R^2. We show in the following *Justification* that:

$$R_{rms} = N^{1/2}l \tag{19.31}$$

We see that, as the number of monomer units increases, the root mean square separation of its end increases as $N^{1/2}$ (Fig. 19.18), and consequently its volume increases as $N^{3/2}$. The result must be multiplied by a factor when the chain is not freely jointed (see below).

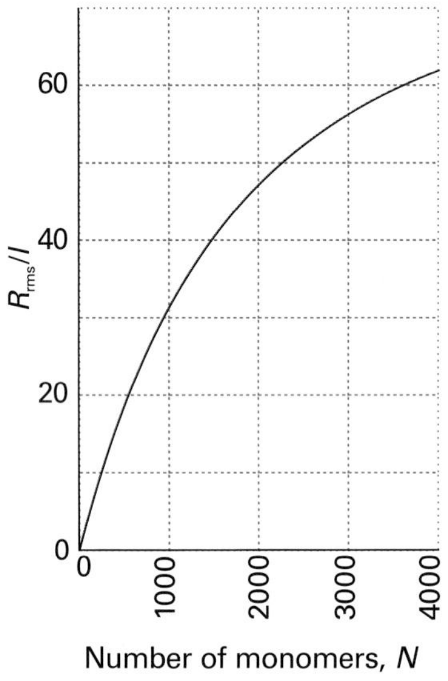

Fig. 19.18 The variation of the root mean square separation of the ends of a three-dimensional random coil, R_{rms}, with the number of monomers.

Justification 19.4 *The root mean square separation of the ends of a freely jointed chain*

In *Appendix* 2 we see that the mean value $\langle X\rangle$ of a variable X with possible values x is given by

$$\langle X\rangle = \int_{-\infty}^{+\infty} xf(x)\mathrm{d}x$$

where the function $f(x)$ is the *probability density*, a measure of the distribution of the probability values over x, and $\mathrm{d}x$ is an infinitesimally small interval of x values. The mean value of a function $g(X)$ can be calculated with a similar formula:

$$\langle g(X)\rangle = \int_{-\infty}^{+\infty} g(x)f(x)\mathrm{d}x$$

To apply these concepts to the calculation of the root mean square separation of the ends of a random coil, we identify $f(r)\mathrm{d}r$ as the probability that the ends of the chain lie in the range $R = r$ to $R = r + \mathrm{d}r$. It follows that the general expression for the mean nth power of the end-to-end separation (a positive quantity that can vary from 0 to $+\infty$) is

$$\langle R^n\rangle = \int_0^{\infty} r^n f(r)\mathrm{d}r$$

To calculate R_{rms}, we first determine $\langle R^2\rangle$ by using $n = 2$ and $f(r)$ from eqn 19.27:

$$\langle R^2\rangle = 4\pi\left(\frac{a}{\pi^{1/2}}\right)^3 \int_0^{\infty} r^4 e^{-a^2r^2}\mathrm{d}r = 4\pi\left(\frac{a}{\pi^{1/2}}\right)^3 \times \frac{3\pi^{1/2}}{8a^5} = \frac{3}{2a^2}$$

where we have used the standard integral

$$\int_0^\infty x^4 e^{-a^2x^2}dx = \frac{3}{2a^2}$$

When we use the expression for a in eqn 19.27 we obtain:

$$\langle R^2\rangle = \frac{3}{2}\times\left(\frac{2Nl^2}{3}\right) = Nl^2$$

The root mean square separation follows from

$$R_{rms} = \langle R^2\rangle^{1/2} = N^{1/2}l$$

Self-test 19.7 Calculate the mean separation of the ends of a freely jointed chain of N bonds of length l. *Hint.* You will need the standard integral $\int_0^\infty x^3 e^{-a^2x^2}dx = \frac{1}{2}a^4$.

$$[\langle R\rangle = \left(\frac{8N}{3\pi}\right)^{1/2} l]$$

Another convenient measure of size is the **radius of gyration**, R_g, which we encountered in Section 19.3a. It is calculated formally from the expression:

$$R_g = \frac{1}{N}\left(\frac{1}{2}\sum_{ij} R_{ij}^2\right)^{1/2} \qquad (19.32)$$

where R_{ij} is the separation of atoms i and j. The radius of gyration of the coil also increases as $N^{1/2}$:

$$R_g = \left(\frac{N}{6}\right)^{1/2} l \qquad (19.33)$$

The radius of gyration may also be calculated for other geometries. For example, a solid uniform sphere of radius R has $R_g = (\frac{3}{5})^{1/2}R$, and a long thin uniform rod of length l has $R_g = l/(12)^{1/2}$ for rotation about an axis perpendicular to the long axis.

The random coil model ignores the role of the solvent: a poor solvent will tend to cause the coil to tighten so that solute–solvent contacts are minimized; a good solvent does the opposite. Therefore, calculations based on this model are better regarded as lower bounds to the dimensions for a polymer in a good solvent and as an upper bound for a polymer in a poor solvent. The model is most reliable for a polymer in a bulk solid sample, where the coil is likely to have its natural dimensions.

(b) Conformational entropy

The random coil is the least structured conformation of a polymer chain and corresponds to the state of greatest entropy. Any stretching of the coil introduces order and reduces the entropy. Conversely, the formation of a random coil from a more extended form is a spontaneous process (provided enthalpy contributions do not interfere). As shown in the *Justification* below, we can use the same model to deduce that the change in **conformational entropy**, the statistical entropy arising from the arrangement of bonds, when a coil containing N bonds of length l is stretched or compressed by nl is

$$\Delta S = -\tfrac{1}{2}kN \ln\{(1+\nu)^{1+\nu}(1-\nu)^{1-\nu}\} \qquad \nu = n/N \qquad (19.34)$$

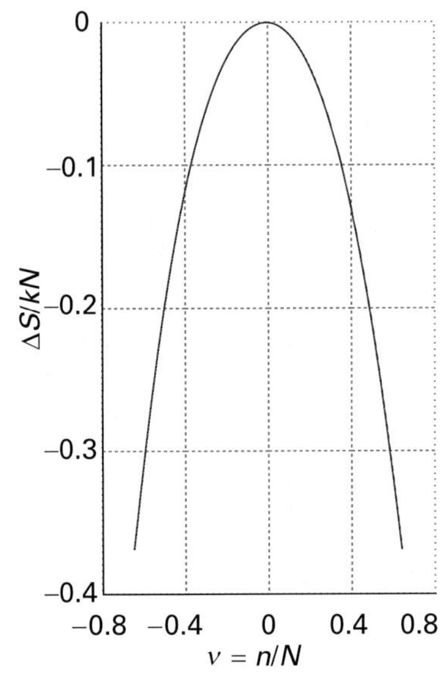

Fig. 19.19 The change in molar entropy of a perfect elastomer as its extension changes; $\nu = 1$ corresponds to complete extension; $\nu = 0$, the conformation of highest entropy, corresponds to the random coil.

This function is plotted in Fig. 19.19, and we see that minimum extension corresponds to maximum entropy.

Justification 19.5 *The conformational entropy of a freely jointed chain*

The conformational entropy of the chain is $S = k \ln W$, where W is given by eqn 19.28. Therefore,

$$S/k = \ln N! - \ln\{\tfrac{1}{2}(N+n)\}! - \ln\{\tfrac{1}{2}(N-n)\}!$$

Because the factorials are large (except for large extensions), we can use Stirling's approximation to obtain

$$S/k = -\ln(2\pi)^{1/2} + (N+1)\ln 2 + (N+\tfrac{1}{2})\ln N - \tfrac{1}{2}\ln\{(N+n)^{N+n+1}(N-n)^{N-n+1}\}$$

The most probable conformation of the chain is the one with the ends close together ($n = 0$), as may be confirmed by differentiation. Therefore, the maximum entropy is

$$S/k = -\ln(2\pi)^{1/2} + (N+1)\ln 2 + \tfrac{1}{2}\ln N$$

The change in entropy when the chain is stretched or compressed by nl is therefore the difference of these two quantities, and the resulting expression is eqn 19.34.

(c) Constrained chains

The freely jointed chain model is improved by removing the freedom of bond angles to take any value. For long chains, we can simply take groups of neighbouring bonds and consider the direction of their resultant. Although each successive individual bond is constrained to a single cone of angle θ relative to its neighbour, the resultant of several bonds lies in a random direction. By concentrating on such groups rather than individuals, it turns out that for long chains the expressions for the root mean square separation and the radius of gyration given above should be multiplied by

$$F = \left(\frac{1-\cos\theta}{1+\cos\theta}\right)^{1/2} \tag{19.35}$$

For tetrahedral bonds, for which $\cos\theta = -\tfrac{1}{3}$ (that is, $\theta = 109.5°$), $F = 2^{1/2}$. Therefore:

$$R_{rms} = (2N)^{1/2} l \qquad R_g = \left(\frac{N}{3}\right)^{1/2} l \tag{19.36}$$

Illustration 19.2 *The dimensions of a polymer chain*

Consider a polyethylene chain with $M = 56$ kg mol^{-1}, corresponding to $N = 4000$. Because $l = 154$ pm for a C—C bond, we find $R_{rms} = 14$ nm and $R_g = 5.6$ nm (Fig. 19.20). This value of R_g means that, on average, the coils rotate like hollow spheres of radius 5.6 nm and mass equal to the molecular mass.

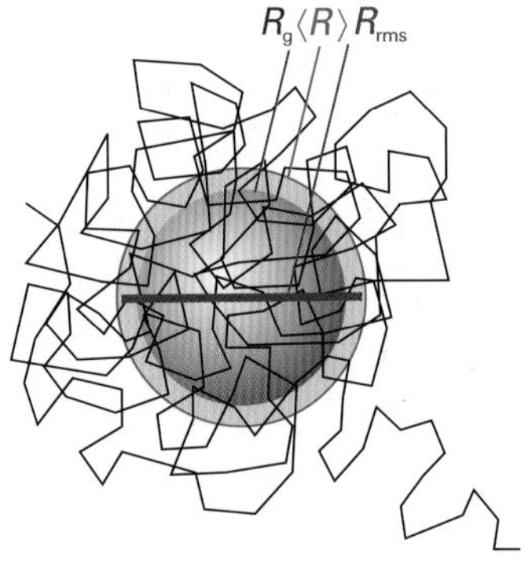

Fig. 19.20 A random coil in three dimensions. This one contains about 200 units. The root mean square distance between the ends (R_{rms}), the mean radius, and the radius of gyration (R_g) are indicated.

The model of a randomly coiled molecule is still an approximation, even after the bond angles have been restricted, because it does not take into account the impossibility of two or more atoms occupying the same place. Such self-avoidance tends to swell the coil, so (in the absence of solvent effects) it is better to regard R_{rms} and R_g as lower bounds to the actual values.

19.9 The structure and stability of synthetic polymers

Synthetic polymers are classified broadly as *elastomers*, *fibres*, and *plastics*, depending on their **crystallinity**, the degree of three-dimensional long-range order attained in the solid state. An **elastomer** is a flexible polymer that can expand or contract easily upon application of an external force. Elastomers are polymers with numerous cross-links that pull them back into their original shape when a stress is removed. A **perfect elastomer**, a polymer in which the internal energy is independent of the extension of the random coil, can be modelled as a freely jointed chain. We saw in Section 19.8b that the contraction of an extended chain to a random coil is spontaneous in the sense that it corresponds to an increase in entropy; the entropy change of the surroundings is zero because no energy is released when the coil forms. In the following *Justification* we also see that the restoring force, F, of a one-dimensional perfect elastomer is

$$F=\frac{kT}{2l}\ln\left(\frac{1+\nu}{1-\nu}\right) \qquad \nu=n/N \tag{19.37a}$$

where N is the total number of bonds of length l and the polymer is stretched or compressed by nl. This function is plotted in Fig. 19.21. At low extensions, when $\nu \ll 1$:

$$F\approx\frac{\nu kT}{l}=\frac{nkT}{Nl} \tag{19.37b}$$

and the sample obeys Hooke's law: the restoring force is proportional to the displacement (which is proportional to n). For small displacements, therefore, the whole coil shakes with simple harmonic motion.

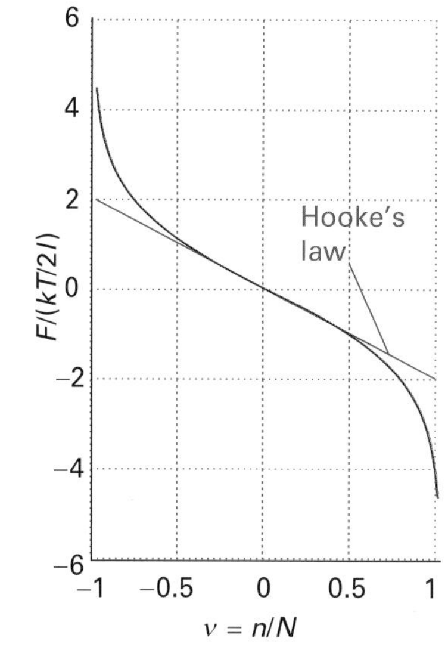

Fig. 19.21 The restoring force, F, of a one-dimensional perfect elastomer. For small strains, F is linearly proportional to the extension, corresponding to Hooke's law.

Justification 19.6 *Hooke's law*

The work done on an elastomer when it is extended through a distance dx is Fdx, where F is the restoring force. The change in internal energy is therefore

$$dU=TdS-pdV+Fdx$$

It follows that

$$\left(\frac{\partial U}{\partial x}\right)_{T,V}=T\left(\frac{\partial S}{\partial x}\right)_{T,V}+F$$

In a perfect elastomer, as in a perfect gas, the internal energy is independent of the dimensions (at constant temperature), so $(\partial U/\partial x)_{T,V}=0$. The restoring force is therefore

$$F=-T\left(\frac{\partial S}{\partial x}\right)_{T,V}$$

If now we substitute eqn 19.34 into this expression (we evade problems arising from the constraint of constant volume by supposing that the sample contracts laterally as it is stretched), we obtain

$$F=-\frac{T}{l}\left(\frac{\partial S}{\partial n}\right)_{T,V}=\frac{T}{Nl}\left(\frac{\partial S}{\partial \nu}\right)_{T,V}=\frac{kT}{2l}\ln\left(\frac{1+\nu}{1-\nu}\right)$$

as in eqn 19.37a.

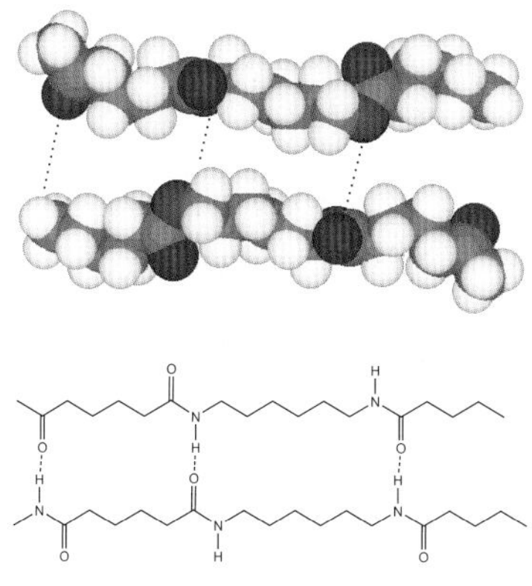

Fig. 19.22 A fragment of two nylon-66 polymer chains showing the pattern of hydrogen bonds that are responsible for the cohesion between the chains.

A **fibre** is a polymeric material that owes its strength to interactions between chains. One example is nylon-66 (Fig. 19.22). Under certain conditions, nylon-66 can be prepared in a state of high crystallinity, in which hydrogen bonding between the amide links of neighbouring chains results in an ordered array.

A **plastic** is a polymer that can attain only a limited degree of crystallinity and as a result is neither as strong as a fibre nor as flexible as an elastomer. Certain materials, such as nylon-66, can be prepared either as a fibre or as a plastic. A sample of plastic nylon-66 may be visualized as consisting of crystalline hydrogen-bonded regions of varying size interspersed amongst amorphous, random coil regions. A single type of polymer may exhibit more than one characteristic, for to display fibrous character, the polymers need to be aligned; if the chains are not aligned, then the substance may be plastic. That is the case with nylon, poly(vinyl chloride), and the siloxanes.

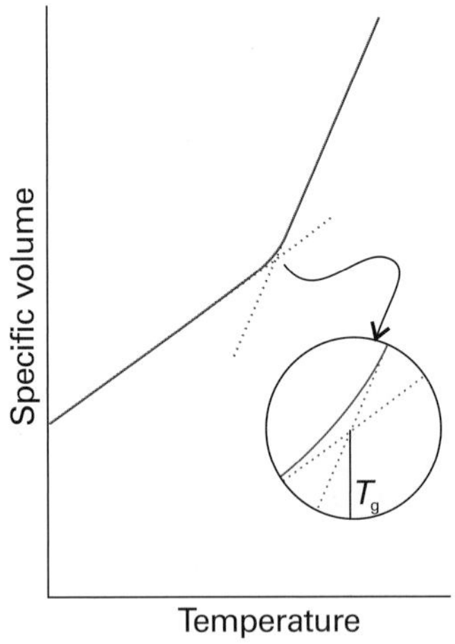

Fig. 19.23 The variation of specific volume with temperature of a synthetic polymer. The glass transition temperature, T_g, is at the point of intersection of extrapolations of the two linear parts of the curve.

The crystallinity of synthetic polymers can be destroyed by thermal motion at sufficiently high temperatures. This change in crystallinity may be thought of as a kind of intramolecular melting from a crystalline solid to a more fluid random coil. Polymer melting also occurs at a specific **melting temperature**, T_m, which increases with the strength and number of intermolecular interactions in the material. Thus, polyethylene, which has chains that interact only weakly in the solid, has $T_m = 414$ K and nylon-66 fibres, in which there are strong hydrogen bonds between chains, has $T_m = 530$ K. High melting temperatures are desirable in most practical applications involving fibres and plastics.

All synthetic polymers undergo a transition from a state of high to low chain mobility at the **glass transition temperature**, T_g. To visualize the glass transition, we consider what happens to an elastomer as we lower its temperature. There is sufficient energy available at normal temperatures for limited bond rotation to occur and the flexible chains writhe. At lower temperatures, the amplitudes of the writhing motion decrease until a specific temperature, T_g, is reached at which motion is frozen completely and the sample forms a glass. Glass transition temperatures well below 300 K are desirable in elastomers that are to be used at normal temperatures. Both the glass transition temperature and the melting temperature of a polymer may be measured by differential scanning calorimetry (*Impact* I2.1). Because the motion of the segments of a polymer chain increase at the glass transition temperature, T_g may also be determined from a plot of the specific volume of a polymer (the reciprocal of its mass density) against temperature (Fig. 19.23).

Comment 19.2

As we shall discuss further in Chapter 20, a metallic conductor is a substance with an electrical conductivity that decreases as the temperature is raised. A semiconductor is a substance with an electrical conductivity that increases as the temperature is raised.

IMPACT ON TECHNOLOGY

I19.2 Conducting polymers

We have just seen how the structure of a polymer chain affects its mechanical and thermal properties. Now we consider the electrical properties of synthetic polymers.

Most of the macromolecules and self-assembled structures considered in this chapter are insulators, or very poor electrical conductors. However, a variety of newly developed macromolecular materials have electrical conductivities that rival those of silicon-based semiconductors and even metallic conductors. We examine one example in detail: **conducting polymers**, in which extensively conjugated double bonds facilitate electron conduction along the polymer chain. The Nobel Prize in chemistry was awarded in 2000 to A.J. Heeger, A.J. McDiarmid, and H. Shirakawa for their pioneering work in the synthesis and characterization of conducting polymers.

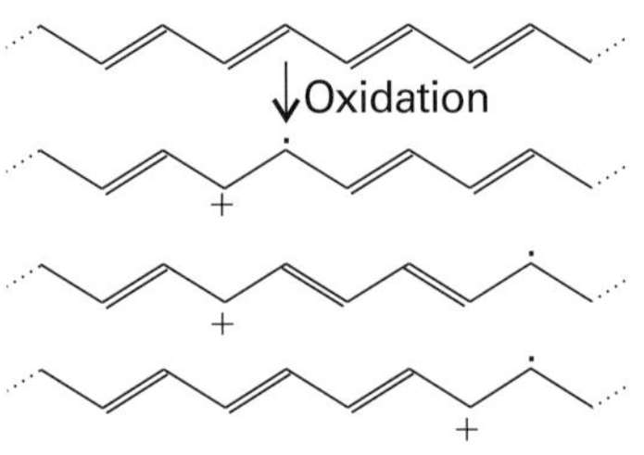

Fig. 19.24 The mechanism of migration of a partially localized cation radical, or polaron, in polyacetylene.

One example of a conducting polymer is polyacetylene (Fig. 19.24). Whereas the delocalized π bonds do suggest that electrons can move up and down the chain, the electrical conductivity of polyacetylene increases significantly when it is partially oxidized by I_2 and other strong oxidants. The product is a **polaron**, a partially localized cation radical that does not delocalize but rather travels through the chain, as shown in Fig. 19.24. Oxidation of the polymer by one more equivalent forms either **bipolarons**, a di-cation that moves as a unit through the chain, or **solitons**, two separate cation radicals that move independently. Polarons and solitons contribute to the mechanism of charge conduction in polyacetylene.

Conducting polymers are slightly better electrical conductors than silicon semiconductors but are far worse than metallic conductors. They are currently used in a number of devices, such as electrodes in batteries, electrolytic capacitors, and sensors. Recent studies of photon emission by conducting polymers may lead to new technologies for light-emitting diodes and flat-panel displays. Conducting polymers also show promise as molecular wires that can be incorporated into nanometre-sized electronic devices.

19.10 The structure of proteins

A protein is a polypeptide composed of linked α-amino acids, $NH_2CHRCOOH$, where R is one of about 20 groups. For a protein to function correctly, it needs to have a well defined conformation. For example, an enzyme has its greatest catalytic efficiency only when it is in a specific conformation. The amino acid sequence of a protein contains the necessary information to create the active conformation of the protein from a newly synthesized random coil. However, the prediction of the conformation from the primary structure, the so-called *protein folding problem*, is extraordinarily difficult and is still the focus of much research.

(a) The Corey–Pauling rules

The origin of the secondary structures of proteins is found in the rules formulated by Linus Pauling and Robert Corey in 1951. The essential feature is the stabilization of structures by hydrogen bonds involving the peptide link. The latter can act both as a donor of the H atom (the NH part of the link) and as an acceptor (the CO part). The **Corey–Pauling rules** are as follows (Fig. 19.25):

1 The four atoms of the peptide link lie in a relatively rigid plane.

The planarity of the link is due to delocalization of π electrons over the O, C, and N atoms and the maintenance of maximum overlap of their p orbitals.

2 The N, H, and O atoms of a hydrogen bond lie in a straight line (with displacements of H tolerated up to not more than 30° from the N—O vector).

3 All NH and CO groups are engaged in hydrogen bonding.

The rules are satisfied by two structures. One, in which hydrogen bonding between peptide links leads to a helical structure, is a *helix*, which can be arranged as either a right- or a left-handed screw. The other, in which hydrogen bonding between peptide links leads to a planar structure, is a *sheet*; this form is the secondary structure of the protein fibroin, the constituent of silk.

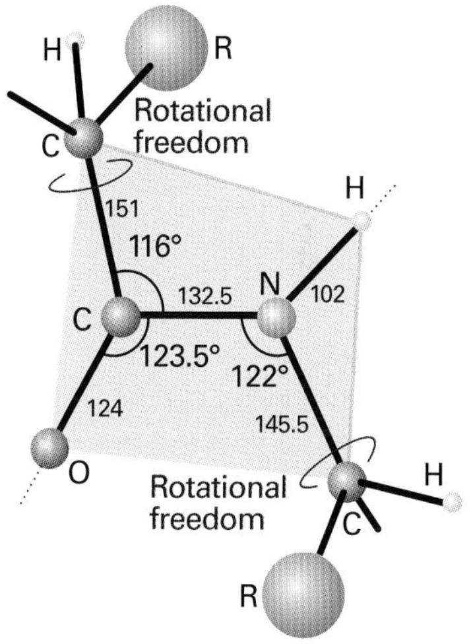

Fig. 19.25 The dimensions that characterize the peptide link (bonds in picometres). The C—NH—CO—C atoms define a plane (the C—N bond has partial double-bond character), but there is rotational freedom around the C—CO and N–C bonds.

(b) Conformational energy

A polypeptide chain adopts a conformation corresponding to a minimum Gibbs energy, which depends on the **conformational energy**, the energy of interaction between different parts of the chain, and the energy of interaction between the chain and surrounding solvent molecules. In the aqueous environment of biological cells, the outer surface of a protein molecule is covered by a mobile sheath of water molecules, and its interior contains pockets of water molecules. These water molecules play an important role in determining the conformation that the chain adopts through hydrophobic interactions and hydrogen bonding to amino acids in the chain.

The simplest calculations of the conformational energy of a polypeptide chain ignore entropy and solvent effects and concentrate on the total potential energy of all the interactions between nonbonded atoms. For example, these calculations predict that a right-handed α helix of L-amino acids is marginally more stable than a left-handed helix of the same amino acids.

To calculate the energy of a conformation, we need to make use of many of the molecular interactions described in Chapter 18, and also of some additional interactions:

1 *Bond stretching*. Bonds are not rigid, and it may be advantageous for some bonds to stretch and others to be compressed slightly as parts of the chain press against one another. If we liken the bond to a spring, then the potential energy takes the form of Hooke's law

$$V_{\text{stretch}} = \tfrac{1}{2}k_{\text{stretch}}(R - R_e)^2 \tag{19.38}$$

where R_e is the equilibrium bond length and k_{stretch} is the force constant, a measure of the stiffness of the bond in question.

2 *Bond bending*. An O—C—H bond angle (or some other angle) may open out or close in slightly to enable the molecule as a whole to fit together better. If the equilibrium bond angle is θ_e, we write

$$V_{\text{bend}} = \tfrac{1}{2}k_{\text{bend}}(\theta - \theta_e)^2 \tag{19.39}$$

where k_{bend} is the force constant, a measure of how difficult it is to change the bond angle.

Self-test 19.8 Theoretical studies have estimated that the lumiflavin isoalloazine ring system has an energy minimum at the bending angle of 15°, but that it requires only 8.5 kJ mol^{-1} to increase the angle to 30°. If there are no other compensating interactions, what is the force constant for lumiflavin bending?

[6.27×10^{-23} J deg^{-2}, equivalent to 37.7 J mol^{-1} deg^{-2}]

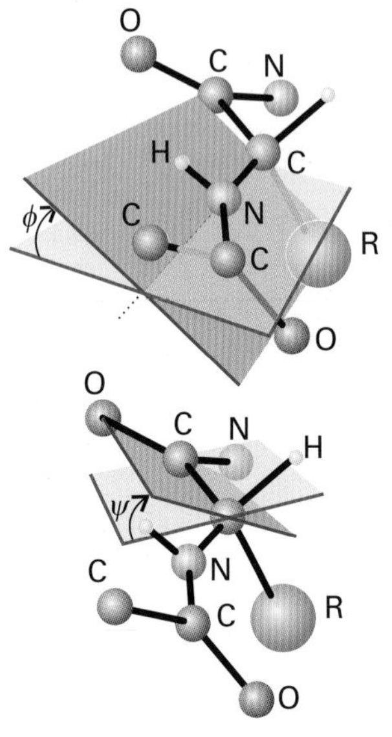

Fig. 19.26 The definition of the torsional angles ψ and ϕ between two peptide units. In this case (an α-L-polypeptide) the chain has been drawn in its all-*trans* form, with $\psi = \phi = 180°$.

3 *Bond torsion*. There is a barrier to internal rotation of one bond relative to another (just like the barrier to internal rotation in ethane). Because the planar peptide link is relatively rigid, the geometry of a polypeptide chain can be specified by the two angles that two neighbouring planar peptide links make to each other. Figure 19.26 shows the two angles ϕ and ψ commonly used to specify this relative orientation. The sign convention is that a positive angle means that the front atom must be rotated clockwise to bring it into an eclipsed position relative to the rear atom. For an all-*trans* form of the chain, all ϕ and ψ are 180°. A helix is obtained when all the ϕ are equal and when all the ψ are equal. For a right-handed helix, all $\phi = -57°$ and all $\psi = -47°$. For a left-handed helix, both angles are positive. The torsional contribution to the total potential energy is

$$V_{\text{torsion}} = A(1 + \cos 3\phi) + B(1 + \cos 3\psi) \tag{19.40}$$

in which A and B are constants of the order of 1 kJ mol^{-1}. Because only two angles are needed to specify the conformation of a helix, and they range from $-180°$ to $+180°$, the torsional potential energy of the entire molecule can be represented on a **Ramachandran plot**, a contour diagram in which one axis represents ϕ and the other represents ψ.

4 *Interaction between partial charges*. If the partial charges q_i and q_j on the atoms i and j are known, a Coulombic contribution of the form $1/r$ can be included (Section 18.3):

$$V_{\text{Coulomb}} = \frac{q_i q_j}{4\pi\varepsilon r} \tag{19.41}$$

where ε is the permittivity of the medium in which the charges are embedded. Charges of $-0.28e$ and $+0.28e$ are assigned to N and H, respectively, and $-0.39e$ and $+0.39e$ to O and C, respectively. The interaction between partial charges does away with the need to take dipole–dipole interactions into account, for they are taken care of by dealing with each partial charge explicitly.

5 *Dispersive and repulsive interactions.* The interaction energy of two atoms separated by a distance r (which we know once ϕ and ψ are specified) can be given by the Lennard-Jones (12,6) form (Section 18.5):

$$V_{\mathrm{LJ}} = \frac{C}{r^{12}} - \frac{D}{r^{6}} \qquad (19.42)$$

6 *Hydrogen bonding.* In some models of structure, the interaction between partial charges is judged to take into account the effect of hydrogen bonding. In other models, hydrogen bonding is added as another interaction of the form

$$V_{\mathrm{H\,bonding}} = \frac{E}{r^{12}} - \frac{F}{r^{10}} \qquad (19.43)$$

The total potential energy of a given conformation (ϕ,ψ) can be calculated by summing the contributions given by eqns 19.38–19.43 for all bond angles (including torsional angles) and pairs of atoms in the molecule. The procedure is known as a **molecular mechanics** simulation and is automated in commercially available molecular modelling software. For large molecules, plots of potential energy against bond distance or bond angle often show several local minima and a global minimum (Fig. 19.27). The software packages include schemes for modifying the locations of the atoms and searching for these minima systematically.

The structure corresponding to the global minimum of a molecular mechanics simulation is a snapshot of the molecule at $T = 0$ because only the potential energy is included in the calculation; contributions to the total energy from kinetic energy are excluded. In a **molecular dynamics** simulation, the molecule is set in motion by heating it to a specified temperature, as described in Section 17.6b. The possible trajectories of all atoms under the influence of the intermolecular potentials correspond to the conformations that the molecule can sample at the temperature of the simulation. At very low temperatures, the molecule cannot overcome some of the potential energy barriers given by eqns 19.38–19.43, atomic motion is restricted, and only a few conformations are possible. At high temperatures, more potential energy barriers can be overcome and more conformations are possible. Therefore, molecular dynamics calculations are useful tools for the visualization of the flexibility of polymers.

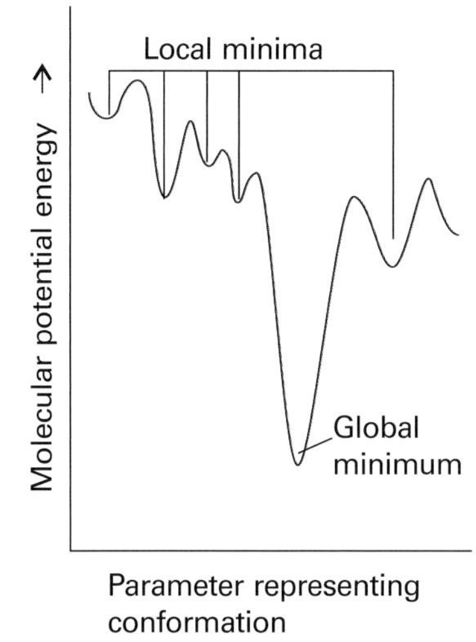

Fig. 19.27 For large molecules, a plot of potential energy against the molecular geometry often shows several local minima and a global minimum.

(c) Helices and sheets

A right-handed **α-helix** is illustrated in Fig. 19.28. Each turn of the helix contains 3.6 amino acid residues, so the period of the helix corresponds to 5 turns (18 residues). The pitch of a single turn (the distance between points separated by 360°) is 544 pm. The N—H···O bonds lie parallel to the axis and link every fourth group (so residue i is linked to residues $i-4$ and $i+4$). All the R groups point away from the major axis of the helix.

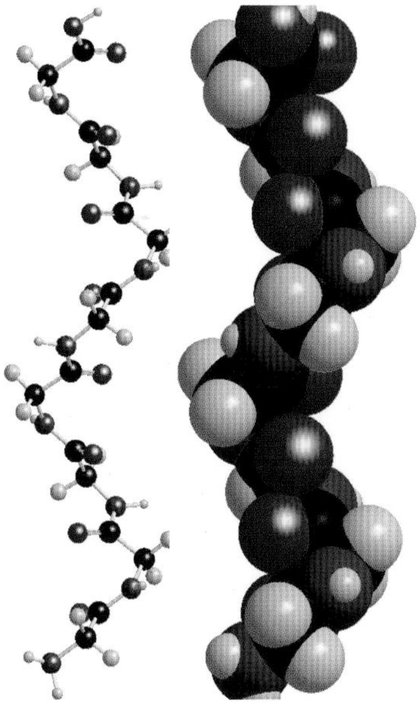

Fig. 19.28 The polypeptide α helix, with poly-L-glycine as an example. Carbon atoms are shown in black, with nitrogen in blue, oxygen in red, and hydrogen atoms in grey. There are 3.6 residues per turn, and a translation along the helix of 150 pm per residue, giving a pitch of 540 pm. The diameter (ignoring side chains) is about 600 pm.

Figure 19.29 shows the Ramachandran plots for the helical form of polypeptide chains formed from the nonchiral amino acid glycine (R = H) and the chiral amino acid L-alanine ($R = CH_3$). The glycine map is symmetrical, with minima of equal depth at $\phi = -80°$, $\psi = +90°$ and at $\phi = +80°$, $\psi = -90°$. In contrast, the map for L-alanine is unsymmetrical, and there are three distinct low-energy conformations (marked I, II, III). The minima of regions I and II lie close to the angles typical of right- and

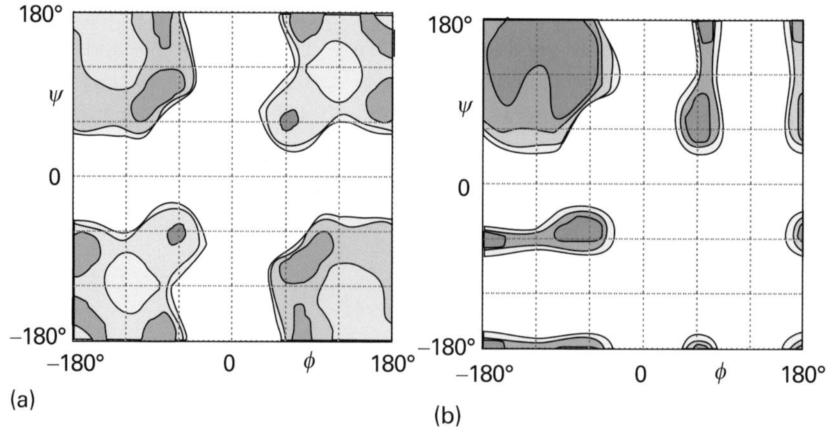

Fig. 19.29 Contour plots of potential energy against the torsional angles ψ and ϕ, also known as Ramachandran plots, for (a) a glycyl residue of a polypeptide chain and (b) an alanyl residue. The darker the shading is, the lower the potential energy. The glycyl diagram is symmetrical, but regions I and II in the correspond to right- and left-handed helices, are unsymmetrical, and the minimum in region I lies lower than that in region II. (After D.A. Brant and P.J. Flory, *J. Mol. Biol.* **23**, 47 (1967).)

left-handed helices, but the former has a lower minimum. This result is consistent with the observation that polypeptides of the naturally occurring L-amino acids tend to form right-handed helices.

A ***β*-sheet** (also called the ***β*-pleated sheet**) is formed by hydrogen bonding between two extended polypeptide chains (large absolute values of the torsion angles ϕ and ψ). Some of the R groups point above and some point below the sheet. Two types of structures can be distinguished from the pattern of hydrogen bonding between the constituent chains.

In an **anti-parallel *β*-sheet** (Fig. 19.30a), $\phi = -139°$, $\psi = 113°$, and the N—H···O atoms of the hydrogen bonds form a straight line. This arrangement is a consequence of the antiparallel arrangement of the chains: every N—H bond on one chain is aligned with a C—O bond from another chain. Antiparallel *β*-sheets are very common in proteins. In a **parallel *β*-sheet** (Fig. 19.30b), $\phi = -119°$, $\psi = 113°$, and the N—H···O atoms of the hydrogen bonds are not perfectly aligned. This arrangement is a result of the parallel arrangement of the chains: each N—H bond on one chain is aligned with a N—H bond of another chain and, as a result, each C—O bond of one chain is aligned with a C—O bond of another chain. These structures are not common in proteins.

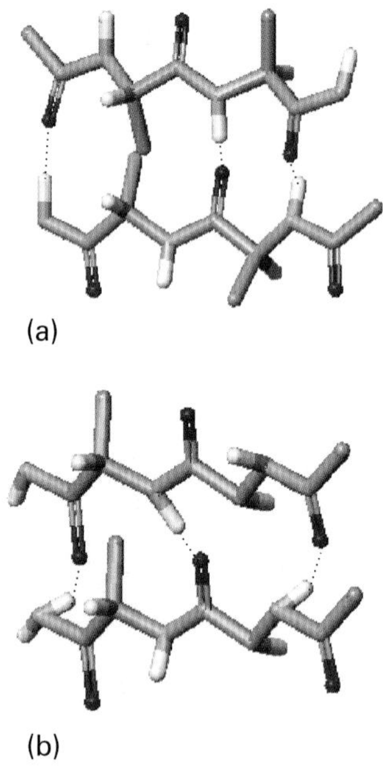

Fig. 19.30 The two types of *β*-sheets: (a) antiparallel ($\phi = -139°$, $\psi = 113°$), in which the N—H—O atoms of the hydrogen bonds form a straight-line; (b) parallel ($\phi = -119°$, $\psi = 113°$ in which the N—H···O atoms of the hydrogen bonds are not perfectly aligned.

Circular dichroism (CD) spectroscopy (Section 14.2) provides a great deal of information about the secondary structure of polypeptides. Consider a helical polypeptide. Not only are the individual monomer units chiral, but so is the helix. Therefore, we expect the *α*-helix to have a unique CD spectrum. Because *β*-sheets and random coils also have distinguishable spectral features (Fig. 19.31), circular dichroism is a very important technique for the study of protein conformation.

(d) Higher-order structures

Covalent and non-covalent interactions may cause polypeptide chains with well defined secondary structures to fold into tertiary structures. Subunits with well defined tertiary structures may interact further to form quaternary structures.

Although we do not know all the rules that govern protein folding, a few general conclusions may be drawn from X-ray diffraction studies of water-soluble natural proteins and synthetic polypeptides. In an aqueous environment, the chains fold in such a way as to place nonpolar R groups in the interior (which is often not very accessible to solvent) and charged R groups on the surface (in direct contact with the polar solvent). A wide variety of structures can result from these broad rules. Among them, a **four-helix bundle** (Fig. 19.32), which is found in proteins such as cytochrome b_{562} (an electron transport protein), forms when each helix has a nonpolar region along its length. The four nonpolar regions pack together to form a nonpolar interior.

Comment 19.3

The web site contains links to sites where you may predict the secondary structure of a polypeptide by molecular mechanics simulations. There are also links to sites where you may visualize the structures of proteins and nucleic acids that have been obtained by experimental and theoretical methods.

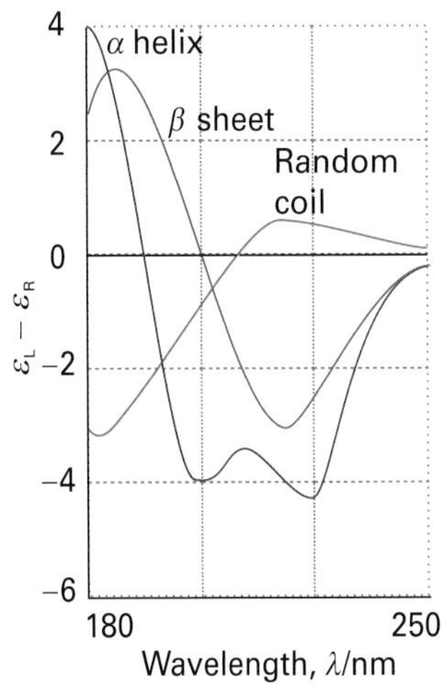

Fig. 19.31 Representative CD spectra of polypeptides. Random coils, α-helices, and β-sheets have different CD features in the spectral region where the peptide link absorbs.

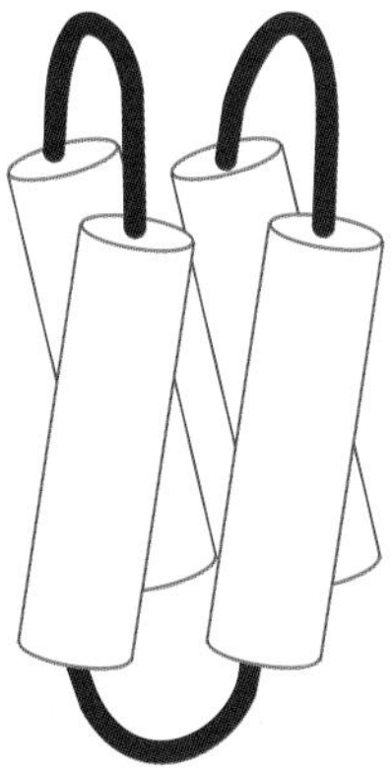

Fig. 19.32 A four-helix bundle forms from the interactions between nonpolar aminoacids on the surfaces of each helix, with the polar aminoacids exposed to the aqueous environment of the solvent.

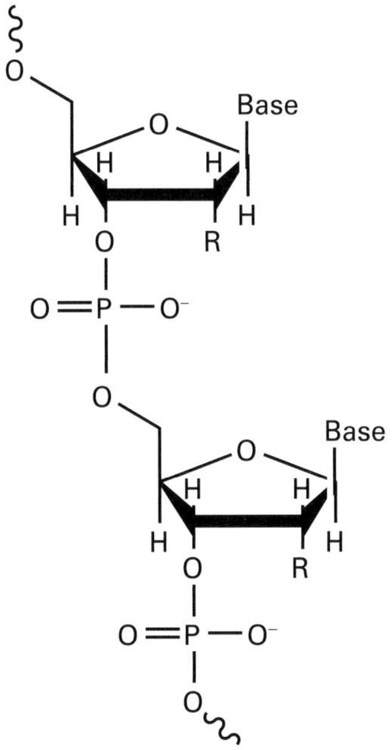

2 D-ribose (R = OH) and 2'-deoxy-D-Ribose (R = H)

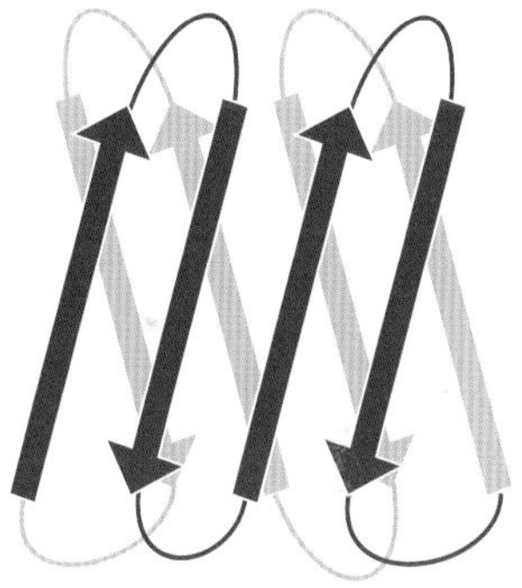

Fig. 19.33 Eight anti-parallel β-sheets, each represented by a purple arrow and linked by short random coils fold together as a β-barrel. Nonpolar aminoacids are in the interior of the barrel.

Similarly, interconnected β-sheets may interact to form a ***β*-barrel** (Fig. 19.33), the interior of which is populated by nonpolar R groups and which has an exterior rich in charged residues. The retinol-binding protein of blood plasma, which is responsible for transporting vitamin A, is an example of a β-barrel structure.

Factors that promote the folding of proteins include covalent disulfide (—S—S—) links, Coulombic interactions between ions (which depend on the degree of protonation of groups and therefore on the pH), hydrogen bonding, van der Waals interactions, and hydrophobic interactions (Section 18.4g). The clustering of nonpolar, hydrophobic, amino acids into the interior of a protein is driven primarily by hydrophobic interactions.

19.11 The structure of nucleic acids

Nucleic acids are key components of the mechanism of storage and transfer of genetic information in biological cells. Deoxyribonucleic acid (DNA) contains the instructions for protein synthesis, which is carried out by different forms of ribonucleic acid (RNA). In this section, we discuss the main structural features of DNA and RNA.

Both DNA and RNA are *polynucleotides* (**2**), in which base–sugar–phosphate units are linked by phosphodiester bonds. In RNA the sugar is β-D-ribose and in DNA it is β-D-2-deoxyribose (as shown in **2**). The most common bases are adenine (A, **3**), cytosine (C, **4**), guanine (G, **5**), thymine (T, found in DNA only, **6**), and uracil (U, found in RNA only, **7**). At physiological pH, each phosphate group of the chain carries a

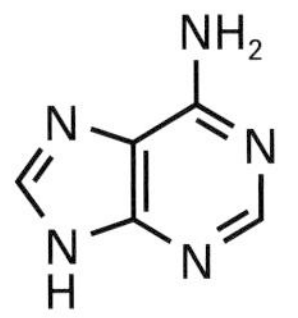

3 Adenine, A

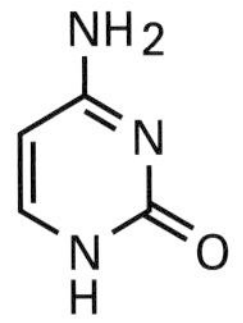

4 Cytosine, C

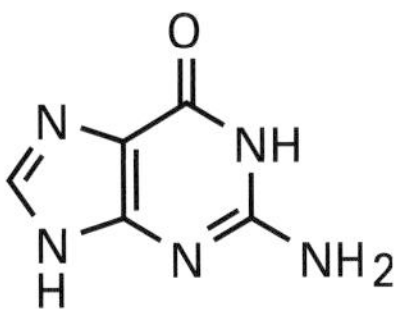

5 Guanine, G

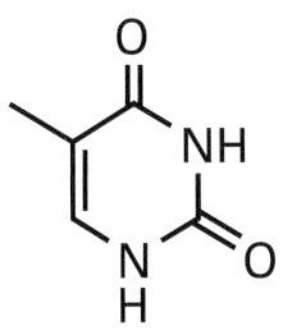

6 Thymine, T

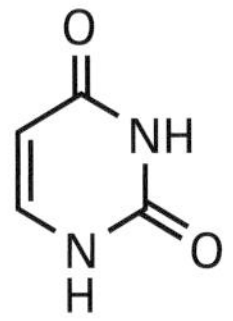

7 Uracil, U

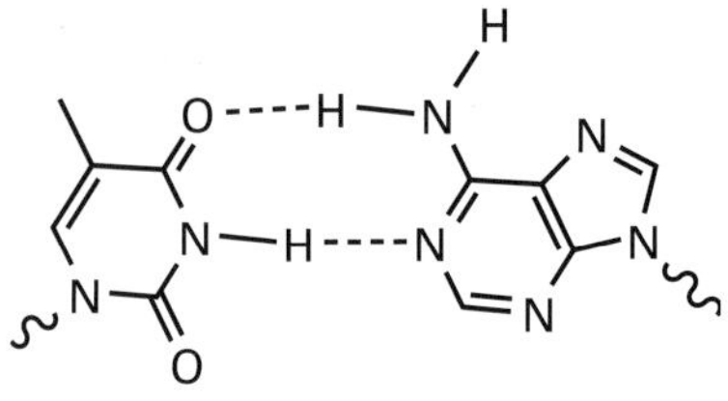

8 A–T base pair

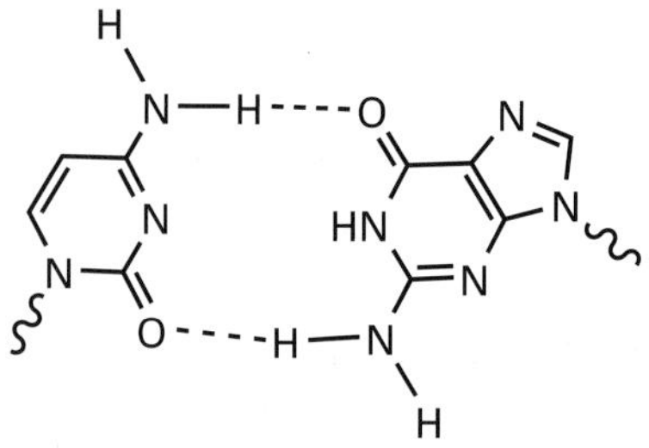

9 C–G base pair

negative charge and the bases are deprotonated and neutral. This charge distribution leads to two important properties. One is that the polynucleotide chain is a **polyelectrolyte**, a macromolecule with many different charged sites, with a large and negative overall surface charge. The second is that the bases can interact by hydrogen bonding, as shown for A—T (**8**) and C—G base pairs (**9**). The secondary and tertiary structures of DNA and RNA arise primarily from the pattern of this hydrogen bonding between bases of one or more chains.

In DNA, two polynucleotide chains wind around each other to form a double helix (Fig. 19.34). The chains are held together by links involving A—T and C—G base pairs that lie parallel to each other and perpendicular to the major axis of the helix. The structure is stabilized further by interactions between the planar π systems of the bases. In B-DNA, the most common form of DNA found in biological cells, the helix is right-handed with a diameter of 2.0 nm and a pitch of 3.4 nm. Long stretches of DNA can fold further into a variety of tertiary structures. Two examples are shown in Fig. 19.35. Supercoiled DNA is found in the chromosome and can be visualized as the twisting of closed circular DNA (ccDNA), much like the twisting of a rubber band.

The extra —OH group in β-D-ribose imparts enough steric strain to a polynucleotide chain so that stable double helices cannot form in RNA. Therefore, RNA exists primarily as single chains that can fold into complex structures by formation of A—U and G—C base pairs. One example of this structural complexity is the structure of transfer RNA (tRNA), shown schematically in Fig. 19.36 in which base-paired regions are connected by loops and coils. Transfer RNAs help assemble polypeptide chains during protein synthesis in the cell.

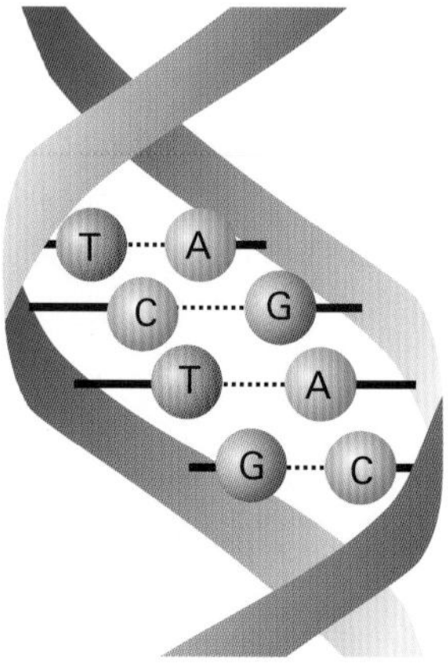

Fig. 19.34 DNA double helix, in which two polynucleotide chains are linked together by hydrogen bonds between adenine (A) and thymine (T) and between cytosine (C) and guanine (G).

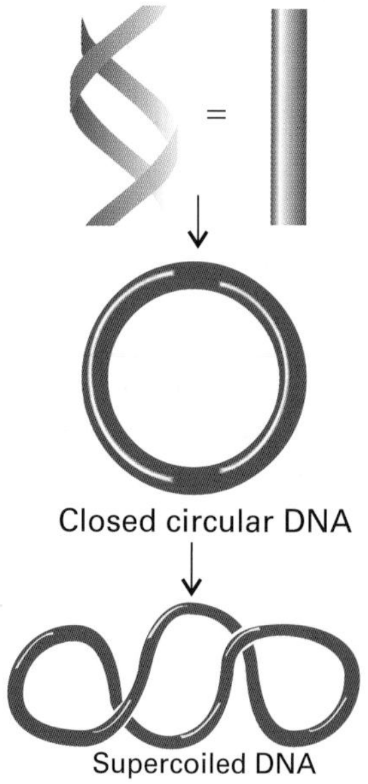

Fig. 19.35 A long section of DNA may form closed circular DNA (ccDNA) by covalent linkage of the two ends of the chain. Twisting of ccDNA leads to the formation of supercoiled DNA.

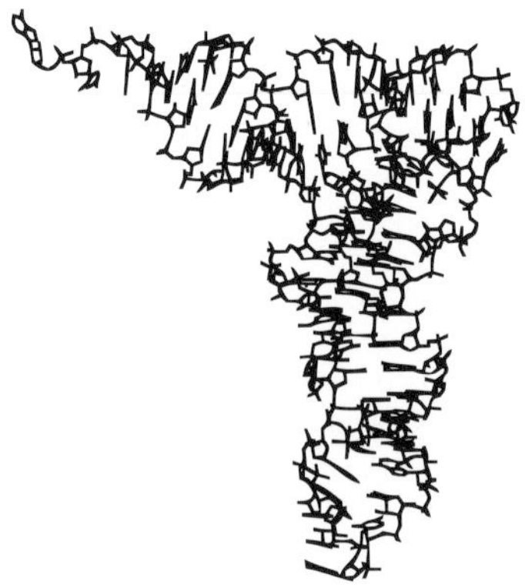

Fig. 19.36 Structure of a transfer RNA (tRNA).

19.12 The stability of proteins and nucleic acids

The loss of their natural conformation by proteins and nucleic acids is called **denaturation**. It can be achieved by changing the temperature or by adding chemical agents. Cooking is an example of thermal denaturation. When eggs are cooked, the protein albumin is denatured irreversibly, collapsing into a structure that resembles a random coil. One example of chemical denaturation is the 'permanent waving' of hair, which is a result of the reorganization of the protein keratin in hair. Disulfide cross-links between the chains of keratin render the protein and, hence hair fibres, inflexible. Chemical reduction of the disulfide bonds unravels keratin, and allows hair to be shaped. Oxidation re-forms the disulfide bonds and sets the new shape. The 'permanence' is only temporary, however, because the structure of newly formed hair is genetically controlled. Other means of chemically denaturing a protein include the addition of compounds that form stronger hydrogen bonds than those within a helix or sheet. One example is urea, which competes for the NH and CO groups of a polypeptide. The action of acids or bases, which can protonate or deprotonate groups involved in hydrogen bonding or change the Coulombic interactions that determine the conformation of a protein, can also result in denaturation.

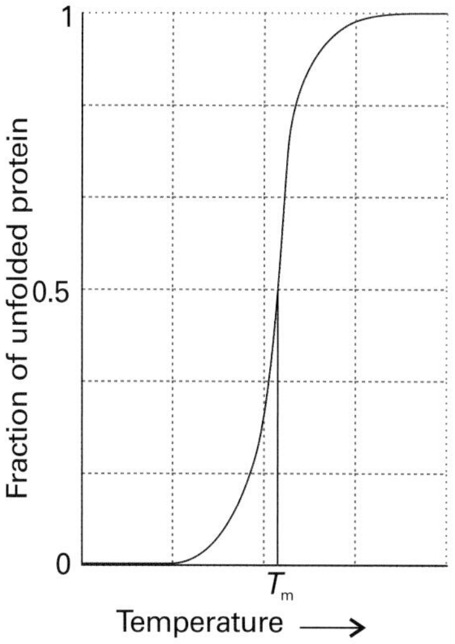

Fig. 19.37 A protein unfolds as the temperature of the sample increases. The sharp step in the plot of fraction of unfolded protein against temperature indicated that the transition is cooperative. The melting temperature, T_m, is the temperature at which the fraction of unfolded polymer is 0.5.

Closer examination of thermal denaturation reveals some of the chemical factors that determine protein and nucleic acid stability. Thermal denaturation is similar to the melting of synthetic polymers (Section 19.9). Denaturation is a **cooperative process** in the sense that the biopolymer becomes increasingly more susceptible to denaturation once the process begins. This cooperativity is observed as a sharp step in a plot of fraction of unfolded polymer versus temperature (*Impact* I16.1). The melting temperature, T_m, is the temperature at which the fraction of unfolded polymer is 0.5 (Fig. 19.37).

A DNA molecule is held together by hydrogen bonding interactions between bases of different chains and by **base-stacking**, in which dispersion interactions bring together the planar π systems of bases. Each G—C base pair has three hydrogen bonds whereas each A—T base pair has only two. Furthermore, experiments show that stacking interactions are stronger between C—G base pairs than between A—T base pairs. It follows that two factors render DNA sequences rich in C—G base pairs more stable than sequences rich in A—T base pairs: more hydrogen bonds between the bases and stronger stacking interactions between base pairs.

Proteins are relatively unstable towards chemical and thermal denaturation. For example, $T_m = 320$ K for ribonuclease T_1 (an enzyme that cleaves RNA in the cell), which is not far above the temperature at which the enzyme must operate (close to body temperature, 310 K). More surprisingly, the Gibbs energy for the unfolding of ribonuclease T_1 at pH 7.0 and 298 K is only 19.5 kJ mol^{-1}, which is comparable to the energy required to break a single hydrogen bond (about 20 kJ mol^{-1}). Therefore, unlike DNA, the stability of a protein does not increase in a simple way with the number of hydrogen bonding interactions. While the reasons for the low stability of proteins are not known, the answer probably lies in a delicate balance of all intra- and intermolecular interactions that allow a protein to fold into its active conformation, as discussed in Section 19.10.

Self-assembly

Much of the material discussed in this chapter also applies to aggregates of particles that form by **self-assembly**, the spontaneous formation of complex structures of molecules or macromolecules held together by molecular interactions, such as

Coulombic, dispersion, hydrogen bonding, or hydrophobic interactions. We have already encountered a few examples of self-assembly, such as the formation of liquid crystals (*Impact I6.1*), of protein quaternary structures from two or more polypeptide chains, and of a DNA double helix from two polynucleotide chains. Now we concentrate on the specific properties of additional self-assembled systems, including small aggregates that are at the heart of detergent action and extended sheets like those forming biological cell membranes.

19.13 **Colloids**

A **colloid**, or **disperse phase**, is a dispersion of small particles of one material in another. In this context, 'small' means something less than about 500 nm in diameter (about the wavelength of visible light). In general, colloidal particles are aggregates of numerous atoms or molecules, but are too small to be seen with an ordinary optical microscope. They pass through most filter papers, but can be detected by light-scattering and sedimentation.

(a) Classification and preparation

The name given to the colloid depends on the two phases involved. A **sol** is a dispersion of a solid in a liquid (such as clusters of gold atoms in water) or of a solid in a solid (such as ruby glass, which is a gold-in-glass sol, and achieves its colour by light scattering). An **aerosol** is a dispersion of a liquid in a gas (like fog and many sprays) or a solid in a gas (such as smoke): the particles are often large enough to be seen with a microscope. An **emulsion** is a dispersion of a liquid in a liquid (such as milk).

A further classification of colloids is as **lyophilic**, or solvent attracting, and **lyophobic**, solvent repelling. If the solvent is water, the terms **hydrophilic** and **hydrophobic**, respectively, are used instead. Lyophobic colloids include the metal sols. Lyophilic colloids generally have some chemical similarity to the solvent, such as —OH groups able to form hydrogen bonds. A **gel** is a semirigid mass of a lyophilic sol in which all the dispersion medium has penetrated into the sol particles.

The preparation of aerosols can be as simple as sneezing (which produces an imperfect aerosol). Laboratory and commercial methods make use of several techniques. Material (for example, quartz) may be ground in the presence of the dispersion medium. Passing a heavy electric current through a cell may lead to the sputtering (crumbling) of an electrode into colloidal particles. Arcing between electrodes immersed in the support medium also produces a colloid. Chemical precipitation sometimes results in a colloid. A precipitate (for example, silver iodide) already formed may be dispersed by the addition of a peptizing agent (for example, potassium iodide). Clays may be peptized by alkalis, the OH^- ion being the active agent.

Emulsions are normally prepared by shaking the two components together vigorously, although some kind of emulsifying agent usually has to be added to stabilize the product. This emulsifying agent may be a soap (the salt of a long-chain carboxylic acid) or other **surfactant** (surface active) species, or a lyophilic sol that forms a protective film around the dispersed phase. In milk, which is an emulsion of fats in water, the emulsifying agent is casein, a protein containing phosphate groups. It is clear from the formation of cream on the surface of milk that casein is not completely successful in stabilizing milk: the dispersed fats coalesce into oily droplets which float to the surface. This coagulation may be prevented by ensuring that the emulsion is dispersed very finely initially: intense agitation with ultrasonics brings this dispersion about, the product being 'homogenized' milk.

One way to form an aerosol is to tear apart a spray of liquid with a jet of gas. The dispersal is aided if a charge is applied to the liquid, for then electrostatic repulsions help to blast it apart into droplets. This procedure may also be used to produce emulsions, for the charged liquid phase may be directed into another liquid.

Colloids are often purified by dialysis (*Impact I5.2*). The aim is to remove much (but not all, for reasons explained later) of the ionic material that may have accompanied their formation. A membrane (for example, cellulose) is selected that is permeable to solvent and ions, but not to the colloid particles. Dialysis is very slow, and is normally accelerated by applying an electric field and making use of the charges carried by many colloidal particles; the technique is then called **electrodialysis**.

(b) Structure and stability

Colloids are thermodynamically unstable with respect to the bulk. This instability can be expressed thermodynamically by noting that because the change in Gibbs energy, dG, when the surface area of the sample changes by $d\sigma$ at constant temperature and pressure is $dG = \gamma d\sigma$, where γ is the interfacial surface tension (Section 18.7a), it follows that $dG < 0$ if $d\sigma < 0$. The survival of colloids must therefore be a consequence of the kinetics of collapse: colloids are thermodynamically unstable but kinetically nonlabile.

At first sight, even the kinetic argument seems to fail: colloidal particles attract each other over large distances, so there is a long-range force that tends to condense them into a single blob. The reasoning behind this remark is as follows. The energy of attraction between two individual atoms i and j separated by a distance R_{ij}, one in each colloidal particle, varies with their separation as $1/R_{ij}^6$ (Section 18.4). The sum of all these pairwise interactions, however, decreases only as approximately $1/R^2$ (the precise variation depending on the shape of the particles and their closeness), where R is the separation of the centres of the particles. The sum has a much longer range than the $1/R^6$ dependence characteristic of individual particles and small molecules.

Several factors oppose the long-range dispersion attraction. For example, there may be a protective film at the surface of the colloid particles that stabilizes the interface and cannot be penetrated when two particles touch. Thus the surface atoms of a platinum sol in water react chemically and are turned into $-Pt(OH)_3H_3$, and this layer encases the particle like a shell. A fat can be emulsified by a soap because the long hydrocarbon tails penetrate the oil droplet but the carboxylate head groups (or other hydrophilic groups in synthetic detergents) surround the surface, form hydrogen bonds with water, and give rise to a shell of negative charge that repels a possible approach from another similarly charged particle.

(c) The electrical double layer

A major source of kinetic nonlability of colloids is the existence of an electric charge on the surfaces of the particles. On account of this charge, ions of opposite charge tend to cluster nearby, and an ionic atmosphere is formed, just as for ions (Section 5.9).

We need to distinguish two regions of charge. First, there is a fairly immobile layer of ions that adhere tightly to the surface of the colloidal particle, and which may include water molecules (if that is the support medium). The radius of the sphere that captures this rigid layer is called the **radius of shear** and is the major factor determining the mobility of the particles. The electric potential at the radius of shear relative to its value in the distant, bulk medium is called the **zeta potential**, ζ, or the **electrokinetic potential.** Second, the charged unit attracts an oppositely charged atmosphere of mobile ions. The inner shell of charge and the outer ionic atmosphere is called the **electrical double layer**.

The theory of the stability of lyophobic dispersions was developed by B. Derjaguin and L. Landau and independently by E. Verwey and J.T.G. Overbeek, and is known as the **DLVO theory**. It assumes that there is a balance between the repulsive interaction between the charges of the electric double layers on neighbouring particles and the attractive interactions arising from van der Waals interactions between the molecules in the particles. The potential energy arising from the repulsion of double layers on particles of radius a has the form

$$V_{\text{repulsion}} = +\frac{Aa^2\zeta^2}{R}e^{-s/r_D} \quad (19.44)$$

where A is a constant, ζ is the **zeta potential**,[1] R is the separation of centres, s is the separation of the surfaces of the two particles ($s = R - 2a$ for spherical particles of radius a), and r_D is the thickness of the double layer. This expression is valid for small particles with a thick double layer ($a \ll r_D$). When the double layer is thin ($r_D \ll a$), the expression is replaced by

$$V_{\text{repulsion}} = \tfrac{1}{2}Aa\zeta^2 \ln(1 + e^{-s/r_D}) \quad (19.45)$$

In each case, the thickness of the double layer can be estimated from an expression like that derived for the thickness of the ionic atmosphere in the Debye–Hückel theory (eqn 5.80):

$$r_D = \left(\frac{\varepsilon RT}{2\rho F^2 I b^{\ominus}}\right)^{1/2} \quad (19.46)$$

where I is the ionic strength of the solution, ρ its mass density, and $b^{\ominus} = 1\ \text{mol kg}^{-1}$. The potential energy arising from the attractive interaction has the form

$$V_{\text{attraction}} = -\frac{B}{s} \quad (19.47)$$

where B is another constant. The variation of the total potential energy with separation is shown in Fig. 19.38.

At high ionic strengths, the ionic atmosphere is dense and the potential shows a secondary minimum at large separations. Aggregation of the particles arising from the stabilizing effect of this secondary minimum is called **flocculation**. The flocculated material can often be redispersed by agitation because the well is so shallow. **Coagulation**, the irreversible aggregation of distinct particles into large particles, occurs when the separation of the particles is so small that they enter the primary minimum of the potential energy curve and van der Waals forces are dominant.

The ionic strength is increased by the addition of ions, particularly those of high charge type, so such ions act as flocculating agents. This increase is the basis of the empirical **Schulze–Hardy rule**, that hydrophobic colloids are flocculated most efficiently by ions of opposite charge type and high charge number. The Al^{3+} ions in alum are very effective, and are used to induce the congealing of blood. When river water containing colloidal clay flows into the sea, the salt water induces flocculation and coagulation, and is a major cause of silting in estuaries. Metal oxide sols tend to be positively charged whereas sulfur and the noble metals tend to be negatively charged.

The primary role of the electric double layer is to confer kinetic non-lability. Colliding colloidal particles break through the double layer and coalesce only if the collision is sufficiently energetic to disrupt the layers of ions and solvating molecules, or if thermal motion has stirred away the surface accumulation of charge. This disruption may occur at high temperatures, which is one reason why sols precipitate

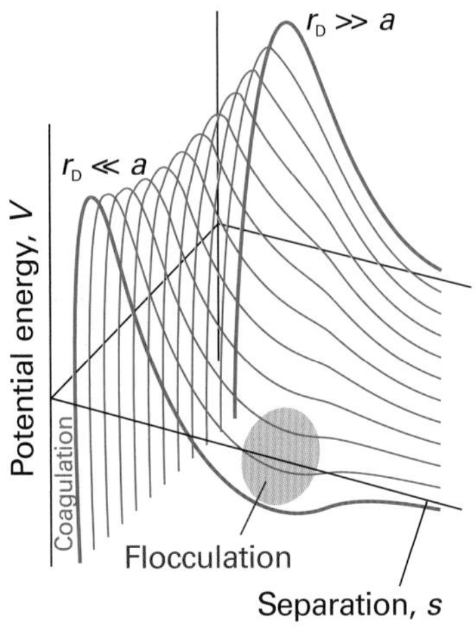

Fig. 19.38 The potential energy of interaction as a function of the separation of the centres of the two particles and its variation with the ratio of the particle size to the thickness a of the electric double layer r_D. The regions labelled coagulation and flocculation show the dips in the potential energy curves where these processes occur.

[1] The actual potential is that of the surface of the particles; there is some danger in identifying it with the zeta potential. See the references in *Further reading*.

when they are heated. The protective role of the double layer is the reason why it is important not to remove all the ions when a colloid is being purified by dialysis, and why proteins coagulate most readily at their isoelectric point.

19.14 Micelles and biological membranes

Surfactant molecules or ions can cluster together as **micelles**, which are colloid-sized clusters of molecules, for their hydrophobic tails tend to congregate, and their hydrophilic heads provide protection (Fig. 19.39).

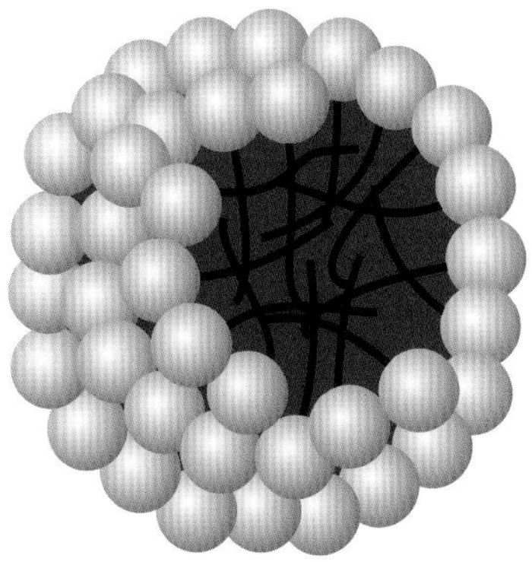

Fig. 19.39 A schematic version of a spherical micelle. The hydrophilic groups are represented by spheres and the hydrophobic hydrocarbon chains are represented by the stalks; these stalks are mobile.

(a) Micelle formation

Micelles form only above the **critical micelle concentration** (CMC) and above the **Krafft temperature**. The CMC is detected by noting a pronounced change in physical properties of the solution, particularly the molar conductivity (Fig. 19.40). There is no abrupt change in properties at the CMC; rather, there is a transition region corresponding to a range of concentrations around the CMC where physical properties vary smoothly but nonlinearly with the concentration. The hydrocarbon interior of a micelle is like a droplet of oil. Nuclear magnetic resonance shows that the hydrocarbon tails are mobile, but slightly more restricted than in the bulk. Micelles are important in industry and biology on account of their solubilizing function: matter can be transported by water after it has been dissolved in their hydrocarbon interiors. For this reason, micellar systems are used as detergents, for organic synthesis, froth flotation, and petroleum recovery.

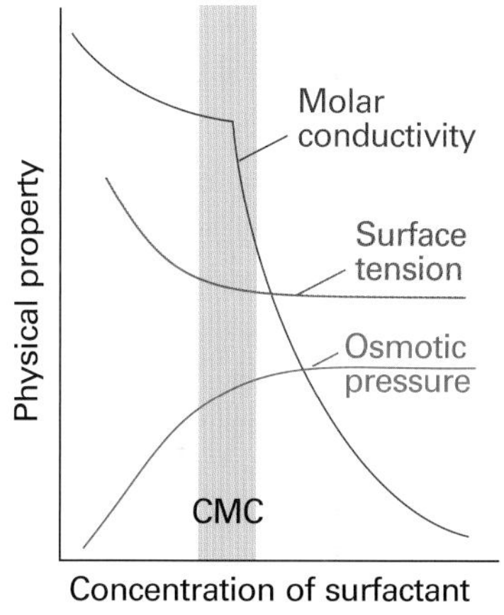

Fig. 19.40 The typical variation of some physical properties of an aqueous solution of sodium dodecylsulfate close to the critical micelle concentration (CMC).

Non-ionic surfactant molecules may cluster together in clumps of 1000 or more, but ionic species tend to be disrupted by the electrostatic repulsions between head groups and are normally limited to groups of less than about 100. The micelle population is often polydisperse, and the shapes of the individual micelles vary with concentration. Spherical micelles do occur, but micelles are more commonly flattened spheres close to the CMC.

Under certain experimental conditions, a **liposome** may form, with an inward pointing inner surface of molecules surrounded by an outward pointing outer layer (Fig. 19.41). Liposomes may be used to carry nonpolar drug molecules in blood. In concentrated solutions micelles formed from surfactant molecules may take the form of long cylinders and stack together in reasonably close-packed (hexagonal) arrays. These orderly arrangements of micelles are called **lyotropic mesomorphs** and, more colloquially, 'liquid crystalline phases'.

The enthalpy of micelle formation reflects the contributions of interactions between micelle chains within the micelles and between the polar head groups and the surrounding medium. Consequently, enthalpies of micelle formation display no readily discernible pattern and may be positive (endothermic) or negative (exothermic). Many non-ionic micelles form endothermically, with ΔH of the order of 10 kJ per mole of surfactant. That such micelles do form above the CMC indicates that the entropy change accompanying their formation must then be positive, and measurements suggest a value of about $+140\ \mathrm{J\,K^{-1}\,mol^{-1}}$ at room temperature. The fact that the entropy change is positive even though the molecules are clustering together shows that hydrophobic interactions (Section 18.*) are important in the formation of micelles.

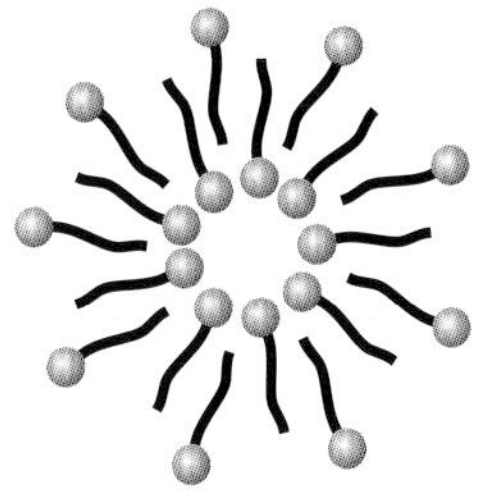

Fig. 19.41 The cross-sectional structure of a spherical liposome.

(b) Membrane formation

Some micelles at concentrations well above the CMC form extended parallel sheets, called **lamellar micelles**, two molecules thick. The individual molecules lie perpendicular to the sheets, with hydrophilic groups on the outside in aqueous solution and

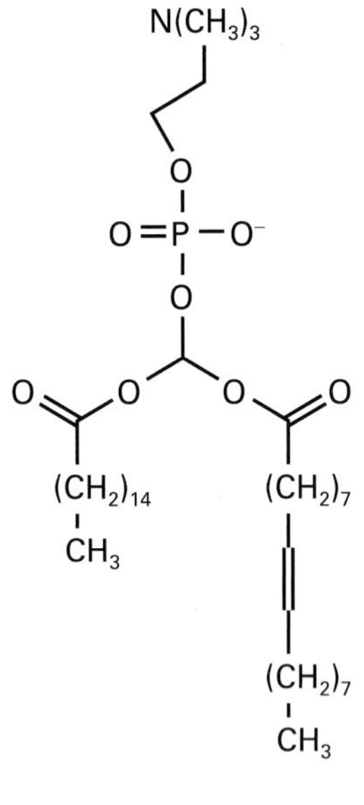

10 Phosphatidyl choline

on the inside in nonpolar media. Such lamellar micelles show a close resemblance to biological membranes, and are often a useful model on which to base investigations of biological structures.

Although lamellar micelles are convenient models of cell membranes, actual membranes are highly sophisticated structures. The basic structural element of a membrane is a phospholipid, such as phosphatidyl choline (**10**), which contains long hydrocarbon chains (typically in the range C_{14}–C_{24}) and a variety of polar groups, such as $—CH_2CH_2N(CH_3)_3^+$ in (**10**). The hydrophobic chains stack together to form an extensive bilayer about 5 nm across. The lipid molecules form layers instead of micelles because the hydrocarbon chains are too bulky to allow packing into nearly spherical clusters.

The bilayer is a highly mobile structure, as shown by EPR studies with spin-labelled phospholipids (*Impact* I15.2). Not only are the hydrocarbon chains ceaselessly twisting and turning in the region between the polar groups, but the phospholipid and cholesterol molecules migrate over the surface. It is better to think of the membrane as a viscous fluid rather than a permanent structure, with a viscosity about 100 times that of water. In common with diffusional behaviour in general (see Section 21.*), the average distance a phospholipid molecule diffuses is proportional to the square-root of the time; more precisely, for a molecule confined to a two-dimensional plane, the average distance travelled in a time t is equal to $(4Dt)^{1/2}$. Typically, a phospholipid molecule migrates through about 1 μm (the diameter of a cell) in about 1 min.

All lipid bilayers undergo a transition from a state of high to low chain mobility at a temperature that depends on the structure of the lipid. To visualize the transition, we consider what happens to a membrane as we lower its temperature (Fig. 19.42). There is sufficient energy available at normal temperatures for limited bond rotation to occur and the flexible chains writhe. However, the membrane is still highly organized in the sense that the bilayer structure does not come apart and the system is best described as a liquid crystal (Fig. 19.42a). At lower temperatures, the amplitudes of the writhing motion decrease until a specific temperature is reached at which motion is largely frozen. The membrane is said to exist as a gel (Fig. 19.42b). Biological membranes exist as liquid crystals at physiological temperatures.

Comment 19.4

The web site contains links to databases of thermodynamic properties of lipids.

Phase transitions in membranes are often observed as 'melting' from gel to liquid crystal by differential scanning calorimetry (*Impact* I2.1). The data show relations between the structure of the lipid and the melting temperature. For example, the melting temperature increases with the length of the hydrophobic chain of the lipid. This correlation is reasonable, as we expect longer chains to be held together more strongly by hydrophobic interactions than shorter chains. It follows that stabilization of the gel phase in membranes of lipids with long chains results in relatively high melting temperatures. On the other hand, any structural elements that prevent alignment of the hydrophobic chains in the gel phase lead to low melting temperatures. Indeed, lipids containing unsaturated chains, those containing some C=C bonds, form membranes

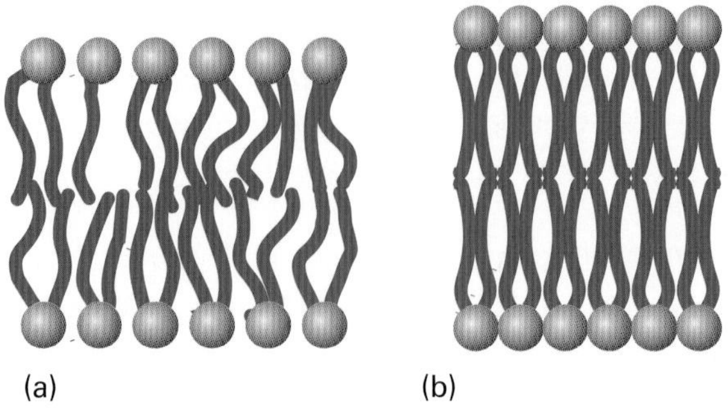

Fig. 19.42 A depiction of the variation with temperature of the flexibility of hydrocarbon chains in a lipid bilayer. (a) At physiological temperature, the bilayer exists as a liquid crystal, in which some order exists but the chains writhe. (b) At a specific temperature, the chains are largely frozen and the bilayer is said to exist as a gel.

with lower melting temperatures than those formed from lipids with fully saturated chains, those consisting of C—C bonds only.

Interspersed among the phospholipids of biological membranes are sterols, such as cholesterol (**11**), which is largely hydrophobic but does contain a hydrophilic —OH group. Sterols, which are present in different proportions in different types of cells, prevent the hydrophobic chains of lipids from 'freezing' into a gel and, by disrupting the packing of the chains, spread the melting point of the membrane over a range of temperatures.

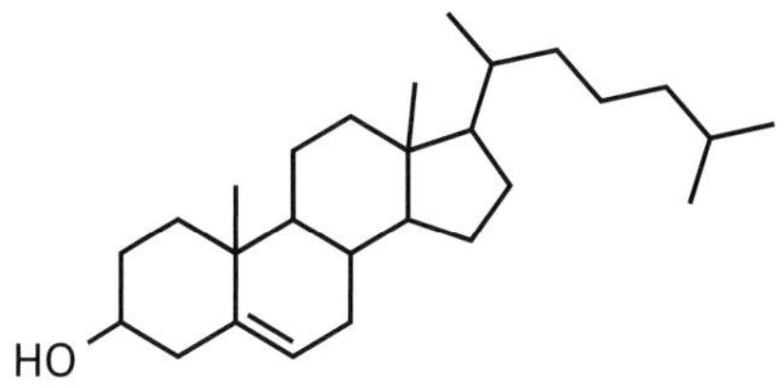

11 Cholesterol

Self-test 19.9 Organisms are capable of biosynthesizing lipids of different composition so that cell membranes have melting temperatures close to the ambient temperature. Why do bacterial and plant cells grown at low temperatures synthesize more phospholipids with unsaturated chains than do cells grown at higher temperatures?

[Insertion of lipids with unsaturated chains lowers the plasma membrane's melting temperature to a value that is close to the lower ambient temperature.]

Peripheral proteins are proteins attached to the bilayer. **Integral proteins** are proteins immersed in the mobile but viscous bilayer. These proteins may span the depth of the bilayer and consist of tightly packed α helices or, in some cases, β sheets containing hydrophobic residues that sit comfortably within the hydrocarbon region of the bilayer. There are two views of the motion of integral proteins in the bilayer. In the **fluid mosaic model** shown in Fig. 19.43 the proteins are mobile, but their diffusion coefficients are much smaller than those of the lipids. In the **lipid raft model**, a number of lipid and cholesterol molecules form ordered structures, or 'rafts', that envelop proteins and help carry them to specific parts of the cell.

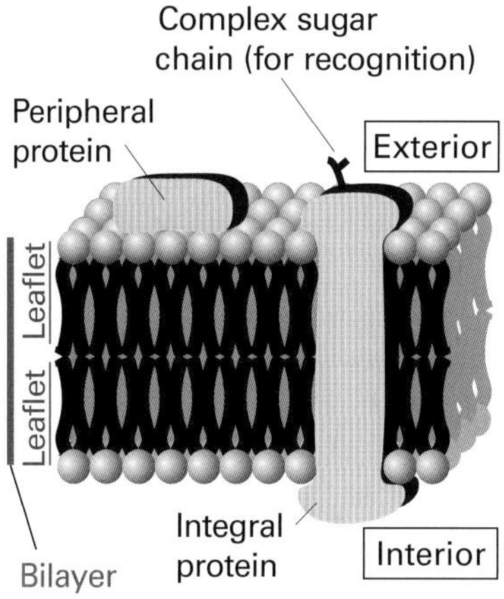

Fig. 19.43 In the fluid mosaic model of a biological cell membrane, integral proteins diffuse through the lipid bilayer. In the alternative lipid raft model, a number of lipid and cholesterol molecules envelop and transport the protein around the membrane.

The mobility of the bilayer enables it to flow round a molecule close to the outer surface, to engulf it, and incorporate it into the cell by the process of *endocytosis*. Alternatively, material from the cell interior wrapped in cell membrane may coalesce with the cell membrane itself, which then withdraws and ejects the material in the process of *exocytosis*. The function of the proteins embedded in the bilayer, though, is to act as devices for transporting matter into and out of the cell in a more subtle manner. By providing hydrophilic channels through an otherwise alien hydrophobic environment, some proteins act as **ion channels** and **ion pumps** (*Impact* I21.2).

19.15 **Surface films**

The compositions of surface layers have been investigated by the simple but technically elegant procedure of slicing thin layers off the surfaces of solutions and analysing their compositions. The physical properties of surface films have also been investigated. Surface films one molecule thick, such as that formed by a surfactant, are called **monolayers**. When a monolayer has been transferred to a solid support, it is called a **Langmuir–Blodgett film**, after Irving Langmuir and Katherine Blodgett, who developed experimental techniques for studying them.

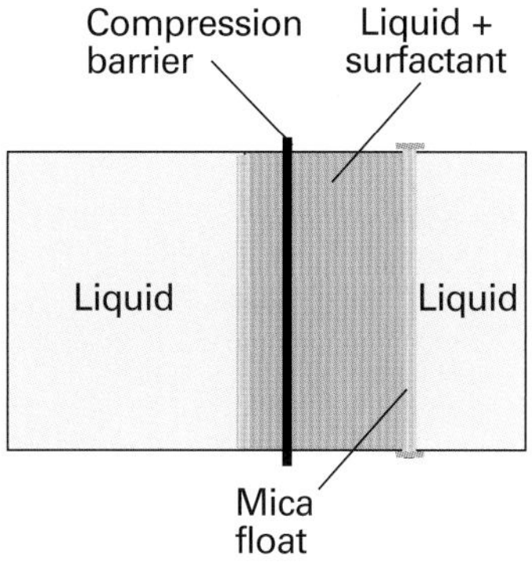

Fig. 19.44 A schematic diagram of the apparatus used to measure the surface pressure and other characteristics of a surface film. The surfactant is spread on the surface of the liquid in the trough, and then compressed horizontally by moving the compression barrier towards the mica float. The latter is connected to a torsion wire, so the difference in force on either side of the float can be monitored.

(a) Surface pressure

The principal apparatus used for the study of surface monolayers is a **surface film balance** (Fig. 19.44). This device consists of a shallow trough and a barrier that can be moved along the surface of the liquid in the trough, and hence compress any monolayer on the surface. The **surface pressure**, π, the difference between the surface

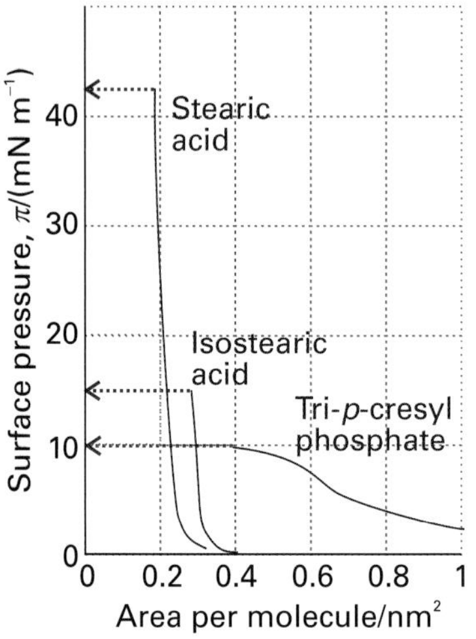

Fig. 19.45 The variation of surface pressure with the area occupied by each surfactant molecule. The collapse pressures are indicated by the horizontal dotted lines.

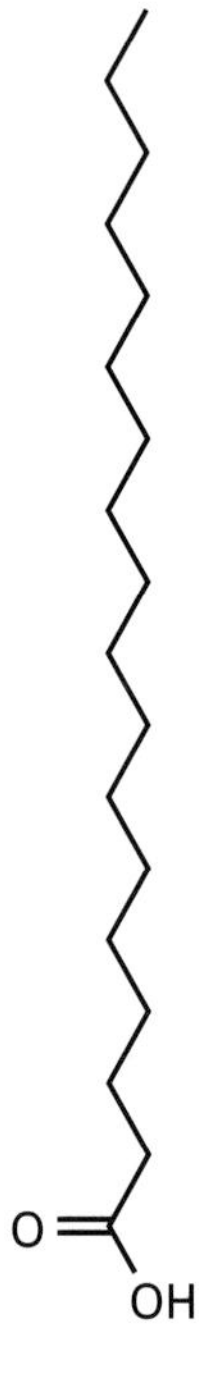

12 Stearic acid, $C_{17}H_{35}COOH$

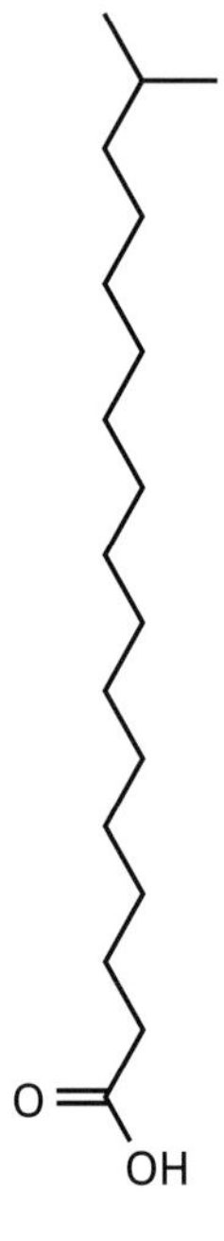

13 Isostearic acid, $C_{17}H_{35}COOH$

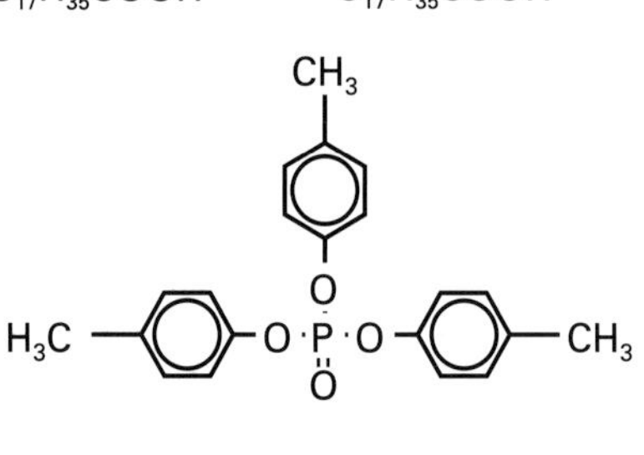

14 Tri-*p*-cresylphosphate

tension of the pure solvent and the solution ($\pi = \gamma^* - \gamma$) is measured by using a torsion wire attached to a strip of mica that rests on the surface and pressing against one edge of the monolayer. The parts of the apparatus that are in touch with liquids are coated in polytetrafluoroethylene to eliminate effects arising from the liquid–solid interface. In an actual experiment, a small amount (about 0.01 mg) of the surfactant under investigation is dissolved in a volatile solvent and then poured on to the surface of the water; the compression barrier is then moved across the surface and the surface pressure exerted on the mica bar is monitored.

Some typical results are shown in Fig. 19.45. One parameter obtained from the isotherms is the area occupied by the molecules when the monolayer is closely packed. This quantity is obtained from the extrapolation of the steepest part of the isotherm to the horizontal axis. As can be seen from the illustration, even though stearic acid (**12**) and isostearic acid (**13**) are chemically very similar (they differ only in the location of a methyl group at the end of a long hydrocarbon chain), they occupy significantly different areas in the monolayer. Neither, though, occupies as much area as the tri-*p*-cresyl phosphate molecule (**14**), which is like a wide bush rather than a lanky tree.

The second feature to note from Fig. 19.45 is that the tri-*p*-cresyl phosphate isotherm is much less steep than the stearic acid isotherms. This difference indicates that the tri-*p*-cresyl phosphate film is more compressible than the stearic acid films, which is consistent with their different molecular structures.

A third feature of the isotherms is the **collapse pressure**, the highest surface pressure. When the monolayer is compressed beyond the point represented by the collapse pressure, the monolayer buckles and collapses into a film several molecules thick. As can be seen from the isotherms in Fig. 19.45, stearic acid has a high collapse pressure, but that of tri-*p*-cresyl phosphate is significantly smaller, indicating a much weaker film.

(b) The thermodynamics of surface layers

A surfactant is active at the interface between two phases, such as at the interface between hydrophilic and hydrophobic phases. A surfactant accumulates at the interface, and modifies its surface tension and hence the surface pressure. To establish the relation between the concentration of surfactant at a surface and the change in surface tension it brings about, we consider two phases α and β in contact and suppose that the system consists of several components J, each one present in an overall amount n_J. If the components were distributed uniformly through the two phases right up to the interface, which is taken to be a plane of surface area σ, the total Gibbs energy, G, would be the sum of the Gibbs energies of both phases, $G = G(\alpha) + G(\beta)$. However, the components are not uniformly distributed because one may accumulate at the interface. As a result, the sum of the two Gibbs energies differs from G by an amount called the **surface Gibbs energy**, $G(\sigma)$:

$$G(\sigma) = G - \{G(\alpha) + G(\beta)\} \tag{19.48}$$

Similarly, if it is supposed that the concentration of a species J is uniform right up to the interface, then from its volume we would conclude that it contains an amount $n_J(\alpha)$ of J in phase α and an amount $n_J(\beta)$ in phase β. However, because a species may accumulate at the interface, the total amount of J differs from the sum of these two amounts by $n_J(\sigma) = n_J - \{n_J(\alpha) + n_J(\beta)\}$. This difference is expressed in terms of the **surface excess**, Γ_J:

$$\Gamma_J = \frac{n_J(\sigma)}{\sigma} \tag{19.49}$$

The surface excess may be either positive (an accumulation of J at the interface) or negative (a deficiency there).

The relation between the change in surface tension and the composition of a surface (as expressed by the surface excess) was derived by Gibbs. In the following *Justification* we derive the **Gibbs isotherm**, between the changes in the chemical potentials of the substances present in the interface and the change in surface tension:

$$d\gamma = -\sum_J \Gamma_J d\mu_J \tag{19.50}$$

Justification 19.7 *The Gibbs isotherm*

A general change in G is brought about by changes in T, p, and the n_J:

$$dG = -SdT + Vdp + \gamma d\sigma + \sum_J \mu_J dn_J$$

When this relation is applied to G, $G(\alpha)$, and $G(\beta)$ we find

$$dG(\sigma) = -S(\sigma)dT + \gamma d\sigma + \sum_J \mu_J dn_J(\sigma)$$

because at equilibrium the chemical potential of each component is the same in every phase, $\mu_J(\alpha) = \mu_J(\beta) = \mu_J(\sigma)$. Just as in the discussion of partial molar quantities (Section 5.1), the last equation integrates at constant temperature to

$$G(\sigma) = \gamma\sigma + \sum_J \mu_J n_J(\sigma)$$

We are seeking a connection between the change of surface tension $d\gamma$ and the change of composition at the interface. Therefore, we use the argument that in Section 5.1 led to the Gibbs–Duhem equation (eqn 5.12b), but this time we compare the expression

$$dG(\sigma) = \gamma d\sigma + \sum_J \mu_J dn_J(\sigma)$$

(which is valid at constant temperature) with the expression for the same quantity but derived from the preceding equation:

$$dG(\sigma) = \gamma d\sigma + \sigma d\gamma + \sum_J \mu_J dn_J(\sigma) + \sum_J n_J(\sigma) d\mu_J$$

The comparison implies that, at constant temperature,

$$\sigma d\gamma + \sum_J n_J(\sigma) d\mu_J = 0$$

Division by σ then gives eqn 19.50.

Now consider a simplified model of the interface in which the 'oil' and 'water' phases are separated by a geometrically flat surface. This approximation implies that only the surfactant, S, accumulates at the surface, and hence that Γ_{oil} and Γ_{water} are both zero. Then the Gibbs isotherm equation becomes

$$d\gamma = -\Gamma_S d\mu_S \tag{19.51}$$

For dilute solutions,

$$d\mu_S = RT \ln c \tag{19.52}$$

where c is the molar concentration of the surfactant. It follows that

$$d\gamma = RT\Gamma_S \frac{dc}{c}$$

at constant temperature, or

$$\left(\frac{\partial \gamma}{\partial c}\right)_T = -\frac{RT\Gamma_S}{c} \tag{19.53}$$

If the surfactant accumulates at the interface, its surface excess is positive and eqn 19.53 implies that $(\partial\gamma/\partial c)_T < 0$. That is, the surface tension decreases when a solute accumulates at a surface. Conversely, if the concentration dependence of γ is known, then the surface excess may be predicted and used to infer the area occupied by each surfactant molecule on the surface.

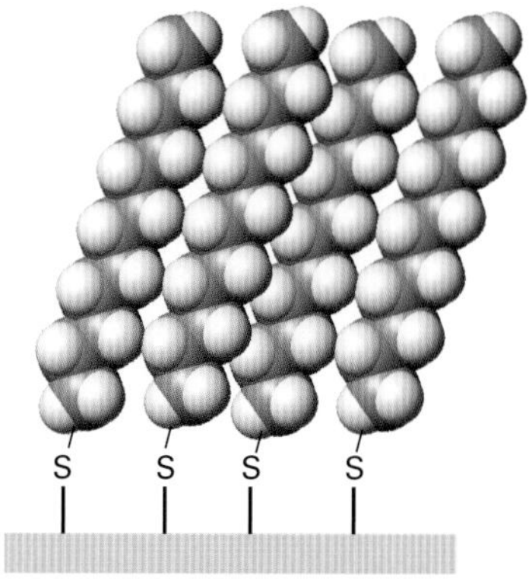

Fig. 19.46 Self-assembled monolayers of alkylthiols formed on to a gold surface by reaction of the thiol groups with the surface and aggregation of the alkyl chains.

IMPACT ON NANOSCIENCE

I19.3 Nanofabrication with self-assembled monolayers

Nanofabrication is the synthesis of *nanodevices*, nanometre-sized assemblies of atoms and molecules that can be used in nanotechnological applications, such as those discussed in *Impact I9.2*. Here we see how molecular self-assembly can be used as the basis for nanofabrication on surfaces . Of current interest are **self-assembled monolayers** (SAMs), ordered molecular aggregates that form a monolayer of material on a surface. To understand the formation of SAMs, consider exposing molecules such as alkyl thiols RSH, where R represents an alkyl chain, to an Au(0) surface. The thiols react with the surface, forming $RS^-Au(I)$ adducts:

$$RSH + Au(0)_n \rightarrow RS^-Au(I) \cdot Au(0)_{n-1} + \tfrac{1}{2}H_2$$

If R is a sufficiently long chain, van der Waals interactions between the adsorbed RS units lead to the formation of a highly ordered monolayer on the surface, as shown in Fig. 19.46. It is observed that the Gibbs energy of formation of SAMs increases with the length of the alkyl chain, with each methylene group contributing 400–4000 J mol^{-1} to the overall Gibbs energy of formation.

The atomic force microscope (*Impact* I9.1) can be used to manipulate SAMs into specific shapes on a surface by digging the microscope tip through the alkyl chains, bringing it into contact with the surface and then moving SAMs around the surface. In one application of the technique, enzymes were bound to patterned SAMs. The experiment shows that it is possible to form nanometre-sized reactors that take advantage of the catalytic properties of enzymes (which we explore in Chapter 23).

Checklist of key ideas

☐ 1. A polymer is a compound formed by linking together small molecules. Many proteins (specifically protein enzymes) are monodisperse; a synthetic polymer is polydisperse.

☐ 2. The number-average molar mass, $\bar{M}_n$, is the value obtained by weighting each molar mass by the number of molecules of that mass present in the sample; the weight-average molar mass, $\bar{M}_w$, is the average calculated by weighting the molar masses of the molecules by the mass of each one present in the sample; the *Z*-average molar mass, $\bar{M}_Z$, is the average molar mass obtained from sedimentation measurements.

☐ 3. The heterogeneity index of a polymer sample is $\bar{M}_w/\bar{M}_n$.

☐ 4. Techniques for the determination of the mean molar masses of macromolecules include mass spectrometry (as MALDI), ultracentrifugation, laser light scattering, and viscometry.

☐ 5. The least structured model of a macromolecule is as a random coil; for a freely jointed random coil of contour length Nl, the root mean square separation is $N^{1/2}l$ and the radius of gyration is $R_g = (N/6)^{1/2}l$.

☐ 6. The conformational entropy is the statistical entropy arising from the arrangement of bonds in a random coil.

☐ 7. The primary structure of a biopolymer is the sequence of its monomer units.

☐ 8. The secondary structure of a protein is the spatial arrangement of the polypeptide chain and includes the α-helix and β-sheet.

☐ 9. Helical and sheet-like polypeptide chains are folded into a tertiary structure by bonding influences between the residues of the chain.

☐ 10. Some macromolecules have a quaternary structure as aggregates of two or more polypeptide chains.

☐ 11. Synthetic polymers are classified as elastomers, fibres, and plastics.

☐ 12. A perfect elastomer is a polymer in which the internal energy is independent of the extension of the random coil; for small extensions a random coil model obeys a Hooke's law restoring force.

☐ 13. Synthetic polymers undergo a transition from a state of high to low chain mobility at the glass transition temperature, T_g.

☐ 14. The melting temperature of a polymer is the temperature at which three-dimensional order is lost.

☐ 15. A mesophase is a bulk phase that is intermediate in character between a solid and a liquid.

☐ 16. A disperse system is a dispersion of small particles of one material in another.

☐ 17. Colloids are classified as lyophilic (solvent attracting, specifically hydrophilic for water) and lyophobic (solvent repelling, specifically hydrophobic).

☐ 18. A surfactant is a species that accumulates at the interface of two phases or substances and modifies the properties of the surface.

☐ 19. The radius of shear is the radius of the sphere that captures the rigid layer of charge attached to a colloid particle.

☐ 20. The zeta potential is the electric potential at the radius of shear relative to its value in the distant, bulk medium.

☐ 21. The inner shell of charge and the outer atmosphere jointly constitute the electric double layer.

☐ 22. Many colloid particles are thermodynamically unstable but kinetically non-labile.

☐ 23. Flocculation is the reversible aggregation of colloidal particles; coagulation is the irreversible aggregation of colloidal particles.

☐ 24. The Schultze–Hardy rule states that hydrophobic colloids are flocculated most efficiently by ions of opposite charge type and high charge number.

☐ 25. A micelle is a colloid-sized cluster of molecules that forms at the critical micelle concentration (CMC) and at the Krafft temperature.

☐ 26. A liposome is a vesicle with an inward pointing inner surface of molecules surrounded by an outward pointing outer layer.

☐ 27. A lamellar micelle is an extended layer of molecules two molecules thick.

☐ 28. A monolayer, a single layer of molecules on a surface. A Langmuir–Blodgett film is a monolayer that has been transferred to a solid support.

☐ 29. Surface pressure is the difference between the surface tension of the pure solvent and the solution: $\pi = \gamma^* - \gamma$.

☐ 30. The collapse pressure is the highest surface pressure sustained by a surface film.

Further reading

Articles and texts

A.W. Adamson and A.P. Gast, *Physical chemistry of surfaces*. Wiley, New York (1997).

J. Barker, *Mass spectrometry*. Wiley, New York (1999).

C.E. Carraher, Jr., *Seymour/Carraher's polymer chemistry*, Marcel Dekker (2000).

D.F. Evans and H. Wennerström, *The colloidal domain : where physics, chemistry, biology, and technology meet*. Wiley-VCH, New York (1999).

P. Flory, *Principles of polymer chemistry*. Cornell University Press, Ithaca (1953).

C.S. Johnson and D.A. Gabriel, *Laser light scattering*. Dover, New York (1995).

A.R. Leach, *Molecular modelling: principles and applications*. Longman, Harlow (1997).

K.E. van Holde, W.C. Johnson, and P.S. Ho, *Principles of physical biochemistry*. Prentice Hall, Upper Saddle River (1998).

P.C. Heimenz and R. Rajagopalan, *Principles of colloid and surface chemistry*. Marcel Dekker, New York (1997).

Sources of data and information

D.R. Lide (ed.), *CRC Handbook of Chemistry and Physics*, Sections 7 and 13. CRC Press, Boca Raton (2000).

Further information

Further information 19.1 *The Rayleigh ratio*

Here we outline the key steps in the derivation of eqn 19.8 for the Rayleigh ratio. You are encouraged to consult sources in the *Further reading* section for additional details.

The ratio of the intensity of the light scattered by a sample to the intensity of the incident light is (see *Further reading*):

$$\frac{I}{I_0} = \frac{\pi^2\alpha^2}{\varepsilon_r^2\lambda^4}\frac{\sin^2\phi}{r^2} \tag{19.54}$$

where r is the distance between the sample and the detector and ϕ is the angle of observation relative to the z-axis ($\phi = 90°$ in Fig. 19.3). It follows from eqn 19.7 that

$$R_\theta = \frac{\pi^2\alpha^2}{\varepsilon_r^2\lambda^4}$$

The relation between the polarizability and the refractive index, n_r, of a solution is (see *Appendix* 3 for a qualitative explanation and *Further reading* for quantitative details)

$$n_r^2 - n_{r,0}^2 = \frac{\mathcal{N}\alpha}{\varepsilon_r} \tag{19.55}$$

where $n_{r,0}$ is the refractive index of the solvent and $\mathcal{N}$ is the number density of polymer molecules. Because $\mathcal{N} = c_P N_A/M$ (where c_P is the mass concentration of the polymer and M is its molar mass), we have:

$$\alpha = \frac{\varepsilon_r M}{c_P N_A}(n_r^2 - n_{r,0}^2)$$

For a dilute solution, n_r differs little from $n_{r,0}$ and we can write:

$$n_r = n_{r,0} + \left(\frac{dn_r}{dc_P}\right)c_P + \cdots$$

It follows that

$$n_r^2 \approx n_{r,0}^2 + 2n_{r,0}\left(\frac{dn_r}{dc_P}\right)c_P$$

and therefore that

$$\alpha = \frac{2\varepsilon_r n_{r,0} M}{N_A}\left(\frac{dn_r}{dc_P}\right) \qquad (19.56)$$

The Rayleigh ratio for scattering by a single molecule now becomes

$$R_\theta = \frac{4\pi^2 n_{r,0}^2 M^2}{\lambda^4 N_A^2}\left(\frac{dn_r}{dc_P}\right)^2 \qquad (19.57)$$

For a sample of N molecules, we multiply the expression above by $N = c_P N_A V/M$ and, after substituting $\bar{M}_w$ for M, obtain

$$R_\theta = Kc_P\bar{M}_w \qquad K = \frac{4\pi^2 n_{r,0}^2 V(dn_r/dc_P)^2}{\lambda^4 N_A} \qquad (19.58)$$

Equation 19.8 follows after we multiply the right-hand side of this equation by the structure factor P_θ.

Now we derive expressions for the structure factor. If the molecule is regarded as composed of a number of atoms i at distances R_i from a convenient point, interference occurs between the radiation scattered by each pair. The scattering from all the particles is then calculated by allowing for contributions from all possible orientations of each pair of atoms in each molecule. If there are N atoms in the macromolecule, and if all are assumed to have the same scattering power, then it is possible to show that (see *Further reading*)

$$P_\theta = \frac{1}{N^2}\sum_{i,j}\frac{\sin sR_{ij}}{sR_{ij}}, \qquad s = \frac{4\pi}{\lambda}\sin\tfrac{1}{2}\theta \qquad (19.59)$$

where R_{ij} is the separation of atoms i and j, and λ is the wavelength of the incident radiation.

When the molecule is much smaller than the wavelength of the incident radiation in the sense that $sR_{ij} \ll 1$ (for example, if $R = 5$ nm, and $\lambda = 500$ nm, all the sR_{ij} are about 0.1), we can use the expansion $\sin x = x - \frac{1}{6}x^3 + \cdots$ to write

$$\sin sR_{ij} = sR_{ij} - \tfrac{1}{6}(sR_{ij})^3 + \cdots$$

and then

$$P_\theta = \frac{1}{N^2}\sum_{i,j}\left\{1 - \tfrac{1}{6}(sR_{ij})^2 + \cdots\right\} = 1 - \frac{s^2}{6N^2}\sum_{i,j}R_{ij}^2 + \cdots$$

The sum over the squares of the separations gives the radius of gyration of the molecule (through eqn 19.32). Therefore

$$P_\theta \approx 1 - \tfrac{1}{3}s^2R_g^2 = 1 - \frac{16\pi^2 R_g^2 \sin^2\frac{1}{2}\theta}{3\lambda^2}$$

which is eqn 19.9.

Discussion questions

19.1 Distinguish between number-average, weight-average, and Z-average molar masses. Discuss experimental techniques that can measure each of these properties.

19.2 Suggest reasons why the techniques described in the preceding question produce different mass averages.

19.3 Distinguish between contour length, root mean square separation, and radius of gyration of a random coil.

19.4 Identify the terms in and limit the generality of the following expressions: (a) $\Delta S = -\frac{1}{2}kN\ln\{(1+\nu)^{1+\nu}(1-\nu)^{1-\nu}\}$, (b) $R_{rms} = (2N)^{1/2}l$, and (c) $R_g = (N/6)^{1/2}l$.

19.5 Distinguish between molecular mechanics and molecular dynamics calculations. Why are these methods generally more popular in the field of polymer chemistry than the quantum mechanical procedures discussed in Chapter 11?

19.6 It is observed that the critical micelle concentration of sodium dodecyl sulfate in aqueous solution decreases as the concentration of added sodium chloride increases. Explain this effect.

19.7 Explain the physical origins of surface activity by surfactant molecules.

19.8 Discuss the physical origins of the surface Gibbs energy.

19.9 Self-assembled monolayers (SAMs) are receiving more attention than Langmuir–Blodgett (LB) films as starting points for nanofabrication. How do SAMs differ from LB films and why are SAMs more useful than LB films in nanofabrication work?

Exercises

19.1a Calculate the number-average molar mass and the mass-average molar mass of a mixture of equal amounts of two polymers, one having $M = 62$ kg mol^{-1} and the other $M = 78$ kg mol^{-1}.

19.1b Calculate the number-average molar mass and the mass-average molar mass of a mixture of two polymers, one having $M = 62$ kg mol^{-1} and the other $M = 78$ kg mol^{-1}, with their amounts (numbers of moles) in the ratio 3:2.

19.2a The radius of gyration of a long chain molecule is found to be 7.3 nm. The chain consists of C—C links. Assume the chain is randomly coiled and estimate the number of links in the chain.

19.2b The radius of gyration of a long chain molecule is found to be 18.9 nm. The chain consists of links of length 450 pm. Assume the chain is randomly coiled and estimate the number of links in the chain.

19.3a A solution consists of solvent, 30 per cent by mass, of a dimer with $M = 30$ kg mol^{-1} and its monomer. What average molar mass would be obtained from measurement of (a) osmotic pressure, (b) light scattering?

19.3b A solution consists of 25 per cent by mass of a trimer with $M = 22$ kg mol^{-1} and its monomer. What average molar mass would be obtained from measurement of: (a) osmotic pressure, (b) light scattering?

19.4a Evaluate the rotational correlation time, $\tau_R = 4\pi a^3\eta/3kT$, for serum albumin in water at 25°C on the basis that it is a sphere of radius 3.0 nm. What is the value for a CCl_4 molecule in carbon tetrachloride at 25°C? (Viscosity data in Table 21.4 in the *Data section* at the end of this volume; take $a(CCl_4) = 250$ pm.)

19.4b Evaluate the rotational correlation time, $\tau_R = 4\pi a^3\eta/3kT$, for a synthetic polymer in water at 20°C on the basis that it is a sphere of radius 4.5 nm.

19.5a What is the relative rate of sedimentation for two spherical particles of the same density, but which differ in radius by a factor of 10?

19.5b What is the relative rate of sedimentation for two spherical particles with densities 1.10 g cm^{-3} and 1.18 g cm^{-3} and which differ in radius by a factor of 8.4, the former being the larger? Use $\rho = 0.794$ g cm^{-3} for the density of the solution.

19.6a Human haemoglobin has a specific volume of 0.749×10^{-3} m^3 kg^{-1}, a sedimentation constant of 4.48 Sv, and a diffusion coefficient of 6.9×10^{-11} m^2 s^{-1}. Determine its molar mass from this information.

19.6b A synthetic polymer has a specific volume of 8.01×10^{-4} m^3 kg^{-1}, a sedimentation constant of 7.46 Sv, and a diffusion coefficient of 7.72×10^{-11} m^2 s^{-1}. Determine its molar mass from this information.

19.7a Find the drift speed of a particle of radius 20 μm and density 1750 kg m^{-3} which is settling from suspension in water (density 1000 kg m^{-3}) under the influence of gravity alone. The viscosity of water is 8.9×10^{-4} kg m^{-1} s^{-1}.

19.7b Find the drift speed of a particle of radius 15.5 μm and density 1250 kg m^{-3} which is settling from suspension in water (density 1000 kg m^{-3}) under the influence of gravity alone. The viscosity of water is 8.9×10^{-4} kg m^{-1} s^{-1}.

19.8a At 20°C the diffusion coefficient of a macromolecule is found to be 8.3×10^{-11} m^2 s^{-1}. Its sedimentation constant is 3.2 Sv in a solution of density 1.06 g cm^{-3}. The specific volume of the macromolecule is 0.656 cm^3 g^{-1}. Determine the molar mass of the macromolecule.

19.8b At 20°C the diffusion coefficient of a macromolecule is found to be 7.9×10^{-11} m^2 s^{-1}. Its sedimentation constant is 5.1 Sv in a solution of density 997 kg m^{-3}. The specific volume of the macromolecule is 0.721 cm^3 g^{-1}. Determine the molar mass of the macromolecule.

19.9a The data from a sedimentation equilibrium experiment performed at 300 K on a macromolecular solute in aqueous solution show that a graph of ln c against r^2 is a straight line with a slope of 729 cm^{-2}. The rotational rate of the centrifuge was 50 000 r.p.m. The specific volume of the solute is 0.61 cm^3 g^{-1}. Calculate the molar mass of the solute.

19.9b The data from a sedimentation equilibrium experiment performed at 293 K on a macromolecular solute in aqueous solution show that a graph of ln c against $(r/\text{cm})^2$ is a straight line with a slope of 821. The rotation rate of the centrifuge was 1080 Hz. The specific volume of the solute is 7.2×10^{-4} m^3 kg^{-1}. Calculate the molar mass of the solute.

19.10a Calculate the radial acceleration (as so many g) in a cell placed at 6.0 cm from the centre of rotation in an ultracentrifuge operating at 80 000 r.p.m.

19.10b Calculate the radial acceleration (as so many g) in a cell placed at 5.50 cm from the centre of rotation in an ultracentrifuge operating at 1.32 kHz.

19.11a A polymer chain consists of 700 segments, each 0.90 nm long. If the chain were ideally flexible, what would be the r.m.s. separation of the ends of the chain?

19.11b A polymer chain consists of 1200 segments, each 1.125 nm long. If the chain were ideally flexible, what would be the r.m.s. separation of the ends of the chain?

19.12a Calculate the contour length (the length of the extended chain) and the root mean square separation (the end-to-end distance) for polyethylene with a molar mass of 280 kg mol^{-1}.

19.12b Calculate the contour length (the length of the extended chain) and the root mean square separation (the end-to-end distance) for polypropylene of molar mass 174 kg mol^{-1}.

Problems*

Numerical problems

19.1 In a sedimentation experiment the position of the boundary as a function of time was found to be as follows:

t/min	15.5	29.1	36.4	58.2
r/cm	5.05	5.09	5.12	5.19

The rotation rate of the centrifuge was 45 000 r.p.m. Calculate the sedimentation constant of the solute.

19.2 Calculate the speed of operation (in r.p.m.) of an ultracentrifuge needed to obtain a readily measurable concentration gradient in a sedimentation equilibrium experiment. Take that gradient to be a concentration at the bottom of the cell about five times greater than at the top. Use $r_{top} = 5.0$ cm, $r_{bott} = 7.0$ cm, $M \approx 10^5$ g mol^{-1}, $\rho v_s \approx 0.75$, $T = 298$ K.

19.3 The concentration dependence of the viscosity of a polymer solution is found to be as follows:

c/(g dm^{-3})	1.32	2.89	5.73	9.17
η/(g m^{-1} s^{-1})	1.08	1.20	1.42	1.73

The viscosity of the solvent is 0.985 g m^{-1} s^{-1}. What is the intrinsic viscosity of the polymer?

19.4 The times of flow of dilute solutions of polystyrene in benzene through a viscometer at 25°C are given in the table below. From these data, calculate the molar mass of the polystyrene samples. Since the solutions are dilute, assume that the densities of the solutions are the same as those of pure benzene. η(benzene) $= 0.601 \times 10^{-3}$ kg m^{-1} s^{-1} (0.601 cP) at 25°C.

c/(g dm^{-3})	0	2.22	5.00	8.00	10.00
t/s	208.2	248.1	303.4	371.8	421.3

19.5 The viscosities of solutions of polyisobutylene in benzene were measured at 24°C (the θ temperature for the system) with the following results:

c/(g/10^2 cm^3)	0	0.2	0.4	0.6	0.8	1.0
η/(10^{-3} kg m^{-1} s^{-1})	0.647	0.690	0.733	0.777	0.821	0.865

Use the information in Table 19.4 to deduce the molar mass of the polymer.

19.6‡ Polystyrene in cyclohexane at 34.5°C forms a θ solution, with an intrinsic viscosity related to the molar mass by $[\eta] = KM^a$. The following data

* Problems denoted with the symbol ‡ were supplied by Charles Trapp and Carmen Giunta.

on polystyrene in cyclohexane are taken from L.J. Fetters, N. Hadjichristidis, J.S. Lindner, and J.W. Mays (*J. Phys. Chem. Ref. Data* **23**, 619 (1994)):

$M/(\text{kg mol}^{-1})$	10.0	19.8	106	249	359	860	1800	5470	9720	56 800
$[\eta]/(\text{cm}^3\,\text{g}^{-1})$	8.90	11.9	28.1	44.0	51.2	77.6	113.9	195	275	667

Determine the parameters K and a. What is the molar mass of a polystyrene that forms a θ solution in cyclohexane with $[\eta] = 100\ \text{cm}^3\,\text{g}^{-1}$?

19.7‡ Standard polystyrene solutions of known average molar masses continue to be used as for the calibration of many methods of characterizing polymer solutions. M. Kolinsky and J. Janca (*J. Polym. Sci., Polym. Chem.* **12**, 1181 (1974)) studied polystyrene in tetrahydrofuran (THF) for use in calibrating a gel permeation chromatograph. Their results for the intrinsic viscosity, $[\eta]$, as a function of average molar mass at 25°C are given in the table below. (a) Obtain the Mark–Houwink constants that fit these data. (b) Compare your values to those in Table 19.4 and Example 19.5. How might you explain the differences?

$\bar{M}_v/(\text{kg mol}^{-1})$	5.0	10.3	19.85	51	98.2	173	411	867
$[\eta]/(\text{cm}^3\,\text{g}^{-1})$	5.2	8.8	14.0	27.6	43.6	67.0	125.0	206.7

19.8 The concentration dependence of the osmotic pressure of solutions of a macromolecule at 20°C was found to be as follows:

$c/(\text{g dm}^{-3})$	1.21	2.72	5.08	6.60
Π/Pa	134	321	655	898

Determine the molar mass of the macromolecule and the osmotic virial coefficient.

19.9 The osmotic pressure of a fraction of poly(vinyl chloride) in a ketone solvent was measured at 25°C. The density of the solvent (which is virtually equal to the density of the solution) was $0.798\ \text{g cm}^{-3}$. Calculate the molar mass and the osmotic virial coefficient, B, of the fraction from the following data:

$c/(\text{g}/10^2\ \text{cm}^3)$	0.200	0.400	0.600	0.088	1.000
h/cm	0.48	1.2	1.86	2.76	3.88

19.10 The following table lists the glass transition temperatures, T_g, of several polymers. Discuss the reasons why the structure of the monomer unit has an effect on the value of T_g.

Polymer	Poly(oxymethylene)	Polyethylene	Poly(vinyl chloride)	Polystyrene
Structure	$-(OCH_2)_n-$	$-(CH_2CH_2)_n-$	$-(CH_2-CHCl)_n-$	$-(CH_2-CH(C_6H_5))_n-$
T_g/K	198	253	354	381

Theoretical problems

19.11 In formamide as solvent, poly(γ-benzyl-L-glutamate) is found by light scattering experiments to have a radius of gyration proportional to M; in contrast, polystyrene in butanone has R_g proportional to $M^{1/2}$. Present arguments to show that the first polymer is a rigid rod whereas the second is a random coil.

19.12 The *kinematic viscosity*, ν, of a fluid is defined as η/ρ, where ρ is the mass density. What are the SI units of kinematic viscosity? Confirm that the drainage times through a narrow tube are governed by the kinematic viscosity by referring to the Poiseuille equation for fluid flow (eqn 21.25) and hence confirm eqn 19.24.

19.13 A polymerization process produced a Gaussian distribution of polymers in the sense that the proportion of molecules having a molar mass in the range M to $M + \text{d}M$ was proportional to $e^{-(M-\bar{M})^2/2\gamma}\text{d}M$. What is the number average molar mass when the distribution is narrow?

19.14 Show how eqn 19.26 for a one-dimensional freely jointed chain can be used to derive eqn 19.27 for a three-dimensional freely-jointed chain. *Hint.* Write the probability that the ends lie in the range n_x to $n_x + \text{d}n_x$ as $\text{d}P_x = P\text{d}n_x$, with P given in eqn 19.26, and similarly for the other two dimensions. Multiply these probabilities together, and integrate $\text{d}n_x\text{d}n_y\text{d}n_z$ over a shell of thickness $\text{d}n$. Don't count negative integers (that is, divide the volume of the shell by 8, corresponding to the all-positive octant of values).

19.15 Use eqn 19.27 to deduce expressions for (a) the root mean square separation of the ends of the chain, (b) the mean separation of the ends, and (c) their most probable separation. Evaluate these three quantities for a fully flexible chain with $N = 4000$ and $l = 154$ pm.

19.16 Construct a two-dimensional random walk by using a random number generating routine with mathematical software or electronic spreadsheet. Construct a walk of 50 and 100 steps. If there are many people working on the problem, investigate the mean and most probable separations in the plots by direct measurement. Do they vary as $N^{1/2}$?

19.17 Evaluate the radius of gyration, R_g, of (a) a solid sphere of radius a, (b) a long straight rod of radius a and length l. Show that in the case of a solid sphere of specific volume v_s, $R_g/\text{nm} \approx 0.056902 \times \{(v_s/\text{cm}^3\,\text{g}^{-1})(M/\text{g mol}^{-1})\}^{1/3}$. Evaluate R_g for a species with $M = 100\ \text{kg mol}^{-1}$, $v_s = 0.750\ \text{cm}^3\,\text{g}^{-1}$, and, in the case of the rod, of radius 0.50 nm.

19.18 The effective radius, a, of a random coil is related to its radius of gyration, R_g, by $a = \gamma R_g$, with $\gamma = 0.85$. Deduce an expression for the osmotic virial coefficient, B, in terms of the number of chain units for (a) a freely jointed chain, (b) a chain with tetrahedral bond angles. Evaluate B for $l =$ 154 pm and $N = 4000$. Estimate B for a randomly coiled polyethylene chain of arbitrary molar mass, M, and evaluate it for $M = 56\ \text{kg mol}^{-1}$. Use $B = \frac{1}{2}N_A v_P$, where v_P is the excluded volume due to a single molecule.

19.19 Radius of gyration is defined in eqn 19.32. Show that an equivalent definition is that R_g is the average root mean square distance of the atoms or groups (all assumed to be of the same mass), that is, that $R_g^2 = (1/N)\sum_j R_j^2$, where R_j is the distance of atom j from the centre of mass.

19.20 Consider the thermodynamic description of stretching rubber. The observables are the tension, t, and length, l (the analogues of p and V for gases). Because $\text{d}w = t\text{d}l$, the basic equation is $\text{d}U = T\text{d}S + t\text{d}l$. (The term $p\text{d}V$ is supposed negligible throughout.) If $G = U - TS - tl$, find expressions for $\text{d}G$ and $\text{d}A$, and deduce the Maxwell relations

$$\left(\frac{\partial S}{\partial l}\right)_T = -\left(\frac{\partial t}{\partial T}\right)_l \qquad \left(\frac{\partial S}{\partial t}\right)_T = -\left(\frac{\partial l}{\partial T}\right)_t$$

Go on to deduce the equation of state for rubber,

$$\left(\frac{\partial U}{\partial l}\right)_T = t - \left(\frac{\partial t}{\partial T}\right)_l$$

19.21 On the assumption that the tension required to keep a sample at a constant length is proportional to the temperature ($t = aT$, the analogue of $p \propto T$), show that the tension can be ascribed to the dependence of the entropy on the length of the sample. Account for this result in terms of the molecular nature of the sample.

Applications: to biochemistry and technology

19.22 In this problem you will use molecular mechanics software of your instructor's choice to gain some appreciation for the complexity of the calculations that lead to plots such as those in Fig. 19.29. Our model for the protein is the dipeptide (**15**) in which the terminal methyl groups replace the rest of the polypeptide chain. (a) Draw three initial conformers of the dipeptide with R = H: one with $\phi = +75°$, $\psi = -65°$, a second with $\phi = \psi =$ $+180°$, and a third with $\phi = +65°$, $\psi = +35°$. Use a molecular mechanics routine to optimize the geometry of each conformer and measure the total potential energy and the final ϕ and ψ angles in each case. (Although any force field will work satisfactorily, the AMBER force field is strongly recommended, as it is optimized for calculations on biopolymers.) Did all of the initial conformers converge to the same final conformation? If not, what do these

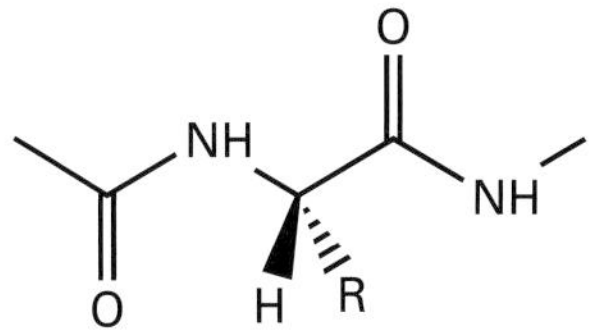

15

final conformers represent? Rationalize any observed differences in total potential energy of the final conformers. (b) Use the approach in part (a) to investigate the case R = CH_3, with the same three initial conformers as starting points for the calculations. Rationalize any similarities and differences between the final conformers of the dipeptides with R = H and R = CH_3.

19.23 Calculate the excluded volume in terms of the molecular volume on the basis that the molecules are spheres of radius a. Evaluate the osmotic virial coefficient in the case of bushy stunt virus, a = 14.0 nm, and haemoglobin, a = 3.2 nm (see Problem 19.18). Evaluate the percentage deviation of the Rayleigh ratios of 1.00 g/(100 cm^3) solutions of bushy stunt virus ($M = 1.07 \times 10^4$ kg mol^{-1}) and haemoglobin (M = 66.5 kg mol^{-1}) from the ideal solution values. In eqn 19.8, let $P_\theta = 1$ and assume that both solutions have the same K value.

19.24 Use the information below and the expression for R_g of a solid sphere quoted in the Problem 19.17, to classify the species below as globular or rod-like.

	M/(g mol^{-1})	v_s/(cm^3 g^{-1})	R_g/nm
Serum albumin	66×10^3	0.752	2.98
Bushy stunt virus	10.6×10^6	0.741	12.0
DNA	4×10^6	0.556	117.0

19.25 Suppose that a rod-like DNA molecule of length 250 nm undergoes a conformational change to a closed-circular (cc) form. (a) Use the information in Problem 19.24 and an incident wavelength λ = 488 nm to calculate the ratio of scattering intensities by each of these conformations, I_{rod}/I_{cc}, when θ = 20°, 45°, and 90°. (b) Suppose that you wish to use light scattering as a technique for the study of conformational changes in DNA molecules. Based on your answer to part (a), at which angle would you conduct the experiments? Justify your choice.

19.26 In an ultracentrifugation experiment at 20°C on bovine serum albumin the following data were obtained: ρ = 1.001 g cm^{-3}, v_s = 1.112 cm^3 g^{-1}, $\omega/2\pi$ = 322 Hz,

r/cm	5.0	5.1	5.2	5.3	5.4
c/(mg cm^{-3})	0.536	0.284	0.148	0.077	0.039

Evaluate the molar mass of the sample.

19.27 Sedimentation studies on haemoglobin in water gave a sedimentation constant S = 4.5 Sv at 20°C. The diffusion coefficient is 6.3×10^{-11} m^2 s^{-1} at the same temperature. Calculate the molar mass of haemoglobin using v_s = 0.75 cm^3 g^{-1} for its partial specific volume and ρ = 0.998 g cm^{-3} for the density of the solution. Estimate the effective radius of the haemoglobin molecule given that the viscosity of the solution is 1.00×10^{-3} kg m^{-1} s^{-1}.

19.28 The rate of sedimentation of a recently isolated protein was monitored at 20°C and with a rotor speed of 50 000 r.p.m. The boundary receded as follows:

t/s	0	300	600	900	1200	1500	1800
r/cm	6.127	6.153	6.179	6.206	6.232	6.258	6.284

Calculate the sedimentation constant and the molar mass of the protein on the basis that its partial specific volume is 0.728 cm^3 g^{-1} and its diffusion coefficient is 7.62×10^{-11} m^2 s^{-1} at 20°C, the density of the solution then being 0.9981 g cm^{-3}. Suggest a shape for the protein given that the viscosity of the solution is 1.00×10^{-3} kg m^{-1} s^{-1} at 20°C.

19.29 For some proteins, the isoelectric point must be obtained by extrapolation because the macromolecule might not be stable over a very wide pH range. Estimate the pH of the isoelectric point from the following data for a protein:

pH	4.5	5.0	5.5	6.0
Drift speed/(μm s^{-1})	−0.10	−0.20	−0.30	−0.35

19.30 Here we use concepts developed in Chapter 16 and this chapter to enhance our understanding of closed-circular and supercoiled DNA. (a) The average end-to-end distance of a flexible polymer (such as a fully denatured polypeptide or a strand of DNA) is $N^{1/2}l$, where N is the number of groups (residues or bases) and l is the length of each group. Initially, therefore, one end of the polymer can be found anywhere within a sphere of radius $N^{1/2}l$ centred on the other end. When the ends join to form a circle, they are confined to a volume of radius l. What is the change in molar entropy? Plot the function you derive as a function of N. (b) The energy necessary to twist ccDNA by i turns is $\varepsilon_i = ki^2$, with k an empirical constant and i being negative or positive depending on the sense of the twist. For example, one twist ($i = \pm 1$) makes ccDNA resemble the number 8. (i) Show that the distribution of the populations $p_i = n_i/N$ of ccDNA molecules with i turns at a specified temperature has the form of a Gaussian function. (ii) Plot the expression you derived in part (a) for several values of the temperature. Does the curve has a maximum? If so, at what value of i? Comment on variations of the shape of the curve with temperature. (iii) Calculate p_0, p_1, p_5, and p_{10} at 298 K.

19.31 The melting temperature of a DNA molecule can be determined by differential scanning calorimetry (*Impact* I2.1). The following data were obtained in aqueous solutions containing the specified concentration c_{salt} of an soluble ionic solid for a series of DNA molecules with varying base pair composition, with f the fraction of G—C base pairs:

$c_{salt} = 1.0 \times 10^{-2}$ mol dm^{-3}

f	0.375	0.509	0.589	0.688	0.750
T_m/K	339	344	348	351	354

c_{salt} = 0.15 mol dm^{-3}

f	0.375	0.509	0.589	0.688	0.750
T_m/K	359	364	368	371	374

(a) Estimate the melting temperature of a DNA molecule containing 40.0 per cent G—C base pairs in both samples. *Hint.* Begin by plotting T_m against fraction of G—C base pairs and examining the shape of the curve. (b) Do the data show an effect of concentration of ions in solution on the melting temperature of DNA? If so, provide a molecular interpretation for the effect you observe.

19.32 The fluidity of a lipid bilayer dispersed in aqueous solution depends on temperature and there are two important melting transitions. One transition is from a 'solid crystalline' state in which the hydrophobic chains are packed together tightly (hence move very little) to a 'liquid crystalline state', in which there is increased but still limited movement of the of the chains. The second transition, which occurs at a higher temperature than the first, is from the liquid crystalline state to a liquid state, in which the hydrophobic interactions holding the aggregate together are largely disrupted. (a) It is observed that the transition temperatures increase with the hydrophobic chain length and decrease with the number of C=C bonds in the chain. Explain these observations. (b) What effect is the inclusion of cholesterol likely to have on the transition temperatures of a lipid bilayer? Justify your answer.

19.33 Polystyrene is a synthetic polymer with the structure —(CH_2—$CH(C_6H_5)$)$_n$—. A batch of polydisperse polystyrene was prepared by initiating the polymerization with t-butyl radicals. As a result, the t-butyl group is expected to be covalently attached to the end of the final products. A sample from this batch was embedded in an organic matrix containing silver trifluoroacetate and the resulting MALDI-TOF spectrum consisted of a large number of peaks separated by 104 g mol^{-1}, with the most intense peak at 25 578 g mol^{-1}. Comment on the purity of this sample and determine the number of (CH_2—$CH(C_6H_5)$) units in the species that gives rise to the most intense peak in the spectrum .

19.34 A manufacturer of polystyrene beads claims that they have an average molar mass of 250 kg mol^{-1}. Solutions of these beads are studied by a physical chemistry student by dilute solution viscometry with an Ostwald viscometer in both the 'good' solvent toluene and the theta solvent cyclohexane. The drainage times, t_D, as a function of concentration for the two solvents are given in the table below. (a) Fit the data to the virial equation for viscosity,

$$\eta = \eta^*(1 + [\eta]c + k'[\eta]^2c^2 + \cdots)$$

where k' is called the *Huggins constant* and is typically in the range 0.35–0.40. From the fit, determine the intrinsic viscosity and the Huggins constant. (b) Use the empirical Mark–Kuhn–Houwink–Sakurada equation (eqn 19.25) to determine the molar mass of polystyrene in the two solvents. For theta solvents, $a = 0.5$ and $K = 8.2 \times 10^{-5}$ dm^3 g^{-1} for cyclohexane; for the good solvent toluene $a = 0.72$ and $K = 1.15 \times 10^{-5}$ dm^3 g^{-1}. (c) According to a general theory proposed by Kirkwood and Riseman, the root mean square end-to-end distance of a polymer chain in solution is related to $[\eta]$ by $[\eta] = \Phi\langle r^2\rangle^{3/2}/M$, where Φ is a universal constant with the value 2.84×10^{26} when $[\eta]$ is expressed in cubic decimetres per gram and the distance is in metres. Calculate this quantity for each solvent. (d) From the molar masses calculate the average number of styrene ($C_6H_5CH{=}CH_2$) monomer units, $\langle n\rangle$, (e) Calculate the length of a fully stretched, planar zigzag configuration, taking the C—C distance as 154 pm and the CCC bond angle to be 109°. (f) Use eqn 19.33 to calculate the radius of gyration, R_g. Also calculate $\langle r^2\rangle^{1/2} = n^{1/2}$. Compare this result with that predicted by the Kirkwood–Riseman theory: which gives the better fit? (g) Compare your values for M to the results of Problem 19.33. Is there any reason why they should or should not agree? Is the manufacturer's claim valid?

c/(g dm^{-3} toluene)	0	1.0	3.0	5.0
t_D/s	8.37	9.11	10.72	12.52
c/(g dm^{-3} cyclohexane)	0	1.0	1.5	2.0
t_D/s	8.32	8.67	8.85	9.03

19.35‡ The determination of the average molar masses of conducting polymers is an important part of their characterization. S. Holdcroft (*J. Polym. Sci., Polym. Phys.* **29**, 1585 (1991)) has determined the molar mases and Mark–Houwink constants for the electronically conducting polymer, poly(3-hexylthiophene) (P3HT) in tetrahydrofuran (THF) at 25°C by methods similar to those used for nonconducting polymers. The values for molar mass and intrinsic viscosity in the table below are adapted from their data. Determine the constants in the Mark–Kuhn–Houwink–Sakurada equation from these results and compare to the values obtained in your solution to Problem 19.7 .

$\bar{M}_v$/(kg mol^{-1})	3.8	11.1	15.3	58.8
$[\eta]$/(cm^3 g^{-1})	6.23	17.44	23.73	85.28

Materials 2: the solid state

20

First, we see how to describe the regular arrangement of atoms in crystals and the symmetry of their arrangement. Then we consider the basic principles of X-ray diffraction and see how the diffraction pattern can be interpreted in terms of the distribution of electron density in a unit cell. X-ray diffraction leads to information about the structures of metallic, ionic, and molecular solids, and we review some typical results and their rationalization in terms of atomic and ionic radii. With structures established, we move on to the properties of solids, and see how their mechanical, electrical, optical, and magnetic properties stem from the properties of their constituent atoms and molecules.

The solid state includes most of the materials that make modern technology possible. It includes the wide varieties of steel that are used in architecture and engineering, the semiconductors and metallic conductors that are used in information technology and power distribution, the ceramics that increasingly are replacing metals, and the synthetic and natural polymers that are used in the textile industry and in the fabrication of many of the common objects of the modern world. The properties of solids stem, of course, from the arrangement and properties of the constituent atoms, and one of the challenges of this chapter is to see how a wide range of bulk properties, including rigidity, electrical conductivity, and optical and magnetic properties stem from the properties of atoms. One crucial aspect of this link is the pattern in which the atoms (and molecules) are stacked together, and we start this chapter with an examination of how the structures of solids are described and determined.

Crystal lattices

Early in the history of modern science it was suggested that the regular external form of crystals implied an internal regularity of their constituents. In this section we see how to describe the arrangement of atoms inside crystals.

20.1 Lattices and unit cells

A crystal is built up from regularly repeating 'structural motifs', which may be atoms, molecules, or groups of atoms, molecules, or ions. A **space lattice** is the pattern formed by points representing the locations of these motifs (Fig. 20.1). The space lattice is, in effect, an abstract scaffolding for the crystal structure. More formally, a space lattice is a three-dimensional, infinite array of points, each of which is surrounded in an identical way by its neighbours, and which defines the basic structure of the crystal. In some cases there may be a structural motif centred on each lattice

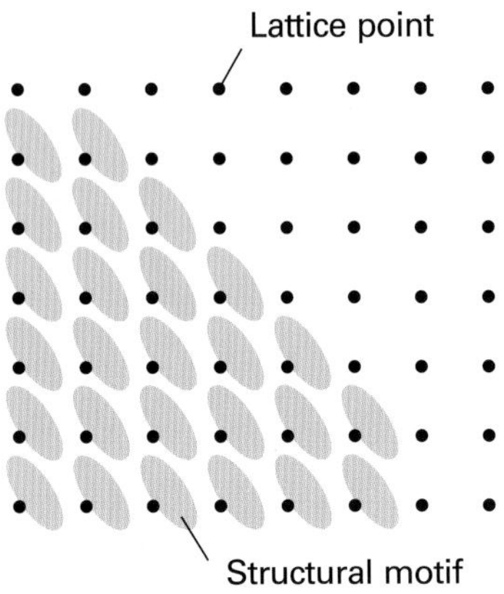

Fig. 20.1 Each lattice point specifies the location of a structural motif (for example, a molecule or a group of molecules). The crystal lattice is the array of lattice points; the crystal structure is the collection of structural motifs arranged according to the lattice.

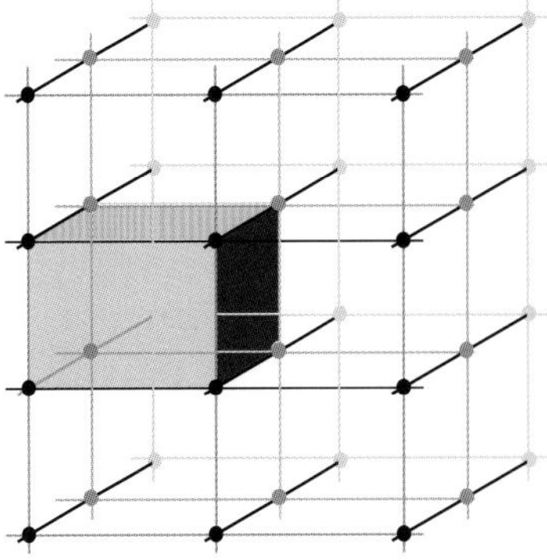

Fig. 20.2 A unit cell is a parallel-sided (but not necessarily rectangular) figure from which the entire crystal structure can be constructed by using only translations (not reflections, rotations, or inversions).

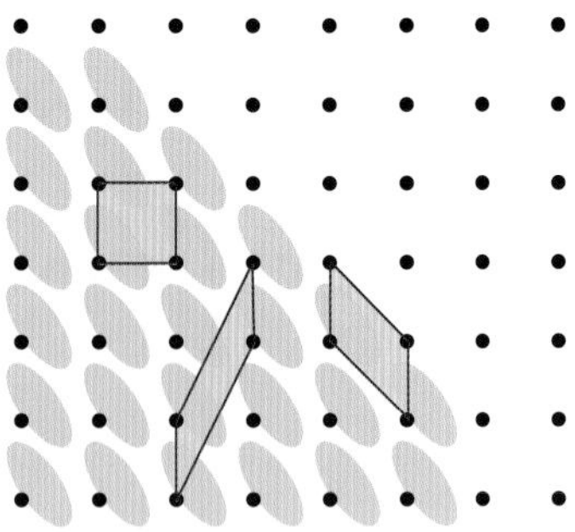

Fig. 20.3 A unit cell can be chosen in a variety of ways, as shown here. It is conventional to choose the cell that represents the full symmetry of the lattice. In this rectangular lattice, the rectangular unit cell would normally be adopted.

point, but that is not necessary. The crystal structure itself is obtained by associating with each lattice point an identical structural motif.

The **unit cell** is an imaginary parallelepiped (parallel-sided figure) that contains one unit of the translationally repeating pattern (Fig. 20.2). A unit cell can be thought of as the fundamental region from which the entire crystal may be constructed by purely translational displacements (like bricks in a wall). A unit cell is commonly formed by joining neighbouring lattice points by straight lines (Fig. 20.3). Such unit cells are called **primitive.** It is sometimes more convenient to draw larger **non-primitive unit cells** that also have lattice points at their centres or on pairs of opposite faces. An infinite number of different unit cells can describe the same lattice, but the one with sides that have the shortest lengths and that are most nearly perpendicular to one another is normally chosen. The lengths of the sides of a unit cell are denoted a, b, and c, and the angles between them are denoted α, β, and γ (Fig. 20.4).

Unit cells are classified into seven **crystal systems** by noting the rotational symmetry elements they possess. A *cubic unit cell*, for example, has four threefold axes in a tetrahedral array (Fig. 20.5). A *monoclinic unit cell* has one twofold axis; the unique

Comment 20.1

A *symmetry operation* is an action (such as a rotation, reflection, or inversion) that leaves an object looking the same after it has been carried out. There is a corresponding *symmetry element* for each symmetry operation, which is the point, line, or plane with respect to which the symmetry operation is performed. For instance, an *n-fold rotation* (the symmetry operation) about an *n-fold axis of symmetry* (the corresponding symmetry element) is a rotation through $360°/n$. See Chapter 12 for a more detailed discussion of symmetry.

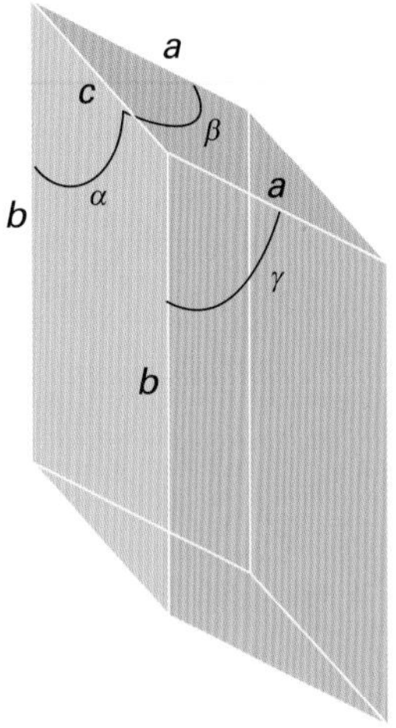

Fig. 20.4 The notation for the sides and angles of a unit cell. Note that the angle α lies in the plane (b,c) and perpendicular to the axis a.

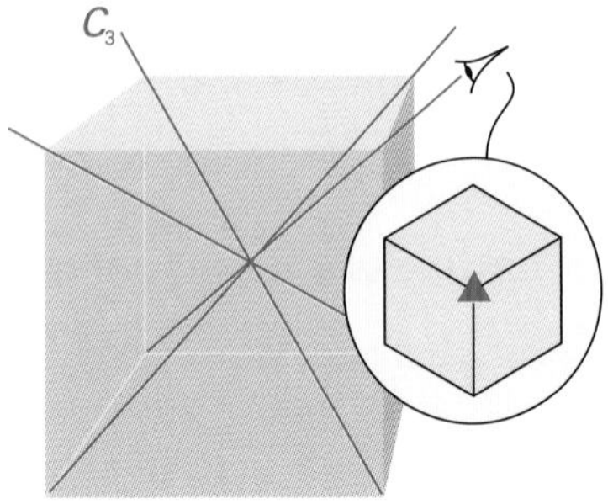

Fig. 20.5 A unit cell belonging to the cubic system has four threefold axes, denoted C_3, arranged tetrahedrally. The insert shows the threefold symmetry.

axis is by convention the b axis (Fig. 20.6). A *triclinic unit cell* has no rotational symmetry, and typically all three sides and angles are different (Fig. 20.7). Table 20.1 lists the **essential symmetries**, the elements that must be present for the unit cell to belong to a particular crystal system.

There are only 14 distinct space lattices in three dimensions. These **Bravais lattices** are illustrated in Fig. 20.8. It is conventional to portray these lattices by primitive unit cells in some cases and by non-primitive unit cells in others. A **primitive unit cell** (with lattice points only at the corners) is denoted P. A **body-centred unit cell** (I) also has a lattice point at its centre. A **face-centred unit cell** (F) has lattice points at its corners and also at the centres of its six faces. A **side-centred unit cell** (A, B, or C) has lattice points at its corners and at the centres of two opposite faces. For simple structures, it is often convenient to choose an atom belonging to the structural motif, or the

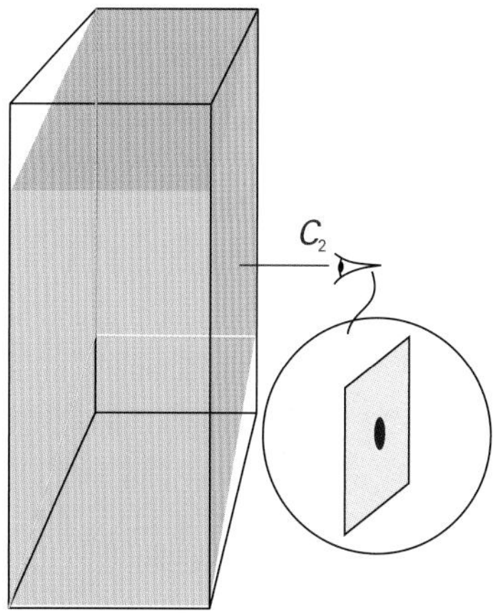

Fig. 20.6 A unit belonging to the monoclinic system has a twofold axis (denoted C_2 and shown in more detail in the insert).

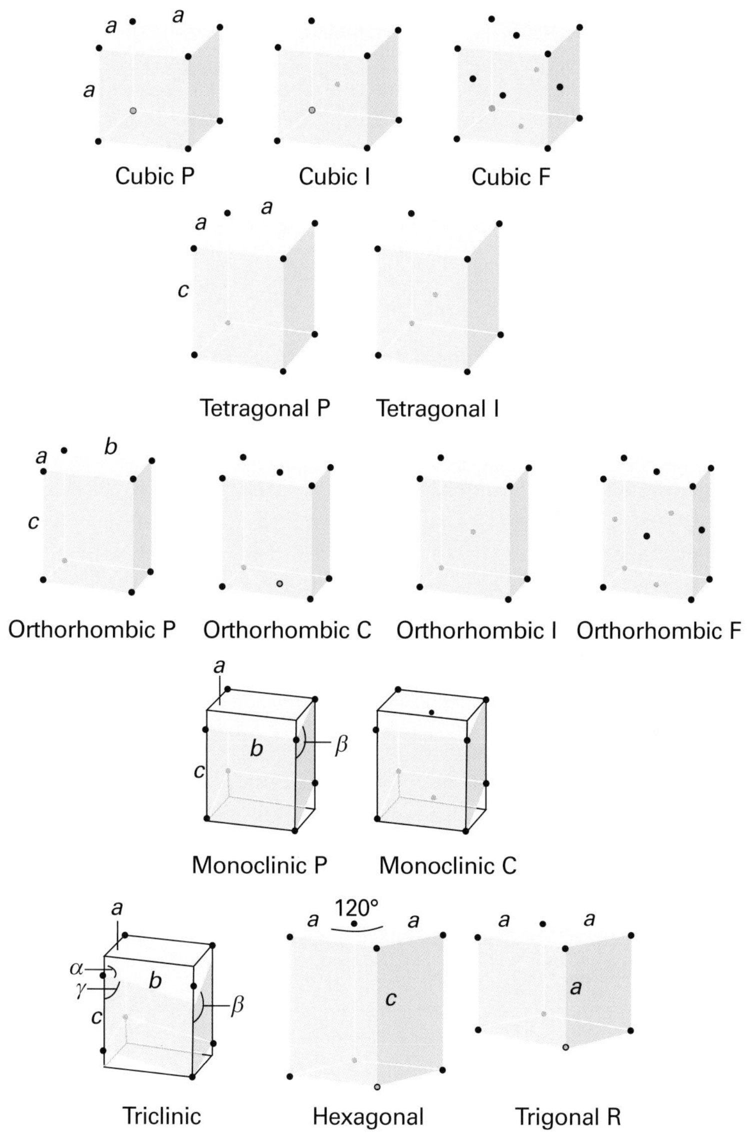

Fig. 20.8 The fourteen Bravais lattices. The points are lattice points, and are not necessarily occupied by atoms. P denotes a primitive unit cell (R is used for a trigonal lattice), I a body-centred unit cell, F a face-centred unit cell, and C (or A or B) a cell with lattice points on two opposite faces.

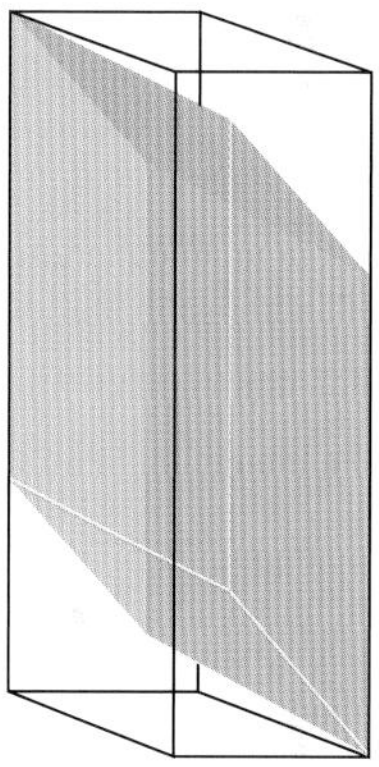

Fig. 20.7 A triclinic unit cell has no axes of rotational symmetry.

Table 20.1 The seven crystal systems

System	Essential symmetries
Triclinic	None
Monoclinic	One C_2 axis
Orthorhombic	Three perpendicular C_2 axes
Rhombohedral	One C_3 axis
Tetragonal	One C_4 axis
Hexagonal	One C_6 axis
Cubic	Four C_3 axes in a tetrahedral arrangement

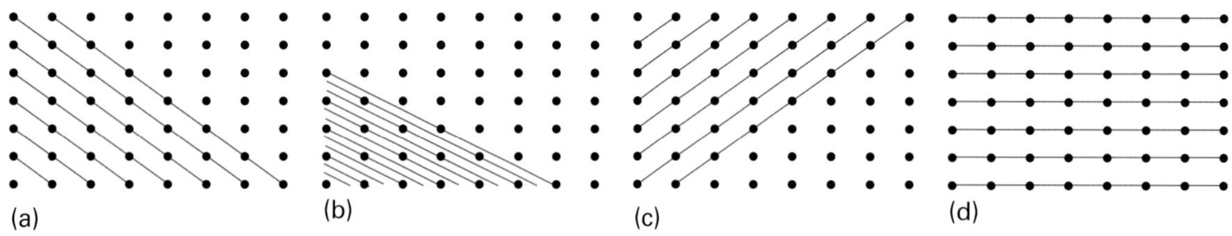

Fig. 20.9 Some of the planes that can be drawn through the points of a rectangular space lattice and their corresponding Miller indices (*hkl*): (a) (110), (b) (230), (c) ($\bar{1}$10), and (d) (010).

centre of a molecule, as the location of a lattice point or the vertex of a unit cell, but that is not a necessary requirement.

20.2 The identification of lattice planes

The spacing of the planes of lattice points in a crystal is an important quantitative aspect of its structure. However, there are many different sets of planes (Fig. 20.9), and we need to be able to label them. Two-dimensional lattices are easier to visualize than three-dimensional lattices, so we shall introduce the concepts involved by referring to two dimensions initially, and then extend the conclusions by analogy to three dimensions.

(a) The Miller indices

Consider a two-dimensional rectangular lattice formed from a unit cell of sides *a*, *b* (as in Fig. 20.9). Each plane in the illustration (except the plane passing through the origin) can be distinguished by the distances at which it intersects the *a* and *b* axes. One way to label each set of parallel planes would therefore be to quote the smallest intersection distances. For example, we could denote the four sets in the illustration as $(1a,1b)$, $(\frac{1}{2}a,\frac{1}{3}b)$, $(-1a,1b)$, and $(\infty a,1b)$. However, if we agree to quote distances along the axes as multiples of the lengths of the unit cell, then we can label the planes more simply as $(1,1)$, $(\frac{1}{2},\frac{1}{3})$, $(-1,1)$, and $(\infty,1)$. If the lattice in Fig. 20.9 is the top view of a three-dimensional orthorhombic lattice in which the unit cell has a length *c* in the *z*-direction, all four sets of planes intersect the *z*-axis at infinity. Therefore, the full labels are $(1,1,\infty)$, $(\frac{1}{2},\frac{1}{3},\infty)$, $(-1,1,\infty)$, and $(\infty,1,\infty)$.

The presence of fractions and infinity in the labels is inconvenient. They can be eliminated by taking the reciprocals of the labels. As we shall see, taking reciprocals turns out to have further advantages. The **Miller indices**, (*hkl*), are the reciprocals of intersection distances (with fractions cleared by multiplying through by an appropriate factor, if taking the reciprocal results in a fraction). For example, the $(1,1,\infty)$ planes in Fig. 20.9a are the (110) planes in the Miller notation. Similarly, the $(\frac{1}{2},\frac{1}{3},\infty)$ planes are denoted (230). Negative indices are written with a bar over the number, and Fig. 20.9c shows the ($\bar{1}$10) planes. The Miller indices for the four sets of planes in Fig. 20.9 are therefore (110), (230), ($\bar{1}$10), and (010). Figure 20.10 shows a three-dimensional representation of a selection of planes, including one in a lattice with non-orthogonal axes.

The notation (*hkl*) denotes an *individual* plane. To specify a *set* of parallel planes we use the notation {*hkl*}. Thus, we speak of the (110) plane in a lattice, and the set of all {110} planes that lie parallel to the (110) plane. A helpful feature to remember is that, the smaller the absolute value of *h* in {*hkl*}, the more nearly parallel the set of planes is to the *a* axis (the {*h*00} planes are an exception). The same is true of *k* and the *b* axis and *l* and the *c* axis. When $h = 0$, the planes intersect the *a* axis at infinity, so the {0*kl*}

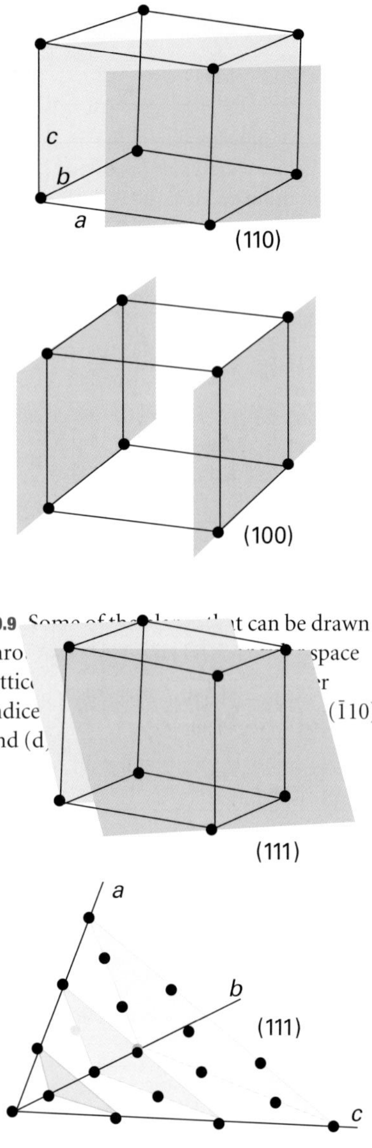

Fig. 20.10 Some representative planes in three dimensions and their Miller indices. Note that a 0 indicates that a plane is parallel to the corresponding axis, and that the indexing may also be used for unit cells with non-orthogonal axes.

planes are parallel to the a axis. Similarly, the $\{h0l\}$ planes are parallel to b and the $\{hk0\}$ planes are parallel to c.

(b) The separation of planes

The Miller indices are very useful for expressing the separation of planes. The separation of the $\{hk0\}$ planes in the square lattice shown in Fig. 20.11 is given by

$$\frac{1}{d_{hk0}^2}=\frac{h^2+k^2}{a^2} \quad \text{or} \quad d_{hk0}=\frac{a}{(h^2+k^2)^{1/2}} \tag{20.1}$$

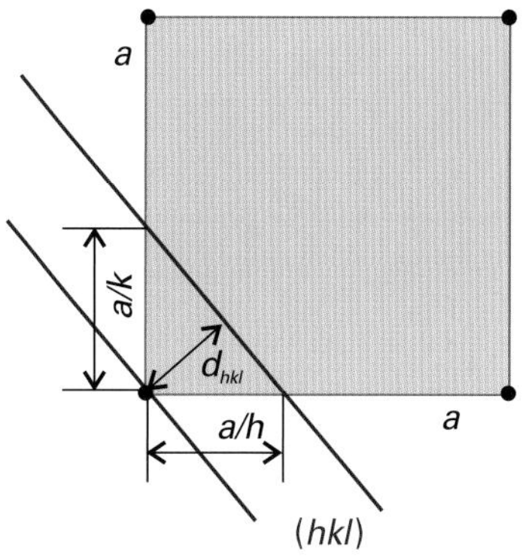

Fig. 20.11 The dimensions of a unit cell and their relation to the plane passing through the lattice points.

By extension to three dimensions, the separation of the $\{hkl\}$ planes of a cubic lattice is given by

$$\frac{1}{d_{hkl}^2}=\frac{h^2+k^2+l^2}{a^2} \quad \text{or} \quad d_{hkl}=\frac{a}{(h^2+k^2+l^2)^{1/2}} \tag{20.2}$$

The corresponding expression for a general orthorhombic lattice is the generalization of this expression:

$$\frac{1}{d_{hkl}^2}=\frac{h^2}{a^2}+\frac{k^2}{b^2}+\frac{l^2}{c^2} \tag{20.3}$$

Example 20.1 *Using the Miller indices*

Calculate the separation of (a) the $\{123\}$ planes and (b) the $\{246\}$ planes of an orthorhombic unit cell with $a=0.82$ nm, $b=0.94$ nm, and $c=0.75$ nm.

Method For the first part, simply substitute the information into eqn 20.3. For the second part, instead of repeating the calculation, note that if all three Miller indices are multiplied by n, then their separation is reduced by that factor (Fig. 20.12):

$$\frac{1}{d_{nh,nk,nl}^2}=\frac{(nh)^2}{a^2}+\frac{(nk)^2}{b^2}+\frac{(nl)^2}{c^2}=n^2\left(\frac{h^2}{a^2}+\frac{k^2}{b^2}+\frac{l^2}{c^2}\right)=\frac{n^2}{d_{hkl}^2}$$

which implies that

$$d_{nh,nk,nl}=\frac{d_{hkl}}{n}$$

Answer Substituting the indices into eqn 20.3 gives

$$\frac{1}{d_{123}^2}=\frac{1^2}{(0.82\text{ nm})^2}+\frac{2^2}{(0.94\text{ nm})^2}+\frac{3^2}{(0.75\text{ nm})^2}=0.22\text{ nm}^{-2}$$

Hence, $d_{123}=0.21$ nm. It then follows immediately that d_{246} is one-half this value, or 0.11 nm.

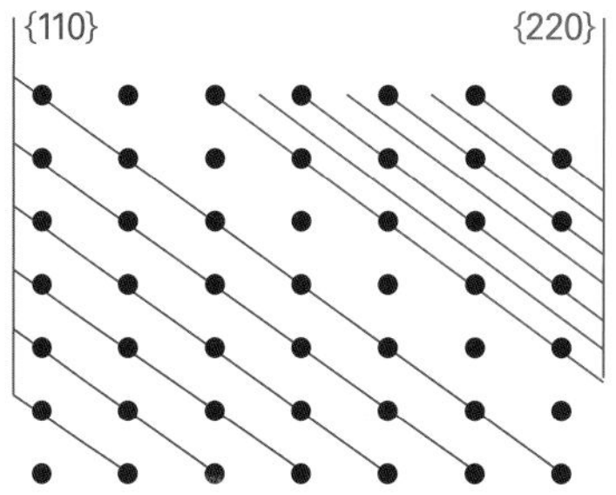

Fig. 20.12 The separation of the $\{220\}$ planes is half that of the $\{110\}$ planes. In general, (the separation of the planes $\{nh,nk,nl\}$ is n times smaller than the separation of the $\{hkl\}$ planes.

A note on good practice It is always sensible to look for analytical relations between quantities rather than to evaluate expressions numerically each time for that emphasizes the relations between quantities (and avoids unnecessary work).

Self-test 20.1 Calculate the separation of (a) the $\{133\}$ planes and (b) the $\{399\}$ planes in the same lattice. [0.19 nm, 0.063 nm]

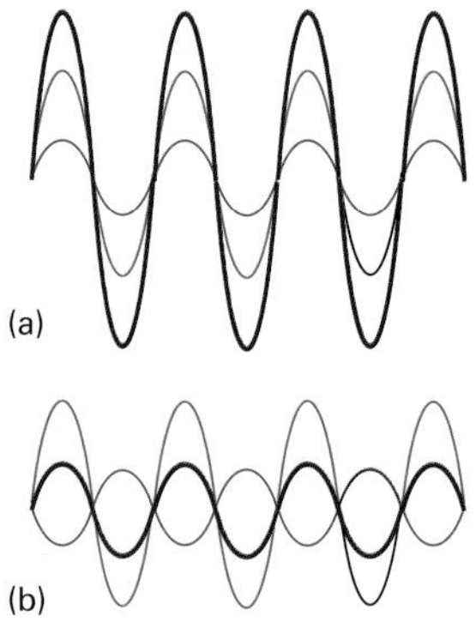

Fig. 20.13 When two waves are in the same region of space they interfere. Depending on their relative phase, they may interfere (a) constructively, to give an enhanced amplitude, or (b) destructively, to give a smaller amplitude. The component waves are shown in green and blue and the resultant in purple.

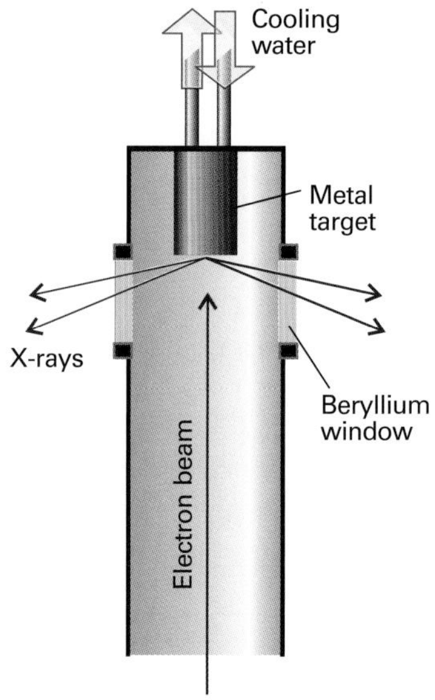

Fig. 20.14 X-rays are generated by directing an electron beam on to a cooled metal target. Beryllium is transparent to X-rays (on account of the small number of electrons in each atom) and is used for the windows.

20.3 The investigation of structure

A characteristic property of waves is that they interfere with one another, giving a greater displacement where peaks or troughs coincide and a smaller displacement where peaks coincide with troughs (Fig. 20.13). According to classical electromagnetic theory, the intensity of electromagnetic radiation is proportional to the square of the amplitude of the waves. Therefore, the regions of constructive or destructive interference show up as regions of enhanced or diminished intensities. The phenomenon of **diffraction** is the interference caused by an object in the path of waves, and the pattern of varying intensity that results is called the **diffraction pattern**. Diffraction occurs when the dimensions of the diffracting object are comparable to the wavelength of the radiation.

(a) X-ray diffraction

Wilhelm Röntgen discovered X-rays in 1895. Seventeen years later, Max von Laue suggested that they might be diffracted when passed through a crystal, for by then he had realized that their wavelengths are comparable to the separation of lattice planes. This suggestion was confirmed almost immediately by Walter Friedrich and Paul Knipping and has grown since then into a technique of extraordinary power. The bulk of this section will deal with the determination of structures using X-ray diffraction. The mathematical procedures necessary for the determination of structure from X-ray diffraction data are enormously complex, but such is the degree of integration of computers into the experimental apparatus that the technique is almost fully automated, even for large molecules and complex solids. The analysis is aided by molecular modelling techniques, which can guide the investigation towards a plausible structure.

X-rays are electromagnetic radiation with wavelengths of the order of 10^{-10} m. They are typically generated by bombarding a metal with high-energy electrons (Fig. 20.14). The electrons decelerate as they plunge into the metal and generate radiation with a continuous range of wavelengths called **Bremsstrahlung**.[1] Superimposed on the

[1] *Bremse* is German for deceleration, *Strahlung* for ray.

continuum are a few high-intensity, sharp peaks (Fig. 20.15). These peaks arise from collisions of the incoming electrons with the electrons in the inner shells of the atoms. A collision expels an electron from an inner shell, and an electron of higher energy drops into the vacancy, emitting the excess energy as an X-ray photon (Fig. 20.16). If the electron falls into a *K* shell (a shell with $n = 1$), the X-rays are classified as ***K*-radiation**, and similarly for transitions into the *L* ($n = 2$) and *M* ($n = 3$) shells. Strong, distinct lines are labelled K_α, K_β, and so on. Increasingly, X-ray diffraction makes use of the radiation available from synchrotron sources (*Further information* 13.1), for its high intensity greatly enhances the sensitivity of the technique.

von Laue's original method consisted of passing a broad-band beam of X-rays into a single crystal, and recording the diffraction pattern photographically. The idea behind the approach was that a crystal might not be suitably orientated to act as a diffraction grating for a single wavelength but, whatever its orientation, diffraction would be achieved for at least one of the wavelengths if a range of wavelengths was used. There is currently a resurgence of interest in this approach because synchrotron radiation spans a range of X-ray wavelengths.

An alternative technique was developed by Peter Debye and Paul Scherrer and independently by Albert Hull. They used monochromatic radiation and a powdered sample. When the sample is a powder, at least some of the crystallites will be orientated so as to give rise to diffraction. In modern powder diffractometers the intensities of the reflections are monitored electronically as the detector is rotated around the sample in a plane containing the incident ray (Fig. 20.17). Powder diffraction techniques are used to identify a sample of a solid substance by comparison of the positions of the diffraction lines and their intensities with diffraction patterns stored in a large data bank. Powder diffraction data are also used to determine phase diagrams, for different solid phases result in different diffraction patterns, and to determine the relative amounts of each phase present in a mixture. The technique is also used for the initial determination of the dimensions and symmetries of unit cells.

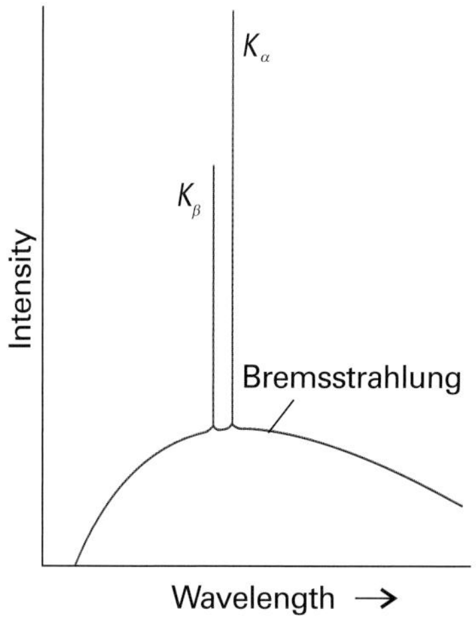

Fig. 20.15 The X-ray emission from a metal consists of a broad, featureless Bremsstrahlung background, with sharp transitions superimposed on it. The label *K* indicates that the radiation comes from a transition in which an electron falls into a vacancy in the *K* shell of the atom.

Comment 20.2

The web site contains links to databases of X-ray diffraction patterns.

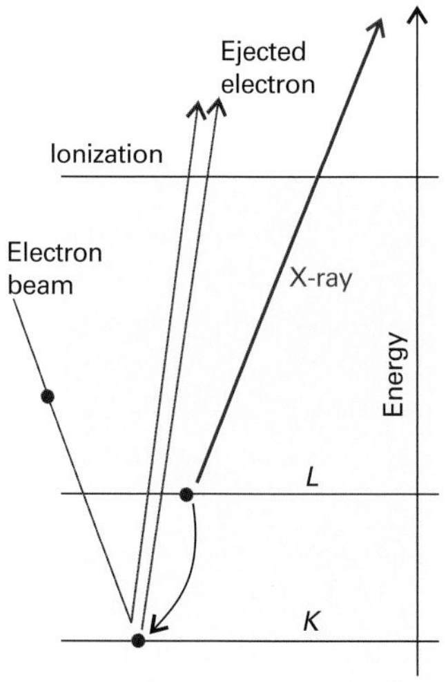

Fig. 20.16 The processes that contribute to the generation of X-rays. An incoming electron collides with an electron (in the *K* shell), and ejects it. Another electron (from the *L* shell in this illustration) falls into the vacancy and emits its excess energy as an X-ray photon.

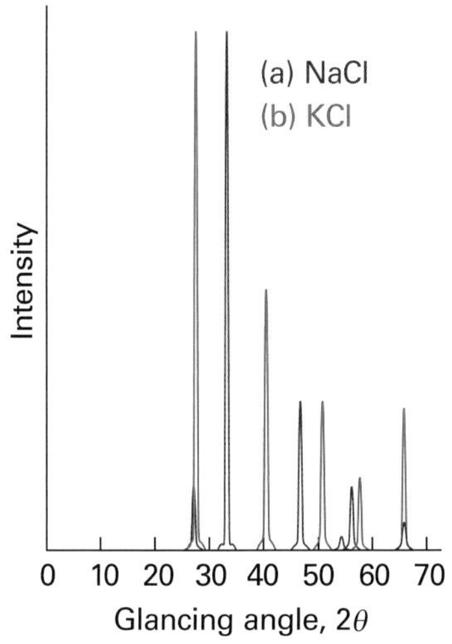

Fig. 20.17 X-ray powder photographs of (a) NaCl, (b) KCl and the indexed reflections. The smaller number of lines in (b) is a consequence of the similarity of the K^+ and Cl^- scattering factors, as discussed later in the chapter.

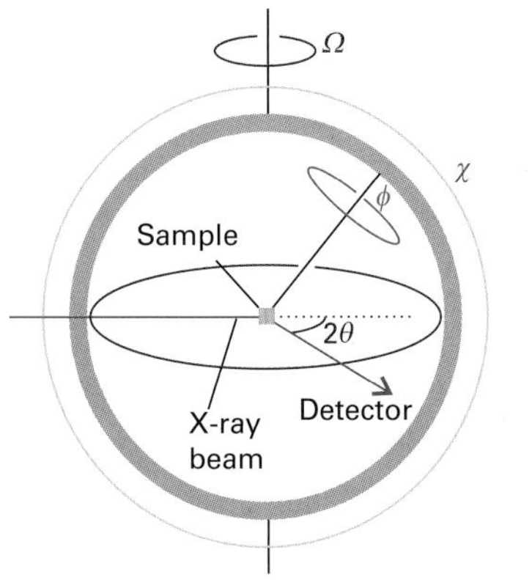

Fig. 20.18 A four-circle diffractometer. The settings of the orientations (ϕ, χ, θ, and Ω) of the components is controlled by computer; each (hkl) reflection is monitored in turn, and their intensities are recorded.

The method developed by the Braggs (William and his son Lawrence, who later jointly won the Nobel Prize) is the foundation of almost all modern work in X-ray crystallography. They used a single crystal and a monochromatic beam of X-rays, and rotated the crystal until a reflection was detected. There are many different sets of planes in a crystal, so there are many angles at which a reflection occurs. The complete set of data consists of the list of angles at which reflections are observed and their intensities.

Single-crystal diffraction patterns are measured by using a **four-circle diffractometer** (Fig. 20.18). The computer linked to the diffractometer determines the unit cell dimensions and the angular settings of the diffractometer's four circles that are needed to observe any particular intensity peak in the diffraction pattern. The computer controls the settings, and moves the crystal and the detector for each one in turn. At each setting, the diffraction intensity is measured, and background intensities are assessed by making measurements at slightly different settings. Computing techniques are now available that lead not only to automatic indexing but also to the automated determination of the shape, symmetry, and size of the unit cell. Moreover, several techniques are now available for sampling large amounts of data, including area detectors and image plates, which sample whole regions of diffraction patterns simultaneously.

(b) Bragg's law

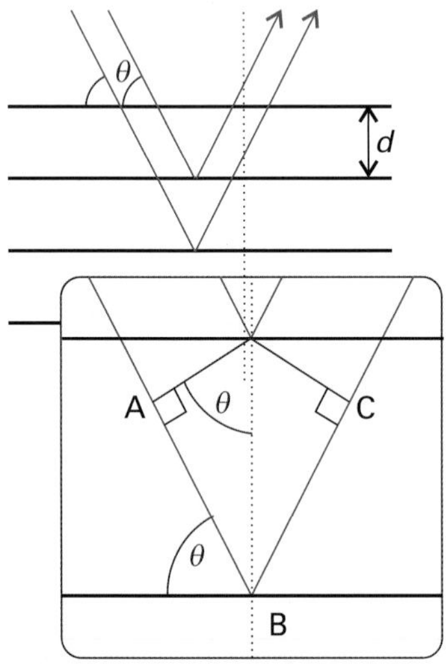

Fig. 20.19 The conventional derivation of Bragg's law treats each lattice plane as a reflecting the incident radiation. The path lengths differ by AB + BC, which depends on the glancing angle, θ. Constructive interference (a 'reflection') occurs when AB + BC is equal to an integer number of wavelengths.

An early approach to the analysis of diffraction patterns produced by crystals was to regard a lattice plane as a semi-transparent mirror, and to model a crystal as stacks of reflecting lattice planes of separation d (Fig. 20.19). The model makes it easy to calculate the angle the crystal must make to the incoming beam of X-rays for constructive interference to occur. It has also given rise to the name **reflection** to denote an intense beam arising from constructive interference.

Consider the reflection of two parallel rays of the same wavelength by two adjacent planes of a lattice, as shown in Fig. 20.19. One ray strikes point D on the upper plane but the other ray must travel an additional distance AB before striking the plane immediately below. Similarly, the reflected rays will differ in path length by a distance BC. The net path length difference of the two rays is then

$$\mathrm{AB} + \mathrm{BC} = 2d \sin \theta$$

where θ is the **glancing angle**. For many glancing angles the path-length difference is not an integer number of wavelengths, and the waves interfere largely destructively. However, when the path-length difference is an integer number of wavelengths ($\mathrm{AB} + \mathrm{BC} = n\lambda$), the reflected waves are in phase and interfere constructively. It follows that a reflection should be observed when the glancing angle satisfies **Bragg's law**:

$$n\lambda = 2d \sin \theta \qquad (20.4)$$

Reflections with $n = 2, 3, \ldots$ are called second-order, third-order, and so on; they correspond to path-length differences of 2, 3, . . . wavelengths. In modern work it is normal to absorb the n into d, to write the Bragg law as

$$\lambda = 2d \sin \theta \qquad (20.5)$$

and to regard the nth-order reflection as arising from the $\{nh,nk,nl\}$ planes (see Example 20.1).

The primary use of Bragg's law is in the determination of the spacing between the layers in the lattice for, once the angle θ corresponding to a reflection has been determined, d may readily be calculated.

Example 20.2 *Using Bragg's law*

A first-order reflection from the {111} planes of a cubic crystal was observed at a glancing angle of 11.2° when Cu(K_α) X-rays of wavelength 154 pm were used. What is the length of the side of the unit cell?

Method The separation of the planes can be determined from Bragg's law. Because the crystal is cubic, the separation is related to the length of the side of the unit cell, a, by eqn 20.2, which may therefore be solved for a.

Answer According to eqn 20.5, the {111} planes responsible for the diffraction have separation

$$d_{111} = \frac{\lambda}{2\sin\theta}$$

The separation of the {111} planes of a cubic lattice of side a is given by eqn 20.2 as

$$d_{111} = \frac{a}{3^{1/2}}$$

Therefore,

$$a = \frac{3^{1/2}\lambda}{2\sin\theta} = \frac{3^{1/2}\times(154\text{ pm})}{2\sin 11.2°} = 687\text{ pm}$$

Self-test 20.2 Calculate the angle at which the same crystal will give a reflection from the {123} planes. [24.8°]

Some types of unit cell give characteristic and easily recognizable patterns of lines. For example, in a cubic lattice of unit cell dimension a the spacing is given by eqn 20.2, so the angles at which the $\{hkl\}$ planes give first-order reflections are given by

$$\sin\theta = (h^2 + k^2 + l^2)^{1/2}\frac{\lambda}{2a}$$

The reflections are then predicted by substituting the values of h, k, and l:

$\{hkl\}$	{100}	{110}	{111}	{200}	{210}	{211}	{220}	{300}	{221}	{310}...
$h^2+k^2+l^2$	1	2	3	4	5	6	8	9	9	10...

Notice that 7 (and 15, . . .) is missing because the sum of the squares of three integers cannot equal 7 (or 15, . . .). Therefore the pattern has absences that are characteristic of the cubic P lattice.

Self-test 20.3 Normally, experimental procedures measure 2θ rather than θ itself. A diffraction examination of the element polonium gave lines at the following values of 2θ (in degrees) when 71.0 pm Mo X-rays were used: 12.1, 17.1, 21.0, 24.3, 27.2, 29.9, 34.7, 36.9, 38.9, 40.9, 42.8. Identify the unit cell and determine its dimensions. [cubic P; a = 337 pm]

(c) Scattering factors

To prepare the way to discussing modern methods of structural analysis we need to note that the scattering of X-rays is caused by the oscillations an incoming electromagnetic

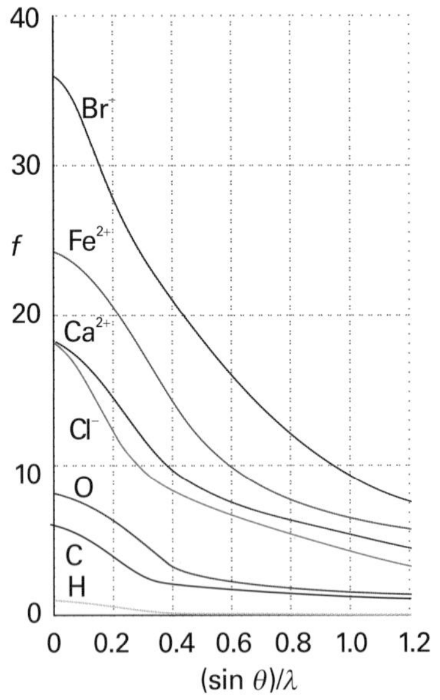

Fig. 20.20 The variation of the scattering factor of atoms and ions with atomic number and angle. The scattering factor in the forward direction (at $\theta = 0$, and hence at $(\sin\theta)/\lambda = 0$) is equal to the number of electrons present in the species.

wave generates in the electrons of atoms, and heavy atoms give rise to stronger scattering than light atoms. This dependence on the number of electrons is expressed in terms of the **scattering factor**, f, of the element. If the scattering factor is large, then the atoms scatter X-rays strongly. The scattering factor of an atom is related to the electron density distribution in the atom, $\rho(r)$, by

$$f = 4\pi\int_0^\infty \rho(r)\frac{\sin kr}{kr}r^2\mathrm{d}r \qquad k = \frac{4\pi}{\lambda}\sin\theta \tag{20.6}$$

The value of f is greatest in the forward direction and smaller for directions away from the forward direction (Fig. 20.20). The detailed analysis of the intensities of reflections must take this dependence on direction into account (in single crystal studies as well as for powders). We show in the *Justification* below that, in the forward direction (for $\theta = 0$), f is equal to the total number of electrons in the atom.

Justification 20.1 *The forward scattering factor*

As $\theta \to 0$, so $k \to 0$. Because $\sin x = x - \frac{1}{6}x^3 + \cdots$,

$$\lim_{x\to 0}\frac{\sin x}{x} = \lim_{x\to 0}\frac{x - \frac{1}{6}x^3 + \cdots}{x} = \lim_{x\to 0}(1 - \tfrac{1}{6}x^2 + \cdots) = 1$$

The factor $(\sin kr)/kr$ is therefore equal to 1 for forward scattering. It follows that in the forward direction

$$f = 4\pi\int_0^\infty \rho(r)r^2\mathrm{d}r$$

The integral over the electron density ρ (the number of electrons in an infinitesimal region divided by the volume of the region) multiplied by the volume element $4\pi r^2\mathrm{d}r$ is the total number of electrons, N_e, in the atom. Hence, in the forward direction, $f = N_e$. For example, the scattering factors of Na^+, K^+, and Cl^- are 8, 18, and 18, respectively.

The scattering factor is smaller in nonforward directions because $(\sin kr)/kr < 1$ for $\theta > 0$, so the integral is smaller than the value calculated above.

(d) The electron density

The problem we now address is how to interpret the data from a diffractometer in terms of the detailed structure of a crystal. To do so, we must go beyond Bragg's law.

If a unit cell contains several atoms with scattering factors f_j and coordinates (x_ja, y_jb, z_jc), then we show in the *Justification* below that the overall amplitude of a wave diffracted by the $\{hkl\}$ planes is given by

$$F_{hkl} = \sum_j f_j e^{i\phi_{hkl}(j)} \qquad \text{where} \quad \phi_{hkl}(j) = 2\pi(hx_j + ky_j + lz_j) \tag{20.7}$$

The sum is over all the atoms in the unit cell. The quantity F_{hkl} is called the **structure factor**.

Justification 20.2 *The structure factor*

We begin by showing that, if in the unit cell there is an A atom at the origin and a B atom at the coordinates (xa,yb,zc), where x, y, and z lie in the range 0 to 1, then the phase difference, ϕ, between the hkl reflections of the A and B atoms is $\phi_{hkl} = 2\pi(hx + ky + lz)$.

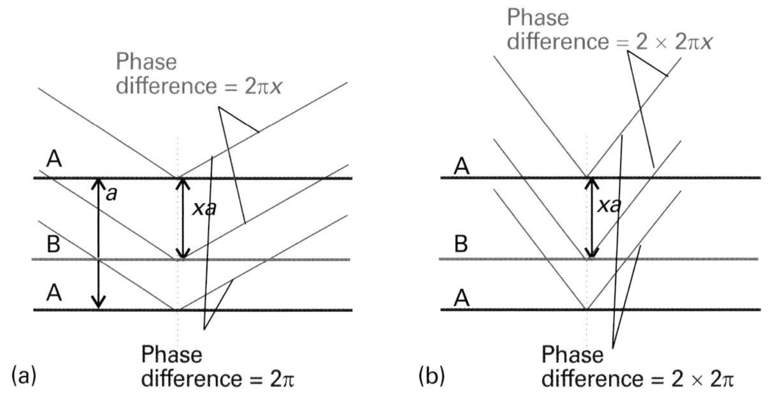

Fig. 20.21 Diffraction from a crystal containing two kinds of atoms. (a) For a (100) reflection from the A planes, there is a phase difference of 2π between waves reflected by neighbouring planes. (b) For a (200) reflection, the phase difference is 4π. The reflection from a B plane at a fractional distance xa from an A plane has a phase that is x times these phase differences.

Consider the crystal shown schematically in Fig. 20.21. The reflection corresponds to two waves from adjacent A planes, the phase difference of the waves being 2π. If there is a B atom at a fraction x of the distance between the two A planes, then it gives rise to a wave with a phase difference $2\pi x$ relative to an A reflection. To see this conclusion, note that, if $x = 0$, there is no phase difference; if $x = \frac{1}{2}$ the phase difference is π; if $x = 1$, the B atom lies where the lower A atom is and the phase difference is 2π. Now consider a (200) reflection. There is now a $2 \times 2\pi$ difference between the waves from the two A layers, and if B were to lie at $x = 0.5$ it would give rise to a wave that differed in phase by 2π from the wave from the upper A layer. Thus, for a general fractional position x, the phase difference for a (200) reflection is $2 \times 2\pi x$. For a general (h00) reflection, the phase difference is therefore $h \times 2\pi x$. For three dimensions, this result generalizes to eqn 20.7.

The A and B reflections interfere destructively when the phase difference is π, and the total intensity is zero if the atoms have the same scattering power. For example, if the unit cells are cubic I with a B atom at $x = y = z = \frac{1}{2}$, then the A,B phase difference is $(h + k + l)\pi$. Therefore, all reflections for odd values of $h + k + l$ vanish because the waves are displaced in phase by π. Hence the diffraction pattern for a cubic I lattice can be constructed from that for the cubic P lattice (a cubic lattice without points at the centre of its unit cells) by striking out all reflections with odd values of $h + k + l$. Recognition of these **systematic absences** in a powder spectrum immediately indicates a cubic I lattice (Fig. 20.22).

If the amplitude of the waves scattered from A is f_A at the detector, that of the waves scattered from B is $f_B e^{i\phi_{hkl}}$, with ϕ_{hkl} the phase difference given in eqn 20.7. The total amplitude at the detector is therefore

$$F_{hkl} = f_A + f_B e^{i\phi_{hkl}}$$

Because the intensity is proportional to the square modulus of the amplitude of the wave, the intensity, I_{hkl}, at the detector is

$$I_{hkl} \propto F^{\star}_{hkl} F_{hkl} = (f_A + f_B e^{-i\phi_{hkl}})(f_A + f_B e^{i\phi_{hkl}})$$

This expression expands to

$$I_{hkl} \propto f_A^2 + f_B^2 + f_A f_B (e^{i\phi_{hkl}} + e^{-i\phi_{hkl}}) = f_A^2 + f_B^2 + 2 f_A f_B \cos \phi_{hkl}$$

The cosine term either adds to or subtracts from $f_A^2 + f_B^2$ depending on the value of ϕ_{hkl}, which in turn depends on h, k, and l and x, y, and z. Hence, there is a variation in the intensities of the lines with different hkl.

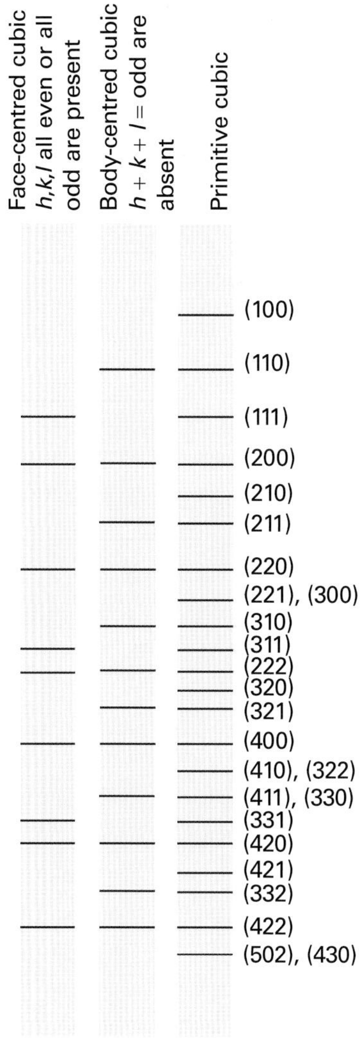

Fig. 20.22 The powder diffraction patterns and the systematic absences of three versions of a cubic cell. Comparison of the observed pattern with patterns like these enables the unit cell to be identified. The locations of the lines give the cell dimensions.

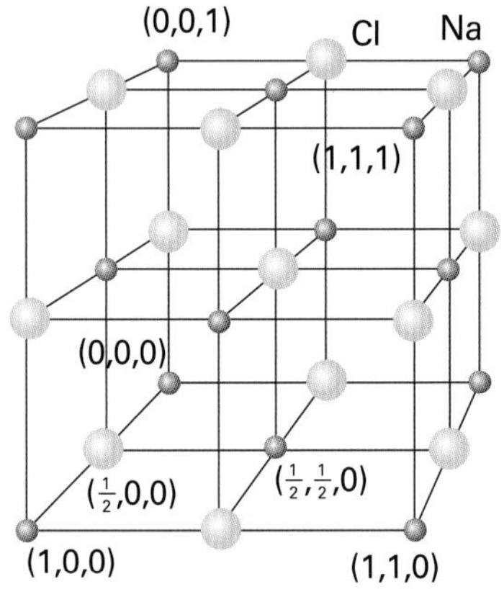

Fig. 20.23 The location of the atoms for the structure factor calculation in Example 20.3. The purple circles are Na^+; the green circles are Cl^-.

Example 20.3 *Calculating a structure factor*

Calculate the structure factors for the unit cell in Fig. 20.23.

Method The structure factor is defined by eqn 20.7. To use this equation, consider the ions at the locations specified in Fig. 20.23. Write f^+ for the Na^+ scattering factor and f^- for the Cl^- scattering factor. Note that ions in the body of the cell contribute to the scattering with a strength f. However, ions on faces are shared between two cells (use $\frac{1}{2}f$), those on edges by four cells (use $\frac{1}{4}f$), and those at corners by eight cells (use $\frac{1}{8}f$). Two useful relations are

$$e^{i\pi} = -1 \qquad \cos\phi = \tfrac{1}{2}(e^{i\phi} + e^{-i\phi})$$

Answer From eqn 20.7, and summing over the coordinates of all 27 atoms in the illustration:

$$\begin{aligned}F_{hkl} = {} & f^+(\tfrac{1}{8} + \tfrac{1}{8}e^{2\pi il} + \cdots + \tfrac{1}{2}e^{2\pi i(\frac{1}{2}h+\frac{1}{2}k+l)}) \\ & + f^-(e^{2\pi i(\frac{1}{2}h+\frac{1}{2}k+\frac{1}{2}l)} + \tfrac{1}{4}e^{2\pi i(\frac{1}{2}h)} + \cdots + \tfrac{1}{4}e^{2\pi i(\frac{1}{2}h+l)})\end{aligned}$$

To simplify this 27-term expression, we use

$$e^{2\pi ih} = e^{2\pi ik} = e^{2\pi il} = 1$$

because h, k, and l are all integers:

$$\begin{aligned}F_{hkl} = {} & f^+\{1 + \cos(h+k)\pi + \cos(h+l)\pi + \cos(k+l)\pi\} \\ & + f^-\{(-1)^{h+k+l} + \cos k\pi + \cos l\pi + \cos h\pi\}\end{aligned}$$

Then, because $\cos h\pi = (-1)^h$,

$$F_{hkl} = f^+\{1 + (-1)^{h+k} + (-1)^{h+l} + (-1)^{l+k}\} + f^-\{(-1)^{h+k+l} + (-1)^h + (-1)^k + (-1)^l\}$$

Now note that:

if h, k, and l are all even, $F_{hkl} = f^+\{1+1+1+1\} + f^-\{1+1+1+1\} = 4(f^+ + f^-)$
if h, k, and l are all odd, $F_{hkl} = 4(f^+ - f^-)$
if one index is odd and two are even, or vice versa, $F_{hkl} = 0$

The hkl all-odd reflections are less intense than the hkl all-even. For $f^+ = f^-$, which is the case for identical atoms in a cubic P arrangement, the hkl all-odd have zero intensity, corresponding to the 'systematic absences' of cubic P unit cells.

Self-test 20.4 Which reflections cannot be observed for a cubic I lattice?

[for $h + k + l$ odd, $F_{hkl} = 0$]

The intensity of the (hkl) reflection is proportional to $|F_{hkl}|^2$, so in principle we can determine the structure factors experimentally by taking the square root of the corresponding intensities (but see below). Then, once we know all the structure factors F_{hkl}, we can calculate the electron density distribution, $\rho(\boldsymbol{r})$, in the unit cell by using the expression

$$\rho(\boldsymbol{r}) = \frac{1}{V}\sum_{hkl} F_{hkl}e^{-2\pi i(hx+ky+lz)} \qquad (20.8)$$

where V is the volume of the unit cell. Equation 20.8 is called a **Fourier synthesis** of the electron density.

Example 20.4 *Calculating an electron density by Fourier synthesis*

Consider the $\{h00\}$ planes of a crystal extending indefinitely in the x-direction. In an X-ray analysis the structure factors were found as follows:

h:	0	1	2	3	4	5	6	7	8	9
F_h	16	−10	2	−1	7	−10	8	−3	2	−3
h:	10	11	12	13	14	15				
F_h	6	−5	3	−2	2	−3				

(and $F_{-h} = F_h$). Construct a plot of the electron density projected on to the x-axis of the unit cell.

Method Because $F_{-h} = F_h$, it follows from eqn 20.8 that

$$V\rho(x) = \sum_{h=-\infty}^{\infty} F_h e^{-2\pi i hx} = F_0 + \sum_{h=1}^{\infty}(F_h e^{-2\pi i hx} + F_{-h}e^{2\pi i hx})$$

$$= F_0 + \sum_{h=1}^{\infty} F_h(e^{-2\pi i hx} + e^{2\pi i hx}) = F_0 + 2\sum_{h=1}^{\infty} F_h \cos 2\pi hx$$

and we evaluate the sum (truncated at $h = 15$) for points $0 \le x \le 1$ using mathematical software.

Answer The results are plotted in Fig. 20.24 (blue line). The positions of three atoms can be discerned very readily. The more terms there are included, the more accurate the density plot. Terms corresponding to high values of h (short wavelength cosine terms in the sum) account for the finer details of the electron density; low values of h account for the broad features.

Self-test 20.5 Use mathematical software to experiment with different structure factors (including changing signs as well as amplitudes). For example, use the same values of F_h as above, but with positive signs for $h \ge 6$.

[Fig. 20.24 (purple line)]

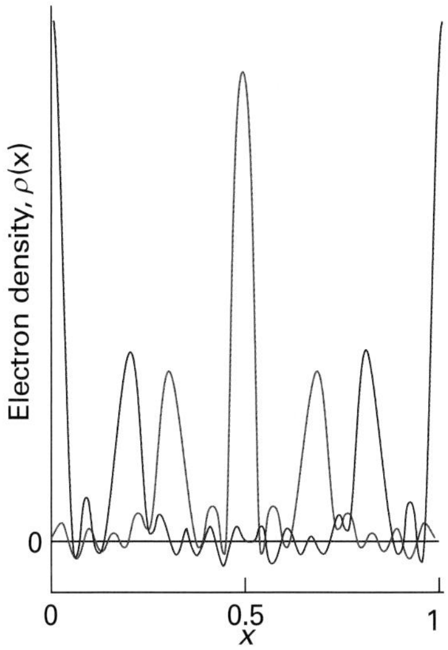

Fig. 20.24 The plot of the electron density calculated in Example 20.4 (blue) and Self-test 20.5 (purple).

Exploration If you do not have access to mathematical software, perform the calculations suggested in Self-test 20.5 by using the interactive applets found in the text's web site.

(e) The phase problem

A problem with the procedure outlined above is that the observed intensity I_{hkl} is proportional to the square modulus $|F_{hkl}|^2$, so we cannot say whether we should use $+|F_{hkl}|$ or $-|F_{hkl}|$ in the sum in eqn 20.8. In fact, the difficulty is more severe for non-centrosymmetric unit cells because, if we write F_{hkl} as the complex number $|F_{hkl}|e^{i\alpha}$, where α is the phase of F_{hkl} and $|F_{hkl}|$ is its magnitude, then the intensity lets us determine $|F_{hkl}|$ but tells us nothing of its phase, which may lie anywhere from 0 to 2π. This ambiguity is called the **phase problem**; its consequences are illustrated by comparing the two plots in Fig. 20.24. Some way must be found to assign phases to the structure factors, for otherwise the sum for ρ cannot be evaluated and the method would be useless.

The phase problem can be overcome to some extent by a variety of methods. One procedure that is widely used for inorganic materials with a reasonably small number of atoms in a unit cell and for organic molecules with a small number of heavy atoms is the **Patterson synthesis**. Instead of the structure factors F_{hkl}, the values of $|F_{hkl}|^2$, which can be obtained without ambiguity from the intensities, are used in an expression that resembles eqn 20.8:

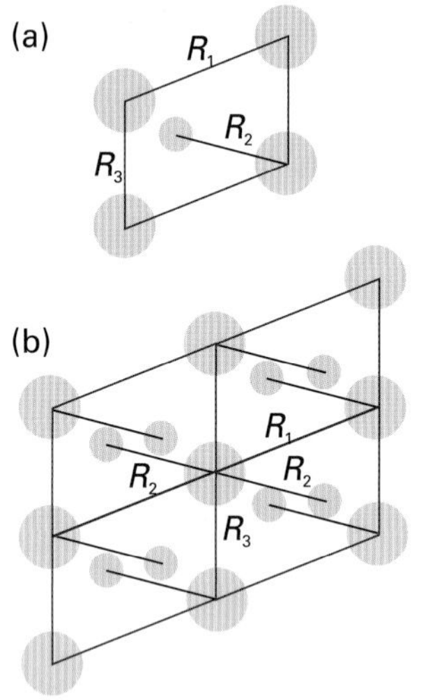

Fig. 20.25 The Patterson synthesis corresponding to the pattern in (a) is the pattern in (b). The distance and orientation of each spot from the origin gives the orientation and separation of one atom–atom separation in (a). Some of the typical distances and their contribution to (b) are shown as R_1, etc.

$$P(\boldsymbol{r}) = \frac{1}{V}\sum_{hkl} |F_{hkl}|^2 e^{-2\pi i(hx+ky+lz)} \tag{20.9}$$

The outcome of a Patterson synthesis is a map of the vector separations of the atoms (the distances and directions between atoms) in the unit cell. Thus, if atom A is at the coordinates (x_A,y_A,z_A) and atom B is at (x_B,y_B,z_B), then there will be a peak at $(x_A - x_B, y_A - y_B, z_A - z_B)$ in the Patterson map. There will also be a peak at the negative of these coordinates, because there is a vector from B to A as well as a vector from A to B. The height of the peak in the map is proportional to the product of the atomic numbers of the two atoms, $Z_A Z_B$. For example, if the unit cell has the structure shown in Fig. 20.25a, the Patterson synthesis would be the map shown in Fig. 20.25b, where the location of each spot relative to the origin gives the separation and relative orientation of each pair of atoms in the original structure.

Heavy atoms dominate the scattering because their scattering factors are large, of the order of their atomic numbers, and their locations may be deduced quite readily. The sign of F_{hkl} can now be calculated from the locations of the heavy atoms in the unit cell, and to a high probability the phase calculated for them will be the same as the phase for the entire unit cell. To see why this is so, we have to note that a structure factor of a centrosymmetric cell has the form

$$F = (\pm)f_{\text{heavy}} + (\pm)f_{\text{light}} + (\pm)f_{\text{light}} + \cdots \tag{20.10}$$

where f_{heavy} is the scattering factor of the heavy atom and f_{light} the scattering factors of the light atoms. The f_{light} are all much smaller than f_{heavy}, and their phases are more or less random if the atoms are distributed throughout the unit cell. Therefore, the net effect of the f_{light} is to change F only slightly from f_{heavy}, and we can be reasonably confident that F will have the same sign as that calculated from the location of the heavy atom. This phase can then be combined with the observed $|F|$ (from the reflection intensity) to perform a Fourier synthesis of the full electron density in the unit cell, and hence to locate the light atoms as well as the heavy atoms.

Modern structural analyses make extensive use of **direct methods**. Direct methods are based on the possibility of treating the atoms in a unit cell as being virtually randomly distributed (from the radiation's point of view), and then using statistical techniques to compute the probabilities that the phases have a particular value. It is possible to deduce relations between some structure factors and sums (and sums of squares) of others, which have the effect of constraining the phases to particular values (with high probability, so long as the structure factors are large). For example, the **Sayre probability relation** has the form

$$\text{sign of } F_{h+h',k+k',l+l'} \text{ is probably equal to } (\text{sign of } F_{hkl}) \times (\text{sign of } F_{h'k'l'}) \tag{20.11}$$

For example, if F_{122} and F_{232} are both large and negative, then it is highly likely that F_{354}, provided it is large, will be positive.

(f) Structure refinement

In the final stages of the determination of a crystal structure, the parameters describing the structure (atom positions, for instance) are adjusted systematically to give the best fit between the observed intensities and those calculated from the model of the structure deduced from the diffraction pattern. This process is called **structure refinement**. Not only does the procedure give accurate positions for all the atoms in the unit cell, but it also gives an estimate of the errors in those positions and in the bond lengths and angles derived from them. The procedure also provides information on the vibrational amplitudes of the atoms.

IMPACT ON BIOCHEMISTRY

I20.1 X-ray crystallography of biological macromolecules

X-ray crystallography is the deployment of X-ray diffraction techniques for the determination of the location of all the atoms in molecules as complicated as biopolymers. Bragg's law helps us understand the features of one of the most seminal X-ray images of all time, the characteristic X-shaped pattern obtained by Rosalind Franklin and Maurice Wilkins from strands of DNA and used by James Watson and Francis Crick in their construction of the double-helix model of DNA (Fig. 20.26). To interpret this image by using the Bragg law we have to be aware that it was obtained by using a fibre consisting of many DNA molecules oriented with their axes parallel to the axis of the fibre, with X-rays incident from a perpendicular direction. All the molecules in the fibre are parallel (or nearly so), but are randomly distributed in the perpendicular directions; as a result, the diffraction pattern exhibits the periodic structure parallel to the fibre axis superimposed on a general background of scattering from the distribution of molecules in the perpendicular directions.

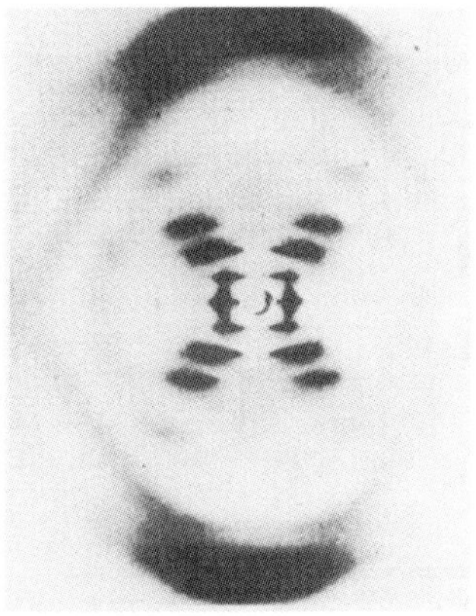

Fig. 20.26 The X-ray diffraction pattern obtained from a fibre of B-DNA. The black dots are the reflections, the points of maximum constructive interference, that are used to determine the structure of the molecule. (Adapted from an illustration that appears in J.P. Glusker and K.N. Trueblood, *Crystal structure analysis: A primer*. Oxford University Press (1972).)

There are two principal features in Fig. 20.26: the strong 'meridional' scattering upward and downward by the fibre and the X-shaped distribution at smaller scattering angles. Because scattering through large angles occurs for closely spaced features (from $\lambda = 2d \sin \theta$, if d is small, then θ must be large to preserve the equality), we can infer that the meridional scattering arises from closely spaced components and that the inner X-shaped pattern arises from features with a longer periodicity. Because the meridional pattern occurs at a distance of about 10 times that of the innermost spots of the X-pattern, the large-scale structure is about 10 times bigger than the small-scale structure. From the geometry of the instrument, the wavelength of the radiation, and Bragg's law, we can infer that the periodicity of the small-scale feature is 340 pm whereas that of the large-scale feature is 3400 pm (that is, 3.4 nm).

To see that the cross is characteristic of a helix, look at Fig. 20.27. Each turn of the helix defines two planes, one orientated at an angle α to the horizontal and the other at $-\alpha$. As a result, to a first approximation, a helix can be thought of as consisting of an array of planes at an angle α together with an array of planes at an angle $-\alpha$ with a separation within each set determined by the pitch of the helix. Thus, a DNA molecule is like two arrays of planes, each set corresponding to those treated in the derivation of Bragg's law, with a perpendicular separation $d = p \cos \alpha$, where p is the pitch of the helix, each canted at the angles $\pm\alpha$ to the horizontal. The diffraction spots from one set of planes therefore occur at an angle α to the vertical, giving one leg of the X, and those of the other set occur at an angle $-\alpha$, giving rise to the other leg of the X. The

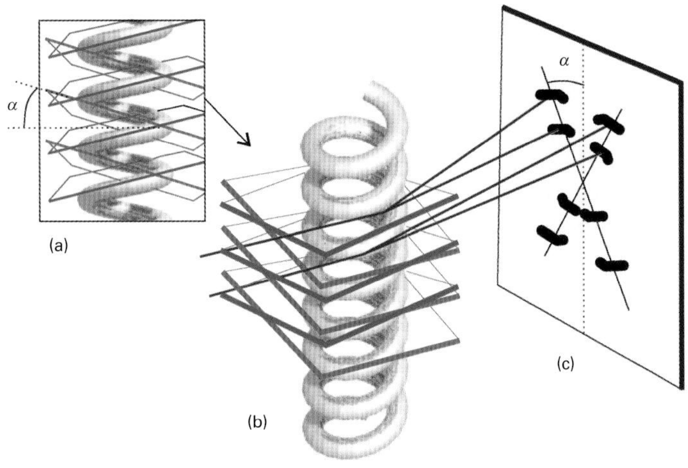

Fig. 20.27 The origin of the X pattern characteristic of diffraction by a helix. (a) A helix can be thought of as consisting of an array of planes at an angle α together with an array of planes at an angle $-\alpha$. (b) The diffraction spots from one set of planes appear at an angle α to the vertical, giving one leg of the X, and those of the other set appear at an angle $-\alpha$, giving rise to the other leg of the X. The lower half of the X appears because the helix has up–down symmetry in this arrangement. (c) The sequence of spots outward along a leg of the X corresponds to first-, second-, . . . order diffraction ($n = 1, 2, \ldots$).

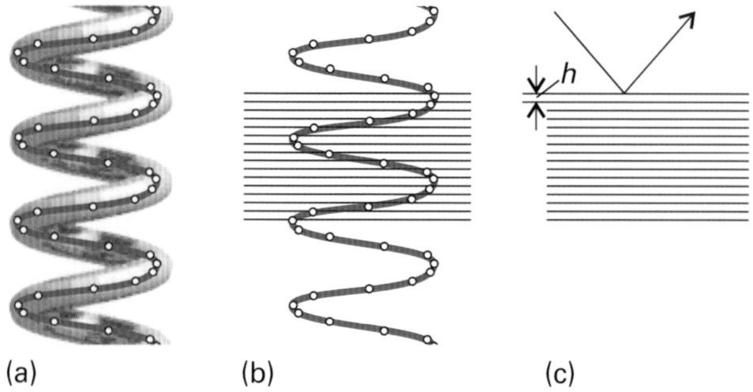

Fig. 20.28 The effect of the internal structure of the helix on the X-ray diffraction pattern. (a) The residues of the macromolecule are represented by points. (b) Parallel planes passing through the residues are perpendicular to the axis of the molecule. (c) The planes give rise to strong diffraction with an angle that allows us to determine the layer spacing h from $\lambda = 2h \sin \theta$.

experimental arrangement has up–down symmetry, so the diffraction pattern repeats to produce the lower half of the X. The sequence of spots outward along a leg corresponds to first-, second-, . . . order diffraction ($n = 1, 2, \ldots$ in eqn 20.4). Therefore from the X-ray pattern, we see at once that the molecule is helical and we can measure the angle α directly, and find $\alpha = 40°$. Finally, with the angle α and the pitch p determined, we can determine the radius r of the helix from $\tan \alpha = p/4r$, from which it follows that $r = (3.4 \text{ nm})/(4 \tan 40°) = 1.0 \text{ nm}$.

To derive the relation between the helix and the cross-like pattern we have ignored the detailed structure of the helix, the fact that it is a periodic array of nucleotide bases, not a smooth wire. In Fig. 20.28 we represent the bases by points, and see that there is an additional periodicity of separation h, forming planes that are perpendicular to the axis to the molecule (and the fibre). These planes give rise to the strong meridional diffraction with an angle that allows us to determine the layer spacing from Bragg's law in the form $\lambda = 2h \sin \theta$ as $h = 340$ pm.

The success of modern biochemistry in explaining such processes as DNA replication, protein biosynthesis, and enzyme catalysis is a direct result of developments in preparatory, instrumental, and computational procedures that have led to the determination of large numbers of structures of biological macromolecules by techniques based on X-ray diffraction. Most work is now done not on fibres but on crystals, in which the large molecules lie in orderly ranks. A technique that works well for charged proteins consists of adding large amounts of a salt, such as $(NH_4)_2SO_4$, to a buffer solution containing the biopolymer. The increase in the ionic strength of the solution decreases the solubility of the protein to such an extent that the protein precipitates, sometimes as crystals that are amenable to analysis by X-ray diffraction. Other common strategies for inducing crystallization involve the gradual removal of solvent from a biopolymer solution, either by *dialysis* (*Impact I5.2*) or *vapour diffusion*. In one implementation of the vapour diffusion method, a single drop of biopolymer solution hangs above an aqueous solution (the reservoir), as shown in Fig. 20.29. If the reservoir solution is more concentrated in a non-volatile solute (for example, a salt) than is the biopolymer solution, then solvent will evaporate slowly from the drop until the vapour pressure of water in the closed container reaches a constant, equilibrium value. At the same time, the concentration of biopolymer in the drop increases gradually until crystals begin to form.

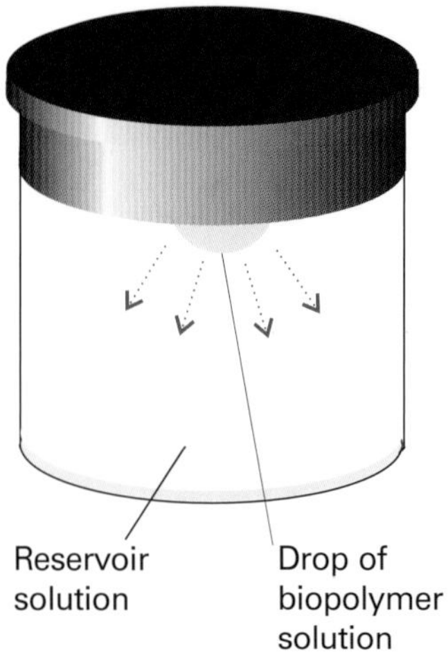

Fig. 20.29 In a common implementation of the vapour diffusion method of biopolymer crystallization, a single drop of biopolymer solution hangs above a reservoir solution that is very concentrated in a non-volatile solute. Solvent evaporates from the more dilute drop until the vapour pressure of water in the closed container reaches a constant equilibrium value. In the course of evaporation (denoted by the downward arrows), the biopolymer solution becomes more concentrated and, at some point, crystals may form.

Special techniques are used to crystallize hydrophobic proteins, such as those spanning the bilayer of a cell membrane. In such cases, surfactant molecules, which like phospholipids contain polar head groups and hydrophobic tails, are used to encase the protein molecules and make them soluble in aqueous buffer solutions. Dialysis or vapour diffusion may then be used to induce crystallization.

After suitable crystals are obtained, X-ray diffraction data are collected and analysed as described in the previous sections. The three-dimensional structures of a very

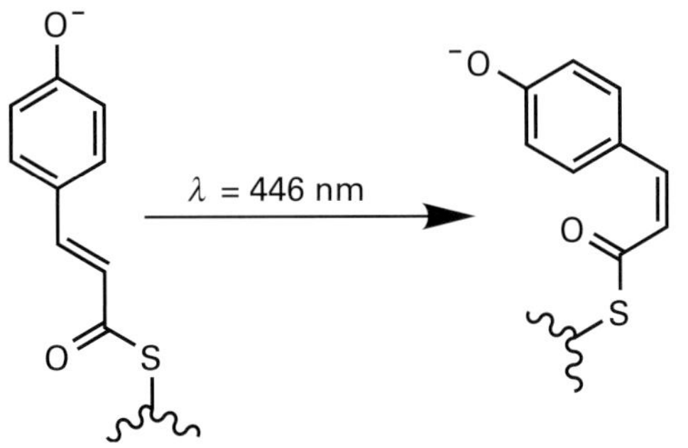

Fig. 20.30 Light-induced isomerization of a protein-bound phenolate ion in the photoactive yellow protein of the bacterium *Ectothiorhodospira halophila*.

large number of biological polymers have been determined in this way. However, the techniques discussed so far give only static pictures and are not useful in studies of dynamics and reactivity. This limitation stems from the fact that the Bragg rotation method requires stable crystals that do not change structure during the lengthy data acquisition times required. However, special time-resolved X-ray diffraction techniques have become available in recent years and it is now possible to make exquisitely detailed measurements of atomic motions during chemical and biochemical reactions.

Comment 20.3

The text's web site contains links to databases of structures of biological macromolecules.

Time-resolved X-ray diffraction techniques make use of synchrotron sources, which can emit intense polychromatic pulses of X-ray radiation with pulse widths varying from 100 ps to 200 ps (1 ps = 10^{-12} s). Instead of the Bragg method, the Laue method is used because many reflections can be collected simultaneously, rotation of the sample is not required, and data acquisition times are short. However, good diffraction data cannot be obtained from a single X-ray pulse and reflections from several pulses must be averaged together. In practice, this averaging dictates the time resolution of the experiment, which is commonly tens of microseconds or less.

An example of the power of time-resolved X-ray crystallography is the elucidation of structural changes that accompany the activation by light of the photoactive yellow protein of the bacterium *Ectothiorhodospira halophila*. Within 1 ns after absorption of a photon of 446 nm light, a protein-bound phenolate ion undergoes *trans–cis* isomerization to form the intermediate shown in Fig. 20.30. A series of rearrangements then follows, which includes the ejection of the ion from its binding site deep in the protein, its return to the site, and re-formation of the *cis* conformation. The physiological outcome of this cycle is a *negative phototactic response*, or movement of the organism away from light. Time-resolved X-ray diffraction studies in the nanosecond to millisecond ranges identified a number of structural changes that follow electronic excitation of the phenolate ion with a laser pulse: isomerization, ejection, protonation of the exposed ion, and a number of amino acid motions.

20.4 Neutron and electron diffraction

According to the de Broglie relation (eqn 8.12, $\lambda = h/p$), particles have wavelengths and may therefore undergo diffraction. Neutrons generated in a nuclear reactor and then slowed to thermal velocities have wavelengths similar to those of X-rays and may also be used for diffraction studies. For instance, a neutron generated in a reactor and slowed to thermal velocities by repeated collisions with a moderator (such as graphite) until it is travelling at about 4 km s^{-1} has a wavelength of about 100 pm. In practice, a range of wavelengths occurs in a neutron beam, but a monochromatic beam can be selected by diffraction from a crystal, such as a single crystal of germanium.

Example 20.5 *Calculating the typical wavelength of thermal neutrons*

Calculate the typical wavelength of neutrons that have reached thermal equilibrium with their surroundings at 373 K.

Method We need to relate the wavelength to the temperature. There are two linking steps. First, the de Broglie relation expresses the wavelength in terms of the linear momentum. Then the linear momentum can be expressed in terms of the kinetic energy, the mean value of which is given in terms of the temperature by the equipartition theorem (see Section 17.3).

Answer From the equipartition principle, we know that the mean translational kinetic energy of a neutron at a temperature T travelling in the x-direction is $E_K = \frac{1}{2}kT$. The kinetic energy is also equal to $p^2/2m$, where p is the momentum of the neutron and m is its mass. Hence, $p = (mkT)^{1/2}$. It follows from the de Broglie relation $\lambda = h/p$ that the neutron's wavelength is

$$\lambda = \frac{h}{(mkT)^{1/2}}$$

Therefore, at 373 K,

$$\lambda = \frac{6.626\times10^{-34}\,\mathrm{J\,s}}{\{(1.675\times10^{-27}\,\mathrm{kg})\times(1.381\times10^{-23}\,\mathrm{J\,K^{-1}})\times(373\,\mathrm{K})\}^{1/2}}$$

$$= \frac{6.626\times10^{-34}}{(1.675\times1.381\times373\times10^{-50})^{1/2}}\,\frac{\mathrm{J\,s}}{(\mathrm{kg^2\,m^2\,s^{-2}})^{1/2}}$$

$$= 2.26\times10^{-10}\,\mathrm{m} = 226\,\mathrm{pm}$$

where we have used $1\,\mathrm{J} = 1\,\mathrm{kg\,m^2\,s^{-2}}$.

Self-test 20.6 Calculate the temperature needed for the average wavelength of the neutrons to be 100 pm. [1.90×10^3 K]

Neutron diffraction differs from X-ray diffraction in two main respects. First, the scattering of neutrons is a nuclear phenomenon. Neutrons pass through the extranuclear electrons of atoms and interact with the nuclei through the 'strong force' that is responsible for binding nucleons together. As a result, the intensity with which neutrons are scattered is independent of the number of electrons and neighbouring elements in the periodic table may scatter neutrons with markedly different intensities. Neutron diffraction can be used to distinguish atoms of elements such as Ni and Co that are present in the same compound and to study order–disorder phase transitions in FeCo. A second difference is that neutrons possess a magnetic moment due to their spin. This magnetic moment can couple to the magnetic fields of atoms or ions in a crystal (if the ions have unpaired electrons) and modify the diffraction pattern. One consequence is that neutron diffraction is well suited to the investigation of magnetically ordered lattices in which neighbouring atoms may be of the same element but have different orientations of their electronic spin (Fig. 20.31).

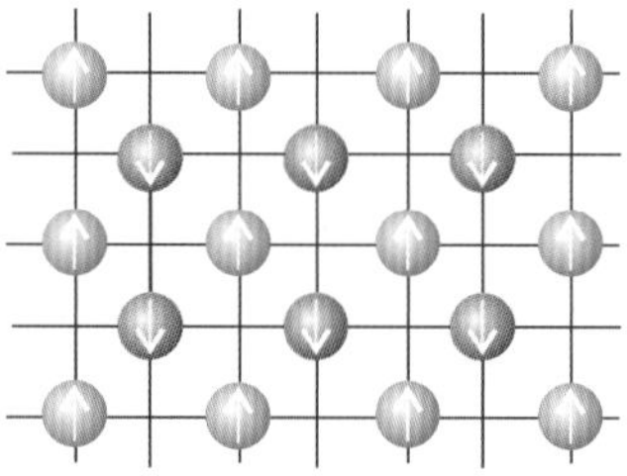

Fig. 20.31 If the spins of atoms at lattice points are orderly, as in this material, where the spins of one set of atoms are aligned antiparallel to those of the other set, neutron diffraction detects two interpenetrating simple cubic lattices on account of the magnetic interaction of the neutron with the atoms, but X-ray diffraction would see only a single bcc lattice.

Electrons accelerated through a potential difference of 40 kV have wavelengths of about 6 pm, and so are also suitable for diffraction studies. However, their main application is to the study of surfaces, and we postpone their discussion until Chapter 25.

Crystal structure

The bonding within a solid may be of various kinds. Simplest of all (in principle) are elemental metals, where electrons are delocalized over arrays of identical cations and bind them together into a rigid but ductile and malleable whole.

20.5 Metallic solids

Most metallic elements crystallize in one of three simple forms, two of which can be explained in terms of hard spheres packing together in the closest possible arrangement.

(a) Close packing

Figure 20.32 shows a **close-packed** layer of identical spheres, one with maximum utilization of space. A close-packed three-dimensional structure is obtained by stacking such close-packed layers on top of one another. However, this stacking can be done in different ways, which result in close-packed **polytypes**, or structures that are identical in two dimensions (the close-packed layers) but differ in the third dimension.

In all polytypes, the spheres of second close-packed layer lie in the depressions of the first layer (Fig. 20.33). The third layer may be added in either of two ways. In one, the spheres are placed so that they reproduce the first layer (Fig. 20.34a), to give an ABA pattern of layers. Alternatively, the spheres may be placed over the gaps in the first layer (Fig. 20.34b), so giving an ABC pattern. Two polytypes are formed if the two stacking patterns are repeated in the vertical direction. If the ABA pattern is repeated, to give the sequence of layers ABABAB . . . , the spheres are **hexagonally close-packed**

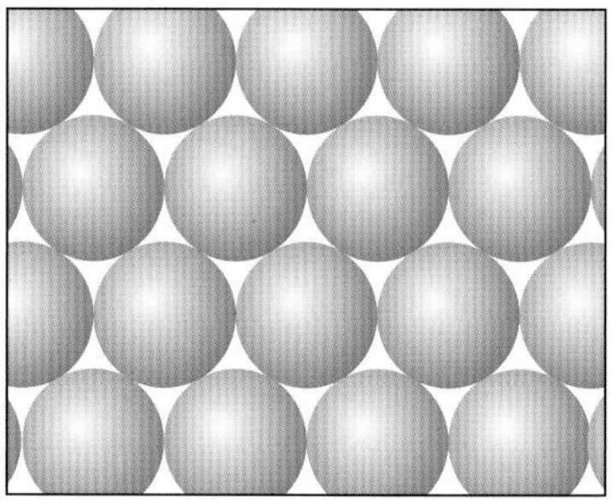

Fig. 20.32 The first layer of close-packed spheres used to build a three-dimensional close-packed structure.

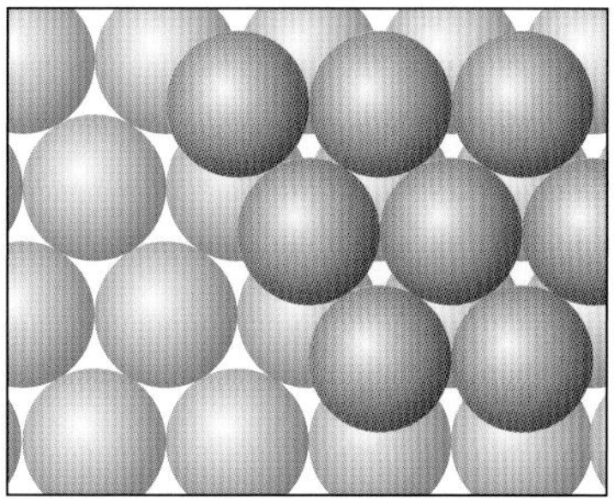

Fig. 20.33 The second layer of close-packed spheres occupies the dips of the first layer. The two layers are the AB component of the close-packed structure.

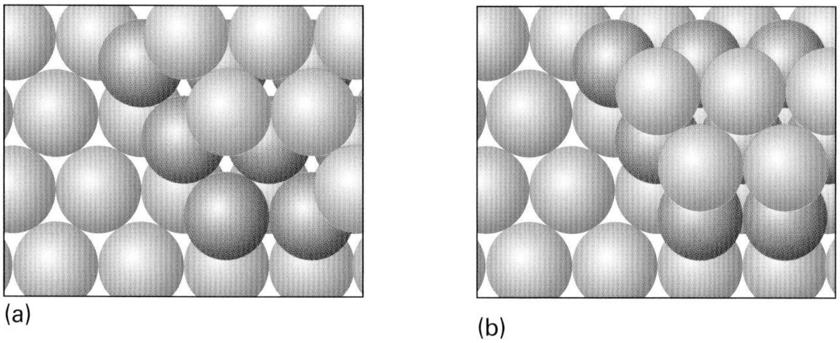

Fig. 20.34 (a) The third layer of close-packed spheres might occupy the dips lying directly above the spheres in the first layer, resulting in an ABA structure, which corresponds to hexagonal close-packing. (b) Alternatively, the third layer might lie in the dips that are not above the spheres in the first layer, resulting in an ABC structure, which corresponds to cubic close-packing

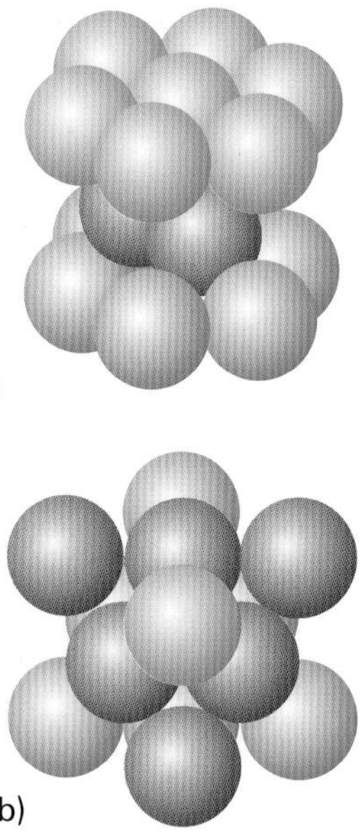
(a)
(b)

Fig. 20.35 A fragment of the structure shown in Fig. 20.34 revealing the (a) hexagonal (b) cubic symmetry. The tints on the spheres are the same as for the layers in Fig. 20.34.

Table 20.2 The crystal structures of some elements

Structure	Element
hcp*	Be, Cd, Co, He, Mg, Sc, Ti, Zn
fcc* (ccp, cubic F)	Ag, Al, Ar, Au, Ca, Cu, Kr, Ne, Ni, Pd, Pb, Pt, Rh, Rn, Sr, Xe
bcc (cubic I)	Ba, Cs, Cr, Fe, K, Li, Mn, Mo, Rb, Na, Ta, W, V
cubic P	Po

* Close-packed structures.

(hcp). Alternatively, if the ABC pattern is repeated, to give the sequence ABCABC . . . , the spheres are **cubic close-packed** (ccp). We can see the origins of these names by referring to Fig. 20.35. The ccp structure gives rise to a face-centred unit cell, so may also be denoted cubic F (or fcc, for face-centred cubic).[2] It is also possible to have random sequences of layers; however, the hcp and ccp polytypes are the most important. Table 20.2 lists some elements possessing these structures.

The compactness of close-packed structures is indicated by their **coordination number**, the number of atoms immediately surrounding any selected atom, which is 12 in all cases. Another measure of their compactness is the **packing fraction**, the fraction of space occupied by the spheres, which is 0.740 (see the following *Justification*). That is, in a close-packed solid of identical hard spheres, only 26.0 per cent of the volume is empty space. The fact that many metals are close-packed accounts for their high densities.

Justification 20.3 *The packing fraction*

To calculate a packing fraction of a ccp structure, we first calculate the volume of a unit cell, and then calculate the total volume of the spheres that fully or partially occupy it. The first part of the calculation is a straightforward exercise in geometry. The second part involves counting the fraction of spheres that occupy the cell.

Refer to Fig. 20.36. Because a diagonal of any face passes completely through one sphere and halfway through two other spheres, its length is $4R$. The length of a side is therefore $8^{1/2}R$ and the volume of the unit cell is $8^{3/2}R^3$. Because each cell contains the equivalent of $6\times\frac{1}{2}+8\times\frac{1}{8}=4$ spheres, and the volume of each sphere is $\frac{4}{3}\pi R^3$, the total occupied volume is $\frac{16}{3}\pi R^3$. The fraction of space occupied is therefore $\frac{16}{3}\pi R^3/8^{3/2}R^3=\frac{16}{3}\pi/8^{3/2}$, or 0.740. Because an hcp structure has the same coordination number, its packing fraction is the same. The packing fractions of structures that are not close-packed are calculated similarly (see Exercises 20.14 and 20.17 and Problem 20.24).

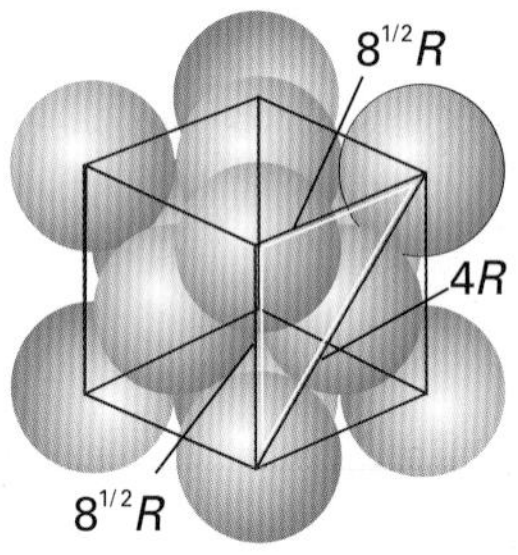

Fig. 20.36 The calculation of the packing fraction of an ccp unit cell.

(b) Less closely packed structures

As shown in Table 20.2, a number of common metals adopt structures that are less than close-packed. The departure from close packing suggests that factors such as specific covalent bonding between neighbouring atoms are beginning to influence the structure and impose a specific geometrical arrangement. One such arrangement results in a cubic I (bcc, for body-centred cubic) structure, with one sphere at the centre of a cube formed by eight others. The coordination number of a bcc structure is

[2] Strictly speaking, ccp refers to a close-packed arrangement whereas fcc refers to the lattice type of the common representation of ccp. However, this distinction is rarely made.

only 8, but there are six more atoms not much further away than the eight nearest neighbours. The packing fraction of 0.68 is not much smaller than the value for a close-packed structure (0.74), and shows that about two-thirds of the available space is actually occupied.

20.6 **Ionic solids**

Two questions arise when we consider ionic solids: the relative locations adopted by the ions and the energetics of the resulting structure.

(a) Structure

When crystals of compounds of monatomic ions (such as NaCl and MgO) are modelled by stacks of hard spheres it is essential to allow for the different ionic radii (typically with the cations smaller than the anions) and different charges. The **coordination number** of an ion is the number of nearest neighbours of opposite charge; the structure itself is characterized as having **(n_+,n_-) coordination**, where n_+ is the coordination number of the cation and n_- that of the anion.

Even if, by chance, the ions have the same size, the problems of ensuring that the unit cells are electrically neutral makes it impossible to achieve 12-coordinate close-packed ionic structures. As a result, ionic solids are generally less dense than metals. The best packing that can be achieved is the (8,8)-coordinate **caesium-chloride structure** in which each cation is surrounded by eight anions and each anion is surrounded by eight cations (Fig. 20.37). In this structure, an ion of one charge occupies the centre of a cubic unit cell with eight counter ions at its corners. The structure is adopted by CsCl itself and also by CaS, CsCN (with some distortion), and CuZn.

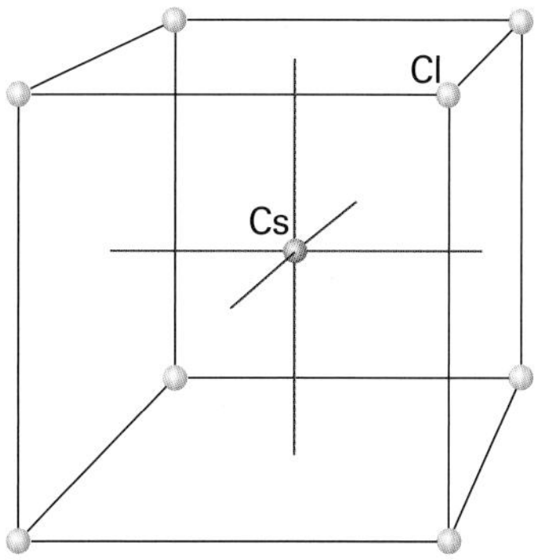

Fig. 20.37 The caesium-chloride structure consists of two interpenetrating simple cubic arrays of ions, one of cations and the other of anions, so that each cube of ions of one kind has a counter-ion at its centre.

When the radii of the ions differ more than in CsCl, even eight-coordinate packing cannot be achieved. One common structure adopted is the (6,6)-coordinate **rock-salt structure** typified by NaCl (Fig. 20.38). In this structure, each cation is surrounded by six anions and each anion is surrounded by six cations. The rock-salt structure can be pictured as consisting of two interpenetrating slightly expanded cubic F (fcc) arrays, one composed of cations and the other of anions. This structure is adopted by NaCl itself and also by several other MX compounds, including KBr, AgCl, MgO, and ScN.

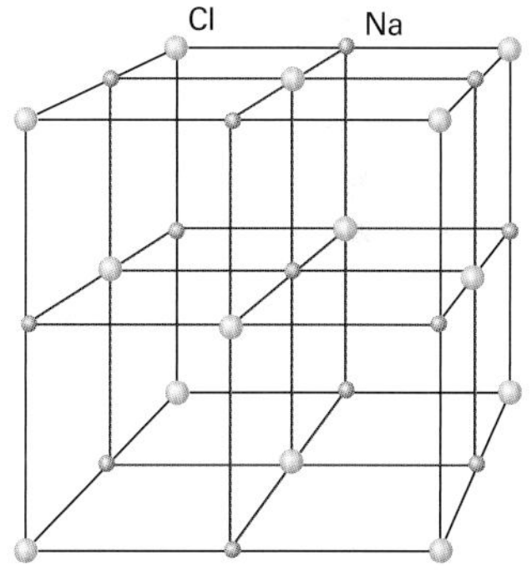

Fig. 20.38 The rock-salt (NaCl) structure consists of two mutually interpenetrating slightly expanded face-centred cubic arrays of ions. The entire assembly shown here is the unit cell.

The switch from the caesium-chloride structure to the rock-salt structure is related to the value of the **radius ratio**, γ:

$$\gamma = \frac{r_{\text{smaller}}}{r_{\text{larger}}} \qquad [20.12]$$

The two radii are those of the larger and smaller ions in the crystal. The **radius-ratio rule** states that the caesium-chloride structure should be expected when $\gamma > 3^{1/2} - 1 = 0.732$ and that the rock-salt structure should be expected when $2^{1/2} - 1 = 0.414 < \gamma < 0.732$. For $\gamma < 0.414$, the most efficient packing leads to four-coordination of the type exhibited by the sphalerite (or zinc blende) form of ZnS (Fig. 20.39). The rule is derived by considering the geometrical problem of packing the maximum number of hard spheres of one radius around a hard sphere of a different radius. The deviation of a structure from that expected on the basis of the radius-ratio rule is often taken to be an indication of a shift from ionic towards covalent bonding; however, a major source of unreliability is the arbitrariness of ionic radii and their variation with coordination number.

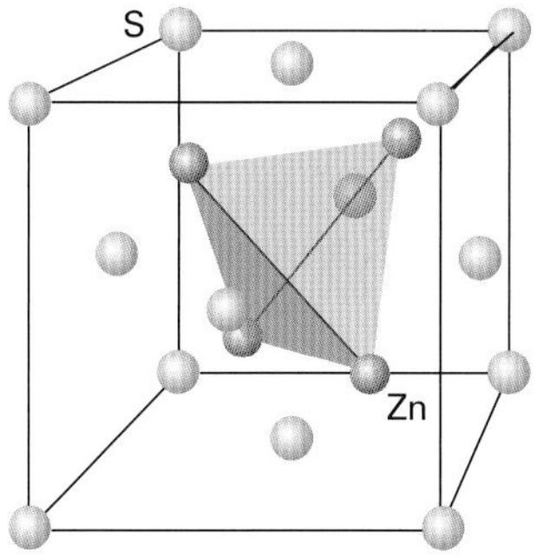

Fig. 20.39 The structure of the sphalerite form of ZnS showing the location of the Zn atoms in the tetrahedral holes formed by the array of S atoms. (There is an S atom at the centre of the cube inside the tetrahedron of Zn atoms.)

Ionic radii are derived from the distance between centres of adjacent ions in a crystal. However, we need to apportion the total distance between the two ions by defining the radius of one ion and then inferring the radius of the other ion. One scale that is widely used is based on the value 140 pm for the radius of the O^{2-} ion (Table 20.3).

Synoptic table 20.3* Ionic radii, r/pm

Na^+	102(6†), 116(8)
K^+	138(6), 151(8)
F^-	128(2), 131(4)
Cl^-	181 (close packing)

* More values are given in the *Data section*.
† Coordination number.

Other scales are also available (such as one based on F^- for discussing halides), and it is essential not to mix values from different scales. Because ionic radii are so arbitrary, predictions based on them must be viewed cautiously.

(b) Energetics

The **lattice energy** of a solid is the difference in potential energy of the ions packed together in a solid and widely separated as a gas. The lattice energy is always positive; a high lattice energy indicates that the ions interact strongly with one another to give a tightly bonded solid. The **lattice enthalpy**, ΔH_L, is the change in standard molar enthalpy for the process

$$MX(s) \rightarrow M^+(g) + X^-(g)$$

and its equivalent for other charge types and stoichiometries. The lattice enthalpy is equal to the lattice energy at $T = 0$; at normal temperatures they differ by only a few kilojoules per mole, and the difference is normally neglected.

Each ion in a solid experiences electrostatic attractions from all the other oppositely charged ions and repulsions from all the other like-charged ions. The total Coulombic potential energy is the sum of all the electrostatic contributions. Each cation is surrounded by anions, and there is a large negative contribution from the attraction of the opposite charges. Beyond those nearest neighbours, there are cations that contribute a positive term to the total potential energy of the central cation. There is also a negative contribution from the anions beyond those cations, a positive contribution from the cations beyond them, and so on to the edge of the solid. These repulsions and attractions become progressively weaker as the distance from the central ion increases, but the net outcome of all these contributions is a lowering of energy.

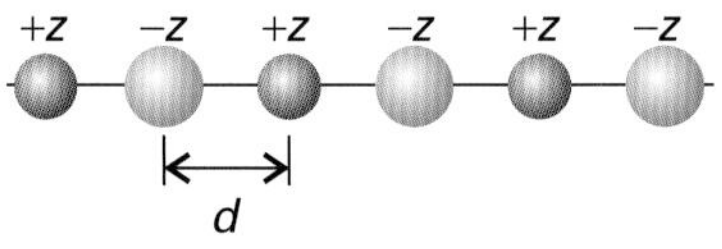

Fig. 20.40 A line of alternating cations and ions used in the calculation of the Madelung constant in one dimension.

First, consider a simple one-dimensional model of a solid consisting of a long line of uniformly spaced alternating cations and anions, with d the distance between their centres, the sum of the ionic radii (Fig. 20.40). If the charge numbers of the ions have the same absolute value (+1 and −1, or +2 and −2, for instance), then $z_1 = +z$, $z_2 = -z$, and $z_1z_2 = -z^2$. The potential energy of the central ion is calculated by summing all the terms, with negative terms representing attractions to oppositely charged ions and positive terms representing repulsions from like-charged ions. For the interaction with ions extending in a line to the right of the central ion, the lattice energy is

$$E_P = \frac{1}{4\pi\varepsilon_0} \times \left(-\frac{z^2e^2}{d} + \frac{z^2e^2}{2d} - \frac{z^2e^2}{3d} + \frac{z^2e^2}{4d} - \cdots \right)$$

$$= -\frac{z^2e^2}{4\pi\varepsilon_0 d}\left(1 - \frac{1}{2} + \frac{1}{3} - \frac{1}{4} + \cdots \right)$$

$$= -\frac{z^2e^2}{4\pi\varepsilon_0 d} \times \ln 2$$

We have used the relation $1 - \frac{1}{2} + \frac{1}{3} - \frac{1}{4} + \cdots = \ln 2$. Finally, we multiply E_P by 2 to obtain the total energy arising from interactions on each side of the ion and then multiply by Avogadro's constant, N_A, to obtain an expression for the lattice energy per mole of ions. The outcome is

$$E_P = -2 \ln 2 \times \frac{z^2 N_A e^2}{4\pi\varepsilon_0 d}$$

with $d = r_{cation} + r_{anion}$. This energy is negative, corresponding to a net attraction. The calculation we have just performed can be extended to three-dimensional arrays of ions with different charges:

$$E_P = -A \times \frac{|z_1 z_2| N_A e^2}{4\pi\varepsilon_0 d} \qquad (20.13)$$

The factor A is a positive numerical constant called the **Madelung constant**; its value depends on how the ions are arranged about one another. For ions arranged in the same way as in sodium chloride, $A = 1.748$. Table 20.4 lists Madelung constants for other common structures.

There are also repulsions arising from the overlap of the atomic orbitals of the ions and the role of the Pauli principle. These repulsions are taken into account by supposing that, because wavefunctions decay exponentially with distance at large distances from the nucleus, and repulsive interactions depend on the overlap of orbitals, the repulsive contribution to the potential energy has the form

$$E_P^{\star} = N_A C' e^{-d/d^{\star}} \qquad (20.14)$$

with C' and $d^{\star}$ constants; the latter is commonly taken to be 34.5 pm. The total potential energy is the sum of E_P and $E_P^{\star}$, and passes through a minimum when $d(E_P + E_P^{\star})/dd = 0$ (Fig. 20.41). A short calculation leads to the following expression for the minimum total potential energy (see Exercise 20.21a):

$$E_{P,\min} = -\frac{N_A |z_A z_B| e^2}{4\pi\varepsilon_0 d}\left(1 - \frac{d^{\star}}{d}\right)A \qquad (20.15)$$

This expression is called the **Born–Mayer equation**. Provided we ignore zero-point contributions to the energy, we can identify the negative of this potential energy with the lattice energy. We see that large lattice energies are expected when the ions are highly charged (so $|z_A z_B|$ is large) and small (so d is small).

Experimental values of the lattice enthalpy (the enthalpy, rather than the energy) are obtained by using a **Born–Haber cycle**, a closed path of transformations starting and ending at the same point, one step of which is the formation of the solid compound from a gas of widely separated ions. A typical cycle, for potassium chloride, is shown in Fig. 20.42. It consists of the following steps (for convenience, starting at the elements):

	$\Delta H/(\text{kJ mol}^{-1})$	
1. Sublimation of K(s)	+89	[dissociation enthalpy of K(s)]
2. Dissociation of $\frac{1}{2}Cl_2(g)$	+122	[$\frac{1}{2}\times$ dissociation enthalpy of $Cl_2(g)$]
3. Ionization of K(g)	+418	[ionization enthalpy of K(g)]
4. Electron attachment to Cl(g)	−349	[electron gain enthalpy of Cl(g)]
5. Formation of solid from gas	$-\Delta H_L/(\text{kJ mol}^{-1})$	
6. Decomposition of compound	+437	[negative of enthalpy of formation of KCl(s)]

Because the sum of these enthalpy changes is equal to zero, we can infer from

$$89 + 122 + 418 - 349 - \Delta H_L/(\text{kJ mol}^{-1}) + 437 = 0$$

that $\Delta H_L = +717 \text{ kJ mol}^{-1}$. Some lattice enthalpies obtained in this way are listed in Table 20.5. As can be seen from the data, the trends in values are in general accord with the predictions of the Born–Mayer equation. Agreement is typically taken to imply that the ionic model of bonding is valid for the substance; disagreement implies that there is a covalent contribution to the bonding.

Table 20.4 Madelung constants

Structural type*	A
Caesium chloride	1.763
Fluorite	2.519
Rock salt	1.748
Rutile	2.408
Sphalerite	1.638
Wurtzite	1.641

* For descriptions of the structural types not covered in this chapter, see references in *Further reading*.

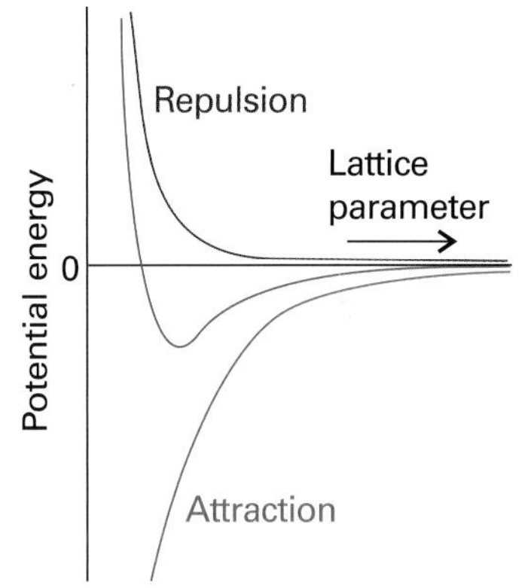

Fig. 20.41 The contributions to the total potential energy of an ionic crystal.

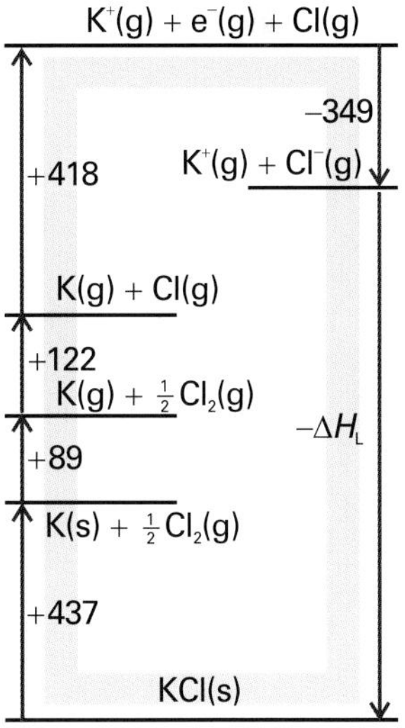

Fig. 20.42 The Born–Haber cycle for KCl at 298 K. Enthalpies changes are in kilojoules per mole.

Synoptic table 20.5* Lattice enthalpies at 298 K, $\Delta H_L/(\text{kJ mol}^{-1})$

NaF	787
NaBr	751
MgO	3850
MgS	3406

* More values are given in the *Data section*.

Comment 20.4

Allotropes are distinct forms of an element that differ in the way that atoms are linked. For example, oxygen has two allotropes: O_2 and O_3 (ozone).

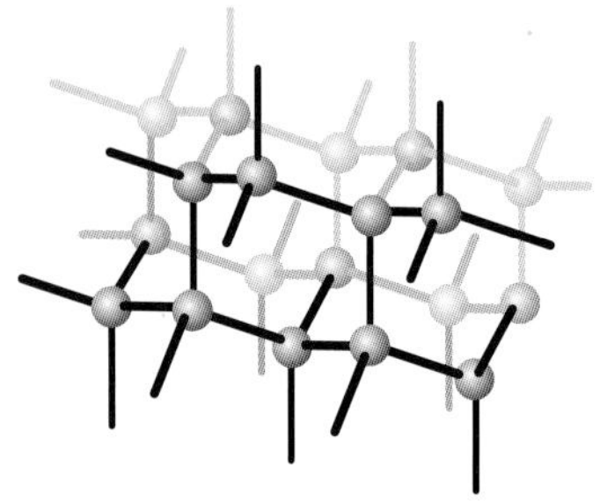

Fig. 20.43 A fragment of the structure of diamond. Each C atom is tetrahedrally bonded to four neighbours. This framework-like structure results in a rigid crystal.

20.7 Molecular solids and covalent networks

X-ray diffraction studies of solids reveal a huge amount of information, including interatomic distances, bond angles, stereochemistry, and vibrational parameters. In this section we can do no more than hint at the diversity of types of solids found when molecules pack together or atoms link together in extended networks.

In **covalent network solids**, covalent bonds in a definite spatial orientation link the atoms in a network extending through the crystal. The demands of directional bonding, which have only a small effect on the structures of many metals, now override the geometrical problem of packing spheres together, and elaborate and extensive structures may be formed. Examples include silicon, red phosphorus, boron nitride, and—very importantly—diamond, graphite, and carbon nanotubes, which we discuss in detail.

Diamond and graphite are two allotropes of carbon. In diamond each sp^3-hybridized carbon is bonded tetrahedrally to its four neighbours (Fig. 20.43). The network of strong C—C bonds is repeated throughout the crystal and, as a result, diamond is the hardest known substance.

In graphite, σ bonds between sp^2-hybridized carbon atoms form hexagonal rings which, when repeated throughout a plane, give rise to sheets (Fig. 20.44). Because the sheets can slide against each other when impurities are present, graphite is used widely as a lubricant.

Carbon nanotubes are thin cylinders of carbon atoms that are both mechanically strong and highly conducting (see *Impact* I20.2). They are synthesized by condensing a carbon plasma either in the presence or absence of a catalyst. The simplest structural motif is called a *single-walled nanotube* (SWNT) and is shown in Fig. 20.45. In a SWNT, sp^2-hybridized carbon atoms form hexagonal rings reminiscent of the structure of the carbon sheets found in graphite. The tubes have diameters between 1 and 2 nm and lengths of several micrometres. The features shown in Fig. 20.45 have been confirmed by direct visualization with scanning tunnelling microscopy (*Impact* I9.1). A *multi-walled nanotube* (MWNT) consists of several concentric SWNTs and its diameter varies between 2 and 25 nm.

Molecular solids, which are the subject of the overwhelming majority of modern structural determinations, are held together by van der Waals interactions (Chapter 18). The observed crystal structure is Nature's solution to the problem of condensing objects of various shapes into an aggregate of minimum energy (actually, for $T > 0$, of

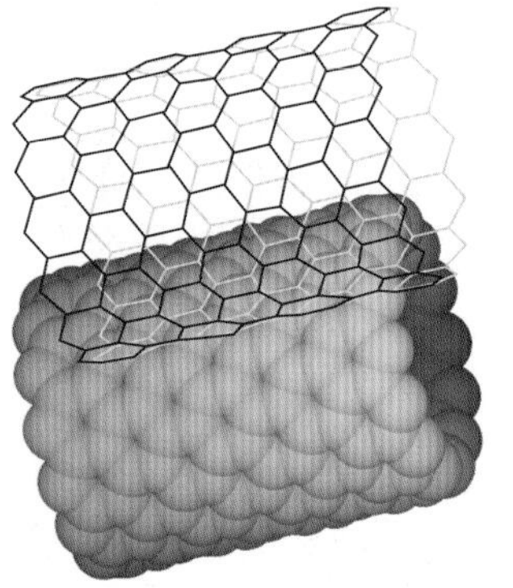

Fig. 20.45 In a single-walled nanotube (SWNT), sp^2-hybridized carbon atoms form hexagonal rings that grow as tubes with diameters between 1 and 2 nm and lengths of several micrometres.

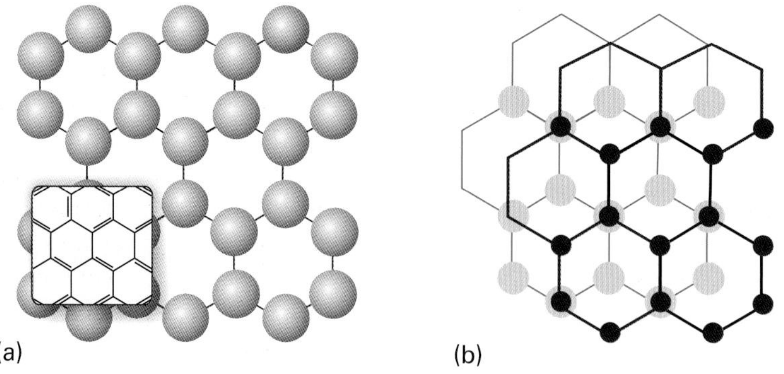

Fig. 20.44 Graphite consists of flat planes of hexagons of carbon atoms lying above one another. (a) The arrangement of carbon atoms in a sheet; (b) the relative arrangement of neighbouring sheets. When impurities are present, the planes can slide over one another easily.

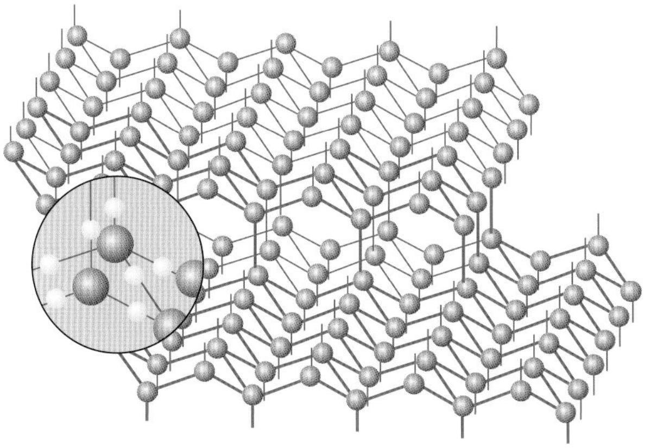

Fig. 20.46 A fragment of the crystal structure of ice (ice-I). Each O atom is at the centre of a tetrahedron of four O atoms at a distance of 276 pm. The central O atom is attached by two short O—H bonds to two H atoms and by two long hydrogen bonds to the H atoms of two of the neighbouring molecules. Overall, the structure consists of planes of hexagonal puckered rings of H_2O molecules (like the chair form of cyclohexane).

minimum Gibbs energy). The prediction of the structure is a very difficult task, but software specifically designed to explore interaction energies can now make reasonably reliable predictions. The problem is made more complicated by the role of hydrogen bonds, which in some cases dominate the crystal structure, as in ice (Fig. 20.46), but in others (for example, in phenol) distort a structure that is determined largely by the van der Waals interactions.

The properties of solids

In this section we consider how the bulk properties of solids, particularly their mechanical, electrical, optical, and magnetic properties, stem from the properties of their constituent atoms. The rational fabrication of modern materials depends crucially on an understanding of this link.

Comment 20.5
The web site contains links to databases of properties of materials, such as metals and polymers.

20.8 Mechanical properties

The fundamental concepts for the discussion of the mechanical properties of solids are stress and strain. The **stress** on an object is the applied force divided by the area to which it is applied. The **strain** is the resulting distortion of the sample. The general field of the relations between stress and strain is called **rheology**.

Stress may be applied in a number of different ways. Thus, **uniaxial stress** is a simple compression or extension in one direction (Fig. 20.47); **hydrostatic stress** is a stress applied simultaneously in all directions, as in a body immersed in a fluid. A **pure shear** is a stress that tends to push opposite faces of the sample in opposite directions. A sample subjected to a small stress typically undergoes **elastic deformation** in the sense that it recovers its original shape when the stress is removed. For low stresses, the strain is linearly proportional to the stress. The response becomes nonlinear at high stresses but may remain elastic. Above a certain threshold, the strain becomes **plastic** in the sense that recovery does not occur when the stress is removed. Plastic deformation occurs when bond breaking takes place and, in pure metals, typically takes place through the agency of dislocations. Brittle solids, such as ionic solids, exhibit sudden fracture as the stress focused by cracks causes them to spread catastrophically.

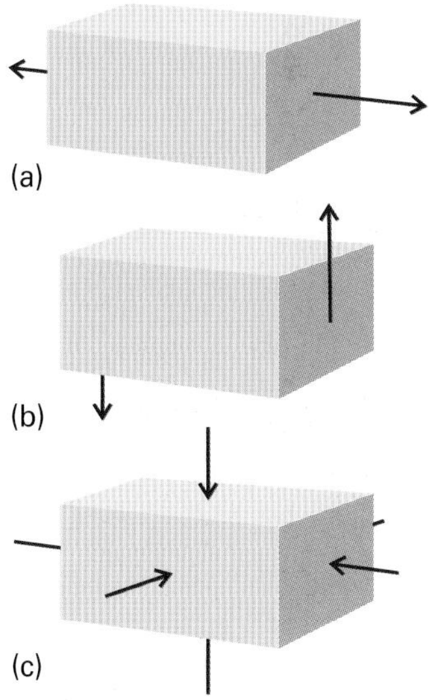

Fig. 20.47 Types of stress applied to a body. (a) Uniaxial stress, (b) shear stress, (c) hydrostatic pressure.

The response of a solid to an applied stress is commonly summarized by a number of coefficients of proportionality known as 'moduli':

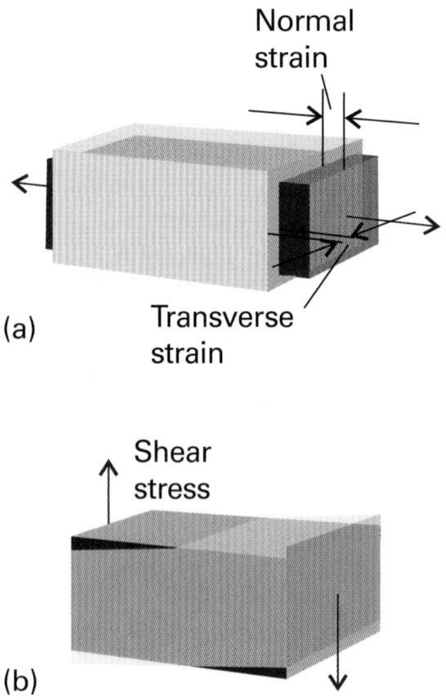

(a) (b)

Fig. 20.48 (a) Normal stress and the resulting strain. (b) Shear stress. Poisson's ratio indicates the extent to which a body changes shape when subjected to a uniaxial stress.

Young's modulus: $$E = \frac{\text{normal stress}}{\text{normal strain}} \quad [20.16a]$$

Bulk modulus: $$K = \frac{\text{pressure}}{\text{fractional change in volume}} \quad [20.16b]$$

Shear modulus: $$G = \frac{\text{shear stress}}{\text{shear strain}} \quad [20.16c]$$

where 'normal stress' refers to stretching and compression of the material, as shown in Fig. 20.48a and 'shear stress' refers to the stress depicted in Fig. 20.48b. The bulk modulus is the inverse of the isothermal compressibility, κ, first encountered in Section 2.11 (eqn 2.44, $\kappa = -(\partial V/\partial p)_T/V$). A third ratio indicates how the sample changes its shape:

Poisson's ratio: $$\nu_P = \frac{\text{transverse strain}}{\text{normal strain}} \quad [20.17]$$

The moduli are interrelated:

$$G = \frac{E}{2(1+\nu_P)} \qquad K = \frac{E}{3(1-2\nu_P)} \quad (20.18)$$

We can use thermodynamic arguments to discover the relation of the moduli to the molecular properties of the solid. Thus, in the *Justification* below, we show that, if neighbouring molecules interact by a Lennard-Jones potential, then the bulk modulus and the compressibility of the solid are related to the Lennard-Jones parameter ε (the depth of the potential well) by

$$K = \frac{8N_A\varepsilon}{V_m} \qquad \kappa = \frac{V_m}{8N_A\varepsilon} \quad (20.19)$$

We see that the bulk modulus is large (the solid stiff) if the potential well represented by the Lennard-Jones potential is deep and the solid is dense (its molar volume small).

Justification 20.4 *The relation between compressibility and molecular interactions*

First, we combine the definition of $K = 1/\kappa$, with the thermodynamic relation $p = -(\partial U/\partial V)_T$ (this is eqn 3.45), to obtain

$$K = V\left(\frac{\partial^2 U}{\partial V^2}\right)_T$$

This expression shows that the bulk modulus (and through eqn 20.18, the other two moduli) depends on the curvature of a plot of the internal energy against volume. To develop this conclusion, we note that the variation of internal energy with volume can be expressed in terms of its variation with a lattice parameter, R, such as the length of the side of a unit cell.

$$\frac{\partial U}{\partial V} = \frac{\partial U}{\partial R}\frac{\partial R}{\partial V}$$

and so

$$\frac{\partial^2 U}{\partial V^2} = \frac{\partial U}{\partial R}\frac{\partial^2 R}{\partial V^2} + \frac{\partial^2 U}{\partial V \partial R}\frac{\partial R}{\partial V} = \frac{\partial U}{\partial R}\frac{\partial^2 R}{\partial V^2} + \frac{\partial^2 U}{\partial R^2}\left(\frac{\partial R}{\partial V}\right)^2$$

To calculate K at the equilibrium volume of the sample, we set $R = R_0$ and recognize that $\partial U/\partial R = 0$ at equilibrium, so

$$K = V\left(\frac{\partial^2 U}{\partial R^2}\right)_{T,0}\left(\frac{\partial R}{\partial V}\right)_{T,0}^2$$

where the 0 denotes that the derivatives are evaluated at the equilibrium dimensions of the unit cell by setting $R = R_0$ after the derivative has been calculated. At this point we can write $V = aR^3$, where a is a constant that depends on the crystal structure, which implies that $\partial R/\partial V = 1/(3aR^2)$. Then, if the internal energy is given by a pairwise Lennard-Jones (12,6)-potential, eqn 18.31, we can write

$$\left(\frac{\partial^2 U}{\partial R^2}\right)_{T,0} = \frac{72nN_A\varepsilon}{R_0^2} \qquad (20.20)$$

where n is the amount of substance in the sample of volume V_0. It then follows that

$$K = \frac{72nN_A\varepsilon}{9aR^3} = \frac{8nN_A\varepsilon}{V_0} = \frac{8N_A\varepsilon}{V_m}$$

where we have used $V_m = V_0/n$, which is the first of eqn 20.19. Its reciprocal is κ.

Comment 20.6

To obtain the result in eqn 20.20, we have used the fact that, at equilibrium, $R = R_0$ and $\sigma^6/R_0^6 = \frac{1}{2}$, where σ is the scale parameter for the intermolecular potential (r_0 in eqn 18.31).

The typical behaviour of a solid under stress is illustrated in Fig. 20.49. For small strains, the stress–strain relation is a Hooke's law of force, with the strain directly proportional to the stress. For larger strains, though, dislocations begin to play a major role and the strain becomes plastic in the sense that the sample does not recover its original shape when the stress is removed.

The differing rheological characteristics of metals can be traced to the presence of **slip planes**, which are planes of atoms that under stress may slip or slide relative to one another. The slip planes of a ccp structure are the close-packed planes, and careful inspection of a unit cell shows that there are eight sets of slip planes in different directions. As a result, metals with cubic close-packed structures, like copper, are malleable: they can easily be bent, flattened, or pounded into shape. In contrast, a hexagonal close-packed structure has only one set of slip planes; and metals with hexagonal close packing, like zinc or cadmium, tend to be brittle.

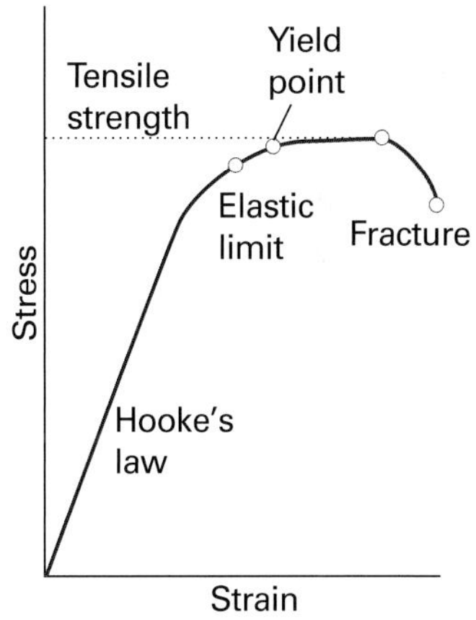

Fig. 20.49 At small strains, a body obeys Hooke's law (stress proportional to strain) and is elastic (recovers its shape when the stress is removed). At high strains, the body is no longer elastic, may yield and become plastic. At even higher strains, the solid fails (at its limiting tensile strength) and finally fractures.

20.9 Electrical properties

We shall confine attention to electronic conductivity, but note that some ionic solids display ionic conductivity. Two types of solid are distinguished by the temperature dependence of their electrical conductivity (Fig. 20.50):

A **metallic conductor** is a substance with a conductivity that decreases as the temperature is raised.

A **semiconductor** is a substance with a conductivity that increases as the temperature is raised.

A semiconductor generally has a lower conductivity than that typical of metals, but the magnitude of the conductivity is not the criterion of the distinction. It is conventional to classify semiconductors with very low electrical conductivities, such as most synthetic polymers, as **insulators**. We shall use this term, but it should be appreciated that it is one of convenience rather than one of fundamental significance. A **superconductor** is a solid that conducts electricity without resistance.

(a) The formation of bands

The central aspect of solids that determines their electrical properties is the distribution of their electrons. There are two models of this distribution. In one, the **nearly**

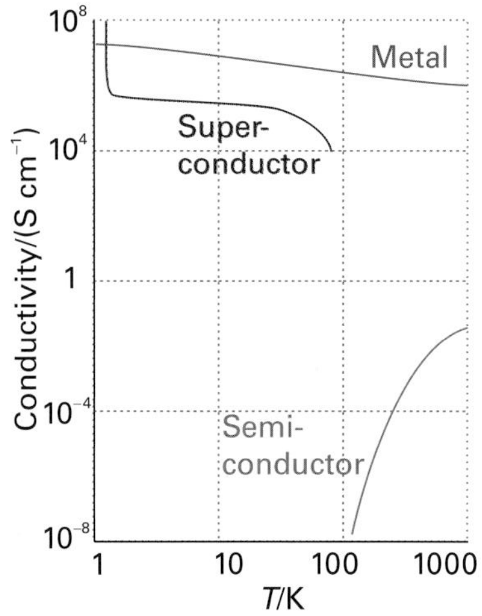

Fig. 20.50 The variation of the electrical conductivity of a substance with temperature is the basis of its classification as a metallic conductor, a semiconductor, or a superconductor. We shall see in Chapter 21 that conductivity is expressed in siemens per metre (S m^{-1} or, as here, S cm^{-1}), where 1 S = 1 Ω^{-1} (the resistance is expressed in ohms, Ω).

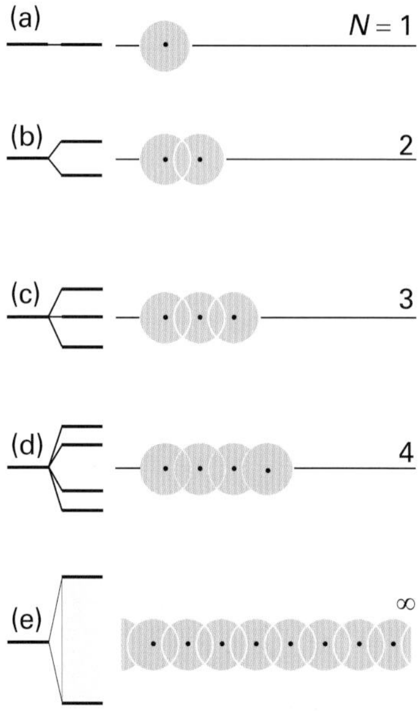

Fig. 20.51 The formation of a band of N molecular orbitals by successive addition of N atoms to a line. Note that the band remains of finite width as N becomes infinite and, although it looks continuous, it consists of N different orbitals.

free-electron approximation, the valence electrons are assumed to be trapped in a box with a periodic potential, with low energy corresponding to the locations of cations. In the **tight-binding approximation**, the valence electrons are assumed to occupy molecular orbitals delocalized throughout the solid. The latter model is more in accord with the discussion in the foregoing chapters, and we confine our attention to it.

We shall consider a one-dimensional solid, which consists of a single, infinitely long line of atoms. At first sight, this model may seem too restrictive and unrealistic. However, not only does it give us the concepts we need to understand conductivity in three-dimensional, macroscopic samples of metals and semiconductors, it is also the starting point for the description of long and thin structures, such as the carbon nanotubes discussed earlier in the chapter.

Suppose that each atom has one *s* orbital available for forming molecular orbitals. We can construct the LCAO-MOs of the solid by adding N atoms in succession to a line, and then infer the electronic structure using the building-up principle. One atom contributes one *s* orbital at a certain energy (Fig. 20.51). When a second atom is brought up it overlaps the first and forms bonding and antibonding orbitals. The third atom overlaps its nearest neighbour (and only slightly the next-nearest), and from these three atomic orbitals, three molecular orbitals are formed: one is fully bonding, one fully antibonding, and the intermediate orbital is nonbonding between neighbours. The fourth atom leads to the formation of a fourth molecular orbital. At this stage, we can begin to see that the general effect of bringing up successive atoms is to spread the range of energies covered by the molecular orbitals, and also to fill in the range of energies with more and more orbitals (one more for each atom). When N atoms have been added to the line, there are N molecular orbitals covering a band of energies of finite width, and the Hückel secular determinant (Section 11.6) is

$$\begin{vmatrix} \alpha-E & \beta & 0 & 0 & 0 & \cdots & 0 \\ \beta & \alpha-E & \beta & 0 & 0 & \cdots & 0 \\ 0 & \beta & \alpha-E & \beta & 0 & \cdots & 0 \\ 0 & 0 & \beta & \alpha-E & \beta & \cdots & 0 \\ 0 & 0 & 0 & \beta & \alpha-E & \cdots & 0 \\ \vdots & \vdots & \vdots & \vdots & \vdots & \cdots & \vdots \\ 0 & 0 & 0 & 0 & 0 & \cdots & \alpha-E \end{vmatrix} = 0$$

where β is now the (*s,s*) resonance integral. The theory of determinants applied to such a symmetrical example as this (technically a 'tridiagonal determinant') leads to the following expression for the roots:

$$E_R = \alpha + 2\beta\cos\frac{k\pi}{N+1} \qquad k = 1, 2, \ldots, N \tag{20.21}$$

When N is infinitely large, the difference between neighbouring energy levels (the energies corresponding to k and $k+1$) is infinitely small, but, as we show in the following *Justification*, the band still has finite width overall:

$$E_N - E_1 \rightarrow 4\beta \quad \text{as} \quad N \rightarrow \infty \tag{20.22}$$

We can think of this band as consisting of N different molecular orbitals, the lowest-energy orbital ($k = 1$) being fully bonding, and the highest-energy orbital ($k = N$) being fully antibonding between adjacent atoms (Fig. 20.52). Similar bands form in three-dimensional solids.

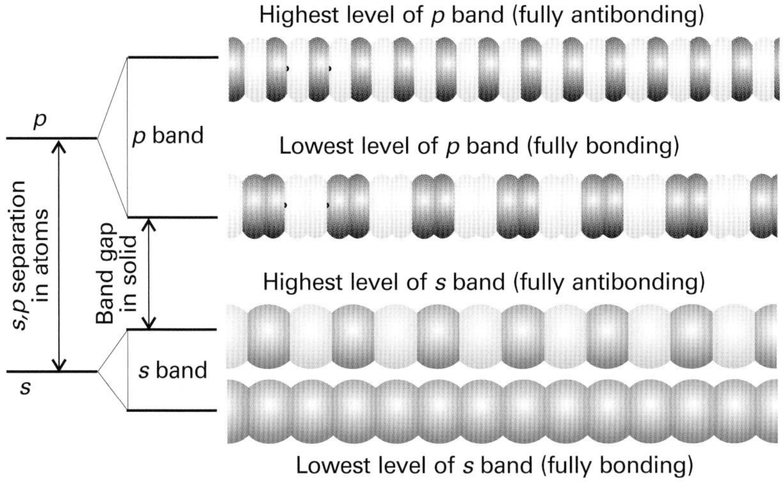

Fig. 20.52 The overlap of *s* orbitals gives rise to an *s* band and the overlap of *p* orbitals gives rise to a *p* band. In this case, the *s* and *p* orbitals of the atoms are so widely spaced that there is a band gap. In many cases the separation is less and the bands overlap.

Justification 20.5 *The width of a band*

The energy of the level with $k = 1$ is

$$E_1 = \alpha + 2\beta \cos\frac{\pi}{N+1}$$

As N becomes infinite, the cosine term becomes $\cos 0 = 1$. Therefore, in this limit

$$E_1 = \alpha + 2\beta$$

When k has its maximum value of N,

$$E_N = \alpha + 2\beta \cos\frac{N\pi}{N+1}$$

As N approaches infinity, we can ignore the 1 in the denominator, and the cosine term becomes $\cos \pi = -1$. Therefore, in this limit

$$E_N = \alpha - 2\beta$$

The difference between the upper and lower energies of the band is therefore 4β.

The band formed from overlap of *s* orbitals is called the ***s* band**. If the atoms have *p* orbitals available, the same procedure leads to a ***p* band** (as shown in the upper half of Fig. 20.52). If the atomic *p* orbitals lie higher in energy than the *s* orbitals, then the *p* band lies higher than the *s* band, and there may be a **band gap**, a range of energies to which no orbital corresponds. However, the *s* and *p* bands may also be contiguous or even overlap (as is the case for the 3*s* and 3*p* bands in magnesium).

(b) The occupation of orbitals

Now consider the electronic structure of a solid formed from atoms each able to contribute one electron (for example, the alkali metals). There are N atomic orbitals and therefore N molecular orbitals packed into an apparently continuous band. There are N electrons to accommodate.

At $T = 0$, only the lowest $\frac{1}{2}N$ molecular orbitals are occupied (Fig. 20.53), and the HOMO is called the **Fermi level**. However, unlike in molecules, there are empty orbitals very close in energy to the Fermi level, so it requires hardly any energy to excite the uppermost electrons. Some of the electrons are therefore very mobile and give rise to electrical conductivity.

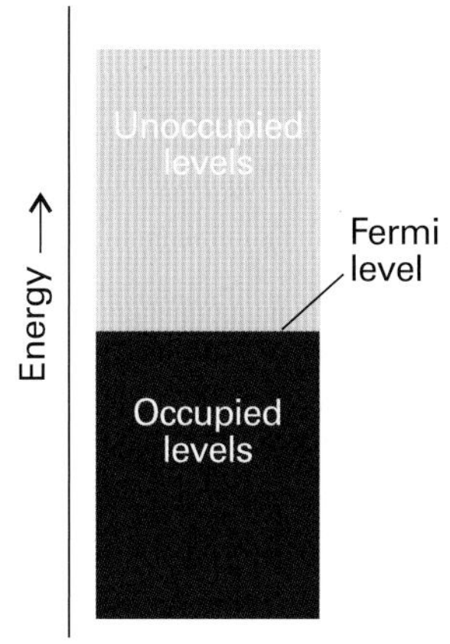

Fig. 20.53 When N electrons occupy a band of N orbitals, it is only half full and the electrons near the Fermi level (the top of the filled levels) are mobile.

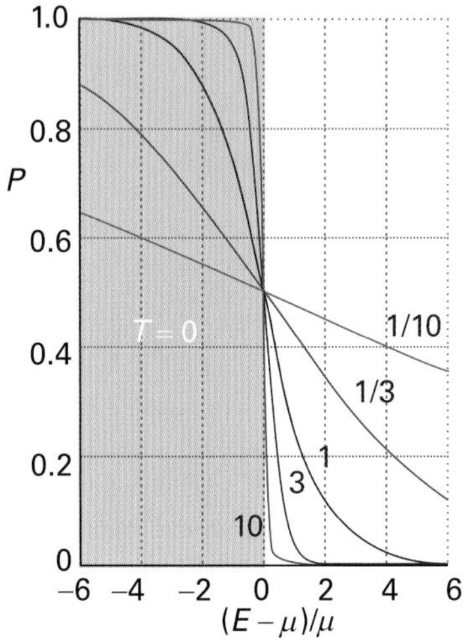

Fig. 20.54 The Fermi–Dirac distribution, which gives the population of the levels at a temperature T. The high-energy tail decays exponentially towards zero. The curves are labelled with the value of μ/kT. The pale green region shows the occupation of levels at $T = 0$.

Exploration Express the population P as a function of the variables $(E - \mu)/\mu$ and μ/kT and then display the set of curves shown in Fig. 20.54 as a single surface.

At temperatures above absolute zero, electrons can be excited by the thermal motion of the atoms. The population, P, of the orbitals is given by the **Fermi–Dirac distribution**, a version of the Boltzmann distribution that takes into account the effect of the Pauli principle:

$$P = \frac{1}{e^{(E-\mu)/kT} + 1} \tag{20.23}$$

The quantity μ is the **chemical potential**, which in this context is the energy of the level for which $P = \frac{1}{2}$ (note that the chemical potential decreases as the temperature increases). The chemical potential in eqn 20.23 has the dimensions of energy, not energy per mole.

The shape of the Fermi–Dirac distribution is shown in Fig. 20.54. For energies well above μ, the 1 in the denominator can be neglected, and then

$$P \approx e^{-(E-\mu)/kT} \tag{20.24}$$

The population now resembles a Boltzmann distribution, decaying exponentially with increasing energy. The higher the temperature, the longer the exponential tail.

The electrical conductivity of a metallic solid decreases with increasing temperature even though more electrons are excited into empty orbitals. This apparent paradox is resolved by noting that the increase in temperature causes more vigorous thermal motion of the atoms, so collisions between the moving electrons and an atom are more likely. That is, the electrons are scattered out of their paths through the solid, and are less efficient at transporting charge.

(c) Insulators and semiconductors

When each atom provides two electrons, the $2N$ electrons fill the N orbitals of the *s* band. The Fermi level now lies at the top of the band (at $T = 0$), and there is a gap before the next band begins (Fig. 20.55). As the temperature is increased, the tail of the Fermi–Dirac distribution extends across the gap, and electrons leave the lower band, which is called the **valence band**, and populate the empty orbitals of the upper band, which is called the **conduction band**. As a consequence of electron promotion, positively charged 'holes' are left in in the valence band. The holes and promoted electrons are now mobile, and the solid is an electrical conductor. In fact, it is a semiconductor, because the electrical conductivity depends on the number of electrons that are promoted across the gap, and that number increases as the temperature is raised. If the gap is large, though, very few electrons will be promoted at ordinary temperatures and the conductivity will remain close to zero, resulting in an insulator. Thus, the conventional distinction between an insulator and a semiconductor is related to the size of the band gap and is not an absolute distinction like that between a metal (incomplete bands at $T = 0$) and a semiconductor (full bands at $T = 0$).

Figure 20.55 depicts conduction in an **intrinsic semiconductor**, in which semiconduction is a property of the band structure of the pure material. Examples of intrinsic semiconductors include silicon and germanium. A **compound semiconductor** is an intrinsic semiconductor that is a combination of different elements, such as GaN, CdS, and many *d*-metal oxides. An **extrinsic semiconductor** is one in which charge carriers are present as a result of the replacement of some atoms (to the extent of about 1 in 10^9) by **dopant** atoms, the atoms of another element. If the dopants can trap electrons, they withdraw electrons from the filled band, leaving holes which allow the remaining electrons to move (Fig. 20.56a). This procedure gives rise to **p-type semiconductivity**, the p indicating that the holes are positive relative to the electrons in the band. An example is silicon doped with indium. We can picture the

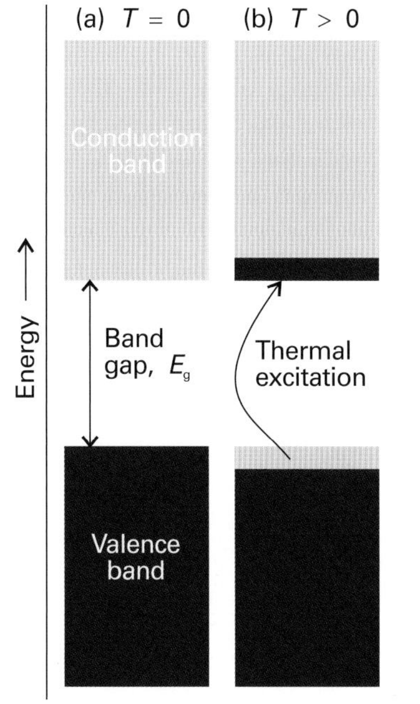

Fig. 20.55 (a) When 2N electrons are present, the band is full and the material is an insulator at $T = 0$. (b) At temperatures above $T = 0$, electrons populate the levels of the upper *conduction band* and the solid is a semiconductor.

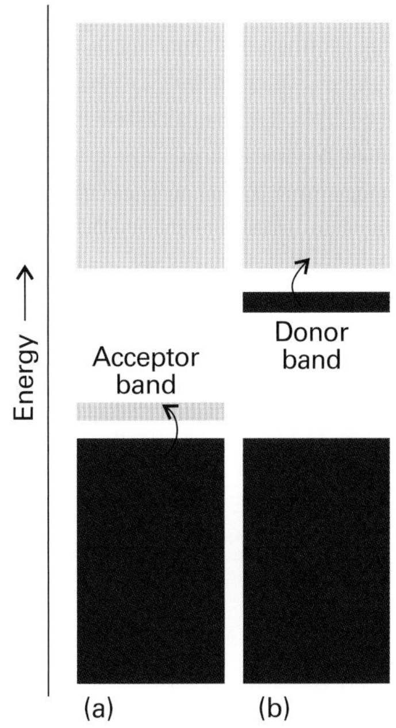

Fig. 20.56 (a) A dopant with fewer electrons than its host can form a narrow band that accepts electrons from the valence band. The holes in the band are mobile and the substance is a p-type semiconductor. (b) A dopant with more electrons than its host forms a narrow band that can supply electrons to the conduction band. The electrons it supplies are mobile and the substance is an n-type semiconductor.

semiconduction as arising from the transfer of an electron from a Si atom to a neighbouring In atom. The electrons at the top of the silicon valence band are now mobile, and carry current through the solid. Alternatively, a dopant might carry excess electrons (for example, phosphorus atoms introduced into germanium), and these additional electrons occupy otherwise empty bands, giving **n-type semiconductivity**, where n denotes the negative charge of the carriers (Fig. 20.56b). The preparation of doped but otherwise ultrapure materials was described in *Impact* I6.2.

Now we consider the properties of a **p–n junction**, the interface of a p-type and n-type semiconductor. Consider the application of a 'reverse bias' to the junction, in the sense that a negative electrode is attached to the p-type semiconductor and a positive electrode is attached to the n-type semiconductor (Fig. 20.57a). Under these conditions, the positively charged holes in p-type semicondutor are attracted to the negative electrode and the negatively charged electrons in the n-type semiconductor are attracted to the positive electrode. As a consequence, charge does not flow across the junction. Now consider the application of a 'forward bias' to the junction, in the sense that the positive electrode is attached to the p-type semiconductor and the negative electrode is attached to the n-type semiconductor (Fig. 20.57b). Now charge flows across the junction, with electrons in the n-type semiconductor moving toward the positive electrode and holes moving in the opposite direction. It follows that a p–n junction affords a great deal of control over the magnitude and direction of current

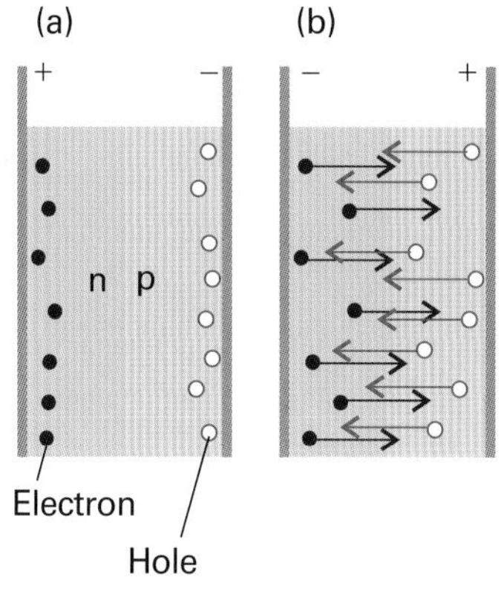

Fig. 20.57 A p–n junction under (a) reverse bias, (b) forward bias.

through a material. This control is essential for the operation of transistors and diodes, which are key components of modern electronic devices.

As electrons and holes move across a p–n junction under forward bias, they recombine and release energy. However, as long as the forward bias continues to be applied, the flow of charge from the electrodes to the semiconductors will replenish them with electrons and holes, so the junction will sustain a current. In some solids, the energy of electron–hole recombination is released as heat and the device becomes warm. This is the case for silicon semiconductors, and is one reason why computers need efficient cooling systems.

IMPACT ON NANOSCIENCE

I20.2 Nanowires

We have already remarked (*Impacts* I9.1, I9.2, and I19.3) that research on nanometre-sized materials is motivated by the possibility that they will form the basis for cheaper and smaller electronic devices. The synthesis of *nanowires*, nanometre-sized atomic assemblies that conduct electricity, is a major step in the fabrication of nanodevices. An important type of nanowire is based on carbon nanotubes, which, like graphite, can conduct electrons through delocalized π molecular orbitals that form from unhybridized $2p$ orbitals on carbon. Recent studies have shown a correlation between structure and conductivity in single-walled nanotubes (SWNTs) that does not occur in graphite. The SWNT in Fig. 20.45 is a semiconductor. If the hexagons are rotated by 90° about their sixfold axis, the resulting SWNT is a metallic conductor.

Carbon nanotubes are promising building blocks not only because they have useful electrical properties but also because they have unusual mechanical properties. For example, an SWNT has a Young's modulus that is approximately five times larger and a tensile strength that is approximately 375 times larger than that of steel.

Silicon nanowires can be made by focusing a pulsed laser beam on to a solid target composed of silicon and iron. The laser ejects Fe and Si atoms from the surface of the target, forming a vapour that can condense into liquid $FeSi_n$ nanoclusters at sufficiently low temperatures. The phase diagram for this complex mixture shows that solid silicon and liquid $FeSi_n$ coexist at temperatures higher than 1473 K. Hence, it is possible to precipitate solid silicon from the mixture if the experimental conditions are controlled to maintain the $FeSi_n$ nanoclusters in a liquid state that is supersaturated with silicon. It is observed that the silicon precipitate consists of nanowires with diameters of about 10 nm and lengths greater than 1 μm.

Nanowires are also fabricated by *molecular beam epitaxy* (MBE), in which gaseous atoms or molecules are sprayed onto a crystalline surface in an ultra-high vacuum chamber. The result is formation of highly ordered structures. Through careful control of the chamber temperature and of the spraying process, it is possible to deposit thin films on to a surface or to create nanometre-sized assemblies with specific shapes. For example, Fig. 20.58 shows an AFM image of germanium nanowires on a silicon surface. The wires are about 2 nm high, 10–32 nm wide, and 10–600 nm long.

Direct manipulation of atoms on a surface also leads to the formation of nanowires. The Coulomb attraction between an atom and the tip of an STM can be exploited to move atoms along a surface, arranging them into patterns, such as wires.

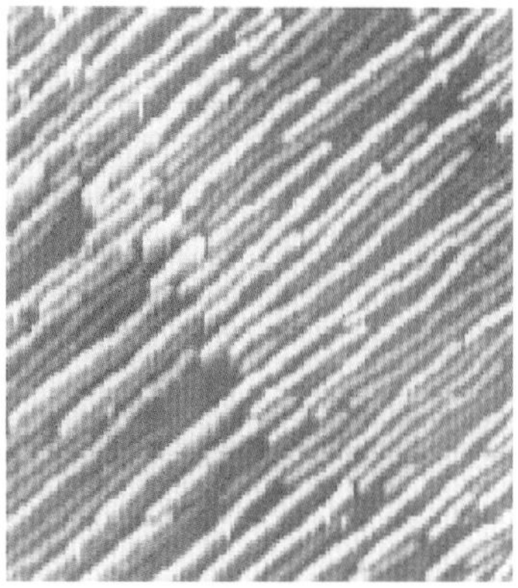

Fig. 20.58 Germanium nanowires fabricated on to a silicon surface by molecular beam epitaxy. (Reproduced with permission from T. Ogino *et al. Acc. Chem. Res.* **32**, 447 (1999).)

20.10 **Optical properties**

In this section, we explore the consequences of interactions between electromagnetic radiation and solids. Our focus will be on the origins of phenomena that inform the design of useful devices, such as lasers and light-emitting diodes.

(a) Light absorption by molecular solids, metallic conductors, and semiconductors

From the discussion in earlier chapters, we are already familiar with the factors that determine the energy and intensity of light absorbed by atoms and molecules in the gas phase and in solution. Now we consider the effects on the electronic absorption spectrum of bringing atoms or molecules together into a solid.

Consider an electronic excitation of a molecule (or an ion) in a crystal. If the excitation corresponds to the removal of an electron from one orbital of a molecule and its elevation to an orbital of higher energy, then the excited state of the molecule can be envisaged as the coexistence of an electron and a hole. This electron–hole pair, the particle-like **exciton**, migrates from molecule to molecule in the crystal (Fig. 20.59). Exciton formation causes spectral lines to shift, split, and change intensity.

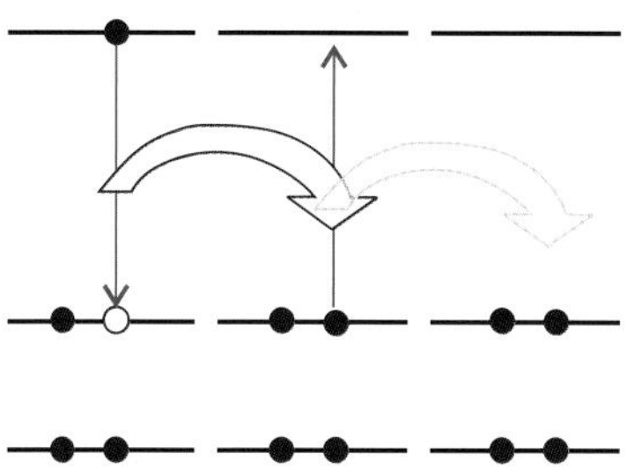

Fig. 20.59 The electron–hole pair shown on the left can migrate through a solid lattice as the excitation hops from molecule to molecule. The mobile excitation is called an exciton.

The electron and the hole jump together from molecule to molecule as they migrate. A migrating excitation of this kind is called a **Frenkel exciton**. The electron and hole can also be on different molecules, but in each other's vicinity. A migrating excitation of this kind, which is now spread over several molecules (more usually ions), is a **Wannier exciton**.

Frenkel excitons are more common in molecular solids. Their migration implies that there is an interaction between the species that constitute the crystal, for otherwise the excitation on one unit could not move to another. This interaction affects the energy levels of the system. The strength of the interaction governs the rate at which an exciton moves through the crystal: a strong interaction results in fast migration, and a vanishingly small interaction leaves the exciton localized on its original molecule. The specific mechanism of interaction that leads to exciton migration is the interaction between the transition dipole moments of the excitation. Thus, an electric dipole transition in a molecule is accompanied by a shift of charge, and the transient dipole exerts a force on an adjacent molecule. The latter responds by shifting its charge. This process continues and the excitation migrates through the crystal.

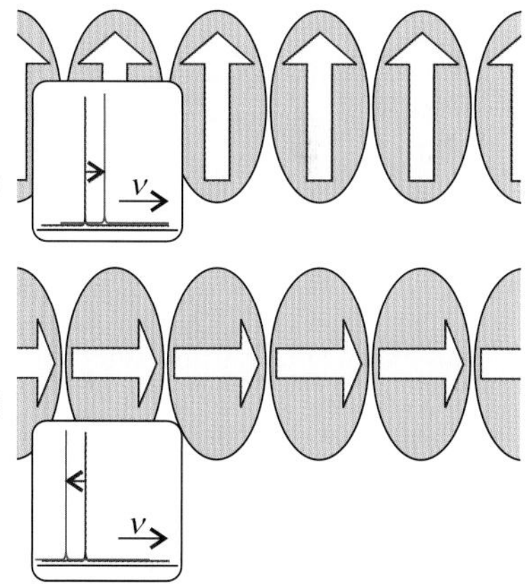

Fig. 20.60 (a) The alignment of transition dipoles (the yellow arrows) is energetically unfavourable, and the exciton absorption is shifted to higher energy (higher frequency). (b) The alignment is energetically favourable for a transition in this orientation, and the exciton band occurs at lower frequency than in the isolated molecules.

The energy shift arising from the interaction between transition dipoles can be understood in terms of their electrostatic interaction. An all-parallel arrangement of the dipoles (Fig. 20.60a) is energetically unfavourable, so the absorption occurs at a higher frequency than in the isolated molecule. Conversely, a head-to-tail alignment of transient dipoles (Fig. 20.60b) is energetically favourable, and the transition occurs at a lower frequency than in the isolated molecules.

Illustration 20.1 *Predicting the frequency of exciton absorption in a molecular solid*

Recall from Section 18.4 that the potential energy of interaction between two parallel dipoles μ_1 and μ_2 separated by a distance r is $V = \mu_1\mu_2(1 - 3\cos^2\theta)/4\pi\varepsilon_0 r^3$, where the angle θ is defined in (**1**). We see that $\theta = 0°$ for a head-to-tail alignment and $\theta = 90°$ for a parallel alignment. It follows that $V < 0$ (an attractive interaction) for $0° \le \theta < 54.74°$, $V = 0$ when $\theta = 54.74°$ (for then $1 - 3\cos^2\theta = 0$), and $V > 0$ (a repulsive interaction) for $54.74° < \theta \le 90°$. This result is expected on the basis of qualitative arguments. In a head-to-tail arrangement, the interaction between the region of partial positive charge in one molecule with the region of partial negative charge in the other molecule is attractive. By contrast, in a parallel arrangement, the molecular interaction is repulsive because of the close approach of regions of partial charge with the same sign.

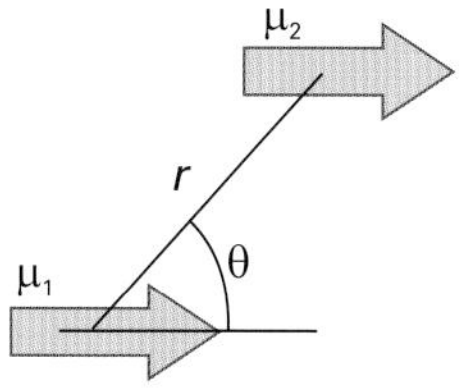

It follows from this discussion that, when $0° \le \theta < 54.74°$, the frequency of exciton absorption is lower than the corresponding absorption frequency for the isolated molecule (a *red shift* in the spectrum of the solid with respect to that of the isolated molecule). Conversely, when $54.74° < \theta \leq 90°$, the frequency of exciton absorption is higher than the corresponding absorption frequency for the isolated molecule (a *blue shift* in the spectrum of the solid with respect to that of the isolated molecule). In the special case $\theta = 54.74°$ the solid and the isolated molecule have absorption lines at the same frequency.

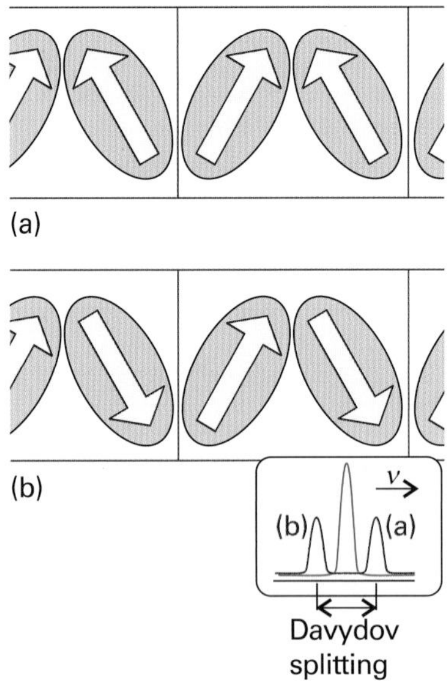

Fig. 20.61 When the transition moments within a unit cell may lie in different relative directions, as depicted in (a) and (b), the energies of the transitions are shifted and give rise to the two bands labelled (a) and (b) in the spectrum. The separation of the bands is the Davydov splitting.

If there are N molecules per unit cell, there are N **exciton bands** in the spectrum (if all of them are allowed). The splitting between the bands is the **Davydov splitting**. To understand the origin of the splitting, consider the case $N = 2$ with the molecules arranged as in Fig. 20.61. Let the transition dipoles be along the length of the molecules. The radiation stimulates the collective excitation of the transition dipoles that are in-phase between neighbouring unit cells. *Within* each unit cell the transition dipoles may be arrayed in the two different ways shown in the illustration. Since the two orientations correspond to different interaction energies, with interaction being repulsive in one and attractive in the other, the two transitions appear in the spectrum at two bands of different frequencies. The Davydov splitting is determined by the energy of interaction between the transition dipoles within the unit cell.

Now we turn our attention to metallic conductors and semiconductors. Again we need to consider the consequences of interactions between particles, in this case atoms, which are now so strong that we need to abandon arguments based primarily on van der Waals interactions in favour of a full molecular orbital treatment, the band model of Section 20.9.

Consider Fig. 20.53, which shows bands in an idealized metallic conductor. The absorption of light can excite electrons from the occupied levels to the unoccupied levels. There is a near continuum of unoccupied energy levels above the Fermi level, so we expect to observe absorption over a wide range of frequencies. In metals, the bands are sufficiently wide that radiation from the radiofrequency to the middle of the ultraviolet region of the electromagnetic spectrum is absorbed (metals are transparent to very high-frequency radiation, such as X-rays and γ-rays). Because this range of absorbed frequencies includes the entire visible spectrum, we expect that all metals should appear black. However, we know that metals are shiny (that is, they reflect light) and some are coloured (that is, they absorb light of only certain wavelengths), so we need to extend our model.

To explain the shiny appearance of a smooth metal surface, we need to realize that the absorbed energy can be re-emitted very efficiently as light, with only a small fraction of the energy being released to the surroundings as heat. Because the atoms near the surface of the material absorb most of the radiation, emission also occurs primarily from the surface. In essence, if the sample is excited with visible light, then visible light will be reflected from the surface, accounting for the lustre of the material.

The perceived colour of a metal depends on the frequency range of reflected light which, in turn, depends on the frequency range of light that can be absorbed and, by extension, on the band structure. Silver reflects light with nearly equal efficiency across the visible spectrum because its band structure has many unoccupied energy levels that can be populated by absorption of, and depopulated by emission of, visible light. On the other hand, copper has its characteristic colour because it has relatively fewer unoccupied energy levels that can be excited with violet, blue, and green light. The material reflects at all wavelengths, but more light is emitted at lower frequencies (corresponding to yellow, orange, and red) Similar arguments account for the colours of other metals, such as the yellow of gold.

Finally, consider semiconductors. We have already seen that promotion of electrons from the valence to the conduction band of a semiconductor can be the result of thermal excitation, if the band gap E_g is comparable to the energy that can be supplied by heating. In some materials, the band gap is very large and electron promotion can occur only by excitation with electromagnetic radiation. However, we see from Fig. 20.55 that there is a frequency $\nu_{min} = E_g/h$ below which light absorption cannot occur. Above this frequency threshold, a wide range of frequencies can be absorbed by the material, as in a metal.

Illustration 20.2 *Predicting the colour of a semiconductor*

The semiconductor cadmium sulfide (CdS) has a band gap energy of 2.4 eV (equivalent to 3.8×10^{-19} J). It follows that the minimum electronic absorption frequency is

$$\nu_{min} = \frac{3.8 \times 10^{-19}\,\text{J}}{6.626 \times 10^{-34}\,\text{J}} = 5.8 \times 10^{14}\,\text{s}^{-1}$$

This frequency, of 5.8×10^{14} Hz, corresponds to a wavelength of 517 nm (green light; see Table 14.1). Lower frequencies, corresponding to yellow, orange, and red, are not absorbed and consequently CdS appears yellow-orange.

Self-test 20.7 Predict the colours of the following materials, given their band-gap energies (in parentheses): GaAs (1.43 eV), HgS (2.1 eV), and ZnS (3.6 eV).

[Black, red, and colourless]

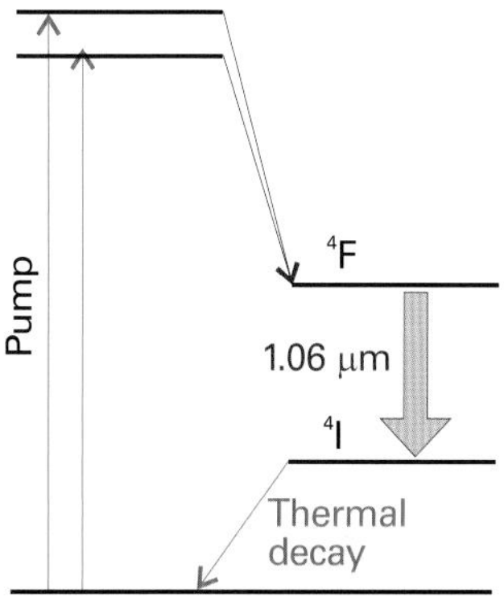

Fig. 20.62 The transitions involved in a neodymium laser. The laser action takes place between the ^{4}F and ^{4}I excited states.

(b) Light emission by solid-state lasers and light-emitting diodes

Here we explore the further consequences of light emission in solids, focusing our attention on ionic crystals and semiconductors used in the design of lasers and light-emitting diodes. In Chapter 14 we discussed the conditions under which a material can become a laser and it would be helpful to review those concepts.

The **neodymium laser** is an example of a four-level laser, in which the laser transition terminates in a state other than the ground state of the laser material (Fig. 20.62). In one form it consists of Nd^{3+} ions at low concentration in yttrium aluminium garnet (YAG, specifically $Y_3Al_5O_{12}$), and is then known as a **Nd–YAG laser**. The population inversion results from pumping a majority of the Nd^{3+} ions into an excited state by using an intense flash from another source, followed by a radiationless transition to another excited state. The pumping flash need not be monochromatic because the upper level actually consists of several states spanning a band of frequencies. A neodymium laser operates at a number of wavelengths in the infrared, the band at 1064 nm being most common. The transition at 1064 nm is very efficient and the laser is capable of substantial power output, either in continuous or pulsed (by Q-switching or mode-locking as discussed in Section 14.5) modes of operation.

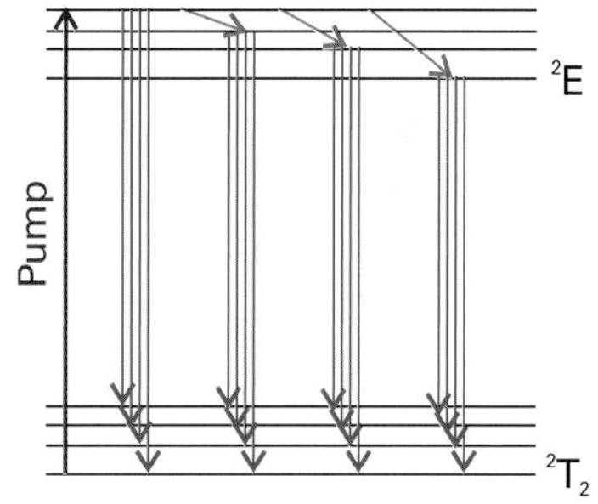

Fig. 20.63 The transitions involved in a titanium sapphire laser. The laser medium consists of sapphire (Al_2O_3) doped with Ti^{3+} ions. Monochromatic light from a pump laser induces a $^2E \leftarrow {}^2T_2$ transition in a Ti^{3+} ion that resides in a site with octahedral symmetry. After radiationless vibrational excitation in the ^{2}E state, laser emission occurs from a very large number of closely spaced vibronic states of the medium. As a result, the titanium sapphire laser emits radiation over a broad spectrum that spans from about 700 nm to about 1000 nm.

The **titanium sapphire laser** consists of Ti^{3+} ions at low concentration in a crystal of sapphire (Al_2O_3). The electronic absorption spectrum of Ti^{3+} ion in sapphire is very similar to that shown in Fig. 14.13, with a broad absorption band centred at around 500 nm that arises from vibronically allowed *d–d* transitions of the Ti^{3+} ion in an octahedral environment provided by oxygen atoms of the host lattice. As a result, the emission spectrum of Ti^{3+} in sapphire is also broad and laser action occurs over a wide range of wavelengths (Fig. 20.63). Therefore, the titanium sapphire laser is an example of a **vibronic laser**, in which the laser transitions originate from vibronic transitions in the laser medium. The titanium sapphire laser is usually pumped by another laser, such as a Nd–YAG laser or an argon-ion laser (*Further information 14.1*), and can be operated in either a continuous or pulsed fashion. Mode-locked titanium sapphire lasers produce energetic (20 mJ to 1 J) and very short (20–100 fs, 1 fs = 10^{-15} s) pulses. When considered together with broad wavelength tunability (700–1000 nm), these features of the titanium sapphire laser justify its wide use in modern spectroscopy and photochemistry.

The unique electrical properties of p–n junctions between semiconductors can be put to good use in optical devices. In some materials, most notably gallium arsenide, GaAs, energy from electron–hole recombination is released not as heat but is carried away by photons as electrons move across the junction under forward bias. Practical **light-emitting diodes** of this kind are widely used in electronic displays. The wavelength of emitted light depends on the band gap of the semiconductor. Gallium arsenide itself emits infrared light, but the band gap is widened by incorporating phosphorus, and a material of composition approximately $GaAs_{0.6}P_{0.4}$ emits light in the red region of the spectrum.

A light-emitting diode is not a laser, because no resonance cavity and stimulated emission are involved. In **diode lasers**, light emission due to electron–hole recombination is employed as the basis of laser action. The population inversion can be sustained by sweeping away the electrons that fall into the holes of the p-type semiconductor, and a resonant cavity can be formed by using the high refractive index of the semiconducting material and cleaving single crystals so that the light is trapped by the abrupt variation of refractive index. One widely used material is $Ga_{1-x}Al_xAs$, which produces infrared laser radiation and is widely used in compact-disc (CD) players.

Comment 20.7

The refractive index, n_r, of the medium, the ratio of the speed of light in a vacuum, c, to its speed c' in the medium: $n_r = c/c'$. A beam of light changes direction ('bends') when it passes from a region of one refractive index to a region with a different refractive index. See *Appendix* 3 for details.

High-power diode lasers are also used to pump other lasers. One example is the pumping of Nd:YAG lasers by $Ga_{0.91}Al_{0.09}As/Ga_{0.7}Al_{0.3}As$ diode lasers. The Nd:YAG laser is often used to pump yet another laser, such as a Ti:sapphire laser. As a result, it is now possible to construct a laser system for steady-state or time-resolved spectroscopy entirely out of solid-state components.

(c) Nonlinear optical phenomena

Nonlinear optical phenomena arise from changes in the optical properties of a material in the presence of an intense electric field from electromagnetic radiation. Here we explore two phenomena that not only can be studied conveniently with intense laser beams but are commonly used in the laboratory to modify the output of lasers for specific experiments, such as those described in Section 14.6.

In **frequency doubling**, or **second harmonic generation**, an intense laser beam is converted to radiation with twice (and in general a multiple) of its initial frequency as it passes though a suitable material. It follows that frequency doubling and tripling of a Nd–YAG laser, which emits radiation at 1064 nm, produce green light at 532 nm and ultraviolet radiation at 355 nm, respectively.

We can account for frequency doubling by examining how a substance responds nonlinearly to incident radiation of frequency $\omega = 2\pi\nu$. Radiation of a particular frequency arises from oscillations of an electric dipole at that frequency and the incident electric field $\mathcal{E}$ induces an electric dipole of magnitude μ, in the substance. At low light intensity, most materials respond linearly, in the sense that $\mu = \alpha\mathcal{E}$, where α is the polarizability (see Section 18.2). To allow for nonlinear response by some materials at high light intensity, we can write

$$\mu = \alpha\mathcal{E} + \tfrac{1}{2}\beta\mathcal{E}^2 + \ldots \tag{20.25}$$

where the coefficient β is the **hyperpolarizability** of the material. The nonlinear term $\beta\mathcal{E}^2$ can be expanded as follows if we suppose that the incident electric field is $\mathcal{E}_0 \cos \omega t$:

$$\beta\mathcal{E}^2 = \beta\mathcal{E}_0^2 \cos^2\omega t = \tfrac{1}{2}\beta\mathcal{E}_0^2(1 + \cos 2\omega t) \tag{20.26}$$

Hence, the nonlinear term contributes an induced electric dipole that oscillates at the frequency 2ω and that can act as a source of radiation of that frequency. Common materials that can be used for frequency doubling in laser systems include crystals of potassium dihydrogenphosphate (KH_2PO_4), lithium niobate ($LiNbO_3$), and β-barium borate (β-BaB_2O_4).

Another important nonlinear optical phenomenon is the **optical Kerr effect**, which arises from a change in refractive index of a well chosen medium, the **Kerr medium**, when it is exposed to intense laser pulses. Because a beam of light changes direction when it passes from a region of one refractive index to a region with a different refractive index, changes in refractive index result in the self-focusing of an intense laser pulse as it travels through the Kerr medium (Fig. 20.64).

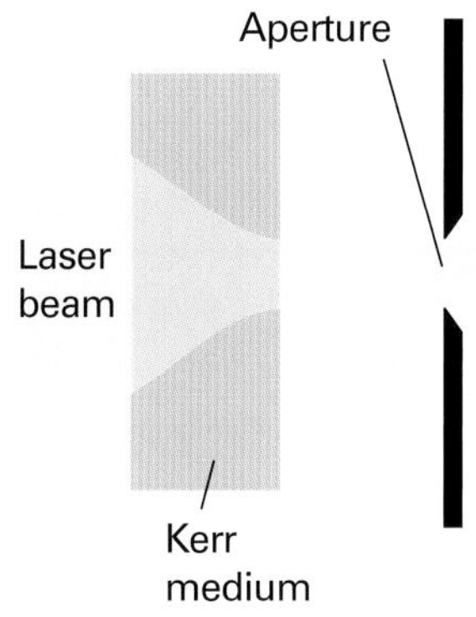

Fig. 20.64 An illustration of the Kerr effect. An intense laser beam is focused inside a Kerr medium and passes through a small aperture in the laser cavity. This effect may be used to mode-lock a laser, as explained in the text.

The optical Kerr effect is used as a mechanism of mode-locking lasers (Section 14.5). A Kerr medium is included in the cavity and next to it is a small aperture. The procedure makes use of the fact that the **gain**, the growth in intensity, of a frequency component of the radiation in the cavity is very sensitive to amplification and, once a particular frequency begins to grow, it can quickly dominate. When the power inside the cavity is low, a portion of the photons will be blocked by the aperture, creating a significant loss. A spontaneous fluctuation in intensity—a bunching of photons—may begin to turn on the optical Kerr effect and the changes in the refractive index of the Kerr medium will result in a **Kerr lens**, which is the self-focusing of the laser beam. The bunch of photons can pass through and travel to the far end of the cavity, amplifying as it goes. The Kerr lens immediately disappears (if the medium is well chosen), but is re-created when the intense pulse returns from the mirror at the far end. In this way, that particular bunch of photons may grow to considerable intensity because it alone is stimulating emission in the cavity. Sapphire is an example of a Kerr medium that facilitates the mode locking of titanium sapphire lasers, resulting in very short laser pulses of duration in the femtosecond range.

In addition to being useful laboratory tools, nonlinear optical materials are also finding many applications in the telecommunications industry, which is becoming ever more reliant on optical signals transmitted through optical fibres to carry voice and data. Judicious use of nonlinear phenomena leads to more ways in which the properties of optical signals, and hence the information they carry, can be manipulated.

20.11 Magnetic properties

The magnetic properties of metallic solids and semiconductors depend strongly on the band structures of the material (see *Further reading*). Here we confine our attention largely to magnetic properties that stem from collections of individual molecules or ions such as *d*-metal complexes. Much of the discussion applies to liquid and gas phase samples as well as to solids.

(a) Magnetic susceptibility

The magnetic and electric properties of molecules and solids are analogous. For instance, some molecules possess permanent magnetic dipole moments, and an applied magnetic field can induce a magnetic moment, with the result that the entire solid sample becomes magnetized. The analogue of the electric polarization, P, is the **magnetization**, $\mathcal{M}$, the average molecular magnetic dipole moment multiplied by the number density of molecules in the sample. The magnetization induced by a field of strength H is proportional to H, and we write

$$\mathcal{M} = \chi H \qquad [20.27]$$

where χ is the dimensionless **volume magnetic susceptibility**. A closely related quantity is the **molar magnetic susceptibility**, χ_m:

$$\chi_m = \chi V_m \qquad [20.28]$$

where V_m is the molar volume of the substance (we shall soon see why it is sensible to introduce this quantity). The **magnetic flux density**, $\mathcal{B}$, is related to the applied field strength and the magnetization by

$$\mathcal{B} = \mu_0(H + \mathcal{M}) = \mu_0(1 + \chi)H \qquad [20.29]$$

where μ_0 is the vacuum permeability, $\mu_0 = 4\pi \times 10^{-7}$ J C^{-2} m^{-1} s^2. The magnetic flux density can be thought of as the density of magnetic lines of force permeating the medium. This density is increased if $\mathcal{M}$ adds to H (when $\chi > 0$), but the density is decreased if $\mathcal{M}$ opposes H (when $\chi < 0$). Materials for which χ is positive are called **paramagnetic**. Those for which χ is negative are called **diamagnetic**.

Just as polar molecules in fluid phases contribute a term proportional to $\mu^2/3kT$ to the electric polarization of a medium (eqn 18.15), so molecules with a permanent magnetic dipole moment of magnitude m contribute to the magnetization an amount proportional to $m^2/3kT$. However, unlike for polar molecules, this contribution to the magnetization is obtained even for paramagnetic species trapped in solids, because the direction of the spin of the electrons is typically not coupled to the orientation of the molecular framework and so contributes even when the nuclei are stationary. An applied field can also induce a magnetic moment by stirring up currents in the electron distribution like those responsible for the chemical shift in NMR (Section 15.5). The constant of proportionality between the induced moment and the applied field is called the **magnetizability**, ξ (xi), and the magnetic analogue of eqn 18.15 is

$$\chi = \mathcal{N}\mu_0\left(\xi + \frac{m^2}{3kT}\right) \qquad (20.30)$$

We can now see why it is convenient to introduce χ_m, because the product of the number density $\mathcal{N}$ and the molar volume is Avogadro's constant, N_A:

$$\mathcal{N}V_m = \frac{NV_m}{V} = \frac{nN_A V_m}{nV_m} = N_A \qquad (20.31)$$

Hence

$$\chi_m = N_A\mu_0\left(\xi + \frac{m^2}{3kT}\right) \qquad (20.32)$$

and the density dependence of the susceptibility (which occurs in eqn 20.30 via $\mathcal{N} = N_A\rho/M$) has been eliminated. The expression for χ_m is in agreement with the empirical **Curie law**:

$$\chi_m = A + \frac{C}{T} \qquad (20.33)$$

with $A = N_A\mu_0\xi$ and $C = N_A\mu_0 m^2/3k$. As indicated above, and in contrast to electric moments, this expression applies to solids as well as fluid phases.

The magnetic susceptibility is traditionally measured with a **Gouy balance**. This instrument consists of a sensitive balance from which the sample hangs in the form of a narrow cylinder and lies between the poles of a magnet. If the sample is paramagnetic, it is drawn into the field, and its apparent weight is greater than when the field is off. A diamagnetic sample tends to be expelled from the field and appears to weigh less when the field is turned on. The balance is normally calibrated against a sample of

Synoptic table 20.6* Magnetic susceptibilities at 298 K

	$\chi/10^{-6}$	$\chi_m/(10^{-5}\ cm^3\ mol^{-1})$
$H_2O(l)$	−9.06	−160
NaCl(s)	−13.9	−38
Cu(s)	−9.6	−6.8
$CuSO_4\cdot 5H_2O(s)$	+176	+1930

* More values are given in the *Data section*.

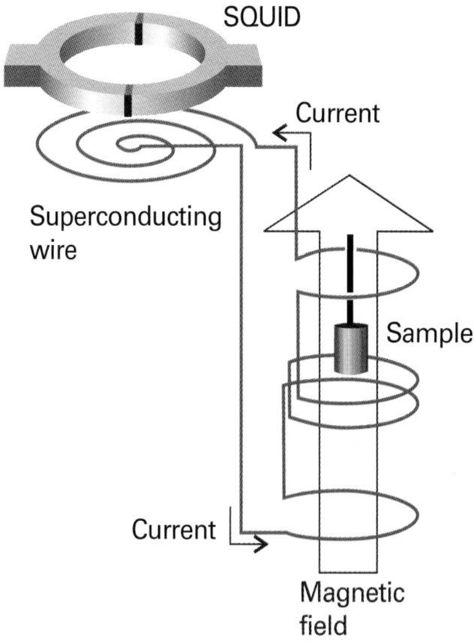

Fig. 20.65 The arrangement used to magnetic susceptibility with a SQUID. The sample is moved upwards in small increments and the potential difference across the SQUID is measured.

known susceptibility. The modern version of the determination makes use of a **superconducting quantum interference device** (SQUID, Fig. 20.65). A SQUID takes advantage of the quantization of magnetic flux and the property of current loops in superconductors that, as part of the circuit, include a weakly conducting link through which electrons must tunnel. The current that flows in the loop in a magnetic field depends on the value of the magnetic flux, and a SQUID can be exploited as a very sensitive magnetometer.

Table 20.6 lists some experimental values. A typical paramagnetic volume susceptibility is about 10^{-3}, and a typical diamagnetic volume susceptibility is about $(-)10^{-5}$. The permanent magnetic moment can be extracted from susceptibility measurements by plotting χ against $1/T$.

(b) The permanent magnetic moment

The permanent magnetic moment of a molecule arises from any unpaired electron spins in the molecule. We saw in Section 10.8 that the magnitude of the magnetic moment of an electron is proportional to the magnitude of the spin angular momentum, $\{s(s+1)\}^{1/2}\hbar$.

$$\mu = g_e\{s(s+1)\}^{1/2}\mu_B \qquad \mu_B = \frac{e\hbar}{2m_e} \tag{20.34}$$

where $g_e = 2.0023$ (see Section 15.1). If there are several electron spins in each molecule, they combine to a total spin S, and then $s(s+1)$ should be replaced by $S(S+1)$. It follows that the spin contribution to the molar magnetic susceptibility is

$$\chi_m = \frac{N_A g_e^2 \mu_0 \mu_B^2 S(S+1)}{3kT} \tag{20.35}$$

This expression shows that the susceptibility is positive, so the spin magnetic moments contribute to the paramagnetic susceptibilities of materials. The contribution decreases with increasing temperature because the thermal motion randomizes the spin orientations. In practice, a contribution to the paramagnetism also arises from the orbital angular momenta of electrons: we have discussed the spin-only contribution.

Illustration 20.3 *Calculating a magnetic susceptibility*

Consider a complex salt with three unpaired electrons per complex cation at 298 K, of mass density 3.24 g cm^{-3}, and molar mass 200 g mol^{-1}. First note that

$$\frac{N_A g_e^2 \mu_0 \mu_B^2}{3k} = 6.3001 \times 10^{-6}\ m^3\ K^{-1}\ mol^{-1}$$

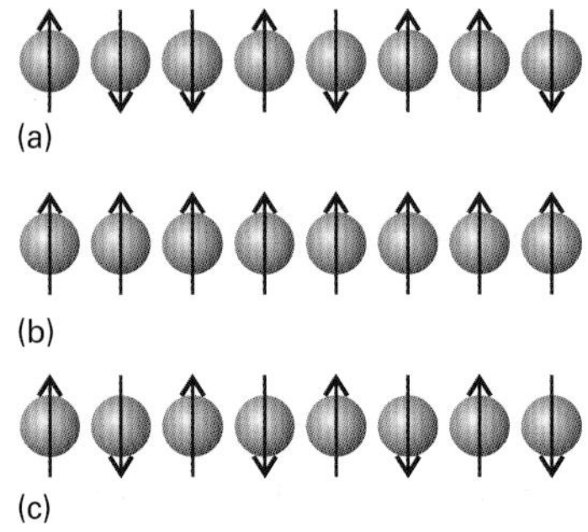

Fig. 20.66 (a) In a paramagnetic material, the electron spins are aligned at random in the absence of an applied magnetic field. (b) In a ferromagnetic material, the electron spins are locked into a parallel alignment over large domains. (c) In an antiferromagnetic material, the electron spins are locked into an antiparallel arrangement. The latter two arrangements survive even in the absence of an applied field.

Consequently,

$$\chi_m = 6.3001 \times 10^{-6} \times \frac{S(S+1)}{T/K} \text{ m}^3 \text{ mol}^{-1}$$

Substitution of the data with $S = \frac{3}{2}$ gives $\chi_m = 7.9 \times 10^{-8}$ m^3 mol^{-1}. Note that the density is not needed at this stage. To obtain the volume magnetic susceptibility, the molar susceptibility is divided by the molar volume $V_m = M/\rho$, where ρ is the mass density. In this illustration, $V_m = 61.7$ cm^3 mol^{-1}, so $\chi = 1.3 \times 10^{-3}$.

At low temperatures, some paramagnetic solids make a phase transition to a state in which large domains of spins align with parallel orientations. This cooperative alignment gives rise to a very strong magnetization and is called **ferromagnetism** (Fig. 20.66). In other cases, the cooperative effect leads to alternating spin orientations: the spins are locked into a low-magnetization arrangement to give an **antiferromagnetic phase**. The ferromagnetic phase has a nonzero magnetization in the absence of an applied field, but the antiferromagnetic phase has a zero magnetization because the spin magnetic moments cancel. The ferromagnetic transition occurs at the **Curie temperature**, and the antiferromagnetic transition occurs at the **Néel temperature**.

(c) Induced magnetic moments

An applied magnetic field induces the circulation of electronic currents. These currents give rise to a magnetic field that usually opposes the applied field, so the substance is diamagnetic. In a few cases the induced field augments the applied field, and the substance is then paramagnetic.

The great majority of molecules with no unpaired electron spins are diamagnetic. In these cases, the induced electron currents occur within the orbitals of the molecule that are occupied in its ground state. In the few cases in which molecules are paramagnetic despite having no unpaired electrons, the induced electron currents flow in the opposite direction because they can make use of unoccupied orbitals that lie close to the HOMO in energy. This orbital paramagnetism can be distinguished from spin paramagnetism by the fact that it is temperature independent: this is why it is called **temperature-independent paramagnetism** (TIP).

We can summarize these remarks as follows. All molecules have a diamagnetic component to their susceptibility, but it is dominated by spin paramagnetism if the molecules have unpaired electrons. In a few cases (where there are low-lying excited states) TIP is strong enough to make the molecules paramagnetic even though their electrons are paired.

20.12 Superconductors

The resistance to flow of electrical current of a normal metallic conductor decreases smoothly with temperature but never vanishes. However, certain solids known as **superconductors** conduct electricity without resistance below a critical temperature, T_c. Following the discovery in 1911 that mercury is a superconductor below 4.2 K, the boiling point of liquid helium, physicists and chemists made slow but steady progress in the discovery of superconductors with higher values of T_c. Metals, such as tungsten, mercury, and lead, tend to have T_c values below about 10 K. Intermetallic compounds, such as Nb_3X (X = Sn, Al, or Ge), and alloys, such as Nb/Ti and Nb/Zr, have intermediate T_c values ranging between 10 K and 23 K. In 1986, **high-temperature superconductors** (HTSC) were discovered. Several **ceramics**, inorganic powders that have

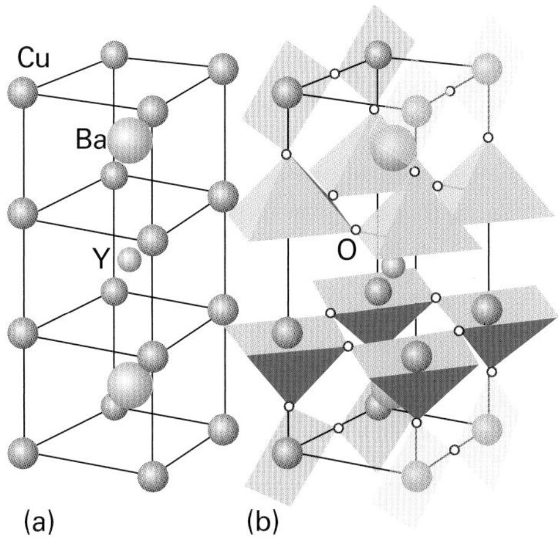

Fig. 20.67 Structure of the $YBa_2Cu_3O_7$ superconductor. (a) Metal atom positions. (b) The polyhedra show the positions of oxygen atoms and indicate that the metal ions are in square-planar and square-pyramidal coordination environments.

been fused and hardened by heating to a high temperature, containing oxocuprate motifs, Cu_mO_n, are now known with T_c values well above 77 K, the boiling point of the inexpensive refrigerant liquid nitrogen. For example, $HgBa_2Ca_2Cu_2O_8$ has $T_c = 153$ K.

Superconductors have unique magnetic properties as well. Some superconductors, classed as **Type I**, show abrupt loss of superconductivity when an applied magnetic field exceeds a critical value H_c characteristic of the material. It is observed that the value of H_c depends on temperature and T_c as

$$H_c(T) = H_c(0)\left(1 - \frac{T^2}{T_c^2}\right) \tag{20.36}$$

where $H_c(0)$ is the value of H_c as $T \to 0$. Type I superconductors are also completely diamagnetic below H_c, meaning that no magnetic field lines penetrate into the material. This complete exclusion of a magnetic field in a material is known as the **Meissner effect**, which can be visualized by the levitation of a superconductor above a magnet. **Type II** superconductors, which include the HTSCs, show a gradual loss of superconductivity and diamagnetism with increasing magnetic field.

There is a degree of periodicity in the elements that exhibit superconductivity. The metals iron, cobalt, nickel, copper, silver, and gold do not display superconductivity, nor do the alkali metals. It is observed that, for simple metals, ferromagnetism and superconductivity never coexist, but in some of the oxocuprate superconductors ferromagnetism and superconductivity can coexist. One of the most widely studied oxocuprate superconductors $YBa_2Cu_3O_7$ (informally known as '123' on account of the proportions of the metal atoms in the compound) has the structure shown in Fig. 20.67. The square-pyramidal CuO_5 units arranged as two-dimensional layers and the square planar CuO_4 units arranged in sheets are common structural features of oxocuprate HTSCs.

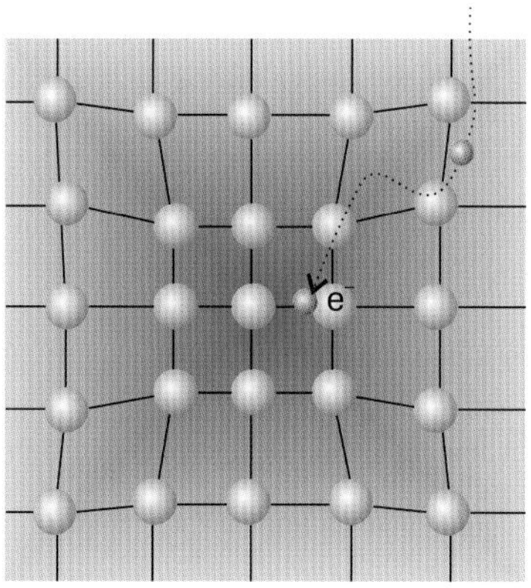

Fig. 20.68 The formation of a Cooper pair. One electron distorts the crystal lattice and the second electron has a lower energy if it goes to that region. These electron–lattice interactions effectively bind the two electrons into a pair.

The mechanism of superconduction is well-understood for low-temperature materials but there is as yet no settled explanation of high-temperature superconductivity. The central concept of low-temperature superconduction is the existence of a **Cooper pair**, a pair of electrons that exists on account of the indirect electron–electron interactions fostered by the nuclei of the atoms in the lattice. Thus, if one electron is in a particular region of a solid, the nuclei there move toward it to give a distorted local structure (Fig. 20.68). Because that local distortion is rich in positive charge, it is favourable for a second electron to join the first. Hence, there is a virtual attraction between the two electrons, and they move together as a pair. The local distortion can be easily disrupted by thermal motion of the ions in the solid, so the virtual attraction occurs only at very low temperatures. A Cooper pair undergoes less scattering than an individual electron as it travels through the solid because the distortion caused by one electron can attract back the other electron should it be scattered out of its path in a collision. Because the Cooper pair is stable against scattering, it can carry charge freely through the solid, and hence give rise to superconduction.

The Cooper pairs responsible for low-temperature superconductivity are likely to be important in HTSCs, but the mechanism for pairing is hotly debated. There is evidence implicating the arrangement of CuO_5 layers and CuO_4 sheets in the mechanism of high-temperature superconduction. It is believed that movement of electrons along the linked CuO_4 units accounts for superconductivity, whereas the linked CuO_5 units act as 'charge reservoirs' that maintain an appropriate number of electrons in the superconducting layers.

Superconductors can sustain large currents and, consequently, are excellent materials for the high-field magnets used in modern NMR spectroscopy (Chapter 15). However, the potential uses of superconducting materials are not limited to the field to chemical instrumentation. For example, HTSCs with T_c values near ambient temperature would be very efficient components of an electrical power transmission system, in which energy loss due to electrical resistance would be minimized. The appropriate technology is not yet available, but research in this area of materials science is active.

Checklist of key ideas

- ☐ 1. Solids are classified as metallic, ionic, covalent, and molecular.
- ☐ 2. A space lattice is the pattern formed by points representing the locations of structural motifs (atoms, molecules, or groups of atoms, molecules, or ions). The Bravais lattices are the 14 distinct space lattices in three dimensions (Fig. 20.8).
- ☐ 3. A unit cell is an imaginary parallelepiped that contains one unit of a translationally repeating pattern. Unit cells are classified into seven crystal systems according to their rotational symmetries.
- ☐ 4. Crystal planes are specified by a set of Miller indices (*hkl*) and the separation of neighbouring planes in a rectangular lattice is given by $1/d_{hkl}^2 = h^2/a^2 + k^2/b^2 + l^2/c^2$.
- ☐ 5. Bragg's law relating the glancing angle θ to the separation of lattice planes is $\lambda = 2d \sin \theta$, where λ is the wavelength of the radiation.
- ☐ 6. The scattering factor is a measure of the ability of an atom to diffract radiation (eqn 20.6).
- ☐ 7. The structure factor is the overall amplitude of a wave diffracted by the $\{hkl\}$ planes (eqn 20.7). Fourier synthesis is the construction of the electron density distribution from structure factors (eqn 20.8).
- ☐ 8. A Patterson synthesis is a map of interatomic vectors obtained by Fourier analysis of diffraction intensities (eqn 20.9).
- ☐ 9. Structure refinement is the adjustment of structural parameters to give the best fit between the observed intensities and those calculated from the model of the structure deduced from the diffraction pattern.
- ☐ 10. Many elemental metals have close-packed structures with coordination number 12; close-packed structures may be either cubic (ccp) or hexagonal (hcp).
- ☐ 11. Representative ionic structures include the caesium-chloride, rock-salt, and zinc-blende structures.
- ☐ 12. The radius-ratio rule may be used cautiously to predict which of these three structures is likely (eqn 20.12).
- ☐ 13. The lattice enthalpy is the change in enthalpy (per mole of formula units) accompanying the complete separation of the components of the solid. The electrostatic contribution to the lattice enthalpy is expressed by the Born–Mayer equation (eqn 20.15).
- ☐ 14. A covalent network solid is a solid in which covalent bonds in a definite spatial orientation link the atoms in a network extending through the crystal. A molecular solid is a solid consisting of discrete molecules held together by van der Waals interactions.
- ☐ 15. The mechanical properties of a solid are discussed in terms of the relationship between stress, the applied force divided by the area to which it is applied, and strain, the distortion of a sample resulting from an applied stress.
- ☐ 16. The response of a solid to an applied stress is summarized by the Young's modulus (eqn 20.16a), the bulk modulus (eqn 20.16b), the shear modulus (eqn 20.16c), and Poisson's ratio (eqn 20.17).
- ☐ 17. Electronic conductors are classified as metallic conductors or semiconductors according to the temperature dependence of their conductivities. An insulator is a semiconductor with a very low electrical conductivity.

☐ 18. According to the band theory, electrons occupy molecular orbitals formed from the overlap of atomic orbitals: full bands are called valence bands and empty bands are called conduction bands. The occupation of the orbitals in a solid is given by the Fermi–Dirac distribution (eqn 20.23).

☐ 19. Semiconductors are classified as p-type or n-type according to whether conduction is due to holes in the valence band or electrons in the conduction band.

☐ 20. The spectroscopic properties of molecular solids can be understood in terms of the formation and migration of excitons, electron–hole pairs, from molecule to molecule.

☐ 21. The spectroscopic properties of metallic conductors and semiconductors can be understood in terms of the light-induced promotion of electrons from valence bands to conduction bands.

☐ 22. Examples of solid state lasers include the neodymium laser, the titanium sapphire laser, and diode lasers.

☐ 23. Nonlinear optical phenomena arise from changes in the optical properties of a material in the presence of an intense field from electromagnetic radiation. Examples include second harmonic generation, and the optical Kerr effect.

☐ 24. A bulk sample exposed to a magnetic field of strength H acquires a magnetization, $\mathcal{M} = \chi H$, where χ is the dimensionless volume magnetic susceptibility. When $\chi < 0$, the material is diamagnetic and moves out of a magnetic field. When $\chi > 0$, the material is paramagnetic and moves into a magnetic field.

☐ 25. The temperature dependence of χ_m is given by the Curie law $\chi_m = A + C/T$, where $A = N_A\mu_0\xi$, $C = N_A\mu_0 m^2/3k$, and ξ is the magnetizability, a measure of the extent to which a magnetic dipole moment may be induced in a molecule.

☐ 26. Ferromagnetism is the cooperative alignment of electron spins in a material and gives rise to strong magnetization. Antiferromagnetism results from alternating spin orientations in a material and leads to weak magnetization.

☐ 27. Temperature-independent paramagnetism arises from induced electron currents within the orbitals of a molecule that are occupied in its ground state.

☐ 28. Superconductors conduct electricity without resistance below a critical temperature T_c. Type I superconductors show abrupt loss of superconductivity when an applied magnetic field exceeds a critical value H_c characteristic of the material. They are also completely diamagnetic below H_c. Type II superconductors show a gradual loss of superconductivity and diamagnetism with increasing magnetic field.

Further reading

Articles and texts

C.D. Graham, Jr., Magnetic materials. In *Encyclopedia of applied physics*, G.L. Trigg (ed.), **9**, 1 VCH, New York (1994).

W.B. Pearson and C. Chieh, Crystallography. In *Encyclopedia of applied physics*, G.L. Trigg (ed.), **4**, 385 VCH, New York (1992).

W.D. Callister, Jr., *Materials science and engineering: an introduction.* Wiley, New York (2000).

P.A. Cox, *The electronic structure and chemistry of solids.* Oxford University Press (1987).

O. Svelto, *Principles of lasers.* Plenum, New York (1998).

M.A. White, *Properties of materials.* Oxford University Press (2000).

Sources of data and information

D.R. Lide (ed.), *CRC Handbook of Chemistry and Physics*, Sections 3, 12, 13, and 15, CRC Press, Boca Raton (2000).

D. Sangeeta and J.R. LaGraff, *Inorganic materials chemistry desk reference*, CRC Press, Boca Raton (2004).

Discussion questions

20.1 Explain how planes of lattice points are labelled.

20.2 Describe the procedure for identifying the type and size of a cubic unit cell.

20.3 What is meant by a systematic absence? How do they arise?

20.4 Describe the phase problem and explain how it may be overcome.

20.5 Describe the structures of elemental metallic solids in terms of the packing of hard spheres. To what extent is the hard-sphere model inaccurate?

20.6 Describe the caesium-chloride and rock-salt structures in terms of the occupation of holes in expanded close-packed lattices.

20.7 Explain how X-ray diffraction can be used to determine the helical configuration of biological molecules.

20.8 Explain how metallic conductors and semiconductors are identified and explain their electrical and optical properties in terms of band theory.

20.9 Describe the characteristics of the Fermi–Dirac distribution. Why is it appropriate to call the parameter μ a chemical potential?

20.10 To what extent are the electric and magnetic properties of molecules analogous? How do they differ?

Exercises

20.1a Equivalent lattice points within the unit cell of a Bravais lattice have identical surroundings. What points within a face-centred cubic unit cell are equivalent to the point $(\frac{1}{2}, 0, 0)$?

20.1b Equivalent lattice points within the unit cell of a Bravais lattice have identical surroundings. What points within a body-centred cubic unit cell are equivalent to the point $(\frac{1}{2}, 0, \frac{1}{2})$?

20.2a Find the Miller indices of the planes that intersect the crystallographic axes at the distances $(2a, 3b, 2c)$ and $(2a, 2b, \infty c)$.

20.2b Find the Miller indices of the planes that intersect the crystallographic axes at the distances $(1a, 3b, -c)$ and $(2a, 3b, 4c)$.

20.3a Calculate the separations of the planes {111}, {211}, and {100} in a crystal in which the cubic unit cell has side 432 pm.

20.3b Calculate the separations of the planes {121}, {221}, and {244} in a crystal in which the cubic unit cell has side 523 pm.

20.4a The glancing angle of a Bragg reflection from a set of crystal planes separated by 99.3 pm is 20.85°. Calculate the wavelength of the X-rays.

20.4b The glancing angle of a Bragg reflection from a set of crystal planes separated by 128.2 pm is 19.76°. Calculate the wavelength of the X-rays.

20.5a What are the values of 2θ of the first three diffraction lines of bcc iron (atomic radius 126 pm) when the X-ray wavelength is 58 pm?

20.5b What are the values of 2θ of the first three diffraction lines of fcc gold {atomic radius 144 pm) when the X-ray wavelength is 154 pm?

20.6a Copper K_α radiation consists of two components of wavelengths 154.433 pm and 154.051 pm. Calculate the separation of the diffraction lines arising from the two components in a powder diffraction pattern recorded in a circular camera of radius 5.74 cm (with the sample at the centre) from planes of separation 77.8 pm.

20.6b A synchrotron source produces X-radiation at a range of wavelengths. Consider two components of wavelengths 95.401 and 96.035 pm. Calculate the separation of the diffraction lines arising from the two components in a powder diffraction pattern recorded in a circular camera of radius 5.74 cm (with the sample at the centre) from planes of separation 82.3 pm.

20.7a The compound Rb_3TlF_6 has a tetragonal unit cell with dimensions $a = 651$ pm and $c = 934$ pm. Calculate the volume of the unit cell.

20.7b Calculate the volume of the hexagonal unit cell of sodium nitrate, for which the dimensions are $a = 1692.9$ pm and $c = 506.96$ pm.

20.8a The orthorhombic unit cell of $NiSO_4$ has the dimensions $a = 634$ pm, $b = 784$ pm, and $c = 516$ pm, and the density of the solid is estimated as 3.9 g cm^{-3}. Determine the number of formula units per unit cell and calculate a more precise value of the density.

20.8b An orthorhombic unit cell of a compound of molar mass 135.01 g mol^{-1} has the dimensions $a = 589$ pm, $b = 822$ pm, and $c = 798$ pm. The density of the solid is estimated as 2.9 g cm^{-3}. Determine the number of formula units per unit cell and calculate a more precise value of the density.

20.9a The unit cells of $SbCl_3$ are orthorhombic with dimensions $a = 812$ pm, $b = 947$ pm, and $c = 637$ pm. Calculate the spacing, d, of the (411) planes.

20.9b An orthorhombic unit cell has dimensions $a = 679$ pm, $b = 879$ pm, and $c = 860$ pm. Calculate the spacing, d, of the (322) planes.

20.10a A substance known to have a cubic unit cell gives reflections with Cu K_α radiation (wavelength 154 pm) at glancing angles 19.4°, 22.5°, 32.6°, and 39.4°. The reflection at 32.6° is known to be due to the (220) planes. Index the other reflections.

20.10b A substance known to have a cubic unit cell gives reflections with radiation of wavelength 137 pm at the glancing angles 10.7°, 13.6°, 17.7°, and 21.9°. The reflection at 17.7° is known to be due to the (111) planes. Index the other reflections.

20.11a Potassium nitrate crystals have orthorhombic unit cells of dimensions $a = 542$ pm, $b = 917$ pm, and $c = 645$ pm. Calculate the glancing angles for the (100), (010), and (111) reflections using CuK_α radiation (154 pm).

20.11b Calcium carbonate crystals in the form of aragonite have orthorhombic unit cells of dimensions $a = 574.1$ pm, $b = 796.8$ pm, and $c = 495.9$ pm. Calculate the glancing angles for the (100), (010), and (111) reflections using radiation of wavelength 83.42 pm (from aluminium).

20.12a Copper(I) chloride forms cubic crystals with four formula units per unit cell. The only reflections present in a powder photograph are those with either all even indices or all odd indices. What is the (Bravais) lattice type of the unit cell?

20.12b A powder diffraction photograph from tungsten shows lines that index as (110), (200), (211), (220), (310), (222), (321), (400), . . . Identify the (Bravais) lattice type of the unit cell.

20.13a The coordinates, in units of a, of the atoms in a body-centred cubic lattice are (0,0,0), (0,1,0), (0,0,1), (0,1,1), (1,0,0), (1,1,0), (1,0,1), and (1,1,1). Calculate the structure factors F_{hkl} when all the atoms are identical.

20.13b The coordinates, in units of a, of the atoms in a body-centred cubic lattice are (0,0,0), (0,1,0), (0,0,1), (0,1,1), (1,0,0), (1,1,0), (1,0,1), (1,1,1), and $(\frac{1}{2},\frac{1}{2},\frac{1}{2})$. Calculate the structure factors F_{hkl} when all the atoms are identical.

20.14a Calculate the packing fraction for close-packed cylinders.

20.14b Calculate the packing fraction for equilateral triangular rods stacked as shown in **2**.

2

20.15a Verify that the radius ratios for sixfold coordination is 0.414.

20.15b Verify that the radius ratios for eightfold coordination is 0.732.

20.16a From the data in Table 20.3 determine the radius of the smallest cation that can have (a) sixfold and (b) eightfold coordination with the O^{2-} ion.

20.16b From the data in Table 20.3 determine the radius of the smallest cation that can have (a) sixfold and (b) eightfold coordination with the K^+ ion.

20.17a Calculate the atomic packing factor for diamond.

20.17b Calculate the atomic packing factor for a side-centred (C) cubic unit cell.

20.18a Is there an expansion or a contraction as titanium transforms from hcp to body-centred cubic? The atomic radius of titanium is 145.8 pm in hcp but 142.5 pm in bcc.

20.18b Is there an expansion or a contraction as iron transforms from hcp to bcc? The atomic radius of iron is 126 pm in hcp but 122 pm in bcc.

20.19a In a Patterson synthesis, the spots correspond to the lengths and directions of the vectors joining the atoms in a unit cell. Sketch the pattern that would be obtained for a planar, triangular isolated BF_3 molecule.

20.19b In a Patterson synthesis, the spots correspond to the lengths and directions of the vectors joining the atoms in a unit cell. Sketch the pattern that would be obtained from the C atoms in an isolated benzene molecule.

20.20a What velocity should neutrons have if they are to have wavelength 50 pm?

20.20b Calculate the wavelength of neutrons that have reached thermal equilibrium by collision with a moderator at 300 K.

20.21a Derive the Born–Mayer equation (eqn 20.15) by calculating the energy at which $d(E_P + E_P^*)/dd = 0$, with E_P and E_P^* given by eqns 20.13 and 20.14, respectively.

20.21b Calculate the lattice enthalpy of $MgBr_2$ from the following data:

	$\Delta H/(\text{kJ mol}^{-1})$
Sublimation of Mg(s)	+148
Ionization of Mg(g) to Mg^{2+}(g)	+2187
Vaporization of Br_2(l)	+31
Dissociation of Br_2(g)	+193
Electron attachment to Br(g)	−331
Formation of $MgBr_2$(s) from Mg(s) and Br_2(l)	−524

20.22a Cotton consists of the polymer cellulose, which is a linear chain of glucose molecules. The chains are held together by hydrogen bonding. When a cotton shirt is ironed, it is first moistened, then heated under pressure. Explain this process.

20.22b Sections of the solid fuel rocket boosters of the space shuttle *Challenger* were sealed together with O-ring rubber seals of circumference 11 m. These seals failed at 0°C, a temperature well above the crystallization temperature of the rubber. Speculate on why the failure occurred.

20.23a Young's modulus for polyethylene at room temperature is 1.2 GPa. What strain will be produced when a mass of 1.0 kg is suspended from a polyethylene thread of diameter 1.0 mm?

20.23b Young's modulus for iron at room temperature is 215 GPa. What strain will be produced when a mass of 10.0 kg is suspended from an iron wire of diameter 0.10 mm?

20.24a Poisson's ratio for polyethylene is 0.45. What change in volume takes place when a cube of polyethylene of volume 1.0 cm^3 is subjected to a uniaxial stress that produces a strain of 1.0 per cent?

20.24b Poisson's ratio for lead is 0.41. What change in volume takes place when a cube of lead of volume 1.0 dm^3 is subjected to a uniaxial stress that produces a strain of 2.0 per cent?

20.25a Is arsenic-doped germanium a p-type or n-type semiconductor?

20.25b Is gallium-doped germanium a p-type or n-type semiconductor?

20.26a The promotion of an electron from the valence band into the conduction band in pure TIO_2 by light absorption requires a wavelength of less than 350 nm. Calculate the energy gap in electronvolts between the valence and conduction bands.

20.26b The band gap in silicon is 1.12 eV. Calculate the minimum frequency of electromagnetic radiation that results in promotion of electrons from the valence to the conduction band.

20.27a The magnetic moment of $CrCl_3$ is $3.81\mu_B$. How many unpaired electrons does the Cr possess?

20.27b The magnetic moment of Mn^{2+} in its complexes is typically $5.3\mu_B$. How many unpaired electrons does the ion possess?

20.28a Calculate the molar susceptibility of benzene given that its volume susceptibility is -7.2×10^{-7} and its density 0.879 g cm^{-3} at 25°C.

20.28b Calculate the molar susceptibility of cyclohexane given that its volume susceptibility is -7.9×10^{-7} and its density 811 kg m^{-3} at 25°C.

20.29a According to Lewis theory, an O_2 molecule should be diamagnetic. However, experimentally it is found that $\chi_m/(\text{m}^3\,\text{mol}^{-1}) = (1.22 \times 10^{-5}\,\text{K})/T$. Determine the number of unpaired spins in O_2. How is the problem of the Lewis structure resolved?

20.29b Predict the molar susceptibility of nitrogen dioxide at 298 K. Why does the molar susceptibility of a sample of nitrogen dioxide gas decrease as it is compressed?

20.30a Data on a single crystal of MnF_2 give $\chi_m = 0.1463\,\text{cm}^3\,\text{mol}^{-1}$ at 294.53 K. Determine the effective number of unpaired electrons in this compound and compare your result with the theoretical value.

20.30b Data on a single crystal of $NiSO_4{\cdot}7H_2O$ give $\chi_m = 6.00 \times 10^{-8}\,\text{m}^3\,\text{mol}^{-1}$ at 298 K. Determine the effective number of unpaired electrons in this compound and compare your result with the theoretical value.

20.31a Estimate the spin-only molar susceptibility of $CuSO_4{\cdot}5H_2O$ at 25°C.

20.31b Estimate the spin-only molar susceptibility of $MnSO_4{\cdot}4H_2O$ at 298 K.

20.32a Approximately how large must the magnetic induction, B, be for the orientational energy of an $S = 1$ system to be comparable to kT at 298 K?

20.32b Estimate the ratio of populations of the M_S states of a system with $S = 1$ in 15.0 T at 298 K.

Problems*

Numerical problems

20.1 In the early days of X-ray crystallography there was an urgent need to know the wavelengths of X-rays. One technique was to measure the diffraction angle from a mechanically ruled grating. Another method was to estimate the separation of lattice planes from the measured density of a crystal. The density of NaCl is 2.17 g cm^{-3} and the (100) reflection using PdK_α radiation occurred at 6.0°. Calculate the wavelength of the X-rays.

20.2 The element polonium crystallizes in a cubic system. Bragg reflections, with X-rays of wavelength 154 pm, occur at sin $\theta = 0.225$, 0.316, and 0.388 from the (100), (110), and (111) sets of planes. The separation between the

* Problems denoted with the symbol ‡ were supplied by Charles Trapp and Carmen Giunta.

sixth and seventh lines observed in the powder diffraction pattern is larger than between the fifth and sixth lines. Is the unit cell simple, body-centred, or face-centred? Calculate the unit cell dimension.

20.3 The unit cell dimensions of NaCl, KCl, NaBr, and KBr, all of which crystallize in face-centred cubic lattices, are 562.8 pm, 627.7 pm, 596.2 pm, and 658.6 pm, respectively. In each case, anion and cation are in contact along an edge of the unit cell. Do the data support the contention that ionic radii are constants independent of the counter-ion?

20.4 The powder diffraction patterns of (a) tungsten, (b) copper obtained in a camera of radius 28.7 mm are shown in Fig. 20.69. Both were obtained with 154 pm X-rays and the scales are marked. Identify the unit cell in each case, and calculate the lattice spacing. Estimate the metallic radii of W and Cu.

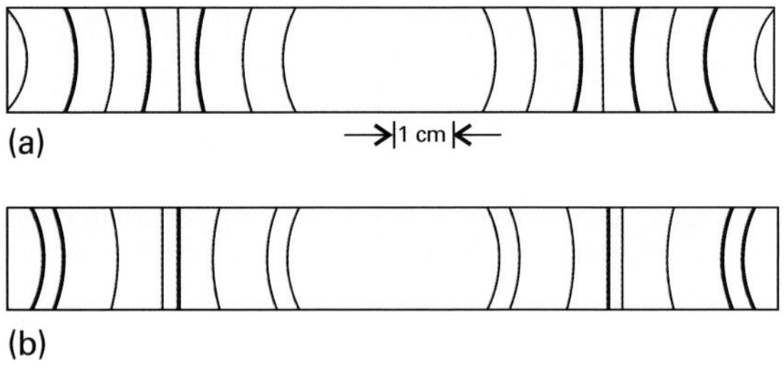

Fig. 20.69

20.5 Elemental silver reflects X-rays of wavelength 154.18 pm at angles of 19.076°, 22.171°, and 32.256°. However, there are no other reflections at angles of less than 33°. Assuming a cubic unit cell, determine its type and dimension. Calculate the density of silver.

20.6 Genuine pearls consist of concentric layers of calcite crystals ($CaCO_3$) in which the trigonal axes are oriented along the radii. The nucleus of a cultured pearl is a piece of mother-of-pearl that has been worked into a sphere on a lathe. The oyster then deposits concentric layers of calcite on the central seed. Suggest an X-ray method for distinguishing between real and cultured pearls.

20.7 In their book *X-rays and crystal structures* (which begins 'It is now two years since Dr. Laue conceived the idea . . . ') the Braggs give a number of simple examples of X-ray analysis. For instance, they report that the reflection from (100) planes in KCl occurs at 5° 23′, but for NaCl it occurs at 6° 0′ for X-rays of the same wavelength. If the side of the NaCl unit cell is 564 pm, what is the side of the KCl unit cell? The densities of KCl and NaCl are 1.99 g cm^{-3} and 2.17 g cm^{-3} respectively. Do these values support the X-ray analysis?

20.8 Calculate the coefficient of thermal expansion of diamond given that the (111) reflection shifts from 22° 2′ 25″ to 21° 57′ 59″ on heating a crystal from 100 K to 300 K and 154.0562 pm X-rays are used.

20.9 The carbon–carbon bond length in diamond is 154.45 pm. If diamond were considered to be a close-packed structure of hard spheres with radii equal to half the bond length, what would be its expected density? The diamond lattice is face-centred cubic and its actual density is 3.516 g cm^{-3}. Can you explain the discrepancy?

20.10 The volume of a monoclinic unit cell is $abc \sin \beta$. Naphthalene has a monoclinic unit cell with two molecules per cell and sides in the ratio 1.377:1:1.436. The angle β is 122° 49′ and the density of the solid is 1.152 g cm^{-3}. Calculate the dimensions of the cell.

20.11 The structures of crystalline macromolecules may be determined by X-ray diffraction techniques by methods similar to those for smaller molecules. Fully crystalline polyethylene has its chains aligned in an orthorhombic unit cell of dimensions 740 pm × 493 pm × 253 pm. There are two repeating CH_2CH_2 units per unit cell. Calculate the theoretical density of fully crystalline polyethylene. The actual density ranges from 0.92 to 0.95 g cm^{-3}.

20.12 Construct the electron density along the *x*-axis of a crystal given the following structure factors:

h	0	1	2	3	4	5	6	7	8	9
F_h	+30.0	+8.2	+6.5	+4.1	+5.5	−2.4	+5.4	+3.2	−1.0	+1.1

h	10	11	12	13	14	15
F_h	+6.5	+5.2	−4.3	−1.2	+0.1	+2.1

20.13 The scattering of electrons or neutrons from a pair of nuclei separated by a distance R_{ij} and orientated at a definite angle to the incident beam can be calculated. When the molecule consists of a number of atoms, we sum over the contribution from all pairs, and find that the total intensity has an angular variation given by the *Wierl equation*:

$$I(\theta) = \sum_{i,j} f_i f_j \frac{\sin sR_{ij}}{sR_{ij}} \qquad s = \frac{4\pi}{\lambda} \sin \tfrac{1}{2}\theta$$

where λ is the wavelength of the electrons in the beam and θ is the scattering angle. The *electron scattering factor*, f, is a measure of the intensity of the electron scattering powers of the atoms. (a) Predict from the Wierl equation the positions of the first maximum and first minimum in the neutron and electron diffraction patterns of a Br_2 molecule obtained with neutrons of wavelength 78 pm wavelength and electrons of wavelength 4.0 pm. (b) Use the Wierl equation to predict the appearance of the 10.0 keV electron diffraction pattern of CCl_4 with an (as yet) undertermined C—Cl bond length but of known tetrahedral symmetry. Take $f_{Cl} = 17f$ and $f_C = 6f$ and note that $R(Cl,Cl) = (8/3)^{1/2}R(C,Cl)$. Plot I/f^2 against positions of the maxima, which occurred at 3° 0′, 5° 22′, and 7° 54′, and minima, which occurred at 1° 46′, 4° 6′, 6° 40′, and 9° 10′. What is the C—Cl bond length in CCl_4?

20.14‡ B.A. Bovenzi and G.A. Pearse, Jr. (*J. Chem. Soc. Dalton Trans.* 2793 (1997)) synthesized coordination compounds of the tridentate ligand pyridine-2,6-diamidoxime ($C_7H_9N_5O_2$). The compound, which they isolated from the reaction of the ligand with $CuSO_4$ (aq), did not contain a $[Cu(C_7H_9N_5O_2)_2]^{2+}$ complex cation as expected. Instead, X-ray diffraction analysis revealed a linear polymer of formula $[Cu(Cu(C_7H_9N_5O_2)(SO_4) \cdot 2H_2O]_n$, which features bridging sulfate groups. The unit cell was primitive monoclinic with $a = 1.0427$ nm, $b = 0.8876$ nm, $c = 1.3777$ nm, and $\beta = 93.254°$. The mass density of the crystals is 2.024 g cm^{-3}. How many monomer units are there per unit cell?

20.15‡ D. Sellmann, M.W. Wemple, W. Donaubauer, and F.W. Heinemann (*Inorg. Chem.* **36**, 1397 (1997)) describe the synthesis and reactivity of the ruthenium nitrido compound $[N(C_4H_9)_4][Ru(N)(S_2C_6H_4)_2]$. The ruthenium complex anion has the two 1,2-benzenedithiolate ligands (**3**) at the base of a rectangular pyramid and the nitrido ligand at the apex. Compute the mass density of the compound given that it crystallizes into an orthorhombic unit cell with $a = 3.6881$ nm, $b = 0.9402$ nm, and $c = 1.7652$ nm and eight formula units per cell. Replacing the ruthenium with an osmium results in a compound with the same crystal structure and a unit cell with a volume less than 1 per cent larger. Estimate the mass density of the osmium analogue.

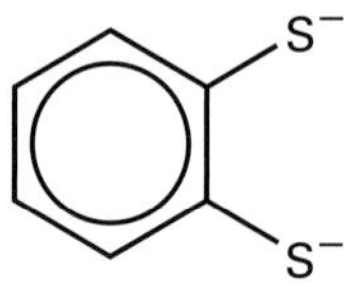

3

20.16 Aided by the Born–Mayer equation for the lattice enthalpy and a Born–Haber cycle, show that formation of CaCl is an exothermic process (the sublimation enthalpy of Ca(s) is 176 kJ mol^{-1}). Show that an explanation for the nonexistence of CaCl can be found in the reaction enthalpy for the reaction $2CaCl(s) \rightarrow Ca(s) + CaCl_2$.

20.17 In an intrinsic semiconductor, the band gap is so small that the Fermi–Dirac distribution results in some electrons populating the conduction band. It follows from the exponential form of the Fermi–Dirac distribution that the conductance G, the inverse of the resistance (with units of siemens, $1\ S = 1\ \Omega^{-1}$), of an intrinsic semiconductor should have an Arrhenius-like temperature dependence, shown in practice to have the form $G = G_0 e^{-E_g/2kT}$, where E_g is the band gap. The conductance of a sample of germanium varied with temperature as indicated below. Estimate the value of E_g.

T/K	312	354	420
G/S	0.0847	0.429	2.86

20.18 Here we investigate quantitatively the spectra of molecular solids. We begin by considering a dimer, with each monomer having a single transition with transition dipole moment μ_{mon} and wavenumber $\tilde{\nu}_{mon}$. We assume that the ground state wavefunctions are not perturbed as a result of dimerization and then write the dimer excited state wavefunctions Ψ_i as linear combinations of the excited state wavefunctions ψ_1 and ψ_2 of the monomer: $\Psi_i = c_j\psi_1 + c_k\psi_2$. Now we write the hamiltonian matrix with diagonal elements set to the energy between the excited and ground state of the monomer (which, expressed as a wavenumber, is simply $\tilde{\nu}_{mon}$), and off-diagonal elements correspond to the energy of interaction between the transition dipoles. Using the arrangement discussed in *Illustration* 20.1, we write this interaction energy (as a wavenumber) as:

$$\beta = \frac{\mu_{mon}^2}{4\pi\varepsilon_0 hcr^3}(1 - 3\cos^2\theta)$$

It follows that the hamiltonian matrix is

$$\hat{H} = \begin{pmatrix} \tilde{\nu}_{mon} & \beta \\ \beta & \tilde{\nu}_{mon} \end{pmatrix}$$

The eigenvalues of the matrix are the dimer transition wavenumbers $\tilde{\nu}_1$ and $\tilde{\nu}_2$. The eigenvectors are the wavefunctions for the excited states of the dimer and have the form $\begin{pmatrix} c_j \\ c_k \end{pmatrix}$. (a) The intensity of absorption of incident radiation is proportional to the square of the transition dipole moment (Section 9.10). The monomer transition dipole moment is $\mu_{mon} = \int \psi_1^* \hat{\mu}\, \psi_0 d\tau = \int \psi_2^* \hat{\mu}\, \psi_0 d\tau$, where ψ_0 is the wavefunction of the monomer ground state. Assume that the dimer ground state may also be described by ψ_0 and show that the transition dipole moment μ_i of each dimer transition is given by $\mu_i = \mu_{mon}(c_j + c_k)$. (b) Consider a dimer of monomers with $\mu_{mon} = 4.00$ D, $\tilde{\nu}_{mon} = 25\,000\ cm^{-1}$, and $r = 0.5$ nm. How do the transition wavenumbers $\tilde{\nu}_1$ and $\tilde{\nu}_2$ vary with the angle θ? The relative intensities of the dimer transitions may be estimated by calculating the ratio μ_2^2/μ_1^2. How does this ratio vary with the angle θ? (c) Now expand the treatment given above to a chain of N monomers ($N = 5$, 10, 15, and 20), with $\mu_{mon} = 4.00$ D, $\tilde{\nu}_{mon} = 25\,000\ cm^{-1}$, and $r = 0.5$ nm. For simplicity, assume that $\theta = 0$ and that only nearest neighbours interact with interaction energy V. For example the hamiltonian matrix for the case $N = 4$ is

$$\hat{H} = \begin{pmatrix} \tilde{\nu}_{mon} & V & 0 & 0 \\ V & \tilde{\nu}_{mon} & V & 0 \\ 0 & V & \tilde{\nu}_{mon} & V \\ 0 & 0 & V & \tilde{\nu}_{mon} \end{pmatrix}$$

How does the wavenumber of the lowest energy transition vary with size of the chain? How does the transition dipole moment of the lowest energy transition vary with the size of the chain?

20.19‡ J.J. Dannenberg, D. Liotard, P. Halvick, and J.C. Rayez (*J. Phys. Chem.* **100**, 9631 (1996)) carried out theoretical studies of organic molecules consisting of chains of unsaturated four-membered rings. The calculations

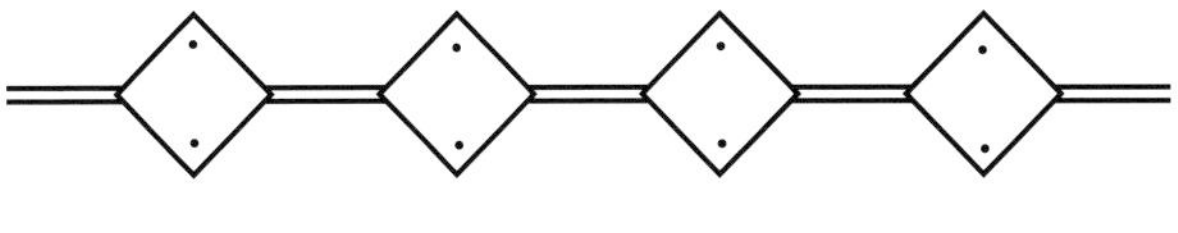

4

suggest that such compounds have large numbers of unpaired spins, and that they should therefore have unusual magnetic properties. For example, the lowest-energy state of the five-ring compound $C_{22}H_{14}$ (**4**) is computed to have $S = 3$, but the energies of $S = 2$ and $S = 4$ structures are each predicted to be $50\ kJ\ mol^{-1}$ higher in energy. Compute the molar magnetic susceptibility of these three low-lying levels at 298 K. Estimate the molar susceptibility at 298 K if each level is present in proportion to its Boltzmann factor (effectively assuming that the degeneracy is the same for all three of these levels).

20.20 Lead has $T_c = 7.19$ K and $H_c = 63\,901\ A\ m^{-1}$. At what temperature does lead become superconducting in a magnetic field of $20\ kA\ m^{-1}$?

20.21‡ P.G. Radaelli, M. Marezio, M. Perroux, S. de Brion, J.L. Tholence, Q. Huang, and A. Santoro (*Science* **265**, 380 (1994)) report the synthesis and structure of a material that becomes superconducting at temperatures below 45 K. The compound is based on a layered compound $Hg_2Ba_2YCu_2O_{8-\delta}$, which has a tetragonal unit cell with $a = 0.38606$ nm and $c = 2.8915$ nm; each unit cell contains two formula units. The compound is made superconducting by partially replacing Y by Ca, accompanied by a change in unit cell volume by less than 1 per cent. Estimate the Ca content x in superconducting $Hg_2Ba_2Y_{1-x}Ca_xCu_2O_{7.55}$ given that the mass density of the compound is $7.651\ g\ cm^{-3}$.

Theoretical problems

20.22 Show that the separation of the (hkl) planes in an orthorhombic crystal with sides a, b, and c is given by eqn 20.3.

20.23 Show that the volume of a triclinic unit cell of sides a, b, and c and angles α, β, and γ is

$$V = abc(1 - \cos^2\alpha - \cos^2\beta - \cos^2\gamma + 2\cos\alpha\cos\beta\cos\gamma)^{1/2}$$

Use this expression to derive expressions for monoclinic and orthorhombic unit cells. For the derivation, it may be helpful to use the result from vector analysis that $V = \boldsymbol{a}\cdot\boldsymbol{b}\times\boldsymbol{c}$ and to calculate V^2 initially.

20.24 Calculate the packing fractions of (a) a primitive cubic lattice, (b) a bcc unit cell, (c) an fcc unit cell.

20.25 The coordinates of the four I atoms in the unit cell of KIO_4 are (0,0,0), $(0,\frac{1}{2},\frac{1}{2})$, $(\frac{1}{2},\frac{1}{2},\frac{1}{2})$, $(\frac{1}{2},0,\frac{3}{4})$. By calculating the phase of the I reflection in the structure factor, show that the I atoms contribute no net intensity to the (114) reflection.

20.26 The coordinates, in units of a, of the A atoms, with scattering factor f_A, in a cubic lattice are (0,0,0), (0,1,0), (0,0,1), (0,1,1), (1,0,0), (1,1,0), (1,0,1), and (1,1,1). There is also a B atom, with scattering factor f_B, at $(\frac{1}{2},\frac{1}{2},\frac{1}{2})$. Calculate the structure factors F_{hkl} and predict the form of the powder diffraction pattern when (a) $f_A = f$, $f_B = 0$, (b) $f_B = \frac{1}{2}f_A$, and (c) $f_A = f_B = f$.

20.27 For an isotropic substance, the moduli and Poisson's ratio may be expressed in terms of two parameters λ and μ called the *Lamé constants*:

$$E = \frac{\mu(3\lambda + 2\mu)}{\lambda + \mu} \qquad K = \frac{3\lambda + 2\mu}{3} \qquad G = \mu \qquad \nu_P = \frac{\lambda}{3(\lambda + \mu)}$$

Use the Lamé constants to confirm the relations between G, K, and E given in eqn 20.18.

20.28 When energy levels in a band form a continuum, the density of states $\rho(E)$, the number of levels in an energy range divided by the width of the range, may be written as $\rho(E) = dk/dE$, where dk is the change in the quantum number k and dE is the energy change. (a) Use eqn 20.21 to show that

$$\rho(E) = -\frac{(N+1)/2\pi\beta}{\left\{1-\left(\frac{E-\alpha}{2\beta}\right)^2\right\}^{1/2}}$$

where k, N, α, and β have the meanings described in Section 20.9. (b) Use the expression above to show that $\rho(E)$ becomes infinite as E approaches $\alpha \pm 2\beta$. That is, show that the density of states increases towards the edges of the bands in a one-dimensional metallic conductor.

20.29 The treatment in Problem 20.28 applies only to one-dimensional solids. In three dimensions, the variation of density of states is more like that shown in Fig. 20.70. Account for the fact that in a three-dimensional solid the greatest density of states lies near the centre of the band and the lowest density at the edges.

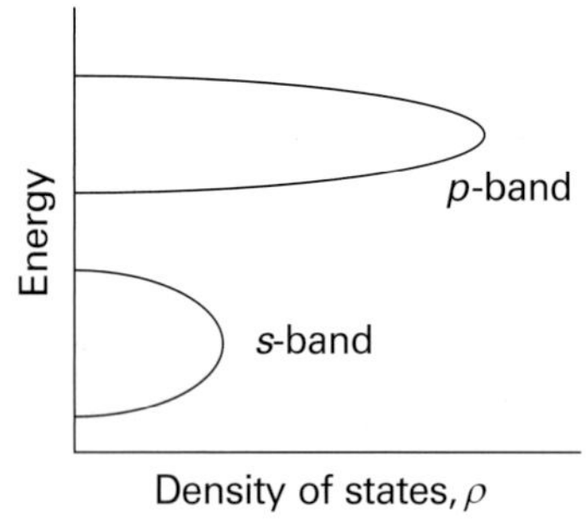

Fig. 20.70

20.30 Show that, if a substance responds nonlinearly to two sources of radiation, one of frequency ω_1 and the other of frequency ω_2, then it may give rise to radiation of the sum and difference of the two frequencies. This nonlinear optical phenomenon is known as *frequency mixing* and is used to expand the wavelength range of lasers in laboratory applications, such as spectroscopy and photochemistry.

20.31 The magnetizability, ξ, and the volume and molar magnetic susceptibilities can be calculated from the wavefunctions of molecules. For instance, the magnetizability of a hydrogenic atom is given by the expression $\xi = -(e^2/6m_e)\langle r^2 \rangle$, where $\langle r^2 \rangle$ is the (expectation) mean value of r^2 in the atom. Calculate ξ and χ_m for the ground state of a hydrogenic atom.

20.32 Nitrogen dioxide, a paramagnetic compound, is in equilibrium with its dimer, dinitrogen tetroxide, a diamagnetic compound. Derive an expression in terms of the equilibrium constant, K, for the dimerization to show how the molar susceptibility varies with the pressure of the sample. Suggest how the susceptibility might be expected to vary as the temperature is changed at constant pressure.

20.33 An NO molecule has thermally accessible electronically excited states. It also has an unpaired electron, and so may be expected to be paramagnetic. However, its ground state is not paramagnetic because the magnetic moment of the orbital motion of the unpaired electron almost exactly cancels the spin magnetic moment. The first excited state (at 121 cm^{-1}) is paramagnetic because the orbital magnetic moment adds to, rather than cancels, the spin magnetic moment. The upper state has a magnetic moment of $2\mu_B$. Because the upper state is thermally accessible, the paramagnetic susceptibility of NO shows a pronounced temperature dependence even near room temperature. Calculate the molar paramagnetic susceptibility of NO and plot it as a function of temperature.

Applications: to biochemistry and nanoscience

20.34 Although the crystallization of large biological molecules may not be as readily accomplished as that of small molecules, their crystal lattices are no different. Tobacco seed globulin forms face-centred cubic crystals with unit cell dimension of 12.3 nm and a density of 1.287 g cm^{-3}. Determine its molar mass.

20.35 What features in an X-ray diffraction pattern suggest a helical conformation for a biological macromolecule? Use Fig. 20.26 to deduce as much quantitative information as you can about the shape and size of a DNA molecule.

20.36 A transistor is a semiconducting device that is commonly used either as a switch or an amplifier of electrical signals. Prepare a brief report on the design of a nanometre-sized transistor that uses a carbon nanotube as a component. A useful starting point is the work summarized by Tans *et al. Nature* **393**, 49 (1998).

20.37 The tip of a scanning tunnelling microscope can be used to move atoms on a surface. The movement of atoms and ions depends on their ability to leave one position and stick to another, and therefore on the energy changes that occur. As an illustration, consider a two-dimensional square lattice of univalent positive and negative ions separated by 200 pm, and consider a cation on top of this array. Calculate, by direct summation, its Coulombic interaction when it is in an empty lattice point directly above an anion.

PART 3 Change

Part 3 considers the processes by which change occurs. We prepare the ground for a discussion of the rates of reactions by considering the motion of molecules in gases and in liquids. Then we establish the precise meaning of reaction rate, and see how the overall rate, and the complex behaviour of some reactions, may be expressed in terms of elementary steps and the atomic events that take place when molecules meet. Characteristic physical and chemical events take place at surfaces, including catalysis, and we see how to describe them. A special type of surface is that of an electrode, and we shall see how to describe and understand the rate at which electrons are transferred between an electrode and species in solution.

Molecules in motion

21

One of the simplest types of molecular motion to describe is the random motion of molecules of a perfect gas. We see that a simple theory accounts for the pressure of a gas and the rates at which molecules and energy migrate through gases. Molecular mobility is particularly important in liquids. Another simple kind of motion is the largely uniform motion of ions in solution in the presence of an electric field. Molecular and ionic motion have common features and, by considering them from a more general viewpoint, we derive expressions that govern the migration of properties through matter. One of the most useful consequences of this general approach is the formulation of the diffusion equation, which is an equation that shows how matter and energy spread through media of various kinds. Finally, we build a simple model for all types of molecular motion, in which the molecules migrate in a series of small steps, and see that it accounts for many of the properties of migrating molecules in both gases and condensed phases.

The general approach we describe in this chapter provides techniques for discussing the motion of all kinds of particles in all kinds of fluids. We set the scene by considering a simple type of motion, that of molecules in a perfect gas, and go on to see that molecular motion in liquids shows a number of similarities. We shall concentrate on the **transport properties** of a substance, its ability to transfer matter, energy, or some other property from one place to another. Four examples of transport properties are

Diffusion, the migration of matter down a concentration gradient.

Thermal conduction, the migration of energy down a temperature gradient.

Electric conduction, the migration of electric charge along an electrical potential gradient.

Viscosity, the migration of linear momentum down a velocity gradient.

It is convenient to include in the discussion **effusion**, the emergence of a gas from a container through a small hole.

Molecular motion in gases

Here we present the kinetic model of a perfect gas as a starting point for the discussion of its transport properties. In the **kinetic model** of gases we assume that the only contribution to the energy of the gas is from the kinetic energies of the molecules. The kinetic model is one of the most remarkable—and arguably most beautiful—models in physical chemistry for, from a set of very slender assumptions, powerful quantitative conclusions can be deduced.

21.1 The kinetic model of gases

The kinetic model is based on three assumptions:

1 The gas consists of molecules of mass m in ceaseless random motion.

2 The size of the molecules is negligible, in the sense that their diameters are much smaller than the average distance travelled between collisions.

3 The molecules interact only through brief, infrequent, and elastic collisions.

An **elastic collision** is a collision in which the total translational kinetic energy of the molecules is conserved.

(a) Pressure and molecular speeds

From the very economical assumptions of the kinetic model, we show in the following *Justification* that the pressure and volume of the gas are related by

$$pV = \tfrac{1}{3}nMc^2 \qquad (21.1)^\circ$$

where $M = mN_A$, the molar mass of the molecules, and c is the **root mean square speed** of the molecules, the square root of the mean of the squares of the speeds, v, of the molecules:

$$c = \langle v^2 \rangle^{1/2} \qquad [21.2]$$

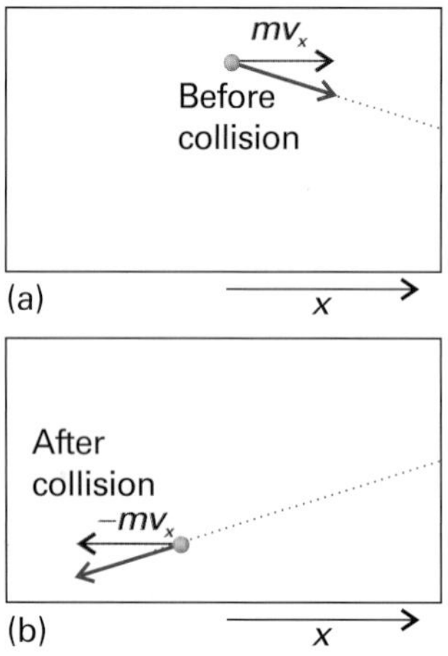

Fig. 21.1 The pressure of a gas arises from the impact of its molecules on the walls. In an elastic collision of a molecule with a wall perpendicular to the x-axis, the x-component of velocity is reversed but the y- and z-components are unchanged.

Justification 21.1 *The pressure of a gas according to the kinetic model*

Consider the arrangement in Fig. 21.1. When a particle of mass m that is travelling with a component of velocity v_x parallel to the x-axis collides with the wall on the right and is reflected, its linear momentum (the product of its mass and its velocity) changes from mv_x before the collision to $-mv_x$ after the collision (when it is travelling in the opposite direction). The x-component of momentum therefore changes by $2mv_x$ on each collision (the y- and z-components are unchanged). Many molecules collide with the wall in an interval Δt, and the total change of momentum is the product of the change in momentum of each molecule multiplied by the number of molecules that reach the wall during the interval.

Because a molecule with velocity component v_x can travel a distance $v_x\Delta t$ along the x-axis in an interval Δt, all the molecules within a distance $v_x\Delta t$ of the wall will strike it if they are travelling towards it (Fig. 21.2). It follows that, if the wall has area A, then all the particles in a volume $A \times v_x\Delta t$ will reach the wall (if they are travelling towards it). The number density of particles is nN_A/V, where n is the total amount of molecules in the container of volume V and N_A is Avogadro's constant, so the number of molecules in the volume $Av_x\Delta t$ is $(nN_A/V) \times Av_x\Delta t$.

At any instant, half the particles are moving to the right and half are moving to the left. Therefore, the average number of collisions with the wall during the interval Δt is $\tfrac{1}{2}nN_AAv_x\Delta t/V$. The total momentum change in that interval is the product of this number and the change $2mv_x$:

$$\text{Momentum change} = \frac{nN_AAv_x\Delta t}{2V} \times 2mv_x = \frac{nmAN_Av_x^2\Delta t}{V} = \frac{nMAv_x^2\Delta t}{V}$$

where $M = mN_A$.

Next, to find the force, we calculate the rate of change of momentum, which is this change of momentum divided by the interval Δt during which it occurs:

$$\text{Rate of change of momentum} = \frac{nMAv_x^2}{V}$$

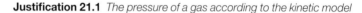

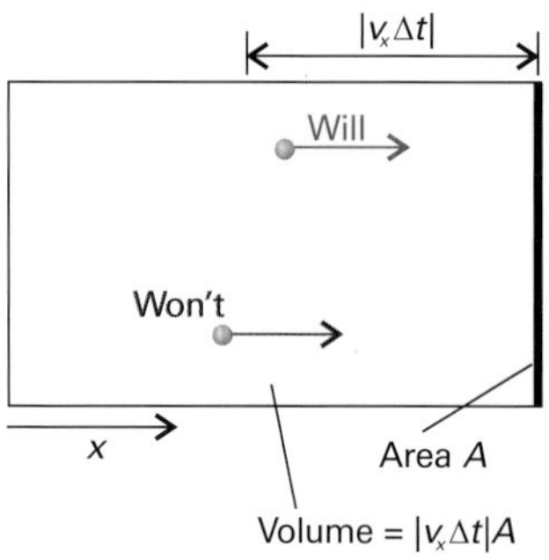

Fig. 21.2 A molecule will reach the wall on the right within an interval Δt if it is within a distance $v_x\Delta t$ of the wall and travelling to the right.

This rate of change of momentum is equal to the force (by Newton's second law of motion). It follows that the pressure, the force divided by the area, is

$$\text{Pressure} = \frac{nMv_x^2}{V}$$

Not all the molecules travel with the same velocity, so the detected pressure, p, is the average (denoted $\langle \ldots \rangle$) of the quantity just calculated:

$$p = \frac{nM\langle v_x^2 \rangle}{V}$$

This expression already resembles the perfect gas equation of state.

To write an expression of the pressure in terms of the root mean square speed, c, we begin by writing the speed of a single molecule, v, as $v^2 = v_x^2 + v_y^2 + v_z^2$. Because the root-mean-square speed, c, is defined as $c = \langle v^2 \rangle^{1/2}$ (eqn 21.2), it follows that

$$c^2 = \langle v^2 \rangle = \langle v_x^2 \rangle + \langle v_y^2 \rangle + \langle v_z^2 \rangle$$

However, because the molecules are moving randomly, all three averages are the same. It follows that $c^2 = 3\langle v_x^2 \rangle$. Equation 21.1 follows immediately by substituting $\langle v_x^2 \rangle = \frac{1}{3}c^2$ into $p = nM\langle v_x^2 \rangle / V$.

Equation 21.1 is one of the key results of the kinetic model. We see that if the root mean square speed of the molecules depends only on the temperature, then at constant temperature

$$pV = \text{constant}$$

which is the content of Boyle's law (Section 1.2). Moreover, for eqn 21.1 to be the equation of state of a perfect gas, its right-hand side must be equal to nRT. It follows that the root mean square speed of the molecules in a gas at a temperature T must be

$$c = \left(\frac{3RT}{M}\right)^{1/2} \qquad (21.3)^\circ$$

We can conclude that *the root mean square speed of the molecules of a gas is proportional to the square root of the temperature and inversely proportional to the square root of the molar mass*. That is, the higher the temperature, the higher the root mean square speed of the molecules, and, at a given temperature, heavy molecules travel more slowly than light molecules. Sound waves are pressure waves, and for them to propagate the molecules of the gas must move to form regions of high and low pressure. Therefore, we should expect the root mean square speeds of molecules to be comparable to the speed of sound in air (340 m s^{-1}). The root mean square speed of N_2 molecules, for instance, is found from eqn 21.3 to be 515 m s^{-1}.

Equation 21.3 is an expression for the mean square speed of molecules. However, in an actual gas the speeds of individual molecules span a wide range, and the collisions in the gas continually redistribute the speeds among the molecules. Before a collision, a molecule may be travelling rapidly, but after a collision it may be accelerated to a very high speed, only to be slowed again by the next collision. The fraction of molecules that have speeds in the range v to $v + dv$ is proportional to the width of the range, and is written $f(v)dv$, where $f(v)$ is called the **distribution of speeds**.

The precise form of f for molecules of a gas at a temperature T was derived by J.C. Maxwell, and is

$$f(v) = 4\pi\left(\frac{M}{2\pi RT}\right)^{3/2} v^2 e^{-Mv^2/2RT} \qquad (21.4)$$

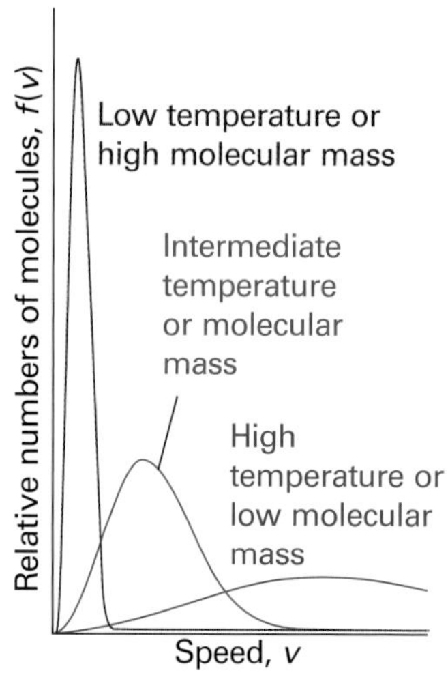

Fig. 21.3 The distribution of molecular speeds with temperature and molar mass. Note that the most probable speed (corresponding to the peak of the distribution) increases with temperature and with decreasing molar mass, and simultaneously the distribution becomes broader.

Exploration (a) Plot different distributions by keeping the molar mass constant at 100 g mol^{-1} and varying the temperature of the sample between 200 K and 2000 K. (b) Use mathematical software or the *Living graph* applet from the text's web site to evaluate numerically the fraction of molecules with speeds in the range 100 m s^{-1} to 200 m s^{-1} at 300 K and 1000 K. (c) Based on your observations, provide a molecular interpretation of temperature.

This expression is called the **Maxwell distribution of speeds** and is derived in the following *Justification*. Let's consider its features, which are also shown pictorially in Fig. 21.3:

1 Equation 21.4 includes a decaying exponential function, the term $e^{-Mv^2/2RT}$. Its presence implies that the fraction of molecules with very high speeds will be very small because e^{-x^2} becomes very small when x^2 is large.

2 The factor $M/2RT$ multiplying v^2 in the exponent is large when the molar mass, M, is large, so the exponential factor goes most rapidly towards zero when M is large. That is, heavy molecules are unlikely to be found with very high speeds.

3 The opposite is true when the temperature, T, is high: then the factor $M/2RT$ in the exponent is small, so the exponential factor falls towards zero relatively slowly as v increases. In other words, a greater fraction of the molecules can be expected to have high speeds at high temperatures than at low temperatures.

4 A factor v^2 (the term before the e) multiplies the exponential. This factor goes to zero as v goes to zero, so the fraction of molecules with very low speeds will also be very small.

5 The remaining factors (the term in parentheses in eqn 21.4 and the 4π) simply ensure that, when we add together the fractions over the entire range of speeds from zero to infinity, then we get 1.

To use eqn 21.4 to calculate the fraction of molecules in a given narrow range of speeds, Δv, we evaluate $f(v)$ at the speed of interest, then multiply it by the width of the range of speeds of interest; that is, we form $f(v)\Delta v$. To use the distribution to calculate the fraction in a range of speeds that is too wide to be treated as infinitesimal, we evaluate the integral:

$$\text{Fraction in the range } v_1 \text{ to } v_2 = \int_{v_1}^{v_2} f(v)\mathrm{d}v \qquad (21.5)$$

This integral is the area under the graph of f as a function of v and, except in special cases, has to be evaluated numerically by using mathematical software (Fig. 21.4).

Justification 21.2 *The Maxwell distribution of speeds*

The Boltzmann distribution is a key result of physical chemistry and is treated fully in Section 16.1. It implies that the fraction of molecules with velocity components v_x, v_y, v_z is proportional to an exponential function of their kinetic energy, which is

$$E = \tfrac{1}{2}mv_x^2 + \tfrac{1}{2}mv_y^2 + \tfrac{1}{2}mv_z^2$$

Therefore, we can use the relation $a^{x+y+z+\cdots} = a^x a^y a^z \cdots$ to write

$$f = Ke^{-E/kT} = Ke^{-(\frac{1}{2}mv_x^2+\frac{1}{2}mv_y^2+\frac{1}{2}mv_z^2)/kT} = Ke^{-mv_x^2/2kT}e^{-mv_y^2/2kT}e^{-mv_z^2/2kT}$$

where K is a constant of proportionality (at constant temperature) and $f\mathrm{d}v_x\mathrm{d}v_y\mathrm{d}v_z$ is the fraction of molecules in the velocity range v_x to $v_x + \mathrm{d}v_x$, v_y to $v_y + \mathrm{d}v_y$, and v_z to $v_z + \mathrm{d}v_z$. We see that the fraction factorizes into three factors, one for each axis, and we can write $f = f(v_x)f(v_y)f(v_z)$ with

$$f(v_x) = K^{1/3}e^{-mv_x^2/2kT}$$

and likewise for the two other directions.

To determine the constant K, we note that a molecule must have a velocity somewhere in the range $-\infty < v_x < \infty$, so

$$\int_{-\infty}^{\infty} f(v_x)\mathrm{d}v_x = 1$$

Substitution of the expression for $f(v_x)$ then gives

$$1 = K^{1/3}\int_{-\infty}^{\infty} e^{-mv_x^2/2kT}dv_x = K^{1/3}\left(\frac{2\pi kT}{m}\right)^{1/2}$$

where we have used the standard integral

$$\int_{-\infty}^{\infty} e^{-ax^2}dx = \frac{\pi}{a}$$

Therefore, $K = (m/2\pi kT)^{3/2} = (M/2\pi RT)^{3/2}$, where M is the molar mass of the molecules. At this stage we know that

$$f(v_x) = \left(\frac{M}{2\pi RT}\right)^{1/2} e^{-Mv_x^2/2RT} \qquad (21.6)$$

The probability that a molecule has a velocity in the range v_x to $v_x + dv_x$, v_y to $v_y + dv_y$, v_z to $v_z + dv_z$ is

$$f(v_x)f(v_y)f(v_z)dv_xdv_ydv_z = \left(\frac{M}{2\pi RT}\right)^{3/2} e^{-Mv_x^2/2RT}dv_xdv_ydv_z$$

where $v^2 = v_x^2 + v_y^2 + v_z^2$. The probability $f(v)dv$ that the molecules have a speed in the range v to $v + dv$ regardless of direction is the sum of the probabilities that the velocity lies in any of the volume elements $dv_xdv_ydv_z$ forming a spherical shell of radius v and thickness dv (Fig. 21.5). The sum of the volume elements on the right-hand side of the last equation is the volume of this shell, $4\pi v^2dv$. Therefore,

$$f(v) = 4\pi\left(\frac{M}{2\pi RT}\right)^{3/2} v^2e^{-Mv_x^2/2RT}$$

as given in eqn 21.4.

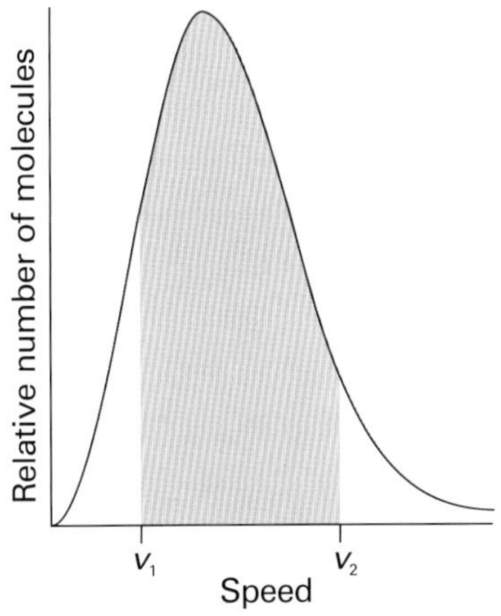

Fig. 21.4 To calculate the probability that a molecule will have a speed in the range v_1 to v_2, we integrate the distribution between those two limits; the integral is equal to the area of the curve between the limits, as shown shaded here.

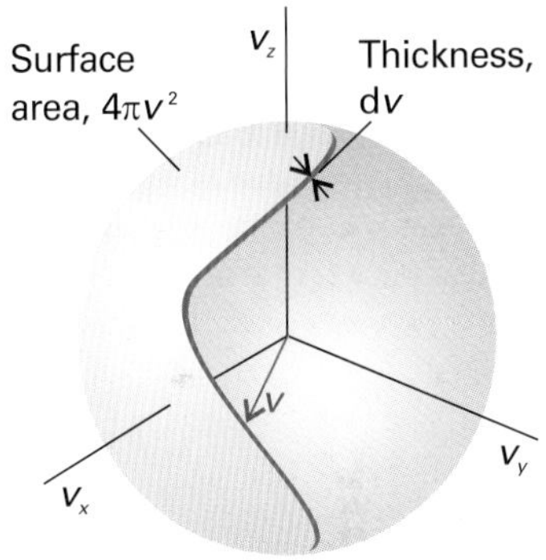

Fig. 21.5 To evaluate the probability that a molecule has a speed in the range v to $v + dv$, we evaluate the total probability that the molecule will have a speed that is anywhere on the surface of a sphere of radius $v = (v_x^2 + v_y^2 + v_z^2)^{1/2}$ by summing the probabilities that it is in a volume element $dv_xdv_ydv_z$ at a distance v from the origin.

Example 21.1 *Calculating the mean speed of molecules in a gas*

What is the mean speed, $\bar{c}$, of N_2 molecules in air at 25°C?

Method We need to use the results of probability theory summarized in *Appendix* 2. In this case, we are asked to calculate the mean speed, not the root mean square speed. A mean speed is calculated by multiplying each speed by the fraction of molecules that have that speed, and then adding all the products together. When the speed varies over a continuous range, the sum is replaced by an integral. To employ this approach here, we note that the fraction of molecules with a speed in the range v to $v + dv$ is $f(v)dv$, so the product of this fraction and the speed is $vf(v)dv$. The mean speed, $\bar{c}$, is obtained by evaluating the integral

$$\bar{c} = \int_0^{\infty} vf(v)dv$$

with f given in eqn 21.4.

Answer The integral required is

$$\bar{c} = 4\pi\left(\frac{M}{2\pi RT}\right)^{3/2}\int_0^{\infty} v^3e^{-Mv^2/2RT}dv$$

$$= 4\pi\left(\frac{M}{2\pi RT}\right)^{3/2} \times \tfrac{1}{2}\left(\frac{2RT}{M}\right)^2 = \left(\frac{8RT}{\pi M}\right)^{1/2}$$

where we have used the standard result from tables of integrals (or software) that

$$\int_0^{\infty} x^3 e^{-ax^2}\,dx = \frac{1}{2a^2}$$

Substitution of the data then gives

$$\bar{c} = \left(\frac{8\times(8.3141\ \mathrm{J\,K^{-1}\,mol^{-1}})\times(298\ \mathrm{K})}{\pi\times(28.02\times10^{-3}\ \mathrm{kg\,mol^{-1}})}\right)^{1/2} = 475\ \mathrm{m\,s^{-1}}$$

where we have used $1\ \mathrm{J} = 1\ \mathrm{kg\,m^2\,s^{-2}}$.

Self-test 21.1 Evaluate the root mean square speed of the molecules by integration. You will need the integral

$$\int_0^{\infty} x^4 e^{-ax^2}\,dx = \tfrac{3}{8}\left(\frac{\pi}{a^5}\right)^{1/2}$$

[$c = (3RT/M)^{1/2}$, $515\ \mathrm{m\,s^{-1}}$]

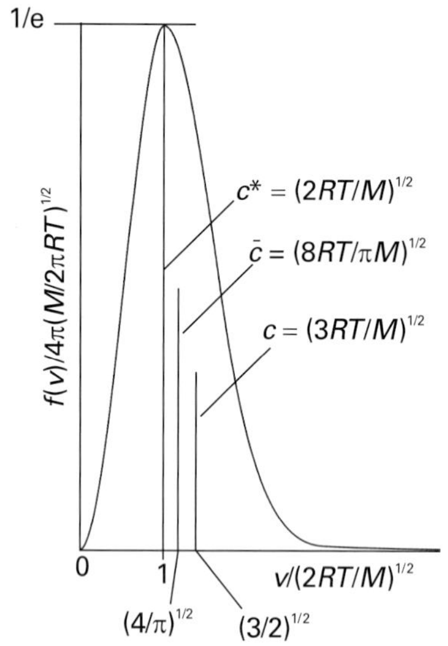

Fig. 21.6 A summary of the conclusions that can be deduced from the Maxwell distribution for molecules of molar mass M at a temperature T: c^* is the most probable speed, $\bar{c}$ is the mean speed, and c is the root mean square speed.

Comment 21.1

To find the location of the peak of the distribution, differentiate f with respect to v and look for the value of v at which the derivative is zero (other than at $v = 0$ and $v = \infty$).

As shown in Example 21.1, we can use the Maxwell distribution to evaluate the **mean speed**, $\bar{c}$, of the molecules in a gas:

$$\bar{c} = \left(\frac{8RT}{\pi M}\right)^{1/2} \tag{21.7}$$

We can identify the **most probable speed**, c^*, from the location of the peak of the distribution:

$$c^* = \left(\frac{2RT}{M}\right)^{1/2} \tag{21.8}$$

Figure 21.6 summarizes these results.

The **relative mean speed**, $\bar{c}_{\mathrm{rel}}$, the mean speed with which one molecule approaches another, can also be calculated from the distribution:

$$\bar{c}_{\mathrm{rel}} = 2^{1/2}\bar{c} \tag{21.9}$$

This result is much harder to derive, but the diagram in Fig. 21.7 should help to show that it is plausible. The last result can also be generalized to the relative mean speed of two dissimilar molecules of masses m_A and m_B:

$$\bar{c}_{\mathrm{rel}} = \left(\frac{8kT}{\pi\mu}\right)^{1/2} \qquad \mu = \frac{m_A m_B}{m_A + m_B} \tag{21.10}$$

Note that the molecular masses (not the molar masses) and Boltzmann's constant, $k = R/N_A$, appear in this expression; the quantity μ is called the **reduced mass** of the molecules. Equation 21.10 turns into eqn 21.7 when the molecules are identical (that is, $m_A = m_B = m$, so $\mu = \tfrac{1}{2}m$).

The Maxwell distribution has been verified experimentally. For example, molecular speeds can be measured directly with a velocity selector (Fig. 21.8). The spinning cylinder has channels that permit the passage of only those molecules moving through them at the appropriate speed, and the number of molecules can be determined by collecting them at a detector.

(b) The collision frequency

A qualitative picture of the events taking place in a gas was first described in Section 1.2. The kinetic model enables us to make that picture more quantitative. In particular,

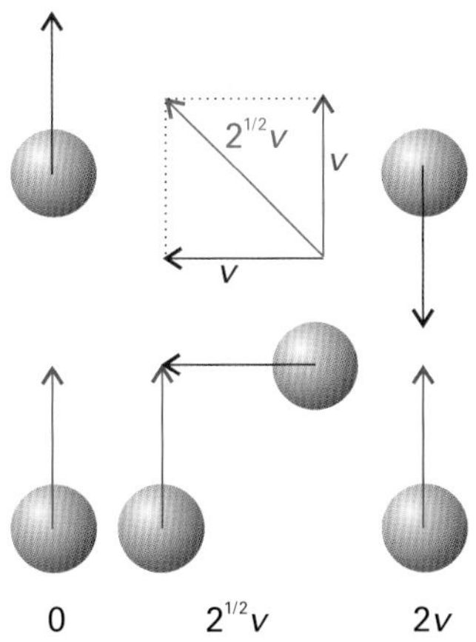

Fig. 21.7 A simplified version of the argument to show that the mean relative speed of molecules in a gas is related to the mean speed. When the molecules are moving in the same direction, the mean relative speed is zero; it is $2v$ when the molecules are approaching each other. A typical mean direction of approach is from the side, and the mean speed of approach is then $2^{1/2}v$. The last direction of approach is the most characteristic, so the mean speed of approach can be expected to be about $2^{1/2}v$. This value is confirmed by more detailed calculation.

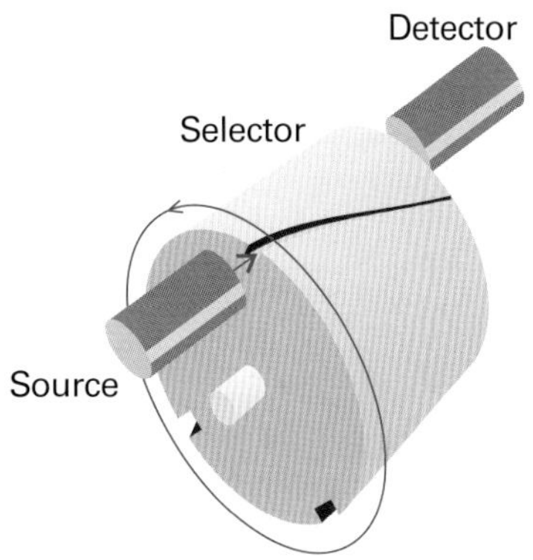

Fig. 21.8 A velocity selector. The molecules are produced in the source (which may be an oven with a small hole in one wall), and travel in a beam towards the rotating channels. Only if the speed of a molecule is such as to carry it along the channel that rotates into its path will it reach the detector. Thus, the number of slow molecules can be counted by rotating the cylinder slowly, and the number of fast molecules counted by rotating the cylinder rapidly.

Synoptic table 21.1* Collision cross-sections

	σ/nm^2
C_6H_6	0.88
CO_2	0.52
He	0.21
N_2	0.43

* More values are given in the *Data section*.

it enables us to calculate the frequency with which molecular collisions occur and the distance a molecule travels on average between collisions.

We count a 'hit' whenever the centres of two molecules come within a distance d of each other, where d, the **collision diameter**, is of the order of the actual diameters of the molecules (for impenetrable hard spheres d is the diameter). As we show in the following *Justification*, we can use kinetic model to deduce that the **collision frequency**, z, the number of collisions made by one molecule divided by the time interval during which the collisions are counted, when there are N molecules in a volume V is

$$z = \sigma\bar{c}_{\text{rel}}\mathcal{N} \qquad (21.11\text{a})^\circ$$

with $\mathcal{N} = N/V$ and $\bar{c}_{\text{rel}}$ given in eqn 21.10. The area $\sigma = \pi d^2$ is called the **collision cross-section** of the molecules. Some typical collision cross-sections are given in Table 21.1 (they are obtained by the techniques described in Section 18.6). In terms of the pressure,

$$z = \frac{\sigma\bar{c}_{\text{rel}}p}{kT} \qquad (21.11\text{b})^\circ$$

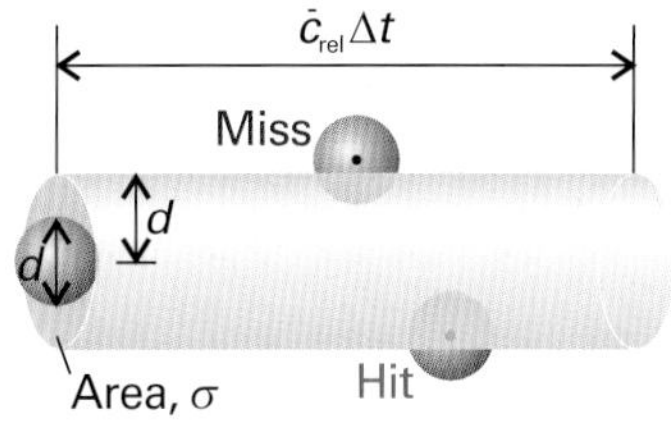

Fig. 21.9 In an interval Δt, a molecule of diameter d sweeps out a tube of radius d and length $\bar{c}_{\text{rel}}\Delta t$. As it does so it encounters other molecules with centres that lie within the tube, and each such encounter counts as one collision. In reality, the tube is not straight, but changes direction at each collision. Nevertheless, the volume swept out is the same, and this straightened version of the tube can be used as a basis of the calculation.

Justification 21.3 *Using the kinetic model to calculate the collision frequency*

We consider the positions of all the molecules except one to be frozen. Then we note what happens as one mobile molecule travels through the gas with a mean relative speed $\bar{c}_{\text{rel}}$ for a time Δt. In doing so it sweeps out a 'collision tube' of cross-sectional area $\sigma = \pi d^2$ and length $\bar{c}_{\text{rel}}\Delta t$, and therefore of volume $\sigma\bar{c}_{\text{rel}}\Delta t$ (Fig. 21.9). The number of stationary molecules with centres inside the collision tube is given by the volume of the tube multiplied by the number density $\mathcal{N} = N/V$, and is $\mathcal{N}\sigma\bar{c}_{\text{rel}}\Delta t$. The number of hits scored in the interval Δt is equal to this number, so the number of collisions divided by the time interval, is $\mathcal{N}\sigma\bar{c}_{\text{rel}}$. The expression in terms of the pressure of the gas is obtained by using the perfect gas equation to write

$$\mathcal{N} = \frac{N}{V} = \frac{nN_{\text{A}}}{V} = \frac{pN_{\text{A}}}{RT} = \frac{p}{kT}$$

Equation 21.11a shows that, at constant volume, the collision frequency increases with increasing temperature. The reason for this increase is that the relative mean speed increases with temperature (eqns 21.9 and 21.10). Equation 21.11b shows that, at constant temperature, the collision frequency is proportional to the pressure. Such a proportionality is plausible, for the greater the pressure, the greater the number density of molecules in the sample, and the rate at which they encounter one another is greater even though their average speed remains the same. For an N_2 molecule in a sample at 1 atm and 25°C, $z \approx 5 \times 10^9\ \text{s}^{-1}$, so a given molecule collides about 5×10^9 times each second. We are beginning to appreciate the timescale of events in gases.

(c) The mean free path

Once we have the collision frequency, we can calculate the **mean free path**, λ (lambda), the average distance a molecule travels between collisions. If a molecule collides with a frequency z, it spends a time $1/z$ in free flight between collisions, and therefore travels a distance $(1/z)\bar{c}$. It follows that the mean free path is

$$\lambda = \frac{\bar{c}}{z} \tag{21.12}$$

Substitution of the expression for z in eqn 21.11b gives

$$\lambda = \frac{kT}{2^{1/2}\sigma p} \tag{21.13}$$

Doubling the pressure reduces the mean free path by half. A typical mean free path in nitrogen gas at 1 atm is 70 nm, or about 10^3 molecular diameters. Although the temperature appears in eqn 21.13, in a sample of constant volume, the pressure is proportional to T, so T/p remains constant when the temperature is increased. Therefore, the mean free path is independent of the temperature in a sample of gas in a container of fixed volume. The distance between collisions is determined by the number of molecules present in the given volume, not by the speed at which they travel.

In summary, a typical gas (N_2 or O_2) at 1 atm and 25°C can be thought of as a collection of molecules travelling with a mean speed of about 500 m s^{-1}. Each molecule makes a collision within about 1 ns, and between collisions it travels about 10^3 molecular diameters. The kinetic model of gases is valid (and the gas behaves nearly perfectly) if the diameter of the molecules is much smaller than the mean free path ($d \ll \lambda$), for then the molecules spend most of their time far from one another.

IMPACT ON ASTROPHYSICS

I21.1 The Sun as a ball of perfect gas

The kinetic model of gases is valid when the size of the particles is negligible compared with their mean free path. It may seem absurd, therefore, to expect the kinetic model and, as a consequence, the perfect gas law, to be applicable to the dense matter of stellar interiors. In the Sun, for instance, the density at its centre is 1.50 times that of liquid water and comparable to that of water about half way to its surface. However, we have to realize that the state of matter is that of a *plasma*, in which the electrons have been stripped from the atoms of hydrogen and helium that make up the bulk of the matter of stars. As a result, the particles making up the plasma have diameters comparable to those of nuclei, or about 10 fm. Therefore, a mean free path of only 0.1 pm satisfies the criterion for the validity of the kinetic theory and the perfect gas law. We can therefore use $pV = nRT$ as the equation of state for the stellar interior.

As for any perfect gas, the pressure in the interior of the Sun is related to the mass density, $\rho = m/V$, by $p = \rho RT/M$. Atoms are stripped of their electrons in the interior of stars so, if we suppose that the interior consists of ionized hydrogen atoms, the mean molar mass is one-half the molar mass of hydrogen, or 0.5 g mol^{-1} (the mean of the molar mass of H^+ and e^-, the latter being virtually 0). Half way to the centre of the Sun, the temperature is 3.6 MK and the mass density is 1.20 g cm^{-3} (slightly denser than water); so the pressure there works out as 7.2×10^{13} Pa, or about 720 million atmospheres.

We can combine this result with the expression for the pressure from the kinetic model (eqn 21.1). Because the total kinetic energy of the particles is $E_K = \frac{1}{2}Nmc^2$, we can write $p = \frac{2}{3}E_K/V$. That is, the pressure of the plasma is related to the *kinetic energy density*, $\rho_K = E_K/V$, the kinetic energy of the molecules in a region divided by the volume of the region, by $p = \frac{2}{3}\rho_K$. It follows that the kinetic energy density half way to the centre of the Sun is about 0.11 GJ cm^{-3}. In contrast, on a warm day (25°C) on Earth, the (translational) kinetic energy density of our atmosphere is only 0.15 J cm^{-3}.

21.2 Collisions with walls and surfaces

The key result for accounting for transport in the gas phase is the rate at which molecules strike an area (which may be an imaginary area embedded in the gas, or part of a real wall). The **collision flux**, Z_W, is the number of collisions with the area in a given time interval divided by the area and the duration of the interval. The **collision frequency**, the number of hits per second, is obtained by multiplication of the collision flux by the area of interest. We show in the *Justification* below that the collision flux is

$$Z_W = \frac{p}{(2\pi mkT)^{1/2}} \qquad (21.14)°$$

When $p = 100$ kPa (1.00 bar) and $T = 300$ K, $Z_W \approx 3 \times 10^{23}$ cm^{-2} s^{-1}.

Justification 21.4 *The collision flux*

Consider a wall of area A perpendicular to the x-axis (as in Fig. 21.2). If a molecule has $v_x > 0$ (that is, it is travelling in the direction of positive x), then it will strike the wall within an interval Δt if it lies within a distance $v_x\Delta t$ of the wall. Therefore, all molecules in the volume $Av_x\Delta t$, and with positive x-component of velocities, will strike the wall in the interval Δt. The total number of collisions in this interval is therefore the volume $Av_x\Delta t$ multiplied by the number density, $\mathcal{N}$, of molecules. However, to take account of the presence of a range of velocities in the sample, we must sum the result over all the positive values of v_x weighted by the probability distribution of velocities (eqn 21.6):

$$\text{Number of collisions} = \mathcal{N}A\Delta t\int_0^\infty v_x f(v_x)\mathrm{d}x$$

The collision flux is the number of collisions divided by A and Δt, so

$$Z_W = \mathcal{N}\int_0^\infty v_x f(v_x)\mathrm{d}x$$

Then, using the velocity distribution in eqn 21.6,

$$\int_0^\infty v_x f(v_x)\mathrm{d}v_x = \left(\frac{m}{2\pi kT}\right)^{1/2}\int_0^\infty v_x \mathrm{e}^{-mv_x^2/2kT}\mathrm{d}v_x = \left(\frac{kT}{2\pi m}\right)^{1/2}$$

where we have used the standard integral

$$\int_0^\infty x\mathrm{e}^{-ax^2}\mathrm{d}x = \frac{1}{2a}$$

Therefore,

$$Z_\mathrm{W} = \mathcal{N}\left(\frac{kT}{2\pi m}\right)^{1/2} = \tfrac{1}{4}\bar{c}\mathcal{N} \qquad (21.15)^\circ$$

where we have used eqn 21.7 in the form $\bar{c} = (8kT/\pi m)^{1/2}$, which implies that $\tfrac{1}{4}\bar{c} = (kT/2\pi m)^{1/2}$. Substitution of $\mathcal{N} = nN_\mathrm{A}/V = p/kT$ gives eqn 21.14.

21.3 The rate of effusion

The essential empirical observations on effusion are summarized by **Graham's law of effusion**, which states that the rate of effusion is inversely proportional to the square root of the molar mass. The basis of this result is that, as remarked above, the mean speed of molecules is inversely proportional to $M^{1/2}$, so the rate at which they strike the area of the hole is also inversely proportional to $M^{1/2}$. However, by using the expression for the rate of collisions, we can obtain a more detailed expression for the rate of effusion and hence use effusion data more effectively.

When a gas at a pressure p and temperature T is separated from a vacuum by a small hole, the rate of escape of its molecules is equal to the rate at which they strike the area of the hole (which is given by eqn 21.14). Therefore, for a hole of area A_0,

$$\text{Rate of effusion} = Z_\mathrm{W}A = \frac{pA_0}{(2\pi mkT)^{1/2}} = \frac{pA_0N_\mathrm{A}}{(2\pi MRT)^{1/2}} \qquad (21.16)^\circ$$

where, in the last step, we have used $R = N_\mathrm{A}k$ and $M = mN_\mathrm{A}$. This rate is inversely proportional to $M^{1/2}$, in accord with Graham's law.

Equation 21.16 is the basis of the **Knudsen method** for the determination of the vapour pressures of liquids and solids, particularly of substances with very low vapour pressures. Thus, if the vapour pressure of a sample is p, and it is enclosed in a cavity with a small hole, then the rate of loss of mass from the container is proportional to p.

Example 21.2 *Calculating the vapour pressure from a mass loss*

Caesium (m.p. 29°C, b.p. 686°C) was introduced into a container and heated to 500°C. When a hole of diameter 0.50 mm was opened in the container for 100 s, a mass loss of 385 mg was measured. Calculate the vapour pressure of liquid caesium at 500 K.

Method The pressure of vapour is constant inside the container despite the effusion of atoms because the hot liquid metal replenishes the vapour. The rate of effusion is therefore constant, and given by eqn 21.16. To express the rate in terms of mass, multiply the number of atoms that escape by the mass of each atom.

Answer The mass loss Δm in an interval Δt is related to the collision flux by

$$\Delta m = Z_\mathrm{W}A_0 m\Delta t$$

where A_0 is the area of the hole and m is the mass of one atom. It follows that

$$Z_W = \frac{\Delta m}{A_0 m \Delta t}$$

Because Z_W is related to the pressure by eqn 21.14, we can write

$$p = \left(\frac{2\pi RT}{M}\right)^{1/2} \frac{\Delta m}{A_0 \Delta t}$$

Because $M = 132.9$ g mol^{-1}, substitution of the data gives $p = 11$ kPa (using 1 Pa = 1 N m^{-2} = 1 J m^{-1}), or 83 Torr.

Self-test 21.2 How long would it take 1.0 g of Cs atoms to effuse out of the oven under the same conditions? [260 s]

21.4 Transport properties of a perfect gas

Transport properties are commonly expressed in terms of a number of 'phenomenological' equations, or equations that are empirical summaries of experimental observations. These phenomenological equations apply to all kinds of properties and media. In the following sections, we introduce the equations for the general case and then show how to calculate the parameters that appear in them.

(a) The phenomenological equations

The rate of migration of a property is measured by its **flux**, J, the quantity of that property passing through a given area in a given time interval divided by the area and the duration of the interval. If matter is flowing (as in diffusion), we speak of a **matter flux** of so many molecules per square metre per second; if the property is energy (as in thermal conduction), then we speak of the **energy flux** and express it in joules per square metre per second, and so on.

Experimental observations on transport properties show that the flux of a property is usually proportional to the first derivative of some other related property. For example, the flux of matter diffusing parallel to the z-axis of a container is found to be proportional to the first derivative of the concentration:

$$J(\text{matter}) \propto \frac{d\mathcal{N}}{dz} \qquad (21.17)$$

where $\mathcal{N}$ is the number density of particles with units number per metre cubed (m^{-3}). The SI units of J are number per metre squared per second (m^{-2} s^{-1}). The proportionality of the flux of matter to the concentration gradient is sometimes called **Fick's first law of diffusion**: the law implies that, if the concentration varies steeply with position, then diffusion will be fast. There is no net flux if the concentration is uniform ($d\mathcal{N}/dz = 0$). Similarly, the rate of thermal conduction (the flux of the energy associated with thermal motion) is found to be proportional to the temperature gradient:

$$J(\text{energy}) \propto \frac{dT}{dz} \qquad (21.18)$$

The SI units of this flux are joules per metre squared per second (J m^{-2} s^{-1}).

A positive value of J signifies a flux towards positive z; a negative value of J signifies a flux towards negative z. Because matter flows down a concentration gradient, from high concentration to low concentration, J is positive if $d\mathcal{N}/dz$ is negative (Fig. 21.10).

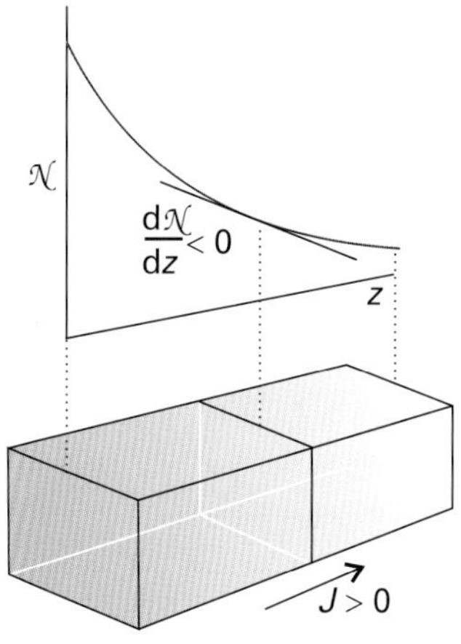

Fig. 21.10 The flux of particles down a concentration gradient. Fick's first law states that the flux of matter (the number of particles passing through an imaginary window in a given interval divided by the area of the window and the duration of the interval) is proportional to the density gradient at that point.

Synoptic table 21.2* Transport properties of gases at 1 atm

	$\kappa/(\mathrm{J\,K^{-1}\,m^{-1}\,s^{-1}})$	$\eta/(\mu\mathrm{P})^{\dagger}$	
	273 K	273 K	293 K
Ar	0.0163	210	223
CO_2	0.0145	136	147
He	0.1442	187	196
N_2	0.0240	166	176

* More values are given in the *Data section*.
† $1\,\mu\mathrm{P} = 10^{-7}\,\mathrm{kg\,m^{-1}\,s^{-1}}$.

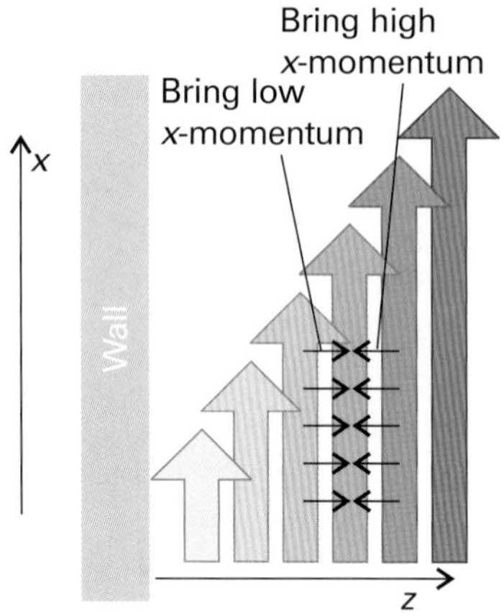

Fig. 21.11 The viscosity of a fluid arises from the transport of linear momentum. In this illustration the fluid is undergoing laminar flow, and particles bring their initial momentum when they enter a new layer. If they arrive with high x-component of momentum they accelerate the layer; if with low x-component of momentum they retard the layer.

Therefore, the coefficient of proportionality in eqn 21.18 must be negative, and we write it $-D$:

$$J(\text{matter}) = -D\frac{\mathrm{d}\mathcal{N}}{\mathrm{d}z} \tag{21.19}$$

The constant D is the called the **diffusion coefficient**; its SI units are metre squared per second ($\mathrm{m^2\,s^{-1}}$). Energy migrates down a temperature gradient, and the same reasoning leads to

$$J(\text{energy}) = -\kappa\frac{\mathrm{d}T}{\mathrm{d}z} \tag{21.20}$$

where κ is the **coefficient of thermal conductivity**. The SI units of κ are joules per kelvin per metre per second ($\mathrm{J\,K^{-1}\,m^{-1}\,s^{-1}}$). Some experimental values are given in Table 21.2.

To see the connection between the flux of momentum and the viscosity, consider a fluid in a state of **Newtonian flow**, which can be imagined as occurring by a series of layers moving past one another (Fig. 21.11). The layer next to the wall of the vessel is stationary, and the velocity of successive layers varies linearly with distance, z, from the wall. Molecules ceaselessly move between the layers and bring with them the x-component of linear momentum they possessed in their original layer. A layer is retarded by molecules arriving from a more slowly moving layer because they have a low momentum in the x-direction. A layer is accelerated by molecules arriving from a more rapidly moving layer. We interpret the net retarding effect as the fluid's viscosity.

Because the retarding effect depends on the transfer of the x-component of linear momentum into the layer of interest, the viscosity depends on the flux of this x-component in the z-direction. The flux of the x-component of momentum is proportional to $\mathrm{d}v_x/\mathrm{d}z$ because there is no net flux when all the layers move at the same velocity. We can therefore write

$$J(x\text{-component of momentum}) = -\eta\frac{\mathrm{d}v_x}{\mathrm{d}z} \tag{21.21}$$

The constant of proportionality, η, is the **coefficient of viscosity** (or simply 'the viscosity'). Its units are kilograms per metre per second ($\mathrm{kg\,m^{-1}\,s^{-1}}$). Viscosities are often reported in poise (P), where $1\,\mathrm{P} = 10^{-1}\,\mathrm{kg\,m^{-1}\,s^{-1}}$. Some experimental values are given in Table 21.2.

(b) The transport parameters

As shown in *Further information* 21.1 and summarized in Table 21.3, the kinetic model leads to expressions for the diffusional parameters of a perfect gas.

Table 21.3 Transport properties of perfect gases

Property	Transported quantity	Simple kinetic theory	Units
Diffusion	Matter	$D = \frac{1}{3}\lambda\bar{c}$	$\mathrm{m^2\,s^{-1}}$
Thermal conductivity	Energy	$\kappa = \frac{1}{3}\lambda\bar{c}C_{V,\mathrm{m}}[\mathrm{A}] = \dfrac{\bar{c}C_{V,\mathrm{m}}}{3\sqrt{2}\sigma N_\mathrm{A}}$	$\mathrm{J\,K^{-1}\,m^{-1}\,s^{-1}}$
Viscosity	Linear momentum	$\eta = \frac{1}{3}\lambda\bar{c}m\mathcal{N} = \dfrac{m\bar{c}}{3\sqrt{2}\sigma}$	$\mathrm{kg\,m^{-1}\,s^{-1}}$

The diffusion coefficient is

$$D = \tfrac{1}{3}\lambda\bar{c} \qquad (21.22)^{\circ}$$

As usual, we need to consider the significance of this expression:

1 The mean free path, λ, decreases as the pressure is increased (eqn 21.13), so D decreases with increasing pressure and, as a result, the gas molecules diffuse more slowly.

2 The mean speed, $\bar{c}$, increases with the temperature (eqn 21.7), so D also increases with temperature. As a result, molecules in a hot sample diffuse more quickly than those in a cool sample (for a given concentration gradient).

3 Because the mean free path increases when the collision cross-section of the molecules decreases (eqn 21.13), the diffusion coefficient is greater for small molecules than for large molecules.

Similarly, according to the kinetic model of gases, the thermal conductivity of a perfect gas A having molar concentration [A] is given by the expression

$$\kappa = \tfrac{1}{3}\lambda\bar{c}C_{V,\mathrm{m}}[\mathrm{A}] \qquad (21.23)^{\circ}$$

where $C_{V,\mathrm{m}}$ is the molar heat capacity at constant volume. To interpret this expression, we note that:

1 Because λ is inversely proportional to the pressure, and hence inversely proportional to the molar concentration of the gas, the thermal conductivity is independent of the pressure.

2 The thermal conductivity is greater for gases with a high heat capacity because a given temperature gradient then corresponds to a greater energy gradient.

The physical reason for the pressure independence of κ is that the thermal conductivity can be expected to be large when many molecules are available to transport the energy, but the presence of so many molecules limits their mean free path and they cannot carry the energy over a great distance. These two effects balance. The thermal conductivity is indeed found experimentally to be independent of the pressure, except when the pressure is very low, when $\kappa \propto p$. At low pressures λ exceeds the dimensions of the apparatus, and the distance over which the energy is transported is determined by the size of the container and not by the other molecules present. The flux is still proportional to the number of carriers, but the length of the journey no longer depends on λ, so $\kappa \propto [\mathrm{A}]$, which implies that $\kappa \propto p$.

Finally, the kinetic model leads to the following expression for the viscosity (see *Further information* 21.1):

$$\eta = \tfrac{1}{3}M\lambda\bar{c}[\mathrm{A}] \qquad (21.24)^{\circ}$$

where [A] is the molar concentration of the gas molecules and M is their molar mass. We can interpret this expression as follows:

1 Because $\lambda \propto 1/p$ (eqn 21.13) and $[\mathrm{A}] \propto p$, it follows that $\eta \propto \bar{c}$, independent of p. That is, the viscosity is independent of the pressure

2 Because $\bar{c} \propto T^{1/2}$ (eqn 21.7), $\eta \propto T^{1/2}$. That is, the viscosity of a gas *increases* with temperature.

The physical reason for the pressure-independence of the viscosity is the same as for the thermal conductivity: more molecules are available to transport the momentum, but they carry it less far on account of the decrease in mean free path. The increase of viscosity with temperature is explained when we remember that at high temperatures the molecules travel more quickly, so the flux of momentum is greater. By contrast, as we shall see in Section 21.6, the viscosity of a liquid *decreases* with increase in temperature because intermolecular interactions must be overcome.

There are two main techniques for measuring viscosities of gases. One technique depends on the rate of damping of the torsional oscillations of a disc hanging in the gas. The half-life of the decay of the oscillation depends on the viscosity and the design of the apparatus, and the apparatus needs to be calibrated. The other method is based on **Poiseuille's formula** for the rate of flow of a fluid through a tube of radius r:

$$\frac{dV}{dt}=\frac{(p_1^2-p_2^2)\pi r^4}{16l\eta p_0} \tag{21.25}$$

where V is the volume flowing, p_1 and p_2 are the pressures at each end of the tube of length l, and p_0 is the pressure at which the volume is measured.

Such measurements confirm that the viscosities of gases are independent of pressure over a wide range. For instance, the results for argon from 10^{-3} atm to 10^2 atm are shown in Fig. 21.12, and we see that η is constant from about 0.01 atm to 20 atm. The measurements also confirm (to a lesser extent) the $T^{1/2}$ dependence. The blue line in the illustration shows the calculated values using $\sigma = 22\times10^{-20}$ m^2, implying a collision diameter of 260 pm, in contrast to the van der Waals diameter of 335 pm obtained from the density of the solid. The agreement is not too bad, considering the simplicity of the model, especially the neglect of intermolecular forces.

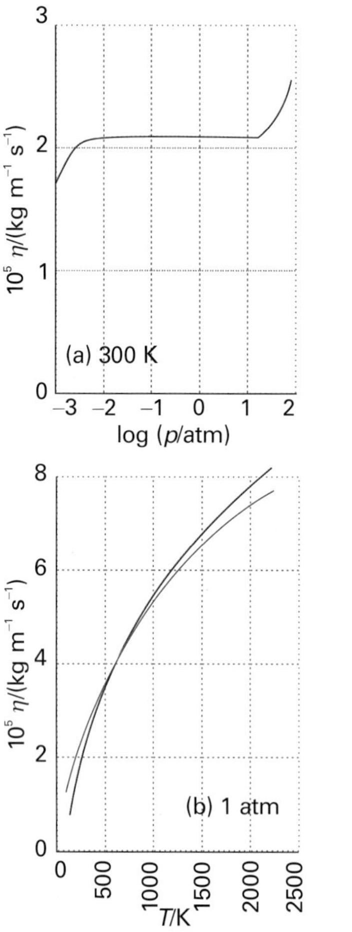

Fig. 21.12 The experimental results for (a) the pressure dependence of the viscosity of argon, and (b) its temperature dependence. The blue line in the latter is the calculated value. Fitting the observed and calculating curves is one way of determining the collision cross-section.

Illustration 21.1 *Using the Poiseuille formula*

In a Poiseuille flow experiment to measure the viscosity of air at 298 K, the sample was allowed to flow through a tube of length 100 cm and internal diameter 1.00 mm. The high-pressure end was at 765 Torr and the low-pressure end was at 760 Torr. The volume was measured at the latter pressure. In 100 s, 90.2 cm^3 of air passed through the tube. The viscosity of air at 298 K is found by reorganizing the Poiseuille formula, eqn 21.25, into

$$\eta=\frac{(p_1^2-p_2^2)\pi r^4}{16lp_0(dV/dt)}$$

and substituting the data (after converting the pressures to pascals by using 1 Torr = 133.3 Pa):

$$\eta=\frac{\{(765\times133.3\ \text{Pa})^2-(760\times133.3\ \text{Pa})^2\}\times\pi\times(5.00\times10^{-4}\ \text{m})^4}{16\times(1.00\times10^{-1}\ \text{m})\times(760\times133.3\ \text{Pa})\times\left(\dfrac{9.02\times10^{-5}\ \text{m}^3}{100\ \text{s}}\right)}$$

$$=1.82\times10^{-4}\ \text{kg m}^{-1}\ \text{s}^{-1}$$

where we have used 1 Pa = 1 kg m^{-1} s^{-2}. The kinetic model expression gives $\eta = 1.4\times10^{-5}$ kg m^{-1} s^{-1}, so the agreement is reasonably good. Viscosities are commonly expressed in centipoise (cP) or (for gases) micropoise (μP), the conversion being 1 cP = 10^{-3} kg m^{-1} s^{-1}; the viscosity of air at 20°C is about 180 μP.

Self-test 21.3 What volume would be collected if the pressure gradient were doubled, other conditions remaining constant? [181 cm^3]

Molecular motion in liquids

We outlined what is currently known about the structure of simple liquids in Section 17.6. Here we consider a particularly simple type of motion through a liquid, that of an ion, and see that the information that motion provides can be used to infer the behaviour of uncharged species too.

21.5 Experimental results

The motion of molecules in liquids can be studied experimentally by a variety of methods. Relaxation time measurements in NMR and EPR (Chapter 15) can be interpreted in terms of the mobilities of the molecules, and have been used to show that big molecules in viscous fluids typically rotate in a series of small (about 5°) steps, whereas small molecules in nonviscous fluids typically jump through about 1 radian (57°) in each step. Another important technique is **inelastic neutron scattering**, in which the energy neutrons collect or discard as they pass through a sample is interpreted in terms of the motion of its particles. The same technique is used to examine the internal dynamics of macromolecules.

More mundane than these experiments are viscosity measurements (Table 21.4). For a molecule to move in a liquid, it must acquire at least a minimum energy to escape from its neighbours. The probability that a molecule has at least an energy E_a is proportional to $e^{-E_a/RT}$, so the mobility of the molecules in the liquid should follow this type of temperature dependence. Because the coefficient of viscosity, η, is inversely proportional to the mobility of the particles, we should expect that

$$\eta \propto e^{E_a/RT} \tag{21.26}$$

(Note the positive sign of the exponent.) This expression implies that the viscosity should decrease sharply with increasing temperature. Such a variation is found experimentally, at least over reasonably small temperature ranges (Fig. 21.13). The activation energy typical of viscosity is comparable to the mean potential energy of intermolecular interactions.

One problem with the interpretation of viscosity measurements is that the change in density of the liquid as it is heated makes a pronounced contribution to the temperature variation of the viscosity. Thus, the temperature dependence of viscosity at constant volume, when the density is constant, is much less than that at constant pressure. The intermolecular interactions between the molecules of the liquid govern the magnitude of E_a, but the problem of calculating it is immensely difficult and still largely unsolved. At low temperatures, the viscosity of water decreases as the pressure is increased. This behaviour is consistent with the rupture of hydrogen bonds.

Synoptic table 21.4* Viscosities of liquids at 298 K

	$\eta/(10^{-3}\ \mathrm{kg\ m^{-1}\ s^{-1}})$
Benzene	0.601
Mercury	1.55
Propane	0.224
Water†	0.891

* More values are given in the *Data section*.
† The viscosity of water corresponds to 0.891 cP.

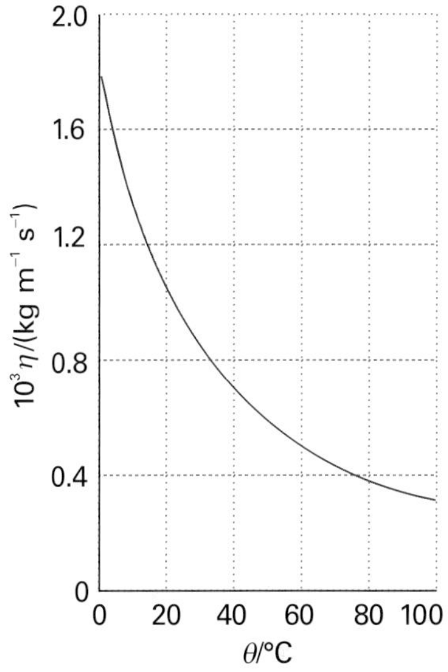

Fig. 21.13 The experimental temperature dependence of the viscosity of water. As the temperature is increased, more molecules are able to escape from the potential wells provided by their neighbours, and so the liquid becomes more fluid. A plot of ln η against $1/T$ is a straight line (over a small range) with positive slope.

21.6 The conductivities of electrolyte solutions

Further insight into the nature of molecular motion can be obtained by studying the motion of ions in solution, for ions can be dragged through the solvent by the application of a potential difference between two electrodes immersed in the sample. By studying the transport of charge through electrolyte solutions it is possible to build up a picture of the events that occur in them and, in some cases, to extrapolate the conclusions to species that have zero charge, that is, to neutral molecules.

(a) Conductance and conductivity

The fundamental measurement used to study the motion of ions is that of the electrical resistance, R, of the solution. The **conductance**, G, of a solution is the inverse of its resistance R: $G = 1/R$. As resistance is expressed in ohms, Ω, the conductance of a sample is expressed in Ω^{-1}. The reciprocal ohm used to be called the mho, but its official designation is now the siemens, S, and $1\ S = 1\ \Omega^{-1} = 1\ C\ V^{-1}\ s^{-1}$. The conductance of a sample decreases with its length l and increases with its cross-sectional area A. We therefore write

$$G = \frac{\kappa A}{l} \tag{21.27}$$

where κ is the **conductivity**. With the conductance in siemens and the dimensions in metres, it follows that the SI units of κ are siemens per metre ($S\ m^{-1}$).

The conductivity of a solution depends on the number of ions present, and it is normal to introduce the **molar conductivity**, Λ_m, which is defined as

$$\Lambda_m = \frac{\kappa}{c} \qquad [21.28]$$

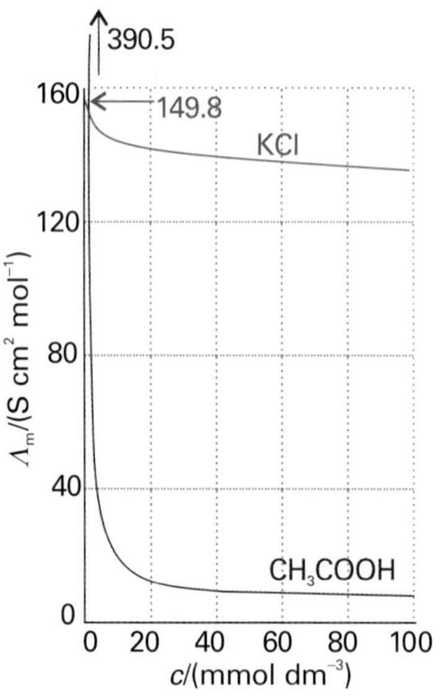

Fig. 21.14 The concentration dependence of the molar conductivities of (a) a typical strong electrolyte (aqueous potassium chloride) and (b) a typical weak electrolyte (aqueous acetic acid).

where c is the molar concentration of the added electrolyte. The SI unit of molar conductivity is siemens metre-squared per mole ($S\ m^2\ mol^{-1}$), and typical values are about $10\ mS\ m^2\ mol^{-1}$ (where $1\ mS = 10^{-3}\ S$).

The molar conductivity is found to vary with the concentration. One reason for this variation is that the number of ions in the solution might not be proportional to the concentration of the electrolyte. For instance, the concentration of ions in a solution of a weak acid depends on the concentration of the acid in a complicated way, and doubling the concentration of the acid added does not double the number of ions. Secondly, because ions interact strongly with one another, the conductivity of a solution is not exactly proportional to the number of ions present.

The concentration dependence of molar conductivities indicates that there are two classes of electrolyte. The characteristic of a **strong electrolyte** is that its molar conductivity depends only slightly on the molar concentration (and in general decreases slightly as the concentration is increased, Fig. 21.14). The characteristic of a **weak electrolyte** is that its molar conductivity is normal at concentrations close to zero, but falls sharply to low values as the concentration increases. The classification depends on the solvent employed as well as the solute: lithium chloride, for example, is a strong electrolyte in water but a weak electrolyte in propanone.

(b) Strong electrolytes

Strong electrolytes are substances that are virtually fully ionized in solution, and include ionic solids and strong acids. As a result of their complete ionization, the concentration of ions in solution is proportional to the concentration of strong electrolyte added.

In an extensive series of measurements during the nineteenth century, Friedrich Kohlrausch showed that at low concentrations the molar conductivities of strong electrolytes vary linearly with the square root of the concentration:

$$\Lambda_m = \Lambda_m^\circ - \mathcal{K} c^{1/2} \tag{21.29}$$

This variation is called **Kohlrausch's law**. The constant Λ_m° is the **limiting molar conductivity**, the molar conductivity in the limit of zero concentration (when the ions are effectively infinitely far apart and do not interact with one another). The constant $\mathcal{K}$ is found to depend more on the stoichiometry of the electrolyte (that is, whether it is of the form MA, or M_2A, etc.) than on its specific identity. In due course we shall see that the $c^{1/2}$ dependence arises from interactions between ions: when charge is conducted ionically, ions of one charge are moving past the ions of interest and retard its progress.

Synoptic table 21.5* Limiting ionic conductivities in water at 298 K

	$\lambda/(\text{mS m}^2\,\text{mol}^{-1})$		$\lambda/(\text{mS m}^2\,\text{mol}^{-1})$
H^+	34.96	OH^-	19.91
Na^+	5.01	Cl^-	7.63
K^+	7.35	Br^-	7.81
Zn^{2+}	10.56	SO_4^{2-}	16.00

* More values are given in the *Data section*.

Kohlrausch was also able to establish experimentally that Λ_m° can be expressed as the sum of contributions from its individual ions. If the limiting molar conductivity of the cations is denoted λ_+ and that of the anions λ_-, then his **law of the independent migration of ions** states that

$$\Lambda_m^\circ = \nu_+\lambda_+ + \nu_-\lambda_- \qquad (21.30)^\circ$$

where ν_+ and ν_- are the numbers of cations and anions per formula unit of electrolyte (for example, $\nu_+ = \nu_- = 1$ for HCl, NaCl, and $CuSO_4$, but $\nu_+ = 1$, $\nu_- = 2$ for $MgCl_2$). This simple result, which can be understood on the grounds that the ions migrate independently in the limit of zero concentration, lets us predict the limiting molar conductivity of any strong electrolyte from the data in Table 21.5.

Illustration 21.2 *Calculating a limiting molar conductivity*

The limiting molar conductivity of $BaCl_2$ in water at 298 K is

$$\Lambda_m^\circ = (12.72 + 2\times 7.63)\ \text{mS m}^2\,\text{mol}^{-1} = 27.98\ \text{mS m}^2\,\text{mol}^{-1}$$

(c) Weak electrolytes

Weak electrolytes are not fully ionized in solution. They include weak Brønsted acids and bases, such as CH_3COOH and NH_3. The marked concentration dependence of their molar conductivities arises from the displacement of the equilibrium

$$\text{HA(aq)} + \text{H}_2\text{O(l)} \rightleftharpoons \text{H}_3\text{O}^+\text{(aq)} + \text{A}^-\text{(aq)} \qquad K_a = \frac{a_{\text{H}_3\text{O}^+}a_{\text{A}^-}}{a_{\text{HA}}} \qquad (21.31)$$

towards products at low molar concentrations.

The conductivity depends on the number of ions in the solution, and therefore on the **degree of ionization**, α, of the electrolyte; when referring to weak acids, we speak instead of the **degree of deprotonation**. It is defined so that, for the acid HA at a molar concentration c, at equilibrium

$$[\text{H}_3\text{O}^+] = \alpha c \qquad [\text{A}^-] = \alpha c \qquad [\text{HA}] = (1-\alpha)c \qquad (21.32)$$

If we ignore activity coefficients, the acidity constant, K_a, is approximately

$$K_a = \frac{\alpha^2 c}{1-\alpha} \qquad (21.33)^\circ$$

from which it follows that

$$\alpha = \frac{K_a}{2c}\left\{\left(1+\frac{4c}{K_a}\right)^{1/2} - 1\right\} \qquad (21.34)^\circ$$

Comment 21.2

It will be familiar from introductory chemistry that a *Brønsted acid* is a proton donor and a *Brønsted base* is a proton acceptor. In eqn 21.31, HA(aq) is a Brønsted acid and H_2O(l) is a Brønsted base. The equilibrium constant for the reaction between HA(aq) and H_2O(l) is the *acidity constant* K_a of HA. Acids with $K_a < 1$, indicating only a small extent of deprotonation in water, are classified as *weak acids*.

The acid is fully deprotonated at infinite dilution, and its molar conductivity is then Λ_m°. Because only a fraction α is actually present as ions in the actual solution, the measured molar conductivity Λ_m is given by

$$\Lambda_m = \alpha\Lambda_m^\circ \qquad (21.35)^\circ$$

with α given by eqn 21.34.

Illustration 21.3 *Using molar conductivity data to calculate an acidity constant*

The molar conductivity of 0.0100 M $CH_3COOH(aq)$ at 298 K is $\Lambda_m = 1.65$ mS m^2 mol^{-1}. The degree of deprotonation, α, is calculated from eqn 21.35 with $\Lambda_m^\circ =$ 39.05 mS cm^2 mol^{-1} (Table 21.5). It follows that $\alpha = 0.0423$. The acidity constant, K_a, is calculated by substitution of α into eqn 21.33, which gives $K_a = 1.9 \times 10^{-5}$.

Self-test 21.4 The molar conductivity of 0.0250 M $HCOOH(aq)$ is 4.61 mS m^2 mol^{-1}. Determine the $pK_a = -\log K_a$ of the acid. [3.44]

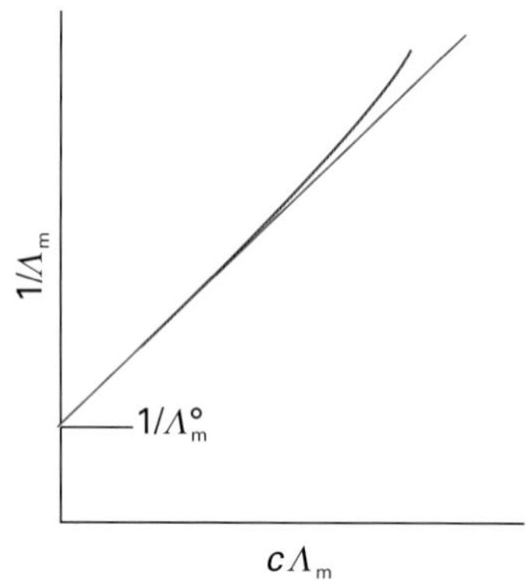

Fig. 21.15 The graph used to determine the limiting value of the molar conductivity of a solution by extrapolation to zero concentration.

Once we know K_a, we can use eqns 21.34 and 21.35 to predict the concentration dependence of the molar conductivity. The result agrees quite well with the experimental curve in Fig. 21.14. More usefully, we can use the concentration dependence of Λ_m in measurements of the limiting molar conductance. First, we rearrange eqn 21.33 into

$$\frac{1}{\alpha} = 1 + \frac{\alpha c}{K_a} \qquad (21.36)^\circ$$

Then, by using eqn 21.35, we obtain **Ostwald's dilution law**:

$$\frac{1}{\Lambda_m} = \frac{1}{\Lambda_m^\circ} + \frac{\Lambda_m c}{K_a(\Lambda_m^\circ)^2} \qquad (21.37)^\circ$$

This equation implies that, if $1/\Lambda_m$ is plotted against $c\Lambda_m$, then the intercept at $c = 0$ will be $1/\Lambda_m^\circ$ (Fig. 21.15).

21.7 The mobilities of ions

To interpret conductivity measurements we need to know why ions move at different rates, why they have different molar conductivities, and why the molar conductivities of strong electrolytes decrease with the square root of the molar concentration. The central idea in this section is that, although the motion of an ion remains largely random, the presence of an electric field biases its motion, and the ion undergoes net migration through the solution.

(a) The drift speed

When the potential difference between two electrodes a distance l apart is $\Delta\phi$, the ions in the solution between them experience a uniform electric field of magnitude

$$\mathcal{E} = \frac{\Delta\phi}{l} \qquad (21.38)$$

In such a field, an ion of charge ze experiences a force of magnitude

$$\mathcal{F} = ze\mathcal{E} = \frac{ze\Delta\phi}{l} \qquad (21.39)$$

Synoptic table 21.6* Ionic mobilities in water at 298 K

	$u/(10^{-8}\ \mathrm{m^2\ s^{-1}\ V^{-1}})$		$u/(10^{-8}\ \mathrm{m^2\ s^{-1}\ V^{-1}})$
H^+	36.23	OH^-	20.64
Na^+	5.19	Cl^-	7.91
K^+	7.62	Br^-	8.09
Zn^{2+}	5.47	SO_4^{2-}	8.29

* More values are given in the *Data section.*

(In this chapter we disregard the sign of the charge number and so avoid notational complications.) A cation responds to the application of the field by accelerating towards the negative electrode and an anion responds by accelerating towards the positive electrode. However, this acceleration is short-lived. As the ion moves through the solvent it experiences a frictional retarding force, $\mathcal{F}_{\text{fric}}$, proportional to its speed. If we assume that the Stokes formula (eqn 19.12) for a sphere of radius a and speed s applies even on a microscopic scale (and independent evidence from magnetic resonance suggests that it often gives at least the right order of magnitude), then we can write this retarding force as

$$\mathcal{F}_{\text{fric}} = fs \qquad f = 6\pi\eta a \tag{21.40}$$

The two forces act in opposite directions, and the ions quickly reach a terminal speed, the **drift speed**, when the accelerating force is balanced by the viscous drag. The net force is zero when

$$s = \frac{ze\mathcal{E}}{f} \tag{21.41}$$

It follows that the drift speed of an ion is proportional to the strength of the applied field. We write

$$s = u\mathcal{E} \qquad [21.42]$$

where u is called the **mobility** of the ion (Table 21.6). Comparison of eqns 21.41 and 21.42 and use of eqn 21.40 shows that

$$u = \frac{ze}{f} = \frac{ze}{6\pi\eta a} \tag{21.43}$$

Illustration 21.4 *Calculating an ionic mobility*

For an order of magnitude estimate we can take $z = 1$ and a the radius of an ion such as Cs^+ (which might be typical of a smaller ion plus its hydration sphere), which is 170 pm. For the viscosity, we use $\eta = 1.0$ cP (1.0×10^{-3} kg m^{-1} s^{-1}, Table 21.4). Then $u \approx 5 \times 10^{-8}$ m^2 V^{-1} s^{-1}. This value means that, when there is a potential difference of 1 V across a solution of length 1 cm (so $\mathcal{E} = 100$ V m^{-1}), the drift speed is typically about 5 μm s^{-1}. That speed might seem slow, but not when expressed on a molecular scale, for it corresponds to an ion passing about 10^4 solvent molecules per second.

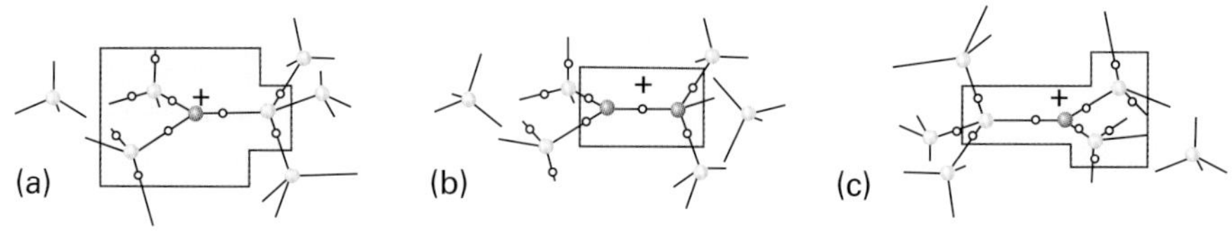

Fig. 21.16 The mechanism of conduction by hydrogen ions in water as proposed by N. Agmon (*Chem. Phys. Letts.* **244**, 456 (1995)). Proton transfer between neighbouring molecules occurs when one molecule rotates into such a position that an O—H···O hydrogen bond can flip into being an O···H—O hydrogen bond. See text for a description of the steps.

Because the drift speed governs the rate at which charge is transported, we might expect the conductivity to decrease with increasing solution viscosity and ion size. Experiments confirm these predictions for bulky ions (such as R_4N^+ and RCO_2^-) but not for small ions. For example, the molar conductivities of the alkali metal ions increase from Li^+ to Cs^+ (Table 21.6) even though the ionic radii increase. The paradox is resolved when we realize that the radius a in the Stokes formula is the **hydrodynamic radius** (or 'Stokes radius') of the ion, its effective radius in the solution taking into account all the H_2O molecules it carries in its hydration sphere. Small ions give rise to stronger electric fields than large ones (the electric field at the surface of a sphere of radius r is proportional to ze/r^2 and it follows that the smaller the radius the stronger the field), so small ions are more extensively solvated than big ions. Thus, an ion of small ionic radius may have a large hydrodynamic radius because it drags many solvent molecules through the solution as it migrates. The hydrating H_2O molecules are often very labile, however, and NMR and isotope studies have shown that the exchange between the coordination sphere of the ion and the bulk solvent is very rapid.

The proton, although it is very small, has a very high molar conductivity (Table 21.6)! Proton and ^{17}O-NMR show that the times characteristic of protons hopping from one molecule to the next are about 1.5 ps, which is comparable to the time that inelastic neutron scattering shows it takes a water molecule to reorientate through about 1 rad (1 to 2 ps). According to the **Grotthuss mechanism**, there is an effective motion of a proton that involves the rearrangement of bonds in a group of water molecules. However, the actual mechanism is still highly contentious. Attention now focuses on the $H_9O_4^+$ unit, in which the nearly trigonal planar H_3O^+ ion is linked to three strongly solvating H_2O molecules. This cluster of atoms is itself hydrated, but the hydrogen bonds in the secondary sphere are weaker than in the primary sphere. It is envisaged that the rate-determining step is the cleavage of one of the weaker hydrogen bonds of this secondary sphere (Fig. 21.16a). After this bond cleavage has taken place, and the released molecule has rotated through a few degrees (a process that takes about 1 ps), there is a rapid adjustment of bond lengths and angles in the remaining cluster, to form an $H_5O_2^+$ cation of structure $H_2O \cdots H^+ \cdots OH_2$ (Fig. 21.16b). Shortly after this reorganization has occurred, a new $H_9O_4^+$ cluster forms as other molecules rotate into a position where they can become members of a secondary hydration sphere, but now the positive charge is located one molecule to the right of its initial location (Fig. 21.16c). According to this model, there is no coordinated motion of a proton along a chain of molecules, simply a very rapid hopping between neighbouring sites, with a low activation energy. The model is consistent with the observation that the molar conductivity of protons increases as the pressure is raised, for increasing pressure ruptures the hydrogen bonds in water. The mobility of NH_4^+ is also anomalous and presumably occurs by an analogous mechanism.

Comment 21.3

The H_3O^+ ion is trigonal pyramidal in the gas phase but nearly planar in water.

(b) Mobility and conductivity

Ionic mobilities provide a link between measurable and theoretical quantities. As a first step we establish in the *Justification* below the following relation between an ion's mobility and its molar conductivity:

$$\lambda = zuF \qquad (21.44)^\circ$$

where F is Faraday's constant ($F = N_A e$).

Justification 21.5 *The relation between ionic mobility and molar conductivity*

To keep the calculation simple, we ignore signs in the following, and concentrate on the magnitudes of quantities: the direction of ion flux can always be decided by common sense.

Consider a solution of a fully dissociated strong electrolyte at a molar concentration c. Let each formula unit give rise to ν_+ cations of charge z_+e and ν_- anions of charge z_-e. The molar concentration of each type of ion is therefore νc (with $\nu = \nu_+$ or ν_-), and the number density of each type is νcN_A. The number of ions of one kind that pass through an imaginary window of area A during an interval Δt is equal to the number within the distance $s\Delta t$ (Fig. 21.17), and therefore to the number in the volume $s\Delta tA$. (The same sort of argument was used in Section 21.1 in the discussion of the pressure of a gas.) The number of ions of that kind in this volume is equal to $s\Delta tA\nu cN_A$. The flux through the window (the number of this type of ion passing through the window divided by the area of the window and the duration of the interval) is therefore

$$J(\text{ions}) = \frac{s\Delta tA\nu cN_A}{A\Delta t} = s\nu cN_A$$

Each ion carries a charge ze, so the flux of charge is

$$J(\text{charge}) = zs\nu ceN_A = zs\nu cF$$

Because $s = u\mathcal{E}$, the flux is

$$J(\text{charge}) = zu\nu cF\mathcal{E}$$

The current, I, through the window due to the ions we are considering is the charge flux times the area:

$$I = JA = zu\nu cF\mathcal{E}A$$

Because the electric field is the potential gradient, $\Delta\phi/l$, we can write

$$I = \frac{zu\nu cFA\Delta\phi}{l} \qquad (21.45)$$

Current and potential difference are related by Ohm's law, $\Delta\phi = IR$, so it follows that

$$I = \frac{\Delta\phi}{R} = G\Delta\phi = \frac{\kappa A\Delta\phi}{l}$$

where we have used eqn 21.27 in the form $\kappa = Gl/A$. Note that the proportionality of current to potential difference ($I \propto \Delta\phi$) is another example of a phenomenological flux equation like those introduced in Section 21.4. Comparison of the last two expressions gives $\kappa = zu\nu cF$. Division by the molar concentration of ions, νc, then results in eqn 21.44.

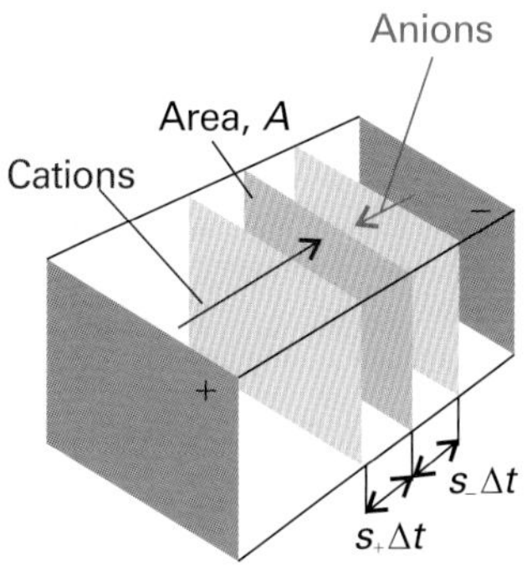

Fig. 21.17 In the calculation of the current, all the cations within a distance $s_+\Delta t$ (that is, those in the volume $s_+A\Delta t$) will pass through the area A. The anions in the corresponding volume the other side of the window will also contribute to the current similarly.

Equation 21.44 applies to the cations and to the anions. Therefore, for the solution itself in the limit of zero concentration (when there are no interionic interactions),

$$\Lambda_m^\circ = (z_+u_+\nu_+ + z_-u_-\nu_-)F \qquad (21.46a)^\circ$$

For a symmetrical z:z electrolyte (for example, $CuSO_4$ with $z = 2$), this equation simplifies to

$$\Lambda_m^\circ = z(u_+ + u_-)F \qquad (21.46b)^\circ$$

Illustration 21.5 *Estimating a limiting molar conductivity*

Earlier, we estimated the typical ionic mobility as $5\times10^{-8}\ \mathrm{m^2\,V^{-1}\,s^{-1}}$; so, with $z=1$ for both the cation and anion, we can estimate that a typical limiting molar conductivity should be about $10\ \mathrm{mS\,m^2\,mol^{-1}}$, in accord with experiment. The experimental value for KCl, for instance, is $15\ \mathrm{mS\,m^2\,mol^{-1}}$.

(c) Transport numbers

The **transport number**, $t_\pm$, is defined as the fraction of total current carried by the ions of a specified type. For a solution of two kinds of ion, the transport numbers of the cations (t_+) and anions (t_-) are

$$t_\pm = \frac{I_\pm}{I} \qquad [21.47]$$

where $I_\pm$ is the current carried by the cation (I_+) or anion (I_-) and I is the total current through the solution. Because the total current is the sum of the cation and anion currents, it follows that

$$t_+ + t_- = 1 \qquad (21.48)$$

The **limiting transport number**, $t^\circ_\pm$, is defined in the same way but for the limit of zero concentration of the electrolyte solution. We shall consider only these limiting values from now on, for that avoids the problem of ionic interactions.

The current that can be ascribed to each type of ion is related to the mobility of the ion by eqn 21.45. Hence the relation between $t^\circ_\pm$ and $u_\pm$ is

$$t^\circ_\pm = \frac{z_\pm \nu_\pm u_\pm}{z_+\nu_+u_+ + z_-\nu_-u_-} \qquad (21.49\text{a})^\circ$$

However, because $z_+\nu_+ = z_-\nu_-$ for any electrolyte, eqn 21.49a simplifies to

$$t^\circ_\pm = \frac{u_\pm}{u_+ + u_-} \qquad (21.49\text{b})^\circ$$

Moreover, because the ionic conductivities are related to the mobilities by eqn 21.44, it also follows from eqn 21.49b that

$$t^\circ_\pm = \frac{\nu_\pm\lambda_\pm}{\nu_+\lambda_+ + \nu_-\lambda_-} = \frac{\nu_\pm\lambda_\pm}{\Lambda^\circ_\mathrm{m}} \qquad (21.50)^\circ$$

and hence, for each type of ion,

$$\nu_\pm\lambda_\pm = t^\circ_\pm \Lambda^\circ_\mathrm{m} \qquad (21.51)^\circ$$

Consequently, because there are independent ways of measuring transport numbers of ions, we can determine the individual ionic conductivities and (through eqn 21.44) the ionic mobilities.

There are several ways to measure transport numbers (see *Further reading*). One of the most accurate (and the only one we describe in detail) is the **moving boundary method**, in which the motion of a boundary between two ionic solutions having a common ion is observed as a current flows.

Let MX be the salt of interest and NX a salt giving a denser solution. The solution of NX is called the **indicator solution**; it occupies the lower part of a vertical tube of cross-sectional area A (Fig. 21.18). The MX solution, which is called the **leading**

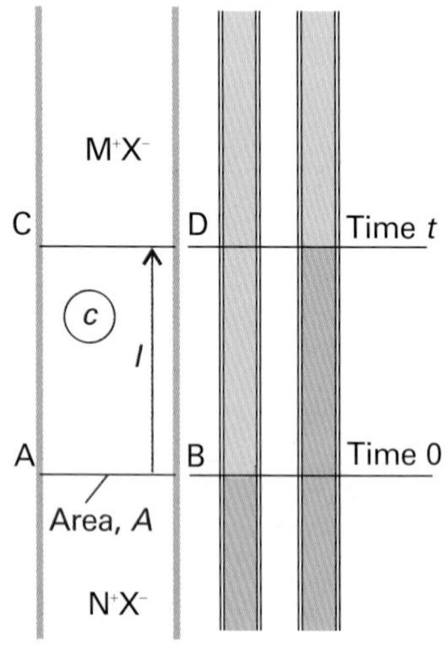

Fig. 21.18 In the moving boundary method for the measurement of transport numbers the distance moved by the boundary is observed as a current is passed. All the M ions in the volume between AB and CD must have passed through CD if the boundary moves from AB to CD. One procedure is to add bromothymol blue indicator to a slightly alkaline solution of the ion of interest and to use a cadmium electrode at the lower end of the vertical tube. The electrode produces Cd^{2+} ions, which are slow moving and slightly acidic (the hydrated ion is a Brønsted acid), and the boundary is revealed by the colour change of the indicator.

solution, occupies the upper part of the tube. There is a sharp boundary between the two solutions. The indicator solution must be denser than the leading solution, and the mobility of the M ions must be greater than that of the N ions. Thus, if any M ions diffuse into the lower solution, they will be pulled upwards more rapidly than the N ions around them, and the boundary will reform. When a current I is passed for a time Δt, the boundary moves from AB to CD, so all the M ions in the volume between AB and CD must have passed through CD. That number is $clAN_A$, so the charge that the M ions transfer through the plane is z_+clAeN_A. However, the *total* charge transferred when a current I flows for an interval Δt is $I\Delta t$. Therefore, the fraction due to the motion of the M ions, which is their transport number, is

$$t_+ = \frac{z_+clAF}{I\Delta t} \tag{21.52}$$

Hence, by measuring the distance moved, the transport number and hence the conductivity and mobility of the ions can be determined.

21.8 Conductivities and ion–ion interactions

The remaining problem is to account for the $c^{1/2}$ dependence of the Kohlrausch law (eqn 21.29). In Section 5.9 we saw something similar: the activity coefficients of ions at low concentrations also depend on $c^{1/2}$ and depend on their charge type rather than their specific identities. That $c^{1/2}$ dependence was explained in terms of the properties of the ionic atmosphere around each ion, and we can suspect that the same explanation applies here too.

To accommodate the effect of motion, we need to modify the picture of an ionic atmosphere as a spherical haze of charge. Because the ions forming the atmosphere do not adjust to the moving ion immediately, the atmosphere is incompletely formed in front of the moving ion and incompletely decayed behind the ion (Fig. 21.19). The overall effect is the displacement of the centre of charge of the atmosphere a short distance behind the moving ion. Because the two charges are opposite, the result is a retardation of the moving ion. This reduction of the ions' mobility is called the **relaxation effect**. A confirmation of the picture is obtained by observing the conductivities of ions at high frequencies, which are greater than at low frequencies: the atmosphere does not have time to follow the rapidly changing direction of motion of the ion, and the effect of the field averages to zero.

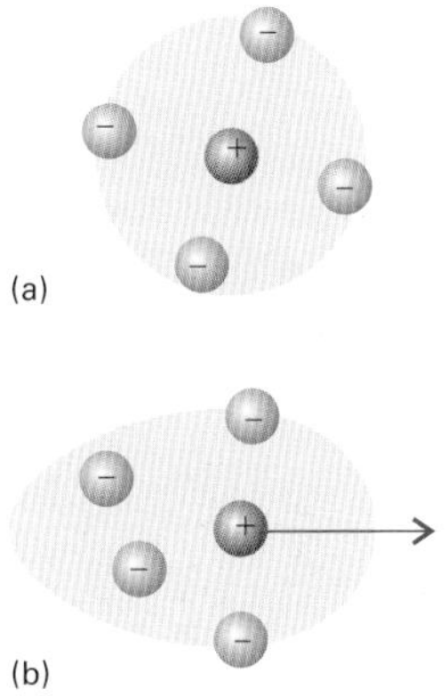

Fig. 21.19 (a) In the absence of an applied field, the ionic atmosphere is spherically symmetric, but (b) when a field is present it is distorted and the centres of negative and positive charge no longer coincide. The attraction between the opposite charges retards the motion of the central ion.

The ionic atmosphere has another effect on the motion of the ions. We have seen that the moving ion experiences a viscous drag. When the ionic atmosphere is present this drag is enhanced because the ionic atmosphere moves in an opposite direction to the central ion. The enhanced viscous drag, which is called the **electrophoretic effect**, reduces the mobility of the ions, and hence also reduces their conductivities.

The quantitative formulation of these effects is far from simple, but the **Debye–Hückel–Onsager theory** is an attempt to obtain quantitative expressions at about the same level of sophistication as the Debye–Hückel theory itself. The theory leads to a Kohlrausch-like expression in which

$$\mathcal{K} = A + B\Lambda_m^\circ \tag{21.53a}$$

with

$$A = \frac{z^2eF^2}{3\pi\eta}\left(\frac{2}{\varepsilon RT}\right)^{1/2} \qquad B = \frac{qz^3eF}{24\pi\varepsilon RT}\left(\frac{2}{\varepsilon RT}\right)^{1/2} \tag{21.53b}$$

where ε is the electric permittivity of the solvent (Section 18.3) and $q = 0.586$ for a 1,1-electrolyte (Table 21.7). The slopes of the conductivity curves are predicted to depend

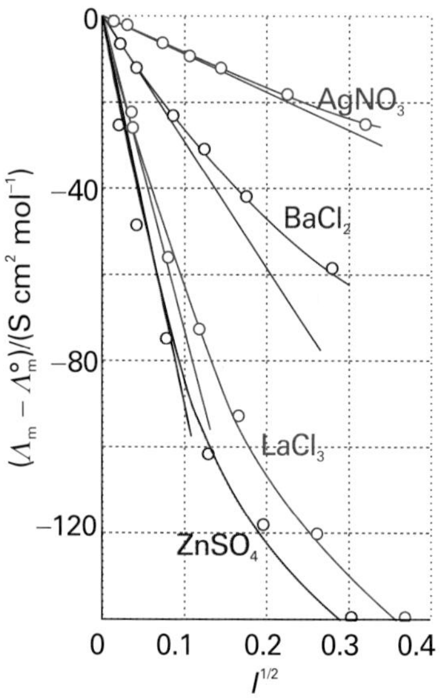

Fig. 21.20 The dependence of molar conductivities on the square root of the ionic strength, and comparison (straight lines) with the dependence predicted by the Debye–Hückel–Onsager theory.

Synoptic table 21.7* Debye–Hückel–Onsager coefficients for (1,1)-electrolytes at 298 K

Solvent	$A/(\text{mS m}^2\text{ mol}^{-1}/(\text{mol dm}^{-3})^{1/2})$	$B/(\text{mol dm}^{-3})^{-1/2}$
Methanol	15.61	0.923
Propanone	32.8	1.63
Water	6.02	0.229

* More values are given in the *Data section*.

on the charge type of the electrolyte, in accord with the Kohlrausch law, and some comparisons between theory and experiment are shown in Fig. 21.20. The agreement is quite good at very low ionic strengths, corresponding to very low molar concentrations (less than about 10^{-3} M, depending on the charge type).

IMPACT ON BIOCHEMISTRY

I21.2 Ion channels and ion pumps

Controlled transport of molecules and ions across biological membranes is at the heart of a number of key cellular processes, such as the transmission of nerve impulses, the transfer of glucose into red blood cells, and the synthesis of ATP by oxidative phosphorylation (*Impact* I7.2). Here we examine in some detail the various ways in which ions cross the alien environment of the lipid bilayer.

Suppose that a membrane provides a barrier that slows down the transfer of molecules or ions into or out of the cell. We saw in *Impact* I7.2 that the thermodynamic tendency to transport an ion through the membrane is partially determined by a concentration gradient (more precisely, an activity gradient) across the membrane, which results in a difference in molar Gibbs energy between the inside and the outside of the cell, and a transmembrane potential gradient, which is due to the different potential energy of the ions on each side of the bilayer. There is a tendency, called **passive transport**, for a species to move down concentration and membrane potential gradients. It is also possible to move a species against these gradients, but now the flow must be driven by an exergonic process, such as the hydrolysis of ATP. This process is called **active transport**.

The transport of ions into or out of a cell needs to be mediated (that is, facilitated by other species) because the hydrophobic environment of the membrane is inhospitable to ions. There are two mechanisms for ion transport: mediation by a carrier molecule and transport through a **channel former**, a protein that creates a hydrophilic pore through which the ion can pass. An example of a channel former is the polypeptide gramicidin A, which increases the membrane permeability to cations such as H^+, K^+, and Na^+.

Ion channels are proteins that effect the movement of specific ions down a membrane potential gradient. They are highly selective, so there is a channel protein for Ca^{2+}, another for Cl^-, and so on. The opening of the gate may be triggered by potential differences between the two sides of the membrane or by the binding of an *effector* molecule to a specific receptor site on the channel.

Ions such as H^+, Na^+, K^+, and Ca^{2+} are often transported actively across membranes by integral proteins called **ion pumps**. Ion pumps are molecular machines that work by adopting conformations that are permeable to one ion but not others depending on the state of phosphorylation of the protein. Because protein phosphorylation requires dephosphorylation of ATP, the conformational change that opens or closes the pump is endergonic and requires the use of energy stored during metabolism.

Let's consider some of the experimental approaches used in the study of ion channels. The structures of a number of channel proteins have been obtained by the now traditional X-ray diffraction techniques described in Chapter 20. Information about the flow of ions across channels and pumps is supplied by the **patch clamp technique**. One of many possible experimental arrangements is shown in Fig. 21.21. With mild suction, a 'patch' of membrane from a whole cell or a small section of a broken cell can be attached tightly to the tip of a micropipet filled with an electrolyte solution and containing an electronic conductor, the so-called *patch electrode*. A potential difference (the 'clamp') is applied between the patch electrode and an intracellular electronic conductor in contact with the cytosol of the cell. If the membrane is permeable to ions at the applied potential difference, a current flows through the completed circuit. Using narrow micropipette tips with diameters of less than 1 μm, ion currents of a few picoamperes (1 pA = 10^{-12} A) have been measured across sections of membranes containing only one ion channel protein.

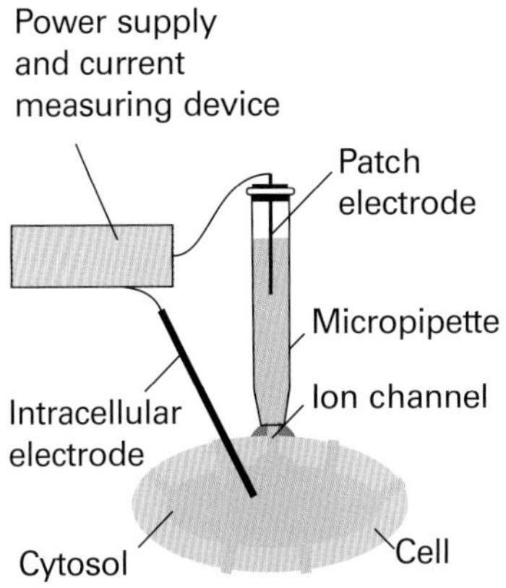

Fig. 21.21 A representation of the patch clamp technique for the measurement of ionic currents through membranes in intact cells. A section of membrane containing an ion channel (shown as a green rectangle) is in tight contact with the tip of a micropipette containing an electrolyte solution and the patch electrode. An intracellular electronic conductor is inserted into the cytosol of the cell and the two conductors are connected to a power supply and current measuring device.

A detailed picture of the mechanism of action of ion channels has emerged from analysis of patch clamp data and structural data. Here we focus on the K^+ ion channel protein, which, like all other mediators of ion transport, spans the membrane bilayer (Fig. 21.22). The pore through which ions move has a length of 3.4 nm and is divided into two regions: a wide region with a length of 2.2 nm and diameter of 1.0 nm and a narrow region with a length of 1.2 nm and diameter of 0.3 nm. The narrow region is called the *selectivity filter* of the K^+ ion channel because it allows only K^+ ions to pass.

Filtering is a subtle process that depends on ionic size and the thermodynamic tendency of an ion to lose its hydrating water molecules. Upon entering the selectivity filter, the K^+ ion is stripped of its hydrating shell and is then gripped by carbonyl groups of the protein. Dehydration of the K^+ ion is endergonic ($\Delta_{\text{dehyd}}G^{\ominus} = +203$ kJ mol^{-1}), but is driven by the energy of interaction between the ion and the protein. The Na^+ ion, though smaller than the K^+ ion, does not pass through the selectivity filter of the K^+ ion channel because interactions with the protein are not sufficient to compensate for the high Gibbs energy of dehydration of Na^+ ($\Delta_{\text{dehyd}}G^{\ominus} = +301$ kJ mol^{-1}). More specifically, a dehydrated Na^+ ion is too small and cannot be held tightly by the protein carbonyl groups, which are positioned for ideal interactions with the larger K^+ ion. In its hydrated form, the Na^+ ion is too large (larger than a dehydrated K^+ ion), does not fit in the selectivity filter, and does not cross the membrane.

Though very selective, a K^+ ion channel can still let other ions pass through. For example, K^+ and Tl^+ ions have similar radii and Gibbs energies of dehydration, so Tl^+ can cross the membrane. As a result, Tl^+ is a neurotoxin because it replaces K^+ in many neuronal functions.

The efficiency of transfer of K^+ ions through the channel can also be explained by structural features of the protein. For efficient transport to occur, a K^+ ion must enter the protein, but then must not be allowed to remain inside for very long so that, as one K^+ ion enters the channel from one side, another K^+ ion leaves from the opposite side. An ion is lured into the channel by water molecules about halfway through the length of the membrane. Consequently, the thermodynamic cost of moving an ion from an aqueous environment to the less hydrophilic interior of the protein is minimized. The ion is 'encouraged' to leave the protein by electrostatic interactions in the selectivity filter, which can bind two K^+ ions simultaneously, usually with a bridging water molecule. Electrostatic repulsion prevents the ions from binding too tightly, minimizing the residence time of an ion in the selectivity filter, and maximizing the transport rate.

Now we turn our attention to a very important ion pump, the H^+-ATPase responsible for coupling of proton flow to synthesis of ATP from ADP and P_i (*Impact* I7.2). Structural studies show that the channel through which the protons flow is linked in

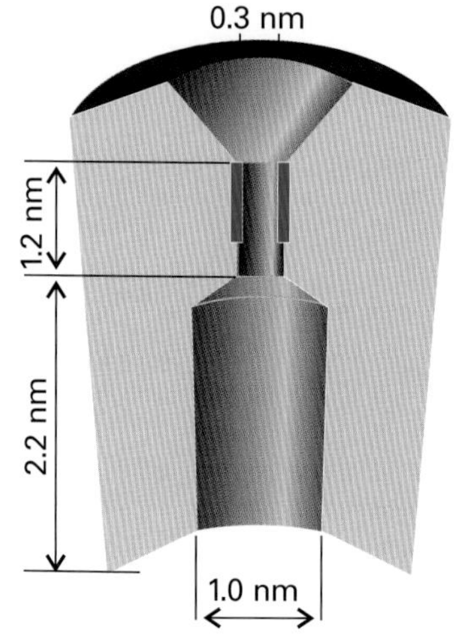

Fig. 21.22 A schematic representation of the cross-section of a membrane-spanning K^+ ion channel protein. The bulk of the protein is shown in beige. The pore through which ions move is divided into two regions: a wide region with a length of 2.2 nm and diameter of 1.0 nm, and a narrow region, the *selectivity filter*, with a length of 1.2 nm and diameter of 0.3 nm. The selectivity filter has a number of carbonyl groups (shown in dark green) that grip K^+ ions. As explained in the text, electrostatic repulsions between two bound K^+ ions 'encourage' ionic movement through the selectivity filter and across the membrane.

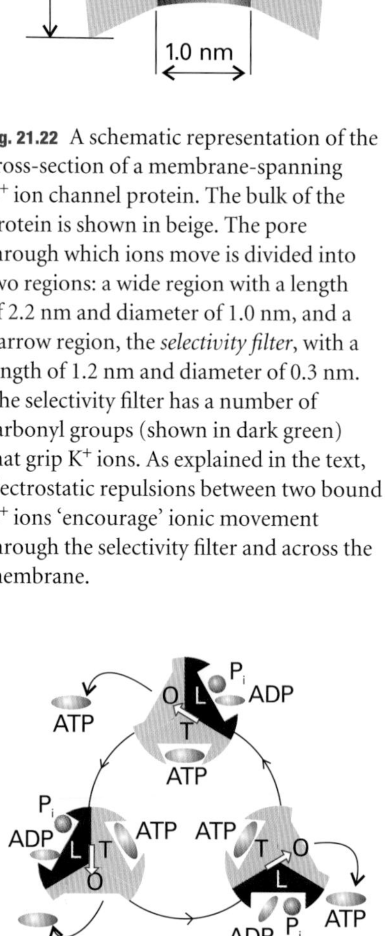

Fig. 21.23 The mechanism of action of H^+-ATPase, a molecular motor that transports protons across the mitochondrial membrane and catalyses either the formation or hydrolysis of ATP.

tandem to a unit composed of six protein molecules arranged in pairs of α and β subunits to form three interlocked $\alpha\beta$ segments (Fig. 21.23). The conformations of the three pairs may be loose (L), tight (T), or open (O), and one of each type is present at each stage. A protein at the centre of the interlocked structure, the γ subunit shown as a white arrow, rotates and induces structural changes that cycle each of the three segments between L, T, and O conformations. At the start of a cycle, a T unit holds an ATP molecule. Then ADP and a P_i group migrate into the L site and, as it closes into T, the earlier T site opens into O and releases its ATP. The ADP and P_i in the T site meanwhile condense into ATP, and the new L site is ready for the cycle to begin again. The proton flux drives the rotation of the γ subunit, and hence the conformational changes of the α/β segments, as well as providing the energy for the condensation reaction itself. Several key aspects of this mechanism have been confirmed experimentally. For example, the rotation of the γ subunit has been observed directly by using single-molecule spectroscopy (Section 14.6).

Diffusion

We are now in a position to extend the discussion of ionic motion to cover the migration of neutral molecules and of ions in the absence of an applied electric field. We shall do this by expressing ion motion in a more general way than hitherto, and will then discover that the same equations apply even when the charge on the particles is zero.

21.9 The thermodynamic view

We saw in Part 1 that, at constant temperature and pressure, the maximum non-expansion work that can be done per mole when a substance moves from a location where its chemical potential is μ to a location where its chemical potential is $\mu + d\mu$ is $dw = d\mu$. In a system in which the chemical potential depends on the position x,

$$dw = d\mu = \left(\frac{\partial \mu}{\partial x}\right)_{p,T} dx \qquad (21.54)$$

We also saw in Chapter 2 (Table 2.1) that in general work can always be expressed in terms of an opposing force (which here we write $\mathcal{F}$), and that

$$dw = -\mathcal{F}\,dx \qquad (21.55)$$

By comparing these two expressions, we see that the slope of the chemical potential can be interpreted as an effective force per mole of molecules. We write this **thermodynamic force** as

$$\mathcal{F} = -\left(\frac{\partial \mu}{\partial x}\right)_{p,T} \qquad [21.56]$$

There is not necessarily a real force pushing the particles down the slope of the chemical potential. As we shall see, the force may represent the spontaneous tendency of the molecules to disperse as a consequence of the Second Law and the hunt for maximum entropy.

(a) The thermodynamic force of a concentration gradient

In a solution in which the activity of the solute is a, the chemical potential is

$$\mu = \mu^{\ominus} + RT \ln a$$

If the solution is not uniform the activity depends on the position and we can write

$$\mathcal{F} = -RT\left(\frac{\partial \ln a}{\partial x}\right)_{p,T} \tag{21.57}$$

If the solution is ideal, a may be replaced by the molar concentration c, and then

$$\mathcal{F} = -\frac{RT}{c}\left(\frac{\partial c}{\partial x}\right)_{p,T} \qquad (21.58)^{\circ}$$

where we have also used the relation $\mathrm{d}\ln y/\mathrm{d}x = (1/y)(\mathrm{d}y/\mathrm{d}x)$.

Example 21.3 *Calculating the thermodynamic force*

Suppose the concentration of a solute decays exponentially along the length of a container. Calculate the thermodynamic force on the solute at 25°C given that the concentration falls to half its value in 10 cm.

Method According to eqn 21.58, the thermodynamic force is calculated by differentiating the concentration with respect to distance. Therefore, write an expression for the variation of the concentration with distance, and then differentiate it.

Answer The concentration varies with position as

$$c = c_0 \mathrm{e}^{-x/\lambda}$$

where λ is the decay constant. Therefore,

$$\frac{\mathrm{d}c}{\mathrm{d}x} = -\frac{c}{\lambda}$$

Equation 21.58 then implies that

$$\mathcal{F} = \frac{RT}{\lambda}$$

We know that the concentration falls to $\frac{1}{2}c_0$ at $x = 10$ cm, so we can find λ from $\frac{1}{2} = \mathrm{e}^{-(10\,\mathrm{cm})/\lambda}$. That is, $\lambda = (10\,\mathrm{cm}/\ln 2)$. It follows that

$$\mathcal{F} = (8.3145\,\mathrm{J\,K^{-1}\,mol^{-1}}) \times (298\,\mathrm{K}) \times \ln 2/(1.0 \times 10^{-1}\,\mathrm{m}) = 17\,\mathrm{kN\,mol^{-1}}$$

where we have used $1\,\mathrm{J} = 1\,\mathrm{N\,m}$.

Self-test 21.5 Calculate the thermodynamic force on the molecules of molar mass M in a vertical tube in a gravitational field on the surface of the Earth, and evaluate $\mathcal{F}$ for molecules of molar mass 100 g mol^{-1}. Comment on its magnitude relative to that just calculated.

[$\mathcal{F} = -Mg$, -0.98 N mol^{-1}; the force arising from the concentration gradient greatly dominates that arising from the gravitational gradient.]

(b) Fick's first law of diffusion

In Section 21.4 we saw that Fick's first law of diffusion (that the particle flux is proportional to the concentration gradient) could be deduced from the kinetic model of gases. We shall now show that it can be deduced more generally and that it applies to the diffusion of species in condensed phases too.

We suppose that the flux of diffusing particles is motion in response to a thermodynamic force arising from a concentration gradient. The particles reach a steady drift speed, s, when the thermodynamic force, $\mathcal{F}$, is matched by the viscous drag. This drift speed is proportional to the thermodynamic force, and we write $s \propto \mathcal{F}$. However, the particle flux, J, is proportional to the drift speed, and the thermodynamic force is proportional to the concentration gradient, $\mathrm{d}c/\mathrm{d}x$. The chain of proportionalities ($J \propto s$, $s \propto \mathcal{F}$, and $\mathcal{F} \propto \mathrm{d}c/\mathrm{d}x$) implies that $J \propto \mathrm{d}c/\mathrm{d}x$, which is the content of Fick's law.

(c) The Einstein relation

If we divide both sides of eqn 21.19 by Avogadro's constant, thereby converting numbers into amounts (numbers of moles), then Fick's law becomes

$$J = -D\frac{\mathrm{d}c}{\mathrm{d}x} \tag{21.59}$$

In this expression, D is the diffusion coefficient and $\mathrm{d}c/\mathrm{d}x$ is the slope of the molar concentration. The flux is related to the drift speed by

$$J = sc \tag{21.60}$$

This relation follows from the argument that we have used several times before. Thus, all particles within a distance $s\Delta t$, and therefore in a volume $s\Delta tA$, can pass through a window of area A in an interval Δt. Hence, the amount of substance that can pass through the window in that interval is $s\Delta tAc$. Therefore,

$$sc = -D\frac{\mathrm{d}c}{\mathrm{d}x}$$

If now we express $\mathrm{d}c/\mathrm{d}x$ in terms of $\mathcal{F}$ by using eqn 21.58, we find

$$s = -\frac{D}{c}\frac{\mathrm{d}c}{\mathrm{d}x} = \frac{D\mathcal{F}}{RT} \tag{21.61}$$

Therefore, once we know the effective force and the diffusion coefficient, D, we can calculate the drift speed of the particles (and vice versa) whatever the origin of the force.

There is one case where we already know the drift speed and the effective force acting on a particle: an ion in solution has a drift speed $s = u\mathcal{E}$ when it experiences a force $ez\mathcal{E}$ from an electric field of strength $\mathcal{E}$ (so $\mathcal{F} = N_A ez\mathcal{E} = zF\mathcal{E}$). Therefore, substituting these known values into eqn 21.61 gives

$$u\mathcal{E} = \frac{zF\mathcal{E}D}{RT}$$

and hence

$$u = \frac{zFD}{RT} \tag{21.62}$$

This equation rearranges into the very important result known as the **Einstein relation** between the diffusion coefficient and the ionic mobility:

$$D = \frac{uRT}{zF} \tag{21.63}^{\circ}$$

On inserting the typical value $u = 5 \times 10^{-8}\ \mathrm{m^2\ s^{-1}\ V^{-1}}$, we find $D \approx 1 \times 10^{-9}\ \mathrm{m^2\ s^{-1}}$ at 25°C as a typical value of the diffusion coefficient of an ion in water.

(d) The Nernst–Einstein equation

The Einstein relation provides a link between the molar conductivity of an electrolyte and the diffusion coefficients of its ions. First, by using eqns 21.44 and 21.63 we write

$$\lambda = zuF = \frac{z^2DF^2}{RT} \qquad (21.64)^\circ$$

for each type of ion. Then, from $\Lambda_m^\circ = \nu_+\lambda_+ + \nu_-\lambda_-$, the limiting molar conductivity is

$$\Lambda_m = (\nu_+z_+^2D_+ + \nu_-z_-^2D_-)\frac{F^2}{RT} \qquad (21.65)^\circ$$

which is the **Nernst–Einstein equation**. One application of this equation is to the determination of ionic diffusion coefficients from conductivity measurements; another is to the prediction of conductivities using models of ionic diffusion (see below).

(e) The Stokes–Einstein equation

Equations 21.43 ($u = ez/f$) and 21.63 relate the mobility of an ion to the frictional force and to the diffusion coefficient, respectively. We can combine the two expressions into the **Stokes–Einstein equation:**

$$D = \frac{kT}{f} \qquad (21.66)$$

If the frictional force is described by Stokes's law, then we also obtain a relation between the diffusion coefficient and the viscosity of the medium:

$$D = \frac{kT}{6\pi\eta a} \qquad (21.67)$$

An important feature of eqn 21.66 (and of its special case, eqn 21.67) is that it makes no reference to the charge of the diffusing species. Therefore, the equation also applies in the limit of vanishingly small charge, that is, it also applies to neutral molecules. Consequently, we may use viscosity measurements to estimate the diffusion coefficients for electrically neutral molecules in solution (Table 21.8). It must not be forgotten, however, that both equations depend on the assumption that the viscous drag is proportional to the speed.

Synoptic table 21.8* Diffusion coefficients at 298 K

	$D/(10^{-9}\ m^2\ s^{-1})$
H^+ in water	9.31
I_2 in hexane	4.05
Na^+ in water	1.33
Sucrose in water	0.522

* More values are given in the *Data section.*

Example 21.4 *Interpreting the mobility of an ion*

Use the experimental value of the mobility to evaluate the diffusion coefficient, the limiting molar conductivity, and the hydrodynamic radius of a sulfate ion in aqueous solution.

Method The starting point is the mobility of the ion, which is given in Table 21.6. The diffusion coefficient can then be determined from the Einstein relation, eqn 21.63. The ionic conductivity is related to the mobility by eqn 21.44. To estimate the hydrodynamic radius, a, of the ion, use the Stokes–Einstein relation to find f and the Stokes law to relate f to a.

Answer From Table 21.6, the mobility of SO_4^{2-} is $8.29 \times 10^{-8}\ m^2\ s^{-1}\ V^{-1}$. It follows from eqn 21.63 that

$$D = \frac{uRT}{zF} = 1.1 \times 10^{-9}\ m^2\ s^{-1}$$

From eqn 21.44 it follows that

$$\lambda = zuF = 16 \text{ mS m}^2 \text{ mol}^{-1}$$

Finally, from $f = 6\pi\eta a$ using 0.891 cP (or 8.91×10^{-4} kg m^{-1} s^{-1}) for the viscosity of water (Table 21.4):

$$a = \frac{kT}{6\pi\eta D} = 220 \text{ pm}$$

The bond length in SO_4^{2-} is 144 pm, so the radius calculated here is plausible and consistent with a small degree of solvation.

Self-test 21.6 Repeat the calculation for the NH_4^+ ion.
[1.96×10^{-9} m^2 s^{-1}, 7.4 mS m^2 mol^{-1}, 125 pm]

Experimental support for the relations derived above comes from conductivity measurements. In particular, **Walden's rule** is the empirical observation that the product $\eta\Lambda_m$ is very approximately constant for the same ions in different solvents (but there are numerous exceptions). Because $\Lambda_m \propto D$, and we have just seen that $D \propto 1/\eta$, we do indeed predict that $\Lambda_m \propto 1/\eta$, as Walden's rule implies. The usefulness of the rule, however, is muddied by the role of solvation: different solvents solvate the same ions to different extents, so both the hydrodynamic radius and the viscosity change with the solvent.

21.10 The diffusion equation

We now turn to the discussion of time-dependent diffusion processes, where we are interested in the spreading of inhomogeneities with time. One example is the temperature of a metal bar that has been heated at one end: if the source of heat is removed, then the bar gradually settles down into a state of uniform temperature. When the source of heat is maintained and the bar can radiate, it settles down into a steady state of nonuniform temperature. Another example (and one more relevant to chemistry) is the concentration distribution in a solvent to which a solute is added. We shall focus on the description of the diffusion of particles, but similar arguments apply to the diffusion of physical properties, such as temperature. Our aim is to obtain an equation for the rate of change of the concentration of particles in an inhomogeneous region.

The central equation of this section is the **diffusion equation**, also called 'Fick's second law of diffusion', which relates the rate of change of concentration at a point to the spatial variation of the concentration at that point:

$$\frac{\partial c}{\partial t} = D\frac{\partial^2 c}{\partial x^2} \qquad (21.68)$$

We show in the following *Justification* that the diffusion equation follows from Fick's first law of diffusion.

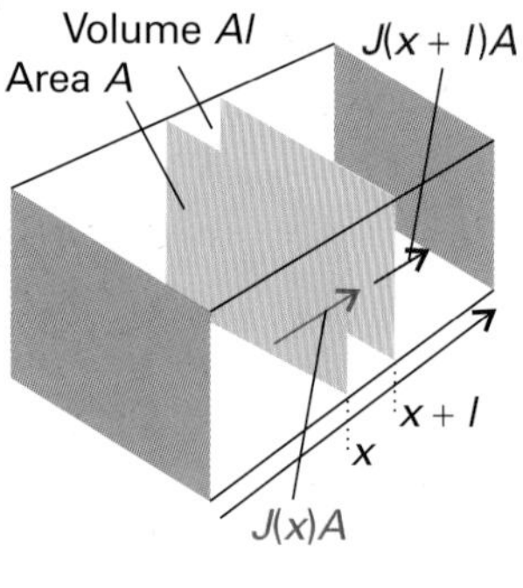

Fig. 21.24 The net flux in a region is the difference between the flux entering from the region of high concentration (on the left) and the flux leaving to the region of low concentration (on the right).

Justification 21.6 *The diffusion equation*

Consider a thin slab of cross-sectional area A that extends from x to $x + l$ (Fig. 21.24). Let the concentration at x be c at the time t. The amount (number of moles) of particles that enter the slab in the infinitesimal interval dt is $JAdt$, so the rate of increase in molar concentration inside the slab (which has volume Al) on account of the flux from the left is

$$\frac{\partial c}{\partial t}=\frac{JA\mathrm{d}t}{Al\mathrm{d}t}=\frac{J}{l}$$

There is also an outflow through the right-hand window. The flux through that window is J', and the rate of change of concentration that results is

$$\frac{\partial c}{\partial t}=-\frac{J'A\mathrm{d}t}{Al\mathrm{d}t}=-\frac{J'}{l}$$

The net rate of change of concentration is therefore

$$\frac{\partial c}{\partial t}=\frac{J-J'}{l}$$

Each flux is proportional to the concentration gradient at the window. So, by using Fick's first law, we can write

$$J-J'=-D\frac{\partial c}{\partial x}+D\frac{\partial c'}{\partial x}=-D\frac{\partial c}{\partial x}+D\frac{\partial}{\partial x}\left\{c+\left(\frac{\partial c}{\partial x}\right)l\right\}=Dl\frac{\partial^2 c}{\partial x^2}$$

When this relation is substituted into the expression for the rate of change of concentration in the slab, we get eqn 21.68.

The diffusion equation shows that the rate of change of concentration is proportional to the curvature (more precisely, to the second derivative) of the concentration with respect to distance. If the concentration changes sharply from point to point (if the distribution is highly wrinkled) then the concentration changes rapidly with time. Where the curvature is positive (a dip, Fig. 21.25), the change in concentration is positive; the dip tends to fill. Where the curvature is negative (a heap), the change in concentration is negative; the heap tends to spread. If the curvature is zero, then the concentration is constant in time. If the concentration decreases linearly with distance, then the concentration at any point is constant because the inflow of particles is exactly balanced by the outflow.

The diffusion equation can be regarded as a mathematical formulation of the intuitive notion that there is a natural tendency for the wrinkles in a distribution to disappear. More succinctly: Nature abhors a wrinkle.

(a) Diffusion with convection

The transport of particles arising from the motion of a streaming fluid is called **convection**. If for the moment we ignore diffusion, then the flux of particles through an area A in an interval Δt when the fluid is flowing at a velocity v can be calculated in the way we have used several times before (by counting the particles within a distance $v\Delta t$), and is

$$J=\frac{cA\Delta vt}{A\Delta t}=cv \qquad (21.69)$$

This J is called the **convective flux**. The rate of change of concentration in a slab of thickness l and area A is, by the same argument as before and assuming that the velocity does not depend on the position,

$$\frac{\partial c}{\partial t}=\frac{J-J'}{l}=\left\{c-\left[c+\left(\frac{\partial c}{\partial x}\right)l\right]\right\}\frac{v}{l}=-v\frac{\partial c}{\partial x} \qquad (21.70)$$

When both diffusion and convection occur, the total change of concentration in a region is the sum of the two effects, and the **generalized diffusion equation** is

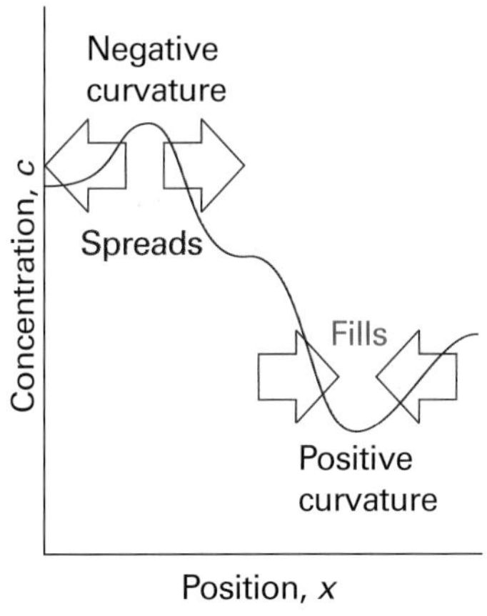

Fig. 21.25 Nature abhors a wrinkle. The diffusion equation tells us that peaks in distribution (regions of negative curvature) spread and troughs (regions of positive curvature) fill in.

$$\frac{\partial c}{\partial t} = D\frac{\partial^2 c}{\partial x^2} - v\frac{\partial c}{\partial x} \qquad (21.71)$$

A further refinement, which is important in chemistry, is the possibility that the concentrations of particles may change as a result of reaction. When reactions are included in eqn 21.71 (Section 24.2), we get a powerful differential equation for discussing the properties of reacting, diffusing, convecting systems and which is the basis of reactor design in chemical industry and of the utilization of resources in living cells.

(b) Solutions of the diffusion equation

The diffusion equation, eqn 21.68, is a second-order differential equation with respect to space and a first-order differential equation with respect to time. Therefore, we must specify two boundary conditions for the spatial dependence and a single initial condition for the time-dependence.

As an illustration, consider a solvent in which the solute is initially coated on one surface of the container (for example, a layer of sugar on the bottom of a deep beaker of water). The single initial condition is that at $t = 0$ all N_0 particles are concentrated on the *yz*-plane (of area A) at $x = 0$. The two boundary conditions are derived from the requirements (1) that the concentration must everywhere be finite and (2) that the total amount (number of moles) of particles present is n_0 (with $n_0 = N_0/N_A$) at all times. These requirements imply that the flux of particles is zero at the top and bottom surfaces of the system. Under these conditions it is found that

$$c(x,t) = \frac{n_0}{A(\pi Dt)^{1/2}} e^{-x^2/4Dt} \qquad (21.72)$$

as may be verified by direct substitution. Figure 21.26 shows the shape of the concentration distribution at various times, and it is clear that the concentration spreads and tends to uniformity.

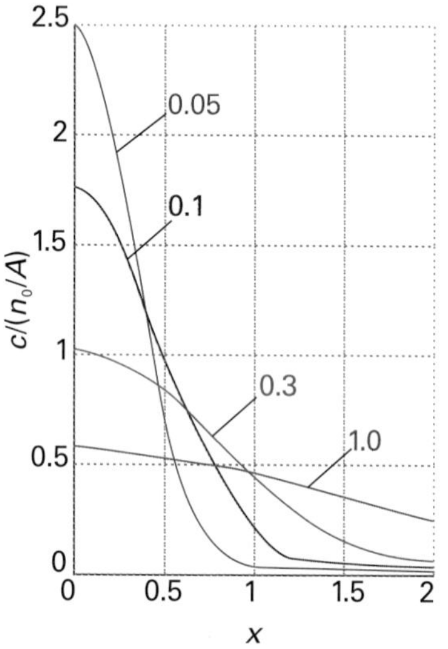

Fig. 21.26 The concentration profiles above a plane from which a solute is diffusing. The curves are plots of eqn 21.72 and are labelled with different values of Dt. The units of Dt and x are arbitrary, but are related so that Dt/x^2 is dimensionless. For example, if x is in metres, Dt would be in metres^2; so, for $D = 10^{-9}$ m^2 s^{-1}, $Dt = 0.1$ m^2 corresponds to $t = 10^8$ s.

Exploration Generate a family of curves similar to that shown in Fig. 21.26 but by using eqn 21.73, which describes diffusion in three dimensions.

Another useful result is for a localized concentration of solute in a three-dimensional solvent (a sugar lump suspended in a large flask of water). The concentration of diffused solute is spherically symmetrical and at a radius r is

$$c(r,t) = \frac{n_0}{8(\pi Dt)^{3/2}} e^{-r^2/4Dt} \qquad (21.73)$$

Other chemically (and physically) interesting arrangements, such as transport of substances across biological membranes can be treated (*Impact I21.3*). In many cases the solutions are more cumbersome.

(c) The measurement of diffusion coefficients

The solutions of the diffusion equation are useful for experimental determinations of diffusion coefficients. In the **capillary technique**, a capillary tube, open at one end and containing a solution, is immersed in a well stirred larger quantity of solvent, and the change of concentration in the tube is monitored. The solute diffuses from the open end of the capillary at a rate that can be calculated by solving the diffusion equation with the appropriate boundary conditions, so D may be determined. In the **diaphragm technique**, the diffusion occurs through the capillary pores of a sintered glass diaphragm separating the well-stirred solution and solvent. The concentrations are monitored and then related to the solutions of the diffusion equation corresponding to this arrangement. Diffusion coefficients may also be measured by the dynamic light scattering technique described in Section 19.3 and by NMR.

IMPACT ON BIOCHEMISTRY

I21.3 Transport of non-electrolytes across biological membranes

We saw in *Impact I21.2* how electrolytes are transported across cell membranes. Here we use the diffusion equation to explore the way in which non-electrolytes cross the lipid bilayer.

Consider the passive transport of an uncharged species A across a lipid bilayer of thickness l. To simplify the problem, we will assume that the concentration of A is always maintained at $[A] = [A]_0$ on one surface of the membrane and at $[A] = 0$ on the other surface, perhaps by a perfect balance between the rate of the process that produces A on one side and the rate of another process that consumes A completely on the other side. This is one example of a steady-state assumption, which will be discussed in more detail in Section 22.7. Then $\partial[A]/\partial t = 0$ and the diffusion equation simplifies to

$$D\frac{d^2[A]}{dx^2} = 0 \qquad (21.74)$$

where D is the diffusion coefficient and the steady-state assumption makes partial derivatives unnecessary. We use the boundary conditions $[A](0) = [A]_0$ and $[A](l) = 0$ to solve eqn 21.74 and the result, which may be verified by differentiation, is

$$[A](x) = [A]_0\left(1 - \frac{x}{l}\right) \qquad (21.75)$$

which implies that the [A] decreases linearly inside the membrane. We now use Fick's first law to calculate the flux J of A through the membrane and the result is

$$J = D\frac{[A]_0}{l} \qquad (21.76)$$

However, we need to modify this equation slightly to account for the fact that the concentration of A on the surface of a membrane is not always equal to the concentration of A measured in the bulk solution, which we assume to be aqueous. This difference arises from the significant difference in the solubility of A in an aqueous environment and in the solution–membrane interface. One way to deal with this problem is to define a **partition ratio**, K_D (D for distribution) as

$$K_D = \frac{[A]_0}{[A]_s} \qquad [21.77]$$

where $[A]_s$ is the concentration of A in the bulk aqueous solution. It follows that

$$J = DK_D\frac{[A]_s}{l} \qquad (21.78)$$

In spite of the assumptions that led to its final form, eqn 21.78 describes adequately the passive transport of many non-electrolytes through membranes of blood cells.

In many cases the flux is underestimated by eqn 21.78 and the implication is that the membrane is more permeable than expected. However, the permeability increases only for certain species and not others and this is evidence that transport can be mediated by carriers. One example is the transporter protein that carries glucose into cells.

A characteristic of a carrier C is that it binds to the transported species A and the dissociation of the AC complex is described by

$$AC \rightleftharpoons A + C \qquad K = \frac{[A][C]}{[AC]} \qquad (21.79)$$

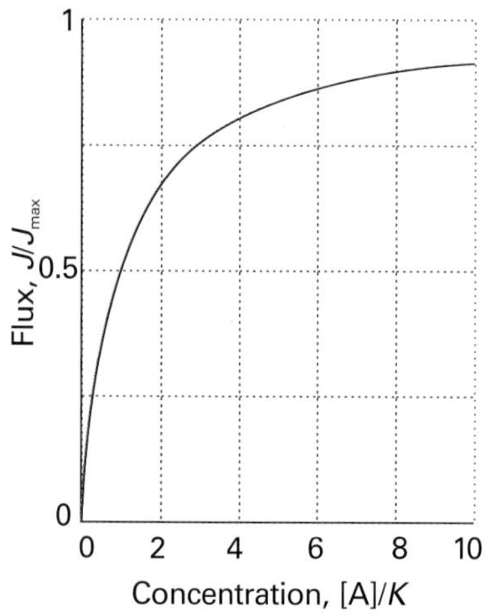

Fig. 21.27 The flux of the species AC through a membrane varies with the concentration of the species A. The behaviour shown in the figure and explained in the text is characteristic of mediated transport of A, with C as a carrier molecule.

where we have used concentrations instead of activities. After writing $[C]_0 = [C] + [AC]$, where $[C]_0$ is the total concentration of carrier, it follows that

$$[AC] = \frac{[A][C]_0}{[A]+K} \tag{21.80}$$

We can now use eqns 21.80 and 21.78 to write an expression for the flux of the species AC through the membrane:

$$J = \frac{DK_D[C]_0}{l}\frac{[A]}{[A]+K} = J_{max}\frac{[A]}{[A]+K} \tag{21.81}$$

where K_D and D are the partition ratio and diffusion coefficient of the species AC. We see from Fig. 21.27 that when $[A] \ll K$ the flux varies linearly with [A] and that the flux reaches a maximum value of $J_{max} = DK_D[C]_0/l$ when $[A] \gg K$. This behaviour is characteristic of mediated transport.

21.11 Diffusion probabilities

The solutions of the diffusion equation can be used to predict the concentration of particles (or the value of some other physical quantity, such as the temperature in a nonuniform system) at any location. We can also use them to calculate the net distance through which the particles diffuse in a given time.

Example 21.5 *Calculating the net distance of diffusion*

Calculate the net distance travelled on average by particles in a time t if they have a diffusion constant D.

Method We need to use the results of probability theory summarized in *Appendix* 2. In this case, we calculate the probability that a particle will be found at a certain distance from the origin, and then calculate the average by weighting each distance by that probability.

Answer The number of particles in a slab of thickness dx and area A at x, where the molar concentration is c, is $cAN_A dx$. The probability that any of the $N_0 = n_0 N_A$ particles is in the slab is therefore $cAN_A dx/N_0$. If the particle is in the slab, it has travelled a distance x from the origin. Therefore, the mean distance travelled by all the particles is the sum of each x weighted by the probability of its occurrence:

$$\langle x\rangle = \int_0^\infty \frac{xcAN_A}{N_0}dx = \frac{1}{(\pi Dt)^{1/2}}\int_0^\infty xe^{-x^2/4Dt}dx = 2\left(\frac{Dt}{\pi}\right)^{1/2}$$

where we have used the same standard integral as that used in Justification 21.4.

The average distance of diffusion varies as the square root of the lapsed time. If we use the Stokes–Einstein relation for the diffusion coefficient, the mean distance travelled by particles of radius a in a solvent of viscosity η is

$$\langle x\rangle = \left(\frac{2kTt}{3\pi^2\eta a}\right)^{1/2}$$

Self-test 21.7 Derive an expression for the root mean square distance travelled by diffusing particles in a time t. $[\langle x^2\rangle^{1/2} = (2Dt)^{1/2}]$

As shown in Example 21.5, the average distance travelled by diffusing particle in a time t is

$$\langle x\rangle = 2\left(\frac{Dt}{\pi}\right)^{1/2} \tag{21.82}$$

and the root mean square distance travelled in the same time is

$$\langle x^2\rangle^{1/2} = (2Dt)^{1/2} \tag{21.83}$$

The latter is a valuable measure of the spread of particles when they can diffuse in both directions from the origin (for then $\langle x\rangle = 0$ at all times). The root mean square distance travelled by particles with a typical diffusion coefficient ($D = 5\times10^{-10}\ \mathrm{m^2\,s^{-1}}$) is illustrated in Fig. 21.28, which shows how long it takes for diffusion to increase the net distance travelled on average to about 1 cm in an unstirred solution. The graph shows that diffusion is a very slow process (which is why solutions are stirred, to encourage mixing by convection).

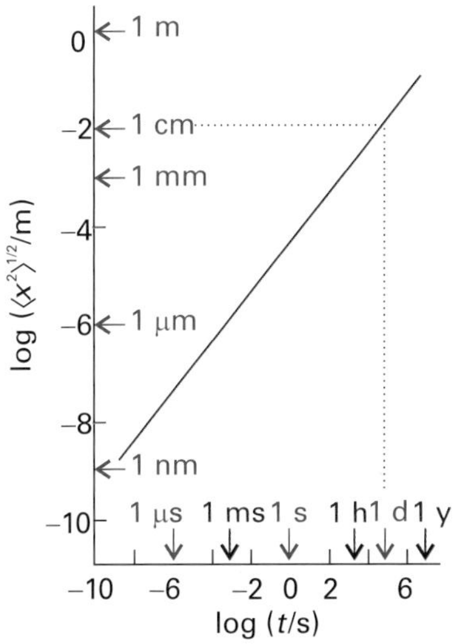

Fig. 21.28 The root mean square distance covered by particles with $D = 5\times10^{-10}\ \mathrm{m^2\,s^{-1}}$. Note the great slowness of diffusion.

21.12 The statistical view

An intuitive picture of diffusion is of the particles moving in a series of small steps and gradually migrating from their original positions. We shall explore this idea using a model in which the particles can jump through a distance λ in a time τ. The total distance travelled by a particle in a time t is therefore $t\lambda/\tau$. However, the particle will not necessarily be found at that distance from the origin. The direction of each step may be different, and the net distance travelled must take the changing directions into account.

If we simplify the discussion by allowing the particles to travel only along a straight line (the x-axis), and for each step (to the left or the right) to be through the same distance λ, then we obtain the **one-dimensional random walk**. The same model was used in the discussion of a one-dimensional random coil in Section 19.8a.

We show in the *Justification* below that the probability of a particle being at a distance x from the origin after a time t is

$$P = \left(\frac{2\tau}{\pi t}\right)^{1/2} \mathrm{e}^{-x^2\tau/2t\lambda^2} \tag{21.84}$$

Justification 21.7 *The one-dimensional random walk*

Consider a one-dimensional random walk in which each step is through a distance λ to the left or right. The net distance travelled after N steps is equal to the difference between the number of steps to the right (N_R) and to the left (N_L), and is $(N_R - N_L)\lambda$. We write $n = N_R - N_L$ and the total number of steps as $N = N_R + N_L$.

The number of ways of performing a walk with a given net distance of travel $n\lambda$ is the number of ways of making N_R steps to the right and N_L steps to the left, and is given by the binomial coefficient

$$W = \frac{N!}{N_L!N_R!} = \frac{N!}{\{\frac{1}{2}(N+n)\}!\{\frac{1}{2}(N-n)\}!}$$

The probability of the net distance walked being $n\lambda$ is

$$P = \frac{\text{number of paths with } N_R \text{ steps to the right}}{\text{total number of steps}}$$

$$= \frac{W}{2^N} = \frac{N!}{\{\frac{1}{2}(N+n)\}!\{\frac{1}{2}(N-n)\}!2^N}$$

The use of Stirling's approximation (Section 16.1a) in the form

$$\ln x! \approx \ln(2\pi)^{1/2} + (x+\tfrac{1}{2})\ln x - x$$

gives (after quite a lot of algebra; see Problem 21.33)

$$\ln P = \ln\left(\frac{2}{\pi N}\right)^{1/2} - \tfrac{1}{2}(N+n+1)\ln\left(1+\frac{n}{N}\right) - \tfrac{1}{2}(N-n+1)\ln\left(1-\frac{n}{N}\right)$$

For small net distances ($n \ll N$) we can use the approximation $\ln(1 \pm x) \approx \pm x - \tfrac{1}{2}x^2$, and so obtain

$$\ln P \approx \ln\left(\frac{2}{\pi N}\right)^{1/2} - \frac{n^2}{2N}$$

At this point, we note that the number of steps taken in a time t is $N = t/\tau$ and the net distance travelled from the origin is $x = n\lambda$. Substitution of these quantities into the expression for $\ln P$ gives

$$\ln P \approx \ln\left(\frac{2\tau}{\pi t}\right)^{1/2} - \frac{x^2\tau}{2t\lambda^2}$$

which, upon using $e^{\ln x} = x$ and $e^{x+y} = e^x e^y$, rearranges into eqn 21.84.

The differences of detail between eqns 21.72 and 21.84 arise from the fact that in the present calculation the particles can migrate in either direction from the origin. Moreover, they can be found only at discrete points separated by λ instead of being anywhere on a continuous line. The fact that the two expressions are so similar suggests that diffusion can indeed be interpreted as the outcome of a large number of steps in random directions.

We can now relate the coefficient D to the step length λ and the rate at which the jumps occur. Thus, by comparing the two exponents in eqns 21.72 and 21.84 we can immediately write down the **Einstein–Smoluchowski equation**:

$$D = \frac{\lambda^2}{2\tau} \qquad (21.85)$$

Illustration 21.6 *Using the Einstein–Smoluchowski equation*

Suppose that a SO_4^{2-} ion jumps through its own diameter each time it makes a move in an aqueous solution; then, because $D = 1.1 \times 10^{-9}$ m^2 s^{-1} and $a = 210$ pm (as deduced from mobility measurements), it follows from $\lambda = 2a$ that $\tau = 80$ ps. Because τ is the time for one jump, the ion makes 1×10^{10} jumps per second.

The Einstein–Smoluchowski equation is the central connection between the microscopic details of particle motion and the macroscopic parameters relating to diffusion (for example, the diffusion coefficient and, through the Stokes–Einstein relation, the viscosity). It also brings us back full circle to the properties of the perfect gas. For if we interpret λ/τ as $\bar{c}$, the mean speed of the molecules, and interpret λ as a mean free path, then we can recognize in the Einstein–Smoluchowski equation exactly the same expression as we obtained from the kinetic model of gases, eqn 21.22. That is, the diffusion of a perfect gas is a random walk with an average step size equal to the mean free path.

Checklist of key ideas

☐ 1. Diffusion is the migration of matter down a concentration gradient; thermal conduction is the migration of energy down a temperature gradient; electric conduction is the migration of electric charge along an electrical potential gradient; viscosity is the migration of linear momentum down a velocity gradient.

☐ 2. The kinetic model of a gas considers only the contribution to the energy from the kinetic energies of the molecules. Important results from the model include expressions for the pressure ($pV = \frac{1}{3}nMc^2$) and the root mean square speed ($c = \langle v^2 \rangle^{1/2} = (3RT/M)^{1/2}$).

☐ 3. The Maxwell distribution of speeds is the function which, through $f(v)\mathrm{d}v$, gives the fraction of molecules that have speeds in the range v to $v + \mathrm{d}v$.

☐ 4. The collision frequency is the number of collisions made by a molecule in an interval divided by the length of the interval: $z = \sigma \bar{c}_{\mathrm{rel}} \mathcal{N}$, where the collision cross-section is $\sigma = \pi d^2$.

☐ 5. The mean free path is the average distance a molecule travels between collisions: $\lambda = \bar{c}/z$.

☐ 6. The collision flux, Z_W, is the number of collisions with an area in a given time interval divided by the area and the duration of the interval: $Z_w = p/(2\pi mkT)^{1/2}$.

☐ 7. Effusion is the emergence of a gas from a container through a small hole. Graham's law of effusion states that the rate of effusion is inversely proportional to the square root of the molar mass.

☐ 8. Flux J is the quantity of a property passing through a given area in a given time interval divided by the area and the duration of the interval.

☐ 9. Fick's first law of diffusion states that the flux of matter is proportional to the concentration gradient, $J(\text{matter}) = -D\mathrm{d}\mathcal{N}/\mathrm{d}z$, where D is the diffusion coefficient.

☐ 10. The conductance, G, is the inverse of resistance. The conductivity is the constant κ in $G = \kappa A/l$ and the molar conductivity is written as $\Lambda_{\mathrm{m}} = \kappa/c$.

☐ 11. A strong electrolyte is an electrolyte with a molar conductivity that varies only slightly with concentration. A weak electrolyte is an electrolyte with a molar conductivity that is normal at concentrations close to zero, but falls sharply to low values as the concentration increases.

☐ 12. Kohlrausch's law for the concentration dependence of the molar conductivity of a strong electrolyte is written as $\Lambda_{\mathrm{m}} = \Lambda_{\mathrm{m}}^{\circ} - \mathcal{K}c^{1/2}$, where the limiting molar conductivity, $\Lambda_{\mathrm{m}}^{\circ}$, is the molar conductivity at zero concentration ($\Lambda_{\mathrm{m}}^{\circ} = \nu_+\lambda_+ + \nu_-\lambda_-$).

☐ 13. The drift speed s is the terminal speed when an accelerating force is balanced by the viscous drag: $s = u\mathcal{E}$, where $u = ze/6\pi\eta a$ is the ionic mobility and a is the hydrodynamic radius (Stokes radius), the effective radius of a particle in solution.

☐ 14. The ionic conductivity is the contribution of ions of one type to the molar conductivity: $\lambda = zuF$.

☐ 15. The transport number is the fraction of total current I carried by the ions of a specified type: $t_{\pm} = I_{\pm}/I$.

☐ 16. The Debye–Hückel–Onsager theory explains the concentration dependence of the molar conductivity of a strong electrolyte in terms of ionic interactions.

☐ 17. The Einstein relation between the diffusion coefficient and the ionic mobility is $D = uRT/zF$.

☐ 18. The Nernst–Einstein relation between the molar conductivity of an electrolyte and the diffusion coefficients of its ions is $\Lambda_{\mathrm{m}} = (\nu_+z_+^2D_+ + \nu_-z_-^2D_-)(F^2/RT)$.

☐ 19. The Stokes–Einstein equation relates the diffusion coefficient to the frictional force: $D = kT/f$.

☐ 20. Walden's rule states that the product $\eta\Lambda_{\mathrm{m}}$ is very approximately constant for the same ions in different solvents.

☐ 21. The diffusion equation is a relation between the rate of change of concentration at a point and the spatial variation of the concentration at that point: $\partial c/\partial t = D\partial^2 c/\partial x^2$.

☐ 22. In a one-dimensional random walk, the probability P that a molecule moves a distance x from the origin for a period t by taking small steps with size λ and time τ is: $P = (2\tau/\pi t)^{1/2}\mathrm{e}^{-x^2\tau/2t\lambda^2}$.

☐ 23. The Einstein–Smoluchowski equation relates the diffusion coefficient to the parameters used in the formulation of the random walk model, $D = \lambda^2/2\tau$.

Further reading

Articles and texts

D.G. Leaist, Diffusion and ionic conduction in liquids. In *Encyclopedia of applied physics* (ed. G.L. Trigg), **5**, 661 VCH, New York (1993).

A.J. Bard and L.R. Faulkner, *Electrochemical methods: fundamentals and applications*. Wiley, New York (2000).

R.B. Bird, W.E. Stewart, and E.N. Lightfoot, *Transport phenomena*. Wiley, New-York (1960).

J.N. Murrell and A.D. Jenkins, *Properties of liquids and solutions*. Wiley-Interscience, New York (1994).

K.E. van Holde, W.C. Johnson, and P.S. Ho, *Principles of physical biochemistry*. Prentice Hall, Upper Saddle River (1998).

A.J. Walton, *Three phases of matter*. Oxford University Press (1983).

Sources of data and information

D.R. Lide (ed.), *CRC Handbook of Chemistry and Physics*, Sections 5 and 6. CRC Press, Boca Raton (2000).

NIST thermodynamic and transport properties of pure fluids database, NIST standard reference database 12, National Institute of Standards and Testing, Gaithersburg (1995). This material is on CD-ROM.

Further information

Further information 21.1 *The transport characteristics of a perfect gas*

In this Further information section, we derive expressions for the diffusion characteristics (specifically, the diffusion coefficient, the thermal conductivity, and the viscosity) of a perfect gas on the basis of the kinetic molecular theory.

The diffusion coefficient, *D*

Consider the arrangement depicted in Fig. 21.29. On average, the molecules passing through the area A at $z=0$ have travelled about one mean free path λ since their last collision. Therefore, the number density where they originated is $\mathcal{N}(z)$ evaluated at $z=-\lambda$. This number density is approximately

$$\mathcal{N}(-\lambda)=\mathcal{N}(0)-\lambda\left(\frac{\mathrm{d}\mathcal{N}}{\mathrm{d}z}\right)_0 \tag{21.86}$$

where we have used a Taylor expansion of the form $f(x)=f(0)+(\mathrm{d}f/\mathrm{d}x)_0x+\cdots$ truncated after the second term (see *Appendix* 2). The average number of impacts on the imaginary window of area A_0 during an interval Δt is $Z_W A_0\Delta t$, with $Z_W=\frac{1}{4}\mathcal{N}\bar{c}$ (eqn 21.15). Therefore, the flux from left to right, $J(\mathrm{L}\rightarrow\mathrm{R})$, arising from the supply of molecules on the left, is

$$J(\mathrm{L}\rightarrow\mathrm{R})=\frac{\frac{1}{4}A_0\mathcal{N}(-\lambda)\bar{c}\Delta t}{A_0\Delta t}=\tfrac{1}{4}\mathcal{N}(-\lambda)\bar{c} \tag{21.87}$$

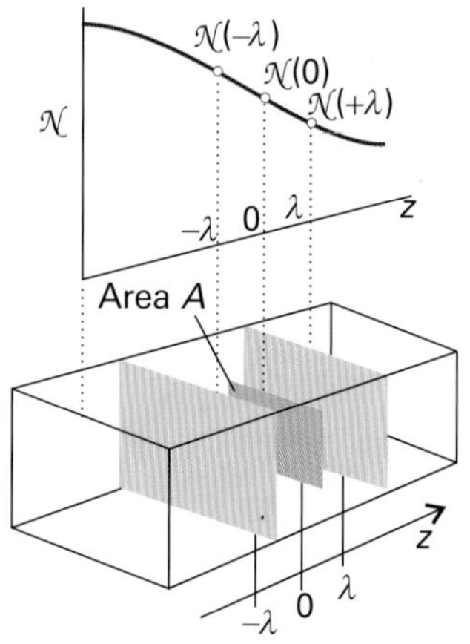

Fig. 21.29 The calculation of the rate of diffusion of a gas considers the net flux of molecules through a plane of area A as a result of arrivals from on average a distance λ away in each direction, where λ is the mean free path.

There is also a flux of molecules from right to left. On average, the molecules making the journey have originated from $z=+\lambda$ where the number density is $\mathcal{N}(\lambda)$. Therefore,

$$J(\mathrm{L}\leftarrow\mathrm{R})=-\tfrac{1}{4}\mathcal{N}(\lambda)\bar{c} \tag{21.88}$$

The average number density at $z=+\lambda$ is approximately

$$\mathcal{N}(\lambda)=\mathcal{N}(0)+\lambda\left(\frac{\mathrm{d}\mathcal{N}}{\mathrm{d}z}\right)_0 \tag{21.89}$$

The net flux is

$$\begin{aligned}J_z&=J(L\rightarrow R)+J(L\leftarrow R)\\&=\tfrac{1}{4}\bar{c}\left\{\left[\mathcal{N}(0)-\lambda\left(\frac{\mathrm{d}\mathcal{N}}{\mathrm{d}z}\right)_0\right]-\left[\mathcal{N}(0)+\lambda\left(\frac{\mathrm{d}\mathcal{N}}{\mathrm{d}z}\right)_0\right]\right\}\\&=-\tfrac{1}{2}\bar{c}\lambda\left(\frac{\mathrm{d}\mathcal{N}}{\mathrm{d}z}\right)_0\end{aligned} \tag{21.90}$$

This equation shows that the flux is proportional to the first derivative of the concentration, in agreement with Fick's law.

At this stage it looks as though we can pick out a value of the diffusion coefficient by comparing eqns 21.19 and 21.90, so obtaining $D=\frac{1}{2}\lambda\bar{c}$. It must be remembered, however, that the calculation is quite crude, and is little more than an assessment of the order of magnitude of D. One aspect that has not been taken into account is illustrated in Fig. 21.30, which shows that, although a molecule may

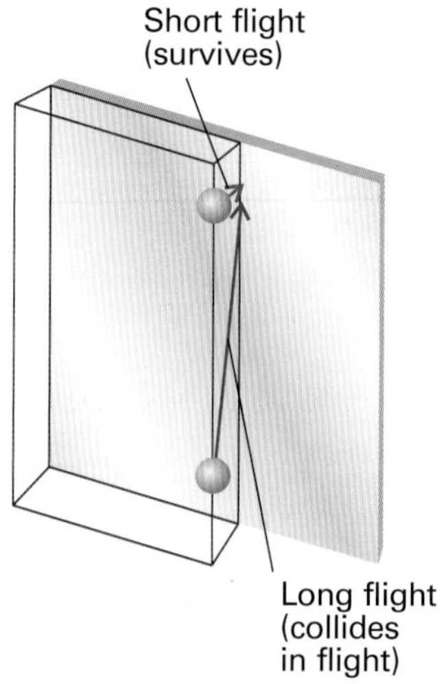

Fig. 21.30 One approximation ignored in the simple treatment is that some particles might make a long flight to the plane even though they are only a short perpendicular distance away, and therefore they have a higher chance of colliding during their journey.

have begun its journey very close to the window, it could have a long flight before it gets there. Because the path is long, the molecule is likely to collide before reaching the window, so it ought to be added to the graveyard of other molecules that have collided. To take this effect into account involves a lot of work, but the end result is the appearance of a factor of $\frac{2}{3}$ representing the lower flux. The modification results in eqn 21.22.

Thermal conductivity

According to the equipartition theorem (Section 17.3), each molecule carries an average energy $\varepsilon = \nu kT$, where ν is a number of the order of 1. For monatomic particles, $\nu = \frac{3}{2}$. When one molecule passes through the imaginary window, it transports that energy on average. We suppose that the number density is uniform but that the temperature is not. On average, molecules arrive from the left after travelling a mean free path from their last collision in a hotter region, and therefore with a higher energy. Molecules also arrive from the right after travelling a mean free path from a cooler region. The two opposing energy fluxes are therefore

$$J(\mathrm{L} \rightarrow \mathrm{R}) = \tfrac{1}{4}\bar{c}\mathcal{N}\varepsilon(-\lambda) \qquad \varepsilon(-\lambda) = \nu k\left\{T - \lambda\left(\frac{\mathrm{d}T}{\mathrm{d}z}\right)_0\right\}$$

$$J(\mathrm{L} \leftarrow \mathrm{R}) = -\tfrac{1}{4}\bar{c}\mathcal{N}\varepsilon(\lambda) \qquad \varepsilon(\lambda) = \nu k\left\{T + \lambda\left(\frac{\mathrm{d}T}{\mathrm{d}z}\right)_0\right\} \qquad (21.91)$$

and the net flux is

$$J_z = J(\mathrm{L} \rightarrow \mathrm{R}) + J(\mathrm{L} \leftarrow \mathrm{R}) = -\tfrac{1}{2}\nu k\lambda\bar{c}\mathcal{N}\left(\frac{\mathrm{d}T}{\mathrm{d}z}\right)_0 \qquad (21.92)$$

As before, we multiply by $\frac{2}{3}$ to take long flight paths into account, and so arrive at

$$J_z = -\tfrac{1}{3}\nu k\lambda\bar{c}\mathcal{N}\left(\frac{\mathrm{d}T}{\mathrm{d}z}\right)_0 \qquad (21.93)$$

The energy flux is proportional to the temperature gradient, as we wanted to show. Comparison of this equation with eqn 21.20 shows that

$$\kappa = \tfrac{1}{3}\nu k\lambda\bar{c}\mathcal{N} \qquad (21.94)$$

Equation 21.23 then follows from $C_{V,\mathrm{m}} = \nu kN_\mathrm{A}$ for a perfect gas, where [A] is the molar concentration of A. For this step, we use $\mathcal{N} = N/V = nN_\mathrm{A}/V = N_\mathrm{A}[\mathrm{A}]$.

Viscosity

Molecules travelling from the right in Fig. 21.31 (from a fast layer to a slower one) transport a momentum $mv_x(\lambda)$ to their new layer at $z = 0$; those travelling from the left transport $mv_x(-\lambda)$ to it. If it is assumed that the density is uniform, the collision flux is $\frac{1}{4}\mathcal{N}\bar{c}$. Those arriving from the right on average carry a momentum

$$mv_x(\lambda) = mv_x(0) + m\lambda\left(\frac{\mathrm{d}v_x}{\mathrm{d}z}\right)_0 \qquad (21.95\mathrm{a})$$

Those arriving from the left bring a momentum

$$mv_x(-\lambda) = mv_x(0) - m\lambda\left(\frac{\mathrm{d}v_x}{\mathrm{d}z}\right)_0 \qquad (21.95\mathrm{b})$$

The net flux of x-momentum in the z-direction is therefore

$$J = \tfrac{1}{4}\mathcal{N}\bar{c}\left\{\left[mv_x(0) - m\lambda\left(\frac{\mathrm{d}v_x}{\mathrm{d}z}\right)_0\right] - \left[mv_x(0) + m\lambda\left(\frac{\mathrm{d}v_x}{\mathrm{d}z}\right)_0\right]\right\}$$

$$= -\tfrac{1}{2}\mathcal{N}m\lambda\bar{c}\left(\frac{\mathrm{d}v_x}{\mathrm{d}z}\right)_0 \qquad (21.96)$$

The flux is proportional to the velocity gradient, as we wished to show. Comparison of this expression with eqn 21.21, and multiplication by $\frac{2}{3}$ in the normal way, leads to

$$\eta = \tfrac{1}{3}\mathcal{N}m\lambda\bar{c} \qquad (21.97)$$

which can easily be converted into eqn 21.24 by using $Nm = nM$ and $[\mathrm{A}] = n/V$.

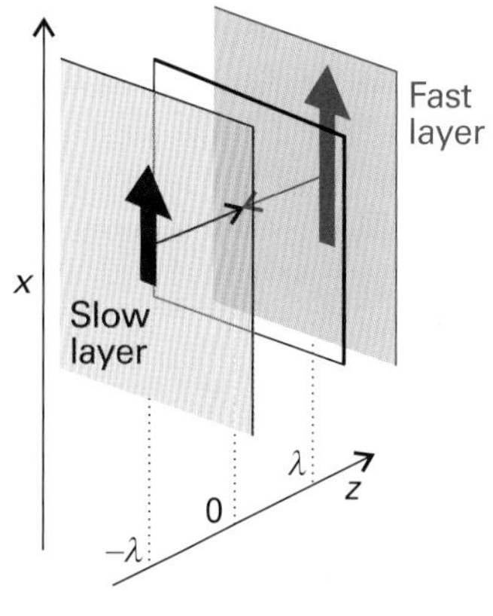

Fig. 21.31 The calculation of the viscosity of a gas examines the net x-component of momentum brought to a plane from faster and slower layers on average a mean free path away in each direction.

Discussion questions

21.1 Use the kinetic theory to justify the following observations: (a) the rate of a reaction in the gas phase depends on the energy with which two molecules collide, which in turn depends on their speeds; (b) in the Earth's atmosphere, light gases, such as H_2 and He, are rare but heavier gases, such as O_2, CO_2, and N_2, are abundant.

21.2 Provide a molecular interpretation for each of the following processes: diffusion, thermal conduction, electric conduction, and viscosity.

21.3 Provide a molecular interpretation for the observation that the viscosity of a gas increases with temperature, whereas the viscosity of a liquid decreases with increasing temperature.

21.4 Discuss the mechanism of proton conduction in liquid water.

21.5 Limit the generality of the following expressions: (a) $J=-D(\mathrm{d}c/\mathrm{d}x)$, (b) $D=kT/f$, and (c) $D=kT/6\pi\eta a$.

21.6 Provide a molecular interpretation for the observation that mediated transport across a biological membrane leads to a maximum flux J_{max} when the concentration of the transported species becomes very large.

21.7 Discuss how nuclear magnetic resonance spectroscopy, inelastic neutron scattering, and dynamic light scattering may be used to measure the mobility of molecules in liquids.

Exercises

21.1a Determine the ratios of (a) the mean speeds, (b) the mean kinetic energies of H_2 molecules and Hg atoms at 20°C.

21.1b Determine the ratios of (a) the mean speeds, (b) the mean kinetic energies of He atoms and Hg atoms at 25°C.

21.2a A 1.0 dm^3 glass bulb contains 1.0×10^{23} H_2 molecules. If the pressure exerted by the gas is 100 kPa, what are (a) the temperature of the gas, (b) the root mean square speeds of the molecules? (c) Would the temperature be different if they were O_2 molecules?

21.2b The best laboratory vacuum pump can generate a vacuum of about 1 nTorr. At 25°C and assuming that air consists of N_2 molecules with a collision diameter of 395 pm, calculate (a) the mean speed of the molecules, (b) the mean free path, (c) the collision frequency in the gas.

21.3a At what pressure does the mean free path of argon at 25°C become comparable to the size of a 1 dm^3 vessel that contains it? Take $\sigma=0.36$ nm^2.

21.3b At what pressure does the mean free path of argon at 25°C become comparable to the diameters of the atoms themselves?

21.4a At an altitude of 20 km the temperature is 217 K and the pressure 0.050 atm. What is the mean free path of N_2 molecules? ($\sigma=0.43$ nm^2.)

21.4b At an altitude of 15 km the temperature is 217 K and the pressure 12.1 kPa. What is the mean free path of N_2 molecules? ($\sigma=0.43$ nm^2.)

21.5a How many collisions does a single Ar atom make in 1.0 s when the temperature is 25°C and the pressure is (a) 10 atm, (b) 1.0 atm, (c) 1.0 μatm?

21.5b How many collisions per second does an N_2 molecule make at an altitude of 15 km? (See Exercise 21.4b for data.)

21.6a Calculate the mean free path of molecules in air using $\sigma=0.43$ nm^2 at 25°C and (a) 10 atm, (b) 1.0 atm, (c) 1.0 μatm.

21.6b Calculate the mean free path of carbon dioxide molecules using $\sigma=0.52$ nm^2 at 25°C and (a) 15 atm, (b) 1.0 bar, (c) 1.0 Torr.

21.7a Use the Maxwell distribution of speeds to estimate the fraction of N_2 molecules at 500 K that have speeds in the range 290 to 300 m s^{-1}.

21.7b Use the Maxwell distribution of speeds to estimate the fraction of CO_2 molecules at 300 K that have speeds in the range 200 to 250 m s^{-1}.

21.8a A solid surface with dimensions 2.5 mm × 3.0 mm is exposed to argon gas at 90 Pa and 500 K. How many collisions do the Ar atoms make with this surface in 15 s?

21.8b A solid surface with dimensions 3.5 mm × 4.0 cm is exposed to helium gas at 111 Pa and 1500 K. How many collisions do the He atoms make with this surface in 10 s?

21.9a An effusion cell has a circular hole of diameter 2.50 mm. If the molar mass of the solid in the cell is 260 g mol^{-1} and its vapour pressure is 0.835 Pa at 400 K, by how much will the mass of the solid decrease in a period of 2.00 h?

21.9b An effusion cell has a circular hole of diameter 3.00 mm. If the molar mass of the solid in the cell is 300 g mol^{-1} and its vapour pressure is 0.224 Pa at 450 K, by how much will the mass of the solid decrease in a period of 24.00 h?

21.10a A manometer was connected to a bulb containing carbon dioxide under slight pressure. The gas was allowed to escape through a small pinhole, and the time for the manometer reading to drop from 75 cm to 50 cm was 52 s. When the experiment was repeated using nitrogen (for which $M=28.02$ g mol^{-1}) the same fall took place in 42 s. Calculate the molar mass of carbon dioxide.

21.10b A manometer was connected to a bulb containing nitrogen under slight pressure. The gas was allowed to escape through a small pinhole, and the time for the manometer reading to drop from 65.1 cm to 42.1 cm was 18.5 s. When the experiment was repeated using a fluorocarbon gas, the same fall took place in 82.3 s. Calculate the molar mass of the fluorocarbon.

21.11a A space vehicle of internal volume 3.0 m^3 is struck by a meteor and a hole of radius 0.10 mm is formed. If the oxygen pressure within the vehicle is initially 80 kPa and its temperature 298 K, how long will the pressure take to fall to 70 kPa?

21.11b A container of internal volume 22.0 m^3 was punctured, and a hole of radius 0.050 mm was formed. If the nitrogen pressure within the vehicle is initially 122 kPa and its temperature 293 K, how long will the pressure take to fall to 105 kPa?

21.12a Calculate the flux of energy arising from a temperature gradient of 2.5 K m^{-1} in a sample of argon in which the mean temperature is 273 K.

21.12b Calculate the flux of energy arising from a temperature gradient of 3.5 K m^{-1} in a sample of hydrogen in which the mean temperature is 260 K.

21.13a Use the experimental value of the thermal conductivity of neon (Table 21.2) to estimate the collision cross-section of Ne atoms at 273 K.

21.13b Use the experimental value of the thermal conductivity of nitrogen (Table 21.2) to estimate the collision cross-section of N_2 molecules at 298 K.

21.14a In a double-glazed window, the panes of glass are separated by 5.0 cm. What is the rate of transfer of heat by conduction from the warm room (25°C) to the cold exterior (−10°C) through a window of area 1.0 m^2? What power of heater is required to make good the loss of heat?

21.14b Two sheets of copper of area 1.50 m^2 are separated by 10.0 cm. What is the rate of transfer of heat by conduction from the warm sheet (50°C) to the cold sheet (−10°C). What is the rate of loss of heat?

21.15a Use the experimental value of the coefficient of viscosity for neon (Table 21.2) to estimate the collision cross-section of Ne atoms at 273 K.

21.15b Use the experimental value of the coefficient of viscosity for nitrogen (Table 21.2) to estimate the collision cross-section of the molecules at 273 K.

21.16a Calculate the inlet pressure required to maintain a flow rate of 9.5×10^5 dm^3 h^{-1} of nitrogen at 293 K flowing through a pipe of length 8.50 m and diameter 1.00 cm. The pressure of gas as it leaves the tube is 1.00 bar. The volume of the gas is measured at that pressure.

21.16b Calculate the inlet pressure required to maintain a flow rate of 8.70 $cm^3\ s^{-1}$ of nitrogen at 300 K flowing through a pipe of length 10.5 m and diameter 15 mm. The pressure of gas as it leaves the tube is 1.00 bar. The volume of the gas is measured at that pressure.

21.17a Calculate the viscosity of air at (a) 273 K, (b) 298 K, (c) 1000 K. Take $\sigma \approx 0.40\ nm^2$. (The experimental values are 173 μP at 273 K, 182 μP at 20°C, and 394 μP at 600°C.)

21.17b Calculate the viscosity of benzene vapour at (a) 273 K, (b) 298 K, (c) 1000 K. Take $\sigma \approx 0.88\ nm^2$.

21.18a Calculate the thermal conductivities of (a) argon, (b) helium at 300 K and 1.0 mbar. Each gas is confined in a cubic vessel of side 10 cm, one wall being at 310 K and the one opposite at 295 K. What is the rate of flow of energy as heat from one wall to the other in each case?

21.18b Calculate the thermal conductivities of (a) neon, (b) nitrogen at 300 K and 15 mbar. Each gas is confined in a cubic vessel of side 15 cm, one wall being at 305 K and the one opposite at 295 K. What is the rate of flow of energy as heat from one wall to the other in each case?

21.19a The viscosity of carbon dioxide was measured by comparing its rate of flow through a long narrow tube (using Poiseuille's formula) with that of argon. For the same pressure differential, the same volume of carbon dioxide passed through the tube in 55 s as argon in 83 s. The viscosity of argon at 25°C is 208 μP; what is the viscosity of carbon dioxide? Estimate the molecular diameter of carbon dioxide.

21.19b The viscosity of a chlorofluorocarbon (CFC) was measured by comparing its rate of flow through a long narrow tube (using Poiseuille's formula) with that of argon. For the same pressure differential, the same volume of the CFC passed through the tube in 72.0 s as argon in 18.0 s. The viscosity of argon at 25°C is 208 μP; what is the viscosity of the CFC? Estimate the molecular diameter of the CFC. Take $M = 200\ g\ mol^{-1}$.

21.20a Calculate the thermal conductivity of argon ($C_{V,m} = 12.5\ J\ K^{-1}\ mol^{-1}$, $\sigma = 0.36\ nm^2$) at room temperature (20°C).

21.20b Calculate the thermal conductivity of nitrogen ($C_{V,m} = 20.8\ J\ K^{-1}\ mol^{-1}$, $\sigma = 0.43\ nm^2$) at room temperature (20°C).

21.21a Calculate the diffusion constant of argon at 25°C and (a) 1.00 Pa, (b) 100 kPa, (c) 10.0 MPa. If a pressure gradient of 0.10 $atm\ cm^{-1}$ is established in a pipe, what is the flow of gas due to diffusion?

21.21b Calculate the diffusion constant of nitrogen at 25°C and (a) 10.0 Pa, (b) 100 kPa, (c) 15.0 MPa. If a pressure gradient of 0.20 $bar\ m^{-1}$ is established in a pipe, what is the flow of gas due to diffusion?

21.22a The mobility of a chloride ion in aqueous solution at 25°C is $7.91 \times 10^{-8}\ m^2\ s^{-1}\ V^{-1}$. Calculate the molar ionic conductivity.

21.22b The mobility of an acetate ion in aqueous solution at 25°C is $4.24 \times 10^{-8}\ m^2\ s^{-1}\ V^{-1}$. Calculate the molar ionic conductivity.

21.23a The mobility of a Rb^+ ion in aqueous solution is $7.92 \times 10^{-8}\ m^2\ s^{-1}\ V^{-1}$ at 25°C. The potential difference between two electrodes placed in the solution is 35.0 V. If the electrodes are 8.00 mm apart, what is the drift speed of the Rb^+ ion?

21.23b The mobility of a Li^+ ion in aqueous solution is $4.01 \times 10^{-8}\ m^2\ s^{-1}\ V^{-1}$ at 25°C. The potential difference between two electrodes placed in the solution is 12.0 V. If the electrodes are 1.00 cm apart, what is the drift speed of the ion?

21.24a What fraction of the total current is carried by Li^+ when current flows through an aqueous solution of LiBr at 25°C?

21.24b What fraction of the total current is carried by Cl^- when current flows through an aqueous solution of NaCl at 25°C?

21.25a The limiting molar conductivities of KCl, KNO_3, and $AgNO_3$ are 14.99 $mS\ m^2\ mol^{-1}$, 14.50 $mS\ m^2\ mol^{-1}$, and 13.34 $mS\ m^2\ mol^{-1}$, respectively (all at 25°C). What is the limiting molar conductivity of AgCl at this temperature?

21.25b The limiting molar conductivities of NaI, $NaCH_3CO_2$, and $Mg(CH_3CO_2)_2$ are 12.69 $mS\ m^2\ mol^{-1}$, 9.10 $mS\ m^2\ mol^{-1}$, and 18.78 $mS\ m^2\ mol^{-1}$, respectively (all at 25°C). What is the limiting molar conductivity of MgI_2 at this temperature?

21.26a At 25°C the molar ionic conductivities of Li^+, Na^+, and K^+ are 3.87 $mS\ m^2\ mol^{-1}$, 5.01 $mS\ m^2\ mol^{-1}$, and 7.35 $mS\ m^2\ mol^{-1}$, respectively. What are their mobilities?

21.26b At 25°C the molar ionic conductivities of F^-, Cl^-, and Br^- are 5.54 $mS\ m^2\ mol^{-1}$, 7.635 $mS\ m^2\ mol^{-1}$, and 7.81 $mS\ m^2\ mol^{-1}$, respectively. What are their mobilities?

21.27a The mobility of a NO_3^- ion in aqueous solution at 25°C is $7.40 \times 10^{-8}\ m^2\ s^{-1}\ V^{-1}$. Calculate its diffusion coefficient in water at 25°C.

21.27b The mobility of a $CH_3CO_2^-$ ion in aqueous solution at 25°C is $4.24 \times 10^{-8}\ m^2\ s^{-1}\ V^{-1}$. Calculate its diffusion coefficient in water at 25°C.

21.28a The diffusion coefficient of CCl_4 in heptane at 25°C is $3.17 \times 10^{-9}\ m^2\ s^{-1}$. Estimate the time required for a CCl_4 molecule to have a root mean square displacement of 5.0 mm.

21.28b The diffusion coefficient of I_2 in hexane at 25°C is $4.05 \times 10^{-9}\ m^2\ s^{-1}$. Estimate the time required for an iodine molecule to have a root mean square displacement of 1.0 cm.

21.29a Estimate the effective radius of a sucrose molecule in water 25°C given that its diffusion coefficient is $5.2 \times 10^{-10}\ m^2\ s^{-1}$ and that the viscosity of water is 1.00 cP.

21.29b Estimate the effective radius of a glycine molecule in water at 25°C given that its diffusion coefficient is $1.055 \times 10^{-9}\ m^2\ s^{-1}$ and that the viscosity of water is 1.00 cP.

21.30a The diffusion coefficient for molecular iodine in benzene is $2.13 \times 10^{-9}\ m^2\ s^{-1}$. How long does a molecule take to jump through about one molecular diameter (approximately the fundamental jump length for translational motion)?

21.30b The diffusion coefficient for CCl_4 in heptane is $3.17 \times 10^{-9}\ m^2\ s^{-1}$. How long does a molecule take to jump through about one molecular diameter (approximately the fundamental jump length for translational motion)?

21.31a What are the root mean square distances travelled by an iodine molecule in benzene and by a sucrose molecule in water at 25°C in 1.0 s?

21.31b About how long, on average, does it take for the molecules in Exercise 21.31a to drift to a point (a) 1.0 mm, (b) 1.0 cm from their starting points?

Problems*

Numerical problems

21.1 Instead of the arrangement in Fig. 21.8, the speed of molecules can also be measured with a rotating slotted-disc apparatus, which consists of five coaxial 5.0 cm diameter discs separated by 1.0 cm, the slots in their rims being displaced by 2.0° between neighbours. The relative intensities, I, of the detected beam of Kr atoms for two different temperatures and at a series of rotation rates were as follows:

ν/Hz	20	40	80	100	120
I (40 K)	0.846	0.513	0.069	0.015	0.002
I (100 K)	0.592	0.485	0.217	0.119	0.057

Find the distributions of molecular velocities, $f(v_x)$, at these temperatures, and check that they conform to the theoretical prediction for a one-dimensional system.

21.2 Cars were timed by police radar as they passed in both directions below a bridge. Their velocities (kilometres per hour, numbers of cars in parentheses) to the east and west were as follows: 80 E (40), 85 E (62), 90 E (53), 95 E (12), 100 E (2); 80 W (38), 85 W (59), 90 W (50), 95 W (10), 100 W (2). What are (a) the mean velocity, (b) the mean speed, (c) the root mean square speed?

21.3 A population consists of people of the following heights (in metres, numbers of individuals in brackets): 1.80 (1), 1.82 (2), 1.84 (4), 1.86 (7), 1.88 (10), 1.90 (15), 1.92 (9), 1.94 (4), 1.96 (0), 1.98 (1). What are (a) the mean height, (b) the root mean square height of the population?

21.4 Calculate the ratio of the thermal conductivities of gaseous hydrogen at 300 K to gaseous hydrogen at 10 K. Be circumspect, and think about the modes of motion that are thermally active at the two temperatures.

21.5 A Knudsen cell was used to determine the vapour pressure of germanium at 1000°C. During an interval of 7200 s the mass loss through a hole of radius 0.50 mm amounted to 43 μg. What is the vapour pressure of germanium at 1000°C? Assume the gas to be monatomic.

21.6 The nuclide ^{244}Bk (berkelium) decays by producing α particles, which capture electrons and form He atoms. Its half-life is 4.4 h. A sample of mass 1.0 mg was placed in a container of volume 1.0 cm^3 that was impermeable to α radiation, but there was also a hole of radius 2.0 μm in the wall. What is the pressure of helium at 298 K, inside the container after (a) 1.0 h, (b) 10 h?

21.7 An atomic beam is designed to function with (a) cadmium, (b) mercury. The source is an oven maintained at 380 K, there being a small slit of dimensions 1.0 cm × 1.0 × 10^{-3} cm. The vapour pressure of cadmium is 0.13 Pa and that of mercury is 12 Pa at this temperature. What is the atomic current (the number of atoms per unit time) in the beams?

21.8 Conductivities are often measured by comparing the resistance of a cell filled with the sample to its resistance when filled with some standard solution, such as aqueous potassium chloride. The conductivity of water is 76 mS m^{-1} at 25°C and the conductivity of 0.100 mol dm^{-3} KCl(aq) is 1.1639 S m^{-1}. A cell had a resistance of 33.21 Ω when filled with 0.100 mol dm^{-3} KCl(aq) and 300.0 Ω when filled with 0.100 mol dm^{-3} CH_3COOH. What is the molar conductivity of acetic acid at that concentration and temperature?

21.9 The resistances of a series of aqueous NaCl solutions, formed by successive dilution of a sample, were measured in a cell with cell constant (the constant C in the relation $\kappa = C/R$) equal to 0.2063 cm^{-1}. The following values were found:

c/(mol dm^{-3})	0.00050	0.0010	0.0050	0.010	0.020	0.050
R/Ω	3314	1669	342.1	174.1	89.08	37.14

Verify that the molar conductivity follows the Kohlrausch law and find the limiting molar conductivity. Determine the coefficient $\mathcal{K}$. Use the value of $\mathcal{K}$ (which should depend only on the nature, not the identity of the ions) and the information that $\lambda(Na^+) = 5.01$ mS m^2 mol^{-1} and $\lambda(I^-) =$ 7.68 mS m^2 mol^{-1} to predict (a) the molar conductivity, (b) the conductivity, (c) the resistance it would show in the cell, of 0.010 mol dm^{-3} NaI(aq) at 25°C.

21.10 After correction for the water conductivity, the conductivity of a saturated aqueous solution of AgCl at 25°C was found to be 0.1887 mS m^{-1}. What is the solubility of silver chloride at this temperature?

21.11 What are the drift speeds of Li^+, Na^+, and K^+ in water when a potential difference of 10 V is applied across a 1.00-cm conductivity cell? How long would it take an ion to move from one electrode to the other? In conductivity measurements it is normal to use alternating current: what are the displacements of the ions in (a) centimetres, (b) solvent diameters, about 300 pm, during a half cycle of 1.0 kHz applied potential?

21.12 The mobilities of H^+ and Cl^- at 25°C in water are 3.623 × 10^{-7} m^2 s^{-1} V^{-1} and 7.91 × 10^{-8} m^2 s^{-1} V^{-1}, respectively. What proportion of the current is carried by the protons in 10^{-3} M HCl(aq)? What fraction do they carry when the NaCl is added to the acid so that the solution is 1.0 mol dm^{-3} in the salt? Note how concentration as well as mobility governs the transport of current.

21.13 In a moving boundary experiment on KCl the apparatus consisted of a tube of internal diameter 4.146 mm, and it contained aqueous KCl at a concentration of 0.021 mol dm^{-3}. A steady current of 18.2 mA was passed, and the boundary advanced as follows:

Δt/s	200	400	600	800	1000
x/mm	64	128	192	254	318

Find the transport number of K^+, its mobility, and its ionic conductivity.

21.14 The proton possesses abnormal mobility in water, but does it behave normally in liquid ammonia? To investigate this question, a moving-boundary technique was used to determine the transport number of NH_4^+ in liquid ammonia (the analogue of H_3O^+ in liquid water) at −40°C (J. Baldwin, J. Evans, and J.B. Gill, *J. Chem. Soc.* A, 3389 (1971)). A steady current of 5.000 mA was passed for 2500 s, during which time the boundary formed between mercury(II) iodide and ammonium iodide solutions in ammonia moved 286.9 mm in a 0.013 65 mol kg^{-1} solution and 92.03 mm in a 0.042 55 mol kg^{-1} solution. Calculate the transport number of NH_4^+ at these concentrations, and comment on the mobility of the proton in liquid ammonia. The bore of the tube is 4.146 mm and the density of liquid ammonia is 0.682 g cm^{-3}.

21.15 A dilute solution of potassium permanganate in water at 25°C was prepared. The solution was in a horizontal tube of length 10 cm, and at first there was a linear gradation of intensity of the purple solution from the left (where the concentration was 0.100 mol dm^{-3}) to the right (where the concentration was 0.050 mol dm^{-3}). What is the magnitude and sign of the thermodynamic force acting on the solute (a) close to the left face of the container, (b) in the middle, (c) close to the right face? Give the force per mole and force per molecule in each case.

* Problems denoted with the symbol ‡ were supplied by Charles Trapp, Carmen Giunta, and Marshall Cady.

21.16 Estimate the diffusion coefficients and the effective hydrodynamic radii of the alkali metal cations in water from their mobilities at 25°C. Estimate the approximate number of water molecules that are dragged along by the cations. Ionic radii are given Table 20.3.

21.17 Nuclear magnetic resonance can be used to determine the mobility of molecules in liquids. A set of measurements on methane in carbon tetrachloride showed that its diffusion coefficient is $2.05 \times 10^{-9}\ m^2\ s^{-1}$ at 0°C and $2.89 \times 10^{-9}\ m^2\ s^{-1}$ at 25°C. Deduce what information you can about the mobility of methane in carbon tetrachloride.

21.18 A concentrated sucrose solution is poured into a cylinder of diameter 5.0 cm. The solution consisted of 10 g of sugar in $5.0\ cm^3$ of water. A further $1.0\ dm^3$ of water is then poured very carefully on top of the layer, without disturbing the layer. Ignore gravitational effects, and pay attention only to diffusional processes. Find the concentration at 5.0 cm above the lower layer after a lapse of (a) 10 s, (b) 1.0 y.

21.19 In a series of observations on the displacement of rubber latex spheres of radius 0.212 μm, the mean square displacements after selected time intervals were on average as follows:

t/s	30	60	90	120
$10^{12}\langle x^2\rangle/m^2$	88.2	113.5	128	144

These results were originally used to find the value of Avogadro's constant, but there are now better ways of determining N_A, so the data can be used to find another quantity. Find the effective viscosity of water at the temperature of this experiment (25°C).

21.20‡ A.K. Srivastava, R.A. Samant, and S.D. Patankar (*J. Chem. Eng. Data* **41**, 431 (1996)) measured the conductance of several salts in a binary solvent mixture of water and a dipolar aprotic solvent 1,3-dioxolan-2-one (ethylene carbonate). They report the following conductances at 25°C in a solvent 80 per cent 1,3-dioxolan-2-one by mass:

NaI

$c/(mmol\ dm^{-3})$	32.02	20.28	12.06	8.64	2.85	1.24	0.83
$\Lambda_m/(S\ cm^2\ mol^{-1})$	50.26	51.99	54.01	55.75	57.99	58.44	58.67

KI

$c/(mmol\ dm^{-3})$	17.68	10.8	87.19	2.67	1.28	0.83	0.19
$\Lambda_m/(S\ cm^2\ mol^{-1})$	42.45	45.91	47.53	51.81	54.09	55.78	57.42

Calculate Λ_m° for NaI and KI in this solvent and $\lambda^\circ(Na) - \lambda^\circ(K)$. Compare your results to the analogous quantities in aqueous solution using Table 21.5 in the *Data section*.

21.21‡ A. Fenghour, W.A. Wakeham, V. Vesovic, J.T.R. Watson, J. Millat, and E. Vogel (*J. Phys. Chem. Ref. Data* **24**, 1649 (1995)) have compiled an extensive table of viscosity coefficients for ammonia in the liquid and vapour phases. Deduce the effective molecular diameter of NH_3 based on each of the following vapour-phase viscosity coefficients: (a) $\eta = 9.08 \times 10^{-6}\ kg\ m^{-1}\ s^{-1}$ at 270 K and 1.00 bar; (b) $\eta = 1.749 \times 10^{-5}\ kg\ m^{-1}\ s^{-1}$ at 490 K and 10.0 bar.

21.22‡ G. Bakale, K. Lacmann, and W.F. Schmidt (*J. Phys. Chem.* **100**, 12477 (1996)) measured the mobility of singly charged C_{60}^- ions in a variety of nonpolar solvents. In cyclohexane at 22°C, the mobility is $1.1\ cm^2\ V^{-1}\ s^{-1}$. Estimate the effective radius of the C_{60} ion. The viscosity of the solvent is $0.93 \times 10^{-3}\ kg\ m^{-1}\ s^{-1}$. *Comment.* The researchers interpreted the substantial difference between this number and the van der Waals radius of neutral C_{60} in terms of a solvation layer around the ion.

Theoretical problems

21.23 Start from the Maxwell–Boltzmann distribution and derive an expression for the most probable speed of a gas of molecules at a temperature T. Go on to demonstrate the validity of the equipartition conclusion that the average translational kinetic energy of molecules free to move in three dimensions is $\frac{3}{2}kT$.

21.24 Consider molecules that are confined to move in a plane (a two-dimensional gas). Calculate the distribution of speeds and determine the mean speed of the molecules at a temperature T.

21.25 A specially constructed velocity-selector accepts a beam of molecules from an oven at a temperature T but blocks the passage of molecules with a speed greater than the mean. What is the mean speed of the emerging beam, relative to the initial value, treated as a one-dimensional problem?

21.26 What is the proportion of gas molecules having (a) more than, (b) less than the root mean square speed? (c) What are the proportions having speeds greater and smaller than the mean speed?

21.27 Calculate the fractions of molecules in a gas that have a speed in a range Δv at the speed nc^* relative to those in the same range at c^* itself? This calculation can be used to estimate the fraction of very energetic molecules (which is important for reactions). Evaluate the ratio for $n = 3$ and $n = 4$.

21.28 Derive an expression that shows how the pressure of a gas inside an effusion oven (a heated chamber with a small hole in one wall) varies with time if the oven is not replenished as the gas escapes. Then show that $t_{1/2}$, the time required for the pressure to decrease to half its initial value, is independent of the initial pressure. *Hint.* Begin by setting up a differential equation relating dp/dt to $p = NkT/V$, and then integrating it.

21.29 Show how the ratio of two transport numbers t' and t'' for two cations in a mixture depends on their concentrations c' and c'' and their mobilities u' and u''.

21.30 Confirm that eqn 21.72 is a solution of the diffusion equation with the correct initial value.

21.31 The diffusion equation is valid when many elementary steps are taken in the time interval of interest, but the random walk calculation lets us discuss distributions for short times as well as for long. Use eqn 21.84 to calculate the probability of being six paces from the origin (that is, at $x = 6\lambda$) after (a) four, (b) six, (c) twelve steps.

21.32 Use mathematical software to calculate P in a one-dimensional random walk, and evaluate the probability of being at $x = n\lambda$ for $n = 6, 10, 14, \ldots, 60$. Compare the numerical value with the analytical value in the limit of a large number of steps. At what value of n is the discrepancy no more than 0.1 per cent?

21.33 Supply the intermediate mathematical steps in *Justification* 21.7.

21.34‡ A dilute solution of a weak (1,1)-electrolyte contains both neutral ion pairs and ions in equilibrium ($AB \rightleftharpoons A^+ + B^-$). Prove that molar conductivities are related to the degree of ionization by the equations:

$$\frac{1}{\Lambda_m} = \frac{1}{\Lambda_m(\alpha)} + \frac{(1-\alpha)\Lambda_m^\circ}{\alpha^2\Lambda_m(\alpha)^2} \qquad \Lambda_m(\alpha) = \lambda_+ + \lambda_- = \Lambda_m^\circ - \mathcal{K}(\alpha c)^{1/2}$$

where Λ_m° is the molar conductivity at infinite dilution and $\mathcal{K}$ is the constant in Kohlrausch's law (eqn 21.29).

Applications: to astrophysics and biochemistry

21.35 Calculate the escape velocity (the minimum initial velocity that will take an object to infinity) from the surface of a planet of radius R. What is the value for (a) the Earth, $R = 6.37 \times 10^6$ m, $g = 9.81\ m\ s^{-2}$, (b) Mars, $R = 3.38 \times 10^6$ m, $m_{Mars}/m_{Earth} = 0.108$. At what temperatures do H_2, He, and O_2 molecules have mean speeds equal to their escape speeds? What proportion of the molecules have enough speed to escape when the temperature is (a) 240 K, (b) 1500 K? Calculations of this kind are very important in considering the composition of planetary atmospheres.

21.36‡ Interstellar space is a medium quite different from the gaseous environments we commonly encounter on Earth. For instance, a typical density of the medium is about 1 atom cm^{-3} and that atom is typically H; the effective temperature due to stellar background radiation is about 10 000 K. Estimate the diffusion coefficient and thermal conductivity of H under these conditions. *Comment.* Energy is in fact transferred much more effectively by radiation.

21.37 The principal components of the atmosphere of the Earth are diatomic molecules, which can rotate as well as translate. Given that the translational kinetic energy density of the atmosphere is 0.15 J cm^{-3}, what is the total kinetic energy density, including rotation?

21.38‡ In the *standard model* of stellar structure (I. Nicholson, *The sun*. Rand McNally, New York (1982)), the interior of the Sun is thought to consist of 36 per cent H and 64 per cent He by mass, at a density of 158 g cm^{-3}. Both atoms are completely ionized. The approximate dimensions of the nuclei can be calculated from the formula $r_{nucleus} = 1.4 \times 10^{-15} A^{1/3}$ m, where A is the mass number. The size of the free electron, $r_e \approx 10^{-18}$ m, is negligible compared to the size of the nuclei. (a) Calculate the excluded volume in 1.0 cm^3 of the stellar interior and on that basis decide upon the applicability of the perfect gas law to this system. (b) The standard model suggests that the pressure in the stellar interior is 2.5×10^{11} atm. Calculate the temperature of the Sun's interior based on the perfect gas model. The generally accepted standard model value is 16 MK. (c) Would a van der Waals type of equation (with $a = 0$) give a better value for T?

21.39 Enrico Fermi, the great Italian scientist, was a master at making good approximate calculations based on little or no actual data. Hence, such calculations are often called 'Fermi calculations'. Do a Fermi calculation on how long it would take for a gaseous air-borne cold virus of molar mass 100 kg mol^{-1} to travel the distance between two conversing people 1.0 m a part by diffusion in still air.

21.40 The diffusion coefficient of a particular kind of t-RNA molecule is $D = 1.0 \times 10^{-11}$ $m^2\,s^{-1}$ in the medium of a cell interior. How long does it take molecules produced in the cell nucleus to reach the walls of the cell at a distance 1.0 μm, corresponding to the radius of the cell?

21.41‡ In this problem, we examine a model for the transport of oxygen from air in the lungs to blood. First, show that, for the initial and boundary conditions $c(x,t) = c(x,0) = c_0$, $(0 < x < \infty)$ and $c(0,t) = c_s$, $(0 \le t \le \infty)$ where c_0 and c_s are constants, the concentration, $c(x,t)$, of a species is given by

$$c(x,t) = c_0 + (c_s - c_0)\{1 - \mathrm{erf}\,\xi\} \qquad \xi(x,t) = \frac{x}{(4Dt)^{1/2}}$$

where $\mathrm{erf}\,\xi$ is the error function (*Justification 9.4*) and the concentration $c(x,t)$ evolves by diffusion from the yz-plane of constant concentration, such as might occur if a condensed phase is absorbing a species from a gas phase. Now draw graphs of concentration profiles at several different times of your choice for the diffusion of oxygen into water at 298 K (when $D = 2.10 \times 10^{-9}$ $m^2\,s^{-1}$) on a spatial scale comparable to passage of oxygen from lungs through alveoli into the blood. Use $c_0 = 0$ and set c_s equal to the solubility of oxygen in water. *Hint.* Use mathematical software.

The rates of chemical reactions

22

This chapter is the first of a sequence that explores the rates of chemical reactions. The chapter begins with a discussion of the definition of reaction rate and outlines the techniques for its measurement. The results of such measurements show that reaction rates depend on the concentration of reactants (and products) in characteristic ways that can be expressed in terms of differential equations known as rate laws. The solutions of these equations are used to predict the concentrations of species at any time after the start of the reaction. The form of the rate law also provides insight into the series of elementary steps by which a reaction takes place. The key task in this connection is the construction of a rate law from a proposed mechanism and its comparison with experiment. Simple elementary steps have simple rate laws, and these rate laws can be combined together by invoking one or more approximations. These approximations include the concept of the rate-determining stage of a reaction, the steady-state concentration of a reaction intermediate, and the existence of a pre-equilibrium.

This chapter introduces the principles of **chemical kinetics**, the study of reaction rates, by showing how the rates of reactions may be measured and interpreted. The remaining chapters of this part of the text then develop this material in more detail and apply it to more complicated or more specialized cases. The rate of a chemical reaction might depend on variables under our control, such as the pressure, the temperature, and the presence of a catalyst, and we may be able to optimize the rate by the appropriate choice of conditions. The study of reaction rates also leads to an understanding of the **mechanisms** of reactions, their analysis into a sequence of elementary steps.

Empirical chemical kinetics

The first steps in the kinetic analysis of reactions are to establish the stoichiometry of the reaction and identify any side reactions. The basic data of chemical kinetics are then the concentrations of the reactants and products at different times after a reaction has been initiated. The rates of most chemical reactions are sensitive to the temperature, so in conventional experiments the temperature of the reaction mixture must be held constant throughout the course of the reaction. This requirement puts severe demands on the design of an experiment. Gas-phase reactions, for instance, are often carried out in a vessel held in contact with a substantial block of metal. Liquid-phase reactions, including flow reactions, must be carried out in an efficient thermostat. Special efforts have to be made to study reactions at low temperatures, as in the study of the kinds of reactions that take place in interstellar clouds. Thus, supersonic expansion of the reaction gas can be used to attain temperatures as low as

10 K. For work in the liquid phase and the solid phase, very low temperatures are often reached by flowing cold liquid or cold gas around the reaction vessel. Alternatively, the entire reaction vessel is immersed in a thermally insulated container filled with a cryogenic liquid, such as liquid helium (for work at around 4 K) or liquid nitrogen (for work at around 77 K). Non-isothermal conditions are sometimes employed. For instance, the shelf-life of an expensive pharmaceutical may be explored by slowly raising the temperature of a single sample.

22.1 Experimental techniques

The method used to monitor concentrations depends on the species involved and the rapidity with which their concentrations change. Many reactions reach equilibrium over periods of minutes or hours, and several techniques may then be used to follow the changing concentrations.

(a) Monitoring the progress of a reaction

A reaction in which at least one component is a gas might result in an overall change in pressure in a system of constant volume, so its progress may be followed by recording the variation of pressure with time.

Example 22.1 *Monitoring the variation in pressure*

Predict how the total pressure varies during the gas-phase decomposition $2\,N_2O_5(g) \rightarrow 4\,NO_2(g) + O_2(g)$ in a constant-volume container.

Method The total pressure (at constant volume and temperature and assuming perfect gas behaviour) is proportional to the number of gas-phase molecules. Therefore, because each mole of N_2O_5 gives rise to $\frac{5}{2}$ mol of gas molecules, we can expect the pressure to rise to $\frac{5}{2}$ times its initial value. To confirm this conclusion, express the progress of the reaction in terms of the fraction, α, of N_2O_5 molecules that have reacted.

Answer Let the initial pressure be p_0 and the initial amount of N_2O_5 molecules present be n. When a fraction α of the N_2O_5 molecules has decomposed, the amounts of the components in the reaction mixture are:

	N_2O_5	NO_2	O_2	Total
Amount:	$n(1-\alpha)$	$2\alpha n$	$\frac{1}{2}\alpha n$	$n(1+\frac{3}{2}\alpha)$

When $\alpha = 0$ the pressure is p_0, so at any stage the total pressure is

$$p = (1 + \tfrac{3}{2}\alpha)p_0$$

When the reaction is complete, the pressure will have risen to $\frac{5}{2}$ times its initial value.

Self-test 22.1 Repeat the calculation for $2\,NOBr(g) \rightarrow 2\,NO(g) + Br_2(g)$.
$[p = (1 + \frac{1}{2}\alpha)p_0]$

Spectrophotometry, the measurement of absorption of radiation in a particular spectral region, is widely applicable, and is especially useful when one substance in the reaction mixture has a strong characteristic absorption in a conveniently accessible region of the electromagnetic spectrum. For example, the progress of the reaction

$$H_2(g) + Br_2(g) \rightarrow 2\,HBr(g)$$

can be followed by measuring the absorption of visible light by bromine. A reaction that changes the number or type of ions present in a solution may be followed by monitoring the electrical conductivity of the solution. The replacement of neutral molecules by ionic products can result in dramatic changes in the conductivity, as in the reaction

$$(CH_3)_3CCl(aq) + H_2O(l) \rightarrow (CH_3)_3COH(aq) + H^+(aq) + Cl^-(aq)$$

If hydrogen ions are produced or consumed, the reaction may be followed by monitoring the pH of the solution.

Other methods of determining composition include emission spectroscopy, mass spectrometry, gas chromatography, nuclear magnetic resonance, and electron paramagnetic resonance (for reactions involving radicals or paramagnetic *d*-metal ions).

(b) Application of the techniques

In a **real-time analysis** the composition of the system is analysed while the reaction is in progress. Either a small sample is withdrawn or the bulk solution is monitored. In the **flow method** the reactants are mixed as they flow together in a chamber (Fig. 22.1). The reaction continues as the thoroughly mixed solutions flow through the outlet tube, and observation of the composition at different positions along the tube is equivalent to the observation of the reaction mixture at different times after mixing. The disadvantage of conventional flow techniques is that a large volume of reactant solution is necessary. This makes the study of fast reactions particularly difficult because to spread the reaction over a length of tube the flow must be rapid. This disadvantage is avoided by the **stopped-flow technique**, in which the reagents are mixed very quickly in a small chamber fitted with a syringe instead of an outlet tube (Fig. 22.2). The flow ceases when the plunger of the syringe reaches a stop, and the reaction continues in the mixed solutions. Observations, commonly using spectroscopic techniques such as ultraviolet–visible absorption, circular dichroism, and fluorescence emission, are made on the sample as a function of time. The technique allows for the study of reactions that occur on the millisecond to second timescale. The suitability of the stopped-flow method to the study of small samples means that it is appropriate for many biochemical reactions, and it has been widely used to study the kinetics of protein folding and enzyme action (see *Impact* I22.1 later in the chapter).

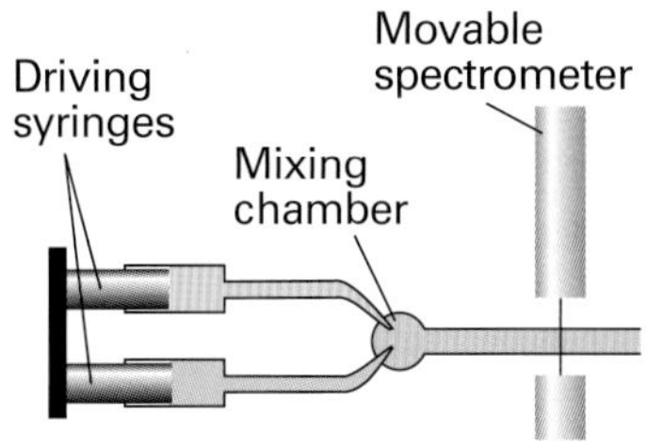

Fig. 22.1 The arrangement used in the flow technique for studying reaction rates. The reactants are injected into the mixing chamber at a steady rate. The location of the spectrometer corresponds to different times after initiation.

Very fast reactions can be studied by **flash photolysis**, in which the sample is exposed to a brief flash of light that initiates the reaction and then the contents of the reaction chamber are monitored. Most work is now done with lasers with photolysis pulse widths that range from femtoseconds to nanoseconds (Section 14.5). The apparatus used for flash photolysis studies is based on the experimental design for time-resolved spectroscopy (Section 14.6). Reactions occurring on a picosecond or femtosecond timescale may be monitored by using electronic absorption or emission, infrared absorption, or Raman scattering. The spectra are recorded at a series of times following laser excitation. The laser pulse can initiate the reaction by forming a reactive species, such as an excited electronic state of a molecule, a radical, or an ion. We discuss examples of excited state reactions in Chapter 23. An example of radical generation is the light-induced dissociation of $Cl_2(g)$ to yield Cl atoms that react with HBr to make HCl and Br according to the following sequence:

$$Cl_2 + h\nu \rightarrow Cl + Cl$$

$$Cl + HBr \rightarrow HCl^* + Br$$

$$HCl^* + M \rightarrow HCl + M$$

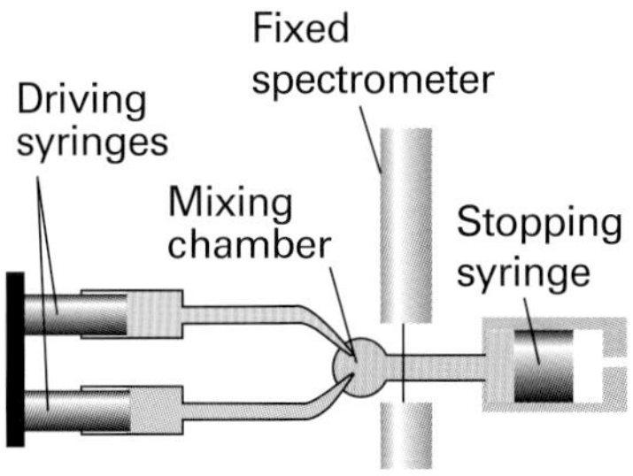

Fig. 22.2 In the stopped-flow technique the reagents are driven quickly into the mixing chamber by the driving syringes and then the time dependence of the concentrations is monitored.

Here HCl* denotes a vibrationally excited HCl molecule and M is a body (an unreactive molecule or the wall of the container) that removes the excess energy stored in HCl. A so-called 'third body' (M) is not always necessary for heteronuclear diatomic molecules because they can discard energy radiatively, but homonuclear diatomic molecules are vibrationally and rotationally inactive, and can discard energy only by collision.

In contrast to real-time analysis, **quenching methods** are based on stopping, or quenching, the reaction after it has been allowed to proceed for a certain time. In this way the composition is analysed at leisure and reaction intermediates may be trapped. These methods are suitable only for reactions that are slow enough for there to be little reaction during the time it takes to quench the mixture. In the **chemical quench flow method**, the reactants are mixed in much the same way as in the flow method but the reaction is quenched by another reagent, such as solution of acid or base, after the mixture has travelled along a fixed length of the outlet tube. Different reaction times can be selected by varying the flow rate along the outlet tube. An advantage of the chemical quench flow method over the stopped-flow method is that spectroscopic fingerprints are not needed in order to measure the concentration of reactants and products. Once the reaction has been quenched, the solution may be examined by 'slow' techniques, such as gel electrophoresis, mass spectrometry, and chromatography. In the **freeze quench method**, the reaction is quenched by cooling the mixture within milliseconds and the concentrations of reactants, intermediates, and products are measured spectroscopically.

22.2 The rates of reactions

Reaction rates depend on the composition and the temperature of the reaction mixture. The next few sections look at these observations in more detail.

(a) The definition of rate

Consider a reaction of the form $A + 2\,B \rightarrow 3\,C + D$, in which at some instant the molar concentration of a participant J is [J] and the volume of the system is constant. The instantaneous **rate of consumption** of one of the reactants at a given time is $-d[R]/dt$, where R is A or B. This rate is a positive quantity (Fig. 22.3). The **rate of formation** of one of the products (C or D, which we denote P) is $d[P]/dt$ (note the difference in sign). This rate is also positive.

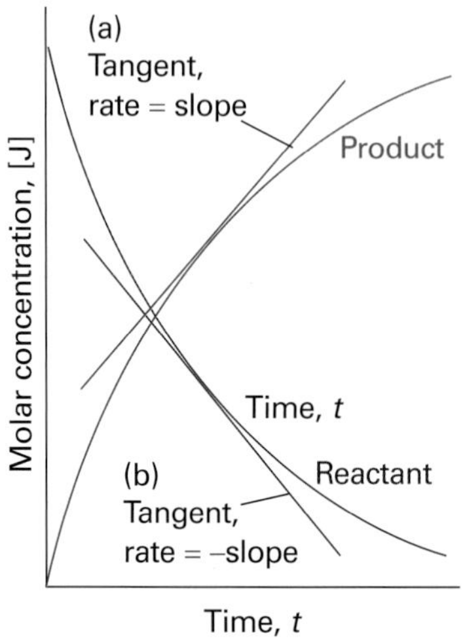

Fig. 22.3 The definition of (instantaneous) rate as the slope of the tangent drawn to the curve showing the variation of concentration with time. For negative slopes, the sign is changed when reporting the rate, so all reaction rates are positive.

It follows from the stoichiometry for the reaction $A + 2\,B \rightarrow 3\,C + D$ that

$$\frac{d[D]}{dt} = \frac{1}{3}\frac{d[C]}{dt} = -\frac{d[A]}{dt} = -\frac{1}{2}\frac{d[B]}{dt}$$

so there are several rates connected with the reaction. The undesirability of having different rates to describe the same reaction is avoided by using the extent of reaction, ξ (xi, the quantity introduced in Section 7.1):

$$\xi = \frac{n_J - n_{J,0}}{\nu_J} \qquad (22.1)$$

where ν_J is the stoichiometric number of species J, and defining the unique **rate of reaction**, v, as the rate of change of the extent of reaction:

$$v = \frac{1}{V}\frac{d\xi}{dt} \qquad [22.2]$$

It follows that

$$v = \frac{1}{\nu_J} \times \frac{1}{V}\frac{dn_J}{dt} \qquad (22.3a)$$

(Remember that ν_J is negative for reactants and positive for products.) For a homogeneous reaction in a constant-volume system the volume V can be taken inside the differential and we use $[J] = n_J/V$ to write

$$v = \frac{1}{\nu_J}\frac{d[J]}{dt} \qquad (22.3b)$$

For a heterogeneous reaction, we use the (constant) surface area, A, occupied by the species in place of V and use $\sigma_J = n_J/A$ to write

$$v = \frac{1}{\nu_J}\frac{d\sigma_J}{dt} \qquad (22.3c)$$

In each case there is now a single rate for the entire reaction (for the chemical equation as written). With molar concentrations in moles per cubic decimetre and time in seconds, reaction rates of homogeneous reactions are reported in moles per cubic decimetre per second (mol dm^{-3} s^{-1}) or related units. For gas-phase reactions, such as those taking place in the atmosphere, concentrations are often expressed in molecules per cubic centimetre (molecules cm^{-3}) and rates in molecules per cubic centimetre per second (molecules cm^{-3} s^{-1}). For heterogeneous reactions, rates are expressed in moles per square metre per second (mol m^{-2} s^{-1}) or related units.

Illustration 22.1 *Rates of formation and consumption*

If the rate of formation of NO in the reaction $2\,NOBr(g) \rightarrow 2\,NO(g) + Br_2(g)$ is reported as 0.16 mmol dm^{-3} s^{-1}, we use $\nu_{NO} = +2$ to report that $v = 0.080$ mmol dm^{-3} s^{-1}. Because $\nu_{NOBr} = -2$ it follows that $d[NOBr]/dt = -0.16$ mmol dm^{-3} s^{-1}. The rate of consumption of NOBr is therefore 0.16 mmol dm^{-3} s^{-1}, or 9.6×10^{16} molecules cm^{-3} s^{-1}.

Self-test 22.2 The rate of change of molar concentration of CH_3 radicals in the reaction $2\,CH_3(g) \rightarrow CH_3CH_3(g)$ was reported as $d[CH_3]/dt = -1.2$ mol dm^{-3} s^{-1} under particular conditions. What is (a) the rate of reaction and (b) the rate of formation of CH_3CH_3? [(a) 0.60 mol dm^{-3} s^{-1}, (b) 0.60 mol dm^{-3} s^{-1}]

(b) Rate laws and rate constants

The rate of reaction is often found to be proportional to the concentrations of the reactants raised to a power. For example, the rate of a reaction may be proportional to the molar concentrations of two reactants A and B, so we write

$$v = k[A][B] \qquad (22.4)$$

with each concentration raised to the first power. The coefficient k is called the **rate constant** for the reaction. The rate constant is independent of the concentrations but depends on the temperature. An experimentally determined equation of this kind is called the **rate law** of the reaction. More formally, a rate law is an equation that expresses the rate of reaction as a function of the concentrations of all the species present in the overall chemical equation for the reaction at some time:

$$v = f([A],[B],\ldots) \qquad [22.5a]$$

For homogeneous gas-phase reactions, it is often more convenient to express the rate law in terms of partial pressures, which are related to molar concentrations by $p_J = RT[J]$. In this case, we write

$$v = f(p_A, p_B, \ldots) \qquad [22.5b]$$

The rate law of a reaction is determined experimentally, and cannot in general be inferred from the chemical equation for the reaction. The reaction of hydrogen and bromine, for example, has a very simple stoichiometry, $H_2(g) + Br_2(g) \rightarrow 2\,HBr(g)$, but its rate law is complicated:

$$v = \frac{k[H_2][Br_2]^{3/2}}{[Br_2] + k'[HBr]} \qquad (22.6)$$

In certain cases the rate law does reflect the stoichiometry of the reaction, but that is either a coincidence or reflects a feature of the underlying reaction mechanism (see later).

A practical application of a rate law is that, once we know the law and the value of the rate constant, we can predict the rate of reaction from the composition of the mixture. Moreover, as we shall see later, by knowing the rate law, we can go on to predict the composition of the reaction mixture at a later stage of the reaction. Moreover, a rate law is a guide to the mechanism of the reaction, for any proposed mechanism must be consistent with the observed rate law.

(c) Reaction order

Many reactions are found to have rate laws of the form

$$v = k[A]^a[B]^b \cdots \qquad (22.7)$$

The power to which the concentration of a species (a product or a reactant) is raised in a rate law of this kind is the **order** of the reaction with respect to that species. A reaction with the rate law in eqn 22.4 is **first-order** in A and first-order in B. The **overall order** of a reaction with a rate law like that in eqn 22.7 is the sum of the individual orders, $a + b + \cdots$. The rate law in eqn 22.4 is therefore second-order overall.

A reaction need not have an integral order, and many gas-phase reactions do not. For example, a reaction having the rate law

$$v = k[A]^{1/2}[B] \qquad (22.8)$$

is half-order in A, first-order in B, and three-halves-order overall. Some reactions obey a **zero-order rate law**, and therefore have a rate that is independent of the concentration of the reactant (so long as some is present). Thus, the catalytic decomposition of phosphine (PH_3) on hot tungsten at high pressures has the rate law

$$v = k \qquad (22.9)$$

The PH_3 decomposes at a constant rate until it has almost entirely disappeared. Zero-order reactions typically occur when there is a bottle-neck of some kind in the mechanism, as in heterogeneous reactions when the surface is saturated and the subsequent reaction slow and in a number of enzyme reactions when there is a large excess of substrate relative to the enzyme.

When a rate law is not of the form in eqn 22.7, the reaction does not have an overall order and may not even have definite orders with respect to each participant. Thus, although eqn 22.6 shows that the reaction of hydrogen and bromine is first-order in H_2, the reaction has an indefinite order with respect to both Br_2 and HBr and has no overall order.

These remarks point to three problems. First, we must see how to identify the rate law and obtain the rate constant from the experimental data. We concentrate on this aspect in this chapter. Second, we must see how to construct reaction mechanisms that are consistent with the rate law. We shall introduce the techniques of doing so in this chapter and develop them further in Chapter 23. Third, we must account for the values of the rate constants and explain their temperature dependence. We shall see a little of what is involved in this chapter, but leave the details until Chapter 24.

(d) The determination of the rate law

The determination of a rate law is simplified by the **isolation method** in which the concentrations of all the reactants except one are in large excess. If B is in large excess, for example, then to a good approximation its concentration is constant throughout the reaction. Although the true rate law might be $v = k[A][B]$, we can approximate [B] by $[B]_0$, its initial value, and write

$$v = k'[A] \qquad k' = k[B]_0 \tag{22.10}$$

which has the form of a first-order rate law. Because the true rate law has been forced into first-order form by assuming that the concentration of B is constant, eqn 22.10 is called a **pseudofirst-order rate law**. The dependence of the rate on the concentration of each of the reactants may be found by isolating them in turn (by having all the other substances present in large excess), and so constructing a picture of the overall rate law.

In the **method of initial rates**, which is often used in conjunction with the isolation method, the rate is measured at the beginning of the reaction for several different initial concentrations of reactants. We shall suppose that the rate law for a reaction with A isolated is $v = k[A]^a$; then its initial rate, v_0, is given by the initial values of the concentration of A, and we write $v_0 = k[A]_0^a$. Taking logarithms gives:

$$\log v_0 = \log k + a \log [A]_0 \tag{22.11}$$

For a series of initial concentrations, a plot of the logarithms of the initial rates against the logarithms of the initial concentrations of A should be a straight line with slope a.

Example 22.2 *Using the method of initial rates*

The recombination of iodine atoms in the gas phase in the presence of argon was investigated and the order of the reaction was determined by the method of initial rates. The initial rates of reaction of $2\,I(g) + Ar(g) \rightarrow I_2(g) + Ar(g)$ were as follows:

$[I]_0/(10^{-5}\ \mathrm{mol\ dm^{-3}})$		1.0	2.0	4.0	6.0
$v_0/(\mathrm{mol\ dm^{-3}\ s^{-1}})$	(a)	8.70×10^{-4}	3.48×10^{-3}	1.39×10^{-2}	3.13×10^{-2}
	(b)	4.35×10^{-3}	1.74×10^{-2}	6.96×10^{-2}	1.57×10^{-1}
	(c)	8.69×10^{-3}	3.47×10^{-2}	1.38×10^{-1}	3.13×10^{-1}

The Ar concentrations are (a) 1.0 mmol dm^{-3}, (b) 5.0 mmol dm^{-3}, and (c) 10.0 mmol dm^{-3}. Determine the orders of reaction with respect to the I and Ar atom concentrations and the rate constant.

Method Plot the logarithm of the initial rate, $\log v_0$, against $\log [I]_0$ for a given concentration of Ar, and, separately, against $\log [Ar]_0$ for a given concentration of I. The slopes of the two lines are the orders of reaction with respect to I and Ar, respectively. The intercepts with the vertical axis give $\log k$.

Answer The plots are shown in Fig. 22.4. The slopes are 2 and 1, respectively, so the (initial) rate law is

$$v_0 = k[I]_0^2[Ar]_0$$

This rate law signifies that the reaction is second-order in [I], first-order in [Ar], and third-order overall. The intercept corresponds to $k = 9\times10^9\ \mathrm{mol^{-2}\ dm^6\ s^{-1}}$.

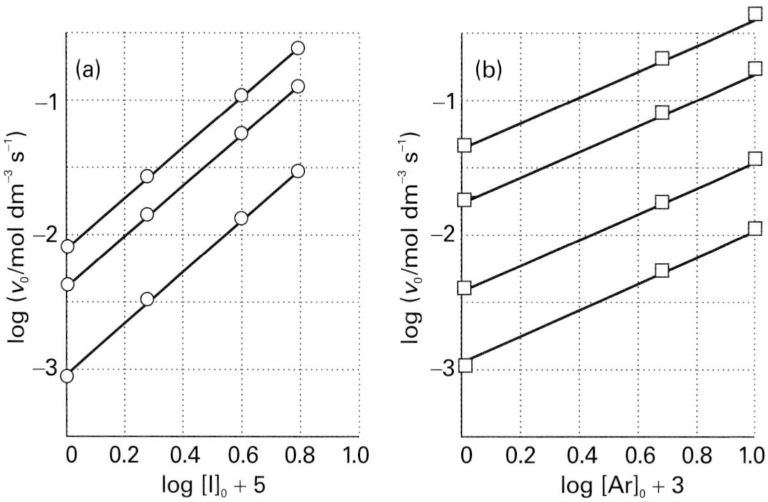

Fig. 22.4 The plot of log v_0 against (a) log $[I]_0$ for a given $[Ar]_0$, and (b) log $[Ar]_0$ for a given $[I]_0$.

A note on good practice The units of k come automatically from the calculation, and are always such as to convert the product of concentrations to a rate in concentration/time (for example, mol dm^{-3} s^{-1}).

Self-test 22.3 The initial rate of a reaction depended on concentration of a substance J as follows:

$[J]_0$/(mmol dm^{-3})	5.0	8.2	17	30
v_0/(10^{-7} mol dm^{-3} s^{-1})	3.6	9.6	41	130

Determine the order of the reaction with respect to J and calculate the rate constant.
[2, 1.4×10^{-2} dm^3 mol^{-1} s^{-1}]

The method of initial rates might not reveal the full rate law, for once the products have been generated they might participate in the reaction and affect its rate. For example, products participate in the synthesis of HBr, because eqn 22.6 shows that the full rate law depends on the concentration of HBr. To avoid this difficulty, the rate law should be fitted to the data throughout the reaction. The fitting may be done, in simple cases at least, by using a proposed rate law to predict the concentration of any component at any time, and comparing it with the data. A law should also be tested by observing whether the addition of products or, for gas-phase reactions, a change in the surface-to-volume ratio in the reaction chamber affects the rate.

22.3 Integrated rate laws

Because rate laws are differential equations, we must integrate them if we want to find the concentrations as a function of time. Even the most complex rate laws may be integrated numerically. However, in a number of simple cases analytical solutions, known as **integrated rate laws**, are easily obtained, and prove to be very useful. We examine a few of these simple cases here.

(a) First-order reactions

As shown in the *Justification* below, the integrated form of the first-order rate law

$$\frac{d[A]}{dt} = -k[A] \qquad (22.12a)$$

Synoptic table 22.1* Kinetic data for first-order reactions

Reaction	Phase	θ/°C	k/s^{-1}	$t_{1/2}$
$2\,N_2O_5 \rightarrow 4\,NO_2 + O_2$	g	25	3.38×10^{-5}	5.70 h
	$Br_2(l)$	25	4.27×10^{-5}	4.51 h
$C_2H_6 \rightarrow 2\,CH_3$	g	700	5.36×10^{-4}	21.6 min

* More values are given in the *Data section*.

Comment 22.1
The web site contains links to databases of rate constants of chemical reactions.

is

$$\ln\left(\frac{[A]}{[A]_0}\right) = -kt \qquad [A] = [A]_0 e^{-kt} \tag{22.12b}$$

where $[A]_0$ is the initial concentration of A (at $t = 0$).

Justification 22.1 *First-order integrated rate law*

First, we rearrange eqn 22.12a into

$$\frac{d[A]}{[A]} = -k\,dt$$

This expression can be integrated directly because k is a constant independent of t. Initially (at $t = 0$) the concentration of A is $[A]_0$, and at a later time t it is $[A]$, so we make these values the limits of the integrals and write

$$\int_{[A]_0}^{[A]} \frac{d[A]}{[A]} = -k\int_0^t dt$$

Because the integral of $1/x$ is $\ln x$, eqn 22.12b is obtained immediately.

A note on good practice To set the limits of integration, identify the start time ($t = 0$) and the corresponding concentration of A ($[A]_0$), and write these quantities as the lower limits of their respective integrals. Then identify the time of interest (t) and the corresponding concentration ($[A]$), and write these quantities as the upper limits of their respective integrals.

Equation 22.12b shows that, if $\ln([A]/[A]_0)$ is plotted against t, then a first-order reaction will give a straight line of slope $-k$. Some rate constants determined in this way are given in Table 22.1. The second expression in eqn 22.12b shows that in a first-order reaction the reactant concentration decreases exponentially with time with a rate determined by k (Fig. 22.5).

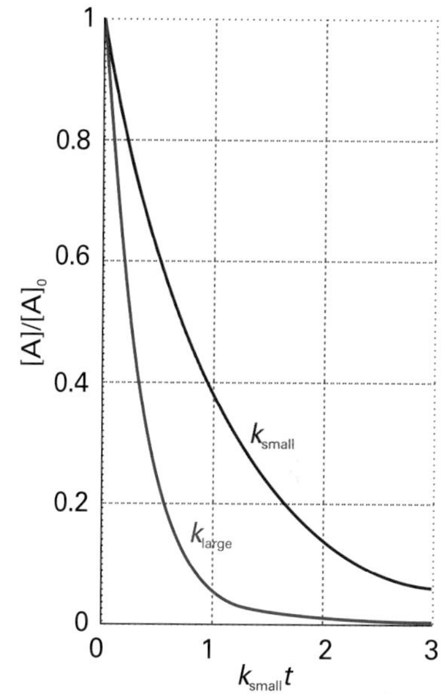

Fig. 22.5 The exponential decay of the reactant in a first-order reaction. The larger the rate constant, the more rapid the decay: here $k_{large} = 3k_{small}$.

Exploration For a first-order reaction of the form $A \rightarrow nB$ (with n possibly fractional), the concentration of the product varies with time as $[B] = n[B]_0(1 - e^{-kt})$. Plot the time dependence of [A] and [B] for the cases $n = 0.5$, 1, and 2.

Example 22.3 *Analysing a first-order reaction*

The variation in the partial pressure of azomethane with time was followed at 600 K, with the results given below. Confirm that the decomposition

$$CH_3N_2CH_3(g) \rightarrow CH_3CH_3(g) + N_2(g)$$

is first-order in azomethane, and find the rate constant at 600 K.

t/s	0	1000	2000	3000	4000
p/Pa	10.9	7.63	5.32	3.71	2.59

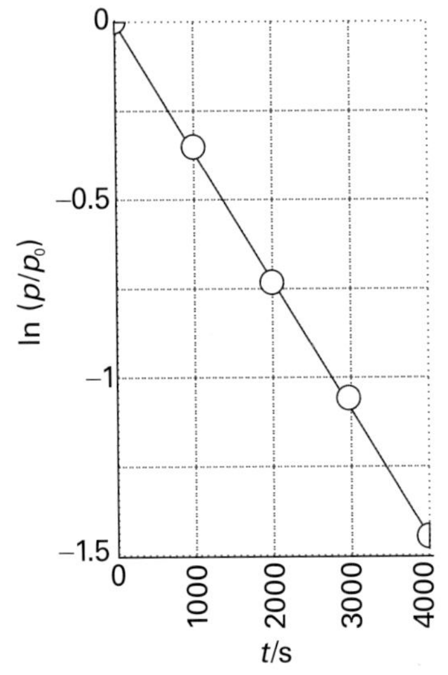

Fig. 22.6 The determination of the rate constant of a first-order reaction: a straight line is obtained when ln [A] (or, as here, ln p) is plotted against t; the slope gives k.

Method As indicated in the text, to confirm that a reaction is first-order, plot $\ln([A]/[A]_0)$ against time and expect a straight line. Because the partial pressure of a gas is proportional to its concentration, an equivalent procedure is to plot $\ln(p/p_0)$ against t. If a straight line is obtained, its slope can be identified with $-k$.

Answer We draw up the following table:

t/s	0	1000	2000	3000	4000
$\ln(p/p_0)$	1	−0.360	−0.720	−1.082	−1.441

Figure 22.6 shows the plot of $\ln(p/p_0)$ against t. The plot is straight, confirming a first-order reaction, and its slope is -3.6×10^{-4}. Therefore, $k=3.6\times10^{-4}\ \mathrm{s}^{-1}$.

A note on good practice Because the horizontal and vertical axes of graphs are labelled with pure numbers, the slope of a graph is always dimensionless. For a graph of the form $y=b+mx$ we can write

$$y=b+(m\ \text{units})(x/\text{units})$$

where 'units' are the units of x, and identify the (dimensionless) slope with 'm units'. Then

$$m=\text{slope/units}$$

In the present case, because the graph shown here is a plot of $\ln(p/p_0)$ against t/s (with 'units' = s) and k is the negative value of the slope of $\ln(p/p_0)$ against t itself,

$$k=-\text{slope/s}$$

Self-test 22.4 In a particular experiment, it was found that the concentration of N_2O_5 in liquid bromine varied with time as follows:

t/s	0	200	400	600	1000
$[N_2O_5]/(\mathrm{mol\ dm^{-3}})$	0.110	0.073	0.048	0.032	0.014

Confirm that the reaction is first-order in N_2O_5 and determine the rate constant.
$[k=2.1\times10^{-3}\ \mathrm{s}^{-1}]$

(b) Half-lives and time constants

A useful indication of the rate of a first-order chemical reaction is the **half-life**, $t_{1/2}$, of a substance, the time taken for the concentration of a reactant to fall to half its initial value. The time for [A] to decrease from $[A]_0$ to $\frac{1}{2}[A]_0$ in a first-order reaction is given by eqn 22.12b as

$$kt_{1/2}=-\ln\left(\frac{\frac{1}{2}[A]_0}{[A]_0}\right)=-\ln\tfrac{1}{2}=\ln 2$$

Hence

$$t_{1/2}=\frac{\ln 2}{k} \qquad (22.13)$$

(ln 2 = 0.693.) The main point to note about this result is that, for a first-order reaction, the half-life of a reactant is independent of its initial concentration. Therefore, if the concentration of A at some *arbitrary* stage of the reaction is [A], then it will have fallen to $\frac{1}{2}$[A] after a further interval of $(\ln 2)/k$. Some half-lives are given in Table 22.1.

Another indication of the rate of a first-order reaction is the **time constant**, τ (tau), the time required for the concentration of a reactant to fall to 1/e of its initial value. From eqn 22.12b it follows that

$$k\tau = -\ln\left(\frac{[\mathrm{A}]_0/\mathrm{e}}{[\mathrm{A}]_0}\right) = -\ln\frac{1}{\mathrm{e}} = 1$$

That is, the time constant of a first-order reaction is the reciprocal of the rate constant:

$$\tau = \frac{1}{k} \qquad (22.14)$$

(c) Second-order reactions

We show in the *Justification* below that the integrated form of the second-order rate law

$$\frac{\mathrm{d}[\mathrm{A}]}{\mathrm{d}t} = -k[\mathrm{A}]^2 \qquad (22.15a)$$

is either of the following two forms:

$$\frac{1}{[\mathrm{A}]} - \frac{1}{[\mathrm{A}]_0} = kt \qquad (22.15b)$$

$$[\mathrm{A}] = \frac{[\mathrm{A}]_0}{1 + kt[\mathrm{A}]_0} \qquad (22.15c)$$

where $[\mathrm{A}]_0$ is the initial concentration of A (at $t = 0$).

Justification 22.2 *Second-order integrated rate law*

To integrate eqn 22.15a we rearrange it into

$$\frac{\mathrm{d}[\mathrm{A}]}{[\mathrm{A}]^2} = -k\mathrm{d}t$$

The concentration of A is $[\mathrm{A}_0]$ at $t = 0$ and [A] at a general time t later. Therefore,

$$-\int_{[\mathrm{A}]_0}^{[\mathrm{A}]} \frac{\mathrm{d}[\mathrm{A}]}{[\mathrm{A}]^2} = k\int_0^t \mathrm{d}t$$

Because the integral of $1/x^2$ is $-1/x$, we obtain eqn 22.15b by substitution of the limits

$$\left.\frac{1}{[\mathrm{A}]}\right|_{[\mathrm{A}]_0}^{[\mathrm{A}]} = \frac{1}{[\mathrm{A}]} - \frac{1}{[\mathrm{A}]_0} = kt$$

We can then rearrange this expression into eqn 22.15c.

Equation 22.15b shows that to test for a second-order reaction we should plot 1/[A] against t and expect a straight line. The slope of the graph is k. Some rate constants determined in this way are given in Table 22.2. The rearranged form, eqn 22.15c, lets us predict the concentration of A at any time after the start of the reaction. It shows that the concentration of A approaches zero more slowly than in a first-order reaction with the same initial rate (Fig. 22.7).

It follows from eqn 22.15b by substituting $t = t_{1/2}$ and $[\mathrm{A}] = \frac{1}{2}[\mathrm{A}]_0$ that the half-life of a species A that is consumed in a second-order reaction is

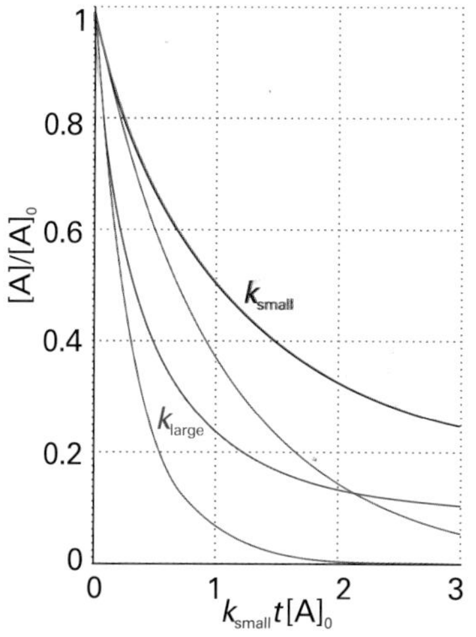

Fig. 22.7 The variation with time of the concentration of a reactant in a second-order reaction. The grey lines are the corresponding decays in a first-order reaction with the same initial rate. For this illustration, $k_{\mathrm{large}} = 3k_{\mathrm{small}}$.

Exploration For a second-order reaction of the form $\mathrm{A} \to n\mathrm{B}$ (with n possibly fractional), the concentration of the product varies with time as $[\mathrm{B}] = nkt[\mathrm{A}]_0^2/(1 + kt[\mathrm{A}]_0)$. Plot the time dependence of [A] and [B] for the cases $n = 0.5$, 1, and 2.

Synoptic table 22.2* Kinetic data for second-order reactions

Reaction	Phase	θ/°C	k/(dm^3 mol^{-1} s^{-1})
$2\,NOBr \rightarrow 2\,NO + Br_2$	g	10	0.80
$2\,I \rightarrow I_2$	g	23	7×10^9
$CH_3Cl + CH_3O^-$	$CH_3OH(l)$	20	2.29×10^{-6}

* More values are given in the *Data section*.

$$t_{1/2} = \frac{1}{k[A]_0} \qquad (22.16)$$

Therefore, unlike a first-order reaction, the half-life of a substance in a second-order reaction varies with the initial concentration. A practical consequence of this dependence is that species that decay by second-order reactions (which includes some environmentally harmful substances) may persist in low concentrations for long periods because their half-lives are long when their concentrations are low. In general, for an nth-order reaction of the form A $\rightarrow$ products, the half-life is related to the rate constant and the initial concentration of A by

$$t_{1/2} = \frac{1}{k[A]^{n-1}} \qquad (22.17)$$

(See Exercise 12.12a.)

Another type of second-order reaction is one that is first-order in each of two reactants A and B:

$$\frac{d[A]}{dt} = -k[A][B] \qquad (22.18)$$

Such a rate law cannot be integrated until we know how the concentration of B is related to that of A. For example, if the reaction is A + B $\rightarrow$ P, where P denotes products, and the initial concentrations are $[A]_0$ and $[B]_0$, then it is shown in the *Justification* below that, at a time t after the start of the reaction, the concentrations satisfy the relation

$$\ln\left(\frac{[B]/[B]_0}{[A]/[A]_0}\right) = ([B]_0 - [A]_0)kt \qquad (22.19)$$

Therefore, a plot of the expression on the left against t should be a straight line from which k can be obtained.

Justification 22.3 *Overall second-order rate law*

It follows from the reaction stoichiometry that, when the concentration of A has fallen to $[A]_0 - x$, the concentration of B will have fallen to $[B]_0 - x$ (because each A that disappears entails the disappearance of one B). It follows that

$$\frac{d[A]}{dt} = -k([A]_0 - x)([B]_0 - x)$$

Because $[A] = [A]_0 - x$, it follows that $d[A]/dt = -dx/dt$ and the rate law may be written as

$$\frac{\mathrm{d}x}{\mathrm{d}t}=k([\mathrm{A}]_0-x)([\mathrm{B}]_0-x)$$

The initial condition is that $x=0$ when $t=0$; so the integration required is

$$\int_0^x \frac{\mathrm{d}x}{([\mathrm{A}]_0-x)([\mathrm{B}]_0-x)}=k\int_0^t \mathrm{d}t$$

The integral on the right is simply kt. The integral on the left is evaluated by using the method of partial fractions and by using $[\mathrm{A}]=[\mathrm{A}]_0$ and $[\mathrm{B}]=[\mathrm{B}]_0$ at $t=0$ to give:

$$\int_0^x \frac{\mathrm{d}x}{([\mathrm{A}]_0-x)([\mathrm{B}]_0-x)}=\frac{1}{[\mathrm{B}]_0-[\mathrm{A}]_0}\left\{\ln\left(\frac{[\mathrm{A}]_0}{[\mathrm{A}]_0-x}\right)-\ln\left(\frac{[\mathrm{B}]_0}{[\mathrm{B}]_0-x}\right)\right\}$$

This expression can be simplified and rearranged into eqn 22.19 by combining the two logarithms by using $\ln y-\ln z=\ln(y/z)$ and noting that $[\mathrm{A}]=[\mathrm{A}]_0-x$ and $[\mathrm{B}]=[\mathrm{B}]_0-x$.

Comment 22.2

To use the method of partial fractions to evaluate an integral of the form $\int \frac{1}{(a-x)(b-x)}\mathrm{d}x$, where a and b are constants, we write

$$\frac{1}{(a-x)(b-x)}=\frac{1}{b-a}\left(\frac{1}{a-x}-\frac{1}{b-x}\right)$$

and integrate the expression on the right. It follows that

$$\int\frac{\mathrm{d}x}{(a-x)(b-x)}=\frac{1}{b-a}\left[\int\frac{\mathrm{d}x}{a-x}-\int\frac{\mathrm{d}x}{b-x}\right]$$

$$=\frac{1}{b-a}\left(\ln\frac{1}{a-x}-\ln\frac{1}{b-x}\right)+\text{constant}$$

Similar calculations may be carried out to find the integrated rate laws for other orders, and some are listed in Table 22.3.

Table 22.3 Integrated rate laws

Order	Reaction	Rate law*	$t_{1/2}$
0	$\mathrm{A}\rightarrow\mathrm{P}$	$v=k$ $kt=x$ for $0\le x\le[\mathrm{A}]_0$	$[\mathrm{A}]_0/2k$
1	$\mathrm{A}\rightarrow\mathrm{P}$	$v=k[\mathrm{A}]$ $kt=\ln\dfrac{[\mathrm{A}]_0}{[\mathrm{A}]_0-x}$	$(\ln 2)/k$
2	$\mathrm{A}\rightarrow\mathrm{P}$	$v=k[\mathrm{A}]^2$ $kt=\dfrac{x}{[\mathrm{A}]_0([\mathrm{A}]_0-x)}$	$1/k[\mathrm{A}]_0$
	$\mathrm{A}+\mathrm{B}\rightarrow\mathrm{P}$	$v=k[\mathrm{A}][\mathrm{B}]$ $kt=\dfrac{1}{[\mathrm{B}]_0-[\mathrm{A}]_0}\ln\dfrac{[\mathrm{A}]_0([\mathrm{B}]_0-x)}{([\mathrm{A}]_0-x)[\mathrm{B}]_0}$	
	$\mathrm{A}+2\,\mathrm{B}\rightarrow\mathrm{P}$	$v=k[\mathrm{A}][\mathrm{B}]$ $kt=\dfrac{1}{[\mathrm{B}]_0-2[\mathrm{A}]_0}\ln\dfrac{[\mathrm{A}]_0([\mathrm{B}]_0-2x)}{([\mathrm{A}]_0-x)[\mathrm{B}]_0}$	
	$\mathrm{A}\rightarrow\mathrm{P}$ with autocatalysis	$v=k[\mathrm{A}][\mathrm{P}]$ $kt=\dfrac{1}{[\mathrm{A}]_0+[\mathrm{P}]_0}\ln\dfrac{[\mathrm{A}]_0([\mathrm{P}]_0+x)}{([\mathrm{A}]_0-x)[\mathrm{P}]_0}$	
3	$\mathrm{A}+2\,\mathrm{B}\rightarrow\mathrm{P}$	$v=k[\mathrm{A}][\mathrm{B}]^2$ $kt=\dfrac{2x}{(2[\mathrm{A}]_0-[\mathrm{B}]_0)([\mathrm{B}]_0-2x)[\mathrm{B}]_0}+\dfrac{1}{(2[\mathrm{A}]_0-[\mathrm{B}]_0)^2}\ln\dfrac{[\mathrm{A}]_0([\mathrm{B}]_0-2x)}{([\mathrm{A}]_0-x)[\mathrm{B}]_0}$	
$n\ge 2$	$\mathrm{A}\rightarrow\mathrm{P}$	$v=k[\mathrm{A}]^n$ $kt=\dfrac{1}{n-1}\left\{\dfrac{1}{([\mathrm{A}]_0-x)^{n-1}}-\dfrac{1}{[\mathrm{A}]_0^{n-1}}\right\}$	$\dfrac{2^{n-1}-1}{(n-1)k[\mathrm{A}]_0^{n-1}}$

* $x=[\mathrm{P}]$ and $v=\mathrm{d}x/\mathrm{d}t$.

22.4 Reactions approaching equilibrium

Because all the laws considered so far disregard the possibility that the reverse reaction is important, none of them describes the overall rate when the reaction is close to equilibrium. At that stage the products may be so abundant that the reverse reaction must be taken into account. In practice, however, most kinetic studies are made on reactions that are far from equilibrium, and the reverse reactions are unimportant.

(a) First-order reactions close to equilibrium

We can explore the variation of the composition with time close to chemical equilibrium by considering the reaction in which A forms B and both forward and reverse reactions are first-order (as in some isomerizations). The scheme we consider is

$$\mathrm{A} \rightarrow \mathrm{B} \qquad v = k[\mathrm{A}] \qquad (22.20)$$

$$\mathrm{B} \rightarrow \mathrm{A} \qquad v = k'[\mathrm{B}]$$

The concentration of A is reduced by the forward reaction (at a rate $k[\mathrm{A}]$) but it is increased by the reverse reaction (at a rate $k'[\mathrm{B}]$). The net rate of change is therefore

$$\frac{\mathrm{d}[\mathrm{A}]}{\mathrm{d}t} = -k[\mathrm{A}] + k'[\mathrm{B}] \qquad (22.21)$$

If the initial concentration of A is $[\mathrm{A}]_0$, and no B is present initially, then at all times $[\mathrm{A}] + [\mathrm{B}] = [\mathrm{A}]_0$. Therefore,

$$\frac{\mathrm{d}[\mathrm{A}]}{\mathrm{d}t} = -k[\mathrm{A}] + k'([\mathrm{A}]_0 - [\mathrm{A}]) = -(k+k')[\mathrm{A}] + k'[\mathrm{A}]_0 \qquad (22.22)$$

The solution of this first-order differential equation (as may be checked by differentiation) is

$$[\mathrm{A}] = \frac{k' + k\mathrm{e}^{-(k+k')t}}{k'+k}[\mathrm{A}]_0 \qquad (22.23)$$

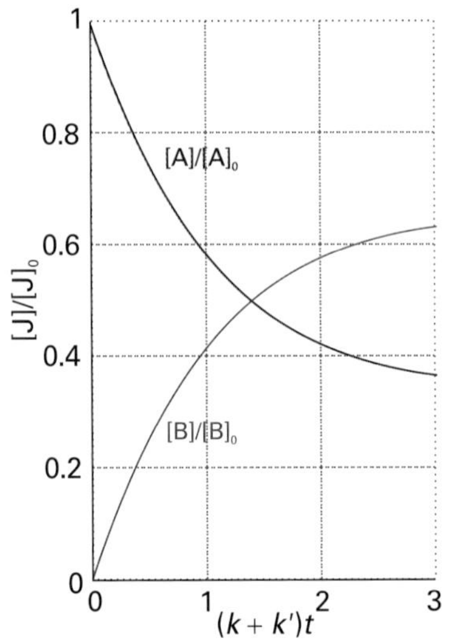

Fig. 22.8 The approach of concentrations to their equilibrium values as predicted by eqn 22.23 for a reaction $\mathrm{A} \rightleftharpoons \mathrm{B}$ that is first-order in each direction, and for which $k = 2k'$.

Exploration Set up the rate equations and plot the corresponding graphs for the approach to and equilibrium of the form $\mathrm{A} \rightleftharpoons 2\,\mathrm{B}$.

Figure 22.8 shows the time dependence predicted by this equation.

As $t \rightarrow \infty$, the concentrations reach their equilibrium values, which are given by eqn 22.23 as:

$$[\mathrm{A}]_{\mathrm{eq}} = \frac{k'[\mathrm{A}]_0}{k+k'} \qquad [\mathrm{B}]_{\mathrm{eq}} = [\mathrm{A}]_0 - [\mathrm{A}]_\infty = \frac{k[\mathrm{A}]_0}{k+k'} \qquad (22.24)$$

It follows that the equilibrium constant of the reaction is

$$K = \frac{[\mathrm{B}]_{\mathrm{eq}}}{[\mathrm{A}]_{\mathrm{eq}}} = \frac{k}{k'} \qquad (22.25)$$

(This expression is only approximate because thermodynamic equilibrium constants are expressed in terms of activities, not concentrations.) Exactly the same conclusion can be reached—more simply, in fact—by noting that, at equilibrium, the forward and reverse rates must be the same, so

$$k[\mathrm{A}]_{\mathrm{eq}} = k'[\mathrm{B}]_{\mathrm{eq}} \qquad (22.26)$$

This relation rearranges into eqn 22.25. The theoretical importance of eqn 22.25 is that it relates a thermodynamic quantity, the equilibrium constant, to quantities relating to rates. Its practical importance is that, if one of the rate constants can be measured, then the other may be obtained if the equilibrium constant is known.

For a more general reaction, the overall equilibrium constant can be expressed in terms of the rate constants for all the intermediate stages of the reaction mechanism:

$$K = \frac{k_a}{k'_a} \times \frac{k_b}{k'_b} \times \cdots \qquad (22.27)$$

where the *k*s are the rate constants for the individual steps and the *k*′s are those for the corresponding reverse steps.

(b) Relaxation methods

The term **relaxation** denotes the return of a system to equilibrium. It is used in chemical kinetics to indicate that an externally applied influence has shifted the equilibrium position of a reaction, normally suddenly, and that the reaction is adjusting to the equilibrium composition characteristic of the new conditions (Fig. 22.9). We shall consider the response of reaction rates to a **temperature jump**, a sudden change in temperature. We know from Section 7.4 that the equilibrium composition of a reaction depends on the temperature (provided $\Delta_r H^{\ominus}$ is nonzero), so a shift in temperature acts as a perturbation on the system. One way of achieving a temperature jump is to discharge a capacitor through a sample made conducting by the addition of ions, but laser or microwave discharges can also be used. Temperature jumps of between 5 and 10 K can be achieved in about 1 μs with electrical discharges. The high energy output of pulsed lasers (Section 14.5) is sufficient to generate temperature jumps of between 10 and 30 K within nanoseconds in aqueous samples. Some equilibria are also sensitive to pressure, and **pressure-jump techniques** may then also be used.

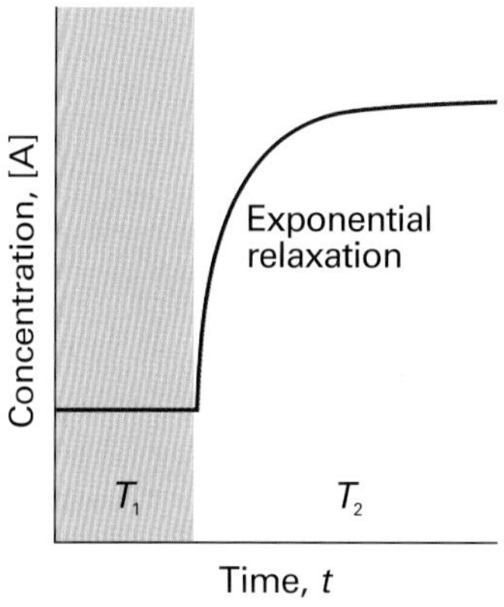

Fig. 22.9 The relaxation to the new equilibrium composition when a reaction initially at equilibrium at a temperature T_1 is subjected to a sudden change of temperature, which takes it to T_2.

When a sudden temperature increase is applied to a simple $A \rightleftharpoons B$ equilibrium that is first-order in each direction, we show in the *Justification* below that the composition relaxes exponentially to the new equilibrium composition:

$$x = x_0 e^{-t/\tau} \qquad \frac{1}{\tau} = k_a + k_b \qquad (22.28)$$

where x_0 is the departure from equilibrium immediately after the temperature jump and x is the departure from equilibrium at the new temperature after a time t.

Justification 22.4 *Relaxation to equilibrium*

When the temperature of a system at equilibrium is increased suddenly, the rate constants change from their earlier values to the new values k_a and k_b characteristic of that temperature, but the concentrations of A and B remain for an instant at their old equilibrium values. As the system is no longer at equilibrium, it readjusts to the new equilibrium concentrations, which are now given by

$$k_a[A]_{eq} = k_b[B]_{eq}$$

and it does so at a rate that depends on the new rate constants. We write the deviation of [A] from its new equilibrium value as x, so $[A] = x + [A]_{eq}$ and $[B] = [B]_{eq} - x$. The concentration of A then changes as follows:

$$\begin{aligned}\frac{d[A]}{dt} &= -k_a[A] + k_b[B] \\ &= -k_a([A]_{eq} + x) + k_b([B]_{eq} - x) \\ &= -(k_a + k_b)x\end{aligned}$$

because the two terms involving the equilibrium concentrations cancel. Because $d[A]/dt = dx/dt$, this equation is a first-order differential equation with the solution that resembles eqn 22.12b and is given in eqn 22.28.

Equation 22.28 shows that the concentrations of A and B relax into the new equilibrium at a rate determined by the sum of the two new rate constants. Because the equilibrium constant under the new conditions is $K \approx k_a/k_b$, its value may be combined with the relaxation time measurement to find the individual k_a and k_b.

Example 22.4 *Analysing a temperature-jump experiment*

The equilibrium constant for the autoprotolysis of water, $H_2O(l) \rightleftharpoons H^+(aq) + OH^-(aq)$, is $K_w = a(H^+)a(OH^-) = 1.008 \times 10^{-14}$ at 298 K. After a temperature-jump, the reaction returns to equilibrium with a relaxation time of 37 μs at 298 K and pH ≈ 7. Given that the forward reaction is first-order and the reverse is second-order overall, calculate the rate constants for the forward and reverse reactions.

Method We need to derive an expression for the relaxation time, τ (the time constant for return to equilibrium), in terms of k_1 (forward, first-order reaction) and k_2 (reverse, second-order reaction). We can proceed as above, but it will be necessary to make the assumption that the deviation from equilibrium (x) is so small that terms in x^2 can be neglected. Relate k_1 and k_2 through the equilibrium constant, but be careful with units because K_w is dimensionless.

Answer The forward rate at the final temperature is $k_1[H_2O]$ and the reverse rate is $k_2[H^+][OH^-]$. The net rate of deprotonation of H_2O is

$$\frac{d[H_2O]}{dt} = -k_1[H_2O] + k_2[H^+][OH^-]$$

We write $[H_2O] = [H_2O]_{eq} + x$, $[H^+] = [H^+]_{eq} - x$, and $[OH^-] = [OH^-]_{eq} - x$, and obtain

$$\frac{dx}{dt} = -\{k_1 + k_2([H^+]_{eq} + [OH^-]_{eq})\}x - k_1[H_2O]_{eq} + k_2[H^+]_{eq}[OH^-]_{eq} + k_2x^2$$

$$\approx -\{k_1 + k_2([H^+]_{eq} + [OH^-]_{eq})\}x$$

where we have neglected the term in x^2 and used the equilibrium condition

$$k_1[H_2O]_{eq} = k_2[H^+]_{eq}[OH^-]_{eq}$$

to eliminate the terms that are independent of x. It follows that

$$\frac{1}{\tau} = k_1 + k_2([H^+]_{eq} + [OH^-]_{eq})$$

At this point we note that

$$K_w = a(H^+)a(OH^-) \approx ([H^+]_{eq}/c^{\ominus})([OH^-]_{eq})/c^{\ominus}) = [H^+]_{eq}[OH^-]_{eq}/c^{\ominus 2}$$

with $c^{\ominus} = 1$ mol dm^{-3}. For this electrically neutral system, $[H^+] = [OH^-]$, so the concentration of each type of ion is $K_w^{1/2}c^{\ominus}$, and hence

$$\frac{1}{\tau} = k_1 + k_2(K_w^{1/2}c^{\ominus} + K_w^{1/2}c^{\ominus}) = k_2\left\{\frac{k_1}{k_2} + 2K_w^{1/2}c^{\ominus}\right\}$$

At this point we note that

$$\frac{k_1}{k_2} = \frac{[H^+]_{eq}[OH^-]_{eq}}{[H_2O]_{eq}} = \frac{K_w c^{\ominus 2}}{[H_2O]_{eq}}$$

The molar concentration of pure water is 55.6 mol dm^{-3}, so $[H_2O]_{eq}/c^{\ominus} = 55.6$. If we write $K = K_w/55.6 = 1.81 \times 10^{-16}$, we obtain

$$\frac{1}{\tau} = k_2\{K + 2K_w^{1/2}\}c^{\ominus}$$

Hence,

$$k_2 = \frac{1}{\tau(K + 2K_w^{1/2})c^{\ominus}}$$

$$= \frac{1}{(3.7 \times 10^{-5}\ \text{s}) \times (2.0 \times 10^{-7}) \times (1\ \text{mol dm}^{-3})} = 1.4 \times 10^{11}\ \text{dm}^3\ \text{mol}^{-1}\ \text{s}^{-1}$$

It follows that

$$k_1 = k_2 K c^{\ominus} = 2.4 \times 10^{-5}\ \text{s}^{-1}$$

The reaction is faster in ice, where $k_2 = 8.6 \times 10^{12}\ \text{dm}^3\ \text{mol}^{-1}\ \text{s}^{-1}$.

A note on good practice Notice how we keep track of units through the use of $c^{\ominus}$: K and K_w are dimensionless; k_2 is expressed in $\text{dm}^3\ \text{mol}^{-1}\ \text{s}^{-1}$ and k_1 is expressed in s^{-1}.

Self-test 22.5 Derive an expression for the relaxation time of a concentration when the reaction $A + B \rightleftharpoons C + D$ is second-order in both directions.
$[1/\tau = k([A] + [B])_{eq} + k'([C] + [D])_{eq}]$

22.5 The temperature dependence of reaction rates

The rate constants of most reactions increase as the temperature is raised. Many reactions in solution fall somewhere in the range spanned by the hydrolysis of methyl ethanoate (where the rate constant at 35°C is 1.82 times that at 25°C) and the hydrolysis of sucrose (where the factor is 4.13).

(a) The Arrhenius parameters

It is found experimentally for many reactions that a plot of $\ln k$ against $1/T$ gives a straight line. This behaviour is normally expressed mathematically by introducing two parameters, one representing the intercept and the other the slope of the straight line, and writing the **Arrhenius equation**

$$\ln k = \ln A - \frac{E_a}{RT} \qquad (22.29)$$

The parameter A, which corresponds to the intercept of the line at $1/T = 0$ (at infinite temperature, Fig. 22.10), is called the **pre-exponential factor** or the 'frequency factor'. The parameter E_a, which is obtained from the slope of the line ($-E_a/R$), is called the **activation energy**. Collectively the two quantities are called the **Arrhenius parameters** (Table 22.4).

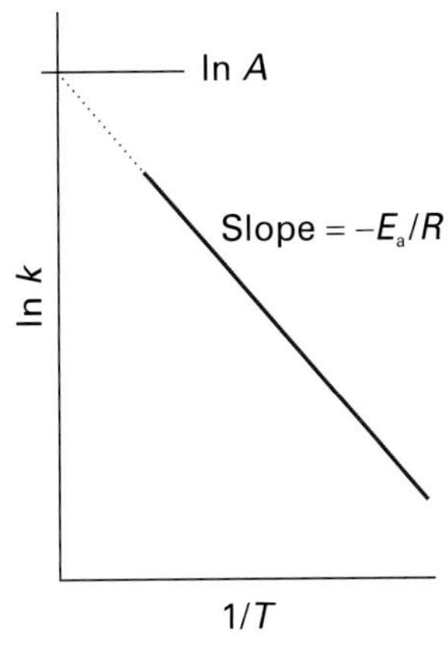

Fig. 22.10 A plot of $\ln k$ against $1/T$ is a straight line when the reaction follows the behaviour described by the Arrhenius equation (eqn 22.29). The slope gives $-E_a/R$ and the intercept at $1/T = 0$ gives $\ln A$.

Synoptic table 22.4* Arrhenius parameters

(1) First-order reactions	A/s^{-1}	$E_a/(\text{kJ mol}^{-1})$
$CH_3NC \rightarrow CH_3CN$	3.98×10^{13}	160
$2\,N_2O_5 \rightarrow 4\,NO_2 + O_2$	4.94×10^{13}	103.4
(2) Second-order reactions	$A/(\text{dm}^3\ \text{mol}^{-1}\ \text{s}^{-1})$	$E_a/(\text{kJ mol}^{-1})$
$OH + H_2 \rightarrow H_2O + H$	8.0×10^{10}	42
$NaC_2H_5O + CH_3I$ in ethanol	2.42×10^{11}	81.6

* More values are given in the *Data section*.

Example 22.5 *Determining the Arrhenius parameters*

The rate of the second-order decomposition of acetaldehyde (ethanal, CH_3CHO) was measured over the temperature range 700–1000 K, and the rate constants are reported below. Find E_a and A.

T/K	700	730	760	790	810	840	910	1000
$k/(\text{dm}^3\,\text{mol}^{-1}\,\text{s}^{-1})$	0.011	0.035	0.105	0.343	0.789	2.17	20.0	145

Method According to eqn 22.29, the data can be analysed by plotting $\ln(k/\text{dm}^3\,\text{mol}^{-1}\,\text{s}^{-1})$ against $1/(T/\text{K})$, or more conveniently $(10^3\,\text{K})/T$, and getting a straight line. As explained in Example 22.3, we obtain the activation energy from the dimensionless slope by writing $-E_a/R$ = slope/units, where in this case 'units' = $1/(10^3\,\text{K})$, so $E_a = -\text{slope} \times R \times 10^3\,\text{K}$. The intercept at $1/T = 0$ is $\ln(A/\text{dm}^3\,\text{mol}^{-1}\,\text{s}^{-1})$.

Answer We draw up the following table:

$(10^3\,\text{K})/T$	1.43	1.37	1.32	1.27	1.23	1.19	1.10	1.00
$\ln(k/\text{dm}^3\,\text{mol}^{-1}\,\text{s}^{-1})$	−4.51	−3.35	−2.25	−1.07	−0.24	0.77	3.00	4.98

Now plot $\ln k$ against $1/T$ (Fig. 22.11). The least-squares fit is to a line with slope −22.7 and intercept 27.7. Therefore,

$$E_a = 22.7 \times (8.3145\,\text{J K}^{-1}\,\text{mol}^{-1}) \times 10^3\,\text{K} = 189\,\text{kJ mol}^{-1}$$

$$A = e^{27.7}\,\text{dm}^3\,\text{mol}^{-1}\,\text{s}^{-1} = 1.1 \times 10^{12}\,\text{dm}^3\,\text{mol}^{-1}\,\text{s}^{-1}$$

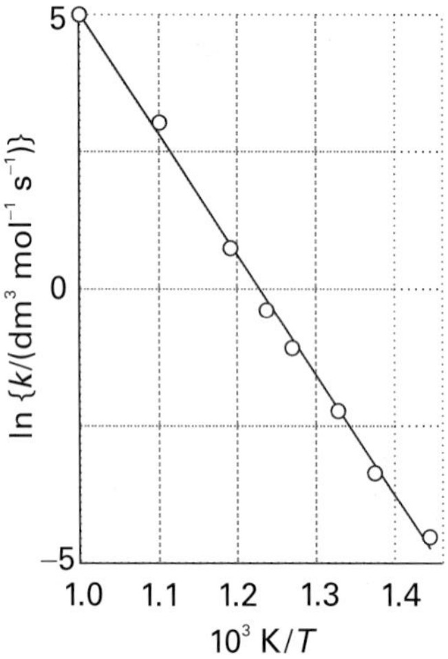

Fig. 22.11 The Arrhenius plot using the data in Example 22.5.

A note on good practice Note that A has the same units as k. In practice, A is obtained from one of the mid-range data values rather than using a lengthy extrapolation.

Self-test 22.6 Determine A and E_a from the following data:

T/K	300	350	400	450	500
$k/(\text{dm}^3\,\text{mol}^{-1}\,\text{s}^{-1})$	7.9×10^6	3.0×10^7	7.9×10^7	1.7×10^8	3.2×10^8

[$8 \times 10^{10}\,\text{dm}^3\,\text{mol}^{-1}\,\text{s}^{-1}$, $23\,\text{kJ mol}^{-1}$]

The fact that E_a is given by the slope of the plot of $\ln k$ against $1/T$ means that, the higher the activation energy, the stronger the temperature dependence of the rate constant (that is, the steeper the slope). *A high activation energy signifies that the rate constant depends strongly on temperature.* If a reaction has zero activation energy, its rate is independent of temperature. In some cases the activation energy is negative, which indicates that the rate decreases as the temperature is raised. We shall see that such behaviour is a signal that the reaction has a complex mechanism.

The temperature dependence of some reactions is non-Arrhenius, in the sense that a straight line is not obtained when $\ln k$ is plotted against $1/T$. However, it is still possible to define an activation energy at any temperature as

$$E_a = RT^2\left(\frac{\mathrm{d}\ln k}{\mathrm{d}T}\right) \qquad [22.30]$$

This definition reduces to the earlier one (as the slope of a straight line) for a temperature-independent activation energy. However, the definition in eqn 22.30 is more general than eqn 22.29, because it allows E_a to be obtained from the slope (at the temperature of interest) of a plot of ln k against $1/T$ even if the Arrhenius plot is not a straight line. Non-Arrhenius behaviour is sometimes a sign that quantum mechanical tunnelling is playing a significant role in the reaction (Section 22.7f).

(b) The interpretation of the parameters

For the present chapter we shall regard the Arrhenius parameters as purely empirical quantities that enable us to discuss the variation of rate constants with temperature; however, it is useful to have an interpretation in mind and write eqn 22.29 as

$$k = A e^{-E_a/RT} \qquad (22.31)$$

To interpret E_a we consider how the molecular potential energy changes in the course of a chemical reaction that begins with a collision between molecules of A and molecules of B (Fig. 22.12).

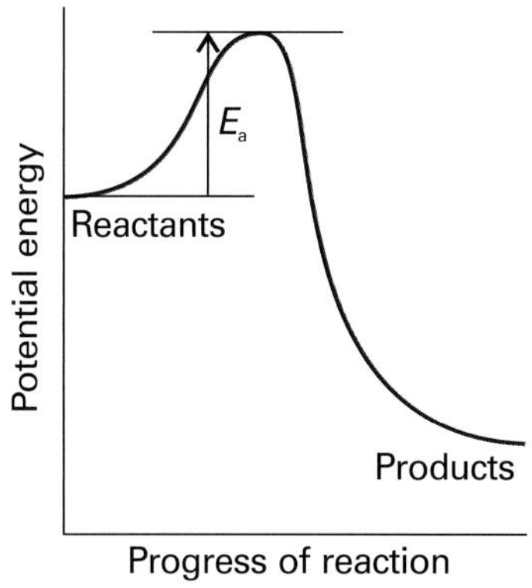

Fig. 22.12 A potential energy profile for an exothermic reaction. The height of the barrier between the reactants and products is the activation energy of the reaction.

As the reaction event proceeds, A and B come into contact, distort, and begin to exchange or discard atoms. The **reaction coordinate** is the collection of motions, such as changes in interatomic distances and bond angles, that are directly involved in the formation of products from reactants. (The reaction coordinate is essentially a geometrical concept and quite distinct from the extent of reaction.) The potential energy rises to a maximum and the cluster of atoms that corresponds to the region close to the maximum is called the **activated complex**. After the maximum, the potential energy falls as the atoms rearrange in the cluster and reaches a value characteristic of the products. The climax of the reaction is at the peak of the potential energy, which corresponds to the activation energy E_a. Here two reactant molecules have come to such a degree of closeness and distortion that a small further distortion will send them in the direction of products. This crucial configuration is called the **transition state** of the reaction. Although some molecules entering the transition state might revert to reactants, if they pass through this configuration then it is inevitable that products will emerge from the encounter.

Comment 22.3

The terms *activated complex* and *transition state* are often used as synonyms; however, we shall preserve a distinction.

We also conclude from the preceding discussion that, for a reaction involving the collision of two molecules, *the activation energy is the minimum kinetic energy that reactants must have in order to form products*. For example, in a gas-phase reaction there are numerous collisions each second, but only a tiny proportion are sufficiently energetic to lead to reaction. The fraction of collisions with a kinetic energy in excess of an energy E_a is given by the Boltzmann distribution as $e^{-E_a/RT}$. Hence, we can interpret the exponential factor in eqn 22.31 as the fraction of collisions that have enough kinetic energy to lead to reaction.

The pre-exponential factor is a measure of the rate at which collisions occur irrespective of their energy. Hence, the product of A and the exponential factor, $e^{-E_a/RT}$, gives the rate of *successful* collisions. We shall develop these remarks in Chapter 24 and see that they have their analogues for reactions that take place in liquids.

Accounting for the rate laws

We now move on to the second stage of the analysis of kinetic data, their explanation in terms of a postulated reaction mechanism.

22.6 Elementary reactions

Most reactions occur in a sequence of steps called **elementary reactions**, each of which involves only a small number of molecules or ions. A typical elementary reaction is

$$H + Br_2 \rightarrow HBr + Br$$

Note that the phase of the species is not specified in the chemical equation for an elementary reaction, and the equation represents the specific process occurring to individual molecules. This equation, for instance, signifies that an H atom attacks a Br_2 molecule to produce an HBr molecule and a Br atom. The **molecularity** of an elementary reaction is the number of molecules coming together to react in an elementary reaction. In a **unimolecular reaction**, a single molecule shakes itself apart or its atoms into a new arrangement, as in the isomerization of cyclopropane to propene. In a **bimolecular reaction**, a pair of molecules collide and exchange energy, atoms, or groups of atoms, or undergo some other kind of change. It is most important to distinguish molecularity from order: reaction order is an empirical quantity, and obtained from the experimental rate law; molecularity refers to an elementary reaction proposed as an individual step in a mechanism.

The rate law of a unimolecular elementary reaction is first-order in the reactant:

$$A \rightarrow P \qquad \frac{d[A]}{dt} = -k[A] \tag{22.32}$$

where P denotes products (several different species may be formed). A unimolecular reaction is first-order because the number of A molecules that decay in a short interval is proportional to the number available to decay. (Ten times as many decay in the same interval when there are initially 1000 A molecules as when there are only 100 present.) Therefore, the rate of decomposition of A is proportional to its molar concentration.

An elementary bimolecular reaction has a second-order rate law:

$$A + B \rightarrow P \qquad \frac{d[A]}{dt} = -k[A][B] \tag{22.33}$$

A bimolecular reaction is second-order because its rate is proportional to the rate at which the reactant species meet, which in turn is proportional to their concentrations. Therefore, if we have evidence that a reaction is a single-step, bimolecular process, we can write down the rate law (and then go on to test it). Bimolecular elementary reactions are believed to account for many homogeneous reactions, such as the dimerizations of alkenes and dienes and reactions such as

$$CH_3I(alc) + CH_3CH_2O^-(alc) \rightarrow CH_3OCH_2CH_3(alc) + I^-(alc)$$

(where 'alc' signifies alcohol solution). There is evidence that the mechanism of this reaction is a single elementary step

$$CH_3I + CH_3CH_2O^- \rightarrow CH_3OCH_2CH_3 + I^-$$

This mechanism is consistent with the observed rate law

$$v = k[CH_3I][CH_3CH_2O^-] \tag{22.34}$$

We shall see below how to combine a series of simple steps together into a mechanism and how to arrive at the corresponding rate law. For the present we emphasize that, *if the reaction is an elementary bimolecular process, then it has second-order kinetics, but if the kinetics are second-order, then the reaction might be complex.* The postulated mechanism can be explored only by detailed detective work on the system, and by investigating whether side products or intermediates appear during the course of the reaction. Detailed analysis of this kind was one of the ways, for example, in which the reaction $H_2(g) + I_2(g) \rightarrow 2\,HI(g)$ was shown to proceed by a complex mechanism. For many years the reaction had been accepted on good, but insufficiently meticulous evidence as a fine example of a simple bimolecular reaction, $H_2 + I_2 \rightarrow HI + HI$, in which atoms exchanged partners during a collision.

22.7 Consecutive elementary reactions

Some reactions proceed through the formation of an intermediate (I), as in the consecutive unimolecular reactions

$$A \xrightarrow{k_a} I \xrightarrow{k_b} P$$

An example is the decay of a radioactive family, such as

$$^{239}U \xrightarrow{23.5\ \text{min}} {}^{239}Np \xrightarrow{2.35\ \text{day}} {}^{239}Pu$$

(The times are half-lives.) We can discover the characteristics of this type of reaction by setting up the rate laws for the net rate of change of the concentration of each substance.

(a) The variation of concentrations with time

The rate of unimolecular decomposition of A is

$$\frac{d[A]}{dt} = -k_a[A] \qquad (22.35)$$

and A is not replenished. The intermediate I is formed from A (at a rate $k_a[A]$) but decays to P (at a rate $k_b[I]$). The net rate of formation of I is therefore

$$\frac{d[I]}{dt} = k_a[A] - k_b[I] \qquad (22.36)$$

The product P is formed by the unimolecular decay of I:

$$\frac{d[P]}{dt} = k_b[I] \qquad (22.37)$$

We suppose that initially only A is present, and that its concentration is $[A]_0$.

The first of the rate laws, eqn 22.35, is an ordinary first-order decay, so we can write

$$[A] = [A]_0 e^{-k_a t} \qquad (22.38)$$

When this equation is substituted into eqn 22.36, we obtain after rearrangement

$$\frac{d[I]}{dt} + k_b[I] = k_a[A]_0 e^{-k_a t} \qquad (22.39)$$

This differential equation has a standard form and, after setting $[I]_0 = 0$, the solution is

$$[I] = \frac{k_a}{k_b - k_a}(e^{-k_a t} - e^{-k_b t})[A]_0 \qquad (22.40)$$

At all times $[A] + [I] + [P] = [A]_0$, so it follows that

$$[P] = \left\{1 + \frac{k_a e^{-k_b t} - k_b e^{-k_a t}}{k_b - k_a}\right\}[A]_0 \qquad (22.41)$$

The concentration of the intermediate I rises to a maximum and then falls to zero (Fig. 22.13). The concentration of the product P rises from zero towards $[A]_0$.

Comment 22.4

The solution of a first-order differential equation with the form

$$\frac{dy}{dx} + yf(x) = g(x)$$

is

$$e^{\int f(x)\,dx} y = \int e^{\int f(x)dx} g(x)dx + \text{constant}$$

Equation 22.39 is a special case of this standard form, with $f(x) = \text{constant}$.

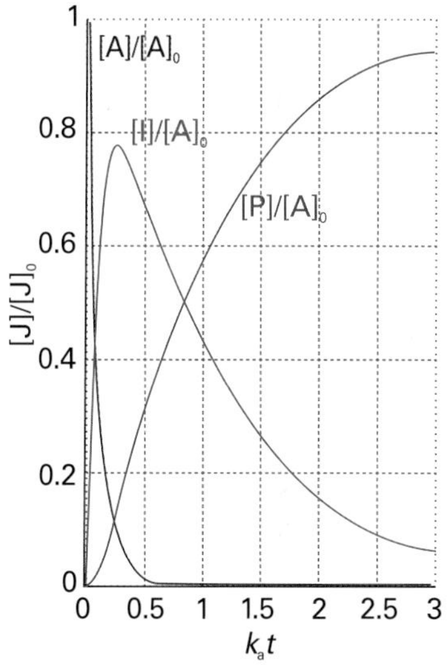

Fig. 22.13 The concentrations of A, I, and P in the consecutive reaction scheme $A \rightarrow I \rightarrow P$. The curves are plots of eqns 22.38, 22.40, and 21.41 with $k_a = 10k_b$. If the intermediate I is in fact the desired product, it is important to be able to predict when its concentration is greatest; see Example 22.6.

Exploration Use mathematical software, an electronic spreadsheet, or the applets found in the *Living graphs* section of the text's web site to investigate the effects on [A], [I], [P], and t_{max} of changing the ratio k_a/k_b from 10 (as in Fig. 22.13) to 0.01. Compare your results with those shown in Fig. 22.15.

Example 22.6 *Analysing consecutive reactions*

Suppose that in an industrial batch process a substance A produces the desired compound I which goes on to decay to a worthless product C, each step of the reaction being first-order. At what time will I be present in greatest concentration?

Method The time-dependence of the concentration of I is given by eqn 22.40. We can find the time at which [I] passes through a maximum, t_{max}, by calculating d[I]/dt and setting the resulting rate equal to zero.

Answer It follows from eqn 22.40 that

$$\frac{d[I]}{dt} = -\frac{k_a[A]_0(k_a e^{-k_a t} - k_b e^{-k_b t})}{k_b - k_a}$$

This rate is equal to zero when

$$k_a e^{-k_a t} = k_b e^{-k_b t}$$

Therefore,

$$t_{max} = \frac{1}{k_a - k_b} \ln \frac{k_a}{k_b}$$

For a given value of k_a, as k_b increases both the time at which [I] is a maximum and the yield of I decrease.

Self-test 22.7 Calculate the maximum concentration of I and justify the last remark.
$[[I]_{max}/[A]_0 = (k_a/k_b)^c, c = k_b/(k_b - k_a)]$

(b) The steady-state approximation

One feature of the calculation so far has probably not gone unnoticed: there is a considerable increase in mathematical complexity as soon as the reaction mechanism has more than a couple of steps. A reaction scheme involving many steps is nearly always unsolvable analytically, and alternative methods of solution are necessary. One approach is to integrate the rate laws numerically (see *Appendix* 2). An alternative approach, which continues to be widely used because it leads to convenient expressions and more readily digestible results, is to make an approximation.

The **steady-state approximation** (which is also widely called the **quasi-steady-state approximation**, QSSA, to distinguish it from a true steady state) assumes that, after an initial **induction period**, an interval during which the concentrations of intermediates, I, rise from zero, and during the major part of the reaction, the rates of change of concentrations of all reaction intermediates are negligibly small (Fig. 22.14):

$$\frac{d[I]}{dt} \approx 0 \qquad (22.42)$$

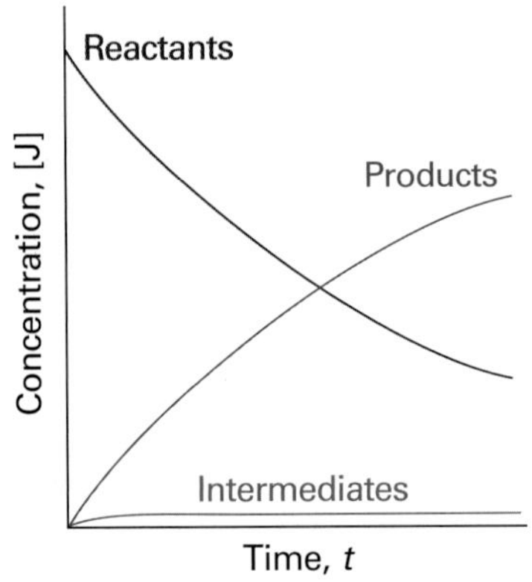

Fig. 22.14 The basis of the steady-state approximation. It is supposed that the concentrations of intermediates remain small and hardly change during most of the course of the reaction.

This approximation greatly simplifies the discussion of reaction schemes. For example, when we apply the approximation to the consecutive first-order mechanism, we set d[I]/dt = 0 in eqn 22.36, which then becomes

$$k_a[A] - k_b[I] \approx 0$$

Then

$$[I] \approx (k_a/k_b)[A] \qquad (22.43)$$

For this expression to be consistent with eqn 22.42, we require $k_a/k_b \ll 1$ (so that, even though [A] does depend on the time, the dependence of [I] on the time is negligible). On substituting this value of [I] into eqn 22.37, that equation becomes

$$\frac{d[P]}{dt} = k_b[I] \approx k_a[A] \qquad (22.44)$$

and we see that P is formed by a first-order decay of A, with a rate constant k_a, the rate-constant of the slower, rate-determining, step. We can write down the solution of this equation at once by substituting the solution for [A], eqn 22.38, and integrating:

$$[P] = k_a[A]_0 \int_0^t e^{-k_a t}dt = (1 - e^{-k_a t})[A]_0 \qquad (22.45)$$

This is the same (approximate) result as before, eqn 22.41, but much more quickly obtained. Figure 22.15 compares the approximate solutions found here with the exact solutions found earlier: k_b does not have to be very much bigger than k_a for the approach to be reasonably accurate.

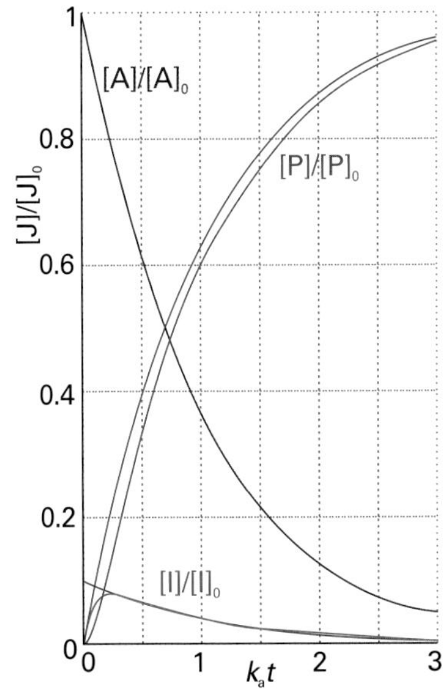

Fig. 22.15 A comparison of the exact result for the concentrations of a consecutive reaction and the concentrations obtained by using the steady-state approximation (red lines) for $k_b = 20k_a$. (The curve for [A] is unchanged.)

Example 22.7 *Using the steady-state approximation*

Devise the rate law for the decomposition of N_2O_5,

$$2\,N_2O_5(g) \rightarrow 4\,NO_2(g) + O_2(g)$$

on the basis of the following mechanism:

$N_2O_5 \rightarrow NO_2 + NO_3$	k_a
$NO_2 + NO_3 \rightarrow N_2O_5$	k_a'
$NO_2 + NO_3 \rightarrow NO_2 + O_2 + NO$	k_b
$NO + N_2O_5 \rightarrow NO_2 + NO_2 + NO_2$	k_c

A note on good practice Note that when writing the equation for an elementary reaction all the species are displayed individually; so we write A → B + B, for instance, not A → 2 B.

Method First identify the intermediates (species that occur in the reaction steps but do not appear in the overall reaction) and write expressions for their net rates of formation. Then, all net rates of change of the concentrations of intermediates are set equal to zero and the resulting equations are solved algebraically.

Answer The intermediates are NO and NO_3; the net rates of change of their concentrations are

$$\frac{d[NO]}{dt} = k_b[NO_2][NO_3] - k_c[NO][N_2O_5] \approx 0$$

$$\frac{d[NO_3]}{dt} = k_a[N_2O_5] - k_a'[NO_2][NO_3] - k_b[NO_2][NO_3] \approx 0$$

The net rate of change of concentration of N_2O_5 is

$$\frac{d[N_2O_5]}{dt} = -k_a[N_2O_5] + k_a'[NO_2][NO_3] - k_c[NO][N_2O_5]$$

and replacing the concentrations of the intermediates by using the equations above gives

$$\frac{d[N_2O_5]}{dt} = -\frac{2k_a k_b [N_2O_5]}{k'_a + k_b}$$

Self-test 22.8 Derive the rate law for the decomposition of ozone in the reaction $2\,O_3(g) \rightarrow 3\,O_2(g)$ on the basis of the (incomplete) mechanism

$$O_3 \rightarrow O_2 + O \qquad k_a$$
$$O_2 + O \rightarrow O_3 \qquad k'_a$$
$$O + O_3 \rightarrow O_2 + O_2 \qquad k_b$$

$[d[O_3]/dt = -k_a k_b [O_3]^2/(k'_a[O_2] + k_b[O_3])]$

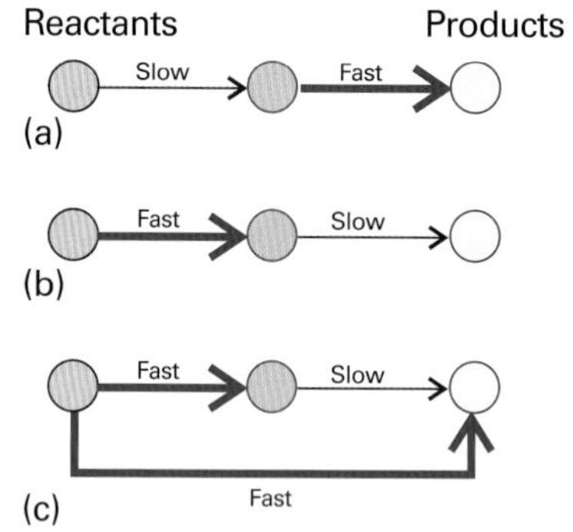

Fig. 22.16 In these diagrams of reaction schemes, heavy arrows represent fast steps and light arrows represent slow steps. (a) The first step is rate-determining; (b) the second step is rate-determining; (c) although one step is slow, it is not rate-determining because there is a fast route that circumvents it.

(c) The rate-determining step

Equation 22.45 shows that, when $k_b \gg k_a$, then the formation of the final product P depends on only the *smaller* of the two rate constants. That is, the rate of formation of P depends on the rate at which I is formed, not on the rate at which I changes into P. For this reason, the step $A \rightarrow I$ is called the 'rate-determining step' of the reaction. Its existence has been likened to building a six-lane highway up to a single-lane bridge: the traffic flow is governed by the rate of crossing the bridge. Similar remarks apply to more complicated reaction mechanisms, and in general the **rate-determining step** is the slowest step in a mechanism and controls the overall rate of the reaction. However, the rate-determining step is not just the slowest step: it must be slow *and* be a crucial gateway for the formation of products. If a faster reaction can also lead to products, then the slowest step is irrelevant because the slow reaction can then be side-stepped (Fig. 22.16).

The rate law of a reaction that has a rate-determining step can often be written down almost by inspection. If the first step in a mechanism is rate-determining, then the rate of the overall reaction is equal to the rate of the first step because all subsequent steps are so fast that once the first intermediate is formed it results immediately in the formation of products. Figure 22.17 shows the reaction profile for a mechanism of this kind in which the slowest step is the one with the highest activation energy. Once over the initial barrier, the intermediates cascade into products. However, a rate-determining step may also stem from the low concentration of a crucial reactant and need not correspond to the step with highest activation barrier.

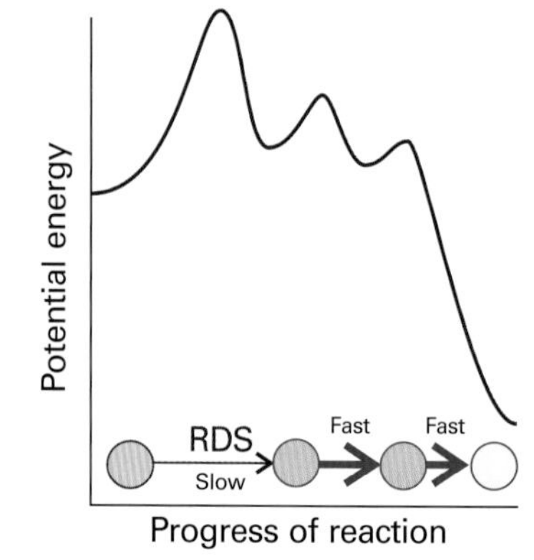

Fig. 22.17 The reaction profile for a mechanism in which the first step (RDS) is rate-determining.

(d) Kinetic and thermodynamic control of reactions

In some cases reactants can give rise to a variety of products, as in nitrations of mono-substituted benzene, when various proportions of the *ortho-*, *meta-*, and *para-*substituted products are obtained, depending on the directing power of the original substituent. Suppose two products, P_1 and P_2, are produced by the following competing reactions:

$$A + B \rightarrow P_1 \qquad \textit{Rate of formation of } P_1 = k_1[A][B]$$
$$A + B \rightarrow P_2 \qquad \textit{Rate of formation of } P_2 = k_2[A][B]$$

The relative proportion in which the two products have been produced at a given stage of the reaction (before it has reached equilibrium) is given by the ratio of the two rates, and therefore of the two rate constants:

$$\frac{[P_2]}{[P_1]}=\frac{k_2}{k_1} \tag{22.46}$$

This ratio represents the **kinetic control** over the proportions of products, and is a common feature of the reactions encountered in organic chemistry where reactants are chosen that facilitate pathways favouring the formation of a desired product. If a reaction is allowed to reach equilibrium, then the proportion of products is determined by thermodynamic rather than kinetic considerations, and the ratio of concentrations is controlled by considerations of the standard Gibbs energies of all the reactants and products.

(e) Pre-equilibria

From a simple sequence of consecutive reactions we now turn to a slightly more complicated mechanism in which an intermediate I reaches an equilibrium with the reactants A and B:

$$A + B \rightleftharpoons I \rightarrow P \tag{22.47}$$

The rate constants are k_a and k_a' for the forward and reverse reactions of the equilibrium and k_b for the final step. This scheme involves a **pre-equilibrium**, in which an intermediate is in equilibrium with the reactants. A pre-equilibrium can arise when the rate of decay of the intermediate back into reactants is much faster than the rate at which it forms products; thus, the condition is possible when $k_a' \gg k_b$ but not when $k_b \gg k_a'$. Because we assume that A, B, and I are in equilibrium, we can write

$$K=\frac{[I]}{[A][B]} \qquad K=\frac{k_a}{k_a'} \tag{22.48}$$

In writing these equations, we are presuming that the rate of reaction of I to form P is too slow to affect the maintenance of the pre-equilibrium (see the example below). The rate of formation of P may now be written:

$$\frac{d[P]}{dt}=k_b[I]=k_bK[A][B] \tag{22.49}$$

This rate law has the form of a second-order rate law with a composite rate constant:

$$\frac{d[P]}{dt}=k[A][B] \qquad k=k_bK=\frac{k_ak_b}{k_a'} \tag{22.50}$$

Example 22.8 *Analysing a pre-equilibrium*

Repeat the pre-equilibrium calculation but without ignoring the fact that I is slowly leaking away as it forms P.

Method Begin by writing the net rates of change of the concentrations of the substances and then invoke the steady-state approximation for the intermediate I. Use the resulting expression to obtain the rate of change of the concentration of P.

Answer The net rates of change of P and I are

$$\frac{d[P]}{dt}=k_b[I]$$

$$\frac{d[I]}{dt}=k_a[A][B]-k_a'[I]-k_b[I]\approx 0$$

The second equation solves to

$$[\mathrm{I}] \approx \frac{k_a[\mathrm{A}][\mathrm{B}]}{k'_a + k_b}$$

When we substitute this result into the expression for the rate of formation of P, we obtain

$$\frac{\mathrm{d}[\mathrm{P}]}{\mathrm{d}t} \approx k[\mathrm{A}][\mathrm{B}] \qquad k = \frac{k_a k_b}{k'_a + k_b}$$

This expression reduces to that in eqn 22.50 when the rate constant for the decay of I into products is much smaller than that for its decay into reactants, $k_b \ll k'_a$.

Self-test 22.9 Show that the pre-equilibrium mechanism in which 2 A $\rightleftharpoons$ I (K) followed by I + B → P (k_b) results in an overall third-order reaction.

[$\mathrm{d}[\mathrm{P}]/\mathrm{d}t = k_b K[\mathrm{A}]^2[\mathrm{B}]$]

(f) The kinetic isotope effect

The postulation of a plausible mechanism requires careful analysis of many experiments designed to determine the fate of atoms during the formation of products. Observation of the **kinetic isotope effect**, a decrease in the rate of a chemical reaction upon replacement of one atom in a reactant by a heavier isotope, facilitates the identification of bond-breaking events in the rate-determining step. A **primary kinetic isotope effect** is observed when the rate-determining step requires the scission of a bond involving the isotope. A **secondary kinetic isotope effect** is the reduction in reaction rate even though the bond involving the isotope is not broken to form product. In both cases, the effect arises from the change in activation energy that accompanies the replacement of an atom by a heavier isotope on account of changes in the zero-point vibrational energies (Section 13.9).

First, we consider the origin of the primary kinetic isotope effect in a reaction in which the rate-determining step is the scission of a C—H bond. The reaction coordinate corresponds to the stretching of the C—H bond and the potential energy profile is shown in Fig. 22.18. On deuteration, the dominant change is the reduction of the zero-point energy of the bond (because the deuterium atom is heavier). The whole reaction profile is not lowered, however, because the relevant vibration in the activated complex has a very low force constant, so there is little zero-point energy associated with the reaction coordinate in either isotopomeric form of the activated complex.

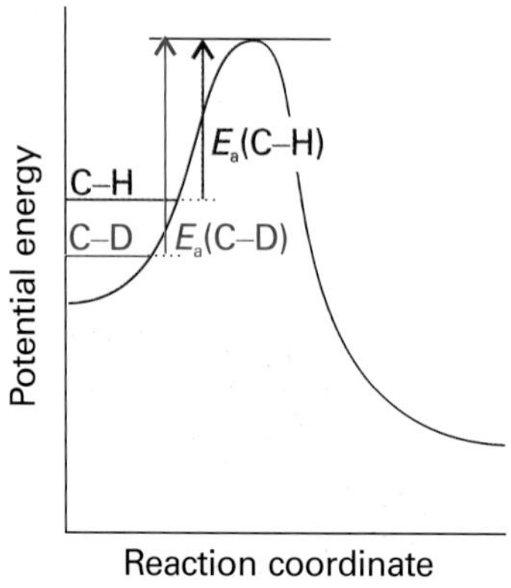

Fig. 22.18 Changes in the reaction profile when a C—H bond undergoing cleavage is deuterated. In this illustration, the C—H and C—D bonds are modelled as simple harmonic oscillators. The only significant change is in the zero-point energy of the reactants, which is lower for C—D than for C—H. As a result, the activation energy is greater for C—D cleavage than for C—H cleavage.

We assume that, to a good approximation, a change in the activation energy arises only from the change in zero-point energy of the stretching vibration, so

$$E_a(\mathrm{C{-}D}) - E_a(\mathrm{C{-}H}) = N_A\{\tfrac{1}{2}hc\tilde{\nu}(\mathrm{C{-}H}) - \tfrac{1}{2}hc\tilde{\nu}(\mathrm{C{-}D})\} \qquad (22.51)$$

where $\tilde{\nu}$ is the relevant vibrational wavenumber. From Section 13.9, we know that $\tilde{\nu}(\mathrm{C{-}D}) = (\mu_{\mathrm{CH}}/\mu_{\mathrm{CD}})^{1/2}\tilde{\nu}(\mathrm{C{-}H})$, where μ is the relevant effective mass. It follows that

$$E_a(\mathrm{C{-}D}) - E_a(\mathrm{C{-}H}) = \tfrac{1}{2}N_A hc\tilde{\nu}(\mathrm{C{-}H})\left\{1 - \left(\frac{\mu_{\mathrm{CH}}}{\mu_{\mathrm{CD}}}\right)^{1/2}\right\} \qquad (22.52)$$

If we assume further that the pre-exponential factor does not change upon deuteration, then the rate constants for the two species should be in the ratio

$$\frac{k(\text{C—D})}{k(\text{C—H})}=e^{-\lambda} \quad \text{with} \quad \lambda=\frac{hc\tilde{\nu}(\text{C—H})}{2kT}\left\{1-\left(\frac{\mu_{\text{CH}}}{\mu_{\text{CD}}}\right)^{1/2}\right\} \tag{22.53}$$

Note that $\lambda > 0$ because $\mu_{\text{CD}} > \mu_{\text{CH}}$ and that $k(\text{C—D})/k(\text{C—H})$ decreases with decreasing temperature.

Illustration 22.2 *Assessing the primary kinetic isotope effect*

From infrared spectra, the fundamental vibrational wavenumber for stretching of a C—H bond is about 3000 cm^{-1}. From $\mu_{\text{CH}}/\mu_{\text{CD}} = 0.538$ and eqn 22.53, it follows that $k(\text{C—D})/k(\text{C—H}) = 0.145$ at 298 K. We predict that at room temperature C—H cleavage should be about seven times faster than C—D cleavage, other conditions being equal. However, experimental values of $k(\text{C—D})/k(\text{C—H})$ can differ significantly from those predicted by eqn 22.53 on account of the severity of the assumptions in the model.

In some cases, substitution of deuterium for hydrogen results in values of $k(\text{C—D})/k(\text{C—H})$ that are too low to be accounted for by eqn 22.53, even when more complete models are used to predict ratios of rate constants. Such abnormal kinetic isotope effects are evidence for a path in which quantum mechanical tunnelling of hydrogen atoms takes place through the activation barrier (Fig. 22.19). We saw in Section 9.3 that the probability of tunnelling through a barrier decreases as the mass of the particle increases, so deuterium tunnels less efficiently through a barrier than hydrogen and its reactions are correspondingly slower. Quantum mechanical tunnelling can be the dominant process in reactions involving hydrogen atom or proton transfer when the temperature is so low that very few reactant molecules can overcome the activation energy barrier. We shall see in Chapter 23 that, because m_e is so small, tunnelling is also a very important contributor to the rates of electron transfer reactions.

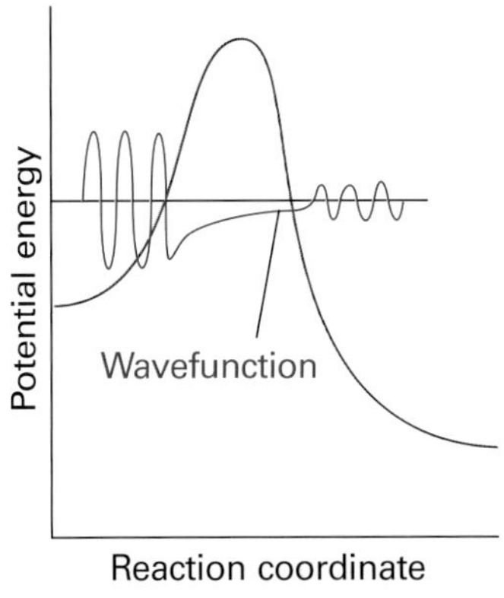

Fig. 22.19 A proton can tunnel through the activation energy barrier that separates reactants from products, so the effective height of the barrier is reduced and the rate of the proton transfer reaction increases. The effect is represented by drawing the wavefunction of the proton near the barrier. Proton tunnelling is important only at low temperatures, when most of the reactants are trapped on the left of the barrier.

Now consider the secondary isotope effect, which arises from differences in the zero-point energies between reactants and an activated complex with a significantly different structure. The activation energy of the undeuterated compound is

$$E_a(\text{H}) = E_a + E^{\ddagger}_{\text{vib},0}(\text{H}) - E_{\text{vib},0}(\text{H})$$

where E_a is the difference between the minima of the molecular potential energy curves of the activated complex and the ground state of the reactant and $E^{\ddagger}_{\text{vib},0}(\text{H})$ and $E_{\text{vib},0}(\text{H})$ are the zero-point vibrational energies of the two states (Fig. 22.20). For the deuterated compound

$$E_a(\text{D}) = E_a + E^{\ddagger}_{\text{vib},0}(\text{D}) - E_{\text{vib},0}(\text{D})$$

The difference in activation energies is therefore

$$E_a(\text{D}) - E_a(\text{H}) = \{E^{\ddagger}_{\text{vib},0}(\text{D}) - E_{\text{vib},0}(\text{D})\} - \{E^{\ddagger}_{\text{vib},0}(\text{H}) - E_{\text{vib},0}(\text{H})\}$$

We now suppose that the difference in zero-point energies is due solely to the vibration of a single C—H (or C—D) bond, and so write

$$\begin{aligned} E_a(\text{D}) - E_a(\text{H}) &= \tfrac{1}{2}N_A hc\{\tilde{\nu}^{\ddagger}(\text{C—D}) - \tilde{\nu}(\text{C—D})\} \\ &\quad - \tfrac{1}{2}N_A hc\{\tilde{\nu}^{\ddagger}(\text{C—H}) - \tilde{\nu}(\text{C—H})\} \end{aligned} \tag{22.54}$$

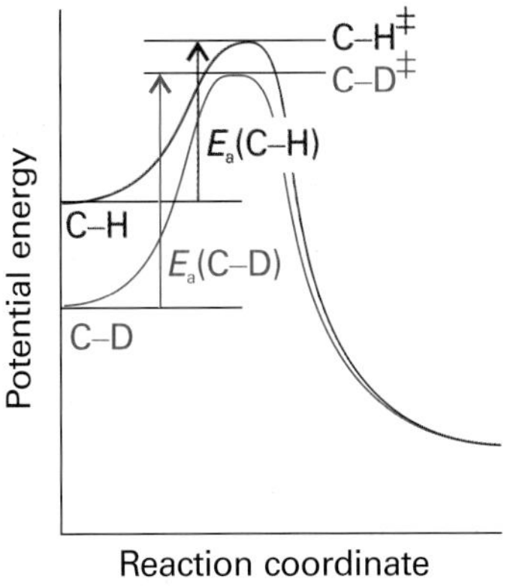

Fig. 22.20 The difference in zero-point vibrational energies used to explain the secondary isotope effect.

where $\tilde{\nu}^{\ddagger}$ and $\tilde{\nu}$ denote vibrational wavenumbers in the activated complex and reactant, respectively. With $\tilde{\nu}^{\ddagger}(\text{C—D}) = (\mu_{\text{CH}}/\mu_{\text{CD}})^{1/2}\tilde{\nu}^{\ddagger}(\text{C—H})$ and $\tilde{\nu}(\text{C—D}) = (\mu_{\text{CH}}/\mu_{\text{CD}})^{1/2}\tilde{\nu}(\text{C—H})$, it follows that

$$E_a(\mathrm{D}) - E_a(\mathrm{H}) = \tfrac{1}{2}N_A hc\{\tilde{\nu}^{\ddagger}(\mathrm{C{-}H}) - \tilde{\nu}(\mathrm{C{-}H})\}\left\{\left(\frac{\mu_{\mathrm{CH}}}{\mu_{\mathrm{CD}}}\right)^{1/2} - 1\right\} \qquad (22.55)$$

and

$$\frac{k(\mathrm{D})}{k(\mathrm{H})} = \mathrm{e}^{-\lambda} \quad \text{with} \quad \lambda = \frac{hc\{\tilde{\nu}^{\ddagger}(\mathrm{C{-}H}) - \tilde{\nu}(\mathrm{C{-}H})\}}{2kT}\left\{\left(\frac{\mu_{\mathrm{CH}}}{\mu_{\mathrm{CD}}}\right)^{1/2} - 1\right\} \qquad (22.56)$$

Because $\mu_{\mathrm{CH}}/\mu_{\mathrm{CD}} < 1$, provided the vibrational wavenumber of the activated complex is less than that of the reactant, $\lambda > 1$ and the deuterated form reacts more slowly than the undeuterated compound.

Illustration 22.3 *Assessing the secondary kinetic isotope effect*

In the heterolytic dissociation $\mathrm{CHCl_3} \rightarrow \mathrm{CHCl_2^+} + \mathrm{Cl^-}$ the activated complex resembles the product $\mathrm{CHCl_2^+}$. From infrared spectra, the fundamental vibrational wavenumber for a bending motion involving the C—H group is about 1350 cm^{-1} in $\mathrm{CHCl_3}$ and about 800 cm^{-1} in $\mathrm{CHCl_2^+}$. Assuming that $\tilde{\nu}^{\ddagger}(\mathrm{C{-}H}) = 800\ \mathrm{cm^{-1}}$ on account of the structural similarity between $\mathrm{CHCl_2^+}$ and the activated complex, it follows from $\mu_{\mathrm{CH}}/\mu_{\mathrm{CD}} = 0.538$ and eqn 22.56 that $k(\mathrm{D})/k(\mathrm{H}) = 0.709$ at 298 K. We predict that at room temperature the dissociation of $\mathrm{CHCl_3}$ should be about 40 per cent faster than dissociation of $\mathrm{CDCl_3}$. Comparison with the result from *Illustration* 22.2 shows that the secondary kinetic isotope effect leads to higher values of $k(\mathrm{D})/k(\mathrm{H})$ than does the primary kinetic isotope effect. This conclusion is supported by a number of experimental observations.

IMPACT ON BIOCHEMISTRY

I22.1 Kinetics of the helix–coil transition in polypeptides

We saw in *Impact* I16.1 that a simple statistical model accounts for the thermodynamic aspects of the helix–coil transition in polypeptides. The unfolding of a helix begins somewhere in the middle of the chain with a nucleation step, which is less favourable than the remaining helix-to-coil conversions, and continues in a cooperative fashion, in which the polymer becomes increasingly more susceptible to structural changes as more conversions occur. Here we examine the kinetics of the helix–coil transition, focusing primarily on experimental strategies and some recent results.

Earlier work on folding and unfolding of small polypeptides and large proteins relied primarily on rapid mixing and stopped-flow techniques. In a typical stopped-flow experiment, a sample of the protein with a high concentration of a chemical denaturant, such as urea or guanidinium hydrochloride, is mixed with a solution containing a much lower concentration of the same denaturant. Upon entering the mixing chamber, the denaturant is diluted and the protein re-folds. Unfolding is observed by mixing a sample of folded protein with a solution containing a high concentration of denaturant. These experiments are ideal for sorting out events in the millisecond timescale, such as the formation of contacts between helical segments in a large protein. However, the available data also indicate that, in a number of proteins, a significant portion of the folding process occurs in less than 1 ms, a time range not accessible by the stopped-flow technique. More recent temperature-jump and flash photolysis experiments have uncovered faster events. For example, at ambient temperature the formation of a loop between helical or sheet segments may be as fast as 1 μs and the formation of tightly packed cores with significant tertiary structure occurs in the

10–100 μs range. Among the fastest events are the formation and denaturation of helices and sheets from fully unfolded peptide chains and we examine how the laser-induced temperature-jump technique has been used in the study of the helix–coil transition.

The laser-induced temperature-jump technique takes advantage of the fact that proteins unfold, or melt, at high temperatures and each protein has a characteristic melting temperature (Section 19.10). Proteins also lose their native structures at very low temperatures, a process known as *cold denaturation*, and re-fold when the temperature is increased but kept significantly below the melting temperature. Hence, a temperature-jump experiment can be configured to monitor either folding or unfolding of a polypeptide, depending on the initial and final temperatures of the sample. The challenge of using melting or cold denaturation as the basis of kinetic measurements lies in increasing the temperature of the sample very quickly so fast relaxation proccess can be monitored. A number of clever strategies have been employed. In one example, a pulsed laser excites dissolved dye molecules that decay largely by internal conversion, or heat transfer to the solution. Another variation makes use of direct excitation of overtones of the O—H or O—D stretching modes of H_2O or D_2O, respectively, with a pulsed infrared laser. The latter strategy leads to temperature jumps in a small irradiated volume of about 20 K in less than 100 ps. Relaxation of the sample can then be probed by a variety of spectroscopic techniques, including absorption, emission, or Raman scattering. For example, the infrared absorption spectrum of a polypeptide is sensitive to polypeptide conformation, as the N—H stretching vibrations in the range 1630–1670 cm^{-1} are significantly different in the helix and coil forms.

Much of the kinetic work on the helix–coil transition has been conducted in small synthetic polypeptides rich in alanine, an aminoacid that is known to stabilize helical structures. Both experimental results and statistical mechanical calculations suggest that the mechanism of unfolding consists of at least two steps: a very fast step in which aminoacids at either end of a helical segment undergo transitions to coil regions and a slower rate-determining step that corresponds to the cooperative melting of the rest of the chain and loss of helical content. Using *h* and *c* to denote an aminoacid residue belonging to a helical and coil region, respectively, the mechanism may be summarized as follows:

hhhh... → *chhh*...	very fast
chhh... → *cccc*...	rate-determining step

The rate-determining step is thought to account for the relaxation time of 160 ns measured with a laser-induced temperature jump between 282.5 K and 300.6 K in an alanine-rich polypeptide containing 21 amino acids. It is thought that the limitation on the rate of the helix–coil transition in this peptide arises from an activation energy barrier of 1.7 kJ mol^{-1} associated with nucleation events of the form ...*hhhh*... → ...*hhch*... in the middle of the chain. Therefore, nucleation is not only thermodynamically unfavourable but also kinetically slow. Models that use concepts of statistical thermodynamics also suggest that a *hhhh*... → *chhh*... transition at either end of a helical segment has a significantly lower activation energy on account of the converting aminoacid not being flanked by *h* regions.

The time constant for the helix–coil transition has also been measured in proteins. In apomyoglobin (myoglobin lacking the haem cofactor), the unfolding of the helices appears to have a relaxation time of about 50 ns, even shorter than in synthetic peptides. It is difficult to interpret these results because we do not yet know how the amino acid sequence or interactions between helices in a folded protein affect the helix–coil relaxation time.

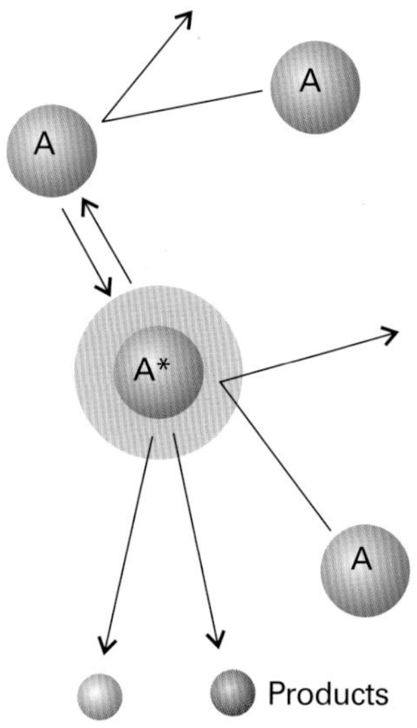

Fig. 22.21 A representation of the Lindemann–Hinshelwood mechanism of unimolecular reactions. The species A is excited by collision with A, and the excited A molecule (A^*) may either be deactivated by a collision with A or go on to decay by a unimolecular process to form products.

22.8 Unimolecular reactions

A number of gas-phase reactions follow first-order kinetics, as in the isomerization of cyclopropane mentioned earlier:

$$cyclo\text{-}C_3H_6 \rightarrow CH_3CH{=}CH_2 \qquad v = k[cyclo\text{-}C_3H_6] \tag{22.57}$$

The problem with the interpretation of first-order rate laws is that presumably a molecule acquires enough energy to react as a result of its collisions with other molecules. However, collisions are simple bimolecular events, so how can they result in a first-order rate law? First-order gas-phase reactions are widely called 'unimolecular reactions' because they also involve an elementary unimolecular step in which the reactant molecule changes into the product. This term must be used with caution, though, because the overall mechanism has bimolecular as well as unimolecular steps.

(a) The Lindemann–Hinshelwood mechanism

The first successful explanation of unimolecular reactions was provided by Frederick Lindemann in 1921 and then elaborated by Cyril Hinshelwood. In the **Lindemann–Hinshelwood mechanism** it is supposed that a reactant molecule A becomes energetically excited by collision with another A molecule (Fig. 22.21):

$$A + A \rightarrow A^* + A \qquad \frac{d[A^*]}{dt} = k_a[A]^2 \tag{22.58}$$

The energized molecule (A^*) might lose its excess energy by collision with another molecule:

$$A + A^* \rightarrow A + A \qquad \frac{d[A^*]}{dt} = -k_a'[A][A^*] \tag{22.59}$$

Alternatively, the excited molecule might shake itself apart and form products P. That is, it might undergo the unimolecular decay

$$A^* \rightarrow P \qquad \frac{d[A^*]}{dt} = -k_b[A^*] \tag{22.60}$$

If the unimolecular step is slow enough to be the rate-determining step, the overall reaction will have first-order kinetics, as observed. This conclusion can be demonstrated explicitly by applying the steady-state approximation to the net rate of formation of A^*:

$$\frac{d[A^*]}{dt} = k_a[A]^2 - k_a'[A][A^*] - k_b[A^*] \approx 0 \tag{22.61}$$

This equation solves to

$$[A^*] = \frac{k_a[A]^2}{k_b + k_a'[A]} \tag{22.62}$$

so the rate law for the formation of P is

$$\frac{d[P]}{dt} = k_b[A^*] = \frac{k_a k_b[A]^2}{k_b + k_a'[A]} \tag{22.63}$$

At this stage the rate law is not first-order. However, if the rate of deactivation by (A^*,A) collisions is much greater than the rate of unimolecular decay, in the sense that

$$k_a'[A^*][A] \gg k_b[A^*] \qquad \text{or} \qquad k_a'[A] \gg k_b$$

then we can neglect k_b in the denominator and obtain

$$\frac{d[P]}{dt} = k[A] \qquad k = \frac{k_a k_b}{k'_a} \tag{22.64}$$

Equation 22.64 is a first-order rate law, as we set out to show.

The Lindemann–Hinshelwood mechanism can be tested because it predicts that, as the concentration (and therefore the partial pressure) of A is reduced, the reaction should switch to overall second-order kinetics. Thus, when $k'_a[A] \ll k_b$, the rate law in eqn 22.63 is

$$\frac{d[P]}{dt} \approx k_a[A]^2 \tag{22.65}$$

The physical reason for the change of order is that at low pressures the rate-determining step is the bimolecular formation of A*. If we write the full rate law in eqn 22.63 as

$$\frac{d[P]}{dt} = k[A] \qquad k = \frac{k_a k_b[A]}{k_b + k'_a[A]} \tag{22.66}$$

then the expression for the effective rate constant, k, can be rearranged to

$$\frac{1}{k} = \frac{k'_a}{k_a k_b} + \frac{1}{k_a[A]} \tag{22.67}$$

Hence, a test of the theory is to plot $1/k$ against $1/[A]$, and to expect a straight line.

(b) The RRK model

Whereas the Lindemann–Hinshelwood mechanism agrees in general with the switch in order of unimolecular reactions, it does not agree in detail. Figure 22.22 shows a typical graph of $1/k$ against $1/[A]$. The graph has a pronounced curvature, corresponding to a larger value of k (a smaller value of $1/k$) at high pressures (low $1/[A]$) than would be expected by extrapolation of the reasonably linear low pressure (high $1/[A]$) data.

An improved model was proposed in 1926 by O.K. Rice and H.C. Ramsperger and almost simultaneously by L.S. Kassel, and is now known as the **Rice–Ramsperger–Kassel model** (RRK model). The model has been elaborated, largely by R.A. Marcus, into the RRKM model. Here we outline Kassel's original approach to the RRK model: the details are set out in *Further information* 22.1 at the end of the chapter. The essential feature of the model is that, although a molecule might have enough energy to react, that energy is distributed over all the modes of motion of the molecule, and reaction will occur only when enough of that energy has migrated into a particular location (such as a bond) in the molecule. Provided the rate constant is proportional to this probability, which we show in *Further information* 22.1 is

$$P = \left(1 - \frac{E^*}{E}\right)^{s-1} \tag{22.68a}$$

where s is the number of modes of motion over which the energy may be dissipated and E^* is the energy required for the bond of interest to break. We can write the **Kassel form** of the unimolecular rate constant for the decay of A* to products as

$$k_b(E) = \left(1 - \frac{E^*}{E}\right)^{s-1} k_b \qquad \text{for} \quad E \geq E^* \tag{22.68b}$$

where k_b is the rate constant used in the original Lindemann theory.

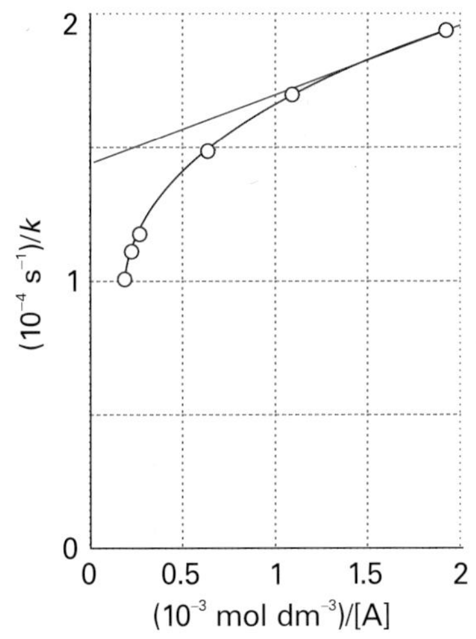

Fig. 22.22 The pressure dependence of the unimolecular isomerization of *trans*-CHD=CHD showing a pronounced departure from the straight line predicted by eqn 22.67 based on the Lindemann–Hinshelwood mechanism.

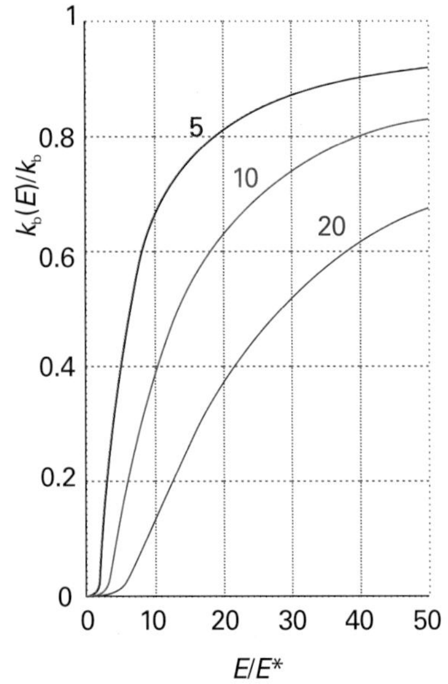

Fig. 22.23 The energy dependence of the rate constant given by eqn 22.68 for three values of *s*.

The energy dependence of the rate constant given by eqn 22.68 is shown in Fig. 22.23 for various values of *s*. We see that the rate constant is smaller at a given excitation energy if *s* is large, as it takes longer for the excitation energy to migrate through all the oscillators of a large molecule and accumulate in the critical mode. As *E* becomes very large, however, the term in parentheses approaches 1, and $k_b(E)$ becomes independent of the energy and the number of oscillators in the molecule, as there is now enough energy to accumulate immediately in the critical mode regardless of the size of the molecule.

(c) The activation energy of a composite reaction

Although the rate of each step of a complex mechanism might increase with temperature and show Arrhenius behaviour, is that true of a composite reaction? To answer this question, we consider the high-pressure limit of the Lindemann–Hinshelwood mechanism as expressed in eqn 22.64. If each of the rate constants has an Arrhenius-like temperature dependence, we can use eqn 22.31 for each of them, and write

$$k=\frac{k_a k_b}{k_a'}=\frac{(A_a e^{-E_a(a)/RT})(A_b e^{-E_a(b)/RT})}{(A_a' e^{-E_a'(a)/RT})} \qquad (22.69)$$

$$=\frac{A_a A_b}{A_a'} e^{-\{E_a(a)+E_a(b)-E_a'(a)\}/RT}$$

That is, the composite rate constant *k* has an Arrhenius-like form with activation energy

$$E_a = E_a(a) + E_a(b) - E_a'(a) \qquad (22.70)$$

Provided $E_a(a) + E_a(b) > E_a'(a)$, the activation energy is positive and the rate increases with temperature. However, it is conceivable that $E_a(a) + E_a(b) < E_a'(a)$ (Fig. 22.24), in which case the activation energy is negative and the rate will *decrease* as the temperature is raised. There is nothing remarkable about this behaviour: all it means is that the reverse reaction (corresponding to the deactivation of A*) is so sensitive to temperature that its rate increases sharply as the temperature is raised, and depletes the steady-state concentration of A*. The Lindemann–Hinshelwood mechanism is an unlikely candidate for this type of behaviour because the deactivation of A* has only a small activation energy, but there are reactions with analogous mechanisms in which a negative activation energy is observed.

When we examine the general rate law given in eqn 22.63, it is clear that the temperature dependence may be difficult to predict because each rate constant in the

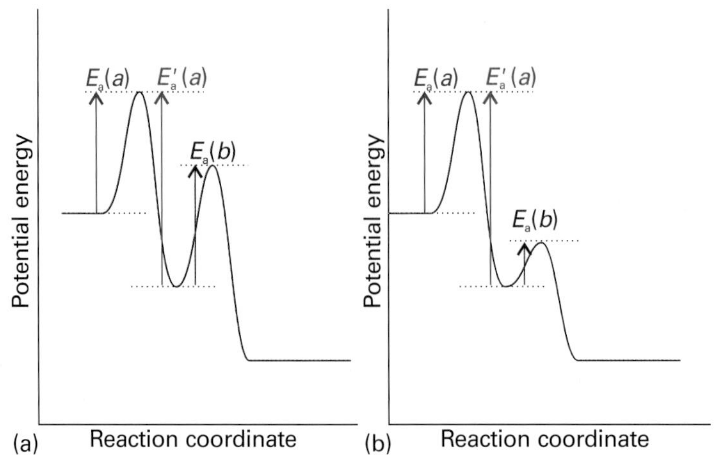

Fig. 22.24 For a reaction with a pre-equilibrium, there are three activation energies to take into account, two referring to the reversible steps of the pre-equilibrium and one for the final step. The relative magnitudes of the activation energies determine whether the overall activation energy is (a) positive or (b) negative.

expression for k increases with temperature, and the outcome depends on whether the terms in the numerator dominate those in the denominator, or vice versa. The fact that so many reactions do show Arrhenius-like behaviour with positive activation energies suggests that their rate laws are in a 'simple' regime, like eqn 22.65 rather than eqn 22.64, and that the temperature dependence is dominated by the activation energy of the rate-determining stage.

Checklist of key ideas

☐ 1. The rates of chemical reactions are measured by using techniques that monitor the concentrations of species present in the reaction mixture. Examples include real-time and quenching procedures, flow and stopped-flow techniques, and flash photolysis.

☐ 2. The instantaneous rate of a reaction is the slope of the tangent ot the graph of concentration against time (expressed as a positive quantity).

☐ 3. A rate law is an expression for the reaction rate in terms of the concentrations of the species that occur in the overall chemical reaction.

☐ 4. For a rate law of the form $v = k[A]^a[B]^b \ldots$, the rate constant is k, the order with respect to A is a, and the overall order is $a + b + \ldots$.

☐ 5. An integrated rate law is an expression for the concentration of a reactant or product as a function of time (Table 22.3).

☐ 6. The half-life $t_{1/2}$ of a reaction is the time it takes for the concentration of a species to fall to half its initial value. The time constant τ is the time required for the concentration of a reactant to fall to $1/e$ of its initial value. For a first-order reaction, $t_{1/2} = (\ln 2)/k$ and $\tau = 1/k$.

☐ 7. The equilibrium constant for a reaction is equal to the ratio of the forward and reverse rate constants, $K = k/k'$.

☐ 8. In relaxation methods of kinetic analysis, the equilibrium position of a reaction is first shifted suddenly and then allowed to readjust the equilibrium composition characteristic of the new conditions.

☐ 9. The temperature dependence of the rate constant of a reaction typically follows the Arrhenius equation, $\ln k = \ln A - E_a/RT$.

☐ 10. The activation energy, the parameter E_a in the Arrhenius equation, is the minimum kinetic energy for reaction during a molecular encounter. The larger the activation energy, the more sensitive the rate constant is to the temperature.

☐ 11. The mechanism of reaction is the sequence of elementary steps involved in a reaction.

☐ 12. The molecularity of an elementary reaction is the number of molecules coming together to react. An elementary unimolecular reaction has first-order kinetics; an elementary bimolecular reaction has second-order kinetics.

☐ 13. The rate-determining step is the slowest step in a reaction mechanism that controls the rate of the overall reaction.

☐ 14. In the steady-state approximation, it is assumed that the concentrations of all reaction intermediates remain constant and small throughout the reaction.

☐ 15. Provided a reaction has not reached equilibrium, the products of competing reactions are controlled by kinetics, with $[P_2]/[P_1] = k_2/k_1$.

☐ 16. Pre-equilibrium is a state in which an intermediate is in equilibrium with the reactants and which arises when the rates of formation of the intermediate and its decay back into reactants are much faster than its rate of formation of products.

☐ 17. The kinetic isotope effect is the decrease in the rate of a chemical reaction upon replacement of one atom in a reactant by a heavier isotope. A primary kinetic isotope effect is observed when the rate-determining step requires the scission of a bond involving the isotope. A secondary kinetic isotope effect is the reduction in reaction rate even though the bond involving the isotope is not broken to form product.

☐ 18. The Lindemann–Hinshelwood mechanism and the RRKM model of 'unimolecular' reactions account for the first-order kinetics of gas-phase reactions.

Further reading

Articles and texts

J. Andraos, A streamlined approach to solving simple and complex kinetic systems analytically. *J. Chem. Educ.* **76**, 1578 (1999).

C.H. Bamford, C.F. Tipper, and R.G. Compton (ed.), *Comprehensive chemical kinetics*. Vols. 1–38, Elsevier, Amsterdam (1969–2001).

M.N. Berberan-Santos and J.M.G. Martinho, Integration of kinetic rate equations by matrix methods. *J. Chem. Educ.* **67**, 375 (1990).

J.M. Goodman, How do approximations affect the solutions to kinetic equations? *J. Chem. Educ.* **76**, 275 (1999).

J.C. Lindon, G.E. Tranter, and J.L. Holmes (ed.), *Encyclopedia of spectroscopy and spectrometry*. Academic Press, San Diego (2000).

S.R. Logan, *Fundamentals of chemical kinetics*. Longman, Harlow (1996).

M.J. Pilling and P.W. Seakins, *Reaction kinetics*. Oxford University Press (1996).

J.I. Steinfeld, J.S. Francisco, and W.L. Hase, *Chemical kinetics and dynamics*. Prentice Hall, Englewood Cliffs (1998).

Sources of data and information

NDRL/NIST solution kinetics database, NIST standard reference database 40, National Institute of Standards and Technology, Gaithersburg (1994). For the URL, see the web site for this book.

NIST chemical kinetics database, NIST standard reference database 17, National Institute of Standards and Technology, Gaithersburg (1998). For the URL, see the web site for this book.

Further information

Further information 22.1 *The RRK model of unimolecular reactions*

To set up the RRK model, we suppose that a molecule consists of s identical harmonic oscillators, each of which has frequency ν. In practice, of course, the vibrational modes of a molecule have different frequencies, but assuming that they are all the same is a good first approximation. Next, we suppose that the vibrations are excited to a total energy $E = nh\nu$ and then set out to calculate the number of ways N in which the energy can be distributed over the oscillators.

We can represent the n quanta as follows:

□□□□□□□□□□□□□□□□□□□□□□□□□□□□□□□
□ . . . □□□

These quanta must be put in s containers (the s oscillators), which can be represented by inserting $s-1$ walls, denoted by |. One such distribution is

□□|□□□□|□□||□□□|□□□□□□□□|□□□□|||□□□□□|□
□□□ . . . □|□□

The total number of arrangements of each quantum and wall (of which there are $n+s-1$ in all) is $(n+s-1)!$ where, as usual, $x! = x(x-1)! \ldots 1$. However the $n!$ arrangements of the n quanta are indistinguishable, as are the $(s-1)!$ arrangements of the $s-1$ walls. Therefore, to find N we must divide $(n+s-1)!$ by these two factorials. It follows that

$$N = \frac{(n+s-1)!}{n!(s-1)!} \tag{22.71}$$

The distribution of the energy throughout the molecule means that it is too sparsely spread over all the modes for any particular bond to be sufficiently highly excited to undergo dissociation. If we suppose that a bond will break if it is excited to at least an energy $E^* = n^*h\nu$, then the number of ways in which at least this energy can be localized in one bond is

$$N^* = \frac{(n-n^*+s-1)!}{(n-n^*)!(s-1)!} \tag{22.72}$$

To obtain this result, we isolate one critical oscillator as the one that undergoes dissociation if it has *at least* n^* of the quanta, leaving up to $n-n^*$ quanta to be accommodated in the remaining $s-1$ oscillators (and therefore with $s-2$ walls in the partition in place of the $s-1$ walls we used above). We suppose that the critical oscillator consists of a single level plus an array of levels like the other oscillators, and that dissociation occurs however many quanta are in this latter array of levels, from 0 upwards. For example, in a system of five oscillators (other than the critical one) we might suppose that at least 6 quanta out of the 28 available must be present in the critical oscillator, then all the following partitions will result in dissociation:

□□□□□□|□□□□□|□□□□□□□□|□□□□||□□□□□□
□□□□□□□|□□□□|□□□□□□□□|□□□□||□□□□□□
□□□□□□□□|□□□|□□□□□□□□|□□□□||□□□□□□
. . .

(The leftmost partition is the critical oscillator.) However, these partitions are equivalent to

□□□□□□ |□□□□□|□□□□□□□□|□□□□||□□□□□
□□□□□□ □|□□□□|□□□□□□□□|□□□□||□□□□□
□□□□□□ □□|□□□|□□□□□□□□|□□□□||□□□□□
. . .

and we see that we have the problem of permuting $28-6=22$ (in general, $n-n^*$) quanta and 5 (in general, $s-1$) walls, and therefore a total of 27 (in general, $n-n^*+s-1$ objects). Therefore, the calculation is exactly like the one above for N, except that we have to find the number of distinguishable permutations of $n-n^*$ quanta in s containers (and therefore $s-1$ walls). The number N^* is therefore obtained from eqn 22.71 by replacing n by $n-n^*$.

From the preceding discussion we conclude that the probability that one specific oscillator will have undergone sufficient excitation to dissociate is the ratio N^*/N, which is

$$P = \frac{N^*}{N} = \frac{n!(n-n^*+s-1)!}{(n-n^*)!(n+s-1)} \tag{22.73}$$

Equation 22.73 is still awkward to use, even when written out in terms of its factors:

$$P = \frac{n(n-1)(n-2)\ldots 1}{(n-n^*)(n-n^*-1)\ldots 1} \times \frac{(n-n^*+s-1)(n-n^*+s-2)\ldots 1}{(n+s-1)(n+s-2)\ldots 1}$$

$$= \frac{(n-n^*+s-1)(n-n^*+s-2)\ldots(n-n^*+1)}{(n+s-1)(n+s-2)\ldots(n+2)(n+1)}$$

However, because $s-1$ is small (in the sense $s-1 \ll n-n^*$), we can approximate this expression by

$$P \approx \frac{(n-n^*)(n-n^*)\ldots(n-n^*)_{s-1\text{ factors}}}{(n)(n)\ldots(n)_{s-1\text{ factors}}} = \left(\frac{n-n^*}{n}\right)^{s-1}$$

Because the energy of the excited molecule is $E = nh\nu$ and the critical energy is $E^* = n^*h\nu$, this expression may be written

$$P = \left(1 - \frac{E^*}{E}\right)^{s-1}$$

as in eqn 22.68a. The dispersal of the energy of the collision reduces the rate constant below its simple 'Lindemann' form, and to obtain the observed rate constant we should multiply the latter by the probability that the energy will in fact be localized in the bond of interest, which gives eqn 22.68b.

Discussion questions

22.1 Consult literature sources and list the observed timescales during which the following processes occur: radiative decay of excited electronic states, molecular rotational motion, molecular vibrational motion, proton transfer reactions, the initial event of vision, energy transfer in photosynthesis, the initial electron transfer events in photosynthesis, the helix-to-coil transition in polypeptides, and collisions in liquids.

22.2 Write a brief report on a recent research article in which at least one of the following techniques was used to study the kinetics of a chemical reaction: stopped-flow techniques, flash photolysis, chemical quench-flow methods, freeze quench methods, temperature-jump methods, or pressure-jump methods. Your report should be similar in content and size to one of the *Impact* sections found throughout this text.

22.3 Describe the main features, including advantages and disadvantages, of the following experimental methods for determining the rate law of a reaction: the isolation method, the method of initial rates, and fitting data to integrated rate law expressions.

22.4 Distinguish between reaction order and molecularity.

22.5 Assess the validity of the following statement: the rate-determining step is the slowest step in a reaction mechanism.

22.6 Distinguish between a pre-equilibrium approximation and a steady-state approximation.

22.7 Distinguish between kinetic and thermodynamic control of a reaction.

22.8 Define the terms in and limit the generality of the expression $\ln k = \ln A - E_a/RT$.

22.9 Distinguish between a primary and a secondary kinetic isotope effect. Discuss how kinetic isotope effects in general can provide insight into the mechanism of a reaction.

22.10 Discuss the limitations of the generality of the expression $k = k_a k_b[A]/(k_b + k_a'[A])$ for the effective rate constant of a unimolecular reaction $A \rightarrow P$ with the following mechanism: $A + A \rightleftharpoons A^* + A$ (k_a, k_a'), $A^* \rightarrow P$ (k_b). Suggest an experimental procedure that may either support or refute the mechanism.

Exercises

22.1a The rate of the reaction $A + 2\,B \rightarrow 3\,C + D$ was reported as $1.0\ \text{mol dm}^{-3}\,\text{s}^{-1}$. State the rates of formation and consumption of the participants.

22.1b The rate of the reaction $A + 3\,B \rightarrow C + 2\,D$ was reported as $1.0\ \text{mol dm}^{-3}\,\text{s}^{-1}$. State the rates of formation and consumption of the participants.

22.2a The rate of formation of C in the reaction $2\,A + B \rightarrow 2\,C + 3\,D$ is $1.0\ \text{mol dm}^{-3}\,\text{s}^{-1}$. State the reaction rate, and the rates of formation or consumption of A, B, and D.

22.2b The rate of consumption of B in the reaction $A + 3\,B \rightarrow C + 2\,D$ is $1.0\ \text{mol dm}^{-3}\,\text{s}^{-1}$. State the reaction rate, and the rates of formation or consumption of A, C, and D.

22.3a The rate law for the reaction in Exercise 22.1a was found to be $v = k[A][B]$. What are the units of k? Express the rate law in terms of the rates of formation and consumption of (a) A, (b) C.

22.3b The rate law for the reaction in Exercise 22.1b was found to be $v = k[A][B]^2$. What are the units of k? Express the rate law in terms of the rates of formation and consumption of (a) A, (b) C.

22.4a The rate law for the reaction in Exercise 22.2a was reported as $d[C]/dt = k[A][B][C]$. Express the rate law in terms of the reaction rate; what are the units for k in each case?

22.4b The rate law for the reaction in Exercise 22.2b was reported as $d[C]/dt = k[A][B][C]^{-1}$. Express the rate law in terms of the reaction rate; what are the units for k in each case?

22.5a At 518°C, the rate of decomposition of a sample of gaseous acetaldehyde, initially at a pressure of 363 Torr, was $1.07\ \text{Torr s}^{-1}$ when 5.0 per cent had reacted and $0.76\ \text{Torr s}^{-1}$ when 20.0 per cent had reacted. Determine the order of the reaction.

22.5b At 400 K, the rate of decomposition of a gaseous compound initially at a pressure of 12.6 kPa, was $9.71\ \text{Pa s}^{-1}$ when 10.0 per cent had reacted and $7.67\ \text{Pa s}^{-1}$ when 20.0 per cent had reacted. Determine the order of the reaction.

22.6a At 518°C, the half-life for the decomposition of a sample of gaseous acetaldehyde (ethanal) initially at 363 Torr was 410 s. When the pressure was 169 Torr, the half-life was 880 s. Determine the order of the reaction.

22.6b At 400 K, the half-life for the decomposition of a sample of a gaseous compound initially at 55.5 kPa was 340 s. When the pressure was 28.9 kPa, the half-life was 178 s. Determine the order of the reaction.

22.7a The rate constant for the first-order decomposition of N_2O_5 in the reaction $2\,N_2O_5(g) \rightarrow 4\,NO_2(g) + O_2(g)$ is $k = 3.38 \times 10^{-5}\ \text{s}^{-1}$ at 25°C. What is the half-life of N_2O_5? What will be the pressure, initially 500 Torr, at (a) 10 s, (b) 10 min after initiation of the reaction?

22.7b The rate constant for the first-order decomposition of a compound A in the reaction $2\,A \rightarrow P$ is $k = 2.78 \times 10^{-7}\ \text{s}^{-1}$ at 25°C. What is the half-life of A? What will be the pressure, initially 32.1 kPa, at (a) 10 s, (b) 10 min after initiation of the reaction?

22.8a A second-order reaction of the type A + B → P was carried out in a solution that was initially 0.050 mol dm^{-3} in A and 0.080 mol dm^{-3} in B. After 1.0 h the concentration of A had fallen to 0.020 mol dm^{-3}. (a) Calculate the rate constant. (b) What is the half-life of the reactants?

22.8b A second-order reaction of the type A + 2 B → P was carried out in a solution that was initially 0.075 mol dm^{-3} in A and 0.030 mol dm^{-3} in B. After 1.0 h the concentration of A had fallen to 0.045 mol dm^{-3}. (a) Calculate the rate constant. (b) What is the half-life of the reactants?

22.9a If the rate laws are expressed with (a) concentrations in moles per decimetre cubed, (b) pressures in kilopascals, what are the units of the second-order and third-order rate constants?

22.9b If the rate laws are expressed with (a) concentrations in molecules per metre cubed, (b) pressures in newtons per metre squared, what are the units of the second-order and third-order rate constants?

22.10a The second-order rate constant for the reaction

$$CH_3COOC_2H_5(aq) + OH^-(aq) \rightarrow CH_3CO_2^-(aq) + CH_3CH_2OH(aq)$$

is 0.11 dm^3 mol^{-1} s^{-1}. What is the concentration of ester after (a) 10 s, (b) 10 min when ethyl acetate is added to sodium hydroxide so that the initial concentrations are [NaOH] = 0.050 mol dm^{-3} and [$CH_3COOC_2H_5$] = 0.100 mol dm^{-3}?

22.10b The second-order rate constant for the reaction A + 2 B → C + D is 0.21 dm^3 mol^{-1} s^{-1}. What is the concentration of C after (a) 10 s, (b) 10 min when the reactants are mixed with initial concentrations of [A] = 0.025 mol dm^{-3} and [B] = 0.150 mol dm^{-3}?

22.11a A reaction 2 A → P has a second-order rate law with $k = 3.50 \times 10^{-4}$ dm^3 mol^{-1} s^{-1}. Calculate the time required for the concentration of A to change from 0.260 mol dm^{-3} to 0.011 mol dm^{-3}.

22.11b A reaction 2 A → P has a third-order rate law with $k = 3.50 \times 10^{-4}$ dm^6 mol^{-2} s^{-1}. Calculate the time required for the concentration of A to change from 0.077 mol dm^{-3} to 0.021 mol dm^{-3}.

22.12a Show that $t_{1/2} \propto 1/[A]^{n-1}$ for a reaction that is nth-order in A.

22.12b Deduce an expression for the time it takes for the concentration of a substance to fall to one-third its initial value in an nth-order reaction.

22.13a The pK_a of NH_4^+ is 9.25 at 25°C. The rate constant at 25°C for the reaction of NH_4^+ and OH^- to form aqueous NH_3 is 4.0×10^{10} dm^3 mol^{-1} s^{-1}. Calculate the rate constant for proton transfer to NH_3. What relaxation time would be observed if a temperature jump were applied to a solution of 0.15 mol dm^{-3} NH_3(aq) at 25°C?

22.13b The equilibrium A ⇌ B + C at 25°C is subjected to a temperature jump that slightly increases the concentrations of B and C. The measured relaxation time is 3.0 μs. The equilibrium constant for the system is 2.0×10^{-16} at 25°C, and the equilibrium concentrations of B and C at 25°C are both 2.0×10^{-4} mol dm^{-3}. Calculate the rate constants for steps (1) and (2).

22.14a The rate constant for the decomposition of a certain substance is 2.80×10^{-3} dm^3 mol^{-1} s^{-1} at 30°C and 1.38×10^{-2} dm^3 mol^{-1} s^{-1} at 50°C. Evaluate the Arrhenius parameters of the reaction.

22.14b The rate constant for the decomposition of a certain substance is 1.70×10^{-2} dm^3 mol^{-1} s^{-1} at 24°C and 2.01×10^{-2} dm^3 mol^{-1} s^{-1} at 37°C. Evaluate the Arrhenius parameters of the reaction.

22.15a The base-catalysed bromination of nitromethane-d$_3$ in water at room temperature (298 K) proceeds 4.3 times more slowly than the bromination of the undeuterated material. Account for this difference. Use k_f(C—H) = 450 N m^{-1}.

22.15b Predict the order of magnitude of the isotope effect on the relative rates of displacement of (a) ^{1}H and ^{3}H, (b) ^{16}O and ^{18}O. Will raising the temperature enhance the difference? Take k_f(C—H) = 450 N m^{-1}, k_f(C—O) = 1750 N m^{-1}.

22.16a The effective rate constant for a gaseous reaction that has a Lindemann–Hinshelwood mechanism is 2.50×10^{-4} s^{-1} at 1.30 kPa and 2.10×10^{-5} s^{-1} at 12 Pa. Calculate the rate constant for the activation step in the mechanism.

22.16b The effective rate constant for a gaseous reaction that has a Lindemann–Hinshelwood mechanism is 1.7×10^{-3} s^{-1} at 1.09 kPa and 2.2×10^{-4} s^{-1} at 25 Pa. Calculate the rate constant for the activation step in the mechanism.

Problems*

Numerical problems

22.1 The data below apply to the formation of urea from ammonium cyanate, $NH_4CNO \rightarrow NH_2CONH_2$. Initially 22.9 g of ammonium cyanate was dissolved in in enough water to prepare 1.00 dm^3 of solution. Determine the order of the reaction, the rate constant, and the mass of ammonium cyanate left after 300 min.

t/min	0	20.0	50.0	65.0	150
m(urea)/g	0	7.0	12.1	13.8	17.7

22.2 The data below apply to the reaction, $(CH_3)_3CBr + H_2O \rightarrow (CH_3)_3COH + HBr$. Determine the order of the reaction, the rate constant, and the molar concentration of $(CH_3)_3CBr$ after 43.8 h.

t/h	0	3.15	6.20	10.00	18.30	30.80
$[(CH_3)_3CBr]/(10^{-2}$ mol dm$^{-3})$	10.39	8.96	7.76	6.39	3.53	2.07

22.3 The thermal decomposition of an organic nitrile produced the following data:

$t/(10^3$ s)	0	2.00	4.00	6.00	8.00	10.00	12.00	∞
[nitrile]/(mol dm^{-3})	1.10	0.86	0.67	0.52	0.41	0.32	0.25	0

Determine the order of the reaction and the rate constant.

22.4 The following data have been obtained for the decomposition of N_2O_5(g) at 67°C according to the reaction $2\,N_2O_5(g) \rightarrow 4\,NO_2(g) + O_2(g)$. Determine the order of the reaction, the rate constant, and the half-life. It is not necessary to obtain the result graphically, you may do a calculation using estimates of the rates of change of concentration.

t/min	0	1	2	3	4	5
$[N_2O_5]$/(mol dm^{-3})	1.000	0.705	0.497	0.349	0.246	0.173

22.5 A first-order decomposition reaction is observed to have the following rate constants at the indicated temperatures. Estimate the activation energy.

$k/(10^{-3}$ s^{-1})	2.46	45.1	576
θ/°C	0	20.0	40.0

* Problems denoted with the symbol ‡ were supplied by Charles Trapp, Carmen Giunta, and Marshall Cady.

22.6 The gas-phase decomposition of acetic acid at 1189 K proceeds by way of two parallel reactions:

(1) $CH_3COOH \rightarrow CH_4 + CO_2$ $\quad k_1 = 3.74\ s^{-1}$

(2) $CH_3COOH \rightarrow H_2C{=}C{=}O + H_2O$ $\quad k_2 = 4.65\ s^{-1}$

What is the maximum percentage yield of the ketene CH_2CO obtainable at this temperature?

22.7 Sucrose is readily hydrolysed to glucose and fructose in acidic solution. The hydrolysis is often monitored by measuring the angle of rotation of plane-polarized light passing through the solution. From the angle of rotation the concentration of sucrose can be determined. An experiment on the hydrolysis of sucrose in 0.50 M HCl(aq) produced the following data:

t/min	0	14	39	60	80	110	140	170	210
[sucrose]/(mol dm^{-3})	0.316	0.300	0.274	0.256	0.238	0.211	0.190	0.170	0.146

Determine the rate constant of the reaction and the half-life of a sucrose molecule. What is the average liftime of a sucrose molecule?

22.8 The composition of a liquid-phase reaction $2\,A \rightarrow B$ was followed by a spectrophotometric method with the following results:

t/min	0	10	20	30	40	∞
[B]/(mol dm^{-3})	0	0.089	0.153	0.200	0.230	0.312

Determine the order of the reaction and its rate constant.

22.9 The ClO radical decays rapidly by way of the reaction, $2\,ClO \rightarrow Cl_2 + O_2$. The following data have been obtained:

t/(10^{-3} s)	0.12	0.62	0.96	1.60	3.20	4.00	5.75
[ClO]/(10^{-6} mol dm^{-3})	8.49	8.09	7.10	5.79	5.20	4.77	3.95

Determine the rate constant of the reaction and the half-life of a ClO radical.

22.10 Cyclopropane isomerizes into propene when heated to 500°C in the gas phase. The extent of conversion for various initial pressures has been followed by gas chromatography by allowing the reaction to proceed for a time with various initial pressures:

p_0/Torr	200	200	400	400	600	600
t/s	100	200	100	200	100	200
p/Torr	186	173	373	347	559	520

where p_0 is the initial pressure and p is the final pressure of cyclopropane. What are the order and rate constant for the reaction under these conditions?

22.11 The addition of hydrogen halides to alkenes has played a fundamental role in the investigation of organic reaction mechanisms. In one study (M.J. Haugh and D.R. Dalton, *J. Amer. Chem. Soc.* **97**, 5674 (1975)), high pressures of hydrogen chloride (up to 25 atm) and propene (up to 5 atm) were examined over a range of temperatures and the amount of 2-chloropropane formed was determined by NMR. Show that, if the reaction $A + B \rightarrow P$ proceeds for a short time δt, the concentration of product follows $[P]/[A] = k[A]^{m-1}[B]^n\delta t$ if the reaction is mth-order in A and nth-order in B. In a series of runs the ratio of [chloropropane] to [propene] was independent of [propene] but the ratio of [chloropropane] to [HCl] for constant amounts of propene depended on [HCl]. For $\delta t \approx 100$ h (which is short on the timescale of the reaction) the latter ratio rose from zero to 0.05, 0.03, 0.01 for $p(HCl) =$ 10 atm, 7.5 atm, 5.0 atm, respectively. What are the orders of the reaction with respect to each reactant?

22.12 Use mathematical software or an electronic spreadsheet to examine the time dependence of [I] in the reaction mechanism $A \rightarrow I \rightarrow P$ (k_1, k_2). You may either integrate eqn 22.39 numerically (see *Appendix* 2) or use eqn 22.40 directly. In all the following calculations, use $[A]_0 = 1$ mol dm^{-3} and a time range of 0 to 5 s. (a) Plot [I] against t for $k_1 = 10\ s^{-1}$ and $k_2 = 1\ s^{-1}$. (b) Increase the ratio k_2/k_1 steadily by decreasing the value of k_1 and examine the plot of [I] against t at each turn. What approximation about d[I]/dt becomes increasingly valid?

22.13 Show that the following mechanism can account for the rate law of the reaction in Problem 22.11:

$HCl + HCl \rightleftharpoons (HCl)_2$ $\quad K_1$

$HCl + CH_3CH{=}CH_2 \rightleftharpoons$ complex $\quad K_2$

$(HCl)_2 +$ complex $\rightarrow CH_3CHClCH_3 + 2\,HCl$ $\quad k$ (slow)

What further tests could you apply to verify this mechanism?

22.14 Consider the dimerization $2\,A \rightleftharpoons A_2$, with forward rate constant k_a and backward rate constant k_b. (a) Derive the following expression for the relaxation time in terms of the total concentration of protein, $[A]_{tot} = [A] + 2[A_2]$:

$$\frac{1}{\tau^2} = k_b^2 + 8k_ak_b[A]_{tot}$$

(b) Describe the computational procedures that lead to the determination of the rate constants k_a and k_b from measurements of τ for different values of $[A]_{tot}$. (c) Use the data provided below and the procedure you outlined in part (b) to calculate the rate constants k_a and k_b, and the equilibrium constant K for formation of hydrogen-bonded dimers of 2-pyridone:

$[A]_{tot}$/(mol dm^{-3})	0.500	0.352	0.251	0.151	0.101
τ/ns	2.3	2.7	3.3	4.0	5.3

22.15 In the experiments described in Problems 22.11 and 22.13 an inverse temperature dependence of the reaction rate was observed, the overall rate of reaction at 70°C being roughly one-third that at 19°C. Estimate the apparent activation energy and the activation energy of the rate-determining step given that the enthalpies of the two equilibria are both of the order of −14 kJ mol^{-1}.

22.16 The second-order rate constants for the reaction of oxygen atoms with aromatic hydrocarbons have been measured (R. Atkinson and J.N. Pitts, *J. Phys. Chem.* **79**, 295 (1975)). In the reaction with benzene the rate constants are 1.44×10^7 dm^3 mol^{-1} s^{-1} at 300.3 K, 3.03×10^7 dm^3 mol^{-1} s^{-1} at 341.2 K, and 6.9×10^7 dm^3 mol^{-1} s^{-1} at 392.2 K. Find the pre-exponential factor and activation energy of the reaction.

22.17 In Problem 22.10 the isomerization of cyclopropane over a limited pressure range was examined. If the Lindemann mechanism of first-order reactions is to be tested we also need data at low pressures. These have been obtained (H.O. Pritchard, R.G. Sowden, and A.F. Trotman-Dickenson, *Proc. R. Soc.* **A217**, 563 (1953)):

p/Torr	84.1	11.0	2.89	0.569	0.120	0.067
$10^4\,k_{eff}$/s^{-1}	2.98	2.23	1.54	0.857	0.392	0.303

Test the Lindemann theory with these data.

22.18‡ P.W. Seakins, M.J. Pilling, L.T. Niiranen, D. Gutman, and L.N. Krasnoperov (*J. Phys. Chem.* **96**, 9847 (1992)) measured the forward and reverse rate constants for the gas-phase reaction $C_2H_5(g) + HBr(g) \rightarrow C_2H_6(g) + Br(g)$ and used their findings to compute thermodynamic parameters for C_2H_5. The reaction is bimolecular in both directions with Arrhenius parameters $A = 1.0 \times 10^9$ dm^3 mol^{-1} s^{-1}, $E_a = -4.2$ kJ mol^{-1} for the forward reaction and $k' = 1.4 \times 10^{11}$ dm^3 mol^{-1} s^{-1}, $E_a = 53.3$ kJ mol^{-1} for the reverse reaction. Compute $\Delta_fH^{\ominus}$, $S_m^{\ominus}$, and $\Delta_fG^{\ominus}$ of C_2H_5 at 298 K.

22.19 Two products are formed in reactions in which there is kinetic control of the ratio of products. The activation energy for the reaction leading to Product 1 is greater than that leading to Product 2. Will the ratio of product concentrations $[P_1]/[P_2]$ increase or decrease if the temperature is raised?

Theoretical problems

22.20 The reaction mechanism

$A_2 \rightleftharpoons A + A$ $\quad$ (fast)

$A + B \rightarrow P$ $\quad$ (slow)

involves an intermediate A. Deduce the rate law for the reaction.

22.21 The equilibrium $\mathrm{A} \rightleftharpoons \mathrm{B}$ is first-order in both directions. Derive an expression for the concentration of A as a function of time when the initial molar concentrations of A and B are $[\mathrm{A}]_0$ and $[\mathrm{B}]_0$. What is the final composition of the system?

22.22 Derive an integrated expression for a second-order rate law $v = k[\mathrm{A}][\mathrm{B}]$ for a reaction of stoichiometry $2\,\mathrm{A} + 3\,\mathrm{B} \rightarrow \mathrm{P}$.

22.23 Derive the integrated form of a third-order rate law $v = k[\mathrm{A}]^2[\mathrm{B}]$ in which the stoichiometry is $2\,\mathrm{A} + \mathrm{B} \rightarrow \mathrm{P}$ and the reactants are initially present in (a) their stoichiometric proportions, (b) with B present initially in twice the amount.

22.24 Set up the rate equations for the reaction mechanism:

$$\mathrm{A} \overset{k_a,\, k_a'}{\rightleftharpoons} \mathrm{B} \overset{k_b,\, k_b'}{\rightleftharpoons} \mathrm{C}$$

Show that the mechanism is equivalent to

$$\mathrm{A} \overset{k_{\mathrm{eff}},\, k_{\mathrm{eff}}'}{\rightleftharpoons} \mathrm{C}$$

under specified circumstances.

22.25 Show that the ratio $t_{1/2}/t_{3/4}$, where $t_{1/2}$ is the half-life and $t_{3/4}$ is the time for the concentration of A to decrease to $\frac{3}{4}$ of its initial value (implying that $t_{3/4} < t_{1/2}$) can be written as a function of n alone, and can therefore be used as a rapid assessment of the order of a reaction.

22.26 Derive an equation for the steady-state rate of the sequence of reactions $\mathrm{A} \rightleftharpoons \mathrm{B} \rightleftharpoons \mathrm{C} \rightleftharpoons \mathrm{D}$, with [A] maintained at a fixed value and the product D removed as soon as it is formed.

22.27‡ For a certain second-order reaction $\mathrm{A} + \mathrm{B} \rightarrow$ Products, the rate of reaction, v, may be written

$$v = \frac{\mathrm{d}x}{\mathrm{d}t} = k([\mathrm{A}]_0 - x)([\mathrm{B}]_0 + x)$$

where x is the decrease in concentration of A or B as a result of reaction. Find an expression for the maximum rate and the conditions under which it applies. Draw a graph of v against x, and noting that v and x cannot be negative, identify the portion of the curve that corresponds to reality.

22.28 Consider the dimerization $\mathrm{A} \Leftrightarrow \mathrm{A}_2$ with forward rate constant k_a and backward rate constant k_b. Show that the relaxation time is:

$$\tau = \frac{1}{k_b + 4k_a[\mathrm{A}]_{\mathrm{eq}}}$$

Applications: to archaeology, biochemistry, and environmental science

22.29 The half-life for the (first-order) radioactive decay of $^{14}\mathrm{C}$ is 5730 y (it emits β rays with an energy of 0.16 MeV). An archaeological sample contained wood that had only 72 per cent of the $^{14}\mathrm{C}$ found in living trees. What is its age?

22.30 One of the hazards of nuclear explosions is the generation of $^{90}\mathrm{Sr}$ and its subsequent incorporation in place of calcium in bones. This nuclide emits β rays of energy 0.55 MeV, and has a half-life of 28.1 y. Suppose 1.00 μg was absorbed by a newly born child. How much will remain after (a) 18 y, (b) 70 y if none is lost metabolically?

22.31 Pharmacokinetics is the study of the rates of absorption and elimination of drugs by organisms. In most cases, elimination is slower than absorption and is a more important determinant of availability of a drug for binding to its target. A drug can be eliminated by many mechanisms, such as metabolism in the liver, intestine, or kidney followed by excretion of breakdown products through urine or faeces. As an example of pharmacokinetic analysis, consider the elimination of beta adrenergic blocking agents (beta blockers), drugs used in the treatment of hypertension. After intravenous administration of a beta blocker, the blood plasma of a patient was analysed for remaining drug and the data are shown below, where c is the drug concentration measured at a time t after the injection.

t/min	30	60	120	150	240	360	480
c/(ng cm^{-3})	699	622	413	292	152	60	24

(a) Is removal of the drug a first- or second-order process? (b) Calculate the rate constant and half-life of the process. *Comment.* An essential aspect of drug development is the optimization of the half-life of elimination, which needs to be long enough to allow the drug to find and act on its target organ but not so long that harmful side-effects become important.

22.32 The absorption and elimination of a drug in the body may be modelled with a mechanism consisting of two consecutive reactions:

A	→	B	→	C
drug at site of administration		drug dispersed in blood		eliminated drug

where the rate constants of absorption ($\mathrm{A} \rightarrow \mathrm{B}$) and elimination are, respectively, k_1 and k_2. (a) Consider a case in which absorption is so fast that it may be regarded as instantaneous so that a dose of A at an initial concentration $[\mathrm{A}]_0$ immediately leads to a drug concentration in blood of $[\mathrm{B}]_0$. Also, assume that elimination follows first-order kinetics. Show that, after the administration of n equal doses separated by a time interval τ, the peak concentration of drug B in the blood, $[\mathrm{P}]_n$, rises beyond the value of $[\mathrm{B}]_0$ and eventually reaches a constant, maximum peak value given by

$$[\mathrm{P}]_\infty = [\mathrm{B}]_0(1 - \mathrm{e}^{-k_2\tau})^{-1}$$

where $[\mathrm{P}]_n$ is the (peak) concentration of B immediately after administration of the nth dose and $[\mathrm{P}]_\infty$ is the value at very large n. Also, write a mathematical expression for the residual concentration of B, $[\mathrm{R}]_n$, which we define to be the concentration of drug B immediately before the administration of the $(n+1)$th dose. $[\mathrm{R}]_n$ is always smaller than $[\mathrm{P}]_n$ on account of drug elimination during the period τ between drug administrations. Show that $[\mathrm{P}]_\infty - [\mathrm{R}]_\infty = [\mathrm{B}]_0$. (b) Consider a drug for which $k_2 = 0.0289\ \mathrm{h}^{-1}$. (i) Calculate the value of τ required to achieve $[\mathrm{P}]_\infty/[\mathrm{B}]_0 = 10$. Prepare a graph that plots both $[\mathrm{P}]_n/[\mathrm{B}]_0$ and $[\mathrm{R}]_n/[\mathrm{B}]_0$ against n. (ii) How many doses must be administered to achieve a $[\mathrm{P}]_n$ value that is 75 per cent of the maximum value? What time has passed during the administration of these doses? (iii) What actions can be taken to reduce the variation $[\mathrm{P}]_\infty - [\mathrm{R}]_\infty$ while maintaining the same value of $[\mathrm{P}]_\infty$? (c) Now consider the administration of a single dose $[\mathrm{A}]_0$ for which absorption follows first-order kinetics and elimination follows zero-order kinetics. Show that with the initial concentration $[\mathrm{B}]_0 = 0$, the concentration of drug in the blood is given by

$$[\mathrm{B}] = [\mathrm{A}]_0(1 - \mathrm{e}^{-k_1t}) - k_2t$$

Plot $[\mathrm{B}]/[\mathrm{A}]_0$ against t for the case $k_1 = 10\ \mathrm{h}^{-1}$, $k_2 = 4.0 \times 10^{-3}\ \mathrm{mmol\ dm^{-3}\ h^{-1}}$, and $[\mathrm{A}]_0 = 0.1\ \mathrm{mmol\ dm^{-3}}$. Comment on the shape of the curve. (d) Using the model from part (c), set $\mathrm{d}[\mathrm{B}]/\mathrm{d}t = 0$ and show that the maximum value of [B] occurs at the time $t_{\max} = \dfrac{1}{k_1}\ln\left(\dfrac{k_1[\mathrm{A}]_0}{k_2}\right)$. Also, show that the maximum concentration of drug in blood is given by $[\mathrm{B}]_{\max} = [\mathrm{A}]_0 - k_2/k_1 - k_2t_{\max}$.

22.33 Consider a mechanism for the helix–coil transition in which nucleation occurs in the middle of the chain:

$$hhhh\ldots \rightleftharpoons hchh\ldots$$

$$hchh\ldots \rightleftharpoons cccc\ldots$$

We saw in *Impact* I22.1 that this type of nucleation is relatively slow, so neither step may be rate-determining. (a) Set up the rate equations for this mechanism. (b) Apply the steady-state approximation and show that, under these circumstances, the mechanism is equivalent to $hhhh\ldots \Leftrightarrow cccc\ldots$.

(c) Use your knowledge of experimental techniques and your results from parts (a) and (b) to support or refute the following statement: It is very difficult to obtain experimental evidence for intermediates in protein folding by performing simple rate measurements and one must resort to special flow, relaxation, or trapping techniques to detect intermediates directly.

22.34 Propose a set of experiments in which analysis of the line-shapes of NMR transitions (Section 15.7) can be used to monitor fast events in protein folding and unfolding. What are the disadvantages and disadvantages of this NMR method over methods that use electronic and vibrational spectroscopy?

22.35 Consider the following mechanism for renaturation of a double helix from its strands A and B:

A + B ⇌ unstable helix (fast)

Unstable helix → stable double helix (slow)

Derive the rate equation for the formation of the double helix and express the rate constant of the renaturation reaction in terms of the rate constants of the individual steps.

22.36‡ *Prebiotic reactions* are reactions that might have occurred under the conditions prevalent on the Earth before the first living creatures emerged and which can lead to analogues of molecules necessary for life as we now know it. To qualify, a reaction must proceed with favourable rates and equilibria. M.P. Robertson and S.I. Miller (*Science* **268**, 702(1995)) have studied the prebiotic synthesis of 5-substituted uracils, among them 5-hydroxymethyluracil (HMU). Amino acid analogues can be formed from HMU under prebiotic conditions by reaction with various nucleophiles, such as H_2S, HCN, indole, imidazole, etc. For the synthesis of HMU (the uracil analogue of serine) from uracil and formaldehyde (HCHO), the rate of addition is given by $\log \{k/(\mathrm{dm^3\,mol^{-1}\,s^{-1}})\} = 11.75 - 5488/(T/\mathrm{K})$ (at pH = 7), and $\log K = -1.36 + 1794/(T/\mathrm{K})$. For this reaction, calculate the rates and equilibrium constants over a range of temperatures corresponding to possible prebiotic conditions, such as 0–50°C, and plot them against temperature. Also, calculate the activation energy and the standard reaction Gibbs energy and enthalpy at 25°C. Prebiotic conditions are not likely to be standard conditions. Speculate about how the actual values of the reaction Gibbs energy and enthalpy might differ from the standard values. Do you expect that the reaction would still be favourable?

22.37‡ Methane is a by-product of a number of natural processes (such as digestion of cellulose in ruminant animals, anaerobic decomposition of organic waste matter) and industrial processes (such as food production and fossil fuel use). Reaction with the hydroxyl radical OH is the main path by which CH_4 is removed from the lower atmosphere. T. Gierczak, R.K. Talukdar, S.C. Herndon, G.L. Vaghjiani, and A.R. Ravishankara (*J. Phys. Chem.* **A 101**, 3125 (1997)) measured the rate constants for the elementary bimolecular gas-phase reaction of methane with the hydroxyl radical over a range of temperatures of importance to atmospheric chemistry. Deduce the Arrhenius parameters A and E_a from the following measurements:

T/K	295	223	218	213	206	200	195
$k/(10^6\,\mathrm{dm^3\,mol^{-1}\,s^{-1}})$	3.55	0.494	0.452	0.379	0.295	0.241	0.217

22.38‡ As we saw in Problem 22.37, reaction with the hydroxyl radical OH is the main path by which CH_4, a by-product of many natural and industrial processes, is removed from the lower atmosphere. T. Gierczak, R.K. Talukdar, S.C. Herndon, G.L. Vaghjiani, and A.R. Ravishankara (*J. Phys. Chem.* **A 101**, 3125 (1997)) measured the rate constants for the bimolecular gas-phase reaction $CH_4(g) + OH(g) \rightarrow CH_3(g) + H_2O(g)$ and found $A = 1.13 \times 10^9\ \mathrm{dm^3\,mol^{-1}\,s^{-1}}$ and $E_a = 14.1\ \mathrm{kJ\,mol^{-1}}$ for the Arrhenius parameters. (a) Estimate the rate of consumption of CH_4. Take the average OH concentration to be $1.5 \times 10^{-21}\ \mathrm{mol\,dm^{-3}}$, that of CH_4 to be $4.0 \times 10^{-8}\ \mathrm{mol\,dm^{-3}}$, and the temperature to be −10°C. (b) Estimate the global annual mass of CH_4 consumed by this reaction (which is slightly less than the amount introduced to the atmosphere) given an effective volume for the Earth's lower atmosphere of $4 \times 10^{21}\ \mathrm{dm^3}$.

22.39‡ T. Gierczak, R.K. Talukdar, S.C. Herndon, G.L. Vaghjiani, and A.R. Ravishankara (*J. Phys. Chem.* **A 101**, 3125 (1997)) measured the rate constants for the bimolecular gas-phase reaction of methane with the hydroxyl radical in several isotopic variations. From their data, the following Arrhenius parameters can be obtained:

	$A/(\mathrm{dm^3\,mol^{-1}\,s^{-1}})$	$E_a/(\mathrm{kJ\,mol^{-1}})$
$CH_4 + OH \rightarrow CH_3 + H_2O$	1.13×10^9	14.1
$CD_4 + OH \rightarrow CD_3 + DOH$	6.0×10^8	17.5
$CH_4 + OD \rightarrow CH_3 + DOH$	1.01×10^9	13.6

Compute the rate constants at 298 K, and interpret the kinetic isotope effects.

22.40‡ The oxidation of HSO_3^- by O_2 in aqueous solution is a reaction of importance to the processes of acid rain formation and flue gas desulfurization. R.E. Connick, Y.-X. Zhang, S. Lee, R. Adamic, and P. Chieng (*Inorg. Chem.* **34**, 4543 (1995)) report that the reaction $2\,HSO_3^- + O_2 \rightarrow 2\,SO_4^{2-} + 2\,H^+$ follows the rate law $v = k[HSO_3^-]^2[H^+]^2$. Given a pH of 5.6 and an oxygen molar concentration of $2.4 \times 10^{-4}\ \mathrm{mol\,dm^{-3}}$ (both presumed constant), an initial HSO_3^- molar concentration of $5 \times 10^{-5}\ \mathrm{mol\,dm^{-3}}$, and a rate constant of $3.6 \times 10^6\ \mathrm{dm^9\,mol^{-3}\,s^{-1}}$, what is the initial rate of reaction? How long would it take for HSO_3^- to reach half its initial concentration?

23 The kinetics of complex reactions

This chapter extends the material introduced in Chapter 22 by showing how to deal with complex reaction mechanisms. First, we consider chain reactions and see that either complicated or simple rate laws can be obtained, depending on the conditions. Under certain circumstances, a chain reaction can become explosive, and we see some of the reasons for this behaviour. An important application of these more complicated techniques is to the kinetics of polymerization reactions. There are two major classes of polymerization process and the average molar mass of the product varies with time in distinctive ways. Second, we describe homogeneous catalysis and apply the associated concepts to enzyme-catalysed reactions. Finally, we describe the principles of photochemistry and apply them to problems in environmental science, biochemistry, and medicine.

Many reactions take place by mechanisms that involve several elementary steps. Some take place at a useful rate only after absorption of light or if a catalyst is present. In this chapter we see how to develop the ideas introduced in Chapter 22 to deal with these special kinds of reactions.

Chain reactions

Many gas-phase reactions and liquid-phase polymerization reactions are **chain reactions**. In a chain reaction, a reaction intermediate produced in one step generates an intermediate in a subsequent step, then that intermediate generates another intermediate, and so on. The intermediates in a chain reaction are called **chain carriers**. In a **radical chain reaction** the chain carriers are radicals (species with unpaired electrons). Ions may also act as chain carriers. In nuclear fission the chain carriers are neutrons.

23.1 The rate laws of chain reactions

A chain reaction can have a simple rate law. As a first example, consider the **pyrolysis**, or thermal decomposition in the absence of air, of acetaldehyde (ethanal, CH_3CHO), which is found to be three-halves order in CH_3CHO:

$$CH_3CHO(g) \rightarrow CH_4(g) + CO(g) \qquad v = k[CH_3CHO]^{3/2} \tag{23.1}$$

Some ethane is also detected. The **Rice–Herzfeld mechanism** for this reaction is as follows (the dot signifies an unpaired electron and marks a radical):

Initiation:	$CH_3CHO \rightarrow \cdot CH_3 + \cdot CHO$	$v = k_i[CH_3CHO]$
Propagation:	$CH_3CHO + \cdot CH_3 \rightarrow CH_3CO\cdot + CH_4$	$v = k_p[CH_3CHO][\cdot CH_3]$
Propagation:	$CH_3CO\cdot \rightarrow \cdot CH_3 + CO$	$v = k_p'[CH_3CO\cdot]$
Termination:	$\cdot CH_3 + \cdot CH_3 \rightarrow CH_3CH_3$	$v = k_t[\cdot CH_3]^2$

The chain carriers $\cdot CH_3$ and $\cdot CHO$ are formed initially in the **initiation step**. To simplify the treatment, we shall ignore the subsequent reactions of $\cdot CHO$, except to note that they give rise to the formation of CO and of the by-product H_2. The chain carrier $\cdot CH_3$ attacks other reactant molecules in the **propagation steps**, and each attack gives rise to a new carrier. Radicals combine and end the chain in the **termination step**.

To test the proposed mechanism we need to show that it leads to the observed rate law. According to the steady-state approximation (Section 22.7b), the net rate of change of the intermediates ($\cdot CH_3$ and $CH_3CO\cdot$) may be set equal to zero:

$$\frac{d[\cdot CH_3]}{dt} = k_i[CH_3CHO] - k_p[\cdot CH_3][CH_3CHO] + k_p'[CH_3CO\cdot] - 2k_t[\cdot CH_3]^2 = 0$$

$$\frac{d[CH_3CO\cdot]}{dt} = k_p[\cdot CH_3][CH_3CHO] - k_p'[CH_3CO\cdot] = 0$$

The sum of the two equations is

$$k_i[CH_3CHO] - 2k_t[\cdot CH_3]^2 = 0$$

which shows that the steady-state approximation also implies that the rate of chain initiation is equal to the rate of chain termination. The steady-state concentration of $\cdot CH_3$ radicals is

$$[\cdot CH_3] = \left(\frac{k_i}{2k_t}\right)^{1/2}[CH_3CHO]^{1/2} \qquad (23.2)$$

It follows that the rate of formation of CH_4 is

$$\frac{d[CH_4]}{dt} = k_p[\cdot CH_3][CH_3CHO] = k_p\left(\frac{k_i}{2k_t}\right)^{1/2}[CH_3CHO]^{3/2} \qquad (23.3)$$

which is in agreement with the three-halves order observed experimentally (eqn 23.1). However, this mechanism does not accommodate the formation of various known reaction by-products, such as propanone (CH_3COCH_3) and propanal (CH_3CH_2CHO).

In many cases, a chain reaction leads to a complicated rate law. An example is the **hydrogen–bromine reaction**:

$$H_2(g) + Br_2(g) \rightarrow 2\,HBr(g) \qquad \frac{d[HBr]}{dt} = \frac{k[H_2][Br_2]^{3/2}}{[Br_2] + k'[HBr]} \qquad (23.4)$$

The following mechanism has been proposed to account for this rate law (Fig. 23.1):

Initiation: $Br_2 + M \rightarrow Br\cdot + Br\cdot + M \qquad v = k_i[Br_2][M]$

where M is either Br_2 or H_2. This step is an example of a **thermolysis**, a reaction initiated by heat, which stimulates vigorous intermolecular collisions.

Propagation:	$Br\cdot + H_2 \rightarrow HBr + H\cdot$	$v = k_p[Br\cdot][H_2]$
	$H\cdot + Br_2 \rightarrow HBr + Br\cdot$	$v = k_p'[H\cdot][Br_2]$
Retardation:	$H\cdot + HBr \rightarrow H_2 + Br\cdot$	$v = k_r[H\cdot][HBr]$
Termination:	$Br\cdot + Br\cdot + M \rightarrow Br_2 + M^*$	$v = k_t[Br\cdot]^2[M]$

A **retardation step** reduces the net rate of formation of product. In this case, the chain carrier $H\cdot$ attacks a molecule of HBr, the product. In the termination step, the third body M removes the energy of recombination. Other possible termination steps include the recombination of H atoms to form H_2 and combination of H and Br atoms. However, it turns out that only Br atom recombination is important because Br atoms propagate the chain more slowly and thus live longer than H atoms. The net rate of formation of the product HBr is

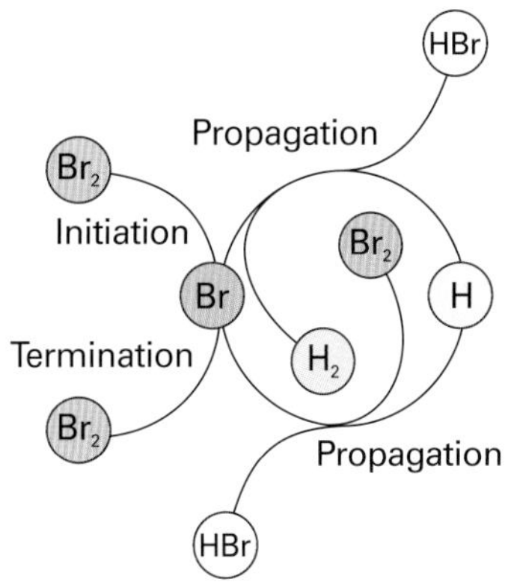

Fig. 23.1 A schematic representation of the mechanism of the reaction between hydrogen and bromine. Note how the reactants and products are shown as arms to the circle, but the intermediates (H and Br) occur only on the circle. Similar diagrams are used to depict the action of catalysts.

$$\frac{d[HBr]}{dt} = k_p[Br\cdot][H_2] + k_p'[H\cdot][Br_2] - k_r[H\cdot][HBr]$$

We can now either analyse the rate equations numerically (see *Appendix* 2 and *Further reading*) or look for approximate solutions and see if they agree with the empirical rate law. The following example illustrates the latter approach.

Example 23.1 *Deriving the rate equation of a chain reaction*

Derive the rate law for the formation of HBr according to the mechanism given above.

Method Make the steady-state approximation for the concentrations of any intermediates (H· and Br· in the present case) by setting the net rates of change of their concentrations equal to zero. Solve the resulting equations for the concentrations of the intermediates, and then use the resulting expressions in the equation for the net rate of formation of HBr.

Answer The net rates of formation of the two intermediates are

$$\frac{d[H\cdot]}{dt} = k_p[Br\cdot][H_2] - k_p'[H\cdot][Br_2] - k_r[H\cdot][HBr] = 0$$

$$\frac{d[Br\cdot]}{dt} = 2k_i[Br_2][M] - k_p[Br\cdot][H_2] + k_p'[H\cdot][Br_2] + k_r[H\cdot][HBr] - 2k_t[Br\cdot]^2[M] = 0$$

The steady-state concentrations of the intermediates are obtained by solving these two simultaneous equations and are

$$[Br\cdot] = \left(\frac{k_i}{k_t}\right)^{1/2}[Br_2]^{1/2} \qquad [H\cdot] = \frac{k_p(k_i/k_t)^{1/2}[H_2][Br_2]^{1/2}}{k_p'[Br_2] + k_r[HBr]}$$

Note that [M] has cancelled. When we substitute these concentrations into the expression for $d[HBr]/dt$, we obtain

$$\frac{d[HBr]}{dt} = \frac{2k_p(k_i/k_t)^{1/2}[H_2][Br_2]^{3/2}}{[Br_2] + (k_r/k_p')[HBr]}$$

This equation has the same form as the empirical rate law (eqn 23.4), so the two empirical rate constants can be identified as

$$k = 2k_p\left(\frac{k_i}{k_t}\right)^{1/2} \qquad k' = \frac{k_r}{k_p'}$$

The rate law shows that the reaction slows down as HBr forms, or as the $[HBr]/[Br_2]$ ratio increases. This effect occurs because Br_2 molecules compete with HBr molecules for H· atoms, with the propagation step $H\cdot + Br_2 \rightarrow HBr + Br\cdot$ yielding product (HBr) and the retardation step $H\cdot + HBr \rightarrow H_2 + Br\cdot$ converting HBr back into reactant (H_2). Numerical integration of the rate law with mathematical software shows the predicted time dependence of the concentration of HBr for this mechanism (Fig. 23.2).

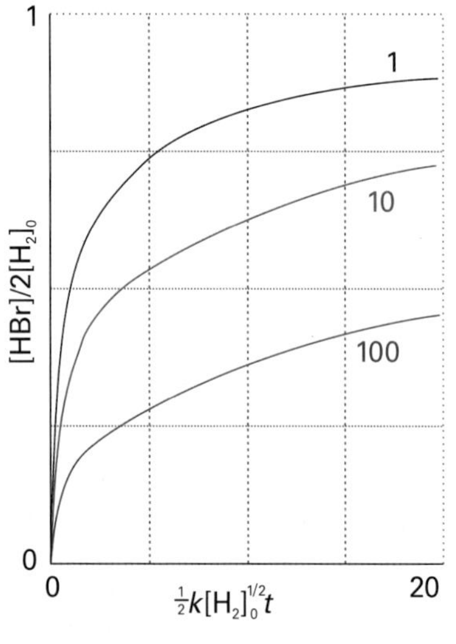

Fig. 23.2 The numerical integration of the HBr rate law, Example 23.1, can be used to explore how the concentration of HBr changes with time. These runs began with stoichiometric proportions of H_2 and Br_2; the curves are labelled with the value of $2k' - 1$.

Exploration Use mathematical software or the interactive applets found in the *Living graphs* section of the text's web site to plot the concentrations of the radicals H· and Br· against time. Find a combination of rate constants that results in steady states for these intermediates.

Self-test 23.1 Deduce the rate law for the production of HBr when the initiation step is the photolysis, or light-induced decomposition, of Br_2 into two bromine atoms, Br·. Let the photolysis rate be $v = I_{abs}$, where I_{abs} is the intensity of absorbed radiation. [See eqn 23.39 below]

23.2 Explosions

A **thermal explosion** is a very rapid reaction arising from a rapid increase of reaction rate with increasing temperature. The temperature of the system rises if the energy released by an exothermic reaction cannot escape, and the reaction goes faster. The acceleration of the rate results in an even faster rise of temperature, so the reaction becomes catastrophically fast. A **chain-branching explosion** occurs when the number of chain centres grows exponentially.

An example of both types of explosion is the reaction between hydrogen and oxygen:

$$2\,H_2(g) + O_2(g) \rightarrow 2\,H_2O(g)$$

Although the net reaction is very simple, the mechanism is very complex and has not yet been fully elucidated. A chain reaction is involved, and the chain carriers include H·, ·O·, and ·OH. Some steps involving H· are:

Initiation:	$H_2 \rightarrow H\cdot + H\cdot$	$v = \text{constant } (v_{\text{init}})$
Propagation:	$H_2 + \cdot OH \rightarrow \cdot H + H_2O$	$v = k_p[H_2][\cdot OH]$
Branching:	$\cdot O_2\cdot + \cdot H \rightarrow \cdot O\cdot + \cdot OH$	$v = k_b[\cdot O_2\cdot][H\cdot]$
	$\cdot O\cdot + H_2 \rightarrow \cdot OH + \cdot H$	$v = k_b'[\cdot O\cdot][H_2]$
Termination:	$H\cdot + \text{wall} \rightarrow \frac{1}{2} H_2$	$v = k_t[H\cdot]$
	$H\cdot + O_2 + M \rightarrow HO_2\cdot + M^*$	$v = k_t'[H\cdot][O_2][M]$

A **branching step** is an elementary reaction that produces more than one chain carrier. Recall that an O atom, with the ground-state configuration $[\text{He}]2s^22p^4$, has two unpaired electrons. The same is true of an O_2 molecule, with 12 valence electrons and a ground-state configuration $1\sigma_g^2 1\sigma_u^2 2\sigma_g^2 1\pi_u^4 1\pi_g^2$.

The occurrence of an explosion depends on the temperature and pressure of the system, and the **explosion regions** for the reaction, the conditions under which explosion occurs, are shown in Fig. 23.3. At very low pressures the system is outside the explosion region and the mixture reacts smoothly. At these pressures the chain carriers produced in the branching steps can reach the walls of the container where they combine. Increasing the pressure along a vertical line in the illustration takes the system through the **first explosion limit** (provided the temperature is greater than about 730 K). The chain carriers react before reaching the walls and the branching reactions are explosively efficient. The reaction is smooth when the pressure is above the **second explosion limit**. The concentration of third-body M molecules is then so high compared to the concentrations of chain carriers that the combination of H· atoms with O_2 molecules to form relatively unreactive $HO_2\cdot$ molecules becomes faster than the branching reaction between H· atoms and O_2 molecules. These long-lived $HO_2\cdot$ molecules then diffuse to the walls and are removed there, in what amounts to another termination step. When the pressure is increased to above the **third explosion limit**, diffusion of $HO_2\cdot$ molecules to the walls becomes so slow that they can react with H_2 molecules (now at very high concentrations) to regenerate H atoms and H_2O_2 molecules.

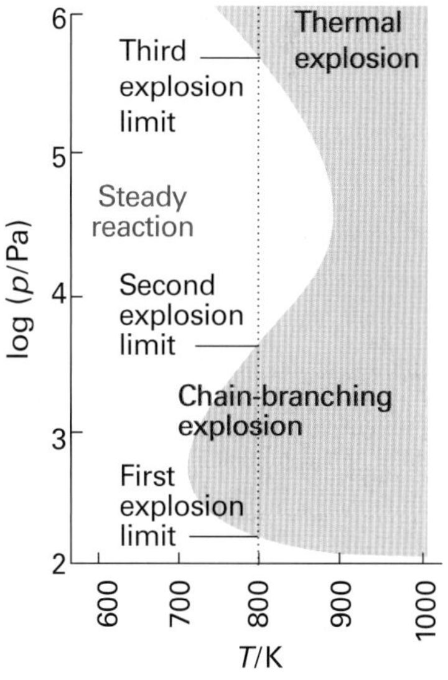

Fig. 23.3 The explosion limits of the $H_2 + O_2$ reaction. In the explosive regions the reaction proceeds explosively when heated homogeneously.

Example 23.2 *Examining the explosion behaviour of a chain reaction*

For the reaction of hydrogen and oxygen described above, show that an explosion occurs when the rate of chain branching exceeds that of chain termination.

Method Identify the onset of explosion with the rapid increase in the concentration of radicals, and for simplicity identify that concentration with the concentration of

H· atoms, which probably outnumber the highly reactive ·OH and ·O· radicals. Set up the corresponding rate laws for the reaction intermediates and then apply the steady-state approximation.

Answer The rate of formation of radicals, v_{rad}, is identified with d[H·]/dt; therefore we write

$$v_{rad} = v_{init} + k_p[\cdot OH][H_2] - k_b[H\cdot][O_2] + k_b'[\cdot O\cdot][H_2] - k_t[H\cdot] - k_t'[H\cdot][O_2][M]$$

Applying the steady-state approximation to ·OH and ·O· gives

$$\frac{d[\cdot OH]}{dt} = -k_p[\cdot OH][H_2] + k_b[H\cdot][O_2] + k_b'[\cdot O\cdot][H_2] = 0$$

$$\frac{d[\cdot O\cdot]}{dt} = k_b[H\cdot][O_2] - k_b'[\cdot O\cdot][H_2] = 0$$

The solutions of these two algebraic equations are

$$[\cdot O\cdot] = \frac{k_b[H\cdot][O_2]}{k_b'[H_2]} \qquad [\cdot OH] = \frac{2k_b[H\cdot][O_2]}{k_p[H_2]}$$

The rate of formation of radicals is therefore

$$v_{rad} = v_{init} + (2k_b[O_2] - k_t - k_t'[O_2][M])[H\cdot]$$

We write $k_{branch} = 2k_b[O_2]$, a measure of the rate of the more important chain-branching step, and $k_{term} = k_t + k_t'[O_2][M]$, a measure of the rate of chain termination. Then,

$$\frac{d[H\cdot]}{dt} = v_{init} + (k_{branch} - k_{term})[H\cdot]$$

There are two solutions. At low O_2 concentrations, termination dominates branching, so $k_{term} > k_{branch}$. Then,

$$[H\cdot] = \frac{v_{init}}{k_{term} - k_{branch}}(1 - e^{-(k_{term}-k_{branch})t})$$

As can be seen from Fig. 23.4a, in this regime there is steady combustion of hydrogen. At high O_2 concentrations, branching dominates termination, or $k_{branch} > k_{term}$. Then,

$$[H\cdot] = \frac{v_{init}}{k_{branch} - k_{term}}(e^{(k_{branch}-k_{term})t} - 1)$$

There is now an explosive increase in the concentration of radicals (Fig. 23.4b).

Although the steady-state approximation does not hold under explosive conditions, the calculation at least gives an indication of the basis for the transition from smooth combustion to explosion.

Self-test 23.2 Calculate the variation in radical composition when the rates of branching and termination are equal. $[[H\cdot] = v_{init}t]$

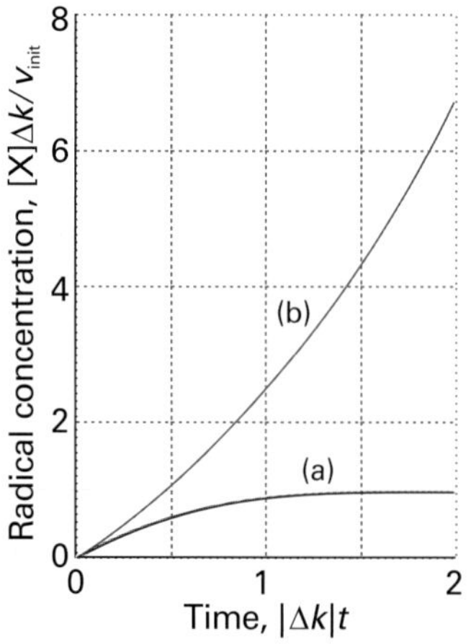

Fig. 23.4 The concentration of radicals in the fuel-rich regime of the hydrogen–oxygen reaction (a) under steady combustion conditions, (b) in the explosive region. For this graph, $\Delta k = k_{branch} - k_{term}$.

Exploration Using mathematical software, an electronic spreadsheet, or the interactive applets found in the *Living graphs* section of the text's web site, explore the effect of changing the parameter $\Delta k = k_{branch} - k_{term}$ on the shapes of the curves in Figs. 23.4a and 23.4b.

Not all explosions are due to chain reactions. Solid-state explosions, such as the explosion of ammonium nitrate or TNT (2,4,6-trinitrotoluene), for instance, are simply decompositions that occur very rapidly with the production of large amounts of gas phase molecules.

Polymerization kinetics

In **stepwise polymerization** any two monomers present in the reaction mixture can link together at any time and growth of the polymer is not confined to chains that are already forming (Fig. 23.5). As a result, monomers are removed early in the reaction and, as we shall see, the average molar mass of the product grows with time. In **chain polymerization** an activated monomer, M, attacks another monomer, links to it, then that unit attacks another monomer, and so on. The monomer is used up as it becomes linked to the growing chains (Fig. 23.6). High polymers are formed rapidly and only the yield, not the average molar mass, of the polymer is increased by allowing long reaction times.

23.3 Stepwise polymerization

Stepwise polymerization commonly proceeds by a condensation reaction, in which a small molecule (typically H_2O) is eliminated in each step. Stepwise polymerization is the mechanism of production of polyamides, as in the formation of nylon-66:

$$\begin{aligned}&H_2N(CH_2)_6NH_2 + HOOC(CH_2)_4COOH\\&\quad \rightarrow H_2N(CH_2)_6NHCO(CH_2)_4COOH + H_2O\\&\quad \rightarrow H{-}[NH(CH_2)_6NHCO(CH_2)_4CO]_n{-}OH\end{aligned}$$

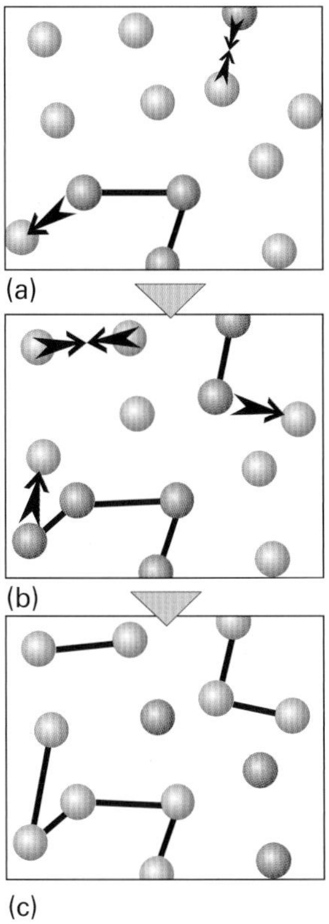

Fig. 23.5 In stepwise polymerization, growth can start at any pair of monomers, and so new chains begin to form throughout the reaction.

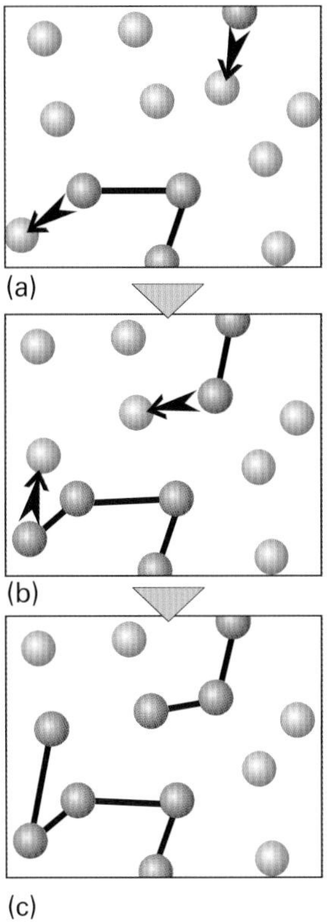

Fig. 23.6 The process of chain polymerization. Chains grow as each chain acquires additional monomers.

Polyesters and polyurethanes are formed similarly (the latter without elimination). A polyester, for example, can be regarded as the outcome of the stepwise condensation of a hydroxyacid HO—M—COOH. We shall consider the formation of a polyester from such a monomer, and measure its progress in terms of the concentration of the —COOH groups in the sample (which we denote A), for these groups gradually disappear as the condensation proceeds. Because the condensation reaction can occur between molecules containing any number of monomer units, chains of many different lengths can grow in the reaction mixture.

In the absence of a catalyst, we can expect the condensation to be overall second-order in the concentration of the —OH and —COOH (or A) groups, and write

$$\frac{d[A]}{dt} = -k[OH][A] \qquad (23.5a)$$

However, because there is one —OH group for each —COOH group, this equation is the same as

$$\frac{d[A]}{dt} = -k[A]^2 \qquad (23.5b)$$

If we assume that the rate constant for the condensation is independent of the chain length, then k remains constant throughout the reaction. The solution of this rate law is given by eqn 22.15, and is

$$[A] = \frac{[A]_0}{1 + kt[A]_0} \qquad (23.6)$$

The fraction, p, of —COOH groups that have condensed at time t is, after application of eqn 23.6:

$$p = \frac{[A]_0 - [A]}{[A]_0} = \frac{kt[A]_0}{1 + kt[A]_0} \qquad (23.7)$$

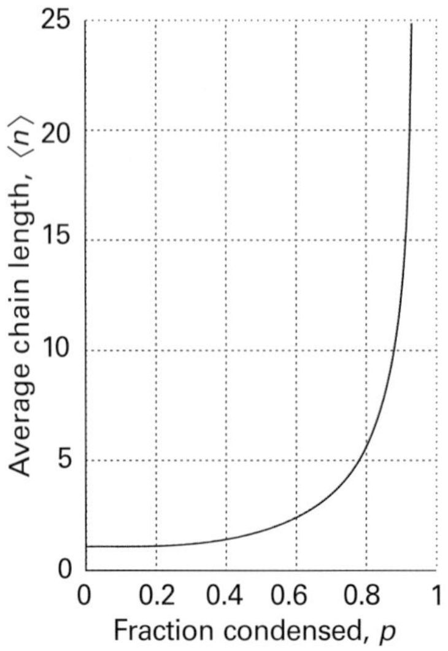

Fig. 23.7 The average chain length of a polymer as a function of the fraction of reacted monomers, p. Note that p must be very close to 1 for the chains to be long.

Exploration Plot the variation of p with time for a range of k values of your choosing (take $[A]_0 = 1.0\ \text{mol dm}^{-3}$).

Next, we calculate the **degree of polymerization**, which is defined as the average number of monomer residues per polymer molecule. This quantity is the ratio of the initial concentration of A, $[A]_0$, to the concentration of end groups, [A], at the time of interest, because there is one —A group per polymer molecule. For example, if there were initially 1000 A groups and there are now only 10, each polymer must be 100 units long on average. Because we can express [A] in terms of p (eqn 23.7), the average number of monomers per polymer molecule, $<n>$, is

$$<n> = \frac{[A]_0}{[A]} = \frac{1}{1-p} \qquad (23.8a)$$

This result is illustrated in Fig. 23.7. When we express p in terms of the rate constant k (eqn 23.7), we find

$$<n> = 1 + kt[A]_0 \qquad (23.8b)$$

The average length grows linearly with time. Therefore, the longer a stepwise polymerization proceeds, the higher the average molar mass of the product.

23.4 Chain polymerization

Chain polymerization occurs by addition of monomers to a growing polymer, often by a radical chain process. It results in the rapid growth of an individual polymer chain for each activated monomer. Examples include the addition polymerizations of ethene, methyl methacrylate, and styrene, as in

$$-CH_2CHX\cdot + CH_2{=}CHX \rightarrow -CH_2CHXCH_2CHX\cdot$$

and subsequent reactions. The central feature of the kinetic analysis (which is summarized in the *Justification* below) is that the rate of polymerization is proportional to the square root of the initiator concentration:

$$v = k[\mathrm{I}]^{1/2}[\mathrm{M}] \qquad (23.9)$$

Justification 23.1 *The rate of chain polymerization*

There are three basic types of reaction step in a chain polymerization process:

(a) Initiation:

$$\mathrm{I} \rightarrow \mathrm{R}\cdot + \mathrm{R}\cdot \qquad v_\mathrm{i} = k_\mathrm{i}[\mathrm{I}]$$

$$\mathrm{M} + \mathrm{R}\cdot \rightarrow \cdot\mathrm{M}_1 \qquad \text{(fast)}$$

where I is the initiator, R· the radical I forms, and $\cdot\mathrm{M}_1$ is a monomer radical. We have shown a reaction in which a radical is produced, but in some polymerizations the initiation step leads to the formation of an ionic chain carrier. The rate-determining step is the formation of the radicals R· by homolysis of the initiator, so the rate of initiation is equal to the v_i given above.

(b) Propagation:

$$\mathrm{M} + \cdot\mathrm{M}_1 \rightarrow \cdot\mathrm{M}_2$$

$$\mathrm{M} + \cdot\mathrm{M}_2 \rightarrow \cdot\mathrm{M}_3$$

$$\vdots$$

$$\mathrm{M} + \cdot\mathrm{M}_{n-1} \rightarrow \cdot\mathrm{M}_n \qquad v_\mathrm{p} = k_\mathrm{p}[\mathrm{M}][\cdot\mathrm{M}]$$

If we assume that the rate of propagation is independent of chain size for sufficiently large chains, then we can use only the equation given above to describe the propagation process. Consequently, for sufficiently large chains, the rate of propagation is equal to the overall rate of polymerization.

Because this chain of reactions propagates quickly, the rate at which the total concentration of radicals grows is equal to the rate of the rate-determining initiation step. It follows that

$$\left(\frac{\mathrm{d}[\cdot\mathrm{M}]}{\mathrm{d}t}\right)_\text{production} = 2fk_\mathrm{i}[\mathrm{I}] \qquad (23.10)$$

where f is the fraction of radicals R· that successfully initiate a chain.

(c) Termination:

$$\cdot\mathrm{M}_n + \cdot\mathrm{M}_m \rightarrow \mathrm{M}_{n+m} \qquad \text{(mutual termination)}$$

$$\cdot\mathrm{M}_n + \cdot\mathrm{M}_m \rightarrow \mathrm{M}_n + \mathrm{M}_m \qquad \text{(disproportionation)}$$

$$\mathrm{M} + \cdot\mathrm{M}_n \rightarrow \cdot\mathrm{M} + \mathrm{M}_n \qquad \text{(chain transfer)}$$

In **mutual termination** two growing radical chains combine. In termination by **disproportionation** a hydrogen atom transfers from one chain to another, corresponding to the oxidation of the donor and the reduction of the acceptor. In **chain transfer**, a new chain initiates at the expense of the one currently growing.

Here we suppose that only mutual termination occurs. If we assume that the rate of termination is independent of the length of the chain, the rate law for termination is

$$v_\mathrm{t} = k_\mathrm{t}[\cdot\mathrm{M}]^2$$

and the rate of change of radical concentration by this process is

$$\left(\frac{d[\cdot M]}{dt}\right)_{\text{depletion}} = -2k_t[\cdot M]^2$$

The steady-state approximation gives:

$$\frac{d[\cdot M]}{dt} = 2fk_i[I] - 2k_t[\cdot M]^2 = 0$$

The steady-state concentration of radical chains is therefore

$$[\cdot M] = \left(\frac{fk_i}{k_t}\right)^{1/2} [I]^{1/2} \qquad (23.11)$$

Because the rate of propagation of the chains is the negative of the rate at which the monomer is consumed, we can write $v_P = -d[M]/dt$ and

$$v_P = k_P[\cdot M][M] = k_P\left(\frac{fk_i}{k_t}\right)^{1/2} [I]^{1/2}[M] \qquad (23.12)$$

This rate is also the rate of polymerization, which has the form of eqn 23.9.

The **kinetic chain length**, ν, is the ratio of the number of monomer units consumed per activated centre produced in the initiation step:

$$\nu = \frac{\text{number of monomer units consumed}}{\text{number of activated centres produced}} \qquad (23.13)$$

The kinetic chain length can be expressed in terms of the rate expressions in *Justification* 23.1. To do so, we recognize that monomers are consumed at the rate that chains propagate. Then,

$$\nu = \frac{\text{rate of propagation of chains}}{\text{rate of production of radicals}}$$

By making the steady-state approximation, we set the rate of production of radicals equal to the termination rate. Therefore, we can write the expression for the kinetic chain length as

$$\nu = \frac{k_p[\cdot M][M]}{2k_t[M\cdot]^2} = \frac{k_p[M]}{2k_t[\cdot M]}$$

When we substitute the steady-state expression, eqn 23.11, for the radical concentration, we obtain

$$\nu = k[M][I]^{-1/2} \qquad k = \tfrac{1}{2}k_p(fk_ik_t)^{-1/2} \qquad (23.14)$$

Consider a polymer produced by a chain mechanism with mutual termination. In this case, the average number of monomers in a polymer molecule, $<n>$, produced by the reaction is the sum of the numbers in the two combining polymer chains. The average number of units in each chain is ν. Therefore,

$$<n> = 2\nu = 2k[M][I]^{-1/2} \qquad (23.15)$$

with k given in eqn 23.14. We see that, the slower the initiation of the chain (the smaller the initiator concentration and the smaller the initiation rate constant), the greater the kinetic chain length, and therefore the higher the average molar mass of the polymer. Some of the consequences of molar mass for polymers were explored in Chapter 19: now we have seen how we can exercise kinetic control over them.

Homogeneous catalysis

A **catalyst** is a substance that accelerates a reaction but undergoes no net chemical change. The catalyst lowers the activation energy of the reaction by providing an alternative path that avoids the slow, rate-determining step of the uncatalysed reaction (Fig. 23.8). Catalysts can be very effective; for instance, the activation energy for the decomposition of hydrogen peroxide in solution is 76 kJ mol^{-1}, and the reaction is slow at room temperature. When a little iodide ion is added, the activation energy falls to 57 kJ mol^{-1} and the rate constant increases by a factor of 2000. **Enzymes**, which are biological catalysts, are very specific and can have a dramatic effect on the reactions they control. For example, the enzyme catalase reduces the activation energy for the decomposition of hydrogen peroxide to 8 kJ mol^{-1}, corresponding to an acceleration of the reaction by a factor of 10^{15} at 298 K.

A **homogeneous catalyst** is a catalyst in the same phase as the reaction mixture. For example, the decomposition of hydrogen peroxide in aqueous solution is catalysed by bromide ion or catalase (Sections 23.5 and 23.6). A **heterogeneous catalyst** is a catalyst in a different phase from the reaction mixture. For example, the hydrogenation of ethene to ethane, a gas-phase reaction, is accelerated in the presence of a solid catalyst such as palladium, platinum, or nickel. The metal provides a surface upon which the reactants bind; this binding facilitates encounters between reactants and increases the rate of the reaction. We examine heterogeneous catalysis in Chapter 25 and consider only homogeneous catalysis here.

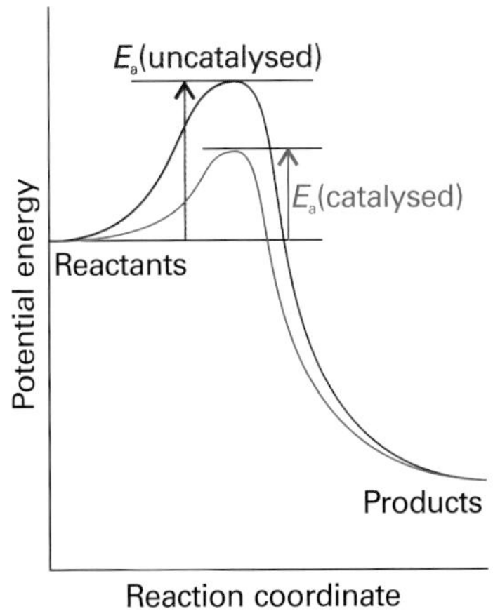

Fig. 23.8 A catalyst provides a different path with a lower activation energy. The result is an increase in the rate of formation of products.

23.5 Features of homogeneous catalysis

We can obtain some idea of the mode of action of homogeneous catalysts by examining the kinetics of the bromide-catalysed decomposition of hydrogen peroxide:

$$2\,H_2O_2(aq) \rightarrow 2\,H_2O(l) + O_2(g)$$

The reaction is believed to proceed through the following pre-equilibrium:

$$H_3O^+ + H_2O_2 \rightleftharpoons H_3O_2^+ + H_2O \qquad K = \frac{[H_3O_2^+]}{[H_2O_2][H_3O^+]}$$

$$H_3O_2^+ + Br^- \rightarrow HOBr + H_2O \qquad v = k[H_3O_2^+][Br^-]$$

$$HOBr + H_2O_2 \rightarrow H_3O^+ + O_2 + Br^- \qquad \text{(fast)}$$

where we have set the activity of H_2O in the equilibrium constant equal to 1 and assumed that the thermodynamic properties of the other substances are ideal. The second step is rate-determining. Therefore, we can obtain the rate law of the overall reaction by setting the overall rate equal to the rate of the second step and using the equilibrium constant to express the concentration of $H_3O_2^+$ in terms of the reactants. The result is

$$\frac{d[O_2]}{dt} = k_{eff}[H_2O_2][H_3O^+][Br^-]$$

with $k_{eff} = kK$, in agreement with the observed dependence of the rate on the Br^- concentration and the pH of the solution. The observed activation energy is that of the effective rate constant kK.

In **acid catalysis** the crucial step is the transfer of a proton to the substrate:

$$X + HA \rightarrow HX^+ + A^- \qquad HX^+ \rightarrow \text{products}$$

Acid catalysis is the primary process in the solvolysis of esters and keto–enol tautomerism. In **base catalysis**, a proton is transferred from the substrate to a base:

$$XH + B \rightarrow X^- + BH^+ \qquad X^- \rightarrow \text{products}$$

Base catalysis is the primary step in the isomerization and halogenation of organic compounds, and of the Claisen and aldol condensation reactions.

23.6 Enzymes

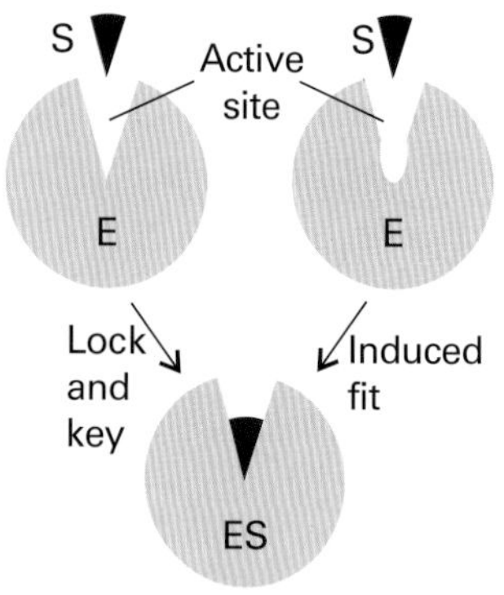

Fig. 23.9 Two models that explain the binding of a substrate to the active site of an enzyme. In the lock-and-key model, the active site and substrate have complementary three-dimensional structures and dock perfectly without the need for major atomic rearrangements. In the induced fit model, binding of the substrate induces a conformational change in the active site. The substrate fits well in the active site after the conformational change has taken place.

Enzymes are homogeneous biological catalysts. These ubiquitous compounds are special proteins or nucleic acids that contain an **active site**, which is responsible for binding the **substrates**, the reactants, and processing them into products. As is true of any catalyst, the active site returns to its original state after the products are released. Many enzymes consist primarily of proteins, some featuring organic or inorganic co-factors in their active sites. However, certain RNA molecules can also be biological catalysts, forming **ribozymes**. A very important example of a ribozyme is the **ribosome**, a large assembly of proteins and catalytically active RNA molecules responsible for the synthesis of proteins in the cell.

The structure of the active site is specific to the reaction that it catalyses, with groups in the substrate interacting with groups in the active site by intermolecular interactions, such as hydrogen bonding, electrostatic, or van der Waals interactions. Figure 23.9 shows two models that explain the binding of a substrate to the active site of an enzyme. In the **lock-and-key model**, the active site and substrate have complementary three-dimensional structures and dock perfectly without the need for major atomic rearrangements. Experimental evidence favours the **induced fit model**, in which binding of the substrate induces a conformational change in the active site. Only after the change does the substrate fit snugly in the active site.

Enzyme-catalysed reactions are prone to inhibition by molecules that interfere with the formation of product. Many drugs for the treatment of disease function by inhibiting enzymes. For example, an important strategy in the treatment of acquired immune deficiency syndrome (AIDS) involves the steady administration of a specially designed protease inhibitor. The drug inhibits an enzyme that is key to the formation of the protein envelope surrounding the genetic material of the human immunodeficiency virus (HIV). Without a properly formed envelope, HIV cannot replicate in the host organism.

(a) The Michaelis–Menten mechanism of enzyme catalysis

Experimental studies of enzyme kinetics are typically conducted by monitoring the initial rate of product formation in a solution in which the enzyme is present at very low concentration. Indeed, enzymes are such efficient catalysts that significant accelerations may be observed even when their concentration is more than three orders of magnitude smaller than that of the substrate.

The principal features of many enzyme-catalysed reactions are as follows:

1 For a given initial concentration of substrate, $[S]_0$, the initial rate of product formation is proportional to the total concentration of enzyme, $[E]_0$.

2 For a given $[E]_0$ and low values of $[S]_0$, the rate of product formation is proportional to $[S]_0$.

3 For a given $[E]_0$ and high values of $[S]_0$, the rate of product formation becomes independent of $[S]_0$, reaching a maximum value known as the **maximum velocity**, v_{max}.

The **Michaelis–Menten mechanism** accounts for these features. According to this mechanism, an enzyme–substrate complex is formed in the first step and either the substrate is released unchanged or after modification to form products:

$$\begin{array}{ll} E + S \rightleftharpoons ES & k_a, k_a' \\ ES \rightarrow P + E & k_b \end{array} \tag{23.16}$$

We show in the following *Justification* that this mechanism leads to the **Michaelis–Menten equation** for the rate of product formation

$$v = \frac{k_b[E]_0}{1 + K_m/[S]_0} \tag{23.17}$$

where $K_M = (k_a' + k_b)/k_a$ is the **Michaelis constant**, characteristic of a given enzyme acting on a given substrate.

Justification 23.2 *The Michaelis–Menten equation*

The rate of product formation according to the Michaelis–Menten mechanism is

$$v = k_b[ES] \tag{23.18}$$

We can obtain the concentration of the enzyme–substrate complex by invoking the steady-state approximation and writing

$$\frac{d[ES]}{dt} = k_a[E][S] - k_a'[ES] - k_b[ES] = 0$$

It follows that

$$[ES] = \left(\frac{k_a}{k_a' + k_b}\right)[E][S] \tag{23.19}$$

where [E] and [S] are the concentrations of *free* enzyme and substrate, respectively. Now we define the Michaelis constant as

$$K_M = \frac{k_a' + k_b}{k_a} = \frac{[E][S]}{[ES]}$$

and note that K_M has the same units as molar concentration. To express the rate law in terms of the concentrations of enzyme and substrate added, we note that $[E]_0 = [E] + [ES]$. Moreover, because the substrate is typically in large excess relative to the enzyme, the free substrate concentration is approximately equal to the initial substrate concentration and we can write $[S] \approx [S]_0$. It then follows that:

$$[ES] = \frac{[E]_0}{1 + K_M/[S]_0}$$

We obtain eqn 23.17 when we substitute this expression for [ES] into that for the rate of product formation ($v = k_b[ES]$).

Equation 23.17 shows that, in accord with experimental observations (Fig. 23.10):

1 When $[S]_0 \ll K_M$, the rate is proportional to $[S]_0$:

$$v = \frac{k_a}{K_M}[S]_0[E]_0 \tag{23.20a}$$

2 When $[S]_0 \gg K_M$, the rate reaches its maximum value and is independent of $[S]_0$:

$$v = v_{max} = k_b[E]_0 \tag{23.20b}$$

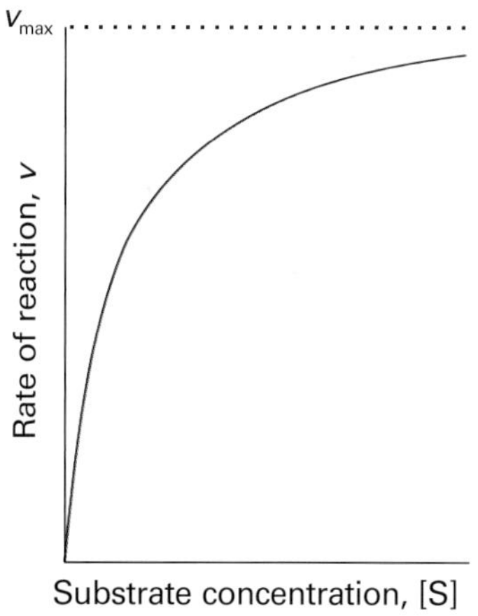

Fig. 23.10 The variation of the rate of an enzyme-catalysed reaction with substrate concentration. The approach to a maximum rate, v_{max}, for large [S] is explained by the Michaelis–Menten mechanism.

Exploration Use the Michaelis–Menten equation to generate two families of curves showing the dependence of v on [S]: one in which K_M varies but v_{max} is constant, and another in which v_{max} varies but K_M is constant.

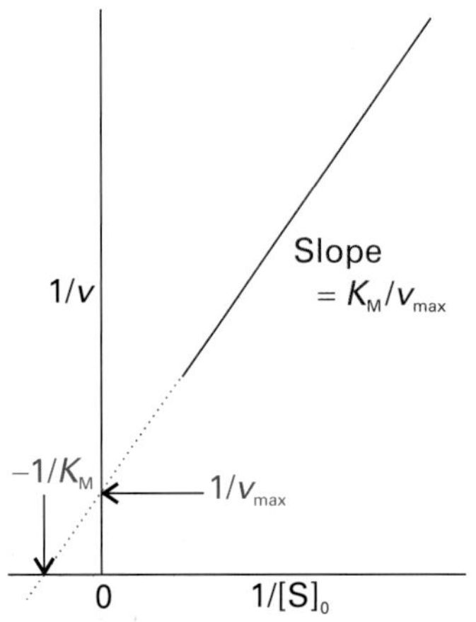

Fig. 23.11 A Lineweaver–Burk plot for the analysis of an enzyme-catalysed reaction that proceeds by a Michaelis–Menten mechanism and the significance of the intercepts and the slope.

Substitution of the definitions of K_M and v_{max} into eqn 23.17 gives:

$$v = \frac{v_{max}}{1 + K_M/[S]_0} \tag{23.21}$$

We can rearrange this expression into a form that is amenable to data analysis by linear regression:

$$\frac{1}{v} = \frac{1}{v_{max}} + \left(\frac{K_M}{v_{max}}\right)\frac{1}{[S]_0} \tag{23.22}$$

A **Lineweaver–Burk plot** is a plot of $1/v$ against $1/[S]_0$, and according to eqn 23.22 it should yield a straight line with slope of K_M/v_{max}, a y-intercept at $1/v_{max}$, and an x-intercept at $-1/K_M$ (Fig. 23.11). The value of k_b is then calculated from the y-intercept and eqn 23.20b. However, the plot cannot give the individual rate constants k_a and k_a' that appear in the expression for K_M. The stopped-flow technique described in Section 22.1b can give the additional data needed, because we can find the rate of formation of the enzyme–substrate complex by monitoring the concentration after mixing the enzyme and substrate. This procedure gives a value for k_a, and k_a' is then found by combining this results with the values of k_b and K_M.

(b) The catalytic efficiency of enzymes

The **turnover frequency**, or **catalytic constant**, of an enzyme, k_{cat}, is the number of catalytic cycles (turnovers) performed by the active site in a given interval divided by the duration of the interval. This quantity has units of a first-order rate constant and, in terms of the Michaelis–Menten mechanism, is numerically equivalent to k_b, the rate constant for release of product from the enzyme–substrate complex. It follows from the identification of k_{cat} with k_b and from eqn 23.20b that

$$k_{cat} = k_b = \frac{v_{max}}{[E]_0} \tag{23.23}$$

The **catalytic efficiency**, ε (epsilon), of an enzyme is the ratio k_{cat}/K_M. The higher the value of ε, the more efficient is the enzyme. We can think of the catalytic activity as the effective rate constant of the enzymatic reaction. From $K_M = (k_a' + k_b)/k_a$ and eqn 23.23, it follows that

$$\varepsilon = \frac{k_{cat}}{K_m} = \frac{k_a k_b}{k_a' + k_b} \tag{23.24}$$

The efficiency reaches its maximum value of k_a when $k_b \gg k_a'$. Because k_a is the rate constant for the formation of a complex from two species that are diffusing freely in solution, the maximum efficiency is related to the maximum rate of diffusion of E and S in solution. This limit (which is discussed further in Section 24.2) leads to rate constants of about 10^8–10^9 dm^3 mol^{-1} s^{-1} for molecules as large as enzymes at room temperature. The enzyme catalase has $\varepsilon = 4.0 \times 10^8$ dm^3 mol^{-1} s^{-1} and is said to have attained 'catalytic perfection', in the sense that the rate of the reaction it catalyses is controlled only by diffusion: it acts as soon as a substrate makes contact.

Comment 23.1

The web site contains links to databases of enzymes.

Example 23.3 *Determining the catalytic efficiency of an enzyme*

The enzyme carbonic anhydrase catalyses the hydration of CO_2 in red blood cells to give bicarbonate (hydrogencarbonate) ion:

$$CO_2(g) + H_2O(l) \rightarrow HCO_3^-(aq) + H^+(aq)$$

The following data were obtained for the reaction at pH = 7.1, 273.5 K, and an enzyme concentration of 2.3 nmol dm^{-3}:

$[CO_2]/(\text{mmol dm}^{-3})$	1.25	2.5	5	20
rate/$(\text{mmol dm}^{-3}\,\text{s}^{-1})$	2.78×10^{-2}	5.00×10^{-2}	8.33×10^{-2}	1.67×10^{-1}

Determine the catalytic efficiency of carbonic anhydrase at 273.5 K.

Method Prepare a Lineweaver–Burk plot and determine the values of K_M and v_{max} by linear regression analysis. From eqn 23.23 and the enzyme concentration, calculate k_{cat} and the catalytic efficiency from eqn 23.24.

Answer We draw up the following table:

$1/([CO_2]/(\text{mmol dm}^{-3}))$	0.800	0.400	0.200	0.0500
$1/(v/(\text{mmol dm}^{-3}\,\text{s}^{-1}))$	36.0	20.0	12.0	6.0

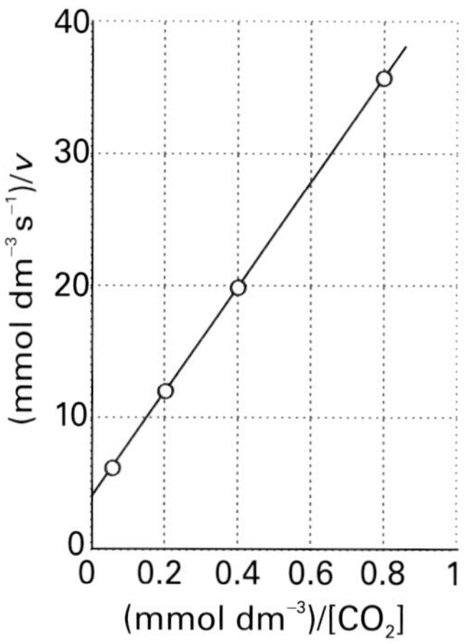

Fig. 23.12 The Lineweaver–Burk plot of the data for Example 23.3.

Figure 23.12 shows the Lineweaver–Burk plot for the data. The slope is 40.0 and the *y*-intercept is 4.00. Hence,

$$v_{max}/(\text{mmol dm}^{-3}\,\text{s}^{-1}) = \frac{1}{\text{intercept}} = \frac{1}{4.00} = 0.250$$

and

$$K_M/(\text{mmol dm}^{-3}) = \frac{\text{slope}}{\text{intercept}} = \frac{40.0}{4.00} = 10.0$$

It follows that

$$k_{cat} = \frac{v_{max}}{[E]_0} = \frac{2.5\times10^{-4}\,\text{mol dm}^{-3}\,\text{s}^{-1}}{2.3\times10^{-9}\,\text{mol dm}^{-3}} = 1.1\times10^{5}\,\text{s}^{-1}$$

and

$$\varepsilon = \frac{k_{cat}}{K_M} = \frac{1.1\times10^{5}\,\text{s}^{-1}}{1.0\times10^{-2}\,\text{mol dm}^{-3}} = 1.1\times10^{7}\,\text{dm}^3\,\text{mol}^{-1}\,\text{s}^{-1}$$

A note on good practice The slope and the intercept are unit-less: we have remarked previously, that all graphs should be plotted as pure numbers.

Self-test 23.3 The enzyme α-chymotrypsin is secreted in the pancreas of mammals and cleaves peptide bonds made between certain amino acids. Several solutions containing the small peptide *N*-glutaryl-L-phenylalanine-*p*-nitroanilide at different concentrations were prepared and the same small amount of α-chymotrypsin was added to each one. The following data were obtained on the initial rates of the formation of product:

$[S]/(\text{mmol dm}^{-3})$	0.334	0.450	0.667	1.00	1.33	1.67
$v/(\text{mmol dm}^{-3}\,\text{s}^{-1})$	0.152	0.201	0.269	0.417	0.505	0.667

Determine the maximum velocity and the Michaelis constant for the reaction.

[$v_{max} = 2.80$ mmol $dm^{-3}\,s^{-1}$, $K_M = 5.89$ mmol dm^{-3}]

(c) Mechanisms of enzyme inhibition

An inhibitor, I, decreases the rate of product formation from the substrate by binding to the enzyme, to the ES complex, or to the enzyme and ES complex simultaneously. The most general kinetic scheme for enzyme inhibition is then:

$$\mathrm{E+S \rightleftharpoons ES} \qquad k_{\rm a}, k'_{\rm a}$$

$$\mathrm{ES \rightarrow E+P} \qquad k_{\rm b}$$

$$\mathrm{EI \rightleftharpoons E+I} \qquad K_{\rm I}=\frac{[\mathrm{E}][\mathrm{I}]}{[\mathrm{EI}]} \tag{23.25a}$$

$$\mathrm{ESI \rightleftharpoons ES+I} \qquad K'_{\rm S}=\frac{[\mathrm{E}][\mathrm{I}]}{[\mathrm{EI}]} \tag{23.25b}$$

The lower the values of $K_{\rm I}$ and $K'_{\rm I}$ the more efficient are the inhibitors. The rate of product formation is always given by $v = k_{\rm b}[\mathrm{ES}]$, because only ES leads to product. As shown in the following *Justification*, the rate of reaction in the presence of an inhibitor is

$$v=\frac{v_{\rm max}}{\alpha' + \alpha K_{\rm M}/[\mathrm{S}]_0} \tag{23.26}$$

where $\alpha = 1 + [\mathrm{I}]/K_{\rm I}$ and $\alpha' = 1 + [\mathrm{I}]/K'_{\rm I}$. This equation is very similar to the Michaelis–Menten equation for the uninhibited enzyme (eqn 23.17) and is also amenable to analysis by a Lineweaver–Burk plot:

$$\frac{1}{v}=\frac{\alpha'}{v_{\rm max}}+\left(\frac{\alpha K_{\rm M}}{v_{\rm max}}\right)\frac{1}{[\mathrm{S}]_0} \tag{23.27}$$

Justification 23.3 *Enzyme inhibition*

By mass balance, the total concentration of enzyme is:

$$[\mathrm{E}]_0=[\mathrm{E}]+[\mathrm{EI}]+[\mathrm{ES}]+[\mathrm{ESI}]$$

By using eqns 23.25a and 23.25b and the definitions

$$\alpha=1+\frac{[\mathrm{I}]}{K_{\rm I}} \quad \text{and} \quad \alpha'=1+\frac{[\mathrm{I}]}{K'_{\rm I}}$$

it follows that

$$[\mathrm{E}]_0=[\mathrm{E}]\alpha+[\mathrm{ES}]\alpha'$$

By using $K_{\rm M} = [\mathrm{E}][\mathrm{S}]/[\mathrm{ES}]$, we can write

$$[\mathrm{E}]_0=\frac{K_{\rm M}[\mathrm{ES}]}{[\mathrm{S}]_0}\alpha+[\mathrm{ES}]\alpha'=[\mathrm{ES}]\left(\frac{\alpha K_{\rm M}}{[\mathrm{S}]_0}+\alpha'\right)$$

The expression for the rate of product formation is then:

$$v=k_{\rm b}[\mathrm{ES}]=\frac{k_{\rm b}[\mathrm{E}]_0}{\alpha K_{\rm M}/[\mathrm{S}]_0+\alpha'}$$

which, upon rearrangement, gives eqn 23.26.

There are three major modes of inhibition that give rise to distinctly different kinetic behaviour (Fig. 23.13). In **competitive inhibition** the inhibitor binds only to the active site of the enzyme and thereby inhibits the attachment of the substrate. This condition corresponds to $\alpha > 1$ and $\alpha' = 1$ (because ESI does not form). The slope of

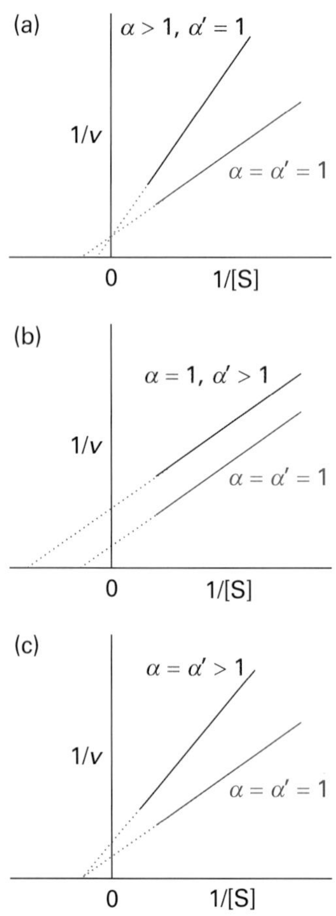

Fig. 23.13 Lineweaver–Burk plots characteristic of the three major modes of enzyme inhibition: (a) competitive inhibition, (b) uncompetitive inhibition, and (c) non-competitive inhibition, showing the special case $\alpha = \alpha' > 1$.

Exploration Use eqn 23.26 to explore the effect of competitive, uncompetitive, and non-competitive inhibition on the shapes of the plots of v against [S] for constant $K_{\rm M}$ and $v_{\rm max}$.

the Lineweaver–Burk plot increases by a factor of α relative to the slope for data on the uninhibited enzyme ($\alpha = \alpha' = 1$). The y-intercept does not change as a result of competitive inhibition (Fig. 23.13a). In **uncompetitive inhibition** the inhibitor binds to a site of the enzyme that is removed from the active site, but only if the substrate is already present. The inhibition occurs because ESI reduces the concentration of ES, the active type of complex. In this case $\alpha = 1$ (because EI does not form) and $\alpha' > 1$. The y-intercept of the Lineweaver–Burk plot increases by a factor of α' relative to the y-intercept for data on the uninhibited enzyme but the slope does not change (Fig. 23.13b). In **non-competitive inhibition** (also called **mixed inhibition**) the inhibitor binds to a site other than the active site, and its presence reduces the ability of the substrate to bind to the active site. Inhibition occurs at both the E and ES sites. This condition corresponds to $\alpha > 1$ and $\alpha' > 1$. Both the slope and y-intercept of the Lineweaver–Burk plot increase upon addition of the inhibitor. Figure 23.13c shows the special case of $K_I = K_I'$ and $\alpha = \alpha'$, which results in intersection of the lines at the x-axis.

In all cases, the efficiency of the inhibitor may be obtained by determining K_M and v_{max} from a control experiment with uninhibited enzyme and then repeating the experiment with a known concentration of inhibitor. From the slope and y-intercept of the Lineweaver–Burk plot for the inhibited enzyme (eqn 23.27), the mode of inhibition, the values of α or α', and the values of K_I or K_I' may be obtained.

Photochemistry

Many reactions can be initiated by the absorption of electromagnetic radiation by one of the mechanisms described in Chapter 14. The most important of all are the photochemical processes that capture the radiant energy of the Sun. Some of these reactions lead to the heating of the atmosphere during the daytime by absorption of ultraviolet radiation (*Impact* I23.1). Others include the absorption of visible radiation during photosynthesis (*Impact* I7.2 and I23.2). Without photochemical processes, the Earth would be simply a warm, sterile, rock. Table 23.1 summarizes common photochemical reactions.

23.7 Kinetics of photophysical and photochemical processes

Photochemical processes are initiated by the absorption of radiation by at least one component of a reaction mixture. In a **primary process**, products are formed directly from the excited state of a reactant. Examples include fluorescence (Section 14.3) and the *cis–trans* photoisomerization of retinal (Table 23.1, see also *Impact* I14.1). Products of a **secondary process** originate from intermediates that are formed directly from the excited state of a reactant. Examples include photosynthesis and photochemical chain reactions (Section 23.8).

Competing with the formation of photochemical products is a host of primary photophysical processes that can deactivate the excited state (Table 23.2). Therefore, it is important to consider the timescales of excited state formation and decay before describing the mechanisms of photochemical reactions.

Comment 23.2
The web site contains links to databases on photochemical reactions.

(a) Timescales of photophysical processes

Electronic transitions caused by absorption of ultraviolet and visible radiation occur within 10^{-16}–10^{-15} s. We expect, then, that the upper limit for the rate constant of a first-order photochemical reaction is about 10^{16} s^{-1}. Fluorescence is slower than

Table 23.1 Examples of photochemical processes

Process	General form	Example
Ionization	$A^* \rightarrow A^+ + e^-$	$NO^* \xrightarrow{134\,\text{nm}} NO^+ + e^-$
Electron transfer	$A^* + B \rightarrow A^+ + B^-$ or $A^- + B^+$	$[Ru(bpy)_3^{2+}]^* + Fe^{3+} \xrightarrow{452\,\text{nm}} Ru(bpy)_3^{3+} + Fe^{2+}$
Dissociation	$A^* \rightarrow B + C$	$O_3^* \xrightarrow{1180\,\text{nm}} O_2 + O$
	$A^* + B{-}C \rightarrow A + B + C$	$Hg^* + CH_4 \xrightarrow{254\,\text{nm}} Hg + CH_3 + H$
Addition	$2\,A^* \rightarrow B$	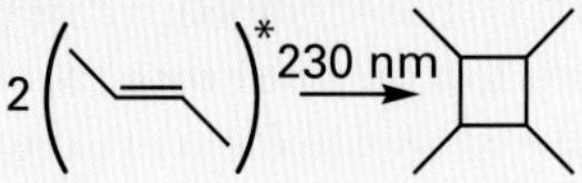
	$A^* + B \rightarrow AB$	
Abstraction	$A^* + B{-}C \rightarrow A{-}B + C$	$Hg^* + H_2 \xrightarrow{254\,\text{nm}} HgH + H$
Isomerization or rearrangement	$A^* \rightarrow A'$	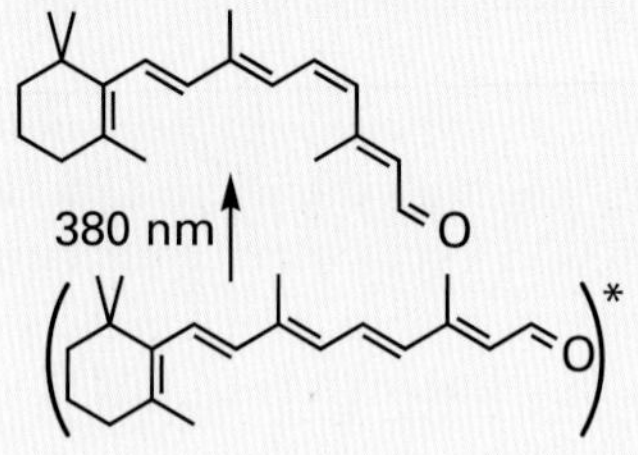

* Excited state.

Table 23.2 Common photophysical processes†

Primary absorption	$S + h\nu \rightarrow S^*$
Excited-state absorption	$S^* + h\nu \rightarrow S^{**}$
	$T^* + h\nu \rightarrow T^{**}$
Fluorescence	$S^* \rightarrow S + h\nu$
Stimulated emission	$S^* + h\nu \rightarrow S + 2h\nu$
Intersystem crossing (ISC)	$S^* \rightarrow T^*$
Phosphorescence	$T^* \rightarrow S + h\nu$
Internal conversion (IC)	$S^* \rightarrow S$
Collision-induced emission	$S^* + M \rightarrow S + M + h\nu$
Collisional deactivation	$S^* + M \rightarrow S + M$
	$T^* + M \rightarrow S + M$
Electronic energy transfer:	
Singlet–singlet	$S^* + S \rightarrow S + S^*$
Triple–triplet	$T^* + T \rightarrow T + T^*$
Excimer formation	$S^* + S \rightarrow (SS)^*$
Energy pooling	
Singlet–singlet	$S^* + S^* \rightarrow S^{**} + S$
Triple–triplet	$T^* + T^* \rightarrow S^* + S$

† S denotes a singlet state, T a triplet state, and M is a third body.

absorption, with typical lifetimes of 10^{-12}–10^{-6} s. Therefore, the excited singlet state can initiate very fast photochemical reactions in the femtosecond (10^{-15} s) to picosecond (10^{-12} s) timescale. Examples of such ultrafast reactions are the initial events of vision (*Impact* I14.1) and of photosynthesis. Typical intersystem crossing (ISC) and phosphorescence times for large organic molecules are 10^{-12}–10^{-4} s and 10^{-6}–10^{-1} s, respectively. As a consequence, excited triplet states are photochemically important. Indeed, because phosphorescence decay is several orders of magnitude slower than most typical reactions, species in excited triplet states can undergo a very large number of collisions with other reactants before deactivation. The interplay between reaction rates and excited state lifetimes is a very important factor in the determination of the kinetic feasibility of a photochemical process.

Illustration 23.1 *Exploring the photochemical roles of excited singlet and triplet states*

To estimate whether the excited singlet or triplet state of the reactant is a suitable product precursor, we compare the emission lifetimes with the relaxation time, τ, of the reactant due to the chemical reaction. As an illustration, consider a unimolecular photochemical reaction with rate constant $k = 1.7 \times 10^4\ \mathrm{s^{-1}}$ and relaxation time $\tau = 1/(1.7 \times 10^4\ \mathrm{s^{-1}}) = 59\ \mu\mathrm{s}$ that involves a reactant with an observed fluorescence lifetime of 1.0 ns and an observed phosphorescence lifetime of 1.0 ms. The excited singlet state is too short-lived and is not expected to be a major source of product in this reaction. On the other hand, the excited triplet state is a good candidate for a precursor.

(b) The primary quantum yield

We shall see that the rates of deactivation of the excited state by radiative, non-radiative, and chemical processes determine the yield of product in a photochemical reaction. The **primary quantum yield**, ϕ, is defined as the number of photophysical or photochemical events that lead to primary products divided by the number of photons absorbed by the molecule in the same interval. It follows that the primary quantum yield is also the rate of radiation-induced primary events divided by the rate of photon absorption. Because the rate of photon absorption is equal to the intensity of light absorbed by the molecule (Section 13.2), we write

$$\phi = \frac{\text{number of events}}{\text{number of photons absorbed}} = \frac{\text{rate of process}}{\text{intensity of light absorbed}} = \frac{v}{I_{\mathrm{abs}}} \qquad [23.28]$$

A molecule in an excited state must either decay to the ground state or form a photochemical product. Therefore, the total number of molecules deactivated by radiative processes, non-radiative processes, and photochemical reactions must be equal to the number of excited species produced by absorption of light. We conclude that the sum of primary quantum yields ϕ_i for *all* photophysical and photochemical events i must be equal to 1, regardless of the number of reactions involving the excited state. It follows that

$$\sum_i \phi_i = \sum_i \frac{v_i}{I_{\mathrm{abs}}} = 1 \qquad (23.29)$$

It follows that for an excited singlet state that decays to the ground state only via the photophysical processes described in Section 23.7(a), we write

$$\phi_{\mathrm{f}} + \phi_{\mathrm{IC}} + \phi_{\mathrm{ISC}} + \phi_{\mathrm{p}} = 1$$

where ϕ_f, ϕ_{IC}, ϕ_{ISC}, and ϕ_p are the quantum yields of fluorescence, internal conversion, intersystem crossing, and phosphorescence, respectively. The quantum yield of photon emission by fluorescence and phosphorescence is $\phi_{emission} = \phi_f + \phi_p$, which is less than 1. If the excited singlet state also participates in a primary photochemical reaction with quantum yield ϕ_R, we write

$$\phi_f + \phi_{IC} + \phi_{ISC} + \phi_p + \phi_R = 1$$

We can now strengthen the link between reaction rates and primary quantum yield already established by eqns 23.28 and 23.29. By taking the constant I_{abs} out of the summation in eqn 23.29 and rearranging, we obtain $I_{abs} = \sum_i v_i$. Substituting this result into eqn 23.29 gives the general result

$$\phi_i = \frac{v_i}{\sum_i v_i} \tag{23.30}$$

Therefore, the primary quantum yield may be determined directly from the experimental rates of *all* photophysical and photochemical processes that deactivate the excited state.

(c) Mechanism of decay of excited singlet states

Consider the formation and decay of an excited singlet state in the absense of a chemical reaction:

Absorption:	$S + h\nu_i \rightarrow S^*$	$v_{abs} = I_{abs}$
Fluorescence:	$S^* \rightarrow S + h\nu_f$	$v_f = k_f[S^*]$
Internal conversion:	$S^* \rightarrow S$	$v_{IC} = k_{IC}[S^*]$
Intersystem crossing:	$S^* \rightarrow T^*$	$v_{ISC} = k_{ISC}[S^*]$

in which S is an absorbing species, S^* an excited singlet state, T^* an excited triplet state, and $h\nu_i$ and $h\nu_f$ are the energies of the incident and fluorescent photons, respectively. From the methods developed in Chapter 22 and the rates of the steps that form and destroy the excited singlet state S^*, we write the rate of formation and decay of S^* as:

Rate of formation of $[S^*] = I_{abs}$

Rate of decay of $[S^*] = -k_f[S^*] - k_{ISC}[S^*] - k_{IC}[S^*] = -(k_f + k_{ISC} + k_{IC})[S^*]$

It follows that the excited state decays by a first-order process so, when the light is turned off, the concentration of S^* varies with time t as:

$$[S^*]_t = [S^*]_0 e^{-t/\tau_0} \tag{23.31}$$

where the **observed fluorescence lifetime**, τ_0, is defined as:

$$\tau_0 = \frac{1}{k_f + k_{ISC} + k_{IC}} \tag{23.32}$$

We show in the following *Justification* that the quantum yield of fluorescence is

$$\phi_f = \frac{k_f}{k_f + k_{ISC} + k_{IC}} \tag{23.33}$$

Justification 23.4 *The quantum yield of fluorescence*

Most fluorescence measurements are conducted by illuminating a relatively dilute sample with a continuous and intense beam of light. It follows that [S*] is small and constant, so we may invoke the steady-state approximation (Section 22.7) and write:

$$\frac{d[S^*]}{dt} = I_{abs} - k_f[S^*] - k_{ISC}[S^*] - k_{IC}[S^*] = I_{abs} - (k_f + k_{ISC} + k_{IC})[S^*] = 0$$

Consequently,

$$I_{abs} = (k_f + k_{ISC} + k_{IC})[S^*]$$

By using this expression and eqn 23.28, the quantum yield of fluorescence is written as:

$$\phi_f = \frac{v_f}{I_{abs}} = \frac{k_f[S^*]}{(k_f + k_{ISC} + k_{IC})[S^*]}$$

which, by cancelling the [S*], simplifies to eqn 23.33.

The observed fluorescence lifetime can be measured with a pulsed laser technique (Section 13.12b). First, the sample is excited with a short light pulse from a laser using a wavelength at which S absorbs strongly. Then, the exponential decay of the fluorescence intensity after the pulse is monitored. From eqn 23.28, it follows that

$$\tau_0 = \frac{1}{k_f + k_{ISC} + k_{IC}} = \left(\frac{k_f}{k_f + k_{ISC} + k_{IC}}\right) \times \frac{1}{k_f} = \frac{\phi_f}{k_f} \qquad (23.34)$$

Illustration 23.2 *Calculating the fluorescence rate constant of tryptophan*

In water, the fluorescence quantum yield and observed fluorescence lifetime of tryptophan are $\phi_f = 0.20$ and $\tau_0 = 2.6$ ns, respectively. It follows from eqn 23.33 that the fluorescence rate constant k_f is

$$k_f = \frac{\phi_f}{\tau_0} = \frac{0.20}{2.6 \times 10^{-9}\,\text{s}} = 7.7 \times 10^7\,\text{s}^{-1}$$

(d) Quenching

The shortening of the lifetime of the excited state is called **quenching**. Quenching may be either a desired process, such as in energy or electron transfer, or an undesired side reaction that can decrease the quantum yield of a desired photochemical process. Quenching effects may be studied by monitoring the emission from the excited state that is involved in the photochemical reaction.

The addition of a quencher, Q, opens an additional channel for deactivation of S*:

Quenching: $S^* + Q \rightarrow S + Q \qquad v_Q = k_Q[Q][S^*]$

The **Stern–Volmer equation**, which is derived in the *Justification* below, relates the fluorescence quantum yields $\phi_{f,0}$ and ϕ_f measured in the absence and presence, respectively, of a quencher Q at a molar concentration [Q]:

$$\frac{\phi_{f,0}}{\phi_f} = 1 + \tau_0 k_Q[Q] \qquad (23.35)$$

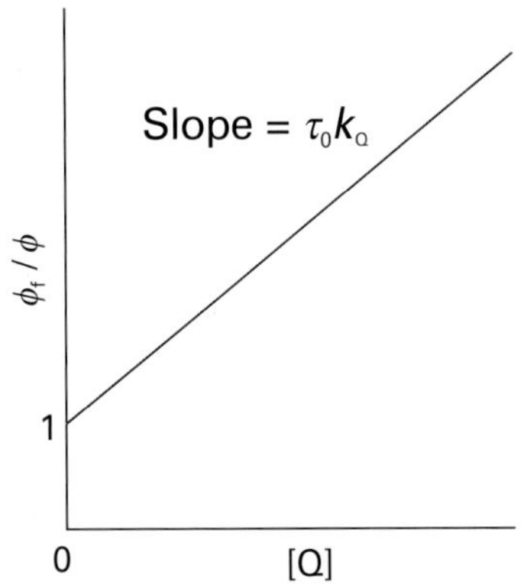

Fig. 23.14 The format of a Stern–Volmer plot and the interpretation of the slope in terms of the rate constant for quenching and the observed fluorescence lifetime in the absence of quenching.

This equation tells us that a plot of $\phi_{f,0}/\phi_f$ against [Q] should be a straight line with slope $\tau_0 k_Q$. Such a plot is called a **Stern–Volmer plot** (Fig. 23.14). The method may also be applied to the quenching of phosphorescence.

Justification 23.5 *The Stern–Volmer equation*

With the addition of quenching, the steady-state approximation for $[S^*]$ now gives:

$$\frac{d[S^*]}{dt} = I_{abs} - (k_f + k_{IC} + k_{ISC} + k_{IC} + k_Q[Q])[S^*] = 0$$

and the fluorescence quantum yield in the presence of the quencher is:

$$\phi_f = \frac{k_f}{k_f + k_{ISC} + k_{IC} + k_Q[Q]}$$

When [Q] = 0, the quantum yield is

$$\phi_{f,0} = \frac{k_f}{k_f + k_{ISC} + k_{IC}}$$

It follows that

$$\begin{aligned}\frac{\phi_{f,0}}{\phi_f} &= \left(\frac{k_f}{k_f + k_{ISC} + k_{IC}}\right) \times \left(\frac{k_f + k_{ISC} + k_{IC} + k_Q[Q]}{k_f}\right) \\ &= \frac{k_f + k_{ISC} + k_{IC} + k_Q[Q]}{k_f + k_{ISC} + k_{IC}} \\ &= 1 + \frac{k_Q}{k_f + k_{ISC} + k_{IC}}[Q]\end{aligned}$$

By using eqn 23.34, this expression simplifies to eqn 23.35.

Because the fluorescence intensity and lifetime are both proportional to the fluorescence quantum yield (specifically, from eqn 23.34, $\tau = \phi_f/k_f$), plots of $I_{f,0}/I_f$ and τ_0/τ (where the subscript 0 indicates a measurement in the absence of quencher) against [Q] should also be linear with the same slope and intercept as those shown for eqn 23.35.

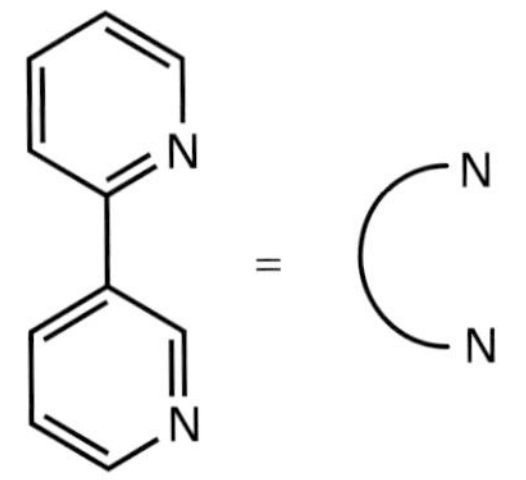

1 2,2′-Bipyridine (bipy)

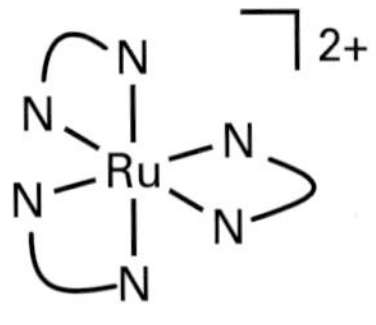

2

Example 23.4 *Determining the quenching rate constant*

The molecule 2,2′-bipyridine (**1**) forms a complex with the Ru^{2+} ion. Ruthenium(II) tris-(2,2′-bipyridyl), $Ru(bipy)_3^{2+}$ (**2**), has a strong metal-to-ligand charge transfer (MLCT) transition (Section 14.2) at 450 nm. The quenching of the $^*Ru(bipy)_3^{2+}$ excited state by $Fe(H_2O)_6^{3+}$ in acidic solution was monitored by measuring emission lifetimes at 600 nm. Determine the quenching rate constant for this reaction from the following data:

$[Fe(H_2O)_6^{3+}]/(10^{-4}\ mol\ dm^{-3})$	0	1.6	4.7	7	9.4
$\tau/(10^{-7}\ s)$	6	4.05	3.37	2.96	2.17

Method Re-write the Stern–Volmer equation (eqn 23.35) for use with lifetime data then fit the data to a straight-line.

Answer Upon substitution of τ_0/τ for $\phi_{0,f}/\phi_f$ in eqn 23.35 and after rearrangement, we obtain:

$$\frac{1}{\tau}=\frac{1}{\tau_0}+k_Q[Q] \qquad (23.36)$$

Figure 23.15 shows a plot of $1/\tau$ versus $[Fe^{3+}]$ and the results of a fit to eqn 23.36. The slope of the line is 2.8×10^9, so $k_Q = 2.8\times10^9$ $dm^3\,mol^{-1}\,s^{-1}$.

This example shows that measurements of emission lifetimes are preferred because they yield the value of k_Q directly. To determine the value of k_Q from intensity or quantum yield measurements, we need to make an independent measurement of τ_0.

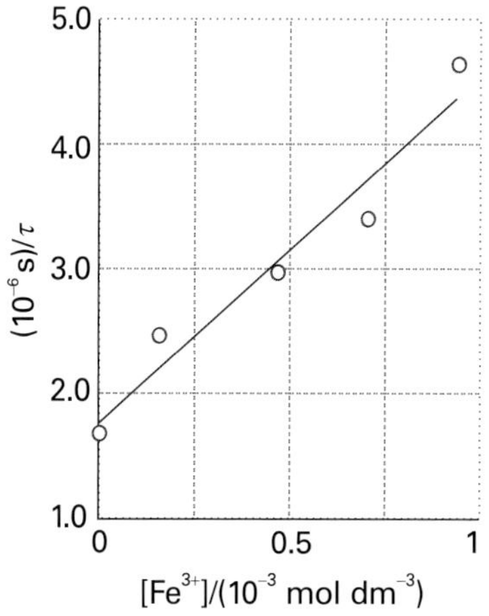

Fig. 23.15 The Stern–Volmer plot of the data for Example 23.4.

Self-test 23.4 The quenching of tryptophan fluorescence by dissolved O_2 gas was monitored by measuring emission lifetimes at 348 nm in aqueous solutions. Determine the quenching rate constant for this process from the following data:

$[O_2]/(10^{-2}\,mol\,dm^{-3})$	0	2.3	5.5	8	10.8
$\tau/(10^{-9}\,s)$	2.6	1.5	0.92	0.71	0.57

$[1.3\times10^{10}\,dm^3\,mol^{-1}\,s^{-1}]$

Three common mechanisms for bimolecular quenching of an excited singlet (or triplet) state are:

Collisional deactivation:	$S^* + Q \rightarrow S + Q$
Resonance energy transfer:	$S^* + Q \rightarrow S + Q^*$
Electron transfer:	$S^* + Q \rightarrow S^+ + Q^-$ or $S^- + Q^+$

Collisional quenching is particularly efficient when Q is a heavy species, such as iodide ion, which receives energy from S^* and then decays primarily by internal conversion to the ground state. This fact may be used to determine the accessibility of amino acid residues of a folded protein to solvent. For example, fluorescence from a tryptophan residue ($\lambda_{abs} \approx 290$ nm, $\lambda_{fluor} \approx 350$ nm) is quenched by iodide ion when the residue is on the surface of the protein and hence accessible to the solvent. Conversely, residues in the hydrophobic interior of the protein are not quenched effectively by I^-.

The quenching rate constant itself does not give much insight into the mechanism of quenching. For the system of Example 23.4, it is known that the quenching of the excited state of $Ru(bipy)_3^{2+}$ is a result of light-induced electron transfer to Fe^{3+}, but the quenching data do not allow us to prove the mechanism. However, there are some criteria that govern the relative efficiencies of energy and electron transfer.

(e) Resonance energy transfer

We visualize the process $S^* + Q \rightarrow S + Q^*$ as follows. The oscillating electric field of the incoming electromagnetic radiation induces an oscillating electric dipole moment in S. Energy is absorbed by S if the frequency of the incident radiation, ν, is such that $\nu = \Delta E_S/h$, where ΔE_S is the energy separation between the ground and excited electronic states of S and h is Planck's constant. This is the 'resonance condition' for absorption of radiation. The oscillating dipole on S now can affect electrons bound to a nearby Q molecule by inducing an oscillating dipole moment in the latter. If the frequency of oscillation of the electric dipole moment in S is such that $\nu = \Delta E_Q/h$ then Q will absorb energy from S.

The efficiency, E_T, of resonance energy transfer is defined as

$$E_T = 1 - \frac{\phi_f}{\phi_{f,0}} \qquad [23.37]$$

Table 23.3 Values of R_0 for some donor–acceptor pairs*

Donor†	Acceptor	R_0/nm
Naphthalene	Dansyl	2.2
Dansyl	ODR	4.3
Pyrene	Coumarin	3.9
IEDANS	FITC	4.9
Tryptophan	IEDANS	2.2
Tryptophan	Haem (heme)	2.9

* Additional values may be found in J.R. Lacowicz in *Principles of fluorescence spectroscopy*, Kluwer Academic/Plenum, New York (1999).
† Abbreviations: Dansyl, 5-dimethylamino-1-naphthalenesulfonic acid; FITC, fluorescein 5-isothiocyanate; IEDANS, 5-((((2-iodoacetyl) amino)ethyl)amino)naphthalene-1-sulfonic acid; ODR, octadecyl-rhodamine.

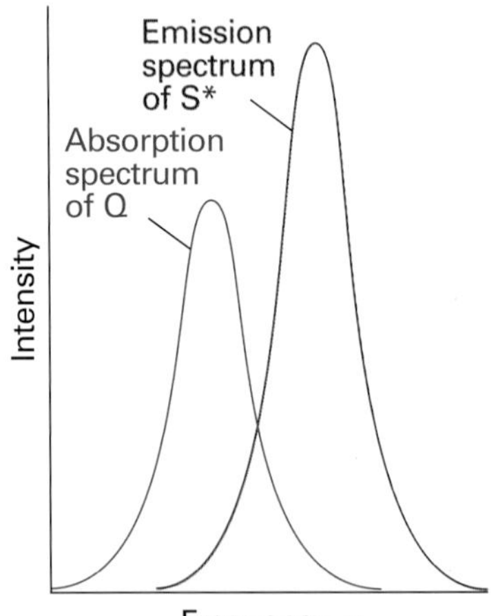

Fig. 23.16 According to the Förster theory, the rate of energy transfer from a molecule S* in an excited state to a quencher molecule Q is optimized at radiation frequencies in which the emission spectrum of S* overlaps with the absorption spectrum of Q, as shown in the shaded region.

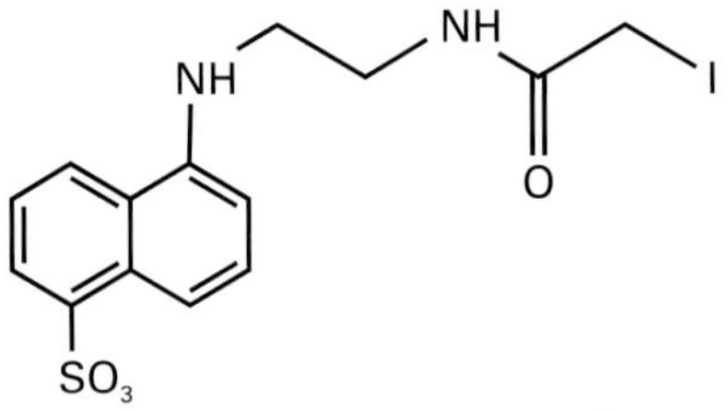

3 1.5-I AEDANS

According to the **Förster theory** of resonance energy transfer, which was proposed by T. Förster in 1959, energy transfer is efficient when:

1 The energy donor and acceptor are separated by a short distance (of the order of nanometres).

2 Photons emitted by the excited state of the donor can be absorbed directly by the acceptor.

We show in *Further information* 23.1 that for donor–acceptor systems that are held rigidly either by covalent bonds or by a protein 'scaffold', E_T increases with decreasing distance, R, according to

$$E_T = \frac{R_0^6}{R_0^6 + R^6} \tag{23.38}$$

where R_0 is a parameter (with units of distance) that is characteristic of each donor–acceptor pair. Equation 23.38 has been verified experimentally and values of R_0 are available for a number of donor–acceptor pairs (Table 23.3).

The emission and absorption spectra of molecules span a range of wavelengths, so the second requirement of the Förster theory is met when the emission spectrum of the donor molecule overlaps significantly with the absorption spectrum of the acceptor. In the overlap region, photons emitted by the donor have the proper energy to be absorbed by the acceptor (Fig. 23.16).

In many cases, it is possible to prove that energy transfer is the predominant mechanism of quenching if the excited state of the acceptor fluoresces or phosphoresces at a characteristic wavelength. In a pulsed laser experiment, the rise in fluorescence intensity from Q* with a characteristic time that is the same as that for the decay of the fluorescence of S* is often taken as indication of energy transfer from S to Q.

Equation 23.38 forms the basis of **fluorescence resonance energy transfer** (FRET), in which the dependence of the energy transfer efficiency, E_T, on the distance, R, between energy donor and acceptor can be used to measure distances in biological systems. In a typical FRET experiment, a site on a biopolymer or membrane is labelled covalently with an energy donor and another site is labelled covalently with an energy acceptor. In certain cases, the donor or acceptor may be natural constituents of the system, such as amino acid groups, cofactors, or enzyme substrates. The distance between the labels is then calculated from the known value of R_0 and eqn 23.38. Several tests have shown that the FRET technique is useful for measuring distances ranging from 1 to 9 nm.

Illustration 23.3 *Using FRET analysis*

As an illustration of the FRET technique, consider a study of the protein rhodopsin (*Impact* I14.1). When an amino acid on the surface of rhodopsin was labelled covalently with the energy donor 1.5-I AEDANS (**3**), the fluorescence quantum yield of the label decreased from 0.75 to 0.68 due to quenching by the visual pigment 11-*cis*-retinal (**4**). From eqn 23.37, we calculate $E_T = 1 - (0.68/0.75) = 0.093$ and from eqn 23.38 and the known value of $R_0 = 5.4$ nm for the 1.5-I AEDANS/11-*cis*-retinal pair we calculate $R = 7.9$ nm. Therefore, we take 7.9 nm to be the distance between the surface of the protein and 11-*cis*-retinal.

If donor and acceptor molecules diffuse in solution or in the gas phase, Förster theory predicts that the efficiency of quenching by energy transfer increases as the average distance travelled between collisions of donor and acceptor decreases. That is, the quenching efficiency increases with concentration of quencher, as predicted by the Stern–Volmer equation.

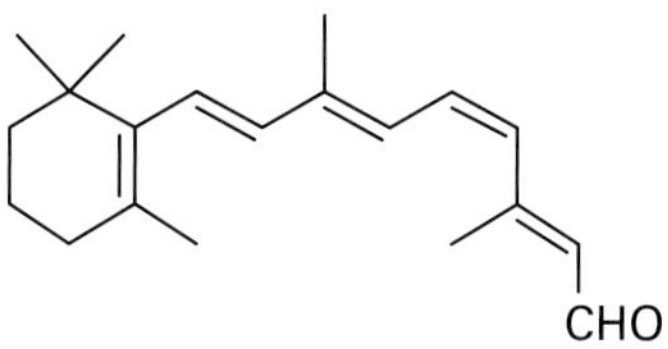

4 11-*cis*-retinal

(f) Electron transfer reactions

According to the **Marcus theory** of electron transfer, which was proposed by R.A. Marcus in 1965 and is discussed fully in Section 24.11, the rates of electron transfer (from ground or excited states) depend on:

1 The distance between the donor and acceptor, with electron transfer becoming more efficient as the distance between donor and acceptor decreases.

2 The reaction Gibbs energy, $\Delta_r G$, with electron transfer becoming more efficient as the reaction becomes more exergonic. For example, efficient photooxidation of S requires that the reduction potential of S* be lower than the reduction potential of Q.

3 The reorganization energy, the energy cost incurred by molecular rearrangements of donor, acceptor, and medium during electron transfer. The electron transfer rate is predicted to increase as this reorganization energy is matched closely by the reaction Gibbs energy.

Electron transfer can also be studied by time-resolved spectroscopy (Section 14.6). The oxidized and reduced products often have electronic absorption spectra distinct from those of their neutral parent compounds. Therefore, the rapid appearance of such known features in the absorption spectrum after excitation by a laser pulse may be taken as indication of quenching by electron transfer.

IMPACT ON ENVIRONMENTAL SCIENCE

I23.1 The chemistry of stratospheric ozone

The Earth's atmosphere contains primarily N_2 and O_2 gas, with low concentrations of a large number of other species of both natural and anthropogenic origins. Indeed, many of the natural trace constituents of our atmosphere participate in complex chemical reactions that have contributed to the proliferation of life on the planet. The development of industrial societies has added new components to the Earth's atmosphere and has led to significant changes in the concentrations of some natural trace species. The negative consequences of these changes for the environment are either already being felt or, more disturbingly, are yet to be felt in the next few decades (see, for example, the discussion of global warming in *Impact* I13.2). Careful kinetic analysis allows us to understand the origins of our complex atmosphere and point to ways in which environmental problems can be solved or avoided.

The Earth's atmosphere consists of layers, as shown in Fig. 23.17. The pressure decreases as altitude increases (see Problems 1.27 and 16.20), but the variation of temperature with altitude is complex, owing to processes that capture radiant energy from the Sun. We focus on the *stratosphere*, a region spanning from 15 km to 50 km above the surface of the Earth, and on the chemistry of the trace component ozone, O_3.

In the *troposphere*, the region between the Earth's surface and the stratosphere, temperature decreases with increasing altitude. This behaviour may be understood in terms of a model in which the boundary between the troposphere and the stratosphere, also called the *tropopause*, is considered adiabatic. Then we know from Section 2.6 that, as atmospheric gases are allowed to expand from layers close to the surface to higher layers, the temperature varies with pressure, and hence height, as

$$\frac{T_{\text{low altitude}}}{T_{\text{high altitude}}} = \left(\frac{p_{\text{low altitude}}}{p_{\text{high altitude}}}\right)^{c} \qquad c = \frac{C_{p,\text{m}}}{C_{V,\text{m}}} - 1$$

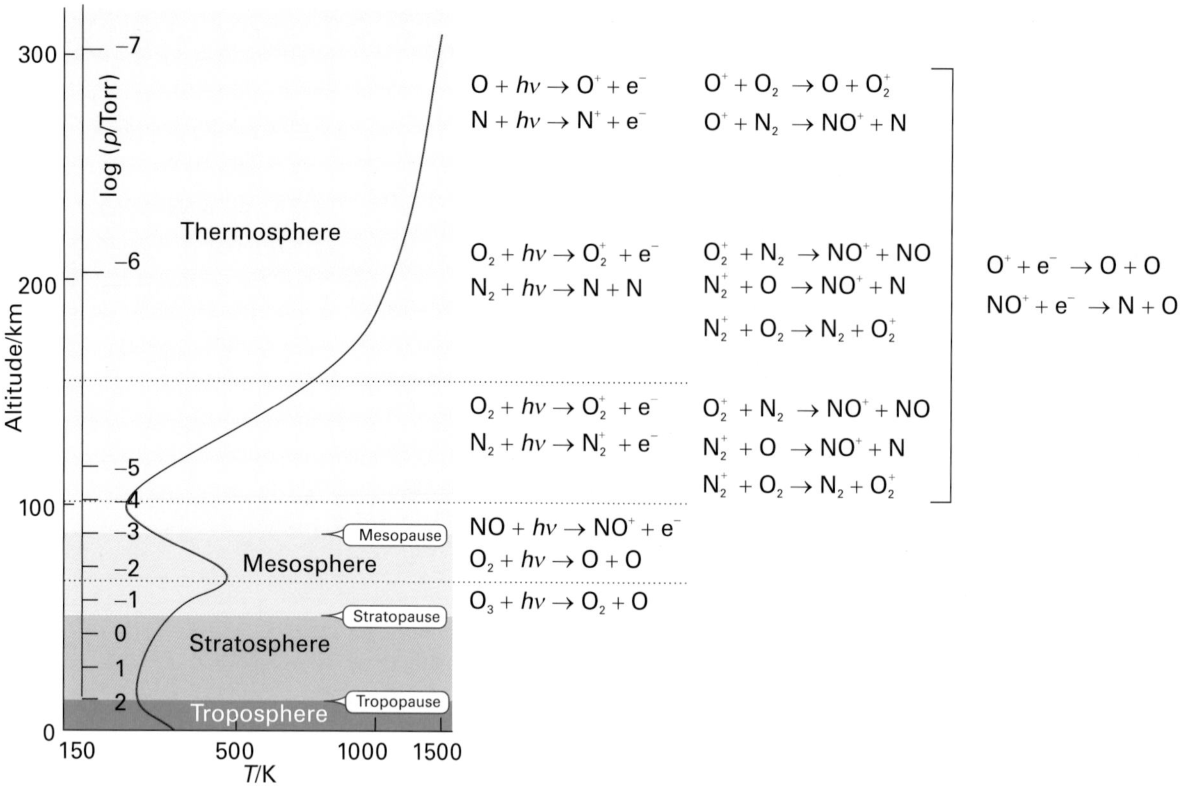

Fig. 23.17 The temperature profile through the atmosphere and some of the reactions that occur in each region.

The model predicts a decrease in temperature with increasing altitude because $C_{p,\mathrm{m}}/C_{V,\mathrm{m}} \approx \frac{7}{5}$ for air. In the stratosphere, a *temperature inversion* is observed because of photochemical chain reactions that produce ozone from O_2. The *Chapman model* accounts for ozone formation and destruction in an atmosphere that contains only O_2:

Initiation:	$O_2 + h\nu \rightarrow O + O$	$185\ \mathrm{nm} < \lambda < 220\ \mathrm{nm}$	$v = k_1[O_2]$
Propagation:	$O + O_2 + M \rightarrow O_3 + M^*$	$\Delta_r H = -106.6\ \mathrm{kJ\ mol^{-1}}$	$v = k_2[O][O_2][M]$
	$O_3 + h\nu \rightarrow O_2 + O$	$210\ \mathrm{nm} < \lambda < 300\ \mathrm{nm}$	$v = k_3[O_3]$
Termination:	$O + O_3 \rightarrow O_2 + O_2$	$\Delta_r H = -391.9\ \mathrm{kJ\ mol^{-1}}$	$v = k_4[O][O_3]$
	$O + O + M \rightarrow O_2 + M^*$		$v = k_5[O]^2[M]$

where M is an arbitrary third body, such as O_2 in an 'oxygen-only' atmosphere, which helps to remove excess energy from the products of combination and recombination reactions. The mechanism shows that absorption of radiation by O_2 and O_3 during the daytime leads to the production of reactive O atoms, which, in turn, participate in exothermic reactions that are responsible for the heating of the stratosphere.

Using values of the rate constants that are applicable to stratospheric conditions, the Chapman model predicts a net formation of trace amounts of ozone, as seen in Fig. 23.18 (see also Problem 23.33). However, the model overestimates the concentration of ozone in the stratosphere because other trace species X contribute to catalytic enhancement of the termination step $O_3 + O \rightarrow O_2 + O_2$ according to

$$X + O_3 \rightarrow XO + O_2$$
$$XO + O \rightarrow X + O_2$$

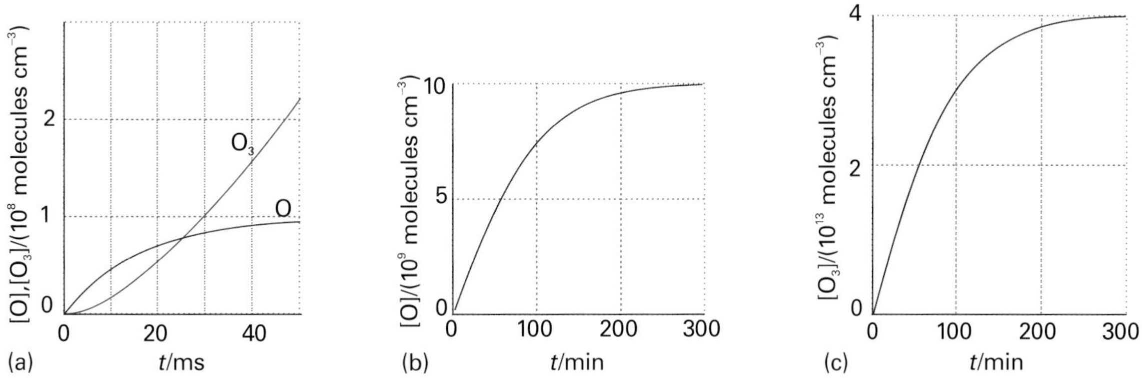

Fig. 23.18 Net formation of ozone via the Chapman model in a stratospheric model containing only O_2, O, andO_3. The rate constants are consistent with reasonable stratospheric conditions. (a) Early reaction period after irradiation begins at $t = 0$. (b) Late reaction period, showing that the concentration of O atoms begins to level off after about 4 hours of continuous irradiation. (c) Late reaction period, showing that the ozone concentration also begins to level off similarly. For details of the calculation, see Problem 23.33 and M.P. Cady and C.A. Trapp, *A Mathcad primer for physical chemistry*. Oxford University Press (1999).

The catalyst X can be H, OH, NO, or Cl. Chlorine atoms are produced by photolysis of CH_3Cl which, in turn, is a by-product of reactions between Cl^- and decaying vegetation in oceans. Nitric oxide, NO, is produced in the stratosphere from reaction between excited oxygen atoms and N_2O, which is formed mainly by microbial denitrification processes in soil. The hydroxyl radical is a product, along with the methyl radical, of the reaction between excited oxygen atoms and methane gas, which is a by-product of a number of natural processes (such as digestion of cellulose in ruminant animals, anaerobic decomposition of organic waste matter) and industrial processes (such as food production and fossil fuel use). In spite of the presence of these catalysts, a natural stratosphere is still capable to maintain a low concentration of ozone.

The chemistry outlined above shows that the photochemical reactions of the Chapman model account for absorption of a significant portion of solar ultraviolet radiation in the stratosphere. Hence, the surface of the Earth is bathed by lower energy radiation, which does not damage biological tissue (see *Impact* I23.3). However, some pollutants can lower the concentration of stratospheric ozone. For example, chlorofluorocarbons (CFCs) have been used as propellants and refrigerants over many years. As CFC molecules diffuse slowly into the middle stratosphere, they are finally photolysed by ultraviolet radiation. For CF_2Cl_2, also known as CFC-12, the reaction is:

$$CF_2Cl_2 + h\nu \rightarrow CF_2Cl + Cl$$

We already know that the resulting Cl atoms can participate in the decomposition of ozone according to the catalytic cycle shown in Fig. 23.19. A number of experimental observations have linked this chemistry of CFCs to a dangerously rapid decline in the concentration of stratospheric ozone over the last three decades.

Ozone depletion has increased the amount of ultraviolet radiation at the Earth's surface, particularly radiation in the 'UVB range', 290–320 nm. The physiological consequences of prolonged exposure to UVB radiation include DNA damage, principally by photodimerization of adjacent thymine bases to yield either a cyclobutane

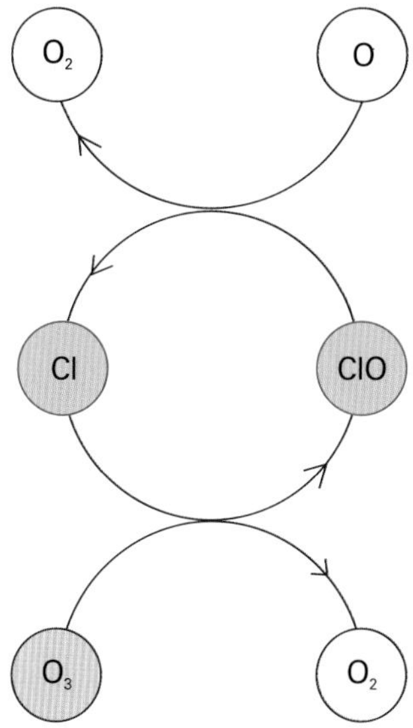

Fig. 23.19 A catalytic cycle showing the propagation of ozone decomposition by chlorine atoms.

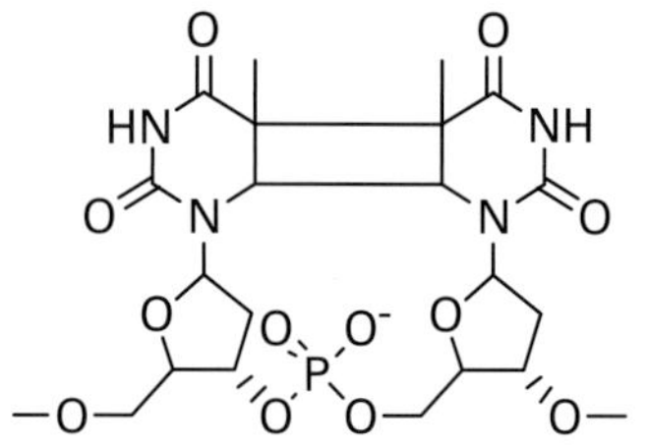

5

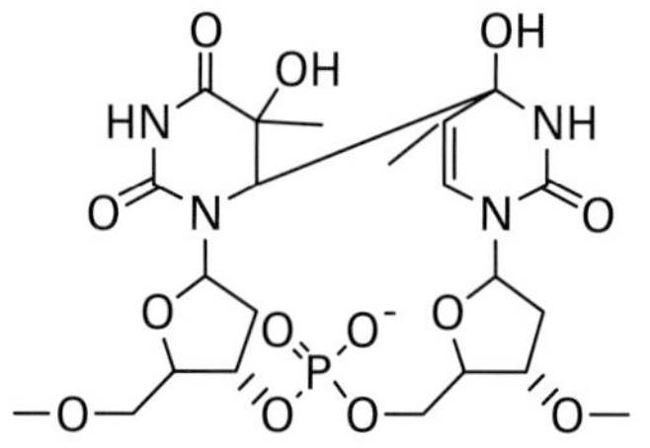

6

thymine dimer (**5**) or a so-called 6–4 photoproduct (**6**). The former has been linked directly to cell death and the latter may lead to DNA mutations and, consequently, to the formation of tumours. There are several natural mechanisms for protection from and repair of photochemical damage. For example, the enzyme DNA photolyase, present in organisms from all kingdoms but not in humans, catalyses the destruction of cyclobutane thymine dimers. Also, ultraviolet radiation can induce the production of the pigment melanin (in a process more commonly known as 'tanning'), which shields the skin from damage. However, repair and protective mechanisms become increasingly less effective with persistent and prolonged exposure to solar radiation. Consequently, there is concern that the depletion of stratospheric ozone may lead to an increase in mortality not only of animals but also the plants and lower organisms that form the base of the food chain.

Chlorofluorocarbons are being phased out according to international agreements and alternatives, such as the hydrofluorocarbon CH_2FCH_3, are already being used. However, the temperature inversion shown in Fig. 23.17 leads to trapping of gases in the troposphere, so CFCs are likely to continue to cause ozone depletion over many decades as the molecules diffuse slowly into the middle stratosphere, where they are photolysed by intense solar UV radiation.

IMPACT ON BIOCHEMISTRY

I23.2 Harvesting of light during plant photosynthesis

A large proportion of solar radiation with wavelengths below 400 nm and above 1000 nm is absorbed by atmospheric gases such as ozone and O_2, which absorb ultraviolet radiation (*Impact* I23.1), and CO_2 and H_2O, which absorb infrared radiation (*Impact* I13.2). As a result, plants, algae, and some species of bacteria evolved photosynthetic apparatus that capture visible and near-infrared radiation. Plants use radiation in the wavelength range of 400–700 nm to drive the endergonic reduction of CO_2 to glucose, with concomitant oxidation of water to O_2 ($\Delta_r G^{\ominus} = +2880\ \text{kJ mol}^{-1}$), in essence the reverse of glycolysis and the citric acid cycle (*Impact* I7.2):

$$6\,CO_2(g) + 6\,H_2O(l) \underset{\text{glycolysis and the citric acid cycle}}{\overset{\text{photosynthesis}}{\rightleftarrows}} C_6H_{12}O_6(s) + 6\,O_2(g)$$

Electrons flow from reductant to oxidant via a series of electrochemical reactions that are coupled to the synthesis of ATP. The process takes place in the *chloroplast*, a special organelle of the plant cell, where chlorophylls *a* and *b* (**7**) and carotenoids (of which *β*-carotene, **8**, is an example) bind to integral proteins called *light harvesting complexes*, which absorb solar energy and transfer it to protein complexes known as *reaction centres*, where light-induced electron transfer reactions occur. The combination of a light harvesting complex and a reaction centre complex is called a photosystem. Plants have two photosystems that drive the reduction of $NADP^+$ (**9**) by water:

$$2\,H_2O + 2\,NADP^+ \xrightarrow{\text{light, photosystems I and II}} O_2 + 2\,NADPH + 2\,H^+$$

It is clear that energy from light is required to drive this reaction because, in the dark, $E^{\ominus} = -1.135$ V and $\Delta_r G^{\ominus} = +438.0\ \text{kJ mol}^{-1}$.

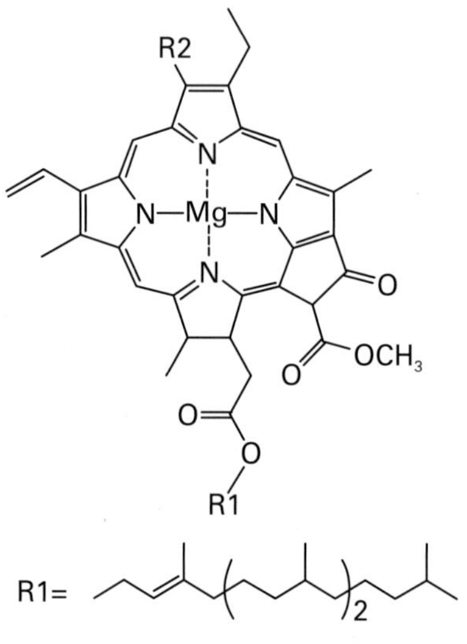

R2 = CH_3 (Chl *a*) or CHO (Chl *b*)

7 Chlorophyll *a* and *b*

Light harvesting complexes bind large numbers of pigments in order to provide a sufficiently large area for capture of radiation. In photosystems I and II, absorption of a photon raises a chlorophyll or carotenoid molecule to an excited singlet state and within 0.1–5 ps the energy hops to a nearby pigment via the Förster mechanism (Section 23.7e). About 100–200 ps later, which corresponds to thousands of hops within the light harvesting complex, more than 90 per cent of the absorbed energy reaches the reaction centre. There, a chlorophyll *a* dimer becomes electronically

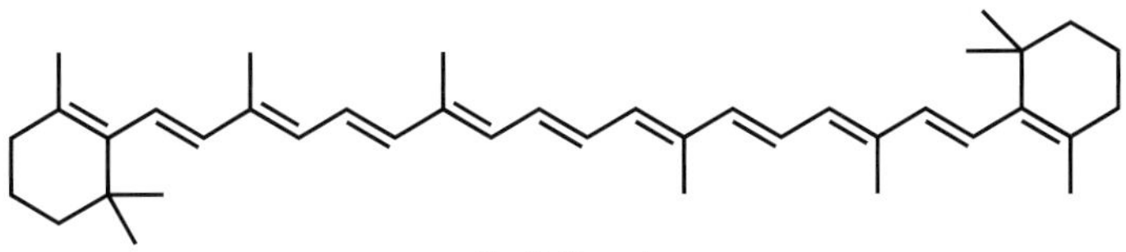

8 *β*-Carotene

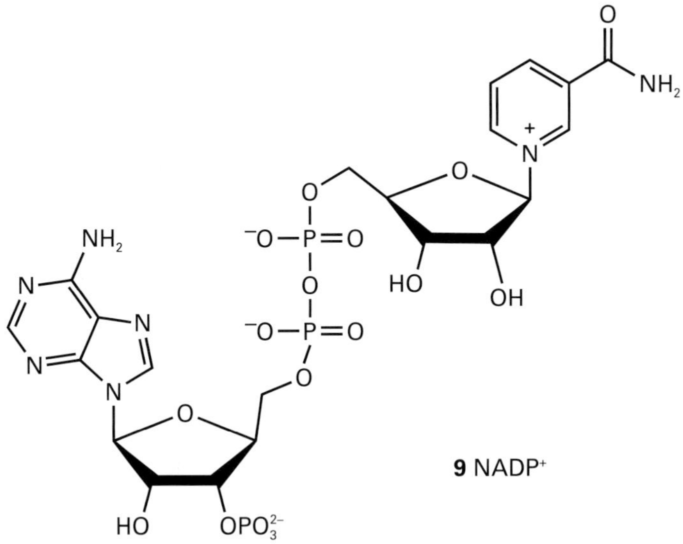

9 NADP⁺

excited and initiates ultrafast electron transfer reactions. For example, the transfer of an electron from the excited singlet state of P680, the chlorophyll dimer of the photosystem II reaction centre, to its immediate electron acceptor, a phaeophytin *a* molecule (a chlorophyll *a* molecule where the central Mg^{2+} ion is replaced by two protons, which are bound to two of the pyrrole nitrogens in the ring), occurs within 3 ps. Once the excited state of P680 has been quenched efficiently by this first reaction, subsequent steps that lead to the oxidation of water occur more slowly, with reaction times varying from 200 ps to 1 ms. The electrochemical reactions within the photosystem I reaction centre also occur in this time interval. We see that the initial energy and electron transfer events of photosynthesis are under tight kinetic control. Photosynthesis captures solar energy efficiently because the excited singlet state of chlorophyll is quenched rapidly by processes that occur with relaxation times that are much shorter than the fluorescence lifetime, which is typically about 1 ns in organic solvents at room temperature.

Working together, photosystem I and the enzyme ferredoxin:$NADP^+$ oxidoreductase catalyse the light-induced oxidation of $NADP^+$ to NADPH. The electrons required for this process come initially from P700 in its excited state. The resulting $P700^+$ is then reduced by the mobile carrier plastocyanin (Pc), a protein in which the bound copper ion can exist in oxidation states +2 and +1. The net reaction is

$$NADP^+ + 2\,Cu^+(Pc) + H^+ \xrightarrow{\text{light, photosystem I}} NADPH + 2\,Cu^{2+}(Pc)$$

Oxidized plastocyanin accepts electrons from reduced plastoquinone (PQ, **10**). The process is catalysed by the cytochrome b_6f complex, a membrane protein complex that resembles complex III of mitochondria (*Impact* I7.2):

$$PQH_2 + 2\,Cu^{2+}(Pc) \xrightarrow{\text{cyt } b_6 f \text{ complex}} PQ + 2\,H^+ + 2\,Cu^+(Pc)$$
$$E^{\ominus} = +0.370\ \text{V},\ \Delta_r G^{\ominus} = -71.4\ \text{kJ mol}^{-1}$$

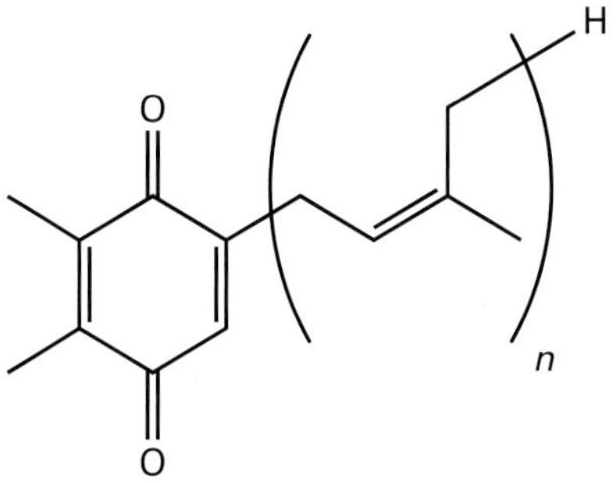

10 Plastoquinone

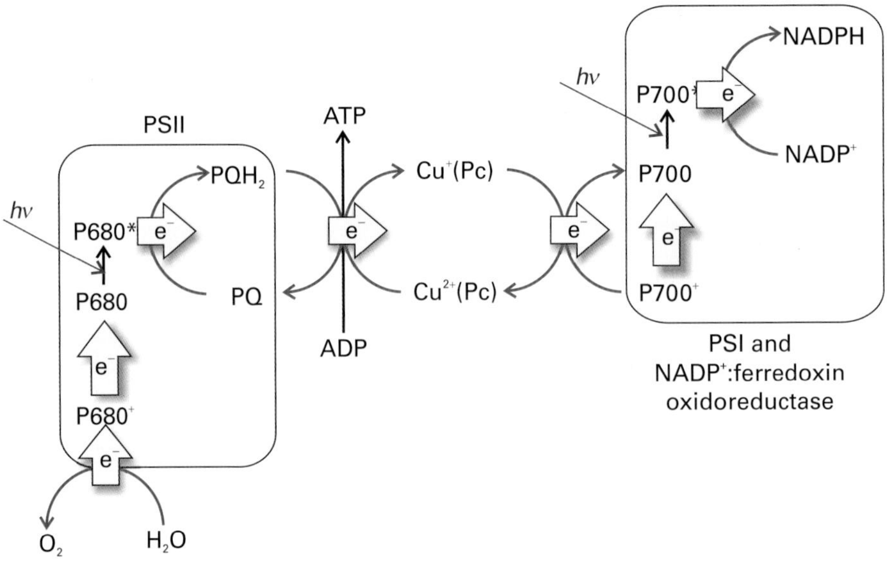

Fig. 23.20 In plant photosynthesis, light-induced electron transfer processes lead to the oxidation of water to O_2 and the reduction of $NADP^+$ to NADPH, with concomitant production of ATP. The energy stored in ATP and NADPH is used to reduce CO_2 to carbohydrate in a separate set of reactions. The scheme summarizes the general patterns of electron flow and does not show all the intermediate electron carriers in photosystems I and II, the cytochrome b_6f complex, and ferredoxin:$NADP^+$ oxidoreductase.

This reaction is sufficiently exergonic to drive the synthesis of ATP in the process known as **photophosphorylation**.

Plastoquinone is reduced by water in a process catalysed by light and photosystem II. The electrons required for the reduction of plastoquinone come initially from P680 in its excited state. The resulting $P680^+$ is then reduced ultimately by water. The net reaction is

$$H_2O + PQ \xrightarrow{\text{light, photosystem II}} \tfrac{1}{2}O_2 + PQH_2$$

In this way, plant photosynthesis uses an abundant source of electrons (water) and of energy (the Sun) to drive the endergonic reduction of $NADP^+$, with concomitant synthesis of ATP (Fig. 23.20). Experiments show that, for each molecule of NADPH formed in the chloroplast of green plants, one molecule of ATP is synthesized.

The ATP and NADPH molecules formed by the light-induced electron transfer reactions of plant photosynthesis participate directly in the reduction of CO_2 to glucose in the chloroplast:

$$6\,CO_2 + 12\,NADPH + 12\,ATP + 12\,H^+ \rightarrow C_6H_{12}O_6 + 12\,NADP^+ + 12\,ADP + 12\,P_i + 6H_2O$$

In summary, plant photosynthesis uses solar energy to transfer electrons from a poor reductant (water) to carbon dioxide. In the process, high energy molecules (carbohydrates, such as glucose) are synthesized in the cell. Animals feed on the carbohydrates derived from photosynthesis. During aerobic metabolism, the O_2 released by photosynthesis as a waste product is used to oxidize carbohydrates to CO_2, driving biological processes, such as biosynthesis, muscle contraction, cell division, and nerve conduction. Hence, the sustenance of life on Earth depends on a tightly regulated carbon–oxygen cycle that is driven by solar energy.

23.8 Complex photochemical processes

Many photochemical processes have complex mechanisms that may be examined by considering the concepts developed above. In this section, we consider two examples: photochemical chain reactions and photosensitization.

(a) The overall quantum yield of a photochemical reaction

For complex reactions involving secondary processes, such as chain reactions initiated by photolysis, many reactant molecules might be consumed as a result of absorption of a single photon. The **overall quantum yield**, Φ, for such reactions, the number of reactant molecules consumed per photon absorbed, might exceed 1. In the photolysis of HI, for example, the processes are

$$HI + h\nu \rightarrow H\cdot + I\cdot$$

$$H\cdot + HI \rightarrow H_2 + I\cdot$$

$$I\cdot + I\cdot + M \rightarrow I_2 + M^*$$

The overall quantum yield is 2 because the absorption of one photon leads to the destruction of two HI molecules. In a chain reaction, Φ may be very large, and values of about 10^4 are common. In such cases the chain acts as a chemical amplifier of the initial absorption step.

Example 23.5 *Determining the quantum yield of a photochemical reaction*

When a sample of 4-heptanone was irradiated for 100 s with 313 nm radiation with a power output of 50 W under conditions of total absorption, it was found that 2.8 mmol C_2H_4 was formed. What is the quantum yield for the formation of ethene?

Method First, calculate the amount of photons generated in an interval Δt: see Example 8.1. Then divide the amount of ethene molecules formed by the amount of photons absorbed.

Answer From Example 8.1, the amount (in moles) of photons absorbed is

$$n = \frac{P\Delta t}{(hc/\lambda)N_A}$$

If $n_{C_2H_4}$ is the amount of ethene formed, the quantum yield is

$$\Phi = \frac{n_{C_2H_4}}{n} = \frac{n_{C_2H_4}N_A hc}{\lambda P\Delta t}$$

$$= \frac{(2.8\times10^{-3}\,\text{mol})\times(6.022\times10^{23}\,\text{mol}^{-1})\times(6.626\times10^{-34}\,\text{J s})\times(2.997\times10^{8}\,\text{m s}^{-1})}{(3.13\times10^{-7}\,\text{m})\times(50\,\text{J s}^{-1})\times(100\,\text{s})}$$

$$= 0.21$$

Self-test 23.5 The overall quantum yield for another reaction at 290 nm is 0.30. For what length of time must irradiation with a 100 W source continue in order to destroy 1.0 mol of molecules? [3.8 h]

(b) Rate laws of complex photochemical reactions

As an example of how to write a rate law for a complex photochemical process, consider the photochemical activation of the reaction

$$H_2(g) + Br_2(g) \rightarrow 2\,HBr(g)$$

In place of the first step in the thermal reaction we have

$$Br_2 + h\nu \rightarrow Br\cdot + Br\cdot \qquad v = I_{abs}$$

where I_{abs} is the number of photons of the appropriate frequency absorbed divided by the volume in which absorption occurs and the time interval. We are assuming a primary quantum yield of unity for the photodissociation of Br_2. It follows that I_{abs} should take the place of $k_i[Br_2][M]$ in the thermal reaction scheme, so from Example 23.1 we can write

$$\frac{d[HBr]}{dt} = \frac{2k_p(1/k_t[M])^{1/2}[H_2][Br_2]I_{abs}^{1/2}}{[Br_2] + (k_r/k'_p)[HBr]} \tag{23.39}$$

We predict that the reaction rate should depend on the square root of the absorbed light intensity, which is confirmed experimentally.

(c) Photosensitization

The reactions of a molecule that does not absorb directly can be made to occur if another absorbing molecule is present, because the latter may be able to transfer its energy to the former during a collision. An example of this **photosensitization** is the reaction used to generate excited state O_2 in a type of treatment known as *photodynamic therapy* (*Impact* I23.3). Another example is the reaction used to generate atomic hydrogen, the irradiation of hydrogen gas containing a trace of mercury vapour using radiation of wavelength 254 nm from a mercury discharge lamp. The Hg atoms are excited (to Hg*) by resonant absorption of the radiation, and then collide with H_2 molecules. Two reactions then take place:

$$Hg^* + H_2 \rightarrow Hg + H\cdot + H\cdot$$
$$Hg^* + H_2 \rightarrow HgH + H\cdot$$

The latter reaction is the initiation step for other mercury photosensitized reactions, such as the synthesis of formaldehyde from carbon monoxide and hydrogen:

$$H\cdot + CO \rightarrow HCO\cdot$$
$$HCO\cdot + H_2 \rightarrow HCHO + H\cdot$$
$$HCO\cdot + HCO\cdot \rightarrow HCHO + CO$$

Note that the last step is termination by disproportionation rather than by combination.

IMPACT ON MEDICINE

I23.3 Photodynamic therapy

In photodynamic therapy (PDT), laser radiation, which is usually delivered to diseased tissue through a fibre optic cable, is absorbed by a drug which, in its first excited triplet state ^3P, photosensitizes the formation of an excited singlet state of O_2, $^1O_2$. The $^1O_2$ molecules are very reactive and destroy cellular components and it is thought that cell membranes are the primary cellular targets. Hence, the photochemical cycle below leads to the shrinkage (and sometimes total destruction) of diseased tissue.

Absorption:	$P + h\nu \rightarrow P^*$
Intersystem crossing:	$P^* \rightarrow {}^3P$
Photosensitization:	${}^3P + {}^3O_2 \rightarrow P + {}^1O_2$
Oxidation reactions:	${}^1O_2$ + reactants $\rightarrow$ products

The photosensitizer is hence a 'photocatalyst' for the production of $^1O_2$. It is common practice to use a porphyrin photosensitizer, such as compounds derived from haematoporphyrin (**11**). However, much effort is being expended to develop better drugs with enhanced photochemical properties.

A potential PDT drug must meet many criteria. From the point of view of pharmacological effectiveness, the drug must be soluble in tissue fluids, so it can be transported to the diseased organ through blood and secreted from the body through

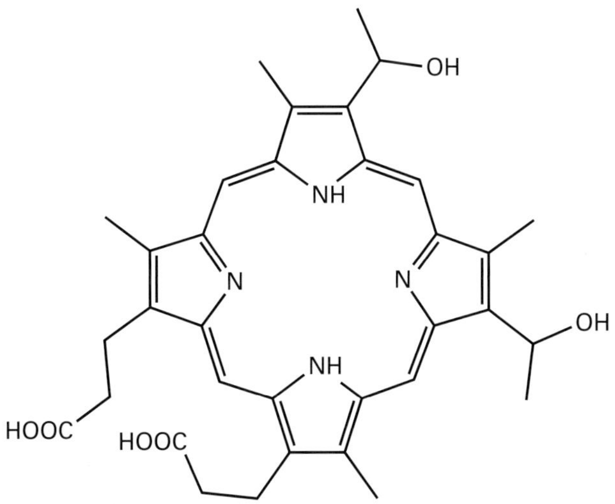

11 Haematoporphyrin

urine. The therapy should also result in very few side effects. The drug must also have unique photochemical properties. It must be activated photochemically at wavelengths that are not absorbed by blood and skin. In practice, this means that the drug should have a strong absorption band at $\lambda > 650$ nm. Drugs based on haematoporphyrin do not meet this criterion very well, so novel porphyrin and related macrocycles with more desirable electronic properties are being synthesized and tested. At the same time, the quantum yield of triplet formation and of $^1O_2$ formation must be high, so many drug molecules can be activated and many oxidation reactions can occur during a short period of laser irradiation. Photodynamic therapy has been used successfully in the treatment of macular degeneration, a disease of the retina that leads to blindness, and in a number of cancers, including those of the lung, bladder, skin, and oesophagus.

Checklist of key ideas

☐ 1. In a chain reaction, an intermediate (the chain carrier) produced in one step (the initiation step) attacks other reactant molecules (in the propagation steps), with each attack giving rise to a new carrier. The chain ends in the termination step. Examples of chain reactions include some explosions.

☐ 2. In stepwise polymerization any two monomers in the reaction mixture can link together at any time and growth of the polymer is not confined to chains that are already forming. The longer a stepwise polymerization proceeds, the higher the average molar mass of the product.

☐ 3. In chain polymerization an activated monomer attacks another monomer and links to it. That unit attacks another monomer, and so on. The slower the initiation of the chain, the higher the average molar mass of the polymer.

☐ 4. Catalysts are substances that accelerate reactions but undergo no net chemical change.

☐ 5. A homogeneous catalyst is a catalyst in the same phase as the reaction mixture. Enzymes are homogeneous, biological catalysts.

☐ 6. The Michaelis–Menten mechanism of enzyme kinetics accounts for the dependence of rate on the concentration of the substrate, $v = v_{max}[S]_0/([S]_0 + K_M)$.

☐ 7. A Lineweaver–Burk plot, based on $1/v = 1/v_{max} + (K_M/v_{max})(1/[S]_0)$, is used to determine the parameters that occur in the Michaelis–Menten mechanism.

☐ 8. In competitive inhibition of an enzyme, the inhibitor binds only to the active site of the enzyme and thereby inhibits the attachment of the substrate.

☐ 9. In uncompetitive inhibition the inhibitor binds to a site of the enzyme that is removed from the active site, but only if the substrate is already present.

☐ 10. In non-competitive inhibition, the inhibitor binds to a site other than the active site, and its presence reduces the ability of the substrate to bind to the active site.

☐ 11. The primary quantum yield of a photochemical reaction is the number of reactant molecules producing specified primary products for each photon absorbed; the overall quantum yield is the number of reactant molecules that react for each photon absorbed.

☐ 12. The observed fluorescence lifetime is related to the quantum yield, ϕ_f, and rate constant, k_f, of fluorescence by $\tau_0 = \phi_f/k_f$.

☐ 13. A Stern–Volmer plot is used to analyse the kinetics of fluorescence quenching in solution. It is based on the Stern–Volmer equation, $\phi_{f,0}/\phi_f = 1 + \tau_0 k_Q[Q]$.

☐ 14. Collisional deactivation, electron transfer, and resonance energy transfer are common fluorescence quenching processes. The rate constants of electron and resonance energy transfer decrease with increasing separation between donor and acceptor molecules.

☐ 15. In photosensitization, the reaction of a molecule that does not absorb radiation directly is made to occur by energy transfer during a collision with a molecule that does absorb radiation.

Further reading

Articles and texts

J. Andraos, How mathematics figures in chemistry: some examples. *J. Chem. Educ.* **76**, 258 (1999).

C.E. Carraher, Jr., *Seymour/Carraher's polymer chemistry*, Marcel Dekker (2000).

A. Fersht, *Structure and mechanism in protein science: a guide to enzyme catalysis and protein folding*. W.H. Freeman, New York (1999).

A.M. Kuznetsov and J. Ulstrup, *Electron transfer in chemistry and biology: an introduction to the theory*. Wiley, New York (1998).

J.I. Steinfeld, J.S. Francisco, and W.L. Hase, *Chemical kinetics and dynamics*. Prentice Hall, Englewood Cliffs (1998).

N.J. Turro, *Modern molecular photochemistry*. University Science Books, Sausalito (1991).

K.E. van Holde, W.C. Johnson and P.H. Ho, *Principles of physical biochemistry*. Prentice Hall, Upper Saddle River (1998).

D. Voet and J.G. Voet, *Biochemistry*. Wiley, New York (2004).

Sources of data and information

D.R. Lide (ed.), *CRC handbook of chemistry and physics*, Section 5. CRC Press, Boca Raton (2000).

S.L. Murov, I. Carmichael, and G.L. Hug, *Handbook of photochemistry*. Marcel Dekker, New York (1993).

NDRL/NIST solution kinetics database, NIST standard reference database 40, National Institute of Standards and Technology, Gaithersburg (1994). The URL is available on the web site for this book.

NIST chemical kinetics database, NIST standard reference database 17, National Institute of Standards and Technology, Gaithersburg (1998). The URL is available on the web site for this book.

Further information

Further information 23.1 *The Förster theory of resonance energy transfer*

From the qualitative description given in Section 23.7e, we conclude that resonance energy transfer arises from the interaction between two oscillating dipoles with moments μ_S and μ_Q. From Section 18.4, the energy of the dipole–dipole interaction, $V_{\text{dipole–dipole}}$, is

$$V_{\text{dipole–dipole}} \propto \frac{\mu_S\mu_Q}{R^3}$$

where R is the distance between the dipoles. We saw in Section 9.10 that the rate of a transition from a state i to a state f at a radiation frequency ν is proportional to the square modulus of the matrix element of the perturbation between the two states:

$$w_{f\leftarrow i} \propto |H^{(1)}_{fi}|^2$$

For energy transfer, the wavefunctions of the initial and final states may be denoted as $\psi_S^*\psi_Q$ and $\psi_S\psi_Q^*$, respectively, and $H^{(1)}$ may be written from $V_{\text{dipole–dipole}}$. It follows that the rate of energy transfer, w_T, at a fixed distance R is given (using notation introduced in *Further information 9.1*) by

$$w_T \propto \frac{1}{R^6}\left\langle \psi_S\psi_{Q^*} \middle| \mu_S\mu_Q \middle| \psi_{S^*}\psi_Q \right\rangle^2$$

$$= \frac{1}{R^6}\left|\left\langle \psi_S \middle| \mu_S \middle| \psi_{S^*} \right\rangle\right|^2 \left\langle \psi_{Q^*} \middle| \mu_Q \middle| \psi_Q \right\rangle^2$$

We have used the fact that the terms related to S are functions of coordinates that are independent of those for the functions related to Q. In the last expression, the integrals are squares of transition dipole moments at the radiation frequency ν, the first corresponding to emission of S* to S and the second to absorption of Q to Q*.

We interpret the expression for w_T as follows. The rate of energy transfer is proportional to R^{-6}, so it decreases sharply with increasing separation between the energy donor and acceptor. Furthermore, the energy transfer rate is optimized when both emission of radiation by S* and absorption of radiation by Q are efficient at the frequency ν. Because the absorption and emission spectra of large molecules in condensed phases are broad, it follows that the energy transfer rate is optimal at radiation frequencies in which the emission spectrum of the donor and the absorption spectrum of the acceptor overlap significantly.

In practice, it is more convenient to measure the efficiency of energy transfer and not the rate itself. In much the same way that we defined the quantum yield as a ratio of rates, we can also define the efficiency of energy transfer, E_T, as the ratio

$$E_T = \frac{w_T}{w_T + w_0} \qquad w_0 = (k_f + k_{IC} + k_{ISC})[S^*] \tag{23.40}$$

where w_0 is the rate of deactivation of S* in the absence of the quencher. The efficiency, E_T, may be expressed in terms of the experimental fluorescence quantum yields $\phi_{f,0}$ and ϕ_f of the donor in the absence and presence of the acceptor, respectively. To proceed, we use eqn 23.30 to write:

$$\phi_{f,0} = \frac{v_f}{w_0} \quad \text{and} \quad \phi_f = \frac{v_f}{w_0 + w_T}$$

where v_f is the rate of fluorescence. Subsituting these results into eqn 23.40 gives, after a little algebra, eqn 23.37.

Alternatively, we can express w_0 in terms of the parameter R_0, the characteristic distance at which $w_T = w_0$ for a specified pair of S and Q (Table 23.3). By using $w_T \propto R^{-6}$ and $w_0 \propto R_0^{-6}$, we can rearrange the expression for E_T into eqn 23.38.

Discussion questions

23.1 Identify any initiation, propagation, retardation, inhibition, and termination steps in the following chain mechanisms:

(a) (1) $AH \rightarrow A\cdot + H\cdot$

(2) $A\cdot \rightarrow B\cdot + C$

(3) $AH + B\cdot \rightarrow A\cdot + D$

(4) $A\cdot + B\cdot \rightarrow P$

(b) (1) $A_2 \rightarrow A\cdot + A\cdot$

(2) $A\cdot \rightarrow B\cdot + C$

(3) $A\cdot + P \rightarrow B\cdot$

(4) $A\cdot + B\cdot \rightarrow P$

23.2 Bearing in mind distinctions between the mechanisms of stepwise and chain polymerization, describe ways in which it is possible to control the molar mass of a polymer by manipulating the kinetic parameters of polymerization.

23.3 Discuss the features, advantages, and limitations of the Michaelis–Menten mechanism of enzyme action.

23.4 Distinguish between competitive, non-competitive, and uncompetitive inhibition of enzymes. Discuss how these modes of inhibition may be detected experimentally.

23.5 Distinguish between the primary quantum yield and overall quantum yield of a chemical reaction. Describe an experimental procedure for the determination of the quantum yield.

23.6 Discuss experimental procedures that make it possible to differentiate between quenching by energy transfer, collisions, or electron transfer.

23.7 Summarize the main features of the Förster theory of resonance energy transfer. Then, discuss FRET in terms of Förster theory.

Exercises

In the following exercises and problems, it is recommended that rate constants are labelled with the number of the step in the proposed reaction mechanism, and any reverse steps are labelled similarly but with a prime.

23.1a Derive the rate law for the decomposition of ozone in the reaction $2\,O_3(g) \rightarrow 3\,O_2(g)$ on the basis of the following proposed mechanism:

(1) $O_3 \rightleftharpoons O_2 + O$ $\quad k_1, k_1'$

(2) $O + O_3 \rightarrow O_2 + O_2$ $\quad k_2$

23.1b On the basis of the following proposed mechanism, account for the experimental fact that the rate law for the decomposition $2\,N_2O_5(g) \rightarrow 4\,NO_2(g) + O_2(g)$ is $v = k[N_2O_5]$.

(1) $N_2O_5 \rightleftharpoons NO_2 + NO_3$ $\quad k_1, k_1'$

(2) $NO_2 + NO_3 \rightarrow NO_2 + O_2 + NO$ $\quad k_2$

(3) $NO + N_2O_5 \rightarrow NO_2 + NO_2 + NO_2$ $\quad k_3$

23.2a A slightly different mechanism for the decomposition of N_2O_5 from that in Exercise 23.1b has also been proposed. It differs only in the last step, which is replaced by

(3) $NO + NO_3 \rightarrow NO_2 + NO_2$ $\quad k_3$

Show that this mechanism leads to the same overall rate law.

23.2b Consider the following mechanism for the thermal decomposition of R_2:

(1) $R_2 \rightarrow R + R$

(2) $R + R_2 \rightarrow P_B + R'$

(3) $R' \rightarrow P_A + R$

(4) $R + R \rightarrow P_A + P_B$

where R_2, P_A, P_B are stable hydrocarbons and R and R′ are radicals. Find the dependence of the rate of decomposition of R_2 on the concentration of R_2.

23.3a Refer to Fig. 23.3 and determine the pressure range for a branching chain explosion in the hydrogen–oxygen reaction at 800 K.

23.3b Refer to Fig. 23.3 and determine the pressure range for a branching chain explosion in the hydrogen–oxygen reaction at (a) 700 K, (b) 900 K.

23.4a The condensation reaction of propanone, $(CH_3)_2CO$, in aqueous solution is catalysed by bases, B, which react reversibly with propanone to form the carbanion $C_3H_5O^-$. The carbanion then reacts with a molecule of propanone to give the product. A simplified version of the mechanism is

(1) $AH + B \rightarrow BH^+ + A^-$

(2) $A^- + BH^+ \rightarrow AH + B$

(3) $A^- + AH \rightarrow$ product

where AH stands for propanone and A^- denotes its carbanion. Use the steady-state approximation to find the concentration of the carbanion and derive the rate equation for the formation of the product.

23.4b Consider the acid-catalysed reaction

(1) $HA + H^+ \rightleftharpoons HAH^+$ $\qquad k_1, k_1'$, both fast

(2) $HAH^+ + B \rightarrow BH^+ + AH$ $\qquad k_2$, slow

Deduce the rate law and show that it can be made independent of the specific term $[H^+]$.

23.5a Consider the following chain mechanism:

(1) $AH \rightarrow A\cdot + H\cdot$

(2) $A\cdot \rightarrow B\cdot + C$

(3) $AH + B\cdot \rightarrow A\cdot + D$

(4) $A\cdot + B\cdot \rightarrow P$

Use the steady-state approximation to deduce that the decomposition of AH is first-order in AH.

23.5b Consider the following chain mechanism:

(1) $A_2 \rightarrow A\cdot + A\cdot$

(2) $A\cdot \rightarrow B\cdot + C$

(3) $A\cdot + P \rightarrow B\cdot$

(4) $A\cdot + B\cdot \rightarrow P$

Use the steady-state approximation to deduce that the rate law for the consumption of A_2.

23.6a The enzyme-catalysed conversion of a substrate at 25°C has a Michaelis constant of 0.035 mol dm^{-3}. The rate of the reaction is 1.15×10^{-3} mol dm^{-3} s^{-1} when the substrate concentration is 0.110 mol dm^{-3}. What is the maximum velocity of this enzymolysis?

23.6b The enzyme-catalysed conversion of a substrate at 25°C has a Michaelis constant of 0.042 mol dm^{-3}. The rate of the reaction is 2.45×10^{-4} mol dm^{-3} s^{-1} when the substrate concentration is 0.890 mol dm^{-3}. What is the maximum velocity of this enzymolysis?

23.7a In a photochemical reaction $A \rightarrow 2\,B + C$, the quantum efficiency with 500 nm light is 2.1×10^2 mol einstein^{-1} (1 einstein = 1 mol photons). After exposure of 300 mmol of A to the light, 2.28 mmol of B is formed. How many photons were absorbed by A?

23.7b In a photochemical reaction $A \rightarrow B + C$, the quantum efficiency with 500 nm light is 1.2×10^2 mol einstein^{-1}. After exposure of 200 mmol A to the light, 1.77 mmol B is formed. How many photons were absorbed by A?

23.8a In an experiment to measure the quantum efficiency of a photochemical reaction, the absorbing substance was exposed to 490 nm light from a 100 W source for 45 min. The intensity of the transmitted light was 40 per cent of the intensity of the incident light. As a result of irradiation, 0.344 mol of the absorbing substance decomposed. Determine the quantum efficiency.

23.8b In an experiment to measure the quantum efficiency of a photochemical reaction, the absorbing substance was exposed to 320 nm radiation from a 87.5 W source for 28.0 min. The intensity of the transmitted light was 0.257 that of the incident light. As a result of irradiation, 0.324 mol of the absorbing substance decomposed. Determine the quantum efficiency.

Problems*

Numerical problems

23.1 Studies of combustion reactions depend on knowing the concentrations of H atoms and HO radicals. Measurements on a flow system using EPR for the detection of radicals gave information on the reactions

(1) $H + NO_2 \rightarrow OH + NO$ $\qquad k_1 = 2.9 \times 10^{10}$ dm^3 mol^{-1} s^{-1}

(2) $OH + OH \rightarrow H_2O + O$ $\qquad k_2 = 1.55 \times 10^9$ dm^3 mol^{-1} s^{-1}

(3) $O + OH \rightarrow O_2 + H$ $\qquad k_3 = 1.1 \times 10^{10}$ dm^3 mol^{-1} s^{-1}

(J.N. Bradley, W. Hack, K. Hoyermann, and H.G. Wagner, *J. Chem. Soc. Faraday Trans.* I, 1889 (1973)). Using initial H atom and NO_2 concentrations of 4.5×10^{-10} mol cm^{-3} and 5.6×10^{-10} mol cm^{-3}, respectively, compute and plot curves showing the O, O_2, and OH concentrations as a function of time in the range 0–10 ns.

23.2 In a flow study of the reaction between O atoms and Cl_2 (J.N. Bradley, D.A. Whytock, and T.A. Zaleski, *J. Chem. Soc. Faraday Trans.* I, 1251 (1973)) at high chlorine pressures, plots of $\ln [O]_0/[O]$ against distances l along the flow tube, where $[O]_0$ is the oxygen concentration at zero chlorine pressure, gave straight lines. Given the flow velocity as 6.66 m s^{-1} and the data below, find the rate coefficient for the reaction $O + Cl_2 \rightarrow ClO + Cl$.

l/cm	0	2	4	6	8	10	12	14	16	18
$\ln([O]_0/[O])$	0.27	0.31	0.34	0.38	0.45	0.46	0.50	0.55	0.56	0.60

with $[O]_0 = 3.3\ 10^{-8}$ mol dm^{-3}, $[Cl_2] = 2.54 \times 10^{-7}$ mol dm^{-3}, $p = 1.70$ Torr.

23.3‡ J.D. Chapple-Sokol, C.J. Giunta, and R.G. Gordon (*J. Electrochem. Soc.* **136**, 2993 (1989)) proposed the following radical chain mechanism for the initial stages of the gas-phase oxidation of silane by nitrous oxide:

(1) $N_2O \rightarrow N_2 + O$

(2) $O + SiH_4 \rightarrow SiH_3 + OH$

(3) $OH + SiH_4 \rightarrow SiH_3 + H_2O$

(4) $SiH_3 + N_2O \rightarrow SiH_3O + N_2$

* Problems denoted with the symbol ‡ were supplied by Charles Trapp, Carmen Giunta, and Marshall Cady.

(5) $SiH_3O + SiH_4 \rightarrow SiH_3OH + SiH_3$

(6) $SiH_3 + SiH_3O \rightarrow (H_3Si)_2O$

Label each step with its role in the chain. Use the steady-state approximation to show that this mechanism predicts the following rate law for SiH_4 consumption (provided k_1 and k_6 are in some sense small):

$$\frac{d[SiH_4]}{dt} = -k[N_2O][SiH_4]^{1/2}$$

23.4‡ The water formation reaction has been studied many times and continues to be of interest. Despite the many studies there is not uniform agreement on the mechanism. But as explosions are known to occur at certain critical values of the pressure, any proposed mechanism to be considered plausible must be consistent with the existence of these critical explosion limits. One such plausible mechanism is that of Example 23.2. Another is the following:

(1) $H_2 \rightarrow H + H$

(2) $H + O_2 \rightarrow OH + O$

(3) $O + H_2 \rightarrow OH + H$

(4) $H + O_2 \rightarrow HO_2$

(5) $HO_2 + H_2 \rightarrow H_2O + OH$

(6) HO_2 + wall → destruction

(7) H + M → destruction

In a manner similar to that in Example 23.2, determine whether or not this mechanism can lead to explosions under appropriate conditions.

23.5‡ For many years the reaction $H_2(g) + I_2(g) \rightarrow 2\,HI(g)$ and its reverse were assumed to be elementary bimolecular reactions. However, J.H. Sullivan (*J. Chem. Phys.* **46**, 73 (1967)) suggested that the following mechanism for the reaction, originally proposed by M. Bodenstein (*Z. Physik. Chem.* **29**, 56 (1898)), provides a better explanation of the experimental results:

(1) $I_2 \rightleftharpoons I + I$ $\quad k_1, k_1'$

(2) $I + I + H_2 \rightarrow HI + HI$ $\quad k_2$

Obtain the expression for the rate of formation of HI based on this mechanism. Under what conditions does this rate law reduce to the one for the originally accepted mechanism?

23.6 The number of photons falling on a sample can be determined by a variety of methods, of which the classical one is chemical actinometry. The decomposition of oxalic acid $(COOH)_2$, in the presence of uranyl sulfate, $(UO_2)SO_4$, proceeds according to the sequence

(1) $UO^{2+} + h\nu \rightarrow (UO^{2+})^*$

(2) $(UO^{2+})^* + (COOH)_2 \rightarrow UO^{2+} + H_2O + CO_2 + CO$

with a quantum efficiency of 0.53 at the wavelength used. The amount of oxalic acid remaining after exposure can be determined by titration (with $KMnO_4$) and the extent of decomposition used to find the number of incident photons. In a particular experiment, the actinometry solution consisted of 5.232 g anhydrous oxalic acid, 25.0 cm^3 water (together with the uranyl salt). After exposure for 300 s the remaining solution was titrated with 0.212 M $KMnO_4$(aq), and 17.0 cm^3 were required for complete oxidation of the remaining oxalic acid. The titration reaction is

$$2\,MnO_4^-(aq) + 5\,(COOH)_2(aq) + 6\,H^+(aq) \rightarrow 10\,CO_2(g) + 8\,H_2O(l) + 2\,Mn^{2+}(aq)$$

What is the rate of incidence of photons at the wavelength of the experiment? Express the answer in photons/second and einstein/second.

23.7 Dansyl chloride, which absorbs maximally at 330 nm and fluoresces maximally at 510 nm, can be used to label aminoacids in fluorescence microscopy and FRET studies. Tabulated below is the variation of the fluorescence intensity of an aqueous solution of dansyl chloride with time after excitation by a short laser pulse (with I_0 the initial fluorescence intensity).

t/ns	5.0	10.0	15.0	20.0
I_f/I_0	0.45	0.21	0.11	0.05

(a) Calculate the observed fluorescence lifetime of dansyl chloride in water. (b) The fluorescence quantum yield of dansyl chloride in water is 0.70. What is the fluorescence rate constant?

23.8 When benzophenone is illuminated with ultraviolet light it is excited into a singlet state. This singlet changes rapidly into a triplet, which phosphoresces. Triethylamine acts as a quencher for the triplet. In an experiment in methanol as solvent, the phosphorescence intensity varied with amine concentration as shown below. A time-resolved laser spectroscopy experiment had also shown that the half-life of the fluorescence in the absence of quencher is 29 μs. What is the value of k_q?

[Q]/(mol dm^{-3})	0.0010	0.0050	0.0100
I_f/(arbitrary units)	0.41	0.25	0.16

23.9 An electronically excited state of Hg can be quenched by N_2 according to

$$Hg^*(g) + N_2(g, v=0) \rightarrow Hg(g) + N_2(g, v=1)$$

in which energy transfer from Hg* excites N_2 vibrationally. Fluorescence lifetime measurements of samples of Hg with and without N_2 present are summarized below ($T = 300$ K):

$p_{N_2} = 0.0$ atm

Relative fluorescence intensity	1.000	0.606	0.360	0.22	0.135
t/μs	0.0	5.0	10.0	15.0	20.0

$p_{N_2} = 9.74 \times 10^{-4}$ atm

Relative fluorescence intensity	1.000	0.585	0.342	0.200	0.117
t/μs	0.0	3.0	6.0	9.0	12.0

You may assume that all gases behave ideally. Determine the rate constant for the energy transfer process.

23.10 The Förster theory of resonance energy transfer and the basis for the FRET technique can be tested by performing fluorescence measurements on a series of compounds in which an energy donor and an energy acceptor are covalently linked by a rigid molecular linker of variable and known length. L. Stryer and R.P. Haugland (*Proc. Natl. Acad. Sci. USA* **58**, 719 (1967)) collected the following data on a family of compounds with the general composition dansyl-(L-prolyl)$_n$-naphthyl, in which the distance R between the naphthyl donor and the dansyl acceptor was varied from 1.2 nm to 4.6 nm by increasing the number of prolyl units in the linker:

R/nm	1.2	1.5	1.8	2.8	3.1	3.4	3.7	4.0	4.3	4.6
$1 - E_T$	0.99	0.94	0.97	0.82	0.74	0.65	0.40	0.28	0.24	0.16

Are the data described adequately by eqn 23.38? If so, what is the value of R_0 for the naphthyl-dansyl pair?

Theoretical problems

23.11 The Rice–Herzfeld mechanism for the dehydrogenation of ethane is specified in Section 23.1, and it was noted there that it led to first-order kinetics. Confirm this remark, and find the approximations that lead to the rate law quoted there. How may the conditions be changed so that the reaction shows different orders?

23.12 The following mechanism has been proposed for the thermal decomposition of acetaldehyde (ethanal):

(1) $CH_3CHO \rightarrow \cdot CH_3 + CHO$

(2) $\cdot CH_3 + CH_3CHO \rightarrow CH_4 + \cdot CH_2CHO$

(3) $\cdot CH_2CHO \rightarrow CO + \cdot CH_3$

(4) $\cdot CH_3 + \cdot CH_3 \rightarrow CH_3CH_3$

Find an expression for the rate of formation of methane and the rate of disappearance of acetaldehyde.

23.13 Express the root mean square deviation $\{\langle M^2\rangle - \langle \bar{M}\rangle^2\}^{1/2}$ of the molar mass of a condensation polymer in terms of p, and deduce its time dependence.

23.14 Calculate the ratio of the mean cube molar mass to the mean square molar mass in terms of (a) the fraction p, (b) the chain length.

23.15 Calculate the average polymer length in a polymer produced by a chain mechanism in which termination occurs by a disproportionation reaction of the form $M\cdot + \cdot M \rightarrow M + :M$.

23.16 Derive an expression for the time dependence of the degree of polymerization for a stepwise polymerization in which the reaction is acid-catalysed by the —COOH acid functional group. The rate law is $d[A]/dt = -k[A]^2[OH]$.

23.17 Autocatalysis is the catalysis of a reaction by the products. For example, for a reaction $A \rightarrow P$ it may be found that the rate law is $v = k[A][P]$ and the reaction rate is proportional to the concentration of P. The reaction gets started because there are usually other reaction routes for the formation of some P initially, which then takes part in the autocatalytic reaction proper. (a) Integrate the rate equation for an autocatalytic reaction of the form $A \rightarrow P$, with rate law $v = k[A][P]$, and show that

$$\frac{[P]}{[P]_0} = (b+1)\frac{e^{at}}{1+be^{at}}$$

where $a = ([A]_0 + [P]_0)k$ and $b = [P]_0/[A]_0$. *Hint.* Starting with the expression $v = -d[A]/dt = k[A][P]$, write $[A] = [A]_0 - x$, $[P] = [P]_0 + x$ and then write the expression for the rate of change of either species in terms of x. To integrate the resulting expression, the following relation will be useful:

$$\frac{1}{([A]_0 - x)([P]_0 + x)} = \frac{1}{[A]_0 + [P]_0}\left(\frac{1}{[A]_0 - x} + \frac{1}{[P]_0 + x}\right)$$

(b) Plot $[P]/[P]_0$ against at for several values of b. Discuss the effect of autocatalysis on the shape of a plot of $[P]/[P]_0$ against t by comparing your results with those for a first-order process, in which $[P]/[P]_0 = 1 - e^{-kt}$. (c) Show that, for the autocatalytic process discussed in parts (a) and (b), the reaction rate reaches a maximum at $t_{max} = -(1/a)\ln b$. (d) An autocatalytic reaction $A \rightarrow P$ is observed to have the rate law $d[P]/dt = k[A]^2[P]$. Solve the rate law for initial concentrations $[A]_0$ and $[P]_0$. Calculate the time at which the rate reaches a maximum. (e) Another reaction with the stoichiometry $A \rightarrow P$ has the rate law $d[P]/dt = k[A][P]^2$; integrate the rate law for initial concentrations $[A]_0$ and $[P]_0$. Calculate the time at which the rate reaches a maximum.

23.18 Conventional equilibrium considerations do not apply when a reaction is being driven by light absorption. Thus the steady-state concentration of products and reactants might differ significantly from equilibrium values. For instance, suppose the reaction $A \rightarrow B$ is driven by light absorption and that its rate is I_a, but that the reverse reaction $B \rightarrow A$ is bimolecular and second-order with a rate $k[B]^2$. What is the stationary state concentration of B? Why does this 'photostationary state' differ from the equilibrium state?

23.19 Derive an expression for the rate of disappearance of a species A in a photochemical reaction for which the mechanism is:

(1) initiation with light of intensity I, $A \rightarrow R\cdot + R\cdot$

(2) propagation, $A + R\cdot \rightarrow R\cdot + B$

(3) termination, $R\cdot + R\cdot \rightarrow R_2$

Hence, show that rate measurements will give only a combination of k_2 and k_3 if a steady state is reached, but that both may be obtained if a steady state is not reached.

23.20 The photochemical chlorination of chloroform in the gas has been found to follow the rate law $d[CCl_4]/dt = k[Cl_2]^{1/2}I_a^{1/2}$. Devise a mechanism that leads to this rate law when the chlorine pressure is high.

23.21 Photolysis of $Cr(CO)_6$ in the presence of certain molecules M, can give rise to the following reaction sequence:

(1) $Cr(CO)_6 + h\nu \rightarrow Cr(CO)_5 + CO$

(2) $Cr(CO)_5 + CO \rightarrow Cr(CO)_6$

(3) $Cr(CO)_5 + M \rightarrow Cr(CO)_5M$

(4) $Cr(CO)_5M \rightarrow Cr(CO)_5 + M$

Suppose that the absorbed light intensity is so weak that $I \ll k_4[Cr(CO)_5M]$. Find the factor f in the equation $d[Cr(CO)_5M]/dt = -f[Cr(CO)_5M]$. Show that a graph of $1/f$ against [M] should be a straight line.

Applications: to biochemistry and environmental science

23.22 Models of population growth are analogous to chemical reaction rate equations. In the model due to Malthus (1798) the rate of change of the population N of the planet is assumed to be given by dN/dt = births − deaths. The numbers of births and deaths are proportional to the population, with proportionality constants b and d. Obtain the integrated rate law. How well does it fit the (very approximate) data below on the population of the planet as a function of time?

Year	1750	1825	1922	1960	1974	1987	2000
$N/10^9$	0.5	1	2	3	4	5	6

23.23 Many enzyme-catalysed reactions are consistent with a modified version of the Michaelis–Menten mechanism in which the second step is also reversible. (a) For this mechanism show that the rate of formation of product is given by

$$v = \frac{(v_{max}/K_M)[S] - (v'_{max}/K'_M)[P]}{1 + [S]/K_M + [P]/K'_M}$$

where $v_{max} = k_b[E]_0$, $v'_{max} = k'_a[E]_0$, $K_M = (k'_a + k_b)/k_a$, and $K'_M = (k'_a + k_b)/k'_b$. (b) Find the limiting behaviour of this expression for large and small concentrations of substrate.

23.24 The following results were obtained for the action of an ATPase on ATP at 20°C, when the concentration of the ATPase was 20 nmol dm^{-3}:

[ATP]/(μmol dm^{-3})	0.60	0.80	1.4	2.0	3.0
v/(μmol dm^{-3} s^{-1})	0.81	0.97	1.30	1.47	1.69

Determine the Michaelis constant, the maximum velocity of the reaction, the turnover number, and the catalytic efficiency of the enzyme.

23.25 Enzyme-catalysed reactions are sometimes analysed by use of the *Eadie–Hofstee plot*, in which v is plotted against $v/[S]_0$. (a) Using the simple Michaelis–Menten mechanism, derive a relation between $v/[S]_0$ and v. (b) Discuss how the values of K_M and v_{max} are obtained from analysis of the Eadie–Hofstee plot. (c) Determine the Michaelis constant and the maximum velocity of the reaction of the reaction from Problem 23.23 by using an Eadie–Hofstee plot to analyse the data.

23.26 In general, the catalytic efficiency of an enzyme depends on the pH of the medium in which it operates. One way to account for this behaviour is to propose that the enzyme and the enzyme–substrate complex are active only in specific protonation states. This situation can be summarized by the following mechanism:

$$EH + S \Leftrightarrow ESH \qquad k_a, k_a'$$

$$ESH \rightarrow E + P \qquad k_b$$

$$EH \rightleftharpoons E^- + H^+ \qquad K_{E,a} = \frac{[E^-][H^+]}{[EH^+]}$$

$$EH_2^+ \rightleftharpoons EH + H^+ \qquad K_{E,b} = \frac{[EH][H^+]}{[EH_2^+]}$$

$$ESH \rightleftharpoons ES^- + H^+ \qquad K_{ES,a} = \frac{[ES^-][H^+]}{[ESH]}$$

$$ESH_2 \rightleftharpoons ESH + H^+ \qquad K_{ES,b} = \frac{[ESH][H^+]}{[ESH_2]}$$

in which only the EH and ESH forms are active. (a) For the mechanism above, show that

$$v = \frac{v'_{max}}{1 + K'_M[S]_0}$$

with

$$v'_{max} = \frac{v_{max}}{1 + \dfrac{[H^+]}{K_{ES,b}} + \dfrac{K_{ES,a}}{[H^+]}}$$

$$K'_M = K_M \frac{1 + \dfrac{[H^+]}{K_{E,b}} + \dfrac{K_{E,a}}{[H^+]}}{1 + \dfrac{[H^+]}{K_{ES,b}} + \dfrac{K_{ES,a}}{[H^+]}}$$

where v_{max} and K_M correspond to the form EH of the enzyme. (b) For pH values ranging from 0 to 14, plot v'_{max} against pH for a hypothetical reaction for which $v_{max} = 1.0 \times 10^{-6}$ mol dm^{-3} s^{-1}, $K_{ES,b} = 1.0 \times 10^{-6}$ mol dm^{-3} and $K_{ES,a} = 1.0 \times 10^{-8}$ mol dm^{-3}. Is there a pH at which v'_{max} reaches a maximum value? If so, determine the pH. (c) Redraw the plot in part (b) by using the same value of v_{max}, but $K_{ES,b} = 1.0 \times 10^{-4}$ mol dm^{-3} and $K_{ES,a} = 1.0 \times 10^{-10}$ mol dm^{-3}. Account for any differences between this plot and the plot from part (b).

23.27 The enzyme carboxypeptidase catalyses the hydrolysis of polypeptides and here we consider its inhibition. The following results were obtained when the rate of the enzymolysis of carbobenzoxy-glycyl-D-phenylalanine (CBGP) was monitored without inhibitor:

$[CBGP]_0/(10^{-2}$ mol dm$^{-3})$	1.25	3.84	5.81	7.13
Relative reaction rate	0.398	0.669	0.859	1.000

(All rates in this problem were measured with the same concentration of enzyme and are relative to the rate measured when $[CBGP]_0 = 0.0713$ mol dm^{-3} in the absence of inhibitor.) When 2.0×10^{-3} mol dm^{-3} phenylbutyrate ion was added to a solution containing the enzyme and substrate, the following results were obtained:

$[CBGP]_0/(10^{-2}$ mol dm$^{-3})$	1.25	2.50	4.00	5.50
Relative reaction rate	0.172	0.301	0.344	0.548

In a separate experiment, the effect of 5.0×10^{-2} mol dm^{-3} benzoate ion was monitored and the results were:

$[CBGP]_0/(10^{-2}$ mol dm$^{-3})$	1.75	2.50	5.00	10.00
Relative reaction rate	0.183	0.201	0.231	0.246

Determine the mode of inhibition of carboxypeptidase by the phenylbutyrate ion and benzoate ion.

23.28 Many biological and biochemical processes involve autocatalytic steps (Problem 23.17). In the SIR model of the spread and decline of infectious diseases the population is divided into three classes; the susceptibles, S, who can catch the disease, the infectives, I, who have the disease and can transmit it, and the removed class, R, who have either had the disease and recovered, are dead, or are immune or isolated. The model mechanism for this process implies the following rate laws:

$$\frac{dS}{dt} = -rSI \qquad \frac{dI}{dt} = rSI - aI \qquad \frac{dR}{dt} = aI$$

What are the autocatalytic steps of this mechanism? Find the conditions on the ratio a/r that decide whether the disease will spread (an epidemic) or die out. Show that a constant population is built into this system, namely, that $S + I + R = N$, meaning that the timescales of births, deaths by other causes, and migration are assumed large compared to that of the spread of the disease.

23.29 In light-harvesting complexes, the fluorescence of a chlorophyll molecule is quenched by nearby chlorophyll molecules. Given that for a pair of chlorophyll *a* molecules $R_0 = 5.6$ nm, by what distance should two chlorophyll *a* molecules be separated to shorten the fluorescence lifetime from 1 ns (a typical value for monomeric chlorophyll *a* in organic solvents) to 10 ps?

23.30 The light-induced electron transfer reactions in photosynthesis occur because chlorophyll molecules (whether in monomeric or dimeric forms) are better reducing agents in their electronic excited states. Justify this observation with the help of molecular orbital theory.

23.31 The emission spectrum of a porphyrin dissolved in O_2-saturated water shows a strong band at 650 nm and a weak band at 1270 nm. In separate experiments, it was observed that the electronic absorption spectrum of the porphyrin sample showed bands at 420 nm and 550 nm, and the electronic absorption spectrum of O_2-saturated water showed no bands in the visible range of the spectrum (and therefore no emission spectrum when excited in the same range). Based on these data alone, make a preliminary assignment of the emission band at 1270 nm. Propose additional experiments that test your hypothesis.

23.32‡ Ultraviolet radiation photolyses O_3 to O_2 and O. Determine the rate at which ozone is consumed by 305 nm radiation in a layer of the stratosphere of thickness 1 km. The quantum efficiency is 0.94 at 220 K, the concentration about 8×10^{-9} mol dm^{-3}, the molar absorption coefficient 260 dm^3 mol^{-1} cm^{-1}, and the flux of 305 nm radiation about 1×10^{14} photons cm^{-2} s^{-1}. Data from W.B. DeMore, S.P. Sander, D.M. Golden, R.F. Hampson, M.J. Kurylo, C.J. Howard, A.R. Ravishankara, C.E. Kolb, and M.J. Molina, *Chemical kinetics and photochemical data for use in stratospheric modeling: Evaluation Number 11*, JPL Publication 94–26 (1994).

23.33‡ Use the Chapman model to explore the behaviour of a model atmosphere consisting of pure O_2 at 10 Torr and 298 K that is exposed to measurable frequencies and intensities of UV radiation. (a) Look up the values of k_2, k_4, and k_5 in a source such as the *CRC Handbook of chemistry and physics* or *Chemical kinetics and photochemical data for use in stratospheric modeling* (the URL is available at the text's web site). The rate constants k_1 and k_3 depend upon the radiation conditions; assume values of 1.0×10^{-8} s^{-1} and 0.016 s^{-1}, respectively. If you cannot find a value for k_5, formulate chemically sound arguments for exclusion of the fifth step from the mechanism.
(b) Write the rate expressions for the concentration of each chemical species.
(c) Assume that the UV radiation is turned on at $t = 0$, and solve the rate expressions for the concentration of all species as a function of time over a period of 4 h. Examine relevant concentrations in the very early time period $t < 0.1$ s. State all assumptions. Is there any ozone present initially? Why must the pressure be low and the UV radiation intensities high for the production of ozone? Draw graphs of the time variations of both atomic oxygen and ozone on both the very short and the long timescales. What is the percentage of ozone after 4.0 h of irradiation? *Hint.* You will need a software package for solving a 'stiff' system of differential equations. Stiff differential equations have at least two rate constants with very different values and result in different behaviours on different timescales, so the solution usually requires that the total time period be broken into two or more periods; one may be

very short and another very long. For help with using mathematical software to solve systems of differential equations, see M.P. Cady and C.A. Trapp, *A Mathcad primer for physical chemistry*. Oxford University Press (1999).

23.34‡ Chlorine atoms react rapidly with ozone in the gas-phase bimolecular reaction $Cl + O_3 \rightarrow ClO + O_2$ with $k = (1.7 \times 10^{10}\ dm^3\ mol^{-1}\ s^{-1})e^{-260/(T/K)}$ (W.B. DeMore, S.P. Sander, D.M. Golden, R.F. Hampson, M.J. Kurylo, C.J. Howard, A.R. Ravishankara, C.E. Kolb, and M.J. Molina, Chemical kinetics and photochemical data for use in stratospheric modeling: Evaluation Number 11, JPL Publication 94–26 (1994)). Estimate the rate of this reaction at (a) 20 km, where $[Cl] = 5 \times 10^{-17}$ mol dm^{-3}, $[O_3] = 8 \times 10^{-9}$ mol dm^{-3}, and $T = 220$ K; (b) 45 km, where $[Cl] = 3 \times 10^{-15}$ mol dm^{-3}, $[O_3] = 8 \times 10^{-11}$ mol dm^{-3}, and $T = 270$ K.

23.35‡ Because of its importance in atmospheric chemistry, the thermal decomposition of nitric oxide, $2\ NO(g) \rightarrow N_2(g) + O_2(g)$, has been amongst the most thoroughly studied of gas-phase reactions. The commonly accepted mechanism has been that of H. Wise and M.F. Freech (*J. Chem. Phys.* **22**, 1724 (1952)):

(1) $NO + NO \rightarrow N_2O + O$ $\quad k_1$

(2) $O + NO \rightarrow O_2 + N$ $\quad k_2$

(3) $N + NO \rightarrow N_2 + O$ $\quad k_3$

(4) $O + O + M \rightarrow O_2 + M$ $\quad k_4$

(5) $O_2 + M \rightarrow O + O + M$ $\quad k_4'$

(a) Label the steps of this mechanism as initiation, propagation, etc. (b) Write down the full expression for the rate of disappearance of NO. What does this expression for the rate become on the basis of the assumptions that $v_2 = v_3$ when [N] reaches its steady state concentration, that the rate of the propagation step is more rapid than the rate of the initiation step, and that oxygen atoms are in equilibrium with oxygen molecules? (c) Find an expression for the effective activation energy, $E_{a,eff}$, for the overall reaction in terms of the activation energies of the individual steps of the reaction. (d) Estimate $E_{a,eff}$ from the bond energies of the species involved. (e) It has been pointed out by R.J. Wu and C.T. Yeh (*Int. J. Chem. Kinet.* **28**, 89 (1996)) that the reported experimental values of $E_{a,eff}$ obtained by different authors have varied from 253 to 357 kJ mol^{-1}. They suggest that the assumption of oxygen atoms and oxygen molecules being in equilibrium is unwarranted and that the steady-state approximation needs to be applied to the entire mechanism. Obtain the overall rate law based on the steady-state approximation and find the forms that it assumes for low NO conversion (low O_2 concentration). (f) When the reaction conversion becomes significant, Wu and Yeh suggest that two additional elementary steps,

(6) $O_2 + M \rightarrow O + O + M$ $\quad k_6$

(7) $NO + O_2 \rightarrow O + NO_2$ $\quad k_7$

start to compete with step (1) as the initiation step. Obtain the rate laws based on these alternative mechanisms and again estimate the apparent activation energies. Is the range of these different theoretically estimated values for $E_{a,eff}$ consistent with the range of values obtained experimentally?

Molecular reaction dynamics

24

The simplest quantitative account of reaction rates is in terms of collision theory, which can be used only for the discussion of reactions between simple species in the gas phase. Reactions in solution are classified into two types: diffusion-controlled and activation-controlled. The former can be expressed quantitatively in terms of the diffusion equation. In transition state theory, it is assumed that the reactant molecules form a complex that can be discussed in terms of the population of its energy levels. Transition state theory inspires a thermodynamic approach to reaction rates, in which the rate constant is expressed in terms of thermodynamic parameters. This approach is useful for parametrizing the rates of reactions in solution. The highest level of sophistication is in terms of potential energy surfaces and the motion of molecules through these surfaces. As we shall see, such an approach gives an intimate picture of the events that occur when reactions occur and is open to experimental study. We also use transition state theory to examine the transfer of electrons in homogeneous systems and see that the rate of the process depends on the distance between electron donor and acceptor, the standard Gibbs energy of reaction, and the energy associated with molecular rearrangements that accompany the transfer of charge.

Now we are at the heart of chemistry. Here we examine the details of what happens to molecules at the climax of reactions. Extensive changes of structure are taking place and energies the size of dissociation energies are being redistributed among bonds: old bonds are being ripped apart and new bonds are being formed.

As may be imagined, the calculation of the rates of such processes from first principles is very difficult. Nevertheless, like so many intricate problems, the broad features can be established quite simply. Only when we enquire more deeply do the complications emerge. In this chapter we look at several approaches to the calculation of a rate constant for elementary bimolecular processes, ranging from electron transfer to chemical reactions involving bond breakage and formation. Although a great deal of information can be obtained from gas-phase reactions, many reactions of interest take place in solution, and we shall also see to what extent their rates can be predicted.

Reactive encounters

In this section we consider two elementary approaches to the calculation of reaction rates, one relating to gas-phase reactions and the other to reactions in solution. Both approaches are based on the view that reactant molecules must meet, and that reaction takes place only if the molecules have a certain minimum energy. In the collision theory of bimolecular gas-phase reactions, which we mentioned briefly in Section 22.5, products are formed only if the collision is sufficiently energetic; otherwise the colliding reactant molecules separate again. In solution, the reactant molecules may simply diffuse together and then acquire energy from their immediate surroundings while they are in contact.

24.1 Collision theory

We shall consider the bimolecular elementary reaction

$$A + B \rightarrow P \qquad v = k_2[A][B] \tag{24.1}$$

where P denotes products, and aim to calculate the second-order rate constant k_2.

We can anticipate the general form of the expression for k_2 by considering the physical requirements for reaction. We expect the rate v to be proportional to the rate of collisions, and therefore to the mean speed of the molecules, $\bar{c} \propto (T/M)^{1/2}$ where M is the molar mass of the molecules, their collision cross-section, σ, and the number densities $\mathcal{N}_A$ and $\mathcal{N}_B$ of A and B:

$$v \propto \sigma(T/M)^{1/2}\mathcal{N}_A\mathcal{N}_B \propto \sigma(T/M)^{1/2}[A][B]$$

However, a collision will be successful only if the kinetic energy exceeds a minimum value, the activation energy, E_a, of the reaction. This requirement suggests that the rate constant should also be proportional to a Boltzmann factor of the form $e^{-E_a/RT}$. So we can anticipate, by writing the reaction rate in the form given in eqn 24.1, that

$$k_2 \propto \sigma(T/M)^{1/2}e^{-E_a/RT}$$

Not every collision will lead to reaction even if the energy requirement is satisfied, because the reactants may need to collide in a certain relative orientation. This 'steric requirement' suggests that a further factor, P, should be introduced, and that

$$k_2 \propto P\sigma(T/M)^{1/2}e^{-E_a/RT} \tag{24.2}$$

As we shall see in detail below, this expression has the form predicted by collision theory. It reflects three aspects of a successful collision:

$$k_2 \propto \text{steric requirement} \times \text{encounter rate} \times \text{minimum energy requirement}$$

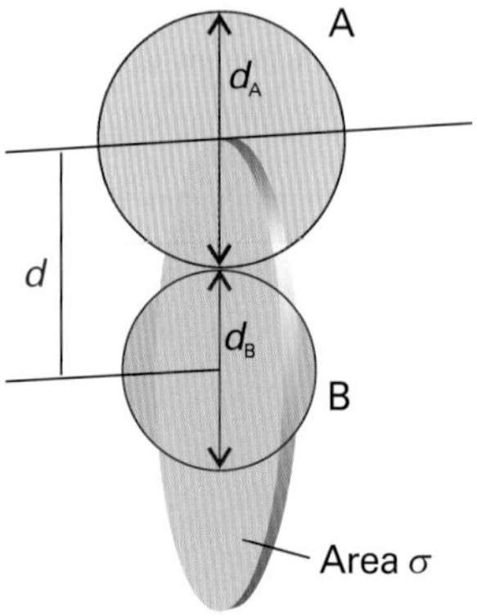

Fig. 24.1 The collision cross-section for two molecules can be regarded to be the area within which the projectile molecule (A) must enter around the target molecule (B) in order for a collision to occur. If the diameters of the two molecules are d_A and d_B, the radius of the target area is $d = \tfrac{1}{2}(d_A + d_B)$ and the cross-section is πd^2.

(a) Collision rates in gases

We have anticipated that the reaction rate, and hence k_2, depends on the frequency with which molecules collide. The **collision density**, Z_{AB}, is the number of (A,B) collisions in a region of the sample in an interval of time divided by the volume of the region and the duration of the interval. The frequency of collisions of a single molecule in a gas was calculated in Section 21.1. As shown in the *Justification* below, that result can be adapted to deduce that

$$Z_{AB} = \sigma\left(\frac{8kT}{\pi\mu}\right)^{1/2} N_A^2[A][B] \tag{24.3a}$$

where σ is the collision cross-section (Fig. 24.1)

$$\sigma = \pi d^2 \qquad d = \tfrac{1}{2}(d_A + d_B) \tag{24.3b}$$

and μ is the reduced mass,

$$\mu = \frac{m_A m_B}{m_A + m_B} \tag{24.3c}$$

Similarly, the collision density for like molecules at a molar concentration [A] is

$$Z_{AA} = \sigma\left(\frac{4kT}{\pi m_A}\right)^{1/2} N_A^2[A]^2 \tag{24.4}$$

Collision densities may be very large. For example, in nitrogen at room temperature and pressure, with $d = 280$ pm, $Z = 5 \times 10^{34}\ m^{-3}\ s^{-1}$.

Justification 24.1 *The collision density*

It follows from eqn 21.11 that the collision frequency, z, for a single A molecule of mass m_A in a gas of other A molecules is

$$z = \sigma\bar{c}_{rel}\mathcal{N}_A \tag{24.5}$$

where $\mathcal{N}_A$ is the number density of A molecules and $\bar{c}_{rel}$ is their relative mean speed. As indicated in Section 21.1a,

$$\bar{c}_{rel} = 2^{1/2}\bar{c} \qquad \bar{c} = \left(\frac{8kT}{\pi m}\right)^{1/2} \tag{24.6}$$

For future convenience, it is sensible to introduce $\mu = \tfrac{1}{2}m$ (for like molecules of mass m), and then to write

$$\bar{c}_{rel} = \left(\frac{8kT}{\pi\mu}\right)^{1/2} \tag{24.7}$$

This expression also applies to the mean relative speed of dissimilar molecules, provided that μ is interpreted as the reduced mass in eqn 24.5.

The total collision density is the collision frequency multiplied by the number density of A molecules:

$$Z_{AA} = \tfrac{1}{2}z\mathcal{N}_A = \tfrac{1}{2}\sigma\bar{c}_{rel}\mathcal{N}_A^2 \tag{24.8}$$

The factor of $\tfrac{1}{2}$ has been introduced to avoid double counting of the collisions (so one A molecule colliding with another A molecule is counted as one collision regardless of their actual identities). For collisions of A and B molecules present at number densities $\mathcal{N}_A$ and $\mathcal{N}_B$, the collision density is

$$Z_{AB} = \sigma\bar{c}_{rel}\mathcal{N}_A\mathcal{N}_B \tag{24.9}$$

Note that we have discarded the factor of $\tfrac{1}{2}$ because now we are considering an A molecule colliding with any of the B molecules as a collision.

The number density of a species J is $\mathcal{N}_J = N_A[J]$, where [J] is their molar concentration and N_A is Avogadro's constant. Equations 24.3 and 24.4 then follow.

(b) The energy requirement

According to collision theory, the rate of change in the molar concentration of A molecules is the product of the collision density and the probability that a collision occurs with sufficient energy. The latter condition can be incorporated by writing the collision cross-section as a function of the kinetic energy of approach of the two colliding species, and setting the cross-section, $\sigma(\varepsilon)$, equal to zero if the kinetic energy of approach is below a certain threshold value, ε_a. Later, we shall identify $N_A\varepsilon_a$ as E_a, the (molar) activation energy of the reaction. Then, for a collision with a specific relative speed of approach v_{rel} (not, at this stage, a mean value),

$$\frac{d[A]}{dt} = -\sigma(\varepsilon)v_{rel}N_A[A][B] \tag{24.10}$$

Comment 24.1

See *Further information* 10.1. The kinetic energy associated with the relative motion of two particles takes the form $\varepsilon = \tfrac{1}{2}\mu v_{rel}^2$ when the centre-of-mass coordinates are separated from the internal coordinates of each particle.

Comment 24.2

To go from eqn 24.10 to eqn 24.11, we need to review concepts of probability theory summarized in *Appendix* 2. Namely, the mean value of a continuous variable X is given by

$$\langle X\rangle = \int x f(x)\mathrm{d}x$$

where the integral is over all values x that X can assume and the probability of finding a value of X between x and $x + \mathrm{d}x$ is $f(x)\mathrm{d}x$, with $f(x)$ a measure of the distribution of the probability values over x. The mean value of a function $g(X)$ is given by

$$\langle g(X)\rangle = \int g(x)f(x)\mathrm{d}x$$

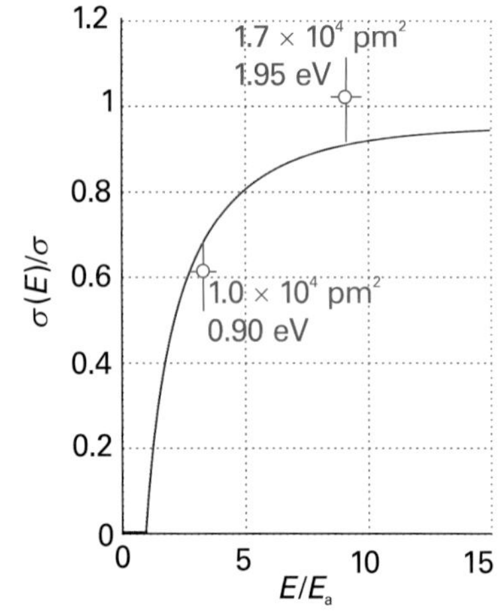

Fig. 24.2 The variation of the reactive cross-section with energy as expressed by eqn 24.13. The data points are from experiments on the reaction $H + D_2 \rightarrow HD + D$ (K. Tsukiyama, B. Katz, and R. Bersohn, *J. Chem. Phys.* **84**, 1934 (1986)).

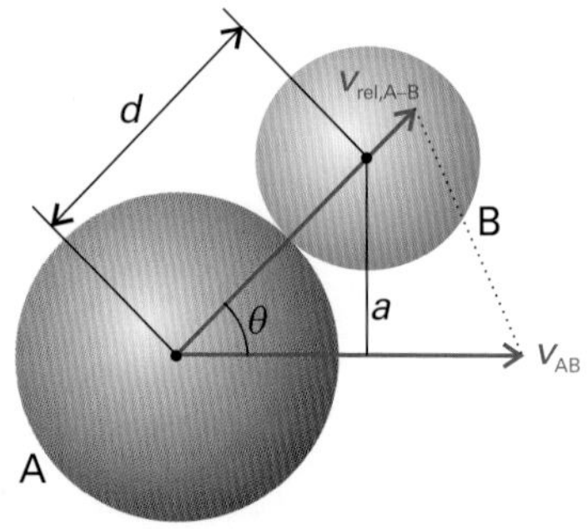

Fig. 24.3 The parameters used in the calculation of the dependence of the collision cross-section on the relative kinetic energy of two molecules A and B.

The relative kinetic energy, ε, and the relative speed are related by $\varepsilon = \frac{1}{2}\mu v_{\mathrm{rel}}^2$, so $v_{\mathrm{rel}} = (2\varepsilon\mu)^{1/2}$. At this point we recognize that a wide range of approach energies is present in a sample, so we should average the expression just derived over a Boltzmann distribution of energies $f(\varepsilon)$, and write (see *Comment* 24.2)

$$\frac{\mathrm{d}[\mathrm{A}]}{\mathrm{d}t} = -\left\{\int_0^\infty \sigma(\varepsilon)v_{\mathrm{rel}}f(\varepsilon)\mathrm{d}\varepsilon\right\}N_{\mathrm{A}}[\mathrm{A}][\mathrm{B}] \qquad (24.11)$$

and hence recognize the rate constant as

$$k_2 = N_{\mathrm{A}}\int_0^\infty \sigma(\varepsilon)v_{\mathrm{rel}}f(\varepsilon)\mathrm{d}\varepsilon \qquad (24.12)$$

Now suppose that the reactive collision cross-section is zero below ε_{a}. We show in the *Justification* below that, above ε_{a}, $\sigma(\varepsilon)$ varies as

$$\sigma(\varepsilon) = \left(1 - \frac{\varepsilon_{\mathrm{a}}}{\varepsilon}\right)\sigma \qquad (24.13)$$

This form of the energy dependence is broadly consistent with experimental determinations of the reaction between H and D_2 as determined by molecular beam measurements of the kind described later (Fig. 24.2). Then, in the *Justification* below, we show that

$$k_2 = N_{\mathrm{A}}\sigma\bar{c}_{\mathrm{rel}}\mathrm{e}^{-E_{\mathrm{a}}/RT} \qquad (24.14)$$

Justification 24.2 *The rate constant*

Consider two colliding molecules A and B with relative speed v_{rel} and relative kinetic energy $\varepsilon = \frac{1}{2}\mu v_{\mathrm{rel}}^2$ (Fig. 24.3). Intuitively we expect that a head-on collision between A and B will be most effective in bringing about a chemical reaction. Therefore, $v_{\mathrm{rel,A-B}}$, the magnitude of the relative velocity component parallel to an axis that contains the vector connecting the centres of A and B, must be large. From trigonometry and the definitions of the distances a and d, and the angle θ given in Fig. 24.3, it follows that

$$v_{\mathrm{rel,A-B}} = v_{\mathrm{rel}}\cos\theta = v_{\mathrm{rel}}\left(\frac{d^2 - a^2}{d^2}\right)^{1/2} \qquad (24.15)$$

We assume that only the kinetic energy associated with the head-on component of the collision, $\varepsilon_{\mathrm{A-B}}$, can lead to a chemical reaction. After squaring both sides of the equation above and multiplying by $\frac{1}{2}\mu$, it follows that

$$\varepsilon_{\mathrm{A-B}} = \varepsilon\frac{d^2 - a^2}{d^2} \qquad (24.16)$$

The existence of an energy threshold, ε_{a}, for the formation of products implies that there is a maximum value of a, a_{max}, above which reactions do not occur. Setting $a = a_{\mathrm{max}}$ and $\varepsilon_{\mathrm{A-B}} = \varepsilon_{\mathrm{a}}$ in the equation above gives

$$a_{\mathrm{max}}^2 = \left(1 - \frac{\varepsilon_{\mathrm{a}}}{\varepsilon}\right)d^2 \qquad (24.17)$$

Substitution of $\sigma(\varepsilon)$ for πa_{max}^2 and σ for πd^2 in the equation above gives eqn 24.13. Note that the equation can be used only when $\varepsilon > \varepsilon_{\mathrm{a}}$.

We proceed with the calculation of the rate constant by considering the Maxwell–Boltzmann distribution of molecular speeds given in Section 21.1. It may be expressed in terms of the kinetic energy, ε, by writing $\varepsilon = \frac{1}{2}\mu v^2$, then $dv = d\varepsilon/(2\mu\varepsilon)^{1/2}$ and eqn 21.4 becomes

$$f(v)dv = 4\pi\left(\frac{\mu}{2\pi kT}\right)^{3/2}\left(\frac{2\varepsilon}{\mu}\right)e^{-\varepsilon/kT}\frac{d\varepsilon}{(2\mu\varepsilon)^{1/2}}$$

$$= 2\pi\left(\frac{1}{\pi kT}\right)^{3/2}\varepsilon^{1/2}e^{-\varepsilon/kT}d\varepsilon = f(\varepsilon)d\varepsilon \qquad (24.18)$$

The integral we need to evaluate is therefore

$$\int_0^\infty \sigma(\varepsilon)v_{rel}f(\varepsilon)d\varepsilon = 2\pi\left(\frac{1}{\pi kT}\right)^{3/2}\int_0^\infty \sigma(\varepsilon)\left(\frac{2\varepsilon}{\mu}\right)^{1/2}\varepsilon^{1/2}e^{-\varepsilon/kT}d\varepsilon$$

$$= \left(\frac{8}{\pi\mu kT}\right)^{1/2}\left(\frac{1}{kT}\right)\int_0^\infty \varepsilon\sigma(\varepsilon)e^{-\varepsilon/kT}d\varepsilon$$

To proceed, we introduce the approximation for $\sigma(\varepsilon)$ in eqn 24.13, and evaluate

$$\int_0^\infty \varepsilon\sigma(\varepsilon)e^{-\varepsilon/kT}d\varepsilon = \sigma\int_{\varepsilon_a}^\infty \varepsilon\left(1-\frac{\varepsilon_a}{\varepsilon}\right)e^{-\varepsilon/kT}d\varepsilon = (kT)^2\sigma e^{-\varepsilon_a/kT}$$

We have made use of the fact that $\sigma = 0$ for $\varepsilon < \varepsilon_a$. It follows that

$$\int_0^\infty \sigma(\varepsilon)v_{rel}f(\varepsilon)d\varepsilon = \sigma\left(\frac{8kT}{\pi\mu}\right)^{1/2}e^{-\varepsilon_a/kT}$$

as in eqn 24.14 (with $\varepsilon_a/kT = E_a/RT$).

Equation 24.14 has the Arrhenius form $k_2 = Ae^{-E_a/RT}$ provided the exponential temperature dependence dominates the weak square-root temperature dependence of the pre-exponential factor. It follows that we can identify the activation energy, E_a, with the minimum kinetic energy along the line of approach that is needed for reaction, and that the pre-exponential factor is a measure of the rate at which collisions occur in the gas.

(c) The steric requirement

The simplest procedure for calculating k_2 is to use for σ the values obtained for non-reactive collisions (for example, typically those obtained from viscosity measurements) or from tables of molecular radii. Table 24.1 compares some values of the pre-exponential factor calculated in this way with values obtained from Arrhenius plots (Section 22.5a). One of the reactions shows fair agreement between theory and

Synoptic table 24.1* Arrhenius parameters for gas-phase reactions

	$A/(dm^3\ mol^{-1}\ s^{-1})$			
	Experiment	Theory	$E_a/(kJ\ mol^{-1})$	P
$2\,NOCl \rightarrow 2\,NO + 2\,Cl$	9.4×10^9	5.9×10^{10}	102	0.16
$2\,ClO \rightarrow Cl_2 + O_2$	6.3×10^7	2.5×10^{10}	0	2.5×10^{-3}
$H_2 + C_2H_4 \rightarrow C_2H_6$	1.24×10^6	7.4×10^{11}	180	1.7×10^{-6}
$K + Br_2 \rightarrow KBr + Br$	1.0×10^{12}	2.1×10^{11}	0	4.8

* More values are given in the *Data section*.

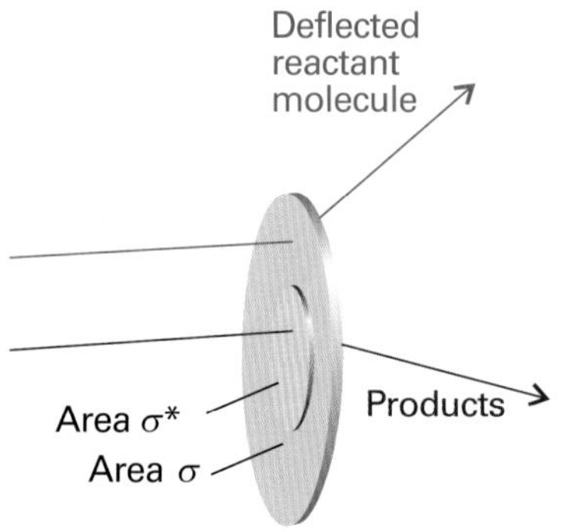

Fig. 24.4 The collision cross-section is the target area that results in simple deflection of the projectile molecule; the reaction cross-section is the corresponding area for chemical change to occur on collision.

experiment, but for others there are major discrepancies. In some cases the experimental values are orders of magnitude smaller than those calculated, which suggests that the collision energy is not the only criterion for reaction and that some other feature, such as the relative orientation of the colliding species, is important. Moreover, one reaction in the table has a pre-exponential factor larger than theory, which seems to indicate that the reaction occurs more quickly than the particles collide!

We can accommodate the disagreement between experiment and theory by introducing a **steric factor**, P, and expressing the **reactive cross-section**, σ^*, as a multiple of the collision cross-section, $\sigma^* = P\sigma$ (Fig. 24.4). Then the rate constant becomes

$$k_2 = P\sigma\left(\frac{8kT}{\pi\mu}\right)^{1/2} N_A e^{-E_a/RT} \tag{24.19}$$

This expression has the form we anticipated in eqn 24.2. The steric factor is normally found to be several orders of magnitude smaller than 1.

Example 24.1 *Estimating a steric factor (1)*

Estimate the steric factor for the reaction $H_2 + C_2H_4 \rightarrow C_2H_6$ at 628 K given that the pre-exponential factor is 1.24×10^6 dm^3 mol^{-1} s^{-1}.

Method To calculate P, we need to calculate the pre-exponential factor, A, by using eqn 24.19 and then compare the answer with experiment: the ratio is P. Table 21.1 lists collision cross-sections for non-reactive encounters. The best way to estimate the collision cross-section for dissimilar spherical species is to calculate the collision diameter for each one (from $\sigma = \pi d^2$), to calculate the mean of the two diameters, and then to calculate the cross-section for that mean diameter. However, as neither species is spherical, a simpler but more approximate procedure is just to take the average of the two collision cross-sections.

Answer The reduced mass of the colliding pair is

$$\mu = \frac{m_1 m_2}{m_1 + m_2} = 3.12 \times 10^{-27}\ \text{kg}$$

because $m_1 = 2.016$ u for H_2 and $m_2 = 28.05$ u for C_2H_4 (the atomic mass unit, 1 u, is defined inside the front cover). Hence

$$\left(\frac{8kT}{\pi\mu}\right)^{1/2} = 2.66 \times 10^3\ \text{m s}^{-1}$$

From Table 21.1, $\sigma(H_2) = 0.27$ nm^2 and $\sigma(C_2H_4) = 0.64$ nm^2, giving a mean collision cross-section of $\sigma = 0.46$ nm^2. Therefore,

$$A = \sigma\left(\frac{8kT}{\pi\mu}\right)^{1/2} N_A = 7.37 \times 10^{11}\ \text{dm}^3\ \text{mol}^{-1}\ \text{s}^{-1}$$

Experimentally $A = 1.24 \times 10^6$ dm^3 mol^{-1} s^{-1}, so it follows that $P = 1.7 \times 10^{-6}$. The very small value of P is one reason why catalysts are needed to bring this reaction about at a reasonable rate. As a general guide, the more complex the molecules, the smaller the value of P.

Self-test 24.1 It is found for the reaction $NO + Cl_2 \rightarrow NOCl + Cl$ that $A = 4.0 \times 10^9$ dm^3 mol^{-1} s^{-1} at 298 K. Use $\sigma(NO) = 0.42$ nm^2 and $\sigma(Cl_2) = 0.93$ nm^2 to estimate the P factor for the reaction. [0.018]

An example of a reaction for which it is possible to estimate the steric factor is $K + Br_2 \rightarrow KBr + Br$, with the experimental value $P = 4.8$. In this reaction, the distance of approach at which reaction occurs appears to be considerably larger than the distance needed for deflection of the path of the approaching molecules in a non-reactive collision. It has been proposed that the reaction proceeds by a **harpoon mechanism**. This brilliant name is based on a model of the reaction that pictures the K atom as approaching a Br_2 molecule, and when the two are close enough an electron (the harpoon) flips across from K to Br_2. In place of two neutral particles there are now two ions, so there is a Coulombic attraction between them: this attraction is the line on the harpoon. Under its influence the ions move together (the line is wound in), the reaction takes place, and KBr + Br emerge. The harpoon extends the cross-section for the reactive encounter, and the reaction rate is greatly underestimated by taking for the collision cross-section the value for simple mechanical contact between $K + Br_2$.

Example 24.2 *Estimating a steric factor (2)*

Estimate the value of P for the harpoon mechanism by calculating the distance at which it becomes energetically favourable for the electron to leap from K to Br_2.

Method We should begin by identifying all the contributions to the energy of interaction between the colliding species. There are three contributions to the energy of the process $K + Br_2 \rightarrow K^+ + Br_2^-$. The first is the ionization energy, I, of K. The second is the electron affinity, E_{ea}, of Br_2. The third is the Coulombic interaction energy between the ions when they have been formed: when their separation is R, this energy is $-e^2/4\pi\varepsilon_0 R$. The electron flips across when the sum of these three contributions changes from positive to negative (that is, when the sum is zero).

Answer The net change in energy when the transfer occurs at a separation R is

$$E = I - E_{ea} - \frac{e^2}{4\pi\varepsilon_0 R}$$

The ionization energy I is larger than E_{ea}, so E becomes negative only when R has decreased to less than some critical value R^* given by

$$\frac{e^2}{4\pi\varepsilon_0 R^*} = I - E_{ea}$$

When the particles are at this separation, the harpoon shoots across from K to Br_2, so we can identify the reactive cross-section as $\sigma^* = \pi R^{*2}$. This value of σ^* implies that the steric factor is

$$P = \frac{\sigma^*}{\sigma} = \frac{R^{*2}}{d^2} = \left\{\frac{e^2}{4\pi\varepsilon_0 d(I - E_{ea})}\right\}^2$$

where $d = R(K) + R(Br_2)$. With $I = 420\ \text{kJ mol}^{-1}$ (corresponding to 7.0×10^{-19} J), $E_{ea} \approx 250\ \text{kJ mol}^{-1}$ (corresponding to 4.2×10^{-19} J), and $d = 400$ pm, we find $P = 4.2$, in good agreement with the experimental value (4.8).

Self-test 24.2 Estimate the value of P for the harpoon reaction between Na and Cl_2 for which $d \approx 350$ pm; take $E_{ea} \approx 230\ \text{kJ mol}^{-1}$. [2.2]

Example 24.2 illustrates two points about steric factors. First, the concept of a steric factor is not wholly useless because in some cases its numerical value can be estimated.

Second (and more pessimistically) most reactions are much more complex than K + Br_2, and we cannot expect to obtain P so easily. What we need is a more powerful theory that lets us calculate, and not merely guess, its value. We go some way to setting up that theory in Section 24.4.

24.2 Diffusion-controlled reactions

Encounters between reactants in solution occur in a very different manner from encounters in gases. Reactant molecules have to jostle their way through the solvent, so their encounter frequency is considerably less than in a gas. However, because a molecule also migrates only slowly away from a location, two reactant molecules that encounter each other stay near each other for much longer than in a gas. This lingering of one molecule near another on account of the hindering presence of solvent molecules is called the **cage effect**. Such an encounter pair may accumulate enough energy to react even though it does not have enough energy to do so when it first forms. The activation energy of a reaction is a much more complicated quantity in solution than in a gas because the encounter pair is surrounded by solvent and we need to consider the energy of the entire local assembly of reactant and solvent molecules.

(a) Classes of reaction

The complicated overall process can be divided into simpler parts by setting up a simple kinetic scheme. We suppose that the rate of formation of an encounter pair AB is first-order in each of the reactants A and B:

$$\mathrm{A} + \mathrm{B} \rightarrow \mathrm{AB} \qquad v = k_\mathrm{d}[\mathrm{A}][\mathrm{B}] \tag{24.20a}$$

As we shall see, k_d (where the d signifies diffusion) is determined by the diffusional characteristics of A and B. The encounter pair can break up without reaction or it can go on to form products P. If we suppose that both processes are pseudofirst-order reactions (with the solvent perhaps playing a role), then we can write

$$\mathrm{AB} \rightarrow \mathrm{A} + \mathrm{B} \qquad v = k_\mathrm{d}'[\mathrm{AB}] \tag{24.20b}$$

and

$$\mathrm{AB} \rightarrow \mathrm{P} \qquad v = k_\mathrm{a}[\mathrm{AB}] \tag{24.20c}$$

The concentration of AB can now be found from the equation for the net rate of change of concentration of AB:

$$\frac{\mathrm{d}[\mathrm{AB}]}{\mathrm{d}t} = k_\mathrm{d}[\mathrm{A}][\mathrm{B}] - k_\mathrm{d}'[\mathrm{AB}] - k_\mathrm{a}[\mathrm{AB}] \approx 0 \tag{24.21}$$

where we have applied the steady-state approximation. This expression solves to

$$[\mathrm{AB}] = \frac{k_\mathrm{d}[\mathrm{A}][\mathrm{B}]}{k_\mathrm{a} + k_\mathrm{d}'} \tag{24.22}$$

The rate of formation of products is therefore

$$\frac{\mathrm{d}[\mathrm{P}]}{\mathrm{d}t} \approx k_\mathrm{a}[\mathrm{AB}] = k_2[\mathrm{A}][\mathrm{B}] \qquad k_2 = \frac{k_\mathrm{a}k_\mathrm{d}}{k_\mathrm{a} + k_\mathrm{d}'} \tag{24.23}$$

Two limits can now be distinguished. If the rate of separation of the unreacted encounter pair is much slower than the rate at which it forms products, then $k_\mathrm{d}' \ll k_\mathrm{a}$ and the effective rate constant is

Synoptic table 24.2* Arrhenius parameters for reactions in solution

	Solvent	$A/(\mathrm{dm^3\,mol^{-1}\,s^{-1}})$	$E_a/(\mathrm{kJ\,mol^{-1}})$
$(CH_3)_3CCl$ solvolysis	Water	7.1×10^{16}	100
	Ethanol	3.0×10^{13}	112
	Chloroform	1.4×10^{4}	45
CH_3CH_2Br	Ethanol	4.3×10^{11}	90

* More values are given in the *Data section*.

$$k_2 \approx \frac{k_a k_d}{k_a} = k_d \qquad (24.24)$$

In this **diffusion-controlled limit**, the rate of reaction is governed by the rate at which the reactant molecules diffuse through the solvent. An indication that a reaction is diffusion-controlled is that its rate constant is of the order of $10^9\ \mathrm{dm^3\,mol^{-1}\,s^{-1}}$ or greater. Because the combination of radicals involves very little activation energy, radical and atom recombination reactions are often diffusion-controlled.

An **activation-controlled reaction** arises when a substantial activation energy is involved in the reaction AB → P. Then $k_a \ll k_d'$ and

$$k_2 \approx \frac{k_a k_d}{k_d'} = k_a K \qquad (24.25)$$

where K is the equilibrium constant for A + B $\rightleftharpoons$ AB. In this limit, the reaction proceeds at the rate at which energy accumulates in the encounter pair from the surrounding solvent. Some experimental data are given in Table 24.2.

(b) Diffusion and reaction

The rate of a diffusion-controlled reaction is calculated by considering the rate at which the reactants diffuse together. As shown in the *Justification* below, the rate constant for a reaction in which the two reactant molecules react if they come within a distance R^* of one another is

$$k_d = 4\pi R^* D N_A \qquad (24.26)$$

where D is the sum of the diffusion coefficients the two reactant species in the solution.

Justification 24.3 *Solution of the radial diffusion equation*

From the form of the diffusion equation (Section 21.10) corresponding to motion in three dimensions, $D_B\nabla^2[B] = \partial[B]/\partial t$, the concentration of B when the system has reached a steady state ($\partial[B]/\partial t = 0$) satisfies $\nabla^2[B]_r = 0$, where the subscript r signifies a quantity that varies with the distance r. For a spherically symmetrical system, ∇^2 can be replaced by radial derivatives alone (see Table 8.1), so the equation satisfied by $[B]_r$ is

$$\frac{d^2[B]_r}{dr^2} + \frac{2}{r}\frac{d[B]_r}{dr} = 0 \qquad (24.27)$$

The general solution of this equation is

$$[B]_r = a + \frac{b}{r} \qquad (24.28)$$

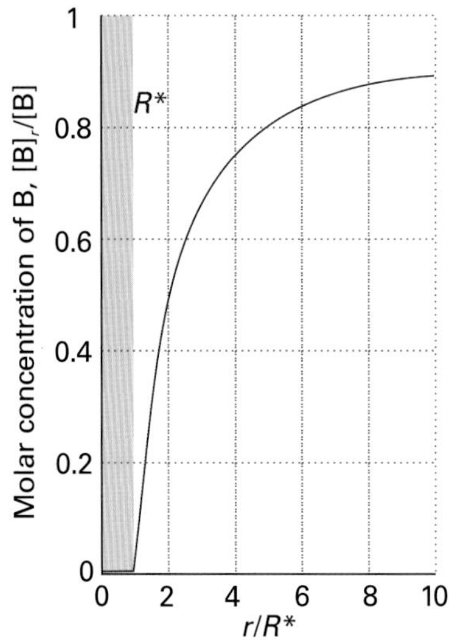

Fig. 24.5 The concentration profile for reaction in solution when a molecule B diffuses towards another reactant molecule and reacts if it reaches R^*.

as may be verified by substitution. We need two boundary conditions to pin down the values of the two constants. One condition is that $[B]_r$ has its bulk value [B] as $r \to \infty$. The second condition is that the concentration of B is zero at $r = R^*$, the distance at which reaction occurs. It follows that $a = [B]$ and $b = -R^*[B]$, and hence that (for $r \geq R^*$)

$$[B]_r = \left(1 + \frac{R^*}{r}\right)[B] \qquad (24.29)$$

Figure 24.5 illustrates the variation of concentration expressed by this equation.

The rate of reaction is the (molar) flux, J, of the reactant B towards A multiplied by the area of the spherical surface of radius R^*:

$$\text{Rate of reaction} = 4\pi R^{*2} J \qquad (24.30)$$

From Fick's first law (eqn 21.17), the flux towards A is proportional to the concentration gradient, so at a radius R^*:

$$J = D_B \left(\frac{d[B]_r}{dr}\right)_{r=R^*} = \frac{D_B[B]}{R^*} \qquad (24.31)$$

(A sign change has been introduced because we are interested in the flux towards decreasing values of r.) When this condition is substituted into the previous equation we obtain

$$\text{Rate of reaction} = 4\pi R^* D_B[B] \qquad (24.32)$$

The rate of the diffusion-controlled reaction is equal to the average flow of B molecules to all the A molecules in the sample. If the bulk concentration of A is [A], the number of A molecules in the sample of volume V is $N_A[A]V$; the global flow of all B to all A is therefore $4\pi R^* D_B N_A[A][B]V$. Because it is unrealistic to suppose that all A are stationary; we replace D_B by the sum of the diffusion coefficients of the two species and write $D = D_A + D_B$. Then the rate of change of concentration of AB is

$$\frac{d[AB]}{dt} = 4\pi R^* D N_A[A][B] \qquad (24.33)$$

Hence, the diffusion-controlled rate constant is as given in eqn 24.26.

We can take eqn 24.26 further by incorporating the Stokes–Einstein equation (eqn 21.66) relating the diffusion constant and the hydrodynamic radius R_A and R_B of each molecule in a medium of viscosity η:

$$D_A = \frac{kT}{6\pi\eta R_A} \qquad D_B = \frac{kT}{6\pi\eta R_B} \qquad (24.34)$$

As these relations are approximate, little extra error is introduced if we write $R_A = R_B = \frac{1}{2}R^*$, which leads to

$$k_d = \frac{8RT}{3\eta} \qquad (24.35)$$

(The R in this equation is the gas constant.) The radii have cancelled because, although the diffusion constants are smaller when the radii are large, the reactive collision radius is larger and the particles need travel a shorter distance to meet. In this approximation, the rate constant is independent of the identities of the reactants, and depends only on the temperature and the viscosity of the solvent.

Illustration 24.1 *Estimating a diffusional rate constant*

The rate constant for the recombination of I atoms in hexane at 298 K, when the viscosity of the solvent is 0.326 cP (with $1\ \mathrm{P} = 10^{-1}\ \mathrm{kg\ m^{-1}\ s^{-1}}$) is

$$k_d = \frac{8\times(8.3145\ \mathrm{J\ K^{-1}\ mol^{-1}})\times(298\ \mathrm{K})}{3\times(3.26\times10^{-4}\ \mathrm{kg\ m^{-1}\ s^{-1}})} = 2.0\times10^{7}\ \mathrm{m^3\ mol^{-1}\ s^{-1}}$$

where we have used $1\ \mathrm{J} = 1\ \mathrm{kg\ m^2\ s^{-2}}$. Because $1\ \mathrm{m^3} = 10^3\ \mathrm{dm^3}$, this result corresponds to $2.0\times10^{10}\ \mathrm{dm^3\ mol^{-1}\ s^{-1}}$. The experimental value is $1.3\times10^{10}\ \mathrm{dm^3\ mol^{-1}\ s^{-1}}$, so the agreement is very good considering the approximations involved.

24.3 The material balance equation

The diffusion of reactants plays an important role in many chemical processes, such as the diffusion of O_2 molecules into red blood corpuscles and the diffusion of a gas towards a catalyst. We can have a glimpse of the kinds of calculations involved by considering the diffusion equation (Section 21.10) generalized to take into account the possibility that the diffusing, convecting molecules are also reacting.

(a) The formulation of the equation

Consider a small volume element in a chemical reactor (or a biological cell). The net rate at which J molecules enter the region by diffusion and convection is given by eqn 21.71:

$$\frac{\partial[\mathrm{J}]}{\partial t} = D\frac{\partial^2[\mathrm{J}]}{\partial x^2} - v\frac{\partial[\mathrm{J}]}{\partial x} \qquad (24.36)$$

The net rate of change of molar concentration due to chemical reaction is

$$\frac{\partial[\mathrm{J}]}{\partial t} = -k[\mathrm{J}] \qquad (24.37)$$

if we suppose that J disappears by a pseudofirst-order reaction. Therefore, the overall rate of change of the concentration of J is

$$\frac{\partial[\mathrm{J}]}{\partial t} = \underbrace{D\frac{\partial^2[\mathrm{J}]}{\partial x^2}}_{\text{Spread due to non-uniform concentration}} - \underbrace{v\frac{\partial[\mathrm{J}]}{\partial x}}_{\text{Change due to convection}} - \underbrace{k[\mathrm{J}]}_{\text{Loss due to reaction}} \qquad (24.38)$$

Equation 24.38 is called the **material balance equation**. If the rate constant is large, then [J] will decline rapidly. However, if the diffusion constant is large, then the decline can be replenished as J diffuses rapidly into the region. The convection term, which may represent the effects of stirring, can sweep material either into or out of the region according to the signs of v and the concentration gradient $\partial[\mathrm{J}]/\partial x$.

(b) Solutions of the equation

The material balance equation is a second-order partial differential equation and is far from easy to solve in general. Some idea of how it is solved can be obtained by considering the special case in which there is no convective motion (as in an unstirred reaction vessel):

$$\frac{\partial[\mathrm{J}]}{\partial t} = D\frac{\partial^2[\mathrm{J}]}{\partial x^2} - k[\mathrm{J}] \qquad (24.39)$$

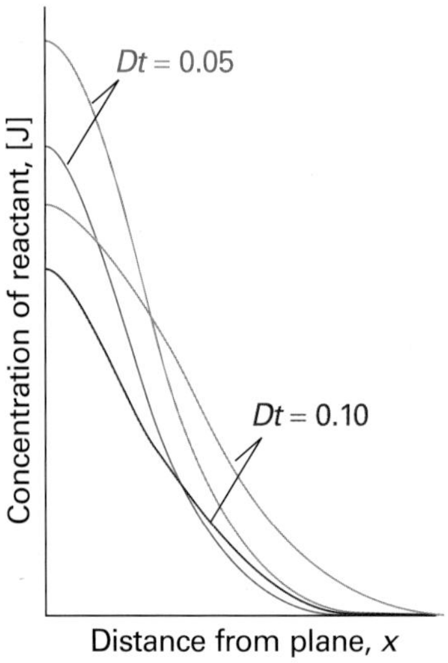

Fig. 24.6 The concentration profiles for a diffusing, reacting system (for example, a column of solution) in which one reactant is initially in a layer at $x = 0$. In the absence of reaction (grey lines) the concentration profiles are the same as in Fig. 21.26.

Exploration Use the interactive applet found in the *Living graphs* section of the text's web site to explore the effect of varying the value of the rate constant k on the spatial variation of [J] for a constant value of the diffusion constant D.

As may be verified by susbtitution, if the solution of this equation in the absence of reaction (that is, for $k = 0$) is [J], then the solution in the presence of reaction ($k > 0$) is

$$[\mathrm{J}]^* = k\int_0^t [\mathrm{J}]\mathrm{e}^{-kt}\,\mathrm{d}t + [\mathrm{J}]\mathrm{e}^{-kt} \qquad (24.40)$$

We have already met one solution of the diffusion equation in the absence of reaction: eqn 21.71 is the solution for a system in which initially a layer of n_0N_A molecules is spread over a plane of area A:

$$[\mathrm{J}] = \frac{n_0\mathrm{e}^{-x^2/4Dt}}{A(\pi Dt)^{1/2}} \qquad (24.41)$$

When this expression is substituted into eqn 24.40 and the integral is evaluated numerically, we obtain the concentration of J as it diffuses away from its initial surface layer and undergoes reaction in the solution above (Fig. 24.6).

Even this relatively simple example has led to an equation that is difficult to solve, and only in some special cases can the full material balance equation be solved analytically. Most modern work on reactor design and cell kinetics uses numerical methods to solve the equation, and detailed solutions for realistic environments, such as vessels of different shapes (which influence the boundary conditions on the solutions) and with a variety of inhomogeneously distributed reactants can be obtained reasonably easily.

Transition state theory

We saw in Section 22.5b that an activated complex forms between reactants as they collide and begin to assume the nuclear and electronic configurations characteristic of products. We also saw that the change in potential energy associated with formation of the activated complex accounts for the activation energy of the reaction. We now consider a more detailed calculation of rate constants which uses the concepts of statistical thermodynamics developed in Chapter 17. The approach we describe, which is called **transition state theory** (also widely referred to as *activated complex theory*), has the advantage that a quantity corresponding to the steric factor appears automatically, and P does not need to be grafted on to an equation as an afterthought. Transition state theory is an attempt to identify the principal features governing the size of a rate constant in terms of a model of the events that take place during the reaction. There are several approaches to the calculation, all of which lead to the same final expression (see *Further reading*); here we present the simplest approach.

Comment 24.3

This chapter inevitably puts heavy demands on the letter K; the various meanings are summarized in Table 24.3 at the end of the chapter.

24.4 The Eyring equation

Transition state theory pictures a reaction between A and B as proceeding through the formation of an activated complex, $C^‡$, in a rapid pre-equilibrium (Fig. 24.7):

$$\mathrm{A} + \mathrm{B} \rightleftharpoons \mathrm{C}^‡ \qquad K^‡ = \frac{p_{\mathrm{C}^‡}p^{\ominus}}{p_\mathrm{A}p_\mathrm{B}} \qquad (24.42)$$

When we express the partial pressures, p_J, in terms of the molar concentrations, [J], by using $p_J = RT[\mathrm{J}]$, the concentration of activated complex is related to the (dimensionless) equilibrium constant by

$$[\mathrm{C}^‡] = \frac{RT}{p^{\ominus}}K^‡[\mathrm{A}][\mathrm{B}] \qquad (24.43)$$

The activated complex falls apart by unimolecular decay into products, P, with a rate constant $k^\ddagger$:

$$C^\ddagger \rightarrow P \qquad v = k^\ddagger[C^\ddagger] \tag{24.44}$$

It follows that

$$v = k_2[A][B] \qquad k_2 = \frac{RT}{p^\ominus} k^\ddagger K^\ddagger \tag{24.45}$$

Our task is to calculate the unimolecular rate constant $k^\ddagger$ and the equilibrium constant $K^\ddagger$.

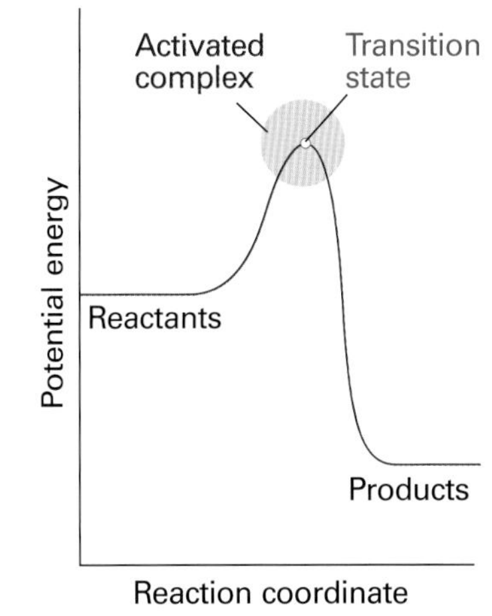

Fig. 24.7 A reaction profile. The horizontal axis is the reaction coordinate, and the vertical axis is potential energy. The activated complex is the region near the potential maximum, and the transition state corresponds to the maximum itself.

(a) The rate of decay of the activated complex

An activated complex can form products if it passes through the **transition state**, the arrangement the atoms must achieve in order to convert to products (Section 22.5b). If its vibration-like motion along the reaction coordinate occurs with a frequency ν, then the frequency with which the cluster of atoms forming the complex approaches the transition state is also ν. However, it is possible that not every oscillation along the reaction coordinate takes the complex through the transition state. For instance, the centrifugal effect of rotations might also be an important contribution to the break-up of the complex, and in some cases the complex might be rotating too slowly, or rotating rapidly but about the wrong axis. Therefore, we suppose that the rate of passage of the complex through the transition state is proportional to the vibrational frequency along the reaction coordinate, and write

$$k^\ddagger = \kappa\nu \tag{24.46}$$

where κ is the **transmission coefficient**. In the absence of information to the contrary, κ is assumed to be about 1.

(b) The concentration of the activated complex

We saw in Section 17.8 how to calculate equilibrium constants from structural data. Equation 17.54 of that section can be used directly, which in this case gives

$$K^\ddagger = \frac{N_A q^\ominus_{C^\ddagger}}{q^\ominus_A q^\ominus_B} e^{-\Delta E_0/RT} \tag{24.47}$$

where $p^\ominus = 1$ bar and

$$\Delta E_0 = E_0(C^\ddagger) - E_0(A) - E_0(B) \tag{24.48}$$

The $q^\ominus_J$ are the standard molar partition functions, as defined in Section 17.1. Note that the units of N_A and the q_J are mol^{-1}, so $K^\ddagger$ is dimensionless (as is appropriate for an equilibrium constant).

In the final step of this part of the calculation, we focus attention on the partition function of the activated complex. We have already assumed that a vibration of the activated complex $C^\ddagger$ tips it through the transition state. The partition function for this vibration is

$$q = \frac{1}{1 - e^{-h\nu/kT}} \tag{24.49a}$$

where ν is its frequency (the same frequency that determines $k^\ddagger$). This frequency is much lower than for an ordinary molecular vibration because the oscillation corresponds to the complex falling apart (Fig. 24.8), so the force constant is very low.

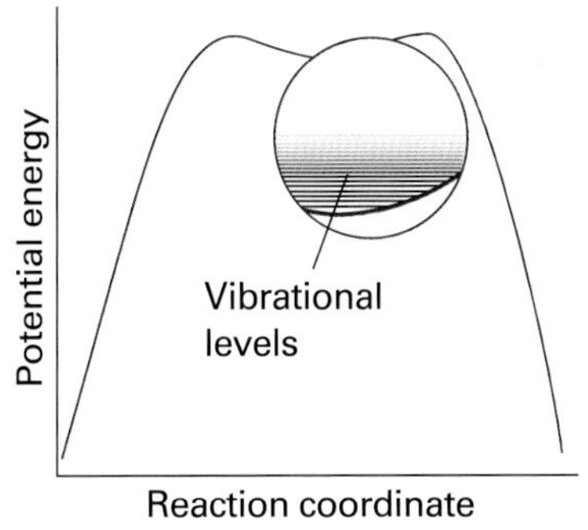

Fig. 24.8 In an elementary depiction of the activated complex close to the transition state, there is a broad, shallow dip in the potential energy surface along the reaction coordinate. The complex vibrates harmonically and almost classically in this well. However, this depiction is an oversimplification, for in many cases there is no dip at the top of the barrier, and the curvature of the potential energy, and therefore the force constant, is negative. Formally, the vibrational frequency is then imaginary. We ignore this problem here, but see *Further reading*.

Therefore, provided that $h\nu/kT \ll 1$, the exponential may be expanded and the partition function reduces to

$$q = \frac{1}{1 - \left(1 - \dfrac{h\nu}{kT} + \cdots\right)} \approx \frac{kT}{h\nu} \tag{24.49b}$$

We can therefore write

$$q_{C^\ddagger} \approx \frac{kT}{h\nu} \bar{q}_{C^\ddagger} \tag{24.50}$$

where $\bar{q}$ denotes the partition function for all the other modes of the complex. The constant $K^\ddagger$ is therefore

$$K^\ddagger = \frac{kT}{h\nu}\bar{K}^\ddagger \qquad \bar{K}^\ddagger = \frac{N_A \bar{q}^{\ominus}_{C^\ddagger}}{q^{\ominus}_A q^{\ominus}_B} e^{-\Delta E_0/RT} \tag{24.51}$$

with $\bar{K}^\ddagger$ a kind of equilibrium constant, but with one vibrational mode of $C^\ddagger$ discarded.

(c) The rate constant

We can now combine all the parts of the calculation into

$$k_2 = k^\ddagger \frac{RT}{p^{\ominus}} K^\ddagger = \kappa\nu \frac{kT}{h\nu}\frac{RT}{p^{\ominus}} \bar{K}^\ddagger \tag{24.52}$$

At this stage the unknown frequencies ν cancel and, after writing $\bar{K}^\ddagger_c = (RT/p^{\ominus})\bar{K}^\ddagger$, we obtain the **Eyring equation**:

$$k_2 = \kappa \frac{kT}{h} \bar{K}^\ddagger_c \tag{24.53}$$

The factor $\bar{K}^\ddagger_c$ is given by eqn 24.51 and the definition $\bar{K}^\ddagger_c = (RT/p^{\ominus})\bar{K}^\ddagger$ in terms of the partition functions of A, B, and $C^\ddagger$, so in principle we now have an explicit expression for calculating the second-order rate constant for a bimolecular reaction in terms of the molecular parameters for the reactants and the activated complex and the quantity κ.

The partition functions for the reactants can normally be calculated quite readily, using either spectroscopic information about their energy levels or the approximate expressions set out in Table 17.3. The difficulty with the Eyring equation, however, lies in the calculation of the partition function of the activated complex: $C^\ddagger$ is difficult to investigate spectroscopically (but see Section 24.9), and in general we need to make assumptions about its size, shape, and structure. We shall illustrate what is involved in one simple but significant case.

(d) The collision of structureless particles

Consider the case of two structureless particles A and B colliding to give an activated complex that resembles a diatomic molecule. Because the reactants J = A, B are structureless 'atoms', the only contributions to their partition functions are the translational terms:

$$q^{\ominus}_J = \frac{V^{\ominus}_m}{\Lambda^3_J} \qquad \Lambda_J = \frac{h}{(2\pi m_J kT)^{1/2}} \qquad V^{\ominus}_m = \frac{RT}{p^{\ominus}} \tag{24.54}$$

The activated complex is a diatomic cluster of mass $m_C = m_A + m_B$ and moment of inertia I. It has one vibrational mode, but that mode corresponds to motion along the reaction coordinate and therefore does not appear in $\bar{q}_{C^\ddagger}$. It follows that the standard molar partition function of the activated complex is

$$q^{\ominus}_{C^\ddagger} = \left(\frac{2IkT}{\hbar^2}\right)\frac{V^{\ominus}_m}{\Lambda^3_{C^\ddagger}} \tag{24.55}$$

The moment of inertia of a diatomic molecule of bond length r is μr^2, where $\mu = m_A m_B/(m_A + m_B)$ is the effective mass, so the expression for the rate constant is

$$\begin{aligned} k_2 &= \kappa\frac{kT}{h}\frac{RT}{p^{\ominus}}\left(\frac{N_A\Lambda^3_A\Lambda^3_B}{\Lambda^3_{C^\ddagger}V^{\ominus}_m}\right)\left(\frac{2IkT}{\hbar^2}\right)e^{-\Delta E_0/RT} \\ &= \kappa\frac{kT}{h}N_A\left(\frac{\Lambda_A\Lambda_B}{\Lambda_{C^\ddagger}}\right)\left(\frac{2IkT}{\hbar^2}\right)e^{-\Delta E_0/RT} \\ &= \kappa N_A\left(\frac{8kT}{\pi\mu}\right)^{1/2}\pi r^2 e^{-\Delta E_0/RT} \end{aligned} \tag{24.56}$$

Finally, by identifying $\kappa\pi r^2$ as the reactive cross-section σ^*, we arrive at precisely the same expression as that obtained from simple collision theory (eqn 24.14).

24.5 Thermodynamic aspects

The statistical thermodynamic version of transition state theory rapidly runs into difficulties because only in some cases is anything known about the structure of the activated complex. However, the concepts that it introduces, principally that of an equilibrium between the reactants and the activated complex, have motivated a more general, empirical approach in which the activation process is expressed in terms of thermodynamic functions.

(a) Activation parameters

If we accept that $\bar{K}^\ddagger$ is an equilibrium constant (despite one mode of $C^\ddagger$ having been discarded), we can express it in terms of a **Gibbs energy of activation**, $\Delta^\ddagger G$, through the definition

$$\Delta^\ddagger G = -RT\ln \bar{K}^\ddagger \tag*{[24.57]}$$

(All the $\Delta^\ddagger X$ in this section are *standard* thermodynamic quantities, $\Delta^\ddagger X^{\ominus}$, but we shall omit the standard state sign to avoid overburdening the notation). Then the rate constant becomes

$$k_2 = \kappa\frac{kT}{h}\frac{RT}{p^{\ominus}}e^{-\Delta^\ddagger G/RT} \tag{24.58}$$

Because $G = H - TS$, the Gibbs energy of activation can be divided into an **entropy of activation**, $\Delta^\ddagger S$, and an **enthalpy of activation**, $\Delta^\ddagger H$, by writing

$$\Delta^\ddagger G = \Delta^\ddagger H - T\Delta^\ddagger S \tag*{[24.59]}$$

When eqn 24.59 is used in eqn 24.58 and κ is absorbed into the entropy term, we obtain

$$k_2 = Be^{\Delta^\ddagger S/R}e^{-\Delta^\ddagger H/RT} \qquad B = \frac{kT}{h}\frac{RT}{p^{\ominus}} \tag{24.60}$$

The formal definition of activation energy, $E_a = RT^2(\partial \ln k/\partial T)$, then gives $E_a = \Delta^\ddagger H + 2RT$, so

$$k_2 = e^2 Be^{\Delta^\ddagger S/R}e^{-E_a/RT} \tag{24.61}$$

Comment 24.4

For reactions of the type $A + B \rightleftharpoons P$ in the gas phase, $E_a = \Delta^\ddagger H + 2RT$. For these reactions in solution, $E_a = \Delta^\ddagger H + RT$.

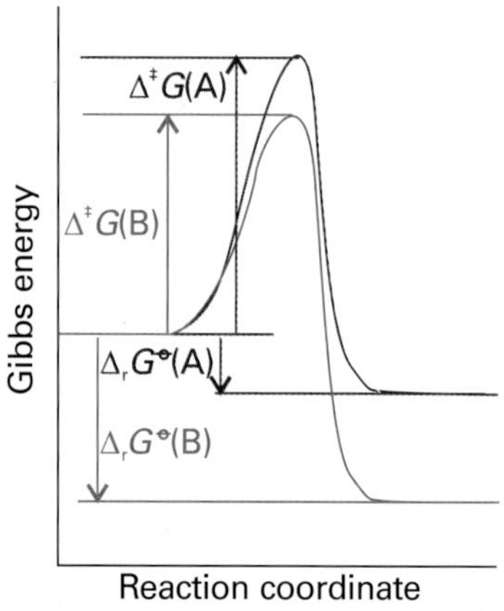

Fig. 24.9 For a related series of reactions, as the magnitude of the standard reaction Gibbs energy increases, so the activation barrier decreases. The approximate linear correlation between $\Delta^{\ddagger}G$ and $\Delta_r G^{\ominus}$ is the origin of linear free energy relations.

from which it follows that the Arrhenius factor A can be identified as

$$A = e^2 B e^{\Delta^{\ddagger}S/R} \tag{24.62}$$

The entropy of activation is negative because two reactant species come together to form one species. However, if there is a reduction in entropy below what would be expected for the simple encounter of A and B, then A will be smaller than that expected on the basis of simple collision theory. Indeed, we can identify that additional reduction in entropy, $\Delta^{\ddagger}S_{\text{steric}}$, as the origin of the steric factor of collision theory, and write

$$P = e^{\Delta^{\ddagger}S_{\text{steric}}/R} \tag{24.63}$$

Thus, the more complex the steric requirements of the encounter, the more negative the value of $\Delta^{\ddagger}S_{\text{steric}}$, and the smaller the value of P.

Gibbs energies, enthalpies, entropies, volumes, and heat capacities of activation are widely used to report experimental reaction rates, especially for organic reactions in solution. They are encountered when relationships between equilibrium constants and rates of reaction are explored using **correlation analysis**, in which ln K (which is equal to $-\Delta_r G^{\ominus}/RT$) is plotted against ln k (which is proportional to $-\Delta^{\ddagger}G/RT$). In many cases the correlation is linear, signifying that, as the reaction becomes thermodynamically more favourable, its rate constant increases (Fig. 24.9). This linear correlation is the origin of the alternative name **linear free energy relation** (LFER; see *Further reading*).

(b) Reactions between ions

The thermodynamic version of transition state theory simplifies the discussion of reactions in solution. The statistical thermodynamic theory is very complicated to apply because the solvent plays a role in the activated complex. In the thermodynamic approach we combine the rate law

$$\frac{d[P]}{dt} = k^{\ddagger}[C^{\ddagger}] \tag{24.64}$$

with the thermodynamic equilibrium constant

$$K = \frac{a_{C^{\ddagger}}}{a_A a_B} = K_{\gamma}\frac{[C^{\ddagger}]}{[A][B]} \qquad K_{\gamma} = \frac{\gamma_{C^{\ddagger}}}{\gamma_A \gamma_B} \tag{24.65}$$

Then

$$\frac{d[P]}{dt} = k_2[A][B] \qquad k_2 = \frac{k^{\ddagger}K}{K_{\gamma}} \tag{24.66}$$

If k_2° is the rate constant when the activity coefficients are 1 (that is, $k_2^{\circ} = k^{\ddagger}K$), we can write

$$k_2 = \frac{k_2^{\circ}}{K_{\gamma}} \tag{24.67}$$

At low concentrations the activity coefficients can be expressed in terms of the ionic strength, I, of the solution by using the Debye–Hückel limiting law (Section 5.9, particularly eqn 5.69) in the form

$$\log \gamma_J = -Az_J^2 I^{1/2} \tag{24.68}$$

with $A = 0.509$ in aqueous solution at 298 K. Then

$$\log k_2 = \log k_2^{\circ} - A\{z_A^2 + z_B^2 - (z_A + z_B)^2\}I^{1/2} = \log k_2^{\circ} + 2Az_A z_B I^{1/2} \tag{24.69}$$

The charge numbers of A and B are z_A and z_B, so the charge number of the activated complex is $z_A + z_B$; the z_J are positive for cations and negative for anions.

Equation 24.69 expresses the **kinetic salt effect**, the variation of the rate constant of a reaction between ions with the ionic strength of the solution (Fig. 24.10). If the reactant ions have the same sign (as in a reaction between cations or between anions), then increasing the ionic strength by the addition of inert ions increases the rate constant. The formation of a single, highly charged ionic complex from two less highly charged ions is favoured by a high ionic strength because the new ion has a denser ionic atmosphere and interacts with that atmosphere more strongly. Conversely, ions of opposite charge react more slowly in solutions of high ionic strength. Now the charges cancel and the complex has a less favourable interaction with its atmosphere than the separated ions.

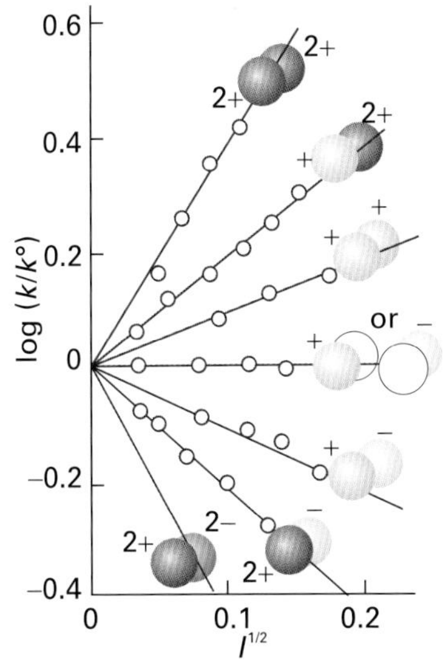

Fig. 24.10 Experimental tests of the kinetic salt effect for reactions in water at 298 K. The ion types are shown as spheres, and the slopes of the lines are those given by the Debye–Hückel limiting law and eqn 24.69.

Example 24.3 *Analysing the kinetic salt effect*

The rate constant for the base hydrolysis of $[CoBr(NH_3)_5]^{2+}$ varies with ionic strength as tabulated below. What can be deduced about the charge of the activated complex in the rate-determining stage?

I	0.0050	0.0100	0.0150	0.0200	0.0250	0.0300
k/k°	0.718	0.631	0.562	0.515	0.475	0.447

Method According to eqn 24.69, plot $\log(k/k^\circ)$ against $I^{1/2}$, when the slope will give $1.02z_Az_B$, from which we can infer the charges of the ions involved in the formation of the activated complex.

Answer Form the following table:

I	0.0050	0.0100	0.0150	0.0200	0.0250	0.0300
$I^{1/2}$	0.071	0.100	0.122	0.141	0.158	0.173
$\log(k/k^\circ)$	−0.14	−0.20	−0.25	−0.29	−0.32	−0.35

These points are plotted in Fig. 24.11. The slope of the (least squares) straight line is −2.04, indicating that $z_Az_B = -2$. Because $z_A = -1$ for the OH^- ion, if that ion is involved in the formation of the activated complex, then the charge number of the second ion is +2. This analysis suggests that the pentaamminebromocobalt(III) cation participates in the formation of the activated complex. The rate constant is also influenced by the relative permittivity of the medium.

Self-test 24.3 An ion of charge number +1 is known to be involved in the activated complex of a reaction. Deduce the charge number of the other ion from the following data:

I	0.0050	0.010	0.015	0.020	0.025	0.030
k/k°	0.930	0.902	0.884	0.867	0.853	0.841

[−1]

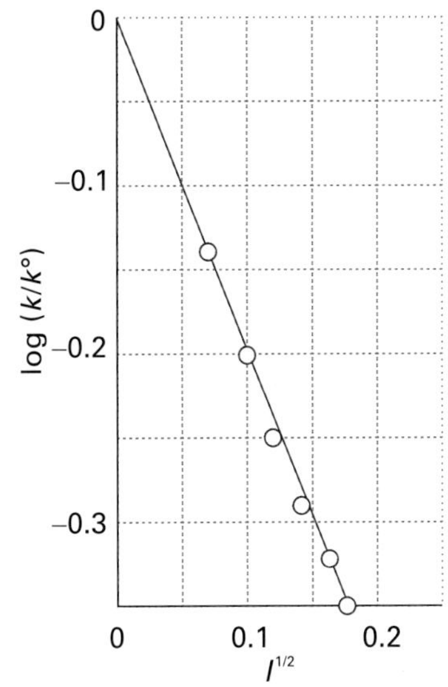

Fig. 24.11 The experimental ionic strength dependence of the rate constant of a hydrolysis reaction: the slope gives information about the charge types involved in the activated complex of the rate-determining step. See Example 24.3.

The dynamics of molecular collisions

We now come to the third and most detailed level of our examination of the factors that govern the rates of reactions.

24.6 Reactive collisions

Molecular beams allow us to study collisions between molecules in preselected energy states, and can be used to determine the states of the products of a reactive collision. Information of this kind is essential if a full picture of the reaction is to be built, because the rate constant is an average over events in which reactants in different initial states evolve into products in their final states.

(a) Experimental probes of reactive collisions

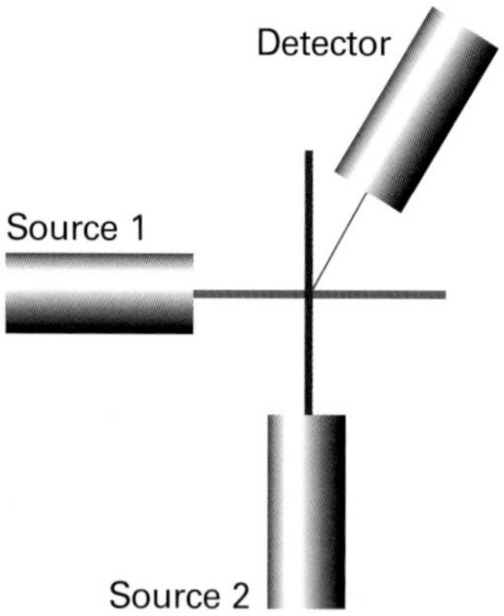

Fig. 24.12 In a crossed-beam experiment, state-selected molecules are generated in two separate sources, and are directed perpendicular to one another. The detector responds to molecules (which may be product molecules if chemical reaction occurs) scattered into a chosen direction.

Detailed experimental information about the intimate processes that occur during reactive encounters comes from molecular beams, especially crossed molecular beams (Fig. 24.12). The detector for the products of the collision of two beams can be moved to different angles, so the angular distribution of the products can be determined. Because the molecules in the incoming beams can be prepared with different energies (for example, with different translational energies by using rotating sectors and supersonic nozzles, with different vibrational energies by using selective excitation with lasers, as shown in Section 24.9b, and with different orientations (by using electric fields), it is possible to study the dependence of the success of collisions on these variables and to study how they affect the properties of the outcoming product molecules.

One method for examining the energy distribution in the products is **infrared chemiluminescence**, in which vibrationally excited molecules emit infrared radiation as they return to their ground states. By studying the intensities of the infrared emission spectrum, the populations of the vibrational states may be determined (Fig. 24.13). Another method makes use of **laser-induced fluorescence.** In this technique, a laser is used to excite a product molecule from a specific vibration-rotation level; the intensity of the fluorescence from the upper state is monitored and interpreted in terms of the population of the initial vibration-rotation state. When the molecules being studied do not fluoresce efficiently, coherent anti-Stokes Raman spectroscopy (CARS, Section 13.16) can be used to monitor the progress of reaction. **Multiphoton ionization** (MPI) techniques are also good alternatives for the study of weakly fluorescing molecules. In MPI, the absorption of several photons by a molecule results in ionization if the total photon energy is greater than the ionization energy of the molecule. One or more pulsed lasers are used to generate the molecular ions, which are commonly detected by time-of-flight mass spectrometry (TOF-MS, Section 19.2). The angular distribution of products can also be determined by **reaction product imaging.** In this technique, product ions are accelerated by an electric field towards a phosphorescent screen and the light emitted from specific spots where the ions struck the screen is imaged by a charge-coupled device (CCD, *Further information* 13.1). An important variant of MPI is **resonant multiphoton ionization** (REMPI), in which one or more photons promote a molecule to an electronically excited state and then additional photons are used to generate ions from the excited state. The power of REMPI lies in the fact that the experimenter can choose which reactant or product to study by tuning the laser frequency to the electronic absorption band of a specific molecule.

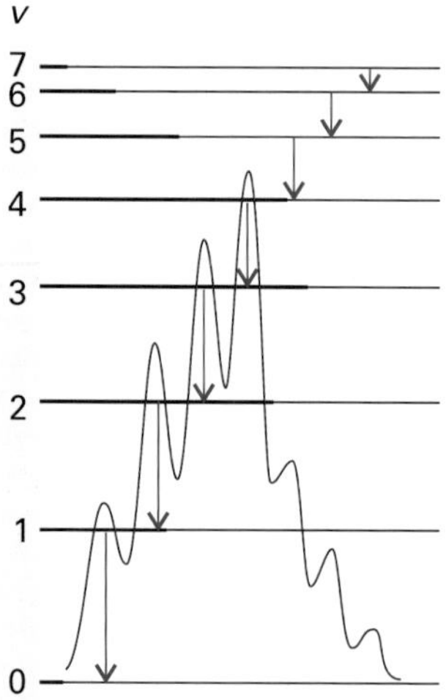

Fig. 24.13 Infrared chemiluminescence from CO produced in the reaction O + CS → CO + S arises from the non-equilibrium populations of the vibrational states of CO and the radiative relaxation to equilibrium.

(b) State-to-state dynamics

The concept of collision cross-section was introduced in connection with collision theory in Section 24.1, where we saw that the second-order rate constant, k_2, can be expressed as a Boltzmann-weighted average of the reactive collision cross-section and the relative speed of approach. We shall write eqn 24.12 as

$$k_2 = \langle \sigma v_{rel} \rangle N_A \qquad (24.70)$$

where the angle brackets denote a Boltzmann average. Molecular beam studies provide a more sophisticated version of this quantity, for they provide the **state-to-state cross-section**, $\sigma_{nn'}$, and hence the **state-to-state rate constant**, $k_{nn'}$:

$$k_{nn'} = \langle \sigma_{nn'} v_{\text{rel}} \rangle N_{\text{A}} \tag{24.71}$$

The rate constant k_2 is the sum of the state-to-state rate constant over all final states (because a reaction is successful whatever the final state of the products) and over a Boltzmann-weighted sum of initial states (because the reactants are initially present with a characteristic distribution of populations at a temperature T):

$$k_2 = \sum_{n,n'} k_{nn'}(T) f_n(T) \tag{24.72}$$

where $f_n(T)$ is the Boltzmann factor at a temperature T. It follows that, if we can determine or calculate the state-to-state cross-sections for a wide range of approach speeds and initial and final states, then we have a route to the calculation of the rate constant for the reaction.

24.7 Potential energy surfaces

One of the most important concepts for discussing beam results and calculating the state-to-state collision cross-section is the **potential energy surface** of a reaction, the potential energy as a function of the relative positions of all the atoms taking part in the reaction. Potential energy surfaces may be constructed from experimental data, with the techniques described in Section 24.6, and from results of quantum chemical calculations (Section 11.7). The theoretical method requires the systematic calculation of the energies of the system in a large number of geometrical arrangements. Special computational techniques are used to take into account electron correlation, which arises from instantaneous interactions between electrons as they move closer to and farther from each other in molecule or molecular cluster. Techniques that incorporate electron correlation are very time-consuming and, consequently, only reactions between relatively small particles, such as the reactions $H + H_2 \rightarrow H_2 + H$ and $H + H_2O \rightarrow OH + H_2$, are amenable to this type of theoretical treatment. An alternative is to use semi-empirical methods, in which results of calculations and experimental parameters are used to construct the potential energy surface.

To illustrate the features of a potential energy surface we consider the collision between an H atom and an H_2 molecule. Detailed calculations show that the approach of an atom along the H—H axis requires less energy for reaction than any other approach, so initially we confine our attention to a collinear approach. Two parameters are required to define the nuclear separations: one is the H_A—H_B separation R_{AB}, and the other is the H_B—H_C separation R_{BC}.

At the start of the encounter R_{AB} is infinite and R_{BC} is the H_2 equilibrium bond length. At the end of a successful reactive encounter R_{AB} is equal to the equilibrium bond length and R_{BC} is infinite. The total energy of the three-atom system depends on their relative separations, and can be found by doing a molecular orbital calculation. The plot of the total energy of the system against R_{AB} and R_{BC} gives the potential energy surface of this collinear reaction (Fig. 24.14). This surface is normally depicted as a contour diagram (Fig. 24.15).

When R_{AB} is very large, the variations in potential energy represented by the surface as R_{BC} changes are those of an isolated H_2 molecule as its bond length is altered. A section through the surface at $R_{AB} = \infty$, for example, is the same as the H_2 bonding potential energy curve drawn in Fig. 11.16. At the edge of the diagram where R_{BC} is very large, a section through the surface is the molecular potential energy curve of an isolated H_AH_B molecule.

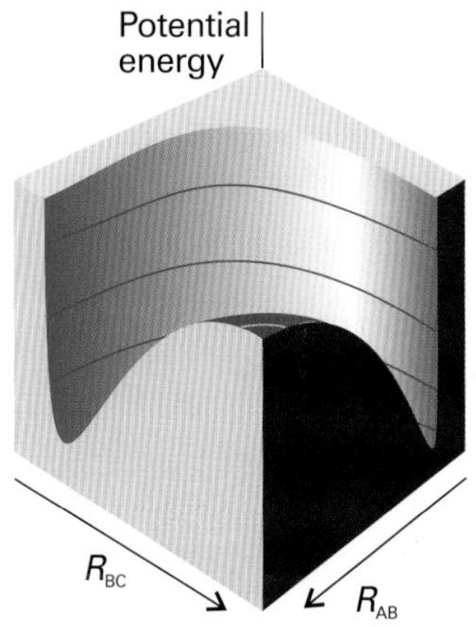

Fig. 24.14 The potential energy surface for the $H + H_2 \rightarrow H_2 + H$ reaction when the atoms are constrained to be collinear.

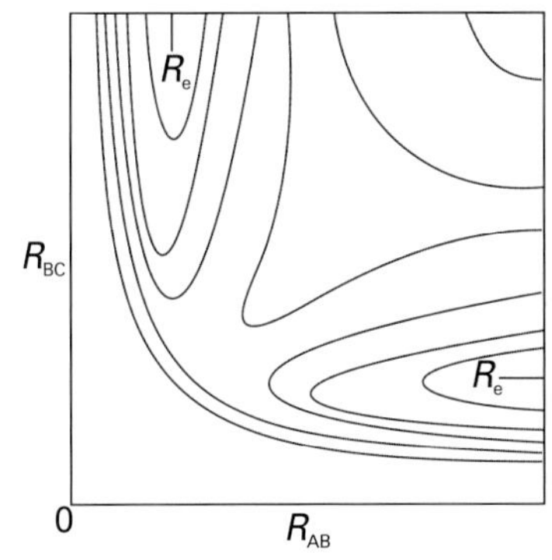

Fig. 24.15 The contour diagram (with contours of equal potential energy) corresponding to the surface in Fig. 24.14. R_e marks the equilibrium bond length of an H_2 molecule (strictly, it relates to the arrangement when the third atom is at infinity).

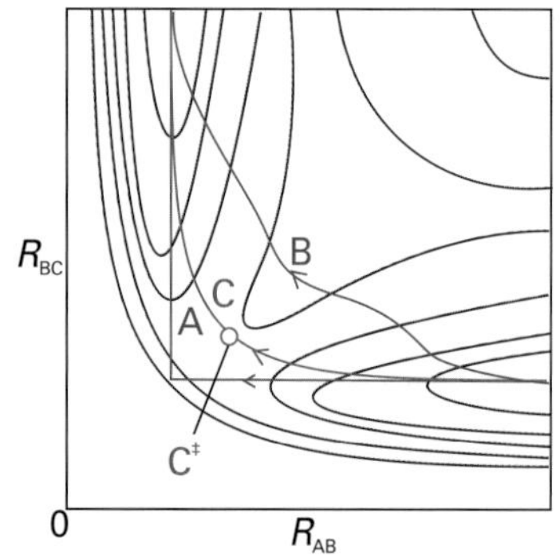

Fig. 24.16 Various trajectories through the potential energy surface shown in Fig. 24.15. Path A corresponds to a route in which R_{BC} is held constant as H_A approaches; path B corresponds to a route in which R_{BC} lengthens at an early stage during the approach of H_A; path C is the route along the floor of the potential valley.

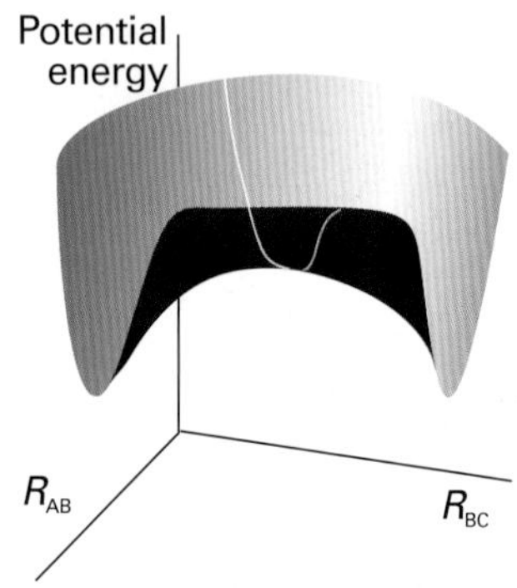

Fig. 24.17 The transition state is a set of configurations (here, marked by the line across the saddle point) through which successful reactive trajectories must pass.

The actual path of the atoms in the course of the encounter depends on their total energy, the sum of their kinetic and potential energies. However, we can obtain an initial idea of the paths available to the system for paths that correspond to least potential energy. For example, consider the changes in potential energy as H_A approaches H_BH_C. If the $H_B—H_C$ bond length is constant during the initial approach of H_A, then the potential energy of the H_3 cluster rises along the path marked A in Fig. 24.16. We see that the potential energy reaches a high value as H_A is pushed into the molecule and then decreases sharply as H_C breaks off and separates to a great distance. An alternative reaction path can be imagined (B) in which the $H_B—H_C$ bond length increases while H_A is still far away. Both paths, although feasible if the molecules have sufficient initial kinetic energy, take the three atoms to regions of high potential energy in the course of the encounter.

The path of least potential energy is the one marked C, corresponding to R_{BC} lengthening as H_A approaches and begins to form a bond with H_B. The $H_B—H_C$ bond relaxes at the demand of the incoming atom, and the potential energy climbs only as far as the saddle-shaped region of the surface, to the **saddle point** marked $C^{\ddagger}$. The encounter of least potential energy is one in which the atoms take route C up the floor of the valley, through the saddle point, and down the floor of the other valley as H_C recedes and the new $H_A—H_B$ bond achieves its equilibrium length. This path is the reaction coordinate we met in Section 24.4.

We can now make contact with the transition state theory of reaction rates. In terms of trajectories on potential surfaces, the transition state can be identified with a critical geometry such that every trajectory that goes through this geometry goes on to react (Fig. 24.17).

24.8 Some results from experiments and calculations

To travel successfully from reactants to products the incoming molecules must possess enough kinetic energy to be able to climb to the saddle point of the potential surface. Therefore, the shape of the surface can be explored experimentally by changing the relative speed of approach (by selecting the beam velocity) and the degree of vibrational excitation and observing whether reaction occurs and whether the products emerge in a vibrationally excited state (Fig. 24.18). For example, one question that can be answered is whether it is better to smash the reactants together with a lot of translational kinetic energy or to ensure instead that they approach in highly excited vibrational states. Thus, is trajectory C_2^*, where the H_BH_C molecule is initially vibrationally excited, more efficient at leading to reaction than the trajectory C_1^*, in which the total energy is the same but has a high translational kinetic energy?

(a) The direction of attack and separation

Figure 24.19 shows the results of a calculation of the potential energy as an H atom approaches an H_2 molecule from different angles, the H_2 bond being allowed to relax to the optimum length in each case. The potential barrier is least for collinear attack, as we assumed earlier. (But we must be aware that other lines of attack are feasible and contribute to the overall rate.) In contrast, Fig. 24.20 shows the potential energy changes that occur as a Cl atom approaches an HI molecule. The lowest barrier occurs for approaches within a cone of half-angle 30° surrounding the H atom. The relevance of this result to the calculation of the steric factor of collision theory should be noted: not every collision is successful, because not every one lies within the reactive cone.

If the collision is sticky, so that when the reactants collide they orbit around each other, the products can be expected to emerge in random directions because all

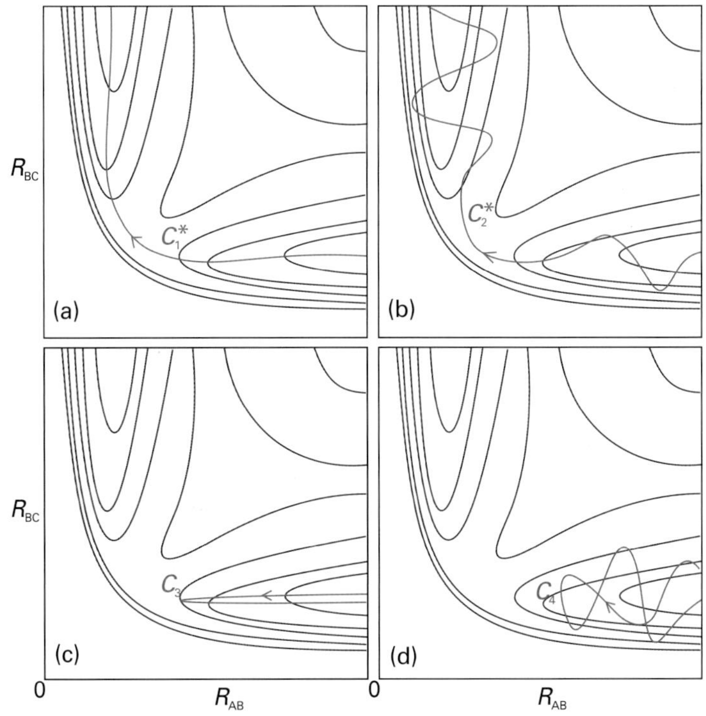

Fig. 24.18 Some successful (*) and unsuccessful encounters. (a) C_1^* corresponds to the path along the foot of the valley. (b) C_2^* corresponds to an approach of A to a vibrating BC molecule, and the formation of a vibrating AB molecule as C departs. (c) C_3 corresponds to A approaching a non-vibrating BC molecule, but with insufficient translational kinetic energy. (d) C_4 corresponds to A approaching a vibrating BC molecule, but still the energy, and the phase of the vibration, is insufficient for reaction.

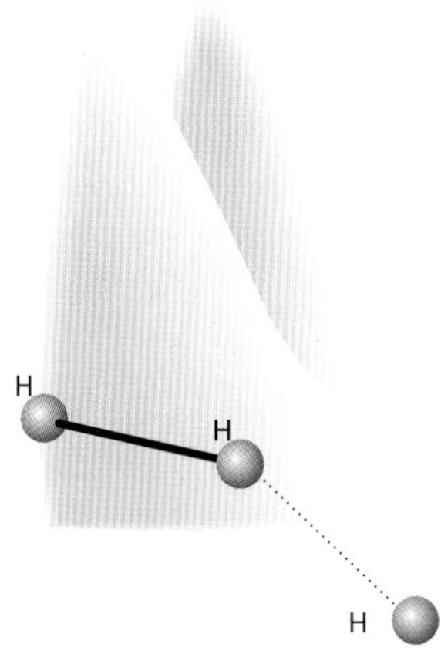

Fig. 24.19 An indication of the anisotropy of the potential energy changes as H approaches H_2 with different angles of attack. The collinear attack has the lowest potential barrier to reaction. The surface indicates the potential energy profile along the reaction coordinate for each configuration.

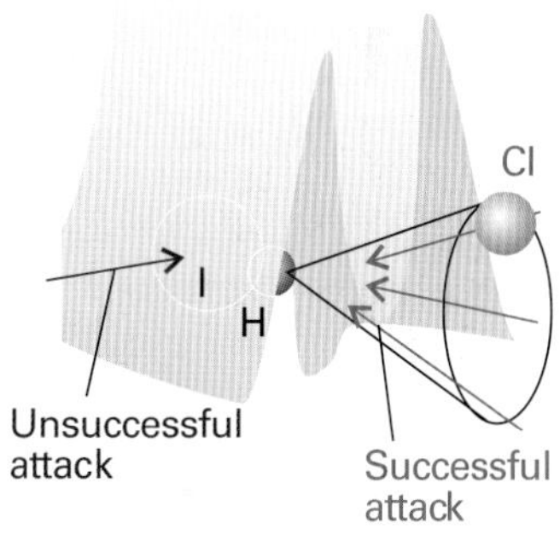

Fig. 24.20 The potential energy barrier for the approach of Cl to HI. In this case, successful encounters occur only when Cl approaches within a cone surrounding the H atom.

memory of the approach direction has been lost. A rotation takes about 1 ps, so if the collision is over in less than that time the complex will not have had time to rotate and the products will be thrown off in a specific direction. In the collision of K and I_2, for example, most of the products are thrown off in the forward direction. This product distribution is consistent with the harpoon mechanism (Section 24.1c) because the transition takes place at long range. In contrast, the collision of K with CH_3I leads to reaction only if the molecules approach each other very closely. In this mechanism, K effectively bumps into a brick wall, and the KI product bounces out in the backward direction. The detection of this anisotropy in the angular distribution of products gives an indication of the distance and orientation of approach needed for reaction, as well as showing that the event is complete in less than 1 ps.

(b) Attractive and repulsive surfaces

Some reactions are very sensitive to whether the energy has been predigested into a vibrational mode or left as the relative translational kinetic energy of the colliding molecules. For example, if two HI molecules are hurled together with more than twice the activation energy of the reaction, then no reaction occurs if all the energy is translational. For F + HCl $\rightarrow$ Cl + HF, for example, the reaction is about five times as efficient when the HCl is in its first vibrational excited state than when, although HCl has the same total energy, it is in its vibrational ground state.

Comment 24.5

In molecular beam work the remarks we make in our discussion normally refer to directions in a centre-of-mass coordinate system. The origin of the coordinates is the centre of mass of the colliding reactants, and the collision takes place when the molecules are at the origin. The way in which centre-of-mass coordinates are constructed and the events in them interpreted involves too much detail for our present purposes, but we should bear in mind that 'forward' and 'backward' have unconventional meanings. The details are explained in the books in *Further reading*.

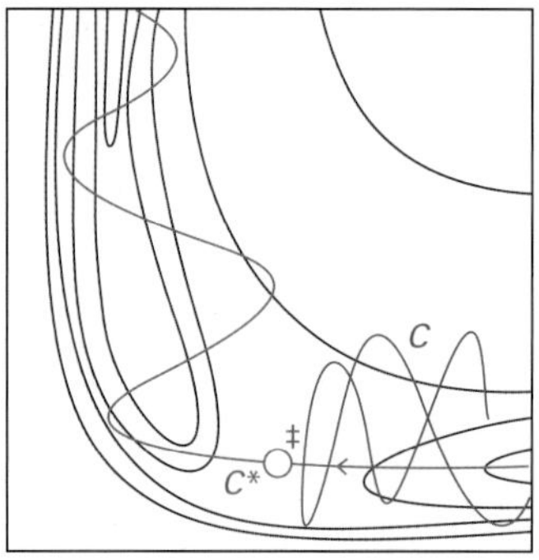

Fig. 24.21 An attractive potential energy surface. A successful encounter (C^*) involves high translational kinetic energy and results in a vibrationally excited product.

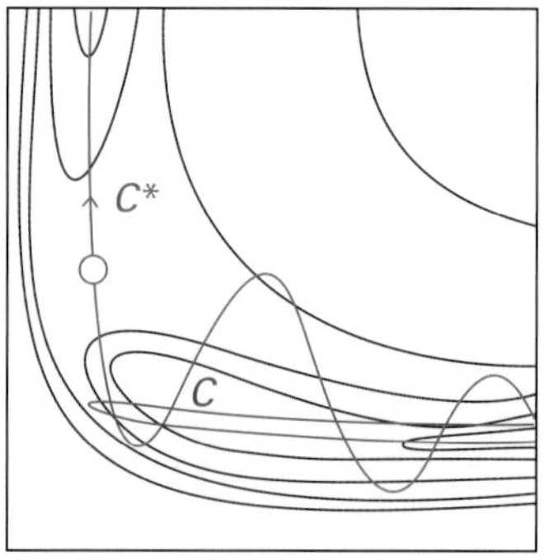

Fig. 24.22 A repulsive potential energy surface. A successful encounter (C^*) involves initial vibrational excitation and the products have high translational kinetic energy. A reaction that is attractive in one direction is repulsive in the reverse direction.

The origin of these requirements can be found by examining the potential energy surface. Figure 24.21 shows an **attractive surface** in which the saddle point occurs early in the reaction coordinate. Figure 24.22 shows a **repulsive surface** in which the saddle point occurs late. A surface that is attractive in one direction is repulsive in the reverse direction.

Consider first the attractive surface. If the original molecule is vibrationally excited, then a collision with an incoming molecule takes the system along C. This path is bottled up in the region of the reactants, and does not take the system to the saddle point. If, however, the same amount of energy is present solely as translational kinetic energy, then the system moves along C^* and travels smoothly over the saddle point into products. We can therefore conclude that reactions with attractive potential energy surfaces proceed more efficiently if the energy is in relative translational motion. Moreover, the potential surface shows that once past the saddle point the trajectory runs up the steep wall of the product valley, and then rolls from side to side as it falls to the foot of the valley as the products separate. In other words, the products emerge in a vibrationally excited state.

Now consider the repulsive surface (Fig. 24.22). On trajectory C the collisional energy is largely in translation. As the reactants approach, the potential energy rises. Their path takes them up the opposing face of the valley, and they are reflected back into the reactant region. This path corresponds to an unsuccessful encounter, even though the energy is sufficient for reaction. On C^* some of the energy is in the vibration of the reactant molecule and the motion causes the trajectory to weave from side to side up the valley as it approaches the saddle point. This motion may be sufficient to tip the system round the corner to the saddle point and then on to products. In this case, the product molecule is expected to be in an unexcited vibrational state. Reactions with repulsive potential surfaces can therefore be expected to proceed more efficiently if the excess energy is present as vibrations. This is the case with the $H + Cl_2 \rightarrow HCl + Cl$ reaction, for instance.

(c) Classical trajectories

A clear picture of the reaction event can be obtained by using classical mechanics to calculate the trajectories of the atoms taking place in a reaction from a set of initial conditions, such as velocities, relative orientations, and internal energies of the reacting particles. The initial values used for the internal energy reflect the quantization of electronic, vibrational, and rotational energies in molecules but the features of quantum mechanics are not used explicitly in the calculation of the trajectory.

Figure 24.23 shows the result of such a calculation of the positions of the three atoms in the reaction $H + H_2 \rightarrow H_2 + H$, the horizontal coordinate now being time and the vertical coordinate the separations. This illustration shows clearly the vibration of the original molecule and the approach of the attacking atom. The reaction itself, the switch of partners, takes place very rapidly and is an example of a **direct mode process.** The newly formed molecule shakes, but quickly settles down to steady, harmonic vibration as the expelled atom departs. In contrast, Fig. 24.24 shows an example of a **complex mode process**, in which the activated complex survives for an extended period. The reaction in the illustration is the exchange reaction $KCl + NaBr \rightarrow KBr + NaCl$. The tetratomic activated complex survives for about 5 ps, during which time the atoms make about 15 oscillations before dissociating into products.

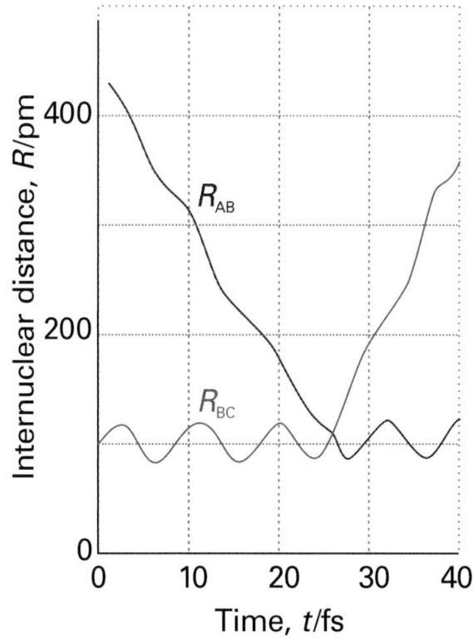

Fig. 24.23 The calculated trajectories for a reactive encounter between A and a vibrating BC molecule leading to the formation of a vibrating AB molecule. This direct-mode reaction is between H and H_2. (M. Karplus, R.N. Porter, and R.D. Sharma, *J. Chem. Phys.*, **43**, 3258 (1965).)

(d) Quantum mechanical scattering theory

Classical trajectory calculations do not recognize the fact that the motion of atoms, electrons, and nuclei is governed by quantum mechanics. The concept of trajectory then fades and is replaced by the unfolding of a wavefunction that represents initially the reactants and finally products.

Complete quantum mechanical calculations of trajectories and rate constants are very onerous because it is necessary to take into account all the allowed electronic, vibrational, and rotational states populated by each atom and molecule in the system at a given temperature. It is common to define a 'channel' as a group of molecules in well-defined quantum mechanically allowed states. Then, at a given temperature, there are many channels that represent the reactants and many channels that represent possible products, with some transitions between channels being allowed but others not allowed. Furthermore, not every transition leads to a chemical reaction. For example, the process $H_2^* + OH \rightarrow H_2 + (OH)^*$, where the asterisk denotes an excited state, amounts to energy transfer between H_2 and OH, whereas the process $H_2^* + OH \rightarrow H_2O + H$ represents a chemical reaction. What complicates a quantum mechanical calculation of trajectories and rate constants even in this simple four-atom system is that many reacting channels present at a given temperature can lead to the desired products $H_2O + H$, which themselves may be formed as many distinct channels. The **cumulative reaction probability**, $N(E)$, at a fixed total energy E is then written as

$$N(E) = \sum_{i,j} P_{ij}(E) \qquad (24.73)$$

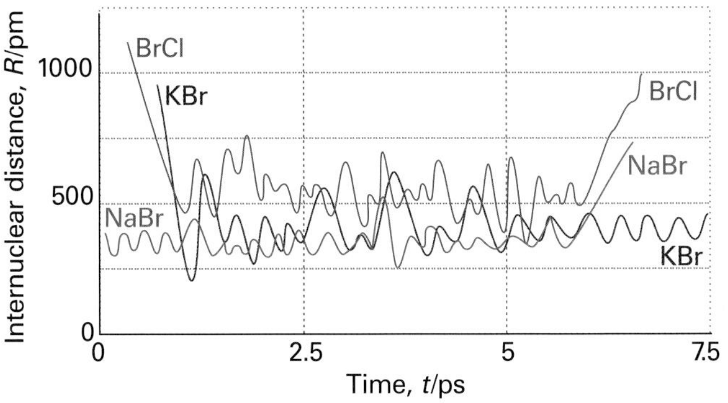

Fig. 24.24 An example of the trajectories calculated for a complex-mode reaction, $KCl + NaBr \rightarrow KBr + NaCl$, in which the collision cluster has a long lifetime. (P. Brumer and M. Karplus, *Faraday Disc. Chem. Soc.*, **55**, 80 (1973).)

where $P_{i,j}(E)$ is the probability for a transition between a reacting channel i and a product channel j and the summation is over all possible transitions that lead to product. It is then possible to show that the rate constant is given by

$$k(T) = \frac{\int_0^{\infty} N(E)e^{-E/kT}\,dE}{hQ_r(T)} \tag{24.74}$$

where $Q_r(T)$ is the partition function density (the partition function divided by the volume) of the reactants at the temperature T. The significance of eqn 24.74 is that it provides a direct connection between an experimental quantity, the rate constant, and a theoretical quantity, $N(E)$.

24.9 The investigation of reaction dynamics with ultrafast laser techniques

The development of femtosecond pulsed lasers (Section 14.6) has made it possible to make observations on species that have such short lifetimes that in a number of respects they resemble an activated complex. Pulsed-laser techniques can also be used to control the outcome of chemical reactions.

(a) Spectroscopic observation of the activated complex

Until very recently there were no direct spectroscopic observations on activated complexes, for they have a very fleeting existence and often survive for only a few picoseconds. In a typical experiment designed to detect an activated complex, a femtosecond laser pulse is used to excite a molecule to a dissociative state, and then a second femtosecond pulse is fired at an interval after the dissociating pulse. The frequency of the second pulse is set at an absorption of one of the free fragmentation products, so its absorption is a measure of the abundance of the dissociation product. For example, when ICN is dissociated by the first pulse, the emergence of CN from the photoactivated state can be monitored by watching the growth of the free CN absorption (or, more commonly, its laser-induced fluorescence). In this way it has been found that the CN signal remains zero until the fragments have separated by about 600 pm, which takes about 205 fs.

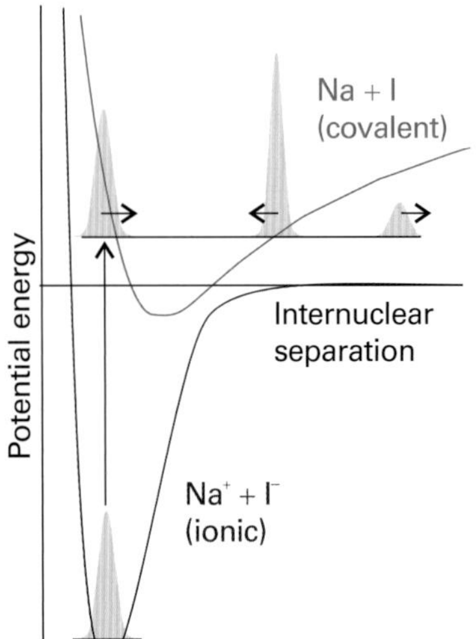

Fig. 24.25 Excitation of the ion pair Na^+I^- forms an excited state with covalent character. Also shown is movement between a 'covalent' surface (in green) and an 'ionic' surface (in purple) of the wavepacket formed by laser excitation.

Some sense of the progress that has been made in the study of the intimate mechanism of chemical reactions can be obtained by considering the decay of the ion pair Na^+I^-. As shown in Fig. 24.25, excitation of the ionic species with a femtosecond laser pulse forms an excited state that corresponds to a covalently bonded NaI molecule. The system can be described with two potential energy surfaces, one largely 'ionic' and another 'covalent', which cross at an internuclear separation of 693 pm. A short laser pulse is composed of a wide range of frequencies, which excite many vibrational states of NaI simultaneously. Consequently, the electronically excited complex exists as a superposition of states, or a localized wavepacket (Section 8.6), which oscillates between the 'covalent' and 'ionic' potential energy surfaces, as shown in Fig. 24.25. The complex can also dissociate, shown as movement of the wavepacket toward very long internuclear separation along the dissociative surface. However, not every outward-going swing leads to dissociation because there is a chance that the I atom can be harpooned again, in which case it fails to make good its escape. The dynamics of the system is probed by a second laser pulse with a frequency that corresponds to the absorption frequency of the free Na product or to the frequency at which Na absorbs when it is a part of the complex. The latter frequency depends on the Na···I distance, so an absorption (in practice, a laser-induced fluorescence) is obtained each time the wavepacket returns to that separation.

A typical set of results is shown in Fig. 24.26. The bound Na absorption intensity shows up as a series of pulses that recur in about 1 ps, showing that the wavepacket oscillates with about that period. The decline in intensity shows the rate at which the complex can dissociate as the two atoms swing away from each other. The free Na absorption also grows in an oscillating manner, showing the periodicity of wavepacket oscillation, each swing of which gives it a chance to dissociate. The precise period of the oscillation in NaI is 1.25 ps, corresponding to a vibrational wavenumber of 27 cm^{-1} (recall that the activated complex theory assumes that such a vibration has a very low frequency). The complex survives for about ten oscillations. In contrast, although the oscillation frequency of NaBr is similar, it barely survives one oscillation.

Femtosecond spectroscopy has also been used to examine analogues of the activated complex involved in bimolecular reactions. Thus, a molecular beam can be used to produce a van der Waals molecule (Section 18.6), such as IH···OCO. The HI bond can be dissociated by a femtosecond pulse, and the H atom is ejected towards the O atom of the neighbouring CO_2 molecule to form HOCO. Hence, the van der Waals molecule is a source of a species that resembles the activated complex of the reaction

$$H + CO_2 \rightarrow [HOCO]^{\ddagger} \rightarrow HO + CO$$

The probe pulse is tuned to the OH radical, which enables the evolution of $[HOCO]^{\ddagger}$ to be studied in real time. Femtosecond transition state spectroscopy has also been used to study more complex reactions, such as the Diels–Alder reaction, nucleophilic substitution reactions, and pericyclic addition and cleavage reactions. Biological process that are open to study by femtosecond spectroscopy include the energy-converting processes of photosynthesis and the photostimulated processes of vision. In other experiments, the photoejection of carbon monoxide from myoglobin and the attachment of O_2 to the exposed site have been studied to obtain rate constants for the two processes.

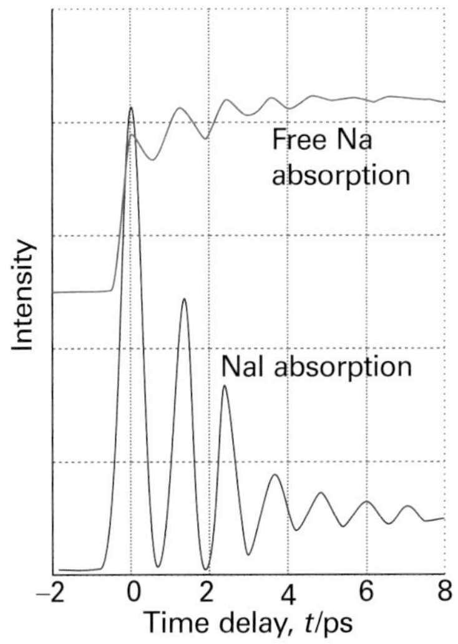

Fig. 24.26 Femtosecond spectroscopic results for the reaction in which sodium iodide separates inot Na and I. The lower curve is the absorption of the electronically excited complex and the upper curve is the absorption of free Na atoms (Adapted from A.H. Zewail, *Science* **242**, 1645 (1988)).

(b) Controlling chemical reactions with lasers

A long-standing goal of chemistry is to control the rate and distribution of products in chemical reactions, with an eye toward minimizing undesirable side reactions and improving the efficiency of industrial processes. We already have at our disposal a number of successful strategies for achieving this goal. For example, it is possible to synthesize a catalyst that accelerates a specific type of reaction but not others, so a desired product may be formed more quickly than an undesired product. However, this strategy is not very general, as a new catalyst needs to be developed for every reaction type of interest. A more ambitious and potentially more powerful strategy consists of using lasers to prepare specific states of reactant molecules that lead to a specific activated complex and hence a specific product, perhaps not even the major product isolated under ordinary laboratory conditions. Here we examine two ways in which the outcome of a chemical reaction can be affected by laser irradiation.

Some reactions may be controlled by exciting the reactants to different vibrational states. Consider the gas-phase reaction between H and HOD. It has been observed that H_2 and OD are the preferred products when thermally equilibrated H atoms react with vibrationally excited HOD molecules prepared by laser irradiation at a wavelength that excites the H—OD stretching mode from the $v = 0$ to the $v = 4$ energy level. On the other hand, when the same stretching mode is excited to the $v = 5$ energy level, HD and OH are the preferred products. This control strategy is commonly referred to as *mode-selective chemistry* and has been used to alter product distributions in a number of bimolecular reactions. However, the technique is limited to those cases in which energy can be deposited and remains localized in the desired vibrational mode of the reactant for a time that is much longer than the reaction time. This is difficult to achieve in large molecules, in which intramolecular vibrational relaxation redistributes energy among the many vibrational modes within a few picoseconds.

A strategy that seeks to avoid the problem of vibrational relaxation uses ultrafast lasers and is related closely to the techniques used for the spectroscopic detection of transition states. Consider the reaction $I_2 + Xe \rightarrow XeI^* + I$, which occurs via a harpoon mechanism with a transition state denoted as $[Xe^+\cdots I^-\cdots I]$. The reaction can be initiated by exciting I_2 to an electronic state at least 52 460 cm^{-1} above the ground state and then followed by measuring the time dependence of the chemiluminescence of XeI^*. To exert control over the yield of the product, a pair of femtosecond pulses can be used to induce the reaction. The first pulse excites the I_2 molecule to a low energy and unreactive electronic state. We already know that excitation by a femtosecond pulse generates a wavepacket that can be treated as a particle travelling across the potential energy surface. In this case, the wavepacket does not have enough energy to react, but excitation by another laser pulse with the proper wavelength can provide the necessary additional energy. It follows that activated complexes with different geometries can be prepared by varying the time delay between the two pulses, as the partially localized wave packet will be at different locations along the potential energy surface as it evolves after being formed by the first pulse. Because the reaction occurs via the harpoon mechanism, the product yield is expected to be optimal if the second pulse is applied when the wave packet is at a point where the $Xe\cdots I_2$ distance is just right for electron transfer from Xe to I_2 to occur (see Example 24.2). This type of control of the $I_2 + Xe$ reaction has been demonstrated.

So far, the control techniques we have discussed have only been applied to reactions between relatively small molecules, with simple and well-understood potential energy surfaces. Extension of these techniques to the controlled synthesis of materials in routine laboratory work will require much more sophisticated knowledge of how laser pulses may be combined to stimulate a specific molecular response in a complex system.

Electron transfer in homogeneous systems

We end the chapter by applying the concepts of transition state theory and quantum theory to the study of a deceptively simple process, electron transfer between molecules in homogeneous systems. We begin by examining the features of a theory that describes the factors governing the rates of electron transfer. Then, we discuss the theory in the light of experimental results on a variety of systems, including protein complexes. We shall see that relatively simple expressions may be used to predict the rates of electron transfer with reasonable accuracy.

24.10 The rates of electron transfer processes

Consider electron transfer from a donor species D to an acceptor species A in solution. The net reaction is

$$D + A \rightarrow D^+ + A^- \qquad v = k_{obs}[D][A] \qquad K = \frac{[D^+][A^-]}{[D][A]} \qquad (24.75)$$

In the first step of the mechanism, D and A must diffuse through the solution and collide to form a complex DA, in which the donor and acceptor are separated by a distance comparable to r, the distance between the edges of each species. We assume that D, A, and DA are in equilibrium:

$$\mathrm{D} + \mathrm{A} \rightleftharpoons \mathrm{DA} \qquad K_{\mathrm{DA}} = \frac{k_\mathrm{a}}{k'_\mathrm{a}} = \frac{[\mathrm{DA}]}{[\mathrm{D}][\mathrm{A}]} \tag{24.76a}$$

where k_a and k'_a are, respectively, the rate constants for the association and dissociation of the DA complex. Next, electron transfer occurs within the DA complex to yield $\mathrm{D^+A^-}$:

$$\mathrm{DA} \rightarrow \mathrm{D^+A^-} \qquad v_\mathrm{et} = k_\mathrm{et}[\mathrm{DA}] \tag{24.76b}$$

where k_et is the first-order rate constant for the forward electron transfer step. The $\mathrm{D^+A^-}$ complex has two possible fates. First, reverse electron transfer with a rate constant k_r can regenerate DA:

$$\mathrm{D^+A^-} \rightarrow \mathrm{DA} \qquad v_\mathrm{r} = k_\mathrm{r}[\mathrm{D^+A^-}] \tag{24.76c}$$

Second, $\mathrm{D^+A^-}$ can break apart and the ions diffuse through the solution:

$$\mathrm{D^+A^-} \rightarrow \mathrm{D^+} + \mathrm{A^-} \qquad v_\mathrm{d} = k_\mathrm{d}[\mathrm{D^+A^-}] \tag{24.76d}$$

We show in the following *Justification* that

$$\frac{1}{k_\mathrm{obs}} = \frac{1}{k_\mathrm{a}} + \frac{k'_\mathrm{a}}{k_\mathrm{a}k_\mathrm{et}}\left(1 + \frac{k_\mathrm{r}}{k_\mathrm{d}}\right) \tag{24.77}$$

Justification 24.4 *The rate constant for electron transfer in solution*

To find an expression for the second-order rate constant k_obs for electron transfer between D and A in solution, we begin by equating the rate of the net reaction (eqn 24.75) to the rate of formation of separated ions, the reaction products (eqn 24.76d):

$$v = k_\mathrm{obs}[\mathrm{D}][\mathrm{A}] = k_\mathrm{d}[\mathrm{D^+A^-}]$$

Now we apply the steady-state approximation to the intermediate $\mathrm{D^+A^-}$:

$$\frac{\mathrm{d}[\mathrm{D^+A^-}]}{\mathrm{d}t} = k_\mathrm{et}[\mathrm{DA}] - k_\mathrm{r}[\mathrm{D^+A^-}] - k_\mathrm{d}[\mathrm{D^+A^-}] = 0$$

It follows that

$$[\mathrm{D^+A^-}] = \frac{k_\mathrm{et}}{k_\mathrm{r} + k_\mathrm{d}}[\mathrm{DA}]$$

However, DA is also an intermediate so we apply the steady-state approximation again

$$\frac{\mathrm{d}[\mathrm{DA}]}{\mathrm{d}t} = k_\mathrm{a}[\mathrm{D}][\mathrm{A}] - k'_\mathrm{a}[\mathrm{DA}] - k_\mathrm{et}[\mathrm{DA}] + k_\mathrm{r}[\mathrm{D^+A^-}] = 0$$

Substitution of the initial expression for the steady-state concentration of $\mathrm{D^+A^-}$ into this expression for [DA] gives, after some algebra, a new expression for $[\mathrm{D^+A^-}]$:

$$[\mathrm{D^+A^-}] = \frac{k_\mathrm{a}k_\mathrm{et}}{k'_\mathrm{a}k_\mathrm{r} + k'_\mathrm{a}k_\mathrm{d} + k_\mathrm{d}k_\mathrm{et}}[\mathrm{D}][\mathrm{A}]$$

When we multiply this expression by k_d, we see that the resulting equation has the form of the rate of electron transfer, $v = k_\mathrm{obs}[\mathrm{D}][\mathrm{A}]$, with k_obs given by

$$k_\mathrm{obs} = \frac{k_\mathrm{d}k_\mathrm{a}k_\mathrm{et}}{k'_\mathrm{a}k_\mathrm{r} + k'_\mathrm{a}k_\mathrm{d} + k_\mathrm{d}k_\mathrm{et}}$$

To obtain eqn 24.77, we divide the numerator and denominator on the right-hand side of this expression by $k_\mathrm{d}k_\mathrm{et}$ and solve for the reciprocal of k_obs.

To gain insight into eqn 24.77 and the factors that determine the rate of electron transfer reactions in solution, we assume that the main decay route for D^+A^- is dissociation of the complex into separated ions, or $k_d \gg k_r$. It follows that

$$\frac{1}{k_{obs}} \approx \frac{1}{k_a}\left(1+\frac{k_a'}{k_{et}}\right)$$

When $k_{et} \gg k_a'$, we see that $k_{obs} \approx k_a$ and the rate of product formation is controlled by diffusion of D and A in solution, which fosters formation of the DA complex. When $k_{et} \ll k_a'$, we see that $k_{obs} \approx (k_a/k_a')k_{et}$ or, after using eqn 24.76a,

$$k_{obs} \approx K_{DA}k_{et} \tag{24.78}$$

and the process is controlled by the activation energy of electron transfer in the DA complex. Using transition state theory (Section 24.4), we write

$$k_{et} = \kappa\nu e^{-\Delta^{\ddagger}G/RT} \tag{24.79}$$

where κ is the transmission coefficient, ν is the vibrational frequency with which the activated complex approaches the transition state, and $\Delta^{\ddagger}G$ is the Gibbs energy of activation. Our first task is to write theoretical expressions for $\kappa\nu$ and $\Delta^{\ddagger}G$ by describing the motions of electrons and nuclei mathematically.

24.11 Theory of electron transfer processes

Our discussion concentrates on the following two key aspects of the theory, which was developed independently by R.A. Marcus, N.S. Hush, V.G. Levich, and R.R. Dogonadze:

1 Electrons are transferred by tunnelling through a potential energy barrier, the height of which is partly determined by the ionization energies of the DA and D^+A^- complexes. Electron tunnelling influences the magnitude of $\kappa\nu$.

2 The complex DA and the solvent molecules surrounding it undergo structural rearrangements prior to electron transfer. The energy associated with these rearrangements and the standard reaction Gibbs energy determine $\Delta^{\ddagger}G$.

(a) Electron tunnelling

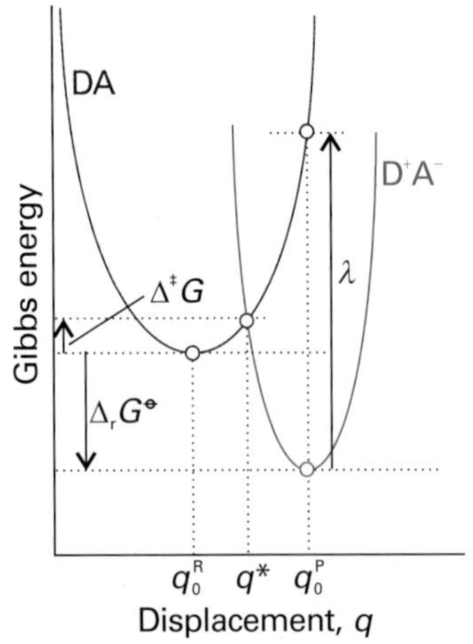

Fig. 24.27 The Gibbs energy surfaces of the complexes DA and D^+A^- involved in an electron transfer process are represented by parabolas characteristic of harmonic oscillators, with the displacement coordinate q corresponding to the changing geometries of the system. In the plot, q_0^R and q_0^P are the values of q at which the minima of the reactant and product parabolas occur, respectively. The parabolas intersect at $q = q^*$. The plots also portray the Gibbs energy of activation, $\Delta^{\ddagger}G$, the standard reaction Gibbs energy, $\Delta_rG^⦵$, and the reorganization energy, λ (discussed in Section 24.11b).

We saw in Section 14.2 that, according to the Franck–Condon principle, electronic transitions are so fast that they can be regarded as taking place in a stationary nuclear framework. This principle also applies to an electron transfer process in which an electron migrates from one energy surface, representing the dependence of the energy of DA on its geometry, to another representing the energy of D^+A^-. We can represent the potential energy (and the Gibbs energy) surfaces of the two complexes (the reactant complex, DA, and the product complex, D^+A^-) by the parabolas characteristic of harmonic oscillators, with the displacement coordinate corresponding to the changing geometries (Fig. 24.27). This coordinate represents a collective mode of the donor, acceptor, and solvent.

According to the Franck–Condon principle, the nuclei do not have time to move when the system passes from the reactant to the product surface as a result of the transfer of an electron. Therefore, electron transfer can occur only after thermal fluctuations bring the geometry of DA to q^* in Fig. 24.27, the value of the nuclear coordinate at which the two parabolas intersect.

The factor $\kappa\nu$ is a measure of the probability that the system will convert from reactants (DA) to products (D^+A^-) at q^* by electron transfer within the thermally excited DA complex. To understand the process, we must turn our attention to the effect that the rearrangement of nuclear coordinates has on electronic energy levels of DA and D^+A^- for a given distance r between D and A (Fig. 24.28). Initially, the electron to be transferred occupies the HOMO of D, and the overall energy of DA is lower than that of D^+A^- (Fig. 24.28a). As the nuclei rearrange to a configuration represented by q^* in Fig. 24.28b, the highest occupied electronic level of DA and the lowest unoccupied electronic level of D^+A^- become degenerate and electron transfer becomes energetically feasible. Over reasonably short distances r, the main mechanism of electron transfer is tunnelling through the potential energy barrier depicted in Fig. 24.28b. The height of the barrier increases with the ionization energies of the DA and D^+A^- complexes. After an electron moves from the HOMO of D to the LUMO of A, the system relaxes to the configuration represented by q_0^P in Fig. 24.28c. As shown in the illustration, now the energy of D^+A^- is lower than that of DA, reflecting the thermodynamic tendency for A to remain reduced and for D to remain oxidized.

The tunnelling event responsible for electron transfer is similar to that described in Section 9.3, except that in this case the electron tunnels from an electronic level of D, with wavefunction ψ_D, to an electronic level of A, with wavefunction ψ_A. We saw in Section 9.3 that the rate of an electronic transition from a level described by the wavefunction ψ_D to a level described by ψ_A is proportional to the square of the integral

$$\langle H_{DA}\rangle = \int \psi_A \hat{H}_{DA} \psi_B \, d\tau$$

where H_{DA} is a hamiltonian that describes the coupling of the electronic wavefunctions. It turns out that in cases where the coupling is relatively weak we may write:

$$\langle H_{DA}\rangle^2 = \langle H_{DA}^{\circ}\rangle^2 e^{-\beta r} \qquad (24.80)$$

where r is the edge-to-edge distance between D and A, β is a parameter that measures the sensitivity of the electronic coupling matrix element to distance, and $\langle H_{DA}^{\circ}\rangle^2$ is the value of the electronic coupling matrix element when D and A are in contact ($r = 0$). The exponential dependence on distance in eqn 24.80 is essentially the same as the exponential decrease in transmission probability through a potential energy barrier described in Section 9.3.

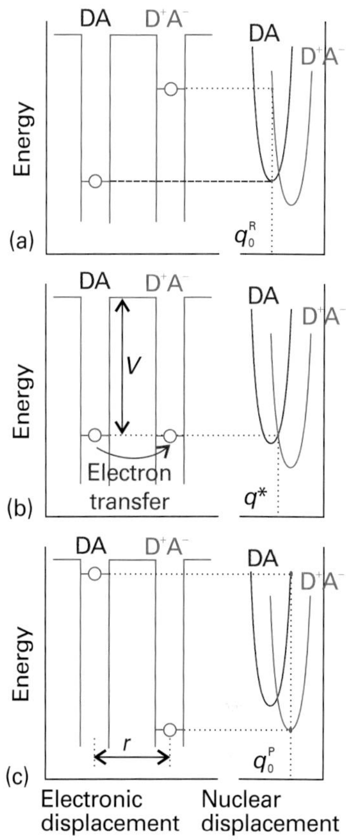

Fig. 24.28 Correspondence between the electronic energy levels (shown on the left) and the nuclear energy levels (shown on the right) for the DA and D^+A^- complexes involved in an electron transfer process. (a) At the nuclear configuration denoted by q_0^R, the electron to be transferred in DA is in an occupied electronic energy level (denoted by a blue circle) and the lowest unoccupied energy level of D^+A^- (denoted by an unfilled circle) is of too high an energy to be a good electron acceptor. (b) As the nuclei rearrange to a configuration represented by q^*, DA and D^+A^- become degenerate and electron transfer occurs by tunnelling through the barrier of height V and width r, the edge-to-edge distance between donor and acceptor. (c) The system relaxes to the equilibrium nuclear configuration of D^+A^- denoted by q_0^P, in which the lowest unoccupied electronic level of DA is higher in energy than the highest occupied electronic level of D^+A^-. (Adapted from R.A. Marcus and N. Sutin, *Biochim. Biophys. Acta* **811**, 265 (1985).)

(b) The expression for the rate of electron transfer

The full expression for k_{et} turns out to be (see *Further reading* for a derivation)

$$k_{et} = \frac{2\langle H_{DA}\rangle^2}{h}\left(\frac{\pi^3}{4\lambda RT}\right)^{1/2} e^{-\Delta^{\ddagger}G/RT} \qquad (24.81)$$

where $\langle H_{DA}\rangle^2$ can be approximated by eqn 24.80. We show in *Further information* 24.1 that the Gibbs energy of activation $\Delta^{\ddagger}G$ is

$$\Delta^{\ddagger}G = \frac{(\Delta_r G^{\ominus} + \lambda)^2}{4\lambda} \qquad (24.82)$$

where $\Delta_r G^{\ominus}$ is the standard reaction Gibbs energy for the electron transfer process $DA \rightarrow D^+A^-$, and λ is the **reorganization energy**, the energy change associated with molecular rearrangements that must take place so that DA can take on the equilibrium geometry of D^+A^-. These molecular rearrangements include the relative reorientation of the D and A molecules in DA and the relative reorientation of the solvent molecules surrounding DA. Equation 24.82 shows that $\Delta^{\ddagger}G = 0$, with the implication that the reaction is not slowed down by an activation barrier, when $\Delta_r G^{\ominus} = -\lambda$, corresponding to the cancellation of the reorganization energy term by the standard reaction Gibbs energy.

Equation 24.81 has some limitations. First, it describes only those processes with weak electronic coupling between donor and acceptor. Weak coupling is observed when the electroactive species are sufficiently far apart that the wavefunctions ψ_A and ψ_D do not overlap extensively. An example of a weakly coupled system is the cytochrome *c*–cytochrome b_5 complex, in which the electroactive haem-bound iron ions shuttle between oxidation states +2 and +3 during electron transfer and are about 1.7 nm apart. Strong coupling is observed when the wavefunctions ψ_A and ψ_D overlap very extensively. Examples of strongly coupled systems are mixed-valence, binuclear *d*-metal complexes with the general structure L_mM^{n+}—B—$M^{p+}L_{m'}$, in which the electroactive metal ions are separated by a bridging ligand B. In these systems, $r < 1.0$ nm. The weak coupling limit applies to a large number of electron transfer reactions, including those between proteins during metabolism (*Impact* I7.2). Second, the term $(\pi^3/4\lambda RT)^{1/2}e^{-\Delta^{\ddagger}G/RT}$ should be used only at high temperatures. At low temperatures, thermal fluctuations alone cannot bring the reactants to the transition state and transition state theory, which is at the heart of the theory presented in this section, fails to account for any observed electron transfer. Electron transfer can still occur, but by *nuclear* tunnelling from the reactant to the product surfaces. We saw in Section 9.5 that the wavefunctions for the lower levels of the quantum mechanical harmonic oscillator extend significantly beyond classically allowed regions, so an oscillator can tunnel into a region of space in which another oscillator may be found. Full quantum mechanical treatments of electron transfer reactions replace the $(\pi^3/4\lambda RT)^{1/2}e^{-\Delta^{\ddagger}G/RT}$ term with Franck–Condon factors similar to those discussed in Section 14.2, which couple the *nuclear* wavefunctions and provide a measure of the contribution of nuclear tunnelling to the rate of electron transfer.

24.12 Experimental results

It is difficult to measure the distance dependence of k_{et} when the reactants are ions or molecules that are free to move in solution. In such cases, electron transfer occurs after a donor–acceptor complex forms and it is not possible to exert control over *r*, the edge-to-edge distance. The most meaningful experimental tests of the dependence of k_{et} on *r* are those in which the same donor and acceptor are positioned at a variety of distances, perhaps by covalent attachment to molecular linkers (see **1** for an example). Under these conditions, the term $e^{-\Delta^{\ddagger}G/RT}$ becomes a constant and, after taking the natural logarithm of eqn 24.81 and using eqn 24.80, we obtain

$$\ln k_{et} = -\beta r + \text{constant} \qquad (24.83)$$

which implies that a plot of $\ln k_{et}$ against *r* should be a straight line with slope $-\beta$. The value of β depends on the medium through which the electron must travel from donor to acceptor. In a vacuum, $28\ \text{nm}^{-1} < \beta < 35\ \text{nm}^{-1}$, whereas $\beta \approx 9\ \text{nm}^{-1}$ when the intervening medium is a molecular link between donor and acceptor.

The dependence of k_{et} on the standard reaction Gibbs energy has been investigated in systems where the edge-to-edge distance, the reorganization energy, and $\kappa\nu$ are constant for a series of reactions. Then eqn 24.81 becomes

$$\ln k_{et} = -\frac{1}{4\lambda}\left(\frac{\Delta_r G^{\ominus}}{RT}\right)^2 - \tfrac{1}{2}\left(\frac{\Delta_r G^{\ominus}}{RT}\right) + \text{constant} \qquad (24.84)$$

and a plot of $\ln k_{et}$ (or $\log k_{et}$) against $\Delta_r G^{\ominus}$ (or $-\Delta_r G^{\ominus}$) is predicted to be shaped like a downward parabola. Equation 24.84 implies that the rate constant increases as $\Delta_r G^{\ominus}$ decreases but only up to $-\Delta_r G^{\ominus} = \lambda$. Beyond that, the reaction enters the **inverted region**, in which the rate constant decreases as the reaction becomes more exergonic ($\Delta_r G^{\ominus}$ becomes more negative). The inverted region has been observed in a series of

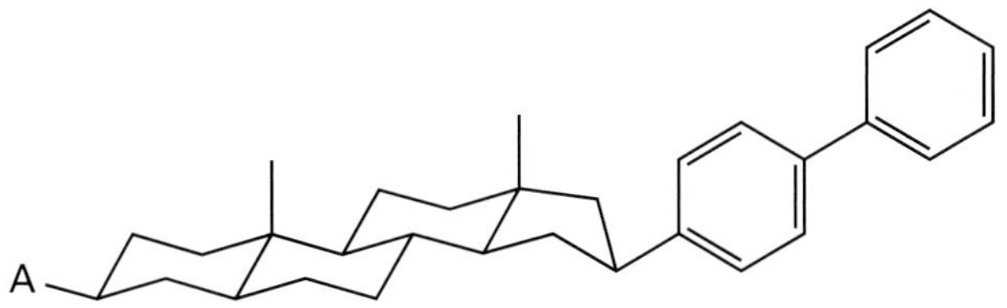

1 An electron donor–acceptor complex

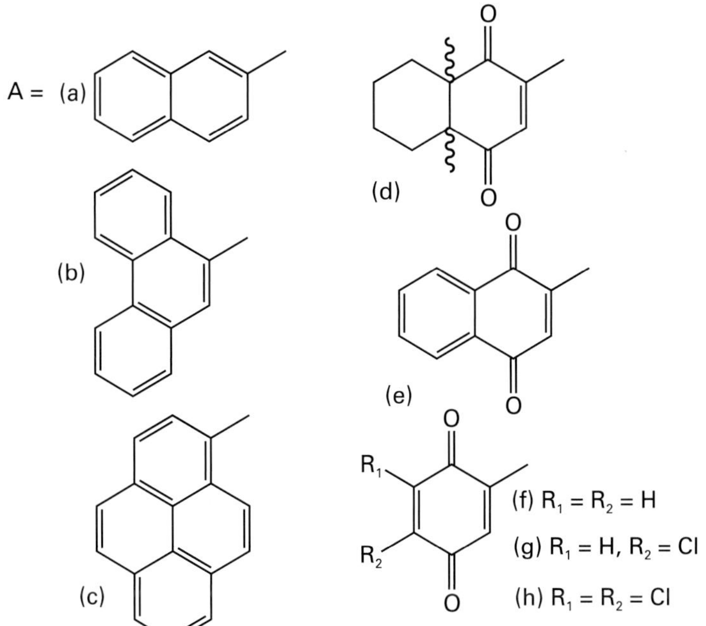

special compounds in which the electron donor and acceptor are linked covalently to a molecular spacer of known and fixed size (Fig. 24.29).

The behaviour predicted by eqn 24.84 and observed experimentally in Fig. 24.29 can be explained by considering the dependence of the activation Gibbs energy on the standard Gibbs energy of electron transfer. We suppose that the energies of the reactant and product complexes can be characterized by parabolas with identical curvatures and fixed but distinct q_0^{R} and q_0^{P}. Now we let the minimum energy of the product complex change while keeping q_0^{P} constant, which corresponds to changing the magnitude of $\Delta_r G^{\ominus}$. Figure 24.30 shows the effect of increasing the exergonicity of the process. In Fig. 24.30a we see that, for a range of values of $\Delta_r G^{\ominus}$, $\Delta^{\ddagger}G > 0$ and the transition state is at $q_a^{*} > q_0^{\mathrm{R}}$. As the process becomes more exergonic, the activation Gibbs energy decreases and the rate constant increases. (This behaviour is another example of a 'linear free-energy relation', first discussed in Section 24.5.) Figure 24.30b shows that, when $\Delta^{\ddagger}G = 0$ and $q_b^{*} = q_0^{\mathrm{R}}$, the rate constant for the process reaches a maximum as there is no activation barrier to overcome. According to eqn 24.81, this condition occurs when $-\Delta_r G^{\ominus} = \lambda$. Finally, Fig. 24.30c shows that, as the process becomes even more exergonic, $\Delta^{\ddagger}G$ becomes positive again but now the transition state is at $q_c^{*} < q_0^{\mathrm{R}}$. The rate constant for the process decreases steadily as the activation barrier for the process increases with decreasing $\Delta_r G^{\ominus}$. This is the explanation for the 'inverted region' observed in Fig. 24.29.

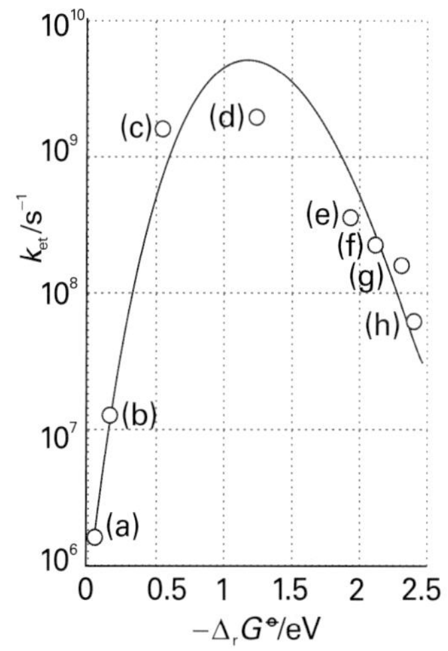

Fig. 24.29 Variation of log k_{et} with $-\Delta_r G^{\ominus}$ for a series of compounds with the structures given in (**1**). Kinetic measurements were conducted in 2-methyltetrahydrofuran and at 296 K. The distance between donor (the reduced biphenyl group) and the acceptor is constant for all compounds in the series because the molecular linker remains the same. Each acceptor has a characteristic standard reduction potential, so it follows that the standard Gibbs energy for the electron transfer process is different for each compound in the series. The line is a fit to a version of eqn 24.84 and the maximum of the parabola occurs at $-\Delta_r G^{\ominus} = \lambda = 1.2$ eV $= 1.2 \times 10^2$ kJ mol^{-1}. (Reproduced with permission from J.R. Miller, L.T. Calcaterra, and G.L. Closs, *J. Amer. Chem. Soc.* **106**, 3047 (1984).)

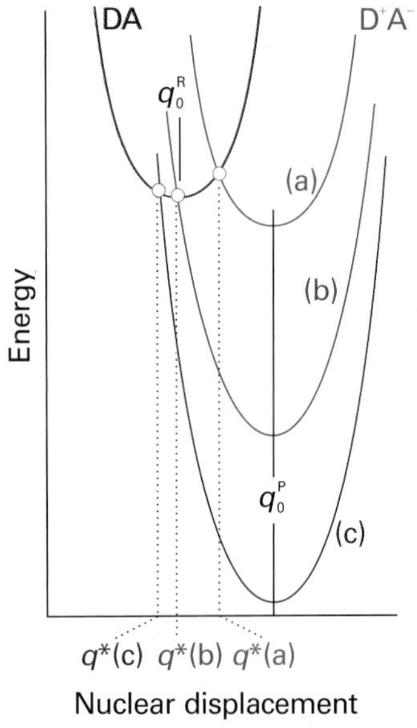

Fig. 24.30 (a) $\Delta^{\ddagger}G > 0$ and the transition state is at $q_a^* > q_0^R$. As the process becomes more exergonic, the activation Gibbs energy decreases and the rate constant increases. (b) When $\Delta^{\ddagger}G = 0$ and $q_b^* = q_0^R$ the rate constant for the process reaches a maximum as there is no activation barrier to overcome. (c) As the process becomes even more exergonic, $\Delta^{\ddagger}G$ becomes positive again but now the transition state is at $q_c^* < q_0^R$. The rate constant for the process decreases steadily as the activation barrier for the process increases with decreasing $\Delta_r G^{\ominus}$.

Some of the key features of electron transfer theory have been tested by experiments, showing in particular the predicted dependence of k_{et} on the standard reaction Gibbs energy and the edge-to-edge distance between electron donor and acceptor. The basic theory presented in Section 24.11 has been extended to transfer of other light particles, such as protons.

IMPACT ON BIOCHEMISTRY

I24.1 Electron transfer in and between proteins

We saw in *Impact* I7.2 and I21.2 that exergonic electron transfer processes drive the synthesis of ATP in the mitochondrion during oxidative phosphorylation. Electron transfer between protein-bound cofactors or between proteins also plays a role in other biological processes, such as photosynthesis (*Impact* I23.2), nitrogen fixation, the reduction of atmospheric N_2 to NH_3 by certain microorganisms, and the mechanisms of action of oxidoreductases, which are enzymes that catalyse redox reactions.

Equation 24.78 applies to a large number of biological systems, such as cytochrome *c* and cytochrome *c* oxidase (*Impact* I7.2), which must form an encounter complex before electron transfer can take place. Electron transfer between protein-bound cofactors can occur at distances of up to about 2.0 nm, a relatively long distance on a molecular scale, with the protein providing an intervening medium between donor and acceptor.

When the electron donor and acceptor are anchored at fixed distances within a single protein, only k_{et} needs to be considered when calculating the rate of electron transfer by using eqn 24.81. Cytochrome *c* oxidase is an example of a system where such intraprotein electron transfer is important. In that enzyme, bound copper ions and haem groups work together to reduce O_2 to water in the final step of respiration. However, there is a great deal of controversy surrounding the interpretation of protein electron transfer data in the light of the theory that leads to eqn 24.81. Much of the available data may be interpreted with $\beta \approx 14\ \text{nm}^{-1}$, a value that appears to be insensitive to the primary and secondary structures of the protein but does depend slightly on the density of atoms in the section of protein that separates donor from acceptor. More detailed work on the specific effect of secondary structure suggests that $12.5\ \text{nm}^{-1} < \beta < 16.0\ \text{nm}^{-1}$ when the intervening medium consists primarily of α helices and $9.0\ \text{nm}^{-1} < \beta < 11.5\ \text{nm}^{-1}$ when the medium is primarily β sheet. Yet another view suggests that the electron takes specific paths through covalent bonds and hydrogen bonds that exist in the protein for the purpose of optimizing the rate of electron transfer.

A value of β is not necessary for the prediction of the rate constants for electron transfer processes between proteins if we take a different approach. It follows from eqns 24.78 and 24.79 that the rate constant k_{obs} may be written as

$$k_{obs} = Ze^{-\Delta^{\ddagger}G/RT} \tag{24.85}$$

where $Z = K_{DA}\kappa\nu$. It is difficult to estimate k_{obs} because we often lack knowledge of β, λ, and $\kappa\nu$. However, when $\lambda \gg |\Delta_r G^{\ominus}|$, k_{obs} may be estimated by a special case of the **Marcus cross-relation**, which we derive in *Further information* 24.1:

$$k_{obs} = (k_{DD}k_{AA}K)^{1/2} \tag{24.86}$$

where K is the equilibrium constant for the net electron transfer reaction (eqn 24.75) and k_{DD} and k_{AA} are the experimental rate constants for the electron self-exchange processes (with the asterisks distinguishing one molecule from another)

$$^{*}D + D^+ \rightarrow {}^{*}D^+ + D \tag{24.87a}$$

$$^{*}A^- + A \rightarrow {}^{*}A + A^- \tag{24.87b}$$

The rate constants estimated by eqn 24.86 agree fairly well with experimental rate constants for electron transfer between proteins, as we see in the following *Example.*

Example 24.4 *Using the Marcus cross-relation*

The following kinetic and thermodynamic data were obtained for cytochrome c and cytochrome c_{551}, two proteins in which haem-bound iron ions shuttle between the oxidation states Fe(II) and Fe(III):

	$k_{ii}/(\mathrm{dm^3\,mol^{-1}\,s^{-1}})$	$E^{\ominus}/\mathrm{V}$
cytochrome c	1.5×10^{2}	0.260
cytochrome c_{551}	4.6×10^{7}	0.286

Estimate the rate constant k_{obs} for the process

$$\text{cytochrome } c_{551}(\text{red}) + \text{cytochrome } c(\text{ox}) \rightarrow \text{cytochrome } c_{551}(\text{ox}) + \text{cytochrome } c(\text{red})$$

Then, compare the estimated value with the observed value of 6.7×10^{4} $\mathrm{dm^3\,mol^{-1}\,s^{-1}}$.

Method We use the standard potentials and eqns 7.30 ($\ln K = \nu FE^{\ominus}/RT$) and 7.37 ($E^{\ominus}=E^{\ominus}_{R}-E^{\ominus}_{L}$) to calculate the equilibrium constant K. Then, we use eqn 24.86, the calculated value of K, and the self-exchange rate constants k_{ii} to calculate the rate constant k_{obs}.

Answer The two reduction half-reactions are

Right: cytochrome $c(\text{ox}) + e^- \rightarrow$ cytochrome $c(\text{red})$ $\quad E^{\ominus}_{R}=+0.260\ \mathrm{V}$

Left: cytochrome $c_{551}(\text{ox}) + e^- \rightarrow$ cytochrome $c_{551}(\text{red})$ $\quad E^{\ominus}_{L}=+0.286\ \mathrm{V}$

The difference is

$$E^{\ominus} = (0.260\ \mathrm{V}) - (0.286\ \mathrm{V}) = -0.026\ \mathrm{V}$$

It then follows from eqn 7.30 with $\nu = 1$ and $RT/F = 25.69$ mV that

$$\ln K = -\frac{0.026\mathrm{V}}{25.69\times10^{-3}\ \mathrm{V}} = -\frac{2.6}{2.569}$$

Therefore, K = 0.36. From eqn 24.76 and the self-exchange rate constants, we calculate

$$k_{obs} = \{(1.5\times10^{2}\ \mathrm{dm^3\,mol^{-1}\,s^{-1}})\times(4.6\times10^{7}\ \mathrm{dm^3\,mol^{-1}\,s^{-1}})\times0.36\}^{1/2}$$
$$= 5.0\times10^{4}\ \mathrm{dm^3\,mol^{-1}\,s^{-1}}$$

The calculated and observed values differ by only 25 per cent, indicating that the Marcus relation can lead to reasonable estimates of rate constants for electron transfer.

Self-test 24.4 Estimate k_{obs} for the reduction by cytochrome c of plastocyanin, a protein containing a copper ion that shuttles between the +2 and +1 oxidation states and for which $k_{AA} = 6.6\times10^{2}$ $\mathrm{dm^3\,mol^{-1}\,s^{-1}}$ and $E^{\ominus} = 0.350$ V.

[1.8×10^{3} $\mathrm{dm^3\,mol^{-1}\,s^{-1}}$]

Table 24.3 Summary of uses of k

Symbol	Significance
k	Boltzmann's constant
k_2	Second-order rate constant
k_2°	Rate constant at zero ionic strength
$k_a, k_b, \ldots$	Rate constants for individual steps
$k'_a, k'_b, \ldots$	Rate constants for individual reverse steps
$k^\ddagger$	Rate constant for unimolecular decay of activated complex
K	Equilibrium constant (dimensionless)
K_γ	Ratio of activity coefficients
$K^\ddagger$	Proportionality constant in transition state theory
κ	Transmission coefficient
$\bar{K}$	Equilibrium constant with one mode discarded
k_f	Force constant

Checklist of key ideas

☐ 1. In collision theory, it is supposed that the rate is proportional to the collision frequency, a steric factor, and the fraction of collisions that occur with at least the kinetic energy E_a along their lines of centres.

☐ 2. A reaction in solution may be diffusion-controlled if its rate is controlled by the rate at which reactant molecules encounter each other in solution. The rate of an activation-controlled reaction is controlled by the rate of accumulating sufficient energy.

☐ 3. In transition state theory, it is supposed that an activated complex is in equilibrium with the reactants, and that the rate at which that complex forms products depends on the rate at which it passes through a transition state. The result is the Eyring equation, $k_2 = \kappa(kT/h)\bar{K}_c^\ddagger$.

☐ 4. The rate constant may be parametrized in terms of the Gibbs energy, entropy, and enthalpy of activation, $k_2 = (kT/h)e^{\Delta^\ddagger S/R}e^{-\Delta^\ddagger H/RT}$.

☐ 5. The kinetic salt effect is the effect of an added inert salt on the rate of a reaction between ions, $\log k_2 = \log k_2^\circ + 2Az_Az_BI^{1/2}$.

☐ 6. Techniques for the study of reactive collisions include infrared chemiluminescence, laser-induced fluorescence, multiphoton ionization (MPI), reaction product imaging, and resonant multiphoton ionization (REMPI).

☐ 7. A potential energy surface maps the potential energy as a function of the relative positions of all the atoms taking part in a reaction. In an attractive surface, the saddle point (the highest point) occurs early on the reaction coordinate. In a repulsive surface, the saddle point occurs late on the reaction coordinate.

☐ 8. Ultrafast laser techniques can be used to probe directly the activated complex and to control the outcome of some chemical reactions.

☐ 9. The rate constant of electron transfer in a donor–acceptor complex depends on the distance between electron donor and acceptor, the standard reaction Gibbs energy, and the reorganization energy, λ: $k_{et} \propto e^{-\beta r}e^{-\Delta^\ddagger G/RT}$ (constant T), with $\Delta^\ddagger G = (\Delta_r G^\ominus + \lambda)^2/4\lambda$.

☐ 10. The Marcus cross-relation predicts the rate constant for electron transfer in solution from the reaction's equilibrium constant K and the self-exchange rate constants k_{ii}: $k_{obs} = (k_{DD}k_{AA}K)^{1/2}$.

Further reading

Articles and texts

A.J. Alexander and R.N. Zare, Anatomy of elementary chemical reactions. *J. Chem. Educ.* **75**, 1105 (1998).

G.D. Billing and K.V. Mikkelsen, *Advanced molecular dynamics and chemical kinetics*. Wiley, New York (1997).

D.C. Clary, Quantum theory of chemical reaction dynamics. *Science* **279**, 1879 (1998).

F.F. Crim, Vibrational state control of bimolecular reactions: discovering and directing the chemistry. *Acc. Chem. Res.* **32**, 877 (1999).

D.M. Hirst, *Potential energy surfaces*. Taylor and Francis, London (1985).

A.M. Kuznetsov and J. Ulstrup, *Electron transfer in chemistry and biology: an introduction to the theory*. Wiley, New York (1998).

R.D. Levine and R.B. Bernstein, *Molecular reaction dynamics and chemical reactivity*. Clarendon Press, Oxford (1987).

J.I. Steinfeld, J.S. Francisco and W.L. Hase, *Chemical kinetics and dynamics*. Prentice Hall, Upper Saddle River (1999).

A.H. Zewail, Femtochemistry: atomic-scale dynamics of the chemical bond using ultrafast lasers. *Angew. Chem. Int. Ed.* **39**, 2586 (2000).

Further information

Further information 24.1 *The Gibbs energy of activation of electron transfer and the Marcus cross-relation*

The simplest way to derive an expression for the Gibbs energy of activation of electron transfer processes is to construct a model in which the surfaces for DA (the 'reactant complex', denoted R) and for D^+A^- (the 'product complex', denoted P) are described by classical harmonic oscillators with identical reduced masses μ and angular frequencies ω, but displaced minima, as shown in Fig. 24.27. The molar Gibbs energies $G_{m,R}(q)$ and $G_{m,P}(q)$ of the reactant and product complexes, respectively, may be written as

$$G_{m,R}(q) = \tfrac{1}{2}N_A\mu\omega^2(q - q_0^R)^2 + G_{m,R}(q_0^R) \qquad (24.88a)$$

$$G_{m,P}(q) = \tfrac{1}{2}N_A\mu\omega^2(q - q_0^P)^2 + G_{m,P}(q_0^P) \qquad (24.88b)$$

where q_0^R and q_0^P are the values of q at which the minima of the reactant and product parabolas occur, respectively. The standard reaction Gibbs energy for the electron transfer process $DA \rightarrow D^+A^-$ is $\Delta_r G^\ominus = G_{m,P}(q_0^P) - G_{m,R}(q_0^R)$, the difference in standard molar Gibbs energy between the minima of the parabolas. In Fig. 24.27, $\Delta_r G^\ominus < 0$.

We also note that q^*, the value of q corresponding to the transition state of the complex, may be written in terms of the parameter α, the fractional change in q:

$$q^* = q_0^R + \alpha(q_0^P - q_0^R) \qquad (24.89)$$

We see from Fig. 24.27 that $\Delta^\ddagger G = G_{m,R}(q^*) - G_{m,R}(q_0^R)$. It then follows from eqns 24.88a, 24.88b, and 24.89 that

$$\Delta^\ddagger G = \tfrac{1}{2}N_A\mu\omega^2(q^* - q_0^R)^2 = \tfrac{1}{2}N_A\mu\omega^2\{\alpha(q_0^P - q_0^R)\}^2 \qquad (24.90)$$

We now define the reorganization energy, λ, as

$$\lambda = \tfrac{1}{2}N_A\mu\omega^2(q_0^P - q_0^R)^2 \qquad (24.91)$$

which can be interpreted as $G_{m,R}(q_0^P) - G_{m,R}(q_0^R)$ and, consequently, as the (Gibbs) energy required to deform the equilibrium configuration of DA to the equilibrium configuration of D^+A^- (as shown in Fig. 24.27). It follows from eqns 24.90 and 24.91 that

$$\Delta^\ddagger G = \alpha^2\lambda \qquad (24.92)$$

Because $G_{m,R}(q^*) = G_{m,P}(q^*)$, it follows from eqns 24.88b, 24.91, and 24.92 that

$$\begin{aligned}\alpha^2\lambda &= \tfrac{1}{2}N_A\mu\omega^2\{(\alpha - 1)(q_0^P - q_0^R)\}^2 + \Delta_r G^\ominus \\ &= (\alpha - 1)\lambda + \Delta_r G^\ominus\end{aligned} \qquad (24.93)$$

which implies that

$$\alpha = \tfrac{1}{2}\left(\frac{\Delta_r G^\ominus}{\lambda} + 1\right) \qquad (24.94)$$

By combining eqns 24.92 and 24.94, we obtain eqn 24.82. We can obtain an identical relation if we allow the harmonic oscillators to have different angular frequencies and hence different curvatures (see *Further reading*).

Equation 24.82 can be used to derive the form of the Marcus cross-relation (eqn 24.86) used in *Impact* I24.1. We begin by using eqn 24.85 to write the rate constants for the self-exchange reactions as

$$k_{DD} = Z_{DD}e^{-\Delta^\ddagger G_{DD}/RT} \qquad k_{AA} = Z_{AA}e^{-\Delta^\ddagger G_{AA}/RT}$$

For the net reaction (also called the 'cross-reaction') and the self-exchange reactions, the Gibbs energy of activation may be written from eqn 24.82 as

$$\Delta^\ddagger G = \frac{\Delta_r G^{\ominus 2}}{4\lambda} + \frac{\Delta_r G^\ominus}{2} + \frac{\lambda}{4}$$

When $\lambda \gg |\Delta_r G^\ominus|$, we obtain

$$\Delta^\ddagger G = \frac{\Delta_r G^\ominus}{2} + \frac{\lambda}{4}$$

This expression can be used without further elaboration to denote the Gibbs energy of activation of the net reaction. For the self-exchange reactions, we set $\Delta_r G^\ominus_{DD} = \Delta_r G^\ominus_{AA} = 0$ and write

$$\Delta^\ddagger G_{DD} = \frac{\lambda_{DD}}{4} \qquad \Delta^\ddagger G_{AA} = \frac{\lambda_{AA}}{4}$$

It follows that

$$k_{DD} = Z_{DD}e^{-\lambda_{DD}/4RT} \qquad k_{AA} = Z_{AA}e^{-\lambda_{AA}/4RT}$$

To make further progress, we assume that the reorganization energy of the net reaction is the arithmetic mean of the reorganization energies of the self-exchange reactions:

$$\lambda_{DA} = \frac{\lambda_{DD} + \lambda_{AA}}{2}$$

It follows that the Gibbs energy of activation of the net reaction is

$$\Delta^{\ddagger}G = \frac{\Delta_r G^{\ominus}}{2} + \frac{\lambda_{DD}}{8} + \frac{\lambda_{AA}}{8}$$

Therefore, the rate constant for the net reaction is

$$k_{obs} = Ze^{-\Delta_r G^{\ominus}/2RT}e^{-\lambda_{DD}/8RT}e^{-\lambda_{AA}/8RT}$$

We can use eqn 7.17 ($\ln K = -\Delta_r G^{\ominus}/RT$) to write

$$K = e^{-\Delta_r G^{\ominus}/RT}$$

Then, by combining this expression with the expressions for k_{DD} and k_{AA}, and using the relations $e^{x+y} = e^x e^y$ and $e^{x/2} = (e^x)^{1/2}$, we obtain the most general case of the Marcus cross-relation:

$$k_{obs} = (k_{DD}k_{AA}K)^{1/2}f$$

where

$$f = \frac{Z}{(Z_{AA}Z_{DD})^{1/2}}$$

In practice, the factor f is usually set to 1 and we obtain eqn 24.86.

Discussion questions

24.1 Describe the essential features of the harpoon mechanism.

24.2 Distinguish between a diffusion-controlled reaction and an activation-controlled reaction.

24.3 Describe the formulation of the Eyring equation.

24.4 Discuss the physical origin of the kinetic salt effect.

24.5 Describe how the following techniques are used in the study of chemical dynamics: infrared chemiluminescence, laser-induced fluorescence, multiphoton ionization, resonant multiphoton ionization, reaction product imaging, and femtosecond spectroscopy.

24.6 Justify the following statements: (a) Reactions with attractive potential energy surfaces proceed more efficiently if the energy is in relative translational motion. (b) Reactions with repulsive potential surfaces proceed more efficiently if the excess energy is present as vibrations.

24.7 A method for directing the outcome of a chemical reaction consists of using molecular beams to control the relative orientations of reactants during a collision. Consider the reaction $Rb + CH_3I \rightarrow RbI + CH_3$. How should CH_3I molecules and Rb atoms be oriented to maximize the production of RbI?

24.8 Discuss how the following factors determine the rate of electron transfer in homogeneous systems: the distance between electron donor and acceptor, and the reorganization energy of redox active species and the surrounding medium.

Exercises

24.1a Calculate the collision frequency, z, and the collision density, Z, in ammonia, $R = 190$ pm, at 25°C and 100 kPa. What is the percentage increase when the temperature is raised by 10 K at constant volume?

24.1b Calculate the collision frequency, z, and the collision density, Z, in carbon monoxide, $R = 180$ pm at 25°C and 100 kPa. What is the percentage increase when the temperature is raised by 10 K at constant volume?

24.2a Collision theory demands knowing the fraction of molecular collisions having at least the kinetic energy E_a along the line of flight. What is this fraction when (a) $E_a = 10$ kJ mol^{-1}, (b) $E_a = 100$ kJ mol^{-1} at (i) 300 K and (ii) 1000 K?

24.2b Collision theory demands knowing the fraction of molecular collisions having at least the kinetic energy E_a along the line of flight. What is this fraction when (a) $E_a = 15$ kJ mol^{-1}, (b) $E_a = 150$ kJ mol^{-1} at (i) 300 K and (ii) 800 K?

24.3a Calculate the percentage increase in the fractions in Exercise 24.2a when the temperature is raised by 10 K.

24.3b Calculate the percentage increase in the fractions in Exercise 24.2b when the temperature is raised by 10 K.

24.4a Use the collision theory of gas-phase reactions to calculate the theoretical value of the second-order rate constant for the reaction $H_2(g) + I_2(g) \rightarrow 2\,HI(g)$ at 650 K, assuming that it is elementary bimolecular. The collision cross-section is 0.36 nm^2, the reduced mass is 3.32×10^{-27} kg, and the activation energy is 171 kJ mol^{-1}.

24.4b Use the collision theory of gas-phase reactions to calculate the theoretical value of the second-order rate constant for the reaction $D_2(g) + Br_2(g) \rightarrow 2\,DBr(g)$ at 450 K, assuming that it is elementary bimolecular. Take the collision cross-section as 0.30 nm^2, the reduced mass as 3.930 u, and the activation energy as 200 kJ mol^{-1}.

24.5a A typical diffusion coefficient for small molecules in aqueous solution at 25°C is 5×10^{-9} m^2 s^{-1}. If the critical reaction distance is 0.4 nm, what value is expected for the second-order rate constant for a diffusion-controlled reaction?

24.5b Suppose that the typical diffusion coefficient for a reactant in aqueous solution at 25°C is 4.2×10^{-9} m^2 s^{-1}. If the critical reaction distance is 0.50 nm, what value is expected for the second-order rate constant for the diffusion-controlled reaction?

24.6a Calculate the magnitude of the diffusion-controlled rate constant at 298 K for a species in (a) water, (b) pentane. The viscosities are 1.00×10^{-3} kg m^{-1} s^{-1}, and 2.2×10^{-4} kg m^{-1} s^{-1}, respectively.

24.6b Calculate the magnitude of the diffusion-controlled rate constant at 298 K for a species in (a) decylbenzene, (b) concentrated sulfuric acid. The viscosities are 3.36 cP and 27 cP, respectively.

24.7a Calculate the magnitude of the diffusion-controlled rate constant at 298 K for the recombination of two atoms in water, for which $\eta = 0.89$ cP. Assuming the concentration of the reacting species is 1.0 mmol dm^{-3} initially, how long does it take for the concentration of the atoms to fall to half that value? Assume the reaction is elementary.

24.7b Calculate the magnitude of the diffusion-controlled rate constant at 298 K for the recombination of two atoms in benzene, for which $\eta = 0.601$ cP. Assuming the concentration of the reacting species is 1.8 mmol dm^{-3} initially, how long does it take for the concentration of the atoms to fall to half that value? Assume the reaction is elementary.

24.8a For the gaseous reaction $A + B \rightarrow P$, the reactive cross-section obtained from the experimental value of the pre-exponential factor is 9.2×10^{-22} m^2. The collision cross-sections of A and B estimated from the transport properties are 0.95 and 0.65 nm^2, respectively. Calculate the P-factor for the reaction.

24.8b For the gaseous reaction $A + B \rightarrow P$, the reactive cross-section obtained from the experimental value of the pre-exponential factor is 8.7×10^{-22} m^2. The collision cross-sections of A and B estimated from the transport properties are 0.88 and 0.40 nm^2, respectively. Calculate the P factor for the reaction.

24.9a Two neutral species, A and B, with diameters 588 pm and 1650 pm, respectively, undergo the diffusion-controlled reaction $A + B \rightarrow P$ in a solvent of viscosity 2.37×10^{-3} kg m^{-1} s^{-1} at 40°C. Calculate the initial rate d[P]/dt if the initial concentrations of A and B are 0.150 mol dm^{-3} and 0.330 mol dm^{-3}, respectively.

24.9b Two neutral species, A and B, with diameters 442 pm and 885 pm, respectively, undergo the diffusion-controlled reaction $A + B \rightarrow P$ in a solvent of viscosity 1.27 cP at 20°C. Calculate the initial rate d[P]/dt if the initial concentrations of A and B are 0.200 mol dm^{-3} and 0.150 mol dm^{-3}, respectively.

24.10a The reaction of propylxanthate ion in acetic acid buffer solutions has the mechanism $A^- + H^+ \rightarrow P$. Near 30°C the rate constant is given by the empirical expression $k_2 = (2.05 \times 10^{13})e^{-(8681\,K)/T}$ dm^3 mol^{-1} s^{-1}. Evaluate the energy and entropy of activation at 30°C.

24.10b The reaction $A^- + H^+ \rightarrow P$ has a rate constant given by the empirical expression $k_2 = (8.72 \times 10^{12})e^{-(6134\,K)/T}$ dm^3 mol^{-1} s^{-1}. Evaluate the energy and entropy of activation at 25°C.

24.11a When the reaction in Exercise 24.10a occurs in a dioxane/water mixture that is 30 per cent dioxane by mass, the rate constant fits $k_2 = (7.78 \times 10^{14})e^{-(9134\,K)/T}$ dm^3 mol^{-1} s^{-1} near 30°C. Calculate $\Delta^{\ddagger}G$ for the reaction at 30°C.

24.11b A rate constant is found to fit the expression $k_2 = (6.45 \times 10^{13})$ $e^{-(5375\,K)/T}$ dm^3 mol^{-1} s^{-1} near 25°C. Calculate $\Delta^{\ddagger}G$ for the reaction at 25°C.

24.12a The gas-phase association reaction between F_2 and IF_5 is first-order in each of the reactants. The energy of activation for the reaction is 58.6 kJ mol^{-1}. At 65°C the rate constant is 7.84×10^{-3} kPa^{-1} s^{-1}. Calculate the entropy of activation at 65°C.

24.12b A gas-phase recombination reaction is first-order in each of the reactants. The energy of activation for the reaction is 49.6 kJ mol^{-1}. At 55°C the rate constant is 0.23 m^3 s^{-1}. Calculate the entropy of activation at 55°C.

24.13a Calculate the entropy of activation for a collision between two structureless particles at 300 K, taking $M = 50$ g mol^{-1} and $\sigma = 0.40$ nm^2.

24.13b Calculate the entropy of activation for a collision between two structureless particles at 500 K, taking $M = 78$ g mol^{-1} and $\sigma = 0.62$ nm^2.

24.14a The pre-exponential factor for the gas-phase decomposition of ozone at low pressures is 4.6×10^{12} dm^3 mol^{-1} s^{-1} and its activation energy is 10.0 kJ mol^{-1}. What are (a) the entropy of activation, (b) the enthalpy of activation, (c) the Gibbs energy of activation at 298 K?

24.14b The pre-exponential factor for a gas-phase decomposition of ozone at low pressures is 2.3×10^{13} dm^3 mol^{-1} s^{-1} and its activation energy is 30.0 kJ mol^{-1}. What are (a) the entropy of activation, (b) the enthalpy of activation, (c) the Gibbs energy of activation at 298 K?

24.15a The rate constant of the reaction $H_2O_2(aq) + I^-(aq) + H^+(aq) \rightarrow H_2O(l) + HIO(aq)$ is sensitive to the ionic strength of the aqueous solution in which the reaction occurs. At 25°C, $k = 12.2$ dm^6 mol^{-2} min^{-1} at an ionic strength of 0.0525. Use the Debye–Hückel limiting law to estimate the rate constant at zero ionic strength.

24.15b At 25°C, $k = 1.55$ dm^6 mol^{-2} min^{-1} at an ionic strength of 0.0241 for a reaction in which the rate-determining step involves the encounter of two singly charged cations. Use the Debye–Hückel limiting law to estimate the rate constant at zero ionic strength.

24.16a For a pair of electron donor and acceptor, $H_{AB} = 0.03$ cm^{-1}, $\Delta_r G^{\ominus} = -0.182$ eV and $k_{et} = 30.5$ s^{-1} at 298 K. Estimate the value of the reorganization energy.

24.16b For a pair of electron donor and acceptor, $k_{et} = 2.02 \times 10^5$ s^{-1} for $\Delta_r G^{\ominus} = -0.665$ eV. The standard reaction Gibbs energy changes to $\Delta_r G^{\ominus} = -0.975$ eV when a substituent is added to the electron acceptor and the rate constant for electron transfer changes to $k_{et} = 3.33 \times 10^6$ s^{-1}. The experiments were conducted at 298 K. Assuming that the distance between donor and acceptor is the same in both experiments, estimate the values of H_{AB} and λ.

24.17a For a pair of electron donor and acceptor, $k_{et} = 2.02 \times 10^5$ s^{-1} when $r = 1.11$ nm and $k_{et} = 4.51 \times 10^5$ s^{-1} when $r = 1.23$ nm. Assuming that $\Delta_r G^{\ominus}$ and λ are the same in both experiments, estimate the value of β.

24.17b Refer to Exercise 24.17a. Estimate the value of k_{et} when $r = 1.48$ nm.

Problems*

Numerical problems

24.1 In the dimerization of methyl radicals at 25°C, the experimental pre-exponential factor is 2.4×10^{10} dm^3 mol^{-1} s^{-1}. What are (a) the reactive cross-section, (b) the P factor for the reaction if the C—H bond length is 154 pm?

24.2 Nitrogen dioxide reacts bimolecularly in the gas phase to give 2 NO + O_2. The temperature dependence of the second-order rate constant for the rate law $d[P]/dt = k[NO_2]^2$ is given below. What are the P factor and the reactive cross-section for the reaction?

* Problems denoted with the symbol ‡ were supplied by Charles Trapp, Carmen Giunta, and Marshall Cady.

T/K	600	700	800	1000
k/(cm^3 mol^{-1} s^{-1})	4.6×10^2	9.7×10^3	1.3×10^5	3.1×10^6

Take $\sigma = 0.60$ nm^2.

24.3 The diameter of the methyl radical is about 308 pm. What is the maximum rate constant in the expression $d[C_2H_6]/dt = k[CH_3]^2$ for second-order recombination of radicals at room temperature? 10 per cent of a 1.0-dm^3 sample of ethane at 298 K and 100 kPa is dissociated into methyl radicals. What is the minimum time for 90 per cent recombination?

24.4 The rates of thermolysis of a variety of *cis*- and *trans*-azoalkanes have been measured over a range of temperatures in order to settle a controversy concerning the mechanism of the reaction. In ethanol an unstable *cis*-azoalkane decomposed at a rate that was followed by observing the N_2 evolution, and this led to the rate constants listed below (P.S. Engel and D.J. Bishop, *J. Amer. Chem. Soc.* **97**, 6754 (1975)). Calculate the enthalpy, entropy, energy, and Gibbs energy of activation at −20°C.

θ/°C	−24.82	−20.73	−17.02	−13.00	−8.95
$10^4\times k$/s^{-1}	1.22	2.31	4.39	8.50	14.3

24.5 In an experimental study of a bimolecular reaction in aqueous solution, the second-order rate constant was measured at 25°C and at a variety of ionic strengths and the results are tabulated below. It is known that a singly charged ion is involved in the rate-determining step. What is the charge on the other ion involved?

I	0.0025	0.0037	0.0045	0.0065	0.0085
k/(dm^3 mol^{-1} s^{-1})	1.05	1.12	1.16	1.18	1.26

24.6 The rate constant of the reaction $I^-(aq) + H_2O_2(aq) \rightarrow H_2O(l) + IO^-(aq)$ varies slowly with ionic strength, even though the Debye–Hückel limiting law predicts no effect. Use the following data from 25°C to find the dependence of $\log k_r$ on the ionic strength:

I/(mol kg^{-1})	0.0207	0.0525	0.0925	0.1575
k_r/(dm^3 mol^{-1} min^{-1})	0.663	0.670	0.679	0.694

Evaluate the limiting value of k_r at zero ionic strength. What does the result suggest for the dependence of $\log\gamma$ on ionic strength for a neutral molecule in an electrolyte solution?

24.7 The total cross-sections for reactions between alkali metal atoms and halogen molecules are given in the table below (R.D. Levine and R.B. Bernstein, *Molecular reaction dynamics*, Clarendon Press, Oxford, 72 (1974)). Assess the data in terms of the harpoon mechanism.

σ^*/nm^2	Cl_2	Br_2	I_2
Na	1.24	1.16	0.97
K	1.54	1.51	1.27
Rb	1.90	1.97	1.67
Cs	1.96	2.04	1.95

Electron affinities are approximately 1.3 eV (Cl_2), 1.2 eV (Br_2), and 1.7 eV (I_2), and ionization energies are 5.1 eV (Na), 4.3 eV (K), 4.2 eV (Rb), and 3.9 eV (Cs).

24.8‡ M. Cyfert, B. Latko, and M. Wawrzeczyk (*Int. J. Chem. Kinet.* **28**, 103 (1996)) examined the oxidation of tris(1,10-phenanthroline)iron(II) by periodate in aqueous solution, a reaction that shows autocatalytic behaviour. To assess the kinetic salt effect, they measured rate constants at a variety of concentrations of Na_2SO_4 far in excess of reactant concentrations and reported the following data:

$[Na_2SO_4]$/(mol kg^{-1})	0.2	0.15	0.1	0.05	0.25	0.0125	0.005
k/(dm$^{3/2}$ mol$^{-1/2}$ s^{-1})	0.462	0.430	0.390	0.321	0.283	0.252	0.224

What can be inferred about the charge of the activated complex of the rate-determining step?

24.9‡ For the thermal decomposition of F_2O by the reaction $2\,F_2O(g) \rightarrow 2\,F_2(g) + O_2(g)$, J. Czarnowski and H.J. Schuhmacher (*Chem. Phys. Lett.* **17**, 235 (1972)) have suggested the following mechanism:

(1) $F_2O + F_2O \rightarrow F + OF + F_2O$ $\quad k_1$

(2) $F + F_2O \rightarrow F_2 + OF$ $\quad k_2$

(3) $OF + OF \rightarrow O_2 + F + F$ $\quad k_3$

(4) $F + F + F_2O \rightarrow F_2 + F_2O$ $\quad k_4$

(a) Using the steady-state approximation, show that this mechanism is consistent with the experimental rate law $-d[F_2O]/dt = k[F_2O]^2 + k'[F_2O]^{3/2}$. (b) The experimentally determined Arrhenius parameters in the range 501–583 K are $A = 7.8\times10^{13}$ dm^3 mol^{-1} s^{-1}, $E_a/R = 1.935\times10^4$ K for k and and $A = 2.3\times10^{10}$ dm^3 mol^{-1} s^{-1}, $E_a/R = 1.691\times10^4$ K for k'. At 540 K, $\Delta_f H^\ominus(F_2O) = +24.41$ kJ mol^{-1}, $D(F{-}F) = 160.6$ kJ mol^{-1}, and $D(O{-}O) = 498.2$ kJ mol^{-1}. Estimate the bond dissociation energies of the first and second F—O bonds and the Arrhenius activation energy of reaction 2.

24.10‡ For the gas-phase reaction $A + A \rightarrow A_2$, the experimental rate constant, k_2, has been fitted to the Arrhenius equation with the pre-exponential factor $A = 4.07\times10^5$ dm^3 mol^{-1} s^{-1} at 300 K and an activation energy of 65.43 kJ mol^{-1}. Calculate $\Delta^\ddagger S$, $\Delta^\ddagger H$, $\Delta^\ddagger U$, and $\Delta^\ddagger G$ for the reaction.

24.11‡ One of the most historically significant studies of chemical reaction rates was that by M. Bodenstein (*Z. physik. Chem.* **29**, 295 (1899)) of the gas-phase reaction $2\,HI(g) \rightarrow H_2(g) + I_2(g)$ and its reverse, with rate constants k and k', respectively. The measured rate constants as a function of temperature are

T/K	647	666	683	700	716	781
k/(22.4 dm^3 mol^{-1} min^{-1})	0.230	0.588	1.37	3.10	6.70	105.9
k'/(22.4 dm^3 mol^{-1} min^{-1})	0.0140	0.0379	0.0659	0.172	0.375	3.58

Demonstrate that these data are consistent with the collision theory of bimolecular gas-phase reactions.

24.12 Use the approximate form of the Marcus relation (eqn 24.86 with $f = 1$) to estimate the rate constant for the reaction $Ru(bpy)_3^{3+} + Fe(H_2O)_6^{2+} \rightarrow Ru(bpy)_3^{2+} + Fe(H_2O)_6^{3+}$, where bpy stands for 4,4′-bipyridine. The following data are useful:

$Ru(bpy)_3^{3+} + e^- \rightarrow Ru(bpy)_3^{2+}$ $\quad E^\ominus = 1.26$ V

$Fe(H_2O)_6^{3+} + e^- \rightarrow Fe(H_2O)_6^{2+}$ $\quad E^\ominus = 0.77$ V

${}^*Ru(bpy)_3^{3+} + Ru(bpy)_3^{2+} \rightarrow {}^*Ru(bpy)_3^{2+} + Ru(bpy)_3^{3+}$
$k_{Ru} = 4.0\times10^8$ dm^3 mol^{-1} s^{-1}

${}^*Fe(H_2O)_6^{3+} + Fe(H_2O)_6^{2+} \rightarrow {}^*Fe(H_2O)_6^{2+} + Fe(H_2O)_6^{3+}$
$k_{Fe} = 4.2$ dm^3 mol^{-1} s^{-1}

Theoretical problems

24.13 Confirm that eqn 24.40 is a solution of eqn 24.39, where $[J]_t$ is a solution of the same equation but with $k = 0$ and for the same initial conditions.

24.14 Evaluate $[J]^*$ numerically using mathematical software for integration in eqn 24.40, and explore the effect of increasing reaction rate constant on the spatial distribution of J.

24.15 Estimate the orders of magnitude of the partition functions involved in a rate expression. State the order of magnitude of q_m^T/N_A, q^R, q^V, q^E for typical molecules. Check that in the collision of two structureless molecules the order of magnitude of the pre-exponential factor is of the same order as that predicted by collision theory. Go on to estimate the P factor for a reaction in which $A + B \rightarrow P$, and A and B are nonlinear triatomic molecules.

24.16 Use the Debye–Hückel limiting law to show that changes in ionic strength can affect the rate of reaction catalysed by H^+ from the deprotonation of a weak acid. Consider the mechanism: $H^+(aq) + B(aq) \rightarrow P$, where H^+ comes from the deprotonation of the weak acid, HA. The weak acid has a fixed concentration. First show that log $[H^+]$, derived from the ionization of HA, depends on the activity coefficients of ions and thus depends on the ionic strength. Then find the relationship between log(rate) and log $[H^+]$ to show that the rate also depends on the ionic strength.

24.17 The Eyring equation can also be applied to physical processes. As an example, consider the rate of diffusion of an atom stuck to the surface of a solid. Suppose that in order to move from one site to another it has to reach the top of the barrier where it can vibrate classically in the vertical direction and in one horizontal direction, but vibration along the other horizontal direction takes it into the neighbouring site. Find an expression for the rate of diffusion, and evaluate it for W atoms on a tungsten surface ($E_a = 60$ kJ mol^{-1}). Suppose that the vibration frequencies at the transition state are (a) the same as, (b) one-half the value for the adsorbed atom. What is the value of the diffusion coefficient D at 500 K? (Take the site separation as 316 pm and $\nu = 1 \times 10^{11}$ Hz.)

24.18 Suppose now that the adsorbed, migrating species treated in Problem 24.17 is a spherical molecule, and that it can rotate classically as well as vibrate at the top of the barrier, but that at the adsorption site itself it can only vibrate. What effect does this have on the diffusion constant? Take the molecule to be methane, for which $B = 5.24$ cm^{-1}.

24.19 Show that the intensities of a molecular beam before and after passing through a chamber of length l containing inert scattering atoms are related by $I = I_0 e^{-\mathcal{N}\sigma l}$, where σ is the collision cross-section and $\mathcal{N}$ the number density of scattering atoms.

24.20 In a molecular beam experiment to measure collision cross-sections it was found that the intensity of a CsCl beam was reduced to 60 per cent of its intensity on passage through CH_2F_2 at 10 μTorr, but that when the target was Ar at the same pressure the intensity was reduced only by 10 per cent. What are the relative cross-sections of the two types of collision? Why is one much larger than the other?

24.21‡ Show that bimolecular reactions between nonlinear molecules are much slower than between atoms even when the activation energies of both reactions are equal. Use transition state theory and make the following assumptions. (1) All vibrational partition functions are close to unity; (2) all rotational partition functions are approximately $1 \times 10^{1.5}$, which is a reasonable order of magnitude number; (3) the translational partition function for each species is 1×10^{26}.

24.22 This exercise gives some familiarity with the difficulties involved in predicting the structure of activated complexes. It also demonstrates the importance of femtosecond spectroscopy to our understanding of chemical dynamics because direct experimental observation of the activated complex removes much of the ambiguity of theoretical predictions. Consider the attack of H on D_2, which is one step in the $H_2 + D_2$ reaction. (a) Suppose that the H approaches D_2 from the side and forms a complex in the form of an isosceles triangle. Take the H—D distance as 30 per cent greater than in H_2 (74 pm) and the D—D distance as 20 per cent greater than in H_2. Let the critical coordinate be the antisymmetric stretching vibration in which one H—D bond stretches as the other shortens. Let all the vibrations be at about 1000 cm^{-1}. Estimate k_2 for this reaction at 400 K using the experimental activation energy of about 35 kJ mol^{-1}. (b) Now change the model of the activated complex in part (a) and make it linear. Use the same estimated molecular bond lengths and vibrational frequencies to calculate k_2 for this choice of model. (c) Clearly, there is much scope for modifying the parameters of the models of the activated complex. Use mathematical software or write and run a program that allows you to vary the structure of the complex and the parameters in a plausible way, and look for a model (or more than one model) that gives a value of k close to the experimental value, 4×10^5 dm^3 mol^{-1} s^{-1}.

Applications: to environmental science and biochemistry

24.23‡ R. Atkinson (*J. Phys. Chem. Ref. Data* **26**, 215 (1997)) has reviewed a large set of rate constants relevant to the atmospheric chemistry of volatile organic compounds. The recommended rate constant for the bimolecular association of O_2 with an alkyl radical R at 298 K is 4.7×10^9 dm^3 mol^{-1} s^{-1} for R = C_2H_5 and 8.4×10^9 dm^3 mol^{-1} s^{-1} for R = cyclohexyl. Assuming no energy barrier, compute the steric factor, P, for each reaction. (*Hint.* Obtain collision diameters from collision cross-sections of similar molecules in the *Data section*.)

24.24‡ The compound α-tocopherol, a form of vitamin E, is a powerful antioxidant that may help to maintain the integrity of biological membranes. R.H. Bisby and A.W. Parker (*J. Amer. Chem. Soc.* **117**, 5664 (1995)) studied the reaction of photochemically excited duroquinone with the antioxidant in ethanol. Once the duroquinone was photochemically excited, a bimolecular reaction took place at a rate described as diffusion-limited. (a) Estimate the rate constant for a diffusion-limited reaction in ethanol. (b) The reported rate constant was 2.77×10^9 dm^3 mol^{-1} s^{-1}; estimate the critical reaction distance if the sum of diffusion constants is 1×10^{-9} m^2 s^{-1}.

24.25 The study of conditions that optimize the association of proteins in solution guides the design of protocols for formation of large crystals that are amenable to analysis by the X-ray diffraction techniques discussed in Chapter 20. It is important to characterize protein dimerization because the process is considered to be the rate-determining step in the growth of crystals of many proteins. Consider the variation with ionic strength of the rate constant of dimerization in aqueous solution of a cationic protein P:

I	0.0100	0.0150	0.0200	0.0250	0.0300	0.0350
$k/k°$	8.10	13.30	20.50	27.80	38.10	52.00

What can be deduced about the charge of P?

24.26 A useful strategy for the study of electron transfer in proteins consists of attaching an electroactive species to the protein's surface and then measuring k_{et} between the attached species and an electroactive protein cofactor. J.W. Winkler and H.B. Gray (*Chem. Rev.* **92**, 369 (1992)) summarize data for cytochrome *c* (*Impact* I7.2) modified by replacement of the haem iron by a zinc ion, resulting in a zinc-porphyrin (ZnP) moiety in the interior of the protein, and by attachment of a ruthenium ion complex to a surface histidine aminoacid. The edge-to-edge distance between the electroactive species was thus fixed at 1.23 nm. A variety of ruthenium ion complexes with different standard reduction potentials were used. For each ruthenium-modified protein, either the $Ru^{2+} \rightarrow ZnP^+$ or the $ZnP^* \rightarrow Ru^{3+}$, in which the electron donor is an electronic excited state of the zinc-porphyrin formed by laser excitation, was monitored. This arrangement leads to different standard reaction Gibbs energies because the redox couples ZnP^+/ZnP and ZnP^+/ZnP^* have different standard potentials, with the electronically excited porphyrin being a more powerful reductant. Use the following data to estimate the reorganization energy for this system:

$-\Delta_r G^\ominus$/eV	0.665	0.705	0.745	0.975	1.015	1.055
$k_{et}/(10^6$ s$^{-1})$	0.657	1.52	1.12	8.99	5.76	10.1

24.27 The photosynthetic reaction centre of the purple photosynthetic bacterium *Rhodopseudomonas viridis* contains a number of bound cofactors that participate in electron transfer reactions. The following table shows data compiled by Moser *et al.* (*Nature* **355**, 796 (1992)) on the rate constants for electron transfer between different cofactors and their edge-to-edge distances:

Reaction	$BChl^- \rightarrow BPh$	$BPh^- \rightarrow BChl_2^+$	$BPh^- \rightarrow Q_A$	cyt $c_{559} \rightarrow BChl_2^+$
r/nm	0.48	0.95	0.96	1.23
k_{et}/s^{-1}	1.58×10^{12}	3.98×10^9	1.00×10^9	1.58×10^8

Reaction	$Q_A^- \rightarrow Q_B$	$Q_A^- \rightarrow BChl_2^+$
r/nm	1.35	2.24
k_{et}/s^{-1}	3.98×10^7	63.1

(BChl, bacteriochlorophyll; $BChl_2$, bacteriochlorophyll dimer, functionally distinct from BChl; BPh, bacteriophaeophytin; Q_A and Q_B, quinone molecules bound to two distinct sites; cyt c_{559}, a cytochrome bound to the reaction centre complex). Are these data in agreement with the behaviour predicted by eqn 24.83? If so, evaluate the value of β.

24.28 The rate constant for electron transfer between a cytochrome *c* and the bacteriochlorophyll dimer of the reaction centre of the purple bacterium *Rhodobacter sphaeroides* (Problem 24.27) decreases with decreasing temperature in the range 300 K to 130 K. Below 130 K, the rate constant becomes independent of temperature. Account for these results.

24.29 Azurin is a protein containing a copper ion that shuttles between the +2 and +1 oxidation states, and cytochrome *c* is a protein in which a haem-bound iron ion shuttles between the +3 and +2 oxidation states. The rate constant for electron transfer from reduced azurin to oxidized cytochrome *c* is 1.6×10^3 $dm^3\ mol^{-1}\ s^{-1}$. Estimate the electron self-exchange rate constant for azurin from the following data:

	$k_{ii}/(dm^3\ mol^{-1}\ s^{-1})$	$E^{\ominus}/V$
cytochrome *c*	1.5×10^2	0.260
azurin	?	0.304

Processes at solid surfaces

25

The growth and structure of solid surfaces

The extent of adsorption

Heterogeneous catalysis

Processes at electrodes

In this chapter we see how solids grow at their surfaces and how the details of the structure and composition of solid surfaces can be determined experimentally. A major part of the material concerns the extent to which a surface is covered and the variation of the extent of coverage with the pressure and temperature. This material is used to discuss how surfaces affect the rate and course of chemical change by acting as the site of catalysis. Reactions at surfaces include the processes that lie at the heart of electrochemistry. Therefore, we revisit in this chapter some of the topics treated in Chapter 7, but focus on the dynamics of electrode processes rather than the equilibrium properties treated there. Finally, we analyse the kinetics of reactions that are responsible for power production in fuel cells and for corrosion.

Processes at solid surfaces govern the viability of industry both constructively, as in catalysis, and destructively, as in corrosion. Chemical reactions at solid surfaces may differ sharply from reactions in the bulk, for reaction pathways of much lower activation energy may be provided, and hence result in catalysis. The concept of a solid surface has been extended in recent years with the availability of microporous materials as catalysts.

An important kinetic problem examined in this chapter is the rate at which oxidizable or reducible species—in short, **electroactive species**—can donate or accept electrons on the surfaces of electrodes. In Chapter 24 we explored the dynamics of electron transfer in homogeneous systems. In heterogeneous systems, the rates of processes that occur at the interface between the phases, such as an electrode immersed in an ionic solution, are very important. A measure of this rate is the **current density**, j, the charge flux through a region (the electric current divided by the area of the region). We shall discuss the properties that control the current density and its consequences.

We shall see that acronyms are widely used in surface studies; for convenience, a list of the acronyms used in this chapter is given in Table 25.7 at the end of the chapter.

The growth and structure of solid surfaces

In this section we see how surfaces are extended and crystals grow. The attachment of particles to a surface is called **adsorption**. The substance that adsorbs is the **adsorbate** and the underlying material that we are concerned with in this section is the **adsorbent** or **substrate**. The reverse of adsorption is **desorption**.

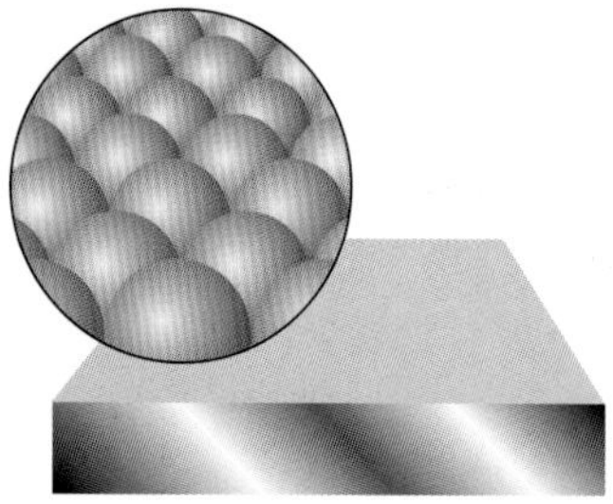

Fig. 25.1 A schematic diagram of the flat surface of a solid. This primitive model is largely supported by scanning tunnelling microscope images (see *Impact* I9.1).

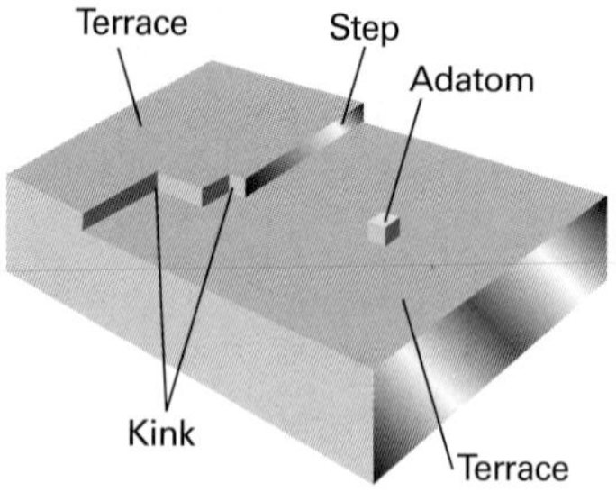

Fig. 25.2 Some of the kinds of defects that may occur on otherwise perfect terraces. Defects play an important role in surface growth and catalysis.

25.1 Surface growth

A simple picture of a perfect crystal surface is as a tray of oranges in a grocery store (Fig. 25.1). A gas molecule that collides with the surface can be imagined as a ping-pong ball bouncing erratically over the oranges. The molecule loses energy as it bounces, but it is likely to escape from the surface before it has lost enough kinetic energy to be trapped. The same is true, to some extent, of an ionic crystal in contact with a solution. There is little energy advantage for an ion in solution to discard some of its solvating molecules and stick at an exposed position on the surface.

The picture changes when the surface has defects, for then there are ridges of incomplete layers of atoms or ions. A common type of surface defect is a **step** between two otherwise flat layers of atoms called **terraces** (Fig. 25.2). A step defect might itself have defects, for it might have kinks. When an atom settles on a terrace it bounces across it under the influence of the intermolecular potential, and might come to a step or a corner formed by a kink. Instead of interacting with a single terrace atom, the molecule now interacts with several, and the interaction may be strong enough to trap it. Likewise, when ions deposit from solution, the loss of the solvation interaction is offset by a strong Coulombic interaction between the arriving ions and several ions at the surface defect.

Not all kinds of defect result in sustained surface growth. As the process of settling into ledges and kinks continues, there comes a stage when an entire lower terrace has been covered. At this stage the surface defects have been eliminated, and growth will cease. For continuing growth, a surface defect is needed that propagates as the crystal grows. We can see what form of defect this must be by considering the types of **dislocations**, or discontinuities in the regularity of the lattice, that exist in the bulk of a crystal. One reason for their formation may be that the crystal grows so quickly that its particles do not have time to settle into states of lowest potential energy before being trapped in position by the deposition of the next layer.

A special kind of dislocation is the **screw dislocation** shown in Fig. 25.3. Imagine a cut in the crystal, with the atoms to the left of the cut pushed up through a distance of one unit cell. The unit cells now form a continuous spiral around the end of the cut, which is called the **screw axis**. A path encircling the screw axis spirals up to the top of the crystal, and where the dislocation breaks through to the surface it takes the form of a spiral ramp.

The surface defect formed by a screw dislocation is a step, possibly with kinks, where growth can occur. The incoming particles lie in ranks on the ramp, and successive ranks reform the step at an angle to its initial position. As deposition continues the step rotates around the screw axis and is not eliminated. Growth may therefore continue indefinitely. Several layers of deposition may occur, and the edges of the spirals might be cliffs several atoms high. Propagating spiral edges can also give rise to flat terraces (Fig. 25.4). Terraces are formed if growth occurs simultaneously at neighbouring left- and right-handed screw dislocations (Fig. 25.5). Successive tables of atoms may form as counter-rotating defects collide on successive circuits, and the terraces formed may then fill up by further deposition at their edges to give flat crystal planes.

The rapidity of growth depends on the crystal plane concerned, and the slowest growing faces dominate the appearance of the crystal. This feature is explained in Fig. 25.6, where we see that, although the horizontal face grows forward most rapidly, it grows itself out of existence, and the slower-growing faces survive.

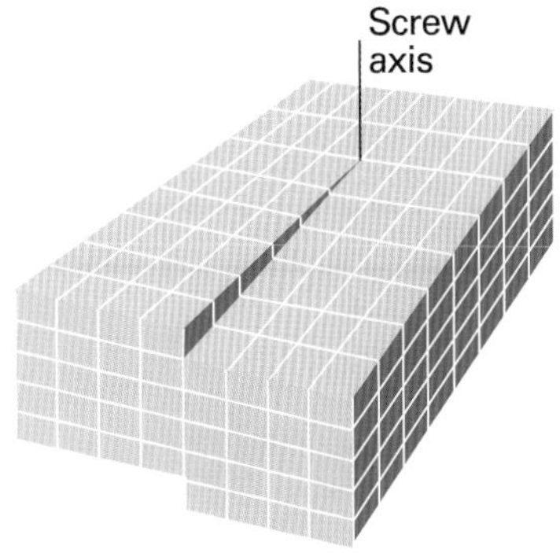

Fig. 25.3 A screw dislocation occurs where one region of the crystal is pushed up through one or more unit cells relative to another region. The cut extends to the screw axis. As atoms lie along the step the dislocation rotates round the screw axis and is not annihilated.

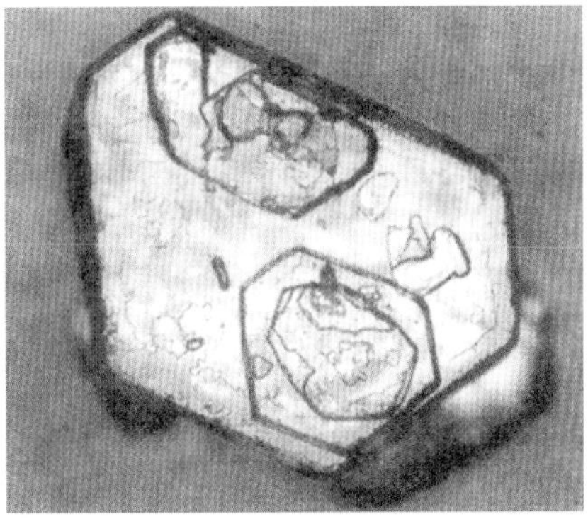

Fig. 25.4 The spiral growth pattern is sometimes concealed because the terraces are subsequently completed by further deposition. This accounts for the appearance of this cadmium iodide crystal. (H.M. Rosenberg, *The solid state*. Clarendon Press, Oxford (1978).)

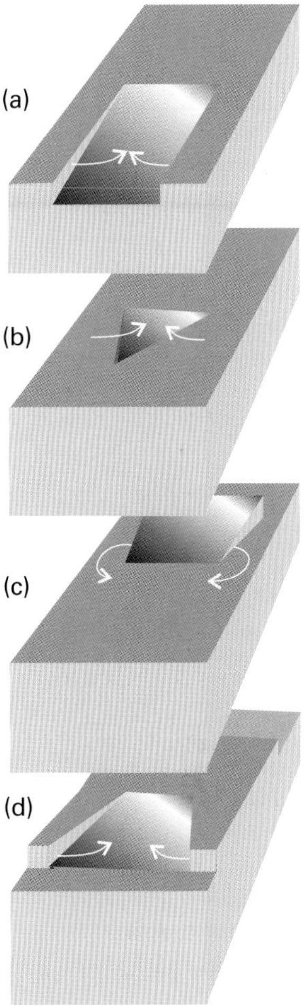

Fig. 25.5 Counter-rotating screw dislocations on the same surface lead to the formation of terraces. Four stages of one cycle of growth are shown here. Subsequent deposition can complete each terrace.

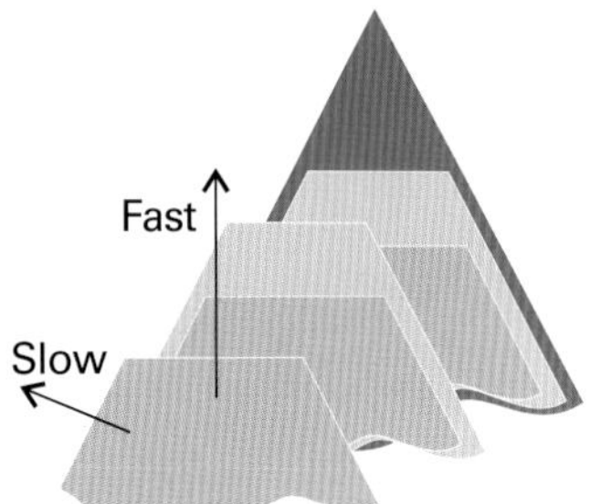

Fig. 25.6 The slower-growing faces of a crystal dominate its final external appearance. Three successive stages of the growth are shown.

25.2 Surface composition

Under normal conditions, a surface exposed to a gas is constantly bombarded with molecules and a freshly prepared surface is covered very quickly. Just how quickly can be estimated using the kinetic model of gases and the expression (eqn 21.14) for the collision flux:

$$Z_W = \frac{p}{(2\pi mkT)^{1/2}} \tag{25.1a}$$

A practical form of this equation is

$$Z_W = \frac{Z_0(p/\mathrm{Pa})}{\{(T/\mathrm{K})(M/(\mathrm{g\ mol^{-1}}))\}^{1/2}} \quad \text{with} \quad Z_0 = 2.63 \times 10^{24}\ \mathrm{m^{-2}\ s^{-1}} \tag{25.1b}$$

where M is the molar mass of the gas. For air ($M \approx 29$ g mol^{-1}) at 1 atm and 25°C the collision flux is 3×10^{27} m^{-2} s^{-1}. Because 1 m^2 of metal surface consists of about 10^{19} atoms, each atom is struck about 10^8 times each second. Even if only a few collisions leave a molecule adsorbed to the surface, the time for which a freshly prepared surface remains clean is very short.

The obvious way to retain cleanliness is to reduce the pressure. When it is reduced to 10^{-4} Pa (as in a simple vacuum system) the collision flux falls to about 10^{18} m^{-2} s^{-1}, corresponding to one hit per surface atom in each 0.1 s. Even that is too brief in most experiments, and in **ultrahigh vacuum** (UHV) techniques pressures as low as 10^{-7} Pa (when $Z_W = 10^{15}$ m^{-2} s^{-1}) are reached on a routine basis and as low as 10^{-9} Pa (when $Z_W = 10^{13}$ m^{-2} s^{-1}) are reached with special care. These collision fluxes correspond to each surface atom being hit once every 10^5 to 10^6 s, or about once a day.

The layout of a typical UHV apparatus is such that the whole of the evacuated part can be heated to 150–250°C for several hours to drive gas molecules from the walls. All the taps and seals are usually of metal so as to avoid contamination from greases. The sample is usually in the form of a thin foil, a filament, or a sharp point. Where there is interest in the role of specific crystal planes the sample is a single crystal with a freshly cleaved face. Initial surface cleaning is achieved either by heating it electrically or by

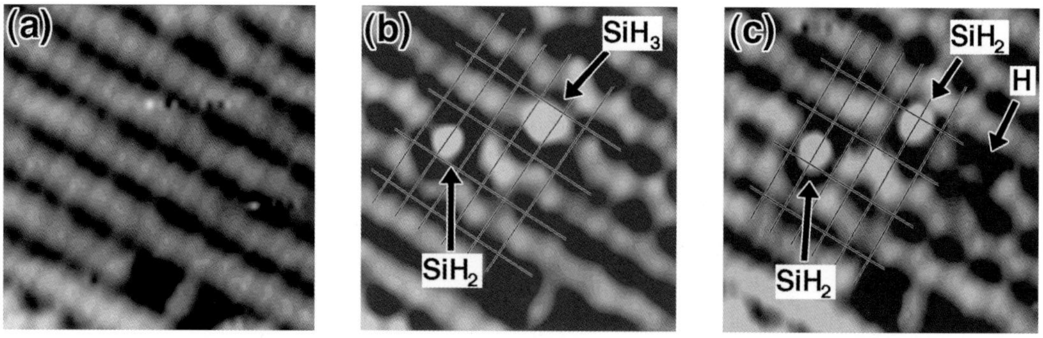

Fig. 25.7 Visualization by STM of the reaction $SiH_3 \rightarrow SiH_2 + H$ on a 4.7 nm × 4.7 nm area of a Si(001) surface. (a) The Si(001) surface before exposure to Si_2H_6(g). (b) Adsorbed Si_2H_6 dissociates into SiH_2(surface), on the left of the image, and SiH_3(surface), on the right. (c) After 8 min, SiH_3(surface) dissociates to SiH_2(surface) and H(surface). (Reproduced with permission from Y. Wang, M.J. Bronikowski, and R.J. Hamers, *Surface Science* **64**, 311 (1994).)

bombarding it with accelerated gaseous ions. The latter procedure demands care because ion bombardment can shatter the surface structure and leave it an amorphous jumble of atoms. High temperature annealing is then required to return the surface to an ordered state.

We have already discussed three important techniques for the characterization of surfaces: scanning electron microscopy (*Impact* I8.1), which is often used to observe terraces, steps, kinks, and dislocations on a surface, and scanning probe microscopy (*Impact* I9.1), which reveals the atomic details of structure of the surface and of adsorbates and can be used to visualize chemical reactions as they happen on surfaces (Fig. 25.7). In the following sections, we describe additional techniques that comprise the toolbox of a surface scientist.

(a) Ionization techniques

Surface composition can be determined by a variety of ionization techniques. The same techniques can be used to detect any remaining contamination after cleaning and to detect layers of material adsorbed later in the experiment. Their common feature is that the **escape depth** of the electrons, the maximum depth from which ejected electrons come, is in the range 0.1–1.0 nm, which ensures that only surface species contribute.

One technique that may be used is photoelectron spectroscopy (Section 11.4), which in surface studies is normally called **photoemission spectroscopy**. X-rays or hard ultraviolet ionizing radiation of energy in the range 5–40 eV may be used, giving rise to the techniques denoted XPS and UPS, respectively.

In XPS, the energy of the incident photon is so great that electrons are ejected from inner cores of atoms. As a first approximation, core ionization energies are insensitive to the bonds between atoms because they are too tightly bound to be greatly affected by the changes that accompany bond formation, so core ionization energies are characteristic of the individual atom. Consequently, XPS gives lines characteristic of the elements present on a surface. For instance, the *K*-shell ionization energies of the second row elements are

Li	Be	B	C	N	O	F
50	110	190	280	400	530	690 eV

Detection of one of these values (and values corresponding to ejection from other inner shells) indicates the presence of the corresponding element (Fig. 25.8). This

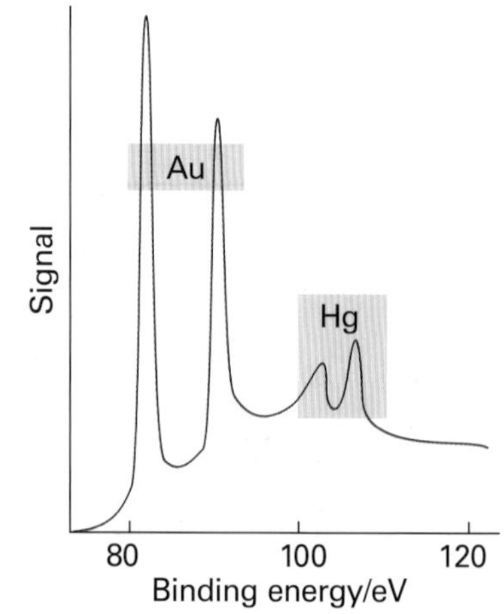

Fig. 25.8 The X-ray photoelectron emission spectrum of a sample of gold contaminated with a surface layer of mercury. (M.W. Roberts and C.S. McKee, *Chemistry of the metal–gas interface*, Oxford (1978).)

application is responsible for the alternative name **electron spectroscopy for chemical analysis** (ESCA). The technique is very useful for studying the surface state of heterogeneous catalysts, the differences between surface and bulk structures, and the processes that can cause damage to high-temperature superconductors and semiconductor wafers.

UPS, which examines electrons ejected from valence shells, is more suited to establishing the bonding characteristics and the details of valence shell electronic structures of substances on the surface. Its usefulness is its ability to reveal which orbitals of the adsorbate are involved in the bond to the substrate. For instance, the principal difference between the photoemission results on free benzene and benzene adsorbed on palladium is in the energies of the π electrons. This difference is interpreted as meaning that the C_6H_6 molecules lie parallel to the surface and are attached to it by their π orbitals.

In **secondary-ion mass spectrometry** (SIMS), the surface is ionized by bombardment with other ions and the secondary ions that emerge from the surface are detected by a mass spectrometer. Among the advantages of SIMS are the ability to detect adsorbed H and He atoms, which are not easily probed by XPS, and the high sensitivity of the mass spectrometer detector. A disadvantage is that SIMS analysis erodes the part of the sample that is bombarded. However, it is possible to control the degree of erosion to one or two monolayers by controlling the bombardment parameters.

(b) Vibrational spectroscopy

Several kinds of vibrational spectroscopy have been developed to study adsorbates and to show whether dissociation has occurred. Measurement of transmitted radiation is not practical in surfaces, which are typically too opaque to infrared or visible radiation. One technique that circumvents this problem is **reflection–absorption infrared spectroscopy** (RAIRS), in which the Fourier-transform IR absorption spectrum of the adsorbate is obtained by comparing the intensity of the incident infrared beam with the intensity of infrared radiation reflected by the surface.

Raman spectroscopy is better suited for studies of surfaces because it involves the detection of scattered radiation, but spectral bands are typically very weak. However, **surface-enhanced Raman scattering** (SERS) is viable for surface studies: the strong enhancement of the Raman spectrum of the adsorbate can increase intensities by a factor as big as 10^6. The effect is due in part to local accumulations of electron density at the features of the roughened surface and at regions where bonding occurs. The SERS effect is also observed when molecules adsorb to colloidal particles of gold and silver, with the surface of the colloid fostering the enhancement of the Raman spectrum. Disadvantages of SERS include weak enhancement observed on flat single crystal surfaces and the fact that the technique works for only certain metals.

(c) Electron spectroscopy

A hybrid version of photoemission spectroscopy and vibrational spectroscopy is **electron energy loss spectroscopy** (EELS, or HREELS, where HR denotes high resolution) in which the energy loss suffered by a beam of electrons is monitored when they are reflected from a surface. As in Raman spectroscopy, the spectrum of energy loss can be interpreted in terms of the vibrational spectrum of the adsorbate. High resolution and sensitivity are attainable, and the technique is sensitive to light elements (to which X-ray techniques are insensitive). Very tiny amounts of adsorbate can be detected, and one report estimated that about 48 atoms of phosphorus were detected in one sample. As an example, Fig. 25.9 shows the EELS result for CO on the (111) face of a platinum crystal as the extent of surface coverage increases. The main peak arises

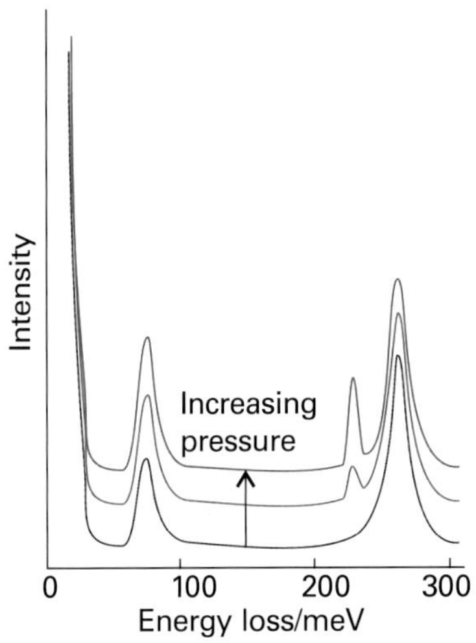

Fig. 25.9 The electron energy loss spectrum of CO adsorbed on Pt(111). The results for three different pressures are shown, and the growth of the additional peak at about 200 meV (1600 cm^{-1}) should be noted. (Based on spectra provided by Professor H. Ibach.)

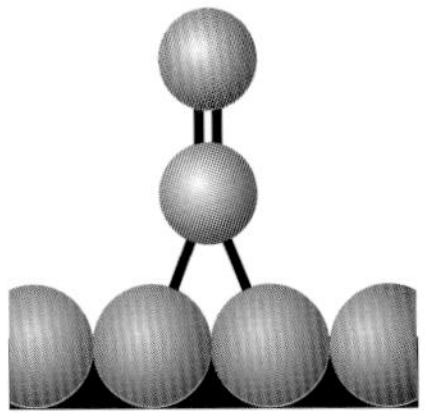

1 Bridge site

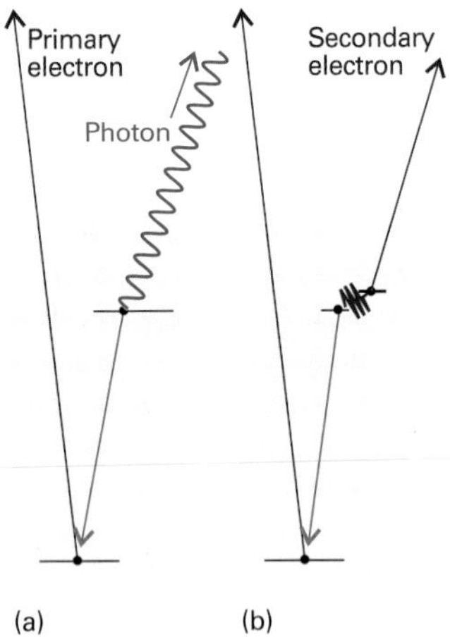

Fig. 25.10 When an electron is expelled from a solid (a) an electron of higher energy may fall into the vacated orbital and emit an X-ray photon to produce X-ray fluorescence. Alternatively, (b) the electron falling into the orbital may give up its energy to another electron, which is ejected in the Auger effect.

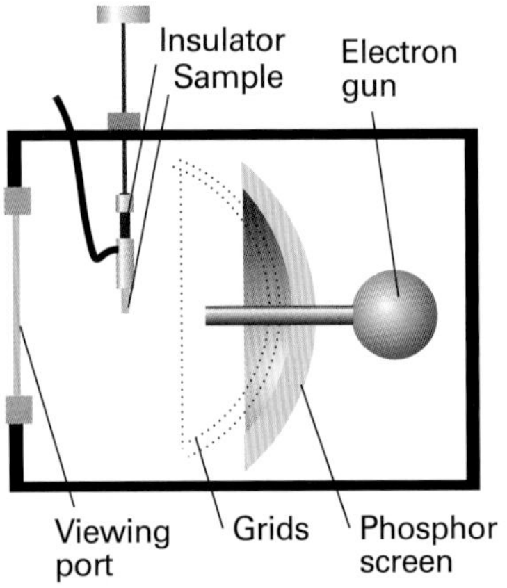

Fig. 25.11 A schematic diagram of the apparatus used for a LEED experiment. The electrons diffracted by the surface layers are detected by the fluorescence they cause on the phosphor screen.

from CO attached perpendicular to the surface by a single Pt atom. As the coverage increases the neighbouring smaller peak increases in intensity. This peak is due to CO at a bridge site, attached to two Pt atoms, as in (**1**).

A very important technique, which is widely used in the microelectronics industry, is **Auger electron spectroscopy** (AES). The **Auger effect** is the emission of a second electron after high energy radiation has expelled another. The first electron to depart leaves a hole in a low-lying orbital, and an upper electron falls into it. The energy this releases may result either in the generation of radiation, which is called **X-ray fluorescence** (Fig. 25.10a) or in the ejection of another electron (Fig. 25.10b). The latter is the secondary electron of the Auger effect. The energies of the secondary electrons are characteristic of the material present, so the Auger effect effectively takes a fingerprint of the sample. In practice, the Auger spectrum is normally obtained by irradiating the sample with an electron beam of energy in the range 1–5 keV rather than electromagnetic radiation. In **scanning Auger electron microscopy** (SAM), the finely focused electron beam is scanned over the surface and a map of composition is compiled; the resolution can reach below about 50 nm.

(d) Surface-extended X-ray absorption fine structure spectroscopy

The technique known as **surface-extended X-ray absorption fine structure spectroscopy** (SEXAFS) uses intense X-radiation from synchrotron sources (*Further information* 13.1). Oscillations in X-ray absorbance are observed on the high-frequency side of the absorption edge (the start of an X-ray absorption band) of a substance. These oscillations arise from a quantum mechanical interference between the wavefunction of a photoejected electron and parts of that electron's wavefunction that are scattered by neighbouring atoms. If the waves interfere destructively, then the photoelectron appears with lower probability and the X-ray absorption is correspondingly less. If the waves interfere constructively, then the photoelectron amplitude is higher, and the photoelectron has a higher probability of appearing; correspondingly, the X-ray absorption is greater. The oscillations therefore contain information about the number and distances of the neighbouring atoms. Such studies show that a solid's surface is much more plastic than had previously been thought, and that it undergoes **reconstruction**, or structural modification, in response to adsorbates that are present.

(e) Low-energy electron diffraction

One of the most informative techniques for determining the arrangement of the atoms close to the surface is **low energy electron diffraction** (LEED). This technique is like X-ray diffraction (Chapter 20) but uses the wave character of electrons, and the sample is now the surface of a solid. The use of low energy electrons (with energies in the range 10–200 eV, corresponding to wavelengths in the range 100–400 pm) ensures that the diffraction is caused only by atoms on and close to the surface. The experimental arrangement is shown in Fig. 25.11, and typical LEED patterns, obtained by photographing the fluorescent screen through the viewing port, are shown in Fig. 25.12.

A LEED pattern portrays the two-dimensional structure of the surface. By studying how the diffraction intensities depend on the energy of the electron beam it is also possible to infer some details about the vertical location of the atoms and to measure the thickness of the surface layer, but the interpretation of LEED data is much more complicated than the interpretation of bulk X-ray data. The pattern is sharp if the surface is well-ordered for distances long compared with the wavelength of the incident electrons. In practice, sharp patterns are obtained for surfaces ordered to depths of about 20 nm and more. Diffuse patterns indicate either a poorly ordered surface or the presence of impurities. If the LEED pattern does not correspond to the pattern

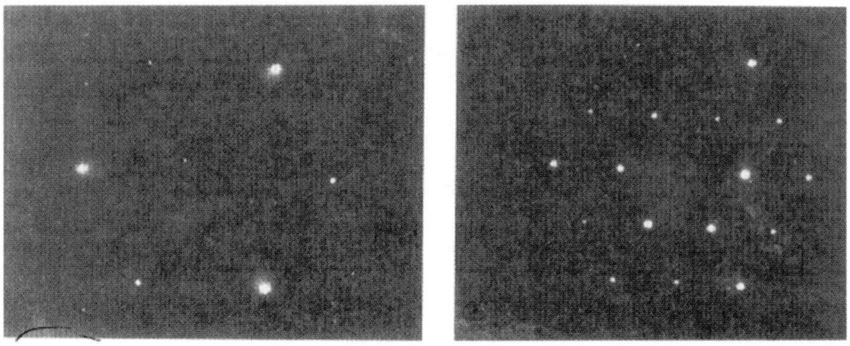

Fig. 25.12 LEED photographs of (a) a clean platinum surface and (b) after its exposure to propyne, $CH_3C{\equiv}CH$. (Photographs provided by Professor G.A. Somorjai.)

expected by extrapolation of the bulk surface to the surface, then either a reconstruction of the surface has occurred or there is order in the arrangement of an adsorbed layer.

The results of LEED experiments show that the surface of a crystal rarely has exactly the same form as a slice through the bulk. As a general rule, it is found that metal surfaces are simply truncations of the bulk lattice, but the distance between the top layer of atoms and the one below is contracted by around 5 per cent. Semiconductors generally have surfaces reconstructed to a depth of several layers. Reconstruction occurs in ionic solids. For example, in lithium fluoride the Li^+ and F^- ions close to the surface apparently lie on slightly different planes. An actual example of the detail that can now be obtained from refined LEED techniques is shown in Fig. 25.13 for CH_3C— adsorbed on a (111) plane of rhodium.

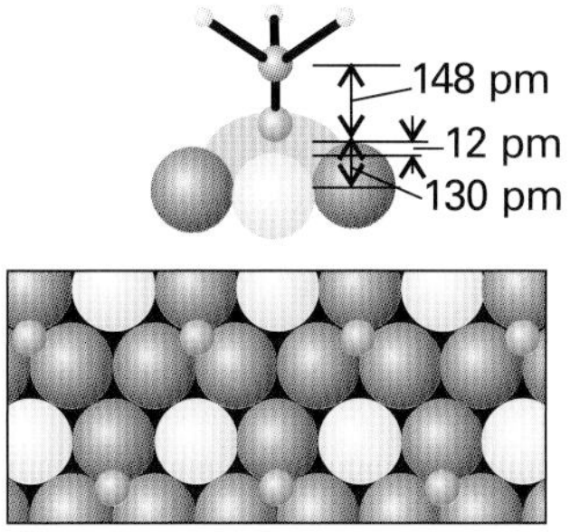

Fig. 25.13 The structure of a surface close to the point of attachment of CH_3C— to the (110) surface of rhodium at 300 K and the changes in positions of the metal atoms that accompany chemisorption.

The presence of terraces, steps, and kinks in a surface shows up in LEED patterns, and their surface density (the number of defects in a region divided by the area of the region) can be estimated. The importance of this type of measurement will emerge later. Three examples of how steps and kinks affect the pattern are shown in Fig. 25.14. The samples used were obtained by cleaving a crystal at different angles to a plane of

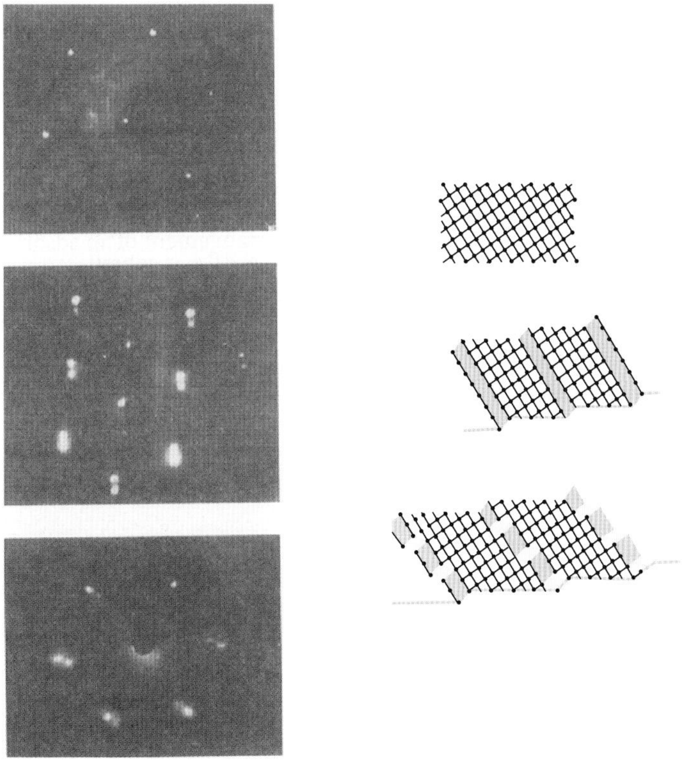

Fig. 25.14 LEED patterns may be used to assess the defect density of a surface. The photographs correspond to a platinum surface with (top) low defect density, (middle) regular steps separated by about six atoms, and (bottom) regular steps with kinks. (Photographs provided by Professor G.A. Samorjai.)

atoms. Only terraces are produced when the cut is parallel to the plane, and the density of steps increases as the angle of the cut increases. The observation of additional structure in the LEED patterns, rather than blurring, shows that the steps are arrayed regularly.

(f) Molecular beam techniques

Whereas many important studies have been carried out simply by exposing a surface to a gas, modern work is increasingly making use of **molecular beam scattering** (MBS). One advantage is that the activity of specific crystal planes can be investigated by directing the beam on to an orientated surface with known step and kink densities (as measured by LEED). Furthermore, if the adsorbate reacts at the surface the products (and their angular distributions) can be analysed as they are ejected from the surface and pass into a mass spectrometer. Another advantage is that the time of flight of a particle may be measured and interpreted in terms of its residence time on the surface. In this way a very detailed picture can be constructed of the events taking place during reactions at surfaces.

The extent of adsorption

The extent of surface coverage is normally expressed as the **fractional coverage**, θ:

$$\theta = \frac{\text{number of adsorption sites occupied}}{\text{number of adsorption sites available}} \qquad [25.2]$$

The fractional coverage is often expressed in terms of the volume of adsorbate adsorbed by $\theta = V/V_\infty$, where V_∞ is the volume of adsorbate corresponding to complete monolayer coverage. The **rate of adsorption**, $d\theta/dt$, is the rate of change of surface coverage, and can be determined by observing the change of fractional coverage with time.

Among the principal techniques for measuring $d\theta/dt$ are flow methods, in which the sample itself acts as a pump because adsorption removes particles from the gas. One commonly used technique is therefore to monitor the rates of flow of gas into and out of the system: the difference is the rate of gas uptake by the sample. Integration of this rate then gives the fractional coverage at any stage. In **flash desorption** the sample is suddenly heated (electrically) and the resulting rise of pressure is interpreted in terms of the amount of adsorbate originally on the sample. The interpretation may be confused by the desorption of a compound (for example, WO_3 from oxygen on tungsten). **Gravimetry**, in which the sample is weighed on a microbalance during the experiment, can also be used. A common instrument for gravimetric measurements is the **quartz crystal microbalance** (QCM), in which the mass of a sample laid on the surface of a quartz crystal is related to changes in the latter's mechanical properties. The key principle behind the operation of a QCM is the ability of a quartz crystal to vibrate at a characteristic frequency when an oscillating electric field is applied. The vibrational frequency decreases when material is spread over the surface of the crystal and the change in frequency is proportional to the mass of material. Masses as small as a few nanograms ($1\ \text{ng} = 10^{-9}\ \text{g}$) can be measured reliably in this way.

25.3 Physisorption and chemisorption

Molecules and atoms can attach to surfaces in two ways. In **physisorption** (an abbreviation of 'physical adsorption'), there is a van der Waals interaction (for example, a dispersion or a dipolar interaction) between the adsorbate and the substrate. Van der Waals interactions have a long range but are weak, and the energy released when a particle is physisorbed is of the same order of magnitude as the enthalpy of condensation. Such small energies can be absorbed as vibrations of the lattice and dissipated as thermal motion, and a molecule bouncing across the surface will gradually lose its energy and finally adsorb to it in the process called **accommodation**. The enthalpy of physisorption can be measured by monitoring the rise in temperature of a sample of known heat capacity, and typical values are in the region of 20 kJ mol^{-1} (Table 25.1). This small enthalpy change is insufficient to lead to bond breaking, so a physisorbed molecule retains its identity, although it might be distorted by the presence of the surface.

In **chemisorption** (an abbreviation of 'chemical adsorption'), the molecules (or atoms) stick to the surface by forming a chemical (usually covalent) bond, and tend to find sites that maximize their coordination number with the substrate. The enthalpy of chemisorption is very much greater than that for physisorption, and typical values are in the region of 200 kJ mol^{-1} (Table 25.2). The distance between the surface and the closest adsorbate atom is also typically shorter for chemisorption than for physisorption. A chemisorbed molecule may be torn apart at the demand of the unsatisfied valencies of the surface atoms, and the existence of molecular fragments on the surface as a result of chemisorption is one reason why solid surfaces catalyse reactions.

Except in special cases, chemisorption must be exothermic. A spontaneous process requires $\Delta G < 0$. Because the translational freedom of the adsorbate is reduced when it is adsorbed, ΔS is negative. Therefore, in order for $\Delta G = \Delta H - T\Delta S$ to be negative, ΔH must be negative (that is, the process is exothermic). Exceptions may occur if the adsorbate dissociates and has high translational mobility on the surface. For example, H_2 adsorbs endothermically on glass because there is a large increase of translational entropy accompanying the dissociation of the molecules into atoms that move quite freely over the surface. In its case, the entropy change in the process $H_2(g) \rightarrow 2\,H(\text{glass})$ is sufficiently positive to overcome the small positive enthalpy change.

The enthalpy of adsorption depends on the extent of surface coverage, mainly because the adsorbate particles interact. If the particles repel each other (as for CO on palladium) the adsorption becomes less exothermic (the enthalpy of adsorption less negative) as coverage increases. Moreover, LEED studies show that such species settle on the surface in a disordered way until packing requirements demand order. If the adsorbate particles attract one another (as for O_2 on tungsten), then they tend to cluster together in islands, and growth occurs at the borders. These adsorbates also show order–disorder transitions when they are heated enough for thermal motion to overcome the particle–particle interactions, but not so much that they are desorbed.

Synoptic table 25.1* Maximum observed enthalpies of physisorption

Adsorbate	$\Delta_{ad}H^{\ominus}/(\text{kJ mol}^{-1})$
CH_4	−21
H_2	−84
H_2O	−59
N_2	−21

* More values are given in the *Data section.*

Synoptic table 25.2* Enthalpies of chemisorption, $\Delta_{ad}H^{\ominus}/(\text{kJ mol}^{-1})$

Adsorbate	Adsorbent (substrate)		
	Cr	Fe	Ni
CH_4	−427	−285	−243
CO		−192	
H_2	−188	−134	
NH_3		−188	−155

* More values are given in the *Data section.*

25.4 Adsorption isotherms

The free gas and the adsorbed gas are in dynamic equilibrium, and the fractional coverage of the surface depends on the pressure of the overlying gas. The variation of θ with pressure at a chosen temperature is called the **adsorption isotherm**.

(a) The Langmuir isotherm

The simplest physically plausible isotherm is based on three assumptions:

1 Adsorption cannot proceed beyond monolayer coverage.

2 All sites are equivalent and the surface is uniform (that is, the surface is perfectly flat on a microscopic scale).

3 The ability of a molecule to adsorb at a given site is independent of the occupation of neighbouring sites (that is, there are no interactions between adsorbed molecules).

The dynamic equilibrium is

$$A(g) + M(surface) \rightleftharpoons AM(surface)$$

with rate constants k_a for adsorption and k_d for desorption. The rate of change of surface coverage due to adsorption is proportional to the partial pressure p of A and the number of vacant sites $N(1-\theta)$, where N is the total number of sites:

$$\frac{d\theta}{dt} = k_a pN(1-\theta) \qquad (25.3a)$$

The rate of change of θ due to desorption is proportional to the number of adsorbed species, $N\theta$:

$$\frac{d\theta}{dt} = -k_d N\theta \qquad (25.3b)$$

At equilibrium there is no net change (that is, the sum of these two rates is zero), and solving for θ gives the **Langmuir isotherm**:

$$\theta = \frac{Kp}{1+Kp} \qquad K = \frac{k_a}{k_d} \qquad (25.4)$$

Example 25.1 *Using the Langmuir isotherm*

The data given below are for the adsorption of CO on charcoal at 273 K. Confirm that they fit the Langmuir isotherm, and find the constant K and the volume corresponding to complete coverage. In each case V has been corrected to 1.00 atm (101.325 kPa).

p/kPa	13.3	26.7	40.0	53.3	66.7	80.0	93.3
V/cm^3	10.2	18.6	25.5	31.5	36.9	41.6	46.1

Method From eqn 25.4,

$$Kp\theta + \theta = Kp$$

With $\theta = V/V_\infty$, where V_∞ is the volume corresponding to complete coverage, this expression can be rearranged into

$$\frac{p}{V} = \frac{p}{V_\infty} + \frac{1}{KV_\infty}$$

Hence, a plot of p/V against p should give a straight line of slope $1/V_\infty$ and intercept $1/KV_\infty$.

Answer The data for the plot are as follows:

p/kPa	13.3	26.7	40.0	53.3	66.7	80.0	93.3
$(p/\text{kPa})/(V/\text{cm}^3)$	1.30	1.44	1.57	1.69	1.81	1.92	2.02

The points are plotted in Fig. 25.15. The (least squares) slope is 0.00900, so $V_\infty = 111\ \text{cm}^3$. The intercept at $p = 0$ is 1.20, so

$$K = \frac{1}{(111\ \text{cm}^3) \times (1.20\ \text{kPa cm}^{-3})} = 7.51 \times 10^{-3}\ \text{kPa}^{-1}$$

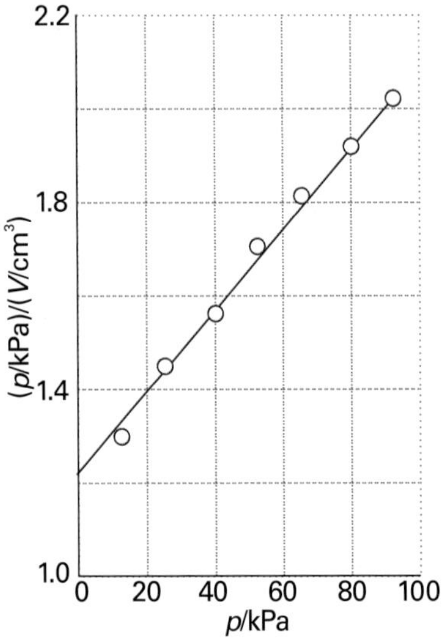

Fig. 25.15 The plot of the data in Example 25.1. As illustrated here, the Langmuir isotherm predicts that a straight line should be obtained when p/V is plotted against p.

Self-test 25.1 Repeat the calculation for the following data:

p/kPa	13.3	26.7	40.0	53.3	66.7	80.0	93.3
V/cm^3	10.3	19.3	27.3	34.1	40.0	45.5	48.0

[128 cm^3, 6.69×10^{-3} kPa^{-1}]

For adsorption with dissociation, the rate of adsorption is proportional to the pressure and to the probability that both atoms will find sites, which is proportional to the square of the number of vacant sites,

$$\frac{d\theta}{dt}=k_a p\{N(1-\theta)\}^2 \tag{25.5a}$$

The rate of desorption is proportional to the frequency of encounters of atoms on the surface, and is therefore second-order in the number of atoms present:

$$\frac{d\theta}{dt}=-k_d(N\theta)^2 \tag{25.5b}$$

The condition for no net change leads to the isotherm

$$\theta=\frac{(Kp)^{1/2}}{1+(Kp)^{1/2}} \tag{25.6}$$

The surface coverage now depends more weakly on pressure than for non-dissociative adsorption.

The shapes of the Langmuir isotherms with and without dissociation are shown in Figs. 25.16 and 25.17. The fractional coverage increases with increasing pressure, and approaches 1 only at very high pressure, when the gas is forced on to every available site of the surface. Different curves (and therefore different values of K) are obtained at different temperatures, and the temperature dependence of K can be used to determine the **isosteric enthalpy of adsorption**, $\Delta_{ad}H^{\ominus}$, the standard enthalpy of adsorption at a fixed surface coverage. To determine this quantity we recognize that K is essentially an equilibrium constant, and then use the van 't Hoff equation (eqn 7.23) to write:

$$\left(\frac{\partial \ln K}{\partial T}\right)_\theta=\frac{\Delta_{ad}H^{\ominus}}{RT^2} \tag{25.7}$$

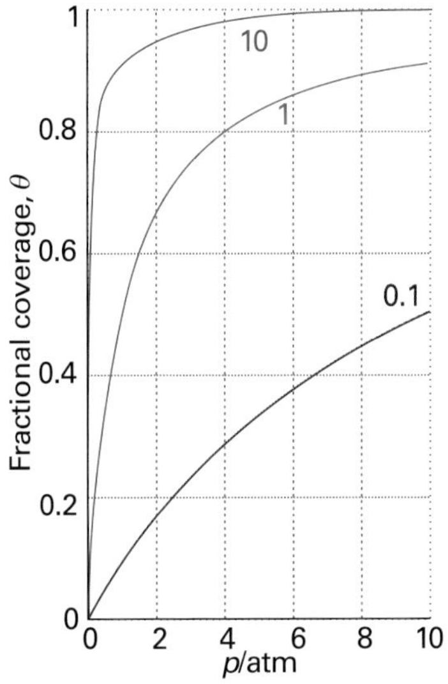

Fig. 25.16 The Langmuir isotherm for dissociative adsorption, $X_2(g) \rightarrow 2\,X(\text{surface})$, for different values of K.

Exploration Using eqn 25.4, generate a family of curves showing the dependence of $1/\theta$ on $1/p$ for several valuesof K.

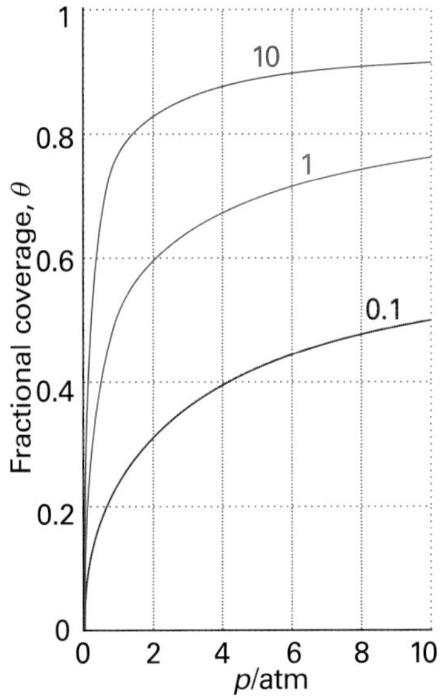

Fig. 25.17 The Langmuir isotherm for non-dissociative adsorption for different values of K.

Exploration Using eqn 25.6, generate a family of curves showing the dependence of $1/\theta$ on $1/p$ for several valuesof K. Taking these results together with those of the previous *Exploration*, discuss how plots of $1/\theta$ against $1/p$ can be used to distinguish between adsorption with and without dissociation.

Example 25.2 *Measuring the isosteric enthalpy of adsorption*

The data below show the pressures of CO needed for the volume of adsorption (corrected to 1.00 atm and 273 K) to be 10.0 cm^3 using the same sample as in Example 25.1. Calculate the adsorption enthalpy at this surface coverage.

T/K	200	210	220	230	240	250
p/kPa	4.00	4.95	6.03	7.20	8.47	9.85

Method The Langmuir isotherm can be rearranged to

$$Kp=\frac{\theta}{1-\theta}$$

Therefore, when θ is constant,

$$\ln K+\ln p=\text{constant}$$

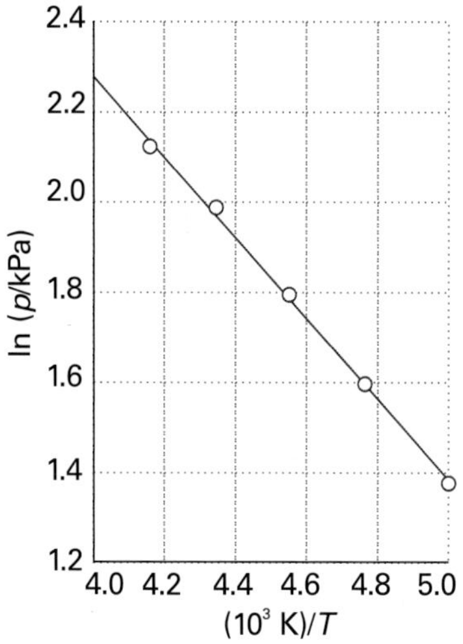

Fig. 25.18 The isosteric enthalpy of adsorption can be obtained from the slope of the plot of $\ln p$ against $1/T$, where p is the pressure needed to achieve the specified surface coverage. The data used are from Example 25.2.

It follows from eqn 25.7 that

$$\left(\frac{\partial \ln p}{\partial T}\right)_\theta = -\left(\frac{\partial \ln K}{\partial T}\right)_\theta = -\frac{\Delta_{\rm ad}H^\ominus}{RT^2}$$

With $\mathrm{d}(1/T)/\mathrm{d}T = -1/T^2$, this expression rearranges to

$$\left(\frac{\partial \ln p}{\partial (1/T)}\right)_\theta = \frac{\Delta_{\rm ad}H^\ominus}{R}$$

Therefore, a plot of $\ln p$ against $1/T$ should be a straight line of slope $\Delta_{\rm ad}H^\ominus/R$.

Answer We draw up the following table:

T/K	200	210	220	230	240	250
$10^3/(T/\mathrm{K})$	5.00	4.76	4.55	4.35	4.17	4.00
$\ln(p/\mathrm{kPa})$	1.39	1.60	1.80	1.97	2.14	2.29

The points are plotted in Fig. 25.18. The slope (of the least squares fitted line) is −0.904, so

$$\Delta_{\rm ad}H^\ominus = -(0.904 \times 10^3\ \mathrm{K}) \times R = -7.52\ \mathrm{kJ\ mol^{-1}}$$

The value of K can be used to obtain a value of $\Delta_{\rm ad}G^\ominus$, and then that value combined with $\Delta_{\rm ad}H^\ominus$ to obtain the standard entropy of adsorption. The expression for $(\partial \ln p/\partial T)_\theta$ in this example is independent of the model for the isotherm.

Self-test 25.2 Repeat the calculation using the following data:

T/K	200	210	220	230	240	250
p/kPa	4.32	5.59	7.07	8.80	10.67	12.80

[$-9.0\ \mathrm{kJ\ mol^{-1}}$]

(b) The BET isotherm

If the initial adsorbed layer can act as a substrate for further (for example, physical) adsorption, then, instead of the isotherm levelling off to some saturated value at high pressures, it can be expected to rise indefinitely. The most widely used isotherm dealing with multilayer adsorption was derived by Stephen Brunauer, Paul Emmett, and Edward Teller, and is called the **BET isotherm**:

$$\frac{V}{V_{\rm mon}} = \frac{cz}{(1-z)\{1-(1-c)z\}} \qquad \text{with} \quad z = \frac{p}{p^*} \tag{25.8}$$

In this expression, p^* is the vapour pressure above a layer of adsorbate that is more than one molecule thick and which resembles a pure bulk liquid, $V_{\rm mon}$ is the volume corresponding to monolayer coverage, and c is a constant which is large when the enthalpy of desorption from a monolayer is large compared with the enthalpy of vaporization of the liquid adsorbate:

$$c = \mathrm{e}^{(\Delta_{\rm des}H^\ominus - \Delta_{\rm vap}H^\ominus)/RT} \tag{25.9}$$

Figure 25.19 illustrates the shapes of BET isotherms. They rise indefinitely as the pressure is increased because there is no limit to the amount of material that may condense when multilayer coverage may occur. A BET isotherm is not accurate at all pressures, but it is widely used in industry to determine the surface areas of solids.

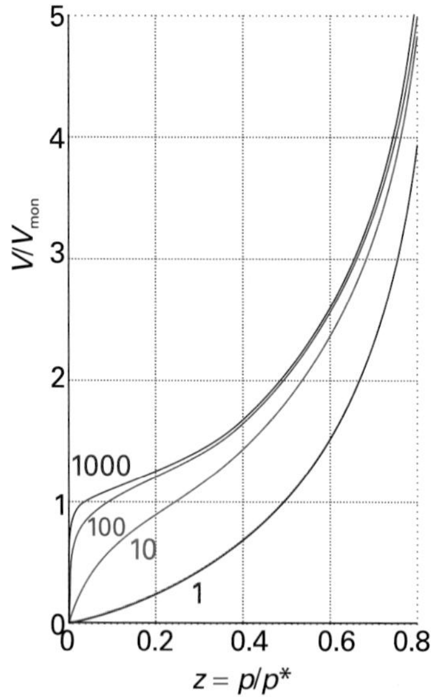

Fig. 25.19 Plots of the BET isotherm for different values of c. The value of $V/V_{\rm mon}$ rises indefinitely because the adsorbate may condense on the covered substrate surface.

Exploration Using eqn 25.8, generate a family of curves showing the dependence of $zV_{\rm mon}/(1-z)V$ on z for different values of c.

Example 25.3 *Using the BET isotherm*

The data below relate to the adsorption of N_2 on rutile (TiO_2) at 75 K. Confirm that they fit a BET isotherm in the range of pressures reported, and determine V_{mon} and c.

p/kPa	0.160	1.87	6.11	11.67	17.02	21.92	27.29
V/mm^3	601	720	822	935	1046	1146	1254

At 75 K, $p^* = 76.0$ kPa. The volumes have been corrected to 1.00 atm and 273 K and refer to 1.00 g of substrate.

Method Equation 25.8 can be reorganized into

$$\frac{z}{(1-z)V} = \frac{1}{cV_{mon}} + \frac{(c-1)z}{cV_{mon}}$$

It follows that $(c-1)/cV_{mon}$ can be obtained from the slope of a plot of the expression on the left against z, and cV_{mon} can be found from the intercept at $z = 0$. The results can then be combined to give c and V_{mon}.

Answer We draw up the following table:

p/kPa	0.160	1.87	6.11	11.67	17.02	21.92	27.29
$10^3 z$	2.11	24.6	80.4	154	224	288	359
$10^4 z/(1-z)(V/\text{mm}^3)$	0.035	0.350	1.06	1.95	2.76	3.53	4.47

These points are plotted in Fig. 25.20. The least squares best line has an intercept at 0.0398, so

$$\frac{1}{cV_{mon}} = 3.98 \times 10^{-6}\ \text{mm}^{-3}$$

The slope of the line is 1.23×10^{-2}, so

$$\frac{c-1}{cV_{mon}} = (1.23 \times 10^{-2}) \times 10^3 \times 10^{-4}\ \text{mm}^{-3} = 1.23 \times 10^{-3}\ \text{mm}^{-3}$$

The solutions of these equations are $c = 310$ and $V_{mon} = 811$ mm^3. At 1.00 atm and 273 K, 811 mm^3 corresponds to 3.6×10^{-5} mol, or 2.2×10^{19} atoms. Because each atom occupies an area of about 0.16 nm^2, the surface area of the sample is about 3.5 m^2.

Self-test 25.3 Repeat the calculation for the following data:

p/kPa	0.160	1.87	6.11	11.67	17.02	21.92	27.29
V/cm^3	235	559	649	719	790	860	950

[370, 615 cm^3]

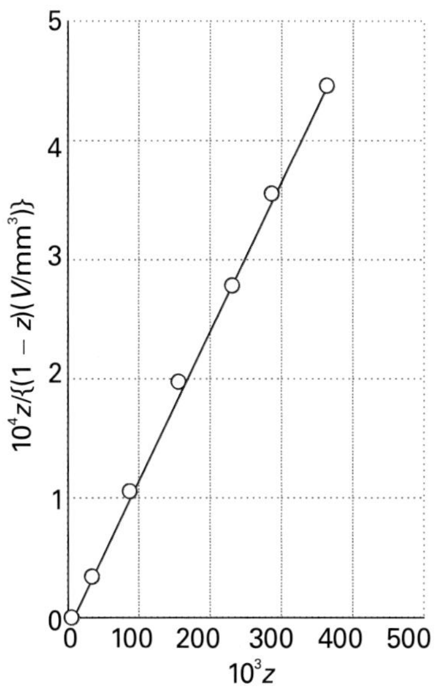

Fig. 25.20 The BET isotherm can be tested, and the parameters determined, by plotting $z/(1-z)V$ against $z = p/p^*$. The data are from Example 25.3.

When $c \gg 1$, the BET isotherm takes the simpler form

$$\frac{V}{V_{mon}} = \frac{1}{1-z} \tag{25.10}$$

This expression is applicable to unreactive gases on polar surfaces, for which $c \approx 10^2$ because $\Delta_{des}H^\ominus$ is then significantly greater than $\Delta_{vap}H^\ominus$ (eqn 25.9). The BET isotherm fits experimental observations moderately well over restricted pressure ranges, but it errs by underestimating the extent of adsorption at low pressures and by overestimating it at high pressures.

(c) Other isotherms

An assumption of the Langmuir isotherm is the independence and equivalence of the adsorption sites. Deviations from the isotherm can often be traced to the failure of these assumptions. For example, the enthalpy of adsorption often becomes less negative as θ increases, which suggests that the energetically most favourable sites are occupied first. Various attempts have been made to take these variations into account. The **Temkin isotherm**,

$$\theta = c_1 \ln(c_2 p) \tag{25.11}$$

where c_1 and c_2 are constants, corresponds to supposing that the adsorption enthalpy changes linearly with pressure. The **Freundlich isotherm**

$$\theta = c_1 p^{1/c_2} \tag{25.12}$$

corresponds to a logarithmic change. This isotherm attempts to incorporate the role of substrate–substrate interactions on the surface (see Problem 25.24).

Different isotherms agree with experiment more or less well over restricted ranges of pressure, but they remain largely empirical. Empirical, however, does not mean useless for, if the parameters of a reasonably reliable isotherm are known, reasonably reliable results can be obtained for the extent of surface coverage under various conditions. This kind of information is essential for any discussion of heterogeneous catalysis.

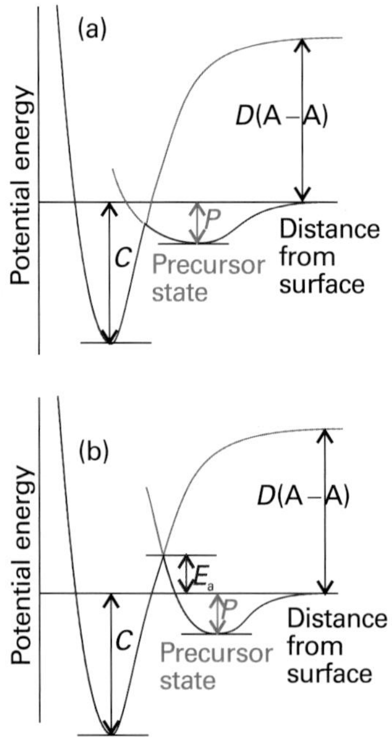

Fig. 25.21 The potential energy profiles for the dissociative chemisorption of an A_2 molecule. In each case, P is the enthalpy of (non-dissociative) physisorption and C that for chemisorption (at $T = 0$). The relative locations of the curves determines whether the chemisorption is (a) not activated or (b) activated.

25.5 The rates of surface processes

The rates of surface processes may be studied by techniques described in Section 25.2 and *Impact* I25.1. Another technique, **second harmonic generation** (SHG), is very important for the study of all types of surfaces, including thin films and liquid–gas interfaces. We saw in Section 20.10 that second harmonic generation is the conversion of an intense, pulsed laser beam to radiation with twice its initial frequency as it passes though a material. In addition to a number of crystals, surfaces are also suitable materials for SHG. Because pulsed lasers are the excitation sources, time-resolved measurements of the kinetics and dynamics of surface processes are possible over timescales as short as femtoseconds.

Figure 25.21 shows how the potential energy of a molecule varies with its distance from the substrate surface. As the molecule approaches the surface its energy falls as it becomes physisorbed into the **precursor state** for chemisorption. Dissociation into fragments often takes place as a molecule moves into its chemisorbed state, and after an initial increase of energy as the bonds stretch there is a sharp decrease as the adsorbate–substrate bonds reach their full strength. Even if the molecule does not fragment, there is likely to be an initial increase of potential energy as the molecule approaches the surface and the bonds adjust.

In most cases, therefore, we can expect there to be a potential energy barrier separating the precursor and chemisorbed states. This barrier, though, might be low, and might not rise above the energy of a distant, stationary particle (as in Fig. 25.21a). In this case, chemisorption is not an activated process and can be expected to be rapid. Many gas adsorptions on clean metals appear to be non-activated. In some cases the barrier rises above the zero axis (as in Fig. 25.21b); such chemisorptions are activated and slower than the non-activated kind. An example is H_2 on copper, which has an activation energy in the region of 20–40 kJ mol^{-1}.

One point that emerges from this discussion is that rates are not good criteria for distinguishing between physisorption and chemisorption. Chemisorption can be fast if the activation energy is small or zero, but it may be slow if the activation energy is large. Physisorption is usually fast, but it can appear to be slow if adsorption is taking place on a porous medium.

(a) The rate of adsorption

The rate at which a surface is covered by adsorbate depends on the ability of the substrate to dissipate the energy of the incoming particle as thermal motion as it crashes on to the surface. If the energy is not dissipated quickly, the particle migrates over the surface until a vibration expels it into the overlying gas or it reaches an edge. The proportion of collisions with the surface that successfully lead to adsorption is called the **sticking probability**, s:

$$s = \frac{\text{rate of adsorption of particles by the surface}}{\text{rate of collision of particles with the surface}} \qquad [25.13]$$

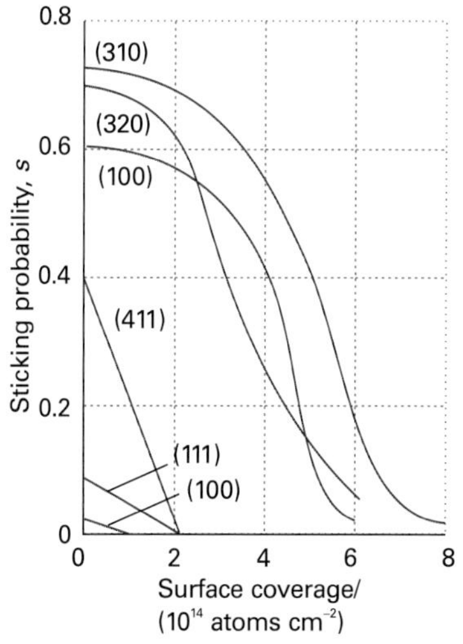

Fig. 25.22 The sticking probability of N_2 on various faces of a tungsten crystal and its dependence on surface coverage. Note the very low sticking probability for the (110) and (111) faces. (Data provided by Professor D.A. King.)

The denominator can be calculated from the kinetic model, and the numerator can be measured by observing the rate of change of pressure.

Values of s vary widely. For example, at room temperature CO has s in the range 0.1–1.0 for several d-metal surfaces, but for N_2 on rhenium $s < 10^{-2}$, indicating that more than a hundred collisions are needed before one molecule sticks successfully. Beam studies on specific crystal planes show a pronounced specificity: for N_2 on tungsten, s ranges from 0.74 on the (320) faces down to less than 0.01 on the (110) faces at room temperature. The sticking probability decreases as the surface coverage increases (Fig. 25.22). A simple assumption is that s is proportional to $1 - \theta$, the fraction uncovered, and it is common to write

$$s = (1 - \theta)s_0 \qquad (25.14)$$

where s_0 is the sticking probability on a perfectly clean surface. The results in the illustration do not fit this expression because they show that s remains close to s_0 until the coverage has risen to about 6×10^{13} molecules cm^{-2}, and then falls steeply. The explanation is probably that the colliding molecule does not enter the chemisorbed state at once, but moves over the surface until it encounters an empty site.

(b) The rate of desorption

Desorption is always activated because the particles have to be lifted from the foot of a potential well. A physisorbed particle vibrates in its shallow potential well, and might shake itself off the surface after a short time. The temperature dependence of the first-order rate of departure can be expected to be Arrhenius-like, with an activation energy for desorption, E_d, comparable to the enthalpy of physisorption:

$$k_d = Ae^{-E_d/RT} \qquad (25.15)$$

Therefore, the half-life for remaining on the surface has a temperature dependence

$$t_{1/2} = \frac{\ln 2}{k_d} = \tau_0 e^{E_d/RT} \qquad \tau_0 = \frac{\ln 2}{A} \qquad (25.16)$$

(Note the positive sign in the exponent.) If we suppose that $1/\tau_0$ is approximately the same as the vibrational frequency of the weak particle–surface bond (about 10^{12} Hz) and $E_d \approx 25$ kJ mol^{-1}, then residence half-lives of around 10 ns are predicted at room temperature. Lifetimes close to 1 s are obtained only by lowering the temperature to about 100 K. For chemisorption, with $E_d = 100$ kJ mol^{-1} and guessing that $\tau_0 = 10^{-14}$ s

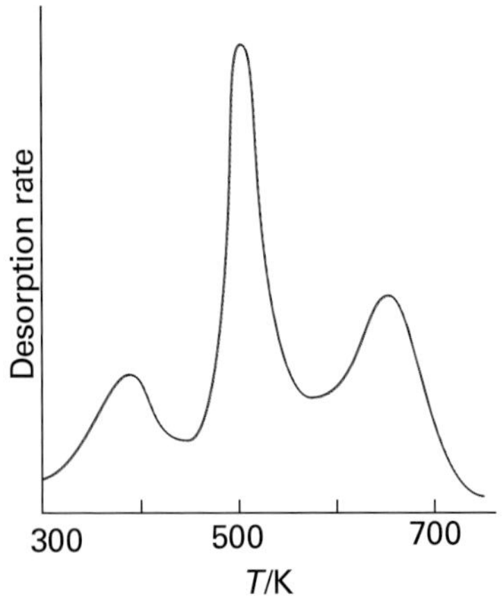

Fig. 25.23 The flash desorption spectrum of H_2 on the (100) face of tungsten. The three peaks indicate the presence of three sites with different adsorption enthalpies and therefore different desorption activation energies. (P.W. Tamm and L.D. Schmidt, *J. Chem. Phys.* **51**, 5352 (1969).)

(because the adsorbate–substrate bond is quite stiff), we expect a residence half-life of about 3×10^3 s (about an hour) at room temperature, decreasing to 1 s at about 350 K.

The desorption activation energy can be measured in several ways. However, we must be guarded in its interpretation because it often depends on the fractional coverage, and so may change as desorption proceeds. Moreover, the transfer of concepts such as 'reaction order' and 'rate constant' from bulk studies to surfaces is hazardous, and there are few examples of strictly first-order or second-order desorption kinetics (just as there are few integral-order reactions in the gas phase too).

If we disregard these complications, one way of measuring the desorption activation energy is to monitor the rate of increase in pressure when the sample is maintained at a series of temperatures, and to attempt to make an Arrhenius plot. A more sophisticated technique is **temperature programmed desorption** (TPD) or **thermal desorption spectroscopy** (TDS). The basic observation is a surge in desorption rate (as monitored by a mass spectrometer) when the temperature is raised linearly to the temperature at which desorption occurs rapidly, but once the desorption has occurred there is no more adsorbate to escape from the surface, so the desorption flux falls again as the temperature continues to rise. The TPD spectrum, the plot of desorption flux against temperature, therefore shows a peak, the location of which depends on the desorption activation energy. There are three maxima in the example shown in Fig. 25.23, indicating the presence of three sites with different activation energies.

In many cases only a single activation energy (and a single peak in the TPD spectrum) is observed. When several peaks are observed they might correspond to adsorption on different crystal planes or to multilayer adsorption. For instance, Cd atoms on tungsten show two activation energies, one of 18 kJ mol^{-1} and the other of 90 kJ mol^{-1}. The explanation is that the more tightly bound Cd atoms are attached directly to the substrate, and the less strongly bound are in a layer (or layers) above the primary overlayer. Another example of a system showing two desorption activation energies is CO on tungsten, the values being 120 kJ mol^{-1} and 300 kJ mol^{-1}. The explanation is believed to be the existence of two types of metal–adsorbate binding site, one involving a simple M—CO bond, the other adsorption with dissociation into individually adsorbed C and O atoms.

(b) Mobility on surfaces

A further aspect of the strength of the interactions between adsorbate and substrate is the mobility of the adsorbate. Mobility is often a vital feature of a catalyst's activity, because a catalyst might be impotent if the reactant molecules adsorb so strongly that they cannot migrate. The activation energy for diffusion over a surface need not be the same as for desorption because the particles may be able to move through valleys between potential peaks without leaving the surface completely. In general, the activation energy for migration is about 10–20 per cent of the energy of the surface–adsorbate bond, but the actual value depends on the extent of coverage. The defect structure of the sample (which depends on the temperature) may also play a dominant role because the adsorbed molecules might find it easier to skip across a terrace than to roll along the foot of a step, and these molecules might become trapped in vacancies in an otherwise flat terrace. Diffusion may also be easier across one crystal face than another, and so the surface mobility depends on which lattice planes are exposed.

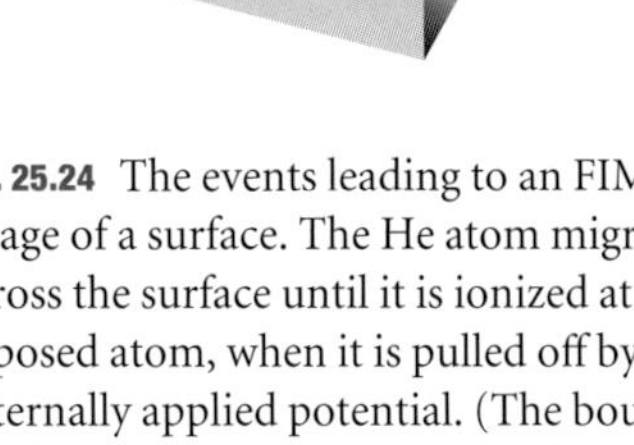

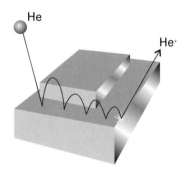

Fig. 25.24 The events leading to an FIM image of a surface. The He atom migrates across the surface until it is ionized at an exposed atom, when it is pulled off by the externally applied potential. (The bouncing motion is due to the intermolecular potential, not gravity!)

Diffusion characteristics of an adsorbate can be examined by using STM to follow the change in surface characteristics or by **field-ionization microscopy** (FIM), which portrays the electrical characteristics of a surface by using the ionization of noble gas atoms to probe the surface (Fig. 25.24). An individual atom is imaged, the temperature is raised, and then lowered after a definite interval. A new image is then recorded, and the new position of the atom measured (Fig. 25.25). A sequence of images shows

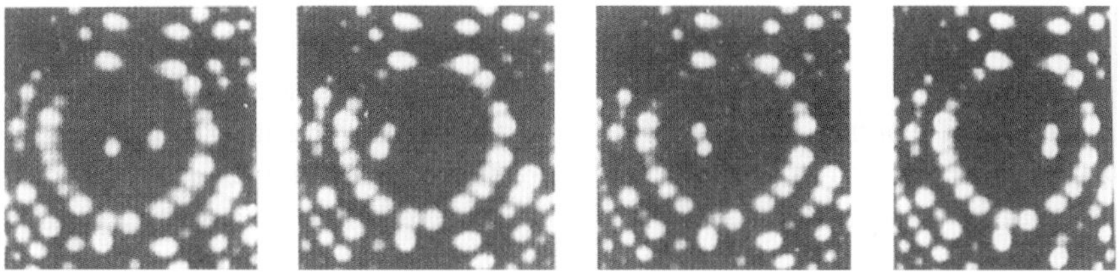

Fig. 25.25 FIM micrographs showing the migration of Re atoms on rhenium during 3 s intervals at 375 K. (Photographs provided by Professor G. Ehrlich.)

that the atom makes a random walk across the surface, and the diffusion coefficient, D, can be inferred from the mean distance, d, travelled in an interval τ by using the two-dimensional random walk expression $d = (D\tau)^{1/2}$. The value of D for different crystal planes at different temperatures can be determined directly in this way, and the activation energy for migration over each plane obtained from the Arrhenius-like expression

$$D = D_0 e^{-E_D/RT} \tag{25.17}$$

where E_D is the activation energy for diffusion. Typical values for W atoms on tungsten have E_D in the range 57–87 kJ mol^{-1} and $D_0 \approx 3.8 \times 10^{-11}$ m^2 s^{-1}. For CO on tungsten, the activation energy falls from 144 kJ mol^{-1} at low surface coverage to 88 kJ mol^{-1} when the coverage is high.

IMPACT ON BIOCHEMISTRY
I25.1 Biosensor analysis

Biosensor analysis is a very sensitive and sophisticated optical technique that is now used routinely to measure the kinetics and thermodynamics of interactions between biopolymers. A biosensor detects changes in the optical properties of a surface in contact with a biopolymer.

The mobility of delocalized valence electrons accounts for the electrical conductivity of metals and these mobile electrons form a **plasma**, a dense gas of charged particles. Bombardment of the plasma by light or an electron beam can cause transient changes in the distribution of electrons, with some regions becoming slightly more dense than others. Coulomb repulsion in the regions of high density causes electrons to move away from each other, so lowering their density. The resulting oscillations in electron density, called **plasmons**, can be excited both in the bulk and on the surface of a metal. Plasmons in the bulk may be visualized as waves that propagate through the solid. A surface plasmon also propagates away from the surface, but the amplitude of the wave, also called an **evanescent wave**, decreases sharply with distance from the surface.

Biosensor analysis is based on the phenomenon of **surface plasmon resonance** (SPR), the absorption of energy from an incident beam of electromagnetic radiation by surface plasmons. Absorption, or 'resonance', can be observed with appropriate choice of the wavelength and angle of incidence of the excitation beam. It is common practice to use a monochromatic beam and to vary the angle of incidence θ (Fig. 25.26). The beam passes through a prism that strikes one side of a thin film of gold or silver. The angle corresponding to light absorption depends on the refractive index of the medium in direct contact with the opposing side of the metallic film. This variation of the resonance angle with the state of the surface arises from the ability of the evanescent wave to interact with material a short distance away from the surface.

As an illustration of biosensor analysis, we consider the association of two polymers, A and B. In a typical experiment, a stream of solution containing a known concentration of A flows above the surface to which B is chemisorbed. Figure 25.27 shows that the kinetics of binding of A to B may be followed by monitoring the time dependence

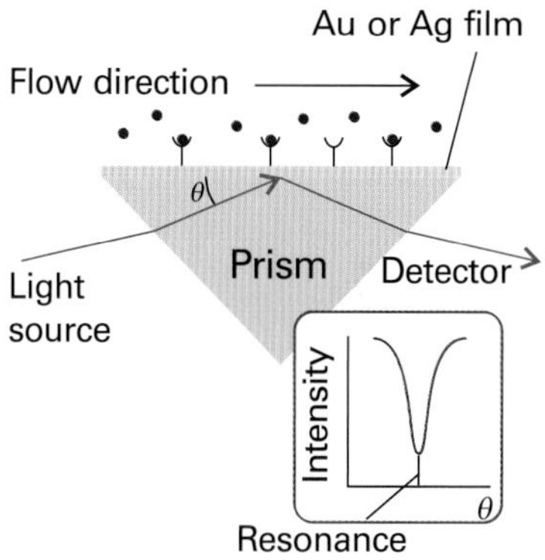

Fig. 25.26 The experimental arrangement for the observation of surface plasmon resonance, as explained in the text.

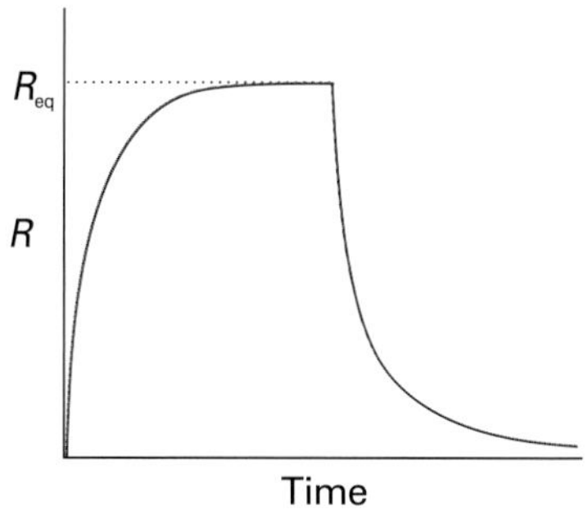

Fig. 25.27 The time dependence of a surface plasmon resonance signal, R, showing the effect of binding of a ligand to a biopolymer adsorbed on to a surface. Binding leads to an increase in R until an equilibrium value, R_{eq}, is obtained. Passing a solution containing no ligand over the surface leads to dissociation and decrease in R.

of the SPR signal, denoted by R, which is typically the shift in resonance angle. The system is normally allowed to reach equilibrium, which is denoted by the plateau in Fig. 25.27. Then, a solution containing no A is flowed above the surface and the AB complex dissociates. Again, analysis of the decay of the SPR signal reveals the kinetics of dissociation of the AB complex.

The equilibrium constant for formation of the AB complex can be measured directly from data of the type displayed in Fig. 25.27. Consider the equilibrium

$$A + B \rightleftharpoons AB \qquad K = \frac{k_{on}}{k_{off}}$$

where k_{on} and k_{off} are the rate constants for formation and dissociation of the AB complex, and K is the equilibrium constant for formation of the AB complex. It follows that

$$\frac{d[AB]}{dt} = k_{on}[A][B] - k_{off}[AB] \tag{25.18}$$

In a typical experiment, the flow rate of A is sufficiently high that $[A] = a_0$ is essentially constant. We can also write $[B] = b_0 - [AB]$ from mass-balance considerations, where b_0 is the total concentration of B. Finally, the SPR signal is often observed to be proportional to [AB]. The maximum value that R can have is $R_{max} \propto b_0$, which would be measured if all B molecules were ligated to A. We may then write

$$\frac{dR}{dt} = k_{on}a_0(R_{max} - R) - k_{off}R = k_{on}a_0R_{max} - (k_{on}a_0 + k_{off})R \tag{25.19}$$

At equilibrium $R = R_{eq}$ and $dR/dt = 0$. It follows that (after some algebra)

$$R_{eq} = R_{max}\left(\frac{a_0K}{a_0K + 1}\right) \tag{25.20}$$

Hence, the value of K can be obtained from measurements of R_{eq} for a series of a_0.

Biosensor analysis has been used in the study of thin films, metal–electrolyte surfaces, Langmuir–Blodgett films, and a number of biopolymer interactions, such as antibody–antigen and protein–DNA interactions. The most important advantage of the technique is its sensitivity; it is possible to measure the adsorption of nanograms of material on to a surface. For biological studies, the main disadvantage is the requirement for immobilization of at least one of the components of the system under study.

Heterogeneous catalysis

A catalyst acts by providing an alternative reaction path with a lower activation energy (Table 25.3). It does not disturb the final equilibrium composition of the system, only the rate at which that equilibrium is approached. In this section we consider heterogeneous catalysis, in which (as mentioned in the introduction to Section 23.5) the catalyst and the reagents are in different phases. For simplicity, we consider only gas/solid systems.

Many catalysts depend on **co-adsorption**, the adsorption of two or more species. One consequence of the presence of a second species may be the modification of the electronic structure at the surface of a metal. For instance, partial coverage of d-metal surfaces by alkali metals has a pronounced effect on the electron distribution and reduces the work function of the metal. Such modifiers can act as promoters (to enhance the action of catalysts) or as poisons (to inhibit catalytic action).

Synoptic table 25.3* Activation energies of catalysed reactions

Reaction	Catalyst	$E_a/(\text{kJ mol}^{-1})$
$2\,HI \rightarrow H_2 + I_2$	None	184
	Au	105
	Pt	59
$2\,NH_3 \rightarrow$	None	350
$N_2 + 3\,H_2$	W	162

* More values are given in the *Data section*.

25.6 Mechanisms of heterogeneous catalysis

Heterogeneous catalysis normally depends on at least one reactant being adsorbed (usually chemisorbed) and modified to a form in which it readily undergoes reaction. This modification often takes the form of a fragmentation of the reactant molecules. In practice, the active phase is dispersed as very small particles of linear dimension less than 2 nm on a porous oxide support. **Shape-selective catalysts**, such as the zeolites (*Impact* I25.2), which have a pore size that can distinguish shapes and sizes at a molecular scale, have high internal specific surface areas, in the range of 100–500 $m^2\,g^{-1}$.

The decomposition of phosphine (PH_3) on tungsten is first-order at low pressures and zeroth-order at high pressures. To account for these observations, we write down a plausible rate law in terms of an adsorption isotherm and explore its form in the limits of high and low pressure. If the rate is supposed to be proportional to the surface coverage and we suppose that θ is given by the Langmuir isotherm, we would write

$$v = k\theta = \frac{kKp}{1+Kp} \qquad (25.21)$$

where p is the pressure of phosphine. When the pressure is so low that $Kp \ll 1$, we can neglect Kp in the denominator and obtain

$$v = kKp \qquad (25.22a)$$

and the decomposition is first-order. When $Kp \gg 1$, we can neglect the 1 in the denominator, whereupon the Kp terms cancel and we are left with

$$v = k \qquad (25.22b)$$

and the decomposition is zeroth-order.

Self-test 25.4 Suggest the form of the rate law for the deuteration of NH_3 in which D_2 adsorbs dissociatively but not extensively (that is, $Kp \ll 1$, with p the partial pressure of D_2), and NH_3 (with partial pressure p') adsorbs at different sites.

[$v = k(Kp)^{1/2}K'p'/(1+K'p')$]

In the **Langmuir–Hinshelwood mechanism** (LH mechanism) of surface-catalysed reactions, the reaction takes place by encounters between molecular fragments and atoms adsorbed on the surface. We therefore expect the rate law to be second-order in the extent of surface coverage:

$$A + B \rightarrow P \qquad v = k\theta_A\theta_B \qquad (25.23)$$

Insertion of the appropriate isotherms for A and B then gives the reaction rate in terms of the partial pressures of the reactants. For example, if A and B follow Langmuir isotherms, and adsorb without dissociation, so that

$$\theta_A = \frac{K_A p_A}{1+K_A p_A + K_B p_B} \qquad \theta_B = \frac{K_B p_B}{1+K_A p_A + K_B p_B} \qquad (25.24)$$

then it follows that the rate law is

$$v = \frac{kK_A K_B p_A p_B}{(1+K_A p_A + K_B p_B)^2} \qquad (25.25)$$

The parameters K in the isotherms and the rate constant k are all temperature-dependent, so the overall temperature dependence of the rate may be strongly non-Arrhenius (in the sense that the reaction rate is unlikely to be proportional to $e^{-E_a/RT}$).

The Langmuir–Hinshelwood mechanism is dominant for the catalytic oxidation of CO to CO_2.

In the **Eley–Rideal mechanism** (ER mechanism) of a surface-catalysed reaction, a gas-phase molecule collides with another molecule already adsorbed on the surface. The rate of formation of product is expected to be proportional to the partial pressure, p_B, of the non-adsorbed gas B and the extent of surface coverage, θ_A, of the adsorbed gas A. It follows that the rate law should be

$$A + B \rightarrow P \qquad v = kp_B\theta_A \tag{25.26}$$

The rate constant, k, might be much larger than for the uncatalysed gas-phase reaction because the reaction on the surface has a low activation energy and the adsorption itself is often not activated.

If we know the adsorption isotherm for A, we can express the rate law in terms of its partial pressure, p_A. For example, if the adsorption of A follows a Langmuir isotherm in the pressure range of interest, then the rate law would be

$$v = \frac{kKp_Ap_B}{1 + Kp_A} \tag{25.27}$$

If A were a diatomic molecule that adsorbed as atoms, we would substitute the isotherm given in eqn 25.6 instead.

According to eqn 25.27, when the partial pressure of A is high (in the sense $Kp_A \gg 1$) there is almost complete surface coverage, and the rate is equal to kp_B. Now the rate-determining step is the collision of B with the adsorbed fragments. When the pressure of A is low ($Kp_A \ll 1$), perhaps because of its reaction, the rate is equal to kKp_Ap_B; now the extent of surface coverage is important in the determination of the rate.

Almost all thermal surface-catalysed reactions are thought to take place by the LH mechanism, but a number of reactions with an ER mechanism have also been identified from molecular beam investigations. For example, the reaction between H(g) and D(ad) to form HD(g) is thought to be by an ER mechanism involving the direct collision and pick-up of the adsorbed D atom by the incident H atom. However, the two mechanisms should really be thought of as ideal limits, and all reactions lie somewhere between the two and show features of each one.

25.7 Catalytic activity at surfaces

Molecular beam reactive scattering (MBRS) studies are able to provide detailed information about catalysed reactions. It has become possible to investigate how the catalytic activity of a surface depends on its structure as well as its composition. For instance, the cleavage of C—H and H—H bonds appears to depend on the presence of steps and kinks, and a terrace often has only minimal catalytic activity. The reaction $H_2 + D_2 \rightarrow 2\,HD$ has been studied in detail. For this reaction, terrace sites are inactive but one molecule in ten reacts when it strikes a step. Although the step itself might be the important feature, it may be that the presence of the step merely exposes a more reactive crystal face (the step face itself). Likewise, the dehydrogenation of hexane to hexene depends strongly on the kink density, and it appears that kinks are needed to cleave C—C bonds. These observations suggest a reason why even small amounts of impurities may poison a catalyst: they are likely to attach to step and kink sites, and so impair the activity of the catalyst entirely. A constructive outcome is that the extent of dehydrogenation may be controlled relative to other types of reactions by seeking impurities that adsorb at kinks and act as specific poisons.

Molecular beam studies can also be used to investigate the details of the reaction process, particularly by using **pulsed beams**, in which the beam is chopped into short

Table 25.4 Chemisorption abilities*

	O_2	C_2H_2	C_2H_4	CO	H_2	CO_2	N_2
Ti, Cr, Mo, Fe	+	+	+	+	+	+	+
Ni, Co	+	+	+	+	+	+	−
Pd, Pt	+	+	+	+	+	−	−
Mn, Cu	+	+	+	+	±	−	−
Al, Au	+	+	+	+	−	−	−
Li, Na, K	+	+	−	−	−	−	−
Mg, Ag, Zn, Pb	+	−	−	−	−	−	−

* +, Strong chemisorption; ±, chemisorption; −, no chemisorption.

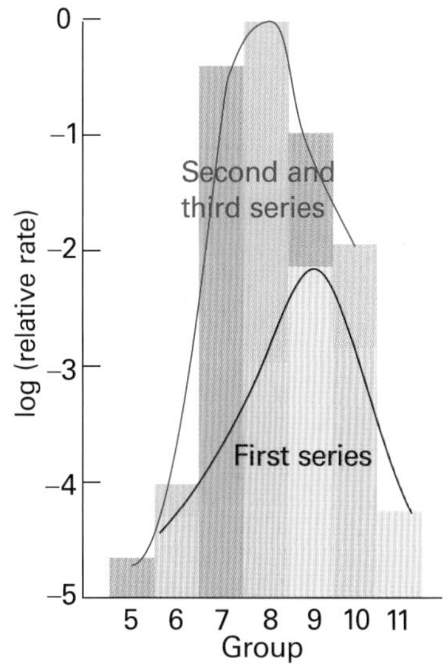

Fig. 25.28 A volcano curve of catalytic activity arises because, although the reactants must adsorb reasonably strongly, they must not adsorb so strongly that they are immobilized. The lower curve refers to the first series of *d*-block metals, the upper curve to the second and third series *d*-block metals. The group numbers relate to the periodic table inside the back cover.

slugs. The angular distribution of the products, for instance, can be used to assess the length of time that a species remains on the surface during the reaction, for a long residence time will result in a loss of memory of the incident beam direction.

The activity of a catalyst depends on the strength of chemisorption as indicated by the 'volcano' curve in Fig. 25.28 (which is so-called on account of its general shape). To be active, the catalyst should be extensively covered by adsorbate, which is the case if chemisorption is strong. On the other hand, if the strength of the substrate–adsorbate bond becomes too great, the activity declines either because the other reactant molecules cannot react with the adsorbate or because the adsorbate molecules are immobilized on the surface. This pattern of behaviour suggests that the activity of a catalyst should initially increase with strength of adsorption (as measured, for instance, by the enthalpy of adsorption) and then decline, and that the most active catalysts should be those lying near the summit of the volcano. Most active metals are those that lie close to the middle of the *d* block.

Many metals are suitable for adsorbing gases, and the general order of adsorption strengths decreases along the series O_2, C_2H_2, C_2H_4, CO, H_2, CO_2, N_2. Some of these molecules adsorb dissociatively (for example, H_2). Elements from the *d* block, such as iron, vanadium, and chromium, show a strong activity towards all these gases, but manganese and copper are unable to adsorb N_2 and CO_2. Metals towards the left of the periodic table (for example, magnesium and lithium) can adsorb (and, in fact, react with) only the most active gas (O_2). These trends are summarized in Table 25.4.

IMPACT ON TECHNOLOGY
I25.2 Catalysis in the chemical industry

Almost the whole of modern chemical industry depends on the development, selection, and application of catalysts (Table 25.5). All we can hope to do in this section is to give a brief indication of some of the problems involved. Other than the ones we consider, these problems include the danger of the catalyst being poisoned by byproducts or impurities, and economic considerations relating to cost and lifetime.

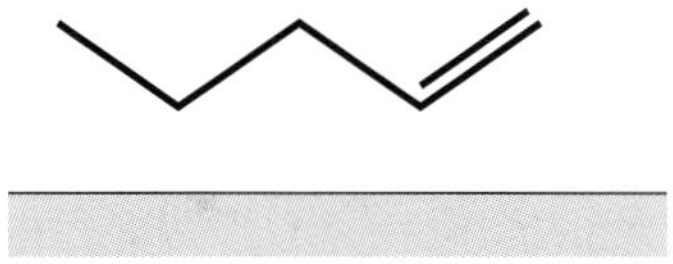

2

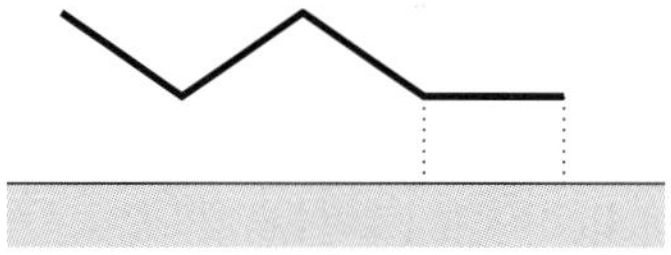

3

An example of catalytic action is found in the hydrogenation of alkenes. The alkene (**2**) adsorbs by forming two bonds with the surface (**3**), and on the same surface there may be adsorbed H atoms. When an encounter occurs, one of the alkene–surface bonds is broken (forming **4** or **5**) and later an encounter with a second H atom releases the fully hydrogenated hydrocarbon, which is the thermodynamically more stable species. The evidence for a two-stage reaction is the appearance of different isomeric alkenes in the mixture. The formation of isomers comes about because, while the hydrocarbon chain is waving about over the surface of the metal, an atom in the chain

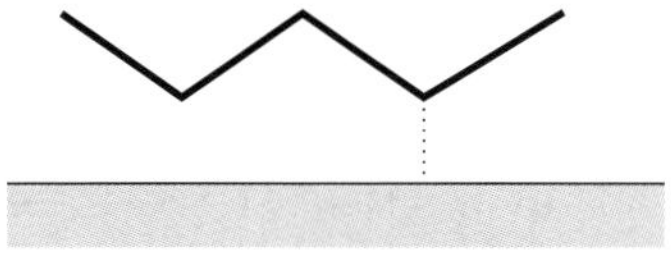

4

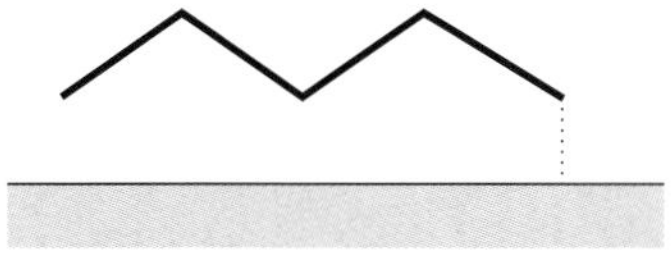

5

Table 25.5 Properties of catalysts

Catalyst	Function	Examples
Metals	Hydrogenation Dehydrogenation	Fe, Ni, Pt, Ag
Semiconducting oxides and sulfides	Oxidation Desulfurization	NiO, ZnO, MgO, Bi_2O_3/MoO_3, MoS_2
Insulating oxides	Dehydration	Al_2O_3, SiO_2, MgO
Acids	Polymerization Isomerization Cracking Alkylation	H_3PO_4, H_2SO_4, SiO_3/Al_2O_3, zeolites

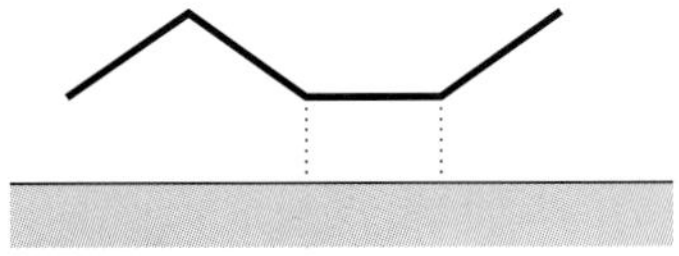

6

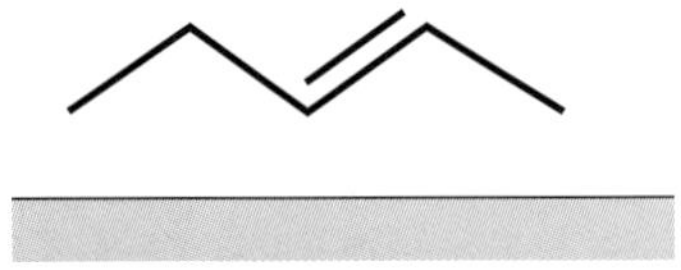

7

might chemisorb again to form (**6**) and then desorb to (**7**), an isomer of the original molecule. The new alkene would not be formed if the two hydrogen atoms attached simultaneously.

A major industrial application of catalytic hydrogenation is to the formation of edible fats from vegetable and animal oils. Raw oils obtained from sources such as the soya bean have the structure $CH_2(OOCR)CH(OOCR')CH_2(OOCR'')$, where R, R′, and R″ are long-chain hydrocarbons with several double bonds. One disadvantage of the presence of many double bonds is that the oils are susceptible to atmospheric oxidation, and therefore are liable to become rancid. The geometrical configuration of the chains is responsible for the liquid nature of the oil, and in many applications a solid fat is at least much better and often necessary. Controlled partial hydrogenation of an oil with a catalyst carefully selected so that hydrogenation is incomplete and so that the chains do not isomerize (finely divided nickel, in fact), is used on a wide scale to produce edible fats. The process, and the industry, is not made any easier by the seasonal variation of the number of double bonds in the oils.

Catalytic oxidation is also widely used in industry and in pollution control. Although in some cases it is desirable to achieve complete oxidation (as in the production of nitric acid from ammonia), in others partial oxidation is the aim. For example, the complete oxidation of propene to carbon dioxide and water is wasteful, but its partial oxidation to propenal (acrolein, $CH_2{=}CHCHO$) is the start of important industrial processes. Likewise, the controlled oxidations of ethene to ethanol, ethanal (acetaldehyde), and (in the presence of acetic acid or chlorine) to chloroethene (vinyl chloride, for the manufacture of PVC), are the initial stages of very important chemical industries.

Some of these oxidation reactions are catalysed by *d*-metal oxides of various kinds. The physical chemistry of oxide surfaces is very complex, as can be appreciated by considering what happens during the oxidation of propene to propenal on bismuth molybdate. The first stage is the adsorption of the propene molecule with loss of a hydrogen to form the propenyl (allyl) radical, $CH_2{=}CHCH_2$. An O atom in the surface can now transfer to this radical, leading to the formation of propenal and its desorption from the surface. The H atom also escapes with a surface O atom, and goes on to form H_2O, which leaves the surface. The surface is left with vacancies and metal ions in lower oxidation states. These vacancies are attacked by O_2 molecules in the overlying gas, which then chemisorb as O_2^- ions, so reforming the catalyst. This sequence of events, which is called the **Mars van Krevelen mechanism**, involves great upheavals of the surface, and some materials break up under the stress.

Many of the small organic molecules used in the preparation of all kinds of chemical products come from oil. These small building blocks of polymers, perfumes, and petrochemicals in general, are usually cut from the long-chain hydrocarbons drawn from the Earth as petroleum. The catalytically induced fragmentation of the long-chain hydrocarbons is called **cracking**, and is often brought about on silica–alumina catalysts. These catalysts act by forming unstable carbocations, which dissociate and rearrange to more highly branched isomers. These branched isomers burn more smoothly and efficiently in internal combustion engines, and are used to produce higher octane fuels.

Catalytic **reforming** uses a dual-function catalyst, such as a dispersion of platinum and acidic alumina. The platinum provides the metal function, and brings about dehydrogenation and hydrogenation. The alumina provides the acidic function, being able to form carbocations from alkenes. The sequence of events in catalytic reforming shows up very clearly the complications that must be unravelled if a reaction as important as this is to be understood and improved. The first step is the attachment of the long-chain hydrocarbon by chemisorption to the platinum. In this process first one and then a second H atom is lost, and an alkene is formed. The alkene migrates to a Brønsted acid site, where it accepts a proton and attaches to the surface as a carbocation. This carbocation can undergo several different reactions. It can break into two, isomerize into a more highly branched form, or undergo varieties of ring-closure. Then the adsorbed molecule loses a proton, escapes from the surface, and migrates (possibly through the gas) as an alkene to a metal part of the catalyst where it is hydrogenated. We end up with a rich selection of smaller molecules which can be withdrawn, fractionated, and then used as raw materials for other products.

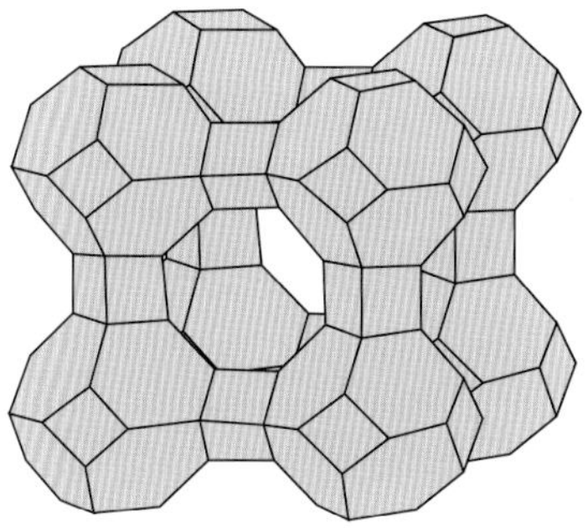

Fig. 25.29 A framework representation of the general layout of the Si, Al, and O atoms in a zeolite material. Each vertex corresponds to a Si or Al atom and each edge corresponds to the approximate location of a O atom. Note the large central pore, which can hold cations, water molecules, or other small molecules.

The concept of a solid surface has been extended with the availability of **microporous materials**, in which the surface effectively extends deep inside the solid. Zeolites are microporous aluminosilicates with the general formula $\{[M^{n+}]_{x/n}\cdot[H_2O]_m\}\{[AlO_2]_x[SiO_2]_y\}^{x-}$, where M^{n+} cations and H_2O molecules bind inside the cavities, or pores, of the Al—O—Si framework (Fig. 25.29). Small neutral molecules, such as CO_2, NH_3, and hydrocarbons (including aromatic compounds), can also adsorb to the internal surfaces and we shall see that this partially accounts for the utility of zeolites as catalysts.

Some zeolites for which $M = H^+$ are very strong acids and catalyse a variety of reactions that are of particular importance to the petrochemical industry. Examples include the dehydration of methanol to form hydrocarbons such as gasoline and other fuels:

$$x\,CH_3OH \xrightarrow{\text{zeolite}} (CH_2)_x + x\,H_2O$$

and the isomerization of *m*-xylene (**8**) to *p*-xylene (**9**). The catalytically important form of these acidic zeolites may be either a Brønsted acid (**10**) or a Lewis acid (**11**). Like enzymes, a zeolite catalyst with a specific compostion and structure is very selective toward certain reactants and products because only molecules of certain sizes can enter

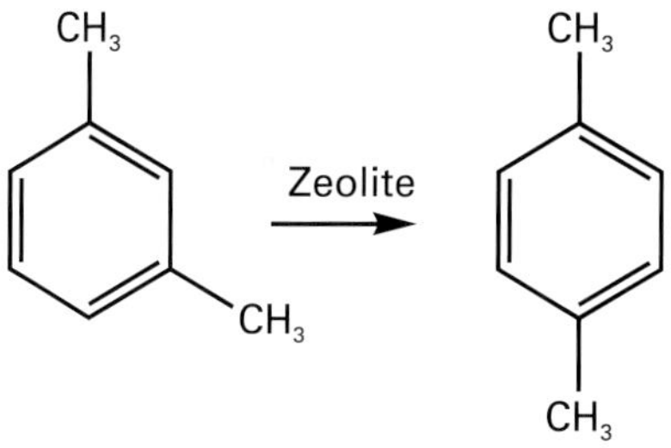

8 **9**

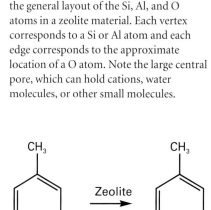

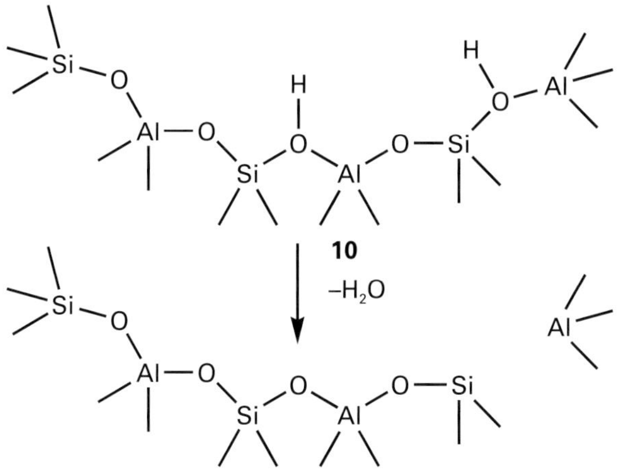

10

11

and exit the pores in which catalysis occurs. It is also possible that zeolites derive their selectivity from the ability to bind and to stabilize only transition states that fit properly in the pores. The analysis of the mechanism of zeolyte catalysis is greatly facilitated by computer simulation of microporous systems, which shows how molecules fit in the pores, migrate through the connecting tunnels, and react at the appropriate active sites.

Processes at electrodes

A special kind of surface is an electrode and the special kind of process that occurs there is the transfer of electrons. Detailed knowledge of the factors that determine the rate of electron transfer at electrodes leads to a better understanding of power production in batteries, and of electron conduction in metals, semiconductors, and nanometre-sized electronic devices. Indeed, the economic consequences of electron transfer reactions are almost incalculable. Most of the modern methods of generating electricity are inefficient, and the development of fuel cells could revolutionize our production and deployment of energy (*Impact* I25.3). Today we produce energy inefficiently to produce goods that then decay by corrosion. Each step of this wasteful sequence could be improved by discovering more about the kinetics of electrochemical processes. Similarly, the techniques of organic and inorganic electrosynthesis, where an electrode is an active component of an industrial process, depend on intimate understanding of the kinetics of electron transfer processes.

As for homogeneous systems (Chapter 24), electron transfer at the surface of an electrode involves electron tunnelling. However, the electrode possesses a nearly infinite number of closely spaced electronic energy levels rather than the small number of discrete levels of a typical complex. Furthermore, specific interactions with the electrode surface give the solute and solvent special properties that can be very different from those observed in the bulk of the solution. For this reason, we begin with a description of the electrode–solution interface. Then, we describe the kinetics of electrode processes by using a largely phenomenological (rather than strictly theoretical) approach that draws on the thermodynamic language inspired by transition state theory.

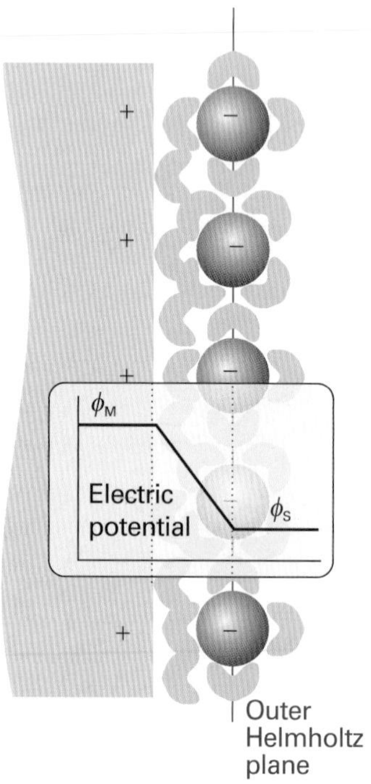

Fig. 25.30 A simple model of the electrode–solution interface treats it as two rigid planes of charge. One plane, the outer Helmholtz plane (OHP), is due to the ions with their solvating molecules and the other plane is that of the electrode itself. The plot shows the dependence of the electric potential with distance from the electrode surface according to this model. Between the electrode surface and the OHP, the potential varies linearly from φ_M, the value in the metal, to φ_S, the value in the bulk of the solution.

25.8 The electrode–solution interface

The most primitive model of the boundary between the solid and liquid phases is as an **electrical double layer**, which consists of a sheet of positive charge at the surface of the electrode and a sheet of negative charge next to it in the solution (or vice versa). We shall see that this arrangement creates an electrical potential difference, called the **Galvani potential difference**, between the bulk of the metal electrode and the bulk of the solution. More sophisticated models for the electrode–solution interface attempt to describe the gradual changes in the structure of the solution between two extremes: the charged electrode surface and the bulk of the solution.

(a) The structure of the interface

A more detailed picture of the interface can be constructed by speculating about the arrangement of ions and electric dipoles in the solution. In the **Helmholtz layer model** of the interface the solvated ions arrange themselves along the surface of the electrode but are held away from it by their hydration spheres (Fig. 25.30). The location of the sheet of ionic charge, which is called the **outer Helmholtz plane** (OHP), is identified as the plane running through the solvated ions. In this simple model, the electrical potential changes linearly within the layer bounded by the electrode surface on one side and the OHP on the other (see Exercise 25.15a). In a refinement of this model, ions that have discarded their solvating molecules and have become attached to the electrode surface by chemical bonds are regarded as forming the **inner Helmholtz plane** (IHP). The Helmholtz layer model ignores the disrupting effect of thermal motion, which tends to break up and disperse the rigid outer plane of charge. In the **Gouy–Chapman model** of the **diffuse double layer**, the disordering effect of thermal motion is taken into account in much the same way as the Debye–Hückel model describes the ionic atmosphere of an ion (Section 5.9) with the latter's single central ion replaced by an infinite, plane electrode.

Figure 25.31 shows how the local concentrations of cations and anions differ in the Gouy–Chapman model from their bulk concentrations. Ions of opposite charge cluster close to the electrode and ions of the same charge are repelled from it. The modification of the local concentrations near an electrode implies that it might be misleading to use activity coefficients characteristic of the bulk to discuss the thermodynamic properties of ions near the interface. This is one of the reasons why measurements of the dynamics of electrode processes are almost always done using a large excess of supporting electrolyte (for example, a 1 M solution of a salt, an acid, or a base). Under such conditions, the activity coefficients are almost constant because the inert ions dominate the effects of local changes caused by any reactions taking place. The use of a concentrated solution also minimizes ion migration effects.

Neither the Helmholtz nor the Gouy–Chapman model is a very good representation of the structure of the double layer. The former overemphasizes the rigidity of the local solution; the latter underemphasizes its structure. The two are combined in the **Stern model**, in which the ions closest to the electrode are constrained into a rigid Helmholtz plane while outside that plane the ions are dispersed as in the Gouy–Chapman model (Fig. 25.32). Yet another level of sophistication is found in the **Grahame model**, which adds an inner Helmholtz plane to the Stern model.

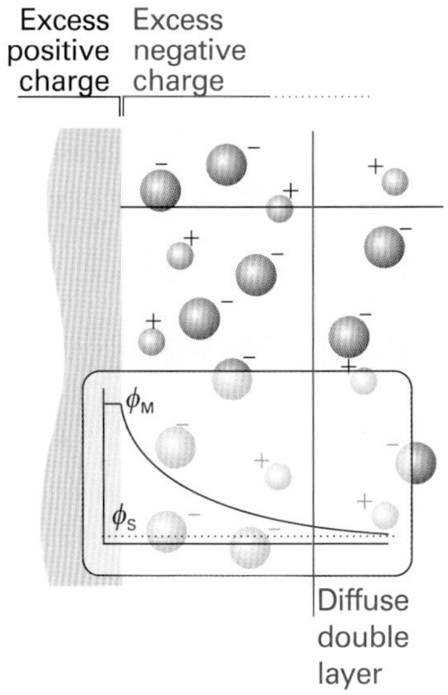

Fig. 25.31 The Gouy–Chapman model of the electrical double layer treats the outer region as an atmosphere of counter-charge, similar to the Debye–Hückel theory of ion atmospheres. The plot of electrical potential against distance from the electrode surface shows the meaning of the diffuse double layer (see text for details).

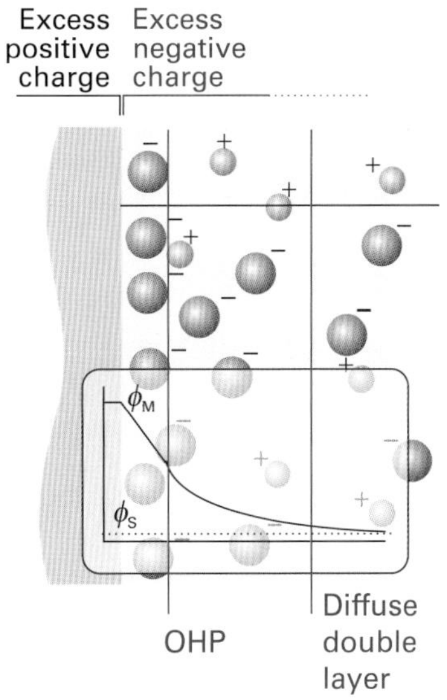

Fig. 25.32 A representation of the Stern model of the electrode–solution interface. The model incorporates the idea of an outer Helmholtz plane near the electrode surface and of a diffuse double layer further away from the surface.

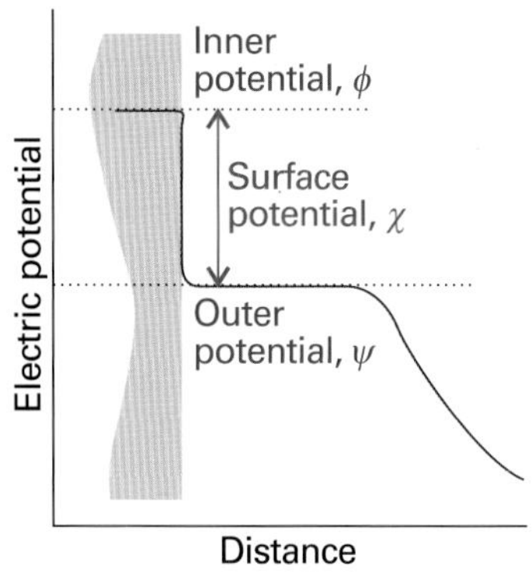

Fig. 25.33 The variation of potential with distance from an electrode that has been separated from the electrolyte solution without there being an adjustment of charge. A similar diagram applies to the separated solution.

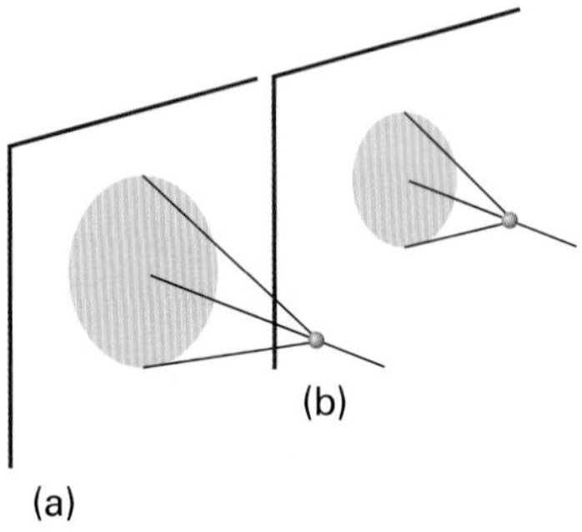

Fig. 25.34 The origin of the distance-independence of the outer potential. (a) Far from the electrode, a point charge experiences a potential arising from a wide area but each contribution is weak. (b) Close to the electrode, the point charge experiences a potential arising from a small area but each contribution is strong. Provided the point charge is in a certain range of values (and, specifically, where image charge effects can be ignored) the potential it experiences is largely independent of distance.

(b) The electric potential at the interface

The potential at the interface can be analysed by imagining the separation of the electrode from the solution, but with the charges of the metal and the solution frozen in position. A positive test charge at great distances from the isolated electrode experiences a Coulomb potential that varies inversely with distance (Fig. 25.33). As the test charge approaches the electrode, which can be a metal or membrane electrode, it enters a region where the potential varies more slowly. This change in behaviour can be traced to the fact that the surface charge is not point-like but is spread over an area. At about 100 nm from the surface the potential varies only slightly with distance because the closer the point of observation is to the surface, although the potential from a given region of charge is stronger, a smaller area of surface is sampled (Fig. 25.34). The potential in this region is called the **outer potential**, ψ. As the test charge is taken through the skin of electrons on the surface of the electrode, the potential it experiences changes until the probe reaches the inner, bulk metal environment, where the potential is called the **inner potential**, ϕ. The difference between the inner and outer potentials is called the **surface potential**, χ.

A similar sequence of changes of potential is observed as a positive test charge is brought up to and through the solution surface. The potential changes to its outer value as the charge approaches the charged medium, then to its inner value as the probe is taken into the bulk.

Now consider bringing the electrode and solution back together again but without any change of charge distribution. The potential difference between points in the bulk metal and the bulk solution is the Galvani potential difference, $\Delta\phi$. Apart from a constant, this Galvani potential difference is the electrode potential that was discussed in Chapter 7. We shall ignore the constant, which cannot be measured anyway, and identify changes in $\Delta\phi$ with changes in electrode potential (see *Further information* 25.1 for a quantitative treatment).

25.9 The rate of charge transfer

Because an electrode reaction is heterogeneous, it is natural to express its rate as the flux of products, the amount of material produced over a region of the electrode surface in an interval of time divided by the area of the region and the duration of the interval.

(a) The rate laws

A first-order heterogeneous rate law has the form

$$\text{Product flux} = k[\text{species}] \quad (25.28)$$

where [species] is the molar concentration of the relevant species in solution close to the electrode, just outside the double layer. The rate constant has dimensions of length/time (with units, for example, of centimetres per second, cm s^{-1}). If the molar concentrations of the oxidized and reduced materials outside the double layer are [Ox] and [Red], respectively, then the rate of reduction of Ox, v_{Ox}, is

$$v_{\text{Ox}} = k_c[\text{Ox}] \quad (25.29\text{a})$$

and the rate of oxidation of Red, v_{Red}, is

$$v_{\text{Red}} = k_a[\text{Red}] \quad (25.29\text{b})$$

(The notation k_c and k_a is justified below.)

Now consider a reaction at the electrode in which an ion is reduced by the transfer of a single electron in the rate-determining step. For instance, in the deposition of cadmium only one electron is transferred in the rate-determining step even though overall the deposition involves the transfer of two electrons.

The net current density at the electrode is the difference between the current densities arising from the reduction of Ox and the oxidation of Red. Because the redox processes at the electrode involve the transfer of one electron per reaction event, the current densities, j, arising from the redox processes are the rates (as expressed above) multiplied by the charge transferred per mole of reaction, which is given by Faraday's constant. Therefore, there is a **cathodic current density** of magnitude

$$j_c = Fk_c[\mathrm{Ox}] \qquad \text{for} \quad \mathrm{Ox} + e^- \rightarrow \mathrm{Red} \tag{25.30a}$$

arising from the reduction (because, as we saw in Chapter 7, the cathode is the site of reduction). There is also an opposing **anodic current density** of magnitude

$$j_a = Fk_a[\mathrm{Red}] \qquad \text{for} \quad \mathrm{Red} \rightarrow \mathrm{Ox} + e^- \tag{25.30b}$$

arising from the oxidation (because the anode is the site of oxidation). The net current density at the electrode is the difference

$$j = j_a - j_c = Fk_a[\mathrm{Red}] - Fk_c[\mathrm{Ox}] \tag{25.30c}$$

Note that, when $j_a > j_c$, so that $j > 0$, the current is anodic (Fig. 25.35a); when $j_c > j_a$, so that $j < 0$, the current is cathodic (Fig. 25.35b).

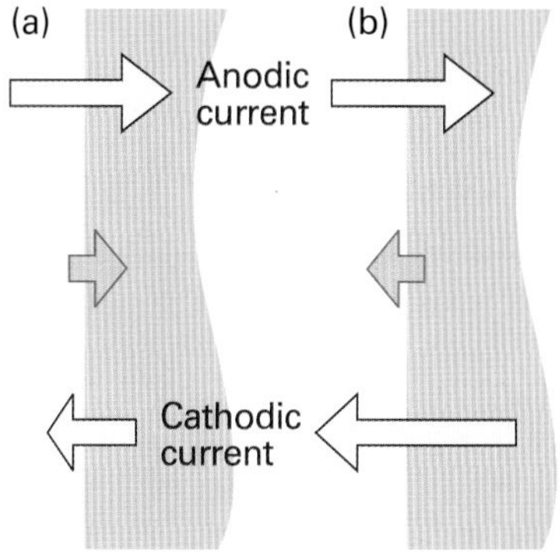

Fig. 25.35 The net current density is defined as the difference between the cathodic and anodic contributions. (a) When $j_a > j_c$, the net current is anodic, and there is a net oxidation of the species in solution. (b) When $j_c > j_a$, the net current is cathodic, and the net process is reduction.

(b) The activation Gibbs energy

If a species is to participate in reduction or oxidation at an electrode, it must discard any solvating molecules, migrate through the electrode–solution interface, and adjust its hydration sphere as it receives or discards electrons. Likewise, a species already at the inner plane must be detached and migrate into the bulk. Because both processes are activated, we can expect to write their rate constants in the form suggested by transition state theory (Section 24.4) as

$$k = B\mathrm{e}^{-\Delta^{\ddagger}G/RT} \tag{25.31}$$

where $\Delta^{\ddagger}G$ is the activation Gibbs energy and B is a constant with the same dimensions as k.

When eqn 25.31 is inserted into eqn 25.30 we obtain

$$j = FB_a[\mathrm{Red}]\mathrm{e}^{-\Delta^{\ddagger}G_a/RT} - FB_c[\mathrm{Ox}]\mathrm{e}^{-\Delta^{\ddagger}G_c/RT} \tag{25.32}$$

This expression allows the activation Gibbs energies to be different for the cathodic and anodic processes. That they are different is the central feature of the remaining discussion.

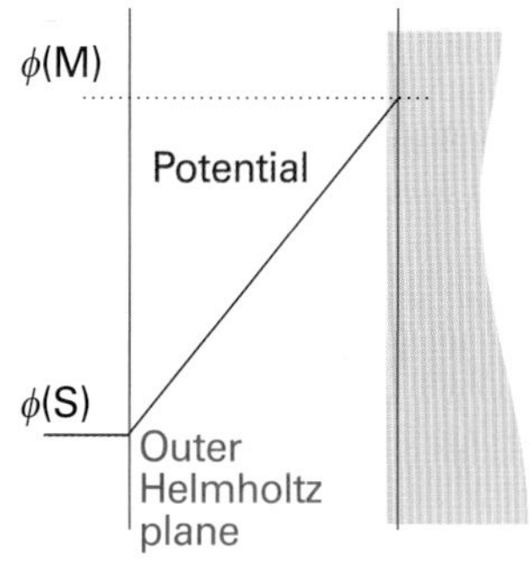

Fig. 25.36 The potential, ϕ, varies linearly between two plane parallel sheets of charge, and its effect on the Gibbs energy of the transition state depends on the extent to which the latter resembles the species at the inner or outer planes.

(c) The Butler–Volmer equation

Now we relate j to the Galvani potential difference, which varies across the electrode–solution interface as shown schematically in Fig. 25.36.

Consider the reduction reaction, $\mathrm{Ox} + e^- \rightarrow \mathrm{Red}$, and the corresponding reaction profile. If the transition state of the activated complex is product-like (as represented by the peak of the reaction profile being close to the electrode in Fig. 25.37), the activation Gibbs energy is changed from $\Delta^{\ddagger}G_c(0)$, the value it has in the absence of a potential difference across the double layer, to

$$\Delta^{\ddagger}G_c = \Delta^{\ddagger}G_c(0) + F\Delta\phi \tag{25.33a}$$

Thus, if the electrode is more positive than the solution, $\Delta\phi > 0$, then more work has to be done to form an activated complex from Ox; in this case the activation Gibbs energy is increased. If the transition state is reactant-like (represented by the peak of

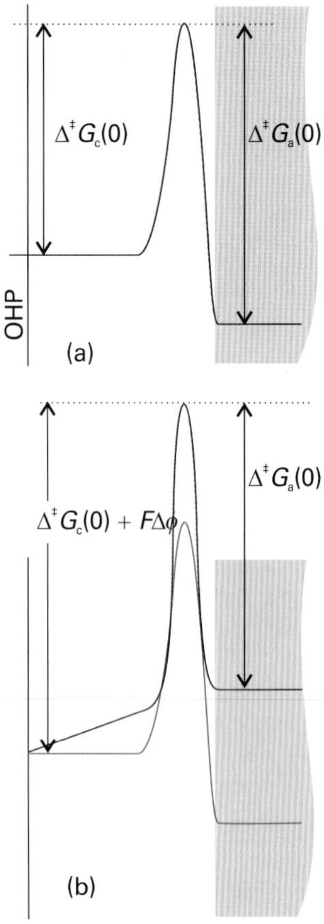

Fig. 25.37 When the transition state resembles a species that has undergone reduction, the activation Gibbs energy for the anodic current is almost unchanged, but the full effect applies to the cathodic current. (a) Zero potential difference; (b) nonzero potential difference.

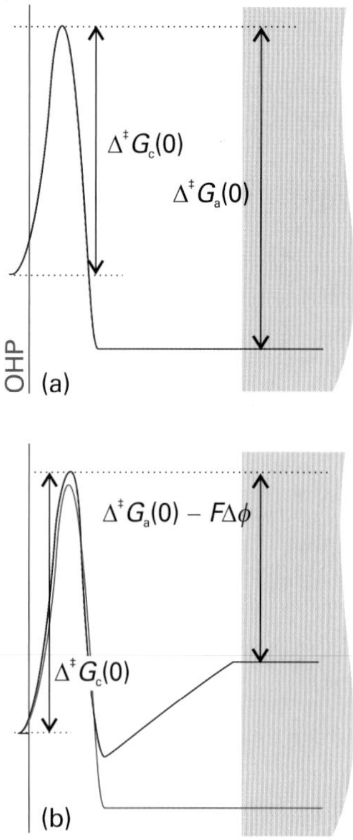

Fig. 25.38 When the transition state resembles a species that has undergone oxidation, the activation Gibbs energy for the cathodic current is almost unchanged but the activation Gibbs energy for the anodic current is strongly affected. (a) Zero potential difference; (b) nonzero potential difference.

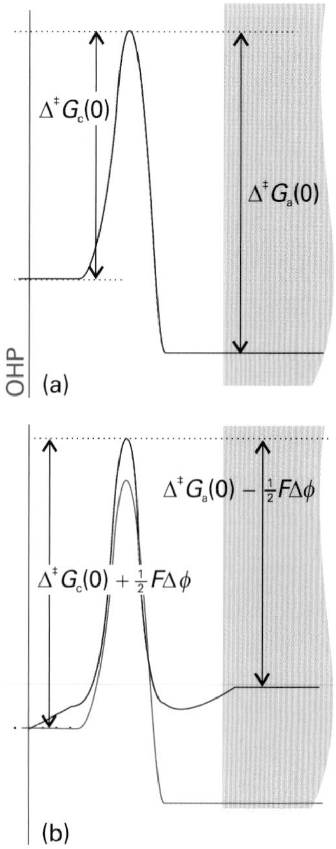

Fig. 25.39 When the transition state is intermediate in its resemblance to reduced and oxidized species, as represented here by a peak located at an intermediate position as measured by α (with $0 < \alpha < 1$), both activation Gibbs energies are affected; here, $\alpha \approx 0.5$. (a) Zero potential difference; (b) nonzero potential difference.

the reaction profile being close to the outer plane of the double-layer in Fig. 25.38), then $\Delta^{\ddagger}G_c$ is independent of $\Delta\phi$. In a real system, the transition state has an intermediate resemblance to these extremes (Fig. 25.39) and the activation Gibbs energy for reduction may be written as

$$\Delta^{\ddagger}G_c = \Delta^{\ddagger}G_c(0) + \alpha F\Delta\phi \qquad (25.33b)$$

The parameter α is called the (cathodic) **transfer coefficient**, and lies in the range 0 to 1. Experimentally, α is often found to be about 0.5.

Now consider the oxidation reaction, Red + e^- → Ox and its reaction profile. Similar remarks apply. In this case, Red discards an electron to the electrode, so the extra work is zero if the transition state is reactant-like (represented by a peak close to the electrode). The extra work is the full $-F\Delta\phi$ if it resembles the product (the peak close to the outer plane). In general, the activation Gibbs energy for this anodic process is

$$\Delta^{\ddagger}G_a = \Delta^{\ddagger}G_a(0) - (1 - \alpha)F\Delta\phi \qquad (25.34)$$

The two activation Gibbs energies can now be inserted in place of the values used in eqn 25.32 with the result that

$$j = FB_a[\text{Red}]e^{-\Delta^\ddagger G_a(0)/RT}e^{(1-\alpha)F\Delta\phi/RT} - FB_c[\text{Ox}]e^{-\Delta^\ddagger G_c(0)/RT}e^{-\alpha F\Delta\phi/RT} \quad (25.35)$$

This is an explicit, if complicated, expression for the net current density in terms of the potential difference.

The appearance of eqn 25.35 can be simplified. First, in a purely cosmetic step we write

$$f = \frac{F}{RT} \quad [25.36]$$

Next, we identify the individual cathodic and anodic current densities:

$$\left.\begin{aligned} j_a &= FB_a[\text{Red}]e^{-\Delta^\ddagger G_a(0)/RT}e^{(1-\alpha)f\Delta\phi} \\ j_c &= FB_c[\text{Ox}]e^{-\Delta^\ddagger G_c(0)/RT}e^{-\alpha f\Delta\phi} \end{aligned}\right\} j = j_a - j_c \quad (25.37)$$

Illustration 25.1 *Calculating the current density 1*

To calculate the change in cathodic current density at an electrode when the potential difference changes from $\Delta\phi'$ and $\Delta\phi$, we use eqn 25.37 to express the ratio of cathodic current densities j_c' and j_c

$$\frac{j_c'}{j_c} = e^{-\alpha f(\Delta\phi' - \Delta\phi)}$$

When $\Delta\phi' - \Delta\phi = 1.0$ V, $T = 298$ K, and $\alpha = \frac{1}{2}$ (a typical value), we obtain

$$\alpha f \times (\Delta\phi' - \Delta\phi) = \frac{\frac{1}{2} \times (9.6485 \times 10^4 \text{ C mol}^{-1}) \times (1.0 \text{ V})}{(8.3145 \text{ J K}^{-1} \text{ mol}^{-1}) \times (298 \text{ K})} = \frac{9.6485 \times 10^4 \times 1.0}{2 \times 8.3145 \times 298}$$

Hence (after using 1 J = 1 VC),

$$\frac{j_c'}{j_c} = e^{-\frac{9.6485\times 10^4 \times 1.0}{2\times 8.3145\times 298}} = 4 \times 10^{-9}$$

This huge change in current density, by a factor of a billion, occurs for a very mild and easily applied change of conditions. We can appreciate why the change is so great by realizing that a change of potential difference by 1 V changes the activation Gibbs energy by (1 V) × F, or about 50 kJ mol^{-1}, which has an enormous effect on the rates.

Self-test 25.5 Calculate the change in anodic current density under the same circumstances. $[j_a'/j_a = 3 \times 10^8]$

If the cell is balanced against an external source, the Galvani potential difference, $\Delta\phi$, can be identified as the (zero-current) electrode potential, E, and we can write

$$\begin{aligned} j_a &= FB_a[\text{Red}]e^{-\Delta^\ddagger G_a(0)/RT}e^{(1-\alpha)fE} \\ j_c &= FB_c[\text{Ox}]e^{-\Delta^\ddagger G_c(0)/RT}e^{-\alpha fE} \end{aligned} \quad (25.38)$$

When these equations apply, there is no net current at the electrode (as the cell is balanced), so the two current densities must be equal. From now on we denote them both as j_0, which is called the **exchange current density**.

When the cell is producing current (that is, when a load is connected between the electrode being studied and a second counter electrode) the electrode potential changes

Comment 25.1

Here we are assuming that we can identify the Galvani potential difference and the zero-current electrode potential. As explained earlier, they differ by a constant amount, which may be regarded as absorbed into the constant B.

from its zero-current value, E, to a new value, E', and the difference is the electrode's **overpotential**, η:

$$\eta = E' - E \qquad [25.39]$$

Hence, $\Delta\phi$ changes to $\Delta\phi = E + \eta$ and the two current densities become

$$j_a = j_0 e^{(1-\alpha)f\eta} \qquad j_c = j_0 e^{-\alpha f\eta} \qquad (25.40)$$

Then from eqn 25.32 we obtain the **Butler–Volmer equation**:

$$j = j_0\{e^{(1-\alpha)f\eta} - e^{-\alpha f\eta}\} \qquad (25.41)$$

This equation is the basis of all that follows.

(d) The low overpotential limit

When the overpotential is so small that $f\eta \ll 1$ (in practice, η less than about 0.01 V) the exponentials in eqn 25.41 can be expanded by using $e^x = 1 + x + \cdots$ to give

$$j = j_0\{1 + (1-\alpha)f\eta + \cdots - (1 - \alpha f\eta + \cdots)\} \approx j_0 f\eta \qquad (25.42)$$

This equation shows that the current density is proportional to the overpotential, so at low overpotentials the interface behaves like a conductor that obeys Ohm's law. When there is a small positive overpotential the current is anodic ($j > 0$ when $\eta > 0$), and when the overpotential is small and negative the current is cathodic ($j < 0$ when $\eta < 0$). The relation can also be reversed to calculate the potential difference that must exist if a current density j has been established by some external circuit:

$$\eta = \frac{RTj}{Fj_0} \qquad (25.43)$$

The importance of this interpretation will become clear below.

Illustration 25.2 *Calculating the current density 2*

The exchange current density of a $Pt(s)|H_2(g)|H^+(aq)$ electrode at 298 K is 0.79 mA cm^{-2}. Therefore, the current density when the overpotential is +5.0 mV is obtained by using eqn 25.42 and $f = F/RT = 1/(25.69\ \text{mV})$:

$$j = j_0 f\eta = \frac{(0.79\ \text{mA cm}^{-2}) \times (5.0\ \text{mV})}{25.69\ \text{mV}} = 0.15\ \text{mA cm}^{-2}$$

The current through an electrode of total area 5.0 cm^2 is therefore 0.75 mA.

Self-test 25.6 What would be the current at pH = 2.0, the other conditions being the same? [−18 mA (cathodic)]

(e) The high overpotential limit

When the overpotential is large and positive (in practice, $\eta \geq 0.12$ V), corresponding to the electrode being the anode in electrolysis, the second exponential in eqn 25.41 is much smaller than the first, and may be neglected. Then

$$j = j_0 e^{(1-\alpha)f\eta} \qquad (25.44)$$

so

$$\ln j = \ln j_0 + (1-\alpha)f\eta \qquad (25.45)$$

When the overpotential is large but negative (in practice, $\eta \le -0.12$ V), corresponding to the cathode in electrolysis, the first exponential in eqn 25.41 may be neglected. Then

$$j = -j_0 e^{-\alpha f \eta} \qquad (25.46)$$

so

$$\ln(-j) = \ln j_0 - \alpha f \eta \qquad (25.47)$$

The plot of the logarithm of the current density against the overpotential is called a **Tafel plot**. The slope gives the value of α and the intercept at $\eta = 0$ gives the exchange current density.

The experimental arrangement used for a Tafel plot is shown in Fig. 25.40. A similar arrangement is typical of all kinds of electrochemical rate measurements. The current-carrying electrodes are the **working electrode**, the electrode of interest, and the **counter electrode**, which is necessary to complete the electrical circuit. The current flowing through them is controlled externally. If the area of the working electrode is A and the current is I, the current density across its surface is I/A. The potential difference across the interface cannot be measured directly, but the potential of the working electrode relative to a third electrode, the **reference electrode**, can be measured with a high impedance voltmeter, and no current flows in that half of the circuit. The reference electrode is in contact with the solution close to the working electrode through a 'Luggin capillary', which helps to eliminate any ohmic potential difference that might arise accidentally. Changing the current flowing through the working circuit causes a change of potential of the working electrode, and that change is measured with the voltmeter. The overpotential is then obtained by taking the difference between the potentials measured with and without a flow of current through the working circuit.

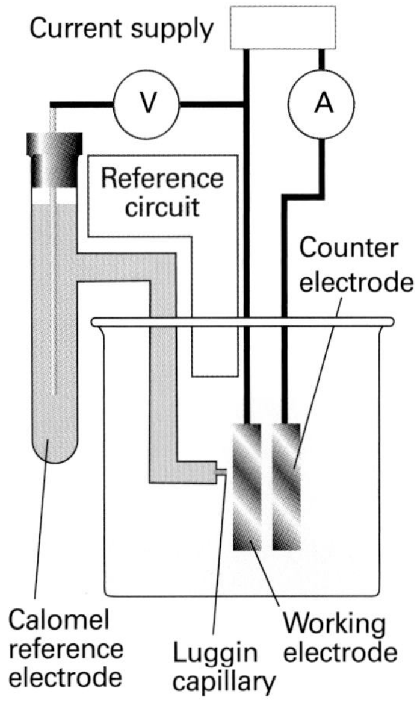

Fig. 25.40 The general arrangement for electrochemical rate measurements. The external source establishes a current between the working electrode and the counter electrode, and its effect on the potential difference of either of them relative to the reference electrode is observed. No current flows in the reference circuit.

Example 25.4 *Interpreting a Tafel plot*

The data below refer to the anodic current through a platinum electrode of area 2.0 cm^2 in contact with an Fe^{3+},Fe^{2+} aqueous solution at 298 K. Calculate the exchange current density and the transfer coefficient for the electrode process.

η/mV	50	100	150	200	250
I/mA	8.8	25.0	58.0	131	298

Method The anodic process is the oxidation $Fe^{2+}(aq) \rightarrow Fe^{3+}(aq) + e^-$. To analyse the data, we make a Tafel plot (of $\ln j$ against η) using the anodic form (eqn 25.45). The intercept at $\eta = 0$ is $\ln j_0$ and the slope is $(1 - \alpha)f$.

Answer Draw up the following table:

η/mV	50	100	150	200	250
$j/(\text{mA cm}^{-2})$	4.4	12.5	29.0	65.5	149
$\ln(j/(\text{mA cm}^{-2}))$	1.48	2.53	3.37	4.18	5.00

The points are plotted in Fig. 25.41. The high overpotential region gives a straight line of intercept 0.88 and slope 0.0165. From the former it follows that $\ln(j_0/(\text{mA cm}^{-2})) = 0.88$, so $j_0 = 2.4$ mA cm^{-2}. From the latter,

$$(1 - \alpha)\frac{F}{RT} = 0.0165 \text{ mV}^{-1}$$

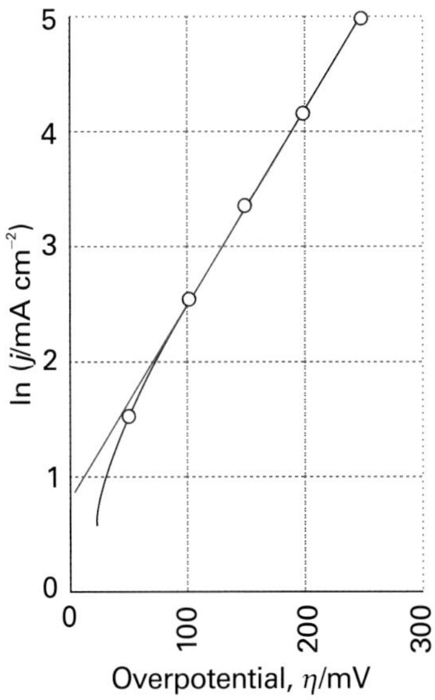

Fig. 25.41 A Tafel plot is used to measure the exchange current density (given by the extrapolated intercept at $\eta = 0$) and the transfer coefficient (from the slope). The data are from Example 25.4.

so $\alpha = 0.58$. Note that the Tafel plot is nonlinear for $\eta < 100$ mV; in this region $\alpha f \eta = 2.3$ and the approximation that $\alpha f \eta \gg 1$ fails.

Self-test 25.7 Repeat the analysis using the following cathodic current data:

η/mV	−50	−100	−150	−200	−250	−300
I/mA	−0.3	−1.5	−6.4	−27.6	−118.6	−510

[$\alpha = 0.75$, $j_0 = 0.041$ mA cm^{-2}]

Some experimental values for the Butler–Volmer parameters are given in Table 25.6. From them we can see that exchange current densities vary over a very wide range. For example, the N_2,N_3^- couple on platinum has $j_0 = 10^{-76}$ A cm^{-2}, whereas the H^+,H_2 couple on platinum has $j_0 = 8 \times 10^{-4}$ A cm^{-2}, a difference of 73 orders of magnitude. Exchange currents are generally large when the redox process involves no bond breaking (as in the $[Fe(CN)_6]^{3-},[Fe(CN)_6]^{4-}$ couple) or if only weak bonds are broken (as in Cl_2,Cl^-). They are generally small when more than one electron needs to be transferred, or when multiple or strong bonds are broken, as in the N_2,N_3^- couple and in redox reactions of organic compounds.

25.10 Voltammetry

The kinetics of electrode processes can be studied by **voltammetry**, in which the current is monitored as the potential of the electrode is changed, and by **chronopotentiometry**, in which the potential is monitored as the current flow is changed. Voltammetry may also be used to identify species present in solution and to determine their concentration.

Before we describe voltammetry in detail, we need to understand how electrode potentials vary with current. Electrodes with potentials that change only slightly when a current passes through them are classified as **non-polarizable**. Those with strongly current-dependent potentials are classified as **polarizable**. From the linearized equation (eqn 25.43) it is clear that the criterion for low polarizability is high exchange current density (so η may be small even though j is large). The calomel and H_2/Pt electrodes are both highly non-polarizable, which is one reason why they are so extensively used as reference electrodes in electrochemistry.

(a) Concentration polarization

One of the assumptions in the derivation of the Butler–Volmer equation is the negligible conversion of the electroactive species at low current densities, resulting in uniformity of concentration near the electrode. This assumption fails at high current

Synoptic table 25.6* Exchange current densities and transfer coefficients at 298 K

Reaction	Electrode	j_0/(A cm^{-2})	α
$2\,H^+ + 2\,e^- \rightarrow H_2$	Pt	7.9×10^{-4}	
	Ni	6.3×10^{-6}	0.58
	Pb	5.0×10^{-12}	
$Fe^{3+} + e^- \rightarrow Fe^{2+}$	Pt	2.5×10^{-3}	0.58

* More values are given in the *Data section*.

densities because the consumption of electroactive species close to the electrode results in a concentration gradient; diffusion of the species towards the electrode from the bulk is slow and may become rate-determining. A larger overpotential is then needed to produce a given current. This effect is called **concentration polarization** and its contribution to the total overpotential is called the **polarization overpotential**, η^c.

Consider a case for which the concentration polarization dominates all the rate processes and a redox couple of the type M^{z+},M with the reduction $M^{z+} + z\,e^- \rightarrow M$. Under zero-current conditions, when the net current density is zero, the electrode potential is related to the activity, a, of the ions in the solution by the Nernst equation (eqn 7.29):

$$E = E^{\ominus} + \frac{RT}{zF}\ln a \qquad (25.48)$$

As remarked earlier, electrode kinetics are normally studied using a large excess of support electrolyte so as to keep the mean activity coefficients approximately constant. Therefore, the constant activity coefficient in $a = \gamma c$ may be absorbed into E, and we write the **formal potential**, E°, of the electrode as

$$E^{\circ} = E^{\ominus} + \frac{RT}{zF}\ln \gamma \qquad [25.49]$$

Then the electrode potential is

$$E = E^{\circ} + \frac{RT}{zF}\ln c \qquad (25.50)$$

When the cell is producing current, the active ion concentration at the OHP changes to c' and the electrode potential changes to

$$E' = E^{\circ} + \frac{RT}{zF}\ln c' \qquad (25.51)$$

The concentration overpotential is therefore

$$\eta^c = E' + E = \frac{RT}{zF}\ln\left(\frac{c'}{c}\right) \qquad (25.52)$$

We now suppose that the solution has its bulk concentration, c, up to a distance δ from the outer Helmholtz plane, and then falls linearly to c' at the plane itself. This **Nernst diffusion layer** is illustrated in Fig. 25.42. The thickness of the Nernst layer (which is typically 0.1 mm, and strongly dependent on the condition of hydrodynamic flow due to any stirring or convective effects) is quite different from that of the electric double layer (which is typically less than 1 nm, and unaffected by stirring). The concentration gradient through the Nernst layer is

$$\frac{\mathrm{d}c}{\mathrm{d}x} = \frac{c' - c}{\delta} \qquad (25.53)$$

This gradient gives rise to a flux of ions towards the electrode, which replenishes the cations as they are reduced. The (molar) flux, J, is proportional to the concentration gradient, and according to Fick's first law (Section 21.4)

$$J = -D\left(\frac{\partial c}{\partial x}\right) \qquad (25.54)$$

Therefore, the particle flux towards the electrode is

$$J = D\frac{c - c'}{\delta} \qquad (25.55)$$

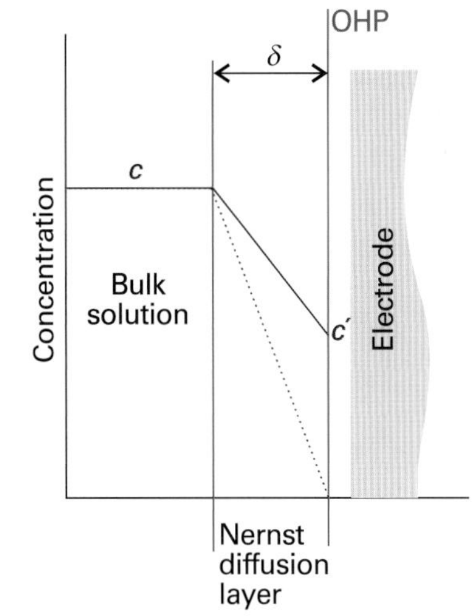

Fig. 25.42 In a simple model of the Nernst diffusion layer there is a linear variation in concentration between the bulk and the outer Helmholtz plane; the thickness of the layer depends strongly on the state of flow of the fluid. Note that the diffusion layer is much thicker relative to the OHP than shown here.

The cathodic current density towards the electrode is the product of the particle flux and the charge transferred per mole of ions, zF:

$$j = zFJ = zFD\frac{c-c'}{\delta} \qquad (25.56)$$

For instance, for the couple $[Fe(CN)_6]^{2-}/[Fe(CN)_6]^{3-}$, $z=1$, but for Fe^{3+}/Fe, $z=3$. The maximum rate of diffusion across the Nernst layer occurs when the gradient is steepest, which is when $c'=0$. This concentration occurs when an electron from an ion that diffuses across the layer is snapped over the activation barrier and on to the electrode. No flow of current can exceed the **limiting current density**, j_{lim}, which is given by

$$j_{\text{lim}} = zFJ_{\text{lim}} = \frac{zFDc}{\delta} \qquad (25.57a)$$

By using the Nernst–Einstein equation (eqn 21.64, written as $D = RT\lambda/z^2F^2$), we can express j_{lim} in terms of the ionic conductivity λ:

$$j_{\text{lim}} = \frac{cRT\lambda}{zF\delta} \qquad (25.57b)$$

Illustration 25.3 *Estimating the limiting current density*

Consider an electrode in a 0.10 M Cu^{2+}(aq) unstirred solution in which the thickness of the diffusion layer is about 0.3 mm. With $\lambda = 107$ S cm^2 mol^{-1} (Table 21.4), $\delta = 0.3$ mm, $c = 0.10$ mol dm^{-3}, $z=2$, and $T=298$ K, it follows from eqn 25.57b that $j_{\text{lim}} = 5$ mA cm^{-2}. The result implies that the current towards an electrode of area 1 cm^2 electrode cannot exceed 5 mA in this (unstirred) solution.

Self-test 25.8 Evaluate the limiting current density for an Ag(s)|Ag^+(aq) electrode in 0.010 mol dm^{-3} Ag^+(aq) at 298 K. Take $\delta = 0.03$ mm. [5 mA cm^{-2}]

It follows from eqn 25.56 that the concentration c' is related to the current density at the double layer by

$$c' = c - \frac{j\delta}{zFD} \qquad (25.58)$$

Hence, as the current density is increased, the concentration falls below the bulk value. However, this decline in concentration is small when the diffusion constant is large, for then the ions are very mobile and can quickly replenish any ions that have been removed.

Finally, we substitute eqn 25.58 into eqn 25.52 and obtain the following expressions for the overpotential in terms of the current density, and vice versa:

$$\eta^{c} = \frac{RT}{zF}\ln\left(1 - \frac{j\delta}{zcFD}\right) \qquad (25.59a)$$

$$j = \frac{zcFD}{\delta}(1 - e^{zf\eta^{c}}) \qquad (25.59b)$$

(b) Experimental techniques

The kind of output from **linear-sweep voltammetry** is illustrated in Fig. 25.43. Initially, the absolute value of the potential is low, and the cathodic current is due to the migration of ions in the solution. However, as the potential approaches the reduction potential of the reducible solute, the cathodic current grows. Soon after the potential exceeds the reduction potential the current rises and reaches a maximum value (as specified in eqn 25.57). This maximum current is proportional to the molar concentration of the species, so that concentration can be determined from the peak height after subtraction of an extrapolated baseline. In **differential pulse voltammetry** the current is monitored before and after a pulse of potential is applied, and the processed output is the slope of a curve like that obtained by linear-sweep voltammetry (Fig. 25.44). The area under the curve (in effect, the integral of the derivative displayed in the illustration) is proportional to the concentration of the species.

In **cyclic voltammetry** the potential is applied in a sawtooth manner to the working electrode and the current is monitored. A typical cyclic voltammogram is shown in Fig. 25.45. The shape of the curve is initially like that of a linear sweep experiment, but after reversal of the sweep there is a rapid change in current on account of the high concentration of oxidizable species close to the electrode that were generated on the reductive sweep. When the potential is close to the value required to oxidize the reduced species, there is a substantial anodic current until all the oxidation is complete, and the current returns to zero.

When the reduction reaction at the electrode can be reversed, as in the case of the $[Fe(CN)_6]^{3-}/[Fe(CN)_6]^{4-}$ couple, the cyclic voltammogram is broadly symmetric about the standard potential of the couple (as in Fig. 25.45b). The scan is initiated with $[Fe(CN)_6]^{3-}$ present in solution and, as the potential approaches $E^{\ominus}$ for the couple, the $[Fe(CN)_6]^{3-}$ near the electrode is reduced and current begins to flow. As the potential continues to change, the cathodic current begins to decline again because all

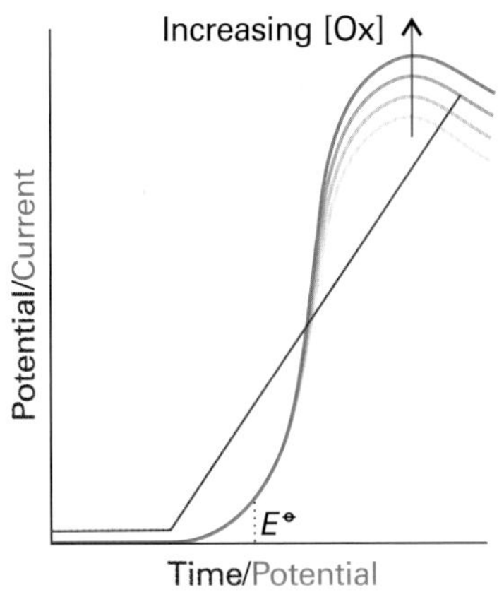

Fig. 25.43 The change of potential with time and the resulting current/potential curve in a voltammetry experiment. The peak value of the current density is proportional to the concentration of electroactive species (for instance, [Ox]) in solution.

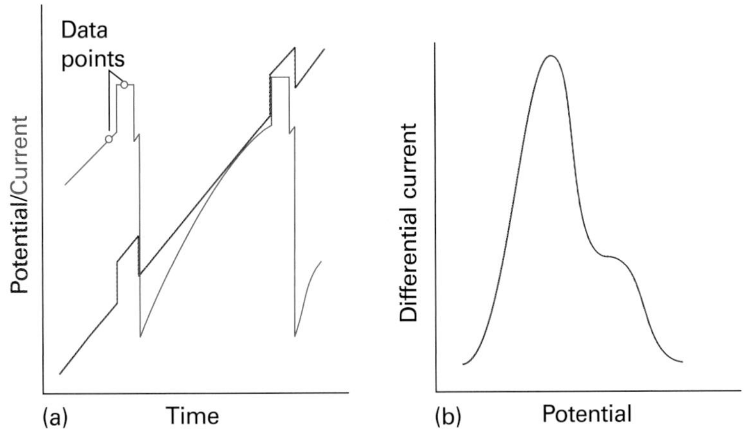

Fig. 25.44 A differential pulse voltammetry experiment. (a) The potential is swept linearly as a mercury droplet grows on the end of a capillary dipping into the sample and then pulsed as shown by the purple line. The resulting current is shown as the blue line and is sampled at the two points shown. (b) The data output is obtained as the difference of the currents at the two sampled points.

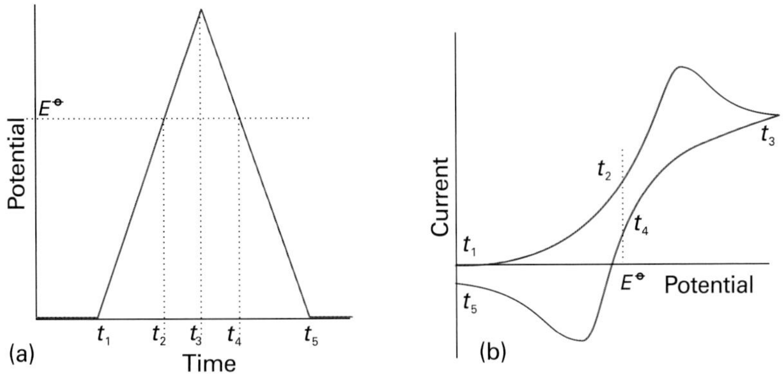

Fig. 25.45 (a) The change of potential with time and (b) the resulting current/potential curve in a cyclic voltammetry experiment.

the $[Fe(CN)_6]^{3-}$ near the electrode has been reduced and the current reaches its limiting value. The potential is now returned linearly to its initial value, and the reverse series of events occurs with the $[Fe(CN)_6]^{4-}$ produced during the forward scan now undergoing oxidation. The peak of current lies on the other side of $E^{\ominus}$, so the species present and its standard potential can be identified, as indicated in the illustration, by noting the locations of the two peaks.

The overall shape of the curve gives details of the kinetics of the electrode process and the change in shape as the rate of change of potential is altered gives information on the rates of the processes involved. For example, the matching peak on the return phase of the sawtooth change of potential may be missing, which indicates that the oxidation (or reduction) is irreversible. The appearance of the curve may also depend on the timescale of the sweep for, if the sweep is too fast, some processes might not have time to occur. This style of analysis is illustrated in the following example.

Example 25.5 *Analysing a cyclic voltammetry experiment*

The electroreduction of *p*-bromonitrobenzene in liquid ammonia is believed to occur by the following mechanism:

$$BrC_6H_4NO_2 + e^- \rightarrow BrC_6H_4NO_2^-$$
$$BrC_6H_4NO_2^- \rightarrow \cdot C_6H_4NO_2 + Br^-$$
$$\cdot C_6H_4NO_2 + e^- \rightarrow C_6H_4NO_2^-$$
$$C_6H_4NO_2^- + H^+ \rightarrow C_6H_5NO_2$$

Suggest the likely form of the cyclic voltammogram expected on the basis of this mechanism.

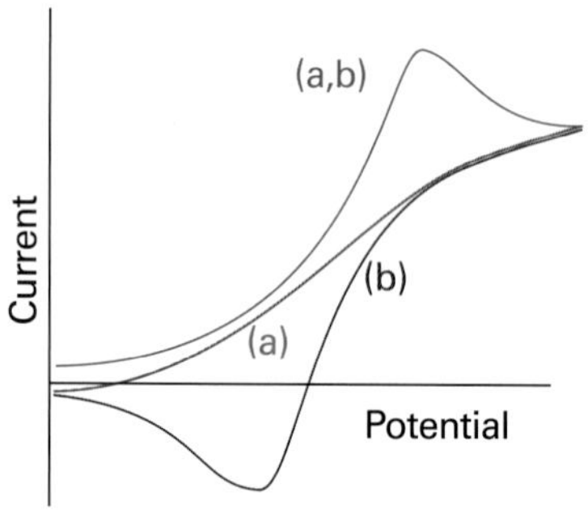

Fig. 25.46 (a) When a non-reversible step in a reaction mechanism has time to occur, the cyclic voltammogram may not show the reverse oxidation or reduction peak. (b) However, if the rate of sweep is increased, the return step may be caused to occur before the irreversible step has had time to intervene, and a typical 'reversible' voltammogram is obtained.

Method Decide which steps are likely to be reversible on the timescale of the potential sweep: such processes will give symmetrical voltammograms. Irreversible processes will give unsymmetrical shapes because reduction (or oxidation) might not occur. However, at fast sweep rates, an intermediate might not have time to react, and a reversible shape will be observed.

Answer At slow sweep rates, the second reaction has time to occur, and a curve typical of a two-electron reduction will be observed, but there will be no oxidation peak on the second half of the cycle because the product, $C_6H_5NO_2$, cannot be oxidized (Fig. 25.46a). At fast sweep rates, the second reaction does not have time to take place before oxidation of the $BrC_6H_4NO_2^-$ intermediate starts to occur during the reverse scan, so the voltammogram will be typical of a reversible one-electron reduction (Fig. 25.46b).

Self-test 25.9 Suggest an interpretation of the cyclic voltammogram shown in Fig. 25.47. The electroactive material is ClC_6H_4CN in acid solution; after reduction to $ClC_6H_4CN^-$, the radical anion may form C_6H_5CN irreversibly.

$$[ClC_6H_4CN + e^- \rightleftharpoons ClC_6H_4CN^-,$$
$$ClC_6H_4CN^- + H^+ + e^- \rightarrow C_6H_5CN + Cl^-, C_6H_5CN + e^- \rightleftharpoons C_6H_5CN^-]$$

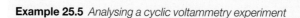

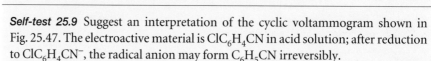

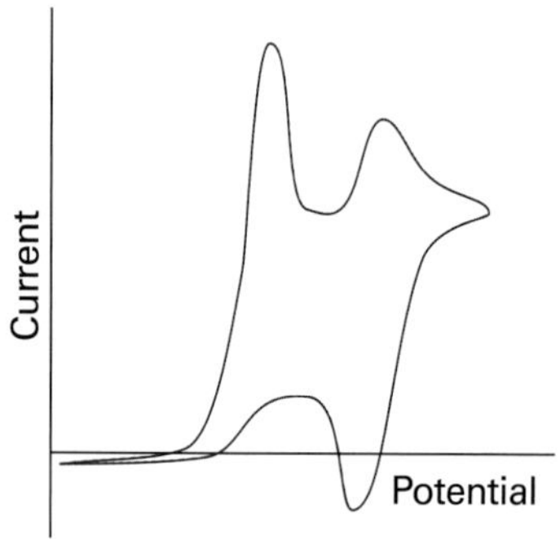

Fig. 25.47 The cyclic voltammogram referred to in Self-test 25.9.

25.11 Electrolysis

To induce current to flow through an electrolytic cell and bring about a nonspontaneous cell reaction, the applied potential difference must exceed the zero-current potential by at least the **cell overpotential**. The cell overpotential is the sum of the overpotentials at the two electrodes and the ohmic drop (IR_s, where R_s is the internal resistance of the cell) due to the current through the electrolyte. The additional potential needed to achieve a detectable rate of reaction may need to be large when the exchange current density at the electrodes is small. For similar reasons, a working galvanic cell generates a smaller potential than under zero-current conditions. In this section we see how to cope with both aspects of the overpotential.

The relative rates of gas evolution or metal deposition during electrolysis can be estimated from the Butler–Volmer equation and tables of exchange current densities. From eqn 25.46 and assuming equal transfer coefficients, we write the ratio of the cathodic currents as

$$\frac{j'}{j} = \frac{j'_0}{j_0} e^{(\eta-\eta')\alpha f} \qquad (25.60)$$

where j' is the current density for electrodeposition and j is that for gas evolution, and j'_0 and j_0 are the corresponding exchange current densities. This equation shows that metal deposition is favoured by a large exchange current density and relatively high gas evolution overpotential (so $\eta - \eta'$ is positive and large). Note that $\eta < 0$ for a cathodic process, so $-\eta' > 0$.

The exchange current density depends strongly on the nature of the electrode surface, and changes in the course of the electrodeposition of one metal on another. A very crude criterion is that significant evolution or deposition occurs only if the overpotential exceeds about 0.6 V.

Self-test 25.10 Deduce an expression for the ratio when the hydrogen evolution is limited by transport across a diffusion layer. [$j'/j = (\delta j'_0/cFD)e^{-\alpha\eta' f}$]

A glance at Table 25.6 shows the wide range of exchange current densities for a metal/hydrogen electrode. The most sluggish exchange currents occur for lead and mercury, and the value of 1 pA cm^{-2} corresponds to a monolayer of atoms being replaced in about 5 years. For such systems, a high overpotential is needed to induce significant hydrogen evolution. In contrast, the value for platinum (1 mA cm^{-2}) corresponds to a monolayer being replaced in 0.1 s, so gas evolution occurs for a much lower overpotential.

The exchange current density also depends on the crystal face exposed. For the deposition of copper on copper, the (100) face has j_0 = 1 mA cm^{-2}, so for the same overpotential the (100) face grows at 2.5 times the rate of the (111) face, for which j_0 = 0.4 mA cm^{-2}.

25.12 **Working galvanic cells**

In working galvanic cells (those not balanced against an external potential), the overpotential leads to a smaller potential than under zero-current conditions. Furthermore, we expect the cell potential to decrease as current is generated because it is then no longer working reversibly and can therefore do less than maximum work.

We shall consider the cell $M|M^+(aq)||M'^+(aq)|M'$ and ignore all the complications arising from liquid junctions. The potential of the cell is $E' = \Delta\phi_R - \Delta\phi_L$. Because the cell potential differences differ from their zero-current values by overpotentials, we can write $\Delta\phi_X = E_X + \eta_X$, where X is L or R for the left or right electrode, respectively. The cell potential is therefore

$$E' = E + \eta_R - \eta_L \qquad (25.61a)$$

To avoid confusion about signs (η_R is negative, η_L is positive) and to emphasize that a working cell has a lower potential than a zero-current cell, we shall write this expression as

$$E' = E - |\eta_R| - |\eta_L| \qquad (25.61b)$$

with E the cell emf. We should also subtract the ohmic potential difference IR_s, where R_s is the cell's internal resistance:

$$E' = E - |\eta_R| - |\eta_L| - IR_s \qquad (25.61c)$$

The ohmic term is a contribution to the cell's irreversibility—it is a thermal dissipation term—so the sign of IR_s is always such as to reduce the potential in the direction of zero.

The overpotentials in eqn 25.61 can be calculated from the Butler–Volmer equation for a given current, I, being drawn. We shall simplify the equations by supposing that the areas, A, of the electrodes are the same, that only one electron is transferred in the rate-determining steps at the electrodes, that the transfer coefficients are both $\frac{1}{2}$, and that the high-overpotential limit of the Butler–Volmer equation may be used. Then from eqns 25.46 and 25.61c we find

$$E' = E - IR_s - \frac{4RT}{F}\ln\left(\frac{I}{A\bar{j}}\right) \qquad \bar{j} = (j_{0L}j_{0R})^{1/2} \qquad (25.62)$$

where j_{0L} and j_{0R} are the exchange current densities for the two electrodes.

The concentration overpotential also reduces the cell potential. If we use the Nernst diffusion layer model for each electrode, the total change of potential arising from concentration polarization is given by eqn 25.59 as

$$E' = E - \frac{RT}{zF}\ln\left\{\left(1 - \frac{I}{Aj_{\text{lim,L}}}\right)\left(1 - \frac{I}{Aj_{\text{lim,R}}}\right)\right\} \qquad (25.63)$$

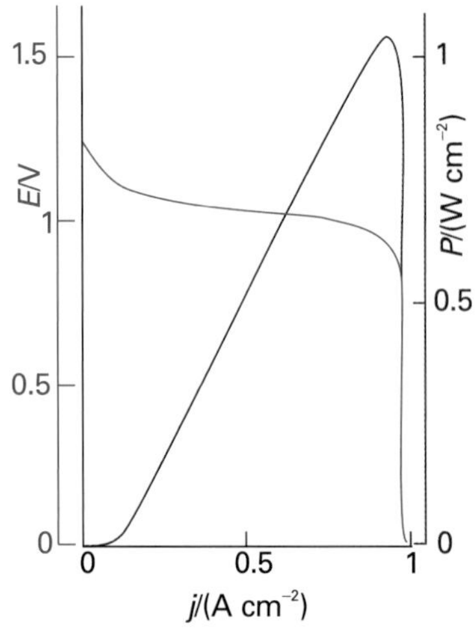

Fig. 25.48 The dependence of the potential of a working galvanic cell on the current density being drawn (blue line) and the corresponding power output (purple line) calculated by using eqns 25.64 and 25.65, respectively. Notice the sharp decline in power just after the maximum.

Exploration Using mathematical software, and electronic spreadsheet, or the interactive applets found in the *Living graphs* section of the text's web site, confirm that the sharp decline in potential and power observed in Fig. 25.48 is true for any value of R_s.

This contribution can be added to the one in eqn 25.62 to obtain a full (but still very approximate) expression for the cell potential when a current I is being drawn:

$$E' = E - IR_s - \frac{2RT}{zF}\ln g(I) \qquad (25.64a)$$

with

$$g(I) = \left(\frac{I}{A\bar{j}}\right)^{2z}\left\{\left(1 - \frac{I}{Aj_{\text{lim,L}}}\right)\left(1 - \frac{I}{Aj_{\text{lim,R}}}\right)\right\}^{1/2} \qquad (25.64b)$$

This equation depends on a lot of parameters, but an example of its general form is given in Fig. 25.48. Notice the very steep decline of working potential when the current is high and close to the limiting value for one of the electrodes.

Because the power, P, supplied by a galvanic cell is IE', from eqn 25.64 we can write

$$P = IE - I^2R_s - \frac{2IRT}{zF}\ln g(I) \qquad (25.65)$$

The first term on the right is the power that would be produced if the cell retained its zero-current potential when delivering current. The second term is the power generated uselessly as heat as a result of the resistance of the electrolyte. The third term is the reduction of the potential at the electrodes as a result of drawing current.

The general dependence of power output on the current drawn is shown in Fig. 25.48 as the purple line. Notice how maximum power is achieved just before the concentration polarization quenches the cell's performance. Information of this kind is essential if the optimum conditions for operating electrochemical devices are to be found and their performance improved.

Electric storage cells operate as galvanic cells while they are producing electricity but as electrolytic cells while they are being charged by an external supply. The lead–acid battery is an old device, but one well suited to the job of starting cars (and the only one available). During charging the cathode reaction is the reduction of Pb^{2+} and its deposition as lead on the lead electrode. Deposition occurs instead of the reduction of the acid to hydrogen because the latter has a low exchange current density on lead. The anode reaction during charging is the oxidation of Pb(II) to Pb(IV), which is deposited as the oxide PbO_2. On discharge, the two reactions run in reverse. Because they have such high exchange current densities the discharge can occur rapidly, which is why the lead battery can produce large currents on demand.

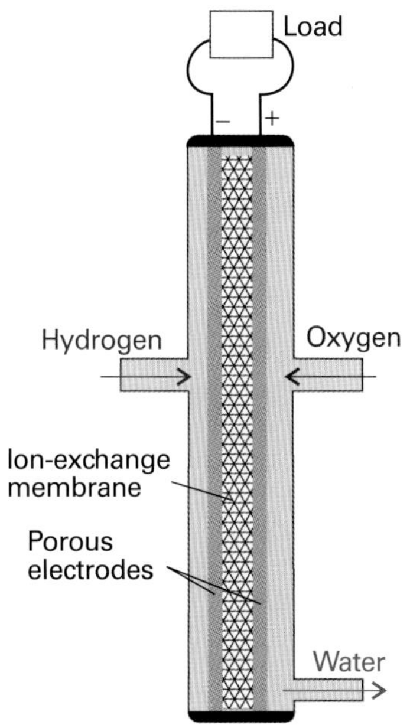

Fig. 25.49 A single cell of a hydrogen/oxygen fuel cell. In practice, a stack of many cells is used.

IMPACT ON TECHNOLOGY
I25.3 Fuel cells

A fuel cell operates like a conventional galvanic cell with the exception that the reactants are supplied from outside rather than forming an integral part of its construction. A fundamental and important example of a fuel cell is the hydrogen/oxygen cell, such as the ones used in space missions (Fig. 25.49). One of the electrolytes used is concentrated aqueous potassium hydroxide maintained at 200°C and 20–40 atm; the electrodes may be porous nickel in the form of sheets of compressed powder. The cathode reaction is the reduction

$$O_2(g) + 2\,H_2O(l) + 4\,e^- \rightarrow 4\,OH^-(aq) \qquad E^\ominus = +0.40\text{ V}$$

and the anode reaction is the oxidation

$$H_2(g) + 2\,OH^-(aq) \rightarrow 2\,H_2O(l) + 2\,e^-$$

For the corresponding reduction, $E^\ominus = -0.83$ V. Because the overall reaction

$$2\,H_2(g) + O_2(g) \rightarrow 2\,H_2O(l) \qquad E^\ominus = +1.23\text{ V}$$

is exothermic as well as spontaneous, it is less favourable thermodynamically at 200°C than at 25°C, so the cell potential is lower at the higher temperature. However, the increased pressure compensates for the increased temperature, and $E \approx +1.2$ V at 200°C and 40 atm.

One advantage of the hydrogen/oxygen system is the large exchange current density of the hydrogen reaction. Unfortunately, the oxygen reaction has an exchange current density of only about 0.1 nA cm^{-2}, which limits the current available from the cell. One way round the difficulty is to use a catalytic surface (to increase j_0) with a large surface area. One type of highly developed fuel cell has phosphoric acid as the electrolyte and operates with hydrogen and air at about 200°C; the hydrogen is obtained from a reforming reaction on natural gas:

Anode: $2\,H_2(g) \rightarrow 4\,H^+(aq) + 4\,e^-$

Cathode: $O_2(g) + 4\,H^+(aq) + 4\,e^- \rightarrow 2\,H_2O(l)$

This fuel cell has shown promise for *combined heat and power systems* (CHP systems). In such systems, the waste heat is used to heat buildings or to do work. Efficiency in a CHP plant can reach 80 per cent. The power output of batteries of such cells has reached the order of 10 MW. Although hydrogen gas is an attractive fuel, it has disadvantages for mobile applications: it is difficult to store and dangerous to handle. One possibility for portable fuel cells is to store the hydrogen in carbon nanotubes (*Impact* 20.2). It has been shown that carbon nanofibres in herringbone patterns can store huge amounts of hydrogen and result in an energy density (the magnitude of the released energy divided by the volume of the material) twice that of gasoline.

Cells with molten carbonate electrolytes at about 600°C can make use of natural gas directly. Solid-state electrolytes are also used. They include one version in which the electrolyte is a solid polymeric ionic conductor at about 100°C, but in current versions it requires very pure hydrogen to operate successfully. Solid ionic conducting oxide cells operate at about 1000°C and can use hydrocarbons directly as fuel. Until these materials have been developed, one attractive fuel is methanol, which is easy to handle and is rich in hydrogen atoms:

Anode: $CH_3OH(l) + 6\,OH^-(aq) \rightarrow 5\,H_2O(l) + CO_2(g) + 6\,e^-$

Cathode: $O_2(g) + 4\,e^- + 2\,H_2O(l) \rightarrow 4\,OH^-(aq)$

One disadvantage of methanol, however, is the phenomenon of 'electro-osmotic drag' in which protons moving through the polymer electrolyte membrane separating the anode and cathode carry water and methanol with them into the cathode compartment where the potential is sufficient to oxidize CH_3OH to CO_2, so reducing the efficiency of the cell.

A *biofuel cell* is like a conventional fuel cell but in place of a platinum catalyst it uses enzymes or even whole organisms. The electricity will be extracted through organic molecules that can support the transfer of electrons. One application will be as the power source for medical implants, such as pacemakers, perhaps using the glucose present in the bloodstream as the fuel.

25.13 **Corrosion**

A thermodynamic warning of the likelihood of corrosion is obtained by comparing the standard potentials of the metal reduction, such as

$$Fe^{2+}(aq) + 2\,e^- \rightarrow Fe(s) \qquad E^\ominus = -0.44\ V$$

with the values for one of the following half-reactions:

In acidic solution:

$$\text{(a)}\ \ 2\,H^+(aq) + 2\,e^- \rightarrow H_2(g) \qquad E^\ominus = 0$$
$$\text{(b)}\ \ 4\,H^+(aq) + O_2(g) + 4\,e^- \rightarrow 2\,H_2O(l) \qquad E^\ominus = +1.23\ V$$

In basic solution:

$$\text{(c)}\ \ 2\,H_2O(l) + O_2(g) + 4\,e^- \rightarrow 4\,OH^-(aq) \qquad E^\ominus = +0.40\ V$$

Because all three redox couples have standard potentials more positive than $E^\ominus(Fe^{2+}/Fe)$, all three can drive the oxidation of iron to iron(II). The electrode potentials we have quoted are standard values, and they change with the pH of the medium. For the first two:

$$E(a) = E^\ominus(a) + (RT/F)\ln a(H^+) = -(0.059\ V)pH$$
$$E(b) = E^\ominus(b) + (RT/F)\ln a(H^+) = 1.23\ V - (0.059\ V)pH$$

These expressions let us judge at what pH the iron will have a tendency to oxidize (see Chapter 7). A thermodynamic discussion of corrosion, however, only indicates whether a tendency to corrode exists. If there is a thermodynamic tendency, we must examine the kinetics of the processes involved to see whether the process occurs at a significant rate.

A model of a corrosion system is shown in Fig. 25.50a. It can be taken to be a drop of slightly acidic (or basic) water containing some dissolved oxygen in contact with the metal. The oxygen at the edges of the droplet, where the O_2 concentration is higher, is reduced by electrons donated by the iron over an area A. Those electrons are replaced by others released elsewhere as $Fe \rightarrow Fe^{2+} + 2\,e^-$. This oxidative release occurs over an area A' under the oxygen-deficient inner region of the droplet. The droplet acts as a short-circuited galvanic cell (Fig. 25.50b).

The rate of corrosion is measured by the current of metal ions leaving the metal surface in the anodic region. This flux of ions gives rise to the **corrosion current**, I_{corr}, which can be identified with the anodic current, I_a. We show in the justification below that the corrosion current is related to the cell potential of the corrosion couple by

$$I_{\text{corr}} = \bar{j}_0 \bar{A} e^{fE/4} \qquad \bar{j}_0 = (j_0 j_0')^{1/2} \qquad \bar{A} = (AA')^{1/2} \tag{25.66}$$

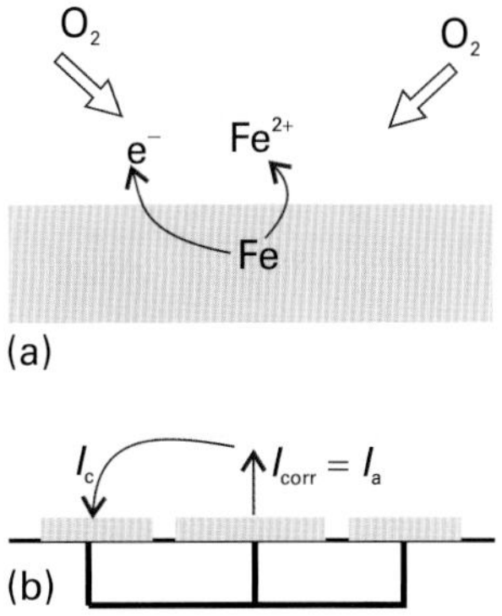

Fig. 25.50 (a) A simple version of the corrosion process is that of a droplet of water, which is oxygen rich near its boundary with air. The oxidation of the iron takes place in the region away from the oxygen because the electrons are transported through the metal. (b) The process may be modelled as a short-circuited electrochemical cell.

Justification 25.1 *The corrosion current*

Because any current emerging from the anodic region must find its way to the cathodic region, the cathodic current, I_c, and the anodic current, I_a, must both be equal to the corrosion current. In terms of the current densities at the oxidation and reduction sites, j and j', respectively, we can write

$$I_{\text{corr}} = jA = j'A' = (jj'AA')^{1/2} = \bar{j}\bar{A} \qquad \bar{j} = (jj')^{1/2} \qquad \bar{A} = (AA')^{1/2} \tag{25.67}$$

The Butler–Volmer equation is now used to express the current densities in terms of overpotentials. For simplicity we assume that the overpotentials are large enough for the high-overpotential limit (eqn 25.46, $j = -j_0 e^{-\alpha f \eta}$) to apply, that polarization overpotential can be neglected, that the rate-determining step is the transfer of a single electron, and that the transfer coefficients are $\frac{1}{2}$. We also assume that, since the droplet is so small, there is negligible potential difference between the cathode and anode regions of the solution. Moreover, because it is short-circuited by the metal, the potential of the metal is the same in both regions, and so the potential difference between the metal and the solution is the same in both regions too; it is denoted $\Delta\phi_{\text{corr}}$. The overpotentials in the two regions are therefore $\eta = \Delta\phi_{\text{corr}} - \Delta\phi$ and $\eta' = \Delta\phi_{\text{corr}} - \Delta\phi'$, and the current densities are

$$j = j_0 e^{\eta f/2} = j_0 e^{f\Delta\phi_{\text{corr}}/2} e^{-f\Delta\phi/2} \qquad j' = j_0' e^{-\eta' f/2} = j_0' e^{-f\Delta\phi_{\text{corr}}/2} e^{f\Delta\phi'/2}$$

These expressions can be substituted into the expression for I_{corr} and $\Delta\phi' - \Delta\phi$ replaced by the difference of electrode potentials E to give eqn 25.66.

The effect of the exchange current density on the corrosion rate can be seen by considering the specific case of iron in contact with acidified water. Thermodynamically, either hydrogen or oxygen reduction reaction (a) or (b) on p. 946 is effective. However, the exchange current density of reaction (b) on iron is only about 10^{-14} A cm^{-2}, whereas for (a) it is 10^{-6} A cm^{-2}. The latter therefore dominates kinetically, and iron corrodes by hydrogen evolution in acidic solution.

For corrosion reactions with similar exchange current densities, eqn 25.66 predicts that the rate of corrosion is high when E is large. That is, rapid corrosion can be expected when the oxidizing and reducing couples have widely differing electrode potentials.

IMPACT ON TECHNOLOGY

I25.4 Protecting materials against corrosion

Several techniques for inhibiting corrosion are available. First, from eqn 25.66 we see that the rate of corrosion depends on the surfaces exposed: if either A or A' is zero, then the corrosion current is zero. This interpretation points to a trivial, yet often effective, method of slowing corrosion: cover the surface with some impermeable layer, such as paint, which prevents access of damp air. Paint also increases the effective solution resistance between the cathode and anode patches on the surface. Unfortunately, this protection fails disastrously if the paint becomes porous. The oxygen then has access to the exposed metal and corrosion continues beneath the paintwork. Another form of surface coating is provided by **galvanizing**, the coating of an iron object with zinc. Because the latter's standard potential is –0.76 V, which is more negative than that of the iron couple, the corrosion of zinc is thermodynamically

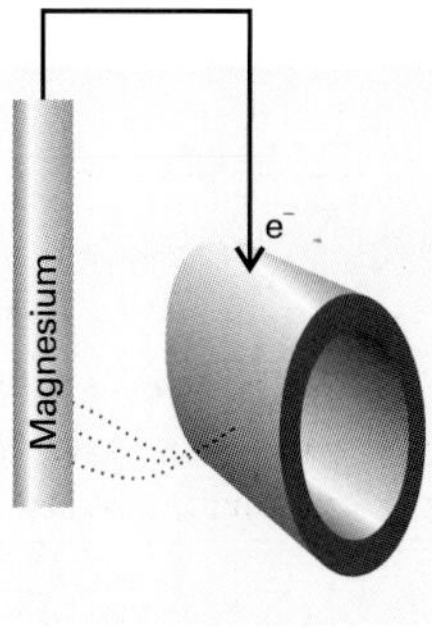

(a)

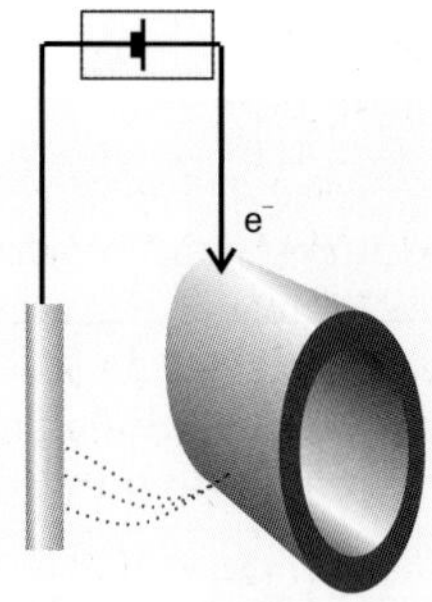

(b)

Fig. 25.51 (a) In cathodic protection an anode of a more strongly reducing metal is sacrificed to maintain the integrity of the protected object (for example, a pipeline, bridge, or boat). (b) In impressed-current cathodic protection electrons are supplied from an external cell so that the object itself is not oxidized. The broken lines depict the completed circuit through the soil.

favoured and the iron survives (the zinc survives because it is protected by a hydrated oxide layer). In contrast, tin plating leads to a very rapid corrosion of the iron once its surface is scratched and the iron exposed because the tin couple ($E^\ominus = -0.14$ V) oxidizes the iron couple ($E^\ominus = -0.44$ V). Some oxides are inert kinetically in the sense that they adhere to the metal surface and form an impermeable layer over a fairly wide pH range. This **passivation**, or kinetic protection, can be seen as a way of decreasing the exchange currents by sealing the surface. Thus, aluminium is inert in air even though its standard potential is strongly negative (−1.66 V).

Another method of protection is to change the electric potential of the object by pumping in electrons that can be used to satisfy the demands of the oxygen reduction without involving the oxidation of the metal. In **cathodic protection**, the object is connected to a metal with a more negative standard potential (such as magnesium, −2.36 V). The magnesium acts as a **sacrificial anode**, supplying its own electrons to the iron and becoming oxidized to Mg^{2+} in the process (Fig. 25.51a). A block of magnesium replaced occasionally is much cheaper than the ship, building, or pipeline for which it is being sacrificed. In **impressed-current cathodic protection** (Fig. 25.51b) an external cell supplies the electrons and eliminates the need for iron to transfer its own.

Table 25.7 Summary of acronyms

AES	Auger electron spectroscopy
AFM	Atomic force microscopy
BET isotherm	Brunauer–Emmett–Teller isotherm
EELS	Electron energy-loss spectroscopy
ER mechanism	Eley–Rideal mechanism
ESCA	Electron spectroscopy for chemical analysis
FIM	Field-ionization microscopy
HREELS	High-resolution electron energy-loss spectroscopy
IHP	Inner Helmholtz plane
LEED	Low-energy electron diffraction
LH mechanism	Langmuir–Hinshelwood mechanism
MBRS	Molecular beam reactive scattering
MBS	Molecular beam scattering
OHP	Outer Helmholtz plane
QCM	Quartz crystal microbalance
RAIRS	Reflection–absorption infrared spectroscopy
SAM	Scanning Auger electron microscopy
SAM	Self-assembled monolayer
SEM	Scanning electron microscopy
SERS	Surface-enhanced Raman scattering
SEXAFS	Surface-extended X-ray absorption fine structure spectroscopy
SHG	Second harmonic generation
SIMS	Secondary ion mass spectrometry
SPM	Scanning probe microscopy
SPR	Surface plasmon resonance
STM	Scanning tunnelling microscopy
TDS	Thermal desorption spectroscopy
TPD	Temperature programmed desorption
UHV	Ultra-high vacuum
UPS	Ultraviolet photoemission spectroscopy
XPS	X-ray photoemission spectroscopy

Checklist of key ideas

☐ 1. Adsorption is the attachment of molecules to a surface; the substance that adsorbs is the adsorbate and the underlying material is the adsorbent or substrate. The reverse of adsorption is desorption.

☐ 2. The collision flux, Z_W, of gas molecules bombarding a solid surface is related to the gas pressure by $Z_W = p/(2\pi mkT)^{1/2}$.

☐ 3. Techniques for studying surface composition and structure include scanning electron microscopy (SEM), scanning probe microscopy (STM), photoemission spectroscopy, sescondary-ion mass spectrometry, surface-enhanced Raman scattering (SERS), Auger electron spectroscopy (AES), low energy electron diffraction (LEED), and molecular beam scattering (MBS).

☐ 4. The fractional coverage, θ, is the ratio of the number of occupied sites to the number of available sites.

☐ 5. Techniques for studying the rates of surface processes include flash desorption, biosensor analysis, second harmonic generation (SHG), gravimetry by using a quartz crystal microbalance (QCM), and molecular beam reactive scattering (MRS).

☐ 6. Physisorption is adsorption by a van der Waals interaction; chemisorption is adsorption by formation of a chemical (usually covalent) bond.

☐ 7. The Langmuir isotherm is a relation between the fractional coverage and the partial pressure of the adsorbate: $\theta = Kp/(1+Kp)$.

☐ 8. The isosteric enthalpy of adsorption is determined from a plot of ln K against $1/T$.

☐ 9. The BET isotherm is an isotherm applicable when multilayer adsorption is possible: $V/V_{mon} = cz/(1-z)\{1-(1-c)z\}$, with $z = p/p^*$.

☐ 10. The sticking probability, s, is the proportion of collisions with the surface that successfully lead to adsorption.

☐ 11. Desorption is an activated process with half-life $t_{1/2} = \tau_0 e^{E_d/RT}$; the desorption activation energy is measured by temperature-programmed desorption (TPD) or thermal desorption spectroscopy (TDS).

☐ 12. In the Langmuir–Hinshelwood mechanism (LH mechanism) of surface-catalysed reactions, the reaction takes place by encounters between molecular fragments and atoms adsorbed on the surface.

☐ 13. In the Eley–Rideal mechanism (ER mechanism) of a surface-catalysed reaction, a gas-phase molecule collides with another molecule already adsorbed on the surface.

☐ 14. An electrical double layer consists of a sheet of positive charge at the surface of the electrode and a sheet of negative charge next to it in the solution (or vice versa).

☐ 15. The Galvani potential difference is the potential difference between the bulk of the metal electrode and the bulk of the solution.

☐ 16. Models of the double layer include the Helmholtz layer model and the Gouy–Chapman model.

☐ 17. The current density, j, at an electrode is expressed by the Butler–Volmer equation, $j = j_0\{e^{(1-\alpha)f\eta} - e^{-\alpha f\eta}\}$, where η is the overpotential, $\eta = E' - E$, α is the transfer coefficient, and j_0 is the exchange-current density.

☐ 18. A Tafel plot is a plot of the logarithm of the current density against the overpotential: the slope gives the value of α and the intercept at $\eta = 0$ gives the exchange-current density.

☐ 19. Voltammetry is the study of the current through an electrode as a function of the applied potential difference. Experimental techniques include linear-sweep voltammetry, differential pulse voltammetry, and cyclic voltammetry.

☐ 20. To induce current to flow through an electrolytic cell and bring about a nonspontaneous cell reaction, the applied potential difference must exceed the cell emf by at least the cell overpotential.

☐ 21. The corrosion current is a current proportional to the rate at which metal ions leave a metal surface in the anodic region during corrosion.

Further reading

Articles and texts

A.W. Adamson and A. Gast, *Physical chemistry of surfaces*. Wiley, New York (1997).

A.J. Bard and L.R. Faulkner, *Electrochemical methods: fundamentals and applications*. Wiley, New York (2000).

J.O'M. Bockris, R.E. White, and B.E. Conway (ed.), *Modern aspects of electrochemistry*. Vol. 33. Plenum, New York (1999).

G. Ertl, H. Knözinger, and J. Weitkamp, *Handbook of heterogeneous catalysis*. VCH, Weinheim (1997).

M.G. Fontanna and R.W. Staehle (ed.), *Advances in corrosion science and technology*. Plenum, New York (1980).

C.H. Hamann, W. Vielstich, and A. Hammett, *Electrochemistry*. Wiley–VCH, New York (1998).

J.C. Lindon, G.E. Tranter, and J.L. Holmes (ed.), *Encyclopedia of spectroscopy and spectrometry*. Academic Press, San Diego (2000).

N. Mizuno and M. Misono, Heterogeneous catalysis. *Chem. Rev.* **98**, 199 (1998).

G.A. Somorjai, Modern surface science and surface technologies: an introduction. *Chem. Rev.* **96**, 1223 (1996).

C.D.S. Tuck, *Modern battery construction*. Ellis Horwood, New York (1991).

J. Vickerman, *Surface analysis: techniques and applications*. Wiley, New York (1997).

Sources of data and information

C.M.A. Brett and A.M.O. Brett, *Electrode potentials*. Oxford Chemistry Primers, Oxford University Press (1998).

D. Linden (ed.), *Handbook of batteries and cells*. McGraw-Hill, New York (1984).

Further information

Further information 25.1 *The relation between electrode potential and the Galvani potential*

To demonstrate the relation between $\Delta\phi$ and E, consider the cell $Pt|H_2(g)|H^+(g)||M^+(aq)|M(s)$ and the half-reactions

$$M^+(aq) + e^- \rightarrow M(s) \qquad H^+(aq) + e^- \rightarrow \tfrac{1}{2} H_2(g)$$

The Gibbs energies of these two half-reactions can be expressed in terms of the chemical potentials, μ, of all the species. However, we must take into account the fact that the species are present in phases with different electric potentials. Thus, a cation in a region of positive potential has a higher chemical potential (is chemically more active in a thermodynamic sense) than in a region of zero potential.

The contribution of an electric potential to the chemical potential is calculated by noting that the electrical work of adding a charge ze to a region where the potential is ϕ is $ze\phi$, and therefore that the work per mole is $zF\phi$, where F is Faraday's constant. Because at constant temperature and pressure the maximum electrical work can be identified with the change in Gibbs energy (Section 7.7), the difference in chemical potential of an ion with and without the electrical potential present is $zF\phi$. The chemical potential of an ion in the presence of an electric potential is called its **electrochemical potential**, $\bar{\mu}$. It follows that

$$\bar{\mu} = \mu + zF\phi \qquad [25.68]$$

where μ is the chemical potential of the species when the electrical potential is zero. When $z = 0$ (a neutral species), the electrochemical potential is equal to the chemical potential.

To express the Gibbs energy for the half-reactions in terms of the electrochemical potentials of the species we note that the cations M^+ are in the solution where the inner potential is ϕ_S and the electrons are in the electrode where it is ϕ_M. It follows that

$$\begin{aligned}\Delta_r G_R &= \bar{\mu}(M) - \{\bar{\mu}(M^+) + \bar{\mu}(e^-)\} \\ &= \mu(M) - \{\mu(M^+) + F\phi_S + \mu(e^-) - F\phi_M\} \\ &= \mu(M) - \mu(M^+) - \mu(e^-) + F\Delta\phi_R\end{aligned}$$

where $\Delta\phi_R = \phi_M - \phi_S$ is the Galvani potential difference at the right-hand electrode. Likewise, in the hydrogen half-reaction, the electrons are in the platinum electrode at a potential ϕ_{Pt} and the H^+ ions are in the solution where the potential is ϕ_S:

$$\begin{aligned}\Delta_r G_L &= \tfrac{1}{2}\bar{\mu}(H_2) - \{\bar{\mu}(H^+) + \bar{\mu}(e^-)\} \\ &= \tfrac{1}{2}\mu(H_2) - \mu(H^+) - \mu(e^-) + F\Delta\phi_L\end{aligned}$$

where $\Delta\phi_L = \phi_{Pt} - \phi_S$ is the Galvani potential difference at the left-hand electrode.

The overall reaction Gibbs energy is

$$\begin{aligned}\Delta_r G_R - \Delta_r G_L &= \mu(M) + \mu(H^+) - \mu(M^+) - \tfrac{1}{2}\mu(H_2) + F(\Delta\phi_R - \Delta\phi_L) \\ &= \Delta_r G + F(\Delta\phi_R - \Delta\phi_L)\end{aligned}$$

where $\Delta_r G$ is the Gibbs energy of the cell reaction. When the cell is balanced against an external source of potential the entire system is at equilibrium. The overall reaction Gibbs energy is then zero (because its tendency to change is balanced against the external source of potential and overall there is stalemate), and the last equation becomes

$$0 = \Delta_r G + F(\Delta\phi_R - \Delta\phi_L)$$

which rearranges to

$$\Delta_r G = -F(\Delta\phi_R - \Delta\phi_L) \qquad (25.69)$$

If we compare this with the result established in Section 7.7 that $\Delta_r G = -FE$ and $E = E_R - E_L$, we can conclude that (ignoring the effects of any metal–platinum and liquid junction potentials that may be present in an actual cell)

$$E_R - E_L = \Delta\phi_R - \Delta\phi_L \qquad (25.70)$$

This is the result we wanted to show, for it implies that the Galvani potential difference at each electrode can differ from the electrode potential by a constant at most; that constant cancels when the difference is taken.

Discussion questions

25.1 (a) Distinguish between a step and a terrace. (b) Describe how steps and terraces can be formed by dislocations.

25.2 (a) Describe the advantages and limitations of each of the spectroscopic techniques designated by the acronyms AES, EELS, HREELS, RAIRS, SERS, SEXAFS, SHG, UPS, and XPS. (b) Describe the advantages and limitations of each of the microscopy, diffraction, and scattering techniques designated by the acronyms AFM, FIM, LEED, MBRS, MBS, SAM, SEM, and STM.

25.3 Distinguish between the following adsorption isotherms: Langmuir, BET, Temkin, and Freundlich.

25.4 Consider the analysis of surface plasmon resonance data (as in biosensor analysis) and discuss how a plot of a_0/R_{eq} against a_0 may be used to evaluate R_{max} and K.

25.5 Describe the essential features of the Langmuir–Hinshelwood, Eley–Rideal, and Mars van Krevelen mechanisms for surface-catalysed reactions.

25.6 Account for the dependence of catalytic activity of a surface on the strength of chemisorption, as shown in Fig. 25.28.

25.7 Discuss the unique physical and chemical properties of zeolites that make them useful heterogeneous catalysts.

25.8 (a) Discuss the main structural features of the electrical double layer. (b) Distinguish between the electrical double layer and the Nernst diffusion layer.

25.9 Define the terms in and limit the generality of the following expressions: (a) $j=j_0 f\eta$, (b) $j=j_0 e^{(1-\alpha)f\eta}$, and (c) $j=-j_0 e^{-\alpha f\eta}$.

25.10 Discuss the technique of cyclic voltammetry and account for the characteristic shape of a cyclic voltammogram, such as those shown in Figs. 25.45 and 25.46.

25.11 Discuss the principles of operation of a fuel cell.

25.12 Discuss the chemical origins of corrosion and useful strategies for preventing it.

Exercises

25.1a Calculate the frequency of molecular collisions per square centimetre of surface in a vessel containing (a) hydrogen, (b) propane at 25°C when the pressure is (i) 100 Pa, (ii) 0.10 μTorr.

25.1b Calculate the frequency of molecular collisions per square centimetre of surface in a vessel containing (a) nitrogen, (b) methane at 25°C when the pressure is (i) 10.0 Pa, (ii) 0.150 μTorr.

25.2a What pressure of argon gas is required to produce a collision rate of $4.5\times10^{20}\ s^{-1}$ at 425 K on a circular surface of diameter 1.5 mm?

25.2b What pressure of nitrogen gas is required to produce a collision rate of $5.00\times10^{19}\ s^{-1}$ at 525 K on a circular surface of diameter 2.0 mm?

25.3a Calculate the average rate at which He atoms strike a Cu atom in a surface formed by exposing a (100) plane in metallic copper to helium gas at 80 K and a pressure of 35 Pa. Crystals of copper are face-centred cubic with a cell edge of 361 pm.

25.3b Calculate the average rate at which He atoms strike an iron atom in a surface formed by exposing a (100) plane in metallic iron to helium gas at 100 K and a pressure of 24 Pa. Crystals of iron are body-centred cubic with a cell edge of 145 pm.

25.4a A monolayer of N_2 molecules (effective area 0.165 nm^2) is adsorbed on the surface of 1.00 g of an Fe/Al_2O_3 catalyst at 77 K, the boiling point of liquid nitrogen. Upon warming, the nitrogen occupies 2.86 cm^3 at 0°C and 760 Torr. What is the surface area of the catalyst?

25.4b A monolayer of CO molecules (effective area 0.165 nm^2) is adsorbed on the surface of 1.00 g of an Fe/Al_2O_3 catalyst at 77 K, the boiling point of liquid nitrogen. Upon warming, the carbon monoxide occupies 4.25 cm^3 at 0°C and 1.00 bar. What is the surface area of the catalyst?

25.5a The volume of oxygen gas at 0°C and 101 kPa adsorbed on the surface of 1.00 g of a sample of silica at 0°C was 0.284 cm^3 at 142.4 Torr and 1.430 cm^3 at 760 Torr. What is the value of V_{mon}?

25.5b The volume of gas at 20°C and 1.00 bar adsorbed on the surface of 1.50 g of a sample of silica at 0°C was 1.60 cm^3 at 52.4 kPa and 2.73 cm^3 at 104 kPa. What is the value of V_{mon}?

25.6a The enthalpy of adsorption of CO on a surface is found to be −120 kJ mol^{-1}. Estimate the mean lifetime of a CO molecule on the surface at 400 K.

25.6b The enthalpy of adsorption of ammonia on a nickel surface is found to be −155 kJ mol^{-1}. Estimate the mean lifetime of an NH_3 molecule on the surface at 500 K.

25.7a The average time for which an oxygen atom remains adsorbed to a tungsten surface is 0.36 s at 2548 K and 3.49 s at 2362 K. Find the activation energy for desorption. What is the pre-exponential factor for these tightly chemisorbed atoms?

25.7b The chemisorption of hydrogen on manganese is activated, but only weakly so. Careful measurements have shown that it proceeds 35 per cent faster at 1000 K than at 600 K. What is the activation energy for chemisorption?

25.8a The adsorption of a gas is described by the Langmuir isotherm with $K=0.85\ kPa^{-1}$ at 25°C. Calculate the pressure at which the fractional surface coverage is (a) 0.15, (b) 0.95.

25.8b The adsorption of a gas is described by the Langmuir isotherm with $K=0.777\ kPa^{-1}$ at 25°C. Calculate the pressure at which the fractional surface coverage is (a) 0.20, (b) 0.75.

25.9a A certain solid sample adsorbs 0.44 mg of CO when the pressure of the gas is 26.0 kPa and the temperature is 300 K. The mass of gas adsorbed when the pressure is 3.0 kPa and the temperature is 300 K is 0.19 mg. The Langmuir isotherm is known to describe the adsorption. Find the fractional coverage of the surface at the two pressures.

25.9b A certain solid sample adsorbs 0.63 mg of CO when the pressure of the gas is 36.0 kPa and the temperature is 300 K. The mass of gas adsorbed when the pressure is 4.0 kPa and the temperature is 300 K is 0.21 mg. The Langmuir isotherm is known to describe the adsorption. Find the fractional coverage of the surface at the two pressures.

25.10a For how long on average would an H atom remain on a surface at 298 K if its desorption activation energy were (a) 15 kJ mol^{-1}, (b) 150 kJ mol^{-1}? Take $\tau_0=0.10$ ps. For how long on average would the same atoms remain at 1000 K?

25.10b For how long on average would an atom remain on a surface at 400 K if its desorption activation energy were (a) 20 kJ mol^{-1}, (b) 200 kJ mol^{-1}? Take $\tau_0=0.12$ ps. For how long on average would the same atoms remain at 800 K?

25.11a A solid in contact with a gas at 12 kPa and 25°C adsorbs 2.5 mg of the gas and obeys the Langmuir isotherm. The enthalpy change when 1.00 mmol of the adsorbed gas is desorbed is +10.2 J. What is the equilibrium pressure for the adsorption of 2.5 mg of gas at 40°C?

25.11b A solid in contact with a gas at 8.86 kPa and 25°C adsorbs 4.67 mg of the gas and obeys the Langmuir isotherm. The enthalpy change when 1.00 mmol of the adsorbed gas is desorbed is +12.2 J. What is the equilibrium pressure for the adsorption of the same mass of gas at 45°C?

25.12a Hydrogen iodide is very strongly adsorbed on gold but only slightly adsorbed on platinum. Assume the adsorption follows the Langmuir isotherm and predict the order of the HI decomposition reaction on each of the two metal surfaces.

25.12b Suppose it is known that ozone adsorbs on a particular surface in accord with a Langmuir isotherm. How could you use the pressure dependence of the fractional coverage to distinguish between adsorption (a) without dissociation, (b) with dissociation into $O + O_2$, (c) with dissociation into $O + O + O$?

25.13a Nitrogen gas adsorbed on charcoal to the extent of $0.921\ cm^3\ g^{-1}$ at 490 kPa and 190 K, but at 250 K the same amount of adsorption was achieved only when the pressure was increased to 3.2 MPa. What is the enthalpy of adsorption of nitrogen on charcoal?

25.13b Nitrogen gas adsorbed on a surface to the extent of $1.242\ cm^3\ g^{-1}$ at 350 kPa and 180 K, but at 240 K the same amount of adsorption was achieved only when the pressure was increased to 1.02 MPa. What is the enthalpy of adsorption of nitrogen on the surface?

25.14a In an experiment on the adsorption of oxygen on tungsten it was found that the same volume of oxygen was desorbed in 27 min at 1856 K and 2.0 min at 1978 K. What is the activation energy of desorption? How long would it take for the same amount to desorb at (a) 298 K, (b) 3000 K?

25.14b In an experiment on the adsorption of ethene on iron it was found that the same volume of the gas was desorbed in 1856 s at 873 K and 8.44 s at 1012 K. What is the activation energy of desorption? How long would it take for the same amount of ethene to desorb at (a) 298 K, (b) 1500 K?

25.15a The Helmholtz model of the electric double layer is equivalent to a parallel plate capacitor. Hence the potential difference across the double layer is given by $\Delta\varphi = \sigma d/\varepsilon$, where d is the distance between the plates and σ is the surface charge density. Assuming that this model holds for concentrated salt solutions calculate the magnitude of the electric field at the surface of silica in 5.0 M NaCl(aq) if the surface charge density is $0.10\ C\ m^{-2}$.

25.15b Refer to the preceding exercise. Calculate the magnitude of the electric field at the surface of silica in 4.5 M NaCl(aq) if the surface charge density is $0.12\ C\ m^{-2}$.

25.16a The transfer coefficient of a certain electrode in contact with M^{3+} and M^{4+} in aqueous solution at 25°C is 0.39. The current density is found to be $55.0\ mA\ cm^{-2}$ when the overvoltage is 125 mV. What is the overvoltage required for a current density of $75\ mA\ cm^{-2}$?

25.16b The transfer coefficient of a certain electrode in contact with M^{2+} and M^{3+} in aqueous solution at 25°C is 0.42. The current density is found to be $17.0\ mA\ cm^{-2}$ when the overvoltage is 105 mV. What is the overvoltage required for a current density of $72\ mA\ cm^{-2}$?

25.17a Determine the exchange current density from the information given in Exercise 25.16a.

25.17b Determine the exchange current density from the information given in Exercise 25.16b.

25.18a To a first approximation, significant evolution or deposition occurs in electrolysis only if the overpotential exceeds about 0.6 V. To illustrate this criterion determine the effect that increasing the overpotential from 0.40 V to 0.60 V has on the current density in the electrolysis of 1.0 M NaOH(aq), which is $1.0\ mA\ cm^{-2}$ at 0.4 V and 25°C. Take $\alpha = 0.5$.

25.18b Determine the effect that increasing the overpotential from 0.50 V to 0.60 V has on the current density in the electrolysis of 1.0 M NaOH(aq), which is $1.22\ mA\ cm^{-2}$ at 0.50 V and 25°C. Take $\alpha = 0.50$.

25.19a Use the data in Table 25.6 for the exchange current density and transfer coefficient for the reaction $2\ H^+ + 2\ e^- \rightarrow H_2$ on nickel at 25°C to determine what current density would be needed to obtain an overpotential of 0.20 V as calculated from (a) the Butler–Volmer equation, and (b) the Tafel equation. Is the validity of the Tafel approximation affected at higher overpotentials (of 0.4 V and more)?

25.19b Use the data in Table 25.6 for the exchange current density and transfer coefficient for the reaction $Fe^{3+} + e^- \rightarrow Fe^{2+}$ on platinum at 25°C to determine what current density would be needed to obtain an overpotential of 0.30 V as calculated from (a) the Butler–Volmer equation, and (b) the Tafel equation. Is the validity of the Tafel approximation affected at higher overpotentials (of 0.4 V and more)?

25.20a Estimate the limiting current density at an electrode in which the concentration of Ag^+ ions is $2.5\ mmol\ dm^{-3}$ at 25°C. The thickness of the Nernst diffusion layer is 0.40 mm. The ionic conductivity of Ag^+ at infinite dilution and 25°C is $6.19\ mS\ m^2\ mol^{-1}$.

25.20b Estimate the limiting current density at an electrode in which the concentration of Mg^{2+} ions is $1.5\ mmol\ dm^{-3}$ at 25°C. The thickness of the Nernst diffusion layer is 0.32 mm. The ionic conductivity of Mg^{2+} at infinite dilution and 25°C is $10.60\ mS\ m^2\ mol^{-1}$.

25.21a A 0.10 M $CdSO_4$(aq) solution is electrolysed between a cadmium cathode and a platinum anode with a current density of $1.00\ mA\ cm^{-2}$. The hydrogen overpotential is 0.60 V. What will be the concentration of Cd^{2+} ions when evolution of H_2 just begins at the cathode? Assume all activity coefficients are unity.

25.21b A 0.10 M $FeSO_4$(aq) solution is electrolysed between a magnesium cathode and a platinum anode with a current density of $1.50\ mA\ cm^{-2}$. The hydrogen overpotential is 0.60 V. What will be the concentration of Fe^{2+} ions when evolution of H_2 just begins at the cathode? Assume all activity coefficients are unity.

25.22a A typical exchange current density, that for H^+ discharge at platinum, is $0.79\ mA\ cm^{-2}$ at 25°C. What is the current density at an electrode when its overpotential is (a) 10 mV, (b) 100 mV, (c) −5.0 V? Take $\alpha = 0.5$.

25.22b The exchange current density for a $Pt|Fe^{3+},Fe^{2+}$ electrode is 2.5 mA cm^{-2}. The standard potential of the electrode is +0.77 V. Calculate the current flowing through an electrode of surface area $1.0\ cm^2$ as a function of the potential of the electrode. Take unit activity for both ions.

25.23a Suppose that the electrode potential is set at 1.00 V. The exchange current density is $6.0 \times 10^{-4}\ A\ cm^{-2}$ and $\alpha = 0.50$. Calculate the current density for the ratio of activities $a(Fe^{3+})/a(Fe^{2+})$ in the range 0.1 to 10.0 and at 25°C.

25.23b Suppose that the electrode potential is set at 0.50 V. Calculate the current density for the ratio of activities $a(Cr^{3+})/a(Cr^{2+})$ in the range 0.1 to 10.0 and at 25°C.

25.24a What overpotential is needed to sustain a current of 20 mA at a $Pt|Fe^{3+},Fe^{2+}$ electrode in which both ions are at a mean activity $a = 0.10$? The surface area of the electrode is $1.0\ cm^2$.

25.24b What overpotential is needed to sustain a current of 15 mA at a $Pt|Ce^{4+},Ce^{3+}$ electrode in which both ions are at a mean activity $a = 0.010$?

25.25a How many electrons or protons are transported through the double layer in each second when the $Pt,H_2|H^+$, $Pt|Fe^{3+},Fe^{2+}$, and $Pb,H_2|H^+$ electrodes are at equilibrium at 25°C? Take the area as $1.0\ cm^2$ in each case. Estimate the number of times each second a single atom on the surface takes part in a electron transfer event, assuming an electrode atom occupies about $(280\ pm)^2$ of the surface.

25.25b How many electrons or protons are transported through the double layer in each second when the $Cu,H_2|H^+$ and $Pt|Ce^{4+},Ce^{3+}$ electrodes are at equilibrium at 25°C? Take the area as $1.0\ cm^2$ in each case. Estimate the number of times each second a single atom on the surface takes part in a electron transfer event, assuming an electrode atom occupies about $(260\ pm)^2$ of the surface.

25.26a What is the effective resistance at 25°C of an electrode interface when the overpotential is small? Evaluate it for $1.0\ cm^2$ (a) $Pt,H_2|H^+$, (b) $Hg,H_2|H^+$ electrodes.

25.26b Evaluate the effective resistance at 25°C of an electrode interface for 1.0 cm^2 (a) Pb,H_2|H^+, (b) Pt|Fe^{2+},Fe^{3+} electrodes.

25.27a State what happens when a platinum electrode in an aqueous solution containing both Cu^{2+} and Zn^{2+} ions at unit activity is made the cathode of an electrolysis cell.

25.27b State what happens when a platinum electrode in an aqueous solution containing both Fe^{2+} and Ni^{2+} ions at unit activity is made the cathode of an electrolysis cell.

25.28a What are the conditions that allow a metal to be deposited from aqueous acidic solution before hydrogen evolution occurs significantly at 293 K? Why may silver be deposited from aqueous silver nitrate?

25.28b The overpotential for hydrogen evolution on cadmium is about 1 V at current densities of 1 mA cm^{-2}. Why may cadmium be deposited from aqueous cadmium sulfate?

25.29a The exchange current density for H^+ discharge at zinc is about 50 pA cm^{-2}. Can zinc be deposited from a unit activity aqueous solution of a zinc salt?

25.29b The standard potential of the Zn^{2+}|Zn electrode is −0.76 V at 25°C. The exchange current density for H^+ discharge at platinum is 0.79 mA cm^{-2}. Can zinc be plated on to platinum at that temperature? (Take unit activities.)

25.30a Can magnesium be deposited on a zinc electrode from a unit activity acid solution at 25°C?

25.30b Can iron be deposited on a copper electrode from a unit activity acid solution at 25°C?

25.31a Calculate the maximum (zero-current) potential difference of a nickel–cadmium cell, and the maximum possible power output when 100 mA is drawn at 25°C.

25.31b Calculate the maximum (zero-current) potential difference of a lead–acid cell, and the maximum possible power output when 100 mA is drawn at 25°C.

25.32a The corrosion current density j_{corr} at an iron anode is 1.0 A m^{-2}. What is the corrosion rate in millimetres per year? Assume uniform corrosion.

25.32b The corrosion current density j_{corr} at a zinc anode is 2.0 A m^{-2}. What is the corrosion rate in millimetres per year? Assume uniform corrosion.

Problems*

Numerical problems

25.1 The movement of atoms and ions on a surface depends on their ability to leave one position and stick to another, and therefore on the energy changes that occur. As an illustration, consider a two-dimensional square lattice of univalent positive and negative ions separated by 200 pm, and consider a cation on the upper terrace of this array. Calculate, by direct summation, its Coulombic interaction when it is in an empty lattice point directly above an anion. Now consider a high step in the same lattice, and let the cation move into the corner formed by the step and the terrace. Calculate the Coulombic energy for this position, and decide on the likely settling point for the cation.

25.2 In a study of the catalytic properties of a titanium surface it was necessary to maintain the surface free from contamination. Calculate the collision frequency per square centimetre of surface made by O_2 molecules at (a) 100 kPa, (b) 1.00 Pa and 300 K. Estimate the number of collisions made with a single surface atom in each second. The conclusions underline the importance of working at very low pressures (much lower than 1 Pa, in fact) in order to study the properties of uncontaminated surfaces. Take the nearest-neighbour distance as 291 pm.

25.3 Nickel is face-centred cubic with a unit cell of side 352 pm. What is the number of atoms per square centimetre exposed on a surface formed by (a) (100), (b) (110), (c) (111) planes? Calculate the frequency of molecular collisions per surface atom in a vessel containing (a) hydrogen, (b) propane at 25°C when the pressure is (i) 100 Pa, (ii) 0.10 μTorr.

25.4 The data below are for the chemisorption of hydrogen on copper powder at 25°C. Confirm that they fit the Langmuir isotherm at low coverages. Then find the value of K for the adsorption equilibrium and the adsorption volume corresponding to complete coverage.

p/Pa	25	129	253	540	1000	1593
V/cm^3	0.042	0.163	0.221	0.321	0.411	0.471

25.5 The data for the adsorption of ammonia on barium fluoride are reported below. Confirm that they fit a BET isotherm and find values of c and V_{mon}.

(a) $\theta = 0°C$, $p^* = 429.6$ kPa:

p/kPa	14.0	37.6	65.6	79.2	82.7	100.7	106.4
V/cm^3	11.1	13.5	14.9	16.0	15.5	17.3	16.5

(b) $\theta = 18.6°C$, $p^* = 819.7$ kPa:

p/kPa	5.3	8.4	14.4	29.2	62.1	74.0	80.1	102.0
V/cm^3	9.2	9.8	10.3	11.3	12.9	13.1	13.4	14.1

25.6 The following data have been obtained for the adsorption of H_2 on the surface of 1.00 g of copper at 0°C. The volume of H_2 below is the volume that the gas would occupy at STP (0°C and 1 atm).

p/atm	0.050	0.100	0.150	0.200	0.250
V/cm^3	1.22	1.33	1.31	1.36	1.40

Determine the volume of H_2 necessary to form a monolayer and estimate the surface area of the copper sample. The density of liquid hydrogen is 0.708 g cm^{-3}.

25.7 The adsorption of solutes on solids from liquids often follows a Freundlich isotherm. Check the applicability of this isotherm to the following data for the adsorption of acetic acid on charcoal at 25°C and find the values of the parameters c_1 and c_2.

[acid]/(mol dm^{-3})	0.05	0.10	0.50	1.0	1.5
w_a/g	0.04	0.06	0.12	0.16	0.19

w_a is the mass adsorbed per unit mass of charcoal.

25.8 In some catalytic reactions the products may adsorb more strongly than the reacting gas. This is the case, for instance, in the catalytic decomposition of ammonia on platinum at 1000°C. As a first step in examining the kinetics of

* Problems denoted with the symbol ‡ were supplied by Charles Trapp, Carmen Giunta, and Marshall Cady.

this type of process, show that the rate of ammonia decomposition should follow

$$\frac{dp_{NH_3}}{dt} = -k_c \frac{p_{NH_3}}{p_{H_2}}$$

in the limit of very strong adsorption of hydrogen. Start by showing that, when a gas J adsorbs very strongly, and its pressure is p_J, the fraction of uncovered sites is approximately $1/Kp_J$. Solve the rate equation for the catalytic decomposition of NH_3 on platinum and show that a plot of $F(t) = (1/t)\ln(p/p_0)$ against $G(t) = (p - p_0)/t$, where p is the pressure of ammonia, should give a straight line from which k_c can be determined. Check the rate law on the basis of the data below, and find k_c for the reaction.

t/s	0	30	60	100	160	200	250
p/kPa	13.3	11.7	11.2	10.7	10.3	9.9	9.6

25.9‡ A. Akgerman and M. Zardkoohi (*J. Chem. Eng. Data* **41**, 185 (1996)) examined the adsorption of phenol from aqueous solution on to fly ash at 20°C. They fitted their observations to a Freundlich isotherm of the form $c_{ads} = Kc_{sol}^{1/n}$, where c_{ads} is the concentration of adsorbed phenol and c_{sol} is the concentration of aqueous phenol. Among the data reported are the following:

$c_{sol}/(\text{mg g}^{-1})$	8.26	15.65	25.43	31.74	40.00
$c_{ads}/(\text{mg g}^{-1})$	4.4	19.2	35.2	52.0	67.2

Determine the constants K and n. What further information would be necessary in order to express the data in terms of fractional coverage, θ?

25.10‡ C. Huang and W.P. Cheng (*J. Colloid Interface Sci.* **188**, 270 (1997)) examined the adsorption of the hexacyanoferrate(III) ion, $[Fe(CN)_6]^{3-}$, on γ-Al_2O_3 from aqueous solution. They modelled the adsorption with a modified Langmuir isotherm, obtaining the following values of K at pH = 6.5:

T/K	283	298	308	318
$10^{-11}K$	2.642	2.078	1.286	1.085

Determine the isosteric enthalpy of adsorption, $\Delta_{ads}H^\ominus$, at this pH. The researchers also reported $\Delta_{ads}S^\ominus = +146\ \text{J mol}^{-1}\ \text{K}^{-1}$ under these conditions. Determine $\Delta_{ads}G^\ominus$.

25.11‡ M.-G. Olivier and R. Jadot (*J. Chem. Eng. Data* **42**, 230 (1997)) studied the adsorption of butane on silica gel. They report the following amounts of absorption (in moles per kilogram of silica gel) at 303 K:

p/kPa	31.00	38.22	53.03	76.38	101.97
$n/(\text{mol kg}^{-1})$	1.00	1.17	1.54	2.04	2.49
p/kPa	130.47	165.06	182.41	205.75	219.91
$n/(\text{mol kg}^{-1})$	2.90	3.22	3.30	3.35	3.36

Fit these data to a Langmuir isotherm, and determine the value of n that corresponds to complete coverage and the constant K.

25.12‡ The following data were obtained for the extent of adsorption, s, of acetone on charcoal from an aqueous solution of molar concentration, c, at 18°C.

$c/(\text{mmol dm}^{-3})$	15.0	23.0	42.0	84.0	165	390	800
s/(mmol acetone/g charcoal)	0.60	0.75	1.05	1.50	2.15	3.50	5.10

Which isotherm fits this data best, Langmuir, Freundlich, or Temkin?

25.13 In an experiment on the Pt|H_2|H^+ electrode in dilute H_2SO_4 the following current densities were observed at 25°C. Evaluate α and j_0 for the electrode.

η/mV	50	100	150	200	250
$j/(\text{mA cm}^{-2})$	2.66	8.91	29.9	100	335

How would the current density at this electrode depend on the overpotential of the same set of magnitudes but of opposite sign?

25.14 The standard potentials of lead and tin are −126 mV and −136 mV respectively at 25°C, and the overvoltage for their deposition are close to zero. What should their relative activities be in order to ensure simultaneous deposition from a mixture?

25.15 The limiting current density for the reaction $I_3^- + 2\,e^- \rightarrow 3\,I^-$ at a platinum electrode is 28.9 μA cm^{-2} when the concentration of KI is 6.6×10^{-4} mol dm^{-3} and the temperature 25°C. The diffusion coefficient of I_3^- is $1.14 \times 10^{-9}\ \text{m}^2\ \text{s}^{-3}$. What is the thickness of the diffusion layer?

25.16 Estimating the power output and potential of a cell under operating conditions is very difficult, but eqn 25.65 summarizes, in an approximate way, some of the parameters involved. As a first step in manipulating this expression, identify all the quantities that depend on the ionic concentrations. Express E in terms of the concentration and conductivities of the ions present in the cell. Estimate the parameters for Zn(s)|$ZnSO_4$(aq)||$CuSO_4$(aq)|Cu(s). Take electrodes of area 5 cm^2 separated by 5 cm. Ignore both potential differences and resistance of the liquid junction. Take the concentration as 1 mol dm^{-3}, the temperature 25°C, and neglect activity coefficients. Plot E as a function of the current drawn. On the same graph, plot the power output of the cell. What current corresponds to maximum power?

25.17 Consider a cell in which the current is activation-controlled. Show that the current for maximum power can be estimated by plotting $\log(I/I_0)$ and $c_1 - c_2 I$ against I (where $I_0 = A^2 j_0 j_0'$ and c_1 and c_2 are constants), and looking for the point of intersection of the curves. Carry through this analysis for the cell in Problem 25.16 ignoring all concentration overpotentials.

25.18‡ The rate of deposition of iron, v, on the surface of an iron electrode from an aqueous solution of Fe^{2+} has been studied as a function of potential, E, relative to the standard hydrogen electrode, by J. Kanya (*J. Electroanal. Chem.* **84**, 83 (1977)). The values in the table below are based on the data obtained with an electrode of surface area 9.1 cm^2 in contact with a solution of concentration 1.70 μmol dm^{-3} in Fe^{2+}. (a) Assuming unit activity coefficients, calculate the zero current potential of the Fe^{2+}/Fe cathode and the overpotential at each value of the working potential. (b) Calculate the cathodic current density, j_c, from the rate of deposition of Fe^{2+} for each value of E. (c) Examine the extent to which the data fit the Tafel equation and calculate the exchange current density.

$v/(\text{pmol s}^{-1})$	1.47	2.18	3.11	7.26
$-E$/mV	702	727	752	812

25.19‡ The thickness of the diffuse double layer according to the Gouy–Chapman model is given by eqn 19.46. Use this equation to calculate and plot the thickness as a function of concentration and electrolyte type at 25°C. For examples, choose aqueous solutions of NaCl and Na_2SO_4 ranging in concentration from 0.1 to 100 mmol dm^{-3}.

25.20‡ V.V. Losev and A.P. Pchel'nikov (*Soviet Electrochem.* **6**, 34 (1970)) obtained the following current–voltage data for an indium anode relative to a standard hydrogen electrode at 293 K:

$-E$/V	0.388	0.365	0.350	0.335
$j/(\text{A m}^{-2})$	0	0.590	1.438	3.507

Use these data to calculate the transfer coefficient and the exchange current density. What is the cathodic current density when the potential is 0.365 V?

25.21‡ The redox reactions of quinones have been the subject of many studies over the years and they continue to be of interest to electrochemists. In a study of methone (1,1-dimethyl-3,5-cyclohexanedione) by E. Kariv, J. Hermolin, and E. Gileadi (*Electrochim. Acta* **16**, 1437 (1971)), the following current–voltage data were obtained for the reduction of the quinone in anhydrous butanol on a mercury electrode:

$-E$/V	1.50	1.58	1.63	1.72	1.87	1.98	≥2.10
$j/(\text{A m}^{-2})$	10	30	50	100	200	250	290

(a) How well do these data fit the empirical Tafel equation? (b) The authors postulate that the reduction product is the dimer HMMH formed by the following mechanism (where the quinone is denoted M):

(1) $M(sol) \rightleftharpoons M(ads)$

(2) $M(ads) + H^+ + e^- \rightarrow MH(ads)$

(3) $MH(ads) + MH(ads) \rightleftharpoons HMMH$

The affixes sol and ads refer to species in solution and on the surface of the electrode, respectively. Does this mechanism help to explain the current–voltage data?

25.22‡ An early study of the hydrogen overpotential is that of H. Bowden and T. Rideal (*Proc. Roy. Soc.* **A120**, 59 (1928)), who measured the overpotential for H_2 evolution with a mercury electrode in dilute aqueous solutions of H_2SO_4 at 25°C. Determine the exchange current density and transfer coefficient, α, from their data:

$j/(\mathrm{mA\ m^{-2}})$	2.9	6.3	28	100	250	630	1650	3300
η/V	0.60	0.65	0.73	0.79	0.84	0.89	0.93	0.96

Explain any deviations from the result expected from the Tafel equation.

Theoretical problems

25.23 Although the attractive van der Waals interaction between individual molecules varies as R^{-6}, the interaction of a molecule with a nearby solid (a homogeneous collection of molecules) varies as R^{-3}, where R is its vertical distance above the surface. Confirm this assertion. Calculate the interaction energy between an Ar atom and the surface of solid argon on the basis of a Lennard-Jones (6,12)-potential. Estimate the equilibrium distance of an atom above the surface.

25.24 Use the Gibbs adsorption isotherm (another name for eqn 19.50), to show that the volume adsorbed per unit area of solid, V_a/σ, is related to the pressure of the gas by $V_a = -(\sigma/RT)(\mathrm{d}\mu/\mathrm{d}\ln p)$, where μ is the chemical potential of the adsorbed gas.

25.25 If the dependence of the chemical potential of the gas on the extent of surface coverage is known, the Gibbs adsorption isotherm, eqn 19.50, can be integrated to give a relation between V_a and p, as in a normal adsorption isotherm. For instance, suppose that the change in the chemical potential of a gas when it adsorbs is of the form $\mathrm{d}\mu = -c_2(RT/\sigma)\mathrm{d}V_a$, where c_2 is a constant of proportionality: show that the Gibbs isotherm leads to the Freundlich isotherm in this case.

25.26 Finally we come full circle and return to the Langmuir isotherm. Find the form of $\mathrm{d}\mu$ that, when inserted in the Gibbs adsorption isotherm, leads to the Langmuir isotherm.

25.27 Show that, for the association part of the surface plasmon resonance experiment in Fig. 25.27, $R(t) = R_{eq}(1 - e^{-k_{obs}t})$ and write an expression for k_{obs}. Then, derive an expression for $R(t)$ that applies to the dissociation part of the surface plasmon resonance experiment in Fig. 25.27.

25.28 If $\alpha = \frac{1}{2}$, an electrode interface is unable to rectify alternating current because the current density curve is symmetrical about $\eta = 0$. When $\alpha \neq \frac{1}{2}$, the magnitude of the current density depends on the sign of the overpotential, and so some degree of 'faradaic rectification' may be obtained. Suppose that the overpotential varies as $\eta = \eta_0 \cos \omega t$. Derive an expression for the mean flow of current (averaged over a cycle) for general α, and confirm that the mean current is zero when $\alpha = \frac{1}{2}$. In each case work in the limit of small η_0 but to second order in $\eta_0 F/RT$. Calculate the mean direct current at 25°C for a 1.0 cm^2 hydrogen–platinum electrode with $\alpha = 0.38$ when the overpotential varies between ±10 mV at 50 Hz.

25.29 Now suppose that the overpotential is in the high overpotential region at all times even though it is oscillating. What waveform will the current across the interface show if it varies linearly and periodically (as a sawtooth waveform) between η_- and η_+ around η_0? Take $\alpha = \frac{1}{2}$.

25.30 Derive an expression for the current density at an electrode where the rate process is diffusion-controlled and η_c is known. Sketch the form of j/j_L as a function of η_c. What changes occur if anion currents are involved?

Applications: to chemical engineering and environmental science

25.31 The designers of a new industrial plant wanted to use a catalyst code-named CR-1 in a step involving the fluorination of butadiene. As a first step in the investigation they determined the form of the adsorption isotherm. The volume of butadiene adsorbed per gram of CR-1 at 15°C varied with pressure as given below. Is the Langmuir isotherm suitable at this pressure?

p/kPa	13.3	26.7	40.0	53.3	66.7	80.0
V/cm^3	17.9	33.0	47.0	60.8	75.3	91.3

Investigate whether the BET isotherm gives a better description of the adsorption of butadiene on CR-1. At 15°C, p^*(butadiene) = 200 kPa. Find V_{mon} and c.

25.32‡ In a study relevant to automobile catalytic converters, C.E. Wartnaby, A. Stuck, Y.Y. Yeo, and D.A. King (*J. Phys. Chem.* **100**, 12483 (1996)) measured the enthalpy of adsorption of CO, NO, and O_2 on initially clean platinum 110 surfaces. They report $\Delta_{ads}H^{\ominus}$ for NO to be -160 kJ mol^{-1}. How much more strongly adsorbed is NO at 500°C than at 400°C?

25.33‡ The removal or recovery of volatile organic compounds (VOCs) from exhaust gas streams is an important process in environmental engineering. Activated carbon has long been used as an adsorbent in this process, but the presence of moisture in the stream reduces its effectiveness. M.-S. Chou and J.-H. Chiou (*J. Envir. Engrg.* ASCE, **123**, 437(1997)) have studied the effect of moisture content on the adsorption capacities of granular activated carbon (GAC) for normal hexane and cyclohexane in air streams. From their data for dry streams containing cyclohexane, shown in the table below, they conclude that GAC obeys a Langmuir type model in which $q_{VOC,RH=0} = abc_{VOC}/(1 + bc_{VOC})$, where $q = m_{VOC}/m_{GAC}$, RH denotes relative humidity, a the maximum adsorption capacity, b is an affinity parameter, and c is the abundance in parts per million (ppm). The following table gives values of $q_{VOC,RH=0}$ for cyclohexane:

c/ppm	33.6°C	41.5°C	57.4°C	76.4°C	99°C
200	0.080	0.069	0.052	0.042	0.027
500	0.093	0.083	0.072	0.056	0.042
1000	0.101	0.088	0.076	0.063	0.045
2000	0.105	0.092	0.083	0.068	0.052
3000	0.112	0.102	0.087	0.072	0.058

(a) By linear regression of $1/q_{VOC,RH=0}$ against $1/c_{VOC}$, test the goodness of fit and determine values of a and b. (b) The parameters a and b can be related to $\Delta_{ads}H$, the enthalpy of adsorption, and $\Delta_b H$, the difference in activation energy for adsorption and desorption of the VOC molecules, through Arrhenius type equations of the form $a = k_a\exp(-\Delta_{ads}H/RT)$ and $b = k_b\exp(-\Delta_b H/RT)$. Test the goodness of fit of the data to these equations and obtain values for k_a, k_b, $\Delta_{ads}H$, and $\Delta_b H$. (c) What interpretation might you give to k_a and k_b?

25.34‡ M.-S. Chou and J.-H. Chiou (*J. Envir. Engrg.*, ASCE, **123**, 437(1997)) have studied the effect of moisture content on the adsorption capacities of granular activated carbon (GAC, Norit PK 1–3) for the volatile organic compounds (VOCs) normal hexane and cyclohexane in air streams. The following table shows the adsorption capacities ($q_{water} = m_{water}/m_{GAC}$) of GAC for pure water from moist air streams as a function of relative humidity (RH) in the absence of VOCs at 41.5°C.

RH	0.00	0.26	0.49	0.57	0.80	1.00
q_{water}	0.00	0.026	0.072	0.091	0.161	0.229

The authors conclude that the data at this and other temperatures obey a Freundlich type isotherm, $q_{\text{water}} = k(\text{RH})^{1/n}$. (a) Test this hypothesis for their data at 41.5°C and determine the constants k and n. (b) Why might VOCs obey the Langmuir model, but water the Freundlich model? (c) When both water vapour and cyclohexane were present in the stream the values given in the table below were determined for the ratio $r_{\text{VOC}} = q_{\text{VOC}}/q_{\text{VOC, RH=0}}$ at 41.5°C.

RH	0.00	0.10	0.25	0.40	0.53	0.76	0.81
r_{VOC}	1.00	0.98	0.91	0.84	0.79	0.67	0.61

The authors propose that these data fit the equation $r_{\text{VOC}} = 1 - q_{\text{water}}$. Test their proposal and determine values for k and n and compare to those obtained in part (b) for pure water. Suggest reasons for any differences.

25.35‡ The release of petroleum products by leaky underground storage tanks is a serious threat to clean ground water. BTEX compounds (benzene, toluene, ethylbenzene, and xylenes) are of primary concern due to their ability to cause health problems at low concentrations. D.S. Kershaw, B.C. Kulik, and S. Pamukcu (*J. Geotech. & Geoenvir. Engrg.* **123**, 324(1997)) have studied the ability of ground tyre rubber to sorb (adsorb and absorb) benzene and *o*-xylene. Though sorption involves more than surface interactions, sorption data are usually found to fit one of the adsorption isotherms. In this study, the authors have tested how well their data fit the linear ($q = Kc_{\text{eq}}$), Freundlich ($q = K_F c_{\text{eq}}^{1/n}$), and Langmuir ($q = K_L M c_{\text{eq}}/(1 + K_L c_{\text{eq}})$) type isotherms, where q is the mass of solvent sorbed per gram of ground rubber (in milligrams per gram), the Ks and M are empirical constants, and c_{eq} the equilibrium concentration of contaminant in solution (in milligrams per litre). (a) Determine the units of the empirical constants. (b) Determine which of the isotherms best fits the data in the table below for the sorption of benzene on ground rubber.

$c_{\text{eq}}/(\text{mg dm}^{-3})$	97.10	36.10	10.40	6.51	6.21	2.48
$q/(\text{mg g}^{-1})$	7.13	4.60	1.80	1.10	0.55	0.31

(c) Compare the sorption efficiency of ground rubber to that of granulated activated charcoal, which for benzene has been shown to obey the Freundlich isotherm in the form $q = 1.0c_{\text{eq}}^{1.6}$ with coefficient of determination $R^2 = 0.94$.

25.36 Calculate the thermodynamic limit to the zero-current potential of fuel cells operating on (a) hydrogen and oxygen, (b) methane and air, and (c) propane and air. Use the Gibbs energy information in the *Data section*, and take the species to be in their standard states at 25°C.

25.37 For each group below, determine which metal has a thermodynamic tendency to corrode in moist air at pH = 7. Take as a criterion of corrosion a metal ion concentration of at least 10^{-6} mol dm^{-3}.

(a) Fe, Cu, Pb, Al, Ag, Cr, Co

(b) Ni, Cd, Mg, Ti, Mn

25.38 Estimate the magnitude of the corrosion current for a patch of zinc of area 0.25 cm^2 in contact with a similar area of iron in an aqueous environment at 25°C. Take the exchange current densities as 1 μA cm^{-2} and the local ion concentrations as 1 μmol dm^{-3}.

25.39 The corrosion potential of iron immersed in a de-aerated acidic solution of pH = 3 is −0.720 V as measured at 25°C relative to the standard calomel electrode with potential 0.2802 V. A Tafel plot of cathodic current density against overpotential yields a slope of 18 V^{-1} and the hydrogen ion exchange current density $j_0 = 0.10$ μA cm^{-2}. Calculate the corrosion rate in milligrams of iron per square centimetre per day (mg cm^{-2} d^{-1}).

Appendix 1 Quantities, units, and notational conventions

A1

The result of a measurement is a **physical quantity** (such as mass or density) that is reported as a numerical multiple of an agreed **unit**:

physical quantity = numerical value × unit

For example, the mass of an object may be reported as $m = 2.5$ kg and its density as $d = 1.010$ kg dm^{-3} where the units are, respectively, 1 kilogram (1 kg) and 1 kilogram per decimetre cubed (1 kg dm^{-3}). Units are treated like algebraic quantities, and may be multiplied, divided, and cancelled. Thus, the expression (physical quantity)/unit is simply the numerical value of the measurement in the specified units, and hence is a dimensionless quantity. For instance, the mass reported above could be denoted m/kg = 2.5 and the density as d/(kg dm^{-3}) = 1.01.

Physical quantities are denoted by italic or (sloping) Greek letters (as in m for mass and Π for osmotic pressure). Units are denoted by Roman letters (as in m for metre).

Names of quantities

A **substance** is a distinct, pure form of matter. The **amount of substance**, n (more colloquially 'number of moles' or 'chemical amount'), in a sample is reported in terms of the **mole** (mol): 1 mol is the amount of substance that contains as many objects (atoms, molecules, ions, or other specified entities) as there are atoms in exactly 12 g of carbon-12. This number is found experimentally to be approximately 6.02×10^{23} (see the endpapers for more precise values). If a sample contains N entities, the amount of substance it contains is $n = N/N_A$, where N_A is the Avogadro constant: $N_A = 6.02 \times 10^{23}$ mol^{-1}. Note that N_A is a quantity with units, not a pure number.

An **extensive property** is a property that depends on the amount of substance in the sample. Two examples are mass and volume. An **intensive property** is a property that is independent of the amount of substance in the sample. Examples are temperature, mass density (mass divided by volume), and pressure.

A **molar property**, X_m, is the value of an extensive property, X, of the sample divided by the amount of substance present in the sample: $X_m = X/n$. A molar property is intensive. An example is the **molar volume**, V_m, the volume of a sample divided by the amount of substance in the sample (the volume per mole). The one exception to the notation X_m is the **molar mass**, which is denoted M. The molar mass of an element is the mass per mole of its atoms. The molar mass of a molecular compound is the mass per mole of molecules, and the molar mass of an ionic compound is the mass per mole of formula units. A **formula unit** of an ionic compound is an assembly of ions corresponding to the chemical formula of the compound; so the formula unit NaCl consists of one Na^+ ion and one Cl^- ion. The names *atomic weight* and *molecular weight* are still widely used in place of molar mass (often with the units omitted), but we shall not use them in this text.

The **molar concentration** ('molarity') of a solute in a solution is the amount ofsubstance of the solute divided by the volume of the solution. Molar concentration is usually expressed in moles per decimetre cubed (mol dm^{-3} or mol L^{-1}; 1 dm^3 is identical to 1 L). A solution in which the molar concentration of the solute is 1 mol dm^{-3} is prepared by dissolving 1 mol of the solute in sufficient solvent to prepare 1 dm^3 of solution. Such a solution is widely called a '1 molar' solution and denoted 1 M. The term **molality** refers to the amount of substance of solute divided by the mass of solvent used to prepare the solution. Its units are typically moles of solute per kilogram of solvent (mol kg^{-1}).

Units

In the **International System** of units (SI, from the French *Système International d'Unités*), the units are formed from seven **base units** listed in Table A1.1. All other physical quantities may be expressed as combinations of these physical quantities and reported in terms of **derived units**. Thus, volume is $(\text{length})^3$ and may be reported as a multiple of 1 metre cubed (1 m^3), and density, which is mass/volume, may be reported as a multiple of 1 kilogram per metre cubed (1 kg m^{-3}).

A number of derived units have special names and symbols. The names of units derived from names of people are lower case (as in torr, joule, pascal, and kelvin), but their symbols are upper case (as in Torr, J, Pa, and K). Among the most important for our purposes are those listed in Table A1.2. In all cases (both for base and derived quantities), the units may be modified by a prefix that denotes a factor of a power of 10. In a perfect world, Greek prefixes of units are upright (as in μm) and sloping for physical properties (as in μ for chemical potential), but available typefaces are not always so obliging. Among the most common prefixes are those listed in Table A1.3. Examples of the use of these prefixes are

$$1\ \text{nm} = 10^{-9}\ \text{m} \qquad 1\ \text{ps} = 10^{-12}\ \text{s} \qquad 1\ \mu\text{mol} = 10^{-6}\ \text{mol}$$

Table A1.1 The SI base units

Physical quantity	Symbol for quantity	Base unit
Length	l	metre, m
Mass	M	kilogram, kg
Time	t	second, s
Electric current	I	ampere, A
Thermodynamic temperature	T	kelvin, K
Amount of substance	n	mole, mol
Luminous intensity	I	candela, cd

Table A1.2 A selection of derived units

Physical quantity	Derived unit*	Name of derived unt
Force	1 kg m s^{-2}	newton, N
Pressure	1 kg m^{-1} s^{-2}	pascal, Pa
	1 N m^{-2}	
Energy	1 kg m^{2} s^{-2}	joule, J
	1 N m	
	1 Pa m^{3}	
Power	kg m^{2} s^{-3}	watt, W
	1 J s^{-1}	

* Equivalent definitions in terms of derived units are given following the definition in terms of base units.

Table A1.3 Common SI prefixes

Prefix	z	a	f	p	n	μ	m	c	d
Name	zepto	atto	femto	pico	nano	micro	milli	centi	deci
Factor	10^{-21}	10^{-18}	10^{-15}	10^{-12}	10^{-9}	10^{-6}	10^{-3}	10^{-2}	10^{-1}
Prefix	k	M	G	T					
Name	kilo	mega	giga	tera					
Factor	10^{3}	10^{6}	10^{9}	10^{12}					

The kilogram (kg) is anomalous: although it is a base unit, it is interpreted as 10^3 g, and prefixes are attached to the gram (as in 1 mg $= 10^{-3}$ g). Powers of units apply to the prefix as well as the unit they modify:

$$1\ \text{cm}^3 = 1\ (\text{cm})^3 = 1\ (10^{-2}\ \text{m})^3 = 10^{-6}\ \text{m}^3$$

Note that 1 cm^3 does not mean 1 c(m^3). When carrying out numerical calculations, it is usually safest to write out the numerical value of an observable as powers of 10.

There are a number of units that are in wide use but are not a part of the International System. Some are exactly equal to multiples of SI units. These include the *litre* (L), which is exactly 10^3 cm^3 (or 1 dm^3) and the *atmosphere* (atm), which is exactly 101.325 kPa. Others rely on the values of fundamental constants, and hence are liable to change when the values of the fundamental constants are modified by more accurate or more precise measurements. Thus, the size of the energy unit *electronvolt* (eV), the energy acquired by an electron that is accelerated through a potential difference of exactly 1 V, depends on the value of the charge of the electron, and the present (2005) conversion factor is 1 eV $= 1.602\ 177\ 33 \times 10^{-19}$ J. Table A1.4 gives the conversion factors for a number of these convenient units.

Notational conventions

We use SI units and IUPAC conventions throughout (see *Further reading*), except in a small number of cases. The default numbering of equations is ($C.n$), where C is the chapter; however, [$C.n$] is used to denote a definition and {$C.n$} is used to indicate that a variable x should be interpreted as $x/x^{\ominus}$, where $x^{\ominus}$ is a standard value. A subscript r

Table A1.4 Some common units

Physical quantity	Name of unit	Symbol for unit	Value*
Time	minute	min	60 s
	hour	h	3600 s
	day	d	86 400 s
Length	ångström	Å	10^{-10} m
Volume	litre	L, l	1 dm^3
Mass	tonne	t	10^3 kg
Pressure	bar	bar	10^5 Pa
	atmosphere	atm	101.325 kPa
Energy	electronvolt	eV	$1.602\ 177\ 33 \times 10^{-19}$ J
			96.485 31 kJ mol^{-1}

* All values in the final column are exact, except for the definition of 1 eV, which depends on the measured value of *e*.

attached to an equation number indicates that the equation is valid only for a reversible change. A superscript ° indicates that the equation is valid for an ideal system, such as a perfect gas or an ideal solution.

We use

$$p^{\ominus} = 1\ \text{bar} \qquad b^{\ominus} = 1\ \text{mol kg}^{-1} \qquad 1\ c^{\ominus} = 1\ \text{mol dm}^{-3}$$

When referring to temperature, T denotes a thermodynamic temperature (for instance, on the Kelvin scale) and θ a temperature on the Celsius scale.

For numerical calculations, we take special care to use the proper number of significant figures. Unless otherwise specified, assume that zeros in data like 10, 100, 1000, etc are significant (that is, interpret such data as 10., 100., 1000., etc).

Further reading

I.M. Mills (ed.), *Quantities, units, and symbols in physical chemistry.* Blackwell Scientific, Oxford (1993).

Appendix 2 Mathematical techniques

A2

Basic procedures

A2.1 Logarithms and exponentials

The **natural logarithm** of a number x is denoted $\ln x$, and is defined as the power to which $e = 2.718\ldots$ must be raised for the result to be equal to x. It follows from the definition of logarithms that

$$\ln x + \ln y + \cdots = \ln xy \cdots \quad \text{(A2.1)}$$

$$\ln x - \ln y = \ln(x/y) \quad \text{(A2.2)}$$

$$a \ln x = \ln x^a \quad \text{(A2.3)}$$

We also encounter the **common logarithm** of a number, $\log x$, the logarithm compiled with 10 in place of e. Common logarithms follow the same rules of addition and subtraction as natural logarithms. Common and natural logarithms are related by

$$\ln x = \ln 10 \log x \approx 2.303 \log x \quad \text{(A2.4)}$$

The **exponential function**, e^x, plays a very special role in the mathematics of chemistry. The following properties are important:

$$e^x e^y e^z \ldots = e^{x+y+z+\cdots} \quad \text{(A2.5)}$$

$$e^x/e^y = e^{x-y} \quad \text{(A2.6)}$$

$$(e^x)^a = e^{ax} \quad \text{(A2.7)}$$

A2.2 Complex numbers and complex functions

Complex numbers have the form

$$z = x + iy \quad \text{(A2.8)}$$

where $i = (-1)^{1/2}$. The real numbers x and y are, respectively, the real and imaginary parts of z, denoted $\mathrm{Re}(z)$ and $\mathrm{Im}(z)$. We write the **complex conjugate** of z, denoted z^*, by replacing i by −i:

$$z^* = x - iy \quad \text{(A2.9)}$$

The **absolute value** or **modulus** of the complex number z is denoted $|z|$ and is given by:

$$|z| = (z^*z)^{1/2} = (x^2 + y^2)^{1/2} \quad \text{(A2.10)}$$

The following rules apply for arithmetic operations involving complex numbers:

1 Addition. If $z = x + iy$ and $z' = x' + iy'$, then

$$z \pm z' = (x \pm x') + i(y \pm y') \quad \text{(A2.11)}$$

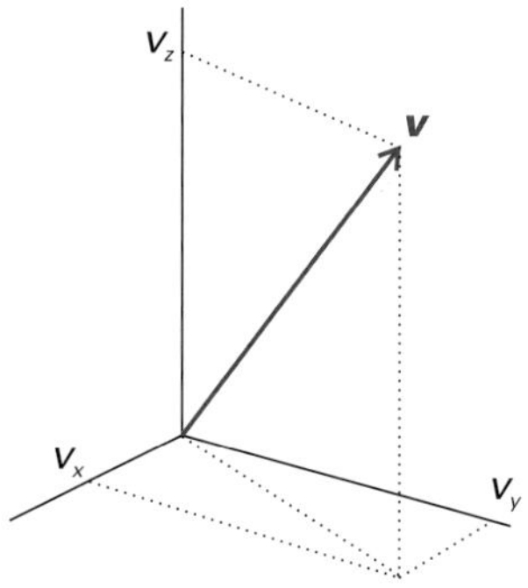

Fig. A2.1 The vector $\boldsymbol{v}$ has components v_x, v_y, and v_z on the x, y, and z axes with magnitudes v_x, v_y, and v_z, respectively.

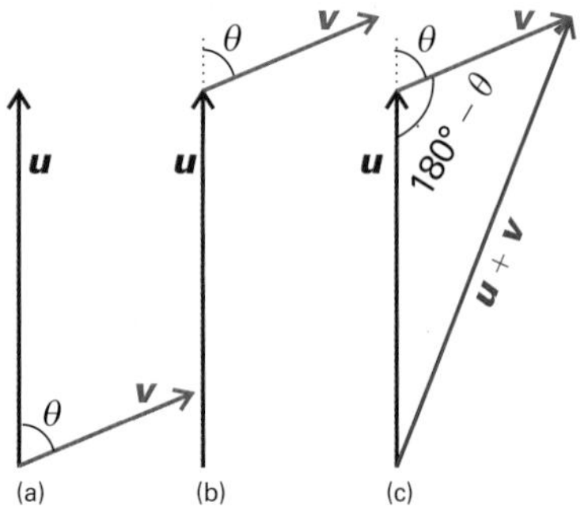

Fig. A2.2 (a) The vectors $\boldsymbol{v}$ and $\boldsymbol{u}$ make an angle θ. (b) To add $\boldsymbol{u}$ to $\boldsymbol{v}$, we first join the tail of $\boldsymbol{u}$ to the head of $\boldsymbol{v}$, making sure that the angle θ between the vectors remains unchanged. (c) To finish the process, we draw the resultant vector by joining the tail of $\boldsymbol{u}$ to the head of $\boldsymbol{v}$.

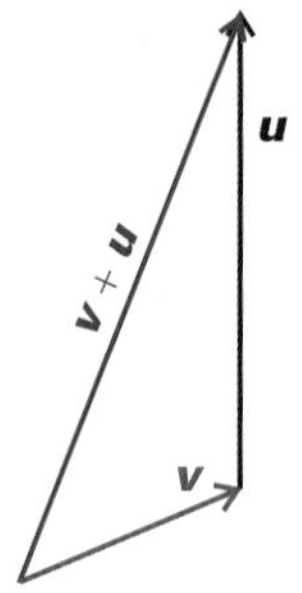

Fig. A2.3 The result of adding the vector $\boldsymbol{v}$ to the vector $\boldsymbol{u}$, with both vectors defined in Fig. A2.2a. Comparison with the result shown in Fig. A2.2c for the addition of $\boldsymbol{u}$ to $\boldsymbol{v}$ shows that reversing the order of vector addition does not affect the result.

2 Multiplication. For z and z' defined above,

$$z \times z' = (x + iy)(x' + iy') = (xx' - yy') + i(xy' + yx') \tag{A2.12}$$

3 Division. For z and z' defined above,

$$\frac{z}{z'} = \frac{z(z')^*}{|z'|^2} \tag{A2.13}$$

Functions of complex arguments are useful in the discussion of wave equations (Chapter 8). We write the complex conjugate, f^*, of a complex function, f, by replacing i wherever it occurs by –i. For instance, the complex conjugate of e^{ix} is e^{-ix}.

Complex exponential functions may be written in terms of trigonometric functions. For example,

$$e^{\pm ix} = \cos x \pm i \sin x \tag{A2.14}$$

which implies that

$$\cos x = \tfrac{1}{2}(e^{ix} + e^{-ix}) \tag{A2.15}$$

$$\sin x = -\tfrac{1}{2}i(e^{ix} - e^{-ix}) \tag{A2.16}$$

A2.3 Vectors

A vector quantity has both magnitude and direction. The vector shown in Fig. A2.1 has components on the x, y, and z axes with magnitudes v_x, v_y, and v_z, respectively. The vector may be represented as

$$\boldsymbol{v} = v_x\boldsymbol{i} + v_y\boldsymbol{j} + v_z\boldsymbol{k} \tag{A2.17}$$

where $\boldsymbol{i}$, $\boldsymbol{j}$, and $\boldsymbol{k}$ are **unit vectors**, vectors of magnitude 1, pointing along the positive directions on the x, y, and z axes. The magnitude of the vector is denoted v or $|\boldsymbol{v}|$ and is given by

$$v = (v_x^2 + v_y^2 + v_z^2)^{1/2} \tag{A2.18}$$

Using this representation, we can define the following vector operations:

1 Addition and subtraction. If $\boldsymbol{v} = v_x\boldsymbol{i} + v_y\boldsymbol{j} + v_z\boldsymbol{k}$ and $\boldsymbol{u} = u_x\boldsymbol{i} + u_y\boldsymbol{j} + u_z\boldsymbol{k}$, then

$$\boldsymbol{v} \pm \boldsymbol{u} = (v_x \pm u_x)\boldsymbol{i} + (v_y \pm u_y)\boldsymbol{j} + (v_z \pm u_z)\boldsymbol{k} \tag{A2.19}$$

A graphical method for adding and subtracting vectors is sometimes desirable, as we saw in Chapters 10 and 18. Consider two vectors $\boldsymbol{v}$ and $\boldsymbol{u}$ making an angle θ (Fig. A2.2a). The first step in the addition of $\boldsymbol{u}$ to $\boldsymbol{v}$ consists of joining the tail of $\boldsymbol{v}$ to the head of $\boldsymbol{u}$, as shown in Fig. A2.2b. In the second step, we draw a vector $\boldsymbol{v}_{res}$, the **resultant vector**, originating from the tail of $\boldsymbol{u}$ to the head of $\boldsymbol{v}$, as shown in Fig. A2.2c. Reversing the order of addition leads to the same result. That is, we obtain the same $\boldsymbol{v}_{res}$ whether we add $\boldsymbol{v}$ to $\boldsymbol{u}$ (Fig. A2.2c) or $\boldsymbol{u}$ to $\boldsymbol{v}$ (Fig. A2.3).

To calculate the magnitude of $\boldsymbol{v}_{res}$, we note that $\boldsymbol{v}$, $\boldsymbol{u}$, and $\boldsymbol{v}_{res}$ form a triangle and that we know the magnitudes of two of its sides (v and u) and of the angle between them (180° – θ; see Fig. A2.2c). To calculate the magnitude of the third side, v_{res}, we make use of the *law of cosines*, which states that:

For a triangle with sides a, b, and c, and angle C facing side c:

$$c^2 = a^2 + b^2 - 2ab \cos C$$

This law is summarized graphically in Fig. A2.4 and its application to the case shown in Fig. A2.2c leads to the expression

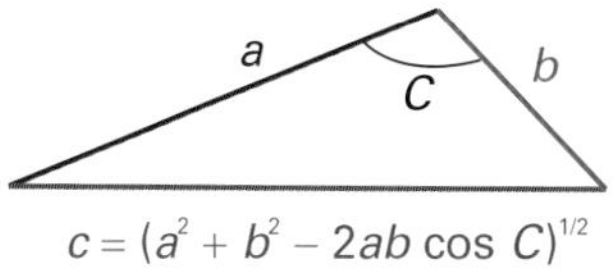

Fig. A2.4 The graphical representation of the law of cosines.

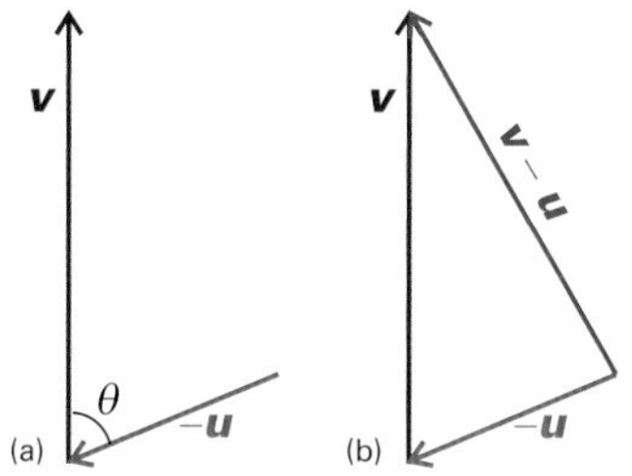

Fig. A2.5 The graphical method for subtraction of the vector $\boldsymbol{u}$ from the vector $\boldsymbol{v}$ (as shown in Fig. A2.2a) consists of two steps: (a) reversing the direction of $\boldsymbol{u}$ to form $-\boldsymbol{u}$, and (b) adding $-\boldsymbol{u}$ to $\boldsymbol{v}$.

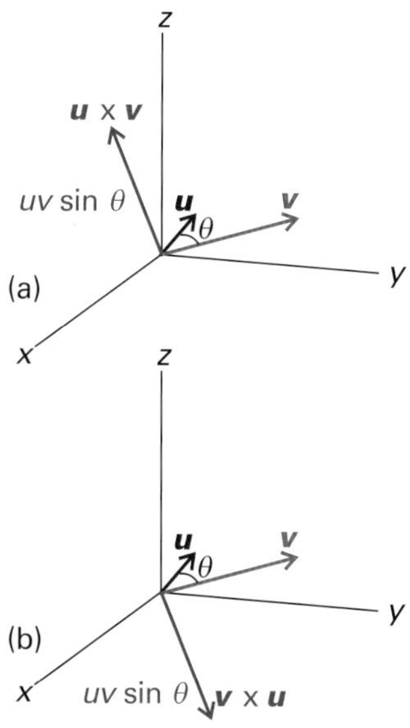

Fig. A2.6 The direction of the cross-products of two vectors $\boldsymbol{u}$ and $\boldsymbol{v}$ with an angle θ between them: (a) $\boldsymbol{u}\times\boldsymbol{v}$ and (b) $\boldsymbol{v}\times\boldsymbol{u}$. Note that the cross-product, and the unit vector $\boldsymbol{l}$ of eqn A2.21, are perpendicular to both $\boldsymbol{u}$ and $\boldsymbol{v}$ but the direction depends on the order in which the product is taken.

$$v_{\text{res}}^2 = v^2 + u^2 - 2vu\cos(180° - \theta)$$

Because $\cos(180° - \theta) = -\cos\theta$, it follows after taking the square-root of both sides of the preceding expression that

$$v_{\text{res}} = (v^2 + u^2 + 2vu\cos\theta)^{1/2} \tag{A2.20}$$

The subtraction of vectors follows the same principles outlined above for addition. Consider again the vectors shown in Fig. A2.2a. We note that subtraction of $\boldsymbol{u}$ from $\boldsymbol{v}$ amounts to addition of $-\boldsymbol{u}$ to $\boldsymbol{v}$. It follows that in the first step of subtraction we draw $-\boldsymbol{u}$ by reversing the direction of $\boldsymbol{u}$ (Fig. A2.5a) Then, the second step consists of adding the $-\boldsymbol{u}$ to $\boldsymbol{v}$ by using the strategy shown in Fig. A2.2c: we draw a resultant vector $\boldsymbol{v}_{\text{res}}$ by joining the tail of $-\boldsymbol{u}$ to the head of $\boldsymbol{v}$.

2 Multiplication. There are two ways to multiply vectors. In one procedure, the **cross-product** of two vectors $\boldsymbol{u}$ and $\boldsymbol{v}$ is a vector defined as

$$\boldsymbol{u}\times\boldsymbol{v} = (uv\sin\theta)\boldsymbol{l} \tag{A2.21a}$$

where θ is the angle between the two vectors and $\boldsymbol{l}$ is a unit vector perpendicular to both $\boldsymbol{u}$ and $\boldsymbol{v}$, with a direction determined as in Fig. A2.6. An equivalent definition is

$$\boldsymbol{u}\times\boldsymbol{v} = \begin{vmatrix} \boldsymbol{i} & \boldsymbol{j} & \boldsymbol{k} \\ u_x & u_y & u_z \\ v_x & v_y & v_z \end{vmatrix} = (u_yv_z - u_zv_y)\boldsymbol{i} - (u_xv_z - u_zv_x)\boldsymbol{j} + (u_xv_y - u_yv_x)\boldsymbol{k} \tag{A2.21b}$$

where the structure in the middle is a determinant (see below). The second type of vector multiplication is the **scalar product** (or **dot product**) of two vectors $\boldsymbol{u}$ and $\boldsymbol{v}$:

$$\boldsymbol{u}\cdot\boldsymbol{v} = uv\cos\theta \tag{A2.22}$$

As its name suggests, the scalar product of two vectors is a scalar.

Calculus

A2.4 Differentiation and integration

Rates of change of functions—slopes of their graphs—are best discussed in terms of the infinitesimal calculus. The slope of a function, like the slope of a hill, is obtained by dividing the rise of the hill by the horizontal distance (Fig. A2.7). However, because

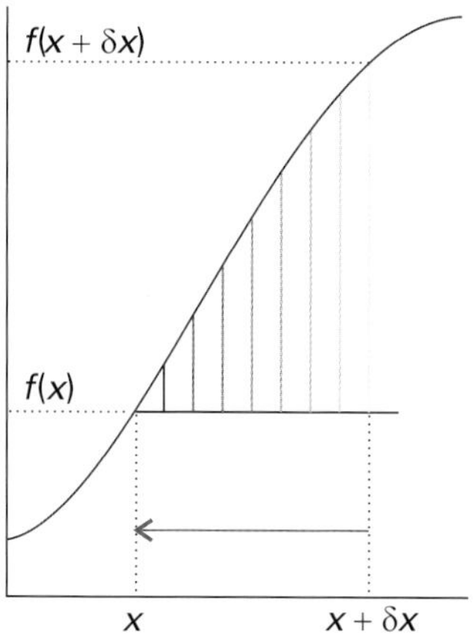

Fig. A2.7 The slope of $f(x)$ at x, $\mathrm{d}f/\mathrm{d}x$, is obtained by making a series of approximations to the value of $f(x+\delta x) - f(x)$ divided by the change in x, denoted δx, and allowing δx to approach 0 (as indicated by the vertical lines getting closer to x).

the slope may vary from point to point, we should make the horizontal distance between the points as small as possible. In fact, we let it become infinitesimally small —hence the name *infinitesimal* calculus. The values of a function f at two locations x and $x+\delta x$ are $f(x)$ and $f(x+\delta x)$, respectively. Therefore, the slope of the function f at x is the vertical distance, which we write δf, divided by the horizontal distance, which we write δx:

$$\text{Slope} = \frac{\text{rise in value}}{\text{horizontal distance}} = \frac{\delta f}{\delta x} = \frac{f(x+\delta x)-f(x)}{\delta x} \qquad \text{(A2.23)}$$

The slope at x itself is obtained by letting the horizontal distance become zero, which we write lim $\delta x \rightarrow 0$. In this limit, the δ is replaced by a d, and we write

$$\text{Slope at } x = \frac{\mathrm{d}f}{\mathrm{d}x} = \lim_{\delta x \to 0} \frac{f(x+\delta x)-f(x)}{\delta x} \qquad \text{(A2.24)}$$

To work out the slope of any function, we work out the expression on the right: this process is called **differentiation** and the expression for $\mathrm{d}f/\mathrm{d}x$ is the **derivative** of the function f with respect to the variable x. Some important derivatives are given inside the front cover of the text. Most of the functions encountered in chemistry can be differentiated by using the following rules (noting that in these expressions, derivatives $\mathrm{d}f/\mathrm{d}x$ are written as $\mathrm{d}f$):

Rule 1 For two functions f and g:

$$\mathrm{d}(f+g) = \mathrm{d}f + \mathrm{d}g \qquad \text{(A2.25)}$$

Rule 2 (the product rule) For two functions f and g:

$$\mathrm{d}(fg) = f\,\mathrm{d}g + g\,\mathrm{d}f \qquad \text{(A2.26)}$$

Rule 3 (the quotient rule) For two functions f and g:

$$\mathrm{d}\frac{f}{g} = \frac{1}{g}\mathrm{d}f - \frac{f}{g^2}\mathrm{d}g \qquad \text{(A2.27)}$$

Rule 4 (the chain rule) For a function $f=f(g)$, where $g=g(t)$,

$$\frac{\mathrm{d}f}{\mathrm{d}t} = \frac{\mathrm{d}f}{\mathrm{d}g}\frac{\mathrm{d}g}{\mathrm{d}t} \qquad \text{(A2.28)}$$

The area under a graph of any function f is found by the techniques of **integration**. For instance, the area under the graph of the function f drawn in Fig. A2.8 can be written as the value of f evaluated at a point multiplied by the width of the region, δx, and then all those products $f(x)\delta x$ summed over all the regions:

$$\text{Area between } a \text{ and } b = \sum f(x)\delta x$$

When we allow δx to become infinitesimally small, written dx, and sum an infinite number of strips, we write:

$$\text{Area between } a \text{ and } b = \int_a^b f(x)\mathrm{d}x \qquad \text{(A2.29)}$$

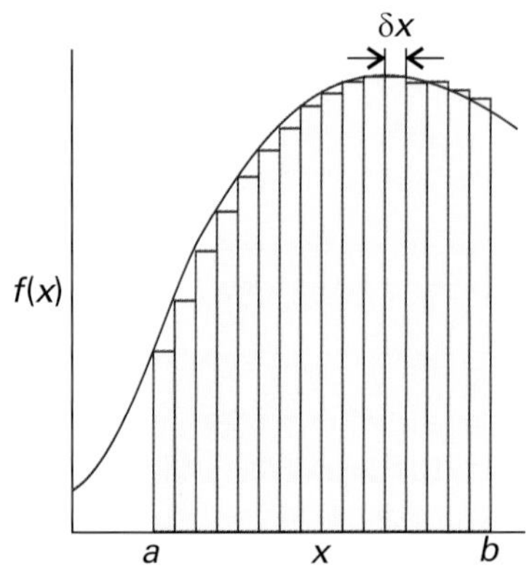

Fig. A2.8 The shaded area is equal to the definite integral of $f(x)$ between the limits a and b.

The elongated S symbol on the right is called the **integral** of the function f. When written as $\int$ alone, it is the **indefinite integral** of the function. When written with limits (as in eqn A2.29), it is the **definite integral** of the function. The definite integral is the indefinite integral evaluated at the upper limit (b) minus the indefinite integral evaluated at the lower limit (a). The **average value** of a function $f(x)$ in the range $x=a$ to $x=b$ is

$$\text{Average value of } f(x) \text{ from } a \text{ to } b = \frac{1}{b-a}\int_a^b f(x)\mathrm{d}x \tag{A2.30}$$

The **mean value theorem** states that a continuous function has its mean value at least once in the range.

Integration is the inverse of differentiation. That is, if we integrate a function and then differentiate the result, we get back the original function. Some important integrals are given on the front back cover of the text. Many other standard forms are found in tables (see *Further reading*) and it is also possible to calculate definite and indefinite integrals with mathematical software. Two integration techniques are useful:

Technique 1 (integration by parts) For two functions f and g:

$$\int f\frac{\mathrm{d}g}{\mathrm{d}x}\mathrm{d}x = fg - \int g\frac{\mathrm{d}f}{\mathrm{d}x}\mathrm{d}x \tag{A2.31}$$

Technique 2 (method of partial fractions) To solve an integral of the form $\int \frac{1}{(a-x)(b-x)}\mathrm{d}x$, where a and b are constants, we write

$$\frac{1}{(a-x)(b-x)} = \frac{1}{b-a}\left(\frac{1}{a-x} - \frac{1}{b-x}\right)$$

and integrate the expression on the right. It follows that

$$\int \frac{\mathrm{d}x}{(a-x)(b-x)} = \frac{1}{b-a}\left[\int \frac{\mathrm{d}x}{a-x} - \int \frac{\mathrm{d}x}{b-x}\right]$$
$$= \frac{1}{b-a}\left(\ln\frac{1}{a-x} - \ln\frac{1}{b-x}\right) + \text{constant} \tag{A2.32}$$

A2.5 Power series and Taylor expansions

A **power series** has the form

$$c_0 + c_1(x-a) + c_2(x-a)^2 + \cdots + c_n(x-a)^n + \cdots = \sum_{n=0}^{\infty} c_n(x-a)^n \tag{A2.33}$$

where c_n and a are constants. It is often useful to express a function $f(x)$ in the vicinity of $x = a$ as a special power series called the **Taylor series**, or **Taylor expansion**, which has the form:

$$f(x) = f(a) + \left(\frac{\mathrm{d}f}{\mathrm{d}x}\right)(x-a) + \frac{1}{2!}\left(\frac{\mathrm{d}^2f}{\mathrm{d}x^2}\right)_a (x-a)^2 + \cdots + \frac{1}{n!}\left(\frac{\mathrm{d}^nf}{\mathrm{d}x^n}\right)_a (x-a)^n$$
$$= \sum_{n=0}^{\infty}\frac{1}{n!}\left(\frac{\mathrm{d}^nf}{\mathrm{d}x^n}\right)_a (x-a)^n \tag{A2.34}$$

where $n!$ denotes a **factorial**, given by

$$n! = n(n-1)(n-2)\ldots 1 \tag{A2.35}$$

By definition $0! = 1$. The following Taylor expansions are often useful:

$$\frac{1}{1+x} = 1 - x + x^2 + \cdots$$
$$\mathrm{e}^x = 1 + x + \tfrac{1}{2}x^2 + \cdots$$
$$\ln x = (x-1) - \tfrac{1}{2}(x-1)^2 + \tfrac{1}{3}(x-1)^3 - \tfrac{1}{4}(x-1)^4 + \cdots$$
$$\ln(1+x) = x - \tfrac{1}{2}x^2 + \tfrac{1}{3}x^3 - \cdots$$

If $x \ll 1$, then $(1+x)^{-1} \approx 1 - x$, $\mathrm{e}^x \approx 1 + x$, and $\ln(1+x) \approx x$.

A2.6 Partial derivatives

A **partial derivative** of a function of more than one variable, such as $f(x,y)$, is the slope of the function with respect to one of the variables, all the other variables being held constant (see Fig. 2.21). Although a partial derivative shows how a function changes when one variable changes, it may be used to determine how the function changes when more than one variable changes by an infinitesimal amount. Thus, if f is a function of x and y, then when x and y change by $\mathrm{d}x$ and $\mathrm{d}y$, respectively, f changes by

$$\mathrm{d}f = \left(\frac{\partial f}{\partial x}\right)_y \mathrm{d}x + \left(\frac{\partial f}{\partial y}\right)_x \mathrm{d}y \qquad \text{(A2.36)}$$

where the symbol ∂ is used (instead of d) to denote a partial derivative. The quantity $\mathrm{d}f$ is also called the **differential** of f. For example, if $f = ax^3y + by^2$, then

$$\left(\frac{\partial f}{\partial x}\right)_y = 3ax^2y \qquad \left(\frac{\partial f}{\partial y}\right)_x = ax^3 + 2by$$

Then, when x and y undergo infinitesimal changes, f changes by

$$\mathrm{d}f = 3ax^2y\mathrm{d}x + (ax^3 + 2by)\mathrm{d}y$$

Partial derivatives may be taken in any order:

$$\frac{\partial^2 f}{\partial x \partial y} = \frac{\partial^2 f}{\partial y \partial x} \qquad \text{(A2.37)}$$

For the function f given above, it is easy to verify that

$$\left(\frac{\partial}{\partial y}\left(\frac{\partial f}{\partial x}\right)_y\right)_x = 3ax^2 \qquad \left(\frac{\partial}{\partial x}\left(\frac{\partial f}{\partial y}\right)_x\right)_y = 3ax^2$$

In the following, z is a variable on which x and y depend (for example, x, y, and z might correspond to p, V, and T).

Relation 1 When x is changed at constant z:

$$\left(\frac{\partial f}{\partial x}\right)_z = \left(\frac{\partial f}{\partial x}\right)_y + \left(\frac{\partial f}{\partial y}\right)_x \left(\frac{\partial y}{\partial x}\right)_z \qquad \text{(A2.38)}$$

Relation 2

$$\left(\frac{\partial y}{\partial x}\right)_z = \frac{1}{(\partial x/\partial y)_z} \qquad \text{(A2.39)}$$

Relation 3

$$\left(\frac{\partial x}{\partial y}\right)_z = -\left(\frac{\partial x}{\partial z}\right)_y \left(\frac{\partial z}{\partial y}\right)_x \qquad \text{(A2.40)}$$

By combining this relation and Relation 2 we obtain the **Euler chain relation**:

$$\left(\frac{\partial y}{\partial x}\right)_z \left(\frac{\partial x}{\partial z}\right)_y \left(\frac{\partial z}{\partial y}\right)_x = -1 \qquad \text{(A2.41)}$$

Relation 4 This relation establishes whether or not $\mathrm{d}f$ is an **exact differential**.

$$\mathrm{d}f = g(x,y)\mathrm{d}x + h(x,y)\mathrm{d}y \quad \text{is exact if} \quad \left(\frac{\partial g}{\partial y}\right)_x = \left(\frac{\partial h}{\partial x}\right)_y \qquad \text{(A2.42)}$$

If $\mathrm{d}f$ is exact, its integral between specified limits is independent of the path.

A2.7 Functionals and functional derivatives

Just as a function f can be regarded as a set of mathematical procedures that associates a number $f(x)$ to a specified value of a variable x, so a **functional** G gives a prescription for associating a number $G[f]$ to a function $f(x)$ over a specified range of the variable x. That is, the functional is a function of a function. Functionals are important in quantum chemistry. We saw in Chapter 11 that the energy of a molecule is a functional of the electron density, which in turn is a function of the position.

To make the following discussion more concrete, consider the functional

$$G[f] = \int_0^1 f(x)^2 \mathrm{d}x \tag{A2.43}$$

If we let $f(x) = x$, then $G[f] = \frac{1}{3}$ over the range $0 \leq x \leq 1$. However, if $f(x) = \sin \pi x$, then $G[f] = \frac{1}{2}$ over the range $0 \leq x \leq 1$.

Just as the derivative of a function $f(x)$ tells us how the function changes with small changes δx in the variable x, so a **functional derivative** tells us about the variation δG of a functional $G[f]$ with small changes δf in the function $f(x)$. By analogy with eqn A2.24, we can write the following definition of the functional derivative as

$$\frac{\delta G}{\delta f} = \lim_{\delta f \to 0} \frac{G[f + \delta f] - G[f]}{\delta f} \tag{A2.44}$$

However, this equation does not give us a simple method for calculating the functional derivative. It can be shown that an alternative definition of $\delta G/\delta f$ is (see *Further reading*):

$$G[f + \delta f] - G[f] = \int_a^b \left(\frac{\delta G}{\delta f}\delta f(x)\right)\mathrm{d}x \tag{A2.45}$$

where the integral is evaluated in the range over which x varies.

To see how eqn A2.45 is used to calculate a functional derivative, consider the functional given by eqn A2.43. We begin by writing

$$\begin{aligned} G[f + \delta f] &= \int_0^1 \{f(x) + \delta f(x)\}^2 \mathrm{d}x = \int_0^1 \{f(x)^2 + 2f(x)\delta f(x) + \delta f(x)^2\}\mathrm{d}x \\ &= \int_0^1 \{f(x)^2 + 2f(x)\delta f(x)\}\mathrm{d}x = G[f] + \int_0^1 2f(x)\delta f(x)\mathrm{d}x \end{aligned}$$

where we have ignored the minute contribution from δf^2 to arrive at the penultimate expression and then used eqn A2.43 to write the final expression. It follows that

$$G[f + \delta f] - G[f] = \int_0^1 2f(x)\delta f(x)\mathrm{d}x$$

By comparing this expression with eqn A2.45, we see that the functional derivative is

$$\frac{\delta G}{\delta f} = 2f(x)$$

A2.8 Undetermined multipliers

Suppose we need to find the maximum (or minimum) value of some function f that depends on several variables $x_1, x_2, \ldots, x_n$. When the variables undergo a small change from x_i to $x_i + \delta x_i$ the function changes from f to $f + \delta f$, where

$$\delta f = \sum_i^n \left(\frac{\partial f}{\partial x_i}\right)\delta x_i \tag{A2.46}$$

At a minimum or maximum, $\delta f = 0$, so then

$$\sum_i^n \left(\frac{\partial f}{\partial x_i}\right)\delta x_i = 0 \tag{A2.47}$$

If the x_i were all independent, all the δx_i would be arbitrary, and this equation could be solved by setting each $(\partial f/\partial x_i) = 0$ individually. When the x_i are not all independent, the δx_i are not all independent, and the simple solution is no longer valid. We proceed as follows.

Let the constraint connecting the variables be an equation of the form $g = 0$. For example, in Chapter 16, one constraint was $n_0 + n_1 + \cdots = N$, which can be written

$$g = 0, \qquad \text{with} \quad g = (n_0 + n_1 + \cdots) - N$$

The constraint $g = 0$ is always valid, so g remains unchanged when the x_i are varied:

$$\delta g = \sum_i \left(\frac{\partial g}{\partial x_i}\right)\delta x_i = 0 \tag{A2.48}$$

Because δg is zero, we can multiply it by a parameter, λ, and add it to eqn A2.47:

$$\sum_i^n \left\{\left(\frac{\partial f}{\partial x_i}\right) + \lambda\left(\frac{\partial g}{\partial x_i}\right)\right\}\delta x_i = 0 \tag{A2.49}$$

This equation can be solved for one of the δx, δx_n for instance, in terms of all the other δx_i. All those other δx_i ($i = 1, 2, \ldots n - 1$) are independent, because there is only one constraint on the system. But here is the trick: λ is arbitrary; therefore we can choose it so that the coefficient of δx_n in eqn A2.49 is zero. That is, we choose λ so that

$$\left(\frac{\partial f}{\partial x_n}\right) + \lambda\left(\frac{\partial g}{\partial x_n}\right) = 0 \tag{A2.50}$$

Then eqn A2.49 becomes

$$\sum_i^{n-1} \left\{\left(\frac{\partial f}{\partial x_i}\right) + \lambda\left(\frac{\partial g}{\partial x_i}\right)\right\}\delta x_i = 0 \tag{A2.51}$$

Now the $n - 1$ variations δx_i *are* independent, so the solution of this equation is

$$\left(\frac{\partial f}{\partial x_i}\right) + \lambda\left(\frac{\partial g}{\partial x_i}\right) = 0 \qquad i = 1, 2, \ldots, n - 1 \tag{A2.52}$$

However, eqn A2.50 has exactly the same form as this equation, so the maximum or minimum of f can be found by solving

$$\left(\frac{\partial f}{\partial x_i}\right) + \lambda\left(\frac{\partial g}{\partial x_i}\right) = 0 \qquad i = 1, 2, \ldots, n \tag{A2.53}$$

The use of this approach was illustrated in the text for two constraints and therefore two undetermined multipliers λ_1 and λ_2 (α and $-\beta$).

The multipliers λ cannot always remain undetermined. One approach is to solve eqn A2.50 instead of incorporating it into the minimization scheme. In Chapter 16 we used the alternative procedure of keeping λ undetermined until a property was calculated for which the value was already known. Thus, we found that $\beta = 1/kT$ by calculating the internal energy of a perfect gas.

A2.9 Differential equations

(a) Ordinary differential equations

An **ordinary differential equation** is a relation between derivatives of a function of one variable and the function itself, as in

$$a\frac{d^2y}{dx^2}+b\frac{dy}{dx}+cy=0 \qquad \text{(A2.54)}$$

The coefficients a, b, etc. may be functions of x. The **order** of the equation is the order of the highest derivative that occurs in it, so eqn A2.54 is a second-order equation. Only rarely in science is a differential equation of order higher than 2 encountered. A solution of a differential equation is an expression for y as a function of x. The process of solving a differential equation is commonly termed 'integration', and in simple cases simple integration can be employed to find $y(x)$. A **general solution** of a differential equation is the most general solution of the equation and is expressed in terms of a number of constants. When the constants are chosen to accord with certain specified **initial conditions** (if one variable is the time) or certain **boundary conditions** (to fulfil certain spatial restrictions on the solutions), we obtain the **particular solution** of the equation. A first–order differential equation requires the specification of *one* boundary (or initial) condition; a second–order differential equation requires the specification of *two* such conditions, and so on.

First-order differential equations may often be solved by direct integration. For example, the equation

$$\frac{dy}{dx}=axy$$

with a constant may be rearranged into

$$\frac{dy}{y}=ax\mathrm{d}x$$

and then integrated to

$$\ln y=\tfrac{1}{2}ax^2+A$$

where A is a constant. If we know that $y=y_0$ when $x=0$ (for instance), then it follows that $A=\ln y_0$, and hence the particular solution of the equation is

$$\ln y=\tfrac{1}{2}ax^2+\ln y_0$$

This expression rearranges to

$$y=y_0e^{ax^2/2}$$

First-order equations of a more complex form can often be solved by the appropriate substitution. For example, it is sensible to try the substitution $y=sx$, and to change the variables from x and y to x and s. An alternative useful transformation is to write $x=u+a$ and $y=v+b$, and then to select a and b to simplify the form of the resulting expression.

Solutions to complicated differential equations may also be found by referring to tables (see *Further reading*). For example, first-order equations of the form

$$\frac{dy}{dx}+yf(x)=g(x) \qquad \text{(A2.55)}$$

appear in the study of chemical kinetics. The solution is given by

$$ye^{\int f(x)\mathrm{d}x} = \int e^{\int f(x)\mathrm{d}x} g(x)\mathrm{d}x + \text{constant} \qquad \text{(A2.56)}$$

Mathematical software is now capable of finding analytical solutions of a wide variety of differential equations.

Second-order differential equations are in general much more difficult to solve than first-order equations. One powerful approach commonly used to lay siege to second-order differential equations is to express the solution as a power series:

$$y = \sum_{n=0}^{\infty} c_n x^n \qquad \text{(A2.57)}$$

and then to use the differential equation to find a relation between the coefficients. This approach results, for instance, in the Hermite polynomials that form part of the solution of the Schrödinger equation for the harmonic oscillator (Section 9.4). All the second-order differential equations that occur in this text are tabulated in compilations of solutions or can be solved with mathematical software, and the specialized techniques that are needed to establish the form of the solutions may be found in mathematical texts.

(b) Numerical integration of differential equations

Many of the differential equations that describe physical phenomena are so complicated that their solutions cannot be cast as functions. In such cases, we resort to numerical methods, in which approximations are made in order to integrate the differential equation. Software packages are now readily available that can be used to solve almost any equation numerically. The general form of such programs to solve $\mathrm{d}f/\mathrm{d}x = g(x)$, for instance, replaces the infinitesimal quantity $\mathrm{d}f = g(x)\mathrm{d}x$ by the small quantity $\Delta f = g(x)\Delta x$, so that

$$f(x+\Delta x) \approx f(x) + g(x)\Delta x$$

and then proceeds numerically to step along the x-axis, generating $f(x)$ as it goes. The actual algorithms adopted are much more sophisticated than this primitive scheme, but stem from it. Among the simple numerical methods, the fourth-order **Runge–Kutta method** is one of the most accurate.

The *Further reading* section lists monographs that discuss the derivation of the fourth-order Runge–Kutta method. Here we illustrate the procedure with a first-order differential equation of the form:

$$\frac{\mathrm{d}y}{\mathrm{d}x} = f(x,y) \qquad \text{(A2.58)}$$

One example of this differential equation in chemical kinetics is eqn 22.36, which describes the time dependence of the concentration of an intermediate I in the reaction sequence $\mathrm{A} \rightarrow \mathrm{I} \rightarrow \mathrm{P}$.

To obtain an approximate value of the integral of eqn A2.58, we proceed by rewriting it in terms of finite differences instead of differentials:

$$\frac{\Delta y}{\Delta x} = f(x,y)$$

where Δy may also be written as $y(x + \Delta x) - y(x)$. The fourth-order Runge–Kutta method is based on the following approximation:

$$y(x+\Delta x) = y(x) + \tfrac{1}{6}(k_1 + 2k_2 + 2k_3 + k_4) \qquad \text{(A2.59)}$$

where

$$k_1 = f(x,y)\Delta x \quad \text{(A2.60a)}$$

$$k_2 = f(x+\tfrac{1}{2}\Delta x, y+\tfrac{1}{2}k_1)\Delta x \quad \text{(A2.60b)}$$

$$k_3 = f(x+\tfrac{1}{2}\Delta x, y+\tfrac{1}{2}k_2)\Delta x \quad \text{(A2.60c)}$$

$$k_4 = f(x+\Delta x, y+k_3)\Delta x \quad \text{(A2.60d)}$$

Therefore, if we know the functional form of $f(x,y)$ and $y(0)$, we can use eqns A2.60(a–d) to calculate values of y for a range of x values. The process can be automated easily with an electronic spreadsheet or with mathematical software. The accuracy of the calculation increases with decreasing values of the increment Δx.

(c) Partial differential equations

A **partial differential equation** is a differential in more than one variable. An example is

$$\frac{\partial^2 y}{\partial t^2} = a\frac{\partial^2 y}{\partial x^2} \quad \text{(A2.61)}$$

with y a function of the two variables x and t. In certain cases, partial differential equations may be separated into ordinary differential equations. Thus, the Schrödinger equation for a particle in a two-dimensional square well (Section 9.2) may be separated by writing the wavefunction, $\psi(x,y)$, as the product $X(x)Y(y)$, which results in the separation of the second-order partial differential equation into two second-order differential equations in the variables x and y. A good guide to the likely success of such a **separation of variables** procedure is the symmetry of the system.

Statistics and probability

Throughout the text, but especially in Chapters 16, 17, 19, and 21, we use several elementary results from two branches of mathematics: **probability theory**, which deals with quantities and events that are distributed randomly, and **statistics**, which provides tools for the analysis of large collections of data. Here we introduce some of the fundamental ideas from these two fields.

A2.10 Random selections

Combinatorial functions allow us to express the number of ways in which a system of particles may be configured; they are especially useful in statistical thermodynamics (Chapters 16 and 17). Consider a simple coin-toss problem. If n coins are tossed, the number $N(n,i)$ of outcomes that have i heads and $(n-i)$ tails, regardless of the order of the results, is given by the coefficients of the **binomial expansion** of $(1+x)^n$:

$$(1+x)^n = 1 + \sum_{i=1}^{n} N(n,i)x^i, \qquad N(n,i) = \frac{n!}{(n-i)!i!} \quad \text{(A2.62)}$$

The numbers $N(n,i)$, which are sometimes denoted $\binom{n}{i}$, are also called **binomial coefficients**.

Suppose that, unlike the coin-toss problem, there are more than two possible results for each event. For example, there are six possible results for the roll of a die. For n rolls of the die, the number of ways, W, that correspond to n_1 occurrences of the number 1, n_2 occurrences of the number 2, and so on, is given by

$$W = \frac{n!}{n_1!n_2!n_3!n_4!n_5!n_6!}, \qquad n = \sum_{i=1}^{6} n_i$$

This is an example of a **multinomial coefficient**, which has the form

$$W = \frac{n!}{n_1!n_2!\ldots n_m!}, \qquad n = \sum_{i=1}^{m} n_i \tag{A2.63}$$

where W is the number of ways of achieving an outcome, n is the number of events, and m is the number of possible results. In Chapter 16 we use the multinomial coefficient to determine the number of ways to configure a system of identical particles given a specific distribution of particles into discrete energy levels.

In chemistry it is common to deal with a very large number of particles and outcomes and it is useful to express factorials in different ways. We can simplify factorials of large numbers by using **Stirling's approximation**:

$$n! \approx (2\pi)^{1/2} n^{n+\frac{1}{2}} e^{-n} \tag{A2.64}$$

The approximation is in error by less than 1 per cent when n is greater than about 10. For very large values of n, it is possible to use another form of the approximation:

$$\ln n! \approx n \ln n - n \tag{A2.65}$$

A2.11 Some results of probability theory

Here we develop two general results of probability theory: the mean value of a variable and the mean value of a function. The calculation of mean values is useful in the description of random coils (Chapter 19) and molecular diffusion (Chapter 21).

The mean value (also called the *expectation value*) $\langle X \rangle$ of a variable X is calculated by first multiplying each discrete value x_i that X can have by the probability p_i that x_i occurs and then summing these products over all possible N values of X:

$$\langle X \rangle = \sum_{i=1}^{N} x_i p_i$$

When N is very large and the x_i values are so closely spaced that X can be regarded as varying continuously, it is useful to express the probability that X can have a value between x and $x + dx$ as

$$\text{Probability of finding a value of } X \text{ between } x \text{ and } x + dx = f(x)dx$$

where the function $f(x)$ is the *probability density*, a measure of the distribution of the probability values over x, and dx is an infinitesimally small interval of x values. It follows that the probability that X has a value between $x = a$ and $x = b$ is the integral of the expression above evaluated between a and b:

$$\text{Probability of finding a value of } X \text{ between } a \text{ and } b = \int_a^b f(x)dx$$

The mean value of the continuously varying X is given by

$$\langle X \rangle = \int_{-\infty}^{+\infty} x f(x)dx \tag{A2.66}$$

This expression is similar to that written for the case of discrete values of X, with $f(x)dx$ as the probability term and integration over the closely spaced x values replacing summation over widely spaced x_i.

The mean value of a function $g(X)$ can be calculated with a formula similar to that for $\langle X\rangle$:

$$\langle g(X)\rangle = \int_{-\infty}^{+\infty} g(x)f(x)\mathrm{d}x \qquad \text{(A2.67)}$$

Matrix algebra

A **matrix** is an array of numbers. Matrices may be combined together by addition or multiplication according to generalizations of the rules for ordinary numbers. Most numerical matrix manipulations are now carried out with mathematical software.

Consider a square matrix $\boldsymbol{M}$ of n^2 numbers arranged in n columns and n rows. These n^2 numbers are the **elements** of the matrix, and may be specified by stating the row, r, and column, c, at which they occur. Each element is therefore denoted M_{rc}. For example, in the matrix

$$\boldsymbol{M} = \begin{pmatrix} 1 & 2 \\ 3 & 4 \end{pmatrix}$$

the elements are $M_{11} = 1$, $M_{12} = 2$, $M_{21} = 3$, and $M_{22} = 4$. This is an example of a 2×2 matrix. The **determinant**, $|\boldsymbol{M}|$, of this matrix is

$$|\boldsymbol{M}| = \begin{vmatrix} 1 & 2 \\ 3 & 4 \end{vmatrix} = 1 \times 4 - 2 \times 3 = -2$$

A **diagonal matrix** is a matrix in which the only nonzero elements lie on the major diagonal (the diagonal from M_{11} to M_{nn}). Thus, the matrix

$$\boldsymbol{D} = \begin{pmatrix} 1 & 0 & 0 \\ 0 & 2 & 0 \\ 0 & 0 & 1 \end{pmatrix}$$

is diagonal. The condition may be written

$$M_{rc} = m_r \delta_{rc} \qquad \text{(A2.68)}$$

where δ_{rc} is the **Kronecker delta**, which is equal to 1 for $r = c$ and to 0 for $r \neq c$. In the above example, $m_1 = 1$, $m_2 = 2$, and $m_3 = 1$. The **unit matrix**, **1** (and occasionally **I**), is a special case of a diagonal matrix in which all nonzero elements are 1.

The **transpose** of a matrix $\boldsymbol{M}$ is denoted $\boldsymbol{M}^{\mathrm{T}}$ and is defined by

$$M^{\mathrm{T}}_{mn} = M_{nm} \qquad \text{(A2.69)}$$

Thus, for the matrix $\boldsymbol{M}$ we have been using,

$$\boldsymbol{M}^{\mathrm{T}} = \begin{pmatrix} 1 & 3 \\ 2 & 4 \end{pmatrix}$$

Matrices are very useful in chemistry. They simplify some mathematical tasks, such as solving systems of simultaneous equations, the treatment of molecular symmetry (Chapter 12), and quantum mechanical calculations (Chapter 11).

A2.12 Matrix addition and multiplication

Two matrices $\boldsymbol{M}$ and $\boldsymbol{N}$ may be added to give the sum $\boldsymbol{S} = \boldsymbol{M} + \boldsymbol{N}$, according to the rule

$$S_{rc} = M_{rc} + N_{rc} \qquad \text{(A2.70)}$$

(that is, corresponding elements are added). Thus, with $\boldsymbol{M}$ given above and

$$\boldsymbol{N}=\begin{pmatrix}5 & 6\\ 7 & 8\end{pmatrix}$$

the sum is

$$\boldsymbol{S}=\begin{pmatrix}1 & 2\\ 3 & 4\end{pmatrix}+\begin{pmatrix}5 & 6\\ 7 & 8\end{pmatrix}=\begin{pmatrix}6 & 8\\ 10 & 12\end{pmatrix}$$

Two matrices may also be multiplied to give the product $\boldsymbol{P}=\boldsymbol{MN}$ according to the rule

$$P_{rc}=\sum_n M_{rn}N_{nc} \tag{A2.71}$$

For example, with the matrices given above,

$$\boldsymbol{P}=\begin{pmatrix}1 & 2\\ 3 & 4\end{pmatrix}\begin{pmatrix}5 & 6\\ 7 & 8\end{pmatrix}=\begin{pmatrix}1\times5+2\times7 & 1\times6+2\times8\\ 3\times5+4\times7 & 3\times6+4\times8\end{pmatrix}=\begin{pmatrix}19 & 22\\ 43 & 50\end{pmatrix}$$

It should be noticed that in general $\boldsymbol{MN}\neq\boldsymbol{NM}$, and matrix multiplication is in general non-commutative.

The **inverse** of a matrix $\boldsymbol{M}$ is denoted $\boldsymbol{M}^{-1}$, and is defined so that

$$\boldsymbol{MM}^{-1}=\boldsymbol{M}^{-1}\boldsymbol{M}=\boldsymbol{1} \tag{A2.72}$$

The inverse of a matrix can be constructed by using mathematical software, but in simple cases the following procedure can be carried through without much effort:

1 Form the determinant of the matrix. For example, for our matrix $\boldsymbol{M}$, $|\boldsymbol{M}|=-2$.

2 Form the transpose of the matrix. For example, $\boldsymbol{M}^{\mathrm{T}}=\begin{pmatrix}1 & 3\\ 2 & 4\end{pmatrix}$.

3 Form $\tilde{\boldsymbol{M}}'$, where $\tilde{\boldsymbol{M}}'_{rc}$ is the **cofactor** of the element M_{rc}, that is, it is the determinant formed from $\boldsymbol{M}$ with the row r and column c struck out. For example,

$$\tilde{\boldsymbol{M}}'=\begin{pmatrix}4 & -2\\ -3 & 1\end{pmatrix}$$

4 Construct the inverse as $\boldsymbol{M}^{-1}=\tilde{\boldsymbol{M}}'/|\boldsymbol{M}|$. For example,

$$\boldsymbol{M}^{-1}=\frac{1}{-2}\begin{pmatrix}4 & -2\\ -3 & 1\end{pmatrix}=\begin{pmatrix}-2 & 1\\ \frac{3}{2} & -\frac{1}{2}\end{pmatrix}$$

A2.13 Simultaneous equations

A set of n **simultaneous equations**

$$\begin{aligned}
&a_{11}x_1+a_{12}x_2+\cdots+a_{1n}x_n=b_1\\
&a_{21}x_1+a_{22}x_2+\cdots+a_{2n}x_n=b_2\\
&\vdots\\
&a_{n1}x_1+a_{n2}x_2+\cdots+a_{nn}x_n=b_n
\end{aligned} \tag{A2.73}$$

can be written in matrix notation if we introduce the **column vectors $\boldsymbol{x}$** and **$\boldsymbol{b}$**:

$$\boldsymbol{x}=\begin{pmatrix}x_1\\ x_2\\ \vdots\\ x_n\end{pmatrix}\qquad \boldsymbol{b}=\begin{pmatrix}b_1\\ b_2\\ \vdots\\ b_n\end{pmatrix}$$

Then, with the $\boldsymbol{a}$ matrix of coefficients a_{rc}, the n equations may be written as

$$\boldsymbol{ax} = \boldsymbol{b} \tag{A2.74}$$

The formal solution is obtained by multiplying both sides of this matrix equation by $\boldsymbol{a}^{-1}$, for then

$$\boldsymbol{x} = \boldsymbol{a}^{-1}\boldsymbol{b} \tag{A2.75}$$

A2.14 Eigenvalue equations

An **eigenvalue equation** is a special case of eqn A2.74 in which

$$\boldsymbol{ax} = \lambda\boldsymbol{x} \tag{A2.76}$$

where λ is a constant, the **eigenvalue**, and $\boldsymbol{x}$ is the **eigenvector**. In general, there are n eigenvalues $\lambda^{(i)}$, and they satisfy the n simultaneous equations

$$(\boldsymbol{a} - \lambda\boldsymbol{1})\boldsymbol{x} = 0 \tag{A2.77}$$

There are n corresponding eigenvectors $\boldsymbol{x}^{(i)}$. Equation A2.77 has a solution only if the determinant of the coefficients is zero. However, this determinant is just $|\boldsymbol{a} - \lambda\boldsymbol{1}|$, so the n eigenvalues may be found from the solution of the **secular equation**:

$$|\boldsymbol{a} - \lambda\boldsymbol{1}| = 0 \tag{A2.78}$$

The n eigenvalues the secular equation yields may be used to find the n eigenvectors. These eigenvectors (which are $n \times 1$ matrices), may be used to form an $n \times n$ matrix $\boldsymbol{X}$. Thus, because

$$\boldsymbol{x}^{(1)} = \begin{pmatrix} x_1^{(1)} \\ x_2^{(1)} \\ \vdots \\ x_n^{(1)} \end{pmatrix} \qquad \boldsymbol{x}^{(2)} = \begin{pmatrix} x_1^{(2)} \\ x_2^{(2)} \\ \vdots \\ x_n^{(2)} \end{pmatrix} \qquad \text{etc.}$$

we may form the matrix

$$\boldsymbol{X} = (\boldsymbol{x}^{(1)}, \boldsymbol{x}^{(2)}, \ldots, \boldsymbol{x}^{(n)}) = \begin{pmatrix} x_1^{(1)} & x_1^{(2)} & \cdots & x_1^{(n)} \\ x_2^{(1)} & x_2^{(2)} & \cdots & x_2^{(n)} \\ \vdots & \vdots & & \vdots \\ x_n^{(1)} & x_n^{(2)} & \cdots & x_n^{(n)} \end{pmatrix}$$

so that $X_{rc} = x_r^{(c)}$. If further we write $\Lambda_{rc} = \lambda_r\delta_{rc}$, so that $\boldsymbol{\Lambda}$ is a diagonal matrix with the elements $\lambda_1, \lambda_2, \ldots, \lambda_n$ along the diagonal, then all the eigenvalue equations $\boldsymbol{ax}^{(i)} = \lambda_i\boldsymbol{x}^{(i)}$ may be confined into the single equation

$$\boldsymbol{aX} = \boldsymbol{X\Lambda} \tag{A2.79}$$

because this expression is equal to

$$\sum_n a_{rn}X_{nc} = \sum_n X_{rn}\Lambda_{nc}$$

or

$$\sum_n a_{rn}x_n^{(c)} = \sum_n x_r^{(n)}\lambda_n\delta_{nc} = \lambda_c x_r^{(c)}$$

as required. Therefore, if we form $\boldsymbol{X}^{-1}$ from $\boldsymbol{X}$, we construct a **similarity transformation**

$$\boldsymbol{\Lambda} = \boldsymbol{X}^{-1}\boldsymbol{aX} \tag{A2.80}$$

that makes $\boldsymbol{a}$ diagonal (because $\boldsymbol{\Lambda}$ is diagonal). It follows that if the matrix $\boldsymbol{X}$ that causes $\boldsymbol{X}^{-1}\boldsymbol{a}\boldsymbol{X}$ to be diagonal is known, then the problem is solved: the diagonal matrix so produced has the eigenvalues as its only nonzero elements, and the matrix $\boldsymbol{X}$ used to bring about the transformation has the corresponding eigenvectors as its columns. The solutions of eigenvalue equations are best found by using mathematical software.

Further reading

P.W. Atkins, J.C. de Paula, and V.A. Walters, *Explorations in physical chemistry.* W.H. Freeman and Company, New York (2005).

J.R. Barrante, *Applied mathematics for physical chemistry.* Prentice Hall, Upper Saddle River (2004).

M.P. Cady and C.A. Trapp, *A Mathcad primer for physical chemistry.* W.H. Freeman and Company, New York (2000).

R.G. Mortimer, *Mathematics for physical chemistry.* Academic Press, San Diego (2005).

D.A. McQuarrie, *Mathematical methods for scientists and engineers.* University Science Books, Sausalito (2003).

D. Zwillinger (ed.), *CRC standard mathematical tables and formulae.* CRC Press, Boca Raton (1996).

Appendix 3 Essential concepts of physics

A3

Energy

The central concept of all explanations in physical chemistry, as in so many other branches of physical science, is that of **energy**, the capacity to do work. We make use of the apparently universal law of nature that *energy is conserved*, that is, energy can neither be created nor destroyed. Although energy can be transferred from one location to another, the total energy is constant.

A3.1 Kinetic and potential energy

The **kinetic energy**, E_K, of a body is the energy the body possesses as a result of its motion. For a body of mass m travelling at a speed v,

$$E_K = \tfrac{1}{2}mv^2 \qquad \text{(A3.1)}$$

The **potential energy**, E_P or V, of a body is the energy it possesses as a result of its position. The zero of potential energy is arbitrary. For example, the gravitational potential energy of a body is often set to zero at the surface of the Earth; the electrical potential energy of two charged particles is set to zero when their separation is infinite. No universal expression for the potential energy can be given because it depends on the type of interaction the body experiences. One example that gives rise to a simple expression is the potential energy of a body of mass m in the gravitational field close to the surface of the Earth (a gravitational field acts on the mass of a body). If the body is at a height h above the surface of the Earth, then its potential energy is mgh, where g is a constant called the **acceleration of free fall**, $g = 9.81\ \text{m s}^{-2}$, and $V = 0$ at $h = 0$ (the arbitrary zero mentioned previously).

The total energy is the sum of the kinetic and potential energies of a particle:

$$E = E_K + E_P \qquad \text{(A3.2)}$$

A3.2 Energy units

The SI unit of energy is the *joule* (J), which is defined as

$$1\ \text{J} = 1\ \text{kg m}^2\ \text{s}^{-2} \qquad \text{(A3.3)}$$

Calories (cal) and kilocalories (kcal) are still encountered in the chemical literature: by definition, 1 cal = 4.184 J. An energy of 1 cal is enough to raise the temperature of 1 g of water by 1°C.

The rate of change of energy is called the **power**, P, expressed as joules per second, or *watt*, W:

$$1\ \text{W} = 1\ \text{J s}^{-1} \qquad \text{(A3.4)}$$

Classical mechanics

Classical mechanics describes the behaviour of objects in terms of two equations. One expresses the fact that the total energy is constant in the absence of external forces; the other expresses the response of particles to the forces acting on them.

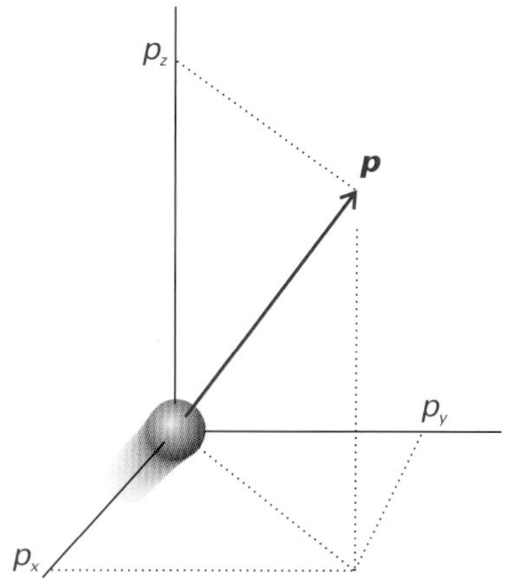

Fig. A3.1 The linear momentum of a particle is a vector property and points in the direction of motion.

A3.3 The trajectory in terms of the energy

The **velocity**, $\boldsymbol{v}$, of a particle is the rate of change of its position:

$$\boldsymbol{v} = \frac{\mathrm{d}\boldsymbol{r}}{\mathrm{d}t} \tag{A3.5}$$

The velocity is a vector, with both direction and magnitude. The magnitude of the velocity is the **speed**, v. The **linear momentum**, $\boldsymbol{p}$, of a particle of mass m is related to its velocity, $\boldsymbol{v}$, by

$$\boldsymbol{p} = m\boldsymbol{v} \tag{A3.6}$$

Like the velocity vector, the linear momentum vector points in the direction of travel of the particle (Fig. A3.1). In terms of the linear momentum, the total energy of a particle is

$$E = \frac{p^2}{2m} + V(x) \tag{A3.7}$$

This equation can be used to show that a particle will have a definite **trajectory**, or definite position and momentum at each instant. For example, consider a particle free to move in one direction (along the x-axis) in a region where $V = 0$ (so the energy is independent of position). Because $v = \mathrm{d}x/\mathrm{d}t$, it follows from eqns A3.6 and A3.7 that

$$\frac{\mathrm{d}x}{\mathrm{d}t} = \left(\frac{2E_{\mathrm{K}}}{m}\right)^{1/2} \tag{A3.8}$$

A solution of this differential equation is

$$x(t) = x(0) + m\left(\frac{2E_{\mathrm{K}}}{m}\right)^{1/2} t \tag{A3.9}$$

The linear momentum is a constant:

$$p(t) = mv(t) = m\frac{\mathrm{d}x}{\mathrm{d}t} = (2mE_{\mathrm{K}})^{1/2} \tag{A3.10}$$

Hence, if we know the initial position and momentum, we can predict all later positions and momenta exactly.

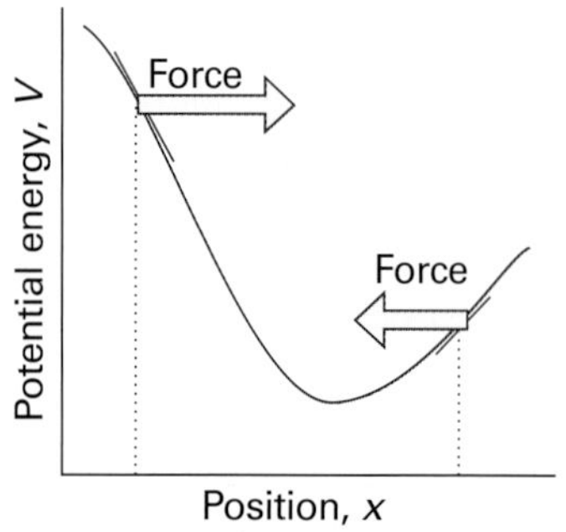

Fig. A3.2 The force acting on a particle is determined by the slope of the potential energy at each point. The force points in the direction of lower potential energy.

A3.4 Newton's second law

The **force**, F, experienced by a particle free to move in one dimension is related to its potential energy, V, by

$$F = -\frac{\mathrm{d}V}{\mathrm{d}x} \tag{A3.11a}$$

This relation implies that the direction of the force is towards decreasing potential energy (Fig. A3.2). In three dimensions,

$$\boldsymbol{F} = -\nabla V \qquad \nabla = \boldsymbol{i}\frac{\partial}{\partial x} + \boldsymbol{j}\frac{\partial}{\partial y} + \boldsymbol{k}\frac{\partial}{\partial z} \tag{A3.11b}$$

Newton's second law of motion states that *the rate of change of momentum is equal to the force acting on the particle*. In one dimension:

$$\frac{\mathrm{d}p}{\mathrm{d}t} = F \tag{A3.12a}$$

Because $p = m(\mathrm{d}x/\mathrm{d}t)$ in one dimension, it is sometimes more convenient to write this equation as

$$m\frac{\mathrm{d}^2x}{\mathrm{d}t^2} = F \tag{A3.12b}$$

The second derivative, $\mathrm{d}^2x/\mathrm{d}t^2$, is the **acceleration** of the particle, its rate of change of velocity (in this instance, along the x-axis). It follows that, if we know the force acting everywhere and at all times, then solving eqn A3.12 will also give the trajectory. This calculation is equivalent to the one based on E, but is more suitable in some applications. For example, it can be used to show that, if a particle of mass m is initially stationary and is subjected to a constant force F for a time τ, then its kinetic energy increases from zero to

$$E_{\mathrm{K}} = \frac{F^2\tau^2}{2m} \tag{A3.13}$$

and then remains at that energy after the force ceases to act. Because the applied force, F, and the time, τ, for which it acts may be varied at will, the solution implies that the energy of the particle may be increased to any value.

A3.5 Rotational motion

The rotational motion of a particle about a central point is described by its **angular momentum**, $\boldsymbol{J}$. The angular momentum is a vector: its magnitude gives the rate at which a particle circulates and its direction indicates the axis of rotation (Fig. A3.3). The magnitude of the angular momentum, J, is given by the expression

$$J = I\omega \tag{A3.14}$$

where ω is the **angular velocity** of the body, its rate of change of angular position (in radians per second), and I is the **moment of inertia**. The analogous roles of m and I, of v and ω, and of p and J in the translational and rotational cases, respectively, should be remembered, because they provide a ready way of constructing and recalling equations. For a point particle of mass m moving in a circle of radius r, the moment of inertia about the axis of rotation is given by the expression

$$I = mr^2 \tag{A3.15}$$

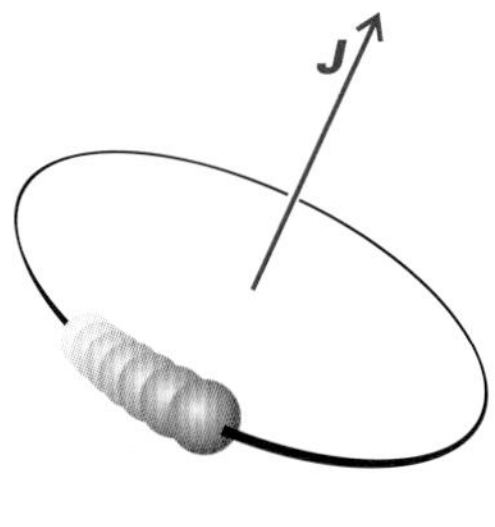

Fig. A3.3 The angular momentum of a particle is represented by a vector along the axis of rotation and perpendicular to the plane of rotation. The length of the vector denotes the magnitude of the angular momentum. The direction of motion is clockwise to an observer looking in the direction of the vector.

To accelerate a rotation it is necessary to apply a **torque**, $\boldsymbol{T}$, a twisting force. Newton's equation is then

$$\frac{\mathrm{d}\boldsymbol{J}}{\mathrm{d}t} = \boldsymbol{T} \tag{A3.16}$$

If a constant torque is applied for a time τ, the rotational energy of an initially stationary body is increased to

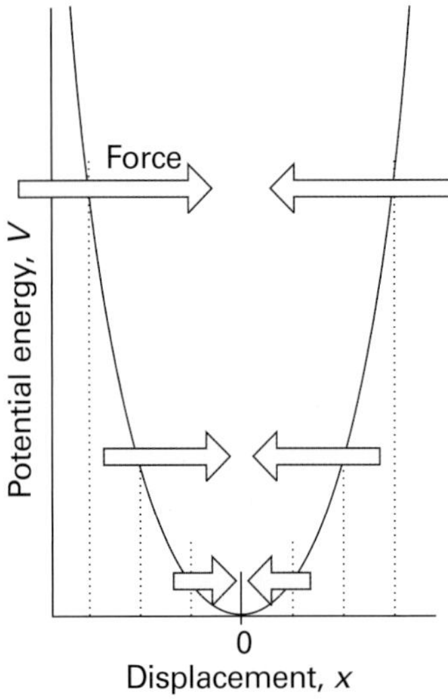

Fig. A3.4 The force acting on a particle that undergoes harmonic motion. The force is directed towards zero displacement and is proportional to the displacement. The corresponding potential energy is parabolic (proportional to x^2).

$$E_K = \frac{T^2\tau^2}{2I} \tag{A3.17}$$

The implication of this equation is that an appropriate torque and period for which it is applied can excite the rotation to an arbitrary energy.

A3.6 The harmonic oscillator

A **harmonic oscillator** consists of a particle that experiences a restoring force proportional to its displacement from its equilibrium position:

$$F = -kx \tag{A3.18}$$

An example is a particle joined to a rigid support by a spring. The constant of proportionality k is called the **force constant**, and the stiffer the spring the greater the force constant. The negative sign in F signifies that the direction of the force is opposite to that of the displacement (Fig. A3.4).

The motion of a particle that undergoes harmonic motion is found by substituting the expression for the force, eqn A3.18, into Newton's equation, eqn A3.12b. The resulting equation is

$$m\frac{d^2x}{dt^2} = -kx$$

A solution is

$$x(t) = A \sin \omega t \qquad p(t) = m\omega A \cos \omega t \qquad \omega = (k/m)^{1/2} \tag{A3.19}$$

These solutions show that the position of the particle varies **harmonically** (that is, as $\sin \omega t$) with a frequency $\nu = \omega/2\pi$. They also show that the particle is stationary ($p = 0$) when the displacement, x, has its maximum value, A, which is called the **amplitude** of the motion.

The total energy of a classical harmonic oscillator is proportional to the square of the amplitude of its motion. To confirm this remark we note that the kinetic energy is

$$E_K = \frac{p^2}{2m} = \frac{(m\omega A \cos \omega t)^2}{2m} = \tfrac{1}{2}m\omega^2 A^2 \cos^2 \omega t \tag{A3.20}$$

Then, because $\omega = (k/m)^{1/2}$, this expression may be written

$$E_K = \tfrac{1}{2}kA^2 \cos^2 \omega t \tag{A3.21}$$

The force on the oscillator is $F = -kx$, so it follows from the relation $F = -dV/dx$ that the potential energy of a harmonic oscillator is

$$V = \tfrac{1}{2}kx^2 = \tfrac{1}{2}kA^2 \sin^2 \omega t \tag{A3.22}$$

The total energy is therefore

$$E = \tfrac{1}{2}kA^2 \cos^2 \omega t + \tfrac{1}{2}kA^2 \sin^2 \omega t = \tfrac{1}{2}kA^2 \tag{A3.23}$$

(We have used $\cos^2 \omega t + \sin^2 \omega t = 1$.) That is, the energy of the oscillator is constant and, for a given force constant, is determined by its maximum displacement. It follows that the energy of an oscillating particle can be raised to any value by stretching the spring to any desired amplitude A. Note that the frequency of the motion depends only on the inherent properties of the oscillator (as represented by k and m) and is independent of the energy; the amplitude governs the energy, through $E = \tfrac{1}{2}kA^2$, and is independent of the frequency. In other words, the particle will oscillate at the same frequency regardless of the amplitude of its motion.

Waves

Waves are disturbances that travel through space with a finite velocity. Examples of disturbances include the collective motion of water molecules in ocean waves and of gas particles in sound waves. Waves can be characterized by a **wave equation**, a differential equation that describes the motion of the wave in space and time. **Harmonic waves** are waves with displacements that can be expressed as sine or cosine functions. These concepts are used in classical physics to describe the wave character of electromagnetic radiation, which is the focus of the following discussion.

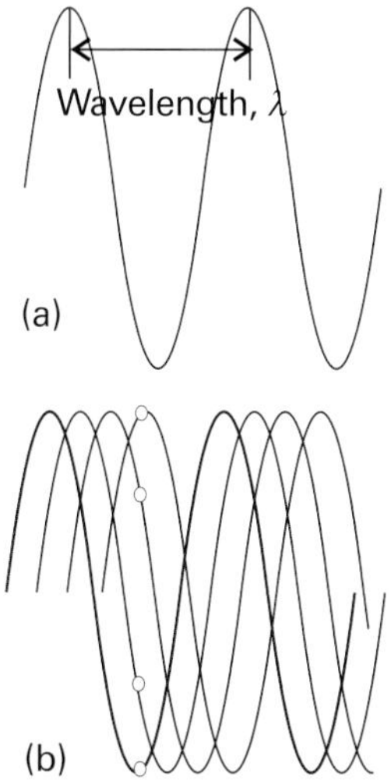

Fig. A3.5 (a) The wavelength, λ, of a wave is the peak-to-peak distance. (b) The wave is shown travelling to the right at a speed c. At a given location, the instantaneous amplitude of the wave changes through a complete cycle (the four dots show half a cycle) as it passes a given point. The frequency, ν, is the number of cycles per second that occur at a given point. Wavelength and frequency are related by $\lambda\nu = c$.

A3.7 The electromagnetic field

In classical physics, electromagnetic radiation is understood in terms of the **electromagnetic field**, an oscillating electric and magnetic disturbance that spreads as a harmonic wave through empty space, the vacuum. The wave travels at a constant speed called the *speed of light*, c, which is about 3×10^8 m s^{-1}. As its name suggests, an electromagnetic field has two components, an **electric field** that acts on charged particles (whether stationary of moving) and a **magnetic field** that acts only on moving charged particles. The electromagnetic field is characterized by a **wavelength**, λ (lambda), the distance between the neighbouring peaks of the wave, and its **frequency**, ν (nu), the number of times per second at which its displacement at a fixed point returns to its original value (Fig. A3.5). The frequency is measured in *hertz*, where 1 Hz = 1 s^{-1}. The wavelength and frequency of an electromagnetic wave are related by

$$\lambda\nu = c \qquad \text{(A3.24)}$$

Therefore, the shorter the wavelength, the higher the frequency. The characteristics of a wave are also reported by giving the **wavenumber**, $\tilde{\nu}$ (nu tilde), of the radiation, where

$$\tilde{\nu} = \frac{\nu}{c} = \frac{1}{\lambda} \qquad \text{(A3.25)}$$

A wavenumber can be interpreted as the number of complete wavelengths in a given length. Wavenumbers are normally reported in reciprocal centimetres (cm^{-1}), so a wavenumber of 5 cm^{-1} indicates that there are 5 complete wavelengths in 1 cm. The classification of the electromagnetic field according to its frequency and wavelength is summarized in Fig. A3.6.

A3.8 Features of electromagnetic radiation

Consider an electromagnetic disturbance travelling along the x direction with wavelength λ and frequency ν. The functions that describe the oscillating electric field, $\mathcal{E}(x,t)$, and magnetic field, $\mathcal{B}(x,t)$, may be written as

$$\mathcal{E}(x,t) = \mathcal{E}_0\cos\{2\pi\nu t - (2\pi/\lambda)x + \phi\} \qquad \text{(A3.26a)}$$

$$\mathcal{B}(x,t) = \mathcal{B}_0\cos\{2\pi\nu t - (2\pi/\lambda)x + \phi\} \qquad \text{(A3.26b)}$$

where $\mathcal{E}_0$ and $\mathcal{B}_0$ are the amplitudes of the electric and magnetic fields, respectively, and the parameter ϕ is the **phase** of the wave, which varies from $-\pi$ to π and gives the relative location of the peaks of two waves. If two waves, in the same region of space, with the same wavelength are shifted by $\phi = \pi$ or $-\pi$ (so the peaks of one wave coincide with the troughs of the other), then the resultant wave will have diminished

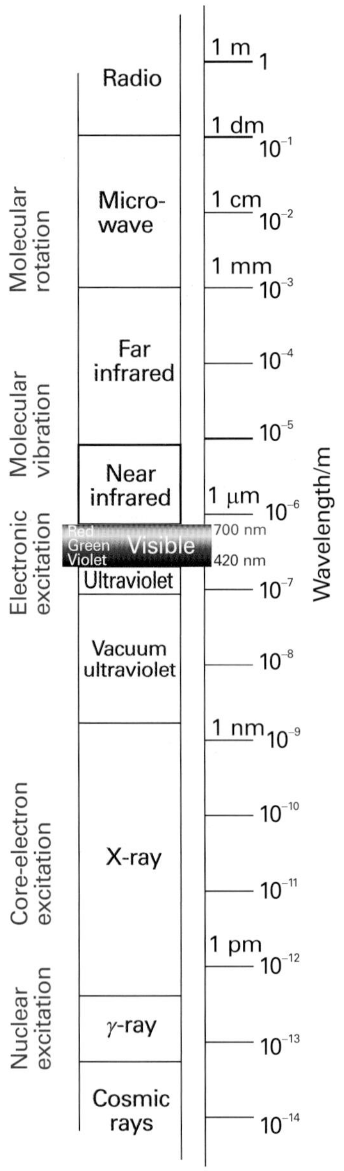

Fig. A3.6 The regions of the electromagnetic spectrum and the types of excitation that give rise to each region.

amplitudes. The waves are said to interfere destructively. A value of $\phi = 0$ (coincident peaks) corresponds to constructive interference, or the enhancement of the amplitudes.

Equations A3.26a and A3.26b represent electromagnetic radiation that is **plane-polarized**; it is so called because the electric and magnetic fields each oscillate in a single plane (in this case the *xy*- and *xz*-planes, respectively, Fig. A3.7). The plane of polarization may be orientated in any direction around the direction of propagation (the *x*-direction in Fig. A3.7), with the electric and magnetic fields perpendicular to that direction (and perpendicular to each other). An alternative mode of polarization is **circular polarization**, in which the electric and magnetic fields rotate around the direction of propagation in either a clockwise or a counterclockwise sense but remain perpendicular to it and each other.

It is easy to show by differentiation that eqns A3.26a and A3.26b satisfy the following equations:

$$\frac{\partial^2}{\partial x^2}\psi(x,t) = -\frac{4\pi^2}{\lambda^2}\psi(x,t) \qquad \frac{\partial^2}{\partial t^2}\psi(x,t) = -4\pi^2\nu^2\psi(x,t) \tag{A3.27}$$

where $\psi(x,t)$ is either $\mathcal{E}(x,t)$ or $\mathcal{B}(x,t)$.

According to classical electromagnetic theory, the intensity of electromagnetic radiation is proportional to the square of the amplitude of the wave. For example, the light detectors discussed in *Further information* 16.1 are based on the interaction between the electric field of the incident radiation and the detecting element, so light intensities are proportional to $\mathcal{E}_0^2$.

A3.9 Refraction

A beam of light changes direction ('bends') when it passes from one transparent medium to another. This effect, called **refraction**, depends on the **refractive index**, n_r, of the medium, the ratio of the speed of light in a vacuum, c, to its speed c' in the medium:

$$n_r = \frac{c}{c'} \qquad [A3.28]$$

It follows from the Maxwell equations (see *Further reading*), that the refractive index at a (visible or ultraviolet) specified frequency is related to the relative permittivity ε_r (discussed in Section 20.10) at that frequency by

$$n_r = \varepsilon_r^{1/2} \tag{A3.29}$$

Table A3.1 lists refractive indices of some materials.

Because the relative permittivity of a medium is related to its polarizability by eqn 20.10, the refractive index is related to the polarizability. To see why this is so, we need to realize that propagation of light through a medium induces an oscillating dipole moment, which then radiates light of the same frequency. The newly generated radiation is delayed slightly by this process, so it propagates more slowly through the medium than through a vacuum. Because photons of high-frequency light carry more energy than those of low-frequency light, they can distort the electronic distributions of the molecules in their path more effectively. Therefore, after allowing for the loss of contributions from low-frequency modes of motion, we can expect the electronic polarizabilities of molecules, and hence the refractive index, to increase as the incident frequency rises towards an absorption frequency. This dependence on frequency is the origin of the dispersion of white light by a prism: the refractive index is greater for blue light than for red, and therefore the blue rays are bent more than the red. The term **dispersion** is a term carried over from this phenomenon to mean the variation of the refractive index, or of any property, with frequency.

A3.10 Optical activity

The concept of refractive index is closely related to the property of optical activity. An **optically active** substance is a substance that rotates the plane of polarization of plane-polarized light. To understand this effect, it is useful to regard the incident plane-polarized beam as a superposition of two oppositely rotating circularly polarized components. By convention, in right-handed circularly polarized light the electric vector rotates clockwise as seen by an observer facing the oncoming beam (Fig. A3.8). On entering the medium, one component propagates faster than the other if their refractive indices are different. If the sample is of length l, the difference in the times of passage is

$$\Delta t = \frac{l}{c_L} - \frac{l}{c_R}$$

where c_R and c_L are the speeds of the two components in the medium. In terms of the refractive indices, the difference is

$$\Delta t = (n_R - n_L)\frac{l}{c}$$

The phase difference between the two components when they emerge from the sample is therefore

$$\Delta\theta = 2\pi\nu\Delta t = \frac{2\pi c\Delta t}{\lambda} = (n_R - n_L)\frac{2\pi l}{\lambda}$$

where λ is the wavelength of the light. The two rotating electric vectors have a different phase when they leave the sample from the value they had initially, so their superposition gives rise to a plane-polarized beam rotated through an angle $\Delta\theta$ relative to the plane of the incoming beam. It follows that the angle of optical rotation is proportional to the difference in refractive index, $n_R - n_L$. A sample in which these two refractive indices are different is said to be **circularly birefringent**.

To explain why the refractive indices depend on the handedness of the light, we must examine why the polarizabilities depend on the handedness. One interpretation is that, if a molecule is helical (such as a polypeptide α-helix described in Section 19.7) or a crystal has molecules in a helical arrangement (as in a cholesteric liquid crystal, as described in *Impact* I6.1), its polarizability depends on whether or not the electric field of the incident radiation rotates in the same sense as the helix.

Associated with the circular birefringence of the medium is a difference in absorption intensities for right- and left-circularly polarized radiation. This difference is known as *circular dichroism*, which is explored in Chapter 14.

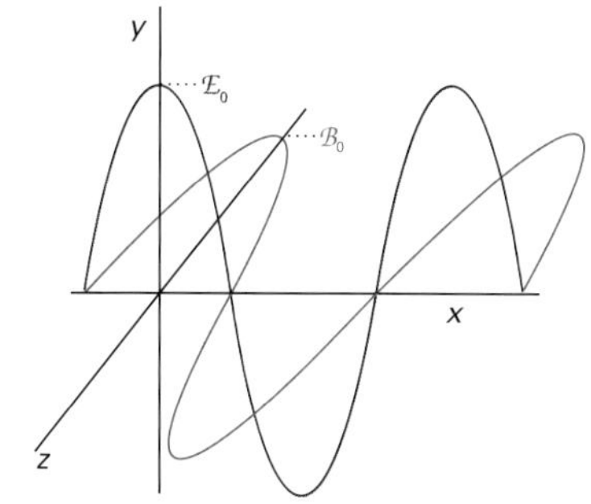

Fig. A3.7 Electromagnetic radiation consists of a wave of electric and magnetic fields perpendicular to the direction of propagation (in this case the x-direction), and mutually perpendicular to each other. This illustration shows a plane-polarized wave, with the electric and magnetic fields oscillating in the xy- and xz-planes, respectively.

Synoptic table A3.1* Refractive indices relative to air at 20°C

	434 nm	589 nm	656 nm
$C_6H_6(l)$	1.524	1.501	1.497
$CS_2(l)$	1.675	1.628	1.618
$H_2O(l)$	1.340	1.333	1.331
KI(s)	1.704	1.666	1.658

* More values are given in the *Data section*.

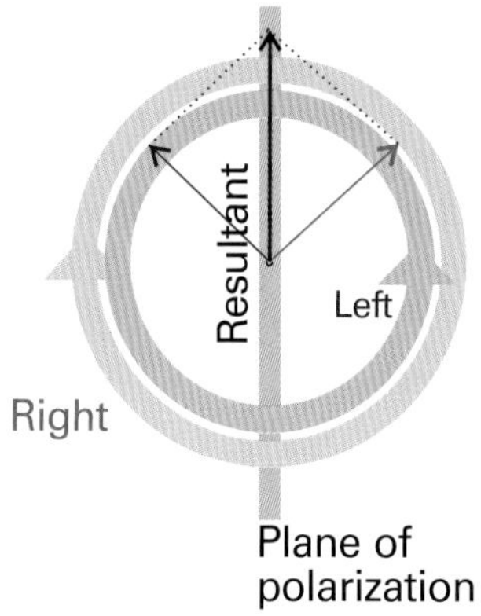

Fig. A3.8 The superposition of left and right circularly polarized light as viewed from an observer facing the oncoming beam.

Electrostatics

Electrostatics is the study of the interactions of stationary electric charges. The elementary charge, the magnitude of charge carried by a single electron or proton, is $e \approx 1.60 \times 10^{-19}$ C. The magnitude of the charge per mole is Faraday's constant: $F = N_A e = 9.65 \times 10^4$ C mol^{-1}.

A3.11 The Coulomb interaction

If a point charge q_1 is at a distance r in a vacuum from another point charge q_2, then their potential energy is

$$V=\frac{q_1q_2}{4\pi\varepsilon_0 r} \qquad \text{(A3.30)}$$

The constant ε_0 is the **vacuum permittivity**, a fundamental constant with the value $8.85\times10^{-12}\ \mathrm{C^2\,J^{-1}\,m^{-1}}$. This very important relation is called the **Coulomb potential energy** and the interaction it describes is called the **Coulomb interaction** of two charges. The Coulomb potential energy is equal to the work that must be done to bring up a charge q_1 from infinity to a distance r from a charge q_2.

It follows from eqns A3.5 and A3.30 that the electrical force, $\boldsymbol{F}$, exerted by a charge q_1 on a second charge q_2 has magnitude

$$F=\frac{q_1q_2}{4\pi\varepsilon_0 r^2} \qquad \text{(A3.31)}$$

The force itself is a vector directed along the line joining the two charges. With charge in coulombs and distance in metres, the force is obtained in newtons.

In a medium other than a vacuum, the potential energy of interaction between two charges is reduced, and the vacuum permittivity is replaced by the **permittivity**, ε, of the medium (see Section 18.3).

A3.12 The Coulomb potential

The potential energy of a charge q_1 in the presence of another charge q_2 can be expressed in terms of the **Coulomb potential**, ϕ:

$$V=q_1\phi \qquad \phi=\frac{q_2}{4\pi\varepsilon_0 r} \qquad \text{(A3.32)}$$

The units of potential are joules per coulomb, $\mathrm{J\,C^{-1}}$, so, when φ is multiplied by a charge in coulombs, the result is in joules. The combination joules per coulomb occurs widely in electrostatics, and is called a *volt*, V:

$$1\ \mathrm{V}=1\ \mathrm{J\,C^{-1}} \qquad \text{(A3.33)}$$

If there are several charges $q_2, q_3, \ldots$ present in the system, the total potential experienced by the charge q_1 is the sum of the potential generated by each charge:

$$\phi=\phi_2+\phi_3+\cdots \qquad \text{(A3.34)}$$

When the charge distribution is more complex than a single point–like object, the Coulomb potential is described in terms of a charge density, ρ. With charge in coulomb and length in metres, the charge density is expressed in coulombs per metre-cubed ($\mathrm{C\,m^{-3}}$). The electric potential arising from a charge distribution with density ρ is the solution to **Poisson's equation**:

$$\nabla^2\phi=-\rho/\varepsilon_0 \qquad \text{(A3.35)}$$

where $\nabla^2=(\partial^2/\partial x^2+\partial^2/\partial y^2+\partial^2/\partial z^2)$. If the distribution is spherically symmetrical, then so too is ϕ and eqn A3.35 reduces to the form used in *Further information* 5.1.

A3.13 The strength of the electric field

Just as the potential energy of a charge q_1 can be written $V=q_1\phi$, so the magnitude of the force on q_1 can be written $F=q_1\mathcal{E}$, where $\mathcal{E}$ is the magnitude of the electric field strength arising from q_2 or from some more general charge distribution. The electric field strength (which, like the force, is actually a vector quantity) is the negative gradient of the electric potential:

$$\mathcal{E} = -\nabla\phi \tag{A3.36}$$

A3.14 Electric current and power

The motion of charge gives rise to an electric **current**, I. Electric current is measured in *ampere*, A, where

$$1\ \text{A} = 1\ \text{C s}^{-1} \tag{A3.37}$$

If the current flows from a region of potential ϕ_i to ϕ_f, through a **potential difference** $\Delta\phi = \phi_f - \phi_i$, the rate of doing work is the current (the rate of transfer of charge) multiplied by the potential difference, $I\Delta\phi$. The rate of doing electrical work is the **electrical power**, P, so

$$P = I\Delta\phi \tag{A3.38}$$

With current in amperes and the potential difference in volts, the power works out in watts. The total energy, E, supplied in an interval Δt is the power (the rate of energy supply) multiplied by the duration of the interval:

$$E = P\Delta t = I\Delta\phi\Delta t \tag{A3.39}$$

The energy is obtained in joules with the current in amperes, the potential difference in volts, and the time in seconds.

Further reading

R.P. Feynman, R.B. Leighton, and M. Sands, *The Feynman lectures on physics*. Vols 1–3. Addison–Wesley, Reading (1966).

G.A.D. Ritchie and D.S. Sivia, *Foundations of physics for chemists*. Oxford Chemistry Primers, Oxford University Press (2000).

W.S. Warren, *The physical basis of chemistry*. Academic Press, San Diego (2000).

R. Wolfson and J.M. Pasachoff, *Physics for scientists and engineers*. Benjamin Cummings, San Francisco (1999).

Data section

Contents

The following is a directory of all tables in the text; those included in this *Data section* are marked with an asterisk. The remainder will be found on the pages indicated.

The following tables reproduce and expand the data given in the short tables in the text, and follow their numbering. Standard states refer to a pressure of $p^{\ominus} = 1$ bar. The general references are as follows:

AIP: D.E. Gray (ed.), *American Institute of Physics handbook*. McGraw Hill, New York (1972).

AS: M. Abramowitz and I.A. Stegun (ed.), *Handbook of mathematical functions*. Dover, New York (1963).

E: J. Emsley, *The elements*. Oxford University Press (1991).

HCP: D.R. Lide (ed.), *Handbook of chemistry and physics*. CRC Press, Boca Raton (2000).

JL: A.M. James and M.P. Lord, *Macmillan's chemical and physical data*. Macmillan, London (1992).

KL: G.W.C. Kaye and T.H. Laby (ed.), *Tables of physical and chemical constants*. Longman, London (1973).

LR: G.N. Lewis and M. Randall, resived by K.S. Pitzer and L. Brewer, *Thermodynamics*. McGraw-Hill, New York (1961).

NBS: *NBS tables of chemical thermodynamic properties*, published as *J. Phys. and Chem. Reference Data*, 11, Supplement 2 (1982).

RS: R.A. Robinson and R.H. Stokes, *Electrolyte solutions*. Butterworth, London (1959).

TDOC: J.B. Pedley, J.D. Naylor, and S.P. Kirby, *Thermochemical data of organic compounds*. Chapman & Hall, London (1986).

Physical properties of selected materials

	ρ/(g cm^{-3}) at 293 K†	T_f/K	T_b/K		ρ/(g cm^{-3}) at 293 K†	T_f/K	T_b/K
Elements				**Inorganic compounds**			
Aluminium(s)	2.698	933.5	2740	$CaCO_3$(s, calcite)	2.71	1612	1171d
Argon(g)	1.381	83.8	87.3	$CuSO_4\cdot 5H_2O$(s)	2.284	383(–H_2O)	423(–$5H_2O$)
Boron(s)	2.340	2573	3931	HBr(g)	2.77	184.3	206.4
Bromine(l)	3.123	265.9	331.9	HCl(g)	1.187	159.0	191.1
Carbon(s, gr)	2.260	3700s		HI(g)	2.85	222.4	237.8
Carbon(s, d)	3.513			H_2O(l)	0.997	273.2	373.2
Chlorine(g)	1.507	172.2	239.2	D_2O(l)	1.104	277.0	374.6
Copper(s)	8.960	1357	2840	NH_3(g)	0.817	195.4	238.8
Fluorine(g)	1.108	53.5	85.0	KBr(s)	2.750	1003	1708
Gold(s)	19.320	1338	3080	KCl(s)	1.984	1049	1773s
Helium(g)	0.125		4.22	NaCl(s)	2.165	1074	1686
Hydrogen(g)	0.071	14.0	20.3	H_2SO_4(l)	1.841	283.5	611.2
Iodine(s)	4.930	386.7	457.5				
Iron(s)	7.874	1808	3023	**Organic compounds**			
Krypton(g)	2.413	116.6	120.8	Acetaldehyde, CH_3CHO(l, g)	0.788	152	293
Lead(s)	11.350	600.6	2013	Acetic acid, CH_3COOH(l)	1.049	289.8	391
Lithium(s)	0.534	453.7	1620	Acetone, $(CH_3)_2CO$(l)	0.787	178	329
Magnesium(s)	1.738	922.0	1363	Aniline, $C_6H_5NH_2$(l)	1.026	267	457
Mercury(l)	13.546	234.3	629.7	Anthracene, $C_{14}H_{10}$(s)	1.243	490	615
Neon(g)	1.207	24.5	27.1	Benzene, C_6H_6(l)	0.879	278.6	353.2
Nitrogen(g)	0.880	63.3	77.4	Carbon tetrachloride, CCl_4(l)	1.63	250	349.9
Oxygen(g)	1.140	54.8	90.2	Chloroform, $CHCl_3$(l)	1.499	209.6	334
Phosphorus(s, wh)	1.820	317.3	553	Ethanol, C_2H_5OH(l)	0.789	156	351.4
Potassium(s)	0.862	336.8	1047	Formaldehyde, HCHO(g)		181	254.0
Silver(s)	10.500	1235	2485	Glucose, $C_6H_{12}O_6$(s)	1.544	415	
Sodium(s)	0.971	371.0	1156	Methane, CH_4(g)		90.6	111.6
Sulfur(s, α)	2.070	386.0	717.8	Methanol, CH_3OH(l)	0.791	179.2	337.6
Uranium(s)	18.950	1406	4018	Naphthalene, $C_{10}H_8$(s)	1.145	353.4	491
Xenon(g)	2.939	161.3	166.1	Octane, C_8H_{18}(l)	0.703	216.4	398.8
Zinc(s)	7.133	692.7	1180	Phenol, C_6H_5OH(s)	1.073	314.1	455.0
				Sucrose, $C_{12}H_{22}O_{11}$(s)	1.588	457d	

d: decomposes; s: sublimes; Data: AIP, E, HCP, KL. † For gases, at their boiling points.

Masses and natural abundances of selected nuclides

Nuclide		m/u	Abundance/%
H	^{1}H	1.0078	99.985
	^{2}H	2.0140	0.015
He	^{3}He	3.0160	0.000 13
	^{4}He	4.0026	100
Li	^{6}Li	6.0151	7.42
	^{7}Li	7.0160	92.58
B	^{10}B	10.0129	19.78
	^{11}B	11.0093	80.22
C	^{12}C	12*	98.89
	^{13}C	13.0034	1.11
N	^{14}N	14.0031	99.63
	^{15}N	15.0001	0.37
O	^{16}O	15.9949	99.76
	^{17}O	16.9991	0.037
	^{18}O	17.9992	0.204
F	^{19}F	18.9984	100
P	^{31}P	30.9738	100
S	^{32}S	31.9721	95.0
	^{33}S	32.9715	0.76
	^{34}S	33.9679	4.22
Cl	^{35}Cl	34.9688	75.53
	^{37}Cl	36.9651	24.4
Br	^{79}Br	78.9183	50.54
	^{81}Br	80.9163	49.46
I	^{127}I	126.9045	100

* Exact value.

Table 1.4 Second virial coefficients, $B/(\text{cm}^3\ \text{mol}^{-1})$

	100 K	273 K	373 K	600 K
Air	−167.3	−13.5	3.4	19.0
Ar	−187.0	−21.7	−4.2	11.9
CH_4		−53.6	−21.2	8.1
CO_2		−142	−72.2	−12.4
H_2	−2.0	13.7	15.6	
He	11.4	12.0	11.3	10.4
Kr		−62.9	−28.7	1.7
N_2	−160.0	−10.5	6.2	21.7
Ne	−6.0	10.4	12.3	13.8
O_2	−197.5	−22.0	−3.7	12.9
Xe		−153.7	−81.7	−19.6

Data: AIP, JL. The values relate to the expansion in eqn 1.22 of Section 1.3b; convert to eqn 1.21 using $B' = B/RT$.
For Ar at 273 K, $C = 1200\ \text{cm}^6\ \text{mol}^{-1}$.

Table 1.5 Critical constants of gases

	p_c/atm	$V_c/(\text{cm}^3\ \text{mol}^{-1})$	T_c/K	Z_c	T_B/K
Ar	48.00	75.25	150.72	0.292	411.5
Br_2	102	135	584	0.287	
C_2H_4	50.50	124	283.1	0.270	
C_2H_6	48.20	148	305.4	0.285	
C_6H_6	48.6	260	562.7	0.274	
CH_4	45.6	98.7	190.6	0.288	510.0
Cl_2	76.1	124	417.2	0.276	
CO_2	72.85	94.0	304.2	0.274	714.8
F_2	55	144			
H_2	12.8	64.99	33.23	0.305	110.0
H_2O	218.3	55.3	647.4	0.227	
HBr	84.0	363.0			
HCl	81.5	81.0	324.7	0.248	
He	2.26	57.76	5.21	0.305	22.64
HI	80.8	423.2			
Kr	54.27	92.24	209.39	0.291	575.0
N_2	33.54	90.10	126.3	0.292	327.2
Ne	26.86	41.74	44.44	0.307	122.1
NH_3	111.3	72.5	405.5	0.242	
O_2	50.14	78.0	154.8	0.308	405.9
Xe	58.0	118.8	289.75	0.290	768.0

Data: AIP, KL.

Table 1.6 van der Waals coefficients

	$a/(\mathrm{atm\ dm^6\ mol^{-2}})$	$b/(10^{-2}\ \mathrm{dm^3\ mol^{-1}})$		$a/(\mathrm{atm\ dm^6\ mol^{-2}})$	$b/(10^{-2}\ \mathrm{dm^3\ mol^{-1}})$
Ar	1.337	3.20	H_2S	4.484	4.34
C_2H_4	4.552	5.82	He	0.0341	2.38
C_2H_6	5.507	6.51	Kr	5.125	1.06
C_6H_6	18.57	11.93	N_2	1.352	3.87
CH_4	2.273	4.31	Ne	0.205	1.67
Cl_2	6.260	5.42	NH_3	4.169	3.71
CO	1.453	3.95	O_2	1.364	3.19
CO_2	3.610	4.29	SO_2	6.775	5.68
H_2	0.2420	2.65	Xe	4.137	5.16
H_2O	5.464	3.05			

Data: HCP.

Table 2.2 Temperature variation of molar heat capacities†

	a	$b/(10^{-3}\ \mathrm{K^{-1}})$	$c/(10^5\ \mathrm{K^2})$
Monatomic gases			
	20.78	0	0
Other gases			
Br_2	37.32	0.50	−1.26
Cl_2	37.03	0.67	−2.85
CO_2	44.22	8.79	−8.62
F_2	34.56	2.51	−3.51
H_2	27.28	3.26	0.50
I_2	37.40	0.59	−0.71
N_2	28.58	3.77	−0.50
NH_3	29.75	25.1	−1.55
O_2	29.96	4.18	−1.67
Liquids (from melting to boiling)			
$C_{10}H_8$, naphthalene	79.5	0.4075	0
I_2	80.33	0	0
H_2O	75.29	0	0
Solids			
Al	20.67	12.38	0
C (graphite)	16.86	4.77	−8.54
$C_{10}H_8$, naphthalene	−115.9	3.920×10^3	0
Cu	22.64	6.28	0
I_2	40.12	49.79	0
NcCl	45.94	16.32	0
Pb	22.13	11.72	0.96

† For $C_{p,\mathrm{m}}/(\mathrm{J\ K^{-1}\ mol^{-1}}) = a + bT + c/T^2$.
Source: LR.

Table 2.3 Standard enthalpies of fusion and vaporization at the transition temperature, $\Delta_{trs}H^{\ominus}/(kJ\ mol^{-1})$

	T_f/K	Fusion	T_b/K	Vaporization
Elements				
Ag	1234	11.30	2436	250.6
Ar	83.81	1.188	87.29	6.506
Br_2	265.9	10.57	332.4	29.45
Cl_2	172.1	6.41	239.1	20.41
F_2	53.6	0.26	85.0	3.16
H_2	13.96	0.117	20.38	0.916
He	3.5	0.021	4.22	0.084
Hg	234.3	2.292	629.7	59.30
I_2	386.8	15.52	458.4	41.80
N_2	63.15	0.719	77.35	5.586
Na	371.0	2.601	1156	98.01
O_2	54.36	0.444	90.18	6.820
Xe	161	2.30	165	12.6
K	336.4	2.35	1031	80.23
Inorganic compounds				
CCl_4	250.3	2.47	349.9	30.00
CO_2	217.0	8.33	194.6	25.23 s
CS_2	161.2	4.39	319.4	26.74
H_2O	273.15	6.008	373.15	40.656
				44.016 at 298 K
H_2S	187.6	2.377	212.8	18.67
H_2SO_4	283.5	2.56		
NH_3	195.4	5.652	239.7	23.35
Organic compounds				
CH_4	90.68	0.941	111.7	8.18
CCl_4	250.3	2.5	350	30.0
C_2H_6	89.85	2.86	184.6	14.7
C_6H_6	278.61	10.59	353.2	30.8
C_6H_{14}	178	13.08	342.1	28.85
$C_{10}H_8$	354	18.80	490.9	51.51
CH_3OH	175.2	3.16	337.2	35.27
				37.99 at 298 K
C_2H_5OH	158.7	4.60	352	43.5

Data: AIP; s denotes sublimation.

Table 2.5 Thermodynamic data for organic compounds (all values are for 298 K)

	$M/(g\ mol^{-1})$	$\Delta_fH^{\ominus}/(kJ\ mol^{-1})$	$\Delta_fG^{\ominus}/(kJ\ mol^{-1})$	$S_m^{\ominus}/(J\ K^{-1}\ mol^{-1})$†	$C_{p,m}^{\ominus}/(J\ K^{-1}\ mol^{-1})$	$\Delta_cH^{\ominus}/(kJ\ mol^{-1})$
C(s) (graphite)	12.011	0	0	5.740	8.527	−393.51
C(s) (diamond)	12.011	+1.895	+2.900	2.377	6.113	−395.40
CO_2(g)	44.040	−393.51	−394.36	213.74	37.11	
Hydrocarbons						
CH_4(g), methane	16.04	−74.81	−50.72	186.26	35.31	−890
CH_3(g), methyl	15.04	+145.69	+147.92	194.2	38.70	
C_2H_2(g), ethyne	26.04	+226.73	+209.20	200.94	43.93	−1300
C_2H_4(g), ethene	28.05	+52.26	+68.15	219.56	43.56	−1411
C_2H_6(g), ethane	30.07	−84.68	−32.82	229.60	52.63	−1560
C_3H_6(g), propene	42.08	+20.42	+62.78	267.05	63.89	−2058
C_3H_6(g), cyclopropane	42.08	+53.30	+104.45	237.55	55.94	−2091
C_3H_8(g), propane	44.10	−103.85	−23.49	269.91	73.5	−2220
C_4H_8(g), 1-butene	56.11	−0.13	+71.39	305.71	85.65	−2717
C_4H_8(g), *cis*-2-butene	56.11	−6.99	+65.95	300.94	78.91	−2710
C_4H_8(g), *trans*-2-butene	56.11	−11.17	+63.06	296.59	87.82	−2707
C_4H_{10}(g), butane	58.13	−126.15	−17.03	310.23	97.45	−2878
C_5H_{12}(g), pentane	72.15	−146.44	−8.20	348.40	120.2	−3537
C_5H_{12}(l)	72.15	−173.1				
C_6H_6(l), benzene	78.12	+49.0	+124.3	173.3	136.1	−3268

Table 2.5 (Continued)

	$M/(\mathrm{g\ mol^{-1}})$	$\Delta_f H^\ominus/(\mathrm{kJ\ mol^{-1}})$	$\Delta_f G^\ominus/(\mathrm{kJ\ mol^{-1}})$	$S_m^\ominus/(\mathrm{J\ K^{-1}\ mol^{-1}})$†	$C_{p,m}^\ominus/(\mathrm{J\ K^{-1}\ mol^{-1}})$	$\Delta_c H^\ominus/(\mathrm{kJ\ mol^{-1}})$
Hydrocarbons (Continued)						
C_6H_6(g)	78.12	+82.93	+129.72	269.31	81.67	−3302
C_6H_{12}(l), cyclohexane	84.16	−156	+26.8	204.4	156.5	−3920
C_6H_{14}(l), hexane	86.18	−198.7		204.3		−4163
$C_6H_5CH_3$(g), methylbenzene (toluene)	92.14	+50.0	+122.0	320.7	103.6	−3953
C_7H_{16}(l), heptane	100.21	−224.4	+1.0	328.6	224.3	
C_8H_{18}(l), octane	114.23	−249.9	+6.4	361.1		−5471
C_8H_{18}(l), iso-octane	114.23	−255.1				−5461
$C_{10}H_8$(s), naphthalene	128.18	+78.53				−5157
Alcohols and phenols						
CH_3OH(l), methanol	32.04	−238.66	−166.27	126.8	81.6	−726
CH_3OH(g)	32.04	−200.66	−161.96	239.81	43.89	−764
C_2H_5OH(l), ethanol	46.07	−277.69	−174.78	160.7	111.46	−1368
C_2H_5OH(g)	46.07	−235.10	−168.49	282.70	65.44	−1409
C_6H_5OH(s), phenol	94.12	−165.0	−50.9	146.0		−3054
Carboxylic acids, hydroxy acids, and esters						
HCOOH(l), formic	46.03	−424.72	−361.35	128.95	99.04	−255
CH_3COOH(l), acetic	60.05	−484.5	−389.9	159.8	124.3	−875
CH_3COOH(aq)	60.05	−485.76	−396.46	178.7		
$CH_3CO_2^-$(aq)	59.05	−486.01	−369.31	+86.6	−6.3	
$(COOH)_2$(s), oxalic	90.04	−827.2			117	−254
C_6H_5COOH(s), benzoic	122.13	−385.1	−245.3	167.6	146.8	−3227
$CH_3CH(OH)COOH$(s), lactic	90.08	−694.0				−1344
$CH_3COOC_2H_5$(l), ethyl acetate	88.11	−479.0	−332.7	259.4	170.1	−2231
Alkanals and alkanones						
HCHO(g), methanal	30.03	−108.57	−102.53	218.77	35.40	−571
CH_3CHO(l), ethanal	44.05	−192.30	−128.12	160.2		−1166
CH_3CHO(g)	44.05	−166.19	−128.86	250.3	57.3	−1192
CH_3COCH_3(l), propanone	58.08	−248.1	−155.4	200.4	124.7	−1790
Sugars						
$C_6H_{12}O_6$(s), α-D-glucose	180.16	−1274				−2808
$C_6H_{12}O_6$(s), β-D-glucose	180.16	−1268	−910	212		
$C_6H_{12}O_6$(s), β-D-fructose	180.16	−1266				−2810
$C_{12}H_{22}O_{11}$(s), sucrose	342.30	−2222	−1543	360.2		−5645
Nitrogen compounds						
$CO(NH_2)_2$(s), urea	60.06	−333.51	−197.33	104.60	93.14	−632
CH_3NH_2(g), methylamine	31.06	−22.97	+32.16	243.41	53.1	−1085
$C_6H_5NH_2$(l), aniline	93.13	+31.1				−3393
$CH_2(NH_2)COOH$(s), glycine	75.07	−532.9	−373.4	103.5	99.2	−969

Data: NBS, TDOC. † Standard entropies of ions may be either positive or negative because the values are relative to the entropy of the hydrogen ion.

Table 2.7 Thermodynamic data for elements and inorganic compounds (all values relate to 298 K)

	M/(g mol^{-1})	$\Delta_f H^\ominus$/(kJ mol^{-1})	$\Delta_f G^\ominus$/(kJ mol^{-1})	$S_m^\ominus$/(J K^{-1} mol^{-1})†	$C_{p,m}^\ominus$/(J K^{-1} mol^{-1})
Aluminium (aluminum)					
Al(s)	26.98	0	0	28.33	24.35
Al(l)	26.98	+10.56	+7.20	39.55	24.21
Al(g)	26.98	+326.4	+285.7	164.54	21.38
Al^{3+}(g)	26.98	+5483.17			
Al^{3+}(aq)	26.98	−531	−485	−321.7	
Al_2O_3(s, α)	101.96	−1675.7	−1582.3	50.92	79.04
$AlCl_3$(s)	133.24	−704.2	−628.8	110.67	91.84
Argon					
Ar(g)	39.95	0	0	154.84	20.786
Antimony					
Sb(s)	121.75	0	0	45.69	25.23
SbH_3(g)	124.77	+145.11	+147.75	232.78	41.05
Arsenic					
As(s, α)	74.92	0	0	35.1	24.64
As(g)	74.92	+302.5	+261.0	174.21	20.79
As_4(g)	299.69	+143.9	+92.4	314	
AsH_3(g)	77.95	+66.44	+68.93	222.78	38.07
Barium					
Ba(s)	137.34	0	0	62.8	28.07
Ba(g)	137.34	+180	+146	170.24	20.79
Ba^{2+}(aq)	137.34	−537.64	−560.77	+9.6	
BaO(s)	153.34	−553.5	−525.1	70.43	47.78
$BaCl_2$(s)	208.25	−858.6	−810.4	123.68	75.14
Beryllium					
Be(s)	9.01	0	0	9.50	16.44
Be(g)	9.01	+324.3	+286.6	136.27	20.79
Bismuth					
Bi(s)	208.98	0	0	56.74	25.52
Bi(g)	208.98	+207.1	+168.2	187.00	20.79
Bromine					
Br_2(l)	159.82	0	0	152.23	75.689
Br_2(g)	159.82	+30.907	+3.110	245.46	36.02
Br(g)	79.91	+111.88	+82.396	175.02	20.786
Br^-(g)	79.91	−219.07			
Br^-(aq)	79.91	−121.55	−103.96	+82.4	−141.8
HBr(g)	90.92	−36.40	−53.45	198.70	29.142
Cadmium					
Cd(s, γ)	112.40	0	0	51.76	25.98
Cd(g)	112.40	+112.01	+77.41	167.75	20.79
Cd^{2+}(aq)	112.40	−75.90	−77.612	−73.2	

Table 2.7 (Continued)

	$M/(\text{g mol}^{-1})$	$\Delta_f H^\ominus/(\text{kJ mol}^{-1})$	$\Delta_f G^\ominus/(\text{kJ mol}^{-1})$	$S_m^\ominus/(\text{J K}^{-1}\text{ mol}^{-1})$†	$C_{p,m}^\ominus/(\text{J K}^{-1}\text{ mol}^{-1})$
Cadmium (Continued)					
CdO(s)	128.40	−258.2	−228.4	54.8	43.43
$CdCO_3$(s)	172.41	−750.6	−669.4	92.5	
Caesium (cesium)					
Cs(s)	132.91	0	0	85.23	32.17
Cs(g)	132.91	+76.06	+49.12	175.60	20.79
Cs^+(aq)	132.91	−258.28	−292.02	+133.05	−10.5
Calcium					
Ca(s)	40.08	0	0	41.42	25.31
Ca(g)	40.08	+178.2	+144.3	154.88	20.786
Ca^{2+}(aq)	40.08	−542.83	−553.58	−53.1	
CaO(s)	56.08	−635.09	−604.03	39.75	42.80
$CaCO_3$(s) (calcite)	100.09	−1206.9	−1128.8	92.9	81.88
$CaCO_3$(s) (aragonite)	100.09	−1207.1	−1127.8	88.7	81.25
CaF_2(s)	78.08	−1219.6	−1167.3	68.87	67.03
$CaCl_2$(s)	110.99	−795.8	−748.1	104.6	72.59
$CaBr_2$(s)	199.90	−682.8	−663.6	130	
Carbon (for 'organic' compounds of carbon, see Table 2.5)					
C(s) (graphite)	12.011	0	0	5.740	8.527
C(s) (diamond)	12.011	+1.895	+2.900	2.377	6.113
C(g)	12.011	+716.68	+671.26	158.10	20.838
C_2(g)	24.022	+831.90	+775.89	199.42	43.21
CO(g)	28.011	−110.53	−137.17	197.67	29.14
CO_2(g)	44.010	−393.51	−394.36	213.74	37.11
CO_2(aq)	44.010	−413.80	−385.98	117.6	
H_2CO_3(aq)	62.03	−699.65	−623.08	187.4	
HCO_3^-(aq)	61.02	−691.99	−586.77	+91.2	
CO_3^{2-}(aq)	60.01	−677.14	−527.81	−56.9	
CCl_4(l)	153.82	−135.44	−65.21	216.40	131.75
CS_2(l)	76.14	+89.70	+65.27	151.34	75.7
HCN(g)	27.03	+135.1	+124.7	201.78	35.86
HCN(l)	27.03	+108.87	+124.97	112.84	70.63
CN^-(aq)	26.02	+150.6	+172.4	+94.1	
Chlorine					
Cl_2(g)	70.91	0	0	223.07	33.91
Cl(g)	35.45	+121.68	+105.68	165.20	21.840
Cl^-(g)	34.45	−233.13			
Cl^-(aq)	35.45	−167.16	−131.23	+56.5	−136.4
HCl(g)	36.46	−92.31	−95.30	186.91	29.12
HCl(aq)	36.46	−167.16	−131.23	56.5	−136.4
Chromium					
Cr(s)	52.00	0	0	23.77	23.35
Cr(g)	52.00	+396.6	+351.8	174.50	20.79

Table 2.7 (Continued)

	M/(g mol^{-1})	$\Delta_f H^\ominus$/(kJ mol^{-1})	$\Delta_f G^\ominus$/(kJ mol^{-1})	$S_m^\ominus$/(J K^{-1} mol^{-1})†	$C_{p,m}^\ominus$/(J K^{-1} mol^{-1})
Chromium (Continued)					
CrO_4^{2-}(aq)	115.99	−881.15	−727.75	+50.21	
$Cr_2O_7^{2-}$(aq)	215.99	−1490.3	−1301.1	+261.9	
Copper					
Cu(s)	63.54	0	0	33.150	24.44
Cu(g)	63.54	+338.32	+298.58	166.38	20.79
Cu^+(aq)	63.54	+71.67	+49.98	+40.6	
Cu^{2+}(aq)	63.54	+64.77	+65.49	−99.6	
Cu_2O(s)	143.08	−168.6	−146.0	93.14	63.64
CuO(s)	79.54	−157.3	−129.7	42.63	42.30
$CuSO_4$(s)	159.60	−771.36	−661.8	109	100.0
$CuSO_4 \cdot H_2O$(s)	177.62	−1085.8	−918.11	146.0	134
$CuSO_4 \cdot 5H_2O$(s)	249.68	−2279.7	−1879.7	300.4	280
Deuterium					
D_2(g)	4.028	0	0	144.96	29.20
HD(g)	3.022	+0.318	−1.464	143.80	29.196
D_2O(g)	20.028	−249.20	−234.54	198.34	34.27
D_2O(l)	20.028	−294.60	−243.44	75.94	84.35
HDO(g)	19.022	−245.30	−233.11	199.51	33.81
HDO(l)	19.022	−289.89	−241.86	79.29	
Fluorine					
F_2(g)	38.00	0	0	202.78	31.30
F(g)	19.00	+78.99	+61.91	158.75	22.74
F^-(aq)	19.00	−332.63	−278.79	−13.8	−106.7
HF(g)	20.01	−271.1	−273.2	173.78	29.13
Gold					
Au(s)	196.97	0	0	47.40	25.42
Au(g)	196.97	+366.1	+326.3	180.50	20.79
Helium					
He(g)	4.003	0	0	126.15	20.786
Hydrogen (see also deuterium)					
H_2(g)	2.016	0	0	130.684	28.824
H(g)	1.008	+217.97	+203.25	114.71	20.784
H^+(aq)	1.008	0	0	0	0
H^+(g)	1.008	+1536.20			
H_2O(s)	18.015			37.99	
H_2O(l)	18.015	−285.83	−237.13	69.91	75.291
H_2O(g)	18.015	−241.82	−228.57	188.83	33.58
H_2O_2(l)	34.015	−187.78	−120.35	109.6	89.1
Iodine					
I_2(s)	253.81	0	0	116.135	54.44
I_2(g)	253.81	+62.44	+19.33	260.69	36.90

Table 2.7 (Continued)

	$M/(\text{g mol}^{-1})$	$\Delta_f H^\ominus/(\text{kJ mol}^{-1})$	$\Delta_f G^\ominus/(\text{kJ mol}^{-1})$	$S_m^\ominus/(\text{J K}^{-1}\text{ mol}^{-1})$†	$C_{p,m}^\ominus/(\text{J K}^{-1}\text{ mol}^{-1})$
Iodine (Continued)					
I(g)	126.90	+106.84	+70.25	180.79	20.786
I^-(aq)	126.90	−55.19	−51.57	+111.3	−142.3
HI(g)	127.91	+26.48	+1.70	206.59	29.158
Iron					
Fe(s)	55.85	0	0	27.28	25.10
Fe(g)	55.85	+416.3	+370.7	180.49	25.68
Fe^{2+}(aq)	55.85	−89.1	−78.90	−137.7	
Fe^{3+}(aq)	55.85	−48.5	−4.7	−315.9	
Fe_3O_4(s) (magnetite)	231.54	−1118.4	−1015.4	146.4	143.43
Fe_2O_3(s) (haematite)	159.69	−824.2	−742.2	87.40	103.85
FeS(s, α)	87.91	−100.0	−100.4	60.29	50.54
FeS_2(s)	119.98	−178.2	−166.9	52.93	62.17
Krypton					
Kr(g)	83.80	0	0	164.08	20.786
Lead					
Pb(s)	207.19	0	0	64.81	26.44
Pb(g)	207.19	+195.0	+161.9	175.37	20.79
Pb^{2+}(aq)	207.19	−1.7	−24.43	+10.5	
PbO(s, yellow)	223.19	−217.32	−187.89	68.70	45.77
PbO(s, red)	223.19	−218.99	−188.93	66.5	45.81
PbO_2(s)	239.19	−277.4	−217.33	68.6	64.64
Lithium					
Li(s)	6.94	0	0	29.12	24.77
Li(g)	6.94	+159.37	+126.66	138.77	20.79
Li^+(aq)	6.94	−278.49	−293.31	+13.4	68.6
Magnesium					
Mg(s)	24.31	0	0	32.68	24.89
Mg(g)	24.31	+147.70	+113.10	148.65	20.786
Mg^{2+}(aq)	24.31	−466.85	−454.8	−138.1	
MgO(s)	40.31	−601.70	−569.43	26.94	37.15
$MgCO_3$(s)	84.32	−1095.8	−1012.1	65.7	75.52
$MgCl_2$(s)	95.22	−641.32	−591.79	89.62	71.38
Mercury					
Hg(l)	200.59	0	0	76.02	27.983
Hg(g)	200.59	+61.32	+31.82	174.96	20.786
Hg^{2+}(aq)	200.59	+171.1	+164.40	−32.2	
Hg_2^{2+}(aq)	401.18	+172.4	+153.52	+84.5	
HgO(s)	216.59	−90.83	−58.54	70.29	44.06
Hg_2Cl_2(s)	472.09	−265.22	−210.75	192.5	102
$HgCl_2$(s)	271.50	−224.3	−178.6	146.0	
HgS(s, black)	232.65	−53.6	−47.7	88.3	

Table 2.7 (Continued)

	M/(g mol^{-1})	$\Delta_f H^\ominus$/(kJ mol^{-1})	$\Delta_f G^\ominus$/(kJ mol^{-1})	$S_m^\ominus$/(J K^{-1} mol^{-1})†	$C_{p,m}^\ominus$/(J K^{-1} mol^{-1})
Neon					
Ne(g)	20.18	0	0	146.33	20.786
Nitrogen					
N_2(g)	28.013	0	0	191.61	29.125
N(g)	14.007	+472.70	+455.56	153.30	20.786
NO(g)	30.01	+90.25	+86.55	210.76	29.844
N_2O(g)	44.01	+82.05	+104.20	219.85	38.45
NO_2(g)	46.01	+33.18	+51.31	240.06	37.20
N_2O_4(g)	92.1	+9.16	+97.89	304.29	77.28
N_2O_5(s)	108.01	−43.1	+113.9	178.2	143.1
N_2O_5(g)	108.01	+11.3	+115.1	355.7	84.5
HNO_3(l)	63.01	−174.10	−80.71	155.60	109.87
HNO_3(aq)	63.01	−207.36	−111.25	146.4	−86.6
NO_3^-(aq)	62.01	−205.0	−108.74	+146.4	−86.6
NH_3(g)	17.03	−46.11	−16.45	192.45	35.06
NH_3(aq)	17.03	−80.29	−26.50	111.3	
NH_4^+(aq)	18.04	−132.51	−79.31	+113.4	79.9
NH_2OH(s)	33.03	−114.2			
HN_3(l)	43.03	+264.0	+327.3	140.6	43.68
HN_3(g)	43.03	+294.1	+328.1	238.97	98.87
N_2H_4(l)	32.05	+50.63	+149.43	121.21	139.3
NH_4NO_3(s)	80.04	−365.56	−183.87	151.08	84.1
NH_4Cl(s)	53.49	−314.43	−202.87	94.6	
Oxygen					
O_2(g)	31.999	0	0	205.138	29.355
O(g)	15.999	+249.17	+231.73	161.06	21.912
O_3(g)	47.998	+142.7	+163.2	238.93	39.20
OH^-(aq)	17.007	−229.99	−157.24	−10.75	−148.5
Phosphorus					
P(s, wh)	30.97	0	0	41.09	23.840
P(g)	30.97	+314.64	+278.25	163.19	20.786
P_2(g)	61.95	+144.3	+103.7	218.13	32.05
P_4(g)	123.90	+58.91	+24.44	279.98	67.15
PH_3(g)	34.00	+5.4	+13.4	210.23	37.11
PCl_3(g)	137.33	−287.0	−267.8	311.78	71.84
PCl_3(l)	137.33	−319.7	−272.3	217.1	
PCl_5(g)	208.24	−374.9	−305.0	364.6	112.8
PCl_5(s)	208.24	−443.5			
H_3PO_3(s)	82.00	−964.4			
H_3PO_3(aq)	82.00	−964.8			
H_3PO_4(s)	94.97	−1279.0	−1119.1	110.50	106.06
H_3PO_4(l)	94.97	−1266.9			
H_3PO_4(aq)	94.97	−1277.4	−1018.7	−222	

Table 2.7 (Continued)

	$M/(\text{g mol}^{-1})$	$\Delta_f H^\ominus/(\text{kJ mol}^{-1})$	$\Delta_f G^\ominus/(\text{kJ mol}^{-1})$	$S_m^\ominus/(\text{J K}^{-1}\,\text{mol}^{-1})$†	$C_{p,m}^\ominus/(\text{J K}^{-1}\,\text{mol}^{-1})$
Phosphorus (Continued)					
PO_4^{3-}(aq)	94.97	−1277.4	−1018.7	−221.8	
P_4O_{10}(s)	283.89	−2984.0	−2697.0	228.86	211.71
P_4O_6(s)	219.89	−1640.1			
Potassium					
K(s)	39.10	0	0	64.18	29.58
K(g)	39.10	+89.24	+60.59	160.336	20.786
K^+(g)	39.10	+514.26			
K^+(aq)	39.10	−252.38	−283.27	+102.5	21.8
KOH(s)	56.11	−424.76	−379.08	78.9	64.9
KF(s)	58.10	−576.27	−537.75	66.57	49.04
KCl(s)	74.56	−436.75	−409.14	82.59	51.30
KBr(s)	119.01	−393.80	−380.66	95.90	52.30
KI(s)	166.01	−327.90	−324.89	106.32	52.93
Silicon					
Si(s)	28.09	0	0	18.83	20.00
Si(g)	28.09	+455.6	+411.3	167.97	22.25
SiO_2(s, α)	60.09	−910.94	−856.64	41.84	44.43
Silver					
Ag(s)	107.87	0	0	42.55	25.351
Ag(g)	107.87	+284.55	+245.65	173.00	20.79
Ag^+(aq)	107.87	+105.58	+77.11	+72.68	21.8
AgBr(s)	187.78	−100.37	−96.90	107.1	52.38
AgCl(s)	143.32	−127.07	−109.79	96.2	50.79
Ag_2O(s)	231.74	−31.05	−11.20	121.3	65.86
$AgNO_3$(s)	169.88	−129.39	−33.41	140.92	93.05
Sodium					
Na(s)	22.99	0	0	51.21	28.24
Na(g)	22.99	+107.32	+76.76	153.71	20.79
Na^+(aq)	22.99	−240.12	−261.91	59.0	46.4
NaOH(s)	40.00	−425.61	−379.49	64.46	59.54
NaCl(s)	58.44	−411.15	−384.14	72.13	50.50
NaBr(s)	102.90	−361.06	−348.98	86.82	51.38
NaI(s)	149.89	−287.78	−286.06	98.53	52.09
Sulfur					
S(s, α) (rhombic)	32.06	0	0	31.80	22.64
S(s, β) (monoclinic)	32.06	+0.33	+0.1	32.6	23.6
S(g)	32.06	+278.81	+238.25	167.82	23.673
S_2(g)	64.13	+128.37	+79.30	228.18	32.47
S^{2-}(aq)	32.06	+33.1	+85.8	−14.6	
SO_2(g)	64.06	−296.83	−300.19	248.22	39.87
SO_3(g)	80.06	−395.72	−371.06	256.76	50.67

Table 2.7 (Continued)

	$M/(\mathrm{g\ mol^{-1}})$	$\Delta_f H^{\ominus}/(\mathrm{kJ\ mol^{-1}})$	$\Delta_f G^{\ominus}/(\mathrm{kJ\ mol^{-1}})$	$S_m^{\ominus}/(\mathrm{J\ K^{-1}\ mol^{-1}})$†	$C_{p,m}^{\ominus}/(\mathrm{J\ K^{-1}\ mol^{-1}})$
Sulfur (Continued)					
H_2SO_4(l)	98.08	−813.99	−690.00	156.90	138.9
H_2SO_4(aq)	98.08	−909.27	−744.53	20.1	−293
SO_4^{2-}(aq)	96.06	−909.27	−744.53	+20.1	−293
HSO_4^-(aq)	97.07	−887.34	−755.91	+131.8	−84
H_2S(g)	34.08	−20.63	−33.56	205.79	34.23
H_2S(aq)	34.08	−39.7	−27.83	121	
HS^-(aq)	33.072	−17.6	+12.08	+62.08	
SF_6(g)	146.05	−1209	−1105.3	291.82	97.28
Tin					
Sn(s, *β*)	118.69	0	0	51.55	26.99
Sn(g)	118.69	+302.1	+267.3	168.49	20.26
Sn^{2+}(aq)	118.69	−8.8	−27.2	−17	
SnO(s)	134.69	−285.8	−256.9	56.5	44.31
SnO_2(s)	150.69	−580.7	−519.6	52.3	52.59
Xenon					
Xe(g)	131.30	0	0	169.68	20.786
Zinc					
Zn(s)	65.37	0	0	41.63	25.40
Zn(g)	65.37	+130.73	+95.14	160.98	20.79
Zn^{2+}(aq)	65.37	−153.89	−147.06	−112.1	46
ZnO(s)	81.37	−348.28	−318.30	43.64	40.25

Source: NBS. † Standard entropies of ions may be either positive or negative because the values are relative to the entropy of the hydrogen ion.

Table 2.7a Standard enthalpies of hydration at infinite dilution, $\Delta_{hyd}H^{\ominus}/(\mathrm{kJ\ mol^{-1}})$

	Li^+	Na^+	K^+	Rb^+	Cs^+
F^-	−1026	−911	−828	−806	−782
Cl^-	−884	−783	−685	−664	−640
Br^-	−856	−742	−658	−637	−613
I^-	−815	−701	−617	−596	−572

Entries refer to $X^+(g) + Y^-(g) \rightarrow X^+(aq) + Y^-(aq)$.
Data: Principally J.O'M. Bockris and A.K.N. Reddy, *Modern electrochemistry*, Vol. 1. Plenum Press, New York (1970).

Table 2.7b Standard ion hydration enthalpies, $\Delta_{hyd}H^{\ominus}/(\mathrm{kJ\ mol^{-1}})$ at 298 K

Cations					
H^+	(−1090)	Ag^+	−464	Mg^{2+}	−1920
Li^+	−520	NH_4^+	−301	Ca^{2+}	−1650
Na^+	−405			Sr^{2+}	−1480
K^+	−321			Ba^{2+}	−1360
Rb^+	−300			Fe^{2+}	−1950
Cs^+	−277			Cu^{2+}	−2100
				Zn^{2+}	−2050
				Al^{3+}	−4690
				Fe^{3+}	−4430
Anions					
OH^-	−460				
F^-	−506	Cl^-	−364	Br^- −337	I^- −296

Entries refer to $X^{\pm}(g) \rightarrow X^{\pm}(aq)$ based on $H^+(g) \rightarrow H^+(aq)$; $\Delta H^{\ominus} = -1090\ \mathrm{kJ\ mol^{-1}}$.
Data: Principally J.O'M. Bockris and A.K.N. Reddy, *Modern electrochemistry*, Vol. 1. Plenum Press, New York (1970).

Table 2.8 Expansion coefficients, α, and isothermal compressibilities, κ_T

	$\alpha/(10^{-4}\ K^{-1})$	$\kappa_T/(10^{-6}\ atm^{-1})$
Liquids		
Benzene	12.4	92.1
Carbon tetrachloride	12.4	90.5
Ethanol	11.2	76.8
Mercury	1.82	38.7
Water	2.1	49.6
Solids		
Copper	0.501	0.735
Diamond	0.030	0.187
Iron	0.354	0.589
Lead	0.861	2.21

The values refer to 20°C.
Data: AIP(α), KL(κ_T).

Table 2.9 Inversion temperatures, normal freezing and boiling points, and Joule–Thomson coefficients at 1 atm and 298 K

	T_I/K	T_f/K	T_b/K	$\mu_{JT}/(K\ atm^{-1})$
Air	603			0.189 at 50°C
Argon	723	83.8	87.3	
Carbon dioxide	1500	194.7s		1.11 at 300 K
Helium	40		4.22	−0.062
Hydrogen	202	14.0	20.3	−0.03
Krypton	1090	116.6	120.8	
Methane	968	90.6	111.6	
Neon	231	24.5	27.1	
Nitrogen	621	63.3	77.4	0.27
Oxygen	764	54.8	90.2	0.31

s: sublimes.
Data: AIP, JL, and M.W. Zemansky, *Heat and thermodynamics*. McGraw-Hill, New York (1957).

Table 3.1 Standard entropies (and temperatures) of phase transitions, $\Delta_{trs}S^{\ominus}/(J\ K^{-1}\ mol^{-1})$

	Fusion (at T_f)	Vaporization (at T_b)
Ar	14.17 (at 83.8 K)	74.53 (at 87.3 K)
Br_2	39.76 (at 265.9 K)	88.61 (at 332.4 K)
C_6H_6	38.00 (at 278.6 K)	87.19 (at 353.2 K)
CH_3COOH	40.4 (at 289.8 K)	61.9 (at 391.4 K)
CH_3OH	18.03 (at 175.2 K)	104.6 (at 337.2 K)
Cl_2	37.22 (at 172.1 K)	85.38 (at 239.0 K)
H_2	8.38 (at 14.0 K)	44.96 (at 20.38 K)
H_2O	22.00 (at 273.2 K)	109.0 (at 373.2 K)
H_2S	12.67 (at 187.6 K)	87.75 (at 212.0 K)
He	4.8 (at 1.8 K and 30 bar)	19.9 (at 4.22 K)
N_2	11.39 (at 63.2 K)	75.22 (at 77.4 K)
NH_3	28.93 (at 195.4 K)	97.41 (at 239.73 K)
O_2	8.17 (at 54.4 K)	75.63 (at 90.2 K)

Data: AIP.

Table 3.2 Standard entropies of vaporization of liquids at their normal boiling point

	$\Delta_{vap}H^{\ominus}/(kJ\ mol^{-1})$	$\theta_b/°C$	$\Delta_{vap}S^{\ominus}/(J\ K^{-1}\ mol^{-1})$
Benzene	30.8	80.1	+87.2
Carbon disulfide	26.74	46.25	+83.7
Carbon tetrachloride	30.00	76.7	+85.8
Cyclohexane	30.1	80.7	+85.1
Decane	38.75	174	+86.7
Dimethyl ether	21.51	−23	+86
Ethanol	38.6	78.3	+110.0
Hydrogen sulfide	18.7	−60.4	+87.9
Mercury	59.3	356.6	+94.2
Methane	8.18	−161.5	+73.2
Methanol	35.21	65.0	+104.1
Water	40.7	100.0	+109.1

Data: JL.

Table 3.3 Standard Third-Law entropies at 298 K: see Tables 2.5 and 2.7

Table 3.4 Standard Gibbs energies of formation at 298 K: see Tables 2.5 and 2.7

Table 3.6 The fugacity coefficient of nitrogen at 273 K

p/atm	ϕ	p/atm	ϕ
1	0.999 55	300	1.0055
10	0.9956	400	1.062
50	0.9912	600	1.239
100	0.9703	800	1.495
150	0.9672	1000	1.839
200	0.9721		

Data: LR.

Table 5.1 Henry's law constants for gases at 298 K, $K/(kPa\ kg\ mol^{-1})$

	Water	Benzene
CH_4	7.55×10^4	44.4×10^3
CO_2	30.1×10^3	8.90×10^2
H_2	1.28×10^5	2.79×10^4
N_2	1.56×10^5	1.87×10^4
O_2	7.92×10^4	

Data: converted from R.J. Silbey and R.A. Alberty, *Physical chemistry*. Wiley, New York (2001).

Table 5.2 Freezing-point and boiling-point constants

	$K_f/(\text{K kg mol}^{-1})$	$K_b/(\text{K kg mol}^{-1})$
Acetic acid	3.90	3.07
Benzene	5.12	2.53
Camphor	40	
Carbon disulfide	3.8	2.37
Carbon tetrachloride	30	4.95
Naphthalene	6.94	5.8
Phenol	7.27	3.04
Water	1.86	0.51

Data: KL.

Table 5.5 Mean activity coefficients in water at 298 K

$b/b^\ominus$	HCl	KCl	$CaCl_2$	H_2SO_4	$LaCl_3$	$In_2(SO_4)_3$
0.001	0.966	0.966	0.888	0.830	0.790	
0.005	0.929	0.927	0.789	0.639	0.636	0.16
0.01	0.905	0.902	0.732	0.544	0.560	0.11
0.05	0.830	0.816	0.584	0.340	0.388	0.035
0.10	0.798	0.770	0.524	0.266	0.356	0.025
0.50	0.769	0.652	0.510	0.155	0.303	0.014
1.00	0.811	0.607	0.725	0.131	0.387	
2.00	1.011	0.577	1.554	0.125	0.954	

Data: RS, HCP, and S. Glasstone, *Introduction to electrochemistry*. Van Nostrand (1942).

Table 5.6 Relative permittivities (dielectric constants) at 293 K

Nonpolar molecules		Polar molecules	
Methane (at −173°C)	1.655	Water	78.54 (at 298 K)
			80.10
Carbon tetrachloride	2.238	Ammonia	16.9 (at 298 K)
			22.4 at −33°C
Cyclohexane	2.024	Hydrogen sulfide	9.26 at −85°C
			5.93 (at 283 K)
Benzene	2.283	Methanol	33.0
		Ethanol	25.3
		Nitrobenzene	35.6

Data: HCP.

Table 7.2 Standard potentials at 298 K. (a) In electrochemical order

Reduction half-reaction	$E^⦵$/V	Reduction half-reaction	$E^⦵$/V
Strongly oxidizing		$Cu^{2+} + e^- \rightarrow Cu^+$	+0.16
$H_4XeO_6 + 2H^+ + 2e^- \rightarrow XeO_3 + 3H_2O$	+3.0	$Sn^{4+} + 2e^- \rightarrow Sn^{2+}$	+0.15
$F_2 + 2e^- \rightarrow 2F^-$	+2.87	$AgBr + e^- \rightarrow Ag + Br^-$	+0.07
$O_3 + 2H^+ + 2e^- \rightarrow O_2 + H_2O$	+2.07	$Ti^{4+} + e^- \rightarrow Ti^{3+}$	0.00
$S_2O_8^{2-} + 2e^- \rightarrow 2SO_4^{2-}$	+2.05	$2H^+ + 2e^- \rightarrow H_2$	0, by definition
$Ag^{2+} + e^- \rightarrow Ag^+$	+1.98	$Fe^{3+} + 3e^- \rightarrow Fe$	−0.04
$Co^{3+} + e^- \rightarrow Co^{2+}$	+1.81	$O_2 + H_2O + 2e^- \rightarrow HO_2^- + OH^-$	−0.08
$H_2O_2 + 2H^+ + 2e^- \rightarrow 2H_2O$	+1.78	$Pb^{2+} + 2e^- \rightarrow Pb$	−0.13
$Au^+ + e^- \rightarrow Au$	+1.69	$In^+ + e^- \rightarrow In$	−0.14
$Pb^{4+} + 2e^- \rightarrow Pb^{2+}$	+1.67	$Sn^{2+} + 2e^- \rightarrow Sn$	−0.14
$2HClO + 2H^+ + 2e^- \rightarrow Cl_2 + 2H_2O$	+1.63	$AgI + e^- \rightarrow Ag + I^-$	−0.15
$Ce^{4+} + e^- \rightarrow Ce^{3+}$	+1.61	$Ni^{2+} + 2e^- \rightarrow Ni$	−0.23
$2HBrO + 2H^+ + 2e^- \rightarrow Br_2 + 2H_2O$	+1.60	$Co^{2+} + 2e^- \rightarrow Co$	−0.28
$MnO_4^- + 8H^+ + 5e^- \rightarrow Mn^{2+} + 4H_2O$	+1.51	$In^{3+} + 3e^- \rightarrow In$	−0.34
$Mn^{3+} + e^- \rightarrow Mn^{2+}$	+1.51	$Tl^+ + e^- \rightarrow Tl$	−0.34
$Au^{3+} + 3e^- \rightarrow Au$	+1.40	$PbSO_4 + 2e^- \rightarrow Pb + SO_4^{2-}$	−0.36
$Cl_2 + 2e^- \rightarrow 2Cl^-$	+1.36	$Ti^{3+} + e^- \rightarrow Ti^{2+}$	−0.37
$Cr_2O_7^{2-} + 14H^+ + 6e^- \rightarrow 2Cr^{3+} + 7H_2O$	+1.33	$Cd^{2+} + 2e^- \rightarrow Cd$	−0.40
$O_3 + H_2O + 2e^- \rightarrow O_2 + 2OH^-$	+1.24	$In^{2+} + e^- \rightarrow In^+$	−0.40
$O_2 + 4H^+ + 4e^- \rightarrow 2H_2O$	+1.23	$Cr^{3+} + e^- \rightarrow Cr^{2+}$	−0.41
$ClO_4^- + 2H^+ + 2e^- \rightarrow ClO_3^- + H_2O$	+1.23	$Fe^{2+} + 2e^- \rightarrow Fe$	−0.44
$MnO_2 + 4H^+ + 2e^- \rightarrow Mn^{2+} + 2H_2O$	+1.23	$In^{3+} + 2e^- \rightarrow In^+$	−0.44
$Br_2 + 2e^- \rightarrow 2Br^-$	+1.09	$S + 2e^- \rightarrow S^{2-}$	−0.48
$Pu^{4+} + e^- \rightarrow Pu^{3+}$	+0.97	$In^{3+} + e^- \rightarrow In^{2+}$	−0.49
$NO_3^- + 4H^+ + 3e^- \rightarrow NO + 2H_2O$	+0.96	$U^{4+} + e^- \rightarrow U^{3+}$	−0.61
$2Hg^{2+} + 2e^- \rightarrow Hg_2^{2+}$	+0.92	$Cr^{3+} + 3e^- \rightarrow Cr$	−0.74
$ClO^- + H_2O + 2e^- \rightarrow Cl^- + 2OH^-$	+0.89	$Zn^{2+} + 2e^- \rightarrow Zn$	−0.76
$Hg^{2+} + 2e^- \rightarrow Hg$	+0.86	$Cd(OH)_2 + 2e^- \rightarrow Cd + 2OH^-$	−0.81
$NO_3^- + 2H^+ + e^- \rightarrow NO_2 + H_2O$	+0.80	$2H_2O + 2e^- \rightarrow H_2 + 2OH^-$	−0.83
$Ag^+ + e^- \rightarrow Ag$	+0.80	$Cr^{2+} + 2e^- \rightarrow Cr$	−0.91
$Hg_2^{2+} + 2e^- \rightarrow 2Hg$	+0.79	$Mn^{2+} + 2e^- \rightarrow Mn$	−1.18
$Fe^{3+} + e^- \rightarrow Fe^{2+}$	+0.77	$V^{2+} + 2e^- \rightarrow V$	−1.19
$BrO^- + H_2O + 2e^- \rightarrow Br^- + 2OH^-$	+0.76	$Ti^{2+} + 2e^- \rightarrow Ti$	−1.63
$Hg_2SO_4 + 2e^- \rightarrow 2Hg + SO_4^{2-}$	+0.62	$Al^{3+} + 3e^- \rightarrow Al$	−1.66
$MnO_4^{2-} + 2H_2O + 2e^- \rightarrow MnO_2 + 4OH^-$	+0.60	$U^{3+} + 3e^- \rightarrow U$	−1.79
$MnO_4^- + e^- \rightarrow MnO_4^{2-}$	+0.56	$Sc^{3+} + 3e^- \rightarrow Sc$	−2.09
$I_2 + 2e^- \rightarrow 2I^-$	+0.54	$Mg^{2+} + 2e^- \rightarrow Mg$	−2.36
$CU^+ + e^- \rightarrow Cu$	+0.52	$Ce^{3+} + 3e^- \rightarrow Ce$	−2.48
$I_3^- + 2e^- \rightarrow 3I^-$	+0.53	$La^{3+} + 3e^- \rightarrow La$	−2.52
$NiOOH + H_2O + e^- \rightarrow Ni(OH)_2 + OH^-$	+0.49	$Na^+ + e^- \rightarrow Na$	−2.71
$Ag_2CrO_4 + 2e^- \rightarrow 2Ag + CrO_4^{2-}$	+0.45	$Ca^{2+} + 2e^- \rightarrow Ca$	−2.87
$O_2 + 2H_2O + 4e^- \rightarrow 4OH^-$	+0.40	$Sr^{2+} + 2e^- \rightarrow Sr$	−2.89
$ClO_4^- + H_2O + 2e^- \rightarrow ClO_3^- + 2OH^-$	+0.36	$Ba^{2+} + 2e^- \rightarrow Ba$	−2.91
$[Fe(CN)_6]^{3-} + e^- \rightarrow [Fe(CN)_6]^{4-}$	+0.36	$Ra^{2+} + 2e^- \rightarrow Ra$	−2.92
$Cu^{2+} + 2e^- \rightarrow Cu$	+0.34	$Cs^+ + e^- \rightarrow Cs$	−2.92
$Hg_2Cl_2 + 2e^- \rightarrow 2Hg + 2Cl^-$	+0.27	$Rb^+ + e^- \rightarrow Rb$	−2.93
$AgCl + e^- \rightarrow Ag + Cl^-$	+0.22	$K^+ + e^- \rightarrow K$	−2.93
$Bi^{3+} + 3e^- \rightarrow Bi$	+0.20	$Li^+ + e^- \rightarrow Li$	−3.05

Table 7.2 Standard potentials at 298 K. (b) In electrochemical order

Reduction half-reaction	$E^{\ominus}$/V	Reduction half-reaction	$E^{\ominus}$/V
$Ag^+ + e^- \rightarrow Ag$	+0.80	$I_2 + 2e^- \rightarrow 2I^-$	+0.54
$Ag^{2+} + e^- \rightarrow Ag^+$	+1.98	$I_3^- + 2e^- \rightarrow 3I^-$	+0.53
$AgBr + e^- \rightarrow Ag + Br^-$	+0.0713	$In^+ + e^- \rightarrow In$	−0.14
$AgCl + e^- \rightarrow Ag + Cl^-$	+0.22	$In^{2+} + e^- \rightarrow In^+$	−0.40
$Ag_2CrO_4 + 2e^- \rightarrow 2Ag + CrO_4^{2-}$	+0.45	$In^{3+} + 2e^- \rightarrow In^+$	−0.44
$AgF + e^- \rightarrow Ag + F^-$	+0.78	$In^{3+} + 3e^- \rightarrow In$	−0.34
$AgI + e^- \rightarrow Ag + I^-$	−0.15	$In^{3+} + e^- \rightarrow In^{2+}$	−0.49
$Al^{3+} + 3e^- \rightarrow Al$	−1.66	$K^+ + e^- \rightarrow K$	−2.93
$Au^+ + e^- \rightarrow Au$	+1.69	$La^{3+} + 3e^- \rightarrow La$	−2.52
$Au^{3+} + 3e^- \rightarrow Au$	+1.40	$Li^+ + e^- \rightarrow Li$	−3.05
$Ba^{2+} + 2e^- \rightarrow Ba$	+2.91	$Mg^{2+} + 2e^- \rightarrow Mg$	−2.36
$Be^{2+} + 2e^- \rightarrow Be$	−1.85	$Mn^{2+} + 2e^- \rightarrow Mn$	−1.18
$Bi^{3+} + 3e^- \rightarrow Bi$	+0.20	$Mn^{3+} + e^- \rightarrow Mn^{2+}$	+1.51
$Br_2 + 2e^- \rightarrow 2Br^-$	+1.09	$MnO_2 + 4H^+ + 2e^- \rightarrow Mn^{2+} + 2H_2O$	+1.23
$BrO^- + H_2O + 2e^- \rightarrow Br^- + 2OH^-$	+0.76	$MnO_4^- + 8H^+ + 5e^- \rightarrow Mn^{2+} + 4H_2O$	+1.51
$Ca^{2+} + 2e^- \rightarrow Ca$	−2.87	$MnO_4^- + e^- \rightarrow MnO_4^{2-}$	+0.56
$Cd(OH)_2 + 2e^- \rightarrow Cd + 2OH^-$	−0.81	$MnO_4^{2-} + 2H_2O + 2e^- \rightarrow MnO_2 + 4OH^-$	+0.60
$Cd^{2+} + 2e^- \rightarrow Cd$	−0.40	$Na^+ + e^- \rightarrow Na$	−2.71
$Ce^{3+} + 3e^- \rightarrow Ce$	−2.48	$Ni^{2+} + 2e^- \rightarrow Ni$	−0.23
$Ce^{4+} + e^- \rightarrow Ce^{3+}$	+1.61	$NiOOH + H_2O + e^- \rightarrow Ni(OH)_2 + OH^-$	+0.49
$Cl_2 + 2e^- \rightarrow 2Cl^-$	+1.36	$NO_3^- + 2H^+ + e^- \rightarrow NO_2 + H_2O$	−0.80
$ClO^- + H_2O + 2e^- \rightarrow Cl^- + 2OH^-$	+0.89	$NO_3^- + 4H^+ + 3e^- \rightarrow NO + 2H_2O$	+0.96
$ClO_4^- + 2H^+ + 2e^- \rightarrow ClO_3^- + H_2O$	+1.23	$NO_3^- + H_2O + 2e^- \rightarrow NO_2^- + 2OH^-$	+0.10
$ClO_4^- + H_2O + 2e^- \rightarrow ClO_3^- + 2OH^-$	+0.36	$O_2 + 2H_2O + 4e^- \rightarrow 4OH^-$	+0.40
$Co^{2+} + 2e^- \rightarrow Co$	−0.28	$O_2 + 4H^+ + 4e^- \rightarrow 2H_2O$	+1.23
$Co^{3+} + e^- \rightarrow Co^{2+}$	+1.81	$O_2 + e^- \rightarrow O_2^-$	−0.56
$Cr^{2+} + 2e^- \rightarrow Cr$	−0.91	$O_2 + H_2O + 2e^- \rightarrow HO_2^- + OH^-$	−0.08
$Cr_2O_7^{2-} + 14H^+ + 6e^- \rightarrow 2Cr^{3+} + 7H_2O$	+1.33	$O_3 + 2H^+ + 2e^- \rightarrow O_2 + H_2O$	+2.07
$Cr^{3+} + 3e^- \rightarrow Cr$	−0.74	$O_3 + H_2O + 2e^- \rightarrow O_2 + 2OH^-$	+1.24
$Cr^{3+} + e^- \rightarrow Cr^{2+}$	−0.41	$Pb^{2+} + 2e^- \rightarrow Pb$	−0.13
$Cs^+ + e^- \rightarrow Cs$	−2.92	$Pb^{4+} + 2e^- \rightarrow Pb^{2+}$	+1.67
$Cu^+ + e^- \rightarrow Cu$	+0.52	$PbSO_4 + 2e^- \rightarrow Pb + SO_4^{2-}$	−0.36
$Cu^{2+} + 2e^- \rightarrow Cu$	+0.34	$Pt^{2+} + 2e^- \rightarrow Pt$	+1.20
$Cu^{2+} + e^- \rightarrow Cu^+$	+0.16	$Pu^{4+} + e^- \rightarrow Pu^{3+}$	+0.97
$F_2 + 2e^- \rightarrow 2F^-$	+2.87	$Ra^{2+} + 2e^- \rightarrow Ra$	−2.92
$Fe^{2+} + 2e^- \rightarrow Fe$	−0.44	$Rb^+ + e^- \rightarrow Rb$	−2.93
$Fe^{3+} + 3e^- \rightarrow Fe$	−0.04	$S + 2e^- \rightarrow S^{2-}$	−0.48
$Fe^{3+} + e^- \rightarrow Fe^{2+}$	+0.77	$S_2O_8^{2-} + 2e^- \rightarrow 2SO_4^{2-}$	+2.05
$[Fe(CN)_6]^{3-} + e^- \rightarrow [Fe(CN)_6]^{4-}$	+0.36	$Sc^{3+} + 3e^- \rightarrow Sc$	−2.09
$2H^+ + 2e^- \rightarrow H_2$	0, by definition	$Sn^{2+} + 2e^- \rightarrow Sn$	−0.14
$2H_2O + 2e^- \rightarrow H_2 + 2OH^-$	−0.83	$Sn^{4+} + 2e^- \rightarrow Sn^{2+}$	+0.15
$2HBrO + 2H^+ + 2e^- \rightarrow Br_2 + 2H_2O$	+1.60	$Sr^{2+} + 2e^- \rightarrow Sr$	−2.89
$2HClO + 2H^+ + 2e^- \rightarrow Cl_2 + 2H_2O$	+1.63	$Ti^{2+} + 2e^- \rightarrow Ti$	−1.63
$H_2O_2 + 2H^+ + 2e^- \rightarrow 2H_2O$	+1.78	$Ti^{3+} + e^- \rightarrow Ti^{2+}$	−0.37
$H_4XeO_6 + 2H^+ + 2e^- \rightarrow XeO_3 + 3H_2O$	+3.0	$Ti^{4+} + e^- \rightarrow Ti^{3+}$	0.00
$Hg_2^{2+} + 2e^- \rightarrow 2Hg$	+0.79	$Tl^+ + e^- \rightarrow Tl$	−0.34
$Hg_2Cl_2 + 2e^- \rightarrow 2Hg + 2Cl^-$	+0.27	$U^{3+} + 3e^- \rightarrow U$	−1.79
$Hg^{2+} + 2e^- \rightarrow Hg$	+0.86	$U^{4+} + e^- \rightarrow U^{3+}$	−0.61
$2Hg^{2+} + 2e^- \rightarrow Hg_2^{2+}$	+0.92	$V^{2+} + 2e^- \rightarrow V$	−1.19
$Hg_2SO_4 + 2e^- \rightarrow 2Hg + SO_4^{2-}$	+0.62	$V^{3+} + e^- \rightarrow V^{2+}$	−0.26
		$Zn^{2+} + 2e^- \rightarrow Zn$	−0.76

Table 7.4 Acidity constants for aqueous solutions at 298 K. (a) In order of acid strength

Acid	HA	A^-	K_a	pK_a
Hydriodic	HI	I^-	10^{11}	−11
Hydrobromic	HBr	Br^-	10^9	−9
Hydrochloric	HCl	Cl^-	10^7	−7
Sulfuric	H_2SO_4	HSO_4^-	10^2	−2
Perchloric*	$HClO_4$	ClO_4^-	4.0×10^1	−1.6
Hydronium ion	H_3O^+	H_2O	1	0.0
Oxalic	$(COOH)_2$	$HOOCCO_2^-$	5.6×10^{-2}	1.25
Sulfurous	H_2SO_3	HSO_3^-	1.4×10^{-2}	1.85
Hydrogensulfate ion	HSO_4^-	SO_4^{2-}	1.0×10^{-2}	1.99
Phosphoric	H_3PO_4	$H_2PO_4^-$	6.9×10^{-3}	2.16
Glycinium ion	$^+NH_3CH_2COOH$	NH_2CH_2COOH	4.5×10^{-3}	2.35
Hydrofluoric	HF	F^-	6.3×10^{-4}	3.20
Formic	HCOOH	HCO_2^-	1.8×10^{-4}	3.75
Hydrogenoxalate ion	$HOOCCO_2^-$	$C_2O_4^{2-}$	1.5×10^{-5}	3.81
Lactic	$CH_3CH(OH)COOH$	$CH_3CH(OH)CO_2^-$	1.4×10^{-4}	3.86
Acetic (ethanoic)	CH_3COOH	$CH_3CO_2^-$	1.4×10^{-5}	4.76
Butanoic	$CH_3CH_2CH_2COOH$	$CH_3CH_2CH_2CO_2^-$	1.5×10^{-5}	4.83
Propanoic	CH_3CH_2COOH	$CH_3CH_2CO_2^-$	1.4×10^{-5}	4.87
Anilinium ion	$C_6H_5NH_3^+$	$C_6H_5NH_2$	1.3×10^{-5}	4.87
Pyridinium ion	$C_5H_5NH^+$	C_6H_5N	5.9×10^{-6}	5.23
Carbonic	H_2CO_3	HCO_3^-	4.5×10^{-7}	6.35
Hydrosulfuric	H_2S	HS^-	8.9×10^{-8}	7.05
Dihydrogenphosphate ion	$H_2PO_4^-$	HPO_4^{2-}	6.2×10^{-8}	7.21
Hypochlorous	HClO	ClO^-	4.0×10^{-8}	7.40
Hydrazinium ion	$NH_2NH_3^+$	NH_2NH_2	8×10^{-9}	8.1
Hypobromous	HBrO	BrO^-	2.8×10^{-9}	8.55
Hydrocyanic	HCN	CN^-	6.2×10^{-10}	9.21
Ammonium ion	NH_4^+	NH_3	5.6×10^{-10}	9.25
Boric*	$B(OH)_3$	$B(OH)_4^-$	5.4×10^{-10}	9.27
Trimethylammonium ion	$(CH_3)_3NH^+$	$(CH_3)_3N$	1.6×10^{-10}	9.80
Phenol	C_6H_5OH	$C_6H_5O^-$	1.0×10^{-10}	9.99
Hydrogencarbonate ion	HCO_3^-	CO_3^{2-}	4.8×10^{-11}	10.33
Hypoiodous	HIO	IO^-	3×10^{-11}	10.5
Ethylammonium ion	$CH_3CH_2NH_3^+$	$CH_3CH_2NH_2$	2.2×10^{-11}	10.65
Methylammonium ion	$CH_3NH_3^+$	CH_3NH_2	2.2×10^{-11}	10.66
Dimethylammonium ion	$(CH_3)_2NH_2^+$	$(CH_3)_2NH$	1.9×10^{-11}	10.73
Triethylammonium ion	$(CH_3CH_2)_3NH^+$	$(CH_3CH_2)_3N$	1.8×10^{-11}	10.75
Diethylammonium ion	$(CH_3CH_2)_2NH_2^+$	$(CH_3CH_2)_2NH$	1.4×10^{-11}	10.84
Hydrogenarsenate ion	$HAsO_4^{2-}$	AsO_4^{3-}	5.1×10^{-12}	11.29
Hydrogenphosphate ion	HPO_4^{2-}	PO_4^{3-}	4.8×10^{-13}	12.32
Hydrogensulfide ion	HS^-	S^{2-}	1.0×10^{-19}	19.00

* At 293 K.

Table 7.4 Acidity constants for aqueous solutions at 298 K. (b) In alphabetical order

Acid	HA	A^-	K_a	pK_a
Acetic (ethanoic)	CH_3COOH	$CH_3CO_2^-$	1.4×10^{-5}	4.76
Ammonium ion	NH_4^+	NH_3	5.6×10^{-10}	9.25
Anilinium ion	$C_6H_5NH_3^+$	$C_6H_5NH_2$	1.3×10^{-5}	4.87
Boric*	$B(OH)_3$	$B(OH)_4^-$	5.4×10^{-10}	9.27
Butanoic	$CH_3CH_2CH_2COOH$	$CH_3CH_2CH_2CO_2^-$	1.5×10^{-5}	4.83
Carbonic	H_2CO_3	HCO_3^-	4.5×10^{-7}	6.35
Diethylammonium ion	$(CH_3CH_2)_2NH_2^+$	$(CH_3CH_2)_2NH$	1.4×10^{-11}	10.84
Dihydrogenphosphate ion	$H_2PO_4^-$	HPO_4^{2-}	6.2×10^{-8}	7.21
Dimethylammonium ion	$(CH_3)_2NH_2^+$	$(CH_3)_2NH$	1.9×10^{-11}	10.73
Ethylammonium ion	$CH_3CH_2NH_3^+$	$CH_3CH_2NH_2$	2.2×10^{-11}	10.65
Formic	$HCOOH$	HCO_2^-	1.8×10^{-4}	3.75
Glycinium ion	$^+NH_3CH_2COOH$	NH_2CH_2COOH	4.5×10^{-3}	2.35
Hydrazinium ion	$NH_2NH_3^+$	NH_2NH_2	8×10^{-9}	8.1
Hydriodic	HI	I^-	10^{11}	−11
Hydrobromic	HBr	Br^-	10^{9}	−9
Hydrochloric	HCl	Cl^-	10^{7}	−7
Hydrocyanic	HCN	CN^-	6.2×10^{-10}	9.21
Hydrofluoric	HF	F^-	6.3×10^{-4}	3.20
Hydrogenarsenate ion	$HAsO_4^{2-}$	AsO_4^{3-}	5.1×10^{-12}	11.29
Hydrogencarbonate ion	HCO_3^-	CO_3^{2-}	4.8×10^{-11}	10.33
Hydrogenoxalate ion	$HOOCCO_2^-$	$C_2O_4^{2-}$	1.5×10^{-5}	3.81
Hydrogenphosphate ion	HPO_4^{2-}	PO_4^{3-}	4.8×10^{-13}	12.32
Hydrogensulfate ion	HSO_4^-	SO_4^{2-}	1.0×10^{-2}	1.99
Hydrogensulfide ion	HS^-	S^{2-}	1.0×10^{-19}	19.00
Hydronium ion	H_3O^+	H_2O	1	0.0
Hydrosulfuric	H_2S	HS^-	8.9×10^{-8}	7.05
Hypobromous	HBrO	BrO^-	2.8×10^{-9}	8.55
Hypochlorous	HClO	ClO^-	4.0×10^{-8}	7.40
Hypoiodous	HIO	IO^-	3×10^{-11}	10.5
Lactic	$CH_3CH(OH)COOH$	$CH_3CH(OH)CO_2^-$	1.4×10^{-4}	3.86
Methylammonium ion	$CH_3NH_3^+$	CH_3NH_2	2.2×10^{-11}	10.66
Oxalic	$(COOH)_2$	$HOOCCO_2^-$	5.6×10^{-2}	1.25
Perchloric*	$HClO_4$	ClO_4^-	4.0×10^{1}	−1.6
Phenol	C_6H_5OH	$C_6H_5O^-$	1.0×10^{-10}	9.99
Phosphoric	H_3PO_4	$H_2PO_4^-$	6.9×10^{-3}	2.16
Propanoic	CH_3CH_2COOH	$CH_3CH_2CO_2^-$	1.4×10^{-5}	4.87
Pyridinim ion	$C_5H_5NH^+$	C_6H_5N	5.9×10^{-6}	5.23
Sulfuric	H_2SO_4	HSO_4^-	10^{2}	−2
Sulfurous	H_2SO_3	HSO_3^-	1.4×10^{-2}	1.85
Triethylammonium ion	$(CH_3CH_2)_3NH^+$	$(CH_3CH_2)_3N$	1.8×10^{-11}	10.75
Trimethylammonium ion	$(CH_3)_3NH^+$	$(CH_3)_3N$	1.6×10^{-10}	9.80

* At 293 K.

Table 9.2 The error function

z	erf z	z	erf z
0	0	0.45	0.475 48
0.01	0.011 28	0.50	0.520 50
0.02	0.022 56	0.55	0.563 32
0.03	0.033 84	0.60	0.603 86
0.04	0.045 11	0.65	0.642 03
0.05	0.056 37	0.70	0.677 80
0.06	0.067 62	0.75	0.711 16
0.07	0.078 86	0.80	0.742 10
0.08	0.090 08	0.85	0.770 67
0.09	0.101 28	0.90	0.796 91
0.10	0.112 46	0.95	0.820 89
0.15	0.168 00	1.00	0.842 70
0.20	0.222 70	1.20	0.910 31
0.25	0.276 32	1.40	0.952 28
0.30	0.328 63	1.60	0.976 35
0.35	0.379 38	1.80	0.989 09
0.40	0.428 39	2.00	0.995 32

Data: AS.

Table 10.2 Screening constants for atoms; values of $Z_{\text{eff}} = Z - \sigma$ for neutral ground-state atoms

	H							He
1*s*	1							1.6875
	Li	Be	B	C	N	O	F	Ne
1*s*	2.6906	3.6848	4.6795	5.6727	6.6651	7.6579	8.6501	9.6421
2*s*	1.2792	1.9120	2.5762	3.2166	3.8474	4.4916	5.1276	5.7584
2*p*			2.4214	3.1358	3.8340	4.4532	5.1000	5.7584
	Na	Mg	Al	Si	P	S	Cl	Ar
1*s*	10.6259	11.6089	12.5910	13.5745	14.5578	15.5409	16.5239	17.5075
2*s*	6.5714	7.3920	8.3736	9.0200	9.8250	10.6288	11.4304	12.2304
2*p*	6.8018	7.8258	8.9634	9.9450	10.9612	11.9770	12.9932	14.0082
3*s*	2.5074	3.3075	4.1172	4.9032	5.6418	6.3669	7.0683	7.7568
3*p*			4.0656	4.2852	4.8864	5.4819	6.1161	6.7641

Data: E. Clementi and D.L. Raimondi, *Atomic screening constants from SCF functions.* IBM Res. Note NJ-27 (1963). *J. chem. Phys.* **38**, 2686 (1963).

Table 10.3 Ionization energies, $I/(\mathrm{kJ\ mol^{-1}})$

H							He
1312.0							2372.3
							5250.4
Li	Be	B	C	N	O	F	Ne
513.3	899.4	800.6	1086.2	1402.3	1313.9	1681	2080.6
7298.0	1757.1	2427	2352	2856.1	3388.2	3374	3952.2
Na	Mg	Al	Si	P	S	Cl	Ar
495.8	737.7	577.4	786.5	1011.7	999.6	1251.1	1520.4
4562.4	1450.7	1816.6	1577.1	1903.2	2251	2297	2665.2
		2744.6		2912			
K	Ca	Ga	Ge	As	Se	Br	Kr
418.8	589.7	578.8	762.1	947.0	940.9	1139.9	1350.7
3051.4	1145	1979	1537	1798	2044	2104	2350
		2963	2735				
Rb	Sr	In	Sn	Sb	Te	I	Xe
403.0	549.5	558.3	708.6	833.7	869.2	1008.4	1170.4
2632	1064.2	1820.6	1411.8	1794	1795	1845.9	2046
		2704	2943.0	2443			
Cs	Ba	Tl	Pb	Bi	Po	At	Rn
375.5	502.8	589.3	715.5	703.2	812	930	1037
2420	965.1	1971.0	1450.4	1610			
		2878	3081.5	2466			

Data: E.

Table 10.4 Electron affinities, $E_{ea}/(\mathrm{kJ\ mol^{-1}})$

H							He
72.8							−21
Li	Be	B	C	N	O	F	Ne
59.8	≤0	23	122.5	−7	141	322	−29
					−844		
Na	Mg	Al	Si	P	S	Cl	Ar
52.9	≤0	44	133.6	71.7	200.4	348.7	−35
					−532		
K	Ca	Ga	Ge	As	Se	Br	Kr
48.3	2.37	36	116	77	195.0	324.5	−39
Rb	Sr	In	Sn	Sb	Te	I	Xe
46.9	5.03	34	121	101	190.2	295.3	−41
Cs	Ba	Tl	Pb	Bi	Po	At	Rn
45.5	13.95	30	35.2	101	186	270	−41

Data: E.

Table 11.2 Bond lengths, R_e/pm

(a) Bond lengths in specific molecules

Br_2	228.3
Cl_2	198.75
CO	112.81
F_2	141.78
H_2^+	106
H_2	74.138
HBr	141.44
HCl	127.45
HF	91.680
HI	160.92
N_2	109.76
O_2	120.75

(b) Mean bond lengths from covalent radii*

H	37						
C	77(1) 67(2) 60(3)	N	74(1) 65(2)	O	66(1) 57(2)	F	64
Si	118	P	110	S	104(1) 95(2)	Cl	99
Ge	122	As	121	Se	104	Br	114
		Sb	141	Te	137	I	133

* Values are for single bonds except where indicated otherwise (values in parentheses). The length of an A—B covalent bond (of given order) is the sum of the corresponding covalent radii.

Table 11.3a Bond dissociation enthalpies, $\Delta H^{\ominus}$(A—B)/(kJ mol^{-1}) at 298 K

Diatomic molecules

H—H	436	F—F	155	Cl—Cl	242	Br—Br	193	I—I	151
O=O	497	C=O	1076	N≡N	945				
H—O	428	H—F	565	H—Cl	431	H—Br	366	H—I	299

Polyatomic molecules

H—CH_3	435	H—NH_2	460	H—OH	492	H—C_6H_5	469
H_3C—CH_3	368	H_2C=CH_2	720	HC≡CH	962		
HO—CH_3	377	Cl—CH_3	352	Br—CH_3	293	I—CH_3	237
O=CO	531	HO—OH	213	O_2N—NO_2	54		

Data: HCP, KL.

Table 11.3b Mean bond enthalpies, $\Delta H^{\ominus}(\text{A}{-}\text{B})/(\text{kJ mol}^{-1})$

	H	C	N	O	F	Cl	Br	I	S	P	Si
H	436										
C	412	348(i)									
		612(ii)									
		838(iii)									
		518(a)									
N	388	305(i)	163(i)								
		613(ii)	409(ii)								
		890(iii)	946(iii)								
O	463	360(i)	157	146(i)							
		743(ii)		497(ii)							
F	565	484	270	185	155						
Cl	431	338	200	203	254	242					
Br	366	276				219	193				
I	299	238				210	178	151			
S	338	259			496	250	212		264		
P	322									201	
Si	318		374	466							226

(i) Single bond, (ii) double bond, (iii) triple bond, (a) aromatic.
Data: HCP and L. Pauling, *The nature of the chemical bond*. Cornell University Press (1960).

Table 11.4 Pauling (*italics*) and Mulliken electronegativities

H							He
2.20							
3.06							
Li	Be	B	C	N	O	F	Ne
0.98	*1.57*	*2.04*	*2.55*	*3.04*	*3.44*	*3.98*	
1.28	1.99	1.83	2.67	3.08	3.22	4.43	4.60
Na	Mg	Al	Si	P	S	Cl	Ar
0.93	*1.31*	*1.61*	*1.90*	*2.19*	*2.58*	*3.16*	
1.21	1.63	1.37	2.03	2.39	2.65	3.54	3.36
K	Ca	Ga	Ge	As	Se	Br	Kr
0.82	*1.00*	*1.81*	*2.01*	*2.18*	*2.55*	*2.96*	*3.0*
1.03	1.30	1.34	1.95	2.26	2.51	3.24	2.98
Rb	Sr	In	Sn	Sb	Te	I	Xe
0.82	*0.95*	*1.78*	*1.96*	*2.05*	*2.10*	*2.66*	*2.6*
0.99	1.21	1.30	1.83	2.06	2.34	2.88	2.59
Cs	Ba	Tl	Pb	Bi			
0.79	*0.89*	*2.04*	*2.33*	*2.02*			

Data: Pauling values: A.L. Allred, *J. Inorg. Nucl. Chem.* **17**, 215 (1961); L.C. Allen and J.E. Huheey, *ibid.*, **42**, 1523 (1980). Mulliken values: L.C. Allen, *J. Am. Chem. Soc.* **111**, 9003 (1989). The Mulliken values have been scaled to the range of the Pauling values.

Table 13.2 Properties of diatomic molecules

	$\tilde{\nu}_0/\text{cm}^{-1}$	θ_V/K	B/cm^{-1}	θ_R/K	r/pm	$k/(\text{N m}^{-1})$	$D/(\text{kJ mol}^{-1})$	σ
$^1H_2^+$	2321.8	3341	29.8	42.9	106	160	255.8	2
$^1H_2$	4400.39	6332	60.864	87.6	74.138	574.9	432.1	2
$^2H_2$	3118.46	4487	30.442	43.8	74.154	577.0	439.6	2
$^1H^{19}F$	4138.32	5955	20.956	30.2	91.680	965.7	564.4	1
$^1H^{35}Cl$	2990.95	4304	10.593	15.2	127.45	516.3	427.7	1
$^1H^{81}Br$	2648.98	3812	8.465	12.2	141.44	411.5	362.7	1
$^1H^{127}I$	2308.09	3321	6.511	9.37	160.92	313.8	294.9	1
$^{14}N_2$	2358.07	3393	1.9987	2.88	109.76	2293.8	941.7	2
$^{16}O_2$	1580.36	2274	1.4457	2.08	120.75	1176.8	493.5	2
$^{19}F_2$	891.8	1283	0.8828	1.27	141.78	445.1	154.4	2
$^{35}Cl_2$	559.71	805	0.2441	0.351	198.75	322.7	239.3	2
$^{12}C^{16}O$	2170.21	3122	1.9313	2.78	112.81	1903.17	1071.8	1
$^{79}Br^{81}Br$	323.2	465	0.0809	10.116	283.3	245.9	190.2	1

Data: AIP.

Table 13.3 Typical vibrational wavenumbers, $\tilde{\nu}/\text{cm}^{-1}$

C—H stretch	2850–2960
C—H bend	1340–1465
C—C stretch, bend	700–1250
C=C stretch	1620–1680
C≡C stretch	2100–2260
O—H stretch	3590–3650
H-bonds	3200–3570
C=O stretch	1640–1780
C≡N stretch	2215–2275
N—H stretch	3200–3500
C—F stretch	1000–1400
C—Cl stretch	600–800
C—Br stretch	500–600
C—I stretch	500
CO_3^{2-}	1410–1450
NO_3^-	1350–1420
NO_2^-	1230–1250
SO_4^{2-}	1080–1130
Silicates	900–1100

Data: L.J. Bellamy, *The infrared spectra of complex molecules* and *Advances in infrared group frequencies*. Chapman and Hall.

Table 14.1 Colour, frequency, and energy of light

Colour	λ/nm	$\nu/(10^{14}\,\text{Hz})$	$\tilde{\nu}/(10^4\,\text{cm}^{-1})$	E/eV	$E/(\text{kJ mol}^{-1})$
Infrared	>1000	<3.00	<1.00	<1.24	<120
Red	700	4.28	1.43	1.77	171
Orange	620	4.84	1.61	2.00	193
Yellow	580	5.17	1.72	2.14	206
Green	530	5.66	1.89	2.34	226
Blue	470	6.38	2.13	2.64	254
Violet	420	7.14	2.38	2.95	285
Near ultraviolet	300	10.0	3.33	4.15	400
Far ultraviolet	<200	>15.0	>5.00	>6.20	>598

Data: J.G. Calvert and J.N. Pitts, *Photochemistry*. Wiley, New York (1966).

Table 14.3 Absorption characteristics of some groups and molecules

Group	$\tilde{\nu}_{max}/(10^4\ cm^{-1})$	λ_{max}/nm	$\varepsilon_{max}/(dm^3\ mol^{-1}\ cm^{-1})$
C=C ($\pi^* \leftarrow \pi$)	6.10	163	1.5×10^4
	5.73	174	5.5×10^3
C=O ($\pi^* \leftarrow n$)	3.7–3.5	270–290	10–20
—N=N—	2.9	350	15
	>3.9	<260	Strong
$—NO_2$	3.6	280	10
	4.8	210	1.0×10^4
$C_6H_5—$	3.9	255	200
	5.0	200	6.3×10^3
	5.5	180	1.0×10^5
$[Cu(OH_2)_6]^{2+}(aq)$	1.2	810	10
$[Cu(NH_3)_4]^{2+}(aq)$	1.7	600	50
H_2O ($\pi^* \leftarrow n$)	6.0	167	7.0×10^3

Table 15.2 Nuclear spin properties

Nuclide	Natural abundance %	Spin I	Magnetic moment μ/μ_N	g-value	$\gamma/(10^7\ T^{-1}\ s^{-1})$	NMR frequency at 1 T, ν/MHz
^1n*		$\frac{1}{2}$	−1.9130	−3.8260	−18.324	29.164
^1H	99.9844	$\frac{1}{2}$	2.792 85	5.5857	26.752	42.576
^2H	0.0156	1	0.857 44	0.857 45	4.1067	6.536
^3H*		$\frac{1}{2}$	2.978 96	−4.2553	−20.380	45.414
^{10}B	19.6	3	1.8006	0.6002	2.875	4.575
^{11}B	80.4	$\frac{3}{2}$	2.6886	1.7923	8.5841	13.663
^{13}C	1.108	$\frac{1}{2}$	0.7024	1.4046	6.7272	10.708
^{14}N	99.635	1	0.403 76	0.403 56	1.9328	3.078
^{17}O	0.037	$\frac{5}{2}$	−1.893 79	−0.7572	−3.627	5.774
^{19}F	100	$\frac{1}{2}$	2.628 87	5.2567	25.177	40.077
^{31}P	100	$\frac{1}{2}$	1.1316	2.2634	10.840	17.251
^{33}S	0.74	$\frac{3}{2}$	0.6438	0.4289	2.054	3.272
^{35}Cl	75.4	$\frac{3}{2}$	0.8219	0.5479	2.624	4.176
^{37}Cl	24.6	$\frac{3}{2}$	0.6841	0.4561	2.184	3.476

* Radioactive.

μ is the magnetic moment of the spin state with the largest value of m_I: $\mu = g_I \mu_N I$ and μ_N is the nuclear magneton (see inside front cover).

Data: KL and HCP.

Table 15.3 Hyperfine coupling constants for atoms, a/mT

Nuclide	Spin	Isotropic coupling	Anisotropic coupling
^{1}H	$\frac{1}{2}$	50.8(1s)	
^{2}H	1	7.8(1s)	
^{13}C	$\frac{1}{2}$	113.0(2s)	6.6(2p)
^{14}N	1	55.2(2s)	4.8(2p)
^{19}F	$\frac{1}{2}$	1720(2s)	108.4(2p)
^{31}P	$\frac{1}{2}$	364(3s)	20.6(3p)
^{35}Cl	$\frac{3}{2}$	168(3s)	10.0(3p)
^{37}Cl	$\frac{3}{2}$	140(3s)	8.4(3p)

Data: P.W. Atkins and M.C.R. Symons, *The structure of inorganic radicals*. Elsevier, Amsterdam (1967).

Table 18.1 Dipole moments, polarizabilities, and polarizability volumes

	$\mu/(10^{-30}\ \mathrm{C\ m})$	μ/D	$\alpha'/(10^{-30}\ \mathrm{m^3})$	$\alpha/(10^{-40}\ \mathrm{J^{-1}\ C^2\ m^2})$
Ar	0	0	1.66	1.85
C_2H_5OH	5.64	1.69		
$C_6H_5CH_3$	1.20	0.36		
C_6H_6	0	0	10.4	11.6
CCl_4	0	0	10.3	11.7
CH_2Cl_2	5.24	1.57	6.80	7.57
CH_3Cl	6.24	1.87	4.53	5.04
CH_3OH	5.70	1.71	3.23	3.59
CH_4	0	0	2.60	2.89
$CHCl_3$	3.37	1.01	8.50	9.46
CO	0.390	0.117	1.98	2.20
CO_2	0	0	2.63	2.93
H_2	0	0	0.819	0.911
H_2O	6.17	1.85	1.48	1.65
HBr	2.67	0.80	3.61	4.01
HCl	3.60	1.08	2.63	2.93
He	0	0	0.20	0.22
HF	6.37	1.91	0.51	0.57
HI	1.40	0.42	5.45	6.06
N_2	0	0	1.77	1.97
NH_3	4.90	1.47	2.22	2.47
1,2-$C_6H_4(CH_3)_2$	2.07	0.62		

Data: HCP and C.J.F. Böttcher and P. Bordewijk, *Theory of electric polarization*. Elsevier, Amsterdam (1978).

Table 18.4 Lennard-Jones (12,6)-potential parameters

	(ε/k)/K	r_0/pm
Ar	111.84	362.3
C_2H_2	209.11	463.5
C_2H_4	200.78	458.9
C_2H_6	216.12	478.2
C_6H_6	377.46	617.4
CCl_4	378.86	624.1
Cl_2	296.27	448.5
CO_2	201.71	444.4
F_2	104.29	357.1
Kr	154.87	389.5
N_2	91.85	391.9
O_2	113.27	365.4
Xe	213.96	426.0

Source: F. Cuadros, I. Cachadiña, and W. Ahamuda, *Molec. Engineering*, **6**, 319 (1996).

Table 18.5 Surface tensions of liquids at 293 K

	γ/(mN m^{-1})
Benzene	28.88
Carbon tetrachloride	27.0
Ethanol	22.8
Hexane	18.4
Mercury	472
Methanol	22.6
Water	72.75
	72.0 at 25°C
	58.0 at 100°C

Data: KL.

Table 19.1 Radius of gyration of some macromolecules

	M/(kg mol^{-1})	R_g/nm
Serum albumin	66	2.98
Myosin	493	46.8
Polystyrene	3.2×10^3	50 (in poor solvent)
DNA	4×10^3	117.0
Tobacco mosaic virus	3.9×10^4	92.4

Data: C. Tanford, *Physical chemistry of macromolecules*. Wiley, New York (1961).

Table 19.2 Diffusion coefficients of macromolecules in water at 20°C

	M/(kg mol^{-1})	D/(10^{-10} m^2 s^{-1})
Sucrose	0.342	4.586
Ribonuclease	13.7	1.19
Lysozyme	14.1	1.04
Serum albumin	65	0.594
Haemoglobin	68	0.69
Urease	480	0.346
Collagen	345	0.069
Myosin	493	0.116

Data: C. Tanford, *Physical chemistry of macromolecules*. Wiley, New York (1961).

Table 19.3 Frictional coefficients and molecular geometry

Major axis/Minor axis	Prolate	Oblate
2	1.04	1.04
3	1.11	1.10
4	1.18	1.17
5	1.25	1.22
6	1.31	1.28
7	1.38	1.33
8	1.43	1.37
9	1.49	1.42
10	1.54	1.46
50	2.95	2.38
100	4.07	2.97

Data: K.E. Van Holde, *Physical biochemistry*. Prentice-Hall, Englewood Cliffs (1971).

Sphere; radius a, $c = af_0$

Prolate ellipsoid; major axis $2a$, minor axis $2b$, $c = (ab^2)^{1/3}$

$$f = \left\{ \frac{(1 - b^2/a^2)^{1/2}}{(b/a)^{2/3} \ln\{[1 + (1 - b^2/a^2)^{1/2}]/(b/a)\}} \right\} f_0$$

Oblate ellipsoid; major axis $2a$, minor axis $2b$, $c = (a^2 b)^{1/3}$

$$f = \left\{ \frac{(a^2/b^2 - 1)^{1/2}}{(a/b)^{2/3} \arctan[(a^2/b^2 - 1)^{1/2}]} \right\} f_0$$

Long rod; length l, radius a, $c = (3a^2/4)^{1/3}$

$$f = \left\{ \frac{(1/2a)^{2/3}}{(3/2)^{1/3}\{2 \ln(l/a) - 0.11\}} \right\} f_0$$

In each case $f_0 = 6\pi\eta c$ with the appropriate value of c.

Table 19.4 Intrinsic viscosity

Macromolecule	Solvent	θ/°C	$K/(10^{-3}\ \mathrm{cm^3\ g^{-1}})$	a
Polystyrene	Benzene	25	9.5	0.74
	Cyclohexane	34†	81	0.50
Polyisobutylene	Benzene	23†	83	0.50
	Cyclohexane	30	26	0.70
Amylose	0.33 M KCl(aq)	25†	113	0.50
Various proteins‡	Guanidine hydrochloride + $HSCH_2CH_2OH$		7.16	0.66

† The θ temperature.

‡ Use $[\eta] = KN^a$; N is the number of amino acid residues.

Data: K.E. Van Holde, *Physical biochemistry*. Prentice-Hall, Englewood Cliffs (1971).

Table 20.3 Ionic radii (r/pm)†

Li^+(4)	Be^{2+}(4)	B^{3+}(4)	N^{3-}	O^{2-}(6)	F^-(6)
59	27	12	171	140	133
Na^+(6)	Mg^{2+}(6)	Al^{3+}(6)	P^{3-}	S^{2-}(6)	Cl^-(6)
102	72	53	212	184	181
K^+(6)	Ca^{2+}(6)	Ga^{3+}(6)	As^{3-}(6)	Se^{2-}(6)	Br^-(6)
138	100	62	222	198	196
Rb^+(6)	Sr^{2+}(6)	In^{3+}(6)		Te^{2-}(6)	I^-(6)
149	116	79		221	220
Cs^+(6)	Ba^{2+}(6)	Tl^{3+}(6)			
167	136	88			

***d*-block elements (high-spin ions)**

Sc^{3+}(6)	Ti^{4+}(6)	Cr^{3+}(6)	Mn^{3+}(6)	Fe^{2+}(6)	Co^{3+}(6)	Cu^{2+}(6)	Zn^{2+}(6)
73	60	61	65	63	61	73	75

† Numbers in parentheses are the coordination numbers of the ions. Values for ions without a coordination number stated are estimates.

Data: R.D. Shannon and C.T. Prewitt, *Acta Cryst.* **B25**, 925 (1969).

Table 20.5 Lattice enthalpies, $\Delta H_L^{\ominus}/(\text{kJ mol}^{-1})$

	F	Cl	Br	I
Halides				
Li	1037	852	815	761
Na	926	787	752	705
K	821	717	689	649
Rb	789	695	668	632
Cs	750	676	654	620
Ag	969	912	900	886
Be		3017		
Mg		2524		
Ca		2255		
Sr		2153		

Oxides							
MgO	3850	CaO	3461	SrO	3283	BaO	3114
Sulfides							
MgS	3406	CaS	3119	SrS	2974	BaS	2832

Entries refer to $MX(s) \rightarrow M^+(g) + X^-(g)$.
Data: Principally D. Cubicciotti, *J. Chem. Phys.* **31**, 1646 (1959).

Table 20.6 Magnetic susceptibilities at 298 K

	$\chi/10^{-6}$	$\chi_m/(10^{-4}\text{ cm}^3\text{ mol}^{-1})$
Water	−90	−16.0
Benzene	−7.2	−6.4
Cyclohexane	−7.9	−8.5
Carbon tetrachloride	−8.9	−8.4
NaCl(s)	−13.9	−3.75
Cu(s)	−96	−6.8
S(s)	−12.9	−2.0
Hg(l)	−28.5	−4.2
$CuSO_4\cdot5H_2O(s)$	+176	+192
$MnSO_4\cdot4H_2O(s)$	+2640	$+2.79\times10^3$
$NiSO_4\cdot7H_2O(s)$	+416	+600
$FeSO_4(NH_4)_2SO_4\cdot6H_2O(s)$	+755	$+1.51\times10^3$
Al(s)	+22	+2.2
Pt(s)	+262	+22.8
Na(s)	+7.3	+1.7
K(s)	+5.6	+2.5

Data: KL and $\chi_m = \chi M/\rho$.

Table 21.1 Collision cross-sections, σ/nm^2

Ar	0.36
C_2H_4	0.64
C_6H_6	0.88
CH_4	0.46
Cl_2	0.93
CO_2	0.52
H_2	0.27
He	0.21
N_2	0.43
Ne	0.24
O_2	0.40
SO_2	0.58

Data: KL.

Table 21.2 Transport properties of gases at 1 atm

	$\kappa/(\text{J K}^{-1}\text{ m}^{-1}\text{ s}^{-1})$	$\eta/\mu\text{P}$	
	273 K	273 K	293 K
Air	0.0241	173	182
Ar	0.0163	210	223
C_2H_4	0.0164	97	103
CH_4	0.0302	103	110
Cl_2	0.079	123	132
CO_2	0.0145	136	147
H_2	0.1682	84	88
He	0.1442	187	196
Kr	0.0087	234	250
N_2	0.0240	166	176
Ne	0.0465	298	313
O_2	0.0245	195	204
Xe	0.0052	212	228

Data: KL.

Table 21.4 Viscosities of liquids at 298 K, $\eta/(10^{-3}\text{ kg m}^{-1}\text{ s}^{-1})$

Benzene	0.601
Carbon tetrachloride	0.880
Ethanol	1.06
Mercury	1.55
Methanol	0.553
Pentane	0.224
Sulfuric acid	27
Water†	0.891

† The viscosity of water over its entire liquid range is represented with less than 1 per cent error by the expression

$$\log(\eta_{20}/\eta) = A/B,$$
$$A = 1.370\,23(t-20) + 8.36\times10^{-4}(t-20)^2$$
$$B = 109 + t \qquad t = \theta/°\text{C}$$

Convert $\text{kg m}^{-1}\text{ s}^{-1}$ to centipoise (cP) by multiplying by 10^3 (so $\eta \approx 1$ cP for water).
Data: AIP, KL.

Table 21.5 Limiting ionic conductivities in water at 298 K, $\lambda/(\text{mS m}^2\text{ mol}^{-1})$

Cations		Anions	
Ba^{2+}	12.72	Br^-	7.81
Ca^{2+}	11.90	$CH_3CO_2^-$	4.09
Cs^+	7.72	Cl^-	7.635
Cu^{2+}	10.72	ClO_4^-	6.73
H^+	34.96	CO_3^{2-}	13.86
K^+	7.350	$(CO_2)_2^{2-}$	14.82
Li^+	3.87	F^-	5.54
Mg^{2+}	10.60	$[Fe(CN)_6]^{3-}$	30.27
Na^+	5.010	$[Fe(CN)_6]^{4-}$	44.20
$[N(C_2H_5)_4]^+$	3.26	HCO_2^-	5.46
$[N(CH_3)_4]^+$	4.49	I^-	7.68
NH_4^+	7.35	NO_3^-	7.146
Rb^+	7.78	OH^-	19.91
Sr^{2+}	11.89	SO_4^{2-}	16.00
Zn^{2+}	10.56		

Data: KL, RS.

Table 21.6 Ionic mobilities in water at 298 K, $u/(10^{-8}\text{ m}^2\text{ s}^{-1}\text{ V}^{-1})$

Cations		Anions	
Ag^+	6.24	Br^-	8.09
Ca^{2+}	6.17	$CH_3CO_2^-$	4.24
Cu^{2+}	5.56	Cl^-	7.91
H^+	36.23	CO_3^{2-}	7.46
K^+	7.62	F^-	5.70
Li^+	4.01	$[Fe(CN)_6]^{3-}$	10.5
Na^+	5.19	$[Fe(CN)_6]^{4-}$	11.4
NH_4^+	7.63	I^-	7.96
$[N(CH_3)_4]^+$	4.65	NO_3^-	7.40
Rb^+	7.92	OH^-	20.64
Zn^{2+}	5.47	SO_4^{2-}	8.29

Data: Principally Table 21.4 and $u = \lambda/zF$.

Table 21.7 Debye–Hückel–Onsager coefficients for (1,1)-electrolytes at 25°C

Solvent	$A/(\text{mS m}^2\text{ mol}^{-1}/(\text{mol dm}^{-3})^{1/2})$	$B/(\text{mol dm}^{-3})^{-1/2}$
Acetone (propanone)	3.28	1.63
Acetonitrile	2.29	0.716
Ethanol	8.97	1.83
Methanol	15.61	0.923
Nitrobenzene	4.42	0.776
Nitromethane	111	0.708
Water	6.020	0.229

Data: J.O'M. Bockris and A.K.N. Reddy, *Modern electrochemistry*. Plenum, New York (1970).

Table 21.8 Diffusion coefficients at 25°C, $D/(10^{-9}\ m^2\ s^{-1})$

Molecules in liquids				Ions in water			
I_2 in hexane	4.05	H_2 in CCl_4(l)	9.75	K^+	1.96	Br^-	2.08
in benzene	2.13	N_2 in CCl_4(l)	3.42	H^+	9.31	Cl^-	2.03
CCl_4 in heptane	3.17	O_2 in CCl_4(l)	3.82	Li^+	1.03	F^-	1.46
Glycine in water	1.055	Ar in CCl_4(l)	3.63	Na^+	1.33	I^-	2.05
Dextrose in water	0.673	CH_4 in CCl_4(l)	2.89			OH^-	5.03
Sucrose in water	0.5216	H_2O in water	2.26				
		CH_3OH in water	1.58				
		C_2H_5OH in water	1.24				

Data: AIP and (for the ions) $\lambda = zuF$ in conjunction with Table 21.5.

Table 22.1 Kinetic data for first-order reactions

	Phase	θ/°C	k/s^{-1}	$t_{1/2}$
$2\ N_2O_5 \rightarrow 4\ NO_2 + O_2$	g	25	3.38×10^{-5}	5.70 h
	HNO_3(l)	25	1.47×10^{-6}	131 h
	Br_2(l)	25	4.27×10^{-5}	4.51 h
$C_2H_6 \rightarrow 2\ CH_3$	g	700	5.36×10^{-4}	21.6 min
Cyclopropane → propene	g	500	6.71×10^{-4}	17.2 min
$CH_3N_2CH_3 \rightarrow C_2H_6 + N_2$	g	327	3.4×10^{-4}	34 min
Sucrose → glucose + fructose	aq(H^+)	25	6.0×10^{-5}	3.2 h

g: High pressure gas-phase limit.
Data: Principally K.J. Laidler, *Chemical kinetics*. Harper & Row, New York (1987); M.J. Pilling and P.W. Seakins, *Reaction kinetics*. Oxford University Press (1995); J. Nicholas, *Chemical kinetics*. Harper & Row, New York (1976). See also JL.

Table 22.2 Kinetic data for second-order reactions

	Phase	θ/°C	$k/(dm^3\ mol^{-1}\ s^{-1})$
$2\ NOBr \rightarrow 2\ NO + Br_2$	g	10	0.80
$2\ NO_2 \rightarrow 2\ NO + O_2$	g	300	0.54
$H_2 + I_2 \rightarrow 2\ HI$	g	400	2.42×10^{-2}
$D_2 + HCl \rightarrow DH + DCl$	g	600	0.141
$2\ I \rightarrow I_2$	g	23	7×10^{9}
	hexane	50	1.8×10^{10}
$CH_3Cl + CH_3O^-$	methanol	20	2.29×10^{-6}
$CH_3Br + CH_3O^-$	methanol	20	9.23×10^{-6}
$H^+ + OH^- \rightarrow H_2O$	water	25	1.35×10^{11}
	ice	−10	8.6×10^{12}

Data: Principally K.J. Laidler, *Chemical kinetics*. Harper & Row, New York (1987); M.J. Pilling and P.W. Seakins, *Reaction kinetics*. Oxford University Press (1995); J. Nicholas, *Chemical kinetics*. Harper & Row, New York (1976).

Table 22.4 Arrhenius parameters

First-order reactions	A/s^{-1}	$E_a/(kJ\ mol^{-1})$
Cyclopropane → propene	1.58×10^{15}	272
$CH_3NC \rightarrow CH_3CN$	3.98×10^{13}	160
cis-CHD=CHD → *trans*-CHD=CHD	3.16×10^{12}	256
Cyclobutane → 2 C_2H_4	3.98×10^{13}	261
$C_2H_5I \rightarrow C_2H_4 + HI$	2.51×10^{17}	209
$C_2H_6 \rightarrow 2\ CH_3$	2.51×10^{7}	384
$2\ N_2O_5 \rightarrow 4\ NO_2 + O_2$	4.94×10^{13}	103
$N_2O \rightarrow N_2 + O$	7.94×10^{11}	250
$C_2H_5 \rightarrow C_2H_4 + H$	1.0×10^{13}	167
Second-order, gas-phase	$A/(dm^3\ mol^{-1}\ s^{-1})$	$E_a/(kJ\ mol^{-1})$
$O + N_2 \rightarrow NO + N$	1×10^{11}	315
$OH + H_2 \rightarrow H_2O + H$	8×10^{10}	42
$Cl + H_2 \rightarrow HCl + H$	8×10^{10}	23
$2\ CH_3 \rightarrow C_2H_6$	2×10^{10}	ca. 0
$NO + Cl_2 \rightarrow NOCl + Cl$	4.0×10^{9}	85
$SO + O_2 \rightarrow SO_2 + O$	3×10^{8}	27
$CH_3 + C_2H_6 \rightarrow CH_4 + C_2H_5$	2×10^{8}	44
$C_6H_5 + H_2 \rightarrow C_6H_6 + H$	1×10^{8}	ca. 25
Second-order, solution	$A/(dm^3\ mol^{-1}\ s^{-1})$	$E_a/(kJ\ mol^{-1})$
$C_2H_5ONa + CH_3I$ in ethanol	2.42×10^{11}	81.6
$C_2H_5Br + OH^-$ in water	4.30×10^{11}	89.5
$C_2H_5I + C_2H_5O^-$ in ethanol	1.49×10^{11}	86.6
$CH_3I + C_2H_5O^-$ in ethanol	2.42×10^{11}	81.6
$C_2H_5Br + OH^-$ in ethanol	4.30×10^{11}	89.5
$CO_2 + OH^-$ in water	1.5×10^{10}	38
$CH_3I + S_2O_3^{2-}$ in water	2.19×10^{12}	78.7
Sucrose + H_2O in acidic water	1.50×10^{15}	107.9
$(CH_3)_3CCl$ solvolysis		
in water	7.1×10^{16}	100
in methanol	2.3×10^{13}	107
in ethanol	3.0×10^{13}	112
in acetic acid	4.3×10^{13}	111
in chloroform	1.4×10^{4}	45
$C_6H_5NH_2 + C_6H_5COCH_2Br$		
in benzene	91	34

Data: Principally J. Nicholas, *Chemical kinetics*. Harper & Row, New York (1976) and A.A. Frost and R.G. Pearson, *Kinetics and mechanism*. Wiley, New York (1961).

Table 24.1 Arrhenius parameters for gas-phase reactions

	$A/(\mathrm{dm^3\,mol^{-1}\,s^{-1}})$		$E_a/(\mathrm{kJ\,mol^{-1}})$	P
	Experiment	Theory		
$2\,NOCl \rightarrow 2\,NO + Cl_2$	9.4×10^{9}	5.9×10^{10}	102.0	0.16
$2\,NO_2 \rightarrow 2\,NO + O_2$	2.0×10^{9}	4.0×10^{10}	111.0	5.0×10^{-2}
$2\,ClO \rightarrow Cl_2 + O_2$	6.3×10^{7}	2.5×10^{10}	0.0	2.5×10^{-3}
$H_2 + C_2H_4 \rightarrow C_2H_6$	1.24×10^{6}	7.4×10^{11}	180	1.7×10^{-6}
$K + Br_2 \rightarrow KBr + Br$	1.0×10^{12}	2.1×10^{11}	0.0	4.8

Data: Principally M.J. Pilling and P.W. Seakins, *Reaction kinetics*. Oxford University Press (1995).

Table 24.2 Arrhenius parameters for reactions in solution. See Table 22.4

Table 25.1 Maximum observed enthalpies of physisorption, $\Delta_{ad}H^{\ominus}/(\mathrm{kJ\,mol^{-1}})$

C_2H_2	−38	H_2	−84
C_2H_4	−34	H_2O	−59
CH_4	−21	N_2	−21
Cl_2	−36	NH_3	−38
CO	−25	O_2	−21
CO_2	−25		

Data: D.O. Haywood and B.M.W. Trapnell, *Chemisorption*. Butterworth (1964).

Table 25.2 Enthalpies of chemisorption, $\Delta_{ad}H^{\ominus}/(\mathrm{kJ\,mol^{-1}})$

Adsorbate	Adsorbent (substrate)										
	Ti	Ta	Nb	W	Cr	Mo	Mn	Fe	Co	Ni	Rh	Pt
H_2		−188			−188	−167	−71	−134			−117	
N_2		−586						−293				
O_2						−720					−494	−293
CO	−640							−192	−176			
CO_2	−682	−703	−552	−456	−339	−372	−222	−225	−146	−184		
NH_3				−301				−188		−155		
C_2H_4		−577		−427	−427			−285		−243	−209	

Data: D.O. Haywood and B.M.W. Trapnell, *Chemisorption*. Butterworth (1964).

Table 25.3 Activation energies of catalysed reactions

	Catalyst	E_a/(kJ mol^{-1})
$2\,HI \rightarrow H_2 + I_2$	None	184
	Au(s)	105
	Pt(s)	59
$2\,NH_3 \rightarrow N_2 + 3\,H_2$	None	350
	W(s)	162
$2\,N_2O \rightarrow 2\,N_2 + O_2$	None	245
	Au(s)	121
	Pt(s)	134
$(C_2H_5)_2O$ pyrolysis	None	224
	I_2(g)	144

Data: G.C. Bond, *Heterogeneous catalysis*. Clarendon Press, Oxford (1986).

Table 25.6 Exchange current densities and transfer coefficients at 298 K

Reaction	Electrode	j_0/(A cm^{-2})	α
$2\,H^+ + 2\,e^- \rightarrow H_2$	Pt	7.9×10^{-4}	
	Cu	1×10^{-6}	
	Ni	6.3×10^{-6}	0.58
	Hg	7.9×10^{-13}	0.50
	Pb	5.0×10^{-12}	
$Fe^{3+} + e^- \rightarrow Fe^{2+}$	Pt	2.5×10^{-3}	0.58
$Ce^{4+} + e^- \rightarrow Ce^{3+}$	Pt	4.0×10^{-5}	0.75

Data: Principally J.O'M. Bockris and A.K.N. Reddy, *Modern electrochemistry*. Plenum, New York (1970).

Table A3.1 Refractive indices relative to air at 20°C

	434 nm	589 nm	656 nm
Benzene	1.5236	1.5012	1.4965
Carbon tetrachloride	1.4729	1.4676	1.4579
Carbon disulfide	1.6748	1.6276	1.6182
Ethanol	1.3700	1.3618	1.3605
KCl(s)	1.5050	1.4904	1.4973
KI(s)	1.7035	1.6664	1.6581
Methanol	1.3362	1.3290	1.3277
Methylbenzene	1.5170	1.4955	1.4911
Water	1.3404	1.3330	1.3312

Data: AIP.

Character tables

The groups C_1, C_s, C_i

C_1 (1)	E	$h = 1$
A	1	

$C_s = C_h$ (m)	E	σ_h	$h = 2$	
A′	1	1	x, y, R_z	x^2, y^2, z^2, xy
A″	1	−1	z, R_x, R_y	yz, xz

$C_i = S_2$ $(\bar{1})$	E	i	$h = 2$	
A_g	1	1	R_x, R_y, R_z	$x^2, y^2, z^2, xy, xz, yz$
A_u	1	−1	x, y, z	

The groups C_{nv}

C_{2v}, $2mm$	E	C_2	σ_v	σ'_v	$h=4$	
A_1	1	1	1	1	z, z^2, x^2, y^2	
A_2	1	1	−1	−1	xy	R_z
B_1	1	−1	1	−1	x, xz	R_y
B_2	1	−1	−1	1	y, yz	R_x

C_{3v}, $3m$	E	$2C_3$	$3\sigma_v$	$h=6$	
A_1	1	1	1	z, z^2, x^2+y^2	
A_2	1	1	−1		R_z
E	2	−1	0	$(x, y), (xy, x^2-y^2)$ (xz, yz)	(R_x, R_y)

C_{4v}, $4mm$	E	C_2	$2C_4$	$2\sigma_v$	$2\sigma_d$	$h=8$	
A_1	1	1	1	1	1	z, z^2, x^2+y^2	
A_2	1	1	1	−1	1		R_z
B_1	1	1	−1	1	−1	x^2-y^2	
B_2	1	1	−1	−1	1	xy	
E	2	−2	0	0	0	$(x, y), (xz, yz)$	(R_x, R_y)

C_{5v}	E	$2C_5$	$2C_5^2$	$5\sigma_v$	$h=10, \alpha=72°$	
A_1	1	1	1	1	z, z^2, x^2+y^2	
A_2	1	1	1	−1		R_z
E_1	2	$2\cos\alpha$	$2\cos 2\alpha$	0	$(x, y), (xz, yz)$	(R_x, R_y)
E_2	2	$2\cos 2\alpha$	$2\cos\alpha$	0	(xy, x^2-y^2)	

C_{6v}, $6mm$	E	C_2	$2C_3$	$2C_6$	$3\sigma_d$	$3\sigma_v$	$h=12$	
A_1	1	1	1	1	1	1	z, z^2, x^2+y^2	
A_2	1	1	1	1	−1	1		R_z
B_1	1	−1	1	−1	−1	1		
B_2	1	−1	1	−1	1	−1		
E_1	2	−2	−1	1	0	0	$(x, y), (xz, yz)$	(R_x, R_y)
E_2	2	2	−1	−1	0	0	(xy, x^2-y^2)	

$C_{\infty v}$	E	$2C_\phi$†	$\infty\sigma_v$	$h=\infty$	
$A_1(\Sigma^+)$	1	1	1	z, z^2, x^2+y^2	
$A_2(\Sigma^-)$	1	1	−1		R_z
$E_1(\Pi)$	2	$2\cos\phi$	0	$(x, y), (xz, yz)$	(R_x, R_y)
$E_2(\Delta)$	2	$2\cos 2\phi$	0	(xy, x^2-y^2)	

† There is only one member of this class if $\phi=\pi$.

The groups D_n

D_2, 222	E	C_2^z	C_2^y	C_2^x	$h=4$	
A_1	1	1	1	1	x^2, y^2, z^2	
B_1	1	1	−1	−1	z, xy	R_z
B_2	1	−1	1	−1	y, xz	R_y
B_3	1	−1	−1	1	x, yz	R_x

D_3, 32	E	$2C_3$	$3C_2'$	$h=6$	
A_1	1	1	1	z^2, x^2+y^2	
A_2	1	1	−1	z	R_z
E	2	−1	0	$(x, y), (xz, yz), (xy, x^2-y^2)$	(R_x, R_y)

D_4, 422	E	C_2	$2C_4$	$2C_2'$	$2C_2''$	$h=8$	
A_1	1	1	1	1	1	z^2, x^2+y^2	
A_2	1	1	1	−1	−1	z	R_z
B_1	1	1	−1	1	−1	x^2-y^2	
B_2	1	1	−1	−1	1	xy	
E	2	−2	0	0	0	$(x, y), (xz, yz)$	(R_x, R_y)

The groups D_{nh}

D_{3h}, $\bar{6}2m$	E	σ_h	$2C_3$	$2S_3$	$3C_2'$	$3\sigma_v$	$h=12$	
A_1'	1	1	1	1	1	1	z^2, x^2+y^2	
A_2'	1	1	1	1	−1	−1		R_z
A_1''	1	−1	1	−1	1	−1		
A_2''	1	−1	1	−1	−1	1	z	
E′	2	2	−1	−1	0	0	$(x, y), (xy, x^2-y^2)$	
E″	2	−2	−1	1	0	0	(xz, yz)	(R_x, R_y)

D_{4h}, 4/mmm	E	$2C_4$	C_2	$2C_2'$	$2C_2''$	i	$2S_4$	σ_h	$2\sigma_v$	$2\sigma_d$	$h=16$	
A_{1g}	1	1	1	1	1	1	1	1	1	1	x^2+y^2, z^2	
A_{2g}	1	1	1	−1	−1	1	1	1	−1	−1		R_z
B_{1g}	1	−1	1	1	−1	1	−1	1	1	−1	x^2-y^2	
B_{2g}	1	−1	1	−1	1	1	−1	1	−1	1	xy	
E_g	2	0	−2	0	0	2	0	−2	0	0	(xz, yz)	(R_x, R_y)
A_{1u}	1	1	1	1	1	−1	−1	−1	−1	−1		
A_{2u}	1	1	1	−1	−1	−1	−1	−1	1	1	z	
B_{1u}	1	−1	1	1	−1	−1	1	−1	−1	1		
B_{2u}	1	−1	1	−1	1	−1	1	−1	1	−1		
E_u	2	0	−2	0	0	−2	0	2	0	0	(x, y)	

D_{5h}	E	$2C_5$	$2C_5^2$	$5C_2$	σ_h	$2S_5$	$2S_5^3$	$5\sigma_v$	$h=20$	$\alpha=72°$
A_1'	1	1	1	1	1	1	1	1	x^2+y^2, z^2	
A_2'	1	1	1	−1	1	1	1	−1		R_z
E_1'	2	$2\cos\alpha$	$2\cos 2\alpha$	0	2	$2\cos\alpha$	$2\cos 2\alpha$	0	(x, y)	
E_2'	2	$2\cos 2\alpha$	$2\cos\alpha$	0	2	$2\cos 2\alpha$	$2\cos\alpha$	0	(x^2-y^2, xy)	
A_1''	1	1	1	1	−1	−1	−1	−1		
A_2''	1	1	1	−1	−1	−1	−1	1	z	
E_1''	2	$2\cos\alpha$	$2\cos 2\alpha$	0	−2	$-2\cos\alpha$	$-2\cos 2\alpha$	0	(xz, yz)	(R_x, R_y)
E_2''	2	$2\cos 2\alpha$	$2\cos\alpha$	0	−2	$-2\cos 2\alpha$	$-2\cos\alpha$	0		

$D_{\infty h}$	E	$2C_\phi$	$\infty C_2'$	i	$2iC_\infty$	iC_2'	$h=\infty$	
$A_{1g}(\Sigma_g^+)$	1	1	1	1	1	1	z^2, x^2+y^2	
$A_{1u}(\Sigma_u^+)$	1	1	1	−1	−1	−1	z	
$A_{2g}(\Sigma_g^-)$	1	1	−1	1	1	−1		R_z
$A_{2u}(\Sigma_u^-)$	1	1	−1	−1	1	1		
$E_{1g}(\Pi_g)$	2	$2\cos\phi$	0	2	$-2\cos\phi$	0	(xz, yz)	(R_x, R_y)
$E_{1u}(\Pi_u)$	2	$2\cos\phi$	0	−2	$2\cos\phi$	0	(x, y)	
$E_{2g}(\Delta_g)$	2	$2\cos 2\phi$	0	2	$2\cos 2\phi$	0	(xy, x^2-y^2)	
$E_{2u}(\Delta_u)$	2	$2\cos 2\phi$	0	−2	$-2\cos 2\phi$	0		
⋮								

The cubic groups

T_d, $\bar{4}3m$	E	$8C_3$	$3C_2$	$6\sigma_d$	$6S_4$	$h=24$	
A_1	1	1	1	1	1	$x^2+y^2+z^2$	
A_2	1	1	1	−1	−1		
E	2	−1	2	0	0	$(3z^2-r^2, x^2-y^2)$	
T_1	3	0	−1	−1	1		(R_x, R_y, R_z)
T_2	3	0	−1	1	−1	$(x, y, z), (xy, xz, yz)$	

O_h $(m3m)$	E	$8C_3$	$6C_2$	$6C_2$	$3C_2$ $(=C_4^2)$	i	$6S_4$	$8S_6$	$3\sigma_h$	$6\sigma_d$	$h=48$	
A_{1g}	1	1	1	1	1	1	1	1	1	1		$x^2+y^2+z^2$
A_{2g}	1	1	−1	−1	1	1	−1	1	1	−1		
E_g	2	−1	0	0	2	2	0	−1	2	0		$(2z^2-x^2-y^2, x^2-y^2)$
T_{1g}	3	0	−1	1	−1	3	1	0	−1	−1	(R_x, R_y, R_z)	
T_{2g}	3	0	1	−1	−1	3	−1	0	−1	1		(xy, yz, xy)
A_{1u}	1	1	1	1	1	−1	−1	−1	−1	−1		
A_{2u}	1	1	−1	−1	1	−1	1	−1	−1	1		
E_u	2	−1	0	0	2	−2	0	1	−2	0		
T_{1u}	3	0	−1	1	−1	−3	−1	0	1	1	(x, y, z)	
T_{2u}	3	0	1	−1	−1	−3	1	0	1	−1		

The icosahedral group

I	E	$12C_5$	$12C_5^2$	$20C_3$	$15C_2$	$h=60$	
A	1	1	1	1	1		$z^2+y^2+z^2$
T_1	3	$\frac{1}{2}(1+\sqrt{5})$	$\frac{1}{2}(1-\sqrt{5})$	0	−1	(x, y, z)	
T_2	3	$\frac{1}{2}(1-\sqrt{5})$	$\frac{1}{2}(1+\sqrt{5})$	0	−1	(R_x, R_y, R_z)	
G	4	−1	−1	1	0		
G	5	0	0	−1	1		$(2z^2-x^2-y^2, x^2-y^2, xy, yz, zx)$

Further information: P.W. Atkins, M.S. Child, and C.S.G. Phillips, *Tables for group theory*. Oxford University Press (1970).

Solutions to a) exercises

Chapter 1

1.1 (a) 24 atm; (b) 22 atm.
1.2 (a) 3.42 bar; (b) 3.38 atm.
1.3 30 lb in^{-2}.
1.4 4.20×10^{-2} atm.
1.5 0.50 m^3.
1.6 104 kPa.
1.7 8.3147 J K^{-1} mol^{-1}.
1.8 256 g mol^{-1}, S_8.
1.9 6.2 kg.
1.10 (a) $x_{N_2} = 0.7583, x_{O_2} = 0.2417, p_{N_2} = 0.748$ bar, $p_{O_2} = 0.239$ bar.
(b) $x_{N_2} = 0.7798, x_{O_2} = 0.2102, p_{N_2} = 0.770$ bar, $p_{O_2} = 0.207$ bar, $p_{Ar} = 0.00987$ bar.
1.11 169 g mol^{-1}.
1.12 −272°C.
1.13 (a) (i) $p = 1.0$ atm, (ii) $p = 8.2 \times 10^2$ atm;
(b) (i) $p = 1.0$ atm, (ii) $p = 1.7 \times 10^3$ atm.
1.14 $a = 7.61 \times 10^{-2}$ kg m^5 s^{-2} mol^{-2}, $b = 2.26 \times 10^{-5}$ m^{-3} mol^{-1}.
1.15 (a) $Z = 0.88$, (b) $V_m = 1.2$ dm^3 mol^{-1}.
1.16 140 atm.
1.17 (a) 50.7 atm; (b) 35.2 atm, $Z = 0.695$.
1.18 (a) $x(H_2) = 0.67, x(N_2) = 0.33$;
(b) $p(H_2) = 2.0$ atm, $p(N_2) = 1.0$ atm;
(c) $p = 3.0$ atm.
1.19 $b = 32.9$ cm^3 mol^{-1}, $a = 1.33$ dm^6 atm mol^{-2}, $v_{mol} = 0.24$ nm.
1.20 (a) 1.41×10^3 K; (b) 0.59 nm.
1.21 (a) $T = 3.64 \times 10^3$ K, $p = 8.7$ atm;
(b) $T = 2.60 \times 10^3$ K, $p = 4.5$ atm;
(c) $T = 46.7$ K, $p = 0.18$ atm.
1.22 $b = 0.46 \times 10^{-4}$ m^3 mol^{-1}, $Z = 0.66$.

Chapter 2

2.1 on earth, 2.6×10^3 J, on the moon, 4.2×10^2 J.
2.2 -1.0×10^2 J.
2.3 (a) $\Delta U = \Delta H = 0, w = -1.57$ kJ, $q = +1.57$ kJ;
(b) $\Delta U = \Delta H = 0, w = -1.13$ kJ, $q = +1.13$ kJ;
(c) $\Delta U = \Delta H = 0, w = 0, q = 0$.
2.4 $p_2 = 1.33$ atm, $\Delta U = +1.25$ kJ, $w = 0$, $q = +1.25$ kJ.
2.5 (a) −88 J; (b) −167 J.
2.6 $\Delta H = -40.656$ kJ, $q = -40.656$ kJ, $w = +3.10$ kJ, $\Delta U = -37.55$ kJ.
2.7 −1.5 kJ.
2.8 (a) $q = \Delta H = +28.3$ kJ, $w = -1.45$ kJ, $\Delta U = +26.8$ kJ;
(b) $\Delta H = +28.3$ kJ, $\Delta U = +26.8$ kJ, $w = 0$, $q = +26.8$ kJ.
2.9 131 K.
2.10 −194 J.
2.11 22 kPa.
2.12 $C_{p,m} = 30$ J K^{-1} mol^{-1}, $C_{V,m} = 22$ J K^{-1} mol^{-1}.
2.13 $q_p = +2.2$ kJ, $\Delta H = +2.2$ kJ, $\Delta U = +1.6$ kJ.
2.14 $q = 0, w = -3.2$ kJ, $\Delta U = -3.2$ kJ, $\Delta T = -38$ K, $\Delta H = -4.5$ kJ.
2.15 $V_f = 0.0113$ m^3, $T_f = 344$ K, $w = 7.1 \times 10^2$ J.
2.16 $q = 13.0$ kJ, $w = -1.0$ kJ, $\Delta U = 12.0$ kJ.
2.17 −4564.7 kJ mol^{-1}
2.18 $\Delta_f H[(CH_2)_3,g] = +53$ kJ mol^{-1}, $\Delta_r H = -33$ kJ mol^{-1}.
2.19 $\Delta_c U^\ominus = -5152$ kJ mol^{-1}, $C = 1.58$ kJ K^{-1}, $\Delta T = +0.205$ K.
2.20 65.49 kJ mol^{-1}.
2.21 −383 kJ mol^{-1}.
2.22 (a) $\Delta_r H^\ominus = -114.40$ kJ mol^{-1}, $\Delta_r U = -111.92$ kJ mol^{-1};
(b) $\Delta_f H^\ominus(HCl,g) = -92.31$ kJ mol^{-1}, $\Delta_f H^\ominus(H_2O,g) = -241.82$ kJ mol^{-1}.
2.23 −1368 kJ mol^{-1}.
2.24 (a) −392.1 kJ mol^{-1}; (b) −946.6 kJ mol^{-1}.
2.25 −56.98 kJ mol^{-1}.
2.26 (a) $\Delta_r H^\ominus(298\ K) = +131.29$ kJ mol^{-1}, $\Delta_r U^\ominus(298\ K) = +128.81$ kJ mol^{-1},
(b) $\Delta_r H^\ominus(378\ K) = +132.56$ kJ mol^{-1}, $\Delta_r U^\ominus(378\ K) = +129.42$ kJ mol^{-1}.
2.27 −218.66 kJ mol^{-1}.
2.28 −1892 kJ mol^{-1}.
2.29 0.71 K atm^{-1}.
2.30 $\Delta U_m = +131$ J mol^{-1}, $q = +8.05 \times 10^3$ J mol^{-1}, $w = -7.92 \times 10^3$ J mol^{-1}.
2.31 1.31×10^{-3} K^{-1}.
2.32 1.1×10^3 atm.
2.33 −7.2 J atm^{-1} mol^{-1}, q (supplied) = +8.1 kJ.

Chapter 3

3.1 (a) 92 J K^{-1}, (b) 67 J K^{-1}.
3.2 152.67 J K^{-1} mol^{-1}.
3.3 −22.1 J K^{-1}.
3.4 $\Delta S = q = 0, w = \Delta U = +4.1$ kJ, $\Delta H = +5.4$ kJ, $\Delta H_{tot} = 0, \Delta S_{tot} = +93.4$ J K^{-1}.
3.6 (a) 0; (b) −20 J; (c) −20 J; (d) −0.347 K; (e) +0.60 J K^{-1}.
3.7 (a) +87.8 J K^{-1} mol^{-1}; (b) −87.8 J K^{-1} mol^{-1}.
3.8 (a) −386.1 J K^{-1} mol^{-1}; (b) +92.6 J K^{-1} mol^{-1}; (c) −153.1 J K^{-1} mol^{-1}.
3.9 (a) −521.5 kJ mol^{-1}; (b) +25.8 kJ mol^{-1}; (c) −178.7 kJ mol^{-1}.
3.10 (a) −522.1 kJ mol^{-1}; (b) +25.78 kJ mol^{-1}; (c) −178.6 kJ mol^{-1}.
3.11 −93.05 kJ mol^{-1}.
3.12 −50 kJ mol^{-1}.
3.13 (a) ΔS(gas) = +2.9 J K^{-1}, ΔS(surroundings) = −2.9 J K^{-1}, ΔS(total) = 0;
(b) ΔS(gas) = +2.9 J K^{-1}, ΔS(surroundings) = 0, ΔS(total) = +2.9 J K^{-1}.
(c) ΔS(gas) = 0, ΔS(surroundings) = 0, ΔS(total) = 0.
3.14 817.90 kJ mol^{-1}.
3.15 (a) 0.11, (b) 0.38.
3.16 −3.8 J.
3.17 −36.5 J K^{-1}.
3.18 10 kJ.
3.19 +7.3 kJ mol^{-1}.
3.20 −0.55 kJ mol^{-1}.
3.21 +10 kJ.
3.22 +11 kJ mol^{-1}.

Chapter 4

4.1 304 K = 31°C.
4.2 $\Delta_{fus} S = +45.23$ J K^{-1} mol^{-1}, $\Delta_{fus} H = +16$ kJ mol^{-1}.
4.3 +20.80 kJ mol^{-1}.
4.4 (a) +34.08 kJ mol^{-1}; (b) $T_b = 350.5$ K.

4.5 281.8 K = 8.7°C.

4.6 25 g s^{-1}.

4.7 (a) 1.7 kg; (b) 31 kg; (c) 1.4 g.

4.8 (a) 49.1 kJ mol^{-1}; (b) 215°C; (c) 101 J K^{-1} mol^{-1}.

4.9 272.80 K.

4.10 0.07630 ≈ 7.6 per cent.

Chapter 5

5.1 886.8 cm^3.

5.2 56.3 cm^3 mol^{-1}.

5.3 6.4×10^3 kPa.

5.4 1.3×10^2 kPa.

5.5 82 g mol^{-1}.

5.6 381 g mol^{-1}.

5.7 −0.09°C.

5.8 $\Delta_{mix}G = -0.35$ kJ, $\Delta_{mix}S = +1.2$ J K^{-1}.

5.9 +4.7 J K^{-1} mol^{-1}.

5.10 (a) 1:1; (b) 0.8600.

5.11 (a) 3.4 mmol kg^{-1}, (b) 34 mmol kg^{-1}.

5.12 0.17 mol kg^{-1}.

5.13 24 g.

5.14 87 kg mol^{-1}.

5.15 $a_A = 0.833$, $\gamma_A = 0.93$, $a_B = 0.125$, $\gamma_B = 1.25$.

5.16 $p(CCl_4) = 32.2$ Torr, $p(Br_2) = 6.1$ Torr, $p(\text{total}) = 38.3$ Torr, $y(CCl_4) = 0.841$, $y(Br_2) = 0.16$.

5.17 $a_A = 0.499$, $a_M = 0.668$, $\gamma_A = 1.25$, $\gamma_M = 1.11$.

5.18 0.90.

5.19 (a) 2.73 g, (b) 2.92 g.

5.20 0.56.

5.21 2.01.

Chapter 6

6.1 $x_A = 0.920$, $x_B = 0.080$, $y_A = 0.968$, $y_B = 0.032$.

6.2 $p = 58.7$ kPa, $x_A = 0.268$, $x_B = 0.732$.

6.3 (a) Yes; (b) $y_A = 0.830$, $y_B = 0.170$.

6.4 (a) $p_{(\text{total})} = 20.5$ kPa; (b) $y_{DE} = 0.67$, $y_{DP} = 0.33$.

6.5 (a) $y_M = 0.36$; (b) $y_M = 0.82$.

6.6 (a) $C = 2$; (b) $C = 2$.

6.7 $P = 2$, $C = 2$.

6.8 (a) $C = 2$, $P = 3$; (b) variance = 1.

6.11 Incongruent melting point (T_1) = 400°C, composition of the eutectic (x_e) ≈ 0.30, melting point (T_2) ≈ 200°C.

6.13 (a) 80 per cent;
(b) Compound Ag_3Sn decomposes;
(c) 80 per cent.

6.14 (b) 620 Torr; (c) 490 Torr; (d) $x(\text{Hex}) = 0.50$, $y(\text{Hex}) \approx 0.72$; (e) $y(\text{Hex}) = 0.50$, $x(\text{Hex}) = 0.30$; (f) $n_{vap} \approx 1.7$ mol, $n_{liq} \approx 0.3$ mol.

Chapter 7

7.1 (a) $K = 2.85 \times 10^{-6}$; (b) $\Delta_r G^\ominus = +240$ kJ mol^{-1}; (c) $\Delta_r G = 0$ (the system is at equilibrium).

7.2 (a) $K = 0.1411$; (b) $\Delta_r G^\ominus = +4.855$ kJ mol^{-1}; (c) K (100°C) = 14.556.

7.3 (a) $\Delta_r G^\ominus = -68.26$ kJ mol^{-1}, $K = 9.2 \times 10^{11}$;
(b) K (400°C) = 1.3×10^9, $\Delta_r G^\ominus$ (400°C) = −69.7 kJ mol^{-1}.

7.4 (a) Mole fractions: A: 0.087, B: 0.370, C: 0.196, D: 0.348, Total: 1.001;
(b) $K_x = 0.33$; (c) $p = 0.33$; (d) $\Delta_r G^\ominus = +2.8 \times 10^3$ J mol^{-1}.

7.5 $T_2 = 1500$ K.

7.6 $\Delta_r H^\ominus = +2.77$ kJ mol^{-1}, $\Delta_r S^\ominus = -16.5$ J K^{-1} mol^{-1}.

7.7 $\Delta G^\ominus = +12.3$ kJ mol^{-1}.

7.8 50 per cent.

7.9 $\dfrac{n_B}{n} = x_B = 0.904$, $x_I = 0.096$.

7.10 (a) $\Delta_r H^\ominus = +53$ kJ mol^{-1}; (b) $\Delta_r H^\ominus = -53$ kJ mol^{-1}.

7.11 $\Delta_r G = -14.38$ kJ mol^{-1}, spontaneous direction of reaction is towards products.

7.12 $T = 1110$ K.

7.13 $\Delta_f G^\ominus = -1108$ kJ mol^{-1}.

7.14

	$E^\ominus$
(a) R: $2Ag^+(aq) + 2e^- \rightarrow 2Ag(s)$	+0.80 V
L: $Zn^+(aq) + 2e^- \rightarrow Zn(s)$	−0.76 V
Overall (R – L): $2Ag^+(aq) + Zns \rightarrow 2Ag(s) + Zn^{2+}(aq)$	+1.56 V
(b) R: $2H^+(aq) + 2e^- \rightarrow H_2(g)$	0
L: $Cd^2 + (aq) + 2e^- \rightarrow Cd(s)$	−0.40 V
Overall (R – L): $Cd(s) + 2H^+(aq) \rightarrow Cd^{2+}(aq) + H_2(g)$	+0.40 V
(c) R: $Cr^{3+}(aq) + 3e^- \rightarrow Cr(s)$	−0.74 V
L: $3[Fe(CN)_6]^{3-}(aq) + 3e^- \rightarrow 3[Fe(CN)_6]^{4-}(aq)$	+0.36 V
Overall (R – L): $Cr^{3+}(aq) + 3[Fe(CN)_6]^{4-}(aq) \rightarrow Cr(s) + 3[Fe(CN)_6]^{3-}(aq)$	−1.10 V

7.15

	$E^\ominus$
(a) R: $2Cu^{2+}(aq) + 2e^- \rightarrow Cu(s)$	+0.34 V
L: $Zn^{2+}(aq) + 2e^- \rightarrow Zn(s)$	−0.76 V
Hence the cell is	
$Zn(s)\vert ZnSO_4(aq)\vert\vert CuSO_4(aq)\vert Cu(s)$	+1.10 V
(b) R: $AgCl(s) + e^- \rightarrow Ag(s) + Cl^-(aq)$	+0.22 V
L: $H^+(aq) + e^- \rightarrow \frac{1}{2}H_2(g)$	0
and the cell is	
$Pt\vert H_2(g)\vert H^+(aq)\vert AgCl(s)\vert Ag(s)$	
or $Pt\vert H_2(g)\vert HCl(aq)\vert AgCl(s)\vert Ag(s)$	+0.22 V
(c) R: $O_2(g) + 4H^+(aq) + 4e^- \rightarrow 2H_2O(l)$	+1.23 V
L: $4H^+(aq) + 4e^- \rightarrow 2H_2(g)$	0
and the cell is	
$Pt\vert H_2(g)\vert H^+(aq), H_2O(l)\vert O_2(g)\vert Pt$	+1.23 V

7.16 $E = -0.62$ V.

7.17 (a) $K = 6.5 \times 10^9$; (b) $K = 1.5 \times 10^{12}$.

7.18 (a) 9.2×10^{-9} M, (b) 8.5×10^{-17} M.

Chapter 8

8.1 $v = 0.0242$ m s^{-1}.

8.2 m = 332 pm.

8.3 $\Delta q = 0.70$ nm.

8.4 (a) $E = 3.31 \times 10^{-19}$ J, E (per mole) = 199 kJ mol^{-1};
(b) $E = 3.61 \times 10^{-19}$ J, E (per mole) = 218 kJ mol^{-1};
(c) $E = 4.97 \times 10^{-19}$ J, E (per mole) = 299 kJ mol^{-1}.

8.5 600 nm: $v = 0.66$ m s^{-1}, 550 nm: $v = 0.72$ m s^{-1}, 400 nm: $v = 0.99$ m s^{-1}.

8.6 $v = 21$ m s^{-1}.

8.7 (a) $N = 2.8 \times 10^{18}$ s^{-1}; (b) $N = 2.8 \times 10^{20}$ s^{-1}.

8.8 (a) no ejection occurs; (b) $E_K = 3.19 \times 10^{-19}$ J, $v = 837$ km s^{-1}.

8.9 (a) $\Delta E = 7 \times 10^{-19}$ J, or 400 kJ mol^{-1};
(b) $\Delta E = 7 \times 10^{-20}$ J, or 40 kJ mol^{-1};
(c) $\Delta E = 7 \times 10^{-34}$ J, or 4×10^{-13} kJ mol^{-1}.

8.10 (a) $\lambda = 6.6\times10^{-29}$ m; (b) $\lambda = 6.6\times10^{-36}$ m; (c) $\lambda = 99.7$ pm.
8.12 $\Delta v_{min} = 1.1\times10^{28}$ m s^{-1}, $\Delta q_{min} = 1\times10^{-27}$ m.
8.13 $I = 1.12\times10^{15}$ J.
8.14 (a) $-\dfrac{1}{x^2}$; (b) $2x$.

Chapter 9

9.1 (a) 1.81×10^{-19} J, 1.13 eV, 9100 cm^{-1}, 109 kJ mol^{-1};
(b) 6.6×10^{-19} J, 4.1 eV, 33 000 cm^{-1}, 400 kJ mol^{-1}.
9.2 $P = 0.04$; (b) $P \approx 0$.
9.3 $p = 0, p^2 = \dfrac{h^2}{4L^2}$.
9.4 $L = \dfrac{h}{8^{1/2}m_e c} = \dfrac{\lambda_c}{8^{1/2}}$.
9.5 $x = \dfrac{L}{6}, \dfrac{L}{2}$ and $\dfrac{5L}{6}$.
9.6 3.
9.7 23 per cent.
9.8 $E_0 = 4.30\times10^{-21}$ J.
9.9 $k = 278$ N m^{-1}.
9.10 $\lambda = 2.63$ µm.
9.11 $\lambda = 3.72$ µm.
9.12 (a) $\Delta E = 3.3\times10^{-34}$ J; (b) $\Delta E = 3.3\times10^{-33}$ J.
9.14 $x = \pm0.525\alpha$ or $\pm1.65\alpha$.
9.15 $E_0 = 5.61\times10^{-21}$ J.
9.16 $N = \left(\dfrac{1}{2\pi}\right)^{1/2}$.
9.17 Magnitude = 1.49×10^{-34} J s; possible projections = 0 or $\pm\eta$, that is, 0 or 1.05×10^{-34} J s.

Chapter 10

10.1 $I = 14.0$ eV.
10.2 $r = 4a_0, r = 0$.
10.3 101 pm and 376 pm.
10.4 $N = \dfrac{2}{a_0^{3/2}}$
10.5 $\langle V\rangle = 2E_{1s}, \langle T\rangle = -E_{1s}$.
10.6 $r^* = 5.24\dfrac{a_0}{Z}$.
10.7 $r = \dfrac{2a_0}{Z}$.
10.8 (a) angular momentum = 0, angular nodes = 0, radial nodes = 0;
(b) angular momentum = 0; angular nodes = 0, radial nodes = 2;
(c) angular momentum = $\sqrt{6}\hbar$, angular nodes = 2, radial nodes = 0.
10.9 (a) $j = \frac{5}{2}, \frac{3}{2}$ (b) $j = \frac{7}{2}, \frac{5}{2}$.
10.10 $l = 1$ in each case.
10.11 (a) $g = 1$; (b) $g = 9$; (c) $g = 25$.
10.12 $L = 2, S = 0, J = 2$.
10.13 $r = 0.35a_0$.
10.14 (a) forbidden; (b) allowed; (c) allowed.
10.15 (a) [Ar]$3d^8$; (b) for $S = 1$, $M_s = -1,0,+1$; for $S = 0$, $M_s = 0$.
10.16 (a) $S = 1,0$; multiplicities = 3,1, respectively. (b) $S' = 1,0$; then $S = \frac{3}{2}, \frac{1}{2}$ [from 1], and $\frac{1}{3}$ [from 0]; multiplicities = 4,2,2 respectively.
10.17 $^3D_3, {}^3D_2, {}^3D_1, {}^1D_2$, the $^3D_3$ set of terms are the lower in energy.
10.18 (a) $J = 0$, only 1 state; (b) $J = \frac{3}{2}, \frac{1}{2}$, with 4, 2 states respectively;
(c) $J = 2,1,0$, with 5, 3, 1 states respectively.
10.19 (a) $^2S_{1/2}$; (b) $^2P_{3/2}$ and $^2P_{1/2}$.

Chapter 11

11.1 (a) $1\sigma^2$, $b = 1$; (b) $1\sigma^2 2\sigma^{*2}$, $b = 0$; (c) $1\sigma^2 2\sigma^{*2} 1\pi^4$, $b = 2$.
11.2 (a) $1\sigma^2 2\sigma^{*2} 1\pi^4 3\sigma^2$; (b) $1\sigma^2 2\sigma^{*2} 1\pi^4 3\sigma^2 2\pi^{*1}$; (c) $1\sigma^2 2\sigma^{*2} 1\pi^4 3\sigma^2$.
11.3 C_2.
11.4 Since the bond order is increased when XeF^+ is formed from XeF (because an electron is removed from an antibonding orbital), XeF^+ will have a shorter bond length than XeF.
11.5 The electron configurations are used to determine the bond orders. Larger bond order corresponds qualitatively to shorter bond length. The bond orders of NO and N_2 are 2.5 and 3, respectively hence N_2 should have the shorter bond length.
11.8 Appropriate. $E_{trial} = \dfrac{5h^2}{mL^2}$, the trial energy is greater than the exact answer by a factor of 1.013.
11.9 $E_{trial} = \dfrac{3a\hbar}{2\mu} - \dfrac{e^2}{\varepsilon_0}\left(\dfrac{a}{2\pi^3}\right)^{1/2}$.
11.10 2.2×10^{-19} J.
11.12 (a) $\begin{vmatrix} \alpha-E & \beta & 0 \\ \beta & \alpha-E & \beta \\ 0 & \beta & \alpha-E \end{vmatrix} = 0$; (b) $\begin{vmatrix} \alpha-E & \beta & \beta \\ \beta & \alpha-E & \beta \\ \beta & \beta & \alpha-E \end{vmatrix} = 0$.

Chapter 12

12.1 The elements, other than the identity E, are a C_3 axis and three vertical mirror planes σ_V. The symmetry axis passes through the C—Cl nuclei. The mirror planes are defined by the three ClCH planes.
12.2 (a) pyridine, (b) nitroethane.
12.3 The integral is necessarily zero.
12.6 D_{2h} contains i and C_{3h} contains σ_h; therefore molecules belonging to these point groups cannot be chiral.
12.7

	First operation	E	C_2	C_2'	C_2''
Second operation	E	E	C_2	C_2'	C_2''
	C_2	C_2	E	C_2''	C_2'
	C_2'	C_2'	C_2''	E	C_2
	C_2''	C_2''	C_2'	C_2	E

12.8 (a) R_3; (b) C_{2V}; (c) D_{3h}; (d) $D_{\infty h}$.
12.9 (a) C_{2V}; (b) $C_{\infty V}$; (c) C_{3V}; (d) D_{2h}; (e) C_{2V}; (f) C_{2h}.
12.10 (a) C_{2V}; (b) C_{2h}.
12.11 (a) NO_2, N_2O, $CHCl_3$ and *cis*-CHBr=CHBr; (b) None are chiral.
12.12 NO_2: no orbitals, SO_2: d_{xy}.
12.13 $B_1(x)$, $B_2(y)$, $A_1(z)$.
12.14 (a) either E_{1U} or A_{2U}; (b) B_{3U} (x-polarized), B_{2U} (y-polarized) B_{1U} (z-polarized).

Chapter 13

13.1 (a) 0.0469 J m^{-3} s; (b) 1.33×10^{-3} J m^{-3} s; (c) 4.50×10^{-16} J m^{-3} s.
13.2 $\lambda_{obs} = 0.999\,999\,925\times660$ nm.

13.3 (a) 53 cm^{-1}; (b) 0.53 cm^{-1}.
13.4 (a) 53 ps; (b) 5 ps.
13.5 $\nu = 4.09 \times 10^{11}$ Hz.
13.6 (a) $\tilde{\nu} = 2.642 \times 10^{-47}$ kg m^2; (b) 127.4 pm.
13.7 $I = 4.442 \times 10^{-47}$ kg m^2, $R = 165.9$ pm.
13.8 $R = 232.1$ pm.
13.9 $R_{HC} = 106.5$ pm, $R_{CN} = 115.6$ pm.
13.10 20 475 cm^{-1}.
13.11 198.9 pm.
13.12 (b) HCl; (d) CH_3Cl; (e) CH_2Cl_2 *(cis)*.
13.13 (a) H_2; (b) HCl; (d) CH_3Cl.
13.14 $k = 1.6 \times 10^2$ N m^{-1}.
13.15 1.089 per cent.
13.16 $k = 327.8$ N m^{-1}.
13.17 (a) 0.067; (b) 0.20.
13.18 HF: $k = 967.0$, HCl: $k = 515.6$, HBr: $k = 411.8$, HI: $k = 314.2$.
13.19 $\tilde{\nu} = 1580.38$ cm^{-1}, $x_e = 7.644 \times 10^{-3}$.
13.20 $D_0 = 41.5 \times 10^3$ cm^{-1} = 5.15 eV
13.21 $\tilde{\nu}_R(2) = 2699.77$ cm^{-1}
13.22 (b) HCl; (c) CO_2; (d) H_2O.
13.23 (a) 3; (b) 6; (c) 12.
13.24 (a) Nonlinear: all modes both infrared and Raman active.
(b) Linear: the symmetric stretch is infrared inactive but Raman active. The antisymmetric stretch is infrared active and (by the exclusive rule) Raman inactive. The two bending modes are infrared active and therefore Raman inactive.
13.25 (a) Raman active.
13.26 $4A_1 + A_2 + 2B_1 + 2B_2$.

Chapter 14

14.1 $S = \frac{1}{2}$, total orbital angular momentum = zero.
14.2 80 per cent.
14.3 $\varepsilon = 6.28 \times 10^3$ dm^3 mol^{-1} cm^{-1}.
14.4 1.5 mmol dm^{-3}.
14.5 5.44×10^7 dm^3 mol^{-1} cm^{-2}.
14.6 192 nm = 2,3-dimethyl-2-butene, 243 = 2,5-dimethyl-2, 4-hexadiene.
14.7 $\varepsilon = 450$ dm^3 mol^{-1} cm^{-1}.
14.8 159 dm^3 mol^{-1} cm^{-1}, 23 per cent.
14.9 (a) 0.9 m; (b) 3 m.
14.10 (a) 5×10^7 dm^3 mol^{-1} cm^{-2}; (b) 25×10^5 dm^3 mol^{-1} cm^{-2}.

Chapter 15

15.1 $\nu = 600$ MHz.
15.2 $E_{m_I} = -1.625 \times 10^{-26}$ J $\times m_I$.
15.3 154 MHz.
15.4 Proton (a) $\Delta E = 3.98 \times 10^{-25}$ J; (b) $\Delta E = 6.11 \times 10^{-26}$ J.
15.5 $\Delta E = 6.116 \times 10^{-26}$ J.
15.6

B_0/T	(a)^{1}H	(b)^{2}H	(c)^{13}C
g_I	5.5857	0.85745	1.4046
(i) 250 MHz	5.87	38.3	23.4
(ii) 500 MHz	11.7	76.6	46.8

15.7 (a) 1×10^{-6}; (b) 5.1×10^{-6}; (c) 3.4×10^{-5}.
15.8 $\dfrac{\delta N(800\text{ MHz})}{\delta N(60\text{ MHz})} = 13$.
15.9 (a) 11 μT; (b) 110 μT.
15.11 6.7×10^2 s^{-1}.
15.14 (a) The H nuclei are both chemically and magnetically equivalent.
(b) The H nuclei are chemically but not magnetically equivalent.
15.15 5.9×10^{-4} T, 20 μs
15.16 2×10^2 T, 10 mT.
15.17 $g = 2.0022$.
15.18 $a = 2.3$ mT; $g = 2.0025$.
15.19 Four lines of equal intensity at the fields: 330.2 mT, 332.2 mT, 332.8 mT, 334.8 mT.
15.21 (a) 331.9 mT; (b) 1.201 T.
15.22 $I = \frac{3}{2}$.

Chapter 16

16.1 1.
16.2 (a) 2.57×10^{27}; 7.26×10^{27}.
16.3 $\dfrac{q(D_2)}{q(H_2)} = 2.83$.
16.4 3.156.
16.5 $E = 2.45$ kJ mol^{-1}.
16.6 $T = 354$ K.
16.7 $q = 1 + e^{-2\mu_B\beta B}$, $\langle\varepsilon\rangle + \dfrac{2\mu_B B e^{-2\mu_B\beta B}}{1 + e^{-2\mu_B\beta B}}$; (a) 0.71; (b) 0.996.
16.8 (a) (i) 5×10^{-5}; (ii) 0.4; (iii) 0.905;
(b) $q = 1.4$;
(c) $E = 22$ J mol^{-1};
(d) $C_v = 1.6$ J K^{-1} mol^{-1};
(e) $S = 4.8$ J K^{-1} mol^{-1}.
16.9 $T = 4303$ K.
16.10 (a) $T = 138$ J K^{-1} mol^{-1}; (b) $T = 146$ J K^{-1} mol^{-1}.
16.11 $S_m = 5.18$ J K^{-1} mol^{-1}.
16.12 (a), (b), (d).

Chapter 17

17.1 (a) I_2: $\frac{7}{2}R$ [experimental = 3.4R];
(b) CH_4: 3R [experimental = 3.2R];
(c) C_2H_6: 7R [experimental = 8.8R].
17.2 (a) NH_3: With vibrations: 1.11, Without vibrations: 1.33; Experimental: 1.31.
(b) CH_4: With vibrations: 1.08, Without vibrations: 1.33; Experimental: 1.31.
17.3 (a) 25°C: 19.6; (b) 250°C: 34.3.
17.4 (a) 1; (b) 2; (c) 2; (d) 12; (e) 3.
17.5 $q^K = 43.1$, $\theta_R = 22.36$ K, $T = 8.79$ K.
17.6 $S_m^R = 43.76$ J K^{-1} mol^{-1}.
17.7 (a) At 298 K, $q^R = 36.95$. At 500 K, $q^R = 80.08$.
(b) At 298 K, $q = 36.7$. At 500 K, $q = 79.7$.
17.8 $q = 72.5$.
17.9 (a) At 298 K: 14.93 J K^{-1} mol^{-1}.
(b) At 500 K: 25.65 J K^{-1} mol^{-1}.
17.10 −13.8 kJ mol^{-1}, −0.20 kJ mol^{-1}.
17.11 (a) At 500 K: 0.236 R. (b) At 900 K: 0.193 R.

17.12 $11.5\ J\ K^{-1}\ mol^{-1}$.

17.13 (a) $S_m = 9.13\ J\ K^{-1}\ mol^{-1}$; (b) $S_m = 13.4\ J\ K^{-1}\ mol^{-1}$; (c) $S_m = 14.9\ J\ K^{-1}\ mol^{-1}$;

17.14 $K = 3.70 \times 10^{-3}$.

Chapter 18

18.1 O_3, H_2O_2.

18.2 *p*-xylene: $\mu = 0$, *o*-xylene: $\mu = 0.7$ D, *m*-xylene: $\mu = 0.4$ D.

18.3 $\mu = 37$ D, $\theta = 11.7°$.

18.4 $\mu = 5.5 \times 10^{-30}$ C m or 1.7 D, $\alpha' = 9.1 \times 10^{-24}\ cm^3$.

18.5 $\varepsilon_r = 4.8$.

18.6 $\mu = 4.9\ \mu$ D.

18.7 $\alpha = 1.42 \times 10^{-39}\ J^{-1}\ C^2\ m^2$, $\alpha' = 1.28 \times 10^{-23}\ cm^3$.

18.8 $n_r = 1.34$.

18.9 $\varepsilon_r = 18$.

18.10 $p = 2.6$ kPa.

18.11 $\gamma = 7.28 \times 10^{-2}\ N\ m^{-1}$.

18.12 $p_{in} - p_{out} = 7.28 \times 10^5$ Pa.

Chapter 19

19.1 Equal amounts imply equal numbers of molecules. $\bar{M}_n = 70\ kg\ mol^{-1}$, $\bar{M}_w = 71\ kg\ mol^{-1}$.

19.2 $R_g = 1.4 \times 10^4$.

19.3 (a) $\bar{M}_n = 18\ kg\ mol^{-1}$; (b) $\bar{M}_w = 20\ kg\ mol^{-1}$.

19.4 $\eta(H_2O)$: $\tau = 2.4 \times 10^{-8}$ s, $\eta(CCl_4)$: $\tau = 1.4 \times 10^{-11}$ s.

19.5 100.

19.6 $\bar{M} = 63\ kg\ mol^{-1}$.

19.7 $s = 7.3 \times 10^{-4}\ m\ s^{-1}$.

19.8 $\bar{M} = 31\ kg\ mol^{-1}$.

19.9 $M = 3.4 \times 10^3\ kg\ mol^{-1}$.

19.10 $a = 4.3 \times 10^5$ g.

19.11 $R_{rms} = 24$ nm.

19.12 $R_c = 3.08 \times 10^{-6}$ m, $R_{rms} = 3.08 \times 10^{-8}$ m.

Chapter 20

20.1 $(1, \frac{1}{2}, 0)$, $(1, 0, \frac{1}{2})$, and $(\frac{1}{2}, \frac{1}{2}, \frac{1}{2})$.

20.2 (3 2 3) and (1 1 0).

20.3 $d_{111} = 249$ pm, $d_{211} = 176$ pm, $d_{100} = 432$ pm.

20.4 $\lambda = 70.7$ pm.

20.5 $2\theta_{110} = 16°$, $2\theta_{200} = 23°$, $2\theta_{211} = 28°$.

20.6 $D = 0.215$ cm.

20.7 $V = 3.96 \times 10^{-28}\ m^3$

20.8 $N = 4$, $\rho = 4.01\ g\ cm^{-3}$.

20.9 $d_{411} = 190$ pm.

20.10

θ	$a^2\left(\frac{2 \sin \theta}{\lambda}\right)^2$	$h^2 + k^2 + l^2$	(hkl)	a/pm
19.4	3.04	3	(111)	402
22.5	4.03	4	(200)	402
32.6	7.99	8	(220)	404
39.4	11.09	11	(311)	402

20.11 $\theta_{100} = 8.17°$, $\theta_{010} = 4.82°$, $\theta_{111} = 11.75°$.

20.12 face-centred cubic.

20.13 $F_{hkl} = f$.

20.14 $f = 0.9069$.

20.15 $\frac{r}{R} = 0.414$.

20.16 (a) 58.0 pm; (b) 102 pm.

20.17 0.340.

20.18 expansion.

20.20 $7.9\ km\ s^{-1}$.

20.23 0.010 or about 1% elongation.

20.24 $9.3 \times 10^{-4}\ cm^3$.

20.25 n-type.

20.26 3.54 eV.

20.27 3.

20.28 $\chi_m = -6.4 \times 10^{-5}\ cm^3\ mol^{-1}$.

20.29 2.

20.30 4.326 effective unpaired spins, theoretical number = 5.

20.31 $\chi_m = +1.6 \times 10^{-8}\ m^3\ mol^{-1}$.

20.32 222 T.

Chapter 21

21.1 (a) 9.975; (b) 1.

21.2 (a) 72 K; (b) $9.5 \times 10^2\ m\ s^{-1}$; (c) No.

21.3 $p = 0.081$ Pa.

21.4 $\lambda = 9.7 \times 10^{-1}$ m.

21.5 (a) $z = 5 \times 10^{10}\ s^{-1}$; (b) $z = 5 \times 10^9\ s^{-1}$; (c) $z = 5 \times 10^3\ s^{-1}$.

21.6 (a) 6.7 nm; (b) 67 nm; (c) 6.7 cm.

21.7 9.06×10^{-3}.

21.8 $N = 1.9 \times 10^{20}$.

21.9 $\Delta m = 104$ mg.

21.10 $M' = 43\ g\ mol^{-1}$.

21.11 $t = 1.1 \times 10^5$ s = 30 h

21.12 $J_z = 4.1 \times 10^{-2}\ J\ m^{-2}\ s^{-1}$.

21.13 $\sigma = 5.6 \times 10^{-20}\ m^2$ or $0.056\ nm^2$.

21.14 17 W.

21.15 $\sigma = 1.42 \times 10^{-19}\ m^2$ or $0.142\ nm^2$.

21.16 205 kPa.

21.17 (a) $T = 273$ K: 130 μP; (b) $T = 298$ K: 130 μP; (c) $T = 1000$ K: 240 μP.

21.18 (a) For Ar, $\kappa = 5.4\ m\ J\ K^{-1}\ m^{-1}\ s^{-1}$, rate of flow = 8.1 m W;
(b) For He, $\kappa = 29\ m\ J\ K^{-1}\ m^{-1}\ s^{-1}$, rate of flow = 44 m W.

21.19 $\eta(CO_2) = 1.38$ μP, $d = 390$ pm.

21.20 $\kappa = 5.4 \times 10^{-3}\ J\ K^{-1}\ m^{-1}\ s^{-1}$.

21.21 (a) At 1 Pa: $D = 1.1\ m^2\ s^{-1}$, $J = 4.4 \times 10^2\ mol\ m^{-2}\ s^{-1}$;
(b) At 100 kPa: $D = 1.1 \times 10^{-5}\ m^2\ s^{-1}$, $J = 4.4 \times 10^{-3}\ mol\ m^{-2}\ s^{-1}$;
(c) At 10 MPa: $D = 1.1 \times 10^{-7}\ m^2\ s^{-1}$, $J = 4.4 \times 10^{-5}\ mol\ m^{-2}\ s^{-1}$.

21.22 $\lambda = 7.63 \times 10^{-3}\ S\ m^2\ mol^{-1}$.

21.23 $s = 3.47 \times 10^{-4}\ m\ s^{-1}$, or $347\ \mu m\ s^{-1}$.

21.24 0.331.

21.25 $\Lambda_m^\circ(AgCl) = 13.83\ mS\ m^2\ mol^{-1}$.

21.26 $u(Li^+) = 4.01 \times 10^{-8}\ m^2\ V^{-1}\ s^{-1}$, $u(Na^+) = 5.19 \times 10^{-8}\ m^2\ V^{-1}\ s^{-1}$, $u(K^+) = 7.62 \times 10^{-6}\ m^2\ V^{-1}\ s^{-1}$.

21.27 $D = 1.90 \times 10^{-9}\ m^2\ s^{-1}$.

21.28 $t = 1.3 \times 10^3$ s.

21.29 $a = 420$ pm.

21.30 $s = 27$ ps.

21.31 For iodine in benzene: $\langle r^2\rangle^{1/2} = 113\ \mu m$;

For sucrose in water: $\sqrt{\langle r^2\rangle} = 5.594 \times 10^{-5}$ m.

Chapter 22

22.1 Rates of formation: $C = 3v = 3.0$ mol dm^{-3} s^{-1}; $D = v = 1.0$ mol dm^{-3} s^{-1}; $A = v = 1.0$ mol dm^{-3} s^{-1}; $B = 2v = 2.0$ mol dm^{-3} s^{-1}.

22.2 $v = 0.50$ mol dm^{-3} s^{-1}; Rates of formation: $D = 3v = 1.5$ mol dm^{-3} s^{-1}; $A = 2v = 1.0$ mol dm^{-3} s^{-1}; $B = v = 0.50$ mol dm^{-3} s^{-1}.

22.3 k: dm^3 mol^{-1} s^{-1};

(a) Rate of formation of $A = v = k[A][B]$;

(b) Rate of consumption of $C = 3v = 3k[A][B]$.

22.4 $v = \dfrac{1}{2}k[A][B][C]$; $[k] = $ dm^6 mol^{-2} s^{-1}.

22.5 Second-order.

22.6 $n = 2$.

22.7 $t_{1/2} = 1.03 \times 10^4$ s; (a) 4.997 Torr; (b) 480 Torr.

22.8 (a) $k = 4.1 \times 10^{-3}$ dm^3 mol^{-1} s^{-1},

(b) $t_{1/2}(B) = 7.4 \times 10^3$ s, $t_{1/2}(A) = 2.6 \times 10^3$ s;

22.9 (a) Second-order: $[k] = $ dm^3 mol^{-1} s^{-1}, Third-order: $[k] = $ dm^6 mol^{-1} s^{-1};

(b) Second-order: $[k] = $ kPa^{-1} s^{-1}, Third-order: $[k] = $ kPa^{-2} s^{-1}.

22.10 (a) NaOH = 0.045 mol dm^{-3}, $CH_3COOC_2H_5 = 0.095$ mol dm^{-3};

(b) NaOH = 0.001 mol dm^{-3}, $CH_3COOC_2H_5 = 0.051$ mol dm^{-3};

22.11 1.24×10^5 s.

22.13 $k = 7.1 \times 10^5$ s^{-1}, $\tau = 7.61$ ns.

22.14 $E_a = 64.9$ kJ mol^{-1}, $A = 4.32 \times 10^8$ mol dm^{-3} s^{-1}.

22.15 0.156.

22.16 1.9 MPa^{-1} s^{-1}.

Chapter 23

23.1 $\text{Rate} = \dfrac{k_1 k_2 [O_3]^2}{k_1[O_2] + k_2[O_3]}$

23.3 0.16 kPa and 4.0 kPa.

23.4 $[A^-] = \dfrac{k_1[AH][B]}{k_2[BH^+] + k_3[AH]}, \quad \dfrac{d[P]}{dt} = \dfrac{k_1 k_3 [AH]^2[B]}{k_2[BH^+] + k_3[AH]}$

23.6 $v_{max} = 1.52 \times 10^{-3}$ mol dm^{-3} s^{-1}.

23.7 Number absorbed $= 3.3 \times 10^{18}$.

23.8 $\phi = 0.518$.

Chapter 24

24.1 $z = 9.5 \times 10^9$ s^{-1}, $Z_{AA} = 1.2 \times 10^{35}$ m^{-3} s^{-1}, 1.7 per cent.

24.2 (a) (i) 0.018, (ii) 0.30; (b) (i) 3.9×10^{-18}, (ii) 6.0×10^{-6}.

24.3 (a) 13 per cent at 300 K, 1.2 per cent at 1000 K;

(b) 130 per cent at 300 K, 12 per cent at 1000 K.

24.4 $k_2 = 1.7 \times 10^{-2}$ dm^3 mol^{-1} s.

24.5 $k_d = 3 \times 10^{10}$ dm^3 mol^{-1} s^{-1}.

24.6 (a) $k_d = 6.61 \times 10^6$ m^3 mol^{-1} s^{-1}, (b) $k_d = 3.0 \times 10^7$ m^3 mol^{-1} s^{-1}.

24.7 $k_d = 7.4 \times 10^9$ dm^3 mol^{-1} s^{-1}, $t_{1/2} = 1.4 \times 10^{-7}$ s.

24.8 $P \approx 1.2 \times 10^{-3}$.

24.9 1.9×10^8 mol dm^{-3} s^{-1}.

24.10 $\Delta^{\ddagger}H = 69.7$ kJ mol^{-1}, $\Delta^{\ddagger}S = -25$ J K^{-1} mol^{-1}.

24.11 $\Delta^{\ddagger}G = +71.9$ kJ mol^{-1}.

24.12 $\Delta^{\ddagger}S = -96.6$ J K^{-1} mol^{-1}.

24.13 $\Delta^{\ddagger}S = -76$ J K^{-1} mol^{-1}.

24.14 (a) $\Delta^{\ddagger}S = -45.8$ J K^{-1} mol^{-1}; (b) $\Delta^{\ddagger}H = +5.0$ kJ mol^{-1}; (c) $\Delta^{\ddagger}G = +18.7$ kJ mol^{-1}.

24.15 $k_2^{\circ} = 20.9$ dm^6 mol^{-2} min^{-1}.

24.16 $\lambda = 1.9 \times 10^{-19}$ J or about 1.2 eV.

24.17 $\beta = 12$ nm^{-1}.

Chapter 25

25.1

	H_2	C_3H_8
M/(g mol^{-1})	2.02	44.09
Z_W (m^{-2} s^{-1})		
(i) 100 Pa	1.07×10^{25}	2.35×10^{24}
(ii) 10^{-7} Torr	1.4×10^{18}	3.1×10^{17}

25.2 $p = 1.3 \times 10^4$ Pa.

25.3 rate per Cu atom $= 3.4 \times 10^5$ s^{-1}.

25.4 $A = 12.7$ m^2.

25.5 $V_{mon} = 20.5$ cm^3.

25.6 Chemisorption, $t_{1/2} = 50$ s.

25.7 $E_d = 610$ kJ mol^{-1}, $\tau_0 = 0.113 \times 10^{-12}$ s, $A = 6.15 \times 10^{12}$ s^{-1}.

25.8 (a) $p = 0.12$ kPa; (b) $p = 22$ kPa.

25.9 $\theta_1 = 0.83$, $\theta_2 = 0.36$.

25.10 (a) $E_d = 4 \times 10^{-11}$ s, $t_{1/2} = 6 \times 10^{-13}$ s; (b) $E_d = 2 \times 10^{13}$ s; $t_{1/2} = 7 \times 10^{-6}$ s.

25.11 $p' = 15$ kPa.

25.12 Zeroth-order on gold; First-order on platinum.

25.13 $\Delta_{ad}H^{\ominus} \approx -13$ kJ mol^{-1}.

25.14 $E_d = 650$ kJ mol^{-1}, (a) At 298 K: $t = 1.1 \times 10^{97}$ min, (b) At 3000 K: $t = 2.9 \times 10^{-6}$ min.

25.15 2.4×10^8 V m^{-1}.

25.16 138 mV.

25.17 $j_0 = 2.8$ mA cm^{-2}.

25.18 The anodic current density increases roughly by a factor of 50 with a corresponding increase in O_2 evolution.

25.19 (a) $j = 17 \times 10^{-4}$ A cm^{-2}; (b) $j = 17 \times 10^{-4}$ A cm^{-2}. The validity of the Tafel equation increases with higher overpotentials, but decreases at lower overpotentials.

25.20 $j_{lim} = 0.99$ A m^{-2}.

25.21 $[Cd^{2+}] = 2 \times 10^{-7}$ mol dm^{-3}.

25.22 (a) $j = 0.31$ mA cm^{-2}; (b) $j = 5.41$ mA cm^{-2}; (c) $j = -2.19$ A cm^{-2}.

25.23

$a(Fe^{3+})$ / $a(Fe^{2+})$	0.1	0.3	0.6	1.0	3.0	6.0	10.0
j/(mA cm^{-2})	684	395	278	215	124	88	68.0

25.24 $\eta = 108$ mV.

25.25 (a) 4.9×10^{15} cm^{-2} s^{-1}; (b) 1.6×10^{16} cm^{-2} s^{-1}; (c) 3.1×10^7 cm^{-2} s^{-1}. The numbers of electrons per atom are therefore 3.8 s^{-1}, 12 s^{-1} and 2.4×10^{-8} s^{-1} respectively.

25.26 (a) 33 Ω; (b) 33 GΩ.

25.29 Zinc may be deposited.

25.30 Zinc will not be plated out.

25.31 $E^{\ominus} = +1.30$ V, $P = 0.13$ W.

25.32 1.2 mm yr^{-1}.

Solutions to b) exercises

Chapter 1

1.1 (a) 10.5 bar, (b) 10.4 bar.
1.2 (a) 1.07 bar; (b) 803 Torr.
1.3 120 kPa.
1.4 2.67×10^3 kg.
1.5 1.5×10^3 Pa.
1.6 115 kPa.
1.7 $R = 0.082\,061\,5\ \mathrm{dm^3\ atm\ K^{-1}\ mol^{-1}}$, $M = 31.9987\ \mathrm{g\ mol^{-1}}$.
1.8 P_4.
1.9 2.61 kg.
1.10 (a) $3.14\ \mathrm{dm^3}$; (b) 28.2 kPa.
1.11 $16.14\ \mathrm{g\ mol^{-1}}$.
1.12 −270°C.
1.13 (a) (i) 1.0 atm, (ii) 270 atm.
(b) (i) 0.99 atm, (ii) 190 atm.
11.4 $a = 1.34 \times 10^{-1}\ \mathrm{kg\ m^5\ s^{-2}\ mol^{-2}}$, $b = 4.36 \times 10^{-5}\ \mathrm{m^3\ mol^{-1}}$.
1.15 (a) 1.12, repulsive; (b) $2.7\ \mathrm{dm^3\ mol^{-1}}$.
1.16 (a) $0.124\ \mathrm{dm^3\ mol^{-1}}$; (b) $0.112\ \mathrm{dm^3\ mol^{-1}}$.
1.17 (a) $8.7\ \mathrm{cm^3}$; (b) $-0.15\ \mathrm{dm^3\ mol^{-1}}$.
1.18 (a) $x_N = 0.63$, $x_H = 0.37$;
(b) $p_N = 2.5$ atm, $p_H = 1.5$ atm;
(c) 4.0 atm.
1.19 $b = 0.0493\ \mathrm{dm^3\ mol^{-1}}$, $r = 1.94 \times 10^{-10}$ m, $a = 3.16\ \mathrm{dm^6\ atm\ mol^{-2}}$.
1.20 (a) 1259 K; (b) 0.129 nm.
1.21 (a) $p = 2.6$ atm, $T = 881$ K;
(b) $p = 2.2$ atm, $T = 718$ K;
(c) $p = 1.4$ atm, $T = 356$ K.
1.22 $b = 1.3 \times 10^{-4}\ \mathrm{m^3\ mol^{-1}}$, $Z = 0.67$.

Chapter 2

2.1 59 J.
2.2 −91 J.
2.3 (a) $\Delta U = \Delta H = 0$, $q = -w = 1.62 \times 10^3$ J;
(b) $\Delta U = \Delta H = 0$, $q = -w = 1.38 \times 10^3$ J;
(c) $\Delta U = \Delta H = 0$, $q = w = 0$.
2.4 $p_2 = 143$ kPa, $w = 0$, $q = 3.28 \times 10^3$ J, $\Delta U = 3.28 \times 10^3$ J.
2.5 (a) −19 J; (b) −52.8 J.
2.6 $\Delta H = q = -70.6$ kJ, $w = 5.60 \times 10^3$ J, $\Delta U = -65.0$ kJ.
2.7 −188 J.
2.8 (a) $\Delta H = q = 14.9 \times 10^3$ J, $w = -831$ J, $\Delta U = 14.1$ kJ.
(b) $\Delta H = 14.9$ kJ, $w = 0$, $\Delta U = q = 14.1$ kJ.
2.9 200 K.
2.10 −325 J.
2.11 8.5 Torr.
2.12 $C_{p,\mathrm{m}} = 53\ \mathrm{J\ K^{-1}\ mol^{-1}}$, $C_{V,\mathrm{m}} = 45\ \mathrm{J\ K^{-1}\ mol^{-1}}$.
2.13 $\Delta H = q = 2.0 \times 10^3\ \mathrm{J\ mol^{-1}}$, $\Delta U = 1.6 \times 10^3\ \mathrm{J\ mol^{-1}}$.
2.14 $q = 0$, $w = -3.5 \times 10^3\ \mathrm{J} = \Delta U$, $\Delta T = -24$ K, $\Delta H = -4.5 \times 10^3$ J.
2.15 $V_f = 0.0205\ \mathrm{m^3}$, $T_f = 279$ K, $w = -6.7 \times 10^2$ J.
2.16 $q = \Delta H = 24.0$ kJ, $w = -1.6$ kJ, $\Delta U = 22.4$ kJ.
2.17 $-3053.6\ \mathrm{kJ\ mol^{-1}}$.
2.18 $-1152\ \mathrm{kJ\ mol^{-1}}$.
2.19 $C = 68.3\ \mathrm{J\ K^{-1}}$, $\Delta T = +64.1$ K.
2.20 $+84.40\ \mathrm{kJ\ mol^{-1}}$.
2.21 $+1.90\ \mathrm{kJ\ mol^{-1}}$.
2.22 (a) $\Delta_r H^\ominus = -589.56\ \mathrm{kJ\ mol^{-1}}$, $\Delta_r U^\ominus = -582.13\ \mathrm{kJ\ mol^{-1}}$.
(b) $\Delta_f H^\ominus(\mathrm{HI}) = 26.48\ \mathrm{kJ\ mol^{-1}}$, $\Delta_f H^\ominus(\mathrm{H_2O}) = -241.82\ \mathrm{kJ\ mol^{-1}}$.
2.23 $-760.3\ \mathrm{kJ\ mol^{-1}}$.
2.24 $+52.5\ \mathrm{kJ\ mol^{-1}}$.
2.25 $-566.93\ \mathrm{kJ\ mol^{-1}}$.
2.26 (a) $\Delta_r H^\ominus(298\ \mathrm{K}) = -175\ \mathrm{kJ\ mol^{-1}}$, $\Delta_r U^\ominus(298\ \mathrm{K}) = -173\ \mathrm{kJ\ mol^{-1}}$;
(b) $\Delta_r H^\ominus(348\ \mathrm{K}) = -176\ \mathrm{mol^{-1}}$.
2.27 $-65.49\ \mathrm{kJ\ mol^{-1}}$.
2.28 $-1587\ \mathrm{kJ\ mol^{-1}}$.
2.29 $0.48\ \mathrm{K\ atm^{-1}}$.
2.30 $\Delta U_\mathrm{m} = +129\ \mathrm{J\ mol^{-1}}$, $q = +7.7465\ \mathrm{kJ\ mol^{-1}}$, $w = -7.62\ \mathrm{kJ\ mol^{-1}}$.
2.31 $1.27 \times 10^{-3}\ \mathrm{K^{-1}}$.
2.32 3.6×10^2 atm.
2.33 $-41.2\ \mathrm{J\ atm^{-1}\ mol^{-1}}$, q (supplied) $= 27.2 \times 10^3$ J.

Chapter 3

3.1 (a) $1.8 \times 10^2\ \mathrm{J\ K^{-1}}$; (b) $1.5 \times 10^2\ \mathrm{J\ K^{-1}}$.
3.2 $152.65\ \mathrm{J\ K^{-1}\ mol^{-1}}$.
3.3 $-7.3\ \mathrm{J\ K^{-1}}$.
3.4 $\Delta S = q = 0$, $w = \Delta U = +2.75$ kJ, $\Delta H = +3.58$ kJ.
3.5 $\Delta H_\mathrm{tot} = 0$, $\Delta S_\mathrm{tot} = 24\ \mathrm{J\ K^{-1}}$.
3.6 (a) 0; (b) −230 J; (c) −230 J; (d) −5.3 K; (e) $3.2\ \mathrm{J\ K^{-1}}$.
3.7 (a) $104.6\ \mathrm{J\ K^{-1}}$; (b) $-104.6\ \mathrm{J\ K^{-1}}$.
3.8 (a) $-21.0\ \mathrm{J\ K^{-1}\ mol^{-1}}$; (b) $+512.0\ \mathrm{J\ K^{-1}\ mol^{-1}}$.
3.9 (a) $-212.40\ \mathrm{kJ\ mol^{-1}}$; (b) $-5798\ \mathrm{kJ\ mol^{-1}}$.
3.10 (a) $-212.55\ \mathrm{kJ\ mol^{-1}}$; (b) $-5798\ \mathrm{kJ\ mol^{-1}}$.
3.11 $-86.2\ \mathrm{kJ\ mol^{-1}}$.
3.12 $-197\ \mathrm{kJ\ mol^{-1}}$.
3.13 (a) $\Delta S(\text{gas}) = +3.0\ \mathrm{J\ K^{-1}}$, $\Delta S(\text{surroundings}) = -3.0\ \mathrm{J\ K^{-1}}$, $\Delta S(\text{total}) = 0$;
(b) $\Delta S(\text{gas}) = +3.0\ \mathrm{J\ K^{-1}}$, $\Delta S(\text{surroundings}) = 0$; $\Delta S(\text{total}) = +3.0\ \mathrm{J\ K^{-1}}$;
(c) $\Delta S(\text{gas}) = 0$, $\Delta S(\text{surroundings}) = 0$, $\Delta S(\text{total}) = 0$.
3.14 $2108.11\ \mathrm{kJ\ mol^{-1}}$.
3.15 (a) 0.500; (b) 0.50 kJ; (c) 0.5 kJ.
3.16 −2.0 J.
3.17 $-42.8\ \mathrm{J\ K^{-1}}$.
3.18 3.0 kJ.
3.19 $2.71\ \mathrm{kJ\ mol^{-1}}$.
3.20 $-0.93\ \mathrm{kJ\ mol^{-1}}$.
3.21 200 J.
3.22 $+2.88\ \mathrm{kJ\ mol^{-1}}$.

Chapter 4

4.1 296 K = 23°C.
4.2 $\Delta_\mathrm{fus} S = +5.5\ \mathrm{J\ K^{-1}\ mol^{-1}}$, $\Delta_\mathrm{fus} H = +2.4\ \mathrm{kJ\ mol^{-1}}$.
4.3 $25.25\ \mathrm{kJ\ mol^{-1}}$.
4.4 (a) $31.11\ \mathrm{kJ\ mol^{-1}}$; (b) 276.9 K.

4.5 272 K.

4.6 $3.6\ kg\ s^{-1}$.

4.7 Frost will sublime, 0.40 kPa or more.

4.8 (a) $29.1\ kJ\ mol^{-1}$; (b) At 25°C, $p_1 = 0.22$ atm = 168 Torr; At 60°C, $p_1 = 0.76$ atm = 576 Torr.

4.9 272.41 K.

4.10 $6.73 \times 10^{-2} = 6.73$ per cent.

Chapter 5

5.1 $843.5\ cm^3$.

5.2 $18\ cm^3$.

5.3 8.2×10^3 kPa.

5.4 1.5×10^2 kPa.

5.5 $270\ g\ mol^{-1}$.

5.6 $178\ g\ mol^{-1}$.

5.7 −0.077°C.

5.8 $\Delta_{mix}G = -17.3$ J, $\Delta_{mix}S = 6.34 \times 10^{-2}\ J\ K^{-1}$.

5.9 $\Delta_{mix}G = -3.43$ kJ, $\Delta_{mix}S = +11.5\ J\ K^{-1}$, $\Delta_{mix}H = 0$.

5.10 (a) 1:1; (b) 0.7358.

5.11 N_2: $0.51\ mmol\ kg^{-1}$, O_2: $0.27\ mmol\ kg^{-1}$.

5.12 $0.067\ mol\ dm^{-3}$.

5.13 11 kg.

5.14 $14.0\ kg\ mol^{-1}$.

5.15 $a_A = 0.9701$, $x_A = 0.980$.

5.16 $-3536\ J\ mol^{-1}$, 212 Torr.

5.17 $a_A = 0.436$, $a_B = 0.755$, $\gamma_A = 1.98$, $\gamma_B = 0.968$.

5.18 0.320.

5.19 (a) 45.0 kg KNO_3; (b) 38.8 g $Ba(NO_3)_2$.

5.20 0.661.

5.21 1.3.

Chapter 6

6.1 $x_A = 0.5$, $y_B = 0.5$.

6.2 $x_A = 0.653$, $x_B = 0.347$, $p = 73.4$ kPa.

6.3 (a) the solution is ideal; (b) $y_A = 0.4582$, $y_B = 0.5418$.

6.4 (a) 6.4 kPa; (b) $y_B = 0.77$, $y_T = 0.23$; (c) $p_{(final)} = 4.5$ kPa.

6.5 (a) $y_A = 0.81$; (b) $x_A = 0.67$, $y_A = 0.925$.

6.6 $C = 3$.

6.7 (a) $C = 1$, $P = 2$; (b) $C = 2$, $P = 2$.

6.8 (a) $C = 2$, $P = 2$; (b) $F = 2$.

6.11 $x_B = 0.53$, $T = T_2$, $x_B = 0.82$, $T = T_3$.

6.13 (a) $x_B \approx 0.53$; (b) $x_{AB_2} \approx 0.8$; (c) $x_{AB_2} \approx 0.6$.

6.14 A solid solution with $x(ZrF_4) = 0.24$ appears at 855°C. The solid solution continues to form, and its ZrF_4 content increases until it reaches $x(ZrF_4) = 0.40$ and 820°C. At that temperature, the entire sample is solid.

6.17 (a) When x_A falls to 0.47, a second liquid phase appears. The amount of new phase increases as x_A falls and the amount of original phase decreases until, at $x_A = 0.314$, only one liquid remains.
(b) The mixture has a single liquid phase at all compositions.

Chapter 7

7.1 (a) $\Delta_r G = 0$; (b) $K = 0.16841$; (c) $\Delta_r G^\ominus = 4.41\ kJ\ mol^{-1}$.

7.2 (a) $K = 0.24$; (b) $\Delta_r G^\ominus = 19\ kJ\ mol^{-1}$; (c) $K = 2.96$.

7.3 (a) $K = 1.3 \times 10^{54}$, $\Delta_r G^\ominus = 308.84\ kJ\ mol^{-1}$;
(b) $K = 3.5 \times 10^{49}$, $\Delta_r G^\ominus = -306.52\ kJ\ mol^{-1}$.

7.4 (a) Mole fractions: A: 0.1782, B: 0.0302, C: 0.1162, 2D: 0.6742, Total: 0.9999; (b) $K_x = 9.6$; (c) $K = 9.6$; (d) $\Delta_r G^\ominus = -5.6\ kJ\ mol^{-1}$.

7.5 $T_2 = 1.4 \times 10^3$ K.

7.6 $\Delta_r H^\ominus = 7.191\ kJ\ mol^{-1}$, $\Delta_r S^\ominus = -21\ J\ K^{-1}\ mol^{-1}$.

7.7 $\Delta G^\ominus = -41.0\ kJ\ mol^{-1}$.

7.9 $x_{NO} = 1.6 \times 10^{-2}$.

7.10 (a) $\Delta_f H^\ominus = 39\ kJ\ mol^{-1}$; (b) $\Delta_f H^\ominus = -39\ kJ\ mol^{-1}$.

7.11 (a) At 427°C, $K = 9.24$, At 459°C, $K = 31.08$;
(b) $\Delta_r G^\ominus = -12.9\ kJ\ mol^{-1}$;
(c) $\Delta_r H^\ominus = +161\ kJ\ mol^{-1}$;
(d) $\Delta_r S^\ominus = +248\ J\ K^{-1}\ mol^{-1}$.

7.12 $T = 397$ K.

7.13 $\Delta_f G^\ominus = -128.8\ kJ\ mol^{-1}$.

7.14 (a) R: $Ag_2CrO_4(s) + 2e^- \rightarrow 2Ag(s) + CrO_4^{2-}(aq)$ +0.45 V
L: $Cl_2(g) + 2e^- \rightarrow 2Cl^-(aq)$ +1.36 V
Overall (R – L): $Ag_2CrO_4(s) + 2Cl^-(aq) \rightarrow 2Ag(s) + CrO_4^{2-}(aq) + Cl_2(g)$ −0.91 V
(b) R: $Sn^{4+}(aq) + 2e^- \rightarrow Sn^{2+}(aq)$ +0.15 V
L: $2Fe^{3+}(aq) + 2e^- \rightarrow 2Fe^{2+}(aq)$ +0.77 V
Overall (R – L): $Sn^{4+}(aq) + 2Fe^{2+}(aq) \rightarrow Sn^{2+}(aq) + 2Fe^{3+}(aq)$ −0.62 V
(c) R: $MnO_2(s) + 4H^+(aq) + 2e^- \rightarrow Mn^{2+}(aq) + 2Fe^{3+}(aq)$ +1.23 V
L: $Cu^{2+}(aq) + 2e^- \rightarrow Cu(s)$ +0.34 V
Overall (R – L): $Cu(s) + MnO_2(s) + 4H^+(aq) \rightarrow Cu^{2+}(aq) + Mn^{2+}(aq) + 2H_2O(l)$ +0.89 V

7.15 (a) R: $2H_2O(l) + 2e^- \rightarrow 2OH^-(aq) + H_2(g)$ −0.83 V
L: $2Na^+(aq) + 2e^- \rightarrow 2Na(s)$ −2.71 V
and the cell is $Na(s)|Na^+(aq), OH^-(aq)|H_2(g)Pt$ +1.88 V
(b) R: $I_2(s) + 2e^- \rightarrow 2I^-(aq)$ +0.54 V
L: $2H^+(aq) + 2e^- \rightarrow H_2(g)$ 0.00 V
and the cell is $Pt|H_2(g)|H^+(aq), I^-(aq)|I_2(s)|Pt$ +0.54 V
(c) R: $2H^+(aq) + 2e^- \rightarrow H_2(g)$ 0.00 V
L: $2H_2O(l) + 2e^- \rightarrow H_2(g) + 2OH^-(aq)$ 0.083 V
and the cell is $Pt|H_2(g)|H^+(aq), OH^-(aq)|H_2(g)|Pt$ 0.083 V

7.16 (a) $E = E^\ominus - \frac{2RT}{F}\ln(\gamma_\pm b)$; (b) $\Delta_r G^\ominus = -89.89\ kJ\ mol^{-1}$;
(c) $E^\ominus = +0.223$ V.

7.17 (a) $K = 1.7 \times 10^{16}$; (b) $K = 8.2 \times 10^{-7}$.

7.18 (a) 1.4×10^{-20}; (b) 5.2×10^{-98}.

Chapter 8

8.1 $v = 1.3 \times 10^{-5}\ m\ s^{-1}$.

8.2 $p = 1.89 \times 10^{-27}\ kg\ m\ s^{-1}$, $v = 0.565\ m\ s^{-1}$.

8.3 $\Delta x = 5.8 \times 10^{-6}$ m.

8.4 (a) $E = 0.93 \times 10^{-19}$ J, E (per mole) = $598\ kJ\ mol^{-1}$;
(b) $E = 1.32 \times 10^{-15}$ J, E (per mole) = $7.98 \times 10^5\ kJ\ mol^{-1}$;
(c) $E = 1.99 \times 10^{-23}$ J, E (per mole) = $0.012\ kJ\ mol^{-1}$.

8.5 (a) $v = 0.499\ m\ s^{-1}$; (b) $v = 665\ m\ s^{-1}$; (c) $v = 9.98 \times 10^{-6}\ m\ s^{-1}$.

8.6 $v = 158\ m\ s^{-1}$.

8.7 (a) $3.52 \times 10^{17}\ s^{-1}$; (b) $3.52 \times 10^{18}\ s^{-1}$.

8.8 (a) 0; (b) $E_K = 6.84 \times 10^{-19}$ J; $v = 1.23 \times 10^6\ m\ s^{-1}$.

8.9 (a) $E = 2.65 \times 10^{-19}$ J, or $160\ kJ\ mol^{-1}$; (b) $E = 3.00 \times 10^{-19}$ J, or $181\ kJ\ mol^{-1}$; (c) $E = 6.62 \times 10^{-31}$ J, or $4.0 \times 10^{-10}\ kJ\ mol^{-1}$.

8.10 (a) $\lambda = 1.23 \times 10^{-10}$ m; (b) $\lambda = 3.9 \times 10^{-11}$ m; (c) $\lambda = 3.88 \times 10^{-1}$ m.

8.12 $\Delta x = 100$ pm, speed (Δv): $5.8 \times 10^5\ m\ s^{-1}$.

8.13 1.67×10^{-16} J.

8.14 $\hbar$.

Chapter 9

9.1 (a) 2.14×10^{-19} J, 1.34 eV, 1.08×10^4 cm^{-1}, 129 kJ mol^{-1}; (b) 3.48×10^{-19} J, 2.17 eV, 1.75×10^4 cm^{-1}, 210 kJ mol^{-1}.

9.2 (a) $P = 0.031$; (b) $P = 0.029$.

9.3 $p = 0,\ p^2 = \dfrac{h^2}{L^2}$.

9.4 $L = \left(\dfrac{3}{8}\right)^{1/2} \dfrac{h}{mc} = \left(\dfrac{3}{8}\right)^{1/2} \lambda_c$.

9.5 $x = \dfrac{L}{10}, \dfrac{3L}{10}, \dfrac{L}{2}, \dfrac{7L}{10}, \dfrac{9L}{10}$.

9.6 6.

9.7 $n = 7.26 \times 10^{10}$, $\Delta E = 1.71 \times 10^{-31}$ J, m = 27.5 pm, the particle behaves classically.

9.8 $E_0 = 3.92 \times 10^{-21}$ J.

9.9 $k = 260$ N m^{-1}.

9.10 $\lambda = 13.2$ μm.

9.11 $\lambda = 18.7$ μm.

9.12 (a) $\Delta E = 2.2 \times 10^{-29}$ J; (b) $\Delta E = 3.14 \times 10^{-20}$ J.

9.14 $0, \pm 0.96\alpha$, or $\pm 2.02\alpha$.

9.15 $E_0 = 2.3421 \times 10^{-20}$ J.

9.17 Magnitude $= 2.58 \times 10^{-34}$ J s; possible projections $= 0, \pm 1.0546 \times 10^{-34}$ J s and $\pm 2.1109 \times 10^{-34}$ J s.

Chapter 10

10.1 $I = 12.1$ eV.

10.2 $r = 11.5a_0/Z,\ r = 3.53a_0/Z,\ r = 0$.

10.3 $r = 0,\ r = 1.382a_0,\ r = 3.618a_0$.

10.4 $N = \dfrac{1}{4\sqrt{2\pi a_0^3}}$.

10.5 $\langle V\rangle = -\dfrac{Z^2e^2}{16\pi\varepsilon_0 a_0},\ \langle E_K\rangle = \dfrac{\hbar^2 Z^2}{8ma_0^2}$.

10.6 $P_{3s} = 4\pi r^2 \left(\dfrac{1}{4\pi}\right) \times \left(\dfrac{1}{243}\right) \times \left(\dfrac{Z}{\alpha_0}\right)^3 \times (6 - 6\rho + \rho^2)^2 e^{-\rho}$,
$r = 0.74\ a_0/Z$, $4.19\ a_0/Z$ and $13.08\ a_0/Z$.

10.7 $r = 1.76\ a_0/Z$.

10.8 (a) angular momentum $= 6\frac{1}{2}h = 2.45 \times 10^{-34}$ J s, angular nodes = 2, radial nodes = 1;
(b) angular momentum $= 2\frac{1}{2}h = 1.49 \times 10^{-34}$ J s, angular nodes = 1, radial nodes = 0;
(c) angular momentum $= 2\frac{1}{2}h = 1.49 \times 10^{-34}$ J s, angular nodes = 1, radial nodes = 1.

10.9 (a) $j = \frac{1}{2}, \frac{3}{2}$; (b) $j = \frac{9}{2}, \frac{11}{2}$.

10.10 $J = 8,7,6,5,4,3,2$.

10.11 (a) $g = 1$; (b) $g = 64$; (c) $g = 25$.

10.12 The letter F indicates that the total orbital angular momentum quantum number L is 3; the superscript 3 is the multiplicity of the term, $2S + 1$, related to the spin quantum number $S = 1$; and the subscript 4 indicates the total angular momentum quantum number J.

10.13 (a) $r = 110$ pm, $r = 20.1$ ppm; (b) $r = 86$ pm, $r = 29.4$ pm.

10.14 (a) forbidden; (b) allowed; (c) forbidden.

10.15 (a) [Ar]$3d^3$; (b) For $S = \frac{3}{2}$, $M_s = \pm\frac{1}{2}$ and $\pm\frac{3}{2}$, for $S = \frac{1}{2}$, $M_s = \pm\frac{1}{2}$.

10.16 (a) $S = 2,1,0$; multiplicities = 5,3,1, respectively. (b) $S' = \frac{5}{2}, \frac{3}{2}, \frac{1}{2}$; multiplicities = 6,4,2 respectively.

10.17 $^1F_3$; $^3F_4$; $^3F_3$; $^3F_2$; $^1D_2$; $^3D_3$; $^3D_2$; $^3D_1$; $^1P_1$; $^3P_2$; $^3P_1$; $^3P_0$, the $^3F_2$ set of terms are the lower in energy.

10.18 (a) $J = 3$, 2 and 1, with 7, 5 and 3 states respectively;
(b) $J = \frac{7}{2}, \frac{5}{2}, \frac{3}{2}, \frac{1}{2}$, with 8, 6, 4 and 2 states respectively;
(c) $J = \frac{9}{2}, \frac{7}{2}$, with 10 and 8 states respectively.

10.19 (a) $^2D_{5/2}$ and $^2D_{3/2}$ (b) $^2P_{3/2}$ and $^2P_{1/2}$.

Chapter 11

11.1 (a) $1\sigma^2\, 2\sigma^{*1}$; (b) $1\sigma^2\, 2\sigma^{*2}\, 1\pi^4\, 3\sigma^2$; (c) $1\sigma^2\, 2\sigma^{*2}\, 3\sigma^2 1\pi^4\, 2\pi^{*2}$.

11.2 (a) $1\sigma^2\, 2\sigma^{*2}\, 3\sigma^2\, 1\pi^4\, 2\pi^{*4}$; (b) $1\sigma^2\, 2\sigma^{*2}\, 1\pi^4\, 3\sigma^2$; (c) $1\sigma^2\, 2\sigma^{*2}\, 3\sigma^2\, 1\pi^4\, 2\pi^{*3}$.

11.3 (a) C_2 and CN; (b) NO, O_2 and F_2.

11.4 BrCl is likely to have a shorter bond length than $BrCl^-$; it has a bond order of 1, while $BrCl^-$ has a bond order of 1/2.

11.5 The sequence O_2^+, O_2, O_2^-, O_2^{2-} has progressively longer bonds.

11.6 $N = \left(\dfrac{1}{1 + 2\lambda S + \lambda^2}\right)^{1/2}$.

11.7 $a = -\dfrac{0.145S + 0.844}{0.145 + 0.844S}b,\ N(0.844A - 0.145B)$.

11.8 Not appropriate.

11.9 $E_{\text{trial}} = \dfrac{-\mu e^4}{12\pi^3\varepsilon_0^2\hbar^2}$.

11.10 3.39×10^{-16} J.

11.12 (a) $a_{2u}^2 e_{1g}^4 e_{2u}^1$, $E = 7\alpha + 7\beta$; (b) $a_{2u}^2 e_{1g}^3$, $E = 5\alpha + 7\beta$.

11.13 (a) 19.31368β; (b) 19.44824β.

Chapter 12

12.1 CCl_4 has 4 C_3 axes (each C—Cl axis), 3 C_2 axes (bisecting Cl—C—Cl angles), 3 S_4 axes (the same as the C_2 axes), and 6 dihedral mirror planes (each Cl—C—Cl plane).

12.2 (a) CH_3Cl.

12.3 Yes, it is zero.

12.4 Forbidden.

12.6 T_d has S_4 axes and mirror planes ($= S_1$), T_h has a centre of inversion ($= S_2$).

12.8 (a) $C_{\infty V}$; (b) D_3; (c) C_{4V}, C_{2V}; (d) C_s.

12.9 (a) D_{2h}; (b) D_{2h}; (c) (i) C_{2V}; (ii) C_{2V}; (iii) D_{2h}.

12.10 (a) $C_{\infty V}$; (b) D_{5h}; (c) C_{2V}; (d) D_{3h}; (e) O_h; (f) T_d.

12.11 (a) *ortho*-dichlorobenzene, *meta*-dichlorobenzene, HF and XeO_2F_2; (b) none are chiral.

12.12 NO_3^-: p_x and p_y, SO_3: all d orbitals except d_z^2.

12.13 A_2.

12.14 (a) B_{3U}(x-polarized), B_{2U}(y-polarized), B_{1U}(z-polarized); (b) A_{1U} or E_{1U}.

12.15 Yes, it is zero.

Chapter 13

13.1 (a) 7.73×10^{-32} J m^{-3} s; (b) $\nu = 6.2 \times 10^{-28}$ J m^{-3} s.

13.2 $s = 6.36 \times 10^7$ m s^{-1}.

13.3 (a) 1.59 ns; (b) 2.48 ps.

13.4 (a) 160 MHz; (b) 16 MHz.
13.5 $\nu = 3.4754 \times 10^{11}\ s^{-1}$.
13.6 (a) $I = 3.307 \times 10^{-47}$ kg m^2; (b) $R = 141$ pm.
13.7 (a) $I = 5.420 \times 10^{-46}$ kg m^2; (b) $R = 162.8$ pm.
13.8 $R = 116.21$ pm.
13.9 $R_{CO} = 116.1$ pm; $R_{CS} = 155.9$ pm.
13.10 $\tilde{\nu}_{Stokes} = 20\,603\ cm^{-1}$.
13.11 $R = 141.78$ pm.
13.12 (a) H_2O, C_{2V}; (b) H_2O_2, C_2; (c) NH_3, C_{3V}; (d) N_2O, $C_{\infty V}$.
13.13 (a) CH_2Cl_2; (b) CH_3CH_3; (d) N_2O.
13.14 $k = 0.71$ N m^{-1}.
13.15 28.4 per cent.
13.16 $k = 245.9$ N m^{-1}.
13.17 (a) 0.212; (b) 0.561.
13.18 DF: $\tilde{\nu} = 3002.3\ cm^{-1}$, DCl: $\tilde{\nu} = 2143.7\ cm^{-1}$, DBr: $\tilde{\nu} = 1885.8\ cm^{-1}$, DI: $\tilde{\nu} = 1640.1\ cm^{-1}$.
13.19 $\tilde{\nu} = 2374.05\ cm^{-1}$, $x_e = 6.087 \times 10^{-3}$.
13.20 $D_0 = 3.235 \times 10^4\ cm^{-1} = 4.01$ eV.
13.21 $\tilde{\nu}_R = 2347.16\ cm^{-1}$.
13.22 (a) CH_3CH_3; (b) CH_4; (c) CH_3Cl.
13.23 (a) 30; (b) 42; (c) 13.
13.24 (a) IR active = $A_2'' + E'$, Raman active = $A_1 + E'$;
(b) IR active = $A_1 + E$, Raman active = $A_1 + E$.
13.25 (a) IR active; (b) Raman active.
13.26 $A_{1g} + A_{2g} + E_{1u}$.

Chapter 14

14.1 multiplicity = 3, parity = u.
14.2 22.2 per cent.
14.3 $\varepsilon = 7.9 \times 10^5$ cm^2 mol^{-1}.
14.4 1.33×10^{-3} mol dm^{-3}.
14.5 $A = 1.56 \times 10^8$ dm^3 mol^{-1} cm^{-2}.
14.6 Rise.
14.7 $\varepsilon = 522$ dm^3 mol^{-1} cm^{-1}.
14.8 $\varepsilon = 128$ dm^3 mol^{-1} cm^{-1}, $T = 0.13$.
14.9 (a) 0.010 cm; (b) 0.033 cm.
14.10 (a) 1.39×10^8 dm^3 mol^{-1} cm^{-2}; (b) 1.39×10^9 m mol^{-1}.
14.11 Stronger.

Chapter 15

15.1 $\nu = 649$ MHz.
15.2 $E_{m_I} = -2.35 \times 10^{-26}$ J, 0, $+2.35 \times 10^{-26}$ J.
15.3 47.3 MHz.
15.4 (a) $\Delta E = 2.88 \times 10^{-26}$ J; (b) $\Delta E = 5.77 \times 10^{-24}$ J.
15.5 3.523 T.
15.6

	B/T	^{14}N	^{19}F	^{31}P
	g_I	0.40356	5.2567	2.2634
(a)	300 MHz	97.5	7.49	17.4
(b)	750 MHz	244	18.7	43.5

15.7 (a) 4.3×10^{-7}; (b) 2.2×10^{-6}; (c) 1.34×10^{-5}.
15.8 (a) δ is independent of both B and υ. (b) $\dfrac{\nu - \nu°(800\ MHz)}{\nu - \nu°(60\ MHz)} = 13$.
15.9 (a) 4.2×10^{-6} T; (b) 3.63×10^{-5} T. Spectrum appears narrower at 650 MHz.
15.11 $2.9 \times 10^3\ s^{-1}$.
15.14 (a) The H and F nuclei are both chemically and magnetically equivalent. (b) The P and H nuclei chemically and magnetically equivalent in both the *cis*- and *trans*-forms.
15.15 $B_1 = 9.40 \times 10^{-4}$ T, 6.25 μs.
15.16 1.3 T.
15.17 $g = 2.0022$.
15.18 2.2 mT, $g = 1.992$.
15.19 Eight equal parts at ±1.445 ± 1.435 ± 1.055 mT from the centre, namely: 328.865, 330.975, 331.735, 331.755, 333.845, 333.865, 334.625 and 336.735 mT.
15.21 (a) 332.3 mT; (b)1209 mT.
15.22 $I = 1$.

Chapter 16

16.1 $T = 623$ K.
16.2 (a) 15.9 pm, 5.04 pm; (b) 2.47×10^{26}, 7.82×10^{27}.
16.3 $\dfrac{q\text{Xe}}{q\text{He}} = 187.9$.
16.4 $q = 4.006$.
16.5 $E = 7.605$ kJ mol^{-1}.
16.6 213 K.
16.7 (a) 0.997, 0.994; (b) 0.999 99, 0.999 98.
16.8 (a) (i) $\dfrac{n_2}{n_1} = 1.39 \times 10^{-11}$, $\dfrac{n_3}{n_1} = 1.93 \times 10^{-22}$;
(ii) $\dfrac{n_2}{n_1} = 0.368$, $\dfrac{n_3}{n_1} = 0.135$;
(iii) $\dfrac{n_2}{n_1} = 0.779$, $\dfrac{n_3}{n_1} = 0.607$;
(b) $q = 1.503$; (c) $U_m = 88.3$ J mol^{-1};
(d) $C_v = 3.53$ J K^{-1} mol^{-1}; (e) $S_m = 6.92$ J K^{-1} mol^{-1}.
16.9 7.26 K.
16.10 (a) 147 J K^{-1} mol^{-1}; (b) 169.6 J K^{-1} mol^{-1}.
16.11 10.7 J K^{-1} mol^{-1}.
16.12 (a)

Chapter 17

17.1 (a) $O_3 : 3R$ [experimental = $3.7R$]
(b) $C_2H_6 : 4R$ [experimental = $6.3R$]
(c) $CO_2 : \frac{5}{2}R$ [experimental = $4.5R$]
17.2 With vibrations: 115, Without vibrations: 140, Experimental: 1.29.
17.3 (a) 143; (b) 251.
17.4 (a) 2; (b) 2; (c) 6; (d) 24; (e) 4.
17.5 $q^K = 5837$, $\theta_R = 0.8479$ K, $T = 0.3335$ K.
17.6 $S_m^R = 84.57$ J K^{-1} mol^{-1}.
17.7 (a) At 298 K, $q^R = 2.50 \times 10^3$. At 500 K, $q^R = 5.43 \times 10^3$;
(b) At 298 K, $q = 2.50 \times 10^3$. At 500 K, $q = 5.43 \times 10^3$.
17.8 (a) At 25°C, $q^R = 7.97 \times 10^3$; (b) At 100°C, $q^R = 1.12 \times 10^4$.
17.9 (a) At 298 K, $S_m = 5.88$ J mol^{-1} K^{-1}.
(b) At 500 K, $S_m = 16.48$ J mol^{-1} K^{-1}.
17.10 $G_m^R - G_m^R(0) = -20.1$ kJ mol^{-1}, $G_m^V - G_m^V(0) = -0.110$ kJ mol^{-1}.
17.11 -3.65 kJ mol^{-1}.
17.12 $S_m = 14.9$ J mol^{-1} K^{-1}.
17.14 $K \approx 0.25$.

Chapter 18

18.1 SF_4.
18.2 $\mu = 1.4$ D.
18.3 $\mu = 9.45\times10^{-29}$ C m, $\theta = 194.0°$.
18.4 $\mu = 3.23\times10^{-30}$ C m, $\alpha = 2.55\times10^{-39}$ C^2 m^2 J^{-1}.
18.5 $\varepsilon_r = 8.97$.
18.6 $\mu^* = 3.71\times10^{-36}$ C m.
18.7 $\alpha = 3.40\times10^{-40}$ C^2 m^2 J^{-1}.
18.8 $n_r = 1.10$.
18.9 $\varepsilon_r = 16$.
18.10 $p = 5.92$ kPa.
18.11 $\gamma = 7.12\times10^{-2}$ N m^{-1}.
18.12 $p_{in} - p_{out} = 2.04\times10^5$ Pa.

Chapter 19

19.1 $\bar{M}_n = 68$ kg mol^{-1}, $\bar{M}_w = 69$ kg mol^{-1}.
19.2 $R_g = 1.06\times10^4$.
19.3 (a) $\bar{M}_n = 8.8$ kg mol^{-1}; (b) $\bar{M}_w = 11$ kg mol^{-1}.
19.4 $\tau = 9.4\times10^{-8}$ s.
19.5 71.
19.6 $\bar{M}_n = 120$ kg mol^{-1}.
19.7 $s = 1.47\times10^{-4}$ m s^{-1}.
19.8 $\bar{M} = 56$ kg mol^{-1}.
19.9 $\bar{M}_w = 3.1\times10^3$ kg mol^{-1}.
19.10 $a/g = 3.86\times10^5$.
19.11 $R_{rms} = 38.97$ nm.
19.12 $R_c = 1.26\times10^{-6}$ m, $R_{rms} = 1.97\times10^{-8}$ m.

Chapter 20

20.1 $(\frac{1}{2},\frac{1}{2},0)$ and $(0,\frac{1}{2},\frac{1}{2})$.
20.2 (3 1 3) and (6 4 3).
20.3 $d_{121} = 214$ pm, $d_{221} = 174$ pm, $d_{244} = 87.2$ pm.
20.4 $\lambda = 86.7$ pm.

20.5

hkl	sin θ	θ/°	2θ/°
111	0.327	19.1	38.2
200	0.378	22.2	44.4
220	0.535	32.3	64.6

20.6 $D = 0.054$ cm.
20.7 $V = 1.2582$ nm^3.
20.8 $d = 5$, $d = 2.90$ g cm^{-3}.
20.9 $d_{322} = 182$ pm.
20.10 (100), (110), (111), and (200).

20.11

hkl	d_{hkl}/pm	θ_{hkl}/°
100	574.1	4.166
010	796.8	3.000
111	339.5	7.057

20.12 body-centred cubic.
20.13 $F_{hkl} = 2f$ for $h+k+l$ even; 0 for $h+k+l$ odd.
20.14 $\frac{2}{3}$.
20.15 $\frac{r}{R} = 0.732$.
20.16 (a) 57 pm; (b) 111 pm.
20.17 $\frac{2V_{atom}}{V_{cell}} = 0.370$.
20.18 contraction.
20.20 $\lambda = 252$ pm.
20.21 $\Delta_L H^{\ominus}(MgBr_2, s) = 2421$ kJ mol^{-1}.
20.23 strain $= 5.8\times10^{-2}$.
20.24 $\Delta V = 0.003$ dm^3.
20.25 p-type.
20.26 2.71×10^4 H.
20.27 5.
20.28 -8.2×10^{-4} cm^3 mol^{-1}.
20.29 $\chi_m = 1.58\times10^{-8}$ m^3 mol^{-1}, dimerization occurs.
20.30 2.52 = effective unpaired spins, theoretical number = 2.
20.31 $\chi_m = 1.85\times10^{-7}$ m^3 mol^{-1}.
20.32 $r = 0.935$.

Chapter 21

21.1 (a) 7.079; (b) 1.
21.2 (a) $c = 4.75\times10^2$ m s^{-1}; (b) $\lambda = 4\times10^4$ m; (c) $z = 0.01$ s^{-1}.
21.3 $p = 2.4\times10^7$ Pa.
21.4 $\lambda = 4.1\times10^{-7}$ m.
21.5 $z = 9.9\times10^8$ s^{-1}.
21.6 (a) $\lambda = 3.7\times10^{-9}$ m; (b) $\lambda = 5.5\times10^{-8}$ m; (c) $\lambda = 4.1\times10^{-5}$ m.
21.7 $F \approx 9.6\times10^{-2}$.
21.8 $N = 5.3\times10^{21}$.
21.9 $\Delta m = 4.98\times10^{-4}$ kg.
21.10 $M_{fluorocarbon} = 554$ g mol^{-1}.
21.11 $t = 1.5\times10^4$ s.
21.12 0.17 J m^{-2} s^{-1}.
21.13 1.61×10^{-19} m^2.
21.14 22 J s^{-1}.
21.15 3.00×10^{-19} m^2.
21.16 1.00×10^5 Pa.
21.17 (a) At 273 K: $\eta = 0.95\times10^{-5}$ kg m^{-1} s^{-1};
(b) At 298 K: $\eta = 0.99\times10^{-5}$ kg m^{-1} s^{-1};
(c) At 1000 K: $\eta = 1.81\times10^{-5}$ kg m^{-1} s^{-1}.
21.18 (a) $\kappa = 0.0114$ J m^{-1} s^{-1} K^{-1}, 0.017 J s^{-1};
(b) $\kappa = 9.0\times10^{-3}$ J m^{-1} s^{-1} K^{-1}, 0.014 J s^{-1}.
21.19 52.0×10^{-7} kg m^{-1} s^{-1}, $d = 923$ pm.
21.20 $\kappa = 9.0\times10^{-3}$ J m^{-1} s^{-1} K^{-1}.
21.21 (a) $D = 0.107$ m^2 s^{-1}, $J = 0.87$ mol m^{-2} s^{-1};
(b) $D = 1.07\times10^{-5}$ m^2 s^{-1}, $J = 8.7\times10^{-5}$ mol m^{-2} s^{-1};
(c) $D = 7.13\times10^{-8}$ m^2 s^{-1}, $J = 5.8\times10^{-7}$ mol m^{-2} s^{-1}.
21.22 4.09×10^{-3} S m^2 mol^{-1}.
21.23 4.81×10^{-5} m V^{-1} s^{-1}.
21.24 0.604.
21.25 $\Lambda_m^{o}(MgI_2) = 25.96$ mS m^2 mol^{-1}.
21.26 F^-: $u = 5.74\times10^{-8}$ m^2 V^{-1} s^{-1};
Cl^-: $u = 7.913\times10^{-8}$ m^2 V^{-1} s^{-1};
Br^-: $u = 8.09\times10^{-8}$ m^2 V^{-1} s^{-1}.
21.27 1.09×10^{-9} m^2 s^{-1}.
21.28 4.1×10^3 s.

21.29 207 pm.

21.30 200×10^{-11} s = 20 ps.

21.31 Iodine: (a) 78 s; (b) 7.8×10^3 s.
Sucrose: (a) 3.2×10^2 s; (b) 3.2×10^4 s.

Chapter 22

22.1 Rates of consumption of A = 1.0 mol dm^{-3} s^{-1}; B = 3.0 mol dm^{-3} s^{-1}; C = 1.0 mol dm^{-3} s^{-1}; D = 2.0 mol dm^{-3} s^{-1}.

22.2 Rate of consumption of B = 1.00 mol dm^{-3} s^{-1}.
Rate of reaction = 0.33 mol dm^{-3} s^{-1}.
Rate of formation of C = 0.33 mol dm^{-3} s^{-1}.
Rate of formation of D = 0.66 mol dm^{-3} s^{-1}.
Rate of consumption of A = 0.33 mol dm^{-3} s^{-1}.

22.3 K: dm^3 mol^{-2} s^{-1}. (a) $v = \frac{d[A]}{dt} = -k[A][B]^2$; (b) $v = \frac{d[C]}{dt} = k[A][B]^2$.

22.4 $v = k[A][B][C]^{-1}$, K: s^{-1}.

22.5 2.00.

22.6 Reaction order = 0.

22.7 $t_{1/2} = 1.80 \times 10^6$ s, (a) $p = 31.5$ kPa; (b) $p = 29.0$ kPa.

22.8 (a) $k = 3.47 \times 10^{-3}$ dm^3 mol^{-1} s^{-1}; (b) $t_{1/2}(A) = 2.4$ h; $t_{1/2}(B) = 0.44$ h.

22.9 (a) Second-order units: m^3 molecule^{-1} s^{-1}, Third-order units: m^6 molecule^{-2} s^{-1}; (b) Second-order units; Pa^{-1} s^{-1}, Third-order units; Pa^{-2} s^{-1}.

22.10 (a) 6.5×10^{-3} mol dm^{-3}; (b) 0.025 mol dm^{-3}.

22.11 1.5×10^6 s.

22.12 $t_{1/3} = \frac{3^{n-1} - 1}{k(n-1)}[A]_0^{1-n}$.

22.13 $K_f = 1.7 \times 10^{-7}$ s^{-1}, $k_r = 8.3 \times 10^8$ dm mol^{-1} s^{-1}.

22.14 $E_a = 9.9$ kJ mol^{-1}, $A = 0.94$ dm^3 mol^{-1} s^{-1}.

22.15 (a) $k_T/k_H \approx 0.06$; (b) $k_{18}/k_{16} \approx 0.89$.

22.16 $k_a = 9.9 \times 10^{-6}$ s^{-1} Pa^{-1} = 9.9 s^{-1} MPa^{-1}.

Chapter 23

23.2 $\frac{d[R_2]}{dt} = -k_1[R_2] - k_2\left(\frac{k_1}{k_4}\right)^{1/2}[R_2]^{3/2}$.

23.3 (a) Does not occur. (b) $p = 1.3 \times 10^2$ Pa to 3×10^4 Pa.

23.4 $\frac{k_1 k_2 K_a^{1/2}}{k_1'}[HA]^{3/2}[B]$.

23.5 $\frac{d[A_2]}{dt} = -k_1[A_2]$.

23.6 $v_{max} = 2.57 \times 10^{-4}$ mol dm^{-3} s^{-1}.

23.7 1.5×10^{-5} moles of photons.

23.8 $\Phi = 1.11$.

Chapter 24

24.1 $z = 6.64 \times 10^9$ s^{-1}, $Z_{AA} = 8.07 \times 10^{34}$ m^{-3} s^{-1}, 1.6 per cent.

24.2 (a) (i) 2.4×10^{-3}, (ii) 0.10; (b) (i) 7.7×10^{-27}, (ii) 1.6×10^{-10}.

24.3 (a) (i) 1.2, (ii) 1.03; (b) (i) 7.4, (ii) 1.3.

24.4 $k = 1.7 \times 10^{-12}$ dm^{-3} mol^{-1} s^{-1}.

24.5 $k_d = 3.2 \times 10^7$ m^3 mol^{-1} s^{-1} or 3.2×10^{10} dm^3 mol^{-1} s^{-1}.

24.6 (a) $k_d = 1.97 \times 10^6$ m^3 mol^{-1} s^{-1}; (b) $k_d = 2.4 \times 10^5$ m^3 mol^{-1} s^{-1}.

24.7 $k_d = 1.10 \times 10^7$ m^3 mol^{-1} s^{-1} or 1.10×10^{10} dm^3 mol^{-1} s^{-1}, $t_{1/2} = 5.05 \times 10^{-8}$ s.

24.8 $P = 1.41 \times 10^{-3}$.

24.9 $v = 1.54 \times 10^8$ mol dm^{-3} s^{-1}.

24.10 $\Delta^{\ddagger}H = 48.52$ kJ mol^{-1}, $\Delta^{\ddagger}S = -32.2$ J K^{-1} mol^{-1}.

24.11 $\Delta^{\ddagger}G = 46.8$ kJ mol^{-1}.

24.12 $\Delta^{\ddagger}S = -93$ J K^{-1} mol^{-1}.

24.13 $\Delta^{\ddagger}S = -80.0$ J K^{-1} mol^{-1}.

24.14 (a) $\Delta^{\ddagger}S = -24.1$ J K^{-1} mol^{-1}; (b) $\Delta^{\ddagger}H = 27.5$ kJ mol^{-1}; (c) $\Delta^{\ddagger}G = 34.7$ kJ mol^{-1}.

24.15 $k_2^{\circ} = 1.08$ dm^6 mol^{-2} min^{-1}.

24.16 $\lambda = 1.531$ eV, $\langle H_{DA}\rangle = 9.39 \times 10^{-24}$ J.

24.17 $k_{et} = 1.4 \times 10^3$ s^{-1}.

Chapter 25

25.1 (a) (i) 2.88×10^{19} cm^{-2} s^{-1}, (ii) 5.75×10^{13} cm^{-2} s^{-1}; (b) (i) 3.81×10^{19} cm^{-2} s^{-1}, (ii) 7.60×10^{13} cm^{-2} s^{-1}.

25.2 $p = 7.3 \times 10^2$ Pa.

25.3 6.6×10^4 s^{-1}.

25.4 $A = 18.8$ m^2.

25.5 $V_{mon} = 9.7$ cm^3.

25.6 $t_{1/2} = 200$ s.

25.7 $E_d = 3.7 \times 10^3$ J mol^{-1}.

25.8 (a) 0.32 kPa; (b) 3.9 kPa.

25.9 $\theta_1 = 0.75$, $\theta_2 = 0.25$.

25.10 (a) At 400 K: 4.9×10^{-11} s, At 800 K: 2.4×10^{-12} s; (b) At 400 K: 1.6×10^{13} s, At 800 K: 1.4 s.

25.11 $p_2 = 6.50$ kPa.

25.12 (a) $\theta = \frac{Kp}{1+Kp}$; (b) $\theta = \frac{(Kp)^{1/2}}{1+(Kp)^{1/2}}$; (c) $\theta = \frac{(Kp)^{1/3}}{1+(Kp)^{1/3}}$. A plot of θ versus p at low pressures (where the denominator is approximately 1) would show progressively weaker dependence on p for dissociation into two or three fragments.

25.13 $\Delta_{ad}H^{\ominus} = -6.40$ kJ mol^{-1}.

25.14 $E_d = 2.85 \times 10^5$ J mol^{-1}. (a) $t = 1.48 \times 10^{36}$ s; (b) $t = 1.38 \times 10^{-4}$ s.

25.15 $\varepsilon = 2.8 \times 10^8$ Vm^{-1}.

25.16 167 mV.

25.17 $j_0 = 1.6$ mA cm^{-2}.

25.18 $j_2 = 8.5$ mA cm^{-2}.

25.19 (a) $j = 0.34$ A cm^{-2};
(b) $j = 0.34$ A cm^{-2}. The validity of the Tafel equation improves as the overpotential increases.

25.20 $j_{lim} = 1.3$ A m^{-2}.

25.21 $[Fe^{2+}] = 4 \times 10^{-6}$ mol dm^{-3}.

25.22 $j = (2.5 \text{ mA cm}^{-2}) \times [(e^{(0.42)E'/f} \times (3.41 \times 10^{-6}) - e^{(-0.58)E'/f} \times (3.55 \times 10^7)]$.

25.23 At $r = 0.1$: $j/j_0 = 1.5$ A cm^{-2},
At $r = 1$: $j/j_0 = 4.8$ A cm^{-2},
At $r = 10$: $j/j_0 = 15$ A cm^{-2}.

25.24 0.61 V.

25.25 For the Cu, $H_2|H^+$ electrode: $N = 6.2 \times 10^{12}$ s^{-1} cm^{-2}, $f = 4.2 \times 10^{-3}$ s^{-1}.
For the Pt|Ce^{4+}, Ce^{3+} electrode: $N = 2.5 \times 10^{14}$ s^{-1} cm^{-2}, $f = 0.17$ s^{-1}.

25.26 (a) 5.1 GΩ; (b) 10 GΩ.

25.29 Deposition would not occur.

25.30 Iron can be deposited.

25.31 $E^{\ominus} = 1.80$ V, $P = 0.180$ W.

25.32 3.0 mm y^{-1}.

Solutions to all problems

Chapter 1

1.1 −233°N.

1.2 $p = \rho \frac{RT}{M}$, $M = 45.9\ \text{g mol}^{-1}$.

1.3 −272.95°C.

1.4 $M \approx 102\ \text{g mol}^{-1}$, CH_2FCF_3 or CHF_2CHF_2.

1.5 (a) $\Delta p = 0.0245$ kPa; (b) $p = 9.14$ kPa; (c) $\Delta p = 0.0245$ kPa.

1.6 $p = 1.66$ atm, $p(H_2) = 0$, $p(N_2) = 0.33$ atm, $p(NH_3) = 1.33$ atm.

1.7 (a) $V_m = 12.5\ \text{dm}^3\ \text{mol}^{-1}$; (b) $V_m = 12.3\ \text{dm}^3\ \text{mol}^{-1}$.

1.8 $Z = 0.927$, $V_m = 0.208\ \text{dm}^3$.

1.9 (a) $0.944\ \text{dm}^3\ \text{mol}^{-1}$;
(b) $2.69\ \text{dm}^3\ \text{mol}^{-1}$, $2.67\ \text{dm}^3\ \text{mol}^{-1}$;
(c) $5.11\ \text{dm}^3\ \text{mol}^{-1}$

1.11 (a) $0.1353\ \text{dm}^3\ \text{mol}^{-1}$; (b) 0.6957; (c) 0.72.

1.12 $T_c = 210$ K, $r = 0.28$ nm.

1.13 $b = 59.4\ \text{cm}^3\ \text{mol}^{-1}$, $a = 5.649\ \text{dm}^6\ \text{atm mol}^{-2}$, $p = 21$ atm.

1.15 $B = b - \frac{a}{RT}$, $C = b^2$, $b = 34.6\ \text{cm}^3\ \text{mol}^{-1}$, $a = 1.26\ \text{dm}^6\ \text{atm mol}^{-2}$.

1.17 $V_c = \frac{3C}{B}$, $T_c = \frac{B^2}{3RC}$, $p_c = \frac{B^3}{27C^2}$, $Z_c = \frac{1}{3}$

1.18 $B' = \frac{B}{RT}$, $C' = \frac{C - B^2}{R^2T^2}$.

1.19 $B' = 0.082\ \text{atm}^{-1}$, $B = 2.0\ \text{dm}^3\ \text{mol}^{-1}$.

1.20 $\left(\frac{dV_m}{dT}\right)_p = \frac{RV_m + b}{2pV_m - RT}$.

1.21 No.

1.22 $\frac{10}{9} = 1.11$.

1.23 0.011.

1.24 (a) $B = -1.32 \times 10^{-2}\ \text{dm}^3\ \text{mol}^{-1}$
(b) $B = -1.51 \times 10^{-2}\ \text{dm}^3\ \text{mol}^{-1}$, $C = 1.07 \times 10^{-3}\ \text{dm}^6\ \text{mol}^{-2}$

1.25 $4.1 \times 10^8\ \text{dm}^3$

1.26 mid–latitude, $n = 1.12 \times 10^{-3}$ mol
$\frac{n}{V} = 2.8 \times 10^{-9}\ \text{mol dm}^{-3}$
ozone hole, $n = 4.46 \times 10^{-4}$ mol
$\frac{n}{V} = 1.1 \times 10^{-9}\ \text{mol dm}^{-3}$

1.27 (a) 0.00; (b) −0.72

1.28 (a) 4.62×10^3 mol; (b) 1.3×10^2 kg; (c) 1.2×10^2 kg.

1.30 (a) $\frac{n(CCl_3F)}{V} = 1.1 \times 10^{-11}\ \text{mol dm}^{-3}$
$\frac{n(CCl_2F_2)}{V} = 2.2 \times 10^{-11}\ \text{mol dm}^{-3}$
(b) $\frac{n(CCl_3F)}{V} = 8.0 \times 10^{-13}\ \text{mol dm}^{-3}$
$\frac{n(CCl_2F_2)}{V} = 1.6 \times 10^{-12}\ \text{mol dm}^{-3}$

1.31 $h = 51.5$ km.
$p = 3.0 \times 10^{-3}$ bar.

Chapter 2

2.1 $T_1 = 273\ \text{K} = T_3$, $T_2 = 546$ K
Step 1 → 2: $w = -2.27 \times 10^3$ J
$\Delta U = +3.40 \times 10^3$ J
$q = +5.67 \times 10^3$ J
$\Delta H = +5.67 \times 10^3$ J
Step 2 → 3: $w = 0$
$q_V = \Delta U = -3.40 \times 10^3$ J
$\Delta H = -5.67 \times 10^3$ J
Step 3 → 1: $\Delta U = \Delta H = 0$
$-q = w = +1.57 \times 10^3$ J
Cycle: $\Delta U = \Delta H = 0$
$q = -w = +0.70 \times 10^3$ J

2.2 $w \approx -8.9$ kJ.

2.3 $w = 0$, $\Delta U = +2.35$ kJ, $\Delta H = +3.03$ kJ,

2.4 $w = -25$ J, $q = +109$ J, $\Delta H = +97$ J.

2.5 (a) $w = 0$, $\Delta U = +6.19$ kJ, $q = +6.19$ kJ, $\Delta H = +8.67$ kJ;
(b) $q = 0$, $\Delta U = -6.19$ kJ, $\Delta H = -8.67$ kJ, $w = -6.19$ kJ;
(c) $\Delta U = \Delta H = 0$, $-q = w = +4.29$ kJ.

2.6 $w = -nRT \ln\left(\frac{V_2 - nb}{V_1 - nb}\right) - n^2a\left(\frac{1}{V_2} - \frac{1}{V_1}\right)$.
(a) $w_0 = -1.7$ kJ; (b) $w = -1.8$ kJ; (c) $w = -1.5$ kJ.

2.7 $-87.33\ \text{kJ mol}^{-1}$.

2.8 $\Delta_f H = -1267\ \text{kJ mol}^{-1}$.

2.9 $\Delta_r H^\ominus = +17.7\ \text{kJ mol}^{-1}$, $\Delta_f H^\ominus(\text{metallocene, 583 K}) = +116.0\ \text{kJ mol}^{-1}$.

2.10 $k = -67.44$, $n = 0.9253$, $\Delta_c H = -6625.5\ \text{kJ mol}^{-1}$, 2.17 per cent error.

2.11 (c) $n = 0.903$, $k = -73.7\ \text{kJ mol}^{-1}$.

2.12 $241\ \text{kJ mol}^{-1}$.

2.13 $\Delta_c H^\ominus = 25\,968\ \text{kJ mol}^{-1}$, $\Delta_f H^\ominus(C_{60}) = 2357\ \text{kJ mol}^{-1}$.

2.14 $-994.30\ \text{kJ mol}^{-1}$.

2.15 (a) $240\ \text{kJ mol}^{-1}$; (b) $228\ \text{kJ mol}^{-1}$.

2.16 (a) $-101.8\ \text{kJ mol}^{-1}$; (b) $-344.2\ \text{kJ mol}^{-1}$; (c) $44.0\ \text{kJ mol}^{-1}$.

2.17 $41.40\ \text{J K}^{-1}\ \text{mol}^{-1}$.

2.18 $-30.5\ \text{J mol}^{-1}$.

2.19 3.60 kJ.

2.21 (a) $dz = (2x - 2y + 2)dx + (4y - 2x - 4)dy$
(c) $dz = \left(y + \frac{1}{x}\right)dx + (x - 1)dy$

2.24 $C_p - C_V = nR$

2.25 (a) $\left(\frac{\partial H}{\partial U}\right)_p = 1 + p\left(\frac{\partial V}{\partial U}\right)_p = 1 + \frac{p}{(\partial U/\partial V)_p}$
(b) $\left(\frac{\partial H}{\partial U}\right)_p = 1 + \frac{p}{(\partial U/\partial V)_p} = 1 + p\left(\frac{\partial V}{\partial U}\right)_p$

2.26 (a) $dV = \left(\frac{\partial V}{\partial p}\right)_T dp + \left(\frac{\partial V}{\partial T}\right)_p dT$
$dp = \left(\frac{\partial p}{\partial V}\right)_T dV + \left(\frac{\partial p}{\partial T}\right)_V dT$

(b) $d\ln V = -\kappa_T dp + \alpha dT$

$$d\ln p = \frac{1}{p\kappa_T}\left(\alpha dT - \frac{dV}{V}\right)$$

2.27 (a) -1.5 kJ
(b) -1.6 kJ

2.28 $$w_r = -\frac{8}{9}nT_r \ln\left(\frac{V_{r,2} - 1/3}{V_{r,1} - 1/3}\right) - n\left(\frac{1}{V_{r,2}} - \frac{1}{V_{r,1}}\right)$$

$$\frac{w_r}{n} = -\frac{8}{9}\ln\left(\frac{3x-1}{2}\right) - \frac{1}{x} + 1$$

2.29 increase

2.32 $$\alpha = \frac{(RV^2)\times(V-nb)}{(RTV^3)-(2na)\times(V-nb)^2}$$

$$\kappa_T = \frac{V^2(V-nb)^2}{nRTV^3 - 2n^2a(V-nb)^2}$$

2.33 $$\mu C_p = \left(\frac{1-\dfrac{nb\zeta}{V}}{\zeta - 1}\right)V$$

$\mu = 1.41$ K atm^{-1}

$$T_1 = \frac{27}{4}T_c\left(1 - \frac{b}{V_m}\right)^2$$

$T_1 = 1946$ K

2.34 $$\left(\frac{\partial H}{\partial p}\right)_T = -T\left(\frac{\partial V}{\partial T}\right)_p + V$$

2.35 $$\frac{1}{\lambda} = 1 - \frac{(3V_r - 1)^2}{4T_rV_r^3}$$

$C_{p,m} - C_{V,m} \approx 1.1R = 9.2$ J K^{-1} mol^{-1}

2.36 $$c = \left(\frac{\gamma p}{\rho}\right)^{1/2}$$

$c = 322$ m s^{-1}

2.37 (a) $\mu = \dfrac{aT^2}{C_p}$ (b) $C_V = C_p - R\left(1 + \dfrac{2apT}{R}\right)^2$

2.38 (a) (ii) $w = \frac{1}{2}k_F x f^2$;
(b) (ii) 9.1×10^{-16} N;
(iv) $w = \dfrac{kNT}{2}[(1+\nu_f)\ln(1+\nu_f) + (1-\nu_f)\ln(1-\nu_f)]$;
(v) $w = kNT\ln 2$.

2.39 7.4%.

2.40 $\Delta T = +37$ K, $m = 4.09$ kg.

2.41 (a) -25 kJ; (b) 9.7 m; (c) 39 kJ; (d) 15 m.

2.42 (a) (i) -2802 kJ mol^{-1}; (ii) -2802 kJ mol^{-1}; (iii) -1274 kJ mol^{-1}.
(b) more exothermic by 2688 kJ mol^{-1}

2.44 (a) 16.2 kJ mol^{-1}; (b) 1146 kJ mol^{-1}; (c) 122.0 kJ mol^{-1}.

2.45 $\Delta T = 2$°C, $\Delta h = 1.6$ m, $\Delta T = 1$°C, $\Delta h = 0.8$ m, $\Delta T = 3.5$°C, $\Delta h = 2.8$ m.

2.46 (a) 29.9 K MPa^{-1}; (b) -2.99 K.

2.47 (a) 23.5 K MPa^{-1}; (b) 14.0 K MPa^{-1}.

Chapter 3

3.1 (a) $\Delta_{trs}S(l\rightarrow s, -5°C) = -21.3$ J K^{-1} mol^{-1}, $\Delta S_{sur} = +21.7$ J K^{-1} mol^{-1}, $\Delta S_{total} = +0.4$ J K^{-1} mol^{-1}.
(b) $\Delta_{trs}S(l\rightarrow g, T) = +109.7$ J K^{-1} mol^{-1}, $\Delta S_{sur} = -111.2$ J K^{-1} mol^{-1}, $\Delta S_{total} = -1.5$ J K^{-1} mol^{-1}.

3.2 10.7 J K^{-1} mol^{-1}

3.3 (a) q(Cu) = 43.9 kJ, $q(H_2O) = -43.9$ kJ, $\Delta S(H_2O) = -118.1$ J K^{-1}, ΔS(Cu) = 145.9 J K^{-1}, ΔS(total) = 28 J K^{-1}.
(b) $\theta = 49.9$°C = 323.1 K, q(Cu) = 38.4 kJ = $-q(H_2O)$, $\Delta S(H_2O) = -119.8$ J K^{-1}, ΔS(Cu) = 129.2 J K^{-1}, ΔS(total) = 9 J K^{-1}.

3.4 (a) $\Delta S_A = 50.7$ J K^{-1}, $\Delta S_B = -11.5$ J K^{-1},
(b) $\Delta A_B = +3.46$ kJ, ΔA_A is indeterminate;
(c) $\Delta G_B = +3.46$ kJ, ΔG_A is indeterminate;
(d) ΔS(total system) = +39.2 J K^{-1} = $-\Delta S$(surroundings).

3.5

	Step 1	Step 2	Step 3	Step 4	Cycle
q	+11.5 kJ	0	−5.74 kJ	0	−5.8 kJ
w	−11.5 kJ	−3.74 kJ	+5.74 kJ	+3.74 kJ	−5.8 kJ
ΔU	0	−3.74 kJ	0	+3.74 kJ	0
ΔH	0	−6.23 kJ	0	+6.23 kJ	0
ΔS	+19.1 J K^{-1}	0	−19.1 J K^{-1}	0	0
ΔS_{tot}	0	0	0	0	0
ΔG	−11.5 kJ	Indeterminate	+11.5 kJ	Indeterminate	0

3.6

	q	w	$\Delta U = \Delta H$	ΔS	ΔS_{sur}	ΔS_{tot}
Path (a)	2.74 kJ	−2.74 kJ	0	9.13 J K^{-1}	−9.13 J K^{-1}	0
Path (b)	1.66 kJ	−1.66 kJ	0	9.13 J K^{-1}	−5.53 J K^{-1}	3.60 J K^{-1}

3.7 (a) 200.7 J K^{-1} mol^{-1};
(b) 232.0 J K^{-1} mol^{-1}.

3.8 1st experiment, $\Delta S = 45.4$ J K^{-1}, 2nd experiment, ΔS(water) = 51.2 J K^{-1}.

3.9 $\Delta S = nC_{p,m}\ln\dfrac{T_f}{T_h} + nC_{p,m}\ln\dfrac{T_f}{T_c}$, $\Delta S = +22.6$ J K^{-1}.

3.10 $\Delta T = -21$ K, $\Delta S_m = 35.9$ J K^{-1} mol^{-1}.

3.11 (a) 63.88 J K^{-1} mol^{-1}; (b) 66.08 J K^{-1} mol^{-1}.

3.12 $\Delta_r H^{\ominus}$ (298 K) = 41.16 kJ mol^{-1},
$\Delta_r S^{\ominus}$ (298 K) = 42.08 J K^{-1} mol^{-1},
$\Delta_r H^{\ominus}$ (398 K) = 40.84 kJ mol^{-1},
$\Delta_r S^{\ominus}$ (398 K) = 41.08 J K^{-1} mol^{-1},

3.13 $H_m^{\ominus}(200\text{ K}) - H_m^{\ominus}(0) = 32.1$ kJ mol^{-1}.

3.14 34.4 kJ mol^{-1}, 243 J K^{-1} mol^{-1}.

3.15 46.60 J K^{-1} mol^{-1}.

3.16 -501 kJ mol^{-1}.

3.17 (a) -7 kJ mol^{-1}; (b) +107 kJ mol^{-1}.

3.18 73 atm.

3.22 $V = \dfrac{RT}{p} + B' + C'p + C'p^2$.

3.28 (a) $\left(\dfrac{\partial H}{\partial p}\right)_T = 0$

(b) $$\left(\frac{\partial H}{\partial p}\right)_T = \frac{nb - \left(\dfrac{2na}{RT}\right)\lambda^2}{1 - \left(\dfrac{2na}{RTV}\right)\lambda^2}, \quad \lambda = 1 - \frac{nb}{V}$$

$$\left(\frac{\partial H}{\partial p}\right)_T \approx -8.3 \text{ J atm}^{-1}$$

$\Delta H \approx -8$ J

3.29 $\pi_T \approx \dfrac{p^2}{R} \times \dfrac{\Delta B}{\Delta T}$

(a) 3.0×10^{-3} atm; (b) 0.30 atm.

3.31 $\pi_T = \dfrac{nap}{RTV}$

3.33 $T\,dS = C_p dT - \alpha TV\,dp$, $q_{rev} = -\alpha TV\,\Delta p$, $q_{rev} = -0.50$ kJ

3.34 (a) $f = pe^{bp/RT}$, 10.2 atm;

(b) $\ln\phi = -\dfrac{ap}{(RT)^2}$

$f = 9.237$ atm

3.35 $\ln\phi = \dfrac{Bp}{RT} + \dfrac{(C-B^2)p^2}{2R^2T^2} + \ldots$

$f = 0.9991$ atm

3.36 477 J K^{-1} mol^{-1}.

3.37 −21 kJ mol^{-1}.

3.38 (a) 35 J K^{-1} mol^{-1}; (b) $P = 12$ W m^{-3}, $P_{battery} = 1.5 \times 10^4$ W m^{-3}.

(c) $\dfrac{0.46 \text{ mol ATP}}{\text{mol glutamine}}$

3.39 13 per cent increase.

3.40 monohydrate, 57.2 kJ mol^{-1}, dihydrate, 85.6 kJ mol^{-1}, trihydrate, 112.8 kJ mol^{-1}.

3.42 7.79 km.

3.43 $\varepsilon = 1 - \left(\dfrac{V_B}{V_A}\right)^{1/c}$

$\varepsilon = 0.47$

$\Delta S_1 = \Delta S_3 = \Delta S_{sur,1} = \Delta S_{sur,3} = 0$ [adiabatic reversible steps]

$\Delta S_2 = \Delta S_{sur,4} = +33$ J K^{-1} $= -\Delta S_{sur,2} = -\Delta S_4$

3.44 (b) $|w|_{total} = 6.86$ kJ, $.t = 68.6$ s

3.45 (a) 1.00 kJ; (b) 8.4 kJ.

Chapter 4

4.1 196.0 K, 11.1 Torr.

4.2 9 atm.

4.3 (a) 5.56 kPa K^{-1}; (b) 2.5 per cent.

4.4 (a) −22.0 J K^{-1} mol^{-1}; (b) −109 J K^{-1} mol^{-1}; (c) +110 J mol^{-1};

4.5 (a) −1.63 cm^3 mol^{-1}; (b) +30.1 dm^3 mol^{-1}, +0.6 kJ mol^{-1}.

4.6 234.4 K.

4.7 22°C.

4.8 (a) $T_b = 357$ K = 84°C; (b) $\Delta_{vap}H = +38$ kJ mol^{-1}.

4.9 (a) $T_b = 227.5$°C; (b) $\Delta_{vap}H = +53$ kJ mol^{-1};

4.11 (b) 171.18 K; (c) $T = 383.6$ K; (d) 33.0 kJ mol^{-1}.

4.12 31.6 kJ mol^{-1}.

4.15 9.8 Torr.

4.16 363 K = 90°C.

4.17 $-\dfrac{1}{T} \times C_{p,m}$.

4.18 The Clapeyron equation cannot apply because both ΔV and ΔS are zero through a second-order transition, resulting in an indeterminate form $\frac{0}{0}$.

4.20 (c) $n \cong 17$.

4.21 (a) $\Delta U_r(T_r,V_r) = -\displaystyle\int_{T_r\text{constant}}^{\infty} \dfrac{2p_r(T_r,V_r)}{T_rV_r}\,dV_r$; (c) 0.85 to 0.90.

4.22 (b) $T_b = 112$ K; (c) 8.07 kJ mol^{-1}.

4.23 $\Delta_{vap}H = 1.60 \times 10^4$ bar.

Chapter 5

5.1 $k_A = 15.58$ kPa, $k_B = 47.03$ kPa

5.2 18.07 cm^3 mol^{-1}.

5.3 $V_{salt} = -1.4$ cm^3 mol^{-1}, $V_{H_2O} = 18.04$ cm^3 mol^{-1}.

5.4 12.0 cm^3 mol^{-1}.

5.5 $V_E = 57.9$ cm^3, $V_W = 45.8$ cm^3, $\Delta V = +0.96$ cm^3.

5.6 36 to 72 g mol^{-1}.

5.7 4 ions.

5.8 $K_A = 60.0$ kPa, $K_I = 62.0$ kPa.

5.10 $V_c = 109.0$ cm^3 mol^{-1}, $V_p = 279.3$ cm^3 mol^{-1}.

5.11 (a) propionic acid: $V_1 = V_{m,1} + a_0x_2^2 + a_1(3x_1 - x_2)x_2^2$,

oxane: $V_2 = V_{m,2} + a_0x_1^2 + a_1(x_1 - 3x_2)x_1^2$,

(b) $V_1 = 75.63$ cm^3 mol^{-1}, $V_2 = 99.06$ cm^3 mol^{-1}.

5.12 For $x_T = 0.228, 0.511, 0.810$, $\gamma_T = 0.490, 0.723, 0.966$, and $\gamma_T = 1.031, 0.920, 0.497$.

5.13 $K_H = 371$ bar, $\gamma_{CO_2} = 1.01$ (at $10p$/bar), 0.99 (at $20p$/bar), 1.00 (at $30p$/bar), 0.99 (at $40p$/bar), 0.98 (at $60p$/bar), 0.94 (at $80p$/bar).

5.14 $S_0 = 19.89$ mol L^{-1}, $\tau = 165$ K; fits well, $R = 0.999$ 78.

5.15 $\Delta_{mix}G = -4.6$ kJ.

5.17 $\mu_A = \mu_A^* + RT\ln x_A + gRTx_B^2$

5.19 80.36 cm^3 mol^{-1}.

5.24 4.5 cm^3.

5.25 p_{N_2} at 4.0 atm = 56 µg, p_{N_2} at 1.0 atm = 14 µg, increase = 1.7×10^2 µg N_2.

5.26 K = 1.167 dm^3 µmol^{-1}, N = 5.24, model is applicable.

5.29 (a) g cm K^{-1} mol^{-1},

(b) $M = 1.1 \times 10^5$ g mol^{-1},

(d) $B' = 21.4$ cm^3 g^{-1}, $C' = 211$ cm^6 g^{-1},

(e) $B' = 28.0$ cm^3 g^{-1}, $C' = 196$ cm^6 g^{-2}.

5.30 $M = 1.26 \times 10^5$ g mol^{-1}, $B = 1.23 \times 10^4$ dm^3 mol^{-1}.

Chapter 6

6.1 (b) 391.0 K; (c) $\dfrac{n_{liq}}{n_{vap}} = 0.532$.

6.2 (b) $n_\alpha/n_\beta = 0.093$, $x = 0.750$ at 302.5 K.

6.3 Temperature (γ_{O_2}): 0.877 (78 K), 1.079 (80 K), 1.039 (82 K), 0.995 (84 K), 0.993 (86 K), 0.990 (88 K), 0.987 (90.2 K).

6.4 (a) The values of x_s corresponding to the three compounds are: (1) P_4S_3: 0.43, (2) P_4S_7: 0.64, (3) P_4S_{10}: 0.71.

6.6 The number of distinct chemical species (as opposed to components) and phases present at the indicated points are, respectively: $b(3,2)$, $d(2,2)$, $e(4,3)$, $f(4,3)$, $g(4,3)$, $d(2,2)$.

6.7 $MgCu_2$: 16 per cent mg by mass, Mg_2Cu: 43 per cent mg by mass.

6.8 (a) (i) At 1500 K, $F = 2$, (ii) At 1100 K, $F = 3$, (b) (i) No phase separation occurs, (ii) $x_{PB}(\alpha) = 0.19$, $x_{PB}(\alpha) = 0.86$, $n_\alpha/n_\beta = 0.36$, (c) solubility of Cu in Pb = 0.050 g Cu/g Pb.

6.9 (a) Eutectic: 40.2 at per cent Si at 1268°C; 69.4 at per cent Si at 1030°C. Congruent melting compounds: Ca_2Si = 1314°C; CaSi = 1324°C. Incongruent melting compounds: $CaSi_2$ = 1040°C;

(b) At 1000°C, the phases at equilibrium will be Ca(s) and liquid (13 at per cent Si). Relative amounts, $n_{Ca}/n_{liq} = 2.86$;

(c) (i) $n_{Si}/n_{liq} = 0.53$ at slightly above 1030°C, (ii) $n_{Si}/n_{CaSi_2} = 0.665$ slightly below 1030°C.

6.10 No phases are in equilibrium as the system cools. At 360°C, K_2FeCl_4(s) appears. The solution becomes richer in $FeCl_2$ until the temperature reaches 351°C at which point $KFeCl_3$(s) also appears. Below 351°C the system is a mixture of K_2FeCl_4(s) and $KFeCl_3$(s).

6.13 (i) Below a denaturant concentration of 0.1 only the native and unfolded forms are stable.

6.14 Above about 33°C the membrane has the highly mobile liquid crystal form. At 33°C the membrane consists of liquid crystal in equilibrium with a relatively small amount of the gel form. Cooling from 33°C to about 20°C, the equilibrium persists but shifts to a greater relative abundance of the gel form. Below 20°C the gel form alone is stable.

6.19 (a) 2150°C, (b) $y(\text{MgO}) = 0.18$, $x(\text{MgO}) = 0.35$, (c) $c = 2640$°C.

6.20 (a) At 460°C, $n_{\text{liq}}/n_{\text{solid}} \approx 5$; (b) At 375°C, there is no liquid.

6.21 (b) $n_{\text{liq}}/n_{\text{vap}} = 10.85$.

Chapter 7

7.1 (a) $\Delta_r G^{\ominus} = +4.48$ kJ mol^{-1}; (b) $p_{\text{IBr}} = 0.101$ atm.

7.2 (a) $K = 1.24 \times 10^{-9}$; (b) $K = 1.29 \times 10^{-8}$; (c) $\alpha_e = 1.8 \times 10^{-4}$.

7.3 $\Delta_f H^{\ominus} = 8.48\ R$.

7.4 $\Delta_r H^{\ominus} = 3.00 \times 10^5$ J mol^{-1}.

7.5 $\Delta_r G^{\ominus}(T)/(\text{kJ mol}^{-1}) = 78 - 0.161 \times (T/\text{K})$.

7.6 $K = 1.69 \times 10^{-5}$.

7.7 First experiment, $K = 0.740$, second experiment, $K = 5.71$, enthalpy of dimerization $= -103$ kJ mol^{-1}.

7.8 H_2: 0.0007 mol, HI: 0.786 mol.

7.9 $\Delta H^{\ominus} = +158$ kJ mol^{-1}.

7.10 $\Delta_f H^{\ominus} = 76.8$ kJ mol^{-1}.

7.11 (a) At 298 K: 1.2×10^8; (b) At 700 K: 2.7×10^3.

7.12 $\Delta_r H^{\ominus} = +18.8$ kJ mol^{-1}.

7.13 (a) $CuSO_4$: 4.0×10^{-3}; $ZnSO_4$: 1.2×10^{-2}; (b) $\gamma_{\pm}(CuSO_4) = 0.74$, $\gamma_{\pm}(ZnSO_4) = 0.60$; (c) $Q = 5.9$; (d) $E^{\ominus} = +1.102$ V; (e) $E = +1.079$ V.

7.14 (a) $E^{\ominus} = +1.23$ V; (b) $E = +1.09$ V.

7.15 pH = 2.0.

7.16 (a) $= E^{\ominus} - (38.54\text{ mV}) \times \ln(4^{1/3}b) - (38.54\text{ mV})\ln(\gamma_{\pm})$;
(b) $E^{\ominus}$(cell) = +1.0304 V;
(c) $\Delta_r G^{\ominus} = -198.84$ J mol^{-1}, $K = 6.84 \times 10^{34}$;
(d) $\gamma_{\pm} = 0.763$; (e) $\gamma_{\pm} = 0.75$ (f) $\Delta_r S = -87.2$ J K^{-1} mol^{-1}, $\Delta H^{\ominus} = -262.4$ kJ mol^{-1}.

7.17 $E^{\ominus} = +0.268\ 43$ V, $\gamma_{\pm} = 0.9659$ (1.6077 mmol kg^{-1}), 0.9509 (3.0769 mmol kg^{-1}), 0.9367 (5.0403 mmol kg^{-1}), 0.9232 (7.6938 mmol kg^{-1}), 0.9094 (10.9474 mmol kg^{-1}),

7.18 $pK_w = 14.23$, $\Delta_r H^{\ominus} = +74.9$ kJ mol^{-1}, $\Delta_r G^{\ominus} = +80.0$ kJ mol^{-1}, $\Delta_r S^{\ominus} = -17.1$ J K^{-1} mol^{-1}.

7.19 $\gamma = 0.533$.

7.21 (a) $\left(\dfrac{\partial E}{\partial p}\right)_{T,n} = -\dfrac{\Delta_r V}{\nu F}$;
(b) 2.80×10^{-3} mV atm^{-1};
(c) the linear fit and constancy of $\left(\dfrac{\partial E}{\partial p}\right)$ are very good;
(d) 3.2×10^{-7} atm^{-1}.

7.22 $pK_a = 6.736$, $B = 1.997$ kg$^{0.5}$ mol$^{-0.5}$, $k = -0.121$ kg mol^{-1}.

7.23 −1.15 V.

7.24 (a) $E = E_{\text{ap}} + 0.059\ 16\text{ V}\log(a_{F^-} + k_{F^-,OH^-}a_{OH^-})$; (b) $5 < \text{pH} < 8$.

7.25 $\xi = 1 - \left(\dfrac{1}{1 + ap/p^{\ominus}}\right)^{1/2}$

7.26 Therefore, $\Delta_r G' = \Delta_r G + (T - T')\Delta_r S + \alpha\Delta a + \beta\Delta b + \gamma\Delta c$, $\Delta_f G^{\ominus} = +225.31$ kJ mol^{-1}.

7.30 (a) 40 per cent; (b) $\Delta_r G' = -2871$ kJ mol^{-1}, 74 per cent.

7.31 Yes.

7.32 2 mol.

7.33 (b) +0.206 V.

7.34 (a) $\Delta_r H^{\ominus} = -72.4$ kJ mol^{-1}, $\Delta_r S^{\ominus} = -144$ J K^{-1} mol^{-1};
(b) $\Delta_r H^{\ominus} = +131.2$ kJ mol^{-1}, $S^{\ominus} = +309.2$ J K^{-1} mol^{-1}.

7.35 (iv) $HNO_3 \cdot 3H_2O$ is most stable.

7.36 (a) (b) Perfect gas: $T = 450$: 156.5 bar, $T = 400$: 81.8 bar; van der Waals gas: $T = 723.25$ K: 132.5 bar, $T = 673.15$ K: 73.7 bar; (c) confirms.

Chapter 8

8.1 (a) $\Delta E = 1.6 \times 10^{33}$ J m^{-3}; (b) $\Delta E = 2.5 \times 10^{-4}$ J m^{-3}.

8.2 $h = 6.54 \times 10^{-34}$ J s.

8.3 (a) $\theta_E = 2231$ K, $\dfrac{C_V}{3R} = 0.031$; (b) $\theta_E = 343$ K, $\dfrac{C_V}{3R} = 0.897$.

8.4 (a) 0.020; (b) 0.007; (c) 7×10^6; (d) 0.5; (e) 0.61.

8.5 (a) 9.0×10^{-6}; (b) 1.2×10^{-6}.

8.6 π.

8.8 $i\hbar$.

8.10 $\lambda_{\max} T = \dfrac{hc}{5k} = \dfrac{c_2}{5}$.

8.12 (a) $a = 8\pi hc$, $b = hc$.

8.13 (a) $N = \left(\dfrac{2}{L}\right)^{1/2}$; (b) $N = \dfrac{1}{c(2L)^{1/2}}$; (c) $N = \dfrac{1}{(\pi a^3)^{1/2}}$; (d) $N = \dfrac{1}{(32\pi a^5)^{1/2}}$.

8.14 (a) (i) $N = \left(\dfrac{1}{32\pi a_0^3}\right)^{1/2}$ (ii) $N = \left(\dfrac{1}{32\pi a_0^5}\right)^{1/2}$.

8.15 (a) Yes, eigenvalue = ik; (b) No; (c) Yes, eigenvalue = 0; (d) No; (e) No.

8.16 (a) Yes, eigenvalue = −1; (b) Yes, eigenvalue = +1; (c) No.

8.17 (a) Yes, eigenvalue = $-k^2$;
(b) Yes, eigenvalue = $-k^2$;
(c) Yes, eigenvalue = 0;
(d) Yes, eigenvalue = 0;
(e) No.

Hence, (a,b,c,d) are eigenfunctions of $\dfrac{d^2}{dx^2}$; (b,d) are eigenfunctions of $\dfrac{d^2}{dx^2}$, but not of $\dfrac{d}{dx}$.

8.18 (a) $\cos^2\chi$ (b) $\sin^2\chi$ (c) $\psi = 0.95e^{ikx} \pm 0.32e^{-ikx}$.

8.19 $\dfrac{\hbar^2 k^2}{2m}$.

8.20 (a) $k\hbar$; (b) 0; (c) 0.

8.21 (a) $r = 6a_0$, $r^2 = 42a_0^2$; (b) $r = 5a_0$, $r^2 = 30a_0^2$.

8.22 (a) $\dfrac{-e^2}{4\pi\varepsilon_0 a_0}$ (b) $\dfrac{\hbar^2}{2m_e a_0^2}$.

8.26 $\lambda_{\max} = 500$ nm, blue-green.

8.27 (a) $\lambda_{\text{relativistic}} = 5.35$ pm.

8.28 $I = 255$ K, $T\lambda_{\max} = 11.3$ µm.

8.29 (a) Methane is unstable above 825 K; (b) $\lambda_{\max}$(1000 K) = 2880 nm; (c) Excitance ratio = 7.7×10^{-4}; Energy density ratio = 8.8×10^{-3}; (d) 2.31×10^{-7}, it hardly shines.

Chapter 9

9.1 $E_2 - E_1 = 1.24 \times 10^{-39}$ J, $n = 2.2 \times 10^9$ J, $E_n - E_{n-1} = 1.8 \times 10^{-30}$ J.

9.2 HI(314 N m^{-1}) < HBr(412 N m^{-1}) < HCl(516 N m^{-1}) < NO(1600 N m^{-1}) < CO(1900 N m^{-1}).

9.3 $E_1 = 1.30 \times 10^{-22}$ J, minimum angular momentum = $\pm\eta$.

9.4 0, 2.62, 7.86, 15.72.

9.5 (a) $E_1^{(1)} = \frac{\varepsilon a}{L} + \frac{\varepsilon}{\pi}\sin\left(\frac{\pi a}{L}\right)$; (b) $E_1^{(1)} = \frac{\varepsilon}{10} + \frac{\varepsilon}{\pi}\sin\left(\frac{\pi}{10}\right) = 0.1984\varepsilon$.

9.6 $\frac{1}{2}mgL$.

9.10 (a) $T = |A_3|^2 = A_3 \times A_3^* = \frac{4k_1^2k_2^2}{(a^2+b^2)\sinh^2(k_2L)+b^2}$.

where $a^2+b^2 = (k_1^2+k_2^2)(k_2^2+k_3^2)$ and $b^2 = k_2^2(k_1+k_3)^2$

9.11 (a) $P = \frac{N^2}{2\kappa}$; (b) $\langle x\rangle = \frac{N^2}{4\kappa^2}$.

9.12 $g = \frac{1}{2}\left(\frac{mk}{\hbar^2}\right)^{1/2}$.

9.13 $\langle T\rangle = \frac{1}{2}\left(\upsilon + \frac{1}{2}\right)\hbar\omega$.

9.14 $\langle x^3\rangle = 0$, $\langle x^4\rangle = \frac{3}{4}(2\upsilon^2+2\upsilon+1)\alpha^4$

9.15 (a) $\delta x = L\left(\frac{1}{12} - \frac{1}{2\pi^2n^2}\right)^{1/2}$, $\delta p = \frac{nh}{2L}$;

(b) $\delta x = \left[\left(\upsilon+\frac{1}{2}\right)\frac{\hbar}{\omega m}\right]^{1/2}$, $\delta p = \left[\left(\upsilon+\frac{1}{2}\right)\hbar\omega m\right]^{1/2}$.

9.16 For $\upsilon' = \upsilon+1$, $\mu = \alpha\left(\frac{\upsilon+1}{2}\right)^{1/2}$, for $\upsilon' = \upsilon-1$, $\mu = \alpha\left(\frac{\upsilon}{2}\right)^{1/2}$.

9.19 $\langle T\rangle = -\frac{1}{2}\langle V\rangle$.

9.20 (a) $l_z = +\eta$, $\langle T\rangle = \frac{\hbar^2}{2I}$; (b) $l_z = -2\eta$, $\langle T\rangle = \frac{2\hbar^2}{I}$; (c) $l_z = 0$, $\langle T\rangle = \frac{\hbar^2}{2I}$;

(d) $l_z = \eta\cos 2\chi$, $\langle T\rangle = \frac{\hbar^2}{2I}$.

9.23 (a) $E = 0$, angular momentum $= 0$; (b) $\frac{3\hbar^2}{I}$, angular momentum $=$ $6^{1/2}\hbar$; (c) $E = \frac{6\hbar^2}{I}$, angular momentum $= 2\sqrt{3}\hbar$.

9.25 $\theta = \arccos\frac{m_l}{\{l(l+1)\}^{1/2}}$, 54°44′.

9.26 eigenvalue $= -(a^2+b^2+c^2)$.

9.31 (a) $\Delta E = 3.3\times10^{-19}$ J; (b) $\upsilon = 4.95\times10^{-14}$ J s^{-1}; (c) lower, increases.

9.32 1.2×10^6.

9.33 $\omega = 2.68\times10^{-14}$ J s^{-1}.

9.35 (a) $l_z = 5.275\times10^{-34}$ J s, $E_{\pm5} = 1.39\times10^{-24}$ J; (b) $\upsilon = 9.2\times10^8$ Hz.

9.36 $\langle x^2\rangle^{1/2} = \left(\frac{\hbar l}{2}\right)^{1/2}\times\left(\frac{1}{mkT}\right)^{1/4}$.

9.37 $F = 5.8\times10^{-11}$ N.

Chapter 10

10.1 $n_2 \to 6$, transitions occur at 12 372 nm, 7503 nm, 5908 nm, 5129 nm, . . . , 3908 nm (at $n_2 = 15$), converging to 3282 nm as $n_2 \to \infty$.

10.2 397.13 nm, 3.40 eV.

10.3 $R_{Li^{2+}} = 987\ 663$ cm^{-1}, the Balmer transitions lie at $\tilde{\nu} = 137\ 175$ cm^{-1}, 85 187 cm^{-1}, 122.5 eV.

10.4 5.39 eV.

10.5 $^2P_{1/2}$ and $^2P_{3/2}$, of which the former has the lower energy, $^2D_{3/2}$ and $^2D_{5/2}$ of which the former has the lower energy, the ground state will be $^2D_{3/2}$.

10.6 A = 38.50 cm^{-1}.

10.7 3.3429×10^{-24} kg, $\frac{I_D}{I_H} = 1.000\ 272$.

10.8 $\tilde{\nu} = 7621$ cm^{-1}, 10 288 cm^{-1}, 11 522 cm^{-1}; $I = 6.80$ eV.

10.9 (a) $\Delta\tilde{\nu} = 0.9$ cm^{-1}, (b) Normal Zeeman splitting is small compared to the difference in energy of the states involved in the transition.

10.10 $r^* = 0.420$ pm.

10.11 ±106 pm.

10.13 (b) For 3*s*, $\rho_{node} = 3+\sqrt{3}$ and $\rho_{node} = 3-\sqrt{3}$, no nodal plane; for $3p_x$, $\rho_{node} = 0$ and $\rho_{node} = 4$, *yz* nodal plane ($\phi = 90°$); for $3d_{xy}$, $\rho_{node} = 0$, *xz* nodal plane ($\phi = 0$) and *yz* nodal plane ($\phi = 90°$); (c) $\langle r\rangle_{3s} = \frac{27a_0}{2}$.

10.14 Neither p_x nor p_y are eigenfunctions of $\hat{l}_z$. However, $p_x + ip_y$ and $p_x - ip_y$ are eigenfunctions.

10.16 $r' = 2.66a_0$.

10.17 $\langle r^{-1}\rangle_{1s} = \frac{Z}{a_0}$; (b) $\langle r^{-1}\rangle_{2s} = \frac{Z}{4a_0}$; (c) $\langle r^{-1}\rangle_{2p} = \frac{Z}{4a_0}$.

10.18 $E = -\frac{Z^2e^4m_e}{32\pi^2\varepsilon_0^2\hbar^2}\times\frac{1}{n^2}$.

10.20 $a_{Ps} = 2a_0$, $E_{1,Ps} = \frac{1}{2}E_{1,H}$

10.22 (b) $\Delta x = 23.8$ T m^{-1}.

10.25 The wavenumbers for $n = 3 \to n = 2$: ^{4}He = 60 957.4 cm^{-1}, ^{3}He = 60 954.7 cm^{-1}. The wavenumbers for $n = 2 \to n = 1$: ^{4}He = 329 170 cm^{-1}, ^{3}He = 329 155 cm^{-1}.

10.26 $\tilde{\nu} = 0.216$ cm^{-1}, $\langle r\rangle_{100} = 529$ nm, $\frac{I_{100}}{hc} = 10.9677$ cm^{-1}, $V_{min} = 511$ m s^{-1}, radius of Bohr orbit $= 8.8\times10^{-13}$ m^2. A neutral H atom in its ground state is likely to pass by the Rydberg atom.

10.27 (a) receeding; $s = 3.381\times10^5$ m s^{-1}.

Chapter 11

11.3 $R = 2.1a_0$.

11.7 (a) $P = 8.6\times10^{-7}$ / $P = 2.0\times10^{-6}$; (b) $P = 8.6\times10^{-7}$ / $P = 2.0\times10^{-6}$; (c) $P = 3.7\times10^{-7}$ / $P = 0$; (d) $P = 4.9\times10^{-7}$/ $P = 5.5\times10^{-7}$.

11.8 Dissociation energy = 1.9 eV, bond length = 130 pm.

11.11 Thermal motion would cause the molecule to break apart; it is not likely to exist for more than one vibrational period.

11.12 $I_1 = 10.20$ eV, $I_2 = 12.98$ eV, $I_3 = 15.99$ eV.

11.13 Delocalization energy $= 2\{E_{\text{with resonance}} - E_{\text{without resonance}}\}$ $= \{(\alpha_O - \alpha_N)^2 + 12\beta^2\}^{1/2} - \{(\alpha_O - \alpha_N)^2 + 4\beta^2\}^{1/2}$

11.14 $\Delta E = 2.7$ eV, $\lambda = 460$ nm, orange.

11.15 (a) C_2H_4: −3.813, C_4H_6: −4.623, C_6H_8: −5.538, C_8H_{10}: −5.873; (b) 8.913 eV.

11.16 (a)

	energy$(E_k - \alpha)/\beta$	
orbital, *k*	C_6H_6	C_8H_8
±4		−2.000
±3	−2.000	−1.414
±2	−1.000	0
±1	1.000	1.414
0	2.000	2.000

(b) Benzene: $E_{\text{delocalisation}} = 2\beta$, aromatic; Hexatriene; $E_{\text{delocal}} = 0.988\beta$.

(c) Cyclooctaene: $E_{\text{delocalisation}} = 1.657\beta$, not aromatic; octatetraene: $E_{\text{deloc}} = 1.518\beta$.

11.18 (b) Linear least-squares best fit is: $\Delta E/\text{eV} = 3.3534 + 1.3791 \times 10^{-4}\,\tilde{\nu}\,\text{cm}^{-1}$ ($r^2 = 0.994$); (c) $\tilde{\nu}\,\text{cm}^{-1} = 30\,937\,\text{cm}^{-1}$

11.24 (a) $E = -hc\mathfrak{R}_H$; (b) $E = -\frac{8}{3\pi} \times hc\mathfrak{R}_H$.

11.26 (a) $E = \alpha - \beta, \alpha - \beta, \alpha + 2\beta$; (b) $\Delta H_2 = -413\,\text{kJ mol}^{-1}$; (c) $\beta = -849\,\text{kJ mol}^{-1} - 2\alpha$; $H_3^+ = -849\,\text{kJ mol}^{-1}$; $H_3 = 3(\alpha/2) - 212\,\text{kJ mol}^{-1}$; $H_3^- = 3\alpha - 425\,\text{kJ mol}^{-1}$.

11.29 (a) linear relationship; (b) $E^{\ominus} = -0.122$ V; (c) $E^{\ominus} = -0.174$ V, ubiquinone a better oxidizing agent than plastiquinone.

Chapter 12

12.1 (a) D_{3d}; (b) chair: D_{3d}, boat: C_{2v}; (c) D_{2h}; (d) D_3; (d) D_{4d}; (i) Polar: Boat C_6H_{12}; (ii) Chiral: $[Co(en)_3]^{3+}$.

12.2

	E	C_2	σ_h	i
E	E	C_2	σ_h	i
C_2	C_2	E	i	σ_h
σ_h	σ_h	i	E	C_2
i	i	σ_h	C_2	E

The *trans*-CHCl=CHCl molecule belongs to the C_{2h}.

12.3 $C_2\sigma_h = i$.

12.6 representation 1: $D(\sigma_v) = D(\sigma_d) = +1$ or -1; representation 2: $D(\sigma_v) = -D(\sigma_d) = +1$ or -1.

12.7

	1	σ_x	σ_y	σ_z
1	1	σ_x	σ_y	σ_z
σ_x	σ_x	1	$i\sigma_x$	$-i\sigma_y$
σ_y	σ_y	$-i\sigma_z$	1	$i\sigma_x$
σ_z	σ_z	$i\sigma_y$	$-i\sigma_z$	1

The matrices do not form a group since the products $i\sigma_z$, $i\sigma_y$, $i\sigma_x$ and their negatives are not among the four given matrices.

12.8 $A_1 + T_2$, the *s* and *p* orbitals of the C atom may form molecular orbitals, the T_2 set (d_{xy}, d_{yx}, d_{zx}) may contribute to molecular orbital formation within the H orbitals.

12.9 All five *d* orbitals may contribute to bonding. (b) All except $A_2(d_{xy})$ may participate in bonding.

12.10 The most distinctive symmetry operation is the S_4 axis through the central atom and aromatic nitrogens on both ligands. That axis is also a C_2 axis. The group is S_4.

12.11 (a) D_{2h}; (b) (i) Staggered: C_{2h}; (ii) Eclipsed: C_{2V}.

12.12 (a) D_{2d}; (b) A_1; (c) C_{4V}; (d) A_1.

12.13 (a) C_{2V}, $f \rightarrow 2A_1 + A_2 + 2B_1 + 2B_2$;
(b) C_{3V}, $f \rightarrow +A_2 + 3E$;
(c) T_d, $f \rightarrow A_1 + T_1 + T_2$;
(d) O_h, $f \rightarrow A_{2U} + T_{1U} + T_{2U}$.
Lanthanide ion (a) tetrahedral complex: $f \rightarrow A_1 + T_1 + T_2$ in T_d symmetry, and there is one nondegenerate orbital and two sets of triply degenerate orbitals. (b) octahedral complex: $f \rightarrow A_{2U} + T_{1U} + T_{2U}$, and the pattern of splitting is the same.

12.14 (a) Must be zero; (b) Need not be zero (c) Must be zero.

12.15 irreducible representations: $4A_1 + 2B_1 + 3B_2 + A_2$

12.18 (a) Not allowed; (b) The transition to T_1 would become allowed however the transition to G would still be forbidden.

12.20 $A_{1g} + B_{1g} + E_U$.

12.22 *z*-polarized transition is not allowed, for *x*- or *y*-polarized transitions the transition is allowed.

Chapter 13

13.1

T/K	E/J m^{-3}	E_{class}/J m^{-3}
(a) 1500	2.136×10^{-6}	2.206
(b) 2500	9.884×10^{-4}	3.676
(c) 5800	3.151×10^{-1}	8.528

13.2 (a) HCl: 2.1×10^{-6}, $\delta\nu(\text{rotation}) = 1.3$ MHz, $\delta\tilde{\nu}(\text{vibration}) = 0.0063\,\text{cm}^{-1}$;
(b) ICl: 9.7×10^{-7}, $\delta\nu(\text{rotation}) = 6.6$ kHz, $\delta\tilde{\nu}(\text{vibration}) = 0.0004\,\text{cm}^{-1}$.

13.3 $\tau = \frac{1}{z} = \frac{kT}{4\sigma p}\left(\frac{\pi m}{kT}\right)^{1/2}$, $\delta\nu \approx 700$ MHz, below 1 Torr.

13.4 Rotational line separations = 596 GHZ, 19.9 cm^{-1}, and 0.503 mm. $B = 9.942\,\text{cm}^{-1}$.

13.5 $R_0 = 112.83$ pm, $R_1 = 123.52$ pm.

13.6 $R_{CC} = 139.6$ pm, $R_{CH} = 108.5$ pm, $R_{CD} = R_{CH}$.

13.7 $I = 2.728 \times 10^{-47}\,\text{kg m}^2$, $R = 129.5$ pm, hence we expect lines at 10.56, 21.11, 31.67, . . . cm^{-1}.

13.8 $R_{HCl} = 128.393$ pm, $R^2_{HCl} = 128.13$ pm.

13.9 218 pm.

13.10 $R_{CO} = 116.28$ pm, $R_{CS} = 155.97$ pm.

13.11 $B = 14.35\,\text{m}^{-1}$, $J_{max} = 26$ at 298 K, $J_{max} = 15$ at 100 K.

13.12 Zero-point level = 142.81 cm^{-1}, $D_0 = 3.36$ eV, $k = 93.8\,\text{N m}^{-1}$.

13.13 linear.

13.14 (a) 2143.26 cm^{-1}; (b) 12.8195 kJ mol^{-1}; (c) $1.855\,63 \times 10^3\,\text{N m}^{-1}$; (d) 1.91 cm^{-1}; (e) 113 pm.

13.15 (a) 5.15 eV; (b) 5.20 eV.

13.17 (a) $\tilde{\nu} = 152\,\text{m}^{-1}$, $k = 2.72 \times 10^{-4}\,\text{kg s}^{-2}$, $I = 2.93 \times 10^{-46}\,\text{kg m}^2$, $B = 95.5\,\text{m}^{-1}$.
(b) $x_e = 0.96$.

13.19 (a) C_{3v}; (b) 9; (c) $2A_1 + A_2 + 3E$. (d) All but the A_2 mode are infrared active. (e) All but the A_2 mode may be Raman active.

13.20 (a) 7; (b) Structure 2: C_{2h}, Structure 3: C_{2v}, Structure 4: C_2; (c) Structure 2 is inconsistent.

13.22 $D = \frac{4B^3}{\tilde{\nu}^2}$.

13.23 (a) spherical rotor; (b) symmetric rotor; (c) linear rotor; (d) asymmetric rotor; (e) symmetric rotor; (f) asymmetric rotor.

13.24 $J_{max} = \left(\frac{kT}{hcB}\right)^{1/2} - \frac{1}{2}$, for ICl $J_{max} = 30$, for CH_4 $J_{max} = 6$.

13.25 $HgCl_2$: 230, $HgBr_2$: 240, HgI_2: 250 pm.

13.26 $\nu = 2D_e/\tilde{\nu} - \frac{1}{2}$

13.27 (a) infrared active; (b) 796 cm^{-1}; (c) O_2: 2, O_2^-: 1.5, O_2^{2-}: 1; (d) $Fe_2^{3+}O_2^{2-}$; (e) Structures 6 and 7 are consistent with this observation, but structures 4 and 5 are not.

13.29 $s = 0.0768\ c$, $T = 8.34 \times 10^5$ K.

13.31 $B = 2.031\ \text{cm}^{-1}$; $T = 2.35$ K.

Chapter 14

14.1 49 364 cm^{-1}.

14.2 5.1147 eV.

14.3 14 874 cm^{-1}.

14.4 $\mathcal{A} = 4.8 \times 10^4\ \text{dm}^3\ \text{mol}^{-1}\ \text{cm}^{-2}$.

14.5 $\mathcal{A} = 1.1 \times 10^6\ \text{dm}^3\ \text{mol}^{-1}\ \text{cm}^{-2}$, Excitations from A_1 to A_1, B_1, and B_2 terms are allowed.

14.6 ${}^2\Sigma_g^+ \leftarrow {}^2\Sigma_u^+$ is allowed.

14.7 5.06 eV.

14.9

Hydrocarbon	E_{HOMO}/eV*
Benzene	−9.7506
Biphenyl	−8.9169
Naphthalene	−8.8352
Phenanthrene	−8.7397
Pyrene	−8.2489
Anthracene	−8.2477

14.10 $t_{1/2}(\text{T} \rightarrow \text{S}) = 2 \times 10^{-4}$ s.

14.11 (a) $\dfrac{n}{V} = 1.7 \times 10^{-9}\ \text{mol dm}^{-3}$, (b) $N = 6.0 \times 10^2$.

14.12 $t = 4 \times 10^{-10}$ so r 0.4 ns.

14.14 (a) allowed; (b) forbidden.

14.15 The transition moves toward the red as the chain lengthens and the apparent color of the dye shifts towards blue.

14.18 (a) lower; (b) nothing can be said about the spacing of the upper state levels (without a detailed analysis of the intensities of the lines).

14.20 (a) $I_{\text{abs}} = I_0 \times (1 - 10^{-\varepsilon[\text{J}]l})$; (b) $I_f(\tilde{\nu}_f) \propto \phi_f I_0(\tilde{\nu})\varepsilon[\text{J}]l \ln 10$.

14.21 (a) 3 + 1, 3 + 3; (b) 4 + 4, 2 + 2.

14.23 4.4×10^3.

14.25 $\mathcal{A} = 1.24 \times 10^5\ \text{dm}^3\ \text{mol}^{-1}\ \text{cm}^{-2}$.

14.26 300 DU, A = 6.37, for 100 DU, A = 2.12.

14.27 (a) $\mathcal{A} = 2.24 \times 10^5\ \text{dm}^3\ \text{mol}^{-1}\ \text{cm}^{-2}$; (b) A = 0.185; (c) $\varepsilon = 135\ \text{dm}^3\ \text{mol}^{-1}\ \text{cm}^{-1}$.

14.28 28 kJ mol^{-1} greater.

14.29 $V_1 - V_0 = 3.1938$ eV, $\tilde{\nu}_1 - \tilde{\nu}_0 = 79.538\ \text{cm}^{-1}$, $\tilde{\nu}_0 = 2034.3\ \text{cm}^{-1}$, $T_{\text{eff}} = 1321$ K.

Chapter 15

15.1 $B_0 = 10.3$ T, $\dfrac{\delta N}{N} \approx 2.42 \times 10^{-5}$, β state lies lower.

15.2 $E_a = 57\ \text{kJ mol}^{-1}$.

15.3 300×10^6 Hz ± 10 Hz, 0.29 s.

15.4 $k = 4 \times 10^2\ \text{s}^{-1}$, $E_{\text{II}} - E_{\text{I}} = 3.7\ \text{KJ mol}^{-1}$, $E_a = 16\ \text{kJ mol}^{-1}$.

15.5 Both fit the data equally well.

15.6 (a) yes; (b) ${}^3J_{\text{SnSn}}/\text{Hz} = 580 - 79 \cos\phi + 395 \cos 2\phi$.

15.8 $g_\parallel = 1.992$, $g_\perp = 2.002$.

15.9 Width of the CH_3 spectrum is $3a_H = 6.9$ mT. The width of the CD_3 spectrum is $6a_D$. The overall width is $6a_D = 2.1$ mT.

15.11 $P(\text{N2}s) = 0.10$, $P(\text{N2}p_z) = 0.38$, total probability: $P(\text{N}) = 0.48$, $P(\text{O}) = 0.52$, hybridization ratio = 3.8, $\Phi = 131°$.

15.13 $\sigma_d = \dfrac{e^2\mu_0 Z}{12\pi m_e a_0} = 1.78 \times 10^{-5}Z$.

15.15 $R = 158$ pm.

15.17 $I(\omega) \approx \dfrac{1}{2}\dfrac{A\tau}{1 + (\omega_0 - \omega)^2\tau^2}$.

15.21 $\dfrac{-g_I\mu_N\mu_0 m_I}{4\pi R^3}(\cos^2\theta_{\max} + \cos\theta_{\max})$, $\langle B_{\text{nucl}}\rangle = 0.58$ mT.

15.26 29 μT m^{-1}.

Chapter 16

16.1 $W = 2 \times 10^{40}$, $S = 1.282 \times 10^{-21}\ \text{J K}^{-1}$, $S_1 = 0.637 \times 10^{-21}\ \text{J K}^{-1}$, $S_2 = 0.645 \times 10^{-21}\ \text{J K}^{-1}$.

16.3 $\dfrac{\Delta W}{W} \approx 2.4 \times 10^{25}$.

16.4 $\dfrac{\Delta W}{W} \approx 4.8 \times 10^{21}$.

16.5 $T = 3.5 \times 10^{-15}$ K, $q = 7.41$.

16.6 Not at equilibrium.

16.7 (a) (i) $q = 5.00$; (ii) $q = 6.26$;
(b) $p_0 = 1.00$ at 298 K, $p_0 = 0.80$ at 5000 K; $p_2 = 6.5 \times 10^{-11}$ at 298 K, $p_2 = 0.12$ at 5000 K.
(c) (i) $S_m = 13.38\ \text{J K}^{-1}\ \text{mol}^{-1}$, (ii) $S_m = 18.07\ \text{J K}^{-1}\ \text{mol}^{-1}$.

16.8 $p({}^3F_2) = 0.257$, $p({}^3F_3) = 0.336$, $p({}^3F_4) = 0.396$, $p({}^4F_1) = 0.011$.

16.9 (a) $p_0 = 0.64$, $p_1 = 0.36$; (b) 0.52 kJ mol^{-1}. At 300 K, $S_m = 11.2\ \text{J K}^{-1}\ \text{mol}^{-1}$, At 500 K, $S_m = 11.4\ \text{J K}^{-1}\ \text{mol}^{-1}$.

16.10 At 298 K, $g = 1.209$. At 1000 K, $g = 3.004$.

16.11 (a) At 100 K: $q = 1.049$, $p_0 = 0.953$, $p_1 = 0.044$, $p_2 = 0.002$; $U_m - U_m(0) = 123\ \text{J mol}^{-1}$, $S_m = 1.63\ \text{J K}^{-1}\ \text{mol}^{-1}$.
(b) At 298 K: $q = 1.55$, $p_0 = 0.64$, $p_1 = 0.230$, $p_2 = 0.083$, $U_m - U_m(0) = 1348\ \text{J mol}^{-1}$, $S_m = 8.17\ \text{J K}^{-1}\ \text{mol}^{-1}$.

16.12 (a) $W = e^{1.254\times10^{24}} = 10^{5.44\times10^{23}}$; (b) $W = e^{1.31\times10^{24}} = 10^{5.69\times10^{23}}$;
(c) At constant internal energy and volume the condition for spontaneity is $\Delta S_{U,V} > 0$. Since $W_{(b)} > W_{(a)}$, the process is spontaneous.

16.13 Most probable configurations are {2, 2, 0, 1, 0, 0} and {2, 1, 2, 0, 0, 0} jointly.

16.14 Most probable configuration is {4, 2, 2, 1, 0, 0, 0, 0, 0}.

16.15 (a) $T = 160$ K.

16.16 (a) $q = 1 + 3e^{-\varepsilon/kT}$; (b) 0.5245 RT, 2.074 J $\text{K}^{-1}\ \text{mol}^{-1}$, 10.55 J $\text{K}^{-1}\ \text{mol}^{-1}$.

16.20 $\dfrac{p(h)}{p_0} = e^{-mgh/kT}$, $\dfrac{N(8.0\ \text{km})}{N(0)}[H_2O] = 0.57$.

Chapter 17

17.1 (a) 0.351; (b) 0.079; (c) 0.029.

17.2 (a) 0.1 per cent; (b) 4×10^{-3} per cent.

17.3 $C_{V,m} = 4.2\ \text{J K}^{-1}\ \text{mol}^{-1}$, $S_m = 15\ \text{J K}^{-1}\ \text{mol}^{-1}$.

17.5 $q = 19.90$.

17.6 $S_m^\ominus = 191.4\ \text{J K}^{-1}\ \text{mol}^{-1}$.

17.7 $S_m^\ominus = 199.4\ \text{J mol}^{-1}\ \text{K}^{-1}$.

17.8 $\nu_V^* = 28$, $C_{V,m} = 258\ \text{J mol}^{-1}\ \text{K}^{-1}$.

17.10

T/K	300	400	500	600	700	800	900	1000
K	945	273	132	83	61	49	42	37

17.11 At 298 K: $K = 3.89$. At 800 K: $K = 2.41$.

17.12 $\Delta S_m = R\ln\left\{\left(\frac{\sigma_m}{V_m}\right)\times\left(\frac{h^2\beta}{2\pi me}\right)^{1/2}\right\}$.

17.13 (a) $\theta_R = 87.55$ K, $\theta_V = 6330$ K.

17.16 (b) $J_{max} = \left(\frac{kT}{2hcB}\right)^{1/2} - \frac{1}{2}$; (c) $T \approx 374$ K.

17.17 (a) $q^R = 660.6$; (b) $q^R = 4.26\times10^4$.

17.18 (a) $U - U(0) = nRT\left(\frac{q}{q}\right)$, $C_V = nR\left\{\frac{q}{q} - \left(\frac{q}{q}\right)^2\right\}$, $S = nR\left(\frac{q}{q} - \ln\frac{eq}{N}\right)$.

(b) $C_{V,m} = 5.41$ J K^{-1} mol^{-1}.

17.22 (a) $c_s = \left(\frac{1.40RT}{M}\right)^{1/2}$; (b) $c_s = \left(\frac{1.40RT}{M}\right)^{1/2}$; (c) $c_s = \left(\frac{4RT}{3M}\right)^{1/2}$.

$c_s = 350$ m s^{-1}.

17.23 $S = 9.57\times10^{-15}$ J K^{-1}

17.25 $G_m^\ominus - G_m^\ominus(0) = 513.5$ k J mol^{-1}.

17.26 $G_m^\ominus - G_m^\ominus(0) = 45.76$ k J mol^{-1}.

17.27 At 10 K, $G_m^\ominus - G_m^\ominus(0) = 660.8$ J mol^{-1}.
At 1000 K, $G_m^\ominus - G_m^\ominus(0) = 241.5$ kJ mol^{-1}.

Chapter 18

18.1 (a) $\varepsilon = 1.1\times10^8$ V m^{-1}; (b) $\varepsilon = 4\times10^9$ V m^{-1}; (c) $\varepsilon = 4$ kV m^{-1}.

18.2 $r = 2.4$ nm.

18.3 $\alpha' = 1.2\times10^{-23}$ cm^3, $\mu = 0.86$ D.

18.4 $\alpha' = 1.38\times10^{-23}$ cm^3, $\mu = 0.34$ D.

18.5 $\alpha' = 2.24\times10^{-24}$ cm^3, $\mu = 1.58$ D, $P'_m = 5.66$ cm^3 mol^{-1}, $\mu = 1.58$ D.

18.6 $\alpha' = 1.36\times10^{-24}$ cm^3, $\mu = 1.85$ D.

18.7 (a) $\varepsilon = 1.51\times10^{-23}$ J, $R_e = 265$ pm.

18.8 $q/e = 0.123$.

18.9 $P_m = 8.14$ cm^3 mol^{-1}, $\varepsilon_r = 1.76$, $n_r = 1.33$.

18.10 (a) $V = \frac{6Q_1Q_2}{\pi\varepsilon_0}\times\frac{1}{r^5}$; (b) $V = V\frac{9Q_1Q_2}{4\pi\varepsilon_0 r^5}$.

18.12 The relative permittivity should decrease as the dilution increases.

18.14 $r = 1.63$.

18.18 The interaction is a dipole-induced-dipole interaction. $V = -1.8\times10^{-27}$ J $= -1.1\times10^{-3}$ J mol^{-1}.

18.19 (a) $V = -39$ J mol^{-1}, (b) The force approaches zero as the distance becomes very large.

18.21 (a) $\mu = 1.03\times10^{-29}$ C m; (b) $V_{max} = 3.55\times10^{-23}$ J.

18.22 (a) $\pi = 3.5$; (b) slope = 1.49, intercept = −1.95, (c) $K = 1.12\times10^{-2}$.

Chapter 19

19.1 $S = 4.97\times10^{-13}$ s or 5.0 Sv.

19.2 $\nu = 58$ Hz, or 3500 r.p.m.

19.3 $[\eta] = 0.0716$ dm^3 g^{-1}.

19.4 $\frac{\bar{M}_v}{\text{g mol}^{-1}} = 2.1\times10^5$.

19.5 $\bar{M} = 158$ kg mol^{-1}.

19.6 $a = 0.500$, $K = 2.73$ cm^3 g^{-1} kg$^{-1/2}$ mol$^{-1/2}$, $\bar{M}_v = 1.34\times10^3$ kg mol^{-1}.

19.7 (a) $K = 0.0117$ cm^3 g^{-1} and $a = 0.717$.

19.8 $\bar{M}_n = 23.1$ kg mol^{-1}, $B = 1.02$ m^3 mol^{-1}.

19.9 $\bar{M}_n = 155$ kg mol^{-1}, $B = 13.7$ m^3 mol^{-1}.

19.13 $\bar{M}_n \approx \bar{M} + \left(\frac{2\gamma}{\pi}\right)^{1/2}$.

19.15 (a) $R_{rms} = lN^{1/2}$, $R_{rms} = 9.74$ nm;

(b) $R_{mean} = \left(\frac{8N}{3\pi}\right)^{1/2} l$, $R_{mean} = 8.97$ nm;

(c) $R^* = l\left(\frac{2}{3}N\right)^{1/2}$, $R^* = 7.95$ nm;

19.17 (a) $R_g = \left(\frac{3}{5}\right)^{1/2} a$; (b) $R_g = \frac{l}{2\sqrt{3}}$. When M = 100 kg mol^{-1},
R_g/nm = 2.40. For a rod of radius 0.50 nm, $R_g = 46$ nm.

19.23 $v_p = 8v_{mol}$
For BSV, $B = 28$ m^3 mol^{-1}. For Hb, $B = 0.33$ m^3 mol^{-1}.
For BSV, $\frac{\Pi - \Pi^\circ}{\Pi^\circ} = 2.6\times10^{-2}$ corresponding to 2.6 per cent.
For Hb, $\frac{\Pi - \Pi^\circ}{\Pi^\circ} = 5.0\times10^{-2}$ corresponding to 5 per cent.

19.24 Globular: serum albumin and bushy stunt virus.

19.25 (a)

θ / °	20	45	90
I_{rod}/I_{cc}	0.976	0.876	0.514

(b) 90°.

19.26 $\bar{M}_w = 65.6$ kg mol^{-1}.

19.27 $\bar{M}_n = 69$ kg mol^{-1}, $a = 3.4$ nm.

19.29 pH = 3.85.

19.32 (a) $c_{salt} = 1.0\times10^{-2}$ mol dm^{-3}, $T_m = 340$ K
(b) $c_{salt} = 0.15$ mol dm^{-3}, $T_m = 360$ K.

19.34 $n = 244$.

19.36 $K = 2.38\times10^{-3}$ cm^3 g^{-1}, $a = 0.955$.

Chapter 20

20.1 $\lambda = 118$ pm.

20.2 Simple (primitive) cubic lattice, $a = 342$ pm.

20.3 Yes, the data support.

20.4 (a) Body-centred cubic, $R = 136$ pm;
(b) Face-centred cubic,
$R = 129$ pm.

20.5 Face-centred cubic, $a = 408.55$ pm, $\rho = 10.507$ g cm^{-3}.

20.6 Very narrow X-ray beam.

20.7 a(KCl) = 628 pm, are broadly conistent.

20.8 $\alpha_{volume} = 4.8\times10^{-5}$ K^{-1}, $\alpha_{volume} = 1.6\times10^{-5}$ K^{-1}.

20.9 $\rho = 7.654$ g cm^{-3}.

20.10 $a = 834$ pm, $b = 606$ pm, $a = 870$ pm.

20.11 $\rho = 1.01$ g cm^{-3}.

20.14 $N = 4$.

20.15 $\rho = 1.385$ g cm^{-3}, $\rho o_s = 1.578$ g cm^{-3}.

20.16 $\Delta_f H^\ominus$(CaCl, s) = −373 kJ mol^{-1}.

20.17 0.736 eV.

20.19 For $S = 2$, $\chi_m = 0.127\times10^{-6}$ m^3 mol^{-1}, For $S = 3$, $\chi_m = 0.254\times10^{-6}$ m^3 mol^{-1}, For $S = 4$, $\chi_m = 0.423\times10^{-6}$ m^3 mol^{-1}.

20.20 $T = 6.0$ K.

20.21 $x = 0.41$.

20.23 For a monoclinic cell, $V = abc\sin\beta$.
For an orthorhombic cell, $V = abc$.

20.24 (a) $f = 0.5236$; (b) $f = 0.6802$; (c) $f = 0.7405$.

20.25 $F_{hkl} \propto 1 + e^{5i\pi} + e^{6i\pi} + e^{7i\pi} = 1 - 1 + 1 - 1 = 0$.

20.26 (a) $F_{hkl} = f$; (b) $F_{hkl} = f_A[1 + \frac{1}{2}(-1)^{(h+k+l)}]$,
(c) $F_{h+k+l} = f\{1 + (-1)^{h+k+l}\} = 0$.

20.31 $\xi = \dfrac{-e^2 a_0^2}{2m_e}, \chi_m = \dfrac{-N_A \mu_0 e^2 a_0^2}{2m_e}$.

20.32 Degree of dimerization $= \left(\dfrac{1}{4(p/K)+1}\right)^{1/2}$. A positive $\Delta_r H^{\ominus}$ indicates that $NO_2(g)$ is favoured as the temperature increases; hence the susceptibility increases with temperature.

20.34 $M = 3.61 \times 10^5$ g mol^{-1}.

Chapter 21

21.2 (a) 2.8 km h^{-1} East; (b) 86 km h^{-1}; (c) 86 km h^{-1}.

21.3 (a) $\langle h \rangle = 1.89$ m; (b) $\sqrt{\langle h^2 \rangle} = 1.89$ m.

21.4 9.1.

21.5 $p = 7.3 \times 10^{-3}$ Pa, or 7.3 mPa.

21.6 (a) 100 Pa; (b) 24 Pa.

21.7 (a) Cadmium: 2×10^{14} s^{-1}; (b) Mercury: 1×10^{17} s^{-1}.

21.8 5.3×10^{-4} mS m^2 mol^{-1}.

21.9 $\Lambda_m^{\circ} = 12.6$ mS m^2 mol^{-1}; $\mathcal{K} = 7.30$ mS m^2 mol^{-1} M$^{-1/2}$.
(a) $\Lambda_m = 11.96$ mS m^2 mol^{-1}; (b) $\kappa = 119.6$ mS m^{-1}; (c) $R = 172.5\ \Omega$.

21.10 1.36×10^{-5} M.

21.11 $s(Li^+) = 4.0 \times 10^{-3}$ cm s^{-1}, $s(Na^+) = 5.2 \times 10^{-3}$ cm s^{-1}; $s(K^+) = 7.6 \times 10^{-3}$ cm s^{-1}. $t(Li^+) = 250$ s, $t(Na^+) = 190$ s, $t(K^+) = 130$ s.
(a) $d(Li^+) = 1.3 \times 10^{-6}$ cm; $d(Na^+) = 1.7 \times 10^{-6}$ cm; $d(K^+) = 2.4 \times 10^{-6}$ cm.
(b) 43, 55 and 81 solvent molecule diameters respectively.

21.12 0.82, 0.0028.

21.13 $t_+ = 0.48$ and $t_- = 0.52$. $u_+ = 7.5 \times 10^{-4}$ cm^2 s^{-1} V^{-1}. $\lambda_+ = 72$ S cm^2 mol^{-1}.

21.14 In the first solution, In the second solution $t_+ = 0.278$. $t_+ = 0.278$.

21.15 (a) 2.1×10^{-20} N molecule^{-1}; (b) 2.8×10^{-20} N molecule^{-1}; (c) 4.1×10^{-20} N molecule^{-1}.

21.17 9.3 kJ mol^{-1}.

21.18 0, 0.063 M.

21.19 1.2×10^{-3} kg m^{-1} s^{-1}.

21.20 In sovlent: $\Lambda_m^{\circ}(NaI) = 60.7$ S cm^2 mol^{-1} and $\Lambda_m^{\circ}(KI) = 58.9$ S cm^2 mol^{-1},
In water: $\Lambda_m^{\circ}(NaI) = 126.9$ S cm^2 mol^{-1} and $\Lambda_m^{\circ}(KI) = 150.3$ S cm^2 mol^{-1}.

21.21 (a) 3.68×10^{-10} m; (b) 3.07×10^{-10} m.

21.22 $a = 830$ pm.

21.25 $\langle v_x \rangle = 0.47 \langle v_x \rangle_{initial}$.

21.26 (a) 39 per cent; (b) 61 per cent; (c) greater: 47 per cent, less than: 53 per cent.

21.27 $\dfrac{f(nc^*)}{f(c^*)} = \dfrac{(nc^*)^2 e^{-mn^2c^{*2}/2kT}}{c^{*2}e^{-mc^{*2}/2kT}}[24.4] = n^2 e^{-(n^2-1)mc^{*2}/2kT} = n^2 e^{(1-n^2)}$,
$\dfrac{f(3c^*)}{f(c^*)} = 3.02 \times 10^{-3}, \dfrac{f(4c^*)}{f(c^*)} = 4.9 \times 10^{-6}$.

21.31 (a) $p = 0$; (b) $p = 0.016$; (c) $p = 0.054$.

21.32 $N > 60$.

21.36 $D = 1.6 \times 10^{16}$ m^2 s^{-1}, $\kappa = 0.34$ J K^{-1} m^{-1} s^{-1}.

21.37 The total energy density (translational plus rotational) $= \rho T = 0.25$ J cm^{-3}.

21.39 $t = 10^8$ s.

21.40 1.7×10^{-2} s.

Chapter 22

22.1 Second-order, $k = 0.0594$ dm^3 mol^{-1} min^{-1}, $m = 2.94$ g.

22.2 First-order, $k = 1.51 \times 10^{-5}$ s^{-1}, $[A] = 9.82 \times 10^{-3}$ mol dm^{-3}.

22.3 First-order, $k = 1.23 \times 10^{-4}$ s^{-1}.

22.4 First-order, $k = 5.84 \times 10^{-3}$ s^{-1}, $t_{1/2} = 1.98$ min.

22.5 9.70×10^4 J mol^{-1}.

22.6 55.4 per cent.

22.7 $k = 3.65 \times 10^{-3}$ min^{-1}, $t_{1/2} = 190$ min.

22.8 First-order, $k = 2.8 \times 10^{-4}$ s^{-1}.

22.9 $k = 2.37 \times 10^7$ dm^3 mol^{-1} s^{-1}, $t_{1/2} = 4.98 \times 10^{-3}$ s.

22.10 First-order, $k = 7.2 \times 10^{-4}$ s^{-1}.

22.11 Propene: first order, HCl: third-order.

22.12 Steady-state approximation.

22.13 rate $= kK_1K_2[HCl]^3[CH_3CH{=}CH_2]$; look for evidence of proposed intermediates, e.g. using infrared spectroscopy to search for $(HCl)_2$.

22.14 (a) $\dfrac{1}{\tau^2} \approx 8k_a k_b [A]_{tot} + k_b^2$; (c) $k_b = 1.7 \times 10^7$ s^{-1},
$k_a = 2.7 \times 10^9$ dm^3 mol^{-1} s^{-1}, $K = 1.6 \times 10^2$.

22.15 $E_{a,eff} = -18$ kJ mol^{-1}, $E_a = +10$ kJ mol^{-1}.

22.16 16.7 kJ mol^{-1}, 1.14×10^{10} dm^3 mol^{-1} s^{-1}.

22.17 There are marked deviations at low pressures, indicating that the Lindemann theory is deficient in that region.

22.18 $\Delta_f H^{\ominus}(C_2H_5) = 121.2$ kJ mol^{-1}, $S_m^{\ominus}(C_2H_5) = 247.0$ J K^{-1} mol^{-1}, $\Delta_f G^{\ominus}(C_2H_5) = 148.3$ kJ mol^{-1}.

22.19 The product concentration ratio increases.

22.20 $v = k_2 K^{1/2}[A_2]^{1/2}[B]$.

22.23 (a) $kt = \dfrac{2x(A_0 - x)}{A_0^2(A_0 - 2x)^2}$; (b) $kt = \left(\dfrac{2x}{A_0^2(A_0 - 2x)}\right) + \left(\dfrac{1}{A_0^2}\right)\ln\left(\dfrac{A_0 - 2x}{A_0 - x}\right)$.

22.27 $v_{max} = k\left(\dfrac{[A]_0 + [B]_0}{2}\right)^2$.

22.29 2720 y.

22.30 (a) 0.642 μg; (b) 0.642 μg.

22.31 (a) First-order, (b) $k = 0.007\ 65$ min^{-1} $= 0.459$ h^{-1}, $t_{1/2} = 1.51$ h $= 91$ min.

22.35 $v = k[A][B]$, $k = \dfrac{k_1 k_2}{k_1}$.

22.36 $E_a = 105$ kJ mol^{-1}, $\Delta_r G^{\ominus} = -26.6$ kJ mol^{-1}, $\Delta_r H^{\ominus} = -34.3$ kJ mol^{-1}.

22.37 $E_a = 13.9$ kJ mol^{-1}, $A = 1.03 \times 10^9$ dm^3 mol^{-1} s^{-1}.

22.38 (a) $v = 1.1 \times 10^{-16}$ mol dm^{-3} s^{-1}; (b) $m = 2.2 \times 10^{11}$ kg or 220 Tg.

22.39 $k_1 = 3.82 \times 10^6$ dm^3 mol^{-1} s^{-1}, $k_2 = 5.1 \times 10^5$ dm^3 mol^{-1} s^{-1},
$k_3 = 4.17 \times 10^6$ dm^3 mol^{-1} s^{-1}, $\dfrac{k_2}{k_1} = 0.13$.

22.40 $v_0 = 9 \times 10^{-10}$ mol dm^{-3} s^{-1}, $t_{1/2} = 2.8 \times 10^5$ s $= 3$ days.

Chapter 23

23.2 $k = 5.0 \times 10^7$ dm^3 mol^{-1} s^{-1}.

23.3 (1) Initiation, (2), (3), (4) and (5) Propagation, (6) Termination.
$\dfrac{d[SiH_4]}{dt} = \left(\dfrac{k_1 k_4 k_5}{k_6}\right)^{1/2}[N_2O][SiH_4]^{1/2}$.

23.4 It is unlikely to lead to explosions.

23.5 $\dfrac{d[HI]}{dt} = \dfrac{2k_b k_a [I_2][H_2]}{k'_a + k_b[H_2]}$. This simple rate law is observed when step (b) is rate-determining so that step (a) is a rapid equilibrium and [I·] is in an approximate steady state. This is equivalent to $k_b[H_2] = k'_a$ and hence, $\dfrac{d[HI]}{dt} = 2k_b K[I_2][H_2]$.

23.6 3.1×10^{-4} einstein s^{-1} or 1.9×10^{20} s^{-1}.

23.7 (a) $\tau_0 = 6.67$ ns; (b) $k_f = 0.105$ ns^{-1}.

23.8 $k_q = 5.1 \times 10^6 \text{ dm}^3 \text{ mol}^{-1} \text{ s}^{-1}$.

23.9 $k_q = 1.98 \times 10^9 \text{ dm}^3 \text{ mol}^{-1} \text{ s}^{-1}$.

23.10 $R_0 = 3.52$ nm.

23.12 $\dfrac{d[CH_4]}{dt} = k_b \left(\dfrac{k_a}{2k_d}\right)^{1/2} [CH_3CHO]^{3/2}$.

$\dfrac{d[CH_3CHO]}{dt} = -k_a[CH_3CHO] - k_b \left(\dfrac{k_a}{2k_d}\right)^{1/2} [CH_3CHO]^{3/2}$.

23.13 $\delta M = \dfrac{p^{1/2}M}{1-p}$, $\delta M = M\{kt[A]_0(1+kt[A]_0)\}^{1/2}$.

23.14 (a) $\dfrac{\overline{M}_n^3}{\overline{M}_n^2} = \dfrac{M(1+4p+p^2)}{1-p^2}$; (b) $\dfrac{\overline{M}_n^3}{\overline{M}_n^2} = (6\langle n\rangle^2 - 6\langle n\rangle + 1)\langle n\rangle$.

23.15 $\langle n\rangle = v = k[\cdot M][I]^{1/2}$.

23.16 $\langle n\rangle = (1+2kt[A]_0)^{1/2}$.

23.18 $[B] = \left(\dfrac{I_a}{k}\right)^{1/2} \propto [A]^{1/2}$.

23.19 $\dfrac{d[A]}{dt} = -I - k_2 \left(\dfrac{I}{k_3}\right)^{1/2} [A]$.

23.21 $f = \dfrac{k_2 k_4 [CO]}{k_2[CO] + k_3[M]}$.

23.22 $N(t) = N_0 e^{(b-d)t} = N_0 e^{kt}$, fits data with $R^2 = 0.983$.

23.24 $v_{max} = 2.31\ \mu\text{mol dm}^{-3} \text{ s}^{-1}$, $k_b = 115 \text{ s}^{-1}$, $k_{cat} = 115 \text{ s}^{-1}$, $K_M = 1.11\ \mu\text{mol dm}^{-3}$, $\varepsilon = 104 \text{ dm}^3\ \mu\text{mol}^{-1} \text{ s}^{-1}$.

23.26 (b) pH = 7.0.

23.27 Uncompetitive.

23.28 Step 1 is autocatalytic. $\dfrac{a}{r} < S_0$: infection spreads, $\dfrac{a}{r} > S_0$: infection dies out.

23.29 $R = 2.6$ nm.

23.32 $v = 5.9 \times 10^{-13} \text{ mol dm}^{-3} \text{ s}^{-1}$.

23.34 (a) $v = 2.1 \times 10^{-15} \text{ mol dm}^{-3} \text{ s}^{-1}$; (b) $v = 1.6 \times 10^{-15} \text{ mol dm}^{-3} \text{ s}^{-1}$.

23.35 (a) Initiation, propagation, propagation, termination, initiation;

(b) $\dfrac{d[NO]}{dt} = -2k_a[NO]^2 - 2k_b[O][NO]$;

(c) $E_{a,eff} = E_b + \tfrac{1}{2}E_{-d} - \tfrac{1}{2}E_d$;

(d) $E_{a,eff} \approx 381.39 \text{ kJ mol}^{-1}$;

(e) $\dfrac{d[NO]}{dt} = -2k_b \left(\dfrac{k_a}{2k_d[M]}\right)^{1/2} [NO]^2$;

(f) $\dfrac{d[NO]}{dt} = -2k_b \left(\dfrac{k_a}{2k_d[M]}\right)^{1/2} [NO]^2$, where k_e is the rat constant for $NO + O_2 \rightarrow O + NO_2$. $E_{a,eff} = 253 \text{ kJ mol}^{-1}$, this value is consistent with the low range of the experimental values of $E_{a,eff}$.

Chapter 24

24.1 (a) $\sigma^* = 4.4 \times 10^{-20} \text{ m}^2$; (b) $P = 0.15$.

24.2 $\sigma^* = 4.0 \times 10^{-21} \text{ m}^2$ or $4.0 \times 10^{-3} \text{ nm}^2$, $P = \dfrac{\sigma^*}{\sigma} = 0.007$.

24.3 $k_2 = 1.7 \times 10^{11} \text{ M}^{-1} \text{ s}^{-1}$, $t = 3.6$ ns.

24.4 $E_a = 85.9 \text{ kJ mol}^{-1}$, $\Delta^\ddagger H = 83.8 \text{ kJ mol}^{-1}$, $\Delta^\ddagger S = +19.1 \text{ JK}^{-} \text{ mol}^{-1}$, $\Delta^\ddagger G = +79.0 \text{ kJ mol}^{-1}$.

24.5 2–.

24.6 $k° = 0.658 \text{ dm}^3 \text{ mol}^{-1} \text{ min}^{-1}$.

24.8 A complex of two univalent ions of the same sign.

24.9 (a) $-\dfrac{d[F_2O]}{dt} = k_1[F_2O]^2 + k_2\left(\dfrac{k_1}{k_4}\right)^{1/2} [F_2O]^{3/2}$; (b) $\Delta H(\text{FO—F}) \approx$

$E_1 = 160.9 \text{ kJ mol}^{-1}$, $\Delta H(\text{O—F}) \approx 224.4 \text{ kJ mol}^{-1}$, $E_2 \approx 60 \text{ kJ mol}^{-1}$.

24.10 $\Delta^\ddagger S = -148 \text{ J K}^{-1} \text{ mol}^{-1}$, $\Delta^\ddagger H = 60.44 \text{ kJ mol}^{-1}$, $\Delta^\ddagger U = 62.9 \text{ kJ mol}^{-1}$, $\Delta^\ddagger G = 104.8 \text{ kJ mol}^{-1}$.

24.11 Linear regression analysis of ln(rate constant) against $1/T$ yields the following results: $R = 0.999\,76$ and $R = 0.998\,48$, which indicate that the data are a good fit.

24.12 $k_{12} \approx 5.6 \times 10^8 \text{ dm}^3 \text{ mol}^{-1} \text{ s}^{-1}$.

24.15 $P = 5.2 \times 10^{-6}$.

24.16 $\log v \propto I^{1/2}$.

24.17 $k_1 = \dfrac{v^3}{v^{\ddagger 2}} e^{-\beta \Delta E_0}$, (a) $D = 2.7 \times 10^{-15} \text{ m}^2 \text{ s}^{-1}$, (b) $D = 1.1 \times 10^{-14} \text{ m}^2 \text{ s}^{-1}$.

24.18 $D \approx 2 \times 10^{-13} \text{ m}^2 \text{ s}^{-1}$, if $v^\ddagger = v$ and $9 \times 10^{-13} \text{ m}^2 \text{ s}^{-1}$ if $v^\ddagger = \tfrac{1}{2}v$.

24.20 5.

24.22 (a) $k_2 = 1.37 \times 10^6 \text{ dm}^3 \text{ mol}^{-1} \text{ s}^{-1}$; (b) $k_2 = 1.16 \times 10^6 \text{ dm}^3 \text{ mol}^{-1} \text{ s}^{-1}$.

24.23 For O_2 with ethyl: $P = 1.6 \times 10^{-3}$; For O_2 with cyclohexyl: $P = 1.8 \times 10^{-3}$.

24.24 (a) $k = 6.23 \times 10^9 \text{ dm}^3 \text{ mol}^{-1} \text{ s}^{-1}$; (b) $R^* = 3.7 \times 10^{-10}$ m or 0.37 nm.

24.25 $z_A = +3.0$.

24.26 $\lambda = 2.30$ eV, $T = 275$ K.

24.27 Yes, the equation appears to apply, $\beta = 13.4 \text{ nm}^{-1}$.

Chapter 25

25.1 For a cation above a flat surface, the energy is 0.11. For a cation at the foot of a high cliff, the energy is –0.51. The latter is the more likely settling point.

25.2 (a) At 100 kPa: $Z_W = 2.69 \times 10^{23} \text{ cm}^{-2} \text{ s}^{-1}$, $Z_{atom} = 2.0 \times 10^8 \text{ s}^{-1}$; (b) At 1000 kPa: $Z_W = 2.69 \times 10^{18} \text{ cm}^{-2} \text{ s}^{-1}$, $Z_{atom} = 2.0 \times 10^3 \text{ s}^{-1}$.

25.3 (a) $1.61 \times 10^{15} \text{ cm}^{-2}$; (b) $1.14 \times 10^{15} \text{ cm}^{-2}$; (c) $1.86 \times 10^{15} \text{ cm}^{-2}$.
For the collision frequencies:

	Hydrogen		Propane	
$Z/(\text{atom}^{-1} \text{ s}^{-1})$	100 Pa	10^{-7} Torr	100 Pa	10^{-7} Torr
(100)	6.8×10^5	8.7×10^{-2}	1.4×10^5	1.9×10^{-2}
(110)	9.6×10^5	1.2×10^{-1}	2.0×10^5	2.7×10^{-2}
(111)	5.9×10^5	7.5×10^{-2}	1.2×10^5	1.7×10^{-2}

25.4 $K = 0.37 \text{ Torr}^{-1}$.

25.5 (a) $c = 164$, V_{mon} 13.1 cm^3; (b) $c = 264$, V_{mon} 12.5 cm^3.

25.6 $V_\infty = 1.39 \text{ cm}^{-3}$, 59 m^2.

25.7 $c_2 = 2.4$, $c_1 = 0.16$.

25.8 $k = 0.02 \text{ Torr s}^{-1}$.

25.9 $K = 0.138 \text{ mg g}^{-1}$, $n = 0.58$.

25.10 $-\Delta_{ad}H^\ominus - 20.1 \text{ kJ mol}^{-1}$, $-\Delta_{ad}G^\ominus = 63.6 \text{ kJ mol}^{-1}$.

25.11 $n_\infty = 5.78 \text{ mol kg}^{-1}$, $K = 7.02 \text{ Pa}^{-1}$.

25.12 Regression analysis provides the following coefficients of correlation: R (Langmuir) = 0.973, R (Freudlich) = 0.999 94, R (Temkin) = 0.9590. The correlation coefficients and standard deviations indicate that the Freudlich isotherm provides the best fit.

25.13 $j_0/(\text{mA cm}^{-2}) = 0.78$, $\alpha = 0.38$.

25.14 $a(Sn^{2+}) \approx 2.2a(Pb^{2+})$.

25.15 $\delta = 2.5 \times 10^{-4}$ m or 0.25 mm.

25.16 87 mA.

25.18 (a) $E_0 = -0.618$ V;

(b)

$-E'/mV$	$jc/(\mu A\ cm^{-12})$
702	0.0324
727	0.0469
752	0.0663
812	0.154

(c) Regression analysis provides the following coefficients of correlation: $R = 0.999\ 94$, suggesting that the data fit the Tafel equation very well. $\alpha = 0.363$.

25.20 $\alpha = 0.50, j_0 = 0.150$ A m^{-2}, $-j_c = 0.0384$ A m^{-2}.

25.21 (a) The Tafel plot of $\ln j$ against E show no region of lineraity so the Tafel equation cannot be used to determine j_0 and α.

25.22 The linear regression explains 99.90 per cent of the variation in a $\ln j$ against η plot and standard deviations are low. There are no deviations from the Tafel equation/plot. $j_0 = 2.00 \times 10^{-5}$ mA m^{-2}, $\alpha = 0.498$.

25.26 $\left(\dfrac{RTV_\infty}{\sigma}\right) \mathrm{d}\ln(1-\theta)$.

25.28 $\langle j \rangle = \frac{1}{4}(1-2\alpha)f^2 j_0 \eta_0^2$, $\langle I \rangle = 7.2$ µA.

25.30 $j = j_L(1 - e^{F\eta^c/RT})$ For the anion current, the sign of η^c is changed, and the current of anions approaches its limiting value as η^c becomes more positive.

25.31 BET isotherm is a much better representation of the data. $V_{mon} =$ 75.4 cm^3, $c = 3.98$.

25.32 $\dfrac{K_1}{K_2} = 40.4$.

25.33 (a) R values in the range 0.975 to 0.991, the fit is good at all temperatures.

(b) $k_a = 3.68 \times 10^{-3}$, $\Delta_{ad}H = -8.67$ kJ mol^{-1}, $k_b = 2.62 \times 10^{-5}$ ppm^{-1}, $\Delta_b H = -15.7$ kJ mol^{-1}.

(c) k_a may be interpreted to be the maximum adsorption capacity at an adsorption enthalpy of zero, while k_b is the maximum affinity in the case for which the adsorbant–surface bonding enthalpy is zero.

25.34 (a) $k = 0.2289$, n = 0.6180; (c) $k = 0.5227$, n = 0.7273;

25.35 (a) K unit: $(g_R\ \mathrm{dm}^{-3})^{-1}$ [g_R = mass (grams) of rubber], K_F unit: $(\mathrm{mg})^{(1-1/n)} g_R^{-1}\ \mathrm{dm}^{-3/n}$, K_L unit: $(\mathrm{mg\ dm}^{-3})^{-1}$, M unit: $(\mathrm{mg}\ g_R^{-1})$.

(b) R (Linear) = 0.9612, R (Freudlich) = 0.9682, R (Langmuir) = 0.9690. on this basis alone, the fits are equally satisfactory, but not good. The Langmuir isotherm can be eliminated as it gives a negative value for K_L: the fit to the Freudlich isotherm has a large standard deviation. Hence the linear isotherm seems the best fit, but the Freudlich isotherm is preferred for this kind of system.

(c) $q_{rubber}/q_{charcoal} = 0.164 c_{eq}^{-0.46}$, hence much worse.

25.36 (a) $E^\ominus = +1.23$ V; (b) $E^\ominus = +1.06$ V.

25.37 (a) Therefore, the metals with a thermodynamic tendency to corrode in moist conditions at pH = 7 are Fe, Al, Co, Cr if oxygen is absent, but, if oxygen is present, all seven elements have a tendency to corrode.

(b) Ni: corrodes, Cd: corrodes, Mg: corrodes, Ti: corrodes, Mn: corrodes.

25.38 $I_{corr} \approx 6$ µA.

25.39 0.28 mg cm^{-2} d^{-1}.

Index

(T) denotes a table in the Data Section.

A

B

C

D

E

F

G

J

K

L

M

N

O

P

Q

R

S

T

Where is the best place to stay in Paris? Azzedine Alaïa 'Trois Chambres'
Where is your favourite place to eat in Paris? Le Voltaire, especially in the Madame Picot space!
Where is your favourite place to shop in Paris? The Bosquet golf shop JEAN TOUITOU

PARIS

TEXT BY TEDDY CZOPP

n 2005, Paris is reinventing and rejuvenating its grand history; within the ity's fashion industry, the legendary luxury brands are leading the way. tronger and more confident, iconic houses like Chanel, Dior, Cartier and Louis 'uitton have lately been revamping their stores and further embellishing their nages. Some might say that one result of the prevailing conservatism in Paris s that now is not a good time for young designers, in comparison with those in Jew York, Milan or London, where emerging talents are proudly celebrated. In ny case, it is the right moment for France to cherish its national treasures. This henomenon also applies to social and night life. Today, it is de rigueur to meet p and go out to old-fashioned cult places such as Maxim's or Castel, two magificent venues in Paris that have recently hosted successful fashion parties. It's ot only tourists that find these kind of places exciting – Parisians, too, are ppreciating their heritage. Looking back to a certain simplicity, clubbing has een new places like Le Baron or Truskel coming out as the coolest places to ang out; cabarets and revues like Le Lido or Le Crazy Horse are also hot again. 'or eating out, typical bistros and brasseries are preferred to fancy restaurants. Classy? Yes, Paris is a classy city but, today, that is exactly what Parisians like bout it.

m Jahr 2005 ist Paris dabei, seine großartige Geschichte neu zu schreiben und u verjüngen. In der Modebranche geben dabei die legendären Luxuslabels den 'on an. Erstarkt und selbstbewusster denn je haben Kultmarken wie Chanel, Dior, Cartier und Louis Vuitton in jüngster Zeit ihre Läden aufpoliert und ihr mage weiter verbessert. Man könnte behaupten, eine der Folgen des in Paris 'orherrschenden Konservativismus sei, dass junge Designer es hier momentan chwerer haben als etwa in New York, Mailand oder London, wo aufstrebende 'alente stolz gefeiert werden. Wie auch immer ist es aber bestimmt der rechte ugenblick für Frankreich, um seine nationalen Aushängeschilder zu hegen. Dieses Phänomen gilt auch für das gesellschaftliche und das Nachtleben. Heute st es Pflicht, gemeinsam altmodische legendäre Lokale wie das Maxim's oder Castel zu besuchen. An diesen beiden prachtvollen Schauplätzen fanden in jüngster Zeit gelungene Modepartys statt. Und es sind nicht nur die Touristen, die solche Orte aufregend finden – auch die Pariser schätzen diese Vermächtnisse. Gemäß dem Trend hin zu einer gewissen Schlichtheit sind beim Clubbing neue Lokale wie Le Baron oder Truskel als die coolsten Orte zum Abhängen gefragt. Aber auch Cabarets und Revue-Lokale wie Le Lido oder Le Crazy Horse sind wieder angesagt. Zum Essengehen gibt man typischen Bistros und Brasserien den Vorzug vor schicken Restaurants. Klassenbewusst? Ja, Paris ist wohl klassenbewusst, aber genau das lieben die Pariser heute an ihrer Stadt.

En 2005, Paris réinvente et rajeunit son illustre passé, et au sein de l'industrie de la mode, les légendaires marques de luxe ouvrent la voie. Plus fortes et plus confiantes, des maisons phares telles que Chanel, Dior, Cartier et Louis Vuitton ont récemment revu la conception de leurs boutiques et retravaillé leur image. D'aucuns diront que face au conservatisme qui prévaut à Paris, la période n'est pas des plus propices pour les jeunes créateurs, du moins par rapport à ceux de New York, Milan ou Londres où les talents émergents sont célébrés avec fierté. En tout cas, la France chérit actuellement ses trésors nationaux, un phénomène qui s'applique aussi à la vie sociale et nocturne. Aujourd'hui, la tendance veut qu'on sorte dans des lieux cultes et démodés comme Maxim's ou Castel, deux superbes adresses parisiennes qui depuis peu organisent avec succès des soirées très fashion. Ces endroits ne séduisent pas que les touristes : les Parisiens aussi savent apprécier leur héritage. Revenant à une certaine simplicité, la scène du clubbing voit de nouveaux rendez-vous, tels que Le Baron ou Le Truskel, s'imposer comme les bars les plus cool du moment ; les revues des cabarets du Lido ou du Crazy Horse bénéficient également d'un retour en grâce. Pour dîner, les Parisiens délaissent les restaurants de luxe et préfèrent se retrouver dans des bistrots et des brasseries plus typiques. Chic ? Oui, Paris reste une ville chic, mais c'est justement la raison pour laquelle ses habitants l'aiment autant aujourd'hui.

Der vormals an der Lower East Side angesiedelte Laden ist kürzlich nach SoHo umgezogen. Man findet dort aber weiterhin Männer- und Damenmode der progressivsten Designer wie Raf Simons, Boudicca, As Four, AF Vandevorst und Cosmic Wonder.

Autrefois niché au milieu des anciens vendeurs de fripes du Lower East Side, Seven vient de déménager à SoHo. La boutique continue de vendre les vêtements pour homme et pour femme des créateurs les plus innovants au monde, tels que Raf Simons, Boudicca, As Four, Preen, AF Vandevorst et Cosmic Wonder.

110 Mercer Street
New York, NY 10012
Tel. +1 212 6466 5400
www.sevennewyork.com

Opening Ceremony
Modelled after the Olympics, each year Opening Ceremony sells designer wares from one spot on the globe, for example, Brazil, Germany, Hong Kong, in addition to local New York favourites such as Cloak, Rachel Comey, Indigo People and Mary Ping. London is the latest to receive the spotlight.

Nach dem Vorbild der Olympischen Spiele verkauft Opening Ceremony jedes Jahr Designerkleidung aus einem bestimmten Land oder einer Stadt wie z. B. Brasilien, Deutschland oder Hongkong. Dazu kommen einheimische New Yorker Lieblingslabels wie Cloak, Rachel Comey, Indigo People und Mary Ping. Als nächstes steht London im Rampenlicht.

Inspirée par les Jeux olympiques, la boutique Opening Ceremony vend chaque année les vêtements de créateurs venus d'un endroit particulier dans le monde, par exemple du Brésil, d'Allemagne ou de Hong-Kong, en plus des chouchous new-yorkais tels que Cloak, Rachel Comey, Indigo People et Mary Ping. Récemment, la ville de Londres a été mise à l'honneur.

35 Howard Street
New York, NY 10013
Tel. +1 212 2219 2688
www.openingceremony.us

Kirna Zabête
This cheery two-storey SoHo boutique carries a wide range of unabashedly girly merchandise – including hard-to-find pieces by Lulu Guinness, Anya Hindmarch and New York's own Alice Roi – and reflects the quirky-cool personalities of its owners, Sarah Easley and Elizabeth Shepherd.

Die fröhliche zweistöckige Boutique in SoHo führt ein breites Sortiment frecher Girly-Mode. Darunter schwer zu findende Teile von Lulu Guinness, Anya Hindmarch und der New Yorkerin Alice Roi. Das Ganze entspricht dem eigenwillig coolen Geschmack der Ladenbesitzerinnen Sarah Easley und Elizabeth Shepherd.

Répartie sur deux niveaux, cette joyeuse boutique de SoHo réalise sans complexe le rêve de toutes les filles avec son large choix de vêtements, notamment des pièces introuvables ailleurs de Lulu Guinness, Anya Hindmarch et de la new-yorkaise Alice Roi, à l'image des personnalités cool et complexes de ses propriétaires Sarah Easley et Elizabeth Shepherd.

96 Greene Street
New York, NY 10012
Tel. +1 212 9419 656
www.kirnazabete.com

Century 21
Situated across from where the World Trade Center once stood, Century 21 was forced to close its doors on 9/11 before making a triumphant return nine months later. The beloved discount department store is where the sartorially obsessed make a beeline when one of its personal shoppers calls with new arrivals from, say, Alexander McQueen, Dolce & Gabbana or Helmut Lang.

Gleich gegenüber vom Standort des ehemaligen World Trade Centers gelegen, war Century 21 gezwungen, seine Tore am 11. September 2002 zu schließen. Die Wiedereröffnung neun Monate später war ein Triumph. In diese heiß geliebte Institution pilgern Fashion-Addict wenn einer ihrer Personal Shoppers Neuzugänge etwa von Alexander McQueen, Dolce & Gabbana oder Helmut Lang vermeldet.

Située en face de l'ex-World Trade Center, la boutique Century 21 a fermé ses portes le 11 septembre avant de faire un retour triomphant neuf mois plus tard. Ce soldeur est particulièrement prisé par les obsédés de la couture qui s'y précipitent dès qu'un des acheteurs de la boutique les prévient de l'arrivée de nouveaux stocks Alexander McQueen, Dolce & Gabbana ou Helmut Lang, entre autres.

22 Cortlandt Street
New York, NY 10007
Tel. +1 212 2279 092

Comme des Garçons
Beyond its funky funnel-like entryway lies a cavernous and curved-wall interior echoing Rei Kawakubo's familiar off-beat approach where all her warped colour and fabric combos can be found. Also available are Comme's popular unisex fragrances.

Hinter dem außergewöhnlichen trichterförmigen Eingang stößt man auf ein höhlenartiges Inneres mit gewölbten Wänden, das Rei Kawakubos bekanntem Stil entspricht. Man findet hier all ihre eigenwilligen Farb- und Materialkombinationen. Auch die beliebten Unisex-Düfte sind erhältlich.

Derrière son entrée funky en forme d'entonnoir, on découvre une caverne toute en courbes qui reflète l'approche décalée familière de Rei Kawakubo et propose toutes ses couleurs et combinaisons de tissus bizarres. On y trouve également les célèbres parfums unisexe de Comme des Garçons.

520 West 22nd Street
New York, NY 10012
Tel. +1 212 6049 200

Prada
Created by design darling Rem Koolhaas, this space is a virtual theme park with not only the complete Prada collections to browse through, but favourite vintage pieces, plus video monitors, digital catalogues and adjustable changing rooms to complete the experience.

Der vom Designer-Liebling Rem Koolhaas entworfene Raum ist genau genommen ein virtueller Vergnügungspark, in dem man nicht nur durch die komplette Prada-Kollektion streifen kann. Begehrte Vintage-Stücke, Videobildschirme, digitale Kataloge und flexible Umkleidekabinen machen das Einkaufserlebnis perfekt.

Conçu par l'architecte chéri Rem Koolhaas, cet espace est une sorte de parc à thème où l'on a non seulement droit au panorama complet des collections Prada, mais aussi à des pièces vintage très recherchées, le tout au milieu d'écrans vidéo, de catalogues numériques et de cabines d'essayage modulables.

575 Broadway
New York, NY 10012
Tel. +1 212 3348 888

Zero
It's all about the pluses at Zero, a sparse shop launched by designer Maria Cornejo after moving to New York from London with her husband, photographer Mark Borthwick. Pieces hang unhindered on well-spaced racks, leaving most of the store open so the mind can roam free.

Diesen schlichten Laden eröffnete die Designerin Maria Cornejo, nachdem sie mit ihrem Mann, dem Fotografen Mark Borthwick, von London nach New York gezogen war. Die Kleider hängen frei auf wohlplatzierten Ständern und lassen den Großteil der Verkaufsfläche frei. So kann man seine Gedanken ungehindert schweifen lassen.

C'est l'idée de plus qui compte pour cette austère boutique lancée par la créatrice Maria Cornejo venue de Londres avec son époux, le photographe Mark Borthwick. Les vêtements sont accrochés librement sur des portants bien espacés, laissant la majeure partie du lieu ouverte à la méditation.

225 Mott Street
New York, NY 10036
Tel. +1 212 9253 849

VINTAGE FASHION SHOPS

Cherry
Cherry boasts an eclectic mix of high-end designer goods with wearable art and expensive vintage leathers. Treasures line the walls, and it isn't uncommon for top designers to pop in for a little secret research every now and then.

Hier findet man eine eklektische Mischung aus hochwertigen Designerwaren, Kunst zum Anziehen und teuren Second-Hand-Stücken aus Leder. Schätze säumen die Wände, und es ist nichts Ungewöhnliches, wenn immer mal wieder Designer von Weltruf reinschauen, um ein wenig Feldforschung zu betreiben.

Cherry affiche un mélange éclectique de pièces de créateurs haut de gamme, d'œuvres d'art portables ou de cuirs vintage à prix d'or. Des trésors s'alignent sur les murs et on y croise de temps en temps des stylistes connus en pleine recherche secrète.

19 Eighth Avenue
New York, NY 10014
Tel. +1 212 9241 410

Resurrection
This secondhand spot is serious about vintage, offering precious pieces by the likes of Ossie Clark, Emilio Pucci, Halston and Rudi Gernreich.

In diesem Second-Hand-Laden gibt es kostbare Stücke von Ossie Clark, Emilio Pucci, Halston und Rudi Gernreich.

Cette boutique de vêtements d'occasion ne rigole pas avec le vintage : elle propose de précieuses créations d'Ossie Clark, Emilio Pucci, Halston et Rudi Gernreich.

217 Mott Street
New York, 10012
Tel. +1 212 6251 374

What Comes Around Goes Around
Flogging more Western looks than high-end designer duds, this SoHo store with antler chandeliers specializes in denim and leather goods.

Der Laden in SoHo mit Lampen aus Geweihen verkauft eher Westernmode als edle Designerklamotten und hat sich auf Jeans und Ledersachen spezialisiert.

Plus axée sur les looks Western que sur les perles des créateurs de luxe, cette boutique de SoHo aux chandeliers en bois se spécialise dans le jean et le cuir.

351 West Broadway
New York, NY 10013
Tel. +1 212 343 9303

FLEA MARKETS

Chelsea Flea Market

(Annex Antique Fair and Flea Market) The sprawling Chelsea Flea Markets are packed with fashion insiders each and every weekend. Among the 10,000 bargain-hunters who flock here, designers are often spotted in shades rummaging for inspiration.

Auf diesem weitläufigen Flohmarkt tummeln sich jedes Wochenende zahllose Insider aus der Modebranche. Unter den 10 000 Schnäppchenjägern entdeckt man oft Designer mit großen Sonnenbrillen, die hier nach neuer Inspiration suchen.

Ces marchés aux puces de Chelsea qui s'étendent sans cesse sont envahis par les groupies de la mode chaque week-end. Parmi les 10 000 farfouilleurs qui s'y précipitent, on croise souvent des créateurs en quête d'inspiration qui espèrent passer incognito sous leurs lunettes noires.

6th Avenue between 24th Street and 26th Street
New York, NY 10116
Tel. +1 212 243 5343

RESTAURANTS

Schiller's Liquor Bar

Restaurateur Keith McNally (Pastis, Balthazar) continues to impress with this atmospheric, low-key bistro in the Lower East Side. French comfort food in the form of steak frites, an iron skillet of garlic shrimps and a penne al' arrabiata can be had here. Perfect for a quick bite any time of the day.

Der Gastronom Keith McNally (Pastis, Balthazar) macht sich auch mit diesem schlichten, gemütlichen Bistro an der Lower East Side alle Ehre. Man serviert hier schnelle französische Küche wie Steak frites oder ein Eisenpfännchen mit Knoblauch-Shrimps, aber auch Penne al'Arrabiata. Perfekt für einen schnellen Happen zu jeder Tageszeit.

Le restaurateur Keith McNally (Pastis, Balthazar) continue à nous impressionner avec ce bistrot discret mais chaleureux du Lower East Side. Il propose de la bonne vieille cuisine française sous forme de steak frites, une poêlée de crevettes à l'ail ou des penne à l'arrabiata. Le lieu idéal pour satisfaire un petit creux à n'importe quelle heure de la journée.

131 Rivington Street
New York, NY 10002
Tel. +1 212 260 4555

Bottino

The art crowd routinely pit stops at Bottino during the synchronized exhibit openings of the flourishing Chelsea gallery scene. With an uncomplicated menu, it cleans the palette of the art patron who just gobbled up a Damien Hirst or a Julian Schnabel.

Die Kunstbranche kehrt während der terminlich aufeinander abgestimmten Vernissagen in der florierenden Galerienszene von Chelsea hier ein. Die schlichte Speisekarte sorgt bei dem Kunstliebhaber, der gerade einen Damien Hirst oder Julian Schnabel gekauft hat, für einen klaren Kopf.

Le milieu de l'art vient par habitude chez Bottino lors des vernissages groupés des galeries d'un Chelsea en plein essor. Grâce à sa carte sans chichi, il réussit à déculpabiliser le mécène qui vient de dépenser une fortune sur un Damien Hirst ou un Julian Schnabel.

246 Tenth Avenue
New York, NY 10001
Tel. +1 212 206 6766
www.bottinonyc.com

Da Silvano

The scenester's Italian restaurant of choice, Da Silvano has been the pinnacle of New York dining (read media-elite ogling) for over twenty years. Anna Wintour and her Voguettes are often spotted preening at outdoor tables while a stream of media giants eagerly put their names on the block-long waiting list.

Der Lieblingsitaliener schlechthin der Szenegänger und seit über zwanzig Jahren der Fels in der New Yorker Brandung (insbesondere für die Medienelite). Man sieht Anna Wintour und ihre Voguetten sich oft an einem der Tische im Freien putzen, während Prominente aus der Medienbranche sich darum reißen, ihre Namen auf die ellenlange Warteliste zu setzen.

Restaurant italien terriblement à la mode, Da Silvano est le lieu préféré des dîneurs new-yorkais (comprenez l'élite médiatique) depuis plus de vingt ans. Anna Wintour et ses Voguettes y sont souvent repérées en train de faire les belles en terrasse, tandis que les célébrités des médias viennent ajouter leur nom sur une liste d'attente assez longue pour faire le tour du bloc.

260 Sixth Avenue
New York, NY 10014
Tel. +1 212 982 2343
www.dasilvano.com

Nobu

New York's fascination with Japanese fusion isn't anywhere close to over. Nobu, with its South American- and Italian-tinted menu, is the latest of these restaurants with a chef, Nobuyuki Matsuhisa, at least as famous as the celebrities who try to slip in for a bit of take-out without being noticed by the gawking diners.

Die Begeisterung der New Yorker für eine Mischung aus italienischer und japanischer Küche ist ungebrochen. Das Nobu mit seiner südamerikanisch und italienisch gefärbten Karte ist das jüngste Restaurant dieser Art. Der Küchenchef Nobuyuki Matsuhisa ist mindestens ebenso berühmt wie die Stars, die versuchen hereinzuschleichen und etwas zum Mitnehmen abzuholen, ohne von den lästigen Gästen bemerkt zu werden.

La fascination de New York pour la cuisine fusion japonaise n'est pas près de passer. Nobu, avec sa carte teintée de cuisine sud-américaine et italienne, est le dernier-né de ces restaurants et bénéficie d'un chef, Nobuyuki Matsuhisa, au moins aussi connu que les innombrables célébrités qui essaient de s'y glisser discrètement pour prendre des plats à emporter sans être remarquées par les clients indiscrets.

105 Hudson Street
New York, NY 10013
Tel. +1 212 219 0500

Café Gitane

A NoLIta institution, little Café Gitane is like the coffeehouses you used to frequent as a student to be seen reading Jean-Paul Sartre while smoking like a chimney. Except the coffee and French-Moroccan fare at this late-night spot are truly excellent, as is the locals-watching.

Als Institution von NoLIta erinnert das kleine Café Gitane an die Lokale der eigenen Studentenzeit. Als man demonstrativ Jean-Paul Sartre las und dazu wie ein Schlot rauchte. Nur dass der Kaffee und die kleinen französisch-marokkanischen Gerichte hier ebenso gut sind wie die Gelegenheit, Einheimische zu beobachten.

Institution de NoLIta, ce petit Café Gitane rappelle ces cafétérias dans lesquelles vous passiez votre temps à l'université pour être vu en train de lire du Jean-Paul Sartre tout en fumant comme un pompier. Sauf que dans ce bar de nuit, le café et la cuisine franco-marocaine sont vraiment excellents. Tout comme le défilé des habitués.

242 Mott Street
New York, NY 10012
Tel. +1 212 334 9552

BARS

Bungalow 8

Bungalow 8's claim to fame is its passion for pampering. Though located in a "developing" part of Chelsea, inside it feels like an Egyptian oasis. You'll imbibe like a pharaoh, and you can even order a helicopter if you've had one too many – just the kind of star treatment that has lured the fashion flock since day one.

Bungalow 8 ist berühmt dafür, seine Gäste zu verwöhnen. Die Bar liegt zwar in einem noch unterentwickelten Teil von Chelsea, doch fühlt man sich darin wie in einer ägyptischen Oase. Sie können bechern wie ein Pharao und sich sogar von einem Helikopter abholen lassen, wenn Sie ein Glas zu viel erwischt haben. Genau die Art von Starbehandlung, die die Modemeute seit dem ersten Tag hierher lockt.

Bungalow 8 prétend être connu pour sa passion du service. Bien que situé dans une zone « en développement » de Chelsea, on s'y sent comme dans une oasis égyptienne. Vous y boirez comme un pharaon, et pourrez même commander un hélicoptère si vous avez bu un verre de trop : voilà le genre de traitement qui attire les gens de la mode depuis le jour de l'ouverture.

515 West 27th Street
New York, NY 10001
Tel. +1 212 629 3333

Dark Room

In the new LES (that's Lower East Side, natch), this stonework basement bar is leading the way. In one room, old-school locals croon to Johnny Cash and his ilk, while in the other room, the glossy fashion set gets down to the sounds of dirty rock.

An der neuen LES (steht natürlich für Lower East Side) begründet diese gemauerte Kellerbar einen Trend. In einem Raum gröhlen Einheimische der alten Garde zu Johnny Cash und Konsorten, während sich in einem anderen die Hochglanz-Modemagazin-Menschen zum Sound von schmutziger Rockmusik entspannen.

Dans le nouveau « L.E.S. » (Lower East Side, bien évidemment), ce bar en sous-sol ouvre la voie. Dans une salle, les vieux habitués reprennent les chansons de Johnny Cash et de sa troupe, tandis que dans une autre, la scène fashion se lâche aux rythmes d'un rock crasseux.

165 Ludlow Street
New York, NY 10002
Tel. +1 212 353 0536

Happy Ending

Once a brothel for those of the Asian persuasion, Happy Ending (a name taken from the way in which the previous tenants would finish a massage) has reemerged as Chinatown's bar of choice for residents of the gentrifying neighbourhood and late-night shoppers of area boutiques.

Das ehemalige asiatische Bordell (Happy Ending nannten die Vormieter das Ende einer Massage) hat sich zu Chinatowns Bar der Wahl sowohl bei den Einheimischen des aufgewerteten Viertels als auch bei den späten Kunden der Boutiquen rundum entwickelt.

Ancien bordel de Chinatown dédié aux amateurs de beautés asiatiques, Happy Ending (un nom inspiré par la façon dont les précédents propriétaires clôturaient leurs massages) renaît comme le bar de choix des habitants de ce quartier en voie d'embourgeoisement et des clients nocturnes des boutiques du coin.

302 Broome Street
New York, NY 10002
Tel. +1 212 334 9676
www.happyendinglounge.com

Passerby

Attached to the Gavin Brown gallery, this insider hang-out that's the size of a postage stamp attracts art students, underground music enthusiasts and the fashion faithful alike, none too cool to kick up their heels on the lit-up dance floor.

Gleich neben der Galerie Gavin Brown lockt dieser briefmarkengroße Insidertreff Kunststudenten, Fans von Underground Music und Modeverrückte an, die sich nicht zu cool sind, auf der beleuchteten Tanzfläche ihre High Heels wegzukicken.

Rattaché à la galerie Gavin Brown, étudiants en art, fans de musique underground et fidèles de la mode n'hésitent pas à faire claquer leurs talons hauts sur le dancefloor lumineux de ce repaire de connaisseurs à peine plus grand qu'un timbre-poste.

436 West 15th Street
New York, NY 10011
Tel. +1 212 206 6847

HOTELS

Chelsea Hotel

The legendary haunt where William S. Burroughs wrote 'Naked Lunch', this one-time home to Jimi Hendrix, Janis Joplin, Andy Warhol, Tennessee Williams, Patti Smith, Allen Ginsberg and, tragically, Sid Vicious and Nancy Spungen still attracts the city's more bohemian passers-through.

Dies ist der legendäre Schlupfwinkel, wo William S. Burroughs „Naked Lunch" schrieb. Nach wie vor zieht das einstige Zuhause von Jimi Hendrix, Janis Joplin, Andy Warhol, Tennessee Williams, Patti Smith, Allen Ginsberg und tragischerweise auch Sid Vicious sowie Nancy Spungen eher die Bohemiens unter den Durchreisenden an.

Lieu légendaire où William Burroughs a écrit « Le Festin Nu », cet hôtel a vu défiler Jimi Hendrix, Janis Joplin, Andy Warhol, Tennessee Williams, Patti Smith et Allen Ginsberg. Sans parler du souvenir tragique laissé par Sid Vicious et Nancy Spungen. Aujourd'hui, il attire toujours les visiteurs les plus bohèmes de la ville.

222 West 23rd Street
New York, NY 10011
Tel. +1 212 243 3700
www.hotelchelsea.com

Maritime Hotel

With its perfectly round windows overlooking the Hudson river, the Maritime Hotel is kitsch raised to an art form. Teak furniture and green-tiled bathrooms typify the over-the-top retro sheen of the rooms. To boot, its two restaurants and its Japanese-themed bar-club, Hiro, are among the city's preferred places to throw back a few.

Mit seinen kreisrunden Fenstern hinaus auf den Hudson hat das Maritime Hotel Kitsch zur Kunstform erhoben. Teakholzmöbel und grün geflieste Bäder kennzeichnen den übersteigerten Retro-Chic der Zimmer. Obendrein zählen die zwei Hotelrestaurants und die Club-Bar im japanischen Stil zu den bevorzugten Retrolokalen der Stadt.

Avec ses fenêtres toutes rondes donnant sur l'Hudson, le Maritime Hotel hisse le kitsch au rang d'art à part entière. Meubles en teck et salles de bains à carrelage vert caractérisent le lustre rétro surchargé de ses chambres. Par-dessus le marché, ses deux restaurants et Hiro, son bar-club japonais, comptent parmi les adresses préférées de la ville.

88 Ninth Avenue
New York, NY 10011
Tel. +1 212 243 3888
www.themaritimehotel.com

Hotel Gansevoort

Named after the old Dutch street onto which it looks, the monolithic Hotel Gansevoort is anything but Old World. The newest resident of the Meatpacking District boasts minimalist-catering features from a metal exterior to granite ceramic floors on the interior. The rooftop swimming pool is a must.

Das monolithische Hotel ist zwar nach der alten holländischen Straße benannt, auf die es blickt, ist aber nichts weniger als ein Vertreter der Alten Welt. Die jüngste Errungenschaft des Meatpacking District ist vielmehr stolz auf ihre Minimalismus, angefangen bei der Metallfassade bis hin zu den Granit-Keramik-Fußböden. Der Pool auf dem Dach ist ein Muss.

Baptisé du nom hollandais de la vieille rue sur laquelle il donne, le monolithique Gansevoort n'a vraiment rien d'un hôtel à l'ancienne. Le tout nouveau résident du quartier des anciens abattoirs affiche tous les attributs du minimalisme hôtelier, de ses extérieurs en métal à ses sols carrelés de granite. A voir absolument : la piscine sur le toit.

18 Ninth Avenue
New York, NY 10014
Tel. +1 212 206 6700
www.hotelgansevoort.com

The Mercer

Built inside a landmark 19th century building in SoHo, this sister hotel to The Standard in Los Angeles serves as the temporary home to models, designers and a slew of Hollywood-variety celebrities. Don't check out before dining at the Mercer Kitchen in the basement.

Das in einem Wahrzeichen SoHos aus dem 19. Jahrhundert untergebrachte Partnerhotel von The Standard in Los Angeles dient Models, Designern und einem Haufen Hollywood-Prominenz als zweites Zuhause. Vor dem Auschecken unbedingt einmal in der Mercer Kitchen im Untergeschoss zu Abend essen.

Aménagé dans un immeuble classé du XIXe siècle en plein SoHo, cet hôtel est le petit frère du Standard de Los Angeles. Mannequins, créateurs de mode et autres célébrités hollywoodiennes en ont fait leur pied-à-terre new-yorkais. Ne partez pas sans avoir dîné au restaurant Mercer Kitchen du sous-sol.

147 Mercer Street
New York, NY 10012
Tel. +1 212 966 6060
www.mercerhotel.com

Soho Grand

This industrial, steel-clad SoHo outpost features spacious rooms and penthouses with possibly the best views in all of Lower Manhattan. Better still, travellers who just can't part with their pets can procure grooming, day care, veterinarian services and limos for their pampered little darlings.

Das stahlverkleidete Haus im Industriedesign ist ein Außenposten in SoHo und präsentiert sich mit geräumigen Zimmern und Penthouses mit der wahrscheinlich besten Aussicht über ganz Lower Manhattan. Noch dazu können Reisende, die sich nic von ihren Haustieren trennen wolle Hundefriseur, Betreuung, Tierarzt u Limousinen für ihre verhätschelten Lieblinge bestellen.

Habillé d'acier, cet hôtel de style industriel à SoHo possède des chambres spacieuses et ses penthouses offrent les plus belles vue sur le sud de Manhattan. Le plus : client qui ne peut laisser son anim domestique chez lui se voit propos divers services de toilettage et soi quotidiens, l'assistance d'un vétérinaire et même des limousines pou promener son toutou adoré.

310 West Broadway
New York, NY 10013
Tel. +1 212 965 3000
www.sohogrand.com

60 Thompson

Designed by Thomas O'Brien, als responsible for homes for Giorgio Armani and Ralph Lauren, this Mediterranean-tinged hotel in SoH has an open and breezy feel that it far apart from the city's other, overly designed mainstays.

Das Design dieses mediterran an hauchten Hotels in SoHo stammt Thomas O'Brien, der auch die Pri häuser von Giorgio Armani und R Lauren eingerichtet hat. Das offer luftige Ambiente grenzt es deutlic von den anderen übertrieben des ten Hauptquartieren der Stadt ab.

Conçu par Thomas O'Brien, à qui doit également les résidences de Giorgio Armani et de Ralph Laure cet hôtel de SoHo au style médit néen dégage un sentiment d'ouv ture et d'élégance éthérée qui le tingue d'autres hôtels trop tape-à de la ville.

60 Thompson Street
New York, NY 10012
Tel. +1 212 431 0400
www.60thompson.com

AIRPORTS

John F. Kennedy Internation Airport

Jamaica, NY 11430
Tel. +1 718 244 4444

Newark International Airpo

Newark, NJ 07114
Tel. +1 973 961 6000

La Guardia Airport

Flushing, NY 11371
Tel. +1 718 533 3400

FASHION SHOWS AND FAIRS

Semaine de la Mode de Paris
Schedules and contacts for menswear, womenswear and haute couture presentations available from www.modeaparis.com

Veranstaltungskalender und Kontaktadressen für Herren- und Damenmode sowie Haute-Couture-Präsentationen findet man unter www.modeaparis.com

Programmes et contacts pour les défilés homme, femme et haute couture disponibles à l'adresse : www.modeaparis.com

Première Classe
Fashion trade show specialising in young emerging designers.

Modemesse speziell für junge, aufstrebende Designer.

Salon dédié aux jeunes créateurs émergents.

www.premiere-classe.com

Salon du Prêt-à-Porter
Major commercial fashion fair.

Wichtige kommerzielle Modemesse.

Principal salon professionnel de la mode.

www.pretparis.com

Tranoï
Fresh accessories and fashion brands.

Frische Accessoires und Modelabels.

Salon consacré aux jeunes griffes de mode et d'accessoires.

www.tranoi.com

Workshop
International fashion and accessories designers.

Internationale Mode- und Accessoire-Designer.

Salon international dédié aux créateurs de mode et d'accessoires.

www.workshopsalons.com

Rendez-vous Paris
Trendy young designer fair with sessions for menswear and womenswear.

Trendige Messe für junge Designer mit Sessions für Herren und Damen.

Salon des jeunes créateurs tendance, avec des défilés pour homme et pour femme.

www.rendez-vous-paris.com

Paris sur Mode
Good commercial young designers fair.

Gute kommerzielle Messe für Jung-Designer.

Salon professionnel intéressant pour les jeunes créateurs.

www.parissurmode.com

19 Vendôme
Luxury trade show.

Luxusmesse für den Handel.

Salon de l'industrie du luxe.

www.19vendome.com

FASHION SCHOOLS

Studio Berçot
29, rue des Petites-Ecuries
75010 Paris
Tel. +33 (0)1 4246 1555
www.studio-bercot.com

Esmod International
16, boulevard Montmartre
75009 Paris
Tel. +33 (0)1 4483 8150
www.esmod.com

Institut Français de la Mode
33, rue Jean-Goujon
75008 Paris
Tel. +33 (0)1 5659 2222
www.ifm-paris.org

Ecole de la Chambre Syndicale de la Couture Parisienne
45, rue Saint-Roch
75001 Paris
Tel. +33 (0)1 4261 0077

FASHION MUSEUMS AND GALLERIES

Musée Galliera
Fashion history (around 100 000 costumes are classified) coexists with modern exhibitions of personal collections from Marlene Dietrich to 1960's singer Sylvie Vartan.

Modegeschichte (etwa 100 000 Exponate), daneben moderne Ausstellungen wie die private Garderobe von Marlene Dietrich oder der französischen Sängerin der 1960er Jahre, Sylvie Vartan.

Les collections historiques (près de 100 000 costumes sont répertoriés) côtoient des expositions plus modernes consacrées, entre autres, aux collections personnelles de Marlène Dietrich ou de Sylvie Vartan.

10, avenue Pierre de Serbie
75016 Paris
Tel. +33 (0)1 5652 8600

Musée de la Mode et du Textile
Following a brilliant retrospective of Dutch duo Viktor & Rolf, the museum exhibited the work of Yohji Yamamoto in summer 2005.

Auf eine brillante Retrospektive über das holländische Designer-Duo Viktor & Rolf folgte im Sommer 2005 eine Ausstellung über das Werk von Yohji Yamamoto.

Après une brillante rétrospective dédiée au duo hollandais Viktor & Rolf, le musée consacre toute une exposition au Yohji Yamamoto à l'été 2005.

107, rue de Rivoli
75001 Paris
Tel. +33 (0)1 4455 5750
www.ucad.fr

Centre Georges Pompidou
The inevitable spot in Paris for contemporary art with a beautiful panoramic view from the terrace of restaurant Georges.

Der obligatorische Pariser Ausstellungsort für moderne Kunst. Von der Terrasse des Restaurants Georges hat man einen wunderbaren Rundblick über die Stadt.

Le rendez-vous incontournable de l'art contemporain à Paris offre également une sublime vue panoramique sur la ville depuis la terrasse du restaurant Georges.

Place Georges-Pompidou
75004 Paris
Tel. +33 (0)1 4478 1233
www.centrepompidou.fr

Le Jeu de Paume
The beautiful space in the Tuileries gardens has become a museum dedicated to photography and the image, and celebrated its change with a long-awaited retrospective of Guy Bourdin's work.

Aus den wunderschönen Räumlichkeiten in den Tuilerien hat man ein Museum für Fotografie gemacht. Diese neue Bestimmung wurde mit einer lang ersehnten Retrospektive über das Werk Guy Bourdins gefeiert.

Situé dans les Jardins des Tuileries, ce magnifique espace a été transformé en musée de la photographie et de l'image. Il a célébré sa reconversion à travers une rétrospective très attendue consacrée à l'œuvre de Guy Bourdin.

1, Place de la Concorde
75008 Paris
Tel. +33 (0)1 4703 1250
www.jeudepaume.org

Fondation Cartier pour l'Art Contemporain
The Fondation is lovated in a magnificent glass building surrounded by a garden and will continue exhibiting contemporary artists in the future.

Die Fondation residiert in einem prächtigen, von einem Garten umgebenen Glasbau und wird dort auch in Zukunft die Werke zeitgenössischer Künstler ausstellen.

Installée dans un superbe immeuble de verre cerné d'un beau jardin, la Fondation continue à exposer les œuvres d'artistes contemporains.

261, Boulevard Raspail
75014 Paris
Tel. +33 (0)1 4218 5650
www.fondation.cartier.fr

Palais de Tokyo
The huge friendly space (open till midnight) showcases experimental young contemporary artists and welcomes you to visit its great library, restaurant TokyoEat and fashion and accessories store BlackBlock.

Das riesige, freundlich wirkende Museum (bis 24 h geöffnet) präsentiert experimentelle junge Künstler und lädt außerdem in seine prächtige Bibliothek, ins Restaurant TokyoEat sowie in den Mode- und Accessoire-Laden BlackBlock ein.

Cet espace immense mais accueillant (ouvert jusqu'à minuit) reçoit de jeunes artistes contemporains expérimentaux. Vous y trouverez également une superbe librairie, le restaurant TokyoEat et la boutique de mode et d'accessoires BlackBlock.

13, avenue du Président Wilson
75016 Paris
Tel. +33 (0)1 4723 5401
www.palaisdetokyo.com

Galerie du Jour Agnès B.
Always on top of fashion, music, cinema and art, Agnès B. has become a highly respected curator exhibiting the works of Martin Parr and Vincent Gallo.

Immer auf dem neuesten Stand, was Mode, Musik, Film und Kunst angeht, ist Agnès B. zu einer angesehenen Kuratorin für die Arbeiten von Martin Parr und Vincent Gallo avanciert.

Toujours à l'avant-garde de la mode, de la musique, du cinéma et de l'art, Agnès B. est aussi une conservatrice respectée dont la galerie expose les œuvres de Martin Parr et de Vincent Gallo.

44, rue Quincampoix
75004 Paris
Tel. +33 (0)1 4454 5590
www.galeriedujour.com

Tools Galerie
This small space is dedicated to innovative product design by artists such as Droog Design, Laurence Brabant or Hella Jongerius.

Diese kleine Galerie hat sich innovativem Produktdesign von Künstlern wie Droog Design, Laurence Brabant oder Hella Jongerius verschrieben.

Ce petit espace consacré aux nouveaux designers expose les créations innovantes d'artistes tels que Droog Design, Laurence Brabant ou Hella Jongerius.

119, rue Vieille-du-Temple
75003 Paris
Tel. + 33 (0)1 4277 3580

Galerie Emmanuel Perrotin
The gallery has moved from the south to the heart of Paris with a brand new space to present its group and solo exhibitions of Sophie Calle or Terry Richardson.

Diese Galerie ist aus dem Süden von Paris in nagelneue Räumlichkeiten im Zentrum umgezogen. Dort präsentiert man Gemeinschafts- oder Einzelausstellungen von Sophie Calle oder Terry Richardson.

La galerie a quitté le sud de la ville pour venir s'installer en plein cœur de Paris, dans un espace flambant neuf qui présente des expositions individuelles ou groupées, notamment de Sophie Calle ou Terry Richardson.

76, rue de Turenne
75003 Paris
Tel. +33 (0)1 4216 7979
www.galerieperrotin.com

Galerie Kamel Mennour
The successful gallery has opened a second space and still rules the photography world in Paris thanks to leading artists such as Nobuyoshi Araki or Stephen Shore.

Die erfolgreiche Galerie hat bereits eine Dependance eröffnet und dominiert nach wie vor die Pariser Szene für Fotografie, was sie Künstlern wie Nobuyoshi Araki oder Stephen Shore verdankt.

Face au succès qu'elle rencontre, la galerie a ouvert un second espace d'exposition et règne toujours en maître sur le petit monde parisien de la photographie en exposant des artistes visionnaires tels que Nobuyoshi Araki ou Stephen Shore.

60 et 72, rue Mazarine
75006 Paris
Tel. +33 (0)1 5624 0363
www.galeriemennour.com

La Maison Rouge
The new all red square building specialising in personal contemporary art collections celebrated its first anniversary in 2005.

Das neue, ganz in rot gehaltene, quadratische Bauwerk ist auf individuelle Kollektionen moderner Kunst spezialisiert. 2005 feierte man das einjährige Bestehen der Galerie.

Exposant les collections d'art contemporain de particuliers, ce nouvel immeuble carré et entièrement rouge fête son tout premier anniversaire en 2005.

10, boulevard de la Bastille
75012 Paris
Tel. +33 (0)1 4001 0881
www.lamaisonrouge.org

FASHION SHOPS

Spree
The best of French designers Isabel Marant and Vanessa Bruno alongside Marc Jacobs or Preen, is housed in these two stores.

In den zwei Läden findet man neben Marc Jacobs oder Preen das Beste der französischen Designerinnen Isabel Marant und Vanessa Bruno.

Dans ces deux boutiques, on trouve le meilleur des créatrices françaises Isabel Marant et Vanessa Bruno, aux côtés de Marc Jacobs ou de Preen.

16, rue de la Vieuxville
75018 Paris
Tel. +33 (0)1 4223 4140
1, rue Saint-Simon
75007 Paris
Tel. +33 (0)1 4223 4140

AB 33
Rue Charlot is becoming a hot spot thanks to stores like AB 33, a new fashion boutique selling cool fashion (Zucca, Isabel Marant) and lingerie (Vanessa Bruno, Fifi Chachnil).

Dank Boutiquen wie der AB 33 wird die rue Charlot zu einem echten Anziehungspunkt. Man verkauft dort coole Mode (Zucca, Isabel Marant) und Dessous (Vanessa Bruno, Fifi Chachnil).

Si la rue Charlot est en train de devenir aussi tendance, c'est en partie grâce à des boutiques telles qu'AB 33, nouvel espace multimarques qui vend des griffes de créateurs (Zucca, Isabel Marant) et de la lingerie (Vanessa Bruno, Fifi Chachnil).

33, rue Charlot
75003 Paris
Tel. +33 (0)1 4271 0282

Culotte
A funny name for a tiny Japanese store that sells the funniest jewellery in town with a selection of fresh fashion brands and, of course, panties.

Ein witziger Name für einen winzigen japanischen Laden, wo man den originellsten Schmuck der Stadt, aber auch eine Auswahl der neuesten Modemarken und natürlich kurze Höschen bekommt.

Drôle de nom pour cette minuscule boutique japonaise qui propose les bijoux les plus rigolos de la ville aux côtés d'une sélection de jeunes griffes et, bien sûr, de petites culottes.

7, rue Mahler
75004 Paris
Tel. +33 (0)1 4271 5889

Colette
Still a leader of its own, Colette has improved its streetwear corner, showing the best in sneakers and T-shirts mixed up with hip young designers like Proenza Schouler or As Four.

Nach wie vor maßgeblich, hat Colette den Bereich Streetwear ausgebaut und präsentiert nun Sneakers und T-Shirts vom Feinsten sowie hippe junge Designer wie Proenza Schouler oder As Four.

Toujours leader dans sa catégorie, Colette a revu son espace streetwear pour y présenter le nec plus ultra de la tendance en matière de baskets et de T-shirts, aux côtés de jeunes créateurs branchés tels que Proenza Schouler ou As Four.

213, rue Saint-Honoré
75001 Paris
Tel. +33 (0)1 5535 3390
www.colette.fr

L'Eclaireur
The amazing selection of designers for men and women (Dries Van Noten, Martin Margiela, Rick Owens etc.) has welcomed a design and furniture store in the beautiful Jardins du Palais-Royal.

Eine beachtliche Auswahl an Designern für Herren- wie Damenmode (Dries van Noten, Martin Margiela, Rick Owens etc.) bereichert diesen Design- und Möbelladen in den wunderschönen Jardins du Palais-Royal.

L'impressionnante sélection de griffes de créateurs pour homme et pour femme (Dries Van Noten, Martin Margiela, Rick Owens etc.) s'enrichit d'une collection de mobilier design avec l'ouverture d'une boutique dans les superbes jardins du Palais-Royal.

Femme: 3 ter, rue des Rosiers
75004 Paris
Tel. +33 (0)1 4887 1022
Homme: 12, rue Mahler
75004 Paris
Tel. +33 (0)1 4454 2211
Mixte: 10, rue Hérold
75001 Paris
Tel. +33 (0)1 4041 0989
Design: 13, Galerie de Valois-Jardins du Palais-Royal
75001 Paris
Tel. +33 (0)1 4020 4252
www.leclaireur.com

Pierre Hardy
We heard they're the most comfortable heels on earth. Today Hardy is showcasing his beautiful sharp designs in his first store recalling a black shoe box.

Man sagt, es seien die bequemsten Absätze des Planeten. Inzwischen präsentiert Hardy seine herrlich klaren Kreationen in seinem ersten eigenen Laden, der an einen schwarzen Schuhkarton erinnert.

On dit qu'il fabrique les talons les plus confortables de la planète. Aujourd'hui, Hardy présente ses belles créations pointues dans une première boutique dont le décor ressemble à une boîte à chaussures toute noire.

156, Galerie de Valois-Jardins du PalaisRoyal
75001 Paris
Tel. +33 (0)1 4260 5975
www.pierrehardy.com

Maria Luisa
If Paris was a woman it would be Maria Luisa Poumaillou. Beautiful and elegant, her four stores will dress you up from head to toe in Zac Posen, Balenciaga, Alexander McQueen and Manolo Blahnik.

Wenn Paris eine Frau wäre, dann Maria Luisa Poumaillou. Schön und elegant kleiden ihre vier Geschäfte Sie von Kopf bis Fuß in Zac Posen, Balenciaga, Alexander McQueen und Manolo Blahnik.

Si Paris était une femme, elle s'appellerait Maria Luisa Poumaillou. Belles et élégantes, ses quatre boutiques vous habilleront de la tête aux pieds en Zac Posen, Balenciaga, Alexander McQueen et Manolo Blahnik.

Femme: 2, rue Cambon
75001 Paris
Tel. +33 (0)1 4703 9615
Homme: 19 bis, rue du Mont-Thabor
75001 Paris
Tel. +33 (0)1 4260 8983
Mixte: 38, rue du Mont-Thabor
75001 Paris
Tel. +33 (0)1 4296 4781
Accessories: 40 rue du Mont-Thabor
75001 Paris
Tel. +33 (0)1 4703 4808

Ofr System'
Still the best bookstore in Paris for fashion, photography and graphics.

Nach wie vor die beste Buchhandlung der Stadt für Mode, Fotografie und Grafikdesign.

C'est toujours la meilleure librairie de Paris pour trouver des livres sur la mode, la photo et le graphisme.

30, rue Beaurepaire
75010 Paris
Tel. +33 (0)1 4245 7288
www.ofrpublications.com

Onward
Thanks to a smart selection of avant-garde designers in fashion and accessories, Onward has never lost its spark in the cosy area of Saint Germain.

Dank einer klugen Auswahl von Avantgarde-Designern im Bereich Mode und Accessoires hat Onward im gemütlichen Saint Germain nie seinen Reiz verloren.

Grâce à une sélection intelligente de créateurs de mode et d'accessoires avant-gardistes, la boutique Onward de Saint Germain des Prés n'a rien perdu de son esprit visionnaire.

147, boulevard Saint-Germain
75006 Paris
Tel. +33 (0)1 5542 7755

Shine
The fashion store of the popular rue de Charonne is on the road to success with a second store opening soon in the Marais and accessible brands like See by Chloé or Cacharel.

Die Modeboutique in der beliebten rue de Charonne ist so erfolgreich, dass man einen zweiten Laden im Marais eröffnen wird. Man führt bezahlbare Marken wie Chloé oder Cacharel.

La boutique de la populaire rue de Charonne poursuit son ascension vers la gloire avec l'ouverture prochaine d'un second espace dans le Marais et une sélection de marques accessibles telles que See by Chloé ou Cacharel.

30, rue de Charonne
75011 Paris
Tel. +33 (0)1 4805 8010

American Apparel
It's like a dream come true: American Apparel opened its first store in Paris where it sells basics like tank tops, polo shirts and hooded sweat-shirts in a large range of plain colours for men, women, babies and even dogs.

Es ist wie ein wahr gewordener Traum: American Apparel hat seinen ersten Laden in Paris eröffnet und verkauft dort Basics wie Tank Tops, Polo Shirts und Kapuzen-Sweater in einer großen Auswahl klarer Farben für Männer, Frauen, Babys und sogar Hunde.

Un rêve devenu réalité : American Apparel a ouvert une première boutique parisienne qui vend des basiques tels que des débardeurs, des polos et des sweats à capuche déclinés dans une large palette de couleurs pour les hommes, les femmes, les bébés et même les chiens.

31, Place du Marché Saint-Honoré
75001 Paris
Tel. + 33 (0)1 4260 0372
www.americanapparel.net

Taschen Book Store
Philippe Starck designed this trendy book store gathering all of TASCHEN's beautiful editions.

Dieser trendige Buchladen, der all die wunderbaren Ausgaben des TASCHEN-Verlags führt, wurde von Philippe Starck gestaltet.

C'est à Philippe Starck que l'on doit le design de cette librairie branchée où vous trouverez la collection complète des éditions TASCHEN.

2, rue de Buci
75006 Paris
Tel. +33 (0)1 4051 7922
www.taschen.com

Galerie 213
One of the best places to find vintage, collectors and limited editions photography books.

Eines der besten Antiquariate für Sammlerwerke und limitierte Auflagen von Fotobüchern.

L'un des meilleurs endroits pour dénicher des livres de photo anciens, de collection ou sortis en édition limitée.

58, rue Charlot
75003 Paris
Tel. +33 (0)1 4322 8323
www.galerie213.com

Madame André
The 10 m^2 concept store is like a mini boudoir with funny accessories, dildos or girly fashion from vintage to Gilles Dufour and is owned by Chloé, the wife of graffiti artist André.

Der 10 Quadratmeter große Concept Store sieht aus wie ein Mini-Boudoir mit witzigen Accessoires, Dildos und Girly-Mode – von Second Hand bis Gilles Dufour. Besitzerin des Ladens ist Chloé, die Ehefrau des Graffiti-Künstlers André.

Appartenant à Chloé, l'épouse de l'artiste graffiti André, ce mini-boudoir concept de 10 m^2 fourmille d'accessoires insolites, de godemichés ou de vêtements très girly, du vintage à Gilles Dufour.

34, rue du Mont-Thabor
75001 Paris
Tel. +33 (0)1 4296 2724

Chantal Thomass
The queen of lingerie makes a brilliant comeback with a brand new shop resembling a candy store and designed by Christian Ghion.

Der Königin der Dessous gelang mit diesem brandneuen Laden, der an ein Bonbongeschäft erinnert und von Christian Ghion designt wurde, ein fulminantes Comeback.

La reine de la lingerie fait un brillant retour avec l'ouverture d'une toute nouvelle boutique ressemblant à une confiserie et conçue par Christian Ghion.

211, rue Saint-Honoré
75001 Paris
Tel. +33 (0)1 4260 4056

VINTAGE FASHION SHOPS

Woch Dom
Two stores fronting each other, one with mixed fashion and one with accessories, will fulfil your love for vintage clothing, it also has a great selection of vintage magazines.

Zwei einander gegenüber liegende Läden. Der eine mit diverser Mode, der andere mit Accessoires. Hier kann man seiner Leidenschaft für Vintage-Mode fröhnen und auch in einer große Sammlung von alten Zeitschriften stöbern.

Ces deux boutiques situées l'une en face de l'autre, l'une dédiée aux vêtements et l'autre aux accessoires, étancheront votre soif de vintage, notamment grâce à une belle sélection de vieux magazines.

72, rue Condorcet
75009 Paris
Tel. +33 (0)1 5321 0972

Nuits de Satin
Cute store specialising in vintage lingerie with corsets and slippers, also has a corner at Le Bon Marché.

Ein süßer Laden, der sich auf Vintage-Dessous, Korsetts und Pantoffeln spezialisiert hat. Es gibt auch eine Verkaufsecke im Le Bon Marché.

Cette adorable boutique spécialisée dans la lingerie vintage vend, entre autres, des corsets et des mules. Elle possède également un stand au Bon Marché.

9, rue Oberkampf
75011 Paris
Tel. +33 (0)1 4357 6505
www.nuitsdesatin.com

Anouschka
Still the most complete vintage corner of Paris, you will find fashion archives classified by genres or designers, ranging from Vionnet to Saint Laurent. By appointment only.

Nach wie vor die umfassendste Vintage-Kollektion von ganz Paris. Hier finden Sie Mode nach Genres oder Designern sortiert – von Vionnet bis Saint Laurent. Nur nach Vereinbarung geöffnet.

C'est toujours la boutique vintage la mieux achalandée de Paris, avec ses archives de la mode classées par genre ou par couturier, de Madeleine Vionnet à Yves Saint Laurent. Uniquement sur rendez-vous.

6, avenue du Coq
75009 Paris
Tel. +33 (0)1 4874 3700

La Bazar de Julie
Brand new small vintage spot owned by lovely costume designer Julie who presents a nice selection of dresses and great accessories, like antique lace collars or feather head pieces.

Brandneuer, kleiner Second-Hand-Laden im Besitz der reizenden Kostümbildnerin Julie, die dort eine hübsche Auswahl von Kleidern und phantastische Accessoires wie alte Spitzenkragen oder Kopfputz aus Federn präsentiert.

Cette petite boutique vintage flambant neuve appartient à l'adorable créatrice de costumes Julie, qui y présente une belle sélection de robes et des accessoires géniaux, tels que des cols en dentelle anciens et autres chapeaux à plumes.

19, rue de Poitou
75009 Paris
Tel. +33 (0)1 4272 7424

Mamie
A chaotic but charming vintage store for men and women where you can meet fashion designers looking for inspiration.

Chaotischer, aber charmanter Vintage-Store für Männer und Frauen, wo man schon mal Modedesignern auf der Suche nach Inspiration begegnen kann.

Boutique vintage chaotique mais charmante pour les hommes et les femmes. On y croise parfois des créateurs en quête d'inspiration.

73, rue de Rochechouart
75009 Paris
Tel. +33 (0)1 4282 0998

Quidam de Revel
Probably the best vintage store in town with a sharp selection of clothing, jewellery and furniture.

Der wahrscheinlich beste Vintage-Laden der Stadt mit einer Spitzenauswahl an Kleidung, Schmuck und Möbeln.

Sans doute la meilleure boutique vintage de Paris grâce à sa sélection pointue de vêtements, de bijoux et de meubles.

24, rue de Poitou
75003 Paris
Tel. +33 (0)1 4271 3707

FLEA MARKETS

Puces de Montreuil
Métro: Porte de Montreuil
75020 Paris

Puces de Saint-Ouen
Métro: Porte de Saint-Ouen
75018 Paris

Puces de Vanves
Métro: Porte de Vanves
75014 Paris
Every Saturday and Sunday.

RESTAURANTS

Anahi
This cosy and elegant restaurant serves the best Argentinian meat in town.

In diesem gemütlichen, eleganten Restaurant wird das beste argentinische Rindfleisch der Stadt serviert.

Ce restaurant cosy et élégant sert la meilleure viande d'Argentine qu'on puisse manger à Paris.

49, rue Volta
75003 Paris
Tel. +33 (0)1 4887 8824

L'ami Louis
Confidential and friendly, L'ami Louis is an inevitable spot for celebrities. French food.

Intim und freundlich. Ein obligatorischer Treffpunkt für Prominente. Französische Küche.

Confidentiel et accueillant, L'ami Louis est particulièrement couru par les célébrités. Cuisine française.

32, rue de Vertbois
75003 Paris
Tel. +33 (0)1 4887 7748

R'Aliment
Nice, modern canteen specialising in bio food and situated in the trendy north of Marais area.

Angenehm moderne Küche, spezialisiert auf Bio-Essen, im trendigen Norden des Marais.

Situé dans la partie nord du Marais, cette cantine moderne et sympa s'est spécialisée dans la cuisine bio.

57, rue Charlot
75003 Paris
Tel. +33 (0)1 4804 8828

Etienne Marcel
Designed by art directors M/M, this psychedelic-style restaurant is a good spot to eat and chill out.

Das von den Art Directors M/M gestaltete Restaurant mit psychedelischem Look ist der ideale Ort zum Essen und Chillen.

Conçu par les directeurs artistiques M/M, ce restaurant au look psychédélique est une adresse idéale pour manger et se détendre.

34, rue Etienne-Marcel
75002 Paris
Tel. +33 (0)1 4508 0103

L'Atelier de Joël Rebuchon
Mister Rebuchon, one of the most famous French chefs, proposes exquisite meals in a 100 % non smoking restaurant.

Joël Rebuchon, einer der berühmtesten Köche Frankreichs, offeriert in diesem reinen Nichtraucher-Restaurant exquisite Menüs.

Monsieur Rebuchon, l'un des plus célèbres chefs français, propose des plats exquis dans son nouveau restaurant 100 % non fumeur.

5, rue Montalembert
75007 Paris
Tel. +33 (0)1 4222 5656

Le Martel
Small French-Moroccan restaurant serving the best tagines in town in a cosy and trendy ambiance.

Kleines, französisch-marokkanisches Restaurant, das die besten Tagines der Stadt in gemütlich-trendigem Ambiente serviert.

Petit restaurant franco-marocain où l'on mange les meilleurs tagines de Paris dans une ambiance feutrée et branchée.

3, rue Martel
75010 Paris
Tel. +33 (0)1 4770 6756

Square Trousseau
One of the best brasseries of the capital, traditional delicious French food.

Eine der besten Brasserien der französischen Hauptstadt mit traditionell köstlicher einheimischer Küche.

L'une des meilleures brasseries de la capitale. Délicieuse cuisine française traditionnelle.

1, rue Antoine Vollon
75012 Paris
Tel. +33 (0)1 4343 0600

Cristal Room Baccarat
The restaurant on the first floor of the magnificent crystal palace was magically decorated by Philippe Starck. A must-see.

Das Restaurant im ersten Stock des prachtvollen Kristallpalasts hat Philippe Starck mit seinem Dekor geradezu verzaubert. Pflichtprogramm für Parisbesucher.

Installé au premier étage de l'étincelant palais du cristal, ce restaurant a bénéficié de la touche déco magique de Philippe Starck. A voir absolument.

11, place des Etats-Unis
75016 Paris
Tel. +33 (0)1 4022 1110

La Famille
Friendly space, the trendiest spot on Montmartre.

Einladendes Lokal. Gegenwärtig das angesagteste von Montmartre.

Restaurant sympa, le spot le plus tendance de Montmartre.

41, rue des Trois-Frères
75018 Paris
Tel. +33 (0)1 4252 1112

Georges
With its magnificent 360° view of Paris, the restaurant on top of the Centre Pompidou is a must-see.

Mit seinem herrlichen 360-Grad-Rundblick ist das Restaurant im obersten Stock des Centre Pompidou für jeden Touristen Pflicht.

Avec son incroyable vue à 360° sur Paris, le restaurant qui trône au dernier étage du Centre Pompidou est aujourd'hui absolument incontournable.

Place Georges-Pompidou
75004 Paris
Tel. +33 (0)1 4478 4799

BARS

Le Baron
Formerly the haunt of hookers, Le Baron has become the hottest bar and club in Paris, thanks to graffiti artist André who revamped it.

Der ehemalige Prostituierten-Treff ist zu einer des heißesten Bars von Paris avanciert. Zu danken ist dies der Neugestaltung durch den Graffiti-Künstler André.

Entièrement redécoré par l'artiste graffiti André, cet ancien bordel est désormais le bar-club le plus branché et le plus couru de Paris.

6, avenue Marceau
75008 Paris
Tel. +33 (0)1 4720 0401

Truskel
The great indie-pop rock pub opens late with a mixed crowd has become inevitable.

Der tolle Pub mit Indie, Pop und Rock öffnet spät und wird von einem gemischten Publikum frequentiert. Pflichtprogramm.

Ouvert tard et désormais incontournable, ce pub indie pop rock accueille une faune bigarrée.

12, rue Feydeau
75002 Paris
Tel. +33 (0)1 4026 5997
www.truskel.com

Le Pop In
Alternative, arty place to listen to pop and rock music.

Alternatives Bohème-Lokal, wo man Pop und Rock hört.

Bar bohème alternatif dédié au pop et rock.

105, rue Amelot
75011 Paris
Tel. +33 (0)1 4805 5611
www.popingays.com

Chez Prune
Based in the arty canal Saint-Martin area, Prune is a famous spot with a cool crowd.

Im Künstlerviertel um den Canal Saint-Martin befindet sich das berühmte Prune mit seinem coolen Publikum.

Le désormais célèbre Chez Prune accueille une clientèle cool dans le quartier bobo du canal Saint-Martin.

36, rue Beaurepaire
75010 Paris
Tel. +33 (0)1 4241 3047

DeLaVille Café
Huge modern space set up in an antique Parisian palace with DJ sets of house and electro music.

Riesige modern eingerichtete Räumlichkeiten in einem alten Pariser Palais. Die DJs legen hier House und Electro Music auf.

Installé dans un ancien hôtel particulier, cet immense espace moderne propose à ses clients les sets de DJ house et électro.

34, boulevard de Bonne-Nouvelle
75010 Paris
Tel. +33 (0)1 4824 4809

Le Café
Cool small café in the heart of the trendy Montorgueuil area.

Cooles, kleines Café im Herzen des Trend-Viertels Montorgueuil.

Petit café cool et sympa en plein cœur du quartier Montorgueuil.

62, rue Tiquetonne
75002 Paris
Tel. +33 (0)1 4039 0800

Pause Café
Still leading the Charonne area, Pause Café has become famous for its enjoyable big terrace.

Das nach wie vor interessanteste Lokal der Charonne-Gegend ist bekannt für seine angenehm große Terrasse.

Indétrônable roi de la rue de Charonne, le Pause Café doit son succès à une grande terrasse très agréable.

41, rue de Charonne
75011 Paris
Tel. +33 (0)1 4806 8033

Maxim's Club
The legendary amazingly beautiful club started hosting really trendy parties.

Der legendäre, unglaublich schöne Club hat begonnen, wirklich angesagte Partys zu veranstalten.

Incroyablement beau, ce club légendaire organise désormais des soirées très branchées.

5, rue Royale
75008 Paris
Tel. +33 (0)1 5305 5640

Castel
Classic night spot of the Saint Germain area on 4 beautiful floors.

Klassisches Nachtlokal von Saint Germain auf vier herrlichen Etagen.

Rendez-vous nocturne classique du quartier de Saint-Germain-des-Prés, 4 étages magnifiquement décorés.

15, rue Princesse
75006 Paris
Tel. +33 (0)1 4051 5280

HOTELS

Cinq rue de Moussy

3 suites split over 3 floors, set up in fashion designer Azzedine Alaïa's beautiful hotel particulier in le Marais.

Drei Suiten auf drei Etagen im hübschen Hotel des Modedesigners Azzedine Alaïa. Noch dazu im Marais.

3 suites réparties sur 3 étages dans le sublime hôtel particulier du Marais appartenant au couturier Azzedine Alaïa.

5, rue de Moussy
75004 Paris
Tel. +33 (0)1 4277 1488
www.3rooms-10corsocomo.com

L'Hotel

Renovated by star decorator Jacques Garcia, this sublime hotel is haunted by the ghosts of mythical figures like Oscar Wilde or Mistinguett.

In diesem erhabenen, vom berühmten Innenarchitekten Jacques Garcia eingerichteten Hotel spuken die Geister so legendärer Gestalten wie Oscar Wilde und Mistinguett.

Rénové par le décorateur des stars Jacques Garcia, ce sublime hôtel reste hanté par les fantômes de personnages mythiques tels qu'Oscar Wilde ou Mistinguett.

13, rue des Beaux-Arts
75006 Paris
Tel. +33 (0)1 4441 9900
www.l-hotel.com

Hotel du Petit Moulin

Built in the oldest boulangerie of the capital and decorated by Christian Lacroix, this small colourful and extravagant hotel is for sure the hottest place to stay in Paris.

In der ältesten Bäckerei von Paris befindet sich dieses von Christian Lacroix dekorierte, kleine, bunte, extravagante Hotel. Zweifellos das heißeste Haus in Paris.

Installé dans la plus ancienne boulangerie de la capitale et décoré par Christian Lacroix, ce petit hôtel extravagant et coloré est sans aucun doute l'endroit le plus à la mode pour séjourner à Paris.

29/31, rue de Poitou
75003 Paris
Tel. +33 (0)1 4274 1010
www.paris-hotel-petitmoulin.com

Murano Urban Resort

A very modern and refined hotel, the 52 white rooms, called after Italian names, are only accessible with fingerprints and the suites have small swimming pools on their terraces.

Sehr modernes, elegantes Hotel. Die 52 ganz in Weiß gehaltenen Zimmer mit italienischen Namen sind nur über Fingerabdruck-Erkennung zugänglich. Die Suiten bieten auf ihren Terrassen sogar eigene kleine Pools.

Cet hôtel raffiné et très moderne propose 52 chambres blanches baptisées de noms italiens et uniquement accessibles sur empreinte digitale, mais aussi des suites avec terrasse et petite piscine privées.

13, boulevard du Temple
75003 Paris
Tel. +33 (0)1 4271 2000
www.muranoresort.com

Le A

Exclusively black and white, Le A is the trendy new spot of the chic 8th arrondissement of Paris designed by architect Frédéric Méchiche with interventions by artist Fabrice Hybert.

Das ganz in Schwarzweiß gehaltene Le A ist ein trendiges neues Hotel im schicken 8. Arrondissement von Paris. Das Design stammt von dem Architekten Frédéric Méchiche, in Zusammenarbeit mit dem Künstler Fabrice Hybert.

Tout en noir et blanc, Le A est le dernier spot à la mode du très chic 8ème arrondissement de Paris. Il a été conçu par l'architecte Frédéric Méchiche en collaboration avec l'artiste Fabrice Hybert.

4, rue d'Artois
75008 Paris
Tel. +33 (0)1 4256 9999
www.hotel-le-a.com

Hotel de Buci

Classy, typical French bourgeois hotel in the heart of Saint Germain, the bedrooms are decorated with charming toile de Jouy.

Stilvolles, typisch französisches bürgerliches Hotel im Herzen von Saint Germain. Die Schlafzimmer sind mit zauberhaftem Toile de Jouy dekoriert.

Hôtel bourgeois classe et typiquement français en plein cœur du quartier Saint Germain. Ses chambres pleines de charme sont tapissées de toile de Jouy.

22, rue de Buci
75006 Paris
Tel. +33 (0)1 5542 7474
www.hotelbuci.fr

La Villa

Very well situated contemporary and comfortable hotel.

Modernes, komfortables Hotel in sehr guter Lage.

Hôtel contemporain et confortable bénéficiant d'un emplacement idéal.

29, rue Jacob
75006 Paris
Tel. + 33 (0)1 43266000
www.villa-saintgermain.com

Le Regina

Home to editors during fashion weeks, Le Regina is classy and charming.

Im noblen und charmanten Le Regina logieren während der Fashion Weeks die Moderedakteurinnen.

Domicile des journalistes pendant les semaines de la mode, Le Regina est un hôtel élégant et plein de charme.

2, Place des Pyramides
75001 Paris
Tel. +33 (0)1 4260 3110
www.regina-hotel.com

Pavillon de la Reine

A discreet palace in the middle of a private garden.

Ein intimer Palast mitten in einem privaten Garten.

Hôtel particulier confidentiel niché au milieu d'un jardin privé.

28, Place des Vosges
75003 Paris
Tel. +33 (0)1 4029 1919
www.pavillon-de-la-reine.com

Hotel La Terrasse

One of the most beautiful views of Paris from the legendary terrace of this cool hotel.

Von der legendären Terrasse dieses coolen Hotels genießt man eine der schönsten Aussichten auf Paris.

La terrasse légendaire de ce bel hôtel offre l'une des plus belles vues sur Paris.

12, rue Joseph de Maistre
75018 Paris
Tel. +33 (0)1 4606 7285

AIRPORTS

Aéroports de Paris information
www.adp.fr

Roissy Charles de Gaulle (north of Paris)

Tel. +33 (0)1 4862 1212

Orly (south of Paris)

Tel. +33 (0)1 4975 1515

Where is the best place to stay in São Paulo? Hotel Unique, designed by the architect Ruy Othake, and the Fasano Hotel, designed by the architect Isay Weinfeld. In São Paulo there aren't many interesting options that are cheap Where is your favourite place to eat in São Paulo? The Ritz at Alameda Franca, Jardins for wonderful rice and beans (typical Brazilian food) and hamburgers. Also a restaurant in the Japanese neighbourhood called Yamamoto. The central city market (mercado municipal) is a cheap option with exotic Brazilian food Where is your favourite place to shop in São Paulo? The flea markets at Benedito Calixto Square (on Saturdays), the flea market at Bexiga neighborhood (the Italian neighbourhood, on Sundays) and 25 de Marco Street where carnival related products can be found What makes São Paulo special to you? São Paulo is the place where I have family and friends ALEXANDRE HERCHCOVITCH

SÃO PAULO

TEXT BY ERIKA PALOMINO

São Paulo, Brazil, is a metropolis that provides the perfect blend of art, fashion, architecture, multicultural events and popular culture. The city is also the proud home to São Paulo Fashion Week, one of the world's largest fashion events, which is held twice a year in the historical, Oscar Niemayer-designed Bienal building. The city caters for distinct tastes, with restaurants, clubs and bars spread across hip districts such as Jardins and Vila Madalena. During the day, a trip to its newer art galleries, such as Leme, Vermelho and Factory, as well as a visit to Oscar Freire's stores, are among the favourite options. São Paulo is also home to Latin America's largest concentration of international fashion labels; its most opulent temple to luxury is Daslu, which sells designer clothes alongside magazines, apartments and cars. However it's not only the big hitters of fashion who score in São Paulo, which has a market that is becoming defined by the increasing number of young Brazilian designers – many of whom are also blossoming on the international scene. Driving the upcoming generation is a desire for music, style and the exploration of different arts and media. The young fashion talents carve out a significant share of the market, showing their collections in two main fashion events: Casa de Criadores and Hot Spot. In terms of education, São Paulo is also part of the Fashion School phenomenon, with a growing number of different courses related to the subject. Senac Moda, Santa Marcelina, Anhembi Morumbi and Faap are now providing background as well as techniques to shape young fashion students. São Paulo's energy and atmosphere are similar to that of New York; its elite glamour can be compared to Milan's; its urban appeal is like Berlin's, but its personality is definitely Brazilian.

Das brasilianische São Paulo ist eine Metropole, die die perfekte Mischung aus Kunst, Mode, Architektur, multikulturellen Events und Popkultur zu bieten hat. Die Stadt ist auch stolze Veranstalterin der São Paulo Fashion Week, einer der größten Modemessen weltweit. Der Event findet zweimal jährlich im historischen, von Oscar Niemayer entworfenen Bienal statt. Mit ihren Restaurants, Clubs und Bars, vor allem in den In-Vierteln Jardins und Vila Madalena, wird die Stadt fast jedem Geschmack gerecht. Tagsüber ist ein Besuch der neuesten Kunstgalerien wie Leme, Vermelho und Factory oder auch ein Trip zu den Läden in der Rua Oscar Freire besonders zu empfehlen. São Paulo ist auch die südamerikanische Stadt mit der höchsten Konzentration internationaler Modelabels. Der opulenteste Luxustempel ist Daslu, wo außer Designermode auch Zeitschriften, Wohnungen und Autos verkauft werden. In São Paulo punkten allerdings nicht nur die großen Namen der Modeszene, sondern der Markt wird zunehmend von jungen brasilianischen Designern bestimmt – von denen wiederum viele auch international Erfolge verzeichnen. Die gerade erwachsen gewordene Generation der Brasilianer interessiert sich für Musik, Stil und die Möglichkeiten diverser Künste und Medien. Junge Modetalente sichern sich beträchtliche Marktanteile und zeigen ihre Kreationen bei zwei wichtigen Events: Casa de Criadores und Hot Spot. Was die Ausbildung angeht, so greift auch in São Paulo das Modeschulen-Syndrom um sich – d. h., es gibt eine ständig wachsende Zahl verschiedener Lehr- und Studiengänge zum Thema. Senac Moda, Santa Marcelina, Anhembi Morumbi und Faap bieten jungen Modestudenten sowohl das nötige Hintergrundwissen als auch die Vermittlung praktisch technischer Kenntnisse. São Paulos Energie und Atmosphäre erinnern stark an New York, der elitäre Glamour lässt sich mit dem Mailands vergleichen und die Urbanität mit der Berlins – die Persönlichkeit dieser Stadt ist allerdings eindeutig brasilianisch.

Au Brésil, la métropole de São Paulo propose la combinaison idéale entre art, mode, architecture, événements multiculturels et culture populaire. La ville accueille fièrement la São Paulo Fashion Week, l'un des rendez-vous les plus importants de la mode internationale, organisé deux fois par an dans l'historique immeuble Bienal conçu par Oscar Niemayer. São Paulo en offre pour tous les goûts, fourmillant de restaurants, de clubs et de bars dans les quartiers branchés comme Jardins et Vila Madalena. En journée, l'une des meilleures options consiste à visiter ses galeries d'art flambant neuves, par exemple Leme, Vermelho et Factory, sans oublier un petit détour par les boutiques de la rue Oscar Freire. São Paulo revendique également l'une des plus fortes concentrations de griffes internationales en Amérique latine ; Daslu, son temple du luxe le plus opulent, vend des vêtements de créateurs, mais aussi des magazines, des appartements et des voitures. Mais les grands noms de la mode ne sont pas les seuls à marquer des points à São Paulo, avec un marché qui a tendance à se transformer face au nombre croissant de jeunes créateurs brésiliens, dont la plupart commencent à se faire reconnaître sur la scène mondiale. Cette nouvelle génération semble motivée par son désir de musique et de style comme par l'exploration des différents arts et médias. Les jeunes talents de la mode, qui s'octroient une part de marché toujours plus importante, présentent leurs collections lors des deux grands rendez-vous de la mode : Casa de Criadores et Hot Spot. En termes d'enseignement, São Paulo n'échappe pas au phénomène de l'école de mode, avec une profusion de formations diverses liées à la mode. Senac Moda, Santa Marcelina, Anhembi Morumbi et Faap dispensent désormais des programmes théoriques et techniques pour former les créateurs de demain. L'énergie et l'atmosphère de São Paulo rappellent celles de New York, son glamour élitiste celui de Milan et son attrait urbain celui de Berlin, mais sa personnalité reste résolument brésilienne.

FASHION SHOWS AND FAIRS

São Paulo Fashion Week

Brazil's biggest and most important fashion event, with 48 participating designers and labels. Held twice a year, in January and June.

Die São Paulo Fashion Week ist das größte und wichtigste Mode-Event Brasiliens, mit 48 teilnehmenden Designern und Labels. Zweimal jährlich, im Januar und im Juni.

Le rendez-vous le plus important de la mode au Brésil, avec la participation de 48 créateurs et autres griffes. Deux fois par an, en janvier et en juin.

www.saopaulofashionweek.com.br

Casa de Criadores

Dedicated to the work of young and upcoming designers, it is the main catwalk for São Paulo's undergound fashion. Its venue is an old hospital.

Kümmert sich um die Arbeiten junger, aufstrebender Designer und ist der wichtigste Laufsteg für São Paulos Underground Fashion. Die Veranstaltung findet in einem ehemaligen Krankenhaus statt.

Organisé dans un ancien hôpital et dédié aux nouveaux créateurs, c'est le principal podium de la mode underground à São Paulo.

FENIT (Feira Internacional da Indústria Têxtil)

Brazil's main textile fair, attended by buyers and exhibitors from all over the country.

Die wichtigste Textilmesse Brasiliens für Einkäufer und Aussteller aus dem ganzen Land.

Principal salon brésilien du textile auquel participent des acheteurs et des exposants venus des quatre coins du pays.

www.fenit.com.br

FASHION SCHOOLS

Senac Moda

Open courses and workshops with high quality teaching.

Bietet offene Kurse und Workshops von hoher Qualität.

Cours et ateliers ouverts à tous, enseignement de grande qualité.

www.sp.senac.br

Faculdade Santa Marcelina

One of São Paulo's best options for studying fashion, in a creative environment.

Eine der besten Optionen, wenn man in São Paulo in kreativer Umgebung Mode studieren möchte.

L'une des meilleures options de São Paulo pour étudier la mode au sein d'un environnement créatif.

www.fasm.com.br

Faculdade Anhembi Morumbi

This is the course to take if you are interested in the commercial aspects of fashion.

Dieser Studiengang ist besonders zu empfehlen, wenn man sich für die kommerziellen Aspekte von Mode interessiert.

L'une des meilleures options de São Paulo pour étudier la mode au sein d'un environnement créatif.

www.anhembimorumbi.com.br

Faap Moda

Best place for postgraduate courses, with a sophisticated environment.

Ist die beste Adresse für ein Postgraduiertenstudium in eleganter Umgebung.

L'école idéale pour les troisièmes cycles, environnement sophistiqué.

www.faap.br

FASHION MUSEUMS AND GALLERIES

Galeria Leme

Beautiful gallery designed by architect Paulo Mendes da Rocha, assembling the work of national and international artists.

Eine wunderschöne, vom Architekten Paulo Mendes da Rocha gestaltete Galerie, die Werke nationaler und internationaler Künstler zeigt.

Belle galerie d'art conçue par l'architecte Paulo Mendes da Rocha. Expose les œuvres d'artistes brésiliens et internationaux.

www.galerialeme.com.br

Galeria Vermelho

Strongly related to fashion, Vermelho's events always promote discussion between art and fashion. Fashion designer Karlla Girotto is amongst its most important names.

Hat einen starken Bezug zur Mode und sucht bei ihren Events stets den Dialog zwischen Kunst und Mode. Zu den wichtigsten hier vertretenen Modedesignern zählt Karlla Girotto.

Avec un accent sur la mode, cette galerie organise des événements qui favorisent toujours les débats entre les mondes de l'art et de la mode. La créatrice Karlla Girotto y figure en très bonne place.

Rua Minas Gerais, 350
Higienópolis
Tel. +55 11 3257 2033
www.galeriavermelho.com.br

Galeria Maria Bonita

Located on the third floor of the Maria Bonita store, one of São Paulo's finest brands.

Befindet sich im dritten Stock des Maria Bonita, einer der besten Einkaufsadressen São Paulos.

Située au troisième étage de la boutique Maria Bonita, l'une des meilleures marques de São Paulo.

Rua Oscar Freire, 702
Tel. +55 11 3062 6433

Factory

Multimedia gallery that blends fashion, art and technology, in São Paulo's hip district of Vila Madalena.

Die Multimedia-Galerie verbindet in São Paulos angesagtem Viertel Vila Madalena Mode, Kunst und Technologie.

Galerie multimédia fusionnant mode, art et technologie à Vila Madalena, le quartier branché de São Paulo.

Rua Mourato Coelho, 790
Vila Madalena
Tel: +55 11 3813 0414

Centro Cultural Itaú

One of the best places for an exhibition, usually showing the work of mainstream artists.

Einer der besten Ausstellungsorte, wo meist die Werke von allgemein bekannten Künstlern präsentiert werden.

L'une des meilleures adresses pour les expositions. Présente généralement des œuvres d'artistes connus.

Av. Paulista, 149
Bela Vista
Tel. +55 11 2168 1700
www.itaucultural.com.br

FASHION SHOPS

Daslu

Chanel, Prada, Dolce & Gabbana as well as other luxury brands can be found in Brazil's most prestigious boutique. From May 2005 the store has been operating from a new and even bigger site, selling everything from stationary to helicopters.

Die berühmteste Boutique ganz Brasiliens. Hier findet man Chanel, Dolce & Gabbana und andere Luxusmarken. Seit Mai 2005 befindet sich der Laden in neuen, noch größeren Räumlichkeiten, wo von Briefpapier bis zum Helikopter praktisch alles verkauft wird.

La boutique la plus prestigieuse du Brésil vend des créations Chanel, Prada et Dolce & Gabbana à côté d'autres griffes de luxe. Depuis mai 2005, elle opère depuis un nouvel espace encore plus grand qui propose de tout, des articles de papeterie aux hélicoptères.

Vila Daslu – Rua Chedid Jafet, 200
Vila Olímpia
Tel. +55 11 04551 065
www.daslu.com.br

Clube Chocolate

São Paulo's main concept store is designed by architect Isay Weinfeld. Four storeys high, the place sells a diversity of different labels such as Marni, Yohji Yamamoto, Dolce & Gabbana, Seven, Paper, Evisu as well as their own label and chic Brazilian designers such as Raia de Goeye and Isabela Capeto. Electronic equipment, books, magazines, CDs, a bakery and even exotic birds can also be found.

São Paulos wichtigster Concept Store wurde von dem Architekten Isay Weinfeld designt. Auf vier Etagen führt man hier so verschiedene Labels wie Marni, Yohji Yamamoto, Dolce & Gabbana, Seven, Paper, Evisu und eine Eigenmarke sowie schicke brasilianische Designer wie Raia de Goeye und Isabela Capeto. Elektrogeräte, Bücher, Zeitschriften, CDs, eine Bäckerei und sogar exotische Vögel findet man hier ebenfalls.

La principale boutique-concept de São Paulo a été conçue par l'architecte Isay Weinfeld. Sur quatre étages, elle vend un large choix de griffes telles que Marni, Yohji Yamamoto, Dolce & Gabbana, Seven, Paper et Evisu, à côté de sa propre collection et de créateurs brésiliens très chic comme Raia de Goeye et Isabela Capeto. Vous y trouverez également du matériel électronique, des livres, des magazines, des CD, une boulangerie et même des oiseaux exotiques.

Rua Oscar Freire, 913
Jardins
Tel. +55 11 3084 1500

Shopping Iguatemi

Iguatemi is São Paulo's oldest mall. With a huge variety of options, it is one of the city's main retail centres.

Iguatemi ist die älteste Mall São Paulos. Mit einer riesigen Auswahl ist sie das wichtigste Einkaufszentrum der Stadt.

Iguatemi est le plus ancien centre commercial de São Paulo. Proposant une large variété de produits, c'est l'une des principales destinations de shopping de la ville.

Av. Brigadeiro Faria Lima, 2232
Pinheiros
www.iguatemisaopaulo.com.br

Complexo de lojas Oscar Freire/ Haddock Lobo

This is the place to shop for Brazil's best brands. A diversity of international labels can also be found, including Bulgari, Armani, Tiffany, Vuitton and Dior. The perfect option for Saturday afternoon, with bars, restaurants and movie theatres around.

Hier kann man die besten Marken Brasiliens kaufen. Zudem findet man eine Vielzahl internationaler Labels wie Bulgari, Armani, Tiffany, Vuitton und Dior. Das perfekte Ziel für einen Samstagnachmittag. Rund herum gibt es auch Bars, Restaurants und Kinos.

C'est le choix qui s'impose pour découvrir les meilleures marques brésiliennes. On y trouve aussi une sélection variée de griffes internationales, dont Bulgari, Armani, Tiffany, Vuitton et Dior. Une option idéale le samedi après-midi, car le quartier recèle de nombreux bars, restaurants et cinémas.

Ellus 2nd Floor
Jeanswear label Ellus replaced itself in the market with the launch of Ellus 2nd Floor, opening its second floor for upcoming designers such as Karlla Girotto and Fabia Bercsek.

Das Jeanslabel Ellus hat sich selbst durch die Einführung von Ellus 2nd Floor ersetzt und seine zweite Etage für aufstrebende Designer wie Karlla Girotto und Fabia Bercsek geöffnet.

La griffe de jeans Ellus a cédé sa place dans le marché avec le lancement d'Ellus 2nd Floor, ouvrant son deuxième étage aux jeunes créateurs qui montent, tels que Karlla Girotto et Fabia Bercsek.

Rua Oscar Freire, 990
Jardins
Tel. +55 11 3061 2900
www.ellus2ndfloor.com.br

Fuxique
Using vintage fabrics, young designers develop feminine and creative pieces with a bohemian appeal.

Für Fuxique entwerfen junge Designer aus Vintage-Stoffen feminine, originelle Kreationen mit Bohème-Touch.

A partir de tissus vintage, de jeunes créateurs produisent de belles pièces féminines et originales à l'esprit bohème.

Rua Oscar Freire, 1119
Jardins
Tel. +55 11 3897 8130

Pelu
Operating as a hair salon, bar and store, this charming venue works as the best option for the young and cool girls.

Der zauberhafte Friseursalon mit Bar und Laden ist die beste Adresse für coole junge Mädchen.

L'endroit fait salon de coiffure, bar et boutique, mais aussi le bonheur des jeunes filles à la mode.

Al. Lorena, 1257, casa 2
Tel. +55 11 3891 1229
www.pelu.com.br

Adriana Barra
Adriana is one of Brazil's most talented designers and leader of a young generation who create clothes which focus on feminine and romantic values.

Adriana ist eine der talentiertesten Designerinnen Brasiliens und führt eine junge Generation von Kollegen an, die Kleider mit feminin-romantischem Schwerpunkt kreiert.

Adriana est l'une des créatrices les plus talentueuses du Brésil, leader d'une jeune génération qui crée des vêtements principalement axés autour de la féminité et du romantisme.

Rua Peixoto Gomide, 1801, Casa 5
Jardins
Tel. +55 11 3062 0387
www.adrianabarra.com.br

adrianabarra@adrianabarra.com.br

Raia de Goeye
Duo Fernanda de Goeye and Paula Raia are São Paulo's Phoebe Philo and Stella McCartney. Good quality clothes, full of personality.

Fernanda de Goeye und Paula Raia sind die Phoebe Philo and Stella McCartney von São Paulo. Kleidung von bester Qualität und mit Persönlichkeit.

Le duo formé par Fernanda de Goeye et Paula Raia est l'équivalent brésilien du tandem Phoebe Philo/Stella McCartney. Vêtements pleins de personnalité et de bonne qualité.

Rua Adolfo Tabacow, 261F
Itaim
Tel. +55 11 3079 3301
www.raiadegoeye.com.br

Theodora
Designer Rita Wainer's streetwear label focuses on T-shirts with political and subversive messages.

Das Streetwear-Label der Designerin Rita Wainer hat sich auf T-Shirts mit politischen und subversiven Messages spezialisiert.

La griffe streetwear lancée par la créatrice Rita Wainer propose des t-shirts imprimés de messages politiques et subversifs.

Rua Augusta, 2690
Loja 206 – Galeria Ouro Fino, 1o andar
Tel. +55 11 3082 9624

ALTERNATIVE SHOPS

Galeria Ouro Fino
Since the beginning of the '90s Galeria Ouro Fino assembles brands that dress the clubbing generation.

Schon seit Beginn der 1990er Jahre führt Galeria Ouro Fino die Marken, in die sich die Clubbing-Generation kleidet.

Depuis le début des années 90, Galeria Ouro Fino vend des griffes qui habillent la nouvelle génération du clubbing.

Rua Augusta, 2690
Jardins

Galeria do Rock
São Paulo's beloved venue sells clothes, vinyl, CDs and all kinds of accessories for rock fans.

Ein Lieblingsladen São Paulos, der Kleidung, Schallplatten, CDs und alle erdenklichen Accessoires für Rockfans verkauft.

Galeria do Rock : l'adresse chérie de São Paulo vend des vêtements, des vinyles, des CD et toutes sortes d'accessoires dédiés aux enfants du rock.

Rua 24 de Maio, 62
Centro

VINTAGE SHOPS

Minha Avó Tinha
Clothes, accessories and furniture from different decades.

Kleidung, Accessoires und Möbel aus verschiedenen Jahrzehnten.

Vêtements, accessoires et meubles de différentes époques.

Rua Dr. Franco da Rocha, 74
Perdizes
Tel. +55 11 3865 1759

Trash Chic
Vintage sunglasses, bags, clothes and accessories from labels such as Prada, Valentino, Dolce & Gabbana and Moschino.

Second-Hand Sonnenbrillen, Taschen, Kleider und Accessoires von Labels wie Prada, Valentino, Dolce & Gabbana und Moschino.

Lunettes de soleil, sacs, vêtements et accessoires vintage griffés Prada, Valentino, Dolce & Gabbana ou Moschino.

Rua Professor Carlos Carvalho, 95
Itaim
Tel. +55 11 3167 4331

RESTAURANTS

Restaurante Hotel Fasano
Expensive place, but worth every real. Architect Isay Weinfeld shines once again with this project.

Teures Etablissement, aber jeden Real wert. Architekt Isay Weinfeld hat sich mit diesem Projekt wieder einmal profiliert.

Prix élevés mais justifiés jusqu'au dernier real. Encore une brillante réalisation de l'architecte Isay Weinfeld.

Rua Vitório Fasano, 88
Cerqueira César
Tel. +55 11 3062 4000
www.fasano.com.br

Spot
Hyped restaurant. Best place to be seen during fashion week season.

Absolutes In-Restaurant. Die beste Adresse, um während der Zeit der Modewoche gesehen zu werden.

Restaurant branché, le meilleur endroit pour être vu pendant les semaines de la mode.

Al. Ministro Rocha Azevedo, 72
Cerqueira César
Tel. +55 11 3283 0946
www.restaurantespot.com.br

Carlota
Charming place, located at residential district Higienópolis, wonderful food by chef Carla Pernambuco, and a good wine selection by her husband, former fashion photographer Nando Pernambuco.

Charmantes Lokal im Wohnviertel Higienópolis. Wunderbares Essen von Küchenchefin Carla Pernambuco und eine hervorragende Weinkarte, für die ihr Ehemann, der ehemalige Modefotograf Nando Pernambuco, verantwortlich ist.

Charmant restaurant situé dans le quartier résidentiel de Higienópolis, merveilleuse cuisine du chef Carla Pernambuco et bonne carte de vins sélectionnés par son mari, l'ex-photographe de mode Nando Pernambuco.

Rua Sergipe, 753
Higienópolis
Tel. +55 11 3661 8670

Prime Burger by Sergio Arno
Great bites and home of São Paulo's famous Nutella milkshake.

Prima Happen und der Geburtsort von São Paulos berühmtem Nutella-Milchshake.

Excellents plats et résidence du fameux milk-shake au Nutella de São Paulo.

Rua Joaquim Floriano, 541
Itaim
Tel. +55 11 3078 7817

Forneria São Paulo
Uptown public meets here after clubbing for some of the best sandwiches in town. The restaurant is coordinated by the Fasano staff.

Das Uptown-Publikum trifft sich hier nach dem Clubbing zu den vielleicht besten Sandwiches der Stadt. Das Lokal wird von den Fasano-Leuten geführt.

Les habitants d'uptown aiment s'y retrouver après une nuit de clubbing pour s'offrir certains des meilleurs sandwiches de la ville. Le restaurant est géré par l'équipe du Fasano.

Rua Amauri, 319
Itaim
Tel. +55 11 3078 0099
www.fasano.com.br

Mercearia do Conde/Condessa
Quiches and salads in a pleasurable environment.

Quiches und Salate in netter Umgebung.

Quiches et salades servies dans un environnement agréable.

Mercearia do Conde
Rua Joaquim Antunes, 217
Pinheiros
Tel. +55 11 3081 7204
www.merceariadoconde.com.br
Condessa
Rua João Lorena, 367
Vila Nova Conceição
Tel. +55 11 3842 5141

Gero
Another treat from the Fasano family, pulling in São Paulo's flashiest crowd. More relaxed than Fasano, with great Italian food.

Ein weiteres Lokal der Fasano-Familie, das das schillerndste Publikum São Paulos anzieht. Etwas legerer als das Fasano und mit tollem italienischen Essen.

Cet autre bijou de la famille Fasano attire la faune la plus fashion de São Paulo. Ambiance plus relax que chez Fasano et sublime cuisine italienne.

Rua Haddock Lobo, 1629
Jardins
Tel. +55 11 3064 0005
www.fasano.com.br

Sushi Jun Sakamoto
São Paulo's equivalent of Nobu. The best sashimi in town. Reservations are strongly recommended.

Quasi das Nobu von São Paulo. Bestes Sashimi der Stadt. Reservierung unbedingt zu empfehlen.

L'équivalent de Nobu à São Paulo. Les meilleurs sashimi de la ville. Réservations vivement recommandées.

Rua Lisboa, 55
Pinheiros
Tel. +55 11 3088 6019

Rodeio
Best place for grills and barbecue, providing comfort, privacy and good service.

Bestes Grillrestaurant. Gemütlich, intim und mit gutem Service.

La meilleure adresse pour les grillades et les barbecues. Service impeccable, ambiance confortable et intimiste.

Rua Haddock Lobo, 1498
Cerqueira César
Tel. +55 11 3083 2322

BARS

Bar 13
Fashionista spot owned by young fashion designer Marcelo Sommer.

Dieser Treffpunkt der Fashionistas gehört dem jungen Modedesigner Marcelo Sommer.

Repaire de fashion victims appartenant au jeune créateur de mode Marcelo Sommer.

Rua Alagoas, 852
Higienópolis
Tel. +55 11 3666 0723

Ritz
Like Spot, Ritz is owned by Sergio Kalil. During lunchtime, the place gets filled with the city's artsy and fashion people. It's also a favourite place to meet at night, when it gets even louder.

Das Ritz befindet sich ebenso wie das Spot im Besitz von Sergio Kalil. Um die Mittagszeit wimmelt es hier von Leuten aus der Kunst- und Modeszene. Auch ein beliebtes Lokal für den Abend. Dann geht es allerdings noch lauter zu.

Comme le Spot, le Ritz appartient à Sergio Kalil. Le midi, ce bar accueille le petit monde de l'art et de la mode. C'est aussi un rendez-vous de prédilection pour le soir, où l'on ne s'entend même plus parler.

Al. Franca, 1088
Jardim Paulista
Tel. +55 11 3088 6808

Skye
One of São Paulo's nicest views, seen from the top floor bar at Hotel Unique.

Die Bar im obersten Stock des Hotel Unique bietet den vielleicht schönsten Blick auf São Paulo.

L'une des plus belles vues sur São Paulo depuis le bar du dernier étage du Hotel Unique.

Av. Brigadeiro Luis Antônio, 4700
Jardim Paulista
Tel. +55 11 3055 4702

Cristallo
Oscar Freire's "official" café, with great patisserie.

Das „offizielle" Café der Rua Oscar Freire mit toller Patisserie.

Café « officiel » de la rue Oscar Freire proposant de délicieuses pâtisseries.

Rua Oscar Freire, 914
Jardim Paulista
Tel. +55 11 3082 1783

Bar da Dida
Best place to have a cold beer (Brazilian style) at the sidewalk. One of the most significant landmarks in São Paulo's bar culture.

Die beste Adresse für ein kaltes Bier (auf brasilianische Art), also auf dem Gehsteig. Eine der Säulen der Barkultur von São Paulo.

La meilleure adresse pour boire une bière bien fraîche sur le trottoir (à la brésilienne). Une véritable institution de la culture des bars à São Paulo.

Rua Dr. Melo Alves, 98
Cerqueira César
Tel. +55 11 3088 7177

Samba
Friendly atmosphere, playing cool samba with a live crowd of musicians. Located at Vila Madalena, district filled with bars and young people.

Freundliche Atmosphäre und coole Samba von Live-Musikern. Das Lokal liegt in Vila Madalena, einem Viertel voller Bars und junger Leute.

Dans une ambiance cordiale, on oscille aux doux rythmes de la samba jouée en direct par des musiciens. Situé dans Vila Madalena, un quartier très jeune et plein de bars.

Rua Fidalga, 308
Vila Madalena
Tel. +55 11 3819 4619

Empanada's
For a quick bite in Vila Madalena, serving the famous snack which gave its name to the bar.

Ideal für einen schnellen Happen in Vila Madalena. Man serviert die berühmten Snacks, nach denen die Bar benannt ist.

Idéal pour manger sur le pouce dans Vila Madalena, on y sert le fameux snack qui a donné son nom à l'établissement.

Rua Wisard, 489
Vila Madalena
Tel. +55 11 3032 2116

D-Edge
Nightclub with a modern and technological atmosphere, by Muti Randolph. Playing different kinds of music, from rock to electro, and presenting a different international attraction every week.

Nachtclub mit moderner High-Tech-Atmosphäre, gestaltet von Muti Randolph. Hier wird verschiedenste Musik, von Rock bis Electro, gespielt. Allwöchentlich eine neue internationale Attraktion.

Boîte de nuit au décor moderne et technologique appartenant à Muti Randolph. Programmation musicale variée, du rock à l'électro, avec une attraction internationale différente chaque semaine.

Al. Olga, 170
Barra Funda
Tel. +55 11 3666 9022
www.d-edge.com.br

Lov.e
São Paulo's oldest and most successful nightclub, home of famous DJs such as Marky, Mau Mau and Renato Cohen.

Ältester und erfolgreichster Nachtclub von São Paulo. Hier sind so berühmte DJs wie Marky, Mau Mau und Renato Cohen zu Hause.

Le club le plus ancien et le plus couru de São Paulo, résidence de célèbres DJ tels que Marky, Mau Mau et Renato Cohen.

Rua Pequetita, 189
Tel. +55 11 3044 1613
www.loveclub.com.br

HOTELS

Hotel Fasano
With an exclusive and traditional atmosphere, the hotel is the favourite of artists and fashionistas who look for privacy. Located at uptown district Jardins.

Exklusive, traditionelle Atmosphäre. Lieblingshotel von Künstlern und Fashion People, die Wert auf ihre Privatsphäre legen. Im Uptown-Viertel Jardins gelegen.

Avec son atmosphère luxueuse à l'ancienne, cet hôtel est le préféré des artistes et des gens de la mode en quête d'un peu d'intimité. Situé uptown dans le quartier Jardins.

Rua Vitório Fasano, 88
Cerqueira César
Tel. +55 11 3062 4000
www.fasano.com.br

Hotel Unique
With a younger and more relaxed atmosphere, Hotel Unique is shaped like a watermelon due to Ruy Ohtake's polemical architectural creation. Don't miss its entrance hall.

Jugendliche, entspannte Atmosphäre. Das Hotel besitzt die Form einer Wassermelone, eine ironische Architektur von Ruy Ohtake. Die Lobby allein ist einen Besuch wert.

Avec son atmosphère plus jeune et plus informelle, l'Hotel Unique ressemble à une pastèque en raison de la création architecturale controversée de Ruy Ohtake. Son hall d'entrée vaut le détour.

Av. Brigadeiro Luis Antônio, 4700
Jardim Paulista
Tel. +55 11 3055 4712

Hotel Emiliano
São Paulo's first design hotel is Gisèle Bündchen's favourite, in the heart of Jardins.

São Paulos erstes Designhotel und Lieblingsquartier von Gisèle Bündchen. Im Herzen von Jardins gelegen.

Situé en plein cœur du quartier Jardins, le premier hôtel design de São Paulo est aussi le préféré de Gisèle Bündchen.

Rua Oscar Freire, 384
Tel. +55 11 3069 4368
www.emiliano.com.br

Pensão da Dna. Zilah
Short of money? This is a cheaper option with a family-like and relaxed atmosphere.

Knapp bei Kasse? Preiswerte Familienpension mit lockerer Atmosphäre.

Vous avez dépensé tout votre argent ? Alors voici une alternative moins chère. Ambiance familiale et décontractée.

Al. Franca, 1621/1633
Jardim Paulista
Tel. +55 11 3062 1444
www.zilah.com.br/portugues

Flat George V
One of São Paulo's best flats. Lots of space and privacy.

Eine der besten Kurzzeit-Mietwohnungen São Paulos. Viel Platz und Privatsphäre.

L'un des plus beaux appartements de São Paulo. Beaucoup d'espace et d'intimité.

Rua José Maria Lisboa, 1000
Jardins
Tel. +55 11 3088 9822
www.george-v.com.br

AIRPORT

Aeroporto internacional de Guarulhos
São Paulo's airport is located one hour from the centre of town and a taxi costs approximately R$60 (U$20).

Der Flughafen von São Paulo liegt eine Autostunde vom Stadtzentrum entfernt. Eine Taxifahrt dorthin kostet etwa 60 Real (20 US$).

L'aéroport de São Paulo se trouve à une heure de route du centre-ville. La course en taxi coûte environ R$60 (20 dollars).

Rod. Hélio Smidt s/n
Ed. Interlagação
São Paulo – SP
Tel. +55 11 6445 2945
Fax +55 11 6488 8476
Distance from centre: 25 km
www.infraero.gov.br

Aeroporto nacional de Congonhas
This is where you catch a flight to Rio de Janeiro, which takes approximately 35 to 50 minutes. Closer to town than the international airport but traffic is always bad so it's recommended that you leave an hour early.

Von hier aus kann man nach Rio de Janeiro fliegen (ca. 35–50 Minuten Flugzeit). Näher zur Stadt als der internationale Flughafen, aber aufgrund der schlechten Verkehrsverhältnisse empfiehlt es sich, auch hier eine Stunde Fahrzeit einzukalkulieren.

C'est ici que vous pouvez prendre un vol pour Rio de Janeiro, qui dure entre 35 et 50 minutes. Plus proche de la ville que l'aéroport international, mais en raison des embouteillages constants, nous vous conseillons de partir avec une heure d'avance.

Av. Washigton Luís s/nº
São Paulo – SP
Tel. +55 11 5090 9000
Fax. +55 11 5531 7718
Distance from centre: 8 km
www.infraero.gov.br

What makes Tokyo special to you? In Tokyo, you can get everything you need from around the world Where is the best place to stay in Tokyo? The Park Hyatt or Grand Hyatt Where is your favorite place to eat in Tokyo? Nobu Tokyo Where is your favourite place to shop in Tokyo? Louis Vuitton at Roppongi Hills and Bapexclusive NIGO

TOKYO

TEXT BY NICOLE FALL

There is no other city in the world like Tokyo for discovering new stores that span the gamut from sublime to absurd and back again. Speciality shops are the current favourite trend and the collaboration between 10 Corso Como and Comme des Garçons is the latest hotspot, as is Norm, both of which are located in designer boutique-land Aoyama, an area seemingly untouched by the decade-long recession. For those shoppers who prefer homegrown Japanese talent to imported designer goods, however, the most fashionable area to shop, drink and eat is Naka Meguro. Low-key, unique, hole-in-the-wall style stores reside by a pretty canal that runs through the centre of the district. It is the antithesis of Shibuya's commercialism and Harajuku's overtly 'too cool to be true' attitude. Naka Meguro's ascent can be attributed to the exorbitant store rents in nearby Daikanayama, an area once popular with young designers just starting out in the fashion world. Talent is not enough when competing against multinational mall developers, many of which have engulfed the former laidback residential neighbourhood and converted traditional Japanese houses into ugly tower blocks. Trends are adopted and discarded in Japan within weeks and stores reflect these changes constantly. The infamous deeply-tanned, micro mini-wearing Gunagaru or Shibuya Girls that made the headlines overseas have disappeared and the shops that catered to this particular style tribe have long since transformed their names, fittings and, of course, products to suit the latest fashion. Cyber trance, street style, gothic, hip hop, Latino teenage prostitute, whatever; there is a store dedicated to it.

Es gibt auf der Welt keine Stadt wie Tokio, um neue Läden aufzuspüren, die das komplette Spektrum vom Erhabenen bis zum Absurden abdecken. Spezialisierte Boutiquen sind der neue Lieblingstrend, und die jüngsten Hot Spots sind das Gemeinschaftsprojekt von 10 Corso Como und Comme des Garçons sowie Norm, die beide in der Gegend der Designerboutiquen liegen, in Aoyama, wo die Dekade der Rezession spurlos vorübergegangen zu sein scheint. Für Käufer, die heimische japanische Talente importierter Designerware vorziehen, ist die hippste Gegend zum Einkaufen, Trinken und Essen Naka Meguro. Unauffällige, schmucklose Geschäfte mit unverwechselbarem Charakter liegen entlang eines hübschen Kanals, der mitten durch das Viertel fließt. Es ist die Antithese zum Kommerz von Shibuya und der übertriebenen „Zu cool, um wahr zu sein"-Pose von Harajuku. Naka Meguros Aufstieg ist wohl den exorbitanten Ladenmieten im benachbarten Daikanayama zuzuschreiben, einer früher bei jungen Designern, die sich gerade erst in der Modewelt etablieren, sehr beliebte Gegend. Talent allein ist nicht genug, wenn man mit den internationalen Betreibergesellschaften von Malls konkurrieren muss, die die früher so stille Wohngegend geradezu eingekreist und Häuser im traditionellen Stil durch hässliche Blocks ersetzt haben. Trends werden in Japan innerhalb weniger Wochen angenommen und wieder verworfen, und in den Geschäften spiegelt sich dieser ständige Wechsel wider. Die berühmt-berüchtigten sonnengebräunten Gunagaru- oder Shibuya-Girls in ihren Mikro-Minis, die auf der anderen Seite der Erde Schlagzeilen machten, sind längst verschwunden, und die zu diesem speziellen Look gehörigen Läden haben mittlerweile ihre Namen, ihre Aufmachung und natürlich ihr Angebot der letzten Mode entsprechend geändert. Cyber Trance, Street Style, Gothic, Hiphop, minderjährige Latino-Prostituierte, was auch immer – für alles gibt es einen entsprechenden Laden.

Tokyo est la seule ville du monde à offrir une telle profusion de nouvelles boutiques, des plus sublimes aux plus absurdes. Aujourd'hui, la tendance est aux boutiques spécialisées. Aux côtés de Norm, l'espace de vente né de la collaboration entre 10 Corso Como et Comme des Garçons est la dernière sensation d'Aoyama, quartier des boutiques de créateurs apparemment épargné par la récession qui sévit depuis dix ans. Pour ceux qui préfèrent les créations 100 % japonaises aux vêtements importés, le coin le plus en vogue pour faire du shopping, boire un verre ou manger un morceau s'appelle Naka Meguro. Des boutiques au chic dépouillé, discrètes et uniques bordent le joli canal qui traverse le centre du quartier. C'est l'antithèse du mercantilisme de Shibuya et de la branchitude ouvertement forcée de Harajuku. L'intérêt croissant de Naka Meguro peut être attribué aux loyers prohibitifs du quartier avoisinant Daikanayama où s'étaient installés nombre de jeunes créateurs qui démarraient dans la mode. Le talent ne suffit pas lorsqu'il s'agit de faire concurrence aux promoteurs des centres commerciaux qui ont envahi cet ancien quartier résidentiel et transformé ses belles maisons traditionnelles japonaises en affreux blocs de béton. Au Japon, les tendances se succèdent en l'espace de quelques semaines et les boutiques reflètent ces changements constants. Avec leur bronzage outrancier et leurs jupes microscopiques, les fameuses filles de Gunagaru ou de Shibuya qui faisaient les gros titres de la presse étrangère ont disparu depuis longtemps. Les boutiques dédiées à cette tribu particulière ont donc changé de nom, de décoration et, bien sûr, de produits pour suivre la dernière mode : cyber trance, streetwear, gothique, hip-hop, pute ado latino... chaque style a son propre magasin.

FASHION SHOWS AND FAIRS

International Fashion Fair (IFF)

Twice yearly mainstream fashion fair organised by one of Japan's largest trade fashion publications.

Zweimal jährlich stattfindende konventionelle Modemesse, die von einem der auflagenstärksten japanischen Modemagazine organisiert wird.

Salon de mode commerciale organisé deux fois par an par une revue spécialisée japonaise à grand tirage.

Pacific Convention Plaza Yokohama

Minato Mirai
Nishi-ku
Yokohama City
Tokyo
Tel. +81 (3) 3263 6881
www.senken.co.jp/iff

Tokyo Collections

Twice yearly fashion collections from the local members of the CFD. This organisation can also provide details of off-schedule shows.

Zweimal im Jahr stattfindende Modenschauen der Tokioter Mitglieder des CFD. Bei dieser Organisation kann man auch mehr über Schauen außerhalb des offiziellen Programms erfahren.

Défilés de mode organisés deux fois par an par les membres tokyoïtes du Council of Fashion Designers. Cette organisation fournit également des informations détaillées sur les défilés « off ».

Council of Fashion Designers
Tel. +81 (3) 3498 5971
www.cdf.or.jp/

FASHION SCHOOLS

Bunka Fashion College

Past alumni of this prestigious school include Yohji Yamamoto, Kenzo and Junko Koshino.

Zu den Absolventen gehören Yohij Yamamoto, Kenzo und Junko Koshino.

Parmi les anciens diplômés de cette prestigieuse école figurent Yohji Yamamoto, Kenzo et Junko Koshino.

3-22-1 Yoyogi
Shibuya-ku
Tokyo
Tel. +81 (3) 3299 2222

Mode Gakuen

This fashion school is connected to the famous Paris college.

Diese Modeschule arbeitet mit dem renommierten Institut in Paris zusammen.

Ecole de mode en relation avec le célèbre institut parisien.

1-6-2 Nishi Shinjuku
Shinjuku-ku
Tokyo
Tel. +81 (3) 3344 6000

FASHION MUSEUMS AND GALLERIES

Mori Art Museum

53F, Mori Tower
6-10-1 Roppongi
Minato-ku
Tokyo
Tel. +81 (3) 6406 6100
www.mori.art.museum/english

Nezu Institute of Fine Arts

6-5-1 Minami Aoyama
Minato-ku
Tokyo
Tel. +81 (3) 3400 2536
www.nezu-muse.or.jp

Watari-Um Museum of Contemporary Art

1-10-10 Jingumae
Shibuya-ku
Tokyo
Tel. +81 (3) 3402 3001
www.watarium.co.jp

Depot

2-43-6 Kamimeguro
Meguro-ku
Tokyo
Tel. +81 (3) 5773 5502

Gallery 360°

5-1-27 Minami Aoyama
Minato-ku
Tokyo
Tel. +81 (3) 3406 5823

Gallery Side 2

Mitsuba Bldg, 1F
2-18-3 Akasaka
Minato-ku
Tokyo
Tel. +81 (3) 6229 3669

National Museum

Ueno Park
Taito-ku
Tokyo
Tel. +81 (3) 3822 1111
www.tnm.go.jp

Tokyo Opera City Art Gallery

3-20-2 Nishi Shinjuku
Shinjuku-ku
Tokyo
Tel. +81 (3) 5353 0756
www.tokyooperacity-cf.or.jp

FASHION SHOPS

Celux

Beg, borrow or steal a Celux entry card to gain access to quite possibly the world's most exclusive boutique located on the top floor of the Louis Vuitton store. Seriously, if you're not a member, you're not getting in….

Erbetteln, borgen oder stehlen Sie sich eine Eintrittskarte, um Zugang zur womöglich exklusivsten Boutique der Welt im obersten Stock des Louis-Vuitton-Ladens zu bekommen. Denn ganz im Ernst, wenn Sie kein Mitglied sind, kommen Sie da nicht rein …

Quémandez, empruntez ou volez une carte Celux pour pouvoir découvrir cette boutique qui est probablement la plus luxueuse du monde, située au dernier étage de l'immeuble Louis Vuitton. Nous sommes sérieux : si vous n'êtes pas membre du club, vous n'entrerez pas…

5-7-5 Jingumae
Shibuya-ku
Tokyo
www.celux.com

LaForet

It would be criminal to leave this influential thirteen-floor fashion building off any guide. Aimed at fashionable teens through to women in their thirties, LaForet has been a Harajuku landmark for over 20 years.

Kein Führer, der den Namen verdient, wäre komplett ohne dieses Trends setzende, dreizehnstöckige Modekaufhaus. LaForet spricht modebewusste Teenies ebenso an wie Frauen in den Dreißigern und ist seit über 20 Jahren eine Institution in Harajuku.

Tout guide digne de ce nom se doit de mentionner cet immeuble de 13 étages entièrement dédié à la mode. Fourmillant d'ados et de trentenaires branchés, LaForet est une référence de Harajuku depuis plus de 20 ans.

1-11-6 Jingumae
Shibuya-ku
Tokyo
Tel. +81 (3) 3475 0411

Norm

Sumptuous surroundings offering carefully selected, difficult-to-source European and local fashion – far from 'norm'.

Eine sorgsam zusammengestellte Auswahl an schwer zu findender europäischer und einheimischer Mode, präsentiert in einem ansprechenden Ambiente – alles andere als die Norm.

Environnement somptueux proposant des griffes japonaises et européennes difficiles à importer et choisies avec le plus grand soin : une sélection complètement hors « norme ».

6-4-6 Minami Aoyama
Minato-ku
Tokyo
Tel. +81 (3) 5467 1055

RealMadHectic

One of the most influential street style stores in Tokyo. An absolute must-see for its industrial concrete interior and the queues of fanatics that form when new stock arrives.

Einer der einflussreichsten Street-Style-Läden in Tokio. Mit seinem Beton-Interieur und den Schlangen der Fans, die anstehen, wenn neue Ware eintrifft, eine absolute Sehenswürdigkeit.

L'une des boutiques streetwear les plus influentes de Tokyo. Une adresse à visiter absolument pour son décor en béton industriel et pour les files d'attente qui se forment instantanément dès l'arrivée d'un nouveau stock.

4-26-21 Jingumae
Shibuya-ku
Tokyo
Tel. +81 (3) 3405 6933

Sou Sou

Sou Sou stocks Tabi or Japanese kimono socks beautifully remade with proper soles into colourful shoes that quite literally fit like a mitten. One-off designs made from traditional fabrics in stunning hues are almost too pretty to wear.

Sou Sou führt Tabi, die japanischen Kimono-Socken, die mit entsprechenden Sohlen zu richtigen Schuhen verarbeitet werden und im wahrsten Sinne des Wortes wie Handschuhe passen. Einzigartige Entwürfe aus traditionellen Stoffen in den erstaunlichsten Farbtönen – zum Anziehen fast zu schade.

Sou Sou vend des Tabi, ces chaussettes japonaises avec séparation du gros orteil, magnifiquement refaçonnées à l'aide de semelles adaptées sous forme de chaussures colorées qui vous iront comme un gant. Ces créations originales et uniques sont réalisées dans des tissus traditionnels et déclinées dans de superbes nuances : elles sont presque trop belles pour être portées.

2F From 1st Bldg
5-3-10 Minami Aoyama
Minato-ku
Tokyo
Tel. +81 (3) 3407 7877
www.sousou.co.jp

Mihara Yasuhiro

Famed for his Puma collaborations, talented cobbler Yasuhiro also produces menswear featuring inimitable detailing and styles that are far from pedestrian.

Der talentierte Autodidakt Yasuhiro wurde durch die Kooperation mit Puma berühmt. Er produziert aber auch Herrenmode mit unnachahmlichen Details in einem Stil, der alles, nur nicht langweilig ist.

Connu pour ses collaborations avec Puma, le talentueux bricoleur Yasuhiro produit également une collection pour homme aux détails inimitables et aux looks très originaux.

SoSu Mihara Yasuhiro
B1 Colonnado Jingumae
Shibuya-ku
Tokyo
Tel. +81 (3) 5778 0675

Naoki Takizawa for Issey Miyake
Issey Miyake's creative director never fails to surprise. The store's interior is as fascinating as the product sold.

Issey Miyakes Creative Director gelingen immer wieder Überraschungen. Die Einrichtung des Ladens ist ebenso faszinierend wie die dort angebotene Produktpalette.

Le directeur de la création d'Issey Miyake ne cessera jamais de nous surprendre. La déco intérieure de sa boutique est aussi fascinante que ses produits.

Issey Miyake Concept Store
Roppongi Hills Keyakizaka Dori
6-12-4 Roppongi
Minato-ku
Tokyo
Tel. +81 (3) 5772 2777

WR
Womenswear store that stocks mainly underground brands – the real draw, however, is its own label apparel and accessories. Check out the Aoyama branch for men's concept apparel.

Laden für Damenmode, der hauptsächlich Underground-Labels führt – die eigentliche Attraktion sind jedoch die Kleider und Accessoires der eigenen Marke. Entsprechende Männermode finden Sie in der Abteilung Aoyama.

Boutique pour femme présentant majoritairement des créateurs underground. Son principal intérêt réside dans sa propre griffe de vêtements et d'accessoires. Visitez aussi le rayon pour homme d'Aoyama.

Vogue II Building
13-4 Daikanyama-cho
Shibuya-ku
Tokyo
Tel. +81 (3) 5489 1457
Also: 3-12-16 Kita Aoyama
Minato-ku
Tokyo
Tel. +81 (3) 5468 2466

VINTAGE FASHION SHOPS

Hanjiro
Supermarket-sized vintage emporium spread over two floors. Original, remade and new items organized by style, design and colour, all within a stripped blond wood floor and chandelier setting.

Vintage-Mekka in Supermarktgröße auf zwei Etagen. Originalteile, Umgearbeitetes und Neues nach Modelstilen, Designs und Farben sortiert. Heller Holzfußboden und Kronleuchter sorgen für Atmosphäre.

Supermarché du vintage réparti sur deux niveaux. Les pièces originales, neuves ou refaçonnées sont classées par style, par marque ou par couleur, le tout dans un décor de chandeliers et de parquet de bois blond.

YM Square Bldg
4-31-10 Jingumae
Shibuya-ku
Tokyo
Tel. +81 (3) 3796 7303

Chicago
Vintage kimonos, yukata, denim and reasonably priced US-sourced street style apparel and accessories

Alte Kimonos, Yukatas (leichte Baumwollkimonos), Jeans und Street-Style-Kleidung sowie Accessoires aus den USA zu moderaten Preisen.

Kimonos, yukatas (kimonos légers en coton) et jeans vintage. Vêtements et accessoires streetwear importés des Etats-Unis mais proposés à des prix raisonnables.

6-31-21 Jingumae, B1F
Shibuya-ku
Tokyo
Tel. +81 (3) 3409 5017

Cocotier
This and sister store Cocotier Deco, selling vintage homewares, are essential for luxurious designer finds.

Hier und im Partnerladen Cocotier Deco wird Vintage-Homeware verkauft. Für luxuriöse Designer-Schnäppchen unvergleichlich.

Cocotier et sa petite sœur Cocotier Deco, qui vend des articles vintage pour la maison, représentent deux destinations incontournables pour faire de luxueuses trouvailles griffées.

5-46-12 Jingumae
Shibuya-ku
Tokyo
Tel. +81 (3) 5466 9055

RT
Five floors of men's and ladies secondhand clothing and accessories from a variety of labels. Mainly local mid-range to high-end brands. Best of all, you can trade in your old gear.

Second-Hand-Kleidung und Accessoires verschiedenster Label für Damen und Herren auf fünf Etagen. Vor allem einheimische Marken der mittleren bis hohen Preisklasse. Und das Beste: Sie können Ihre alten Sachen gleich eintauschen.

Cinq étages de vêtements et d'accessoires d'occasion pour homme et pour femme dans un large choix de marques. Du milieu de gamme japonais aux plus grandes griffes de luxe. Cerise sur le gâteau, la boutique reprend aussi les vêtements dont vous ne voulez plus.

3-3-15 Ginza
Chuo-ku
Tokyo
Tel. +81 (3) 3535 4100

Santa Monica
Ladies, men's and some kids' wear imported from the US. Many of the vintage items are reworked into new designs that follow seasonal trends.

Damen-, Herren- und Kindermode aus den USA. Viele Vintage-Sachen sind gemäß aktuellen Trends zu neuen Kreationen umgearbeitet.

Vêtements pour femme, pour homme et parfois pour enfant importés des Etats-Unis. La plupart des pièces vintage sont refaçonnées et transformées en de nouveaux modèles inspirés par la mode du moment.

5-8-5 Jingumae
Shibuya-ku
Tokyo
Tel. +81 (3) 3498 3260

FLEA MARKETS

Roi Building Flea Market
A tiny market held on the steps outside of Roppongi's Roi Building on the last Thursday and Friday of the month, this is the perfect destination for those searching for inexpensive vintage kimonos and shop fixtures from the 1950s and 1960s.

Winziger Markt auf den Stufen des Roi Building im Stadtteil Roppongi, jeden letzten Donnerstag und Freitag im Monat. Die ideale Anlaufstelle für alle, die preiswerte Second-Hand-Kimonos und Ladeninventar aus den 1950ern und 1960ern suchen.

Les derniers jeudi et vendredi de chaque mois, ce petit marché s'installe sur les marches du Roi Building dans le quartier Roppongi. Idéal pour dégotter des kimonos vintage à bas prix, du mobilier et de la PLV des années 50 et 60.

Salvation Army Bazaar
Weekly Saturday bazaar offering secondhand apparel, household goods and accessories of differing degrees of quality.

Jeden Samstag werden hier Second-Hand-Kleidung, Hausrat und Accessoires unterschiedlichster Qualität verkauft.

Tous les samedis, grand bazar de vêtements, d'objets et d'accessoires d'occasion de qualité variée.

2-21-2 Wada
Suginami-ku
Tokyo
Tel. +81 (3) 3384 3769

Yoyogi Koen Flea market
Tokyo's largest flea market, with around eight hundred stalls. Held every second Sunday, it's a definite must-visit, even if just to check out the crowds of cool kids, bands and food stalls.

Tokios größter Flohmarkt mit etwa 800 Ständen. Muss man gesehen haben, und sei es nur, um die vielen coolen Kids, Bands und Essensstände zu erleben. Jeden zweiten Sonntag, neben der NKH Hall, gegenüber vom Yoyogi Park.

Les plus grandes puces de Tokyo, avec près de 800 stands. Un samedi sur deux. A voir absolument, ne serait-ce que pour sa foule de kids branchés, ses groupes de musique et ses stands de nourriture. Près du NHK Hall, en face du Yoyogi Park.

Next to NHK Hall, opposite Yoyogi Park
Tel. +81 (3) 3226 6800

RESTAURANTS

Beige
The staff wear skinny black ties à la Karl Lagerfeld and the soft furnishings are Chanel's signature tweed. The ultimate fashion/food crossover.

Das Personal trägt figurbetonte Smokings à la Lagerfeld, und die Sessel sind mit typischem Chanel-Tweed überzogen. Das ultimative Crossover von Fashion und Food.

Le personnel porte des cravates étroites à la Karl Lagerfeld, les nappes et les rideaux sont taillés dans le tweed signature de Chanel : le meilleur de la fusion mode et cuisine.

10F Chanel Ginza Bldg
3-5-3 Ginza
Chuo-ku
Tokyo
Tel. +81 (3) 5159 5501
www.beige-tokyo.com

Café@Idée
Funky, lived-in joint offering cheap and tasty Japanese standards by day, cool local DJ sounds by night. Run by influential design group Idée.

Funkiges, gemütliches Lokal mit preiswerten, aber schmackhaften japanischen Standardgerichten am Tag und coolen einheimischen DJ-Sounds bei Nacht. Geführt wird das Ganze von der einflussreichen Designertruppe Idée.

Adresse funky et vivante proposant des classiques japonais savoureux et bon marché le jour, et les sets des DJ locaux le soir. Géré par l'influent groupe de designers Idée.

6-1-16 Minami-Aoyama
Tel. +81 (3) 3409 6744

Casita
A rooftop hideaway with laid-back Balinese interior. Tasty Asian fusion, located seconds from Roppingi Hills. Reserve the popular outdoor terrace during summer.

Verstecktes Dachgarten-Lokal mit lässiger balinesischer Einrichtung. Leckere asiatische Fusion-Küche, nur ein paar Schritte vom Roppingi Hills entfernt. Im Sommer ist die Terrasse im Freien heiß begehrt.

Un havre de paix sur le toit d'un immeuble, avec une décoration intérieure balinaise des plus apaisantes. Délicieuse cuisine fusion asiatique, à quelques minutes de Roppingi Hills. En été, mieux vaut réserver si vous voulez une table en terrasse.

3F Roppongi Place
5-10-25 Roppongi
Minato-ku
Tokyo
Tel. +81 (3) 5414 3190

Higashi-yama
Beautifully prepared Japanese cuisine in traditional-style surroundings with a modern twist.

Wunderschön angerichtete japanische Gerichte in traditioneller Umgebung mit modernem Touch.

Cuisine japonaise magnifiquement préparée. Décor traditionnel avec une touche de modernité.

1-21-25 Higashiyama
Meguro-ku
Tokyo
Tel. +81 (3) 5720 1300

Maisen
This former public bathhouse is the stage for Tokyo's tastiest Tonkatsu (deep fried pork chops). Retro interior, cheap and simple comfort cuisine.

In diesem ehemaligen öffentlichen Badehaus serviert man die köstlichsten Tonkatsu (frittierte Schweinekoteletts) von ganz Tokio. Retro-Design und preisgünstige Hausmannskost.

Cet ancien établissement de bains publics offre aujourd'hui les meilleurs tonkatsus de Tokyo (côtelettes de porc panées et frites). Décoration rétro, cuisine simple et bon marché.

4-8-5 Jingumae
Shibuya-ku
Tokyo
Tel. +81 (3) 3470 0071

Organic Café
Parisian-style café that doubles as a late night bar, lunch joint and dinner haunt. Permanently packed with fashionable locals.

Café im Pariser Stil, serviert Lunch und Dinner und hat bis spät in die Nacht geöffnet. Immer brechend voll mit hippen Leuten aus dem Viertel.

Café de style parisien qui fait également déjeuner, dîner et bar de nuit. Peuplé en permanence par la faune branchée du quartier.

1-24-1 Kamimeguro
Meguro-ku
Tokyo
Tel. +81 (3) 3791 5151

Shibuya Underpass Society
Located beneath the train tracks close to Shibuya station, SUS (as it is known by locals) serves simple fish and salad with a Japanese flair.

Das SUS (wie es die Einheimischen nennen) unter den Gleisen in der Nähe des Shibuya-Bahnhofs serviert einfache Fischgerichte und Salate mit japanischem Flair.

Situé sous une voie ferrée près de la gare de Shibuya, SUS (comme l'appellent les Tokyoïtes) propose des poissons et des salades simples de style japonais.

3-23-1 Shibuya
Shibuya-ku
Tokyo
Tel. +81 (3) 5778 4103

Zipangu
Lively atmosphere, impressive smart and modern interior with good city views. Offers a variety of reasonably priced Japanese staples and lengthy sake and Sho-chu lists.

Stark frequentiertes Lokal mit eindrucksvoll intelligentem, modernem Interieur und schönem Blick auf die Stadt. Gute Auswahl an japanischen Standardgerichten zu moderaten Preisen. Große Sake- und Sho-chu-Karte.

Atmosphère vivante, décoration classe, moderne et impressionnante avec de belles vues sur la ville. Propose toute une variété de classiques japonais à prix raisonnables, ainsi que du saké chaud et différentes sortes de Sho-chu.

14F Akasaka Excel Hotel
2-14-3 Nagatacho
Chiyoda-Ku
Tel. +81 (3) 3580 3661

BARS

Bape Café
A Bathing Ape and its subsidiary brands have become iconic in Japan. The Bape Café takes the empire one step further towards ape domination. This place offers a limited choice of just beer, water and coffee, but only for a few hours every afternoon.

A Bathing Ape und seine Ableger sind in Japan mittlerweile Kult. Das Bape Café ist ein weiterer Schritt auf dem Weg zur Weltherrschaft der Affen. Die Auswahl ist auf wenige Sorten Bier, Mineralwasser und Kaffee begrenzt, und auch das nur für einige Stunden am Nachmittag.

A Bathing Ape et ses marques dérivées sont devenus des symboles du Japon et le Bape Café ne fait qu'asseoir la domination de l'empire du singe. L'endroit propose un choix limité de bières, d'eaux minérales et de cafés, seulement l'après-midi.

5-3-18, B1F Minami Aoyama
Minato-ku
Tokyo
Tel. +81 (3) 5778 9726
www.apeshallneverkillape.com

Montoak
Cool, polished, boutique style bar popular with design elite. Located in the heart of Harajuku, ideal for daytime coffee and evening cocktails.

Coole Hochglanz-Bar im Boutiquenstil, die sich bei der Designelite großer Beliebtheit erfreut. Mitten in Harajuku und ideal für einen Kaffee am Tag und abendliche Cocktails.

Petit bar concept à l'ambiance cool et chic particulièrement prisé par l'élite créative. Situé en plein cœur de Harajuku, idéal pour boire un café en journée ou siroter un cocktail le soir venu.

6-1-9 Jingumae
Shibuya-ku
Tokyo
Tel. +81 (3) 5468 5928

Office
What a concept.... A bar that resembles an office, right down to the photocopier. No, we're not joking.

Was für ein Konzept ... Eine Bar, die wie ein Büro aussieht, perfekt bis zum Kopierer. Nein, das ist kein Witz.

Quel concept... Un bar qui ressemble à un bureau jusque dans les moindres détails, avec même son coin photocopieuse. Non, ce n'est pas une blague.

Yamazaki Bldg, 5F
2-7-18 Kita Aoyama
Minato-ku
Tokyo
Tel. +81 (3) 5786 1052

Shida
Chinese tea salon, stockist of accessories and difficult-to-locate denim brand Somét, Shida is the antithesis of mass market retailing and is located in Tokyo's most up-and-coming area, Naka Meguro. In case you're wondering why a denim store and Chinese tea salon is in the bar guide, it's worth knowing that tea is the new beer in Japan.

Chinesisches Teehaus, in dem auch Accessoires und die schwer erhältliche Jeansmarke Somét verkauft werden. Shida ist das absolute Gegenteil von Massengeschmack und liegt in Tokios angesagtestem Viertel Naka Meguro. Wer fragt, warum ein Jeansladen und Teehaus in der Rubrik Bars steht, muss wissen, dass Tee in Japan das neue Bier ist.

Salon de thé chinois, stockiste d'accessoires et de la griffe de jeans Somét, introuvable ailleurs. Situé dans le quartier le plus branché de Tokyo, Naka Meguro, Shida est l'antithèse de la distribution de masse. Au cas où vous vous demanderiez pourquoi une boutique de jeans et un salon de thé chinois se trouvent sous la rubrique Bars, sachez que le thé est la nouvelle bière du Japon.

1-13-7 Kamimeguro
Meguro-ku
Tokyo
Tel. +81 (3) 5428 0804

/SO/RA/SI/O
Incredible views of Tokyo's harbour, a heart-stopping 46 floors above Shiodome. Visit between 4 – 7 pm for half-price drinks – a rare bargain in this town.

Unglaubliche Aussicht auf den Tokioter Hafen, atemberaubende 46 Stockwerke über dem Viertel Shiodome. Zwischen 16 und 19 Uhr alle Drinks zum halben Preis – in dieser Stadt eine Seltenheit.

Vues à couper le souffle sur le Port de Tokyo, au 46ème étage de l'immeuble du Shiodome. Allez-y entre 16 et 19 heures, les verres sont à moitié prix : une vraie rareté dans cette ville.

46F Caretta Shiodome
1-8-1 Higashi-Shinbashi
Minato-ku
Tokyo
Tel.+81 (3) 6215 8055

The Hanazawa Garden
Former stately home turned into stylish drinking spot. Stunning Japanese gardens set off the low key Asian décor. If you only visit one bar in Tokyo make it this one.

Stilvolle Bar in einem ehemaligen vornehmen Wohnhaus. Eindrucksvolle japanische Gärten bringen die dezente asiatische Einrichtung perfekt zur Geltung. Falls Sie in Tokio nur eine einzige Bar besuchen, dann sollte es diese sein.

Ancienne propriété de l'Etat transformée en bar très élégant. Des jardins japonais d'une beauté renversante s'harmonisent avec un décor asiatique plutôt sobre. Si vous ne devez aller que dans un bar à Tokyo, choisissez celui-ci.

3-12-15 Hiroo
Shibuya-ku
Tokyo
Tel. +81 (3) 3400 2013

Unit
The city's hottest club – yes, people still go clubbing in Tokyo – located over three subterranean floors. Attracts local and international DJs with a bias towards dub and global beats.

Heißester Club der Stadt – ja, die Leute gehen in Tokio nach wie vor zum Clubbing – auf drei unterirdischen Ebenen. Anziehungspunkt für in- und ausländische DJs mit einem Faible für Dub und Global Beats.

Situé au troisième sous-sol d'un immeuble, c'est le club le plus hype de la ville : et oui, les gens sortent encore en boîte à Tokyo. Attire des DJ japonais et internationaux, avec un penchant pour le dub et l'électro fusion.

Za House Bull
1-34-17 Ebisu-Nishi
Shibuya-ku
Tokyo
www.unit-tokyo.com

HOTELS

Four Seasons Hotel

Everything you would expect from the understated yet glamorous Four Seasons chain, including a fabulous Marunouchi location close to Ginza's shopping and the capital's main train station.

Alles, was man von der glamourösen und zugleich auf Understatement setzenden Kette Four Seasons erwarten kann, inklusive traumhafter Lage im Marunouchi-Viertel, nahe der Einkaufsgegend Ginza und dem Hauptbahnhof.

Tout ce que vous êtes en droit d'attendre de la discrète mais néanmoins glamour chaîne hôtelière Four Seasons, y compris un emplacement idéal sur Marunouchi, à deux pas du quartier des boutiques de Ginza et de la principale gare ferroviaire de la capitale.

1-11-1 Marunouchi
Chiyoda-ku
Tokyo
Tel. +81 (3) 5222 7222
www.fourseasons.com

Park Hyatt Tokyo

One of Tokyo's most exclusive hotels and also one of its most expensive. VIPs and celebrities are drawn to the Japanese touches that are incorporated into each room.

Eines der exklusivsten und teuersten Hotels von Tokio. VIPs und Prominente schätzen die japanischen Elemente, die in jedes der Zimmer integriert sind.

L'un des hôtels les plus luxueux de Tokyo mais aussi l'un des plus chers. Les VIP et les célébrités y sont attirés par la touche déco japonaise intégrée à chacune des chambres.

3-7-1-2 Nishi Shinjuku
Shinjuku-ku
Tokyo
Tel. +81 (3) 5322 1234
www.parkhyatttokyo.com

The Claska

Tokyo's only boutique hotel offers just nine rooms – no two are alike, decorated minimally in neutral tones. Located close to the capital's best concentration of interior stores, but otherwise a twenty-minute taxi ride from anything else notable.

Einziges Boutique-Hotel Tokios mit nur neun Zimmern, die alle individuell, minimalistisch und in neutralen Farbtönen eingerichtet sind. Ganz nah zu den besten Einrichtungsläden der Stadt, aber zwanzig Taxi-Minuten von allen anderen Attraktionen entfernt.

Le seul « boutique hotel » de Tokyo ne propose que neuf chambres : toutes différentes, déco minimaliste et couleurs neutres. A deux pas de la meilleure concentration de boutiques de design d'intérieur de la capitale, mais compter vingt minutes en taxi pour accéder aux autres attractions touristiques.

1-3-18 Chuo-ku
Meguro-ku
Tokyo
Tel. +81 (3) 3719 8121
www.claska.com

AIRPORT

Narita International Airport

Narita City
Tel. +81 (476) 3450 73
www.narita-airport.or.jp/airport/

To stay informed about upcoming TASCHEN titles, please
request our magazine at www.taschen.com/magazine or write to
TASCHEN, Hohenzollernring 53, D-50672 Cologne, Germany,
contact@taschen.com, Fax +49-221-254919. We will be happy
to send you a free copy of our magazine which is filled
with information about all of our books.

Editors: Terry Jones & Susie Rushton
Production director: Matthew Hawker
Fashion director: Edward Enninful
Fashion editors: David Lamb, Erika Kurihara
Production manager: Karen Leong
Managing editor: Eloise Alemany
Sub-editor: Richard Hodkinson
Design co-ordinator: Aya Naito
Assistant: Patrick Waugh
Editorial assistance: Glenn Waldron
Executive director: Tricia Jones

Writers: Aimee Farrell, Ben Reardon, Dan Jones, David Lamb, Ed Perry, Elena Moretti, Erika Palomino, Gianluca Longo, Glenn Waldron, Hellin Kay, Holly Shackleton, James Anderson, James Sherwood, Jamie Huckbody, Jo-Ann Furniss, Jörg Koch, Josh Sims, Karen Leong, Lauren Cochrane, Lee Carter, Lesley Arfin, Liz Hancock, Marcus Ross, Mark Hooper, Nancy Waters, Nicole Fall, Penny Martin, Peter de Potter, Simon Chilvers, Skye Sherwin, Steve Cook, Steven Taylor, Susie Rushton, Teddy Czopp, Terry Newman, Will Fairman, Zoë Jackson

Photographers: Alasdair McLellan, Alex Hoerner, Alexia Silvagni, Ali Mahdavi, Ali Peck, Amanda DeCadenet, Anette Aurell, Bianca Pilet, Carter Smith, Cellina Von Mannstein, Christophe Cufos, Corinne Day, David Armstrong, David Bailey, David Slijper, Dennis Schoenberg, Gerald Jenkins, Derek Henderson, Desmond Muckian, Dirk Seiden Schwan, Donald Christie, Donald Milne, Duc Liao, Dusan Reljin, Ed Sykes, Ellen Nolan, Ellen Von Unwerth, Ellin Wasson, Gemma Booth, Gerrard Needham, Gleb Kosorukov, Greg Lotus, Gustavo Ten Hoever, Henrique Gendré, Hiroshi Kutomi, Horst Diekgerdes, Immo Klink, Isabel Asha Penzlien, Idesu Ohayo, Jason Evans, Jennifer Tzar, Jerome Albertini, Jesse Shadoan, Jessica Craig Martin, Jo Barker, Jo Metson-Scott, James Mountford, Jess Shadoan, Jérome Albertini, Johnny Giunta, Karina Taira, Kate Garner, Kayt Jones, Kevin Davies, Keiron O' Connor, Kerry Hallihan, Laetitia Negre, Lena Modigh, Manuela Pavesi, Manuel Vason, Marcus Tomlinson, Marius Hansen, Mark Lebon, Mark Segal, Matt Black, Matt Jones, Matthias Vriens, Mauro Cocilio, Max Zambelli, Max Vadukul, Mert Alas & Marcus Piggott, Michael Roberts, Michael Sanders, Michel Momy, Mikael Jansson, Mitchell Sams, Mischa Richter, Patrick Demarchelier, Paul Wetherell, Randall Mesdon, Rebecca Pierce, Rebecca Lewis, Richard Burbridge, Richard Bush, Sachiko Okada, Satoshi Saikusa, Schohaja, Sean Cunningham, Sean Ellis, Shawn Mortensen, Shiro, Sophie Delaporte, Steen Sundland, Steve Smith, Takay, Tesh, Thomas Schenk, Timur Çelikdag, Todd Cole, Vanina Sorrenti, Vava Ribeiro, Viviane Sassen, Will Sanders, Willy Vanderperre, Wing Shya, Xevi, Yelena Yemchuk

Stylists: Aaron Sharif, Adam Howe, Nadine Sanders, Alastair Mckimm, Allan Kennedy, Anna Burns, Anna Foster, Annett Monheim, Belen Casadevall, Bill Mullen, Cathy Dixon, Charles Adesanya, Cher Coulter, Christine Fortune, Claudia Carretti, David Lamb, Dean Voykovich, Dudu Bertholini, Edward Enninful, Erika Kurihara, Elisa Nalin, Gabriel Feliciano, Gareth Griffiths, Gaspard Yurkievich, Giannie Couji, Harris Elliott, Havana Laffitte, Heathermary Jackson, Heidi Bivens, Hortense Manga, Jani Savolainen, Jo Barker, Jodie Barnes, John Hullum, June Nakamoto, K8 Hardy, Kanako B Koga, Karina Givargisoff, Karl Plewka, Karl Templer, Katie Mossman, Lucy Ewing, Marcia Taylor, Marcus Ross, Mark Anthony, Mark Morrison, Merryn Leslie, Michelle Cameron, Mika Mizutani, Nadine Shaw, Neil Stuart, Olivier Rizzo, Panos Yiapanis, Patti Wilson, Rushka Bergman, Sarah Richardson, Sean Kunjambu, Simon Foxton, Soraya Dayani, Tamara Rothstein, Tiffany Pentz, Venetia Scott, William Baker, Yoshiyuki Shimizu

Editorial co-ordination: Simone Philippi, Cologne
Production co-ordinator: Ute Wachendorf, Cologne
German translation: Henriette Zeltner, Munich
French translation: Claire Le Breton, Paris

Printed in Italy
ISBN 3-8228-4241-9